a

...omia

Deuterostomia

Ecdysozoa

1 Annelida

Mollusca

Sipuncula

Nematoda

Nematomorpha

Loricifera

Kinorhyncha

Priapulida

Onychophora

Tardigrada

2 Arthropoda

3 Chaetognatha

Xenoturbellida

Echinodermata

Hemichordata

Chordata

Cuticle molted

Radial cleavage

Mouth formed from blastopore

ia

1 Annelida includes pogonophorans and echiurans

2 Arthropoda includes pentastomids

3 Chaetognatha is discussed with deuterostomes

INTEGRATED PRINCIPLES OF
ZOOLOGY

SEVENTEENTH EDITION

Cleveland P. Hickman, Jr.
WASHINGTON AND LEE UNIVERSITY

Susan L. Keen
UNIVERSITY OF CALIFORNIA–DAVIS

David J. Eisenhour
MOREHEAD STATE UNIVERSITY

Allan Larson
WASHINGTON UNIVERSITY

Helen I'Anson
WASHINGTON AND LEE UNIVERSITY

Original Artwork by
WILLIAM C. OBER, M.D.
WASHINGTON AND LEE UNIVERSITY
and
CLAIRE W. OBER, B.A.
WASHINGTON AND LEE UNIVERSITY

Mc
Graw
Hill
Education

INTEGRATED PRINCIPLES OF ZOOLOGY, SEVENTEENTH EDITION

Published by McGraw-Hill Education, 2 Penn Plaza, New York, NY 10121. Copyright © 2017 by McGraw-Hill
Education. All rights reserved. Printed in the United States of America. Previous editions © 2014, 2011, and
2008. No part of this publication may be reproduced or distributed in any form or by any means, or stored in a
database or retrieval system, without the prior written consent of McGraw-Hill Education, including, but not
limited to, in any network or other electronic storage or transmission, or broadcast for distance learning.

Some ancillaries, including electronic and print components, may not be available to customers outside the
United States.

This book is printed on acid-free paper.

1 2 3 4 5 6 7 8 9 DOW 21 20 19 18 17 16

ISBN 978-1-259-56231-0
MHID 1-259-56231-X

Chief Product Officer, SVP Products & Markets: *G. Scott Virkler*
Vice President, Content Design and Delivery: *Betsy Whalen*
Vice President, General Manager, Products & Markets: *Marty Lange*
Brand Manager: *Justin K. Wyatt, PhD*
Director, Product Development: *Rose Koos*
Product Developer: *Elizabeth Sievers*
Marketing Manager: *Patrick Reidy*
Director of Digital Content: *Michael G. Koot, PhD*
Director, Content Design & Delivery: *Linda Avenarius*
Program Manager: *Angela FitzPatrick*
Content Project Managers: *Jessica Portz, Evan Roberts, Sandra Schnee*
Buyer: *Laura M. Fuller*
Design: *Tara McDermott*
Content Licensing Specialists: *Lori Hancock, Lorraine Buczek*
Cover Image: *© Kevin Schafer/Getty Images*
Compositor: *Lumina Datamatics, Inc.*
Printer: *R. R. Donnelley*

All credits appearing on page or at the end of the book are considered to be an extension of the copyright page.

Library of Congress Cataloging-in-Publication Data

Names: Hickman, Cleveland P., Jr., author.
Title: Integrated principles of zoology / Cleveland P. Hickman, Jr.,
    Washington and Lee University, Susan L. Keen, University of
    California-Davis, David J. Eisenhour, Morehead State University, Allan
    Larson, Washington University, Helen I'Anson, Washington and Lee
    University; original artwork by William C. Ober, M.D. Washington and Lee
    University, and Claire W. Ober, B.A., Washington and Lee University.
Description: Seventeenth edition. | New York, NY: McGraw-Hill Education,
    2016. | Includes bibliographical references and index.
Identifiers: LCCN 2016026850 | ISBN 9781259562310 (alk. paper)
Subjects: LCSH: Zoology.
Classification: LCC QL47.2 .H54 2016 | DDC 590—dc23 LC record available at https://lccn.loc.gov/2016026850

The Internet addresses listed in the text were accurate at the time of publication. The inclusion of a website does
not indicate an endorsement by the authors or McGraw-Hill Education, and McGraw-Hill Education does not
guarantee the accuracy of the information presented at these sites.

mheducation.com/highered

CONTENTS IN BRIEF

TABLE OF CONTENTS

PART ONE

Introduction to Living Animals

PART TWO

Continuity and Evolution of Animal Life

PART THREE

Diversity of Animal Life

PART FOUR

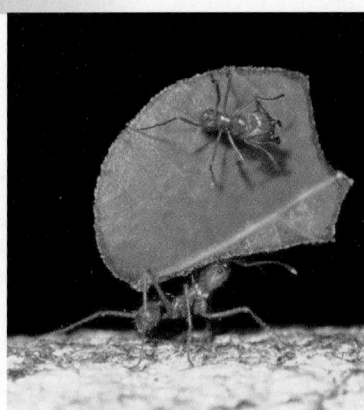

Activity of Life

PART FIVE

Animals and Their Environments

ABOUT THE AUTHORS

CLEVELAND P. HICKMAN, JR.

Cleveland P. Hickman, Jr., Professor Emeritus of Biology at Washington and Lee University in Lexington, Virginia, has taught zoology and animal physiology for more than 30 years. He received his Ph.D. in comparative physiology from the University of British Columbia, Vancouver, B.C., in 1958 and taught animal physiology at the University of Alberta before moving to Washington and Lee University in 1967. He has published numerous articles and research papers in fish physiology, in addition to co-authoring these highly successful texts: *Integrated Principles of Zoology, Biology of Animals, Animal Diversity, Laboratory Studies in Animal Diversity,* and *Laboratory Studies in Integrated Principles of Zoology.*

Over the years Dr. Hickman has led many field trips to the Galápagos Islands. His research is on intertidal zonation and marine invertebrate systematics in the Galápagos. He has published three field guides in the Galápagos Marine Life Series for the identification of echinoderms, marine molluscs, and marine crustaceans.

His interests include scuba diving, woodworking, and participating in chamber music ensembles.

Dr. Hickman can be contacted at:
hickmanc@wlu.edu

SUSAN KEEN

Susan Keen is Associate Dean for Undergraduate Academic Programs, College of Biological Sciences at the University of California–Davis. She received her Ph.D. in zoology from the University of California–Davis, following a M.Sc. from the University of Michigan at Ann Arbor. She is a native of Canada and obtained her undergraduate education at the University of British Columbia in Vancouver.

Dr. Keen is an invertebrate zoologist fascinated with sea jelly life histories. She has a particular interest in life cycles where both asexual and sexual phases of organisms are present, as they are in most jellies. Her other research has included work on sessile marine invertebrate communities, spider populations, and Andean potato evolution.

Dr. Keen has been teaching evolution and animal diversity within the Introductory Biology series for 22 years. She enjoys all facets of the teaching process, from lectures and discussions to the design of effective laboratory exercises. In addition to her work with introductory biology, she offers seminars on teaching methods in biology, and works with an animator to create self-guided teaching modules on animal development. She has won awards for Excellence in Education from the Associated Students group at Davis. She attended the National Academies Summer Institute on Undergraduate Education in Biology, and was a National Academies Education Fellow in the Life Sciences for 2005–2006.

Her interests include horseback riding, gardening, travel, and mystery novels.

Dr. Keen can be contacted at:
slkeen@ucdavis.edu

DAVID J. EISENHOUR

David J. Eisenhour is professor of biology at Morehead State University in Morehead, Kentucky. He received his Ph.D. in zoology from Southern Illinois University, Carbondale. He teaches courses in environmental science, human anatomy, mammalogy, general zoology, comparative anatomy, ichthyology, and vertebrate zoology. David has an active research program that focuses on systematics, conservation biology, and natural history of North American freshwater fishes. He has a particular interest in the diversity of Kentucky's fishes and is writing a book about that subject. He and his students have authored several publications. David serves as an academic advisor to prepharmacy students.

His interests include fishing, landscaping, softball, hiking, and entertaining his three children, who, along with his wife, are enthusiastic participants in fieldwork.

Dr. Eisenhour can be contacted at:
d.eisenhour@morehead-st.edu

ALLAN LARSON

Allan Larson is a professor at Washington University, St. Louis, MO. He received his Ph.D. in genetics at the University of California–Berkeley, following B.S. and M.S. degrees in zoology from the University of Maryland. His fields of specialization include evolutionary biology, molecular population genetics and systematics, and amphibian systematics. He teaches courses in introductory genetics, zoology, macroevolution, molecular evolution, and the history of evolutionary theory, and has organized and taught a special course in evolutionary biology for high-school teachers. Dr. Larson's students have participated in zoological field studies around the world, including projects in Africa, Asia, Australia, Madagascar, North America, South America, the Indo-Pacific Ocean, and the Caribbean Islands. Dr. Larson has authored numerous scientific publications, and has edited for the journals *The American Naturalist, Evolution, Journal of Experimental Zoology, Molecular Phylogenetics and Evolution,* and *Systematic Biology.* Dr. Larson serves as an academic advisor to undergraduate students and supervises the undergraduate biology curriculum at Washington University.

Dr. Larson can be contacted at:
larson@wustl.edu

HELEN I'ANSON

Helen I'Anson, a native of England, is the John T. Perry Professor of Biology and Research Science at Washington and Lee University in Lexington, Virginia. She received her Ph.D. in physiology at the University of Kentucky, Lexington, and postdoctoral training at the University of Michigan, Ann Arbor. She teaches courses in animal physiology, microanatomy, neuroendocrinology, general biology, and reproductive physiology. She has an active research program that focuses on the neural regulation of energy partitioning during development using the female weanling rat. In particular, she is interested in mechanisms regulating the development of childhood obesity. Currently, she is studying the role of snacking from weaning to adulthood in the onset of childhood obesity.

Her interests include gardening, hiking, fishing, aromatherapy, music, and participating in choral ensembles.

Dr. I'Anson can be contacted at:
iansonh@wlu.edu

Integrated Principles of Zoology continues to be the leading text for the introductory zoology course. With the seventeenth edition, the authors bring a wealth of real experience as they describe the diversity of animal life and the fascinating adaptations that enable animals to inhabit so many ecological niches.

The overall organization of this text has proven to work well to help students understand the content. Distinctive features, especially the emphasis on principles of evolution and zoological science, have been strengthened. To aid in student learning, several pedagogical features have been retained: opening chapter prologues drawn from the chapter's theme; chapter summaries and review questions to aid in comprehension and study; concise and visually appealing illustrations; chapter notes and essays that offer interesting sidelights to the narrative; literature citations; and an extensive glossary providing pronunciations, derivations, and definitions of terms used in the text.

NEW TO THE SEVENTEENTH EDITION

Revisions for the seventeenth edition are primarily for improved pedagogy, and guided by an electronic tabulation of student responses to questions. All parts of the sixteenth edition were linked to electronic questions designed to measure students' comprehension of their reading. Authors received a heat map of the sixteenth edition, showing for each paragraph the percentage of correct student responses for that material. We focused our revisions on improving explanations wherever the heat map showed less than 50% of the students responding correctly. With this detailed and insightful guidance, we have made our text more accessible to its readers.

In addition to focused revision of the writing, we have replaced many photographs and diagrams throughout the book to improve clarity and vibrancy.

We summarize our major revisions in order by the book's five major parts.

Part One, Introduction to Living Animals

In Chapter 1, a new section, *Historical Continuity of Life,* replaces the section formerly titled *Does Life Have Defining Properties?* A new figure illustrating protein structure strengthens the visual correspondence between the bonding of amino acids and the three-dimensional macromolecular structure of a protein. Regarding general properties of living systems, we expand the explanation of speciation as part of the biological hierarchy, and extend the principle of movement to the macromolecular level, emphasizing reversible conformational changes in proteins. In Chapter 2, expanded captions for Figures 2.12–2.15 make the diagrams more independent of the text in conveying the fundamental structures of biological macromolecules. The important role of ATP in cellular metabolism and as a precursor to RNA gets expanded treatment, as does the role of membrane-bound vesicles in the evolution of biological polymers. Chapter 3 features a new explanation of electron microscopy, more information on the Golgi apparatus and on how cells adapt to temperature, and a more detailed description of how spindle fibers organize chromosomes for cellular division.

Part Two, Continuity and Evolution of Animal Life

Chapter 5 features greater detail on Mendel's investigations, plus a more detailed and precise explanation of how geneticists use the term "chromosome." Also given expanded explanations are the topics of chromosomal sex determination, how a new allele arises by mutation, occurrence of multiple alleles in a population, autosomal linkage and crossing over, DNA structure, and gene expression. Chapter 6 contains a more complete explanation of geological dating, a reorganized section, *Forces of Evolutionary Change,* and a new section describing studies of Caribbean ectoprocts as the best-documented case of punctuated equilibrium. New information on the CRISPR tool used to splice and to manipulate genes appears in Chapter 8. Chapter 8 includes updates on mesoderm and coelom formation, plus new and improved diagrams of development.

Part Three, Diversity of Animal Life

Chapters 9 and 11 update taxonomy of unicellular eukaryotes. Chapter 10 features more precise explanation of cladistic terminology, phylogenetic reconstruction, the conflicts among the major schools of taxonomy, and the major divisions of life. Chapter 11 emphasizes symbiogenesis as a major theme emerging from studies of eukaryotic origins. Chapter 11 also includes clarifications on saprozoic feeding and sexual reproduction in unicellular eukaryotes, and updates on recent U.S. outbreaks of *Cryptosporidium parvum.* Chapter 12 updates sponge taxonomy to use Calcispongiae instead of Calcarea, provides greater detail on Homoscleromorpha, and updates the debate on whether sponges have true tissue epithelia, including classification of pinacoderm as an incipient tissue. Chapter 13 is retitled *Cnidarians and Ctenophores.* "Jellyfishes" are now consistently called "jellies." The section on sea anemones is revised with greater detail, and the section on coral reefs now includes an example of coral bleaching in the Maldives. The section on phylogeny and adaptive diversification is substantially revised to include detail on the evolutionary biology of muscle cells in cnidarians and ctenophores. In Chapter 14, the section on Cestoda is updated with greater detail on the life cycle of pork tapeworms. New phylogenetic evidence groups acoelomorphs and xenoturbellids as a clade. The introduction to Chapter 15 now mentions controversy concerning homology of the lophophore across the three lophophorate phyla. The section on Phylum Ectoprocta now includes detail on *Tricellaria inopinate* as an invasive species.

Chapter 16 features a new description of evolution of terrestrial slugs, new information on how ocean acidification threatens marine molluscs, and the economic importance of molluscs. Chapter 17 presents a completely revised introductory section on annelids with updated information on the fluid-filled coelomic cavity, and more background on annelid evolutionary history. The section describing earthworms' peristaltic movement is augmented and improved. Chapter 18 reports that all plant parts have specialized nematode parasites. The section on form and function of nematodes is augmented and improved with updated information on the life cycle of *Ascaris* worms. Information from the CDC on filarial worms is updated. The evolutionary placement of the fossil *Hallucigenia* sheds light on the evolution of teeth in ecdysozoans and their loss in velvet worms. Additional detail appears on mating in nematomorphs. Chapter 19 features greater detail on spider respiration. Updates include the role of the *Varroa* mite in the dramatic loss of U.S. honeybees. Chapter 20 reports the first venomous crustacean and includes updates on the symbiotic feeding habits of yeti crabs. Chapter 21 features new phylogenetic data suggesting that insects arose far earlier than the fossil record indicates. The section on insect metamorphosis and growth includes new information on molting and ecdysis. The defense section includes a new box on luna moths. The section on insects and human welfare continues from Chapter 19 the discussion of colony collapse disorder in honeybees, plus updates on defoliation and the Zika virus. Chapter 22 adds information on species diversity of Chaetognatha, echinoderm larvae and nervous systems, and echinoderms of Class Ophiuroidea.

In Chapters 23–24, we adopt unranked nested clade names for vertebrate taxonomy. There is no longer a consensus system for placing the higher groups of vertebrates into a Linnaean system. We update hagfish and lamprey relationships; these are now considered sister groups, forming the clade Cyclostomata. Also presented is new information about hagfish development and reproduction, which includes the first documentation that at least some hagfishes do have vertebrae. Extensive clarification of wording throughout these chapters follows the guidance of the heat map. The fish cladogram was heavily modified, incorporating new ideas about relationships of placoderms and acanthodians. There is new lungfish art depicting aestivation behavior. An updated boxed essay describes the impact of dams, and their removals, on salmon populations. Chapter 25 presents more precise dating information on the Devonian origin of tetrapods in the text and accompanying figures. Chapter 26 features updated information on turtle evolution, highlighting evolutionary developmental biology of the shell. The opening section on enclosing the pond has been substantially revised for context and clarity, now specifying the four living reptile groups and noting that dinosaurs form a clade with birds. A new boxed essay describes the Burmese python as an invasive species in southern Florida. Chapter 27 features more precise terminology and description of the categories of bird flight. The boxed essay on flightless birds includes an update on the height of the largest bird, and the essay on Himalayan migration by bar-headed geese is updated. Chapter 28 presents an expanded section on mammalian omnivory. Updates to human evolution include new information from recent South African and European fossils and from DNA studies of Neanderthal and modern humans. The boxed essay on trade in rhinoceros horns is updated.

Part Four, Activity of Life

The section on integument in Chapter 29 includes a new footnote directing readers to the Annual Report to the Nation on the Status of Cancer 1975–2011. The section on skeletal systems presents an update on the running capability and speed of *Tyrannosaurus*. The Chapter 31 subsection on excitation and control of the heart now features greater detail on how sympathetic nerves increase stroke volume and heart rate when activated. Chapter 32 presents additional detail on the functionality of pepsin. There are updated statistics and a new footnote on the prevalence of childhood and adult obesity in the United States, plus more detail on secretions of fat cells and inflammatory responses. The section on nutritional requirements includes updated statistics on meat consumption. Coverage of hearing in Chapter 33 includes new material on luna moth defenses. In the section on prostoglandins and cytokines, Chapter 34 features more information on secretion of these hormones and inflammatory responses. In the section on hormones of metabolism, more detail appears on possible consequences of defects in thyroid hormone production or receptors, along with more detail on the evolution of bone and parathyroid glands in fishes. A boxed essay on anabolic steroids has updated statistics. The Chapter 35 section on innate defense mechanisms adds current studies on the gut microbiome, which may be altered by obesity. In Chapter 36, the subsection on genetics of behavior includes greater detail on W. C. Dilger's hybridization of lovebirds and its disruption of genetically programmed behavior. The subsection on agonistic or competitive behavior features added detail on peck order. The subsection on cooperative behavior, altruism, and kin selection describes W. D. Hamilton's modification of Ronald Fisher's concept of the fitness of an allele and self-sacrifice. The subsection on the language of honeybees features a more detailed description.

Part Five, Animals and Their Environments

The Chapter 37 discussion of distribution by dispersal and vicariance includes more information on the geographic distribution of ratite birds. There is more information on glacial cycles in the section on climatic cycles and vicariance, and greater detail on the relationships between upwellings of phytoplankton in the Antarctic sea and krill. The Chapter 38 subsection on populations includes a new paragraph on the metapopulation model, plus greater detail and clarity on the three principal types of theoretical survivorship. Figure 38.3 has a revised and more detailed caption. This section also now includes a note on the United Nations' position that a world population of 7–11 billion people would reach the point of being unsustainable agriculturally. The section on community ecology is significantly revised and now includes coverage of ecological character displacement, microhabitat selection, and the theory of competitive exclusion. It also features new coverage of ectoparasites and endoparasites and the complexity of parasite/host relationships. The section on ecosystems is revised for clarity and to provide greater detail. The caption for Figure 38.13 is revised for better context.

OUR APPROACH
TO THE REVISION

Our goal now is, as it has always been, to ensure student success. To that end, we approached this revision differently. To help guide our revision for this seventeenth edition, we were able to incorporate student usage data and input, derived from thousands of our Smart-Book® users. SmartBook "heat maps" provided a quick visual snapshot of chapter usage data and the relative difficulty students experienced in mastering the content. With these data, we were able to hone not only our text content but also the SmartBook probes.

- If the data indicated that the subject was more difficult than other parts of the chapter, as evidenced by a high proportion of students responding incorrectly to the probes, we revised or reorganized the content to be as clear and illustrative as possible.
- In other cases, if one or more of the SmartBook probes for a section was not as clear as it might be or did not appropriately reflect the content, the probes were edited, rather than the text.

Below is an example of one of the heat maps from Chapter 13. The color-coding in highlighted sections indicates the various levels of difficulty students experienced in learning the material, topics highlighted in red being the most challenging for students.

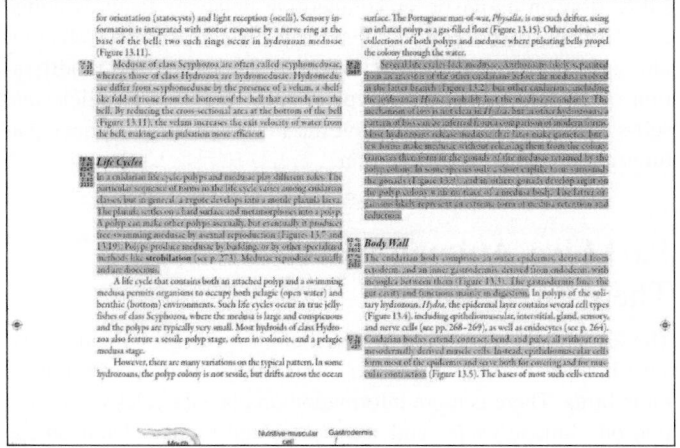

TEACHING AND LEARNING AIDS

To help students in **vocabulary development,** key words are bold-faced and derivations of technical and zoological terms are provided, along with generic names of animals where they first appear in the text. In this way students gradually become familiar with the more common roots that form many technical terms. An extensive **glossary** provides pronunciation, derivation, and definition for many terms, including new ones added to the glossary or existing ones rewritten for this edition.

A distinctive feature of this text is a **prologue** for each chapter that highlights a theme or fact relating to the chapter. Some prologues present biological, particularly evolutionary, principles; those in Part Three on animal diversity illuminate distinguishing characteristics of the group presented in the chapter.

Chapter notes, which appear throughout the book, augment the text material and offer interesting sidelights without interrupting the narrative. We prepared many new notes for this edition and revised several existing notes.

To assist students in chapter review, each chapter ends with a **concise summary,** a list of **review questions,** and **annotated selected references.** The review questions enable a student to self-test retention and understanding of the more important chapter material.

Again, William C. Ober and Claire W. Ober have strengthened the art program for this text with many new full-color paintings that replace older art, or that illustrate new material. Bill's artistic skills, knowledge of biology, and experience gained from an earlier career as a practicing physician have enriched this text through 10 of its editions. Claire practiced pediatric and obstetric nursing before turning to scientific illustration as a full-time career. Texts illustrated by Bill and Claire have received national recognition and won awards from the Association of Medical Illustrators, American Institute of Graphic Arts, Chicago Book Clinic, Printing Industries of America, and Bookbuilders West. They are also recipients of the Art Directors Award.

For the Zoology Lab

Laboratory Studies in Integrated Principles of Zoology **by Cleveland Hickman, Jr., Susan Keen, and Lee B. Kats**

Now in its seventeenth edition, this lab manual was written to accompany *Integrated Principles of Zoology,* and can be easily adapted to fit a variety of course plans.

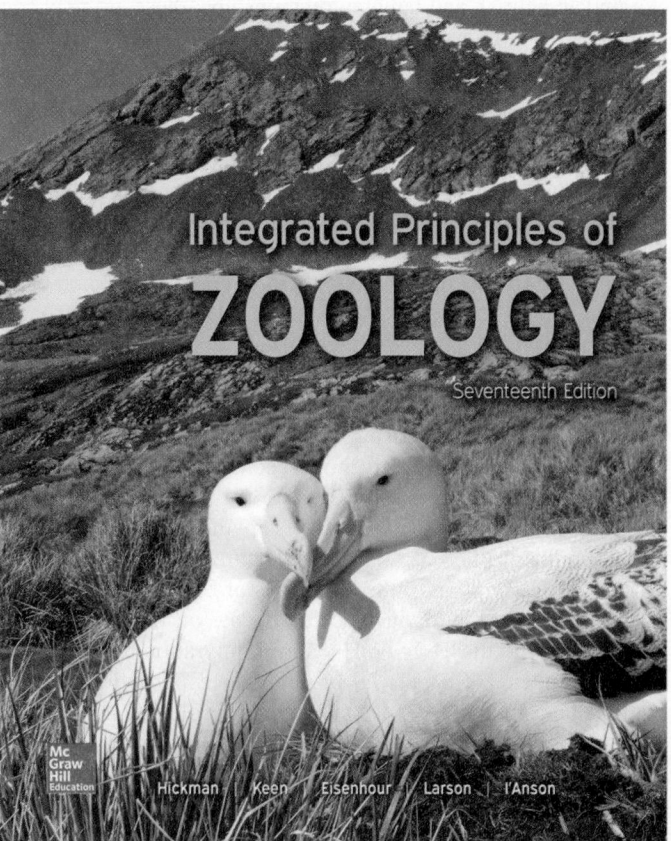

ACKNOWLEDGMENTS

The authors express their appreciation to the editors and support staff at McGraw-Hill Higher Education who made this project possible. Special thanks are due Justin Wyatt, Brand Manager, and Elizabeth Sievers, Lead Product Developer who were the driving forces in piloting this text throughout its development. Jessica Portz, Content Project Manager, somehow kept authors, text, art, and production programs on schedule. Lori Hancock oversaw the extensive photographic program and Tara McDermott managed the book's interior and cover design. We are indebted to them for their talents and dedication.

Although we make every effort to bring to you an error-free text, errors of many kinds inevitably find their way into a textbook of this scope and complexity. We will be grateful to readers who have comments or suggestions concerning content and send their remarks to Patrick Reidy, Executive Marketing Manager, at patrick.reidy@mheducation.com

Cleveland P. Hickman, Jr.
Susan Keen
David J. Eisenhour
Allan Larson
Helen I'Anson

McGraw-Hill Connect®
Learn Without Limits

Connect is a teaching and learning platform that is proven to deliver better results for students and instructors.

Connect empowers students by continually adapting to deliver precisely what they need, when they need it, and how they need it, so your class time is more engaging and effective.

> 73% of instructors who use **Connect** require it; instructor satisfaction **increases** by 28% when **Connect** is required.

Connect's Impact on Retention Rates, Pass Rates, and Average Exam Scores

| | 70.1% | 89.9% | 72.5% | 85.2% | 71.0% | 80.1% |

Retention Rates   Course Pass Rates   Average Exam Scores

without Connect   with Connect

> Using **Connect** improves passing rates by **12.7%** and retention by **19.8%**.

Analytics

Connect Insight®

Connect Insight is Connect's new one-of-a-kind visual analytics dashboard—now available for both instructors and students—that provides at-a-glance information regarding student performance, which is immediately actionable. By presenting assignment, assessment, and topical performance results together with a time metric that is easily visible for aggregate or individual results, Connect Insight gives the user the ability to take a just-in-time approach to teaching and learning, which was never before available. Connect Insight presents data that empowers students and helps instructors improve class performance in a way that is efficient and effective.

Impact on Final Course Grade Distribution

without Connect		with Connect
22.9%	A	31.0%
27.4%	B	34.3%
22.9%	C	18.7%
11.5%	D	6.1%
15.4%	F	9.9%

> Students can view their results for any **Connect** course.

Mobile

Connect's new, intuitive mobile interface gives students and instructors flexible and convenient, anytime–anywhere access to all components of the Connect platform.

Adaptive

©Getty Images/iStockphoto

THE **ADAPTIVE** **READING EXPERIENCE** DESIGNED TO TRANSFORM THE WAY STUDENTS READ

More students earn **A's** and **B's** when they use McGraw-Hill Education **Adaptive** products.

SmartBook®

Proven to help students improve grades and study more efficiently, SmartBook contains the same content within the print book, but actively tailors that content to the needs of the individual. SmartBook's adaptive technology provides precise, personalized instruction on what the student should do next, guiding the student to master and remember key concepts, targeting gaps in knowledge and offering customized feedback, and driving the student toward comprehension and retention of the subject matter. Available on smartphones and tablets, SmartBook puts learning at the student's fingertips—anywhere, anytime.

Over **5.7 billion questions** have been answered, making McGraw-Hill Education products more intelligent, reliable, and precise.

STUDENTS WANT

SMARTBOOK®

95% of students reported **SmartBook** to be a more effective way of reading material

100% of students want to use the Practice Quiz feature available within **SmartBook** to help them study

100% of students reported having reliable access to off-campus wifi

90% of students say they would purchase **SmartBook** over print alone

95% reported that **SmartBook** would impact their study skills in a positive way

Mc Graw Hill Education

*Findings based on a 2015 focus group survey at Pellissippi State Community College administered by McGraw-Hill Education

www.mheducation.com

Life: Biological Principles and the Science of Zoology

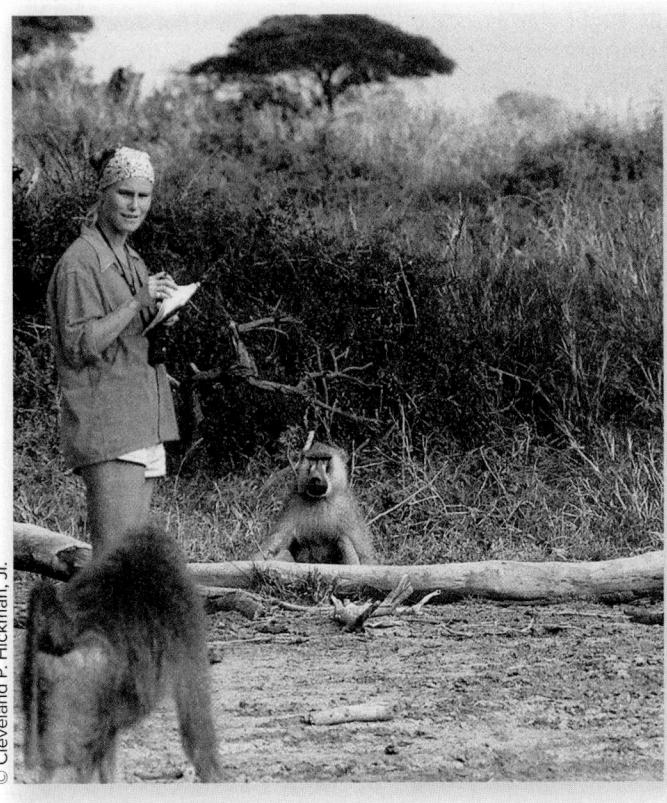

Zoologist studying the behavior of yellow baboons (Papio cynocephalus) *in the Amboseli Reserve, Kenya.*

The Uses of Principles

We explore the animal world by actively applying important guiding principles to our investigations. Just as the exploration of outer space is both guided and limited by available technologies, exploration of the animal world depends critically on our questions, methods, and principles. Zoology makes sense to us only when we understand the principles used to construct this knowledge.

The principles of modern zoology trace their long history to many sources. Some principles come from laws of physics and chemistry, which all living systems obey. Others come from the scientific method, which tells us that our hypothetical explanations of the animal world must guide us to gather data that potentially can refute these explanations.

Many important principles come from previous studies of the living world, of which animals are one part.

Principles of heredity, variation, and organic evolution guide the study of life from the simplest unicellular forms to the most complex animals, fungi, and plants. Because life shares a common evolutionary origin, principles learned from the study of one group often provide insights into other groups as well. By tracing the origins of our operating principles, we see that zoologists are not an island unto themselves but part of a larger scientific community.

We begin our study of zoology by searching broadly for our most basic principles and their diverse sources. These principles simultaneously guide our studies of animals and integrate those studies into the broader context of human knowledge.

Zoology, the scientific study of animal life, builds on centuries of human observations of the animal world. Mythologies of nearly every human culture reveal early attempts to solve the mysteries of animal life and its origin. Zoologists now confront these same mysteries with the most advanced methods and technologies developed by all branches of science. We document the diversity of animal life and organize it in a systematic way. This complex and exciting process builds on the contributions of thousands of zoologists working in all dimensions of the biosphere (Figure 1.1). We strive to explain how animal diversity originated and how animals perform the basic processes of life that permit them to inhabit diverse environments.

This chapter introduces the fundamental properties of animal life, the methodological principles that govern their study, and two important theories that guide our research: (1) the theory of evolution, which is the central organizing principle of biology, and (2) the chromosomal theory of inheritance, which explains heredity and variation in animals. These theories unify our knowledge of the animal world.

1.1 FUNDAMENTAL PROPERTIES OF LIFE

Historical Continuity of Life

We begin by asking, *What is life?* Our definition lies in the historical continuity of life on earth. Life's history of common descent with modification gives it an identity separate from the nonliving world. We trace this common history backward through time from the diverse forms observed today and in the fossil record to a common ancestor that must have arisen almost 4 billion years ago (see Chapter 2). There are no traces in the fossil record or on earth's surface of what we postulate to have been the incipient stages of life, those that predate cells. Replicating molecular systems, which could not have produced fossils, must have preceded and given rise to cellular life, whose history appears in the fossil record. All descendants of life's common ancestor, past and present, lie within our concept of life.

Life's most fundamental attribute is its reproduction of individuals with heredity and variation. Replication of large molecules that

A

C

B

Figure 1.1 Examples of observation in zoological research. **A,** Observing a coral reef. **B,** Observing nematocyst discharge from cnidarian tentacles (**C,** see p. 260)

store information is unique to life and must trace to life's origin. These properties establish a temporal continuity of ancestral and descendant populations showing extensive and ongoing change, which we call **evolution**. Through its evolution, life has generated many spectacular features that have no counterparts in the nonliving world. Novel properties emerge at all levels of life's hierarchical systems, from molecules and cells to organismal form and behavior.

We must resist giving life a definition based in essential characteristics that must occur in all living forms past and present. Such a definition would be particularly difficult for our theories of the origin of life from nonliving matter. Nonetheless, all living cells share metabolic processes and genetic information that reveal unmistakably their hereditary descent from life's common ancestor.

General Properties of Living Systems

Life's most outstanding general features include chemical uniqueness; complexity and hierarchical organization; reproduction (heredity and variation); possession of a genetic program; metabolism; development; environmental interaction; and movement.

1. **Chemical uniqueness.** *Living systems demonstrate a unique and complex molecular organization.* Living systems assemble large molecules, called macromolecules, that greatly exceed in complexity the small molecules of nonliving matter. Macromolecules contain the same kinds of atoms and chemical bonds that occur in nonliving matter and obey all fundamental laws of chemistry; it is only the complex organizational structure of these macromolecules that makes them unique to life. We recognize four major categories of biological macromolecules: nucleic acids, proteins, carbohydrates, and lipids (see Chapter 2). These categories differ in the structures of their component parts, the kinds of chemical bonds that link their subunits together, and their roles in living systems.

    The general structures of these macromolecules evolved and stabilized early in the history of life. With some modifications, these same general structures occur in every form of life today.

Proteins, for example, are built from 20 specific kinds of amino acid subunits linked together by peptide bonds in a linear sequence (Figure 1.2). Additional bonds occurring between amino acids that are not adjacent to each other in the protein chain give the protein a complex, three-dimensional structure (see Figures 1.2 and 2.15). A typical protein contains several hundred amino acid subunits. Despite the stability of this basic protein structure, the ordering of the different amino acids in a protein molecule shows enormous variation. This variation underlies much of the diversity that we observe among different kinds of living organisms. The nucleic acids, carbohydrates, and lipids likewise contain characteristic bonds that link variable subunits (see Chapter 2). This organization gives living systems a common biochemical theme with great potential diversity.

2. **Complexity and hierarchical organization.** *Living systems demonstrate a unique and complex hierarchical organization.* Nonliving matter is organized at least into atoms and molecules and often has a higher degree of organization as well. However, atoms and molecules are combined into patterns in the living world that do not exist in nonliving matter. In living systems, we find a hierarchy of levels that includes, in ascending order of complexity, macromolecules, cells, organisms, populations, and species (Figure 1.3). Each level builds on the level below it and has its own internal structure, which is also often hierarchical. Within a cell, for example, macromolecules are assembled into structures such as ribosomes, chromosomes, and membranes, and these are likewise combined in various ways to form even more complex subcellular structures called organelles, such as mitochondria (see Chapters 3 and Chapter 4). The organismal level also has a hierarchical substructure; cells combine to form tissues, which combine to form organs, which likewise combine to form organ systems (see Chapter 9).

    Cells (Figure 1.4) are the smallest units of the biological hierarchy that are semiautonomous in their ability to conduct basic functions, including reproduction. Replication of molecules and subcellular components occurs only within a cellular

© Laguna Design/Science Source

A   **B**

Amino acid   Amino acid   Amino acid

Figure 1.2   A computer simulation of the three-dimensional structure of the human endothelin-1 protein **(A),** which can constrict blood vessels to increase blood pressure. The protein is a linear string of molecular subunits called amino acids, connected as shown in **B,** which fold in a three dimensional pattern to form the active protein. The white balls correspond to carbon atoms, the red balls to oxygen, the blue balls to nitrogen, the yellow balls to sulfur, the green balls to hydrogen, and the black balls **(B)** to molecular groups formed by various combinations of carbon, oxygen, nitrogen, hydrogen, and sulfur atoms that differ among amino acids. Hydrogen atoms are not shown in **A.**

context, not independently. Cells are therefore considered the basic units of living systems (see Chapter 3). We can isolate cells from an organism and cause them to grow and to multiply under laboratory conditions in the presence of nutrients alone. This semiautonomous replication cannot occur for any individual molecules or subcellular components, which require additional cellular constituents for their reproduction.

Each successively higher level of the biological hierarchy is composed of units of the preceding lower level in the hierarchy. An important consequence of this hierarchy is that we cannot infer the properties at any given level even from the most complete knowledge of the properties of its component parts. A physiological feature, such as blood pressure, is a property of the organismal level; it is impossible to predict someone's blood pressure simply by knowing the physical characteristics of individual cells of the body. Likewise, systems of social interaction, as seen in bees, appear at the populational level; one cannot infer properties of this social system by studying individual bees in isolation.

Appearance of new characteristics at a given level of organization is called **emergence,** and these characteristics are called **emergent properties.** These properties arise from interactions among the component parts of a system. For this reason, we must study all levels directly, each one being the focus of a different subfield of biology (molecular biology; cell biology; organismal anatomy, physiology, and genetics; population biology; Table 1.1). Emergent properties expressed at a particular level of the biological hierarchy are certainly influenced and restricted by properties of the lower-level components. For example, a population of organisms that lack hearing could not

Figure 1.4 Electron micrograph of ciliated epithelial cells and mucus-secreting cells (see pp. 189–191) lining the interior of a rat oviduct. Cells are the basic building blocks of living organisms.

develop a spoken language. Nonetheless, properties of parts of a living system do not rigidly determine properties of the whole. Many different spoken languages have emerged in human culture from the same basic anatomical structures that permit hearing and speech. The freedom of the parts to interact in different ways makes possible a great diversity of potential emergent properties at each level of the biological hierarchy.

Different levels of the biological hierarchy and their particular emergent properties are built by evolution. Before multicellular organisms evolved, there was no distinction between the organismal and cellular levels, and this distinction remains absent from single-celled organisms (see Chapter 11). The diversity of emergent properties that we see at all levels of the biological hierarchy contributes to the difficulty of giving life a simple definition or description.

3. **Reproduction.** *Living systems can reproduce themselves.* Life does not arise spontaneously but comes only from prior life, through reproduction. Although life certainly originated from nonliving matter at least once (see Chapter 2), this origin featured enormously long periods of time and conditions very different from the current biosphere. At each level of the biological hierarchy, living forms reproduce to generate others like themselves (Figure 1.5). Genes replicate to produce new genes. Cells divide to produce new cells. Organisms reproduce, sexually or asexually, to produce new organisms (see Chapter 7). Populations reproduce themselves through time to form lineages of ancestral-descendant populations. Should a geographic barrier split a population into spatially isolated parts, multiple population lineages can emerge from a common ancestral one. Evolutionary divergence of character among separated population lineages can produce a multiplication of species, in a process called speciation. Reproduction at any hierarchical level usually features an increase in numbers. Individual genes, cells, organisms, populations, or species may fail to reproduce themselves, but reproduction is nonetheless an expected property of these individuals.

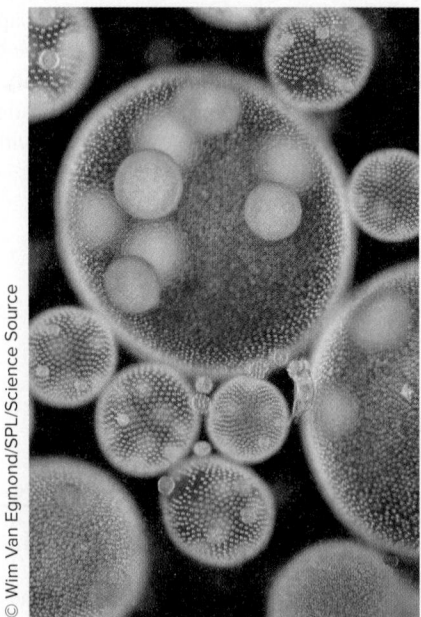

Figure 1.3 *Volvox globator* (see pp. 239–240) is a multicellular flagellate that illustrates three different levels of the biological hierarchy: cellular, organismal, and populational. Each individual spheroid (organism) contains cells embedded in a gelatinous matrix. The larger cells function in reproduction, and the smaller ones perform the general metabolic functions of the organism. The individual spheroids together form a population.

TABLE 1.1

Different Hierarchical Levels of Biological Complexity That Display Reproduction, Variation, and Heredity

Level	Timescale of Reproduction	Fields of Study	Methods of Study	Some Emergent Properties
Cell	Hours (mammalian cell = ~16 hours)	Cell biology, molecular biology	Microscopy (light, electron), biochemistry	Chromosomal replication (meiosis, mitosis), synthesis of macromolecules (DNA, RNA, proteins, lipids, polysaccharides)
Organism	Hours to days (unicellular); days to years (multicellular)	Organismal anatomy, physiology, genetics	Dissection, genetic crosses, clinical studies, physiological experimentation	Structure, functions and coordination of tissues, organs and organ systems (blood pressure, body temperature, sensory perception, feeding)
Population	Up to thousands of years	Population biology, population genetics, ecology	Statistical analysis of variation, abundance, geographical distribution	Social structures, systems of mating, age distribution of organisms, levels of variation, action of natural selection
Species	Thousands to millions of years	Systematics and evolutionary biology, community ecology	Study of reproductive barriers, phylogeny, paleontology, ecological interactions	Method of reproduction, reproductive Barriers

Reproduction at each of these levels shows the complementary, and yet apparently contradictory, phenomena of **heredity** and **variation.** Heredity is the faithful transmission of traits from parents to offspring, usually (but not necessarily) observed at the organismal level. Variation is the production of *differences* among the traits of different individuals. In a reproductive process, properties of descendants resemble those of their parents to varying degrees but usually are not identical to them. Replication of deoxyribonucleic acid (DNA) occurs with high fidelity, but errors occur at repeatable rates. Cell division is exceptionally precise, especially with regard to nuclear material, but chromosomal changes occur nonetheless at measurable rates. Organismal reproduction likewise demonstrates both heredity and variation, the latter most obvious in sexually reproducing forms. Production of new populations and species also demonstrates conservation of some properties and changes of others. Two closely related frog species may have similar mating calls but differ in the rhythms of repeated sounds.

Interaction of heredity and variation in the reproductive process makes organic evolution possible and inevitable (see Chapter 6). If heredity were perfect, living systems would never change; if variation were uncontrolled by heredity, biological systems would lack the stability that allows them to persist through time.

4. **Possession of a genetic program.** *A genetic program provides fidelity of inheritance* (Figure 1.6). **Nucleic acids** encode structures of the protein molecules needed for organismal develop-

ment and functioning (see Chapter 5). For animals and most other organisms, **DNA** stores genetic information. DNA is a very long, linear chain of subunits called nucleotides, each of which contains a sugar phosphate (deoxyribose phosphate) and one of four nitrogenous bases (adenine, cytosine, guanine, or thymine, abbreviated A, C, G, and T, respectively). The sequence of nucleotide bases contains a code for the order of amino acids in the protein specified by the DNA molecule. The correspondence between the sequence of bases in DNA and the sequence of amino acids in a protein is the **genetic code**.

The genetic code arose early in the evolutionary history of life, and the same code occurs in bacteria and in the nuclear genomes of almost all animals and plants. The near constancy of this code among living forms provides strong evidence for a single origin of life. The genetic code has undergone very little evolutionary change since its origin because an alteration would disrupt the structure of nearly every protein, which would in turn severely disrupt cellular functions that require very specific protein structures. Only in the rare instance that the altered protein structures maintained their cellular functions would such a change possibly survive and be reproduced. Evolutionary change in the genetic code has occurred in the DNA contained in animal mitochondria, the organelles that regulate cellular energy. The genetic code in animal mitochondrial DNA therefore is slightly different from the standard code of nuclear and bacterial DNA. Because mitochondrial DNA specifies far fewer proteins than does nuclear DNA, the

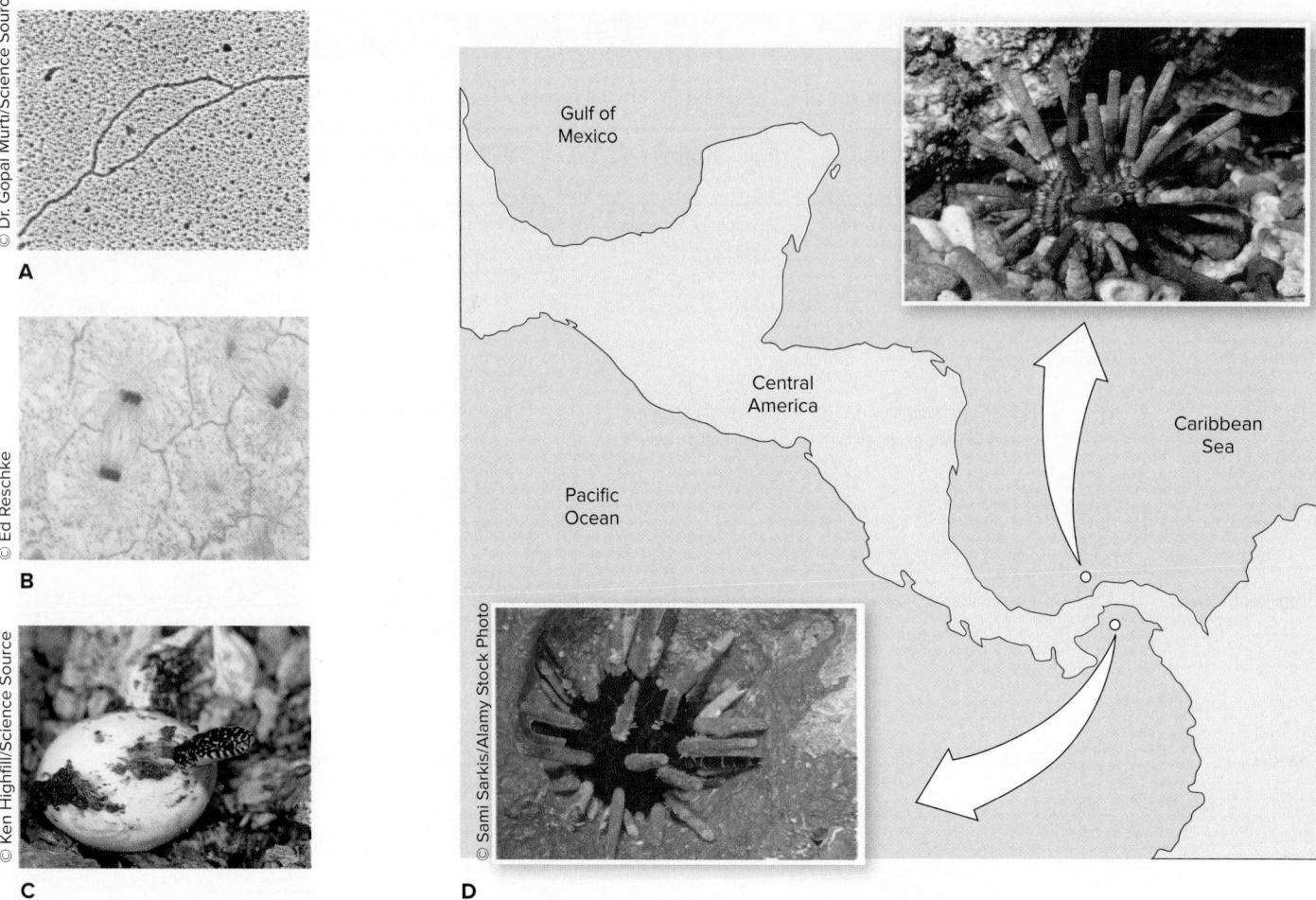

Figure 1.5 Reproductive processes observed at four different levels of biological complexity. **A,** Molecular level—electron micrograph of a replicating DNA molecule. **B,** Cellular level—micrograph of cell division at mitotic telophase. **C,** Organismal level—a king snake hatching. **D,** Species level—formation of new species in sea urchins (*Eucidaris*) after geographic separation of Caribbean (*E. tribuloides*) and Pacific (*E. thouarsi*) populations by a land bridge that formed approximately 3.5 million years ago.

likelihood of getting a change in the code that maintains cellular functions is greater there than in the nucleus.

5. **Metabolism.** *Living organisms maintain themselves by acquiring nutrients from their environments* (Figure 1.7). Nutrients supply the chemical energy and molecular components for building and maintaining a living system (see Chapter 4). We call these essential chemical processes **metabolism.** They include digestion, acquisition of energy (respiration), and synthesis of molecules and structures. Metabolism is an interaction of destructive (catabolic) and constructive (anabolic) reactions. The most fundamental anabolic and catabolic chemical processes used by living systems arose early in the evolutionary history of life, and all living forms share them. These reactions include synthesis of carbohydrates, lipids, nucleic acids, and proteins and their constituent parts and cleavage of chemical bonds to recover energy stored in them. In animals, many fundamental metabolic reactions occur at the cellular level, often in specific organelles present throughout the animal kingdom. Cellular respiration occurs, for example, in mitochondria. Cellular and nuclear membranes regulate metabolism by controlling the

movement of molecules across the cellular and nuclear boundaries, respectively. The study of metabolic functions from the biochemical to the organismal levels is called **physiology.** We devote a large portion of this book to describing and comparing the diverse tissues, organs, and organ systems that different groups of animals have evolved to perform basic physiological functions of life (see Chapters 11 through Chapter 36).

6. **Development.** *All organisms pass through a characteristic life cycle.* Development describes the characteristic changes that an organism undergoes from its origin (usually the fertilization of an egg by sperm) to its final adult form (see Chapter 8). Development usually features changes in size and shape, and differentiation of structures within an organism. Even the simplest one-celled organisms grow in size and replicate their component parts until they divide into two or more cells. Multicellular organisms undergo more dramatic changes during their lives. Different developmental stages of some multicellular forms are so dissimilar that they are hardly recognizable as belonging to the same species. Embryos are distinctly different from juvenile and adult forms into which they develop. Even

Figure 1.7   Feeding processes illustrated by **(A)** an ameba surrounding food and **(B)** a chameleon capturing insect prey with its projectile tongue.

Figure 1.6   James Watson and Francis Crick with a model of the DNA double helix **(A)**. The nucleotide base sequence inside the DNA molecule encodes genetic information. Genetic variation is shown **(B)** in DNA molecules that are similar in base sequence but differ from each other at four positions. Such differences can specify alternative traits, such as different eye colors.

postembryonic development of some organisms includes stages dramatically different from each other. The transformation that occurs from one stage to another is called **metamorphosis.** There is little resemblance, for example, among the egg, larval, pupal, and adult stages of metamorphic insects (Figure 1.8). Early stages of development are often more similar among organisms of different species than are later developmental stages. In our survey of animal diversity, we describe all stages of observed life histories but concentrate on adult stages, in which diversity tends to be most obvious.

7. **Environmental interaction.** *All animals interact with their environments.* The study of organismal interaction with an environment is called **ecology.** Of special interest are the factors that influence geographic distribution and abundance of animals (see Chapters 37 and 38). The science of ecology reveals how an organism perceives environmental stimuli and responds

© McGraw-Hill Education/Ken Cavanagh Photographer

© McGraw-Hill Education/Ken Cavanagh Photographer

A   B

Figure 1.8 **A,** Adult monarch butterfly emerging from its pupal case. **B,** Fully formed adult monarch butterfly.

in appropriate ways by adjusting its metabolism and physiology (Figure 1.9). All organisms respond to environmental stimuli, a property called **irritability.** The stimulus and response may be simple, such as a unicellular organism moving from or toward a light source or away from a noxious substance, or it may be quite complex, such as a bird responding to a complicated series of signals in a mating ritual (see Chapter 36). Life and environment are inseparable. We cannot isolate the evolutionary history of a lineage of populations from the environments in which it occurred.

8. **Movement.** *Living systems and their parts show precise and controlled movements arising from within the system.* The energy that living systems extract from their environments permits them to initiate controlled movements. Such movements at the cellular level are essential for reproduction, growth, and many responses to stimuli in all living forms and for development in multicellular ones. Semiautonomous movement occurs even in some biological macromolecules. An enzymatic protein undergoes characteristic and reversible changes in shape as it binds a substrate, catalyzes a reaction, and releases a product. These characteristic molecular movements occur even when the enzyme is removed from its cellular context and used as a reagent to catalyze reactions in a laboratory. Autonomous movement reaches great diversity in animals, and much of this book comprises descriptions of animal movement and the many adaptations that animals have evolved for locomotion. On a larger scale, entire populations or species may disperse from one geographic location to another one over time through their powers of movement. Movement characteristic of nonliving matter, such as that of particles in solution, radioactive decay of nuclei, and eruption of volcanoes is not precisely controlled by the moving objects themselves and often involves forces entirely external to them. The adaptive and often purposeful movements initiated by living systems are absent from the nonliving world.

Life Obeys Physical Laws

To untrained observers, these eight properties of life might appear to violate basic laws of physics. Vitalism, the idea that life has a mystical vital force that violates physical and chemical laws, was once widely advocated. Biological research has consistently rejected vitalism, showing instead that all living systems obey basic laws of physics and chemistry. Laws governing energy and its transformations (thermodynamics) are particularly important for understanding life

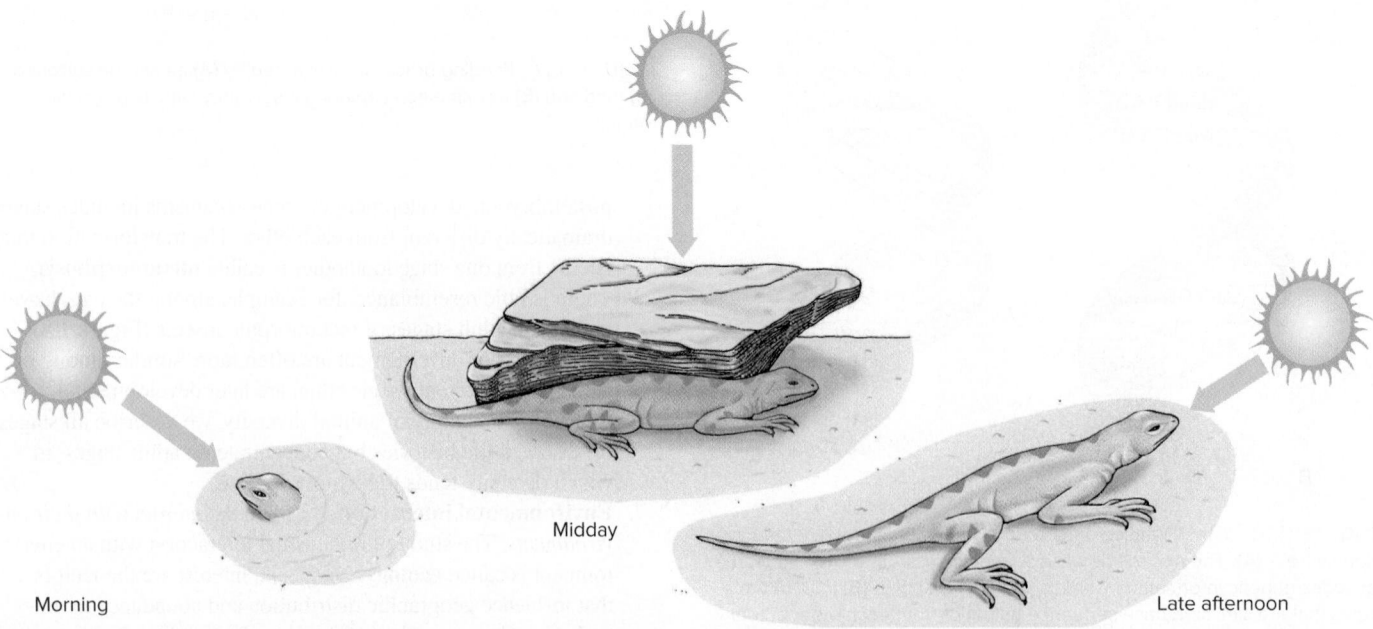

Morning   Midday   Late afternoon

Figure 1.9 A lizard regulates its body temperature by choosing different locations (microhabitats) at different times of day.

(see Chapter 4). The **first law of thermodynamics** is the law of conservation of energy. Energy is neither created nor destroyed but can be transformed from one form to another. All aspects of life require energy and its transformation. The energy to support life on earth flows from the fusion reactions in our sun and reaches the earth as light and heat. Photosynthesis in green plants and cyanobacteria transforms energy captured as sunlight into chemical bonds. Energy in chemical bonds is a form of potential energy released when the bond is broken; the energy is used to perform numerous cellular tasks. Energy transformed and stored in plants is then used by animals that eat the plants, and these animals may in turn provide energy for predators.

The **second law of thermodynamics** states that physical systems tend to proceed toward a state of greater disorder, or **entropy.** Energy obtained and stored by plants is subsequently released by various mechanisms and finally dissipated as heat. Living cells maintain complex molecular organization only as long as energy fuels the organization. The ultimate fate of materials in the cells is degradation and dissipation of their chemical-bond energy as heat. An evolutionary increase over time in organismal complexity may appear at first to violate the second law of thermodynamics, but it does not. Organismal complexity is achieved and maintained only by the perpetual use and dissipation of energy flowing into the biosphere from the sun. Survival, growth, and reproduction of animals require energy that comes from breaking complex food molecules into simple organic waste. The processes by which animals acquire energy through nutrition and respiration reveal themselves to us through the many physiological sciences.

1.2 ZOOLOGY AS A PART OF BIOLOGY

Animals form a distinct branch on the evolutionary tree of life. It is a large and old branch that originated in the Precambrian seas over 600 million years ago. Animals form part of an even larger limb called **eukaryotes,** organisms whose cells contain membrane-enclosed nuclei. This larger limb includes plants, fungi, and numerous unicellular forms. Perhaps the most distinctive characteristic of animals as a group is their means of nutrition, which consists in eating other organisms. Animal evolution has elaborated this basic way of life through diverse systems for capturing and processing a wide array of food items and for locomotion.

We distinguish animals also by the absence of characteristics that have evolved in other eukaryotes but not in animals. Plants, for example, use light energy to produce organic compounds (photosynthesis), and they have evolved rigid cell walls that surround their cell membranes; photosynthesis and cell walls do not occur in animals. Fungi acquire nutrition by absorption of small organic molecules from their environments, and their body plan contains tubular filaments called *hyphae;* these structures do not occur in the animal kingdom.

Some organisms that are neither animals nor plants combine properties of animals and plants. For example, *Euglena* (Figure 1.10) is a motile, single-celled organism that resembles plants in being photosynthetic, but resembles animals in its ability to eat food particles. *Euglena* is part of a separate eukaryotic lineage that diverged from those of plants and animals early in the evolutionary history of eukaryotes. *Euglena* and other unicellular eukaryotes formerly were grouped as the kingdom Protista, although this kingdom is an arbi-

Chloroplasts (photosynthesis)

Flagellum (motility)

Figure 1.10 Some organisms that are neither animals nor plants, such as single-celled *Euglena* (shown here) and *Volvox* (see Figure 1.3), combine properties that distinguish animals (locomotion) from plants (photosynthetic ability).

trary grouping of taxa that are not each other's closest relatives and thus violates taxonomic principles (see Chapter 10).

We summarize in Chapters 8 and 9 the fundamental structural and developmental features evolved by the animal kingdom.

1.3 PRINCIPLES OF SCIENCE

Nature of Science

A basic understanding of zoology requires understanding what science is, what it is not, and how we gain knowledge using the scientific method.

Science is a way of asking questions about the natural world and sometimes obtaining precise answers to them. Although science, in the modern sense, arose within the last 200 years or so, a tradition of asking questions about the natural world is ancient. In this section, we examine the methodology that zoology shares with science as a whole. These procedures for constructing data-based explanations of natural phenomena distinguish sciences from activities that we exclude from the realm of science, such as art and religion.

Despite an enormous impact of science on our lives, many people have only a minimal understanding of science. For example, on March 19, 1981, the governor of Arkansas signed into law the Balanced Treatment for Creation-Science and Evolution-Science Act (Act 590 of 1981). This act falsely presented "creation-science" as a valid scientific endeavor. "Creation-science" is instead a religious position advocated by a minority of the American religious community, and it does not qualify as science. Enactment of this law led to a historic lawsuit tried in December 1981 in the court of Judge William R. Overton, U.S. District Court, Eastern District of Arkansas. The American Civil Liberties Union brought this suit on behalf of 23 plaintiffs, including religious leaders and groups representing several denominations, individual parents, and educational associations. The plaintiffs contended that the law violated the First Amendment to the U.S. Constitution, which prohibits "establishment of religion" by government. This prohibition precludes passing a law that would aid one religion or prefer one religion over another. On January 5, 1982, Judge Overton permanently stopped the State of Arkansas from enforcing Act 590.

Considerable testimony during the trial described the process of science. Some witnesses defined science simply, if not very informatively, as "what is accepted by the scientific community" and "what scientists do." However, on the basis of other testimony by scientists, Judge Overton stated explicitly these essential characteristics of science:

1. It is guided by natural law.
2. It has to be explanatory by reference to natural law.
3. It is testable against the observable world.
4. Its conclusions are tentative and therefore not necessarily the final word.
5. It is falsifiable.

Pursuit of scientific knowledge must be guided by the physical and chemical laws that govern the state of existence. Scientific knowledge must explain what is observed by reference to natural law without requiring intervention of a supernatural being or force. We must be able to observe events in the real world, directly or indirectly, to test hypotheses about nature. If we draw a conclusion relative to some event, we must be ready always to discard or to modify our conclusion if further observations contradict it. As Judge Overton stated, "While anybody is free to approach a scientific inquiry in any fashion they choose, they cannot properly describe the methodology used as scientific if they start with a conclusion and refuse to change it regardless of the evidence developed during the course of the investigation." Science lies outside religion, and the results of science do not favor one religious position over another.

Unfortunately, the religious position formerly called "creation-science" has reappeared in American politics with the name "intelligent-design theory." We are forced once again to defend the teaching of science against this scientifically meaningless dogma. On December 20, 2005, Judge John E. Jones III of the U.S. District Court for the Middle District of Pennsylvania ruled unconstitutional the teaching of intelligent design, which had been mandated by the Dover school board. The local voters already had rejected the eight board members who supported the intelligent-design requirement, replacing them with candidates who actively opposed teaching intelligent design as science.

Scientific Method

These essential criteria of science form the **hypothetico-deductive method.** This method requires us to generate hypotheses or potential answers to a question being asked. These hypotheses are usually based on prior observations of nature or derived from theories based on such observations. Scientific hypotheses often constitute general statements about nature that may explain a large number of diverse observations. Darwin's hypothesis of natural selection, for example, explains the observations that many different species have properties that adapt them to their environments. From the hypothesis, a scientist must predict future observations. The scientist must say, "If my hypothesis is a valid explanation of past observations, then future observations ought to have certain characteristics." The best hypotheses are those that make many predictions that, if found erroneous, cause rejection, or falsification, of the hypothesis.

The scientific method is summarized as a series of steps:

1. Observation
2. Question
3. Hypothesis
4. Empirical test
5. Conclusions
6. Publication

Observations are a critical first step in evaluating the life histories of natural populations. For example, observations of moth populations in industrial areas of England for more than a century have revealed that moths in polluted areas have tended primarily to have darkly colored wings and body, whereas moths of the same species in unpolluted areas are more lightly colored. This observation pertains to multiple moth species, but we focus here on the species *Biston betularia.*

Our question is "Why do pigmentation patterns vary according to habitat?" With no prior knowledge of the biology of these moth populations, one might hypothesize that coloration is influenced somehow by a direct action of the environment. Does consumption of soot by caterpillars somehow darken pigmentation of the adult moth? One could test this hypothesis by rearing moths under artificial conditions. If darkly pigmented moths and lightly pigmented moths are allowed to reproduce in unpolluted conditions, our hypothesis predicts that offspring of both will be lightly pigmented; by contrast, offspring of both groups would be darkly pigmented if raised in polluted conditions.

To test our hypothesis, we construct a null hypothesis. A null hypothesis is one that permits a statistical test of our data to reject its predictions if the hypothesis is false. We can choose as our null hypothesis the prediction that moths reared in unpolluted conditions should be lightly pigmented regardless of whether their parents were from light or dark populations, and that offspring from both populations reared in polluted conditions should be dark. This experiment is a special case of a "common garden" experiment as used in agriculture. Do contrasting populations from different habitats retain their contrasting characteristics when reared in a common garden?

For *Biston betularia,* a common garden experiment reveals that the contrasting wing colors of populations from polluted and unpolluted environments are maintained in the common garden. Offspring of moths from polluted populations retain the dark pigmentation of their parents, whereas offspring of lightly pigmented moths are lightly colored like their parents. We thereby reject the hypothesis that the color contrasts represent a direct action of environmental conditions.

We have gained important knowledge by rejecting our initial hypothesis, and we now test an alternative hypothesis, that pigmentation is a genetic trait in *Biston betularia.* Using the genetic methodology of Gregor Mendel (p. 72), we cross the darkly and lightly colored populations and trace the inheritance of pigmentation in subsequent generations. Experimental results reveal that the offspring produced by crossing light and dark populations have dark pigmentation, and that the second-generation progeny include both dark and light moths in the 3:1 ratio predicted by the null hypothesis for a single-gene Mendelian trait in which dark color is genetically dominant.

We still have not answered our initial question, why pigmentation differs between populations in polluted versus unpolluted environments. We have learned, however, that the critical question is why different forms of a single gene have contrasting frequencies in these two areas. We know that the moth populations have inhabited England since well before the introduction of industrial pollution. The lightly pigmented populations most likely resemble the ancestral condition, so why have darkly pigmented moths accumulated in the polluted environments? The simplest hypothesis is that darkly pigmented individuals are more likely to survive and to reproduce in polluted environments.

Further observations reveal that *Biston betularia* is typical of moths in being active at night and inactive during the day, resting on the bark of trees. Contrasting photographs of light and dark moths resting on unpolluted, lichen-covered tree bark versus sooty tree bark lead us to a hypothesis that might explain why dark moths predominate in polluted areas. Figure 1.11 shows that the lightly colored moth is camouflaged against the unpolluted substrate whereas the dark moth is highly visible; by contrast, the dark moth is camouflaged against the sooty bark whereas the light moth is highly visible. Camouflage suggests that a predator using vision to find its prey preferentially kills moths that contrast with the background color of their diurnal resting place. How can we test this hypothesis?

Many birds are diurnal predators guided to their prey by visual cues. Many experiments have revealed that birds will attack clay models that closely resemble their favored prey items. We can test our hypothesis by constructing clay models of light and dark moths. We place equal numbers of the light and dark models against the bark of unpolluted trees and equal numbers of light and dark models against sooty tree bark. When a bird attacks a clay model, it typically leaves an imprint of its beak in the clay. Because beak shape varies among bird species, the beak shape marked in the clay often reveals which species attacked the model. Our null hypothesis is that equal numbers of dark and light models have beak impressions on both the unpolluted and the polluted substrates. We reject this hypothesis if we find a large excess of beak marks in the uncamouflaged models relative to the camouflaged ones; dark models should be attacked preferentially in unpolluted conditions and light models attacked preferentially in polluted conditions. Note that in this case, we use a null hypothesis that is the *opposite* of our favored explanation, that birds preferentially destroy uncamouflaged moths. In this case, data that reject the null hypothesis serve to verify our favored explanation.

Experiments of this kind have rejected the null hypothesis as expected, verifying our explanation that dark moths prevail in polluted environments because the dark color protects them from predation by birds during the day. Note that our experiments have led us to a strong, specific explanation for the initial observations. It is a strong working hypothesis, but our experiments have not proven the correctness of this hypothesis. We can test it further in various ways. For example, we might raise light and dark moths in equal numbers in an outdoor enclosure that excludes birds; our null hypothesis is then that the dark and light forms should persist in equal numbers regardless of whether the tree bark is polluted or unpolluted. Rejection of this null hypothesis would tell us that our favored explanation was not the full answer to our original question.

We publish our results and conclusions to guide other researchers further to test our hypotheses. Over the past century, many research papers have reported results and conclusions to explain "industrial melanism" in moths. With some ambiguities, the favored explanation is that differential bird predation on uncamouflaged moths best explains industrial melanism. These studies have drawn much attention because this explanation illustrates a mechanism of Darwin's theory of natural selection (p. 14).

Note that a hypothesis cannot be proved correct using the scientific method. If the available data are compatible with it, the hypothesis serves as a guide for collecting additional data that potentially might reject it. Our most successful hypotheses are the ones that make specific predictions confirmed by large numbers of empirical tests.

If a hypothesis is very powerful in explaining a wide variety of related phenomena, it attains the status of a **theory.** Natural selection is a good example. Our example of the use of natural selection to explain observed pigmentation patterns in moth populations is only one of many phenomena to which natural selection applies. Natural selection provides a potential explanation for the occurrence of many different traits distributed among virtually all animal species. Each of these instances constitutes a specific hypothesis generated from the theory of natural selection. Note, however, that falsification of a specific hypothesis does not necessarily lead to rejection of the theory as a whole. Natural selection may fail to explain origins of a human behavior, for example, but it provides an excellent explanation for many structural modifications of the pentadactyl (five-fingered) vertebrate limb for diverse functions. Scientists test many subsidiary hypotheses of their major theories to ask whether their theories are generally applicable. Most useful are theories that explain the largest array of different natural phenomena.

We emphasize that "theory," when used by scientists, does not mean "speculation" as it does in ordinary English usage. Failure to

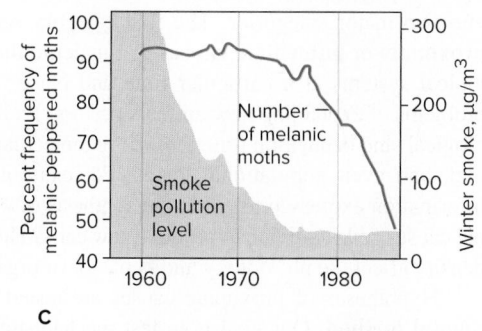

A   **B**   **C**

Figure 1.11 Light and melanic forms of the peppered moth, *Biston betularia,* on **A,** a lichen-covered tree in unpolluted countryside and **B,** a soot-covered tree near industrial Birmingham, England. These color variants have a simple genetic basis. **C,** Recent decline in the frequency of the melanic form of the peppered moth with falling air pollution in industrial areas of England. The frequency of the melanic form still exceeded 90% in 1960, when smoke and sulfur dioxide emissions were still high. Later, as emissions fell and light-colored lichens began to grow again on the tree trunks, the melanic form became more conspicuous to predators. By 1986, only 50% of the moths were still of the melanic form, the rest having been replaced by the light form.

make this distinction is prominent in creationist challenges to evolution. The creationists call evolution "only a theory," as if it were little better than a rough guess. In fact, the theory of evolution is supported by such massive evidence that biologists view repudiation of evolution as a repudiation of rational thinking. Nonetheless, evolution, along with all other theories in science, is not proven in a mathematical sense, but it is testable, tentative, and falsifiable.

Powerful theories that guide extensive research are called **paradigms.** The history of science shows that even major paradigms are subject to refutation and replacement when they fail to explain our observations of the natural world. They are then replaced by new paradigms in a process called a **scientific revolution.** For example, prior to the 1800s, animal species were studied as if they were historically unrelated and unchanging entities. Darwin's theories led to a scientific revolution that replaced these views with the evolutionary paradigm. The evolutionary paradigm has guided biological research for more than 150 years, and to date there is no scientific evidence that falsifies it; it has strong explanatory power and continues to guide active inquiry into the natural world. Evolutionary theory is generally accepted as the cornerstone of biology.

Chemists and physicists often use the term "law" to denote highly corroborated theories that appear to apply without exception to the physical world. Such laws are considered uniform throughout time and space. Because the biological world is temporally and spatially bounded, and because evolutionary change has produced an enormous diversity of forms with different emergent properties at multiple levels (Table 1.1), biologists now avoid using the term "law" for their theories. Nearly all of the biological laws proposed in the past have been found to apply only to some of life's diverse forms and not to all. Mendel's laws of inheritance, for example, do not apply to bacteria and often are violated even in animal and plant species that usually follow them. Darwin's theories of perpetual change and common descent of living forms (p. 13) are perhaps the only statements that one meaningfully might call laws of biology.

Experimental Versus Comparative Methods

The many hypotheses tested about the animal world can be grouped into two major categories. The first category seeks to explain the **proximate** or **immediate causes** that underlie the operation of biological systems at a particular time and place. These include the problems of explaining how animals perform their metabolic, physiological, and behavioral functions at the molecular, cellular, organismal, and even populational levels. For example, how is genetic information expressed to guide the synthesis of proteins? What signal causes cells to divide to produce new cells? How does population density affect the physiology and behavior of organisms?

Hypotheses of proximate causes are tested using the **experimental method.** Our goal is to test mechanistic explanations of a biological system. We predict the results of an experimental treatment of the system based on our tentative explanation of it. If our explanation is correct, then the predicted outcome should occur. If, after the experimental treatment, we see an unexpected outcome, we then discover that our explanation is incorrect or incomplete. Exper-

imental conditions are repeated to eliminate chance occurrences that might produce erroneous conclusions. **Controls**—repetitions of the experimental procedure that lack the treatment—eliminate unknown factors that might bias the outcome of the experiment. The processes by which animals maintain a body temperature under different environmental conditions, digest their food, migrate to new habitats, or store energy are some additional examples of physiological phenomena studied by experiment (see Chapters 29 through 36). Experimentation is important at all levels of biological complexity, including molecular biology, cell biology, endocrinology, developmental biology, animal behavior, and community ecology.

Our example in the preceding section of using clay models of moths to test avian predation on differently colored forms illustrates experimental testing of a hypothesis. By placing darkly colored models on both light and dark backgrounds, we see that birds attack the ones on light backgrounds much more frequently than they do dark models on dark backgrounds. Our interpretation that dark moths on dark backgrounds avoid predation by camouflage requires a control. Perhaps birds choose to feed only on light, unpolluted branches. Our control is to place light moths on both light and dark backgrounds. When we observe that birds preferentially attack the light models placed on dark backgrounds, we reject the hypothesis that birds choose not to feed on dark, polluted substrates. The simplest interpretation of the results as described here is that birds will eat both dark and light moths that fail to match their backgrounds, and that camouflage conceals potential prey items from avian predators.

In contrast to questions concerning the proximate causes of biological systems are questions of the **ultimate causes** that have produced these systems and their distinctive characteristics through evolutionary time. For example, what are the evolutionary factors that caused some birds to acquire complex patterns of seasonal migration between temperate and tropical areas? Why do different species of animals have different numbers of chromosomes in their cells? Why do some animal species maintain complex social systems, whereas other species have solitary individuals?

Tests of hypotheses of ultimate causality require the **comparative method.** We compare characteristics of molecular biology, cell biology, organismal structure, development, and ecology among related species to identify their patterns of variation. We then use patterns of similarity and dissimilarity to test hypotheses of relatedness and thereby to reconstruct the evolutionary tree that relates the species being studied. Advances in DNA sequencing technology permit precise tests of relationships among all animal species.

We then use the evolutionary tree to examine hypotheses of the evolutionary origins of the diverse molecular, cellular, organismal, and populational properties observed in the animal world. For example, comparative methodology rejects the hypothesis of a common origin for flight in bats and birds. Comparative morphology of vertebrates and comparisons of DNA sequences from living species clearly place bats within the mammals (Chapter 28) and birds within a separate group that also includes crocodilians, lizards, snakes, and turtles (see Figure 23.2). The most recent common ancestor of these vertebrates clearly could not fly, and close inspection reveals that bats and birds evolved flight via very different modifications of their bodies and forelimbs (pp. 591, 623). The ultimate causes of flight in bats and birds thus require separate explanations, not a shared one. The comparative method likewise reveals that homeothermy (p. 673) evolved in a lineage ancestral to birds and separately in a lineage ancestral to mammals. Furthermore, comparative studies of fossil birds reject the hypothesis that feathers arose for the purpose of flight, because feathers preceded evolution of the flight apparatus in avian ancestry. Feathers most likely served initially primarily for insulation and only later acquired a role in aerodynamics. It should be clear that none of these important historical questions could have been answered by experiment.

Clearly, the comparative method often relies on results of experimental sciences as a starting point. The comparative method likewise applies to all levels of biological complexity, as illustrated by the fields of molecular evolution, comparative cell biology, comparative anatomy, comparative physiology, and phylogenetic systematics.

Figure 1.12 Modern evolutionary theory is strongly identified with Charles Robert Darwin, who, with Alfred Russel Wallace, provided the first credible explanation of evolution. This photograph of Darwin was taken in 1854 when he was 45 years old. His most famous book, *On the Origin of Species*, appeared five years later.

Courtesy of the National Library of Medicine

> A scientist's use of the phrase "ultimate cause," unlike Aristotle's usage, does not imply a preconceived goal for natural phenomena. An argument that nature has a predetermined goal, such as evolution of the human mind, is termed teleological. **Teleology** is the mistaken notion that evolution of living organisms is guided by purpose toward an optimal design. A major success of Darwinian evolutionary theory is its rejection of teleology in explaining biological diversification.

1.4 THEORIES OF EVOLUTION AND HEREDITY

We turn now to the two major paradigms that guide zoological research today: Darwin's theory of evolution and the chromosomal theory of inheritance.

Darwin's Theory of Evolution

Darwin's theory of evolution is now over 150 years old (see Chapter 6). Darwin articulated the complete theory when he published his famous book, *On the Origin of Species by Means of Natural Selection,* in England in 1859 (Figure 1.12). Biologists today often hear the questions, "What is Darwinism?" and "Do biologists still accept Darwin's theory of evolution?" These questions cannot be given simple answers, because Darwinism encompasses several different, although mutually compatible, theories. Professor Ernst Mayr of Harvard University argued that Darwinism should be viewed as five major theories. These five theories have somewhat different origins and different fates and cannot be treated as only a single statement. The theories are (1) perpetual change, (2) common descent, (3) multiplication of species, (4) gradualism, and (5) natural selection. The first three theories are generally accepted as having universal application throughout the living world. Gradualism and natural selection clearly occur during evolution, but their explanatory power as theories might not be as widespread as Darwin intended. Legitimate scientific controversies regarding gradualism and natural selection often are misrepresented by creationists as challenges to the first three theories listed, although the validity of those first three theories is strongly supported by all relevant observations.

1. **Perpetual change.** This is the basic theory of evolution on which the others depend. It states that the living world is neither constant nor perpetually cycling, but is always changing, with continuity between past and present forms of life. The

varying forms of organisms undergo measurable changes across generations throughout time. This theory originated in antiquity but did not gain widespread acceptance until Darwin advocated it in the context of his other four theories. "Perpetual change" is documented by the fossil record, which clearly refutes creationists' claims for a recent origin of all living forms. Because it has withstood repeated testing and is supported by an overwhelming number of observations, we now regard "perpetual change" as a scientific fact; evidence has rejected all rational alternatives to this theory.

2. **Common descent.** The second Darwinian theory, "common descent," states that all forms of life descend from a common ancestor through a branching of lineages (Figure 1.13). The opposing argument, that the different forms of life arose independently and descended to the present in linear, unbranched genealogies, is refuted by comparative studies of organismal form, cell structure, and macromolecular structures (including those of the genetic material, DNA). All of these studies confirm the theory that life's history has the structure of a branching evolutionary tree, called a **phylogeny.** Species that share recent common ancestry have more similar features at all levels than do species whose most recent common ancestor occurred early in the history of life. Much current research is guided by Darwin's theory of common descent toward reconstructing life's phylogeny using the patterns of similarity and dissimilarity observed among species. The resulting phylogeny serves as the basis for our taxonomic classification of animals (see Chapter 10).

3. **Multiplication of species.** Darwin's third theory states that the evolutionary process produces new species by splitting and transforming older ones. Species are now generally viewed as reproductively distinct populations of organisms that usually but not always differ from each other in organismal form. Once species are fully formed, interbreeding among members of different species does not occur or is too restricted to permit the species' lineages to merge. Evolutionists generally agree that the splitting and transformation of lineages produces new species, although there is still much controversy concerning details of this process (see Chapter 6) and the precise meaning of

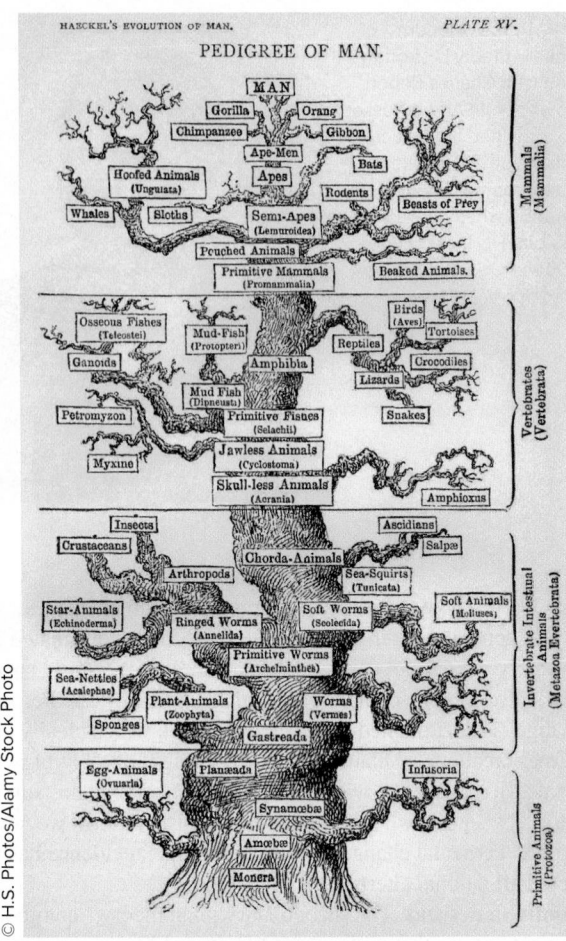

Figure 1.13 An early tree of life drawn in 1874 by the German biologist, Ernst Haeckel, who was strongly influenced by Darwin's theory of common descent. Although it was a good hypothesis in its time, many of the specific groupings shown in this tree have been refuted. (*Source:* S. Gould, *Ontogeny and Phylogeny*, Harvard University Press, 1977.)

the term "species" (see Chapter 10). Much active scientific research examines historical processes that generate new species.

4. **Gradualism.** Gradualism states that the large differences in anatomical traits that characterize disparate species originate through the accumulation of many small incremental changes over very long periods of time. This theory is important because genetic changes that have very large effects on organismal form are usually harmful to an organism. It is possible, however, that some genetic variants whose expression causes large changes in organismal form are nonetheless sufficiently beneficial to be favored by natural selection. For example, some members of the fish species *Cichlasoma minckleyi* have greatly enlarged jaws, muscles, and teeth capable of crushing snails. These individuals look so different from other members of their species that they were once mistakenly considered a separate species. A simple genetic or developmental switch nonetheless appears to produce this "molariform" morph in one large evolutionary step rather than through a graded series of intermediate forms. Therefore, although gradual evolution is known to occur, it may not explain the origins of all structural differences that we observe among species (Figure 1.14). Scientists are actively studying this question.

5. **Natural selection.** Natural selection, Darwin's most famous theory, rests on three propositions. First, there is variation among organisms (within populations) for anatomical, behavioral, and physiological traits. Second, the variation is at least partly heritable so that offspring tend to resemble their parents. Third, organisms with different variant forms are expected to leave different numbers of offspring to future generations. Variants that permit their possessors most effectively to exploit their environments will preferentially survive and be transmitted to future generations. Over many generations, favorable new traits will spread throughout a population. Accumulation of such changes leads, over long periods of time, to production of new organismal characteristics and new species. Natural selection is therefore a creative process that generates novel forms from the small individual variations that occur among organisms within a population.

Natural selection explains why organisms are constructed to meet the demands of their environments, a phenomenon called **adaptation** (Figure 1.15). Adaptation is the expected result of a process that accumulates the most favorable variants occurring in a population throughout long periods of evolutionary time. Adaptation was viewed previously as strong evidence against evolution, and Darwin's theory of natural selection was therefore important for convincing people that a natural process, capable of being studied scientifically, could produce new species. The demonstration that natural processes could produce adaptation was important to the eventual success of all five Darwinian theories.

Darwin's theory of natural selection faced a major obstacle when it was first proposed: it lacked a successful theory of heredity. People assumed incorrectly that heredity was a blending process, and that any favorable new variant appearing in a population therefore would be lost. The new variant arises initially in a single organism, and that organism therefore must mate with one lacking the favorable new trait. Under blending inheritance, the organism's offspring would then have only a diluted form of the favorable trait. These offspring likewise would mate with others that lacked the favorable trait. With its effects diluted by half each generation, the trait eventually would cease to exist. Natural selection would be completely ineffective in this situation.

Darwin was never able to counter this criticism successfully. It did not occur to Darwin that hereditary factors could be discrete and nonblending and that a new genetic variant therefore could persist unaltered from one generation to the next. This principle is called **particulate inheritance.** It was established after 1900 with the discovery of Gregor Mendel's genetic experiments, and it was eventually incorporated into what we now call the **chromosomal theory of inheritance.** We use the term **neo-Darwinism** to describe Darwin's theories as modified by incorporating this theory of inheritance.

Mendelian Heredity and the Chromosomal Theory of Inheritance

The chromosomal theory of inheritance is the foundation for current studies of genetics and evolution in animals (see Chapters 5 and 6). This theory comes from the consolidation of research done in the

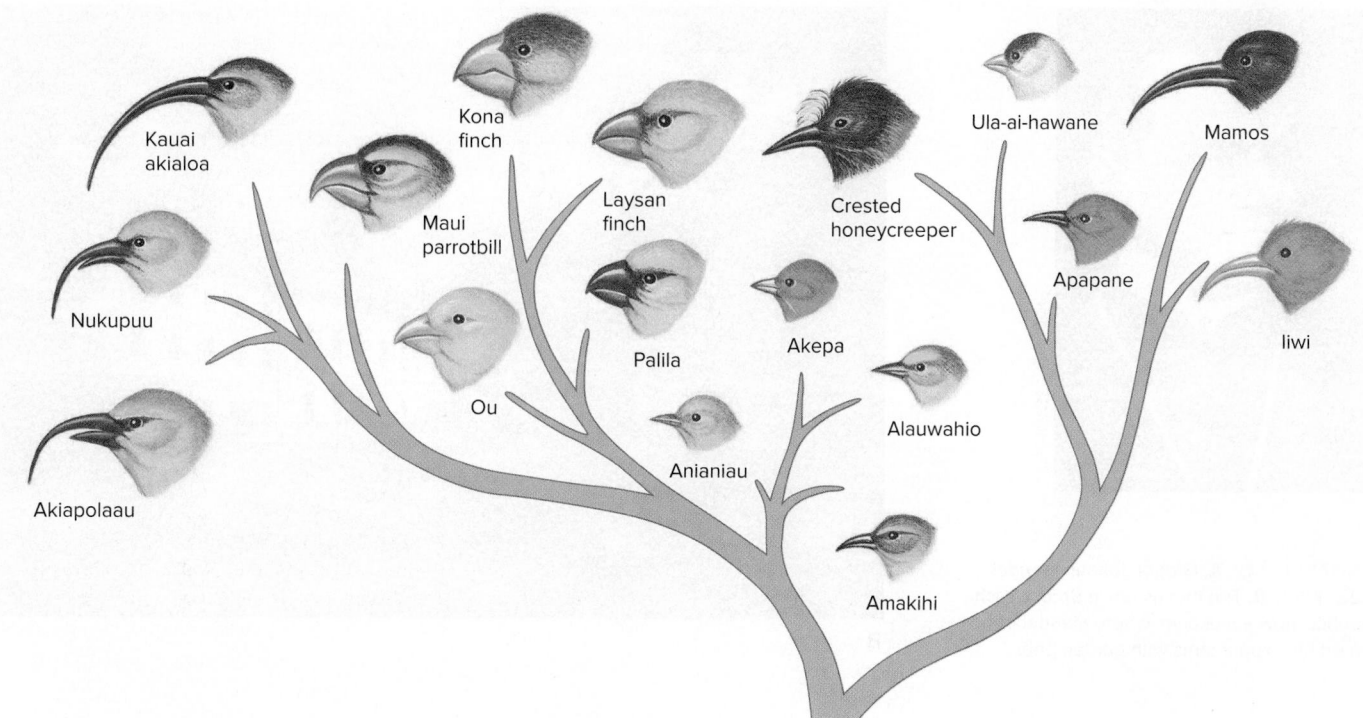

Figure 1.14 Gradualism provides a plausible explanation for the origins of different beak shapes in the Hawaiian honeycreepers shown here. This theory has been challenged, however, as an explanation of the evolution of such structures as vertebrate scales, feathers, and hair from a common ancestral structure. The geneticist Richard Goldschmidt viewed the latter forms as unbridgeable by any gradual transformation series, thus requiring mutations of large phenotypic effect, which he called "hopeful monsters."

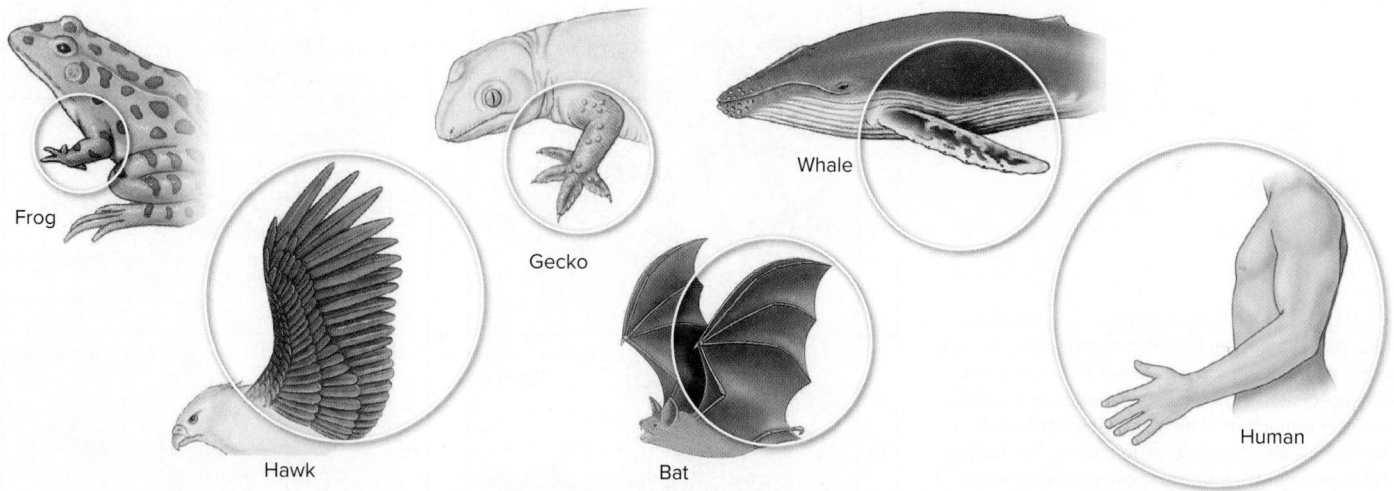

Figure 1.15 According to Darwinian evolutionary theory, the different forms of these vertebrate forelimbs were molded by natural selection to adapt them for different functions. We show in later chapters that, despite these adaptive differences, these limbs share basic structural similarities.

fields of genetics, which was founded by the experimental work of Gregor Mendel (Figure 1.16), and cell biology.

Genetic Approach

The genetic approach consists of mating or "crossing" populations of organisms that are true-breeding for contrasting traits, and then following hereditary transmission of those traits through subsequent generations. "True-breeding" means that a population maintains across generations only one of the contrasting traits when propagated in isolation from other populations. For example, most populations of fruit flies produce only red-eyed individuals, generation after generation, regardless of the environments in which they are raised; such strains are true-breeding for red eyes. Some laboratory strains of fruit flies produce only white-eyed individuals and are therefore true-breeding for white eyes (pp. 82–83).

A

Figure 1.16 **A,** Gregor Johann Mendel (1822–1884). **B,** The monastery in Brno, Czech Republic, now a museum, where Mendel performed his experiments with garden peas.

B

Figure 1.17 Different predictions of particulate versus blending inheritance regarding the outcome of Mendel's crosses of tall and short plants. The prediction of particulate inheritance is upheld and the prediction of blending inheritance is falsified by the results of the experiments. The reciprocal experiments (crossing short female parents with tall male parents) produced similar results. (P_1 = parental generation; F_1 = first filial generation; F_2 = second filial generation.)

Gregor Mendel studied the transmission of seven variable features in garden peas, crossing populations that were true-breeding for contrasting traits (for example, tall versus short plants). In the first generation (called the F_1 generation, for "filial"), only one of the contrasting parental traits was observed; there was no indication of blending of the parental traits. In the example, the offspring (called F_1 *hybrids* because they represent a cross between two different forms) formed by crossing the tall and short plants were tall, regardless of whether the tall trait was inherited from the male or the female parent. These F_1 hybrids were allowed to self-pollinate, and both parental traits were found among their offspring (called the F_2 generation), although the trait observed in the F_1 hybrids (tall plants in this example) was three times more common than the other trait. Again, there was no indication of blending of the parental traits (Figure 1.17).

Mendel's experiments showed that the effects of a genetic factor can be masked in a hybrid individual, but that these factors are not physically altered during the transmission process. He postulated that variable traits are specified by paired hereditary factors, which we now call "genes." When **gametes** (eggs or sperm) are produced, the paired genes controlling a particular feature are segregated from each other and each gamete receives only one of them. Fertilization restores the paired condition. If an organism possesses different forms of the paired genes for a feature, only one of them is expressed in its appearance, but both genes nonetheless are transmitted unaltered in equal numbers to the gametes produced. Transmission of these genes is particulate, not blending. Mendel observed that inheritance of one pair of traits is independent of inheritance of other paired traits. We now know, however, that not all pairs of traits are inherited independently of each other; different traits that tend to be inherited together are called genetically linked (p. 83). Numerous studies, particularly of the fruit fly, *Drosophila melanogaster,* have shown that principles of inheritance discovered initially in plants apply also to animals.

Figure 1.19 Paired chromosomes being separated before nuclear division in the process of forming gametes.

© Alvin Telser, PhD/Cultura RM Exclusive/Getty Images

Contributions of Cell Biology

Improvements in microscopes during the 1800s permitted cytologists to study the production of gametes by direct observation of reproductive tissues. Interpreting the observations was initially difficult, however. Some prominent biologists hypothesized, for example, that sperm were parasitic worms in semen (Figure 1.18). This hypothesis was soon falsified, and the true nature of gametes was clarified. As the precursors of gametes prepare to divide early in gamete production, the nuclear material condenses to reveal discrete, elongate structures called **chromosomes.** Chromosomes occur in pairs that are usually similar but not identical in appearance and informational content. The number of chromosomal pairs varies among species. One member of each pair is derived from the female parent and the other from the male parent. Paired chromosomes are physically associated and then segregated into different daughter cells during cell division prior to gamete formation (Figure 1.19). Each resulting gamete receives one chromosome from each pair. Different pairs of chromosomes are sorted into gametes independently of each other.

Because the behavior of chromosomal material during gamete formation parallels that postulated for Mendel's genes, Sutton and Boveri in 1903 through 1904 hypothesized that chromosomes were the physical bearers of genetic material. This hypothesis met with extreme skepticism when first proposed. A long series of tests designed to falsify it nonetheless showed that its predictions were upheld. The chromosomal theory of inheritance is now well established.

SUMMARY

Zoology is the scientific study of animals, and it is part of biology, the scientific study of life. Animals and life in general can be identified by attributes that they have acquired over their long evolutionary histories. The most outstanding attributes of life include chemical uniqueness, complexity and hierarchical organization, reproduction, possession of a genetic program, metabolism, development, interaction with the environment, and movement. Biological systems comprise a hierarchy of integrative levels (molecular, cellular, organismal, populational, and species levels), each of which demonstrates specific emergent properties.

Scientists construct knowledge by formulating and then testing hypotheses through observations of the natural world. Science is guided by natural law, and its hypotheses are testable, tentative, and falsifiable. Zoological sciences are subdivided into two categories, experimental sciences and evolutionary sciences. Experimental sciences use the experimental method to ask how animals perform their basic metabolic, developmental, behavioral, and reproductive functions, including investigations of their molecular,

Pre'vost and Dumas

Figure 1.18 An early nineteenth-century micrographic drawing of sperm from (1) guinea pig, (2) white mouse, (3) hedgehog, (4) horse, (5) cat, (6) ram, and (7) dog. Some biologists initially interpreted these as parasitic worms in the semen, but in 1824, Jean Prévost and Jean Dumas correctly identified their role in egg fertilization.

cellular, and populational systems. Evolutionary sciences use the comparative method to reconstruct the history of life, and then use that history to explain how diverse species and their molecular, cellular, organismal, and populational properties arose through evolutionary time.

A hypothesis that withstands repeated testing and therefore explains many diverse phenomena gains the status of a theory. Powerful theories that guide extensive research are called "paradigms." The major paradigms that guide the study of zoology are Darwin's theory of evolution and the chromosomal theory of inheritance.

The principles given in this chapter illustrate the unity of biological science. All components of biological systems are guided by natural laws and are constrained by those laws. Living organisms arise only from other living organisms, just as new cells can be produced only from preexisting cells. Reproductive processes occur at all levels of the biological hierarchy and demonstrate both heredity and variation. Interaction of heredity and variation at all levels of the biological hierarchy produces evolutionary change and has generated the great diversity of animal life documented in this book.

 ## REVIEW QUESTIONS

1. Why is life difficult to define?
2. What are the basic chemical differences that distinguish living from nonliving systems?
3. Describe the hierarchical organization of life. How does this organization lead to the emergence of new properties at different levels of biological complexity?
4. What is the relationship between heredity and variation in reproducing biological systems?
5. Describe how evolution of complex organisms is compatible with the second law of thermodynamics.
6. What are the essential characteristics of science? Describe how evolutionary studies fit these characteristics whereas "scientific creationism" or "intelligent-design theory" does not.
7. Use studies of natural selection in British moth populations to illustrate the hypothetico-deductive method of science.
8. How do we distinguish the terms hypothesis, theory, paradigm, and scientific fact?
9. Contrast hypotheses of proximate causality with those of ultimate causality, including the general methods of testing these hypotheses.
10. What are Darwin's five theories of evolution (as identified by Ernst Mayr)? Which are accepted as fact and which continue to stir controversy among biologists?

11. What major obstacle confronted Darwin's theory of natural selection when it was first proposed? How was this obstacle overcome?
12. How does neo-Darwinism differ from Darwinism?
13. Describe the respective contributions of the genetic approach and cell biology to formulating the chromosomal theory of inheritance.

For Further Thought Explain Darwinian evolution as an emergent property of the population level of biological organization.

 ## SELECTED REFERENCES

Futuyma, D. J. 1995. Science on trial: the case for evolution. Sunderland, Massachusetts, Sinauer Associates, Inc. *A defense of evolutionary biology as the exclusive scientific approach to the study of life's diversity.*

Kitcher, P. 1982. Abusing science: the case against creationism. Cambridge, Massachusetts, MIT Press. *A treatise on how knowledge is gained in science and why creationism does not qualify as science. Note that the position refuted as "scientific creationism" in this book is equivalent in content to the position more recently termed "intelligent-design theory."*

Kuhn, T. S. 1970. The structure of scientific revolutions, ed. 2, enlarged. Chicago, University of Chicago Press. *An influential and controversial commentary on the process of science.*

Lebo, L. 2008. The devil in Dover. New York, New Press / W. W. Norton. *A detailed and insightful account of the legal trial that revealed "intelligent design" to be anti-science propaganda.*

Mayr, E. 1982. The growth of biological thought: diversity, evolution and inheritance. Cambridge, Massachusetts, The Belknap Press of Harvard University Press. *An interpretive history of biology with special reference to genetics and evolution.*

Medawar, P. B. 1989. Induction and intuition in scientific thought. London, Methuen & Company. *A commentary on the basic philosophy and methodology of science.*

Moore, J. A. 1993. Science as a way of knowing: the foundations of modern biology. Cambridge, Massachusetts, Harvard University Press. *A lively, wide-ranging account of the history of biological thought and the workings of life.*

Perutz, M. F. 1989. Is science necessary? Essays on science and scientists. New York, E. P. Dutton. *A general discussion of the utility of science.*

Pigliucci, M. 2002. Denying evolution: creationism, scientism, and the nature of science. Sunderland, Massachusetts, Sinauer Associates, Inc. *A critique of science education and the public perception of science.*

Rennie, J. 2002. 15 answers to creationist nonsense. Sci. Am. **287:** 78–85 (July). *A guide to the most common arguments used by creationists against evolutionary biology, with concise explanations of the scientific flaws of creationists' claims.*

2
CHAPTER

The Origin and Chemistry of Life

Earth's abundant supply of water was critical for the origin of life.

Spontaneous Generation of Life?

In ancient times, people commonly thought that life arose repeatedly by spontaneous generation from nonliving material in addition to parental reproduction. For example, frogs appeared to arise from damp earth, mice from putrefied matter, insects from dew, and maggots from decaying meat. Warmth, moisture, sunlight, and even starlight often were considered factors that encouraged spontaneous generation of living organisms.

Among the efforts to synthesize organisms in the laboratory is a recipe for making mice, given by the Belgian plant nutritionist Jean Baptiste van Helmont (1648). "If you press a piece of underwear soiled with sweat together with some wheat in an open jar, after about 21 days the odor changes and the ferment … changes the wheat into mice. But what is more remarkable is that the mice which came out of the wheat

and underwear were not small mice, not even miniature adults or aborted mice, but adult mice emerge!"

In 1861, the great French scientist Louis Pasteur convinced scientists that living organisms do not arise spontaneously from nonliving matter. In his famous experiments, Pasteur introduced fermentable material into a flask with a long S-shaped neck that was open to air. The flask and its contents were then boiled for a long time to kill any microorganisms that might be present. Afterward the flask was cooled and left undisturbed. No fermentation occurred because all organisms that entered the open end were deposited in the neck and did not reach the fermentable material. When the neck of the flask was removed, microorganisms in the air promptly entered the fermentable material and proliferated. Pasteur concluded that life could not originate in the absence of previously existing organisms and their reproductive elements, such as eggs and spores. Announcing his results to the French Academy, Pasteur proclaimed, "Never will the doctrine of spontaneous generation arise from this mortal blow."

All living organisms share a common ancestor; most likely, a population of microorganisms that lived almost 4 billion years ago was the last universal common ancestor (LUCA) of life on earth. This common ancestor was itself the product of a long period of prebiotic assembly of nonliving matter, including organic molecules and water, to form self-replicating units. All living organisms retain a fundamental chemical composition inherited from their ancient common ancestor.

ccording to the big-bang model, the universe originated from a primeval fireball and has been expanding and cooling since its inception 10 to 20 billion years ago. The sun and planets formed approximately 4.6 billion years ago from a spherical cloud of cosmic dust and gases. The cloud collapsed under its own gravity into a rotating disc. As material in the central part of the disc condensed to form the sun, gravitational energy was released as radiation. The pressure of this outwardly directed radiation prevented a collapse of the nebula into the sun. The material left behind cooled and eventually produced the planets, including earth (Figure 2.1).

In the 1920s, Russian biochemist Alexander I. Oparin and British biologist J. B. S. Haldane independently proposed that life originated on earth after an inconceivably long period of "abiogenic molecular evolution." Rather than arguing that the first living organisms miraculously originated all at once, a notion that had discouraged scientific inquiry, Oparin and Haldane argued that the simplest form of life arose gradually by the progressive assembly of small molecules into more complex organic molecules. Molecules capable of self-replication eventually would be produced, ultimately leading to assembly of living microorganisms.

2.1 WATER AND LIFE

The origin and maintenance of life on earth depend critically upon water. Water is the most abundant of all compounds in cells, forming 60% to 90% of most living organisms. Water has several extraordinary properties that explain its essential role in living systems and their origin. Hydrogen bonds that form between adjacent water molecules underlie these properties (Figure 2.2).

Water has a **high specific heat capacity:** 1 calorie * is required to elevate the temperature of 1 g of water 1° C, a higher thermal

---

*A calorie is defined as the amount of heat required to heat 1 g of water from 14.5° to 15.5° C. Although the calorie is the traditional unit of heat widely used in publications and tables, it is not part of the International System of Units (the SI system) which uses the joule (J) as the energy unit (1 cal = 4.184 J).

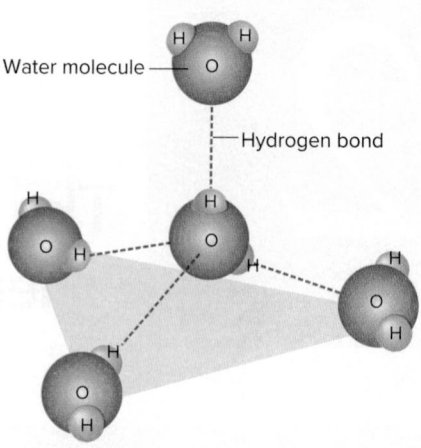

Figure 2.2 Geometry of water molecules. Each water molecule is linked by hydrogen bonds (*dashed lines*) to four other molecules. Imaginary lines connecting the water molecules as shown form a tetrahedron.

capacity than any other liquid except ammonia. Much of this heat energy is used to rupture some hydrogen bonds in addition to increasing the kinetic energy (molecular movement), and thus the temperature, of the water. Water's high thermal capacity greatly moderates environmental temperature changes, thereby protecting living organisms from extreme thermal fluctuation. Water also has a **high heat of vaporization,** requiring more than 500 calories to convert 1 g of liquid water to water vapor. All hydrogen bonds between a water molecule and its neighbors must be ruptured before that water molecule can escape the surface and enter the air. For terrestrial animals (and plants), cooling produced by evaporation of water is important for expelling excess heat.

Another property of water important for life is its **unique density behavior** during changes of temperature. Most liquids become denser with decreasing temperature. Water, however, reaches its maximum density at 4° C *while still a liquid,* then

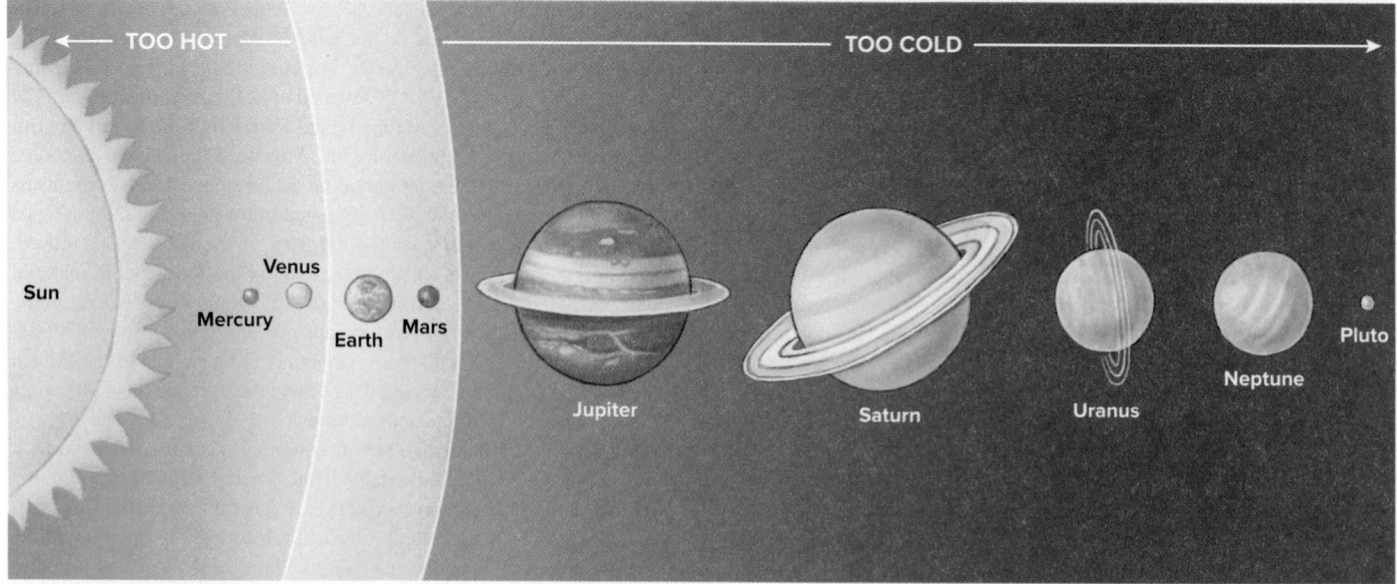

Figure 2.1 Solar system showing narrow range of thermal conditions suitable for life.

Figure 2.3   When water freezes at 0° C, the four partial charges of each atom in the molecule interact with the opposite charges of atoms in other water molecules. The hydrogen bonds between all the molecules form a crystal-like lattice structure, and the molecules are farther apart (and thus less dense) than when some of the molecules have not formed hydrogen bonds at 4° C.

Figure 2.4   Because of hydrogen bonds between water molecules at the water-air interface, the water molecules cling together and create a high surface tension. Thus some insects, such as this water strider, can literally walk on water.

becomes less dense with further cooling (Figure 2.3). Therefore, ice *floats* rather than sinking to the bottoms of lakes and ponds. If ice were denser than liquid water, bodies of water would freeze solid from the bottom upward in winter and might not melt completely in summer. Such conditions would severely limit aquatic life. In ice, water molecules form an extensive, open, crystal-like network supported by hydrogen bonds that connect all molecules. The molecules in this lattice are farther apart, and thus less dense, than in liquid water at 4° C.

Water has **high surface tension,** exceeding that of any other liquid but mercury. Hydrogen bonding among water molecules produces a cohesiveness important for maintaining protoplasmic form and movement. The resulting surface tension creates an ecological niche (see p. 817) for insects, such as water striders and whirligig beetles, that skate on the surfaces of ponds (Figure 2.4). Despite its high surface tension, water has **low viscosity,** permitting movement of blood through minute capillaries and of cytoplasm inside cellular boundaries.

Water is an excellent **solvent.** Salts dissolve more extensively in water than in any other solvent. The dipolar nature of water produces this property, which causes water molecules to orient around charged dissolved particles. When, for example, crystalline NaCl dissolves in water, the Na$^+$ and Cl$^-$ ions separate (Figure 2.5). The negative zones of the water dipoles attract the Na$^+$ ions while the positive zones attract the Cl$^-$ ions. This orientation keeps the ions separated, promoting their dissociation. Solvents lacking this dipolar character are less effective at keeping the ions separated. Binding of water to dissolved protein molecules is essential to the proper functioning of many proteins.

Water also participates in many chemical reactions in living organisms. Many compounds are split into smaller pieces by the addition of a molecule of water, a process called **hydrolysis.** Likewise, larger compounds may be synthesized from smaller

components by the reverse of hydrolysis, called **condensation reactions.**

$$R - R + H_2O \xrightarrow[\text{Hydrolysis}]{} R - OH + H - R$$

$$R - OH + H - R \xrightarrow[\text{Condensation}]{} R - R + H_2O$$

Because water is critical to the support of life, the continuing search for extraterrestrial life usually begins with a search for water. Plans for a human outpost on the moon likewise depend upon finding water there. NASA estimates that at least 600 metric tons of water ice lie in permanently shadowed craters near the moon's north

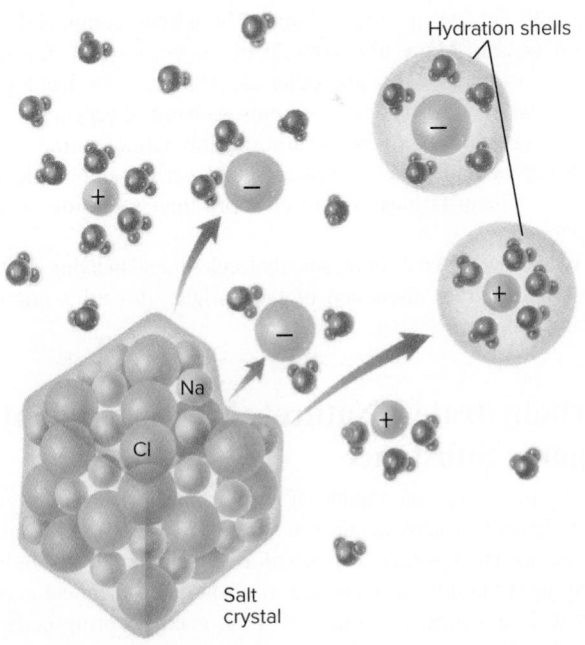

Figure 2.5   When a crystal of sodium chloride dissolves in water, the negative ends of the dipolar molecules of water surround the Na$^+$ ions, while the positive ends of water molecules face the Cl$^-$ ions. The ions are thus separated and do not reenter the salt lattice.

---

pH of Water Solutions

In pure liquid water (= distilled water), a small fraction of the water molecules split into ions of hydrogen (H⁺) and hydroxide (OH⁻); the concentration of both ions is 10^{-7} moles/liter. An acidic substance, when dissolved in water, contributes H⁺ ions to solution, thereby increasing their concentration and causing an excess of H⁺ ions over OH⁻ ions in solution. A basic substance does the reverse, contributing OH⁻ ions to the solution and making OH⁻ ions more common than H⁺ ions. The degree to which a solution is acidic or basic is critical for most cellular processes and requires precise quantification and control; the structure and function of dissolved proteins, for example, depend critically on the concentration of H⁺ in the solution.

The **pH** scale quantifies the degree to which a solution is acidic or basic. The scale ranges from 0 to 14 and represents the additive inverse of the logarithm (base 10) of the H⁺ concentration (in moles/liter) of the solution. Pure liquid water therefore has a pH of 7 (H⁺ concentration = 10^{-7} moles/liter). A solution with pH = 6.0 has an H⁺ concentration ten times higher than that of pure water and is acidic, whereas a solution with pH = 8.0 has an H⁺ concentration ten times lower than pure water and is basic. A concentrated strong acid, such

as hydrochloric acid (HCl, known commercially as "muriatic acid" used to clean masonry), has an H⁺ concentration of ~10^0 mole/liter, giving a pH of 0 (a concentration of H⁺ 10,000,000 times that of pure water). A concentrated base, such as sodium hydroxide (NaOH, used commercially in liquid drain cleaners), has an H⁺ concentration of approximately 10^{-14} mole/liter, giving a pH of 14.

A **buffer** is a dissolved substance (solute) that causes a solution to resist changes in pH because the buffer can remove added H⁺ and OH⁻ ions from solution by binding them into compounds. Dissolved carbon dioxide in the form of bicarbonate (HCO_3^-) is a buffer that helps to protect human blood (pH = 7.3 to 7.5) from changes in pH. H⁺ ions are removed from solution when they react with bicarbonate ions to form carbonic acid, which then dissociates into carbon dioxide and water. The excess carbon dioxide is removed during exhalation (pp. 696–697). OH⁻ ions are removed from solution when this reaction is reversed, forming bicarbonate and hydrogen ions. The excess bicarbonate ions are secreted in the urine (pp. 669–670), and the hydrogen ions serve to increase blood pH back to normal levels. Severe health problems occur if the pH of blood drops to 7 or rises to 7.8.

---

pole. This ice could provide a "lunar freezer" for future human explorers.

2.2 ORGANIC MOLECULAR STRUCTURE OF LIVING SYSTEMS

Chemical evolution in a prebiotic environment produced simple organic compounds that ultimately formed the building blocks of living cells. The term "organic" refers broadly to compounds that contain carbon. Many also contain hydrogen, oxygen, nitrogen, sulfur, phosphorus, salts, and other elements. Carbon has a great ability to bond with other carbon atoms in chains of varying lengths and configurations. Carbon-to-carbon combinations introduce the possibility of enormous complexity and variety into molecular structure. Chemists have identified more than a million organic compounds.

We review the kinds of organic molecules found in living systems, followed by further discussion of their origins in earth's primitive reducing atmosphere.

Carbohydrates: Nature's Most Abundant Organic Substance

Carbohydrates are compounds of carbon, hydrogen, and oxygen. These elements usually occur in the ratio of 1 C: 2 H: 1 O and are grouped as H—C—OH. Carbohydrates function in protoplasm mainly as structural elements and as a source of chemical energy. Glucose is the most important of these energy-storing carbohydrates. Familiar examples of carbohydrates include sugars, starches, and cellulose (the woody structure of plants). Cellulose occurs on earth in greater quantities than all other organic materials combined. Carbohydrates are synthesized by green plants from water and carbon dioxide, with the aid of solar energy. This process,

called **photosynthesis,** is a reaction upon which all life depends, for it is the starting point in the formation of food.

Carbohydrates are usually grouped into the following three classes: (1) **monosaccharides,** or simple sugars; (2) **disaccharides,** or double sugars; and (3) **polysaccharides,** or complex sugars. Simple sugars have a single carbon chain containing 4 carbons (tetroses), 5 carbons (pentoses), or 6 carbons (hexoses). Other simple sugars have up to 10 carbons, but these sugars are not biologically important. Simple sugars, such as glucose, galactose, and fructose, all contain a free sugar group,

$$\begin{array}{c} \text{OH} \quad \text{O} \\ | \qquad || \\ -\text{C}-\text{C}- \\ | \\ \text{H} \end{array}$$

in which the double-bonded O may be attached to the terminal or nonterminal carbons of a chain. The hexose **glucose** (also called dextrose) is particularly important to the living world. Glucose is often shown as a straight chain (Figure 2.6A), but in water it forms a cyclic compound (Figure 2.6B). The "chair" diagram (Figure 2.7) of glucose best represents its true configuration, but all forms of glucose, however represented, are chemically equivalent. Other hexoses of biological significance include galactose and fructose, which are compared with glucose in Figure 2.8.

Disaccharides are double sugars formed by bonding two simple sugars. An example is maltose (malt sugar), composed of two glucose molecules. As shown in Figure 2.9, the two glucose molecules are joined by removing a molecule of water, causing the sharing of an oxygen atom by the two sugars. All disaccharides are formed in this manner. Two other common disaccharides are sucrose (ordinary cane, or table, sugar), formed by the linkage of glucose and fructose, and lactose (milk sugar), composed of glucose and galactose.

Figure 2.6   Two ways of depicting the simple sugar glucose. In **A,** the carbon atoms are shown in open-chain form. When dissolved in water, glucose tends to assume a ring form as in **B.** In this ring model the carbon atoms located at each turn in the ring are usually not shown.

Figure 2.7   "Chair" representation of a glucose molecule.

Glucose   **Galactose**   **Fructose**

Figure 2.8   These three hexoses are the most common Monosaccharides.

Figure 2.9   Formation of a double sugar (disaccharide maltose) from two glucose molecules with the removal of one molecule of water.

Polysaccharides contain many molecules of simple sugars (usually glucose) linked in long chains called polymers. Their empirical formula is usually written $(C_6H_{10}O_5)_n$, where n designates the number of simple-sugar subunits in the polymer. Starch is the common polymer in which plants store sugar, and it is an important food for animals. **Chitin** is an important structural polysaccharide in the exoskeletons of insects and other arthropods (p. 404). **Glycogen,** a multibranched polymer of glucose, is an important polymer for storing sugar in animals. It is stored mainly in liver and muscle cells in vertebrates as globular granules, each one containing around 30,000 connected glucose subunits. When needed, glucose molecules are cleaved from glycogen and delivered by blood to the tissues. Another polymer is **cellulose,** the principal structural carbohydrate of plants.

Lipids: Fuel Storage and Building Material

Lipids are fats and fatlike substances. They are molecules of low polarity; consequently, they are virtually insoluble in water but are soluble in organic solvents, such as acetone and ether. The three principal groups of lipids are triglycerides, phospholipids, and steroids.

Triglycerides

The triglycerides or "true" fats are major fuels of animals. Stored fat is derived either directly from dietary fat or indirectly from dietary carbohydrates that the body has converted to fat for storage. Fats are oxidized and released into the bloodstream as needed to meet tissue demands, especially those of active muscle.

Triglycerides contain glycerol and three molecules of fatty acids. Triglycerides are therefore esters, a combination of an alcohol (glycerol) and an acid. Fatty acids in triglycerides are simply long-chain monocarboxylic acids; they vary in size but are commonly 14 to 24 carbons long. Figure 2.10A shows the production of a typical fat by the union of glycerol and stearic acid. In this reaction, three fatty acid molecules have united with OH groups of glycerol to form stearin plus three molecules of water.

Most triglycerides contain two or three different fatty acids attached to glycerol, and bear weighty names such as myristoyl palmitoyl stearoyl glycerol (Figure 2.10B). The fatty acids in this triglyceride are **saturated;** every carbon within the chain holds two hydrogen atoms.

Figure 2.10 Triglycerides. **A,** Formation of a triglyceride from three molecules of stearic acid (a fatty acid) and glycerol. **B,** A triglyceride bearing three different fatty acids.

Saturated fats, more common in animals than in plants, are usually solid at room temperature. **Unsaturated fatty acids,** typical of plant oils, have two or more carbon atoms joined by double bonds; the carbons are not "saturated" with hydrogen atoms and are available to form bonds with other atoms. Two common unsaturated fatty acids are oleic acid and linoleic acid (Figure 2.11). Plant fats, such as peanut oil and corn oil, tend to be liquid at room temperature.

Phospholipids

Unlike fats that are fuels and serve no structural roles in the cell, phospholipids are important components of the molecular organization of tissues, especially membranes. They resemble triglycerides in structure, except that one of the three fatty acids is replaced by phosphoric acid and an organic base. An example is lecithin, an important phospholipid of nerve membranes (Figure 2.12). Because the phosphate group on phospholipids is charged and polar and therefore soluble in water, and the remainder of the molecule is nonpolar, phospholipids can bridge two environments and bind water-soluble molecules, such as proteins, to water-insoluble materials.

The term **amphiphilic** describes compounds, like phospholipids (Figure 2.12), that are polar and water-soluble on one end and nonpolar on the other end. Laboratory experiments show that amphiphilic compounds have a natural tendency to assemble themselves into semipermeable membranes.

Steroids

Steroids are complex alcohols. Although they are structurally unlike fats, they have fatlike properties, including low polarity. The steroids are a large group of biologically important molecules, including cholesterol (Figure 2.13), vitamin D3, many adrenocortical hormones, and sex hormones.

Figure 2.11 Unsaturated fatty acids. Oleic acid has one double bond and linoleic acid has two double bonds. The remainder of both acids is saturated.

Figure 2.12 Lecithin, an important phospholipid of nerve membranes, in which phosphatidyl choline replaces one of the fatty acids in the typical structure of a triglyceride (Figure 2.10).

Amino Acids and Proteins

Proteins are large, complex molecules composed of 20 kinds of amino acids (Figure 2.14). The amino acids are linked by **peptide bonds** to form long, chainlike polymers. In the formation of a peptide bond, the carboxyl group of one amino acid is linked by a covalent bond to the amino group of another, with elimination of water, as shown here:

Figure 2.13 Cholesterol is a steroid. All steroids have a basic skeleton of four rings (three 6-carbon rings and one 5-carbon ring) with various side groups attached. The carbon chain of the side group is the main structural similarity between a steroid and a fatty acid.

Figure 2.14 Five of the 20 kinds of amino acids. Glycine is the simplest amino acid. Moving from right to left in its structure, the first carbon is in a carboxyl group and the second one attaches to an amino group. This characteristic is repeated in the other amino acids, which differ from one another by the "R" group shown to the left of the second carbon. In glycine, the "R" is simply a hydrogen atom. In proline, the characteristic amino group is bound to the "R" group to form a ring structure.

The combination of two amino acids by a peptide bond forms a dipeptide having a free amino group on one end and a free carboxyl group on the other; therefore, additional amino acids can be joined to produce a long chain. The 20 different kinds of amino acids can be arranged in an enormous variety of sequences of up to several hundred amino acid units, explaining the large diversity of proteins found among living organisms.

A protein is not just a long string of amino acids; it is a highly organized molecule. For convenience, biochemists recognize four levels of protein organization, called primary, secondary, tertiary, and quaternary structures.

The **primary structure** of a protein is the sequence of amino acids composing the polypeptide chain. Because bonds between the amino acids in the chain can form only a limited number of stable angles, the chain assumes certain recurring structural patterns. These bond angles generate the **secondary structure,** such as the **alpha-helix,** which makes helical turns in a clockwise direction like a screw (Figure 2.15). Hydrogen bonds stabilize the spirals of the chains, usually by connecting a hydrogen atom of one amino acid and the peptide-bond oxygen of another amino acid from an adjacent turn of the helix.

The helical and other configurations formed by the polypeptide chain bend and fold, giving the protein its complex, yet stable,

three-dimensional **tertiary structure** (Figure 2.15). Chemical bonds between pairs of amino acids from different parts of the polypeptide chain stabilize tertiary structure. These bonds form between "side groups," parts of the amino acid not involved in a peptide bond. An example is the **disulfide bond,** a covalent bond between the sulfur atoms in two cysteine amino acids that are brought together by folds in the polypeptide chain. Also stabilizing the tertiary structure of proteins are hydrogen bonds, ionic bonds, and hydrophobic bonds.

The term **quaternary structure** describes proteins that contain more than one polypeptide chain. For example, hemoglobin (the oxygen-carrying substance in blood) of jawed vertebrates comprises four polypeptide subunits held together in a single protein molecule (Figure 2.15).

Proteins perform many functions in living organisms. They form the structural framework of protoplasm and many cellular components. Many proteins function as **enzymes,** the biological catalysts required for almost every reaction in the body. Enzymes lower the activation energy required for specific reactions and enable life processes to proceed at moderate temperatures rather than requiring high temperatures. Enzymes control the reactions by which food is digested, absorbed, and metabolized. They promote the synthesis of structural materials for growth and to replace those lost by wear. They determine the release of energy used in respiration, growth, muscle contraction, physical and mental activities, and many other activities. Enzyme action is described in Chapter 4 (p. 57).

A **prion** is an infectious protein particle in which a protein of the host organism is contorted into an abnormal three-dimensional structure. Upon infection, the prion causes its host's normal copies of the protein to be refolded into the abnormal form, with pathological results. In "mad cow disease," a prion infection severely damages brain tissues and is fatal. Fatal neurological diseases associated with transmissible prions occur also in people (for example, kuru), and in sheep and goats (scrapie).

Nucleic Acids

Nucleic acids are complex polymeric molecules whose sequence of nitrogenous bases encodes the genetic information necessary for biological inheritance. They store directions for the synthesis of enzymes and other proteins, and are the only molecules that can (with the help of the right enzymes) replicate themselves. The two kinds of nucleic acids in cells are **deoxyribonucleic acid (DNA)** and **ribonucleic acid (RNA).** They are polymers of repeated units called **nucleotides,** each of which contains a sugar, a nitrogenous base, and a phosphate group. In addition to its role as one of the four precursors of RNA, adenosine triphosphate (ATP) is the primary transporter of chemical energy in cellular metabolism (p. 60). Because the structure of nucleic acids is crucial to the mechanism of inheritance and protein synthesis, detailed information on nucleic acids is presented in Chapter 5 (p. 85).

2.3 CHEMICAL EVOLUTION

Both Oparin and Haldane proposed that earth's primitive atmosphere consisted of simple compounds such as water, molecular hydrogen, methane, and ammonia, but lacked oxygen gas (O_2, also called "molecular oxygen"). The nature of the primeval atmosphere is critical for understanding life's origin. The organic compounds that compose living organisms are neither synthesized outside cells nor stable in the presence of molecular oxygen, which is abundant in the atmosphere today. The best evidence indicates, however, that the primitive atmosphere contained not more than a trace of molecular oxygen. The primeval atmosphere therefore was a reducing one, consisting primarily of molecules in which hydrogen exceeds oxygen; methane (CH_4) and ammonia (NH_3), for example, constitute fully reduced compounds. Such compounds are called "reducing" because they tend to donate electrons to other compounds, thereby "reducing" those compounds (p. 61). During this time, the earth was bombarded by large (100 km diameter) comets and meteorites, generating heat that repeatedly vaporized seawater.

This reducing atmosphere was conducive to the prebiotic synthesis that led to life's beginnings, although totally unsuited for organisms alive today. Oparin and Haldane proposed that ultraviolet radiation of such a gas mixture caused many organic substances, such as sugars and amino acids, to form. Haldane proposed that the early organic molecules accumulated in the primitive oceans to form a "hot dilute soup." In this primordial broth, carbohydrates, fats, proteins, and nucleic acids could have assembled to form the earliest structures capable of guiding their own replication.

If the simple gaseous compounds present in the early atmosphere are mixed with methane and ammonia in a closed glass system and kept at room temperature, they never react chemically with each other. To produce a chemical reaction, a continuous source of **free energy** sufficient to overcome reaction-activation barriers must be supplied. Ultraviolet light from the sun must have been intense on earth before the accumulation of atmospheric oxygen; ozone, a three-atom form of oxygen located high in the atmosphere, now blocks much of the ultraviolet radiation from reaching the earth's surface. Electrical discharges could have provided further energy for chemical evolution. Although the total amount of electrical energy released by lightning during a thunderstorm is small compared with solar energy, nearly all of the energy of lightning is effective in synthesizing organic compounds in a reducing atmosphere. A single flash of lightning through a reducing atmosphere generates a large amount of organic matter. Lightning from thunderstorms was probably the most important source of energy for organic synthesis.

Widespread volcanic activity is another possible source of energy. One hypothesis maintains, for example, that life originated not on the surface of the earth, but deep beneath the sea in or around **hydrothermal vents** (p. 830). Hydrothermal vents are submarine hot springs; seawater seeps through cracks in the seafloor until the water comes close to hot magma. The water is then superheated and expelled forcibly, carrying various dissolved molecules from the superheated rocks. These molecules include hydrogen sulfide, methane, iron ions, and sulfide ions. Hydrothermal vents have been discovered in several locations beneath the deep sea, and they would have been much more widely prevalent on the early earth. Interestingly, many heat- and sulfur-loving bacteria grow in hot springs today. Volcanic vents with unique biotic communities occur in Yellowstone Lake.

Figure 2.15 Structure of proteins. The *primary structure* of a protein is the ordering of amino acids, shown as circles in the figure, from the carboxyl end of the molecule to the amino end. Proteins differ in primary structure by the number and sequence of amino acids present. The amino acid sequence of a protein often encourages the formation of hydrogen bonds between nearby amino acids, producing coils and foldbacks (the *secondary structure*). Bonds between R groups of amino acids cause the chain to fold back on itself in a complex manner (*tertiary structure*). Amino acids whose R groups interact to form the tertiary structure usually are far apart in the amino acid sequence of the primary structure, but their R groups meet in the folded three-dimensional structure of the protein. Individual polypeptide chains of some proteins aggregate to form a functional molecule composed of several subunits (*quaternary structure*). The quaternary structure shown is that of hemoglobin, which comprises two alpha polypeptides (blue) and two beta polypeptides (brown), each of which also binds a heme molecule (red).

Prebiotic Synthesis of Small Organic Molecules

The Oparin-Haldane hypothesis stimulated experimental work to test the hypothesis that organic compounds characteristic of life could be formed from the simpler molecules present in the prebiotic

Figure 2.16 Dr. S. L. Miller with a replica of the apparatus used in his historic 1953 experiment on the synthesis of amino acids with an electric spark in a strongly reducing atmosphere.

environment. In 1953, Stanley Miller and Harold Urey in Chicago successfully simulated the conditions thought to prevail on the primitive earth. Miller built an apparatus designed to circulate a mixture of methane, hydrogen, ammonia, and water past an electric spark (Figure 2.16). Water in the flask was boiled to produce steam that helped to circulate the gases. The products formed in the electrical discharge (representing lightning) were condensed in the condenser and collected in the U-tube and small flask (representing an ocean).

After a week of continuous sparking, approximately 15% of the carbon from the reducing "atmosphere" had been converted into organic compounds that collected in the "ocean." The most striking finding was that many compounds related to life were synthesized. These compounds included four of the amino acids commonly found in proteins, urea, and several simple fatty acids. We can appreciate the astonishing nature of this synthesis when we consider that there are thousands of known organic compounds with structures no more complex than those of the amino acids formed. Yet in Miller's synthesis, most of the relatively few substances formed were compounds found in living organisms. This result is surely no coincidence, and it suggests that prebiotic synthesis on the primitive earth may have occurred under conditions not greatly different from those that Miller simulated.

Miller's experiments have been criticized in light of current opinion that the early atmosphere on earth was quite different from Miller's strongly reducing simulated atmosphere. Nevertheless, Miller's work stimulated many other investigators to repeat and to extend his experiment. Amino acids are synthesized in many different kinds of gas mixtures that are heated (volcanic heat), irradiated with ultraviolet light (solar radiation), or subjected to electrical discharge (lightning). The only conditions required to produce amino acids are that the gas mixture be reducing and that it be subjected violently to a source of energy. In other experiments, electrical discharges passed through mixtures of carbon monoxide, nitrogen, and water yield amino acids and nitrogenous bases. Although reaction rates are much slower than in atmospheres containing methane and ammonia, and yields poor in comparison, these experiments support the hypothesis that the chemical beginnings of life can occur in atmospheres that are only mildly reducing. The need for methane and ammonia, however, leads to proposals that these substances might have been introduced by comets or meteorites, or that they were synthesized near the hydrothermal vents.

Thus the experiments of many scientists have shown that highly reactive intermediate molecules such as hydrogen cyanide, formaldehyde, and cyanoacetylene are formed when a reducing mixture of gases is subjected to a violent energy source. These molecules react with water and ammonia or nitrogen to form more complex organic molecules, including amino acids, fatty acids, urea, aldehydes, sugars, and nitrogenous bases (purines and pyrimidines), all of the building blocks required for the synthesis of the most complex organic compounds of living matter. Further evidence for the natural abiotic synthesis of amino acids comes from finding amino acids in meteorites, such as the Murchison meteorite that hit Australia in 1969.

Formation of Polymers

The next stage in chemical evolution involved the joining of amino acids, nitrogenous bases, and sugars to yield larger molecules, such as proteins and nucleic acids. These polymerizations are condensation (dehydration) reactions, in which monomers are linked together by the removal of water (p. 21). Such synthesis does not occur easily

NASA

Figure 2.17 Micrograph showing membranous vesicular structures formed by amphiphilic molecules extracted from the Murchison meteorite.

in dilute solutions, because excess water drives reactions toward decomposition (hydrolysis). In living systems, condensation reactions always occur in an aqueous (cellular) environment containing appropriate enzymes. Without enzymes and energy supplied by ATP, macromolecules (proteins and nucleic acids) of living systems soon decompose into their constituent monomers. How then could the subunits of biological macromolecules become sufficiently concentrated to form polymers prior to evolution of cellular life?

Amphiphilic molecules in aqueous solution spontaneously form membrane-bound vesicles, which tend to concentrate inside them the organic subunits of biological macromolecules. Reactions catalyzed by metal or mineral surfaces produce organic amphiphilic molecules. Work by David Deamer and colleagues shows that membranes can self-assemble from aqueous mixtures of organic amphiphiles. They propose that extraterrestrial material falling on the primordial earth was an important source of such compounds, which are common in meteorites. Amphiphiles extracted from the Murchison meteorite form membranous vesicles in aqueous solutions (Figure 2.17). The fatty acids and long-chain alcohols that form the amphiphilic components of familiar biological membranes occur in meteorites and have been synthesized under simulated prebiotic conditions. They are thus possible components of prebiotic membranes.

In liquid water, small amphiphiles associate spontaneously through hydrophobic interactions to form monolayered and bilayered membranes organized as hollow vesicles (Figure 2.17). The semipermeable membranes of these vesicles concentrate solutes inside the vesicle, thereby providing conditions conducive to dehydration synthesis of polymers from amino acids or nucleotides. Such vesicles internalize and encapsulate double-stranded DNA molecules under laboratory conditions.

Earlier work by Sydney Fox shows that polypeptides in aqueous solution spontaneously form microspheres, and that these microspheres are similar in size and shape to bacteria and can proliferate by

budding. Deamer and colleagues nonetheless argue that vesicles formed from fatty acids or long-chain alcohols are a better model for pre-cellular origins of cell membranes than are proteinoid microspheres. Proteinoid microspheres appear not to provide the semipermeable barrier needed for effective concentration of amino acids and nucleotides for polymerization. Semipermeable membranes formed from fatty acids and long-chain alcohols cause the interior of a vesicle to accumulate amino acids and nucleotides and thereby to promote dehydration synthesis of polypeptides and nucleic acids.

2.4 ORIGIN OF LIVING SYSTEMS

The fossil record reveals that life existed 3.8 billion years ago; therefore, the origin of the earliest form of life can be estimated at approximately 4 billion years ago. The first living organisms were protocells, autonomous membrane-bound units with a complex functional organization that permitted the essential activity of self-reproduction. The primitive chemical systems that we have described lack this essential property. The principal problem in understanding the origin of life is explaining how primitive chemical systems could have become organized into living, autonomous, self-reproducing cells.

As we have seen, a lengthy chemical evolution on the primitive earth produced several molecular components of living forms. In a later stage of evolution, nucleic acids (DNA and RNA) began to behave as simple genetic systems that directed the synthesis of proteins, especially enzymes. However, this conclusion leads to a troublesome chicken-egg paradox: (1) How could nucleic acids have appeared without the enzymes that catalyze their synthesis? (2) How could enzymes have evolved without nucleic acids to encode their amino acid sequence? These questions come from a long-accepted notion that only proteins could act as enzymes. Startling evidence presented in the 1980s shows that RNA has catalytic activity.

Catalytic RNA (ribozymes) can mediate processing of messenger RNA (removal of introns, p. 90), and can catalyze formation of peptide bonds. Strong evidence suggests that translation of mRNA by ribosomes (p. 90) is catalyzed by their RNA, not protein, content.

Therefore the earliest enzymes could have been RNA, and the earliest self-replicating molecules could have been RNA. Investigators are now calling this stage the "RNA world." The "RNA world" would have been encapsulated by membranous vesicular structures possibly resembling those shown in Figure 2.17. Tests of this hypothesis include attempts to assemble a working ribozyme system inside a membranous vesicle in the laboratory. Because proteins have several important advantages over RNA as catalysts, and DNA is a more stable carrier of genetic information than RNA, the first protocells containing protein enzymes and DNA should have been more stable than those with only RNA.

Once this protocellular stage of organization was reached, natural selection (pp. 119–121) would have acted on these primitive self-replicating systems. This stage was critical. Before this stage, biogenesis was shaped by the favorable environmental conditions on the primitive earth and by the nature of the reacting elements themselves. When self-replicating systems became responsive to natural selection, they began to evolve. The more rapidly replicating and more successful systems were favored, thereby gradually evolving efficient replicators. Evolution of the genetic code and fully directed protein synthesis followed. The system now meets

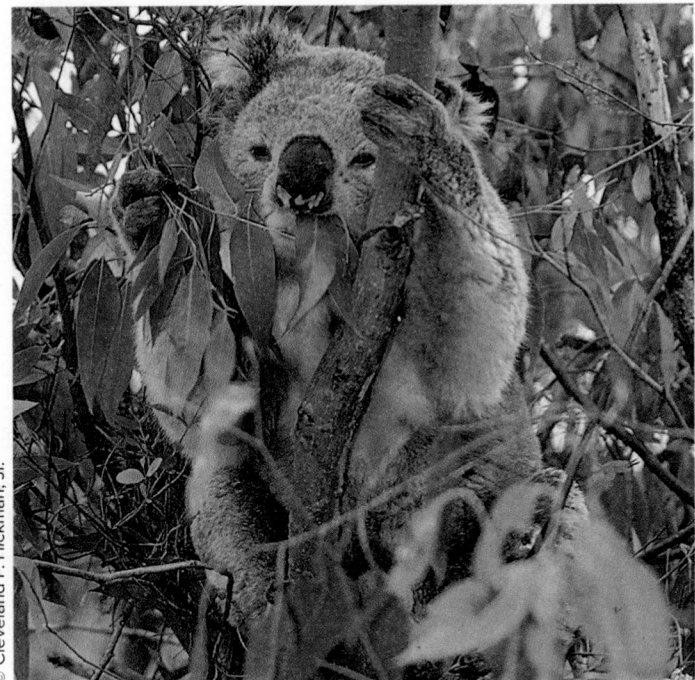

Figure 2.18   Koala, a heterotroph, feeding on a eucalyptus tree, an autotroph. All heterotrophs depend for their nutrients directly or indirectly on autotrophs, which capture the sun's energy to synthesize their own nutrients.

the requirements for being the common ancestor of all living organisms.

Origin of Metabolism

Living cells today are organized systems with complex and highly ordered sequences of enzyme-mediated reactions. How did such vastly complex metabolic schemes develop? The exact history of this phase of life's evolution is unknown. We present here a model of the simplest sequence of events that could explain the origin of observed metabolic properties of living systems.

Organisms that can synthesize their food from inorganic sources using light or another source of energy are called **autotrophs** (Gr. *autos,* self, + *trophos,* feeder) (Figure 2.18). Organisms lacking this ability must obtain their food supplies directly from the environment and are called **heterotrophs** (Gr. *heteros,* another, + *trophos,* feeder). The earliest postulated microorganisms are sometimes called **primary heterotrophs** because they relied on environmental sources for their food and existed prior to the evolution of any autotrophs. They were probably anaerobic organisms similar to bacteria of the genus *Clostridium.* Because chemical evolution had supplied generous stores of organic nutrients in the prebiotic soup, the earliest organisms would not have been required to synthesize their own food.

Autotrophs would have had a tremendous selective advantage over the primary heterotrophs in areas where organic nutrients became depleted. Evolution of autotrophic organisms most likely required acquisition of enzymatic activities to catalyze conversion of inorganic molecules to more complex ones, such as carbohydrates.

The numerous enzymes of cellular metabolism appeared when cells started to utilize proteins for catalytic functions.

Carl Woese challenged the traditional view that the first organisms were primary heterotrophs. He found it easier to visualize the earliest living forms as membrane-associated molecular aggregates that absorbed visible light and converted it with some efficiency into chemical energy. Thus the first organisms would have been autotrophs. Woese also suggested that the earliest "metabolism" comprised numerous chemical reactions catalyzed by nonprotein cofactors (substances necessary for the function of many of the protein enzymes in living cells). These cofactors would have been associated with membranes.

Appearance of Photosynthesis and Oxidative Metabolism

Autotrophy evolved in the form of photosynthesis, which arose separately in multiple evolutionarily distinct groups of bacteria. In the earliest forms of bacterial photosynthesis, hydrogen atoms were derived from hydrogen sulfide or hydrogen gas. Thus, the first form of photosynthesis occurred in bacteria and did not generate oxygen. Cyanobacteria later evolved a water-dependent photosynthesis, which was transferred to an ancestor of plants by endosymbiosis (p. 31). In this most familiar form of photosynthesis, hydrogen atoms obtained from water react with carbon dioxide obtained from the atmosphere to generate sugars and molecular oxygen. Energy is stored in the form of covalent bonds between carbon atoms in the sugar molecule. Sugars provide nutrition to the organism and molecular oxygen is released into the atmosphere.

$$6 CO_2 + 6 H_2O \xrightarrow[\text{light}]{} C_6H_{12}O_6 + 6 O_2$$

This equation summarizes the many reactions now known to occur in photosynthesis. Undoubtedly these reactions did not appear simultaneously, and other reduced compounds, such as hydrogen sulfide (H_2S), probably were the early sources of hydrogen.

Gradually, oxygen produced by photosynthesis accumulated in the atmosphere. When atmospheric oxygen reached approximately 1% of its current level, ozone began to accumulate and to absorb ultraviolet radiation, thereby greatly restricting the amount of ultraviolet light that reached the earth. Land and surface waters then were occupied by photosynthetic organisms, thereby increasing oxygen production.

As the atmosphere slowly accumulated oxygen gas (O_2), a new and highly efficient kind of metabolism appeared: **oxidative (aerobic) metabolism.** By using available oxygen as a terminal electron acceptor (p. 65) and completely oxidizing glucose to carbon dioxide and water, much of the bond energy stored by photosynthesis could be recovered. Most living forms became completely dependent upon oxidative metabolism.

Our atmosphere today is strongly oxidizing. It contains 78% molecular nitrogen, approximately 21% free oxygen, 1% argon, and 0.03% carbon dioxide. Although the time course for production of atmospheric oxygen is much debated, the most important source of oxygen is photosynthesis. Almost all oxygen currently produced comes from cyanobacteria (blue-green algae), eukaryotic algae, and plants. Each day these organisms combine approximately 400 million tons of carbon dioxide with 70 million tons of hydrogen to produce 1.1 billion tons of oxygen. Oceans are a major source of oxygen. Almost all oxygen produced today is consumed by organisms for respiration; otherwise, the amount of oxygen in the atmosphere would double in approximately 3000 years. Because Precambrian fossil cyanobacteria resemble modern cyanobacteria, it is reasonable to suppose that oxygen entering the early atmosphere came from their photosynthesis.

2.5 PRECAMBRIAN LIFE

The Precambrian period covers the geological time before the beginning of the Cambrian period some 542 million years BP. Most major animal phyla appear in the fossil record within a few million years at the beginning of the Cambrian period. This appearance has been called the "Cambrian explosion" because before this time, fossil deposits are mostly devoid of any organisms more complex than single-celled bacteria. Comparative molecular studies (p. 206) now suggest that the rarity of Precambrian fossils may represent poor fossilization rather than absence of animal diversity from the Precambrian period. Nonetheless, animals make a relatively late appearance in the history of life on earth. What were the early forms of life that generated both the oxidizing atmosphere critical for animal evolution and the evolutionary lineage from which animals would arise?

Prokaryotes and the Age of Cyanobacteria (Blue-Green Algae)

The earliest bacterium-like organisms proliferated, generating a great variety of forms, some of which were capable of photosynthesis. From these arose the oxygen-producing **cyanobacteria** approximately 3 billion years ago.

Bacteria and Archaea are called **prokaryotes,** meaning literally "before the nucleus." They contain a single, large molecule of DNA not located in a membrane-bound nucleus, but found in a nuclear region, or **nucleoid.** The DNA is not complexed with histone proteins, and prokaryotes lack membranous organelles such as mitochondria, plastids, Golgi apparatus, and endoplasmic reticulum (see Chapter 3). During cell division, the nucleoid divides and

replicates of the cell's DNA are distributed to the daughter cells. Prokaryotes lack the chromosomal organization and chromosomal (mitotic) division seen in animals, fungi, and plants.

Carl Woese and his colleagues at the University of Illinois discovered the evolutionary distinctness of the Bacteria ("true" bacteria) and the Archaebacteria, also called Archaea (p. 213). Although specimens of these two groups look very much alike when viewed with the electron microscope, they are biochemically distinct. Archaebacteria differ fundamentally from bacteria in cellular metabolism, and their cell walls lack muramic acid, which is present in the cell walls of all Bacteria. The most compelling evidence for differentiating these two groups comes from the use of one of the newest and most powerful tools at the disposal of the evolutionist, sequencing of nucleic acids (p. 85). Woese found that Archaebacteria differ fundamentally from bacteria in the sequence of bases in ribosomal RNA (p. 90). Woese considers the Archaebacteria so distinct from the true bacteria that they should be considered a separate taxonomic domain, Archaea.

Fossil data indicate that oxygen-producing cyanobacteria became prominent in the world's oceans approximately 2.5 billion years ago, causing the atmosphere to become oxygenated within the next 100 million years. Oxygen reacts with water to produce caustic substances like superoxide and hydrogen peroxide, which challenged the ecological tolerances of the existing life forms, including many cyanobacteria responsible for producing the molecular oxygen. Several competing names describe the resulting massive destruction of life forms that were poisoned by oxygen, to be replaced by ones adapted to an oxygenated environment: oxygen revolution, oxygen catastrophe, oxygen crisis, and great oxygenation event (GOE). An important product of evolution in an oxygenated atmosphere was the eukaryotic cell, whose metabolism usually depends upon reacting organic compounds with molecular oxygen.

Appearance of Eukaryotes

Eukaryotes ("true nucleus"; Figure 2.19) have cells with membranebound nuclei containing **chromosomes** composed of **chromatin.** Constituents of eukaryotic chromatin include proteins called **histones** and RNA, in addition to DNA. Some nonhistone proteins

Cell wall
Cell membrane
DNA
Nucleoid
Cytoplasm with ribosomes

Prokaryote

Golgi apparatus
Lysosome
Nucleus
Endoplasmic reticulum
Mitochondrion
Cell membrane

Eukaryote

Figure 2.19 Comparison of prokaryotic and eukaryotic cells. Prokaryotic cells are about one-tenth the size of eukaryotic cells.

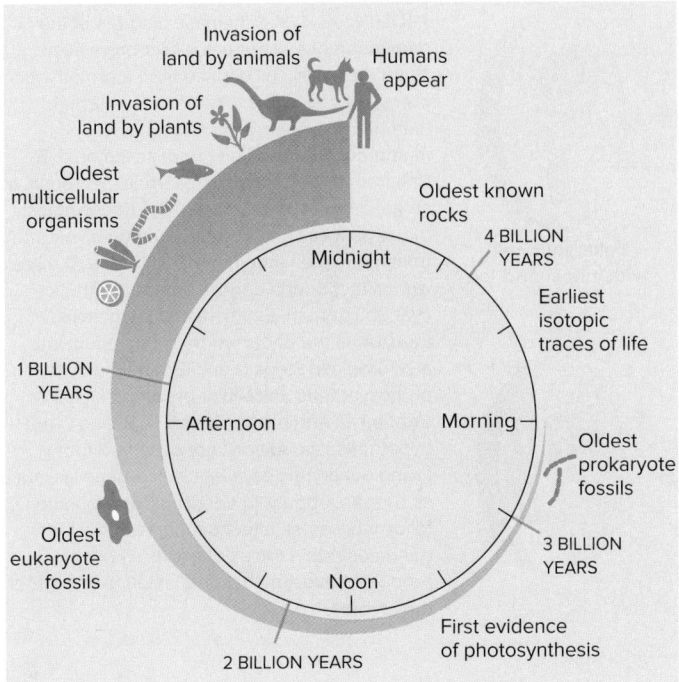

Figure 2.20 The clock of biological time. A billion seconds ago it was 1985, and most students using this text had not yet been born. A billion minutes ago the Roman empire was at its zenith. A billion hours ago Neanderthals were alive. A billion days ago the first bipedal hominids walked the earth. A billion months ago the dinosaurs were at the climax of their radiation. A billion years ago no animal had ever walked on the surface of the earth.

Molecular sequencing has emerged as a very successful approach to unraveling the ancient genealogies of living forms. The sequences of nucleotides in the DNA of an organism's genes are a record of evolutionary relationships, because every gene that exists today is an evolved copy of a gene that existed millions or even billions of years ago. Genes become altered by mutations through the course of time, but vestiges of the original gene usually persist. With modern techniques, one can determine the sequence of nucleotides in an entire molecule of DNA or in short segments of the molecule. When corresponding genes are compared between two different organisms, the extent to which the genes differ can be correlated with the time elapsed since the two organisms diverged from a common ancestor. Similar comparisons are made with RNA and proteins. These methods also permit scientists to synthesize long-extinct genes and proteins and to measure biochemical properties of the extinct proteins.

Figure 2.21 Dr. Lynn Margulis, whose endosymbiotic theory of the origins of mitochondria and chloroplasts is strongly supported by molecular evolutionary studies.

are associated with both prokaryotic DNA and eukaryotic chromosomes. Eukaryotes are generally larger than prokaryotes and contain much more DNA. Cellular division usually is by some form of mitosis (p. 50). Within their cells are numerous membranous organelles, including mitochondria, in which the enzymes for oxidative metabolism are packaged. Eukaryotes include animals, fungi, plants, and numerous single-celled forms formerly called "protozoans" or "protists." Fossil evidence suggests that single-celled eukaryotes arose at least 1.5 billion years ago (Figure 2.20).

Because a eukaryotic cell has an abundance of diverse subcellular structures, called organelles, that have no counterparts within prokaryotic cells, it is difficult to visualize how a eukaryote could have arisen from any single known prokaryote. The American biologist Lynn Margulis (Figure 2.21) and others have proposed that eukaryotes did not arise from any single prokaryote but were derived from a **symbiosis** ("life together") of two or more types of bacteria. Mitochondria and plastids (photosynthetic organelles found only in plant cells), for example, each contain their own DNA (apart from the nucleus of the cell), which has some prokaryotic characteristics.

Nuclei, plastids, and mitochondria each contain genes encoding ribosomal RNA. Comparisons of the sequences of bases of these genes show that the nuclear, plastid, and mitochondrial DNAs represent distinct evolutionary lineages. Plastid and mitochondrial DNAs are closer in their evolutionary history to bacterial DNAs than to the eukaryotic nuclear DNA (p. 213). Plastids are closest evolutionarily to cyanobacteria, and mitochondria are closest to another group of bacteria (alpha-proteobacteria), consistent with the symbiotic hypothesis of eukaryotic origins. Mitochondria contain the enzymes of oxidative metabolism, and plastids (a plastid with chlorophyll is a chloroplast) conduct photosynthesis. It is easy to see how a host cell able to accommodate such guests in its cytoplasm would have gained enormous evolutionary success.

The endosymbiotic theory proposes that a population ancestral to eukaryotic cells, derived from and resembling **anaerobic** (lacking oxidative metabolism) bacteria, evolved a nucleus and other intracellular membranes (Figure 2.22) from infoldings of the cell membrane. Cells of this population acquired, by ingestion or parasitism, aerobic bacteria that avoided digestion and came to reside in the host cell's

Figure 2.22 Schematic diagram of the origination of eukaryotic cellular organelles. **A,** An ancestral cell derived from a prokaryotic precursor cell with evolution of an internal membrane system, including a nuclear membrane surrounding genomic material. **B,** Cell from part A engulfs an aerobic bacterium in an ameboid manner (p. 219). **C,** Endosymbiosis of the host cell and aerobic bacterium transforms the latter into mitochondria. **D,** A cell with mitochondria engulfs a photosynthetic cyanobacterium. **E,** Stable endosymbiosis transforms the photosynthetic bacterium into a chloroplast. Steps D and E pertain only to photosynthetic unicellular eukaryotes (see Chapter 11) and plants. Stages A, B, and D are hypothetical conditions not observed today. Living eukaryotic cells that lack organelles, such as *Giardia*, appear to descend from condition C following loss of mitochondria; their nuclear genomes retain traces of genes whose products have a mitochondrial function in other eukaryotes.

cytoplasm (Figure 2.22). The endosymbiotic aerobic bacteria would have metabolized oxygen, which is toxic for their anaerobic host, and the anaerobic host cell would have given its aerobic residents food and physical protection. This mutually beneficial relationship would produce selection for the host cells and their residents to evolve a means of making their relationship a permanent one. Among the evolutionary outcomes of this selection would be compactness of the endosymbiotic aerobic residents and loss from them of genes whose functions are redundant with those of the host's nuclear genome (or the reverse).

Data collected to test this proposed mechanism show that its conditions are reasonable ones. Fossil data show that both aerobic and anaerobic bacteria were well established by 2.5 billion years ago, and that cells containing nuclei and internal membranes first appeared at this time. Some anaerobic, nucleated forms that lack mitochondria are alive today, including the human parasite *Giardia intestinalis,* although these forms probably represent descendants of lineages that formerly had mitochondria and lost them rather than lineages whose ancestry never featured mitochondria. Eukaryotic cells containing mitochondria are evident approximately 1.2 billion years ago. Bacteria have been introduced experimentally into single-celled eukaryotes and propagated as a symbiotic unit for many generations. Such experiments have shown further that the host cell can become dependent upon its resident bacteria for proteins whose functions formerly were performed by the host population prior to the experimental endosymbiosis.

In addition to claiming that mitochondria and plastids originated as bacterial symbionts, Lynn Margulis argued that eukaryote flagella, cilia (locomotory structures), and even the spindle of mitosis came from a kind of bacterium like a spirochete. Indeed, she suggested that this association (the spirochete with its new host cell) made evolution of mitosis possible. Margulis's evidence that organelles are former partners of an ancestral cell is now accepted by most biologists. Such merging of disparate organisms to produce evolutionarily novel forms is called symbiogenesis.

The first eukaryotes were undoubtedly unicellular, and many were photosynthetic autotrophs. Some of these forms lost their photosynthetic ability and became heterotrophs, feeding on eukaryotic autotrophs and prokaryotes. As cyanobacteria were cropped, their dense filamentous mats began to thin, providing space for other organisms. Carnivores appeared and fed on herbivores. Soon a balanced ecosystem of carnivores, herbivores, and primary producers appeared. By freeing space, cropping herbivores encouraged a greater diversity of producers, which in turn promoted evolution of new and more specialized croppers. An ecological pyramid developed with carnivores at the top of the food chain (p. 829).

The burst of evolutionary activity that followed at the end of the Precambrian period and beginning of the Cambrian period was unprecedented. Some investigators hypothesize that the explanation for the "Cambrian explosion" lies in the accumulation of oxygen in the atmosphere to a critical threshold level. Larger, multicellular animals required the increased efficiency of oxidative metabolism; these pathways could not be supported under conditions of limiting oxygen concentration.

SUMMARY

Living organisms show a remarkable uniformity in their chemical constituents and metabolism, reflecting their common descent from an ancient ancestor.

Life on earth could not have appeared without water, the primary component of living cells. The unique structure of water and its ability to form hydrogen bonds between adjacent water molecules are responsible for its special properties: solvency, high heat capacity, boiling point, surface tension, and lower density as a solid than as a liquid.

Life also depends critically on the chemistry of carbon. Carbon is especially versatile in bonding with itself and with other atoms, and it is the only element capable of forming the large molecules found in living organisms. Carbohydrates are composed primarily of carbon, hydrogen, and oxygen grouped as H—C—OH. The simplest carbohydrates are sugars, which serve as immediate sources of energy in living systems. Monosaccharides, or simple sugars, may bond together to form disaccharides or polysaccharides, which serve as storage forms of sugar or perform structural roles. Lipids constitute another class of large molecules featuring chains of carbon compounds; fats exist principally as triglycerides, phospholipids, and steroids. Proteins are large molecules composed of amino acids linked by peptide bonds. Many proteins function as enzymes that catalyze biological reactions. Each kind of protein has a characteristic primary, secondary, tertiary, and often, quaternary structure critical for its functioning. Nucleic acids are polymers of nucleotide units, each composed of a sugar, a nitrogenous base, and a phosphate group. They contain the material of inheritance and function in protein synthesis.

Experiments by Louis Pasteur in the 1860s convinced scientists that organisms do not arise repeatedly from inorganic matter. About 60 years later, A. I. Oparin and J. B. S. Haldane provided an explanation for how a common ancestor of all living forms could have arisen from nonliving matter almost 4 billion years ago. The origin of life followed a long period of "abiogenic molecular evolution" on earth in which organic molecules slowly accumulated in a "primordial soup." The atmosphere of the primitive earth was reducing, with little or no free oxygen present. Ultraviolet radiation, electrical discharges of lightning, or energy from hydrothermal vents could have provided energy for early formation of organic molecules. Stanley Miller and Harold Urey demonstrated the plausibility of the Oparin-Haldane hypothesis by simple but ingenious experiments. The concentration of reactants necessary for early synthesis of organic molecules might have been provided inside vesicles formed by aggregations of fatty acids and long-chain alcohols into precellular membranes. RNA might have been the primordial biomolecule, performing the functions of both genetic coding of information and catalysis. When self-replicating systems became established, evolution by natural selection could have increased their diversity and complexity.

The first organisms are hypothesized to have been primary heterotrophs, living on energy stored in molecules dissolved in a primordial soup. Later evolution produced autotrophic organisms, which can synthesize their own organic nutrients (carbohydrates) from inorganic materials. Autotrophs are better protected than heterotrophs from depletion of organic compounds from their environments. Molecular oxygen began to accumulate in the atmosphere as an end product of photosynthesis, an autotrophic process that produces sugars and oxygen by reacting water and carbon dioxide. Cyanobacteria appear to be primarily responsible for generation of atmospheric oxygen early in life's history. Accumulation of atmospheric oxygen produced a revolution in earth's biota. It poisoned many susceptible forms and favored ones that could at least tolerate if not use oxygen. Oxygen-based metabolism would give rise to eukaryotic cells.

Archaea and Bacteria lack a membrane-bound nucleus and other organelles in their cytoplasm. The term "prokaryote" traditionally designates this grade of cellular organization, but the "prokaryotes" do not form a monophyletic group.

The eukaryotes apparently arose from symbiotic unions of two or more types of prokaryotes. The genetic material (DNA) of eukaryotes is borne in a membrane-bound nucleus, and also in mitochondria and sometimes plastids. Mitochondria and plastids have resemblances to bacteria, and their DNA is more closely allied to that of certain bacteria than to eukaryotic nuclear genomes.

REVIEW QUESTIONS

1. Explain each of these properties of water, and describe how each is conferred by the dipolar nature of a water molecule: high specific heat capacity; high heat of vaporization; unique density behavior; high surface tension; capacity to be a good solvent for ions of salts.

2. What was the composition of the earth's atmosphere at the time of the origin of life, and how did it differ from the atmosphere of today?

3. Regarding the experiments of Miller and Urey described in this chapter, explain what constituted the following in each case: observations, hypothesis, deduction, prediction, data, control. (The scientific method is described on pp. 10–12.)

4. Explain the significance of the Miller-Urey experiments.

5. Name three different sources of energy that could have powered reactions on early earth to form organic compounds.

6. By what mechanism might organic molecules have been concentrated within a semipermeable membrane in a prebiotic world so that polymerization reactions could occur?

7. Name two simple carbohydrates, two storage carbohydrates, and a structural carbohydrate.

8. What characteristic differences in molecular structure distinguish lipids and carbohydrates?

9. Explain the difference between the primary, secondary, tertiary, and quaternary structures of a protein.

10. What are the important nucleic acids in a cell, and of what units are they constructed?

11. Distinguish among the following: primary heterotroph, autotroph, secondary heterotroph.

12. What is the source of oxygen in the present-day atmosphere, and what is its metabolic significance to most organisms living today?

13. Distinguish prokaryotes from eukaryotes.

14. Describe Margulis's view on the origin of eukaryotes from prokaryotes.

15. What was the "Cambrian explosion" and how might you explain it?

For Further Thought   Why will our knowledge of the origin of life always be more tentative than our knowledge of the subsequent evolution of life's diversity?

SELECTED REFERENCES

Berg, J. M., T. L. Tymoczko, G. J. Gatto, Jr., and L. Stryer. 2015. Biochemistry, ed. 8. New York, W. H. Freeman. *A current and thorough textbook of biochemistry.*

Deamer, D. 2012. First life: discovering the connections between stars, cells, and how life began. Berkeley, University of California Press. *Book-length summary of the latest research and theory of life's origins by a leading contributor.*

Deamer, D., J. P. Dworkin, S. A. Sanford, M. P. Bernstein, and L. J. Allamandola. 2002. The first cell membranes. Astrobiology **2:** 371–381. *A good explanation of how amphiphilic molecules can self-assemble vesicles proposed as the origin of cell membranes.*

Falkowski, P. G., and A. Knoll (eds.). 2007. Evolution of the Primary Producers in the Sea. Amsterdam, The Netherlands, Elsevier Academic Press. *Chapters cover origins of oxygenic photosynthesis, origins of eukaryotic cells and mitochondria, and the geological succession of oceanic life.*

Fenchel, T. 2003. Origin and early evolution of life. Oxford, Oxford Univ. Press. *A review of current theories of the origin and early diversification of living forms.*

Gesteland, R. F., and J. F. Atkins (eds.) 1999. The RNA world. Cold Spring Harbor, New York, Cold Spring Harbor Laboratory Press. *Evidence that there was a period when RNA served in both catalysis and transmission of genetic information.*

Glansdorff, N., Y. Xu, and B. Labedan. 2008. The last universal common ancestor: emergence, constitution and genetic legacy of an elusive forerunner. Biology Direct **3:** 29. *Argues that the most recent common ancestor of life was a community of protoeukaryotes with an RNA genome.*

Kasting, J. F. 1993. Earth's early atmosphere. Science **259:** 920–926. *Most investigators agree that there was little or no oxygen in the atmosphere of early earth and that there was a significant increase about 2 billion years ago.*

Lehninger, A. L., D. L. Nelson, and M. M. Cox. 2012. Lehninger principles of biochemistry, ed. 6. New York, W. H. Freeman & Co. *Clearly presented advanced textbook in biochemistry.*

Lodish, H., A. Berk, C. A. Kaiser, M. Krieger, A. Bretscher, H. Ploegh, A. Amon, and M. P. Scott. 2012. Molecular cell biology, ed. 7. New York, W. H. Freeman &

Co. *Thorough treatment; begins with fundamentals such as energy, chemical reactions, bonds, pH, and biomolecules, then proceeds to advanced molecular biology.*

Margulis, L. 1998. Symbiotic planet: a new look at evolution. New York, Basic Books. *An important discussion of symbiogenesis in evolution.*

Morris, S. C. 1998. The crucible of creation: the Burgess shale and the rise of animals. Oxford, Oxford Univ. Press. *An important summary correlating fossil and molecular evidence.*

CHAPTER 3

Cells as Units of Life

Fluorescence microscopy of fibroblasts in culture reveal cellular detail. Nucleus (purple), actin filaments (pale blue) and microtubules (yellow).

The Fabric of Life

It is remarkable that living forms, from amebas and unicellular algae to whales and giant redwood trees, share a single type of building unit: cells. All animals and plants are composed of cells and cell products. New cells come from division of preexisting cells, and the activity of a multicellular organism as a whole is the sum of activities and interactions of its constituent cells. Thus the cell theory, originally proposed by Schleiden and Schwann, is another great unifying concept of biology.

Energy is required to support life's activities, and virtually all of it flows from sunlight that is captured by green plants and algae and transformed by photosynthesis into chemical bond energy. Chemical

bond energy is a form of potential energy that can be released when a bond is broken and used to perform electrical, mechanical, and osmotic tasks in the cell. Ultimately, all energy is dissipated into heat in accord with the second law of thermodynamics, which states that nature proceeds toward a state of greater molecular disorder, or entropy. Thus the high degree of molecular organization in living cells is attained and maintained only as long as energy fuels the organization. See Chapter 4, opening essay.

Techniques in fluorescence microscopy reveal cellular components in unprecedented detail, as shown in the accompanying photograph.

3.1 CELL CONCEPT

More than 300 years ago the English scientist and inventor Robert Hooke, using a primitive compound microscope, observed boxlike cavities in slices of cork and leaves. He called these compartments "little boxes or cells." In the years that followed Hooke's first demonstration of the remarkable powers of the microscope to the Royal Society of London in 1663, biologists gradually learned that cells were far more than simple containers filled with "juices."

Cells are the fabric of life (Figure 3.1) and are complex structures that form the basic units of all living organisms. In single-celled organisms all functions of life are performed within the confines of one microscopic package. In multicellular organisms, such as humans, cells interact, each performing its specialized role in an organized partnership. There is no life without cells. The idea that a cell represents the basic structural and functional unit of life is an important unifying concept of biology.

With the exception of some egg cells, which are the largest cells (in volume) known, cells are small and mostly invisible to the unaided eye. Consequently, our understanding of cells paralleled technical advances in the resolving power of microscopes. The Dutch microscopist Antoni van Leeuwenhoek sent letters to the Royal Society of London containing detailed descriptions of the numerous organisms he had observed using high-quality single lenses that he

had made (1673 to 1723). In the early nineteenth century, improved design of microscopes permitted biologists to see separate objects only 1 μm apart, and laid the groundwork for the **cell theory**—a theory stating that all living organisms are composed of cells.

In 1838 Matthias Schleiden, a German botanist, announced that all plant tissue was composed of cells. A year later one of his countrymen, Theodor Schwann, described animal cells as being similar to plant cells, an understanding that had been long delayed because animal cells are bounded only by a nearly invisible plasma membrane rather than a distinct cell wall characteristic of plant cells. Schleiden and Schwann are thus credited with the unifying cell theory that guided a new era of productive exploration in cell biology. Another German, Rudolf Virchow, recognized that all cells came from preexisting cells (1858).

In 1840 J. Purkinje introduced the term **protoplasm** to describe cell contents. Protoplasm was at first described as a granular, gel-like mixture with special and elusive life properties of its own; cells were viewed as bags of thick soup containing a nucleus. Later the interior of cells became increasingly visible as microscopes were improved and better tissue-sectioning and staining techniques were introduced. Rather than being a uniform granular soup, a cell's interior is composed of numerous cellular **organelles** within a network of membranes. The components of a cell are so highly organized, structurally

CELLULAR LEVEL

Atoms   Molecule   Macromolecule   Organelle   Cell

ORGANISMAL LEVEL

Tissue   Organ   Organ system   Organism

© Russell Illig/Getty Images

Figure 3.1 Biological organization from simple atoms to complex organisms. The atoms of molecules and macromolecules are assembled into organelles within each cell. Cells are grouped into tissues, organs, and organ systems to form a complex multicellular organism.

and functionally, that describing its contents as "protoplasm" is like describing the contents of an automobile engine as "autoplasm."

How Cells Are Studied

Light microscopes, with all their variations and modifications, have, for the past 300 years, contributed more to biological investigation than any other instrument. They continue to do so more than 50 years after invention of the electron microscope (Figure 3.2).

Electron microscopy has vastly enhanced our appreciation of the internal organization of cells, and modern biochemical, immunological, physical, and molecular techniques have contributed enormously to our understanding of cell structure and function. Electron microscopes use magnets to direct a beam of electrons through or at the surface of objects examined (Figure 3.2). The wavelength of the electron beam is approximately 0.00001 that of ordinary white light, thus permitting far greater magnification and resolution.

In preparation for viewing using the transmission electron microscope, specimens are cut into extremely thin sections (10 nm to 100 nm thick) and treated with "electron stains" (ions of elements such as osmium, lead, and uranium) to increase contrast between different structures. Electrons pass through a specimen where cellular structures are less dense and are deflected where cellular structures are "stained." The images are seen on a fluorescent screen and photographed (Figure 3.2).

In contrast, specimens prepared for scanning electron microscopy are not sectioned, and electrons do not pass through them. The whole specimen is coated with an electron-dense material, such as gold or platinum, and bombarded with electrons, causing some electrons to be reflected back and secondary electrons to be emitted. An apparent three-dimensional image is recorded in the photograph (Figure 3.2). Although the magnification capability of scanning instruments is not as great as that of transmission microscopes, much has been learned about surface features of organisms and cells, as well as internal, membrane-bound structures. Examples of scanning electron micrographs are shown on pages 139, 158 and 680.

A still greater level of resolution can be achieved with X-ray crystallography and nuclear magnetic resonance (NMR) spectroscopy. These techniques reveal the shapes of biomolecules and the relationships among atoms within them. Both techniques are laborious, but NMR spectroscopy does not require purification and crystallization of a substance, and molecules can be observed in solution.

Advances in techniques of cell study (cytology) are not limited to improvements in microscopes but include new methods of tissue preparation and staining for microscopic study, as well as contributions from modern biochemistry and molecular biology. Cells can be disrupted with most of the organelles remaining intact, then centrifuged in a density gradient (Figure 3.3), and relatively pure preparations of each organelle may be recovered. Thus the biochemical functions of various organelles may be studied separately.

DNA and various types of RNA can be extracted and studied. Many enzymes can be purified and their characteristics determined. We use radioactive isotopes to study many metabolic reactions and pathways in cells. Modern chromatographic techniques can separate chemically similar intermediates and products. A particular cellular protein can be extracted and purified, and specific antibodies (see p. 770) against the protein can be prepared. When the

Figure 3.2 Comparison of optical paths of light, a transmission, and scanning electron microscopes. To facilitate comparison, the scheme of the light microscope has been inverted from its usual orientation with light source below and image above. In electron microscopes the lenses are magnets to focus the beam of electrons.

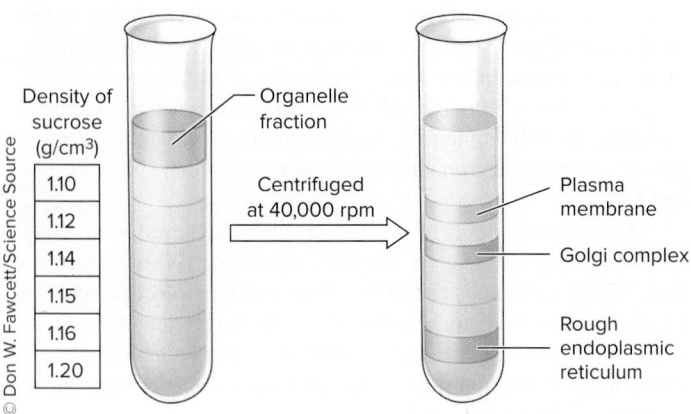

© Don W. Fawcett/Science Source

Figure 3.3 Separation of cell organelles in a density gradient by ultracentrifugation. The gradient is formed by layering sucrose solutions in a centrifuge tube, then carefully placing a preparation of mixed organelles on top. The tube is centrifuged at about 40,000 revolutions per minute for several hours, and organelles become separated down the tube according to their density.

antibody is complexed with a fluorescent substance and the complex is used to "stain" cells, the complex binds to the protein of interest, and its precise cellular location can be determined (see chapter opener figure and Figure 3.13A, p. 43). With the development of such fluorescence stains and confocal microscopy, we are now able to view cells in 3-D and cellular processes, such as cell division, can be observed in real time.

3.2   ORGANIZATION OF CELLS

If we were to restrict our study of cells to fixed and sectioned tissues, we would be left with the erroneous impression that cells are static, quiescent, rigid structures. In fact, a cell's interior is in a state of constant flux. Most cells are continually changing shape; their organelles twist and regroup in a cytoplasm teeming with starch granules, fat droplets, and vesicles of various kinds. This description is derived from studies of living cell cultures with time-lapse photography and video. At the submicroscopic level, the swift shuttling of molecular traffic through protein channels or carriers (see pp. 45 and 46) in the plasma membrane, the movement of molecules and organelles within the cytoplasm, and the metabolic energy transformations within cell organelles are representative of dynamic phenomena that occur in a highly ordered and regulated manner during cell functioning.

Prokaryotic and Eukaryotic Cells

We already have described the radically different cell plan of prokaryotes and eukaryotes (pp. 30–31). A fundamental distinction, expressed in their names, is that prokaryotes lack the membrane-bound nucleus present in all eukaryotic cells. Among other differences, eukaryotic cells contain many membrane-bound organelles (Table 3.1).

Despite these differences, prokaryotes and eukaryotes have much in common. Both have DNA, use the same genetic code, and synthesize proteins. Many specific molecules such as ATP (adenosine triphosphate) perform similar roles in both. These fundamental similarities imply common ancestry. The upcoming discussion is restricted to eukaryotic cells, of which all animals are composed.

Components of Eukaryotic Cells and Their Functions

Typically, eukaryotic cells are enclosed within a thin, selectively permeable **plasma membrane** (Figure 3.4). The most prominent organelle is a spherical or ovoid **nucleus,** enclosed within *two*

TABLE 3.1

Comparison of Prokaryotic and Eukaryotic Cells

Characteristic	Prokaryotic Cell	Eukaryotic Cell
Cell size	Mostly small (1–10 µm)	Mostly large (10–100 µm)
Genetic system	DNA with some DNA-binding protein; simple, circular DNA molecule in nucleoid; nucleoid is not membrane bound	DNA complexed with DNA-binding proteins in complex linear chromosomes within nucleus with membranous envelope; circular mitochondrial and chloroplast DNA
Cell division	Direct by binary fission or budding; no mitosis	Some form of mitosis; centrioles in many; mitotic spindle present
Sexual system	Absent in most; highly modified if present	Present in most; male and female partners; gametes that fuse to form zygote
Nutrition	Absorption by most; photosynthesis by some	Absorption, ingestion; photosynthesis by some
Energy metabolism	No mitochondria; oxidative enzymes bound to cell membrane, not packaged separately; great variation in metabolic pattern	Mitochondria present; oxidative enzymes packaged therein; more unified pattern of oxidative metabolism
Intracellular movement	None	Cytoplasmic streaming, phagocytosis, pinocytosis
Flagella/cilia	If present, not with "9 + 2" microtubular pattern	If present, with "9 + 2" microtubular pattern
Cell wall	Contains disaccharide chains cross-linked with peptides in Bacteria, but not in Archaea. Archeons have ester-linked membrane lipids.	If present, not with disaccharide polymers linked with peptides

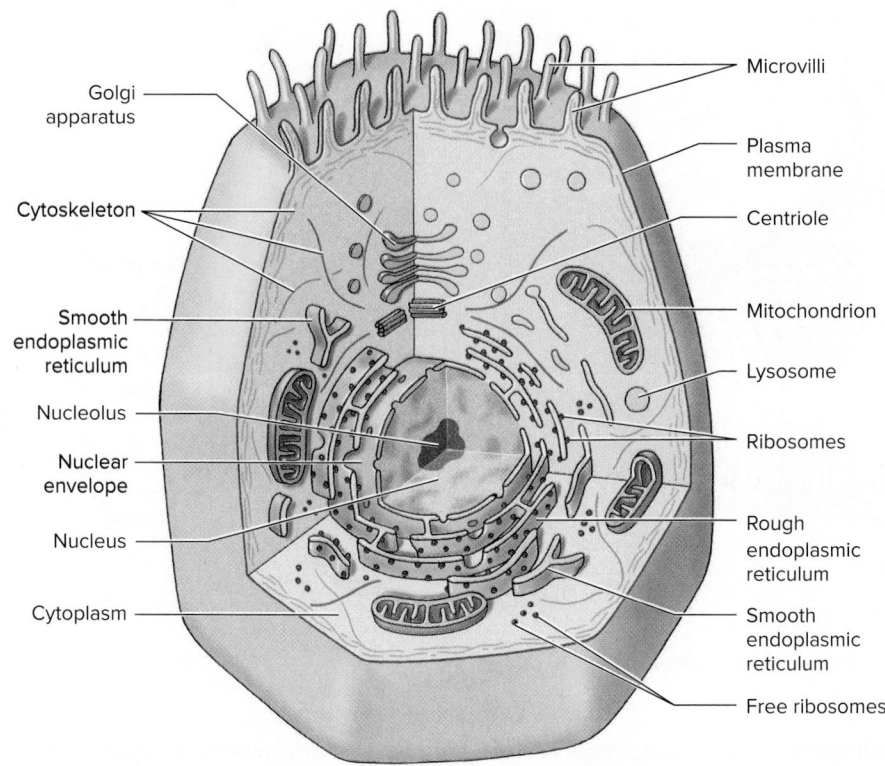

Golgi apparatus

Cytoskeleton

Smooth endoplasmic reticulum

Nucleolus

Nuclear envelope

Nucleus

Cytoplasm

Microvilli

Plasma membrane

Centriole

Mitochondrion

Lysosome

Ribosomes

Rough endoplasmic reticulum

Smooth endoplasmic reticulum

Free ribosomes

Figure 3.4 Generalized cell with principal organelles, as might be seen with the electron microscope.

© Don W. Fawcett/Science Source

Figure 3.5 Plasma membranes of two adjacent cells. Each membrane (*between arrows*) shows a typical dark-light-dark staining pattern.

membranes to form a double-layered **nuclear envelope** (Figure 3.4). Cellular material located between the plasma membrane and the nuclear envelope is collectively called **cytoplasm.** Within the cytoplasm are many organelles, such as mitochondria, Golgi complexes, centrioles, and endoplasmic reticulum. In addition, plant cells typically contain **plastids,** some of which are photosynthetic organelles, and plant cells bear a cell wall containing cellulose outside the plasma membrane.

The **fluid-mosaic model** is the currently accepted concept describing plasma membrane structure. By electron microscopy, a plasma membrane appears as two dark lines, each approximately 3 nm thick, at each side of a light zone (Figure 3.5). The entire membrane is 8 to 10 nm thick. This image is the result of a phospholipid bilayer, two layers of phospholipid molecules, all oriented with their water-soluble (hydrophilic) head regions toward the outside and their fatsoluble (hydrophobic) tail regions toward the inside of the membrane (Figure 3.6). An important characteristic of the phospholipid bilayer is that it is fluidlike, giving the membrane flexibility and allowing the phospholipid molecules to move sideways freely within their own monolayer. Molecules of cholesterol are interspersed in the lipid portion of the bilayer (Figure 3.6). They make the membrane even less permeable to water-soluble ions and molecules and decrease membrane flexibility. One way that cells acclimate to changes in temperature is to add or remove cholesterol to make the membrane less or more flexible, respectively, in an effort to maintain function.

Glycoproteins (proteins with carbohydrates attached) are essential components of plasma membranes (Figure 3.6). Some of

these proteins act as transporters of substances such as polar molecules or charged ions across the membrane (see p. 44, Membrane Function). Others act as specific receptors for various molecules or as highly specific cell markers. For example, the self/nonself recognition that enables the immune system to react to invaders and not to our own cells (see Chapter 35) is based on proteins of this type. Some aggregations of protein molecules form pores or channels through which charged ions or small polar molecules may enter (see gap junctions, p. 44, and channels, p. 45). Like the phospholipid molecules, most glycoproteins can move laterally in the membrane, most often associated with surrounding phospholipid molecules to form moving lipid rafts.

Nuclear envelopes contain less cholesterol than plasma membranes, and complex protein pores in the envelope (Figure 3.7) allow selected molecules to move between nucleus and cytoplasm and vice versa. Nuclei contain linear **chromosomes** suspended in **nucleoplasm.** The chromosomes are normally loosely condensed, flexible strands of **chromatin,** composed of a complex of DNA, and DNA-binding proteins. Chromosomal DNA carries the genetic information encoding cellular RNA and protein molecules (see Chapter 5). The linear chromosomes become condensed and visible as discrete structures only during cell division (see p. 50 for mitosis and p. 72 for meiosis). The **nucleolus** is composed of specialized parts of certain chromosomes that stain in a characteristically dark manner. Nucleoli carry multiple copies of DNA information used to synthesize ribosomal RNA. After transcription from DNA, ribosomal RNA combines with protein to form the two subunits of **ribosomes,** which leave the nucleolus and pass to the cytoplasm

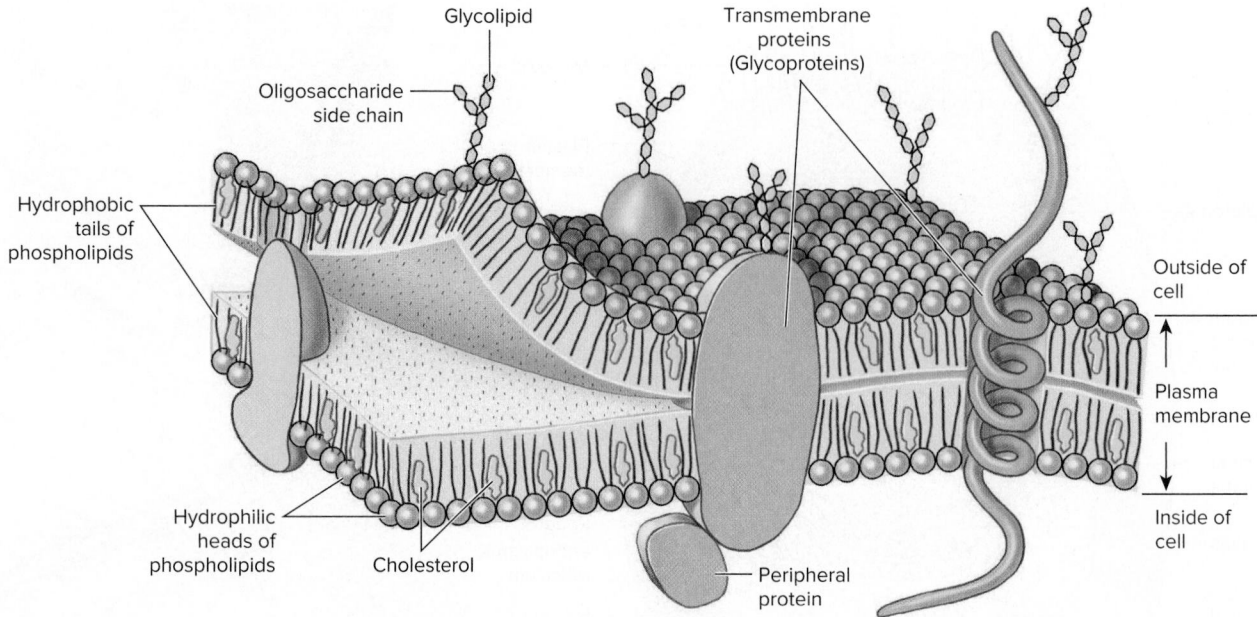

Figure 3.6   Diagram illustrating the fluid-mosaic model of a plasma membrane.

Figure 3.7   Transmission electron micrograph of a portion of a human kidney epithelial cell. Mitochondria and glycogen granules are visible in the cytoplasm, and pores (arrows) can be seen in the nuclear envelope. (×34000)

through pores in the nuclear envelope. Ribosomes are sites of polypeptide or protein synthesis. They perform this function free within the cytoplasm when manufacturing polypeptides for use in the cytoplasm, organelles, or nucleus. Alternatively, they become attached to **endoplasmic reticulum (ER)** when manufacturing polypeptides destined for the plasma membrane, for lysosomes, or for export from the cell.

The outer membrane of the nuclear envelope is continuous with a cytoplasmic **endomembrane system** composed of the ER (Figure 3.7 and Figure 3.8), the **Golgi complex** or **apparatus, lysosomes,** the plasma membrane, and the vesicles that pass between them. The space between the membranes of the nuclear envelope communicates with the space between the ER membranes (**cisterna,** pl., **cisternae**). The ER membranes may be covered on their outer surfaces with ribosomes and are thus designated **rough ER.** This is one of the sites of polypeptide synthesis, mentioned on page 39. Polypeptides synthesized on rough ER enter the ER cisternae or membrane and are destined for incorporation into the plasma membrane (Figure 3.9), for export from the cell, or for use by the lysosomes. When ER lacks a ribosomal covering, it is called **smooth ER.** Smooth ER functions in synthesis of lipids and phospholipids and as a detoxification site within cells.

The Golgi complex or apparatus (Figure 3.9 and Figure 3.10) is composed of a stack of membranous cisternae that function in modification and packaging of polypeptide and protein products produced by rough ER. Small vesicles of ER membrane containing polypeptide or protein detach and then fuse with sacs on the *cis* or "forming face" of a Golgi complex. During the modification process, polypeptides or proteins may be cleaved to form smaller polypeptides or proteins, or they may have complex carbohydrates added to them as they move through the Golgi cisternae until they reach the *trans* or "maturing face" of the complex (Figure 3.9 and Figure 3.10). Finally, vesicles detach from the *trans* face of the complex and their contents may be expelled to the outside of the cell, as secretory products such as from a glandular cell. Some vesicles may carry transmembrane polypeptides or proteins for incorporation into the plasma membrane, such as receptors or carrier proteins. Others may contain enzymes that remain in the same cell that produced them. Such vesicles are called **lysosomes** (literally "loosening body," a body capable

Ribosomes

Rough
endoplasmic
reticulum

Smooth
endoplasmic
reticulum

A   B

© Don W. Fawcett/Science Source

Figure 3.8   Endoplasmic reticulum. **A,** Endoplasmic reticulum is continuous with the nuclear envelope. It may have associated ribosomes (rough endoplasmic reticulum) or not (smooth endoplasmic reticulum). **B,** Electron micrograph showing rough and smooth endoplasmic reticulum.

of causing lysis, or disintegration) (Figure 3.10). Enzymes that they contain are involved in the breakdown of foreign material, including bacteria engulfed by a cell. Lysosomes also destroy injured or diseased cells and worn-out cellular components. Their enzymes are so powerful that they kill the cell that formed them if enough of the lysosome membranes rupture. In normal cells the enzymes remain safely enclosed within the protective vesicle membranes. Lysosomal vesicles may pour their enzymes into a larger membrane-bound body containing an ingested food particle, a **food vacuole** or **phagosome** (see Figure 3.20), as is the case in unicellular eukaryotes (see Chapter 11, p. 224).

Mitochondria (sing., **mitochondrion**) (Figure 3.11) are conspicuous organelles present in nearly all eukaryotic cells. They are diverse in size, number, and shape; some are rodlike, and others are nearly spherical. They may be scattered uniformly throughout the cytoplasm or localized near regions of high metabolic activity. A mitochondrion is composed of a double membrane. The outer membrane is smooth, whereas the inner membrane is folded into numerous platelike or fingerlike projections called **cristae** (sing., **crista**; Figure 3.11), which increase the internal surface area where chemical reactions occur. These characteristic features make mitochondria easy to identify among organelles. Mitochondria are often called "powerhouses of cells," because enzymes located on the cristae catalyze the energy-yielding steps of aerobic metabolism (see Figure 4.14, p. 64). Most of a cell's ATP, the most important energy-transfer molecule of all cells, is produced in this organelle. Mitochondria are self-replicating. They have a tiny, circular genome, much like genomes of prokaryotes except much smaller, which contains DNA specifying some, but not all, proteins of a mitochondrion.

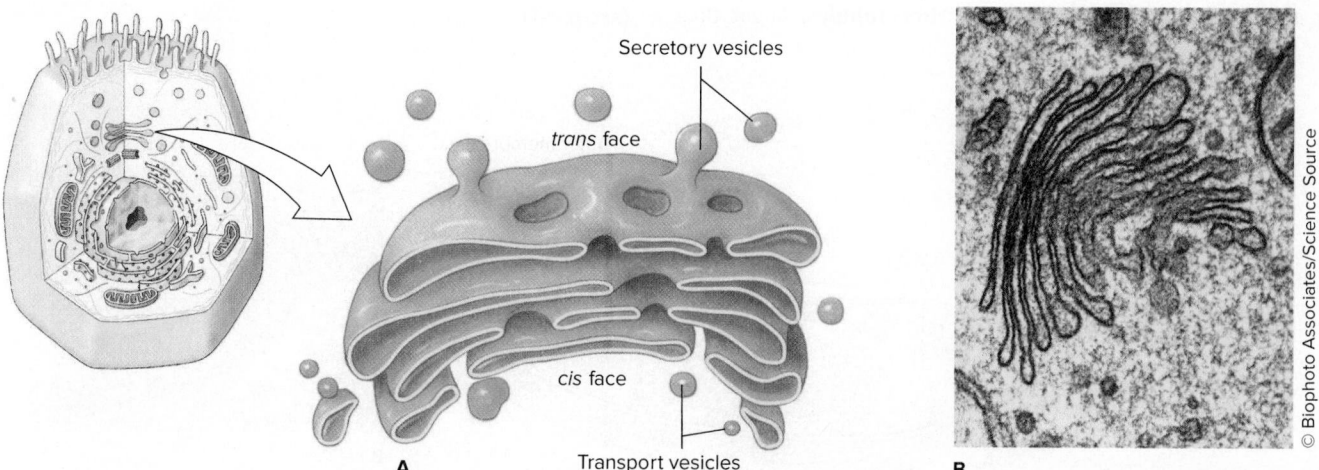

Secretory vesicles

trans face

cis face

Transport vesicles

A   B

© Biophoto Associates/Science Source

Figure 3.9   Golgi complex (= Golgi body, Golgi apparatus). **A,** The smooth cisternae of the Golgi complex have enzymes that modify polypeptides or proteins synthesized by the rough endoplasmic reticulum. **B,** Transmission electron micrograph of a Golgi complex. (×96,000)

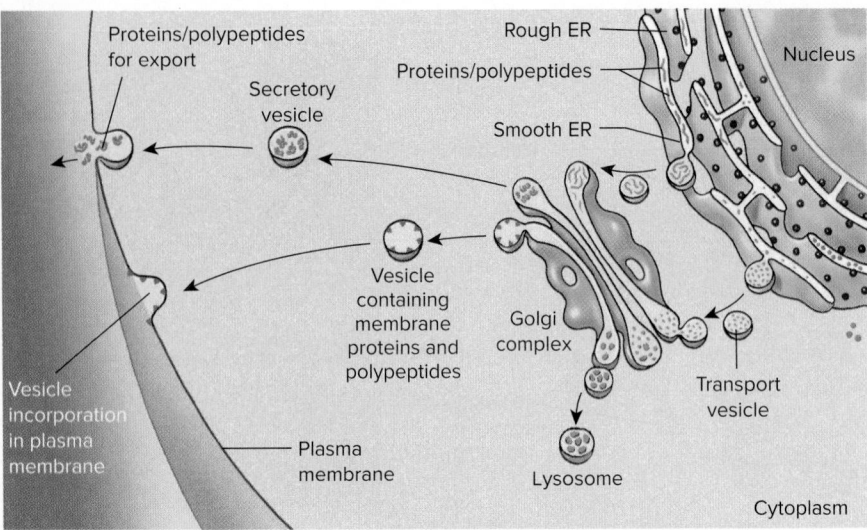

Figure 3.10   The endomembrane system—a system in eukaryotic cells for assembling, isolating, and secreting polypeptides and proteins for export, for lysosomes, or for incorporation into the plasma membrane.

Eukaryotic cells characteristically have a system of tubules and filaments that form a **cytoskeleton** (Figure 3.12 and Figure 3.13). These provide support and maintain the form of cells, as well as a means of cellular locomotion and movement of macromolecules and organelles within a cell. The cytoskeleton is composed of microfilaments, microtubules, and intermediate filaments. **Microfilaments** are thin, linear structures, observed distinctly in some protozoan groups and macrophages of the immune system, where they facilitate cellular locomotion (see Chapters 9 and Chapter 35, respectively), and in some cell types, such as muscle cells, where they cause cellular contraction (see Chapter 29). They are made of a protein called **actin.** Several dozen other proteins (called actin-binding proteins or ABPs) bind actin and determine its configuration and behavior in particular cell types. One ABP is **myosin,** whose interaction with actin causes contraction in muscle and other cells (p. 653). Actin is also instrumental in cytokinesis, the process of cytoplasmic division that takes place at the end of mitosis (p. 51) and meiosis (see Chapter 5), as well as in endocytosis and exocytosis (pp. 48 and 49). **Microtubules,** larger than

microfilaments, are hollow, tubular structures composed of a protein called **tubulin** (Figure 3.13). Each tubulin molecule is actually a doublet, or dimer, composed of two globular proteins. The molecules are attached head-to-tail to form a strand, and 13 strands aggregate to form a microtubule. Because the tubulin subunits in a microtubule are always attached head-to-tail, the ends of the microtubule differ chemically and functionally. One end (called the plus end) both adds and deletes tubulin subunits more rapidly than the other end (the minus end). Microtubules play a vital role in moving chromosomes toward daughter cells during cell division (p. 50), and they are important in intracellular architecture, organization, and transport. They provide a means for movement of molecules and organelles through the cytoplasm, as well as movement of messenger RNA (p. 88) from the nucleus to particular positions within the cytoplasm. Microtubules and associated motor proteins are also important in movement of vesicles between the ER, the Golgi complex, and the plasma membrane or lysosomes. In addition, microtubules form essential parts of the structures of cilia and flagella (see next section, Surfaces of Cells and Their Specializations, and Chapter 11, Unicellular Eukaryotes). Microtubules radiate from a microtubule organizing center, the **centrosome,** near the nucleus. Centrosomes are not membrane bound. Within centrosomes are found a pair of **centrioles** (Figure 3.4 and Figure 3.14), which are themselves composed of microtubules. Each centriole of a pair lies at right angles to the other and is a short cylinder of nine triplets of microtubules. They replicate before cell division. Although cells of higher plants do not have centrioles, a microtubule organizing center is present. **Intermediate filaments** are larger than microfilaments but smaller than microtubules. There are six biochemically distinct subtypes of intermediate filaments, and their composition and arrangement depend on the cell type in which they occur. These filaments resist cell stretching, and help to hold adjacent cells together. They are particularly prominent in skin cells associated with desmosomes (see p. 44).

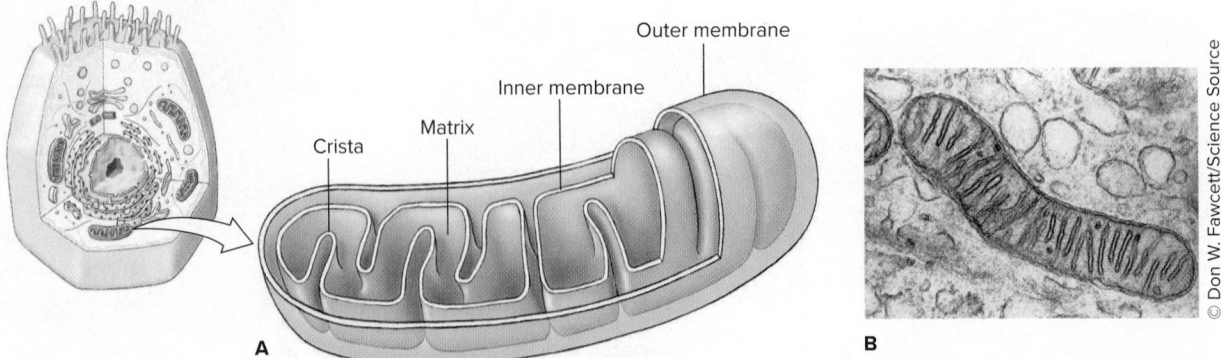

Figure 3.11   Mitochondria. **A,** Structure of a typical mitochondrion. **B,** Electron micrograph of mitochondria in cross section (top right) and longitudinal section (lower organelle).

Figure 3.12 Cytoskeleton of a cell, showing its complex nature. Three visible cytoskeletal elements, in order of increasing diameter, are microfilaments, intermediate filaments, and microtubules. (×66,600)

Figure 3.13 **A,** The microtubules in kidney cells of a kangaroo rat have been rendered visible by treatment with a preparation of fluorescent proteins that specifically bind to tubulin (green), actin filaments (red) and nuclei (blue). **B,** A microtubule is composed of 13 strands of tubulin molecules, and each molecule is a dimer. Tubulin dimers are added to and removed from the (+) end of the microtubule more rapidly than at the (−) end.

Surfaces of Cells and Their Specializations

The free surface of epithelial cells (cells that cover the surface of a structure or line a tube or cavity; see p. 191) sometimes bears either **cilia** or **flagella** (sing., **cilium, flagellum**). These are motile extensions of the cell surface that may be used to sweep materials past the cell. This technique is used during feeding in some unicellular eukaryotes (p. 224) and in sponges (p. 250). In many single-celled eukaryotes and some small multicellular organisms, they propel the entire organism through a liquid medium (see pp. 230, 283). Flagella provide the means of locomotion for male reproductive cells (sperm) of most animals (see p. 141) and many plants. In addition to their function in cell movement and fluid movement around cells, cilia are proposed to play an integral role in cell signaling, both during development and in the adult organism, from simple eukaryotes to mammals.

Cilia and flagella have different beating patterns (see p. 649 and Figure 29.12), but their internal structure is the same. With few exceptions, the internal structures of locomotory cilia and flagella are composed of a long cylinder of nine pairs of microtubules enclosing a central pair (see Figure 29.11). At the base of each cilium or flagellum is a **basal body (kinetosome)**, which is identical in structure to a single centriole. Cilia and flagella movement mechanisms are described in Chapter 29, p. 649.

Many cells move neither by cilia nor by flagella but by **ameboid movement** using **pseudopodia**. Some groups of unicellular eukaryotes (p. 222), migrating cells in embryos of multicellular animals, and some cells of adult multicellular animals, such as lymphocytes and macrophages (see Chapter 35), show ameboid movement. Cytoplasmic streaming through the assembly and disassembly of actin microfilaments extends a cytoplasmic process (pseudopodium) outward from the surface of the cell. Continued streaming in the direction of a pseudopodium brings cytoplasmic organelles into the process, followed by the rest of the cell, and accomplishes movement of the entire cell. Some specialized pseudopodia have cores of microtubules (p. 221), and movement is effected by assembly and disassembly of the tubulin subunits.

Cells covering the surface of a structure (epithelial cells; see p. 289) or cells packed together in a tissue may have specialized junctional complexes between them. Nearest the free or exposed surface, the membranes of two cells next to each other appear to fuse, forming a **tight junction** (Figure 3.15A). They are formed from rows of transmembrane proteins that bind tightly between adjacent cells. There is usually a space of about 20 nm between the plasma membranes of

Microtubule
triplet

A

B

© Don W. Fawcett/Science Source

Figure 3.14 The centrosome. **A,** Each centrosome contains a pair of centrioles and each centriole is composed of nine triplets of microtubules arranged as a cylinder. **B,** Electron micrograph of a pair of centrioles, one in longitudinal (*right*) and one in cross section (*left*). The normal orientation of centrioles is at right angles to each other.

adjacent cells; however, tight junctions close this gap and function as seals to prevent the passage of molecules between cells from one side of a layer of cells to another. The number of rows of transmembrane proteins in the tight junction determines how tightly adjacent cells are sealed together. Tight junctions between intestinal cells, for example, do not allow passage of molecules between the cells and thus enhance regulated absorption of molecules from intestinal contents (see Chapter 32, p. 713). **Adhesion junctions** (Figure 3.15A) occur just beneath tight junctions. These anchoring junctions are similar to tight junctions in that they encircle the cell. They are different from tight junctions in that they do not seal adjacent cells to each other. Rather, the transmembrane proteins link together across a small intercellular space. Inside adjacent cells, the transmembrane proteins are attached to actin microfilaments and thus indirectly attach the cytoskeletons of adjacent cells to each other. Modified adhesion junctions occur between cardiac muscle cells, and these hold the cells together as the heart beats throughout the life of an organism (p. 688). At various points beneath tight junctions and adhesion junctions, in epithelial cells, small ellipsoid discs occur, within the plasma membrane in each adjacent cell. These appear to act as "spot-welds" and are called **desmosomes** (Figure 3.15A). From each desmosome a tuft of

intermediate filaments extends into the cytoplasm connecting desmosomes within a cell, and transmembrane linker proteins extend through the plasma membrane into the intercellular space to bind the desmosomal discs of adjacent cells together. Desmosomes are not seals but seem to increase the strength of the tissue. Many are found between the cells of the skin in vertebrates (p. 640). **Hemidesmosomes** (Figure 3.15A) occur at the base of cells and anchor them to underlying connective tissue layers. **Gap junctions** (Figure 3.15A), rather than serving as points of attachment, provide a means of intercellular communication. They form tiny canals between cells, so that their cytoplasm becomes continuous, and small molecules and ions can pass from one cell to the other. Gap junctions may occur between cells of epithelial, nervous, and muscle tissues (see Chapter 9, p. 190).

Another specialization of cell surfaces occurs where plasma membranes of adjacent cells infold and interdigitate very much like a zipper. These infoldings are especially common in epithelial cells of kidney tubules (see Chapter 30, p. 668) and serve to increase the surface area of the cells for absorption or secretion. The distal or apical boundaries of some epithelial cells, as seen by electron microscopy, show regularly arranged **microvilli** (sing., **microvillus**). They are small, fingerlike projections consisting of tubelike evaginations of the plasma membrane with a core of cytoplasm containing bundles of actin microfilaments (Figure 3.15A and B). They are seen clearly in the lining of the intestine where they greatly increase the absorptive and digestive surface (see Chapter 32, p. 711). Such specializations are often termed brush borders due to their appearance when seen by light microscopy.

Membrane Function

The incredibly thin, yet sturdy, plasma membrane that encloses every cell is vitally important in maintaining cellular integrity. Plasma membranes (also called the plasmalemma) form dynamic structures having remarkable activity and selectivity. They are a permeability barrier that separates the interior from the external environment. They regulate the flow of molecules into and out of the cell, and provide many of the unique functional properties of specialized cells, such as allowing communication with the surrounding extracellular fluid and other cells.

Membranes within a cell surround a variety of organelles (see p. 39), and thus divide the cell into numerous compartments. If all membranes present in one gram of liver tissue were spread out flat, they would cover 30 square meters! Internal membranes share many structural features of plasma membranes and are the site for many of a cell's enzymatic reactions and internal communication systems.

A plasma membrane acts as a selective gatekeeper for entrance and exit of many substances involved in cell metabolism. Some substances can pass through with ease, others enter slowly and with difficulty, and still others cannot enter at all. Because conditions outside a cell are different from and more variable than conditions within a cell, it is necessary that passage of substances across the membrane be rigorously controlled.

We recognize three principal ways that a substance may enter across a cell membrane: (1) by **diffusion** along a concentration gradient; (2) by a **mediated transport** system, in which the substance uses a specific transmembrane protein that assists it across the membrane; and (3) by **endocytosis,** in which the substance is enclosed within a vesicle that forms from the membrane surface and detaches inside the cell.

Figure 3.15 **A,** Junction types and locations are shown in columnar epithelial cells. Actin microfilaments (shown in green) and intermediate filaments (shown in orange) attach adhesion junctions and desmosomes, respectively, to the cytoskeleton. **B,** Electron micrograph of microvilli. (×59,000)

Diffusion

Diffusion is a movement of particles or molecules from an area of higher concentration to an area of lower concentration of the particles or molecules, thus tending to equalize the concentration throughout the area of diffusion. If a living cell surrounded by a membrane is immersed in a solution having a higher concentration of solute molecules than the fluid inside the cell, a concentration gradient instantly exists between the two fluids across the membrane. If the membrane is permeable to the solute, there is a net movement of solute toward the inside of the cell, the side having the lower concentration. The solute diffuses "downhill" across the membrane until its concentrations on each side are equal.

Most cell membranes are **selectively permeable,** often, permeable to water but variably permeable or impermeable to solutes. In free diffusion it is this selectiveness that regulates molecular traffic. As a rule, gases (such as oxygen and carbon dioxide), urea, and lipid-soluble solutes (such as fats, fatlike substances, and alcohol; see p. 24) are the only solutes that can diffuse through biological membranes with any degree of freedom. Water and many water-soluble molecules readily pass through membranes, but movements cannot be explained by simple diffusion. Sugars, water, many electrolytes, and macromolecules are moved across membranes by mediated transport systems.

Diffusion Through Channels

Water and dissolved ions, since they are charged, cannot diffuse through the phospholipid component of the plasma membrane. Instead, they utilize a mediated transport system composed of specialized channels created by transmembrane proteins. Ions and water move through these channels by diffusion. Ion channels allow specific ions of a certain size and charge to diffuse through them. They may allow ion diffusion at all times or they may be **gated channels,** requiring a signal to open or close them. Gated ion channels may open or close when a signaling molecule binds to a specific binding site on the transmembrane protein (**chemically-gated ion channels;** Figure 3.16A), when the ionic charge across a plasma membrane changes (**voltage-gated ion channels;** Figure 3.16B), or when the membrane is distorted (**mechanically-gated ion channels).** Ion diffusion through channels is the basis of signaling mechanisms in the nervous system (see Chapter 33, p. 723) and in muscles (see Chapter 29, p. 654). Water channels are called **aquaporins,** and several different types have been discovered. They are especially important in the digestive system for absorption of water from food (see Chapter 32, p. 713), and in the kidney for water reabsorption during urine formation (see Chapter 30, p. 671).

Osmosis

If we place a membrane between two unequal concentrations of solutes to which the membrane is impermeable, water flows through the membrane from the more dilute to the more concentrated solution. The water molecules move across the membrane down a concentration gradient from an area where the *water* molecules are more concentrated to an area on the other side of the membrane where they are less concentrated. This is **osmosis**—diffusion of water molecules across a membrane.

Water flows across the plasma membrane via osmosis because the cytoplasm and surrounding external environment are often at differing concentrations. The process of osmosis can be readily demonstrated using red blood cells (see p. 47). If red blood cells are placed in a container of pure water, after a short time the red blood cells will swell and burst (lyse). This occurs because of a pressure buildup inside the cell due to a net movement of water into the cells—the cell membrane is unable to withstand this high pressure and ruptures. Inside the cells the cytoplasm contains large macromolecules, as well as salts and water molecules, while the container contains only water molecules. Thus, the concentration of water is less inside the cells, because some of the available space is occupied by the nondiffusible macromolecules and salt ions. A concentration gradient therefore exists for water molecules in the system. Water diffuses from the region of greater concentration of water (pure water in the container) to the region of lesser water concentration inside the cells (Figure 3.17A). In the experiment just described, the region of highest water concentration (in the container) is termed **hypotonic** to the cytoplasm due to low levels of (or no) salt ions or macromolecules in the container, while the cytoplasm is **hyperosmotic** to the water in the container due to the presence of high levels of macromolecules and salt ions. The pressure that resists the flow of water into the cytoplasm is termed the **osmotic pressure**.

If red blood cells were placed in a solution of water and salts that was similar in concentration to the cytoplasm, then the solution would be **isotonic** or **isoosmotic** to the cytoplasm and there would be no net movement of water (Figure 3.17B). If red blood cells were placed in a solution of water and salt that was at a higher concentration (**hypertonic**) than that of the red blood cell cytoplasm, then water molecules would flow out of the cytoplasm and the cell would shrink (Figure 3.17C). The cytoplasm would be **hypoosmotic** to the solution in the container due to the presence of fewer macromolecules and salt ions than in the container.

The concept of osmosis is very important in understanding how animals control their internal fluid and solute environment (see Chapter 30). For example, unicellular eukaryotes possess a contractile vacuole that functions in osmoregulation (see p. 225). In particular, freshwater forms accumulate water via osmosis because their cytoplasm is hyperosmotic (a higher solute concentration) compared with the surrounding environment. The contractile vacuole rapidly fills with this excess water and expels it through the plasma membrane by exocytosis (see p. 48 and p. 225). Osmoregulation is critical for marine bony fishes that maintain a solute concentration in their blood about one-third of that in seawater; they are hypoosmotic to seawater. If a fish swims into a river mouth and then up a freshwater stream, as salmon do, it passes through a region where its blood solutes are equal in concentration to those in its environment (isoosmotic), then enters freshwater, where its blood solutes are hyperosmotic to those in its environment. It must have physiological mechanisms to avoid net loss of water in the sea and gain of water in the river (see p. 662).

A Chemically-gated ion channel

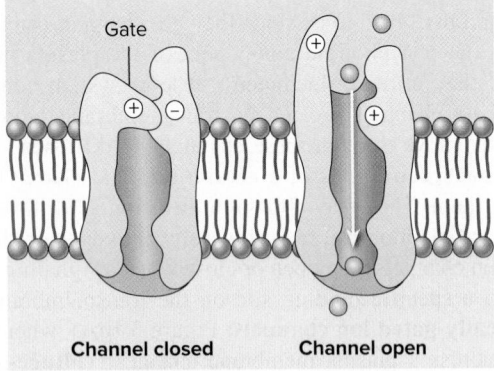

Channel closed   **Channel open**

B Voltage-gated ion channel

Figure 3.16 Gated channels require a signal to open (or close) them. **A,** Chemically-gated ion channels open (or close) when a signaling molecule binds to a specific binding site on the transmembrane protein. In this figure, binding of signaling molecules opens the channel to allow passage of ions. **B,** Voltage-gated ion channels open (or close) when the ionic charge across the membrane changes. In this figure, the ionic charge change across the membrane opens the channel

Carrier-Mediated Transport

The plasma membrane is an effective barrier to free diffusion of most molecules of biological significance, yet it is essential that such materials enter and leave a cell. Nutrients such as sugars and materials for growth such as amino acids must enter a cell, and wastes of metabolism must leave. Such molecules utilize a mediated transport system composed of transmembrane proteins called **transporters,** or carriers. Transporters enable solute molecules to cross the phospholipid bilayer (Figure 3.18A). They are usually quite specific, recognizing and transporting only a limited group of chemical substances or perhaps even a single substance.

At high concentrations of solute, mediated transport systems show a saturation effect. This means simply that the rate of influx reaches a plateau beyond which increasing the solute concentration has no further effect on influx rate (Figure 3.18B). This is evidence that the number of transporters available in a membrane is limited.

Figure 3.17 Red blood cell experiment showing the process of osmosis. **A,** Red blood cells are placed in a beaker of pure water (a hypotonic solution). Water molecules move into the red blood cells through the plasma membrane from an area of high concentration to an area of lower concentration. The red blood cells swell and lyse. **B,** Red blood cells are placed in a beaker of an isotonic solution. Because the concentration of water is equal on each side of the membrane, there is no net water movement. **C,** Red blood cells are placed in a hypertonic solution. The concentration of water molecules is now higher inside the cells, so water moves out of the cells into the beaker and the cells shrink.

A Red blood cells in hypotonic solution

B Red blood cells in isotonic solution

No net water movement

C Red blood cells in hypertonic solution

Net water movement

When all transporters become occupied by solutes, the rate of transport is at a maximum and it cannot be increased. Simple diffusion shows no such limitation; the greater the difference in solute concentrations on the two sides of the membrane, the faster the influx.

Two distinctly different kinds of carrier-mediated transport mechanisms are recognized: (1) **facilitated diffusion** or **facilitated transport,** in which a transporter assists a molecule to diffuse through the membrane that it cannot otherwise penetrate, and (2) **active transport,** in which energy from ATP is supplied to the transporter system to transport molecules in the direction opposite to a concentration gradient (Figure 3.19). Facilitated diffusion or transport therefore differs from active transport in that it supports movement only in

a downhill direction (in the direction of a concentration gradient) and requires no metabolic energy from ATP to drive the transport system.

In many animals facilitated diffusion aids transport of glucose (blood sugar) into body cells that oxidize it as a principal energy source for the synthesis of ATP (see Chapter 4, p. 60). The concentration of glucose is greater in blood than in the cells that consume it, favoring inward diffusion, but glucose is a water-soluble molecule that does not, by itself, penetrate cell membranes rapidly enough to support the metabolism of many cells; the carrier-mediated transport system increases the inward flow of glucose. In active transport, molecules are moved against the forces of passive diffusion. Active transport always

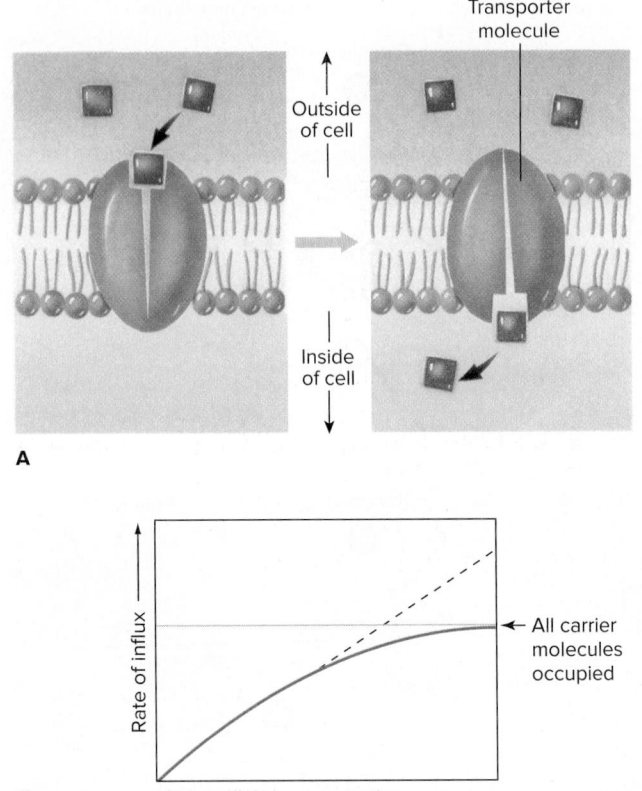

A

B

Figure 3.18   Facilitated diffusion or transport. **A,** A transporter protein molecule binds with a molecule to be transported (substrate) on one side of a plasma membrane, changes shape, and releases the molecule on the other side. Facilitated transport occurs in the direction of a concentration gradient. **B,** Rate of transport increases with increasing substrate concentration until all transporter molecules are occupied.

involves an expenditure of energy (from ATP) because materials are transported against a concentration gradient. Among the most important active-transport systems in all animals are those that maintain sodium and potassium ion gradients between cells and the surrounding extracellular fluid or external environment. Most animal cells require a high internal concentration of potassium ions for protein synthesis at the ribosome and for certain enzymatic functions. The potassium ion concentration may be 20 to 50 times greater inside a cell than outside. Sodium ions, on the other hand, may be 10 times more concentrated outside a cell than inside. This sodium gradient forms the basis for electrical signal generation within the nervous system of animals (see Chapter 33, p. 722). Both of these ionic gradients are maintained by active transport of potassium ions into and sodium ions out of the cell. In many cells outward transport of sodium is linked to inward transport of potassium; the same transporter protein does both. As much as 10% to 40% of all energy produced by cells is consumed by the **sodium-potassium pump** (Figure 3.19).

Endocytosis

Endocytosis, the ingestion of material by cells, is a collective term that describes three similar processes: phagocytosis, pinocytosis, and receptor-mediated endocytosis (Figure 3.20). They are pathways for specifically internalizing solid particles, small molecules and ions, and macromolecules, respectively. All require energy and thus are forms of active transport.

Phagocytosis, which literally means "cell eating," is a common method of feeding among unicellular forms (see Chapter 11, p. 224): sponges (see Chapter 12, p. 249), cnidaria (see Chapter 13, p. 269), and flatworms (see Chapter 14, p. 296). It is also the way in which white blood cells (leukocytes) and macrophages engulf cellular debris and invading microbes or other pathogens in the blood (see Chapter 35, p. 766). During phagocytosis, an area of the plasma membrane, coated

Figure 3.19   Sodium-potassium pump, powered by bond energy of ATP, maintains the normal gradients of these ions across the cell membrane. The pump works by a series of conformational changes in the transporter molecule: *Step 1.* Three ions of Na$^+$ bind to the interior end of the transporter, producing a conformational (shape) change in the protein complex. *Step 2.* The complex binds a molecule of ATP, and cleaves it, and phosphate binds to the complex. *Step 3.* The binding of the phosphate group induces a second conformational change, passing the three Na$^+$ ions across the membrane, where they are now positioned facing the exterior. This new conformation has a very low affinity for the Na$^+$ ions, which dissociate and diffuse away, but it has a high affinity for K$^+$ ions and binds two of them as soon as it is free of the Na$^+$ ions. *Step 4.* Binding of the K$^+$ ions leads to another conformational change in the complex, this time leading to dissociation of the bound phosphate. Freed of the phosphate, the complex reverts to its original conformation, with the two K$^+$ ions exposed on the interior side of the membrane. This conformation has a low affinity for K$^+$ ions so that they are now released, and the complex has the conformation it started with (step 1), having a high affinity for Na$^+$ ions.

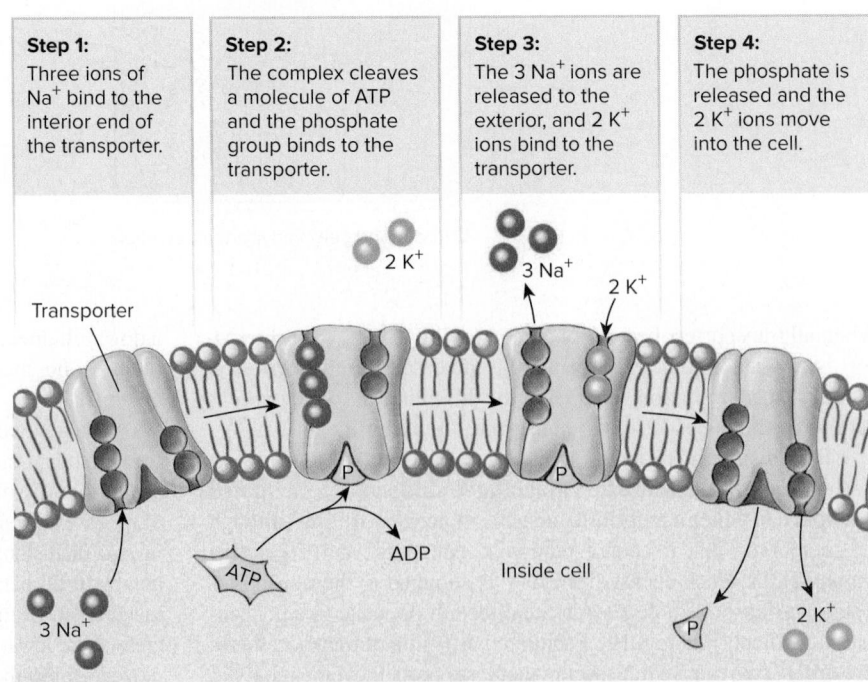

Step 1: Three ions of Na$^+$ bind to the interior end of the transporter.

Step 2: The complex cleaves a molecule of ATP and the phosphate group binds to the transporter.

Step 3: The 3 Na$^+$ ions are released to the exterior, and 2 K$^+$ ions bind to the transporter.

Step 4: The phosphate is released and the 2 K$^+$ ions move into the cell.

externally with specific receptors and internally with actin and actin-binding proteins, forms a pocket that engulfs the solid material. The membrane-enclosed vesicle, a food vacuole or phagosome, then detaches from the cell surface and moves into the cytoplasm where it fuses with lysosomes, its contents are digested by lysosomal enzymes, and useful products are absorbed from the lysosome across its membrane by diffusion or carrier-mediated transport.

Pinocytosis is similar to phagocytosis except that small areas of the surface membrane are invaginated into cells to form tiny vesicles. The invaginated pits and vesicles are called **caveolae** (ka-vēə-lē). Specific binding receptors for the molecule or ion to be internalized are concentrated on the cell surface of caveolae. Pinocytosis apparently functions for intake of at least some vitamins, hormones, and growth factors. Lysosomes fuse with these caveolae, causing digestion of their contents prior to absorption into the cytoplasm. Similar mechanisms may be important in translocating substances from one side of a cell to the other (see Exocytosis, next section), as occurs during some capillary exchange (see p. 690). This variation of pinocytosis is often called **transcytosis**. In this case, contents of the caveolae would remain largely unchanged as they are translocated across the cell.

Receptor-mediated endocytosis is a specific mechanism for bringing large molecules within the cell. Proteins of the plasma membrane specifically bind particular molecules (termed **ligands**), which may be present in the extracellular fluid in very low concentrations. The invaginations of the cell surface that bear the receptors are coated on the cytoplasmic side of the membrane with a protein. One example of a coating protein is called **clathrin;** hence, they are called **clathrin-coated pits.** As a clathrincoated pit with its receptor-bound ligand invaginates and is brought within the cell, it is uncoated, the receptor and the ligand are dissociated, and the receptor and surrounding membrane are recycled back to the surface membrane. Lysosomes fuse with the remaining vesicle, now called an **endosome,** and its contents are digested and absorbed into the cytoplasm. Some important proteins, peptide hormones, and cholesterol are brought into cells in this manner.

In phagocytosis, pinocytosis, and receptor-mediated endocytosis, some amount of extracellular fluid is necessarily trapped in the vesicle and nonspecifically brought within the cell. We call this **bulk-phase endocytosis.**

Exocytosis

Just as materials can be brought into a cell by membrane invagination and formation of a vesicle, the membrane of a vesicle within a cell can fuse with the plasma membrane and extrude its contents to the surrounding medium. This is the process of **exocytosis.** This process occurs in various cells to remove undigestible residues of substances brought in by endocytosis, to secrete substances such as hormones (see Figure 3.10), to recycle membrane receptors and membranes, as mentioned in receptor-mediated endocytosis (Figure 3.20), and to transport a substance completely across a cellular barrier (transcytosis).

Actin and actin-binding proteins are essential cytoskeletal components in the processes of endocytosis and exocytosis.

3.3 MITOSIS AND CELL DIVISION

All cells arise from the division of preexisting cells. In most multicellular organisms, all cells originate from the division of a single

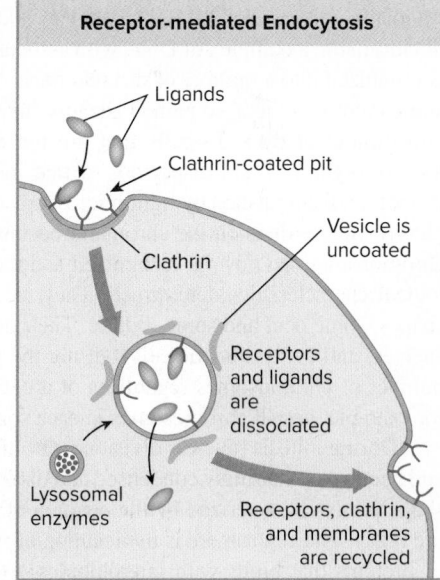

Figure 3.20   Three types of endocytosis. In phagocytosis the cell membrane binds to a large particle and extends to engulf it, forming a membrane-enclosed vesicle, a food vacuole or phagosome. In pinocytosis small areas of cell membrane, bearing specific receptors for a small molecule or ion, invaginate to form caveolae. Receptor-mediated endocytosis is a mechanism for selective uptake of large molecules in clathrincoated pits. Binding of the ligand to the receptor on the surface membrane stimulates invagination of pits. Lysosomes fuse with the vesicle created during phagocytosis and receptor-mediated endocytosis, and during pinocytosis if the caveolae contents are not translocated across the cell. Lysosomal enzymes digest vesicle contents that are then absorbed into the cytoplasm by diffusion or carrier-mediated transport.

cell, a **zygote,** which is the product of union (fertilization) between an **egg** and a **sperm (gametes or germ cells).** Cell division provides the basis for growth in multicellular organisms and for transmission of hereditary information from one cell generation to the next for both sexual and asexual reproduction.

In the formation of body cells or **somatic** cells the process of nuclear division is called **mitosis.** By mitosis each "daughter cell" is ensured a complete set of genetic instructions. Mitosis is a delivery system for distributing the chromosomes and the DNA they contain to subsequent cell generations. Thus, a single zygote divides by mitosis to produce a multicellular organism, and damaged cells are replaced by mitosis during wound healing. As an animal grows, its somatic cells differentiate and assume different functions and appearances because of differential gene action. Although most of the genes in specialized cells remain silent and unexpressed throughout the lives of those cells, every cell possesses a complete genetic complement. Mitosis ensures equality of genetic potential; later, other processes direct the orderly expression of genes during embryonic development by selecting from the genetic instructions that each cell contains. (These fundamental properties of cells of multicellular organisms are discussed further in Chapter 8.)

In animals that reproduce **asexually** (see Chapter 7), mitosis is the only mechanism for the transfer of genetic information from parent to progeny, and thus the progeny are genetically identical to the parents in this case. In animals that reproduce **sexually** (see Chapter 7), the two gametes that fuse to form the zygote must contain only half the usual number of chromosomes, so that progeny formed by the union of gametes will not contain double the parental number of chromosomes. Production of these gametes requires a special type of *reductional* division called **meiosis,** described in Chapter 5 (p. 72).

Structure of Chromosomes

As mentioned on page 39, DNA in eukaryotic cells occurs in strands of chromatin, a complex of DNA with associated protein. Chromatin is organized into a number of discrete linear bodies called **chromosomes** (color bodies), so named because they stain deeply with certain biological dyes. In cells that are not dividing, chromatin is loosely organized and dispersed, so that individual chromosomes cannot be distinguished by light microscopy (see Figure 3.23, Interphase). Before division the chromatin becomes more compact, and chromosomes can now be recognized and their individual morphological characteristics determined. They are of varied lengths and shapes, some bent and some rodlike. Their number is constant for a species, and every somatic cell (but not the gametes) has the same number of chromosomes regardless of a cell's function. A human, for example, has 46 chromosomes in each somatic cell.

During mitosis (nuclear division), chromosomes shorten further and become increasingly condensed and distinct, and each assumes a shape partly characterized by the position of, the **centromere** (Figure 3.21). The centromere is the location of the **kinetochore,** a disc of proteins that binds with microtubules of the spindle that forms during mitosis.

When chromosomes become condensed, the DNA is inaccessible, such that transcription (see Chapter 5, p. 88) cannot occur. Chromosomal condensation does, however, enable a cell to distribute chromosomal material efficiently and equally to daughter cells during cell division.

Phases in Mitosis

There are two distinct stages of cell division: nuclear division of chromosomes **(mitosis)** and division of cytoplasm **(cytokinesis).** Mitosis is certainly the most obvious and complex part of cell division and that of greatest interest to cytologists. Cytokinesis normally immediately follows mitosis, although occasionally the nucleus may divide a number of times without a corresponding division of the cytoplasm. In such a case the resulting mass of protoplasm containing many nuclei is called a **multinucleate** cell. An example is the giant cell found in bone marrow (megakaryocyte) that produces blood platelets (see p. 683), which may contain 24 to 32 nuclei. Sometimes a multinucleate mass is formed by cell fusion rather than nuclear proliferation. This arrangement is called a **syncytium.** An example is vertebrate skeletal muscle (see p. 651), which is composed of multinucleate fibers formed by fusion of numerous embryonic cells.

Mitosis is artificially divided into four successive stages or phases, although in reality one stage merges into the next without sharp lines of transition. These phases are prophase, metaphase, anaphase, and telophase (Figure 3.22 and Figure 3.23). When cells are not actively dividing, they are in interphase, a major part of the cell cycle described on page 51.

Prophase

At the beginning of prophase, centrosomes (and their centrioles) replicate, the nuclear envelope disintegrates, and the two centrosomes migrate to opposite poles of the cell (Figure 3.22). At the same time, microtubules are manufactured between the two centrosomes to form an oval-shaped **spindle** that provides a framework for aligning the chromosomes during division. Other microtubules radiate outward from each centrosome to form **asters.** The asters develop into the microtubular portion of the cytoskeleton in each new daughter cell formed during cell division.

At this time the diffuse nuclear chromatin condenses to form visible chromosomes. These actually consist of two identical sister

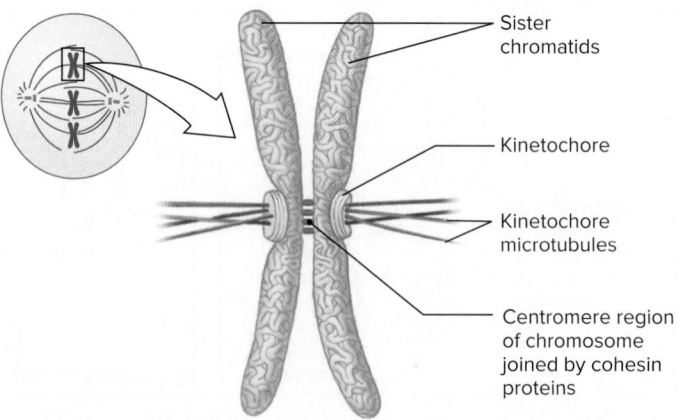

Figure 3.21 Structure of a metaphase chromosome. Sister chromatids are attached at their centromere by cohesin proteins. Each chromatid has a kinetochore, to which the kinetochore microtubules are attached. Kinetochore microtubules from each chromatid run to one of the centrosomes, which are located at opposite poles.

Sister chromatids

Kinetochore

Kinetochore microtubules

Centromere region of chromosome joined by cohesin proteins

Figure 3.22   Stages of mitosis, showing division of a cell with two pairs of chromosomes. One chromosome of each pair is shown in red.

chromosomes toward each pole of the cell (Figure 3.22 and Figure 3.23). Present evidence indicates that the force moving the chromosomes is disassembly of the tubulin subunits at the kinetochore end of each microtubule. As chromosomes are pulled apart, the centrosomes are moved farther apart, so the cell becomes elongated.

Telophase

When daughter chromosomes reach their respective centrosomes or poles, telophase has begun (Figure 3.22 and Figure 3.23). At this time, the daughter chromosomes are crowded together and stain intensely with histological stains. Spindle fibers disappear as the microtubules are disassembled and the chromosomes lose their identity, reverting to a diffuse chromatin network characteristic of an interphase nucleus. Finally, nuclear membranes reappear around the two daughter nuclei.

Cytokinesis: Cytoplasmic Division

During the final stages of nuclear division a **cleavage** furrow appears on the surface of a dividing cell and encircles it at the midline of the spindle (Figure 3.22 and Figure 3.23). The cleavage furrow deepens and constricts the plasma membrane as though it were being tightened by an invisible rubber band. The constriction is due to the presence of actin microfilaments just beneath the surface in the furrow between the cells. Interaction with myosin and other actin-binding proteins, similar to muscle contraction mechanisms (p. 654), draws the furrow inward. Finally, the infolding edges of the plasma membrane meet and fuse, completing cell division.

chromatids (Figure 3.21) formed by DNA replication (see Chapter 5, p. 87) during interphase and joined at their centromere. Dynamic spindle microtubules repeatedly extend from and retract toward each centrosome. When a microtubule encounters a kinetochore, it binds to the kinetochore, ceases extending and retracting, and is now called a **kinetochore microtubule.** Thus, centrosomes send out microtubules that behave like "feelers" to find chromosomes.

Metaphase

The centromere region of each chromosome has two kinetochores, and each of the kinetochores is attached to one of the centrosomes by kinetochore microtubules. As tubulin dimers are added to the kinetochore microtubules during metaphase, the condensed sister chromatids are moved to the middle of the nuclear region, called the **metaphase plate** (Figure 3.22 and Figure 3.23). The centromeres line up precisely in this region with the arms of the sister chromatids trailing off randomly in various directions.

Anaphase

The cohesin proteins that hold the two sister chromatids together at the centromere region are now removed, so that the two sister chromatids separate to become two independent chromosomes. The chromosomes move toward their respective poles, pulled by their kinetochore microtubules. The arms of each chromosome trail behind as the microtubules shorten to drag a complete set of

Cell Cycle

Cycles are conspicuous attributes of life. The progression of a species through time is in a very real sense a sequence of life cycles. Similarly, cells undergo cycles of growth and replication as they repeatedly divide. A cell cycle is the interval between one cell division and the next (Figure 3.24).

Actual nuclear division, or mitosis, occupies only about 5% to 10% of the cell cycle; a cell spends its remaining time in **interphase**, the stage between nuclear divisions. For many years it was thought that interphase was a period of rest, because nuclei appeared inactive when observed by ordinary light microscopy. In the early 1950s new techniques allowed investigation of DNA replication in nuclei, and it was discovered that replication occurred during the interphase stage. Further studies revealed that many other protein and nucleic acid components essential to normal cell function, growth, and division were synthesized during the seemingly quiescent interphase period.

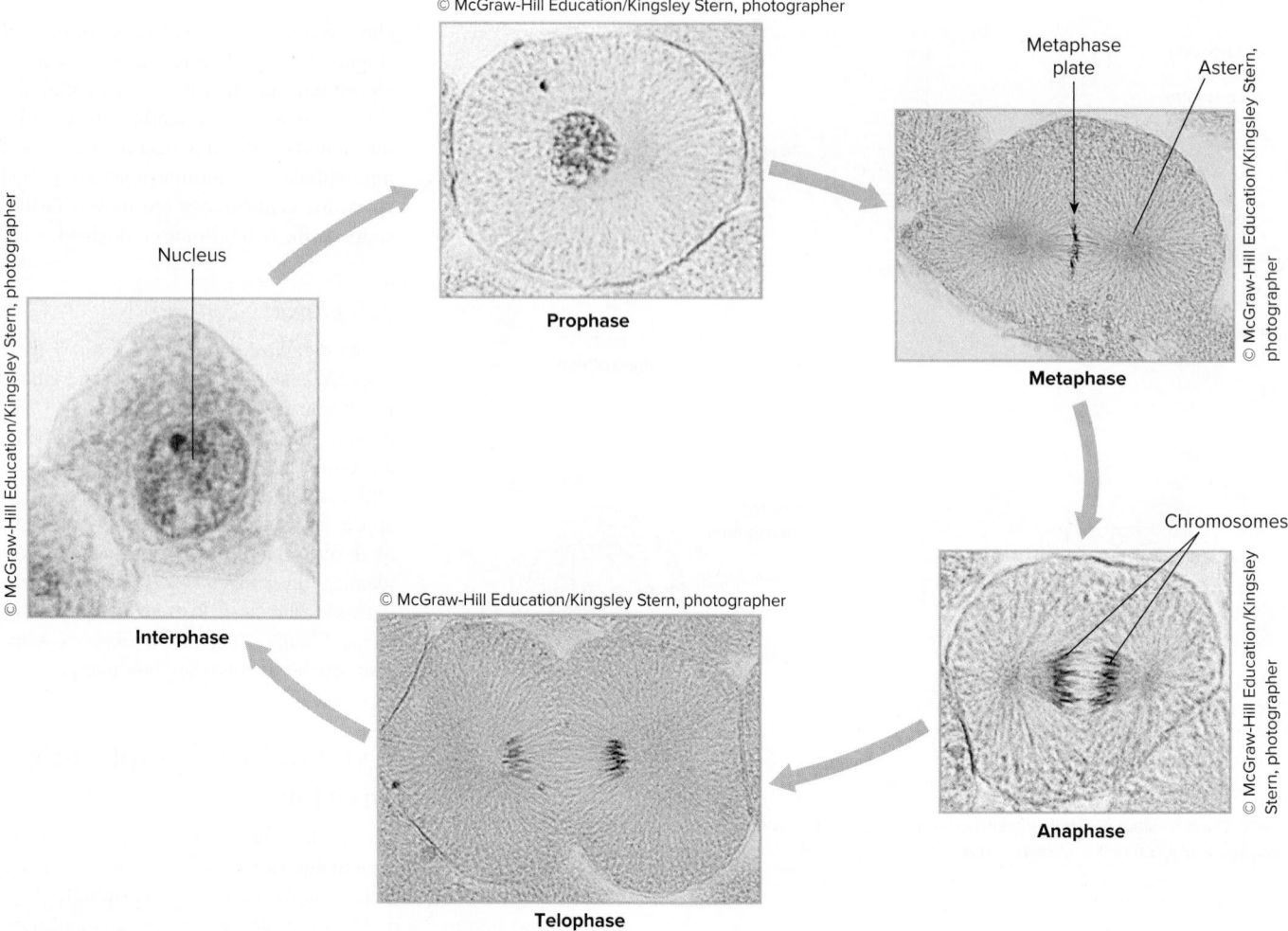

Figure 3.23 Stages of mitosis in whitefish.

Replication of DNA occurs during a phase called the S phase (period of synthesis). In mammalian cells in tissue culture, the S phase lasts about six of the 18 to 24 hours required to complete one cell cycle. In this phase both strands of a DNA molecule must replicate; new complementary partners are synthesized for each strand so that two identical DNA molecules are produced from the original strand (see Chapter 5, p. 87). These complementary partners are the sister chromatids that are separated during the next mitosis.

The S phase is preceded and succeeded by G_1 and G_2 phases, respectively (G stands for "gap"), during which no DNA synthesis is occurring. For most cells, G_1 is an important preparatory stage for the replication of DNA that follows. During G_1, transfer RNA, ribosomes, messenger RNA, and many enzymes are synthesized. During G_2, spindle and aster proteins are synthesized in preparation for chromosome separation during mitosis. G_1 is typically of longer duration than G_2, although there is much variation in different cell types. Embryonic cells divide very rapidly because there is no cell growth between divisions, only subdivision of mass. DNA synthesis may proceed a hundred times more rapidly in embryonic cells than in adult cells, and the G_1 phase is greatly shortened. As an organism develops, the cycles of most of its cells lengthen and many cells may be arrested for long periods of time in G_1 and enter a nonproliferative

or quiescent phase called G_0. Most neurons or nerve cells, for example, divide no further and are essentially in a permanent G_0.

The events in cell cycles are exquisitely regulated. Transitions during cell cycles are mediated by enzymes called **cyclin-dependent kinases (cdk's)** and regulatory protein subunits that activate them, called **cyclins.** In general, kinases are enzymes that add phosphate groups to other proteins to activate or inactivate them, and kinases themselves may require activation. Cdk's become active only when they are bound to the appropriate cyclin, together called MPFs (Mitosis-Promoting Factors), and cyclins are synthesized and degraded during each cell cycle (Figure 3.25). It seems likely that phosphorylation and dephosphorylation of specific cdk's and their interaction with phase-specific cyclins regulates the passage from one cell cycle phase to the next. Research focuses on the checkpoints that regulate this phase-to-phase passage since disregulation of these mechanisms has been implicated in cancer.

Cell Turnover

Cell division is important for growth, for replacement of cells lost to natural attrition, and for wound healing. Cell division is especially rapid during early development of an organism. At birth a human

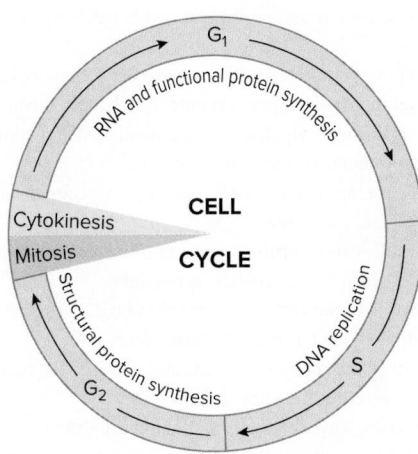

Figure 3.24   Cell cycle, showing relative duration of recognized phases. S, G_1, and G_2 are phases within interphase: S, synthesis of DNA; G_1, presynthetic phase; G_2, postsynthetic phase. After mitosis and cytokinesis the cell may go into an arrested, quiescent stage known as G_0. Actual duration of the cycle and the different phases varies considerably in different cell types.

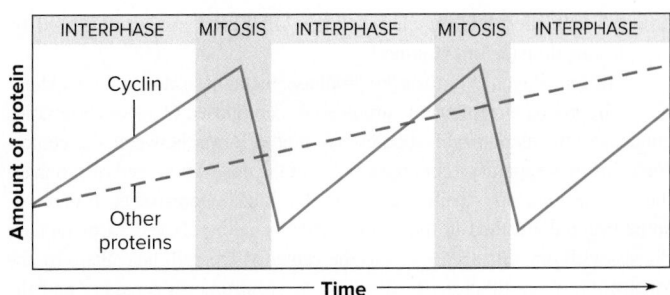

Figure 3.25   Variations in the level of cyclin in dividing cells of early sea urchin embryos. Cyclin binds with its cyclin-dependent kinase to activate the enzyme.

infant has about 2 trillion cells from repeated division of a single fertilized egg or zygote. This immense number could be attained by just 42 cell divisions, with each generation dividing once every six to seven days. With only five more cell divisions, the cell number would increase to approximately 60 trillion, the number of cells in a human adult weighing 75 kg. Of course no organism develops in this machinelike manner. Cell division is rapid during early embryonic development, then slows with age. Furthermore, different cell populations divide at widely different rates. In some the average period between divisions is measured in hours, whereas in others it is measured in days, months, or even years. Some cells in the central nervous system stop dividing after the early months of fetal development and generally persist without further division for the life of the individual. Muscle cells also stop dividing during the third month of fetal development, and most future growth depends on enlargement of fibers already present, although some cell division occurs after muscle injury.

In other tissues that are subject to abrasion, lost cells must be constantly replaced. It is estimated that in humans about 1% to 2% of all body cells—a total of 100 billion—are shed daily. Mechanical rubbing wears away the outer cells of the skin, while digestive enzymes and food in the alimentary canal removes lining epithelial cells. In addition, the restricted life cycle of blood cells involves enormous numbers of replacements. Such lost cells are replaced by mitosis.

Normal development, however, does entail cell death in which cells are not replaced. As cells age, they accumulate damage from destructive oxidizing agents and eventually die. Other cells undergo a programmed cell death, or **apoptosis** (a-puh-TOE-sis) (Gr. *apo-*, from, away from; + *ptosis*, a falling), which is in many cases necessary for continued health and development of an organism. For example, during embryonic development of vertebrates, fingers and toes develop as tissues between them die, inappropriately programmed immune cells that would attack the body's own tissues are

destroyed during their maturation process, and nerve cells die to create cerebral convolutions. Apoptosis consists of a well-coordinated and predictable series of events: the cells shrink, they disintegrate, and their components are recycled by surrounding cells.

SUMMARY

Cells are the basic structural and functional units of all living organisms. Eukaryotic cells differ from the prokaryotic cells of Bacteria and Archaea in several respects, the most distinctive of which is presence of a membrane-bound nucleus containing hereditary material composed of DNA bound to proteins to form chromatin. Chromatin consists of flexible, linear chromosomes, which become condensed and visible only during cell division.

Cells are surrounded by a plasma membrane that regulates the flow of molecules between the cell and its surroundings. The nucleus, enclosed by a double membrane, contains chromatin, associated proteins, and one or more nucleoli. Outside the nuclear envelope is cell cytoplasm, subdivided by a membranous network, the endoplasmic reticulum. Among the organelles within cells are the Golgi complex, mitochondria, lysosomes, and other membrane-bound vesicles. The cytoskeleton is composed of microfilaments (actin), microtubules (tubulin), and intermediate filaments (several types). Cilia and flagella are hairlike, motile appendages that contain microtubules. Ameboid movement by pseudopodia operates by means of the assembly and disassembly of actin microfilaments. Tight junctions, adhesion junctions, desmosomes, and gap junctions are structurally and functionally distinct connections between cells.

Membranes in a cell are composed of a phospholipid bilayer and other materials, including cholesterol and transmembrane proteins. Hydrophilic ends of the phospholipid molecules are on the outer and inner surfaces of membranes, and the fatty acid portions are directed inward, toward each other, to form a hydrophobic core.

Substances can enter cells by diffusion, mediated transport, and endocytosis. Solutes to which the membrane is impermeable require channels or a transporter molecule to traverse the membrane. Water and ions move through open channels by diffusion (in the direction of a concentration gradient). Osmosis is diffusion of water through channels. Transport-mediated systems include facilitated diffusion (with a concentration gradient, not requiring energy) and active transport (against a concentration gradient, which requires energy). Endocytosis includes bringing droplets (pinocytosis) or particles (phagocytosis) into a cell. In exocytosis the process of endocytosis is reversed.

Cell division is necessary for the production of new cells from preexisting cells and is the basis for growth in multicellular organisms. During this process, replicated nuclear chromosomes divide by mitosis followed by cytoplasmic division or cytokinesis.

The four stages of mitosis are prophase, metaphase, anaphase, and telophase. In prophase, replicated chromosomes composed of sister chromatids condense into recognizable bodies. A spindle forms between the centrosomes as they separate to opposite poles of the cell. At the end of prophase the nuclear envelope disintegrates, and the kinetochores of each chromosome become attached to the centrosomes by microtubules. At metaphase the sister chromatids are moved to the center of the cell, held there by the kinetochore microtubules. At anaphase the centromeres separate, and the sister chromatids are pulled apart by the attached kinetochore microtubules of the mitotic spindle. At telophase the sister chromatids, now called chromosomes, gather in the position of the nucleus in each cell and revert to a diffuse chromatin network. A nuclear membrane reappears, and cytokinesis occurs. At the end of mitosis and cytokinesis, two cells genetically identical to the parent cell have been produced.

The cell cycle in eukaryotes includes mitosis, cytokinesis, and interphase. During interphase, G_1, S, and G_2 phases are recognized, and the S phase is the time when DNA is synthesized (the chromosomes are replicated).

Cells divide rapidly during embryonic development, then more slowly with age. Some cells continue to divide throughout the life of an animal to replace cells lost by attrition and wear, whereas others, such as nerve and muscle cells, complete their division during early development and many do not divide again. Some cells undergo a programmed cell death, or apoptosis, especially during embryonic development.

REVIEW QUESTIONS

1. Explain the difference (in principle) between a light microscope and a transmission electron microscope.
2. Briefly describe the structure and function of each of the following: plasma membrane, chromatin, nucleus, nucleolus, rough endoplasmic reticulum (rough ER), Golgi complex, lysosomes, mitochondria, microfilaments, microtubules, intermediate filaments, centrioles, basal body (kinetosome), tight junction, gap junction, desmosome, glycoprotein, microvilli.
3. Name two functions each for actin and tubulin.
4. Distinguish among cilia, flagella, and pseudopodia.
5. What are the functions of each of the main constituents of the plasma membrane?
6. Our current concept of the plasma membrane is called the fluid-mosaic model. Why?
7. You place some red blood cells in a solution and observe that they swell and burst. You place some cells in another solution, and they

shrink and become wrinkled. Explain what has happened in each case.
8. The plasma membrane is an effective barrier to molecular movement across it, yet many substances do enter and leave the cell. Explain the mechanisms through which this is accomplished and comment on the energy requirements of these mechanisms.
9. Distinguish among phagocytosis, pinocytosis, receptor-mediated endocytosis, and exocytosis.
10. Define the following: chromosome, centromere, centrosome, kinetochore, mitosis, cytokinesis, syncytium.
11. Explain phases of the cell cycle, and comment on important cellular processes that characterize each phase. What is G_0?
12. Name the stages of mitosis in order, and describe the behavior and structure of the chromosomes at each stage.
13. Briefly describe ways that cells may die during the normal life of a multicellular organism.

For Further Thought Cystic fibrosis is the most common recessively inherited disease among Caucasians. It occurs primarily because of the absence or malfunction of a transmembrane protein channel (CFTR) that allows regulated flow of chloride ions in the normal human. From your knowledge of cell structure and function and of membrane transport, propose some reasons why the CFTR channel may not be present in the plasma membrane of some cystic fibrosis patients, or may not function correctly in other patients.

SELECTED REFERENCES

Ainsworth, C. 2007. Tails of the unexpected. Nature **4 48:** 638–641. *Discusses research into the role of cilia in cell signaling of single-celled and multicellular organisms. Clues to their wide-ranging functions come from the study of cilia dysfunction in a variety of diseases.*

Alberts, B., D. Bray. 2013. Essential cell biology, ed. 4. New York, Garland Science Publishing. *A well-written text describing more detailed cellular mechanisms.*

Kaksonen, M., C. P. Toret, and D. G. Drubin. 2006. Harnessing action dynamics for clathrin-coated endocytosis. Nature Reviews, Molecular Cell Biology **7:** 404–414. *While receptor-mediated endocytosis is the focus of this article, the role of actin in phagocytosis and pinocytosis is discussed.*

Pollard, T. D., and W. C. Earnshaw. 2008. Cell Biology, ed. 2. Philadelphia, Saunder/Elsevier. *A thorough textbook of cell biology.*

Roth, R. 2006. Clathrin-mediated endocytosis before fluorescent proteins. Nature Reviews, Molecular Cell Biology **7:** 63–68. *An excellent discussion of the history of the scientific timeline underlying the dynamic process of receptor-mediated endocytosis.*

CHAPTER 4

Cellular Metabolism

White-tailed deer (Odocoileus virginianus) *foraging* on fallen apples.

Deferring the Second Law

Living systems appear to contradict the second law of thermodynamics, which states that energy in the universe has direction and that it has been, and always will be, degraded to heat. This increase in disorder, or randomness, is termed **entropy.** Living systems, however, *decrease* their entropy by *increasing* the molecular orderliness of their structure. An organism becomes vastly more complex and organized during its development from fertilized egg to adult.

The second law of thermodynamics, however, applies to closed systems, and living organisms are not closed systems. Animals grow and maintain themselves by borrowing free energy from the environment. When a deer feasts on the acorns and beechnuts at the end of summer, it transfers potential energy, stored as chemical bond energy in acorn and

beechnut tissues, to its own body. Then, in step-by-step sequences called biochemical pathways, this energy is gradually released to fuel the deer's many activities. In effect, the deer decreases its own internal entropy by increasing the entropy of its food. The orderly structure of the deer is not permanent, however, but will be dissipated when it dies.

The ultimate source of this energy for the deer—and for almost all life on earth—is the sun (Figure 4.1). Sunlight is captured by green plants, which fortunately accumulate enough chemical bond energy to sustain both themselves and animals that feed on them. Thus the second law is not violated; it is simply held at bay by life on earth, which uses the continuous flow of solar energy to maintain a biosphere of high internal order, at least for the period of time that life exists.

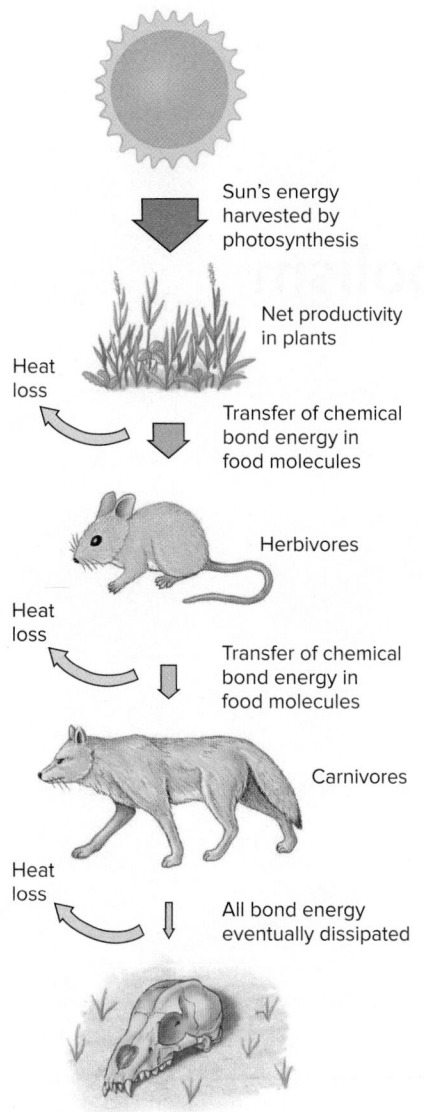

Sun's energy
harvested by
photosynthesis

Net productivity
in plants

Heat
loss

Transfer of chemical
bond energy in
food molecules

Herbivores

Heat
loss

Transfer of chemical
bond energy in
food molecules

Carnivores

Heat
loss

All bond energy
eventually dissipated

Figure 4.1 Solar energy sustains virtually all life on earth. With each energy transfer, however, about 90% of the energy is lost as heat energy.

Lump
of sugar

A   B   C   D

All cells must obtain energy, synthesize their own internal structure, control much of their own activity, and guard their boundaries. **Cellular metabolism** refers to the collective chemical processes that occur within living cells to accomplish these activities. Although the enormous number of reactions involved are extremely complex, the central metabolic processes routes through which matter and energy are channeled are conserved among living organisms.

4.1 ENERGY AND THE LAWS OF THERMODYNAMICS

The concept of energy is fundamental to all life processes. We usually express energy as the capacity to do work, to cause change. Yet energy is a somewhat abstract quantity that is difficult to define and elusive to measure. Energy cannot be seen; it can be identified only by how it affects matter.

Energy can exist in either of two states: kinetic or potential. **Kinetic energy** is the energy of motion. **Potential energy** is stored energy, energy that is not doing work but has the capacity to do so. Energy can be transformed from one state to another. Especially important for living organisms is chemical energy, a form of potential energy stored in the chemical bonds of molecules. Chemical energy can be used when bonds are rearranged to release kinetic energy. Much of the work done by living organisms involves conversion of potential energy to kinetic energy.

Conversion of one form of energy to another is governed by the two laws of thermodynamics. The **first law of thermodynamics** states that energy cannot be created or destroyed. It can change from one form to another (Figure 4.1), but the total amount of energy remains the same. In short, energy is conserved. If we burn gasoline in an engine, we do not create new energy but merely convert the chemical energy in gasoline to another form, in this example, mechanical energy and heat energy. The **second law of thermodynamics,** introduced in the prologue to this chapter, concerns the transformation of energy. This fundamental law states that a closed system moves toward increasing disorder, or entropy, as energy is dissipated from the system (Figure 4.2). Living systems, however, are open systems that not only maintain their organization

Figure 4.2 Diffusion of a solute through a solution, an example of entropy. When the solute (sugar molecules) is first introduced into a solution, the system is ordered and unstable **(B).** Without energy to maintain this order, the solute particles become distributed into solution, reaching a state of disorder (equilibrium) **(D).** Entropy has increased from left diagram to right diagram.

but also increase it, as during the development of an animal from egg to adult.

Free Energy

To describe the energy changes of chemical reactions, biochemists use the concept of **free energy.** Free energy is simply the energy in a system available for doing work. In a molecule, free energy equals the energy present in chemical bonds minus the energy that cannot be used. Most reactions in cells release free energy and are called **exergonic** (Gr. *ex,* out, + *ergon,* work). Such reactions are spontaneous, often proceed slowly, and are said to proceed "downhill" since free energy is lost from the system. Thus:

Exergonic reaction

However, many important reactions in cells require the addition of free energy and are said to be **endergonic** (Gr. *endon,* within, + *ergon,* work). Such reactions are said to be "pushed uphill" because the products store more energy than the reactants.

Endergonic reaction

A ubiquitous, energy-rich intermediate, called ATP and described on page 60, is used by organisms to power important endergonic reactions such as those required for active transport of molecules across membranes (see Chapter 3, p. 47) and cellular synthesis.

4.2 THE ROLE OF ENZYMES

Enzymes and Activation Energy

For any reaction to occur, even exergonic ones that tend to proceed spontaneously, chemical bonds first must be destabilized. Some energy, termed the **activation energy,** must be supplied before the bond is stressed enough to break. Only then will formation of reaction products occur, with overall loss of free energy, if the reaction is exergonic. This requirement of activation energy in chemical reactions can be likened to the energy needed to push a ball over the crest of a hill before it will roll spontaneously down the other side, releasing energy as it descends (Figure 4.3, top panel).

One way to provide the activation energy to chemical reactants is to raise the temperature to increase the rate of molecular collisions and to disrupt chemical bonds. Thus, heat can impart the necessary activation energy to make a reaction proceed. However, metabolic reactions must occur at biologically tolerable temperatures, which are usually too low for reactions to proceed fast enough to sustain life. Instead, living systems have evolved a different strategy: they employ **catalysts.**

Catalysts are chemical substances that accelerate reaction rates without affecting the products of the reaction and without being altered or destroyed by the reaction. A catalyst cannot make an energetically impossible reaction happen; it simply accelerates a reaction that would proceed at a very slow rate otherwise.

Enzymes are catalysts of the living world. They reduce the amount of activation energy required for a reaction. In effect, an enzyme steers the reaction through one or more intermediate steps, each of which requires much less activation energy than that required for a single-step reaction (Figure 4.3, bottom panel). Note that enzymes do not supply the activation energy. Instead they lower the activation energy barrier, making a reaction more likely to proceed. Enzymes affect only the reaction rate. They do not in any way alter the free energy change of a reaction (energy released or consumed).

Figure 4.3 Activation energy supplies enough energy for a reaction to proceed at biologically relevant temperatures. Enzymes lower the activation energy. Once the activation energy is supplied, the overall reaction proceeds with a net release of energy (exergonic). In the absence of an enzyme, substrate is stable because of the large amount of activation energy needed to disrupt strong chemical bonds. The enzyme reduces the energy barrier by forming a chemical intermediate with a much lower internal energy state.

Nature of Enzymes

Enzymes are complex molecules that vary in size from small, simple molecules with a molecular weight of 10,000 to highly complex molecules with molecular weights up to 1 million. Most enzymes are proteins—highly folded and interlinked chains of amino acids. Some require participation of small nonprotein groups called **cofactors** to perform their enzymatic function. In some cases these cofactors are metallic ions (such as ions of iron, copper, zinc, magnesium, potassium, and calcium) that form a functional part of the enzyme. Examples are carbonic anhydrase (see Chapter 31, p. 699), which contains zinc; the cytochromes (some enzymes of the electron transport chain, p. 64), which contain iron; and troponin (a muscle contraction enzyme, see Chapter 29, p. 653), which requires calcium to perform its function. Another class of cofactors, called **coenzymes,** is organic. Coenzymes contain groups derived from vitamins, most of which must be supplied in the diet. All B-complex vitamins are coenzymatic compounds. Since animals have lost the ability to synthesize the vitamin components of coenzymes, it is obvious that a vitamin deficiency can be serious. However, unlike dietary fuels and nutrients that must be replaced after they are burned or assembled into structural materials, vitamin components of coenzymes are recovered in their original form and are used repeatedly. Examples of coenzymes that contain vitamins are nicotinamide adenine dinucleotide (NAD), which contains the vitamin nicotinic acid (niacin); coenzyme A, which contains the vitamin pantothenic acid; and flavin adenine dinucleotide (FAD), which contains riboflavin (vitamin B_2). Another type of molecule, ribonucleic acid (RNA), is now known to possess enzymatic activity. Specifically, ribosomal RNA (rRNA, p. 90), a major component of ribosomes (p. 90), provides the activation energy that enables amino acids to assemble into polypeptide chains during the process of translation (see Chapter 5, p. 90).

Action of Enzymes

An enzyme functions by associating in a highly specific way with its **substrate,** the molecule whose reaction it catalyzes. Enzymes have an active site located within a cleft or pocket in a unique molecular configuration. The active site has a flexible surface that enfolds and conforms to the substrate (Figure 4.4). The binding of enzyme to substrate forms an **enzyme-substrate complex (ES complex),** in which the substrate is secured by covalent bonds to one or more points in the active site of the enzyme. The ES complex is not strong and will quickly dissociate, but during this fleeting moment the enzyme provides a unique chemical environment that stresses certain chemical bonds in the substrate so that much less energy is required to complete the reaction.

If the formation of an enzyme-substrate complex is so rapidly followed by dissociation, how can biochemists be certain that an ES complex exists? The original evidence offered by Leonor Michaelis in 1913 is that, when the substrate concentration is increased while the enzyme concentration is held constant, the reaction rate reaches a maximum velocity, much like that observed during carrier-mediated transport (see p. 46). This *saturation effect* is interpreted to mean that all catalytic sites become filled at high substrate concentration. A saturation effect is not seen in uncatalyzed reactions. Other evidence includes the observation that the ES complex displays unique spectroscopic characteristics not displayed by either the enzyme or the substrate alone. Furthermore, some ES complexes can be isolated in pure form, and at least one kind (nucleic acids and their polymerase enzymes) has been directly visualized by electron microscopy.

Enzymes that engage in crucial energy-providing reactions of cells often proceed constantly and usually operate in sets rather than in isolation. For example, conversion of glucose to carbon dioxide

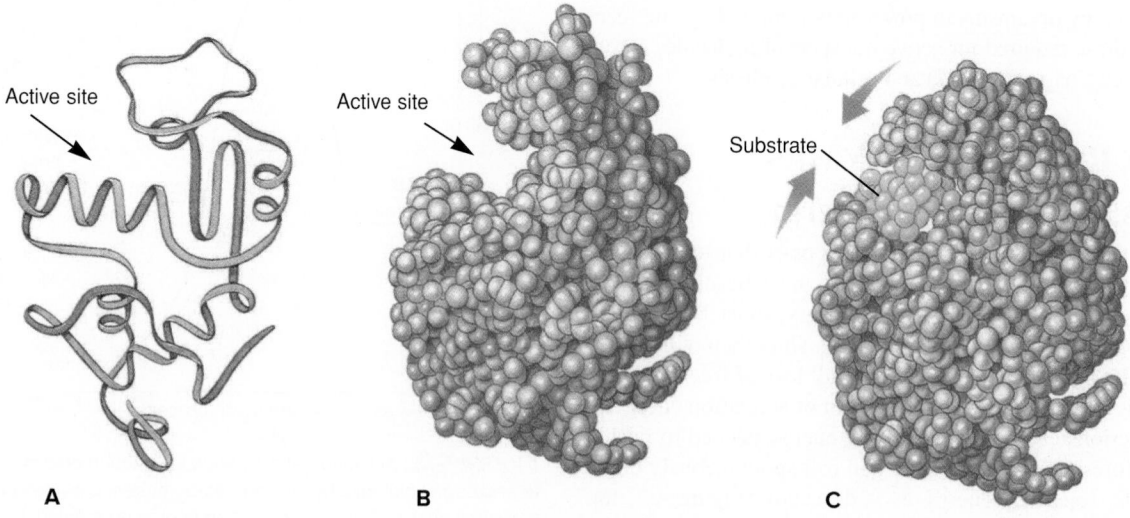

Figure 4.4   How an enzyme works. The ribbon model **(A)** and the space-filling model **(B)** show that the enzyme lysozyme bears a pocket containing the active site. When a chain of sugars (substrate) enters the pocket **(C),** the protein enzyme changes shape slightly so that the pocket enfolds the substrate and conforms to its shape to form the enzyme-substrate complex (ES complex). This positions the active site (amino acids in the protein) next to a bond between adjacent sugars in the chain, causing the sugar chain to break.

and water proceeds through 19 reactions, each requiring a specific enzyme (see p. 62). Such crucial enzymes occur in relatively high concentrations in cells, and they may implement quite complex and highly integrated enzymatic sequences. One enzyme performs the first step, then another enzyme binds to the product and catalyzes the next step. This process continues until the end of the enzymatic pathway is reached. The reactions are said to be coupled. Coupled reactions are explained in a section on chemical energy transfer by ATP (see p. 60).

Specificity of Enzymes

One of the most distinctive attributes of enzymes is their high specificity. Specificity is a consequence of the exact molecular fit required between enzyme and substrate, so that both are specific to each other. Furthermore, an enzyme catalyzes only one reaction. Unlike reactions performed in an organic chemist's laboratory, no side reactions or by-products result.

However, there is some variation in degree of specificity. Some enzymes catalyze the oxidation (dehydrogenation or removal of hydrogen) of only one substrate. For example, during cellular respiration succinic dehydrogenase catalyzes the oxidation of succinic acid only (see the Krebs cycle, p. 63). Others, such as proteases (for example, pepsin and trypsin, released into the digestive tract during digestion, pp. 710 and 712), act on almost any protein, although each protease has its particular point of attack in the protein (Figure 4.5). Usually an enzyme binds one substrate molecule at a time, catalyzes its chemical change, releases the product, and then repeats the process with another substrate molecule. An enzyme may repeat this process billions of times until it is finally dysfunctional (after a few hours to several years) and is degraded by scavenger enzymes in the cell. Some enzymes undergo successive catalytic cycles at speeds of up to a million cycles per minute, but most operate at slower rates. Many enzymes are repeatedly activated and inactivated; several mechanisms for regulating enzyme activity are well known.

Figure 4.5 Substrate specificity of trypsin. It splits only peptide bonds adjacent to lysine or arginine, but does so in many different proteins.

Enzyme-Catalyzed Reactions

Enzyme-catalyzed reactions are reversible, which is signified by double arrows between substrate and products, as in this example:

$$\text{Fumaric acid} + H_2O \rightleftharpoons \text{Malic acid}$$

However, for various reasons reactions catalyzed by most enzymes tend to go predominantly in one direction. For example, the proteolytic enzyme pepsin degrades proteins into amino acids (a **catabolic** reaction), but it does not accelerate the rebuilding of amino acids into any significant amount of protein (an **anabolic** reaction). The same is true of most enzymes that catalyze the cleavage of large molecules such as nucleic acids, polysaccharides, lipids, and proteins. There is usually one set of reactions with specific enzymes that degrade them (catabolism; Gr. *kata,* down, + *bole,* throw), but they must be resynthesized by a different set of reactions catalyzed by different enzymes (anabolism; Gr. *ana,* up, + *bole,* throw).

The net *direction* of any chemical reaction depends on the relative energy contents of the substances involved. If there is little change in chemical bond energy of substrate and products, the reaction is more easily reversible. However, if large quantities of energy are released as the reaction proceeds in one direction, more energy must be provided to drive the reaction in the reverse direction. For this reason, many if not most enzyme-catalyzed reactions are in practice irreversible unless the reaction is coupled to another one that makes enough energy available to reverse the reaction coupled to it. In cells both reversible and irreversible reactions are combined in complex ways to make possible both synthesis and degradation.

Hydrolysis literally means "breaking with water." In hydrolysis reactions, a molecule is cleaved by the addition of water at the cleavage site. A hydrogen atom is attached to one subunit and a hydroxyl (—OH) unit is attached to another (Figure 4.5). This breaks the covalent bond between subunits. Hydrolysis is the opposite of condensation (water-losing) reactions, in which subunits of molecules are linked together by removal of water. Macromolecules are built by condensation reactions.

4.3 ENZYME REGULATION

Although some enzymes appear to function automatically, the activity of others is rigidly controlled. In the former case, suppose that the function of an enzyme is to convert A to B. If B is removed by conversion into another compound, the enzyme tends to restore the original ratio of B to A. If the enzyme acts reversibly, either synthesis or degradation may result. For example, an excess of an intermediate in the Krebs cycle (p. 63) would contribute to glycogen synthesis; a depletion of such a metabolite would lead to glycogen breakdown. This automatic compensation (equilibration) is not, however, sufficient to explain regulation of cellular metabolism.

Mechanisms exist for critically regulating enzymes in both *quantity* and *activity.* Genes leading to synthesis of an enzyme may be switched on or off, depending on the presence or absence of a substrate molecule. In this way the *quantity* of an enzyme is controlled.

Mechanisms that alter activity of enzymes can quickly and finely adjust metabolic pathways to changing conditions in a cell. The presence or increase in concentration of some molecules can alter the shape (conformation) of particular enzymes, thus activating or inhibiting the enzyme (Figure 4.6 shows an example of an enzyme activator). For example, phosphofructokinase, which catalyzes phosphorylation of glucose-6-phosphate to fructose-1, 6-bisphosphate (see Figure 4.16, p. 66), is inhibited by high concentrations of ATP or citric acid. Their presence means that a sufficient amount of

Figure 4.6 Enzyme regulation. **A,** The active site of an enzyme may only loosely fit its substrate in the absence of an activator. **B,** With the regulatory site of the enzyme occupied by an activator, the enzyme binds the substrate, and the site becomes catalytically active.

energy has been produced and at that moment additional glucose metabolism is not needed. In some cases, the final end product of a particular metabolic pathway inhibits the first enzyme in the pathway. This method is termed **feedback inhibition.**

As well as being subject to alteration in physical shape, many enzymes exist in both an active and an inactive form. These forms may be chemically different. For example, one common way to activate or inactivate an enzyme is to add a phosphate group to the molecule, thus changing its conformational shape and either exposing or blocking the enzyme's active site. Enzymes that degrade glycogen (phosphorylase) and synthesize it (synthase) are found in both active and inactive forms. Conditions that activate phosphorylase tend to inactivate synthase and vice versa.

4.4 CHEMICAL ENERGY TRANSFER BY ATP

Endergonic reactions are those that do not proceed spontaneously because their products require an input of free energy. However, an endergonic reaction may be driven by coupling the energy-requiring reaction with an energy-producing reaction. ATP is one of the most common intermediates in **coupled reactions,** and because it can drive such energetically unfavorable reactions, it is of central importance in metabolic processes.

An ATP molecule consists of adenosine (the purine adenine and the 5-carbon sugar ribose) and a triphosphate group. Most free energy in ATP resides in the triphosphate group, especially in two **phosphoanhydride bonds** between the phosphate groups called **"high-energy bonds"** (Figures 4.7 and 4.8). Usually, only the most exposed high-energy bond is hydrolyzed to release free energy when ATP is converted to adenosine diphosphate (ADP) and inorganic phosphate:

$$ATP + H_2O \rightarrow ADP + P_i$$

where P_i represents inorganic phosphate (i = inorganic). The high-energy groups in ATP are often designated by the "tilde" symbol ~ (Figure 4.7). A high-energy phosphate bond is shown as ~P and a low-energy bond (such as the bond linking the triphosphate group to adenosine) as —P. Thus, ATP may be symbolized as A—P~P~P and ADP as A—P~P.

The way in which ATP can drive a coupled reaction is shown in Figure 4.9. A coupled reaction is really a system involving two reactions linked by an energy shuttle (such as ATP). The conversion of substrate A to product A is endergonic because the product contains

Figure 4.7 **A,** Structure of ATP. **B,** ATP formation from ADP and AMP. ATP: adenosine triphosphate; ADP: adenosine diphosphate; AMP: adenosine monophosphate.

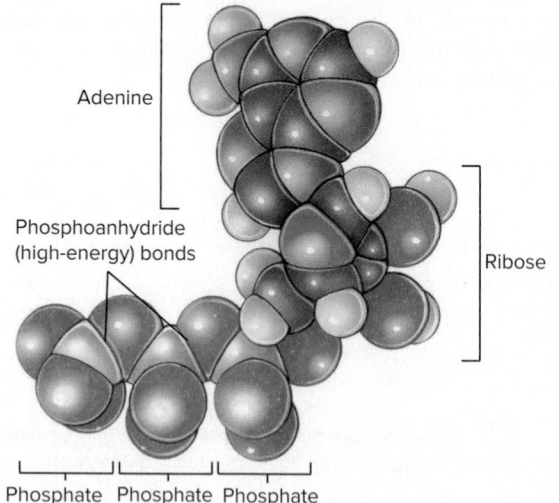

Figure 4.8 Space-filling model of ATP. In this model, carbon is shown in black; nitrogen in blue; oxygen in red; and phosphorus in orange.

more free energy than the substrate. Therefore energy must be supplied by coupling the reaction to an exergonic one, the conversion of substrate B to product B. Substrate B in this reaction is commonly called a **fuel** (for example, glucose or a lipid). Bond energy released

Figure 4.9 A coupled reaction. The endergonic conversion of substrate A to product A will not occur spontaneously but requires an input of energy from another reaction involving a large release of energy. ATP is the intermediate through which the energy is shuttled.

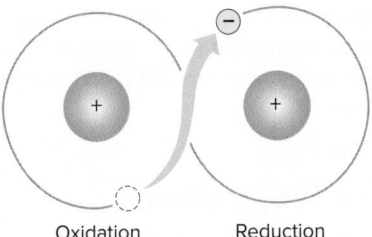

Figure 4.10 A redox pair. The molecule at left is oxidized by the loss of an electron. The molecule at right is reduced by gaining an electron.

in reaction B is transferred to ADP, which in turn is converted to ATP. ATP now contributes its phosphate-bond energy to reaction A, and ADP and P_i are produced again.

The high-energy bonds of ATP are actually rather weak, unstable bonds. Because they are unstable, the energy of ATP is readily released when ATP is hydrolyzed in cellular reactions. Note that ATP is an **energy-coupling agent** and *not* a fuel. It is not a storehouse of energy set aside for some future need. Rather, it is produced by one set of reactions and is almost immediately consumed by another set. ATP is formed as it is needed, primarily by oxidative processes in mitochondria (see p. 41). Oxygen is not consumed unless ADP and phosphate molecules are available, and these do not become available until ATP is hydrolyzed by some energy-consuming process. *Metabolism is therefore mostly self-regulating.*

4.5 CELLULAR RESPIRATION

How Electron Transport Is Used to Trap Chemical Bond Energy

Having seen that ATP is the one common energy denominator by which most cellular machines are powered, we must ask how this energy is captured from fuel substrates. This question directs us to an important generalization: *all cells obtain their chemical energy requirements from oxidation-reduction reactions.* This means that in the degradation of fuel molecules, hydrogen atoms (electrons and protons) are passed from electron donors to electron acceptors with a release of energy. A portion of this energy can be trapped and used to form the high-energy bonds of molecules such as ATP.

An oxidation-reduction ("redox") reaction involves a transfer of electrons from an electron donor (the reducing agent) to an electron acceptor (the oxidizing agent). As soon as the electron donor loses its electrons, it becomes oxidized. As soon as the electron acceptor accepts electrons, it becomes reduced (Figure 4.10). In other words, a reducing agent becomes oxidized when it reduces another

compound, and an oxidizing agent becomes reduced when it oxidizes another compound. Thus for every oxidation there must be a corresponding reduction.

In an oxidation-reduction reaction the electron donor and electron acceptor form a redox pair:

$$\text{Electron donor} \rightleftharpoons e^- + \text{Electron acceptor} + \text{Energy}$$
$$\text{(reducing agent;} \qquad\qquad \text{(oxidizing agent;}$$
$$\text{becomes oxidized)} \qquad\qquad \text{becomes reduced)}$$

When electrons are accepted by the oxidizing agent, energy is liberated because the electrons move to a more stable position.

ATP may be produced in a cell when electrons flow through a series of carrier molecules. Each carrier is reduced as it accepts electrons and then is reoxidized when it passes electrons to the next carrier in the series. By transferring electrons stepwise in this manner, energy is gradually released, and ATP is produced. Ultimately, the electrons are transferred to a **final electron acceptor.** The nature of this final acceptor is the key that determines the overall efficiency of cellular metabolism.

Aerobic versus Anaerobic Metabolism

Heterotrophs are organisms that cannot synthesize their own food but must obtain nutrients from the environment, including animals, fungi, and many single-celled organisms. They can be divided into two groups based on their overall efficiency of energy production during cellular metabolism: **aerobes,** those that use molecular oxygen as the final electron acceptor, and **anaerobes,** those that employ another molecule as the final electron acceptor.

As discussed in Chapter 2, life originated in the absence of oxygen, and an abundance of atmospheric oxygen was produced only after photosynthetic organisms (autotrophs) evolved. Some strictly anaerobic organisms still exist and indeed play important roles in specialized habitats. However, evolution has favored aerobic metabolism, not only because oxygen became available, but also because aerobic metabolism is vastly more efficient in useful energy production than anaerobic metabolism. In the absence of oxygen, only a very small fraction of the bond energy present in nutrients can be released. For example, when an anaerobic microorganism degrades glucose, the final electron acceptor (such as pyruvic acid) still contains most of the energy of the original glucose molecule so it cannot be used by the organism. An aerobic organism, on the other hand, using oxygen as the final electron acceptor, completely oxidizes glucose to carbon dioxide and water. Almost 20 times as much energy is released when glucose is completely oxidized as when it is degraded only to the stage of pyruvic acid. Thus, an obvious advantage of aerobic metabolism is that a much smaller quantity of food is required to maintain a given rate of metabolism.

Overview of Respiration

Aerobic metabolism is usually called **cellular respiration,** and is defined as the oxidation of fuel molecules to produce energy with molecular oxygen as the final electron acceptor. Oxidation of fuel molecules describes the *removal of electrons* from fuel molecules and *not* the direct combination of molecular oxygen with fuel molecules. Let us look at this process in general before considering it in more detail.

Hans Krebs, the British biochemist who contributed so much to our understanding of respiration, described three stages in the complete oxidation of fuel molecules to carbon dioxide and water (Figure 4.11). In stage I, food passing through the intestinal tract is digested into small molecules that can be absorbed into the circulation (see Chapter 32). There is no useful energy yield during digestion. In stage II, also called **glycolysis,** most of the glucose is converted into two 3-carbon units (pyruvic acid) in the cell cytoplasm. The pyruvic acid molecules then enter mitochondria, where in another reaction they join with a coenzyme (coenzyme A or CoA) to form acetyl coenzyme A or acetyl-CoA. Some ATP is generated in stage II, but the yield is small compared with that obtained in stage III of respiration. In stage III the final oxidation of fuel molecules occurs, with a large yield of ATP. This stage occurs entirely in mitochondria. Acetyl-CoA is channeled into the Krebs cycle, where the acetyl group is completely oxidized to carbon dioxide. Electrons released from acetyl groups are transferred to special carriers that pass them to electron acceptor compounds in the electron transport chain. At the end of the electron transport chain, the electrons (and the protons accompanying them) are accepted by molecular oxygen to form water.

Glycolysis

We begin our journey through the stages of respiration with stage II or glycolysis, a nearly universal pathway in living organisms that converts glucose into pyruvic acid. In a series of reactions occurring in the cell cytoplasm, glucose and other 6-carbon monosaccharides are split into 3-carbon molecules of **pyruvic acid** (Figure 4.12). A

Figure 4.11 Overview of cellular respiration, showing the three stages in the complete oxidation of food molecules to carbon dioxide and water.

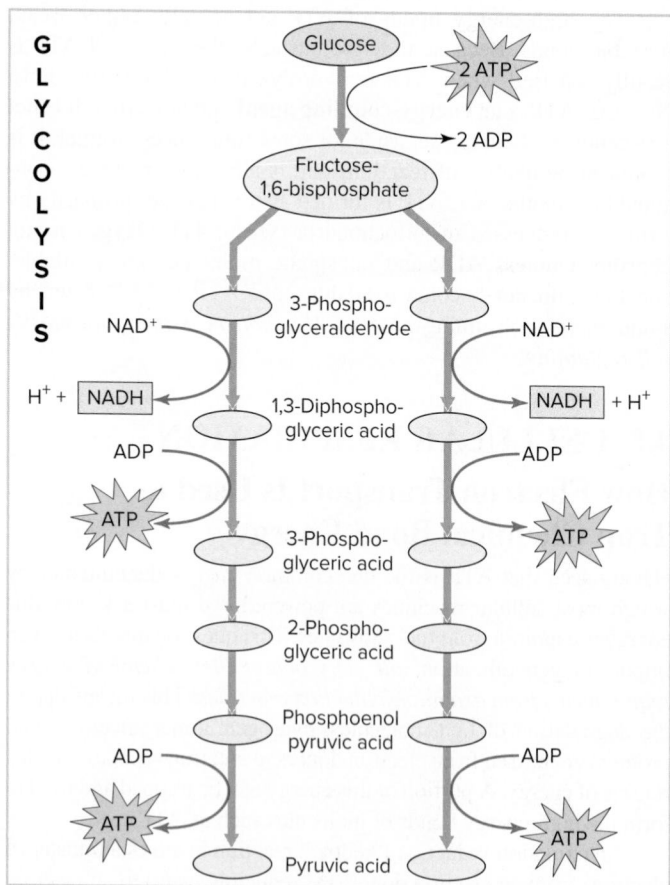

Figure 4.12 Glycolysis. Glucose is phosphorylated in two steps and raised to a higher energy level. High-energy fructose-1, 6-bisphosphate is split into 3-carbon sugar phosphates that are oxidized exergonically to pyruvic acid, yielding ATP and NADH.

single oxidation occurs during glycolysis, and each molecule of glucose has a net yield of two molecules of ATP. In this pathway the carbohydrate molecule is phosphorylated twice by ATP, first to glucose-6-phosphate (not shown in Figure 4.12) and then to fructose-1,6-bisphosphate. The fuel has now been "primed" with phosphate groups in this, energy-consuming portion of glycolysis and is sufficiently reactive to enable subsequent reactions to proceed. This is a kind of energy input required for an ultimate energy return many times greater than the original energy investment.

In the energy-releasing portion of glycolysis, fructose-1,6-bisphosphate is cleaved into two 3-carbon sugars, which undergo an oxidation (electrons are removed), with the electrons and one of the hydrogen ions being accepted by nicotinamide adenine dinucleotide (NAD$^+$, a derivative of the vitamin niacin) to produce a reduced form called **NADH + H$^+$**. NADH serves as a carrier molecule to convey high-energy electrons to the final electron transport chain, where ATP is produced.

$$NAD^+ + 2H \longrightarrow NADH + H^+$$

The two 3-carbon sugars next undergo a series of reactions, ending with the formation of two molecules of pyruvic acid (Figure 4.12). In two of these steps, a molecule of ATP is produced. In other words, each 3-carbon sugar yields two ATP molecules, and since there are two 3-carbon sugars, four ATP molecules are generated. Two ATP molecules were used to phosphorylate the glucose initially; therefore, the net yield at this point is two ATP molecules. The 10 enzymatically catalyzed reactions in glycolysis can be summarized as follows:

Glucose + 2 ADP + 2 P$_i$ + 2 NAD$^+$ → 2 pyruvic acid + 2 NADH + 2 ATP

Acetyl-CoA: Strategic Intermediate in Respiration

In aerobic metabolism the two molecules of pyruvic acid formed during glycolysis enter a mitochondrion. There, each molecule is oxidized, and one of the carbons is released as carbon dioxide (Figure 4.13). The 2-carbon residue reacts with **coenzyme A (CoA)** to form **acetyl coenzyme A,** or **acetyl-CoA,** and an NADH molecule is also produced.

Pyruvic acid is the undissociated form of the acid:

$$\begin{array}{c} O \\ \| \\ CH_3-C-COOH \end{array}$$

Under physiological conditions pyruvic acid typically dissociates

$$\begin{array}{c} O \\ \| \\ CH_3-C-COO^- \end{array}$$

into pyruvate (CH$_3$—C—COO$^-$) and H$^+$. It is correct to use either term in describing this and other organic acids (such as lactic acid or lactate) in metabolism.

Acetyl-CoA is a critically important intermediate compound. Its oxidation in the Krebs cycle provides energized electrons to generate ATP, and it is also a crucial intermediate in lipid metabolism (p. 67).

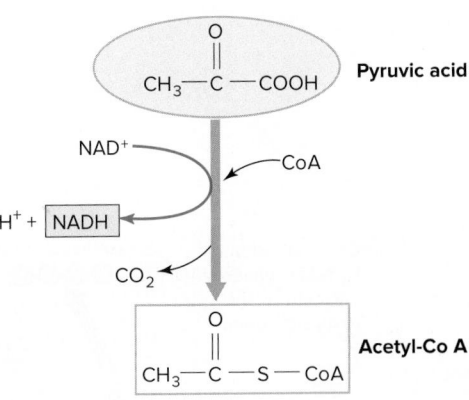

Figure 4.13   Formation of acetyl-CoA from pyruvic acid.

Krebs Cycle: Oxidation of Acetyl-CoA

Oxidation of the 2-carbon acetyl group of acetyl-CoA occurs within the mitochondrial matrix in a cyclic sequence called the **Krebs cycle** (also called the citric acid cycle and tricarboxylic acid cycle [TCA cycle]) (Figure 4.14). Acetyl-CoA reacts with a 4-carbon acid (oxaloacetic acid), releasing CoA to react with more pyruvic acid (see Figure 4.13). Through a cyclic series of reactions in the Krebs cycle, the two carbons from acetyl-CoA are released as carbon dioxide, and oxaloacetic acid is regenerated. Hydrogen ions and electrons in the oxidations transfer to NAD$^+$ and to FAD (flavine adenine dinucleotide, another electron acceptor), and a pyrophosphate bond is generated in the form of guanosine triphosphate (GTP). This high-energy phosphate readily transfers to ADP to form ATP. Thus, the overall products of the Krebs cycle are CO$_2$, ATP, NADH, and FADH$_2$:

$$Acetyl\ CoA + 3\ NAD^+ + FAD + ADP + P_I \rightarrow$$
$$2\ CO_2 + 3\ NADH + FADH_2 + ATP$$

The molecules of NADH and FADH$_2$ formed yield 11 molecules of ATP when oxidized in the electron transport chain. The other molecules in the cycle behave as intermediate reactants and products, which are continuously regenerated as the cycle turns.

Aerobic cellular respiration uses oxygen as the final electron acceptor and releases carbon dioxide and water from the complete oxidation of fuels. The carbon dioxide that we, and other aerobic organisms, produce is removed from our bodies into the atmosphere during external respiration (see Chapter 31, p. 692). Fortunately for us and other aerobes, oxygen is continuously produced by cyanobacteria (blue-green algae), eukaryotic algae, and plants by the process of photosynthesis. In this process, hydrogen atoms obtained from water react with carbon dioxide from the atmosphere to generate sugars and molecular oxygen. Thus, a balance between oxygen used and produced, and carbon dioxide produced and used, is obtained across our planet. Unfortunately, excessive production of carbon dioxide due to human industrialization, and decreased production of oxygen due to our continuous removal of the world's forests, are threatening this delicate balance. Carbon dioxide levels continue to rise, leading to global atmospheric warming caused by the "greenhouse effect" (see Chapter 37, p. 805).

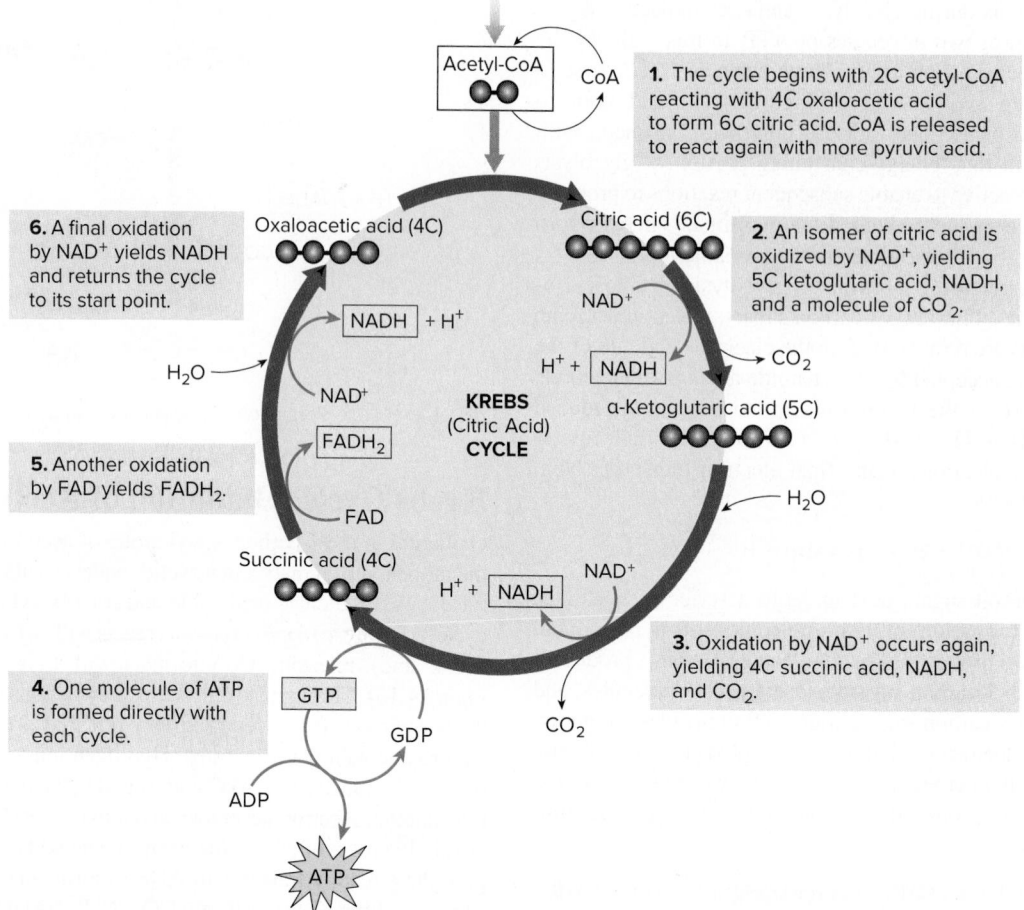

1. The cycle begins with 2C acetyl-CoA reacting with 4C oxaloacetic acid to form 6C citric acid. CoA is released to react again with more pyruvic acid.

2. An isomer of citric acid is oxidized by NAD$^+$, yielding 5C ketoglutaric acid, NADH, and a molecule of CO_2.

3. Oxidation by NAD$^+$ occurs again, yielding 4C succinic acid, NADH, and CO_2.

4. One molecule of ATP is formed directly with each cycle.

5. Another oxidation by FAD yields FADH$_2$.

6. A final oxidation by NAD$^+$ yields NADH and returns the cycle to its start point.

Figure 4.14 The Krebs cycle in outline form, showing production of three molecules of reduced NAD, one molecule of reduced FAD, one molecule of ATP, and two molecules of carbon dioxide. The molecules of NADH and FADH$_2$ will yield 11 molecules of ATP when oxidized in the electron transport system.

Electron Transport Chain

Transfer of hydrogen ions and electrons from NADH and FADH$_2$ to the final electron acceptor, molecular oxygen, is accomplished in an elaborate electron transport chain embedded in the inner membrane of mitochondria (Figure 4.15; see also p. 41). Each carrier molecule in the chain (labeled I to IV in Figure 4.15) is a large transmembrane protein-based complex that accepts and releases electrons at lower energy levels than the carrier preceding it in the chain. Oxygen is the final electron acceptor, combining with them and protons to produce water. Thus, as electrons pass from one carrier molecule to the next, free energy is released. Some of this energy is used to transport H$^+$ ions across the inner mitochondrial membrane and in this way creates a H$^+$ gradient across the membrane. The H$^+$ gradient produced drives the synthesis of ATP. This process is called chemiosmotic coupling (Figure 4.15). According to this model, as electrons are carried along the electron transport chain, they activate proton-transporting molecules, which move protons (hydrogen ions) outward and into the space between the two mitochondrial membranes. This causes the proton concentration in this intermembrane space to rise, producing a diffusion gradient that is used to drive the protons back into the mitochondrial matrix through special proton channels. These channels are ATP-forming transmembrane protein complexes (ATP synthase) that use the inward movement of protons to induce the formation of ATP. By this means, oxidation of one NADH yields three ATP molecules. FADH$_2$ from the Krebs cycle enters the electron transport chain at a lower energy level than NADH and so yields two ATP molecules. This method of energy capture is called **oxidative phosphorylation** because the formation of high-energy phosphate is coupled to oxygen consumption, and these reactions depend on demand for ATP by other metabolic activities within the cell.

Efficiency of Oxidative Phosphorylation

We can now calculate the ATP yield from the complete oxidation of glucose (Figure 4.16). The overall reaction is

$$\text{Glucose} + 2\,\text{ATP} + 36\,\text{ADP} + 36\,\text{P} + 6\,O_2 \rightarrow$$
$$6\,CO_2 + 2\,\text{ADP} + 36\,\text{ATP} + 6\,H_2O$$

ATP has been generated at several points along the way (Table 4.1). The cytoplasmic NADH generated in glycolysis requires a molecule

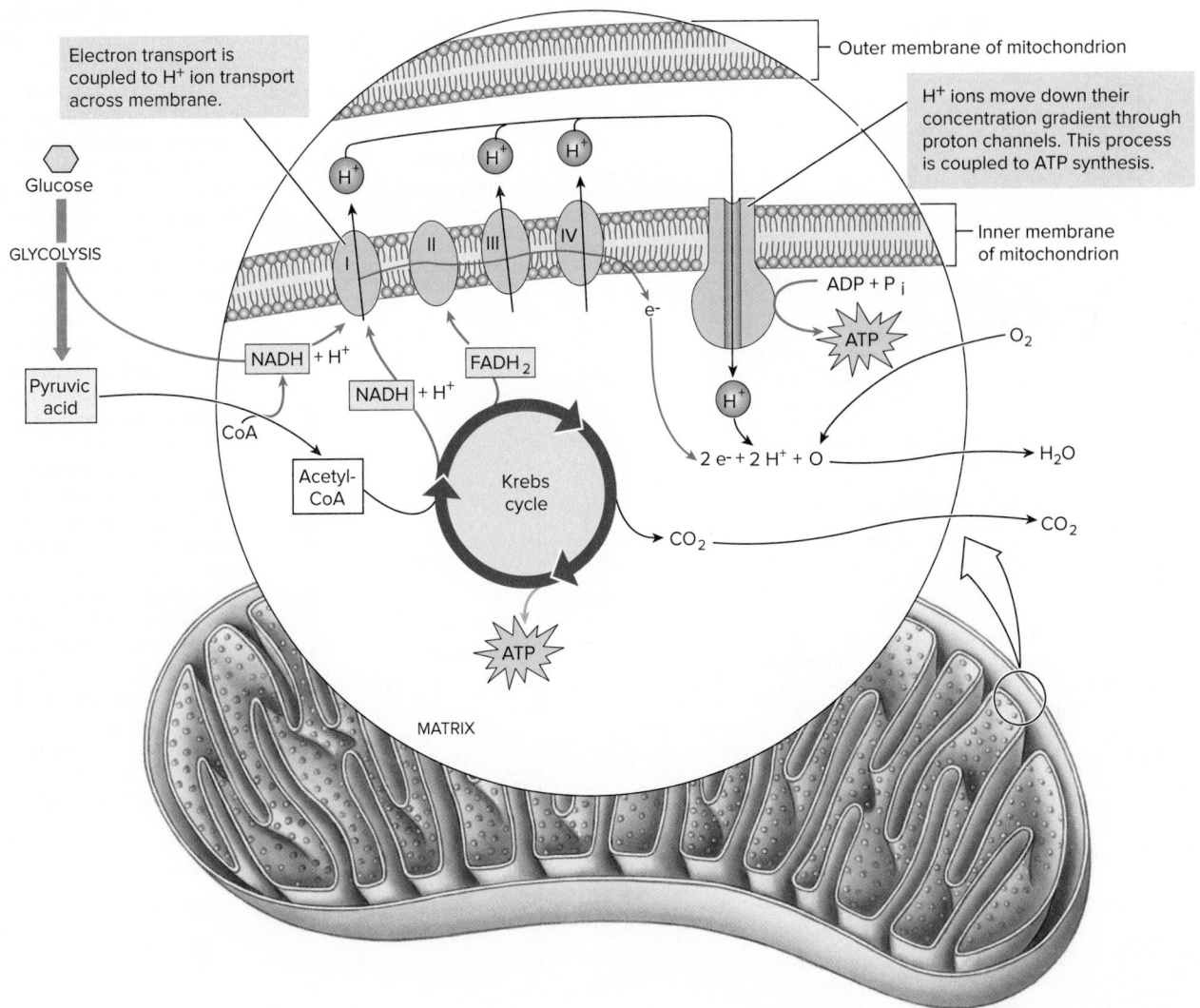

Figure 4.15 Oxidative phosphorylation. Most of the ATP in living organisms is produced in the electron transport chain. Electrons removed from fuel molecules in cellular oxidations (glycolysis and the Krebs cycle) flow through the electron transport chain, the major components of which are four transmembrane protein complexes (I, II, III, and IV). Electron energy is tapped by the major complexes and used to push H^+ outward across the inner mitochondrial membrane. The H^+ gradient created drives H^+ inward through proton channels (ATP synthase) that couple H^+ movement to ATP synthesis.

of ATP to fuel transport of each molecule of NADH into a mitochondrion; therefore, each NADH from glycolysis yields only two ATP (total of four), compared with the three ATP per NADH (total of six) formed within mitochondria. Accounting for the two ATP used in the phosphorylation reactions in glycolysis, the net yield may be as high as 36 molecules of ATP per molecule of glucose. The yield of 36 ATP is a theoretical maximum because some of the H^+ gradient produced by electron transport may be used for other functions, such as transporting substances in and out of the mitochondrion. The overall efficiency of aerobic oxidation of glucose is about 38%, comparing very favorably with human-designed energy conversion systems, which seldom exceed 5% to 10% efficiency.

The capacity for oxidative phosphorylation is also increased by the elaborate folding of the inner mitochondrial membrane (the cristae shown in Figure 4.15 and labeled in Figure 3.11, p. 42), providing a much greater surface area for more electron-transport-chain and ATP-synthase proteins.

TABLE 4.1

Calculation of Total ATP Molecules Generated in Respiration

ATP Generated	Source
4	Directly in glycolysis
2	As GTP (→ ATP) in Krebs cycle
4	From NADH in glycolysis
6	From NADH produced in pyruvic acid to acetyl-CoA reaction
4	From reduced FAD in Krebs cycle
18	From NADH produced in Krebs cycle
38 Total	
−2	Used in phosphorylation reactions in glycolysis
36 Net	

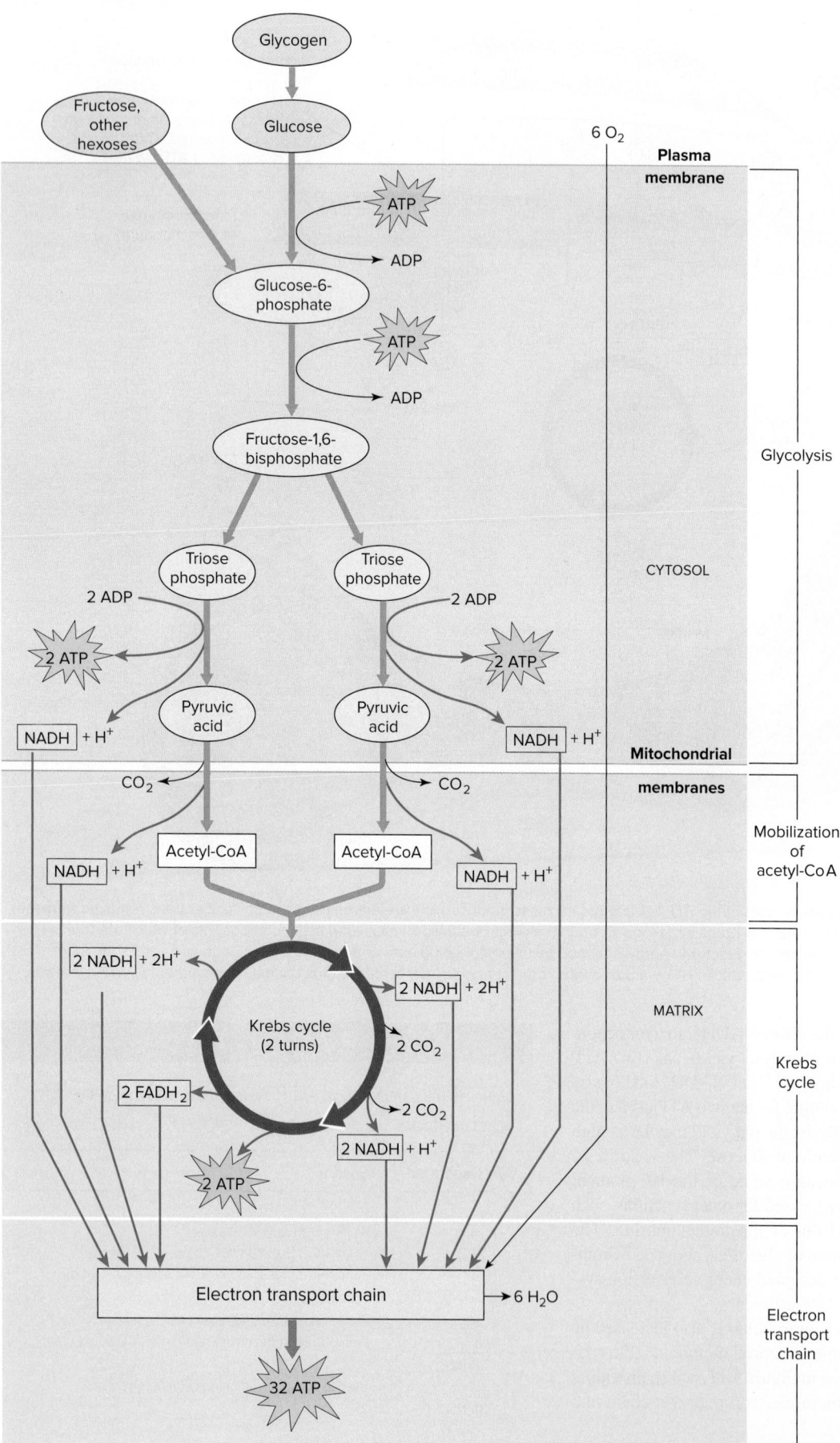

Figure 4.16 Pathway for oxidation of glucose and other carbohydrates. Glucose is degraded to pyruvic acid by cytoplasmic enzymes (glycolytic pathway). Acetyl-CoA is formed from pyruvic acid and is fed into the Krebs cycle. An acetyl-CoA molecule (two carbons) is oxidized to two molecules of carbon dioxide with each turn of the cycle. Pairs of electrons are removed from the carbon skeleton of the substrate at several points in the pathway and are carried by oxidizing agents NADH or FADH$_2$ to the electron transport chain where 32 molecules of ATP are generated. Four molecules of ATP are also generated by substrate phosphorylation in the glycolytic pathway, and two molecules of ATP (initially GTP) are formed in the Krebs cycle. This yields a total of 38 molecules of ATP (36 molecules net) per glucose molecule. Molecular oxygen is involved only at the very end of the pathway as the final electron acceptor at the end of the electron transport chain to yield water.

Anaerobic Glycolysis: Generating ATP Without Oxygen

Under anaerobic conditions, glucose and other 6-carbon sugars are first converted stepwise to a pair of 3-carbon pyruvic acid molecules during glycolysis, described on page 62 (see also Figure 4.12). This series of reactions yields two molecules of ATP and two molecules of NADH. In the absence of molecular oxygen, further oxidation of pyruvic acid cannot occur because without oxygen as the final electron acceptor in the electron transport chain, the Krebs cycle and electron transport chain cannot operate and cannot, therefore, reoxidize the NADH produced in glycolysis to generate more ATP molecules. The problem is neatly solved in most animal cells by reducing pyruvic acid to lactic acid (Figure 4.17). Pyruvic acid becomes the final electron acceptor and lactic acid the end product of anaerobic glycolysis. This step converts NADH to NAD$^+$, effectively freeing it to recycle and pick up more H$^+$ and electrons. In **alcoholic fermentation** (as in yeast, for example), the steps are identical from glycolysis down to pyruvic acid. One of its carbons is then released as carbon dioxide, and the resulting 2-carbon compound is reduced to ethanol, thus regenerating the NAD$^+$.

Anaerobic glycolysis is only one-eighteenth as efficient as complete oxidation of glucose to carbon dioxide and water, but its key virtues are that it provides *some* high-energy phosphate in situations in which oxygen is absent or in short supply, and that it does so at a higher rate than during oxidative phosphorylation. Many microorganisms live in places where oxygen is severely depleted, such as in waterlogged soil, in mud of a lake or seafloor, or within a decaying carcass. Vertebrate skeletal muscle may rely heavily on glycolysis during short bursts of activity when contraction is so rapid and powerful that oxygen delivery to tissues cannot supply energy demands by oxidative phosphorylation alone. At such times anaerobic glycolysis is used to supplement oxidative phosphorylation. One kind of muscle fiber (white muscle) has few mitochondria and primarily uses anaerobic glycolysis for ATP production (see Chapter 29, p. 657). In all muscle types, intense or strenuous activity is followed by a period of increased oxygen consumption as lactic acid, the end product of anaerobic glycolysis, diffuses from muscle to the liver, where it is metabolized. Because oxygen consumption increases following heavy activity, the animal is said to have acquired an **oxygen debt** during such activity, which is "repaid" when activity ceases, and accumulated lactic acid is metabolized—it can be converted back into pyruvic acid and used in aerobic respiration to yield more ATP.

Some animals rely heavily on anaerobic glycolysis during normal activities. For example, diving birds and mammals use glycolysis almost entirely to give them the energy needed to sustain long dives without breathing air. Salmon would never reach their spawning grounds were it not for anaerobic glycolysis providing almost all of the ATP used in the powerful muscular bursts needed to carry them up rapids and waterfalls (see Chapter 24, p. 538). Many parasitic animals have dispensed with oxidative phosphorylation entirely at some stages of their life cycles. They secrete relatively reduced end products of their energy metabolism, such as succinic acid, acetic acid, and propionic acid. These compounds are produced in mitochondrial reactions that derive several more molecules of ATP than does the path from glycolysis to lactic acid, although such sequences are still far less efficient than the aerobic electron-transport chain.

4.6 METABOLISM OF LIPIDS

The first step in the breakdown of a triglyceride (see p. 23) is its hydrolysis to glycerol and three fatty acid molecules (Figure 4.18). Glycerol is phosphorylated and enters the glycolytic pathway (see Figure 4.11).

The remainder of the triglyceride molecule consists of fatty acids (see p. 23). For example, an abundant naturally occurring fatty acid is **stearic acid.**

H$_3$C   CH$_2$ CH$_2$ CH$_2$ CH$_2$ CH$_2$ CH$_2$ CH$_2$ CH$_2$ O
\ / \ / \ / \ / \ / \ / \ / \ / \ //
CH$_2$ CH$_2$ CH$_2$ CH$_2$ CH$_2$ CH$_2$ CH$_2$ CH$_2$ C

Stearic acid   OH

The long hydrocarbon chain of a fatty acid is broken down by oxidation, two carbons at a time; these are released from the end of the molecule as acetyl-CoA (see Figure 4.11). Although two high-energy phosphate bonds are required to phosphorylate each 2-carbon fragment, energy is derived both from the reduction of NAD$^+$ and FAD to NADH + H$^+$ and FADH$_2$, respectively, and from each acetyl-CoA as it is degraded in the Krebs cycle. The complete oxidation of one molecule of 18-carbon stearic acid yields 146 ATP molecules. By comparison, three molecules of glucose (also totaling 18 carbons) yield 108 ATP molecules. Since there are three fatty acids in each triglyceride molecule, a total of 440 ATP molecules are formed. An additional 22 molecules of ATP are generated in the breakdown of glycerol, giving a grand total of 462 molecules of ATP—little wonder that fat is considered the king of animal fuels! Fats are more concentrated fuels than carbohydrates. Because fats are almost pure hydrocarbons, they contain more hydrogen per carbon atom than sugars do (see pp. 23 and 24), and it is the energized electrons of hydrogen that generate high-energy bonds when they are carried through the mitochondrial electron-transport chain.

Figure 4.17   Anaerobic glycolysis, a process that proceeds in the absence of oxygen. Glucose is broken down to two molecules of pyruvic acid, with a net production of two molecules of ATP. Pyruvic acid, the final electron acceptor for the hydrogen ions and electrons released during pyruvic acid formation, is converted to lactic acid. Hydrogen and electrons are recycled through the carrier, NAD$^+$.

Figure 4.18 Hydrolysis of a triglyceride (neutral fat) by intracellular lipase. The R groups of each fatty acid represent a hydrocarbon chain as seen in the stearic acid example.

Fat stores are derived principally from surplus fats and carbohydrates in the diet. Acetyl-CoA is the source of carbon atoms used to build fatty acids. Because all major classes of organic molecules (carbohydrates, fats, and proteins) can be degraded to acetyl-CoA (see Figure 4.11), all can be converted into stored fat. The biosynthetic pathway for fatty acids resembles a reversal of the catabolic pathway already described, but it requires an entirely different set of enzymes. From acetyl-CoA, the fatty acid chain is assembled two carbons at a time. Because fatty acids release energy when they are oxidized, they obviously require an input of energy for their synthesis. This energy is provided principally by energy from glucose degradation. Thus the total ATP derived from oxidation of one molecule of triglyceride is not as great as calculated, because varying amounts of energy are required for synthesis and storage.

Stored fats are the greatest reserve fuel in the body. Most usable fat resides in white adipose tissue composed of specialized cells packed with globules of triglycerides. White adipose tissue is widely distributed in the abdominal cavity, in muscles, around deep blood vessels and large organs (for example, heart and kidneys), and especially under the skin. Women average about 30% more fat than men, which is largely responsible for differences in shape between males and females. Humans can only too easily deposit large quantities of fat, generating hazards to health.

Physiological and psychological aspects of obesity are now being investigated by many scientists. There is increasing evidence that food intake, and therefore the amount of fat deposition, is regulated by feeding centers located in the brain (lateral and ventral hypothalamus and brain stem—see Chapter 33, pp. 730 and 731). The set point of these regions determines normal food intake and body weight for an individual, which may be maintained above or below what is considered normal for humans. Although evidence is accumulating for a genetic component to obesity, the epidemic proportions of obesity in the United States are more easily explained by lifestyle and feeding habits. Other industrialized countries show a similar trend toward development of an obesity problem.

Research also reveals that lipid metabolism in obese individuals appears to be abnormal compared to lean individuals. This research has resulted in the development of drugs that act at various stages of lipid metabolism, such as decreasing lipid digestion and absorption from the digestive tract, or increasing metabolism of lipids once they have been absorbed into the body.

4.7 METABOLISM OF PROTEINS

Since proteins are composed of amino acids, of which 20 kinds commonly occur (p. 24), the central topic of our consideration is amino acid metabolism. Amino acid metabolism is complex, because each of the 20 amino acids requires a separate pathway for biosynthesis and degradation. Amino acids are precursors to tissue proteins, enzymes, nucleic acids, and other nitrogenous constituents that form cellular components. The central purpose of carbohydrate and fat oxidation is to provide energy, much of which is needed to construct and to maintain these vital macromolecules.

When animals eat proteins, most are digested in the digestive tract, releasing their constituent amino acids, which are then absorbed (Figure 4.19). Tissue proteins also are hydrolyzed during normal growth, repair, and tissue restructuring; their amino acids join those derived from protein found in food to enter the **amino acid pool.** A portion of the amino acid pool is used to rebuild tissue proteins, but most animals ingest a surplus of protein. Since amino acids are not excreted as such in any significant amounts, they must be metabolized in some other way. In fact, amino acids can be and are metabolized through oxidative pathways to yield high-energy phosphate. In short, excess proteins serve as fuel as do carbohydrates and fats. Their importance as fuel obviously depends on the nature of the diet. In carnivores that ingest a diet of almost pure protein and fat (see Chapter 28, p. 620), nearly half of their ATP comes from amino acid oxidation.

Figure 4.19 Fate of dietary protein.

Before an amino acid molecule may enter the fuel depot, nitrogen must be removed by deamination (the amino group splits to form ammonia and a keto acid) or by transamination (the amino group is transferred to a keto acid to yield a new amino acid). Thus amino acid degradation yields two main products, carbon skeletons and ammonia, which are handled in different ways. Once nitrogen atoms are removed, the carbon skeletons of amino acids can be completely oxidized, usually by way of pyruvic acid or acetic acid. These residues then enter the final common pathway for fuel oxidation used by carbohydrate and fat metabolism (see Figure 4.11).

The other product of amino acid degradation is ammonia. Ammonia is highly toxic because it inhibits respiration by reacting with α-ketoglutaric acid to form glutamic acid (an amino acid), and effectively removes α-ketoglutarate from the Krebs cycle (see Figure 4.14). Disposal of ammonia offers little problem to aquatic animals because it is soluble and readily diffuses into the surrounding medium, often through respiratory surfaces. Terrestrial animals cannot get rid of ammonia so conveniently and must detoxify it by converting it to a relatively nontoxic compound. The two principal compounds formed are **urea** and **uric acid,** although a variety of other detoxified forms of ammonia are excreted by different animals. Among vertebrates, amphibians and especially mammals produce mainly urea. Reptiles and birds, as well as many terrestrial invertebrates, produce mainly uric acid (the excretion of uric acid by insects and birds is described on pp. 452 and 594, respectively).

The key feature that determines choice of nitrogenous waste is availability of water in the environment. When water is abundant, the chief nitrogenous waste is ammonia. When water is restricted, it is urea. Animals living in truly arid habitats use uric acid. Uric acid is highly insoluble and easily precipitates from solution, allowing its removal in solid form. The production of uric acid requires greater energy expenditure, but the benefit is water conservation. Embryos of birds and reptiles benefit greatly from excretion of nitrogenous waste as uric acid, because waste cannot be eliminated through their eggshells (see Chapter 26, p. 563). During embryonic development, harmless, solid uric acid is retained in one of the extraembryonic membranes. When a hatchling emerges into its new world, accumulated uric acid, along with the shell and membranes that supported development, is discarded.

SUMMARY

Living systems are subject to the same laws of thermodynamics that govern nonliving systems. The first law states that energy cannot be destroyed, although it may change form. The second law states that the structure of systems proceeds toward total randomness, or increasing entropy, as energy is dissipated. Solar energy trapped by photosynthesis as chemical bond energy is passed through the food chain, where it is used for biosynthesis, active transport, and motion, before finally being dissipated as heat. Living organisms are able to decrease their entropy and to maintain high internal order because the biosphere is an open system from which energy can be captured and used. Energy available for use in biochemical reactions is termed "free energy."

Enzymes are usually proteins, often associated with nonprotein cofactors, that vastly accelerate rates of chemical reactions in living systems. An enzyme acts by temporarily binding its reactant (substrate) onto an active site in a highly specific fit. In this configuration, internal activation energy

barriers are lowered enough to modify the substrate, and the enzyme is restored to its original form.

Integration of metabolic pathways is finely regulated by mechanisms that control both amount and activity of enzymes. The quantity of some enzymes is regulated by certain molecules that switch enzyme synthesis on or off. Enzyme activity may be altered by the presence or absence of metabolites that cause conformational changes in enzymes and thus improve or diminish their effectiveness as catalysts.

Cells use the energy stored in chemical bonds of organic fuels by degrading fuels through a series of enzymatically controlled steps. This bond energy is transferred to ATP and packaged in the form of "high-energy" phosphate bonds. ATP is produced as it is required in cells to power various synthetic, secretory, and mechanical processes.

Glucose is an important source of energy for cells. In aerobic metabolism (respiration), the 6-carbon glucose is split into two 3-carbon molecules of pyruvic acid. Pyruvic acid is decarboxylated to form 2-carbon acetyl-CoA, a strategic intermediate that enters the Krebs cycle. Acetyl-CoA can also be derived from breakdown of fat. In the Krebs cycle, acetyl-CoA is oxidized in a series of reactions to carbon dioxide, yielding, in the course of the reactions, energized electrons that are passed to electron acceptor molecules (NAD^+ and FAD). In the final stage, the energized electrons are passed along an electron transport chain consisting of a series of electron carriers located in the inner membranes of mitochondria. A hydrogen gradient is produced as electrons are passed from carrier to carrier and finally to oxygen, and ATP is generated as the hydrogen ions flow down their electrochemical gradient through ATP synthase molecules located in the inner mitochondrial membrane. A net total of 36 molecules of ATP may be generated from one molecule of glucose.

In the absence of oxygen (anaerobic glycolysis), glucose is degraded to two 3-carbon molecules of lactic acid, yielding two molecules of ATP. Although anaerobic glycolysis is vastly less efficient than aerobic metabolism, it provides essential energy for muscle contraction when heavy energy expenditure outstrips the oxygen-delivery system of an animal; it also is the only source of energy generation for microorganisms living in oxygen-free environments.

Triglycerides (neutral fats) are especially rich depots of metabolic energy because the fatty acids of which they are composed are highly reduced and free of water. Fatty acids are degraded by sequential removal of 2-carbon units, which enter the Krebs cycle through acetyl-CoA.

Amino acids in excess of requirements for synthesis of proteins and other biomolecules are used as fuel. They are degraded by deamination or transamination to yield ammonia and carbon skeletons. The latter enter the Krebs cycle to be oxidized. Ammonia is a highly toxic waste product that aquatic animals quickly expel, often through respiratory surfaces. Terrestrial animals, however, convert ammonia into much less toxic compounds, urea or uric acid, for disposal.

REVIEW QUESTIONS

1. State the first and second laws of thermodynamics. Living systems may appear to violate the second law of thermodynamics because living things maintain a high degree of organization despite a universal trend toward increasing disorganization. What is the explanation for this apparent paradox?
2. Explain what is meant by "free energy" in a system. Will a reaction that proceeds spontaneously have a positive or negative change in free energy?
3. Many biochemical reactions proceed slowly unless the energy barrier to the reaction is lowered. How is this accomplished in living systems?

4. What happens in the formation of an enzyme-substrate complex that favors the disruption of substrate bonds?

5. Explain three ways that enzymes may be regulated in cells.

6. What is meant by a "high-energy bond," and why might the production of molecules with such bonds be useful to living organisms?

7. Although ATP supplies energy to an endergonic reaction, why is it not considered a fuel?

8. What is an oxidation-reduction reaction and why are such reactions considered so important in cellular metabolism?

9. Give an example of a final electron acceptor found in aerobic and anaerobic organisms. Why is aerobic metabolism more efficient than anaerobic metabolism?

10. Why must glucose be "primed" with a high-energy phosphate bond before it can be degraded in the glycolytic pathway?

11. What happens to the electrons removed during the oxidation of triose phosphates during glycolysis?

12. Why is acetyl-CoA considered a "strategic intermediate" in respiration?

13. Why are oxygen molecules important in oxidative phosphorylation? What are the consequences if they are absent for a short period of time in tissues that routinely use oxidative phosphorylation to produce useful energy?

14. Explain how animals can generate ATP *without* oxygen. Given that anaerobic glycolysis is much less efficient than oxidative phosphorylation, why has anaerobic glycolysis not been discarded during animal evolution?

15. Why are animal fats sometimes called "the king of fuels"? What is the significance of acetyl-CoA to lipid metabolism?

16. The breakdown of amino acids yields two products: ammonia and carbon skeletons. What happens to these products?

17. Explain the relationship between the amount of water in an animal's environment and the kind of nitrogenous waste it produces.

For Further Thought Diet foods often are advertised as "low-fat" foods, but they contain sugars instead. Explain why they still cause weight gain.

SELECTED REFERENCES

Alberts, B., D. Bray, K. Hopkin, A. Johnson, J. Lewis, M. Raff, K. Roberts, and P. Walter. 2009. Essential cell biology, ed. 3. New York, Garland Science Publishing. *Provides a more in-depth and well-written description of cellular metabolism.*

Berg, J., J. Tymoczko, and L. Stryer. 2012. Biochemistry, ed. 7. San Francisco, W. H. Freeman & Company. *One of the best undergraduate biochemistry texts.*

Lodish, H., A. Berk, C. A. Kaiser, M. Krieger, A. Bretscher, H. Ploegh, A. Amon, and M. P. Scott 2012. Molecular cell biology, ed. 7. San Francisco, W. H. Freeman & Company. *Chapter 12 is a comprehensive, well-illustrated treatment of energy metabolism.*

PART TWO

5

Genetics: A Review

The site of Gregor Mendel's experimental garden, Brno, Czech Republic.

A Code for All Life

The principle of hereditary transmission is a central tenet of life on earth: all organisms inherit a structural and functional organization from their progenitors. What is transmitted from parents to offspring is a functioning cell (zygote) containing a set of coded instructions that it uses to guide its cell divisions, growth, and development to construct a body resembling those of its parents. These instructions are in the form of genes, the fundamental units of inheritance. A great triumph of modern biology was the discovery in 1953 by James Watson and Francis Crick of the nature of the coded instructions in genes. The genetic material (deoxyribonucleic acid, DNA) is composed of nitrogenous bases arranged on a chemical chain of sugar-phosphate units. The genetic code lies in the linear order or sequence of bases in the DNA strand.

Because DNA molecules replicate and pass from generation to generation, genetic variations can persist and spread in a population. Molecular alterations, called mutations, are the ultimate source of biological variation and the raw material of evolution.

A basic principle of modern evolutionary theory is that organisms attain their diversity through hereditary modifications of populations. All known lineages of plants and animals are related by descent from common ancestral populations.

Heredity establishes the continuity of living forms. Although offspring and parents in a particular generation may look different, there is nonetheless a genetic continuity that runs from generation to generation for any species of plant or animal. An offspring inherits from its parents a set of coded information (**genes**), which a fertilized egg uses, together with environmental factors, to guide its development into an adult bearing unique physical characteristics. Each generation passes to the next the instructions required for maintaining continuity of life.

The gene is the unit entity of hereditary function. The study of what genes are, how they are transmitted, and how they work is the science of genetics. It is a science that reveals the underlying causes of *resemblance,* as seen in the remarkable fidelity of reproduction, and of *variation,* the working material for organic evolution. All living forms use the same information storage, transfer, and translation system, which explains the stability of all life and reveals its descent from a common ancestral form. This is one of the most important unifying concepts of biology.

5.1 MENDEL'S INVESTIGATIONS

The first person to formulate the principles of heredity was Gregor Johann Mendel (1822 to 1884) (Figure 5.1 and p. 16), an Augustinian monk living in Brünn (Brno), Moravia. Brünn was then part of Austria but now lies in the eastern part of the Czech Republic. While breeding peas in a small monastery garden from 1856 to 1864, Mendel examined with great care the progeny of many thousands of plants. He presented in elegant simplicity the laws governing transmission of characters from parents to offspring. His discoveries, published in 1866, were of great significance, coming just after Darwin's publication of *On the Origin of Species by Means of Natural Selection.* The modes of inheritance that Mendel described were initially considered peculiar to garden peas. Beginning in 1900, 35 years after the completion of the work and 16 years after Mendel's death, biologists showed that Mendel's principles illustrated the basic hereditary systems of most plants, animals, and fungi.

Mendel chose garden peas for his classic experiments because they had pure strains differing from each other by sharply contrasting paired traits. For example, some varieties had dwarf plants and others were tall; some strains produced smooth seeds and others wrinkled seeds (Figure 5.1). Mendel carefully avoided more quantitative, continuously varying characteristics. A second reason for selecting peas was that they were self-fertilizing but subject to experimental cross-fertilization. The flower of a garden pea normally does not open, guaranteeing that the male and female gametes united in fertilization are from the same flower. This system of mating produces strains that are "true breeding": offspring match the parents' genetic characteristics.

A giant advance in chromosomal genetics was made when the American geneticist Thomas Hunt Morgan and his colleagues selected a species of fruit fly, *Drosophila melanogaster,* for their studies (1910–1920). Flies were cheaply and easily reared in bottles in the laboratory, fed on a simple medium of bananas and yeast. Most important, they produced a new generation every 10 days, enabling Morgan to collect data at least 25 times faster than with organisms that take longer to mature, such as garden peas. Morgan's work led to the mapping of genes on chromosomes and founded the discipline of cytogenetics.

Mendel crossed varieties having contrasting traits, making crosses for each of the seven characters shown in Figure 5.1. He removed the stamens (male part, containing the pollen) from a flower to prevent self-fertilization and then placed on the stigma (female part of flower) pollen from the flower of a plant true-breeding for the contrasting trait. Pollination from other sources such as wind and insects was rare and did not affect his results. Offspring from these crosses are called hybrids, meaning that they contain genetic information from two different parental strains. He collected seeds from the cross-fertilized flowers, planted these hybrid seeds, and examined the resulting plants for the contrasting traits being studied. These hybrid plants then produced offspring by self-pollination within unopened flowers.

Mendel knew nothing of the cellular basis of heredity, since chromosomes and genes were not yet discovered. Although we can admire Mendel's power of intellect in his discovery of the principles of inheritance without knowledge of chromosomes, these principles are easier to understand if we first review chromosomal behavior in the production of **germ cells**.

5.2 CHROMOSOMAL BASIS OF INHERITANCE

In sexually reproducing organisms, special germ cells, or **gametes** (ova and sperm), transmit genetic information from parents to offspring. A scientific explanation of genetic principles required microscopic study of germ cells and their behavior, and correlations between their transmission and certain visible results of inheritance. Nuclei of germ cells, especially the chromosomes, were early suspected of furnishing the real answer to the hereditary mechanism. Chromosomes are apparently the only entities transmitted in equal quantities from both parents to offspring.

When Mendel's laws were rediscovered in 1900, their correspondence with the cytological behavior of chromosomes was obvious. Later experiments showed that chromosomes carried hereditary information.

Meiosis: Reduction Division of Gametes

Although animal species differ greatly in the characteristic numbers, sizes, and shapes of chromosomes present in their body cells, a common feature is that chromosomes occur in pairs. The two members of a chromosomal pair contain similar genes encoding the same set of characteristics and usually, but not always, have the same size and shape. The members of such a pair are called **homologous** chromosomes; each individual member of a pair is called a **homolog**. One homolog comes from the mother and the other from the father. A **set** of chromosomes comprises one chromosome from each pair. Mendel's garden peas contain 14 chromosomes, two sets with 7 chromosomes per set. The number of chromosomes per set varies greatly among species.

Whenever an animal or plant produces gametes, it must produce haploid daughter cells from a diploid progenitor. **Meiosis** is

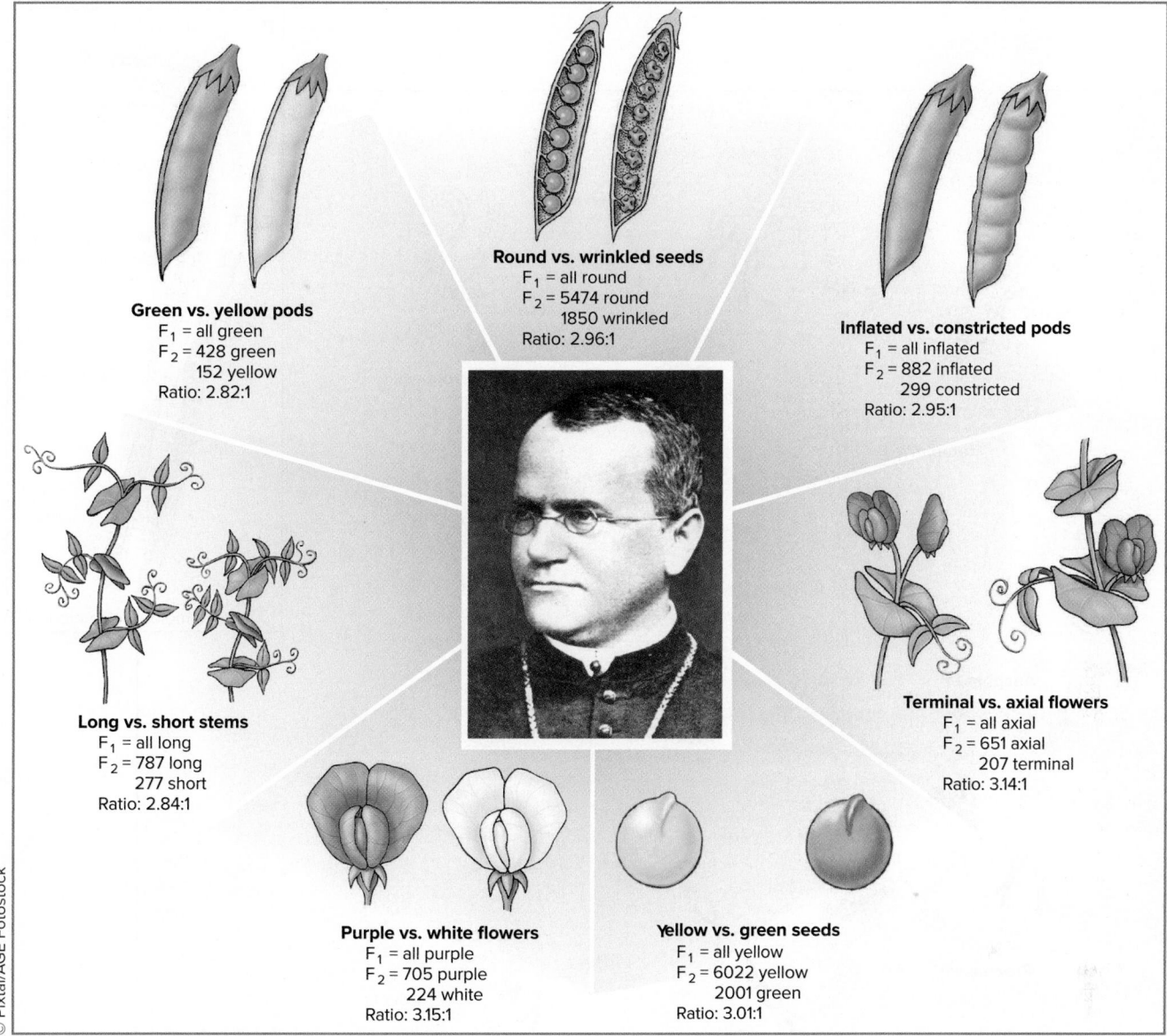

© Pixtal/AGE Fotostock

Green vs. yellow pods
F_1 = all green
F_2 = 428 green
152 yellow
Ratio: 2.82:1

Round vs. wrinkled seeds
F_1 = all round
F_2 = 5474 round
1850 wrinkled
Ratio: 2.96:1

Inflated vs. constricted pods
F_1 = all inflated
F_2 = 882 inflated
299 constricted
Ratio: 2.95:1

Long vs. short stems
F_1 = all long
F_2 = 787 long
277 short
Ratio: 2.84:1

Terminal vs. axial flowers
F_1 = all axial
F_2 = 651 axial
207 terminal
Ratio: 3.14:1

Purple vs. white flowers
F_1 = all purple
F_2 = 705 purple
224 white
Ratio: 3.15:1

Yellow vs. green seeds
F_1 = all yellow
F_2 = 6022 yellow
2001 green
Ratio: 3.01:1

Figure 5.1   Seven experiments on which Gregor Mendel based his postulates. These are the results of monohybrid crosses for first and second generations.

a special pair of cell divisions in which a cell's genetic material replicates once followed by two rounds of cell division (Figure 5.2). The result is a set of four daughter cells, each of which has only *one* member of each homologous chromosome pair. The chromosomes present in a meiotic daughter cell or gamete collectively form a single set of chromosomes. The number of chromosomes in a single set, which varies among species, is called the **haploid** (n) number of chromosomes. When a pair of gametes unites in fertilization, each gamete contributes its set of chromosomes to the newly formed cell, called a **zygote**, which has two complete sets of chromosomes. The number of chromosomes in two complete sets is called the **diploid** ($2n$) number. In humans the zygotes and all body cells normally have a diploid number ($2n$) of 46 chromosomes; the gametes have the haploid number (n), or 23, and meiosis reduces the number of chromosomes per cell from diploid to haploid.

Thus each cell normally has two copies of each gene coding for a given trait, one on each of the homologous chromosomes. Alternative forms of genes for the same trait are **allelic** forms, or **alleles**. Sometimes only one of the alleles has a visible effect on the organism, although both are present in each cell, and either may be passed to progeny by meiosis and subsequent fertilization.

Alleles are alternative forms of the same gene that have arisen by mutation of the DNA sequence. Like a baseball team with several pitchers, only one of whom can occupy the pitcher's mound at a time, only one allele can occupy a chromosomal locus (position). Alternative alleles for the locus may be on homologous chromosomes of a single individual, making that individual heterozygous for the gene in question. Numerous allelic forms of a gene may be found among different individuals in a population, a condition called "multiple alleles" (p. 80).

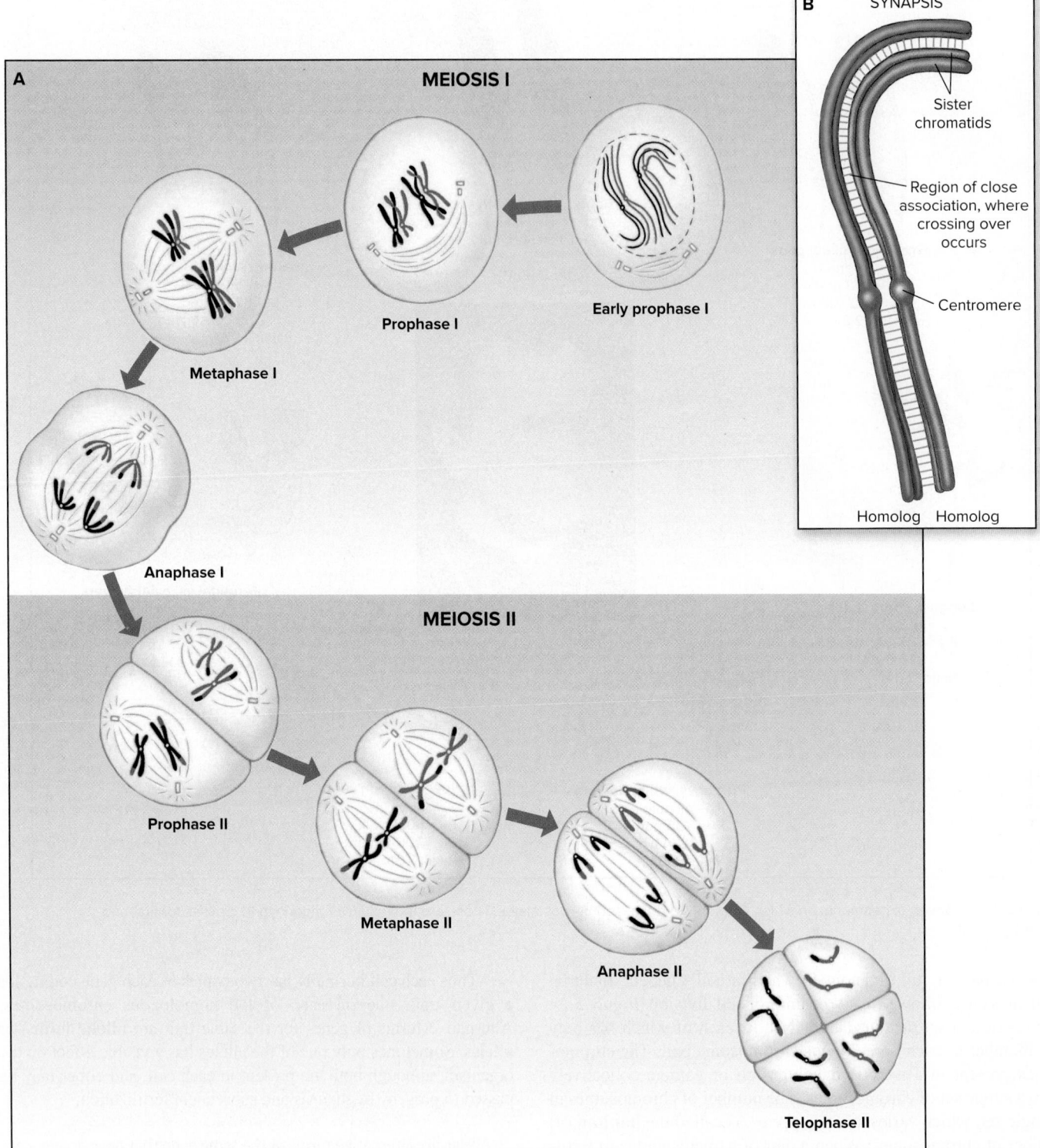

Figure 5.2 **A,** Meiosis in a sex cell with two pairs of chromosomes. Prophase I, homologous chromosomes come to lie with side-to-side contact, or synapsis, forming bivalents. A bivalent comprises a pair of homologous chromosomes, with each of the chromosomes initially containing a pair of identical chromatids joined by a centromere. Metaphase I, bivalents align at the spindle equator. Anaphase I, chromosomes of former bivalents are pulled toward opposite poles. Prophase II, daughter cells contain one of each homologous chromosome (haploid) but each chromosome is in replicated form (two chromatids attached at a centromere). Metaphase II, chromosomes move to the spindle equator. Anaphase II, chromatids of each chromosome separate. Telophase II, four haploid cells (gametes) formed, each with unreplicated chromosomes (one chromatid per chromosome). **B,** Synapsis occurs in prophase I, in which homologous chromosomes can break and exchange corresponding portions (material exchanged between chromatids of homologous chromosomes is shown by color coding in metaphase I of part **A**). The labeled sister chromatids and region of close association extend the full length of the bivalent. Meiosis-specific cohesion proteins called the synaptonemal complex hold the four strands of the bivalent in close association during synapsis.

During an individual's growth, all dividing cells contain the double set of chromosomes (mitosis is described on p. 50). In the reproductive organs, gametes (germ cells) are formed after meiosis, which *separates* the chromosomes of each homologous pair. Without this reductional division, the union of ovum (egg) and sperm would produce an individual with twice as many chromosomes as the parents.

Geneticists use a precise, standard definition of a chromosome: all material attached to a single centromere. *Understanding this definition is critical to interpreting the following description of meiosis.* Immediately following a mitotic cell division (p. 50), each chromosome contains a single long double-stranded DNA molecule bound to chromosomal proteins, which is called a chromatid. DNA and its associated chromosomal material are replicated during an "interphase" of the cell cycle between cell divisions. The replicated material is attached to the same centromere as the chromatid from which it was replicated, yielding two identical chromatids **in a single chromosome**. According to the definition of a chromosome, replication of the chromosomal material does not change the number of chromosomes present in a cell because the number of centromeres has not changed. Each centromere nonetheless carries two chromatids. A chromosome having a single chromatid is called an "unreplicated chromosome," and one carrying two chromatids is a "replicated chromosome." A replicated chromosome is a single chromosome until a subsequent cell division disrupts its centromere to produce two daughter centromeres, each of which carries a single chromatid to opposite poles of the cell. Each chromatid is then a separate, unreplicated chromosome. Cell division thus breaks each replicated chromosome to form two unreplicated chromosomes. Although the terminology is counterintuitive to many students, replication of chromosomal material in interphase doubles the amount of chromosomal material in the cell without changing the number of chromosomes present; cell division doubles the number of chromosomes present without changing the amount of chromosomal material. Cell division, of course, also doubles the number of cells present, thus keeping constant the number of chromosomes per cell. A replicated chromosome entering meiosis is often called a **dyad** because it contains a pair of identical chromatids.

Most unique features of meiosis occur during prophase of the first meiotic division (Figure 5.2). Prior to meiosis, each chromosome has already replicated to form two chromatids joined at one point, the centromere. The two members of each pair of homologous chromosomes make side-by-side contact (**synapsis**) to form a **bivalent**, which permits genetic recombination between the paired homologous chromosomes (p. 83). A meiosis-specific lattice of cohesion proteins called the **synaptonemal complex** holds the four strands of the bivalent in close association during synapsis. Each bivalent is composed of two pairs of chromatids (each pair is a dyad, sister chromatids held together at their centromere), or *four* future chromosomes, and is thus called a **tetrad**. The position or location of any gene on a chromosome is the gene **locus** (pl., **loci**), and in synapsis all gene loci on a chromatid normally lie exactly opposite the corresponding loci on the sister chromatid and both chromatids of the homologous chromosome. Toward the end of prophase, the chromosomes shorten and thicken and then enter the first meiotic division.

In contrast to mitosis, the centromeres holding the chromatids together *do not divide* at anaphase. As a result, each of the dyads is pulled toward one of the opposite poles of the cell by microtubules of the division spindle. At telophase of the first meiotic division, each pole of the cell has one dyad from each tetrad formed at prophase. Therefore, at the end of the first meiotic division, the daughter cells contain *one* chromosome of *each* homologous pair from the parent cell, so that the total chromosome number is reduced to haploid. However, because each chromosome contains two chromatids joined at a centromere, each cell contains twice the amount of DNA present in a gamete.

The second meiotic division more closely resembles events in mitosis. The dyads are split at the beginning of anaphase by division of their centromeres, and single-stranded chromosomes move toward each pole. Thus, by the end of the second meiotic division, the cells have the haploid number of chromosomes, and each chromatid of the original tetrad exists in a separate nucleus. Four cells, each containing one complete haploid set of chromosomes and only one copy of each gene, result from each cell that enters meiosis. In female gametogenesis, only one of the four resulting cells becomes a functional gamete (p. 141).

Sex Determination

Before the importance of chromosomes in heredity was realized in the early 1900s, genetic control of gender was totally unknown. The first scientific clue to chromosomal determination of sex came in 1902 when C. McClung observed that bugs (Hemiptera) produced two kinds of sperm in approximately equal numbers. One kind contained among its regular set of chromosomes a so-called accessory chromosome lacking in the other kind of sperm. Since all eggs of these species had the same number of haploid chromosomes, half the sperm would have the same number of chromosomes as the eggs, and half of them would have one chromosome less. When an egg was fertilized by a spermatozoon carrying the accessory (sex) chromosome, the offspring was a female; when fertilized by a spermatozoon without an accessory chromosome, the offspring was a male. Therefore a distinction was made between sex chromosomes, which determine sex (and sex-linked traits), and **autosomes**, the remaining chromosomes, which do not influence sex. The particular type of sex determination just described is often called the XX-XO type, which indicates that females have two X chromosomes and males only one X chromosome (the O indicates absence of the chromosome; see Figure 5.3).

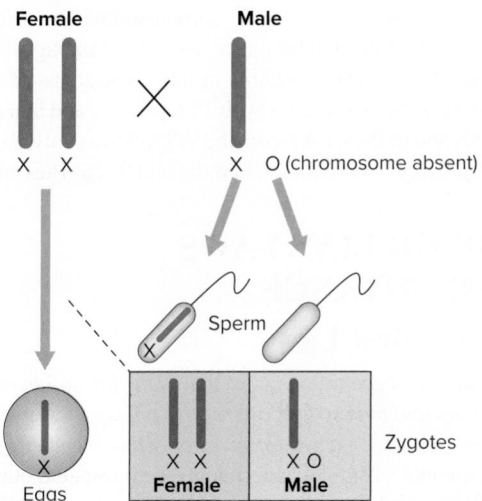

Figure 5.3 XX-XO sex determination. Only the sex chromosomes are shown.

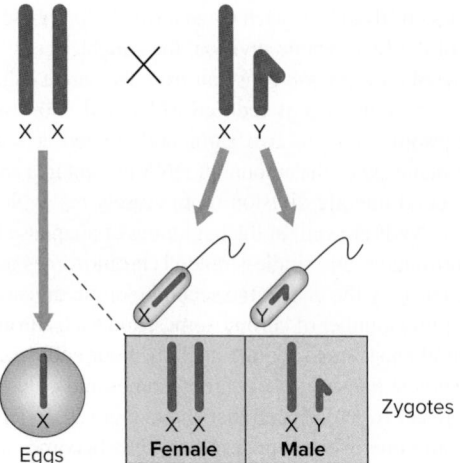

Figure 5.4 XX-XY sex determination. Only the sex chromosomes are shown.

Later, other types of sex determination were discovered. In humans and many other animals each sex contains the same number of chromosomes; however, the sex chromosomes (XX) are alike in females but unlike (XY) in males. Hence a human egg contains 22 autosomes + 1X chromosome. Sperm are of two kinds; half carry 22 autosomes +1X and half bear 22 autosomes +1Y. The Y chromosome is much smaller than the X and carries very little genetic information. At fertilization, when the 2X chromosomes come together, offspring are female; when X and Y come together, offspring are male (Figure 5.4).

A third type of sex determination occurs in birds, moths, butterflies, and some fishes, in which the male has 2 X (or sometimes called ZZ) chromosomes and the female an X and Y (or ZW). Finally, there are both invertebrates (p. 373) and vertebrates (p. 571) in which sex is determined by environmental or behavioral conditions rather than by sex chromosomes, or by genetic loci whose variation is not associated with visible difference in chromosomal structure.

In the case of X and Y chromosomes, homologous chromosomes are unlike in size and shape. Therefore, they do not both carry the same genes. Genes of the X chromosome often do not have allelic counterparts on the diminutive Y chromosome. This fact is very important in sex-linked inheritance (p. 82). Although sex-linked inheritance follows similar patterns in the XY systems of mammals and *Drosophila*, the XY system evolved separately and independently in mammals and in flies. Likewise, the WZ system evolved separately and independently in birds and in Lepidoptera (butterflies and moths).

5.3 MENDELIAN LAWS OF INHERITANCE

Mendel's First Law

Mendel's **law of segregation** states that *in the formation of gametes, paired factors that may specify alternative phenotypes (visible traits) separate so that each gamete receives only one member of the pair.* In one of Mendel's original experiments, he pollinated pure-line tall plants with the pollen of pure-line dwarf plants. Thus the visible characteristics, or **phenotypes**, of the parents were tall and dwarf. Mendel found that all progeny in the first generation (F_1) were tall,

just as tall as the tall parents of the cross. The reciprocal cross—dwarf plants pollinated with tall plants—gave the same result. The tall phenotype appeared in all progeny no matter which way the cross was made. Obviously, this kind of inheritance was not a blending of two traits, because none of the progeny was intermediate in size.

Next Mendel self-fertilized ("selfed") the tall F_1 plants and raised several hundred progeny, the second (F_2) generation. This time, *both* tall and dwarf plants appeared. Again, there was no blending (no plants of intermediate size), but the appearance of dwarf plants from all tall F_1 parents was surprising. The dwarf trait, seen in half of the grandparents but not in the parents, had reappeared. When he counted the actual number of tall and dwarf plants in the F_2 generation, he discovered that there were almost exactly three times more tall plants than dwarf ones.

Mendel then repeated this experiment for the six other contrasting traits that he had chosen, and in every case he obtained ratios very close to 3:1 (see Figure 5.1). These results made clear to Mendel that hereditary determinants for the contrasting traits did not blend when brought together in the same organism. Although the dwarf trait disappeared in the F_1 generation, it reappeared fully expressed in the F_2 generation. He realized that the F_1 generation plants carried determinants (which he called "factors") of both tall and dwarf parents, even though only the tall trait was visible.

One must distinguish the ability of Mendel's paired hereditary factors to influence phenotype from their ability to be transmitted through gametes to the next generation. When factors for tall and dwarf plants occur together in Mendel's F_1 generation, only the tall factor influences phenotype. Both factors, however, are transmitted through gametes to the next generation; half of the gametes produced by F_1 plants carry the tall factor, and the other half carry the dwarf factor.

Mendel called the tall factor **dominant** and the short **recessive**. Similarly, the other pairs of traits that he studied showed dominance and recessiveness. Whenever a dominant factor is present, the recessive one is not visible. The recessive trait appears only when both paired factors are recessive.

In representing his crosses, Mendel used letters as symbols; a capital letter denotes a dominant trait, and the corresponding lowercase letter denotes its recessive alternative. Modern geneticists still often follow this custom. Thus the factors for pure tall plants might be represented by *T/T*, the pure recessive by *t/t*, and the mix, or hybrid, of the two plants by *T/t*. The slash mark indicates that the alleles are on homologous chromosomes. The zygote bears the complete genetic constitution of the organism. All gametes produced by *T/T* must necessarily be *T*, whereas those produced by *t/t* must be *t*. Therefore a zygote produced by union of the two must be *T/t*, or a **heterozygote**. The pure tall plants (*T/T*) and pure dwarf plants (*t/t*) are **homozygotes**, meaning that the paired factors are alike on the homologous chromosomes and represent copies of the same allele. A cross involving variation at only a single locus is called a **monohybrid cross**.

In the cross between tall and dwarf plants there were two phenotypes: tall and dwarf. On the basis of genetic formulas there are three *hereditary* types: *T/T*, *T/t*, and *t/t*. These are called **genotypes**. A genotype is an allelic combination present in a diploid organism (*T/T*, *T/t*, or *t/t*), and the phenotype is the corresponding appearance of the organism (tall or dwarf).

One of Mendel's original crosses (tall plant and dwarf plant) could be represented as follows:

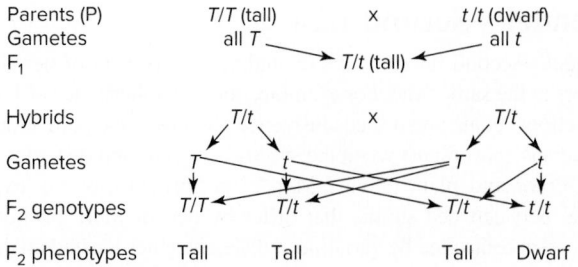

All possible combinations of F_1 gametes in the F_2 zygotes yield a 3:1 phenotypic ratio and a 1:2:1 genotypic ratio. It is convenient in such crosses to use the checkerboard method devised by Punnett (Punnett square) for representing the various combinations produced by a cross. In the F_2 cross this scheme would apply:

Pollen \ Ova	½ T	½ t
½ T	¼ T/T (homozygous tall)	¼ T/t (hybrid tall)
½ t	¼ T/t (hybrid tall)	¼ t/t (homozygous dwarf)

Ratio: 3 tall to 1 dwarf

The next step was an important one because it enabled Mendel to test his hypothesis that every plant contained nonblending factors from both parents. He self-fertilized the plants in the F_2 generation; the pollen of a flower fertilized the stigma of the same flower. The results showed that self-pollinated F_2 dwarf plants produced only dwarf plants, whereas one-third of the F_2 tall plants produced tall and the other two-thirds produced both tall and dwarf in the ratio of 3:1, just as the F_1 plants had done. Genotypes and phenotypes were as follows:

F_2 plants: Tall {
  ¼ T/T —Selfed→ all T/T (homozygous tall)
  ½ T/t —Selfed→ 1 T/T: 2 T/t: 1 t/t (3 tall: 1 dwarf)

Dwarf ¼ t/t —Selfed→ all t/t (homozygous dwarf)

This experiment showed that the dwarf plants were pure because they at all times gave rise to short plants when self-pollinated; the tall plants contained both pure tall and hybrid tall. It also demonstrated that, although the dwarf trait was not expressed in the F_1 plants, which were all tall, hereditary factors for dwarfness were transmitted in unaltered form to the F_2 plants.

Mendel's law of segregation states that whenever two factors are brought together in a hybrid, they segregate into separate gametes produced by the hybrid. The paired factors of the parent pass with equal frequency to the gametes. We now understand that the factors segregate because they occur on different chromosomes of a homologous pair, but the gametes receive only one chromosome of each pair in meiosis. Thus in current usage the law of segregation denotes the parting of homologous chromosomes during meiosis.

Mendel's great contribution was his quantitative approach to inheritance. His approach marks the birth of genetics, because before Mendel, people assumed that traits were blended like mixing together two colors of paint, a notion that unfortunately still lingers in the minds of many and was a problem for Darwin's theory of natural selection when he first proposed it (p. 14). If traits blended,

variability would be lost by hybridization. With particulate inheritance, different alleles remain intact through the hereditary process and can be resorted like particles.

Testcross

A testcross is used to determine whether an individual having a genetically dominant phenotype has a homozygous or heterozygous genotype for the trait in question. When an allele is dominant, heterozygous individuals containing that allele are identical in phenotype to individuals homozygous for it. Therefore one cannot determine the genotypes of these individuals just by observing their phenotypes. For instance, in Mendel's experiment of tall and dwarf traits, it is impossible to determine the genetic constitution of the tall plants of the F_2 generation by mere inspection of the tall plants. Three-fourths of this generation are tall, but which ones are heterozygotes?

As Mendel reasoned, the test is to cross the questionable individuals with pure recessives. If the tall plant is homozygous, all offspring in such a testcross are tall, thus:

Parents   T/T (tall) x t/t (dwarf)

Pollen \ Ova	T	T
t	T/t (hybrid tall)	T/t (hybrid tall)
t	T/t (hybrid tall)	T/t (hybrid tall)

All of the offspring are T/t (hybrid tall). If the tall plant is heterozygous, half of the offspring are tall and half dwarf, thus:

Parents   T/t (hybrid tall) x t/t (dwarf)

Pollen \ Ova	T	t
t	T/t (hybrid tall)	t/t (homozygous dwarf)
t	T/t (hybrid tall)	t/t (homozygous dwarf)

The **testcross** is often used in modern genetics to assess the genetic constitution of offspring and to make desirable homozygous stocks of animals and plants.

Intermediate Inheritance

In some cases neither allele is completely dominant over the other, and the heterozygous phenotype is distinct from those of the parents, often intermediate between them. This is called **intermediate inheritance**, or **incomplete dominance**. In the four-o'clock flower (*Mirabilis*), two allelic variants determine red versus pink or white flowers; homozygotes are red or white flowered, but heterozygotes have pink flowers. In a certain strain of chickens, a cross between those with black and those with splashed white feathers produces offspring that are not gray but a distinctive color called Andalusian blue (Figure 5.5). In each case, if the F_1s are crossed, the F_2s have a ratio of 1:2:1 in colors, or 1 red: 2 pink: 1 white in four-o'clock

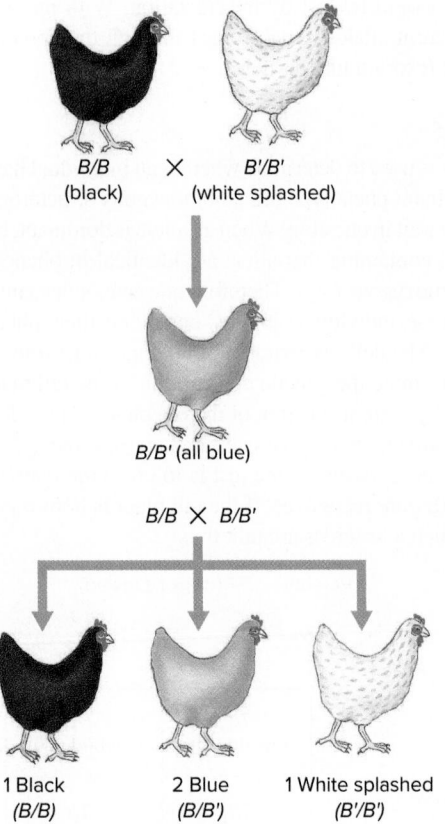

Figure 5.5 Cross between chickens with black and splashed white feathers. Black and white are homozygous; Andalusian blue is heterozygous.

flowers and 1 black: 2 blue: 1 white for Andalusian chickens. This phenomenon is illustrated for the chickens as follows:

Parents	*B/B*	(black feathers)	χ	*B'/B'*	(white feathers)
Gametes	all *B*			all *B'*	
F₁			*B/B'* (all blue)		
Crossing hybrids		*B/B'*	χ	*B/B'*	
Gametes		*B,B'*		*B,B'*	

F₂ genotypes	*B/B*	*B/B'*	*B/B'*	*B'/B'*
F₂ phenotypes	Black	Blue	Blue	White

When neither of the alleles is recessive, one can represent both by capital letters and distinguish them by the addition of a "prime" sign (B') or by superscript letters, for example, B^b (equals black feathers) and B^w (equals white feathers). This system for designating alleles is not universal, especially for loci having multiple alleles with complex patterns of dominance (p. 80).

In this kind of cross, the heterozygous *phenotype* is indeed a blending of both parental types. It is easy to see how such observations would encourage the notion of blending inheritance. However, in the cross of black and white chickens or red and white flowers, only the hybrid phenotype is a blend; its hereditary factors do *not* blend and homozygous offspring breed true to the parental phenotypes.

Mendel's Second Law

Mendel's second law pertains to studies of two pairs of hereditary factors at the same time. For example, does the inheritance of factors for yellow versus green seeds influence the inheritance of factors for tall versus short plants when the strains being crossed differ for both seed color and plant height? Mendel performed crossing experiments between pea strains that differ by two or more phenotypic characters controlled by variation at different genes located on different chromosomes. According to Mendel's **law of independent assortment**, *genes located on different pairs of homologous chromosomes assort independently during meiosis.*

Mendel had already established that tall plants were dominant to dwarf. He also noted that crosses between plants bearing yellow seeds and plants bearing green seeds produced plants with yellow seeds in the F₁ generation; therefore yellow was dominant to green. The next step was to make a cross between plants differing in these two characteristics. When a tall plant with yellow seeds (*T/T Y/Y*) was crossed with a dwarf plant with green seeds (*t/t y/y*), the F₁ plants were tall and yellow as expected (*T/t Y/y*).

Parents	*T/T Y/Y*	χ	*t/t y/y*
	(tall, yellow)		(dwarf, green)
Gametes	all *TY*		all *ty*
F₁		*T/t Y/y*	
		(tall, yellow)	

The F₁ hybrids were then self-fertilized, giving the F₂ results shown in Figure 5.6.

Mendel already knew that a cross between two plants bearing a single pair of alleles of the genotype *T/t* would yield a 3:1 ratio. Similarly, a cross between two plants with the genotypes *Y/y* would yield the same 3:1 ratio. If we examine *only* the tall and dwarf phenotypes expected in the outcome of the dihybrid experiment, they produce a ratio of 12 tall to 4 dwarf, which reduces to a ratio of 3:1. Likewise, a total of 12 plants have yellow seeds for every 4 plants that have green—again a 3:1 ratio. Thus the monohybrid ratio prevails for both traits when they are considered independently. The 9:3:3:1 ratio is nothing more than a combination of the two 3:1 ratios.

$$3: 1 \times 3:1 = 9:3:3:1$$

When one of the alleles is unknown, it can be designated by a dash (*T/—*). This designation is used also when we are concerned only with phenotypic ratios. The dash could be either *T* or *t*.

The F₂ genotypes and phenotypes are as follows:

1	*T/T Y/Y*	⎫		
2	*T/t Y/Y*	⎬ 9 *T/—Y/—*	9 Tall yellow
2	*T/T Y/y*			
4	*T/t Y/y*	⎭		
1	*T/T y/y*	⎫ 3 *T/—y/y*	3 Tall green
2	*T/t y/y*	⎭		
1	*t/t Y/Y*	⎫ 3 *t/t—Y/—*	3 Dwarf yellow
2	*t/t Y/y*	⎭		
1	*t/t y/y*	1 *t/t y/y*	1 Dwarf green

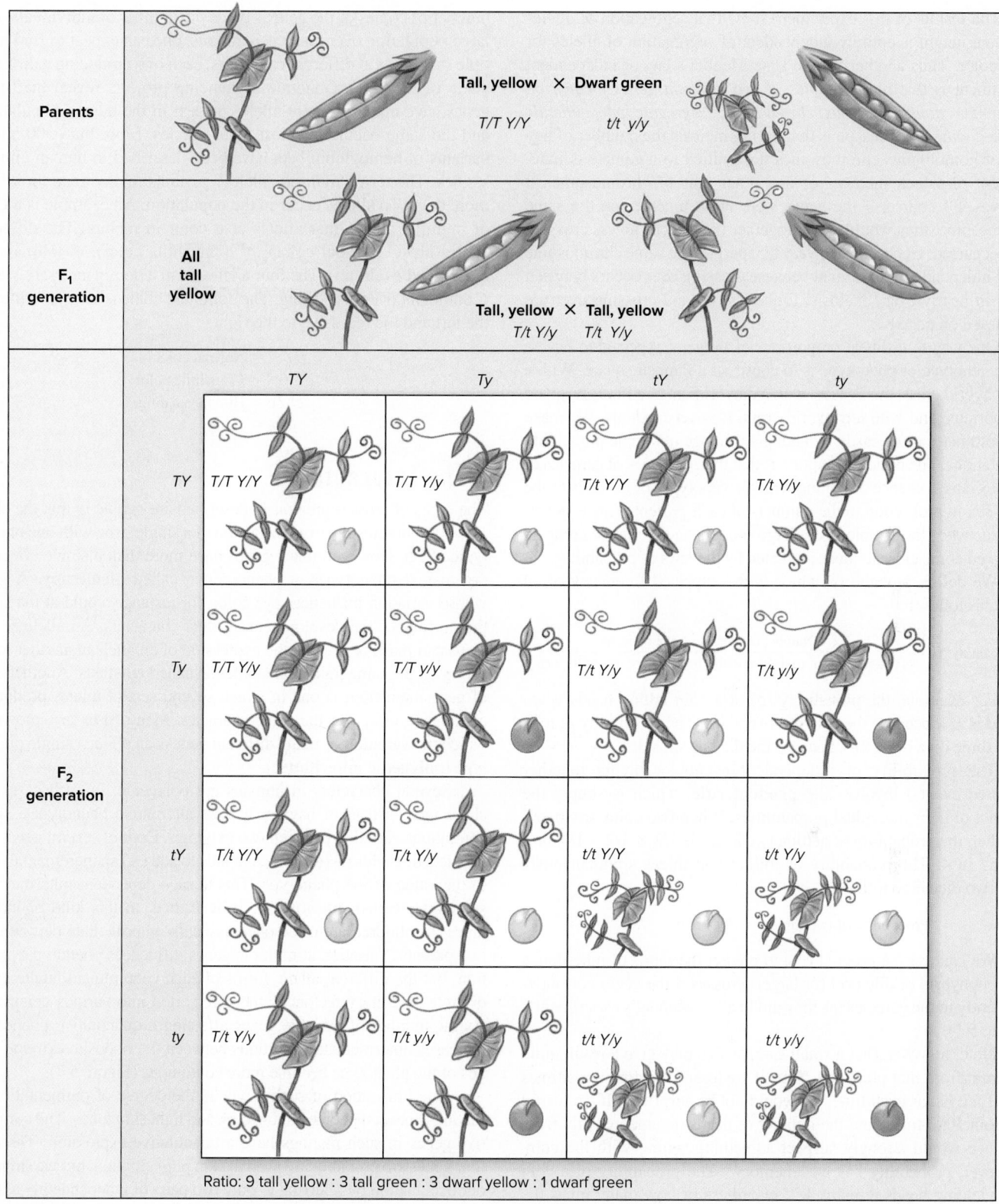

Figure 5.6 Punnett square method for determining ratios of genotypes and phenotypes expected in a dihybrid cross for independently assorting genes.

The results of this experiment show that segregation of alleles for plant height is entirely independent of segregation of alleles for seed color. Thus another way to state Mendel's law of independent assortment is that *paired copies of two different genes located on different (= nonhomologous) chromosomes segregate independently of one another.* The reason is that during meiosis the member of any pair of homologous chromosomes transmitted to a gamete is independent of which member of any other pair of chromosomes it receives. Of course, if the genes were close together on the same chromosome, they would assort together (be linked) unless crossing over occurred. Genes located very far apart on the same chromosome show independent assortment because crossing over occurs between them in nearly every meiosis. Linked genes and crossing over are discussed on page 83.

One way to estimate proportions of progeny expected to have a given genotype or phenotype is to construct a Punnett square. With a monohybrid cross, this is easy; with a dihybrid cross, a Punnett square is laborious; and with a trihybrid cross, it is very tedious. We make such estimates more easily by taking advantage of simple probability calculations. The basic assumption is that the genotypes of gametes of one sex have a chance of uniting with the genotypes of gametes of the other sex in proportion to the numbers of each present. This is generally true when the sample size is large enough, and the actual numbers observed come close to those predicted by the laws of probability.

We define probability, which is the expected frequency of an event, as follows:

$$\text{Probability (p)} = \frac{\text{Number of times an event happens}}{\text{Total number of trials or possibilities for the event to happen}}$$

For example, the probability (p) of a coin falling heads when tossed is 1/2, because the coin has two sides. The probability of rolling a three on a die is 1/6, because the die has six sides.

The probability of independent events occurring together (ordered events) involves the **product rule,** which is simply the product of their individual probabilities. When two coins are tossed together, the probability of getting two heads is 1/2 × 1/2 = 1/4, or 1 chance in 4. The probability of rolling two threes simultaneously with two dice is as follows:

$$\text{Probability of two threes} = 1/6 \times 1/6 = 1/36$$

We can use the product rule to predict the ratios of inheritance in monohybrid or dihybrid (or larger) crosses if the genes sort independently in the gametes (as they did in all of Mendel's experiments) (Table 5.1).

Note, however, that a small sample size may give a result quite different from that predicted. Thus if we tossed the coin three times and it fell heads each time, we would not be surprised. If we tossed the coin 1000 times and the number of heads diverged greatly from 500, we would strongly suspect something wrong with the coin. However, probability has no "memory." The probability of a coin toss yielding heads remains 1/2, regardless of how many times the coin was tossed previously or results of the tosses.

Multiple Alleles

On page 73 we defined alleles as alternative forms of a gene. A new allele arises by mutation of one copy of a preexisting allele. A gene may contain thousands of base pairs and, in a diploid population, the number of copies of the gene is twice the number of individuals. In a large population over many generations, one can expect to find multiple mutations at different base pairs, each one producing a different allele of the gene. Genomic sequencing projects reveal that most genes have many different alleles present in the human population, and the same occurs for most other species. More than 500 allelic variants of hemoglobin beta have been described in human clinical studies. The term "multiple alleles" pertains to any gene for which more than two alleles occur in the population. An example is the set of multiple alleles that affects coat color in rabbits. The different alleles are C (normal color), c^{ch} (chinchilla color), c^h (Himalayan color), and c (albino). The four alleles form a dominance series with C dominant over everything. The dominant allele is always written to the left and the recessive to the right:

$$C / c^h = \text{Normal color}$$
$$c^{ch} / c^h = \text{Chinchilla color}$$
$$c^h / c = \text{Himalayan color}$$
$$c / c = \text{albino}$$

Gene Interaction

The types of crosses previously described are simple in that the character variation results from the action of a single gene with one phenotypic effect. However, many genes have more than a single effect on organismal phenotypes, a phenomenon called **pleiotropy.** A gene whose variation influences eye color, for instance, could at the same time influence the development of other characters. An allele at one locus can mask or prevent the expression of an allele at another locus acting on the same trait, a phenomenon called **epistasis.** Another case of gene interaction is one in which several sets of alleles produce a cumulative effect on the same character. Many cases are known in which the variation of many different genes can affect a single phenotype (**polygenic inheritance).**

Several characters in humans are polygenic. In such cases the characters, instead of having discrete alternative phenotypes, show continuous variation between two extremes. Each of several genes has an allele that adds (+) and one that fails to add (−) an incremental dose to the value of the phenotype. This dosage-dependent inheritance is sometimes called **quantitative inheritance.** In this kind of inheritance the children are often phenotypically intermediate between the two parents. Variation at multiple genes influences phenotypic variation, but the different allelic forms of each gene remain unaltered as discrete hereditary factors as they are sorted into various genotypes. As the number of variable genes affecting a quantitative phenotype increases, intermediate conditions between the opposite extreme values of the phenotype become more continuous (Figure 5.7).

One illustration of such a type is the degree of pigmentation in matings between people having dark and light skin tones. The cumulative genes in such matings have a quantitative expression. Three or four genes are probably involved in skin pigmentation, but we simplify our explanation by referring to only two pairs of independently assorting genes. Thus a person with very dark pigment has two genes for pigmentation on separate chromosomes (*A/A B/B*). Each copy of a capital allele contributes one unit of pigment. A person with very light pigment has alleles (*a/a b/b*) that contribute no color. (Freckles that commonly appear in the skin of very light people represent pigment contributed by entirely separate genes.) The offspring of very dark and very light parents would have an intermediate skin color (*A/a B/b*).

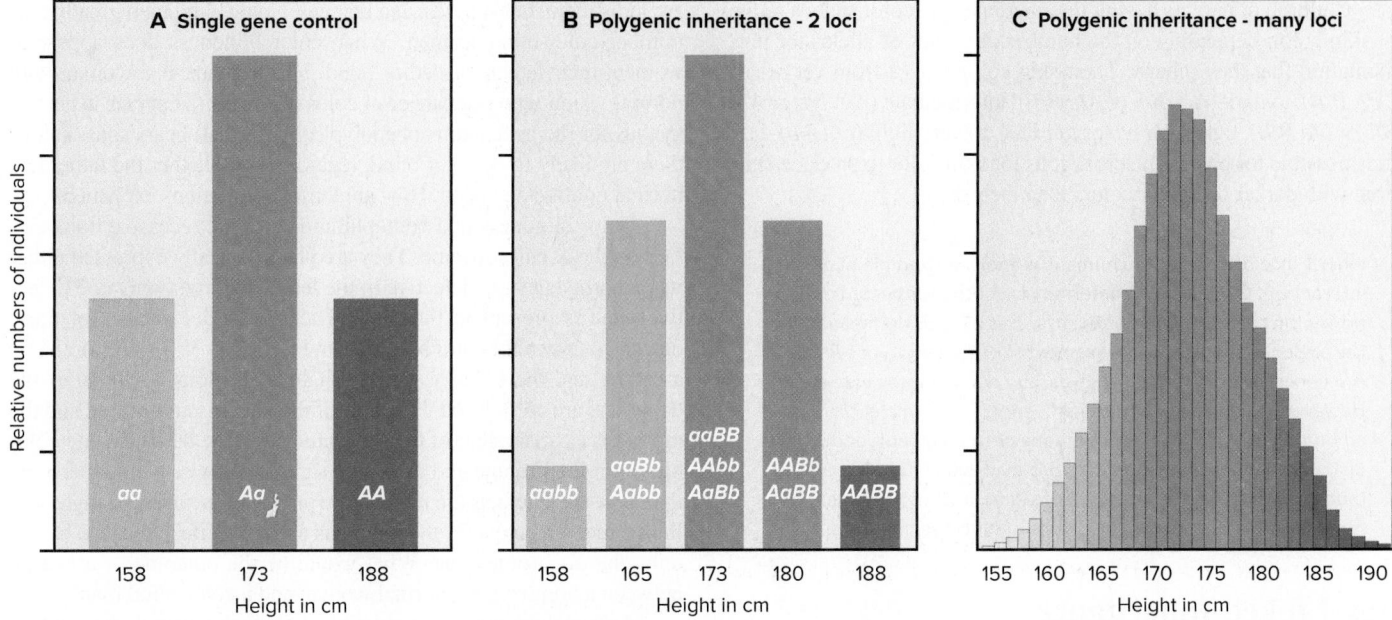

Figure 5.7 Polygenic inheritance and continuous variation. Suppose that variation in height between 158 and 188 cm is controlled by genic variation with intermediate inheritance (heterozygous genotype is intermediate in height between the alternative homozygous genotypes). **A,** If a single gene controls this variation, then there are three discrete height classes in the population, with adjacent classes differing by 15 cm: 158 cm, 173 cm, and 188 cm. **B,** If two genes control the variation, and each capital allele contributes an increment of height, we have five height classes with adjacent classes differing by only 7 or 8 cm: 158 cm, 165 cm, 173 cm, 180 cm, and 188 cm. **C,** As the number of genes influencing the variation increases, the number of size classes increases and the difference in size between adjacent classes decreases. The limit as the number of genes becomes very large is a continuous distribution of heights from 158 to 188. These graphs illustrate the special case in which the contrasting alleles at each locus each have a frequency of 0.5 in the population (p. 122).

TABLE 5.1

Use of Product Rule for Determining Genotypic and Phenotypic Ratios in a Dihybrid Cross for Independently Assorting Genes

Parents' genotypes	*T/t Y/y*	×	*T/t Y/y*
Equivalent monohybrid crosses	*T/t × T/t*	and	*Y/y × Y/y*
Genotype ratios in F₁s of monohybrid crosses		1/4 *T/T*	1/4 *Y/Y*
		2/4 *T/t*	2/4 *Y/y*
		1/4 *t/t*	1/4 *y/y*
Combine two monohybrid ratios to determine dihybrid genotype ratios	1/4 *T/T*	×	⎰ 1/4 *Y/Y* = 1/16 *T/T Y/Y* 2/4 *Y/y* = 2/16 *T/T Y/y* ⎱ 1/4 *y/y* = 1/16 *T/T y/y*
	2/4 *T/t*	×	⎰ 1/4 *Y/Y* = 2/16 *T/t Y/Y* 2/4 *Y/y* = 4/16 *T/t Y/y* ⎱ 1/4 *y/y* = 2/16 *T/t y/y*
	1/4 *t/t*	×	⎰ 1/4 *Y/Y* = 1/16 *t/t Y/Y* 2/4 *Y/y* = 2/16 *t/t Y/y* ⎱ 1/4 *y/y* = 1/16 *t/t y/y*
Phenotype ratios in F₁s of monohybrid crosses			3/4 *T/—*(tall), 1/4 *t/t* (dwarf) 3/4 *Y/—*(yellow), 1/4 *y/y* (green)
Combine two monohybrid ratios to determine phenotype ratios	3/4 *T/—*	×	⎰ 3/4 *Y/—* = 9/16 *T/—Y/—*(tall, yellow) ⎱ 1/4 *y/y* = 3/16 *T/—y/y* (tall, green)
	1/4 *t/t*	×	⎰ 3/4 *Y/—*=3/16 *t/t Y/—*(dwarf, yellow) ⎱ 1/4 *y/y*=1/16 *t/t y/y* (dwarf, green)

Therefore phenotype ratios = 9 tall, yellow: 3 tall, green: 3 dwarf, yellow: 1 dwarf, green

Children of parents having intermediate skin color show a range of skin color, depending on the number of copies of alleles for pigmentation that they inherit. Their skin color ranges from very dark (*A/A B/B*), to dark (*A/A B/b* or *A/a B/B*), intermediate (*A/A b/b* or *A/a B/b* or *a/a B/B*), light (*A/a b/b* or *a/a B/b*), to very light (*a/a b/b*). It is thus possible for parents heterozygous for skin color to produce children with darker or lighter colors than their own.

Inheritance of eye color in humans is another example of gene interaction. One allele (*B*) determines whether pigment is present in the front layer of the iris. This allele is dominant over the allele for the absence of pigment (*b*). The genotypes *B/B* and *B/b* generally produce brown eyes, and *b/b* produces blue eyes. However, many modifier genes influence, for example, the amount of pigment present, the tone of the pigment, and its distribution. Thus a person with *B/b* may even have blue eyes if modifier genes determine a lack of pigment, thus explaining the rare instances of a brown-eyed child of blue-eyed parents.

Sex-Linked Inheritance

Inheritance of some characters depends on the sex of the parent carrying the gene and the sex of the offspring. One of the best-known sex-linked traits of humans is hemophilia (see Chapter 31, p. 684). Another example is red-green color blindness, in which red and green colors are indistinguishable to varying degrees. Color-blind men greatly outnumber color-blind women. When color blindness does appear in women, their fathers are color blind. Furthermore, if a woman with normal vision who is a carrier of color blindness (a **carrier** is heterozygous for the gene and is phenotypically normal) bears sons, half of them are likely to be color blind, regardless of whether the father had normal or affected vision. How are these observations explained?

Color blindness and hemophilia defects are recessive traits carried on the X chromosome. They are phenotypically expressed either when both genes are defective in the female or when only one defective gene is present in the male. The inheritance pattern of these defects is shown for color blindness in Figure 5.8. When the mother is a carrier and the father is normal, half of the sons but none of the daughters are color blind. However, if the father is color blind and the mother is a carrier, half of the sons *and* half of the daughters are color blind (on the average and in a large sample). It is easy to understand then why such defects are much more prevalent in males: a single sex-linked recessive gene in the male has a visible effect because he has only one X chromosome. What would be the outcome of a mating between a homozygous normal woman and a color-blind man?

Another example of a sex-linked character was discovered by Thomas Hunt Morgan (1910) in *Drosophila*. Normal eye color of this fly is red, but mutations for white eyes do occur (Figure 5.9). The X chromosome carries a gene for eye color. If true-breeding white-eyed males and red-eyed females are crossed, all F₁ offspring have red eyes

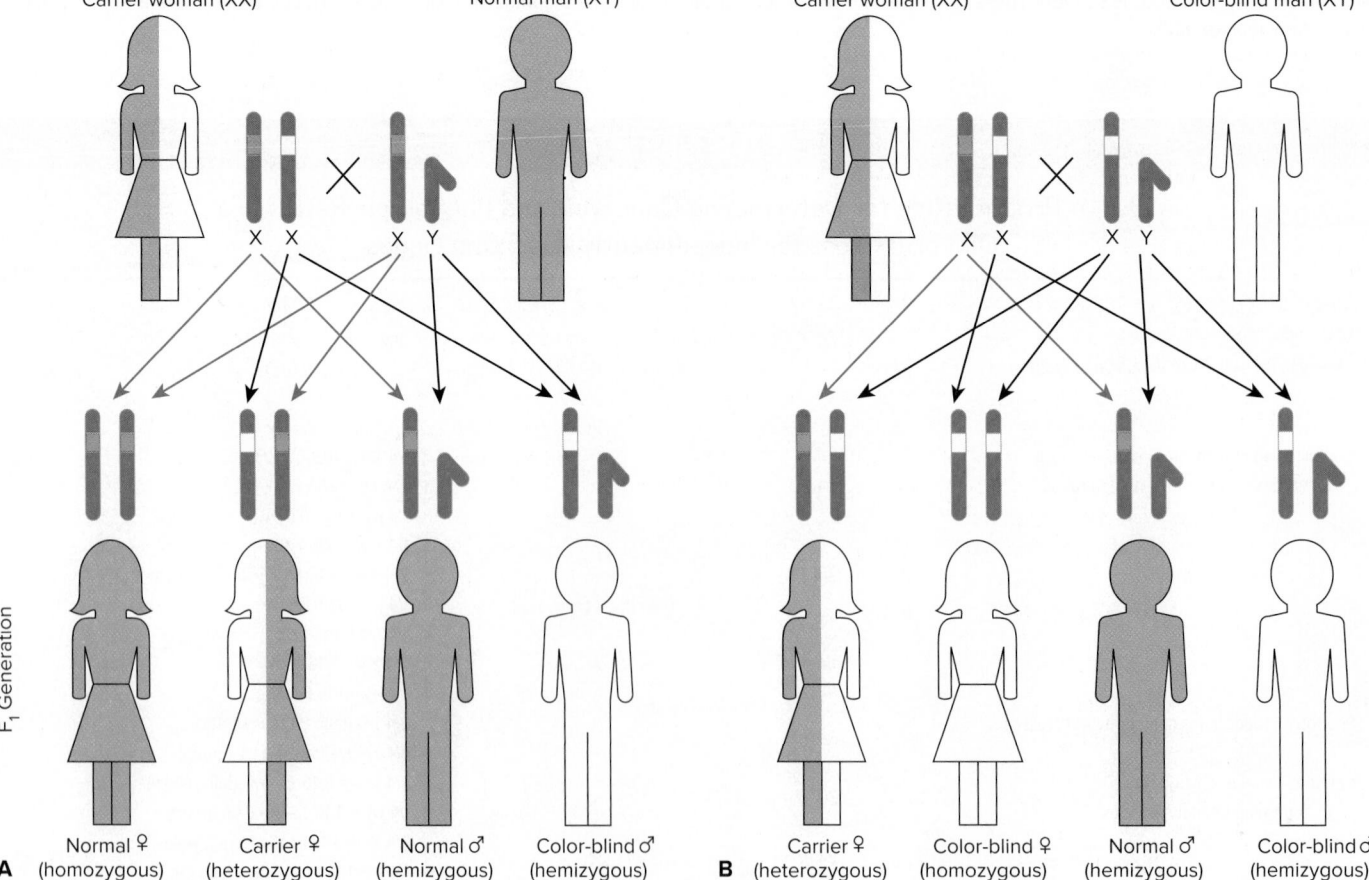

Figure 5.8   Sex-linked inheritance of red-green color blindness in humans. **A,** Carrier mother and normal father produce color blindness in one-half of their sons but in none of their daughters. **B,** Half of both sons and daughters of carrier mother and color-blind father are color blind.

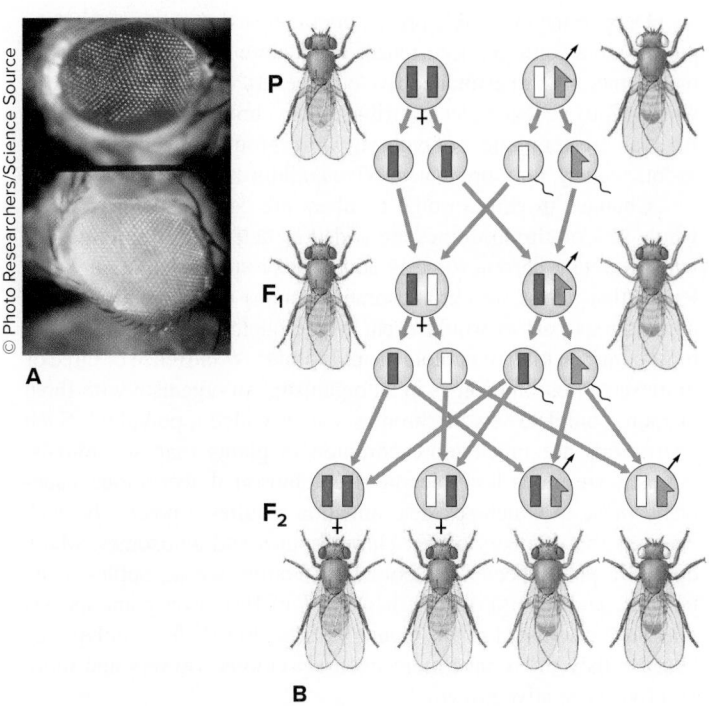

Figure 5.9   Sex-linked inheritance of eye color in fruit fly *Drosophila melanogaster*. **A,** Red and white eyes of *D. melanogaster.* **B,** Genes for eye color are carried on X chromosome; Y carries no genes for eye color. Normal red is dominant to white. Homozygous red-eyed females mated with white-eyed males give all red-eyed in F$_1$. F$_2$ ratios from F$_1$ cross are one homozygous red-eyed female and one heterozygous red-eyed female to one red-eyed male and one white-eyed male.

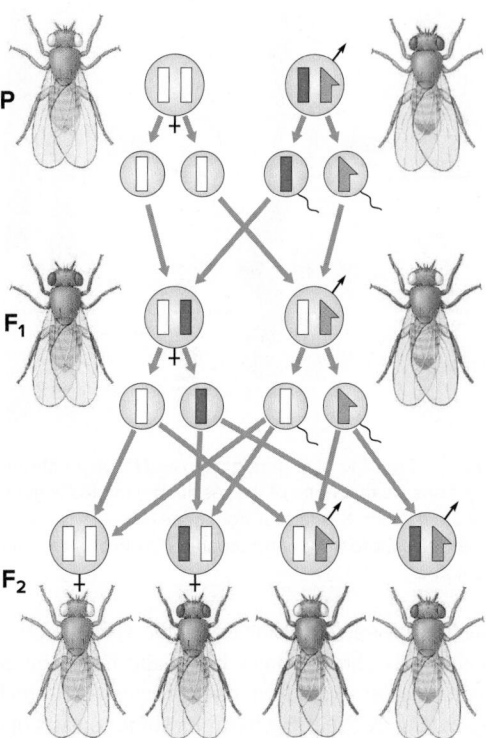

Figure 5.10   Reciprocal cross of Figure 5.9 (homozygous white-eyed female with red-eyed male) gives white-eyed males and red-eyed females in F$_1$. F$_2$ shows equal numbers of red-eyed and white-eyed females and red-eyed and white-eyed males.

because this trait is dominant (Figure 5.9). If these F$_1$ offspring are interbred, all F$_2$ females have red eyes; half of the males have red eyes and the other half have white eyes. No white-eyed females appear in this generation; only males have the recessive character (white eyes). The allele for white eyes is recessive and should affect eye color only in a homozygous condition. However, since the male has only one X chromosome (the Y does not carry a gene for eye color), white eyes appear whenever the X chromosome carries the allele for this trait. Males are **hemizygous** (only one copy of a genetic locus is present) for traits carried on the X chromosome.

If the reciprocal cross is made in which females are white eyed and males red eyed, all F$_1$ females are red eyed and all males are white eyed (Figure 5.10). If these F$_1$ offspring are interbred, the F$_2$ generation shows equal numbers of red-eyed and white-eyed males and females.

Autosomal Linkage and Crossing Over

Linkage

After Mendel's laws were rediscovered in 1900, it became clear that, contrary to Mendel's second law, not all factors segregate independently. Indeed, many traits are inherited together. Since the number of chromosomes in any organism is relatively small compared with the number of traits, each chromosome must contain many genes. All genes present on a chromosome are **linked.** Linkage specifies that the genes are on the same chromosome, and all genes present on homologous chromosomes belong to the same linkage group. Therefore there should be as many linkage groups as there are chromosome pairs.

Geneticists commonly use the word "linkage" in two somewhat different meanings. Sex linkage denotes inheritance of a trait on the sex chromosomes, and thus its phenotypic expression depends on the sex of the organism and the factors already discussed. Autosomal linkage, or simply, linkage, denotes inheritance of genes on a given autosomal chromosome. Letters used to represent such genes are normally written without a slash mark between them, indicating that they are on the same chromosome. For example, *AB/ab* shows that genes *A* and *B* are on the same chromosome. Interestingly, Mendel studied seven characteristics of garden peas that assorted independently because they are on seven different chromosomes. If he had studied eight characteristics, he would not have found independent assortment in two of the traits because garden peas have only seven pairs of homologous chromosomes.

In *Drosophila,* there are four linkage groups that correspond to the four pairs of chromosomes present. Usually, small chromosomes have small linkage groups, and large chromosomes have large groups.

Crossing Over

Linkage, however, is usually not complete. If we perform a dihybrid cross in animals such as *Drosophila,* we find that linked traits separate in some percentage of the offspring. Separation of traits linked on the same chromosome occurs because of **crossing over.**

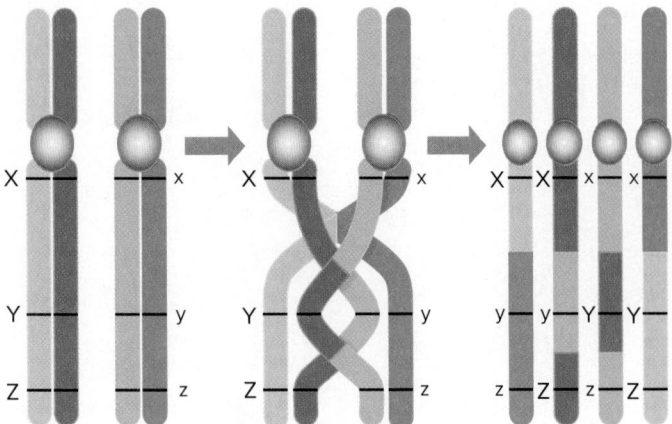

Figure 5.11 Crossing over during meiosis. Nonsister chromatids exchange portions, so that none of the resulting gametes is genetically the same as any other. Gene X is farther from gene Y than Y is from Z; therefore, gene X is more frequently separated from Y in crossing over than Y is from Z.

During the protracted prophase of the first meiotic division, paired homologous chromosomes break and exchange equivalent portions; genes "cross over" from one chromosome to its homolog, and vice versa (Figure 5.11). Each chromosome consists of two sister chromatids, and the paired chromosomes (four chromatids total) are held together by means of a proteinaceous structure called a synaptonemal complex. Breaks and exchanges occur at corresponding points on nonsister chromatids. (Breaks and exchanges also occur between sister chromatids but usually have no genetic significance because sister chromatids are identical.) Crossing over is a means for exchanging genes between homologous chromosomes and thus greatly increases the amount of genetic recombination. The frequency of crossing over varies depending on the species and chromosomal region, but usually at least one and often several crossovers occur per bivalent each time chromosomes pair.

Because the frequency of recombination is proportional to the distance between loci, the relative linear position of each locus can be determined. *Genes located far apart on very large chromosomes may assort independently because the probability of a crossover occurring between them in each meiosis is close to 100%.* We discover that such genes are on the same chromosome if they share genetic linkage to a third gene located physically between them on the chromosome. For example, in Figure 5.11, if recombination between genes X and Y is 30% and recombination between genes Y and Z is 20%, genes X and Z would assort independently of each other (50% recombination), but their common genetic linkage to gene Y would reveal that they are on the same chromosome. Laborious genetic experiments followed by genome sequencing projects have mapped the many thousands of genes distributed on the four chromosomes of *Drosophila melanogaster*.

Chromosomal Aberrations

Structural and numerical deviations from the norm that affect many genes at once are called chromosomal aberrations. They are sometimes called chromosomal mutations, but the term "mutation" usually refers to qualitative changes within a gene; gene mutations are discussed on page 95.

Despite the incredible precision of meiosis, chromosomal aberrations are a common occurrence. They cause many human genetic malformations. An estimated five out of every 1000 people are born with *serious* genetic defects attributable to chromosomal anomalies. An even greater number of embryos with chromosomal defects abort spontaneously, far more than survive until birth.

Changes in chromosome numbers are called **euploidy** when whole sets of chromosomes are added or deleted, and **aneuploidy** when a single chromosome is added to or subtracted from a set. Recall that a "set" of chromosomes contains one member of each homologous pair as would occur in the nucleus of a gamete. The most common kind of euploidy is **polyploidy,** occurrence of three or more sets of chromosomes in an organism. An organism with three or more complete sets of chromosomes is called a polyploid. Such aberrations are much more common in plants than in animals. Animals are much less tolerant of chromosomal aberrations, especially those in which sex determination requires a precise balance between the numbers of sex chromosomes and autosomes. Many domestic plant species are polyploid (cotton, wheat, apples, oats, tobacco, and others), and perhaps 40% of flowering plant species may have originated in this manner. Horticulturists favor polyploids because they often have more intensely colored flowers and more vigorous vegetative growth.

Aneuploidy is usually caused by failure of a pair of homologous chromosomes to separate during meiosis (**nondisjunction**). If a pair of chromosomes fails to separate during the first or second meiotic divisions, both members go to one pole and none to the other. This condition causes at least one gamete or polar body to have $n - 1$ chromosomes and another to have $n + 1$ chromosomes. If an $n - 1$ gamete is fertilized by a normal n gamete, a **monosomic** individual results. Survival is rare because the lack of one chromosome gives an uneven balance of genetic instructions. **Trisomic** individuals, the result of the fusion of a normal n gamete and an $n + 1$ gamete, are much more likely to survive, especially trisomics for small chromosomes. Several kinds of trisomic conditions occur in humans. Perhaps the most familiar is **trisomy 21, or Down syndrome.** As the name indicates, it involves an extra chromosome 21 combined with the chromosome pair 21, and it is caused by nondisjunction of that pair during meiosis. It occurs spontaneously, and there is seldom any family history of the abnormality. However, the risk of its appearance rises dramatically with increasing age of the mother; it occurs 40 times more often in women over 40 years old than among women between the ages of 20 and 30. In cases where maternal age is not a factor, 20% to 25% of trisomy 21 results from nondisjunction during spermatogenesis; it is paternal in origin and is apparently independent of the father's age.

A *syndrome* is a group of symptoms associated with a particular disease or abnormality, although every symptom is not necessarily shown by every patient with the condition. An English physician, John Langdon Down, described in 1866 the syndrome whose underlying cause is trisomy 21. Because Down judged facial features of affected individuals to be mongoloid in appearance, the condition has been called mongolism. The resemblances are superficial, however, and currently accepted names are trisomy 21 and Down syndrome. Among the numerous characteristics of the condition, the most disabling is cognitive impairment. This and other chromosomal aberrations are diagnosed *prenatally* by a procedure called *chorionic villus sampling* (CVS). The physician inserts a catheter or hypodermic needle through the abdominal wall of the mother and into tiny outgrowths of the placenta called chorionic villi. Cells of the chorionic villi are genetically identical to cells of the fetus. The cells are grown in culture, their chromosomes are examined, and other tests are done. If a severe birth defect is found, the mother has the option of having an abortion performed. As an extra "bonus," the sex of the fetus is learned by observing the sex chromosomes. Alternatively, measurements of certain substances in maternal serum can detect about 60% of Down syndrome fetuses. Ultrasound scanning is more than 80% accurate.

In all diploid species, normal development requires exactly two of each kind of autosome (not sex chromosomes). Nondisjunction can cause trisomies of other chromosomes, but because these lead to imbalances of many gene products, they almost always cause death before or soon after birth. However, each cell requires only one functional X chromosome (the other is inactivated in females). Nondisjunction of sex chromosomes is better tolerated but usually causes sterility and abnormalities of sex organs. For example, a human with XXY (Klinefelter syndrome) is a phenotypic male, usually infertile and with some female sexual characteristics. Presence of only one X (and no Y) is usually lethal in embryos, but the occasional live birth produces a phenotypic female with a variety of developmental abnormalities (Turner syndrome). The only aneuploid condition that has no abnormal phenotypic effects is presence of an extra Y chromosome in males (XYY). Because it carries few genes, an extra Y chromosome does not impair normal development, although its possessors on average experience a higher growth rate and a greater-than-average adult height.

Structural aberrations involve multiple genes within a chromosome. A portion of a chromosome may be reversed, placing the linear arrangement of genes in reverse order (**inversion**); nonhomologous chromosomes may exchange sections (**translocation**); entire blocks of genes may be lost (**deletion**), usually causing serious developmental defects; or an extra section of chromosome may attach to a normal chromosome (**duplication**). These structural changes often produce phenotypic changes. Duplications, although rare, are important for evolution because they supply additional genetic information that may enable new functions.

5.4 GENE THEORY

Gene Concept

Wilhelm Johannsen in 1909 coined the term "gene" (Gr. *genos,* descent) for the hereditary factors of Mendel. Initially, genes were considered indivisible subunits of the chromosomes on which they occurred. Later studies with multiple mutant alleles demonstrated that alleles are in fact divisible by recombination; *portions* of a gene are separable. Furthermore, parts of many genes in eukaryotes are separated by sections of DNA that do not specify a part of the finished product (**introns**).

As the chief unit of genetic information, a gene encodes products essential for the structure and metabolism of every cell. Because of their ability to mutate and to be assorted in different combinations,

genes are important units of variation in evolution. Genes maintain their identities for many generations despite mutational changes in some parts of their structure.

One Gene–One Polypeptide Hypothesis

Since genes act to influence variable phenotypes, we may infer that their action follows this scheme: gene → gene product → phenotypic expression. Many genes encode in their sequence of bases the sequence of amino acids in a polypeptide that results from the gene's expression. Polypeptides form proteins, which act as enzymes, antibodies, hormones, and structural elements throughout the body.

The first clear, well-documented study to link genes and enzymes was performed on the common bread mold *Neurospora* by Beadle and Tatum in the early 1940s. This organism was ideally suited to a study of gene function for several reasons: these molds are much simpler to handle than fruit flies, they grow readily in well-defined chemical media, and they are haploid organisms unencumbered with dominance relationships among alleles. Furthermore, mutations are readily induced by irradiation with ultraviolet light. Each mutant strain used by Beadle and Tatum was defective in one enzyme, which prevented that strain from synthesizing one or more complex molecules.

From these experiments Beadle and Tatum made an important and exciting formulation: *one gene produces one enzyme.* For this work they received the Nobel Prize for Physiology or Medicine in 1958. The new hypothesis was soon validated by research on many biosynthetic pathways. Hundreds of inherited disorders, including dozens of human hereditary diseases, are caused by single mutant genes causing loss of a specific enzyme. We now know that a particular protein may contain several chains of amino acids (polypeptides), each one often specified by a different gene, and not all proteins specified by genes are enzymes (for example, structural proteins, antibodies, transport proteins, and hormones). Furthermore, genes directing the synthesis of various kinds of RNA were not included in Beadle and Tatum's formulation. Therefore, we now define a gene more inclusively as *a nucleic acid sequence (usually DNA) that encodes a functional polypeptide or RNA sequence.*

5.5 STORAGE AND TRANSFER OF GENETIC INFORMATION

Nucleic Acids: Molecular Basis of Inheritance

Cells contain two kinds of nucleic acids: deoxyribonucleic acid (DNA), which is the genetic material, and ribonucleic acid (RNA), which functions in protein synthesis. Both DNA and RNA are polymers built of repeated units called **nucleotides.** Each nucleotide contains three parts: a **sugar,** a **nitrogenous base,** and a **phosphate group.** The sugar is a pentose (5-carbon) sugar; in DNA it is **deoxyribose** and in RNA it is **ribose** (Figure 5.12).

Nitrogenous bases of nucleotides are also of two types: pyrimidines, whose characteristic structure is a single, six-membered ring of carbon and nitrogen atoms, and purines, which contain two fused rings. Presence of nitrogen as well as carbon in their rings is one obvious way to distinguish "nitrogenous" bases from the rings of sugars, which contain carbon atoms and one oxygen atom. The

Figure 5.12 Ribose and deoxyribose, the pentose sugars of nucleic acids. A carbon atom lies in each of the four corners of the pentagon (labeled 1' to 4'). Ribose has a hydroxyl group (—OH) and a hydrogen on the number 2' carbon; deoxyribose has two hydrogens at this position. Carbon atoms in ribose or deoxyribose are assigned primed numbers as shown to denote their positions in the sugar; unprimed numbers denote the positions of carbons or nitrogens within the purine or pyrimidine rings of the nucleotide bases (see Figure 5.13).

Figure 5.13 Purines and pyrimidines of DNA and RNA. These are the "nitrogenous" bases of nucleic acids, so called because their rings include nitrogen atoms as well as carbon atoms.

	DNA	RNA
Purines	Adenine	Adenine
	Guanine	Guanine
Pyrimidines	Cytosine	Cytosine
	Thymine	Uracil
Sugar	2-Deoxyribose	Ribose
Phosphate	Phosphoric acid	Phosphoric acid

TABLE 5.2

Chemical Components of DNA and RNA

Figure 5.14 Section of a strand of DNA polynucleotide chain. At the top of the figure, the 5' end of the chain has a phosphate group bonded to the 5' carbon of the terminal nucleotide. A phosphate group connects the 3' carbon of the first nucleotide to the 5' carbon of the next one, and this pattern continues to the 3' end of the chain. The 3' end (not shown) has a hydroxyl group on the 3' carbon of the terminal nucleotide. Nitrogenous bases attach to the 1' carbon of the sugar in each nucleotide.

purines in both RNA and DNA are adenine and guanine (Table 5.2). The pyrimidines in DNA are thymine and cytosine, and in RNA they are uracil and cytosine. Carbon atoms in the bases are numbered (for identification) according to standard biochemical notation (Figure 5.13). Carbons in ribose and deoxyribose are also numbered, but to distinguish them from the carbons in the bases, numbers for carbons in the sugars are given prime signs (see Figure 5.12).

Bases attach to the 1' carbon of the sugar (deoxyribose in DNA, ribose in RNA) portion the nucleotide. There is no bonding between adjacent bases in a polynucleotide chain. The chain is formed by phosphate groups connecting the 3' carbon of one sugar to the 5' carbon of

Hydrogen bond

To deoxyribose of chain

To deoxyribose of chain

Adenine — Thymine

Hydrogen bond

To deoxyribose of chain

To deoxyribose of chain

Guanine — Cytosine

Figure 5.15   Positions of hydrogen bonds between thymine and adenine and between cytosine and guanine in DNA

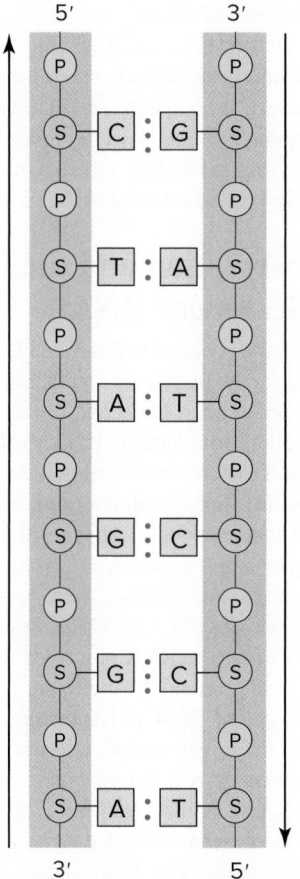

Figure 5.16   DNA, showing how the complementary pairing of bases between the sugar-phosphate "backbones" keeps the double helix at a constant diameter for the entire length of the molecule. Dots represent the three hydrogen bonds between each cytosine and guanine and the two hydrogen bonds between each adenine and thymine.

the next sugar in the nucleotide chain (Figure 5.14). Each end of a polynucleotide chain contains a nucleotide whose sugar connects to the chain through either its 3' or its 5' carbon, but not both. At the **3' end** of the polynucleotide chain, the sugar connects to the chain through its 5' carbon, and has a hydroxyl group at its "free" 3' carbon. The sugar of the nucleotide at the opposite **5' end** of the polynucleotide chain connects to the chain through its 3' carbon, and has a phosphate group at its "free" 5' carbon. A growing nucleotide chain receives new nucleotides at its free 3' end, whose hydroxyl group reacts with a phosphate attached to the 5' carbon of the incoming nucleotide (Figure 5.12).

DNA is not a single polynucleotide chain; it has *two* chains that are precisely cross-linked by specific hydrogen bonding between purine and pyrimidine bases. Adenine pairs with thymine, and guanine with cytosine (Figure 5.15).

Thus, the number of adenines equals the number of thymines, and the number of guanines equals the number of cytosines in the double-stranded DNA molecule. The paired strands of DNA are called **complementary** because the sequence of bases along one strand specifies the sequence of bases along the other strand. The hydrogen-bonded bases of the complementary strands form the central portion of double-stranded DNA (Figure 5.16), and the sugar-phosphate links form its outer rims, analogous in structure to a spiral staircase. DNA twists into a double helix with approximately 10 base pairs for each complete turn of the helix (Figure 5.17). The two DNA strands run in opposite directions (**antiparallel**), meaning that the 5' end of one strand is opposite the 3' end of the other (Figure 5.17).

The structure of DNA is widely considered the single most important biological discovery of the twentieth century. It was based on X-ray diffraction studies of Maurice H. F. Wilkins and Rosalind Franklin and on ingenious proposals of Francis H. C. Crick and James D. Watson published in 1953. Watson, Crick, and Wilkins were later awarded the Nobel Prize for Physiology or Medicine for their work. Rosalind Franklin was not included because she died prior to the award.

RNA is similar to DNA in structure except that it consists of a *single* polynucleotide chain (except in some viruses), has ribose instead of deoxyribose, and has uracil instead of thymine. Ribosomal, transfer, and messenger RNAs are the most abundant and well-known types (function described on pp. 88–91), but many structural and regulatory RNAs, such as microRNAs, are known.

Before a cell divides, the structure of DNA must be precisely copied so that identical copies can be distributed to the daughter cells. This is called **replication** (Figure 5.18). During replication, the two strands of the double helix unwind, and each separated strand serves as a **template** against which a complementary strand is synthesized. An enzyme (DNA polymerase) catalyzes assembly of a new strand of polynucleotides with a thymine group going opposite the adenine group in the template strand, a guanine group opposite the cytosine group, and the two reverse conditions. DNA polymerase synthesizes new strands only in the direction of 5' to 3'. Because the parent DNA strands are antiparallel, one of which runs 5' to 3' and the other running 3' to 5', synthesis along one of the strands is continuous, and the other must be formed in a series of fragments, each of which begins with a 5' end running toward a 3' end (Figure 5.18).

A triplet code permits considerable redundancy of triplets (codons), because DNA encodes just 20 amino acids. Later work confirmed that nearly all amino acids are specified by more than one triplet code.

DNA shows surprising stability, both in prokaryotes and in eukaryotes. Interestingly, it is susceptible to damage by harmful chemicals in the environment and by radiation. Such damage is usually not permanent, because cells have an efficient repair system. Various types of damage and repair are known, one of which is **excision repair.** Ultraviolet irradiation often damages DNA by linking adjacent pyrimidines by covalent bonds (dimerize), preventing transcription and replication. A series of several enzymes "recognizes" the damaged strand and excises the pair of dimerized pyrimidines and several bases following them. DNA polymerase then synthesizes the missing strand using the remaining one as a template, according to the base-pairing rules, and the enzyme **DNA ligase** joins the end of the new strand to the old one.

Transcription and the Role of Messenger RNA

Information is coded in DNA, but DNA does not participate directly in protein synthesis. The intermediary molecule between DNA and protein is another nucleic acid called **messenger RNA (mRNA).** The triplet codes in DNA are **transcribed** into mRNA, with uracil substituting for thymine (Table 5.3). Each triplet in mRNA that encodes an amino acid in the corresponding protein is called a **codon.**

Ribosomal, transfer, and messenger RNAs are transcribed directly from DNA, each encoded by different sets of genes. RNA is formed as a complementary copy of one strand of the appropriate gene using an enzyme called **RNA polymerase.** (In eukaryotes each type of RNA [ribosomal, transfer, and messenger] is transcribed by a different type of RNA polymerase.) The RNA contains a sequence of bases that complements the bases in one of the two DNA strands, just as the DNA strands complement each other. Thus A in the template DNA strand is replaced by U in RNA; C is replaced by G; G is replaced by C; and T is replaced by A. Only one of the two chains is used as the template for RNA synthesis (Figure 5.19). A codon is referenced as the sequence of bases present in a mRNA molecule (Table 5.3), which is complementary and antiparallel to the template DNA strand (often called the "sense" strand) from which it is made. The DNA strand not used as a template during transcription of a gene is called the "antisense" strand.

A bacterial gene is encoded on a continuous stretch of DNA, transcribed into mRNA, and then translated (see the next section, Translation: Final Stage in Information Transfer). The hypothesis that eukaryotic genes had a similar structure was rejected by the surprising discovery that some stretches of DNA are transcribed in the nucleus but are not found in the corresponding mRNA in the cytoplasm. Some pieces of the initial transcript containing the mRNA were removed in the nucleus before the finished mRNA was transported to the

Sugar-phosphate "backbone"

Hydrogen bonds between nitrogenous bases

Phosphodiester bond

OH
3' end

5' end

Figure 5.17 DNA molecule.

DNA Coding by Base Sequence

Because DNA is the genetic material and contains a linear sequence of base pairs, an obvious extension of the Watson-Crick model is that the sequence of base pairs in DNA codes for, and is colinear with, the sequence of amino acids in a protein. The coding hypothesis must explain how a string of four different bases—a four-letter alphabet— could specify the sequence of 20 different amino acids.

In the coding procedure, obviously there cannot be a 1:1 correspondence between four bases and 20 amino acids. If a coding unit were two bases, only 16 amino acids (4^2) could be encoded, which could not specify all 20 amino acids found in proteins. Therefore the coding unit must contain at least three bases or three letters, because 64 possible amino acids (4^3) could be encoded by four bases when taken as triplets.

Figure 5.19 Transcription of mRNA from a DNA template. Transcription is similar for mRNA, rRNA, and tRNA except that each kind of RNA uses a different form of the enzyme, RNA polymerase. This diagram shows transcription midway to completion. Transcription began by unwinding of the DNA helix, annealing of an RNA primer to the template strand of DNA, and extension of the primer at its 3' end by adding nucleotides (not shown) complementary to the sequence of bases in the template DNA strand. The primer is at the 5' end of the mRNA, which continues to grow in length by adding nucleotides at its annealed 3' end. When transcription is finished, the pre-mRNA will detach completely from the DNA template. If the gene contains introns, the pre-mRNA transcript must be processed by deleting the introns before it becomes mRNA.

Figure 5.18 Replication of DNA. Parent strands of DNA part, and DNA polymerase synthesizes daughter strands using the base sequence of parent strands as a template. Because synthesis always proceeds in a 5' to 3' direction, synthesis of one strand is continuous, and the other strand must be synthesized as a series of fragments.

TABLE 5.3

The Genetic Code: Amino Acids Specified by Codons of Messenger RNA

First Letter		Second Letter				Third letter
	U	**C**	**A**	**G**	
U	UUU UUC } Phenylalanine UUA UUG } Leucine	UCU UCC UCA UCG } Serine	UAU UAC } Tyrosine UAA UAG } End chain	UGU UGC } Cysteine UGA End chain UGG Tryptophan	U C A G
C	CUU CUC CUA CUG } Leucine	CCU CCC CCA CCG } Proline	CAU CAC } Histidine CAA CAG } Glutamine	CGU CGC CGA CGG } Arginine	U C A G
A	AUU AUC AUA } Isoleucine AUG } Methionine*	ACU ACC ACA ACG } Threonine	AAU AAC } Asparagine AAA AAG } Lysine	AGU AGC } Serine AGA AGG } Arginine	U C A G
G	GUU GUC GUA GUG } Valine	GCU GCC} GCA GCG } Alanine	GAU GAC } Aspartic acid GAA GAG } Glutamic acid	GGU GGC GGA GGG } Glycine	U C A G

*Also, begin polypeptide chain.

Figure 5.20 Expression of ovalbumin gene of chicken. The entire gene of 7700 base pairs is transcribed to form the primary pre-mRNA, then the 5' cap of methyl guanine and the 3' polyadenylate tail are added. After the introns are spliced out, the mature mRNA is transferred to the cytoplasm.

cytoplasm (Figure 5.20). Thus many genes are interrupted by sequences of bases that do not code for the final product, and the precursor of mRNA (pre-mRNA) transcribed from them must be edited or "matured" before translation in the cytoplasm. The intervening segments of DNA are called **introns**, and those that encode part of the mature RNA and are translated into protein are called **exons**. The pre-mRNA has its introns removed and the exons spliced together to form the mRNA. Before mRNA leaves the nucleus, a methylated guanine "cap" is added at the 5' end, and a tail of adenine nucleotides (poly-*A*) is often added at the 3' end (Figure 5.20). The cap and the poly-*A* tail distinguish mRNA from other kinds of RNA molecules.

In mammals, genes coding for histones and for interferons have no introns. However, we now know that genes coding for many proteins contain one or more introns. In lymphocyte differentiation the parts of the split genes coding for immunoglobulins are actually *rearranged* during development, so that different proteins result from subsequent transcription and translation. This rearrangement partly explains the enormous diversity of antibodies manufactured by descendants of the lymphocytes (p. 770). Presence of introns is thought to have facilitated evolution of such rearrangements of gene structure during development.

Base sequences in some introns are complementary to other base sequences in the intron, suggesting that the intron could fold so that complementary sequences would pair. This folding may be necessary to control proper alignment of intron boundaries before splicing. Most surprising of all is the discovery that, in some cases, RNA can "self-catalyze" the excision of introns. The ends of the intron join; the intron thus becomes a small circle of RNA, and the exons are spliced together. This process does not fit the classical definition of an enzyme or other catalyst because the molecule itself is changed by the reaction. In other cases, removal of introns from a pre-mRNA transcript requires a complex of protein catalysts and other small RNAs; this splicing machinery is called the spliceosome, and introns that require it for their excision are called spliceosomal introns.

Translation: Final Stage in Information Transfer

The **translation** process occurs on **ribosomes**, granular cytoplasmic structures composed of protein and **ribosomal RNA (rRNA).**

Figure 5.21 How the polypeptide chain is formed. As ribosomes move along messenger RNA in a 5' to 3' direction, amino acids are added stepwise to form the polypeptide chain.

Ribosomal RNA contains a large and a small subunit, and the small subunit comes to lie in a depression of a large subunit to form a functional ribosome (Figure 5.21). Messenger RNA molecules attach themselves to ribosomes to form a messenger RNA-ribosome complex. Because only a short section of mRNA makes contact with a single ribosome, the mRNA usually has several ribosomes attached along its length, each one at a different stage of synthesizing the encoded polypeptide. The entire complex, called a **polyribosome** or **polysome,** allows several molecules of the same kind of polypeptide to be synthesized concurrently, one on each ribosome of the polysome (Figure 5.21).

Assembly of polypeptides on the mRNA-ribosome complex requires another kind of RNA called **transfer RNA (tRNA).** Transfer RNAs have a complex secondary structure of folded stems and loops, often illustrated in the form of a cloverleaf (Figure 5.22), although the three-dimensional shape is somewhat different. Molecules of tRNA collect free amino acids from the cytoplasm and deliver them to the polysome, where they are assembled into a polypeptide. There are special tRNA molecules for every amino acid. Furthermore, each tRNA is accompanied by a specific tRNA synthetase. Transfer RNA synthetases are enzymes that attach the correct amino acid through its carboxyl group to the terminal adenine on the 3' end of each tRNA by a process called **charging.**

On the cloverleaf-shaped molecule of tRNA, a special sequence of three bases (the **anticodon**) is exposed in just the right way to

Figure 5.22   Diagram of a tRNA molecule. The anticodon loop bears bases complementary to those in the mRNA codon. The other two loops function in binding to the ribosome in polypeptide synthesis. The amino acid is added through its carboxyl group to the free single-stranded 3' end by tRNA synthetase.

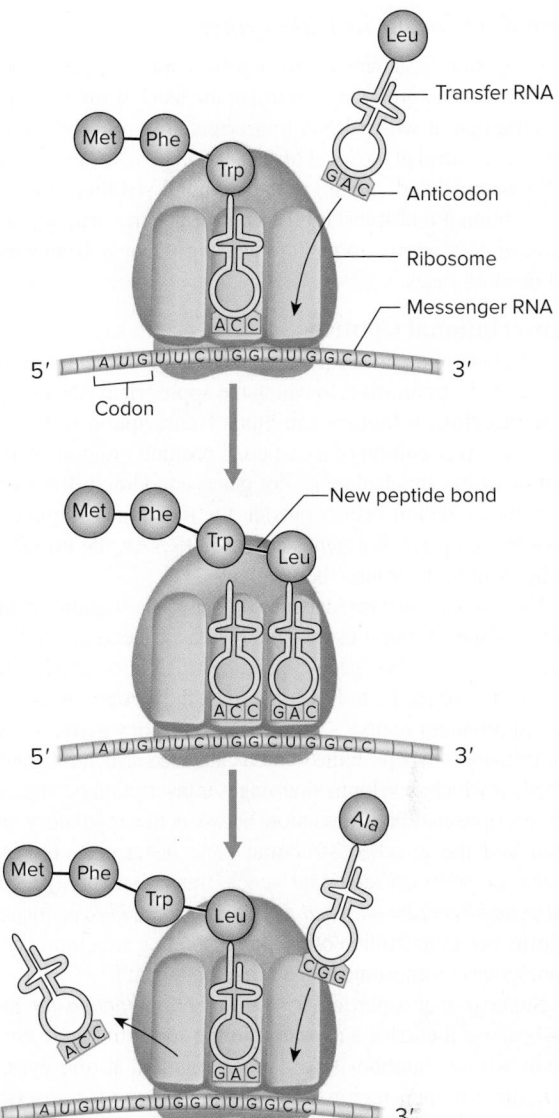

Figure 5.23   Formation of polypeptide chain on messenger RNA. As a ribosome moves down the messenger RNA molecule, transfer RNA molecules with attached amino acids enter the ribosome (top). Amino acids are joined together into a polypeptide chain, and transfer RNA molecules leave the ribosome (bottom).

form base pairs with complementary bases (the codon) in the mRNA. The codons are read along the mRNA in a 5' to 3' direction starting at an AUG codon, specifying methionine (Table 5.3). The anticodon of each tRNA is the key to the correct ordering of amino acids in the polypeptide being assembled. As each tRNA brings a new amino acid to the growing polypeptide chain, the free amino group on the newly arriving amino acid reacts with the carboxyl group of the preceding amino acid already in place, thereby detaching the previous amino acid from its own tRNA and linking the amino acids with a peptide bond (Figure 5.23).

For example, tryptophan is assembled into a polypeptide when signaled by the codon UGG in an mRNA. The translation is accomplished by a charged tryptophan tRNA, in which the anticodon is CCA. The tryptophan-tRNA complex enters the ribosome where it fits precisely into the right place on the mRNA strand. Formation of a peptide bond then links tryptophan to the preceding amino acid (phenylalanine in Figure 5.23) in the growing polypeptide chain. Then the next charged tRNA specified by the mRNA code (leucine tRNA in Figure 5.23) enters the ribosome and attaches itself beside the tryptophan tRNA. The two amino acids are united by a peptide bond, and the tryptophan tRNA then detaches from the ribosome. The process continues stepwise as the polypeptide chain is built (Figure 5.23). A polypeptide of 500 amino acids can be assembled in less than 30 seconds. The polypeptide has a free amino group at the end where its synthesis began and a free carboxyl group where its synthesis ended. Synthesis ends when the ribosome reaches a codon specifying "end chain" (Table 5.3).

Regulation of Gene Expression

In Chapter 8 we show how the orderly differentiation of an organism from fertilized ovum to adult requires expression of genetic material at every stage of development. Developmental biologists have provided convincing evidence that every cell in a developing embryo is genetically equivalent. Thus as tissues differentiate (change developmentally), each one uses only a part of the genetic instruction present in every cell. The stages of development and locations within the body at which a particular gene is active vary greatly among genes. A gene encoding one of the enzymes of cellular metabolism would be active in all cells, whereas one encoding hemoglobin might be active only in precursors of red blood cells. Most of the genes are inactive at any given moment in a particular cell or tissue. The problem in development is to explain how, if every cell has a full gene complement, certain genes are "turned on" to produce proteins required for a particular developmental stage while other genes remain silent.

Gene Regulation in Eukaryotes

Several metabolic stages in eukaryotic cells may serve as control points for gene expression. Control at the level of transcription modulates the rate at which RNA transcripts are produced from a gene sequence. Control at the level of translation involves the rate at which mRNA is translated, including the temporal stability of the mRNA. Transcriptional and translational control are the primary stages for control of gene expression in animals, with gene rearrangement also used in some cases.

Transcriptional Control

Transcription of a gene requires that the gene contain a sequence of bases, called a **promoter,** to which the appropriate RNA polymerase and **transcription factors** can bind. Transcription factors are proteins whose recognition of a particular promoter region enables transcription of the attached gene. For genes encoding mRNA or rRNA, the promoter region occurs outside the transcribed region at the 3' end of the template. For genes encoding tRNAs, the promoter function lies within the transcribed region.

Variation in the base sequence of a promoter region can influence binding of transcription factors and thereby affect transcription of the attached template. For genes encoding mRNA or rRNA, the DNA sequence transcribed into RNA is termed the **structural gene** and the attached promoter region is called the **regulatory gene.** A regulatory gene influences where in the body its attached structural gene is transcribed, at which developmental stages transcription occurs, and rates of transcription. This interaction between the regulatory promoter region and the attached structural gene is termed **cis-regulation** because the promoter region influences transcription *only of the structural gene physically attached to it on the same chromosome;* it does *not* influence transcription of the corresponding structural gene on the homologous chromosome in the same diploid cell.

Suppose that a particular strain of *D. melanogaster* has white eyes because it carries a mutation in the structural gene encoding a protein whose function is to deposit pigment in the eyes. In this strain, the promoter region is intact and the gene is transcribed normally, but the transcript specifies a nonfunctional protein. Suppose that another strain of flies has white eyes because a mutation in the promoter region prevents proper binding by transcription factors, although the attached structural gene encodes a functional protein. Progeny of a genetic cross between these strains would have white eyes. Although the hybrid progeny have one normal promoter and one normal structural gene, the normal genetic functions occur on different homologous chromosomes and therefore fail to restore normal pigmentation. On one homolog, a normal promoter stimulates transcription of a structural gene encoding an inactive protein, and on the other homolog, a defective promoter precludes transcription of the normal structural gene attached to it.

Genes encoding transcription factors influence the transcription of other genes. Transcription factors move from the cytoplasm into the nucleus and can bind any promoters that they recognize on any of the cell's chromosomes. Genes that encode transcription factors are often called **trans-regulatory.** Suppose that a strain of *D. melanogaster* has white eyes because a mutation in the gene encoding a critical transcription factor makes that factor unable to recognize the promoter of a normal copy of the pigmentation gene described in the preceding paragraph. If we cross this strain to the one from the preceding paragraph whose only genetic defect is in the structural gene

for the pigment-depositing protein, the resulting progeny have normally pigmented eyes. The trans-regulating transcription factor inherited from this latter strain binds the promoter of the normal target gene inherited from the other strain. In this case, the mutations of the two strains **complement** each other, each one supplying to their hybrid progeny a critical function absent from the other strain. By contrast, the mutations in the strains of the preceding paragraph fail to complement each other. Cis-regulation of transcription is distinguished experimentally from trans-regulation by testing for complementation of mutations in this manner.

Hormones influence gene expression by binding receptor proteins and thereby activating those proteins as transcription factors. Steroid hormones produced by endocrine glands elsewhere in the body enter a target cell and bind with a receptor protein in the nucleus. The steroid-receptor complex then binds with DNA near the target gene (p. 748). Progesterone, for example, binds with a nuclear receptor protein in cells of the chicken oviduct; the hormone-receptor complex then activates transcription of genes encoding egg albumin and other substances.

An important mechanism for silencing genes is methylation of cytosine bases; a methyl group (CH_3—) binds the carbon in the 5 position in the cytosine ring (Figure 5.24A). This usually happens when the cytosine is next to a guanine base; thus, the bases in the complementary DNA strand would also be a cytosine and a guanine (Figure 5.24B). When the DNA is replicated, an enzyme recognizes the CG sequence and quickly methylates the daughter strand, keeping the gene inactive.

Translational Control

The mRNA from a particular gene can be sequestered so that translation is delayed. Development of eggs of many animals commonly uses this mechanism. Oocytes accumulate large quantities of messenger RNA as they develop in the mother; these maternal mRNAs

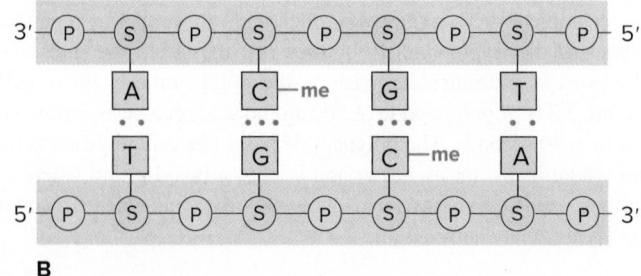

Figure 5.24 Some genes in eukaryotes are turned off by methylation of some cytosine residues in the chain. **A,** Structure of 5-methyl cytosine. **B,** Cytosine residues next to guanine are those that are methylated in a strand, thus allowing both strands to be symmetrically methylated.

are sequestered from translation until the egg is fertilized, and then become actively translated to generate proteins needed for zygotic development. The mother thus provides the means for a zygote to begin development before its own nuclear genes are activated to guide further development of the embryo.

Short (~22 nucleotides long), noncoding RNAs called micro-RNAs (miRNAs) or small interfering RNAs (siRNAs) can inhibit translation of target mRNAs or even destroy the target mRNA. The enzyme called Dicer plays an important role in generating miRNAs or siRNAs in the cytoplasm by cleaving them from longer RNA precursors. MiRNAs are cut from single-stranded RNA precursors that fold back on themselves to form stems and loops in a manner similar to that of a transfer RNA (Figure 5.22), whereas siRNAs are cleaved from a longer, double-stranded RNA molecule.

Regulation of gene expression by these small RNAs is critically important in animal development; experimental inactivation of the gene encoding Dicer in mice causes embryonic lethality. The miRNAs or siRNAs are packaged into a ribonucleoprotein complex called the RNA-induced silencing complex (RISC), which binds specific mRNAs having sequences at least partially complementary to the miRNA or siRNA. Different organs, tissue types, and even stages of development have their own particular miRNAs or siRNAs not found elsewhere in the body. Much remains to be learned about the roles of these small RNAs in regulating gene expression. MiRNAs and siRNAs hold much promise for medical applications in gene therapies.

Gene Rearrangement

Vertebrates contain cells called lymphocytes that bear genes encoding proteins called antibodies (p. 770). Each type of antibody binds only a particular foreign substance (antigen). Because the number of different antigens is enormous, diversity of antibody genes must be equally great. One source of this diversity is rearrangement of DNA sequences coding for antibodies during development of lymphocytes. A shuffling of exons encoding modules of protein function permits assembly of new antibodies.

Molecular Genetics

We describe briefly the most important biochemical techniques that make possible our studies of gene structure and function.

Recombinant DNA

An important tool in this technology is a series of enzymes called **restriction endonucleases**. Each of these enzymes, derived from bacteria, cleaves double-stranded DNA at particular sites determined by their base sequences. Many of these endonucleases cut the DNA strands so that one has several bases projecting farther than the other strand (Figure 5.25), leaving what are called "sticky ends." When these DNA fragments are mixed with others that have been cleaved by the same endonuclease, their sticky ends tend to anneal (join) by the rules of complementary base pairing. The ends are sealed into their new position by the enzyme DNA ligase in a process called **ligation**.

Bacteria often have small circles of double-stranded DNA called *plasmids*. Plasmids are distinct from and much smaller than a bacterial cell's primary genome and can occur in multiple copies per

Figure 5.25   Action of restriction endonuclease, *Eco* RI. Such enzymes recognize specific base sequences that are palindromic (a palindrome is a word spelled the same backward and forward). *Eco* RI leaves "sticky ends," which anneal to other DNA fragments cleaved by the same enzyme. The strands are joined by DNA ligase.

cell. Although constituting only 1% to 3% of bacterial DNA, plasmids may carry important genetic information, for example, resistance to an antibiotic.

If DNA from a foreign source (such as a mammal) is ligated into a plasmid, the product is **recombinant DNA**. To produce the recombinant DNA in large quantities, the modified plasmid must be cloned in bacteria. The bacteria are treated with dilute calcium chloride to make them more susceptible to entry by the recombinant DNA, but plasmids do not enter most cells within the bacterial population. To select the cells that have acquired the recombinant DNA one uses a plasmid that carries an easily selectable marker, for example, resistance to an antibiotic. Then, the only bacteria that can grow in the presence of the antibiotic are ones that have absorbed the recombinant DNA. Some bacteriophages (bacterial viruses) also are used as carriers for recombinant DNA. Plasmids and bacteriophages that carry recombinant DNA are called **vectors**. The vectors retain an ability to replicate in the bacterial cells; therefore, the recombinant insert is produced in large quantities, a process called amplification.

A clone is a collection of individuals or cells all derived by asexual reproduction from a single individual. When we speak of cloning a gene or plasmid in bacteria, we mean that we isolate a colony or group of bacteria derived from a single ancestor into which the gene or plasmid was inserted. Cloning is used to obtain large quantities of a gene that has been ligated into a bacterial plasmid.

Polymerase Chain Reaction

A researcher can clone a specific gene enzymatically from any organism as long as part of the sequence of that gene is known. The technique is called the **polymerase chain reaction (PCR).** To describe this technique, we denote the DNA sequences that occur adjacent to the opposite ends of a gene as being "upstream" versus "downstream" of the target gene. The investigator synthesizes two short chains of nucleotides called primers, each of which is about 15 bases long. The first primer used for PCR is complementary to one strand of the double helix in the "upstream" region; the second primer is complementary to the opposite strand of the double helix in the "downstream" region. Each primer is designed so that its 3' end faces toward the target gene when the primer anneals to its complementary sequence either upstream or downstream of that gene. A large excess of each primer is added to a sample of DNA from the organism, and the mixture is heated to separate the double helix into single strands. When the mixture is cooled, each strand of the gene of interest anneals to a primer rather than to the other strand of the gene—because the primer occurs in a much higher concentration. A heat-stable DNA polymerase (Taq polymerase, originally from the bacterium *Thermus aquaticus*) and the four deoxyribonucleotide triphosphates are added to the reaction mixture. DNA synthesis proceeds from the 3' end of each primer, extending the primer in the 5' to 3' direction through the target gene. Entire new complementary strands are synthesized, and the number of copies of the gene has doubled (Figure 5.26). The reaction mixture is then reheated and cooled again to allow more primers to bind original and new copies of each strand. With each cycle of DNA synthesis, the number of copies of the gene doubles. Since each cycle can take less than five minutes, the number of copies of a gene can increase from one to over 1 million in less than two hours! The PCR allows cloning a known gene from an individual patient, identification of a drop of dried blood at a crime scene, or cloning DNA of a 40,000-year-old woolly mammoth.

Recombinant DNA technology and PCR are currently being used to engineer crop plants, including soybeans, cotton, rice, corn, and tomato. Transgenic mice are commonly used in research, and gene therapy for human genetic diseases is being developed.

Genomics and Proteomics

The scientific field of mapping, sequencing, and analyzing genomes is now called **genomics**. Some researchers divide genomic analysis into "structural genomics" (mapping and sequencing) and "functional genomics" (development of genome-wide or system-wide experimental approaches to understand gene function).

In the 1970s Allan Maxam and Walter Gilbert in the United States and Frederick Sanger in England reported practical techniques for identifying the sequence of bases in DNA. By 1984 and 1985 scientists proposed to sequence and to map the entire human genome, an effort called the Human Genome Project. It was a most ambitious undertaking: the genome was estimated at 50,000 to 100,000 genes and regulatory subunits encoded in a linear sequence of about 3 to 6 billion pairs of bases. Using techniques available in 1988, it would have taken until 2700 to sequence the genome completely, but biologists then expected technical improvements to make it possible by the twenty-second century. In fact, development and improvement of automated sequencers, as well as competition between the publicly supported Human Genome Sequencing Consortium and a large group of privately supported scientists (Celera Genomics and collaborators) led to publication of draft sequences in 2001!

Whether determination of the draft sequence was "the greatest scientific discovery of our time," as claimed by Davies' book

Figure 5.26   Steps in the polymerase chain reaction (PCR). Note that two different primers are required, one for each end of the target sequence.

Cracking the Genome (in Selected References), is debatable. Nevertheless, it was very exciting and yielded many surprises. For example, the human genome has fewer genes than thought previously, with 21,724 genes currently known. Only 5% of the 28% of the genome that is actually transcribed into RNA encodes protein. More than half the DNA present is repeated sequences of several types, including 45% in parasitic DNA elements. Parasitic DNA (also called "selfish" and "junk" DNA) is DNA that seems to serve no cellular or organismal function except its own propagation, but evolution sometimes co-opts it for new biological roles.

Animal species vary by several orders of magnitude in the total amount of DNA in their nuclear genomes (from 10^8 to 10^{11} base pairs in a haploid gamete nucleus). At the low end are sponges (10^8; see Chapter 12). Insects (see Chapter 21) range from slightly less than 10^9 through 10^{10}. Most vertebrates have genomes of approximately 10^9 base pairs, but salamanders (p. 548), caecilians (p. 548), and lungfishes (p. 517) have genome sizes exceeding 10^{10} base pairs, with some salamanders reaching 10^{11} base pairs. Large genomes should not be considered advantageous, however, because most of the difference in genome size is contributed by accumulation of large amounts of the "selfish" or "junk" DNA in the larger genomes rather than DNA sequences useful to cellular metabolism and organismal function. The metabolic demands of replicating large amounts of DNA and the physical demands of housing it within the cell nucleus produce selection against accumulation of too much parasitic DNA in the genome. Animal groups with the largest genomes are likely those most able to tolerate accumulation of large amounts of parasitic DNA in their nuclear genomes without harming cellular and organismal functions. Because the emphasis of our book is on organismal biology, our review of genetics concentrates on genes that have clear roles in cellular and organismal functions, although these genes are a small minority of the DNA sequences present in animal nuclear genomes. Some DNA sequences considered useless to the organism exhibit variation that is useful in studies of population genetics (p. 122) and evolutionary relationships among species.

Defects in single genes explain many human diseases, including cystic fibrosis and Huntington's disease. Almost 300 disease-associated genes are known. Information developed from knowledge of gene sequences permits new diagnostic tests, treatments, possible preventive strategies, and advances in molecular understanding of genetic diseases.

The human genome is responsible for hundreds of thousands of different proteins (**proteome**). Because of alternative gene splicing and post-translational modifications of polypeptides, the number of proteins in an animal's proteome exceeds the number of genes in its genome. The polypeptide encoded by a gene may be cleaved into separate functional parts or associated with polypeptides coded by other genes to produce diverse protein functions. Many scientists are now engaged in the difficult field of **proteomics**: to identify all the proteins in a cell, tissue, or organism; to determine how the proteins interact to accomplish their functions; and to outline the folding structures of the proteins.

5.6 GENE MUTATIONS

The creative force of evolution is natural selection acting on biological variation. Without variability among individuals, there could be no continued adaptation to a stable or changing environment and no evolution (see Chapter 6). Although natural selection acts on varying organismal phenotypes, phenotypic variation within a population in a particular environment is often caused by variation in genotype. Mutation at individual genes introduces allelic diversity into a population, and genetic recombination amplifies the diversity of genotypes in a population.

Gene mutations are chemicophysical changes that alter the sequence of bases in DNA. Once a gene is mutated, it faithfully reproduces its new form. These mutations are studied directly by examining the DNA sequence and indirectly through their effects on organismal phenotype, if such effects occur.

Most mutations likely have no consequences for organismal phenotype. Mutations that occur in non-genic regions of the genome would typically have no phenotypic consequences.

Mutations in a gene's introns in most cases would not influence phenotype unless they altered processing of the RNA transcript. Even within an exon, many base substitutions are "silent" because they do not cause an amino-acid substitution in the protein. Because most amino acids have multiple codons (Table 5.3), there are many cases where a base substitution would produce a different codon for the same amino acid. For example, mutations changing the base present at the third position of a leucine codon all would be silent. Because most proteins can withstand amino-acid substitutions at some portion of their sites without changing protein function, even some base substitutions that are not "silent" would nonetheless have no effect on organismal phenotype.

Some mutations produce a codon substitution, as in the human condition called **sickle cell anemia**. A single base substitution in the gene encoding beta-hemoglobin (see Figure 2.15) causes the sixth codon of the mRNA to change from GAG (specifying glutamic acid in the normal allele) to GUG (specifying valine in the sickle cell allele). This single amino-acid substitution causes the hemoglobin molecule to depart from its normal globular structure (Figure 2.15) when deprived of oxygen. The altered structure has a cavity in the center of the molecule and a protrusion of polypeptides from the margins, forming a wedge. The protruding part of one hemoglobin molecule becomes wedged into the cavity of another one, beginning a chain reaction in which many hemoglobin molecules become strung together into long, sickle-shaped aggregates that distort the shape of the cell.

People who are homozygous for the sickle cell allele often die before the age of 30 because the ability of their red blood cells to carry oxygen is greatly impaired by substitution of only a single amino acid in their hemoglobin. Individuals heterozygous for the sickle cell and normal alleles have normal respiration except under conditions of oxygen stress and thereby acquire some resistance to malarial infection. When a malarial parasite enters a red blood cell of a heterozygous individual, it produces an oxygen stress that distorts the cell into a sickle shape; such cells are destroyed by the spleen before the parasite has completed its life cycle. This example shows how consequences of a single mutation can be harmful under some conditions and favorable in others, depending in this case on environment (malarial versus nonmalarial) and genotype (homozygous versus heterozygous for the sickle cell allele). See p. 125 for further discussion of the action of natural selection on this variation.

Other mutations involve deletion of one or more bases or insertion of additional bases into a DNA chain. Translation of mRNA is thus shifted, producing codons that specify incorrect amino acids and usually a nonfunctional or dysfunctional protein product. Insertion of a transposable element into a gene likewise can disrupt the gene's ability to specify a functional protein product.

When a mutation causes a loss of gene function, the new form often is recessive, and its effects are normally masked when heterozygous with a functional allele. Only in the homozygous condition can such mutant alleles influence phenotype. Thus a population carries a reservoir of mutant recessive alleles, some of which would be lethal if homozygous but rarely occur in the homozygous condition. Inbreeding encourages formation of homozygotes and increases the probability of phenotypic expression of recessive mutants.

In the well-studied fruit fly, *Drosophila melanogaster,* there is approximately one detectable mutation per 10,000 loci (rate of 0.01%

per locus per generation). The rate for humans is one per 10,000 to one per 100,000 loci per generation. If we accept the latter, more conservative figure, then a single normal allele is expected to undergo 100,000 generations before it is mutated. However, since human chromosomes contain approximately 21,724 loci, about every third person carries a new mutation. Similarly, each ovum or spermatozoon contains, on average, one mutant allele.

Genomic surveys have identified millions of single nucleotide polymorphisms (SNP) in the human population. These are base positions in the genome for which two of the four possible bases occur at substantial frequencies in the population. Mutation is thus very effective in producing large amounts of genetic variation in humans and most other species. Most of this variation occurs outside the coding regions of genes, however, and makes no contribution to phenotypic variation in organismal traits.

5.7 MOLECULAR GENETICS OF CANCER

The crucial defect in cancer cells is that they proliferate in an unrestrained manner (**neoplastic growth**). The mechanism that controls the rate of division of normal cells has somehow been lost, and cancer cells multiply much more rapidly, invading other tissues in the body. Cancer cells originate from normal cells that lose their constraint on division and thus become dedifferentiated (less specialized) to some degree. There are many kinds of cancer, depending on the original founder cells of the tumor. The change in many cancerous cells, perhaps all, has a genetic basis, and investigation of the genetic damage that causes cancer is now a major thrust of cancer research.

Cancer cells typically reveal a genetic change in one of numerous genes of two types, **oncogenes** and **tumor-suppressor genes.**

Oncogenes (Gr. *onkos,* bulk, mass, + *genos,* descent) occur normally in cells, and in their normal form they are called **proto-oncogenes.** One of these encodes a protein called **Ras.** Ras protein is a guanosine triphosphatase (GTPase) located just beneath the cell membrane. When a receptor on the cell surface binds a growth factor, Ras is activated and initiates a cascade of reactions causing cell division. The oncogene form encodes a protein that initiates the cell-division cascade even when the growth factor is absent from the surface receptor.

Of the many ways that cellular DNA can sustain damage, the three most important are ionizing radiation, ultraviolet radiation, and chemical mutagens. The high energy of ionizing radiation (X-rays and gamma rays) causes electrons to be ejected from the atoms it encounters, producing ionized atoms with unpaired electrons (free radicals). The free radicals (principally from water) are highly reactive chemically, and they react with molecules in the cell, including DNA. Some damaged DNA is repaired, but if the repair is inaccurate, a mutation results. Ultraviolet radiation is of much lower energy than ionizing radiation and does not produce free radicals; it is absorbed by pyrimidines in DNA and causes formation of a double covalent bond between the adjacent pyrimidines. UV repair mechanisms can also be inaccurate. Chemical mutagens react with the DNA bases and cause mispairing during replication.

Gene products of tumor-suppressor genes act as a constraint on cell proliferation. One such product is called **p53** (for "53-kilodalton protein," a reference to its molecular weight). Mutations in the gene encoding p53 occur in about half of the 6.5 million cases of human cancer diagnosed each year. Normal p53 has several crucial functions, depending on the circumstances of the cell. It can trigger apoptosis (p. 53), act as a transcription activator or repressor (turning genes on or off), control progression from G_1 to S phase in the cell cycle, and promote repair of damaged DNA. Many of the mutations known in p53 interfere with its binding to DNA and thus its function.

SUMMARY

In sexual animals genetic material is distributed to offspring via gametes (ova and sperm) produced by meiosis. Each somatic cell in an organism has two chromosomes of each kind (homologous chromosomes) and is thus diploid.

Meiosis separates homologous chromosomes, so that each gamete has half the somatic chromosome number (haploid). In the first meiotic division, centromeres do not divide, and each daughter cell receives one of each pair of replicated homologous chromosomes with sister chromatids still attached to the centromere. At the beginning of the first meiotic division, replicated homologous chromosomes come to lie alongside each other (synapsis), forming a bivalent. The gene loci on one set of chromatids lie opposite the corresponding loci on the homologous chromatids. Portions of adjacent chromatids can exchange with the nonsister chromatids (crossing over) to produce new genetic combinations. At the second meiotic division, the centromeres divide, completing the reduction in chromosome number and amount of DNA. The diploid number is restored when male and female gametes fuse to form a zygote.

Gender is determined in many animals by the sex chromosomes; in humans, fruit flies, and many other animals, females have two X chromosomes, and males have an X and a Y.

Genes are the unit entities that influence all characteristics of an organism and are inherited by offspring from their parents. Allelic variants of genes might be dominant, recessive, or intermediate; a recessive allele in the heterozygous genotype will not be expressed in the phenotype but requires the homozygous condition for overt expression. In a monohybrid cross involving a dominant allele and a recessive alternative allele (both parents homozygous), the F_1 generation is all heterozygous, whereas F_2 genotypes occur in a 1:2:1 ratio, and phenotypes in a 3:1 ratio. This result demonstrates Mendel's law of segregation. Heterozygotes in intermediate inheritance show phenotypes distinct from homozygous phenotypes, sometimes intermediate forms, with corresponding alterations in phenotypic ratios.

Dihybrid crosses (in which genes for two different characteristics occur on separate pairs of homologous chromosomes) demonstrate Mendel's law of independent assortment, and phenotypic ratios are 9:3:3:1 with dominant and recessive characters. Expected ratios in crosses of two or more characters are calculated from laws of probability.

Genes can have more than two alleles in a population, and different combinations of alleles can produce different phenotypic effects. Alleles of different genes can interact in producing a phenotype, as in polygenic inheritance and epistasis, in which one gene affects the expression of another gene.

A gene on the X chromosome shows sex-linked inheritance and produces an effect in males, even if a recessive allele is present, because the Y chromosome does not carry a corresponding allele. All genes on a given autosomal chromosome are linked, and their variants do not assort independently unless they are very far apart on the chromosome, in which case crossing over occurs between them in nearly every meiosis. Crossing over increases the amount of genetic recombination in a population.

Occasionally, a pair of homologous chromosomes fails to separate in meiosis, causing the gametes to get one chromosome too many or too few. Resulting zygotes usually do not survive; humans with $2n + 1$ chromosomes sometimes live, but they have serious abnormalities, such as Down syndrome.

Nucleic acids in the cell are DNA and RNA, which are large polymers of nucleotides composed of a nitrogenous base, pentose sugar, and phosphate group. The nitrogenous bases in DNA are adenine (A), guanine (G), thymine (T), and cytosine (C), and those in RNA are the same except that uracil (U) is substituted for thymine. DNA is a double-stranded, helical molecule in which the bases extend toward each other from the sugar-phosphate backbone: A always pairs with T and G with C. The strands are antiparallel and complementary, being held in place by hydrogen bonds between the paired bases. In DNA replication the strands part, and the enzyme DNA polymerase synthesizes a new strand along each parental strand, using the parental strand as a template.

A gene can encode either a ribosomal RNA (rRNA), a transfer RNA (tRNA), or a messenger RNA (mRNA); a gene of the latter kind specifies the sequence of amino acids in a polypeptide (one gene–one polypeptide hypothesis). In mRNA, each triplet of three bases specifies a particular amino acid.

Proteins are synthesized by transcription of DNA into the base sequence of a molecule of messenger RNA (mRNA), which functions in concert with ribosomes (containing ribosomal RNA [rRNA] and protein) and transfer RNA (tRNA). Ribosomes attach to the strand of mRNA and move along it, assembling the amino acid sequence of the protein. Each amino acid is brought into position for assembly by a molecule of tRNA, which itself bears a base sequence (anticodon) complementary to the respective codons of the mRNA. In eukaryotic nuclear DNA the sequences of bases in DNA coding for amino acids in a protein (exons) are interrupted by intervening sequences (introns). The introns are removed from the primary mRNA before it leaves the nucleus, and the protein is synthesized in the cytoplasm.

Genes, and the synthesis of the products for which they are responsible, must be regulated: turned on or off in response to varying environmental conditions or cell differentiation. Gene regulation in eukaryotes occurs at several levels, with control of transcription and translation of particular mRNAs being particularly important.

Molecular genetic methods have made spectacular advances possible. Restriction endonucleases cleave DNA at specific base sequences, and DNA from different sources can be rejoined to form recombinant DNA. By combining mammalian with plasmid or viral DNA, a mammalian gene can be introduced into bacterial cells, which then multiply and produce many copies of the mammalian gene. The polymerase chain reaction (PCR) is used to clone specific genes if the base sequences of short pieces of DNA surrounding the gene are known. Draft sequences of the human genome were published in 2001. Among many exciting results was a revision of the number of genes to 21,724, down from previous estimates of 100,000. These genes are responsible for hundreds of thousands of proteins in a typical cell.

A mutation is a physicochemical alteration in the bases of DNA that may change the phenotypic effect of a gene. Although rare and usually detrimental to survival and reproduction of an organism, mutations are occasionally beneficial, and natural selection acts to accumulate favorable mutations in a population.

Genetic changes that permit cells to undergo unrestrained proliferation cause cancer (neoplastic growth). Oncogenes (such as the gene coding for Ras protein) and inactivation of tumor-suppressor genes (such as that coding for p53 protein) are implicated in many cancers.

REVIEW QUESTIONS

1. What is the relationship between homologous chromosomes, copies of a gene, and alleles?
2. Describe or diagram the sequence of events in meiosis (both divisions).
3. What are the designations of the sex chromosomes in males of bugs, humans, and butterflies?
4. How do the chromosomal mechanisms determining sex differ in the three taxa in question 3?
5. Diagram by Punnett square a cross between individuals with the following genotypes: $A/a \times A/a$; $A/a\ B/b \times A/a\ B/b$.
6. Concisely state Mendel's law of segregation and his law of independent assortment.
7. Assuming brown eyes (B) are dominant over blue eyes (b), determine the genotypes of all the following individuals. The blue-eyed son of two brown-eyed parents marries a brown-eyed woman whose mother was brown eyed and whose father was blue eyed. Their child is blue eyed.
8. Recall that red color (R) in four-o'clock flowers is incompletely dominant over white ($R\,'$). In the following crosses, give the genotypes of the gametes produced by each parent and the flower color of the offspring: $R/R\,' \times R/R\,'$; $R'/R\,' \times R/R\,'$; $R/R \times R/R\,'$; $R/R \times R'/R\,'$.
9. A brown male mouse is mated with two female black mice. In several litters of young, the first female has had 48 black and the second female has had 14 black and 11 brown young. Can you deduce the pattern of inheritance of coat color and the genotypes of the parents?
10. Rough coat (R) is dominant over smooth coat (r) in guinea pigs, and black coat (B) is dominant over white (b). This pair of contrasting traits obeys both of Mendel's laws. If a homozygous rough black is mated with a homozygous smooth white, give the appearance of each of the following: F_1; F_2; offspring of F_1 mated with smooth, white parent; offspring of F_1 mated with rough, black parent.
11. Assume that right-handedness (R) is genetically dominant over left-handedness (r) in humans, and that brown eyes (B) are genetically dominant over blue (b). A right-handed, blue-eyed man marries a right-handed, brown-eyed woman. Their two children are (1) right handed, blue eyed and (2) left handed, brown eyed. The man marries again, and this time the woman is right handed and brown eyed. They have 10 children, all right handed and brown eyed. What are the probable genotypes of the man and his two wives?
12. In *Drosophila melanogaster,* red eyes are dominant over white and the variation for this characteristic is on the X chromosome. Vestigial wings (v) are recessive to normal (V) for an autosomal gene. Predict the appearance of offspring of the following crosses: $X^w/X^w\ V/v \times X^w/Y\ v/v$, $X^w/X^w\ V/v \times X^W/Y\ V/v$.
13. Assume that color blindness is a recessive character on the X chromosome. A man and woman with normal vision have the following offspring: a daughter with normal vision who has one color-blind son and one normal son; a daughter with normal vision who has six normal sons; and a color-blind son who has a daughter with normal vision. What are the probable genotypes of all individuals?
14. Distinguish the following: euploidy, aneuploidy, and polyploidy; monosomy and trisomy.
15. Name the purines and pyrimidines in DNA and tell which pairs occur in the double helix. What are the purines and pyrimidines in RNA and to what are they complementary in DNA?
16. Explain how DNA is replicated.
17. Why could a codon not consist of only two bases?
18. Explain the transcription and processing of mRNA in the nucleus.

19. Explain the role of mRNA, tRNA, and rRNA in polypeptide synthesis.
20. Describe three ways that gene expression can be regulated in eukaryotes.
21. In modern molecular genetics, what is recombinant DNA, and how is it prepared?
22. Why do many mutations have no detectable effects on organismal phenotype?
23. Distinguish between proto-oncogene and oncogene. Describe two mechanisms by which genetic change causes cancer.
24. What are Ras protein and p53? How can mutations in the genes for these proteins contribute to cancer?
25. Outline the essential steps in the polymerase chain reaction.
26. What are some general conclusions from human genomic studies regarding the number of genes present and the relative proportion of the genome that performs genic functions?
27. What is the proteome? Why is genomic information alone insufficient to characterize the proteome?

For Further Thought How does the chromosomal genetic system in animals permit both precise control of gene content and expression in cells and an opportunity for enormous amounts of genetic variation?

SELECTED REFERENCES

Burt, A., and R. Trivers. 2006. Genes in conflict: the biology of selfish genetic elements. Cambridge, Massachusetts, Belknap Press of Harvard Univ. Press. *A thorough discussion of the means by which some genes acquire an ability to promote their own transmission or expression at the expense of others.*

Conery, J. S., and M. Lynch. 2000. The evolutionary fate and consequences of duplicate genes. Science **290:** 1151–1155. *Gene duplications are an important source of genetic variation.*

Davies, K. 2001. Cracking the genome: inside the race to unlock human DNA. Craig Venter, Francis Collins, James Watson, and the story of the greatest scientific discovery of our time. New York, The Free Press. *Fascinating story of competition between the Human Genome Sequencing Project and Craig Venter's Celera Genomics. Of course, the genome is not "cracked" until its meaning is deciphered, and we are a long way from that.*

Ezzell, C. 2002. Proteins rule. Sci. Am. **286:** 40–47 (April). *An excellent explanation of proteomics.*

Futreal, P. A., A. Kasprzyk, E. Birney, J. C. Mullikin, R. Wooster, and M. R. Stratton. 2001. Cancer and genomics. Nature **409:** 850–852. *Good list of cancer genes.*

Griffiths, A. J. F., S. R. Wessler, S. B. Carroll, and J. Doebley. 2015. Introduction to genetic analysis, ed. 11. New York, W. H. Freeman & Company. *A good general genetics text.*

International Human Genome Mapping Consortium. 2001. A physical map of the human genome. Nature **409:** 934–941. *The draft sequence published by the publicly supported consortium*

Jimenez-Sanchez, G., B. Childs, and D. Valle. 2001. Human disease genes. Nature **409:** 853–855. *They found "striking correlation between the function of the gene product and features of disease...."*

Johns Hopkins Medicine. 2015. Online Mendelian inheritance in man: an online catalog of human genes and genetic disorders. Baltimore, MD, Johns Hopkins University. www.omim.org. *A resource for physicians and the general public documenting human genetic variation.*

Korf, B. R. 2006. Human genetics and genomics, ed. 3. Hoboken, New Jersey, John Wiley and Sons. *A problem-based approach to medical applications of genetics and genomics.*

Krebs, J. E., E. S. Goldstein, and S. T. Kilpatrick. 2013. Lewin's genes XI. Burlington, MA, Jones and Bartlett Learning. *A thorough coverage of the molecular biology of genes.*

Mullis, K. B. 1990. The unusual origin of the polymerase chain reaction. Sci. Am. **262:** 56–65 (April). *How the author had the idea for the simple production of unlimited copies of DNA while driving through the mountains of California.*

Pennisi, E. 2000. Genomics comes of age. Science **290:** 2220–2221. *Determination of genomes of a number of organisms is "Breakthrough of the Year."*

Pevsner, J. 2009. Bioinformatics and functional genomics, ed. 2. Hoboken, NJ, Wiley-Blackwell. *A major text of functional genomics.*

Roberts, L. 2001. A history of the Human Genome Project. Science **291:** chart. *Each major discovery from double helix to full genome (1953–2001). Has glossary.*

Russell, P. J. 2009. iGenetics: a molecular approach, ed. 3. San Francisco, Benjamin Cummings. *A thorough introductory textbook of genetics.*

Venter, J. C., M. D. Adams, E. W. Myers, P. W. Li, R. J. Mural, and 265 others. 2001. The sequence of the human genome. Science **291:** 1304–1351. *The draft sequence published by Celera Genomics and its collaborators.*

Watson, J. D., and A. Berry. 2003. DNA: the secret of life. New York, Alfred A. Knopf. *An exciting account of the history and applications of genetics.*

Wu, L., and J. G. Belasco. 2008. Let me count the ways: mechanisms of gene regulation by miRNAs and siRNAs. Molecular Cell **29:** 1–7. *A short summary of post-transcriptional regulation of gene expression by an RNA-induced silencing complex.*

CHAPTER 6

Organic Evolution

A trilobite fossilized in Paleozoic rock.

A Legacy of Change

Life's history is a legacy of perpetual change. Despite the apparent permanence of the natural world, change characterizes all things on earth and in the universe. Earth's rock strata record the irreversible, historical change that we call organic evolution. Countless kinds of animals and plants have flourished and disappeared, leaving behind a sparse fossil record of their existence. Many, but not all, have left living descendants that bear some resemblance to them.

We observe life's changes and measure them in many ways. On a short evolutionary timescale, we see changes in the frequencies of different genetic traits within populations. Evolutionary changes in the relative frequencies of light- and dark-colored moths occurred within a single human lifetime in the polluted towns of industrial England. The formation of new species and dramatic changes in organismal form, as illustrated by the evolutionary diversification of Hawaiian birds, requires longer timescales covering 100,000 to 1 million years. Major evolutionary trends and episodic mass extinctions occur on even larger timescales, covering tens of millions of years. The fossil record of horses through the past 50 million years shows a series of different species replacing older ones through time and ending with the horses alive today. The fossil record of marine invertebrates shows episodic mass extinctions separated by intervals of approximately 26 million years.

Because every feature of life as we know it today is a product of evolution, biologists consider organic evolution the keystone of all biological knowledge.

n Chapter 1, we introduced Darwinian evolutionary theory as the dominant paradigm of biology. Charles Robert Darwin and Alfred Russel Wallace (Figure 6.1) first established evolution as a powerful scientific theory. Today the reality of organic evolution can be denied only by abandoning reason. As the noted English biologist Sir Julian Huxley wrote, "Charles Darwin effected the greatest of all revolutions in human thought, greater than Einstein's or Freud's or even Newton's, by simultaneously establishing the fact and discovering the mechanism of organic evolution." Darwinian theory helps us to understand both the genetics of populations and long-term trends in the fossil record. Darwin and Wallace did not originate the basic idea of organic evolution, which has an ancient history. We review the history of evolutionary thinking as it led to Darwin's theory, evidence supporting it, and changes to the theory that have produced our modern Synthetic Theory of evolution.

6.1 ORIGINS OF DARWINIAN EVOLUTIONARY THEORY

Pre-Darwinian Evolutionary Ideas

Early Greek philosophers, notably Xenophanes, Empedocles, and Aristotle, recorded an early idea of evolutionary change. They recognized fossils as evidence for former life that they believed had been destroyed by natural catastrophe. Despite their inquiry, the Greeks failed to establish an evolutionary concept. Opportunity for evolutionary thinking became even more restricted as a literally interpreted biblical account of the earth's creation became accepted as a tenet of faith. Archbishop James Ussher (mid-seventeenth century) declared the year 4004 BCE as the date of life's creation. Evolutionary views were considered rebellious and heretical, but they refused to die. The French naturalist Georges Louis Buffon (1707 to 1788) stressed the influence of environment on the

modifications of animal form. He also extended the age of the earth to 70,000 years.

Lamarckism: The First Scientific Explanation of Evolution

French biologist Jean Baptiste de Lamarck (1744 to 1829; Figure 6.2) authored the first complete explanation of evolution in 1809, the year of Darwin's birth. He made a convincing case that fossils were remains of extinct animals. Lamarck's proposed evolutionary mechanism, **inheritance of acquired characteristics,** was engagingly simple: organisms, by striving to meet the demands of their environments, acquire adaptations and pass them by heredity to their offspring. According to Lamarck, the giraffe evolved its long neck because its ancestors lengthened their necks by stretching to obtain food and then passed the lengthened neck to their offspring. Over many generations, these changes accumulated to produce the long necks of modern giraffes.

We call Lamarck's concept of evolution **transformational,** because it claims that as individual organisms transform their characteristics through the use and disuse of body parts, heredity makes corresponding adjustments to produce evolution. We now reject transformational theories because genetic studies show that traits acquired by an organism during its lifetime, such as strengthened muscles, are not inherited by offspring. Darwin's evolutionary theory differs from Lamarck's in being a **variational** theory, based in genetic differences that occur among organisms within a population. Evolution occurs at the level of the population, and it includes changes across generations in the organismal characteristics that prevail in the population. Darwin argued that organisms possessing hereditary characteristics that conferred an advantage for survival or reproduction would leave more offspring than would other organisms, thereby causing the characteristics most favorable to the survival and reproductive success of their bearers to accumulate in populations across generations.

Charles Lyell and Uniformitarianism

The geologist Sir Charles Lyell (1797 to 1875; Figure 6.3) established in his *Principles of Geology* (1830 to 1833) the principle of uniformitarianism. Uniformitarianism encompasses two important principles that guide scientific study of the history of nature: (1) that the laws of physics and chemistry have not changed throughout the history of the earth, and (2) that past geological events occurred by natural processes similar to those observed today. Lyell showed that natural forces, acting over long periods of time, could explain the formation of fossil-bearing rocks. For example, as skeletal remains of corals (p. 277), foraminiferans (p. 238), and molluscs (p. 333) accumulate on the sea floor, they form sediments of calcium carbonate that eventually become compressed into limestone. Lyell's geological studies led him to conclude that the earth's age must be measured in hundreds of millions of years. Measured rates of sedimentation are much too slow to have produced earth's sedimentary rock formations in a shorter period of time. These

National Library of Medicine

© Biophoto Associates/Science Source

Figure 6.1   Founders of the theory of evolution by natural selection. **A,** Charles Robert Darwin (1809 to 1882). **B,** Alfred Russel Wallace (1823 to 1913) in 1895. Darwin and Wallace independently developed the same theory. A letter and essay from Wallace written to Darwin in 1858 spurred Darwin into writing *On the Origin of Species*, published in 1859.

Figure 6.2   Jean Baptiste de Lamarck (1744 to 1829), French naturalist who offered the first scientific explanation of evolution. Lamarck's hypothesis that evolution proceeds by inheritance of acquired characteristics has been rejected and replaced by neo-Darwinian theories.

Figure 6.3   Sir Charles Lyell (1797 to 1875), English geologist and friend of Darwin. His book *Principles of Geology* greatly influenced Darwin during Darwin's formative period.

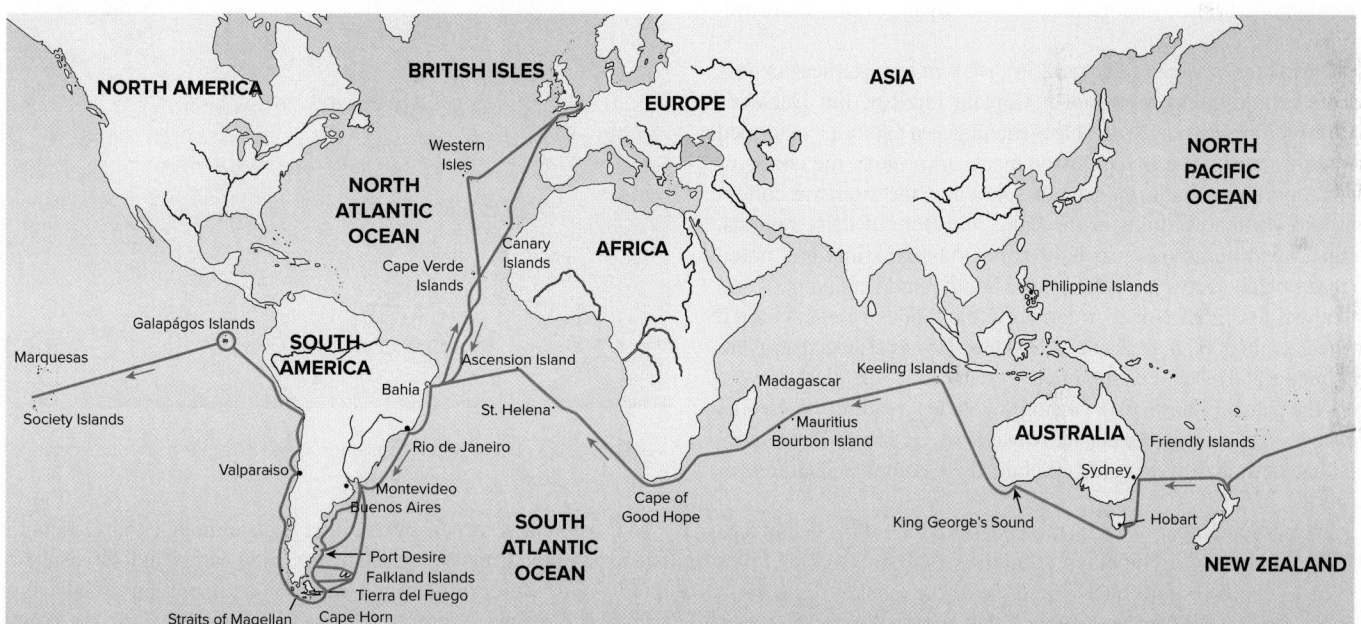

Figure 6.4   Five-year voyage of H.M.S. *Beagle.*

principles were important for discrediting miraculous and supernatural explanations of the history of nature and replacing them with scientific explanations. Lyell also stressed that geological changes occur primarily in small increments whose gradual accumulation through time builds the earth's major geological formations, and he argued further that such changes have no inherent tendency to occur in any particular direction. Positions of mountains and seas would gradually change through time, for example, but the earth's surface would have no directional tendency toward becoming more mountainous or more inundated. Both of these claims left important marks on Darwin's evolutionary theory.

Darwin's Great Voyage of Discovery

"After having been twice driven back by heavy southwestern gales, Her Majesty's ship *Beagle,* a ten-gun brig, under the command of Captain Robert FitzRoy, R.N., sailed from Devonport on the 27th of December, 1831." Thus began Charles Darwin's account of the historic five-year voyage of the *Beagle* around the world (Figure 6.4). Darwin, not quite 23 years old, had been asked to accompany Captain FitzRoy on the *Beagle,* a small vessel only 90 feet in length, which was about to depart on an extensive surveying voyage to South America and the Pacific Ocean (Figure 6.5). It was the

A   B

Figure 6.5 Charles Darwin and H.M.S. *Beagle*. **A,** Darwin in 1840, four years after the *Beagle* returned to England, and a year after his marriage to his cousin, Emma Wedgwood. **B,** The H.M.S. *Beagle* sails in Beagle Channel, Tierra del Fuego, on the southern tip of South America in 1833. The watercolor was painted by Conrad Martens, one of two official artists on the voyage of the *Beagle*.

beginning of the most important scientific voyage of the nineteenth century.

During the voyage (1831 to 1836), Darwin endured seasickness and the erratic companionship of Captain FitzRoy, but Darwin's youthful physical strength and early training as a naturalist equipped him for his work. The *Beagle* made many stops along the coasts of South America and adjacent islands. Darwin made extensive collections and written accounts of the fauna and flora of these regions. He unearthed numerous fossils of animals long extinct and noted the resemblance between fossils of the South American pampas and the known fossils of North America. In the Andes he encountered seashells embedded in rocks at 13,000 feet. He experienced a severe earthquake and watched mountain torrents that relentlessly wore away the earth. These observations, and his reading of Lyell's *Principles of Geology* during the voyage, strengthened his conviction that natural forces could explain the geological features of the earth.

In mid-September of 1835, the *Beagle* arrived at the Galápagos Islands, a volcanic archipelago straddling the equator 600 miles west of Ecuador (Figure 6.6). The fame of the islands stems from their oceanic isolation and rugged volcanic terrain. Circled by capricious currents, surrounded by shores of twisted lava bearing skeletal brushwood baked by the equatorial sun, inhabited by strange reptiles and by convicts stranded by the Ecuadorian government, the islands had few admirers among mariners. By the middle of the seventeenth century, the Spaniards called these islands "Las Islas Galápagos"—the tortoise islands. The giant tortoises, used for food first by buccaneers and later by American and British whalers, sealers, and ships of war, were the islands' principal attraction. At the time of Darwin's visit, the tortoises already were heavily exploited.

During the *Beagle*'s five-week visit to the Galápagos, Darwin documented the unique character of the Galápagos plants and animals, including the giant tortoises, marine iguanas, mockingbirds, and ground finches. Darwin later described these studies as the "origin of all my views."

Figure 6.6 The Galápagos Islands viewed from the rim of a volcano.

Darwin was struck by the fact that, although the Galápagos Islands and the Cape Verde Islands (visited earlier in this voyage of the *Beagle*) were similar in climate and topography, Galápagos plants and animals resembled most closely those of the South American mainland, and they were entirely different from the African-derived forms of the Cape Verde Islands. Each Galápagos Island often contained a unique species that nonetheless resembled forms on other Galápagos Islands. In short, Galápagos life must have originated in continental South America and then undergone modification in the various environmental conditions of the different islands. He concluded that living forms were neither divinely created nor immutable; they were, in fact, products of a long history of evolutionary change.

On October 2, 1836, the *Beagle* returned to England, where Darwin conducted most of his scientific work (Figure 6.7). Most of Darwin's extensive collections had preceded him there, as had notebooks and diaries kept during the cruise. Darwin's journal, published three years after the *Beagle*'s return to England, was an instant

© Cleveland P. Hickman, Jr.

Figure 6.7   Darwin's study at Down House in Kent, England, is preserved today much as it was when Darwin wrote *On the Origin of Species*.

success and required two additional printings within the first year. Darwin later revised his journal as *The Voyage of the Beagle,* a travelogue of continuing popularity.

The main product of Darwin's voyage, his theory of evolution, would continue to develop for more than 20 years after the *Beagle*'s return. In 1838, he "happened to read for amusement" an essay on populations by T. R. Malthus (1766 to 1834), who stated that animal and plant populations, including human populations, have the reproductive capacity to increase beyond the capacity of environmental resources to support them. Darwin already had been gathering information on artificial selection of animals under domestication. Darwin was especially fascinated by artificial breeds of pigeons. Many pigeon breeds differed so much in appearance and behavior that they would be considered different species if found in nature, yet all had clearly been derived from a single wild species, the rock pigeon (*Columba livia*). After reading Malthus's article, Darwin realized that a process of selection in nature, a "struggle for existence" because of overpopulation, could be a powerful force for evolution of wild species.

Darwin allowed the idea to develop in his own mind, writing private, trial essays in 1842 and 1844. Finally in 1856, he began to assemble his voluminous data into a work on the origin of species. He expected to write four volumes, a very big book, "as perfect as I can make it." However, his plans took an unexpected turn.

In 1858, he received a manuscript from Alfred Russel Wallace (1823 to 1913), an English naturalist in Malaya with whom he corresponded. Darwin was stunned to find that in a few pages, Wallace summarized the main points of the natural selection theory on which Darwin had been working for two decades. Rather than withholding his own work in favor of Wallace's as he was inclined to do, Darwin was persuaded by two close friends, the geologist Lyell and the botanist Hooker, to publish his views in a brief statement that would appear together with Wallace's paper in the *Journal of the Linnean Society*. Portions of both papers were read to an unresponsive audience on July 1, 1858.

For the next year, Darwin worked urgently to prepare an "abstract" of the planned four-volume work. This book was published in November 1859, with the title *On the Origin of Species by Means of Natural Selection, or the Preservation of Favoured Races*

in the Struggle for Life. The 1250 copies of the first printing sold the first day! The book instantly generated a storm that has never abated. Darwin's views were to have extraordinary consequences on scientific and religious beliefs and remain among the greatest intellectual achievements of all time.

> "Whenever I have found that I have blundered, or that my work has been imperfect, and when I have been contemptuously criticized, and even when I have been overpraised, so that I have felt mortified, it has been my greatest comfort to say hundreds of times to myself that 'I have worked as hard and as well as I could, and no man can do more than this.'" *Charles Darwin, in his autobiography, 1876*

Once Darwin's caution had been swept away by the publication of *On the Origin of Species,* he entered an incredibly productive period of evolutionary thinking for the next 23 years, producing five revisions of *On the Origin of Species* and a dozen new books. He continued his scientific interactions with Wallace, who carefully documented the geographic distributions of animal and plant species and thereby founded the field of historical biogeography (p. 798). Darwin died on April 19, 1882, and was buried in Westminster Abbey. The little *Beagle* had already disappeared, having been retired in 1870 and sold for scrap.

6.2 DARWINIAN EVOLUTIONARY THEORY: THE EVIDENCE

Perpetual Change

The main premise underlying Darwinian evolution is that the living world is neither constant nor perpetually cycling, but always changing and with hereditary continuity from past to present life. Perpetual change in the form and diversity of animal life throughout its 600- to 700-million-year history is seen most directly in the fossil record. A **fossil** is a remnant of past life uncovered from the crust of the earth (Figure 6.8). Some fossils constitute complete remains (insects in amber and mammoths), actual hard parts (teeth and bones), and petrified skeletal parts infiltrated with silica or other minerals (ostracoderms and molluscs). Other fossils include footprints or other impressions, burrows of marine worms in sediment on the seafloor, and fossil excrement (coprolites). In addition to documenting organismal evolution, fossils reveal profound changes in the earth's environment, including major changes in the distributions of lands and seas. Fossils formed on the floors of ancient seas can be quarried high atop current mountains (p. 105).

Fossil remains may on rare occasions include soft tissues preserved so well that recognizable cellular organelles are revealed by electron microscopy! Insects are frequently found entombed in amber, the fossilized resin of trees. One study of a fly entombed in 40-million-year-old amber revealed structures corresponding to muscle fibers, nuclei, ribosomes, lipid droplets, endoplasmic reticulum, and mitochondria (Figure 6.8D). This extreme case of mummification probably occurred because chemicals in the plant sap diffused into and embalmed the insect's tissues.

Interpreting the Fossil Record

The fossil record is biased because preservation is selective. Vertebrate skeletal parts and invertebrates with shells and other hard structures left the best records (Figure 6.8). Soft-bodied animals, including jellyfishes and most worms, are fossilized only under very unusual circumstances, such as those that formed the Burgess Shale of British Columbia (Figure 6.9). Exceptionally favorable conditions for fossilization produced the Precambrian fossil bed of South Australia, the tar pits of Rancho La Brea (Hancock Park, Los Angeles), the great dinosaur beds (Alberta, Canada, and Jensen, Utah; Figure 6.10), and fossil beds of the Yunnan and Lianoning provinces of China.

Fossils form in stratified layers with new deposits lying on top of older ones. If left undisturbed, which is rare, the ages of fossils in a preserved sequence are directly proportional to depth in the stratified layers. Characteristic fossils often serve to identify particular layers. Certain widespread marine invertebrate fossils, including various foraminiferans (p. 238) and echinoderms (p. 474), are such good indicators of specific geological periods that they are called "index," or "guide," fossils. Unfortunately, the layers are usually tilted or show faults (cracks). Old deposits exposed by erosion may be covered with new deposits in a different plane. When exposed to tremendous pressures or heat, stratified sedimentary rock metamorphoses into crystalline quartzite, slate, or marble, which destroys fossils.

Stratigraphy of fossils for two major groups of African antelopes is shown on Figure 6.11. Species in this group have different characteristic sizes and shapes of horns, which form much of the fossil record of this group. Solid vertical lines in Figure 6.11 show the temporal distributions of species determined by presence of their characteristic horns in rock strata of various ages. Red lines denote the fossil records of living species, and gray lines denote the fossil records of extinct species. The dotted gray lines show the inferred relationships among living and fossil species based on their sharing of homologous structural features.

Geological Time

Long before the earth's age was known, geologists divided its history into a table of succeeding events based on the ordered layers of sedimentary rock. The "law of stratigraphy" produced a relative dating with the oldest layers at the bottom and the youngest at the top of the sequence. Time was divided into eons, eras, periods, and epochs as shown in the Timeline of Major Biological Events on the endsheet inside the back cover of this book. Time during the last eon (Phanerozoic) is expressed in eras (for example, Cenozoic), periods (for example, Tertiary), epochs (for example, Paleocene), and sometimes smaller divisions of an epoch.

In the late 1940s, radiometric dating methods were developed for determining the absolute age in years of rock formations. Several

Figure 6.8 Four examples of fossil material. **A,** Fish fossil from rocks of the Green River Formation, Wyoming. Such fish swam here during the Eocene epoch of the Tertiary period, approximately 50 million years ago. **B,** Stalked crinoids (class Crinoidea, p. 489) from Devonian rocks. The fossil record of these echinoderms shows that they reached their peak millions of years earlier and began a slow decline to the present. **C,** An insect fossil that got stuck in the resin of a tree approximately 25 million years ago and that has since hardened into amber. **D,** Electron micrograph of tissue from a fly fossilized as shown in **C;** the nucleus of a cell is marked in red.

Figure 6.9 **A,** Fossil trilobite visible at the Burgess Shale Quarry, in the Rocky Mountains of British Columbia, Canada. **B,** Animals of the Cambrian period, approximately 505 million years ago, as reconstructed from fossils preserved in the Burgess Shale. The main new body plans that appeared rather abruptly at this time established the body plans of animals familiar to us today.

© Kevin Schafer/Alamy Stock Photo

A

Amiskwia, from an extinct phylum

Odontogriphus, from an extinct phylum

Eldonia, a possible echinoderm

Pikaia, an early chordate

Anomalocaris canadensis, from an extinct phylum

Aysheaia, an onychophoran

Halichondrites, a sponge

Xianguangia, an anenome-like animal

Opabinia, from an extinct phylum

Yohoia, a unique arthropod

Canadia, a polychaete

Sidneyia, a unique arthropod

Marrella splendens, a unique arthropod

Wiwaxia, from an extinct phylum

Ottoia, a priapulid

Dinomischus, from an extinct phylum

Hallucigenia, from an extinct phylum

B

Figure 6.10 A dinosaur skeleton partially excavated from rock at Dinosaur Provincial Park, Alberta.

independent methods are now used, all based on the radioactive decay of naturally occurring elements into other elements.

These "radioactive clocks" are independent of pressure and temperature changes and therefore are not affected by often violent earth-building activities.

One method, potassium-argon dating, uses the decay of potassium-40 (^{40}K) to argon-40 (^{40}Ar) (12%) and calcium-40 (^{40}Ca) (88%). Argon is a noble gas that evaporates from liquid media. It accumulates in the crystal structure of rock only after the rock has solidified and the nuclear decay of potassium-40 produces a trapped atom of argon. The half-life of potassium-40 is 1.3 billion years; half of the original atoms will decay in 1.3 billion years, and half of the remaining atoms will be gone at the end of the next 1.3 billion years. This decay continues until all radioactive potassium-40 atoms are gone. To measure the age of the rock, one calculates the ratio of remaining potassium-40 atoms to the amount of potassium-40 originally there (the remaining potassium-40 atoms plus the argon-40 and calcium-40 into which other potassium-40 atoms have decayed). A standard equation converts these data to time elapsed since formation of the rock as a function of the half-life of potassium-40. Several

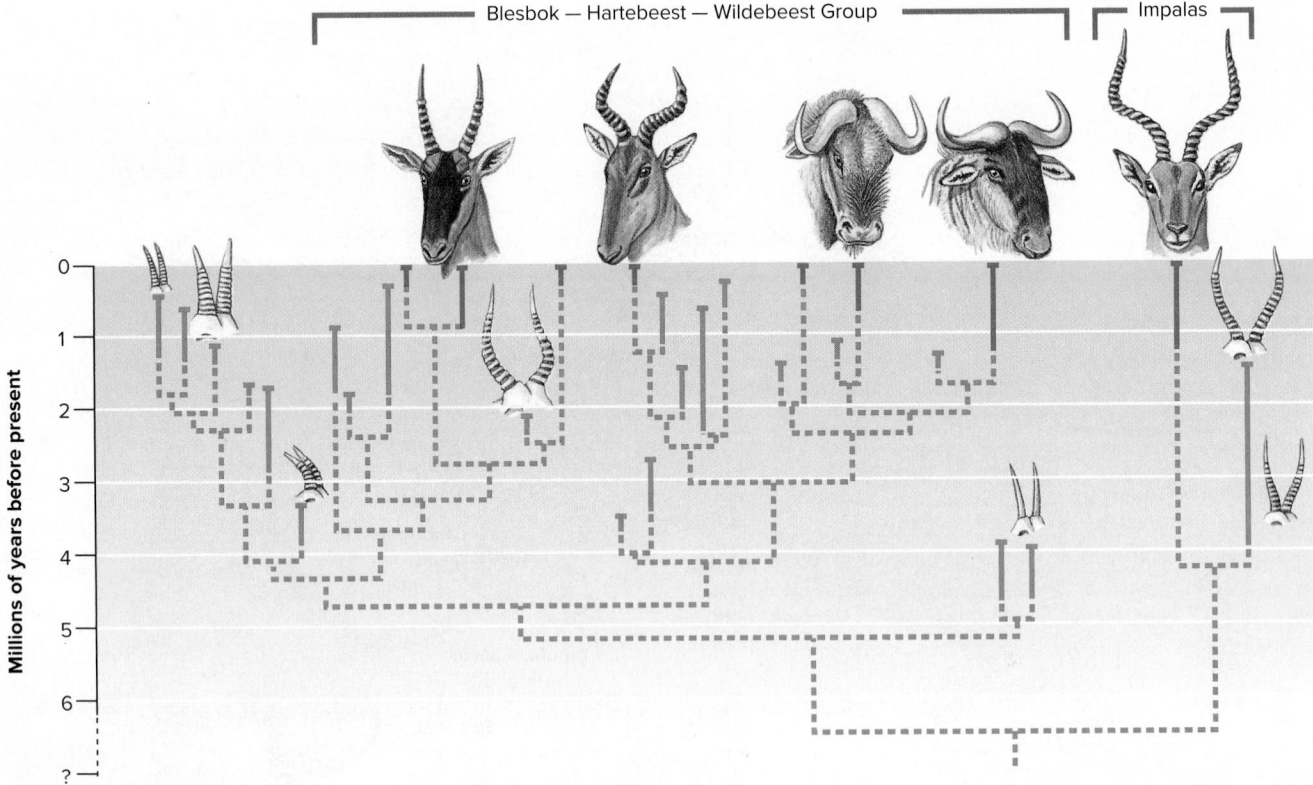

Figure 6.11 Stratigraphic record and inferred evolutionary relationships among alcelaphine (blesboks, hartebeests, wildebeests) and aepycerotine (impalas) antelopes in Africa. Species in this group are identified by characteristic sizes and shapes of horns found in rock strata of various ages. Solid vertical lines show the temporal distribution of species in rock strata whose ages are shown on the scale at the left side of the figure. Red lines show the temporal distributions of living species, and gray lines show the temporal distributions of extinct species in rock strata. Dotted gray lines show the inferred relationships among species based on their sharing of homologous structural features. The relative constancy of horn structure within species through geological time is consistent with the theory of punctuated equilibrium (p. 119). This fossil record shows that rates of speciation and extinction are higher for alcelaphine antelopes than for impalas.

other isotopes can be used in a similar manner to date the ages of rock formations, some for dating the age of the earth itself. One of the most useful radioactive clocks depends on the decay of uranium into lead. With this method, rocks over 2 billion years old can be dated with a probable error of less than 1%.

The fossil record of macroscopic organisms begins near the start of the Cambrian period of the Paleozoic era, approximately 542 million years BP. Geological time before the Cambrian is called the Precambrian era or Proterozoic eon. Although the Precambrian era occupies 85% of all geological time, it has received much less attention than later eras, partly because oil, which provides the commercial incentive for much geological work, seldom exists in Precambrian formations. The Precambrian era contains well-preserved fossils of bacteria and algae, and casts of jellyfishes, sponge spicules, soft corals, segmented flatworms, and worm trails. Most, but not all, are microscopic fossils.

Evolutionary Trends

The fossil record allows us to view evolutionary change across the broadest scale of time. Species arise and go extinct repeatedly throughout the geological history recorded by the fossil record. Animal species typically survive approximately 1 to 10 million years, although their durations are highly variable. We can summarize patterns of species or taxon replacement through time as trends. Trends are directional changes in the characteristic features or patterns of diversity in a group of organisms. Fossil trends clearly demonstrate Darwin's principle of perpetual change.

Our use of the phrase "evolutionary trend" does not imply that more recent forms are superior to older ones or that the changes represent progress in adaptation or organismal complexity. Although Darwin predicted that such trends would show progressive adaptation, many contemporary paleontologists consider progressive adaptation rare among evolutionary trends. Observed trends in the evolution of horses do not imply that contemporary horses are superior in any general sense to their Eocene ancestors.

A well-studied fossil trend is the evolution of horses from the Eocene epoch to the present. Looking back at the Eocene epoch, we see many different genera and species of horses replaced by others through time (Figure 6.12). George Gaylord Simpson (p. 202) showed that this trend is compatible with Darwinian evolutionary theory. The three characteristics that show the clearest trends in horse evolution are body size, foot structure, and tooth structure. Compared to modern horses, those of extinct genera were small; their teeth had a relatively small grinding surface, and their feet had a relatively large number of toes (four). Throughout the subsequent Oligocene, Miocene, Pliocene, and Pleistocene epochs, new genera arose and old ones went extinct. In each case, a net increase in body size, expansion of the grinding surface of the teeth, and reduction in the number of toes occurred. As the number of toes declined, the central digit became increasingly more prominent in the foot, and eventually only this central digit remained.

The fossil record shows not only a net change in the characteristics of horses but also variation in the numbers of different horse genera (and numbers of species) through time. The many horse genera of past epochs have been lost to extinction, leaving only a single survivor, *Equus*. Evolutionary trends in diversity are observed in fossils of many different groups of animals (Figure 6.13).

Trends in fossil diversity through time are produced by different rates of species formation versus extinction through time. Why do some lineages generate large numbers of new species whereas others generate relatively few? Why do different lineages undergo higher or lower rates of extinction (of species, genera, or taxonomic families) throughout evolutionary time? To answer these questions, we must turn to Darwin's other four theories of evolution. Regardless of how we answer these questions, however, the observed trends in animal diversity clearly illustrate Darwin's principle of perpetual change. Because the remaining four theories of Darwinism rely on the theory of perpetual change, evidence supporting these theories strengthens Darwin's theory of perpetual change.

Common Descent

Darwin proposed that all plants and animals have descended from an ancestral form into which life was first breathed. Life's history is depicted as a branching tree, called a **phylogeny.** Pre-Darwinian evolutionists, including Lamarck, advocated multiple independent origins of life, each of which gave rise to lineages that changed through time without extensive branching. Like all good scientific theories, common descent makes several important predictions that can be tested and potentially used to reject it. According to this theory, we should be able to trace the genealogies of all modern species backward until they converge on ancestral lineages shared with other species, both living and extinct.

We should be able to continue this process, moving farther backward through evolutionary time, until we reach the primordial ancestor of all life on earth. All forms of life, including many extinct forms that represent dead branches, connect to this tree somewhere. Although reconstructing the history of life in this manner may seem almost impossible, phylogenetic research has been extraordinarily successful. How has this difficult task been accomplished?

Homology and Phylogenetic Reconstruction

Darwin recognized the major source of evidence for common descent in the concept of **homology.** Darwin's contemporary, Richard Owen (1804 to 1892), used this term to denote "the same organ in different organisms under every variety of form and function." A classic example of homology is the limb skeleton of vertebrates. Bones of vertebrate limbs maintain characteristic structures and patterns of connection despite diverse modifications for different functions (Figure 6.14). According to Darwin's theory of common descent, the structures that we call homologies represent characteristics inherited with some modification from a corresponding feature in a common ancestor.

Darwin devoted an entire book, *The Descent of Man and Selection in Relation to Sex,* largely to the idea that humans share common descent with apes and other animals. This idea was repugnant to many Victorians, who responded with outrage (Figure 6.15). Darwin built his case mostly on anatomical comparisons revealing homology between humans and apes. To Darwin, the close resemblances between apes and humans could be explained only by common descent.

Figure 6.12 Stratigraphy of genera of horses from Eocene to present. Evolutionary trends toward increased size, elaboration of molars, and loss of toes are shown together with bars denoting temporal durations and continental locations of genera.

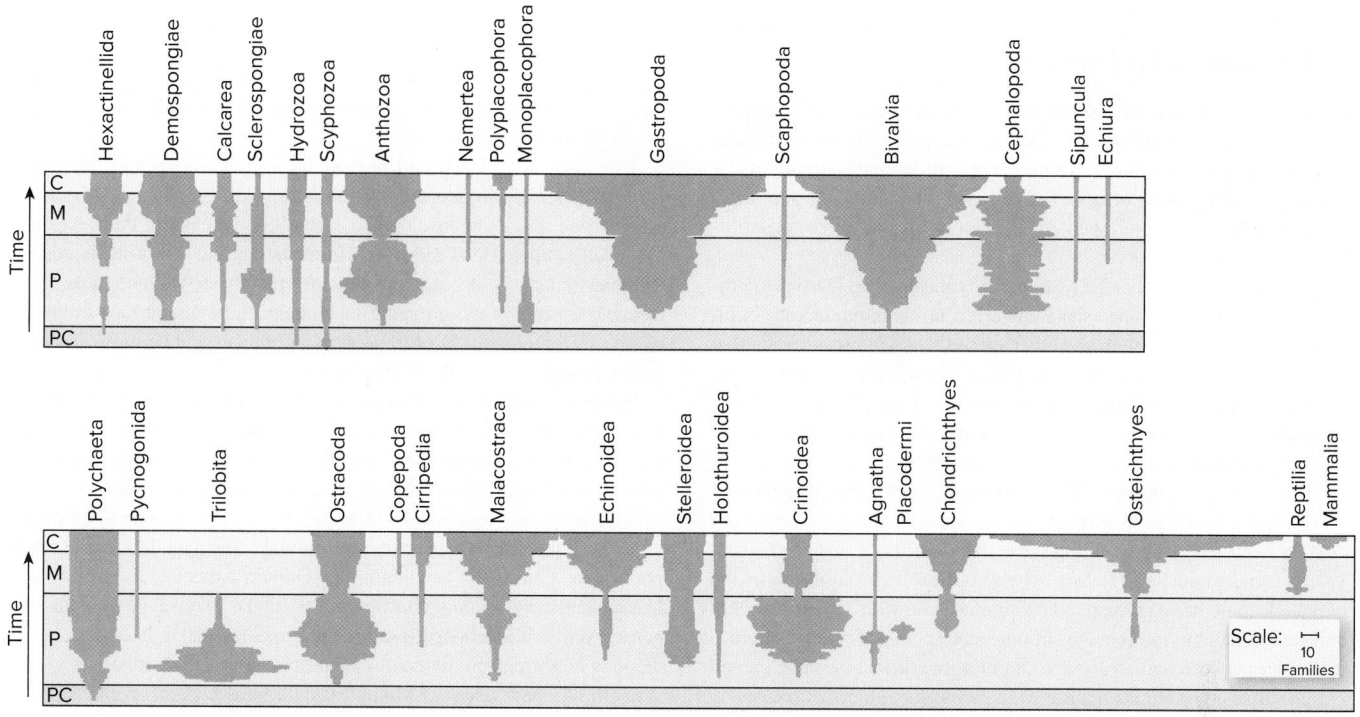

Figure 6.13   Diversity profiles of taxonomic families from different animal groups in the fossil record. The scale marks the Precambrian (PC), Paleozoic (P), Mesozoic (M), and Cenozoic (C) eras. The relative number of families is indicated from the width of the profile.

Figure 6.14   Forelimbs of five vertebrates show skeletal homologies. Clear homologies of bones and patterns of connection are evident despite evolutionary modification for different functions.

Figure 6.15   This 1873 advertisement for Merchant's Gargling Oil ridicules Darwin's theory of the common descent of humans and apes, which received only limited acceptance by the general public during Darwin's lifetime.

The Power of a Theory

Darwin's theory of common descent illustrates the scientific importance of general theories that give unified explanations to diverse kinds of data. Darwin proposed his theory of descent with modification of all living forms because it explained the patterns of similarity and dissimilarity among organisms in anatomical structures and cellular organization.

Anatomical similarities between humans and apes led Darwin to propose that humans and apes share more recent common ancestry with each other than they do with any other species. Darwin was unaware that his theory, a century later, would provide the primary explanation for similarities and dissimilarities among species in the structures of their chromosomes, sequences of amino acids in homologous proteins, and sequences of bases in homologous genomic DNA.

The accompanying figure shows photographs of a complete haploid set of chromosomes from each of four ape species: human (*Homo sapiens*), bonobo (the pygmy chimpanzee, *Pan paniscus*), gorilla (*Gorilla gorilla*), and orangutan (*Pongo pygmaeus*). Each chromosome in the human genome has a corresponding chromosome with similar structure and gene content in the genomes of other ape species. The most obvious difference between human and ape chromosomes is that the large second chromosome in the human nuclear genome was formed evolutionarily by a fusion of two smaller chromosomes characteristic of the ape genomes. Detailed study of the human and other ape chromosomes shows remarkable correspondence between them in genic content and organization.

Ape chromosomes are more similar to each other than they are to chromosomes of any other animals.

Comparison of DNA and protein sequences among apes likewise confirms their close genetic relationships, with humans and the two chimpanzee species being closer to each other than any of these species are to other apes. DNA sequences from the nuclear and mitochondrial genomes independently support the close relationships among ape species and especially the grouping of humans and chimpanzees as close relatives. Homologous DNA sequences of humans and chimpanzees are approximately 99% similar in base sequence.

Studies of variation in chromosomal structure, mitochondrial DNA sequences, and nuclear DNA sequences produced multiple independent data sets, each one potentially capable of rejecting Darwin's theory of common descent. Darwin's theory would be rejected, for example, if the chromosomal structures and DNA sequences of apes were no more similar to each other than to those of other animals. The data in this case support rather than reject predictions of Darwin's theory. The ability of Darwin's theory of common descent to make precise predictions of genetic similarities among these and other species, and to have those predictions confirmed by numerous empirical studies, illustrates its great strength. As new kinds of biological data have become available, the scope and strength of Darwin's theory of common descent have increased enormously. Indeed, nothing in biology makes sense in the absence of this powerful explanatory theory.

Comparative Karyotype of Great Apes

© Dr. Marano Rocchi

The human haploid genome contains 22 autosomes (I–XXII) and a sex chromosome (X or Y). The human chromosome is shown first in each group of four, followed by the corresponding chromosomes of bonobo, gorilla, and orangutan, in that order. Note that the chromatin of human chromosome II corresponds to that of two smaller chromosomes (marked p and q) in other apes.

Throughout the history of all forms of life, evolutionary processes generate new characteristics that are then inherited by subsequent generations. Every time a new feature arises on an evolving lineage, we see the origin of a new homology. That homology gets transmitted to all descendant lineages unless it is subsequently lost. The sharing of homologies among species provides evidence for common descent and allows us to reconstruct the branching evolutionary history of life.

We illustrate such evidence using a phylogenetic tree for a group of large, ground-dwelling birds (Figure 6.16). A new skeletal homology arises on each of the lineages shown (descriptions of specific homologies are not included because they are highly technical). The different groups of species located at the tips of the branches contain different combinations of these homologies, which reflect ancestry. For example, ostriches show homologies 1 through 5 and 8, whereas kiwis show homologies 1, 2, 13, and 15. Branches of the tree combine these species into a **nested hierarchy** of groups within groups (see Chapter 10). Smaller groups (species grouped near terminal branches) are contained within larger ones (species grouped by basal branches, including the trunk of the tree). If we erase the tree structure but retain patterns of homology observed in the living species, we are able to reconstruct the branching structure of the entire tree. Evolutionists test the theory of common descent by observing patterns of homology present within all groups of organisms. The pattern formed by all homologies taken together should specify a single branching tree that represents the evolutionary genealogy of all living organisms.

The tree structure inferred from analysis of skeletal structures of flightless birds can be tested by data gathered independently from DNA-sequence information (see Chapter 10). The phylogeny of flightless birds inferred from DNA-sequence data does not agree completely with the one inferred from skeletal structures (Figure 6.16); if we choose the hypothesis favored by DNA-sequence data, then we must hypothesize that some of the skeletal structures either arose multiple times or were lost on some lineages as shown on Figure 6.16B. Conflict between the phylogenetic hypotheses derived from skeletal structures and from DNA

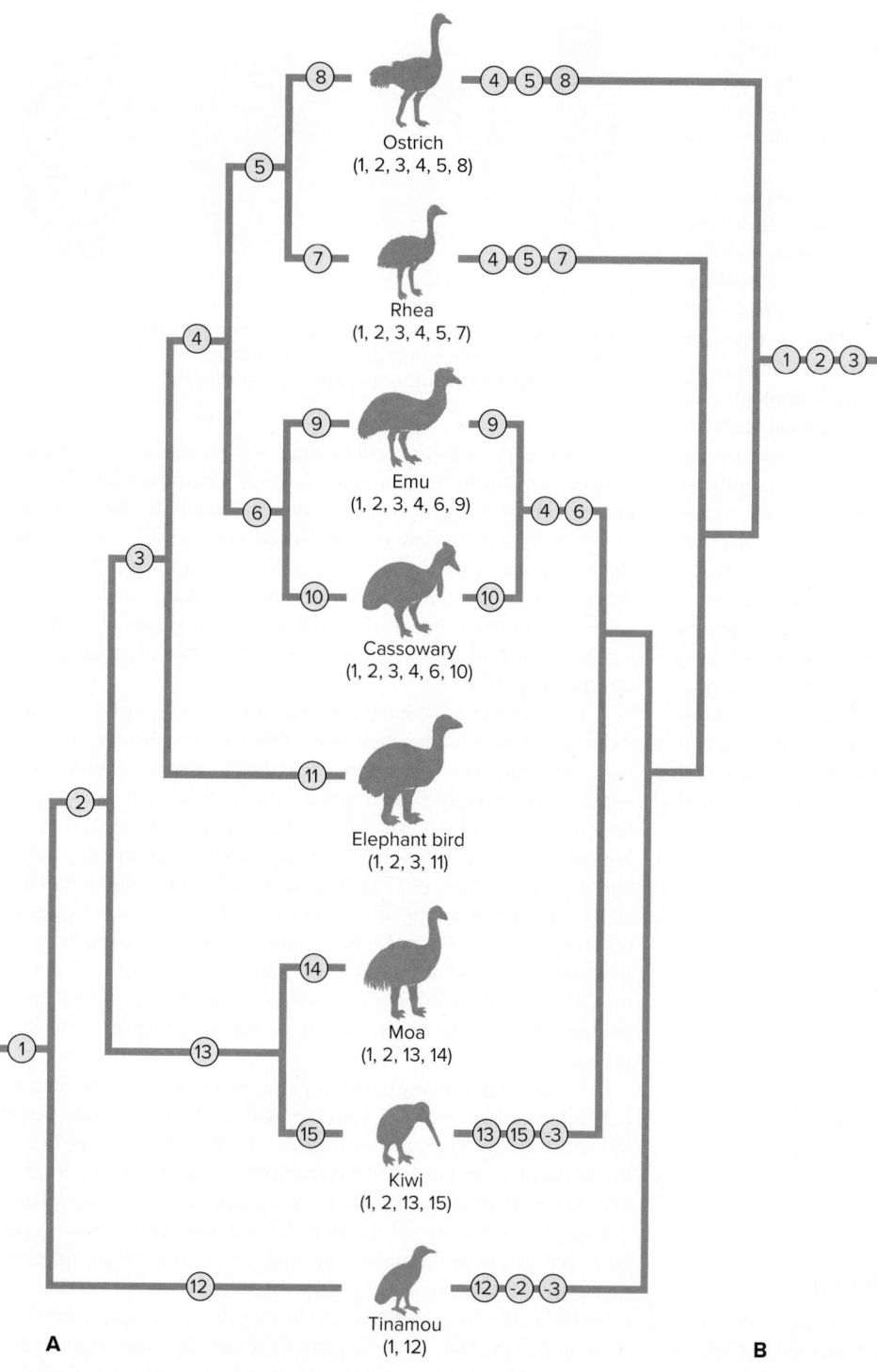

Figure 6.16 **A,** The phylogenetic pattern specified by 15 homologous structures in the skeletons of a group of flightless birds. Homologous features are numbered 1 through 15 and are marked both on the branches of the tree on which they arose and on the birds that have them. If you were to erase the tree structure, you would be able to reconstruct it without error from the distributions of homologous features shown for the birds at the terminal branches. **B,** An analysis of molecular data suggests a different pattern of relationships among the living flightless birds (all except moas and elephant birds). If the molecular analysis is correct, then one must reinterpret evolution of skeletal characters 2, 3, 4, and 5. In the interpretation shown, character 2 is lost by tinamous (−2), and character 3 is lost by both kiwis and tinamous (−3). Character 4 arises independently in ostriches, rheas, and a common ancestor of cassowaries and emus. Character 5 originates separately in ostriches and rheas. Multiple origins and losses complicate phylogenetic analysis, as explained in Chapter 10.

sequences guides systematists to examine their phylogenetic characters and analyses for sources of error in inferring the detailed phylogenetic relationships among these species. All of the phylogenetic data support the hypothesis that these flightless birds are more closely related to each other than they are to any other living species.

The nested hierarchical structure of homology is so pervasive in the living world that it forms the basis for our systematic classification of all forms of life (genera grouped into families, families grouped into orders, and other categories). Hierarchical classification even preceded Darwin's theory because this pattern is so evident, but it was not explained scientifically before Darwin. Once the idea of common descent was accepted, biologists began investigating the structural, molecular, and chromosomal homologies of animal groups. Taken together, the nested hierarchical patterns uncovered by these studies have permitted us to reconstruct evolutionary trees of many groups and to continue investigating others. Use of Darwin's theory of common descent to reconstruct the evolutionary history of life and to classify animals is the subject of Chapter 10.

Note that the earlier evolutionary hypothesis that life arose many times, forming unbranched lineages, predicts linear sequences of evolutionary change with no nested hierarchy of homologies among species. Because we do observe nested hierarchies of homologies, that hypothesis is rejected. Note also that because the creationist argument of intelligent design is not a scientific hypothesis, it can make no testable predictions about any pattern of homology and therefore fails to meet the criteria of a scientific theory of animal diversity.

Characters of different organisms that perform similar functions are not necessarily homologous. The wings of bats and birds, although homologous as vertebrate forelimbs, are not homologous as wings. The most recent common ancestor of bats and birds had forelimbs, but the forelimbs were not in the form of wings. Wings of bats and birds evolved independently and have only superficial similarity in their flight structures.

Bat wings are formed by skin stretched over elongated digits, whereas bird wings are formed by feathers attached along the forelimb. Such functionally similar but nonhomologous structures are often called analogues.

Ontogeny, Phylogeny, and Recapitulation

Ontogeny is the development of an organism through its entire life, from its origin as a fertilized egg or bud throughout adulthood to death. Early developmental and embryological features contribute greatly to our knowledge of homology and common descent. Comparative studies of ontogeny show how evolutionary changes in developmental timing generate new body forms and structures.

The German zoologist Ernst Haeckel, a contemporary of Darwin, proposed that each successive stage in the development of an individual represented one of the adult forms that occurred in its evolutionary history. The human embryo with groovelike pharyngeal slits in the neck corresponded, for example, to the adult appearance of gills in a fishlike ancestor. On this basis Haeckel gave his

Ray-finned fish   Lizard   Bird   Human

Figure 6.17 Comparison of pharyngeal arches of different embryos. All are shown separated from the yolk sac. Note the remarkable similarity of the four embryos at this early stage in development.

generalization: *ontogeny (individual development) recapitulates (repeats) phylogeny (evolutionary descent).* This notion later became known simply as **recapitulation** or the **biogenetic law.** Haeckel based his biogenetic law on the flawed premise that evolutionary change occurs primarily by successively adding new features onto the end of an unaltered ancestral ontogeny while condensing the ancestral ontogeny into earlier developmental stages. This notion was based on Lamarck's concept of the inheritance of acquired characteristics (p. 100).

The nineteenth-century embryologist, K. E. von Baer, gave a more satisfactory explanation of the relationship between ontogeny and phylogeny. He argued that early developmental features were simply more widely shared among different animal groups than later ones were. Figure 6.17 shows, for example, the early embryological similarities of organisms whose adult forms are very different (see Figure 8.22, p. 175). The adults of animals with relatively short and simple ontogenies often resemble pre-adult stages of other animals whose ontogeny is more elaborate, but embryos of descendants do not necessarily resemble the adults of their ancestors. Even early development undergoes evolutionary divergence among lineages, however, and it is not as stable as von Baer proposed.

We now know many parallels between ontogeny and phylogeny, but features of an ancestral ontogeny can be shifted to either earlier or later stages in descendant ontogenies. Evolutionary change in timing of development is called **heterochrony,** a term initially used by Haeckel to denote exceptions to recapitulation. If a descendant's ontogeny extends beyond its ancestral one, new characteristics can be added late in development, beyond the point at which development would have ended in an evolutionary ancestor. Features observed in the ancestor often are moved to earlier stages of development in this process, and ontogeny therefore does recapitulate phylogeny to some degree. Ontogeny also can be shortened during evolution, however. Terminal stages of the ancestor's ontogeny can be deleted, causing adults of descendants to resemble pre-adult stages of their ancestors (Figure 6.18). This outcome reverses the parallel between ontogeny and phylogeny (reverse recapitulation) producing **paedomorphosis** (retention of ancestral juvenile characters by descendant adults). Because lengthening or shortening of ontogeny can change different parts of the body independently, we often see a mosaic of different kinds of developmental evolutionary change occurring concurrently. Therefore, cases in which an entire ontogeny recapitulates phylogeny are rare.

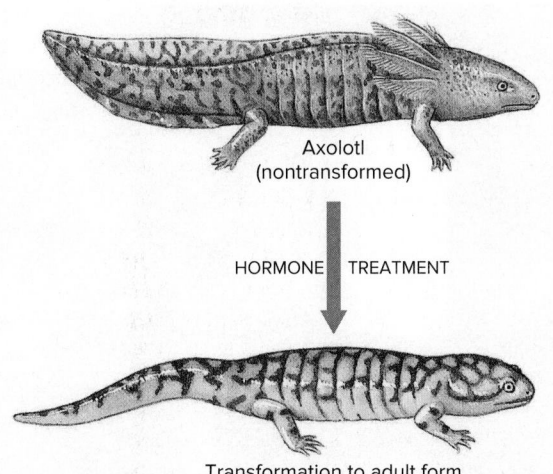

Figure 6.18 Aquatic and terrestrial forms of axolotls. Axolotls retain the juvenile, aquatic morphology (*top*) throughout their lives unless forced to metamorphose (*bottom*) by hormone treatment. Axolotls evolved from metamorphosing ancestors, an example of paedomorphosis.

Developmental Modularity and Evolvability

Evolutionary innovations occur not only by simple changes in rates of developmental processes but by changes in the physical location in the body where a process is activated. **Heterotopy** is the term traditionally used to describe a change in the physical location of a developmental process in an organism's body. To move the development of a structure from one part of the body to a different one, the developmental process must form a semiautonomous module that can be activated in new locations.

An interesting example of modularity and heterotopy occurs in some geckos. Geckos as a group typically have toepads, adhesive structures on the ventral side of the toes that permit climbing and clinging to smooth surfaces. Toepads consist of modified scales containing long protrusions, called setae, which can be molded to the surface of a substrate. A module responsible for toepad development is expressed in an unusual gecko species not only on the toes but also on the ventral side of the tip of the tail. This species has thereby acquired an additional adhesive appendage by ectopic expression of a standard developmental module.

Modularity is evident also in the homeotic mutations of the fruit fly, *Drosophila melanogaster.* Such mutations can substitute a developmental module for a leg in place of one normally specifying an antenna, thereby producing a fly with a pair of legs on its head. Another homeotic mutation in fruit flies transforms the balancer organs on the thorax into a second pair of wings; the balancer module is replaced by activation of the wing module, which in flies is activated normally only in a more anterior portion of the thorax.

Modularity is important in explaining some major evolutionary changes, such as evolution of tetrapod limbs (p. 544). The evolutionary transition from finlike limbs to the standard tetrapod limbs occurred by activating at the site of limb formation a set of homeobox genes (p. 167) whose expression pattern evolved initially as a module for forming part of the vertebral column. Patterns of gene expression shared by the vertebral column and the forelimbs and hindlimbs of tetrapods revealed the genetic and developmental mechanics of this module.

The term **evolvability** has been introduced to denote the great evolutionary opportunities created by having semiautonomous developmental modules whose expression can be moved from one part of the body to another. An evolving lineage that contains a large modular developmental toolkit can "experiment" with the construction of many new structures, some of which will persist and give rise to new homologies.

Multiplication of Species

A branch point on the evolutionary tree means that an ancestral species has split into two different ones. Darwin's theory postulates that genetic variation present within a species, especially variation that occurs between geographically separated populations, provides the material from which new species arise. Because evolution is a branching process, the total number of species produced by evolution increases through time, although most of these species eventually become extinct without leaving descendant species. A major challenge for evolutionists is to discover the process by which an ancestral species "branches" to form two or more descendant species.

This theory adds a spatial dimension to evolutionary processes. When populations of a species become isolated from each other by geographic barriers, the isolated populations undergo separate evolutionary change and diverge from each other. For example, when sea level was higher in the past than it is now, low areas of Cuba were inundated, dividing its land area into multiple isolates. Lizard populations that were formerly parts of a single species evolved species-level differences in isolation before another lowering of sea level reconsolidated Cuba as we know it today.

Before we explore multiplication of species, we must decide what we mean by "species." As explained in Chapter 10, no consensus exists regarding definition of species. Most biologists agree, however, that important criteria for recognizing species include (1) descent of all members from a common ancestral population forming a **lineage** of ancestor-descendant populations, (2) reproductive compatibility (ability to interbreed) within and reproductive incompatibility between species for sexually reproducing animals, and (3) maintenance within species of genotypic and phenotypic cohesion (lack of abrupt differences among populations in allelic frequencies and organismal characteristics). The criterion of reproductive compatibility has received the greatest attention in studies of species formation, also called **speciation.**

Biological features that prevent different species from interbreeding are called **reproductive barriers.** The primary problem of speciation is to discover how two initially compatible populations evolve reproductive barriers that cause them to become distinct, separately evolving lineages. How do populations diverge from each other in their reproductive properties while maintaining complete reproductive compatibility within each population?

Reproductive barriers between populations usually evolve gradually. Evolution of reproductive barriers requires that diverging populations must be kept physically separate for long periods of time. If diverging populations reunite before reproductive barriers have evolved, interbreeding occurs between the populations and they merge. Speciation by gradual divergence in animals may require extraordinarily long periods of time, perhaps 10,000 to 100,000 years or more. Geographical isolation followed by gradual divergence is the most effective way for reproductive barriers to evolve, and many evolutionists consider geographical separation a prerequisite for branching speciation.

> Geographical barriers between populations are not the same thing as reproductive barriers. Geographical barriers refer to spatial separation of two populations. They prevent gene exchange and are usually a precondition for speciation. Reproductive barriers result from evolution and refer to various behavioral, physical, physiological, and ecological factors that prevent interbreeding between different species. Behavioral barriers often evolve faster than other kinds of reproductive barriers. Geographical barriers do not guarantee that reproductive barriers will evolve. Reproductive barriers are most likely to evolve under conditions that include a generation of small population size, a favorable combination of selective factors, and long periods of geographical isolation. One or both of a pair of geographically isolated populations may become extinct prior to evolution of reproductive barriers between them. Over the vast span of geological time, however, conditions sufficient for speciation have occurred millions of times.

Allopatric Speciation

Allopatric ("in another land") populations of a species are those that occupy separate geographical areas. Because of their geographical separation, they cannot interbreed, but would be expected to do so if the geographic barriers between them were removed. If populations are allopatric immediately preceding and during evolution of reproductive barriers between them, the resulting speciation is called **allopatric speciation** or geographic speciation. The separated populations evolve independently and adapt to their respective environments, generating reproductive barriers between them as a result of their separate evolutionary paths. Because their genetic variation arises and changes independently, physically separated populations will diverge genetically even if their environments remain very similar. Environmental change between populations also can promote genetic divergence between them by favoring different phenotypes in the separated populations. Ernst Mayr (Figure 6.19) has contributed greatly to our knowledge of allopatric speciation through his studies of speciation in birds.

Allopatric speciation begins when a species splits into two or more geographically separated populations. This splitting can happen in either of two ways: by **vicariant speciation** or by a **founder event.** Vicariant speciation is initiated when climatic or geological changes fragment a species' habitat, producing impenetrable barriers that separate different populations geographically. For example, a mammalian species inhabiting a lowland forest could be divided by uplifting of a mountain barrier, sinking and flooding of a geological fault, or climatic changes that cause prairie or desert conditions to encroach on the forest. Formation of the isthmus of Panama separated populations of the sea urchin genus *Eucidaris* into separate Caribbean and Pacific Ocean isolates, leading to formation of the pair of species shown in Figure 1.5D (p. 6).

Vicariant speciation has two important consequences. Although the ancestral population is fragmented, individual fragments are

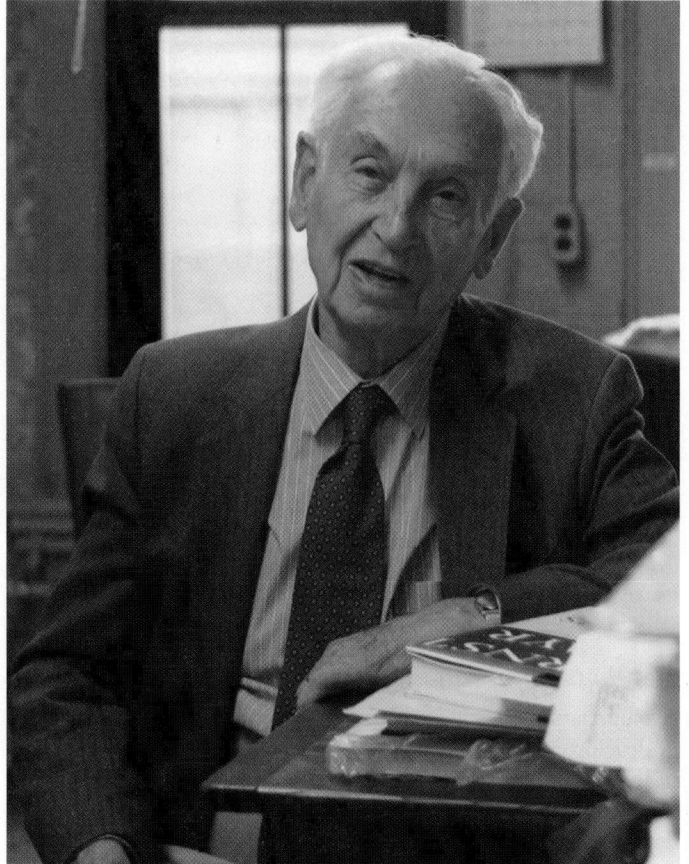

Figure 6.19 Ernst Mayr (1904 to 2005), a major contributor to our knowledge of speciation and of evolution in general.

usually left fairly intact. The vicariant process itself does not induce genetic change by reducing populations to a small size or by transporting them to unfamiliar environments. Another important consequence is that the same vicariant events may fragment several different species simultaneously. For example, fragmentation of a lowland forest most likely would disrupt species from numerous and diverse taxonomic groups, including salamanders, frogs, snails, and many other forest dwellers. Indeed, the same geographic patterns are observed among closely related species in different groups of organisms whose habitats are similar. Such patterns provide strong evidence for vicariant speciation.

An alternative means of initiating allopatric speciation is for a small number of individuals to disperse to a distant place where no other members of their species occur. The dispersing individuals may establish a new population in what is called a founder event. Allopatric speciation caused by founder events has been observed, for example, in the native fruit flies of Hawaii. Hawaii contains numerous patches of forest separated by volcanic lava flows. On rare occasions, strong winds can transport a few flies from one forest to another, geographically isolated forest where the flies are able to start a new population. Sometimes, a single fertilized female may found a new population. Unlike what happens in vicariant speciation, the new population initially has a very small size, which can cause its genetic structure to change quickly and dramatically from that of its ancestral population (see p. 123). When this event happens, phenotypic characteristics that were stable in the ancestral population often reveal unprecedented variation in the new population. As the newly expressed variation is sorted by natural selection, large changes in phenotype and reproductive properties occur, hastening evolution of reproductive barriers between the ancestral and newly founded populations.

> The term *founder event* in its most general usage denotes a dispersal of organisms from an ancestral population across a geographic barrier to start a new, allopatric population. Origins of Galápagos finches by South American emigrants provide good examples (p. 117). A founder event does not always directly cause important changes in the genetic constitution of the new population relative to the old one, although such changes are expected when the number of founding individuals is very small (less than 5 to 10 individuals, for example) and the ancestral population has a large amount of genetic variation. A change in the genetic constitution of a newly founded population because of a small number of founders is termed a *founder effect,* which includes a population bottleneck (p. 125). If a founder effect is so profound that selection acts in new ways on reproductively important characters, the founder event can induce speciation. *Founder-induced speciation* describes the subset of founder events in which a founder effect hastens species-level divergence of the newly founded population. Speciation in Hawaiian *Drosophila* fruit flies as described in the text illustrates founder-induced speciation. Excluded from founder-induced speciation are founder events whose role in speciation is strictly to establish a new allopatric population capable of independent evolutionary change.

Surprisingly, we often learn most about the genetics of allopatric speciation from cases in which formerly separated populations regain geographic contact following evolution of incipient reproductive barriers that are not absolute. The occurrence of mating between divergent populations is called **hybridization** and offspring of these matings are called **hybrids** (Figure 6.20). By studying the genetics of hybrid populations, we can identify the genetic bases of reproductive barriers.

Biologists often distinguish between reproductive barriers that impair fertilization (premating barriers) and those that impair growth and development, survival, or reproduction of hybrid individuals (postmating barriers). Premating barriers cause members of divergent populations either not to recognize each other as potential mates or not to complete the mating ritual successfully. Details of the horn structures of African antelopes (see Figure 6.11) are important for recognizing members of the same species as potential mates. In some other cases, female and male genitalia of the different populations may be incompatible or the gametes may be incapable of fusing to form a zygote. In others, premating barriers may be strictly behavioral, with members of different species being otherwise nearly identical in phenotype. Different species that are indistinguishable in organismal appearance are called **sibling species.** Sibling species arise when allopatric populations diverge in the seasonal timing of reproduction or in auditory, behavioral, or chemical signals required for mating. Evolutionary divergence in these features can produce effective premating barriers without obvious changes in organismal

A   **B**   **C**

Figure 6.20 Pure and hybrid salamanders. Hybrids are intermediate in appearance between parental populations. **A,** Pure white-spotted *Plethodon teyahalee;* **B,** a hybrid between white-spotted *P. teyahalee* and red-legged *P. shermani,* intermediate in appearance for both spotting and leg color; **C,** pure red-legged *P. shermani.*

appearance. Sibling species occur in groups as diverse as ciliates, flies, and salamanders.

Nonallopatric Speciation

Can speciation ever occur without prior geographic separation of populations? Allopatric speciation may seem an unlikely explanation for situations where many closely related species occur together in restricted areas that have no traces of physical barriers to animal dispersal. For example, several large lakes around the world contain very large numbers of closely related species of fish. The great lakes of Africa (Lake Malawi, Lake Tanganyika, and Lake Victoria) each contain many species of cichlid fishes that are found nowhere else. Likewise, Lake Baikal in Siberia contains many different species of sculpins that occur nowhere else in the world (Figure 6.21). It is difficult to conclude that these species arose anywhere other than in the lakes they inhabit, and yet those lakes are young on an evolutionary timescale and have no obvious environmental barriers that would fragment fish populations.

To explain speciation of fish in freshwater lakes and other examples like these, **sympatric** ("same land") **speciation** has been hypothesized. According to this hypothesis, different individuals within a species become specialized for occupying different components of the environment. By seeking and using very specific habitats in a single geographic area, different populations achieve sufficient physical and adaptive separation to evolve reproductive barriers. For example, cichlid species of African lakes are very different from each other in their feeding specializations. In many parasitic organisms, particularly parasitic insects, different populations may use different host species, thereby providing the physical separation necessary for reproductive barriers to evolve. Supposed cases of sympatric speciation have been criticized, however, because the reproductive distinctness of the different populations often is not well demonstrated, so that we may not be observing formation of distinct evolutionary lineages that will become different species. Furthermore, climatic cycles are likely to produce geographically

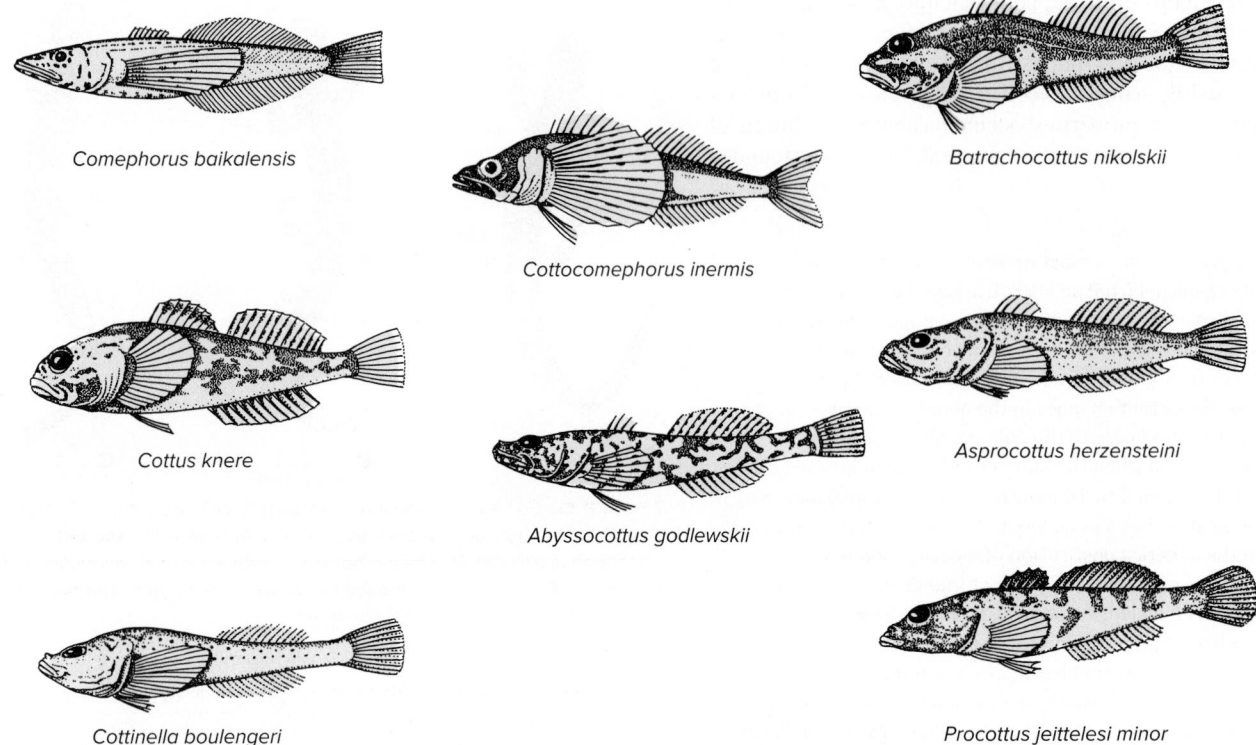

Comephorus baikalensis

Cottocomephorus inermis

Batrachocottus nikolskii

Cottus knere

Abyssocottus godlewskii

Asprocottus herzensteini

Cottinella boulengeri

Procottus jeittelesi minor

Figure 6.21   The sculpins of Lake Baikal, products of speciation that occurred within a single lake.

isolated lakes in dry periods, followed by consolidation into larger lakes during wet periods, thus producing an allopatric model for evolution of fish species flocks.

Another possible mode of speciation, termed **parapatric speciation,** is geographically intermediate between allopatric and sympatric speciation. Two species are called parapatric with respect to each other if their geographic ranges are primarily allopatric but make contact along a borderline that neither species successfully crosses. In parapatric speciation, a geographically continuous ancestral species evolves within its range a borderline across which populations evolve species-level differences while maintaining geographic contact along the border.

The simplest model of parapatric speciation is one in which a change in environmental conditions splits the geographic range of a species into two environmentally distinct but geographically adjoining parts. An increase in temperature on a Caribbean island, for example, might cause part of a wet tropical forest to become a dry, sandy one. A lizard species occupying the formerly wet forest might become divided into geographically adjacent wet and dry forest populations. Unlike vicariant allopatric speciation, however, the populations in the different habitat types are not isolated by a physical barrier but maintain genetic interactions along the geographic borderline between the different habitat types. The disparity in environmental conditions across the border, nonetheless, causes the populations to evolve as separate lineages adapted to different environments despite some gene exchange between them.

A parapatric distribution of species does not necessarily imply that speciation occurred parapatrically. Most cases of parapatrically distributed species show evidence of past allopatry, with subsequent removal of a geographic barrier permitting the two species to make geographic contact but with each species excluding the other one from its geographic territory.

The prevalence of parapatric speciation is controversial. This model of speciation predicts that parapatrically distributed populations differ mainly in adaptive features associated with the observed environmental differences, but show relative homogeneity for other genetic variation. Comparisons of parapatrically distributed populations, including those of lizards occupying different forest types on Caribbean islands, often show extensive divergence for molecular variation unrelated to adaptive differentiation of the populations; such results are better explained by vicariant allopatric speciation than by parapatric speciation. In some cases, geological evidence shows that what is now a single island was physically fragmented into separate islands during warm periods when the sea level was higher than it is now; such evidence likewise favors the interpretation of allopatric speciation for parapatrically distributed species whose geographic contact occurs in formerly inundated areas.

Adaptive Radiation

Evolution of several ecologically diverse species from a common ancestral species is called **adaptive radiation,** especially when many disparate species arise within a short interval of geological time (a few million years). Some of our best examples of adaptive radiation are associated with lakes and young islands, which provide new evolutionary opportunities for aquatic and terrestrial organisms, respectively. Oceanic islands formed by volcanoes are initially devoid of life. They are gradually colonized by plants and animals from a continent or from other islands in separate founder events. These colonizations involve floating of seeds across an expanse of ocean followed by germination on the island. Birds or insects might be carried by wind from a continent to the island, or animals might be carried by flotation in sea-swept debris. The probability that a particular species will make a successful colonization is low, but given large numbers of continental species and millions of years of evolutionary time, such colonizations occur and eventually establish an oceanic island biota. Archipelagoes, such as the Galápagos Islands, greatly increase opportunities for both founder events and ecological diversification. The entire archipelago is isolated from the continent and each island is geographically isolated from the others by sea; moreover, each island is different from every other one in its physical, climatic, and biotic characteristics.

Galápagos finches illustrate adaptive radiation on an oceanic archipelago (Figures 6.22 and 6.23). Galápagos finches (the name "Darwin's finches" was popularized in the 1940s by the British ornithologist David Lack) are closely related to each other, but each of the contrasting ecological forms differs from others in size and shape of the beak and in feeding habits. If the finches were specially created, it would require the strangest kind of coincidence for 13 similar kinds of finches to be created on the Galápagos Islands and nowhere else. Darwin's finches descended from a single ancestral population that arrived from the mainland and subsequently colonized the different islands of the Galápagos archipelago. The finches underwent adaptive radiation, occupying habitats that on the mainland were denied to them by other species better able to exploit those habitats. Galápagos finches thus assumed characteristics of mainland birds as diverse and unfinchlike as warblers and woodpeckers. A fourteenth Darwin's finch, found on isolated Cocos Island far north of the Galápagos archipelago, is similar in appearance to the Galápagos finches and almost certainly descended from the same ancestral founder. Although long considered different species, the contrasting ecological and morphological forms of Galápagos finches do not form distinct evolutionary lineages in molecular genetic studies. Their adaptive radiation is perhaps even more impressive knowing that it evolved as varieties within a species rather than by multiplication of species.

Gradualism

Darwin's theory of gradualism opposes arguments for the sudden origin of species. Small differences, resembling those that we observe among organisms within populations today, are the raw material from which the different major forms of life evolved. This theory shares with Lyell's uniformitarianism the notion that we must not explain past changes by invoking unusual catastrophic events that are not observed today. If new species originated in single, catastrophic events, we should be able to see such events happening today and we do not. Instead, what we usually observe in natural populations are small, continuous changes in phenotypes. Such continuous changes can produce major differences among species only by accumulating over many thousands to millions of years. A simple statement of Darwin's theory of gradualism is that accumulation of quantitative changes leads to qualitative change.

Mayr (see Figure 6.19) makes an important distinction between populational gradualism and phenotypic gradualism. **Populational gradualism** states that new traits become established in a population by increasing their frequency initially from a small fraction of the population to a majority of the population. Populational gradualism is not controversial. **Phenotypic gradualism** states that new traits, even those that are strikingly different from ancestral ones, are produced in a series of small, incremental steps accumulating over hundreds to thousands of generations.

Phenotypic Gradualism

Phenotypic gradualism was controversial when Darwin first proposed it, and it is still controversial. Not all phenotypic changes are small, incremental ones. Some mutations that appear during artificial breeding change the phenotype substantially in a single mutational step. Such mutations traditionally are called "sports."

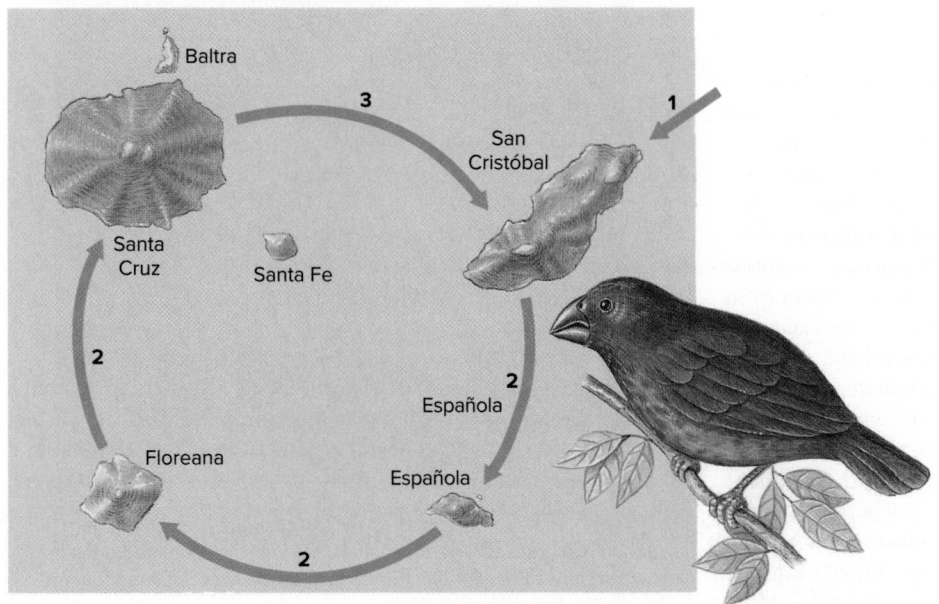

Figure 6.22 Model for evolution of the 13 Darwin's finches on the Galápagos Islands. The model postulates three steps: (1) immigrant finches from the South American mainland reach the Galápagos and colonize an island; (2) once the population becomes established, finches disperse to other islands where they adapt to new conditions and change genetically; and (3) after a period of isolation, secondary contact is established between different populations. The two populations are then recognized as separate species if they cannot interbreed successfully. Molecular genetic studies reveal that speciation has not reached completion in Darwin's finches, and that the contrasting ecological forms occur within a genetically continuous species.

Warbler finch

Cactus ground finch

Woodpecker finch

Sharp-beaked ground finch

Small insectivorous tree finch

Small ground finch

Warbler finch

Probing bills

Cactus eaters

Ground finches

Large insectivorous tree finch

Tree finches

Grasping bills

Insect eaters

Crushing bills

Medium ground finch

Seed eaters

Parrotlike bill

Fruit eaters

Vegetarian tree finch

Large ground finch

A

Figure 6.23 **A,** Adaptive radiation in 10 contrasting forms of Darwin's finches from Santa Cruz, one of the Galápagos Islands. Differences in bills and feeding habits are shown. All apparently descend from a single common ancestral finch from the South American continent. **B,** Woodpecker finch using a slender twig as a tool for feeding. This finch worked for about 15 minutes before spearing a wood roach and removing it from a break in the tree.

© Cleveland P. Hickman, Jr.

B

Sports that produce dwarfing are observed in many species, including humans, dogs, and sheep, and have been used by animal breeders to achieve desired results; for example, a sport that deforms the limbs was used to produce Ancon sheep, which cannot jump hedges and are therefore easily contained (Figure 6.24). Many colleagues of Darwin who accepted his other theories considered phenotypic gradualism too extreme. If sporting mutations can be used in animal breeding, why must we exclude them from our evolutionary theory? In favor of gradualism, Darwin and others have replied that sporting mutations always have negative side-effects that would cause selection to eliminate them from natural populations. Indeed, it is questionable whether Ancon sheep, despite their attractiveness to farmers, would propagate successfully in the presence of their long-legged relatives without human intervention. A mutation of large effect appears responsible, however, for an adaptive bill size polymorphism in an African finch species (*Pyrenestes ostrinus*) in which large-billed forms eat hard seeds and small-billed forms eat softer ones. Recent work in

Figure 6.24 The Ancon breed of sheep arose from a "sporting mutation" that caused dwarfing of legs. Many of his contemporaries criticized Darwin for his claim that such mutations are not important for evolution by natural selection.

© Herbert Gehr/Time Life Pictures/Getty Images

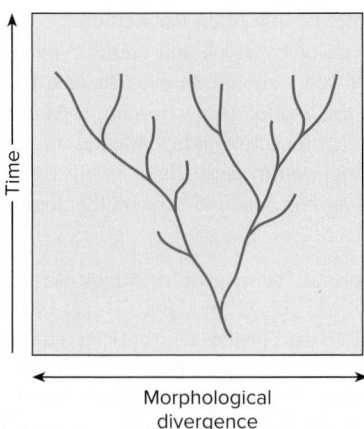

Figure 6.25 The phyletic gradualist model of evolutionary change in morphology, viewed as proceeding more or less steadily through geological time (millions of years). Bifurcations followed by gradual divergence lead to speciation. Note that most morphological change accumulates incrementally within species lineages between branch points, which are not accompanied by unusually large amounts of morphological change.

Figure 6.26 The punctuated equilibrium model sees morphological evolutionary change being concentrated in relatively rapid bursts of branching speciation (lateral lines) followed by prolonged periods of no cumulative change throughout geological time (millions of years).

evolutionary developmental genetics (p. 174) illustrates the continuing controversy surrounding phenotypic gradualism.

Punctuated Equilibrium

When we view Darwinian gradualism on a geological timescale, we may expect to find in the fossil record a long series of intermediate forms connecting the phenotypes of ancestral and descendant populations (Figure 6.25). This predicted pattern is called **phyletic gradualism.** Darwin recognized that phyletic gradualism is not often revealed by the fossil record. Studies conducted since Darwin's time generally have not revealed the continuous series of fossils predicted by phyletic gradualism. Is the theory of gradualism therefore refuted by the fossil record? Darwin and others claim that it is not, because the fossil record is too imperfect to preserve transitional series. Although evolution is a slow process by our standards, it is rapid relative to the rate at which good fossil deposits accumulate. Others have argued, however, that abrupt origins and extinctions of species in the fossil record force us to conclude that phyletic gradualism is rare.

American evolutionary paleontologists Niles Eldredge and Stephen Jay Gould proposed **punctuated equilibrium** in 1972 to explain the discontinuous evolutionary changes observed throughout geological time. Punctuated equilibrium states that phenotypic evolution is concentrated in relatively brief events of branching speciation, followed by much longer intervals of morphological evolutionary stasis (Figure 6.26). Speciation is an episodic event, having a duration of approximately 10,000 to 100,000 years. Because species may survive for 5 to 10 million years, the speciation event is a "geological instant," representing 1% or less of a species' lifespan. Ten thousand years is plenty of time, however, for Darwinian gradual evolution to accomplish dramatic changes. A small fraction of the evolutionary history of a group therefore contributes most of the morphological evolutionary change that we observe. Punctuated equilibrium contrasts with the views of paleontologist George Simpson, who attributed only moderate rates of morphological evolution to branching speciation and expected most morphological change to accumulate gradually in the "phyletic" mode between events of branching speciation.

Founder-induced allopatric speciation provides a possible explanation for punctuated equilibria. Remember that founder-induced speciation requires the breaking of genetic equilibrium in a small, geographically isolated population. Such small populations have very little chance of being preserved in the fossil record. After a new genetic equilibrium forms and stabilizes, the new population may increase in size, thereby increasing the chance that some of its members will form fossils. Founder-induced speciation cannot be the exclusive cause of punctuated equilibrium, however, because punctuated equilibrium characterizes groups where speciation by founder events is unlikely.

Probably the best-documented case of punctuated equilibrium is for ectoprocts (p. 322) of the genera *Metrarabdotos* and *Stylopoma* of the Caribbean Sea. These genera originated 15 to 25 million years ago and have an excellent fossil record as well as living species. Durations of morphologically static species range from 2 to 16 million years, whereas events of branching speciation can be narrowed to no more than 0.86 million years and often less than 0.16 million years in the fossil record. Molecular genetic studies of living populations confirm that the morphological data accurately diagnose biological species. Morphological change is episodic and concentrated in events of branching speciation. Interestingly, rates of morphological divergence during speciation are no greater than would be expected by genetic drift (p. 123). The persistent morphological stability of species is the phenomenon that challenges evolutionary explanations. Evolution by natural selection appears to have created effective mechanisms for maintaining morphological developmental stability within these species of ectoprocts.

Natural Selection

Natural selection is the major process by which evolution occurs in Darwin's theory of evolution. It gives us a natural explanation for the origins of **adaptation,** including all developmental, behavioral, anatomical, and physiological attributes that enhance an organism's ability to use environmental resources to survive and to reproduce. Evolution of color patterns that conceal moths from predators (see Figure 1.11, p. 11), and of bills adapted to different modes of feeding in finches (see Figure 6.23), illustrate natural selection leading to adaptation. Darwin developed his theory of natural selection as a series of five observations and three inferences drawn from them:

Observation 1—Organisms have great potential fertility.

All populations produce large numbers of gametes and potentially large numbers of offspring each generation. Population size would increase exponentially at an enormous rate if all fertilized eggs that are produced each generation successfully completed their development to form reproductively active adult individuals. Darwin calculated that even in slow-breeding animals such as elephants, a single pair breeding from ages 30 to 90 and having only six young could produce 19 million descendants in 750 years.

Observation 2—Natural populations normally remain constant in size, except for minor fluctuations.

Natural populations fluctuate in size across generations and sometimes go extinct, but no natural populations show the continued exponential growth that their reproductive biology theoretically could sustain.

Observation 3—Natural resources are limited.

Exponential growth of a natural population would require unlimited natural resources to provide food and habitat for the expanding population, but natural resources are finite.

Inference 1—A continuing *struggle for existence* exists among members of a population.

Survivors represent only a part, usually a very small part, of the individuals produced each generation. Darwin wrote in *On the Origin of Species* that "it is the doctrine of Malthus applied with manifold force to the whole animal and vegetable kingdoms." The struggle for food, shelter, and space becomes increasingly severe as overpopulation develops.

Observation 4—Populations show *variation* among organisms.

No two individuals are exactly alike. They differ in size, color, physiology, behavior, and many other ways.

Observation 5—Some variation is *heritable*.

Darwin noted that offspring tend to resemble their parents, although he did not understand how. The hereditary mechanism discovered by Gregor Mendel would be applied to Darwin's theory many years later.

Inference 2—Varying organisms show *differential survival and reproduction* favoring advantageous traits (= natural selection).

Survival in the struggle for existence is not random with respect to contrasting hereditary traits present in the population. Some traits give their possessors an advantage in using the environment for effective survival and reproduction. Survivors transmit their favored traits to offspring, thereby causing those traits to accumulate in the population.

Inference 3—Over many generations, natural selection generates new adaptations and new species.

The preferential propagation of favorable traits across generations gradually transforms species and causes long-term "improvement" of populations.

Darwin knew that people often use hereditary variation to produce useful new breeds of livestock and plants. *Natural* selection acting over millions of years should be even more effective in producing new types than the *artificial* selection imposed during a human lifetime. Natural selection acting independently on geographically separated populations would cause them to diverge from each other, thereby generating reproductive barriers that lead to speciation.

> The popular phrase "survival of the fittest" was not originated by Darwin but was coined a few years earlier by the British philosopher Herbert Spencer, who anticipated some of Darwin's principles of evolution. Unfortunately the phrase later came to be coupled with unbridled aggression and violence in a bloody, competitive world. In fact, natural selection operates through many other characteristics of living organisms. The fittest animal may be one that enhances the living conditions of its population. Fighting prowess is only one of several means toward survival and reproductive advantage.

Natural selection may be considered a two-step process with a random component and a nonrandom component. Production of variation among organisms is the random component. The mutational process does not preferentially generate traits that are favorable to the organism; new variants are probably more likely to be unfavorable. The nonrandom component is the survival of different traits. This differential survival is determined by the effectiveness of different traits in permitting their possessors to use environmental resources to survive and to reproduce. Differential survival and reproduction among varying organisms is now called **sorting** and should not be equated with natural selection. We now know that even random processes (genetic drift, p. 123) can produce sorting among varying organisms. When selection operates, sorting occurs *because certain traits give their possessors advantages in survival and reproduction* relative to others that lack those traits. Selection is therefore a specific cause of sorting.

Darwin's theory of natural selection has been challenged repeatedly. One challenge claims that directed (nonrandom) variation governs evolutionary change. In the decades around 1900, diverse evolutionary hypotheses collectively called **orthogenesis** proposed that variation has momentum that forces a lineage to evolve in a particular direction that is not always adaptive. The extinct Irish elk was a popular example of orthogenesis. Newly produced variation was considered biased toward enlarging their antlers, thereby generating an evolutionary momentum for producing larger antlers. Natural selection was considered ineffective at stopping the antlers eventually from becoming so large and cumbersome that they forced the Irish elk into extinction (Figure 6.27). Orthogenesis explained apparently nonadaptive evolutionary trends that supposedly forced species into decline. Because extinction is the expected evolutionary fate of most species, disappearance of the Irish elk is not extraordinary and probably not related to large antlers. Subsequent genetic research on the nature of variation clearly has rejected the genetic predictions of orthogenesis.

Another recurring criticism of natural selection is that it cannot generate new structures or species but can only modify old ones. Most structures in their early evolutionary stages could not have performed the biological roles that the fully formed structures perform, and it is therefore unclear how natural selection could have favored

Darwin's Explanatory Model of Evolution by Natural Selection

Observation 1
Organisms have great potential fertility, which permits exponential growth of populations.
(*Source*: Thomas Malthus)

Observation 2
Natural populations normally do not increase exponentially but remain fairly constant in size.
(*Source*: Charles Darwin and many others)

Observation 3
Natural resources are limited.
(*Source*: Thomas Malthus)

Inference 1
A struggle for existence occurs among organisms in a population.
(*Source*: Thomas Malthus)

Observation 4
Variation occurs among organisms within populations.
(*Source*: Animal breeding and systematics)

Observation 5
Variation is heritable.
(*Source*: Animal breeding)

Inference 2
Varying organisms show differential survival and reproduction, favoring advantageous traits (= natural selection).
(*Source*: Charles Darwin)

Inference 3
Natural selection, acting over many generations, gradually produces new adaptations and new species.
(*Source*: Charles Darwin)

© Biosphoto/Superstock

Figure 6.27 Irish elk, a fossil species that once was used to support the orthogenetic idea that momentum in variation caused the antlers to become so large that the species was forced into extinction.

them. What use is half a wing or the rudiment of a feather for a flying bird? To answer this criticism, we propose that many structures evolved initially for purposes different from the ones they have today. Rudimentary feathers would have been useful in thermoregulation, for example. The feathers later became useful for flying after they incidentally acquired aerodynamic properties. Natural selection then could act to improve the usefulness of feathers for flying. **Exaptation** denotes the utility of a structure for a biological role that was not part of the structure's evolutionary origin. Exaptation contrasts with adaptation, which implies that a structure arose by natural selection for a particular biological role. Bird feathers are therefore adaptations for thermoregulation but exaptations for flight. Because structural changes that separate members of different species are similar in kind to variation that we observe within species, it is reasonable to propose that selection can produce new species.

6.3 REVISIONS OF DARWIN'S THEORY

Neo-Darwinism

The most serious weakness in Darwin's theory was his failure to identify correctly the mechanism of inheritance. Darwin saw heredity as a blending phenomenon in which the hereditary factors of parents melded together in their offspring. Darwin also invoked the Lamarckian hypothesis that an organism could alter its heredity through use and disuse of body parts and through the direct influence of the environment. German developmental biologist August Weismann rejected Lamarckian inheritance in the late 1800s by showing experimentally that modifications of an organism during its lifetime do not change its heredity (see Chapter 5), and he revised Darwin's theory accordingly. We now use the term **neo-Darwinism** to denote Darwin's theory as revised by Weismann.

Mendelian genetics eventually clarified the particulate inheritance that Darwin's theory of natural selection required (p. 76). Ironically, when Mendel's work was rediscovered in 1900, it was considered antagonistic to Darwin's theory of natural selection. When mutations were discovered in the early 1900s, most geneticists thought that they produced new species in single large steps. These geneticists relegated natural selection to the role of executioner, a negative force that merely eliminated the obviously unfit.

Emergence of Modern Darwinism: The Synthetic Theory

In the 1930s a new generation of geneticists began to reevaluate Darwin's theory from a mathematical perspective. These were population geneticists, scientists who studied variation in natural populations using statistics and mathematical models. Gradually, a new comprehensive theory emerged that brought together population genetics, paleontology, biogeography, embryology, systematics, and animal behavior in a Darwinian framework.

6.4 MICROEVOLUTION: GENETIC VARIATION AND CHANGE WITHIN SPECIES

Microevolution is the study of genetic change occurring within natural populations. Occurrence of different allelic forms of a gene in a population is called **polymorphism.** All alleles of all genes possessed by members of a population collectively form the **gene pool** of that population. The amount of polymorphism present in large populations is potentially enormous, because at observed mutation rates, many different alleles are expected for all genes.

Population geneticists study polymorphism by identifying the different allelic forms of a gene present in a population and then measuring the relative frequencies of the different alleles in the population. The relative frequency of a particular allelic form of a gene in a population is called its **allelic frequency.** For example, in the human population, there are three different allelic forms of the gene encoding the ABO blood types (p. 775). Using the symbol I to denote the gene encoding the ABO blood types, I^A and I^B denote genetically codominant alleles encoding blood types A and B, respectively. Allele i is a recessive allele encoding blood group O. Therefore genotypes $I^A I^A$ and $I^A i$ produce type A blood, genotypes $I^B I^B$ and $I^B i$ produce type B blood, genotype $I^A I^B$ produces type AB blood, and genotype ii produces type O blood. Because each individual contains two copies of this gene, the total number of copies present in the population is twice the number of individuals. What fraction of this total is represented by each of the three different allelic forms? In France, we find the following allelic frequencies: I^A =.46, I^B =.14, and i =.40. In Russia, the corresponding allelic frequencies differ (I^A =.38, I^B =.28, and i =.34), demonstrating microevolutionary divergence between these populations (Figure 6.28). Although alleles I^A and I^B are dominant to i, i is nearly as frequent as I^A and exceeds the frequency of I^B in both populations. Dominance describes the *phenotypic effect* of an allele in heterozygous individuals, not its relative abundance in a population of individuals. In many human populations, genetically recessive traits, including the O blood type, blond hair, and blue eyes, are very common. We demonstrate that Mendelian inheritance and dominance do not alter allelic frequencies directly or produce evolutionary change in a population.

Genetic Equilibrium

A mathematical theorem called **Hardy-Weinberg equilibrium** (see box, p. 124) permits us to estimate the relationship between the frequencies of alleles at a genetic locus and the frequencies of genotypes formed by random fertilization of gametes with respect to this variation. The frequency of an allele in the gametes produced by a population equals its frequency in the gene pool as defined in the preceding paragraph. Random mating consists mathematically of drawing pairs of gametes at random from the gene pool, fertilizing each random pair, and then quantifying the frequencies of the resulting diploid genotypes. We can estimate from Hardy-Weinberg equilibrium the ratios of genotypes and phenotypes that should occur in the absence of evolution in the following generation.

Consider a human population that contains a dominant allele for normal pigmentation (A) and a recessive allele for albinism (a), which is rare in the gene pool. Figure 6.29 shows the expected frequency of the dominant phenotype in a population as a function of the frequency of the dominant allele in the gene pool. Note that when an allele is rare, copies of that allele occur almost entirely in heterozygous genotypes (Aa in our example), which express the dominant allele phenotypically; only a tiny fraction of the copies of a rare allele occur in homozygous form (aa). The frequency of the recessive phenotype in the population is thus much less than the frequency of the recessive allele in the gene pool. The frequency of albinism in humans is approximately 1/20,000. Assuming that mating is random with respect to genotype at this locus, we calculate using Hardy-Weinberg equilibrium:

$$q^2 = 1/20,000$$
$$q = (1/20,000)^{1/2} = 1/141$$
$$p = 1 - q = 140/141$$

The frequency of carriers is

$$A/a = 2pq = 2 \times 140/141 \times 1/141 = 1/70$$

Figure 6.28 Frequencies of the blood-type B allele among humans in Europe. The allele is more common in the east and rarer in the west. The allele may have arisen in the east and gradually diffused westward through the genetic continuity of human populations. This allele has no known selective advantage; its changing frequency probably represents the effects of random genetic drift.

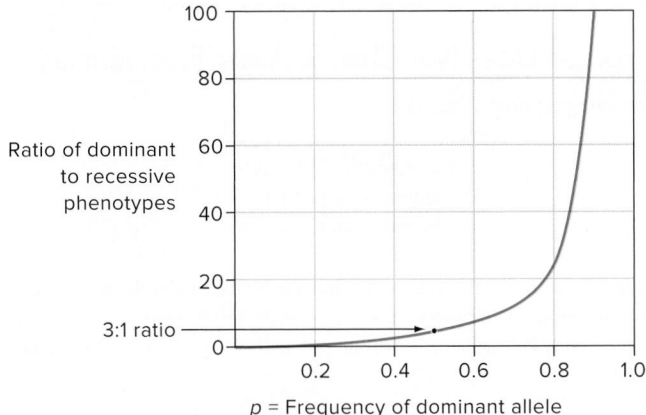

Figure 6.29   The ratio of dominant to recessive phenotypes in a randomly mating population (in Hardy-Weinberg equilibrium) plotted as a function of the frequency of the dominant allele. Frequency of the recessive allele equals $1 - p$. The 3:1 ratio of dominant to recessive phenotypes in the second generation of Mendel's crosses occurred because all individuals of the preceding generation were heterozygous; thus, $p = q = 0.5$, a condition not common in natural populations. As the dominant allele approaches a frequency of one, virtually no individuals express the recessive phenotype, thus permitting recessive lethal alleles to persist in a population at very low frequencies (0.001).

One person in every 70 is a carrier! Tay-Sachs disease in humans is associated with homozygosity for a recessive lethal allele; individuals homozygous for the lethal allele die early in childhood. Natural selection keeps recessive lethal alleles rare in the population, because individuals homozygous for such alleles never reproduce. Natural selection does not eliminate recessive lethal alleles from the population, however, because virtually all copies of those alleles occur in heterozygous genotypes, which are phenotypically normal. Mating is thus random with respect to whether individuals carry the lethal allele. For a recessive lethal allele present in 2 of every 100 persons (but homozygous in only 1 in 1000 fertilizations), 50 generations of selection are required to reduce the frequency of the allele to 1 of every 100 persons.

Nonrandom Mating

If mating is nonrandom, genotypic frequencies will deviate from Hardy-Weinberg expectations. For example, if two different alleles of a gene are equally frequent ($p = q = .5$), we expect half of the genotypes to be heterozygous ($2pq = 2[.5] [5] = .5$) and one-quarter to be homozygous for each of the respective alleles ($p^2 = q^2 = [.5]^2 = .25$). If we have **positive assortative mating,** individuals mate preferentially with others of the same genotype, such as albinos mating with other albinos. Matings among individuals homozygous for the same allele generate offspring that are homozygous like themselves. Matings among individuals heterozygous for the same pair of alleles produce on average 50% heterozygous offspring and 50% homozygous offspring (25% of each alternative type) each generation. Positive assortative mating increases the frequency of homozygous genotypes and decreases the frequency of heterozygous genotypes in a population but does not change allelic frequencies.

Preferential mating among close relatives also increases homozygosity and is called **inbreeding**. Whereas positive assortative mating usually affects one or a few traits, inbreeding simultaneously affects all variable traits. Strong inbreeding greatly increases chances that rare recessive alleles will become homozygous and be expressed.

Inbreeding has surfaced as a serious problem in zoos holding small populations of rare mammals. Matings of close relatives tend to bring together genes from a common ancestor and increase the probability that two copies of a deleterious gene will come together in the same organism. The result is "inbreeding depression." Our management solution is to enlarge genetic diversity by bringing together captive animals from different zoos or by introducing new stock from wild populations if possible. Paradoxically, where zoo populations are extremely small and no wild stock can be obtained, deliberate inbreeding is recommended. This procedure selects for genes that tolerate inbreeding; deleterious genes disappear if they kill animals homozygous for them.

Forces of Evolutionary Change

Population geneticists measure evolutionary change as a change in the frequency of an allele in the gene pool of a population. "Forces" of evolutionary change are factors that can alter the frequency of an allele in the gene pool, and include (1) recurring mutation, (2) genetic drift, (3) migration, (4) natural selection, and (5) interactions among these factors. Recurring mutation is the ultimate source of variability in all populations, but it usually requires interaction with one or more of the other factors to cause a noteworthy change in allelic frequencies. For diploid individuals, a new mutation occurs in one copy in heterozygous condition in a single individual. Its frequency is 1/2N where N is the number of individuals in the population. The total change in allelic frequencies caused by a single mutation is vanishingly small unless the new allele increases in frequency by one of the other forces of evolutionary change. We consider these other forces individually.

Genetic Drift

Some species, such as cheetahs (Figure 6.30), contain very little genetic variation, probably because their ancestral lineages passed through periods when the total number of individuals in the population was very small. A small population clearly cannot contain large amounts of genetic variation. Each individual organism has at most

Figure 6.30   Cheetahs, a species whose genetic variability has been depleted to very low levels because of small population size in the past.

© Jack Hollingsworth/Getty Images

Hardy-Weinberg Equilibrium: Why the Hereditary Process Does Not Change Allelic Frequencies

The Hardy-Weinberg law is a logical consequence of Mendel's first law of segregation and expresses the tendency toward equilibrium inherent in Mendelian heredity.

Let us select for our example a population having a single locus bearing just two alleles T and t. The phenotypic expression of this gene might be, for example, the ability to taste a chemical compound called phenylthiocarbamide. Individuals in the population will be of three genotypes for this locus, T/T, T/t (both tasters), and t/t (nontasters). In a sample of 100 individuals, let us suppose that we have 20 of T/T genotype, 40 of T/t genotype, and 40 of t/t genotype. We could then make a table showing the allelic frequencies (remember that every individual has two copies of the gene):

Genotype	Number of Individuals	Copies of the T Allele	Copies of the t Allele
T/T	20	40	0
T/t	40	40	40
t/t	40	0	80
Total	100	80	120

Of the 200 copies, the proportion of the T allele is $80/200 = 0.4$ (40%), and the proportion of the t allele is $120/200 = 0.6$ (60%). It is customary to use "p" and "q" to represent the two allelic frequencies. The genetically dominant allele is represented by p, and the genetically recessive by q. Thus:

$$P = \text{frequency of } T = 0.4$$
$$q = \text{frequency of } t = 0.6$$
$$\text{Therefore } p + q = 1$$

Having calculated allelic frequencies in the sample, let us determine whether these frequencies will change spontaneously in a new generation of the population. Assuming that mating is random (gametes are sampled independently in pairs), each individual contributes an equal number of gametes to the "common pool" from which the next generation is formed. Frequencies of gametes in the "pool" then equal the allelic frequencies in the sample: 40% of the gametes are T, and 60% are t (ratio of 0.4:0.6). Both ova and sperm, of course, show the same frequencies. The next generation is formed:

	Ova	
Sperm	$T = 0.4$	$t = 0.6$
$T = 0.4$	$T/T = 0.16$	$T/t = 0.24$
$t = 0.6$	$T/t = 0.24$	$t/t = 0.36$

Collecting genotypes, we have

$$\text{frequency of } T/T = 0.16$$
$$\text{requency of } T/t = 0.48$$
$$\text{frequency of } t/t = 0.36$$

Next, we determine the values of p and q from the randomly mated populations. From the table above, we see that the frequency of T is the sum of genotypes T/T, which is 0.16, and one-half of the genotype T/t, which is 0.24:

$$T(p) = 0.16 + .5(0.48) = 0.4$$

Similarly, the frequency of t is the sum of genotypes t/t, which is 0.36, and one-half the genotype T/t, which is 0.24:

$$t(p) = 0.36 + .5(0.48) = 0.6$$

The new generation bears exactly the same allelic frequencies as the parent population! Note that there has been no increase in the frequency of the genetically dominant allele T. Thus, *in a freely interbreeding, sexually reproducing population, the frequency of each allele would remain constant generation after generation in the absence of natural selection, migration, recurring mutation, and genetic drift* (see text). A mathematically minded reader should recognize that the genotype frequencies T/T, T/t, and t/t are actually a binomial expansion of $(p + q)^2$:

$$(p + q)^2 = p^2 + 2pq + q^2 = 1$$

Note that the equilibrium calculations give *expected* frequencies, which are unlikely to be realized exactly in a population of finite size. For this reason, finite population size is a cause of evolutionary change.

Most genes have more than just a single pair of alleles, especially when we measure genetic variation at the DNA sequence level. The binomial expansion shown above can be used for any number of alleles. Suppose that we have three alleles (T_1, T_2, T_3) whose frequencies are denoted p, q, and r, respectively. We now have six possible genotypes with the following Hardy-Weinberg equilibrium frequencies:

$$
\begin{array}{ccccccc}
 & T_1/T_1 & T_1/T_2 & T_2/T_2 & T_1/T_3 & T_2/T_3 & T_3/T_3 \\
(p + q + r)^2 = & p^2 & + 2pq & + q^2 & + 2pr & + 2qr & + r^2
\end{array}
$$

As the number of alleles at a gene increases, the proportion of the population having heterozygous genotypes also increases.

---

two different allelic forms of each gene, and a single breeding pair contains at most four different allelic forms of each gene. Suppose that we have such a breeding pair. We know from Mendelian genetics (see Chapter 5) that chance decides which of the different allelic forms of a gene gets passed to offspring. It is therefore possible by chance alone that one or two of the parental alleles in this example will not be passed to any offspring. It is highly unlikely that the different alleles present in a small ancestral population are all passed to

descendants without any change of allelic frequency. This chance fluctuation in allelic frequency from one generation to the next, including loss of alleles from the population, is called **genetic drift.**

Genetic drift occurs to some degree in all populations of finite size. Perfect constancy of allelic frequencies, as predicted by Hardy-Weinberg equilibrium, occurs only in populations whose size approaches infinity, and such populations occur only in mathematical models. All populations of animals are finite and therefore experience some effect of genetic drift, which becomes greater, on average, as population size declines.

Genetic drift erodes genetic variability of a population. If population size remains small for many generations in a row, genetic variation can be greatly depleted. Alleles are lost by chance at a much higher rate than new ones are introduced by mutation. This loss is harmful to a species' evolutionary success because it restricts potential genetic responses to environmental change. Indeed, biologists are concerned that cheetah populations may have insufficient variation for continued survival.

A large reduction in the size of a population that increases evolutionary change by genetic drift is commonly called a bottleneck. A bottleneck associated with the founding of a new geographic population is called a founder effect and may be associated with formation of a new species (p. 114).

Migration

Migration is the movement of individuals from one population to another one prior to mating. Movement of alleles between populations in this manner is called "gene flow." Migration is an evolutionary force only in species that are geographically subdivided into two or more separate populations; otherwise, the species evolves as a single undivided gene pool. Migration prevents different populations of a species from diverging. If a large species is divided into many small populations, genetic drift and selection acting separately in the different populations can produce evolutionary divergence among them. A small amount of migration in each generation keeps the different populations from becoming too distinct genetically. For example, the French and Russian populations whose ABO allele frequencies were discussed previously show some genetic divergence, but their genetic connection through intervening populations by continuing migration prevents them from becoming completely distinct.

Natural Selection

Natural selection can change both allelic frequencies and genotypic frequencies in a population. Although the effects of selection are often reported for particular polymorphic genes, we must stress that natural selection acts on the whole animal, not on isolated traits. An organism that possesses a superior combination of traits is favored. An animal may have traits that confer no advantage or even a disadvantage, but it is successful overall if its combination of traits is favorable. When we claim that a genotype at a particular gene has a higher **relative fitness** than others, we state that on average that genotype confers an advantage in survival and reproduction in the population. If alternative genotypes have unequal probabilities of survival and reproduction, Hardy-Weinberg equilibrium is upset.

Using the genetic theory of natural selection, one can measure relative **fitness** values associated with different genotypes in a population. Geneticists often use W to denote the expected average fitness of a genotype in a population, with the genotype of highest fitness given a value of one and fitnesses of other genotypes indicated as fractions.

We illustrate measurement of fitness using genetic variation associated with the disease sickle cell anemia in human populations. Considering only the alleles for normal hemoglobin (A) and

sickle cell hemoglobin (S) for the beta-hemoglobin gene in human populations (p. 95), the possible genotypes are AA, AS, and SS. Measurements of viability of individuals of these three genotypes in nonmalarial environments give a fitness value of 1 to genotypes AA and AS and a fitness of 0.2 to genotype SS. People having the SS genotype, who are susceptible to severe anemia, are expected to contribute only 20% as many offspring to the next generation on average as are individuals having the AA or AS genotypes. In malarial environments, genotype AS has the highest fitness (=1); genotype AA has a slightly decreased fitness (=0.9) because these individuals have a greater incidence of malaria than AS individuals, and SS has a low fitness (=0.2) because of anemia. From these measured fitness values and knowledge of the frequencies of alleles in a population and its system of mating, one can calculate the **average effect** that an allele has on the phenotype of relative fitness in that population. In the example of sickle cell anemia, the average effect of allele S on fitness in a malarial environment is a balance between the strongly negative effect it has when homozygous and the positive effect that it has when heterozygous with allele A.

In Chapter 36, we discuss the related concept of **inclusive fitness.** The average effect of an allele on fitness is expressed not only by its direct contribution to the fitness of its possessors but by aid that its possessors give to close relatives, who are likely also to contain copies of the allele. The term "inclusive fitness" pertains to cases where the average effect of an allele on its own propagation in the gene pool would be calculated incorrectly if its effects on survival of its possessors' relatives were ignored. For example, natural selection might favor an allele for a behavior through which an individual dies in saving the lives of many relatives.

Some traits and combinations of traits are advantageous for certain aspects of an organism's survival or reproduction and disadvantageous for others. Darwin used the term **sexual selection** to denote the selection of traits that are advantageous for obtaining mates but not for survival. Bright colors and elaborate feathers can enhance a male bird's competitive ability in obtaining mates while simultaneously increasing his visibility to predators (Figure 6.31). Environmental changes, such as extinction of a predator population, can alter the selective values of alternative traits. The action of selection on character variation is therefore very complex.

Interactions of Selection, Drift, and Migration

Subdivision of a species geographically into small populations that exchange migrants is an optimal situation for promoting rapid adaptive evolution of a species. Interaction of genetic drift and selection in different populations permits many different genetic combinations of many polymorphic genes to be tested against natural selection. Migration among populations permits particularly favorable new genetic combinations to spread throughout the species as a whole. Interaction of selection, genetic drift, and migration in this example produces evolutionary change qualitatively different from what would result if any of these three factors acted alone. Geneticist Sewall Wright called this interaction *shifting balance* because it permits a population to explore different adaptive combinations of variable traits. Natural selection, genetic drift, mutation, nonrandom mating, and migration interact in natural populations to create an enormous opportunity for

© Tim Zurowski/Getty Images

Figure 6.31 A pair of wood ducks. Brightly colored feathers of male birds probably confer no survival advantage and might even be harmful by alerting predators. Such colors nonetheless confer advantage in attracting mates, which overcomes, on average, the negative consequences of these colors for survival. Darwin used the term "sexual selection" to denote evolution of traits that give an individual an advantage in reproduction, even if the traits are neutral or harmful for survival.

evolutionary change; perpetual stability, as predicted by Hardy-Weinberg equilibrium, almost never occurs across any significant amount of evolutionary time.

The importance of interactions between natural selection and genetic drift in adaptive evolution is illustrated by the beta-hemoglobin variation discussed in the preceding section, Natural Selection. We noted that in malarial environments, natural selection retains both the *A* and *S* alleles in the population because individuals with the *AS* genotype uniquely combine the benefits of malarial resistance and normal respiration. A rare third allele called hemoglobin *C* also occurs in some malarial regions of western Africa. Like hemoglobin *S*, allele *C* was derived from allele A by a single mutation changing the sixth codon (p. 95); in allele *C*, lysine (AAG) substitutes for glutamic acid (GAG). In malarial areas, natural selection acts against allele *C* in heterozygous genotypes because *AC* individuals are susceptible to malaria (as are *AA* individuals), and *CS* individuals have severe anemia. Natural selection favors the *C* allele in homozygous form; *CC* individuals have malarial resistance, normal respiration, and much higher fitness than do *AS* individuals in malarial areas. If the population were fixed for the *C* allele, all individuals would benefit from both malarial resistance and normal respiration.

Why has the *C* allele not become the most frequent one in malarial regions of western Africa? We know from Hardy-Weinberg equilibrium that in a randomly mating population, a rare allele occurs almost exclusively in heterozygous genotypes with the more common alleles. Because selection favors *AS* individuals over *AC* and *CS* individuals in malarial areas, selection acts to eliminate allele *C* from those populations.

In a few local populations of western Africa, genetic drift caused the *C* allele to reach a relatively high frequency prior to the onset of malaria. In these populations alone, the *CC* genotype occurs with sufficient frequency that the positive action of natural selection on

these individuals exceeds the negative effects of selection on the *AC* and *CS* individuals in the same population. The *C* allele has increased in frequency by natural selection in these local populations. One expects that natural selection eventually would cause the *C* allele to become fixed in these local populations, and that gene flow from them to others would allow the *C* allele to become fixed across malarial Africa. One hopes, of course, that effective treatment of malaria will overcome the need for an evolutionary solution to this problem, as the latter would require many generations and much intervening illness. Nonetheless, this example shows how interactions between genetic drift, natural selection, and gene flow can shift a population's response to malarial selection from maintaining a polymorphism of the *A* and *S* alleles to fixing the *C* allele.

Measuring Genetic Variation Within Populations

How do we measure the genetic variation that occurs in natural populations? Genetic dominance, interactions between alleles of different genes, and environmental effects on a phenotype make it difficult to quantify genetic variation indirectly by observing organismal phenotypes. Variability can be quantified, however, at the molecular level.

Beginning in the 1960s, studies of protein variation provided the first unambiguous evidence that animal populations typically contain large amounts of genetic variation. Studies of protein polymorphism are now largely superseded by studies of variation in DNA sequences sampled from both the nuclear and mitochondrial genomes. DNA studies reveal even larger amounts of variation than did studies of proteins. We focus here on protein variation, for both its historical importance and its more straightforward interpretation using the principles of Hardy-Weinberg equilibrium (p. 124).

Protein Polymorphism

Different allelic forms of genes encode proteins that often differ slightly in their amino acid sequence. This phenomenon is called **protein polymorphism.** If these differences affect the protein's net electric charge, the different allelic forms can be separated in a laboratory experiment using protein electrophoresis (Figure 6.32). We can identify the genotypes of particular individuals for protein-coding genes and measure allelic frequencies in a population.

Over the last 45 years, geneticists using this approach have discovered far more variation than was previously expected. Despite the high levels of polymorphism discovered using protein electrophoresis (Table 6.1), these studies underestimate both protein polymorphism and the total genetic variation present in a population. For example, protein polymorphism that does not involve charge differences is not detected. Furthermore, because the genetic code is degenerate (more than one codon for most amino acids, p. 89), protein polymorphism does not reveal all of the genetic variation present in protein-coding genes. Genetic changes that do not alter protein structure sometimes alter patterns of protein synthesis during development and can be very important to an organism. When all kinds of variation are considered, it is evident

Figure 6.32 Study of genetic variation in proteins using gel electrophoresis. **A,** An electrophoretic apparatus separates allelic variants of proteins that differ in charge because of differences in their sequence of amino acids. **B,** Genetic variation in the protein leucine aminopeptidase for nine brown snails, *Helix aspersa.* Two different sets of allelic variants are revealed. The top set contains two alleles [denoted fast (F) and slow (S) according to their relative movement in the electric field]. Individuals homozygous for the fast allele show only a single fast band on the gel (FF), those homozygous for the slow allele show only a single slow band (SS), and heterozygous individuals have both bands (FS). The lower set contains three different alleles denoted fast (F), medium (M), and slow (S). Note that no individuals shown are homozygous for the medium (M) allele.

TABLE 6.1

Values of Polymorphism (P) and Heterozygosity (H) for Various Animals and Plants as Measured Using Protein Electrophoresis

(a) Species	Number of Proteins	P*	H*
Humans	71	0.28	0.067
Northern elephant seal	24	0.0	0.0
Horseshoe crab	25	0.25	0.057
Elephant	32	0.29	0.089
Drosophila pseudoobscura	24	0.42	0.12
Barley	28	0.30	0.003
Tree frog	27	0.41	0.074

(b) Taxa	Number of Species	P*	H*
Plants	—	0.31	0.10
Insects (excluding *Drosophila*)	23	0.33	0.074
Drosophila	43	0.43	0.14
Amphibians	13	0.27	0.079
Reptiles	17	0.22	0.047
Birds	7	0.15	0.047
Mammals	46	0.15	0.036
Average		0.27	0.078

Source: Data from P. W. Hedrick, *Population biology.* Jones and Bartlett, Boston, 1984.

*P, the average number of alleles per gene per species; H, the proportion of heterozygous genes per individual.

that most species have an enormous potential for further evolutionary change.

Quantitative Variation

Quantitative traits are those that show continuous variation in a population. The values of the trait in offspring often are intermediate between the values in the parents. Such traits are influenced by variation at many genes, each of which follows Mendelian inheritance and contributes a small, incremental amount to the total phenotype. Examples of traits that show quantitative variation include tail length in mice, length of a leg segment in grasshoppers, number of gill rakers in sunfishes, number of peas in pods, and height of adult males of the human species. When trait values are graphed with respect to frequency distribution, they often approximate a normal, or bell-shaped, probability curve (Figure 6.33A). Most individuals fall near the average; fewer fall somewhat above or below the average, and extremes form the "tails" of the frequency curve with increasing rarity. Usually, the larger the population sample, the more closely the frequency distribution resembles a normal curve.

Selection can act on quantitative traits to produce three different kinds of evolutionary response (Figure 6.33B, C, and D). One outcome is to favor average values of the trait and to disfavor extreme ones; this kind of selection is called **stabilizing selection** (Figure 6.33B). **Directional selection** favors a phenotypic value either above or below the average and causes the population average to shift toward the favored value over time (Figure 6.33C). When we think about natural selection producing evolutionary change, it is usually directional selection that we have in mind, although we must remember that this is not the only possibility. A third alternative is **disruptive selection,** in which two different extreme phenotypes are simultaneously favored, but their average is disfavored (Figure 6.33D). The population then becomes bimodal, meaning that two very different phenotypic values predominate.

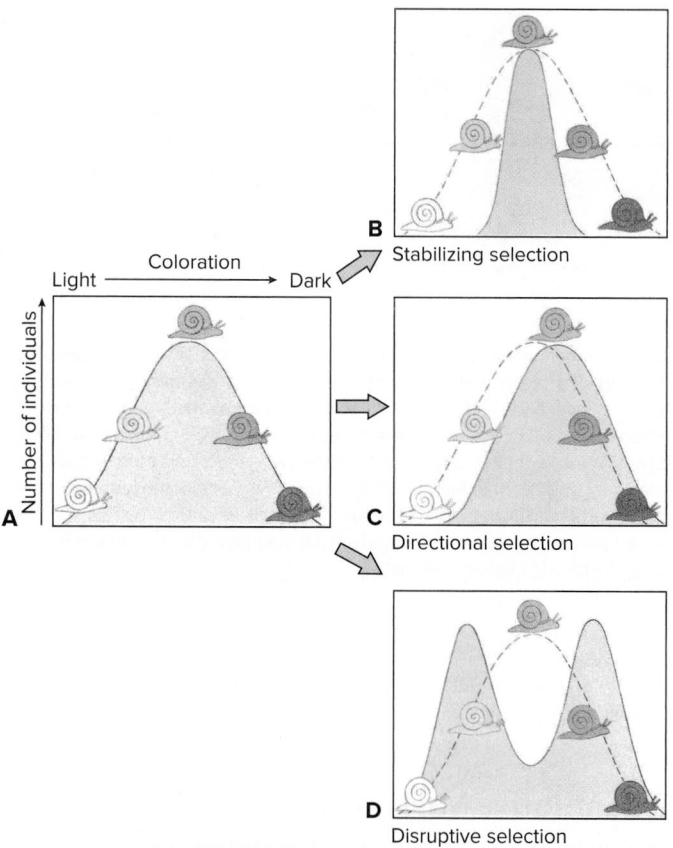

Figure 6.33 Responses to selection on a continuous (polygenic) character, coloration in a snail. **A,** The frequency distribution of coloration before selection. **B,** Stabilizing selection culls extreme variants from the population, in this case eliminating individuals that are unusually light or dark, thereby stabilizing the mean. **C,** Directional selection shifts the population mean, in this case by favoring darkly colored variants. **D,** Disruptive selection favors both extremes but not the mean; the mean is unchanged but the population no longer has a bell-shaped distribution of phenotypes.

6.5 MACROEVOLUTION: MAJOR EVOLUTIONARY EVENTS

Macroevolution describes large-scale events in organic evolution. Speciation links macroevolution and microevolution. Major trends in the fossil record (see Figures 6.11, 6.12, and 6.13) are clearly within the realm of macroevolution. Patterns and processes of macroevolutionary change emerge from those of microevolution, but they acquire some degree of autonomy in doing so. The emergence of new adaptations and species, and the varying rates of speciation and extinction observed in the fossil record, go beyond the fluctuations of allelic frequencies within populations.

Stephen Jay Gould recognized three different "tiers" of time at which we observe distinct evolutionary processes. The first tier constitutes the timescale of population genetic processes, from tens to thousands of years. The second tier covers millions of years, the scale on which rates of speciation and extinction are measured and compared among different groups of organisms. Punctuated equilibrium is a theory of the second tier, explaining the occurrence of

speciation and morphological change and their association over millions of years. The third tier covers tens to hundreds of millions of years, and is marked by occurrence of episodic mass extinctions. In the fossil record of marine organisms, mass extinctions recur at intervals of approximately 26 million years. Five of these mass extinctions have been particularly disastrous (Figure 6.34). The study of long-term changes in animal diversity focuses on the third-tier timescale (see Figures 6.13 and 6.34).

Speciation and Extinction Through Geological Time

Evolutionary change at the second tier provides a new perspective on Darwin's theory of natural selection. Although a species may persist for many millions of years, it ultimately has two possible evolutionary fates: It may give rise to new species or become extinct without leaving descendants. Rates of speciation and extinction vary among lineages, and lineages that have the highest speciation rates and lowest extinction rates produce the greatest number of living species. The characteristics of a species may make it more or less likely than others to undergo speciation or extinction events. Because many characteristics are passed from ancestral to descendant species (analogous to heredity at the organismal level), lineages whose characteristics increase the probability of speciation and confer resistance to extinction should dominate the living world. This species-level process that produces differential rates of speciation and extinction among lineages is analogous in many ways to natural selection. It represents an expansion of Darwin's theory of natural selection. This expansion is particularly important for macroevolution if one accepts the theory of punctuated equilibrium, which states that the evolutionarily important variation occurs primarily among rather than within species.

Species selection encompasses the differential survival and multiplication of species through geological time based on variation among lineages, especially in emergent, species-level properties. These species-level properties include mating rituals, social structuring, migration patterns, geographic distribution, and all other properties that emerge at the species level (see p. 5). Descendant species usually resemble their ancestors in these properties. For example, a "harem" system of mating in which a single male and several females compose a breeding unit characterizes some mammalian lineages but not others. We expect speciation rates to be enhanced by social systems that promote founding of new populations by small numbers of individuals. Certain social systems may increase the likelihood that a species will survive environmental challenges through cooperative action. Such properties would be favored by species selection over geological time.

Differential speciation and extinction among lineages also can be caused by variation in organismal-level properties (such as specialized versus generalized feeding) rather than species-level properties (see p. 5). Organisms that specialize in eating a restricted range of foods, for example, may be subjected more readily than generalized feeders to geographic isolation among populations, because areas where their preferred food is scarce or absent will function as geographic barriers to dispersal. Such geographic isolation could generate more frequent opportunities for speciation to occur throughout geological time. The fossil records of two major groups of

Figure 6.34 Changes in numbers of taxonomic families (p. 109) of marine animals through time from the Cambrian period to the present. Sharp drops represent five major extinctions of skeletonized marine animals. Note that despite the extinctions, the overall number of marine families has increased to the present.

African antelopes suggest this result (see Figure 6.11). A group of specialized grazers that contains blesboks, hartebeests, and wildebeests shows high speciation and extinction rates. Since the late Miocene, 33 extinct and 7 living species are known, representing at least 18 events of branching speciation and 12 terminal extinctions. In contrast, a group of generalist grazers and browsers that contains impalas shows little branching speciation or extinction during this same interval of time. Interestingly, although these two groups differ greatly in speciation rates, extinction rates, and species diversity, they do not differ significantly in total number of individual animals alive today.

Paleontologist Elisabeth Vrba, whose research produced the results in Figure 6.11, uses the term **effect macroevolution** to describe differential speciation and extinction rates among lineages caused by organismal-level properties. She reserves the term "species selection" for cases where species-level emergent properties are of primary importance. Some other evolutionary paleontologists consider effect macroevolution a subset of species selection because fitness differences occur among different species lineages rather than among varying organisms within species.

Mass Extinctions

When we study evolutionary change on an even larger timescale, we observe episodic events in which large numbers of taxa go extinct simultaneously. These events are called **mass extinctions** (see Figure 6.34). The most cataclysmic of these extinction episodes happened about 245 million years ago, when at least half of the families of shallow-water marine invertebrates and fully 90% of marine invertebrate species disappeared within a few million years. This event was the **Permian extinction.** The **Cretaceous extinction,** which occurred about 65 million years ago, marked the end of dinosaurs, as well as numerous marine invertebrates and many small reptilian taxa.

Causes of mass extinctions and evolutionary timing of mass extinctions at intervals of approximately 26 million years are difficult to explain. Some people have proposed biological explanations for these episodic mass extinctions and others consider many mass extinctions artifacts of our statistical and taxonomic analyses. Walter Alvarez proposed that the earth was occasionally bombarded by asteroids, causing these mass extinctions (Figure 6.35). The drastic effects of such bombardment of a planet were observed in July 1994,

Figure 6.35 Meteor crater of the Arizona desert is among the youngest and best-preserved impact craters. Geologists estimate that it formed approximately 50,000 years ago when a meteor measuring 30 meters wide and weighing 100,000 tons landed at a speed of 20 kilometers per second. Asteroid impacts associated with the End Cretaceous mass extinctions would have greatly exceeded the power of this one, but geological changes accumulating over the past 65 million years have lessened the preservation of those craters.

when fragments of Comet Shoemaker-Levy 9 bombarded Jupiter. The first fragment to hit Jupiter was estimated to have the force of 10 million hydrogen bombs. Twenty additional fragments hit Jupiter within the following week, one of which was 25 times more powerful than the first fragment. This bombardment was the most violent event in the recorded history of the solar system. A similar bombardment on earth would send debris into the atmosphere, blocking sunlight and causing drastic changes of climate. Temperature changes would challenge ecological tolerances of many species. Alvarez's hypothesis is being tested in several ways, including a search for impact craters left by asteroids and for altered mineral content of rock strata where mass extinctions occurred. Atypical concentrations of the rare-earth element iridium in strata at the Cretaceous-Tertiary boundary imply that this element entered the earth's atmosphere through asteroid bombardment. Although multiple kinds of geological data support a massive asteroid impact at the Cretaceous-Tertiary boundary, such evidence is lacking for the other major extinctions shown on Figure 6.34.

Sometimes, lineages favored by species selection are unusually susceptible to mass extinction. Climatic changes produced by the hypothesized asteroid bombardments could produce selective challenges very different from those encountered at other times in the earth's history. Selective discrimination of particular biological traits by events of mass extinction is termed catastrophic species selection. For example, mammals survived the End Cretaceous mass extinction that destroyed the dinosaurs and other prominent vertebrate and invertebrate groups. Following this event, mammals were able to use environmental resources that previously had been denied them, leading to their adaptive radiation.

Natural selection, species selection, and catastrophic species selection interact to produce the macroevolutionary trends seen in the fossil record. Studies of these interacting causal processes have made modern evolutionary paleontology an active and exciting field.

SUMMARY

Organic evolution explains the diversity of living organisms as the historical outcome of gradual change from previously existing forms. Evolutionary theory is strongly identified with Charles Robert Darwin, who presented the first credible explanation for evolutionary change. Darwin derived much of the material used to construct his theory from his experiences on a five-year voyage around the world aboard the H.M.S. *Beagle.*

Darwin's evolutionary theory has five major components. Its most basic proposition is perpetual change, the theory that the living world is neither constant nor perpetually cycling but is steadily undergoing irreversible change with continuity from past to present life. The fossil record amply demonstrates perpetual change in the continuing fluctuation of animal form and diversity following the Cambrian explosion 600 million years ago.

Darwin's theory of common descent states that all organisms descend from a common ancestor through a branching of genealogical lineages. This theory explains morphological homologies among organisms as characteristics inherited with modification from a corresponding feature in their common evolutionary ancestor. Patterns of homology formed by common descent with modification permit us to classify organisms according to their evolutionary relationships.

Changes in the timing of developmental processes, termed heterochrony, and changes in their physical location within the body plan, termed heterotopy, explain the evolution of new morphological homologies. A developmental evolutionary module is a set of developmental processes and associated genes that can be expressed as a unit at different parts of the body to produce different structures with some shared developmental properties. Evolution of limbs in terrestrial vertebrates occurred by expressing at the limb bud a set of developmental processes that evolved initially to construct part of the vertebral column. Evolvability denotes the potential of a lineage to evolve new morphological features by using a set of developmental modules as an evolutionary toolkit.

A corollary of common descent is the *multiplication of species* through evolutionary time. Allopatric speciation denotes the evolution of reproductive barriers between geographically separated populations to generate new species. In some animals, especially parasitic insects that specialize on different host species, speciation may occur without geographical isolation and is called sympatric speciation. Intermediate between allopatric speciation and sympatric speciation is a third mode, parapatric speciation, in which an environmental change splits a species into two environmentally distinct parts that maintain contact along a geographic borderline as they diverge to become separate species.

Adaptive radiation is the proliferation of many adaptively disparate species from a single ancestral lineage within a relatively short period of evolutionary time, such as a few million years. Oceanic archipelagoes, such as the Galápagos Islands, are particularly conducive to adaptive radiation of terrestrial organisms.

Darwin's theory of *gradualism* states that large phenotypic differences between species are produced by accumulation through evolutionary time of many individually small changes. Gradualism is still controversial. Mutations that have large effects on an organism have been useful in animal breeding, leading some to dispute Darwin's claim that such mutations are not important in evolution. On a macroevolutionary timescale, punctuated equilibrium states that most evolutionary change occurs in relatively brief events of branching speciation, separated by long intervals in which little phenotypic change accumulates.

Darwin's fifth major statement is that *natural selection* is the guiding force of evolution. This principle is founded on observations that all species overreproduce their kind, causing a struggle for the limited resources that support existence. Because no two organisms are exactly alike, and because variable traits are at least partially heritable, those organisms whose hereditary endowment enhances their use of resources for survival and reproduction contribute disproportionately to the next generation. Over many generations, the sorting of variation by selection produces new species and new adaptations.

Mutations are the ultimate source of all new variation on which selection acts. Darwin's theory emphasizes that variation is produced at random with respect to an organism's needs and that differential survival and reproduction provide the direction for evolutionary change. Darwin's theory of natural selection was modified around 1900 and in subsequent decades by correction of his genetic errors. This modified theory is called neo-Darwinism.

Population geneticists discovered the principles by which genetic properties of populations change through time. A particularly important discovery, called Hardy-Weinberg equilibrium, showed that the hereditary process itself does not change the genetic compositions of populations. Important sources of evolutionary change include mutation, genetic drift, nonrandom mating, migration, natural selection, and their interactions.

Neo-Darwinism, as elaborated by population genetics, formed the basis for the synthetic theory of the 1930s and 1940s. Genetics, natural history, paleobiology, and systematics were unified by the common goal of expanding our knowledge of Darwinian evolution. Microevolution comprises studies of genetic change within contemporary populations. These studies show that most natural populations contain enormous amounts of variation.

Macroevolution comprises studies of evolutionary change on a geological timescale. Macroevolutionary studies measure rates of speciation, extinction, and changes of diversity through time. These studies have expanded Darwinian evolutionary theory to include higher-level processes that regulate rates of speciation and extinction among lineages, including species selection and catastrophic species selection.

REVIEW QUESTIONS

1. Briefly summarize Lamarck's concept of the evolutionary process. What is wrong with this concept?
2. What is "uniformitarianism"? How did it influence Darwin's evolutionary theory?
3. Why was the *Beagle's* journey so important to Darwin's thinking?
4. What was the key idea contained in Malthus's essay on populations that was to help Darwin formulate his theory of natural selection?
5. Explain how each of the following contribute(s) to Darwin's evolutionary theory: fossils; geographic distributions of closely related animals; homology; animal classification.
6. How do modern evolutionists view the relationship between ontogeny and phylogeny? Explain how the observation of paedomorphosis conflicts with Haeckel's "biogenetic law."
7. What are the important differences between the vicariant and founder-event modes of allopatric speciation?
8. What are reproductive barriers? How do premating and postmating barriers differ?
9. Under what conditions is sympatric speciation proposed?
10. What is the main evolutionary lesson provided by Darwin's finches on the Galápagos Islands?
11. How is the observation of "sporting mutations" in animal breeding used to challenge Darwin's theory of gradualism? Why did Darwin reject such mutations as having little evolutionary importance?
12. What does the theory of punctuated equilibrium state about the occurrence of speciation throughout geological time? What observation led to this theory?
13. Describe the observations and inferences that compose Darwin's theory of natural selection.
14. Identify the random and nonrandom components of Darwin's theory of natural selection.
15. Describe some recurring criticisms of Darwin's theory of natural selection. How can these criticisms be refuted?
16. Using the data shown in Figure 6.32, calculate the frequencies of the alleles in the top set of bands and in the lower set of bands.
17. Assume that you are sampling a trait in animal populations; the trait is controlled by a single allelic pair *A* and *a*, and you can distinguish all three phenotypes *AA, Aa,* and *aa* (intermediate inheritance). Your sample includes:

Population	*AA*	*Aa*	*aa*	Total
I	300	500	200	1000
II	400	400	200	1000

Calculate the distribution of phenotypes in each population as expected under Hardy-Weinberg equilibrium. Is population I in equilibrium? Is population II in equilibrium?

18. If after studying a population for a trait determined by a single pair of alleles you find that the population is not in equilibrium, what possible reasons might explain the lack of equilibrium?
19. Explain why genetic drift is more powerful in small populations.
20. Describe how the effects of genetic drift and natural selection can interact in a subdivided species.
21. Is it easier for selection to remove a deleterious recessive allele from a randomly mating population or from a highly inbred population? Why?
22. Distinguish between microevolution and macroevolution, and describe some evolutionary processes evident only at the macroevolutionary level.

For Further Thought Explain why the evidence supporting Darwin's theory of common descent does not depend on the validity of specific hypotheses of evolutionary gradualism or natural selection.

SELECTED REFERENCES

Avise, J. C. 2004. Molecular markers, natural history, and evolution, ed. 2. Sunderland, Massachusetts, Sinauer Associates. *An exciting and readable account of using molecular studies to help us understand evolution.*

Browne, E. J. 2002. Charles Darwin: a biography. New York, Knopf. *A comprehensive two-volume biography of Darwin completed in 2002.*

Conner, J. K., and D. L. Hartl. 2004. A primer of ecological genetics. Sunderland, Massachusetts, Sinauer Associates. *An introductory text on population genetics.*

Coyne, J. A., and H. A. Orr. 2004. Speciation. Sunderland, Massachusetts, Sinauer Associates. *A detailed coverage of speciation with emphasis on controversies in this field.*

Darwin, C. 1859. On the origin of species by means of natural selection, or the preservation of favoured races in the struggle for life. London, John Murray. *There were five subsequent editions by the author.*

Freeman, S., and J. C. Herron. 2014. Evolutionary analysis, ed. 5. Upper Saddle River, New Jersey, Pearson/Prentice Hall. *An introductory textbook on evolutionary biology designed for undergraduate biology majors.*

Futuyma, D. J. 2013. Evolution, ed. 3. Sunderland, Massachusetts, Sinauer Associates. *A very thorough introductory textbook on evolution.*

Gould, S. J. 2002. The structure of evolutionary theory. Cambridge, Massachusetts, Belknap Press of Harvard University Press. *A provocative discussion of what fossils tell us about the nature of life's evolutionary history.*

Hall, B. K., and W. M. Olson (eds.). 2003. Keywords and concepts in evolutionary developmental biology. Cambridge, Massachusetts, Harvard Univ. Press. *A thorough explanation of the major concepts of evolutionary developmental biology by leading researchers.*

Hartl, D. L., and A. G. Clark. 2007. Principles of population genetics, ed. 4. Sunderland, Massachusetts, Sinauer Associates. *A current textbook on population genetics.*

Jablonski, D. 2005. Mass extinctions and macroevolution. Paleobiology **31:** S192–210. *A review of extinction theory.*

Kliman, R. (ed.). 2016. Encyclopedia of evolutionary biology. San Diego, California, Academic Press. *An authoritative and comprehensive review of evolutionary biology authored by many evolutionary biologists and historians of science.*

Levinton, J. S. 2001. Genetics, paleontology and macroevolution, ed. 2. Cambridge, U.K., Cambridge University Press. *A provocative discussion on the Darwinian basis of macroevolutionary theory.*

Losos, J., D. J. Futuyma, and H. E. Hoekstra (eds). 2013. The Princeton guide to evolution. Princeton, New Jersey, Princeton University Press. *Contributions from many evolutionary biologists provide a good entry into the evolutionary literature.*

Mayr, E. 2001. What evolution is. New York, Basic Books. *A general survey of evolution by a leading evolutionary biologist.*

McCalman, I. 2009. Darwin's armada: four voyages and the battle for the theory of evolution. New York, W. W. Norton and Co. *An entertaining history of oceanic* voyages taken by Charles Darwin, Joseph Hooker, Thomas Henry Huxley, and Alfred Russel Wallace, and how these voyages influenced their evolutionary theories.

McKay, B. D., and R. M. Zink. 2015. Sisyphean evolution in Darwin's finches. Biological Reviews **90**:689–698. *Reviews molecular population-genetic evidence that the "adaptive radiation" of Galápagos finches occurred as intraspecific variation without speciation.*

Mousseau, T. A., B. Sinervo, and J. Endler (eds.). 2000. Adaptive genetic variation in the wild. Oxford, U.K., Oxford University Press. *Detailed examples of adaptively important genetic variation in natural populations.*

Nielsen, R., and M. Slatkin. 2013. An introduction to population genetics: theory and applications. Sunderland, Massachusetts, Sinauer Associates. *A thorough introductory coverage of population genetics intended for advanced undergraduates or graduate students.*

Ruse, M. 1998. Philosophy of biology. Amherst, New York, Prometheus Books. *A collection of essays on evolutionary biology, including information on the Arkansas Balanced Treatment for Creation-Science and Evolution-Science Act.*

Sloan, P. R., G. McKenny, and K. Eggleston (eds.). 2015. Darwin in the twenty-first century. Notre Dame, Indiana, University of Notre Dame Press. *Explores changes in Darwinian evolutionary theory made necessary by recent discoveries in evolutionary developmental biology, and implications of evolutionary theory for social and religious institutions. Some essays explain how evolutionary theory in the Darwinian tradition is compatible with Catholic theology, whereas notions of "creation science" and "intelligent design" are incompatible with Catholic theology.*

Stokstad, E. 2001. Exquisite Chinese fossils add new pages to book of life. Science **291:** 232–236. *Exciting fossil discoveries help to complete our understanding of life's evolutionary history. Some related articles immediately follow this one.*

Templeton, A. R. 2006. Population genetics and microevolutionary theory. Hoboken, New Jersey, Wiley-Liss. *An insightful treatment of evolutionary theory at the population level.*

Wagner, G. P. 2014. Homology, genes and evolutionary innovation. Princeton, New Jersey, Princeton University Press. *A new synthetic theory of the molecular and developmental bases of evolutionary novelty.*

West-Eberhard, M. J. 2003. Developmental plasticity and evolution. Oxford, U.K., Oxford University Press. *A provocative discussion of the role of development and modularity in evolution.*

CHAPTER 7

The Reproductive Process

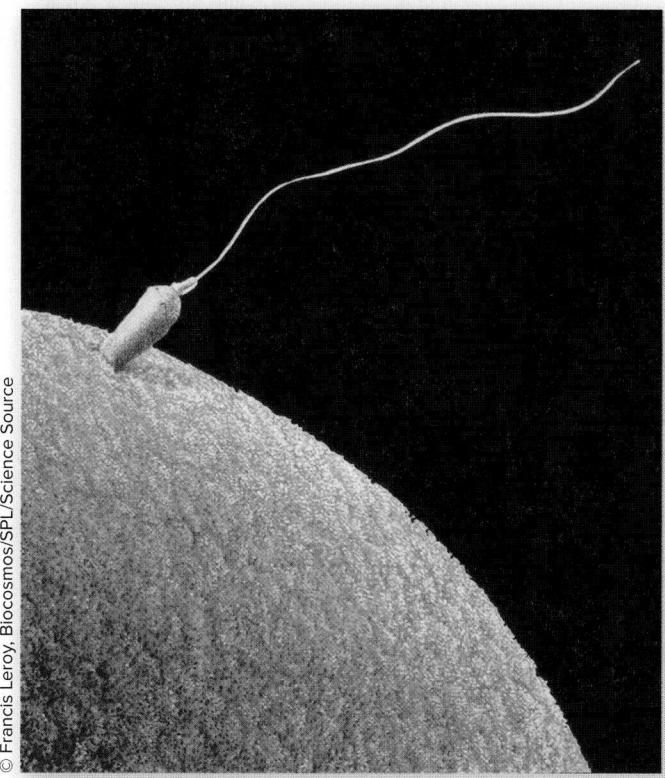

Human egg and sperm at the moment of fertilization

"Omne vivum ex ovo"

In 1651, William Harvey, the English physiologist who earlier had founded experimental physiology by explaining the circuit of blood, published a treatise on reproduction. He asserted that all life developed from the egg—*omne vivum ex ovo*. This was insightful, since Harvey had no means for visualizing eggs of many animals, in particular the microscopic mammalian egg, which is no larger than a speck of dust to the unaided eye. Further, argued Harvey, eggs begin their developmental course by some influence from semen, a conclusion that was either remarkably perceptive or a lucky guess, since sperm also were invisible to Harvey. Such ideas differed sharply from existing notions of biogenesis, which saw life springing from many sources of which eggs were but one. Harvey was describing characteristics of sexual reproduction in which two parents, male and female, must produce gametes that fuse to become a new individual.

Despite the importance of Harvey's assertion that all life arises from eggs, it was not wholly correct. Life springs from reproduction of preexisting life, and reproduction may not be restricted to eggs and sperm. Asexual reproduction, the creation of new, genetically identical individuals by budding or fragmentation or fission from a single parent, is common among some phyla. Nevertheless, most animals have found sex the winning strategy, probably because sexual reproduction promotes diversity, enhancing long-term survival of the lineage in a world of perpetual change.

Reproduction is one of the ubiquitous properties of life. Evolution is inextricably linked to reproduction, because the ceaseless replacement of predecessors with new life gives animal populations the means to adapt to a changing environment. In this chapter we distinguish asexual and sexual reproduction and explore the reasons why, for multicellular animals at least, sexual reproduction is most prevalent. We then consider, in turn, the origin and maturation of germ cells; reproductive patterns in animals; plans of reproductive systems; and, finally, the endocrine events that orchestrate reproduction.

7.1 NATURE OF THE REPRODUCTIVE PROCESS

Two modes of reproduction are recognized: asexual and sexual. In **asexual** reproduction (Figure 7.1A and B) there is only one parent with no special reproductive organs or cells. Each organism can produce genetically identical copies of itself as soon as it becomes an adult. **Sexual** reproduction (Figure 7.1C and D) as a rule involves two parents, each of which contributes special **germ cells (gametes** or **sex cells)** that, once united during the process of fertilization, develop into a new individual. The **zygote** formed from this union receives genetic material from both parents, and the combination of genes (p. 72) produces a genetically unique individual, bearing characteristics of the species but also bearing traits that distinguish it from its parents. Sexual reproduction, by recombining parental characters, multiplies variations and makes possible evolution of more diverse forms. Mechanisms for interchange of genes between individuals are more limited in organisms with only asexual reproduction.

Asexual Reproduction: Reproduction Without Gametes

Asexual reproduction (Figure 7.1A and B; see also Chapter 13, p. 270, and Chapter 17, p. 368) is the production of individuals without gametes (eggs or sperm). It includes a number of distinct processes, all without involving sex or a second parent. Offspring produced by asexual reproduction all have the same genotype (unless mutations occur) and are **clones** of the parent.

Asexual reproduction appears in bacteria and unicellular eukaryotes and in many invertebrate phyla, such as cnidarians (see Chapter 13, p. 270), bryozoans (see Chapter 15, p. 324), annelids (see Chapter 17, p. 368), echinoderms (see Chapter 22, p. 480), and hemichordates (see Chapter 22, p. 494). Animals of some taxa can reproduce both asexually and sexually. In these groups, asexual reproduction ensures rapid increase in numbers prior to the individual's attaining sexual maturity. Asexual reproduction is rare among vertebrates. There are at least 50 species that reproduce by parthenogenesis (see p. 136); some do not involve meiosis or involve only the early stages of meiosis before chromosome reduction occurs. Thus, they are essentially clones and could be categorized as asexual reproduction.

The basic forms of asexual reproduction are fission (binary and multiple), budding, gemmulation, and fragmentation.

Binary fission is common among bacteria and protozoa or unicellular eukaryotes (Figure 7.1A). In binary fission the body of

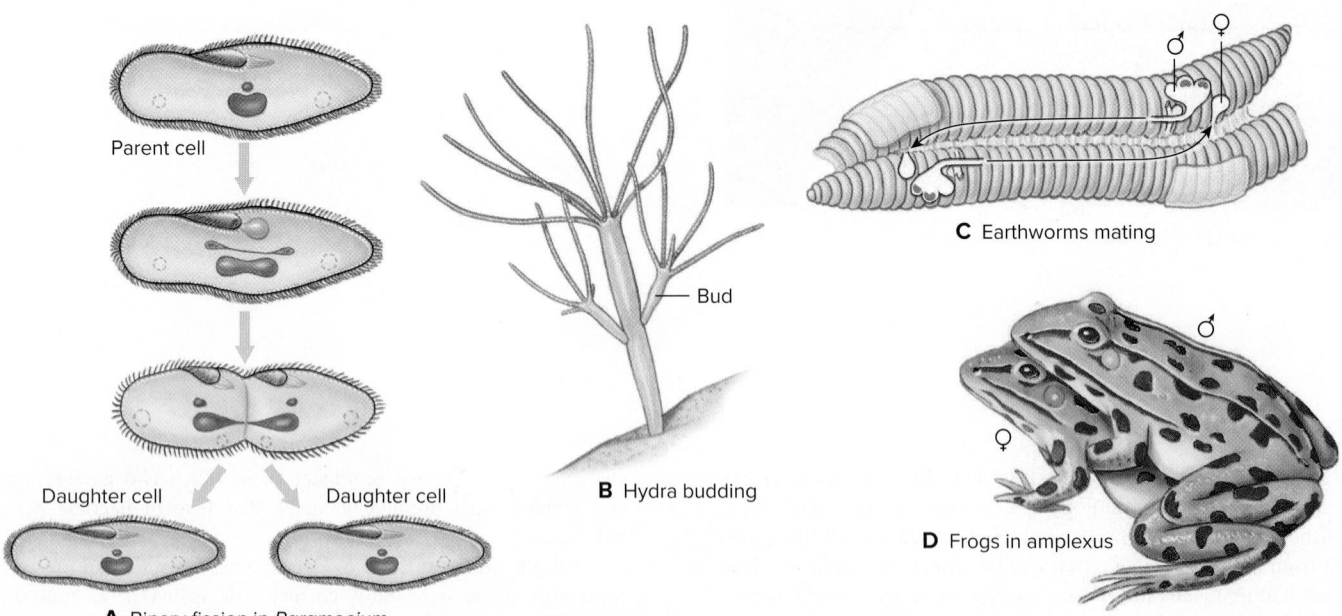

A Binary fission in *Paramecium*

Parent cell

Daughter cell   Daughter cell

B Hydra budding

Bud

C Earthworms mating

D Frogs in amplexus

Figure 7.1 Examples of asexual and sexual reproduction in animals. **A,** Binary fission in *Paramecium,* a single-celled eukaryote, produces two individuals from one parent. **B,** Budding, a simple form of asexual reproduction as shown in a hydra, a radiate animal. The buds, shown growing out of the central, parent hydra, eventually detach themselves and grow into fully formed individuals. **C,** Earthworms reproduce sexually, but are hermaphroditic, with each individual bearing both male and female organs. Each earthworm passes sperm from genital pores along grooves to seminal receptacles of its mate. **D,** Frogs, here in mating position (amplexus), represent bisexual reproduction, the most common form of sexual reproduction involving separate male and female individuals.

the unicellular parent divides by mitosis (p. 49) into two approxi-mately equal parts, each of which grows into an individual similar to the parent. Binary fission may be lengthwise, as in flagellate unicel-lular eukaryotes, or transverse, as in ciliate unicellular eukaryotes (see Chapter 11, p. 233). In **multiple fission,** or **schizogony,** the nucleus divides repeatedly before division of the cytoplasm, pro-ducing many daughter cells simultaneously. Spore formation, called sporogony, is a form of multiple fission common among some par-asitic unicellular eukaryotes, for example, malarial parasites (see Figure 11.30, p. 236).

Budding is an unequal division of an organism. A new individ-ual arises as an outgrowth (bud) from its parent, develops organs like those of the parent, and then detaches itself. Budding occurs in sev-eral animal phyla and is especially prominent in cnidarians (Figure 7.1B; also see Chapter 17, p. 368).

Gemmulation is the formation of a new individual from an aggregation of cells surrounded by a resistant capsule, called a gem-mule (see Figure 12.11, p. 253). In many freshwater sponges, gem-mules develop in the fall and survive the winter in the dried or frozen body of the parent. In spring, the enclosed cells become active, emerge from the capsule, and grow into a new sponge.

In **fragmentation** a multicellular animal breaks into two or more parts, with each fragment capable of becoming a complete individual. Many invertebrates, for example, most anemones and many hydroids, can reproduce asexually by simply breaking into two parts and then regenerating the missing parts of the fragments (see Chapter 13, p. 270). Many echinoderms can regenerate lost parts, but this is not the same as reproduction by fragmentation.

Sexual Reproduction: Reproduction with Gametes

Sexual reproduction is the production of individuals from gametes. It includes **bisexual** (or **biparental**) reproduction as the most common form, involving two separate individuals. **Hermaphroditism** and **parthenogenesis** are less common forms of sexual reproduction.

Bisexual Reproduction

Bisexual reproduction is the *production of offspring formed by the union of gametes from two genetically different parents* (Figures 7.1C and D, and 7.2). Offspring will thus have a new gen-otype different from that of either parent (see Chapter 5, p. 76). Individuals sharing parenthood are characteristically of different **sexes,** male and female (there are exceptions among sexually reproducing organisms, such as bacteria and some unicellular

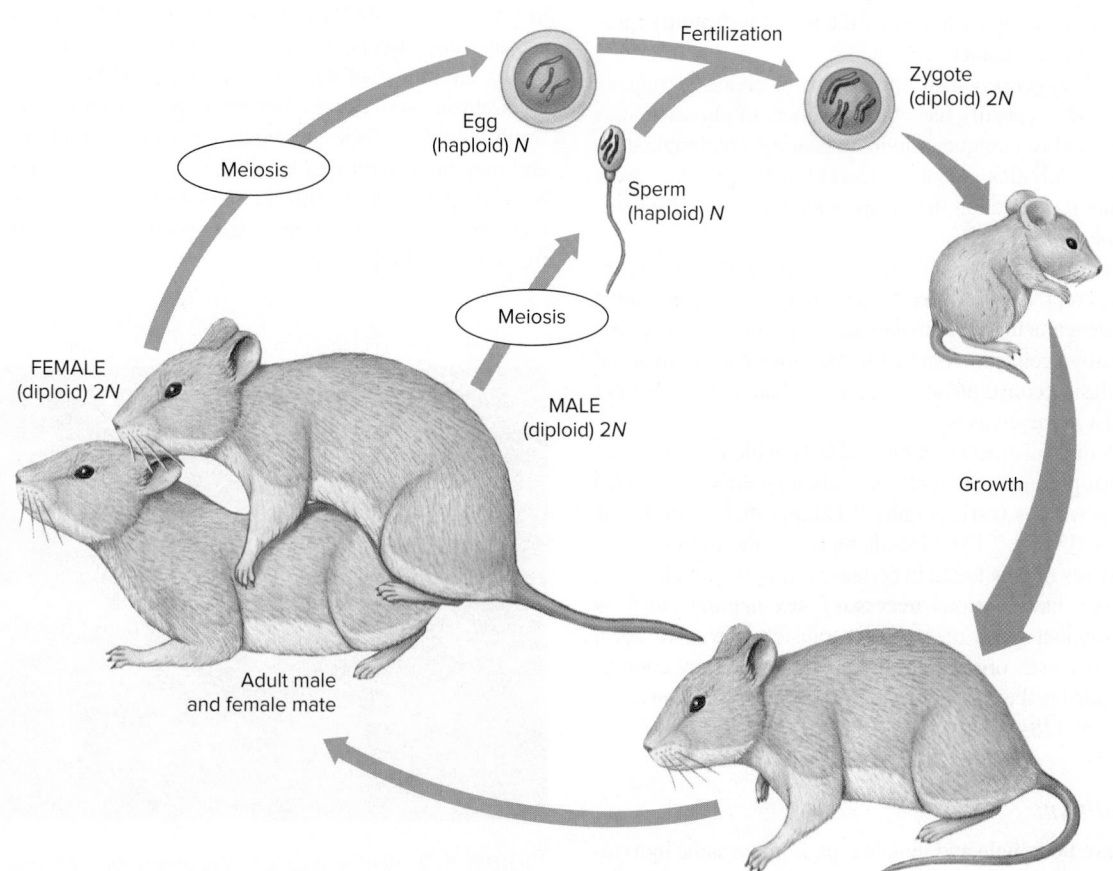

Figure 7.2 A sexual life cycle. The life cycle begins with haploid germ cells, formed by meiosis, combining to form a diploid zygote, which grows by mitosis to an adult. Most of the life cycle is spent as a diploid organism.

eukaryotes in which sexes are lacking). Each parent has its own reproductive system and produces only one kind of germ cell, spermatozoon or ovum, rarely both. Nearly all vertebrates and many invertebrates have separate sexes, and such a condition is called **dioecious** (Gr. *di,* two, + *oikos,* house). Individual animals that have both male and female reproductive organs are called **monoecious** (Gr. *monos,* single, + *oikos,* house). These animals are called **hermaphrodites** (from a combination of the names of the Greek god Hermes and the goddess Aphrodite); this form of reproduction is described below.

Distinctions between male and female are based not on any differences in parental size or appearance, but on the size and mobility of the gametes they produce. The **ovum** (egg), produced by the female, is large (because it is the primary source of cytoplasm for the zygote and often contains stored yolk to sustain early development), nonmotile, and produced in relatively small numbers. The **spermatozoon** (sperm), produced by the male, is small, motile, and produced in enormous numbers. Each sperm is a stripped-down package of highly condensed genetic material produced for the single purpose of reaching and fertilizing an egg.

There is another crucial event that distinguishes sexual from asexual reproduction: **meiosis,** a distinctive type of gamete-producing nuclear division (described in detail on p. 72). Meiosis differs from ordinary cell division (mitosis) in being a double division. Chromosomes split once, but the cell divides *twice,* producing four cells, each with half the original number of chromosomes (the **haploid** number). Meiosis is followed by **fertilization,** in which two haploid gametes are combined to restore the normal **(diploid)** chromosomal number of the species.

The new cell (zygote), which now begins to divide by mitosis (described on p. 49), typically has equal numbers of chromosomes from each parent and is a unique individual bearing a recombination of parental characteristics. Genetic recombination is the great strength of sexual reproduction that keeps adding new genetic combinations into the population.

Many unicellular organisms reproduce both sexually and asexually. When sexual reproduction does occur, it may or may not involve male and female gametes. When gametes are not involved, two mature sexual parent cells join together to exchange nuclear material or merge cytoplasm (**conjugation,** p. 233 in Chapter 11). Distinct sexes do not exist in these cases.

The male-female distinction is more clearly evident in most animals. Organs that produce germ cells are called **gonads.** The gonad that produces sperm is a **testis** (Figure 7.12) and that which forms eggs is an **ovary** (Figure 7.13). Gonads represent the **primary sex organs,** the only sex organs found in certain groups of animals. Most animals, however, have various **accessory sex organs** (such as penis, vagina, oviducts, and uterus) that transfer or receive germ cells. In the primary sex organs, germ cells undergo many complicated changes during their development, the details of which are described on pages 138–142.

Hermaphroditism

Animals that have both male and female organs in the same individual are called **hermaphrodites,** and the condition is called **hermaphroditism.** In contrast to the dioecious state of separate sexes, hermaphrodites are monoecious, meaning that the same organism bears both male and female organs. Many sessile, burrowing, or endoparasitic invertebrate animals are hermaphoditic, for example, most flatworms (see Chapter 14, p. 297), some hydroids and annelids, and all barnacles (see Chapter 20, p. 435) and pulmonate snails (see Chapter 16, p. 347), as well as a few vertebrates (some fishes). Some hermaphrodites fertilize themselves, but most avoid self-fertilization by exchanging germ cells with another member of the same species (Figures 7.1C and 7.3). An advantage is that with every individual producing eggs, a hermaphroditic species could potentially produce twice as many offspring as could a dioecious species in which half the individuals are nonproductive males. Some fishes are **sequential hermaphrodites,** in which a genetically programmed sex change occurs within an individual organism in contrast to the aforementioned simultaneous hermaphrodites. In many species of reef fishes, such as wrasses, an animal begins life as either a female or a male (depending on the species) but later becomes the opposite sex (see Chapter 24, p. 540).

Parthenogenesis

Parthenogenesis ("virgin origin") is the development of an embryo from an unfertilized egg or one in which the male and female nuclei fail to unite following fertilization. Classification of parthenogenesis as either asexual or sexual reproduction is difficult, since there are many different patterns. In one type, called **ameiotic** or **diploid parthenogenesis,** no meiosis occurs, and the egg is formed by mitotic cell division. This "asexual" form of parthenogenesis occurs in some species of flatworms, rotifers (see Figure 14.32, p. 312), crustaceans, insects, and probably others. In these cases, the offspring are clones of the parent because, without meiosis, the parent's chromosomal complement is passed intact to offspring. In some species of fishes, a female produces diploid or triploid eggs, and may be inseminated by a male of the same or related species, but the sperm serves only to activate the egg; the male's genetic material is rejected before it can penetrate the egg (**gynogenesis**) (see Chapter 24, p. 540).

© Premaphotos/Alamy Stock Photo

Figure 7.3 Hermaphroditic earthworms mating. Earthworms are "simultaneous" hermaphrodites; during mating each partner passes sperm from genital pores along grooves to seminal receptacles of its mate. Mucous secretions hold mating earthworms together during this process.

In **meiotic parthenogenesis** a haploid ovum is formed by meiosis, and it may or may not be activated by the influence of a male's sperm. In several species of flatworms, rotifers (see Figure 14.32, p. 312), annelids, mites, and insects, the haploid egg begins development spontaneously; no males are required to stimulate activation of an ovum. The diploid condition may be restored by chromosomal duplication or by autogamy (rejoining of haploid nuclei). A variant of this type of parthenogenesis occurs in many bees, wasps, and ants. In honey bees, for example, the inseminated queen bee can either fertilize eggs as she lays them or allow them to pass unfertilized. Fertilized eggs become diploid females (queens or workers), and unfertilized eggs develop parthenogenetically to become haploid males (drones); this type of sex determination is called **haplodiploidy** (see Chapter 21, p. 461). In some animals meiosis may be so severely modified that offspring are clones of the parent. Certain populations of whiptail lizards of the American southwest are clones consisting solely of females (Figure 7.4). In these lizards, the chromosomes are duplicated to 4 n before meiosis like divisions reduce it to 2 $n;$ however, each egg is a clone of the mother.

Parthenogenesis is surprisingly widespread in animals. It is an abbreviation of the usual steps of bisexual reproduction. It may have evolved to avoid the problem—which may be great in some animals—of bringing together males and females at the right moment for successful fertilization. The disadvantage of parthenogenesis is that if the environment should suddenly change, parthenogenetic species have limited capacity to shift gene combinations to adapt to any new conditions. Bisexual species, by recombining parental characteristics, have a better chance of producing variant offspring that can utilize new environments.

Studies of ovarian development in mammals suggest that parthenogenesis does occur, but that the resulting cells **(parthenotes)** fail to develop into offspring. Mouse and rabbit parthenotes form a small group of pluripotent stem cells—cells that are capable of developing into any cell type. Similar findings in the human may ultimately provide stem cells for therapeutic use to replace defective or dying cells in such diseases as heart failure, diabetes, Parkinson's disease, and spinal cord injury.

Why Do So Many Animals Reproduce Sexually Rather than Asexually?

Because sexual reproduction is so nearly universal among animals, it might be inferred to be highly advantageous. Yet it is easier to list disadvantages to sex than to list advantages. Sexual reproduction is complicated, requires more time, and uses much more energy than asexual reproduction. Males may waste valuable energy in competition for a mate and often possess sexual characteristics that can be detrimental to survival—for example, the elongated tail feathers of peacocks. Mating partners must come together, and this can be a disadvantage in sparsely populated areas for some species. Males and females must also coordinate their activities to produce young. Many biologists believe that an even more troublesome problem is the "cost of meiosis." A female that reproduces asexually passes all of her genes to each offspring, but when she reproduces sexually the genome is divided

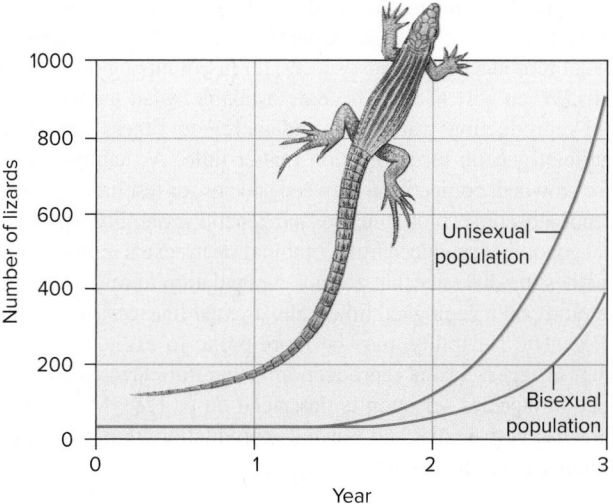

Figure 7.4   Comparison of the growth of a population of unisexual whiptail lizards with that of a population of bisexual lizards. Because all individuals of the unisexual population are females, all produce eggs, whereas only half the bisexual population are egg-producing females. By the end of the third year the unisexual lizards are more than twice as numerous as the bisexual ones.

during meiosis and each offspring receives only half her genes. Another cost is wastage in production of males, many of which fail to reproduce and thus consume resources that could be applied to production of females. Whiptail lizards of the American southwest offer a fascinating example of the potential advantage of parthenogenesis. When unisexual and bisexual species of the same genus are reared under similar conditions in the laboratory, the population of the unisexual species grows more quickly because all unisexual lizards (all females) deposit eggs, whereas only 50% of the bisexual lizards do so (Figure 7.4). Yet another cost is that due to separation and recombination of genetic material, sexual reproduction may break up very favorable gene combinations that could be perpetuated in clones.

Clearly, the costs of sexual reproduction are substantial. How are they offset? Biologists have disputed this question for years. One hypothesis suggests that sexual reproduction, with its separation and recombination of genetic material, enriches a species gene pool by producing novel genotypes that *in times of environmental change* may be advantageous to survival and thus the organism may live to reproduce, whereas most others die. An often-quoted example is the rapidly changing environment produced by parasites continuously evolving new mechanisms to attack a host and therefore favoring recombination of their hosts. Variability, advocates of this viewpoint argue, is sexual reproduction's trump card. Another hypothesis suggests that sexual recombination provides a means for the spread of beneficial mutations without a population's being held back by deleterious ones. Experimental support for this hypothesis has been provided using the fruit fly, *Drosophila,* in which beneficial mutations increased to a greater degree in sexual populations compared with clonal (asexual) ones. These hypotheses are not mutually exclusive, however, and both provide possible explanations for the evolution of sexual reproduction.

There still remains the question of why sexual reproduction is maintained despite its costs. Considerable evidence suggests that asexual reproduction is most successful in colonizing new environments. When new habitats become available, what matters most is rapid reproduction; variability and increased fitness provided by beneficial genetic recombination matter little. As habitats become more crowded, competition between species for resources increases. Selection becomes more intense, and genetic variability—new beneficial genotypes produced by recombination in sexual reproduction—furnishes the diversity that permits a population to resist extinction. Therefore, on a geological timescale, asexual lineages, because they lack genetic flexibility, may be more prone to extinction than are sexual lineages. Sexual reproduction is therefore favored by species selection (species selection is described on p. 128). Many invertebrates use both sexual and asexual reproduction, thus enjoying the advantages each has to offer.

7.2 THE ORIGIN AND MATURATION OF GERM CELLS

Many sexually reproducing organisms are composed of non-reproductive **somatic cells,** which become differentiated for specialized functions and die with the individual, and **germ cells,** which form the gametes: eggs and sperm. Germ cells provide continuity of life between generations, the **germ cell line.**

A traceable germ cell line, as is present in vertebrates, is also distinguishable in some invertebrates, such as nematodes and arthropods. In many invertebrates, however, germ cells develop directly from somatic cells at some period in the life of an individual.

Migration of Germ Cells

In vertebrates, the actual tissue from which gonads arise appears in early embryonic development (described in Chapter 8) as a pair of **genital ridges,** growing into the coelom from the dorsal coelomic lining on each side of the hindgut near the anterior end of the kidney (mesonephros).

Surprisingly perhaps, germ cells, or their precursors, **primordial germ cells,** arise not in the developing gonad, but in the yolk-sac endoderm (p. 172). From studies with frogs and toads, one can trace the germ cell line back to the fertilized egg, in which a localized area of germinal cytoplasm (called **germ plasm**) is identified in the vegetal pole of the uncleaved egg mass (see Figure 8.8B, p. 160). This material is followed through subsequent cell divisions of the embryo until it becomes situated in primordial germ cells in gut endoderm. From here the cells migrate by ameboid movement to the genital ridges, located on either side of the hindgut. A similar migration of primordial germ cells occurs in mammals (Figure 7.5). Primordial germ cells are the future stock of gametes for an animal. Once in the genital ridges and during subsequent gonadal development, germ cells begin to divide by mitosis, increasing their numbers from a few dozen to several thousand.

The other cells of the gonads are somatic cells. They cannot form eggs or sperm, but they are necessary for support, protection, and nourishment of the germ cells during their development **(gametogenesis).**

Sex Determination

At first gonads are sexually indifferent. In mammalian males, a "male determining gene" on the Y chromosome called **SRY (sex-determining region Y)** organizes the developing gonad into a testis instead of an ovary. *DMRT1* may function in a similar way in birds. *SRY* appears to activate another gene, *SOX9,* that drives production of Sertoli cells (p. 139) necessary for the support, protection, and nourishment of developing sperm. Once formed, the testis secretes the steroid **testosterone.** This hormone, and its metabolite, **dihydrotestosterone (DHT),** masculinizes the fetus, causing the differentiation of penis, scrotum, and the male ducts and glands. It also destroys the incipient breast primordia, but leaves behind the nipples that are a reminder of the indifferent ground plan from which both sexes develop. Testosterone is also responsible for the masculinization of the brain, but it does so indirectly. Surprisingly, testosterone is enzymatically converted to estrogen in the brain, and it is

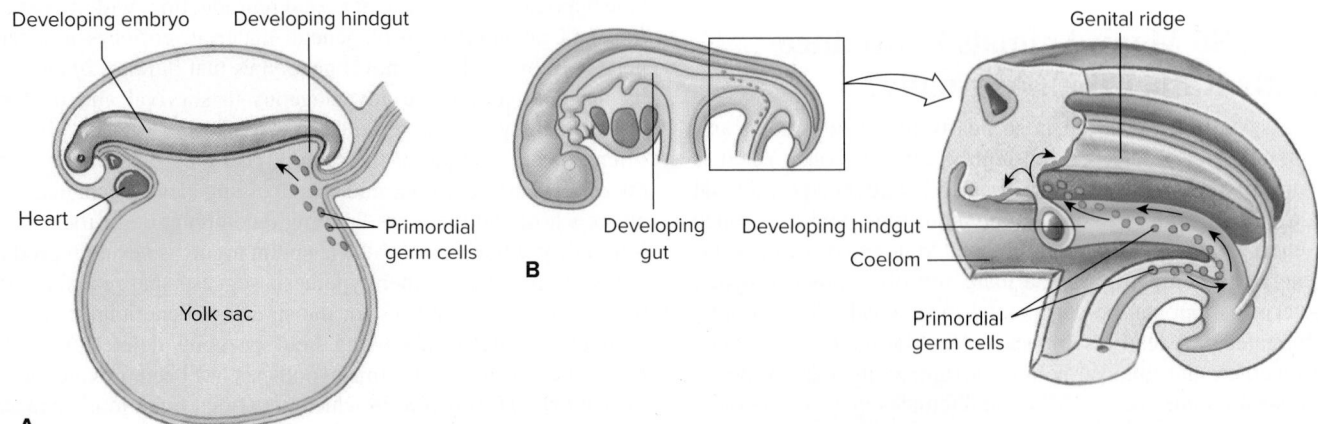

Figure 7.5 Migration of mammalian primordial germ cells. **A,** From the yolk sac the primordial germ cells migrate toward the region where the hindgut develops. **B,** Later-stage embryo in which the hindgut is more developed. Enlarged figure shows the germ cells migrating through the hindgut and into the genital ridges. In human embryos, migration is complete by the end of the fifth week of gestation.

estrogen that determines the organization of the brain for male-typical behavior.

Biologists have often stated that in mammals the indifferent gonad has an inherent tendency to become an ovary. Classic experiments performed in rabbits provide support for the idea that the female is the default sex during development. Removal of the fetal gonads before they have differentiated invariably produces a female with uterine tubes, uterus, and vagina, even if the rabbit is a genetic male. Recent molecular evidence indicates that the X chromosome expresses ovary-determining genes, such as *WNT4* and *DAX1,* that repress one or more of the genes involved in testicular development. Furthermore, sex determination appears to be gene-dosage dependent, so the presence of the Y chromosome in the male overrides these X-chromosome genes when only one X chromosome is present. The developing female brain does require special protection from the effects of estrogen because, as mentioned earlier, estrogen causes masculinization of the brain. In rats, a blood protein (alpha-fetoprotein) binds to estrogen and keeps estrogen from reaching the developing female brain. This does not appear to be the case in humans, however, and even though circulating fetal estrogen levels can be quite high, the developing female brain does not become masculinized. One possible explanation is that the level of brain estrogen receptors in the developing female brain is low, and therefore, high levels of circulating estrogen would have no effect.

The genetics of sex determination are treated in Chapter 5 (p. 75). Gonadal sex is chromosomal in mammals, birds, most amphibians, most reptiles, and probably most fishes, while phenotypic and behavioral sex are determined by secretion of the appropriate hormones (see p. 145). In some fishes and reptiles, gender is determined by nongenetic factors such as temperature or behavior.

In crocodilians, many turtles, and some lizards, the incubation temperature of the nest determines the sex ratio, probably by indirectly activating and/or suppressing genes that direct development of the animals' sex organs. Evidence suggests that temperature regulates expression of the *DMRT1* gene, which is expressed in higher doses in male embryonic testes than in female embryonic ovaries. In turtles *DMRT1* gene expression is higher at low temperatures and promotes development of males. In contrast, alligator eggs incubated at low temperature all become females; those incubated at higher temperature all become males (Figure 7.6), so there is a large degree of variability, particularly in reptiles. Sex determination of many fishes is behavior dependent. Most of these species are hermaphroditic, possessing both male and female gonads. Sensory stimuli from the animal's social environment determine whether it becomes male or female.

Gametogenesis

Mature gametes are produced by a process called gametogenesis. Although the same essential processes are involved in maturation of both sperm and eggs in vertebrates, there are some important differences. Gametogenesis in testes is called **spermatogenesis,** and in ovaries, **oogenesis.**

Spermatogenesis

Walls of the seminiferous tubules contain differentiating germ cells arranged in a stratified layer five to eight cells deep (Figure 7.7). Germ cells develop in close contact with large **Sertoli** (sustentacular) **cells,** which extend from the periphery of the seminiferous tubules to the lumen and provide nourishment during germ-cell development and differentiation (Figure 7.8). The outermost layers contain **spermatogonia,** diploid cells that have increased in number by mitosis. Each spermatogonium increases in size and becomes a **primary spermatocyte.** Each primary spermatocyte then

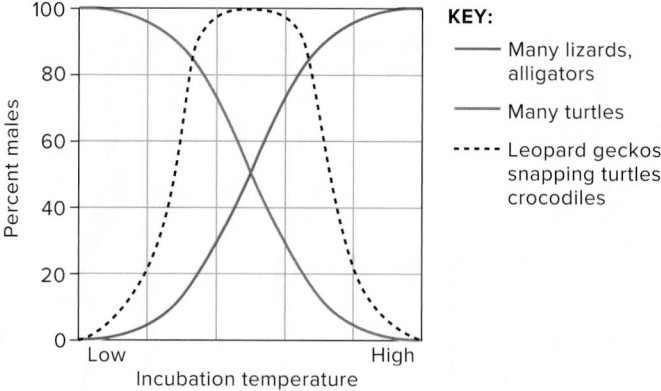

Figure 7.6 Temperature-dependent sex determination. In many reptiles that lack sex chromosomes, incubation temperature of the nest determines gender. The graph shows that embryos of many turtles develop into males at low temperature, whereas embryos of many lizards and alligators become males at high temperatures. Embryos of crocodiles, leopard geckos, and snapping turtles become males at intermediate temperatures, and become females at higher or lower temperatures.

Source: Data from David Crews, "Animal Sexuality," *Scientific American* 270(1):108–114, January 1994.

KEY:
— Many lizards, alligators
— Many turtles
---- Leopard geckos, snapping turtles, crocodiles

(y-axis: Percent males, 0–100; x-axis: Incubation temperature, Low to High)

© Susumu Nishinaga/Science Source

Figure 7.7 Section of a seminiferous tubule containing male germ cells. More than 200 meters long, highly coiled seminiferous tubules are packed in each human testis. This colored scanning electron micrograph reveals numerous spermatozoa within the seminiferous tubules. Each consisting of a head (green), which contains the genetic material that fertilizes the female egg cell, and a tail (blue), which propels the sperm. The heads of the sperm are buried in Sertoli cells (yellow and orange), which nourish the developing sperm.

undergoes the first meiotic division, as described in Chapter 5 (p. 74), to become two **secondary spermatocytes** (Figure 7.8).

For every structure in the reproductive system of males or females, there is a homologous structure in the other. This happens because during early development male and female characteristics begin to differentiate from the embryonic genital ridge (Figure 7.5), and two duct systems develop, which at first are identical in both sexes. Under the influence of sex hormones, the genital ridge develops into the testes of males and the ovaries of females. One duct system (mesonephric or Wolffian) becomes ducts of the testes in males and regresses in females. The other duct system (paramesonephric or Müllerian) develops into the oviducts, uterus, and vagina of females and regresses in males. Similarly, the clitoris and labia of females are homologous to the penis and scrotum of males, because they develop from the same embryonic structures.

Each secondary spermatocyte enters the second meiotic division and produces two **spermatids,** each containing the haploid number (23 in humans) of chromosomes. Thus, in the two steps of meiosis each primary spermatocyte gives rise to four spermatids. A spermatid usually contains a combination of its parents' chromosomes but may contain all chromosomes that the male inherited from his mother or from his father. Without further divisions the spermatids transform into mature **spermatozoa** (or **sperm,** Figure 7.8). Modifications include great reduction of cytoplasm, condensation of the nucleus into a head, formation of a middle piece containing mitochondria, and a whiplike, flagellar tail for locomotion (Figures 7.8 and 7.9). The head consists of a nucleus containing the chromosomes for heredity and an **acrosome,** a distinctive feature of nearly all metazoa (exceptions include teleost fishes and certain invertebrates). In many species, both invertebrate and vertebrate, the acrosome contains enzymes that are released to clear a path through the cellular layers and matrix that surround an egg. In mammals at least, one of the enzymes is hyaluronidase, which allows a sperm to penetrate between the follicular cells surrounding an egg. A striking feature of many invertebrate spermatozoa is the acrosome filament, an extension of varying length in different species that projects suddenly from the sperm

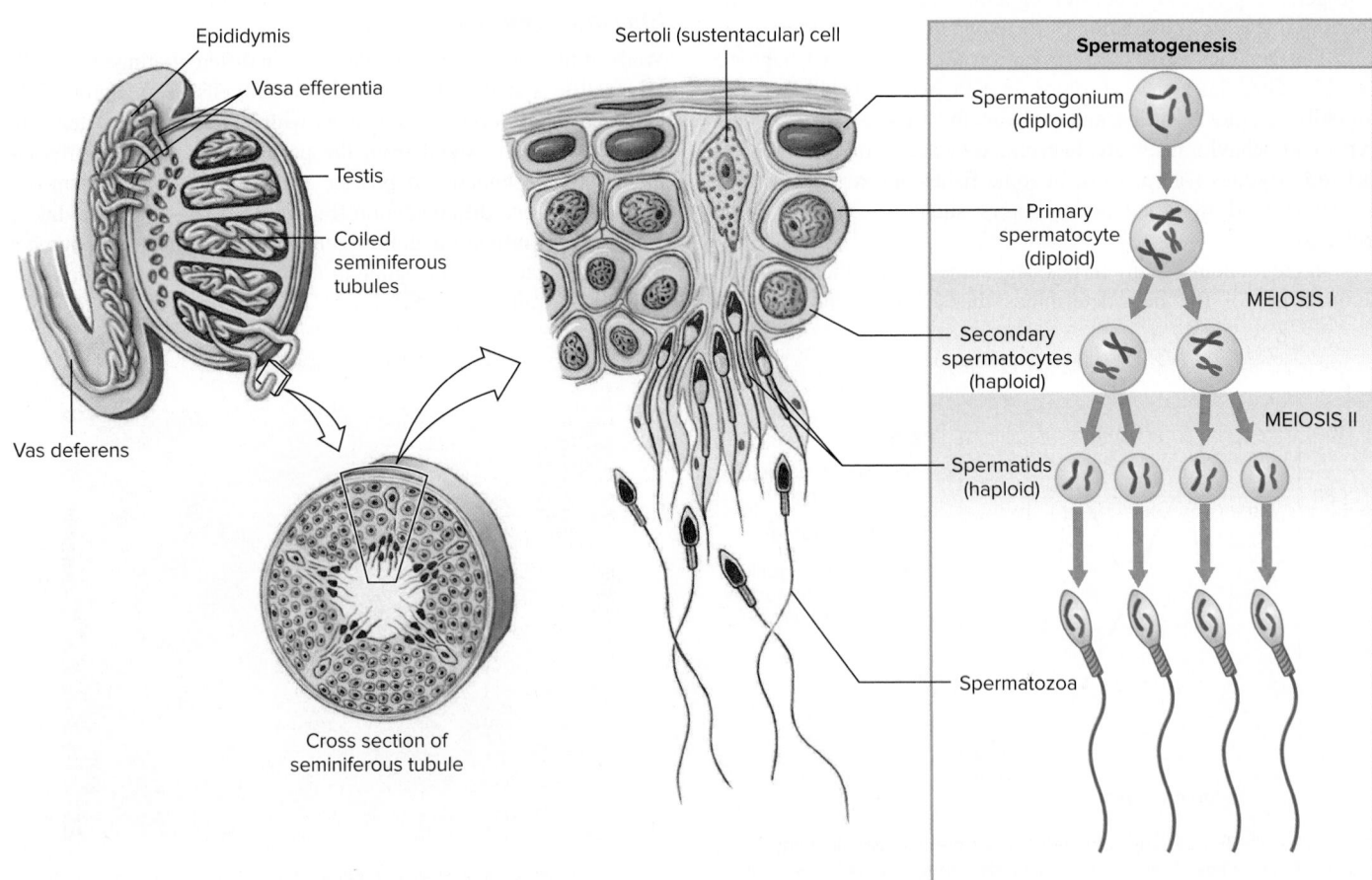

Figure 7.8 Spermatogenesis. Section of seminiferous tubule showing spermatogenesis. Germ cells develop within the recesses of large Sertoli (sustentacular) cells, which extend from the periphery of seminiferous tubules to their lumen, and which provide nourishment to the germ cells. Stem germ cells from which sperm differentiate are the spermatogonia, diploid cells located peripherally in the tubule. These divide by mitosis to produce either more spermatogonia or primary spermatocytes. Meiosis begins when primary spermatocytes divide to produce haploid secondary spermatocytes with replicated chromosomes. The second meiotic division forms four haploid spermatids with nonreplicated chromosomes. As sperm develop, they are gradually pushed toward the lumen of the seminiferous tubule.

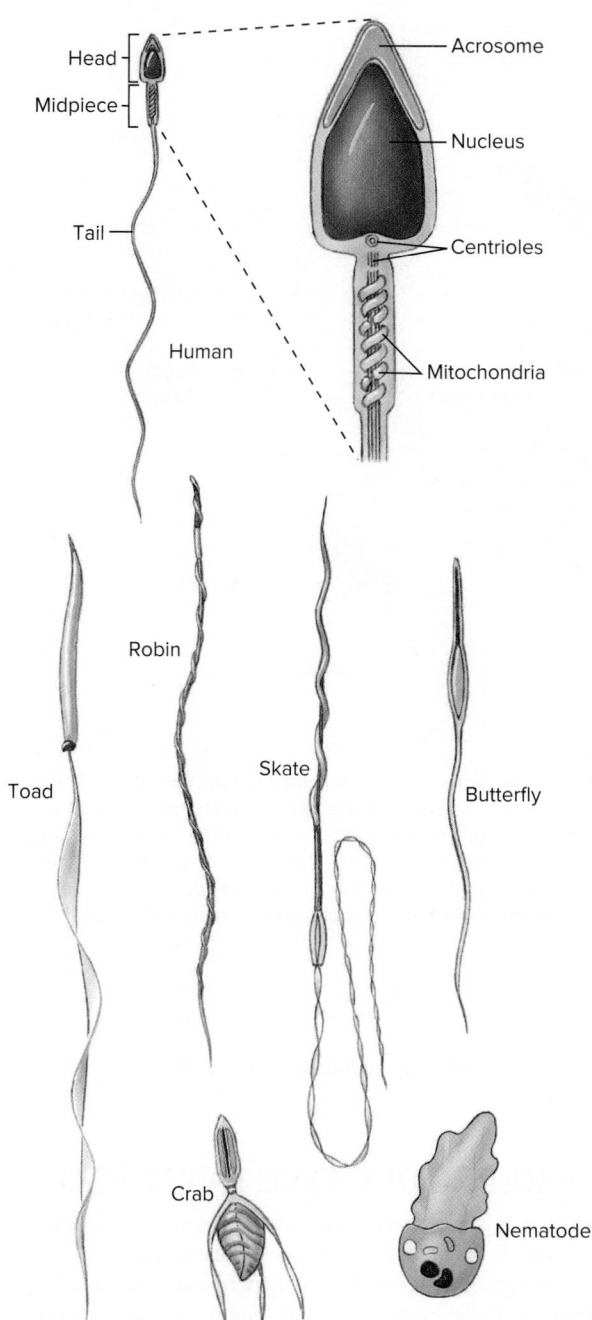

Figure 7.9 Examples of vertebrate and invertebrate sperm. The head and midpiece region of the human sperm is shown in more detail.

head when the latter first contacts the surface of an egg. Fusion of the egg and sperm plasma membranes is the initial event of fertilization (see Contact and Recognition Between Egg and Sperm, p. 157).

The total length of a human sperm is 50 to 70 μm. Some toads have sperm that exceed 2 mm (2000 μm) in length (Figure 7.9) and are easily visible to the unaided eye. Most sperm, however, are microscopic in size (see Figure 8.1 for an early seventeenth-century drawing of a mammalian sperm). In all sexually reproducing animals the number of sperm in males is far greater than the number

of eggs in corresponding females. The number of eggs produced is correlated with the chances of young to hatch and to reach maturity.

Oogenesis

Early germ cells in the ovary, called **oogonia,** increase in number by mitosis. Each oogonium contains the diploid number of chromosomes. After the oogonia cease to increase in number, they grow in size and become **primary oocytes** (Figure 7.10). Before the first meiotic division, the chromosomes in each primary oocyte meet in pairs, paternal and maternal homologues, just as in spermatogenesis. When the first maturation (reduction) division occurs, the cytoplasm is divided unequally. One of the two daughter cells, the **secondary oocyte,** is large and receives most of the cytoplasm; the other is very small and is called the **first polar body** (Figure 7.10). Each of these daughter cells, however, has received half of the chromosomes.

In the second meiotic division, the secondary oocyte divides into a large **ootid** and another small polar body is formed. If the first polar body also divides in this division, which sometimes happens, there are three polar bodies and one ootid (Figure 7.10). The ootid develops into a functional, haploid **ovum.** Polar bodies are nonfunctional, and they disintegrate. Formation of these polar bodies is necessary to dispose of excess chromosomes that result from each nuclear division during meiosis. In addition, the unequal cytoplasmic division makes possible a large cell with the cytoplasm containing a full set of cytoplasmic components needed for early development. Thus a mature ovum or egg has N (haploid) number of chromosomes, the same as a sperm. However, each primary oocyte gives rise to only *one* functional gamete instead of four as in spermatogenesis.

In most vertebrates and many invertebrates, the egg does not actually complete meiotic division before fertilization occurs. The general rule is that development is arrested during prophase I of the first meiotic division (in the primary oocyte phase). Meiosis resumes and is completed either at the time of ovulation (birds and most mammals) or shortly after fertilization (many invertebrates, teleost fishes, amphibians, and reptiles). In humans, the ova begin the first meiotic division at about the thirteenth week of fetal development. Then their development arrests in prophase I as the primary oocyte until puberty, at which time one of these primary oocytes typically develops into a secondary oocyte each menstrual month. In humans, meiosis II is completed only when the secondary oocyte is penetrated by a spermatozoon.

In many animals, the most obvious feature of egg maturation is deposition of yolk. In oviparous animals (see next section, Reproductive Patterns), yolk is composed mainly of proteins and lipids that can contain some carbohydrate and phosphate groups. In addition, some yolk proteins and lipids may also bind minerals that are important for oocyte maturation. Yolk may be synthesized within an egg from raw materials supplied by surrounding follicle cells, or preformed yolk proteins and lipids may be manufactured in the liver and transferred by receptor-mediated endocytosis (see p. 49) from the systemic circulation (see Chapter 31, p. 687) to the oocyte. Yolk is usually stored as granules or vesicles in the oocyte cytoplasm.

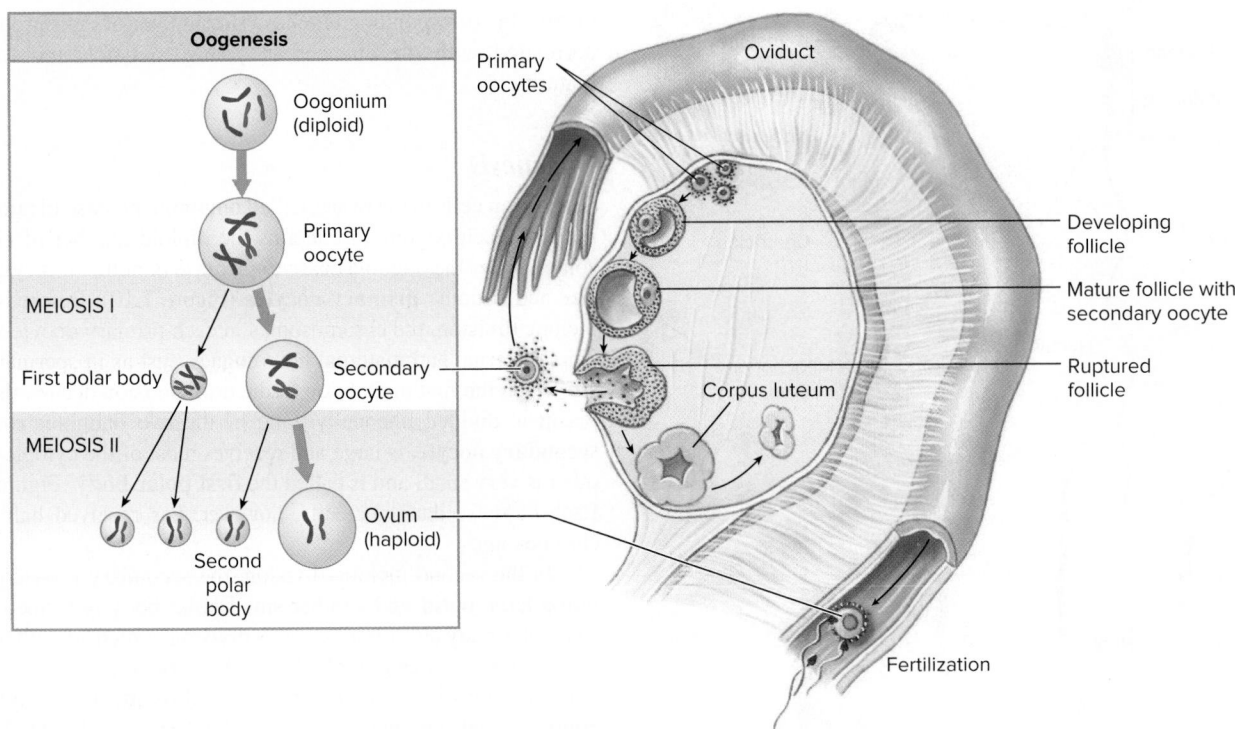

Figure 7.10 Oogenesis in humans. Early germ cells (oogonia) increase by mitosis during embryonic development to form diploid primary oocytes. Prior to birth, each primary oocyte enters meiosis and proceeds to prophase of the first meiotic division. Oocytes remain arrested in this state until after puberty, when each menstrual month, a diploid primary oocyte divides in the first meiotic division into a haploid secondary oocyte and a haploid polar body. The haploid secondary oocyte and attached polar body are released from the ovary during ovulation, and if the secondary oocyte is fertilized, it enters the second meiotic division. The double-stranded chromosomes separate into a large ootid and small second polar body. The ootid develops into an ovum. Both ovum and second polar body now contain *N* number of chromosomes. Fusion of the haploid egg nucleus with a haploid sperm nucleus produces a diploid (2 *N*) zygote.

Eggs also contain a large amount of mRNA that is not translated (p. 90) into polypeptides/proteins until fertilization triggers activation of these previously quiescent mRNA molecules. At this time the newly formed polypeptides/proteins begin to orchestrate the developmental process (see Chapter 8, p. 165).

Enormous accumulation of yolk granules, other nutrients (glycogen and lipid droplets), and quiescent mRNA cause an egg to grow well beyond the normal limits that force ordinary body (somatic) cells to divide. A young frog oocyte 50 μm in diameter, for example, grows to 1500 μm in diameter when mature after 3 years of growth in the ovary, and its volume is increased by a factor of 27,000. Bird eggs attain even greater absolute size; a hen egg increases 200 times in volume in only the last 6 to 14 days of rapid growth preceding ovulation.

Thus eggs are remarkable exceptions to the otherwise universal rule that organisms are composed of relatively minute cellular units. An egg's large size creates a problematic surface area-to-cell volume ratio (see p. 193), because everything that enters and leaves the ovum (nutrients, respiratory gases, wastes, and so on) must pass through the plasma membrane. As the egg becomes larger, the available surface per unit of cytoplasmic volume (mass) becomes smaller. As we would anticipate, the metabolic rate of an egg gradually diminishes until a secondary oocyte or ovum (depending on the species) is in suspended animation until fertilization (the relationship between metabolic rate and cell size is described on p. 193).

7.3 REPRODUCTIVE PATTERNS

A great majority of invertebrates, as well as many vertebrates, lay their eggs outside the body for development; these animals are called **oviparous** ("egg-birth"). Fertilization may be either internal (eggs are fertilized inside the body of a female before she lays them) or external (eggs are fertilized by a male after a female lays them). While many oviparous animals simply abandon their eggs rather indiscriminately, others display extreme care in finding places that provide immediate and suitable sources of food for the young when they hatch.

Fertilized eggs of some animals remain in the oviduct or uterus while they develop, with embryos deriving all their nourishment from yolk stored within the egg. These animals are called **ovoviviparous** ("egg-live-birth"). Ovoviviparity occurs in several invertebrate groups (for example, various annelids, brachiopods, insects, and gastropod molluscs) and is common among certain fishes (p. 538) and reptiles (p. 577).

In the third pattern, **viviparous** ("live-birth"), fertilized eggs develop in the oviduct or uterus with embryos deriving their nourishment directly from the mother. Usually some kind of intimate anatomical relationship is established between developing embryos and their mother. In both ovoviviparity and viviparity, fertilization must be internal (within the body of the female) and the mother gives birth to young usually in a more advanced stage of development. Viviparity is confined mostly to lizards, snakes, mammals, and elasmobranch fishes, although viviparous invertebrates (scorpions, for example) and amphibians are known. Development of embryos within a mother's body, whether ovoviviparous or viviparous, obviously affords more protection to the offspring than does egg-laying. Some physiologists consider ovoviviparity as a special kind of viviparity—**lecithotrophic** viviparity.

7.4 STRUCTURE OF REPRODUCTIVE SYSTEMS

The basic components of reproductive systems are similar in sexual animals, although differences in reproductive habits and methods of fertilization have produced many variations. Sexual systems consist of two components: (1) **primary organs,** which are the gonads that produce sperm and eggs and sex hormones; and (2) **accessory organs,** which assist the gonads in formation and delivery of gametes, and may also serve to support the embryo. They are of great variety, and include gonoducts (sperm ducts and oviducts), accessory organs for transferring spermatozoa into the female, storage organs for spermatozoa or yolk, packaging systems for eggs, and nutritional organs such as yolk glands and placenta.

Invertebrate Reproductive Systems

Invertebrates that transfer sperm from male to female require organs and plumbing for internal fertilization that may be as complex as those of any vertebrate (see Figure 7.11 and Figure 20.6). In contrast, reproductive systems of invertebrates that simply release their gametes into water for external fertilization may be little more than centers for gametogenesis (see Figures 16.10, 16.31, and 22.2). Polychaete annelids, for example, have no permanent reproductive organs. Gametes arise by proliferation of cells lining the body cavity. When mature, the gametes are released through coelomic or nephridial ducts or, in some species, may exit through ruptures in the body wall (see p. 368 and Figures 17.6 and Figures 17.7).

Insects have separate sexes (dioecious), practice internal fertilization by copulation and insemination, and consequently have complex reproductive systems (Figure 7.11). Sperm from the testes pass through sperm ducts to seminal vesicles (where the sperm are stored) and then through a single ejaculatory duct to a penis. Seminal fluid from one or more accessory glands is added to the semen in the ejaculatory duct. Females have a pair of ovaries formed from a series of egg tubes (ovarioles). Mature ova pass through oviducts to a common genital chamber and then to a short copulatory bursa (vagina). In many more highly evolved insects, the male transfers sperm by inserting the penis directly into the female's genital bursa (vagina); from here they migrate, and are stored in a seminal receptacle (see p. 455 and Figure 21.21). Spiders, scorpions, and some insects package sperm into a spermatophore that is deposited directly into the vagina of the female or close to the female for her to deposit into her vagina. Scorpions have spring-loaded spermatophores that they guide the females across during a complex mating dance. The spermatophore may also contain nutrients. Often a single mating provides sufficient sperm to last the reproductive life of a female.

Vertebrate Reproductive Systems

In vertebrates the reproductive and excretory systems are together called the **urogenital system** because of their close anatomical connection, especially in males. This association is very striking during

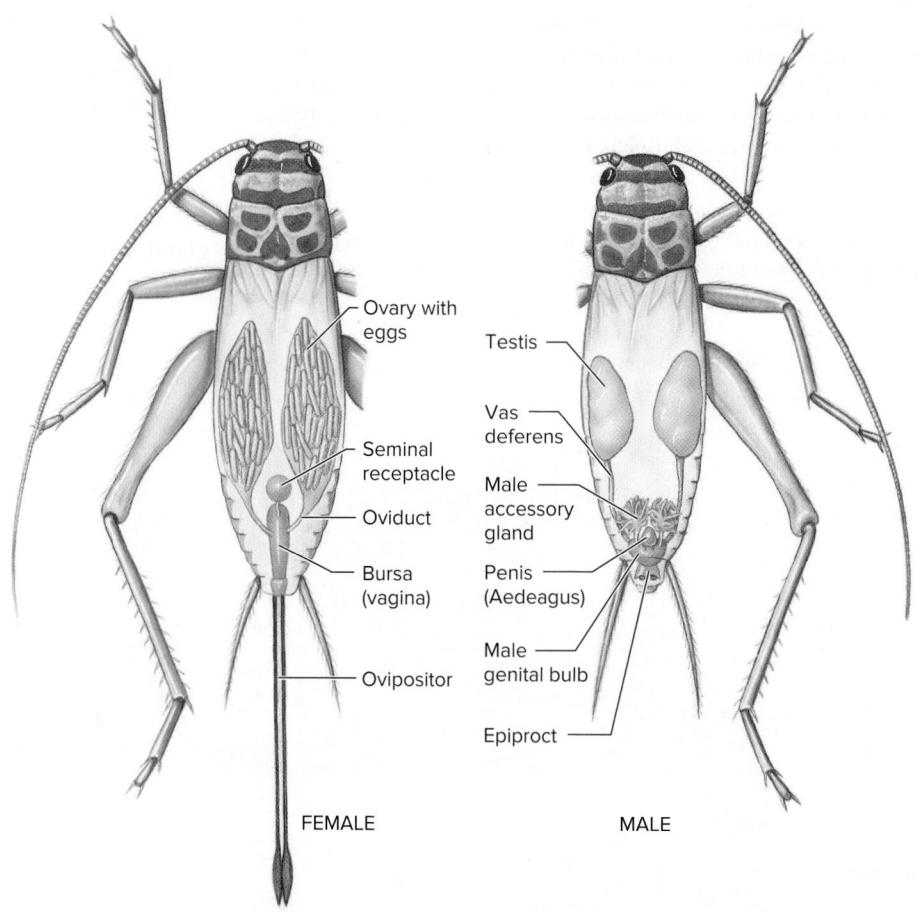

Figure 7.11 Reproductive system of crickets. Sperm from the paired testes of males pass through sperm tubes (vas deferens) to an ejaculatory duct housed in the penis. In females, eggs from the ovaries pass through oviducts to the genital bursa. At mating, sperm enclosed in a membranous sac (spermatophore) formed by the secretions of the accessory gland are deposited in the genital bursa of the female, then migrate to her seminal receptacle, where they are stored. The female controls the release of a few sperm to fertilize her eggs at the moment they are laid, using the needlelike ovipositor to deposit eggs in the soil.

embryonic development. In male fishes and amphibians, the duct that drains the kidney (**opisthonephric duct** or **Wolffian duct**) also serves as a sperm duct (see Figure 30.9 for an evolutionary perspective of kidney and male reproductive duct development). In male reptiles, birds, and mammals in which the kidney develops its own independent duct (**ureter**) to carry away waste, the old **mesonephric duct** becomes exclusively a sperm duct or **vas deferens.** In all these forms, with the exception of most mammals, the ducts open into a **cloaca** (derived, appropriately, from the Latin meaning "sewer"), a common chamber into which intestinal, reproductive, and excretory canals empty. Almost all placental mammals have no cloaca; instead the urogenital system has its own opening separate from the anal opening. In females, the **oviduct** or **uterine duct** is an independent duct that opens to the outside or into the cloaca in animals having a cloaca.

Male Reproductive System

The male reproductive system of vertebrates, such as that of human males (Figure 7.12), includes testes, vasa efferentia, vas deferens, accessory glands, and (in some birds and reptiles, and all mammals) a penis.

Paired **testes** are the sites of sperm production. Each testis is composed of numerous **seminiferous tubules,** in which the sperm develop (Figure 7.8). The sperm are surrounded by Sertoli cells (or **sustentacular cells**), which nourish the developing sperm. Between the tubules are **interstitial cells** (or **Leydig cells**), which produce the male sex hormone (**testosterone**). In most mammals the two testes are housed permanently in a saclike scrotum suspended outside the abdominal cavity, or the testes descend into the scrotum during the breeding season. This odd arrangement provides an environment of slightly lower temperature, since in most mammals (including humans) viable sperm do not form at the higher temperatures maintained within the body. In marine mammals and all other vertebrates the testes are positioned permanently within the abdomen.

The sperm travel from the seminiferous tubules to the **vasa efferentia,** small tubes connected to a coiled **epididymis** (one for each testis), where final sperm maturation occurs and then to a **vas deferens,** the ejaculatory duct (Figures 7.8 and 7.12). In mammals the vas deferens joins the **urethra,** a duct that carries both sperm and urinary products through the **penis,** or external intromittent organ.

Most aquatic vertebrates have no need for a penis, because sperm and eggs are liberated into the water in close proximity to each other. However, in terrestrial (and some aquatic) vertebrates that bear their young alive or enclose the egg within a shell or egg case, sperm must be transferred to the female. Few birds have a true penis (examples of exceptions are the ostrich and the Argentine lake duck) and the mating process simply involves pressing cloaca to cloaca. Most reptiles and mammals have a true penis. In mammals the normally flaccid organ becomes erect when engorged with blood. Some mammals possess a bone in the penis (baculum), which presumably helps with rigidity and therefore sperm transfer.

In most mammals three sets of accessory glands open into the reproductive channels: a pair of **seminal vesicles,** a single **prostate gland,** and a pair of **bulbourethral glands** (Figure 7.12). Fluid secreted by these glands furnishes food to the sperm, lubricates the female reproductive tract for sperm, and counteracts the acidity of the vagina so that the sperm retain their viability longer after being deposited in the female.

Female Reproductive System

The ovaries of female vertebrates produce both ova and female sex hormones (estrogens and progesterone). In all jawed vertebrates (see p. 512) mature ova from each ovary enter the funnel-like opening of an **oviduct** (also termed Fallopian tube or uterine tube), which typically has a fringed margin (fimbriae) that envelops the ovary at the time of ovulation. The terminal end of the uterine tube is unspecialized in most fishes and amphibians, but in cartilaginous fishes, reptiles, and birds that produce a large, shelled egg, special regions have developed for production of albumin and shell. In amniotes (reptiles, birds, and mammals; see Amniotes and the Amniotic Egg, p. 174), the terminal portion of the uterine tube is expanded into a muscular **uterus** in which shelled eggs are held before laying or in which embryos complete their development. In placental mammals, the walls of the uterus

Figure 7.12 Human male reproductive system showing the reproductive structures in sagittal view.

Labels: Peritoneum, Vas deferens, Urinary bladder, Ureter, Seminal vesicle, Vertebral column, Pubic bone, Prostate gland, Urethra, Penis, Glans penis, Prepuce, External urinary meatus, Testicle, Scrotum, Bulbourethral gland, Epididymis, Anus, Ejaculatory duct, Rectum

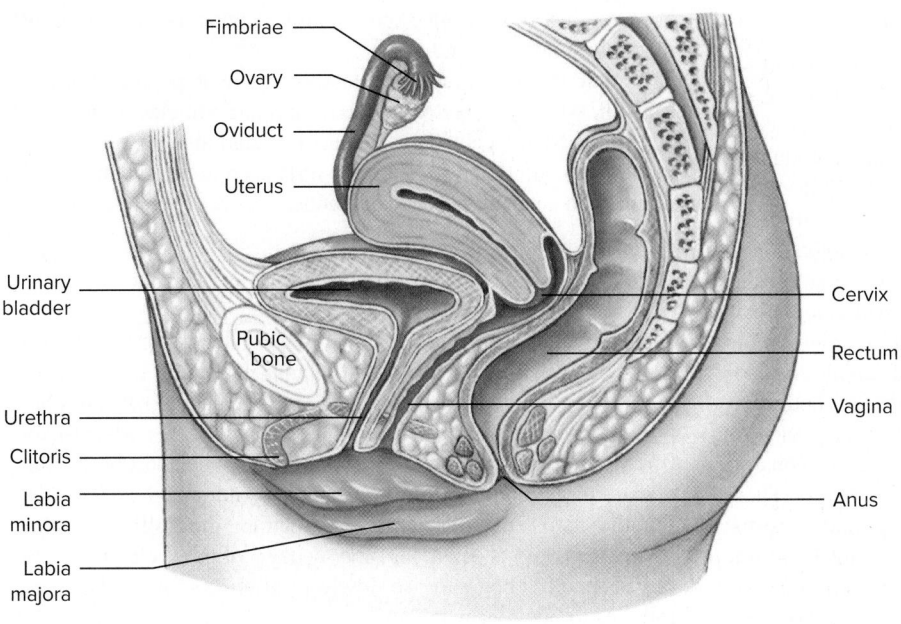

Figure 7.13 Human female reproductive system showing the pelvis in sagittal section

corners of the uterus, or womb, which is specialized for housing the embryo during its intrauterine existence. It consists of thick muscular walls, many blood vessels, and a specialized lining: the **endometrium.** The uterus varies among different mammals, and in many it is designed to hold more than one developing embryo. Ancestrally it was paired but is fused to form one large chamber in some eutherian mammals.

The **vagina** is a muscular tube adapted to receive the male's penis and serves as the birth canal during expulsion of a fetus from the uterus. Where vagina and uterus meet, the uterus projects down into the vagina to form a **cervix.**

The external genitalia of human females, or **vulva,** include folds of skin, the **labia majora** and **labia minora,** and a small erectile organ, the **clitoris** (the female homolog of the glans penis of males). The opening into the vagina is often reduced in size in the virgin state by a membrane, the **hymen,** although in today's more physically active females, this membrane may be much reduced in extent.

establish a close vascular association with the embryonic membranes through a **placenta** (see p. 175).

The paired ovaries of the human female (Figure 7.13), slightly smaller than the male testes, contain many thousands of oocytes. Each oocyte develops within a **follicle** that enlarges and finally ruptures to release a secondary oocyte (see Figure 7.10). During a woman's fertile years, except following fertilization, approximately 13 oocytes mature each year, and usually the ovaries alternate in releasing oocytes. Because a woman is fertile for only about 30 years, of the approximately 400,000 primary oocytes in her ovaries at birth, only 300 to 400 have a chance to reach maturity; the others degenerate and are resorbed.

A long-held principle of mammalian reproductive biology has been that in the male, germ-cell lines continue to remain functional and form sperm throughout adult life, while females possess a finite number of germ cells, and oocyte production ceases at birth. Indeed, we have just described human follicular development as such a case, where the primary oocytes present at birth provide her only source of follicles. An exciting discovery in mice has challenged this reproductive dogma. Juvenile and adult mouse ovaries have been shown to possess actively dividing germ cells that replenish the oocyte pool. In addition, germline stem cells have now been identified in both aged mouse ovaries and postmenopausal human ovaries. If this evidence can be extended to other mammalian species, there could be significant implications in the management of endangered species, where assisted reproductive techniques could be used to expand pools of oocytes that might mean the difference between extinction and survival.

The **oviducts** are lined with cilia for propelling the egg away from the ovary from which it was released. The two ducts open into the upper

7.5 ENDOCRINE EVENTS THAT ORCHESTRATE REPRODUCTION
Hormonal Control of Timing of Reproductive Cycles

From fishes to mammals, reproduction in vertebrates is usually a seasonal or cyclic activity. Timing is crucial, because offspring should appear when food is available and other environmental conditions are optimal for survival. The sexual reproductive process is controlled by hormones, which are regulated by environmental cues, such as food intake, and seasonal changes in photoperiod, rainfall, or temperature, and by social cues. A region within the forebrain called the hypothalamus (p. 730) regulates the release of anterior pituitary gland hormones, some of which stimulate tissues of the gonads (neurosecretion and the pituitary gland are described in Chapter 34). This hormonal system controls development of the gonads, accessory sex structures, and secondary sexual characteristics (see next section, Gonadal Steroids and Their Control), as well as timing of reproduction.

The cyclic reproductive patterns of female mammals are of two types: **estrous cycle,** characteristic of most mammals, and **menstrual cycle,** characteristic only of the anthropoid primates (monkeys, apes, and humans). These two cycles differ in two important ways. First, in estrous cycles, females are receptive to males only during brief periods of **estrus,** or "heat," whereas in the menstrual cycle receptivity may occur throughout the cycle (although some data indicate an increase in receptivity around the time of ovulation). Second, a menstrual cycle, but not an estrous cycle, ends with breakdown and discharge of the inner portion of the uterus (endometrium). In an estrous cycle, each cycle ends with the endometrium simply reverting to its original state, without the discharge characteristic of the menstrual cycle.

Gonadal Steroids and Their Control

The ovaries of female vertebrates produce two kinds of steroid sex hormones—**estrogens** and **progesterone** (Figure 7.14). There are three kinds of estrogens: estradiol, estrone, and estriol, of which estradiol is secreted in the highest amounts during reproductive cycles. Estrogens are responsible for development of female accessory sex structures (oviducts, uterus, and vagina) and for stimulating female reproductive activity. Secondary sex characters, those characteristics that are not primarily involved in formation and delivery of ova (or sperm in males), but that are essential for behavioral and functional success of reproduction, are also controlled or maintained by estrogens. Secondary sexual characteristics include distinctive skin or feather coloration, bone development, body size, and, in mammals, initial development of the mammary glands. In female mammals, both estrogen and progesterone are responsible for preparing the uterus to receive a developing embryo. These hormones are controlled by **anterior pituitary gonadotropins: follicle-stimulating hormone (FSH)** and **luteinizing hormone (LH)** (Figure 7.15). The release of these two gonadotropins is in turn governed by **gonadotropinreleasing hormone (GnRH),** produced by neurosecretory cells in the **hypothalamus** (see p. 730 and Table 34.1). Through this control system environmental factors such as light, nutrition, and stress may influence reproductive cycles. Estrogens and progesterone feed back to the hypothalamus and anterior pituitary to keep secretion of GnRH, FSH, and LH in check (see Chapter 34, for a discussion of negative feedback of hormones).

The male sex steroid, **testosterone** (Figure 7.14), is manufactured by the **interstitial cells** of the testes. Testosterone and its metabolite, **dihydrotestosterone (DHT),** are necessary for the growth and development of the male accessory sex structures (penis, sperm ducts, and glands), development of secondary male sex characters (such as bone and muscle growth, male plumage or pelage coloration, antlers in deer, and, in humans, voice quality), and male sexual behavior. Development of the testes and secretion of testosterone is controlled by FSH and LH, the same anterior pituitary hormones that regulate the female reproductive cycle, and ultimately by GnRH from the hypothalamus. Like estrogens and progesterone in the female, testosterone and DHT feed back to the hypothalamus and anterior pituitary to regulate secretion of GnRH, FSH, and LH.

Identification of a peptide in the hypothalamus of birds and mammals that inhibits the secretion of GnRH and LH has led some scientists to believe that a **gonadotropin-inhibiting hormone (GnIH)** has finally been discovered. Further study is necessary, however, before we can be sure that this peptide antagonizes GnRH in all physiological conditions.

Both the ovary and the testes secrete a peptide hormone, **inhibin,** which is secreted by the developing follicles in the female and by the Sertoli cells (or sustentacular cells) in the male. This hormone is an additional regulator of FSH secretion from the anterior pituitary in a negative feedback manner.

The Menstrual Cycle

The human menstrual cycle (L. *mensis,* month) consists of two distinct phases within the ovary—follicular phase and luteal phase—and three distinct phases within the uterus: menstrual phase, proliferative phase, and secretory phase (Figure 7.15). Menstruation (the "period") signals the **menstrual phase,** when part of the lining of the uterus (endometrium) degenerates and sloughs off, producing the menstrual discharge. Meanwhile, the **follicular phase** within the ovary is occurring, and by day 3 of the cycle blood levels of FSH and LH begin to rise slowly, prompting some of the ovarian follicles to begin growing and to secrete estrogen. As estrogen levels in the blood increase, the uterine endometrium heals and begins to thicken, and uterine glands within the endometrium enlarge **(proliferative phase).** By day 10 most of the ovarian follicles that began to develop at day 3 now degenerate (become **atretic),** leaving only one (sometimes two or three) to continue developing until it appears as a bulge on the surface of the ovary. This is a mature follicle or **graafian follicle.** During the latter part of the follicular phase, the graafian follicle secretes more estrogen, and also inhibin. As the levels of inhibin rise, the levels of FSH fall.

At day 13 or 14 in the cycle, the now high levels of estrogen from the graafian follicle stimulate a surge of GnRH from the hypothalamus, which induces a surge of LH (and to a lesser extent, FSH) from the anterior pituitary. The LH surge causes the graafian follicle to rupture **(ovulation),** releasing an oocyte from the ovary. The oocyte remains viable for approximately 12 hours, during which time it may be fertilized by a sperm. During the ovarian **luteal phase,** a **corpus luteum** ("yellow body" for its appearance in cow ovaries, where it was first described) forms from the remains of the ruptured follicle that released the oocyte at ovulation (Figures 7.10 and 7.15). The corpus luteum, responding to continued stimulation by LH, becomes a transitory endocrine gland that secretes progesterone (and estrogen in primates). Progesterone ("before carrying [gestation]"), as its name implies, stimulates the uterus to undergo final maturational changes that prepare it for gestation **(secretory phase).** The uterus is now fully

Figure 7.14 Sex hormones. These three sex hormones show the basic four-ring steroid structure. The main female sex hormone, estradiol (an estrogen), is a C_{18} (18-carbon) steroid with an aromatic A ring (first ring to left). The main male sex hormone, testosterone (an androgen), is a C 19 steroid with a carbonyl group (C=O) on the A ring. The female sex hormone progesterone is a C 21 steroid, also bearing a carbonyl group on the A ring.

Testosterone

Progesterone

Estradiol-17β

Figure 7.15 Human menstrual cycle, showing changes in blood-hormone levels and uterine endometrium during the 28-day ovarian cycle. FSH promotes maturation of ovarian follicles, which secrete estrogen. Estrogen prepares the uterine endometrium and causes a surge in LH, which in turn causes ovulation and stimulates the corpus luteum to secrete progesterone and estrogen. Progesterone and estrogen production persist only if an ovum is fertilized; without pregnancy progesterone and estrogen levels decline and menstruation follows.

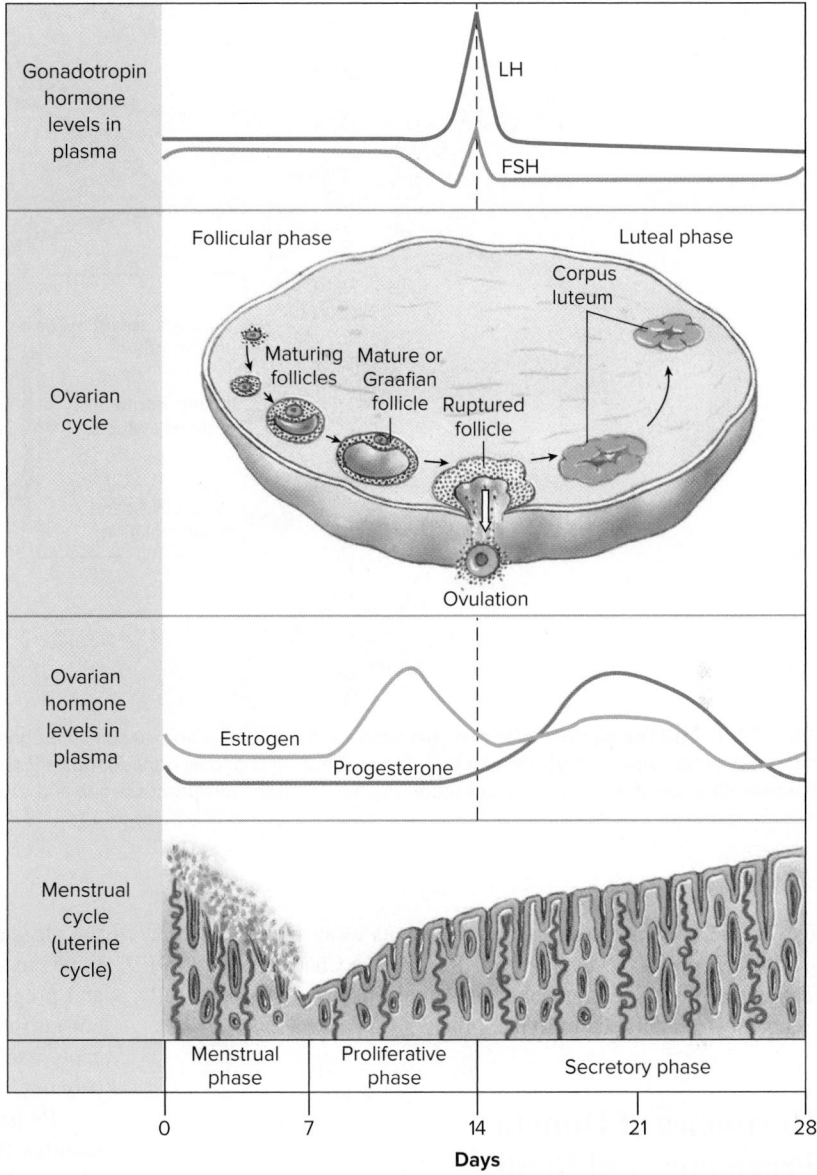

ready to house and to nourish an embryo. If fertilization has *not* occurred, the corpus luteum degenerates, and its hormones are no longer secreted. Because the uterine lining (endometrium) depends on progesterone and estrogen for its maintenance, their declining levels cause the uterine lining to deteriorate, leading to menstrual discharge of the next cycle.

Oral contraceptives (the "Pill") usually are combined preparations of estrogen and progesterone that act to decrease the output of pituitary gonadotropins FSH and LH. This prevents the ovarian follicles from ripening fully and usually prevents ovulation from occurring. Oral contraceptives are highly effective, with a failure rate of less than 1% if the treatment procedure is followed properly. Estrogen and progesterone also can be administered as a skin patch (Ortho

Evra), or as a vaginal ring (NuvaRing). Progesterone acts on the reproductive tract as a whole, making it inhospitable for sperm and any fertilized oocyte. This mechanism has been exploited in progesterone-only contraceptives ("mini-pill," Depo-Provera, Implanon), which may not block follicular development or ovulation, and in emergency contraception (Plan B One-Step or the "morning after pill") that is taken postcoitus and is now available over the counter in the United States.

GnRH from the hypothalamus, and LH and FSH from the anterior pituitary, are controlled by **negative feedback** of ovarian steroids (and inhibin). This negative feedback occurs throughout the menstrual cycle, except for a few days before ovulation. As just mentioned, ovulation is due to the *high levels of estrogen* causing a surge of GnRH, LH (and FSH). Such **positive feedback** mechanisms are

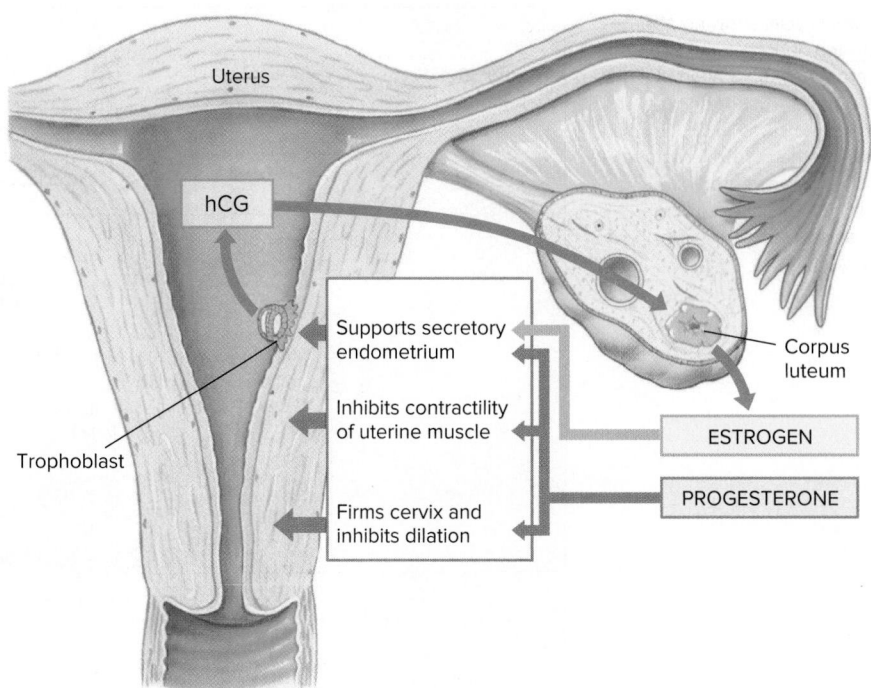

Figure 7.16 The multiple roles of progesterone and estrogen in normal human pregnancy. After implantation of an embryo in the uterus, the trophoblast (the future embryo and placenta) secretes human chorionic gonadotropin (hCG), which maintains the corpus luteum until the placenta, at about the seventh week of pregnancy, begins producing the sex hormones progesterone and estrogen.

rare in the body, because they move events away from stable set points. (Feedback mechanisms are described in Chapter 34, p. 749) This surge event is terminated by ovulation when estrogen levels fall as an oocyte is released from the follicle.

Hormones of Human Pregnancy and Birth

If fertilization occurs, it normally does so in the first third of the oviduct **(ampulla).** The **zygote** travels from here to the uterus, dividing by mitosis to form a **blastocyst** (see Chapter 8, p. 161) by the time it reaches the uterus. The developing blastocyst adheres to the uterine surface after about 6 days and embeds itself in the endometrium. This process is called **implantation.** Growth of the embryo continues, producing a spherically shaped **trophoblast.** This embryonic stage contains three distinct tissue layers, the amnion, chorion, and embryo proper, the inner cell mass (see Figure 8.26, p. 176). The **chorion** secretes **human chorionic gonadotropin (hCG),** which appears in the bloodstream soon after implantation. hCG stimulates the corpus luteum to continue to synthesize and to release both estrogen and progesterone (Figure 7.16).

The placenta forms the point of attachment between trophoblast and uterus (evolution and development of the placenta are described in Chapter 8, p. 175). Besides serving as a medium for the transfer of materials between maternal and fetal bloodstreams,

the placenta also serves as an endocrine gland. The placenta continues to secrete hCG and also produces estrogen (mainly estriol) and progesterone. After about the third month of pregnancy, the corpus luteum degenerates in some mammals, but by then the placenta itself is the main source of both progesterone and estrogen (Figure 7.17).

Preparation of the mammary glands for secretion of milk requires two additional hormones, **prolactin (PRL)** and **human placental lactogen (hPL)** (or **human chorionic somatomammotropin**). PRL is produced by the anterior pituitary, but in nonpregnant women its secretion is inhibited. During pregnancy, elevated levels of progesterone and estrogen depress the inhibitory signal, and PRL begins to appear in the blood. PRL is also produced by the placenta during pregnancy. PRL, in combination with hPL, prepares the mammary glands for milk secretion. hPL, together with **human placental growth hormone (hPGH)** and maternal growth hormone, also stimulates an increase in available nutrients in the mother, so that more are provided to the developing embryo. The placenta also secretes-endorphin and other endogenous opioids (see Chapter 33, p. 737) that regulate appetite and mood during pregnancy. Opioids may also contribute to a sense of well-being and help to alleviate some of the discomfort associated with the later months of pregnancy. Later the placenta begins to synthesize a peptide hormone called **relaxin;** this hormone allows some expansion of the pelvis by increasing the flexibility of the pubic symphysis (see Figure 29.9), and also dilates the cervix in preparation for delivery.

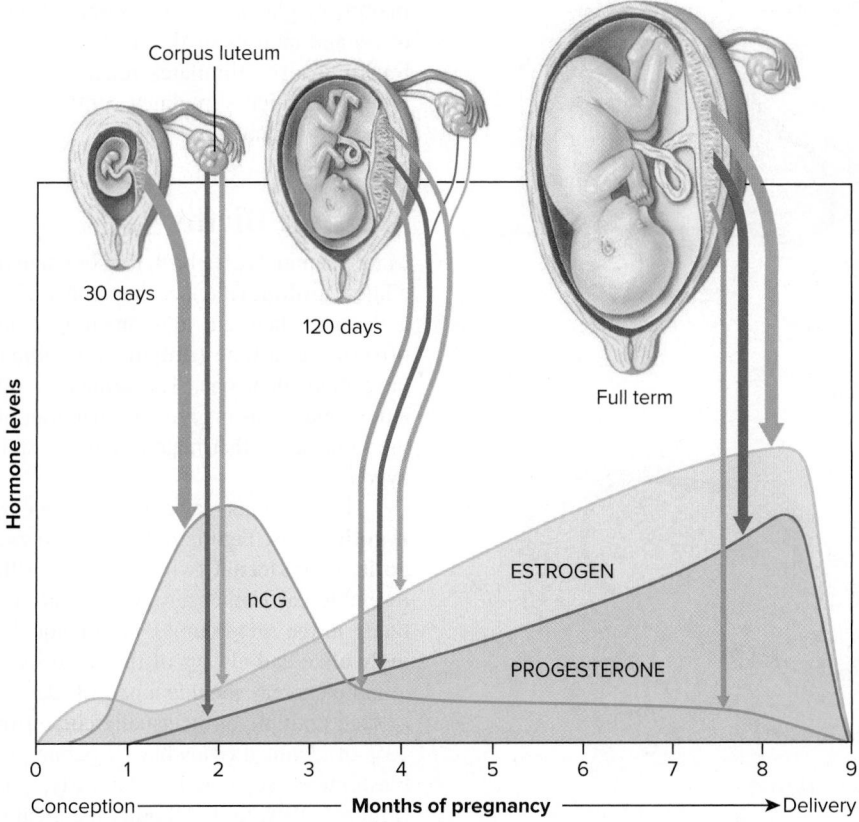

Figure 7.17 Hormone levels released from the corpus luteum and placenta during pregnancy. The width of the arrows suggests the relative amounts of hormone released; hCG (human chorionic gonadotropin) is produced solely by the placenta. Synthesis of progesterone and estrogen shifts during pregnancy from the corpus luteum to the placenta.

Birth, or **parturition,** occurs after approximately 9 months in humans and begins with a series of strong, rhythmic contractions of the uterine musculature, called **labor.** The exact signal that triggers birth is not fully understood in humans, but **placental corticotropinreleasing hormone (CRH)** appears to initiate the birth process. Just before birth, and locally within the placenta and uterus, secretion of estrogen, which stimulates uterine contractions, rises sharply, while the level of progesterone, which inhibits uterine contractions, declines (Figure 7.17). This removes the "progesterone block" that keeps the uterus quiescent throughout pregnancy. **Prostaglandins,** a large group of hormones (long-chain fatty acid derivatives), also increase at this time, making the uterus more responsive (see Chapter 34, p. 754, for more on prostaglandins). Finally, stretching of the cervix sets in motion neural reflexes that stimulate secretion of **oxytocin** from the posterior pituitary. Oxytocin stimulates uterine smooth muscle, leading to stronger and more frequent labor contractions. Secretion of oxytocin during childbirth is another example of **positive feedback.** This time the event is terminated by birth of the baby.

Childbirth occurs in three stages. In the first stage the cervix is enlarged by pressure from the baby in its bag of amniotic fluid, which may be ruptured at this time (**dilation;** Figure 7.18B). In the second stage, the baby is forced out of the uterus and through the vagina to the outside (**expulsion;** Figure 7.18C). In the third stage, the placenta, or **afterbirth,** is expelled from the mother's body, usually within 10 minutes after the baby is born (**placental delivery;** Figure 7.18D).

Miscarriages during pregnancy, or spontaneous abortions, are quite common and appear to serve as a mechanism to reject prenatal abnormalities caused by chromosomal damage and other genetic errors. Exposure to drugs or toxins, immune irregularities, or improper hormonal priming of the uterus also may cause miscarriage. Modern hormonal tests show that about 30% of fertile zygotes are spontaneously aborted before or right after implantation; such miscarriages are unknown to the mother or are expressed as a slightly late menstrual period. Another 20% of established pregnancies end in miscarriage (those known to the mother), giving a spontaneous abortion rate of about 50%.

After birth, secretion of milk is triggered when the infant sucks on its mother's nipple. This leads to a reflex release of oxytocin from the posterior pituitary; when oxytocin reaches the

A Human fetus just before birth

B First stage of labor dilation

C Second stage of labor: expulsion

D Third stage of labor: placental delivery

Figure 7.18 Birth, or parturition, in humans.

mammary glands it causes contraction of smooth muscles lining ducts and sinuses of the mammary glands and ejection of milk. Suckling also stimulates release of prolactin from the anterior pituitary, which stimulates continued production of milk by the mammary glands.

Multiple Births

Many mammals give birth to more than one offspring at a time or to a litter (**multiparous**), each member of which has come from a separate egg. There are some mammals, however, that have only one offspring at a time (**uniparous**), although occasionally they may have more than one. The armadillo (*Dasypus*) is almost unique among mammals in giving birth to four offspring at one time—all of the same sex, either male or female, and all derived from the same zygote.

Human twins may come from one zygote (**identical,** or **monozygotic** twins; Figure 7.19A) or two zygotes (**nonidentical, dizygotic,** or **fraternal** twins; Figure 7.19B). Fraternal twins do not resemble each other any more than do other children born separately in the same family, but identical twins are, of course, strikingly alike and always of the same sex. Triplets, quadruplets, and quintuplets may include a pair of identical twins. The other babies in such multiple births usually come from separate zygotes. About 33% of identical twins have separate placentas, indicating that the blastomeres separated at an early, possibly the two-cell, stage (Figure 7.19A, *top*). All other identical twins share a common placenta, indicating that splitting occurred after formation of the inner cell mass (see Figure 8.26 on p. 177). If splitting were to happen after placenta formation, but before the amnion forms, the twins would have individual amniotic sacs (Figure 7.19A, *middle*), as observed in the great majority of identical twins. Finally, a very small percentage of identical twins share one amniotic sac and a single placenta (Figure 7.19A, *bottom*), indicating that separation occurred after day 9 of pregnancy, by which time the amnion has formed. In these cases, the twins are at risk of becoming conjoined, a condition known as Siamese twinning. Embryologically, each member of fraternal twins has its own placenta and amnion (Figure 7.19B).

The frequency of twin births in comparison to single births is approximately 1 in 86, that of triplets 1 in 86^2, and that of quadruplets approximately 1 in 86^3. The ratio of identical twin births to all births is about the same throughout the world (1 in 250–300). A striking exception occurs in the village of Mohammad Pur Umri in India, where 1 in 10 births are identical twins. Scientists have yet to agree on the explanation for this high twinning rate. The frequency of fraternal births varies with race and country. In the United States, three-fourths of all twin births are dizygotic (fraternal), whereas in Japan only about one-fourth are dizygotic. The tendency for fraternal twinning (but apparently not identical twinning) seems to run in family lines; fraternal twinning (but not identical twinning) also increases in frequency as mothers get older.

A MONOZYGOTIC (IDENTICAL) TWINS

Inner cell mass

Two placentas

Blastomere

Splitting at 2-cell stage

Blastocoel

Fertilization of single oocyte

Complete split of inner cell mass

One placenta

Two amnions

Split of inner cell mass late in development

One placenta

One amnion

B DIZYGOTIC (FRATERNAL) TWINS

Blastomere

Placenta

Embryo

Yolk sac

Fertilization of 2 different oocytes

Figure 7.19 Formation of human twins. **A,** Monozygotic (identical) twin formation. **B,** Dizygotic (fraternal) twin formation. See text for explanation.

SUMMARY

Reproduction is the production of new life and provides an opportunity for evolution to occur. Asexual reproduction is a rapid and direct process by which a single organism produces genetically identical copies of itself. It may occur by fission, budding, gemmulation, or fragmentation. Sexual reproduction involves production of gametes (sex cells), usually by two parents (bisexual reproduction), which combine by fertilization to form a zygote that develops into a new individual. Gametes are formed by meiosis, reducing the number of chromosomes to haploid, and the diploid chromosome number is restored at fertilization. Sexual reproduction recombines parental characters and thus reshuffles and amplifies genetic diversity. Genetic recombination is important for evolution. Two alternatives to typical bisexual reproduction are hermaphroditism, the presence of both male and female organs in the same individual, and parthenogenesis, the development of an unfertilized egg.

Sexual reproduction exacts heavy costs in time and energy, requires cooperative investments in mating, and causes a 50% loss of genetic representation of each parent in the offspring. The classical view of why sex is needed is that it maintains variable offspring within the population, which may help the population to survive environmental change.

In vertebrates the primordial germ cells arise in the yolk-sac endoderm, then migrate to the gonad. In mammals, a gonad becomes a testis in response to masculinizing signals encoded on the Y chromosome of the male, and the reproductive tract masculinizes in response to circulating male sex steroids. Female reproductive structures (ovary, oviducts, uterus, and vagina) develop in the absence of signals encoded on the Y chromosome, and in the presence of two X chromosomes expressing ovary-determining genes that are gene-dosage dependent.

Germ cells mature in the gonads by a process called gametogenesis (spermatogenesis in males and oogenesis in females), involving both mitosis and meiosis. In spermatogenesis, each primary spermatocyte gives rise by meiosis and growth to four motile sperm, each bearing the haploid number of chromosomes. In oogenesis, each primary oocyte gives rise to only one mature, nonmotile, haploid ovum. The remaining nuclear material is discarded in polar bodies. During oogenesis an egg accumulates large food reserves as yolk within its cytoplasm.

Sexual reproductive systems vary enormously in complexity, ranging from some invertebrates, such as polychaete worms, that lack any permanent reproductive structures to the complex systems of vertebrates and many invertebrates consisting of permanent gonads and various accessory structures for transferring, packaging, and nourishing gametes and embryos.

The male reproductive system of humans includes testes, composed of seminiferous tubules in which millions of sperm develop, and a duct system (vasa efferentia and vas deferens) that joins the urethra, glands (seminal vesicles, prostate, bulbourethral), and penis. The human female system includes ovaries, containing thousands of eggs within follicles; oviducts; uterus; and vagina.

The seasonal or cyclic nature of reproduction in vertebrates has required evolution of precise hormonal mechanisms that control production of germ cells, signal readiness for mating, and prepare ducts and glands for successful fertilization of eggs. Neurosecretory centers within the hypothalamus of the brain secrete gonadotropin-releasing hormone (GnRH), which stimulates endocrine cells of the anterior pituitary to release follicle-stimulating hormone (FSH) and luteinizing hormone (LH), which in turn stimulate the gonads. Estrogens and progesterone in females, and testosterone and dihydrotestosterone (DHT) in males, control the growth of accessory sex structures and secondary sex characteristics, in addition to feeding back to the hypothalamus and anterior pituitary to regulate GnRH, FSH, and LH secretion.

In the human menstrual cycle, estrogen induces the initial proliferation of uterine endometrium. A surge in GnRH and LH, induced by rising estrogen levels from the developing follicle(s), midway in the cycle causes ovulation and the corpus luteum to secrete progesterone (and estrogen in humans), which completes preparation of the uterus for implantation. If an egg is fertilized, pregnancy is maintained by hormones produced by the placenta and mother. Human chorionic gonadotropin (hCG) maintains secretion of progesterone and estrogen from the corpus luteum, while the placenta grows and eventually secretes estrogen, progesterone, hCG, human placental lactogen (hPL), human placental growth hormone (hPGH), prolactin (PRL), endogenous opioids, placental corticotropin-releasing hormone (CRH), and relaxin. Estrogen, progesterone, PRL, and hPL, as well as maternal prolactin, induce development of the mammary glands in preparation for lactation. hPL, hPGH, and maternal growth hormone also increase nutrient availability for the developing embryo.

Birth or parturition (at least in most mammals) appears to be initiated by release of placental CRH. In addition, a decrease in progesterone and an increase in estrogen levels occur so that the uterine muscle begins to contract. Oxytocin (from the posterior pituitary) and uterine prostaglandins continue this process until the fetus (followed by the placenta) is expelled. Placental relaxin makes the birth process easier by enabling expansion of the pelvis and dilation of the cervix.

Multiple births in mammals may result from division of one zygote, producing identical, monozygotic twins, or from separate zygotes, producing fraternal, dizygotic twins. Identical twins in humans may have separate placentas, or (most commonly) they may share a common placenta but have individual amniotic sacs.

REVIEW QUESTIONS

1. Define asexual reproduction, and describe four forms of asexual reproduction in invertebrates.
2. Define sexual reproduction and explain why meiosis contributes to one of its great strengths.
3. Explain why genetic mutations in asexual organisms lead to much more rapid evolutionary change than do genetic mutations in sexual forms. Why might harmful mutations be more deleterious to asexual organisms compared with sexual organisms?
4. Define two alternatives to bisexual reproduction—hermaphroditism and parthenogenesis—and offer a specific example of each from the animal kingdom. What is the difference between ameiotic and meiotic parthenogenesis?
5. Define the terms dioecious and monoecious. Can either of these terms be used to describe a hermaphrodite?
6. A paradox of sexual reproduction is that despite its being widespread in nature, the question of why it exists at all is still unresolved. What are some disadvantages of sex? What are some consequences of sex that make it so important?
7. What is a germ-cell line? How do germ cells pass from one generation to the next?
8. Explain how a spermatogonium, containing a diploid number of chromosomes, develops into four functional sperm, each containing a haploid number of chromosomes. In what significant way(s) does oogenesis differ from spermatogenesis?
9. Define, and distinguish among, the terms oviparous, ovoviviparous, and viviparous.
10. Name the general location and give the function of the following reproductive structures: seminiferous tubules, vas deferens, urethra,

seminal vesicles, prostate gland, bulbourethral glands, mature follicle, oviducts, uterus, vagina, endometrium.

11. How do the two kinds of mammalian reproductive cycles—estrous and menstrual—differ from each other?

12. What are the male sex hormones and what are their functions?

13. Explain how the female hormones GnRH, FSH, LH, and estrogen interact during the menstrual cycle to induce ovulation and, subsequently, formation of the corpus luteum.

14. Explain the function of the corpus luteum in the menstrual cycle. If an ovulated egg is fertilized, what endocrine events occur to support pregnancy?

15. Describe the role of pregnancy hormones during human pregnancy. What hormones prepare the mammary glands for lactation and what hormones continue to be important during this process?

16. If identical human twins develop from separate placentas, when must the embryo have separated? When must separation have occurred if the twins share a common placenta but develop within separate amnions?

For Further Thought Fish and amphibians tend to have less complex reproductive systems than those of some reptiles, birds, and mammals. Why do you think this might be the case?

SELECTED REFERENCES

Arukwe, A., and A. Goksoyr. 2003. Eggshell and egg yolk proteins in fish: hepatic proteins for the next generation: oogenetic, population, and evolutionary implications of endocrine disruption. Comp. Hepatology **2:** 4–25. *This review discusses the synthesis of yolk proteins in teleost fish and the ability of environmental estrogens to mimic this process.*

Crow, J. F. 1994. Advantages of sexual reproduction. Developmental Genetics **15:** 205–213. *A n excellent discussion of the advantages and disadvantages of sexual reproduction with a critique of the various hypotheses presented on this issue. Very readable.*

Cyranoski, D. 2009. Two by two. Nature **4 58:** 826–829. *A discussion of the remarkable twinning rate observed in a small village in India.*

Ferguson-Smith, M. 2007. The evolution of sex chromosomes and sex determination in vertebrates and the key role of *DMRT1.* Sex Development **1:** 2–11. *This review article describes the variety of mechanisms that determine sex in vertebrates from an evolutionary perspective.*

Johnson, M. H., and B. J. Everitt. 2012. Essential reproduction, ed. 7. Hoboken, NJ. Wiley & Sons, Inc.. *Excellent coverage of reproductive physiology with emphasis on humans.*

Jones, R. E., and K. H. Lopez. 2013. Human reproductive biology, ed. 4. San Diego, Academic Press. *Thorough treatment of human reproductive physiology.*

Kiessling, A. A. 2005. Eggs alone. Human parthenotes: an ethical source of stem cells for therapies? Nature **434:** 145. *A very short summary of the pros and cons for therapeutic use of using human ovarian stem cells that have recently been discovered in adult ovaries.*

Kinsley, C. H., and K. G. Lambert. 2006. The maternal brain. Sci. Am. **294:** 72–79. *This excellent review discusses how the hormones secreted during pregnancy and lactation in mammals appear to confer long-lasting benefits to the brain that alter skills and behavior associated with better parental care.*

Kriegsfeld, L. J., D. F. Mei, G. E. Bentley, Y. Ubuka, A. O. Mason, K. Inoue, K. Ukena, K. Tsutsui, and R. Silver. 2006. Identification and characterization of a gonadotropin-inhibitory system in the brains of mammals. Proceedings of the National Academy of Science **103:** 2410–2415. *An original research paper that presents evidence of a gonadotropin-inhibiting hormone that suppresses the reproductive axis.*

Lee, D. M., R. R. Yeoman, D. E. Battaglia, R. L. Stouffer, M. B. Zelinski-Wooten, J. W. Fanton, and D. P. Wolf. 2004. Live birth after ovarian tissue transplant. Nature **428:** 137–138. *New hope in the future for cancer patients that are made prematurely sterile is provided by the news of successful ovarian tissue transplants in monkeys.*

Lombardi, J. 1998. Comparative vertebrate reproduction. Boston, Kluwer Academic Publishers. *Comprehensive coverage of vertebrate reproductive physiology.*

Niikura, Y., T. Niikura, and J. L. Tilly. 2009. Aged mouse ovaries possess rare premeiotic germ cells that can generate oocytes following transplantation into a young host environment. Aging **1** (12) **:** 971–978. *A research paper that presents evidence of germline stem cells in the adult female mouse.*

Piñón, R. 2002. Biology of human reproduction. Sausalito, University Science Books. *An updated examination of human reproductive physiology.*

Ridley, M. 2001. The advantage of sex. www.pbs.org/wgbh/evolution/sex/advantage/ *An essay adapted from a* New Scientist *publication (4 Dec, 1993) summarizing hypotheses proposed for the evolution of sex.*

Sekido, R., and R. Lovell-Badge. 2008. Sex determination and SRY: down to a wink and a nudge? Trends in Genetics **25:** 19–29. *An excellent current review of sex determination and its evolution.*

Tilly, J. L., Y. Niikura, and B. R. Rueda. 2009. The current status of evidence for and against postnatal oogenesis in mammals: a case of ovarian optimism versus pessimism? Biol. Reprod. **80:** 2–12. *An excellent mini-review that reports experimental evidence from studies that provide support for or argue against the ability of adult female mammals to replenish their oocyte pool.*

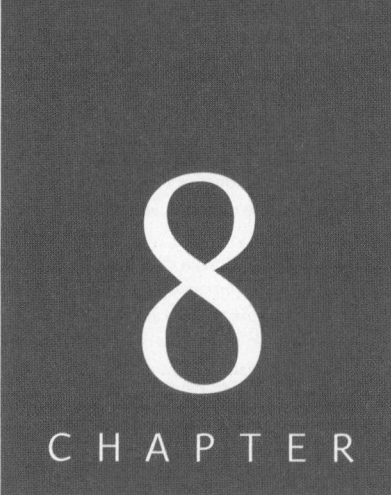

8

CHAPTER

Principles of Development

In a modern reconstruction of a classic experiment, a twinned frog develops after the Spemann organizer region from one frog embryo is grafted into another frog embryo.

The Primary Organizer

During the first half of the twentieth century, experiments by the German embryologist Hans Spemann (1869 to 1941) and his student, Hilde Pröscholdt Mangold (1898 to 1924), started the first of two golden ages of embryology. Working with salamanders, they found that tissue transplanted from one embryo into another could induce development of a complete organ, such as an eyeball, at the site of the transplant. This phenomenon is called embryonic induction. Mangold later discovered that one particular tissue, the dorsal lip from an embryonic stage called the gastrula, could induce the development of an entirely new salamander joined to the host salamander at the site of the transplant. (This work earned Spemann the Nobel Prize in Physiology or Medicine in 1935, but Hilde Mangold had died in a household accident only a few weeks after her research was published.) Spemann designated this dorsal lip tissue the **primary organizer,** now often called the **Spemann organizer.** Recent advances in molecular

biology have inaugurated the second golden age of embryology, by revealing that induction occurs by secretion of certain molecules that trigger or repress the activity of combinations of genes in nearby cells. For example, cells of the Spemann organizer migrate over the dorsal midline, secreting proteins with names like noggin, chordin, and folli-statin. These proteins allow nearby cells to develop the nervous system and other tissues along the middle of the back, and those tissues in turn release other proteins that induce development of other parts of the body. Such organizer proteins have been characterized for many other vertebrates and even invertebrates. Because all animals appear to share similar molecular mechanisms for development, it may now be possi-ble to understand how changes in such developmental controls led to the evolution of the great variety of animals. Research in this area has given rise to the exciting new field called evolutionary developmental biology.

How can a tiny, spherical fertilized egg, scarcely visible to the naked eye, develop into a fully formed, unique organism, consisting of thousands of billions of cells, each cell performing a specific functional or structural role? How is this marvelous unfolding controlled? Clearly, all information needed must originate from the nucleus and in the surrounding cytoplasm. But how does this information guide the conversion of a fertilized egg into a fully differentiated animal? Despite intense scrutiny by thousands of scientists over many decades, it seemed until very recently that developmental biology, almost alone among the biological sciences, lacked a satisfactory explanatory theory. This now has changed. During the last two decades the combination of genetics and evolution with modern techniques of cellular and molecular biology has provided the long-sought explanation of animal development. Biologists are excited by the very recent discovery of a genetic tool, RNA-guided nucleases (CRISPR-associated nuclease 9 or Cas9), that makes it possible to manipulate almost any set of genes to determine what they do during development. Causal relationships between development and evolution have also become the focus of research. We do at last appear to have a conceptual framework to explain development.

8.1 EARLY CONCEPTS: PREFORMATION VERSUS EPIGENESIS

Early scientists and laypeople alike speculated about the mystery of development long before the process was submitted to modern techniques of biochemistry, molecular biology, tissue culture, and electron microscopy. An early and persistent idea was that young animals were preformed in eggs and that development was simply a matter of unfolding what was already there. Some claimed to see a miniature adult in the egg or sperm (Figure 8.1). Even the more cautious argued that all parts of the embryo were in the egg, but too small and transparent to be seen. This concept of **preformation** was strongly advocated by most seventeenth- and eighteenth-century naturalist-philosophers.

In 1759 German embryologist Kaspar Friedrich Wolff clearly showed that in the earliest developmental stages of chickens, there was no preformed individual, only undifferentiated granular material that became arranged into layers. These layers continued to thicken in some areas, to become thinner in others, to fold, and to segment, until the body of the embryo appeared. Wolff called this process **epigenesis** ("origin upon or after"), an idea that a fertilized egg contains building material only, somehow assembled by an unknown directing force. Current ideas of development are essentially epigenetic in concept, although we know far more about what directs growth and differentiation.

Development describes the progressive changes in an individual from its beginning to maturity (Figure 8.2). In sexual multicellular organisms, development usually begins with a fertilized egg that divides mitotically to produce a many-celled embryo. These cells then undergo extensive rearrangements and interact with one another to generate an animal's body plan and all of the many kinds of specialized cells in its body. This generation of cellular diversity does not occur all at once, but emerges sequentially by a **hierarchy of developmental decisions.** The many familiar cell types that form the body do not simply "unfold" at some point, but arise from conditions created in preceding stages. At each stage of development new structures arise from the interactions among groups of cells whose

Figure 8.1 Preformed human infant in sperm as imagined by seventeenth-century Dutch histologist Niklaas Hartsoeker, one of the first to observe sperm, using a microscope of his own construction. Other remarkable pictures published during this period depicted the figure sometimes wearing a nightcap! (Source: N. Hartsoeker, *Essai de deoprique*, 1964.)

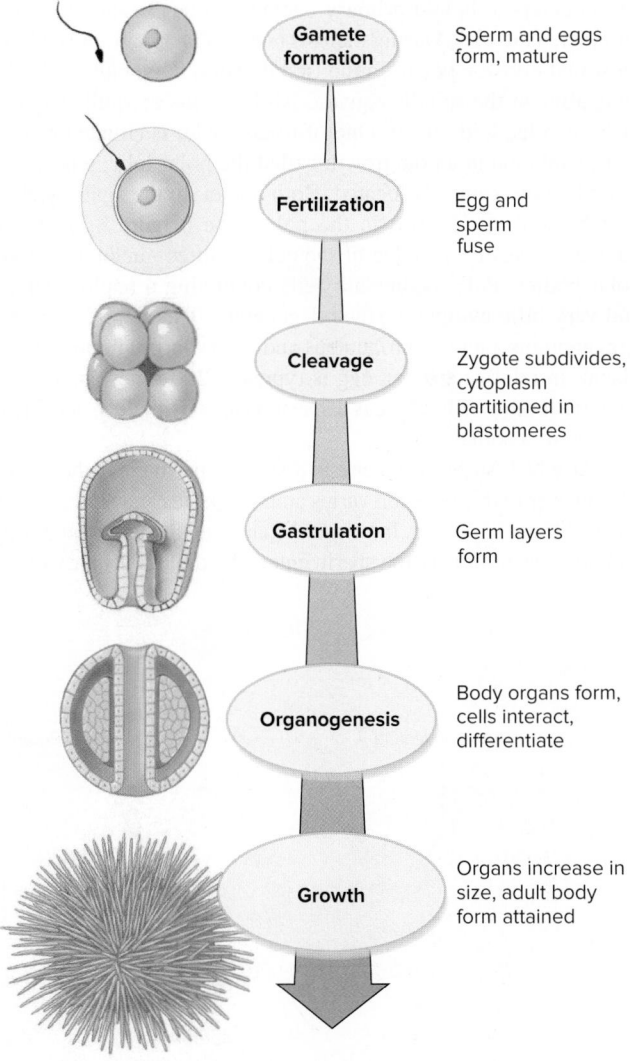

Figure 8.2 Key events in animal development.

fate is not yet determined. However, once new structures are formed, the options for future development are limited. As development proceeds, cell fate becomes more restricted. Once cells embark on a course of differentiation, they become irrevocably committed to that course. They no longer depend on the stage that preceded them, nor do they have the option of becoming something different. Once a structure becomes committed it is said to be **determined.** Thus the hierarchy of commitment is progressive and usually irreversible. The two basic processes responsible for this progressive subdivision are **cytoplasmic specification** and **induction.** We discuss both processes as we proceed through this chapter.

8.2 FERTILIZATION

The initial event in development is **fertilization,** the union of male and female gametes to form a **zygote.** Fertilization accomplishes two things: it brings haploid genomes from the mother and father together in one nucleus, thus restoring the original diploid number of chromosomes characteristic of a species, and it activates the egg to begin development.

Readers may be surprised to learn that in many organisms, the sperm enters a diploid primary oocyte that has begun but not yet completed meiosis. During meiosis (see p. 72), two rounds of chromosomal division occur. In the first division at metaphase I, bivalents align at the spindle equator. All bivalents are pulled apart to make two haploid nuclei. One of these nuclei is situated near the cell membrane in an outgrowth called the polar lobe, whereas the other is more centrally located. Both nuclei divide again, yielding four haploid nuclei, two in the polar lobe (Figure 8.3). A third nucleus moves to the edge of the cell, so three nuclei are shed as polar bodies. Polar bodies are cells containing a haploid nucleus and very little cytoplasm; they degenerate. The mature oocyte, or egg, contains a female pronucleus and a very large amount of cytoplasm. In relative size, an egg is typically 200 times as large as a somatic (body) cell, whereas a sperm is approximately one-fiftieth the size of a somatic cell.

At which stage of oogenesis does the sperm enter the oocyte? The answer to this question varies among species and shows no simple evolutionary pattern. In taxa as disparate as sponges, clams, dogs, and polychaete worms, the sperm enters the diploid primary oocyte.

The oocyte nucleus undergoes meiosis as the sperm pronucleus approaches it. In other taxa as disparate as insects, sea stars, fishes, amphibians, and some mammals, the sperm enters the oocyte after the first or second meiotic division but before polar bodies are released. Sea anemones and sea urchins are among the comparatively few taxa in which a sperm enters a haploid egg with a prepared female pronucleus.

Sperm are not always required for development. Eggs of some species can be artificially induced to initiate development without sperm fertilization (artificial parthenogenesis), but in most cases such an embryo cannot progress very far before lethal developmental abnormalities arise. However, some species have natural parthenogenesis (p. 136). Of these, some have eggs that develop normally in the absence of sperm. In other species (some fishes and salamanders), sperm is required for egg activation, but the sperm contributes no genetic material. Sometimes, during oogenesis, haploid nuclei fuse to restore the diploid condition instead of forming all three polar bodies. Neither sperm contact nor the paternal genome is always essential for egg activation.

Oocyte Maturation

During oogenesis, described in Chapter 7, an egg prepares itself for fertilization and for the beginning of development. Whereas a sperm eliminates all its cytoplasm and condenses its nucleus to the smallest possible dimensions, an egg grows in size by accumulating yolk reserves to support future growth. An egg's cytoplasm also contains vast amounts of messenger RNA, ribosomes, transfer RNA, and other elements required for protein synthesis. In addition, eggs of most species contain **morphogenetic determinants,** such as transcription factors and inducing factors, that switch genes on or off following fertilization. The nucleus grows rapidly in size during egg maturation, becoming bloated with RNA and so changed in appearance that it is given a special name, the **germinal vesicle.** Most of this intense preparation occurs during an arrested stage of meiosis. In mammals, for example, it occurs during the prolonged prophase of the first meiotic division. The oocyte has become a highly structured system, provided with a dowry that, after fertilization, nourishes the embryo and directs its development through cleavage.

Figure 8.3 Process of maturation for the primary oocyte. Sperm may enter the oocyte at any of the stages shown depending upon the species studied. (After Austin, C. R. 1965. *Fertilization.* Prentice Hall, Englewood Cliffs, N.J.)

Fertilization and Activation

Our current understanding of fertilization and activation derives in large part from more than a century of research on marine invertebrates, especially sea urchins. Sea urchins produce large numbers of eggs and sperm, which can be combined in the laboratory for study. Fertilization also has been studied in many vertebrates and, more recently, in mammals, using sperm and eggs of mice, hamsters, and rabbits. We describe fertilization and activation using a sea urchin model system.

Contact and Recognition Between Egg and Sperm

Most marine invertebrates and many marine fishes simply release their gametes into the ocean. Although an egg is a large target for a sperm, the enormous dispersing effect of the ocean and limited swimming range of a spermatozoon conspire against an egg and a sperm meeting by chance encounter. To improve likelihood of contact, eggs of numerous marine species release a chemotactic factor that attracts sperm to eggs. The chemotactic molecule is species-specific, attracting to eggs only sperm of the same species.

In sea urchin eggs, sperm first penetrate a jelly layer surrounding the egg, then contact an egg's vitelline envelope, a thin membrane lying just above the egg plasma membrane (Figure 8.4). At this point, egg-recognition proteins on the acrosomal process of the sperm (Figure 8.5) bind to species-specific sperm receptors on the vitelline envelope. This mechanism ensures that an egg recognizes only sperm of the same species. This is important in the marine environment, where many closely related species may be spawning at the same time. Species-specific recognition proteins have been found on sperm of vertebrate species (including mammals) and presumably are a universal property of animals.

In animal eggs, fertilization induces an increase in the number of free calcium ions inside the egg cytoplasm. This increase in intracellular free calcium regulates later developmental events and is essential for normal development to occur in all taxa studied, but the mechanisms controlling calcium levels vary. In some taxa, calcium ions are released from intracellular stores such as the endoplasmic reticulum, whereas in others calcium enters the egg from outside via voltage-gated calcium channels (see Chapter 3, Figure 3.16; p. 46). Some organisms combine both mechanisms. The calcium signal can occur in a single pulse, as it does in jellyfish, starfish, and frog zygotes, or in a series of closely spaced pulses identified in ribbon worms, polychaetes, and mammals. Researchers once thought that the calcium signaling pattern might vary as part of the developmental dichotomy between protostomes and deuterostomes, but this is not the case. In even the short list of taxa just given, the two chordate deuterostomes exhibit different calcium release patterns, suggesting that the differing patterns are more likely related to the number and duration of developmental events that require calcium signaling.

Prevention of Polyspermy

In marine invertebrates, at the point of sperm contact with the egg vitelline envelope, a **fertilization cone** appears into which the sperm head is later drawn (Figure 8.5). This event is followed immediately

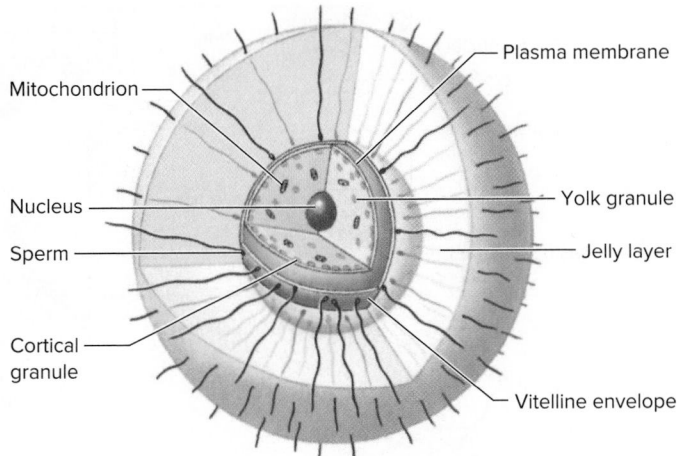

Figure 8.4   Structure of sea urchin egg during fertilization.

by important changes in the egg surface that block entrance of additional sperm, which, in marine eggs especially, may quickly surround the egg in swarming numbers (Figure 8.6). Entrance of more than one sperm, called **polyspermy,** must be prevented because union of more than two haploid nuclei would be ruinous for normal development. Entrance of more than one sperm produces a polyploid egg nucleus, which cannot undergo normal division: more than two mitotic spindles form, so chromosomes are unevenly divided among daughter cells.

In a sea urchin egg, contact of the first sperm with the egg membrane is instantly followed by an electrical potential change in the egg membrane that acts as a transient electrical barrier to prevent additional sperm from fusing with the membrane. This event, called the **fast block,** is followed immediately by a second event called the slow block, a **cortical reaction** in which thousands of enzyme-rich cortical granules located just beneath the egg membrane fuse with the membrane and release their contents into the space between the egg membrane and the overlying vitelline envelope (Figure 8.5). The cortical reaction creates an osmotic gradient, causing water to rush into this space, elevating the envelope and lifting away all sperm bound to it, except the one sperm that has successfully fused with the egg membrane. One of the cortical-granule enzymes causes the vitelline envelope to harden, and it is now called a **fertilization membrane.** The fertilization membrane acts as a permanent physical barrier to sperm. The block to polyspermy is complete. Figure 8.7 summarizes timing of these early events. Mammals have a similar security system that is erected within seconds after the first sperm fuses with the egg membrane—although there is no fertilization membrane, the cortical reaction releases enzymes that modify the egg surface to prevent other sperm from binding.

Fusion of Pronuclei and Egg Activation

Once sperm and egg membranes have fused, the sperm loses its flagellum, which disintegrates. Its nuclear envelope then breaks apart, allowing the sperm chromatin to expand from its extremely condensed state.

Figure 8.5 Sequence of events during sperm contact and penetration of a sea urchin egg.

Figure 8.6 Binding of sperm to the surface of a sea urchin egg. Only one sperm penetrates the egg surface, the others being blocked from entrance by rapid changes in the egg membranes. Unsuccessful sperm are soon lifted away from the egg surface by a newly formed fertilization membrane.

The enlarged sperm nucleus, now called a **pronucleus,** migrates inward to contact the female pronucleus. Their fusion forms the diploid **zygote nucleus.** Nuclear fusion takes only about 12 minutes in sea urchin eggs (Figure 8.7), but requires about 12 hours in mammals.

Fertilization sets in motion several important changes in the cytoplasm of an egg—now called a zygote—that prepare it for cleavage. Inhibitors that had blocked metabolism and kept the egg quiescent, in suspended animation, are removed. Fertilization is immediately followed by a burst of DNA and protein synthesis, the latter utilizing the abundant supply of messenger RNA previously stored in the egg cytoplasm. Messenger RNAs encode proteins such as actins and tubulins, necessary for cell division. Fertilization also initiates an almost complete reorganization of the cytoplasm within which are morphogenetic determinants, for example, transcription factors, that activate or repress specific genes as development proceeds. Movement of cytoplasm repositions the determinants, into new and correct spatial arrangements that are essential for proper development.

The relative positions of transcription factors and inducing factors within the cytoplasm are important because the giant zygote cell is about to undergo a sequence of mitotic cell divisions called **cleavage.** During cleavage, both the nucleus and the cytoplasm divide, so the cytoplasm is partitioned with each division. If certain mRNAs, transcription factors, and other cytoplasmic components are placed in some cells and not in others, then the later fates of these cells may be restricted by what they possess. Cleavage starts the zygote along the path to multicellularity, but it does not produce a uniform mass of cells.

What Can We Learn from Development?

Biologists study development for different reasons. Some studies focus on understanding how the zygote, a single large cell, can produce the multitude of body parts in an organism. Understanding the

Figure 8.7 Timing of events during fertilization and early development in a sea urchin.

characters. Therefore, an understanding of cleavage and an overview of a developmental sequence following cleavage is needed before the developmental patterns of particular animal groups can be described.

8.3 CLEAVAGE AND EARLY DEVELOPMENT

During cleavage, the embryo divides repeatedly, converting a single large cell into many smaller cells called **blastomeres.** No growth occurs during this period, only subdivision of mass, which continues until normal **somatic** cell size is attained. In essence, the fertilized egg divides the cytoplasm present at fertilization over and over, so cells become smaller and smaller as the divisions proceed. At the end of cleavage the zygote has been divided into many hundreds or thousands of cells and the blastula stage is formed.

Before cleavage begins, an animal-vegetal axis is visible on the embryo. This axis exists because yolk, nutrition for the developing embryo, occurs only at one end, establishing **polarity** in the embryo. The yolk-rich end is the **vegetal pole** and the other end is the **animal pole** (Figure 8.8); the animal pole contains mostly cytoplasm and very little yolk. The animal-vegetal axis provides a landmark or reference point on the embryo. Cleavage is generally an orderly sequence of cell divisions so that one cell divides to form two cells, these each divide to form four cells, the four make eight cells, and the process continues. During each division, a distinct cleavage furrow is visible in the cell. This **cleavage furrow** can be parallel or perpendicular to the animal-vegetal axis.

How Amount and Distribution of Yolk Affect Cleavage

The amount of yolk at the vegetal pole varies among taxa. Four terms are used to describe the amount and placement of the yolk in the egg. Eggs with very little yolk, evenly distributed throughout the egg (Figure 8.8A, C, and E), are called **isolecithal** (Gr. *isos,* equal, + *lekithos,* yolk). **Mesolecithal** (Gr. *mesos,* middle, + *lekithos,* yolk) eggs have a moderate amount of yolk concentrated at the vegetal pole (Figure 8.8B), whereas **telolecithal** eggs (Gr. *telos,* end, + *lekithos,* yolk) contain an abundance of yolk densely concentrated at the vegetal pole of the egg (Figure 8.8D). **Centrolecithal** eggs have a large, centrally located mass of yolk (Figure 8.9).

Why does the position and amount of yolk matter for cleavage? Yolk is a mixture of proteins that provides nutrition for a developing embryo; it may be very dense. When a cleavage furrow forms, a cell membrane divides the cytoplasm of one cell into two cells. The furrow has difficulty forming when the cytoplasm is dense with yolk. Of the four yolk distribution patterns just described, which would be expected to interfere least with cleavage? Isolecithal eggs have only a small amount of yolk evenly distributed throughout the cytoplasm, so they should cleave most easily.

In some situations, the cleavage furrow does not completely divide the cytoplasm of the egg at each cell division. When much yolk is present, cleavage is **meroblastic** (Gr. *meros,* part, + *blastos,* germ), with cells sitting atop a mass of undivided yolk (Figure 8.8D). Meroblastic cleavage is incomplete because cleavage furrows cannot cut through the heavy concentration of yolk, but instead stop at the

mechanisms of development requires knowledge of how cleavage partitions cytoplasm, how different cells interact, and how gene expression proceeds. These topics are covered in the next section, Cleavage and Early Development.

Another reason to study development is to search for commonalities among organisms. Commonalities in mechanisms of development are discussed on page 169, but there are also commonalities among organisms in the sequence of developmental events. All multicellular animals begin as zygotes and all go through cleavage and some subsequent developmental stages. Embryos of sponges, snails, and frogs diverge at some point to produce different adults. When does this divergence occur? Developmental variation among animals begins with patterns of zygotic cleavage. Types of cleavage characterize particular groups of animals, but cleavage type varies with three other developmental features to form a suite of

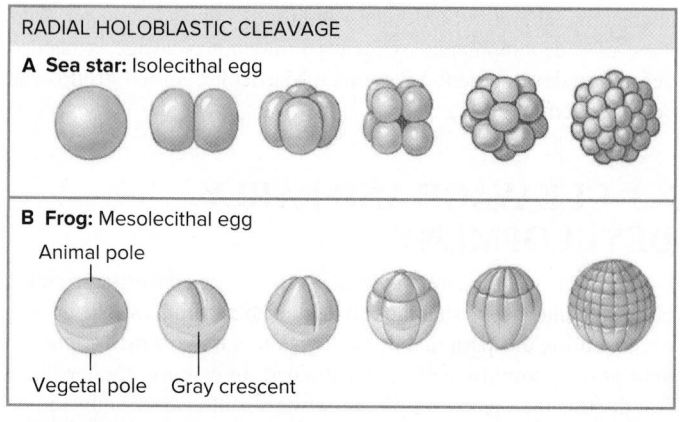

RADIAL HOLOBLASTIC CLEAVAGE

A Sea star: Isolecithal egg

B Frog: Mesolecithal egg

Animal pole

Vegetal pole   Gray crescent

SPIRAL HOLOBLASTIC CLEAVAGE

C Nemertean worm: Isolecithal egg

DISCOIDAL MEROBLASTIC CLEAVAGE

D Chick: Telolecithal egg

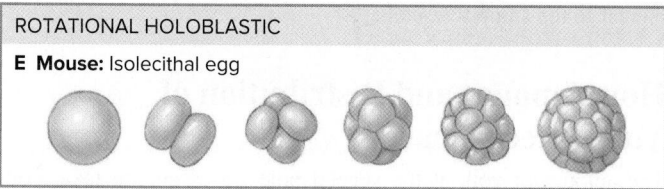

ROTATIONAL HOLOBLASTIC

E Mouse: Isolecithal egg

Figure 8.8 Cleavage stages in sea star, frog, nemertean worm, chick, and mouse. Yellow regions represent yolk in each diagram.

border between the cytoplasm and the yolk below. When little yolk is present, cleavage furrows extend completely through the egg in **holoblastic** (Gr. *holo,* whole, + *blastos,* germ) **cleavage** (Figure 8.8A, B, C, and E).

Cleavage is a critical process for development, so a variety of terms are used to describe contrasting patterns. The yolk distribution prior to cleavage is isolecithal, mesolecithal, telolecithal, or centrolecithal. The extent to which cleavage is able to bisect cell cytoplasm is either holoblastic or meroblastic. Combining these two descriptors makes clear that a complete cleavage furrow is likely to occur in cells with little yolk or evenly distributed yolk; one expects holoblastic cleavage in isolecithal or mesolecithal eggs, but not in telolecithal or centrolecithal eggs.

Another set of terms is used to discuss the angles and directions from which the cleavage furrows cut into the cytoplasm. We illustrate four descriptors for this aspect of cleavage: radial, spiral, discoidal, and rotational. Rotational cleavage can be distinguished from the other patterns of cleavage by comparing the cleavage furrow in the two-cell stage for a mouse embryo (Figure 8.8E) with the two-cell stages for the other embryos (Figure 8.8A–D). Discoidal cleavage occurs when the cleaving cells form a flat disc atop a mass of yolk as in Figure 8.8D. Radial cleavage can be distinguished from spiral cleavage by comparing the eight-cell stages in sea stars and frogs shown in Figure 8.8A and B, respectively, with that stage in Figure 8.8C, a nemertean worm. Notice that in radial cleavage the upper layer of cells sits directly atop the lower layer of cells, whereas in spiral cleavage, the upper layer of cells is packed into the spaces between the cells of the lower layer. These aspects of cleavage are discussed in more detail on page 170.

The reader may wonder how cleavage patterns are distributed among animal taxa. Holoblastic cleavage occurs in isolecithal eggs and is present in echinoderms, tunicates, cephalochordates, nemerteans, and most molluscs, as well as in marsupial and placental mammals, including humans (Figure 8.8A, C, and E). Mesolecithal eggs also cleave holoblastically, but cleavage proceeds more slowly in the presence of yolk, leaving the vegetal region with a few large, yolk-filled cells, whereas the animal region has many small cells. Amphibian eggs (Figure 8.8B) illustrate this process.

Meroblastic cleavage occurs in telolecithal and centrolecithal eggs. In telolecithal eggs of birds, reptiles, most fishes, a few amphibians, cephalopod molluscs, and monotreme mammals, cleavage is

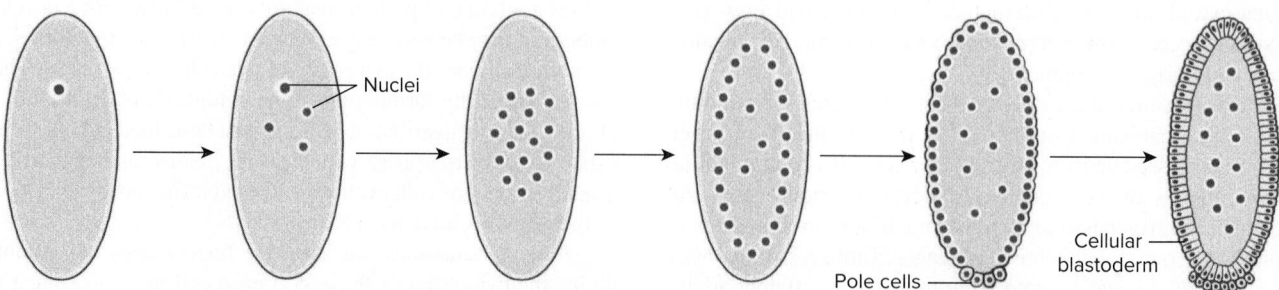

Nuclei

Pole cells

Cellular blastoderm

Figure 8.9 Superficial cleavage in a *Drosophila* embryo. The zygote nucleus at first divides repeatedly in the yolk-rich endoplasm by mitosis without cytokinesis. After several rounds of mitosis, most nuclei migrate to the surface where they are separated by cytokinesis into separate cells. Some nuclei migrate to the posterior pole to form the primordial germ cells, called pole cells. Several nuclei remain in the endoplasm where they will regulate breakdown of yolk products. The cellular blastoderm stage corresponds to the blastula stage of other embryos.

restricted to cytoplasm in a narrow disc on top of the yolk and is therefore called discoidol meroblastic cleavage (see chick development in Figure 8.8D).

Centrolecithal eggs of *Drosophila* undergo **superficial cleavage** (Figure 8.9) where the centrally located mass of yolk restricts cleavage to the cytoplasmic rim of the egg. This pattern is highly unusual because cytoplasmic cleavage (cytokinesis) does not occur until after many rounds of nuclear division. After roughly eight rounds of mitosis in the absence of cytoplasmic division (yielding 256 nuclei), the nuclei migrate to the yolk-free periphery of the egg. A few nuclei at the posterior end of the egg become surrounded by cytoplasm to form pole cells which give rise to germ cells of the adult. Next, the entire egg cell membrane folds inward, partitioning each nucleus into a single cell, and yielding a layer of cells at the periphery surrounding the mass of yolk (Figure 8.9). Because yolk is an impediment to cleavage, this pattern avoids cleaving the yolk and instead confines early cytoplasmic division to small regions of yolk-free cytoplasm.

The function of yolk is to nourish the embryo. When much yolk is present, as in telolecithal eggs, young exhibit **direct development,** going from an embryo to a miniature adult. When little yolk is present, as in isolecithal or mesolecithal eggs, young develop into various larval stages capable of feeding themselves. In this **indirect development,** larvae differ from adults and must metamorphose to produce an adult body form (Figure 8.10). There is another way to compensate for absence of yolk: in most mammals, the mother nourishes the embryos by means of a placenta.

8.4 AN OVERVIEW OF DEVELOPMENT FOLLOWING CLEAVAGE

Blastulation

Cleavage subdivides the mass of the zygote until a cluster of cells called a **blastula** (Gr. *blastos,* germ, + *ule,* little) is formed (Figure 8.11). In mammals, the cluster of cells is called a blastocyst (see Figure 8.20E). In most animals, the blastula cells form a layer around a central fluid-filled cavity (Figure 8.11) called a **blastocoel** (Gr. *blastos,* germ, + *koilos,* cavity). (A hollow blastula can be called a coeloblastula to distinguish it from a solid stereoblastula; the general account here assumes the blastula is hollow.) In the blastula stage, the embryo consists of a few hundred to several thousand cells poised for further development. There has been a great increase in total DNA content because each of the many daughter cell nuclei formed by chromosomal replication contains as much DNA as the original zygote nucleus. The whole embryo, however, is no larger than the zygote. The egg cytoplasm has been partitioned among the cells as previously described, potentially limiting the fates of certain cells.

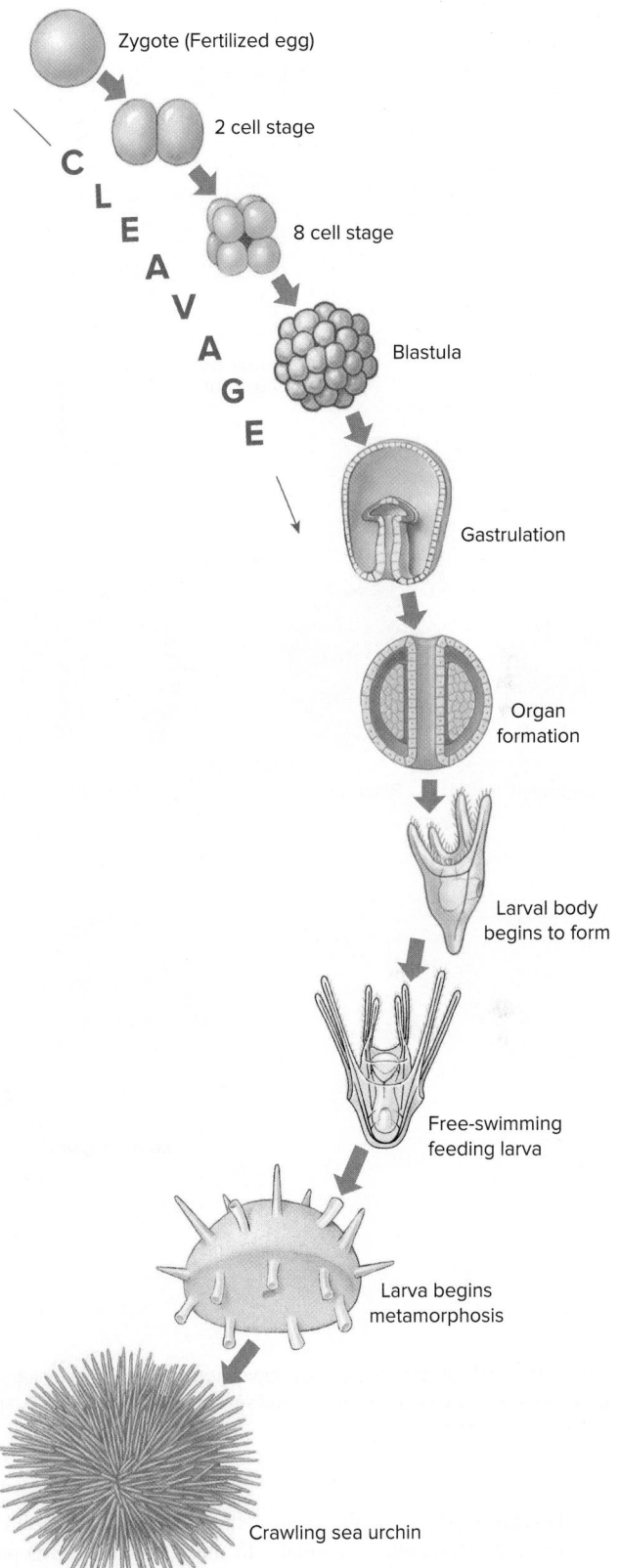

Figure 8.10   Indirect development in a sea urchin. Fertilization of an urchin egg is followed by cleavage, producing a mass of cells. This mass rearranges to form a single cell layer surrounding a cavity (blastula stage). In the next stage, a gut and more tissue layers form (gastrulation). Once a gut is present, the urchin embryo develops a larval body. The free-swimming larva feeds and grows in ocean surface waters. The larva metamorphoses into a tiny bottom-dwelling sea urchin; the urchin feeds and grows, reaching sexual maturity in this body form.

Zygote (Fertilized egg)

CLEAVAGE

2 cell stage

8 cell stage

Blastula

Gastrulation

Organ formation

Larval body begins to form

Free-swimming feeding larva

Larva begins metamorphosis

Crawling sea urchin

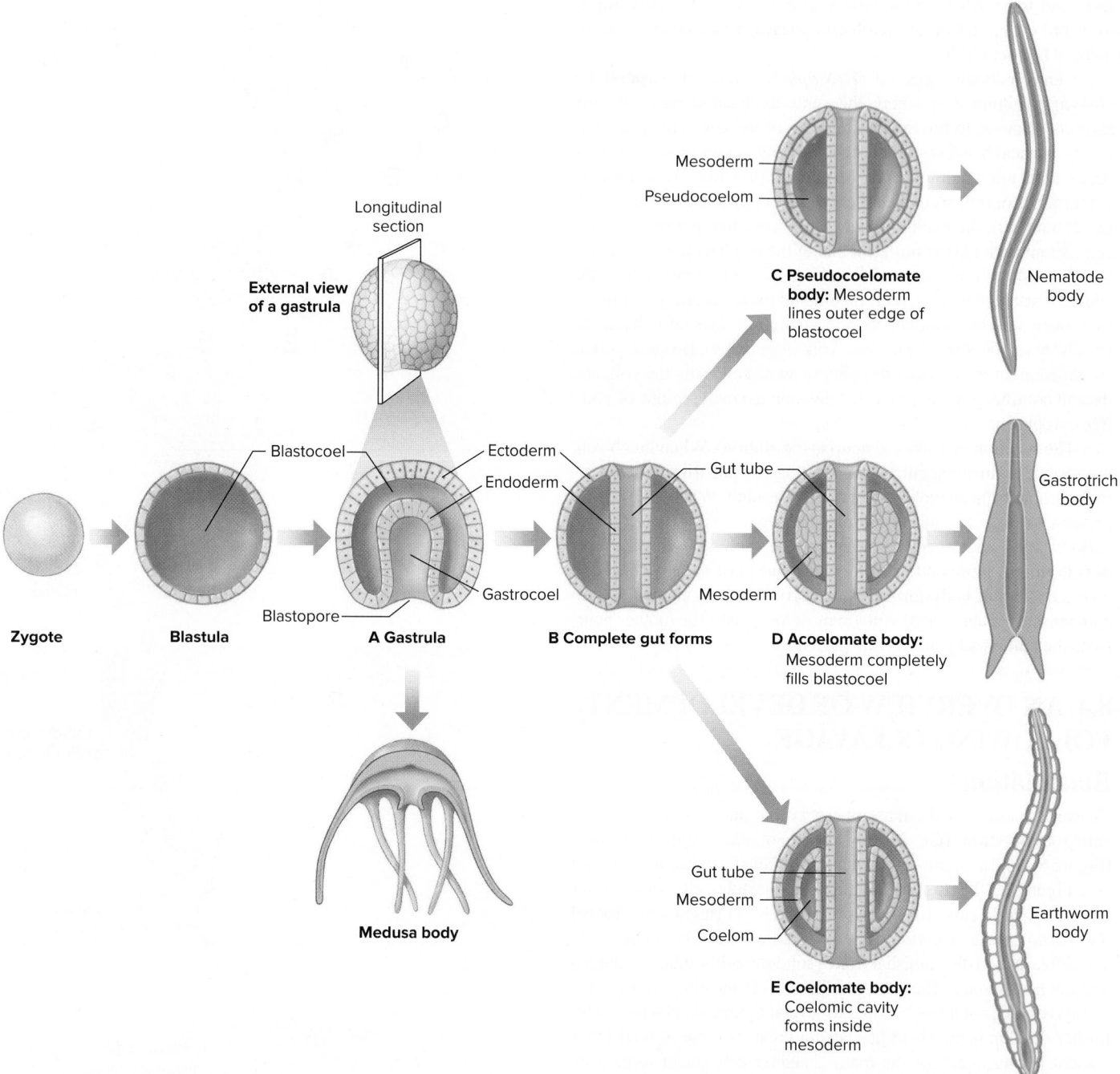

Figure 8.11   A generalized developmental sequence beginning with a zygote. All animals develop to a blastula stage; arrows between stages **A** and **E** illustrate possible later developmental sequences. Stages **A, C, D,** and **E** are paired with generalized examples of adult animals resulting from each developmental sequence.

Formation of a blastula stage, with its one layer of germ cells, occurs in all multicellular animals. In most animals, development continues beyond the blastula to form one or two more germ layers in a gastrula stage. The germ layers ultimately produce all structures of the adult body; the germ layer derivatives for vertebrates are shown in Figure 8.27.

Gastrulation and Formation of Two Germ Layers

Gastrulation converts the spherical blastula into a two- or three-layered embryo. The layers are called germ layers and all subsequent body parts develop from one or more germ layers. We begin with a general account of how the germ layers form. To form a second germ layer (Figure 8.11), one side of the blastula bends inward in a process called invagination. This inward bending continues until the surface of the bending region extends about one-third of the way into the blastocoel, forming a new internal cavity (Figure 8.11). Picture a spherical balloon being pushed inward on one side—the inward region forms a pouch. The internal pouch is the gut cavity, called an **archenteron** (Gr. *archae,* old, + *enteron,* gut) or a **gastrocoel** (Gr. *gaster,* stomach, + *koilos,* cavity). It sits inside the now-reduced blastocoel. The opening to the gut, where the inward bending began, is the **blastopore** (Gr. *blastos,* germ, + *poros,* hole).

The **gastrula** (Gr. *gaster,* stomach, + *ule,* little) stage has two layers: an outer layer of cells surrounding the blastocoel, called **ectoderm** (Gr. *ecto,* out, + *deros,* skin), and an inner layer of cells lining the gut, called **endoderm** (Gr. *endon,* inside, + *deros,* skin). In forming a mental picture of the developmental process, remember that cavities or spaces can be defined only by their boundaries. Thus, the gut cavity is a space defined by the layer of cells surrounding it (Figure 8.11). The colors of the layers are standard within developmental biology: ectoderm is always blue and endoderm is always yellow. Use this convention to follow the developmental process.

When the gut opens only at the blastopore, it is called a blind or **incomplete gut.** Anything consumed by an animal with a blind gut must either be completely digested or have the undigested parts egested through the mouth. Certain animals, sea anemones and flatworms, for example, have a blind gut, sometimes called a gastrovascular cavity. However, most animals have a **complete gut** with a second opening, the anus (Figure 8.11). The blastopore becomes the mouth in some organisms, but it becomes the anus in others. These differences are discussed in the sections called Protostome Development and Deuterostome Development (see pp. 170 and 171).

Formation of a Complete Gut

When a complete gut forms, the inward movement of the archenteron continues until the end of the archenteron meets the ectodermal wall of the gastrula. The archenteron cavity extends through the animal, and the ectoderm and endoderm layers join together. This joining produces an endodermal tube, the gut, surrounded by the blastocoel, inside an ectodermal tube, the body wall (Figure 8.11). The endodermal tube now has two openings, the blastopore and a second, unnamed, opening that formed when the archenteron tube merged with the ectoderm (Figure 8.11).

Formation of Mesoderm, a Third Germ Layer

Most animals proceed from a blastula to a gastrula, producing two germ layers. In one of many quirks of biological terminology, there is no term for organisms with only a single germ cell layer. Many sponges appear to develop to the blastula stage, and not beyond, and thus have a single germ cell layer. However, recent research in some sponges indicates that they develop two germ layers. Such animals are called **diploblastic** (Gr. *diploos,* twofold, + *blastos,* germ).

Diploblastic animals include sea anemones and comb jellies. Most animals have a third germ layer and are **triploblastic** (L. *tres,* three, + *blastos,* germ).

The third layer, **mesoderm** (Gr. *mesos,* middle, + *deros,* skin), eventually lies between the ectoderm and the endoderm (Figure 8.11). It is always colored red in developmental diagrams. Mesoderm can form in two ways: cells arise from a ventral area near the lip of the blastopore and proliferate into the space between the archenteron and outer body wall (see Figure 8.20C), or the central region of the archenteron wall pushes outward into the space between the archenteron and the outer body wall (see Figure 8.20A). Regardless of method, initial cells of mesoderm come from endoderm. (In a few groups, such as amphibians, part of a third layer of cells is made from ectoderm; this is called **ectomesoderm** [Gr. *ecto,* out, + *mesos,* middle, + *deros,* skin] to distinguish it from true, endodermally derived, mesoderm).

At the end of gastrulation, ectoderm covers the embryo, and mesoderm and endoderm have been brought inside (Figure 8.11). As a result, cells have new positions and new neighbors, so interactions among cells and germ layers then generate more of the body plan.

Formation of the Coelom

A **coelom** (Gr. *koilos,* cavity) is a body cavity completely surrounded by mesoderm; the band of mesoderm with its internal coelom lies inside the space previously occupied by the blastocoel (Figure 8.11). How did this happen? During gastrulation, the blastocoel is filled, partially or completely, with mesoderm. The coelomic cavity appears inside the mesoderm by one of two methods: **schizocoely** or **enterocoely.** These methods are discussed on pages 171 and 172. A coelom made by schizocoely is functionally equivalent to a coelom made by enterocoely. The method by which the coelom forms is an inherited character, so it can be used as evidence of shared ancestry (see p. 174).

When coelom formation is complete, the body has three germ layers and two cavities (Figure 8.11). One cavity is the gut cavity and the other is the fluid-filled coelomic cavity. The coelom, surrounded by its mesodermal walls, has completely filled the blastocoel. Mesoderm around the coelom eventually produces layers of muscles, among other structures. All structures in the adult derive from one of the three germ layers.

8.5 MECHANISMS OF DEVELOPMENT
Nuclear Equivalence

How does a developing embryo generate a multitude of cell types of a complete multicellular organism from the starting point of a single diploid nucleus of a zygote? To many nineteenth-century embryologists there seemed only one acceptable answer: as cell division ensued, hereditary material had to be parceled unequally to daughter cells. In this view, the genome gradually became broken into smaller and smaller units until finally only the information required to impart the characteristics of a single cell type remained. This is called the Roux-Weismann hypothesis, after the two German embryologists who developed the concept.

However, in 1892 Hans Driesch discovered that if he mechanically shook apart a two-celled sea urchin into separate cells, both half-embryos developed into normal larvae. Driesch concluded that both cells contained all genetic information of the original zygote. Still, this experiment did not settle the argument, because many embryologists thought that even if all cells contained complete genomes, the nuclei might become progressively modified in some way to dispense with the information not used in forming differentiated cells.

The efforts of Hans Driesch to disrupt egg development are poetically described by Peattie: "Behold Driesch grinding the eggs of Loeb's favorite sea urchin up between plates of glass, pounding and breaking and deforming them in every way. And when he ceased from thus abusing them, they proceeded with their orderly and normal development. Is any machine conceivable, Driesch asks, which could thus be torn down . . . have its parts all disarranged and transposed, and still have them act normally? One cannot imagine it. But of the living egg, fertilized or not, we can say that there lie latent within it all the potentialities presumed by Aristotle, and all of the sculptor's dream of form, yes, and the very power in the sculptor's arm." From Peattie, D. C. 1935. *An Almanac for Moderns.* New York, G. P. Putnam's Sons.

Around the turn of the century Hans Spemann introduced a new approach to testing the Roux-Weismann hypothesis. Spemann placed minute ligatures of human hair around salamander zygotes just as they were about to divide, constricting them until they were almost, but not quite, separated into two halves. The nucleus lay in one half of the partially divided zygote; the other side was anucleate, containing only cytoplasm. The zygote then completed its first cleavage division on the side containing the nucleus; the anucleate side remained undivided. Eventually, when the nucleated side had divided into about 16 cells, one of the cleavage nuclei would wander across the narrow cytoplasmic bridge to the anucleate side. Immediately this side began to divide and developed normally.

Sometimes, however, Spemann observed that the nucleated half of the embryo developed only into an abnormal ball of "belly" tissue. The explanation, Spemann discovered, depended on the presence of the gray crescent, a pigment-free area shown in Figure 8.8B. The gray crescent is required for normal development because it is the precursor of the Spemann organizer discussed in the opening essay on page 154.

Spemann's experiment demonstrated that every blastomere contains sufficient genetic information for the development of a complete animal. In 1938 he suggested another experiment that would demonstrate that even somatic cells of an adult contain a complete genome. The experiment, which Spemann characterized as being "somewhat fantastical" at that time, would be to remove the nucleus of an egg cell and replace it with the nucleus from a somatic cell from a different individual. We now call this process "somatic cell nuclear transplant," or SCNT. If all cells contained the same genetic information as a zygote, then the embryo should develop into an individual that was genetically identical to the animal from which the nucleus was obtained. It took several decades to solve the technical difficulties, but the experiment was successfully performed on amphibians, and today it is done in a variety of mammals. The procedure is now familiarly called **cloning** (Gr. *Klon,* twig). One of the most famous

cloned mammals, Dolly the sheep, got the genetic material in her nuclei from the mammary glands of a six-year-old ewe.

Mammalian clones often fail to develop and, when development is successful, adult clones often tend to be unhealthy because of abnormal gene expression. These problems limit the potential usefulness of mammalian cloning. However, therapeutic cloning has many more potential medical applications. Therapeutic cloning aims only to produce a lineage of undifferentiated cells capable of differentiating into many different adult cell types. These undifferentiated cells, called **embryonic stem cells,** can be used to replace damaged tissues in gravely ill patients. Stem cells made by following the SCNT method are genetically identical to other cells in the patient, thereby avoiding any tissue rejection. Therapeutic cloning is an area of very active research because undifferentiated cells occur naturally only for a limited time during development. As will be clear shortly, once the developmental fate of cells is fixed, the type of cells into which they can develop is quite limited.

Relatively recently, scientists have discovered that adult tissues such as the brain, liver, and reproductive organs contain undifferentiated cells; these are called adult stem cells. The potential therapeutic use of these cells is limited by their relative rarity and by what appears to be a small number of future divisions possible for these cells. Researchers are asking whether the fates of these cells are restricted to tissues from which they came or whether the cells can be transdifferentiated into other cell types, for example, a liver cell into a heart cell. The use of **adult stem cells** or stem cells produced by SCNT avoids ethical controversy associated with the use of embryonic stem cells. Embryonic stem cells are typically cultured from human eggs donated for research and fertilized in the laboratory.

If all nuclei are equivalent, what causes some cells to develop into neurons while others develop into skeletal muscle? In most animals (excluding insects), there are two major ways by which cells become committed to particular developmental fates: (1) cytoplasmic partitioning of determinative molecules during cleavage and (2) interaction with neighboring cells (inductive interactions). All animals use both of these mechanisms to some extent to specify different cell types. However, in some animals cytoplasmic specification is dominant, whereas others rely predominantly on inductive interactions.

The fate of most cells is determined sequentially: a cell becomes ectoderm, endoderm, or mesoderm early in development of an embryo, and then each of these germ layers makes particular derivatives. For example, skin, eye, and nerve cells are derived only from ectodermal cells. Skin cells must make different proteins than eye cells or nerve cells, and certain proteins must be made at particular times.

How is the production of particular proteins limited to certain cell types at certain times, ensuring, for example, that eye cells make corneal proteins and not liver proteins? One means of controlling production is to have only certain mRNAs or proteins present in the cytoplasm. Another way is to regulate the timing of gene expression. Genes are regulated by transcription factors binding to promoter regions and adjacent enhancers when RNA polymerase initiates transcription. Thus, transcription factors and enhancers determine where, when, and how much protein is made—by acting on transcription.

A gene may have more than one enhancer, so that one enhancer works in one kind of cell and another works in a different cell type.

For example, there is one enhancer for the *Pax6* gene in mouse pancreas cells and another for this gene in mouse eye cells. In certain cell types, a combination of enhancers may be required; both of the mouse *Pax6* examples just given require more than one enhancer. Enhancers may activate a promoter or repress it, so enhancers may be "negative enhancers," or silencers.

Transcription factors act on a larger scale than enhancers and may bind to enhancers or directly to promoters to control transcription. Like enhancers, transcription factors may repress or activate transcription. Thus, the fates of cells may be determined by multiple interacting factors such as promoters, enhancers, and transcription factors, turning on and off production of particular proteins in highly structured patterns.

Cytoplasmic Specification

The cytoplasm of a fertilized egg is not homogeneous. It contains unequally distributed morphogenetic components such as mRNAs and proteins, which act as enzymes or as transcription factors. These components may be tethered to the cytoskeleton. As cleavage proceeds, the mRNAs and proteins are unequally partitioned among the resulting new cells (blastomeres). Thus, the fate of each cell is specified by the type of cytoplasm it acquires in cleavage, and even isolated cells differentiate along the path dictated by the cytoplasmic components. Cytoplasmic specification, sometimes called autonomous specification, causes mosaic development of the embryo. The term "mosaic" is used because the embryo appears to be a composite of independently developing parts rather than of interacting parts (see p. 170).

Cytoplasmic specification is especially striking (and easily visualized) in some tunicate species in which the fertilized egg contains as many as five differently colored types of cytoplasm (Figure 8.12). These differently pigmented cytoplasms are segregated into different blastomeres, which then proceed to form distinct tissues or organs. For example, yellow cytoplasm gives rise to muscle cells, while gray equatorial cytoplasm produces notochord and neural tube. Clear cytoplasm produces larval epidermis and gray vegetal cytoplasm (not visible in Figure 8.12) gives rise to the gut.

We know now that yellow cytoplasm gives rise to muscle cells because it contains macho-1 mRNA. Macho-1 mRNA encodes a transcription factor that activates other genes leading to muscle formation in cells descended from yellow cytoplasm cells. However, not all descendant cells make muscles; those in contact with cells whose descendants will make the gut are directed or induced to a different fate by a process called conditional specification.

Conditional Specification

Conditional specification differs from cytoplasmic or autonomous specification because the fate of a particular cell is not fixed until it receives positional information from its neighbors. The cell is induced to a particular fate by the interaction between populations of cells. **Induction** is the capacity of some cells to evoke a developmental response, such as a change in cell shape or cell fate, in other cells. For example, formation of hair or feathers occurs when the ectodermally derived epidermis responds to inducers from the underlying mesodermally derived dermis. While the interaction requires an inducer and a responder, the interactions may be reciprocal in that the tissues influence each other. Inducers act through contacts between the surface proteins of adjacent cells, or through diffusible molecules that travel between cells. Such molecules are called growth and differentiating factors. Diffusion of the molecules away from the source cell population produces a gradient in the strength of the signal available to the responding cell. In some cases, a threshold level of signal strength is required for a response.

The classic experiments in induction, cited in the opening essay on page 154, were reported by Hans Spemann and Hilde Mangold in 1924. A key feature of these experiments was the use of salamanders with different-colored body pigments so that donor and recipient (host) tissues could be distinguished. When a piece of dorsal blastopore lip from a salamander gastrula was transplanted into a ventral or lateral position of another, differently pigmented, salamander gastrula, it invaginated and developed a notochord and somites. It also induced the *host* ectoderm to form a neural tube. Eventually a whole system of organs developed where the graft was placed, and then grew into a nearly complete secondary embryo (Figure 8.13). This creature was composed partly of grafted tissue and partly of induced host tissue.

It was soon found that *only* grafts from the dorsal lip of the blastopore were capable of inducing the formation of a complete or nearly complete secondary embryo. This area corresponds to the presumptive areas of notochord, somites, and prechordal mesoderm (see p. 502). It was also found that only ectoderm of the host would develop a nervous system in the graft and that the reactive ability was greatest at the early gastrula stage and declined as the recipient embryo got older.

Spemann designated the dorsal lip area the **primary organizer** because it was the only tissue capable of inducing the development of a secondary embryo in the host. It is now often called the Spemann organizer. Spemann also termed this inductive event **primary induction** because he considered it the first inductive event in development. Subsequent studies showed that many other cell types originate by later inductive interactions, a process called **secondary induction.**

Usually cells that have differentiated act as inductors for adjacent undifferentiated cells. Timing is important. Once a primary inductor sets in motion a specific developmental pattern in some cells, numerous secondary inductions follow. What emerges is a sequential pattern of development involving not only inductions but cell movement, changes in adhesive properties of cells, and cell proliferation. There is no "hard-wired" master control panel directing development, but rather a

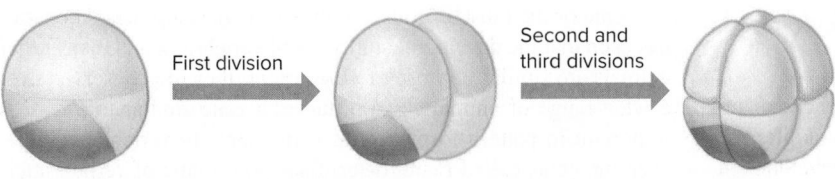

First division

Second and third divisions

Figure 8.12 Bilateral cleavage in tunicate embryos. The first cleavage division divides the asymmetrically distributed cytoplasm evenly between the first two blastomeres, establishing the future right and left sides of the adult animal. Bilateral symmetry of the embryo is maintained through subsequent cleavage divisions.

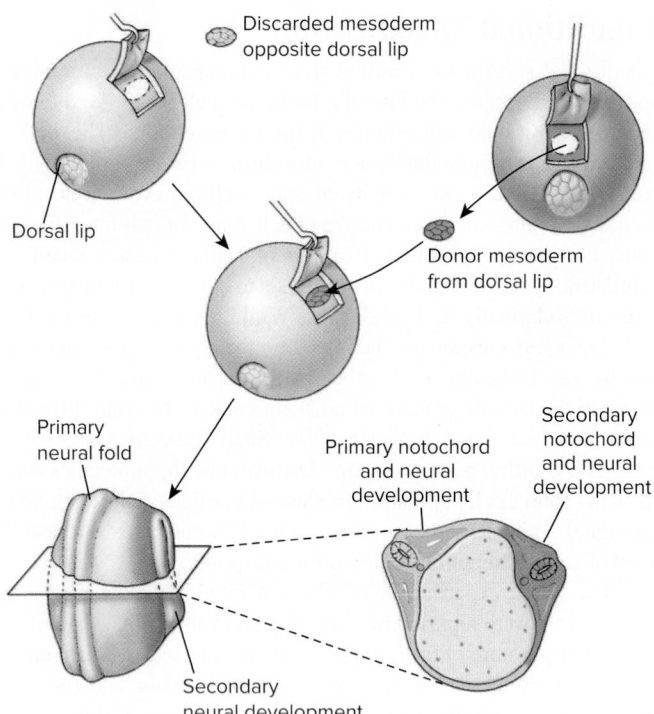

Discarded mesoderm opposite dorsal lip

Dorsal lip

Donor mesoderm from dorsal lip

Primary neural fold

Primary notochord and neural development

Secondary notochord and neural development

Secondary neural development

Figure 8.13 The Spemann-Mangold primary organizer experiment. The host animal was not pigmented, but the donor animal was darkly pigmented, so the tissue source could be determined by color.

sequence of local patterns in which one step in development is a subunit of another. In showing that each step in the developmental hierarchy is a necessary preliminary for the next, Hans Spemann's induction experiments were among the most significant events in experimental embryology.

Syncytial Specification

A syncytium occurs when a single cell membrane surrounds multiple nuclei. A familiar example of a syncytium is the centrolecithal egg typical of the fruit fly genus *Drosophila* (see Figure 8.9). In insect development, the syncytium is eventually cellularized, but some developmental processes occur before this point. Syncytial specification is similar to conditional specification, but the molecules that influence cell fate diffuse *within* the cytoplasm of a single large cell and *not among* cells. Species of *Drosophila* are model organisms for studies of genetics and development, so many principles of embryonic pattern formation were discovered in syncytial embryos.

8.6 GENE EXPRESSION DURING DEVELOPMENT

Since every cell with few exceptions receives the same genetic material, specification of cell fates must involve the activation of different combinations of genes in different cells. Understanding development is therefore ultimately a problem of understanding the genetics involved, so it is not surprising that developmental genetics was first studied in *Drosophila*. These studies have been repeated in several other model animals, such as the nematode worm *Caenorhabditis*

elegans, zebrafish *Danio rerio*, frog *Xenopus laevis*, chick *Gallus gallus*, and mouse *Mus musculus*. Research suggests that epigenesis proceeds in three general stages: pattern formation, determination of position in the body, and induction of limbs and organs appropriate for that position. Each stage is guided by gradients of gene products that function as **morphogens.** Morphogens are diffusible molecules that create a concentration gradient as they move away from a source. A cell's position in the concentration gradient of one or more morphogens determines its fate. Cells respond only to morphogen concentrations above a threshold level.

Pattern Formation

The first step in organizing development of an embryo is pattern formation by specification of the body axes: front-to-rear (anteroposterior), left-to-right, and back-to-front (dorsoventral) axes. As Spemann demonstrated in salamanders, the anteroposterior axis of the embryo is determined by the Spemann organizer, located in the gray crescent of a zygote. In *Drosophila* the anteroposterior axis is determined even before an egg is fertilized. Christiane Nüsslein-Volhard and her colleagues in Germany found that this determination is due to a gradient of mRNA that is secreted into the egg by nurse cells in the mother. To make nurse cells, the oocyte divides mitotically several times, all prior to fertilization. The cells produced are genetically identical to the oocyte and to each other. They remain interconnected and eventually contribute all their cytoplasm and its constituents to the oocyte. In nurse cells, one mRNA involved in specifying the anterior region of the embryo, which will eventually form the head region, is transcribed from a gene called *bicoid* (pronounced BICK-oyd). After an egg is fertilized, *bicoid* mRNA is translated into a protein morphogen called bicoid (not italicized) that binds to certain other genes. The products of these genes in turn activate others in a cascade that ultimately causes the production of an anteroposterior gradient.

Another gene, *nanos,* is involved in specifying the posterior region of the embryo that is fated to become the larva's abdominal region. *Bicoid* and *nanos* are two of about 30 maternal genes that control pattern formation in an embryo and allow a fly to distinguish its head from its tail. *Bicoid*-mutant embryos that do not produce the morphogen develop without a head; those that do not produce *nanos* lack an abdomen. Some of the maternal genes also specify a second dorsoventral axis that allows the fly to distinguish its back and belly. For instance, the gene *short gastrulation* leads to development of ventral structures, such as the nerve cord. Clearly, the heterogeneous nature of the egg is critical for development: the gradients in morphogens provided by the mother define the axes for subsequent development, at which time the zygotic genes in the embryonic nuclei are activated.

One of the most exciting discoveries in developmental genetics has been that the developmental genes of vertebrates and many other animals are similar to those of *Drosophila;* they are conserved over a wide range of animals. For instance, a gene similar to *bicoid* is important in pattern formation in vertebrates. In vertebrates, however, the gene, called *Pitx2,* determines positioning of certain internal organs to either the left or the right side of the body. Mutations in *Pitx2* in frogs, chicks, and mice can place the heart and stomach on the right instead of the left side. Such mutations may explain a reversal of organ position that sometimes occurs in humans. *Pitx2* is in turn activated by a protein produced by the gene *sonic hedgehog (Shh)*, which is similar to a *Drosophila* gene called *hedgehog*. (The name *hedgehog* refers to the bristly appearance of fruit flies lacking the gene. The *"sonic"* comes from the video-game character Sonic the Hedgehog.) In vertebrates, *sonic hedgehog* is active in the left side only at the anterior end of the primitive streak, which determines the anteroposterior axis (see Figure 8.21). *Short gastrulation* also has a counterpart in vertebrates—the gene *chordin*, which produces one of the proteins from the Spemann organizer. Genes critical for development in a wide range of organisms are sometimes called "toolkit genes."

In *Drosophila*, as well as other arthropods, annelid worms, chordates, and a few other groups, one important aspect of pattern formation along the anteroposterior axis is **segmentation,** also called **metamerism.** Segmentation is a division of the body into discrete segments or metameres (see Figure 9.6, p. 189). The segments are identical early in development, but later activation of different combinations of genes causes each segment to form different structures. For example, the anterior segment of insect embryos forms antennae, eyes, and mouthparts, while segments farther back form legs. Segments are obvious in insects, but in fishes segmentation is apparent only in somites that produce such structures as vertebrae and repeated muscle bands (myomeres) (see Figure 24.23, p. 532). In *Drosophila* the number and orientation of segments is controlled by **segmentation genes.** There are three classes of segmentation genes: gap, pair-rule, and segment-polarity. **Gap genes** are activated first and divide an embryo into regions such as head, thorax, and abdomen. **Pair-rule genes** divide these regions into segments. Finally, **segment-polarity genes,** such as *hedgehog,* organize the anterior-to-posterior structures within each segment.

Homeotic and *Hox* Genes

Segmentation genes apparently regulate expression of other genes, ensuring that they are active only in appropriate segments. Such segment-specific genes are called homeotic genes. Mutations in homeotic genes, called **homeotic mutations,** place appendages or other structures in the wrong part of the body. For example, in *Drosophila* the homeotic gene *Antennapedia,* which helps trigger development of legs, is normally active only in the thorax. If the *Antennapedia* gene is activated by a homeotic mutation in the head of a larval fly, or maggot, the adult fly will have legs in place of antennae (Figure 8.14). *Antennapedia* and some other homeotic genes, as well as many other genes involved in development, include a sequence of 180 DNA base pairs, called the **homeobox.** The homeobox produces the part of a protein that attaches to the DNA of other genes, activating or blocking their expression.

Several other homeotic and nonhomeotic genes that are clustered close to *Antennapedia* on the same chromosome in *Drosophila* also include a homeobox. Genes in this cluster are called *Hom* genes. *Hom* genes do not encode specific limbs and organs. Instead, they function by specifying the location in the body along the anteroposterior axis. Intriguingly, the order of the *Hom* genes within the cluster on the chromosome is the same as the order in which they are expressed along the length of the body (Figure 8.15). One of the

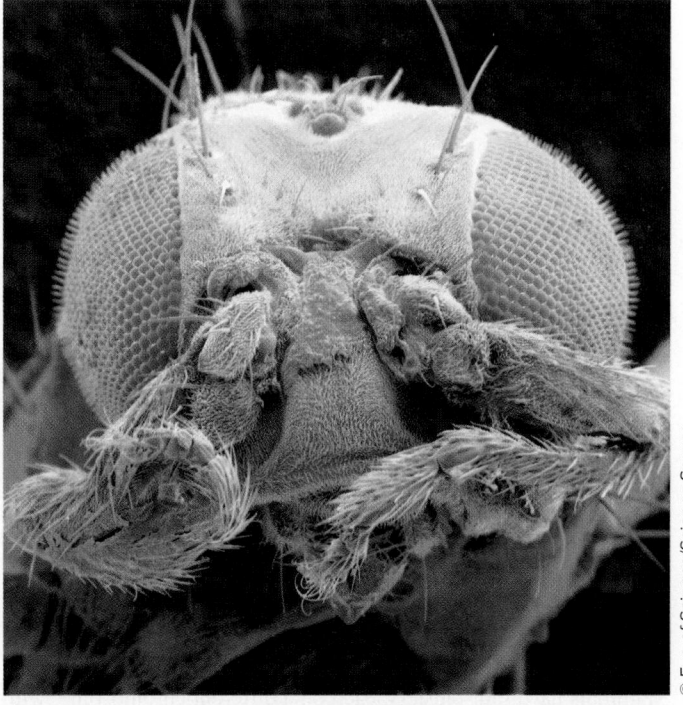

A **B**

Figure 8.14 **A,** Head of a normal fruit fly with two antennae. **B,** Head of a fruit fly with a pair of legs growing out of head sockets where antennae normally grow. The *Antennapedia* homeotic gene normally specifies the second thoracic segment (with legs), but the dominant mutation of this gene leads to this bizarre phenotype.

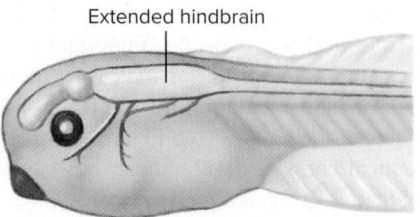

Figure 8.16 How the inhibition of a homeodomain regulatory protein alters normal development of the central nervous system of a frog tadpole. When the protein (encoded by a homeobox DNA sequence known as *X1Hbox 1*) was inactivated by antibodies directed against it, the area that should have become anterior spinal cord transformed into hindbrain instead.

Figure 8.15 Homology of *Hox* genes in insects and mammals. These genes in both insects (fruit fly) and mammals (mouse) control the subdivision of the embryo into regions of different developmental fates along the anterior-posterior axis. The homeobox-containing genes lie on a single chromosome of the fruit fly and on four separate chromosomes in the mouse. Clearly defined homologies between the two, and the parts of the body in which they are expressed, are shown in color. The open boxes denote areas where it is difficult to identify specific homologies between the two. The *Hox* genes shown here are only a small subset of all the homeobox genes.

Morphogenesis of Limbs and Organs

Hox and other homeobox genes also play a role in shaping individual organs and limbs. As shown in Figure 8.15 and 8.16, for example, regions of the brain and identity of somites are specified by particular *Hox* and homeobox genes. Many other developmental genes that are also involved in pattern formation for the entire body also help shape individual limbs and organs by producing gradients of morphogens. One example, which has been studied by Cheryll Tickle and her coworkers at University College in London, is formation and development of limb buds in chicks. They have found that a new limb bud can be induced to grow from the side of a chick by implanting a bead soaked in fibroblast growth factor (FGF). This result implies that limbs are normally induced to develop by activation of the gene for FGF in appropriate parts of the body. Whether the limb bud develops into a wing or a leg depends on whether the FGF is applied toward the front or the rear of the chick.

FGF also plays a role in shaping the limb. It is secreted by cells in an **apical ectodermal ridge** at the end of the limb bud. FGF acts as a morphogen that forms a gradient from the apical ectodermal ridge to the base of a limb bud. This gradient helps establish a proximodistal axis—one of three axes that guide development of a limb (Figure 8.17). Fingers or toes develop at the end of the proximodistal axis with the highest level of FGF. An anteroposterior axis is established by a gradient of sonic hedgehog and ensures that fingers or toes develop in the appropriate order. Finally, Wnt7a, a protein produced by a gene that is similar to the segment-polarity gene *wingless* in *Drosophila*, helps determine the dorsoventral axis. Wnt7a makes the dorsal side of the wing or foot different from the ventral side.

most exciting discoveries of the late twentieth century was that genes similar to *Hom* genes of *Drosophila* occur in other insects, as well as in chordates and unsegmented animals such as hydra and nematode worms. They also occur in plants and yeasts, and perhaps in all eukaryotes. These genes in organisms other than *Drosophila* were called *Hox* genes, but now all such genes are usually called *Hox* genes. Most *Hox* genes occur in a cluster on one chromosome. Mammals have four clusters, each on a different chromosome, with from 9 to 11 *Hox* genes each (Figure 8.15). As in *Drosophila,* the sequence of *Hox* genes within a cluster is the same as the front-to-rear order in which they are expressed in the body.

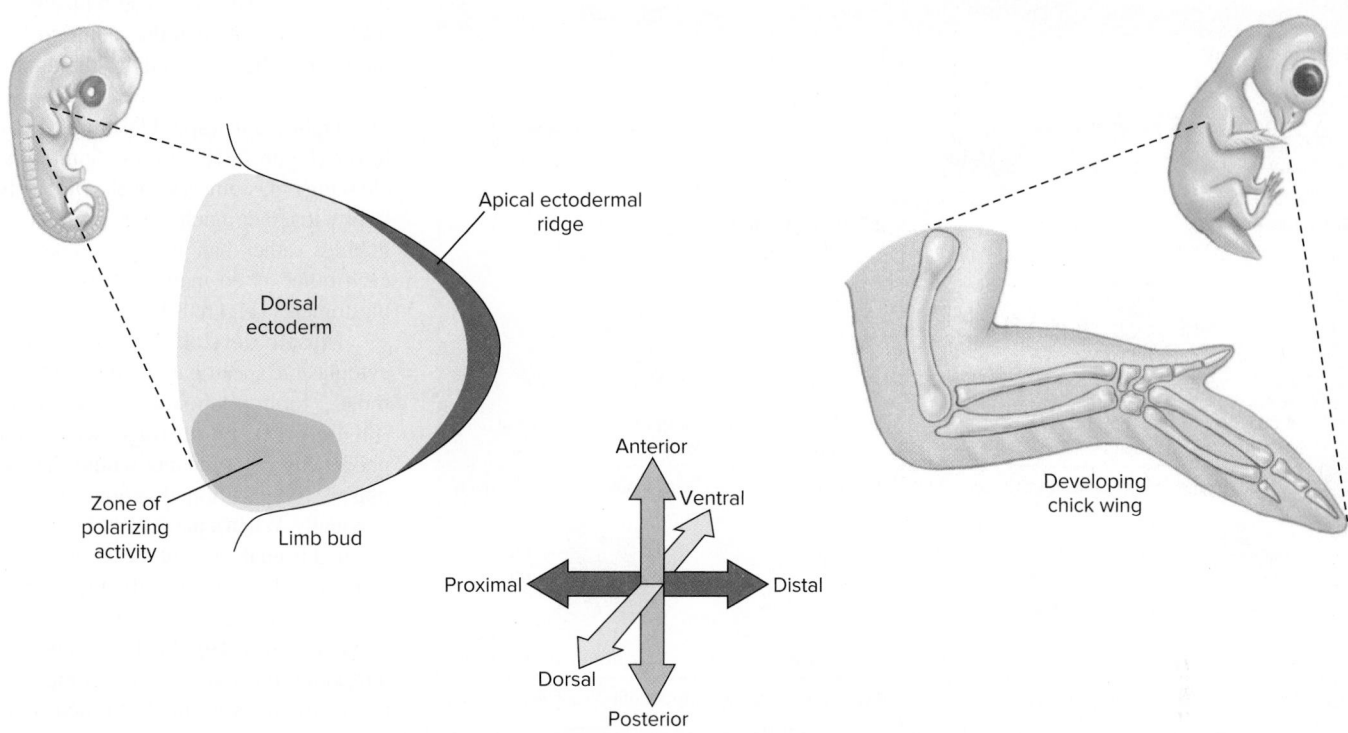

Figure 8.17 Morphogenesis in a vertebrate limb bud. The skeleton of a mature chicken limb is shown for orientation. Three axes are established in the limb bud: a proximal-distal axis by fibroblast growth factor (FGF) from the apical ectodermal ridge; an anterior-posterior axis by sonic hedgehog protein from the zone of polarizing activity; and a dorsal-ventral axis by Wnt7a protein from dorsal ectoderm.

8.7 DEVELOPMENTAL PATTERNS IN ANIMALS

There are 34 phyla of animals, as shown on the Cladogram of All Animal Phyla, inside the front cover. Because these taxa descended from a common multicellular ancestor, we expect some aspects of development to be shared among them. Which aspects of development are shared? The presence of a blastula stage in development is a fundamental animal homology (see the Cladogram of All Animal Phyla). All animals (metazoans), from sponges to chordates, produce some version of a blastula following cleavage.

Examine the base of the cladogram inside the front cover to find the character "diploblasts." This term refers to formation of two germ layers, ectoderm and endoderm, during development. Placozoans, cnidarians, and ctenophores make only these two germ layers. Cnidarians, such as sea anemones, corals, and jellies, are familiar to most people, but placozoans and ctenophores are less well known. We describe the biology of members of all animal phyla in Chapters 12 to 28.

Examine the base of the Cladogram of All Animal Phyla (inside the front cover) to find the character "triploblastic Bilateria." The term "triploblastic" refers to the formation of three germ layers during development: ectoderm, endoderm, and mesoderm. Thirty animal phyla share this developmental feature. These phyla are also bilaterally symmetrical, which means that only one plane of symmetry divides them into two halves that are mirror images of each other. We discuss symmetry and other aspects of body architecture in Chapter 9.

Examine the brackets at the top of the Cladogram of All Animal Phyla to discover that the triploblastic bilaterally symmetrical animals are divided into two major clades: Protostomia and Deuterostomia. Organisms belonging to these clades are called **protostomes** and **deuterostomes,** respectively. The names of these two groups refer to another developmental feature: the origin of the embryonic opening that becomes the adult mouth (Gr. *stoma,* mouth). The first embryonic opening, the blastopore, becomes the mouth in protostomes (Gr. *protos,* first). The second embryonic opening becomes the mouth in deuterostomes (Gr. *deuteros,* second), whereas the blastopore becomes the anus. Snails and earthworms, among many others, are protostomes. Sea stars, fishes, frogs, humans, and other mammals are representative deuterostomes.

In general, protostomes and deuterostomes are distinguished by four developmental questions: (1) as the embryo cleaves, do the cells form a spiral or a radial pattern? (2) after cleavage, is the fate of each cell decided primarily by cytoplasmic specification, producing mosaic cleavage, or primarily by conditional specification, producing regulative cleavage? (3) what is the fate of the embryonic blastopore—does it become the mouth or the anus in the adult animal? and (4) if a coelom is present, does it develop by schizocoelous formation or enterocoelous formation? Most protostomes share a suite of developmental characteristics that includes spiral mosaic cleavage, formation of the mouth from the blastopore, and development of a coelom by schizocoely (Figure 8.18). Most deuterostomes share a suite of developmental characters that includes radial regulative cleavage, formation of the anus from the blastopore, and development of a coelom by enterocoely (Figure 8.18).

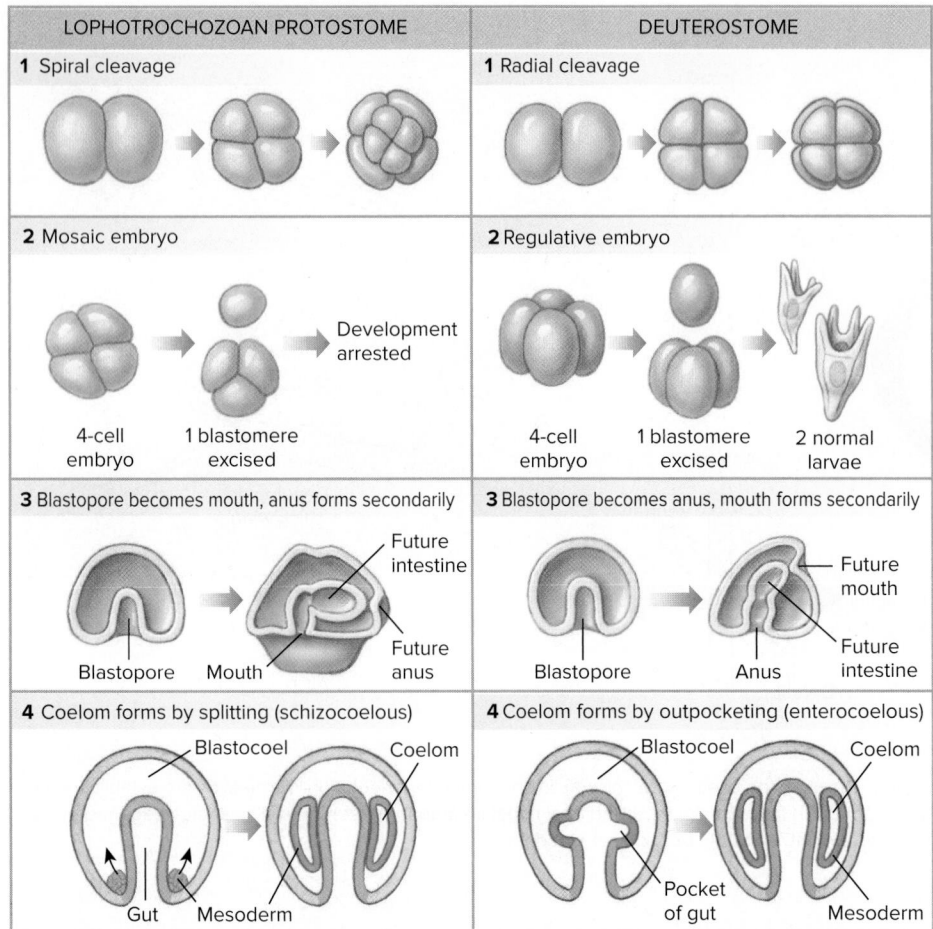

LOPHOTROCHOZOAN PROTOSTOME	DEUTEROSTOME
1 Spiral cleavage	**1** Radial cleavage
2 Mosaic embryo	**2** Regulative embryo
4-cell embryo 1 blastomere excised Development arrested	4-cell embryo 1 blastomere excised 2 normal larvae
3 Blastopore becomes mouth, anus forms secondarily	**3** Blastopore becomes anus, mouth forms secondarily
Blastopore Mouth Future intestine Future anus	Blastopore Anus Future mouth Future intestine
4 Coelom forms by splitting (schizocoelous)	**4** Coelom forms by outpocketing (enterocoelous)
Blastocoel Coelom Gut Mesoderm	Blastocoel Coelom Pocket of gut Mesoderm

Figure 8.18 Developmental tendencies of lophotrochozoan protostomes (flatworms, annelids, molluscs, etc.) and deuterostomes. These tendencies are much modified in some groups, for example, vertebrates. Cleavage in mammals is rotational rather than radial; in reptiles, birds, and many fishes, cleavage is discoidal. Vertebrates have also evolved a derived form of coelom formation that is basically schizocoelous.

the animal-vegetal axis and typically produce quartets of cells that come to lie in the furrows between cells of the underlying layer. The upper layer of cells appears offset (shifted in a spiral fashion) from the lower (Figure 8.18). In addition, spirally cleaving blastomeres pack themselves tightly together much like a group of soap bubbles, rather than just lightly contacting each other as do many radially cleaving blastomeres (Figure 8.18).

Mosaic development as a result of cytoplasmic specification predominates in most lophotrochozoan protostomes (Figure 8.18). As cleavage occurs, morphogenetic determinants within the cytoplasm are partitioned among the cells unequally. When a particular blastomere is isolated from the rest of the embryo, it still forms the characteristic structures decided by the morphogenetic determinants it contains (Figure 8.19). In the absence of a particular blastomere, the animal lacks those structures normally formed by that blastomere, so it cannot develop normally. This pattern is called mosaic development because the embryo seems to be a mosaic of self-differentiating parts.

Many ecdysozoans do not exhibit spiral cleavage; in some, cleavage appears radial, and in others, such as insects, cleavage is neither spiral nor radial, but is superficial (see Figure 8.9). Superficial cleavage is accompanied by syncytial specification of cell fate.

Protostome Development

Examine the brackets beneath the heading "Protostomia" at the top of the Cladogram of All Animal Phyla to discover that protostomes are divided into two clades: Lophotrochozoa and Ecdysozoa. The **lophotrochozoan** clade of protostomes contains segmented worms, molluscs (snails, slugs, clams, octopuses, and their kin), and several less familiar taxa. The name of this clade denotes two features present in some members of the group: a horseshoe-shaped whorl of tentacles called a **lophophore** (see pp. 320; 323–326) and a **trochophore** larva (see pp. 337–338). Lophotrochozoans typically develop via the four protostome features described previously.

Members of the other clade, **ecdysozoan** protostomes, include arthropods (insects, spiders, crabs, and their kin), roundworms, and other taxa that also molt their exoskeletons. The name of this clade refers to the shedding of the cuticle, **ecdysis** (Gr. *ekdyo,* take off, strip).

Cleavage Patterns

Spiral cleavage (Figure 8.18) occurs in most protostomes. Blastomeres cleave obliquely at an approximately 45-degree angle to

Fate of Blastopore

A **protostome** (Gr. *protos,* first, + *stoma,* mouth) is so named because the blastopore becomes the mouth, and the second, unnamed, opening becomes the anus.

Mesoderm and Coelom Formation

To form mesoderm, endodermal cells sitting at the edge of the blastopore opening (the ventral lip of the blastopore) migrate into the space between the walls of the archenteron (endoderm) and the outer body wall (ectoderm). (Figure 8.18, step 4). The process of migration is called ingression. Following migration, the cells divide and the new cells, called mesodermal precursors, are deposited between the two existing endodermal and ectodermal cell layers (Figure 8.20C). The proliferating cells become the mesoderm. Meticulous cell-lineage studies by embryologists established that in many organisms with spiral cleavage, for example, flatworms, snails, and related organisms, these mesodermal precursors arise from one large blastomere, called the 4d cell, that is present in a 29- to 64-cell embryo.

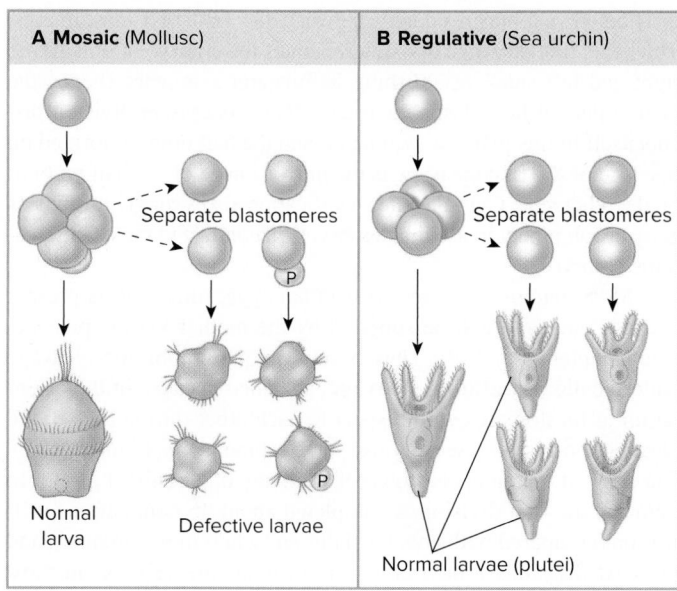

Figure 8.19 Regulative and mosaic cleavage. **A,** Mosaic cleavage. In a mollusc, when blastomeres are separated, each gives rise to only a part of an embryo. The larger size of one defective larva is the result of the formation of a polar lobe (P) composed of clear cytoplasm of the vegetal pole, which this blastomere alone receives. **B,** Regulative cleavage. Each of the early blastomeres (such as that of a sea urchin) when separated from the others develops into a small pluteus larva.

Protostomes that do not form a coelom, such as flatworms like *Planaria,* develop to an early gastrula stage, then form a mesodermal layer as just described. Mesoderm completely fills the blastocoel and a coelom never forms (see Figure 9.3). Animals without a coelom are called **acoelomate.** In other protostomes, mesoderm is deposited along the outermost edge of the blastocoel, leaving a fluid-filled blastocoel next to the gut (see Figure 9.3). The fluid-filled cavity surrounding the gut is called a **pseudocoelom** (Gr. *pseudés,* false, + *koilos,* cavity); it is bordered on the inner edge by the endodermal gut lining, and on the outer edge by a layer of mesoderm next to the ectoderm. Thus, a pseudocoelom has mesoderm on only one side, whereas a true **coelom** is a fluidfilled cavity completely surrounded by mesoderm (see Figure 9.3). We discuss acoelomate and pseudocoelomate body plans in more detail in Chapter 9.

For **coelomate** protostomes, like earthworms and snails, the mesodermal layer forms as just described, and a coelom is made by **schizocoely** (Gr. *schizein,* to split, + *koilos,* cavity). A coelom arises, as the name suggests, when the mesodermal band around the gut splits open centrally (Figure 8.18). The cells in the center of the mesoderm undergo programmed cell death to create a cavity, the coelom. Fluid collects in the coelom.

Deuterostome Development

Cleavage Patterns

Radial cleavage (Figure 8.18) is so named because the embryonic cells are arranged in radial symmetry around the animal-vegetal axis. In radial cleavage of sea stars, the first cleavage plane passes right

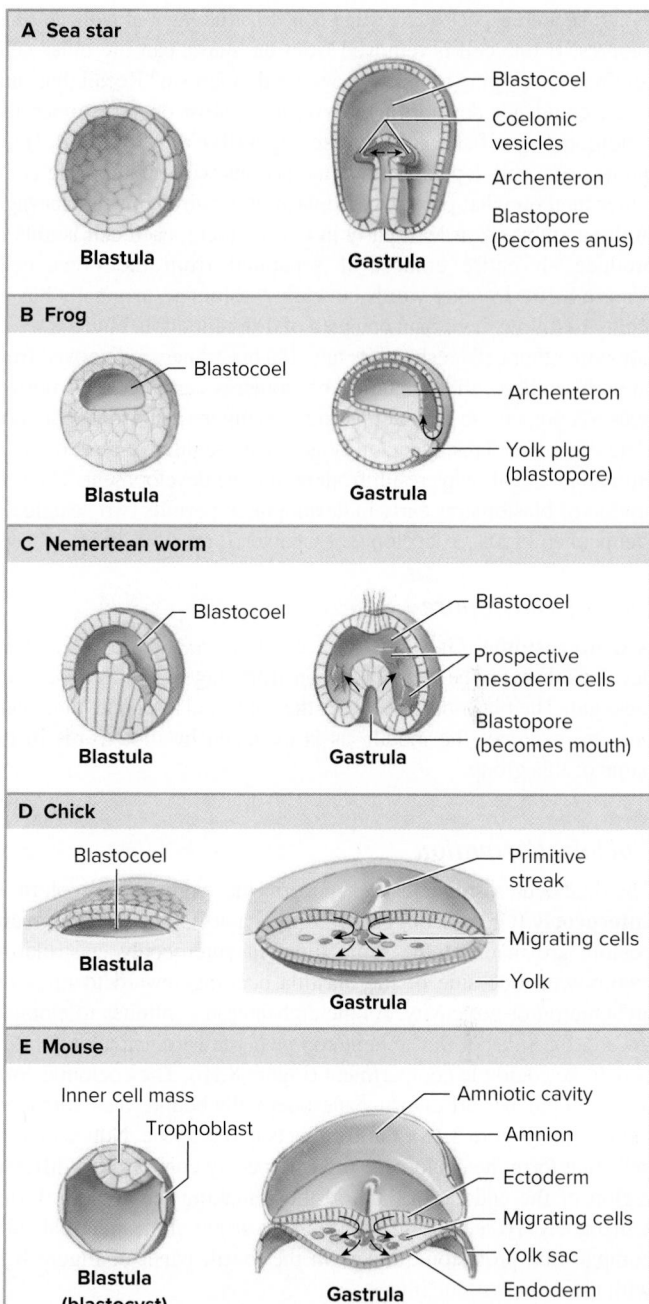

Figure 8.20 Blastula and gastrula stages in embryos of sea star, frog, nemertean worm, chick, and mouse. Arrows indicate the direction of cell movements to form mesoderm.

through the animal-vegetal axis, yielding two identical daughter cells (blastomeres). For the second cleavage division, furrows form simultaneously in both blastomeres, and these are oriented parallel to the animalvegetal axis (but perpendicular to the first cleavage furrow). Cleavage furrows next form simultaneously in the four daughter blastomeres, this time oriented perpendicular to the animal-vegetal axis, yielding two tiers of four cells each. An upper tier of cells sits directly atop the tier of cells below it (Figure 8.18). Subsequent cleavages yield an embryo composed of several tiers of cells.

A second aspect of cleavage concerns the fates of isolated blastomeres. If one cell is removed from the mass, can the other cells continue developing to produce a normal organism? Recall that such a cell could not develop in protostomes. Most deuterostomes use conditional specification leading to **regulative development.** Here, the fate of a cell depends on its interactions with neighboring cells, rather than on what piece of cytoplasm it acquired during cleavage. In these embryos, at least early in development, each cell is able to produce an entire embryo if separated from the other cells (Figure 8.19). In other words, an early blastomere originally has an ability to follow more than one path of differentiation, but its interaction with other cells restricts its fate. If a blastomere is removed from an early embryo, the remaining blastomeres can alter their normal fates to compensate for the missing blastomere and produce a complete organism. This adaptability is termed regulative development. Human identical twins result from regulative development. The separation of blastomeres early in development permits two genetically identical embryos to develop (see Chapter 7, pp. 150–151).

Fate of Blastopore

A **deuterostome** (Gr. *deuteros,* second, + *stoma,* mouth) embryo develops through the blastula and gastrula stages, and forms a complete gut. The blastopore becomes the anus, and a second, unnamed, opening becomes the mouth, as is indicated by root words in the name of this group.

Coelom Formation

The final deuterostome feature concerns the origin of the coelom. In **enterocoely** (Gr. *enteron,* gut, + *koilos,* cavity), both mesoderm and coelom are made at the same time. In enterocoely, gastrulation begins with one side of the blastula bending inward to form the archenteron or gut cavity. As the archenteron continues to elongate inward, the sides of the archenteron push outward, expanding into a pouchlike coelomic compartment (Figure 8.18). The coelomic compartment pinches off to form a mesodermally bound space surrounding the gut (Figure 8.18). Fluid collects in this space. Notice that the cells that form the coelom during enterocoely come from a different region of the endoderm than do those making the coelom during schizocoely. To see this difference in the origin of mesodermal cells, compare the protostome image in the fourth panel of Figure 8.18 with the deuterostome image.

Examples of Deuterostome Development

The general outline of deuterostome development just given varies in some of its details depending upon the animal being studied and the abundance of yolk. A few examples of specific developmental sequences illustrate this variation.

Variations in Deuterostome Cleavage   The typical deuterostome pattern is radial cleavage, but ascidian chordates (also called tunicates) exhibit **bilateral cleavage.** In ascidian eggs, the anteroposterior axis is established prior to fertilization by asymmetrical distribution of several cytoplasmic components (see Figure 8.12). The first cleavage furrow passes through the animal-vegetal axis defined by the yolk-rich vegetal pole, dividing the asymmetrically

distributed cytoplasm equally between the first two blastomeres. Thus, this first cleavage division separates the embryo into its future right and left sides, establishing its bilateral symmetry (hence the name bilateral holoblastic cleavage). Each successive division orients itself to this plane of symmetry, and the half embryo formed on one side of the first cleavage is the mirror image of the half embryo on the other side. Cytoplasmic specification also strongly influences early development in ascidian embryos, in contrast to most deuterostome embryos.

Most mammals possess isolecithal eggs; little yolk is present because nutrients will be supplied by the mother via the placenta (see Chapter 7, p. 148). They have a unique holoblastic cleavage pattern called **rotational cleavage,** so called because of the orientation of blastomeres with respect to each other during the second cleavage division (see mouse development in Figure 8.8E). Cleavage in mammals is slower than in any other animal group. In humans, the first division is completed about 36 hours after fertilization (compared with about an hour and a half in sea urchins), and the next divisions follow at 12- to 24-hour intervals. As in most other animals, the first cleavage plane runs through the animal-vegetal axis to yield a two-cell embryo. However, during the second cleavage one of these blastomeres divides meridionally (through the animal-vegetal axis) while the other divides equatorially (perpendicular to the animal-vegetal axis). Thus, the cleavage plane in one blastomere is rotated 90 degrees with respect to the cleavage plane of the other blastomere (hence the name rotational cleavage). Furthermore, early divisions are asynchronous; not all blastomeres divide at the same time. Thus, mammalian embryos may not increase regularly from two to four to eight blastomeres, but often contain odd numbers of cells. After the third division, the cells suddenly close into a tightly packed configuration, which is stabilized by tight junctions that form between outermost cells of the embryo. These outer cells form the **trophoblast.** The trophoblast is not part of the embryo proper but will form the embryonic portion of the placenta when the embryo implants in the uterine wall. Cells that actually give rise to the embryo proper form from the inner cells, called the **inner cell mass** (see blastula stage in Figure 8.20E). These cells are undifferentiated, so their fates are not restricted; they are also called "stem cells."

Telolecithal eggs of reptiles, birds, and most fish divide by **discoidal cleavage.** Because of the great mass of yolk in these eggs, cleavage is confined to a small disc of cytoplasm lying atop a mound of yolk (see chick development in Figure 8.8D). Early cleavage furrows carve this cytoplasmic disc to yield a single layer of cells called the blastoderm. Further cleavages divide the blastoderm into five to six layers of cells (Figure 8.20D).

Variations in Deuterostome Gastrulation   In sea stars, gastrulation begins when the entire vegetal area of the blastula flattens to form a **vegetal plate** (a sheet of epithelial tissue). This event is followed by a process called **invagination,** in which the vegetal plate bends inward and extends about one-third of the way into the blastocoel, forming the archenteron (Figure 8.20A). Coelomic formation is typical of enterocoely. As the archenteron continues to elongate toward the animal pole, its anterior end expands into two pouchlike **coelomic vesicles.** These vesicles pinch off to form left and right coelomic compartments (Figure 8.20A).

The **ectoderm** gives rise to the epithelium of the body surface and to the nervous system. The **endoderm** gives rise to the epithelial lining of the digestive tube. The outpocketing of the archenteron is the origin of **mesoderm.** This third germ layer will form the muscular system, the reproductive system, the peritoneum (lining of the coelomic compartments), and the calcareous plates of the sea star's endoskeleton.

Frogs are deuterostomes with radial cleavage, but morphogenetic movements of gastrulation are greatly influenced by the mass of inert yolk in the vegetal half of the embryo. Cleavage divisions are slowed in this half so that the resulting blastula consists of many small cells in the animal half and a few large cells in the vegetal half (Figures 8.8B and 8.20B). Gastrulation in amphibians begins when cells located at the future dorsal side of the embryo invaginate to form a slitlike blastopore. Thus, as in sea stars, invagination initiates archenteron formation, but amphibian gastrulation begins in the marginal zone of the blastula, where animal and vegetal hemispheres come together, and where there is less yolk than in the vegetal region. Gastrulation progresses as sheets of cells in the marginal zone turn inward over the blastopore lip and move inside the gastrula to form mesoderm and endoderm. The three germ layers now formed are the primary structural layers that play crucial roles in further differentiation of the embryo. Fates of cells in each germ layer are restricted to particular adult derivatives as discussed later (Figure 8.27).

In bird and other reptile embryos (Figure 8.20D), gastrulation begins with a thickening of the blastoderm at the caudal end of the embryo. This thicker region develops a central depression, which migrates forward to form a **primitive streak** (Figure 8.21). The primitive streak becomes the anteroposterior axis of the embryo and the center of early growth. The primitive streak is homologous to the blastopore of frog embryos, but in chicks it does not open into the gut cavity because of the obstructing mass of yolk that lies beneath the blastoderm. The blastoderm consists of two layers (epiblast and hypoblast) with a blastocoel between them. Cells of the epiblast move as a sheet toward the primitive streak, then roll over the edge and migrate as individual cells into the blastocoel. These migrating cells separate into two streams. One stream of cells moves deeper (displacing the hypoblast along the midline) and forms endoderm. The other stream moves between the epiblast and the hypoblast to form mesoderm. Cells on the surface of the embryo compose the ectoderm. The embryo now has three germ layers, at this point arranged as sheetlike layers with ectoderm on top and endoderm at the bottom. This arrangement changes, however, when all three germ layers lift from the underlying yolk (Figure 8.21), then fold under to form a three-layered embryo that is pinched off from the yolk except for a stalk attachment to the yolk at midbody (see Figure 8.23).

Gastrulation in mammals is remarkably similar to gastrulation in reptiles, including birds (Figure 8.20E). Gastrulation movements in the inner cell mass produce a primitive streak. Epiblast cells move medially through the primitive streak into the blastocoel, and individual cells then migrate laterally through the blastocoel to form mesoderm and endoderm. Endoderm cells (derived from the hypoblast) form a yolk sac devoid of yolk (since mammalian embryos derive nutrients directly from the mother via the placenta).

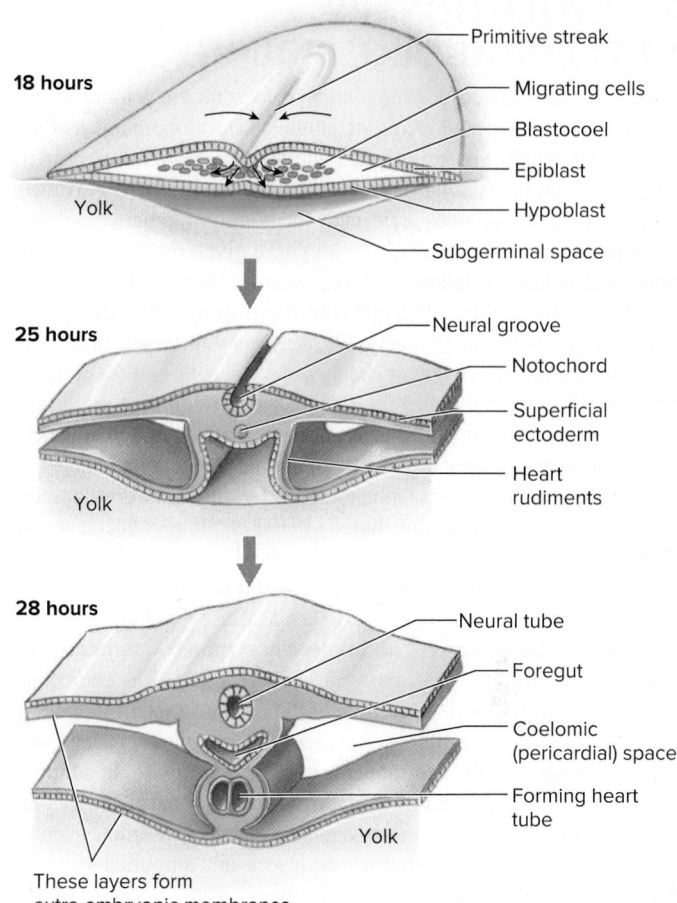

18 hours — Primitive streak, Migrating cells, Blastocoel, Epiblast, Hypoblast, Subgerminal space, Yolk

25 hours — Neural groove, Notochord, Superficial ectoderm, Heart rudiments, Yolk

28 hours — Neural tube, Foregut, Coelomic (pericardial) space, Forming heart tube, Yolk

These layers form extra-embryonic membranes

Figure 8.21   Gastrulation in a chick. Transverse sections through the heart-forming region of the chick show development at 18, 25, and 28 hours of incubation. Ectoderm is blue, mesoderm is red, and endoderm is yellow.

Amphibians and reptiles (including birds), which have moderate to large amounts of yolk concentrated in the vegetal region of the egg, have evolved derived gastrulation patterns in which the yolk does not participate in gastrulation. Yolk is an impediment to gastrulation and consequently the gastrulation process occurs around (amphibians) or on top of (reptiles and birds) the vegetal yolk. Mammalian eggs are isolecithal, and thus one might expect them to have a gastrulation pattern similar to that of sea stars. Instead they have a pattern more suited to telolecithal eggs. The best explanation for this feature of mammalian egg development is common ancestry with birds and other reptiles. Reptiles, birds, and mammals share a common ancestor whose eggs were telolecithal. Thus, all inherited their gastrulation patterns from this common ancestor, and mammals subsequently evolved isolecithal eggs but retained the telolecithal gastrulation pattern.

A further developmental complication in vertebrates is that coelom formation occurs by a modified form of schizocoely (see Figure 8.18), not enterocoely. The nonvertebrate chordates form the coelom by enterocoely, as is typical of deuterostomes. More details of vertebrate development are provided on pages 174–180.

8.8 EVOLUTIONARY DEVELOPMENTAL BIOLOGY

Zoologists have always looked to embryology for clues to the evolutionary history, or phylogeny, of animals. Developmental features such as the number of germ layers and the fate of the blastopore do suggest evolutionary relationships among different phyla. Advances in developmental genetics have made the relationship between development and evolution even closer and have given rise to an exciting new field called evolutionary developmental biology. Evolutionary developmental biology, often nicknamed evo-devo, is based on a realization that evolution is essentially a process in which organisms become different as a result of changes in the genetic control of development. The fact that the genes that control development are similar in animals as different as fruit flies and mice offers hope that we can reconstruct the evolutionary history of animals by understanding how functioning of those genes came to differ. Evolutionary developmental biology has already contributed several exciting concepts to our thinking about animal evolution, but the field is so new that it would be premature to accept these concepts as established. It is best to state them as questions for further study, such as those that follow.

Are the body plans of all bilaterally symmetric animals fundamentally similar? As noted on page 167, *chordin,* one of the genes responsible for development of the nervous system in the dorsal part of a frog, is similar to *short gastrulation,* which is necessary for development of the ventral nerve cord in *Drosophila.* In addition, the gene *decapentaplegic* promotes dorsal development in *Drosophila,* and the similar gene *bone morphogenetic protein-4* promotes ventral development in frogs. In other words, insects and amphibians, whose body plans look so different, actually share a similar control of dorsoventral patterning, except that one is upside down compared with the other. This finding has prompted a reappraisal of an idea first proposed by the French naturalist Etienne Geoffroy St. Hilaire in 1822 after he noticed that in a lobster dissected ventrally and lying on its back the nerve cord was above the gut, and the heart was below it, exactly the opposite of a ventrally dissected vertebrate. The idea that a vertebrate is like an inverted invertebrate was quickly rejected, but now biologists are once more considering whether the body plans of protostomes and deuterostomes are perhaps simply inverted dorsoventrally, but otherwise similar.

Can the anatomy of extinct ancestral species be inferred from the developmental genes shared by their descendants? The fact that dorsoventral patterning is similar in protostomes and deuterostomes suggests that the most recent common ancestor of these two branches had a similar dorsoventral patterning with a heart and nervous system separated by the gut. It may also have had at least rudimentary eyes, judging from the fact that similar genes, *eyeless/ Pax-6,* are used in eye formation in a wide range of protostomes and deuterostomes.

Instead of evolution proceeding by the gradual accumulation of numerous small mutations, could it proceed by relatively few mutations in a few developmental genes? The fact that formation of legs or eyes can be induced by a mutation in one gene suggests that these and other organs develop as modules (see p. 167). If so, then entire limbs and organs could have been lost or acquired during evolution as a result of one or a few mutations. If this is correct, then the apparently rapid evolution of numerous groups of animals during the few million years of the Cambrian explosion and at other times is more easily explained. Instead of requiring mutations in numerous genes, each with a small effect, evolution of different groups could be a result of changes in timing, number, or expression of relatively few developmental genes.

8.9 VERTEBRATE DEVELOPMENT

The Common Vertebrate Heritage

A prominent outcome of shared ancestry of vertebrates is their common pattern of development. This common pattern is best seen in the remarkable similarity of postgastrula vertebrate embryos (Figure 8.22). The likeness occurs at a brief moment in the development of vertebrates when shared chordate hallmarks of dorsal neural tube, notochord, pharyngeal gill pouches with aortic arches, ventral heart, and postanal tail are present at about the same stage of development. Their moment of similarity—when the embryos seem almost interchangeable—is all the more extraordinary considering the great variety of eggs and widely different types of early development that have converged toward a common design. Then, as development continues, the embryos diverge in pace and direction, becoming recognizable as members of their class, then their order, then their family, and finally their species. The important contribution of early vertebrate development to our understanding of homology and evolutionary common descent is described in Chapter 6 in the section on Ontogeny, Phylogeny, and Recapitulation, page 112.

Amniotes and the Amniotic Egg

Reptiles (including birds) and mammals form a monophyletic grouping of vertebrates called **amniotes,** so named because their embryos develop within a membranous sac, the **amnion.** The amnion is one of four **extraembryonic membranes** that compose a sophisticated support system within the **amniotic egg** (Figure 8.23), which evolved when the first amniotes appeared in the late Paleozoic era.

The amnion is a fluid-filled sac that encloses the embryo and provides an aqueous environment in which the embryo floats, protected from mechanical shock and adhesions.

Evolution of the second extraembryonic membrane, the **yolk sac,** actually predates appearance of amniotes by many millions of years. The yolk sac with its enclosed yolk is a conspicuous feature of all fish embryos. After hatching, a growing fish larva depends on the remaining yolk provisions to sustain it until it can begin to feed itself (Figure 8.24). The yolk sac functions differently in animals that give live birth. In many viviparous vertebrates from diverse groups, the yolk sac becomes vascular and intimately associated with the mother's reproductive tract, allowing transfer of nutrients and respiratory gases between mother and fetus. Thus, a yolk sac placenta is formed. The mass of yolk is an extraembryonic structure because it is not a part of the embryo proper, and the yolk sac is an extraembryonic membrane because it is an accessory structure that develops outside the embryo and is discarded after the yolk is consumed.

The **allantois** is a sac that grows from the embryo's hindgut and serves as a repository for metabolic wastes during development. It also functions as a respiratory surface for exchange of oxygen and carbon dioxide.

The **chorion** lies just beneath the eggshell and completely encloses the rest of the embryonic system. As the embryo grows and

Figure 8.22 Early vertebrate embryos drawn from photographs. Embryos as diverse as fish, salamander, tortoise, bird, and human show remarkable similarity following gastrulation. At this stage (top row) they reveal features common to the entire subphylum Vertebrata. As development proceeds they diverge, each becoming increasingly recognizable as belonging to a specific class, order, family, and finally, species.

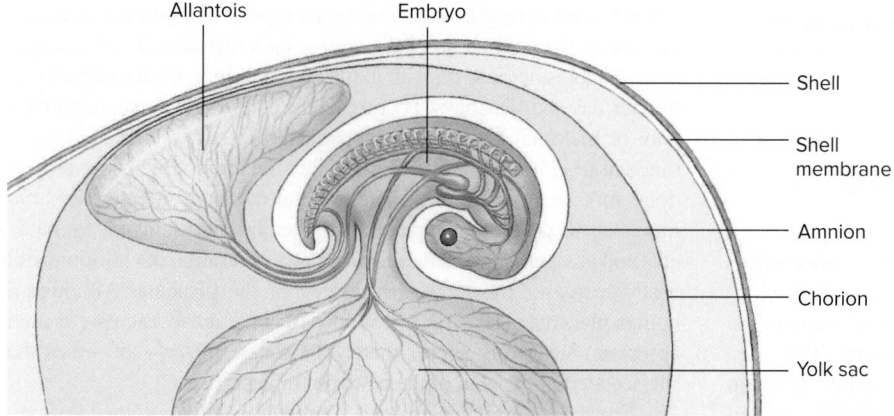

Figure 8.23 Amniotic egg at an early stage of development showing a chick embryo and its extraembryonic membranes.

its need for oxygen increases, the allantois and chorion fuse to form the **chorioallantoic membrane.** This double membrane has a rich vascular network connected to the embryonic circulation. Lying just beneath the porous shell, the vascular chorioallantois serves as a provisional "lung" across which oxygen and carbon dioxide can freely exchange. Thus an amniotic egg provides a complete life-support system for the embryo, enclosed by a tough outer shell. The amniotic egg is one of the most important adaptations to have evolved in vertebrates.

The evolution of a shelled amniotic egg made internal fertilization a reproductive requirement. A male must introduce sperm directly into the female reproductive tract, since sperm must reach and fertilize the egg before the eggshell is wrapped around it.

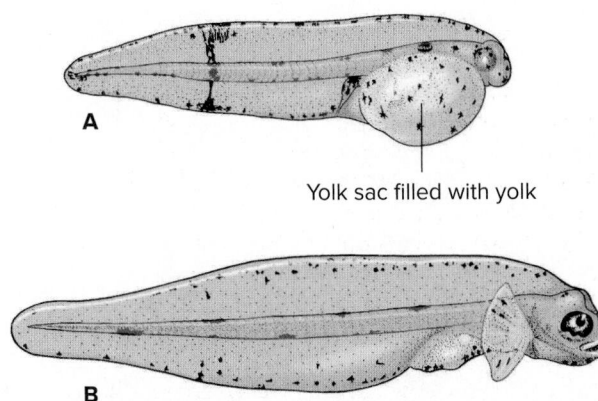

Figure 8.24 Fish larvae showing yolk sac. **A,** The one-day-old larva of a marine flounder has a large yolk sac. **B,** After 10 days of growth the larva has developed mouth, sensory organs, and a primitive digestive tract. With its yolk supply now exhausted, it must capture food to grow and survive.

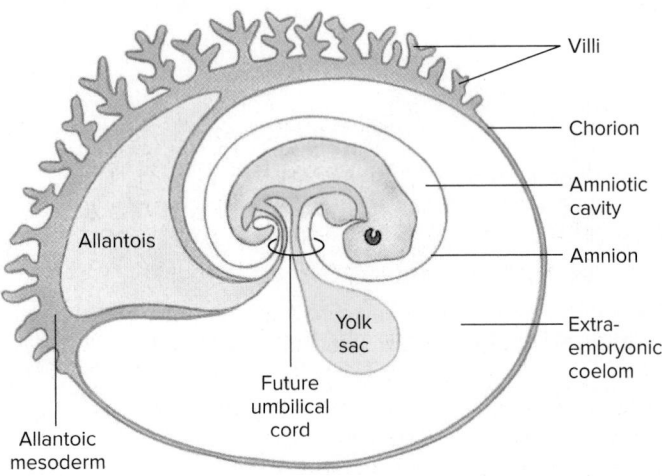

Figure 8.25 Generalized diagram of extraembryonic membranes of a mammal, showing how its development parallels that of a chick (compare with Figure 8.23). Most extraembryonic membranes of mammals have been redirected to new functions.

The Mammalian Placenta and Early Mammalian Development

Rather than developing within an eggshell like most other vertebrates, most mammalian embryos evolved the strategy of developing within the mother's body. We have already seen that mammalian gastrulation closely parallels that of egg-laying amniotes. The earliest mammals were egg layers, and even today some mammals retain this primitive character; **monotremes** (duck-billed platypus and spiny anteater) lay large, yolky eggs that closely resemble bird eggs. In **marsupials** (pouched mammals such as opossums and kangaroos), embryos develop for a time within the mother's uterus, but an embryo does not "take root" in the uterine wall, and consequently it receives little nourishment from the mother before birth. The young of marsupials are born at an early stage of development and

continue developing sheltered in a pouch in the mother's abdominal wall, nourished with milk (we describe reproduction in marsupials on pp. 624-626).

All other mammals, composing 94% of class Mammalia, are **placental mammals.** These mammals have evolved a **placenta,** a remarkable fetal structure through which an embryo is nourished. Evolution of this fetal organ required substantial restructuring, not only of extraembryonic membranes to form the placenta, but also of the maternal oviduct, part of which had to expand into long-term housing for embryos, the **uterus** (see Chapter 7, p. 144). Despite these modifications, development of extraembryonic membranes in placental mammals is remarkably similar to their development in egg-laying amniotes (compare Figure 8.23 and 8.25). In fact, for some nonmammalian vertebrates that give live birth, the extraembryonic membranes form a placenta. Some viviparous lizards and snakes have either a yolk sac placenta or a chorioallantoic placenta or both.

One of the most intriguing questions the placenta presents is this: Why is it not immunologically rejected by the mother? Both placenta and embryo are genetically alien to the mother because they contain proteins (called major histocompatibility proteins, p. 770) that differ from those of the mother. We would expect uterine tissues to reject the embryo just as the mother would reject an organ transplanted from her own child. The placenta is a uniquely successful foreign transplant, or **allograft,** because it has evolved measures for suppressing the immune response that normally would be mounted against it and the fetus by the mother. Experiments suggest that the chorion produces proteins and lymphocytes that block the normal immune response by suppressing formation of specific antibodies by the mother.

Early stages of mammalian cleavage, shown in Figure 8.20E, occur while a **blastocyst** is traveling down the oviduct toward the uterus, propelled by ciliary action and muscular peristalsis. When a human blastocyst is about six days old and composed of about 100 cells, it contacts the uterine endometrium (uterine lining) (Figure 8.26). On contact, the trophoblast cells proliferate rapidly and produce enzymes that digest the epithelium of the uterine endometrium. These changes allow the blastocyst to implant in the endometrium. By the eleventh or twelfth day the blastocyst is completely buried and surrounded by a pool of maternal blood. The trophoblast thickens, sending out thousands of tiny, fingerlike projections, the **chorionic villi.** These projections sink like roots into the uterine endometrium after the embryo implants. As development proceeds and embryonic demands for nutrients and gas exchange increase, the great proliferation of chorionic villi vastly increases the total surface area of the placenta. Although a human placenta at term measures only 18 cm (7 inches) across, its total absorbing surface is approximately 13 square meters—50 times the surface area of the skin of the newborn infant.

Since a mammalian embryo is protected and nourished through the placenta rather than with stored yolk, what happens to the four extraembryonic membranes it has inherited from early amniotes? The amnion remains unchanged, a protective water jacket in which the embryo floats. A fluid-filled yolk sac is also retained, although it contains no yolk. It has acquired a new function: during early development it is the source of stem cells that give rise to blood, lymphoid

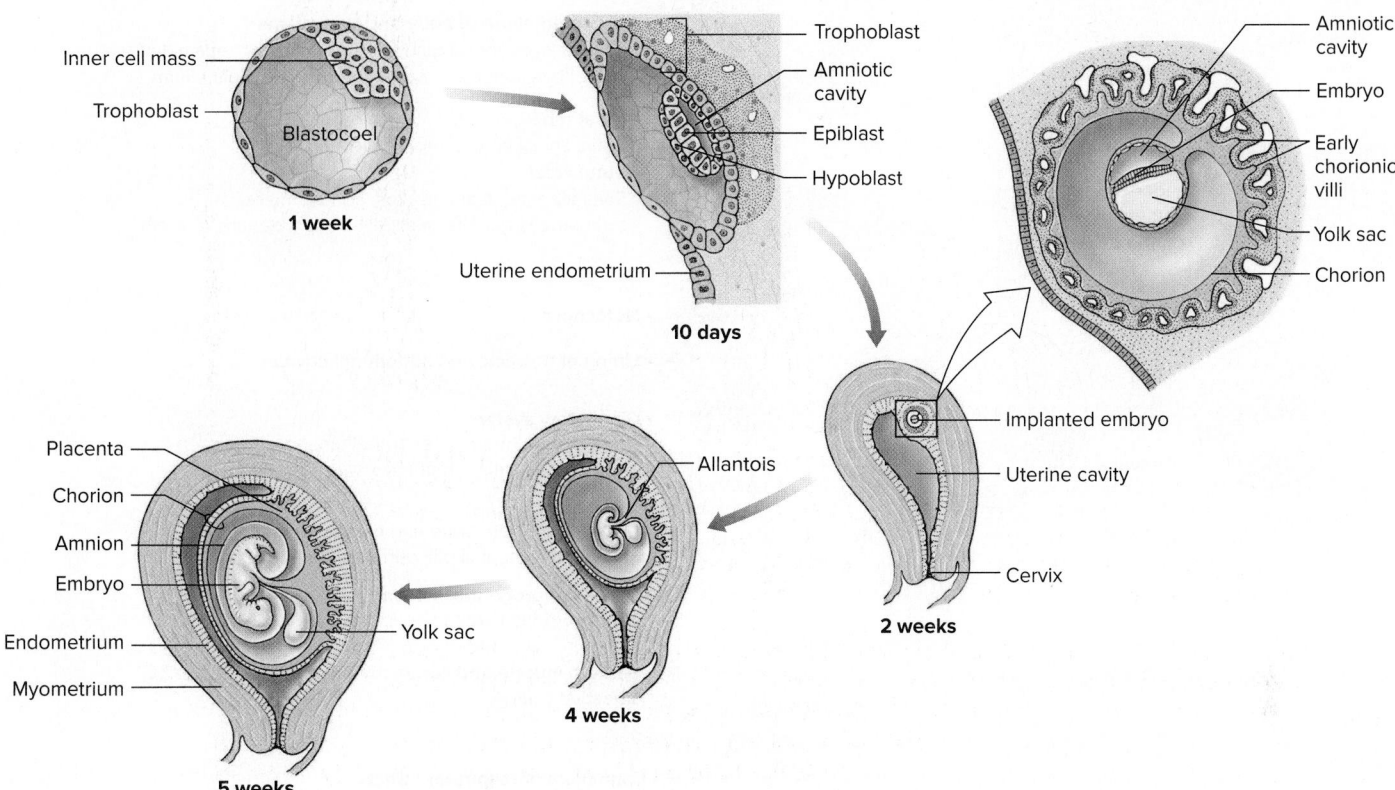

Figure 8.26 Early development of the human embryo and its four extraembryonic membranes: amnion, yolk sac without yolk, allantois, and chorion.

cells, and gametes. These stem cells later migrate into the developing embryo. In some mammals, including raccoons and mice, a heavily vascularized yolk sac implants in the uterus, along with the typical placenta. The two remaining extraembryonic membranes, allantois and the chorion, are recommitted to new functions. The allantois is no longer needed for storage of metabolic wastes. Instead it contributes to the **umbilical cord,** which links the embryo physically and functionally with the placenta (see Chapter 7, Figure 7.18). The chorion, the outermost membrane, forms most of the placenta itself. The rest of the placenta is formed by the adjacent uterine endometrium.

The embryo grows rapidly, and in humans all major organs of the body have begun their formation by the end of the fourth week of development. The embryo is now about 5 mm in length and weighs approximately 0.02 g. During the first two weeks of development (**germinal period**) the embryo is quite resistant to outside influences. However, during the next eight weeks, when all major organs are being established and body shape is forming (**embryonic period**), an embryo is more sensitive to disturbances that might cause malformations (such as exposure to alcohol or drugs taken by the mother) than at any other time in its development. The embryo becomes a **fetus** at approximately two months after fertilization. The **fetal period** is primarily a growth phase, although organ systems (especially the nervous and endocrine systems) continue to differentiate. The fetus grows from approximately 28 mm and 2.7 g at 60 days to approximately 350 mm and 3000 g at term (nine months).

8.10 DEVELOPMENT OF SYSTEMS AND ORGANS

During vertebrate gastrulation the three germ layers are formed. These differentiate, as we have seen, first into primordial cell masses and then into specific organs and tissues. During this process, cells become increasingly committed to specific directions of differentiation. Derivatives of the three germ layers are diagrammed in Figure 8.27.

> Assignment of early embryonic layers to specific "germ layers" (not to be confused with "germ cells," which are the eggs and sperm) is for the convenience of embryologists and is of no concern to the embryo. Whereas the three germ layers normally differentiate to form the tissue and organs described here, it is not the germ layer itself that determines differentiation, but rather the precise position of an embryonic cell with relation to other cells.

Derivatives of Ectoderm: Nervous System and Nerve Growth

The brain, spinal cord, and nearly all outer epithelial structures of the body develop from primitive ectoderm. They are among the earliest organs to appear. Just above the notochord, the ectoderm thickens to form a **neural plate.** The edges of this plate rise up, fold, and join together at the top to create an elongated, hollow **neural tube.** The

Figure 8.27 Derivatives of the primary germ layers in mammals.

neural tube gives rise to most of the nervous system: anteriorly it enlarges and differentiates into the brain and cranial nerves; posteriorly it forms the spinal cord and spinal motor nerves. Much of the rest of the peripheral nervous system is derived from **neural crest** cells, which pinch off from the neural tube before it closes (Figure 8.28). Some of the many different cell types and structures that originate with the neural crest are itemized in Figure 8.27 and Figure 8.28. Neural crest tissue is unique to vertebrates and was probably of prime importance in evolution of the vertebrate head and jaws.

How are the billions of nerve axons in the body formed? What directs their growth? Biologists were intrigued with these questions, which seemed to have no easy solutions. Because a single nerve axon may be more than a meter in length (for example, motor nerves running from the spinal cord to the toes in humans are about 1 meter long), it seemed impossible that a single cell could reach out so far. The answer had to await the development of one of the most powerful tools available to biologists, the cell culture technique.

In 1907 embryologist Ross G. Harrison discovered that he could culture living neuroblasts (embryonic nerve cells) for weeks outside the body by placing them in a drop of frog lymph hung from the underside of a cover slip. Watching nerves grow for periods of days, he saw that each axon was an outgrowth of a single cell. As the axon extended outward, materials for growth flowed down the axon center to the growing tip (growth cone), where they were incorporated into new protoplasm (Figure 8.29).

The second question—what directs nerve growth—has taken longer to unravel. The major hypothesis of the 1940s proposed nerve growth as a random, diffuse process: the nervous system developed as an equipotential network, or blank slate, shaped by usage into a functional system. We now understand that nerve fibers find their way selectively to many destinations by following external cues. Research with invertebrate nervous systems indicated that each of the billions of nerve cell axons acquires a distinct identity that somehow directs it along a specific pathway to its destination. Harrison observed that a growing nerve axon terminated in a growth cone, from which extend numerous tiny threadlike pseudopodial processes (filopodia) (Figure 8.29). Santiago Ramón y Cajal had inferred the dynamic nature of this growth cone in 1890, even though he studied only fixed tissues, not live cells. Research has shown that the growth cone is steered by an array of guidance molecules secreted along the

Figure 8.28 Development of neural tube and neural crest cells from neural plate ectoderm.

pathway and by the axon's target. This chemical guidance system, which requires particular patterns of gene expression, is just one example of the amazing flexibility that characterizes the entire process of differentiation.

The tissue culture technique developed by Ross G. Harrison is now used extensively by scientists in all fields of active biomedical research, not just by developmental biologists. The great impact of the technique has been felt only in recent years. Harrison was twice considered for the Nobel Prize (1917 and 1933), but he failed ever to receive the award because, ironically, the tissue culture method was then considered "of rather limited value."

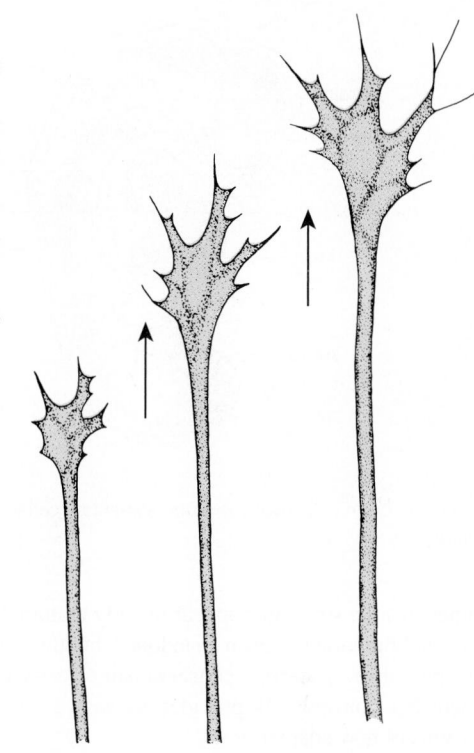

Figure 8.29
Growth cone at the growing tip of a nerve axon. Materials for growth flow down the axon to the growth cone from which numerous threadlike filopodia extend. These serve as a pioneering guidance system for the developing axon. Direction of growth is shown by arrows.

Derivatives of Endoderm: Digestive Tube and Survival of Gill Arches

In frog embryos the primitive gut makes its appearance during gastrulation with the formation of the archenteron. From this simple endodermal cavity develop the lining of the digestive tract, the lining of the pharynx and lungs, most of the liver and pancreas, the thyroid and parathyroid glands, and the thymus (see Figure 8.27).

In other vertebrates the **alimentary canal** develops from the primitive gut and is folded off from the yolk sac by growth and folding of the body wall (Figure 8.30). The ends of the tube open to the exterior and are lined with ectoderm, whereas the rest of the tube is lined with endoderm. **Lungs, liver,** and **pancreas** arise from the foregut.

Among the most intriguing derivatives of the digestive tract are the pharyngeal pouches, which make their appearance in the early embryonic stages of all vertebrates (see Figure 8.22). During development the endodermally-lined pharyngeal pouches interact with overlying ectoderm to form gill arches. In fishes, gill arches develop into gills and supportive structures and serve as respiratory organs. When early vertebrates moved onto land, gills were unsuitable for aerial respiration, and respiratory function was performed by independently evolved lungs.

Why then do gill arches persist in embryos of terrestrial vertebrates? Although gill arches serve no respiratory function in either embryos or adults of terrestrial vertebrates, they are necessary starting points for a variety of other structures. For example, the first arch and its endoderm-lined pouch (the space between adjacent arches) form the upper and lower jaws and inner ear of vertebrates. The second, third, and fourth gill pouches contribute to the tonsils, parathyroid glands, and thymus. We can understand then why gill arches and

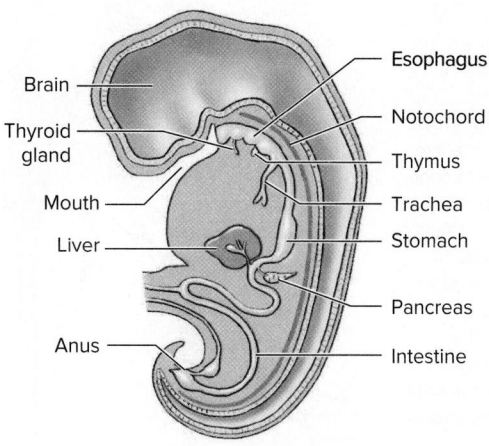

Brain — Esophagus
Thyroid gland — Notochord
Mouth — Thymus
Liver — Trachea
Anus — Stomach
— Pancreas
— Intestine

Figure 8.30 Derivatives of the alimentary canal of a human embryo.

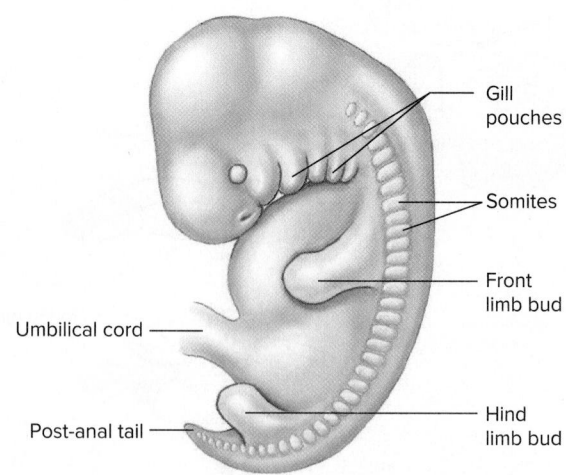

Gill pouches
Somites
Umbilical cord — Front limb bud
Post-anal tail — Hind limb bud

Figure 8.31 Human embryo showing somites, which differentiate into skeletal muscles and axial skeleton.

other fishlike structures appear in early mammalian embryos. Their original function has been abandoned, but the structures are retained for new uses. The great conservatism of early embryonic development has conveniently provided us with a telescoped view of the origins of new adaptations.

Derivatives of Mesoderm: Support, Movement, and Beating Heart

The mesoderm forms most skeletal and muscular tissues, the circulatory system, and urinary and reproductive organs (see Figure 8.27). As vertebrates have increased in size and complexity, mesodermally derived supportive, movement, and transport structures have become an even greater proportion of the body.

Most **muscles** arise from the mesoderm along each side of the neural tube (Figure 8.31). This mesoderm divides into a linear series of blocklike somites (38 in humans), which by splitting, fusion, and migration become the axial skeleton, dermis of the dorsal skin, and muscles of the back, body wall, and limbs.

Mesoderm gives rise to the first functional organ, the embryonic heart. Guided by the underlying endoderm, two clusters of precardiac mesodermal cells move amebalike into position on either side of the developing gut. These clusters of cells differentiate into a pair of double-walled tubes, which later fuse to form a single, thin tube (Figure 8.21, p. 173).

As the cells group together, the first twitchings are evident. In a chick embryo, a favorite animal for experimental embryological studies, the primitive heart begins to beat on the second day of the 21-day incubation period; it begins beating before any true blood vessels have formed and before there is any blood to pump. As the ventricle primordium develops, the spontaneous cellular twitchings become coordinated into a feeble but rhythmical beat. New heart chambers, each with a beat faster than its predecessor, then develop.

Finally a specialized area of heart muscle called the **sinoatrial (SA) node** develops and takes command of the entire heartbeat (the role of the SA node in the excitation of the heart is described on p. 688). The SA node becomes the heart's primary **pacemaker.** As

the heart builds a strong and efficient beat, vascular channels open within the embryo and across the yolk. Within the vessels are the first primitive blood cells suspended in plasma.

Early development of the heart and circulation is crucial to continued embryonic development, because without circulation an embryo could not obtain materials for growth. Food is absorbed from the yolk and carried to the embryonic body, oxygen is delivered to all tissues, and carbon dioxide and other wastes are carried away. An embryo is totally dependent on these extraembryonic support systems, and the circulation is the vital link between them.

SUMMARY

Developmental biology encompasses the emergence of order and complexity during the development of a new individual from a fertilized egg, and the control of this process. The early preformation concept of development gave way in the eighteenth century to the theory of epigenesis, which holds that development is the progressive appearance of new structures that arise as the products of antecedent development. Fertilization of an egg by a sperm restores the diploid number of chromosomes and activates the egg for development. Both sperm and egg have evolved devices to promote efficient fertilization. The sperm is a highly condensed haploid nucleus provided with a locomotory flagellum. Many eggs release chemical sperm attractants, most have surface receptors that recognize and bind only with sperm of their own species, and all have developed devices to prevent polyspermy. In many cases sperm enter the oocyte prior to the completion of meiosis.

During cleavage an embryo divides rapidly and usually synchronously, producing a multicellular blastula. Cleavage is greatly influenced by quantity and distribution of yolk in the egg. Eggs with little yolk, such as those of many marine invertebrates, divide completely (holoblastic) and usually have indirect development with a larval stage interposed between the embryo and the adult. Eggs having an abundance of yolk, such as those of birds, other reptiles, and most arthropods, divide only partially (meroblastic), and birds and other reptiles have no larval stage.

At gastrulation, cells on an embryo's surface move inward to form germ layers (endoderm, ectoderm, mesoderm) and the embryonic body plan. Like cleavage, gastrulation is much influenced by the quantity of yolk.

Despite the different developmental fates of embryonic cells, every cell contains a complete genome and thus the same nuclear information. Early development through cleavage is governed by cytoplasmic determinants derived from the maternal genome and placed in the egg cortex. As gastrulation approaches, control gradually shifts from maternal to embryonic as an embryo's own nuclear genes begin transcribing mRNA.

Harmonious differentiation of tissues proceeds in three general stages: pattern formation, determination of position in the body, and induction of limbs and organs appropriate for each position. Each stage is guided by morphogens. Pattern formation refers to determination of the anteroposterior, dorsoventral, and left-to-right body axes. In amphibians the anteroposterior axis is established by morphogens such as chordin from the Spemann organizer in the gray crescent of the zygote. In *Drosophila* the anterior-posterior axis is determined by morphogens such as *bicoid* and *nanos,* which are transcribed from maternal mRNA deposited at the anterior of the egg. Morphogen gradients also specify other axes, such as the dorsoventral, or back and belly, axis. In flies and other segmented animals, such morphogens activate genes that divide the body into head, thorax, and abdomen, and then into correctly oriented segments. The structures appropriate to each segment are then induced by homeotic genes, which include a particular sequence of DNA bases called the homeobox. Mutations in homeotic genes can cause the development of inappropriate structures on a segment: legs on the head, for example.

Based on several developmental characteristics, bilateral metazoan animals are divided into two major groups. The Protostomia have cytoplasmic specification resulting in spiral mosaic cleavage for lophotrochozoan protostomes and superficial cleavage coupled with syncytial specification for many ecdysozoan protostomes. The mouth forms at or near the embryonic blastopore. The Deuterostomia use conditional specification resulting in regulative cleavage; the mouth forms secondarily and not from the blastopore.

The anteroposterior axis of an embryo is determined by homeotic and other homeobox-containing genes contained in one or more clusters on particular chromosomes. These genes, called *Hox* genes, occur not only in *Drosophila* and amphibians, but apparently in all animals. Each *Hox* gene is active in a particular region of the body, depending on its position within the cluster. Dorsoventral and left-right axes are similarly determined by morphogens that are produced only in the appropriate regions of the embryo. Similarly, morphogens guide the development of limbs along three body axes. Morphogens have been found to be remarkably similar in animals as different as *Drosophila* and amphibians. This realization has given rise to the field of evolutionary developmental biology, which is based on the idea that the evolution of the enormous variety of animals is the result of changes in the position and timing of relatively few genes that control development.

The postgastrula stage of vertebrate development represents a remarkable conservation of morphology when jawed vertebrates from fish to humans exhibit features common to all. As development proceeds, species-specific characteristics are formed.

Amniotes are terrestrial vertebrates that develop extraembryonic membranes during embryonic life. The four membranes are amnion, allantois, chorion, and yolk sac, each serving a specific life-support function for the embryo that develops within a self-contained egg (as in birds and most reptiles) or within the maternal uterus (mammals).

Mammalian embryos are nourished by a placenta, a complex fetal-maternal structure that develops in the uterine wall. During pregnancy the placenta becomes an independent nutritive, endocrine, and regulatory organ for the embryo.

Germ layers formed at gastrulation differentiate to form tissues and organs. The ectoderm gives rise to skin and nervous system; endoderm gives rise to alimentary canal, pharynx, lungs, and certain glands; and mesoderm forms muscular, skeletal, circulatory, reproductive, and excretory organs.

REVIEW QUESTIONS

1. What is meant by epigenesis? How did Kaspar Friedrich Wolff's concept of epigenesis differ from the early notion of preformation?
2. How is an egg (oocyte) prepared during oogenesis for fertilization? Why is preparation essential to development?
3. Describe events that follow contact of a spermatozoon with an egg. What is polyspermy and how is it prevented?
4. What is meant by the term "activation" in embryology?
5. How does amount of yolk affect cleavage? Compare cleavage in a sea star with that in a bird.
6. What is the difference between radial and spiral cleavage?
7. What other developmental hallmarks are often associated with spiral or radial cleavage?
8. What is indirect development?
9. Using sea star embryos as an example, describe gastrulation. Explain how the mass of inert yolk affects gastrulation in frog and bird embryos.
10. What is the difference between schizocoelous and enterocoelous origins of a coelom?
11. Describe two different experimental approaches that provide evidence for nuclear equivalence in animal embryos.
12. What is meant by "induction" in embryology? Describe the famous organizer experiment of Spemann and Mangold and explain its significance.
13. What are homeotic genes and what is the "homeobox" contained in such genes? What is the function of the homeobox? What are *Hox* genes? What is the significance of their apparently universal occurrence in animals?
14. What is the embryological evidence that vertebrates form a monophyletic group?
15. What are the four extraembryonic membranes of amniotic eggs of birds and other reptiles and what is the function of each membrane?
16. What is the fate of the four extraembryonic membranes in embryos of placental mammals?
17. Explain what the "growth cone" that Ross Harrison observed at the ends of growing nerve fibers does to influence direction of nerve growth.
18. Name two organ-system derivatives of each of the three germ layers.
19. What developmental characters are used to diagnose protostome and deuterostome groups (clades)?

For Further Thought SCNT makes possible therapeutic cloning. What are the likely benefits of this technique for medical science?

SELECTED REFERENCES

Carroll, S. B. 2008. Evo-devo and an expanding evolutionary synthesis: a genetic theory of morphological evolution. Cell **134:** 25–36. *Eight principles of morphological evolution are explained.*

Carroll, S. B., J. K. Grenier, and S. D. Weatherbee. 2005. From DNA to diversity: molecular genetics and the evolution of animal design, ed. 2. Malden, Massachusetts, Blackwell Publishing. *Animal body plans develop through a hierarchy of gene interactions. As these interactions are understood, biologists seek commonalities in the "genetic toolkit" across a wide variety of taxa.*

Carroll, S. B., B. Prud'homme, and N. Gompel. 2008. Regulating evolution. Sci. Am. **298:** 60–67 (May). *A general explanation of how gene regulation generates a variety of animal body forms.*

Cibelli, J. B., R. P. Lanza, and M. D. West. 2002. The first human cloned embryo. Sci. Am. **286:** 44–51 (Jan.). *Describes the first cloning of human embryos—but only to the 6-cell stage. Many scientists remain skeptical.*

Degnan, B. M., S. P. Leys, and C. Larroux. 2005. Sponge development and antiquity of animal pattern formation. Integr. Comp. Biol. **45:** 335–341. *After blastula formation in a demosponge embryo, cell migration produces a two-layered gastrula that develops a third layer before becoming a free-swimming larva. If this pattern is typical, it suggests that both blastula and gastrula stages were present in ancestral metazoans.*

Gilbert, S. F. 2006. Developmental biology, ed. 8. Sunderland, Massachusetts, Sinauer Associates. *An excellent text that combines descriptive and mechanistic aspects; good selection of examples from many animal groups.*

Godwin, J. 2014. The promise of perfect adult tissue repair and regeneration in mammals: learning from regenerative amphibians and fish. Bioessays **36:** 861–871. *An increased understanding of how regeneration is controlled in a range of vertebrates offers new research avenues for the repair of human organs and limbs.*

Goodman, C. S., and M. J. Bastiani. 1984. How embryonic nerve cells recognize one another. Sci. Am. **2 51:** 58–66 (Dec.). *Research with insect larvae shows that developing neurons follow pathways having specific molecular labels.*

Harrison, M. M., B. V. Jenkins, K. M. O'Connor-Giles, and J. Wildonger. 2015. A CRISPR view of development. Genes & Development **28:**1859–1872. *A review paper illustrating the power of a new system for manipulating the genes that influence development.*

Leys, S. P., and D. Eerkes-Medrano. 2005. Gastrulation in calcareous sponges: in search of Haeckel's gastraea. Integr. Comp. Biol. **45:** 342–351. *Ingression of cells during embryogenesis produces two germ layers (interpreted as gastrulation) in Sycon, a calcareous sponge. The ancestral pattern of gastrulation may be via ingression rather than by invagination, which occurs much later, during larval metamorphosis.*

Nüsslein-Volhard, C. 1996. Gradients that organize embryo development. Sci. Am. **275:** 54–61 (Aug.). *An account of the author's Nobel Prize-winning research.*

Rosenberg, K. R., and W. R. Trevathan. 2001. The evolution of human birth. Sci. Am. **285:** 72–77 (Nov.). *Examines reasons why humans are the only primates to seek assistance during childbirth.*

PART THREE

9

Architectural Pattern of an Animal

Cnidarian polyps have radial symmetry and cell-tissue grade of organization (Dendronephthya *sp.*).

© Larry S. Roberts

New Designs for Living

Zoologists today recognize 34 phyla of multicellular animals, each phylum characterized by a distinctive body plan and biological properties that set it apart from all other phyla. All are survivors of perhaps 100 phyla that appeared 600 million years ago during the Cambrian explosion, the most important evolutionary event in the geological history of life. Within the space of a few million years, virtually all major body plans that we see today, together with many other novel plans that we know only from the fossil record, evolved. Entering a world sparse in species and mostly free of competition, these new life-forms diversified, producing new themes in animal architecture. Later bursts of diversification that followed major extinction events produced mainly variations on established themes.

Established themes, in the form of distinctive body plans, are passed from an ancestral population to its descendants; molluscs usually carry a hard shell, bird forelimbs make wings. These ancestral traits limit the morphological scope of descendants regardless of their lifestyle. Although penguin bodies are modified for an aquatic life, their wings and feathers are clearly remnants of an ancestral flight apparatus. Despite structural and functional evolution, new forms are often constrained by the architecture of their ancestors.

The basic uniformity of biological organization derives from the common ancestry of animals and from their cellular construction. Despite vast differences of structural complexity of organisms ranging from unicellular forms to humans, all share an intrinsic material design and fundamental functional plan. In this introduction to the diversity chapters (Chapters 11 through 28), we consider the most basic body plans that underlie the apparent diversity of animal form and examine some of the common architectural themes that animals share.

9.1 HIERARCHICAL ORGANIZATION OF ANIMAL COMPLEXITY

Among the different unicellular and metazoan groups, we recognize five major grades of organization (Table 9.1). Each grade is more complex than the one preceding, and builds on it in a hierarchical manner.

The unicellular groups are the simplest eukaryotic organisms and represent the *protoplasmic* grade of organization. They are nonetheless complete organisms that perform all functions of life seen in more complex animals. Within the confines of their cell, they show remarkable organization and division of labor, possessing distinct supportive structures, locomotor devices, and simple sensory structures. The diversity observed among unicellular organisms is achieved by varying the architectural patterns of subcellular structures, organelles, and the cell as a whole (see Chapter 11).

The multicellular animals evolved greater structural complexity by combining cells into larger units. An animal cell is a specialized part of the whole organism and, unlike a unicellular organism, is not capable of independent existence. Cells of a multicellular organism are specialized for performing the various tasks accomplished by subcellular elements in unicellular forms. The simplest metazoans show the *cellular* grade of organization, in which cells demonstrate division of labor but are not strongly associated to perform a specific collective function (Table 9.1).

In the more complex *cell-tissue* grade of organization, cells are grouped together and perform their common functions as a highly coordinated unit called a **tissue** (see p. 189). Animals at or beyond the cell-tissue grade of organization are termed **eumetazoans.** In animals of the *tissue-organ* grade of organization, tissues are assembled into still larger functional units called **organs.** Usually one type of tissue performs an organ's chief function, as muscle tissue does in the heart; other tissues—epithelial and connective—perform supportive roles. The chief functional cells of an organ are called **parenchyma** (pa-ren'ka-ma; Gr. *para,* beside, + *enchyma,* infusion). The supportive tissues are its **stroma** (Gr. bedding). For instance, in the vertebrate pancreas (see p. 759), the secreting cells are the parenchyma; capsule and connective tissues represent stroma.

Most animals have an additional level of complexity in which different organs operate together as **organ systems.** Eleven different

TABLE 9.1

Grades of Organization in Organismal Complexity

1. *Protoplasmic grade of organization*—characterizes unicellular organisms. All life functions are confined within the boundaries of a single cell, the fundamental unit of life. Within a cell, protoplasm is differentiated into organelles capable of performing specialized functions.

2. *Cellular grade of organization*—an aggregation of cells that are functionally differentiated. A division of labor is evident, so that some cells are concerned with, for example, reproduction, and others with nutrition. Some flagellates, such as *Volvox,* that have distinct somatic and reproductive cells are placed at the cellular level of organization. Many authorities also place sponges at this level.

3. *Cell-tissue grade of organization*—an aggregation of similar cells into definite patterns or layers and organized to perform a common function, to form a tissue. Sponges are considered by some authorities to belong to this grade, although jellyfishes and their relatives (Cnidaria) more clearly demonstrate the tissue plan. Both groups are still largely of the cellular grade of organization because most cells are scattered and not organized into tissues. An excellent example of a tissue in cnidarians is the nerve net, in which nerve cells and their processes form a definite tissue structure, with the function of coordination.

4. *Tissue-organ grade of organization*—an aggregation of tissues into organs is a further step in complexity. Organs are usually composed of more than one kind of tissue and have a more specialized function than tissues. This is the organizational level of flatworms (Platyhelminthes), in which well-defined organs such as eyespots, proboscis, and reproductive organs occur. In flatworms, the reproductive organs transcend the tissue-organ grade and are organized into a reproductive system.

5. *Organ-system grade of organization*—organs work together to perform some function, producing the highest level of organization—an organ system. Systems are associated with basic body functions such as circulation, respiration, and digestion. The simplest animals having this type of organization are nemertean worms, which have a complete digestive system distinct from the circulatory system. Most animal phyla demonstrate this type of organization.

kinds of organ systems are described in metazoans: skeletal, muscular, integumentary, digestive, respiratory, circulatory, excretory, nervous, endocrine, immune, and reproductive. The great evolutionary diversity of these organ systems is covered in Chapters 14 through 28, and their structures are related to their functions in Chapters 29 through 35.

9.2 ANIMAL BODY PLANS

Animal body plans differ in the grade of organization, in body symmetry, in the number of embryonic germ layers, and in the number of body cavities. Body symmetry can be generally determined from the external appearance of an animal, but determining other features of a body plan typically requires dissection and sometimes microscopy.

Animal Symmetry

Symmetry refers to balanced proportions, or correspondence in size and shape of parts on opposite sides of a median plane.

Spherical symmetry means that any plane passing through the center divides a body into equivalent, or mirrored, halves (Figure 9.1, *top left).* This type of symmetry occurs among some unicellular forms and is rare in animals. Spherical forms are best suited for floating and rolling.

Radial symmetry (Figure 9.1, *top right)* applies to forms that can be divided into similar halves by more than two planes passing through the longitudinal axis. These are tubular, vase, or bowl shapes found in some sponges and in hydras, jellyfish, sea urchins, and related groups, in which one end of the longitudinal axis is usually the mouth (the **oral** surface). In sessile forms, such as hydra and sea anemones, the basal attachment disc is the **aboral** surface. A variant form is **biradial symmetry** in which, because of some part that is single or paired rather than radial, only two planes passing through the longitudinal axis produce mirrored halves. Comb jellies (phylum Ctenophora, p. 282), which are globular but have a pair of tentacles, are an example. Radial and biradial animals are usually sessile, freely floating, or weakly swimming. Radial animals, with no anterior or posterior end, can interact with their environment in all directions—an advantage to sessile or free-floating forms with feeding structures arranged to snare prey approaching from any direction.

The two phyla that are primarily radial as adults, Cnidaria and Ctenophora, have been called the **Radiata,** although phylogenetic results suggest that this group is not monophyletic (p. 286). Echinoderms (sea stars and their kin) are primarily bilateral animals (their larvae are bilateral) that have become secondarily radial as adults.

Bilateral symmetry applies to animals that can be divided along a sagittal plane into two mirrored portions—right and left halves (Figure 9.1, *bottom).* The appearance of bilateral symmetry in animal evolution was a major innovation, because bilateral animals are much better fitted for directional (forward) movement than are radially symmetrical animals. Bilateral animals form a monophyletic group of phyla called the **Bilateria.**

Bilateral symmetry is strongly associated with **cephalization,** differentiation of a head. The concentration of nervous tissue and

Spherical symmetry

Radial symmetry

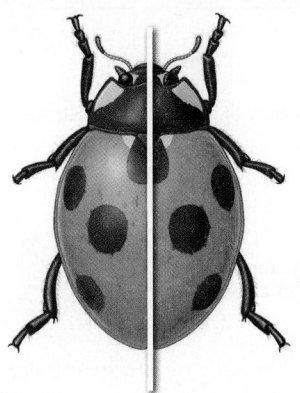
Bilateral symmetry

Figure 9.1   Animal symmetry. Illustrated are animals showing spherical, radial, and bilateral symmetry.

sense organs in a head bestows obvious advantages to an animal moving through its environment head first—it produces an efficient positioning of organs for sensing the environment and responding to it. Usually the mouth of an animal is located on the head as well, since so much of an animal's activity is concerned with procuring food. Cephalization is always accompanied by differentiation along an anteroposterior axis, although the evolution of this axis preceded cephalization.

Some convenient terms used for locating regions of bilaterally symmetrical animals (Figure 9.2) are **anterior,** used to designate the head end; **posterior,** the opposite or tail end; **dorsal,** the back or upper side; and **ventral,** the front or belly side. **Medial** refers to the midline of the body; **lateral,** to the sides. **Distal** parts are farther from the middle of the body; **proximal** parts are nearer. A **frontal** plane (sometimes called coronal plane) divides a bilateral body into dorsal and ventral halves by running through the anteroposterior axis and the right-left axis at right angles to the **sagittal** plane, the plane dividing an animal into right and left parts. One that divides exactly down the middle is sometimes termed the **midsagittal** plane. A **transverse** plane (also called a cross section) cuts through a dorsoventral and a right-left axis at right angles to both the sagittal and frontal planes and would separate anterior and posterior portions (Figure 9.2). In vertebrates **pectoral** denotes the chest region or area associated with the anterior pair of appendages, and **pelvic** denotes the hip region or area associated with the posterior pair of appendages.

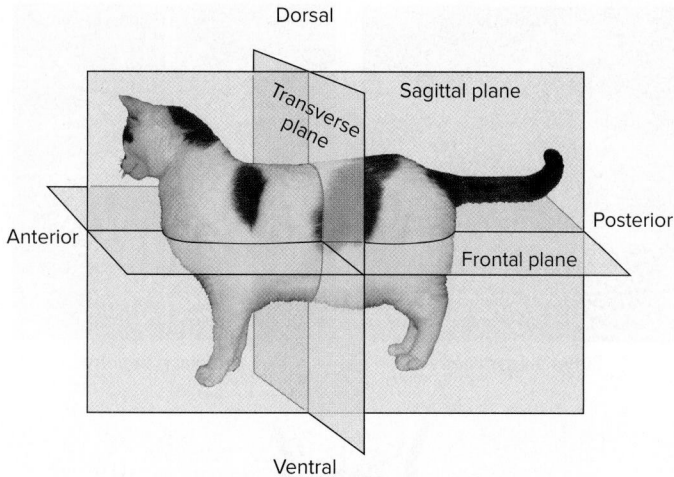

Figure 9.2 The planes of symmetry as illustrated by a bilaterally symmetrical animal.

Body Cavities and Germ Layers

A body cavity is an internal space. The most obvious example is a gut cavity or digestive tract, but most animals have an additional, less obvious, cavity. When this second cavity is fluid-filled, it may cushion and protect the gut from forces exerted on the body. In some animals, such as an earthworm, it also forms part of a hydrostatic skeleton used in locomotion (see Chapter 17, p. 374, and Chapter 29, pp. 643–644).

Sponges, at the cellular grade of organization, have no body cavities, not even a gut cavity. If sponges share the same developmental sequence as other metazoans, why do they lack a gut cavity? Where in the developmental sequence does a gut form? Sponges, like all animals, develop from a zygote to a blastula stage. A typical spherical blastula is composed of a layer of cells surrounding a fluid-filled cavity (see Figure 8.11). This cavity, a **blastocoel,** has no external opening, so it could not serve as a gut. In sponges, after the formation of a blastula, the cells reorganize to form an adult animal

External view of a gastrula

Longitudinal section

Longitudinal section

Cross section

ACOELOMATE PLAN

Mesoderm fills blastocoel

Gut cavity

Ectoderm
Blastocoel
Endoderm
Gut cavity
Early mesoderm cells
Blastopore

A   **Gastrula**

PSEUDOCOELOMATE PLAN

Mesoderm lines one side of blastocoel

Pseudocoelom

Band of mesoderm surrounds gut and then splits open

SCHIZOCOELOUS COELOMATE PLAN

Coelom

ENTEROCOELOUS COELOMATE PLAN

Mesodermal pouches surround gut

Early mesoderm cells

Coelom

B   **Gastrula**

Figure 9.3 Mesoderm resides in different parts of the gastrula during formation of acoelomate, pseudocoelomate, and schizocoelous body plans **(A)**. Mesoderm and a coelom form together in the enterocoelus plan **(B)**.

in which the cells become embedded in an extracellular matrix and surround a chamber, called a spongocoel, through which water and nutrients flow. (See Figure 9.5, *upper pathways,* and Figures 12.5 and 12.10.)

In animals other than sponges, development proceeds from a blastula to a **gastrula** stage, as one side of the blastula pushes inward, making a depression (Figure 9.3). The depression becomes a gut cavity, also called a **gastrocoel** or **archenteron.** The external opening to the depression is the **blastopore;** it typically becomes the adult mouth or anus. The gut lining is **endoderm** and the outer layer of cells, surrounding the blastocoel, is **ectoderm** (Figure 9.3). The embryo now has two cavities, a gut cavity and a blastocoel. Animals such as sea anemones and jellyfish develop from these two germ layers and are called **diploblastic** (see Figure 9.5, *upper pathway).* They typically have radial symmetry as adult animals. The fluid-filled blastocoel persists in diploblasts, but in others it is filled with a third germ layer, **mesoderm.** Animals that possess ectoderm, mesoderm, and endoderm are termed **triploblastic,** and most are bilaterally symmetrical.

Methods of Mesoderm Formation

Cells forming mesoderm are derived from endoderm, but there are two ways that a middle tissue layer of mesoderm can form. In protostomes, mesoderm forms as endodermal cells from near the blastopore migrate into the blastocoel (Figure 9.3A). Following this event, three different body plans—acoelomate, pseudocoelomate, and coelomate—are possible (Figure 9.3A).

In the **acoelomate** plan, mesodermal cells completely fill the blastocoel, leaving a gut cavity as the only body cavity (Figure 9.3A). The region between the ectodermal epidermis and the endodermal digestive tract is filled with a spongy mass of space-filling cells, the **parenchyma** (Figure 9.4). Parenchyma is derived from embryonic connective tissue and is important in assimilation and transport of food and in disposal of metabolic wastes.

In the **pseudocoelomate** plan, mesodermal cells line the outer edge of the blastocoel, leaving two body cavities: a persistent blastocoel and a gut cavity (Figures 9.3 A and 9.4). The blastocoel is now called a **pseudocoelom;** the name means false coelom in reference to mesoderm only partially surrounding the cavity, instead of completely surrounding it, as in a true **coelom.**

The coelomate body plan, by which a true coelom is formed, develops via a **schizocoelous** or an **enterocoelus** plan. In the schizocoelous plan, mesodermal cells fill the blastocoel, forming a solid band of tissue around the gut cavity. Then, through programmed cell death, space opens *inside* the mesodermal band (Figure 9.3A). This new space is a coelom. The embryo has two body cavities, a gut cavity and a coelom.

In deuterostomes, mesoderm forms by an enterocoelous plan, where cells from the central portion of the gut lining begin to grow outward as pouches, expanding into the blastocoel (Figure 9.3B). The expanding pouch walls form a mesodermal ring. As the pouches move outward, they enclose a space. The space becomes a coelomic cavity or coelom. Eventually the pouches pinch off from the gut lining, completely enclosing a coelom bounded by mesoderm on all sides. The embryo has two body cavities, a gut cavity and a coelom.

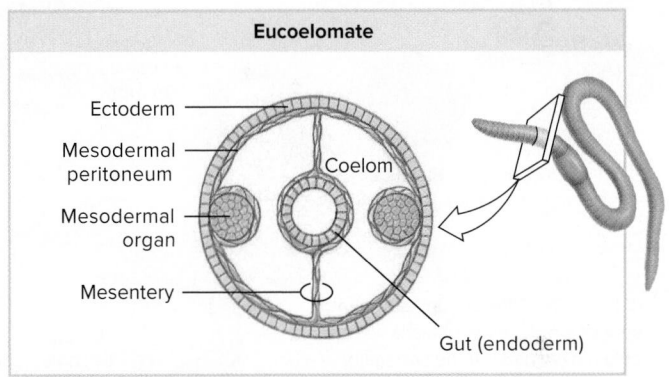

Figure 9.4 Acoelomate, pseudocoelomate, and eucoelomate body plans are shown as cross sections of representative animals. Note the relative positions of parenchyma, peritoneum, and body organs.

A coelom made by **enterocoely** is functionally equivalent to a coelom made by **schizocoely,** and both are represented as such in the **eucoelomate,** or simply, **coelomate,** body plan (Figure 9.4). Both kinds of coelomic cavities are bounded by mesoderm and lined with a **peritoneum,** a thin cellular membrane derived from mesoderm (Figure 9.4). Mesodermal **mesenteries** suspend organs in the coelom (Figure 9.4). A pseudocoelom lacks a peritoneum.

Developmental Origins of Body Plans in Triploblasts

Triploblastic animals follow one of several major developmental pathways to form a blastula from a zygote (Figure 9.5). The most common pathways are by spiral or radial cleavage (see Figure 8.18, p. 169).

Radial cleavage is typically accompanied by three other traits: the blastopore becomes an anus and a new opening makes the mouth,

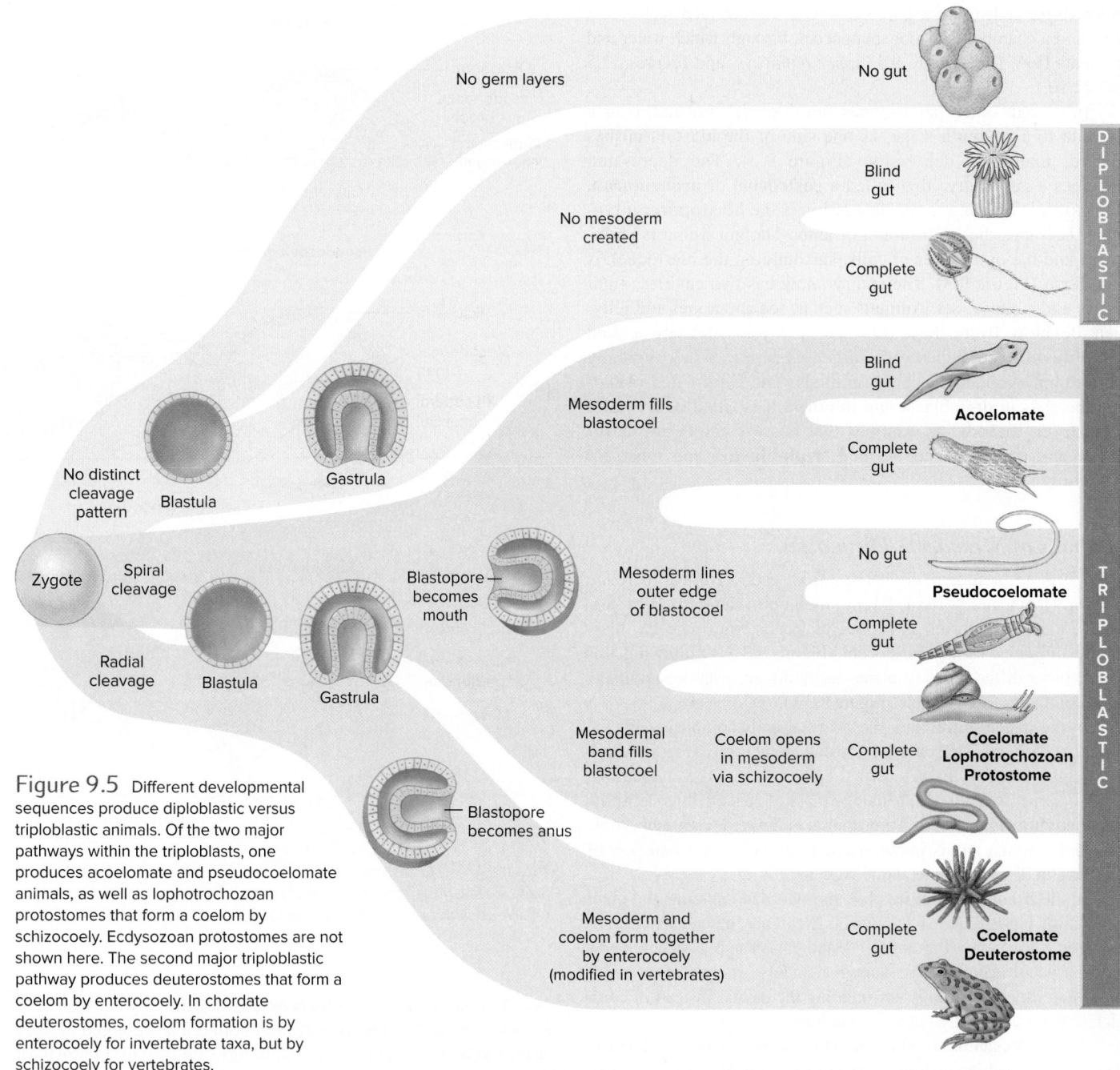

Figure 9.5 Different developmental sequences produce diploblastic versus triploblastic animals. Of the two major pathways within the triploblasts, one produces acoelomate and pseudocoelomate animals, as well as lophotrochozoan protostomes that form a coelom by schizocoely. Ecdysozoan protostomes are not shown here. The second major triploblastic pathway produces deuterostomes that form a coelom by enterocoely. In chordate deuterostomes, coelom formation is by enterocoely for invertebrate taxa, but by schizocoely for vertebrates.

the coelom forms via enterocoely, and cleavage is regulative (see Figure 8.18, p. 169). Animals with these features are called deuterostomes (Figure 9.5, *lower pathway*); this group includes sea urchins and chordates.

Spiral cleavage produces an embryo whose developmental patterns contrast with those described for deuterostomes: the blastopore becomes the mouth, and cleavage is mosaic (see Figure 8.18, p. 169). The body may become acoelomate, pseudocoelomate, or coelomate, depending on the taxa (Figure 9.5, *central pathway*). If a coelom is present, it is made via schizocoely. Animals in this latter group are

called lophotrochozoan protostomes, and include molluscs, segmented worms, and other taxa (Figure 9.5).

Lophotrochozoans are distinguished from ecdysozoan protostomes (not shown in Figure 9.5), for which a range of cleavage patterns have been described. These include spiral cleavage, a superficial cleavage pattern in which nuclei proliferate within common cytoplasm prior to separation by multiple cytoplasmic divisions (see Figure 8.9, p. 160), and a pattern initially resembling radial cleavage. Ecdysozoans may be coelomate or pseudocoelomate. Insects, crabs, and nematodes are among the ecdysozoans.

A Complete Gut and Segmentation

A few diploblasts and triploblasts have a blind or incomplete gut cavity where food must enter and exit the same opening, but most forms possess a complete gut (Figure 9.5). A complete gut makes possible a one-way flow of food from mouth to anus. A body constructed in this way is essentially a gut tube within another body tube. A tube-within-a-tube design appears to be very adaptive, as members of the most common animal phyla, both invertebrate and vertebrate, have this plan.

Segmentation, also called metamerism, is another common feature of metazoans. Segmentation is a serial repetition of similar body segments along the longitudinal axis of the body. Each segment is called a metamere, or somite. In forms such as earthworms and other annelids (Figure 9.6), in which metamerism is most clearly represented, the segmental arrangement includes both external and internal structures of several systems. There is repetition of muscles, blood vessels, nerves, and setae of locomotion (see Figures 17.9–17.22, pp. 375–376). Some other organs, such as those of sex, may be repeated in only a few segments (see Figure 17.19, p. 375). Evolutionary changes have obscured much of the segmentation in many animals, including humans, although it is often observed during development.

Segmentation permits greater body mobility and complexity of structure and function. Its potential is amply displayed in phylum Arthropoda, the largest assemblage of animals on earth. Segmentation occurs in phylum Chordata in addition to Annelida and Arthropoda (Figure 9.6), although superficial segmentation of ectoderm and body wall may appear among diverse groups of animals. The importance and potential of segmentation is discussed in Chapters 17 and 18.

9.3 COMPONENTS OF ANIMAL BODIES

Animal bodies consist of cellular components, derived from the three embryonic germ layers—ectoderm, mesoderm, and endoderm—as well as extracellular components.

Extracellular Components

Metazoan animals contain two important noncellular components: body fluids and extracellular structural elements. In all eumetazoans, body fluids are subdivided into two fluid "compartments": those that occupy intracellular space, within the body's cells, and those that occupy extracellular space, outside the cells. In animals with closed vascular systems (such as segmented worms and vertebrates), the extracellular fluids are subdivided further into blood plasma (the fluid portion of blood) and interstitial fluid (see Figure 31.1). Interstitial fluid, also called tissue fluid, occupies the space surrounding cells (intercellular space). Many invertebrates have open blood systems, however, with no true separation of blood plasma from interstitial fluid. We explore these relationships further in Chapter 31.

Extracellular structural elements are the supportive material of the organism, including connective tissue (especially well developed in vertebrates but present in all metazoa), cartilage (molluscs and chordates), bone (vertebrates), and cuticle (arthropods, nematodes, annelids, and others). These elements provide mechanical stability and protection (see Chapter 29). In some instances, they act also as a storage depot of materials for exchange between the cells and interstitial fluid, and serve as a medium for extracellular reactions. We describe diversity of extracellular structural elements characteristic of different groups of animals in Chapters 12 through 28.

Cellular Components: Tissues

A tissue is a group of similar cells (together with associated cell products) specialized for performance of a common function. The study of tissues is called histology (Gr. histos, tissue, + logos, discourse) or microanatomy. All cells in metazoan animals form tissues. Sometimes cells of a tissue may be of several kinds, and some tissues have much extracellular material.

During embryonic development, the germ layers become differentiated into four kinds of tissues. These are epithelial, connective, muscular, and nervous tissues (Figure 9.7). These four basic tissue types are able to meet the diverse requirements of animal life.

Epithelial Tissue

An epithelium (pl., epithelia) is a sheet of cells that covers an external or internal surface. Outside the body, epithelium forms a protective covering. Inside, epithelium lines all organs of the body cavity, as well as ducts or passageways through which various materials and secretions move. Thus, ions and molecules are likely to pass through

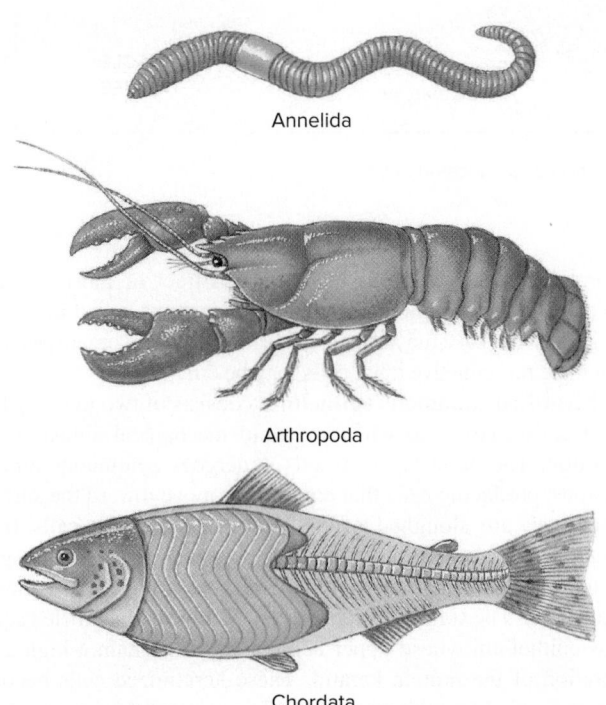

Annelida

Arthropoda

Chordata

Figure 9.6　Segmented phyla. Segmentation (also called metamerism), or repetition of structural units, brings more varied specialization because segments, especially in arthropods, have become modified for different functions.

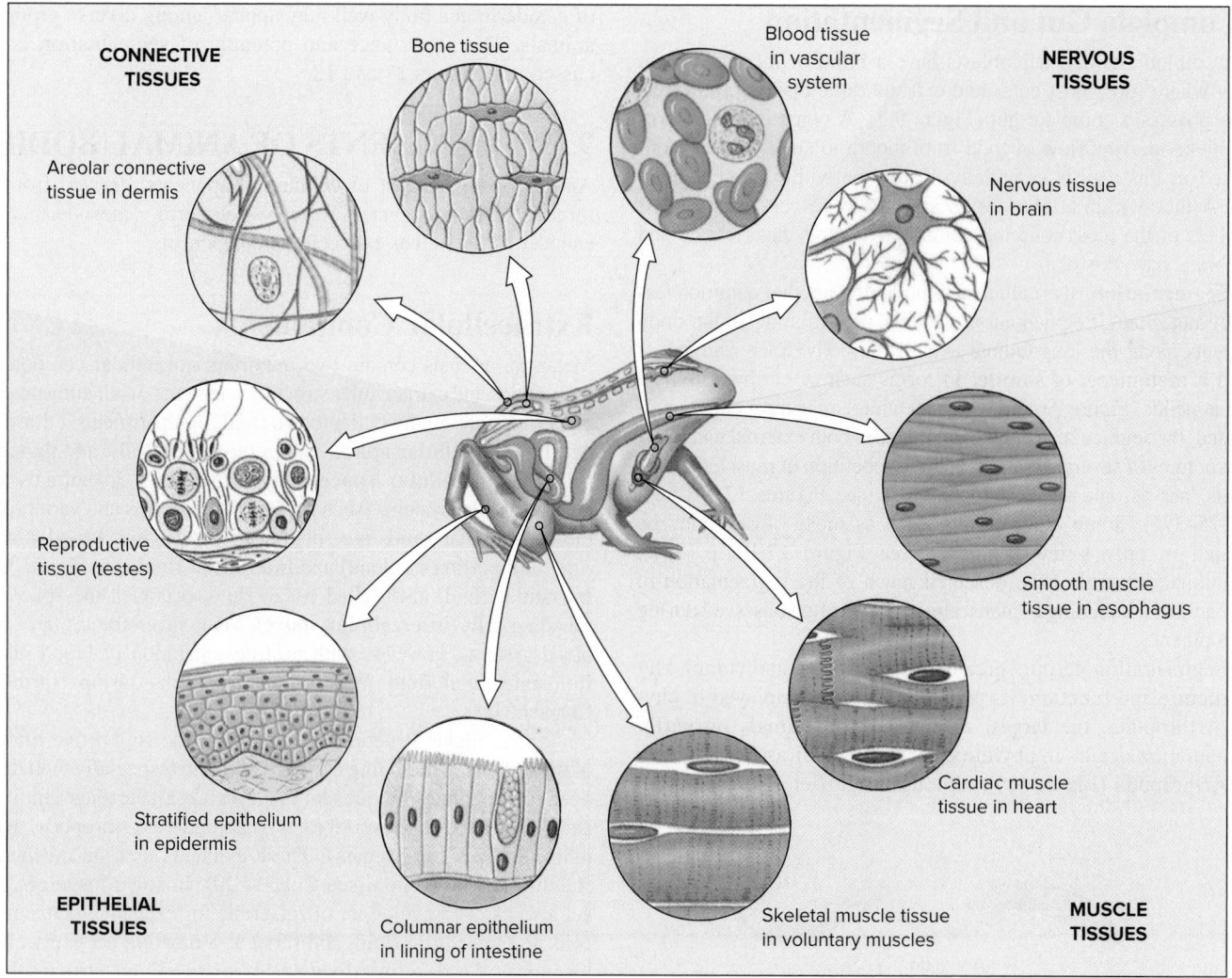

Figure 9.7 Types of tissues in a vertebrate, showing examples of where different tissues are located in a frog.

epithelial cells as they move to and from all other cells of the body. Consequently a large variety of transport molecules occur on epithelial cell membranes (see Chapter 3). Epithelial cells are also modified into glands that produce lubricating mucus or specialized products such as hormones or enzymes.

Epithelia are classified by cell form and number of cell layers. Simple epithelia (a single layer of cells; Figure 9.8) occur in all metazoan animals, while stratified epithelia (many cell layers; Figures 9.9 and 9.10) are mostly restricted to vertebrates.

Simple squamous epithelium is composed of flattened cells that form a continuous lining of blood capillaries, lungs, and other surfaces where it permits diffusion of gases and transport of other molecules into and out of cavities (Figure 9.8A). **Simple cuboidal epithelium** is composed of short, boxlike cells. Cuboidal epithelium often lines small ducts and tubules, such as those of the kidney and salivary glands, and may have active secretory or absorptive functions (Figure 9.8B). **Simple columnar epithelium** resembles cuboidal epithelium, but the cells are taller and usually have elongate nuclei (Figure 9.8C). This type of epithelium occurs on highly

absorptive surfaces such as the intestinal tract of most animals. The cells often bear minute, fingerlike projections called microvilli that greatly increase the absorptive surface. In some organs, such as the female reproductive tract, cells may be ciliated.

Stratified squamous epithelium consists of two to many layers of cells adapted to withstand mild mechanical abrasion and distortion. The basal layer of cells undergoes continuous mitotic divisions, producing cells that continually move toward the surface where they are sloughed off and replaced by new cells from beneath. This type of epithelium lines the oral cavity, esophagus, and anal canal of many vertebrates, and the vagina of mammals (Figure 9.9). The skin of vertebrates is composed of stratified squamous epithelium whose upper cellular layers contain a high concentration of the protein keratin. These keratinized cells become dead and scalelike with no nuclei. They protect the lower, living cellular layers and are waterproof in reptiles, birds, and mammals (see Figure 29.1 and p. 640).

Transitional epithelium is a type of stratified epithelium specialized to accommodate great stretching. This type of epithelium

Figure 9.9 Stratified squamous epithelium

Simple squamous epithelium

A

Simple cuboidal epithelium

B

Simple columnar epithelium

C

Figure 9.8 Types of simple epithelium.

occurs in the urinary tract and bladder of vertebrates. In the relaxed state it appears to be four or five cell layers thick, but when stretched it appears to have only two or three layers of extremely flattened cells (Figure 9.10).

All types of epithelia are supported by an underlying basement membrane, which is a condensed region of connective tissue matrix, secreted by both epithelial and connective tissue cells.

Blood vessels never penetrate into epithelial tissues, which depend on diffusion of oxygen and nutrients from underlying connective tissue proper.

Connective Tissue

Connective tissues are a diverse group of tissues that serve various binding and supportive functions. They are so widespread in the body that removal of other tissues would still leave the complete form of the body clearly apparent. Connective tissue is composed of relatively few cells, a great many extracellular fibers, and a **ground substance,** in which the fibers are suspended (together called **matrix**). We recognize several different types of connective tissue. Two kinds of **connective tissue proper** occur in vertebrates: **loose** and **dense connective tissue.**

Loose connective tissue, also called areolar connective tissue, is the "packing material" of the body that anchors blood vessels, nerves, and body organs. It contains fibroblasts that synthesize the fibers and ground substance of connective tissue and wandering macrophages that phagocytize pathogens or damaged cells. The different fiber types include collagen fibers (purple and thick in Figure 9.11A) and thin elastic fibers (black and branching in Figure 9.11A) formed of the protein elastin.

Dense connective tissue forms tendons, ligaments, and fasciae (fa'sha), the latter arranged as sheets or bands of tissue surrounding skeletal muscle. In a tendon (Figure 9.11B) the collagenous fibers are extremely long and tightly packed together with little ground substance. Many of the fibers of connective tissue are composed of **collagen** (Gr. *kolla,* glue, + *genos,* descent), a protein of great tensile strength. Collagen is the most abundant protein in the animal kingdom, found in animal bodies wherever both flexibility and resistance to stretching are required. Connective tissue of invertebrates, as in vertebrates, consists of cells, fibers, and ground substance, and shows a wide diversity of structure ranging from highly cellular to acellular histologies.

Other types of specialized connective tissue include **blood, lymph** (collectively considered vascular tissue), **adipose** (fat)

Figure 9.10 Transitional epithelium—a type of stratified epithelium that allows great stretching.

Figure 9.11 Types of connective tissue.

tissue, **cartilage,** and **bone.** Vascular tissue is composed of distinctive cells in a fluid ground substance, the plasma. Vascular tissue lacks fibers under normal conditions. Blood composition is discussed in Chapter 31. Adipose tissue is discussed in Chapter 32.

Cartilage is a semirigid form of connective tissue composed of a firm matrix containing cells (chondrocytes) located in pockets called lacunae, and collagen and/or elastic fibers (depending on the type of cartilage). Hyaline cartilage (Figure 9.11C) is the most common type. Because cartilage lacks a blood supply, all nutrients and

Figure 9.12 Types of muscle tissue.

waste materials must diffuse through the ground substance from surrounding tissues. Due to the lack of blood supply, cartilage heals slowly after it is damaged.

Bone is the strongest of vertebrate connective tissues, composed of a calcified matrix containing salts organized around collagen fibers (Figure 9.11D). Small pockets (lacunae) within the matrix contain bone cells, called osteocytes. The osteocytes communicate with each other by means of a tiny network of channels called canaliculi. Blood vessels, extensive in bone, are located in larger channels, including central canals. Bone undergoes continuous remodeling during an animal's life, and can repair itself following even extensive damage. Structure of cartilage and bone is discussed in the section on skeletons in Chapter 29.

Muscular Tissue

Muscle is the most abundant tissue in the body of most animals. It originates (with few exceptions) from mesoderm, and its unit is the **muscle** cell or **fiber,** specialized for contraction. When viewed with a light microscope, **striated muscle** appears transversely striped (striated), with alternating dark and light bands (Figure 9.12). In vertebrates we recognize two types of striated muscle: **skeletal** and **cardiac muscle.** Skeletal muscle occurs in both invertebrates and vertebrates. It is composed of extremely long, cylindrical fibers, which are multinucleate cells that may reach from one end of the whole muscle to the other. Skeletal muscle is called voluntary muscle (in vertebrates) because it contracts when stimulated by nerves under conscious central nervous system control (Figure 9.12A). In invertebrates, skeletal muscle may be controlled by both stimulatory and inhibitory nerve activity. Cardiac muscle is found only in the vertebrate heart. The cells are much shorter than those of skeletal muscle and have only one nucleus per cell (uninucleate). Cardiac muscle tissue is a branching network of fibers with individual cells interconnected by junctional complexes called intercalated discs. Cardiac muscle is considered involuntary muscle because it does not respond to nerves under conscious control of the central nervous system. In fact, heart rate is controlled by specialized pacemaker cells located in the heart itself. However,

autonomic nerves from the brain may alter pacemaker activity (Figure 9.12B). In invertebrates, a third striated muscle type called **obliquely striated muscle** has been described.

Smooth (or visceral) **muscle,** which lacks the characteristic alternating bands of the striated type, occurs in both invertebrates and vertebrates, although major ultrastructural differences have been described between them (Figure 9.12C). Smooth muscle cells are long and tapering, each containing a single, central nucleus. Smooth muscle is the most common type of muscle in invertebrates, in which it serves as body-wall musculature and surrounds ducts and sphincters. In vertebrates, smooth muscle surrounds blood vessels and internal organs such as intestine and uterus. It is called involuntary muscle in vertebrates because its contraction is usually not consciously controlled; it is controlled by autonomic nerves as well as by local signals originating from within the tissue and surrounding regions.

Cytoplasm of muscle fibers is called **sarcoplasm,** and contractile proteins within the fiber are **myofibrils.** Muscular movement is covered in Chapter 29.

Nervous Tissue

Nervous tissue is specialized for reception of stimuli and conduction of impulses from one region to another. Two basic types of cells in nervous tissue are **neurons** (Gr. nerve), the basic functional unit of nervous systems, and **neuroglia** (nu-rog′le-a; Gr. nerve, + *glia,* glue), a nonneuronal cell type that insulates neuron membranes and serves various supportive functions. Figure 9.13 shows the functional anatomy of a typical nerve cell. From the nucleated cell body, or **soma,** extend one or more **dendrites** (Gr. *dendron,* tree), which receive electrical signals from receptors or other nerve cells, and a single **axon** that carries signals away from the cell body to other nerve cells or to an effector organ. The axon is often called a **nerve fiber.** The insulating sheath or **myelin** sheath increases the speed at which electrical signals are transmitted along an axon. Neurons are separated from other neurons or from effector organs by specialized junctions called synapses. The functional roles of nervous tissue are treated in Chapter 33.

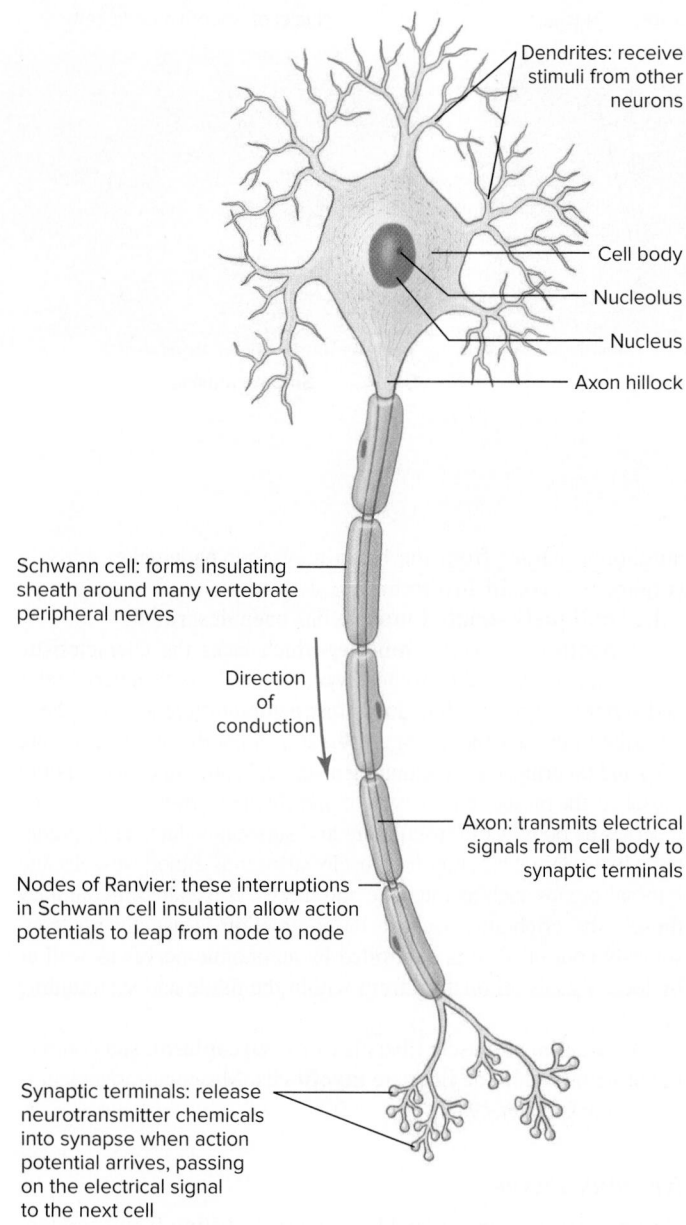

Figure 9.13 Functional anatomy of a neuron.

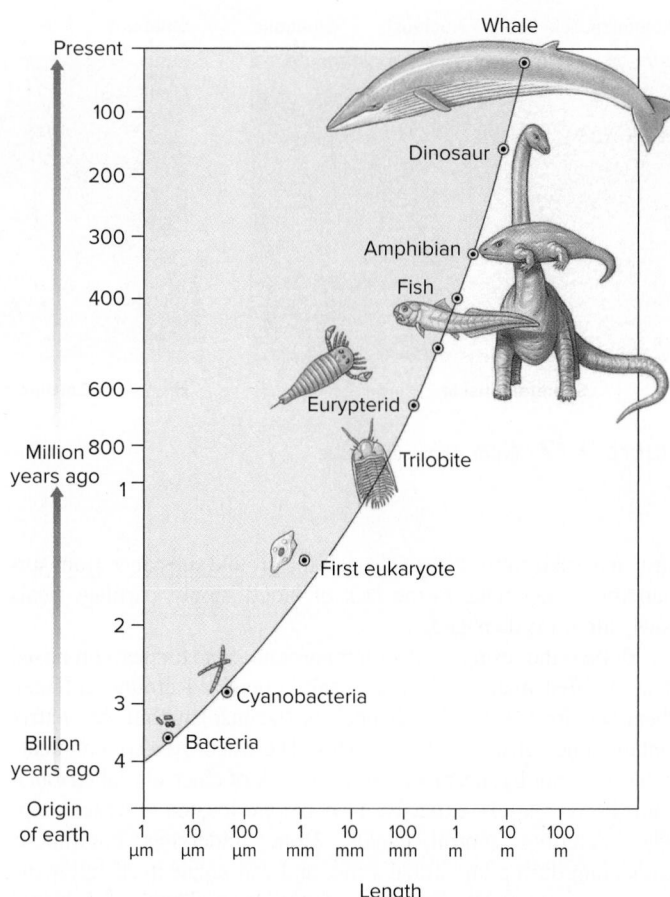

Figure 9.14 Graph showing an increase in maximum sizes of organisms throughout the history of life on earth. Note that both scales are logarithmic.

9.4 COMPLEXITY AND BODY SIZE

The most complex grades of animal organization permit, and to some extent even promote, evolution of large body size (Figure 9.14). Large size confers several important physical and ecological consequences for an organism. As animals become larger, the body surface increases much more slowly than does body volume because surface area increases as the square of body length (length^2), whereas volume (and therefore mass) increases as the cube of body length (length^3). In other words, a large animal has less surface area relative to its volume than does a small animal of the same shape. The surface area of a large animal may be inadequate to provide enough respiratory gas exchange and nutrients for cells located deep within its body. There are two possible solutions to this problem. One

solution is to fold or to invaginate the body surface to increase the surface area or, as exploited by flatworms, to flatten the body into a ribbon or disc so that no internal space is far from the surface (see Chapter 14). This solution allows a body to become large without an increase in internal complexity. However, most large animals adopted a second solution; they developed internal transport systems to shuttle nutrients, gases, and waste products between cells and the external environment.

Larger size buffers an animal against environmental fluctuations; it provides greater protection against predation and enhances offensive tactics; and it permits a more efficient use of metabolic energy. A large mammal uses more oxygen than does a small mammal, but the cost of maintaining its body temperature is less per gram of weight for a large mammal than for a small one because less heat energy is lost from the body surface relative to heat production in a larger mammal. Large animals also can move at less energy cost than can small animals. For example, a large mammal uses more oxygen in running than does a small mammal, but the energy cost of moving 1 g of its body over a given distance is much less for a large mammal than for a small one (Figure 9.15). For all of these reasons, ecological opportunities of larger animals are very different from those of small ones. In subsequent chapters we describe the extensive adaptive diversifications observed in taxa of large animals.

Figure 9.15   Net cost of running for mammals of various sizes. Each point represents the cost (measured in rate of oxygen consumption) of moving 1 g of body over 1 km. Cost decreases with increasing body size.

SUMMARY

From the relatively simple organisms that mark the beginnings of life on earth, animal evolution has produced more intricately organized forms. Whereas a unicellular organism performs all life functions within the confines of a single cell, a multicellular animal is an organization of subordinate units united as a hierarchical system.

Every organism has an inherited body plan described in terms of body symmetry, number of embryonic germ layers, grade of organization, and number of body cavities. Most animals exhibit bilateral symmetry, but spherical and radial symmetry occur in some groups. Most animals are triploblastic and develop from three embryonic germ layers, but cnidarians and a few other forms are diploblastic. Sponges lack germ layers and possess a cellular grade of organization. Most animals have the tissue grade of organization.

All animals other than sponges have a gut cavity. Most animals have a second cavity that surrounds the gut cavity. The second cavity may be a pseudocoelom or a coelom. There are two taxon-specific patterns of coelom formation, schizocoely and enterocoely.

Triploblastic animals are divided among deuterostomes and protostomes according to their particular developmental sequence. Protostomes are further divided into lophotrochozoan and ecdysozoan forms, whose developmental differences include alternative cleavage patterns in the blastula.

An animal body consists of functionally specialized cells; body fluids, divided into intracellular and extracellular fluid compartments; and extracellular structural elements, which are fibers or formless materials that serve various structural functions in the extracellular space. The cells of animals develop into various tissues; basic types are epithelial, connective, muscular, and nervous. Tissues are organized into larger functional units called organs, and organs are associated to form systems.

One correlate of increased anatomical complexity is an increase in body size, which offers certain advantages such as more effective predation, reduced energy cost of locomotion, and improved homeostasis.

REVIEW QUESTIONS

1. Name the five grades of organization in organismal complexity and explain how each successive grade is more complex than the one preceding it.
2. Can you suggest why, during the evolutionary history of animals, there has been a tendency for maximum body size to increase? Do you think it inevitable that complexity should increase along with body size? Why or why not?
3. What is the meaning of the terms parenchyma and stroma as they relate to body organs?
4. Body fluids of eumetazoan animals are separated into fluid "compartments." Name these compartments and explain how compartmentalization may differ in animals with open and closed circulatory systems.
5. What are the four major types of tissues in animals?
6. How would you distinguish simple and stratified epithelium? What characteristic of stratified epithelium might explain why it, rather than simple epithelium, lines the oral cavity, esophagus, and vagina?
7. What three elements occur in all connective tissue? Give some examples of different types of connective tissue.
8. What are three muscle tissue types found among animals? Explain how each is specialized for particular functions.
9. Describe the principal structural and functional features of a neuron.
10. Match the animal group with its body plan:

    _____ Unicellular   a. Nematode
    _____ Cell aggregate   b. Vertebrate
    _____ Blind sac, acoelomate c. Unicellular eukaryote
    _____ Tube-within-a-tube,   d. Flatworm
          pseudocoelomate   e. Sponge
    _____ Tube-within-a-tube,   f. Arthropod
          eucoelomate   g. Nemertean

11. Distinguish spherical, radial, biradial, and bilateral symmetry.
12. Use the following terms to identify regions on your body and on the body of a frog: anterior, posterior, dorsal, ventral, lateral, distal, proximal.
13. How would frontal, sagittal, and transverse planes divide your body?
14. What is meant by segmentation? Name three phyla showing segmentation.

For Further Thought   Calculate the surface area and volume of two spherical organisms, one with a radius of 1 mm and the other with a radius of 10 mm. Now determine the surface area-to-volume ratio for each organism. What problems might occur for each organism because of its size? How might they be solved?

SELECTED REFERENCES

Arthur, W. 1997. The origin of animal body plans. Cambridge, U.K., Cambridge University Press. *Explores genetic, developmental, and population-level processes involved in the evolution of the 35 or so body plans that arose in the geological past.*

Baguna, J., and M. Ruitort. 2004. The dawn of bilaterian animals: the case of acoelomorph flatworms. Bioessays **26:** 1046–1057. *Goes through the hypotheses for the transition between radial and bilateral body plans and the evidence for them.*

Cole, A. G., and B. K. Hall. 2004. The nature and significance of invertebrate cartilages revisited: distribution and histology of cartilage and cartilage-like tissues within the Metazoa. Zoology **107**: 261–273. *Beautifully illustrated and well-written discussion of these tissues.*

Martindale, M. Q., J. R. Finnerty, and J. Q. Henry. 2002. The Radiata and the evolutionary origins of the bilaterian body plan. Molecular Phylogenetics and Evolution **24**: 358–365. *Examines germ layers and symmetry in cnidarians and ctenophores to reconstruct the origin of the bilaterally symmetrical animals.*

McGowan, C. 1999. A practical guide to vertebrate mechanics. New York, Cambridge University Press. *Using many examples from his earlier book,* Diatoms to Dinosaurs, *the author describes principles of biomechanics that underlie functional anatomy. Includes practical experiments and laboratory exercises.*

Mescher, A. 2013. Junqueira's Basic Histology: Text and Atlas, ed. 13. New York, McGraw-Hill. *A comprehensive textbook of animal histology.*

Royuela, M., B. Fraile, M. I. Arenas, and R. Paniagua. 2000. Characterization of several invertebrate muscle cell types: a comparison with vertebrate muscles. Microsc. Res. Tech. **41**: 107–115. *Examines the ultrastructure of smooth, striated, and oblique striated muscle in order to establish clearer classification criteria.*

Willmer, P. 2003. Convergence and homoplasy in the evolution of organismal form. Pp. 33–49 in Müller, G. B., and S. A. Newman (eds.), Origination of organismal form. Cambridge, MA, MIT Press. *Discusses evolutionary origins of animal body plans in light of molecular phylogenetic evidence.*

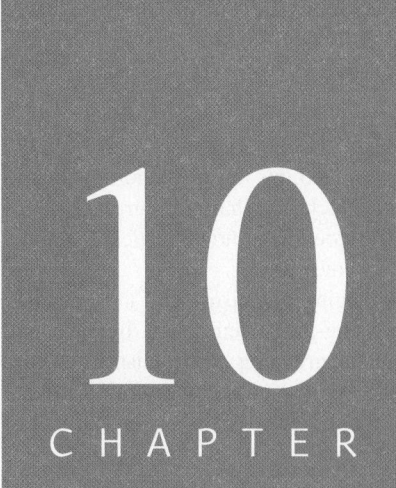

CHAPTER 10

Taxonomy and Phylogeny of Animals

Molluscan shells from the collection of Jean Baptiste de Lamarck (1744 to 1829)

© Cleveland P. Hickman, Jr.

Order in Diversity

Evolution has produced a great diversity of species in the animal kingdom. Zoologists have named more than 1.5 million species of animals, and thousands more are described each year. Some zoologists estimate that species named so far constitute less than 20% of all living animals and less than 1% of all those that have existed.

Despite its magnitude, the diversity of animals is not without limits. Many conceivable forms do not exist in nature, as our myths of minotaurs and winged horses show. Animal diversity is not random but has definite order. Characteristic features of humans and cattle never occur together in a single organism as they do in the mythical minotaurs; nor do characteristic wings of birds and bodies of horses occur together naturally as they do in the mythical horse, Pegasus. Humans, cattle, birds, and horses are distinct groups of animals, yet they do share some important features, including vertebrae and

homeothermy, that separate them from even more dissimilar forms such as insects and flatworms.

All human cultures classify familiar animals according to patterns in animal diversity. These classifications have many purposes. Some societies classify animals according to their usefulness or destructiveness to human endeavors; others may group animals according to their roles in mythology. Biologists organize animal diversity in a nested hierarchy of groups within groups according to evolutionary relationships as revealed by ordered patterns in their sharing of homologous features. This ordering is called a "natural system" because it reflects relationships that exist among animals in nature, outside the context of human activity. A systematic zoologist has three major goals: to discover all species of animals, to reconstruct their evolutionary relationships, and to communicate those relationships by constructing an informative taxonomic system.

Darwin's theory of common descent (Chapters 1 and 6) is the underlying principle that guides our search for order in the diversity of animal life. Our science of **taxonomy** ("arrangement law") produces a formal system for naming and grouping species to communicate this order. Animals that have very recent common ancestry share many features in common and are grouped most closely in our taxonomic system.

Taxonomy is part of the broader science of **systematics,** or comparative biology, in which studies of variation among animal populations are used to reveal their evolutionary relationships. The study of taxonomy predates evolutionary biology, however, and many taxonomic practices are remnants of a pre-evolutionary world view. Adjusting our taxonomic system to accommodate evolution has produced many problems and controversies. Taxonomy has reached an unusually active and controversial point in its development in which several alternative taxonomic systems are competing for use. To explain this controversy, one must first review the history of animal taxonomy.

10.1 LINNAEUS AND TAXONOMY

The Greek philosopher and biologist Aristotle (384 to 332 BC) was the first to group organisms according to their structural similarities. A unified taxonomic system for all animals and plants appeared for the first time more than two millennia later, in the work of Carolus Linnaeus (Figure 10.1).

Linnaeus was a Swedish botanist at the University of Uppsala. Linnaeus produced an extensive system of taxonomy for both plants and animals. This scheme, published in his great work, *Systema Naturae,* used morphology (the comparative study of organismal form) for arranging specimens in collections. He divided the animal kingdom into species and gave each one a distinctive name. He grouped species into genera, genera into orders, and orders into "classes" (we use quotation marks or a capital letter to distinguish "class" as a formal taxonomic rank from its broader meaning as a group of organisms that share a common essential property). Because

his knowledge of animals was limited, his lower categories, such as genera, often were much more inclusive than are current uses of these categories. Much of his taxonomy is now drastically altered, but the basic principle of his scheme remains.

Linnaeus's scheme of arranging organisms into an ascending series of groups of ever-increasing inclusiveness is a **hierarchical system** of taxonomy. Each major grouping of organisms, called a **taxon** (pl., **taxa**), took one of several standard **taxonomic ranks** to indicate the general degree of inclusiveness of the group. The hierarchy of taxonomic ranks has been expanded considerably since Linnaeus's time (Table 10.1). It now includes seven mandatory ranks for the animal kingdom, in descending series: kingdom, phylum, "class," order, family, genus, and species. All organisms must be placed into at least seven taxa, one at each of the mandatory ranks. Taxonomists have the option of subdividing these seven ranks further to recognize more than seven taxa (superfamily, subfamily, superorder, suborder, etc.) for any particular group of organisms. More than 30 taxonomic ranks are recognized. For very large and complex groups, such as fishes and insects, these additional ranks are needed to express different degrees of evolutionary divergence.

A taxonomist's choice of a group of species for recognition as a formally ranked taxon always has some arbitrariness. For example, should the taxonomic family Hominidae be restricted to genus *Homo* (humans) and all fossil genera that are closer to *Homo* than to genus *Pan* (bonobos and chimpanzees), or should it comprise the genera *Homo, Pan, Gorilla,* and *Pongo* (orangutans) plus fossils closer to these genera than to gibbons? Within the last few decades, anthropologists have shifted Hominidae from the former usage primarily to the latter one (p. 208). Does the confusing arbitrariness of ranking taxa outweigh the usefulness of ranks for reminding us which taxa are more inclusive than others? Could we devise a rank-free taxonomy that encodes the positions of species on an evolutionary tree of common descent? As we write, taxonomists are actively constructing both ranked and rank-free taxonomies of animals. Rank-free taxonomies often use indentations to specify levels of inclusiveness of each taxon as illustrated by the rank-free taxonomic ordering of bilaterally symmetrical animals shown in the last section of this chapter.

Systematization versus Classification

Introduction of evolutionary theory into animal taxonomy has changed the taxonomist's role from one of classification to one of **systematization.** Classification denotes the construction of classes, groupings of organisms that possess a common feature, called an essence, used to define the class. Organisms that possess the essential feature are members of the class by definition, and those that lack it are excluded. Because evolving species are subject always to change, the static nature of classes makes them a poor basis for a taxonomy of living systems. The activity of a taxonomist whose groupings of species represent units of common evolutionary descent is systematization, not classification. Species placed into a taxonomic group include the most recent common ancestor of the group and its descendants and thus form a branch of the phylogenetic tree of life. The species of a group thus formed constitute a system of common descent, not a class defined by possession of an essential characteristic. It remains common, although technically erroneous, for systematists to call their taxonomic systems classifications.

Figure 10.1   Carolus Linnaeus (1707 to 1778). This portrait was made of Linnaeus at age 68, three years before his death.

TABLE 10.1

Examples of Taxonomic Categories to Which Representative Animals Belong

Linnaean Rank	Human	Gorilla	Southern Leopard Frog	Fork-Tailed Bush Katydid
Kingdom	Animalia	Animalia	Animalia	Animalia
Phylum	Chordata	Chordata	Chordata	Arthropoda
Subphylum	Vertebrata	Vertebrata	Vertebrata	Uniramia
Class	Mammalia	Mammalia	Amphibia	Insecta
Subclass	Eutheria	Eutheria	—	Pterygota
Order	Primates	Primates	Anura	Orthoptera
Suborder	Anthropoidea	Anthropoidea	—	Ensifera
Family	Hominidae	Hominidae	Ranidae	Tettigoniidae
Subfamily	—	—	Raninae	Phaneropterinae
Genus	*Homo*	*Gorilla*	*Lithobates*	*Scudderia*
Species	*Homo sapiens*	*Gorilla gorilla*	*Lithobates sphenocephala*	*Scudderia furcate*
Subspecies	—	—	—	*Scudderia furcata furcate*

The hierarchical taxonomy of four species (human, gorilla, Southern leopard frog, and fork-tailed bush katydid). Higher taxa generally are more inclusive than lower-level taxa, although taxa at two different levels may be equivalent in content. Closely related species are united at a lower point in the hierarchy than are distantly related species. For example, humans and gorillas are united at the level of the family (Hominidae) and above; they are united with the Southern leopard frog at the subphylum level (Vertebrata) and with the katydid at the kingdom level (Animalia).

Because organismal characteristics are inherited from ancestral to descendant species, character variation is used to diagnose systems of common descent, but there is no requirement that an essential character be maintained throughout the system for its recognition as a taxon. The role of morphological or other features in systematization is therefore fundamentally different from the role of such characters in classification. In classification, a taxonomist asks whether a species being classified contains the defining feature(s) of a particular taxonomic class; in systematization, a taxonomist asks whether the characteristics of a species confirm or reject the hypothesis that it descends from the most recent common ancestor of a particular taxon. For example, tetrapod vertebrates descend from a common ancestor that had four limbs, a condition retained in most but not all of its descendants. Although they lack limbs, caecilians (p. 548) and snakes (p. 574) are tetrapods because they are parts of this system of common descent; other morphological and molecular characters group them respectively with living amphibians and lizards.

Although the hierarchical structure of Linnaean classification is retained in current taxonomy, the taxa are groupings of species related by evolutionary descent with modification, as diagnosed by sharing of homologous characters. As one moves up the taxonomic hierarchy from a species toward more inclusive groups, each taxon represents the descendants of an earlier ancestor, a larger branch of the tree of life.

Binomial Species Nomenclature

Linnaeus's system for naming species is called **binomial nomenclature.** Each species has a latinized name composed of two words (hence binomial) printed in italics (or underlined if handwritten or typed). The first word names the **genus,** which is capitalized; the second word is the **species epithet,** which identifies the species within the genus and is written in lowercase (Table 10.1). The great communicative value of Latin species names is that they are used consistently by scientists in all countries and languages; they are much more precise than are "common names," which vary culturally and geographically.

The genus name is always a noun, and the species epithet is usually an adjective that must agree in gender with the genus. For instance, the scientific name of the common robin is *Turdus migratorius* (L. *turdus*, thrush; *migratorius*, of migratory habit). The species epithet never stands alone; the complete binomial must be used to name a species. Names of genera must refer only to single groups of organisms; the same name cannot be given to two different genera of animals. The same species epithet may be used in different genera, however, to denote different species. For example, the scientific name of the whitebreasted nuthatch is *Sitta carolinensis*. The species epithet *"carolinensis"* is used in other genera for the species *Poecile carolinensis* (Carolina chickadee) and *Anolis carolinensis* (green anole, a lizard) to mean "of Carolina." All ranks above the species are designated using uninomial nouns, written with a capital initial letter.

Sometimes a species is divided into subspecies using a trinomial nomenclature (see katydid example, Table 10.1, and salamander example, Figure 10.2); such species are called **polytypic.** The generic, specific, and subspecific names are printed in italics (underlined if handwritten or typed). A polytypic species contains one subspecies whose subspecific name is a repetition of the species epithet and one or more additional subspecies whose names differ. Thus, to distinguish geographic variants of *Ensatina eschscholtzii*, one subspecies is named *Ensatina eschscholtzii eschscholtzii*, and different subspecies names are used for each of six other subspecies (Figure 10.2). Both the genus name and the species epithet may be abbreviated as shown in Figure 10.2. Formal recognition of subspecies has lost popularity among taxonomists because subspecies are often based on minor differences in appearance that do not necessarily diagnose evolutionarily distinct units. When further study reveals that named subspecies are distinct evolutionary lineages, the subspecies are then often recognized as full species; indeed, many authors argue that the subspecies of *Ensatina eschscholtzii* are in fact separate species. Subspecies designations, therefore, should be viewed as tentative statements indicating that the species status of the populations needs further investigation.

10.2 SPECIES

While discussing Darwin's book, *On the Origin of Species,* in 1859, Thomas Henry Huxley asked, "In the first place, what is a species? The question is a simple one, but the right answer to it is hard to find, even if we appeal to those who should know most about it." We have used the term "species" so far as if it had a simple and unambiguous meaning. Actually, Huxley's commentary is as valid today as it was in 1859. Our concepts of species have become more sophisticated, but the diversity of different concepts and disagreements surrounding their use are as evident now as in Darwin's time.

Despite widespread disagreement about the nature of species, biologists have repeatedly used certain criteria for identifying species. First, **common descent** is central to nearly all modern concepts of species. Members of a species must trace their ancestry to a common ancestral population, although not necessarily to a single pair of parents. Species are thus historical entities. A second criterion is that species must be the **smallest distinct groupings** of organisms sharing patterns of ancestry and descent; otherwise, it would be difficult to separate species from higher taxa, whose members also share common descent. Morphological characters traditionally have been important in identifying such groupings, but chromosomal and molecular characters now are used extensively for this purpose. A third important criterion is that of **reproductive community.** Members of a species must form a reproductive community that excludes members of other species. For sexually reproducing populations, interbreeding is critical for maintaining a reproductive community. For organisms whose reproduction is strictly asexual, reproductive community entails occupation of a particular ecological

habitat in a particular place so that a reproducing population responds as a unit to evolutionary forces such as natural selection and genetic drift (pp. 123–126).

Any species has a distribution through space, its **geographic range,** and a distribution through time, its **evolutionary duration.** Species differ greatly from each other in both dimensions. Species having very large geographic ranges or worldwide distributions are called **cosmopolitan,** whereas those with very restricted geographic distributions are called **endemic.** If a species were restricted to a single point in space and time, we would have little difficulty recognizing it, and nearly every species concept would lead us to the same decision. We have little difficulty distinguishing from each other the different species of animals that we can find living in our local park or woods. However, when we compare a local population to similar but not identical populations located hundreds of miles away, it may be hard to determine whether these populations represent a single species or multiple species (Figure 10.2).

Throughout the evolutionary duration of a species, its geographic range can change many times. A geographic range may be either continuous or disjunct, the latter having breaks within it where the species is absent. Suppose that we find two similar but not

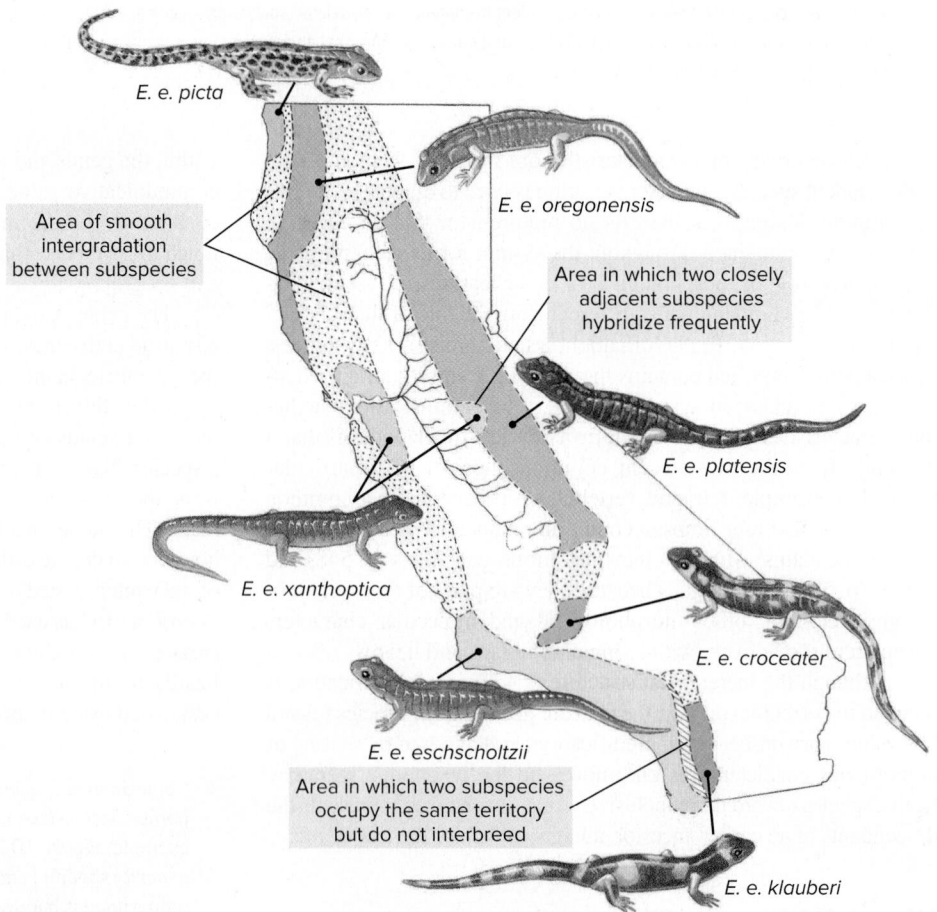

Figure 10.2 Geographic variation of color patterns in the salamander genus *Ensatina.* The species status of these populations has puzzled taxonomists for generations and continues to do so. Current taxonomy recognizes only a single species (*Ensatina eschscholtzii*) divided into subspecies as shown. Hybridization is evident between most adjacent populations, but studies of variation in proteins and DNA show large amounts of genetic divergence among populations. Furthermore, populations of the subspecies *E. e. eschscholtzii* and *E. e. klauberi* can overlap geographically without interbreeding.

identical populations living 300 miles apart with no related populations between them. Are we observing a single species with a disjunct distribution or two different but closely related species? Suppose that these populations have been separated historically for 50,000 years. Is this enough time for them to have evolved separate reproductive communities, or can we still view them as parts of the same reproductive community? Clear answers to such questions are very hard to find. Differences among species concepts pertain to solving these problems.

Typological Species Concept

Before Darwin, a species was considered a distinct and immutable entity. Species were defined by fixed, essential features (usually morphological) considered a divinely created pattern or archetype. This practice constitutes the **typological** (or **morphological**) **species concept.** Scientists recognized species formally by designating a **type specimen** that was labeled and deposited in a museum to represent the ideal form or morphology for the species (Figure 10.3). When scientists obtained additional specimens and wanted to assign them to a species, the type specimens of described species were consulted. The new specimens were assigned to a previously described species if they possessed the essential features of its type specimen. Small differences from the type specimen were considered accidental imperfections. Large differences from existing type specimens would lead a scientist to describe a new species with its own type specimen. In this manner, the living world was categorized into species.

Evolutionists discarded the typological species concept, but some of its traditions remain. Scientists still name species by describing type specimens deposited in museums, and the type specimen formally bears the name of the species. Organismal morphology is likewise still important in recognizing species; however, species are no longer viewed as classes of organisms defined by possession of certain morphological features. The basis of the evolutionary world view is that species are historical entities whose properties are subject always to change. Variation that we observe among organisms within a species is not an imperfect manifestation of an eternal "type"; the type itself is only an abstraction taken from the very real and important variation present within the species. A type is at best an average form that changes as organismal variation is sorted through time by natural selection. A type specimen serves only as a guide to the general morphological features that one might expect to find in a particular species as we observe it today.

The person who first describes a type specimen and publishes the name of a species is called the authority. This person's name and date of publication are often written after the species name. Thus, *Didelphis marsupialis* Linnaeus, 1758, tells us that Linnaeus was the first person to publish the species name of the opossum. Sometimes, the generic status of a species is revised following its initial description. In this case, the name of the authority is presented in parentheses. The Nile monitor lizard is denoted *Varanus niloticus* (Linnaeus, 1766) because the species originally was named by Linnaeus as *Lacerta nilotica*, and subsequently placed into a different genus.

Biological Species Concept

The most influential concept of species inspired by Darwinian evolutionary theory is the **biological species concept** formulated by Theodosius Dobzhansky and Ernst Mayr. This concept emerged during the evolutionary synthesis of the 1930s and 1940s from earlier ideas, and it has been refined and reworded several times since then. In 1982, Mayr stated the biological species concept as follows: *"A species is a reproductive community of populations (reproductively isolated from others) that occupies a specific niche in nature."* Note that a species is identified here according to reproductive properties of populations, not according to possession of any specific organismal characteristics. A species is an **interbreeding population** of individuals having common descent and sharing intergrading characteristics. Studies of populational variation in organismal morphology, chromosomal structure, and molecular genetic features are very useful for evaluating the geographical boundaries of interbreeding populations in nature. The criterion of the "niche" (see Chapter 38) recognizes that members of a reproductive community are expected also to have common ecological properties.

Because a reproductive community should maintain genetic cohesiveness, we expect organismal variation to be relatively smooth and continuous within species and discontinuous between them. Although the biological species is based on reproductive properties of populations rather than organismal morphology, morphology nonetheless can help us to diagnose biological species. Sometimes species status can be evaluated directly by conducting breeding experiments. Controlled breeding is practical only in a minority of cases, however, and our decisions regarding species membership usually are made by studying character variation. Variation in molecular characters is very useful for identifying geographical boundaries of reproductive communities. Molecular studies have revealed the occurrence of cryptic or **sibling species** (p. 115), which are too similar in morphology to be diagnosed as separate species by morphological characters alone.

Criticism of the biological species concept reveals several perceived problems. First, the concept lacks an explicit temporal dimension. It provides a means for diagnosing species status of contemporary populations but little guidance for tracing the temporal duration of a species lineage through its past history. For example,

Figure 10.3 Specimens of birds from the Smithsonian Institution (Washington, D.C.), including birds originally collected by John J. Audubon, Theodore Roosevelt, John Gould, and Charles Darwin.

humans constitute one biological species relative to other living populations. Human populations worldwide form a single reproductive community, reproductively isolated from all other populations. We have no guidance from the biological species concept regarding whether we should place human fossils of increasing evolutionary age into the current human species or into separate species. Should all fossil populations that are closer to us than to other living apes be judged part of our species or a series of temporal species that replaced each other through the past 5 million years of human evolution?

Second, proponents of the biological species concept often disagree on the degree of reproductive isolation necessary for considering two populations separate species, thereby revealing further ambiguity in the concept. For example, should occurrence of limited hybridization between populations in a small geographic area cause them to be considered a single species despite evolutionary differences between them? Hybrids between populations of *Ensatina eschscholtzii platensis* and *E. e. xanthoptica* (see Figure 10.2) occur geographically only at their point of contact in the Sierra Nevada mountains; molecular genetic distinctness between these taxonomic subspecies is otherwise so large that many authorities consider them separate reproductive communities and thus separate species under the biological species concept. Different authors invoking the biological species concept have interpreted the same data as demonstrating that the *Ensatina* populations of Figure 10.2 constitute one species versus demonstrating at least 11 distinct species.

A third problem is that because the biological species concept emphasizes interbreeding as the criterion of reproductive community, it denies the existence of species in groups of organisms that reproduce only asexually. It is common systematic practice, however, to describe species in all groups of organisms, regardless of whether reproduction is sexual or asexual.

Evolutionary and Cohesion Species Concepts

The time dimension creates obvious problems for the biological species concept. How do we assign fossil specimens to biological species that are recognized today? If we trace a lineage backward through time, how far must we go before we have crossed a species boundary? If we could follow the unbroken genealogical chain of populations backward through time to the point where two sister species converge on their common ancestor, we would need to cross at least one species boundary somewhere. It would be very hard to decide, however, where to draw a sharp line between the two species.

To address this problem, mammalian palaeontologist George Gaylord Simpson (Figure 10.4) proposed the **evolutionary species concept** to add an evolutionary time dimension to the biological species concept. This concept persists in a modified form today. A current definition of the evolutionary species is *a single lineage of ancestor-descendant populations that maintains its identity from other such lineages and that has its own evolutionary tendencies and historical fate*. Note that the criterion of common descent is retained here in the need for a lineage to have a distinct historical identity. Reproductive cohesion is the means by which a species maintains its identity from other such lineages and keeps its evolutionary fate separate from other species. The same kinds of diagnostic features discussed for the biological species concept are relevant for identifying

Figure 10.4   George Gaylord Simpson (1902 to 1984) formulated the principles of evolutionary taxonomy.

evolutionary species, although in most cases only morphological features are available from fossils. Unlike the biological species concept, the evolutionary species concept applies to both sexually and asexually reproducing forms. As long as continuity of diagnostic features is maintained by the evolving lineage, it is recognized as a species. Abrupt changes in diagnostic features mark the boundaries of different species in evolutionary time.

An ability of geographic populations to evolve collectively as a single, genetically cohesive unit through evolutionary time is critical to the evolutionary species concept. Population geneticist Alan Templeton updated this concept in 1989 to make explicit the expectation that populations of a species evolve as a genetically cohesive unit by natural selection and genetic drift. Using his **cohesion species concept,** Templeton defined a species as follows: *the most inclusive population of individuals having the potential for phenotypic cohesion through intrinsic cohesion mechanisms.* The cohesion mechanisms include gene flow across the geographic expanse of the species, shared loss of alleles through genetic drift, and shared genetic changes caused by natural selection. Another way of stating the cohesion criterion is that any individual in a species is a possible common ancestor of the entire species at some future time. For example, a new allele that arises by mutation in a single person could spread throughout the human population over many generations, ultimately to become shared by all members of the species at some future time. New mutations arising in other species, even in our closest relatives of the genus *Pan,* could not enter the human gene pool.

Phylogenetic Species Concept

Ornithologist Joel Cracraft defined the last concept that we present: the **phylogenetic species concept.** The phylogenetic species concept is an *irreducible (basal) grouping of organisms diagnosably distinct from other such groupings and within which there is a parental pattern of ancestry and descent.* This concept emphasizes most strongly the criterion of common descent. Both asexual and sexual groups are covered.

Species Concepts in Practice

These hypothetical examples illustrate conditions that challenge taxonomists in judging the numbers of species represented by a group of populations being studied. Assume that the individual populations discussed are internally genetically homogeneous based on molecular genetic data, but that they are diagnosably distinct from others using those same data. I. Two geographic populations of sexually reproducing forms make geographic contact along a borderline at which hybrids (offspring whose parents come from different geographic populations) occur (I. A. natural hybrids are fertile; I. B. natural hybrids are infertile). II. Two geographically allopatric populations of sexually reproducing forms appear morphologically and ecologically equivalent (II. A. artificial crosses produce fertile hybrids; II. B. artificial crosses produce infertile hybrids). III. Six genetically divergent populations of sexually reproducing forms have geographic and phylogenetic relationships as shown with hybrids produced between populations B and D. Populations A–C (group 1)

are morphologically indistinguishable from each other but differ from populations D–F (group 2), which are morphologically indistinguishable from each other (III. A. groups 1 and 2 are ecologically equivalent; III. B. groups 1 and 2 are ecologically distinct). IV. Sexually reproducing populations A–E are genetically divergent but similar in ecology and morphology; population F was derived from population E by a founder event and is reproductively isolated from the others. V. Two allopatric populations of animals that reproduce only asexually are ecologically and morphologically distinct from each other. Each box indicates the number of different species that a taxonomist would recognize using the species concept(s) at the head of the column. Some concepts have inherent ambiguities permitting some room for individual judgment separating taxonomic "lumpers" from "splitters"; such cases are indicated by "1-2" in the appropriate column with the source of ambiguity indicated in column 5.

Geography / Phylogeny		Biological Species	Evolutionary and Cohesion Species	Phylogenetic Species	Source of Ambiguity
I. Hybrid zone	A. hybrids fertile	1 - 2	1 - 2	2	Will 1 & 2 eventually merge or remain distinct with a small hybrid zone?
	B. hybrids infertile	2	2	2	None
II. Artificial cross	A. hybrids fertile	1	1	2	None
	B. hybrids infertile	2	2	2	None
III. Hybrid zone (fertile)	A. 1 & 2 are ecologically similar	1 - 2	1 - 2	6	Will 1 & 2 eventually merge or remain distinct with a small hybrid zone?
	B. 1 & 2 are ecologically distinct	1 - 2	2	6	Is ecological isolation sufficient for separate species status without reproductive isolation?
IV.		2	2	6	None
V.		0	2	2	None

A phylogenetic species is a single population lineage with no detectable branching. The main difference in practice between the evolutionary/cohesion and phylogenetic species concepts is that the latter emphasizes recognizing as separate species the smallest groupings of organisms that have undergone independent evolutionary change. The evolutionary and cohesion species concepts place greater emphasis on whether historically separated populations have the biological potential to merge into a single lineage in the future. The evolutionary or cohesion species concepts would group into a single species geographically disjunct populations that demonstrate some phylogenetic divergence but are judged similar in their "evolutionary tendencies," thus permitting future gene exchange and possible merging, whereas the phylogenetic species concept would treat them as separate species. In general, a greater number of species would be described using the phylogenetic species concept than by any other species concept, and many taxonomists consider it impractical for this reason. For strict adherence to cladistic systematics (p. 209), the phylogenetic species concept is ideal because only this concept guarantees strictly monophyletic units at the species level.

The phylogenetic species concept intentionally disregards details of evolutionary process and gives us a criterion that allows us to describe species without first needing to conduct detailed studies on evolutionary processes. Advocates of the phylogenetic species concept do not necessarily disregard the importance of studying evolutionary process. They argue, however, that the first step in studying evolutionary process is to have a clear picture of life's history. To accomplish this task, the pattern of common descent must be reconstructed in the greatest detail possible by starting with the smallest taxonomic units that have a history of common descent distinct from other such units. How would the phylogenetic species concept resolve the controversy over how many species exist in *Ensatina eschscholtzii* (see Figure 10.2)? The molecular genetic data reveal at least 11 diagnosably distinct lineages, each of which would constitute a separate phylogenetic species. Grouping these lineages as a single species because they retain a potential for future genetic exchanges is not an option under the phylogenetic species concept.

Dynamism of Species Concepts

Herpetologist Kevin de Queiroz argues that the various competing concepts of species have a common underlying principle despite their differences. In each case, a species constitutes a segment of a population-level lineage, what de Queiroz calls the **general lineage concept of species**. For the biological species concept, the segment is a temporally short one with reproductive community among sexually reproducing populations being the critical secondary attribute that separates the biological species concept from alternatives. For the phylogenetic species concept, a population lineage diagnosable as having evolved independently since its evolutionary separation from another such lineage provides the secondary attribute that distinguishes this concept from others. The general lineage concept of species has gained popularity among systematists because it emphasizes the common goal of identifying the phylogenetic history of population-level lineages in detail. It does not solve the problem, however, that taxonomists using contrasting species concepts may differ greatly in how many species they judge worthy of a Linnaean Latin binomial (see boxed essay Species Concepts in Practice).

Taxonomists agree that historically distinct population lineages, the species of the phylogenetic species concept, are real entities in nature. Such entities exist as the indivisible units of evolutionary process and change independent of our knowledge of them. Advocates of the other species concepts do not deny these claims, but they consider such lineages too numerous and too ephemeral for each one to deserve recognition with a Latin species binomial, especially when the biological differences among lineages are judged superficial. Given the power of molecular genetic data to diagnose species lineages, it is perhaps impractical to expect each one to be given formal species status. A taxonomic system must be practical to serve us well, but when we defer to practicality we risk making our recognized species arbitrary constructs that lose their integrity as natural individuals.

Current disagreements concerning concepts of species should not be considered discouraging. Whenever a field of scientific investigation enters a phase of dynamic growth, old concepts are reevaluated and either refined or replaced with newer, more progressive ones. The active debate occurring within systematics shows that this field has acquired unprecedented activity and importance in biology. Just as Thomas Henry Huxley's time was one of enormous advances in biology, so is the present time. Both times are marked by fundamental reconsiderations of the meaning of species. Researchers whose main interests are branching of evolutionary lineages, evolution of reproductive barriers among populations (p. 113), or ecological properties of species may favor different species concepts. The conflicts among the current concepts lead us into the future. In many cases, different concepts agree on the locations of species boundaries, and disagreements identify particularly interesting cases of evolution in action. Understanding the conflicting perspectives, rather than learning a single species concept, is therefore of greatest importance for people now entering the study of zoology.

DNA Barcoding of Species

DNA barcoding is a technique for identifying organisms to species using sequence information from a standard gene present in all animals. The mitochondrial gene encoding cytochrome *c* oxidase subunit 1 (*COI*), which contains about 650 nucleotide base pairs, is a standard "barcode" region for animals. DNA sequences of *COI* usually vary among individuals of the same species but not extensively, so that variation within a species is much smaller than differences among species. DNA barcoding is applied to specimens in nature by taking a small DNA sample from blood or another expendable tissue. The method is useful also for specimens in natural-history museums, zoos, aquaria, and frozen-tissue collections. DNA sequences from such sources are checked against a public reference library of species identifiers to assign unknown specimens to known species. DNA barcoding does not solve the controversies regarding use of different species concepts, but it often permits the origin of a specimen to be identified to a particular local population, which is valuable information regardless of the species status that a taxonomist assigns to that population.

10.3 TAXONOMIC CHARACTERS AND PHYLOGENETIC RECONSTRUCTION

A major goal of systematics is to infer the evolutionary tree or **phylogeny** that relates all extant and extinct species. This task is accomplished by identifying organismal features, formally called **characters,** that vary among species. A character is any feature that a taxonomist uses to study variation within and among species. Taxonomists find characters by observing patterns of similarity among organisms in morphological, chromosomal, and molecular features (see p. 206), and less frequently in behavioral and ecological ones. Phylogenetic analysis depends upon finding features that species share because the features are inherited from a common ancestor. Character sharing that results from common ancestry is called **homology** (see Chapter 6). Character similarity does not always reflect common ancestry, however. Independent evolutionary origin of similar features on different lineages produces patterns of similarity among organisms that misrepresent common descent; this occurrence complicates the work of taxonomists. Character similarity that misrepresents common descent is called nonhomologous similarity or **homoplasy.** Endothermy of birds and mammals illustrates homoplasy; this condition arose separately in an ancestor of birds and in an ancestor of mammals. Variation in other characters shows that birds and mammals are not each other's closest relatives (p. 565). For an example of molecular homoplasy, see the interpretation of character 41 (p. 211) in the boxed essay, Phylogenies from DNA Sequences.

Using Character Variation to Reconstruct Phylogeny

To infer the phylogeny of a taxon using characters that vary among its species, the first step is to ask, for each character, which character state was present in the most recent common ancestor of the taxon. This character state is called **ancestral** for the taxon as a whole. We presume that all other contrasting states of the character arose later within the group, and these are called evolutionarily **derived character states.** Determining the **polarity** of a character refers to identifying which one of its contrasting states is ancestral and which one(s) derived. For example, if we consider as a character the dentition of amniotic vertebrates (reptiles, birds, and mammals), presence versus absence of teeth in the jaws constitute alternative character states. Teeth are absent from modern birds but present in most other amniotes. To evaluate the polarity of this character, we must determine which character state, presence or absence of teeth, characterized the most recent common ancestor of amniotes and which state was derived subsequently within amniotes.

The method used to examine the polarity of a variable character is **outgroup comparison.** We consult an additional group of organisms, called an **outgroup,** that is phylogenetically close but not within the taxon being studied, which we call the "ingroup." A character state shared by ingroup and outgroup species is judged ancestral for the ingroup. Amphibians and different groups of bony fishes constitute appropriate outgroups to the amniotes for polarizing variation in dentition of amniotes. Teeth are usually present in amphibians and bony fishes; therefore, we infer that presence of teeth is ancestral for amniotes and absence of teeth is derived. The observation of abundant teeth in crocodilians, the closest living relatives of birds,

strengthens this inference. The polarity of this character indicates that teeth were lost in the ancestral lineage of all modern birds. Polarity of characters is evaluated most effectively when several different outgroups are used. All ingroup character states that are absent from appropriate outgroups are considered derived.

A **clade** (Gr. *klados,* branch) is the fundamental unit of phylogenetic grouping of species; it comprises an ancestral lineage and all species descended from that lineage. A derived character shared uniquely by all members of a clade is formally called a **synapomorphy** (Gr. *synapsis,* joining together, + *morphē,* form) of that clade. The derived character arose in the common ancestral lineage of the clade and was inherited by all descendants of that ancestor. Taxonomists use shared derived characters as evidence of homology to infer that a particular group of species forms a clade. Among living amniotes, jaws forming a beak without teeth and presence of feathers are synapomorphies that identify birds as a clade. Within birds, the derived morphological character of a flexible palate marks Neognathae as a nested clade separate from Paleognathae. Within Neognathae, derived, swimming-adapted wings mark the penguin order Sphenisciformes as a nested clade. The pattern formed by the derived states of all characters within our ingroup reveals a **nested hierarchy** of clades within clades. Our goal is to identify all of the different clades nested within the ingroup, which would reveal the structure of common descent among its species.

Character states ancestral for a taxon are often called **plesiomorphic** for that taxon, and the sharing of ancestral states among species is termed **symplesiomorphy.** Unlike synapomorphies, however, symplesiomorphies do not provide useful information on nesting of clades within clades. In the example just given, we found that presence of teeth in jaws was plesiomorphic for amniotes. If we grouped together mammalian and reptilian groups, which possess teeth, to the exclusion of modern birds, we would not obtain a valid clade. Birds also descend from all common ancestors of reptiles and mammals and therefore lie within any clade that includes all reptiles and mammals. Errors in determining polarity of characters therefore clearly can produce errors in inference of phylogeny. It is important to note, however, that character states that are plesiomorphic at one taxonomic level can be synapomorphies at a more inclusive level. For example, the presence of jaws bearing teeth is a synapomorphy of gnathostome vertebrates (p. 500), a group that includes amniotes plus amphibians, bony fishes, and cartilaginous fishes, although teeth have been lost in birds and some other gnathostomes. The goal of phylogenetic analysis therefore can be restated as one of finding the appropriate taxonomic level at which any given character state is a synapomorphy. The character state is then used at that level to identify a clade.

A nested hierarchy of clades is presented as a branching diagram called a **cladogram** (Figure 10.5; see also Figure 6.16, and try to reconstruct this cladogram using only the sharing of numbered synapomorphies among the bird species). Taxonomists often make a technical distinction between a cladogram and a **phylogenetic tree.** The branches of a cladogram are only a formal device for indicating the nested hierarchy of clades within clades. The cladogram is not strictly equivalent to a phylogenetic tree, whose branches represent real lineages that occurred in the evolutionary past. To obtain a phylogenetic tree, we must add to the cladogram important additional interpretations concerning ancestors, the durations of

Figure 10.5 A cladogram as a nested hierarchy of taxa among five sampled chordate groups (amphioxus, bass, lizard, horse, monkey). Amphioxus is the outgroup, and the study group comprises the four vertebrates. We generate a simple cladogram from four characters that vary among vertebrates: presence versus absence of four legs, amniotic eggs, hair, and mammary glands. For all four characters, absence is the ancestral state in vertebrates because this is the condition found in the outgroup, amphioxus; for each character, presence is the derived state in vertebrates. Because they share presence of four legs and amniotic eggs as synapomorphies, the lizard, horse, and monkey form a clade relative to the bass. This clade is subdivided further by two synapomorphies (presence of hair and mammary glands) that unite the horse and monkey relative to the lizard. We know from comparisons involving even more distantly related animals that vertebrae and jaws constitute synapomorphies of vertebrates and that amphioxus, which lacks these features, falls outside the vertebrate clade.

evolutionary lineages, or the amounts of evolutionary change that occurred on the lineages. A cladogram is often used, however, as a first approximation of the branching structure of the corresponding phylogenetic tree.

Sources of Phylogenetic Information

We find characters that are useful for constructing cladograms in comparative morphology (including embryology), comparative cytology, and comparative biochemistry. **Comparative morphology** examines the varying shapes and sizes of organismal structures, including their developmental origins. Both macroscopic and microscopic characters are used, including details of cellular structure revealed by histology. As seen in Chapters 23 through 28, the variable structures of skull bones, limb bones, and integument (scales, hair, feathers) are particularly important for reconstructing the phylogeny of vertebrates. Comparative morphology uses specimens obtained from both living organisms and fossilized remains. **Comparative biochemistry** uses sequences of amino acids in proteins and the sequences of nucleotides in nucleic acids (see Chapter 5) to identify variable characters for constructing a cladogram (Figure 10.6). Direct sequencing of DNA is regularly applied to phylogenetic studies; however, comparisons of protein sequences are usually indirect, involving immunological or allozymic (see Figure 6.32) methods, or inferences from DNA sequences of protein-coding genes. Recent studies show that comparative biochemistry can be applied to some fossils in addition to living organisms. **Comparative cytology** (also called karyology) uses variation in the numbers, shapes, and sizes of chromosomes and their parts (see Chapter 3 and p. 110) to obtain variable characters

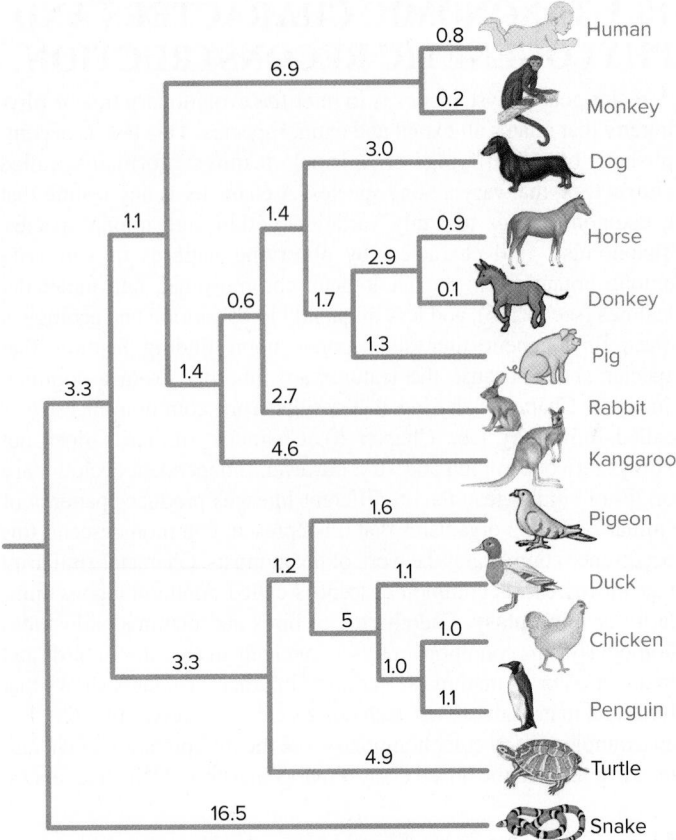

Figure 10.6 An early phylogenetic tree of representative amniotes based on amino acid sequences of the respiratory protein, cytochrome **c**. Numbers on the branches are the expected minimum numbers of mutational changes needed to explain amino acid substitutions along the different evolutionary lineages. Publication of this tree by Fitch and Margoliash in 1967 convinced systematists that molecular sequences contain phylogenetic information. Subsequent work confirms some hypotheses, including the hypothesis that mammals (blue) and birds (red) form nonoverlapping clades, while rejecting others; kangaroo, for example, should be outside a branch containing all other mammals shown.

for constructing cladograms. Comparative cytology is used almost exclusively on living rather than fossilized organisms.

To add an evolutionary timescale necessary for producing a phylogenetic tree, we must consult the fossil record. We can look for the earliest appearance in fossils of derived morphological characters to estimate the ages of clades distinguished by those characters. The age of a fossil showing the derived characters of a particular clade is determined by radioactive dating (p. 106). An example of a phylogenetic tree constructed using these methods is Figure 25.1, page 545.

We can use comparative biochemical data to estimate the ages of different lineages on a phylogenetic tree. Some protein and DNA sequences undergo approximately linear rates of divergence through evolutionary time. The age of the most recent common ancestor of two species is therefore proportional to the differences measured between their proteins and DNA sequences. We calibrate evolution of proteins and DNA sequences by measuring their divergence between species whose most recent common ancestor has been dated using fossils. We then use the molecular evolutionary calibration to estimate ages of other branches on the phylogenetic tree.

10.4 THEORIES OF TAXONOMY

A theory of taxonomy establishes the principles that we use to recognize and to rank taxonomic groups. There are two currently popular theories of taxonomy: (1) evolutionary taxonomy and (2) phylogenetic systematics (cladistics). Both are based on evolutionary principles. These two theories differ, however, on how evolutionary principles are used. These differences have important implications for how we use a taxonomy to study evolutionary processes. Evolutionary taxonomy predates phylogenetic systematics and retains many aspects of Linnaean taxonomy; for this reason, it is sometimes called "traditional evolutionary taxonomy." Evolutionary taxonomy was well established by the 1940s; phylogenetic systematics arose in the 1960s as a replacement for evolutionary taxonomy, which some systematists considered arbitrary and misleading.

The relationship between a taxonomic group and a phylogenetic tree or cladogram is important for both theories. This relationship can take one of three forms: **monophyly, paraphyly, or polyphyly** (Figure 10.7). A taxon is monophyletic if it includes the most recent common ancestor of the group and all descendants of that ancestor (Figure 10.7A). The terms "monophyletic group" and "clade" are synonymous. A taxon is paraphyletic if it includes the most recent common ancestor of all members of a group and some but not all descendants of that ancestor (Figure 10.7B). A taxon is polyphyletic if it does not include the most recent common ancestor of all members of a group; this condition requires that the group has had at least two separate evolutionary origins, usually requiring independent evolutionary acquisition of similar features (Figure 10.7C).

Monophyletic and paraphyletic groups share the property of **convexity**, which distinguishes them from polyphyletic groups. A group is convex if you can trace a path between any two members of the group on a cladogram or phylogenetic tree without leaving the group. For example, on Figure 10.7 you could trace a connection between any pair of points in the blue areas of parts A or B without leaving the blue area. For the polyphyletic grouping in part C of Figure 10.7, one cannot trace the path between species C and E without leaving the group designated by blue shading. In Figure 10.7C, if a systematist added the full path connecting species C and E to the group shown but continued to omit the paths leading to species A, B, and H, then the new grouping thus formed would be convex and paraphyletic rather than polyphyletic. Demonstration that a group is

not convex is the formal criterion for considering the group polyphyletic.

Both evolutionary and cladistic taxonomy accept monophyletic groups and reject polyphyletic groups. They differ on acceptance of paraphyletic groups, however, and this difference has important evolutionary implications.

Evolutionary Taxonomy

Evolutionary taxonomy retains the basic structure of Linnaean taxonomy; species are grouped in a nested hierarchy of increasingly more inclusive higher taxa, each of which receives one of the mandatory or optional Linnaean ranks. The taxonomy is evolutionary because all taxa must have a single evolutionary origin, must include the most recent common ancestor of all members of the taxon, and must be convex on the phylogenetic tree. Unlike pre-evolutionary uses of Linnaean taxonomy, evolutionary taxonomy explicitly excludes polyphyletic groups.

George Gaylord Simpson (see Figure 10.4) and Ernst Mayr (see Figure 6.19) were highly influential in developing and formalizing the procedures of evolutionary taxonomy. According to Simpson and Mayr, a particular branch on an evolutionary tree is considered a higher taxon if it represents a distinct **adaptive zone.** Simpson describes an adaptive zone as "a characteristic reaction and mutual relationship between environment and organism, a way of life and not a place where life is led." By entering a new adaptive zone through a fundamental change in organismal structure and behavior, an evolving population can use environmental resources in a new way.

A taxon that constitutes a distinct adaptive zone is termed a **grade.** Simpson gives the example of penguins as a distinct adaptive zone within birds. The lineage immediately ancestral to all penguins underwent fundamental changes in the form of the body and wings to switch from aerial to aquatic locomotion (Figure 10.8). Aquatic birds that use their flight apparatus to propel themselves through air and water are somewhat intermediate in habitat, morphology, and behavior between aerial and aquatic adaptive zones. Nonetheless, the obvious modifications of the wings and body of penguins for swimming represent a new grade of organization. Penguins are therefore recognized as a distinct taxon within birds, the order Sphenisciformes. The broader the adaptive zone when fully occupied by a group of organisms, the higher the rank given to the corresponding taxon.

Figure 10.7　Relationships between phylogeny and taxonomic groups illustrated for a hypothetical phylogeny of eight species (A through H). **A,** *Monophyly*—a monophyletic group contains the most recent common ancestor of all members of the group and all of its descendants. **B,** *Paraphyly*—a paraphyletic group contains the most recent common ancestor of all members of the group and some but not all of its descendants. **C,** *Polyphyly*—a polyphyletic group typically does not contain the most recent common ancestor of all members of the group, thereby requiring that the group have at least two separate phylogenetic origins. Monophyletic and paraphyletic groups are *convex*, meaning that one can trace a path from any member of the group to any other member without leaving the group; any group that fails the convexity criterion is considered polyphyletic.

A   **B**

Figure 10.8 **A,** Penguin. **B,** Diving petrel. Penguins (avian order Sphenisciformes) were recognized by George G. Simpson as a distinct adaptive zone within birds because of their adaptations for submarine flight. Simpson believed that the adaptive zone ancestral to penguins resembled that of diving petrels, which display adaptations for combined aerial and aquatic flight. Adaptive zones of penguins and diving petrels are distinct enough to be recognized taxonomically as different orders.

Evolutionary taxa may be either monophyletic or paraphyletic. Recognition of paraphyletic taxa requires, however, that our taxonomies distort patterns of common descent. An evolutionary taxonomy of the anthropoid primates provides a good example (Figure 10.9). This taxonomy places humans (genus *Homo*) and their closest fossil relatives in the family Hominidae. It groups the bonobos and chimpanzees (genus *Pan*), gorillas (genus *Gorilla*), and orangutans (genus *Pongo*) in the family Pongidae. However, the pongid genera *Pan* and *Gorilla* share more recent common ancestry with the Hominidae

than they do with the remaining pongid genus, *Pongo*. This arrangement makes the family Pongidae paraphyletic because it excludes humans, who also descend from the most recent common ancestor of all pongids (Figure 10.9). Evolutionary taxonomists nonetheless consider the pongid genera a single, family-level grade of arboreal, herbivorous primates having no language; in other words, they show the same family-level adaptive zone. Humans are terrestrial, omnivorous primates who have language and cultures, thereby forming a distinct adaptive zone at the taxonomic level of the family. Unfortunately, if we want our taxa to constitute adaptive zones, our taxonomy will distort common descent.

Evolutionary taxonomy has faced two opposite challenges. **Phenetic taxonomy** sought to group species into higher taxa according to quantitative measurements of overall similarity among species, without regard to phylogeny. Phylogeny was considered too difficult to measure to be the basis of our taxonomic system. Later work showed that phylogeny could be measured precisely and that the notion of overall similarity was highly ambiguous. Phenetic taxonomy contributed some useful analytical methods but did not have a strong impact on animal taxonomy, and scientific interest in this approach has declined.

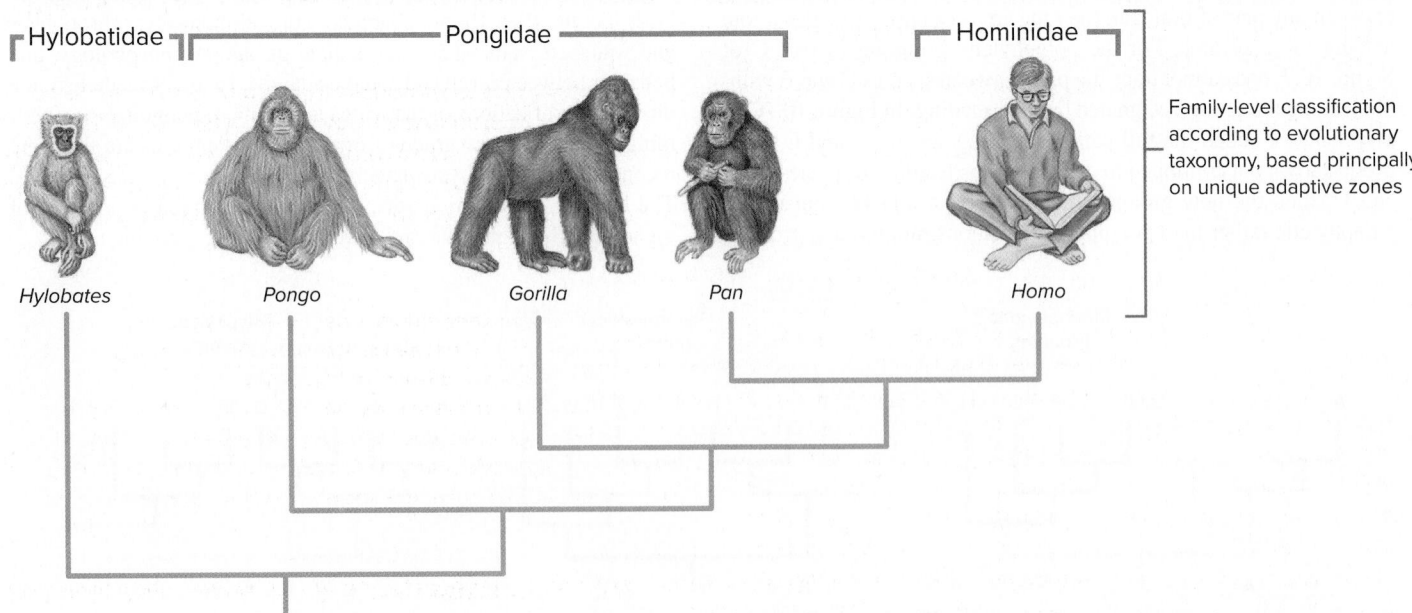

Family-level classification according to evolutionary taxonomy, based principally on unique adaptive zones

Hylobates   *Pongo*   *Gorilla*   *Pan*   *Homo*

Figure 10.9 Phylogeny and family-level classification of anthropoid primates. Evolutionary taxonomy groups the genera *Gorilla*, *Pan*, and *Pongo* into a paraphyletic family Pongidae because they share the same adaptive zone or grade of organization. Humans (genus *Homo*) are phylogenetically closer to *Gorilla* and *Pan* than any of these genera are to *Pongo*, but humans are placed in a separate family (Hominidae) because they represent a new grade of organization. Cladistic taxonomy discontinues recognition of paraphyletic family Pongidae, consolidating *Pongo*, *Gorilla*, *Pan*, and *Homo* in family Hominidae.

Figure 10.10   Willi Hennig (1913 to 1976), German entomologist who formulated the principles of phylogenetic systematics/cladistics in 1950.

Phylogenetic Systematics/Cladistics

A second and stronger challenge to evolutionary taxonomy is one called **phylogenetic systematics** or **cladistics**. As the first name implies, this approach emphasizes the criterion of common descent and, as the second name implies, it is based on the cladogram of the group being classified. Because the German entomologist Willi Hennig (Figure 10.10) formulated cladistic principles, phylogenetic systematics is sometimes called "Hennigian systematics." All taxa recognized by Hennig's cladistic system must be monophyletic. We saw in Figure 10.9 how evolutionary taxonomists' recognition of the primate families Hominidae and Pongidae distorts genealogical relationships to emphasize adaptive uniqueness of the Hominidae. Because the most recent common ancestor of the paraphyletic family Pongidae is also an ancestor of the Hominidae, recognition of the Pongidae is incompatible with cladistic taxonomy. To avoid paraphyly, cladistic taxonomists have discontinued use of the traditional family Pongidae, placing chimpanzees, gorillas, and orangutans with humans in the family Hominidae. We adopt the cladistic classification in this book.

Disagreement on the validity of paraphyletic groups may seem trivial at first, but its important consequences become clear when we discuss evolution. For example, claims that amphibians evolved from bony fish, that birds evolved from reptiles, or that humans evolved from apes might be made by an evolutionary taxonomist but are meaningless to a cladist. We imply by these statements that a descendant group (amphibians, birds, or humans) evolved from part of an ancestral group (bony fish, reptiles, and apes, respectively) to which the descendant does not belong. This usage automatically makes the ancestral group paraphyletic, and indeed bony fish, reptiles, and apes as traditionally recognized are paraphyletic groups. How are such paraphyletic groups recognized? Do they share distinguishing features not shared by the descendant group?

Paraphyletic groups are usually defined in a negative manner. They are distinguished only by lacking features found in a particular descendant group, because any traits that they share from their common ancestry are symplesiomorphies present also in the excluded descendants (unless secondarily lost). For example, apes are those "higher" primates that are not humans. Likewise, fish are those vertebrates that lack the distinguishing characteristics of tetrapods (amphibians and amniotes). What does it mean then to say that humans evolved from apes? To an evolutionary taxonomist, apes and humans are different adaptive zones or grades of organization; to say that humans evolved from apes states that bipedal organisms of large brain capacity evolved from arboreal organisms of smaller brain capacity. To a cladist, however, the statement that humans evolved from apes says essentially that humans evolved from an arbitrary grouping of species that lack the distinctive characteristics of humans, a trivial statement that conveys no useful information. To a cladist, any statement that a particular monophyletic group descends from a paraphyletic one is nothing more than a claim that the descendant group evolved from something that it is not. Extinct ancestral groups are always paraphyletic because they exclude a descendant that shares their most recent common ancestor. Although many such groups have been recognized by evolutionary taxonomists, none are recognized by cladists.

Zoologists often construct paraphyletic groups because they are interested in a terminal, monophyletic group (such as humans), and they want to ask questions about its ancestry. It is often convenient to lump together species whose features are considered approximately equally distant from the group of interest and to ignore the unique features of those species. It is significant in this regard that humans have never been placed in a paraphyletic group, whereas most other organisms have been. Apes, reptiles, fishes, and invertebrates are all terms that traditionally designate paraphyletic groups formed by combining various "side branches" found when tracing human ancestry backward through the tree of life. Such a taxonomy can give the erroneous impression that all of evolution is a progressive march toward humanity or, within other groups, a progressive march toward whatever species humans designate most "advanced." Such thinking is a relic of pre-Darwinian views that there is a linear scale of nature having "primitive" creatures at the bottom and humans near the top just below angels. Darwin's theory of common descent states, however, that evolution is a branching process with no linear scale of increasing perfection along a single branch. Nearly every branch contains its own combination of ancestral and derived features. In cladistics, this perspective is emphasized by recognizing taxa only by their own unique properties and not grouping organisms only because they lack the unique properties found in related groups.

Fortunately, there is a convenient way to express the common descent of groups without constructing paraphyletic taxa. It is done by finding what is called the **sister group** of the taxon of interest to us. Two different monophyletic taxa are each other's sister group if they share common ancestry with each other more recently than either one does with any other taxa. The sister group of humans appears to be genus *Pan* (bonobos and chimpanzees), with gorillas forming the sister group to humans, bonobos, and chimpanzees combined. Orangutans are the sister group of a clade that comprises humans, bonobos, chimpanzees, and gorillas; gibbons form the sister group of the clade that comprises orangutans, bonobos, chimpanzees, gorillas, and humans (Figure 10.9).

Current State of Animal Taxonomy

The formal taxonomy of animals that we use today was established using the principles of evolutionary systematics and has been revised recently in part using the principles of cladistics. Introduction of cladistic principles initially replaces paraphyletic groups with monophyletic subgroups while leaving the remaining taxonomy mostly unchanged. A thorough revision of taxonomy along cladistic principles, however, would require profound changes, one of which is abandonment of Linnaean ranks. A new taxonomic system called PhyloCode is being developed as an alternative to Linnaean taxonomy; this system replaces Linnaean ranks with codes that denote the nested hierarchy of monophyletic groups conveyed by a cladogram. In our coverage of animal taxonomy, we try to use taxa that are monophyletic and therefore consistent with criteria of both evolutionary and cladistic taxonomy. We use Linnaean ranks where this system still prevails in current taxonomy of a group of animals.

Where unranked taxa have gained common usage, we present a rank-free taxonomy. For familiar taxa that are clearly paraphyletic grades, we note this fact and suggest alternative taxonomic schemes that contain only monophyletic taxa.

When discussing patterns of descent, we avoid statements such as "mammals evolved from reptiles" that imply paraphyly and instead specify appropriate sister-group relationships. We avoid calling groups of living organisms primitive, advanced, specialized, or generalized because all groups of animals contain combinations of primitive, advanced, specialized, and generalized features; these terms are best restricted to describing specific characteristics and not an entire group. We likewise avoid calling a living species or group of living species "basal," because no species or group is more basal than its sister taxon. The term "basal" is best reserved for describing branch points or "nodes" on a phylogenetic tree; the most basal node is the one closest to the root of the tree.

Phylogenies from DNA Sequences

Most phylogenetic reconstruction currently underway comes from collecting genomic DNA sequences for the species of interest and analyzing those data using statistical approaches, especially maximum-likelihood and Bayesian methods. Such analysis requires sophisticated computer algorithms and cannot be done by hand. Nonetheless, statistical inference of phylogenies is designed to separate patterns of homology from homoplasy in the data, and to use the former to trace common descent of the species being studied. To explain statistical phylogenetics, we first must examine the basic phylogenetic principles of character analysis that underlie statistical phylogenetics.

A simple example illustrates cladistic analysis of DNA sequence data to examine phylogenetic relationships among species. The study group in this example contains three species of chameleons, two from the island of Madagascar (*Brookesia theili* and *B. brygooi*) and one from Equatorial Guinea (*Chamaeleo feae*). The outgroup is a lizard of genus *Uromastyx*, which is a distant relative of chameleons. Do the molecular data in this example confirm or reject the prior taxonomic hypothesis that the two Madagascan chameleons are more closely related to each other than either one is to the Equatorial Guinean species?

The molecular information in this example comes from a piece of the mitochondrial DNA sequence (57 bases) for each species. Each sequence encodes amino acids 221–239 of a protein called "NADH dehydrogenase subunit 2" in the species from which it was obtained. These DNA base sequences are aligned and numbered as follows:

Each column in the aligned sequences constitutes a character that takes one of four states: A, C, G, or T (a fifth possible state, absence of a base, is not observed in this example). Only characters that vary among the three chameleon species potentially contain information on which pair of species is most closely related. Twenty-three of the 57 aligned bases show variation *among chameleons*, as shown here in bold letters:

To be useful for constructing a cladogram, a character must demonstrate sharing of derived states (=synapomorphy). Which of these 23 characters demonstrate synapomorphies for chameleons? For each of the 23 variable characters, we must ask whether one of the states observed in chameleons is shared with the outgroup, *Uromastyx*. If so, this state is judged ancestral for chameleons and the alternative state(s) derived.

Derived states are identified for 21 of the 23 characters just identified; derived states are shown in blue:

Note that polarity is ambiguous for two variable characters (at positions 23 and 54) whose alternative states in chameleons are not observed in the outgroup.

Of the characters showing derived states, 10 of them show synapomorphies among chameleons. These characters are marked here with numbers 1, 2, or 3 below the appropriate column.

The eight characters marked 1 show synapomorphies grouping the two Madagascan species (*Brookesia theili* and *B. brygooi*) to the exclusion of the Equatorial Guinean species, *Chamaeleo feae*. We can represent these relationships as a cladogram:

We can explain evolution of all characters favoring this cladogram by placing a single mutational change on the branch ancestral to the two *Brookesia* species. This is the simplest explanation for evolutionary change of these characters.

Characters marked 2 and 3 disagree with our cladogram and favor alternative relationships as shown here:

To explain evolutionary changes in characters favoring cladograms 2 or 3 using cladogram 1, we need at least two changes per character. Likewise, if we try to explain evolution of characters favoring cladogram 1 on cladograms 2 or 3, we need at least two changes for each of these characters. These two diagrams show the minimum numbers of changes required for character 5 (which favors cladogram 1) and character 41 (which favors cladogram 3) on cladogram 1; the ancestral state of each character is shown at the root of the tree and the states observed in each species at the tips of the branches:

character 5 (1 change)

character 41 (2 changes)

A principle called **parsimony** can be used to resolve conflicts among taxonomic characters, as seen here. Parsimony represents the simplest limiting case of the maximum-likelihood and Bayesian methods for statistical inference of phylogeny, so we present it first and then show how the statistical methods build on this framework. We choose as our best working hypothesis the cladogram that requires the smallest total amount of

(continued on next page)

character change. In our example, parsimony favors cladogram 1. For all 10 phylogenetically informative characters, cladogram 1 requires a total of 12 changes of character state (one for each of the 8 characters favoring it and two for each of the other 2 characters). Cladograms 2 and 3 each require at least 19 character-state changes, 7 steps longer than cladogram 1. By choosing cladogram 1, we claim that characters favoring cladograms 2 and 3 show homoplasy in their evolution.

The molecular sequences shown in this example therefore confirm predictions of the prior hypothesis, based on appearance and geography of these chameleons, that the *Brookesia* species shared a common ancestor with each other more recently than either one did with *Chamaeleo feae*.

As a further exercise, you should convince yourself that the 12 characters that vary among chameleons but which do not demonstrate unambiguous sharing of derived states are equally compatible with each of the three possible cladograms. For each character, find the minimum total number of changes that must occur to explain its evolution on each cladogram. You will see, if you do this exercise correctly, that the three cladograms do not differ in minimum numbers of changes required for each of these characters. For this reason, the characters are phylogenetically uninformative by the parsimony criterion.

The parsimony method just illustrated makes some assumptions: that base substitutions are equally likely to occur at any of the 57 sites, that all 3 possible substitutions are equally likely at any site (for example, at site 5 one assumes that C would be equally likely to change to A, G, or T given that a substitution occurs), and that the expected amount of molecular evolution on any branch is proportional to its temporal duration. Suppose that site 16 is unusually subject to mutational change, that a change from C to T is ten times more likely than one from C to A or G, and that the internal branch on the tree is very short compared to the tip branches (those that terminate in a living species whose DNA sequence was determined). The likelihood of getting the observed data for that site by parallel changes C to T in the tip branches of the two *Brookesia* species might equal or exceed the likelihood of a single change C to T in the internal branch of hypothesis 1. The data at site 16 then would not necessarily favor hypothesis 1 over the alternatives. If we can obtain detailed knowledge of the evolutionary properties of the sites in this DNA sequence (for example, by studying its evolution in a large number of lizards), we might favor a phylogenetic method that incorporates a detailed model of DNA-sequence evolution.

Maximum-likelihood and Bayesian approaches to phylogenetic inference are efficient means for using detailed models of molecular evolution to test phylogenetic hypotheses from aligned DNA sequences. First, we analyze the aligned sequences to estimate an evolutionary model: how much do sites differ from each other in their tendencies to vary, and which kinds of substitutions (C to A, C to G, C to T, etc.) are most likely to occur? Second, we evaluate each site with respect to each of the alternative possible trees to determine which tree has the highest likelihood of producing the observed data (such as the CTTC pattern shown for site 16). In this second step, branches of the tree can vary in length, relaxing the assumption that expected amounts of substitution are proportional to the temporal duration of the branch. For our data, the maximum-likelihood method would evaluate the probability of observing the results in each of the 57 columns considering all possible trees, and measure the likelihood of the entire data set for each contrasting tree. The tree with the highest likelihood of producing the observed data is the favored phylogenetic hypothesis. We can reject contrasting trees whose likelihoods are judged much lower than would be expected by chance alone. The Bayesian approaches operate in a similar manner, but they permit an investigator to evaluate the contributions of a new data set relative to prior results. For example, we could evaluate the probability that the data favor hypothesis 1 after incorporating phylogenetic results for these same species obtained from another data set. Calculating likelihoods for contrasting trees is part of the Bayesian operation, and in most phylogenetic applications this is the critical factor in favoring one tree over alternatives. For this reason, results of maximum-likelihood and Bayesian analyses are usually the same, with Bayesian analyses currently more popular because of computational efficiency.

Note that the maximum-likelihood and Bayesian approaches use more of the data set than our parsimony analysis did. To estimate branch lengths (numbers of substitutions occurring on them), sites whose derived states arose on a single terminal branch contribute useful information. Given hypothesis 1, the terminal branch connecting *C. feae* to the tree is longer than the terminal branches connecting the *Brookesia* species to their most recent common ancestor with each other. Using sites for which we identified derived states in the preceding exercise but which were not parsimony informative, we see that the terminal branch leading to *C. feae* requires 7 substitutions (sites 10, 15, 30, 32, 38, 56, 57), whereas the terminal branch leading to *B. brygooi* requires only 3 changes (sites 32, 50, 55), and the terminal branch leading to *B. theili* requires only two changes (sites 12, 14). Using the maximum-likelihood method, these sites collectively would favor hypothesis 1 even though none of the sites is parsimony informative. We thus expect parallel substitutions to occur more frequently in the *C. feae* lineage and one of the two *Brookesia* lineages (as probably occurred at sites 35 and 41) than in both *Brookesia* lineages (a pair of parallel changes on the *Brookesia* lineages would produce a parsimony-informative site favoring hypothesis 1, indistinguishable in our data from sites undergoing one change in the lineage immediately ancestral to the two *Brookesia* species).

The previous two paragraphs should make clear why inferring phylogenies using maximum-likelihood and Bayesian approaches would be very hard to do by hand, even for the data in our example. Finding the optimal tree topologies, branch lengths, relative probabilities of substitution at different sites, and relative probabilities of different kinds of substitution requires testing many alternative conditions and comparing their likelihoods. Computer algorithms can explore this parameter space in an efficient manner, but it would be prohibitively tedious to do by hand. Nonetheless, sharing of derived states at a site as predicted by hypotheses of homology remains the primary reason why the tree that requires only a single evolutionary change to explain the observed variation has a higher likelihood than do trees that require two or more parallel substitutions. Statistical inference of phylogeny using maximum-likelihood and Bayesian methods thus remains grounded in the basic cladistic principles. When parameters of the maximum-likelihood and Bayesian analyses are set to the simple conditions used for parsimony analysis, their results correspond to those that we obtained using parsimony.

Data from Townsend, T., and A. Larson. 2002. Molecular phylogenetics and mitochondrial genomic evolution in the Chamaeleonidae (Reptilia, Squamata). *Molecular Phylogenetics and Evolution* **23**:22–36.

Revision of taxonomy according to cladistic principles can cause confusion. In addition to new taxonomic names, we see old ones used in unfamiliar ways. For example, cladistic use of "bony fishes" includes amphibians and amniotes (including non-avian reptilian groups, birds, and mammals) in addition to finned, aquatic animals that we normally term "fish." Cladistic use of "reptiles" includes birds in addition to snakes, lizards, turtles, and crocodilians; however, it excludes some fossil forms, such as synapsids, that were traditionally placed in Reptilia (see Chapters 26 through 28). Taxonomists must be very careful to specify when using these seemingly familiar terms whether the traditional evolutionary taxa or newer cladistic taxa are being referenced.

10.5 MAJOR DIVISIONS OF LIFE

The oldest phylogenetic events in the history of life have been obscure because the different forms of life share very few characters that can be compared among them to reconstruct phylogeny. More recently, however, a cladistic classification of all life-forms has been proposed based on phylogenetic information obtained from molecular data (the nucleotide base sequence of DNA encoding ribosomal RNA). According to this tree (Figure 10.11), Woese, Kandler, and Wheelis (1990) recognized three monophyletic **domains** above the kingdom level: Eucarya (all eukaryotes), Bacteria (the true bacteria), and Archaea (prokaryotes differing from bacteria in membrane structure and ribosomal RNA sequences). Multicellular animals, plants, and fungi are each traditionally given the Linnaean rank of kingdom. Because each of these taxa is a monophyletic group, continued recognition of these three kingdoms is compatible with phylogenetic systematics. Prior to the 1990s, it was common also to group unicellular eukaryotes as kingdom Protista and all prokaryotes as kingdom Monera in a five-kingdom Linnaean classification. Monera and Protista are paraphyletic groups, and their recognition as taxa is now discontinued. Unicellular eukaryotes were sometimes placed within the animal kingdom as phylum Protozoa, a hypothesis clearly rejected by all current phylogenetic analyses of eukaryotes (Chapter 11). Calling unicellular eukaryotes "protozoans" or "protistans" is archaic and should be resisted. Likewise, "metazoan" is archaic because this term formerly denoted all nonprotozoan members of the animal kingdom. Under current taxonomy, "metazoan" is a redundant synonym for "animal."

10.6 MAJOR SUBDIVISIONS OF THE ANIMAL KINGDOM

All animal species occupy one of 34 currently recognized phyla. Each of these phyla is a monophyletic group. The deepest branches in animal phylogeny combine two or more phyla into more inclusive monophyletic groups. Phylogenetic relationships among the animal phyla are very difficult ones to resolve. Nonetheless, some groupings of phyla are sufficiently repeatable in phylogenetic studies that they receive common usage.

Taxon Bilateria comprises all phyla except Placozoa, Porifera, Cnidaria, and Ctenophora. All bilaterians are bilaterally symmetrical at least in their early postembryonic stages (Chapter 9).

The earliest phylogenetic split within taxon Bilateria separates it into two taxa, Deuterostomia and Protostomia. These taxa differ in the fate of the embryonic blastopore, which becomes the anus in

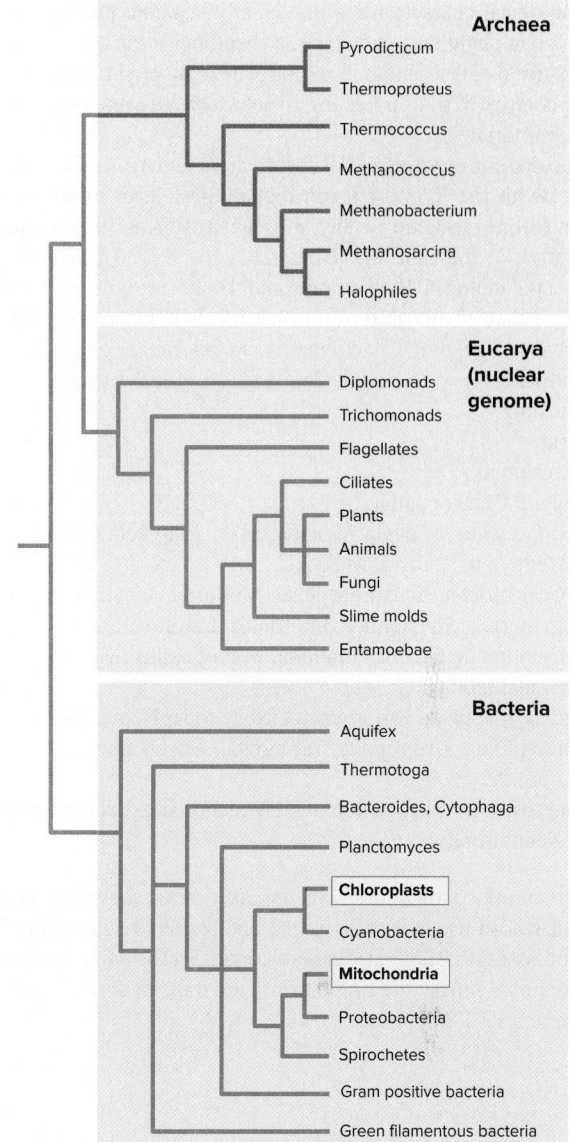

Figure 10.11 Phylogenetic overview of the three domains of life, Archaea, Eucarya, and Bacteria, based on analysis of genes encoding ribosomal RNA. Because of their endosymbiotic origin (p. 31), organellar genomes of domain Eucarya (mitochondria, chloroplasts) are phylogenetically within the Bacteria rather than the clade that includes all eukaryotic nuclear genomes. Organisms of domain Eucarya therefore include cellular components of disparate evolutionary origins. See p. 217 for a more detailed phylogeny of eukaryotes.

Deuterostomia and the mouth in Protostomia (Chapter 8). Deuterostomia comprises phyla Chordata, Echinodermata, Hemichordata, and Xenoturbellida; all remaining bilaterian phyla are in Protostomia.

Molecular phylogenetic studies reveal the earliest phylogenetic split within Protostomia to separate taxon Ecdysozoa from taxon Lophotrochozoa, with phylum Chaetognatha being of uncertain placement (see Chapter 14 for details). The names Ecdysozoa and Lophotrochozoa were chosen following molecular phylogenetic evidence that each one is a monophyletic group in an attempt to find morphological characteristics that clearly separate the two groups. Ecdysozoan phyla feature ecdysis, a shedding of the outer body surface, during development. Lophotrochozoan phyla have either a lophophore (p. 320) as a feeding structure or a trochophore (p. 337) as a larval form.

Taxonomic ordering of Bilateria can be shown using an indented system, with the leftmost taxon including all taxa listed below it farther to the right, up to any subsequently listed taxon equal in indentation to the first one. Bilateria is the leftmost taxon in this listing and includes Protostomia and Deuterostomia as mutually exclusive taxa at the next rightward level of indentation. Within Protostomia, phylum Chaetognatha, taxon Ecdysozoa and taxon Lophotrochozoa appear equivalent to each other at the most rightward nesting.

Bilateria
  Protostomia:
    Phylum Chaetognatha
    Lophotrochozoa: phyla Acoelomorpha, Platyhelminthes, Nemertea, Rotifera, Mesozoa, Gastrotricha, Acanthocephala, Mollusca, Annelida, Echiurida, Sipunculida, Phoronida, Ectoprocta, Entoprocta, Gnathostomulida, Micrognathozoa, Brachiopoda
    Ecdysozoa: phyla Kinorhyncha, Nematoda, Nematomorpha, Priapulida, Arthropoda, Tardigrada, Onychophora, Loricifera
  Deuterostomia: phyla Chordata, Hemichordata, Echinodermata, Xenoturbellida

In summary, there are five unranked taxa above the level of phylum whose usage is widespread and should be familiar to students of zoology: Bilateria, Deuterostomia, Protostomia, Ecdysozoa, and Lophotrochozoa. We present further details of animal taxonomy in Chapters 12–28.

SUMMARY

Animal systematics has three major goals: (1) to identify all species of animals, (2) to evaluate evolutionary relationships among animal species, and (3) to group animal species in a hierarchy of taxonomic groups (taxa) that conveys evolutionary relationships. Taxa are ranked to denote increasing inclusiveness as follows: species, genus, family, order, "class," phylum, and kingdom. All of these ranks can be subdivided to signify taxa that are intermediate between them. Names of species are binomial, with the first name designating the genus to which the species belongs (capitalized) followed by a species epithet (lowercase), both written in italics. Taxa at all other ranks are given single capitalized but nonitalicized names.

The biological species concept has guided the recognition of most animal species. A biological species is defined as a reproductive community of populations (reproductively isolated from others) that occupies a specific niche in nature. A biological species is not immutable through time but changes during the course of evolution. Because the biological species concept may be difficult to apply in spatial and temporal dimensions, and because it excludes asexually reproducing forms, alternative concepts have been proposed. These alternatives include the evolutionary species concept, the cohesion species concept, and the phylogenetic species concept. No single concept of species is universally accepted by all

zoologists, but zoologists agree that a species should constitute a population lineage with a history of evolutionary descent separate from other such lineages. Because species lineages are expected to differ from each other in the DNA sequence of the rapidly evolving mitochondrial gene *COI,* this gene sequence is used as a diagnostic "barcode" to assign specimens to species.

Two major schools of taxonomy are currently active. Evolutionary taxonomy groups species into higher taxa according to the joint criteria of common descent and adaptive evolution; such taxa have a single evolutionary origin and occupy a distinctive adaptive zone. A second approach, called phylogenetic systematics or cladistics, emphasizes common descent exclusively in grouping species into higher taxa. Only monophyletic taxa (those having a single evolutionary origin and containing all descendants of the group's most recent common ancestor) are used in cladistics. In addition to monophyletic taxa, evolutionary taxonomy recognizes some taxa that are paraphyletic (having a single evolutionary origin but excluding some descendants of the most recent common ancestor of the group). Both schools of taxonomy exclude polyphyletic taxa (those having more than one evolutionary origin).

Both evolutionary taxonomy and cladistics require that patterns of common descent among species be assessed before higher taxa are recognized. Comparative morphology (including development), cytology, and biochemistry are used to reconstruct nested hierarchical relationships among taxa that reflect the branching of evolutionary lineages through time. The fossil record provides estimates of the ages of evolutionary lineages. We diagnose clades by identifying shared derived characters, formally called synapomorphies, that distinguish members of the clade from all other taxa. We hypothesize that such synapomorphies represent homologies that arose in the clade's most recent common ancestor. Comparative studies of living species and the fossil record jointly permit us to reconstruct a phylogenetic tree representing the evolutionary history of the animal kingdom. Using a simple example, we illustrate how a systematist infers a phylogenetic tree from aligned DNA sequences using the principles of maximum-parsimony, maximum-likelihood, and Bayesian statistics.

The three most inclusive taxa of living organisms consistent with cladistic taxonomy are domains Archaea, Bacteria, and Eukarya, which includes animals. The most inclusive formal taxon within animals is the phylum, but zoologists commonly use some more inclusive taxa above the phylum level. Bilaterians comprise clades Deuterostomia and Protostomia, distinguished by the fate of the embryonic blastopore to become the anus versus the mouth, respectively. Protostomia comprises clades Ecdysozoa and Lophotrochozoa, diagnosed respectively by the presence of molting versus a trochophore larva or lophophore.

REVIEW QUESTIONS

1. List in order, from most inclusive to least inclusive, the principal categories (taxa) in Linnaean classification as currently applied to animals.
2. Explain why the system for naming species that originated with Linnaeus is "binomial."
3. How does the biological species concept differ from earlier typological concepts of a species? Why do evolutionary biologists prefer it to typological species concepts?
4. What problems have been identified with the biological species concept? How do other species concepts attempt to overcome these problems?
5. How are taxonomic characters recognized? How are such characters used to construct a cladogram?

6. How do monophyletic, paraphyletic, and polyphyletic taxa differ? How do these differences affect the validity of such taxa for both evolutionary and cladistic taxonomies?

7. How many different clades of two or more species are possible for species A–H shown in Figure 10.7 A?

8. What is the difference between a cladogram and a phylogenetic tree? Given a cladogram for a group of species, what additional interpretation is needed to obtain a phylogenetic tree?

9. How would cladists and evolutionary taxonomists differ in their interpretations of the statement that humans evolved from apes, which evolved from monkeys?

10. What taxonomic practices based on the typological species concept are retained in systematics today? How has their interpretation changed?

11. What are the five major unranked taxa above the phylum level in animal taxonomy?

For Further Thought If a taxonomist constructs a rooted phylogenetic tree for a group of living species, the structure of the tree alone can be used to distinguish hypotheses of monophyly versus nonmonophyly of a particular subgroup. If monophyly is rejected for a particular subgroup, tree topology alone cannot distinguish paraphyly from polyphyly. What additional information is needed to distinguish paraphyly from polyphyly?

SELECTED REFERENCES

Avise, J. C. 2006. Evolutionary pathways in nature: a phylogenetic approach. Cambridge, U.K., Cambridge University Press. *A synthetic treatment of phylogenetic knowledge.*

Baum, D. A., and S. D. Smith. 2013. Tree thinking: an introduction to phylogenetic biology. Greenwood Village, CO, Roberts and Company Publishers. *A thorough coverage of phylogenetic principles for beginners.*

de Queiroz, K. 1988. Systematics and the Darwinian revolution. Philosophy of Science **55:** 238–259. *Criticism of evolutionary taxonomy by a phylogenetic systematist, including the contrast between classification and systematization.*

Ereshefsky, M. (ed.). 1992. The units of evolution. Cambridge, Massachusetts, MIT Press. *A thorough coverage of concepts of species, including reprints of important papers on the subject.*

Ereshefsky, M. 2001. The poverty of the Linnaean hierarchy. Cambridge, U.K., Cambridge University Press. *A philosophical critique of Linnaean taxonomy illustrating its problems with cladistic taxonomy.*

Felsenstein, J. 2004. Inferring phylogenies. Sunderland, Massachusetts, Sinauer Associates, Inc. *A thorough coverage of phylogenetic methods.*

Hall, B. K. 1994. Homology: the hierarchical basis of comparative biology. San Diego, Academic Press. *A collection of papers discussing the many dimensions of homology, the central concept of comparative biology, and systematics.*

Hull, D. L. 1988. Science as a process. Chicago, University of Chicago Press. *A study of the working methods and interactions of systematists, containing a thorough review of the principles of evolutionary, phenetic, and cladistic taxonomy.*

Lemey, P., M. Salemi, and A.-M. Vandamme (eds.). 2009. The phylogenetic handbook, ed. 2. Cambridge, U.K., Cambridge University Press. *A thorough and technically detailed summary of how to use standard methods and computer programs to construct phylogenetic trees.*

Maddison, W. P., and D. R. Maddison. 2008. MacClade version 4.08. Sunderland, Massachusetts, Sinauer Associates, Inc. *A computer program for the MacIntosh that conducts phylogenetic analyses of systematic characters. The instruction manual stands alone as an excellent introduction to phylogenetic procedures. The computer program is user-friendly and excellent for instruction in addition to serving as a tool for analyzing real data. This published version of MacClade is superseded by updated versions freely available at macclade.org.*

Swofford, D. 2002. Phylogenetic analysis using parsimony (and other methods) PAUP* version 4. Sunderland, Massachusetts, Sinauer Associates, Inc. *A powerful computer package for constructing phylogenetic trees from data.*

Valentine, J. W. 2004. On the origin of phyla. Chicago, University of Chicago Press. *Phylogenetic analysis of animal phyla in a paleontological perspective.*

Wagner, G. P. (ed.). 2001. The character concept in evolutionary biology. San Diego, Academic Press. *A thorough coverage of evolutionary character concepts.*

Woese, C. R., O. Kandler, and M. L. Wheelis. 1990. Towards a natural system of organisms: proposal for the domains Archaea, Bacteria, and Eucarya. Proceedings of the National Academy of Sciences, USA **87:** 4576–4579. *Proposed cladistic classification for the major taxonomic divisions of life.*

11
CHAPTER

Unicellular Eukaryotes
Protozoan Groups

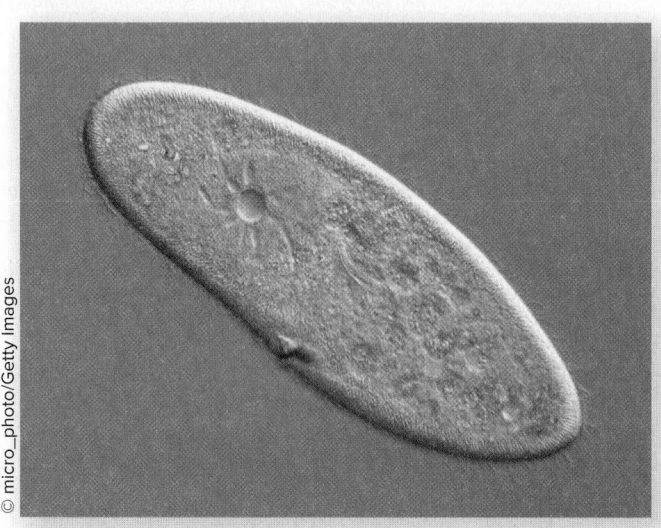

A paramecium.

Emergence of Eukaryotes and a New Life Pattern

The first reasonable evidence for life on earth dates from approximately 3.5 billion years ago. The first cells were prokaryotes whose descendants diversified greatly over an enormous time span. We know them as Bacteria and Archaea. The common ancestor of eukaryotes was formed by merging cells through symbiogenesis. The key element of this merger is that one cell engulfed another in much the same way that an ameba takes in a prey item: by surrounding it with a host plasma membrane. However, in this case, the engulfed organism was not digested; instead the two organisms became interdependent. The engulfed cell maintained its own genetic material, but lost some vital functions and is now an organelle in the host cell. The organelles produced by symbiogenesis include mitochondria and plastids.

A mitochondrion originated from an alpha-proteobacterium capable of deriving energy from carbon compounds using environmental oxygen. An anaerobic host cell that engulfed such an aerobic form gained the capacity to grow in an oxygen-rich environment at a time when cyanobacteria were transforming the world by releasing oxygen as a by-product of photosynthesis. We know the engulfed bacterium as a

mitochondrion. Over evolutionary time, most, but not all, genes from the mitochondrion came to reside in the host cell nucleus. Almost all present-day eukaryotes have mitochondria and are aerobic.

Some eukaryotes have both mitochondria and photosynthetic organelles called plastids. One type, the chloroplast, originated from an engulfed cyanobacterium. When a prokaryote becomes a eukaryotic organelle, we say that the organelle developed via primary endosymbiosis. The chloroplasts in red algae, green algae, and multicellular plants arose this way. However, in some cases, a eukaryotic cell may obtain plastids from another eukaryote. This is secondary endosymbiosis. Two similar cells may have formed very differently, so it is not easy to untangle the evolutionary relationships among the diverse array of unicellular forms.

The assemblage of eukaryotic unicellular organisms was called protozoa. The inclusion of "zoa" in the name referred to two animal-like features: the absence of a cell wall, and the presence of at least one motile stage in the life cycle. However, the unicellular forms are not animals. You will see that the myriad of ways to live as a unicellular organism is fascinating, beguiling, and a little bewildering.

unicellular eukaryote is a complete organism in which all life's activities occur within the limits of a single plasma membrane. Unicellular eukaryotes are found wherever life exists. They are highly adaptable and easily moved from place to place. They require moisture, whether they live in marine or freshwater habitats, soil, decaying organic matter, or plants and animals. They may be sessile or free swimming, and they form a large part of the suspended plankton. The same species are often found widely separated in time as well as in space. Some forms span geological eras exceeding 100 million years.

Despite their wide distribution, many can live successfully only within narrow environmental ranges. Species adaptations vary greatly, and successions of species frequently occur as environmental conditions change.

Unicellular eukaryotes play an enormous role in the economy of nature. Their fantastic numbers are witnessed by the gigantic ocean soil deposits formed over millions of years by their skeletons. About 10,000 species of unicellular eukaryotes are symbiotic in or on animals or plants, sometimes even other unicellular eukaryotes. The relationship may be **mutualistic** (both partners benefit), **commensalistic** (one partner benefits, no effect on the other), or **parasitic** (one partner benefits at the expense of the other), depending on the species involved. Parasitic forms cause some of the most important diseases of humans and domestic animals.

11.1 NAMING AND IDENTIFYING UNICELLULAR EUKARYOTIC TAXA

For many years, all unicellular eukaryotes were placed within a single phylum, but phylogenetic studies showed this group was not monophyletic. Evidence suggests that the origin of the first eukaryote was followed by great diversification, leading some biologists to predict that more than 60 exclusive eukaryotic clades will eventually emerge. The well-supported clade Opisthokonta (Figure 11.1) includes unicellular choanoflagellates, multicellular animals, and fungi, among others (see p. 241). Like the Opisthokonta, the clade Plantae has both unicellular and multicellular members; this group contains red algae, green algae, bryophytes, and vascular plants. The remaining eukaryotic clades contain less well-known organisms only distantly related to plants or animals (Figure 11.1).

The two concepts, plantlike and animal-like, refer in part to the way that food is gathered. Plants are typically **autotrophic,** meaning that they synthesize their own organic constituents from inorganic substrates. Photosynthesis is one kind of autotrophy. Animals are typically **heterotrophic,** meaning that they obtain organic molecules synthesized by other organisms. Heterotrophic unicellular eukaryotes may ingest their food in a soluble form or in a particulate form. Particulate food is acquired by **phagocytosis** (see Chapter 3, pp. 47–48) via an

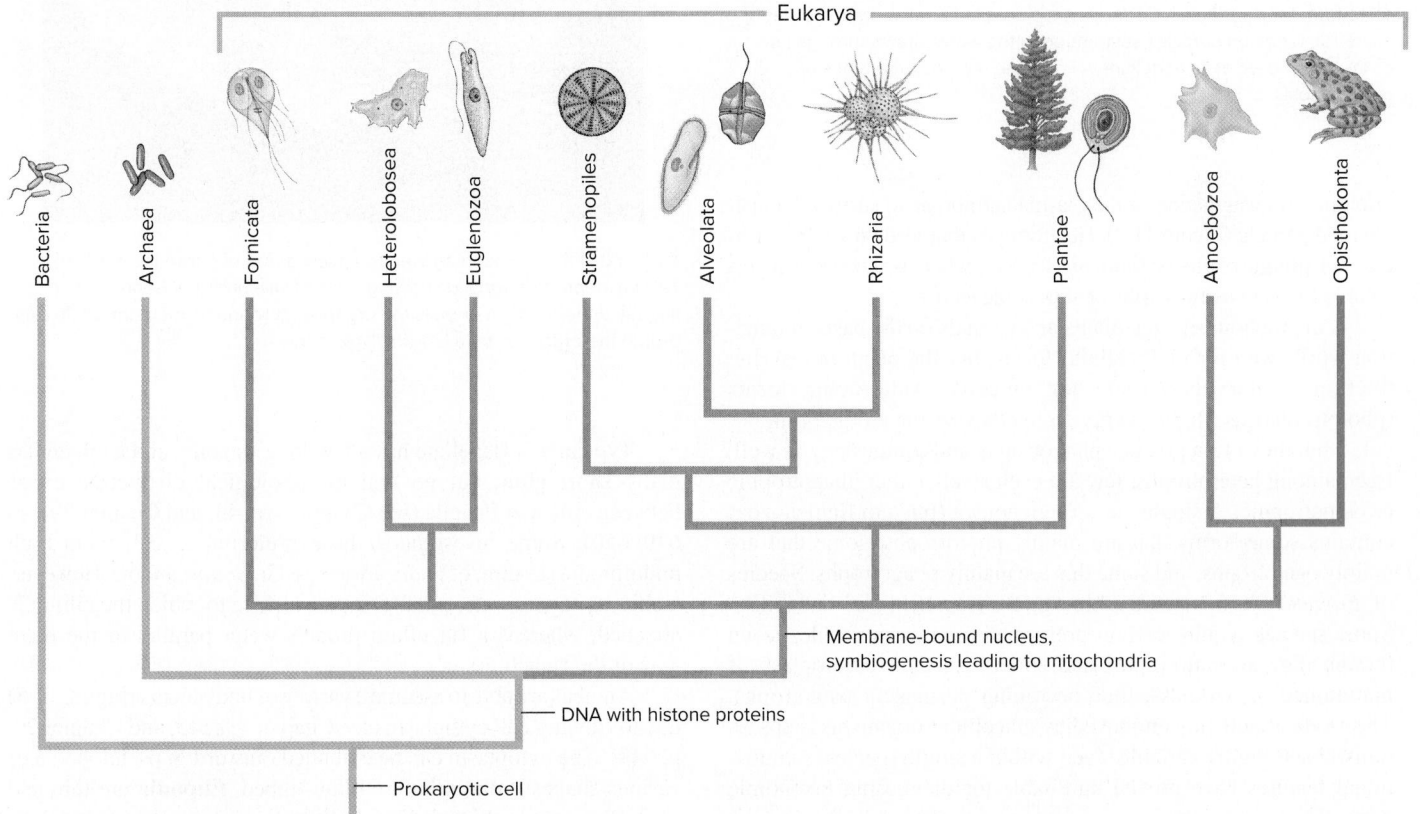

Figure 11.1 Cladogram showing two major prokaryotic branches and diversification of the eukaryotes. Only a few of the eukaryotic clades to be discussed are shown here; more detail is provided in Figure 11.17. The order of branching remains to be determined for most clades. The animals, choanoflagellates, and fungi are within the Opisthokonta.

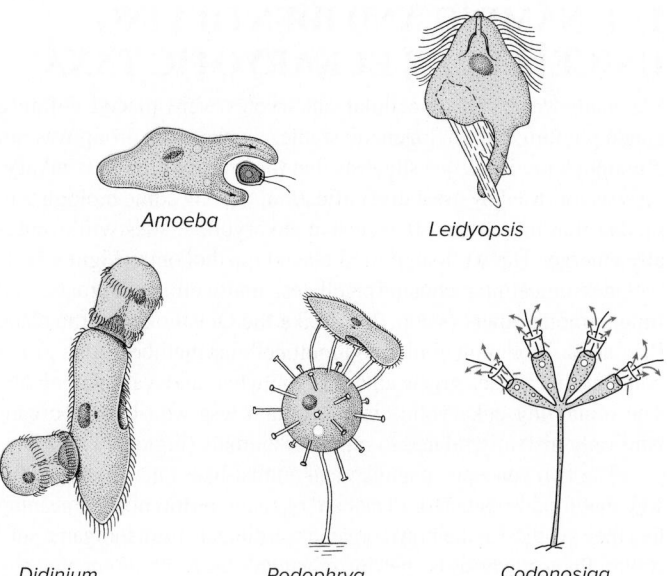

Amoeba

Leidyopsis

Didinium

Podophrya

Codonosiga

Figure 11.2 Some feeding methods among protozoa. *Amoeba* surrounds a small flagellate with pseudopodia. *Leidyopsis,* a flagellate living in the intestine of termites, forms pseudopodia and ingests wood chips. *Didinium,* a ciliate, feeds only on *Paramecium,* which it swallows through a temporary cytostome in its anterior end. Sometimes more than one *Didinium* feed on the same *Paramecium;* two are figured here. *Podophrya* is a suctorian ciliophoran. Its tentacles attach to its prey and suck prey cytoplasm into the body of the *Podophrya,* where it is pinched off to form food vacuoles. *Codonosiga,* a sessile flagellate with a collar of microvilli, feeds on particles suspended in the water drawn through its collar by the beat of its flagellum. All of these methods are types of phagocytosis.

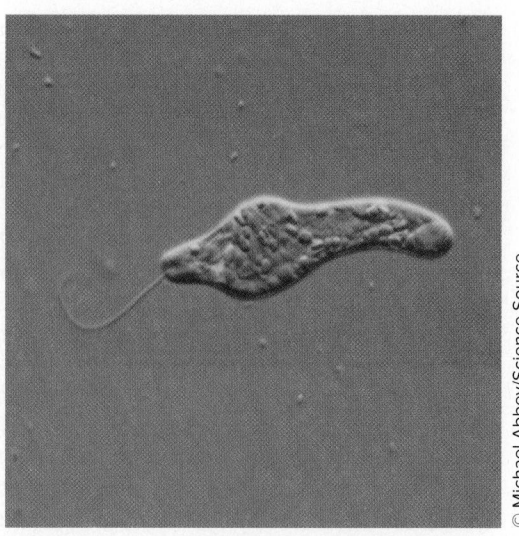

Figure 11.3 One flagellum is clearly visible in the center left of this photograph of *Euglena.*

© Michael Abbey/Science Source

© Aaron J. Bell/Science Source

Figure 11.4 Scanning electron micrograph of a free-living ciliate *Tetrahymena thermophila* showing rows of cilia (×2000). Beating of flagella either pushes or pulls the organism through its medium, while cilia propel the organism by a "rowing" mechanism.

infolding or invagination of the plasma membrane to surround a visible food particle (Figure 11.2). Heterotrophs that feed on visible particles are **phagotrophs** or **holozoic** feeders, whereas those that ingest soluble food are **osmotrophs** or **saprozoic** feeders.

A distinction between plants and animals on the basis of nutrition works well for multicellular forms, but the plant-animal distinction is not appropriate for unicells. Autotrophic forms (phototrophs) use light energy to synthesize their organic molecules, but they often practice phagotrophy and osmotrophy as well. Even among heterotrophs, few are exclusively either phagotrophic or osmotrophic. A single class Euglenoidea (phylum Euglenozoa) contains some forms that are mainly phototrophs, some that are mainly osmotrophs, and some that are mainly phagotrophs. Species of *Euglena* show considerable variety in nutritional capability. Some species require certain preformed organic molecules, even though they are autotrophs, and some lose their chloroplasts if maintained in darkness, thus becoming permanent osmotrophs. The mode of nutrition employed by unicellular organisms is opportunistic and highly variable, even within a single species, so nutritional features have proved unreliable for diagnosing taxonomic groups.

Originally, the means of locomotion was used to distinguish unicellular eukaryotes: flagellates (Figure 11.3) use **flagella,** ciliates (Figure 11.4) travel via a ciliated body surface, and amebas extend their **pseudopodia** (Figure 11.5) to move.

Typically, a flagellate has a few long flagella, and a ciliate has many short **cilia,** but no real morphological distinction exists between cilia and flagella (see Chapter 3, p. 43, and Chapter 29, pp. 649–650). Some investigators have preferred to call them both undulipodia (L. dim. of *unda,* a wave, + Gr. *podos,* a foot). However, a cilium propels water parallel to the surface to which the cilium is attached, whereas a flagellum propels water parallel to the main axis of the flagellum.

Amebas are able to assume a variety of body forms (Figure 11.5) due to flowing cell cytoplasm (see Chapter 3, p. 43, and Chapter 29, p. 649). The cytoplasm can be extended outward in pseudopodia of various shapes: **lobopodia** are blunt-tipped, **filipodia** are thin and sharply pointed, **rhizopodia** are branched filaments, and **reticulopodia** are branched filaments, that merge to form a netlike structure (Figure 11.6). **Axopodia** are thin, pointed pseudopodia that contain a central longitudinal (axial) filament of microtubules called an axoneme (Figure 11.8).

© Eric V. Grave/Science Source

© Eric V. Grave/Science Source

© Eric V. Grave/Science Source

Figure 11.5 Ameboid movement. At top and center, the ameba extends a pseudopodium toward a *paramecium*. At the bottom, the ameba surrounds the *paramecium*, engulfing it by phagocytosis.

on unicellular forms uses the old names, but retaining these old names makes an informed reading of new research impossible. So, we therefore cover both old and new names, making links between the two wherever possible. Some traditional names no longer represent monophyletic groups. Molecular analyses show that the ameboid body form has evolved independently several times, so many amebic taxa were shifted among old groups or moved to newly erected taxa. Amebas are now distributed among eight of the eukaryotic branches that we discuss here. Some group names in common parlance, such as radiolarians or heliozoans, are still in use, but some members of these groups have been moved to other taxa. The classification of unicellular eukaryotes is an area of very active research.

Over 64,000 species have been named, and over half of these are fossils. Some workers estimate there may be 250,000 unicellular eukaryotic species. They are functionally complete organisms with many complicated, microanatomical structures. Particular organelles may perform skeletal, sensory, locomotory, and other functions. These organelles bear closer scrutiny because of their functional importance and because differences in organelle structure can provide homologous characters on which to base taxonomic categories.

Amebas that make shells are called **testate** (Figure 11.6). *Arcella* and *Difflugia* have their delicate plasma membrane covered with a protective **test** or shell of secreted siliceous or chitinoid material that may be reinforced with grains of sand. They move by means of pseudopodia that project from openings in the shell (Figure 11.6). Some very abundant shelled amebas include foraminiferans (*Globigerina,* Figure 11.6), radiolarians (Figure 11.7), and heliozoans. Amebas without shells are called naked amebas.

Relationships among the unicellular eukaryotes are difficult to discern due to the limited number of characters available for morphological analysis and the likelihood of convergent evolution. However, the recent application of molecular analyses, using sequences of bases from multiple genes (see Chapter 10, pp. 210–211, for an example of molecular phylogenetic analysis), has revolutionized our concepts of phylogenetic affinities for all eukaryotes. It has also led to the discovery of many more groups that could be considered phyla.

We now have a far better understanding of the relationships among unicellular or microbial eukaryotes, but these relationships can be difficult to convey because of the proliferation of new clade names arising from each new analysis. Some scientific literature

Characteristics of Unicellular Eukaryotes

1. **Unicellular;** some colonial, and some with multicellular stages in their life cycles. Coloniality occurs when cells divide, but remain together. The cells may specialize on particular functions, but all cells are capable of reproduction. Multicellularity requires the conditions of coloniality plus the additional criterion that only some cells are capable of reproduction.
2. **Mostly microscopic,** although some are large enough to be seen with the unaided eye
3. All symmetries represented in the group; shape variable or constant (oval, spherical, or other)
4. **No germ layer present**
5. No organs or tissues, but **specialized organelles** are found; nucleus single or multiple
6. Free-living, mutualism, commensalism, parasitism all represented in the groups
7. Locomotion by **pseudopodia, flagella, cilia,** and direct cell movements; some sessile
8. Some provided with a **simple endoskeleton** or **exoskeleton,** but most are naked
9. **Nutrition of all types:** autotrophic (manufacturing own nutrients by photosynthesis), heterotrophic (depending on other plants or animals for food), saprozoic (using nutrients dissolved in the surrounding medium)
10. Aquatic or terrestrial habitat; free-living or symbiotic mode of life
11. Reproduction **asexually** by fission, budding, and cysts and **sexually** by conjugation or by syngamy (union of male and female gametes to form a zygote)

Difflugia

Amoeba

Arcella

Clathrulina

Actinophrys

Globigerina

Chlamydophrys

Actinosphaerium

Figure 11.6 The amebas are a very diverse group distributed across many clades: *Arcella, Difflugia,* and *Amoeba* have lobopodia and are amoebozoans, *Globigerina* has a testate body with netlike reticulopodia and is a foraminiferan, and *Chlamydophrys* and *Clathrulina* are cercozoans, whereas *Actinosphaerium* and *Actinophrys* are now stramenopiles.

© Alfred Pasieka/Science Source

Figure 11.7 Some amebas, like those whose shells are shown here, are commonly called radiolarians.

11.2 FORM AND FUNCTION

Locomotion

Cilia and Flagella

A cilium or flagellum has considerable internal structure. Each flagellum or cilium contains nine pairs of longitudinal microtubules arranged in a circle around a central pair (Figure 11.8), and this is true for all motile flagella and cilia in the animal kingdom, with a few notable exceptions. This "9 + 2" tube of microtubules in a flagellum or cilium is its **axoneme;** an axoneme is covered by a

Figure 11.8 **A,** A flagellum illustrating the central axoneme, which is composed of nine pairs of microtubules plus a central pair. The axoneme is enclosed within the plasma membrane. The central pair of microtubules ends near the level of the cell surface in a basal plate (axosome). The peripheral microtubules continue inward for a short distance to compose two of each of the triplets in the kinetosome (basal body) (at level y in **A**). **B,** Electron micrograph of a section through several cilia, corresponding to level x in **A**. (×133,000)

membrane continuous with the plasma membrane covering the rest of the organism. At about the point where an axoneme enters the cell proper, the central pair of microtubules ends at a small plate within the circle of nine pairs (Figure 11.8A). Also at about that point, another microtubule joins each of the nine pairs, so that these form a short tube extending from the base of the flagellum into the cell. The tube consists of nine *triplets* of microtubules and is called a **kinetosome** (or **basal body**). Kinetosomes are identical in structure to **centrioles** that organize mitotic spindles during cell division (see Figure 3.14, p. 44). Centrioles of some flagellates may give rise to kinetosomes, or kinetosomes may function as centrioles. All typical eukaryotic flagella and cilia have a kinetosome at their base. Many small metazoans use cilia not only for locomotion but also to create water currents for their feeding and respiration. Ciliary movement is vital to many species in such functions as handling food, reproduction, excretion, and osmoregulation (as in flame cells, p. 296).

The current explanation for ciliary and flagellar movement is the **sliding-microtubule hypothesis.** The movement is powered by a release of chemical bond energy in ATP (p. 60). Two "arms" of dynein proteins are visible in electron micrographs on each of the pairs of peripheral microtubules in the axoneme (level *x* in Figure 11.8), and these bear the enzyme adenosine triphosphatase (ATPase), which cleaves the ATP. When bond energy in ATP is released, the dynein proteins "walk along" one of the microtubules in the adjacent pair, causing it to slide relative to the other microtubule in the pair. Shear resistance, causing the axoneme to bend when the filaments slide past each other, is provided by radial "spokes" of another microtubule-associated protein from each doublet to the central pair of fibrils. These spokes are visible in electron micrographs and are composed of multiple proteins. Direct evidence for the sliding microtubule hypothesis was obtained by attaching tiny gold beads to axonemal microtubules and observing their movement microscopically.

Pseudopodia

Pseudopodia are extensions of the cell cytoplasm used in locomotion and feeding. (Figure 11.9). The cytoplasm is not homogeneous; sometimes peripheral and central areas of cytoplasm can be distinguished as **ectoplasm** and **endoplasm** (Figure 11.9). Endoplasm appears more granular and contains the nucleus and cytoplasmic organelles. Ectoplasm appears more transparent (hyaline) by light microscopy, and it bears the bases of the cilia or flagella. Ectoplasm is often more rigid and is in the gel state of a colloid, whereas the more fluid endoplasm is in the **sol** state (see box below).

Colloidal systems are permanent suspensions of finely divided particles that do not precipitate, such as milk, blood, starch, soap, ink, and gelatin. Colloids in living systems are commonly proteins, lipids, and polysaccharides suspended in the watery fluid of cells (cytoplasm). Such systems may undergo sol-gel transformations, depending on whether the fluid or particulate components become continuous. In the sol state of cytoplasm, solids are suspended in a liquid, and in the semisolid gel state, liquid is suspended in a solid.

Pseudopodia vary in composition and are of several types. The most familiar are **lobopodia** (Figure 11.6 and 11.9), which are rather large, blunt extensions of the cell body containing both endoplasm and ectoplasm. **Filopodia** are thin extensions, usually branching, and containing only ectoplasm. They occur in some amebas, such as *Euglypha* (see Figure 11.16B). **Reticulopodia** (see Figure 11.6) are distinguished from filopodia in that reticulopodia repeatedly rejoin to form a netlike mesh, although some biologists consider the distinction between filopodia and reticulopodia artificial. **Axopodia** (Figure 11.10) are long, thin pseudopodia supported by axial rods of microtubules (Figure 11.10). The microtubules are arranged in a definite spiral or geometrical array, depending on the species, and constitute the axoneme of the axopod. Axopodia can be extended or retracted, apparently by addition or removal of microtubular material. Since the tips can adhere to the substrate, the organism can progress by a rolling motion, shortening the axonemes in front and extending those in the rear. Cytoplasm can flow along the axonemes, toward the body on one side and in the reverse direction on the other.

When a typical lobopodium begins to form, an extension of ectoplasm called a **hyaline** cap appears, and endoplasm begins to flow toward and into the hyaline cap (Figure 11.9 and 11.11). The flowing endoplasm contains actin subunits attached to regulatory, actin-binding proteins (ABPs) that prevent actin from polymerizing. As endoplasm flows into the hyaline cap, it spreads to the periphery. Interaction with phospholipids in the plasma membrane releases the actin subunits from their regulatory binding proteins and allows them to polymerize into actin filaments. The actin filaments become cross-linked to each other by another ABP to form a semisolid gel, transforming the ectoplasm into a tube through which the fluid endoplasm flows as the pseudopodium extends. Near the trailing edge of the gel, calcium ions activate an ABP that releases actin filaments from the gel and permits myosin to associate with and to pull these actin filaments, much in the same way skeletal muscle contraction occurs in vertebrates (see Chapter 29, p. 654). Thus contraction at the trailing edge creates a pressure that forces the fluid endoplasm, along with its now-dissociated actin subunits, back toward the hyaline cap.

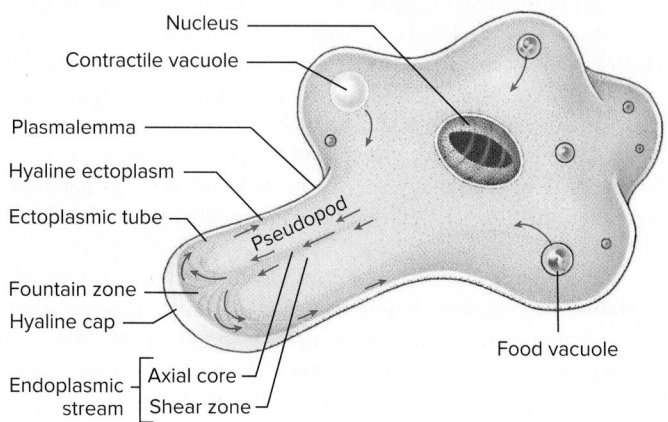

Figure 11.9 Ameba in active locomotion. Arrows indicate the direction of streaming endoplasm. The first sign of a new pseudopodium is thickening of the ectoplasm to form a clear hyaline cap, into which the fluid endoplasm flows. As the endoplasm reaches the forward tip, it fountains out and is converted into ectoplasm, forming a stiff outer tube that lengthens as the forward flow continues. Posteriorly the ectoplasm is converted into fluid endoplasm, replenishing the flow. Substratum is necessary for ameboid movement.

Figure 11.10 **A,** Diagram of an axopodium to show orientation of **B.** The axoneme of an axopodium is composed of an array of microtubules, which may vary from three to many in number depending on the species. Some species can extend or retract their axopodia quite rapidly. **B,** Electron micrograph of axopodium (from *Actinosphaerium nucleofilum*) in cross section (×99,000).

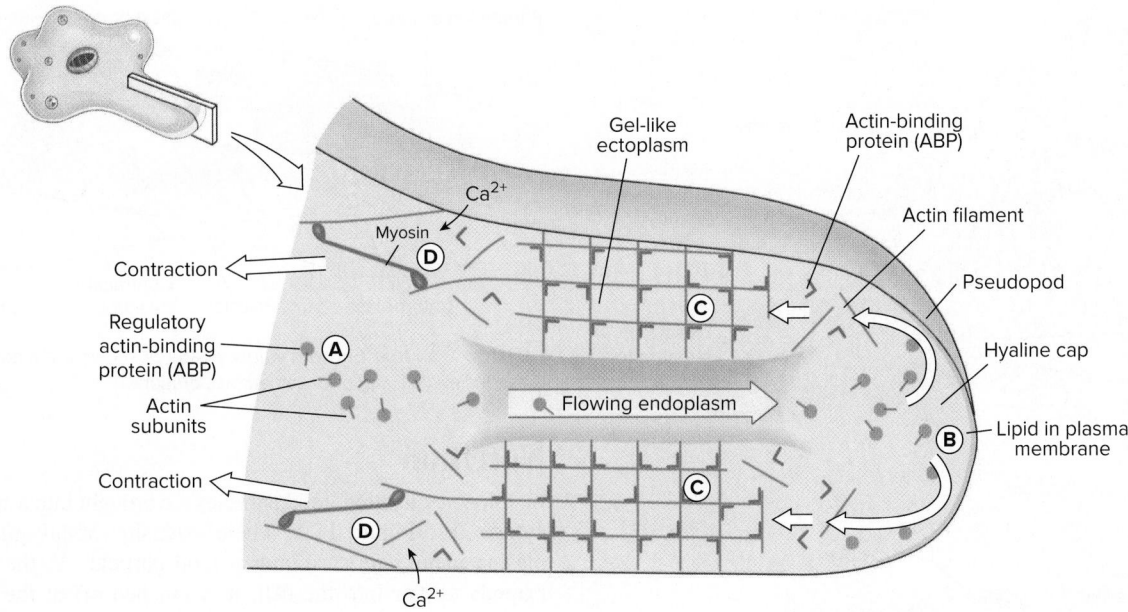

Figure 11.11 Mechanism of pseudopodial movement. In endoplasm, actin subunits are bound to regulatory actin-binding proteins that keep them from assembling **(A).** Upon stimulation, hydrostatic force carries the subunits through a weakened gel to the hyaline cap. The actin subunits are freed from the regulatory proteins by lipids in the plasma membrane **(B).** Subunits quickly assemble into filaments and, upon interaction with actin-binding protein (ABP), form gel-like ectoplasm **(C).** At the trailing edge, calcium ions activate an ABP that releases actin filaments from the gel, loosening the network enough that myosin molecules can pull on it **(D).** Subunits pass through the tube of ectoplasm to be reused.

Pseudopodia occur in free-living amebas, in a variety of flagellated unicellular eukaryotes, and in ameboid cells of many animals. The mammalian body depends on ameboid white blood cells for defense against disease, and similar roles are played by ameboid cells in many other animals. However, movement by pseudopodia occurs over relatively short distances, so to travel longer distances, the ameboid stages of one free-living slime mold aggregate to form a large slug-like body (a limax form) that glides across surfaces. This transition to a multicellular body is quite unusual.

Functional Components of Cells of Unicellular Eukaryotes

Nucleus

As in other eukaryotes, the nucleus is a membrane-bound structure whose interior communicates with the cytoplasm by small pores. Within the nucleus the genetic material (DNA) is borne on chromosomes (see Chapter 3, p. 39, for detailed information on nuclear structure). Except during cell division, chromosomes are not usually condensed in a form that can be distinguished, although during fixation of the cells for light microscopy, chromosomal material (chromatin) often clumps together irregularly, leaving some areas within the nucleus relatively clear. This appearance is described as **vesicular** and is characteristic of the nuclei of many unicellular eukaryotes (Figure 11.12). Condensations of chromatin may be distributed around the periphery of the nucleus or internally in distinct patterns. In most dinoflagellates (p. 233) chromosomes are visible through interphase as they would appear during prophase of mitosis (see Chapter 3, p. 50).

Also within the nucleus, one or more **nucleoli** representing active transcription of ribosomal RNA are often present (Figure 11.12). Characters such as the persistence of nucleoli during mitosis are useful in identifying protozoan clades.

Ciliates possess two kinds of nuclei: a germ line nucleus called the micronucleus and a somatic nucleus called the macronucleus. Only genes of the macronucleus are transcribed (see Chapter 5, p. 88). Macronuclei of ciliates are described as compact or condensed because the chromatin material is more finely dispersed and clear areas cannot be observed with the light microscope.

Mitochondria

A mitochondrion is an organelle used in recovering energy from carbon bonds of fuel molecules where oxygen serves as the terminal electron acceptor (see p. 64). It contains DNA. Cristae, the internal membranes of a mitochondrion (Figure 11.12), are of variable form, being flat, tubular, discoid, or branched (ramifying). The form of cristae is considered a homologous character and, in conjunction with other morphological features, is used to describe particular taxa. In cells without mitochondria, **hydrogenosomes** may be present. Hydrogenosomes are organelles that perform a respiratory function in the absence of oxygen and are assumed to have evolved from mitochondria. Hydrogenosomes and other mitochondrial derivatives that lack DNA are collectively called mitosomes. **Kinetoplasts** are also assumed to be mitochondrial derivatives. They contain masses of circular DNA molecules and work in association with a kinetosome, an organelle at the base of a flagellum.

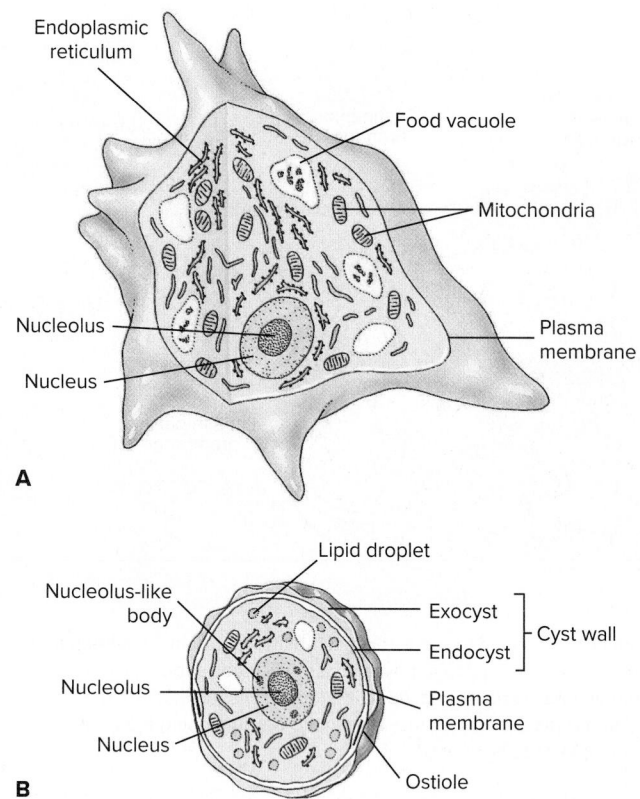

A

B

Figure 11.12   Structure of *Acanthamoeba palestinensis*. **A,** Active, feeding form. **B,** Cyst.

Golgi Complex

The Golgi complex is part of the endomembrane system that participates in cellular secretory processes and intracellular digestion of ingested food vacuoles (see Figure 3.10 and 3.20, p. 42 and p. 49 respectively). Golgi complexes are also called **dictyosomes.** Parabasal bodies are similar structures with potentially similar functions.

Plastids

Plastids are organelles containing a variety of photosynthetic pigments. The original addition of a plastid to eukaryotic cells likely occurred when a cyanobacterium was engulfed and not digested. **Chloroplasts** (Figure 11.13) contain different versions of chlorophylls (*a, b,* or *c*), but other kinds of plastids contain other pigments. For example, red algal plastids contain phycobilins. Particular pigments shared among unicellular eukaryotes may indicate shared ancestry, but plastids also could have been gained by secondary endosymbiosis.

Extrusomes

This general term refers to membrane-bound organelles in unicellular eukaryotes that are used to extrude something from the cell. The wide variety of structures extruded suggests that not all extrusomes are homologous. The ciliate **trichocyst** (p. 231) is an extrusome.

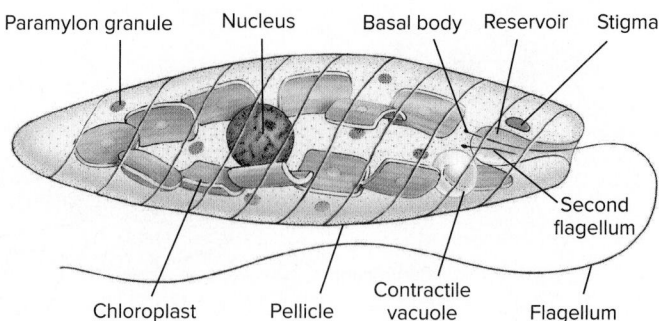

Figure 11.13   *Euglena viridis.* Features shown are a combination of those visible in living and stained preparations.

Nutrition

In holozoic nutrition, food particles are brought into a cell by phagocytosis (see Figure 11.5) where extension or invagination of the plasma membrane surrounds a food particle. As the invagination extends farther into the cell, it is pinched off at the surface (see Figure 3.20). The food particle thus is contained in an intracellular, membrane-bound vesicle, a **food vacuole** or **phagosome.** Lysosomes, small vesicles containing digestive enzymes, fuse with the phagosome and pour their contents into it, where digestion begins (see Figure 3.20). As digested products are absorbed across the vacuolar membrane, the phagosome becomes smaller. Any undigestible material may be released to the outside by exocytosis, the vacuole again fusing with the cell-surface membrane. In most ciliates, many flagellates, and many apicomplexans, the site of phagocytosis is a definite mouth structure, the **cytostome** (Figure 11.14). In amebas, phagocytosis can occur at almost any point by envelopment of a particle with pseudopodia. Particles must be ingested through the opening of the test, or shell, in amebas that have tests. Flagellates may form a temporary cytostome, usually in a characteristic position, or they may have a permanent cytostome with specialized structure. Many ciliates have a characteristic structure for expulsion of waste matter, the **cytopyge** or **cytoproct,** found in a characteristic location. In some, the cytopyge also serves as the site for expulsion of the contents of the contractile vacuole. (See next section on excretion and osmoregulation.)

Uptake of dissolved salts or nutrients, called saprozoic feeding, may be by pinocytosis or by transport of solutes directly across the outer plasma membrane (see Chapter 3, p. 49). Direct transport across a membrane may be by diffusion, facilitated transport, or active transport (see Chapter 3, pp. 46–48). Diffusion is probably of little or no importance in nutrition of unicells, except possibly in some endosymbiotic species. Some important food molecules, such as glucose and amino acids, can enter a cell by facilitated diffusion and active transport.

It has been shown that a stimulatory substance, or "inducer," must be present in the surrounding medium for many unicellular eukaryotes to initiate pinocytosis. Several proteins act as inducers, as can some salts and other substances; it appears that the inducer must be a positively charged molecule. Pinocytosis takes place at the inner end of the cytopharynx in unicells possessing that structure.

Figure 11.14 *Left,* enlarged section of a contractile vacuole (water expulsion vesicle) of *Paramecium.* Water is apparently collected by the endomembrane system, emptied into feeder canals and then into the vesicle. The vesicle contracts to empty its contents to the outside, thus serving as an osmoregulatory organelle. *Right, Paramecium,* showing cytopharynx, food vacuoles, and nuclei.

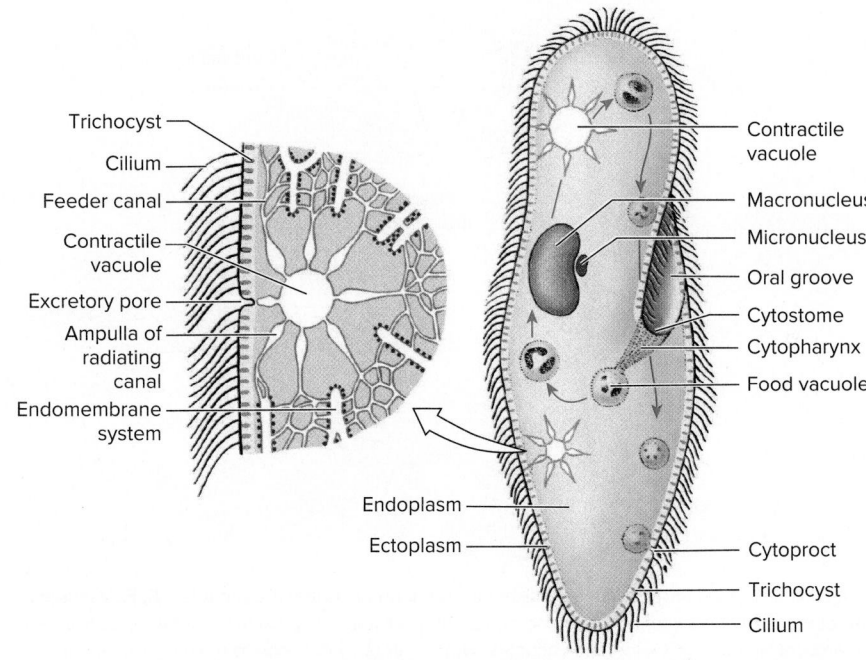

Trichocyst
Cilium
Feeder canal
Contractile vacuole
Excretory pore
Ampulla of radiating canal
Endomembrane system

Contractile vacuole
Macronucleus
Micronucleus
Oral groove
Cytostome
Cytopharynx
Food vacuole

Endoplasm
Ectoplasm

Cytoproct
Trichocyst
Cilium

Excretion and Osmoregulation

Vacuoles can be seen by light microscopy in the cytoplasm of many unicellular eukaryotes. Some of these vacuoles periodically fill with a fluid substance that is then expelled. Evidence is strong that these **contractile vacuoles** (Figures 11.9, 11.13, and 11.14) function principally in osmoregulation. They are more prevalent and fill and empty more frequently in freshwater unicellular eukaryotes than in marine and endosymbiotic species, where their surrounding medium would be more nearly isoosmotic (having the same osmotic pressure) to their cytoplasm. Smaller species, which have a greater surface-to-volume ratio, generally have more rapid filling and expulsion rates in their contractile vacuoles. Excretion of metabolic wastes, on the other hand, is almost entirely by diffusion. The main end product of nitrogen metabolism is ammonia, which readily diffuses from the small bodies of protozoa.

Although it seems clear that contractile vacuoles remove excess water that has entered cytoplasm by osmosis, a reasonable mechanism for such removal has been elusive. One hypothesis suggests that proton transporters (p. 64) on the vacuolar surface and on tubules radiating from it actively transport H^+ and cotransport bicarbonate (HCO_3-) ions (Figure 11.15), which are osmotically active particles. As these particles accumulated within a vacuole, water would be drawn into the vacuole. Fluid within the vacuole would remain isoosmotic to the cytoplasm. Then as the vacuole finally joined its membrane to the surface membrane and emptied its contents to the outside, it would expel water, H^+, and HCO_3-. These ions can be replaced readily in amebas by action of cytoplasmic carbonic anhydrase on CO_2 and H_2O.

Some ciliates, such as *Blepharisma,* have contractile vacuoles with structure and filling mechanisms apparently similar to those described for amebas. Others, such as *Paramecium,* have more complex contractile vacuoles. Such vacuoles occupy a specific position beneath the plasma membrane, with an "excretory" pore leading to the outside, and surrounded by ampullae of about six feeder canals (Figure 11.14).

Feeder canals, in turn, are surrounded by fine tubules about 20 nm in diameter, which connect with the canals during filling of ampullae and at their lower ends connect with the tubular endomembrane system. Ampullae and contractile vacuoles are surrounded by bundles of fibrils, which may function in contraction of these structures. Contraction of ampullae fills the vacuole. When the vacuole contracts to discharge its contents to the outside, the ampullae become disconnected from the vacuole, so that backflow is prevented. Tubules, ampullae, or vacuoles may be supplied with proton transporters to draw water into their lumens by the mechanism already described.

Reproduction

Sexual reproduction does occur in most unicellular eukaryotes. Diploid cells undergo reduction division (meiosis) to make haploid gametes or simply haploid gamete nuclei (p. 73). The haploid gametes or nuclei fuse to make a diploid stage or a zygote, but there are no embryos. The life cycles vary greatly in how much time is spent in haploid and diploid stages, and in the presence or absence of asexual multiplication.

Fission

The cell multiplication process that produces more individuals in unicellular eukaryotes is called fission. The most common type of fission is **binary,** in which two essentially identical individuals result (Figure 11.16). When a progeny cell is considerably smaller than the parent and then grows to adult size, the process is called **budding.** Budding occurs in some ciliates. In **multiple fission,** division of the cytoplasm (cytokinesis) is preceded by several nuclear divisions, so that a number of individuals are produced almost simultaneously (see Figure 11.29). Multiple fission, or **schizogony,** is common among the Apicomplexa and some amebas. If the multiple fission is preceded by or associated with union of gametes, it is called **sporogony.**

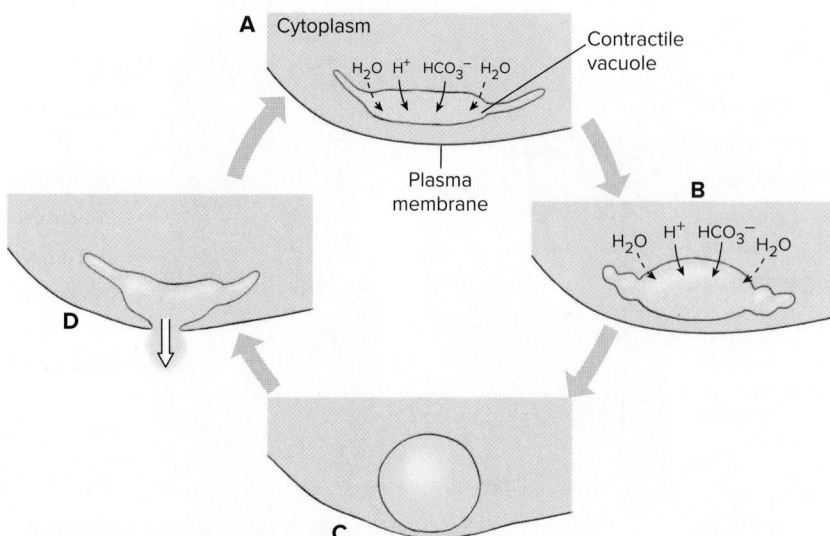

Figure 11.15   Proposed mechanism for operation of contractile vacuoles. **A, B,** Vacuoles are composed of a system of cisternae and tubules. Proton pumps in their membranes transport H^+ and cotransport HCO_3 into the vacuoles. Water enters by diffusion to maintain an osmotic pressure equal to that in the cytoplasm. When the vacuole fills, **C,** its membrane fuses with the cell's surface membrane, expelling water, H^+, and HCO_3 **D,** Protons and bicarbonate ions are replaced readily by action of carbonic anhydrase on carbon dioxide and water.

Fertilization of an individual gamete by another is **syngamy,** but not all sexual phenomena involve syngamy. Examples are **autogamy,** in which gametic nuclei arise by meiosis and fuse to form a zygote within the same organism that produced them, and **conjugation,** in which an exchange of gametic nuclei occurs between paired organisms (conjugants). We describe conjugation further in the discussion of *Paramecium.*

Encystment and Excystment

Although separated from their external environment only by their delicate plasma membrane, unicellular forms are amazingly successful in habitats frequently subjected to extremely harsh conditions. Survival under harsh conditions surely is related to the ability to form **cysts,** dormant forms marked by resistant external coverings and a complete shutdown of metabolic machinery. Cyst formation is important also to many parasitic forms that must survive a harsh environment between hosts (see Figure 11.12). However, some parasites do not form cysts,

The foregoing types of division are accompanied by some form of mitosis (p. 49). However, this mitosis is often somewhat unlike that found in animals. For example, the nuclear membrane often persists through mitosis, and the microtubular spindle may be formed within the nuclear membrane. Centrioles have not been observed in nuclear division of ciliates; the nuclear membrane persists in micronuclear mitosis, with the spindle within the nucleus. The macronucleus of ciliates seems simply to elongate, to constrict, and to divide without any recognizable mitotic phenomena; thus this division process is amitotic.

Sexual Processes

Although all unicellular eukaryotes reproduce asexually, and some are apparently exclusively asexual, the widespread occurrence of sex testifies to its importance as a means of genetic recombination. Gamete nuclei, or pronuclei, which fuse in fertilization to restore the diploid number of chromosomes, are usually borne in special gametic cells. When gametes all look alike, they are called **isogametes,** but most species have two dissimilar types, or **anisogametes.**

In animals meiosis usually occurs during or just before gamete formation (meiosis is discussed in detail in Chapter 5, pp. 72–75). The same process occurs in some unicellular eukaryotes such as Ciliophora and some flagellated and amebic groups. However, in other flagellated groups and in Apicomplexa, the first divisions *after* fertilization are meiotic **(zygotic meiosis),** and all individuals produced asexually (mitotically) in the life cycle up to the next zygote are haploid. Most unicellular eukaryotes that do not reproduce sexually probably are haploid, although demonstration of ploidy is difficult in the absence of meiosis. In some amebas (foraminiferans), haploid and diploid generations alternate **(intermediary meiosis),** a phenomenon widespread among plants.

Figure 11.16   Binary fission in some amebas and flagellates. **A,** The two nuclei of *Arcella* divide as some of its cytoplasm is extruded and begins to secrete a new test for the daughter cell. **B,** The test of another ameba, *Euglypha,* is constructed of secreted platelets. Secretion of platelets for the daughter cell is begun before cytoplasm begins to move out of the aperture. As these are used to construct the test of the daughter cell, the nucleus divides. **C,** *Trypanosoma* has a kinetoplast near the kinetosome of its flagellum close to its posterior end in the stage shown. All of these parts must be replicated before the cell divides. **D,** Division of *Euglena.* Compare C and D with Figure 11.25, fission in a ciliophoran.

apparently depending on direct transfer from one host to another. Reproductive phases such as fission, budding, and syngamy may occur in cysts of some species. Encystment has not been found in *Paramecium,* and it is rare or absent in marine forms.

> Cysts of some soil-inhabiting and freshwater protozoa have amazing durability. Cysts of the soil ciliate *Colpoda* can survive 12 days in liquid nitrogen and 3 hours at 100° C. Survival of *Colpoda* cysts in dried soil has been shown for up to 38 years, and those of a certain small flagellate (*Podo*) can survive up to 49 years! Not all cysts are so sturdy, however. Those of *Entamoeba histolytica,* which causes amebic dysentery, tolerate gastric acidity but not desiccation, temperature above 50° C, or sunlight.

Conditions stimulating encystment are incompletely understood, although in some cases cyst formation is cyclic, occurring at a certain stage in the life cycle. In most free-living forms, adverse environmental change favors encystment. Such conditions include food deficiency, desiccation, increased environmental osmotic pressure, decreased oxygen concentration, and change in pH or temperature.

During encystment some organelles, such as cilia or flagella, are resorbed, and the Golgi complex secretes cyst wall material, which is carried to the surface in vesicles and extruded.

Although the exact stimulus for excystation (escape from cysts) is usually unknown, a return of favorable conditions initiates excystment for those unicellular eukaryotes in which the cysts are a resistant stage. In parasitic forms the excystment stimulus may be more specific, requiring conditions similar to those found in the host.

11.3 MAJOR UNICELLULAR EUKARYOTIC TAXA

Evolution of a eukaryotic cell was followed by diversification into many clades (Figure 11.17), some of which contain both unicellular and multicellular forms. Clades of this type include the Opisthokonta, the Stramenopiles, the Viridiplantae, and the red algal clade, traditionally the phylum Rhodophyta. Rhodophytans have plastids, are not heterotrophic, and lack flagellated stages (motile sperm) in the life cycle; these characteristics identify them as part of the plant clade (Figure 11.17). We discuss Viridiplantae and Opisthokonta, but not Rhodophyta. We discuss these clades in

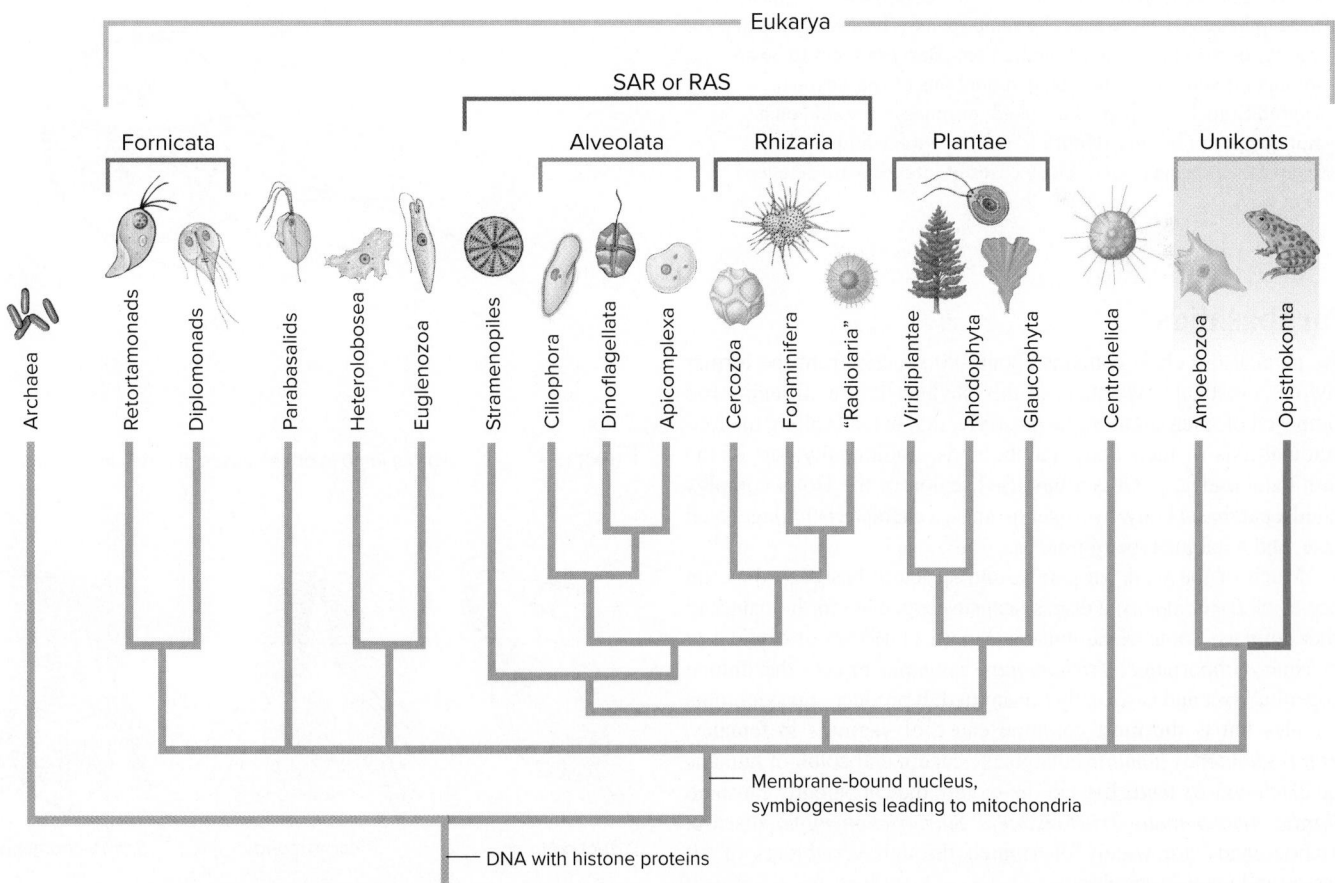

Figure 11.17 Cladogram showing the major eukaryotic clades; in many cases the order of branching remains to be determined. Opisthokonta is a very large clade comprising choanoflagellates, fungi, and multicellular animals. The terms "SAR" and "RAS" are interchangeable; SAR represents the first letter of the names Stramenopiles, Alveolates, and Rhizaria, whereas RAS represents the same taxa listed in the reverse order.

the order in which they appear in Figures 11.1 and 11.17, moving from left to right.

Phylum Retortamonada and the Diplomonads

This phylum is divided into two exclusive clades: retortamonads and diplomonads. Retortamonads include commensal and parasitic unicells, such as *Chilomastix* and *Retortamonas.* They lack mitochondria and Golgi bodies, so biologists wondered whether their ancestors separated from the eukaryotic lineage before the mitochondrial symbiosis evolved. Diplomonads, once a subgroup of retortamonads, also lack mitochondria, and were likewise proposed to lie outside the clade of mitochondria-containing eukaryotes. However, work showing that mitochondrial genes occur in the cell nucleus[1] makes it much more likely that the absence of mitochondria is a secondary loss, not a primary absence.

Giardia, a diplomonad, is a well-studied parasite (Figure 11.18). Some species inhabit the human digestive tract, but others occur in birds or amphibians. It is often asymptomatic but may cause a rather discomfiting, but not fatal, diarrhea. Cysts are passed in the feces, and new hosts are infected by ingestion of cysts, often in contaminated water.

> *Giardia lamblia* is commonly transmitted through water supplies contaminated with sewage. The same species, however, lives in a variety of mammals other than humans. Beavers seem to be an important source of infection in mountains of the western United States. When one has hiked for miles in the wild on a hot day, it can be very tempting to fill a canteen and drink from a clear beaver pond. Many cases of infection are acquired that way.

Parabasalids

The parabasalid clade contains about 400 species from the former phylum Axostylata. Members of this phylum have a stiffening rod composed of microtubules, the **axostyle,** that extends along the longitudinal axis of their body. Parabasalids, traditionally part of the class Parabasalea, possess a modified region of the Golgi complex called a parabasal body, hydrogenosomes, a characteristic flagellated stage, and a unique type of mitosis.

Much of the work on parabasalid structure has been done on species of *Trichomonas,* a disease-causing organism for humans and other animals. Some trichomonads (Figure 11.19) are of medical or veterinary importance. *Trichomonas vaginalis* infects the human urogenital tract and is sexually transmitted. It produces no symptoms in males but is the most common cause of vaginitis in females. *Pentatrichomonas hominis* inhabits the cecum and colon of humans and *Trichomonas tenax* lives in the mouth; they apparently cause no *Giardia Trichomonas Trichonympha Spirotrichonympha* disease. Trichomonads are widely distributed through vertebrates of all classes and many invertebrates.

Heterolobosea

Heteroloboseans are naked amebas whose pseudopodia form abruptly, in what is sometimes called an "eruptive" manner. The life cycle of many heteroloboseans includes both amebic and flagellated stages, so group members are sometimes called amoeboflagellates or schizopyrenids. In *Naegleria gruberi,* the amebic stage feeds on bacteria, but once all local food is exhausted, the ameba completely alters its cytoskeleton, transforming into a flagellated cell within 90 minutes. The flagellated stage is better able to seek distant food sources than the amebic stage. Most heteroloboseans feed on bacteria and are harmless, but *Naegleria fowleri* causes a few deaths each year. It lives in hot pools and causes primary amebic meningoencephalitis in humans when water containing amebas is inhaled. The amebas enter through the nasal passages and migrate along olfactory nerves to the brain, where tissue is destroyed.

Phylum Euglenozoa

The Euglenozoa (Figure 11.20) is generally considered a monophyletic group, based on the shared persistence of the nucleoli during mitosis, and the presence of discoid mitochondrial cristae. Members of this phylum have a series of longitudinal microtubules just beneath the plasma membrane that help to stiffen the membrane into a

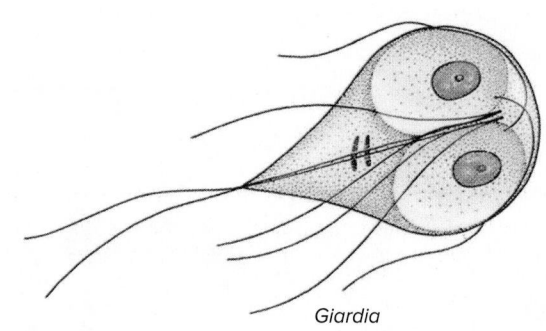

Giardia

Figure 11.18 *Giardia lamblia* often causes diarrhea in humans.

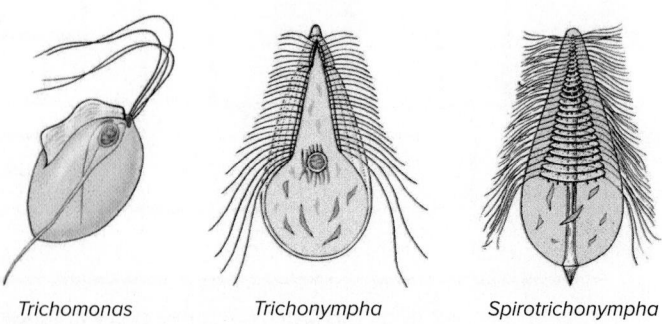

Trichomonas   Trichonympha   Spirotrichonympha

Figure 11.19 These three animals belong to the parabasalid clade. *Trichomonas vaginalis* is transmitted sexually and is a frequent cause of vaginitis in humans. *Trichonympha* and *Spirotrichonympha* are mutualistic symbionts in termites.

[1]Roger, A. J. 1999. Amer. Nat. 154 (supplement):S146–S163.

Peranema

Phacus

Trypanosoma

Leishmania

Figure 11.20   Examples of phylum Euglenozoa. *Peranema* is a colorless, free-living phagotroph, and *Phacus* is a green, free-living photoautotroph. *Trypanosoma* and *Leishmania* are parasitic, and some species cause serious diseases of humans and domestic animals. *Leishmania* is shown as its intracellular form, without an external flagellum.

pellicle. The phylum is divided into two subphyla, the Euglenida and the Kinetoplasta. Kinetoplastans are named for the presence of a unique organelle, the kinetoplast. This modified mitochondrion, associated with a kinetosome, carries large discs of DNA. Kinetoplastans are all parasites of plants or animals.

Subphylum Euglenida

Euglenids have chloroplasts with chlorophyll *b*. These chloroplasts are surrounded by a double membrane and are likely to have arisen via secondary endosymbiosis.

Euglena viridis (Figure 11.13) is a representative flagellate commonly studied in introductory zoology courses. Its natural habitat is freshwater streams and ponds where there is considerable vegetation. The organisms are spindle shaped and about 60µm long, but some species of *Euglena* are smaller and some larger (*E. oxyuris* is 500µm long). Just beneath the outer membrane of *Euglena* are proteinaceous strips and microtubules that form a pellicle. In *Euglena* the pellicle is flexible enough to permit bending, but in other euglenids it may be more rigid. A flagellum extends from a flask-shaped **reservoir** at the anterior end, and another, short flagellum ends within the reservoir. A kinetosome occurs at the base of each flagellum, and a **contractile vacuole** empties into the reservoir. A red eyespot, or stigma, apparently functions in orientation to light.

Within the cytoplasm are oval chloroplasts that bear chlorophyll and give the organism its greenish color. **Paramylon granules** of various shapes are masses of a starchlike food-storage material.

Nutrition of *Euglena* is normally autotrophic (holophytic), but if kept in the dark the organism uses saprozoic nutrition, absorbing nutrients through its plasma membrane. Mutants of *Euglena* can be produced that have permanently lost their photosynthetic ability. Although *Euglena* does not ingest solid food, some euglenids are phagotrophic. *Peranema* has a cytostome that opens alongside its flagellar reservoir.

Euglena reproduces by binary fission and can encyst to survive adverse environmental conditions.

Subphylum Kinetoplasta

Some of the most important unicellular eukaryotic parasites are kinetoplastans. Many of them belong to the genus *Trypanosoma* (Gr. *trypanon*, auger, + *soma*, body; Figure 11.20) and inhabit the blood of fish, amphibians, reptiles, birds, and mammals. Some are nonpathogenic, but others produce severe diseases in humans and domestic animals. *Trypanosoma brucei gambiense* and *T. brucei rhodesiense* cause African sleeping sickness in humans, and *T. brucei brucei* causes a related disease in domestic animals. Trypanosomes are transmitted by tsetse flies (*Glossina* spp.). *Trypanosoma b. rhodesiense*, the more virulent of the sleeping sickness trypanosomes, and *T. b. brucei* have natural reservoirs (antelope and other wild mammals) that are apparently not harmed by the parasites. Some 10,000 new cases of human sleeping sickness are diagnosed each year, of which about half are fatal, and many of the remainder sustain permanent brain damage.

Trypanosoma cruzi causes Chagas' disease in humans in Central America and South America. It is transmitted by "kissing bugs" (Triatominae), a name arising from the bug's habit of biting its sleeping victim on the face. Acute Chagas' disease is most common and severe among children less than five years old, while the chronic disease occurs most often in adults. Symptoms are primarily a result of central and peripheral nervous dysfunction. Two to 3 million people in South and Central America show chronic Chagas' disease, and 45,000 of these die each year.

Several species of *Leishmania* (Figure 11.20) cause disease in humans. Infection with some species may cause a serious visceral disease affecting especially the liver and spleen; others can cause disfiguring lesions in the mucous membranes of the nose and throat, and the least serious result is a skin ulcer. *Leishmania* spp. are transmitted by sand flies. Visceral leishmaniasis and cutaneous leishmaniasis are common in parts of Africa and Asia, and the mucocutaneous form occurs in Central America and South America.

Stramenopiles

Members of the clade Stramenopiles have tubular mitochondrial cristae. They may have flagellated cells, but stramenopiles are heterokont (Gr. *hetero*, different, + *kontos*, pole) flagellates. They have two different flagella, both inserted at the cell anterior. In heterokonts, the forward directed flagellum is long and hairy, whereas the other is short and smooth, and trails behind the cell. This clade is sometimes called Heterokonta; the name stramenopile (L. *stramen*, straw, + *pile*, hair) refers to three-part tubular hairs covering the flagellum. This clade contains brown algae, yellow algae, and diatoms, all collecting energy with plastids, as well as forms that lack plastids and are heterotrophs.

The opalinids, commensals in animals such as frogs, are now placed in Stramenopiles. They were once thought to be modified ciliates. Stramenopiles is also home to labyrinthulids, Oomycetes, and some organisms previously called heliozoans. Labyrinthulids, sometimes called slime nets, may be commensals or mutualists with plants, but some are parasitic on aquatic eel grasses and terrestrial turf grasses, including those on golf courses. Members of Oomycetes were once presumed to be fungi. Perhaps the most well-known Oomycetes belong to the genus *Phytophthora*. *P. infestans* causes potato late blight and was responsible for the Irish potato famine. Another species of *Phytophthora* causes "sudden oak death," which is currently damaging large areas of California. The name heliozoan refers to testate amebas with axopodia (see Figure 11.10), but molecular phylogenetics has shown that the heliozoan body form developed independently several times. Heliozoans in the group Actinophryida, including the genera, *Actinosphaerium* and *Actinophrys* (see Figure 11.6), are now placed with Stramenopiles.

Alveolata

The alveolate clade, sometimes called a superphylum, contains three phyla united by the shared presence of **alveoli,** membrane-bound sacs that lie beneath the plasma membrane. In the Ciliophora (Figure 11.21), the alveoli produce pellicles; in the Dinoflagellata, a group of armored flagellates (see Figure 11.27), the alveoli produce thecal plates; and in the Apicomplexa, containing intracellular parasitic species previously called sporozoans (see Figure 11.29), the alveoli have structural functions.

Phylum Ciliophora

Ciliates (Figure 11.21) are so named because the body surface is covered with cilia that beat in a coordinated, rhythmical manner. The arrangement of cilia varies within the phylum and some ciliates lack cilia as adults, although cilia occur at other stages in the life cycle. In general, ciliates are larger than most other unicellular eukaryotes, but they range from 10 μm to 3 mm in length. Most ciliates are free-living in freshwater or marine habitats, but commensal and parasitic forms do occur. They are usually solitary and motile, but some are sessile and others are colonial. Ciliates are very structurally complex, exhibiting a wide range of specializations.

The pellicle of ciliates may consist of only a plasma membrane or in some species may form a thickened armor. Cilia are short and usually arranged in longitudinal or diagonal rows. Cilia may cover the surface of the organism or may be restricted to the oral region or to certain bands. In some forms cilia are fused into a sheet called an **undulating membrane** or into smaller **membranelles,** both used to propel food into the **cytopharynx** (gullet). In other forms there may be fused cilia forming stiffened tufts called **cirri,** often used in locomotion by the creeping ciliates (Figure 11.21).

An apparently structural system of fibers, in addition to the kinetosomes (basal body), forms the **infraciliature,** just beneath the pellicle (Figure 11.22). Each cilium terminates beneath the pellicle

Figure 11.21 Some representative ciliates. *Euplotes* have stiff cirri used for crawling. Contractile fibrils in ectoplasm of *Stentor* and in stalks of *Vorticella* allow great expansion and contraction. Note the macronuclei, long and curved in *Euplotes* and *Vorticella,* shaped like a string of beads in *Stentor.*

in its kinetosome, and from each kinetosome a fibril arises and passes beneath the row of cilia, joining the other fibrils of that row. The cilia, kinetosomes, and other fibrils of that ciliary row form what is called a **kinety** (Figure 11.22). All ciliates seem to have kinety systems, even those that lack cilia at some stage. The infraciliature apparently does not coordinate ciliary beat, as formerly thought. Coordination of ciliary movement seems to be by waves of depolarization of the plasma membrane moving down the organism, similar to a nerve impulse.

Ciliates are always multinucleate, possessing at least one **macronucleus** and one **micronucleus,** but varying from one to many of either type. Genes in the micronucleus are never transcribed, unlike those in the macronucleus. The macronuclei are

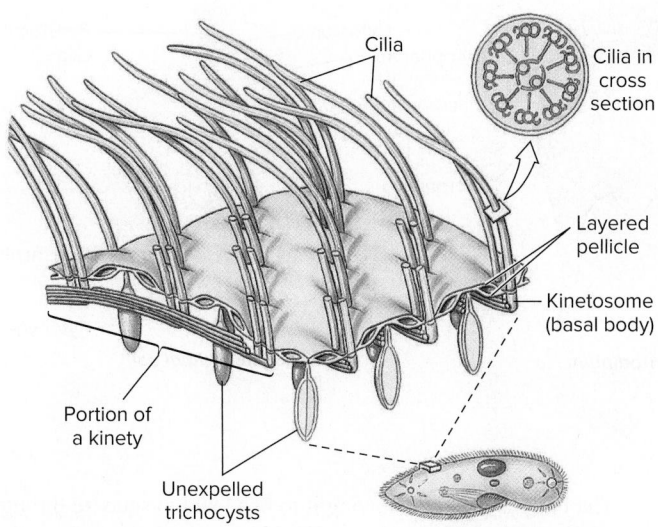

Cilia

Cilia in cross section

Layered pellicle

Kinetosome (basal body)

Portion of a kinety

Unexpelled trichocysts

A Structure of pellicle

B Expelled trichocyst

Figure 11.22 Infraciliature and associated structures in ciliates. **A,** Structure of the pellicle and its relation to the infraciliature system. **B,** Expelled trichocyst.

responsible for metabolic and developmental functions and for maintaining all the visible traits, such as the pellicular apparatus. Macronuclei vary in shape among the different species (Figures 11.14 and 11.21). Micronuclei participate in sexual reproduction and give rise to macronuclei after exchange of micronuclear material between individuals. Micronuclei divide mitotically, and macronuclei divide amitotically (see pp. 225–226).

Some ciliates have small bodies in their ectoplasm between the bases of the cilia. Examples are **trichocysts** (Figures 11.14 and 11.22) and **toxicysts.** Upon mechanical or chemical stimulation, these bodies explosively expel a long, threadlike structure. The mechanism of expulsion is unknown. The function of trichocysts is thought to be defensive. When attacked by a *Didinium,* a paramecium expels its trichocysts but to no avail. Toxicysts, however, release a poison that paralyzes the prey of carnivorous ciliates. Toxicysts are structurally quite distinct from trichocysts. Many dinoflagellates also have trichocysts.

Most ciliates are holozoic. Most of them possess a cytostome (mouth) which in some forms is a simple opening and in others is connected to a gullet or ciliated groove. The mouth in some is strengthened with stiff, rodlike trichites for swallowing larger prey; in others, such as paramecia, ciliary water currents carry microscopic food particles toward the mouth. *Didinium* has a proboscis for

engulfing paramecia on which it feeds (see Figure 11.2). Suctorians paralyze their prey and then ingest the contents through tubelike tentacles by a complex feeding mechanism that apparently combines phagocytosis with a sliding filament action of microtubules in the tentacles (see Figure 11.2).

Suctorians

Suctorians are ciliates in which the young possess cilia and are free-swimming, and the adults grow a stalk for attachment, become sessile, and lose their cilia. They have no cytostome but feed by long, slender, tubelike tentacles. The suctorian captures living prey, usually a ciliate, by the tip of one or more tentacles and paralyzes it. The cytoplasm of the prey then flows through the attached tentacles, by a complex feeding mechanism that apparently combines phagocytosis with a sliding filament action of microtubules in the tentacles (see Figure 11.2). Food vacuoles form in the feeding suctorian.

One of the best places to find freshwater suctorians is in algae that grow on the carapace of turtles. Common genera of suctorians found there are *Anarma* (without stalk or test) and *Squalorophrya* (with stalk and test). Other freshwater representatives are *Podophrya* (see Figure 11.2) and *Dendrosoma. Acinetopsis* and *Ephelota* are saltwater forms.

Suctorian parasites include *Trichophrya,* whose species occur on various invertebrates and freshwater fish; *Allantosoma,* which inhabits the intestine of certain mammals; and *Sphaerophrya,* which is found in *Stentor.*

Symbiotic Ciliates

Many symbiotic ciliates live as commensals, but some can be harmful to their hosts. *Balantidium coli* lives in the large intestine of humans, pigs, rats, and many other mammals (Figure 11.23). Some strains seem to be host-specific, so the organism is not easily transmitted from one species to another. Transmission is by fecal contamination of food or water. Usually the organisms are not pathogenic, but in humans they sometimes invade the intestinal lining and cause a dysentery similar to that caused by *Entamoeba histolytica* (p. 241). The disease can be serious and even fatal. Infections are common in parts of Europe, Asia, and Africa but are rare in the United States.

Other species of ciliates live in other hosts. *Entodinium* (Figure 11.23) belongs to a group that has very complex structure and lives in the digestive tract of ruminants, sometimes in great abundance. *Nyctotherus* inhabits the colon of frogs and toads. In aquarium and wild freshwater fishes, *Ichthyophthirius* causes a disease known to many fish culturists as "ick." Untreated, it can cause much loss of exotic fishes.

Free-Living Ciliates

Among the more striking and familiar ciliates are *Stentor* (Gr. herald with a loud voice), trumpet shaped and solitary, with a bead-shaped macronucleus (see Figure 11.21); *Vorticella* (L. dim. of *vortex,* a whirlpool), bell shaped and attached by a contractile stalk (see Figure 11.21); and *Euplotes* (Gr. *eu,* true, good, + *ploter,* swimmer), with a flattened body and groups of fused cilia (cirri) that function as legs. Paramecia are usually abundant in ponds or

Figure 11.23

Some symbiotic ciliates. *Ichthyophthirius* causes a common disease in aquarium and wild freshwater fishes. *Entodinium* is found in the rumen of cows and sheep. *Balantidium coli* is a parasite of humans and other mammals.

Cytostome

Macronucleus

Ichthyophthirius

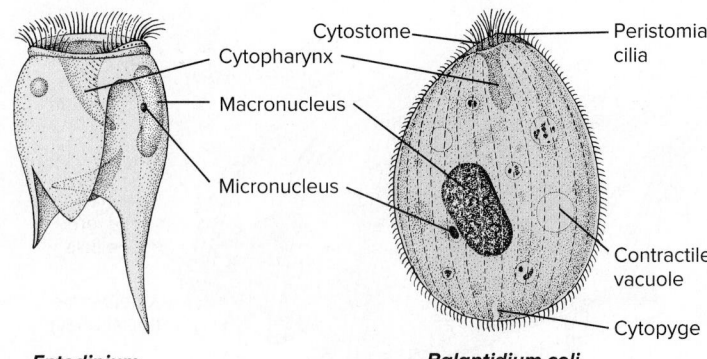

Cytostome
Cytopharynx
Macronucleus

Micronucleus

Entodinium

Peristomial cilia

Contractile vacuole

Cytopyge

Balantidium coli

sluggish streams containing aquatic plants and decaying organic matter. We discuss *Paramecium* in more detail, as a representative free-living ciliate.

Form and Function in Paramecium

Paramecia are often described as slipper shaped. *Paramecium caudatum* is 150 to 300μm in length and is blunt anteriorly and somewhat pointed posteriorly (see Figure 11.14). The organism has an asymmetrical appearance because of the **oral groove,** a depression that runs obliquely backward on the ventral side.

The **pellicle** is a clear, elastic membrane that may be ornamented by ridges or papilla-like projections (see Figure 11.22), and its entire surface is covered with cilia arranged in lengthwise rows. Just below the pellicle is the thin, clear **ectoplasm** that surrounds the larger mass of granular **endoplasm.** Embedded in ectoplasm just below the surface are spindle-shaped **trichocysts** (see Figure 11.22), which alternate with the bases of cilia. The infraciliature can be seen only with special fixing and staining methods.

A **cytostome** at the end of the oral groove leads into a tubular **cytopharynx,** or **gullet.** Along the gullet an undulating membrane of modified cilia keeps food moving. Fecal material is discharged through a **cytoproct** posterior to the oral groove (see Figure 11.14). Within the endoplasm are food vacuoles containing food in various stages of digestion. There are two **contractile vacuoles,** each consisting of a central space surrounded by several **radiating canals** (see Figure 11.14) that collect fluid and empty it into the central vacuole. We describe excretion and osmoregulation on page 225.

Paramecium caudatum has two nuclei: a large, kidney-shaped macronucleus and a smaller micronucleus fitted into the depression of the former. These can usually be seen only in stained specimens. The number of micronuclei varies in different species; for example, *P. multimicronucleatum* may have as many as seven.

Paramecia are holozoic, living on bacteria, algae, and other small organisms. Cilia in the oral groove sweep food particles in the water into the cytostome, from which they are carried into the cytopharynx by the undulating membrane. From the cytopharynx food is collected into a food vacuole that is constricted into the endoplasm. Food vacuoles circulate in a definite course through the cytoplasm while the food is being digested. Indigestible parts of the food are ejected through the cytoproct.

The body is elastic, allowing it to bend and to squeeze through narrow places. Its cilia can beat either forward or backward, so that the organism can swim in either direction. The cilia beat obliquely, causing the organism to rotate on its long axis. In the oral groove the cilia are longer and beat more vigorously than the others so that the anterior end swerves aborally. As a result of these factors, the organism moves forward in a spiral path (Figure 11.24A).

When a ciliate, such as a paramecium, contacts a barrier or a disturbing chemical stimulus, it reverses its cilia, backs up a short distance, and swerves the anterior end as it pivots on its posterior end. This behavior is called an **avoiding reaction** (Figure 11.24B). A paramecium may continue to change its direction to keep itself

A

B

Figure 11.24 **A,** Spiral path of swimming *Paramecium.* **B,** Avoidance reaction of *Paramecium.*

away from a noxious stimulus, and it may react in a similar fashion to keep itself within the zone of an attractant. A paramecium may also change its swimming speed. How does a paramecium "know" when to change directions or swimming speed? Interestingly, reactions of the organism depend on effects of the stimulus on the electrical potential difference across its plasma membrane (see Chapter 33, pp. 722–723). Paramecia slightly hyperpolarize in attractants and depolarize in repellents that produce the avoiding reaction. Hyperpolarization increases the rate of the forward ciliary beat, and depolarization induces ciliary reversal and backward swimming.

Locomotor responses, by which an organism more or less continuously orients itself with respect to a stimulus, are called *taxes* (sing. *taxis*). Movement toward the stimulus is a positive taxis; movement away is a negative taxis. Some examples are thermotaxis, response to heat; phototaxis, response to light; thigmotaxis, response to contact; chemotaxis, response to chemical substances; rheotaxis, response to currents of air or water; galvanotaxis, response to constant electric current; and geotaxis, response to gravity. Some stimuli cause not an orienting response, but simply a change in movement: more rapid movement, more frequent random turning, or slowing or cessation of movement. Such responses are called kineses. Is the avoiding reaction of a paramecium a taxis or a kinesis?

Reproduction in Paramecium

Paramecia reproduce only by binary fission across kineties (ciliary rows) but have certain forms of sexual phenomena called conjugation and autogamy.

In **binary fission** the micronucleus divides mitotically into two daughter micronuclei, which move to opposite ends of the cell (Figure 11.25). The macronucleus elongates and divides amitotically.

Conjugation occurs at intervals in ciliates. Conjugation is the temporary union of two individuals to exchange chromosomal material (Figure 11.26). During the union the macronucleus disintegrates and the micronucleus of each individual undergoes meiosis, giving rise to four haploid micronuclei, three of which degenerate (Figure 11.26A–C). The remaining micronucleus then divides into two haploid pronuclei, one of which is exchanged with the other conjugant. The pronuclei fuse to restore the diploid number of chromosomes, followed by several more nuclear events detailed in Figure 11.26. Following this complicated process, the organisms may continue to reproduce by binary fission without conjugation.

The result of conjugation is similar to that of zygote formation, for each exconjugant contains hereditary material from two individuals. Sexual reproduction permits gene recombinations, thus increasing genetic variation in the population. Although ciliates in clone cultures can apparently reproduce repeatedly and indefinitely without conjugation, the stock seems eventually to lose vigor. Conjugation restores vitality to a stock. Seasonal changes or a deteriorating environment usually stimulate sexual reproduction.

Autogamy is a process of self-fertilization similar to conjugation except that there is no exchange of nuclei. After the

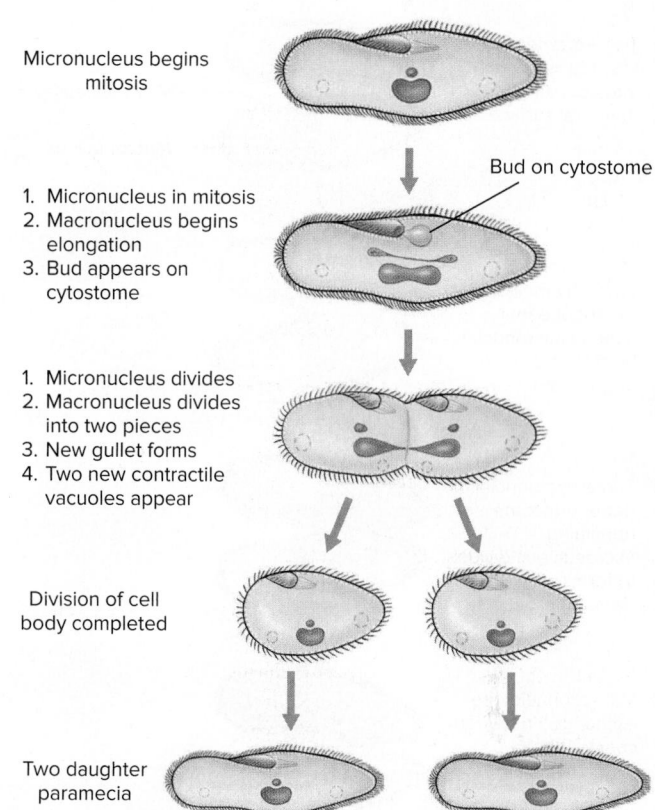

Micronucleus begins mitosis

Bud on cytostome

1. Micronucleus in mitosis
2. Macronucleus begins elongation
3. Bud appears on cytostome

1. Micronucleus divides
2. Macronucleus divides into two pieces
3. New gullet forms
4. Two new contractile vacuoles appear

Division of cell body completed

Two daughter paramecia

Figure 11.25 Binary fission in a ciliophoran (*Paramecium*). Division is across rows of cilia.

disintegration of the macronucleus and the meiotic divisions of the micronucleus, two haploid pronuclei fuse to form a synkaryon that is completely homozygous (see Chapter 5, p. 76).

Phylum Dinoflagellata

Approximately half of the dinoflagellate species are photoautotrophic with chromoplasts bearing chlorophyll. The rest are colorless and heterotrophic. Ancestral dinoflagellates probably were heterotrophic, and some acquired chloroplasts by endosymbiosis from a variety of algal sources. Ecologically, some species are among the most important primary producers in marine environments. They commonly have two flagella, one equatorial and one longitudinal, each borne at least partially in grooves on the body (Figure 11.27). The body may be naked or covered by cellulose plates or valves. Many species can ingest prey through a mouth region between the plates near the posterior area of the body. *Ceratium* (Figure 11.27), for example, has a thick covering with long spines, into which the body extends, but it can catch food with posterior pseudopodia and ingest it between the flexible plates in the posterior groove. *Noctiluca* (Figure 11.27), a colorless dinoflagellate, is a voracious predator and has a long, motile tentacle, near the base of which its single, short flagellum emerges. *Noctiluca* is one of many marine organisms that can produce light (bioluminescence).

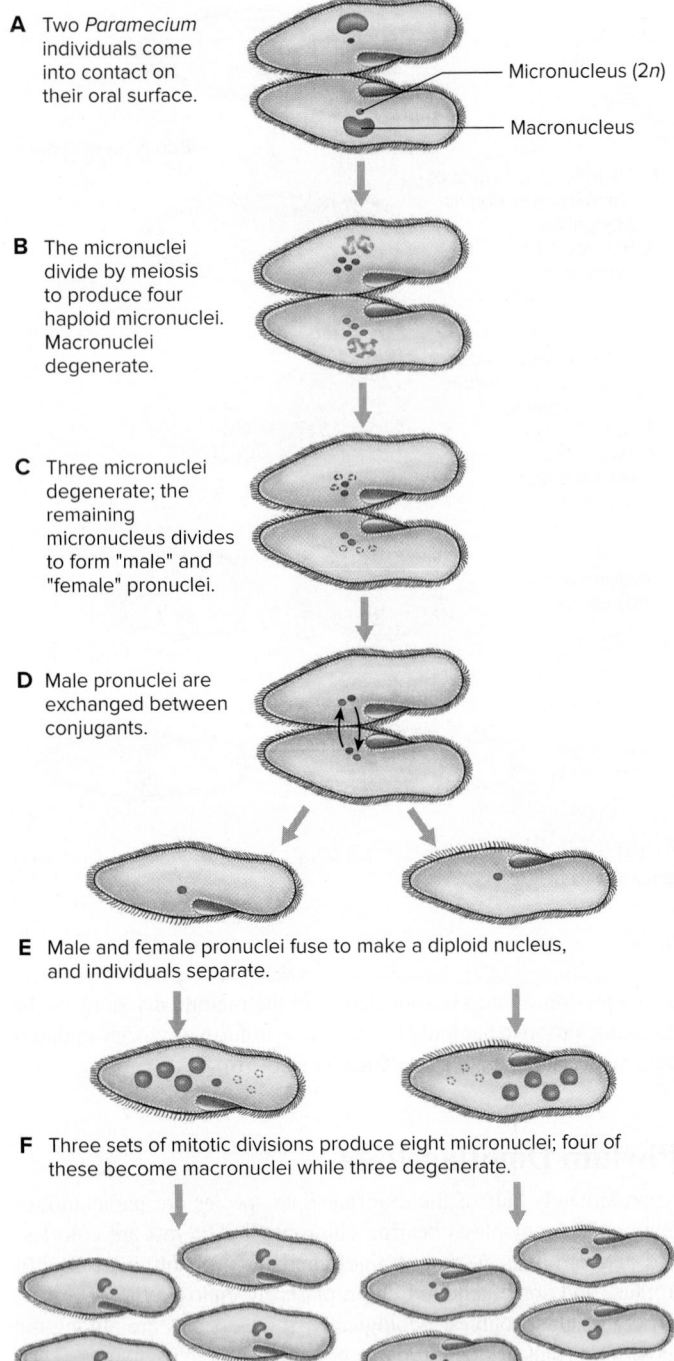

A Two *Paramecium* individuals come into contact on their oral surface.

Micronucleus (2*n*)

Macronucleus

B The micronuclei divide by meiosis to produce four haploid micronuclei. Macronuclei degenerate.

C Three micronuclei degenerate; the remaining micronucleus divides to form "male" and "female" pronuclei.

D Male pronuclei are exchanged between conjugants.

E Male and female pronuclei fuse to make a diploid nucleus, and individuals separate.

F Three sets of mitotic divisions produce eight micronuclei; four of these become macronuclei while three degenerate.

G The remaining micronucleus divides twice as does the cell, producing four daughter cells.

Figure 11.26   Scheme of conjugation in *Paramecium*.

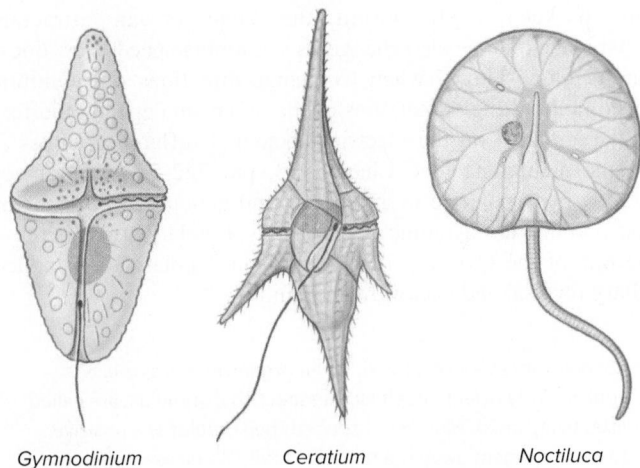

Gymnodinium   Ceratium   Noctiluca

Figure 11.27   Examples of phylum Dinoflagellata. *Gymnodinium* bears no cellulose plates. Some members of its family are autotrophic and some phagotrophic. *Ceratium* bears plates and is both autotrophic and phagotrophic. *Noctiluca* is entirely phagotrophic, can be very large (more than 1 mm wide), and has a large tentacle involved in feeding.

and stony corals, and clams. The association with stony corals is of ecological and economic importance because only corals with symbiotic zooxanthellae can form coral reefs (see Chapter 13).

Dinoflagellates can damage other organisms, such as when they produce a "red tide" (Figure 11.28). Although this name originally was applied to situations in which the organisms reproduced in such profusion (producing a "bloom") that the water turned red from their color, any instance of a bloom producing detectable levels of toxic substances is now called a red tide. The water may be red, brown, yellow, or not remarkably colored at all. The toxic substances are apparently not harmful to the organisms that produce them, but they may be highly poisonous to fish and other marine life. Several different types of dinoflagellates and one species of cyanobacterium have been responsible for red tides. Red tides have caused considerable economic losses to the shellfish industry. Another dinoflagellate produces a toxin concentrated in the food chain, especially in large, coral reef fishes. The illness produced in humans after eating such fish is called ciguatera.

Pfiesteria piscicida is one of several related dinoflagellate species that may affect fish in brackish waters along the Atlantic coast, south of North Carolina. Much of the time *Pfiesteria* feeds on algae and bacteria, but something in the excreta of large fish schools causes it to release a powerful, short-lived toxin. The toxin may stun or kill fish, often creating skin lesions. *Pfiesteria* has flagellated and ameboid forms among its more than 20 body types; some forms feed on fish tissues and blood. Although it does not have chloroplasts, it may sequester choroplasts from its algal prey and gain energy from them for a short time. This fascinating group of species was discovered in 1988.

Several groups of autotrophic flagellates are planktonic primary producers (p. 826) in freshwater and marine environments; however, dinoflagellates are the most important, particularly in the sea. Zooxanthellae are dinoflagellates that live in mutualistic association in tissues of certain invertebrates, including sea anemones, horny

Figure 11.28 "Red tide" makes a striking appearance on the coastline. The water appears colored because it contains a high concentration of dinoflagellates.

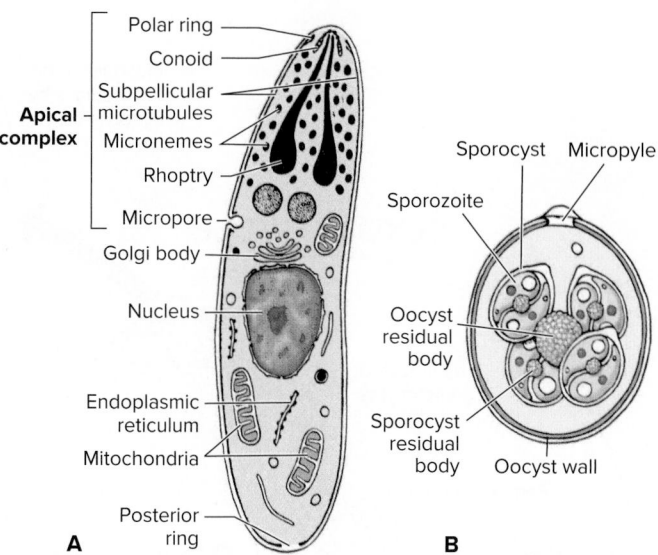

Figure 11.29 **A,** Diagram of an apicomplexan sporozoite or merozoite at the electron-microscope level, illustrating the apical complex. The polar ring, conoid, micronemes, rhoptries, subpellicular microtubules, and micropore (cytostome) are all considered components of the apical complex. **B,** Infective oocyst of *Eimeria*. The oocyst is the resistant stage and has undergone multiple fission after zygote formation (sporogony).

Phylum Apicomplexa

Almost all apicomplexans are endoparasites, and their hosts include many animal phyla. The presence of a certain combination of organelles, the **apical complex,** distinguishes this phylum (Figure 11.29A). The apical complex usually occurs only in certain developmental stages of the organisms; for example, **merozoites** and **sporozoites** (Figure 11.30). Some structures, especially the **rhoptries** and **micronemes,** apparently aid in penetrating the host's cells or tissues.

Locomotor organelles are less obvious in this group than in other protozoa. Pseudopodia occur in some intracellular stages, and gametes of some species are flagellated. Tiny contractile fibrils can form waves of contraction across the body surfaces to propel the organism through a liquid medium.

The life cycle usually includes both asexual and sexual reproduction, and sometimes an invertebrate intermediate host. At some point in the life cycle, the organisms develop a **spore (oocyst),** which is infective for the next host and is often protected by a resistant coat.

Class Coccidea

Coccidia are intracellular parasites in invertebrates and vertebrates, and the group includes species of very great medical and veterinary importance. We discuss three examples: *Eimeria,* which generally affects birds; *Toxoplasma,* which causes toxoplasmosis, a disease affecting cats and humans; and *Plasmodium,* the organism that causes malaria.

Eimeria *Species*

The name "coccidiosis" is generally applied only to infections with *Eimeria* or *Isospora.* Humans can be infected with species of *Isospora,* but there is usually little disease. However, *Isospora* infections can be very serious in AIDS patients. Some species of *Eimeria* may cause serious disease in some domestic animals. Symptoms usually include severe diarrhea or dysentery.

Eimeria tenella is often fatal to young fowl, producing severe pathogenesis in the intestine. The organisms undergo schizogony (p. 225) in the intestinal cells, finally producing gametes. After fertilization the zygote forms an oocyst that exits its host via the feces (Figure 11.29B). Sporogony occurs within the oocyst outside the host, producing eight sporozoites in each oocyst. Infection occurs when a new host accidentally ingests a sporulated oocyst and the sporozoites are released by digestive enzymes.

Toxoplasma gondii

A similar life cycle occurs in *Toxoplasma gondii,* a parasite of cats, but this species produces extraintestinal stages as well. When rodents, cattle, sheep, humans, many other mammals, or even birds ingest sporozoites, the sporozoites cross from the intestine and begin rapid, asexual reproduction in a variety of tissues. As the host mounts an immune response, reproduction of the zoites slows, and they become enclosed in tough **tissue cysts.** The zoites, now called **bradyzoites,** accumulate in large numbers in each tissue cyst. Bradyzoites are infective for other hosts, including cats, where they can initiate the intestinal cycle in a cat that eats infected prey. Bradyzoites can remain viable and infective for months or years, and it is estimated that one-third of the world's human population carries tissue cysts containing bradyzoites. The normal route of infection for

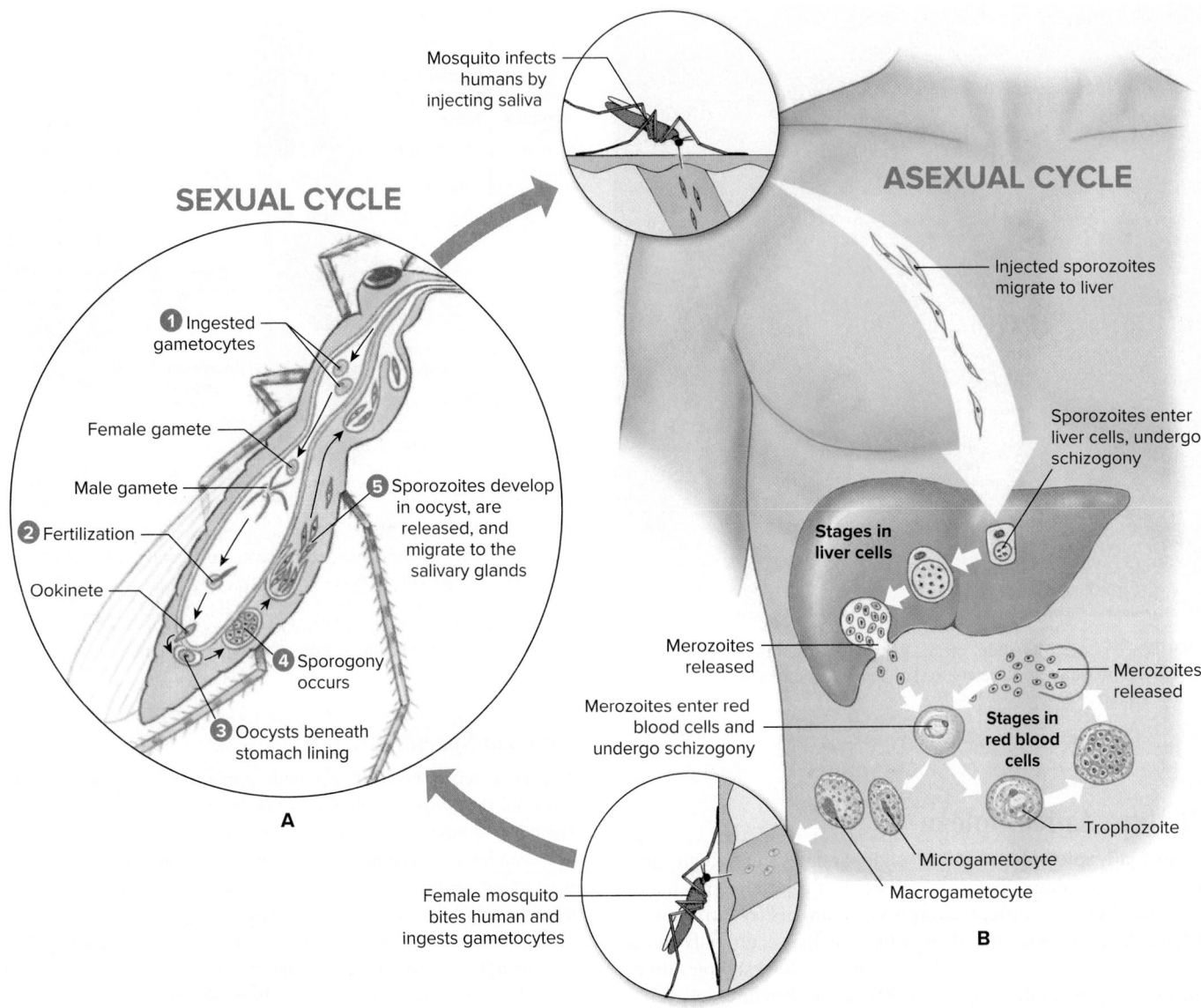

SEXUAL CYCLE

Mosquito infects humans by injecting saliva

ASEXUAL CYCLE

❶ Ingested gametocytes

Female gamete

Male gamete

❷ Fertilization

Ookinete

❺ Sporozoites develop in oocyst, are released, and migrate to the salivary glands

❹ Sporogony occurs

❸ Oocysts beneath stomach lining

A

Female mosquito bites human and ingests gametocytes

Injected sporozoites migrate to liver

Sporozoites enter liver cells, undergo schizogony

Stages in liver cells

Merozoites released

Merozoites enter red blood cells and undergo schizogony

Merozoites released

Stages in red blood cells

Trophozoite

Microgametocyte

Macrogametocyte

B

Figure 11.30 Life cycle of *Plasmodium vivax,* one of the protozoa (class Coccidia) that causes malaria in humans. **A,** Sexual cycle produces sporozoites in a mosquito. Meiosis occurs just after zygote formation (zygotic meiosis). **B,** Sporozoites infect a person and reproduce asexually, first in liver cells and then in red blood cells. Malaria is spread by *Anopheles* mosquito, which ingests gametocytes along with human blood, then, when biting another victim, leaves sporozoites in new wound.

humans is apparently consumption of infected meat that is insufficiently cooked.

Some 16% or more of adults in the United States are infected with *Toxoplasma gondii;* we have no symptoms because the parasite is held in check by our immune systems. However, *T. gondii* is one of the most important opportunistic infections in AIDS patients. The latent infection is activated in between 5% and 15% of AIDS patients, often in the brain, with serious consequences.

In women infected with *Toxoplasma* during pregnancy, especially in the first trimester, such infection greatly increases the chances of a birth defect in the baby; perhaps 2% of all mental retardation in the United States results from congenital toxoplasmosis.

The coccidian *Cryptosporidium parvum* first was reported in people in 1976. We now recognize it as a major cause of diarrheal disease worldwide, especially in children in tropical countries. Waterborne outbreaks have occurred in the United States, and the diarrhea can be life threatening in immunocompromised patients (such as those with AIDS). Infection rates for 2005 were about 3 cases per 100,000 persons. The latest coccidian pathogen to emerge has been *Cyclospora cayetanensis.* In the United States, outbreaks occurred annually from 2013 to 2015 with 546 people infected in 2015. Diarrhea is the most common symptom of infection. Infection usually occurs via ingestion of contaminated food or water.

Plasmodium: **The Malarial Organism**

The best-known coccidians are *Plasmodium* spp., causative organisms of the most important infectious disease of humans: **malaria.** Malaria is a very serious disease, difficult to control and widespread, particularly in tropical and subtropical countries. Four species of *Plasmodium* infect humans: *P. falciparum, P. vivax, P. malariae,* and *P. ovale.* Although each species produces its own peculiar clinical picture, all four have similar cycles of development in their hosts (Figure 11.30).

The parasite is carried by mosquitoes (*Anopheles*), and sporozoites are injected into a person with the insect's saliva during its bite. Sporozoites penetrate liver cells and initiate schizogony. In *P. falciparum* a single sporozoite produces up to 40,000 merozoites by schizogony. The products of this division then enter other liver cells to repeat the schizogonous cycle, or in *P. falciparum* they penetrate red blood cells after only one cycle in the liver. The period when the parasites are in the liver is the **incubation period,** and it lasts from 6 to 15 days, depending on the species of *Plasmodium.*

Merozoites released as a result of liver schizogony enter red blood cells, where they begin a series of schizogonous cycles. When they enter red blood cells, they become ameboid **trophozoites,** feeding on hemoglobin. The end product of the parasite's digestion of hemoglobin is a dark, insoluble pigment: **hemozoin.** Hemozoin accumulates in the host cell, is released when the next generation of merozoites is produced, and eventually accumulates in the liver, spleen, or other organs. A trophozoite within a red blood cell grows and undergoes schizogony, producing 6 to 36 merozoites, which, depending on the species, burst forth to infect new red blood cells. When a red blood cell containing merozoites bursts, it releases the parasite's metabolic products, which have accumulated there. Release of these foreign substances into the patient's circulation causes the chills and fever characteristic of malaria.

Since the populations of schizonts maturing in red blood cells are synchronized to some degree, the episodes of chills and fever have a periodicity characteristic of the particular species of *Plasmodium.* In *P. vivax* (benign tertian) malaria and *P. ovale* malaria, episodes occur every 48 hours; in *P. malariae* (quartan) malaria, every 72 hours; and in *P. falciparum* (malignant tertian) malaria, about every 48 hours, although synchrony is less well defined in this species. People usually recover from infections with the first three species, but mortality is high in untreated cases of *P. falciparum* infection. Sometimes grave complications, such as **cerebral malaria,** occur. Unfortunately, *P. falciparum* is the most common species, accounting for 50% of all malaria in the world. Certain genes, for example, the gene for sickle cell hemoglobin (p. 95 and p. 125), confer some resistance to malaria on people who carry these genes.

After some cycles of schizogony in red blood cells, infection of new cells by some of the merozoites causes production of **microgametocytes** and **macrogametocytes** rather than another generation of merozoites. When gametocytes are ingested by a mosquito feeding on a patient's blood, they mature into **gametes,** and fertilization occurs. The zygote becomes a motile **ookinete,** which penetrates the stomach wall of the mosquito and becomes an **oocyst.** Within the oocyst, sporogony occurs, and thousands of **sporozoites** are produced. The oocyst ruptures, and the sporozoites migrate to the salivary glands, from which they are transferred to a human by a bite of the mosquito. Development in a mosquito requires 7 to 18 days but may be longer in cool weather.

Forty-one percent of the earth's people live in malarial regions. Elimination of mosquitoes and their breeding places by insecticides, drainage, and other methods has been effective in controlling malaria in some areas. However, difficulties in performing such activities in remote areas and areas suffering civil unrest, and acquisition of resistance to insecticides by mosquitoes and to antimalarial drugs by *Plasmodium* (especially *P. falciparum),* mean that malaria will be a serious disease for a long time to come. Global estimates of infection were 247 million people in 2006, with an estimated 1 million deaths, most being African children.

Other species of *Plasmodium* parasitize birds, reptiles, and mammals. Those of birds are transmitted chiefly by *Culex* mosquitoes.

A *disease* is any illness or disorder that can be recognized by a given set of signs and symptoms. *Epidemiology* is the study of all factors that influence transmission, geographic distribution, incidence, and prevalence of a disease. Epidemiology of parasitic diseases often involves poor sanitation and contamination of water or food with infectious stages. That is not the case with arthropod-borne diseases, such as malaria. Transmission and distribution of malaria depend on the presence of a suitable *Anopheles* species, as well as its breeding, feeding, and resting habits. The climate (whether the mosquito can breed and feed throughout the year) is important, as is the prevalence of infected humans (especially asymptomatic individuals). It has nothing to do with improper waste disposal or poverty.

Cercozoa

Members of phylum Cercozoa do not share a common body plan. There are flagellated and ameboid members. The ameboid members of the group may be naked or testate. The testate amebas include *Euglypha,* which makes a test from collected particles (see Figure 11.16), *Clathrulina,* which has a siliceous capsule (see Figure 11.6), and some former members of Radiolaria. Radiolarians are marine amebas with silica skeletons. Phaeodarians, formerly considered radiolarians and now placed in Cercozoa, have an amorphous silica skeleton with magnesium, calcium, and copper added. The skeleton of these unusual unicells contains hollow spines and is known from microfossils dating back to the Cambrian. *Clathrulina* belongs to a group of organisms, called desmothoracids, that was placed within the Heliozoa (see p. 240). Ameboid cercozoans vary in the types of pseudopodia formed: axopodia are made in phaeodarians and desmothoracids, but other members of the group make filopodia.

The clade Cercozoa emerged from molecular phylogenetic studies. From the brief description above it is clear that the group is heterogeneous in terms of morphology. The group is equally heterogeneous in lifestyle. There are photosynthetic members such as the chlorarachniophytes, which are naked green amebas with filopodia. Other group members are free-living heterotrophs, and still others, such as plasmodiophorids and haplosporidians, are parasites. Plasmodiophorids are obligate intracellular parasites responsible for crop damage—they were once thought to be fungi. Haplosporidians are parasites of marine invertebrates; one species causes MSX disease, which adversely affects commercial populations of the eastern oyster, *Crassostrea virginica,* along the Atlantic coast of the United States.

Foraminifera

In this clade of amebas, slender pseudopodia extend through openings in the test, then branch and run together to form a protoplasmic net (reticulopodia) in which they ensnare their prey. Here captured prey is digested, and digested products are carried into the interior by flowing cytoplasm. Cytoplasmic flow is bidirectional throughout the reticulum.

Foraminiferans, or forams, are an ancient group of shelled amebas found in all oceans, with a few in fresh and brackish water. Most foraminiferans live on the ocean floor in incredible numbers, having perhaps the largest biomass of any animal group on earth. Their tests are of numerous types (Figures 11.6 and 11.31). Most tests are many chambered and are made of calcium carbonate, although they sometimes accrete silica, silt, and other foreign materials. Life cycles of foraminiferans are complex, for they have multiple fission and alternation of haploid and diploid generations (intermediary meiosis).

Foraminiferans have existed since Precambrian times and have left excellent fossil records. In many instances their hard shells are preserved unaltered. Many extinct species closely resemble those of the present day. They were especially abundant during the Cretaceous and Tertiary periods. Some measured up to 100 mm (4 in.) in diameter, but they were small in comparison to the extant deep-sea xenophyophorans. These giant, multinucleate benthic forms reach 20 cm in diameter. They cement a wide variety of particles into a fragile test. They may be filter or deposit feeders, and are unusual in that their cytoplasm often contains barium sulfate crystals and their fecal pellets, retained in the body, concentrate heavy metals such as lead or mercury.

For untold millions of years tests of dead foraminiferans have been sinking to the bottom of the ocean, building up a characteristic ooze rich in lime and silica. About one-third of the seafloor is covered with shells of the genus *Globigerina.* This ooze is especially abundant in the Atlantic Ocean.

Of equal interest and of greater practical importance are the limestone and chalk deposits that were laid down by the accumulation of foraminiferans when sea covered what is now land. Later, through a rise in the ocean floor and other geological changes, this sedimentary rock emerged as dry land. The chalk deposits of many areas of England, including the White Cliffs of Dover, formed this way. The great pyramids of Egypt were made from stone quarried from limestone beds that were formed by a very large foraminiferan population that flourished during the early Tertiary period.

Since fossil foraminiferans can be found in well drillings, their identification is often important to oil geologists for identifying rock strata.

"Radiolaria"

Radiolarians are marine testate amebas with axopodia (see Figure 11.7). With the exception of the phaeodarians, which are now placed in Cercozoa (p. 237), most organisms described as radiolarians before the advent of molecular phylogenetics are still considered radiolarians. Many radiolarians live in surface waters, but there are taxa living at depths up to 1000 meters. Radiolarians have a body divided by a central capsule (the test or skeleton) that separates inner and outer zones of the cytoplasm. The central capsule, which may be spherical, ovoid, or branched, is perforated to allow cytoplasmic continuity (Figure 11.32). Around the capsule is a frothy mass of cytoplasm (see Figure 11.7) from which the axopodia arise (see Figure 11.10). Sticky axopodia capture prey items, which are carried by the streaming protoplasm through the capsule. The ectoplasm on one side of the axial rod moves outward toward the axial tip, while that on the other side moves inward toward the test. Solitary radiolarians feed on bacteria, micro-algae and micro-flagellates, whereas colonial radiolarians collect nutrients from symbiotic algae.

Radiolarian cells contain one or many nuclei. Life histories are poorly known, but binary fission, budding, and sporulation have all been observed in radiolarians.

The chemical composition and complexity of the radiolarian skeleton varies: members of subgroup Acantharea have a strontium sulfate skeleton, whereas members of the non-monophyletic subgroup Polycistinea possess an amorphous silica skeleton, which may consist of spicules or develop as a helmet shape. The fossil record of radiolarians dates from the Jurassic because the relatively insoluble siliceous test makes a durable fossil. Fossils are usually found at great depths (4600 to 6100 meters)

A **B**

© Claude Carre/Science Source

© Comstock Images/Getty Images

Figure 11.31 **A,** Air-dried foraminiferan, showing thin pseudopodia extending from test. **B,** Test of foraminiferan. Foraminiferans are ameboid marine unicellular eukaryotes that secrete a calcareous, many-chambered test in which to live and then extrude protoplasm through pores to form a layer over the outside. The animal begins with one chamber, and as it grows, it secretes a succession of new and larger chambers, continuing this process throughout life. Many foraminiferans are planktonic, and when they die, their shells are added to the ooze on the ocean's floor.

Figure 11.32   Types of radiolarian tests. In his study of these beautiful forms collected on the famous *Challenger* expedition of 1872 to 1876, Haeckel proposed our present concepts of symmetry.

© Mehau Kulyk/Science Source

products of cell division maintain cytoplasmic bridges between cells and cells sit within an extra-cellular matrix (ECM). True multicellularity requires a division of labor among cells: some cells become the germ line and others become the somatic line. The genetic changes necessary to set up this division have been elucidated in the green algal lineage that produced a multicellular species, *Volvox carteri*. A colonial body plan is visible in *Gonium, Eudorina,* and *Pandorina* (Figure 11.33), but *Volvox* was one of nine green algal lineages in which multicellularity evolved independently. There are over 25 such instances in the history of life.

Volvox (Figure 11.34) is a green, hollow sphere that may reach a diameter of 0.5 to 1 mm. A single organism contains many thousands of cells (up to 50,000) embedded in the gelatinous surface of a jelly ball. Each cell is much like a euglenid (p. 229), with a nucleus, a pair of fla-gella, a large chloroplast, and a red **stigma.** A stigma is a shallow pigment cup that allows light from only one direction to strike a light-sensitive receptor. Adjacent cells are connected with each

mainly in the Pacific and Indian oceans. Radiolarian ooze covers 5 to 8 million square kilometers to a thickness of 700 to 4000 meters. Under certain conditions, radiolarian ooze forms rocks (chert). Most fossil radiolarians occur in Tertiary rocks of California. The identifi-cation of particular species is important to oil geologists interested in determining the age of particular strata.

Plantae

The clade Plantae comprises three photosynthetic lineages: glauco-phytes, rhodophytes (red algae), and Viridiplantae. Plantae is some-times called Archaeplastida in reference to the ancient primary symbiosis with a cyanobacterium that produced the chloroplasts of photosynthetic eukaryotes. Viridiplantae contains the unicellular, colonial, and multicellular green algae, which were once in phylum Chlorophyta, as well as the nonvascular and vascular plants. The phylum Chlorophyta is a non-monophyletic group containing only the green algae. We now know that the vascular and nonvascular plants (jointly called "land" plants) shared a common ancestor with the green algae. Speaking evolutionarily, organisms that most people would call plants are, in fact, terrestrial green algae. Viridiplantae includes the former members of Chlorophyta and the nonvascular and vascular plants. The chloroplasts of group members contain chlorophylls *a* and *b*.

Viridiplantae

We discuss only a few flagellated taxa, which are typically considered microbial eukaryotes. *Chlamydomonas* (Figure 11.33) is a unicellular biflagellate cell. Colony formation in green algae occurs when the

Chlamydomonas

Gonium

Eudorina

Pandorina

Figure 11.33   Examples of Viridiplantae. They are all photoautotrophs.

other by cytoplasmic strands. At one pole (usually in front as the colony moves), the stigmata are a little larger. Coordinated action of the flagella causes the sphaeroid to move by rolling over and over.

Volvox carteri has a complete division of labor: small, haploid, motile, photosynthetic, somatic cells are embedded in an extracellular matrix (ECM) to form a sphaeroid body and 16 large, haploid, nonmotile reproductive cells called gonidia are embedded beneath the somatic cells. Gonidia undergo cleavage and inversion (see insert below) to form new sphaeroids with both somatic and reproductive cells present. As the 16 new sphaeroids that developed from gonidia leave by digesting through the ECM, the sphaeroids of somatic cells that once housed them die (Figure 11.34).

Male and female strains of V. *carteri* are not distinguishable in the asexual phase, but when exposed to a particular pheromone, the sexes become distinct. In female sphaeroids, gonidia develop into sphaeroids containing about 48 eggs, whereas the gonidia of male sphaeroids develop into organisms containing 64–128 sperm packets. Sperm packets are motile. After fertilization, a dormant diploid resting stage is made. This resting stage tolerates freezing and dessication. Upon germination, the resting stage undergoes meiosis to produce a single viable haploid "germling" that undergoes cleavage to make a new sphaeroid. Sexual reproduction takes place only periodically.

The original polarity of cells in *Volvox* is such that their flagella are protruding into the interior cavity of the developing organism. To move the flagella to the outside so that locomotion is possible, the entire sphaeroid must turn itself inside out. This process, called inversion, is *very unusual*. Of all other living organisms, only the sponges (phylum Porifera) have a comparable developmental process.

Centrohelida (Centroheliozoa, "Heliozoa")

Prior to the widespread use of molecular methods in phylogenetics, the Heliozoa was a larger group of testate amebas forming axopodia. All species formerly placed in Heliozoa, except members of the order Centrohelida, have been moved to other taxa; hence the name change. Former members of Heliozoa have been moved to Stramenopiles (actinophryids) and to Cercozoa (desmothoracids). Centrohelids are amebas with flattened mitochondrial cristae. The structure of their axopodia is distinctive: microtubules (see Figure 11.10) within the axoneme are arranged in hexagons or triangles in this group. In most centrohelids, the axopodia extend through a coat of variously shaped silica scales,

Figure 11.34 The life cycle of *Volvox carteri* showing both the asexual and the sexual phases of the life cycle. New sphaeroids develop from specialized reproductive cells called gonidia. Somatic cells in sphaeroids live only 48 hours.

but this coat may be absent and is replaced in some taxa by a mucous layer. Most centrohelids live in freshwater, but a few clades have colonized brackish or marine habitats. These lovely unicells are predators.

Amoebozoa

Amoebozoans include naked and testate amebas, as well as amebas with flagellated stages in the life cycle. Amoebozoans typically have branched tubular mitochondrial cristae, but this feature is not unique to the group. Ameboid forms within the group include the fascinating plasmodial and cellular slime molds in the subgroup Mycetozoa (e.g., *Physarum* and *Dictyostelium,* respectively); testate amebas with lobose pseudopodia, such as *Arcella* (see Figures 11.6 and 11.16); and naked lobose amebas, such as *Chaos carolinense,* *Amoeba proteus,* or members of genus *Acanthamoeba* (see Figures 11.12 and 11.35). *C. carolinense* or *A. proteus* is sometimes used in biology classrooms, but *Acanthamoeba castellani* has gained notoriety because of its impact on human health. *A. castellani* kills cells of the human cornea and is spread by contact lenses that are not properly disinfected.

Amoebozoa also houses entozoic amebas—those that live inside humans or other animals. *Entamoeba histolytica* is an important human parasite. It lives in the large intestine and on occasion can invade the intestinal wall by secreting enzymes that attack the intestinal lining. If this occurs, a serious and sometimes fatal amebic dysentery may result. The organisms may be carried by the blood to the liver and other organs and cause abscesses there. Many infected persons show few or no symptoms but are carriers, passing cysts in their feces. Diagnosis is complicated by the existence of a nonpathogenic species, *E. dispar,* which is morphologically identical to *E. histolytica.* Infection is spread by contaminated water or food containing cysts. *E. histolytica* occurs throughout the world, but clinical amebiasis is most prevalent in tropical and subtropical areas.

Other species of *Entamoeba* found in humans are *E. coli* in the intestine and *E. gingivalis* in the mouth. Neither of these species is known to cause disease.

Opisthokonta

The Opisthokonta is a clade characterized by a combination of flattened mitochondrial cristae and one posterior flagellum on flagellated cells, if such cells exist. Recent protein sequence comparisons among taxa have also identified a short sequence of amino acids from one protein (elongation factor 1-alpha) that is shared by both unicellular and multicellular clade members. Relationships among clade members as suggested from sequence data from several proteins are shown in Figure 11.36.

The Opisthokonta contains animals, fungi, and some unicellular taxa. The best known unicells in this group are the microsporidians and choanoflagellates. Microsporidians are intracellular parasites now recognized as specialized fungi. Choanoflagellates (Figure 11.36) are solitary or colonial eukaryotes considered the most likely sister taxon to the animals. They are used to test hypotheses of how animal multicellularity arose, specifically to identify features of the most recent common ancestor of animals

and their closest unicellular relatives. We discuss them with sponges (phylum Porifera; pp. 247–257) because of the strong resemblance between choanoflagellate cells and sponge choanocytes.

The Opisthokonta also contains less well-known unicells such as ichthyosporeans (animal parasites sometimes called DRIPs), nucleariid amebas, corallochytreans, and ministeriid amebas.

11.4 PHYLOGENY AND ADAPTIVE DIVERSIFICATION

Phylogeny

Molecular evidence has almost completely revised our phylogeny of unicellular eukaryotes. It now seems that the ancestral eukaryote diversified into many morphologically distinct clades, although the branching order for diversification is still poorly understood. Many characters for phylogenetic analyses come from structural features of organelles. However, one must be able to distinguish an ancient organelle, formed through symbioses among prokaryotes, from a more recently acquired organelle formed through secondary symbioses. The absence of an organelle such as a mitochondrion can be informative, but only if we have a way to distinguish whether mitochondria were present and later lost, or never present at all. Detailed studies of nuclear genomes and gene products—for example, mitochondrial enzymes encoded by the nuclear genes—can distinguish between the primary absence of a structure and its secondary loss. It is now assumed that all amitochondriate unicellular eukaryotes had ancestors with mitochondria.

Plastids were another variable character that held promise for elucidating phylogenetic relationships. However, the presence of particular plastids in a wide variety of seemingly unrelated unicellular and multicellular eukaryotes created confusion until it eventually became clear that the primary endosymbiotic event with a cyanobacterium was followed by secondary and tertiary endosymbiotic events that transferred plastids among eukaryotic lineages. A disentangled pathway of endosymbiont transfers, in combination with results from new molecular data sets, suggests that many eukaryotic lineages now can be combined into a few eukaryotic supergroups.

Figure 11.35
Acanthamoeba polyphaga is free-living species, but it may cause a serious disease of the nervous system in people with damaged immune systems. It is shown dividing.

10 μm

© Scott Camazine/Science Source

Figure 11.36 One hypothesis regarding relationships among some members of Opisthokonta: Choanoflagellates are shown as the sister taxon to Metazoa. Choanoflagellates shown are *Codonosiga* on the left and *Proterospongia* on the right.

The membership of some new supergroups is shown in Figure 11.17. Retortamonads and Diplomonads are united in clade Fornicata. Some research supports the union of Fornicata with Parabasalea, Heterolobosea, and Euglenozoa, as well as other taxa that we have not discussed, to form the supergroup Excavata. Excavate cells share an unusual feeding groove. However, support for Excavata varies with the genes under analysis, so we do not discuss it. Members of the clade Alveolata share a morphological feature, the presence of alveoli, and this group is well supported in molecular phylogenies. Most research places Alveolata as the sister taxon to the Stramenopiles.

The supergroup Rhizaria unites Cercozoa, Foraminifera, "Radiolaria," and some other taxa into a clade where many members are testate amebas, some with rhizopodia. However, members of Cercozoa, in particular, have a range of other body forms, and there are testate amebas outside the Rhizaria. Rhizaria also includes a clade of testate amebas in genus *Gromia*. *Gromia sphaerica* made headlines when it was discovered rolling across the seafloor off the Bahamas at a depth of 2000 feet. This ameba is several centimeters in diameter, so the size is striking, but the really exciting part is that its track on the seafloor is much like fossil tracks attributed to bilaterally symmetrical organisms. Paleontologists are now reevaluating some trace fossil evidence, wondering about the role of giant amebas.

The union of Stramenopiles, Alveolata, and Rhizaria into the group SAR has been proposed. This group takes its name from the first letter of each of the member superclades; it is sometimes called RAS because the taxa are listed in the reverse order. The validity of the supergroup Plantae is not in doubt—this group is supported by both molecular and morphological characters, although some studies include or exclude minor taxa not discussed here. One recent study placed Plantae as the sister taxon to SAR. There is also strong support for the Unikonts. This group unites Amoebozoa and Opisthokonta on the grounds that when flagellated cells are present, there is only one flagellum.

Some phylogenies based on molecular data place the root of the eukaryote tree between the Unikonts and all other unicellular taxa, which are grouped as "Bikonts" due to the shared presence of two flagella, when flagella are present. The phylogeny of the eukaryotes is of interest to a great many biologists, and research continues on these groups at a rapid pace. Some interest lies in finding phylogenetic homes for taxa that cause harm to humans, because treatments effective against particular pathogens often work on related organisms. For example, there were no effective treatments against microsporidians until it was discovered that microsporidians were highly specialized fungi and antifungal drugs were applied. Perhaps the greater interest in eukaryote phylogeny lies in being able to describe the entire pathway of evolution for this enormous group, but much work remains before this goal will be achieved.

Adaptive Diversification

We have described some of the wide range of adaptations in this chapter. Amebas range from seafloor-dwelling, naked species to planktonic forms such as foraminiferans and radiolarians with beautiful, intricate tests. There are many symbiotic species of amebas. Flagellated forms likewise show adaptations for a similarly wide range of habitats, with the added variation of photosynthetic ability in many groups.

Within a single-cell body plan, the division of labor and specialization of organelles are carried furthest by structurally complex ciliates. Specializations for intracellular parasitism have been adopted by Apicomplexa.

Taxonomy of Unicellular Eukaryotes

The taxonomy below is not exhaustive, and with few exceptions, we list only taxa discussed in this chapter.

Phylum Retortamonada (re-tor'ta-mo'nad-a) (L. *retorqueo,* to twist back, + *monas,* single, unit). Mitochondria and Golgi bodies lacking; three anterior and one recurrent (running toward posterior) flagellum lying in a groove; intestinal parasites or free-living in anoxic environments. Example: *Retortamonas.*

Phylum Diplomonada (di'plo-mon-a'da) (Gr. *diploos,* double, + L. *monas,* unit). One or two karyomastigonts (group of kinetosomes with a nucleus); individual mastigonts with one to four flagella; mitotic spindle within nucleus; cysts present; free-living or parasitic. Example: *Giardia.*

Phylum Parabasala (par'a-bas'a-la) (Gr. *para,* beside, + *basis,* base). With very large Golgi bodies associated with karymastigont; up to thousands of flagella. *Trichomonas* and two other forms comprise clade Parabasalids.

> **Order Trichomonadida** (tri'ko-mon-a'di-da) (Gr. *trichos,* hair, + *monas,* unit). Typically at least some kinetosomes associated with rootlet filaments characteristic of trichomonads; parabasal body present; division spindle extranuclear; hydrogenosomes present; no sexual reproduction; true cysts rare; all parasitic. Examples: *Dientamoeba, Trichomonas.*

Phylum Heterolobosea (het'ə-rō-lō bō'se-ə) (Gr. *heteros,* different, + *lobos,* lobe). Naked amebas with eruptive pseudopodia; a typical life cycle includes both amebic and flagellated stages. Group members sometimes called amoeboflagellates or schizopyrenids. Examples: *Naegleria fowleri, Naegleria gruberi.*

Phylum Euglenozoa (yu-glen-a-zo'a) (Gr. *eu* -, good, true, + *glēnē,* cavity, socket, + *zöon,* animal). With cortical microtubules; flagella often with paraxial rod (rodlike structure accompanying axoneme in flagellum); mitochondria with discoid cristae; nucleoli persist during mitosis. This phylum is synonymous with clade Euglenozoa. Classification of Unicellular Eukaryotes

> **Subphylum Euglenida** (yu-glen'i-da) With pellicular microtubules that stiffen pellicle.

>> **Class Euglenoidea** (yu-glen-oyd'e-a) (Gr. *eu-,* good, true, + *glēnē,* cavity, socket, + - *ōideos,* form of, type of). Two heterokont flagella (flagella with different structures) arising from apical reservoir; some species with lightsensitive stigma and chloroplasts. Example: *Euglena.*

> **Subphylum Kinetoplasta** (ky-neet'o-plas'ta) (Gr. *kinētos,* to move, + *plastos,* molded, formed). With a unique mitochondrion containing a large disc of DNA; paraxial rod.

>> **Class Trypanosomatidea** (try-pan'o-som-a-tid'e-a) (Gr. *trypanon,* a borer, + *sōma,* the body). One or two flagella arising from pocket; flagella typically with paraxial rod that parallels axoneme; single mitochondrion (nonfunctional in some forms) extending length of body as tube, hoop, or network of branching tubes, usually with single conspicuous DNA-containing kinetoplast located near flagellar kinetosomes; Golgi body typically in region of flagellar pocket, not connected to kinetosomes and flagella; all parasitic. Examples: *Leishmania, Trypanosoma.*

Phylum Stramenopiles (stra-men' M-piles) (L. *stramen,* straw, + *pile,* hair). Flagellates with two different flagella, one long and one short; forward flagellum is covered with tubular three-part hairs; mitochondria with tubular cristae. Free-living and parasitic plantlike and animal-like forms. The Oomycetes belong to this group. Examples: *Phytophthora infestans, Actinosphaerium, Actinophrys.*

Phylum Ciliophora (sil-i-of'-or-a) (L. *cilium,* eyelash, + Gr. *phora,* bearing). Cilia or ciliary organelles in at least one stage of life cycle; two types of nuclei, with rare exception; binary fission across rows of cilia, budding and multiple fission also occur; sexuality involving conjugation, autogamy, and cytogamy; nutrition heterotrophic; contractile vacuole typically present; most species free-living, but many commensal, some parasitic. (This is a very large group, now divided by the Society of Protozoologists into three classes and numerous orders and suborders. The classes are separated on the basis of technical characteristics of the ciliary patterns, especially around the cytostome, the development of the cytostome, and other characteristics.) Examples: *Paramecium, Colpoda, Tetrahymena, Balantidium, Stentor, Blepharisma, Epidinium, Euplotes, Vorticella, Carchesium, Trichodina, Podophrya, Ephelota.* This phylum is within clade Alveolata.

Phylum Dinoflagellata (dy'no-fla-jel-at'a) (Gr. *dinos,* whirling, + *flagellum,* little whip). Typically with two flagella, one transverse and one trailing; body usually grooved transversely and longitudinally, each groove containing a flagellum; chromoplasts usually yellow or dark brown, occasionally green or blue-green, bearing chlorophylls a and c; nucleus unique among eukaryotes in having chromosomes that lack or have low levels of histones; mitosis intranuclear; body form sometimes of spherical unicells, colonies, or simple filaments; sexual reproduction present; members free-living, planktonic, parasitic, or mutualistic. Examples: *Zooxanthella, Ceratium, Noctiluca, Ptychodiscus.* This phylum is within clade Alveolata.

Phylum Apicomplexa (ap'i-compleks'a) (L. *apex,* tip or summit, + *complex,* twisted around). Characteristic set of organelles (apical complex) associated with anterior end present in some developmental stages; cilia and flagella absent except for flagellated microgametes in some groups; cysts often present; all parasitic. This phylum is within clade Alveolata.

> **Class Gregarinea** (gre-ga-ryn'e-a) (L. *gregarius,* belonging to a herd or flock). Mature gamonts (individuals that produce gametes) large, extracellular; gametes usually alike in shape and size; zygotes forming oocysts within gametocysts; parasites of digestive tract or body cavity of invertebrates; life cycle usually with one host. Examples: *Monocystis, Gregarina.*

> **Class Coccidea** (kok-sid'e-a) (Gr. *kokkos,* kernel, grain). Mature gamonts small, typically intracellular; life cycle typically with merogony, gametogony, and sporogony; most species live inside vertebrates. Examples: *Cryptosporidium, Cyclospora, Eimeria, Toxoplasma, Plasmodium, Babesia.*

Phylum Cercozoa (ser-kō -zo'a) (Gr. *kerkos,* tail, + *zöon,* animal). A diverse group of unicells, heterogeneous in lifestyle and morphology; monophyly supported by molecular data. Most are free-living, some parasitic. Examples: *Euglypha, Clathrulina.*

Phylum Foraminifera (for'a-min-if'er-a) (L. *foramin,* hole, + *fero,* to bear). Shelled amebas bearing slender pseudopodia that extend through many openings in the test, forming a net that ensnares prey; includes xenophyophores. Examples: *Vertebralina, Globigerina.*

Phylum "Radiolaria" (rā̆de-a-laˊre-a) (L. *radiolus,* small sunbeam). Most are amebas with a well-developed internal skeleton of strontium sulfate or silica, forming beautiful tests. Axopodia are present. Examples: *Tetrapyle, Pterocorys.*

Phylum Viridiplantae (virˊi-di-planˊtee) (L. *viridis,* green, + *planto,* to set, plant). Unicellular and multicellular photoautotrophs with chlorophylls a and b; reserve material is starch. Examples: *Chlamydomonas, Volvox, Zea mays.*

Phylum Centrohelida (cen-tro-helˊi-da) (Gr. *kentron,* center of a circle, + *hēlios,* the sun). Amebas with flattened mitochondrial cristae; axoneme of axopodia with microtubules arranged in hexagons or triangles; in most, axopodia extend through a coat of variously shaped silica

scales. Mostly freshwater, some marine. Examples: *Acanthocystis, Pterocystis, Heterophrys.*

Phylum Amoebozoa (uh-meeˊbo-zoˊa) (Gr. *amoibē,* to change, + *zoon,* animal). Naked and shelled amebas, many with flagellated stages in life cycle; cellular and plasmodial slime molds; mitochondria, when present, have tubular and branched cristae. Free-living and parasitic. Examples: *Entamoeba, Dictyostelium, Chaos.*

Phylum Opisthokonta (o-pistˊtho-konˊta) (Gr., *opisthen,* behind, at the back, + *kontos,* a pole, referring to a flagellum). Many are flagellates with one posterior flagellum; the group includes nucleariid amebas, choanoflagellates, fungi, and the animals. Examples: *Codonosiga, Penicillium,* animals.

SUMMARY

"Animal-like," single-celled eukaryotes were formerly assigned to the phylum Protozoa. Molecular phylogenetic studies show that the "phylum" was composed of numerous taxa that do not form a monophyletic group. The terms *protozoa* and *protozoan* are still used informally to refer to all these highly diverse unicellular eukaryotes. They are also sometimes called microbial eukaryotes. They demonstrate the great adaptive potential of the basic body plan: a single eukaryotic cell. They occupy a vast array of niches and habitats. Many species have complex and specialized organelles.

All unicellular eukaryotes have one or more nuclei, and these often appear vesicular with light microscopy. Macronuclei of ciliates are compact. Nucleoli are often evident in the nuclei. Many unicellular eukaryotes have organelles similar to those of animal cells.

Pseudopodial or ameboid movement is a locomotory and food-gathering mechanism in unicellular eukaryotes and plays a vital role as a defense mechanism in animals. It is accomplished by assembly of actin subunits into filaments and interaction of actin filaments with actin-binding proteins and myosin, and it requires expenditure of energy from ATP. Ciliary movement is likewise important in both unicellular eukaryotes and animals. The most widely accepted mechanism to account for ciliary movement is the sliding-microtubule hypothesis.

Various unicellular eukaryotes feed by holophytic, holozoic, or saprozoic means. The excess water that enters their bodies is expelled by contractile vacuoles (water-expulsion vesicles). Respiration and waste elimination are through the body surface. Protozoa can reproduce asexually by binary fission, multiple fission, and budding; sexual processes are common. Cyst formation to withstand adverse environmental conditions is an important adaptation in many protozoa.

The evolution of a eukaryotic cell was followed by diversification of lineages to form morphologically disparate clades, some of which contain both unicellular and multicellular forms. Major taxa discussed are identified partly on the basis of molecular characters and contain subsets of species from traditional phyla. Photoautotrophic species occur in several phyla, including Viridiplantae, Euglenozoa, and Dinoflagellata. Some of these are very important planktonic organisms. Euglenozoa includes many nonphotosynthetic species, and some of these cause serious diseases of humans, such as African sleeping sickness and Chagas' disease. Apicomplexa are parasitic, including *Plasmodium,* which causes malaria. Ciliophora move by means of cilia or ciliary organelles. They are a large and diverse group, and

many are complex in structure. Amebas move by pseudopodia and are now assigned to a number of phyla.

REVIEW QUESTIONS

1. How does a unicellular eukaryote acquire considerable structural complexity within the constraints of a single cell?
2. Distinguish among the following groups: Euglenozoa, Apicomplexa, Ciliophora, Dinoflagellata.
3. Distinguish vesicular and compact nuclei.
4. Explain the transitions of endoplasm and ectoplasm in ameboid movement. What is a current hypothesis regarding the role of actin in ameboid movement?
5. Distinguish lobopodia, filipodia, reticulopodia, and axopodia.
6. Contrast the structure of an axoneme of a cilium with that of a kinetosome.
7. What is the sliding-microtubule hypothesis?
8. Describe the different ways that unicellular eukaryotes feed.
9. Distinguish the following: binary fission, budding, multiple fission, and sexual and asexual reproduction.
10. What is the survival value of encystment?
11. Contrast and give an example of autotrophic and heterotrophic unicellular eukaryotes.
12. Name three kinds of amebas, and describe their habitats.
13. Outline the general life cycle of malarial organisms. What explains the resurgence of malaria in recent years?
14. What is the public health importance of *Toxoplasma,* and how do humans become infected with it? What is the public health importance of *Cryptosporidium* and *Cyclospora?*
15. Define the following with reference to ciliates: macronucleus, micronucleus, pellicle, undulating membrane, cirri, infraciliature, trichocysts, conjugation.
16. Outline the steps in conjugation of ciliates.
17. Explain why unicellular eukaryotes are neither plants nor animals.
18. Distinguish primary endosymbiogenesis from secondary endosymbiogenesis.

For Further Thought Mark the distribution of photoautotrophic taxa on Figure 11.17. Research suggests that members of Ciliophora contain genes of algal origin. From a "tree-thinking" perspective, how would you explain the presence of these genes?

SELECTED REFERENCES

Baldauf, S. L. 2008. An overview of the phylogeny and diversity of eukaryotes. J. Syst. Evol. **46:** 263–273. *A nice summary of the current status of eukaryote phylogeny.*

Burkholder, J. M. 2002. *Pfiesteria:* the toxic *Pfiesteria* complex. In G. Bitton (ed.), Encyclopedia of environmental microbiology, pp. 2431–2447. New York, Wiley Publishers. *A nice summary of work on habitat and life cycles of* Pfiesteria, *including its effects on fish, shellfish, and humans.*

Cavalier-Smith, T. 1999. Principles of protein and lipid targeting in secondary symbiogenesis: euglenoid, dinoflagellate, and sporozoan plastid origins and the eukaryote family tree. J. Euk. Microbiol. **46:** 347–366. *A eukaryotic cell arose by primary symbiogenesis, but some eukaryotes are the products of secondary symbiogenesis, where one eukaryote engulfed another and the engulfed cell became a symbiont. Tertiary symbiogenesis also has occurred.*

Geo, F., and L. A. Katz. 2014. Phylogenomic analyses support the bifurcation of ciliates into two major clades that differ in properties of nuclear division. Molecular Phylogenetics and Evolution **70:** 240–243. *The very large ciliate phylum is divided into two clades based on the details of how the somatic macronuclei divide or, in some cases, fail to divide.*

Harper, J. T., E. Waanders, and P. J. Keeling. 2005. On the monophyly of chromalveolates using a six-protein phylogeny of eukaryotes. Int. J. Syst. Evol. Microbiol. **55:** 487–496. *Outlines support for a large clade uniting stramenopiles and alveolates.*

Keeling, P. J. 2004. Diversity and evolutionary history of plastids and their hosts. Am. J. Bot. **91:** 1481–1493. *A lucid description of plastid evolution and the evidence for primary, secondary, and tertiary endosymbioses.*

Keeling, P. J., G. Burger, D. J. Durnford, B. F. Lang, R. W. Lee, R. E. Pearlman, A. J. Roger, and M. W. Gray. 2005. The tree of eukaryotes. Trends Ecol. Evol. **20:** 670–676. *Support for five eukaryotic supergroups is presented.*

Kirk, D. L. 2001. Germ-soma differentiation in *Volvox*. Developmental Biology **238:** 213–223. *A description of sexual and asexual reproduction in the multicellular* Volvox *species.*

Lecroq, B., A. J. Gooday, M. Tsuchiya, and J. Pawlowski. 2009. A new genus of xenophyophores (Foraminifera) from Japan Trench: morphological description, molecular phylogeny, and elemental analysis. Zoological Journal of the Linnean Society **156:** 455–464. *Photographs and text descriptions of the structure and function of these giant unicells.*

Lee, J. J., G. F. Leedale, and P. Bradbury (eds). 2000. An illustrated guide to the protozoa, ed. 2, 1432 pp., 2 vols. Lawrence, Kansas, Society of Protozoologists. *This long-awaited guide appeared in 2002. It is an essential reference for students of unicellular eukaryotes.*

Margulis, L., and M. J. Chapman. 2009. Kingdoms and domains: an illustrated guide to life on earth, ed. 4. Boston, Academic Press/Elsevier. *This book has good descriptions of many taxa, clear descriptions of basic morphology, and useful photographs and drawings.*

Pawlowski, J., and F. Burki. 2009. Untangling the phylogeny of amoeboid protists. J. Eukaryot. Microbiol. **56:** 16–25. *Many amebas have found new phylogenetic homes, as shown on detailed trees included here.*

Pawlowski, J., and A. J. Gooday. 2008. Precambrian biota: protistan origin of trace fossils? Curr. Biol. **19:** R28–R30. *A discussion of the impact of the giant rolling ameba on paleobiology.*

Roberts, L. S., and J. J. Janovy, Jr. 2009. Foundations of parasitology, ed. 8. Dubuque, Iowa, McGraw-Hill Higher Education. *Up-to-date and readable information on parasitic protozoa.*

Roger, A. J. 1999. Reconstructing early events in eukaryotic evolution. Amer. Nat. **154** (supplement): S146–S163. *Methods in determining whether absence of mitochondria is primary or due to a secondary loss are discussed here.*

Sachs, J. L. 2008. Resolving the first steps to multicellularity. Trends Ecol. Evol. **23:** 245–248. *An interesting summary of work on Volvox evolution covering morphological change and selection pressures.*

Savory, F., G. Leonard, and T. A. Richards. 2015. The role of horizontal gene transfer in the evolution of the Oomycetes. PLoS Pathog 11(5): e1004805. doi:10.1371/journal.ppat.1004805. *The oomycetes were once thought to be fungi because of their morphological similarity to fungi. Both organisms live in the same habitats, but it was still surprising to discover that some similarities are due to gene transfer from fungi to the oomycetes rather than due to convergent evolution.*

Sebé-Pedrós, A., A. J. Roger, F. B. Lang, N. King, and I. Ruiz-Trillo. 2010. Ancient origin of integrin-mediated adhesion and signaling machinery. Proceedings of the National Academy of Sciences, **107:** 10142–10147. *Cells of multicellular organisms must interact with the extracellular matrix (ECM). In animals, integrins mediate these interactions, but this research shows that parts of the integrin protein complex were present before the split between fungi and animals. Some parts predate the Amoebozoa. Ancient genes were modified for new functions in the evolution of multicellularity.*

Sierra, R., M. V. Matz, G. Aglyamova, L. Pillet, J. Decele, F. Not, C. de Vargas, and J. Pawlowski. 2013. Deep relationships of Rhizaria revealed by phylogenomics: a farewell to Haeckel's Radolaria. Molecular Phylogenetics and Evolution **67:** 53–59. *Radiolarians are not a monophyletic group, although the clade* Acantharea *is monophyletic. Foraminifera branches within the former radiolarian phylum.*

Steenkamp, E. T., J. Wright, and S. L. Baldauf. 2006. The protistan origins of animals and fungi. Mol. Biol. Evol. **23:** 93–106. *Opisthokonta is a well supported clade whose members share a short amino acid sequence in elongation factor 1-alpha.*

12 CHAPTER

Sponges and Placozoans

- PHYLUM PORIFERA: SPONGES
- PHYLUM PLACOZOA

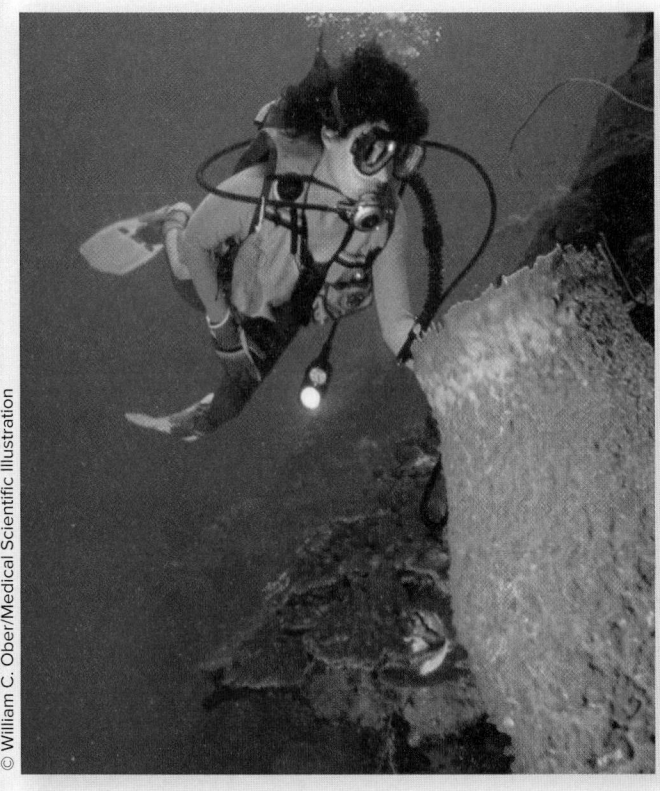

A Caribbean demosponge, *Aplysina fistularis*.

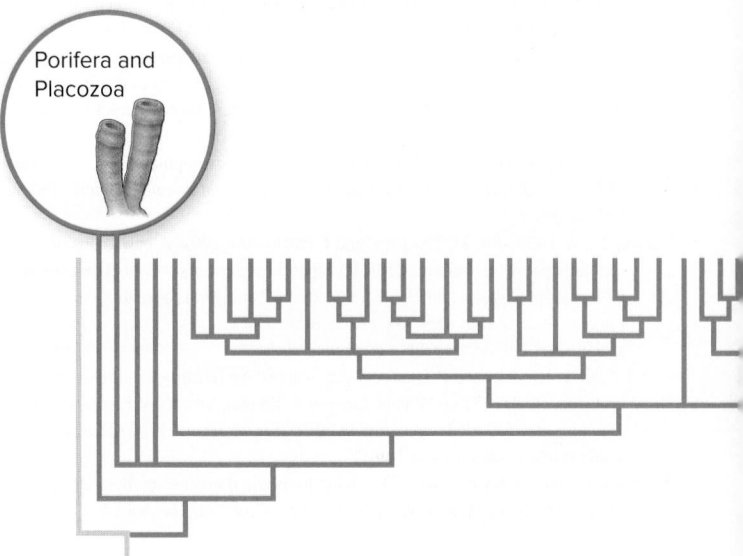

Porifera and Placozoa

The Origins of Multicellularity

Sponges are the simplest multicellular animals. Because cells are the elementary units of life, organisms larger than unicellular organisms arose as aggregates of such building units. Nature experimented with producing larger organisms without cellular differentiation—certain large, single-celled marine algae, for example—but such examples are rarities. Multicellularity has many advantages over increasing the mass of a single cell. Because cell surfaces exchange molecules with the environment, dividing a mass into smaller units greatly increases the surface area available for metabolic activities. A workable surface-to-mass ratio cannot be maintained by increasing the size of a single-celled

organism. Thus multicellularity is a highly adaptive path toward a larger body size.

Strangely, while sponges are multicellular, their organization is quite distinct from that of other metazoans. A sponge body is an assemblage of cells embedded in an extracellular matrix and supported by a skeleton of minute needlelike spicules and protein. Because sponges neither look nor behave like other animals, they were not completely accepted as animals by zoologists until well into the nineteenth century. Nonetheless, molecular evidence demonstrates that sponges are phylogenetically grouped with animals.

12.1 ORIGIN OF ANIMALS (METAZOA)

Evolution of a eukaryotic cell was followed by diversification into many lineages (see Figure 11.1). Modern descendants of these lineages include unicellular taxa (see Chapter 11), as well as colonial and multicellular plants, animals, and fungi. Animals traditionally were divided into single-celled "animals," or protozoans, and multicellular "animals," or metazoans." Because protozoans belong outside the animal kingdom (see Chapter 11), metazoan is now synonymous with "animal." Animals fall within the Opisthokont clade (see Figure 11.17) along with fungi, choanoflagellates, and a few other groups. The sister group to the animals is the choanoflagellates according to several phylogenies using molecular characters.

Choanoflagellates are solitary or colonial aquatic eukaryotes, with each cell carrying a flagellum surrounded by a collar of microvilli. Beating of the flagellum draws water into the collar, where microvilli collect tiny particles, typically bacteria. Many choanoflagellates are sessile and attached to hard surfaces, although one species attaches to floating diatom colonies allowing it to feed in midwater, even though it does not swim. Swimming does occur in *Proterospongia,* an unusual colonial form that propels itself through water using flagella.

Choanoflagellate cells are noteworthy because they strongly resemble sponge feeding cells called choanocytes (see p. 252). It is very interesting to find a collared cell used in filter-feeding in a colonial eukaryote and in a sponge, whose ancestral lineage represents an early divergence from the lineage of all other animals (see the Cladogram of All Animal Phyla on inside front cover). Was a sponge choanocyte inherited from a common ancestor with choanoflagellates? Arguments against this hypothesis include the observation that choanocytes occur only in adult sponges and are not part of the early developmental sequence. Instead, flagellated cells without collars develop into choanocytes after larval metamorphosis. Collar cells also occur in certain corals and some echinoderms, so if they were part of the earliest animal lineage, this morphology has been lost or suppressed in most taxa. Despite these objections, there is another clear link between choanoflagellates and animals: proteins used by colonial choanoflagellates for cell communication and adhesion are homologous to those that animals use in cell-to-cell signaling.[1]

Was the common ancestor of choanoflagellates and animals a colonial organism? Researchers approach this question by asking whether the ancestral choanoflagellate was solitary or colonial. Intriguingly, the most recent molecular phylogeny shows that colonial bodies evolved very early in the choanoflagellate lineage, but we are not yet sure about the ancestral state for the group.

Recent approaches to the problem of animals origins involve inferring the regulatory components of the first animal genome. As already mentioned, the genetic instructions for cell-signaling proteins predate the transition from unicellular to multicellular forms. What other cell transmitters or morphogens did the first animal possess? One way to discover the answer is to compare the genomes or proteomes of sponges with those of other animals, because their shared characteristics would have been inherited from the most recent common ancestor of animals.

Adult sponges have deceptively simple bodies; they are aggregations of several different cell types, including choanocytes, held together by an extracellular matrix. Most sponge bodies are not symmetrical, but some appear radial. A sponge body has neither a mouth nor a digestive tract. Thus, we expect it to have a simple genetic architecture, perhaps reminiscent of the first animals. Surprisingly, the sponge genome contains many elements that code for parts of the regulatory pathways of more complex animals, including proteins involved in spatial patterning, like those that specify an anterior and posterior pole in the larva. This discovery has led some biologists to hypothesize that modern sponges are less morphologically complex than were their ancestors.

Similar hypotheses have been applied to another phylum of simple animals discussed in this chapter. Members of Placozoa (see p. 257) have the smallest nuclear genome, and the largest mitochondrial genome, of any known animal. Their circular mitochondrial genome shares some features with mitochondrial genomes from animal outgroups, including chytrid fungi and choanoflagellates, but also has derived animal features.

Remember that sponges alive today are products of millions of years of evolution since their ancestors diverged from those of other animals. We expect to find genes unique to each phylum, along with those homologous to genes in animal outgroups, and to genes shared with more complex animals. Gene functions may have changed as new morphologies evolved, and there is much yet to be understood about the modern forms of the two phyla we discuss here. The body of a placozoan is at least as puzzling as that of a sponge; one cannot find heads or tails in either of them.

12.2 PHYLUM PORIFERA: SPONGES

Most animals move to search for food, but a sessile sponge (Figure 12.1) draws food and water into its body instead. The entrance of water through myriads of tiny pores is reflected in the phylum name, Porifera (po-rif′-er-a) (L. *porus,* pore, + *fera,* bearing). A sponge uses a flagellated "collar cell," the **choanocyte,** to move water (Figure 12.2). The beating of many tiny flagella, one per choanocyte, draws water past each cell, bringing in food and oxygen, as well as carrying away wastes. The sponge body is designed as an efficient aquatic filter for removing

Red boring sponge | Encrusting sponge | Finger sponge | Variable sponge | Tube sponge

Coral head

Figure 12.1 Some growth habits and forms of sponges.

[1]Abedin, M., and N. King. 2010. Trends in Cell Biology 20:732–742.

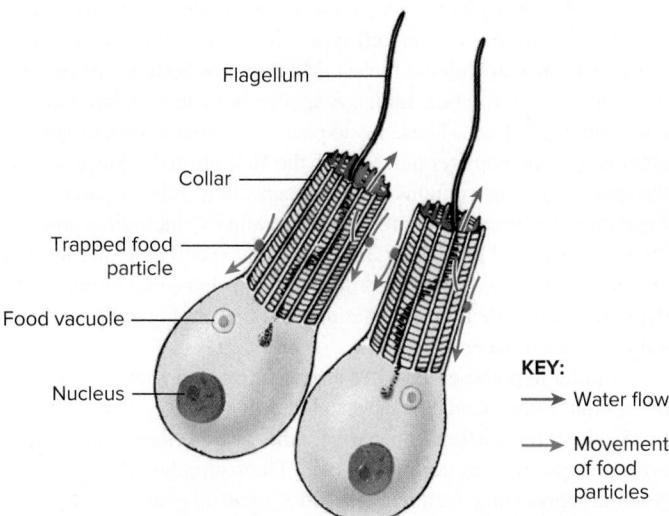

Figure 12.2 Sponge choanocytes have a collar of microvilli surrounding a flagellum. Beating of the flagellum draws water through the collar (blue arrows) where food is trapped on microvilli (red arrows).

Characteristics of Phylum Porifera

1. Multicellular; body an aggregation of several types of cells differentiated for various functions, some of which are organized into **incipient tissues** with some integration. However, the pinacoderm approaches a true tissue epithelium in homoscleromorph sponges.
2. Body with pores (ostia), canals, and chambers that form a unique system of **water currents** on which sponges depend for food and oxygen
3. Mostly marine; all aquatic
4. Radial symmetry or none
5. Outer surface of flat pinacocytes; most interior surfaces lined with flagellated collar cells (choanocytes) that create water currents; a gelatinous protein matrix called mesohyl contains amebocytes of various types and skeletal elements
6. Skeletal structure of fibrillar collagen (a protein) and calcareous or siliceous crystalline spicules, often combined with variously modified collagen (spongin); type IV collagen, characteristic of other animals, occurs only in homoscleromorph sponges.
7. No organs or true tissues; digestion intracellular; excretion and respiration by diffusion
8. Reactions to stimuli apparently local and independent in cellular sponges, but electrical signals in syncytial glass sponges; nervous system probably absent
9. All adults sessile and attached to substratum
10. Asexual reproduction by buds or gemmules and sexual reproduction by eggs and sperm; free-swimming flagellated larvae in most

suspended particles from the surrounding water. Sponges clear many liters of water each day and are important primary consumers within their ecosystems. Biologists are often fascinated by sponges because they function so well with so few body parts.

Most of the approximately 8600 sponge species are marine; a few inhabit **brackish** water, and some 150 species live in fresh water. Marine sponges are abundant in all seas and at all depths. Sponges

vary in size from a few millimeters to over 2 m in diameter; the latter size characterizes the great loggerhead sponges. Recent studies of the size and growth rate of the Caribbean reef sponge *Xestospongia muta* suggest that this sponge may be an astonishing 2300 years old. Many sponge species are brightly colored because of pigments in their dermal cells. Red, yellow, orange, green, and purple sponges are not uncommon.

Although their embryos are free-swimming, adult sponges are always attached, usually to rocks, shells, corals, or other submerged objects. Some bore holes into shells or rocks; others even grow on sand or mud. Some sponges, including the simplest, appear radially symmetrical, but many are quite irregular in shape. Some stand erect, some are branched or lobed, and others are low, even encrusting, in form (Figure 12.1). Their growth patterns often depend on shape of the substratum, direction and speed of water currents, and availability of space, so that the same species may differ markedly in appearance under different environmental conditions. Sponges in calm waters may grow taller and straighter than those in rapidly moving waters.

Many animals, such as crabs, nudibranchs, mites, bryozoans, and fishes, live as commensals or parasites in or on sponges. Larger sponges particularly tend to harbor a great variety of invertebrate **commensals.** Sponges also grow on many other living animals, such as molluscs, barnacles, brachiopods, corals, or hydroids. Some crabs attach pieces of sponge to their carapace for camouflage and for protection against predators. Some reef fishes do graze shallow-water sponges and sponges are an important part of the diet of hawksbill turtles. Surprisingly, dorid nudibranchs (see Figure 16.1C) eat glass sponges.

Sponges, and the microorganisms that live in or on them, produce a wide variety of bioactive chemicals. An extract from a marine sponge appears effective against leishmaniasis, a disease caused by a kinetoplastan parasite (see Chapter 11, p. 229), and another shows promise for treating herpetic infections. Many bacteria isolated from marine taxa also have antimicrobial or antiviral effects: for example, some inhibit *Staphylococcus aureus* infections and others are active against *Escherichia coli*, some strains of which may cause sickness due to food poisoning. These and other results have increased interest in sponge culturing as a source of valuable pharmaceuticals.

The skeletal framework of a sponge can be fibrous and/or rigid. When present, the rigid skeleton consists of calcareous or siliceous support structures called **spicules** (Figure 12.3). The fibrous part of the skeleton comes from collagen protein fibrils in the intercellular matrix of all sponges. Collagen comes in several types differing in chemical composition and form (for example, fibers, filaments, or masses surrounding spicules). One form of collagen is traditionally called **spongin** (Figure 12.3).

Sponges harbor microalgae and cyanobacteria on the body surface and deep inside the body. The presence of photosynthetic organisms inside the sponge led some to propose that spicules were able to transmit light into the body. The fiber optics of siliceous spicules have now been confirmed. This has sparked interest among materials scientists and engineers in the enzymatic machinery needed to form silica nanoparticles and to fuse these particles into spicules inside and outside sponge cells. In many cases, the exterior simplicity of a sponge masks chemical and functional sophistication.

Sponges are an ancient group, with an abundant fossil record extending back to the early Cambrian period and even, according to some claims, the Precambrian. Classification is based on spicule form

Siliceous spicules
(Hexactinellida)

Siliceous spicules
(Demospongiae)

Spongin

Calcareous
spicules

Figure 12.3 Diverse forms of spicules, many amazingly complex and beautiful, support a sponge's body. Spongin fibers provide support in some sponges.

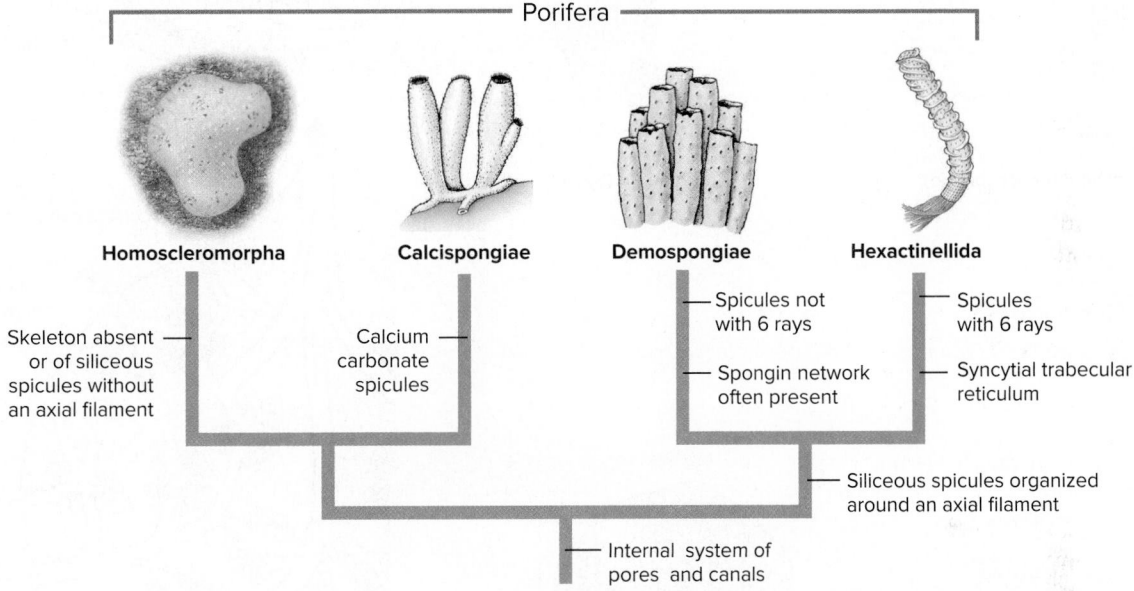

Figure 12.4 Cladogram depicting evolutionary relationships among the four classes of sponges with living representatives.

and chemical composition. Living poriferans traditionally have been assigned to three classes: Calcispongiae, Hexactinellida, and Demospongiae (Figure 12.4). Members of Calcispongiae typically have spicules of crystalline calcium carbonate with one, three, or four rays. Hexactinellids are glass sponges with six-rayed siliceous spicules, where the six rays are arranged in three planes at right angles to each other. Members of Demospongiae have a skeleton of siliceous spicules that develop around an axial filament, or spongin fibers, or both. A fourth clade, Homoscleromorpha, contains sponges that lack a skeleton or have siliceous spicules without an axial filament.

Form and Function

Sponges feed primarily by collecting suspended particles from water pumped through internal canal systems. Water enters canals through a multitude of tiny incurrent pores in the outer layer of cells, a **pinacoderm.** Incurrent pores, called **dermal ostia** (Figure 12.5), have an average diameter of 50 μm. Inside the body, water is directed past the choanocytes, where food particles are collected on the choanocyte collar (Figure 12.2). The collar comprises many fingerlike projections, called microvilli, spaced about 0.1 μm apart. The use of the

collar as a filter is one form of **suspension feeding.** One may be surprised to discover that a large sponge can filter up to 1500 liters of water each day.

Sponges nonselectively consume food particles (bits of detritus, planktonic organisms, and bacteria) sized between 0.1 μm and 50 μm. The smallest particles, accounting for about 80% of the particulate organic carbon, are taken into choanocytes by **phagocytosis** (see Chapter 3, p. 48). Choanocytes may acquire protein molecules by **pinocytosis** (see Chapter 3, p. 49). Two other cell types, **pinacocytes** and **archaeocytes,** play a role in sponge feeding (see p. 253). Sponges may also absorb dissolved nutrients from the water.

The capture of food depends on the movement of water through the body. How is water flow controlled in an animal with such a simple body? There are three main designs for a sponge body, differing in placement of the choanocytes. In the simplest **asconoid** system, the choanocytes lie in a large chamber called the **spongocoel;** in the **syconoid** system, the choanocytes lie in canals; and, in the **leuconoid** system, the choanocytes occupy distinct chambers (Figure 12.5). These three designs demonstrate an increase in complexity and efficiency of the waterpumping system, but they do not imply an evolutionary sequence. The leuconoid grade of

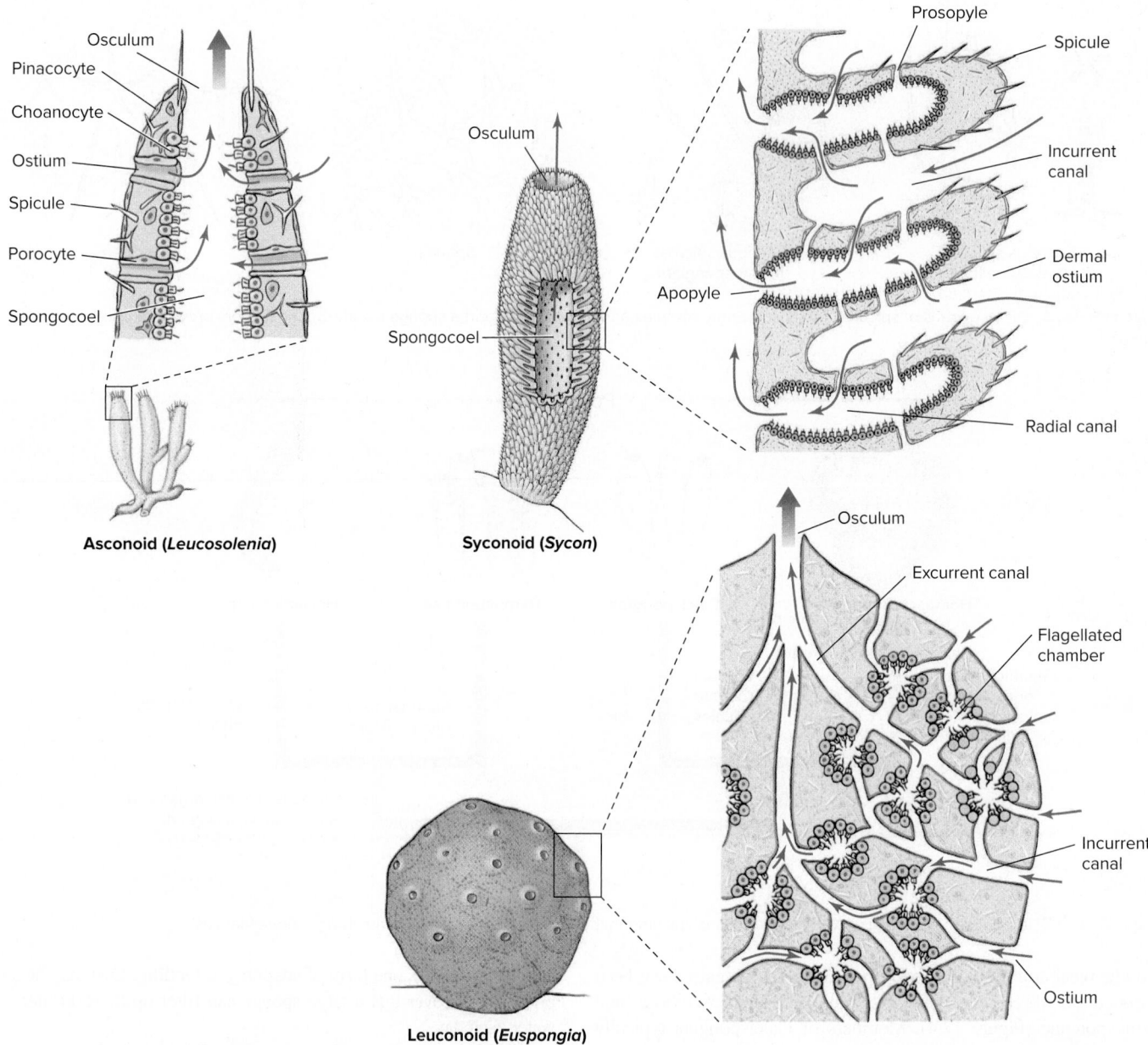

Figure 12.5 Three types of sponge structure. The degree of complexity from simple asconoid to complex leuconoid type has involved mainly the water-canal and skeletal systems, accompanied by outfolding and branching of the collar-cell layer. The leuconoid type is considered the major plan for sponges, for it permits greater size and more efficient water circulation.

construction is of clear adaptive value; it has the highest proportion of flagellated surface area for a given volume of cell tissue, so it efficiently meets food demands. This leuconoid grade has evolved independently many times in sponges.

Types of Canal Systems

Asconoids   Asconoid sponges have the simplest organization. The sponge draws water inside through microscopic dermal pores by the beating of large numbers of flagella on the choanocytes. These choanocytes line the internal cavity called the spongocoel. As the choanocytes filter the water and extract food particles from it, used water is expelled through a single large osculum (Figure 12.5). This design

has distinct limitations because choanocytes line the spongocoel and can collect food only from water directly adjacent to the spongocoel wall. Were the spongocoel large, most of the water and food in its central cavity would lie in "dead space," inaccessible to choanocytes. Thus, asconoid sponges are small and tube-shaped. As an example, examine *Leucosolenia* (Gr. *leukos,* white, + *solen,* pipe), where slender, tubular individuals grow in groups attached by a common stolon, or stem, to objects in shallow seawater (Figure 12.5). *Clathrina* (L. *clathri,* latticework), another asconoid, has bright yellow, intertwined tubes (Figure 12.6). Asconoids occur only in class Calcispongiae.

Syconoids   Syconoid sponges look somewhat like larger asconoids. They have a tubular body and single osculum, but the body wall, which is really the spongocoel lining, is thicker and more complex than that of asconoids. The lining has been folded outward to make choanocyte-lined canals (Figure 12.5). Folding the body wall into canals increases the surface area of the wall and thus increases the surface area covered by choanocytes. The canals are of small diameter compared with an asconoid spongocoel, so most of the water in a canal is accessible to choanocytes.

Water enters the syconoid body through dermal ostia that lead into incurrent canals. It then filters through tiny openings, or **prosopyles,** into the **radial canals** (Figure 12.7). Here food is ingested by the choanocytes. The beating of the choanocytes' flagella forces the used water through internal pores, or **apopyles,** into the spongocoel. Notice that food capture does not occur in the syconoid spongocoel,

so it is lined with epithelial-type cells rather than the flagellated cells present in asconoids. After the used water reaches the spongocoel, it exits the body through an **osculum.** As an example, examine *Sycon* (Gr. *sykon,* a fig) in Figure 12.5.

During development, syconoid sponges pass through an asconoid stage, following which flagellated canals form by evagination of the body wall. This developmental pattern provides evidence that syconoid sponges were derived from an ancestor with an asconoid body plan, but the syconoid condition is not homologous among all the sponges that possess it. Syconoids occur in class Calcispongiae and in some members of class Hexactinellida.

Leuconoids   Leuconoid organization is the most complex of the sponge types and permits an increase in sponge size. In the leuconoid design, the surface area of the food-collecting regions with choanocytes is greatly increased; here the choanocytes line the walls of small chambers where they can effectively filter all the water present (Figure 12.5). The sponge body comprises an enormous number of these tiny chambers. Clusters of flagellated chambers are filled from incurrent canals and discharge water into excurrent canals that eventually lead to an osculum (Figure 12.8).

A sponge pumps a remarkable amount of water. *Leuconia* (Gr. *leukos,* white), for example, is a small leuconoid sponge about 10 cm tall and 1 cm in diameter. It is estimated that water enters through some 81,000 incurrent canals at a velocity of 0.1 cm/second in each canal. However, because water passes into flagellated chambers with a greater cross-sectional area than those of the entry canals, water flow through the chambers slows to 0.001 cm/second. Such a flow rate allows ample opportunity for food capture by choanocytes. *Leuconia* has more than 2 million flagellated chambers where food collection occurs.

Figure 12.6   *Clathrina canariensis* (class Calcispongiae) is common on Caribbean reefs in caves and under ledges.

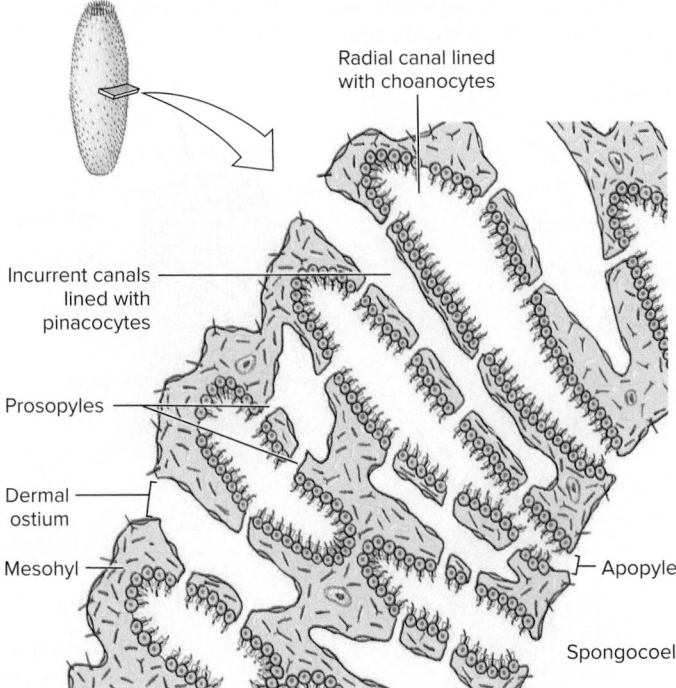

Figure 12.7   Cross section through wall of sponge *Sycon,* showing choanocytes in canals within the wall. Notice choanocytes do not line spongocoel.

Radial canal lined with choanocytes

Incurrent canals lined with pinacocytes

Prosopyles

Dermal ostium

Mesohyl

Apopyle

Spongocoel

Figure 12.8   This orange demosponge, *Mycale laevis,* often grows beneath platelike colonies of the stony coral *Montastrea annularis.* The large oscula of the leuconoid canal system are seen at the edges of the plates. Unlike some other sponges, *Mycale* does not burrow into the coral skeleton and may actually protect coral from invasion by more destructive species. Pinkish radioles of a Christmas tree worm, *Spirobranchus giganteus* (phylum Annelida, class Polychaeta) also project from the coral colony. An unidentified reddish sponge can be seen to the right of the Christmas tree worm.

After food is removed, the used water is pooled to form an exit stream. The exit stream, containing the entire volume of water that entered the sponge over the myriad incurrent canals, leaves the sponge through an exit pore whose cross-sectional area is many times less than the total cross-sectional area of all the incurrent canals. The relatively small size of the exit pore, together with the large volume of used water, produces a very high exit velocity. In *Leuconia,* all water is expelled through a single osculum at a striking velocity of 8.5 cm/second—a jet force capable of carrying used water and wastes far enough from the sponge to avoid refiltering.

Some large sponges can filter 1500 liters of water a day, but unlike *Leuconia,* most leuconoids form large masses with numerous oscula (Figures 12.5 and 12.8), so that water exits from many local sites on the sponge. Most sponges are of the leuconoid type; leuconoid bodies account for most species within class Calcispongiae and are the most common types in other classes.

Types of Cells in the Sponge Body

Sponge cells are arranged in a gelatinous extracellular matrix called **mesohyl,** or **mesenchyme** (Figure 12.9). The mesohyl is the connective tissue of sponges; in it are found various fibrils, skeletal elements, and ameboid cells. The absence of organs means that all fundamental processes must occur at the level of individual cells. Respiration and excretion occur by diffusion in each cell and, in freshwater sponges, excess water is expelled via contractile vacuoles in archaeocytes and choanocytes.

The visible activities and responses in sponges, other than propulsion of water, are alterations in shape, local contractions, propagating contractions, and closing and opening of incurrent and excurrent pores. Incurrent pores can close in response to heavy sediment in the water or other conditions that reduce efficiency of feeding. The most common response is closure of the oscula. These movements are very slow, but the fact that there are responses of the whole body in animals that lack organization above the cellular level

is puzzling. Apparently excitation spreads from cell to cell by an unknown mechanism; suggested mechanisms include mechanical stimuli and signaling molecules, possibly hormones. Electrical communication across the syncytial tissue of hexactinellid sponges (see p. 255) has been demonstrated, but nothing similar has been found in cellular sponges. Some zoologists point to the possibility of coordination by means of substances carried in the water currents, and others have tried, not very successfully, to demonstrate the presence of nerve cells. Although nerve cells have not been found, several other types of cells do occur.

Choanocytes   Choanocytes, which line flagellated canals and chambers, are ovoid cells with one end embedded in mesohyl and the other exposed. The exposed end bears a flagellum surrounded by a collar (Figures 12.9 and 12.10). The collar comprises adjacent

Figure 12.9   Small section through sponge wall, showing four types of sponge cells. Pinacocytes are protective and contractile; choanocytes create water currents and engulf food particles; archaeocytes have a variety of functions; collencytes secrete collagen.

Figure 12.10   Food trapping by sponge cells. **A,** Cutaway section of canals showing cellular structure and direction of water flow. **B,** Two choanocytes and **C,** structure of the collar. In part C, the small red arrows indicate movement of food particles.

microvilli, connected to each other by delicate microfibrils, forming a fine filtering device for straining food particles from water (Figure 12.10B). The beating flagellum pulls water through the sieve-like collar and forces water out through the open top of the collar. Particles too large to enter the collar become trapped in secreted mucus and slide down the collar to the base, where they are phagocytized by the cell body. Larger particles already have been excluded by the small size of the dermal pores and prosopyles. Food engulfed by the cells is passed to a neighboring archaeocyte for digestion. Thus, digestion is entirely **intracellular** (see Chapter 32, p. 707), so there is no extracellular gut cavity. Choanocytes also have a role in sexual reproduction.

Archaeocytes   **Archaeocytes** are ameboid cells that move in the mesohyl (Figures 12.9 and 12.10) and perform a number of functions. They can phagocytize particles at the pinacoderm and receive particles for digestion from choanocytes. Archaeocytes apparently can differentiate into any of the other types of more specialized cells in the sponge. Some, called **sclerocytes,** secrete spicules. Others, called **spongocytes,** secrete the spongin fibers of the skeleton, and **collencytes** secrete fibrillar collagen (p. 191). **Lophocytes** secrete large quantities of collagen but are distinguishable morphologically from collencytes.

Pinacocytes   The nearest approach to a true tissue in sponges is arrangement of the **pinacocyte** cells of the **pinacoderm** (Figures 12.9 and 12.10). A true tissue is a grouping of cells specialized for one function; a true tissue epithelium consists of a layer of specialized cells with particular intercellular junctions resting on a fibrous basal membrane. Pinacocytes are thin, flat, epithelial-like cells that cover the exterior surface and some interior surfaces of a sponge. Some are T-shaped with their cell bodies extending into the mesohyl. A layer of pinacocytes does not constitute an epithelium because a basal membrane is lacking in most sponges. However, a basal lamina does occur in homoscleromorph sponges, along with a distinct type of collagen found in other metazoans. Pinacoderm cells also have a variety of intercellular junctions, making the pinacoderm sufficiently specialized to be called an incipient tissue by some (see box, p. 248), and a true tissue epithelium by others.

Pinacocytes can ingest food particles by phagocytosis at the sponge surface. Pinacocytes are somewhat contractile and help to regulate surface area of a sponge. Some pinacocytes are modified as contractile **myocytes,** which are usually arranged in circular bands around oscula or pores, where they help to regulate rate of water flow.

Cell Independence: Regeneration and Somatic Embryogenesis

Sponges have a tremendous ability to repair injuries and to restore lost parts, a process called regeneration. Regeneration implies reorganization not of the entire animal, but only of the wounded portion. However, a complete reorganization of the structure and function of participating cells or bits of tissue does occur in somatic embryogenesis. If a sponge is cut into small fragments, or if the cells of a sponge are entirely dissociated and are allowed to fall into small groups, or aggregates, entire new sponges can develop from these fragments or aggregates of cells. This process has been termed **somatic embryogenesis.** Somatic embryogenesis involves a complete reorganization of the structure and functions of participating cells or bits of tissue. Isolated from influence

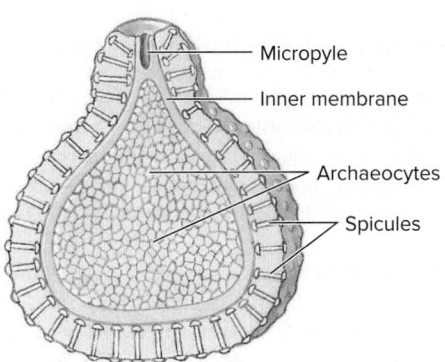

Figure 12.11   Section through a gemmule of a freshwater sponge (Spongillidae). Gemmules are a mechanism for survival of the harsh conditions of winter. On return of favorable conditions, the archaeocytes exit through the micropyle to form a new sponge. The archaeocytes of the gemmule give rise to all cell types of the new sponge structure.

of adjoining cells, they can realize their own potential to change in shape or function as they develop into a new organism.

Much experimental work has been done in this field. The process of reorganization appears to differ in sponges of differing complexity. There is still some controversy concerning just what mechanisms cause adhesion of the cells and the role that each type of cell plays in the formative process.

Regeneration following fragmentation is one means of asexual reproduction: in fragmentation a sponge breaks into parts that are each capable of becoming a new sponge. Asexual reproduction can also occur by bud formation. **External buds,** after reaching a certain size, may become detached from the parent and float away to form new sponges, or they may remain to form colonies. **Internal buds,** or **gemmules** (Figure 12.11), are formed in freshwater sponges and some marine sponges. Here, archaeocytes collect in the mesohyl and become surrounded by a tough spongin coat incorporating siliceous spicules. When the parent animal dies, the gemmules survive and remain dormant, preserving the species during periods of freezing or severe drought. Later, cells in the gemmules escape through special openings, the **micropyles,** and develop into new sponges. Gemmulation in freshwater sponges (Spongillidae) is thus an adaptation to changing seasons. Gemmules are also a means of colonizing new habitats, since they can spread by streams or animal carriers. What prevents gemmules from hatching during the season of formation rather than remaining dormant? Some species secrete a substance that inhibits early germination of gemmules, and gemmules do not germinate as long as they are held in the body of the parent. Other species undergo maturation at low temperatures (as in winter) before they germinate. Gemmules in marine sponges also seem to be an adaptation to pass the cold of winter; they are the only form in which *Haliclona loosanoffi* exists during the colder parts of the year in the northern part of its range.

Sexual Reproduction

In **sexual reproduction** most sponges are **monoecious** (have both male and female sex cells in one individual). Sperm sometimes arise from transformation of choanocytes. In Calcispongiae and at least some

Taxonomy of Phylum Porifera

Class Calcispongiae (kal-se-spun′je-e) (L. *calcis*, lime). Have spicules of calcium carbonate that often form a fringe around the osculum (main water outlet); spicules needle-shaped or three or four rayed; all three types of canal systems (asconoid, syconoid, leuconoid) represented; all marine. Examples: *Sycon, Leucosolenia, Clathrina.*

Class Hexactinellida (hex-ak-tin-el′i-da) (Gr. *hex*, six, + *aktis*, ray, + L. *-ellus*, dim. suffix) **(Hyalospongiae).** Have six-rayed, siliceous spicules extending at right angles from a central point; spicules often united to form network; body often cylindrical or funnel-shaped; flagellated chambers in simple syconoid or leuconoid arrangement; habitat mostly deep water; all marine. Examples: Venus' flower basket *(Euplectella), Hyalonema.*

Class Demospongiae (de-mo-spun′je-e) (Gr. *demos*, people, + *spongos*, sponge). Have siliceous spicules that are not six rayed, or spongin, or both; leuconoid-type canal systems; one family found in freshwater; all others marine. Examples: *Amphimedon, Cliona, Spongilla, Myenia, Poterion, Callyspongia,* and all bath sponges.

Class Homoscleromorpha (hō-mō-′skle-rō-mor-f∂) (Gr. *homos*, same, + *skleros*, hard, + *morphe*, form). Previously a subgroup of Demospongiae; spicules may be absent as in *Oscarella;* if present, spicules are small, are simple in shape, and do not form around an axial filament; pinacoderm with a distinct basement membrane. Examples: *Oscarella, Corticium.*

such sponges, sperm are released into the water by one individual and taken into the canal system of another. There choanocytes phagocytize the sperm; then the choanocytes transform into carrier cells, which carry the sperm through the mesohyl to oocytes. Other sponges are oviparous, and both oocytes and sperm are expelled into the surrounding water. The free-swimming larva of most sponges is a solid-bodied **parenchymula** (Figure 12.12A), although six other larval types exist, and some sponges exhibit direct development. The outwardly directed, flagellated cells of the parenchymula migrate to the interior after the larva settles and become choanocytes in the flagellated chambers.

Calcispongiae and a few Demospongiae have a very strange developmental pattern. A hollow blastula, called an stomoblastula (Figure 12.12B), develops, with flagellated cells toward the interior. The blastula then turns *inside out* **(inversion),** the flagellated ends of the cells becoming directed to the outside! Flagellated cells **(micromeres)** of the amphiblastula larva are at the anterior end, and larger, nonflagellated cells **(macromeres)** are at the posterior end. In contrast to other metazoan embryos, the micromeres invaginate into and are overgrown by the macromeres at metamorphosis during settlement. The flagellated micromeres become choanocytes, archaeocytes, and collencytes of the new sponge, and the nonflagellated cells give rise to pinacoderm and sclerocytes.

Class Calcispongiae

Calcispongiae are calcareous sponges, so called because their spicules are composed of calcium carbonate. Spicules are straight (monaxons) or have three or four rays. These sponges tend to be small—10 cm or less in height—and tubular or vase-shaped. They may be asconoid, syconoid, or leuconoid in structure. Although many are drab in color, some are bright yellow, red, green, or

Demospongiae, oocytes also develop from choanocytes; in other demosponges gametes apparently are derived from archaeocytes. Most sponges are viviparous; after fertilization the zygote is retained in and derives nourishment from its parent, and a ciliated larva is released. In

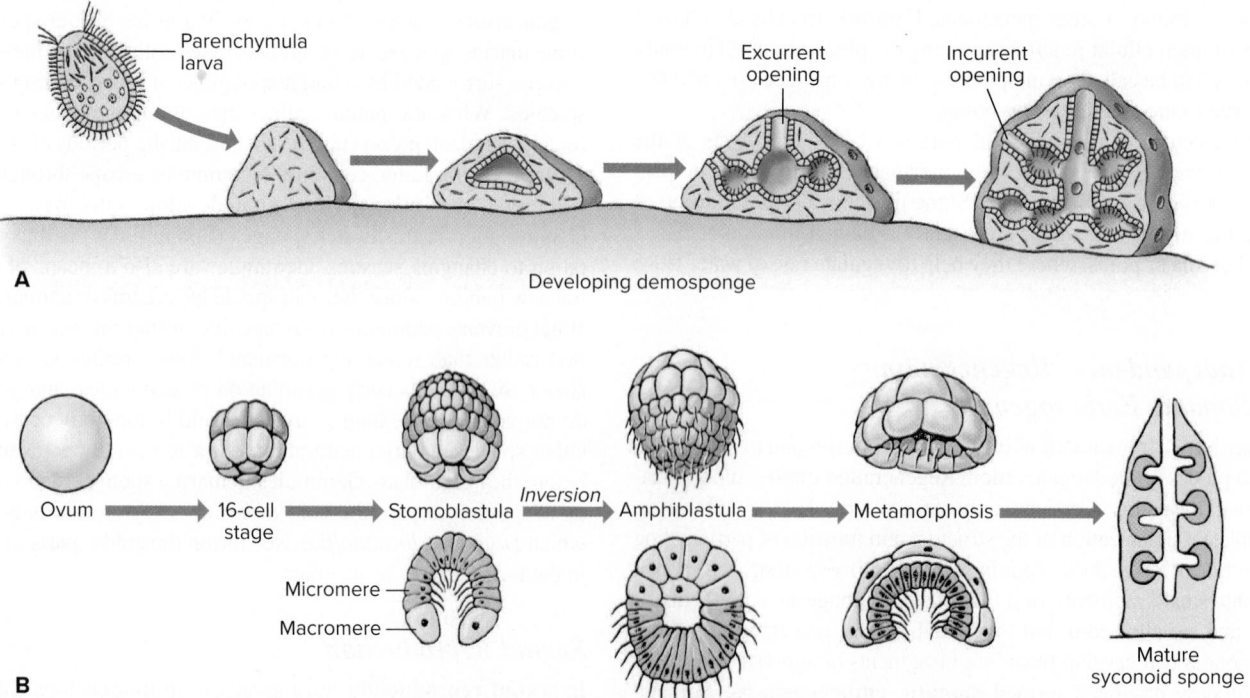

Figure 12.12   **A,** Development of demosponges. **B,** Development of the calcareous syconoid sponge *Sycon.*

lavender. *Leucosolenia* and *Sycon* (often called *Scypha* or *Grantia* by biological. supply companies) are marine shallow-water forms commonly studied in the laboratory (see Figure 12.5). *Leucosolenia* is a small asconoid sponge that grows in branching colonies,

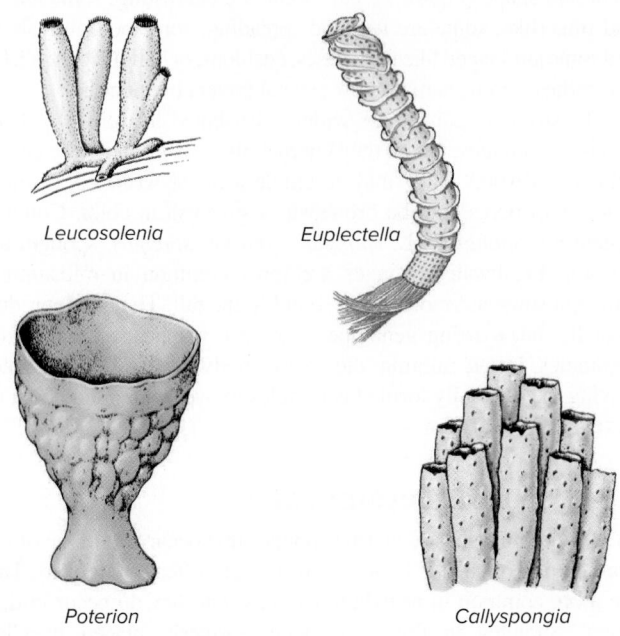

Figure 12.13 Some sponge body forms. *Leucosolenia* is in Calcispongiae, *Euplectella* is in Hexactinellida, *Poterion* and *Callyspongia* are members of Demospongiae.

Leucosolenia

Euplectella

Poterion

Callyspongia

usually arising from a network of horizontal, stolonlike tubes (Figure 12.13). *Clathrina* is small with intertwined tubes (see Figure 12.6). *Sycon* is a solitary sponge that may live singly or form clusters by budding. The vase-shaped, typically syconoid animal is 1 to 3 cm long, with a fringe of straight spicules around the osculum to discourage small animals from entering.

Class Hexactinellida (Hyalospongiae): Glass Sponges

Glass sponges form class Hexactinellida (or Hyalospongiae). Nearly all are deep-sea forms that are collected by dredging. Most are radially symmetrical, with vase- or funnel-shaped bodies usually attached by stalks of root spicules to a substratum (Figure 12.13, *Euplectella*) (N. L. from Gr. *euplektos,* well plaited). They range from 7.5 cm to more than 1.3 m in length. Their distinguishing features include a skeleton of six-rayed siliceous spicules that are commonly bound together into a network forming a glasslike structure.

Their tissue structure differs dramatically from that of other sponges in that it is syncytial—there are many nuclei inside a single very large cell. Syncytia are produced by fusion of many cells, or by repeated divisions of a cell nucleus without division of the cell cytoplasm. The body of a hexactinellid sponge is composed of a single, continuous syncytial tissue called a **trabecular reticulum.** A glass sponge 1 m in diameter makes the largest syncytium reported for the animal kingdom.

The trabecular reticulum is bilayered and can be sheetlike or tubular. Between the layers of the sheet, or inside the tubes, is a thin,

Figure 12.14 Diagram of part of a flagellated chamber of hexactinellids. The primary and secondary reticula are branches of the trabecular reticulum, which is syncytial. Cell bodies of the choanoblasts and their processes are borne by the primary reticulum and are embedded in a thin, collagenous mesohyl. Processes of the choanoblasts end in collar bodies, whose collars extend up through the secondary reticulum. Flagellar action propels water *(arrows)* to be filtered through the mesh of collar microvilli (see Figure 12.10).

collagenous mesohyl in which cells, such as archaeocytes or choanoblasts, occur (Figure 12.14). Choanoblasts and other cells are connected to each other, and to the trabecular reticulum, by cytoplasmic bridges. Choanoblasts are unusual cells that make two or more flagellated outgrowths called **collar bodies.** The flagellum on a collar body beats to drive water flow in the same way that it would on a choanocyte.

An assemblage of collar bodies forms a flagellated chamber. Here, the trabecular reticulum branches to become two distinct bilayered sheets: a primary reticulum and a thin, secondary reticulum that lacks mesohyl. The two sheets make a sandwich around the center of a collar body (Figure 12.14). Collar bodies extend through openings in both sheets, but the openings surround the collar bodies tightly. There is a space between the two sheets. To collect food, the incurrent water is directed to the primary reticulum, where pores lead into the space between the primary and secondary reticular sheets. Water entering this space must leave by moving through the mesh of microvilli on collar bodies; water cannot go anywhere else because of the presence of the secondary reticulum. Food particles captured on microvilli are shared throughout the syncytium.

The syncytial nature of these unusual sponges might suggest a syncytial origin for metazoans, but the details of development refute this idea. The tissue of the reticulum forms after typical embryonic cleavage and blastula formation. Following the 32-cell cleavage stage, new cells remain connected via cytoplasmic bridges, and the syncytium forms through a combination of cell fusion and envelopment. Thus, the animal is initially cellular.

The latticelike network of spicules found in many glass sponges is of exquisite beauty, such as that of *Euplectella,* or Venus' flower basket (Figure 12.13), a classic example of Hexactinellida.

Class Demospongiae

This group contains 95% of living sponge species, including most large sponges. Spicules are siliceous, but are not six rayed. Spicules may be bound together by spongin, or may be absent. So-called bath sponges, *Spongia* and *Hippospongia,* belong to the group called horny sponges, which have spongin skeletons and lack siliceous spicules entirely. All members of the class are leuconoid, and all are marine, except for members of the freshwater family Spongillidae.

Marine Demospongiae are quite varied and may be striking in color and shape (Figure 12.15). Some are encrusting; some are tall and fingerlike; some are low and spreading; some bore into shells; and some are shaped like fans, vases, cushions, or balls (Figure 12.15). Loggerhead sponges may grow several meters in diameter.

Freshwater sponges are widely distributed in well-oxygenated ponds and streams, where they encrust plant stems and old pieces of submerged wood. They may resemble a bit of wrinkled scum, be pitted with pores, and be brownish or greenish in color. Common genera are *Spongilla* (L. *spongia,* from Gr. *spongos,* sponge) and *Myenia.* Freshwater sponges are most common in midsummer, although some are more easily found in the fall. They do reproduce sexually, but existing genotypes may also reappear annually from gemmules. In late autumn, the sponge body dies and disintegrates, leaving the asexually formed gemmules to overwinter and begin the next year's population.

Class Homoscleromorpha

Homoscleromorphs are marine sponges that occur in a range of colors, but live in cryptic habitats, so they are often overlooked. They are more common in nearshore habitats, but they do occur in deep water. Sponges in this class were formerly placed in Class Demospongiae but have been separated because they possess unique features such as a pinacoderm layer with a true basement membrane or ECM (see p. 253). The cells of this layer also differ from those in cells of other sponge classes because they connect not only to each other, but also to the ECM with special adherens cell junctions. Thus some researchers consider the pinacoderm a true tissue. However, in a true tissue the cells connect to each other via proteins called cadherins in particular junctions called desmosomes. There are no desmosomes in the homoscleromorph pinacoderm, so it may be more correct to call the pinacoderm an incipient tissue, not a true tissue. Homoscleromorpha is divided into two clades, one whose members

A **B** **C**

Figure 12.15 Marine Demospongiae on Caribbean coral reefs. **A,** *Pseudoceratina crassa* is a colorful sponge growing at moderate depths. **B,** *Aplysina fistularis* is tall and tubular. **C,** *Monanchora unguifera* with commensal brittle star, *Ophiothrix suensoni* (phylum Echinodermata, class Ophiuroidea).

lack spicules entirely, and the other with spicules that do not form around a central longitudinal (axial) filament. Representative genera are *Plakina*, *Oscarella*, and *Corticium*.

Phylogeny and Adaptive Diversification

Phylogeny

Sponges originated before the Cambrian period. Two groups of calcareous spongelike organisms occupied early Paleozoic reefs. The Devonian period saw rapid development of many glass sponges. Sponges are the sister taxon to a group composed of all other animal phyla, as shown on the Cladogram of All Animal Phyla inside the front cover. The simple body plans of most sponges, aside from hexactinellids, might suggest that sponges share few features with other animals, but this is not the case. To form a multicellular body, cells must adhere to each other in a stable fashion. The proteins used in cell adhesion and cell signaling in sponges are homologous to those in other animals; in fact many of these are present in choanoflagellates, evolving before the last common ancestor of all animals. The animal epithelium is a unique structure where cells adhere to each other in a single layer attached to a basement membrane that contains collagen type IV. Adherens junctions in which cadherin molecules function as adhesives[2] connect cells in an epithelium to the basement membrane, but a true epithelium also requires the cells to link to each other using particular junctions called desmosomes. Because the homoscleromorphs do not have these latter junctions, but do have the former adhesive junctions, there is debate as to whether sponges have the true tissue epithelia that are elaborated in other animals. Sponge development includes the characteristic animal blastula stage, and some sponges actually develop to the two-layered gastrula stage before reorganizing their bodies into asymmetrical adults. It is possible that the sessile lifestyle of sponges favored a deceptively simple body in most species and that a closer look at sponges will reveal more features typical of animals.

Phylogenetic studies[3] using sequence data from large subunit rRNA, small subunit rRNA, and protein kinase C indicate that sponges with calcareous spicules in the class Calcispongiae belong to a separate clade from those with spicules made of silica in classes Demospongiae and Hexactinellida. Two potential placements emerge for calcareous sponges: in one, calcareous sponges are the sister taxon to a clade of siliceous sponges, as we show in Figure 12.4, and in another, the phylum Porifera is paraphyletic because the calcareous "sponges" are more closely related to other animal taxa than they are to siliceous sponges.

Adaptive Diversification

Porifera are a highly successful group that includes several thousand species in a variety of marine and freshwater habitats. Their diversification centers largely on their unique water-current system and its various degrees of complexity. However, within the siliceous Demospongiae, a new feeding mode has evolved for a family of sponges inhabiting

Figure 12.16 The carnivorous sponge, *Chondrocladia lyra*, is commonly called a "harp sponge." It lives off the coast of California at depths down to 3500 meters, where it was discovered in 2012 by researchers using undersea robots operated by the Monterey Bay Aquarium Research Institute (MBARI).

nutrient-poor deep-water caves. These deep-water sponges have a fine coating of tiny hooklike spicules over their highly branched bodies. The spicule layer entangles the appendages of tiny crustaceans swimming near the surface of the sponge. Later, filaments of the sponge body grow over prey, enveloping and digesting them. These sponges are carnivores, not suspension feeders, although some of them may augment their diets with nutrients obtained from symbiotic methanotrophic bacteria. The presence of the typical siliceous spicules clearly identifies these animals as sponges, but they lack choanocytes and internal canals.

The loss of choanocytes in these species is doubtless disturbing for students learning to identify sponges, but students of evolution should be fascinated by it. The convoluted pathway taken by one branch of the sponge lineage clearly illustrates the nondirectional nature of evolution. To colonize such a nutrient-poor habitat initially, the ancestors of this group must have had at least one alternative feeding system, either carnivory or chemoautotrophy, already in place. Presumably, after the alternative method of food capture was in use, the choanocytes and internal canals were no longer formed. If there are further body modifications in this lineage, we might eventually not recognize the descendants as sponges. Imagine how the lineage would look if spicules were lost in favor of a greater reliance on the bacterial symbionts, and you will begin to understand why it is sometimes hard to trace morphological evolution or to identify the closest relatives of certain animals.

In fact, the discovery of the deep sea harp sponge, *Chondrocladia lyra* (Figure 12.16) shows the astonishing morphology of a predatory sponge. This animal has multiple vanes, each with a basal stolon close to 40 cm in length, anchored by rhizoids. Vertical branches extend nearly 20 cm upward from stolons. Small prey, often copepods (see p. 434), are captured on the branches and consumed by phagocytosis. Terminal balls on each branch contain spermatophores; eggs are held midway up the branches. Body form varies from stalked and spherical to branching in the 36 other species in this genus.

12.3 PHYLUM PLACOZOA

The phylum Placozoa (Gr. *plax, plakos*, tablet, plate, + *zōon*, animal) was proposed in 1971 by K. G. Grell to contain a single species, *Trichoplax adhaerens* (Figure 12.17A), a tiny (2 to 3 mm) marine form. The body is platelike and has no symmetry, no organs, and no muscular or nervous system. It also lacks both a basal lamina beneath

[2]Abedin, M., and N. King. 2010. Diverse evolutionary paths to cell adhesion. Trends in Cell Biology **20**: 734–742.

[3]Borchiellini, C., M. Manuel, E. Alivon, N. Boury-Esnault, J. Vacelet, and Y. Le Parco. 2001. Sponge paraphyly and the origin of Metazoa. J. Evol. Biol. **14**: 171–179; Medina, M., A. G. Collins, J. D. Silberman, and M. L. Sogin. 2001. Evaluating hypotheses of basal animal phylogeny using complete sequences of large and small subunit rRNA. Proc. Nat. Acad. Sci., USA **98**: 9707–9712.

Figure 12.17

A, *Trichoplax adhaerens* is a marine, platelike animal only 2 to 3 mm in diameter. **B,** Section through *Trichoplax adhaerens,* showing histological structure.

the epidermis and an extracellular matrix (ECM), two features that were considered animal hallmarks. However, the genome contains many genes that encode predicted ECM proteins. The body of a placozoan is composed of a dorsal epithelium of cover cells and shiny spheres and a thick ventral epithelium containing monociliated cells (cylinder cells) and nonciliated gland cells (Figure 12.17B). The space between the epithelia contains multinucleate fibrous contractile cells. There are four cell types distinguished morphologically, but gene-expression studies suggest the presence of a fifth type.

Placozoans glide over their food, secrete digestive enzymes on it, and then absorb the products. In the laboratory, they feed on organic matter and small algae.

The life cycle of placozoans is not completely known. They divide asexually and produce "swarmer" stages by budding. Although sexual reproduction has not been observed, eggs occur in laboratory animals. Genetic studies of placozoans from around the world show that eight distinct lineages equivalent to species exist, although they cannot be distinguished morphologically. Sexual reproduction has been inferred from molecular evidence of genetic diversity within a clade.

Grell considers *Trichoplax* diploblastic (see p. 187), with dorsal epithelium representing ectoderm and ventral epithelium representing endoderm because of its nutritive function. Gene-expression studies support these homologies. The origin of the middle fibrous layer is currently under study. As this group becomes better understood, the branching order for Placozoa and the two diploblastic phyla (see Chapter 13) may become clear relatively soon. At present we depict branching of placozoans, cnidarians, and ctenophorans as a polytomy (see the Cladogram of All Animal Phyla on inside front cover).

SUMMARY

Sponges (phylum Porifera) are an abundant marine group with some freshwater representatives. They have various specialized cells, but these do not form organs. They depend on the flagellar beat of their choanocytes to circulate water through their bodies for food gathering and respiratory gas exchange. They are supported by secreted skeletons of fibrillar collagen, collagen in the form of large fibers or filaments (spongin), calcareous or siliceous spicules, or a combination of spicules and spongin in most species.

Sponges reproduce asexually by budding, fragmentation, and gemmules (internal buds). Most sponges are monoecious but produce sperm and oocytes at different times. Embryogenesis is unusual, with a migration of flagellated cells at the surface to the interior (in parenchymella larvae) or the production of an amphiblastula larva with inversion and growth of macromeres over micromeres. Sponges have great regenerative abilities.

Sponges are an ancient group, seemingly remote phylogenetically from other metazoa, but molecular evidence suggests that they are the sister group to Eumetazoa. Their adaptive diversification is centered on elaboration of the water circulation and filter-feeding system, except for one family of sponges where filter-feeding has been replaced by carnivory and reliance on bacterial symbionts for extra nutrition.

Phylum Placozoa is represented by a small platelike marine organism. It has only two cell layers with a layer of fibrous syncytial cells between them. Some workers hypothesize that these layers are homologous to ectoderm and endoderm of more complex animals. Genetic studies indicate that there are eight species of placozoans.

REVIEW QUESTIONS

1. The term "metazoan" is synonymous with "multicellular animal." Are there any single-celled animals? How are single-celled eukaryotes classified?
2. Give eight characteristics of sponges.
3. Briefly describe asconoid, syconoid, and leuconoid body types in sponges.
4. What sponge body type appears most efficient and makes possible the largest body size?
5. Define the following: ostia, osculum, spongocoel, apopyles, prosopyles, spicules.
6. Define the following: pinacocytes, choanocytes, archaeocytes, sclerocytes, spongocytes, collencytes.
7. What material is found in the skeleton of all sponges?
8. Describe the skeletons of each class of sponges.
9. Describe how sponges feed, respire, and excrete.
10. What is a gemmule?
11. Why are glass sponges distinguished from sponges with cellular bodies?
12. Describe possible ancestors to sponges. Justify your answer.
13. Describe the body plan of Placozoa.

14. What features make placozoans interesting from a phylogenetic perspective?

For Further Thought Mutualisms are interactions between two species where both participants benefit. How might the relationship between a sponge with glassy spicules and microalgae or cyanobacteria be a mutualism?

SELECTED REFERENCES

Abedin, M., and N. King. 2010. Diverse evolutionary paths to cell adhesion. Trends in Cell Biology **20:** 734–742. *A clear description of the ways cells attach to each other.*

Borchiellini, C., M. Manuel, E. Alivon, N. Boury-Esnault, J. Vacelet, and Y. Le Parco. 2001. Sponge paraphyly and the origin of Metazoa. J. Evol . Biol. **14:** 171–179. *Results of this study suggest that members of Class Calcispongiae (Calcarea) are more closely related to other animals than to siliceous sponges.*

Brummer, F., M. Pfannkuchen, A. Baltz, T. Hauser, and V. Thiel. 2008. Light inside sponges. J. Exp. Mar. Biol. Ecol. **367:** 61–64. *Measurement of light transmission through spicules in sponges.*

Carr, M., B. S. C. Leadbeater, R. Hassan, M. Nelson, and S. L. Baldauf. 2008. Molecular phylogeny of choanoflagellates, the sister group to Metazoa. PNAS **105:** 16641–16646. *Discussion of coloniality and the nature of the common ancestor of choanoflagellates and metazoans.*

Dohrmann, M., D. Janussen, J. Reitner, A. G. Collins, and G. Worheide. 2008. Phylogeny and evolution of glass sponges (Porifera, Hexactinellida). Syst. Biol. **57:** 388–405. *Homoscleromorpha forms a clade within the monophyletic sponges.*

Grell, K. G. 1982. Placozoa. In S. P. Parker (ed.), Synopsis and classification of living organisms, vol. 1. New York, McGraw-Hill Book Company. *Synopsis of placozoan characteristics.*

Hooper, J. N. A., and R. W. M. van Soest (eds.). 2002. Systema Porifera: a guide to the classification of sponges. New York, Kluwer Academic/Plenum. *A large and comprehensive work on sponge systematics and biology.*

King, N., C. T. Hittinger, and S. B. Carroll. 2003. Evolution of key cell signaling and adhesion protein families predates the origin of animals. Science **301:** 361–363. *Cells in multicellular animals must aggregate and communicate. Proteins responsible for these functions in metazoans are homologous to those in choanoflagellates.*

Lee, W. M., H. M. Reiswig, W. C. Austin, and I. Lundsen. 2012. An extraordinary new carnivorous sponge, *Chondrocladia lyra,* in the new subgenus, Symmetrocladia (Demospongiae, Cladorhizidae), from off of northern California, USA. Invertebrate Biology **131:** 259–284. *A fascinating description of the morphology and biology of this animal.*

Leys, S. P., and A. E. Ereskovsky. 2006. Embryogenesis and larval differentiation in sponges. Can. J. Zool. **84:** 262–287. *A review of sponge development with clearly explained terms and excellent photomicrographs.*

Leys, S. P., G. O. Mackie, and H. M. Reiswig. 2007. The biology of glass sponges. Advances in Marine Biology **52:** 1–145. *Provides descriptions, photos, and drawings of these fascinating animals.*

Leys, S. P., and R. W. Meech. 2006. Physiology of coordination in sponges. Can. J. Zool. **84:** 288–306. *Research shows how sponge cells communicate.*

Leys, S. P., S. A. Nichols, and E. D. M. Adams. 2009. Epithelia and integration in sponges. Integ. Comp. Biol. **49:** 167–177. *A nice discussion of the nature of epithelia and their existence in sponges.*

McMurray, S. E., J. E. Blum, and J. R. Pawlik. 2008. Redwood of the reef: growth and age of the giant barrel sponge *Xestospongia muta* in the Florida Keys. Marine Biology **155:** 159–171. *Some giant sponges may be 2300 years old.*

Medina, M., A. G. Collins, J. D. Silberman, and M. L. Sogin. 2001. Evaluating hypotheses of basal animal phylogeny using complete sequences of large and small subunit rRNA. Proc. Nat. Acad. Sci., USA **98:** 9707–9712. *Molecular data support placement of calcareous sponges in a clade distinct from siliceous sponges.*

Reiswig, H. M., and D. Mehl. 1991. Tissue organization of *Farrea occa* (Porifera, Hexactinellida). Zoomorphology **110:** 301–311. *A description of the whole sponge body, as well as the syncytium and spicules, for a glass sponge.*

Schierwater, B. 2005. My favorite animal, *Trichoplax adhaerens.* Bioessays **27:** 1294–1302. *A personal description of the author's fascination with this animal.*

Srivastava, M., E. Begovic, J. Chapman, N. H. Putman, U. Hellsten, T. Kawashima, A. Kuo, T. Mitros, A. Salamov, M. L. Carpenter, A. Y. Signorovitch, M. A. Moreno, K. Kamm, J. Grimwood, J. Schmutz, H. Shapiro, I. V. Grigoriev, L. W. Buss, B. Schierwater, S. L. Dellaporta, and D. S. Rokhsar. 2008. The *Trichoplax* genome and the nature of placozoans. Nature **454:** 955–960. *The gene content of placozoans is surprisingly diverse for an animal with such simple morphology. Genes for many kinds of transcription factors and signaling pathways are present, but the function of their products is not known in* Trichoplax. Trichoplax *is placed as the sister taxon to cnidarians and bilaterians here, but another recent phylogeny shows* Trichoplax *as the earliest diverging metazoan.*

Vogel, S. 1981. Life in moving fluids: the physical biology of flow. Princeton, Princeton University Press. *A clear general discussion of how water flow influences animal design, with specific reference to water movement in the sponge body.*

CHAPTER

Cnidarians and Ctenophores

- **PHYLUM CNIDARIA**
- **PHYLUM CTENOPHORA**

The coral Tubastraea coccinea from the Caribbean.

© William C. Ober/Medical Scientific Illustration

A Fearsome Tiny Weapon

Most cnidarians, aside from corals, have soft bodies. Sea anemones and many others are attached to hard surfaces and thus unable to move away from danger. The jellyfishes or jellies that swim are not able to swim against ocean currents. Indeed, we might easily get the false impression that cnidarians provide easy meals for other animals. The truth is, however, that many cnidarians are very effective predators able to kill and eat prey much more highly organized, swift, and intelligent than themselves. They manage these feats because they possess tentacles that bristle with tiny, remarkably sophisticated weapons called nematocysts.

As it is secreted within the cell that contains it, a nematocyst is endowed with potential energy to power its discharge. It is as though a factory manufactured a gun, cocked and ready with a bullet in its chamber, as it rolled off the assembly line. Like a cocked gun, a completed nematocyst requires only a small stimulus to make it fire. Rather than a bullet, a tiny thread bursts from a nematocyst. Achieving a velocity of 2 meters/sec and an acceleration of 40,000 × gravity, it instantly penetrates its prey and injects a paralyzing toxin. A small animal unlucky enough to brush against one of the tentacles is suddenly speared with hundreds or even thousands of nematocysts and quickly immobilized. Some nematocyst threads can penetrate human skin, causing sensations ranging from minor irritation to great pain, even death, depending on the species. A nematocyst is a fearsome, but wondrous, tiny weapon.

The two phyla discussed in this chapter are diploblastic (see The Cladogram of All Animal Phyla on front inside cover), meaning that they have two embryonic cell layers, ectoderm and endoderm, from which adult structures develop. Two layers are produced as the embryo develops from a single-layered blastula to a gastrula (see Chapters 8 and 9). In adult diploblasts, the epidermis develops from ectoderm, and the gut cavity lining or gastrodermis develops from endoderm; this body plan is in marked contrast to that of adult sponges, where there are neither cell layers nor a gut cavity.

A new stage of development, gastrulation, characterizes diploblasts and produces the cell layers of adult animals. Sponges and placozoans usually are excluded from diploblasts. However, as mentioned in Chapter 12, work on sponge development suggests that cell layers do develop in sponge larvae, but disappear as adults become a nonlayered aggregate of different cell types. The developmental sequence for placozoans is not known, but some biologists consider the two adult layers equivalent to derivatives of ectoderm and endoderm. Thus, it may be appropriate to add more phyla to the diploblast category if stages other than the adult are considered, or new homologies are established. Currently, the diploblastic phyla are Cnidaria and Ctenophora, although some might argue that these groups contain triploblastic members (see p. 286). Adult organisms of both groups exhibit radial or biradial symmetry (see p. 185) and are not cephalized.

Familiar cnidarians are sea anemones and jellies (jellyfishes), and some readers may know ctenophores as comb jellies or sea walnuts. Gelatinous animals are making news more often in recent years as their populations impact humans. Problems caused by abundant jellies include beaches closed to swimming, reduced fish harvests, clogged fishing nets and intake valves for desalination and power plants, to name just a few such effects. Abundance has been correlated with abiotic factors such as solar cycles, as well as with anthropogenic effects such as warming oceans, nutrient additions, overfishing that reduces the abundance of competitors, and direct reductions in the populations of predators such as leatherback turtles.

13.1 PHYLUM CNIDARIA

Phylum Cnidaria (ni-da′-rē-ə) (Gr. *knide,* nettle, + L. *aria* [pl. suffix], like or connected with) is an interesting group of more than 9000 species. It includes some of nature's strangest and loveliest creatures: branching, plantlike hydroids; flowerlike sea anemones; jellyfishes; and those architects of the ocean floor, horny corals (sea whips, sea fans, and others) and stony corals whose thousands of years of calcareous house-building have produced great reefs and coral islands (pp. 277–282).

The phylum takes its name from cells called **cnidocytes,** which contain organelles (cnidae) characteristic of the phylum. The most common type of cnida is the **nematocyst** described in the opening essay. Only cnidarians produce cnidocytes, but some ctenophores, molluscs, and flatworms eat hydroids bearing nematocysts, then store and use these stinging structures for their own defense.

Cnidarians are an ancient group with the longest fossil history of any animal, reaching back more than 700 million years. They are widespread in marine habitats, and a few inhabit freshwater. Cnidarians are most abundant in shallow marine habitats, especially in warm temperatures and tropical regions. There are no terrestrial species. Colonial hydroids are usually found attached to mollusc shells, rocks, wharves, and other animals in shallow coastal water, but some species live at great depths. Floating and free-swimming medusae occur in open seas and lakes, often far from shore. Animals such as the Portuguese man-of-war and *Velella* (L. *velum,* veil, + *ellus,* dim. suffix) have floats or sails by which the wind carries them. Although they are mostly sessile, or at best, fairly slow moving or slow swimming, cnidarians are quite efficient predators of organisms that are much swifter and more complex.

Cnidarians sometimes live symbiotically with other animals, often as commensals on the shell or other surface of their host. Certain hydroids (Figure 13.1) and sea anemones commonly live on snail shells inhabited by hermit crabs, providing the crabs some protection from predators. Algal cells frequently live as mutuals in the

A   **B**

Figure 13.1 **A,** A hermit crab with its cnidarian mutuals. The host snail shell is blanketed with polyps of the hydrozoan *Hydractinia symbiolongicarpus.* The crab gets some protection from predation by the cnidarians, and the cnidarians get a free ride and bits of food from their host's meals. **B,** Portion of a colony of *Hydractinia,* showing the types of zooids and the stolon (hydrorhiza) from which they grow.

© William C. Ober/Medical Scientific Illustration

tissues of cnidarians, notably in some freshwater hydras and in reef-building corals. The presence of the algae in reef-building corals limits the occurrence of coral reefs to relatively shallow, clear water where there is sufficient light for the photosynthetic requirements of the algae. These kinds of corals are an essential component of coral reefs, and reefs are extremely important habitats for many other species of invertebrates and vertebrates in tropical waters. We discuss coral reefs further on pages 281 and 282.

Although many cnidarians have little positive economic importance, reef-building corals are an important exception. Fish and other animals associated with reefs provide substantial amounts of food for people, and reefs are of economic value as tourist attractions. Precious coral is used for making jewelry and ornaments, and coral rock for constructing buildings.

Four classes of Cnidaria were traditionally recognized (Figure 13.2): Hydrozoa (most variable class, including hydroids, fire corals, Portuguese man-of-war, and others), Scyphozoa ("true" jellyfishes), Cubozoa (cube jellyfishes), and Anthozoa (largest

class, including sea anemones, stony corals, soft corals, and others). A fifth class, Staurozoa, has been proposed because recent phylogenies show that stauromedusans do not belong within the Scyphozoa. These odd animals do not make medusae, but the polyp body is topped by a medusa-like region (see p. 274).

Form and Function
Dimorphism and Polymorphism in Cnidarians

One of the most interesting—and sometimes puzzling—aspects of this phylum is the dimorphism displayed by many of its members. All cnidarian forms fit into one of two morphological types **(dimorphism):** a **polyp,** or hydroid form, which is adapted to a sedentary or sessile life, and a **medusa,** or jellyfish form, which is adapted for a floating or free-swimming existence (Figure 13.3). The common name "jellyfish" is largely being replaced by the more accurate term "jelly" or "sea jelly."

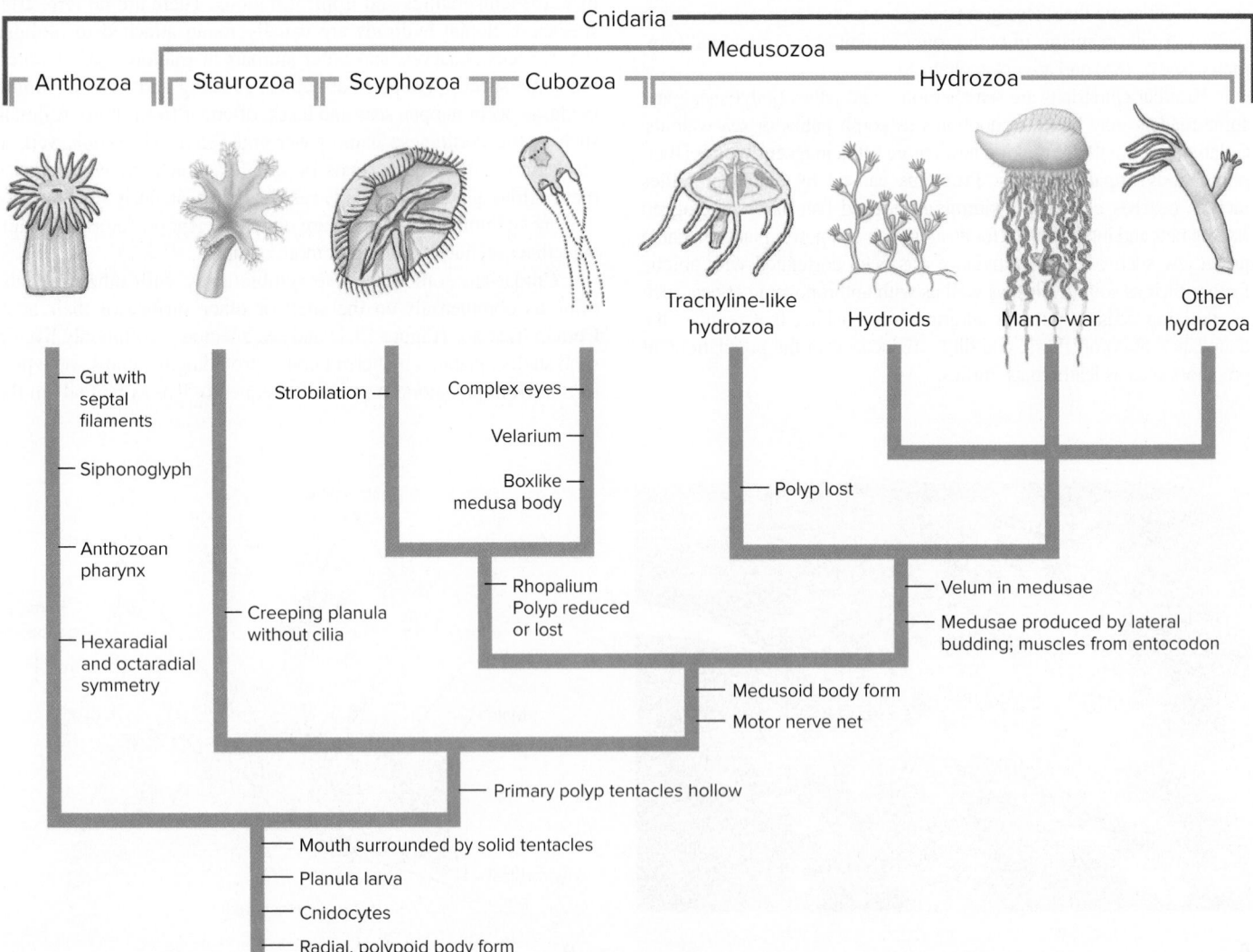

Figure 13.2 Cladogram showing hypothetical relationships of cnidarian classes with some shared derived characters (synapomorphies) indicated. Relationships are according to data of Collins et al. (2006, Syst. Biol. **55:** 97–115). Synapomorphies are adapted from Brusca and Brusca (2003, Invertebrates, ed. 2. Sunderland, Massachusetts, Sinauer Associates, Inc.).

Characteristics of Phylum Cnidaria

1. **Cnidocytes** present, typically housing stinging organelles called **nematocysts**
2. Entirely aquatic, some in freshwater, but most marine
3. **Radial symmetry** or biradial symmetry around a longitudinal axis with oral and aboral ends; no definite head
4. Two types of individuals: **polyps** and **medusae**
5. Adult body two-layered (**diploblastic**) with epidermis and gastrodermis derived from embryonic ectoderm and endoderm, respectively
6. Mesoglea, an extracellular matrix ("jelly"), lies between body layers; amount of mesoglea variable; mesoglea with cells and connective tissue from ectoderm in some
7. Incomplete gut called **gastrovascular cavity;** often branched or divided with septa
8. **Extracellular digestion** in gastrovascular cavity and intracellular digestion in gastrodermal cells
9. Extensible tentacles usually encircle mouth or oral region
10. Muscular contractions via **epitheliomuscular cells,** which form an outer layer of longitudinal fibers at base of epidermis and an inner layer of circular fibers at base of gastrodermis; modifications of plan in hydrozoan medusa (independent ectodermal muscle fibers) and other complex cnidarians
11. Sense organs include well-developed statocysts (organs of balance) and ocelli (photosensitive organs); complex eyes in members of Cubozoa
12. **Nerve net** with symmetrical and asymmetrical synapses; diffuse conduction; two nerve rings in hydrozoan medusae
13. Asexual reproduction by budding in polyps forms clones and colonies; some colonies exhibit **polymorphism**[1] (different polyp types within a colony)
14. Sexual reproduction by gametes in all medusae and some polyps; monoecious or dioecious; holoblastic indeterminate cleavage; planula larval form
15. No excretory or respiratory system
16. No coelomic cavity

[1]Note that polymorphism here refers to more than one structural form of individual within a species, as contrasted with the use of the word in genetics (p. 122), in which it refers to different allelic forms of a gene in a population.

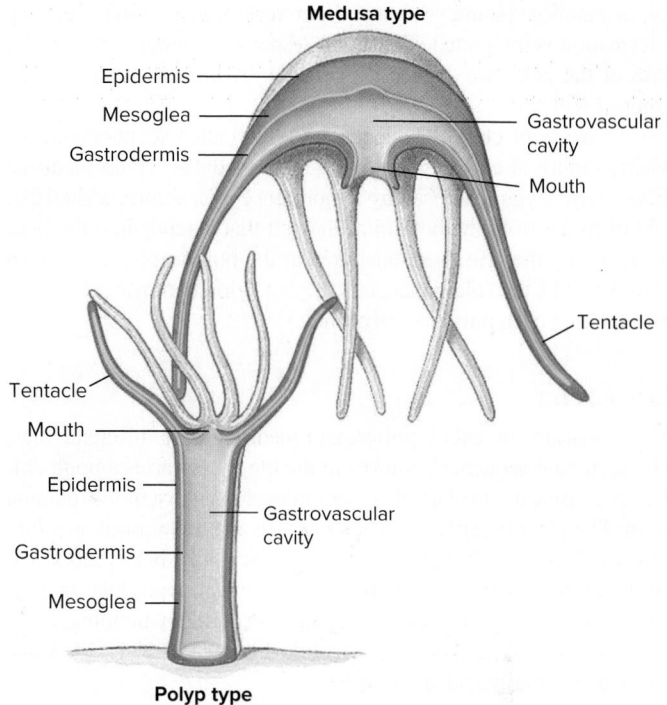

Figure 13.3 Comparison between polyp and medusa types of individuals.

cavity (Figures 13.1B and 13.7). Polyps that do not bud are solitary; polyps that do bud are clonal or colonial. The distinction between colonies and clones is sometimes blurred when a colony fragments.

A shared gastrovascular cavity permits polyp specialization. Many colonies include several morphologically distinct polyps, each specialized for a certain function, such as feeding, reproduction, or defense (Figure 13.1). Such colonies exhibit **polymorphism** (not to be confused with the population-genetic use of this term introduced in Chapter 6). Polymorphism occurs when a single genotype can express more than one body form, for example, when an individual asexually produces other individuals with different morphologies. In class Hydrozoa, feeding polyps, or **hydranths,** are easily distinguished from reproductive polyps, or **gonangia,** by the absence of tentacles in gonangia. Gonangia typically make medusae. Individual polyps within a colony are called **zooids** and may be specialized for particular functions such as feeding (gastrozooids) or reproduction (gonozooids).

Other methods of asexual reproduction in polyps are fission, where an individual divides in half as one side of the polyp pulls away from the other side, and pedal laceration, where tissue torn from the pedal disc develops into tiny new polyps. Pedal laceration and fission are common in sea anemones in class Anthozoa.

Medusae Medusae are usually free-swimming and have bell- or umbrella-shaped bodies (Figures 13.3 and 13.10). They often exhibit tetramerous symmetry where body parts are arranged in fours. The mouth is usually centered on the concave (subumbrellar) side, and it may be pulled downward into frilly lobes that extend a long way beneath the umbrella or bell (see Figure 13.17). Tentacles extend outward from the rim of the umbrella. Medusae have sensory structures

Superficially the polyp and medusa seem very different, but actually each has retained the saclike body plan characteristic of the phylum (Figure 13.3). A medusa is essentially an unattached polyp with the tubular portion widened and flattened into a bell shape.

Polyps Most polyps have tubular bodies. A mouth surrounded by tentacles defines the oral end of the body. The mouth leads into a blind gut or **gastrovascular cavity** (Figure 13.3). The aboral end of the polyp is usually attached to a substratum by a pedal disc or other device.

Polyps may reproduce asexually by budding, fission, or pedal laceration. In **budding,** a knob of tissue forms on the side of an existing polyp and develops a functional mouth and tentacles (see Figure 13.14). If a bud detaches from the polyp that made it, a clone is formed. If a bud stays attached to the polyp that made it, a colony will form and food may be shared through a common gastrovascular for orientation (statocysts) and light reception (ocelli). Sensory information is integrated with motor response by a nerve ring at the base of the bell; two such rings occur in hydrozoan medusae (see Figure 13.11).

Medusae of class Scyphozoa are often called scyphomedusae, whereas those of class Hydrozoa are hydromedusae. Hydromedusae differ from scyphomedusae by the presence of a velum, a shelflike fold of tissue from the bottom of the bell that extends into the bell. By reducing the cross-sectional area at the bottom of the bell (see Figure 13.11), the velum increases the exit velocity of water from the bell, making each pulsation efficient.

Life Cycles

In a cnidarian life cycle, polyps and medusae play different roles. The particular sequence of forms in the life cycle varies among cnidarian classes, but in general, a zygote develops into a motile planula larva. The planula settles on a hard surface and metamorphoses into a polyp. A polyp can make other polyps asexually, but eventually it produces free-swimming medusae by asexual reproduction (see Figures 13.7 and 13.19). Polyps produce medusae by budding, or by other specialized methods like **strobilation** (see p. 274). Medusae reproduce sexually and are dioecious.

A life cycle that contains both an attached polyp and a swimming medusa permits organisms to occupy both pelagic (open water) and benthic (bottom) environments. Such life cycles occur in true jellies of class Scyphozoa, where the medusa is large and conspicuous and the polyps are typically very small. Most hydroids of class Hydrozoa also feature a sessile polyp stage, often in colonies, and a pelagic medusa stage.

However, there are many variations on the typical pattern. In some hydrozoans, the polyp colony is not sessile, but drifts across the ocean surface. The Portuguese man-of-war, *Physalia,* is one such drifter, using an inflated polyp as a gas-filled float (see Figure 13.15). Other colonies are collections of both polyps and medusae where pulsating bells propel the colony through the water.

Several life cycles lack medusae. Anthozoans likely separated from an ancestor of the other cnidarians before the medusa evolved in the latter branch (see Figure 13.2), but other cnidarians, including the hydrozoan *Hydra,* probably lost the medusa secondarily. The mechanism of loss is not clear in *Hydra,* but in other hydrozoans, a pattern of loss can be inferred from a comparison of modern forms. Most hydrozoans release medusae that later make gametes, but a few forms make medusae without releasing them from the colony. Gametes then form in the gonads of the medusae retained by the polyp colony. In some species only a short, cuplike form surrounds the gonads (see Figure 13.9), and in others gonads develop right on the polyp colony with no trace of a medusa body. The latter organisms likely represent an extreme form of medusa retention and reduction.

Body Wall

The cnidarian body comprises an outer epidermis, derived from ectoderm, and an inner gastrodermis, derived from endoderm, with mesoglea between them (Figure 13.3). The gastrodermis lines the gut cavity and functions mainly in digestion. In polyps of the solitary hydrozoan, *Hydra,* the epidermal layer contains several cell types (Figure 13.4), including epitheliomuscular, interstitial, gland, sensory, and nerve cells (see pp. 269–270), as well as cnidocytes (see p. 265). Cnidarian bodies extend, contract, bend, and pulse, all without true mesodermally derived muscle cells. Instead, epitheliomuscular cells form most of the epidermis and serve both for covering and for muscular contraction (Figure 13.5). The bases of most such

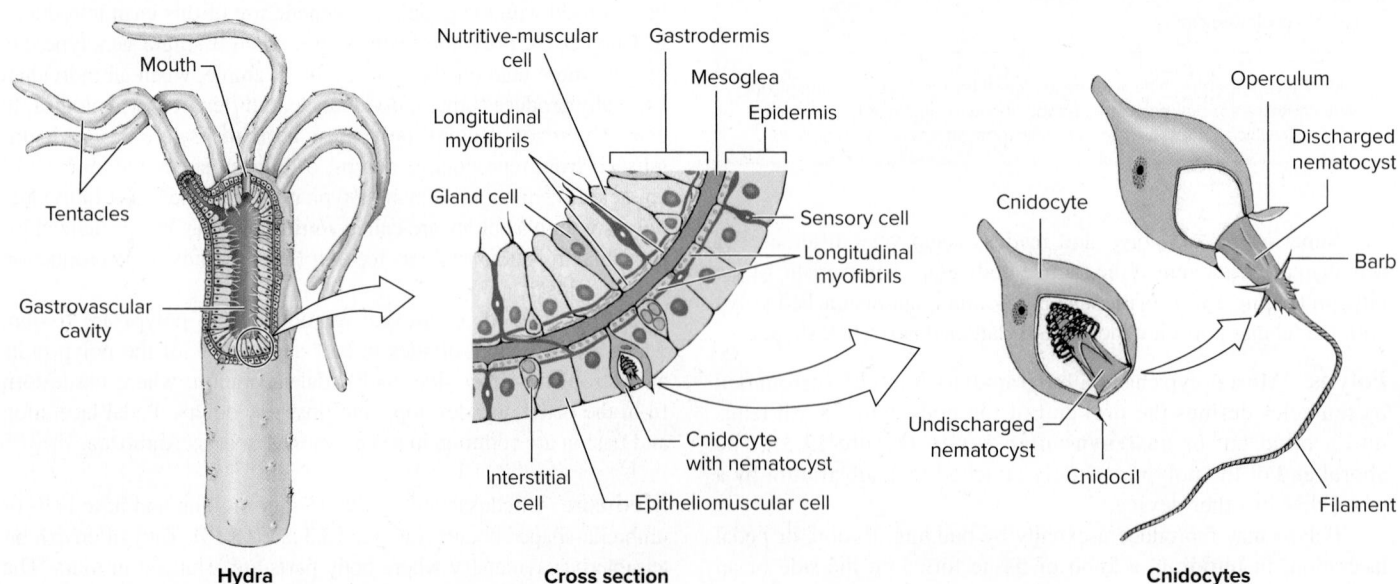

Figure 13.4 At right, structure of a stinging cell. Center, portion of the body wall of a hydra. Cnidocytes, which contain the nematocysts, arise in the epidermis from interstitial cells. At left, a *Hydra* polyp showing the gastrovascular lining.

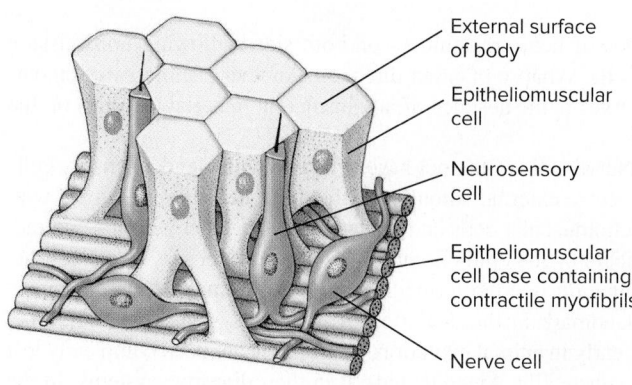

Figure 13.5   Epitheliomuscular and nerve cells in hydra.

Figure 13.6   **A,** Several types of cnidae shown after discharge. At bottom are two views of a type that does not impale prey; it recoils like a spring, catching any small part of the prey in the path of the recoiling thread. **B,** Fired and unfired cnidae from *Corynactis californica.*

cells extend parallel to the tentacle or body axis and contain myofibrils; they form the functional equivalent of a layer of longitudinal muscle next to the mesoglea. Contraction of these fibrils shortens the body or tentacles.

The mesoglea lies between the epidermis and the gastrodermis and is attached to both layers (see Figure 13.3). It is gelatinous, or jellylike, and both epidermal and gastrodermal cells send processes into it. In polyps, it is a continuous layer extending over both body and tentacles, thinnest in the tentacles and thickest in the stalk. This arrangement allows the pedal region at the base of the animal to withstand great mechanical strain and gives the tentacles more flexibility.

The mesoglea helps to support the body and acts as a type of elastic skeleton. In class Anthozoa, the mesoglea is substantial and possesses ameboid cells. The mesogleal layer is also very thick in scyphozoan medusae and contains ameboid cells and fibers. The medusa bell has a fairly firm consistency, despite the fact that mesogleal jelly is 95% to 96% water. The buoyant mass of mesogleal "jelly" gives medusae the common names jellyfishes, or jellies. The mesoglea is much thinner in the bells of hydromedusae and lacks ameboid cells or fibers.

Cnidocytes

As the opening essay attests, many cnidarians are very effective predators on prey larger and more intelligent than themselves. Such efficient predation is made possible by amply arming the tentacles with a unique cell type, the cnidocyte (see Figure 13.4). Cnidocytes are borne in invaginations of ectodermal cells (see Figure 13.4) and, in some forms, in endodermal cells. Each cnidocyte produces one of over 20 kinds of distinctive organelles called **cnidae** (Figure 13.6) that are discharged from the cell. During its development, a cnidocyte is properly called a **cnidoblast.** Once its cnida has been discharged, a cnidocyte is absorbed and replaced.

One type of cnida, the **nematocyst** (see Figure 13.4), is used to inject a toxin for prey capture and defense. Nematocysts are tiny capsules composed of material similar to chitin and containing a coiled tubular "thread" or filament, which is a continuation of the narrowed end of the capsule. This end of the capsule is covered by a little lid, or **operculum.** The inside of the undischarged thread may bear tiny barbs, or spines. Not all cnidae have barbs or inject poison. Some kinds, for example, do not penetrate prey but rapidly recoil like a spring after discharge, grasping and holding any part of the prey caught in the coil (Figure 13.6). Adhesive cnidae usually do not discharge in food capture, but are used in attachment and locomotion.

Except in Anthozoa, cnidocytes are equipped with a triggerlike **cnidocil,** which is a modified cilium. Anthozoan cnidocytes have a somewhat different ciliary mechanoreceptor. In some sea anemones, and perhaps other cnidarians, small organic molecules from the prey "tune" the mechanoreceptors, sensitizing them to the frequency of vibration caused by prey swimming. Tactile stimulation causes the nematocyst to discharge.

The mechanism of nematocyst discharge is remarkable. Evidence indicates that discharge is due to a combination of tensional forces generated during nematocyst formation and to an astonishingly high osmotic pressure within the nematocyst: 140 atmospheres. When stimulated to discharge, the high internal osmotic pressure causes water to rush into the capsule. The operculum opens, and the rapidly increasing *hydrostatic pressure* within the capsule forces the thread out with great force, turning inside out as it goes. At the everting end of the thread, the barbs flick to the outside like tiny switchblades. This minute but awesome weapon then injects toxin when it penetrates prey.

Note again the distinction between osmotic and hydrostatic pressure (p. 46). The nematocyst is never required actually to contain 140 atmospheres of hydrostatic pressure within itself; such a hydrostatic pressure would doubtless cause it to explode. As the water rushes inside during discharge, the osmotic pressure falls rapidly, while the hydrostatic pressure rapidly increases.

Nematocysts of most cnidarians are not harmful to humans and are a nuisance at worst. However, the stings of a Portuguese man-of-war (see Figure 13.15) and certain jellies are quite painful and sometimes dangerous (see boxed note, p. 275).

Feeding and Digestion

Polyps are typically carnivorous, catching prey with their tentacles, and passing them through the mouth into the gastrovascular cavity for digestion. In *Hydra,* the tentacles are hollow and the tentacle cavity communicates with the gastrovascular cavity. Inside the gastrovascular cavity, gland cells discharge enzymes on the food to begin extracellular digestion, but intracellular digestion occurs in the cells of the gastrodermis (see p. 269).

The polyps of a hydrozoan colony capture prey and begin digestion in the mouth. They then pass a digestive broth into the common gastrovascular cavity, where intracellular digestion proceeds in cells lining the gastrovascular cavity (see p. 269). In hydromedusae, both food type and digestive system are similar to that of the polyp. However, the body is oriented with the mouth facing downward in the center of the bell; the mouth is at the end of a tube called the **manubrium** (see Figure 13.11).

Scyphomedusae are typically larger than hydromedusae, but their basic form is similar. The mouth edge is extended as a manubrium, often with four frilly oral arms, sometimes called mouth lobes, used in capturing and ingesting prey (see Figure 13.19).

Anthozoan polyps, such as sea anemones, are carnivorous, feeding on fish or almost any animals of suitable size. They can expand and stretch their tentacles in search of small vertebrates and invertebrates, which they overpower with tentacles and nematocysts and carry to the mouth. A few species feed on minute forms caught by ciliary currents instead of eating large prey. Corals supplement their nutrition by collecting carbohydrates from their algal symbionts (see pp. 280–281).

Nerve Net

The nerve net of cnidarians is one of the best examples of a diffuse nervous system. This plexus of nerve cells occurs both at the base of the epidermis and at the base of the gastrodermis, forming two interconnected nerve nets. Nerve processes (axons) end on other nerve cells at synapses with sensory cells or effector organs (nematocysts or epitheliomuscular cells). Nerve action potentials move from one cell to another by release of a neurotransmitter from small vesicles on one side of the synapse or junction (p. 725). One-way transmission between nerve cells in higher animals is ensured because the vesicles are located on only one side of the synapse. However, cnidarian nerve nets are peculiar in that many of the synapses have vesicles of neurotransmitters on both sides, allowing transmission across the synapse in either direction. Another peculiarity of cnidarian nerves is the absence of any insulating material (myelin) on the axons.

Nerve cells of the net have synapses with slender sensory cells that receive external stimuli, and the nerve cells have synapses with epitheliomuscular cells and nematocysts. Together with the contractile fibers of the epitheliomuscular cells, the sensory-nerve cell net combination is often termed a **neuromuscular system,** an important landmark in the evolution of nervous systems. The nerve net arose early in animal evolution, and it has never been completely lost phylogenetically. Annelids have it in their digestive systems. In the human digestive system it appears as nerve plexuses in the musculature. This counterpart of the cnidarian nerve net coordinates rhythmical peristaltic movements of the stomach and intestine (see Chapter 32, p. 708).

Cnidarians do not have a local concentration of nerve cells that would approximate a central nervous system. However, some have argued that the nerve net and ring system in cnidarian medusae is as effective as a central nervous system when processing and responding to stimuli arriving from all directions. In scyphomedusae and the medusae of cubozoans, nerves are grouped in marginal sense organs, called **rhopalia,** that house chemoreceptors, statocysts, and often ocelli. The nerve nets form two or more systems, including a fast-conducting system to coordinate swimming movements and a slower one to coordinate movements of tentacles. In hydromedusae, two nerve rings that lie at the margin of the bell are formed by condensing the epidermal nerve net. Nerve rings process information from the sense organs and respond by changing swimming direction, pulsation rate, and position of tentacles.

Class Hydrozoa

Most Hydrozoa are marine and colonial in form, and a typical life cycle includes both an asexual polyp and a sexual medusa stage, as exemplified by a colonial marine hydroid such as *Obelia* (Gr. *obelias,* round cake).

Hydroid Colonies

A typical hydroid has a base, a stalk, and one or more terminal zooids. The base by which colonial hydroids attach to the substratum is a rootlike stolon, or **hydrorhiza** (see Figure 13.1), which gives rise to one or more stalks called **hydrocauli.** The living cellular part of the hydrocaulus is a tubular **coenosarc** (Figure 13.7), composed of the three typical cnidarian layers surrounding the gastrovascular cavity. The protective covering of the hydrocaulus is a nonliving chitinous sheath, or **perisarc.** Attached to the hydrocaulus are individual polyp animals, or zooids. Most zooids are feeding polyps called **hydranths,** or **gastrozooids.** They may be tubular, bottle-shaped, or vaselike, but all have a terminal mouth and a circle of tentacles. In **thecate** forms, such as *Obelia,* the perisarc continues as a protective cup around the polyp into which it can withdraw for protection (Figure 13.7). In others the polyp is **athecate** (naked) (Figure 13.8). In some forms the perisarc is an inconspicuous, thin film.

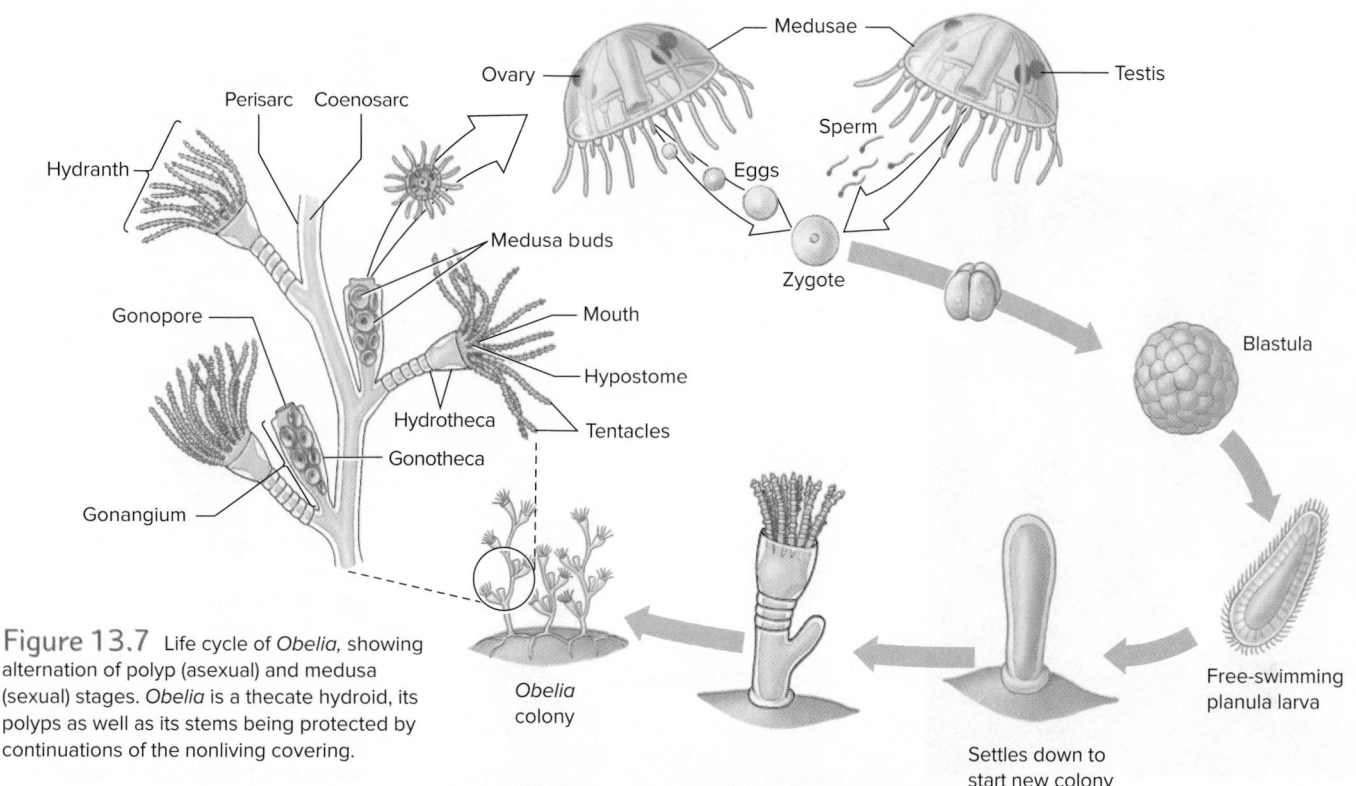

Figure 13.7 Life cycle of *Obelia,* showing alternation of polyp (asexual) and medusa (sexual) stages. *Obelia* is a thecate hydroid, its polyps as well as its stems being protected by continuations of the nonliving covering.

Hydranths capture and ingest prey, such as tiny crustaceans, worms, and larvae, thus providing nutrition for the entire colony. After partial extracellular digestion in a hydranth, the digestive broth passes along the common gastrovascular cavity where it is absorbed by gastrodermal cells, and intracellular digestion occurs.

Circulation within the gastrovascular cavity is a function of the ciliated gastrodermis but is also aided by rhythmical contractions and pulsations of the body.

Colonial hydroids bud off new individuals, thus increasing the size of the colony. New feeding polyps arise by budding, and medusa buds also arise on the colony. In *Obelia* these medusae bud from a reproductive polyp called a **gonangium.** Young medusae leave the colony as free-swimming individuals that mature and produce gametes (eggs and sperm) (Figure 13.7). In some species medusae remain attached to the colony and shed their gametes there. In other species medusae never develop and gametes are shed by male and female gonophores (Figure 13.9). Embryonation of the zygote produces a ciliated planula larva that swims freely for a time. Then it attaches to a substratum to develop into a minute polyp that gives rise, by asexual budding, to the hydroid colony, thus completing the life cycle.

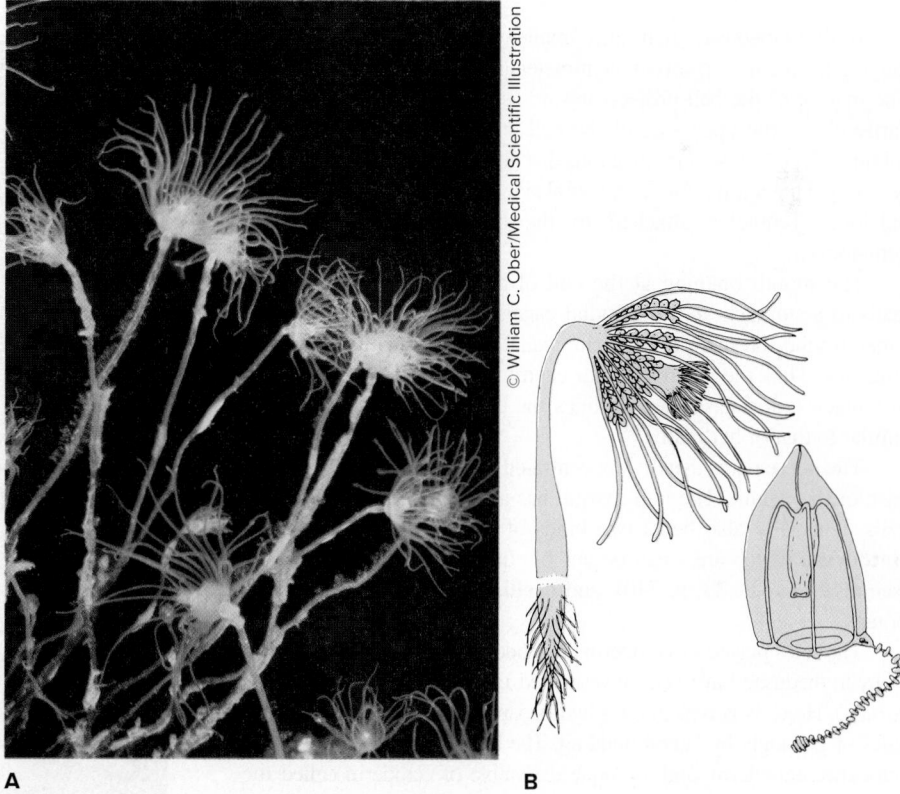

© William C. Ober/Medical Scientific Illustration

A   **B**

Figure 13.8 Athecate hydroids. **A,** *Ectopleura integra,* a solitary polyp with naked hydranths and gonophores. **B,** *Corymorpha* is a solitary hydroid that produces free-swimming medusae, each with a single trailing tentacle.

Reduced medusae
(gonophores)

Gonophore

Planula larva

Planula develops
into an actinula
larva

Emerging
actinula
larva

Actinula larva
is released

Free-swimming
actinula larva

Actinula larva
becomes a polyp

Polyp makes a colony by budding

© William C. Ober/Medical Scientific Illustration

Figure 13.9 In some hydroids, such as this *Tubularia crocea*, medusae are reduced to gonadal tissue and do not detach. These reduced medusae are called gonophores.

Hydroid medusae are usually smaller than scyphozoan medusae, ranging from 2 mm to several centimeters in diameter (Figure 13.10). The margin of the bell projects inward as a shelflike **velum,** which partly closes the open side of the bell and functions in swimming (Figure 13.11). Muscular pulsations that alternately fill and empty the bell propel the animal forward, aboral side first, with a weak "jet propulsion." Tentacles attached to the bell margin are rich in nematocysts.

The mouth opening at the end of a suspended **manubrium** leads to a stomach and four radial canals that connect with a ring canal around the margin. This ring canal connects with the hollow tentacles. Thus the gastrovascular cavity is continuous from mouth to tentacles, and gastrodermis lines the entire system. Nutrition is similar to that of hydranths.

The nerve net is usually concentrated into two nerve rings at the base of the velum. The bell margin has a liberal supply of sensory cells. It usually also bears two kinds of specialized sense organs: **statocysts,** which are small organs of equilibrium (Figure 13.11B) (see also Chapter 33, p. 740), and **ocelli,** which are light-sensitive organs.

The roles played by ectoderm and endoderm during the formation of hydromedusae have been investigated in one species (*Podocoryne carnea*). Here, as is typical for a hydrozoan, medusa buds arise on the sides of gonangia by lateral budding. The buds have three cell layers: ectoderm, endoderm, and a unique derivative of ectoderm called the **entocodon.** Portions of the entocodon differentiate into smooth and striated muscles. Further smooth muscles in the velum and tentacles originate from ectoderm. The reader may recall that cnidarians lack true mesodermally derived muscle, using epitheliomuscular cells for contraction of polyps and of nonhydrozoan medusae. Thus the presence of

© Daniel W. Gotshall

Figure 13.10 Bell medusa, *Polyorchis penicillatus,* medusa stage of an unknown attached polyp.

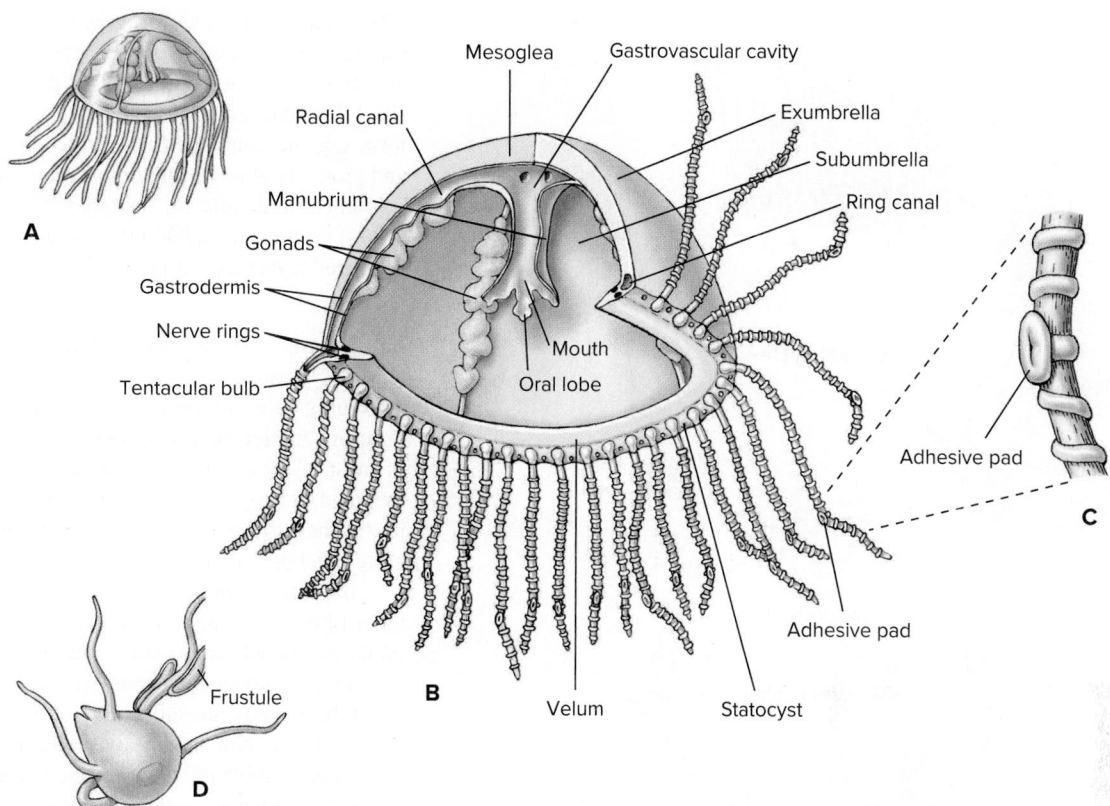

Figure 13.11 Structure of *Gonionemus*. **A,** Medusa with typical tetramerous arrangement. **B,** Cutaway view showing morphology. **C,** Portion of a tentacle with its adhesive pad and ridges of nematocysts. **D,** Tiny polyp, or hydroid stage, that develops from the planula larva. It can produce more polyps by budding (frustules) or produce medusa buds.

smooth and striated muscles (see pp. 651–653) in hydrozoan medusae is surprising, as is the ectodermal origin of these muscles. We discuss the potential significance of this finding on page 286.

Freshwater Medusae

The freshwater medusa *Craspedacusta sowberii* (Figure 13.12) (order Hydroida) most likely evolved from marine ancestors in the Yangtze River of China. Probably introduced with shipments of aquatic plants, this interesting form now occurs in many parts of Europe, throughout the United States, and in parts of Canada. Medusae may attain a diameter of 20 mm.

The polyp phase of this animal is tiny (2 mm) and has a very simple form with no perisarc and no tentacles. It occurs in colonies of a few polyps. For a long time its relation to the medusa was not recognized, and thus the polyp was given a name of its own, *Microhydra ryderi.* On the basis of its relationship to the jellies and the law of priority, both polyp and medusa should be called *Craspedacusta* (N. L. *craspedon,* velum, + Gr. *kystis,* bladder).

The polyp has three methods of asexual reproduction, as shown in Figure 13.12.

Hydra: A Solitary Freshwater Hydrozoan

Common freshwater hydras (Figure 13.13) live on the underside of aquatic leaves and lily pads in cool, clean freshwater of pools and streams. The hydra family occurs throughout the world, with 16 species in North America. Members of this family have been well studied, and much is known about their habits and body plan.

The body of a hydra can extend to a length of 25 to 30 mm or can contract to a tiny, gelatinous mass. It is a cylindrical tube with the aboral end drawn out into a slender stalk, ending in a **basal (or pedal) disc** for attachment.

Feeding and Digestion   Hydras feed on a variety of small crustaceans, insect larvae, and annelid worms. The mouth, located on a conical elevation called the **hypostome,** is encircled by 6 to 10 hollow tentacles that, like the body, can greatly extend when the animal is hungry.

The mouth opens into the **gastrovascular cavity,** which communicates with cavities in the tentacles. A hydra awaits its prey with tentacles extended (Figure 13.13). The food organism that brushes against its tentacles might find itself harpooned by scores of nematocysts that render it helpless, even though it may be larger than the hydra. The tentacles move toward the mouth, which slowly widens. Well moistened with mucous secretions, the mouth glides over and around the prey, totally engulfing it.

Inside the gastrovascular cavity, gland cells discharge enzymes on the food. Digestion is extracellular, but many food particles are drawn by pseudopodia into nutritive-muscular cells of the gastrodermis, where intracellular digestion occurs.

Nutritive-muscular cells are usually tall columnar cells and have laterally extended bases containing myofibrils. The myofibrils run at right angles to the body or tentacle axis and so form a circular muscle layer. However, this muscle layer in hydras is very weak, and longitudinal extension of the body and tentacles is achieved mostly by increasing the volume of water in the gastrovascular cavity. Water

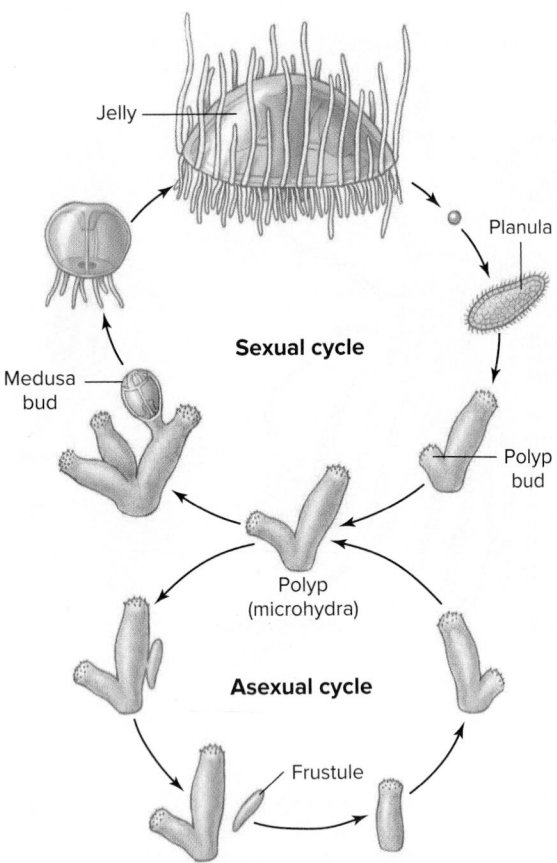

Figure 13.12 Life cycle of *Craspedacusta,* a freshwater hydrozoan. The polyp has three methods of asexual reproduction: by budding off new individuals, which may remain attached to the parent (colony formation); by constricting off nonciliated planula-like larvae (frustules), which can move around and give rise to new polyps; and by producing medusa buds, which develop into sexual jellies.

Figure 13.13 Hydra catches an unwary crustacean with the nematocysts of its tentacles. This animal has two buds with tentacles on the left and right sides near the substrate. Above the buds are two ovaries, one on each side of the hydra.

is brought in through the mouth by beating of cilia on the nutritive-muscular cells. Thus, water in the gastrovascular cavity serves as a **hydrostatic skeleton.** Gastrodermal cells in green hydras

(*Chlorohydra*) (Gr. *chloros,* green, + *hydra,* a mythical nine-headed monster slain by Hercules) bear green algae (zoo-chlorellae), which give the hydras their color. This is probably a symbiotic mutualism, since the algae use the respiratory carbon dioxide from the hydra to form organic compounds useful to the host. Algae receive shelter and probably other physiological requirements in return.

Interstitial cells are scattered among the bases of the nutritive cells. They transform into other types of cells when the need arises.

Cnidocytes are not present in the gastrodermis.

Epidermis   The epidermal layer contains epitheliomuscular, interstitial, gland, cnidocyte, and sensory and nerve cells. **Epitheliomuscular cells** compose most of the epidermis and serve both for covering and for muscular contraction (See Figure 13.5). The bases of most of these cells are extended parallel to the tentacle or body axis and contain myofibrils, thus forming a layer of longitudinal muscle next to the mesoglea. Contraction of these fibrils shortens the body or tentacles.

Interstitial cells are undifferentiated stem cells found among the bases of the epitheliomuscular cells. Differentiation of interstitial cells produces cnidoblasts, sex cells, buds, nerve cells, and others, but generally not epitheliomuscular cells (which reproduce themselves).

Gland cells are tall cells, located around the basal disc and mouth, that secrete an adhesive substance for attachment and sometimes a gas bubble for floating (see Figure 13.4).

Cnidocytes occur throughout the epidermis. Hydras have three functional types of cnidae: those that penetrate prey and inject toxin (penetrants; see Figure 13.4), those that recoil and entangle prey (volvents), and those that secrete an adhesive substance used in locomotion and attachment (glutinants).

Sensory cells are scattered among the other epidermal cells, especially near the mouth and tentacles and on the basal disc. The free end of each sensory cell bears a flagellum, which is the sensory receptor for chemical and tactile stimuli. The other end branches into fine processes that synapse with nerve cells.

Nerve cells of the epidermis are generally multipolar (have many processes), although in some cnidarians the cells may be bipolar (with two processes). Their processes (axons) form synapses with sensory cells, other nerve cells, epitheliomuscular cells, and cnidocytes. There are both one-way (morphologically asymmetrical) and two-way synapses with other nerve cells.

Reproduction   Hydras reproduce sexually and asexually. In asexual reproduction, buds appear as outpocketings of the body wall (Figures 13.13 and 13.14) and develop into young hydras that eventually detach from the parent. Most species are dioecious. Temporary gonads (Figure 13.13) usually appear in autumn, stimulated by lower temperatures and perhaps also by reduced aeration of stagnant waters. Testes or ovaries, when present, appear as rounded projections on the surface of the body (Figure 13.13). Eggs in the ovary usually mature one at a time and are fertilized by sperm shed into the water.

Zygotes undergo holoblastic cleavage to form a hollow blastula. The inner part of the blastula delaminates to form the endoderm, and the mesoglea is laid down between ectoderm and endoderm. A cyst forms around the embryo before it breaks loose from the parent, enabling it to survive winter. Young hydras hatch in spring when weather is favorable.

Figure 13.14   Hydra with developing bud.

Other Hydrozoans

Members of orders Siphonophora and Chondrophora are among the most specialized Hydrozoa. They usually form polymorphic swimming or floating colonies containing modified medusae and polyps.

There are several types of polyp individuals. Gastrozooids are feeding polyps with a single long tentacle arising from the base of each. Some of these long, stinging tentacles become separated from the feeding polyp and are called **dactylozooids,** or fishing tentacles. These tentacles sting prey and lift it to the feeding polyps. Among the modified medusoid individuals are the **gonophores,** which are little more than sacs containing either ovaries or testes.

Physalia (Gr. *physallis,* bladder), the Portuguese man-of-war (Figure 13.15), is a colony with a rainbow-hued float of blues and pinks that carries it along the surface waters of tropical seas. Many are blown to shore on the eastern coast of the United States. The long, graceful tentacles, actually zooids, are laden with nematocysts and are capable of inflicting painful stings. The float, called a **pneumatophore,** is hypothesized to have expanded from the original larval polyp. It contains a sac arising from the body wall and is filled with a gas similar to air. The float acts as a type of "nurse carrier" for future generations of individuals that bud from it and hang suspended in the water. Some siphonophores, such as *Stephalia* and *Nectalia,* possess swimming bells as well as a float.

Invasive species create problems for existing ecosystems throughout the world. Aquatic invaders often reach new habitats through the discharge of ballast water from transoceanic ships. Biologists thought that few pelagic cnidarians would invade in this way because ballast water pickup and discharge would damage fragile medusae. However, *Turritopsis dohrnii,* a hydrozoan with a world-wide distribution, demonstrates the astonishing potential of adaptive life cycle evolution. Damaged or starved medusae of this species do not die; instead, they transdifferentiate into highly resistant benthic cysts containing living tissue. Cysts deposited in new localities hatch to release small polyps, which begin the life cycle anew. Genetic sampling shows that very closely related populations of this species now occur in widely separated parts of the world.

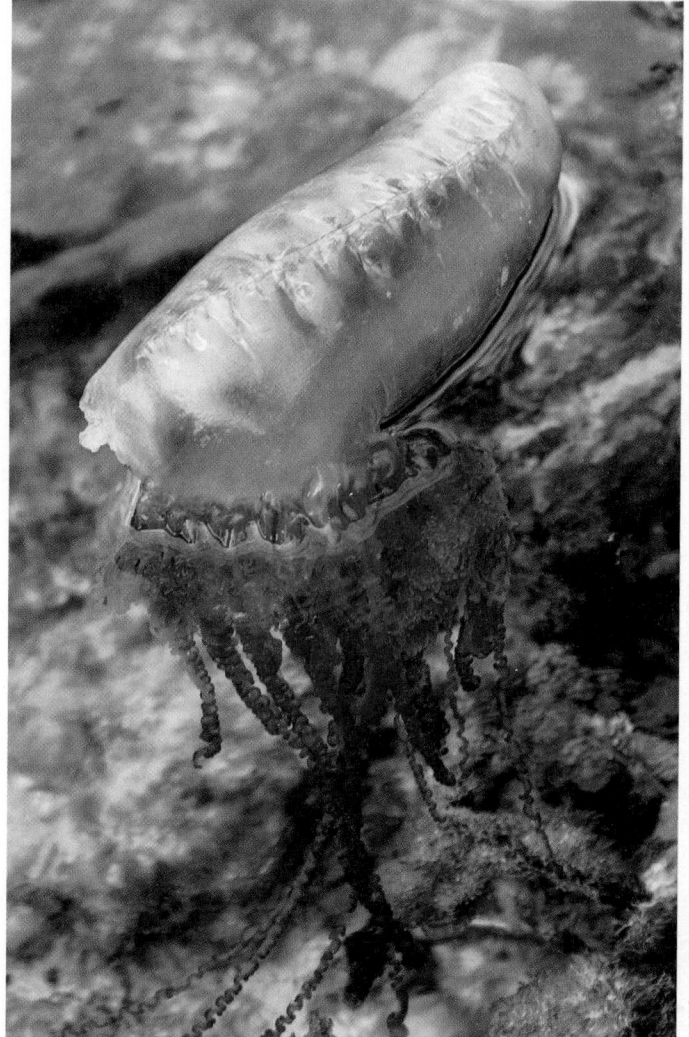

Figure 13.15   A Portuguese man-of-war colony, *Physalia physalis* (order Siphonophora, class Hydrozoa). Colonies often drift onto southern ocean beaches, where they are a hazard to bathers. Each colony of medusa and polyp types is integrated to act as one individual. As many as 1000 zooids may be found in one colony. The nematocysts secrete a powerful neurotoxin.

Other hydrozoans secrete massive calcareous skeletons that resemble true corals (Figure 13.16). They are sometimes called **hydrocorals.**

Class Scyphozoa

Class Scyphozoa (si-fo-zo′a) (Gr. *skyphos,* cup) includes most of the larger jellies, or "cup animals." A few scyphozoans, such as *Cyanea* (Gr. *kyanos,* dark-blue substance), can attain a bell diameter exceeding 2 m and tentacles 60 to 70 m long (Figure 13.17), but most range from 2 to 40 cm in diameter. Most drift or swim in the open sea, some even at depths of 3000 m. Movement is by rhythmical pulsations of the bell.

Bells of different species vary in depth from a shallow saucer shape to a deep helmet or goblet shape, but a velum is never present. Tentacles around the bell, or umbrella, may be numerous or few, and they may be short as in *Aurelia* (L. *aurum,* gold;

A

B

Figure 13.16   These hydrozoans form calcareous skeletons that resemble true coral. **A,** *Stylaster roseus* (order Stylasterina) occurs commonly in caves and crevices in coral reefs. These fragile colonies branch in only a single plane and may be white, pink, purple, red, or red with white tips. **B,** Species of *Millepora* (order Milleporina) form branching or platelike colonies and often grow over the horny skeleton of gorgonians (see Figure 13.34), as is shown here. They have a generous supply of powerful nematocysts that produce a burning sensation on human skin, justly earning the common name fire coral. The inset photo shows the extended tentacles.

Figure 13.18) or long as in *Cyanea*. The margin of the umbrella is scalloped, usually with each indentation, or notch, bearing a pair of **lappets,** and between them is a sense organ called a **rhopalium** (tentaculocyst). *Aurelia* has eight such notches. Some scyphozoans have 4, others 16. Each rhopalium is club-shaped and contains a hollow statocyst for equilibrium and one or two pits lined with sensory epithelium. In some species the rhopalia also bear ocelli.

Figure 13.17   Giant jelly, *Cyanea capillata* (order Semaeostomeae, class Scyphozoa). A North Atlantic species of *Cyanea* reaches a bell diameter exceeding 2 m. It is called the "sea blubber" by fishermen.

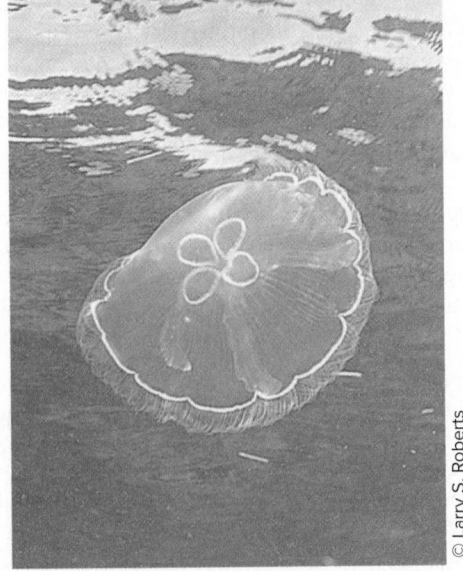

Figure 13.18   Moon jelly *Aurelia aurita* (class Scyphozoa) is cosmopolitan in distribution. It feeds on planktonic organisms caught in mucus on its umbrella.

The **nervous system** in scyphozoans is a nerve net, with a subumbrellar net that controls bell pulsations and another, more diffuse net that controls local reactions such as feeding.

Tentacles, manubrium, and often the entire body surface are well supplied with nematocysts that can deliver painful stings. However, the primary function of scyphozoan nematocysts is not to attack humans but to paralyze prey animals, which are conveyed to the mouth lobes by other tentacles or by bending of the umbrella margin.

The mouth is centered on the subumbrellar side. The manubrium usually forms four frilly **oral arms** that are used in capturing and ingesting prey. The mouth leads into a stomach.

Internally, extending out from the stomach of scyphozoans are four **gastric pouches** in which gastrodermis extends down in little tentacle-like projections called **gastric filaments.** These filaments are covered with nematocysts to quiet any prey that may still be struggling. Gastric filaments are lacking in hydromedusae. A complex system of

radial canals branches outward from the pouches to a **ring canal** in the margin and forms a part of the gastrovascular cavity.

Aurelia, the familiar moon jelly (Figure 13.18), feeds on small planktonic animals. Its medusae, 7 to 10 cm in diameter, commonly occur in waters off both the east and west coasts of the United States. The bell has relatively short tentacles, not used in prey capture. Food items are caught in mucus on the umbrella surface, and are carried to "food pockets" on the umbrella margin by cilia. From there, ciliated oral lobes move food to the gastrovascular cavity. Cilia in the gastrodermis layer keep a current of water moving to bring food and oxygen into the stomach and to expel wastes.

Sexes are separate, with gonads located in the gastric pouches. Fertilization is internal, with sperm being carried by ciliary currents

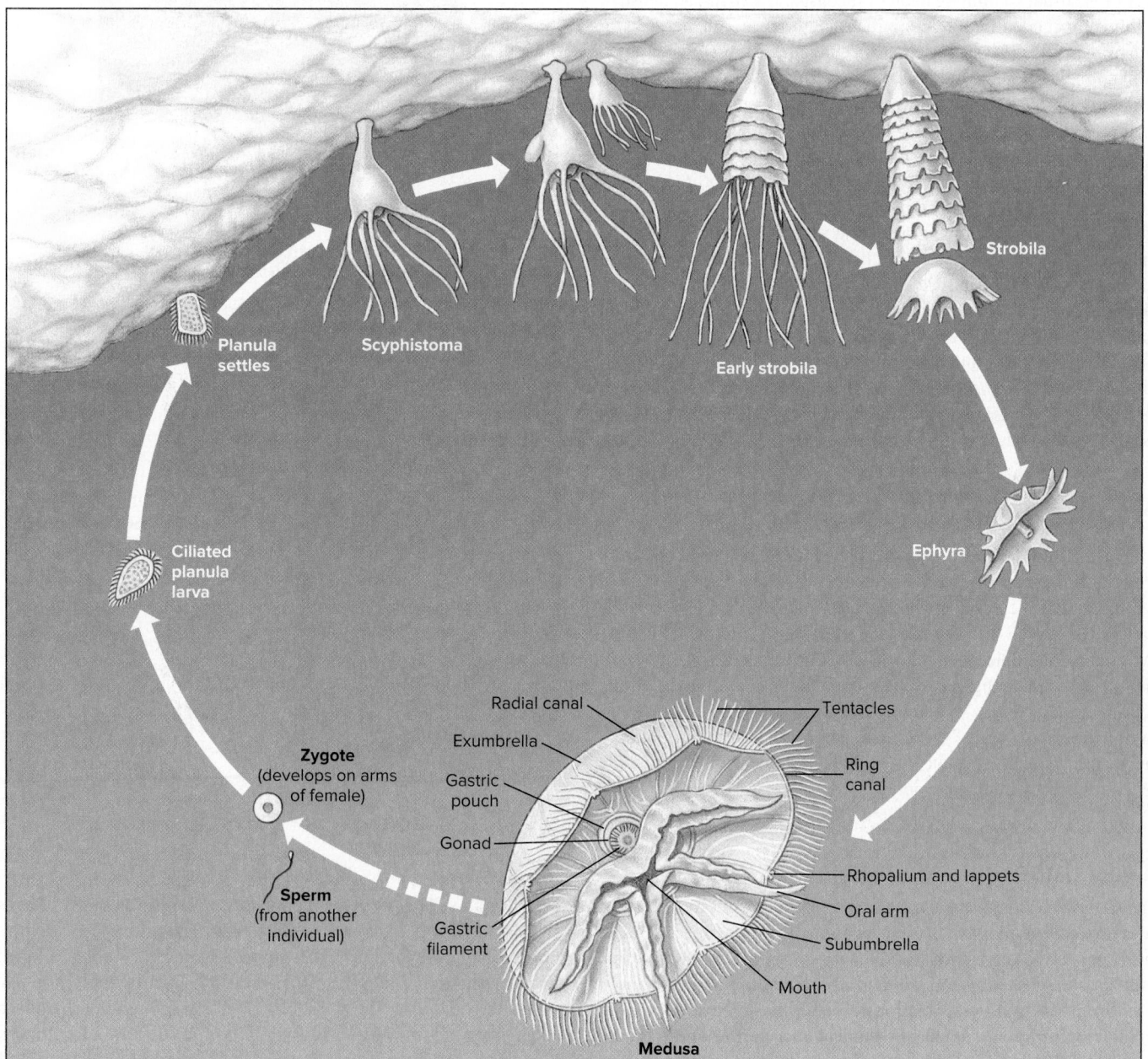

Figure 13.19   Life cycle of *Aurelia,* a marine scyphozoan medusa.

into the gastric pouch of females. Zygotes may develop in seawater or may be brooded in folds of the oral arms. The ciliated planula larva becomes attached and develops into a **scyphistoma,** a hydralike form (Figure 13.19) that may bud to produce a polyp clone. By a process of **strobilation** the scyphistoma of *Aurelia* forms a series of saucer-like buds, **ephyrae,** and is now called a **strobila** (Figure 13.19). When ephyrae break loose, they grow into mature jellies.

The life cycle just described is typical of scyphozoans, but there is variation within this class. In a few species, a larva develops directly into a medusa, and the polyp stage is absent.

The scyphozoans *Cassiopeia* and *Rhizostoma* also exhibit odd body forms. Visitors to Florida often notice the "upside-down jelly," *Cassiopeia* (L. mythical queen of Ethiopia), because it is typically found lying on its "back" in strongly sunlit shallow lagoons, in contrast to the usual swimming habit of medusae. It also has an unusual, highly branched, mouth. A similar mouth form can be seen in *Rhizostoma* (Gr. *rhiza,* root, + *stoma,* mouth), from colder waters. Both animals belong to a group of scyphozoans without tentacles on the umbrella margin and with a characteristic oral arm structure. During development, edges of the oral lobes fold over and fuse, forming canals (**arms** or **brachial canals**) that become highly branched. These canals open to the surface at frequent intervals by pores called "mouths"; the original mouth is obliterated in the fusion of the oral lobes. Planktonic organisms caught in the mucus of the frilly oral arms are transported by cilia to the mouths and then up the brachial canals to the gastric cavity. *Cassiopeia's* umbrella margin contracts about 20 times a minute, creating water currents to bring plankton into contact with the mucus and nematocysts of its oral lobes. Its tissues are abundantly supplied with symbiotic dinoflagellates (p. 233) (**zooxanthellae**). As they lie sunning themselves in shallow water, *Cassiopeia* are thus reminiscent of large flowers in more ways than one.

Class Staurozoa

Animals in this class are commonly called stauromedusans and were previously considered unusual scyphozoans, even though their life cycle does not include a medusa phase. The solitary polyp body is stalked (Figure 13.20) and uses an adhesive disc to attach to seaweeds and other objects on the seafloor. The top of the polyp resembles a medusa, although previous interpretations have noted that the bottom of the "medusa" resembles a polyp. The top of the polyp has eight extensions ("arms"), ending in tentacle clusters, surrounding the mouth. Polyps reproduce sexually. The nonswimming planula develops directly into a new polyp.

Class Cubozoa

The Cubozoa once were considered an order (Cubomedusae) of Scyphozoa. The medusa is the predominant form (Figure 13.21); the polyp is inconspicuous and in most cases unknown. Some cubozoan medusae may range up to 25 cm tall, but most are about 2 to 3 cm. In transverse section the bells are almost square. A tentacle or group of tentacles is found at each corner of the square at the umbrella margin. The base of each tentacle is differentiated into a flattened, tough blade called a **pedalium** (Figure 13.21). Rhopalia are present, each housing six eyes in addition to other

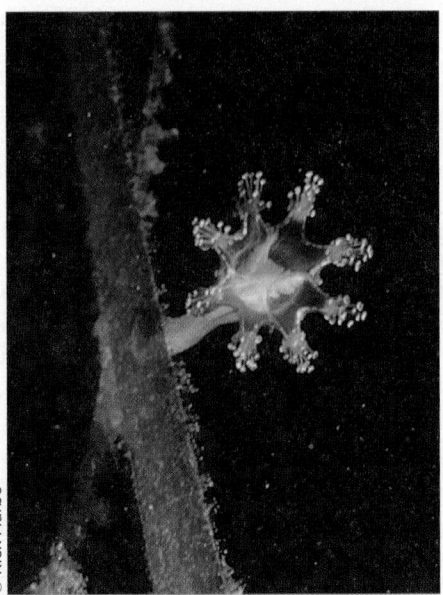

© Rick Harbo

Figure 13.20 *Thaumatoscyphus hexaradiatus* are an example of class Staurozoa.

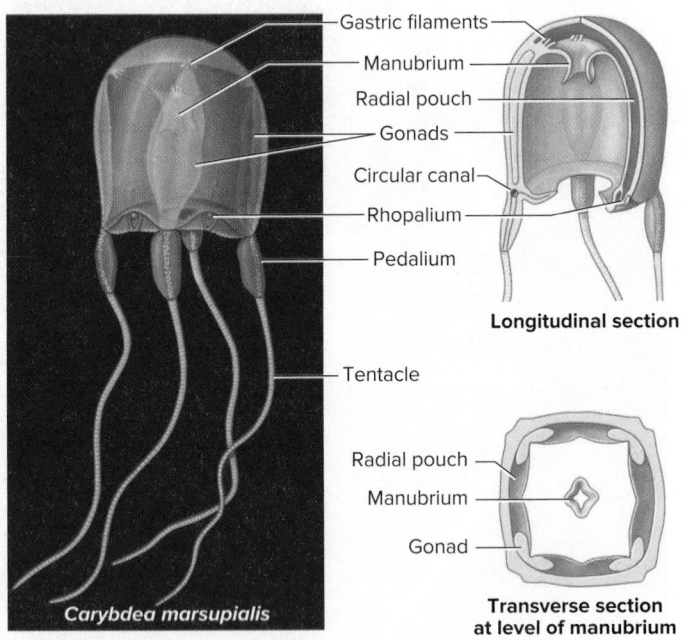

Gastric filaments
Manubrium
Radial pouch
Gonads
Circular canal
Rhopalium
Pedalium

Longitudinal section

Tentacle

Radial pouch
Manubrium
Gonad

Transverse section at level of manubrium

Carybdea marsupialis

Figure 13.21 *Carybdea,* a cubozoan medusa.

sense organs. There are two copies of each of three kinds of eyes: two forms of ocelli, and a sophisticated camera-type eye with a cornea and cellular lens. The umbrella margin is not scalloped, and the subumbrella edge turns inward to form a **velarium.** The velarium functions as a velum does in hydrozoan medusae, increasing swimming efficiency, but it differs structurally. Cubomedusae are strong swimmers and voracious predators, feeding mostly on fish in near-shore areas, such as mangrove swamps. Stings of some species can be fatal to people.

The complete life cycle is known for only one species, *Tripedalia cystophora* (L. *tri,* three, + Gr. *pedalion,* rudder). The polyp is tiny (1 mm tall), solitary, and sessile. New polyps bud laterally, detach, and creep away. Polyps metamorphose directly into medusae.

Chironex fleckeri (Gr. *cheir,* hand, + *nexis,* swimming) is a large cubomedusa called the sea wasp. Its stings are quite dangerous and sometimes fatal. Most fatalities have been reported from tropical Australian waters, usually following quite massive stings. Witnesses have described victims as being covered with "yards and yards of sticky wet string." Stings are very painful, and death, if it is to occur, ensues within a matter of minutes. If death does not occur within 20 minutes after stinging, complete recovery is likely.

Class Anthozoa

Anthozoans, or "flower animals," are polyps with a flowerlike appearance (Figure 13.22). There is no medusa stage. Anthozoa are all marine and occur in both deep and shallow water and in polar as well as tropical seas. They vary greatly in size and can be solitary or colonial. Many forms are supported by skeletons.

The class has three subclasses: **Hexacorallia** (or **Zoantharia),** containing sea anemones, hard corals, and others; **Ceriantipatharia,** containing only tube anemones and thorny corals; and **Octocorallia** (or **Alcyonaria),** containing soft and horny corals, such as sea fans, sea pens, sea pansies, and others. Zoantharians and ceriantipatharians have a **hexamerous** plan (of six or multiples of six) or polymerous symmetry and have simple tubular tentacles arranged in one or more circlets on the oral disc. Octocorallians are **octomerous** (built on a plan of eight) and always have eight pinnate (featherlike) tentacles arranged around the margin of the oral disc (Figure 13.23).

The gastrovascular cavity is large and partitioned by septa, or mesenteries, which are inward extensions of the body wall. Where one septum extends into the gastrovascular cavity from the body wall, another extends from the diametrically opposite side; thus, they are said to be **coupled.** In Hexacorallia, the septa are not only coupled, they are also **paired** (Figure 13.24). The muscular arrangement varies among different groups, but usually features circular muscles in the body wall and longitudinal and transverse muscles in the septa.

The mesoglea is a mesenchyme containing ameboid cells. A general tendency toward biradial symmetry in the septal arrangement occurs in the shape of the mouth and pharynx. There are no special organs for respiration or excretion.

Sea Anemones

Sea anemone (order Actiniaria) polyps are larger and heavier than hydrozoan polyps (Figure 13.22). Most range from 5 mm or less to 100 mm in diameter, and from 5 mm to 200 mm long, but some grow much larger. Some sea anemones are quite colorful. Anemones occur in coastal areas all over the world, especially in warmer waters. They attach by their pedal discs to shells, rocks, timber, or whatever submerged substrata they can find. Some burrow in mud or sand.

Sea anemones are cylindrical in form with a crown of tentacles arranged in one or more circles around the mouth of the flat **oral disc** (Figure 13.24). The slit-shaped mouth leads into a **pharynx.** At one or both ends of the mouth is a ciliated groove called a **siphonoglyph,** which extends into the pharynx. The siphonoglyph creates a water current directed into the pharynx. Cilia elsewhere on the pharynx direct water outward. Currents thus created supply oxygen and remove wastes. They also help to maintain an internal fluid pressure, providing a hydrostatic skeleton that serves in lieu of a true skeleton as a support for opposing muscles.

The pharynx leads into a large **gastrovascular cavity** that is divided into six radial chambers by means of six pairs of **primary (complete) septa,** or **mesenteries,** extending vertically from the body

Figure 13.22 Sea anemones are the familiar and colorful "flower animals" of tide pools, rocks, and pilings of the intertidal zone. Most, however, are subtidal, their beauty seldom revealed to human eyes. These are rose anemones, *Tealia piscivora* (subclass Hexacorallia, class Anthozoa).

A   **B**

Figure 13.23 **A,** White sea pen *Pteroeides* sp. from the Solomon Islands. Sea pens are colonial forms that inhabit soft substrates. The base of the fleshy body of the primary polyp is buried in the substrate. It gives rise to numerous secondary, branching polyps. **B,** Close-up of a gorgonian. The pinnate tentacles characteristic of subclass Octocorallia are apparent.I wall to the pharynx (Figure 13.24). Openings between chambers (septal perforations) in the upper part of the pharyngeal region aid water circulation. Smaller **(incomplete)** septa partially subdivide the large chambers and provide a means of increasing the surface area of the gastrovascular cavity. The free edge of each incomplete septum forms a type of sinuous cord called a **septal filament,** which contains nematocysts and gland cells for digestion. In some anemones (such as *Metridium)* the lower ends of the septal filaments are prolonged into **acontia threads,** also provided with nematocysts and gland cells, which are protruded through the mouth or through pores in the body wall to aid prey capture or defense. The pores also aid in rapid discharge of water from the body when the animal is endangered and contracts to a small size.

Sea anemones are carnivorous, feeding on fish or almost any live (and sometimes dead) animals of suitable size. Some species live on minute forms caught by ciliary currents.

Feeding behavior in many zoantharians is chemically controlled. Some respond to reduced glutathione. In certain others two compounds are involved: asparagine, the feeding activator, causes a bending of tentacles toward the mouth; then reduced glutathione induces swallowing of food.

Muscles are well developed in sea anemones, but the arrangement is quite different from that in Hydrozoa. Longitudinal fibers of the epidermis occur only in the tentacles and oral disc of most species.

The strong longitudinal muscles of the column are gastrodermal and located in the septa (Figure 13.24). Gastrodermal circular muscles in the column are well developed.

Most anemones can glide slowly along a substrate on their pedal discs. They can expand and stretch their tentacles in search of small vertebrates and invertebrates, which they overpower with tentacles and nematocysts and carry to their mouth. When disturbed, sea anemones contract and withdraw their tentacles and oral discs. Some anemones swim to a limited extent by rhythmical bending movements, which may permit escape from enemies such as sea stars and nudibranchs. *Stomphia,* for example, at the touch of a predatory sea star, detaches its pedal disc and creeps or swims to escape (Figure 13.25). This escape reaction is elicited not only by the touch of the star but also by exposure to liquids exuded by the star or to crude extracts made from its tissues. The sea star exudes steroid saponins that are toxic and irritating to most invertebrates. Extracts from nudibranchs also can provoke this reaction in some sea anemones.

Anemones form some interesting mutualistic relationships with other organisms. Many species harbor symbiotic dinoflagellates (zooxanthellae) within their tissues, similar to the hard coral–zooxanthellae association (p. 280), and the anemones profit from the product of algal photosynthesis. Some anemones habitually attach to the shells occupied by certain hermit crabs. The hermit encourages

Figure 13.24 Structure of a sea anemone. Free edges of septa and acontia threads are equipped with nematocysts to complete paralyzation of prey begun by the tentacles.

A   **B**

Figure 13.25 **A,** A sea anemone that swims. **B,** When attacked by a predatory sea star *Dermasterias,* the anemone *Stomphia didemon* (subclass Hexacorallia, class Anthozoa) detaches from the bottom and rolls or swims spasmodically to a safer location.

Figure 13.26 Orangefin anemone fish (*Amphiprion chrysopterus*) nestles in the tentacles of its sea anemone host. Anemone fishes do not elicit stings from their hosts but may lure unsuspecting other fish to become meals for the anemone.

the relationship and, finding its favorite species, which it recognizes by touch, it massages the anemone until it detaches from its original substrate and moves to the hermit crab's shell. The hermit crab holds the anemone against its own shell until the anemone is firmly attached. The crab derives some protection against predators by the anemone. The anemone gets free transportation and particles of food dropped by the hermit crab.

Certain damselfishes (anemone fishes) (family Pomacentridae) form associations with large anemones, especially in tropical Indo-Pacific waters (Figure 13.26). Mucus on the skin of the fish prevents an anemone's nematocysts from discharging, but if some other fish is so unfortunate as to brush the anemone's tentacles, it is likely to become a meal. Experiments show that shelter in an anemone is essential for protecting anemone fish from predators. Swimming fish may help to ventilate the anemone, and keep the anemone free of sediment. New research shows that damselfishes depend on fire corals (reef-forming hydrozoans; see Figure 13.16)

for food, shelter, and protection from predators in much the same way that anemone fish depend on anemones.

Sexes are separate in some sea anemones, whereas others are hermaphroditic. Monoecious species are **protandrous** (produce sperm first, then eggs). Gonads are arranged on the margins of the septa, and fertilization occurs externally or in the gastrovascular cavity. The zygote develops into a ciliated larva. Asexual reproduction commonly occurs by **pedal laceration** or by longitudinal fission, occasionally by transverse fission or by budding. In pedal laceration, small pieces of the pedal disc break off as the animal moves, and each of these regenerates a small anemone.

Hexacorallian Corals

Hexacorallian corals belong to the order Scleractinia, sometimes called true or stony corals. Stony corals might be described as miniature sea anemones that live in calcareous cups that they have secreted (Figures 13.27 and 13.28). Like that of anemones, a coral polyp's gastrovascular cavity is subdivided by septa arranged in multiples of six (hexamerous) and its hollow tentacles surround the mouth, but there is no siphonoglyph.

Instead of a pedal disc, the epidermis at the base of the column secretes a limy skeletal cup, including sclerosepta, which project into the polyp between its true septa (Figure 13.28). Living polyps can retract into the safety of their cup when not feeding. Since the skeleton is secreted below the living tissue rather than within it, the calcareous material is an exoskeleton. In many colonial corals, the skeleton may become massive, building up over many years, with the living coral forming a sheet of tissue over the surface (Figure 13.29). The gastrovascular cavities of the polyps are all connected through this sheet of tissue.

Three other small orders of Zoantharia are recognized.

Tube Anemones and Thorny Corals

Members of subclass Ceriantipatharia have unpaired septa. Tube anemones (order Ceriantharia) (Figure 13.30) are solitary and live buried to the level of the oral disc in soft sediments. They occupy

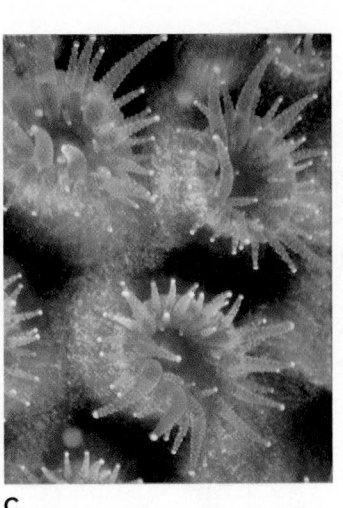

A   **B**   **C**

Figure 13.27   **A,** Cup coral *Tubastrea* sp. Its polyps form clumps resembling groups of sea anemones. Although often found on coral reefs, *Tubastrea* is not a reef-building coral (ahermatypic) and has no symbiotic zooxanthellae in its tissues. **B,** Polyps of *Montastrea cavernosa* are tightly withdrawn during daytime but open to feed at night, as in **C** (subclass Hexacorallia).

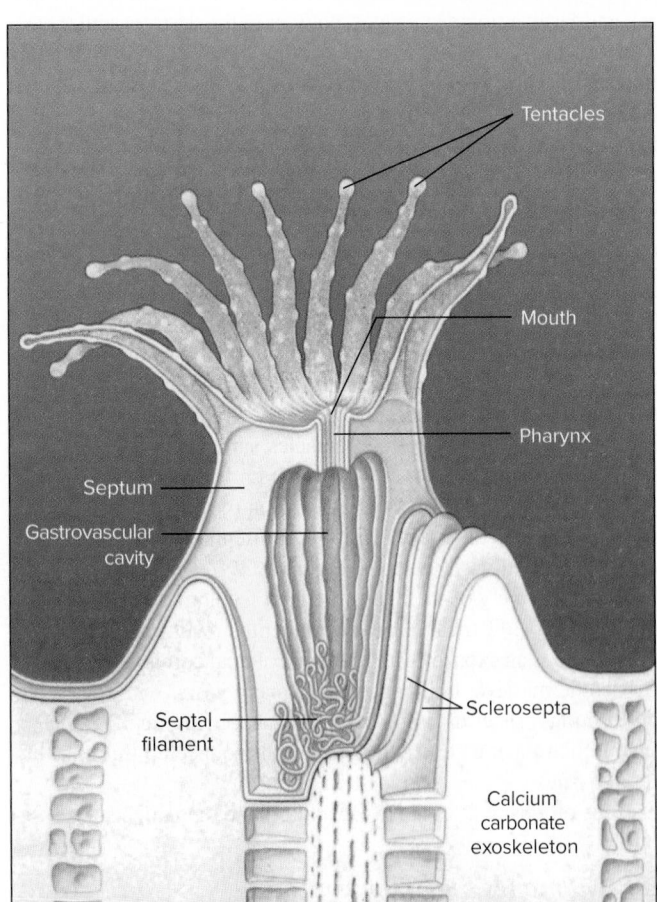

Figure 13.28   Polyp of a hexacorallian coral (order Scleractinia) showing calcareous cup (exoskeleton), gastrovascular cavity, sclerosepta, septa, and septal filaments.

Figure 13.29   Boulder star coral, *Montastrea annularis* (subclass Hexacorallia, class Anthozoa). Colonies can grow up to 10 feet (3 m) high.

Figure 13.30 A tube anemone (subclass Ceriantipatharia, order Ceriantharia) extends from its tube at night. Its oral disc bears long tentacles around the margin and short tentacles around the mouth.

Figure 13.31 **A,** Colony of *Antipathes,* a black or thorny coral (order Antipatharia, subclass Ceriantipatharia, class Anthozoa). Most abundant in deep waters in the tropics, black corals secrete a tough, proteinaceous skeleton that can be worked into jewelry. **B,** The polyps of Antipatharia have six simple, nonretractile tentacles. The spiny processes in the skeleton are the origin of the common name thorny corals.

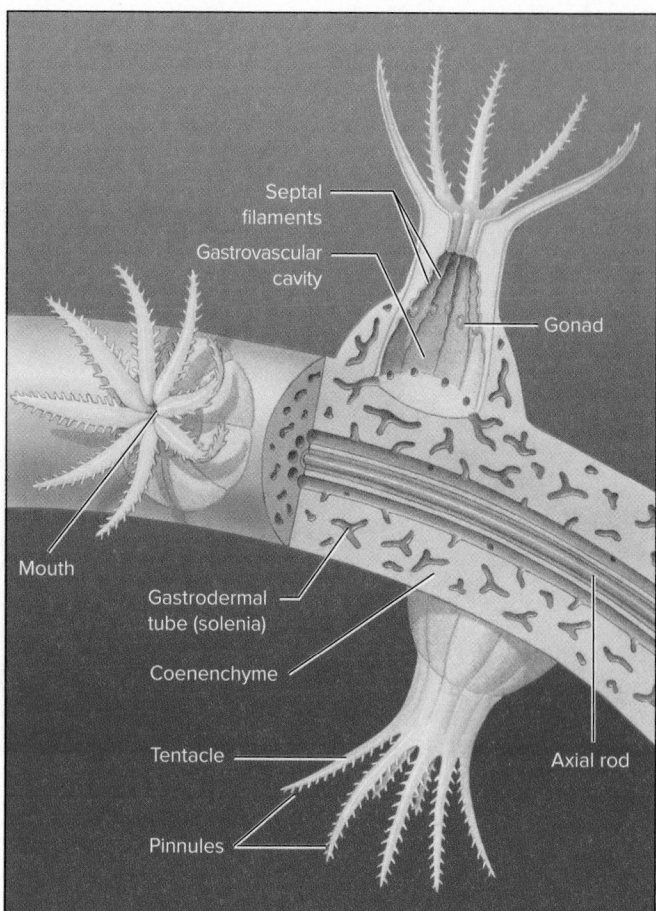

Figure 13.32 Polyps of an octocorallian coral. Note the eight pinnate tentacles, coenenchyme, and solenia. They have an endoskeleton of limy spicules often with a horny protein, which may be in the form of an axial rod.

tubes constructed of secreted mucus and threads of nematocyst-like organelles, into which they can withdraw. Thorny or black corals (order Antipatharia) (Figure 13.31) are colonial and attached to a firm substratum. Their skeleton is of a horny material and has thorns. Both of these orders have few species and are limited to warmer waters of the sea.

Octocorallian Corals

Octocorallia (Alcyonaria) have strict octomerous symmetry, with eight pinnate tentacles and eight unpaired, complete septa (see Figure 13.23). They are all colonial, and gastrovascular cavities of the polyps communicate through a system of gastrodermal tubes called **solenia** (Figure 13.32). The solenia run through an extensive mesoglea (**coenenchyme**) in most octocorallians, and the surface of the colony is covered by epidermis. The skeleton is secreted in the coenenchyme and contains limy spicules, fused spicules, or a horny protein, often in combination. Thus the skeletal support of most alcyonarians is an endoskeleton. The variation in pattern among the species of octocorallians lends great variety to the form of the colonies: from soft corals such as *Dendronephthya* (Figure 13.33), with their spicules scattered through the coenenchyme, to the tough, axial supports of sea fans and other gorgonian corals (Figure 13.34), to the fused spicules of organ-pipe coral. *Renilla* (L. *ren,* kidney, + *illa,* suffix), the sea pansy, is a colony reminiscent of a pansy flower. Its polyps are embedded in the fleshy upper side, and a short stalk that supports the colony is embedded in the seafloor. *Ptilosarcus* (Gr. *ptilon,* feather, + *sarkos,* flesh), a sea pen, belongs to the same order and may reach a length of 50 cm (see Figure 13.23).

The graceful beauty of octocorallians—in hues of yellow, red, orange, and purple—helps to create the "submarine gardens" of the coral reefs.

Coral Reefs

Coral reefs are among the most productive of all ecosystems, and they have a diversity of beautiful and colorful life-forms rivaled only by tropical rainforests. They are large formations of calcium carbonate (limestone) in shallow tropical seas deposited by living organisms over thousands of years; living plants and animals are confined to the top layer of reefs, where they add calcium carbonate to that deposited by their predecessors. The most important organisms that precipitate

Figure 13.33   A soft coral, *Dendronephthya* sp. (order Alcyonacea, subclass Octocorallia, class Anthozoa), on a Pacific coral reef. The showy hues of this soft coral vary from pink and yellow to bright red and contribute much color to Indo-Pacific reefs.

calcium carbonate from seawater to form reefs are the scleractinian, **hermatypic** (reef-building) **corals** (see Figure 13.28) and **coralline algae.** Not only do coralline algae contribute to the total mass of calcium carbonate, but their precipitation of the substance helps to hold the reef together. Some octocorallians and hydrozoa (especially *Millepora* [L. *mille,* thousand, + *porus,* pore] spp., "fire coral," Figure 13.16B) contribute in some measure to the calcareous material, and an enormous variety of other organisms contributes small amounts. However, hermatypic (Gr. *herma,* support, mound, + *typos,* type) corals seem essential to formation of large reefs, since such reefs do not occur where these corals cannot live.

Hermatypic corals require warmth, light, and the salinity of undiluted seawater. These requirements limit coral reefs to shallow waters between 30 degrees north and 30 degrees south latitude and excludes them from areas with upwelling of cold water or areas near major river outflows with attendant low salinity and high turbidity. These corals require light because they have mutualistic dinoflagellates (zooxanthellae) living in their tissues. The microscopic zooxanthellae are very important to the corals; their photosynthesis and fixation of carbon dioxide furnish food molecules for their hosts, they recycle phosphorus and nitrogenous waste compounds that otherwise would be lost, and they enhance the ability of the coral to deposit calcium carbonate.

This highly beneficial symbiosis between corals and zooxanthellae is threatened by **coral bleaching** (Figure 13.35). Coral bleaching occurs when corals lose their zooxanthellae and become white and brittle. The loss of zooxanthellae is

Figure 13.34   Colonial gorgonian, or horny, corals (order Gorgonacea, subclass Octocorallia, class Anthozoa) are conspicuous components of reef faunas. These examples are from the western Pacific. **A,** Red gorgonian *Melithaea* sp. **B,** A sea fan, *Subergorgia mollis.***C,** Red whip coral, *Ellisella* sp.

correlated with global warming and the resultant increase in ocean temperature. As the water warms, heat damages part of the photosynthetic mechanism in the zooxanthellae, leading to the buildup of harmful oxidants.

The oxidants diffuse into coral tissues, destroying the finely tuned relationship between the mutualists. Zooxanthellae die or are expelled in what appears to be an immune response from the corals. An initial reduction in the numbers of symbionts worsens the problem because the strongly reflective coral skeleton brings more light into the damaged photosynthetic pathway. There are at least eight clades of symbionts that vary in thermal tolerance, but none appears able to survive continued warming or to prevent bleaching. Bleaching occurred prior to global warming, but never at the intensity or on the scale seen now. Bleaching levels in 2002 on the Great Barrier Reef in Australia were the worst in recorded history, with 60% of the entire reef showing bleaching; in some areas bleaching was visible in 90% of the coral present. Caribbean reefs also showed 90% bleaching, which was accompanied by death of half of the affected corals. Reefs in the Maldives underwent a severe bleaching event in 1998. Recruitment to the site was noticeable after 6 years, and coral cover returned to pre-bleaching levels after 15 years. However, not all species recovered and species with zooxanthellae did not return, so recovery is actually a community restructuring.

Figure 13.35 A comparison of healthy and bleached polyps within a colony of the zoanthid *Palythoa caribbaeorum* in La Parguera, Puerto Rico.
Photo courtesy of Dr. Ernesto Weil, Department of Marine Sciences, University of Puerto Rico.

© Ernesto Weil

Because zooxanthellae are vital to hermatypic corals, and water absorbs light, hermatypic corals rarely live at depths greater than 100 feet (30 m). Interestingly, some deposits of coral reef limestone, particularly around Pacific islands and atolls, reach great thickness—even thousands of feet. Clearly, corals and other organisms could not have grown from the bottom in the abyssal blackness of deep sea and reached shallow water where light could penetrate. Charles Darwin was the first to realize that such reefs began their growth in *shallow* water around volcanic islands; then as the islands slowly sank beneath the sea, the growth of the reefs kept pace with the rate of sinking, thus producing deep deposits.

Taxonomy of Phylum Cnidaria

Strong molecular and morphological evidence now indicates that fish parasites of the former phylum Myxozoa are highly derived cnidarians.[1] At this time, we cannot place them with confidence in the following classification; they could be hydrozoans or a separate class.

Class Hydrozoa (hi-dro-zōa) (Gr. *hydra*, water serpent, + *zōon*, animal). Solitary or colonial; asexual polyps and sexual medusae, although one type may be suppressed; hydranths with no mesenteries; medusae (when present) with a velum; both freshwater and marine. Examples: *Hydra, Obelia, Physalia, Tubularia.*

Class Scyphozoa (si-fo-zōa) (Gr. *skyphos*, cup, + *zōon*, animal). Solitary; polyp stage reduced or absent; bell-shaped medusae without velum; gelatinous mesoglea much enlarged; margin of bell or umbrella typically with eight notches that are provided with sense organs; all marine. Examples: *Aurelia, Cassiopeia, Rhizostoma.*

Class Staurozoa (sto.-ro-zōa) (Gr. *stauros*, a cross, + *zōon*, animal). Solitary; polyps only; medusae absent; polyp surface extended into eight clusters of tentacles surrounding mouth; attachment via adhesive disc; all marine. Examples: *Haliclystus, Lucernaria.*

Class Cubozoa (ku′ bo-zo′a) (Gr. *kybos*, a cube, + *zōon*, animal). Solitary; polyp stage reduced; bell-shaped medusae square in cross section, with tentacle or group of tentacles hanging from a bladelike pedalium at each corner of the umbrella; margin of umbrella entire, without velum but with velarium; all marine. Examples: *Tripedalia, Carybdea, Chironex, Chiropsalmus.*

Class Anthozoa (an-tho-zo′a) (Gr. *anthos*, flower, + *zōon*, animal). All polyps; no medusae; solitary or colonial; gastrovascular cavity subdivided by at least eight mesenteries or septa bearing nematocysts; gonads endodermal; all marine.

    Subclass Hexacorallia (heks′a-ko-ral′e-a) (Gr. *hex*, six, + *korallion*, coral) (**Zoantharia**). With simple unbranched tentacles; mesenteries in pairs; sea anemones, hard corals, and others. Examples: *Metridium, Anthopleura, Tealia, Astrangia, Acropora.*

    Subclass Ceriantipatharia (se-re-an-tip′a-tha′re-a) (N. L. combination of Ceriantharia and Antipatharia). With simple unbranched tentacles; mesenteries unpaired; tube anemones and black or thorny corals. Examples: *Cerianthus, Antipathes, Stichopathes.*

    Subclass Octocorallia (ok′to-ko-ral′e-a) (L. *octo*, eight, + Gr. *korallion*, coral) (**Alcyonaria**). With eight pinnate tentacles; eight complete, unpaired mesenteries; soft and horny corals. Examples: *Tubipora, Alcyonium, Gorgonia, Plexaura, Renilla.*

[1]Siddall, M. E., et al. 1995. J. Parasitol. **81**: 961–967.

A

© Cleveland P. Hickman, Jr.

B

Figure 13.36 **A,** Profile of a barrier reef. **B,** Portion of an atoll from the air. Reef slope plunges into deep water at left (dark blue); lagoon at right.

Several types of reefs are commonly recognized. **Fringing reefs** are close to a landmass with either no lagoon or a narrow lagoon between reef and shore. A **barrier reef** (Figure 13.36) runs roughly parallel to shore and has a wider and deeper lagoon than does a fringing reef. **Atolls** are reefs that encircle a lagoon but not an island. These types of reefs typically slope rather steeply into deep water at their seaward edge. **Patch** or **bank reefs** occur some distance back from the steep, seaward slope in lagoons of barrier reefs or atolls. The so-called Great Barrier Reef, extending 2027 km long and up to 145 km from shore off the northeast coast of Australia, is actually a complex of reef types.

Fringing, barrier, and atoll reefs all have distinguishable zones characterized by different groups of corals and other animals. The side of the reef facing the sea is the **reef front** or **fore reef slope** (Figure 13.36). The reef front is parallel to the shore and perpendicular to the predominant direction of wave travel. It slopes downward into deeper water, sometimes gently at first, then precipitously. Characteristic assemblages of scleractinian corals grow deep on the slope, high near the crest, and in intermediate zones. In shallow water or slightly emergent at the top of the reef front is a **reef crest.** The upper front and the crest bear the greatest force of waves and must absorb great energy during storms. Pieces of coral and other organisms are broken off at such times and thrown shoreward onto the **reef flat,** which slopes down into the lagoon. The reef flat thus accumulates calcareous material eventually ground into coral sand. The sand is stabilized by growth of plants such as turtle grass and

coralline algae and ultimately becomes cemented into the mass of the reef by precipitation of carbonates. A reef is not an unbroken wall facing the sea but is highly irregular, with grooves, caves, crevices, channels through from the flat to the front, and deep, cup-shaped holes ("blue holes"). Octocorallians grow in areas more protected from the full force of waves, as well as on the flat and the deeper areas of the fore reef slope. Many other kinds of organisms inhabit cryptic locations such as caves and crevices.

Enormous numbers of species and individuals of invertebrate groups and fishes populate the reef ecosystem. For example, there are 300 *common* species of fishes on Caribbean reefs and more than 1200 on the Great Barrier Reef complex of Australia. It is marvelous that such diversity and productivity can be maintained, since reefs are washed by nutrient-poor waves of the open ocean. Although relatively little nutrient enters the ecosystem, little is lost because the interacting organisms are so efficient in recycling. The corals even feed on feces of fish swimming over them! Because corals structure the reef ecosystem, their loss is catastrophic in many dimensions, including fishing, tourism, and the resulting economic impacts.

Coral reefs in many areas are threatened by a variety of factors, mostly of human origin. These include enrichment with nutrients (from sewage and runoff of agricultural fertilizer from nearby land) and overfishing of herbivorous fishes, both of which contribute to overgrowth of multicellular algae. Agricultural pesticides, sediment from tilled fields and dredging, and oil spills degrade reefs. Such environmental stresses kill corals directly, or make them more susceptible to the numerous coral diseases that have been observed in recent years, as well as to bleaching. Furthermore, higher atmospheric concentrations of carbon dioxide (from burning hydrocarbon fuels) tend to acidify ocean water, which makes precipitation of $CaCO_3$ by corals more difficult metabolically.

13.2 PHYLUM CTENOPHORA

Ctenophora (te-nof´o-ra) (Gr. *kteis, ktenos,* comb, + *phora,* pl. of bearing) is composed of about 150 species. All are marine forms occurring in all seas but especially in warm waters. They take their name from eight rows of comblike plates of cilia used for locomotion. Common names for ctenophores are "sea walnuts" and "comb jellies."

Except for a few creeping and sessile forms, ctenophores are free-swimming. Although these feeble swimmers are more common in surface waters, ctenophores sometimes occur at considerable depths. They are often at the mercy of tides and strong currents, but they avoid storms by swimming into deeper water. In calm water they often rest vertically with little movement, but when moving they use their ciliated comb plates to propel themselves mouth-end forward.

From an examination of *Pleurobrachia* (Figure 13.37), it is clear that the presence of two tentacles on the body produces biradial symmetry. There is no head, but an oral-aboral axis is present. The mouth leads from a pharynx into a branched digestive tract ending with an anal pore. The body is transparent and has a gelatinous layer, derived from embryonic ectoderm and endoderm, between the two adult tissue layers. The gelatinous layer contains an extensive set of muscle fibers; fiber patterns are radial, as well as in meridional and latitudinal bands around the body. Muscle fibers are also present in the extensible tentacles.

A *Pleurobrachia*   **B** *Mnemiopsis*

© William C. Ober/Medical Scientific Illustration

© Gregory G. Dimijian/Science Source

Figure 13.37   **A,** Comb jelly *Pleurobrachia* sp. (order Cydippida). Its fragile beauty is especially evident at night when it luminesces from its comb rows. **B,** *Mnemiopsis* sp. (order Lobata).

Ctenophore tentacles capture small planktonic organisms, typically crustaceans such as copepods, from the surrounding waters. Extended tentacles trail in the water, and passing prey are caught by epidermal glue cells called **colloblasts** (Figure 13.38C). Colloblasts contain an adhesive material discharged on contact with prey; the adhesive binds to the prey and the rest of the colloblast cell remains attached to the tentacle. Food-laden tentacles are wiped across the mouth.

Ctenophores with short tentacles may collect food on the ciliated body surface. Ctenophores without tentacles may feed on other gelatinous animals such as medusae, salps (see p. 505), or other ctenophores. Entire prey may be consumed or small parts, such as tentacles, removed. Some ctenophores that feed on cnidarians collect undischarged cnidocytes from their prey and incorporate them into epidermal tissue as a defense mechanism. The ctenophore *Haeckelia rubra* (named after Ernst Haeckel, nineteenth-century German zoologist) consumes hydromedusae tentacles in this way.

Ctenophores were previously divided between two classes: Tentaculata and Nuda. Based on evidence that the classes are not monophyletic groups, most biologists discuss ctenophore diversity using seven orders below the class level. Morphological and molecular evidence suggest that one common order (Cydippida) is polyphyletic. One family within Cydippida appears related to members of the order Beroida (see Figure 13.40), whereas others

Tentacle

Comb row

A

Adhesive granule

Nucleus

Helical thread

Nerve cell

Root

C

Tentacle

Tentacle sheath

Statocyst

Anal canal

Aboral canal

Gastrovascular canals

Stomach

Collenchyme

Paragastric canal

Pharynx

Mouth

B

Statolith

Comb row

Comb plates

D

Figure 13.38   Comb jelly *Pleurobrachia,* a ctenophore. **A,** External view. **B,** Hemisection. **C,** Colloblast, an adhesive cell characteristic of ctenophores. **D,** Portion of comb rows showing comb plates, each composed of transverse rows of long, fused cilia.

cannot be placed unambiguously into currently recognized classes. Thus, it is not useful to discuss ctenophore subgroups.

A fundamental understanding of the ctenophore body plan can be gained from consideration of *Pleurobrachia* and a few other examples.

Representative Type: *Pleurobrachia*

Pleurobrachia (Gr. *pleuron*, side, + L. *brachia*, arms) is about 1.5 to 2 cm in diameter (Figure 13.37). The oral pole bears the mouth opening, and the aboral pole has a sensory organ, the **statocyst.**

Comb Plates

On the surface are eight equally spaced bands called **comb rows,** which extend as meridians from the aboral pole and end before reaching the oral pole (Figure 13.38). Each band consists of transverse plates of long, fused cilia called **comb plates** (Figure 13.38D). Ctenophores are propelled by beating of cilia on the comb plates (see Figure 29.12, p. 650). The beat in each row starts at the aboral end and proceeds successively along the combs to the oral end. All eight rows normally beat in unison. The animal is thus driven forward with the mouth in advance. The animal can swim backward by reversing the direction of the wave.

Tentacles

The two **tentacles** are long, solid, and very extensible, and they can be retracted into a pair of **tentacle sheaths.** When completely extended, they may measure 15 cm in length. The surface of the tentacles bears **colloblasts,** or glue cells (Figure 13.38C), which secrete a sticky substance for catching and holding small animals.

Body Wall

The cellular layers of ctenophores are generally similar to those of cnidarians. Between the epidermis and the gastrodermis is a gelatinous **collenchyme** that fills most of the interior of the body and contains muscle cells and ameboid cells. Muscle cells are distinct and are not contractile portions of epitheliomuscular cells (in contrast to Cnidaria).

Feeding, Digestion, and Respiration

The **gastrovascular system** comprises a mouth, a pharynx, a stomach, and a system of gastrovascular canals that branch through the jelly to extend to the comb plates, tentacular sheaths, and elsewhere (Figure 13.38). Two blind canals terminate near the mouth, and an aboral canal passes near the statocyst and then divides into two small **anal canals** through which undigested material is expelled. Digestion is both extracellular and intracellular.

Respiration and excretion occur by diffusion across the body surface.

Nervous and Sensory Structures

Ctenophores have a nerve network similar to those of cnidarians. It features a subepidermal plexus, which is concentrated under each comb plate, but no central control.

Characteristics of Phylum Ctenophora

1. Eight rows of combs (ctenes) arranged radially around body
2. **Colloblasts,** adhesive cells used in prey capture, present in most
3. Entirely marine
4. Symmetry **biradial;** arrangement of internal canals and position of the paired tentacles change the radial symmetry into a combination of radial and bilateral
5. Body ellipsoidal or spherical in shape with oral and aboral ends; no definite head
6. Adult body with gelatinous middle layer containing muscle cells; derivation of middle cellular layer controversial (ectodermal vs. endodermal) affecting status as diploblastic or triploblastic
7. Complete gut; mouth opens into pharynx; gut with a series of branching **gastrovascular canals;** gut terminates at **anal pore;** wastes exit via anal pore and mouth
8. **Extracellular digestion** in pharynx
9. Two extensible tentacles occur in most
10. Muscular contractions via **muscle fibers (cells),** not epitheliomuscular cells
11. Nervous system consisting of a subepidermal plexus concentrated around the mouth and beneath the comb plate rows; an **aboral sense organ** (statocyst)
12. Reproduction monoecious in most; gonads (endodermal origin) on the walls of the digestive canals, which are under the rows of comb plates; mosaic or regulative cleavage within embryos; cydippid larva
13. No respiratory system
14. No coelomic cavity

The sense organ at the aboral pole is a statocyst (Figure 13.38B and D). Tufts of cilia support a calcareous statolith, with the whole being enclosed in a bell-like container. Alterations in the position of the animal change the pressure of the statolith on the tufts of cilia (see also Chapter 33, p. 740). The sense organ coordinates beating of the comb rows but does not trigger their beat.

The epidermis of ctenophores bears abundant sensory cells to detect chemical and other stimuli. When a ctenophore contacts an unfavorable stimulus, it often reverses the beat of its comb plates and moves backward. Comb plates are very sensitive to touch, which often causes them to withdraw into the animal.

Reproduction and Development

Pleurobrachia, like most other ctenophores, is monoecious. Gonads are located on the lining of the gastrovascular canals under the comb plates. Fertilized eggs are discharged through the epidermis into the water.

Cleavage in ctenophores varies among cell lineages. Some lineages are determinate (mosaic), since the parts of the animal to be formed by each blastomere are determined early in embryogenesis. If one of these blastomeres is removed in the early stages, the resulting embryo is deficient. Other cell lineages are like those of cnidarians where development is regulative (indeterminate). The free-swimming **cydippid larva** (Figure 13.39) develops gradually into an adult without metamorphosis.

Figure 13.39   A cydippid larva.

© Ronald L. Shimek, 2004

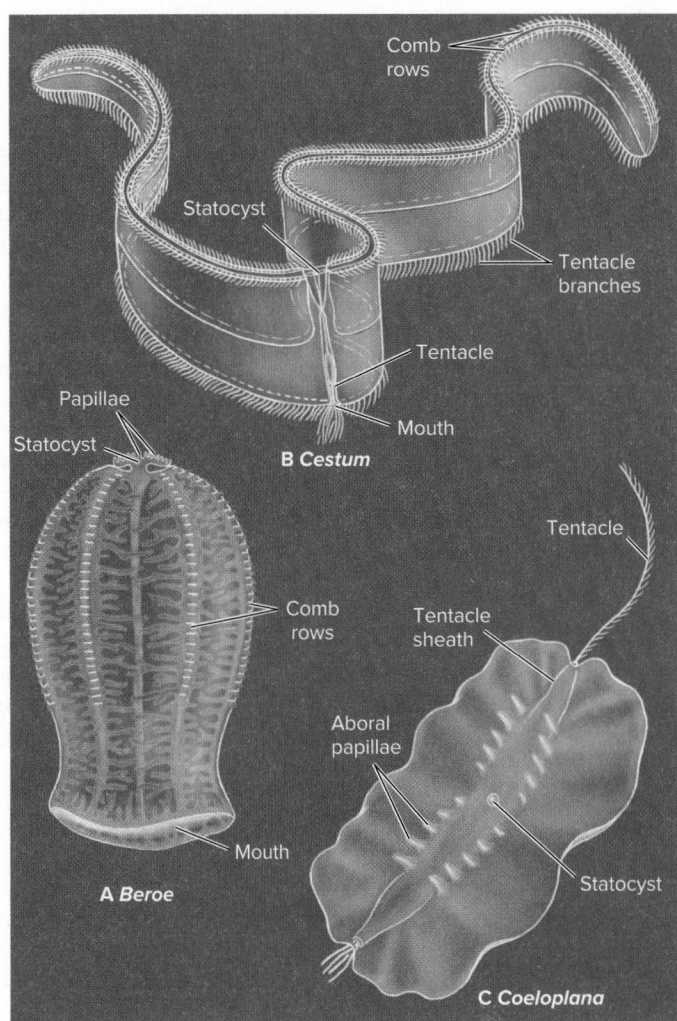

Figure 13.40   Diversity among phylum Ctenophora. **A,** *Beroe* sp. (order Beroida). **B,** *Cestum* sp. (order Cestida). **C,** *Coeloplana* sp. (order Platyctenea).

Other Ctenophores

Ctenophores are fragile and beautiful creatures. Their transparent bodies glisten like fine glass, brilliantly iridescent during the day and luminescent at night.

Among the most striking ctenophores is *Beroe* (L. a nymph), which may be more than 100 mm in length and 50 mm in breadth (Figure 13.40A). It is conical or thimble-shaped and flattened in the tentacular plane. The tentacular plane in *Beroe* is defined as where the tentacles would have been, because it has a large mouth but no tentacles. The animal is pink or rusty brown. Its body wall is covered with an extensive network of canals formed by union of the paragastric and meridional canals.

Highly modified ctenophores such as *Cestum* (L. *cestus,* girdle) use sinuous body movements as well as their comb plates in locomotion. Venus' girdle (*Cestum,* Figure 13.40B) is highly compressed in the tentacular plane. Bandlike, it may be more than 1 m long and presents a graceful appearance as it swims in the oral direction. The highly modified *Ctenoplana* (Gr. *ktenos,* comb, + L. *planus,* flat) and *Coeloplana* (Gr. *koilos,* hollow, + L. *planus,* flat) (Figure 13.40C) are rare but interesting because they have disc-shaped bodies flattened in the oral-aboral axis and are adapted for creeping rather than swimming. A common ctenophore along the Atlantic and Gulf coasts is *Mnemiopsis* (Gr. *mneme,* memory, + *opsis,* appearance) (see Figure 13.37B), which has a laterally compressed body with two large oral lobes and unsheathed tentacles.

Nearly all ctenophores emit flashes of luminescence at night, especially such forms as *Mnemiopsis* (Figure 13.37B). The vivid flashes of light seen at night in southern seas are often caused by members of this phylum.

In the 1980s population explosions of *Mnemiopsis leidyi* in the Black and Azov seas led to catastrophic declines in fisheries there. Inadvertently introduced from the coast of the Americas with ballast water of ships, the ctenophores fed on zooplankton, including small crustaceans and eggs and larvae of fish. The normally inoffensive *M. leidyi* was kept in check in the Atlantic by certain specialized predators. However, accidental introduction of the predatory ctenophore *Beroe ovata* into the Black Sea seems to have caused a decline of *M. leidyi* and then loss of *B. ovata.*

13.3 PHYLOGENY AND ADAPTIVE DIVERSIFICATION

Phylogeny of the Diploblasts

Cnidarians and ctenophores were traditionally considered diploblastic, radially symmetrical animals whose body plans were distinct from both the sponges and the triploblastic, bilaterally symmetrical animals that form the rest of the animal kingdom. The distinctions between the diploblastic and triploblastic conditions are increasingly blurred by detailed morphological studies and by studies of gene expression. Both cnidarians and ctenophores have a gelatinous middle layer surrounded by an outer (epidermal) layer derived from ectoderm and by an inner gut lining derived from endoderm. This body plan is clearly diploblastic, but the presence of cells within the middle gelatinous layer is problematic. If cells of the middle layer derive from endoderm, then they represent a true mesodermal layer of the type seen in triploblasts. If cells in the middle layer derive from ectoderm, then the middle layer is not the same as in the majority of triploblasts; some workers call this layer ectomesoderm.

Sponges, placozoans, cnidarians, and ctenophores are not bilaterally symmetrical. Which of these taxa are most closely related to members of the Bilateria? The answer is not yet clear, but if some of these four taxa shared features, such as muscles, with the bilaterians to the exclusion of other taxa, we could infer a branching order. Sponges and placozoans lack muscles, whereas cnidarians and ctenophores have a range of contractile elements that includes epitheliomuscular cells, smooth muscles, and striated muscles. Epitheliomuscular cells (see Figure 13.5) contain myosin thick filaments and actin thin filaments, are contained within a tissue (epithelium), and anchor to the extracellular matrix (ECM). True muscle cells, whether smooth or striated, contain the same two filaments, but do not form a tissue epithelium. In striated muscles, striations result from repeated units of the two kinds of filaments (called sarcomeres), bordered by Z-discs (see p. 653). In smooth muscle cells, the two kinds of filaments are not in repeated units. If cnidarians, ctenophores, and bilaterians have smooth and striated muscles, did they inherit them from a common ancestor? Researchers identified a key set of muscle motor proteins and decided to see how these proteins were distributed among taxa inside and outside animals. To their surprise they found that these motor proteins were present outside animals—they occur in other members of Opisthokonta and in other unicellular eukaryotes such as amoebozoans and heteroloboseans. In unicells, the contractile motor proteins likely function in cell movement and shape change. Thus, the essential parts of muscle function predate muscles. The researchers also found differences in the ways that striated muscle cells were assembled in cnidarians and bilaterians, suggesting convergent evolution, not homology.

Bilaterians are triploblastic animals (see Chapter 14) whose muscles are derived from the mesodermal layer. Cnidarians and ctenophores are diploblastic animals lacking a mesodermal layer, so biologists wondered whether their muscle cells arose from embryonic ectoderm or endoderm. Recall that mesoderm normally originates from endoderm, although a middle layer called ectomesoderm does sometimes occur. Most ctenophores have only smooth muscle, but one species has striated muscle in its tentacles. Ctenophore muscle cells arise from endoderm. The cells lie within the mesoglea and

are not epithelial, so they are "true" muscle cells. However, there is no sign of a mesodermal layer in ctenophores, so it appears that ctenophores independently evolved striated muscles, as did cnidarians. Cnidarian muscle cells develop from the ectoderm; in medusae the ectoderm gives rise to a tissue layer called the entocodon that is separated from the ectoderm and endoderm by ECM. The entocodon layer produces striated muscle cells, but there is no evidence that this layer is homologous to mesoderm. It appears only in the medusa phase of the life cycle. Striated muscle cells also appear in a highly derived group of swimming sea anemones (anthozoans), but in this case the cells are epithelial and lie within the epidermis. The most recent studies suggest that striated muscles evolved independently at least four times: in highly derived swimming anemones, in medusozoans, in ctenophores, and in triploblasts. Some results even suggest that the striated muscles of triploblasts evolved independently in protostomes and deuterostomes, but that is another story.

It might seem that the designation of body symmetry as radial or bilateral would be more straightforward than the issue of number of embryonic layers. However, this too is much debated. An adult cnidarian is clearly radially symmetrical, and adult ctenophores are biradially symmetrical. However, studies of the cnidarian planula larva show that it swims with one end consistently moving forward. If the forward end is designated "anterior," then the larva has a distinct anterior-posterior axis. The planula larva settles onto a hard substratum with the forward-swimming end serving as the attached end. The trailing end of the larva becomes the oral end of the developing polyp.

Recall that *Hox* genes are highly conserved throughout almost all animals and control expression of other genes determining body axis and morphogenesis along the body axis (see p. 167). Cnidarians do not have as many anterior, central, and posterior *Hox* genes as do most triploblasts, but they do have some genes homologous to anterior and posterior *Hox* genes (central *Hox* genes are lacking). Genes homologous to the anterior *Hox* genes of triploblasts are expressed in the oral end of the polyp. These results are puzzling; did the radially symmetrical cnidarians have a bilaterally symmetrical ancestor, or does the genetic potential for bilateral symmetry predate the bilateral body plan? At present, the answer is not clear.

The reader may have noticed another curiosity in the above account: the forward-swimming end of the larva attaches to the substratum at metamorphosis and becomes the aboral end of the polyp. The aboral end of the polyp is where the posterior *Hox* gene expression occurs. Does this mean that larval orientation is inversely related to polyp orientation? No one knows, but in sponges, where the adult animal has no distinct body axis at all, the larva also has one end that swims forward. With which end does it attach to the substratum? In the sponge *Sycon raphanus,* larvae usually attach with the forward-swimming end, but sometimes with the trailing end, and occasionally with the side of the larva. In most of the triploblastic, bilaterally symmetrical animals, the anteroposterior axis of the adult is already obvious in the larval stage, so there is little basis for comparison with sponges and cnidarians.

Given the preceding discussion, it is perhaps not surprising that we depict a polytomy for cnidarian, ctenophoran, and placozoan branches. Several recent phylogenies place ctenophores as the sister taxon to all other animals, including sponges, which would mean that radial symmetry evolved independently in cnidarians and ctenophores[1].

Cnidarian Phylogeny

Potential antecedents of those hallmark organelles of cnidarians, nematocysts, occur among some single-celled groups, for example, trichocysts and toxicysts in ciliates and trichocysts in dinoflagellates (see p. 231). Some dinoflagellates have organelles that are strikingly similar in structure to nematocysts.

Relationships among cnidarian classes are still controversial. A fascinating area for speculation is the structure of an ancestral cnidarian life cycle: which came first, the polyp or the medusa? Of two important hypotheses, one postulates that the ancestral cnidarian was a trachyline-like hydrozoan with a medusa stage, the other that the ancestral cnidarian was an anthozoan polyp without a medusa in the life cycle.

If the ancestral cnidarians had life cycles similar to those of trachyline-like hydrozoans, a larval form would metamorphose directly into a medusa without an intervening polyp. Under this hypothesis, a polyp phase was added later in evolutionary history, explaining why some biologists consider the polyp a second larval stage. However, molecular evidence suggests that Anthozoa is the sister taxon to the rest of phylum Cnidaria (see Figure 13.2). Development of medusae would then become a synapomorphy of the other classes, with a subsequent loss of a polyp stage in ancestors of Trachylina. One feature that fits well with this hypothesis is the shared possession of a linear mitochondrial genome in groups with medusae: anthozoans and all the other metazoans have a circular mitochondrial genome, considered the ancestral condition. The taxon Medusozoa combines all classes with medusae in the life cycle.

Adaptive Diversification

In Cnidaria, both polyp and medusa are constructed similarly, but medusae have expanded sensory and locomotor capacities. Cnidarians have achieved large numbers of individuals and species, demonstrating a surprising degree of diversity. They are efficient predators, many feeding on prey quite large in relation to themselves. Some are adapted for feeding on small particles. The colonial form of life is well explored, with some colonies growing to great size among corals, and others, such as siphonophores, showing astonishing polymorphism and specialization of individuals within a colony.

Ctenophores all possess eight comb plates and biradial symmetry, but they vary in body shape, and in presence or absence of tentacles. A few have adopted a creeping or sessile habit.

SUMMARY

Phyla Cnidaria and Ctenophora have radially or biradially symmetrical adults; radial symmetry is an advantage for sessile or free-floating organisms because environmental stimuli come from all directions equally. Cnidaria are surprisingly efficient predators because they possess stinging organelles called nematocysts. Both phyla are essentially diploblastic (some triploblastic, depending on the definition of mesoderm), with a body wall composed of epidermis and gastrodermis separated by a mesoglea. The digestive-circulatory (gastrovascular) cavity has a mouth and no anus in cnidarians, but an anal pore is present in ctenophores. Cnidarians have two basic body types (polypoid and medusoid), and in many hydrozoans and scyphozoans the life cycle involves both an asexually reproducing polyp and a sexually reproducing medusa.

That unique organelle, the cnida, is produced by a cnidoblast (which becomes the cnidocyte) and is coiled within a capsule. When discharged, some types of cnidae called nematocysts penetrate prey and inject poison. Discharge occurs by a change in permeability of the capsule and an increase in internal hydrostatic pressure because of the high osmotic pressure within the capsule.

Most hydrozoans are colonial and marine, but the freshwater hydras are commonly demonstrated in class laboratories. They have a typical polypoid form but are not colonial and have no medusoid stage. Most marine hydrozoans form a branching colony of many polyps (hydranths). Hydrozoan medusae can be free-swimming or remain attached to their colony.

Scyphozoans are typical jellies, in which the medusa is the dominant body form, and many have an inconspicuous polypoid stage. A new class, Staurozoa, contains stauromedusans, formerly part of Scyphozoa. Cubozoans are predominantly medusoid. They include the dangerous sea wasps.

Anthozoans are all marine and polypoid; there is no medusoid stage. The most important subclasses are Hexacorallia (with hexamerous or polymerous symmetry) and Octocorallia (with octomerous symmetry). The largest hexacorallian orders contain sea anemones, which are solitary and do not have a skeleton, and stony corals, which are mostly colonial and secrete a calcareous exoskeleton. Stony corals are a critical component in coral reefs, which are habitats of great beauty, productivity, and ecological and economic value. Octocorallia contain the soft and horny corals, many of which are important and beautiful components of coral reefs.

Ctenophora are biradial and swim using eight comb rows. Colloblasts, with which they capture small prey, characterize the phylum.

REVIEW QUESTIONS

1. Explain the utility of radial symmetry for sessile and free-floating animals.
2. What characteristics of phylum Cnidaria are most important in distinguishing it from other phyla?
3. Name and distinguish the taxonomic classes in phylum Cnidaria.
4. Distinguish between polyp and medusa forms.
5. Explain the mechanism of nematocyst discharge. How can a hydrostatic pressure of one atmosphere be maintained within the nematocyst until it receives an expulsion stimulus?
6. What is an unusual feature of the nervous network of cnidarians?
7. In what way is a hydra atypical as a hydrozoan?
8. Name and give functions of the main cell types in the epidermis and in the gastrodermis of hydra.
9. What stimulates feeding behavior in hydras?
10. Define the following with regard to hydroids: hydrorhiza, hydrocaulus, coenosarc, perisarc, hydranth, gonangium, manubrium.
11. Give an example of a highly polymorphic, floating, colonial hydrozoan.
12. Distinguish the following from each other: statocyst and rhopalium; scyphomedusae and hydromedusae; scyphistoma, strobila, and ephyrae; velum, velarium, and pedalium; Hexacorallia and Octocorallia.
13. Define the following with regard to sea anemones: siphonoglyph; primary septa or mesenteries; incomplete septa; septal filaments; acontia threads; pedal laceration.

[1]Dohrmann, M and G. Wörheide. 2013. Integr. Comp. Biol. **53**: 503–511.

14. Describe three specific interactions of anemones with nonprey organisms.
15. Contrast the skeletons of hexacorallian and alcyonarian corals.
16. Coral reefs generally are limited in geographic distribution to shallow marine waters. How do you explain this observation?
17. Specifically, what kinds of organisms are most important in deposition of calcium carbonate on coral reefs?
18. How do zooxanthellae contribute to the welfare of hermatypic corals?
19. Distinguish each of the following from each other: fringing reefs; barrier reefs; atolls; patch or bank reefs.
20. What characteristics of Ctenophora are most important in distinguishing it from other phyla?
21. How do ctenophores swim, and how do they obtain food?
22. Compare cnidarians and ctenophores, giving five ways in which they resemble each other and five ways in which they differ.
23. Cnidarians and ctenophores are considered diploblastic, but why might some biologists label them triploblastic?

For Further Thought What would be the best way to make non-biologists aware of the ecological and economic costs of global warming as it affects coral reefs and jelly populations?

SELECTED REFERENCES

Burton, P. M. 2008. Insights from Diploblasts: the Evolution of Mesoderm and Muscle. Journal of Experimental Zoology (MOL DEV EVOL) **310B:** 5–14. *A very clearly written discussion of muscle structure in cnidarians, ctenophores, and bilaterally symmetrical animals. Coates, M. M. 2003. Visual ecology and functional morphology of Cubozoa (Cnidaria). Integr. Comp. Biol.* **43:** 542–548. *Collected information on the three different kinds of eyes in cubozoans.*

Coates , M. M. 2003. Visual ecology and functional morphology of Cubozoa (Cnidaria). Integr. Comp. Biol. **43:** 542–548. *Collected information on the three different kinds of eyes in cubozoans.*

Collins, A. G., P. Schuchert, A.C. Marques, T. Jankowski, M. Medina, and B. Schierwater. 2006. Medusozoan phylogeny and character evolution clarified by new large and small subunit rDNA and an assessment of the utility of phylogenetic mixture models. Syst. Biol. **55:** 97–115. *Authors produce a working cladogram of major cnidarian taxa, including class Staurozoa.*

Crossland, C. J., B. G. Hatcher, and S. V. Smith. 1991. Role of coral reefs in global ocean production. Coral Reefs **10:** 55–64. *Because of extensive recycling of nutrients within reefs, their net energy production for export is relatively minor. However, they play a major role in inorganic carbon precipitation by biologically mediated processes.*

Dohrmann, M., and G. Wörheide. 2013. Novel Scenarios of Early Animal Evolution—Is It Time to Rewrite Textbooks? Integr. Comp. Biol. **53:** 503–511. *These authors review the evidence for major changes in branching pattern at the base of the animal tree and advocate for collection of more data before changes are accepted.*

Finnerty, J. R., K. Pang, P. Burton, D. Paulson, and M. Q. Martindale. 2004. Origins of bilateral symmetry: *Hox* and Dpp expression in a sea anemone. Science **304:** 1335–1337. *Suggested homology between sea anemone oral end and anterior region of triploblasts based on Hox gene expression. Letters and comments that followed this article offer a full discussion of issues.*

Hejnol, A. 2012. Evolutionary biology: Muscle's dual origins. Nature **487:** 181–182. *This article for general readers is the companion piece to the more technical article on muscle evolution in the same issue of this journal.*

Leal, I. C. S., M. E. de Araujo, S. Rabela da Cunha, and P. H. Cipresso Pereira. 2015. The influence of fire-coral colony size and agonistic behaviour of territorial damselfish on associated coral reef fish communities. Marine Environmental Research **108:** 45–54. *Damselfish depend on fire-coral colonies for food and refuge from predators.*

Martindale, M. Q., K. Pang, and J. R. Finnerty. 2004. Investigating the origins of triploblasty: "mesodermal" gene expression in a diploblastic animal, the sea anemone *Nematostella vectensis* (phylum Cnidaria; class Anthozoa). Development **131:** 2463–2474. *A discussion of the entocodon problem and putative mesoderm in cnidarians with illustrations and photos showing gene expression during development*

Miglietta, M. P., and H. A. Lessios. 2009. A silent invasion. Biol. Invasions **11:** 825–834. *Medusae of an unusual hydrozoan transform into cysts when damaged; cysts are easily transported, making it likely that this species invades via the ballast water released from ships.*

Morri, C., M. Montefalcone, R. Lasagna, G. Gatti, A. Rovere, V. Parravicini, G. Baldelli, P. Colantoni, and C. Nike Bianchi. 2015. Through bleaching and tsunami: coral reef recovery in the Maldives. Marine Pollution Bulletin **98:**188–200. *Although the coral community recovered from a major bleaching event, species with zooxanthellae were among those that did not return.*

Podar, M., S. H. D. Haddock, M. I. Sogin, and G. R. Harbison. 2001. A molecular phylogenetic framework for the phylum Ctenophora using 18S rRNA genes. Mol. Phylogen. Evol. **21:** 218–230. *Similar evolutionary relationships among ctenophore orders appear from morphological and molecular studies.*

Purcell, J. E. 2012. Jellyfish and ctenophore blooms coincide with human proliferations and environmental perturbations. Annual Review of Marine Science (Carlson, C. A., and S. J. Giovannoni, eds.) **4:** 209–235. *Human population increases are associated with abundance of jellies.*

Seipel, K., and V. Schmid. 2005. Evolution of striated muscle: Jellyfish and the origin of triploblasty. Developmental Biology **282:** 14–26. *This article explains why the cnidarian entocodon, which makes striated muscle cells, is not homologous to mesoderm.*

Steinmetz, Patrick R. H., J. E. M. Kraus, C. Larroux, J. U. Hammel, A. Amon-Hassenzahl, E. Houliston, G. Wörheide, M. Nickel, B. M. Degnan, and U. Technau. 2012. Independent evolution of striated muscles in cnidarians and bilaterians. Nature **487:** 231–234. *The common ancestor of animals possessed a core set of contractile proteins that was independently augmented by new and different muscle proteins in different types of animals.*

Weis, V. M. 2008. Cellular mechanisms of cnidarian bleaching: stress causes the collapse of symbiosis. J. Exp. Biol. **211:** 3059–3066. *Clear description of our current understanding of the mechanisms underlying coral bleaching.*

Acoelomorpha, Platyzoa, and Mesozoa

Prosthecereaus vittatus, the candy-striped flatworm, lives in the intertidal zone under stones or in mud in Western Europe.

- **ACOELOMORPHA**
  - Phylum Acoelomorpha
- **PLATYZOA**
  - Phylum Platyhelminthes
  - Phylum Gastrotricha
  - Phylum Gnathostomulida
  - Phylum Micrognathozoa
  - Phylum Rotifera
  - Phylum Acanthocephala
- **MESOZOA**
  - Phylum Mesozoa

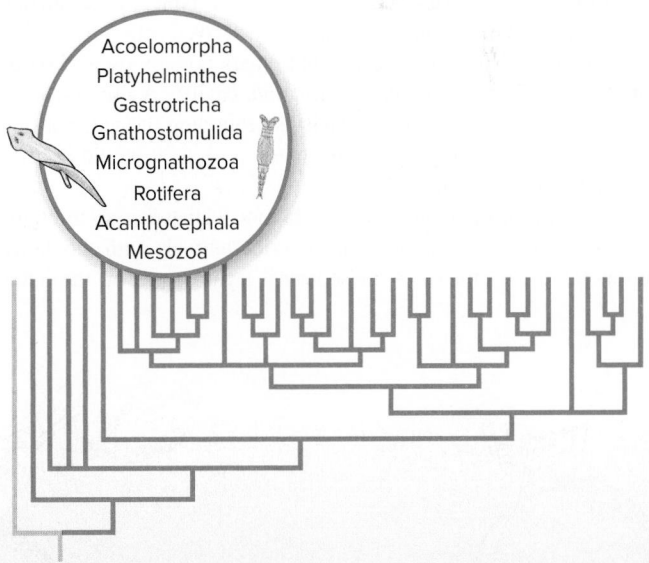

Acoelomorpha
Platyhelminthes
Gastrotricha
Gnathostomulida
Micrognathozoa
Rotifera
Acanthocephala
Mesozoa

Getting Ahead

The radially symmetrical cnidarians and ctenophores snare prey approaching from any direction, but an animal that actively seeks food, shelter, or reproductive mates is not well served by radial symmetry. Directed movement is most efficient with an elongated body form with head (anterior) and tail (posterior) ends. In addition, one side of the body is kept up (dorsal) and the other side, specialized for locomotion, is kept down (ventral). The result is a bilaterally symmetrical animal in which the body can be divided along only one plane of symmetry to yield two halves that are mirror images of each other. Furthermore, because it is better to determine where one is going rather than where one has been, sense organs and centers for nervous control have come to be concentrated on the head. This process is called cephalization. Cephalization and primary bilateral symmetry occur together in almost all triploblastic animals.

In this chapter we introduce two phyla of flatworms and six other phyla that lack a mesodermally lined body cavity. All but one phylum contains animals with bilaterally symmetrical, triploblastic bodies. This body plan occurs in modern or ancestral forms of all metazoans yet to be discussed. The body is triploblastic because it contains a middle germ layer, mesoderm, derived from endoderm. Three germ layers—ectoderm, endoderm, and mesoderm—produce all adult body structures (see p. 178 for typical derivatives of these layers).

Members of two of the bilaterally symmetrical triploblastic phyla discussed here have **acoelomate** (Gr. *a,* not, + *koilōma,* cavity) bodies. A coelom is a cavity that develops *entirely within* mesoderm (see p. 163). Acoelomate bodies do not have a coelom. Readers may recall that diploblasts also lack a coelom but were not called acoelomate; the term is used only for animals possessing mesoderm. Acoelomate taxa do not form a monophyletic group in most analyses, so we use the term only to describe a particular body plan.

Typical acoelomates have only one internal space, the digestive cavity (Figure 14.1). The region between the epidermis and the digestive cavity lining is filled with a cellular, mesodermally derived **parenchyma.** Parenchyma is a form of packing tissue containing more cells and fibers, and less extracellular matrix (ECM), than the mesoglea of cnidarians. Organs constitute another derivative of mesoderm that increases internal complexity in triploblasts. We see this complexity in members of the acoelomate phyla Acoelomorpha and Platyhelminthes.

Some members of Acoelomorpha are atypical acoelomates because they lack a digestive cavity. In these small worms food particles enter through the mouth and move into a cellular or syncytial mass derived from endoderm. A temporary digestive cavity may form within the endoderm.

Some phyla described in this chapter are typical acoelomates with a gut cavity lined with endodermally derived cells surrounded by tissue derived from mesoderm, but others possess pseudocoelomate bodies (Gr. *pseudo,* false, + *koilōma,* cavity). A pseudocoelomate body contains an internal cavity surrounding the gut, but this cavity is not completely lined with mesoderm, as it would be in a coelomate animal (Figure 14.1).

A pseudocoelom is an embryonic blastocoel that persists throughout development, leading some to describe animals with this body plan as blastocoelomates. A mesodermal layer occurs on the outer edge of the cavity, but that layer does not extend around the gut so it does not form a mesentery. The endodermal gut lining forms the inner boundary of the pseudocoelom (see p. 171). The pseudocoel may be filled with fluid, or it may contain a gelatinous matrix with some mesenchymal cells. It shares some functions of a coelom: space for development and differentiation of digestive, excretory, and reproductive systems, a simple means of circulation or distribution of materials throughout the body, a storage place for waste products to be discharged to the outside by excretory ducts, and a hydrostatic support.

Many pseudocoelomate animals are quite small, so the most likely function of the pseudocoel in these animals is to permit internal circulation in the absence of a true circulatory system.

Animals sharing a particular body plan do not necessarily form a monophyletic group. Pseudocoelomate groups include rotifers and acanthocephalans, among others discussed in later chapters, whereas acoelomate taxa include acoelomorphs, platyhelminths, gastrotrichs, gnathostomulids, and probably micrognathozoans and mesozoans. Members of the Acoelomorpha are hypothesized to have branched from a lineage ancestral to all other bilateral animals before that lineage split to form protostomes and deuterostomes. Examine the Cladogram of All Animal Phyla, inside the front cover, to see this branching pattern. Notice that the other phyla discussed in this chapter belong to Protostomia and are placed in the subgroup Lophotrochozoa (see p. 170).

14.1 PHYLUM ACOELOMORPHA

Acoelomorphs (Figure 14.2) are small, flat worms less than 5 mm in length. The word "worm" is loosely applied to elongate bilateral invertebrate animals without appendages. At one time zoologists considered worms (Vermes) a single taxon. This taxon included a highly diverse assortment of forms that were eventually distributed among various phyla, commonly called flatworms, ribbon worms, roundworms, and segmented worms.

Acoelomorph flatworms typically live in marine sediments, although a few forms are pelagic. Some species inhabit brackish water. Most acoelomorphs are free-living, but some are symbiotic and others parasitic. The group contains approximately 350 species.

Members of phylum Acoelomorpha were formerly placed in class Turbellaria within phylum Platyhelminthes (see p. 293). Two turbellarian orders, Acoela and Nemertodermatida, now represent

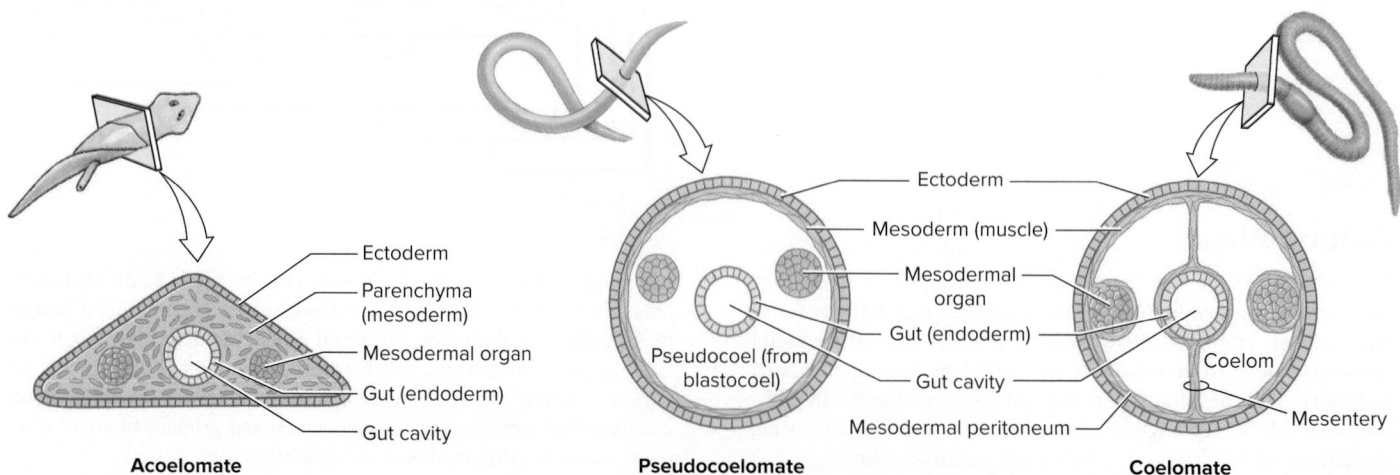

Figure 14.1 Acoelomate, pseudocoelomate, and coelomate body plans.

Figure 14.2 Acoelomorph worms, *Waminoa* sp., on a bubble coral, *Plerogyra sinuosa*.

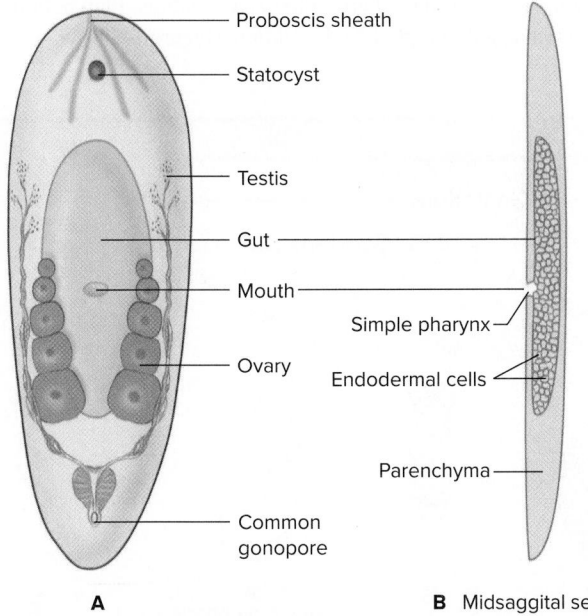

A   **B** Midsaggital section

Figure 14.3 **A,** Generalized acoelomorph flatworm. **B,** Midsagittal section showing gut cavity filled with endodermal cells.

---

two subgroups in phylum Acoelomorpha, although some researchers consider each former order to be a new phylum.

Acoelomorphs have a cellular ciliated epidermis. The parenchyma layer contains a small amount of ECM and circular, longitudinal, and diagonal muscles.

The digestive system of some acoelomorphs opens from a mouth to a tubelike pharynx followed by a saclike gut. There is no anus. In many acoels, the gut and pharynx are absent, so the mouth leads into a mass of endodermally derived cells or an endodermally derived syncytial mass (Figure 14.3). When food is passed into temporary spaces, gastrodermal phagocytotic cells digest food intracellularly.

Acoelomorphs are monoecious. The female reproductive organ produces gametes and nutrition for the young at the same time; the resultant yolk-filled eggs are called **endolecithal** eggs. After

---

Characteristics of Phylum Acoelomorpha

1. Rootlets of epidermal cilia form an interconnecting network
2. Entirely aquatic, some in brackish water, but most in marine sediments
3. Most free-living, some commensal, some parasitic
4. **Bilateral symmetry;** anterior concentration of nerve cells; **body flattened dorsoventrally**
5. Adult body three-layered (**triploblastic**)
6. Body acoelomate, ECM reduced
7. Epidermis cellular
8. Gut absent, or if present, gut incomplete and saclike
9. **Mesodermal** muscle cells produce longitudinal, circular, and diagonal muscles
10. Diffuse system of anterior neurons connected to radially arranged nerve cords
11. Sense organs include statocysts (organs of balance) and ocelli
12. Asexual reproduction by fragmentation
13. Monoecious sexual reproduction via well-developed gonads, ducts, and accessory organs; internal fertilization; duet-spiral cleavage
14. No excretory or respiratory system

---

fertilization, some or all cleavage events produce an unusual duet-spiral pattern of new cells. The duet-spiral pattern may be a defining morphological feature for acoelomorphs, but further study is required for confirmation.

Other defining features proposed for acoelomorphs are biochemical (patterns of neurotransmitters) or rely on details of cellular ultrastructure such as formation of a network of interconnecting rootlets from epidermal cilia.

Acoelomorphs have a distinct anteroposterior axis, but the diffuse collection of nerve cells at the anterior end of the body lacks ganglia typical of a "true" brain. Acoelomorphs have a radial arrangement of nerves in the body, instead of a ladderlike pattern seen in flatworms within phylum Platyhelminthes. Acoelomorph statocysts differ in structure from those of platyhelminths.

Phylogeny of Acoelomorpha

Several phylogenetic studies using molecular characters (for example, mitochondrial genome and myosin II genes) describe acoelomorphs as early diverging, bilaterally symmetrical triploblasts. Acoelomorphs have only four or five *Hox* genes (see Chapter 8, p. 167), unlike free-living members of Platyhelminthes, which have seven or eight such genes.

14.2 CLADES WITHIN PROTOSTOMIA

Most triploblastic metazoans are divided among two large clades or superphyla: Protostomia and Deuterostomia (see the Cladogram of All Animal Phyla, inside the front cover). Division into these two groups is based largely on features of development (see p. 169), although the two groups are also recovered in most phylogenies using molecular characters.

The Protostomia is divided into two large clades: **Lophotrochozoa** and **Ecdysozoa.** Platyhelminthes is the first protostome phylum discussed; it and the remaining phyla included in this chapter belong to Lophotrochozoa. The set of phyla now considered to be lophotrochozoans first appeared as a clade in molecular phylogenies. Prior to the construction of these phylogenies, the primary taxonomic division within protostomes placed all acoelomate phyla in one group and all coelomate phyla in a separate group.

Molecular phylogenies rejected the hypothesis that the acoelomates and the coelomates formed separate monophyletic groups, instead dividing protostomes into two subsets with distinctive molecular signatures. Some morphological characters were shared by members of each subset. Members of Ecdysozoa possess a cuticle that is molted as their bodies grow. Members of Lophotrochozoa share either an odd horseshoe-shaped feeding structure, the **lophophore** (see p. 320, pp. 323–326), or a particular larval form called the **trochophore** (see pp. 337–338).

Trochophore larvae are minute, translucent, and roughly top-shaped (see Figure 16.7, p. 338). They have a prominent circlet of cilia and sometimes one or two accessory circlets. Trochophores occur in the early development of many marine members of Annelida and Mollusca and are assumed ancestral for these groups. Trochophore-like larvae occur in some marine members of Platyhelminthes, Nemertea, Echiura, and Sipunculida, among others.

Clade Platyzoa

Platyzoa is a group of lophotrochozoan protostome phyla. The evolutionary relationships among lophotrochozoans are still under debate because different combinations of molecular and morphological characters yield different cladograms. When gene sequences for certain taxa are not available, these taxa are excluded from the study. However, recent efforts to broaden the number and type of characters as well as the number of taxa in a phylogenetic analysis (see p. 205) have enhanced our understanding of the protostomes. The Platyzoa emerged as a clade in several phylogenies, but readers should be aware that contrasting relationships have been proposed for the phyla forming Platyzoa. As described here, Platyzoa contains Platyhelminthes, Gastrotricha, and four phyla in Gnathifera (Figure 14.4).

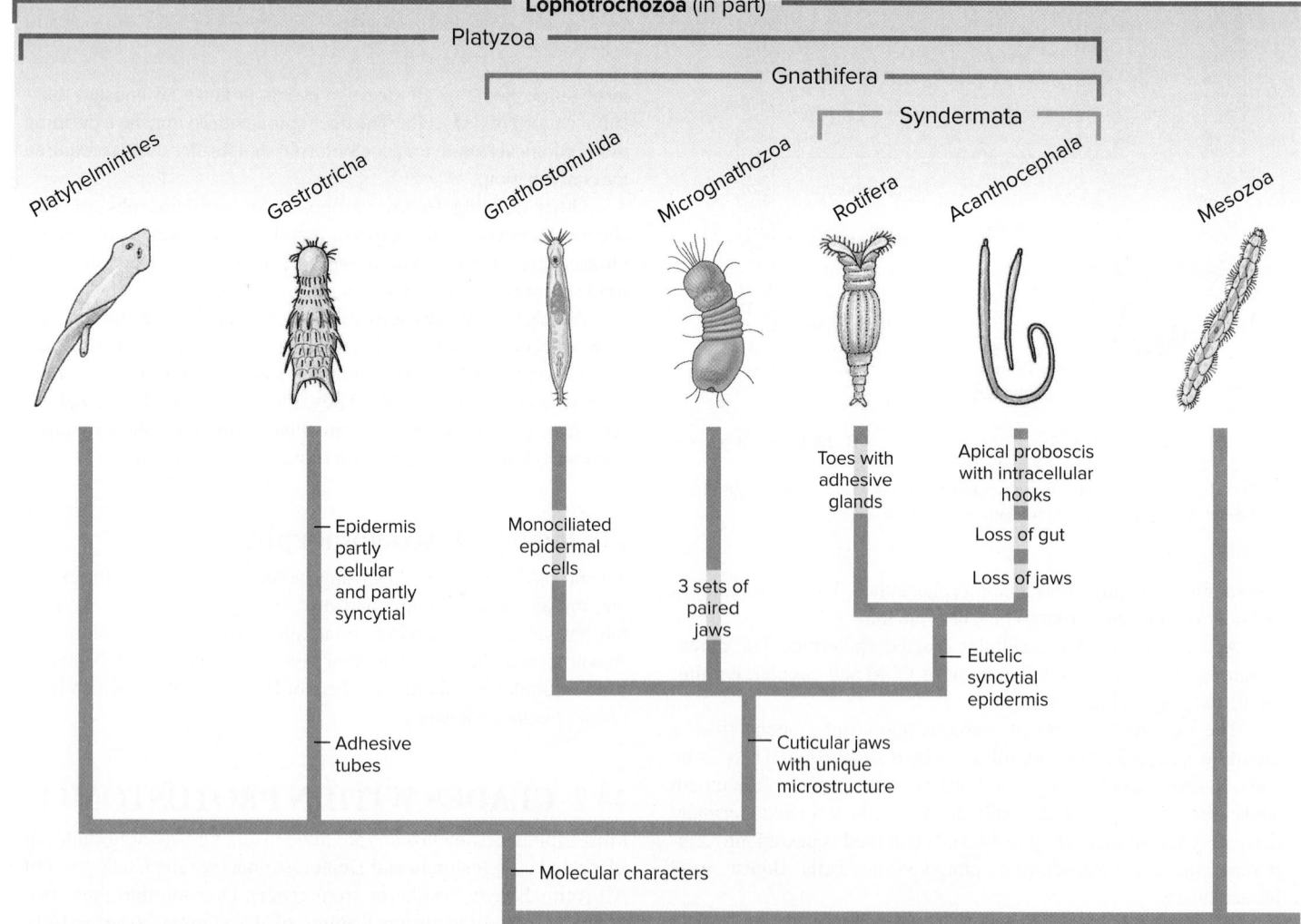

Figure 14.4 Hypothetical relationships among members of Platyzoa. Characters are modified subsets of those in Kristensen (2002; see p. 318 for citation) and Brusca and Brusca (2003; see p. 330 for citation).

14.3 PHYLUM PLATYHELMINTHES

Members of phylum Platyhelminthes (Gr. *platys*, flat, + *helmins*, worm) are commonly called flatworms. They range in size from a millimeter or less to many meters in length (for some tapeworms), although most are 1 to 3 cm. Their bodies may be slender, broadly leaflike, or long and ribbonlike. The phylum contains free-living forms, such as the common planarian (Figure 14.5), as well as parasitic flukes and tapeworms.

Because there is no single unique characteristic (synapomorphy) for the phylum as a whole, some authorities argue that the phylum Platyhelminthes is not a valid monophyletic group. However, Platyhelminthes emerges as a clade in recent molecular phylogenies, and there is a defining feature for a large parasitic clade within Platyhelminthes. The parasites share an external body covering, called a syncytial tegument, or **neodermis,** which contrasts with the cellular ciliated epidermis of most free-living forms. Some morphological features of the free-living flatworms suggest shared ancestry with the parasitic forms. Pending resolution of the intense debate on the nature of this group of worms, we continue to present it as a phylum.

Platyhelminthes is divided into four classes (Figure 14.6): Turbellaria, Trematoda, Monogenea, and Cestoda. Class Turbellaria contains the free-living flatworms, along with some symbiotic and parasitic forms. Most turbellarians are bottom dwellers in marine or freshwater, living under stones or other hard objects. Freshwater planarians inhabit streams, pools, and even hot springs. Terrestrial flatworms are limited to moist places under stones or logs. Class Turbellaria is depicted as a paraphyletic taxon (see p. 207) but a large-scale revision has recently been published.

All members of classes Monogenea, Trematoda (flukes), and Cestoda (tapeworms) are parasitic. Most Monogenea are ectoparasites, but all trematodes and cestodes are endoparasitic. Many species have indirect life cycles with more than one host; the first host is often an invertebrate, and the final host is usually a vertebrate. Humans serve as hosts for a number of species.

Many species covered in this chapter and Chapters 11, 17, 18, 19, 20, and 21 are parasites. People have suffered greatly through the centuries from their parasites and those of their domestic animals. Fleas and bacteria conspired to destroy a third of the European population in the seventeenth century, and malaria, schistosomiasis, and African sleeping sickness have sent millions to their graves. Even today, after successful campaigns against yellow fever, malaria, and hookworm infections in many parts of the world, parasitic diseases in association with nutritional deficiencies are the primary killers of people. Civil wars and environmental changes have led to resurgences in malaria, trypanosomiasis, and leishmaniasis, and global prevalences of intestinal roundworms are unchanged in the last 50 years.

Forms and Function

Epidermis, Muscles

Most turbellarians have a cellular, ciliated epidermis resting on a basement membrane. It contains rod-shaped **rhabdites,** which swell and form a protective mucous sheath around the body when discharged. Single-cell mucous glands open on the surface of the epidermis (Figure 14.7). Most orders of turbellarians have **dual-gland adhesive organs** in the epidermis. These organs consist of three cell types: viscid and releasing gland cells and anchor cells (Figure 14.8). Secretions of the viscid-gland cells apparently fasten microvilli of the anchor cells to the substrate, and secretions of the releasing-gland cells provide a quick, chemical detaching mechanism.

In contrast to the ciliated cellular epidermis of most turbellarians, adult members of the three parasitic classes have a nonciliated body covering called a **syncytial tegument** (Figure 14.9). The term **syncytial** means that many nuclei are enclosed within a single cell membrane. It might appear that a completely new body covering appeared in the parasitic classes, but there are some free-living turbellarians with an atypical epidermis that is syncytial, or "insunk."

Some turbellarians have a syncytial epidermis and others have a syncytial "insunk" epidermis where cell bodies (containing nuclei) are located beneath the basement membrane of the epidermis. Cell bodies communicate with the surface cytoplasm (distal cytoplasm) by sending extensions upward. These extensions fuse to form the syncytial covering, much as they do in syncytial tegument formation. The term "insunk" is a misnomer because the surface cytoplasm arises from upward extensions of cell bodies, not from cell bodies sinking beneath the basement membrane.

Adults of all members of Trematoda, Monogenea, and Cestoda possess a syncytial covering that entirely lacks cilia and is designated a tegument (Figure 14.9). The larval forms of many of these groups

A

B

Figure 14.5 **A,** Stained planarian. **B,** *Bipalium*, a terrestrial flatworm.

© Michael Abbey/Science Source

Platyhelminthes

Neodermata

Turbellaria (in part)

Trematoda   Monogenea   Cestoda

Macrostomida   Polycladida   Tricladida   Aspidogastrea   Digenetic flukes   Monogenetic flukes   Tapeworms

— Loss of digestive tract
— Sporocyst in snail host
— Oncomiracidium larval stage
— Scolex
— Anterior adhesive organ
— Posterior adhesive organ with hooks
— Intestine with many lateral branches
— 3 branched intestine
— Loss of rhabdites
— Syncytial tegument
— Ectolecithal eggs
— Lamellate rhabdites
— Endolecithal eggs
— Mehlis's gland

Figure 14.6   Hypothetical relationships among parasitic Platyhelminthes. The traditionally accepted class Turbellaria is paraphyletic. Some turbellarians have ectolecithal development and, together with the Trematoda, Monogenea, and Cestoda, form a clade to the exclusion of the endolecithal turbellarians. For the sake of simplicity, the synapomorphies of those turbellarians and of the Aspidogastrea, as well as many others given by Brooks (1989), are omitted. Brooks further defines a clade called Cercomeria that includes all members of Neodermata plus two turbellarian taxa not shown here. Members of Cercomeria possess a posterior adhesive organ. Hooks are present on this organ in monogeneans and cestodes.

Source: Modified from D. R. Brooks. The phylogeny of the Cercomeria (Platyhelminthes: Rhabdocoela) and general evolutionary principles. Journal of Parasitology 75: 606–616, 1989.

Ocellus

Pharynx

Intestine

Gland cell   Epidermis   Circular muscles   Pharynx   Pharyngeal cavity   Columnar epithelium

Rhabdite cell

Rhabdite

Dual-gland adhesive organ   Parenchyma   Intestine   Pharyngeal muscles   Nerve cord   Cilia

Parenchymal muscles

Longitudinal muscles

A

B

Figure 14.7   **A,** Whole planarian. **B,** Cross section of planarian through pharyngeal region, showing relationships of body structures.

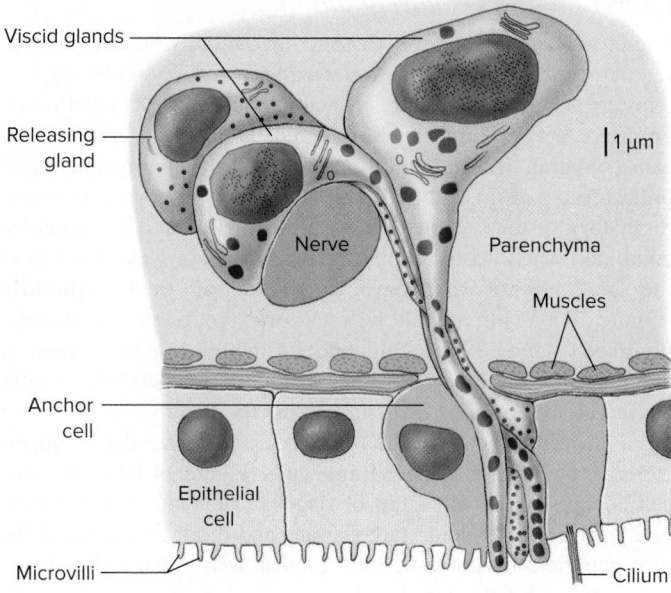

Figure 14.8 Reconstruction of dual-gland adhesive organ of the turbellarian *Haplopharynx* sp. There are two viscid glands and one releasing gland, which lie beneath the body wall. The anchor cell lies within the epidermis, and one of the viscid glands and the releasing gland are in contact with a nerve.

Characteristics of Phylum Platyhelminthes

1. No clear defining feature
2. In marine, freshwater, and moist terrestrial habitats
3. Turbellarian flatworms are mostly free-living; classes Monogenea, Trematoda, and Cestoda entirely parasitic
4. **Bilateral symmetry;** definite polarity of anterior and posterior ends; **body flattened dorsoventrally**
5. Adult body three-layered (**triploblastic**)
6. Body acoelomate
7. Epidermis may be cellular or syncytial (ciliated in some); **rhabdites** in epidermis of most Turbellaria; epidermis a syncytial **tegument** in Monogenea, Trematoda, Cestoda, and some Turbellaria
8. Gut incomplete, may be branched, absent in cestodes
9. Muscular system primarily of a sheath form and of mesodermal origin; layers of circular, longitudinal, and sometimes oblique fibers beneath the epidermis
10. Nervous system consisting of a **pair of anterior ganglia** with **longitudinal nerve cords** connected by transverse nerves and located in the mesenchyme in most forms
11. Sense organs include statocysts (organs of balance) and ocelli
12. Asexual reproduction by fragmentation and other methods as part of complex parasitic life cycles
13. Most forms monoecious; reproductive system complex, usually with well-developed gonads, ducts, and accessory organs; internal fertilization; development direct in free-swimming forms and those with single hosts; complicated life cycle often involving several hosts in many internal parasites
14. Excretory system of two lateral canals with branches bearing **flame cells (protonephridia);** lacking in some forms
15. Respiratory, circulatory, and skeletal systems lacking; lymph channels with free cells in some trematodes

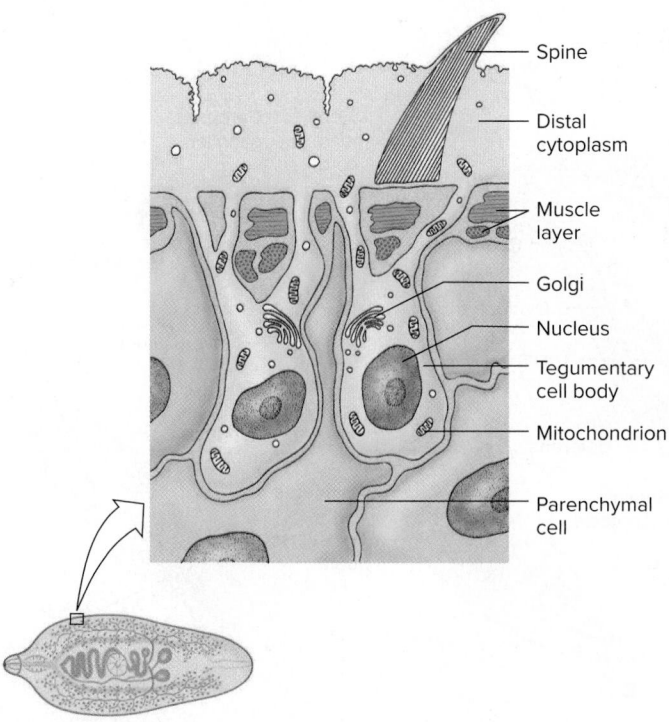

Fasciola hepatica

Figure 14.9 Diagrammatic drawing of the structure of the tegument of a trematode *Fasciola hepatica*.

are ciliated, but the ciliated covering is shed once a host is contacted. Epidermal shedding has been suggested as a means of avoiding host immune response. Development of tegument occurs as several surface layers of the epidermis are shed; eventually fused cytoplasmic extensions from cell bodies below the basement membrane become the surface covering of the body. The tegument is sometimes called the **neodermis,** and its shared presence among the parasites is the basis for uniting trematodes, monogeneans, and cestodes in clade **Neodermata** (Figure 14.6).

The tegument is resistant to the immune system of the host in endoparasites, and it resists digestive juices in tapeworms and others that dwell in a host gut. The syncytial nature of the tegument might render it more resistant because there are no penetrable junctions between cells. The tegument can be both absorptive and secretory. The tegument of one tapeworm has been shown to release enzymes that reduce the effectiveness of the host's digestive system. Tapeworm tegument absorbs nutrients from the host's digestive cavity—tapeworms have neither mouth nor gut.

In the body wall below the basement membrane of flatworms are layers of **muscle fibers** that run circularly, longitudinally, and diagonally. A meshwork of **parenchyma** cells, developed from mesoderm, fills the spaces between muscles and visceral organs. Parenchyma cells in some, perhaps all, flatworms are not a separate cell type but are the noncontractile portions of muscle cells.

Nutrition and Digestion

In general, platyhelminth digestive systems include a mouth, a pharynx, and an intestine (Figure 14.10). In turbellarians, such as the planarian *Dugesia,* the pharynx is enclosed in a **pharyngeal sheath** (Figure 14.10) and opens posteriorly just inside the mouth, through which it can extend. The intestine has three many-branched trunks, one anterior and two posterior. The whole forms a **gastrovascular cavity** lined with columnar epithelium (Figure 14.10).

Planarians are mainly carnivorous, feeding largely on small crustaceans, nematodes, rotifers, and insects. They can detect food from some distance by means of chemoreceptors. They entangle prey in mucous secretions from the mucous glands and rhabdites. A planarian grips prey with its anterior end, wraps its body around the prey item, extends its pharynx, and sucks up food in small amounts.

Intestinal secretions contain proteolytic enzymes for some **extracellular digestion.** Bits of food are sucked into the intestine, where phagocytic cells of the gastrodermis complete digestion **(intracellular).** Undigested food is egested through the pharynx. Monogeneans and trematodes graze on host cells, feeding on cellular debris and body fluids. The mouth of trematodes and monogeneans usually opens at or near the anterior end of their body into a muscular, nonextensible pharynx (Figures 14.11 and 14.19). Posteriorly, their esophagus opens into a blindly ending intestine, which is commonly Y-shaped but may be highly branched or unbranched, depending on the species.

Because cestodes have no digestive tract, they must depend on host digestion, and absorption is confined to small molecules from the host's digestive tract.

Excretion and Osmoregulation

Excretory systems remove wastes from the body, whereas osmoregulatory systems control water balance. Osmoregulatory systems are very common in freshwater animals where concentration gradients between internal fluids and the external environment cause bloating as water enters across the body's permeable membranes (see p. 662). Excess water is typically removed via osmoregulatory systems. Sometimes osmoregulation and excretion are combined when waste products are dissolved in water shed from the body. Flatworms have a system of **protonephridia** (Figure 14.10A) that could be used for excretion or osmoregulation (see Chapter 30, pp. 665–666). Although a small amount of ammonia is excreted via protonephridia, most metabolic wastes are largely removed by diffusion across the body wall.

Flatworm protonephridia (excretory or osmoregulatory organs closed at the inner end) have **flame cells** (Figure 14.10A). A flame cell is cup-shaped with a tuft of flagella extending from the inner face of the cup. In some turbellarians and in all Neodermata, the protonephridia form a **weir** (Old English *wer,* a fence placed in a stream to catch fish); the rim of the cup is elongated into fingerlike projections that extend between similar projections of a tubule cell. The space (lumen) enclosed by the tubule cell continues into collecting ducts that finally open to the outside by pores. Beating flagella (resembling a flickering flame) drive fluid down the collecting ducts and provide a negative pressure to draw fluid through the delicate interlacing projections of the weir. The wall of the duct beyond the flame cell commonly bears folds or microvilli that probably function in resorption of certain ions or molecules.

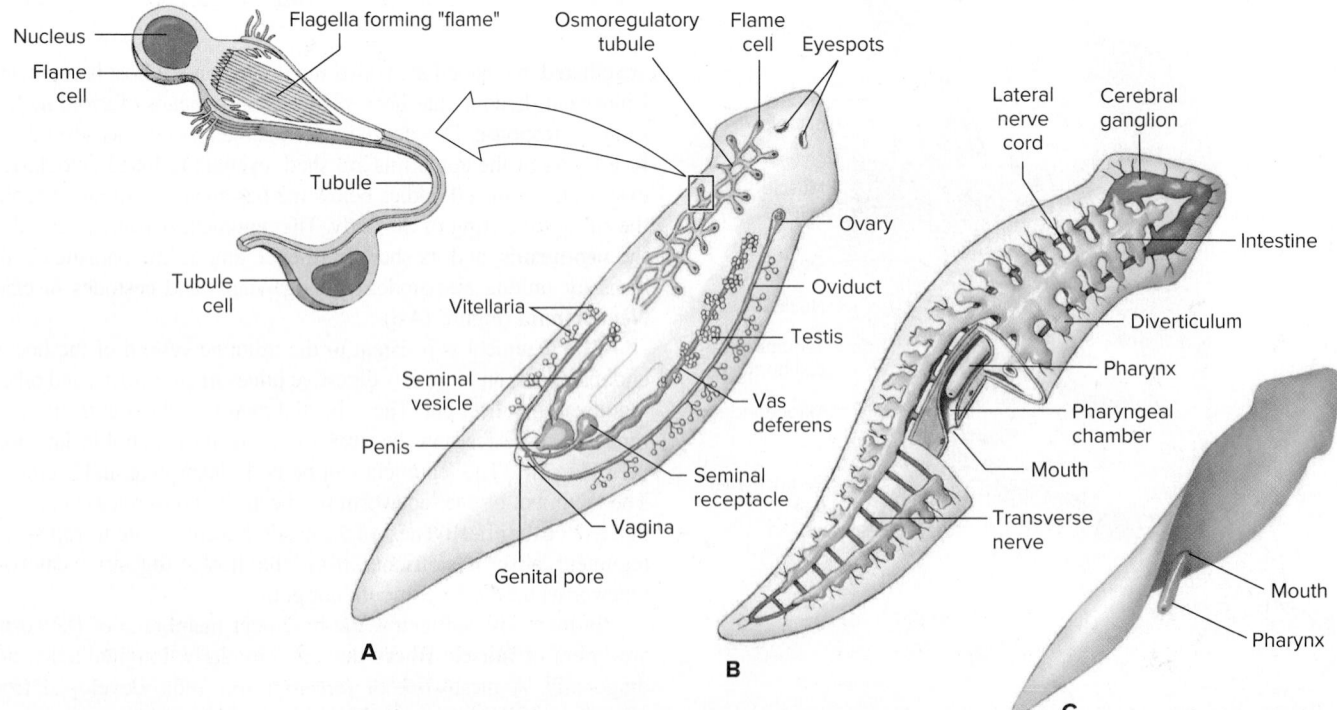

Figure 14.10 Structure of a planarian. **A,** Reproductive and osmoregulatory systems, shown in part. Inset at left is enlargement of flame cell. **B,** Digestive tract and ladder-type nervous system. Pharynx is shown in resting position. **C,** Pharynx extended through ventral mouth.

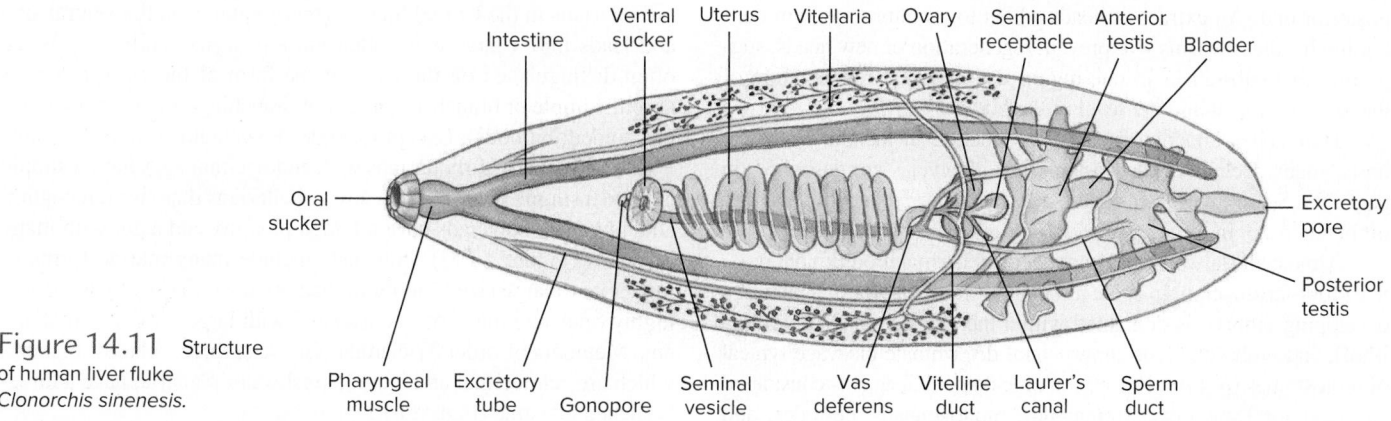

Figure 14.11   Structure of human liver fluke *Clonorchis sinenesis*.

In planarians, collecting ducts join and rejoin into a network along each side of the animal (Figure 14.10) and may empty through many nephridiopores. This system is mainly osmoregulatory because it is reduced or absent in marine turbellarians, which do not have to expel excess water.

Flame-cell protonephridia are present also in the parasitic taxa. Monogeneans usually have two excretory pores opening laterally, near the anterior. Collecting ducts of trematodes empty into an excretory bladder that opens to the exterior by a terminal pore (Figure 14.11). In cestodes there are two main excretory canals on each side that are continuous through the entire length of the worm (see Figure 14.23). They join in the last segment (proglottid, see p. 303) to form an excretory bladder that opens by a terminal pore. When the terminal proglottid is shed, the two canals open separately.

Nervous System

The most primitive flatworm nervous system, found in some turbellarians, is a **subepidermal nerve plexus** resembling the nerve net of cnidarians. Other flatworms have, in addition to a nerve plexus, one to five pairs of **longitudinal nerve cords** lying under the muscle layer (see Chapter 33, p. 727). Freshwater planarians have one ventral pair (Figure 14.10B). Connecting nerves form a "ladder-type" pattern. Their brain is a bilobed mass of ganglion cells arising anteriorly from the ventral nerve cords. Neurons are organized into sensory, motor, and association types—an important development in evolution of nervous systems.

Sense Organs

Active locomotion in flatworms has favored not only cephalization in the nervous system but also further evolution of sense organs. **Ocelli**, or light-sensitive eyespots, are common in turbellarians (Figure 14.10A), monogeneans, and larval trematodes.

Tactile cells and chemoreceptive cells are abundant over the body, and in planarians they form distinct organs on the auricles (the earlike lobes on the sides of the head). Some species also have statocysts for equilibrium and rheoreceptors for sensing direction of the water current. Sensory endings are abundant around the oral sucker of trematodes and holdfast organ (scolex, p. 303) of cestodes and around genital pores in both groups.

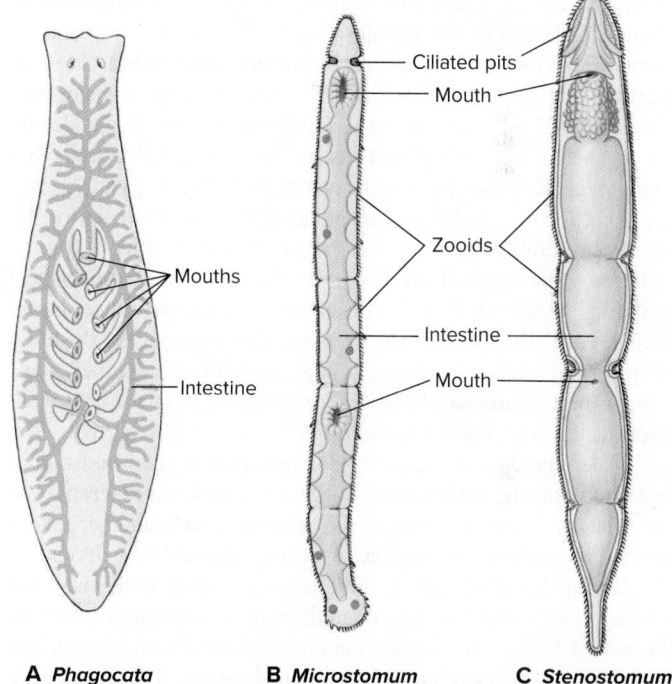

A *Phagocata*   **B** *Microstomum*   **C** *Stenostomum*

Figure 14.12   Some small freshwater turbellarians. **A,** *Phagocata* has numerous pharynges. **B** and **C,** Incomplete fission results for a time in a series of attached zooids.

Reproduction and Regeneration

Many turbellarians reproduce both asexually (by fission) and sexually. Asexually, freshwater planarians merely constrict behind the pharynx and separate into two animals, each of which regenerates the missing parts—a quick means of population increase. Reduced population density may cause an increase in the rate of fissioning. In some fissioning forms individuals may remain temporarily attached, forming chains of zooids (Figure 14.12).

The considerable powers of regeneration in planarians have provided an interesting system for experimental studies of development. For example, a piece excised from the middle of a planarian can regenerate both a new head and a new tail. However, the piece retains its original polarity: a head grows at the anterior end and a tail at the posterior end. An extract of heads added to a culture medium containing headless worms will prevent regeneration of new heads, suggesting that substances in one region will suppress regeneration of the same region at another level of the body.

Trematodes undergo asexual reproduction in their intermediate hosts, snails. Details of their astonishing life cycles are described on page 299. Some juvenile cestodes show asexual reproduction, budding off hundreds, or in some cases, even millions, of offspring (p. 306).

Almost all flatworms are monoecious (hermaphroditic) but practice cross-fertilization. In some turbellarians the yolk for nutrition of a developing embryo is contained within the egg cell itself (**endolecithal**), and embryogenesis shows spiral determinate cleavage typical of protostomes (p. 170). Possession of endolecithal eggs is considered ancestral for flatworms. Trematodes, monogeneans, cestodes, and many of the turbellarian groups share a derived condition in which female gametes contain little or no yolk, and yolk is contributed by cells released from separate organs called **vitellaria**. Yolk cells are conducted toward a juncture with the **oviduct** by **vitelline ducts** (Figures 14.10 and 14.11). Usually a number of yolk cells surround the zygote within the eggshell; thus development is **ectolecithal**. Cleavage is affected in such a way that a spiral pattern cannot be distinguished. The entire package consisting of yolk cells and zygote, surrounded by the eggshell, moves into the **uterus** and finally is released through a common genital pore or a separate uterine pore (Figure 14.11).

Access to the yolk in ectolecithal eggs is problematic for the developing embryo, but the outermost epidermal layers of some embryos grow outward to encompass the yolk. As the outermost epidermal layer is shed during development, successive inner layers enclose and utilize yolk. It has been hypothesized that the shedding of epidermal layers to permit yolk intake in ectolecithal turbellarians formed the evolutionary basis for the shedding of larval epidermal layers as the syncytial tegument forms.

Male reproductive organs include one, two, or more **testes** connected to **vasa efferentia** that join to become a single **vas deferens**. The vas deferens commonly leads into a **seminal vesicle** and hence to a papilla-like **penis** or an extensible copulatory organ called a **cirrus**.

During breeding season turbellarians develop both male and female organs, which usually open through a common genital pore (Figure 14.10A). After copulation one or more fertilized eggs and some yolk cells become enclosed in a small cocoon. The cocoons are attached by little stalks to the underside of stones or plants. Embryos emerge as juveniles that resemble mature adults. In some marine forms embryos develop into ciliated free-swimming larvae very similar to the trochophore larvae of other members of Lophotrochozoa.

Monogeneans hatch as free-swimming larvae that attach to the next host and develop into juveniles. Larval trematodes emerge from the eggshell as ciliated larvae that penetrate a snail intermediate host, or they may hatch only after being eaten by a snail. Most cestodes hatch only after being consumed by an intermediate host. Many different animals serve as intermediate hosts, and depending on the species, a tapeworm may require one or more specific intermediate hosts to complete the life cycle.

Class Turbellaria

Turbellarians are mostly free-living worms that range in length from 5 mm or less to 50 cm. They live under objects in marine, freshwater, and terrestrial habitats. There are about six species of terrestrial turbellarians in the United States. Their mouth is on the ventral side and leads into a gut cavity, often via a pharynx. Turbellarians are often distinguished on the basis of the form of the gut (present or absent; simple or branched; pattern of branching) and pharynx (simple; folded; bulbous). Except for order Polycladida (Gr. *poly,* many, + *klados,* branch), turbellarians with endolecithal eggs have a simple gut and a simple pharynx. In a few turbellarians there is no recognizable pharynx. Polyclads have a folded pharynx and a gut with many branches (Figure 14.13). Polyclads include many marine forms of moderate to large size (3 to more than 40 mm) (Figure 14.14), and a highly branched intestine is correlated with larger size in turbellarians. Members of order Tricladida (Gr. treis, three, klados, branch), which are ectolecithal and include freshwater planaria, have a three-branched intestine (Figure 14.13).

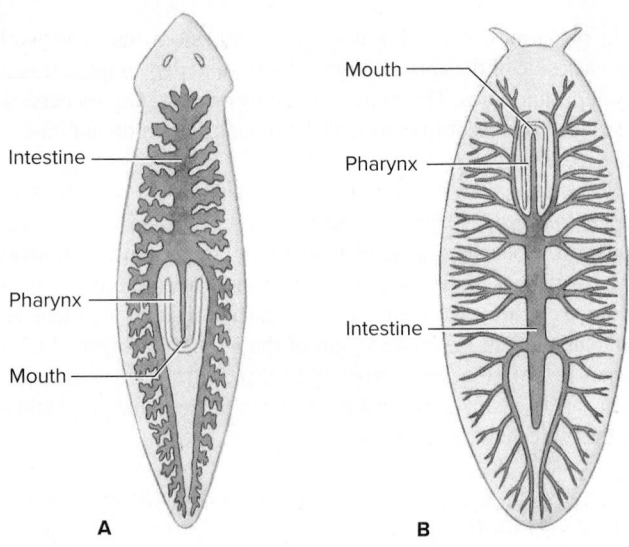

Figure 14.13   Intestinal pattern of two orders of turbellarians. **A,** Tricladida. **B,** Polycladida.

Figure 14.14   *Pseudobiceros hancockanus,* a marine polyclad turbellarian. Marine polyclads are often large and beautifully colored. This Australian species was photographed on the Great Barrier Reef.

Turbellarians typically are creeping forms that combine muscular with ciliary movements to achieve locomotion. Very small planaria swim by means of their cilia. Others move by gliding, head slightly raised, over a slime track secreted by the marginal adhesive glands. Beating of epidermal cilia in the slime track moves the animal forward, while rhythmical muscular waves can be seen passing backward from the head. Large polyclads and terrestrial turbellarians crawl by muscular undulations, much in the manner of a snail.

Class Trematoda

Trematodes are all parasitic flukes, and as adults they are almost all found as endoparasites of vertebrates. They are chiefly leaflike in form with one or more suckers but lack the opisthaptor present in monogenean flukes (p. 303).

Other structural adaptations for parasitism are apparent: various penetration glands or glands to produce cyst material, organs for adhesion such as suckers and hooks, and increased reproductive capacity. Otherwise, trematodes share several characteristics with ectolecithal turbellarians, such as a well-developed alimentary canal (but with the mouth at the anterior, or cephalic, end) and similar reproductive, excretory, and nervous systems, as well as a musculature and parenchyma that are only slightly modified from those of turbellarians. Sense organs are poorly developed.

Of the subclasses of Trematoda, subclass Aspidogastrea is the least well known. Most parasites in this group have only a single host, usually a mollusc. If there is a second host, it is usually a fish or turtle. Subclass Digenea (Gr. *dis,* double, + *genos,* race) is the largest and best known, with many species of medical and economic importance.

Subclass Digenea

With rare exceptions, digeneans have a complex life cycle, the first **(intermediate) host** being a mollusc and the **definitive host** (the host in which sexual reproduction occurs, sometimes called the **final** host) being a vertebrate. In some species a second, and sometimes even a third, intermediate host intervenes. The group has radiated greatly, and its members parasitize almost all kinds of vertebrate hosts. Digeneans inhabit, according to species, a wide variety of sites in their hosts: all parts of the digestive tract, respiratory tract, circulatory system, urinary tract, and reproductive tract.

Among the world's most amazing biological phenomena are digenean life cycles. Although cycles of different species vary widely in detail, a typical example would include adult, egg (shelled embryo), miracidium, sporocyst, redia, cercaria, and metacercaria stages (Figure 14.15). The shelled embryo or larva usually passes from the definitive host in excreta and must reach water to develop further. There, it hatches to a free-swimming, ciliated larva, the **miracidium.** The miracidium penetrates the tissues of a snail, where it transforms into a **sporocyst.** Sporocysts reproduce asexually to yield either more sporocysts or a number of **rediae.** Rediae, in turn, reproduce asexually to produce more rediae or to produce **cercariae.** In this way a single egg can give rise to an enormous number of progeny. Cercariae emerge from the snail and can either penetrate the final host directly (for example, the blood fluke *Schistosoma mansoni*), penetrate a second intermediate host (for example, the lung fluke *Paragonimus westermani*), or encyst on aquatic vegetation (for

example, the intestinal fluke *Fasciolopsis buski).* At this point, cercariae develop into **metacercariae,** which are essentially juvenile flukes. When the metacercariae are eaten by the final host, the juveniles migrate to the site of final infection and grow into adults.

Some of the most serious parasites of humans and domestic animals belong to Digenea (Table 14.1). The first digenean life cycle revealed was that of *Fasciola hepatica* (L. *fasciola,* a small bundle, band), which causes "liver rot" in sheep and other ruminants. Adult flukes live in the bile passage of the liver, and eggs are passed in feces. After hatching, a miracidium penetrates a snail to become a sporocyst. There are two generations of rediae, and the cercaria encysts on vegetation. When infested vegetation is eaten by a sheep or other ruminant (or sometimes humans), the metacercariae excyst and grow into young flukes.

Clonorchis sinensis: **Liver Fluke in Humans**   *Clonorchis* (Gr. *clon,* branch, + *orchis,* testis) is the most important liver fluke of humans and is common in many regions of eastern Asia, especially in China, Southeast Asia, and Japan. Cats, dogs, and pigs are also often infected.

Structure   The worms vary from 10 to 20 mm in length (see Figure 14.11). Their structure is typical of many trematodes in most respects. They have an **oral sucker** and a **ventral sucker.** The **digestive system** consists of a pharynx, a muscular esophagus, and two long, unbranched intestinal ceca. The **excretory system** consists of two protonephridial tubules, with branches lined with flame cells. The two tubules unite to form a single median bladder that opens to the outside. The nervous system, like that of other flatworms, consists of two cerebral ganglia connected to longitudinal cords that have transverse connectives.

As is common in flukes, about 80% of the body is devoted to reproduction. The **reproductive system** is hermaphroditic and complex. They have two branched **testes** that unite to form a single **vas deferens,** which widens into a **seminal vesicle.** The seminal vesicle leads into an **ejaculatory duct,** which terminates at the genital opening. The female system contains a branched **ovary** with a short **oviduct,** which is joined by ducts from the **seminal receptacle** and **vitellaria** at an **ootype.** The ootype is surrounded by a glandular mass, Mehlis's gland, of uncertain function. From Mehlis's gland the much-convoluted **uterus** runs to the genital pore. Cross-fertilization between individuals is usual, and sperm are stored in the seminal receptacle. When an oocyte is released from the ovary, it is joined by a sperm and a group of vitelline cells and is fertilized. The vitelline cells release a proteinaceous shell material, which is stabilized by a chemical reaction; the Mehlis's gland secretions are added, and the egg passes into the uterus.

Life Cycle   The normal habitat of the adults is in the bile passageways of humans and other fish-eating mammals (Figure 14.15). Eggs, each containing a complete miracidium, are shed into water with the feces but do not hatch until they are ingested by the snail *Parafossarulus* or related genera. Eggs, however, may live for some weeks in water. In a snail a miracidium enters the tissues and transforms into a sporocyst, which produces one generation of rediae. A redia is elongated, with an alimentary canal, a nervous system, an excretory system, and many germ cells in the process of development. Rediae pass into the

Figure 14.15   Life cycle of *Clonorchis sinensis*.

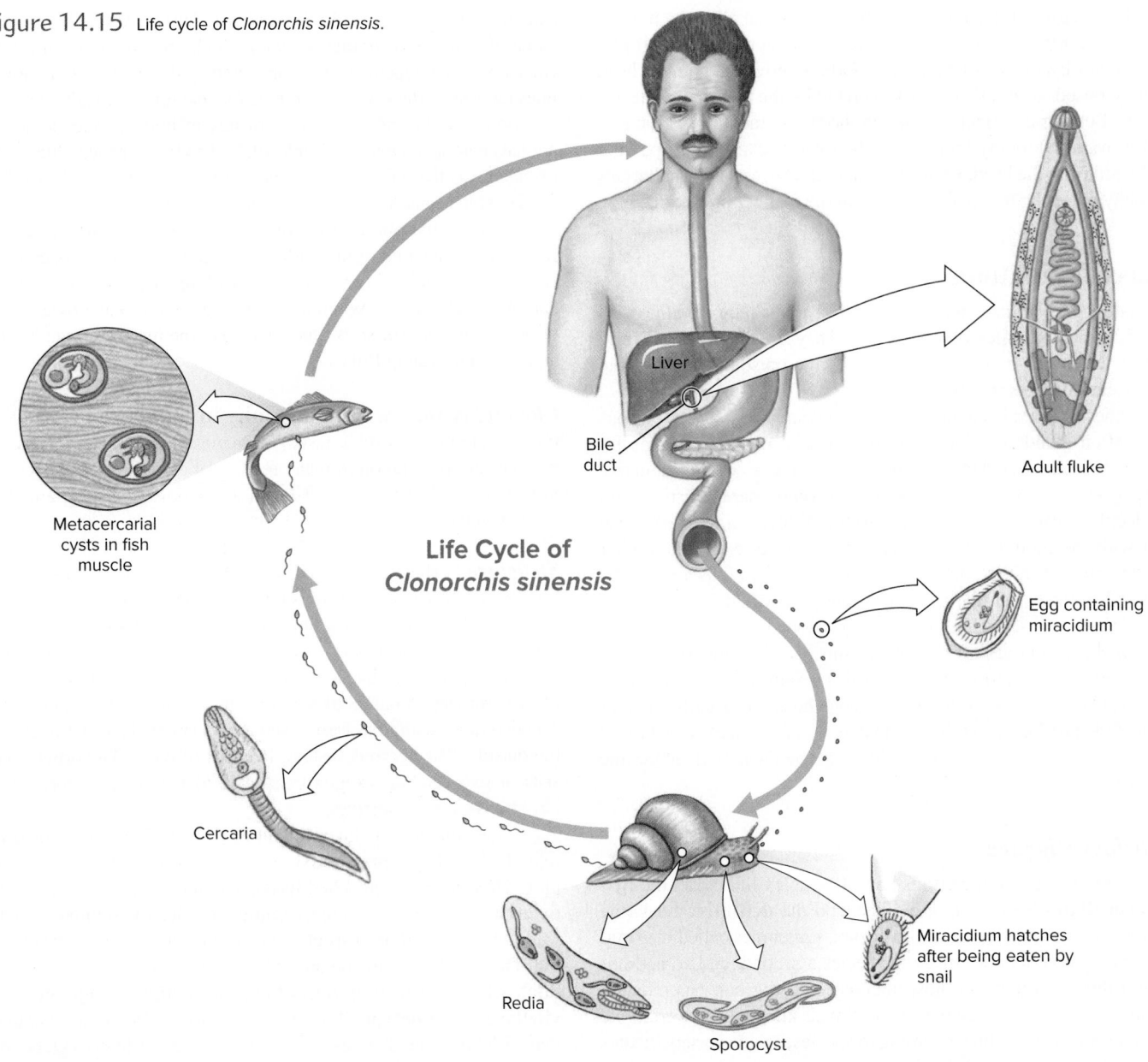

Life Cycle of *Clonorchis sinensis*

Liver

Bile duct

Adult fluke

Metacercarial cysts in fish muscle

Egg containing miracidium

Cercaria

Miracidium hatches after being eaten by snail

Redia

Sporocyst

TABLE 14.1

Examples of Flukes Infecting Humans

Common and Scientific Names	Means of Infection; Distribution and Prevalence in Humans
Blood flukes (*Schistosoma* spp.); three widely prevalent species, others reported	Cercariae in water penetrate skin; 200 million people infected with one or more species
S. mansoni	Africa, South and Central America
S. haematobium	Africa
S. japonicum	Eastern Asia
Chinese liver flukes (*Clonorchis sinensis*)	Eating metacercariae in raw fish; about 30 million cases in eastern Asia
Lung flukes (*Paragonimus* spp.), seven species, most prevalent is *P. westermani*	Eating metacercariae in raw freshwater crabs, crayfish; Asia and Oceania, sub-Saharan Africa, South and Central America; several million cases in Asia
Intestinal fluke (*Fasciolopsis buski*)	Eating metacercariae on aquatic vegetation; 10 million cases in eastern Asia
Sheep liver fluke (*Fasciola hepatica*)	Eating metacercariae on aquatic vegetation; widely prevalent in sheep and cattle, occasional in humans

liver of the snail, where the germ cells continue embryonation and give rise to tadpolelike cercariae. These two asexual stages in the intermediate host allow a single miracidium to produce as many as 250,000 infective cercariae.

Cercariae escape into the water and swim until they encounter a fish of family Cyprinidae. They then bore under scales into the fish's muscles. Here cercariae lose their tails and encyst as metacercariae. If a mammal eats raw or undercooked infected fish, the metacercarial cyst dissolves in the intestine, and young flukes apparently migrate up the bile duct, where they become adults. There the flukes may live for 15 to 30 years.

The effect of these flukes on a person depends mainly on the extent of the infection but includes abdominal pain and other abdominal symptoms. A heavy infection can cause a pronounced cirrhosis of the liver and death. Cases are diagnosed through fecal examinations. Destruction of snails that carry larval stages has been used as a method of control. However, the simplest method to avoid infection is to make sure that all fish used as food is thoroughly cooked.

Schistosoma: Blood Flukes

Schistosomiasis, infection with blood flukes of genus *Schistosoma* (Gr. *schistos,* divided, + *soma,* body), ranks among the major infectious diseases in the world, with 200 million people infected. The disease is widely prevalent over much of Africa and parts of South America, the West Indies, the Middle East, and the Far East. The old generic name for the worms was *Bilharzia* (from Theodor Bilharz, German parasitologist who discovered *Schistosoma haematobium*), and the infection was called bilharziasis, a name still used in many areas.

Blood flukes differ from most other flukes in being dioecious and having the two branches of the digestive tube united into a single tube in the posterior part of the body. Males are broader and heavier and have a large, ventral groove, the **gynecophoric canal,** posterior to the ventral sucker. The gynecophoric canal embraces the long, slender female (Figure 14.16).

Three species account for most schistosomiasis in humans: *S. mansoni,* which lives primarily in veins draining the large intestine; *S. japonicum,* which occurs mostly in veins of the small intestine; and *S. haematobium,* which lives in veins of the urinary bladder. *Schistosoma mansoni* is common in parts of Africa, Brazil, northern South America, and the West Indies; species of *Biomphalaria* are the principal snail intermediate hosts. *Schistosoma haematobium* is widely prevalent in Africa, using snails of the genera *Bulinus* and *Physopsis* as the main intermediate hosts. *Schistosoma japonicum* is confined to the Far East, and its hosts are several species of *Oncomelania* (see Chapter 35, p. 769, for a discussion of immunity in snails).

Unfortunately, some projects intended to raise the standard of living in some tropical countries, such as the Aswan High Dam in Egypt, have increased the prevalence of schistosomiasis by creating more habitats for the snail intermediate hosts. Before the dam was constructed, the 500 miles of the Nile River between Aswan and Cairo was subjected to annual floods; alternate flooding and drying killed many snails. Four years after dam completion, prevalence of schistosomiasis had increased sevenfold along that segment of the river. Prevalence in fishermen around the lake above the dam increased from a very low level to 76%.

The life cycle of blood flukes is similar in all species. Eggs are discharged in human feces or urine; if they get into water, they hatch as ciliated miracidia, which must contact the required kind of snail within a few hours to survive. In the snail, they transform into sporocysts, which produce another generation of sporocysts. Daughter sporocysts give rise to cercariae directly, without formation of rediae. Cercariae escape from the snail and swim until they contact bare human skin. They penetrate the skin, shedding their tails in the process, and reach a blood vessel where they enter the circulatory system. There is no metacercarial stage. The young schistosomes make their way to the hepatic portal system of blood vessels and undergo a period of development in the liver before migrating to their characteristic sites. As eggs are released by adult females, they are somehow extruded through the wall of veins and through the gut or bladder lining, to be voided with feces or urine, according to species. Many eggs do not make this difficult transit and are swept by blood flow back to the liver or other areas, where they become centers of inflammation and tissue reaction.

The parasite's eggs produce the main ill effects of schistosomiasis. With *S. mansoni* and *S. japonicum,* eggs in the intestinal wall cause ulceration, abscesses, and bloody diarrhea with abdominal pain. Similarly, *S. haematobium* causes ulceration of the bladder wall with bloody urine and pain on urination. Eggs swept to the liver or other sites cause symptoms associated with the organs where they lodge. When they are caught in the capillary bed of the liver, they impede circulation and cause cirrhosis, a fibrotic reaction that interferes with liver function (Figure 14.17). Of the three species, *S. haematobium* is considered least serious and *S. japonicum* most severe. The prognosis is poor in heavy infections of *S. japonicum* without early treatment.

Control is best achieved by educating people to dispose of body wastes hygienically and to avoid exposure to contaminated water. These are difficult problems for poverty-stricken people living under unsanitary, crowded conditions.

Although proper disposal of body wastes is the best control for schistosomiasis, other strategies are being pursued with varying success: chemotherapy, vector control, and vaccination. Development of a vaccine is the subject of much research, but an effective vaccine is not yet available. Vector control by environmental management and by biological means appears promising. Biological controls include introduction of species of snails, crayfish, and fish that prey on the snail vectors. However, biological control attempts for other species have often been fraught with unexpected ecological impacts. In some cases, the biological control has been more of a problem in the long run than the pest species it was supposed to control. Many biologists consider such introductions an extreme risk that should be avoided.

Schistosome Dermatitis (Swimmer's Itch)

Various species of schistosomes in several genera cause a rash or dermatitis when their cercariae penetrate hosts that are unsuitable for further development. Cercariae of several genera whose normal hosts are North American birds cause dermatitis in bathers in northern lakes. Severity of the rash increases with an increasing number of contacts with the organisms, or sensitization. After penetration, cercariae are attacked and killed by the host's immune mechanisms, and they release allergenic substances, causing itching. The condition is more an annoyance than a serious threat to health, but there may be economic losses to vacation trade around infested lakes.

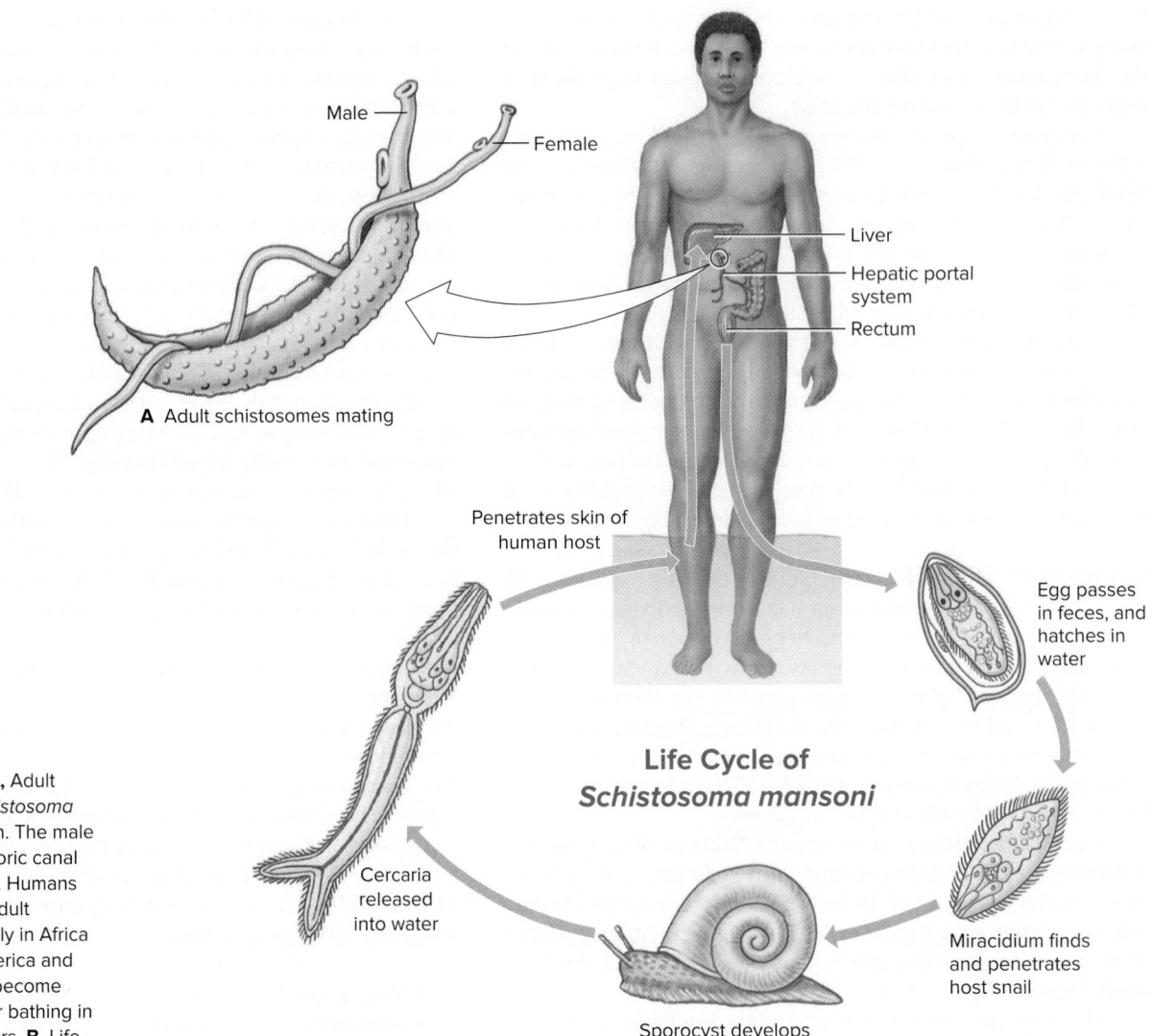

Figure 14.16 **A,** Adult male and female *Schistosoma mansoni* in copulation. The male has a long gynecophoric canal that holds the female. Humans are usually hosts of adult parasites, found mainly in Africa but also in South America and elsewhere. Humans become infected by wading or bathing in cercariainfested waters. **B,** Life cycle of *Schistosoma mansoni*.

Labels in figure:
Male
Female
Liver
Hepatic portal system
Rectum
A Adult schistosomes mating
Penetrates skin of human host
Egg passes in feces, and hatches in water
Life Cycle of *Schistosoma mansoni*
Cercaria released into water
Miracidium finds and penetrates host snail
Sporocyst develops in snail host
B

Paragonimus: **Lung Flukes**   Several species of *Paragonimus* (Gr. *para,* beside, + *gonimos,* generative), a fluke that lives in the lungs of its host, are known from a variety of mammals. *Paragonimus westermani* (Figure 14.18) from East Asia and the southwest Pacific parasitizes a number of wild carnivores, humans, pigs, and rodents. Its eggs are coughed up in the sputum, swallowed, then eliminated with feces. Zygotes develop in the water, and the miracidium penetrates a snail host. Within the snail, miracidia give rise to sporocysts, which in turn develop into rediae. Cercariae form in rediae and are then shed into the water or ingested directly by freshwater crabs that prey on infected snails. Metacercariae develop in the crabs, and the infection is acquired by eating raw or undercooked crab meat. The infection causes respiratory symptoms, with breathing difficulties and chronic cough. Fatal cases are common. A closely related species, *P. kellicotti,* occurs in mink and similar animals in North America, but only one human case has been recorded. Its metacercariae live in crayfish.

Some Other Trematodes   *Fasciolopsis buski* (L. *fasciola,* small bundle, + Gr. *opsis,* appearance) parasitizes the intestine of humans and

pigs in India and China. Larval stages occur in several species of planorbid snails, and cercariae encyst on water chestnuts, an aquatic plant eaten raw by both humans and pigs.

Leucochloridium is noted for its remarkable sporocysts. Snails (*Succinea*) eat vegetation infected with eggs from bird droppings. Sporocysts become much enlarged and branched, and cercariae encyst within the sporocyst. Sporocysts enter the snail's head and tentacles, become brightly striped with orange and green bands, and pulsate at frequent intervals. Birds are attracted by the enlarged and pulsating tentacles, eat the snails, and so complete the life cycle.

Class Monogenea

Monogenetic flukes were traditionally placed as an order of Trematoda, but morphological and molecular data support their placement in a different class. Cladistic analysis places them closer to the Cestoda, and some authorities now argue that cestodes and monogenean flukes are sister taxa, both having a posterior attachment organ with hooks. Monogeneans are all parasites, primarily of gills and external surfaces of fishes. A few inhabit the urinary bladder of frogs and turtles, and one parasitizes the eye of a

Figure 14.17 This cut surface of a liver shows schistosomal hepatic fibrosis associated with the deposition of many schistosome eggs along the portal vein. Fibrosis leads to vascular obstruction, but it develops in only 5% to 10% of infected people. Researchers suspect a genetic component to the severity of response to the presence of schistosome eggs.

Courtesy A. W. Cheever/From H. Zaiman, A Pictorial Presentation of Parasites.

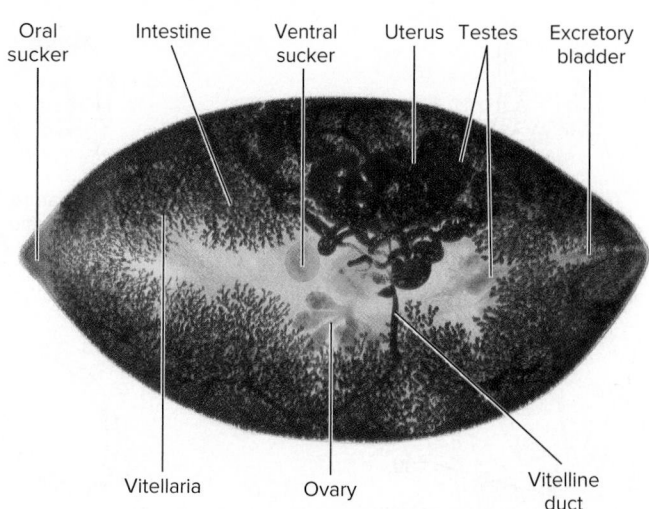

Figure 14.18 Lung fluke *Paragonimus westermani.* Adults are up to 2 cm long. Eggs discharged in sputum or feces hatch into freeswimming miracidia that enter snails. Cercariae from snails enter freshwater crabs and encyst in soft tissues. Humans are infected by eating poorly cooked crabs or by drinking water containing larvae freed from dead crabs.

hippopotamus. Although widespread and common, monogeneans seem to cause little damage to their hosts under natural conditions. However, like numerous other fish pathogens, they become a serious threat when their hosts are crowded together, as, for example, in fish farming.

Life cycles of monogeneans are direct, with a single host. The egg hatches to produce a ciliated larva, called an **oncomiracidium**, that attaches to its host. The oncomiracidium bears hooks on its posterior, which in many species become the hooks on the large posterior attachment organ (**opisthaptor**) of the adult (Figure 14.19). Because monogeneans must cling to the host and withstand the force of water flow over the gills or skin, adaptive diversification has produced a wide array of opisthaptors in different species. Opisthaptors may bear large and small hooks, suckers, and clamps, often in combination with each other.

Class Cestoda

Cestodes, or tapeworms, differ in many respects from the preceding classes. They usually have long, flat bodies composed of a **scolex**, for attachment to the host, followed by a linear series of reproductive units or **proglottids** (Figure 14.20). The scolex, or holdfast, is usually provided with suckers or suckerlike organs and often with hooks or spiny tentacles as well (Figure 14.20).

Tapeworms entirely lack a digestive system but do have well-developed muscles, and their excretory system and nervous system are somewhat similar to those of other flatworms. They have no special sense organs but do have sensory endings in the tegument that are modified cilia (Figure 14.21).

As in Monogenea and Trematoda, no external, motile cilia occur in adults, and the tegument is of a distal cytoplasm with sunken cell bodies beneath the superficial muscle layer (Figure 14.21). In contrast to monogeneans and trematodes, however, the entire surface of cestodes is covered with minute projections similar to microvilli of the vertebrate small intestine (p. 44). These **microtriches** (sing. **microthrix**) greatly enlarge the surface area of the tegument, which

is a vital adaptation for a tapeworm since it must absorb all its nutrients across its tegument.

Subclass Eucestoda contains the great majority of species of Cestoda. The main body of tapeworms, the chain of proglottids, is called a **strobila** (Figure 14.20). Typically, there is a **germinative zone** just behind the scolex where new proglottids are formed. As younger proglottids are differentiated in front of it, each individual proglottid moves outward in the strobila, and its gonads mature.

Unlike most other flatworms, many eucestodes are known to self-fertilize, although mutual cross-fertilization remains the norm when mates are available. Each proglottid contains a complete male and female reproductive system, and during mutual cross-fertilization, sperm from each strobila is transferred to the other. However, many tapeworms are known to double back upon themselves so that two proglottids from the same individual may fertilize one another. The shelled embryos form in the uterus of the proglottid, and they are expelled through a uterine pore, or the entire proglottid is shed from the worm as it breaks free at zones of muscle weakness between each proglottid.

The tapeworm body is unusual because of the absence of many typical landmarks. There is no head. The scolex, used for attachment, is a remnant of the *posterior* part of the ancestral body. Cestodes and monogeneans thus share a posterior attachment organ with hooks.

Some zoologists regard proglottid formation in cestodes as "true" segmentation (metamerism), but we do not support this view. Segmentation of tapeworms is best considered a replication of sex organs to increase reproductive capacity (pseudometamerism) and is not homologous to the metamerism found in Annelida, Arthropoda, and Chordata (see p. 189 and p. 382).

More than 1000 species of tapeworms are known to parasitologists. With rare exceptions, cestodes require at least two hosts, and adults are parasites in the digestive tract of vertebrates. Often their intermediate host is an invertebrate. Collectively these animals are

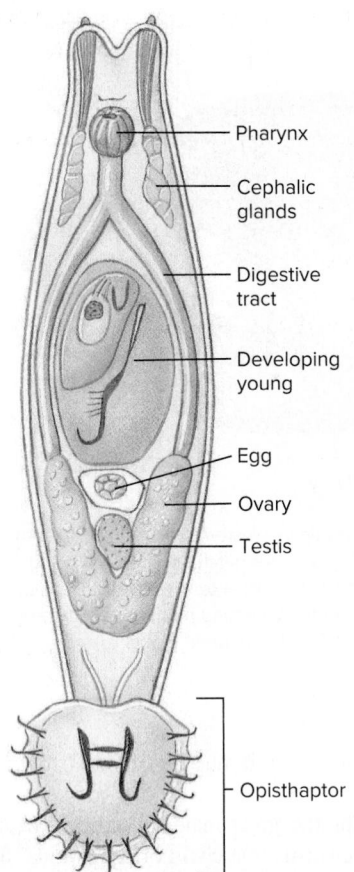

Figure 14.19 A monogenetic fluke *Gyrodactylus cylindriformis,* ventral view.

Labels: Pharynx, Cephalic glands, Digestive tract, Developing young, Egg, Ovary, Testis, Opisthaptor

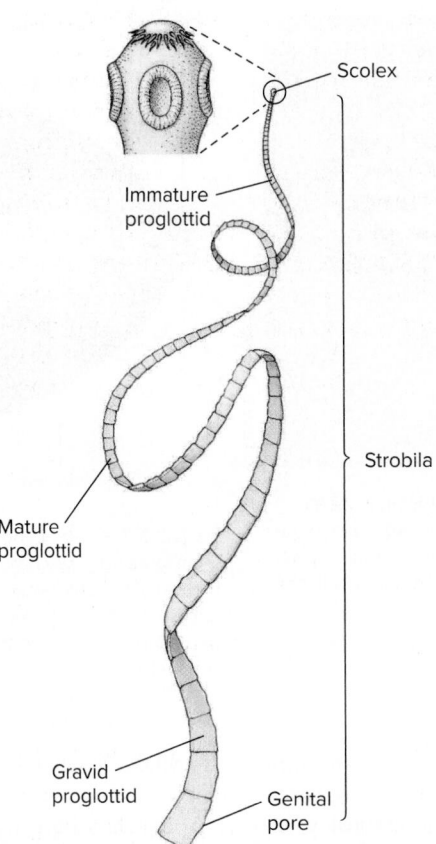

Figure 14.20 A tapeworm, showing strobila and scolex. The scolex is the organ of attachment.

Labels: Scolex, Immature proglottid, Strobila, Mature proglottid, Gravid proglottid, Genital pore

Figure 14.21 Schematic drawing of a longitudinal section through a sensory ending in the tegument of *Echinococcus granulosus.*

Labels: Microthrix, Distal cytoplasm of tegument, Mitochondria, Nerve process, Longitudinal muscle, Circular muscle

capable of infecting almost all vertebrate species. Normally, adult tapeworms do little harm to their hosts. The most common tapeworms found in humans are listed in Table 14.2.

Taenia saginata: Beef Tapeworm

Structure   *Taenia saginata* (Gr. *tainia*, band, ribbon) is called the beef tapeworm, but it lives as an adult in the human intestine. Juvenile forms occur primarily in intermuscular tissue of cattle. A mature adult may reach a length of 10 m or more. Its scolex has four suckers for attachment to the intestinal wall, but no hooks. A short neck connects the scolex to the strobila, which may have as many as 2000 proglottids. Gravid proglottids bear shelled, infective larvae (Figure 14.22), which become detached and pass in feces.

Although tapeworms lack true metamerism, there is repetition of the reproductive and excretory systems in each proglottid. Excretory canals in the scolex are continued the length of the body by a pair of dorsolateral, and a pair of ventrolateral, excretory canals. These paired canals are connected by a transverse canal near the distal end of each proglottid. Two longitudinal **nerve cords** from a **nerve ring** in the scolex also run back through each proglottid (Figure 14.23). Attached to the excretory ducts are flame cells. Each mature proglottid also contains muscles and parenchyma as well as a complete set of male and female organs similar to those of a trematode.

TABLE 14.2
Common Cestodes of Humans

Common and Scientific Name	Means of Infection; Prevalence in Humans
Beef tapeworm (*Taenia saginata*)	Eating rare beef; most common of all tapeworms in humans
Pork tapeworm (*Taenia solium*)	Eating rare pork; less common than *T. saginata*
Fish tapeworm (*Diphyllobothrium latum*)	Eating rare or poorly cooked fish; fairly common in Great Lakes region of United States, and other areas of world where raw fish is eaten
Dog tapeworm (*Dipylidium caninum*)	Unhygienic habits of children (juveniles in flea and louse); moderate frequency
Dwarf tapeworm (*Hymenolepis nana*)	Juveniles in flour beetles; common
Unilocular hydatid (*Echinococcus granulosus*)	Cysts of juveniles in humans; infection by contact with dogs; common wherever humans are in close relationship with dogs and ruminants
Multilocular hydatid (*Echinococcus multilocularis*)	Cysts of juveniles in humans; infection by contact with foxes; less common than unilocular hydatid

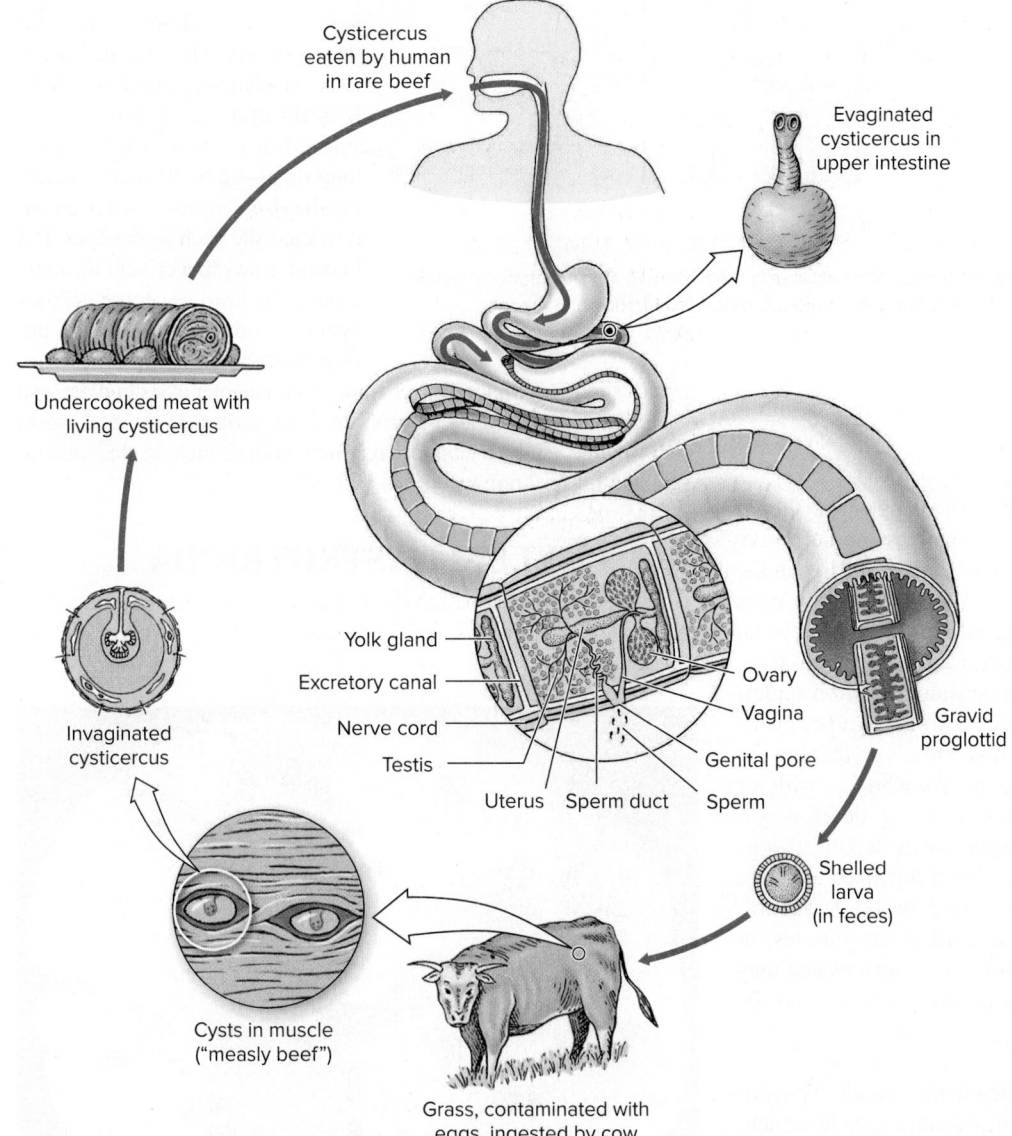

Figure 14.22 Life cycle of beef tapeworm, *Taenia saginata*. Ripe proglottids detach in the human intestine, leave the body in feces, crawl onto grass, and are ingested by cattle. Eggs hatch in the cow's intestine, freeing oncospheres, which penetrate into muscles and encyst, developing into "bladder worms." A human eats infected rare beef, and cysticercus is freed in the intestine where it attaches to the intestinal wall, forms a strobila, and matures.

In this group of tapeworms, vitellaria are typically a single, compact **vitelline gland** located just posterior to the ovaries. When gravid proglottids break off and pass out with the feces, they usually crawl out of the fecal mass and onto vegetation nearby. There they may be eaten by grazing cattle. A proglottid ruptures as it dries, further scattering the embryos on soil and grass. Embryos may remain viable on grass for as long as 5 months.

Life Cycle Shelled larvae (**onco-spheres**) swallowed by cattle hatch and use their hooks to burrow through the intestinal wall into blood or lymph vessels and finally reach voluntary muscle, where they encyst to become **bladder worms** (juveniles called **cysticerci**). There the juveniles develop an invaginated scolex but remain quiescent. When raw or undercooked "measly" meat is eaten by a suitable host, the cyst wall dissolves, the scolex evaginates and attaches to the intestinal mucosa, and new proglottids begin to develop. It takes 2 to 3 weeks for a mature worm to form. When a person is infected with one of these tapeworms, numerous gravid proglottids are expelled daily, sometimes even crawling out the anus. Humans become infected by eating rare beef, steaks, and barbecues. Considering that about 1% of American cattle are infected, that 20% of all cattle slaughtered are not federally inspected, and that even in inspected meat one-fourth of infections are missed, it is not surprising that tapeworm infection is fairly common. Infection is easily avoided when meat is thoroughly cooked.

M. I. Walker/NHPA

Figure 14.23   A, Two mature proglottids of *Taenia pisiformis,* a dog tapeworm. **B,** More anatomical details are marked in a drawing of two complete proglottids, two other partial proglottids are shown.

Some Other Tapeworms

***Taenia solium:* Pork Tapeworm**   The tapeworm *Taenia solium* (Gr. *tainia,* band, ribbon) uses humans as the definitive host and pigs as the intermediate host. Adult tapeworms live in the human gut, and infected humans shed fertilized tapeworm eggs with their feces. Pigs that consume infected human fecal material are colonized by the worms. Inside a pig, tapeworm larvae form cysts in muscle tissue. When humans consume infected undercooked pork, the cysts hatch and develop into adult tapeworms; this is the most common mode of infection. However, there is another way that humans become infected: people who live with, or contact, other people infected with tapeworms may develop cysticercosis, a serious disease caused by ingesting the fertilized eggs directly, without any intermediate host. When fertilized eggs are ingested by a human, instead of by a pig, the resulting larvae migrate to tissues such as the brain, spinal cord, liver, muscles, or eyes. Infection of the brain or spinal cord is very serious and may cause death (Figure 14.24). Infection of other organs may be treatable.

***Diphyllobothrium latum:* Fish Tapeworm**   Adult *Diphyllobothrium* (Gr. *dis,* double, + *phyllon,* leaf, + *bothrion,* hole, trench) live in the intestines of humans, dogs, cats, and other mammals; immature stages develop in crustaceans and fish. With a length up to 20 m, it is the largest cestode that infects humans. Fish tapeworm infections can occur anywhere in the world where people commonly eat raw fish; in the United States infections are most common in the

Great Lakes region. In Finland the worm can cause a serious anemia not apparent in other areas.

***Echinococcus granulosus:* Unilocular Hydatid**   *Echinococcus granulosus* (Gr. *echinos,* hedgehog, + *kokkos,* kernel) (Figure 14.25B), a dog tapeworm, causes hydatidosis, a very serious human disease in many parts of the world. Adult worms develop in canines, and juveniles grow in more than 40 species of mammals, including humans, monkeys, sheep, reindeer, and cattle. Thus humans may serve as a dead-end host for this tapeworm. The juvenile stage is a special kind of cysticercus called a **hydatid cyst** (Gr. *hydatis,* watery vesicle). It grows slowly but for a long time—up to 20 years—reaching the size of a basketball in an unrestricted site such as the liver. If a hydatid grows in a critical location, such as the heart or central nervous system, serious symptoms may appear sooner. The main cyst maintains a single (or unilocular) chamber, but daughter cysts bud off, and each contains thousands of scolices. Each scolex produces a worm only when eaten by a canine. Treatment options include chemotherapies and surgical removal of the hydatid.

14.4 PHYLUM GASTROTRICHA

Gastrotricha (gas-tro-tri′ka) (N. L. fr. Gr. *gaster, gastros,* stomach or belly, + *thrix, trichos,* hair) includes small, ventrally flattened

© Ana Flisser

Figure 14.24   Section through the brain of a person who died of cerebral cysticercosis, an infection with cysticerci of *Taenia solium.*

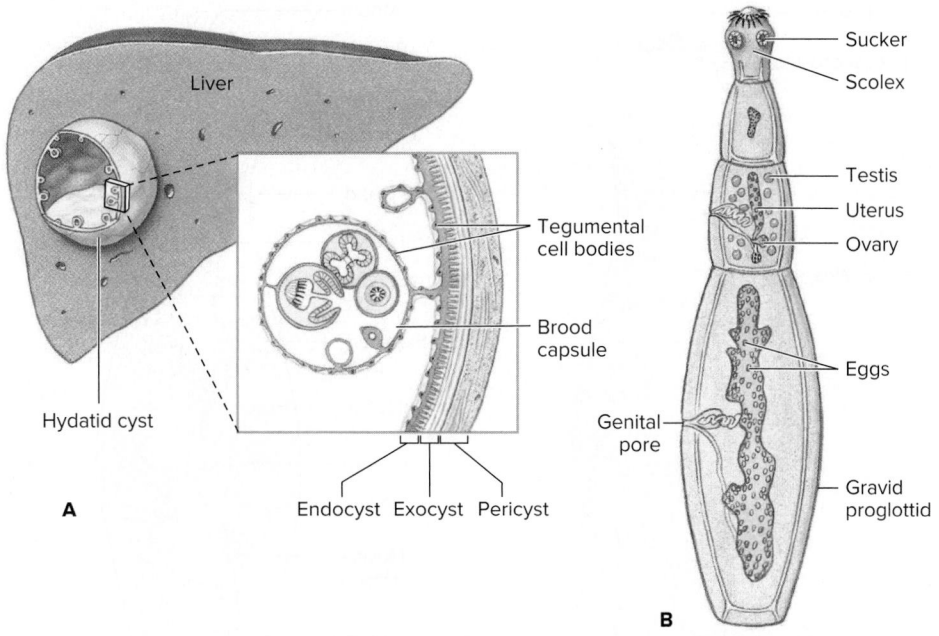

Figure 14.25 *Echinococcus granulosus,* a dog tapeworm, which may be dangerous to humans. **A,** Early hydatid cyst or bladder-worm stage found in cattle, sheep, hogs, and sometimes humans produces hydatid disease. Humans acquire disease by unsanitary habits in association with dogs. When eggs are ingested, liberated larvae encyst in the liver, lungs, or other organs. Brood capsules containing scolices are formed from the inner layer of each cyst. The cyst enlarges, developing other cysts with brood pouches. It may grow for years to the size of a basketball, necessitating surgery. **B,** The adult tapeworm lives in the intestine of a dog or other carnivore.

Many species are cosmopolitan, but only a few occur in both freshwater and the sea. Much is yet to be learned about their distribution and biology.

Form and Function

A gastrotrich (Figures 14.26 and 14.27) is usually elongated, with a convex dorsal surface bearing a pattern of bristles, spines, or scales, and a flattened, ciliated ventral surface. Cells on the ventral surface may be monociliated or multiciliated. The head is often lobed and ciliated, and the tail end may be greatly elongated or forked in some species.

A partially syncytial epidermis is found beneath the cuticle; it has some cellular regions. Longitudinal muscles are better developed than are circular ones, and in most cases they are unstriated. Adhesive tubes secrete a substance for attachment. A dual-gland system for attachment and release resembles that described for Turbellaria (p. 293 and p. 295).

No specialized respiratory or circulatory structures occur in gastrotrichs; gas exchange is by simple diffusion in these tiny animals. At least some species appear capable of anaerobic respiration. Their digestive system is complete and comprises a mouth, a muscular pharynx, a stomach-intestine, and an anus (Figure 14.26 C). Food is largely algae, protozoa, bacteria, and detritus, which are directed to their mouth by their head cilia. Digestion appears to be extracellular, although little is known about the exact mechanisms of digestion and nutrient absorption. Protonephridia are equipped with **solenocytes** rather than flame cells (see p. 296). Solenocytes have a single flagellum enclosed in a cylinder of cytoplasmic rods as opposed to the many flagella found

animals usually less than 1 mm in length. The largest species of gastrotrichs can reach lengths of about 3 mm. Superficially, gastrotrichs may appear somewhat like rotifers but lacking a corona and mastax and having a characteristically bristly or scaly body. They are usually found gliding on the substrate, or the surface of an aquatic plant or animal, by means of their ventral cilia, or they compose part of the meiofauna in interstitial spaces between substrate particles.

Gastrotrichs occur in fresh, brackish, and salt water. The 450 or so species are about equally divided between these environments.

Taxonomy of Phylum Platyhelminthes

Class Turbellaria (tur'bel-lar'e-a) (L. *turbellae* [pl.], stir, bustle, + *aria,* like or connected with): **turbellarians.** Usually free-living forms with soft, flattened bodies; covered with ciliated epidermis containing secreting cells and rodlike bodies (rhabdites); mouth usually on ventral surface sometimes near center of body; no body cavity except intercellular lacunae in parenchyma; mostly hermaphroditic, but some have asexual fission. A paraphyletic taxon awaiting revision. Examples: *Dugesia* (planaria), *Microstomum, Planocera.*

Class Trematoda (trem'a-to'da) (Gr. *trema,* with holes, + *eidos,* form): **digenetic flukes.** Body of adults covered with a syncytial tegument without cilia; leaflike or cylindrical in shape; usually with oral and ventral suckers, no hooks; alimentary canal usually with two main branches; mostly monoecious; development indirect, with first host a mollusc, final host usually a vertebrate; parasitic in all classes of vertebrates. Examples: *Fasciola, Clonorchis, Schistosoma.*

Class Monogenea (mon'o-gen'ē-a) (Gr. *mono,* single, + *gene,* origin, birth): **monogenetic flukes.** Body of adults covered with a syncytial tegument without cilia; body usually oval to cylindrical in shape; posterior attachment organ with hooks, suckers, or clamps, usually in combination; monoecious; development direct, with single host and usually with free-swimming, ciliated larva; all parasitic, mostly on skin or gills of fish. Examples: *Dactylogyrus, Polystoma, Gyrodactylus.*

Class Cestoda (ses-to'da) (Gr. *kestos,* girdle, + *eidos,* form): **tapeworms.** Body of adults covered with nonciliated, syncytial tegument; general form of body tapelike; scolex with suckers or hooks, sometimes both, for attachment; body usually divided into series of proglottids; no digestive organs; usually monoecious; larvae with hooks; parasitic in digestive tract of all classes of vertebrates; development indirect with two or more hosts; first host may be vertebrate or invertebrate. Examples: *Diphyllobothrium, Hymenolepis, Taenia.*

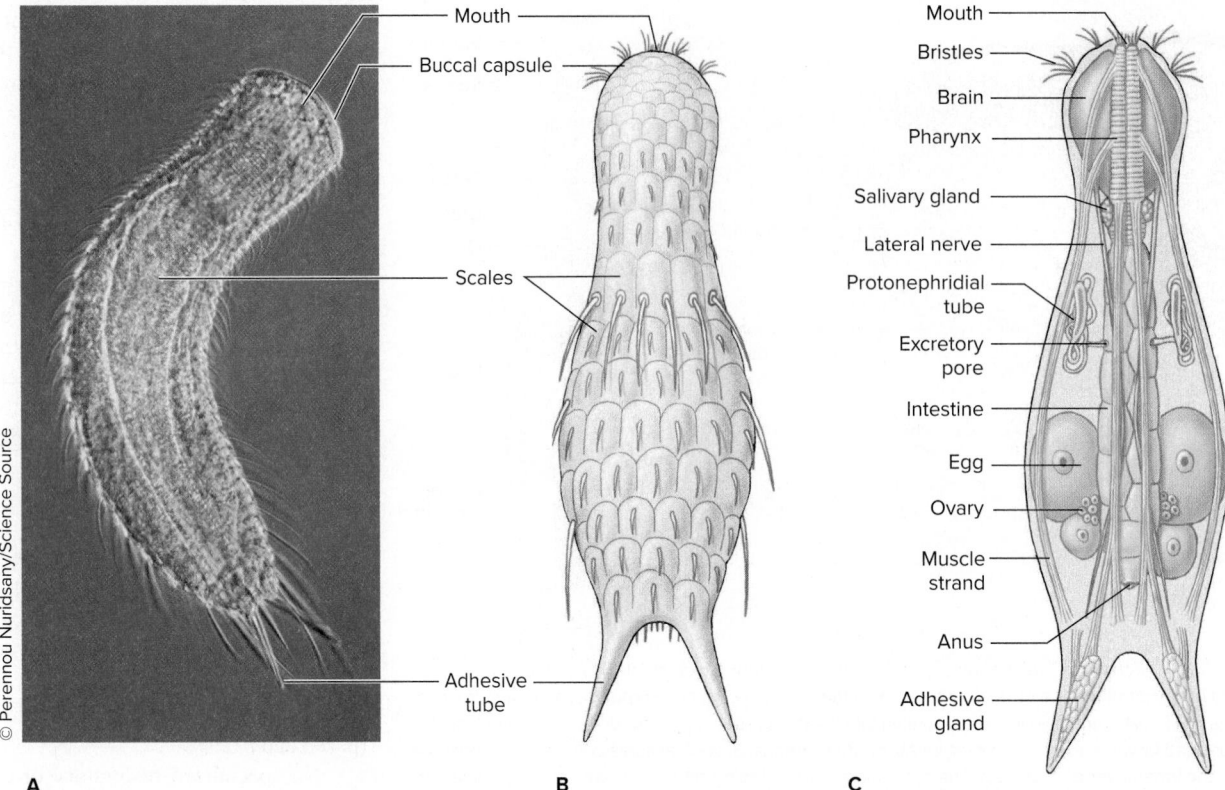

Figure 14.26 **A,** Live *Chaetonotus simrothic,* a common gastrotrich. **B,** Dorsal surface. **C,** Internal structure, ventral view.

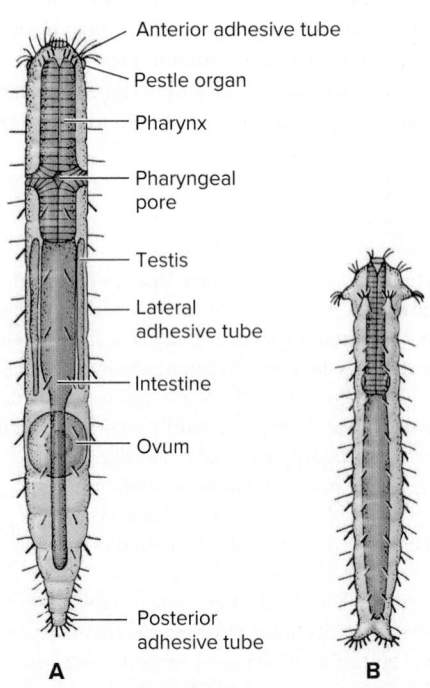

Figure 14.27 Gastrotrichs in order Macrodasyida. **A,** *Macrodasys.* **B,** *Turbanella.*

in flame bulbs. There is no body cavity in gastrotrichs, and the internal organs are all packed tightly into the compact body.

Their nervous system includes a brain near the pharynx and a pair of lateral nerve trunks. Sensory structures are similar to those in rotifers, except that eyespots are generally lacking, although some species have pigmented eyespots (ocelli) in the brain. Sensory bristles, often concentrated on the head, are modified from cilia and are primarily tactile. The pestle organ on the head (Figure 14.27A) is thought to be a chemoreceptor.

Gastrotrichs are typically hermaphroditic, although the male system of some is so rudimentary that they are functionally parthenogenetic females. Like rotifers, some gastrotrichs produce thin-walled, rapidly developing eggs and thick-shelled, dormant eggs. The thick-shelled eggs can withstand harsh environments and may survive dormancy for some years. Cleavage is not well studied but appears to be radial. Development is direct, and juveniles have the same form as adults. Growth and maturation are often rapid, and newly hatched juveniles usually reach sexual maturity within just a few days.

14.5 CLADE GNATHIFERA

Four lophotrochozoan phyla belong to a small clade whose ancestors possessed complex cuticular jaws with a homologous microstructure. The clade is called Gnathifera, and its members are phyla Gnathostomulida, Micrognathozoa, Rotifera, and Acanthocephala (see Figure 14.4). Living gnathiferans, other than acanthocephalans, possess jaws, but the number of pairs of such jaws varies within the clade. Members of Gnathostomulida, Micrognathozoa, and Rotifera are tiny, free-living, aquatic animals. Acanthocephalans are worm-like endoparasites living as adults in fishes or other vertebrates.

Rotifera and Acanthocephala are presumed sister taxa, together forming a clade called Syndermata. Their close relationship first appeared in molecular phylogenies and led morphologists to examine acanthocephalans anew, searching for evidence that these parasites were highly derived rotifers. There is little external similarity between free-swimming rotifers and endoparasitic worms, but members of both groups have a **eutelic** syncytial epidermis. Eutely refers to constancy in the numbers of nuclei present, as illustrated by the constant numbers of nuclei in various organs of one species of rotifer: E. Martini (1912) reported that he always found 183 nuclei in the brain, 39 in the stomach, and 172 in the coronal epithelium.

14.6   PHYLUM GNATHOSTOMULIDA

Gnathostomulids are delicate wormlike animals less than 2 mm long (Figure 14.28). The first known species of Gnathostomulida (nāth′-ō-sto-myu′lid-a) (Gr. *gnathos,* jaw, + *stoma,* mouth, + L. *ulus,* dim. suffix) was observed in 1928 in the Baltic, but its description was not published until 1956. Since then jaw worms have been found in many parts of the world, including the Atlantic coast of the United States, and over 80 species in 18 genera have been described.

Gnathostomulids live in interstitial spaces of very fine sandy coastal sediments and silt, from the intertidal to depths of several hundred meters. They can endure conditions of very low oxygen. They often occur in large numbers and frequently in association with gastrotrichs, nematodes, ciliates, tardigrades, and other small forms.

Gnathostomulids can glide, swim in loops and spirals, and bend the head from side to side. The epidermis is ciliated, but each epidermal cell has only one cilium, a condition rarely found in lophotrochozoans other than some gastrotrichs (p. 307). The nervous system is only partially described, but it appears to be primarily a host of sensory cilia and ciliary pits on the head.

Gnathostomulids feed by scraping bacteria and fungi from the substratum with a pair of jaws on the pharynx. The pharynx leads into a simple blind gut. Some morphologists have suggested that a tissue strand connecting the posterior gut to the epidermis is a remnant of an ancestral complete gut, but this conjecture requires more support.

The body is acoelomate with a poorly developed parenchyma layer. There is no circulatory system, so gnathostomulids probably rely on diffusion for circulation, excretion, and gas exchange.

Description of the reproductive systems and mating behavior of these worms is far from complete. Gnathostomulids are primarily protandric or simultaneous hermaphrodites engaging in mutual cross-fertilization that occurs internally. Fertilized animals each appear to produce a single zygote developing via spiral cleavage.

14.7   PHYLUM MICROGNATHOZOA

The first and only micrognathozoan species, *Limnognathia maerski,* was collected from Greenland in 1994 but not formally described until 2000. Micrognathozoans are tiny animals that are interstitial (living between sand grains) and about 142 μm long. The body

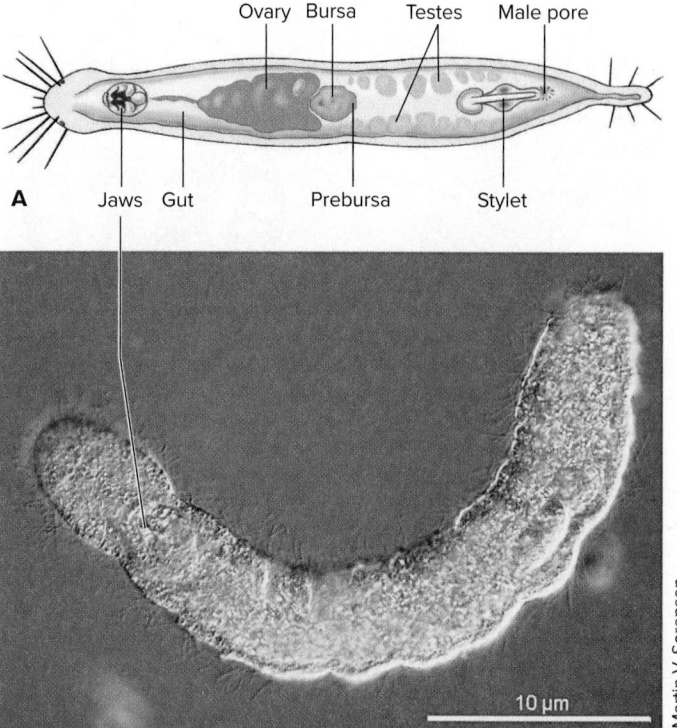

Ovary   Bursa   Testes   Male pore

A   Jaws Gut   Prebursa   Stylet

10 μm

© Martin V. Sorensen

B

Figure 14.28   **A,** *Gnathostomula jenneri* (phylum Gnathostomulida) is a tiny member of the interstitial fauna between grains of sand or mud. Species in this family are among the most commonly encountered jaw worms, found in shallow water and down to depths of several hundred meters. **B,** *Gnathostomula paradoxa* is abundant in sediments near burrows of marine polychaetes in the North Sea. Its ecology is very similar to that of *G. jenneri* from the North American Atlantic coast.

consists of a two-part head, a thorax, and an abdomen with a short tail (Figure 14.29). The cellular epidermis has dorsal plates but lacks plates ventrally. These animals move using cilia and also possess a unique ventral ciliary adhesive pad that produces glue.

There are three pairs of complex jaws. The mouth leads into a relatively simple gut. An anus opens to the outside only periodically. There are two pairs of protonephridia.

The reproductive system is not well understood. Only female reproductive organs have been found, so perhaps the animals reproduce parthenogenetically. Cleavage and subsequent development have not been studied.

14.8   PHYLUM ROTIFERA

Rotifera (ro-tif′e-ra) (L. *rota,* wheel, + *fera,* those that bear) derive their name from the characteristic ciliated crown, or **corona,** that, when beating, often gives the impression of rotating wheels (Figure 14.30). Rotifers range from 40 μm to 3 mm in length, but most are between 100 and 500 μm long. There are about 2000 species of rotifers.

Rotifers are common inhabitants of freshwater lakes and ponds. They are adapted to many ecological conditions. Most freshwater species are benthic, living on the substrate, in vegetation, or along the shore. They can swim or crawl on the vegetation. Some species

C

Figure 14.29 **A,** *Limnognathia maerski,* a micrognathozoan. **B,** Detail of complex jaws. **C,** A living specimen. This animal was found on moss in a freshwater spring on Disko Island, Greenland. It swims, or crawls, consuming bacteria, blue-green algae, and diatoms.

© Martin V. Sorensen

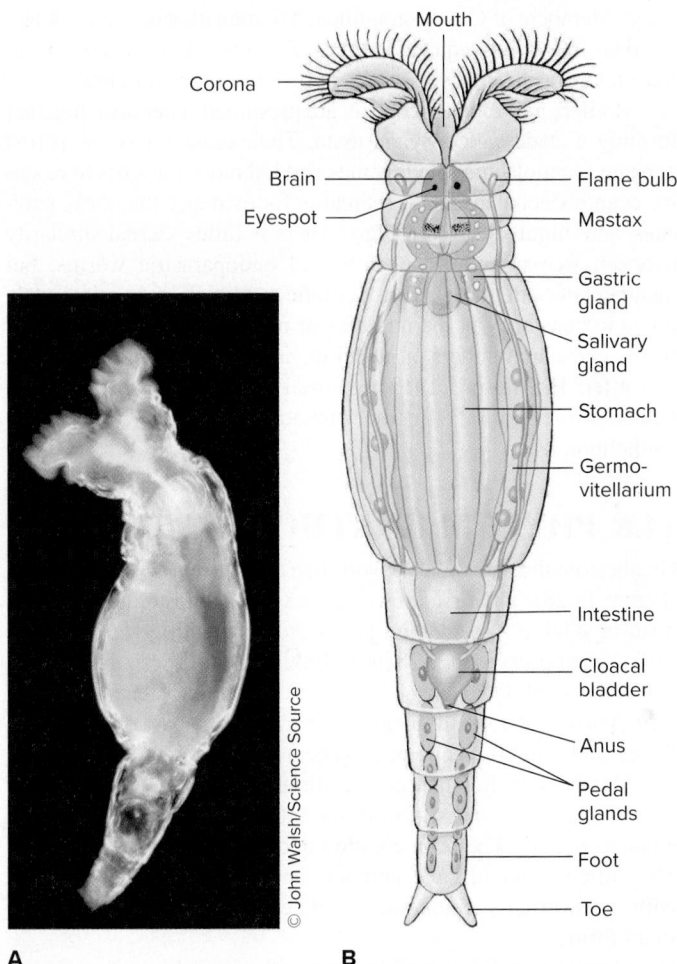

Figure 14.30 **A,** A live *Philodina,* a common rotifer. **B,** Structure of *Philodina.*

© John Walsh/Science Source

live in the water film between sand grains of beaches (meiofauna). Pelagic forms (Figure 14.31B) are common in surface waters of freshwater lakes and ponds. Some rotifers are epizoic (live on the body of another animal) or parasitic.

Some rotifers have bizarre shapes (Figure 14.31). Their shapes are often correlated with their mode of life. Floaters are usually globular and saclike; creepers and swimmers are somewhat elongated and wormlike; and sessile types are commonly vaselike, with a thickened outer epidermis (lorica). Some are colonial.

Many species of rotifers can endure long periods of desiccation, during which they resemble grains of sand. Desiccated rotifers are very tolerant of environmental extremes. For example, some moss-dwelling species have been dried for up to 4 years before reviving upon addition of water. Other rotifers have survived temperatures as cold as −272° C before being successfully revived.

Form and Function

External Features

A rotifer's body comprises a head bearing a ciliated corona, a trunk, and a posterior tail, or foot. Except for the corona, the body is nonciliated and covered with a cuticle. One of the best-known genera is *Philodina* (Gr. *philos,* fond of, + *dinos,* whirling) (Figure 14.30).

Their ciliated **corona,** or crown, surrounds a nonciliated central area of their head, which may bear sensory bristles or papillae. A head's appearance depends on which of several types of corona it has—usually a circlet of some sort, or a pair of trochal (coronal) discs (the term *trochal* comes from a Greek word meaning wheel). Cilia on the corona beat in succession, giving the appearance of a revolving wheel or pair of wheels. Their mouth is located in the corona on the midventral side. Coronal cilia function in both locomotion and feeding.

The trunk may be elongated, as in *Philodina* (Figure 14.30), or saccular in shape (Figure 14.31). The trunk contains visceral organs and often bears sensory antennae. The body wall of many species is superficially ringed, giving the appearance of segmentation. Although some rotifers have a true, secreted cuticle, all have a fibrous layer within their epidermis. The fibrous layer in some is quite thick and forms a caselike **lorica,** which is often arranged in plates or rings.

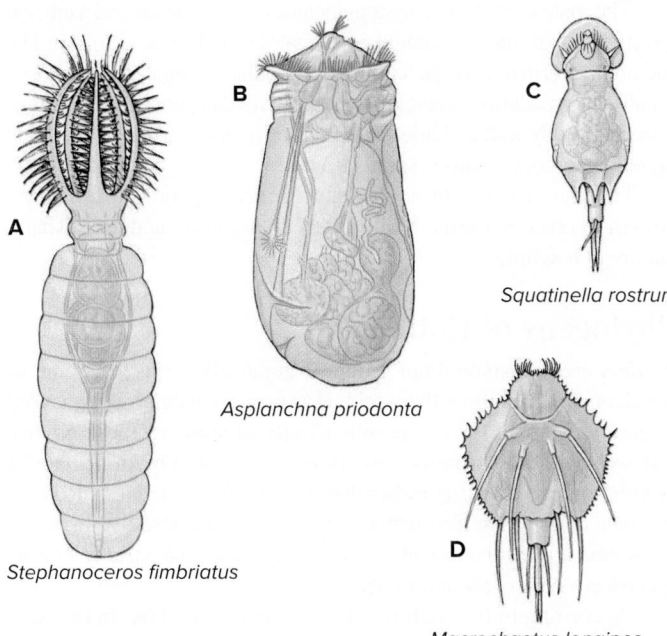

Figure 14.31   Variety of form in rotifers. **A,** *Stephanoceros* has five long, fingerlike coronal lobes with whorls of short bristles. It catches its prey by closing its funnel when food organisms swim into it, and the bristly lobes prevent prey from escaping. **B,** *Asplanchna* is a pelagic, predatory genus with no foot. **C,** *Squatinella* has a semicircular nonretractable, transparent hoodlike extension covering the head. **D,** *Macrochaetus* is dorsoventrally flattened.

Squatinella rostrum

Asplanchna priodonta

Stephanoceros fimbriatus

Macrochaetus longipes

Their **foot** is narrower and usually bears one to four **toes**. The cuticle of the foot may be ringed so that it is telescopically retractile. The foot is an attachment organ and contains **pedal glands** that secrete an adhesive material used by both sessile and creeping forms. It is tapered gradually in some forms (Figure 14.30) and sharply set off in others (Figure 14.31). In swimming pelagic forms, the foot is usually reduced. Rotifers can move by creeping with leechlike movements aided by the foot, or by swimming with the coronal cilia, or both.

Internal Features

Underneath the cuticle is a **syncytial epidermis**, which secretes the cuticle, and bands of **subepidermal muscles**, some circular, some longitudinal, and some running through the pseudocoel to visceral organs. The **pseudocoel** is large, occupying the space between body wall and viscera. It is filled with fluid, some of the muscle bands, and a network of mesenchymal ameboid cells.

The digestive system is complete. Some rotifers feed by sweeping minute organic particles or algae toward the mouth by beating the coronal cilia. Cilia discard larger unsuitable particles. Their pharynx (**mastax**) is fitted with a muscular portion equipped with hard jaws (**trophi**) for sucking in and grinding food particles. The mastax can be a crushing and grinding form among suspension feeders or a grasping and piercing form in predatory species. The constantly chewing mastax is often a distinguishing feature of these tiny animals. Carnivorous species feed on unicellular forms and small animals, which they capture by trapping or grasping. Trappers have a funnel-shaped area around the mouth. When small prey swim into the funnel, the lobes fold inward to capture and to hold them until they are drawn into the mouth and pharynx. Hunters have trophi that are projected and used like forceps to seize prey, bring it back into the pharynx, and then pierce it or break it so that edible parts may be recovered and the rest discarded. **Salivary** and **gastric glands** probably secrete enzymes for extracellular digestion. Absorption occurs in the stomach.

The excretory system typically consists of a pair of **protonephridial tubules**, each with several **flame cells** (see p. 296) that empty into a common bladder. The bladder, by pulsating, empties into a **cloaca**—into which the intestine and oviducts also empty. The fairly rapid pulsation of the protonephridia—one to four times per minute—would indicate that protonephridia are important osmoregulatory organs. Water apparently enters through the mouth rather than across the epidermis; even marine species empty their bladder at frequent intervals.

A bilobed **brain**, dorsal to the mastax in the "neck" region of the body, sends paired nerves to the sense organs, mastax, muscles, and viscera. Sensory organs include paired **eyespots** (in some species such as *Philodina*), sensory bristles and papillae, and ciliated pits and dorsal antennae.

Reproduction

Rotifers are dioecious, and males are usually smaller than females. However, despite having separate sexes, males are entirely unknown in the class Bdelloidea, and in the Monogononta they seem to occur only for a few weeks of the year.

The female reproductive system in the Bdelloidea and Monogononta consists of combined ovaries and yolk glands (**germovitellaria**) and oviducts that open into the cloaca. Yolk flows into developing ova through cytoplasmic bridges, rather than being stored as separate yolk cells as in ectolecithal Platyhelminthes.

In Bdelloidea (*Philodina,* for example), all females are parthenogenetic and produce diploid eggs that hatch into diploid females. These females reach maturity in a few days. In class Seisonidea females produce haploid eggs that must be fertilized and that develop into either males or females. In Monogononta, however, females produce two kinds of eggs (Figure 14.32). During most of the year diploid females produce thin-shelled, **diploid amictic eggs**. Amictic eggs develop parthenogenetically into diploid (amictic) females. However, such rotifers often live in temporary ponds or streams and are cyclic in their reproductive patterns. Any one of several environmental factors—for example, crowding, diet, or photoperiod (according to species)—may induce amictic eggs to develop into diploid mictic females that produce thin-shelled **haploid eggs**. If these eggs are not fertilized, they develop into haploid males. But if fertilized, the eggs, called **mictic** eggs, develop a thick, resistant shell and become dormant. They survive over winter ("winter eggs") or until environmental conditions are again suitable, at which time they hatch into amictic females. Dormant eggs are often dispersed by winds or birds.

Mictic (Gr. *miktos,* mixed, blended) refers to the capacity of haploid eggs to be fertilized (that is, "mixed") with the male's sperm nucleus to form a diploid embryo. Amictic ("without mixing") eggs are diploid and develop parthenogenetically (without fertilization).

The male reproductive system includes a single testis and a ciliated sperm duct that runs to a genital pore (males usually lack a cloaca). The end of the sperm duct is specialized as a copulatory organ. Copulation is usually by hypodermic impregnation; the penis can penetrate any part of a female's body wall and inject sperm directly into her pseudocoel. The zygote undergoes modified spiral cleavage.

Females hatch with adult features, needing only a few days' growth to reach maturity. Males often do not grow and are sexually mature at hatching.

Phylogeny of Rotifera

Rotifers are a cosmopolitan group of about 2000 species, some of which occur throughout the world. However, molecular work has led to questions about the taxonomic affinity of some of these groups, and worldwide distributions of some are apparently an artifact of morphological similarity rather than taxonomic relatedness. Rotifers are most common in freshwater, but many species also live in brackish water or even damp soils or mosses. In contrast, strictly marine species are rather few in number.

According to the traditional taxonomy in the box, in this section, Rotifera has three classes, but some authorities demote Seisonidea and Bdelloidea to orders within a class called Digonata. Others divide the phylum into two classes: one containing the seisonids, and the other containing bdelloids and monogononts under the name Eurotatoria.

In some molecular phylogenies, spiny-headed worms called acanthocephalans (see the next section, Phylum Acathocephala) arise within Rotifera. The idea that these specialized endoparasites are highly derived rotifers[1] is controversial, but if this placement is substantiated by other data sets, phylum Acanthocephala will be demoted to a class within Rotifera. At present, we depict Acanthocephala as the sister taxon to Rotifera.

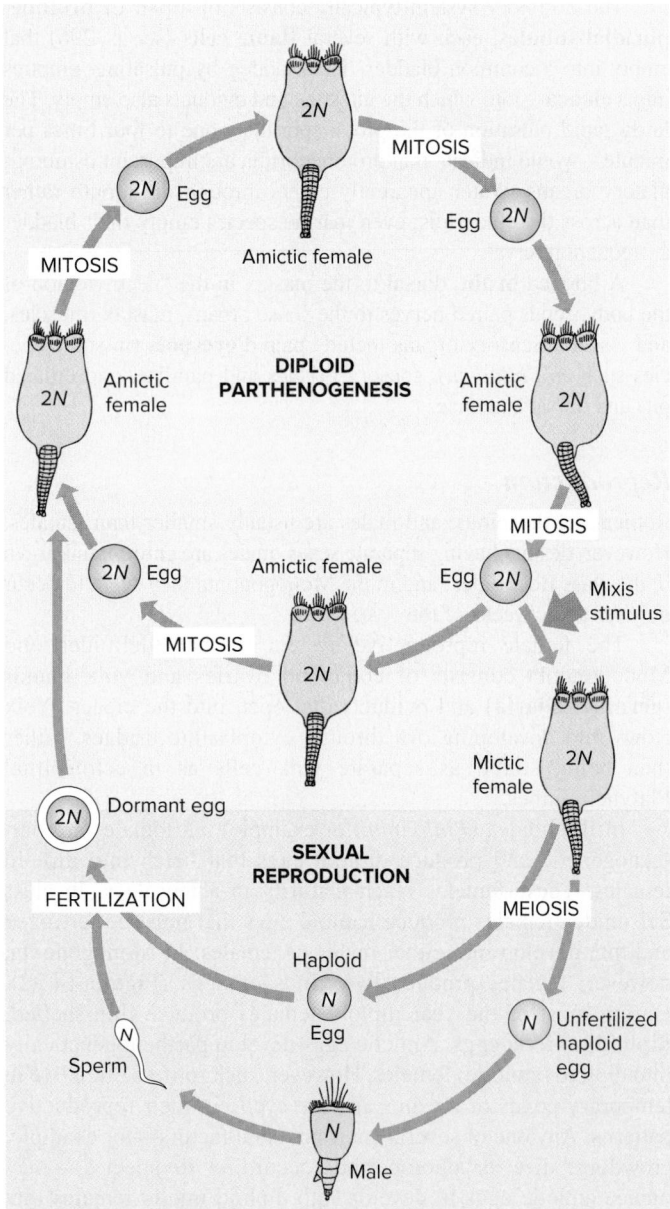

Figure 14.32 Reproduction of some rotifers (class Monogononta) is parthenogenetic during the part of the year when environmental conditions are suitable. In response to certain stimuli, females begin to produce haploid (*N*) eggs. If haploid eggs are not fertilized, they hatch into haploid males. Males provide sperm to fertilize other haploid eggs, which then develop into diploid (2 *N*), dormant eggs that can resist the rigors of winter. When suitable conditions return, dormant eggs resume development, and a female hatches.

Taxonomy of Phylum Rotifera

Classification of the rotifers remains a subject of debate. Some authorities demote Seisonidea and Bdelloidea to orders within the class Digonata. Others demote the phylum Acanthocephala to a class within the phylum Rotifera. Until this debate is resolved, we continue to present the traditional taxonomy here.

Class Seisonidea (sy'son-id'e-a) (Gr. *seison,* earthen vessel, + *eidos,* form). Marine; elongated form; corona vestigial; sexes similar in size and form; females with pair of ovaries and no vitellaria; single genus (*Seison*) with two species; epizoic on gills of a crustacean (*Nebalia*).

Class Bdelloidea (del-oyd'e-a) (Gr. *bdella,* leech, + *eidos,* form). Swimming or creeping forms; anterior end retractile; corona usually with pair of trochal discs; males unknown; parthenogenetic; two germovitellaria. Examples: *Philodina* (Figure 14.30), *Rotaria.*

Class Monogononta (mon'o-go-non'ta) (Gr. *monos,* one, + *gonos,* primary sex gland). Swimming or sessile forms; single germovitellarium; males reduced in size; eggs of three types (amictic, mictic, dormant). Examples: *Asplanchna* (Figure 14.31), *Epiphanes.*

[1]Welch, M. D. B. 2000. Invert. Biol. **119:** 17–26.

14.9 PHYLUM ACANTHOCEPHALA

Members of phylum Acanthocephala (a-kan'tho-sef'a-la) (Gr. *akantha,* spine or thorn, + *kephalē,* head) are commonly called "spiny-headed worms." The phylum derives its name from one of its most distinctive features, a cylindrical, invaginable proboscis bearing rows of recurved spines, by which it attaches itself to the intestine of its host (Figure 14.33). The phylum is cosmopolitan, and more than 1100 species are known, most of which parasitize fish, birds, and mammals. All acanthocephalans are endoparasitic, living as adults in the intestine of vertebrates. Larvae of spiny-headed worms develop in arthropods, either crustaceans or insects, depending on the species.

Various species range in size from less than 2 mm to more than 1 m in length. Females of a species are usually larger than males. Their body is usually bilaterally flattened, with numerous transverse wrinkles. Worms are typically cream colored but may absorb yellow or brown pigments from the intestinal contents.

Form and Function

In life the body is somewhat flattened, although students may see turgid and cylindrical specimens that were treated with tap water before fixation (Figure 14.33).

The body wall is syncytial, and its surface is covered by minute crypts 4 to 6 μm deep, which greatly increase the surface area of the tegument. About 80% of the thickness of the tegument is the radial fiber zone, which contains a **lacunar system** of ramifying fluid-filled canals (Figure 14.33D and E). Diffusion across the body wall exchanges gases, nutrients, and wastes between the circulating lacunar fluid and the external water. There is no heart, but muscles of the body wall form tubes connected to the lacunar system. Both longitudinal and circular body-wall muscles are present. By forming a system of contractable tubes, the body-wall muscles collectively function as a heart. Thus, the lacunar fluid, which also permeates most tissues of the body, appears to be an unusual circulatory system.

The proboscis, which bears rows of recurved hooks, is attached to the neck region (Figure 14.33) and can be inverted into a **proboscis receptacle** by retractor muscles. Attached to the neck region (but not within the proboscis) are two elongated hydraulic sacs (lemnisci) whose exact function is unknown. They might aid in gas exchange between the body and proboscis, or perhaps serve as reservoirs of lacunar fluid from the proboscis when it is evaginated.

There is no respiratory system. When present, the excretory system consists of a pair of **protonephridia** with flame cells. These unite to form a common tube opening into the sperm duct or uterus.

The nervous system has a central ganglion within the proboscis receptacle and nerves to the proboscis and body. There are sensory endings on the proboscis and genital bursa. However, as in many obligate endoparasites, both the nervous system and the sense organs of these animals are greatly reduced.

Acanthocephalans have no digestive tract, and they must absorb all nutrients through their tegument. They can absorb various molecules by specific membrane-transport mechanisms, and other substances can cross the tegumental membrane by pinocytosis. Their tegument bears some enzymes, such as peptidases, which can cleave several dipeptides, and the amino acids are then absorbed by the worm. Like cestodes

(p. 303), acanthocephalans require host dietary carbohydrate, but their mechanism for absorption of glucose is different. As glucose is absorbed, it is rapidly phosphorylated and compartmentalized, forming a metabolic "sink" into which glucose from the surrounding medium can flow. Glucose diffuses down the concentration gradient into the worm because it is constantly removed as soon as it enters.

Acanthocephalans are dioecious. Males have a pair of testes, each with a vas deferens, and a common ejaculatory duct that ends in a small penis. During copulation sperm are ejected into the vagina, travel up the genital duct, and escape into the pseudocoel of the female.

In females the ovarian tissue in the ligament sac breaks into **ovarian balls** that are released from the genital ligaments, or ligament sacs, and float free in the pseudocoel. One of the ligament sacs leads to a funnel-shaped **uterine bell** that receives the developing shelled embryos and passes them to the uterus (Figure 14.33). An interesting and unique selective apparatus operates here. Fully developed embryos are slightly longer than immature ones, and they are actively selected and passed into the uterus, while immature eggs are rejected and retained for further maturation. The shelled embryos, which are discharged in feces of their vertebrate host, do not hatch until eaten by an intermediate host.

No species is normally a parasite of humans, although species that usually occur in other hosts infect humans occasionally. *Macracanthorhynchus hirudinaceus* (Gr. *makros,* long, large, + *akantha,* spine, thorn, + *rhynchos,* beak) occurs throughout the world in the small intestine of pigs and sometimes in other mammals. For *M. hirudinaceus* the intermediate host is any of several species of soil-inhabiting beetle larvae, especially scarabeids. Grubs of the June beetle (*Phyllophaga*) are frequent hosts. Here the larva (**acanthor**) burrows through the intestine and develops into a juvenile (**cystacanth**) in the insect's hemocoel. Pigs become infected by eating the grubs. Acanthocephalans penetrate the intestinal wall with their spiny proboscis to attach to the host. In many cases there is remarkably little inflammation, but in some species the inflammatory response of the host is intense. Infection with these worms can cause great pain, particularly if the gut wall is completely perforated. Multiple infections may do considerable damage to a pig's intestine, and perforations can occur.

Phylogeny of Acanthocephala

Based largely on the shape and organization of spines on the proboscis, acanthocephalans are traditionally divided into three classes: Archiacanthocephala, Eoacanthocephala, and Palaeacanthocephala. Molecular work suggests that the phylum status of this group is unwarranted, and in fact the acanthocephalans are a class of highly derived rotifers, possibly the sister taxon of Bdelloidea. This finding has stirred considerable debate among invertebrate zoologists, but recent work[2] outlines some steps in the evolution of parasitism from a rotifer ancestor to an acanthocephalan.

[2] Wey-Fabrizius, A., H. Holger, B. Rieger, D. Rosenkranz, A. Witek, D. M. Welch, I. Ebersberger, and T. Hankeln. 2014. PLoS ONE **9:** e88618. doi:10.1371/journal.pone.0088618.

Figure 14.33 **A,** A lethal number of acanthocephalan worms, *Polymorphus botulus,* fill the intestine of an eider duck. Prior to death, the duck fed on shore crabs, an intermediate host for the endoparasites, due to absence of blue mussels, the preferred prey for eider ducks on Cape Cod. **B,** An adult worm. **C,** An everted proboscis, showing the trait for which spiny-headed worms are named. The proboscis attaches to the intestinal wall; nutrients are absorbed across the tegument. **D,** Diagrammatic longitudinal section of an everted proboscis showing muscles. **E,** Diagrammatic longitudinal section of an inverted proboscis within the pseudocoelom. **F,** Male is typically smaller than female. **G,** Scheme of the genital selective apparatus of a female acanthocephalan. It is a unique device for separating immature from mature fertilized eggs. Eggs containing larvae enter the uterine bell and pass. Immature eggs are shunted into the ventral ligament sac or into the pseudocoel to undergo further development.

14.10 PHYLUM MESOZOA

The name Mesozoa (mes-o-zo′a) (Gr. *mesos,* in the middle, + *zōon,* animal) was coined by an early investigator (van Beneden, 1876) who considered the group a "missing link" between unicellular eukaryotes and animals; these groups were formerly "protozoa" and "metazoa," respectively, so the middle group was called "mesozoa." These minute, ciliated, wormlike animals represent an extremely simple level of organization. All mesozoans live as parasites or symbionts in marine invertebrates, and most of them are only 0.5 to 7 mm in length and composed of only 20 to 30 cells arranged basically in two layers. The layers are not homologous to the germ layers of other metazoans.

The two classes of mesozoans, Rhombozoa and Orthonectida, differ so much from each other that some authorities place them in separate phyla.

Figure 14.34 Two methods of reproduction by mesozoans. **A,** Asexual development of vermiform larvae from reproductive cells in the axial cell of an adult. **B,** Under crowded conditions in the host kidney, reproductive cells develop into gonads with gametes that produce infusoriform dispersal larvae that emerge in the host urine.

Jacket cell

Axial cell

Axial cell nucleus

Vermiform embryos

Reproductive cell

A

Vermiform larva— parasitic in kidney

Axial cell

Infusoriform larva

Hermaphroditic gonad

Infusoriform larva (enlarged)

Excreted in urine

B

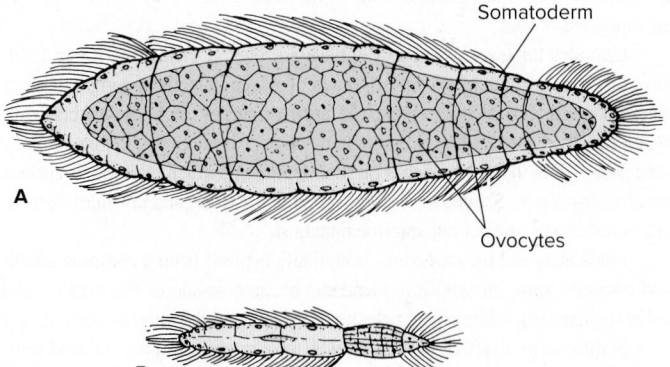

A

Somatoderm

Ovocytes

B

Figure 14.35 **A,** Female and, **B,** male orthonectid (*Rhopalura*). This mesozoan parasitizes such forms as flatworms, molluscs, annelids, and brittle stars. The structure is a single layer of ciliated epithelial cells surrounding an inner mass of sex cells.

Rhombozoans (Gr. *rhombos,* a spinning top, + *zōon,* animal) live in kidneys of benthic cephalopods (bottom-dwelling octopuses, cuttlefishes, and squids), where they apparently cause no harm to the host. Adults, called **vermiforms** (or nematogens), are long and slender (Figure 14.34). Their inner, reproductive cells give rise to vermiform larvae that grow and then reproduce. When a population becomes crowded, reproductive cells of some adults develop into gonadlike structures producing male and female gametes. Zygotes grow into minute (0.04 mm) ciliated infusoriform larvae (Figure 14.34B), quite unlike the parent. These larvae are shed with host urine into the seawater. The next part of the life cycle is unknown because infusoriform larvae are not immediately infective to a new host.

Orthonectids (Gr. *orthos,* straight, + *nektos,* swimming) (Figure 14.35) parasitize a variety of invertebrates, such as brittle stars, bivalve molluscs, polychaetes, and nemerteans. Their life cycles involve sexual and asexual phases, and the asexual stage is quite different from that of rhombozoans. It consists of a multinucleated mass called a **plasmodium,** which by division ultimately gives rise to males and females.

14.11 PHYLOGENY

Examination of the Cladogram of All Animal Phyla, inside the front cover, shows a large clade called Bilateria that contains almost all the triploblastic animals. Most current phylogenies place members of phylum Acoelomorpha as the sister taxon to all other Bilateria. Acoelomorphs differ from flatworms in phylum Platyhelminthes in their embryonic cleavage patterns, in the way that their mesoderm forms, and in nervous system structure. Given that acoelomorphs are not flatworms, some very recent phylogenies have identified a clade consisting of acoelomorphs and strange wormlike animals called xenoturbellids that are typically classified as deuterostomes (see Chapter 22, p. 474). The evolutionary position of this clade is highly variable. It is sometimes placed as the sister taxon to the Bilateria or within the deuterostomes.[3]

Within Bilateria, evidence from sequence analysis of ribosomal genes suggests that ancestral protostomes split from ancestral deuterostomes in the Precambrian. Protostomes later split into two large groups, sometimes called superphyla, Ecdysozoa and Lophotrochozoa. Evolutionary relationships within Lophotrochozoa are still in flux. We depict a clade called Platyzoa, but not all phylogenies support this grouping. Platyzoa contains Platyhelminthes, Gastrotricha, and Gnathifera. Some phylogenies place Platyhelminthes as the sister taxon to Gastrotricha, but others place Gnathostomulida as the sister taxon of Gastrotricha, so for now we depict a polytomy within Platyzoa.

Flatworms in phylum Platyhelminthes are provisional members of Platyzoa. Within Platyhelminthes, class Turbellaria is clearly paraphyletic, but we are retaining the taxon for the present because thorough cladistic analysis requires use of taxa and characteristics that are not yet common in the zoological literature. For example, ectolecithal turbellarians should be allied with trematodes, monogeneans, and cestodes in the sister group to endolecithal turbellarians. Some ectolecithal turbellarians share a number of other derived characteristics with trematodes and cestodes and have been placed by Brooks in a group called Cercomeria. Several synapomorphies, such as the unique structure of the tegument and molecular sequence features, indicate that neodermatans (trematodes, monogeneans, and cestodes) form a monophyletic group.

The group Gnathifera emerges in several different phylogenies, but most molecular studies do not include sequences from Micrognathozoa. Within Gnathifera, the sister taxon relationship between Acanthocephala and Rotifera is surprising for reasons already discussed (see p. 312). Both taxa are united as Syndermata, sharing a eutelic syncytial epidermis. Syndermata emerges from phylogenetic studies repeatedly and several recent studies show that acanthocephalans belong within Rotifera. The taxonomy of rotifer taxa is in flux, but a group called bdelloids is of interest. These rotifers are swimming or creeping forms, most with trochal discs at the top (see *Philodina,* Figure 14.30, p. 310). There are no known male bdelloids, and reproduction is by parthenogenesis. Several recent phylogenetic studies place acanthocephalans as the sister taxon to bdelloids, making the Rotifera paraphyletic as presently conceived. Acanthocephala will be subsumed within phylum Rotifera if this result is confirmed. The group name Syndermata would no longer be necessary.

Mesozoans are identified as lophotrochozoan protostomes based on molecular data, but are not placed in Platyzoa. They do not undergo gastrulation, but the absence of both typical developmental stages and the complex body plan associated with lophotrochozoans may be due to the endosymbiotic and parasitic lifestyles of modern mesozoans.

 SUMMARY

Acoelomorpha and Platyhelminthes are among the simplest bilaterally symmetrical forms, a condition of adaptive value for actively crawling or swimming animals. They are acoelomate. They are triploblastic and at the organ-system level of organization. Members of Acoelomorpha have very simple nervous and digestive systems; some lack a gut entirely.

The body surface of turbellarians is usually a cellular epithelium, at least in part ciliated, containing mucous cells and rod-shaped rhabdites that function together in locomotion. Members of all other classes of flatworms are covered by a nonciliated, syncytial tegument with cell bodies beneath superficial muscle layers. Digestion is extracellular and intracellular in most; cestodes must absorb predigested nutrients across their tegument because they have no digestive tract. Osmoregulation is accomplished by flame-cell protonephridia, and removal of metabolic wastes and respiration occur by diffusion across the body wall. Flatworms have a ladder-type nervous system with motor, sensory, and association neurons. Most flatworms are hermaphroditic, and asexual reproduction occurs in some groups.

Class Turbellaria is a paraphyletic group that includes mostly free-living and carnivorous members. Digenetic trematodes have a mollusc intermediate host and almost always a vertebrate definitive host. The extensive asexual reproduction that occurs in their intermediate host helps to increase the chances that some of their offspring will reach a definitive host. Aside from the tegument, digeneans share many basic structural characteristics with turbellarians. Digenea includes a number of important parasites of humans and domestic animals. Digeneans contrast with Monogenea, which are important ectoparasites of fishes and have a direct life cycle (without intermediate hosts).

Cestodes (tapeworms) generally have a scolex at their posterior end, followed by a long chain of proglottids, each of which contains a complete set of reproductive organs of both sexes. The anterior end of the body has been lost evolutionarily. Cestodes live as adults in the digestive tracts of vertebrates. They have microvillus-like microtriches on their tegument, which increase its surface area for absorption. Shelled larvae are passed in the feces, and juveniles develop in a vertebrate or invertebrate intermediate host.

Flatworms and the cnidarians both likely evolved from a common planuloid ancestor, some of whose descendants became sessile or freefloating and radial (cnidarians), while others became creeping and bilateral (flatworms).

Sequence analysis of rDNA, as well as some developmental and morphological criteria, suggest that Acoelomorpha, heretofore considered an order of turbellarians, diverged from a common ancestor shared by other Bilateria and is the sister group of all other bilateral phyla.

Gastrotrichs are also tiny aquatic animals. They have ventrally flattened bodies with bristles or scales. They move by cilia or adhesive glands.

[3] Achatz, J. G., M. Chiodin, W. Salvenmoser, S. Tyler, and P. Martinez. 2013. Organisms, Diversity and Evolution **13:** 267–286.

Clade Gnathifera contains four phyla whose common ancestor is hypothesized to possess cuticular jaws with a unique microstructure. Included phyla are Gnathostomulida, Micrognathozoa, Rotifera, and Acanthocephala.

Gnathostomulida is a curious phylum containing tiny wormlike animals living among sand grains and silt. The animals do not have an anus.

Micrognathozoa consists of a single species of tiny animals living between sand grains. These animals have three pairs of complex jaws similar to those of rotifers and gnathostomulids.

Phylum Rotifera is composed of small, mostly freshwater species with a ciliated corona, which creates currents of water to draw planktonic food toward the mouth. Their mouth opens into a muscular pharynx, or mastax, which is equipped with jaws. The Bdelloidea are obligate parthenogens, and males appear not to exist in this group.

Acanthocephalans are all parasitic in the intestine of vertebrates as adults, and their juvenile stages develop in arthropods. They have an anterior, invaginable proboscis armed with spines, which they embed in the intestinal wall of their host. They do not have a digestive tract and so must absorb all nutrients across their tegument.

Molecular evidence and a shared eutelic syncytial epidermis suggest a phylogenetic affinity of acanthocephalans and rotifers, and therefore a gnathiferan origin of acanthocephalans, which requires an evolutionary loss of jaws in an ancestral acanthocephalan lineage.

Members of phylum Mesozoa are very simply organized animals that are parasitic in kidneys of cephalopod molluscs (class Rhombozoa) and in several other invertebrate groups (class Orthonectida). They have only two cell layers, but these are not homologous to the germ layers of higher metazoans. They have a complicated life history that is still incompletely known. Their simple organization may have been derived from a more complex platyhelminth-like ancestor.

REVIEW QUESTIONS

1. Why is bilateral symmetry of adaptive value for actively motile animals?
2. Match the terms in the right column with the classes in the left column:

   _____ Turbellaria   a. Endoparasitic
   _____ Monogenea   b. Free-living and commensal
   _____ Trematoda   c. Ectoparasitic
   _____ Cestoda

3. Describe the body plan of a typical turbellarian.
4. Distinguish two mechanisms by which flatworms supply yolk for their embryos. Which system is evolutionarily ancestral for flatworms and which one is derived?
5. What do planarians (triclad flatworms) eat, and how do they digest it?
6. Briefly describe the osmoregulatory system, the nervous system, and the sense organs of turbellarians, trematodes, and cestodes.
7. Contrast asexual reproduction in triclad turbellarians, Trematoda, and Cestoda.
8. Contrast a typical life cycle of a monogenean with that of a digenetic trematode.
9. Describe and contrast the tegument of most turbellarians and the other classes of platyhelminths. Does the tegument provide evidence that trematodes, monogeneans, and cestodes form a clade within Platyhelminthes? Why?
10. Answer the following questions with respect to both *Clonorchis* and *Schistosoma*: (a) how do humans become infected? (b) what is the general geographical distribution? (c) what are the main disease conditions produced?
11. Why is *Taenia solium* a more dangerous infection than *Taenia saginata?*
12. What are two cestodes for which humans can serve as intermediate hosts?
13. Define each of the following with reference to cestodes: scolex, microtriches, proglottids, strobila.
14. What are some adaptive advantages of a pseudocoel compared to the acoelomate condition?
15. Some evidence suggests that members of Acoelomorpha constitute the sister group for all other Bilateria. How do members of this group differ from typical protostomes?
16. What character unites members of clade Gnathifera?
17. What characters unite rotifers and acanthocephalans as members of clade Syndermata?
18. What habitat is shared by micrognathozoans and gnathostomulids?
19. What is the normal size of a rotifer, where is it found, and what are its major features?
20. Explain the difference between mictic and amictic eggs of rotifers. What is the adaptive value of each?
21. What is eutely?
22. Describe the major features of the acanthocephalan body.
23. How do acanthocephalans get food?
24. The evolutionary ancestry of acanthocephalans is particularly obscure. Describe some characters of acanthocephalans that make it surprising that they could be highly derived rotifers.
25. About how big are gastrotrichs, gnathostomulids, and micrognathozoans?
26. How many kinds of mesozoans exist and where would you search for them?

For Further Thought How could an endoparasite evolve from a free-living ancestor, and what other ways of living (e.g., symbiont, commensal, ectoparasite) would you predict as part of the transition?

SELECTED REFERENCES

Achatz, J. G., M. Chiodin, W. Salvenmoser, S. Tyler, and P. Martinez. 2013. The Acoela: on their kind and kinships, especially with nemertodermatids and xenoturbellids (Bilateria incertae sedis). Organisms, Diversity and Evolution **13**: 267–286.

Baguñà, J., and M. Ruitort. 2004. The dawn of bilaterian animals: the case of acoelomorph flatworms. Bioessays **26**: 1046–1057. *Genetic and morphological evidence for acoelomorphs as early diverging bilaterally symmetrical animals.*

Brooks, D. R. 1989. The phylogeny of the Cercomeria (Platyhelminthes: Rhabdocoela) and general evolutionary principles. J. Parasitol. **75**: 606–616. *Cladistic analysis of parasitic flatworms.*

Giribet, G., M. V. Sorenson, P. Funch, R. M. Kristensen, and W. Sterrer. 2004. Investigations into the phylogenetic position of Micrognathozoa using four molecular loci. Cladistics **20**: 1–13. *Research supports the placement of micrognathozoans outside any known phylum.*

Hanelt, B., D. Van Schyndel, C. M. Adema, L. L. Lewis, and E. S. Loker. 1996. The phylogenetic position of *Rhopalura ophiocomae* (Orthonectida) based on 18S ribosomal DNA sequence analysis. Mol. Biol. Evol. **13**: 1187–1191. *Orthonectid mesozoans align with triploblastic animals and do not form the sister taxon to rhombozoans.*

Kobayashi, M., H. Furuya, and P. W. H. Holland. 1999. Dicyemids are higher animals. Nature **401**: 762. *Sequence analysis of the gene for Hox protein is evidence that mesozoans are members of superphylum Lophotrochozoa and* that they are derived from a more complex ancestor that has undergone simplification during its parasitic evolution.

Kristensen, R. M. 2002. An introduction to Loricifera, Cycliophora, and Micrognathozoa. Integ. and Comp. Biol. **42:** 641–651. *A clear and informative description of these little-known animal groups.*

Laumer, C. E., A. Hejol, and G. Giribet. 2015. Nuclear genomic signals of the "microturbellarian" roots of platyhelminth evolutionary innovation. eLife **4:** e05503 *A detailed presentation of turbellarian phylogeny and the evolution of the parasitic taxa.*

Roberts, L. S., and J. Janovy, Jr. 2005. G. D. Schmidt and L. S. Roberts's foundations of parasitology, ed. 7. Dubuque, Iowa, McGraw-Hill Higher Education. *Readable accounts of parasitic flatworms.*

Ruiz-Trillo, I., M. Ruitort, H. M. Fourcade, J. Baguñà, and J. L. Boore. 2004. Mitochondrial genome data support the basal position of Acoelomorpha and the polyphyly of the Platyhelminthes. Mol. Phylogen. Evol. **33:** 321–332. *Evidence that Acoelomorpha is the sister taxon to the remaining Bilateria.*

Schmidt-Rhaesa, A. 2015. Gastrotricha, Cycloneuralia and Gnathifera, Vol 3: Gastrotricha and Gnathifera Book Series: Handbook of Zoology SchmidtRhaesa, A., ed. Walter dy Gruyter GmbH, Berlin/Munich/Boston. *A richly illustrated and detailed compilation of biology and morphology for these lesser-known animals.*

Strickland, G. T. 2000. Hunter's tropical medicine and emerging infectious diseases, ed. 8. Philadelphia, W. B. Saunders Company. *A valuable source of information on parasites of medical importance.*

Telford, M. J., A. E. Lockyear, C. Cartwright-Finch, and D. T. J. Littlewood. 2003. Combined large and small subunit ribosomal RNA phylogenies support a basal position of the acoelomate flatworms. Proc. R. Soc. London B **270:** 1077–1083. *Review of the molecular evidence that led to removal of the Acoela from Platyhelminthes.*

Tyler, S., and M. S. Tyler. 1997. Origin of the epidermis in parasitic Platyhelminthes. Int. J. Parasit. **27:** 715–738. *Descriptions of epidermal replacement as tegument forms in Neodermata.*

Wallace, R. L. 2002. Rotifers: exquisite metazoans. Integ. and Comp. Biol. **42:** 660–667. *This paper summarizes recent work on rotifers, but assumes basic knowledge of the group.*

Witek, A., H. Herlyn, A. Meyer, L. Boell, G. Bucher, and T. Hankeln. 2008. EST-based phylogenomics of Syndermata questions monophyly of Eurotatoria. BMC Evolutionary Biology **8:** 345. *New sequence data suggest that acanthocephalans are the sister taxon to bdelloid rotifers.*

CHAPTER 15

Polyzoa and Kryptrochozoa

Ectoprocts (Bugula neritina) *grow on hard substrates in the ocean.*

- **CLADE POLYZOA**
  - Phylum Cycliophora
  - Phylum Entoprocta
  - Phylum Ectoprocta (Bryozoa)
- **CLADE KRYPTROCHOZOA**
- **CLADE BRACHIOZOA**
  - Phylum Brachiopoda
  - Phylum Phoronida
  - Phylum Nemertea (Rhynchocoela)

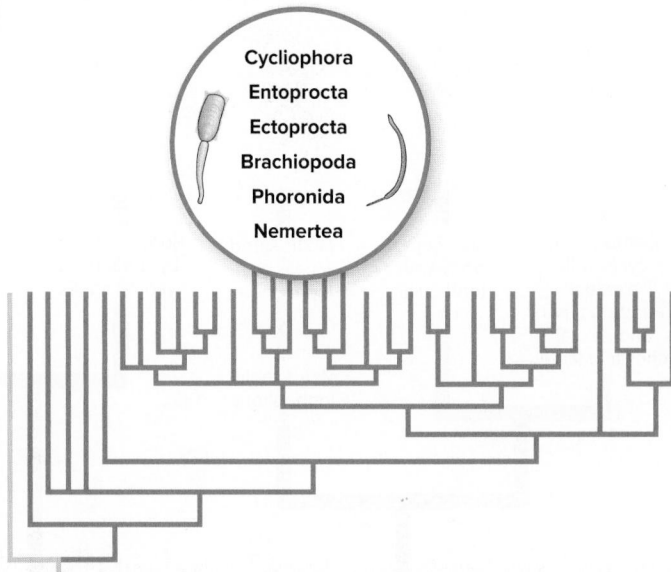

Cycliophora
Entoprocta
Ectoprocta
Brachiopoda
Phoronida
Nemertea

Some Evolutionary Experiments

The early Cambrian period, about 570 million years ago, was the most fertile time in evolutionary history. For 3 billion years before this period, evolution had forged little more than bacteria and blue-green algae. Then, within the space of a few million years, all major phyla, and probably all smaller phyla, became established. This was the Cambrian explosion, the greatest evolutionary "bang" the world has known. In fact, the fossil record suggests that more phyla existed in the Paleozoic era than exist now, but some disappeared during major extinction events that punctuated evolution of life on earth. The greatest of these disruptions was the Permian extinction about 230 million years ago. Thus, evolution has led to many "experimental models." Some of these models failed because they were unable to survive in changing conditions. Others gave rise to abundant and dominant species and individuals that inhabit the world today. Some persisted with small numbers of species, living in very specialized habitats—for example, on the lips of lobsters. The odd shapes and unusual feeding structures of many such animals seem to belong in the realm of science fiction.

Three phyla—Phoronida, Ectoprocta, and Brachiopoda—possess a crown of ciliated tentacles, called a lophophore, that they use in food capture and respiration. Brachiopods were abundant in the Paleozoic era but began to decline thereafter. Phylum Ectoprocta arose in the Cambrian period, became widespread during the Paleozoic era, and remains a prevalent group today. Homology of the lophophore across the three lophophorate phyla is the subject of intense debate and conflicting research.

The six phyla described in this chapter are lophotrochozoan protostomes. The name Lophotrochozoa was developed by merging terms for two features present in many animals in this group: the lophophore and the trochophore. A **lophophore** (Gr. *lophas,* crest or tuft, + *phorein,* to bear) is a crown of tentacles covered with cilia that is borne on a ridge or fold of the body wall. It is an efficient feeding device. The cavity inside the lophophore is part of the coelom and is filled with coelomic fluid. The thin, ciliated walls of the lophophore act as a respiratory surface for exchange of gases between environmental water and coelomic fluid. A lophophore can usually be extended for feeding and respiration and withdrawn for protection.

Three animal phyla possess a lophophore: Ectoprocta, Brachiopoda, and Phoronida. Does the common possession of this feature indicate shared ancestry? A detailed analysis of lophophore structure and function[1] indicates that the lophophore evolved twice,

once in Ectoprocta and once in the common ancestor of Brachiopoda and Phoronida, but some very recent phylogenies unite the lophophorates as a monophyletic group. Here, we place the Ectoprocta in a clade with two other taxa that possess ciliated tentacles (Cycliophora and Entoprocta). These three taxa form a group called Polyzoa (Figure 15.1). Brachiopoda and Phoronida are united in a clade called Brachiozoa (Figure 15.1). The sister taxon to Brachiozoa differs among phylogenetic analyses, but we depict Nemertea in this position. Nemerteans are very thin, stretchy, marine worms with an unusual way of capturing prey (see p. 328). The three taxa form a clade called Kryptrochozoa (Gr. *kryptos,* hidden, + *trochos,* wheel, + *zoa,* animal) that is nested in a larger clade called Trochozoa (Figure 15.1). The remaining members of Trochozoa, Mollusca, Annelida, and Sipuncula, are discussed in Chapters 16 and 17. The term "troch" in the name Trochozoa refers to part of a trochophore larva. A trochophore is a free-swimming,

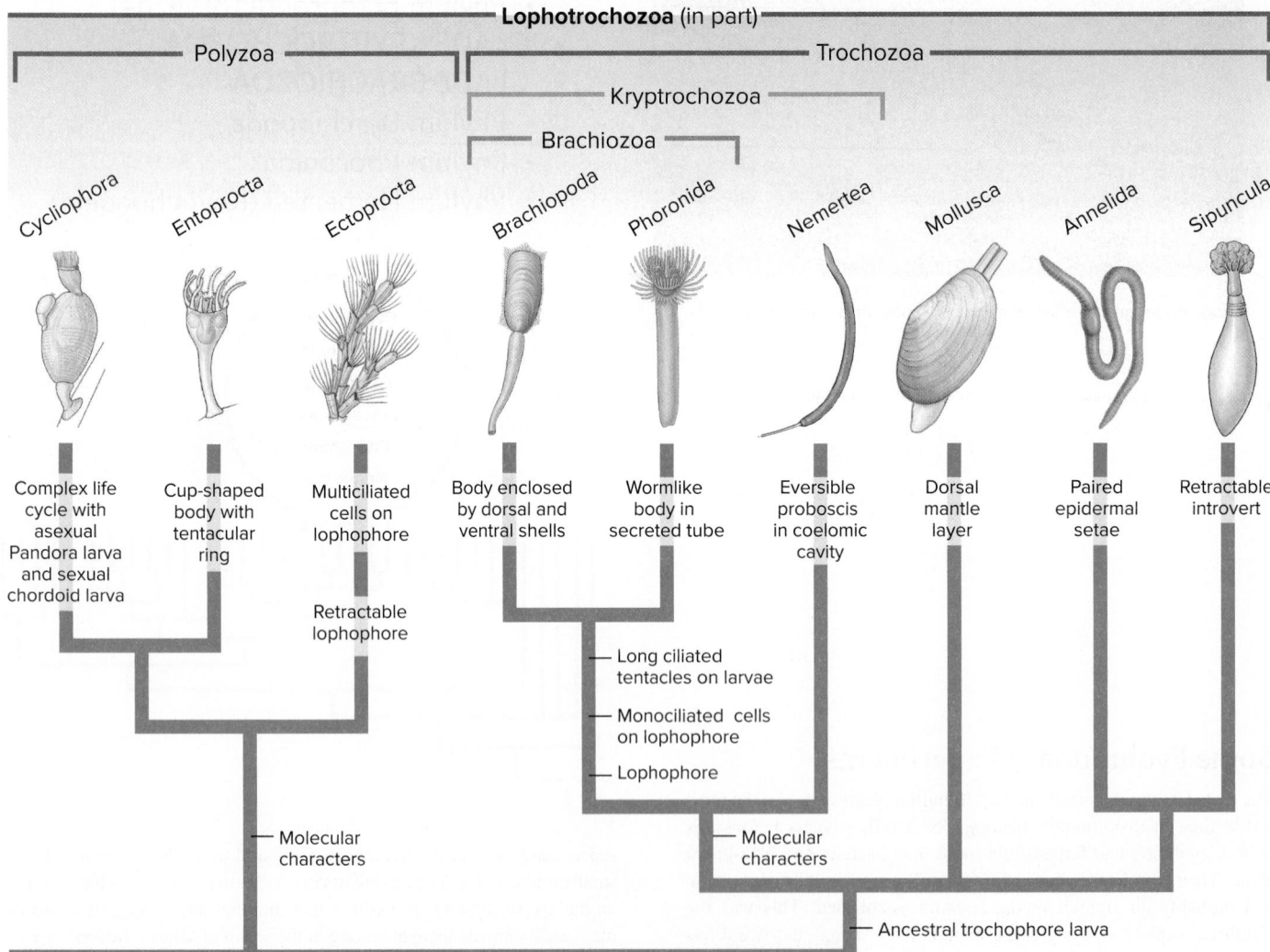

Figure 15.1 Proposed relationships among polyzoans and kryptrochozoans. Characters are modified from those in Nielson (2002) and Brusca and Brusca (2003).

[1]Nielsen, C. 2002. Integ. and Comp. Biol. **42:** 685–691.

feeding, larval stage having a ring of large ciliated cells in front of the mouth (see Figure 16.7, p. 338). This ring of cells is called a prototroch and is used for locomotion primarily, although it may play a role in feeding. The trochophore stage is obvious in the development of molluscs, annelids, and sipunculans, among other taxa. Is there a trochophore in the life cycles of kryptrochozoans? As the name implies, the larval forms in this group may be modified trochophores, carrying a band of cilia or a "hidden" (cryptic) prototroch. Although a detailed discussion of larval forms is not covered here, there are undoubtedly biologists who consider some of these larvae well beyond "modified." The developmental patterns, molecular characteristics, and morphologies of the animals covered here are all under intense study, so the evolutionary relationships of lophotrochozoans may change.

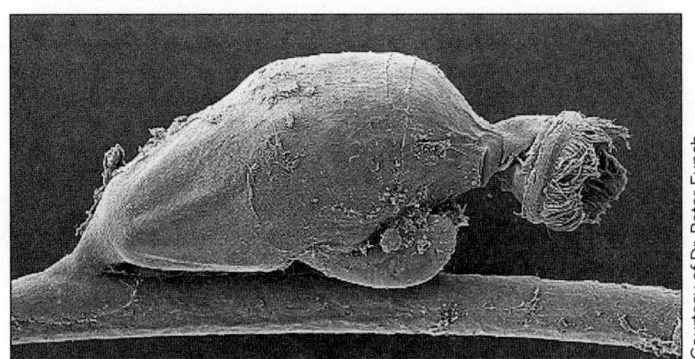

Figure 15.2 *Symbion pandora,* a cycliophoran living on setae on the mouthparts of lobsters.

There are many ways to value biological diversity, but those looking for practical or economic value need go no further than the anticancer properties of compounds from a wide range of marine invertebrates. The bryozoan *Bugula neritina,* pictured on page 323, is the source of bryostatin-1, a treatment for tumors and lymphoma. Bryostatin is actually made by a bacterial symbiont of *B. neritina* called *Candidatus endobugula sertula,* but the bacterium cannot be cultured outside the host. Other anticancer compounds are derived from sponges (e.g., discodermolide), corals (e.g., eleutherobin analogues), tunicates (e.g., ecteinascidin), clams (e.g., spisulosine), and sea cucumbers (e.g., calcigeroside B). The organisms just named are among the best-known marine invertebrates, so we wonder which unique compounds will be discovered within the animals discussed in this chapter. Many animals covered here are studied only by specialists, so their potential may be untapped. We know as little about their role in maintaining a healthy ocean (ecosystem services) as we do about their biochemical novelty.

15.1   CLADE POLYZOA

The clade Polyzoa unites cycliophorans with entoprocts and ectoprocts, also called bryozoans (Figure 15.1). It is now supported by phylogenetic studies using multiple genes, but a close relationship between Ectoprocta and Entoprocta was proposed on morphological grounds 40 years ago. Cycliophora was not discovered until 1995, so it could not have been included in those early studies, but similarities to entoprocts were noted when the animals were first described. The tiny animals in these three phyla have fascinating body plans and life cycles.

15.2   PHYLUM CYCLIOPHORA

In December 1995, P. Funch and R. M. Kristensen reported their discovery of some very strange little creatures clinging to the mouthparts of the Norway lobster (*Nephrops norvegicus*). The animals were tiny, only 0.35 mm long and 0.10 mm wide, and did not fit into any known phylum. They were named *Symbion pandora,* the first members of phylum Cycliophora (Figure 15.2).

Cycliophorans have a very specialized habitat: they live on mouthparts of marine decapod crustaceans in the Northern Hemisphere. They attach to bristles on the mouthparts with an adhesive disc on the end of an acellular stalk. They feed by collecting

bacteria, or bits of food dropped from their lobster host, on a ring of compound cilia that surrounds the mouth.

The body plan is relatively simple. The mouth leads into a U-shaped gut ending with an anus that opens outside the ciliated ring. The body is acoelomate. The epidermis is cellular and surrounded by a cuticle.

The life cycle has sexual and asexual phases. Feeding animals make internal buds, called Pandora larvae, which become new feeding individuals upon release. Clone-members quickly occupy vacant areas on the lobster mouthparts. Internal budding is also used to make a new feeding and digestive system for a feeding animal—the existing system degenerates and is replaced by one from the internal bud.

As a prelude to sexual reproduction, male or female larvae are made. A male larva is released from a feeding individual and settles atop another animal housing a female larva. A male larva produces secondary males with reproductive organs; internal fertilization occurs as one secondary male mates with a female larva leaving the body of a feeding animal. Once the egg in the female is fertilized, a chordoid larva develops inside the body of its mother, consuming it. The chordoid larva swims to a new lobster host, where it makes a feeding animal by internal budding. The feeding animal then makes a clone of feeders by internal budding.

15.3   PHYLUM ENTOPROCTA

Entoprocta (en'to-prok'ta) (Gr. *entos,* within, + *proktos,* anus) is a small phylum of about 150 species of tiny, sessile animals that superficially resemble hydroid cnidarians but have ciliated tentacles that tend to roll inward (Figure 15.3B and C). Most entoprocts are microscopic, and none is more than 5 mm long. They may be solitary or colonial, but all are stalked and sessile. All are ciliary feeders.

With the exception of *Urnatella* (L. *urna,* urn, + *ellus,* dim. suffix), which occurs in freshwater, all entoprocts are marine forms that have a wide distribution from polar regions to tropics. Most marine species are restricted to coastal and brackish waters and often grow on shells and algae. Some are commensals on marine annelid worms. Entoprocts occur from the intertidal to depths of around 500 m. Freshwater entoprocts occur on the undersides of rocks in running water. *Urnatella gracilis* is the only common freshwater species in North America (Figure 15.3A).

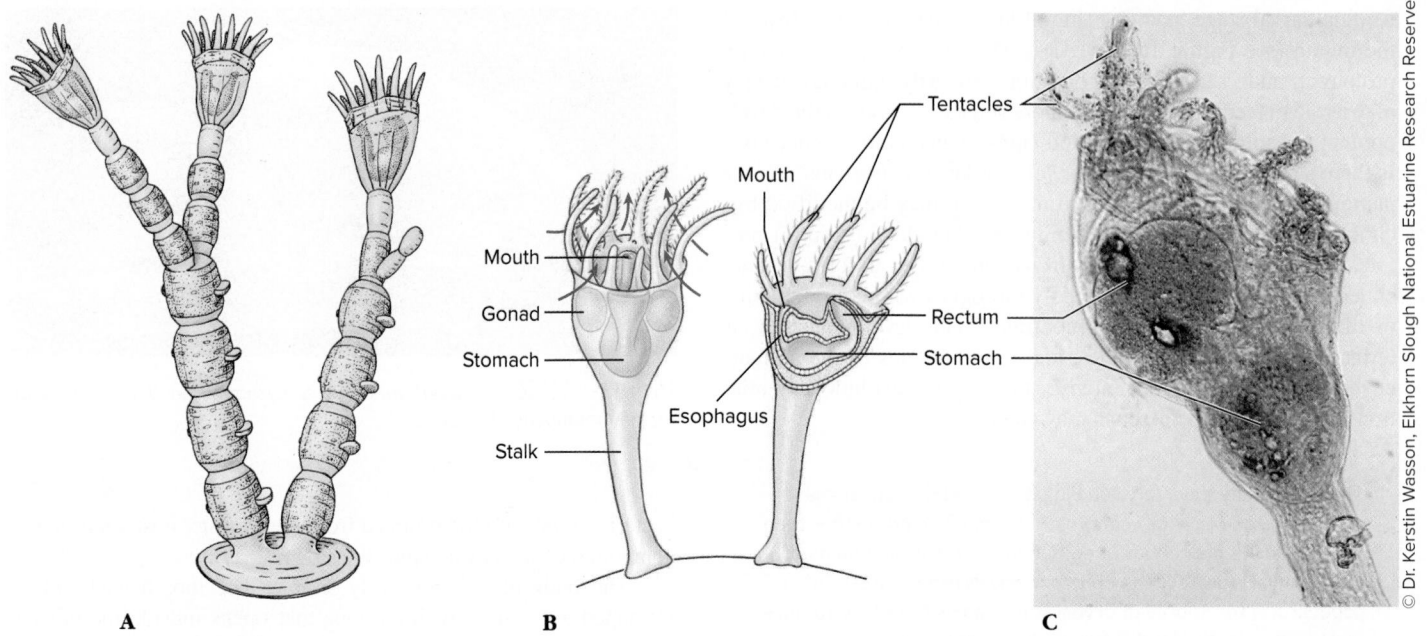

Figure 15.3 **A,** *Urnatella,* a freshwater entoproct, forms small colonies of two or three stalks from a basal plate. **B,** *Loxosomella,* a solitary entoproct. Blue arrows indicate the direction of water movement; flow occurs in a direction opposite to that in lophophorates, such as bryozoans. Both solitary and colonial entoprocts can reproduce asexually by budding, as well as sexually. **C,** A live *Loxosomella.*

Form and Function

The body, or **calyx,** of an entoproct is cup-shaped, bears a crown, or circle, of ciliated tentacles, and may be attached to a substratum by a single stalk and an attachment disc with adhesive glands, as in the solitary *Loxosoma* and *Loxosomella* (Gr. *loxos,* crooked, + *soma,* body) (Figure 15.3B), or by two or more stalks in colonial forms. Movement is usually restricted in entoprocts, but *Loxosoma,* which lives in the tubes of marine annelids, is quite active, moving over the annelid and its tube freely.

The body wall comprises a cuticle, cellular epidermis, and longitudinal muscles. The tentacles and stalk are continuations of the body wall. The 8 to 30 tentacles forming the crown are ciliated on their lateral and inner surfaces, and each can move individually. Tentacles can roll inward to cover and to protect the mouth and anus but cannot be retracted into the calyx. The gut is U-shaped and ciliated, and both the mouth and the anus open within the circle of tentacles. Entoprocts are ciliary suspension feeders. Long cilia on the sides of the tentacles keep a current of water containing protozoa, diatoms, and particles of detritus moving inward between the tentacles. Short cilia on the inner surfaces of the tentacles capture food and direct it downward toward the mouth. Digestion and absorption occur within the stomach and intestine before wastes are discharged from the anus.

The pseudocoel is largely filled with a gelatinous parenchyma in which is embedded a pair of flame bulb protonephridia and their ducts, which unite and empty near the mouth. A well-developed **nerve ganglion** occurs on the ventral side of the stomach, and the body surface bears sensory bristles and pits. Circulatory and respiratory organs are absent. Exchange of gases occurs across the body surface, probably much of it through the tentacles.

Some species are dioecious, but many are monoecious, most often protandrous hermaphrodites, where the gonad at first produces sperm and later eggs. Colonial forms may have monoecious or dioecious zooids, and colonies can even contain zooids of both sexes. The gonoducts open within the circle of tentacles.

Fertilized eggs develop in a depression, or brood pouch, between the gonopore and the anus. Entoprocts have a modified spiral cleavage pattern with mosaic blastomeres. The embryo gastrulates by invagination. Mesoderm develops from the 4d cell (see p. 170). The trochophore-like larva (see p. 338) is ciliated and free-swimming. It has an apical tuft of cilia at the anterior end and a ciliated girdle around the ventral margin of the body. Eventually the larva settles onto a substratum and metamorphoses into an adult zooid.

15.4 PHYLUM ECTOPROCTA (BRYOZOA)

Ectoprocta (ek-to-prok′ta) (Gr. *ektos,* outside, + *proktos,* anus) contains aquatic animals that often encrust hard surfaces. Most species are sessile, but some slide slowly, and others crawl actively, across the surfaces they inhabit. With very few exceptions, they are colony builders. Each member of a colony is small, typically less than 0.5 mm. Colony members, called **zooids,** feed by extending their lophophores into surrounding water to collect tiny particles. Zooids secrete small containers in which they live and thus form an exoskeleton (Figure 15.4). The exoskeleton, or **zoecium,** may, according to species, be gelatinous, chitinous, or stiffened with calcium and possibly also impregnated with sand. Its shape may be boxlike, vaselike, oval, or tubular.

Ectoprocts have left a rich fossil record since the Ordovician period and are diverse and abundant today. There are about

Figure 15.4 **A,** A colony of *Membranipora,* a marine encrusting bryozoan (Ectoprocta). Each little oblong zoecium is the calcareous home of a zooid. **B,** Portion of a colony of an encrusting bryozoan. Two zooids are shown with lophophores extended from their chambers, the zoecia. The tiny zooids pop up to feed with their tentacular crown, then quickly withdraw at the slightest disturbance. The mouth is inside the lophophore ring, but the anus lies outside.

Figure 15.5 Colonies of marine ectoprocts. **A,** The zooids are encrusting in this lacy colony of *Membranipora tuberculata.* **B,** *Bugula neritina* has upright, branching colonies.

4500 living species of ectoprocts. They occupy both freshwater and marine habitats but largely live in shallow waters.

Some colonies form limy encrustations on seaweed, shells, and rocks; others form fuzzy or shrubby growths or erect, branching colonies that look like seaweed (Figure 15.5). Some ectoprocts might easily be mistaken for hydroids but can be distinguished under a microscope by following the movement of particles or dye through the lophophore. Unlike cnidarian tentacles, the lophophore has cilia which generate a current. Although zooids are minute, the colonies are often several centimeters in diameter; some encrusting colonies may be a meter or more in width, and erect forms may reach 30 cm or more in height. Marine forms exploit all kinds of firm surfaces, such as shells, rocks, large brown algae, mangrove roots, ship bottoms, and even the bottoms of icebergs! One species,

Tricellaria inopinate, is spreading through the northeastern Atlantic Ocean and the Mediterranean, displacing native bryozoans. Although it began to diminish naturally in Venice, native byrozoans did not re-establish. Like many invasive species, it reproduces throughout the year and disperses well. Cold temperatures appear to limit its northernmost distribution.

Freshwater ectoprocts may form mosslike colonies on stems of plants or on rocks, usually in shallow ponds or pools. In some freshwater forms individuals are borne on finely branching stolons that trace delicate patterns onto the undersides of rocks or plants. Other freshwater ectoprocts are embedded in large masses of gelatinous material.

Ectoprocts have long been called bryozoans, or moss animals (Gr. *bryon,* moss, + *zōon,* animal), a term that originally included Entoprocta also. However, because entoprocts have their anus located within the tentacular crown, they are commonly separated from ectoprocts, which, like other lophophorates, have their anus outside the circle of tentacles. Many authors continue to use the name "Bryozoa" but exclude entoprocts from the group.

Form and Function

Each member of a colony lives in a tiny chamber, called a zoecium, which is secreted by its epidermis (Figure 15.4). Each **zooid** consists of a feeding polypide and a case-forming cystid. A **polypide** includes the lophophore, digestive tract, muscles, and nerve centers. A **cystid** includes the body wall of an animal, together with its secreted exoskeleton or zoecium.

Polypides live a type of jack-in-the-box existence, popping up to feed but, at the slightest disturbance, quickly withdrawing into their little chamber, which often has a tiny trapdoor (operculum) that shuts to conceal its inhabitant (Figure 15.4). To extend its tentacular crown, certain muscles contract, which increases hydrostatic pressure within the body cavity and pushes the lophophore out by a hydraulic mechanism. Other muscles can contract to withdraw the crown to safety with great speed.

When feeding, an animal extends its lophophore and spreads its tentacles to form a funnel. Cilia on the tentacles draw water into the funnel and out between the tentacles. Food particles caught by cilia in the funnel are drawn into the mouth, both by a pumping action of the muscular pharynx and by action of cilia along the length of the tentacles and in the pharynx itself. Undesirable particles can be rejected by reversing the ciliary action, by drawing the tentacles close together, or by retracting the whole lophophore into the zoecium. The lophophore ridge tends to be circular in marine ectoprocts and U-shaped in freshwater species (Figure 15.6). A septum divides the mesocoel in the lophophore from the larger posterior metacoel. A protocoel and epistome (a flap over the mouth; see Figure 15.10) occur only in freshwater ectoprocts.

Digestion in the ciliated, U-shaped digestive tract begins extracellularly in the stomach and is completed intracellularly in the intestine. The gut is complete. Respiratory, vascular, and excretory organs are absent. Gaseous exchange is through the body surface, and since ectoprocts are small, coelomic fluid is adequate for internal transport. In some, pores in the walls between adjoining zooids permit exchange of materials throughout the colony by way of the coelomic fluid, but in others a funicular system transports lipids and other nutritive substances throughout the colony. Coelomocytes engulf and store waste materials. A ganglionic mass and a nerve ring surround the pharynx, but no specialized sense organs are present.

Most colonies contain only feeding individuals, but specialized zooids incapable of feeding (collectively called heterozooids) occur in some species. One type of modified zooid (called an *avicularium*) resembles a bird beak that snaps at small invading organisms that might foul a colony. Another type (called a *vibraculum*) has a long bristle that apparently helps to sweep away foreign particles.

Most ectoprocts are hermaphroditic. Some species shed eggs into seawater, but most brood their eggs, some within the coelom and some externally in a special brood chamber called an ovicell, which is a modified zoecium in which an embryo develops. In some cases many embryos proliferate asexually from that initial embryo in a process called **polyembryony.** Cleavage is radial but apparently mosaic. Little is known of mesoderm derivation. Larvae of nonbrooding species have a functional gut and swim for a few months before settling; larvae of brooding species do not feed and settle after a brief free-swimming existence. They attach to the substratum by secretions from an **adhesive sac,** then metamorphose to their adult form.

Each colony begins from this single metamorphosed primary zooid, which is called an **ancestrula.** The ancestrula then undergoes asexual budding to produce the many zooids of a colony. Freshwater ectoprocts have another type of budding that produces **statoblasts** (Figure 15.7), which are hard, resistant capsules containing a mass of germinative cells. Statoblasts form during summer and fall. When a colony dies in late autumn, statoblasts remain, and in spring they give rise to new polypides and eventually to new colonies.

Figure 15.6 *Plumatella repens,* a freshwater bryozoan (phylum Ectoprocta). It grows on the underside of rocks and on vegetation in lakes, ponds, and streams.

© blickwinkel/Alamy Stock Photo

Figure 15.7
Statoblast of a freshwater ectoproct *Cristatella.* This statoblast is about 1 mm in diameter and bears hooked spines.

© Larry S. Roberts

15.5 CLADE KRYPTROCHOZOA

The taxon Trochozoa consists of animals with a trochophore larva, as discussed in the opening section of this chapter (pp. 320–321). Within Trochozoa, three taxa are united as Kryptrochozoa (see Figure 15.1); the name refers to a highly modified or "hidden" trochophore larval stage, also described at the beginning of this chapter. Kryptrochozoa comprises a group of thin marine worms called nemerteans, and two other taxa that form Clade Brachiozoa.

15.6 CLADE BRACHIOZOA

The clade Brachiozoa unites brachiopods with phoronids; both molecular characters and morphological features support this pairing (see Figure 15.1). Both taxa possess a lophophore, although one group is shelled and the other is wormlike.

15.7 PHYLUM BRACHIOPODA

Brachiopoda (brak-i-op′o-da) (Gr. *brachio-n,* arm, + *pous, podos,* foot), or lamp shells, are an ancient group. Although about 325 species are now living, some 12,000 fossil species, which once flourished in Paleozoic and Mesozoic seas, have been described. Modern forms have changed little from early ones. Genus *Lingula* (L. tongue) (Figure 15.8A) is considered a "living fossil," having existed virtually unchanged since Ordovician times. Most modern brachiopod shells range between 5 and 80 mm in length, but some fossil forms reached 30 cm.

Brachiopods are attached, bottom-dwelling, marine forms that mostly prefer shallow water, although they occur at nearly all ocean depths. Externally brachiopods resemble bivalved molluscs in having two calcareous shell valves secreted by a mantle. They were, in fact, classified with molluscs until the middle of the nineteenth century, and their name refers to the arms of the lophophore, which were thought homologous to the mollusc foot. Brachiopods, however, have dorsal and ventral valves instead of right and left lateral valves as do bivalve molluscs and, unlike bivalves, most of them are attached to a substrate either directly or by means of a fleshy stalk called a **pedicel.** Some, such as *Lingula,* live in vertical burrows in sand or mud. Muscles both open and close the valves and provide movement for the stalk and tentacles.

In most brachiopods the ventral (pedicel) valve is slightly larger than the dorsal (brachial) valve, and one end projects in the form of a short, pointed beak perforated where the fleshy

stalk passes through the shell to attach to the substratum (Figure 15.8B). In many the pedicel valve is shaped like a classic oil lamp of ancient Greece and Rome, so that brachiopods came to be called "lamp shells."

Shell structure distinguishes the two classes of brachiopods. Shell valves of Articulata have a connecting hinge with an interlocking tooth-and-socket arrangement, as in *Terebratella* (L. *terebratus,* a boring, + *ella,* dim. suffix); those of Inarticulata lack the hinge and are held together by muscles only, as in *Lingula* and *Glottidia* (Gr. *gl ottidos,* mouth of windpipe).

Their body occupies only the posterior part of the space between the valves (Figure 15.9), and extensions of the body wall form mantle lobes that line and secrete the shell. Their large horseshoe-shaped lophophore in the anterior mantle cavity bears long, ciliated tentacles used in respiration and feeding. Ciliary water currents carry food particles between the gaping valves and over the lophophore. Tentacles catch food particles, and ciliated grooves carry the particles along the arm of the lophophore to their mouth. Rejection tracts carry unwanted particles to the mantle lobe, where they are swept out in ciliary currents. Organic detritus and some algae are apparently primary food sources. A brachiopod's lophophore not only creates food currents, as it does in other lophophorates, but also seems to absorb dissolved nutrients directly from environmental seawater.

There are three coelomic cavities, called the protocoel, mesocoel, and metacoel, but the posterior cavity (metacoel) bears the viscera. One or two pairs of nephridia open into the coelom and empty into the mantle cavity. Coelomocytes, which ingest particulate wastes, are expelled by nephridia. There is an open circulatory system with a contractile heart. Lophophore and mantle are probably the chief sites of gaseous exchange. There is a nerve ring with a small dorsal and a larger ventral ganglion.

Most species have separate sexes, and temporary gonads discharge gametes through the nephridia. Most fertilization is external, but a few species brood their eggs and young.

Cleavage is radial, and coelom and mesoderm formation in at least some brachiopods is enterocoelic. The blastopore closes, but its relationship to the mouth is uncertain. In articulates, metamorphosis of larvae occurs after they have attached by a pedicel. In inarticulates, juveniles resemble a minute brachiopod with a

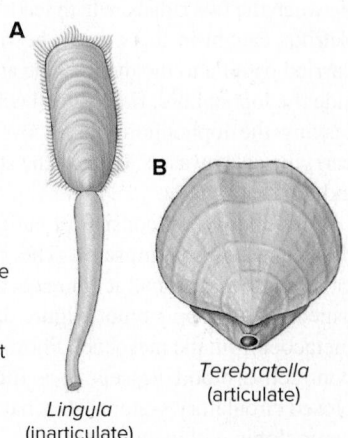

Figure 15.8
Brachiopods. **A,** *Lingula,* an inarticulate brachiopod that normally occupies a burrow. The contractile pedicel can withdraw the body into the burrow. **B,** An articulate brachiopod, *Terebratella.* The valves have a tooth-and-socket articulation, and a short pedicel projects through the pedicel valve to attach to the substratum.

Lingula
(inarticulate)

Terebratella
(articulate)

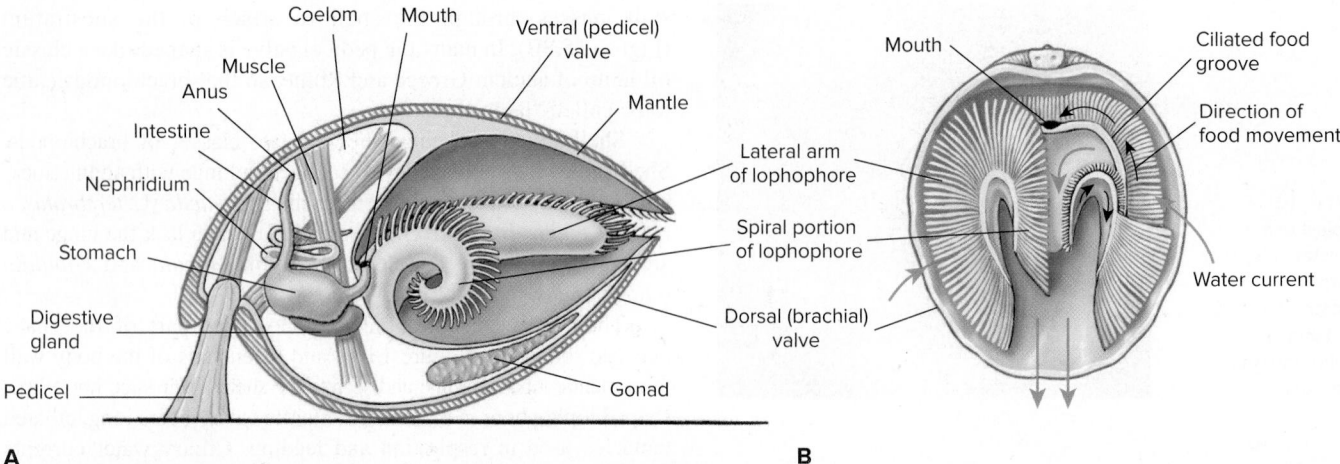

Figure 15.9 Phylum Brachiopoda. **A,** An articulate brachiopod (longitudinal section). Note that its pedicel emerges from the ventral valve, so that when it is attached to a substrate, an articulate brachiopod is "upside down," with its ventral valve on top and its dorsal valve below. **B,** Feeding and respiratory currents. Large (blue) arrows show water flow over lophophore; small (black) arrows indicate food movement toward mouth in ciliated food groove.

coiled pedicel in the mantle cavity. There is no metamorphosis. As a larva settles, its pedicel attaches to the substratum, and adult existence begins.

15.8 PHYLUM PHORONIDA

Phylum Phoronida (fo-ron′i-da) (L. *Phoronis,* in mythology, surname of Io, who was turned into a white heifer) contains approximately 20 species of small, wormlike animals. Most live on the substrate below shallow coastal waters, especially in temperate seas. They range from a few millimeters to 30 cm in length. Each worm secretes a leathery or chitinous tube in which it lies free, but which it never leaves. The tubes may be anchored singly or in a tangled mass on rocks, shells, or pilings or buried in sand. They thrust out the tentacles on the lophophore for feeding, but a disturbed animal can withdraw completely into its tube.

A lophophore has two parallel ridges curved in a horseshoe shape, the bend located ventrally and the mouth lying between the two ridges (Figure 15.10). Horns of the ridges often coil into twin spirals. Each ridge carries hollow ciliated tentacles, which, like the ridges themselves, are extensions of the body wall.

Cilia on the tentacles direct a water current toward a groove between the two ridges, which leads toward the mouth. Plankton and detritus caught in this current become entangled in mucus and are carried by cilia to the mouth. The anus lies dorsal to the mouth, outside the lophophore, flanked on each side by a nephridiopore. Water leaving the lophophore passes over the anus and nephridiopores, carrying away wastes. Cilia in the stomach area of the U-shaped gut aid food movement.

The body wall consists of cuticle, epidermis, and both longitudinal and circular muscles. The protocoel is present as a small cavity in the epistome; it connects to the mesocoel along the lateral aspects of the epistome (Figure 15.10). A septum separates the metacoel from the mesocoel. Phoronids have an extensive system of contractile blood vessels in a functionally, but not technically, closed circulatory system. They have no heart. Their blood contains hemoglobin within nucleated cells. There is a pair of metanephridia.

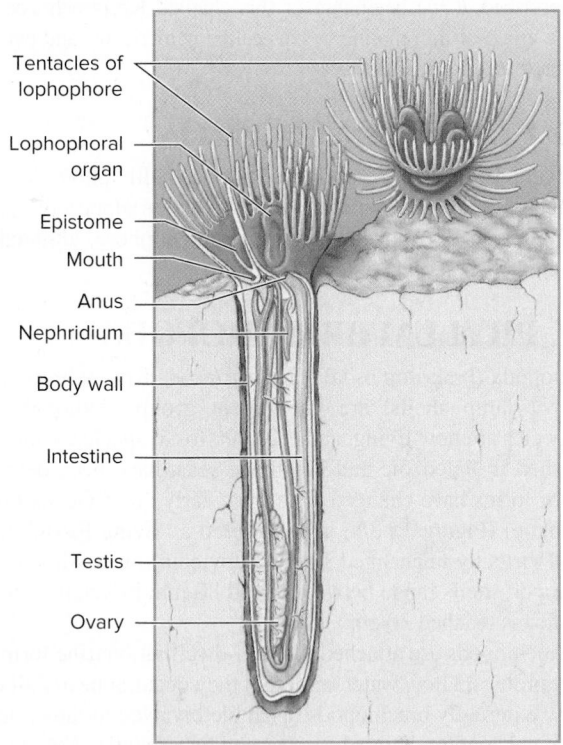

Figure 15.10 Internal structure of *Phoronis* (phylum Phoronida), in diagrammatic vertical section.

A nerve ring sends nerves to tentacles and body wall, but the system is diffuse and lacks a distinct ganglion that could be called a brain. A single giant motor fiber lies in the epidermis, and an epidermal nerve plexus supplies the body wall and epidermis.

There are both monoecious (the majority) and dioecious species of Phoronida, and at least two species reproduce asexually. Fertilization may be internal or external, but contrary to early reports, cleavage is radial. Coelom formation is by a highly modified enterocoelous route, but the blastopore becomes the mouth. A free-swimming ciliated larva, called an actinotroch, sinks to the sea floor where it metamorphoses into an adult animal. It secretes a tube and becomes sessile.

15.9 PHYLUM NEMERTEA (RHYNCHOCOELA)

Nemerteans (nēm-er′tē-ans) are often called ribbon worms (Figure 15.11). Their name (Gr. *Nemertes,* one of the Nereids, unerring one) refers to the unerring aim of the proboscis, a long muscular tube (Figure 15.12) section. that can be thrust out swiftly to grasp their prey. The phylum is also called Rhynchocoela (ring′ko-se′la) (Gr. *rhynchos,* beak, + *koilos,* hollow), which also refers to the proboscis. They are thread-shaped or ribbon-shaped bilaterally symmetrical, triploblastic worms. There are about 1000 species in the group; nearly all are marine.

Nemertean worms are usually less than 20 cm long, although a few are several meters in length (Figure 15.13). *Lineus longissimus* (L. *linea,* line) is the longest animal in the world, reaching 60 m in

Figure 15.12   **A,** Structure of female nemertean worm *Amphiporus* (diagrammatic). Dorsal view to show proboscis. **B,** *Amphiporus,* with proboscis extended to catch prey.

Figure 15.13   *Baseodiscus* is a genus of nemerteans whose members typically measure several meters in length. This *B. mexicanus* is from the Galápagos Islands.

length! This worm is only 5 to 10 mm wide. Nemerteans can be brightly colored, although most are dull or pallid. Some live in secreted gelatinous tubes.

With few exceptions, the general body plan of nemerteans is similar to that of turbellarians (Chapter 14, p. 298). The epidermis of nemerteans is ciliated and has many gland cells. The excretory system contains flame cells. Rhabdites occur in several nemerteans, including *Lineus,* but some work indicates that they are not homologous to rhabdites in flatworms. Nemerteans also differ from flatworms in their reproductive system. Ribbon worms are mostly dioecious. The ciliated larva of marine forms has some resemblance to trochophore larvae of annelids and molluscs.

Nemerteans show some derived features absent from flatworms. The most obvious of these is the eversible **proboscis** and its sheath, for which there are no counterparts within any other phylum. In the odd genus *Gorgonorhynchus* (Gr. *Gorgo,* name of a female monster of terrible aspect, + *rhynchos,* beak, snout), the proboscis is divided into many proboscides, which appear as a mass of wormlike structures when everted. Another difference is the presence of an **anus** in adults, producing a **complete digestive system.** A digestive system with an anus is more efficient because ejection of waste materials back through the mouth is not necessary. Ingestion and defecation can occur simultaneously. Nemerteans are also the simplest animals to have a closed-loop **blood-vascular system.**

A few nemerteans occur in moist soil and freshwater. *Prostoma rubrum* (Gr. *pro,* before, in front of, + *stoma,* month), which is 20 mm or less in length, is a well-known freshwater species. A larger number of nemertean species are marine; at low tide they are often coiled under stones. It seems probable that they are active at high tide and quiescent at low tide. Some nemerteans, such as *Cerebratulus* (L. *cerebrum,* brain, + *ulus,* dim. suffix), often live in empty mollusc shells. Small species often live among seaweed, or they may be found swimming near the surface of the water.

Nemerteans are often secured by dredging at depths of 5 to 8 m or deeper.

Although a few species are commensals or scavengers, most ribbon worms are active predators on small invertebrates. A few species are specialized egg predators (considered ectoparasites) on brachyuran crabs, and in high numbers can consume all the embryos in their host's clutch.

Form and Function

Many nemerteans are difficult to examine because they are so long and fragile. *Amphiporus* (Gr. *amphi,* on both sides, + *poros,* pore), a genus of smaller forms that ranges from 2 to 10 cm in length, is fairly typical of nemertean structure (Figure 15.12). Its body wall consists of ciliated epidermis and layers of circular and longitudinal muscles (Figure 15.14). Locomotion consists largely of gliding over a slime track, although larger species move by muscular contractions. Some large species are even capable of undulatory swimming when threatened.

The mouth is anterior and ventral, and the digestive tract is complete, extending the full length of the body and ending at an anus. There are usually no muscles in the gut wall itself; instead cilia move food through the intestine. Digestion is largely extracellular in the gut lumen.

The favorite prey of most nemerteans is annelids and other small invertebrates. Their diets may be highly specialized or extremely varied, depending on the species. Some species appear able to detect prey only when they physically bump into it, whereas others are capable of tracking prey over great distances. When prey is encountered, nemerteans seize it with a proboscis that lies in an interior cavity of its own, the **rhynchocoel,** above the digestive tract (but not connected with it). The proboscis itself is a long, blind muscular tube that opens at the anterior end at a

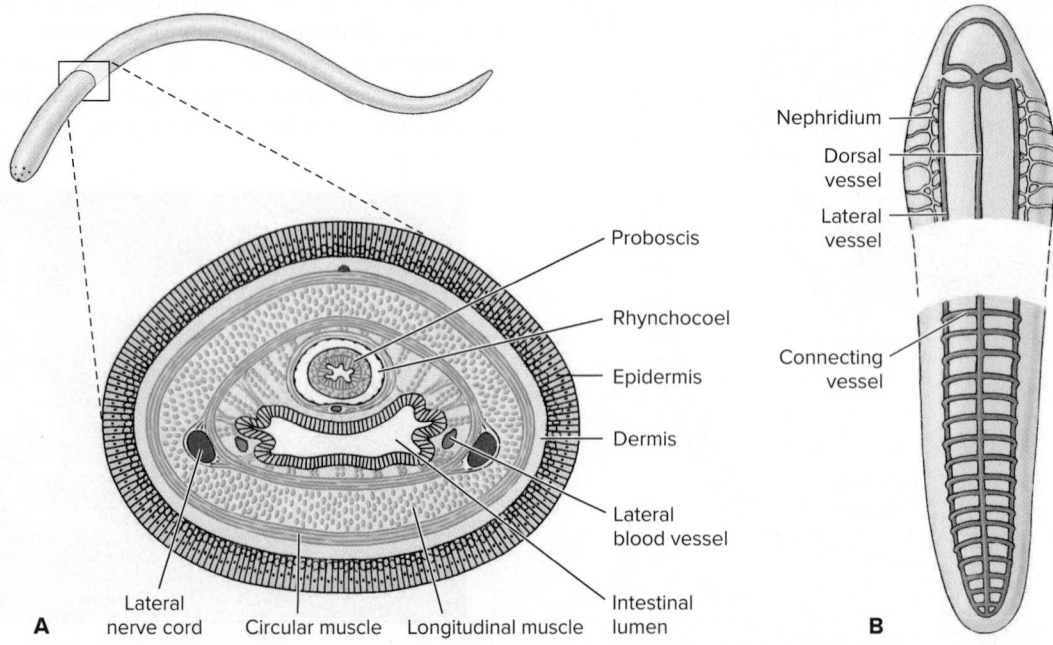

Figure 15.14 **A,** Diagrammatic cross section of female nemertean worm. **B,** Excretory and circulatory systems of nemertean worm. Note how the canals connecting one nephridium to the next are closely associated with lateral blood vessels.

proboscis pore above the mouth (Figure 15.12). Muscular pressure on fluid in the rhynchocoel causes the long, tubular proboscis to be everted rapidly through the proboscis pore. Eversion of the proboscis exposes a sharp barb, called a stylet (absent in some nemerteans). The sticky, slime-covered proboscis coils around the prey and stabs it (often repeatedly) with the stylet, while pouring a toxic secretion on the prey (Figure 15.12). The neurotoxin in some species has recently been identified as tetrodotoxin, commonly known as the poison in puffer fishes. Upon retracting its proboscis, a nemertean draws the prey near its mouth and the subdued prey is swallowed whole.

Characteristics of Phylum Nemertea

1. An **eversible proboscis,** unique to nemerteans, lies free in a cavity (rhynchocoel) above the alimentary canal
2. In marine, freshwater, and moist terrestrial habitats
3. Nemerteans are mostly free-living, with a few parasitic species
4. Bilateral symmetry; highly contractile body that is cylindrical anteriorly and flattened posteriorly
5. Body **triploblastic;** adult parenchyma partly gelatinous
6. Rhynchocoel is a true coelomic cavity, but its unusual position and function as part of the proboscis mechanism leads many to question whether it is homologous to the coelom of other protostomes
7. Epidermis with cilia and gland cells; rhabdites in some
8. **Complete digestive system** (mouth to anus)
9. Body-wall musculature of outer circular and inner longitudinal layers with diagonal fibers between the two; sometimes another circular layer inside the longitudinal layer
10. Nervous system usually a four-lobed brain connected to paired longitudinal nerve trunks or, in some, middorsal and midventral trunks
11. Sensory **ciliated pits** or **head slits** on each side of head, which communicate between the outside and the brain; tactile organs and ocelli (in some)
12. Asexual reproduction by fragmentation
13. Sexes separate with simple gonads; few hermaphrodites; **pilidium larvae** in some
14. Excretory system of two coiled canals, which are branched with **flame cells**
15. **Blood-vascular system** with two or three longitudinal trunks
16. No respiratory system

Nemerteans have a true circulatory system, and blood flow is maintained by a combination of the contractile walls of the vessels and general body movements. As a result the flow is irregular and often reverses direction in the vessels. Two to many flame-bulb protonephridia are closely associated with the circulatory system, so that their function appears to be truly excretory (for disposal of metabolic wastes), in contrast to their apparently osmoregulatory role in Platyhelminthes.

Nemerteans have a pair of nerve ganglia, and one or more pairs of longitudinal nerve cords are connected by transverse nerves.

Some species reproduce asexually by fragmentation and regeneration. Nemerteans show a surprising range of sexual reproductive

strategies. Most species are dioecious and fertilization is often external, although many exceptions are known: some species are hermaphroditic, some have internal fertilization, and some even have ovoviviparous development.

Phylogeny of Nemertea

Like many other lophotrochozoan protostomes, nemerteans exhibit spiral cleavage. Later development varies across the phylum, with some work showing typical formation of mesoderm from endoderm, as well as instances of mesodermal formation from ectoderm.

Nemerteans produce a variety of larval forms, and in some species all stages of development occur within an egg case. The evolutionary relationship between the various larval forms and the typical trochophore has been much discussed. Some similarities exist, but a distinctive central ring of cilia (a prototroch) is not present. However, a new developmental study of an early diverging species revealed that a band of cilia does form around the larva, although it later degenerates. The brief existence of a ciliary band provides evidence that a trochophore larva was present in ancestral nemerteans, and supports their placement in Lophotrochozoa.

A second area of controversy is the nature of the nemertean body plan. Are nemerteans acoelomate or coelomate animals? The rhynchocoel is an internal cavity lined by mesoderm that forms by schizocoely, making it a true coelom. However, a typical coelom (p. 163) forms a fluid-filled cavity around the digestive tract. The rhynchocoel instead lies above the digestive tract, extending about three-quarters the length of the body from the anterior end. The rhynchocoel differs from a typical coelom in both position and function. A typical coelom surrounds, cushions, and protects the gut, but it also forms part of the hydrostatic skeleton, stiffening when surrounding muscles contract. A rhynchocoel is fluid-filled and surrounded by muscles; muscular contraction increases hydrostatic pressure and eventually everts the proboscis. We leave the reader, in the company of future generations of biologists, to ponder whether the protostome coelom and the rhynchocoel are homologous structures.

Taxonomy of Phylum Nemertea

Class Enopla (en'o-pla) (Gr. *enoplos,* armed). Proboscis usually armed with stylets; mouth opens in front of brain. Examples: *Amphiporus, Prostoma.*
Class Anopla (an'o-pla) (Gr. *anoplos,* unarmed). Proboscis lacks stylets; mouth opens below or posterior to brain.
Examples: *Cerebratulus, Tubulanus, Lineus.* Class Anopla is contentious because some authorities consider it a paraphyletic group.

15.10 PHYLOGENY AND ADAPTIVE DIVERSIFICATION

Phylogeny

Since Lophotrochozoa first emerged from phylogenies based on molecular characters, biologists have struggled to understand patterns of morphological change across this diverse group of phyla. The suite of developmental characters associated with spiral cleavage is presumed to be ancestral for the clade. This suite includes spirally cleaving embryos, mosaic cleavage where cell fates are determined by the cytoplasmic factors they contain, and formation of mesoderm from derivatives of a particular cell, called the 4d cell, that is present at about the 64-cell stage of cleavage. Mesoderm is derived from both endoderm (via the 4d cell) and ectoderm in most taxa with spiral cleavage. If some or all of these characters are ancestral, then there must be character loss in several taxa. For example, spiral cleavage does not occur in rotifers, gastrotrichs, bryozoans, phoronids, or brachiopods; nor does mesoderm come from the 4d cell in most of these taxa. The likelihood that these characters were lost over evolutionary time, as opposed to never having been present, cannot be determined until there is a good understanding of the branching pattern within Lophotrochozoa. Phylogenetic analysis has proved challenging because the results vary depending upon which genes are examined. The next few years are likely to yield resolution of this problem as large multigene datasets are developed for most or all taxa. Several recent studies support a lophophorate clade.

Placement of Nemertea is a contentious issue. Nemerteans were once allied with Platyhelminthes because they share a ciliated epidermis and flame-cell excretory structures, but the presence of a complete digestive tract and an eversible proboscis in a unique coelomic cavity in nemerteans makes a close relationship unlikely. However, the position of the coelomic cavity above the digestive tract, rather than surrounding it, sets nemerteans apart from other coelomate animals.

Adaptive Diversification

The boxlike modular unit of ectoprocts has proved very flexible. Zooids bud to form colonies, but a colony may be encrusting or tree-like, to name just two extreme forms. A colony may be soft or calcified, and the shape of the zooids varies greatly, as does the surface ornamentation. Ectoprocts have colonized freshwater habitats, although they are far less widespread in freshwater as compared to marine habitats.

SUMMARY

Cycliophorans are very tiny animals living on the setae of mouthparts of lobsters. They have complex life cycles with sexual and asexual phases.

Entoprocts are small, sessile aquatic animals with a cup-shaped body on a small stalk. They have a crown of ciliated feeding tentacles encircling both the mouth and the anus.

Ectoprocta, Brachiopoda, and Phoronida all bear a lophophore, which is a crown of ciliated tentacles surrounding the mouth but not the anus and containing an extension of the mesocoel. They are sessile as adults, have a U-shaped digestive tract, and have free-swimming larvae. The lophophore functions as both a respiratory and a feeding structure, its cilia creating water currents from which food particles are filtered.

Ectoprocts are abundant in marine habitats, living on a variety of submerged substrata, and a number of species are common in freshwater. Ectoprocts are colonial, and although each individual is quite small,

colonies are commonly several centimeters or more in width or height. Each individual lives in a chamber (zoecium), which is a secreted exoskeleton of chitinous, calcium carbonate, or gelatinous material.

Brachiopods were very abundant in the Paleozoic era but have been declining in numbers and species since the early Mesozoic era. Their bodies and lophophores are covered by a mantle, which secretes a dorsal and a ventral valve (shell). They are usually attached to the substrate directly or by means of a pedicel.

Phoronida are the least common lophophorates, living in tubes mostly in shallow coastal waters. They thrust the lophophore out of the tube for feeding.

REVIEW QUESTIONS

1. Where would you look if you had to find a cycliophoran?
2. How does an entoproct differ from an ectoproct?
3. What are distinguishing characteristics of entoprocts?
4. Examine Figure 15.1. How many times did the lophophore evolve and how does the lophophore differ among taxa?
5. Describe how an ectoproct feeds.
6. Define each of the following: lophophore, zoecium, zooid, polypide, cystid, statoblasts.
7. How are the valves of a brachiopod oriented in terms of the dorsalventral axis?
8. Brachiopods superficially resemble bivalve molluscs, such as clams. How would you explain the differences in symmetry and internal structure to a layperson?
9. Where would you find phoronids in marine ecosystems? What role do they play in the food web?
10. What factors support the placement of phoronids and brachiopods as sister taxa?
11. What morphological evidence suggests that nemerteans are typical coelomates?
12. How does a nemertean catch and consume prey?
13. How are animals as delicate as nemerteans able to function as predators?

For Further Thought How would you advocate for more research on the animals discussed in this chapter? Consider economics, practical use, aesthetics, and the role of a diverse and healthy ecosystem in your answer.

SELECTED REFERENCES

Brusca, R. C., and G. J. Brusca. 2003. Invertebrates, ed. 2. Sunderland, Massachusetts, Sinauer Associates, Inc. *A comprehensive invertebrate text*

Cohen, B. L., and A. Weydmann. 2005. Molecular evidence that phoronids are a subtaxon of brachiopods (Brachiopoda: Phoronata) and that genetic divergence of metazoan phyla began long before the early Cambrian. Org. Divers. Evol. **5:** 253–273. *Indicates that phoronids arose within Brachiopoda and should no longer be considered a phylum.*

Cook, E. J., J. Stehlíková, C. M. Beveridge, M. T. Burrows, H. De Blauwe, and M. Faasse. 2013. Distribution of the invasive bryozoan *Tricellaria inopinata* in Scotland and a review of its European expansion. Aquatic Invasions **8:** 281–288. *Spread of this species is described along with its impacts on native species.*

Funch, P., and R. M. Kristensen. 1995. Cycliophora is a new phylum with affinities to Entoprocta and Ectoprocta. Nature **378:** 711–714. *The first description of Symbion pandora.*

Gittenberger, A., and C. Schipper. 2008. Long live Linnaeus, *Lineus longissimus* (Gunnerus, 1770) (Vermes, Nemertea: Anopla: Heteronemertea: Lineidae), the longest animal worldwide and its relatives occurring in The Netherlands. Zool. Med. Leiden **82:** 59–63. *Lovely photographs of nemerteans in nature.*

Halanych, K. M., J. D. Bacheller, A. M. A. Aguinaldo, S. M. Liva, D. M. Hillis, and J. A. Lake. 1995. Evidence from 18S ribosomal DNA that lophophorates are protostome animals. Science **267:** 1641–1643. *Despite much morphological and developmental evidence that lophophorates are deuterostomes, they are clustered with annelids and molluscs in this analysis. The authors proposed the group called Lophotrochozoa, defined as the last common ancestor of lophophorate taxa, annelids, and molluscs, and all descendants of that ancestor.*

Hejnol, A. 2010. A twist in time—the evolution of spiral cleavage in the light of animal phylogeny. Integ. Comp. Biol. **50:** 695–706. *Interesting review article on the extent to which spiralian taxa are lophotrochozoan taxa.*

Helfenbein, K. G., and J. L. Boore. 2004. The mitochondrial genome of *Phoronis architecta*—Comparisons demonstrate that phoronids are Lophotrochozoan protostomes. Mol. Biol. Evol. **21:** 153–157. *Analysis of the mitochondrial DNA sequence shows a gene arrangement very similar to that of a chiton.*

Lambert, J. D. 2010. Developmental patterns in spiralian embryos. Current Biology **20:** R72–R77. *A discussion of the spiralian characters as ancestral features for Lophotrochozoa.*

Mayer, A. M. S., and K. R. Gustafson. 2003. Marine pharmacology in 2000: antitumor and cytotoxic compounds. Int. J. Cancer **105:** 291–299. *A concise summary of results from marine natural products chemistry.*

Neilsen, C. 2002. The phylogenetic position of Entoprocta, Ectoprocta, Phoronida, and Brachiopoda. Integ. and Comp. Biol. **42:** 685–691. *Presents evidence that lophophorates do not form a monophyletic group and that phoronids and brachiopods are deuterostomes.*

Nesnidal, M. P., M. Helmkampf, I. Bruchhaus, I. Ebersberger, and B. Hausdorf. 2014. Lophophorata monophyletic—after all. In J. W. Wägele and T. Bartolomaeus (eds.), Deep metazoan phylogeny: backbone of the Tree of Life: new insights from analyses of molecules, morphology, and theory of data analysis. Berlin, Germany, Walter de Gruyter. pp.127–142. *New data to support a return to the old idea that the lophophore evolved only once.*

Podsiadlowski, L., A. Braband, T. H. Struck, J. von Doehren, and T. Bartolomaeus. 2009. Phylogeny and mitochondrial gene order in Lophotrochozoa in the light of new mitogenomic data from Nemertea. BMC Genomics **10:** 364–377. *A subset of lophotrochozoan taxa often emerges as a clade, but the relative relationships of phyla change depending upon the study; mitochondrial gene order data are not helpful.*

Turbeville, J. M. 2002. Progress in nemertean biology: development and phylogeny. Integ. and Comp. Biol. **42:** 692–703. *A discussion of nemertean features that suggest shared ancestry with flatworms and other lophotrochozoan phyla.*

16
CHAPTER

Molluscs

- PHYLUM MOLLUSCA

Fluted giant clam, Tridacna maxima.

A Shell-Collector's Delight

Molluscs are astonishingly diverse. The group includes wormlike animals and giant squids, as well as animals with a single shell, two shells, eight shell plates, and no shell at all. Some have wondered whether the molluscs really form a clade, but there are several features present in almost all molluscs that suggest shared ancestry. Most molluscs have an unusual ribbon of teeth called a radula that they use to feed, often scraping algae from hard surfaces. Most molluscs have a large muscular foot used in locomotion and a unique tissue layer called a mantle. The mantle secretes the shell and makes the respiratory and sensory organs, among other things. These common molluscan features are employed in very different ways across the eight classes of molluscs: for example, the radular teeth are used to inject a paralyzing venom in some species, whereas another group has lost the radula, along with its head. Snails crawl on the foot, whereas octopi divide it into muscular, prehensile arms.

Humans have exploited this diversity in many ways. Shells were used as money on almost every continent—the money cowry is a snail whose shell was widely used as currency. There is debate as to whether the first use of shells was as currency or adornment, but shells are still used in jewelry, buttons, and decoration throughout the world. Pearls are another important product of bivalves. By far the most common use of molluscs is as food. We eat clams, oysters, scallops, mussels, snails, abalone, squids, and octopi, to name just a few of the molluscs exploited commercially. Sometimes we eat the mantle, sometimes the foot, and other times the entire body. Our dependence on these animals behooves us to be good stewards of the environment for molluscs by maintaining unpolluted riverine, coastal, and oceanic habitats and by developing sustainable harvest practices.

16.1 MOLLUSCS

Mollusca (mol-lus′ka) (L. *molluscus,* soft) is one of the largest animal phyla after Arthropoda. There are over 90,000 living species and some 70,000 fossil species. Molluscs are coelomate lophotrochozoan protostomes, and as such they develop via spiral mosaic cleavage and make a coelom by schizocoely. The ancestral larval stage is a trochophore, but development is variously modified within the classes.

The name Mollusca indicates one of their distinctive characteristics, a soft body. This very diverse group (Figure 16.1) includes chitons, tusk shells, snails, slugs, nudibranchs, sea butterflies, clams, mussels, oysters, squids, octopuses, and nautiluses. The group ranges from fairly simple organisms to some of the most complex of invertebrates; sizes range from almost microscopic to the giant squid *Architeuthis.* These huge molluscs may grow to nearly 20 m long, including their tentacles. They may weigh up to 900 kg (1980 pounds). The shells of some giant clams, *Tridacna gigas,* which inhabit Indo-Pacific coral reefs, reach 1.5 m in length and weigh more than 250 kg. These are extremes, however, for probably 80% of all molluscs are less than 10 cm in maximum shell size. The phylum includes some of the most sluggish and some of the swiftest and most active invertebrates. It includes herbivorous grazers, predaceous carnivores, filter feeders, detritus feeders, and parasites.

Molluscs occupy a great range of habitats, from the tropics to polar seas, at altitudes exceeding 7000 m, in ponds, lakes, and streams, on mud flats, in pounding surf, and in open ocean from the surface to abyssal depths. They represent a variety of lifestyles, including bottom feeders, burrowers, borers, and pelagic forms.

According to fossil evidence, molluscs originated in the sea, and most of them have remained there. Much of their evolution occurred along the shores, where food was abundant and habitats were varied. Only bivalves and gastropods moved into brackish and freshwater habitats. As filter feeders, bivalves were unable to leave aquatic surroundings. Only slugs and snails (gastropods) actually invaded the land. Terrestrial snails are limited in their range by their need for humidity, shelter, and presence of calcium in the soil.

Molluscs are an extremely important food source for people around the world; in 2006, 450,687,000 tonnes of molluscs were harvested commercially on the eastern seaboard of the US, with the western seaboard and the Gulf of Mexico adding 3,870,000 and 44,537,000 tonnes, respectively. The value of the eastern seaboard catch alone was nearly $6.28 billion dollars[1], so the economic importance of molluscs cannot be overstated. A healthy mollusc "fishery" depends on healthy oceans, but recent threats to mollusc populations come from an unexpected direction: ocean acidification.

Increased amounts of CO_2 in the atmosphere initiate a set of chemical reactions in the oceans that lower pH. As the ocean becomes more acidic, levels of biologically available calcium decline, making it more difficult for marine organisms to form calcium skeletons. Calcium is essential for the foundation beneath living coral tissue (see p. 280), and is critical to healthy mollusc shells. Clams, oysters, mussels, and scallops produce thinner and weaker shells when reared at low pH. Similarly, larval abalone fail to develop normally, and many cannot create normal shells. Larval survivorship for oysters is reduced when ocean acidity increases, so the oyster industry in the Pacific Northwest is concerned about its

Figure 16.1 Molluscs: a diversity of lifeforms. The basic body plan of this ancient group has become variously adapted for different habitats. **A,** A chiton (*Tonicella lineata*), class Polyplacophora. **B,** A marine snail (*Calliostoma*), class Gastropoda. **C,** A nudibranch (*Chromodoris* sp.), class Gastropoda. **D,** Geoduck clams from Puget Sound, Washington, extend their large siphons, class Bivalvia. **E,** An octopus (*Octopus briareus*), class Cephalopoda, forages at night on a Caribbean coral reef.

© Kjell Sandved/Science Source
© Carver Mostardi/Alamy Stock Photo
© Larry S. Roberts
© Doug Willson/Alamy Stock Photo
© Larry S. Roberts

[1]*Source:* http://www.seaaroundus.org

$278 million dollar revenue (2009 data). One Oregon oyster farm adds calcium chloride and sodium carbonate to seawater to increase the amount of biologically available calcium for young animals. The one bright spot in the future of mollusc harvesting lies with cephalopods such as squid and octopus. They appear to be much less affected by ocean acidification than their kin, presumably because they do not have shells.

Research has shown direct adverse effects of ocean acidification on coral settlement and growth, and on the neurological development of larval tropical reef fishes, but indirect damage to the coral reef ecosystem is equally worrying. The complex food webs (see p. 280) present on the reef will change with the distribution and abundance of corals and coralline algae. If acidification leads to increased abundance of algal turfs and sea grass beds, community changes may act on those molluscs and other taxa not directly impacted by the acidification of our oceans.

In this chapter we explore the various major groups of molluscs (Figure 16.2), including those groups having limited diversity (classes

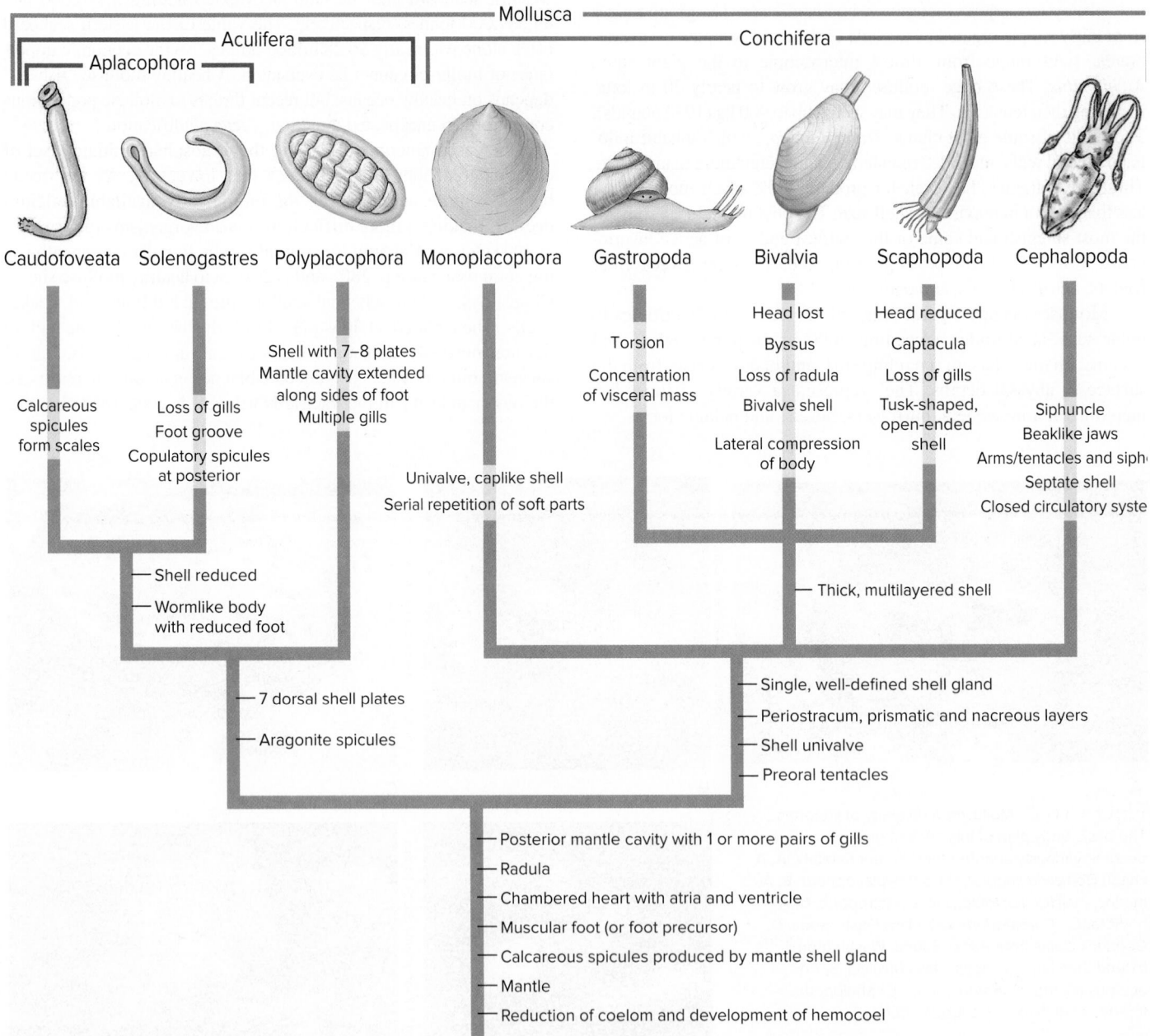

Figure 16.2 Cladogram showing hypothetical relationships among classes of Mollusca. Synapomorphies that identify the various clades are shown, although a number of these have been modified or lost in some descendants. For example, the univalve shell (as well as shell coiling) has been reduced or lost in many gastropods and cephalopods, and many gastropods have undergone detorsion. The bivalve shell of the Bivalvia was derived from an ancestral univalve shell. The byssus is not present in most adult bivalves but functions in larval attachment in many; therefore the byssus is considered a synapomorphy of Bivalvia.

Caudofoveata, Solenogastres, Monoplacophora, and Scaphopoda). Members of class Polyplacophora (chitons) are common to abundant marine animals, especially in the intertidal zone. Bivalves (class Bivalvia) have evolved many species, both marine and freshwater. Class Cephalopoda (squids, cuttlefish, octopuses, and their kin) contains the largest and most intelligent of all invertebrates. Most abundant and widespread of molluscs, however, are snails and their relatives (class Gastropoda). Although enormously diverse, molluscs have in common a basic body plan (pp. 335–337). The coelom in molluscs is limited to a space around the heart, and perhaps around the gonads and part of the kidneys. Although it develops embryonically in a manner similar to the coelom of annelids (see Chapter 17), the functional consequences of this space are quite different because it is not used in locomotion.

16.2 FORM AND FUNCTION

The enormous variety, great beauty, and easy availability of shells of molluscs have made shell collecting a popular pastime. However, many amateur shell collectors, even though able to name hundreds of the shells that grace our beaches, know very little about the living animals that created those shells and once occupied them. Reduced to its simplest dimensions, the mollusc body plan has a **head-foot** portion and a **visceral mass** portion (Figure 16.3). The head-foot is the more active area, containing the feeding, cephalic sensory, and locomotor organs. It depends primarily on muscular action for its function. The visceral mass is the portion containing digestive, circulatory, respiratory, and reproductive organs, and it depends primarily on ciliary tracts for its functioning. Two folds of skin, outgrowths of the dorsal body wall, form a protective **mantle,** which encloses a space between the mantle and body wall called the **mantle cavity.** The mantle cavity houses **gills (ctenidia)** or a lung, and in some molluscs the mantle secretes a protective **shell** over the visceral mass. Modifications of the structures that form the head-foot and the visceral mass produce the great diversity of patterns observed in Mollusca. Greater emphasis on either the headfoot portion or the visceral mass portion can be observed in various classes of molluscs.

Head-Foot

Most molluscs have well-developed heads, which bear their mouth and some specialized structures such as sensory tentacles and photoreceptors. Photoreceptors may be simple, but some, such as the eyes of cephalopods, are quite complex. Within the mouth is a structure unique to molluscs, the radula, and usually posterior to the mouth is the chief locomotor organ, or foot.

Radula

The radula is a rasping, protrusible, tonguelike organ found in all molluscs except bivalves and most

---

Characteristics of Phylum Mollusca

1. Dorsal body wall forms pair of folds called the **mantle,** which encloses the **mantle cavity,** is modified into **gills** or **lungs,** and secretes the **shell** (shell absent in some); ventral body wall specialized as a muscular **foot,** variously modified but used chiefly for locomotion; radula in mouth
2. Live in marine, freshwater, and terrestrial habitats
3. Free-living or occasionally parasitic
4. Body bilaterally symmetrical (bilateral asymmetry in some); unsegmented; often with definite head
5. Triploblastic body
6. **Coelom** limited mainly to area around heart, and perhaps lumen of gonads, part of kidneys, and occasionally part of the intestine
7. Surface epithelium usually ciliated and bearing mucous glands and sensory nerve endings
8. Complex digestive system; rasping organ **(radula)** usually present; anus usually emptying into mantle cavity; internal and external **ciliary tracts** often of great functional importance
9. Circular, diagonal, and longitudinal muscles in the body wall; mantle and foot highly muscular in some classes (for example, cephalopods and gastropods)
10. Nervous system of paired cerebral, pleural, pedal, and visceral ganglia, with nerve cords and subepidermal plexus; ganglia centralized in nerve ring in gastropods and cephalopods
11. Sensory organs of touch, smell, taste, equilibrium, and vision (in some); the highly developed direct **eye** (photosensitive cells in retina face light source) of cephalopods is similar to the indirect eye (photosensitive cells face away from light source) of vertebrates but arises as a skin derivative in contrast to the brain eye of vertebrates
12. No asexual reproduction
13. Both **monoecious** and **dioecious** forms; **spiral cleavage;** ancestral larva a **trochophore,** many with a **veliger** larva, some with direct development
14. One or two kidneys **(metanephridia)** opening into the pericardial cavity and usually emptying into the mantle cavity
15. Gaseous exchange by **gills, lungs, mantle,** or **body surface**
16. **Open circulatory system** (secondarily closed in cephalopods) of heart (usually three chambered), blood vessels, and sinuses; respiratory pigments in blood

Figure 16.3 Generalized mollusc. Although this construct is often presented as a "hypothetical ancestral mollusc (HAM)," most experts now reject this interpretation. Such a diagram is useful, however, to facilitate description of the general body plan of molluscs.

solenogasters. It is a ribbonlike membrane bearing a surface of tiny, backward-pointing teeth used for feeding (Figure 16.4). Complex muscles move the radula and its supporting cartilages (**odontophore**) in and out of the mouth while the membrane is partly rotated over the tips of the cartilages. There may be a few or as many as 250,000 teeth, which, when protruded, can scrape, pierce, tear, or cut. The usual function of the radula is twofold: to rasp fine particles of food material from hard surfaces and to serve as a conveyor belt for carrying particles in a continuous stream toward the digestive tract. As the radula wears away anteriorly, new rows of teeth are continuously replaced by secretion at its posterior end. The pattern and number of teeth in a row are specific for each species and are used in the classification of molluscs. Very interesting radular specializations, such as for boring through hard materials or for harpooning prey, occur in some forms.

Foot

The molluscan foot (Figure 16.3) may be variously adapted for locomotion, for attachment to a substratum, or for a combination of functions. It is usually a ventral, sole-like structure in which waves of muscular contraction effect a creeping locomotion. However, there are many modifications, such as the attachment disc of limpets, the laterally compressed "hatchet foot" of bivalves, or the siphon for jet propulsion in squids and octopuses. Secreted mucus is often used as an aid to adhesion or as a slime tract by small molluscs that glide on cilia.

In snails and bivalves the foot extends from the body hydraulically, by engorgement with blood. Burrowing forms can extend the foot into the mud or sand, enlarge it with blood pressure, then use the engorged foot as an anchor to draw the body forward. In pelagic (free-swimming) forms the foot may be modified into winglike parapodia, or thin, mobile fins for swimming.

Visceral Mass

Mantle and Mantle Cavity

The mantle is a sheath of skin, extending from the visceral mass, that hangs down on each side of the body, protecting the soft parts and creating between itself and the visceral mass a space called the mantle cavity. The outer surface of the mantle secretes the shell.

The mantle cavity (Figure 16.3) plays an enormous role in the life of a mollusc. It usually houses respiratory organs (gills or lung), which develop from the mantle, and the mantle's own exposed surface serves also for gaseous exchange. Products from the digestive, excretory, and reproductive systems are emptied into the mantle cavity. In aquatic molluscs a continuous current of water, kept moving by surface cilia or by muscular pumping, brings in oxygen and, in some forms, food. This same water current also flushes out wastes and carries reproductive products out to the environment. In aquatic forms the mantle is usually equipped with sensory receptors for sampling environmental water. In cephalopods (squids and octopuses) the muscular mantle and its cavity create jet propulsion used in locomotion. Many molluscs can withdraw their head or foot into the mantle cavity, which is surrounded by the shell, for protection.

In the simplest form, a mollusc ctenidium (gill) consists of a long, flattened axis extending from the wall of the mantle cavity (Figure 16.5). Many leaflike gill filaments project from the central axis. Water is propelled by cilia between gill filaments, and blood diffuses from an afferent vessel in the central axis through the filament to an efferent vessel. Direction of blood movement is opposite to the direction of water movement, thus establishing a countercurrent

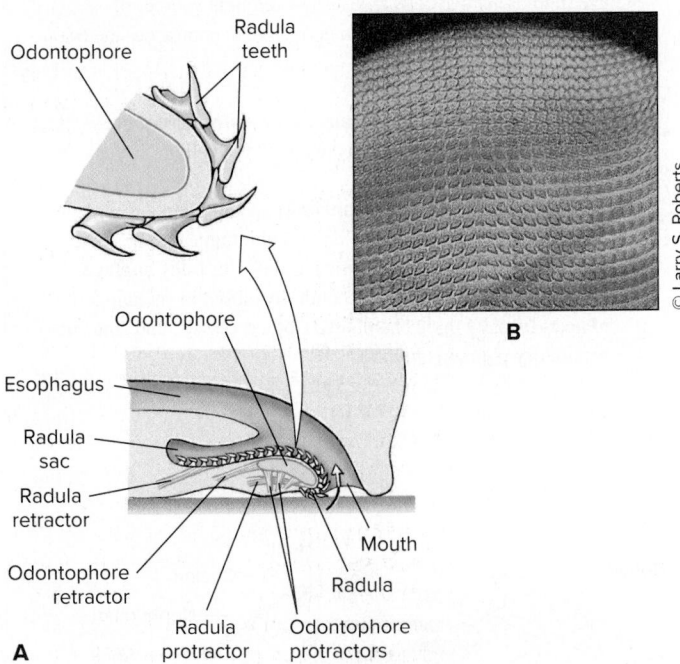

© Larry S. Roberts

Figure 16.4 **A,** Diagrammatic longitudinal section of a gastropod head showing a radula and radula sac. The radula moves back and forth over the odontophore cartilage. As the animal grazes, the mouth opens, the odontophore is thrust forward, the radula gives a strong scrape backward bringing food into the pharynx, and the mouth closes. The sequence is repeated rhythmically. As the radula ribbon wears out anteriorly, it is continually replaced posteriorly. **B,** Radula of a snail prepared for microscopic examination.

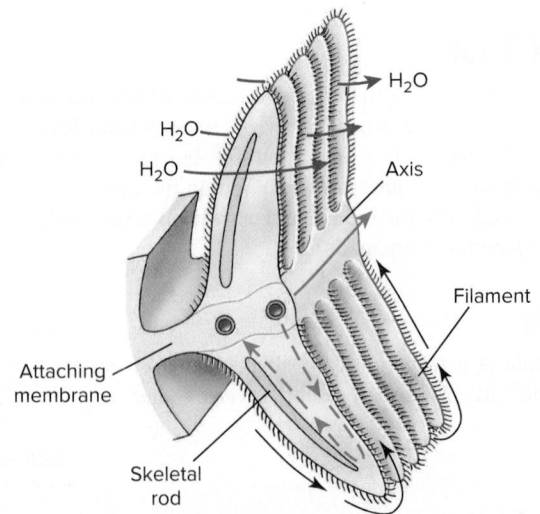

Figure 16.5 Primitive condition of mollusc gill (ctenidium). Circulation of water between gill filaments is by cilia, and blood diffuses through the filament from the afferent vessel to the efferent vessel. Black arrows are ciliary cleansing currents. Red arrows indicate blood flow.

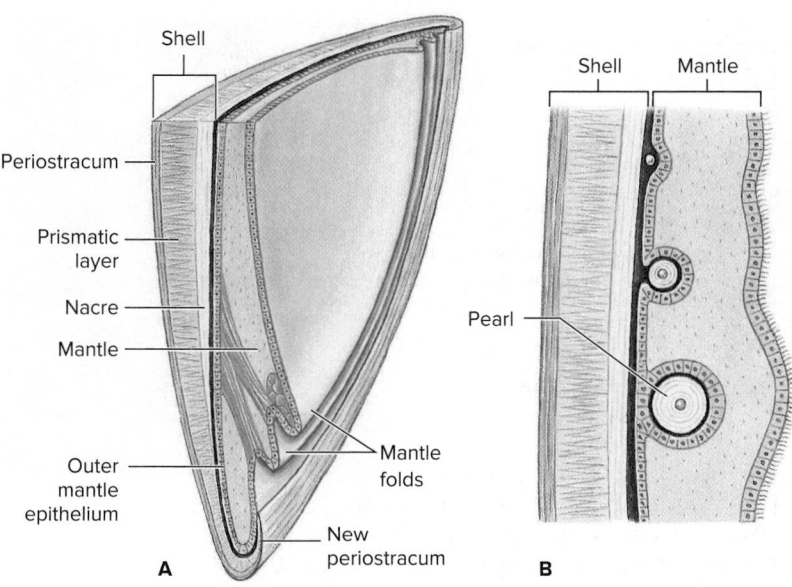

Figure 16.6 A, Diagrammatic vertical section of shell and mantle of a bivalve. The outer mantle epithelium secretes the shell; the inner epithelium is usually ciliated. **B,** Formation of pearl between mantle and shell as a parasite or bit of sand under the mantle becomes covered with nacre.

exchange mechanism (see p. 533). The two ctenidia are located on opposite sides of the mantle cavity and are arranged so that the cavity is functionally divided into an incurrent chamber and an excurrent chamber. The basic arrangement of gills is variously modified in many molluscs.

Shell

The shell of a mollusc, when present, is secreted by the mantle and is lined by it. Typically there are three layers (Figure 16.6A). The **perio-stracum** is the outer organic layer, composed of an organic substance called conchiolin, which consists of quinone-tanned protein. It helps to protect underlying calcareous layers from erosion by boring organisms. It is secreted by a fold of the mantle edge, and growth occurs only at the margin of the shell. On the older parts of the shell, periostracum often becomes worn away. The middle **prismatic layer** is composed of densely packed prisms of calcium carbonate (either aragonite or calcite) laid down in a protein matrix. It is secreted by the glandular margin of the mantle, and increase in shell size occurs at the shell margin as the animal grows. The inner **nacreous layer** of the shell lies next to the mantle and is secreted continuously by the mantle surface, so that it increases in thickness during the life of the animal. The calcareous nacre is laid down in thin layers. Very thin and wavy layers produce the iridescent mother-of-pearl found in abalones (*Haliotis*), chambered nautiluses (*Nautilus*), and many bivalves. Such shells may have 450 to 5000 fine parallel layers of crystalline calcium carbonate for each centimeter of thickness.

There is great variation in shell structure among molluscs. Freshwater molluscs usually have a thick periostracum that gives some protection against acids produced in the water by decay of leaf litter. In many marine molluscs the periostracum is relatively thin, and in some it is absent. Calcium for the shell comes from environmental water or soil or from food. The first shell appears during the larval period and grows continuously throughout life.

Internal Structure and Function

Gas exchange occurs in specialized respiratory organs such as ctenidia, secondary gills and lungs, as well as the body surface, particularly the mantle. There is an **open circulatory system** with a pumping heart, blood vessels, and blood sinuses. In an open circulatory system blood is not entirely contained within blood vessels; rather, it flows through vessels in some parts of the body and enters open sinuses in other parts (see p. 685). An open circulatory system is less efficient at supplying oxygen to all tissues in the body, so it is common in slow-moving animals. Insects are a notable exception, but in these animals oxygen is distributed by the tracheal system, not by the circulatory system. In a closed circulatory system, blood moves to and from tissues within blood vessels. Most cephalopods have a closed circulatory system with vessels and capillaries.

The digestive tract is complex and highly specialized, according to feeding habits of the various molluscs, and is usually equipped with extensive ciliary tracts. Most molluscs have a pair of kidneys (**metanephridia,** a type of nephridium in which the inner end opens into the coelom by a **nephrostome;** see p. 665). Ducts of the kidneys in many forms also serve for discharge of eggs and sperm.

The **nervous system** consists of several pairs of ganglia with connecting nerve cords, and it is generally simpler than that of annelids and arthropods. The nervous system contains neurosecretory cells that, at least in certain air-breathing snails, produce a growth hormone and function in osmoregulation. There are various types of highly specialized sense organs.

Reproduction and Life History

Most molluscs are dioecious, although some are hermaphroditic. The free-swimming **trochophore** larva that emerges from the egg in many molluscs (Figure 16.7) is remarkably similar to that seen in annelids. Direct metamorphosis of a trochophore into a small juvenile, as in chitons, is considered ancestral for molluscs. However, in many molluscan groups (especially gastropods and bivalves) the trochophore stage gives rise to a uniquely molluscan larval stage called a **veliger.** The free-swimming veliger (Figure 16.8) has the beginnings of a foot, shell, and mantle. In many molluscs the trochophore stage occurs in the egg, and a veliger hatches to become the only free-swimming stage. Cephalopods, some freshwater bivalves, and freshwater and some marine snails have no free-swimming larvae; instead, juveniles hatch directly from eggs.

Trochophore larvae (Figure 16.7) are minute, translucent, and more or less top-shaped, and have a prominent circlet of cilia (prototroch) and sometimes one or two accessory circlets. They characterize molluscs and annelids exhibiting the ancestral embryonic development pattern and are usually considered homologous between the two phyla. Some form of trochophorelike larva also characterizes marine turbellarians, nemertines, brachiopods, phoronids, sipunculids, and echiurids, and together with molecular evidence, it supports a taxonomic group called Trochozoa within superphylum Lophotrochozoa.

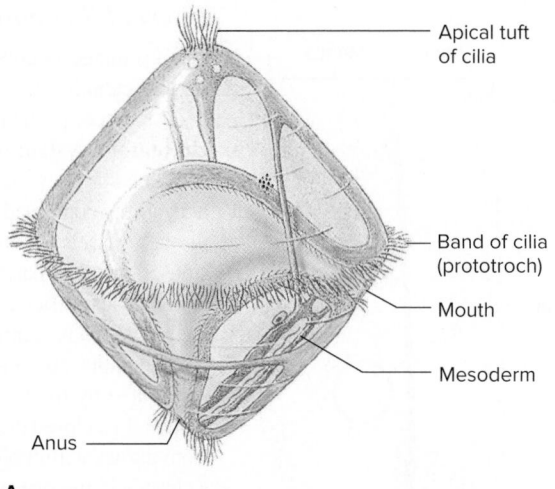

Figure 16.7 **A,** Generalized trochophore larva. Molluscs and annelids with an ancestral pattern of embryonic development have trochophore larvae, as do several other phyla. **B,** Trochophore of a Christmas tree worm, *Spirobranches spinosus* (Annelida).

Apical tuft of cilia

Band of cilia (prototroch)

Mouth

Mesoderm

Anus

A

B

© Dr. Thurston C. Lacalli/University of Victoria

16.3 CLASSES OF MOLLUSCS

For more than 50 years taxonomists recognized five classes of living molluscs: Amphineura, Gastropoda, Scaphopoda, Bivalvia (also called Pelecypoda), and Cephalopoda. Discovery of *Neopilina* in the 1950s added another class (Monoplacophora), and Hyman[2] contended that solenogasters and chitons were separate classes (Aplacophora and Polyplacophora), lapsing the name Amphineura. Subsequently, Aplacophora was divided into the sister groups Caudofoveata and Solenogastres.[3] Members of both groups are wormlike and shell-less with calcareous **sclerites** or spicules in their integument. They have reduced heads and lack nephridia. In spite of these similarities, there are important differences between groups.

Class Caudofoveata

Members of class Caudofoveata comprise about 120 species of wormlike, marine organisms ranging from 2 to 140 mm in length (Figure 16.9). They are mostly burrowers and orient themselves vertically, with the terminal mantle cavity and gills at the entrance of the burrow. They feed primarily on microorganisms and detritus. They possess an oral shield, an organ apparently associated with food selection and intake, as well as a radula. They have one pair of gills. They are dioecious. The body plan of caudofoveates may have more features in common with the ancestor of molluscs than any other living group. This class is sometimes called Chaetodermomorpha.

Class Solenogastres

Solenogasters (Figure 16.9) are a small group of about 250 species of marine animals similar to caudofoveates. Solenogasters, however, usually have no radula and no gills (although

Figure 16.8 Veliger of a snail, *Pedicularia,* swimming. The adults are parasitic on corals. The ciliated process (velum) develops from the prototroch of the trochophore (Figure 16.7).

© Kjell Sandved/Science Source

secondary respiratory structures may be present). Their foot is represented by a midventral, narrow furrow, the pedal groove. They are hermaphroditic. Solenogasters are bottom-dwellers, and often live and feed on cnidarians. This class is sometimes called Neomeniomorpha.

Class Polyplacophora: Chitons

Chitons (Gr. coat of mail, tunic) (Figure 16.10) represent a somewhat more diverse molluscan group with about 1000 currently described species. They are rather flattened dorsoventrally and have a convex dorsal surface that bears seven or eight articulating limy plates, or valves; hence their name Polyplacophora ("many plate bearers"). The plates overlap posteriorly and are usually dull colored to match the rocks to which chitons cling. Their head and cephalic sensory organs are

---

[2]Hyman, L. H. 1967. The Invertebrates, vol. VI. New York, McGraw-Hill Book Company.

[3]Hyman, L. H. 1967. The Invertebrates, vol. VI. New York, McGraw-Hill Book Company.

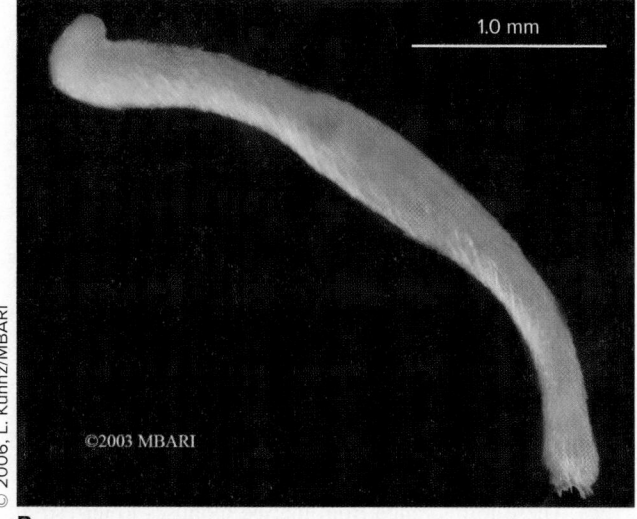

Figure 16.9 Spicules are clearly visible on the skin of solenogasters and caudofoveates. **A,** *Neomeniomorpha,* a solenogaster. **B,** *Chaetoderma elegans,* a caudofoveate.

Figure 16.10 Mossy chiton, *Mopalia muscosa.* The upper surface of the mantle, or "girdle," is covered with hairs and bristles, an adaptation for defense.

reduced, but photosensitive structures **(esthetes),** which have the form of eyes in some chitons, pierce the plates.

Most chitons are small (2 to 5 cm); the largest, *Cryptochiton* (Gr. *crypto,* hidden, + *chiton,* coat of mail), rarely exceeds 30 cm. They prefer rocky surfaces in intertidal regions, although some live at great depths. Most chitons are stay-at-home organisms, straying only very short distances for feeding. Most feed by projecting the radula outward from the mouth to scrape algae from rocks. Scraping is aided by radular teeth reinforced with the iron-containing mineral, magnetite. However, the chiton *Placiphorella velata* is an unusual predatory species that uses a specialized head flap to capture small invertebrate prey. A chiton clings tenaciously to its rock with its broad, flat foot. If detached, it can roll up like an armadillo for protection.

The mantle forms a girdle around the margin of the plates, and in some species mantle folds cover part or all of the plates. Compared with other molluscan classes, the mantle cavity of polyplacophorans is extended along the side of the foot, and the gills are more numerous. The gills are suspended from the roof of the mantle cavity along each side of the broad ventral foot (Figure 16.11). With the foot and the mantle margin adhering tightly to the substrate, these grooves become closed chambers, open only at the ends. Water enters the grooves anteriorly, flows across the gills bringing a continuous supply of oxygen, and leaves posteriorly. At low tide the margins of the mantle can be tightly pressed to the substratum to diminish water

loss, but in some circumstances, the mantle margins can be held open for limited air breathing. A pair of **osphradia** (chemoreceptive sense organs for sampling water) occupy the mantle grooves near the anus of many chitons.

Blood pumped by the three-chambered heart reaches the gills by way of an aorta and sinuses. A pair of kidneys (metanephridia) carries waste from the pericardial cavity to the exterior. Two pairs of longitudinal nerve cords connect in the buccal region.

Sexes are separate in most chitons, and trochophore larvae metamorphose directly into juveniles, without an intervening veliger stage.

Class Monoplacophora

Monoplacophora were long thought to be extinct; they were known only from Paleozoic shells. However, in 1952 living specimens of *Neopilina* (Gr. *neo,* new, + *pilos,* felt cap) were dredged up from the seafloor near the west coast of Costa Rica. About 25 species of monoplacophorans are now known. These molluscs are small and have a low, rounded shell and a creeping foot (Figure 16.12). The mouth bears a characteristic radula. They superficially resemble limpets, but unlike most other molluscs, have some serially repeated organs. These animals have three to six pairs of gills, two pairs of heart atria, three to seven pairs of metanephridia, one or two pairs of gonads, and a ladderlike nervous system with 10 pairs of pedal nerves. Such serial repetition occurs to a more limited extent in chitons. Why should there be repeated sets of body structures in these animals? Body structures repeat in each segment of an annelid worm (see p. 363). Are the repeated structures indications that molluscs had a segmented (metameric) ancestor? Most current research indicates that *Neopilina* shows pseudometamerism, and that molluscs did not have a metameric ancestor. The Serialia hypothesis (see p. 358) proposes monoplacophorans and polyplacophorans as sister taxa, where serially repeated structures evolved in the ancestor of these two groups.

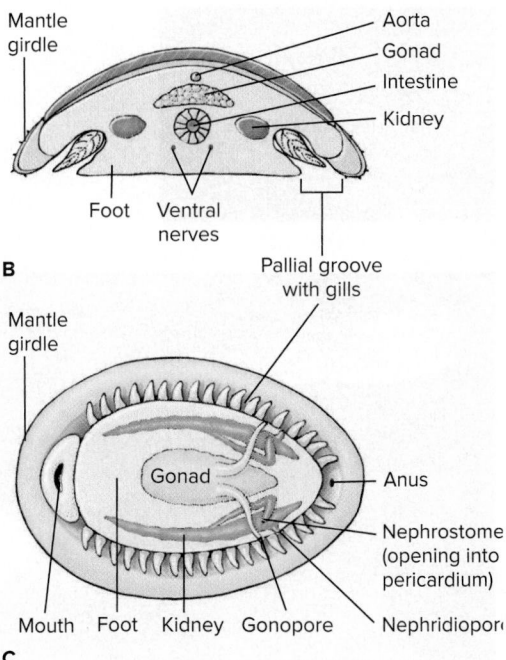

Figure 16.11   Anatomy of a chiton (class Polyplacophora).
A, Longitudinal section. **B,** Transverse section. **C,** External ventral view.

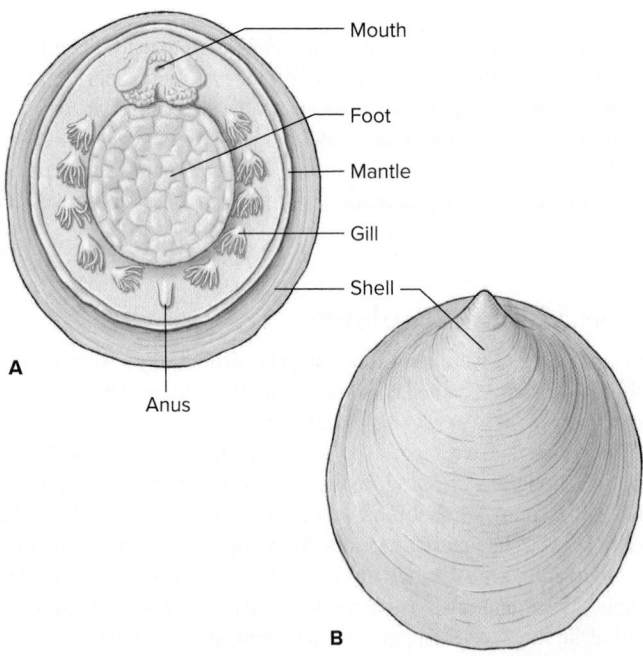

Figure 16.12   *Neopilina,* class Monoplacophora. Living specimens range from 3 mm to about 3 cm in length. **A,** Ventral view. **B,** Dorsal view.

Class Gastropoda

Among molluscs, class Gastropoda is by far the largest and most diverse, containing over 70,000 living and more than 15,000 fossil species. It contains so much diversity that there is no single general term in our language that can apply to it. It contains snails, limpets, slugs, whelks, conchs, periwinkles, sea slugs, sea hares, and sea butterflies. These forms range from marine molluscs to terrestrial, air-breathing snails and slugs. Gastropods are usually sluggish, sedentary animals because most of them have heavy shells and slow locomotion. Some are specialized for climbing, swimming, or burrowing. Shells are their chief defense.

The shell, when present, is always of one piece (**univalve**) and may be coiled or uncoiled. Starting at the **apex,** which contains the oldest and smallest **whorl,** the whorls become successively larger and spiral around the central axis, or **columella** (Figure 16.13). The shell may be right handed (**dextral**) or left handed (**sinistral**), depending on the direction of coiling. Direction of coiling is genetically controlled and dextral shells are far more common. Many snails have an **operculum,** a plate made of tanned protein that covers the shell **aperture** when the body is withdrawn into the shell.

Gastropods range from microscopic forms to giant marine forms such as *Pleuroploca gigantea,* a snail with a shell up to 60 cm long, and sea hares, *Aplysia* (see Figure 16.22), some species of which reach 1 m in length. Most gastropods, however, are between 1 and 8 cm in length. Some fossil species are as much as 2 m long.

The range of gastropod habitats is large. Marine gastropods are common both in littoral zones and at great depths, and some are even pelagic. Some are adapted to brackish water and others to fresh water. On land they are restricted by such factors as mineral content of the soil and extremes of temperature, dryness, and acidity. Even so, they are widespread; some occur at great altitudes and some even in polar regions. Snails occupy all kinds of habitats: in small pools or large bodies of water, in woodlands, in pastures, under rocks, in mosses, on cliffs, in trees, underground, and on the bodies of other animals. They have successfully undertaken every mode of life except aerial locomotion.

Gastropods may be protected by shells, by distasteful or toxic secretions, and/or by secretive habits. Some species are even capable of deploying the stinging cells of their cnidarian prey for their own defense.

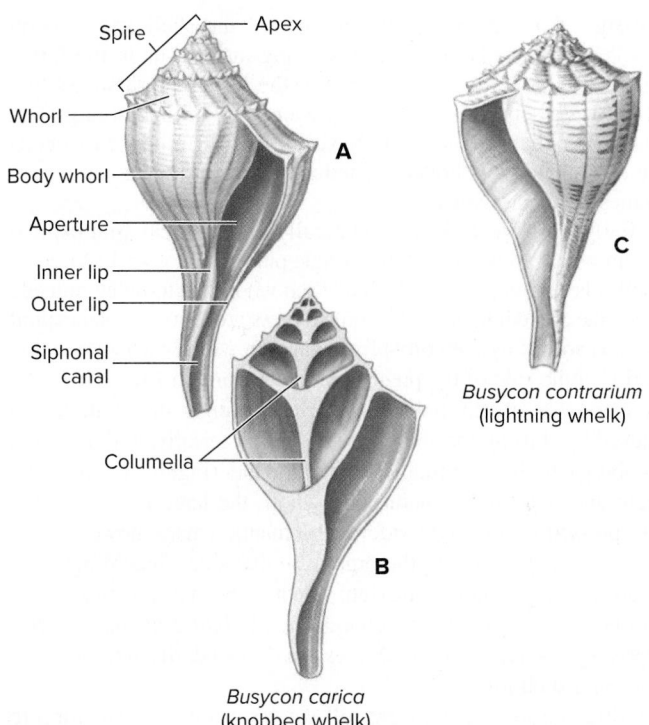

Figure 16.13   Shell of the whelk *Busycon*. **A** and **B,** *Busycon carica,* a dextral, or right-handed, shell. A dextral shell has the aperture on the right side when the shell is held with the apex up and the aperture facing the observer. **C,** *B. contrarium,* a sinistral, or left-handed, shell.

A few, such as *Strombus,* can deal an active blow with their foot, which bears a sharp operculum. Nevertheless, they are eaten by birds, beetles, small mammals, fish, and other predators. Serving as intermediate hosts for many kinds of parasites, especially trematodes (p. 299), snails are often harmed by larval stages of parasites. There are three gastropod subclasses: Prosobranchia, Opisthobranchia, and Pulmonata. They are described on pp. 345–348, but subclass names are commonly used to discuss particular animals. Familiar prosobranchs include periwinkles, limpets, whelks, conchs, abalones, and cowries. Marine sea slugs, sea hares, and nudibranchs are often called opisthobranchs. Pulmonates include most land and freshwater snails and slugs.

Form and Function

Torsion   Gastropod development varies with the particular group under discussion, but in general there is a trochophore larval stage followed by a veliger larval stage where the shell first forms. The veliger has two ciliated velar lobes, used in swimming, and the developing foot is visible (Figure 16.14). The mouth is anterior and the anus is posterior initially, but the relative positions of the shell, the digestive tract and anus, the nerves that lie along both sides of the digestive tract, and the mantle cavity containing the gills all change in a process called torsion.

Torsion is usually described as a two-step process. In the first step, an asymmetrical foot retractor muscle running along the side of the gut and body organs contracts and pulls the shell and enclosed viscera (containing organs of the body) 90 degrees counterclockwise, relative to the head. This movement brings the anus from the posterior to the right side

ONTOGENETIC TORSION

Lateral View

Shell   Mouth

Velar lobe

Foot

Anus

Anus
Mantle cavity now anterior

Dorsal View

Mouth

Velar lobe   Velar lobe

Anus

Developing mantle cavity

In **veliger** larva, anus is posterior; no mantle cavity is visible

Shell rotates to adult position; viscera rotates 90°; mantle cavity visible

Anus and mantle continue to rotate; mantle cavity expands

Anus rotated 180°; mantle epithelium reorganizes to enlarge cavity

Adult snail

Figure 16.14   Ontogenetic torsion in a gastropod veliger larva.

of the body (Figure 16.14). Typical descriptions state that movement of the shell accompanies visceral movement, but detailed studies show that movement of the shell is independent of visceral movements. The first movements of the shell rotate it between 90 and 180 degrees into a position that will persist into adulthood. It was previously assumed that the mantle cavity, which houses both the gills and the anus in adult animals, moved with the anus in the first 90 degrees of torsion. However, studies show that the mantle cavity develops on the right side of the body near the anus, but is initially separate from it. The anus and mantle cavity usually move farther to the right and the mantle cavity is remodeled to encompass the anus. In a slower and more variable series of changes, the digestive tract moves both laterally and dorsally so that the anus lies above the head within the mantle cavity (Figure 16.14).

After torsion, the anus and mantle cavity open above the mouth and head. The left gill, kidney, and heart atrium are now on the right side, whereas the original right gill, kidney, and heart atrium are now on the left, and the nerve cords form a figure eight. Because of the space available in the mantle cavity, the animal's sensitive head end can now be withdrawn into the protection of the shell, with the tougher foot, and when present the operculum, forming a barrier to the outside.

The developmental sequence just described is called ontogenetic torsion. Evolutionary torsion is the series of changes that produced the modern torted gastropod body from the ancestral untorted form. The hypothetical ancestral gastropod was assumed to have a posterior mantle cavity like the hypothetical ancestral mollusc (see Figure 16.3). It has long been assumed that morphological changes in ontogenetic torsion represent the sequence of evolutionary changes. However, new studies of development in several kinds of gastropods suggest a different scenario; researchers hypothesize that the ancestral gastropod had two lateral mantle cavities, much like those in *Neopilina* (see Figure 16.12) and in chitons (see Figure 16.11). A single mantle cavity over the head may have arisen when the left lateral mantle cavity was lost and the right cavity expanded toward the middle of the body after the first 90 degrees of torsion. Careful study of ontogenetic torsion shows that asynchronous displacements of the shell, the visceral mass and anus, and the mantle cavity are possible, although some features move together in some taxa. Torsion has been reinterpreted as a conserved anatomical stage, where the shell has moved to the adult position and the anus and mantle cavity are on the right side of the body, rather than a conserved process of change.[4]

Varying degrees of **detorsion** occur in opisthobranchs and pulmonates, and the anus opens to the right side or even to the posterior (see Figure 16.22). However, both of these groups were derived from torted ancestors.

The curious arrangement that results from torsion, where the anus opens over the head and gills, creates the possibility of wastes being washed back over the gills (**fouling**) and causes us to wonder what strong evolutionary pressures selected for such a strange realignment of body structures. Several explanations have been proposed, none entirely satisfying. For example, sense organs of the mantle cavity (osphradia) would better sample water when turned in the direction of travel. Certainly the consequences of torsion and the resulting need to avoid fouling have been very important in the subsequent evolution of gastropods. These consequences cannot be explored, however, until we describe another unusual feature of gastropods—coiling.

[4]Page, L. R. 2003. J. Exp. Zool. Part B: **297B:** 11–26.

Coiling   Coiling, or spiral winding, of the shell and visceral mass is not the same as torsion. Coiling may occur in the larval stage at the same time as torsion, but the fossil record shows that coiling was a separate evolutionary event and originated in gastropods earlier than did torsion. Nevertheless, all living gastropods have descended from coiled, torted ancestors, although some have lost these characteristics.

Early gastropods had a bilaterally symmetrical **planospiral** shell, in which all whorls lay in a single plane (Figure 16.15A). Such a shell was not very compact, since each whorl had to lie completely outside the preceding one. The compactness problem of a planospiral shell was solved by the **conispiral** shape, in which each succeeding whorl is at the side of the preceding one (Figure 16.15B). However, this shape was clearly unbalanced. Better weight distribution was achieved by shifting the shell upward and posteriorly, with the shell axis oblique to the longitudinal axis of the foot (Figure 16.15C). The weight and bulk of the main body whorl, the largest whorl of the shell, pressed on the right side of the mantle cavity, however, and apparently interfered with the organs on that side. Accordingly, the gill, atrium, and kidney of the right side have been lost in most living gastropods, leading to a condition of *bilateral asymmetry*. Surprisingly, a few modern species have secondarily returned to a planospiral shell form.

Although loss of the right gill was probably an adaptation to the mechanics of carrying a coiled shell, that condition, displayed

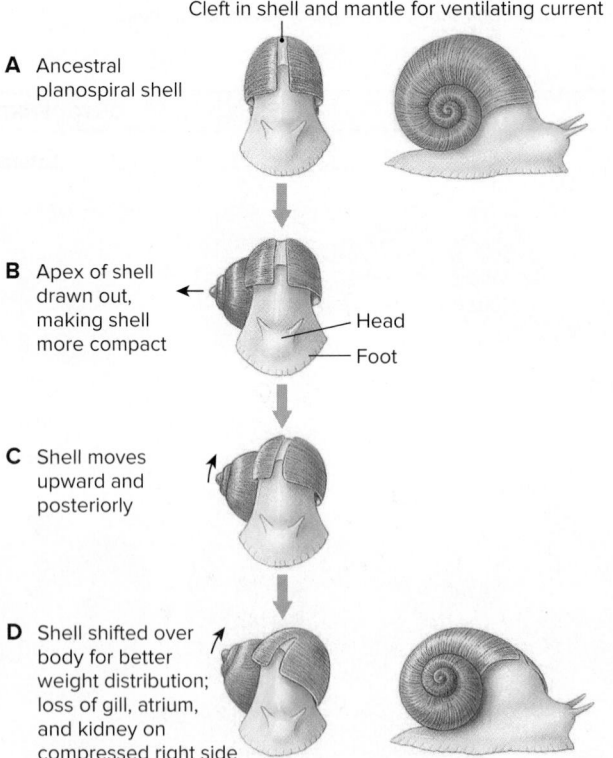

Figure 16.15 Evolution of shell in gastropods. **A,** Earliest coiled shells were planospiral, each whorl lying completely outside the preceding whorl. **B,** Better compactness was achieved by snails in which each whorl lay partially to the side of the preceding whorl. **C** and **D,** Better weight distribution resulted when shell was moved upward and posteriorly. However, some modern forms have reevolved the planospiral form.

A

B

Figure 16.16 **A,** Red abalone, *Haliotis rufescens.* These large limpetlike snails are prized as food and extensively marketed. Abalones are strict vegetarians, feeding especially on sea lettuce and kelp. **B,** Moon snail, *Polinices lewisii.* A common inhabitant of West Coast sand flats, the moon snail is a predator of clams and mussels. It uses its radula to drill neat holes through its victim's shell, through which the proboscis is then extended to eat the bivalve's fleshy body.

in most modern prosobranchs, made possible a way to avoid the fouling problem caused by torsion. Water is brought into the left side of the mantle cavity and out the right side, carrying with it wastes from the anus and nephridiopore, which lie near the right side. Ways in which fouling is avoided in other gastropods are mentioned on pages 345–348.

Feeding Habits   Feeding habits of gastropods are as varied as their shapes and habitats, but all include use of some adaptation of the radula. Most gastropods are herbivorous, rasping particles of algae from hard surfaces. Some herbivores are grazers, some are browsers, and some are planktonic feeders. *Haliotis,* the abalone (Figure 16.16A), holds seaweed with its foot and breaks off pieces with its radula. Land snails forage at night for green vegetation.

Some snails, such as *Bullia* and *Buccinum,* are scavengers living on dead and decaying flesh; others are carnivores that tear their prey with radular teeth. *Melongena* feeds on clams, especially *Tagelus,* the razor clam, thrusting its proboscis between the gaping shell valves. *Fasciolaria* and *Polinices* (Figure 16.16B) feed on a variety of molluscs, preferably bivalves. *Urosalpinx cinerea,* oyster borers, drill holes through the shells of oysters. The radula, bearing three longitudinal rows of teeth, begins the drilling action; then the snail glides forward, everts an accessory boring organ through a pore in the anterior sole of the foot, and holds it against the oyster's shell, using a chemical agent to soften the shell. Short periods of rasping alternate with long periods of chemical activity until a neat round hole is completed. With its proboscis inserted through the hole, a snail may feed continuously for hours or days, using its radula to tear away the soft flesh. *Urosalpinx* is attracted to its prey at some distance by sensing some chemical, probably one released in metabolic wastes of the prey.

Cyphoma gibbosum (see Figure 16.21B) and related species live and feed on gorgonians (phylum Cnidaria, Chapter 13) in shallow, tropical coral reefs. These snails are commonly called flamingo tongues. During normal activity their brightly colored mantle entirely envelops the shell, but it can be quickly withdrawn into the shell aperture when the animal is disturbed.

Members of genus *Conus* (Figure 16.17) feed on fish, worms, and molluscs. Their radula is highly modified for prey capture. A gland charges the radular teeth with a highly toxic venom. When *Conus* senses the presence of its prey, a single radular tooth slides into position at the tip of the proboscis. Upon striking the prey, the proboscis expels a tooth like a harpoon, and the venom immediately paralyzes the prey. This is an effective adaptation for a slowly moving predator to prevent escape of a swiftly moving prey. Some species of *Conus* can deliver very painful stings, and in several species the sting is lethal to humans. Each *Conus* species carries a venom of toxic peptides (**conotoxins**) specific for its preferred prey. Conotoxins are valuable tools in research on the various receptors and ion channels of nerve cells.

Some gastropods, such as the queen conch (*Strombus gigas*), feed on organic deposits on the sand or mud. Others collect the same sort of organic debris but can digest only microorganisms contained in it. Some sessile gastropods, such as some limpets, are ciliary feeders that use gill cilia to draw in particulate matter, roll it into a mucous ball, and carry it to their mouth. Some sea butterflies secrete a mucous net to catch small planktonic forms; then they draw the web into their mouth.

In ciliary feeders the stomachs are sorting regions, and most digestion is intracellular in digestive glands. Other forms macerate food using the radula and sometimes a gizzard, and then digest it extracellularly in the lumen of the stomach or digestive glands.

Internal Form and Function   Respiration in most gastropods is performed by a **ctenidium** (two **gills** is the primitive condition, found in some prosobranchs) located in the mantle cavity, though some aquatic forms lack gills and instead depend on the mantle and skin. After some prosobranchs lost one gill, most of them lost half of the remaining one, and the central axis became attached to the wall of the mantle cavity (Figure 16.18). Thus they attained the most efficient gill arrangement for the way the water circulated through the mantle cavity (in one side and out the other).

Figure 16.17 **A,** *Conus* extends its long, wormlike proboscis. When a fish attempts to consume this tasty morsel, the *Conus* stings it in the mouth and kills it. **B,** The snail engulfs the fish with its distensible stomach then regurgitates the scales and bones some hours later.

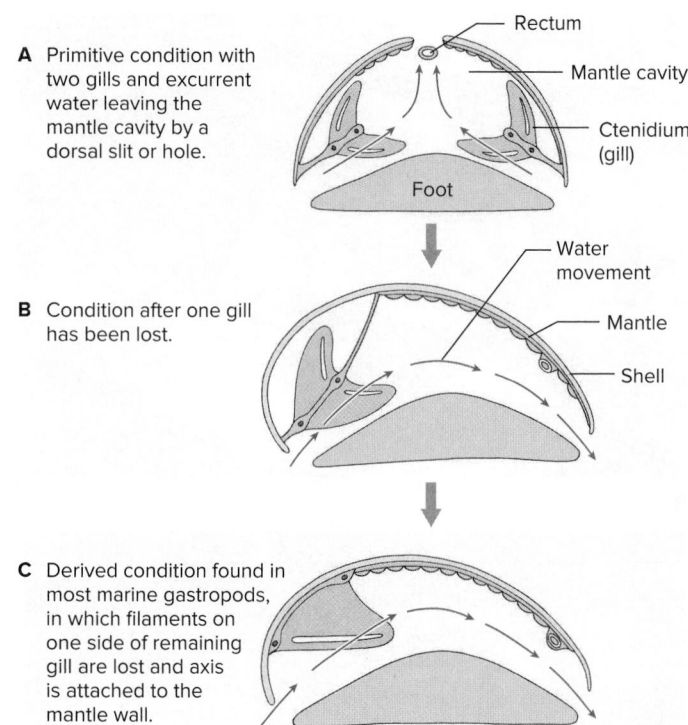

A Primitive condition with two gills and excurrent water leaving the mantle cavity by a dorsal slit or hole.

B Condition after one gill has been lost.

C Derived condition found in most marine gastropods, in which filaments on one side of remaining gill are lost and axis is attached to the mantle wall.

Rectum

Mantle cavity

Ctenidium (gill)

Foot

Water movement

Mantle

Shell

Figure 16.18 Evolution of gills in gastropods. **A,** Primitive condition with two gills and excurrent water leaving the mantle cavity by a dorsal slit or hole. **B,** Condition after one gill had been lost. **C,** Derived condition found in most marine gastropods, in which filaments on one side of remaining gill are lost, and axis is attached to the mantle wall. *Note:* The terms *ctenidium* and *gill* both refer to the respiratory structure, but most biologists limit the use of ctenidium to refer to the greatly expanded gill in bivalves.

Pulmonates lack gills, but have a highly vascular area in their mantle that serves as a **lung** (Figure 16.19). Most of the mantle margin seals to the back of the animal, and the lung opens to the outside by a small opening called a **pneumostome.** The mantle cavity fills with air by contraction of the mantle floor. Many aquatic pulmonates must surface to expel a bubble of gas from their lung. To inhale, they curl the edge of the mantle around the pneumostome to form a siphon.

Most gastropods have a single nephridium (kidney). The circulatory and nervous systems are well developed (Figure 16.19). The latter incorporates three pairs of ganglia connected by nerves. Sense organs include eyes or simple photoreceptors, statocysts, tactile organs, and chemoreceptors. The simplest type of gastropod eye is simply a cuplike indentation in the skin lined with pigmented photoreceptor cells. In many gastropods the eyecup contains a lens covered with a cornea. A sensory area called an **osphradium,** located at the base of the incurrent siphon of most gastropods, is chemosensory in some forms, although its function may be mechanoreceptive in some and remains unknown in others.

There are both dioecious and monoecious gastropods. Many gastropods perform courtship ceremonies. During copulation in monoecious species there is an exchange of spermatozoa or spermatophores (bundles of sperm). Many terrestrial pulmonates eject a dart from a dart sac (Figure 16.19) into the partner's body to heighten excitement before copulation. After copulation each partner deposits its eggs in shallow burrows in the ground. Gastropods with the most primitive reproductive characteristics discharge ova and sperm into seawater where fertilization occurs, and embryos soon hatch as free-swimming trochophore larvae. In most gastropods fertilization is internal.

Fertilized eggs encased in transparent shells may be emitted singly to float among the plankton or may be laid in gelatinous layers attached to a substratum. Some marine forms enclose their eggs, either in small groups or in large numbers, in tough egg capsules, or in a wide variety of egg cases (Figure 16.20). Offspring generally emerge as veliger larvae (see Figure 16.8), or they may spend the veliger stage in the case or capsule and emerge as young snails. Some species, including many freshwater snails, are ovoviviparous, brooding their eggs and young in their oviduct

Major Groups of Gastropod

Traditional taxonomy of class Gastropoda recognizes three subclasses: Prosobranchia, the largest subclass, almost all of which are marine; Opisthobranchia, an assemblage including sea slugs, sea hares, nudibranchs, and canoe shells, all marine; and Pulmonata, containing most freshwater and terrestrial species. Currently, gastropod taxonomy is in flux. Evidence suggests that Prosobranchia is

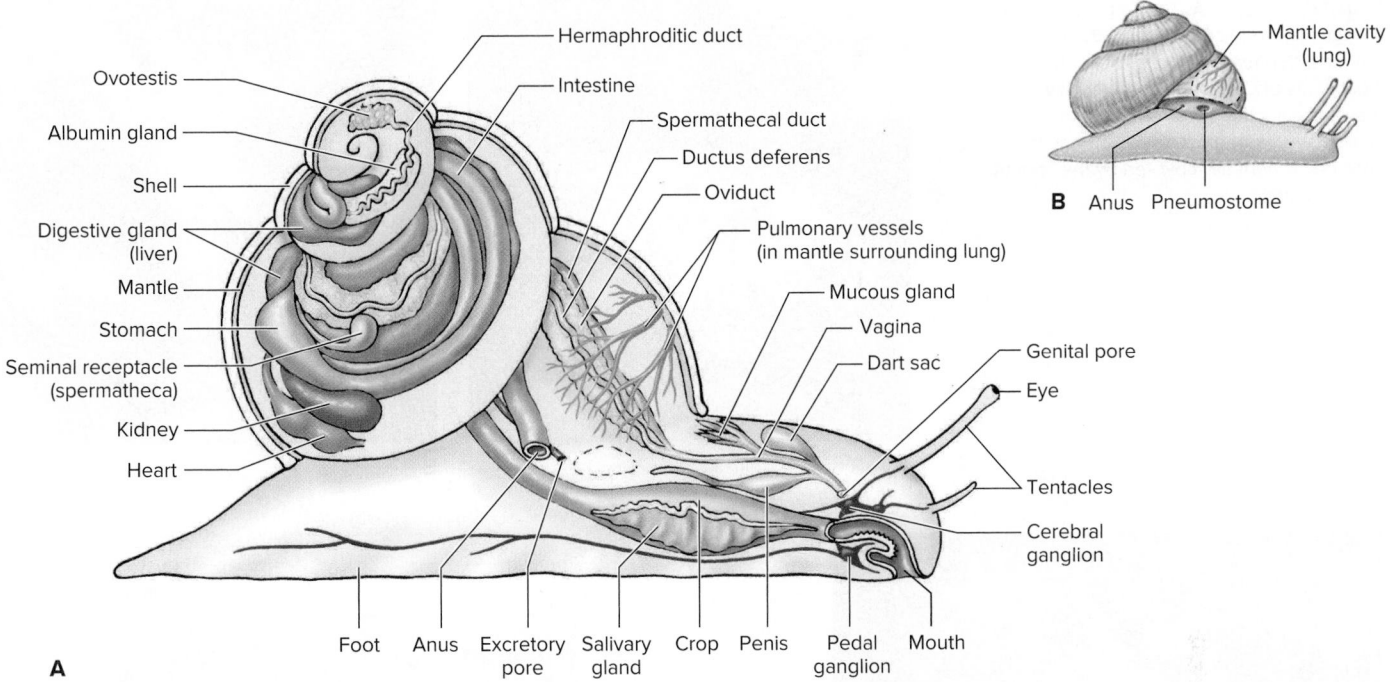

Figure 16.19 **A,** Anatomy of a pulmonate snail. **B,** Position of the mantle cavity serving as a lung. Air enters and exits via the pneumostome.

Figure 16.20 Eggs of marine gastropods. **A,** The dog whelk, *Nucella lapillus*, lays yellow urn-shaped egg cases on barnacles. Each case contains hundreds of embryos. **B,** Egg ribbon of a nudibranch.

paraphyletic. Opisthobranchia may be paraphyletic, but Opisthobranchia and Pulmonata together apparently form a monophyletic grouping. The number of subclasses within the Gastropoda and the relationships among them remain subjects of considerable controversy. For convenience and organization, we continue to use the words "prosobranchs" and "opisthobranchs," recognizing that they may not represent valid taxa.

Prosobranchs   This group contains most marine snails and some freshwater and terrestrial gastropods. The mantle cavity is anterior as a

result of torsion, with the gill or gills lying in front of the heart. Water enters the left side and exits from the right side, and the edge of the mantle often extends into a long siphon to separate incurrent from excurrent flow. In prosobranchs with two gills (for example, the abalone *Haliotis*, Figure 16.16A and keyhole limpet *Diodora*), fouling (see p. 342) is avoided by having the excurrent water go up and out through one or more holes in the shell above the mantle cavity.

Prosobranchs have one pair of tentacles. Sexes are usually separate. An operculum is often present. They range in size from periwinkles and small limpets (*Patella* and *Diodora*) (Figure 16.21A)

Figure 16.21 **A,** The limpet *Patella* has a single flattened caplike shell. **B,** Flamingo tongues, *Cyphoma gibbosum,* are showy inhabitants of Caribbean coral reefs, where they are associated with gorgonians. These snails have a smooth, creamy, orange to pink shell that is normally covered by the brightly marked mantle.

A   B

A   B

Figure 16.22 **A,** A sea hare, *Aplysia dactylomela,* crawls and swims across a coral reef, assisted by large, winglike parapodia, here curled above the body. **B,** When attacked, sea hares squirt a copious protective secretion derived from their red algal food source. Substances from the algae pass from the digestive gland to a "purple gland" where they are modified. Secretions from the purple gland exit the body through the mantle cavity, as do wastes from the anus. The anus and mantle cavity are posterior in sea hares because these animals undergo detorsion during development.

to horse conchs (*Pleuroploca*), which grow shells up to 60 cm in length, making them the largest gastropods in the Atlantic Ocean. Familiar examples of prosobranchs include abalones (*Haliotis*), which have an ear-shaped shell; whelks (*Busycon*), which lay their eggs in double-edged, disc-shaped capsules attached to a cord a meter long; common periwinkles (*Littorina*); moon snails (*Polinices*); oyster borers (*Urosalpinx*), which bore into oysters and suck their juices; rock shells (*Murex*), a European species that was used to make the royal purple of the ancient Romans; and some freshwater forms (*Goniobasis* and *Viviparus*).

Opisthobranchs   Opisthobranchs are an odd assemblage of molluscs that include sea slugs, sea hares, sea butterflies, and bubble shells. They are nearly all marine; most are shallow-water forms, hiding under stones and seaweed; a few are pelagic. Currently nine or more orders of opisthobranchs are recognized. Opisthobranchs show partial or complete detorsion (Figure 16.22); thus, the anus and gill (if present) are displaced to the right side or rear of the body. Clearly, the fouling problem is obviated if the anus is moved away from the head toward the posterior. Two pairs of tentacles are usually found, and the second pair is often further modified (**rhinophores,** Figure 16.22), with platelike folds that apparently increase the area for chemoreception. Their shell is typically reduced or absent. All are monoecious.

Sea hares (*Aplysia,* Figure 16.22) have large, earlike anterior tentacles and vestigial shells. In pteropods or sea butterflies (*Corolla* and *Clione*) the foot is modified into fins for swimming; thus, they are pelagic.

Nudibranchs are carnivorous and often brightly colored (Figure 16.23). Plumed sea slugs (Aeolidae), which feed primarily on sea anemones and hydroids, have elongate papillae (**cerata**) covering their back. They ingest their prey's nematocysts and transport the nematocysts undischarged to the tips of their cerata. There the nematocysts are placed in cnidosacs that open to the outside, and the aeolid can use these high-jacked nematocysts for its own defense. *Hermissenda* is one of the more common West Coast nudibranchs.

Sacoglossan sea slugs are characterized by a radula with a single tooth per row that is used to pierce algal cells, allowing the slug to suck the contents. Similar to their aeolid cousins, some sacoglossans can steal functional organelles from their prey for their own benefit. In fact, many species have evolved special branches of the gut that run throughout the body; photosynthetic plastids from their algal food are directed into these branches rather than being digested, and they continue to function for quite some time. Some carnivorous nudibranchs likewise take advantage of intact zooxanthellae from their cnidarian prey (p. 234). This ability to usurp the photosynthetic machinery of their prey has led to the nickname "solar-powered sea slugs" for some species (for example, *Elysia crispata*).

Pulmonates   Pulmonates include land and most freshwater snails and slugs (and a few brackish and saltwater forms). They have lost their ancestral ctenidia, but their vascularized mantle wall has become a lung, which fills with air by contraction of the mantle floor (some aquatic species have developed secondary gills in the mantle cavity). The anus and nephridiopore open near the pneumostome, and waste is expelled forcibly with air or water from the lung. Pulmonates show some detorsion. They are monoecious. Aquatic species have one pair of nonretractile tentacles, at the base of which are eyes; land forms have two pairs of tentacles, with the posterior pair bearing eyes (Figure 16.24).

Until recently, all terrestrial slugs were presumed to have evolved from terrestrial snail ancestors whose shells prevented

Figure 16.24   **A,** Pulmonate land snail. Note two pairs of tentacles; the second, larger pair bears the eyes. **B,** Banana slug, *Ariolimax columbianus*. Note pneumostome.

Pneumostome

A   **B**

dessication and permitted the colonization of land. These early snail colonists belonged to several different lineages, each independently losing the shell as they adapted to moist habitats. However, research in Palau, Western Pacific, has identified a new terrestrial slug, Aiteng marefugitus, whose intertidal marine ancestors did not have shells. Ancestral intertidal slugs were able to invade the rainforest habitat in Palau because of its high humidity and lack of terrestrial competitors.

Class Bivalvia (Pelecypoda)

Bivalvia are also called Pelecypoda (pel-e-sip'o-da), or "hatchet-footed" animals, as their name implies (Gr. *pelekys*, hatchet, + *pous, podos,* foot). They are bivalved molluscs that include mussels, clams, scallops, oysters, and shipworms (Figures 16.25 to 16.36) and they range in size from tiny seed shells 1 to 2 mm in length to giant South Pacific clams, *Tridacna,* which may reach more than 1 m in length and as much as 225 kg (500 pounds) in weight (see Figure 16.33). Most bivalves are sedentary **filter feeders** that depend on currents produced by cilia on their gills to gather food materials. Unlike gastropods, they have no head, no radula, and very little cephalization.

Most bivalves are marine, but many live in brackish water and in streams, ponds, and lakes.

Freshwater bivalves were once abundant and diverse in streams throughout the eastern United States, but they are now easily the most jeopardized group of animals in the country. Of more than 300 species once present, nearly two dozen are extinct, more than 60 are considered endangered, and as many as 100 more are threatened. Damming and impoundment of rivers are likely the most important threats to this species. Pollution and sedimentation from mining, industry, and agriculture are other important culprits. The cultured pearl industry uses ground shells of freshwater bivalves to stimulate pearl formation in oysters, leading to poaching. In addition, introductions of exotic species make the problem worse. For example, the prolific zebra mussels (see note, p. 350) attach in great numbers to the native clams, exhausting food supplies (phytoplankton) in the surrounding water.

Figure 16.23   *Phyllidia ocellata*, a nudibranch. Like other *Phyllidia* sp. it has a firm body with dense calcareous spicules and bears its gills along the sides, between its mantle and foot.

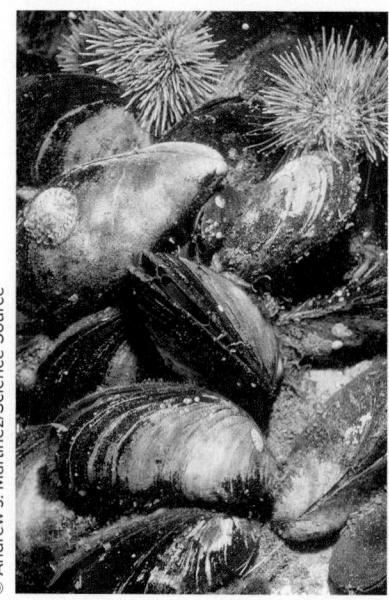

© Andrew J. Martinez/Science Source

A

Figure 16.25 Bivalve molluscs. **A,** Mussels, *Mytilus edulis,* occur in northern oceans around the world; they form dense beds in the intertidal zone. A host of marine creatures live protected beneath attached mussels. **B,** Scallops (*Chlamys opercularis)* swim to escape attack by starfish (*Asterias rubens).* When alarmed, these most agile of bivalves swim by clapping the two shell valves together.

© D.P. Wilson/FLPA/Science Source

B

Form and Function

Shell   Bivalves are laterally compressed, and their two shells **(valves)** are held together dorsally by a hinge ligament that causes the valves to gape ventrally. The valves are drawn together by adductor muscles that work in opposition to the hinge ligament (Figure 16.26C and D). The **umbo** is the oldest part of the shell, and growth occurs in concentric lines around it (Figure 16.26A).

Pearl production is a by-product of a protective device used by the animals when a foreign object (grain of sand, parasite, or other) becomes lodged between the shell and mantle (see Figure 16.6). The mantle secretes many layers of nacre around the irritating object. Pearls are cultured by inserting particles of nacre, usually taken from the shells of freshwater bivalves, between the shell and mantle of a certain species of oyster and by keeping the oysters in enclosures for several years. *Meleagrina* is an oyster used extensively by the Japanese for pearl culture.

Body and Mantle   The **visceral mass** is suspended from the dorsal midline, and the muscular foot is attached to the visceral mass anteroventrally (Figure 16.27). The ctenidia hang down on each side, each covered by a fold of the mantle. The posterior edges of the mantle folds are modified to form dorsal excurrent and ventral incurrent openings (Figure 16.28A). In some marine bivalves the mantle is drawn out into long muscular siphons that allow the clam to burrow into the mud or sand and to extend the siphons to the water above (Figure 16.28A to D).

Locomotion   Bivalves initiate movement by extending a slender muscular foot between the valves (Figure 16.28D). They pump blood into their foot, causing it to swell and to act as an anchor in the mud or sand; then longitudinal muscles contract to shorten the foot and pull the animal forward.

Scallops and file shells swim with a jerky motion by clapping their valves together to create a sort of jet propulsion. The mantle edges can direct the stream of expelled water, so that the animals can swim in virtually any direction (Figures 16.25B and 16.29).

Gills   Gaseous exchange occurs through both mantle and gills. Gills of most bivalves are highly modified for filter-feeding; they are derived from primitive ctenidia by a great lengthening of filaments on each side of the central axis (Figure 16.27). As ends of long filaments became folded back toward the central axis, ctenidial filaments developed a somewhat elongated W shape. Filaments lying beside each other became joined by ciliary junctions or tissue fusions, forming platelike **lamellae** with many vertical water tubes inside. Thus water enters the incurrent siphon, propelled by ciliary action, then enters the water tubes through pores between the filaments in the lamellae, proceeds dorsally into a common **suprabranchial chamber** (Figure 16.30), and then leaves through the excurrent aperture.

Feeding   Most bivalves are filter feeders (see Chapter 32, p. 703). Respiratory currents bring both oxygen and organic materials to the gills, where ciliary tracts direct currents to the tiny pores of the gills. Gland cells on the gills and labial palps secrete copious amounts of mucus, which entangles particles suspended in water going through gill pores. These mucous masses slide down the outside of the gills toward food grooves at the lower edge of the gills (Figure 16.31). Heavier particles of sediment fall from the ctenidia, but smaller particles travel along the food grooves toward the labial palps. The palps, being also grooved and ciliated, sort the particles and direct tasty ones encased in the mucous mass into the mouth.

Some bivalves, such as *Nucula* and *Yoldia,* are deposit feeders and have long proboscides attached to the labial palps (Figure 16.28C). These can be protruded onto sand or mud to collect food particles, in addition to particles attracted by gill currents.

Shipworms (Figure 16.32) burrow in wood and feed on particles they excavate. Symbiotic bacteria live in a special organ in the bivalve and produce cellulase to digest wood (see p. 709). Other bivalves, such as giant clams, gain much of their nutrition as adults from the photosynthetic products of symbiotic dinoflagellates living in their mantle tissue (Figure 16.33).

Septibranchs, another group of bivalves, draw small crustaceans or bits of organic debris into the mantle cavity by sudden inflow of water created by the pumping action of a muscular septum in the mantle cavity.

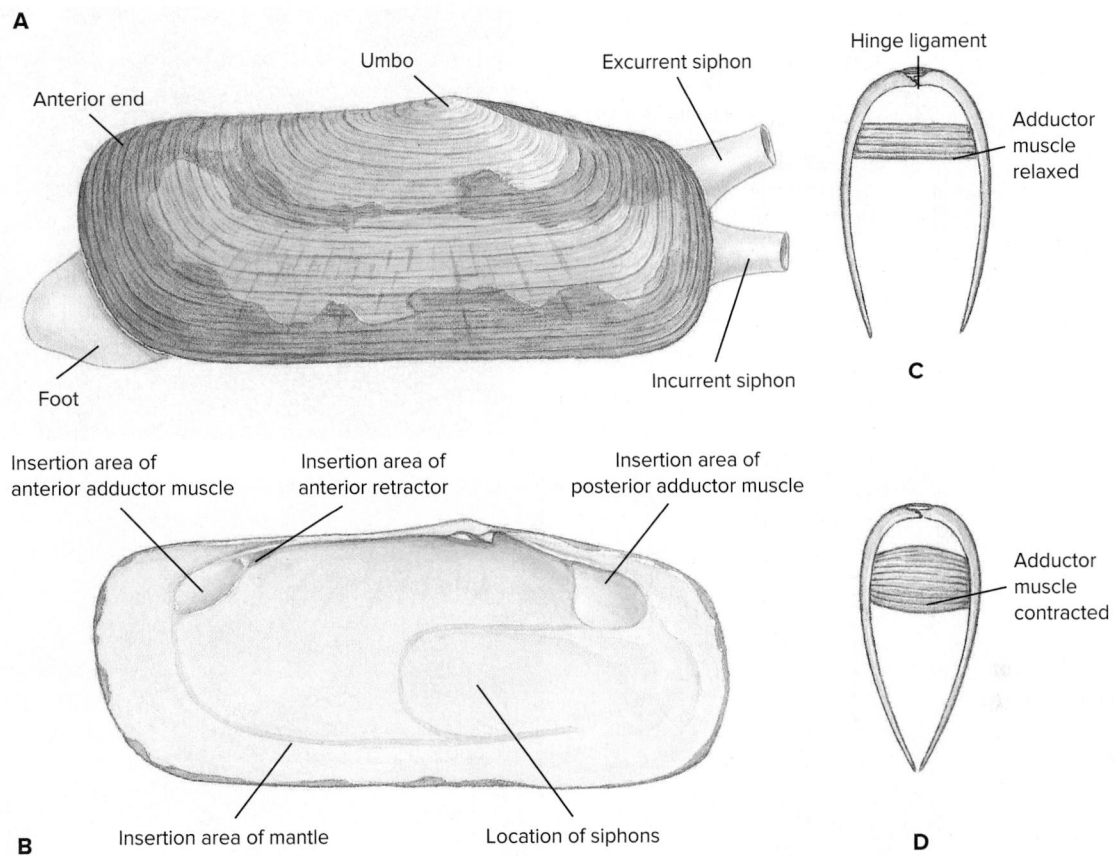

Figure 16.26 *Tagelus plebius,* stubby razor clam (class Bivalvia). **A,** External view of left valve. **B,** Inside of right shell showing scars where muscles were attached. The mantle was attached at its insertion area. **C** and **D,** Sections showing function of adductor muscles and hinge ligament. In **C,** the adductor muscle is relaxed, allowing the hinge ligament to pull the valves apart. In **D,** the adductor muscle is contracted, pulling the valves together.

Internal Structure and Function The floor of the stomach of filter-feeding bivalves is folded into ciliary tracts for sorting a continuous stream of particles. In most bivalves a cylindrical **style sac** opening into the stomach secretes a gelatinous rod called a **crystalline style,** which projects into the stomach and is kept whirling by means of cilia in the style sac (Figure 16.34). Rotation of the style helps to dissolve its surface layers, freeing digestive enzymes (especially amylase) that it contains, and to roll the mucous food mass. Dislodged particles are sorted, and suitable ones are directed to the digestive gland or engulfed by amebocytes. Further digestion is intracellular.

The three-chambered heart, which lies in the pericardial cavity (Figure 16.31), has two atria and a ventricle and beats slowly, ranging from 0.2 to 30 times per minute. Part of the blood is oxygenated in the mantle and returns to the ventricle through the atria; the rest circulates through sinuses and passes in a vein to the kidneys, from there to the gills for oxygenation, and back to the atria.

A pair of U-shaped kidneys (nephridial tubules) lies just ventral and posterior to the heart (Figure 16.31B). The glandular portion of each tubule opens into the pericardium; the bladder portion empties into the suprabranchial chamber.

Zebra mussels, *Dreissena polymorpha,* are a disastrous biological introduction into North America. They were apparently picked up as veligers with ballast water by one or more ships in freshwater ports in northern Europe and then expelled between Lake Huron and Lake Erie in 1986. This 4 cm bivalve spread throughout the Great Lakes by 1990, and by 1994 it was as far south on the Mississippi River as New Orleans, as far north as Duluth, Minnesota, and as far east as the Hudson River in New York. It attaches to any firm surface and filter-feeds on phytoplankton. Populations rapidly increase in size. They foul water-intake pipes of municipal and industrial plants, impede intake of water for municipal supplies, and have far-reaching effects on the ecosystem (see note, p. 348). Zebra mussels will cost billions of dollars to control if they can be controlled at all.

Another freshwater bivalve, *Corbicula fluminea,* was introduced into the United States from Asia more than 50 years ago by unknown means. Despite efforts to control *Corbicula* that cost over a billion dollars per year, it is now a pest throughout most of the continental United States, infesting water systems and clogging pipes.

Figure 16.27 Sections through a bivalve shell and body, showing relative positions of visceral mass and foot. Evolution of bivalve ctenidia: by a great lengthening of individual gill filaments, ctenidia became adapted for filter-feeding and separated the incurrent chamber from the excurrent, suprabranchial chamber.

Shell
Mantle
Gill
Visceral mass
Foot

© Steve Trewhella/FLPA/Science Source

A

B   C   D

Figure 16.28 Adaptations of siphons in bivalves. **A,** In the razor clam, *Ensis,* incurrent and excurrent siphons are clearly visible above the sediment surface. **B** to **D,** In many marine forms the mantle is drawn out into long siphons. In **A, B,** and **D,** the incurrent siphon brings in both food and oxygen. In **C,** *Yoldia,* the siphons are respiratory; long ciliated palps explore the mud surface and convey food to the mouth.

The nervous system consists of three pairs of widely separated ganglia connected by commissures and a system of nerves. Sense organs are poorly developed. They include a pair of statocysts in the foot, a pair of osphradia of uncertain function in the mantle cavity, tactile cells, and sometimes simple pigment cells on the mantle. Scallops (*Aequipecten, Chlamys*) have a row of small blue eyes along each mantle edge (see Figure 16.29). Each eye has a cornea, lens, retina, and pigmented layer. Tentacles on the margin of the mantle of *Aequipecten* and *Lima* have tactile and chemoreceptor cells.

Reproduction and Development   Sexes are usually separate. Gametes are discharged into the suprabranchial chamber to be carried out with the excurrent flow. An oyster may produce 50 million eggs in a single season. In most bivalves fertilization is external. The embryo develops into trochophore, veliger, and spat (juvenile shelled form) stages (Figure 16.35).

In most freshwater clams fertilization is internal. Eggs drop into the water tubes of the ctenidia, where they are fertilized by sperm entering with the incurrent flow (Figure 16.31). They develop there into a bivalved **glochidium larva** stage, which is a specialized veliger (Figure 16.36A). Glochidia need to attach themselves to specific fish hosts and live parasitically for several weeks to complete their development. Various mussel species have unique tactics for getting their larvae in contact with a suitable fish host species. Some simply

© Larry S. Roberts

Figure 16.29 Representing a group that has evolved from burrowing ancestors, the surface-dwelling scallop *Aequipecten irradians* has developed sensory organs along its mantle edges (tentacles and a series of blue eyes).

Figure 16.30 Section through heart region of a freshwater clam to show relation of circulatory and respiratory systems. Respiratory water currents: water is drawn in by cilia, enters gill pores, and then passes up water tubes to suprabranchial chambers and out excurrent aperture. Blood in gills exchanges carbon dioxide for oxygen. Blood circulation: ventricle pumps blood forward to sinuses of foot and viscera, and posteriorly to mantle sinuses. Blood returns from mantle to atria; it returns from viscera to the kidney, and then goes to the gills, and finally to the atria.

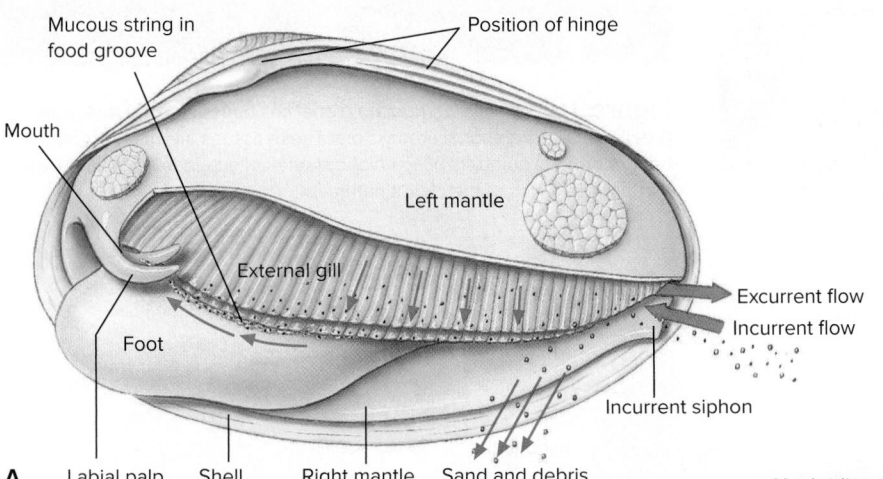

Figure 16.31 **A,** Feeding mechanism of freshwater clam. Left valve and mantle are removed. Water enters the mantle cavity posteriorly and is drawn forward by ciliary action to the gills and palps. As water enters the tiny openings of the gills, food particles are sieved out and caught in strings of mucus that are carried by cilia to the palps and directed to the mouth. Sand and debris drop into the mantle cavity and are removed by cilia. **B,** Clam anatomy.

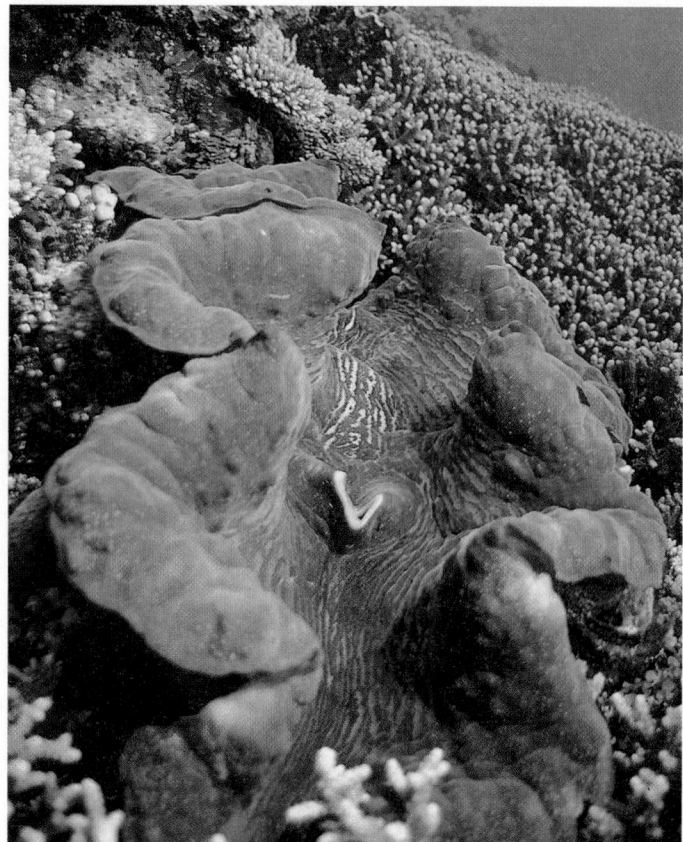

Figure 16.33 Clam (*Tridacna gigas*) lies buried in coral rock with greatly enlarged siphonal area visible. These tissues are richly colored and bear enormous numbers of symbiotic single-celled algae (zooxanthellae) that provide much of the clam's nutriment.

Figure 16.32 **A,** Shipworms are bivalves that burrow in wood, causing great damage to unprotected wooden hulls and piers. They are nicknamed "termites of the sea." **B,** The two small, anterior valves, seen at left, are used as rasping organs to extend the burrow.

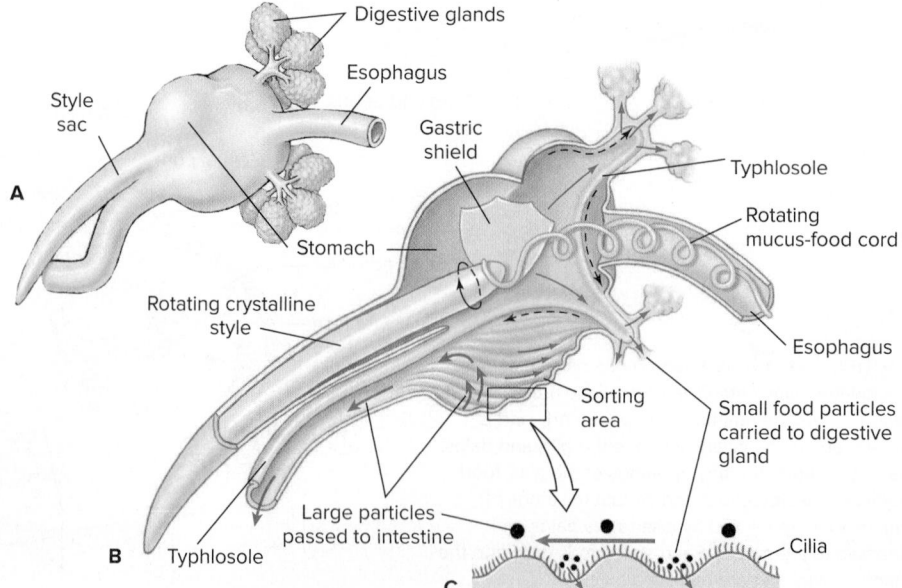

Figure 16.34 Stomach and crystalline style of ciliary-feeding clam. **A,** External view of stomach and style sac. **B,** Transverse section showing direction of food movements. Food particles in incoming water are caught in a cord of mucus that is kept rotating by the crystalline style. Ridged sorting areas direct large particles to the intestine and small food particles to digestive glands. **C,** Sorting action of cilia.

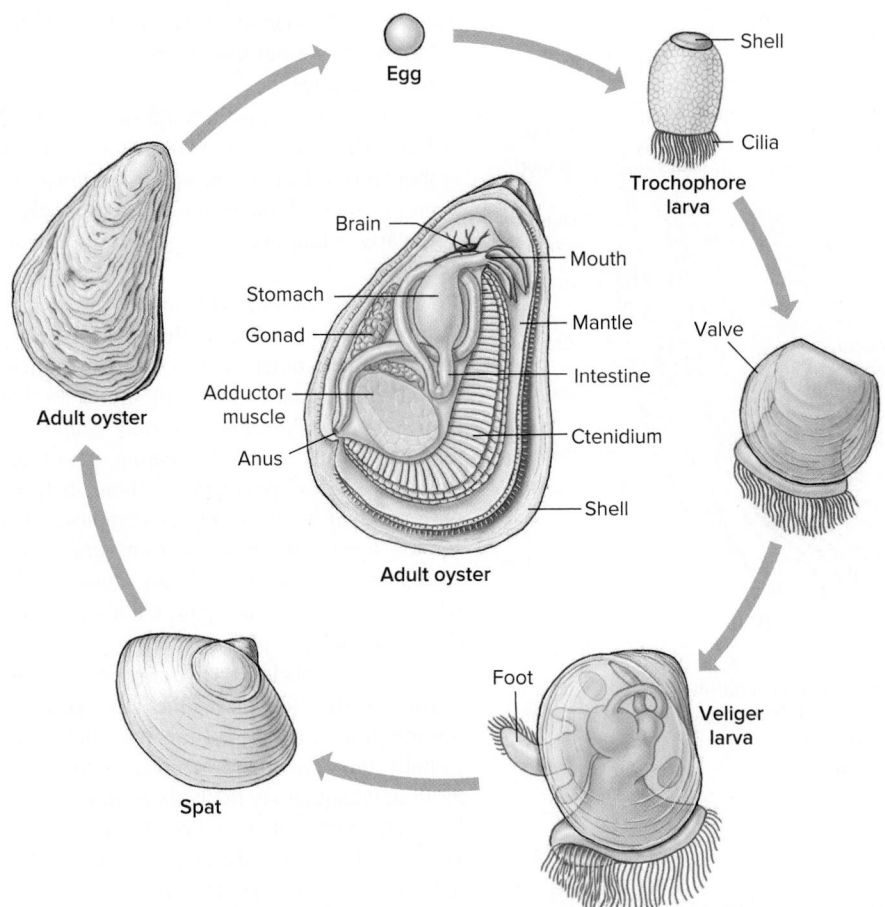

Figure 16.35   Life cycle of oysters. Oyster larvae swim about for approximately 2 weeks before settling down for attachment to become spats. Oysters take about 4 years to grow to commercial size.

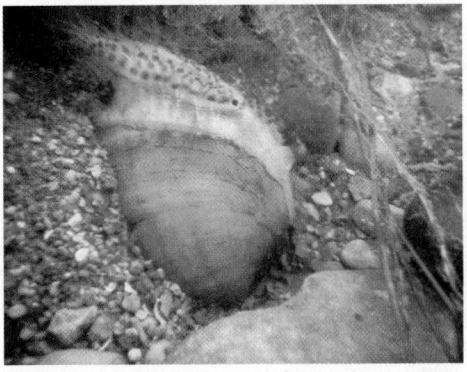

Figure 16.36   **A,** Glochidium, or larval form, for some freshwater clams. When larvae are released from the mother's brood pouch, they may become attached to a fish's gill by clamping their valves closed. They remain as parasites on the fish for several weeks. Their size is approximately 0.3 mm. **B,** Some clams have adaptations that help their glochidia find a host. The mantle edge of this female pocketbook mussel (*Lampsilis ovata*) mimics a small minnow, complete with eye. When a smallmouth bass comes to dine, it gets doused with glochidia.

discharge glochidia into the water column; if they come into contact with a suitable passing fish or amphibian, they can attach to the gills or skin and complete their development. In other species, the mantle flap of brooding females, in which the glochidia are held in a gelatinous packet called a **conglutinate,** has a size and shape unique to each mussel species. This mantle flap is often used as a lure to bring potential host species into contact with the glochidia. For example, the conglutinate of a gravid female pocketbook mussel (*Lampsilis ovata*) grows to resemble a small fish (Figure 16.36B). This mantle flap is then wriggled like a fishing lure to attract nearby bass, which serve as host for the glochidia. When a hungry bass strikes the mantle, instead of a meal it gets a mouthful of glochidia, which promptly attach to the fish's gills.

After encysting on a suitable host to complete development, the juveniles detach and sink to the substratum to begin independent lives. Larval "hitchhiking" helps to distribute organisms whose locomotion is very limited while also preventing larvae from being swept downstream out of lakes.

Boring   Many bivalves can burrow into mud or sand, but some have evolved a mechanism for burrowing into much harder substances, such as wood or stone.

Teredo, Bankia, and some other genera are called shipworms. They can be very destructive to wooden ships and wharves. These strange little clams have a long, wormlike appearance, with a pair of slender posterior siphons that keep water flowing over the gills, and a pair of small globular valves on the anterior end with which they burrow (Figure 16.32). The valves have microscopic teeth that function as very effective wood rasps. The animals extend their burrows with an unceasing rasping motion of the valves. This motion sends a continuous flow of fine wood particles into the digestive tract, where they are attacked by cellulase produced by symbiotic bacteria. Interestingly, these bacteria also fix nitrogen, an important property for their hosts, which live on a diet (wood) high in carbon but deficient in nitrogen.

Some clams bore into rock. The piddock (*Pholas*) bores into limestone, shale, sandstone, and sometimes wood or peat. It has strong valves that bear spines, which it uses to cut away the rock gradually while anchoring itself with its foot. *Pholas* may grow to 15 cm long and make rock burrows up to 30 cm long.

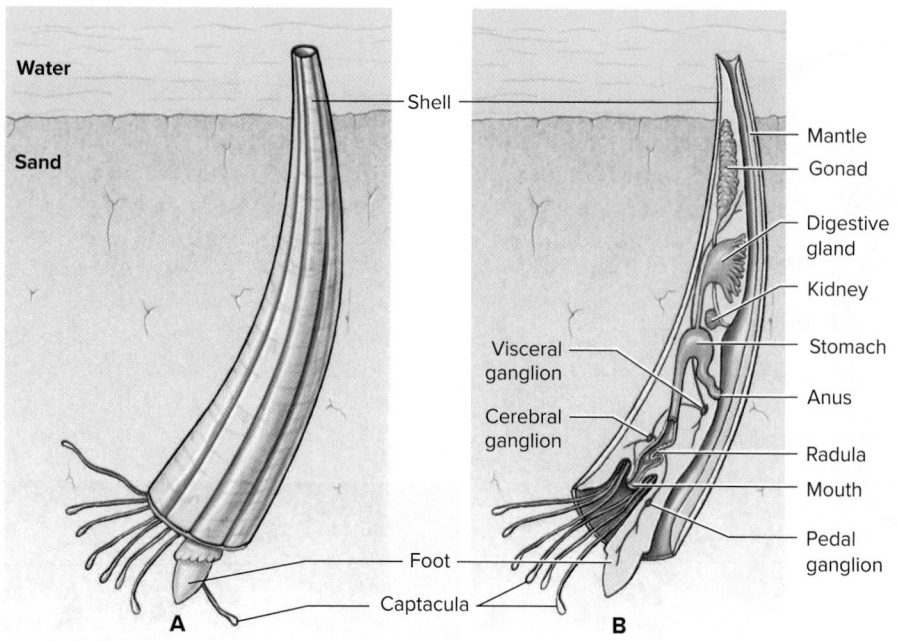

Figure 16.37 The tusk shell, *Dentalium* (class Scaphopoda). **A,** It burrows into soft mud or sand and feeds by means of its prehensile tentacles (captacula). Respiratory currents of water are drawn in by ciliary action through the small open end of the shell, then expelled through the same opening by muscular action. **B,** Internal anatomy of *Dentalium*.

Class Scaphopoda

Scaphopoda, commonly called tusk shells or tooth shells, are benthic marine molluscs found from the subtidal zone to over 6000 m in depth. They have a slender body covered with a mantle and a tubular shell open at both ends. In scaphopods the molluscan body plan has taken a new direction, with the mantle wrapped around the viscera and fused to form a tube. There are about 900 living species of scaphopods, and most are 2.5 to 5 cm long, although they range from 4 mm to 25 cm long.

The foot, which protrudes through the larger end of the shell, is used to burrow into mud or sand, always leaving the small end of the shell exposed to the water above (Figure 16.37). Respiratory water circulates through the mantle cavity by both movements of the foot and ciliary action. Gills are absent, and gaseous exchange therefore occurs in the mantle. Most food is detritus and protozoa from the substratum. It is caught on cilia of the foot or on the mucus-covered, ciliated adhesive knobs of the long tentacles (**captacula**) extending from the head and is conveyed to the nearby mouth. A radula carries food to a crushing gizzard. The captacula may serve some sensory function, but eyes, tentacles, and osphradia typical of many other molluscs are lacking.

Sexes are separate, and the larva is a trochophore.

Class Cephalopoda

Cephalopoda (Gr. *kephalē*, head, + *pous, podos,* foot) include squids, octopuses, nautiluses, devilfish, and cuttlefish. All are marine, and all are active predators. Their modified foot is merged with the head region. It takes the form of a funnel for expelling water from the mantle cavity, and the anterior margin is drawn out into a circle or crown of arms or tentacles.

Cephalopods range upward in size from 2 or 3 cm. The common squid of markets, *Loligo,* is about 30 cm long. Giant squids, *Architeuthis,* at up to almost 19 meters in length and weighing nearly a ton, are the largest invertebrates known.

Fossil records of cephalopods go back to Cambrian times. The earliest shells were straight cones; others were curved or coiled, culminating in the coiled shell similar to that of the modern *Nautilus,* the only remaining member of the once flourishing nautiloids (Figure 16.38). Cephalopods without shells or with internal shells (such as octopuses and squids) apparently evolved from some early straightshelled ancestor. Many ammonoids, which are extinct, had quite elaborate shells (Figure 16.38C).

Modern cephalopods live at various depths. Octopuses are often seen in the intertidal zone, lurking among rocks and crevices, but occasionally they are found at great depths. The more active squids are rarely found in very shallow water, and some have been taken at depths of 5000 m. *Nautilus* is usually found near the bottom in water 50 to 560 m deep, near islands in the southwestern Pacific.

The giant squid, *Architeuthis,* is very poorly known because few specimens have been studied. The anatomy has been described from stranded animals, from those captured in nets of fishermen, and from specimens found in the stomach of sperm whales. The mantle length is 5 to 6 m, and the head is up to 1 m.

Squid are an important food item for sperm whales who detect their prey using sonar. Squid have the largest eyes of any animal, reaching 27 cm in diameter with a 9 cm pupil, and recent work on their eye structure suggests that it is adapted for long-range vision. So although whale sonar would reveal the presence of squid before a squid could see a whale, long-range vision could provide the squid with sufficient time for evasive action.

Form and Function

Shell   Although early nautiloid and ammonoid shells were heavy, they were made buoyant by a series of **gas chambers,** as is that of *Nautilus* (Figure 16.38B), enabling the animal to maintain neutral buoyancy. The shell of *Nautilus,* although coiled, is quite different from that of a gastropod. The shell is divided by transverse septa into internal chambers (Figure 16.38B), only the last inhabited by the living animal. As it grows, it moves forward, secreting behind its body a new septum. The chambers are connected by a cord of living

Figure 16.38 *Nautilus,* a cephalopod. **A,** Live *Nautilus.*
B, Longitudinal section, showing gas-filled chambers of shell, and diagram of body
structure. **C,** Longitudinal section through shell of an ammonoid.

Figure 16.39 Cuttlefish, *Sepia latimanus,* has an internal shell
familiar to keepers of caged birds as "cuttlebone."

After a member of genus *Nautilus* secretes a new septum, the
new chamber is filled with fluid similar in ionic composition to
that of *Nautilus* blood (and of seawater). Fluid removal involves
active secretion of ions into tiny intercellular spaces in the
siphuncular epithelium, so that a very high local osmotic
pressure is produced, and osmosis draws water from the
chamber. The gas in the chamber is just the respiratory gas from
the siphuncle tissue that diffuses into the chamber as fluid is
removed. Thus gas pressure in the chamber is 1 atmosphere or
less because it is in equilibrium with gases dissolved in the
seawater surrounding the *Nautilus,* which are in turn in
equilibrium with air at the surface of the sea, despite the fact
that the *Nautilus* may be swimming at 400 m beneath the
surface. That the shell can withstand implosion by the
surrounding 41 atmospheres (about 600 pounds per square
inch) and that the siphuncle can remove water against this
pressure are marvelous feats of natural engineering!

tissue called a **siphuncle,** which extends from the visceral mass.
Cuttlefishes (Figure 16.39) also have a small, curved shell, but it is
entirely enclosed by the mantle. In squids most of the shell has disappeared, leaving only a thin, proteinaceous strip called a pen, which is
enclosed by the mantle. In *Octopus* (Gr. *oktos,* eight, + *pous, podos,*
foot) the shell has disappeared entirely.

Locomotion   Cephalopods swim by forcefully expelling water
from the mantle cavity through a ventral **funnel** (or **siphon**)—a sort
of jet propulsion. The funnel is mobile and can be pointed forward
or backward to control direction; the force of water expulsion controls speed.

Squids and cuttlefishes are excellent swimmers. The squid body
is streamlined and built for speed (Figure 16.40). Cuttlefishes swim

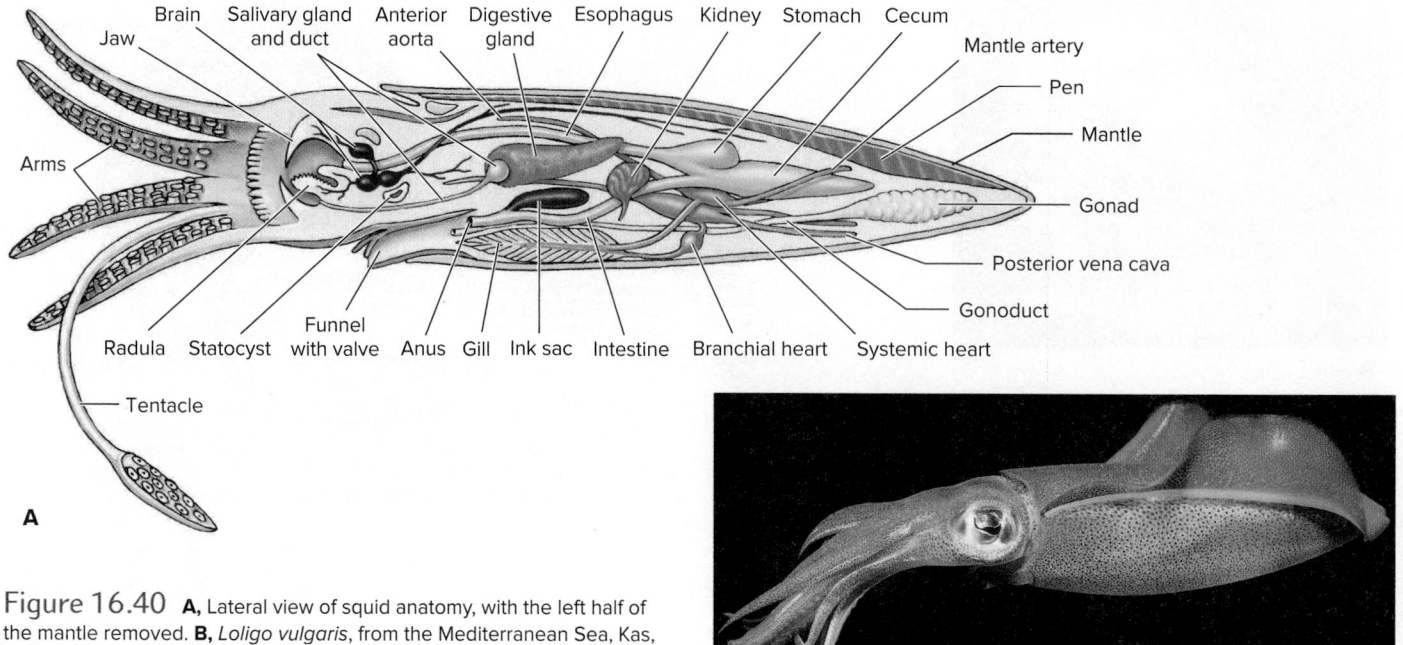

Figure 16.40 **A,** Lateral view of squid anatomy, with the left half of the mantle removed. **B,** *Loligo vulgaris*, from the Mediterranean Sea, Kas, Turkey Figure.

more slowly. The lateral fins of squids and cuttlefishes serve as stabilizers, but they are held close to the body for rapid swimming.

Nautilus is active at night; its gas-filled chambers keep the shell upright. Although not as fast as squids, it moves surprisingly well.

Octopus has a rather globular body and no fins (see Figure 16.1E). An octopus can swim backward by spurting jets of water from its funnel, but it is better adapted to crawling over rocks and coral, using suction discs on its arms to pull or to anchor itself. Some deepwater octopods have the arms webbed like an umbrella and swim in a medusa-like fashion (p. 268).

Internal Features   The active habits of cephalopods are reflected in their internal anatomy, particularly their respiratory, circulatory, and nervous systems.

Respiration and Circulation   Except for nautiloids, cephalopods have one pair of gills. Because ciliary propulsion would not circulate enough water for their high oxygen requirements, there are no cilia on the gills. Instead, radial muscles in the mantle wall compress the wall and enlarge the mantle cavity, drawing water inside. Strong circular muscles contract and expel water forcibly through the funnel. A system of one-way valves prevents water from being taken in through the funnel or expelled around the mantle margin.

Likewise, the open circulatory system of ancestral molluscs would be inadequate for cephalopods. Their circulatory system has evolved into a closed network of vessels, and capillaries conduct blood through the gill filaments. Furthermore, the molluscan plan of circulation places the entire systemic circulation before the blood reaches the gills (in contrast to vertebrates, in which the blood leaves the heart and goes directly to the gills or lungs; see p. 686). This functional problem was solved by the development of **accessory** or **branchial hearts** (Figure 16.40A) at the base of each gill to increase the pressure of blood going through the capillaries there.

Nervous and Sensory Systems   Nervous and sensory systems are more elaborate in cephalopods than in other molluscs. The brain, the largest in any invertebrate, consists of several lobes with millions of nerve cells. Squids have giant nerve fibers (among the largest known in the animal kingdom), which are activated when the animal is alarmed and initiate maximal contractions of the mantle muscles for a speedy escape.

Squid nerves played an important role in early biophysical studies. Our current understanding of transmission of action potentials along and between nerve fibers (see Chapter 33) is based primarily on work performed using the giant nerve fibers of squids, *Loligo* sp. A. Hodgkin and A. Huxley received the Nobel Prize in Physiology or Medicine, 1963, for their achievements in this field.

Sense organs are well developed. Except for *Nautilus*, which has relatively simple eyes, cephalopods have highly complex eyes with cornea, lens, chambers, and retina (Figure 16.41). Orientation of the eyes is controlled by the statocysts, which are larger and more complex than in other molluscs. The eyes are held in a constant relation to gravity, so that the slit-shaped pupils are always in a horizontal position.

Almost all cephalopods are colorblind, but their eyesight is excellent and their visual acuity underwater far surpasses our own. They can also be taught to discriminate between shapes—for example, a square and a rectangle—and to remember such a discrimination for a considerable time. Experimenters find it easy to modify octopod behavior patterns by devices of reward and punishment. Octopods are capable of observational learning; when one octopus observes another being rewarded by making a correct choice, the observer learns which choice is rewarded and consistently makes the same selection when given the opportunity.

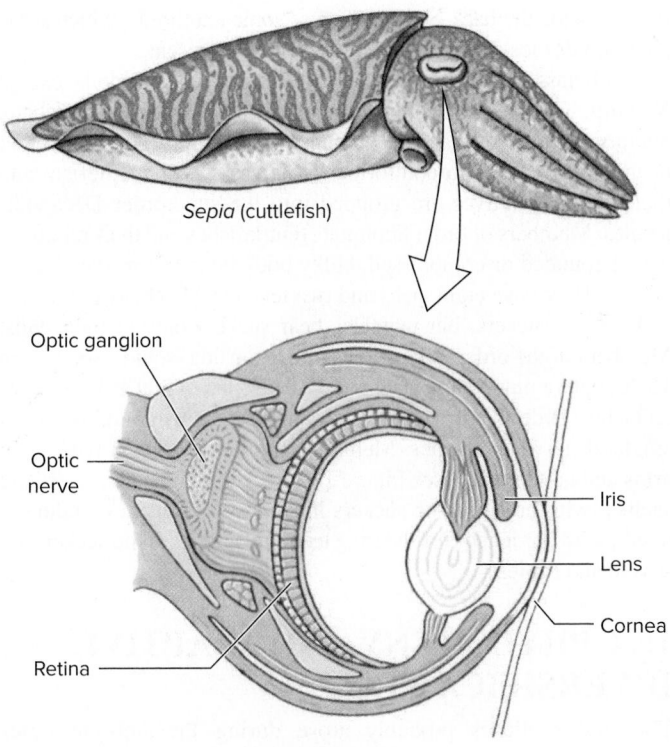

Sepia (cuttlefish)

Optic ganglion

Optic
nerve

Iris

Lens

Cornea

Retina

Figure 16.41 Eye of a cuttlefish (*Sepia*). The structure of cephalopod eyes is very similar to that of eyes of vertebrates (see item 11, p. 335).

When similar structures that are not inherited from a common ancestor evolve in different ways in unrelated animals, we call it **convergence, or convergent evolution.** For many years cephalopod eyes and vertebrate eyes have been cited as a marvelous example of convergent evolution. Cephalopod and vertebrate eyes are similar in many details of structure but differ in development. Compound eyes of arthropods (pp. 428, 741), differing in both structure and development, were viewed as examples of other, independently derived eyes in animals. Now we recognize that all triploblastic animals with eyes, even those with the most simple eyespots, such as platyhelminths, share at least two conserved genes: that for rhodopsin, a visual pigment, and that for *Pax 6*, now sometimes called the "master control gene for eye morphogenesis." Once these two genes originated, natural selection eventually produced the specialized organs of vertebrates, molluscs, and arthropods.

Octopods use their arms for tactile exploration and can discriminate between textures by feel but apparently not between shapes. Their arms are well supplied with both tactile and chemoreceptor cells. Cephalopods seem to lack a sense of hearing.

Communication   Little is known of social behavior of nautiloids or deep-water cephalopods, but inshore and littoral forms such as *Sepia, Sepioteuthis, Loligo,* and *Octopus* have been studied extensively. Although their tactile sense is well developed and they have some chemical sensitivity, visual signals are the predominant means of communication. These signals consist of

a host of movements of the arms, fins, and body, as well as many color changes. Movements may range from minor body motions to exaggerated spreading, curling, raising, or lowering of some or all of the arms.

Color changes are effected by **chromatophores,** cells in the skin that contain pigment granules (see Chapter 29, p. 642). Tiny muscle cells surround each elastic chromatophore, whose contractions pull the cell boundary of the chromatophore outward, causing it to expand greatly. As the cell expands, the pigment becomes dispersed, changing the color pattern of the animal. When the muscles relax, chromatophores return to their original size, and pigment becomes concentrated again. By means of the chromatophores, which are under nervous and probably hormonal control, an elaborate system of changes in color and pattern is possible, including general darkening or lightening; flushes of pink, yellow, or lavender; and formation of bars, stripes, spots, or irregular blotches. These colors may be used variously as danger signals, as protective coloring, in courtship rituals, and probably in other ways. By assuming different color patterns of different parts of the body, a squid can transmit three or four different messages *simultaneously* to different individuals and in different directions, and it can instantaneously change any or all of the messages. Probably no other system of communication in invertebrates can convey so much information so rapidly.

The degree to which cephalopods are able to match colorful backgrounds makes it difficult to believe that they lack color vision. However, in an experiment where background colors were equal in intensity, cephalopods were unable to match colors. This result, coupled with the physiology of their eyes, makes color vision unlikely. Cephalopod photoreceptors are rhabdomere cells, not rods and cones, as in vertebrates. Their ability to see color depends on the number of visual pigments present; almost all cephalopods have only one such pigment and cannot detect color. An exception is the bioluminescent firefly squid, which has three visual pigments, the same number found in humans. Many deep-water cephalopods have evolved elaborate luminescent organs.

It appears that the wonderful color patterns we see on cephalopod skin are visible to other cephalopods as polarized light patterns; all cephalopods detect differences in polarized light. They also use this ability to enhance discrimination of translucent prey and fish with silvery reflective scales in much the same way fishermen use polarized glasses to reduce glare from water.

Most cephalopods other than nautiloids use ink as a protective device. An ink sac that empties into the rectum contains an **ink gland** that secretes **sepia,** a dark fluid containing the pigment melanin, into the sac. When the animal is alarmed, it releases a cloud of ink, which may hang in the water as a blob or be contorted by water currents. The animal quickly departs from the scene, leaving the ink as a decoy to the predator.

Reproduction   Sexes are separate in cephalopods. Spermatozoa are encased in spermatophores and stored in a sac that opens into the mantle cavity. One arm of adult males is modified as an intromittent organ, called a **hectocotylus,** used to pluck a spermatophore from his own mantle cavity and insert it into the mantle cavity of a female

A

B

Figure 16.42 Copulation in cephalopods. **A,** Mating cuttlefishes.
B, Male octopus uses modified arm to deposit spermatophores in female mantle cavity to fertilize her eggs. Octopuses often tend their eggs during development.

near the oviduct opening (Figure 16.42). Before copulation males often undergo color displays, apparently directed against rival males. Eggs are fertilized as they leave the oviduct and are then usually attached to stones or other objects. Some octopods tend their eggs. Females of *Argonauta,* the paper nautilus, secrete a fluted "shell," or capsule, in which eggs develop.

The large, yolky eggs undergo meroblastic cleavage. During embryonic development, the head and foot become indistinguishable. The ring around the mouth, which bears the arms, or tentacles, may be derived from the anterior part of the foot. Juveniles hatch from eggs; no free-swimming larva exists in cephalopods.

Major Groups of Cephalopods

There are three subclasses of cephalopods: Nautiloidea, which have two pairs of gills; the entirely extinct Ammonoidea; and Coleoidea, which have one pair of gills. Nautiloidea populated Paleozoic and Mesozoic seas, but there survives only one genus, *Nautilus* (see Figure 16.38), of which there are five species. The head of *Nautilus,* with its 60 to 90 or more tentacles, can be extended from the opening of the body compartment of its shell. Its tentacles have no suckers but are made adhesive by secretions. They are used in searching for, sensing, and grasping food.

Beneath its head is the funnel. Mantle, mantle cavity, and visceral mass are sheltered by the shell.

Ammonoids were widely prevalent in the Mesozoic era but became extinct by the end of the Cretaceous period. They had chambered shells analogous to those of nautiloids, but the septa were more complex, and the septal sutures (where septa contact the inside of the shell) were frilled (compare shells in Figure 16.38B and C). The reasons for their extinction remain a mystery. Present evidence suggests that they were gone before the asteroid bombardment at the end of the Cretaceous period (see the Timeline of Major Biological

Events inside the back cover), whereas some nautiloids, which some ammonoids closely resembled, survive to the present.

Subclass Coleoidea includes all living cephalopods except *Nautilus.* The taxonomy of living cephalopods is a subject of debate, but most authorities place the octopuses and vampire squids together in the superorder Octopodiformes, whereas squids, cuttlefish, and their closest relatives are grouped into the superorder Decapodiformes. Members of order Sepioidea (cuttlefishes and their relatives) have a rounded or compressed, bulky body bearing fins (see Figure 16.39). They have eight arms and two tentacles. Both arms and tentacles have suckers, but tentacles bear suckers only at their ends. Members of the orders Myopsida and Degopsida (squids, see Figure 16.40) have a more cylindrical body but also have eight arms and two tentacles. Order Vampyromorpha (vampire squid) contains only a single, deep-water species. Members of order Octopoda have eight arms and no tentacles (see Figure 16.1E). Their bodies are short and saclike, with no fins. The suckers in squids are stalked (pedunculated), with hardened rims bearing teeth; in octopuses the suckers are sessile and rimless.

16.4 PHYLOGENY AND ADAPTIVE DIVERSIFICATION

The first molluscs probably arose during Precambrian times (Figure 16.43), because fossils attributed to Mollusca appear in geological strata as old as the early Cambrian period. A Cambrian fossil radula was recently found in Alberta, Canada. On the basis of such shared features as spiral cleavage, mesoderm from the 4d blastomere, and trochophore larva, molluscs are protostomes, allied with the annelids in the lophotrochozoan subgroup Trochozoa (see p. 320). Opinions differ, however, on the exact relationships among lophotrochozoans. Some characters suggest that molluscs and annelids are sister taxa, but we do not depict a branching order for these taxa.

Annelid worms have a developmental pattern very similar to that of molluscs, but the annelid body is metameric, composed of serially repeated segments, whereas there are no true segments in molluscs. Both annelids and molluscs are coelomate protostomes, but the coelom is greatly reduced in molluscs as compared with annelids. Opinions differ as to whether molluscs were derived from a wormlike ancestor independent of annelids, share an ancestor with annelids after the advent of the coelom, or share a segmented common ancestor with annelids.

The hypothesis that annelids and molluscs shared a segmented ancestor is strengthened if the repeated body parts present in *Neopilina* (class Monoplacophora), and in some chitons, can be considered evidence of metamerism. However, morphological and developmental studies indicate that these parts are not remnants of an ancestral metameric body. One perspective on the evolution of repeated parts (gills and muscles) comes from analysis of molecular characters from a wide range of molluscs, including a monoplacophoran[5]. The cladogram from this study places monoplacophorans as the sister taxon to chitons, uniting the two taxa with repeated body parts in a clade called Serialia. Further, clade Serialia is not the sister taxon to other molluscs, as would be expected if the ancestral mollusc were segmented. Instead, clade Serialia is

[5]Giribet, G., et al. 2006. Proc. Natl. Acad. Sci. USA **103:** 7723–7728.

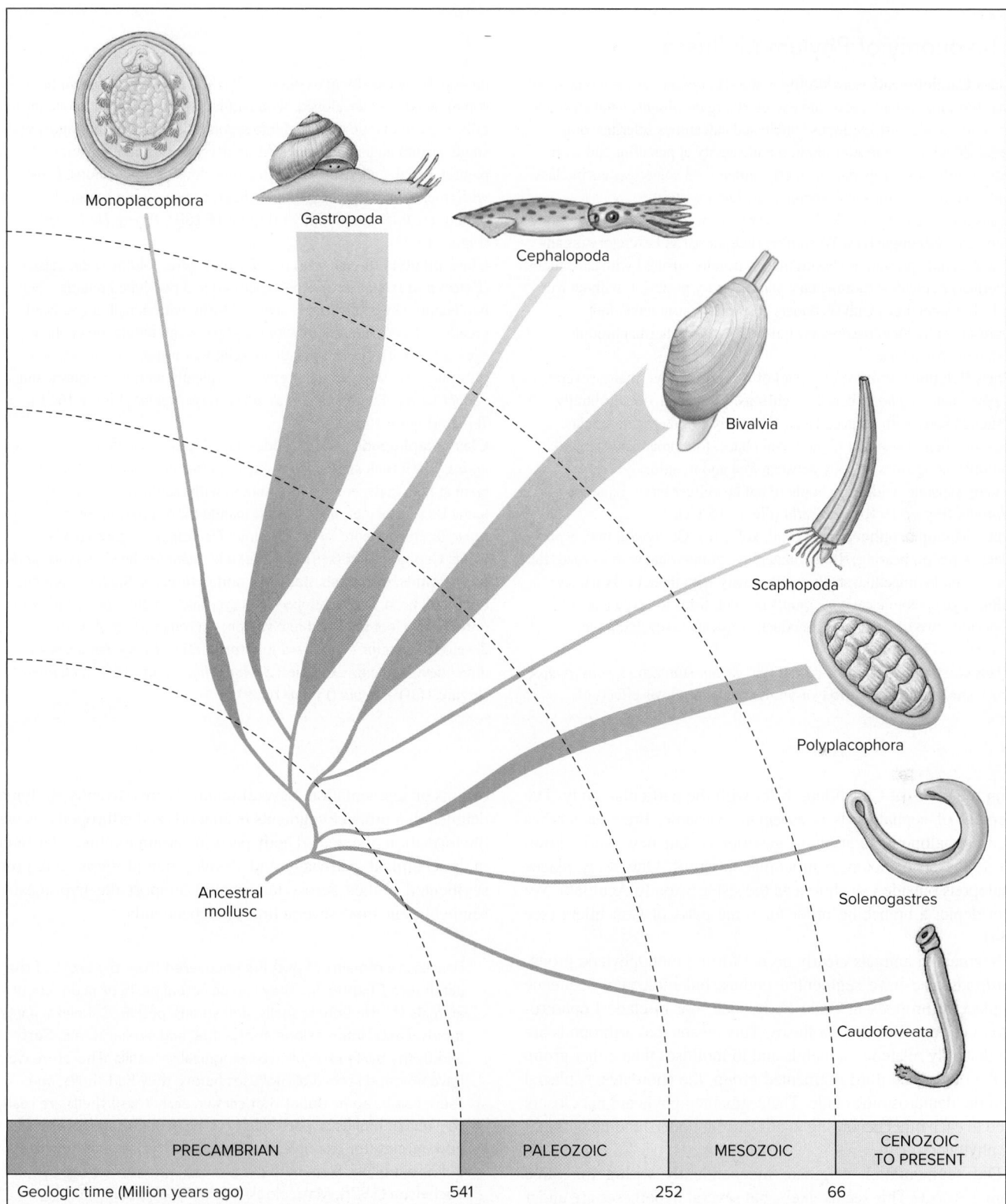

Figure 16.43 Classes of Mollusca, showing their derivations and relative abundance.

	PRECAMBRIAN	PALEOZOIC	MESOZOIC	CENOZOIC TO PRESENT
Geologic time (Million years ago)		541	252	66

phylogenetically nested within a clade of unsegmented molluscs, indicating that the repeated structures are derived molluscan features, not ancestral features. However, several studies using molecular characters do not support the Serialia hypothesis. These studies support clade Aculifera, uniting caudofoveates, solenogasters, and chitons. The two wormlike taxa appear to be derived forms that lost the shell plates and other features. The sister taxon to Aculifera is Conchifera, a clade of shelled molluscs. However, relationships

Taxonomy of Phylum Mollusca

Class Caudofoveata (kaw′dō-fō-ve-at′a) (L. *cauda*, tail, + *fovea*, small pit). Wormlike; shell, head, and excretory organs absent; radula usually present; mantle with chitinous cuticle and calcareous sclerites; oral pedal shield near anterior mouth; mantle cavity at posterior end with pair of gills; sexes separate; formerly united with solenogasters in class Aplacophora. Examples: *Chaetoderma, Limifossor.*

Class Solenogastres (so-len′ō-gas′trez) (Gr. *solen*, pipe, + *gaster*, stomach): **solenogasters.** Wormlike; shell, head, and excretory organs absent; radula present or absent; mantle usually covered with calcareous sclerites or spicules; rudimentary mantle cavity posterior, without true gills, but sometimes with secondary respiratory structures; foot represented by long, narrow, ventral pedal groove; hermaphroditic. Example: *Neomenia.*

Class Polyplacophora (pol′y-pla-kōf′o-ra) (Gr. *polys*, many, several, + *plax*, plate, + *phora*, bearing): **chitons.** Elongated, dorsoventrally flattened body with reduced head; bilaterally symmetrical; radula present; shell of seven or eight dorsal plates; foot broad and flat; gills multiple along sides of body between foot and mantle edge; sexes usually separate, with a trochophore but no veliger larva. Examples: *Mopalia* (Figure 16.10), *Tonicella* (Figure 16.1A).

Class Monoplacophora (mon′ō-pla-kōf′o-ra) (Gr. *monos*, one, + *plax*, plate, + *phora*, bearing). Body bilaterally symmetrical with a broad, flat foot; a single limpetlike shell; mantle cavity with three to six pairs of gills; large coelomic cavities; radula present; three to seven pairs of nephridia, two of which are gonoducts; separate sexes. Example: *Neopilina* (Figure 16.12).

Class Gastropoda (gas-trō-p′ōda) (Gr. *gaster*, stomach, + *pous, podos*, foot): **snails and slugs.** Body asymmetrical and shows effects of torsion; body usually in a coiled shell (shell uncoiled or absent in some); head well developed, with radula; foot large and flat; one or two gills, or with mantle modified into secondary gills or a lung; most with single atrium and single nephridium; nervous system with cerebral, pleural, pedal, and visceral ganglia; dioecious or monoecious, some with trochophore, typically with veliger, some without pelagic larva. Examples: *Busycon, Polinices* (Figure 16.16B), *Physa, Helix, Aplysia* (Figure 16.22).

Class Bivalvia (bi-val′ve-a) (L. *bi*, two,+*valva*, folding door, valve) **(Pelecypoda): bivalves.** Body enclosed in a two-lobed mantle; shell of two lateral valves of variable size and form, with dorsal hinge; head greatly reduced, but mouth with labial palps; no radula; no cephalic eyes, a few with eyes on mantle margin; foot usually wedge shaped; gills platelike; sexes usually separate, typically with trochophore and veliger larvae. Examples: *Anodonta, Venus, Tagelus* (Figure 16.26), *Teredo* (Figure 16.32).

Class Scaphopoda (ska-fō-p′ōda) (Gr. *skaphe-*, trough, boat, + *pous, podos*, foot): **tusk shells.** Body enclosed in a one-piece tubular shell open at both ends; conical foot; mouth with radula and contractile tentacles (captacula); head absent; mantle for respiration; sexes separate; trochophore larva. Example: *Dentalium* (Figure 16.37).

Class Cephalopoda (sef′ālo-p′ōda) (Gr. *kephale-*, head, + *pous, podos*, foot): **squids, cuttlefish, nautilus, and octopuses.** Shell often reduced or absent; head well developed with eyes and a radula; head with arms or tentacles; foot modified into siphon; nervous system of well-developed ganglia, centralized to form a brain; sexes separate, with direct development. Examples: *Sepioteuthis* (Figure 16.40), *Octopus* (Figure 16.1E), *Sepia* (Figure 16.39).

---

among members of Conchifera differ with the particular study. The placement of cephalopods is especially variable. Previous studies placed cephalopods as sister to gastropods, but new work shows them as the sister taxon to monoplacophorans. One study places cephalopods outside Conchifera as the sister taxon to Aculifera. We do not depict a branching order for some parts of Conchifera (see p. 334).

Segmented animals clearly do not form a monophyletic group. Arthropods also have segmented bodies, but molecular sequence data place arthropods in clade Ecdysozoa, not in clade Lophotrochozoa with annelids and molluscs. This means that arthropods are more distantly related to annelids and to molluscs than either group is to the other. The third segmented group, the chordates, is placed within the deuterostome clade. The segmented phyla are not closely related to each other according to our current understanding of metazoan phylogeny.

Did segmentation originate independently within the three metameric taxa? This seems likely, but several hypotheses are under consideration. One hypothesis suggests that the two segmented protostomes, annelids and arthropods, are sister taxa, but this hypothesis conflicts with the recent placement of annelids and arthropods in different clades within Protostomia. Another hypothesis is that genes whose products are needed for segmentation arose early in bilaterian evolutionary history, but served different functions. Expression of these genes was later recruited in multiple separate evolutionary origins of segmentation. Several scientists are currently studying the detailed that produce segments in annelids and arthropods, as well as the formation of repeated body parts in certain molluscs. Differences in biochemical pathways and developmental steps that produce segmented bodies across taxa would support the hypothesis that segmentation arose several times independently.

Fossils are remains of past life uncovered from the crust of the earth (see Chapter 6). They can be actual parts or products of animals (teeth, bones, shells, and so on), petrified skeletal parts, molds, casts, impressions, footprints, and similar items. Soft and fleshy parts rarely leave recognizable fossils. Therefore we have no good record of molluscs before they had shells, and there can be some doubt that certain early fossil shells are really remains of molluscs, particularly if the group they represent is now extinct (for example, the Hyolitha). The issue of how to define a mollusc from hard parts alone was emphasized by Yochelson (1978, Malacologia **17:** 165), who said, "If scaphopods were extinct and soft parts were unknown, would they be called mollusks? I think not."

Most diversity among molluscs is related to their adaptation to different habitats and modes of life and to a wide variety of feeding methods, ranging from sedentary filter-feeding to active predation. There are many adaptations for food gathering within the phylum and an enormous variety in radular structure and function, particularly among gastropods.

The versatile glandular mantle has probably shown more plastic adaptive capacity than any other molluscan structure. Besides secreting the shell and forming the mantle cavity, it is variously modified into gills, lungs, siphons, and apertures, and it sometimes functions in locomotion, in the feeding processes, or in a sensory capacity. The shell, too, has undergone a variety of evolutionary adaptations, making molluscs one of the most successful groups in existence today.

SUMMARY

Mollusca is the largest lophotrochozoan phylum and one of the largest and most diverse of all phyla, its members ranging in size from very small organisms to the largest of invertebrates. The basic body divisions of molluscs are headfoot and visceral mass, which is usually covered by a shell. Most are marine, but some are freshwater, and a few are terrestrial. They occupy a wide variety of niches. A number are economically important, and a few are medically important as hosts of parasites.

Molluscs are coelomate, although their coelom is limited to the area around the heart, the gonads, and occasionally part of the intestine. Evolutionary development of a coelom was important because it enabled better organization of visceral organs and, in many of the animals that have it, an efficient hydrostatic skeleton. However, because most molluscs have a hard external shell, they cannot use the coelom for shape change or locomotion via a hydrostatic skeleton.

The mantle and mantle cavity are important characteristics of molluscs. The mantle secretes the shell and overlies a part of the visceral mass to form a cavity housing the gills. The mantle cavity has been modified into a lung in some molluscs. The foot is usually a ventral, sole-like, locomotory organ, but it may be variously modified, as in cephalopods, where it has become arms and a funnel. The radula occurs in all molluscs except bivalves and many solenogasters and is a protrusible, tonguelike organ with teeth used in feeding. Except in cephalopods, which have secondarily developed a closed circulatory system, the circulatory system of molluscs is open, with a heart and blood sinuses. Molluscs usually have a pair of nephridia connecting with the coelom and a complex nervous system with a variety of sense organs. The primitive larva of molluscs is the trochophore, and in most marine molluscs the trochophore develops into a second larval stage, the veliger.

Classes Caudofoveata and Solenogastres are small groups of wormlike molluscs with no shell, but which possess sclerites or spicules in bands across the larval body. These bands may be homologous to the body regions that produce the shell plates of chitons.

Class Monoplacophora is a tiny, univalve marine group showing pseudometamerism. Polyplacophora are more common, marine organisms with shells in the form of a series of seven or eight plates. They are rather sedentary animals with a row of gills along each side of their foot.

Gastropoda are the most successful and largest class of molluscs. Their interesting evolutionary history includes a torted stage where anus and head are at the same end of the body, as well as coiling, which represents an elongation and spiraling of the visceral mass. Torsion has led to the problem of fouling, the release of excreta over the head and in front of the gills, which is avoided in various ways among different gastropods. Among the solutions to fouling are bringing water into one side of the mantle cavity and out the other (many gastropods), some degree of detorsion (opisthobranchs), and conversion of the mantle cavity into a lung (pulmonates).

Class Bivalvia are marine and freshwater, and they have their shell divided into two valves joined by a dorsal ligament and held together by an adductor muscle. Most of them are filter feeders, drawing water through their gills by ciliary action.

Scaphopoda is a small class whose members have a tubular shell, open at both ends, and the mantle wrapped around the body.

Members of class Cephalopoda are all predators and many can swim rapidly. Their arms or tentacles capture prey by adhesive secretions or by suckers. They swim by forcefully expelling water from their mantle cavity through a funnel, which was derived from the foot.

There is both embryological and molecular evidence that molluscs share a common ancestor with annelids more recently than either of these phyla do with arthropods or deuterostome phyla. However, there remains considerable debate as to where the molluscs arose within the Lophotrochozoa and their relationship to other protostome phyla. Possibly, molluscs with repeated body parts are derived from unsegmented molluscan ancestors.

REVIEW QUESTIONS

1. Explain how ocean acidification is likely to affect molluscs. Describe the long-term effects of this process on human economies.
2. How does a molluscan coelom develop embryologically? Why was the evolutionary development of a coelom important?
3. What characteristics of Mollusca distinguish it from other phyla?
4. Briefly describe characteristics of the hypothetical ancestral mollusc, and tell how each class of molluscs (Caudofoveata, Solenogastres, Polyplacophora, Monoplacophora, Gastropoda, Cephalopoda, Bivalvia, Scaphopoda) differs from the ancestral condition with respect to each of the following: shell, radula, foot, mantle cavity and gills, circulatory system, and head.
5. Briefly describe the habitat and habits of a typical chiton.
6. Define the following with respect to gastropods: operculum, torsion, fouling, bilateral asymmetry, rhinophore, pneumostome.
7. What functional problem results from torsion? How have gastropods evolved to avoid this problem?
8. Gastropods have diversified enormously. Illustrate this statement by describing variations in feeding habits found in gastropods.
9. Distinguish between opisthobranchs and pulmonates.
10. Use examples to explain how the external morphology of cephalopods has changed over evolutionary time.
11. Describe how cephalopods swim and how they eat.
12. Describe adaptations in the circulatory and neurosensory systems of cephalopods that are particularly valuable for actively swimming, predaceous animals.
13. Distinguish between ammonoids and nautiloids.
14. Briefly describe how a typical bivalve feeds and how it burrows.
15. How is the ctenidium modified from the ancestral form in a typical bivalve?
16. Define the following: ctenidia, odontophore, periostracum, prismatic layer, nacreous layer, metanephridia, nephrostome, trochophore, veliger, glochidium, osphridium.
17. Which other invertebrate groups are likely to be the closest relatives of molluscs? What evidence supports and contradicts these relationships?

For Further Thought Read descriptions of the Serialia hypothesis and compare the molluscan tree developed under this hypothesis with the one shown in Figure 16.1. Explain how the ancestral molluscan characters and the evolutionary pathway for molluscs differ depending on the cladogram.

SELECTED REFERENCES

Abbott, R. T., and P. A. Morris. 2001. R. T. Peterson (ed.). A field guide to shells: Atlantic coasts and the West Indies, ed. 5. Boston, Houghton Mifflin Company. *An excellent revision of a popular handbook.*

Barinaga, M. 1990. Science digests the secrets of voracious killer snails. Science **249:** 250–251. *Describes research on the toxins produced by cone snails.*

Bergström, J. 1989. The origin of animal phyla and the new phylum Procoelomata. Lethaia **22:** 259–269. *Argues that Caudofoveata are the only surviving members of Procoelomata, putative ancestral sclerite-bearing early Cambrian metazoan.*

Branch, T. A., B. M. DeJoseph, L. J. Ray, and C. A. Wagner. 2013. Impacts of ocean acidification on marine seafood. Trends in Ecology & Evolution, **28:**178–186. *Excellent summary of the causes and consequences of oceanic change.*

Fleischman, J. 1997. Mass extinctions come to Ohio. Discover **18** (5):84–90. *Of the 300 species of freshwater bivalves in the Mississippi River basin, 161 are extinct or endangered.*

Gehring, W. J., and I. Kazuho. 1999. *Pax 6:* mastering eye morphogenesis and eye evolution. Trends Genet. **15:** 371–377. *The authors discuss morphogenetic pathways by which various animal eyes could have evolved from a common ancestral type of photoreceptive cell.*

Giribet, G., A. Okusu, A. R. Lindgren, S. W. Huff, M. Schrödl, and M. Nishiguchi. 2006. Evidence for a clade composed of molluscs with serially repeated structures: monoplacophorans are related to chitons. Proc. Natl. Acad. Sci. USA **103:** 7723–7728. *Molluscs with repeated structures are derived from an unsegmented molluscan ancestor.*

Gosline, J. M., and M. D. DeMont. 1985. Jet-propelled swimming in squids. Sci. Am. **252:** 96–103 (Jan.). *Mechanics of swimming in squid are analyzed; elasticity of collagen in mantle increases efficiency*

Hanlon, R. 2007. Cephalopod dynamic camouflage. Curr. Biol. **17:** 400–404. *Experiments clearly demonstrate that typical cephalopods cannot see background colors and that they match backgrounds by matching intensity.*

Haszprunar, G. 2000. Is the Aplacophora monophyletic? A cladistic point of view. Amer. Malac. Bull. **15:** 115–130. *Asserts that solenogasters are the sister group to all extant molluscan groups, including the caudofaveates.*

Holloway, M. 2000. Cuttlefish say it with skin. Nat. Hist. **109** (3):70–76. *Cuttlefish and other cephalopods can change texture and color of their skin with astonishing speed. Fifty-four components of cuttlefish "vocabulary" have been described, including color display, skin texture, and a variety of arm and fin signals.*

Kano, Y., T. P. Neusser, H. Fukumori, K.M. Joerger, and M. Schroedl. 2015. Sea-slug invasion of the land. Biological Journal of the Linnean Society Volume: **116:** 253–259 *The evolution and ecology of terrestrial sea-slugs in Palau.*

Page, L. R. 2003. Gastropod ontogenetic torsion: developmental remnants of an ancient evolutionary change in body plan. J. Exp. Zool. Part B: **297B:** 11–26. *A torted body is the result of asynchronous movements of the shell and viscera, accompanied by remodeling of mantle epithelium.*

Smith, S. A., N. G. Wilson, F. E. Goetz, C. Feehery, S. C. S. Andrade, G. W. Rouse, G. Giribet, and C. W. Dunn. 2011. Resolving the evolutionary relationships of molluscs with phylogenomic tools. Nature **480:** 364–367. *This study supports clades Aculifera and Conchifera, and fails to support the Serialia hypothesis.*

Vinther, J., E. A. Sperling, D. E. G. Briggs, and K. J. Peterson. 2012. A molecular palaeobiological hypothesis for the origin of aplacophoran molluscs and their derivation from chitonlike ancestors. Proc. R. Soc. B. **279:** 1259–1268. *A nice discussion of fossil molluscs in light of a new phylogeny.*

Ward, P. D. 1998. Coils of time. Discover **19** (3):100–106. *Present* Nautilus *has apparently existed essentially unchanged for 100 million years, and all other species known were derived from it, including king nautilus (Allonautilus), a recent derivation.*

Woodruff, D. S., and M. Mulvey. 1997. Neotropical schistosomiasis: African affinities of the host snail *Biomphalaria glabrata* (Gastropoda: Planorbidae). Biol. J. Linn. Soc. **60:** 505–516. *The pulmonate snail* Biomphalaria glabrata *is the intermediate host in the New World for* Schistosoma mansoni, *an important trematode of humans (p. 301). Allozyme analysis shows that* B. glabrata *clusters with African species rather than the neotropical ones. Thus, when* S. mansoni *was brought to the New World in African slaves, it found a compatible host.*

Annelids and Allied Taxa

Chloeia sp., *an errant polychaete with large bundles of setae.*

- PHYLUM ANNELIDA, INCLUDING POGONOPHORANS (SIBOGLINIDS) AND ECHIURANS
- PHYLUM SIPUNCULA

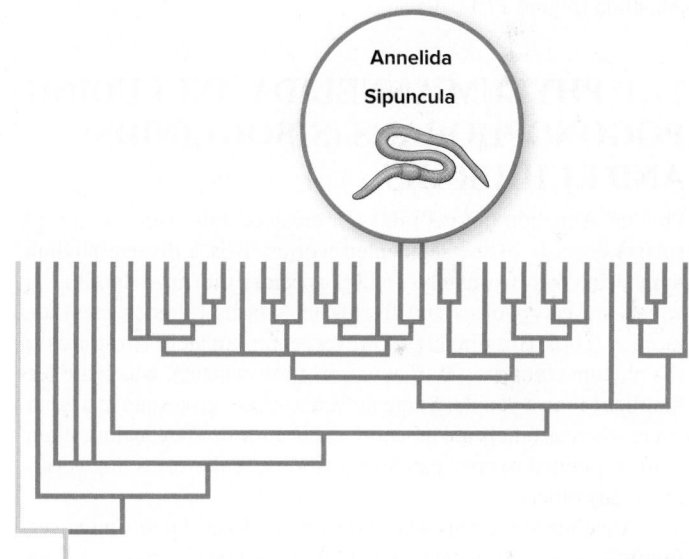

Annelida

Sipuncula

Annelids Illustrate Metamerism

Each annelid body segment is a serially repeated unit that contains components of circulatory, nervous, and excretory organ systems. The metameric body plan occurs in two other animal phyla: Arthropoda and Chordata. If our current understanding of animal phylogeny is correct, each of these phyla evolved metamerism independently. Why is metamerism a successful body plan?

A dominant feature of modern annelids is the fluid-filled coelomic cavity in every segment. The cavity is surrounded by longitudinal and circular muscles. Circular muscle contractions act on the fluid in the closed coelomic space to make the segment long and thin. In contrast, when longitudinal muscles contract, the segment becomes shorter and wider. Partitions, called septa, prevent fluid from moving from one segment to the next, so muscle contractions change the shape of a segment, but not its volume. To understand why this shape change is important, one has only to watch an earthworm crawl or burrow. The worm uses short wide segments as anchors and extends its body

forward from an anchor point by becoming long and thin. The front of the worm then anchors in its new position and the rest of the body is pulled up behind the anchor point, after which forward extension begins again. A burrower requires a compartmentalized coelom.

Was the ancestral annelid a burrower? Looking at an earthworm, one is tempted to picture the ancestral annelid as a streamlined metameric animal with minimal sensory structures. One might assume that enhanced sensory structures and expanded bodies are derived features. However, recent phylogenies indicate that the earthworm morphology is derived. The ancestral annelid had flap-like extensions on the sides of its body and sensory palps on its head that were used also to collect food. The ancestral annelid was a less-elaborate version of some of the marine worms you will see in this chapter. Descendant annelids diverged, with some evolving characters suited to an active crawling or swimming lifestyle, and others adapting to a sessile or burrowing lifestyle. Descendants of the latter group colonized freshwater and land.

The wormlike animal phyla described in this chapter are coelomate protostomes belonging to subgroup Lophotrochozoa. They develop by spiral mosaic cleavage, form mesoderm from derivatives of the 4d cell (see p. 170), make a coelom by schizocoely, and share a trochophore as the ancestral larval form. Two phyla are discussed: Annelida and Sipuncula.

Members of phylum Annelida are segmented worms living in marine, freshwater, and moist terrestrial habitats. Marine bristle worms, leeches, and the familiar earthworms belong to this group. Annelida now includes pogonophoran and vestimentiferan worms, formerly either placed together in phylum Pogonophora, or placed in distinct phyla: Pogonophora and Vestimentifera. These deep-ocean worms belong in clade Siboglinidae. Members of the former phylum Echiura are placed within Annelida. These sausage-shaped "spoon worms" are not segmented, but do contain repeated body parts that suggest a segmented ancestor.

Worms in phylum Sipuncula are benthic marine animals with unsegmented bodies. Several studies using molecular sequence data place sipunculans within phylum Annelida, but not all data sets support this placement. We depict Sipuncula as the sister taxon to Annelida (Figure 17.1).

17.1 PHYLUM ANNELIDA, INCLUDING POGONOPHORANS (SIBOGLINIDS) AND ECHIURANS

Phylum Annelida (an-nel'i-da) (L. *annelus,* little ring, + *ida,* pl. suffix) consists of the segmented worms. It is a diverse phylum, numbering approximately 15,000 species, the most familiar of which are earthworms and freshwater worms (oligochaetes) and leeches (class Hirudinida). However, approximately two-thirds of the phylum comprises marine worms (polychaetes), which are less familiar to most people. Some polychaetes are grotesque in appearance, whereas others are graceful and beautiful. They include clamworms, plumed worms, parchment worms, scaleworms, lugworms, and many others.

Annelida are worms whose bodies are divided into similar **segments** (also called metameres) arranged in linear series and externally marked by circular rings called **annuli** (the name of the phylum refers to this characteristic). Body segmentation (**metamerism**) is a division of the body into a series of segments, each of which contains similar components of all major organ systems. In annelids the segments are delimited internally by septa.

Annelids are sometimes called "bristle worms" because, with the exception of leeches, most annelids bear tiny chitinous bristles called **setae** (L. *seta,* hair or bristle). Short, needlelike setae help anchor segments during locomotion and long, hairlike setae aid aquatic forms in swimming. Since many annelids burrow or live in secreted tubes, stiff setae also aid in preventing the worm from being pulled out or washed out of its home.

Annelids have a worldwide distribution, and a few species are cosmopolitan. Polychaetes are chiefly marine forms. Most are benthic, but some live pelagic lives in the open seas. Oligochaetes and leeches occur predominantly in freshwater or terrestrial soils. Some freshwater species burrow in mud and sand and others among submerged vegetation. Many leeches are predators, specialized for piercing their prey and feeding on blood or soft tissues. A few leeches are marine, but most live in freshwater or in damp regions. Suckers are typically found at both ends of the body for attachment to the substratum or to their prey.

Body Plans

The annelid body typically has a two-part head, composed of a **prostomium** and a **peristomium,** followed by a segmented body and a terminal portion called the **pygidium** bearing an anus (Figure 17.2). The head and pygidium are not considered to be segments. New segments differentiate during development just in front of the pygidium; thus the oldest segments are at the anterior end and the youngest segments are at the posterior end. Each segment typically contains respiratory, nervous, and excretory structures, as well as a coelom. Flaps called **parapodia** may be present on each segment.

In most annelids the coelom develops embryonically as a split in the mesoderm on each side of the gut (**schizocoel**), forming a pair of coelomic compartments in each segment. **Peritoneum** (a layer of mesodermal epithelium) lines the body wall of each compartment, forming dorsal and ventral **mesenteries** that cover all organs (Figure 17.3). Peritonea of adjacent segments meet to form **septa,** which are perforated by the gut and longitudinal blood vessels. The body wall surrounding the peritoneum and coelom contains strong circular and longitudinal muscles adapted for swimming, crawling, and burrowing (Figure 17.3).

Except in leeches, the coelom of most annelids is filled with fluid and serves as a **hydrostatic skeleton** as described in the opening essay. Crawling motions are produced by alternating waves of contraction by longitudinal and circular muscles passing down the body (peristaltic contractions). Forces powerful enough for rapid burrowing as well as locomotion can thus be generated. Swimming forms use undulatory rather than peristaltic movements in locomotion.

An annelid body has a thin outer layer of nonchitinous cuticle (see Chapter 29, p. 640) surrounding the epidermis (Figure 17.3). Paired epidermal setae (Figures 17.2 and 17.20) are ancestral for annelids, although they have been reduced or lost in some. The annelid digestive system is not segmented: the gut runs the length of the body perforating each septum (Figure 17.3). Longitudinal dorsal and ventral blood vessels follow the same path, as does the ventral nerve cord.

Historically, annelids were divided among three classes: Polychaeta, Oligochaeta, and Hirudinida. Phylogenetic analyses show polychaetes and oligochaetes to be paraphyletic groups; leeches form a clade. Leeches arose within Oligochaeta; together the two groups form a clade called Clitellata, characterized by presence of a reproductive structure called a **clitellum** (see p. 374). Oligochaetes arose within Polychaeta, so the term "polychaete" is descriptive rather than taxonomic.

Polychaete is a term used to denote any of 80 morphologically distinct families of worms, typically those with many setae. The evolutionary relationships among these families have not been easy to discern, but recent phylogenies based on molecular characters have supported two main groups of annelids: Errantia and Sedentaria. The division of annelids into two such groups, one errant (freely moving; Figure 17.2) and the other sedentary (spending most of their lives in tubes or burrows; Figure 17.4) was proposed long ago on

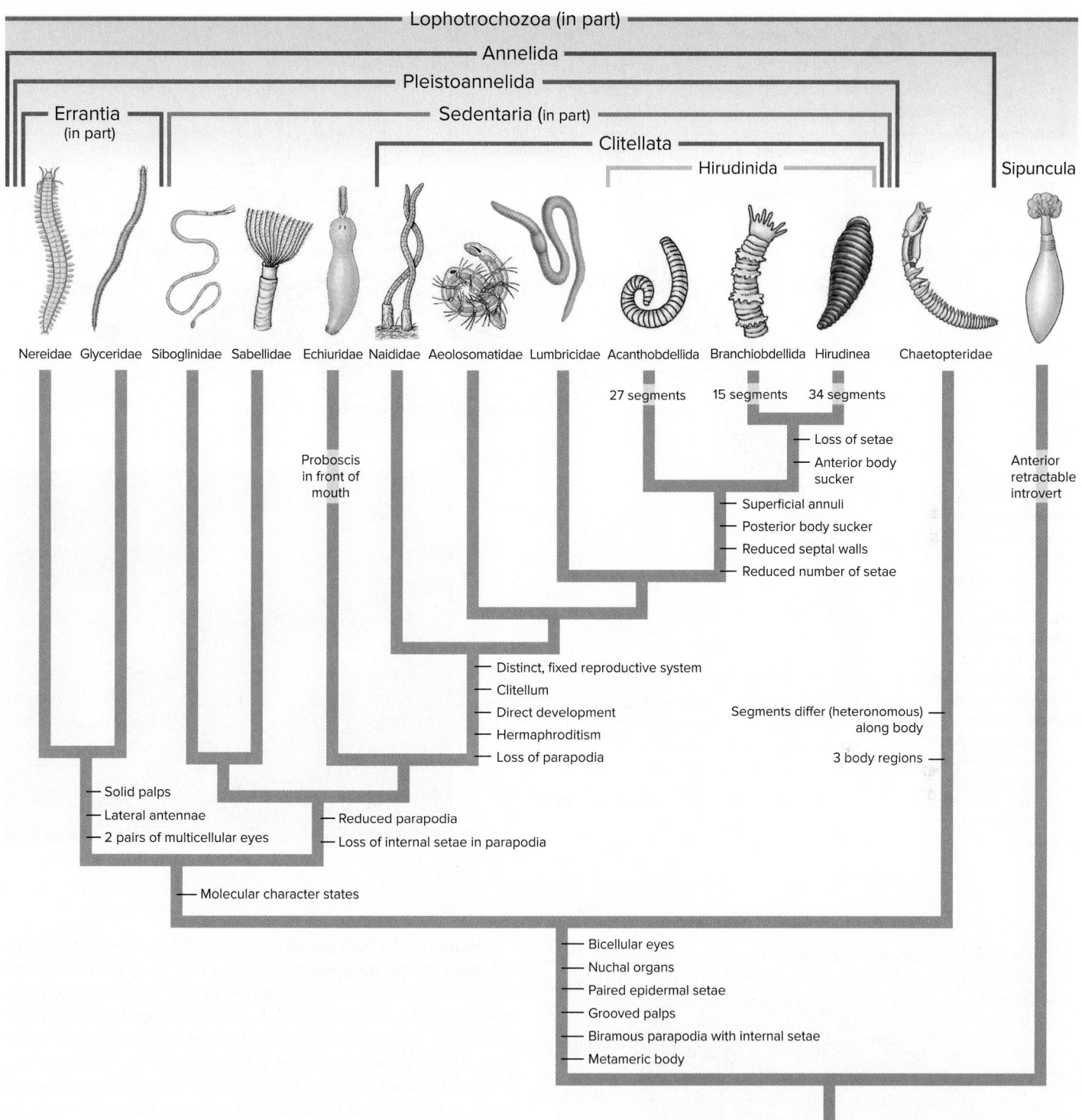

Figure 17.1 Cladogram of annelids, based on morphological characters and a molecular phylogeny by Struck et al. 2011 (see p. 384 for a reference). Most annelids belong to two large clades: Errantia and Sedentaria. Annelids with a clitellum are nested within Sedentaria.

morphological grounds and later rejected. Recently these two groups have been supported by phylogenies based on molecular characters. Sedentaria contains some polychaetes, as well as oligochaetes and leeches (Clitellata; Figure 17.1). Most other polychaetes are in Errantia (Latin *errare*, to wander). Many of these, like the clam worm, *Nereis* (Greek, name of a sea nymph), are predatory. They

have an eversible muscular pharynx armed with jaws or teeth that can be thrust out with surprising speed to capture prey.

Errantia and Sedentaria together form the new group Pleistoannelida. There are a few polychaetes placed outside Pleistoannelida (Figure 17.1); of these excluded annelids we discuss only Chaetopteridae.

Figure 17.2
Nereis virens (A–D) and *Nereis diversicolor* (E) are errant polychaetes. **A,** Anterior end, with pharynx everted. **B,** External structure. **C,** Posterior end. **D,** Generalized transverse section through region of the intestine. **E,** In this photo of a live *N. diversicolor,* note the well-defined segments, the lobed parapodia, and the prostomium with tentacles in the upper center.

A (diagram labels): Jaw, Prostomial tentacles, Palp, Prostomium, Peristomium, Parapodia, Everted pharynx, Eyes, Tentacles (cirri)

B (diagram label): Parapodium

C (diagram labels): Parapodia, Pygidium, Anus, Cirrus

D (diagram labels): Dorsal cirrus, Oblique muscle, Dorsal vessel, Intestine, Coelomic epithelium, Longitudinal muscle, Circular muscle, Notopodium, Respiratory capillaries, Eggs, Parapodium, Setae, Neuropodium, Epidermis, Aciculum, Ventral cirrus, Nephridium, Nerve cord, Ventral vessel

E: © Jeff Goodman/NHPA//Newscom

Figure 17.3 Annelid body plan.

Diagram labels: Circular muscle, Epidermis, Cuticle, Segment, Intestine, Longitudinal muscle, Parietal peritoneum, Septum, Ventral blood vessel, Dorsal mesentery, Intestine, Septum, Visceral peritoneum, Ventral mesentery, Ventral blood vessel, Dorsal blood vessel, Circular muscle, Longitudinal muscle, Ventral mesentery

Members of Chaetopteridae are unusual tube-dwelling annelids with three distinct body regions (Figure 17.5). The parchment worm, *Chaetopterus* (Gr. *chaitē,* long hair, + *pteron,* wing), feeds on suspended particles (Figure 17.5). It lives in a U-shaped, parchmentlike tube buried, except for the tapered ends, in sand or mud along the shore. The worm attaches to the side of the tube by ventral suckers. Fans (modified parapodia on segments 14 to 16) pump water through the tube by rhythmical movements. A pair of enlarged parapodia on segment 12 secretes a long mucous net that reaches back to a small food cup just in front of the fans. All water passing through the tube is filtered through this mucous net, the end of which is rolled into a ball by cilia in the cup. When the ball is about the size of a BB shot (about 3 mm diameter), the fans stop beating and the ball of food and mucus is rolled forward by ciliary action to the mouth and swallowed.

Pleistoannelida

Pleistoannelida comprises two large clades, as mentioned previously. We will provide a brief description of the errant polychaete body

A

B

Figure 17.4   Tube-dwelling sedentary polychaetes. **A,** Christmas-tree worm, *Spirobranchus giganteus,* lives in a calcareous tube. On its head are two whorls of modified tentacles (radioles) used to collect suspended food particles from the surrounding water. Notice the finely branched filters visible on the edge of one radiole. **B,** Sabellid polychaetes, *Bispira brunnea,* live in leathery tubes.

plan and then discuss representative errant worms. Descriptions of the major types of sedentary annelids, including members of Clitellata, follow the discussion on errant taxa.

Errantia

Errantia is one of two large subgroups within Pleistoannelida. It comprises the motile polychaetes.

Most errant polychaetes (Gr. *polys,* many, + *chaitē,* long hair) are marine, and while most are 5 to 10 cm long, some are less than 1 mm, and others may be as long as 3 m. They may be brightly colored in reds and greens, iridescent, or dull.

Many polychaetes are euryhaline and can tolerate a wide range of environmental salinity. The freshwater polychaete fauna are more diversified in warmer regions than in temperate zones.

Many polychaetes live under rocks, in coral crevices, or in abandoned shells. Some are planktonic. They play a significant part in marine food chains because they are eaten by fish, crustaceans, hydroids, and many other predators. Polychaetes have a well-differentiated head with specialized sense organs; paired appendages, called **parapodia,** on most segments; and no clitellum (Figure 17.2). As their name implies, they have many setae, usually arranged in bundles on the parapodia. They exhibit the most pronounced specialization of sensory organs found in annelids (see p. 368).

Form and Function

A polychaete typically has a **prostomium,** which may or may not be retractile and which often bears eyes, tentacles, and sensory palps (Figure 17.2). The **peristomium** surrounds the mouth and may bear setae, palps, or, in predatory forms, chitinous jaws (Figure 17.4).

The polychaete trunk is segmented, and most segments bear parapodia, which may have lobes, cirri, setae, and other parts on them (Figure 17.2). Parapodia typically are used in crawling. They usually serve as the chief respiratory organs, although some polychaetes also have gills.

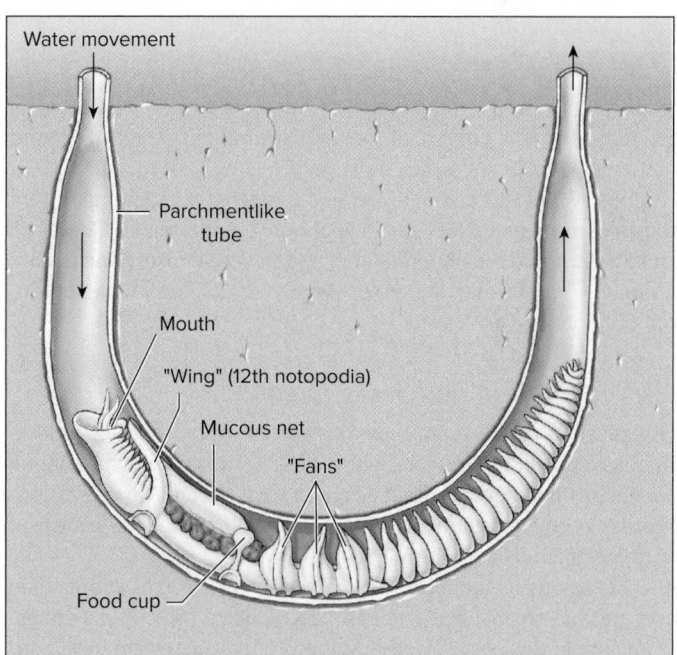

Figure 17.5   *Chaetopterus,* a sedentary polychaete, lives in a U-shaped tube in the sea bottom. It pumps water through the parchmentlike tube (of which one-half has been cut away here) with its three pistonlike fans. The fans beat 60 times per minute to keep water currents moving. The winglike notopodia of the twelfth segment continuously secrete a mucous net that strains out food particles. As the net fills with food, the food cup rolls it into a ball, and when the ball is large enough (about 3 mm), the food cup bends forward and deposits the ball in a ciliated groove to be carried to the mouth and swallowed.

Nutrition

Errant polychaetes are usually predators and scavengers. A polychaete's digestive system consists of a foregut, a midgut, and a hindgut. The foregut includes a stomodeum, a pharynx, and an anterior esophagus. It is lined with cuticle, and the jaws, where present, are constructed of cuticular protein. The more anterior portions of the midgut secrete digestive enzymes but absorption occurs toward the posterior end. A short hindgut connects the midgut to the exterior via the anus, which is on the pygidium.

Circulation and Respiration

Polychaetes show considerable diversity in both circulatory and respiratory structures. As previously mentioned, parapodia and gills serve for gaseous exchange in various species. However, in some polychaetes there are no special organs for respiration, and gaseous exchange occurs across the body surface.

The circulatory pattern varies greatly. In *Nereis* a dorsal longitudinal vessel carries blood anteriorly, and a ventral longitudinal vessel conducts it posteriorly (Figure 17.2D). Blood flows between these two vessels via segmental networks in the parapodia and septa, and around the intestine. In the burrowing predatory worm *Glycera* (Gr. *Glykera,* a feminine proper name) the circulatory system is reduced and joins directly with the coelom. Septa are incomplete, and thus the coelomic fluid assumes the function of circulation.

Many polychaetes have respiratory pigments such as hemoglobin, chlorocruorin, or hemerythrin (pp. 698–699).

Excretion

Excretory organs consist of protonephridia and mixed protoand metanephridia in some, but most polychaetes have metanephridia (Figure 17.2). There is one pair per segment, each ending with a **nephrostome** that opens into a coelomic compartment. Coelomic fluid passes into the nephrostome, and selective resorption occurs along the nephridial duct (see Figure 17.21; see Chapter 30, pp. 665–666).

Nervous System and Sense Organs

Organization of the central nervous system in polychaetes follows the basic annelid plan (see Figure 17.22). Dorsal cerebral ganglia connect with a subpharyngeal ganglion via a circumpharyngeal connective. A double ventral nerve cord courses the length of the worm, with metamerically arranged ganglia.

Sense organs are highly developed in polychaetes and include eyes, nuchal organs, and statocysts. Eyes, when present, may range from simple eyespots to well-developed organs. Eyes are most conspicuous in errant worms. Usually the eyes are retinal cups, with rodlike photoreceptor cells (lining the cup wall) directed toward the lumen of the cup. The highest degree of eye development occurs in the family Alciopidae, which has large, image-resolving eyes similar in structure to those of some cephalopod molluscs (see Figure 16.41, p. 357), with cornea, lens, retina, and retinal pigments. Alciopid eyes also have accessory retinas, a characteristic independently evolved by deep-sea fishes and some deep-sea cephalopods. The accessory retinas of alciopids are sensitive to different wavelengths. The eyes of these pelagic animals may be well adapted to function because penetration by the different wavelengths of light varies with depth. Studies with electroencephalograms show that they are sensitive to dim light of the deep sea. Nuchal organs are ciliated sensory pits or slits that appear to be chemoreceptive, an important factor in food gathering. Some burrowing and tube-building polychaetes have statocysts that function in body orientation.

Reproduction and Development

Polychaetes have no permanent sex organs, and they usually have separate sexes. Reproductive systems are simple: gonads appear as temporary swellings of the peritoneum and shed their gametes into the coelom. The gametes are then carried to the outside through gonoducts, through the metanephridia, or by rupture of the body wall. Fertilization is external, and the early larva is a trochophore (see Figure 16.7, p. 338).

Some polychaetes live most of the year as sexually immature animals called atokes, but during the breeding season a portion of the body becomes sexually mature and swollen with gametes (Figure 17.6). An example is the palolo worm, which lives in burrows among coral reefs. During the swarming period, the sexually mature portions, now called epitokes, break off and swim to the surface. Just before sunrise, the sea is literally covered with them, and at sunrise they burst, freeing eggs and sperm for fertilization. Anterior portions of the worms regenerate new posterior sections. Swarming is of great adaptive value because the synchronous maturation of all the epitokes ensures the maximum number of fertilized eggs. However, this reproductive strategy is very hazardous; many types of predators have a feast on the swarming worms. In the meantime, the atoke remains safely in its burrow to produce another epitoke at the next cycle. In some polychaetes, epitokes arise from atokes by asexual budding (Figure 17.7) and become complete worms.

Representative Members of Errantia

Clam Worms: *Nereis* Clam worms (Figure 17.2), or sand worms as they are sometimes called, are errant polychaetes that live in mucus-lined burrows in or near low tide. Sometimes they occupy temporary hiding places, such as under stones, where they stay with their bodies covered and their heads protruding. They are most active at night, when they wiggle out of their hiding places and swim or crawl over the sand in search of food.

The body, containing about 200 segments, may grow to 30 or 40 cm in length. The head is composed of a prostomium and a peristomium. The prostomium bears a pair of stubby palps, sensitive to touch and taste; a pair of short sensory tentacles; and two pairs of small dorsal eyes that are light sensitive. The peristomium bears the ventral mouth, a pair of chitinous jaws, and four pairs of sensory tentacles (Figure 17.2A).

Each parapodium has two lobes: a dorsal **notopodium** and a ventral **neuropodium** (Figure 17.2D) that bear setae with many blood vessels. Parapodia are used for both creeping and swimming and are controlled by oblique muscles that run from the midventral line to the parapodia in each segment. The worm swims by lateral undulatory movement of the body. It can dart through the water with considerable speed. These undulatory movements can also be used to suck water into or pump it out of the burrow.

Clam worms feed on small animals, other worms, and a variety of larval forms. They seize food with their chitinous jaws, which they

Figure 17.6 *Eunice viridis,* the Samoan palolo worm. The posterior segments form the epitokal region, consisting of segments packed with gametes. Each segment has an eyespot on the ventral side. Once a year the worms swarm, and the epitokes detach, rise to the surface, and discharge their ripe gametes, leaving the water milky. By the next breeding season, the epitokes are regenerated.

Figure 17.8 A scale worm, *Hesperonoe adventor,* normally lives as a commensal in the tubes of *Urechis* (family Echiuridae, p. 372).

protrude through the mouth when they evert their pharynx. Food is swallowed as the worm withdraws its pharynx. Movement of food through the alimentary canal is by peristalsis.

Scale Worms   Scale worms (Figure 17.8) are members of the family Polynoidae (Gr. *polynoē,* daughter of Nereus and Doris, a sea god and goddess), one of the most diverse, abundant, and widespread of polychaete families. Their flattened bodies are covered with broad scales, modified from dorsal parts of the parapodia. Most species are of modest size, but some are enormous (up to 190 mm long and 100 mm wide). They are carnivorous and eat a wide variety of animals. Many are commensal, living in burrows of other polychaetes or in association with cnidarians, molluscs, or echinoderms.

Fireworms   *Hermodice carunculata* (Gr. *herma,* reef, + *dex,* a worm found in wood) (Figure 17.9) and related species are called fireworms because their hollow, brittle setae contain a poisonous secretion. The setae puncture a hand that touches them, and then break off in the wound to cause skin irritation. Fireworms feed on corals, gorgonians, and other cnidarians.

Figure 17.9 A fireworm, *Hermodice carunculata,* feeds on gorgonians and stony corals. Its setae are like tiny glass fibers and serve to ward off predators.

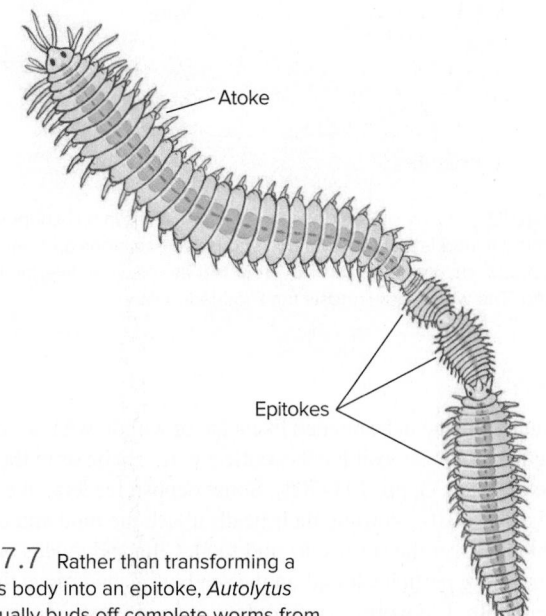

Figure 17.7 Rather than transforming a portion of its body into an epitoke, *Autolytus prolife* asexually buds off complete worms from its posterior end that become sexual epitokes.

Sedentaria

Sedentaria contains many polychaetes and oligochaetes that live in tubes or burrows, including members of the former phyla Pogonophora and Echiura. It also includes members of Clitellata.

The body plan of sedentary polychaetes is much like that of errant polychaetes, except that that the head is often modified by the addition of tentacles used for food capture. Parapodia are often small and sometimes are modified to help anchor the worm in the tube; setae may be hooklike to attach to the tube wall. Parapodia may function in respiration, but many tube-dwellers also have gills. *Amphitrite* (Gr., a mythical sea nymph), for example, has three pairs of branched gills and long extensible tentacles (Figure 17.10). *Arenicola*

Characteristics of Phylum Annelida

1. Unique annelid head and paired epidermal setae present (lost in leeches and some echiurans); parapodia present in the ancestral condition
2. Marine, freshwater, and terrestrial
3. Most free-living, some symbiotic, some ectoparasitic
4. Body bilaterally symmetrical, **metameric,** often with distinct head; metamerism reduced or lost in some, especially echiurans
5. Triploblastic body
6. Coelom (schizocoel) well developed and divided by septa, except in leeches; coelomic fluid functions as hydrostatic skeleton
7. Epithelium secretes outer transparent, moist cuticle
8. Digestive system complete and not segmentally arranged
9. Body wall with outer circular and inner longitudinal muscle layers
10. Nervous system with a double ventral nerve cord and a pair of ganglia with lateral nerves in each segment; brain a pair of dorsal cerebral ganglia with connectives to ventral nerve cord
11. Sensory system of tactile organs, taste buds, statocysts (in some), photoreceptor cells, and eyes with lenses (in some); specialization of head region into differentiated organs, such as tentacles, palps, and eyespots of polychaetes
12. Asexual reproduction by fission and fragmentation; capable of complete regeneration
13. Hermaphroditic or separate sexes; larvae, if present, are trochophore type; asexual reproduction by budding in some; spiral cleavage and mosaic development
14. Excretory system typically a **pair of nephridia for each segment;** nephridia remove waste from blood as well as from coelom
15. Respiratory gas exchange through skin, **gills,** or **parapodia**
16. **Circulatory system closed** with muscular blood vessels and aortic arches ("hearts") for pumping blood, segmentally arranged; respiratory pigments (hemoglobin, hemerythrin, or chlorocruorin) often present; amebocytes in blood plasma

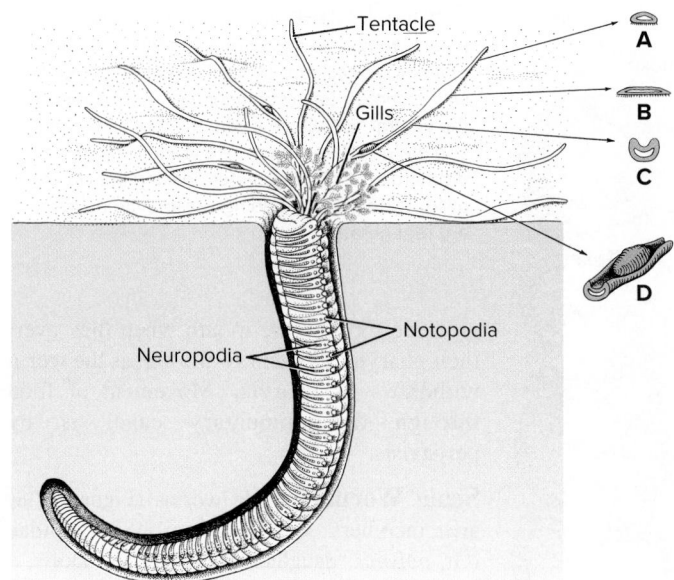

Figure 17.10 *Amphitrite,* which builds its tubes in mud or sand, extends long, grooved tentacles out over the mud to pick up bits of organic matter. The smallest particles are moved along food grooves by cilia, larger particles by peristaltic movement. Its plumelike gills are blood red. **A,** Section through exploratory end of tentacle. **B,** Section through tentacle in area adhering to substratum. **C,** Section showing ciliary groove. **D,** Particle being carried toward mouth.

Figure 17.11 *Arenicola,* the lugworm, lives in a U-shaped burrow in intertidal mudflats. It burrows by successive eversions and retractions of its proboscis. By peristaltic movements it keeps water filtering through the sand. The worm then ingests the food-laden sand.

(L. *arena,* sand, + *colo,* inhabit), the burrowing lugworm (Figure 17.11), has paired gills on certain segments. Another modification is the dual use of tentacles on the head in food capture and in respiration for some tube-dwellers (see Figure 17.4). The expanded surface area afforded by branching within the tentacles is useful for both functions.

Representative Members of Sedentaria

Tubeworms   Polychaete tube-dwellers secrete many types of tubes. Some are parchmentlike or leathery (see Figure 17.4B); some are firm, calcareous tubes attached to rocks or other surfaces (see Figure 17.4A); and some are simply grains of sand or bits of shell or seaweed cemented together with mucous secretions. Many species burrow in sand or mud, lining their burrows with mucus (Figure 17.11).

Most sedentary tube- and burrow-dwellers are particle feeders, using cilia or mucus to obtain food, typically plankton and detritus. Ciliary feeders feed on suspended particles and may bear a crown of

tentacles that can be opened like a fan or withdrawn into the tube (see Figure 17.4). Deposit feeders collect particles on or in the sediments see Chapter 32, pp. 703–705). Some deposit feeders, like *Amphitrite* (Figure 17.10), protrude their heads above the mud and extend long tentacles over the surface to find food. Cilia and mucus on the tentacles entrap particles found on the sea bottom and move them toward the mouth. Lugworms, *Arenicola,* use an interesting combination of suspension and deposit feeding. They live in a U-shaped burrow through which, by peristaltic movements, they cause water to flow. Food particles are trapped by the sand at the front of the burrow, and *Arenicola* then ingests the food-laden sand (Figure 17.11).

Fanworms, or "featherduster" worms, are beautiful tubeworms, fascinating to watch as they emerge from their secreted tubes and unfurl their lovely tentacular crowns to feed (see Figure 17.4). A slight disturbance, sometimes even a passing shadow, causes them to duck back quickly into the safety of the homes. Food drawn to the feathery arms, or **radioles,** by ciliary action is trapped in mucus and carried down ciliated food grooves to the mouth (Figure 17.12). Particles too large for the food grooves pass along the margins of the food grooves and fall away before they reach the mouth. Only small particles of food enter the mouth; sand grains are stored in a sac to be used later in enlarging the tube.

Family Siboglinidae (Pogonophorans)   Members of former phylum Pogonophora (pō′gō-nof′e-ra) (Gr. *pōgōn,* beard, + *phora,* bearing), or beardworms, were entirely unknown before the twentieth century. The first specimens to be described were collected from deep-sea dredgings off the coast of Indonesia in 1900. They have since been discovered in several seas, including the western Atlantic off the U.S. eastern coast. Some 150 species have been described so far. Most species are less than 1 mm in diameter but they can be 10 to 75 cm in length.

Most siboglinids live in mud on the ocean floor, at depths of 100 to 10,000 m. This location accounts for their delayed discovery, for they are obtained only by dredging. They are sessile animals that secrete very long chitinous tubes in which they live, and probably only extend the anterior end of their body for absorbing nutrients. The tubes are generally oriented upright in bottom sediments. A tube can be three to four times the length of the animal, which can move up or down inside its tube but cannot turn around.

Beardworms have a long, cylindrical body covered with cuticle. Cuticle, epidermis, and circular and longitudinal muscles compose the body wall. The body is divided into a short anterior **forepart;** a long, very slender **trunk;** and a small, segmented **opisthosoma** (Figure 17.13). Paired epidermal setae are present on the trunk and opisthosoma. At the anterior end of the body, a cephalic lobe bears

from 1 to 260 long tentacles (the "beard" that gives this phylum its name), depending on species. Tentacles are hollow extensions of the coelom and bear minute pinnules. For a part or all of their length, tentacles lie parallel with each other, enclosing a cylindrical intertentacular space into which the pinnules project (Figure 17.14).

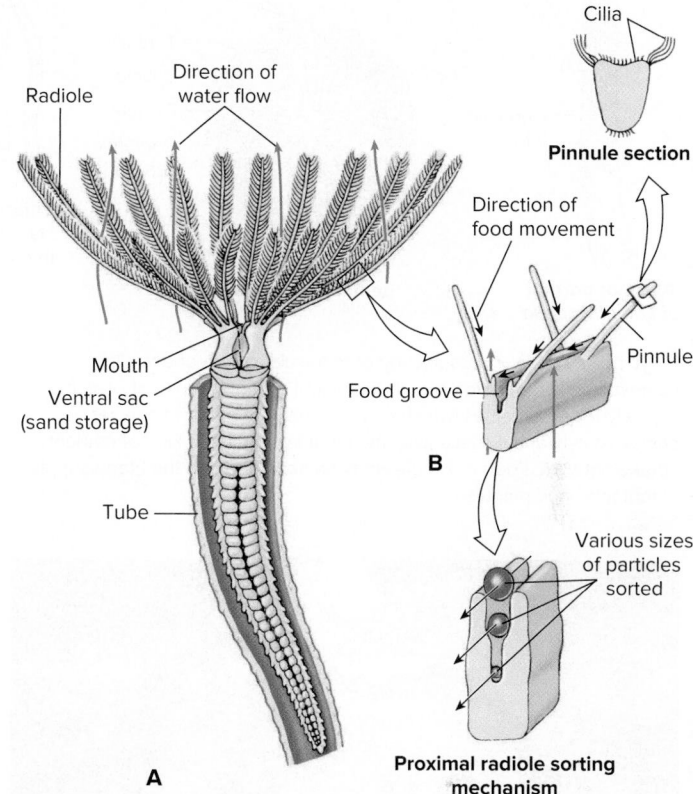

Figure 17.12 *Sabella,* a polychaete ciliary feeder, extends its crown of feeding radioles from its leathery secreted tube, reinforced with sand and debris. **A,** Anterior view of the crown. Cilia direct small food particles along grooved radioles to mouth and discard larger particles. Sand grains are directed to storage sacs and later are used in tube building. **B,** Distal portion of radiole showing ciliary tracts of pinnules and food grooves.

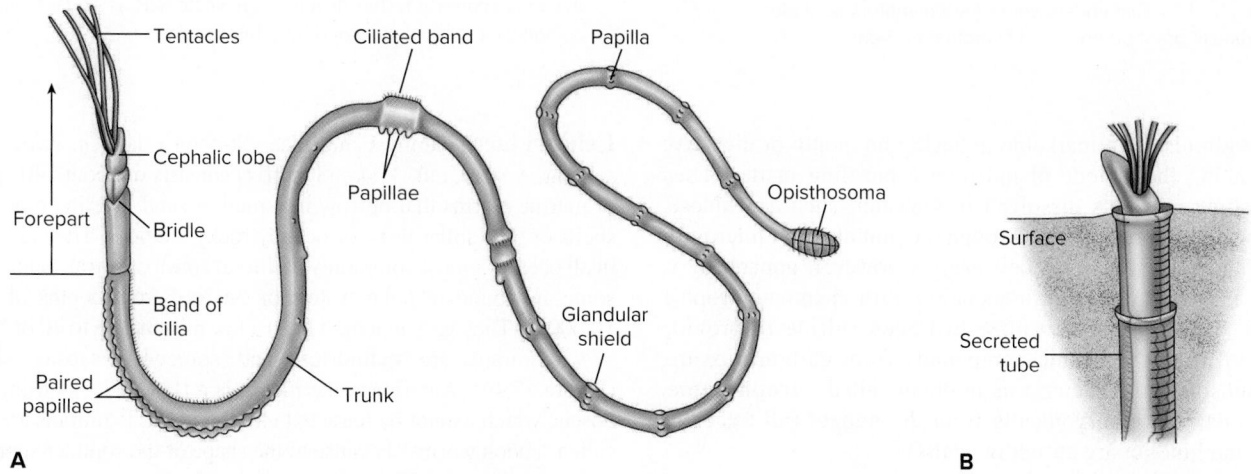

Figure 17.13 Diagram of a typical siboglinid. **A,** External features. In life, the body is much more elongated than shown in this diagram. **B,** Position in tube.

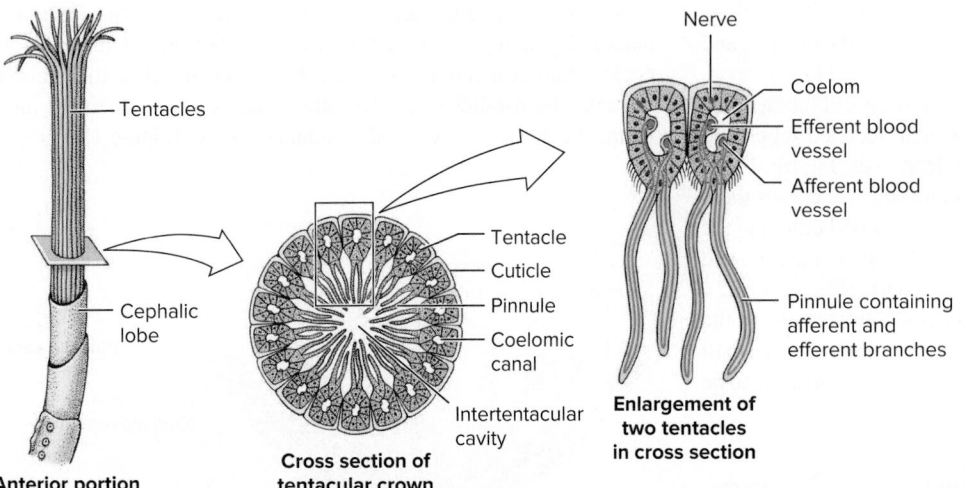

Anterior portion of *Lamellisabella*

Tentacles

Cephalic lobe

Cross section of tentacular crown

Tentacle
Cuticle
Pinnule
Coelomic canal
Intertentacular cavity

Nerve

Coelom
Efferent blood vessel
Afferent blood vessel

Pinnule containing afferent and efferent branches

Enlargement of two tentacles in cross section

Figure 17.14   Cross section of tentacular crown of siboglinid *Lamellisabella*. Tentacles arise from ventral side of forepart at base of cephalic lobe. Tentacles (which vary in number in different species) enclose a cylindrical space, with the pinnules forming a kind of nutrient uptake network. Food molecules may be absorbed into the blood supply of tentacles and pinnules.

OAR/NURP/College of William Mary/NOAA

Figure 17.15   Giant beardworms (vestimentiferans, clade Siboglinidae) at great depth near a hydrothermal vent.

Among the most amazing animals found in deep-water, Pacific rift communities (see Chapter 38, p. 830) are vestimentiferans, *Riftia pachyptila*. These giant beardworms live around deep-water hydrothermal vents and grow up to 3 m long and 5 cm in diameter (Figure 17.15). The trophosome of other siboglinids is confined to the posterior part of the trunk, which is buried in sulfide-rich sediments, but the trophosome of *Riftia* occupies most of its large trunk. It has a much larger supply of hydrogen sulfide, enough to nourish its large body, in the effluent of the hydrothermal vents.

Sexes are separate, with a pair of gonads and a pair of gonoducts in the trunk section. Little developmental work has been done on these deep-sea worms, but research suggests that cleavage is unequal and atypical. It seems to be closer to spiral than to radial. Development of the apparent coelom is schizocoelic, not enterocoelic as was originally described. The worm-shaped embryo is ciliated but a poor swimmer. It is probably carried by water currents until it settles.

The discovery of bone-eating marine worms in genus *Osedax* was very exciting. These siboglinid polychaetes bore into bones of whale carcasses through a "root" system and use endosymbiotic bacteria to digest the bone. The green, branching, vascularized "root" system develops from the posterior part of the egg sac and invades bone marrow. One might wonder if whale falls (carcasses) are sufficiently abundant for this lifestyle to succeed—surprisingly, they are typically only 5 to 15 km apart in coastal regions where the worms occur. Female worms take no chances in finding mates; dwarf males share the tubes of females. How do males and females find each other? Sex determination seems to be environmental, so larvae that land on females become male. Bone-eating worms are only one component of a rich and dynamic ecosystem that develops on whale falls as species that colonize early are succeeded by other species.

Siboglinids are remarkable in having no mouth or digestive tract, making their mode of nutrition a puzzling matter. They absorb some nutrients dissolved in seawater, such as glucose, amino acids, and fatty acids, through the pinnules and microvilli of their tentacles. Most of their energy, however, apparently is derived from a mutualistic association with chemoautotrophic bacteria. These bacteria oxidize hydrogen sulfide to provide energy to produce organic compounds from carbon dioxide. Siboglinids bear the bacteria in an organ called a **trophosome,** which is derived embryonically from the midgut (all traces of foregut and hindgut are absent in adults).

There is a well-developed closed circulatory system. Photoreceptors are similar to those of other annelids.

Echiuridae   Family Echiuridae (ek-ē-yur′i-da) (Gr. *echis*, viper, serpent, + *oura*, tail, + *ida,* pl. suffix) consists of about 140 species of marine worms that burrow into mud or sand, live in empty snail shells or sanddollar tests, or occupy rocky crevices. They are found in all oceans—most commonly in littoral zones of warm waters—but some are found in polar waters or dredged from depths of up to 10,000 m. They vary in length from a few millimeters to 40 or 50 cm.

Echiurans are cylindrical and somewhat sausage-shaped (Figure 17.16). Anterior to the mouth is a flattened, extensible proboscis, which cannot be retracted into the trunk. Echiurans are often called "spoon worms" because of the shape of the contracted proboscis in some species. The nervous system of echiurans is fairly simple, with a ventral nerve cord that runs the length of the trunk and

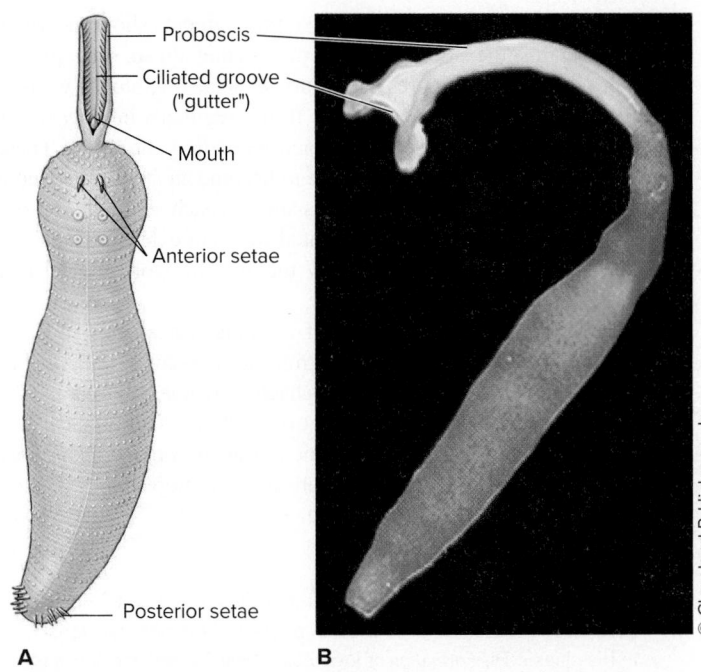

Figure 17.16 **A,** *Echiurus,* an echiuran common on both Atlantic and Pacific coasts of North America. **B,** *Anelassorhynchus,* an echiuran of the tropical Pacific. The shape of their proboscis lends them the common name of "spoon worms."

© Cleveland P. Hickman, Jr.

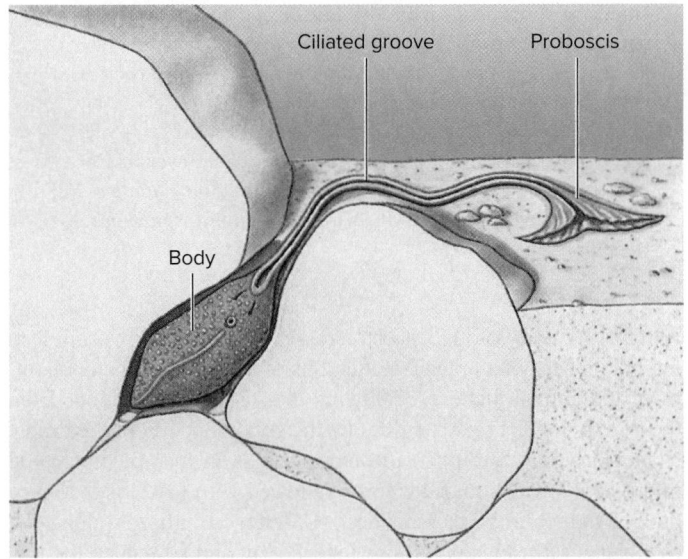

Figure 17.17 *Bonellia* is a detritus feeder. Lying in its burrow, it explores the surface with its long proboscis, which picks up organic particles and carries them along a ciliated groove to the mouth.

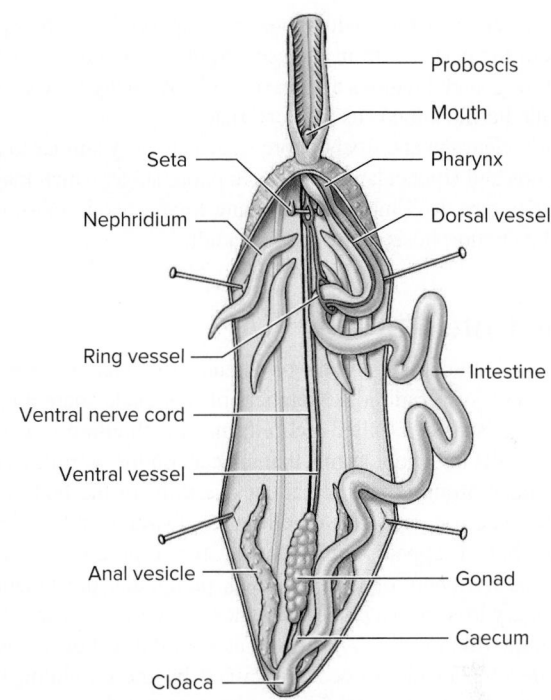

Figure 17.18 Internal anatomy of an echiuran worm.

One common form, *Urechis* (Gr. *oura,* tail, + *echis,* viper, serpent), has a very short proboscis and lives in a U-shaped burrow in which it secretes a funnel-shaped mucous net. It pumps water through the net, capturing bacteria and fine particulate material in it. *Urechis* periodically swallows the food-laden net. *Lissomyema* (Gr. *lissos,* smooth, + *mys,* muscle) lives in empty gastropod shells in which it constructs galleries irrigated by rhythmical pumping of water and feeds on detritus and the organic coating of sand and mud gathered by this process.

Cuticle and epithelium, which may be smooth or ornamented with papillae, cover the muscular body wall. The coelom is large. The digestive tract is long and coiled and terminates at the posterior end (Figure 17.18). A pair of anal sacs may have an excretory and osmoregulatory function. Most echiurans, with the exception of *Urechis,* have a closed circulatory system with colorless blood but contain hemoglobin in coelomic corpuscles and certain body cells. There are one to many pairs of nephridia, which serve mainly as gonoducts in some species. Gas exchange probably occurs primarily in the hindgut, which is continually filled and emptied by cloacal irrigation. Echiurans are not segmented, but they presumably had a segmented ancestor. Many species retain the ancestral paired epidermal setae, either anteriorly or as a row of bristles around the posterior end (Figure 17.16).

In some species sexual dimorphism is pronounced, with the female being much the larger of the two. *Bonellia* has an extreme sexual dimorphism, and tiny males live on the body of the female or in her nephridia. Determination of sex in *Bonellia* is very interesting. Free-swimming larvae are sexually undifferentiated. Those that settle on the proboscis of a female become males (1 to 3 mm long). About 20 males are usually found in a single female. Larvae that do not contact a female proboscis metamorphose into females. The stimulus for development into males is apparently a hormone produced by a female's proboscis.

continues dorsally into the proboscis. The proboscis has a ciliated groove leading to the mouth. While they lie buried, the proboscis can extend out over the mud for exploration and deposit feeding (Figure 17.17). Most species gather very small particles of detritus and move them along the proboscis by cilia; larger particles are moved by a combination of cilia and muscular action or by muscular action alone. Unwanted particles can be rejected along the route to the mouth. The proboscis is short in some forms and long in others. *Bonellia,* which is only 8 cm long, can extend its proboscis up to 2 m.

Sexes are separate, with gonads being produced by special regions of the peritoneum in each sex. Mature sex cells break loose from these gonadal regions and leave the body cavity by way of the nephridia. Fertilization is usually external.

Early cleavage and trochophore stages are very similar to those of annelids and sipunculans. The trochophore stage, which may last from a few days to 3 months, according to species, is followed by gradual metamorphosis to a wormlike adult.

Clade Clitellata

Clade Clitellata contains earthworms and their relatives as well as leeches in class Hirudinida. Members of this clade share a unique reproductive structure called a **clitellum.** The clitellum is a ring of secretory cells in the epidermis that appears on the worm's exterior as a fat band around the body about one-third of the body length from the anterior end. The clitellum is always visible in oligochaetes, but it appears only during the reproductive season in leeches. Members of Clitellata lack parapodia, presumably an evolutionary loss from a polychaete ancestor. Clitellates are all hermaphroditic (monoecious) animals that exhibit direct development: young develop inside a cocoon secreted by the clitellum, so no trochophore larva is visible. Small worms emerge from cocoons.

Oligochaetes

Oligochaetes do not form a monophyletic group. More than 3000 species of worms with an oligochaete body plan occur in a great variety of sizes and habitats. They include the familiar earthworms and many species that live in freshwater. Most are terrestrial or freshwater forms, but some are parasitic, and a few live in marine or brackish water.

With few exceptions, oligochaetes bear setae, which may be long or short, straight or curved, blunt or needlelike, or arranged singly or in bundles. Whatever the type, setae are less numerous in oligochaetes than in polychaetes, as is implied by the class name, which means "few long hairs." Aquatic forms usually have longer setae than do earthworms.

Form and Function

The main features of an oligochaete body are described with reference to the familiar earthworm. The circulatory system and excretory structures described in earthworms are typical of annelids in general, but the digestive and nervous systems have aspects specific to oligochaetes.

Earthworms, sometimes called "night crawlers," burrow in moist, rich soil, and usually live in branched, interconnected tunnels. The species commonly studied in laboratories is *Lumbricus terrestris* (L. *lubricum,* earthworm). It ranges in size from 12 to 30 cm long (Figure 17.19), but is small in comparison to giant tropical forms whose 4 m long bodies may comprise 150 to upward of 250 segments.

Earthworms normally emerge at night, but in damp, rainy weather they stay near the surface, often with mouth or anus protruding from the burrow. In very dry weather they may burrow several feet underground, coil in a slime chamber, and become dormant.

Earthworms use peristaltic movement as described in the essay that opens this chapter. In the central segments of the body, the longitudinal muscles contract to make the segments short and wide, pushing against the sides of the burrow. As they do so, setae project outward through small pores in the cuticle. Setae dig into the walls of the burrow to anchor the segments. In the segments in front of the anchor, the circular muscles contract, extending the body. These anterior segments then change shape to become anchors. When contractions of the longitudinal muscles shorten each segment, the rest of the body is drawn forward. Repeated waves of extension and contraction pass along the entire body moving the worm ahead (see Figure 29.5, p. 643).

The paired epidermal setae of oligochaetes are set in a sac within the body wall and moved by muscles (Figure 17.20), as they are in polychaetes. However, oligochaetes do not have parapodia; instead, the setae extend directly out of the body wall on each segment. In most earthworms each segment bears four pairs of chitinous setae (Figure 17.19C), although there may be more than 100 such setae per segment in some oligochaetes.

Aristotle called earthworms the "intestines of the soil." Some 22 centuries later Charles Darwin published his observations in his classic *The Formation of Vegetable Mould Through the Action of Worms.* He showed how worms enrich soil by bringing subsoil to the surface and mixing it with topsoil. An earthworm can ingest its own weight in soil every 24 hours, and Darwin estimated that from 10 to 18 tons of dry earth per acre pass through earthworm intestines annually, thus bringing potassium and phosphorus from the subsoil and also adding nitrogenous products to the soil from their own metabolism. They also drag leaves, twigs, and organic substances into their burrows closer to the roots of plants. Their activities are vitally important in aerating soil. Darwin's views were at odds with those of his contemporaries, who thought earthworms were harmful to plants. All relevant research has amply confirmed Darwin's findings, and earthworm management is now practiced in many countries.

Nutrition   Most oligochaetes are scavengers. Earthworms feed mainly on decaying organic matter, bits of leaves and vegetation, refuse, and animal matter. After being moistened by secretions from the mouth, food is drawn inward by the sucking action of their muscular pharynx. The liplike prostomium aids in manipulating food into position. Calcium from soil swallowed with food tends to produce a high blood calcium level. **Calciferous glands** along the esophagus secrete calcium ions into the gut and so reduce the calcium ion concentration of their blood. Calciferous glands also function in regulating acid-base balance of body fluids.

Leaving the esophagus, food is stored temporarily in the thin-walled **crop** before being passed on to the **gizzard,** which grinds food into small pieces. Digestion and absorption occur in the **intestine.** The wall of the intestine is infolded dorsally to form a **typhlosole,** which greatly increases the absorptive and digestive surface (Figure 17.19C).

Surrounding the intestine and dorsal vessel and filling much of the typhlosole is a layer of yellowish **chloragogen cells** (Gr. *chlōros,* green, + *agōgē,* a carrying away). This tissue serves as a center for synthesis of glycogen and fat, a function roughly equivalent to that of liver cells. When full of fat, chloragogen cells are

Figure 17.19 Earthworm anatomy. **A,** External features, lateral view. **B,** Internal structure of anterior portion of worm. **C,** Generalized transverse section through region posterior to clitellum. **D,** Portion of epidermis showing sensory, glandular, and epithelial cells.

released into the coelom where they float freely as cells called **eleo-cytes** (Gr. *elaio,* oil, + *kytos,* hollow vessel [cell]), which transport materials to the body tissues. Eleocytes can pass from segment to segment and may accumulate around wounds and regenerating areas, where they break down and release their contents into the coelom. Chloragogen cells also function in excretion.

Circulation and Respiration   Annelids have a double trans-port system: coelomic fluid and a closed circulatory system. Food, wastes, and respiratory gases are carried by both coelomic fluid and blood in varying degrees. Blood circulates in a closed system of vessels, which includes capillary systems in the tissues. Five main blood trunks run lengthwise through the body.

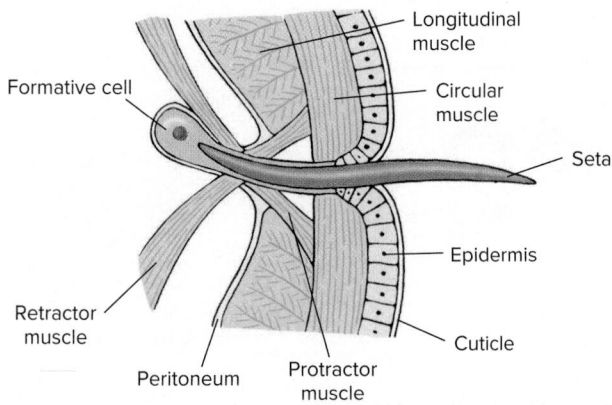

Figure 17.20   Seta with its muscle attachments showing relation to adjacent structures. Setae lost by wear are replaced by new ones, which develop from formative cells.

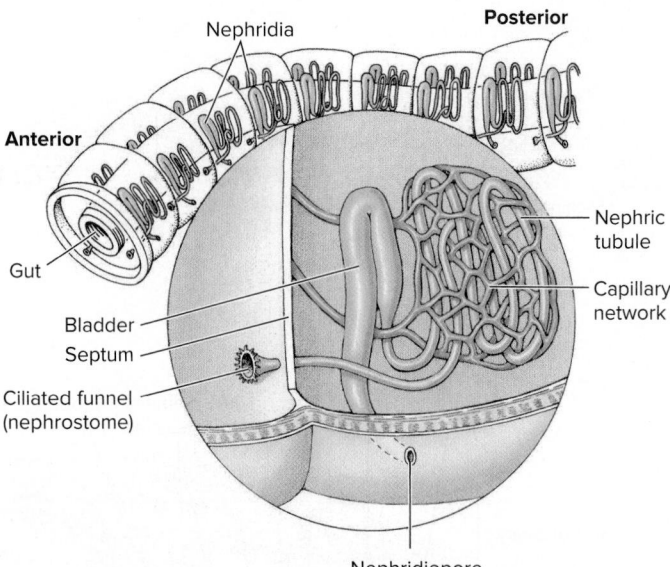

Figure 17.21   Nephridium of earthworm. Wastes are drawn into the ciliated nephrostome in one segment, then passed through the loops of the nephridium, and expelled through the nephridiopore of the next segment.

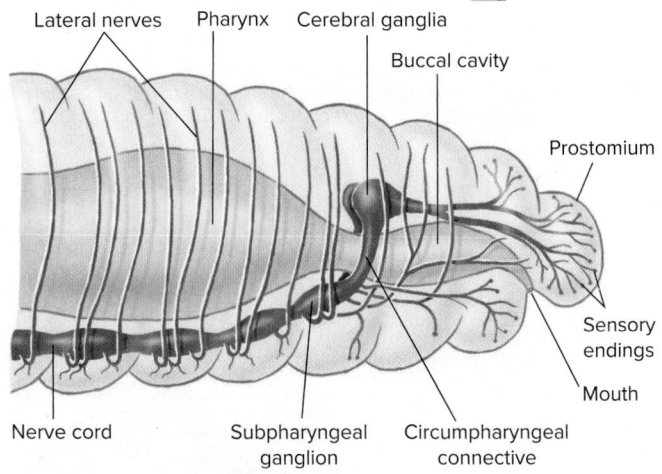

Figure 17.22   Anterior portion of earthworm and its nervous system. Note concentration of sensory endings in this region.

A single **dorsal vessel** runs above the alimentary canal from the pharynx to the anus. It is a pumping organ, provided with valves, and it functions as a true heart. This vessel receives blood from vessels of the body wall and digestive tract and pumps it anteriorly into five pairs of **aortic arches.** The function of aortic arches is to maintain a steady pressure of blood in the ventral vessel.

A single **ventral vessel** serves as an aorta. It receives blood from the aortic arches and delivers it to the brain and rest of the body, providing segmental vessels to the walls, nephridia, and digestive tract.

Their blood contains colorless ameboid cells and a dissolved respiratory pigment, **hemoglobin** (pp. 698–699). The blood of some annelids may have respiratory pigments other than hemoglobin, as noted on page 368.

Earthworms have no special respiratory organs, but gaseous exchange occurs across their moist skin.

Excretion   Each segment (except the first three and the last one) bears a pair of **metanephridia.** Each metanephridium occupies parts of two successive segments (Figure 17.21). A ciliated funnel, the **nephrostome,** lies just anterior to an intersegmental septum and leads by a small ciliated tubule through the septum into the segment behind, where it connects with the main part of the nephridium. Several complex loops of increasing size compose the nephridial duct, which terminates in a bladderlike structure leading to an opening, the **nephridiopore.** The nephridiopore opens to the outside near the ventral row of setae. By means of cilia, wastes from the coelom are drawn into the nephrostome and tubule, where they are joined by salts and organic wastes transported from blood capillaries in the glandular part of the nephridium. Waste is discharged to the outside through a nephridiopore.

Aquatic oligochaetes excrete ammonia; terrestrial oligochaetes usually excrete the much less toxic urea. *Lumbricus* produces both, the level of urea depending somewhat on environmental conditions. Both urea and ammonia are produced by chloragogen cells, which may break off and enter the metanephridia directly; or, their products may be carried by the blood. Some nitrogenous waste is eliminated through the body surface.

Oligochaetes are largely freshwater animals, and even such terrestrial forms as earthworms must exist in a moist environment. Osmoregulation is a function of the body surface and the nephridia, as well as the gut and dorsal pores. *Lumbricus* will gain weight when placed in tap water and lose it when returned to soil. Salts as well as water can pass across the integument, salts apparently being actively transported.

Nervous System and Sense Organs   The nervous system in earthworms (Figure 17.22) consists of a central system and peripheral nerves. The central system reflects the typical annelid pattern: a pair of **cerebral ganglia** (the "brain") above the pharynx, a pair of **connectives** passing around the pharynx connecting the brain with

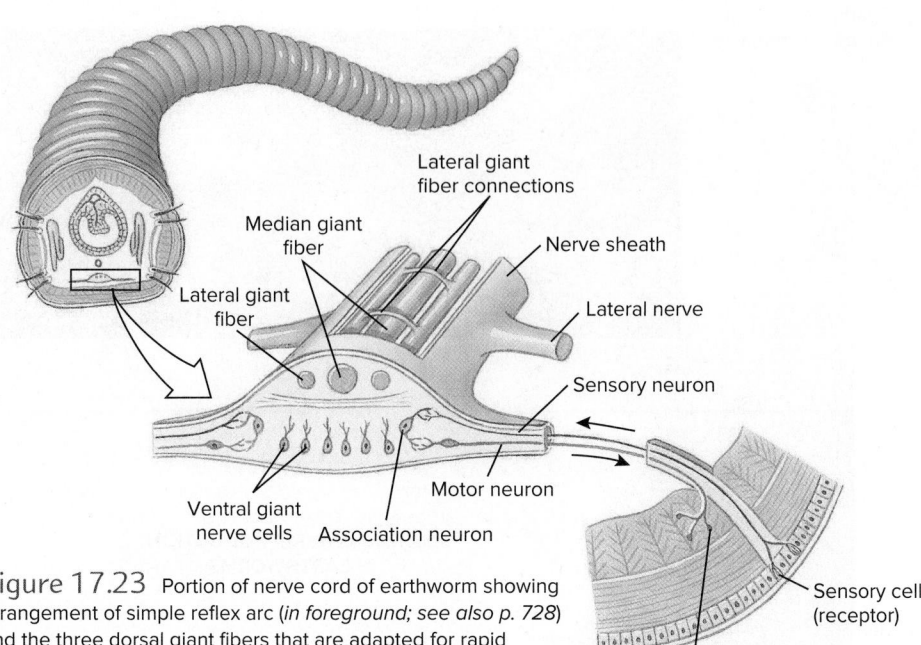

Figure 17.23 Portion of nerve cord of earthworm showing arrangement of simple reflex arc (*in foreground; see also p. 728*) and the three dorsal giant fibers that are adapted for rapid reflexes and escape movements. Ordinary crawling involves a succession of reflex acts, the stretching of one segment stimulating the next to stretch. Impulses are transmitted much faster in giant fibers than in regular nerves so that all segments can contract simultaneously when quick withdrawal into a burrow is necessary.

the first pair of ganglia in the nerve cord; a solid **ventral nerve cord,** really double, running along the floor of the coelom to the last segment; and a pair of fused ganglia on the nerve cord in each segment. Each pair of fused ganglia provides nerves to the body structures, which contain both sensory and motor fibers.

Neurosecretory cells have been found in the brain and ganglia of both oligochaetes and polychaetes. They are endocrine in function and secrete neurohormones concerned with the regulation of reproduction, secondary sex characteristics, and regeneration.

For rapid escape movements most annelids have from one to several very large axons commonly called **giant axons** (Figure 17.23), or giant fibers, located in the ventral nerve cord. Their large diameter increases rate of conduction (see pp. 723–724) and makes possible simultaneous contractions of muscles in many segments.

In the dorsal median giant fiber of *Lumbricus,* which is 90 to 160 μm in diameter, speed of conduction has been estimated at 20 to 45 m/second, several times faster than in ordinary neurons of this species. This is also much faster than in polychaete giant fibers, probably because in earthworms the giant fibers are enclosed in myelinated sheaths, which insulate them.

Simple sense organs are distributed all over the body. Earthworms have no eyes but do have many lens-shaped photoreceptors in their epidermis. Most oligochaetes are negatively phototactic to strong light but positively phototactic to weak light. Many single-celled sense organs are widely distributed in the epidermis. What are presumably chemoreceptors are most numerous on the prostomium. In the integument are many free nerve endings which are probably tactile in nature.

General Behavior Earthworms are among the most defenseless of creatures, yet their abundance and wide distribution indicate their ability to thrive. Although they have no specialized sense organs, they are sensitive to many stimuli. They react positively to mechanical stimuli when such stimuli are moderate and negatively to a strong stimulus (such as a footfall near them), which causes them to retire quickly into their burrows. They react to light, which they avoid unless it is very weak. Chemical responses aid them in the choice of food.

Chemical as well as tactile responses are very important to earthworms. They not only must sample the organic content of soil to find food, but also must sense its texture, acidity, and calcium content.

Experiments show that earthworms have some learning ability. They can be taught to avoid an electric shock, and thus can develop an association reflex. Darwin credited earthworms with a great deal of intelligence because they pulled leaves into their burrows by the narrow end, the easiest way for drawing a leaf-shaped object into a small hole. Darwin assumed that seizure of leaves by worms did not result from random handling or from chance but was deliberate. However, investigations since Darwin's time have shown that the process is mainly one of trial and error, for earthworms often seize a leaf several times before getting it right.

Reproduction and Development Earthworms are monoecious (hermaphroditic); each animal has both male and female organs (see Figure 17.19B). In *Lumbricus* reproductive systems are found in segments 9 to 15. Two pairs of small testes and two pairs of sperm funnels are surrounded by three pairs of large seminal vesicles. Immature sperm from the testes mature in seminal vesicles, then pass into sperm funnels and down sperm ducts to the male genital pores in segment 15, where they are expelled during copulation. Eggs are discharged by a pair of small ovaries into the coelomic cavity, where ciliated funnels of the oviducts carry them outside through female genital pores on segment 14. Two pairs of seminal receptacles in segments 9 and 10 receive and store sperm from the mate during copulation.

Reproduction in earthworms may occur throughout the year as long as warm, moist weather prevails at night (Figure 17.24). When mating, worms extend their anterior ends from their burrows and bring their ventral surfaces together (Figure 17.24). Their surfaces are held together by mucus secreted by the **clitellum** (L. *clitellae,* packsaddle) and by special ventral setae, which penetrate each other's bodies in the regions of contact. After discharge, sperm travel to seminal receptacles of the other worm via its seminal grooves. After copulation each worm secretes first a mucous tube and then a tough, chitinlike band that forms a **cocoon** around its clitellum. As the cocoon passes forward, eggs from the oviducts, albumin from skin glands, and sperm from the mate (stored in the seminal receptacles) pour into it. Fertilization of eggs then occurs within the cocoon. When the cocoon slips past the anterior end of the worm,

Figure 17.24 Earthworm copulation and formation of egg cocoons. **A,** Mutual insemination; sperm from genital pore (segment 15) pass along seminal grooves to seminal receptacles (segments 9 and 10) of each mate. **B** and **C,** After worms separate, the clitellum secretes first a mucous tube and then a tough band that forms a cocoon. The developing cocoon passes forward to receive eggs from oviducts and sperm from seminal receptacles. **D,** As cocoon slips off over anterior end, its ends close and seal. **E,** Cocoon is deposited near burrow entrance. **F,** Young worms emerge in 2 to 3 weeks. **G,** Two earthworms in copulation. Their anterior ends point in opposite directions as their ventral surfaces are held together by mucous bands secreted by the clitella.

its ends close, producing a sealed, lemon-shaped body. Embryogenesis occurs within the cocoon, and the form that hatches from the egg is a young worm similar to the adult. Thus development is direct with no metamorphosis. Juveniles do not develop a clitellum until they are sexually mature.

Representative Oligochaetes Freshwater oligochaetes usually are smaller and have more conspicuous setae than earthworms. They are more mobile than earthworms and tend to have better-developed sense organs. Most are benthic forms that crawl on the substrate or burrow in soft mud. Aquatic oligochaetes are an important food source for fishes. A few are ectoparasitic.

Some of the more common freshwater oligochaetes are the 1 mm long *Aeolosoma* (Gr. *aiolos,* quick-moving, + *soma,* body) (Figure 17.25B); the 10 to 25 mm long *Stylaria* (Gr. *stylos,* pillar) (Figure 17.25A); and the 5 to 10 mm long *Dero* (Gr. *dere,* neck or throat) (Figure 17.25D). The common 30 to 40 mm long *Tubifex* (L. *tubus,* tube, + *faciens,* to make or do) (Figure 17.25C) is reddish and lives with its head in mud at the bottom of ponds and its tail waving in the water. *Tubifex* is an alternate host necessary in the life cycle of *Myxobolus cerebralis,* a parasite that causes a very serious condition called whirling disease in rainbow trout in North America. Some oligochaetes, such as *Aeolosoma,* may asexually form chains of zooids by transverse fission (Figure 17.25B).

Figure 17.25
Some freshwater
oligochaetes.
A, *Stylaria* has the
prostomium drawn out
into a long snout.
B, *Aeolosoma* uses cilia
around the mouth to
sweep in food particles,
and it buds off new
individuals asexually.
C, *Tubifex* lives head
down in long tubes.
D, *Dero* has ciliated
anal gills.

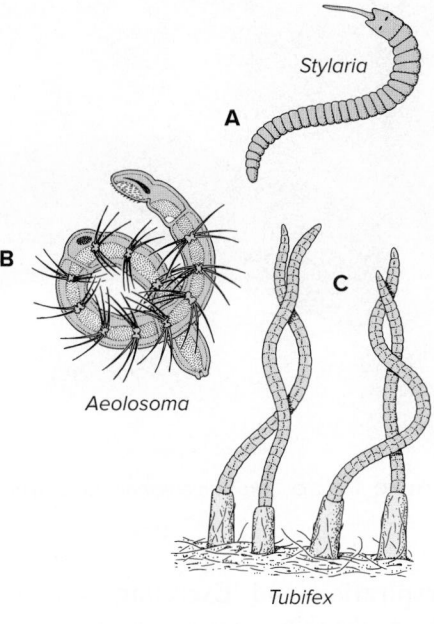

Stylaria

A

B

Aeolosoma

C

Tubifex

D

Dero

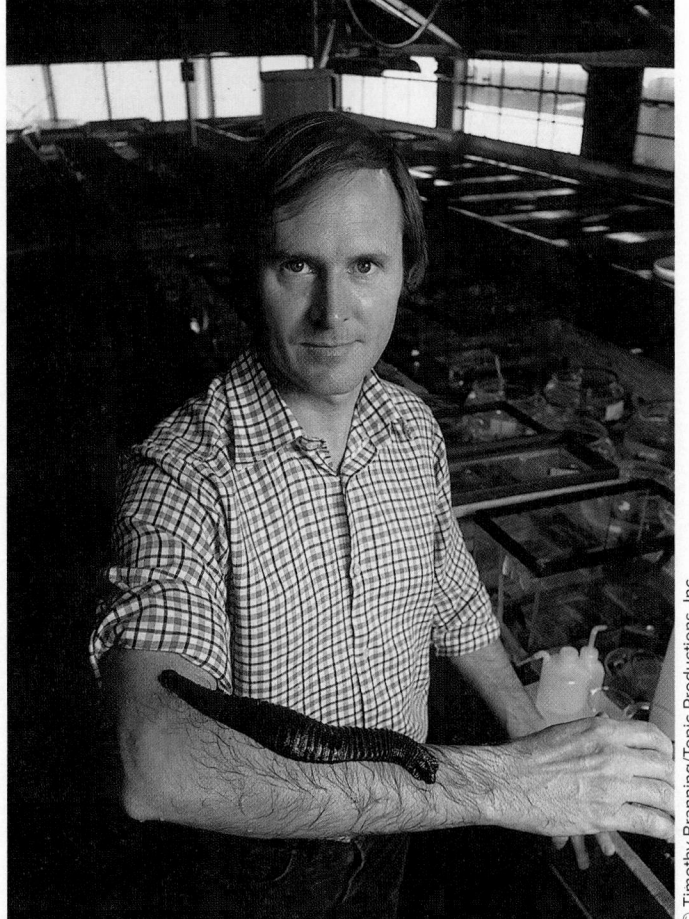

© Timothy Branning/Topic Productions, Inc.

Figure 17.26   The world's largest leech, *Haementeria ghilianii*, on
the arm of Dr. Roy K. Sawyer, who found it in French Guiana, South
America.

Class Hirudinida: Leeches   Class Hirudinida is divided into
three orders, Hirudinea, the "true" leeches, and two others that
merit mention here because their members are morphological intermediates between oligochaetes and true leeches (see Figure 17.1).
Oligochaetes have variable numbers of segments, segments bear
setae, and there are no suckers on the body. True leeches have
34 segments, entirely lack setae, and possess anterior and posterior
suckers. Members of order Acanthobdellida have 27 segments, bear
setae on the first five segments, and have a posterior sucker. Members of order Branchiobdellida have 14 or 15 segments, no setae,
and an anterior sucker. Branchiobdellids are commensal or parasitic
on crayfish. Hereafter, "leech" refers to members of order Hirudinea.

Leeches occur predominantly in freshwater habitats, but a
few are marine, and some have even adapted to terrestrial life in
warm, moist places. They are more abundant in tropical countries
than in temperate zones.

Most leeches are between 2 and 6 cm in length, but some,
including "medicinal" leeches, reach 20 cm. The giant of all is the
Amazonian *Haementeria* (Gr. *haimateros,* bloody) (Figure 17.26),
which reaches 30 cm.

Leeches are usually flattened dorsoventrally and exhibit a variety of patterns and colors: black, brown, red, or olive green. Many
leeches live as carnivores on small invertebrates; some are temporary
parasites; and some are permanent parasites, never leaving their host.
Some leeches attack human beings and are a nuisance to outdoor
enthusiasts.

Leeches are hermaphroditic and have a clitellum, which appears
only during breeding season. The clitellum secretes a cocoon for
reception of eggs.

Form and Function   Unlike other annelids, leeches have a fixed
number of segments but they appear to have many more because
each segment is marked by transverse grooves to form superficial
rings (Figure 17.27).

Leeches lack distinct coelomic compartments. In all but one species the septa have disappeared, and the coelomic cavity is filled with
connective tissue and a system of spaces called **lacunae.** The coelomic
lacunae form a regular system of channels filled with coelomic fluid,
which in some leeches serves as an auxiliary circulatory system.

Leeches are more highly specialized than oligochaetes. They have
lost the setae used by oligochaetes in locomotion and have developed
suckers for attachment while sucking blood (their gut is specialized for
storage of large quantities of blood). Most leeches crawl with looping

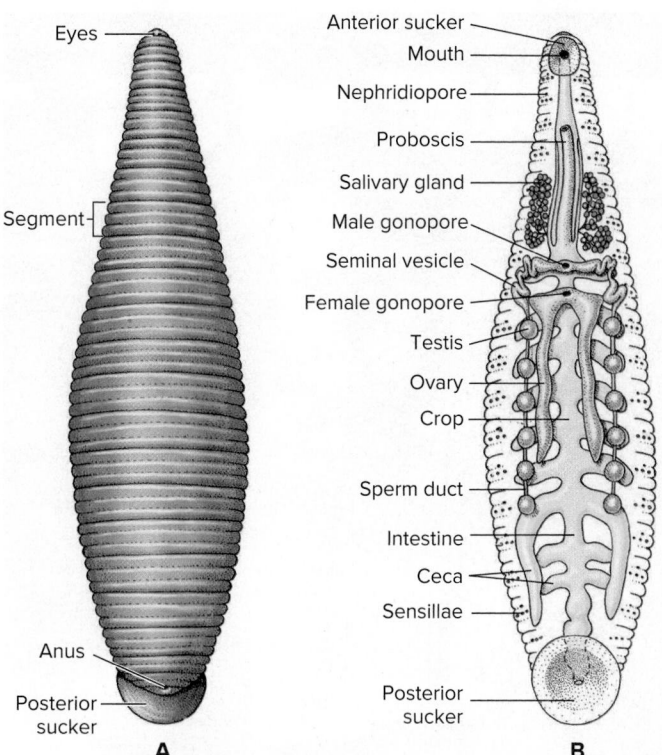

Figure 17.27   Structure of a leech, *Placobdella* **A,** External appearance, dorsal view. **B,** Internal structure, ventral view.

Figure 17.28   *Hirudo medicinalis* feeding on blood from human arm.

movements of the body, by attaching first one sucker and then the other and pulling the body along the surface. Aquatic leeches swim with a graceful undulatory movement.

Nutrition   Leeches are popularly considered parasitic, but many are predaceous. Most freshwater leeches are active predators or scavengers equipped with a proboscis that can be extended to ingest small invertebrates or to take blood from cold-blooded vertebrates. Some can force their pharynx or proboscis into soft tissues such as the gills of fish. Some terrestrial leeches feed on insect larvae, earthworms, and slugs, which they hold by an oral sucker while using a strong sucking pharynx to ingest food. Other terrestrial forms climb bushes or trees to reach warm-blooded vertebrates such as birds or mammals.

Most leeches are fluid feeders. Many prefer to feed on tissue fluids and blood pumped from open wounds. Some fresh-water leeches are true bloodsuckers, preying on cattle, horses, humans, and other mammals. True bloodsuckers, which include the so-called medicinal leech, *Hirudo medicinalis* (L. *hirudo,* a leech) (Figure 17.28), have cutting plates, or chitinous "jaws," for cutting through tough skin. Salivary glands secrete an anesthetic, as well as anticoagulant enzymes (see Chapter 32, p. 708). Some parasitic leeches leave their hosts only during the breeding season, and certain fish parasites are permanently parasitic, depositing their cocoons on their host fish. However, even the true bloodsuckers rarely remain on the host for a long period of time.

Respiration and Excretion   Gas exchange occurs only through the skin except in some fish leeches, which have gills. There are 10 to 17 pairs of nephridia, in addition to coelomocytes and certain other specialized cells that also may be involved in excretory functions.

Nervous and Sensory Systems   Leeches have two "brains": one is anterior and composed of six pairs of fused ganglia (forming a ring around the pharynx); the other is posterior and composed of seven pairs of fused ganglia. An additional 21 pairs of segmental ganglia occur along the double nerve cord. In addition to free sensory nerve endings and photoreceptor cells in the epidermis, there is a row of sense organs, called **sensilla,** in the central annulus of each segment. Pigment-cup ocelli also are present in many species.

Leeches are highly sensitive to stimuli associated with the presence of a prey or host. They are attracted by and will attempt to attach to an object smeared with appropriate host substances, such as fish scales, oil secretions, or sweat. Those that feed on the blood of mammals are attracted by warmth; terrestrial haemadipsids of the tropics will converge on a person standing in one place.

For centuries "medicinal leeches" (*Hirudo medicinalis*) were used for bloodletting because of the mistaken idea that a host of bodily disorders and fevers were caused by an excess of blood. A 10- to 12-cm-long leech can extend to a much greater length when distended with blood, and the amount of blood it can suck is considerable. Leech collecting and leech culture in ponds were practiced in Europe on a commercial scale during the nineteenth century. Wordsworth's poem "The Leech-Gatherer" was based on this use of leeches.

Leeches are once again being used medically. When fingers, toes, or ears are severed, microsurgeons can reconnect arteries but not all the more delicate veins. Leeches are used to relieve congestion until the veins can grow back into the healing appendage.

Taxonomy of Phylum Annelida

Annelids are wormlike forms sharing a segmented ancestor with paired epidermal setae. Taxonomy based on morphology focused on the presence of parapodia and many setae in polychaetes and on the absence of parapodia and a reduction in setae in oligochaetes and leeches. Phylogenies using molecular and morphological characters distinguished a large clade called Pleistoannelida and several lineages, such as Chaetopteridae (example: *Chaetopterus*) that lie outside of Pleistoannelida. Errantia and Sedentaria are within Pleistoannelida. Sedentaria includes worms with polychaete and oligochaete body plans. Because both oligochaetes and hirudineans (leeches) bear a clitellum, these two groups are united under the heading Clitellata (cli-tel-la′ta) and members are called clitellates.

Pleistoannelida comprises Errantia and Sedentaria; Marine, freshwater and terrestrial annelids, most with segmented bodies.

Errantia Freely moving polychaetes (pol′ē-ke′ta) (Gr. *polys*, many, + *chaitē*, long hair). Mostly marine; head distinct and bearing eyes and tentacles; most segments with parapodia (lateral appendages) bearing tufts of many setae; clitellum absent; sexes usually separate; gonads transitory; asexual budding in some; trochophore larva usually present. Examples: *Nereis, Aphrodita, Glycera.*

Sedentaria Sedentary annelids including tube-dwelling polychaetes and those living in burrows, as well as members of Clitellata (oligochaetes and leeches with a clitellum at some phase of the life cycle). Examples with a polychaete body plan: *Arenicola, Amphitrite,* and *Riftia;* examples with an unsegmented body plan: *Urechis* and *Bonellia.* Clitellate animals with an oligochaete body plan have conspicuous segmentation; number of segments variable; setae few per segment; no parapodia; head absent; coelom spacious and usually divided by intersegmental septa; hermaphroditic; development direct, no larva; chiefly terrestrial and freshwater. Examples: *Lumbricus, Stylaria, Aeolosoma, Tubifex.*

Class Hirudinida (hir′u-din′i-da) (L. *hirudo,* leech, + *ida,* pl. suffix): leeches. Sedentary clitellate annelids with fixed number of segments (normally 34; 15 or 27 in some groups) with many annuli; oral and posterior suckers usually present; clitellum present; no parapodia; setae absent (except in Acanthobdellida); coelom closely packed with connective tissue and muscle; development direct; hermaphroditic; terrestrial, freshwater, and marine. Examples: *Hirudo, Placobdella, Macrobdella.*

Reproduction   Leeches are hermaphroditic but cross-fertilize during copulation. Sperm are transferred by a penis or by hypodermic impregnation (a spermatophore is expelled from one worm and penetrates the integument of the other). After copulation their clitellum secretes a cocoon that receives eggs and sperm. Leeches may bury their cocoons in mud, attach them to submerged objects, or, in terrestrial species, place them in damp soil. Development is similar to that of oligochaetes.

Circulation   The coelom of leeches has been reduced by the invasion of connective tissue and, in some, by a proliferation of chloragogen tissue, to a system of coelomic sinuses and channels. Some orders of leeches retain a typical oligochaete circulatory system, and in these the coelomic sinuses act as an auxiliary blood-vascular system. In other orders the traditional blood vessels are lacking and the system of coelomic sinuses forms the only blood-vascular system. In those orders contractions of certain longitudinal channels provide propulsion for the blood (the equivalent of coelomic fluid).

17.2 PHYLUM SIPUNCULA

Phylum Sipuncula (sī-pun′kyū-la) (L. *sipunculus,* little siphon) consists of about 250 species of benthic marine worms, living at depths ranging from the intertidal to over 5000 m. They live sedentary lives in burrows in mud or sand, occupy borrowed snail shells, or live in coral crevices or among vegetation. Some species construct their own rock burrows by chemical and perhaps mechanical means. Most of the species are restricted to tropical zones. Some are tiny, slender worms, but most range from 3 to 10 cm in length. Some are commonly called "peanut worms" because, when disturbed, they can contract to a peanut shape (Figure 17.29).

Sipunculans have no segmentation or setae. They are most easily recognized by a slender, retractile **introvert,** or **proboscis,** which is continually and rapidly being run in and out of the anterior end. Walls of the **trunk** are muscular. When the introvert is everted, the mouth can be seen at its tip surrounded by a crown of ciliated tentacles. Little is known about the details of sipunculan feeding. Some species appear to be deposit feeders or detritivores, whereas others appear to be suspension feeders. Some nutrition may also come from uptake of dissolved organic compounds directly from the water column. Undisturbed sipunculans usually extend the anterior end from their burrow or hiding place and stretch out their tentacles to explore and to feed. Organic matter collected in mucus on the tentacles is moved to the mouth by ciliary action. The introvert is extended by hydrostatic pressure produced by contraction of bodywall muscles against the coelomic fluid. The lumen of the hollow tentacles is connected not to the coelom but rather to one or two blind, tubular compensation sacs that lie along their

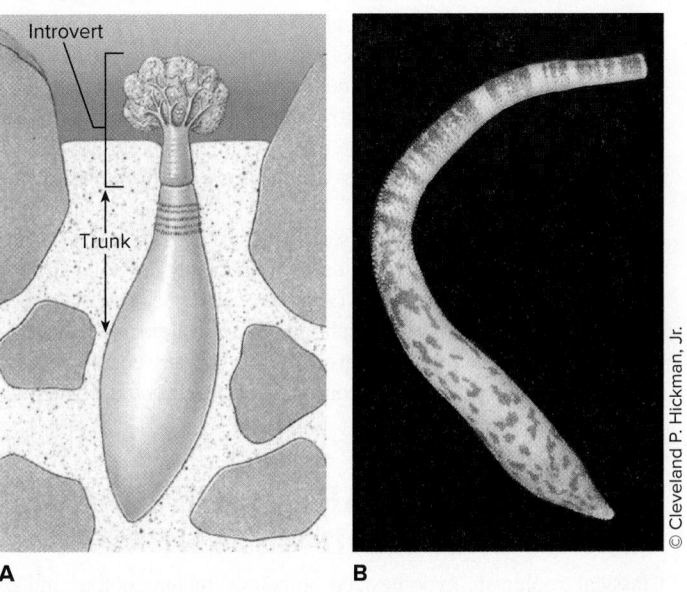

A   **B**

Figure 17.29   Sipunculans. *Themiste* (**A**) and *Phascolosoma* (**B**) are both burrowing genera of cosmopolitan distribution.

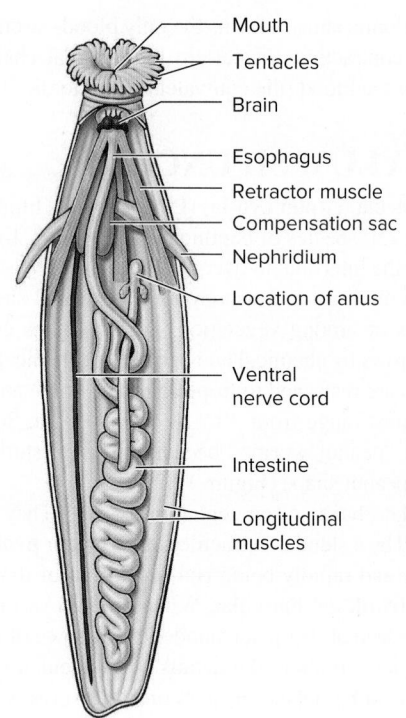

Mouth
Tentacles
Brain
Esophagus
Retractor muscle
Compensation sac
Nephridium
Location of anus
Ventral
nerve cord
Intestine
Longitudinal
muscles

Figure 17.30   Internal structure of *Sipunculus.*

esophagus (Figure 17.30). These sacs receive fluid from the tentacles when the introvert is retracted. Retraction is effected by special retractor muscles. The surface of the introvert is often rough because of surface spines, hooks, or papillae.

There is a large, fluid-filled coelom traversed by muscle and connective-tissue fibers. Their digestive tract is a long tube that doubles back on itself to form a U-shape and ends in an anus near the base of the introvert (Figure 17.30). A pair of large nephridia opens to the outside to expel waste-filled coelomic amebocytes; the nephridia also serve as gonoducts. Circulatory and respiratory systems are lacking, but coelomic fluid contains red corpuscles that have a respiratory pigment, hemerythrin, used in transportation of oxygen. Gas exchange appears to occur largely across the tentacles and introvert. Their nervous system has a bilobed cerebral ganglion just behind the tentacles and a ventral nerve cord extending the length of the body.

With only a few exceptions, sexes are separate. Permanent gonads are lacking, and ovaries or testes develop seasonally in the connective tissue covering the origins of one or more of the retractor muscles. Sex cells are released through the nephridia. The larval form is usually a trochophore. Asexual reproduction also occurs by transverse fission, the posterior one-fifth of the parent constricting off to become a new individual in some species.

17.3 EVOLUTIONARY SIGNIFICANCE OF A COELOM AND METAMERISM

Classical zoologists hypothesized homology of the coelom and of metamerism in animals and sought general explanations for each character. Current systematics rejects the hypothesis of homology in both cases. The coelom appears to have evolved independently in

protostomes and deuterostomes—we presume this because it forms by different methods in each group. Metamerism apparently has arisen separately at least three times, once in deuterostomes, once in ecdysozoan protostomes, and once in lophotrochozoan protostomes. We discuss here the significance of evolution of a coelom and of metamerism in lophotrochozoans. No truly satisfactory explanation has yet been given for the origins of segmentation and the coelom, although the subject has stimulated much speculation and debate. All classical explanations have had important arguments leveled against them, and more than one may be correct, or none, as suggested by R. B. Clark.[1] Clark stressed the functional and evolutionary significance of these features. He argued forcefully that the adaptive value of a coelom was as a **hydrostatic skeleton** in a burrowing animal. Thus contraction of muscles in one part of the animal could act antagonistically on muscles in another part by transmission of the force of contraction through the enclosed constant volume of fluid in the coelom (see the essay that opens this chapter).

Although the original function of the coelom may have served to facilitate burrowing in the substrate, certain other advantages accrued to its possessors. For example, coelomic fluid would have acted as a circulatory fluid for nutrients and wastes, making large numbers of flame cells distributed throughout the tissues unnecessary. Gametes could be stored in the spacious coelom for simultaneous release by all individuals in the population (thus enhancing chances of fertilization). Such a synchronous release of gametes would have selected for greater nervous and endocrine control. Within lophotrochozoans, the size of the coelom varies and many taxa are not segmented. Annelids and molluscs have very similar developmental programs leading to a trochophore larva, but the annelid trochophore develops a series of segments as it grows, whereas the mollusc trochophore does not grow in this way (see discussion, p. 337).

It is possible that all bilaterally symmetrical animals shared a segmented ancestor and that segmentation genes were suppressed in most lineages, but preliminary studies of the details of how segments form (genetic control and chemical signaling) in different phyla do not support this hypothesis.[2] Instead, current evidence supports the hypothesis that segmentation arose independently multiple times.

The selective advantage of a segmented body for annelids appears to lie in the efficiency of burrowing made possible by shape change in individual coelomic compartments of the hydrostatic skeleton.

17.4 PHYLOGENY AND ADAPTIVE DIVERSIFICATION

Phylogeny

Annelids and molluscs share many developmental features, so they were presumed by many biologists to be very closely related, perhaps sister taxa. However, the shared developmental features are likely to be retained ancestral features for lophotrochozoan protostomes.

[1]Clark, R. B. 1964. Dynamics in metazoan evolution. The origin of the coelom and segments. Oxford, U.K., Clarendon Press.

[2]Seaver, E. C. 2003. Int. J. Dev. Biol. **47:** 583–595.

Pogonophoran and vestimentiferan worms were once placed outside phylum Annelida, but they have been reinterpreted as members of Sedentaria and are now placed in clade Siboglinidae within this group. Only a small portion of the siboglinid body is segmented. Like siboglinids, echiurans are now considered members of Sedentaria. Setae are present, but segmentation has been lost. There are serially repeated structures, such as nerve-cord ganglia and mucous glands in echiuran larvae, and serially repeated nephridia in echiuran adults. Some species have metameric ganglia as adults. The presence of paired epidermal setae in some echiuran species provides strong support for placing echiuran within Annelida. A recent phylogenetic analysis places echiuran near capitellid polychaetes; both taxa dwell in sediments.

The placement of Sipuncula is contentious. Sipunculans are not metameric and they do not have setae. Larval development is similar to that of annelids, molluscs, and echiurans. Nerve formation follows a metameric pattern. For the present, we depict sipunculans as the sister taxon to annelids, but several recent phylogenies using data from mitochondrial genomes placed Sipuncula within Annelida.

Adaptive Diversification

Annelids are an ancient group that has undergone extensive adaptive diversification. The basic body structure, particularly of polychaetes, lends itself to almost endless modification. As marine worms, polychaetes occupy a wide range of habitats.

An adaptive feature in evolution of annelids is their septal arrangement of fluid-filled coelomic compartments. Fluid pressure in these compartments is used to create a hydrostatic skeleton, which in turn permits precise movements such as burrowing and swimming. Powerful circular and longitudinal muscles can flex, shorten, and lengthen the body.

Feeding adaptations show great variation, from the sucking pharynx of oligochaetes and the chitinous jaws of carnivorous polychaetes to the specialized tentacles and radioles of particle feeders. The evolution of a trophosome to house the chemoautotrophic bacteria that provide nutrients to siboglinids is an adaptation to deepsea life.

In polychaetes the parapodia have been adapted in many ways and for a variety of functions, chiefly locomotion and respiration.

In leeches many adaptations (such as suckers, cutting jaws, pumping pharynx, and distensible gut) relate to their predatory and bloodsucking habits.

SUMMARY

Phylum Annelida is a large, cosmopolitan group containing marine polychaetes, earthworms and freshwater oligochaetes, and leeches. Certainly the most important structural innovation underlying diversification of this group is metamerism (segmentation), a division of the body into a series of similar segments, each of which contains a repeated arrangement of many organs and systems. The coelom also is highly developed in annelids, and this, together with the septal arrangement of fluid-filled compartments and a well-developed body-wall musculature, is an effective hydrostatic skeleton for precise burrowing and swimming movements.

Most annelids are in a large group called Pleistoannelida. This group includes animals with several different body plans. Annelids with a polychaete body plan have segments with many setae, which are borne on paired parapodia. Parapodia show a wide variety of adaptations among polychaetes, including specialization for swimming, respiration, crawling, maintaining position in a burrow, pumping water through a burrow, and accessory feeding. Some polychaetes are mostly predaceous and have an eversible pharynx with jaws. Other polychaetes rarely leave the burrows or tubes in which they live. Several types of deposit- and filter-feeding are known among members of this group. Polychaetes are dioecious and have a reproductive system lacking a clitellum, external fertilization, and a trochophore larva.

Pleistoannelida is divided into two clades, both of which include animals with a polychaete body plan. Errantia contains freely moving polychaetes. Sedentaria contains animals with polychaete body plans that have been modified for life in tubes or burrows, such as siboglinids and echiurans.

Siboglinids live in tubes on the deep-ocean floor, and they are metameric. They have no mouth or digestive tract but apparently absorb some nutrient by the crown of tentacles at their anterior end. Much of their energy is due to chemoautotrophy of bacteria in their trophosome. Echiurans are burrowing marine worms, and most are deposit feeders, with a proboscis anterior to their mouth. Some species bear epidermal setae. They lack segmentation.

Sedentaria also contains annelids with oligochaete and leech body plans; these two groups fall within Clitellata. Earthworms and many freshwater annelids are oligochaetes; they have a small number of setae per segment (as compared with polychaetes) and no parapodia. They have a closed circulatory system, and the dorsal blood vessel is the main pumping organ. Paired nephridia occur in most segments. Earthworms contain the typical annelid nervous system: dorsal cerebral ganglia connected to a double, ventral nerve cord with segmental ganglia running the length of the worm. Oligochaetes are hermaphroditic and practice cross-fertilization. The clitellum plays an important role in reproduction, including secretion of mucus to surround the worms during copulation and secretion of a cocoon to receive eggs and sperm and in which embryonation occurs. A small, juvenile worm hatches from the cocoon.

Leeches (class Hirudinida) are mostly freshwater, although a few are marine and a few are terrestrial. They feed mostly on fluids; many are predators, some are temporary parasites, and a few are permanent parasites. The hermaphroditic leeches reproduce in a fashion similar to that of oligochaetes, with cross-fertilization and cocoon formation by the clitellum.

Sipunculans are small, burrowing marine worms with an eversible introvert at their anterior end. The introvert bears tentacles used for deposit feeding. Sipunculans are not segmented.

Embryological evidence places annelids with molluscs and arthropods in the Protostomia. Molecular evidence suggests that annelids and molluscs are more closely related to each other (in Lophotrochozoa) than either phylum is to arthropods (in Ecdysozoa). Sipunculans are allied to annelids, but do share certain features with molluscs.

REVIEW QUESTIONS

1. What characteristics of phylum Annelida distinguish it from other phyla?
2. How are members of clade Clitellata distinguished from other annelids?
3. Describe the annelid body plan, including body wall, segments, coelom and its compartments, and coelomic lining.

4. Explain how the hydrostatic skeleton of annelids helps them to burrow.
   How is the efficiency for burrowing increased by segmentation?
5. Describe three ways that various polychaetes obtain food.
6. Define each of the following: prostomium, peristomium, pygidium, radioles, parapodium.
7. Explain functions of each of the following in earthworms: pharynx, calciferous glands, crop, gizzard, typhlosole, chloragogen tissue.
8. Describe the function of the clitellum and cocoon.
9. How are freshwater oligochaetes generally different from earthworms?
10. Describe the ways in which leeches obtain food.
11. What is the largest siboglinid known, and how is it nourished?
12. What features made it difficult for biologists to recognize echiurans and siboglinids as members of Annelida?
13. Where does a sipunculan live, and how does it collect food?
14. What was the evolutionary significance of segmentation and the coelom to its earliest lophotrochozoan possessors?

For Further Thought Review the extent of segmentation in polychaetes, leeches, siboglinids, echiurans, and sipunculans. Which morphological features must change to produce an unsegmented body cavity? Is there evidence of transitional stages for these features?

SELECTED REFERENCES

Clark, R. B. 1964. Dynamics in metazoan evolution. The origin of the coelom and segments. Oxford, O.K., Clarendon Press. *This classic work described the functional role of the coelom in burrowing.*

Cutler, E. B. 1995. The Sipuncula. Their systematics, biology, and evolution. Ithaca, New York, Cornell University Press. *The author tried to "bring together everything known about" sipunculans.*

Fischer, A., and U. Fischer. 1995. On the life-style and life-cycle of the luminescent polychaete *Odontosyllis enopla* (Annelida: Polychaeta). Invert. Biol. **114:** 236–247. *If epitokes of this species survive their spawning swarm, they can return to a benthic existence.*

Halanych, K. M., T. D. Dahlgren, and D. McHugh. 2002. Unsegmented annelids? Possible origins of four lophotrochozoan worm taxa. Integ. and Comp. Biol. **42:** 678–684. *A nice summary of current morphological and molecular studies on classification of pogonophorans, echiurans, myzostomids, and sipunculans.*

McClintock, J. 2001. Blood suckers. Discover **22:** 56–61 (Dec.). *Describes modern medical uses for leeches.*

McHugh, D. 2000. Molecular phylogeny of Annelida. Can. J. Zool. **78:** 1873–1884. *Descriptions of monophyletic groups within Annelida supported by molecular data.*

Menon, J., and A. J. Arp. 1998. Ultrastructural evidence of detoxification in the alimentary canal of *Urechis caupo.* Invert. Biol. **117:** 307–317. *This curious echiuran has detoxification bodies in its gut cells and epithelial cells that allow it to live in a highly toxic sulfide environment.*

Mirsky, S. 2000. When good hippos go bad. Sci. Am. **282:** 28 (Jan.). Placobdelloides jaegerskioeldi *is a parasitic leech that breeds only in the rectum of hippopotamuses.*

Patel, N. H. 2003. The ancestry of segmentation. Dev. Cell **5:** 2–4. *Explores the idea that segmentation is an ancestral feature of all bilaterally symmetrical animals.*

Pernet, B. 2000. A scaleworm's setal snorkel. Invert. Biol. **119:** 147–151. Sthenelais berkeleyi *is an apparently rare but large (20 cm) polychaete that buries its body in sediment and communicates with water above just by its anterior end. Ciliary movement on parapodia pumps water into the burrow for ventilation. The worm remains immobile for long periods, except when prey comes near; it then rapidly everts its pharynx to capture prey.*

Rouse, G. W. 2001. A cladistic analysis of Siboglinidae Caullery, 1914 (Polychaeta: Annelida): Formerly the phyla Pogonophora and Vestimentifera. Zool. J. Linn. Soc. **132:** 55–80. *Diagnostic features of Siboglinidae and its subgroups are provided.*

Rouse, G. W., S. K. Gofferedi, and R. C. Vrijenhoek. 2004. Osedax: Bone-eating marine worms with dwarf males. Science **305:** 668–671. *Fascinating worms that feed on carcasses of whales.*

Seaver, E. C. 2003. Segmentation: mono- or polyphyletic. Int. J. Dev. Biol. **47:** 583–595. *Preliminary comparisons of the segmentation process in annelids, arthropods, and chordates suggest that annelids and arthropods do not share mechanisms of segmentation, but vertebrates and arthropods may share some mechanisms.*

Shen, X., X. Ma, J. Ren, and F. Zhao. 2009. A close phylogenetic relationship between Sipuncula and Annelida evidenced from the complete mitochondrial genome sequence of *Phascolosoma esculenta.* BMC Genomics **10:** Article 136. *Both echiurans and sipunculans are placed with Annelida, suggesting that segmentation was lost, perhaps as long ago as the Cambrian.*

Smith, C. R., A. G. Glover, T. Treude, N. D. Higgs, and D. J. Amon. 2015. Whale-fall ecosystems: recent insights into ecology, paleoecology, and evolution. Annual Review of Marine Science **7:** 571–596. *A fascinating description of the rich community of organisms living on whale falls; includes photographs.*

Struck, T. H. 2011. Direction of evolution within Annelida and the definition of Pleistoannelida. J Zool Syst Evol Res **49:** 340–345. *Description of the ancestral annelid body plan and lifestyle.*

Struck, T. H., C. Paul, N. Hill, S. Hartmann, C. Hösell, M. Kube, B. Lieb, A. Meyer, R. Tiedemann, G. Purschke, and C. Bleidorn. 2011. Phylogenomic analyses unravel annelid evolution. Nature **471:** 95–98. *Descriptions of annelid clades Sedentaria and Errantia.*

Weigert, A., A. Golombek, M. Gerth, F. Schwarz, T. Struck, and C. Bleidorn. 2016. Evolution of mitochondrial gene order in Annelida. Molecular Phylogenetics and Evolution **94:**196–206. *Mitochondrial gene order differs between annelids within Pleistoannelida and those excluded from this group.*

18
CHAPTER

Smaller Ecdysozoans

Male *Trichinella spiralis, a nematode.*

A World of Nematodes

Without any doubt, nematodes are the most important pseudocoelomate animals, in terms of both numbers and their impact on humans. Nematodes are abundant over most of the world, yet most people are only occasionally aware of them as parasites of humans or of their pets. We are not aware of the millions of these worms in the soil, in ocean and freshwater habitats, in plants, and in all kinds of animals. Their dramatic abundance moved N. A. Cobb[1] to write in 1914:

> If all the matter in the universe except the nematodes were swept away, our world would still be dimly recognizable, and if, as

disembodied spirits, we could then investigate it, we should find its mountains, hills, vales, rivers, lakes, and oceans represented by a thin film of nematodes. The location of towns would be decipherable, since for every massing of human beings there would be a corresponding massing of certain nematodes. Trees would still stand in ghostly rows representing our streets and highways. The location of the various plants and animals would still be decipherable, and, had we sufficient knowledge, in many cases even their species could be determined by an examination of their erstwhile nematode parasites.

[1]From N. A. Cobb. 1914. Yearbook of the United States Department of Agriculture, p. 472.

Many protostomes, such as annelids, roundworms, and arthropods, possess a **cuticle,** a nonliving external layer secreted by the epidermis. A stiff or hardened cuticle surrounding the body wall, like that present in roundworms and arthropods, restricts growth. In such animals, the cuticle is molted, and the outer layer shed via **ecdysis,** as the body increases in size. Ecdysozoa (Figure 18.1) comprises those taxa that molt the cuticle as they grow. Where it has been studied, molting is regulated by the hormone **ecdysone;** biologists assume that a homologous set of biochemical steps regulates molting among all ecdysozoans. Ecdysozoan taxa, other than loriciferans, were first united as a clade in phylogenies based on molecular characters.

As was the case with lophotrochozoan phyla, ecdysozoans do not share a common body plan. Members of Nematoda, Nematomorpha, and Kinorhyncha have pseudocoelomate bodies. Members of Priapulida have not been carefully studied, but are assumed to be pseudocoelomate. The pseudocoelom is used as a hydrostatic skeleton in nematodes, kinorhynchs, and priapulids. Within Loricifera, species apparently vary in body plan: some are described as pseudocoelomate and others appear acoelomate. Members of clade Panarthropoda have coelomate bodies, but their coeloms are quite reduced in size as compared with those of

annelids. Panarthropoda is an enormous group of animals, containing three phyla: Onychophora, Tardigrada, and Arthropoda.

Arthropoda is the largest phylum in terms of numbers of described species and forms the subject of Chapters 19, 20, and 21. The present chapter describes all other ecdysozoan phyla.

18.1 PHYLUM NEMATODA: ROUNDWORMS

Approximately 25,000 species of Nematoda (nem-a-tō′da) (Gr., *nematos,* thread) have been named, but many authorities now prefer Nemata for the name of this phylum. It has been estimated that if all species were known, the number might be nearer 500,000. They live in the sea, in freshwater, and in soil, from polar regions to the tropics, and from mountaintops to the depths of the sea. Good topsoil may contain billions of nematodes per acre. Nematodes also parasitize virtually every type of animal and many plants. Effects of nematode infestation on crops, domestic animals, and humans make this phylum one of the most important of all parasitic animal groups.

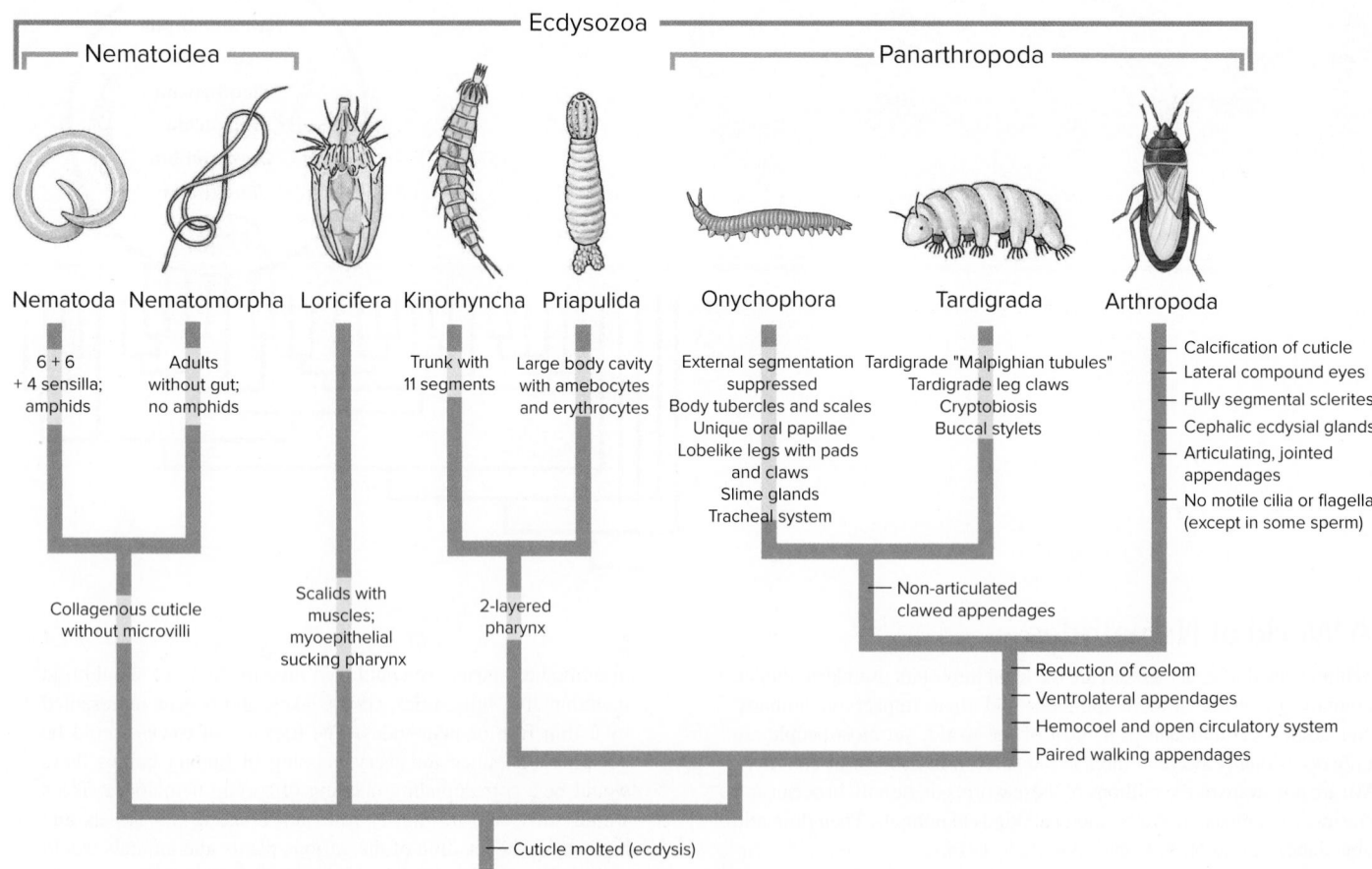

Figure 18.1 Cladogram depicting one hypothesis for relationships among ecdysozoan phyla. Characters shown are subsets of those in Nielsen (1995), Neuhaus and Higgins (2002), and Brusca and Brusca (2003); the nematode character "6 + 6 + 4 sensilla" refers to the anterior rings of sensory papillae.

Free-living nematodes feed on bacteria, yeasts, fungal hyphae, and algae. They may be saprozoic or coprozoic (live in fecal material). Predatory species may eat rotifers, tardigrades, small annelids, and other nematodes. All plant parts have specialized nematode parasites which vary in lifestyle from ectoparasitic forms to migratory or sedentary endoparasites. Sedentary endoparasites in roots cause agricultural damage of great proportions. Nematodes themselves may be prey for mites, insect larvae, and even nematode-capturing fungi. *Caenorhabditis elegans,* a free-living nematode, is easy to culture in the laboratory and has become an invaluable model for studies of developmental biology.

In 1963 Sydney Brenner started studying a free-living nematode, *Caenorhabditis elegans,* the beginning of some extremely fruitful research. Now this small worm has become one of the most important experimental models in biology. The origin and lineage of all the cells in its body (959) have been traced from zygote to adult, and the complete "wiring diagram" of its nervous system is known—all neurons and all connections between them. Its genome has been completely mapped, and scientists have sequenced its entire genome of 3 million bases comprising 19,820 genes. Many basic discoveries of gene function, such as genes encoding proteins essential for programmed cell death, have been made and will be made using *C. elegans.*

Virtually every species of vertebrate and many invertebrates serve as hosts for one or more types of parasitic nematodes. Nematode parasites in humans cause much discomfort, disease, and death, and in domestic animals they are a source of great economic loss.

Form and Function

Distinguishing characteristics of this large group of animals are their eutely; their cylindrical shape; their flexible, nonliving cuticle; their lack of motile cilia or flagella (except in one species); and the muscles of their body wall, which have several unusual features, such as running in a longitudinal direction only. Correlated with their lack of cilia, nematodes do not have protonephridia; their excretory system consists of one or more large gland cells opening by an excretory pore, or a canal system without gland cells, or both cells and canals together. Their pharynx is characteristically muscular with a triradiate lumen and resembles the pharynx of gastrotrichs and of kinorhynchs.

Most nematode worms are less than 5 cm long, and many are microscopic, but some parasitic nematodes are more than 1 m in length.

Their outer body covering is a relatively thick, noncellular **cuticle** secreted by the underlying epidermis (**hypodermis).** This cuticle is shed during juvenile growth stages, which is one of the characters that places nematodes in the Ecdysozoa. The hypodermis is syncytial, and its nuclei are located in four **hypodermal cords** that project inward (Figure 18.2). Dorsal and ventral hypodermal cords bear longitudinal dorsal and ventral nerves, and the lateral cords bear excretory canals. The cuticle is of great functional importance to the worm, serving to contain the high **hydrostatic pressure** (turgor) exerted by fluid in the pseudocoel and protecting the worm from hostile environments such as dry soils or the digestive tracts of their

hosts. The several layers of the cuticle are primarily of **collagen,** a structural protein also abundant in vertebrate connective tissue. Three of the layers are composed of crisscrossing fibers, which confer some longitudinal elasticity on the worm but severely limit its capacity for lateral expansion.

Body-wall muscles of nematodes are very unusual. They lie beneath the hypodermis (epidermal syncytium) and contract longitudinally only. There are no circular muscles in the body wall. The muscles are arranged in four bands, or quadrants, separated by the four hypodermal cords (Figure 18.2). Each muscle cell has a contractile **fibrillar** portion (or **spindle)** and a noncontractile **sarcoplasmic** portion (cell body). The spindle is distal and abuts the hypodermis, and the cell body projects into the pseudocoel. The spindle is striated with bands of actin and myosin, reminiscent of vertebrate skeletal muscle (see Figure 9.12, p. 193 and p. 651). The cell bodies contain the nuclei and are a major depot for glycogen storage in the worm. From each cell body a process or **muscle arm** extends to either the ventral or the dorsal nerve. Although not unique to nematodes, this arrangement is very unusual; in most animals nerve processes (axons, p. 721) extend to the muscle, rather than the other way around.

The fluid-filled pseudocoel, in which the internal organs lie, constitutes a hydrostatic skeleton. Hydrostatic skeletons, found in many invertebrates, lend support by transmitting the force of muscle contraction to the enclosed, noncompressible fluid. Normally, muscles are arranged antagonistically, so that movement is effected in one direction by contraction of one group of muscles, and movement back in the opposite direction is effected by the antagonistic set of muscles. Recall how the longitudinal and circular muscles operate antagonistically in each annelid segment (p. 363). However, nematodes do not have circular body-wall muscles to antagonize the longitudinal muscles; therefore the cuticle must serve that function. As muscles on one side of the body contract, they compress the cuticle on that side, and the force of the contraction is transmitted (by the fluid in the pseudocoel) to the other side of the nematode, stretching the cuticle on that side. When the muscles relax, compression and stretching of the cuticle antagonize the muscle and return the body to resting position; this action produces the characteristic thrashing motion seen when nematodes move through fluids. However, nematode movements are much more directed when they move through sediments or between small objects. An increase in efficiency of the nematode muscular system can be achieved only by an increase in hydrostatic pressure. Consequently, hydrostatic pressure in the nematode pseudocoel is much higher than is usually found in other kinds of animals that have hydrostatic skeletons but that also have antagonistic muscle groups.

The alimentary canal of nematodes consists of a mouth (Figure 18.2), a muscular pharynx, a long nonmuscular intestine, a short rectum, and a terminal anus. Nematodes parasitic in plants have a stylet to pierce cells. Food is sucked into the pharynx when the muscles in its anterior portion contract rapidly and open the lumen. Relaxation of the muscles anterior to the food mass closes the lumen of the pharynx, forcing the food posteriorly toward the intestine. The intestine is one cell-layer thick. Food matter moves posteriorly by body movements and by additional food being passed into the intestine from the pharynx. Defecation is accomplished by muscles that simply pull the anus open, and expulsive force is provided by the high pseudocoelomic pressure that surrounds the gut.

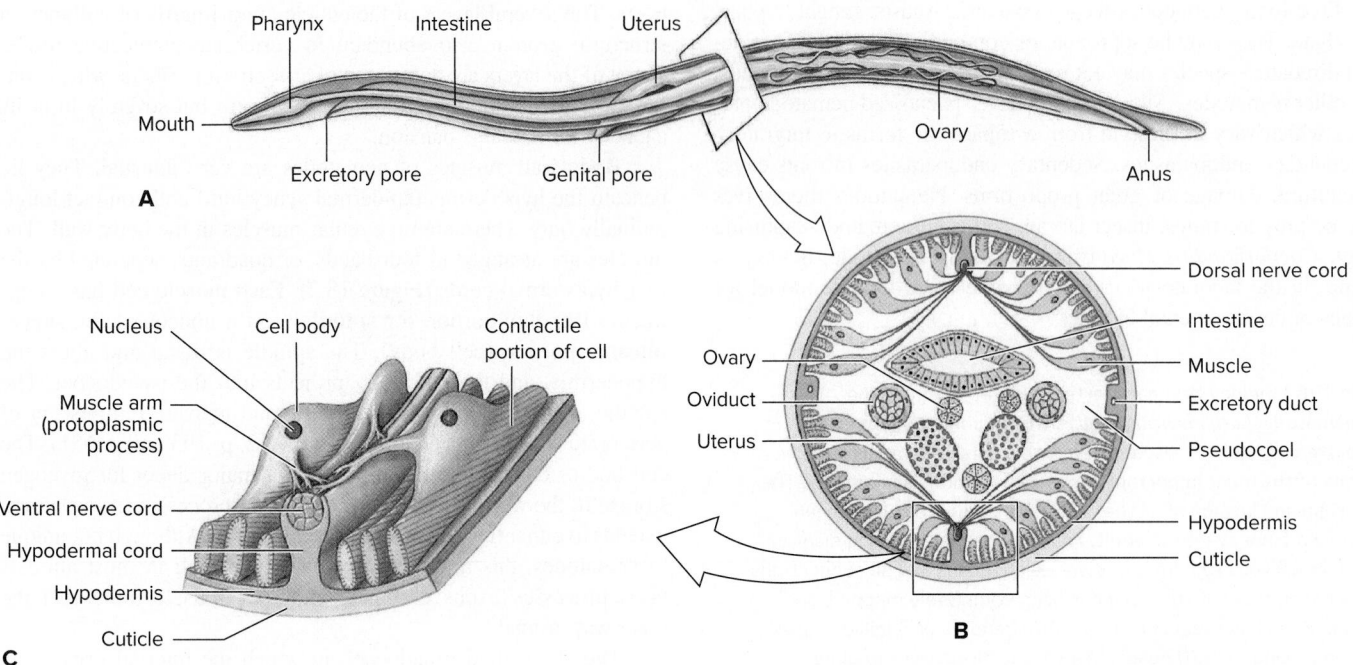

Figure 18.2 **A**, Structure of a nematode as illustrated by *Ascaris* female. *Ascaris* has two ovaries and uteri, which open to the outside by a common genital pore. **B**, Cross section. **C**, Single muscle cell; spindle abuts hypodermis, muscle arm extends to dorsal or ventral nerve.

Adults of many parasitic nematodes have an anaerobic energy metabolism; thus, a Krebs cycle and cytochrome system characteristic of aerobic metabolism are absent. They derive energy through glycolysis and probably through some incompletely known electron-transport sequences. Interestingly, some free-living nematodes and free-living stages of parasitic nematodes are obligate aerobes and have a Krebs cycle and cytochrome system.

A ring of nerve tissue and ganglia around the pharynx gives rise to small nerves at the anterior end and to two **nerve cords**, one dorsal and one ventral. **Sensory papillae** are concentrated around the head and tail. The **amphids** (Figure 18.3) are a pair of somewhat more complex sensory organs that open on each side of the head at about the same level as the cephalic circle of papillae. The amphidial opening leads into a deep cuticular pit with sensory endings of modified cilia. Amphids are usually reduced in nematode parasites of animals, but most parasitic nematodes bear a bilateral pair of **phasmids** near the posterior end. They are similar in structure to amphids.

Most nematodes are dioecious. Males are smaller than females, and their posterior end usually bears a pair of **copulatory spicules** (Figure 18.4). Fertilization is internal, and eggs are usually stored in the uterus until deposition. Development among free-living forms is typically direct. The four juvenile stages are each separated by a molt, or shedding, of the cuticle. Many parasitic nematodes have free-living juvenile stages. Others require an intermediate host to complete their life cycles.

Representative Nematode Parasites

As mentioned on page 387, nearly all vertebrates and many invertebrates are parasitized by nematodes. A number of these are very

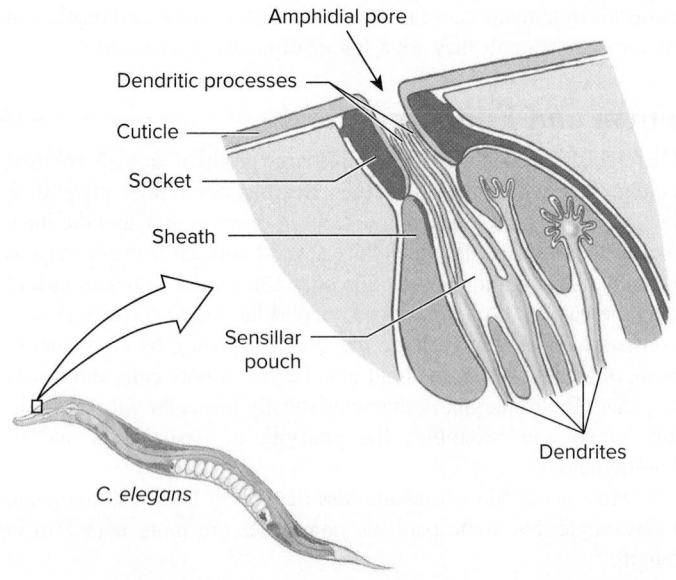

Figure 18.3 Diagram of an amphid in *Caenorhabditis elegans*. Redrawn from Wright, K. A. 1980. Nematode sense organs . In B. M. Zuckerman (ed.), *Nematodes as Biological Models, Vol. 2, Aging and Other Model Systems*. Copyright © Academic Press, New York.

important pathogens of humans and domestic animals. A few nematodes are common in humans in North America (Table 18.1), but they and many others usually abound in tropical countries. Space permits mention of only a few in this discussion.

Copulatory spicules of male nematodes are not true intromittent organs, since they do not conduct sperm, but are another adaptation to cope with high internal hydrostatic pressure. Spicules must hold the vulva of a female open while the ejaculatory muscles overcome the hydrostatic pressure in the female and rapidly inject sperm into her reproductive tract. Furthermore, nematode spermatozoa are unique among those studied in the animal kingdom in that they lack a flagellum and an acrosome. Within a female's reproductive tract, sperm become ameboid and move by pseudopods. Could this be another adaptation to the high hydrostatic pressure in the pseudocoel?

Figure 18.4 **A**, Cross section of a male nematode. **B**, Posterior end of a male nematode.

Ascaris lumbricoides:
The Large Roundworm of Humans

Because of its size and availability, *Ascaris* (Gr. *askaris,* intestinal worm) is usually studied in zoology, as well as in experimental work. Thus it is probable that parasitologists know more about the structure, physiology, and biochemistry of *Ascaris* than of any other nematode. This genus includes several species. One of the most common, *A. megalocephala,* lives in the intestines of horses. *Ascaris lumbricoides* (Figure 18.5) is one of the most common nematode parasites found in humans; it occurs in warm, humid regions of the earth, and about one billion people worldwide are likely to have this worm. It is now uncommon in the United States. The large roundworm of pigs, *A. suum,* is morphologically similar to *A. lumbricoides,* and they were long considered the same species.

Adult *Ascaris* worms live in the small intestine of their host. *Ascaris* females may reach 30 cm in length and can produce 200,000 eggs a day. Fertilized eggs leave the host's body with the feces and are amazingly tolerant to dessication or lack of oxygen, but are killed by direct sunlight and high temperature. In the absence of a sewage system, defecation onto soil allows the life cycle to continue. Given suitable soil conditions, embryos develop into infective juveniles within two weeks. Shelled juveniles can remain viable for many months or even years in soil, so they may be transmitted to a new host by consumption of contaminated vegetables or ingestion of soil.

Tiny juveniles hatch from the shelled eggs once they are swallowed by a host. They burrow through the host's intestinal wall, entering the blood or lymph and then travelling to the heart. From the heart, they migrate to the lungs, moving from the alveoli to the trachea. If abundant, they may cause a serious pneumonia at this stage. The worms, still juveniles, pass from the trachea to the pharynx where they are swallowed, bringing them back into the stomach to complete development. The worms feed on intestinal contents. Two months after being ingested as shelled juveniles, they reach maturity. The intestine, worms cause abdominal symptoms and allergic reactions, and in large numbers they may cause intestinal blockage. Parasitism by *Ascaris* is rarely fatal unless the intestine is blocked by a heavy infestation. Perforation of the intestine with resultant peritonitis is not uncommon, and wandering worms may occasionally emerge from the anus or throat or may enter the trachea or eustachian tubes and middle ears.

TABLE 18.1
Common Parasitic Nematodes of Humans in North America

Common and Scientific Names	Mode of Infection; Prevalence
Hookworm (*Ancylostoma duodenale* and *Necator americanus*)	Contact in soil with juveniles that burrow into skin; occurs in the southern United States, but less prevalent than previously. An estimated 650 million people are infected worldwide.
Pinworm (*Enterobius vermicularis*)	Inhalation of dust with ova and by contamination with fingers; most common worm parasite in United States
Intestinal roundworm (*Ascaris lumbricoides*)	Ingestion of embryonated ova in contaminated food
Trichina worm (*Trichinella* spp.)	Ingestion of infected muscle; occasional in humans throughout North America
Whipworm (*Trichuris trichiura*)	Ingestion of contaminated food or by unhygienic habits; usually common wherever ***Ascaris*** is found

A

© Cleveland P. Hickman, Jr.

B

© Dr. M.A. Ansary/SPL/Science Source

Figure 18.5   **A,** Intestinal roundworm *Ascaris lumbricoides,* male and female. Male, *top,* is smaller and has characteristic sharp kink in the end of the tail. Females of this large nematode may be over 30 cm long. **B,** Intestine of a pig, nearly completely blocked by *Ascaris suum.* Such heavy infections are also fairly common with *A. lumbricoides* in humans.

Figure 18.6
A, Mouth of hookworm displaying cutting plates. **B**, Section through anterior end of hookworm attached to dog intestine. Note cutting plates pinching off mucosa from which the thick muscular pharynx sucks blood. Esophageal glands secrete anticoagulant to prevent blood from clotting.

Plates

Pharynx

© Eye of Science/Science Source

From Herman Zaiman, 'A Presentation of Pictorial Parasites'

A

B

Other ascarids are common in wild and domestic animals. Species of *Toxocara,* for example, parasitize dogs and cats. Their life cycle is generally similar to that of *Ascaris,* but juveniles often do not complete their tissue migration in adult dogs, remaining in the host's body in a stage of arrested development. Pregnancy in a female dog, however, stimulates juvenile worms to wander, and they infect the embryos in the uterus. Puppies are then born with worms. These ascarids also survive in humans but do not complete their development, leading to an occasionally serious condition in children called *visceral larva migrans.* This is a good argument for pet owners to practice immediate hygienic disposal of canine wastes!

Hookworms

Hookworms are so named because the anterior end curves dorsally, suggesting a hook. The most common species is *Necator americanus* (L. *necator,* killer), whose females are up to 11 mm long. Males can reach 9 mm in length. Large plates in their mouths (Figure 18.6) cut into the intestinal mucosa of the host, where they suck blood and pump it through their intestine, partially digesting it and absorbing the nutrients. They suck much more blood than they need for food, and heavy infections cause anemia in patients. Hookworm disease in children may retard mental and physical growth and deplete energy.

Eggs pass in the feces, and juveniles hatch in the soil, where they live on bacteria (Figure 18.7). When human skin comes in contact with infected soil, infective juveniles burrow through the skin to the blood, then reach the lungs and finally the intestine in a manner similar to that described for *Ascaris.*

Trichina Worm

Trichinella spiralis (Gr. *trichinos,* of hair, + *-ella,* diminutive) is one of several species of tiny nematodes responsible for the potentially lethal disease trichinosis. Adult worms burrow in the mucosa of the small intestine where females produce living young. Juveniles penetrate blood vessels and are carried throughout the body, where they may occupy almost any tissue or body space. Eventually, they penetrate skeletal muscle cells, becoming one of the largest known intracellular parasites. Juveniles cause astonishing redirection of gene expression in their host cell, which loses its striations and becomes a **nurse cell** that nourishes the worm (Figure 18.8). When raw or poorly cooked meat containing encysted juveniles is swallowed, the worms are liberated into the intestine, where they mature.

Trichinella spp. can infect a wide variety of mammals in addition to humans, including hogs, rats, cats, and dogs. Hogs become infected by eating garbage containing pork scraps with juveniles or by eating infected rats.

E *Ancylostoma duodenale* can also infect a new individual by by oral route

G Juveniles break out of circulatory system into alveoli and then migrate to small intestine via the trachea

F Juveniles migrate through circulatory system to lungs

C Infective juvenile develops in soil

D A new individual is infected when juvenile penetrates human skin

H Adult worms develop in small intestine, mate and produce eggs

B First-stage juvenile hatches

A Eggs passed in feces

Embryo develops

Start

Drawing by William Ober and Claire Garrison

Figure 18.7 The life cycle of hookworms: a shelled embryo develops into a first-stage juvenile followed by two molts. The resulting third-stage juvenile enters developmental arrest until it reaches a new host (**A** to **C**). Human infection may be via the skin (**D**) or mouth (**E**). Juveniles migrate through the circulatory system to lungs (**F**), enter alveoli (**G**), and then reach the intestine where they mate (**H**).

Figure 18.8 Muscle infected with trichina worm *Trichinella spiralis.* The juveniles lie within muscle cells that the worms have induced to transform into nurse cells (commonly called cysts). An inflammatory reaction occurs around the nurse cells. Juveniles may live 10 to 20 years, and nurse cells eventually may calcify.

© Larry S. Roberts

In addition to *T. spiralis,* we now know there are four other sibling species in the genus. They differ in geographic distribution, infectivity to different host species, and freezing resistance.

Heavy infections may cause death, but lighter infections are more common—recent cases of infection come from wild boar or from hogs raised in close proximity to wild animals. In 2008, 30 people in Northern California were infected after consuming black bear meat.

Guinea worm (*dracunculiasis*) is a nematode parasite of mammals. It has a larval stage in a copepod (an aquatic arthropod) and an adult stage in a mammal. Mammal hosts, such as humans, dogs, cats, and horses, are infected by drinking water containing parasitized copepods. Worms mate in the mammal, then pregnant females, up to one meter in length, emerge from blisters. When the painful blister is bathed, larvae are released into the water. Clean drinking water prevents human infection, so eradication is possible. However, war and social turmoil damage public health programs—96% of the 521 new cases for 2012 were in South Sudan.

Pinworms

Pinworms, *Enterobius vermicularis* (Gr. *enteron*, intestine, + *bios*, life), cause relatively little disease, but they are the most common nematode parasites in the United States, estimated at 30% of all children and 16% of adults. Adult parasites (Figure 18.9A) live in the large intestine and cecum. Females, up to about 12 mm in length, migrate to the anal region at night to lay their eggs (Figure 18.9B). Scratching the resultant itch effectively contaminates hands and bedclothes. Eggs develop rapidly and become infective within 6 hours at body temperature. When they are swallowed, they hatch in the duodenum, and the worms mature in the large intestine.

> Diagnosis of most intestinal roundworms is usually made by examination of a small bit of feces under the microscope and finding characteristic eggs. However, pinworm eggs are not often found in the feces because the female deposits them on the skin around the anus. The "Scotch tape method" is more effective. The sticky side of cellulose tape is applied around the anus to collect the eggs, then the tape is placed on a glass slide and examined under a microscope. Several drugs are effective against this parasite, but all members of a family should be treated at the same time because the worms easily spread through a household.

Taxonomy of Phylum Nematoda

The traditional taxonomy is based on the work of Kampfer, et al.

Class Secernentea (= Phasmida) Amphids ventrally coiled or derived therefrom; three esophageal glands; some with phasmids; both freeliving and parasitic forms. Examples: *Caenorhabditis, Ascaris, Enterobius, Necator, Wuchereria.*

Class Adenophorea (= Aphasmida) Amphids generally well-developed, pocketlike; five or more esophageal glands; phasmids absent; excretory system lacking lateral canals, formed of single, ventral, glandular cells, or entirely absent; mostly free-living, but includes some parasites. Examples: *Dioctophyme, Trichinella, Trichuris.* Taxonomy of nematodes is somewhat more satisfactory at the order and superfamily level; division into classes relies on characteristics that are not striking and that are difficult for novices to distinguish. Argument exists about monophyly of the nematodes, but some molecular work supports the traditional classes. A recent molecular phylogeny divides nematodes among 12 clades.[1]

---

[1]Holterman, T., et al. 2006. *Mol. Biol. Evol.* **23:** 1792–1800.

Filarial Worms

At least eight species of filarial nematodes infect humans, and some of these are major causes of disease. Some 120 million people in tropical countries are infected with *Wuchereria bancrofti* (named for Otto Wucherer) or *Brugia malayi* (named for S. L. Brug), which places these species among the scourges of humanity. The worms live in the lymphatic system, and females are as long as 10 cm. Disease symptoms are associated with inflammation and obstruction of the lymphatic system. Females release live young, tiny **microfilariae** into the blood and lymphatic system (Figure 18.10). As they feed, mosquitos ingest microfilariae, which develop inside the mosquitos to the infective stage. They escape from the mosquito when it is feeding again on a human and penetrate the wound made by the mosquito bite.

The dramatic manifestations of elephantiasis are produced occasionally after long and repeated exposure to reinfection. The condition is marked by an excessive growth of connective tissue and enormous swelling of affected parts, such as the scrotum, legs, arms, and more rarely, the vulva and breasts (Figure 18.11).

Another filarial worm causes river blindness (onchocerciasis) and is carried by black flies. It infects more than 37 million people in parts of Africa, Arabia, Central America, and South America. Diseases caused by filarial worms fall into the Center for Disease Control's (CDC's) "Neglected Tropical Disease" category. Although largely eliminated in the developed world, these diseases cause enormous suffering in the world's poorest countries where people cannot pay for treatment.

The most common filarial worm in the United States is probably the dog heartworm, *Dirofilaria immitis* (Figure 18.12). Carried by mosquitoes, it also can infect other canids, cats, ferrets, sea lions, and occasionally humans. Along the Atlantic and Gulf

A

B

Figure 18.9   Pinworms, *Enterobius vermicularis*. **A**, A female worm from a human large intestine. The inset shows a closer view of the head. **B**, Eight pinworm eggs. Female worms normally discharge their eggs at night around the anus of the host, irritating the host skin at the same time. When the host scratches, fingernails and clothing are contaminated.

B Ingested microfilariae pass through mosquito gut into hemocoel and eventually develop into infective juveniles

Start **A** Mosquito ingests microfilariae when biting human

C Infected mosquito transmits infective juveniles, which enter through wound puncture

Figure 18.10 Life cycle of *Wuchereria bancrofti:* mosquito ingests microfilariae, which penetrate mosquito gut wall and develop to infective juveniles. Juveniles escape from mosquito's proboscis when the insect is feeding and then penetrate wound. (**A** to **C**), Juveniles migrate to regional lymph nodes and develop to sexual maturity in afferent lymphatic vessels. Adult worms produce microfilariae, which enter blood circulation (**D** to **G**).

Blood vessel

D Juveniles migrate via lymphatics to regional lymph nodes

G Microfilariae migrate to bloodstream

Lymph node

Afferent lymphatic vessel

F Adult worms mate and female gives birth to microfilariae

E Adult worms develop to sexual maturity in afferent lymphatic vessels

Figure 18.11 Elephantiasis of leg caused by adult filarial worms of *Wuchereria bancrofti,* which live in lymph passages and block the flow of lymph. Tiny juveniles, called microfilariae, are ingested when a mosquito feeds on the blood of an infected human. In the mosquito, they develop to infective stage and are transmitted to a new host.

CDC

© Nancy Nehring/Getty Images

Figure 18.12 Heartworms, *Dirofilaria immitis,* extending from the heart of a dog.

Coast states and northward along the Mississippi River throughout the Midwestern states, prevalence in dogs is up to 45%. It occurs in other states at a lower prevalence. This worm causes a very serious disease among dogs, and no responsible owner should fail to provide "heartworm pills" for a dog.

18.2 PHYLUM NEMATOMORPHA

The popular name for Nematomorpha (nēm´a-tō-mor´fa) (Gr. *nema, nematos,* thread, + *morphē,* form) is "horsehair worms," based on an old superstition that the worms arise from horsehairs that happen to fall into water. The worms do resemble hairs from a horse's tail. They were long included within Nematoda because both groups share the structure of the cuticle, presence of epidermal cords, longitudinal muscles only, and a similar pattern of the nervous system. They are currently placed as the sister taxon to nematodes.

About 320 species of horsehair worms have been named. Worldwide in distribution, they are free-living as adults and parasitic in arthropods as juveniles. Adults live almost anywhere in wet to moist surroundings if oxygen is adequate.

Form and Function

Horsehair worms are extremely long and slender, with a cylindrical body. They are generally about 0.5 to 3 mm in diameter but can be up to 1 m in overall length. Their anterior end is usually rounded, and their posterior end is rounded or has two or three caudal lobes (Figure 18.13).

Their body wall is much like that of nematodes: a secreted cuticle, a hypodermis, and musculature of **longitudinal muscles** only.

Their digestive system is vestigial. Larval forms absorb food from their arthropod hosts through their body wall. Adults were thought to live entirely on stored nutrients, but in fact they absorb organic molecules through their vestigial gut and body wall in much the same way as juveniles.

Circulatory, respiratory, and excretory systems are lacking and these functions probably occur on a primarily cellular level. However, very little is known about the physiology of these worms. There is a nerve ring around the pharynx and a midventral nerve cord.

Life cycles of nematomorphs are poorly known. In the cosmopolitan genus, *Gordius* (named for an ancient king who tied an intricate knot), juveniles may encyst on vegetation likely to be eaten by a grasshopper or other arthropod. Gordiid larval stages also have hooks or stylets that may be used to bore into a host, perhaps via the integument or the gut lining. In other cases, the gordiid may infect the host via its drinking water. Larvae encyst in the host; in some cases, development continues after the first host is eaten by a second host. In the marine nematomorph, *Nectonema* (Gr. *nektos,* swimming, + *nema,* a thread), juveniles occur in hermit crabs and other crabs.

After several months in the hemocoel (p. 397) of an arthropod host, juveniles complete a single molt and emerge into water as mature adults. If the host is a terrestrial insect, the parasite somehow stimulates the host to seek water; worms will not emerge unless water is nearby.

Nematomorphs are dioecious. Adults wriggle slowly once in water, often making dense mating aggregations, whose tangled forms reminded some biologists of the famed "Gordian Knot" and led to the genus name *Gordius.* As males enwrap females, sperm is deposited near the genital pore. Females discharge long strings of fertilized eggs into water.

© Dr. Andreas Schmidt-Rhaesa

Figure 18.13   Structure of *Paragordius,* a nematomorph. **A,** Longitudinal section through the anterior end. **B,** Transverse section. **C,** Posterior end of male and female worms. Nematomorphs, or "horsehair worms," are very long and very thin. Their pharynx is usually a solid cord of cells and appears nonfunctional. *Paragordius,* whose pharynx opens through to the intestine, is unusual in this respect and also in the possession of a photosensory organ ("eye"). **D,** *Paragordius tricuspidatus* emerges from the body of a European cricket, *Nemobius sylvestris.*

Figure 18.14 **A,** Dorsal view of adult loriciferan, *Nanaloricus mysticus,* showing internal features. **B,** Live animal, 0.3 mm.

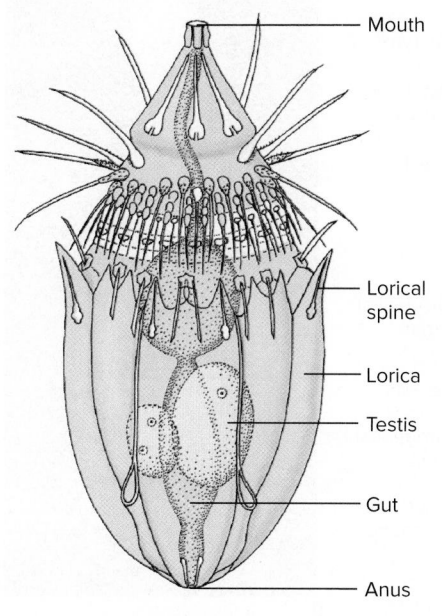

- Mouth
- Lorical spine
- Lorica
- Testis
- Gut
- Anus

A

B

which molts into an adult; and in a third species the life cycle is more complex, as a Higgins larva is followed by parthenogenetic stages. Higgins larvae themselves also differ in form, with benthic larvae having toes and pelagic larvae lacking toes.

18.4 PHYLUM KINORHYNCHA

Kinorhyncha (kīn´ō-ring´ka) (Gr. *kinein,* to move, + *rhynchos,* beak) are marine worms a little larger than rotifers and gastrotrichs but usually not more than 1 mm long. This phylum has also been called Echinodera, meaning spiny necked. About 179 species have been described to date.

Kinorhynchs are cosmopolitan, living from pole to pole, from intertidal areas to 8000 m in depth. Most live in mud or sandy mud, but some have been found in algal holdfasts, sponges, or other invertebrates.

18.3 PHYLUM LORICIFERA

Loricifera (L., *lorica,* corselet, + Gr., *phora,* bearing) was identified in 1983 and now has only 11 currently described species and approximately 80 undescribed species. The tiny animals (ranging from 0.1 to 0.5 mm long) have a protective external case (lorica) and live in spaces between grains of marine gravel, to which they cling tightly. Although they were first described from specimens collected off the coast of France, they are apparently distributed worldwide. Most species occupy coarse marine sediments at depths of 300 to 450 m, although one species was collected from 8000 m.

Form and Function

The loriciferan body has five regions: mouth cone, head or introvert, neck, thorax, and abdomen. There are nine circlets of scalids on the introvert. Scalids are curved spines with sensory and locomotory functions. The covering on the abdomen, a lorica, may have thick cuticular plates, or it may be thin and folded. The entire forepart of their body can be retracted into the circular lorica (Figure 18.14). Their diet is unknown but there is speculation that they feed on bacteria. Their brain fills most of the head, and nerves from the brain and other ganglia innervate oral spines (scalids). The body cavity has been described as a pseudocoel in some species, but other species are considered acoelomate.

Loriciferans are dioecious, with dimorphic males and females. Copulation occurs, but life cycles are not well known. There is a distinct larval phase called a Higgins larva. Three species within genus *Rugiloricus* have life cycles that differ in the number of larval stages. In one species, a Higgins larva molts into an adult; in another a Higgins larva molts into a second stage,

Form and Function

The body of kinorhynchs is divided into head, neck, and trunk regions. The trunk has 11 segments, marked externally by spines and cuticular plates (Figure 18.15). The retractile head, sometimes called an introvert, has five to seven circlets of spines with a small retractile proboscis. The spines, called scalids, function in locomotion, chemoreception, and mechanoreception. Each contains 10 or fewer monociliary sensory cells. The body is flat ventrally and arched dorsally. The body wall is composed of a chitinous cuticle, a cellular epidermis, and longitudinal epidermal cords, much like those of nematodes. The arrangement of muscles is correlated with the segments, and unlike the nematodes, kinorhynchs have circular, longitudinal, and diagonal muscle bands.

A kinorhynch cannot swim. In silt and mud where it commonly lives, it burrows by extending the head into the mud and anchoring it with spines. Extension of the head takes place as trunk muscles increase hydrostatic pressure on the small amount of fluid in the pseudocoel. After extension it draws its body forward until its head is retracted into its body. When disturbed, a kinorhynch draws in its head and protects it with a closing apparatus of cuticular plates on the neck, or on the neck and trunk.

Their digestive system is complete, with a mouth at the tip of a proboscis, followed by a pharynx, an esophagus, a nonciliated midgut, and a cuticle-lined hindgut, as well as an anus. Kinorhynchs feed on diatoms or by digesting organic material from the surface of mud particles through which they burrow.

Their pseudocoel is filled with amebocytes and organs, leaving little fluid space. A multinucleated solenocyte protonephridium (see p. 307) on each side of the gut between the eighth and ninth segments serves as their excretory system.

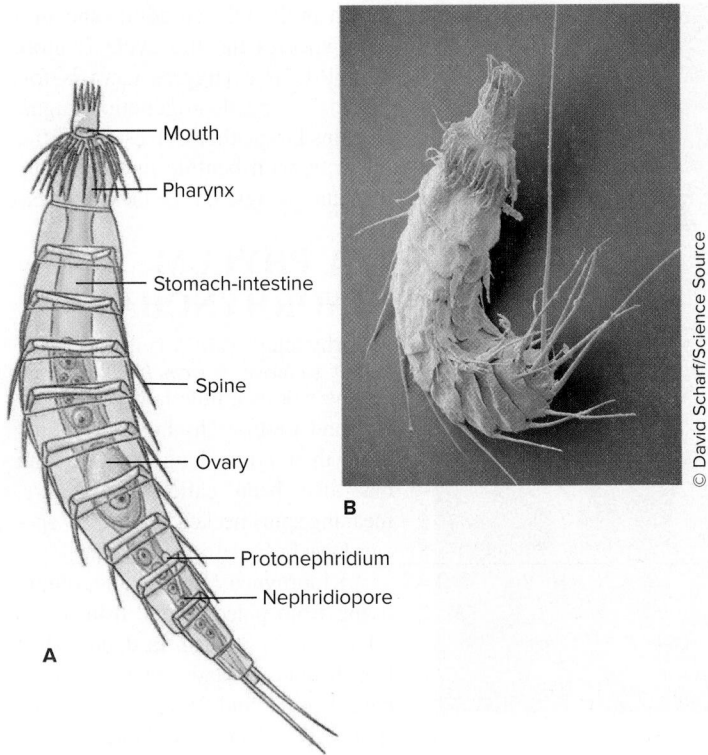

© David Scharf/Science Source

Figure 18.15 **A**, *Echinoderes*, a kinorhynch, is a minute marine worm. Segmentation is superficial. The head, with its circle of spines, is retractile. **B**, Colored scanning electron micrograph (SEM) of kinorhynch *Antigomonas sp.*

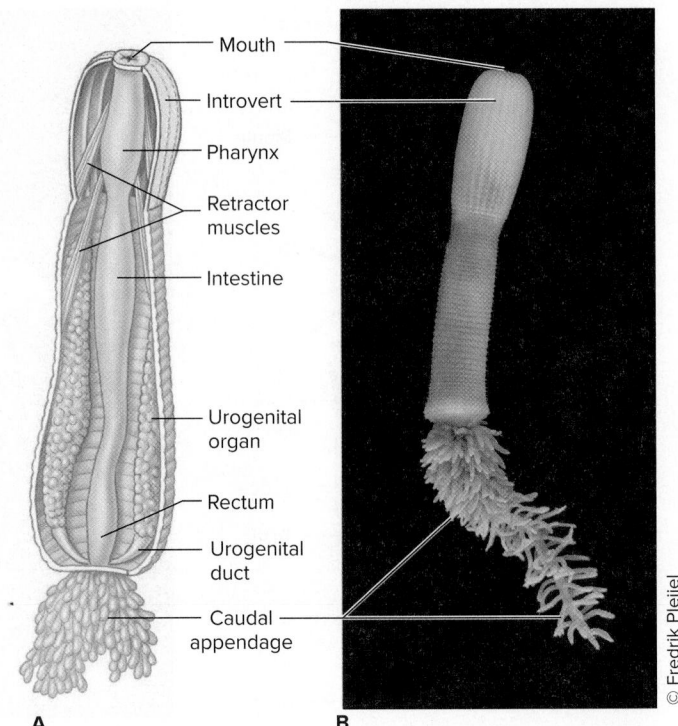

© Fredrik Pleijel

Figure 18.16 **A**, Major internal structure of *Priapulus*. **B**, *Priapulus caudatus* from Lurefjord, Norway.

The nervous system is in contact with the epidermis, with a multilobed brain encircling their pharynx, and with a ventral ganglionated nerve cord extending throughout the body. Sense organs are represented by sensory bristles and by eyespots in some.

Sexes are separate, with paired gonads and gonoducts. There is a series of about six juvenile stages and a definitive, nonmolting adult. No asexual reproduction has been found.

18.5 PHYLUM PRIAPULIDA

Priapulida (pri´a-pyu´li-da) (Gr. *priapos*, phallus, + *ida*, pl. suffix) is a small group (only 16 species) of marine worms found chiefly in colder waters of both hemispheres. They have been reported along the Atlantic coast from Massachusetts to Greenland and along the Pacific coast from California to Alaska. They live in mud and sand of the seafloor and range from intertidal zones to depths of several thousand meters. *Tubiluchus* (L. *tubulus*, dim. of *tubus*, waterpipe) is a minute detritus feeder adapted to interstitial life in warm coralline sediments. *Maccabeus* (named for a Judean patriot who died in 160 B.C.) is a tiny tube-dweller discovered in muddy Mediterranean substrates.

Form and Function

Priapulids have cylindrical bodies, most less than 12 to 15 cm long, but *Halicryptus higginsi* is up to 39 cm in length. Most are burrowing predaceous animals that feed on soft-bodied invertebrates such as polychaete worms (p. 364). They usually orient themselves upright

in mud with their mouth at the surface. However, *Tubiluchus* feeds on organic detritus in the sediments around coral reefs. They are adapted for burrowing by body contractions.

The body includes an introvert, a trunk, and usually one or two caudal appendages (Figure 18.16). Their retractable introvert is ornamented with papillae and ends with rows of scalids that surround the mouth. Extension of the introvert occurs as circular muscles increase hydrostatic pressure on the internal fluid-filled cavity. Derivation of the cavity is not clear. The eversible pharynx is used for capturing small, soft-bodied prey. *Maccabeus* has a crown of brachial tentacles around its mouth.

Their trunk is not metameric but is superficially divided into 30 to 100 rings and is covered with tubercles and spines. The tubercles are probably sensory in function. The anus and urogenital pores are located at the posterior end of the trunk. Caudal appendages are hollow stems believed to be respiratory and probably chemoreceptive in function. A chitinous cuticle, molted periodically throughout life, covers their body.

Their digestive system contains a muscular pharynx and a straight intestine and rectum (Figure 18.16). There is a nerve ring around the pharynx and a midventral nerve cord. Amebocytes occur in the fluids of the body cavity and, at least in some species, have corpuscles containing a respiratory pigment called hemerythrin.

Sexes are separate, although males have never been discovered for *Maccabeus*. Paired urogenital organs each contain a gonad and clusters of solenocytes, both connected to a protonephridial tubule that carries both gametes and excretory products outside the body. Fertilization is external. Embryology is poorly known. In

Meiopriapulus development is direct, and females brood their developing embryos. In most species, the zygote appears to undergo radial cleavage and to develop into a loricate larva. Larvae of *Priapulus* dig into mud and become detritus feeders.

18.6 CLADE PANARTHROPODA

Panarthropoda contains Arthropoda and two allied phyla, Onychophora and Tardigrada. In these taxa, the coelom is reduced and a hemocoel develops. In onychophorans and arthropods, a coelom develops by schizocoely, but coelom formation has been described as enterocoelic in tardigrades. In all three phyla, the main coelomic cavity later fuses with the blastocoel to form a new cavity called a **hemocoel,** or a mixocoel. The hemocoel is lined by an extracellular matrix, not the mesodermal peritoneum that originally lined the coelom. Blood from the open circulatory system enters the hemocoel and surrounds the internal organs. A muscular heart is present, but tubular blood vessels occur in only part of the body; blood enters and leaves the hemocoel through blood vessels. There may be small coelomic cavities surrounding a few organs in other parts of the body.

18.7 PHYLUM ONYCHOPHORA

Members of phylum Onychophora (on-y-kof´o-ra) (Gr. *onyx,* claw, + *pherein,* to bear) are commonly called "velvet worms," or "walking worms." There are approximately 70 species of caterpillar-like animals, ranging from about 0.5 to 15 cm in length. They live in rainforests and other moist, leafy habitats in tropical and subtropical regions and in some temperate regions of the Southern Hemisphere. Most velvet worms are predaceous, feeding on caterpillars, insects, snails, and worms. Some onychophorans live in termite nests and feed on termites.

Their fossil record shows that they have changed little in their 500-million-year history. A fossil form, *Aysheaia* (see Figure 6.9, p. 105), discovered in the Burgess Shale deposit of British Columbia and dating back to mid-Cambrian times, is very much like modern onychophorans. Onychophorans were probably far more common at one time than they are now. Today they are terrestrial and extremely retiring, becoming active only at night or when the air is nearly saturated with moisture.

Form and Function

External Features Onychophorans are more or less cylindrical and show no external segmentation except for the paired appendages (Figure 18.17). The skin is soft, velvety, and covered with a thin, flexible cuticle that contains protein and chitin. In structure and chemical composition it resembles arthropod cuticle; however, it never hardens like arthropod cuticle, and it is molted in patches rather than all at one time. The body is studded with tiny **tubercles,** some of which bear sensory bristles. The color may be green, blue, orange, dark gray, or black, and minute scales on the tubercles give the body an iridescent and velvety appearance. The head bears a pair of large **antennae,** each with an annelid-like eye at the base. The ventral mouth has a pair of clawlike **mandibles** and is flanked by a pair of **oral papillae,** which can expel a slimy defensive secretion (Figure 18.17).

Their 14 to 43 pairs of **unjointed legs** are short, stubby, and clawed. Onychophorans crawl by passing waves of contraction from anterior to posterior. When a body region extends, the legs lift up and move forward. The legs are more ventrally located than are parapodia of annelids.

Internal Features The body wall is muscular like that of annelids. The body cavity is a hemocoel, imperfectly divided into

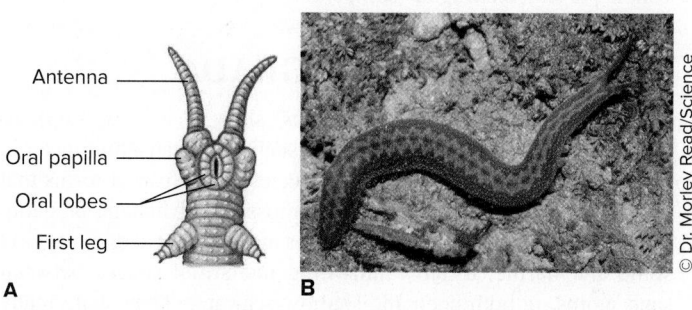

A   **B**

Figure 18.17 *Peripatus,* a caterpillar-like onychophoran that has characteristics in common with both annelids and arthropods. **A,** Ventral view of head. **B,** In natural habitat.

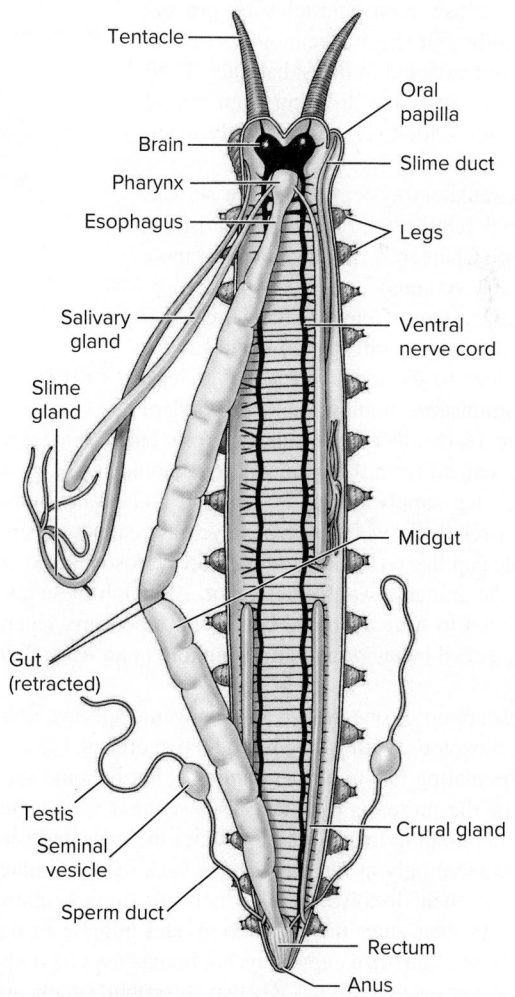

Figure 18.18 Internal anatomy of an onychophoran.

compartments, or sinuses, much like those of arthropods (see p. 425). **Slime glands** on each side of the body cavity open on the oral papillae. When disturbed by a predator, the animal can launch two streams of sticky fluid from these glands up to 30 cm. Hardening rapidly, this adhesive can entangle the would-be predator and hold it firmly for the onychophoran to consume at its leisure.

The mouth, surrounded by lobes of skin, contains a dorsal tooth and a pair of lateral mandibles for grasping and cutting prey. There is a muscular pharynx and a straight digestive tract (Figure 18.18).

Each leg-bearing body segment contains a pair of **nephridia**, each nephridium with a vesicle, ciliated funnel and duct, and nephridiopore opening at the base of a leg. Absorptive cells in the midgut excrete crystalline uric acid, and certain pericardial cells function as nephrocytes, storing excretory products taken from the blood.

For respiration there is a **tracheal system** (sets of branching internal tubes that pipe air directly to cells) that connects to all parts of the body and communicates with the outside by many openings, or **spiracles**, scattered all over the body. Onychophorans cannot close their spiracles to prevent water loss, so although the tracheae are efficient, these animals are restricted to moist habitats. Their tracheal system is somewhat different from that of arthropods (see pp. 450–452), and probably originated independently.

The open circulatory system has, in the pericardial sinus, a dorsal, tubular heart with a pair of ostia in each segment (see Chapter 31, pp. 685–686, for more on open circulatory systems).

The nervous system of onychophorans is organized much like a ladder, with paired ventral nerve cords running close to the top of each row of legs, connected by commisures running across the width of the body (Figure 18.18). Nerves to antennae (also called tentacles) and head region extend from the brain, and ganglionic swellings at the base of each leg supply nerves to legs and body wall. Sense organs include a relatively well-developed eye, taste spines around the mouth, tactile papillae on the integument, and hygroscopic receptors that orient the animal toward water vapor. Although these animals were assumed to have a limited behavioral repertoire, recent work has shown social behavior and group hunting in an Australian species.

With the exception of one known parthenogenic species, onychophorans are dioecious, with paired reproductive organs. Little is known about the mating habits of these animals, but in some species a portion of the uterus is expanded as a seminal receptacle, presumably for copulation. In at least one species the male deposits spermatophores, seemingly at random, on the back of the female. White blood cells then dissolve the skin beneath the spermatophores. Sperm can then enter the body cavity and migrate in the blood to the ovaries to fertilize eggs. Onychophorans may be oviparous, ovoviviparous, or viviparous. Only two Australian genera are oviparous, laying shell-covered eggs in moist places. In all other

A   B

Figure 18.19 **A and B,** Color enhanced scanning electron micrographs (SEM) of an active tardigrade, *Echiniscus granulatus*, from Tubingen Germany. It feeds on moss cells. It clings to moss or water plants with its claws. During a drought it exhibits anhydrobiosis, a stable state of suspended animation, until there is sufficient water for rehydration. This animal ranges in size from 300 to 500 µm.

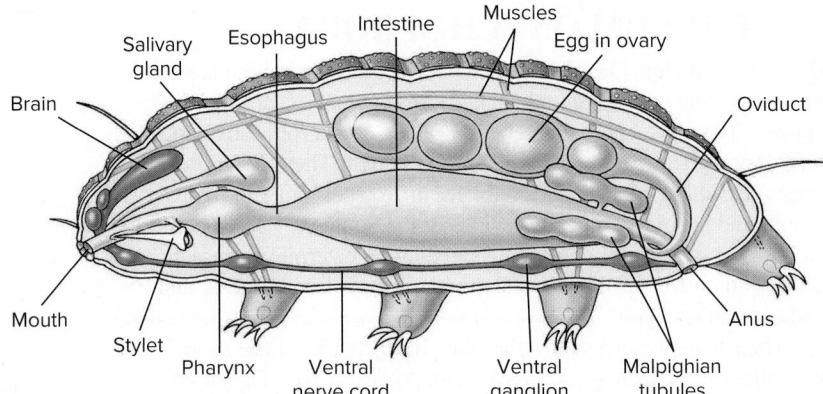

Figure 18.20 Internal anatomy of a tardigrade.

onychophorans eggs develop in the uterus, and live young leave the mother's body. In some species there is a placental attachment between mother and young (viviparous); in others young develop in the uterus without attachment (ovoviviparous). Nonplacental species typically have eggs with a large amount of yolk; the eggs cleave superficially, in a manner similar to that in arthropods. When little yolk is present, cleavage is complete.

18.8 PHYLUM TARDIGRADA

Tardigrada (tar-di-grā-´da) (L. *tardus,* slow, + *gradus,* step), or "water bears," are minute organisms usually less than a millimeter in length. Most of the 900 described species are terrestrial forms that live in the water film surrounding mosses and lichens or damp soils. Some live in freshwater algae or mosses or in sediments, and some are marine, usually inhabiting interstitial spaces between sand grains, in both deep and shallow seawater. They share many characteristics with arthropods.

They have an elongated, a cylindrical, or a long oval body that is unsegmented. The head is merely the anterior part of the trunk.

Figure 18.22 Molted cuticle of a tardigrade, containing a number of fertilized eggs.

Figure 18.21 Color-enhanced scanning electron micrograph of the very ornate surface of a tardigrade egg. This egg of *Macrobiotus sapiens* was found on moss samples from Croatia. Magnification: ×4800 when printed 10cm wide.

The trunk bears four pairs of short, stubby, unjointed legs, each armed with four to eight claws (Figure 18.19). They are covered by a nonchitinous cuticle that is molted along with claws and buccal apparatus four or more times in the life history. Some, such as *Echiniscus* (Figure 18.19B), expel feces when molting, leaving the feces in the discarded cuticle.

The mouth in tardigrades opens into a buccal tube that empties into a muscular pharynx adapted for sucking (Figure 18.20). Two needlelike stylets flanking the buccal tube can be protruded through the mouth. These stylets pierce plant or animal cells, and the pharynx sucks in the liquid contents. Some tardigrades suck body juices of nematodes, rotifers, and other small animals, while others are parasitic on larger animals such as sea cucumbers or barnacles. At the junction of intestine and rectum, three glands, thought to be excretory and often called Malpighian tubules, empty into the digestive system. Cilia are absent.

Most of the body cavity is a hemocoel, with their true coelom restricted to the gonadal cavity. There are no circulatory or respiratory systems, gaseous exchange occurring by diffusion through the body surface.

Their muscular system consists of a number of long muscle bands, usually comprising one or a few large muscle cells each. Circular muscles are absent, but hydrostatic pressure of the body fluid may act as a skeleton. Being unable to swim (with one exception), water bears crawl awkwardly, clinging to the substrate with their claws.

Their brain is relatively large and covers most of the dorsal surface of the pharynx. Circumpharyngeal connectives link it to the subpharyngeal ganglion, from which the double ventral nerve cord extends posteriorly as a chain of four ganglia that appear to control the four pairs of legs. Eyes, inverse pigment-cup ocelli, are present.

Sexes are separate in tardigrades. In some freshwater and moss-dwelling species, males are unknown and parthenogenesis seems to be the rule. Some species also have dwarf males, but in most tardigrades that have been studied males and females occur with approximately equal frequency. In some species sperm is deposited directly into the female's seminal receptacle or cloaca during copulation; in others sperm is injected into the body cavity by piercing the cuticle. Eggs of some species are highly ornate (Figure 18.21). Egg-laying, like defecation, apparently occurs primarily at molting, when the volume of coelomic fluid is reduced. Females of some species cement their eggs to a submerged object, whereas others deposit eggs in the molted cuticle (Figure 18.22). In some such cases, fertilization is indirect and males gather around the old cuticle containing unfertilized eggs and shed sperm into it.

Detailed research on the development of tardigrades is lacking, but cleavage is apparently complete. A stereogastrula is formed. Five pairs of coelomic pouches appear, reminiscent of the enterocoelous development of many deuterostomes. However, all but the last pair, which fuse to form the gonad, disappear during development, and the gonocoel is the only true coelomic space left in adults. Development is direct and rapid. After about 14 days, juveniles use their claws to break free of the egg. At this time the number of cells in the body is relatively fixed, and growth occurs primarily by increase in cell size rather than cell number.

One of the most intriguing features of terrestrial tardigrades is their capacity to enter a state of suspended animation, called cryptobiosis, during which metabolism is virtually imperceptible; such an organism can withstand prolonged harsh environmental conditions. Under gradual drying conditions, water content of the body decreases from 85% to only 3%, movement ceases, and the body becomes barrel-shaped. In a cryptobiotic state tardigrades can resist temperature extremes from +149° C to –272° C, ionizing radiation, oxygen deficiency, preservatives such as ether and absolute alcohol, and other adverse conditions, and may survive for years. Activity resumes when moisture is again available. Some nematodes and rotifers also can undergo cryptobiosis.

18.9 PHYLOGENY AND ADAPTIVE DIVERSIFICATION

Phylogeny

Evolutionary relationships among ecdysozoans are not well understood. Members of this clade do not share a common cleavage pattern. Cleavage in nematodes and nematomorphs is described as unique, or not obviously spiral or radial. In priapulids, it is somewhat similar to radial cleavage. Cleavage has not been studied in kinorhynchs, loriciferans, and tardigrades. In onychophoran eggs containing large amounts of yolk, the cytoplasm does not cleave, but nuclei do divide. Development is similar to that in arthropods with centrolecithal eggs (see pp. 159–160). In onychophoran eggs with little yolk, cleavage is complete (holoblastic), but the cleavage pattern varies, appearing spiral in some taxa and radial in others.

In the absence of developmental characters, branch order is not defined for all ecdysozoans, but roundworms, phylum Nematoda, are united with horsehair worms, phylum Nematomorpha, in clade Nematoidea (see Figure 18.1). Recent phylogenies place the two phyla as sister taxa sharing a collagenus cuticle, but other work suggests that nematomorphs are related to loriciferans.

Phylum Kinorhyncha is shown as the sister taxon to phylum Priapulida based on the shared two-layered pharynx. Kinorhynchs have mouthparts (oral styles on a noninversible mouth cone) similar to those of loriciferans, but loriciferans also share some morphological features with larval nematomorphs and with priapulids. Some workers erect clade Scalidophora to contain kinorhynchs, priapulids, and loriciferans, but more work is needed on these animals before a branch order is specified. Mitochondrial genome sequence data support a sister-taxon relationship between kinorhynchs and priapulids and between this clade and Panarthropoda.

Clade Panarthropoda unites three phyla: velvet worms, phylum Onychophora; water bears, phylum Tardigrada; and Arthropoda.

Onychophoran characteristics shared with arthropods include a tubular heart and hemocoel with open circulatory system, presence of tracheae (probably not homologous), absence of ectodermal cilia, and large size of their brain. Onychophorans also share a number of characteristics with annelids: metamerically arranged nephridia, muscular body wall, pigment cup ocelli, and ciliated reproductive ducts. Unique characteristics include oral papillae, slime glands, body tubercles, and suppression of external segmentation.

Tardigrades have some similarities to rotifers, particularly in their reproduction and their cryptobiotic tendencies, and some authors have called them pseudocoelomates. Their embryogenesis, however, would seem to put them among coelomates. The enterocoelic origin of the mesoderm is a deuterostome characteristic, but five pouches form, some of which fuse while others disappear, unlike the pattern in typical deuterostomes Other authors identify several important synapomorphies that suggest grouping them with arthropods (Figure 18.1). DNA sequence analysis supports alignment with arthropods in Ecdysozoa. Tardigrades and arthropods also share two morphological features: arthropod-type setae and muscles that insert on the cuticle (Figure 18.1). However, mitochondrial genome sequence data indicate that onychophorans and tardigrades are sister taxa, and this pair is the sister taxon to Arthropoda.

Reconstructing the evolutionary history of life is a fascinating pursuit, but biologists lack developmental and morphological information for many taxa, as is clear from comments presented here. However, a recent study of the pharynx and mouthparts of an ancient onychophoran called *Hallucigenia* led to the discovery of an ecdysozoan morphological synapomorphy: hardened tooth-like elements around the mouth and a pharynx lined with teeth. These features were present in early arthropods and in the other ecdysozoans just described, but they are not present in modern onychophorans. Researchers wondered whether modern onychophorans lost these key features or never had them. From the morphology of *Hallucigenia* they infer a loss in modern velvet worms and conclude that the features are ancestral for Ecdysozoa.

Adaptive Diversification

Arthropods aside, certainly the most impressive adaptive diversification in this group of phyla is shown by nematodes. They are by far the most numerous in terms of both individuals and species, and they have been able to adapt to almost every habitat available to animal life. Their basic pseudocoelomate body plan, with the cuticle, hydrostatic skeleton, and longitudinal muscles, has proved generalized and plastic enough to adapt to an enormous variety of physical conditions. Free-living lineages gave rise to parasitic forms on at least several occasions, and virtually all potential hosts have been exploited. One phylogeny indicates that plant-parasitic forms have arisen from fungal-feeding ancestors in three independent evolutionary events. All types of life cycle occur: from simple and direct to complex, with intermediate hosts; from normal dioecious reproduction to parthenogenesis, hermaphroditism, and alternation of free-living and parasitic generations. A major factor contributing to evolutionary opportunism of nematodes has been their extraordinary capacity to survive suboptimal conditions, for example, developmental arrests in many free-living and animal parasitic species and ability to undergo cryptobiosis (survival in harsh conditions by assuming a very low metabolic rate) in many free-living and plant parasitic species.

SUMMARY

Phyla covered in this chapter possess a range of body plans. Analysis of nucleotide similarities in the gene for 18S small-subunit rDNA provides evidence that they belong to superphylum Ecdysozoa. All members of this clade molt their cuticles.

Other than Arthropoda, Nematoda is the largest and most important of these phyla, and although only 25,000 species are described currently, estimates suggest there may be as many as 500,000 species alive today. They are more or less cylindrical, tapering at the ends, and covered with a tough, secreted cuticle. Their body-wall muscles are longitudinal only, and to function well in locomotion, such an arrangement must enclose a volume of fluid in the pseudocoel at high hydrostatic pressure. This fact of nematode life has a profound effect on most of their other physiological functions, for example, ingestion of food, egestion of feces, excretion, copulation, and others. Most nematodes are dioecious, and there are four juvenile stages, each separated by a molt of the cuticle. Almost all invertebrate and vertebrate animals and many plants have nematode parasites, and many other nematodes are free-living in soil and aquatic habitats. Some parasitic nematodes have part of their life cycle free-living, some undergo a tissue migration in their host, and some have an intermediate host in their life cycle. Some parasitic nematodes cause severe diseases in humans and other animals.

Nematomorpha or horsehair worms superficially resemble nematodes and have parasitic juvenile stages in arthropods, followed by a free-living, aquatic adult stage.

Loriciferans can withdraw their bodies into their lorica. Kinorhynchs anchor and then pull themselves by spines on their head. Loricifera and Kinorhyncha are small phyla of tiny, aquatic pseudocoelomates. Priapulids are marine burrowing worms of moderate size.

Clade Panarthropoda contains onychophorans, tardigrades, and arthropods. They have open circulatory systems with a hemocoel.

Onychophora are caterpillar-like animals found in humid, mostly tropical habitats. They are segmented and crawl by means of a series of unjointed, clawed appendages.

Tardigrades are minute animals, mostly terrestrial, living in the water film that surrounds mosses and lichens. They have eight unjointed legs and a nonchitinous cuticle. They can undergo cryptobiosis, withstanding adverse conditions for long periods.

REVIEW QUESTIONS

1. What is a cuticle?
2. Define ecdysis.
3. What is a hydrostatic skeleton?
4. Distinguish a solenocyte from a flame-cell protonephridium.
5. Explain two peculiar features of the body-wall muscles in nematodes.
6. What feature of body-wall muscles in nematodes requires a high hydrostatic pressure in the pseudocoelomic fluid for efficient function?
7. Explain the interaction of cuticle, body-wall muscles, and pseudocoelomic fluid in locomotion of nematodes.
8. Explain how the high pseudocoelomic pressure affects feeding and defecation in nematodes.
9. Outline the life cycle of each of the following: *Ascaris lumbricoides,* hookworm, *Enterobius vermicularis, Trichinella spiralis, Wuchereria bancrofti.*
10. Where in the human body are adults of each species in question 9 found?
11. Outline the life cycle of a gordiid nematomorph.
12. How are nematodes and nematomorphs alike, and how are they different?
13. Where do kinorhynchs live?
14. Describe the introvert of a loriciferan and a priapulid.
15. How is a hemocoel different from a true coelom?
16. In what sense is a hemocoel part of the circulatory system?
17. In what habitats would you encounter tardigrades?
18. How does cryptobiosis in tardigrades increase the likelihood of survival?
19. Describe the two major protostome clades and give a diagnostic feature for each.
20. List the predominant body plan (acoelomate, pseudocoelomate, or coelomate) for members of each protostome phylum and discuss how our picture of protostome evolution would change if each body plan were a homologous character.

For Further Thought The comments of N. A. Cobb quoted in the opening essay for this chapter make clear just how successful are nematodes. What features of these animals account for their astonishing abundance?

SELECTED REFERENCES

Aguinaldo, A. M. A., J. M. Turbeville, L. S. Linford, M. C. Rivera, J. J. F. R. Garey, R. A. Raff, and J. A. Lake. 1997. Evidence for a clade of nematodes, arthropods and other moulting animals. Nature **387:** 489–493. *Sequence analysis to support a superphylum Ecdysozoa.*

Balavoine, G., and A. Adoutte. 1998. One or three Cambrian radiations? Science **280:** 397–398. *Discusses radiation into superphyla Ecdysozoa, Lophotrochozoa, and Deuterostomia.*

Bird, A. F., and J. Bird. 1991. The structure of nematodes, ed. 2. New York, Academic Press. *The most authoritative reference available on nematode morphology. Highly recommended.*

Brusca, R. C., and G. J. Brusca. 2003. Invertebrates, ed. 2. Sunderland, Massachusetts, Sinauer Associates, Inc. *A comprehensive invertebrate text.*

Chan, M.-S. 1997. The global burden of intestinal nematode infections—fifty years on. Parasitol. Today **13:** 438–443. *According to this author, estimates are 1.273 billion infections (24% prevalence) with* Ascaris, *0.902 billion (17% prevalence) with* Trichuris, *and 1.277 billion (24% prevalence) with hookworms. Worldwide prevalence of these nematodes has remained essentially unchanged in 50 years!*

Despommier, D. D. 1990. *Trichinella spiralis:* the worm that would be virus. Parasitol. Today **6:** 193–196. *Juveniles of* Trichinella *are among the largest of all intracellular parasites.*

Dopazo, H., and J. Dopazo. 2005. Genome-scale evidence of the nematode-arthropod clade. Genome Biol. **6:** R41. *Phylogenetic trees constructed under the hypothesis that a coelomate body is homologous are compared with those assuming ecdysis is a homologous trait.*

Duke, B. O. L. 1990. Onchocerciasis (river blindness)—can it be eradicated? Parasitol. Today **6:** 82–84. *Despite the introduction of a very effective drug, the author predicts that this parasite will not be eradicated in the foreseeable future.*

Gould, S. J. 1995. Of tongue worms, velvet worms, and water bears. Natural History **104** (1):6–15. *Intriguing essay on affinities of Pentastomida, Onychophora, and Tardigrada and how they, along with larger phyla, were products of the Cambrian explosion.*

Halanych, K. M., and Y. Passamaneck. 2001. A brief review of metazoan phylogeny and future prospects in *Hox*-research. Am. Zool. **41:** 629–639. *A good review of the arguments for and against the lophotrochozoa and ecdysozoa hypotheses.*

Holterman, M., A. van der Wurff, S. van den Elsen, H. van Megen, T. Bongers, O. Holovachov, J. Bakker, and J. Helder. 2006. Phylum-wide analysis of SSU rDNA reveals deep phylogenetic relationships among nematodes and accelerated evolution toward crown clades. Mol. Biol. Evol. **23:** 1792–1800. *A nematode phylogeny shows that plant-parasitic species arose from fungivorous ancestors in three lineages.*

Neuhaus, B., and R. P. Higgins. 2002. Ultrastructure, biology, and phylogenetic relationships of Kinorhyncha. Integ. and Comp. Biol. **42:** 619–632. *A detailed summary of the biology and morphology of these animals.*

Nielsen, C. 1995. Animal evolution: interrelationships of the living phyla. Oxford University Press, New York. *The author proposes homologous features for many lesser-known taxa.*

Ogilvie, B. M., M. E. Selkirk, and R. M. Maizels. 1990. The molecular revolution and nematode parasitology: yesterday, today, and tomorrow. J. Parasitol. **76:** 607–618. *Modern molecular biology has wrought enormous changes in investigations on nematodes.*

Reinhard, J., and D. M. Rowell. 2005. Social behaviour in an Australian velvet worm, *Euperipatoides rowelli* (Onychophora: Peripatopsidae). J. Zool. **267:** 1–7. *This velvet worm hunts collectively and has an organized social structure where one female is dominant.*

Rota-Stabelli, O., E. Kayal, D. Gleeson, J. Daub, J. L. Boore, M. J. Telford, D. Pisani, M. Blaxter, and D. V. Lavrov. 2010. Ecdysozoan mitogenomics: evidence for a common origin of the legged invertebrates, the Panarthropoda. Genome Biol. Evol. **2:** 425–440. *Support for the sister-taxon relationships between onychophorans and tardigrades, and between this pair and arthropods. The mandibulate hypothesis for arthropod relationships .*

Smith, M. R., and J. Caron. 2015. *Hallucigenia*'s head and the pharyngeal armature of early ecdysozoans. Nature **523:**75–78. *An onychophoran fossil sheds light on the features of the ancestral ecdysozoan.*

Taylor, M. J., and A. Hoerauf. 1999. *Wolbachia* bacteria of filarial nematodes. Parasitol. Today **15:** 437–442. *All filarial parasites of humans have endosymbiotic* Wolbachia, *and most filarial nematodes of all kinds are infected. Nematodes can be "cured" by treatment with the antibiotic tetracycline. If cured, they cannot reproduce. Bacteria are apparently passed vertically from females to offspring.*

19
CHAPTER

Trilobites, Chelicerates, and Myriapods

A scorpion.

- **PHYLUM ARTHROPODA**
  - Subphylum Trilobita
  - Subphylum Chelicerata
  - Subphylum Myriapoda

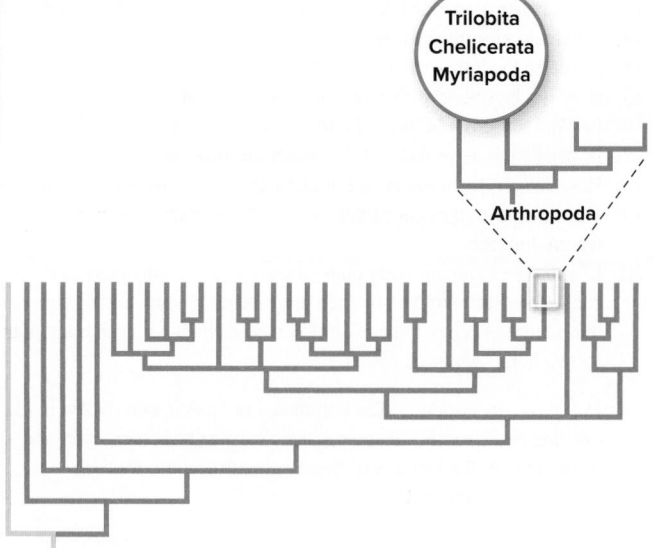

A Suit of Armor

Sometime, somewhere in the Precambrian era, a major milestone occurred in the evolution of life on earth. The soft cuticle in the segmented ancestor of animals we now call arthropods was hardened by deposition of additional protein and an inert polysaccharide called chitin. The cuticular exoskeleton offered some protection against predators and other environmental hazards, and it conferred on its possessors a formidable array of other selective advantages. For example, a hardened cuticle provided a more secure site for muscle attachment, allowed adjacent segments and joints to function as levers, and vastly improved the potential for rapid locomotion, including flight. Of course, a suit of armor could not be uniformly hard; the animal would be as immobile as the rusted Tin Woodsman in *The Wizard of Oz*. Stiff sections of cuticle were separated from each other by thin,

flexible sections, which formed sutures and joints. The cuticular exoskeleton had enormous evolutionary potential. Jointed extensions on each segment became appendages.

As the hardened cuticle evolved, or perhaps concurrently with it, many other changes took place in the bodies and life cycles of protoarthropods. Growth required a sequence of cuticular molts controlled by hormones. Coelomic compartments reduced their hydrostatic skeletal function, associated with a regression of the coelom and its replacement with an open system of sinuses (hemocoel). Motile cilia were lost. These changes and others are called "arthropodization." They produced animals with astonishing diversity and abundance, able to colonize almost every habitat on earth.

19.1 PHYLUM ARTHROPODA

Phylum Arthropoda (ar-thrō'pōda) (Gr. *arthron,* joint, + *pous, podos,* foot) is currently the most species-diverse phylum in the animal kingdom, composed of more than three-fourths of all known species. Approximately 1,100,000 species of arthropods are recorded, and it is likely that as many more remain to be classified. (In fact, based on surveys of insect faunas in the canopies of rainforests, many estimates of yet undescribed species are much higher.) Arthropods include spiders, scorpions, ticks, mites, crustaceans, millipedes, centipedes, insects, and other less well-known groups. In addition, there is a rich fossil record extending to the very late Precambrian period.

Few arthropods exceed 60 cm in length, and most are much smaller. However, Paleozoic eurypterids reached up to 3 m and some ancient dragonfly-like insects (Protodonata) had wingspans approaching 1 m. Currently, the largest arthropod, a Japanese crab *Macrocheira* (Gr. *makros,* large, + *cheir,* hand), spans approximately 4 m; the smallest is a parasitic mite *Demodex* (Gr. *dēmos,* people, + *dex,* a wood worm), which is less than 0.1 mm long.

Arthropods are usually active, energetic animals. They utilize all modes of feeding—carnivory, herbivory, and omnivory—although most are herbivorous. Many aquatic arthropods are omnivorous or depend on algae for nourishment, and most land forms live chiefly on plants. In diversity of ecological distribution, arthropods have no rivals.

Although many terrestrial arthropods compete with humans for food and spread serious vertebrate diseases, they are essential for the pollination of many food plants, and they also serve as food in the ecosystem, yield drugs, and generate products such as silk, honey, beeswax, and dyes.

Arthropods are more widely distributed throughout all regions of the earth's biosphere than are members of any other eukaryotic phylum. They occur in all types of environment from the deepest ocean depths to very high elevations, and from the tropics far into both northern and southern polar regions. Different species are adapted for life in the air; on land; in fresh, brackish, and marine waters; and in or on the bodies of plants and other animals. Some species live in places where no other animal could survive.

Relationships Among Arthropod Subgroups

Arthropods are ecdysozoan protostomes belonging to clade Panarthropoda (see Figure 18.1, p. 386). They have segmented bodies, a chitinous cuticle often containing calcium, and jointed appendages. The critical stiffening of the cuticle to form a jointed exoskeleton is sometimes called "arthropodization."

Arthropods diversified greatly, but it is relatively easy to identify particular body plans characterizing arthropod subgroups. For example, centipedes and millipedes have trunks composed of repeated similar segments, whereas spiders have two distinct body regions and lack repeated segments. Arthropoda is divided into several **subphyla** based on our current understanding of the relationships among subgroups.

Traditionally, centipedes, millipedes, and related forms, called pauropods and symphylans, were grouped with the insects in subphylum Uniramia. Members of Uniramia all possessed **uniramous** appendages—those with a single branch—as opposed to **biramous**

appendages, which have two branches (Figure 19.1). Phylogenies constructed using molecular data did not support Uniramia as a monophyletic group. Further, as the genetic basis for the uniramian versus biramian appendage was better understood (see pp. 437–439), it became increasingly unlikely that all uniramous appendages were inherited from a single common ancestor with such appendages.

Currently, five arthropod subphyla are defined. Centipedes, millipedes, pauropods, and symphylans are placed in subphylum **Myriapoda.** Insects are placed in subphylum **Hexapoda.** Spiders, ticks, horseshoe crabs, and their relatives form subphylum **Chelicerata.** Lobsters, crabs, barnacles, and many others form subphylum **Crustacea.** We include tongue worms, members of former phylum Pentastomida, in Crustacea. The extinct trilobites are placed in subphylum **Trilobita.**

Relationships among the subphyla are controversial. One hypothesis assumes that all arthropods possessing a particular mouthpart, called a **mandible** (Figure 19.1), form a clade called Mandibulata. This clade includes members of Myriapoda, Hexapoda, and Crustacea. Arthropods that do not have mandibles possess **chelicerae** (Figure 19.1), as exemplified by spiders. Thus, according to the "mandibulate hypothesis," myriapods, hexapods, and crustaceans are more closely related to each other than are any of them to chelicerates. Critics of the mandibulate hypothesis argue that the mandibles in each group are so different from each other that they could not be homologous. Mandibles of crustaceans are multijointed with chewing or biting surfaces on the mandible bases (gnathobasic mandible), whereas those of myriapods and hexapods have a single joint with the biting surface on the distal edge (entire-limb mandible). There are also some differences in the muscles controlling the two·types. Proponents of the mandibulate hypothesis respond that the 550-million-year history of the mandibulates makes possible the evolution of diverse mandibles from an ancestral type. Several recent phylogenies, including those using mitochondrial genome characters, support the "mandibulate hypothesis."

We assume that subphylum Trilobita is the sister taxon to all other arthropods, and that the earliest split within the latter group separates chelicerates from a common ancestor of myriapods, hexapods, and crustaceans. However, one phylogeny places trilobites as the sister taxon to Chelicerata, and shows the mandibulates forming a clade.[1] We depict subphylum Crustacea as the sister taxon of subphylum Hexapoda (Figure 19.2). Evidence of a close relationship between hexapods and crustaceans emerged from several phylogenetic studies using molecular characters; these studies prompted a reevaluation of the morphological characters in members of both taxa. We unite subphylum Crustacea with subphylum Hexapoda in clade Pancrustacea. The exact nature of the close relationship between these two subphyla is at issue and is discussed in Chapters 20 and 21.

Following a general introduction to the arthropods in this chapter, we cover three subphyla: Trilobita, Chelicerata, and Myriapoda. Chapter 20 is devoted to subphylum Crustacea and Chapter 21 to subphylum Hexapoda.

---

[1]Legg, D. A., M. D. Sutton, and G. D. Edgecombe. 2013. Arthropod fossil data increase congruence of morphological and molecular phylogenies. *Nature Communications* **4**:2485.

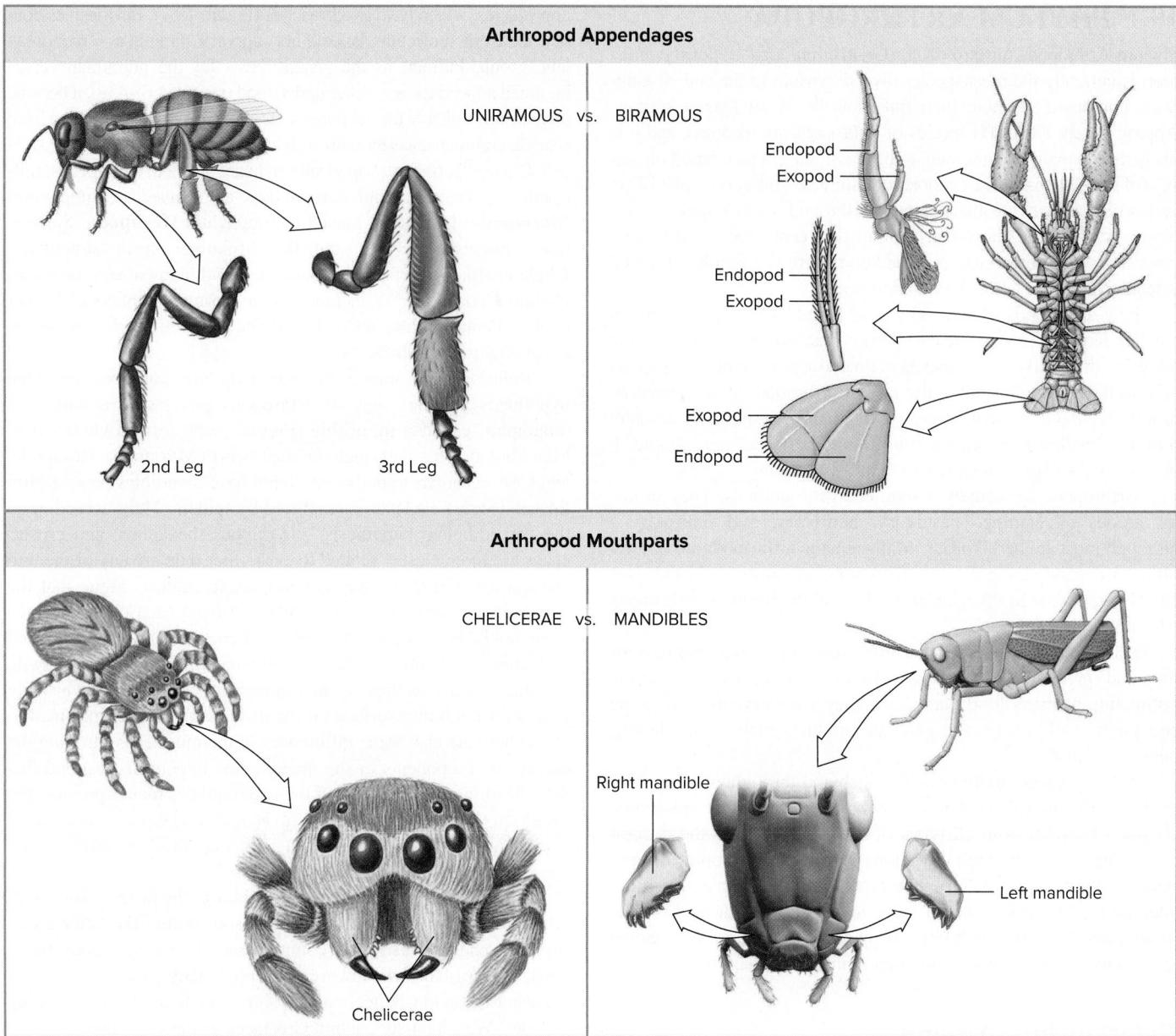

Figure 19.1 Two important arthropod characters: appendages may be uniramous (honey bee leg) or biramous (lobster limbs); mouthparts may include chelicerae (spider) or mandibles (grasshopper). Note that presence or absence of gills is unrelated to appendage form.

Why Have Arthropods Achieved Such Great Diversity and Abundance?

Arthropods exhibit great diversity (number of species), wide distribution, and variety of habitats and feeding habits, and have an uncanny genetic predisposition for adaptation to changing conditions. In our discussion we briefly summarize some structural and physiological patterns that have aided their rise to dominance.

1. **A versatile exoskeleton**. Arthropods possess an exoskeleton that is highly protective without sacrificing flexibility or mobility. This skeleton is the **cuticle,** an outer covering secreted by the underlying epidermis. The cuticle consists of an inner, relatively thick **procuticle** and an outer, relatively thin **epicuticle** (Figure 19.3). The procuticle and the epicuticle each consist of several layers (lamina). The outer epicuticle is composed of protein, often with lipids. The protein is stabilized and hardened by chemical cross-linking, called **sclerotization,** which increases its protective ability. In many insects the outermost layer of epicuticle is composed of waxes that reduce water loss.

The procuticle is divided into an **exocuticle,** which is secreted before a molt, and an **endocuticle,** which is secreted after molting. Both of these layers are composed of **chitin** bound with protein. Chitin is a tough, nitrogenous polysaccharide that is insoluble in water, alkalis, and weak acids. The procuticle not only is flexible and lightweight, but also affords protection against dehydration and other biological and physical stresses. In insects chitin forms about 50% of the procuticle, with the

Arthropoda
Mandibulata
Pancrustacea

Chelicerata	Myriapoda	Crustacea	Hexapoda

Crustacea
- Biramous 2nd antennae
- 2 pairs maxillae
- Nauplius larva
- 2nd pair of antennae

Hexapoda
- Unique trachael system
- 6 legs
- Tagmata = head, thorax and abdomen

Chelicerata
- Loss of antennae
- 1st pair of appendages = chelicerae
- Tagmata = cephalothorax + abdomen

Myriapoda
- Tagmata = head + trunk
- Loss of compound eye

- All head appendages except 1st antennae used for feeding sometime in life
- Tripartite brain
- Shared derived DNA sequences

- Mandibles

- Compound eye
- 1 pair of antennae
- Chitinous exoskeleton with articulated appendages

Figure 19.2 Cladogram of extant arthropods showing probable relationships of the four subphyla. Only a few synapomorphies are included here. Crustaceans and hexapods are shown as sister taxa, in clade Pancrustacea. A sister taxon relationship between Pancrustacea and Myriapoda is based on shared possession of mandibles and data from molecular phylogenies.

EPICUTICLE

PROCUTICLE

Exocuticle

Endocuticle

Principal layer

Membranous layer

- Seta
- Opening of duct to tegumental gland

- Epidermis
- Basement membrane
- Tegumental gland

Figure 19.3 Structure of crustacean cuticle.

remainder being protein. In some crustaceans, chitin is 60% to 80% of the procuticle; additionally, most crustaceans possess some areas of the procuticle impregnated with **calcium salts**. The addition of calcium salts reduces flexibility, but increases strength. In the hard shells of lobsters and crabs, for example, this calcification is extreme.

The cuticle may be soft and permeable or may form a veritable coat of armor. Between body segments and between the segments of appendages the cuticle is thin and flexible, creating movable joints and permitting free movements. In crustaceans and insects the cuticle forms ingrowths (**apodemes**) to which muscles attach. Cuticle may also line foregut and hindgut, line and support the tracheae, and be adapted for biting mouthparts, sensory organs, copulatory organs, and ornamental purposes. It is indeed a versatile material.

The nonexpansible cuticular exoskeleton does, however, impose important restrictions on growth. To grow, an arthropod must shed its outer covering at intervals and grow a larger one—a process called **molting**. The process of molting terminates in the actual shedding of the skin, or **ecdysis**. Arthropods may molt many times before reaching adulthood, and some continue to molt after that. More details of the molting process are given for crustaceans (pp. 429–430) and for insects (pp. 456–457).

2. **Segmentation and appendages provide for more efficient locomotion.** The ancestral arthropod body plan was likely a linear series of similar segments, each with a pair of jointed appendages. However, extant groups exhibit a wide variety of segments and appendages. There has been a tendency for segments to combine or to fuse into functional groups, called **tagmata** (sing., **tagma**), which have specialized purposes. Spider bodies, for example, have two tagmata. Appendages are frequently differentiated and specialized for pronounced division of labor. Limb segments are essentially hollow levers moved by internal muscles, most of which are striated, providing rapid action. The appendages have sensory hairs (as well as bristles and spines) and may be modified and adapted for sensory functions, food handling, swift and efficient walking, and swimming.

3. **Air piped directly to cells.** Most terrestrial arthropods have a highly efficient tracheal system of air tubes, which delivers oxygen directly to the tissues and cells and makes a high metabolic rate possible during periods of intense activity. This system also tends to limit body size. Aquatic arthropods breathe mainly by some form of internal or external gill system.

4. **Highly developed sensory organs.** Sensory organs occur in great variety, from the compound (mosaic) eye to organs of touch, smell, hearing, balancing, and chemical reception. Arthropods are keenly alert to what happens in their environment.

5. **Complex behavior patterns.** Arthropods exceed most other invertebrates in complexity and organization of their activities. Innate (unlearned) behavior unquestionably controls much of what they do, but learning also plays an important part in the lives of many species.

6. **Use of diverse resources through metamorphosis.** Many arthropods pass through metamorphic changes, including a larval form quite different from the adult in structure. Larval forms often are adapted for eating food different from that of adults and occupy a different space. The use of diverse resources seems more noteworthy because intraspecific competition will occur when population numbers are large; regardless of the resources used, one resource will become limiting.

19.2 SUBPHYLUM TRILOBITA

Trilobites probably arose before the Cambrian period, in which they flourished. They have been extinct for 245 million years, but were abundant during the Cambrian and Ordovician periods. Their name refers to the trilobed shape of the body in cross section, caused by a pair of longitudinal grooves. They were dorsoventrally flattened bottom dwellers and probably scavengers (Figure 19.4A). Most of them could roll up like

---

Characteristics of Phylum Arthropoda

1. **Jointed appendages;** ancestrally, one pair to each segment, but number often reduced; appendages often modified for specialized functions
2. Living in marine, freshwater, and terrestrial habitats; many capable of flight
3. Free-living and parasitic taxa
4. Bilateral symmetry; **segmented body** divided into functional groups called **tagmata:** head and trunk; head, thorax, and abdomen or cephalothorax and abdomen; definite head
5. Triploblastic body
6. **Reduced coelom** in adult; most of body cavity consisting of hemocoel (sinuses, or spaces, in the tissues) filled with blood
7. **Cuticular exoskeleton;** containing protein, lipid, **chitin,** and often calcium carbonate secreted by underlying epidermis and shed (molted) at intervals; chitin occurs less pervasively in some other groups
8. **Complete digestive system;** mouthparts modified from ancestral appendages and adapted for different methods of feeding; alimentary canal shows great specialization by having, in various arthropods, chitinous teeth, compartments, and gastric ossicles
9. **Complex muscular system,** with exoskeleton for attachment, **striated muscles** for rapid actions, smooth muscles for visceral organs; no cilia
10. **Nervous system** similar to that of annelids, with dorsal brain connected by a ring around the gullet to a double nerve chain of ventral ganglia; fusion of ganglia in some species
11. Well-developed sensory organs; behavioral patterns much more complex than those of most invertebrates, with wider occurrence of **social organization**
12. Parthenogenesis in some taxa
13. **Sexes usually separate,** with paired reproductive organs and ducts; usually internal fertilization; oviparous, ovoviviparous, or viviparous; often with **metamorphosis**
14. Paired excretory glands called **coxal, antennal,** or **maxillary glands** present in some; others with excretory organs called **Malpighian tubules**
15. Respiration by **body surface, gills, tracheae** (air tubes), or **book lungs**
16. **Open circulatory system,** with dorsal **contractile heart,** arteries, and hemocoel (blood sinuses)

---

pill bugs (isopods), and they ranged from 2 to 67 cm in length. Despite their antiquity, they were highly specialized arthropods.

Their exoskeleton contained chitin, strengthened in some areas by calcium carbonate. There were three tagmata in the body: head (also called cephalon), trunk, and pygidium. Their cephalon was one piece but showed signs of ancestral segmentation; their trunk had a variable number of segments; and segments of the pygidium, at the posterior end, were fused into a plate. Their cephalon bore a pair of antennae, compound eyes, a mouth, and four pairs of leglike appendages. There were no true mouthparts, but a hypostome (p. 413) likely served in feeding. Each body segment except the last also bore a pair

A

B

Figure 19.4 Fossils of early arthropods. **A,** Trilobite fossils, dorsal view. These animals were abundant in the mid-Cambrian period. **B,** Eurypterid fossil from the Silurian period, *Eurypteris remipes.* Eurypterids flourished in Europe and North America from the Ordovician to the Permian periods.

of biramous appendages. One of the branches had a fringe of filaments that may have served as gills.

Trilobite fossils are identified to species by their morphology. Calmoniid trilobites are an extremely diverse group. Members of the calmoniid genus *Metacryphaeus* were abundant in the Devonian and so well preserved in Bolivian strata that the evolution and extinction of taxa can be followed. An excellent fossil record and good information on biogeography and relative sea levels permits the study of evolutionary diversification in relation to abiotic factors. It appears that calmoniid speciation occurred after populations were isolated by falling sea level. Later, rising sea level permitted dispersal of new species. Extinction rates were low for much of the time, but the entire genus disappeared when extinction outpaced speciation. Speciation slowed when sea level was high, presumably because geographic isolation of populations did not occur.

19.3 SUBPHYLUM CHELICERATA

Chelicerate arthropods are known from the Ordovician period over 445 million years ago and include eurypterids (extinct), horseshoe crabs, spiders, ticks and mites, scorpions, sea spiders, and other less well-known groups such as sun scorpions and whip scorpions. Their bodies are composed of two tagmata: a cephalothorax, or prosoma, and an abdomen, or opisthosoma. They are characterized by six pairs of cephalothoracic appendages that include a pair of chelicerae (mouthparts), a pair of pedipalps, and four pairs of walking legs. They have no antennae. Most chelicerates suck liquid food from their prey. There are three chelicerate classes (Figure 19.5).

Class Merostomata

Class Merostomata contains eurypterids, all now extinct, and xiphosurids, or horseshoe crabs, which are sometimes called "living fossils" because extant forms resemble Ordovician fossils.

Subclass Eurypterida

The eurypterids, or giant water scorpions (Figure 19.4B), were the largest of all fossil arthropods, some reaching a length of 3 m. Their fossils occur in rocks from the Ordovician to the Permian periods. They had many features resembling those of marine horseshoe crabs (Figure 19.6) as well as those of scorpions. Their heads had six fused segments and bore both simple and compound eyes as well as chelicerae and pedipalps. They also had four pairs of walking legs, and their abdomen had 12 segments and a spikelike telson.

Eurypterids were the dominant predators of their time and some had anterior appendages modified into large, crushing claws. It is possible that development of dermal armor in early fishes (pp. 511–512) resulted from selection pressure of eurypterid predation.

Subclass Xiphosurida: Horseshoe Crabs

Xiphosurids are an ancient marine group that dates from the Ordovician period. Our common horseshoe crab *Limulus* (L. *limus,* sidelong, askew) (Figure 19.6) goes back practically unchanged to the Triassic period. Only three genera (four species) survive today: *Limulus,* which lives in shallow water along the North American Atlantic coast, including the Gulf coast down through Texas and Mexico; *Carcinoscorpius* (Gr. *karkinos,* crab, + *skorpiōn,* scorpion), along the southern shore of Japan; and *Tachypleus* (Gr. *tachys,* swift, + *pleutēs,* sailor), in the East Indies and along the coast of southern Asia. They usually live in shallow water.

Xiphosurids have an unsegmented, horseshoe-shaped **carapace** (hard dorsal shield) and a broad abdomen, which has a long **telson,** or tailpiece. Their cephalothorax bears a pair of chelicerae, one pair of pedipalps, and four pairs of walking legs, whereas their abdomen has six pairs of broad, thin appendages that are fused in the median line (Figure 19.6). On five abdominal appendages, **book gills** (flat, leaflike gills) occur under the gill opercula. There are two lateral, rudimentary eyes and two simple eyes on the carapace. A horseshoe crab swims by means of its abdominal plates and can walk with its walking legs. It feeds at night on worms and small molluscs, which it seizes with its chelicerae and walking legs.

During the mating season horseshoe crabs come to shore by the thousands at high tide to mate. A female burrows into sand where she

Chelicerata

Pycnogonida

Merostomata

Arachnida

— Multiple gonopores

— Abdomen reduced

— Preoral proboscis

— Ovigers

— Telson

— Book gills

— Slit sensilla

— Abdominal appendages
reduced, lost, or
modified

— 1st or 2nd abdominal somite
modified as genital somite

— Two median eyes

— Cephalothorax a carapace-like shield

— Loss of antennae

— Four pairs of legs

— Chelicerae and pedipalps

— Cephalothorax and abdomen

Figure 19.5 Cladogram of chelicerates, showing one proposed relationship within the chelicerate clade.

Source: Modified from R. C. Brusca and G. J. Brusca, Invertebrates, Sinauer Associates, Inc., Sunderland, MA, 1990.

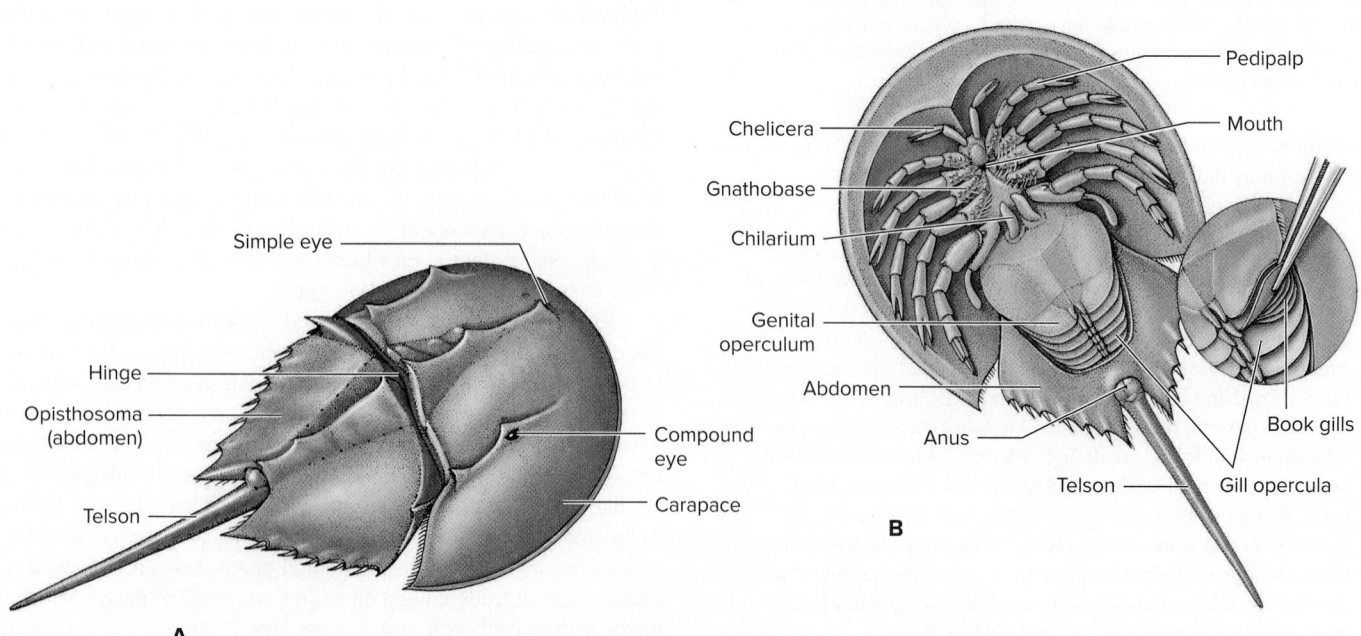

Simple eye

Hinge

Opisthosoma
(abdomen)

Telson

Compound
eye

Carapace

A

Pedipalp

Chelicera

Mouth

Gnathobase

Chilarium

Genital
operculum

Abdomen

Anus

Telson

Book gills

Gill opercula

B

Figure 19.6 **A,** Dorsal view of horseshoe crab *Limulus* (class Merostomata). They grow to 0.5 m in length. **B,** Ventral view of female.

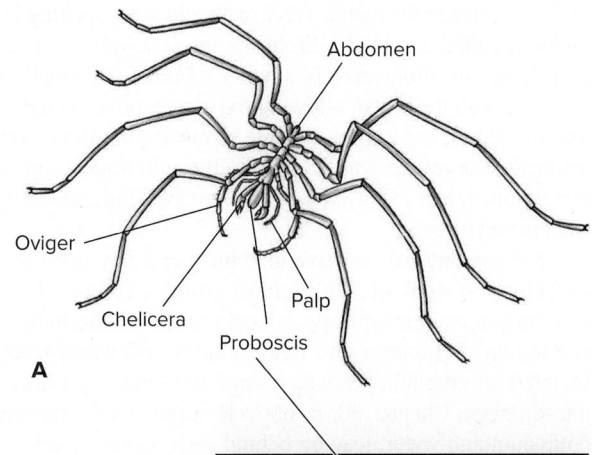

A, Pycnogonid, *Nymphon* sp. In this genus all anterior appendages (chelicerae, palps, and ovigers) are present in both sexes, although ovigers are often not present in females of other genera.

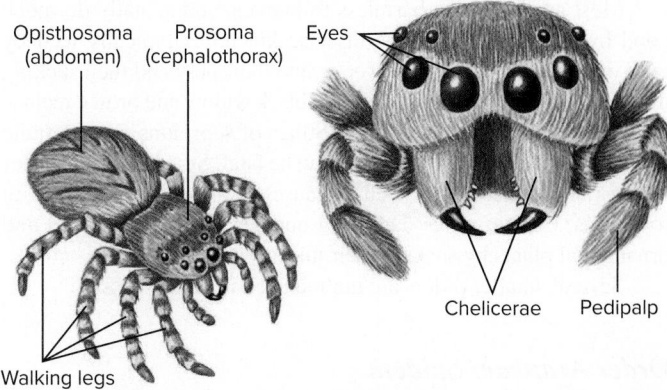

Figure 19.8 External anatomy of a jumping spider, with anterior view of head (at *right*).

Figure 19.7

A, Pycnogonid, *Nymphon* sp. In this genus all anterior appendages (chelicerae, palps, and ovigers) are present in both sexes, although ovigers are often not present in females of other genera. **B,** *Pycnogonum hancockii,* a pycnogonid with relatively short legs. Females of this genus have neither chelicerae nor ovigers and males have ovigers.

© Cleveland P. Hickman, Jr.

lays eggs, with one or more smaller males following closely to add sperm to the nest before the female covers it with sand. American *Limulus* mate and lay eggs during high tides of full and new moons in spring and summer. Eggs are warmed by the sun and protected from waves until young larvae hatch and return to the sea, carried by another high tide. Larvae are segmented and are often called "trilobite larvae" because they resemble trilobites.

Class Pycnogonida: Sea Spiders

About 1000 species of sea spiders occupy marine habitats ranging from shallow, coastal waters to deep-ocean basins. Some sea spiders are only a few millimeters long, but others are much larger, with legspans up to nearly 0.75 m. They have small, thin bodies and usually four pairs of long, thin walking legs. In addition, they have a feature unique among arthropods: segments are duplicated in some groups, so that they possess five or six pairs of legs instead of the four pairs normally characteristic of chelicerates. Males of many species bear a subsidiary pair of legs (**ovigers**) (Figure 19.7) on which they carry developing eggs, and ovigers are often absent in females. Many species also are equipped with chelicerae and palps. Chelicerae are sometimes called chelifores in this group.

The small head (cephalon) has a raised projection with two pairs of simple eyes. The mouth is at the tip of a long **proboscis,** which sucks juices from cnidarians and soft-bodied animals. Their circulatory system is limited to a simple dorsal heart, and excretory and respiratory systems are absent. The long, thin body and legs provide

a large surface area, in proportion to body volume, that is evidently sufficient for diffusion of gases and wastes. Because of the small size of the body, the digestive system and gonads have branches that extend into the legs.

Sea spiders occur in all oceans, but they are most abundant in polar waters. *Pycnogonum* (Figure 19.7B) is a common intertidal genus on both Atlantic and Pacific coasts of the United States; it has relatively short, heavy legs. *Nymphon* (Figure 19.7A) is the largest genus of pycnogonids, with over 200 species. It occurs from subtidal depths to 6800 m in all oceans except the Black and Baltic seas.

Some research suggests that pycnogonids belonged to an early diverging arthropod lineage outside any of the subphyla, but morphological and molecular evidence strongly supports the placement of pycnogonids in the Chelicerata (see Phylogeny section, pp. 417–418).

Class Arachnida

Arachnids (Gr. *arachnē,* spider) exhibit enormous anatomical variation. In addition to spiders, the group includes scorpions, pseudoscorpions, whip scorpions, ticks, mites, daddy longlegs (harvestmen), and others. There are many differences among these taxa with respect to form and appendages. They are mostly free-living and are most common in warm, dry regions.

Arachnids have become extremely diverse: more than 80,000 species have been described to date. They were among the first arthropods to move into terrestrial habitats. For example, scorpions are among Silurian fossils, and by the end of the Paleozoic era mites and spiders had appeared.

All arachnids have two tagmata: a cephalothorax (head and thorax) and an abdomen, which may or may not be segmented. The abdomen houses the reproductive organs and respiratory organs such as tracheae and book lungs. The cephalothorax usually bears a pair of chelicerae, a pair of pedipalps, and four pairs of walking legs (Figure 19.8). Most arachnids are predaceous and have fangs, claws, venom glands, or stingers; fangs are modified chelicerae, whereas claws (chelae) are modified pedipalps. They usually have a strong sucking pharynx with which they ingest the fluids and soft tissues from the bodies of their prey. Among their interesting adaptations are spinning glands of spiders.

Most arachnids are harmless to humans and actually do much good by destroying injurious insects. Arachnids typically feed by releasing digestive enzymes over or into their prey and then sucking the predigested liquid. A few, such as black widow and brown recluse spiders, can give dangerous bites. Stings of scorpions may be quite painful, and those of a few species can be fatal. Some ticks and mites are carriers of diseases as well as causes of annoyance and painful irritations. Certain mites damage a number of important food and ornamental plants by sucking their juices.

Several smaller orders are not included in our discussion.

Order Araneae: Spiders

Spiders are a large group of arachnids comprising about 40,000 species distributed throughout the world and known from the Silurian period approximately 420 million years ago. The spider body is compact: a **cephalothorax (prosoma)** and **abdomen (opisthosoma),** both unsegmented and joined by a slender pedicel. A few spiders have a segmented abdomen, which is considered an ancestral character.

Anterior appendages include a pair of **chelicerae** (Figure 19.8), which have terminal **fangs** through which run ducts from venom glands, and a pair of leglike **pedipalps,** which have sensory function and are also used by males to transfer sperm. The basal parts of pedipalps may be used to manipulate food (Figure 19.8). Four pairs of **walking legs** terminate in claws.

All spiders are predaceous, feeding largely on insects, which they effectively dispatch with venom from their fangs. Some spiders chase prey; others ambush them; and many trap them in a net of silk. After a spider seizes prey with its chelicerae and injects venom, it liquefies the prey's tissues with digestive fluid and sucks the resulting broth into its stomach. Spiders with teeth at the bases of chelicerae crush or chew prey, aiding digestion by enzymes from their mouth.

Spiders respire using book lungs or a system of internal tubes called tracheae. Some spiders have tracheae and book lungs. Book lungs open to the outside through a slit in the exoskeleton. Internal to the slit are many thin-walled chambers filled with blood (Figure 19.9). Between a pair of chambers is a cavity where air enters the body. Oxygen moves across the thin chamber walls to enter the blood, whereas carbon dioxide leaves the blood. A stack of thin-walled chambers resembles the pages of a book; hence the name. Tracheae begin at an opening in the exoskeleton called a spiracle. From the spiracle, tubes extend inward, branching into progressively smaller tubules. The smallest tubules interface with the blood, allowing oxygen to move from the tracheae into the blood, and carbon dioxide to move from the blood into the tracheae. The tracheae are similar to those in insects (pp. 450–452) but are much less extensive and have evolved independently in both spiders and insects.

Spiders and insects have also independently evolved a unique **excretory system of Malpighian tubules** (Figure 19.9), which work in conjunction with specialized resorptive cells in the intestinal epithelium. Potassium and other solutes and waste materials are secreted into the tubules, which drain the fluid, or "urine," into the intestine (see Chapter 30, p. 666). Resorptive cells recapture most potassium and water, leaving behind such wastes as uric acid. This recycling of water and potassium allows species living in dry environments to conserve body fluids by producing a nearly dry mixture of urine and feces. Many spiders also have **coxal glands,** which are modified nephridia that open at the coxa, or base, of the first and third walking legs.

Spiders usually have eight **simple eyes,** each with a lens, optic rods, and a retina (Figure 19.8). They are used chiefly to perceive moving objects, but some, such as those of hunting and jumping spiders, may form images. Since a spider's vision is often poor, its awareness of its environment depends largely on cuticular mechanoreceptors, such as **sensory setae** (sensilla). Fine setae covering the legs can detect vibrations in the web, struggling prey, or even air movements.

Web-Spinning Habits   An ability to spin silk is central to a spider's life, as it is in some other arachnids such as tetranychid spider mites. Two or three pairs of spinnerets containing hundreds of microscopic tubes run to special abdominal **silk glands** (Figure 19.9). A scleroprotein secretion emitted as a liquid from the spinnerets hardens to form a silk thread. Threads of spider silk are stronger than steel threads of the same diameter and second in strength only to fused quartz fibers.

Many species of spiders spin silk webs. The kind of net varies among species. Some webs are simple and consist merely of a few strands of silk radiating out from a spider's burrow or place of retreat.

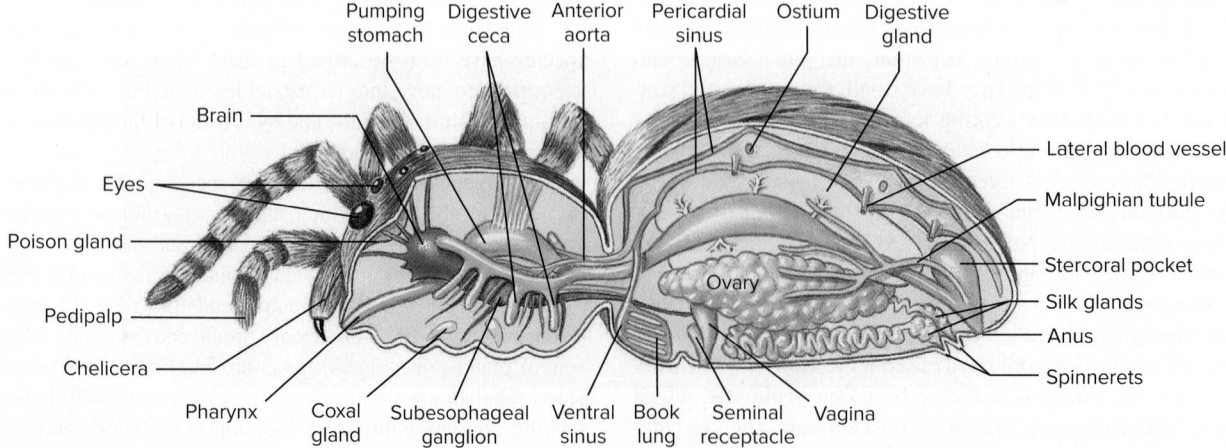

Figure 19.9 Spider, internal anatomy.

Figure 19.10 Grasshopper, snared and helpless in the web of a golden garden spider (*Argiope aurantia*), is wrapped in silk while still alive. If the spider is not hungry, the prize is saved for a later meal.

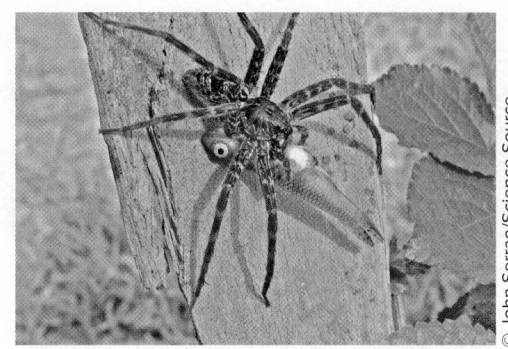

Figure 19.11 A dark fishing spider, *Dolomedes tenebrosus*, feeds on a fish. This handsome spider feeds mostly on aquatic and terrestrial insects but occasionally captures small fishes and tadpoles. It pulls its paralyzed victim from the water, pumps in digestive enzymes, then sucks out the predigested contents.

Others spin beautiful, geometrical orb webs. However, spiders use silk threads for many additional purposes: nest lining, sperm webs or egg sacs, bridge lines, draglines, warning threads, molting threads, attachment discs, nursery webs, and the wrapping of prey items (Figure 19.10). Not all spiders spin webs for traps. Some spiders throw a sticky bolus of silk to capture their prey. Others, such as wolf spiders, jumping spiders (Figure 19.8), and fisher spiders (Figure 19.11), simply chase and catch their prey. These spiders likely lost the ability to produce silk for prey capture.

Reproduction   A courtship ritual precedes mating. Before mating, a male spins a small web, deposits a drop of sperm on it, and then picks up the sperm to be stored in special cavities of his pedipalps. When he mates, he inserts his pedipalps into the female genital opening to store the sperm in his mate's seminal receptacles. A female lays her eggs in a silken net, which she may carry or attach to a web or plant. A cocoon may contain hundreds of eggs, which hatch in approximately two weeks. Young usually remain in the egg sac for a few weeks and molt once before leaving it. The number of molts may vary, but typically ranges between 4 and 12 before adulthood is reached.

Are Spiders Really Dangerous?   It is amazing that such small and innocuous creatures have generated so much unreasonable fear in human minds. Spiders are timid creatures that, rather than being dangerous enemies to humans, are actually allies in the continuing battle with insects and other arthropod pests. Venom produced to kill prey is usually harmless to humans. The most venomous spiders bite only when threatened or when defending their eggs or young. Even American tarantulas (Figure 19.12), despite their fearsome size, are *not* dangerous. They rarely bite, and their bite is about as serious as a bee sting.

There are, however, two genera in the United States that can give severe or even fatal bites: *Latrodectus* (L. *latro*, robber, + *dectes*, biter; **black widow,** five species) and *Loxosceles* (Gr. *loxos*, crooked, + *skelos*, leg; **brown recluse,** 13 species). Black widows are moderate to

Figure 19.12 A tarantula in the family Theraphosidae.

small in size and shiny black, usually with a bright orange or red spot, commonly in the shape of an hourglass, on the underside of their abdomen (Figure 19.13A). Their venom is neurotoxic, acting on the nervous system. About four or five of every 1000 reported bites are fatal.

Brown recluse spiders are brown and bear a violin-shaped dorsal stripe on their cephalothorax (Figure 19.13B). Their venom is hemolytic rather than neurotoxic, producing death of tissues and skin surrounding the bite. Their bite can be mild to serious and there are a few unconfirmed reports of deaths of small children and older individuals.

Some spiders in other parts of the world are also dangerous, for example, funnelweb spiders *Atrax* spp. in Australia. Most dangerous of all are spiders in the South and Central American genus *Phoneutria*. They are large (10 to 12 cm leg span) and quite aggressive. Their venom is among the most pharmacologically toxic of spider venoms, and their bites cause intense pain, neurotoxic effects, sweating, acute allergic reaction, and nonsexual enlargement of the penis. However, most spiders simply use their venom in prey capture (Figures 19.13 C and D).

Figure 19.13 **A,** Black widow spider, *Latrodectus mactans.* Note the red "hourglass" on the ventral side of her abdomen. **B,** Brown recluse, *Loxosceles reclusa,* is a small spider whose venom is hemolytic and dangerous. There is a small violin-shaped marking on its cephalothorax. **C,** A camouflaged crab spider, *Misumenoides formosipes,* awaits its insect prey. Its coloration permits it to blend in among the flowers as it awaits an insect in search of pollen or nectar. **D,** A female regal jumping spider, *Phidippus regius.* This species has excellent vision and stalks an insect until it is close enough to leap with unerring precision, fixing its chelicerae into its prey.

W. S. Bristowe (*The World of Spiders.* 1971. Rev. ed. London, Collins) estimated that at certain seasons a field in Sussex, England (that had been undisturbed for several years) had a population of 2 million spiders to the acre. He concluded that so many spiders could not successfully compete except for the many specialized adaptations they had evolved. These include adaptations to cold and heat, wet and dry conditions, and light and darkness.

Some spiders capture large insects, some only small ones; web-builders snare mostly flying insects, whereas hunters seek those that live on the ground. Some lay eggs in the spring, others in the late summer. Some feed by day, others by night, and some have developed flavors that are distasteful to birds or to certain predatory insects. As it is with spiders, so has it been with other arthropods; their adaptations are many and diverse and contribute in no small way to their long success.

Order Scorpiones: Scorpions

Scorpions comprise about 1400 species worldwide and, like spiders, are known from the Silurian period. Although scorpions are more common in tropical and subtropical regions, some occur in temperate zones. Scorpions are generally secretive, hiding in burrows or under objects by day and feeding at night. They feed largely on insects and spiders, which they seize with their pedipalps and shred with their chelicerae.

Sand-dwelling scorpions locate prey by sensing surface waves generated by the movements of insects on or in the sand. These waves are detected by compound slit sensilla located on the last segment of the legs. A scorpion can locate a burrowing cockroach 50 cm away and reach it in three or four quick movements.

Scorpion tagmata are a rather short **cephalothorax,** which bears chelicerae, pedipalps, legs, a pair of large median eyes, and usually two to five pairs of small lateral eyes; a **preabdomen** (or mesosoma) of seven segments; and a long slender **postabdomen** (or metasoma) of five segments, which ends in a stinging apparatus (Figure 19.14A). Their chelicerae are small, their pedipalps are large and chelate (pincerlike), and the four pairs of walking legs are long and eight-jointed.

On the ventral side of the abdomen are curious comblike **pectines,** which serve as tactile organs for exploring the ground and for

sex recognition. The stinger on the last segment consists of a bulbous base and a curved barb that injects venom. Venom of most species is not harmful to humans but may produce a painful swelling. However, the sting of certain species of *Androctonus* in Africa and *Centruroides* (Gr. *kenteō,* to prick, + *oura,* tail, + *oides,* form) in Mexico can be fatal unless antivenom is administered. In general, larger species tend to be less venomous than smaller species and rely on their greater strength to overpower prey.

Scorpions perform a complex mating dance, the male holding the female's chelae as he steps back and forth. He kneads her chelicerae with his own and, in some species, stings her on her pedipalp or on the edge of her cephalothorax. The stinging action is slow and deliberate, and the stinger remains in the female's body for several minutes. Both individuals remain motionless during that time. Finally, the male deposits a spermatophore and pulls the female over it until the sperm mass is taken into the female orifice. Scorpions are truly viviparous; females brood their young within their reproductive tract. After several months to a year of development anywhere from 1 to over 100 young are produced, depending on the species. The young, only a few millimeters long, crawl onto their mother's back until after their first molt (Figure 19.14A). They mature in 1 to 8 years and may live for as long as 15 years.

Order Solpugida: Sun or Camel Spiders

Solpugids, also called solfugids, and by such common names as sun, camel, or wind spiders, are nonvenomous arachnids that shred prey with their large chelicerae (Figure 19.14B). They range in size from 1 cm to nearly 15 cm. They are common in the tropical and subtropical deserts in America, the Middle East, Asia, and Africa.

Order Opiliones: Harvestmen

Harvestmen (Figure 19.14C), often called "daddy longlegs," are common throughout the world and comprise about 5000 species. They are easily distinguished from spiders: their abdomen and cephalothorax are rounded and broadly joined, without the constriction of a pedicel; their abdomen shows external segmentation; and they have only two eyes, mounted on a tubercle on their cephalothorax. They have four pairs of long, spindly legs that end in tiny claws. They can cast off one or more of these legs without apparent ill effect if they are

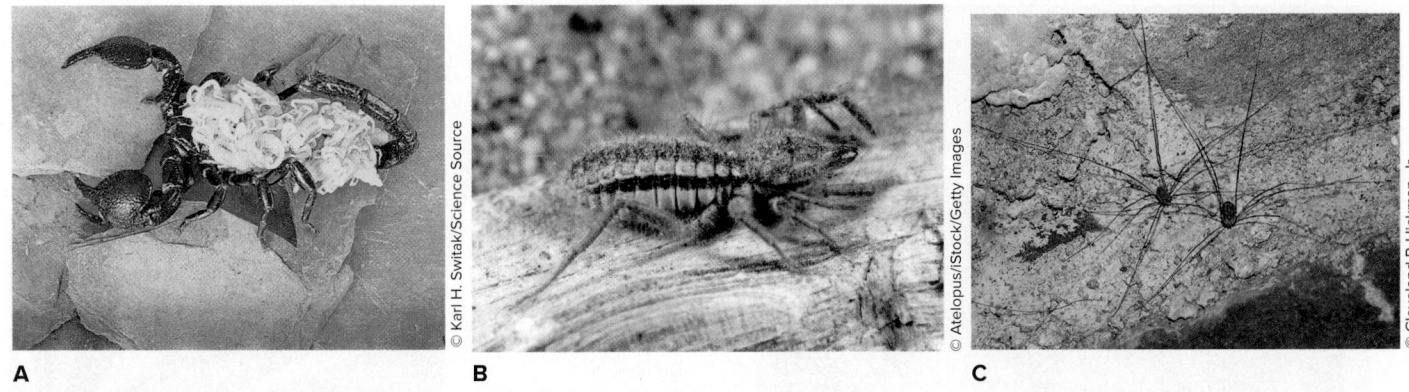

Figure 19.14 **A,** An emperor scorpion (order Scorpiones), *Paninus imperator,* with young, which stay with the mother until their first molt. **B,** A camel spider. **C,** Harvestmen, *Mitopus* sp. (order Opiliones). Harvestmen run rapidly on their stiltlike legs. They are especially noticeable during the harvesting season; hence the common name.

grasped by a predator (or human hand). The ends of their chelicerae are pincerlike, and, while carnivorous, they are often scavengers as well.

Harvestmen are not venomous and are harmless to humans. Odoriferous glands that open on the cephalothorax deter some predators with their noxious secretions. Other than some mites, opilionids are unique among arachnids in having a penis for direct sperm transfer; all are oviparous.

Traditionally allied with Acari, some studies indicate that Opiliones forms a clade with scorpions and two smaller orders. They are the sister group of scorpions.

Order Acari: Ticks and Mites

Members of order Acari are without doubt the most medically and economically important group of arachnids. They far exceed other orders in numbers of individuals and species. Although about 40,000 species have been described, some authorities estimate that from 500,000 to 1 million species exist. Hundreds of individuals of several species of mites live in a few square meters of leaf mold in forests. They occur throughout the world in both terrestrial and aquatic habitats, even extending into such inhospitable regions as deserts, polar areas, and hot springs. Many acarines are parasitic during one or more stages of their life cycle.

Most mites are 1 mm or less in length. Ticks, which are only one suborder of Acari, range from a few millimeters to occasionally 3 cm. A tick may become enormously distended with blood after feeding on its host.

Acarines differ from all other arachnids in having complete fusion of the cephalothorax and abdomen, with no sign of external division or segmentation (Figure 19.15). They carry their mouthparts on a little anterior projection, the **capitulum,** which consists mainly of the feeding appendages surrounding the mouth. On each side of their mouth is a chelicera, which functions in piercing, tearing, or gripping food. The form of the chelicerae varies greatly in different families. Lateral to the chelicerae is a pair of segmented pedipalps, which also vary greatly in form and function related to feeding. Ventrally the bases of the pedipalps fuse to form a **hypostome,** whereas a **rostrum,** or **tectum,** extends dorsally over their mouth. Adult mites and ticks usually have four pairs of legs, although there may be only one to three in some specialized forms.

Figure 19.15 **A,** The Western blacklegged tick, *Ixodes pacificus,* (order Acari) is a vector for the bacterium *Borrelia burgdorferi* which causes Lyme disease. **B,** Velvet mite, *Dinothrombium* sp. from South Africa. Adults are free-living predators, while certain immature instars are often ectoparasites.

Most acarines transfer sperm directly, but many species use a spermatophore. A larva with six legs hatches from the egg, and one or more eight-legged nymphal stages follow before the adult stage is reached.

© Eye of Science/Science Source

Figure 19.16   Color enhanced scanning electron micrograph of a house dust mite.

Figure 19.17
Damage to *Chamaedorea* sp. Palm caused by mites of the family etranychidae (order Acari). Over 130 species of this family occur in North America, and some are serious agricultural pests. Mites pierce plant cells and suck out contents, giving leaves the mottled appearance shown here.

© Larry S. Roberts

Many species of mites are entirely free-living. *Dermatophagoides farinae* (Gr. *dermatos,* skin, + *phag,* to eat, + *eidos,* likeness of form) (Figure 19.16) and related species are denizens of house dust all over the world, sometimes causing allergies and dermatoses. There are some marine mites, but most aquatic species live in freshwater. They have long, hairlike setae on their legs for swimming, and their larvae may be parasitic on aquatic invertebrates. Such abundant organisms must be important ecologically, but many acarines have more direct effects on our food supply and health. Spider mites (family Tetranychidae) are serious agricultural pests on fruit trees, cotton, clover, and many other plants. They suck the contents of plant cells, producing a mottled appearance on the leaves (Figure 19.17), and construct a protective web from silk glands opening near the base of the chelicerae. Larvae of genus *Trombicula* are called **chiggers** or **redbugs**. They feed on the dermal tissues of terrestrial vertebrates, including humans, and may cause an irritating dermatitis, but they do not burrow or remain attached to the host. Some species of chiggers transmit a disease called Asiatic scrub typhus. Hair follicle mites, *Demodex* (Figure 19.18), are apparently nonpathogenic in humans; they infect most of us although we are unaware of them. In some cases they may produce a mild dermatitis. Other species of *Demodex* and other genera of mites cause mange in domestic animals. Human itch mites, *Sarcoptes scabiei* (Figure 19.19), cause intense itching as they burrow beneath the skin. The *Varroa* mite was implicated in the dramatic loss of honey bees that began in the United States in 2006 (see the discussion of colony collapse disorder, p. 462). The mite carries "deformed wing virus" or DWV. Global reemergence of this virus is linked to viral evolution, new hosts for the mite, and human movement of infected bees.

The inflamed welt and intense itching that follows a chigger bite is not the result of the chigger burrowing into the skin, as is popularly believed. Rather, a chigger bites through the skin with its chelicerae and injects a salivary secretion containing powerful enzymes that liquefy skin cells. Human skin responds defensively by forming a hardened tube that the larva uses as a drinking straw and through which it gorges itself with host cells and fluid. Scratching usually removes the chigger but leaves the tube, which is a source of irritation for several days.

In addition to disease conditions that they themselves cause, ticks are among the world's premier disease vectors, ranking second only to mosquitos. They surpass other arthropods in carrying a great variety of infectious agents, including apicomplexans, rickettsial, viral, bacterial, and fungal organisms. Species of *Ixodes* carry the most common arthropod-borne infection in the United States, Lyme disease (see note). Species of *Dermacentor* and other ticks transmit Rocky Mountain spotted fever, a poorly named disease because most cases occur in the eastern United States. *Dermacentor* also transmits tularemia and agents of several other diseases. Texas cattle fever, also called red-water fever, is caused by an apicomplexan parasite transmitted by cattle ticks, *Rhipicephalus annulatus.* The brown dog tick is a disease-vector for domestic animals and humans (Figure 19.20). Many more examples could be cited.

An epidemic of arthritis occurred in the 1970s in the town of Lyme, Connecticut. Subsequently called Lyme disease, it is caused by a bacterium and carried by ticks of the genus *Ixodes.* There are now thousands of cases a year in Europe and North America, and other cases have been reported from Japan, Australia, and South Africa. Many people bitten by infected ticks recover spontaneously or do not get the disease. Others, if not treated at an early stage with appropriate antibiotics, develop a chronic, disabling disease.

19.4 SUBPHYLUM MYRIAPODA

The term "myriapod," meaning "many footed," describes members of four classes in subphylum Myriapoda that have evolved a pattern of two tagmata—head and trunk—with paired appendages on most or all trunk segments. Myriapods include Chilopoda (centipedes), Diplopoda, (millipedes), Pauropoda (pauropods), and Symphyla (symphylans) (Figure 19.21).

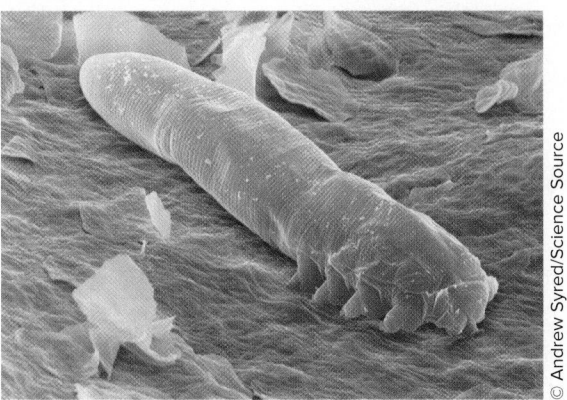

Figure 19.18 Colored scanning electron micrograph of *Demodex folliculorum,* a human follicle mite.

Figure 19.19 *Sarcoptes scabiei,* a human itch mite.

Figure 19.20 *Rhipicephalus sanguine,* the brown dog tick, occurs throughout the world.

Myriapods use tracheae to carry respiratory gases directly to and from all body cells in a manner similar to that of onychophorans (p. 398) and some arachnids, but tracheal systems have likely evolved independently in each group.

Excretion is usually by Malpighian tubules, but these have evolved independently of Malpighian tubules found in Chelicerata.

Class Chilopoda

Chilopoda (ki-lop′o-da) (Gr. *cheilos,* margin, lip, + *pous, podos,* foot), or centipedes, are land forms with somewhat flattened bodies. Centipedes prefer moist places such as under logs, bark, and stones. They are very agile carnivores, living on cockroaches and other insects, and on earthworms. They kill their prey with their venom claws and then chew it with their mandibles. The largest centipede in the world, *Scolopendra gigantea,* is nearly 30 cm in length. Common house centipedes *Scutigera* (L. *scutum,* shield, + *gera,* bearing), which have 15 pairs of legs, are much smaller and often seen scurrying around bathrooms and damp cellars, where they catch insects. Most species of centipedes are harmless to humans, although many tropical centipedes are dangerous. There are about 3000 species worldwide.

Centipede bodies may contain from a few to 177 segments (Figure 19.22). Each segment, except the one behind the head and the last two in the body, bears a pair of jointed legs. Appendages of the first body segment are modified to form venom claws. The last pair of legs is longer than the others and serves a sensory function.

The head appendages are similar to those of an insect (Figure 19.22B). There are a pair of antennae, a pair of mandibles, and one or two pairs of maxillae. A pair of eyes on the dorsal side of the head consists of groups of ocelli.

The digestive system is a straight tube into which salivary glands empty at the anterior end. Two pairs of Malpighian tubules empty into the hind part of the intestine. There is an elongated heart with a pair of arteries to each segment. The heart has a series of ostia to provide for return of blood to the heart from the hemocoel. Respiration is by means of a tracheal system of branched air tubes that come from a pair of spiracles in each segment. The nervous system is typically arthropodan, and there is also a visceral nervous system.

Sexes are separate, with unpaired gonads and paired ducts. Some centipedes lay eggs and others are viviparous. The young are similar in form to adults and do not undergo metamorphosis.

Class Diplopoda

Diplopoda (Gr. *diploo,* double, two, + *pous, podos,* foot) are commonly called millipedes, which literally means "thousand feet" (Figure 19.23A). Millipedes are not as active as centipedes: they walk with a slow, graceful motion, not wriggling as centipedes do. They prefer dark, moist places under logs or stones. Most millipedes are herbivorous, feeding on decayed plant matter, although sometimes they eat living plants. Millipedes are slow-moving animals and may roll into a coil when disturbed. Many millipedes also protect themselves from predation by secreting toxic or repellent fluids from special glands (**repugnatorial glands**) positioned along the sides of

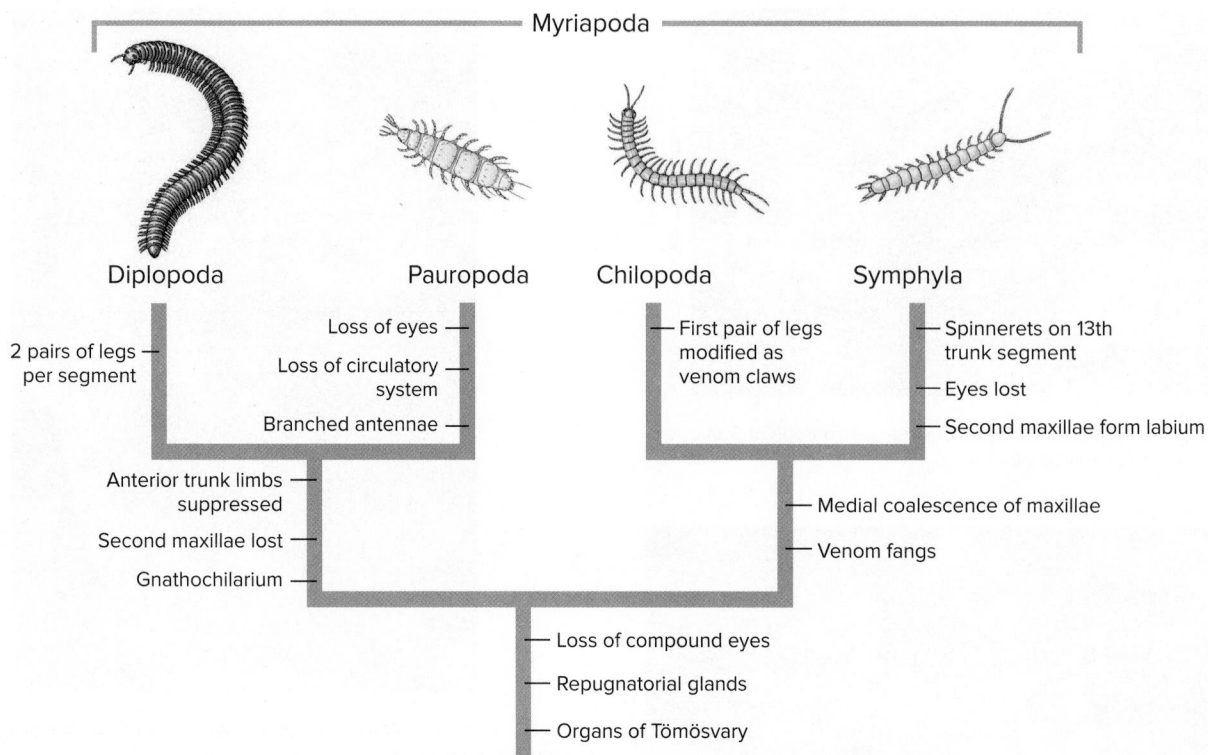

Figure 19.21 Cladogram showing hypothetical relationships of myriapods. Organs of Tömösvary are unique sensory organs opening at the bases of the antennae, and repugnatorial glands, located on certain segments or legs, secrete an obnoxious substance for defense. The gnathochilarium is formed in diplopods and pauropods by fusion of the first maxillae, and the collum is the collarlike tergite of the first trunk segment.

Source: Modified from R. C. Brusca and G. J. Brusca, Invertebrates, *Sinauer Associates, Inc., Sunderland, MA, 1990.*

Figure 19.22 A, The centipede *Scolopendra heros* sits atop a lichen-covered rock. It occurs in the central United States from Texas to Colorado and may be more than 15 cm (6 inches) long. Like all centipedes, it is a carnivore and uses the paired venom claws on the first segment to kill prey. **B,** Head of centipede.

the body. Common examples of this class are *Spirobolus* and *Julus,* both of which have wide distribution. There are more than 10,000 species of millipedes worldwide.

The cylindrical body of a millipede is formed by 25 to more than 100 segments. Their short thorax consists of four segments, each bearing one pair of legs. Each abdominal segment has two pairs

of legs, leading to the impression of a thousand feet. The millipede exoskeleton is reinforced with calcium carbonate.

Their head bears two clumps of simple eyes and a pair each of antennae, mandibles, and maxillae (Figure 19.23B). The general body structures are similar to those of centipedes. Two pairs of spiracles on each abdominal segment open into air chambers that connect to tracheal air tubes. There are two genital apertures toward the anterior end.

In most millipedes the appendages of the seventh segment are specialized as copulatory organs. After millipedes copulate, females lay eggs in a nest and guard them carefully. Interestingly, larval forms have only one pair of legs to each segment.

Class Pauropoda

Pauropoda (Gr. *pauros,* small, + *pous, podos,* foot) are a group of minute (2 mm or less), soft-bodied myriapods, numbering almost 500 species. Although widely distributed, pauropods are the least well-known myriapods. They live in moist soil, leaf litter, or decaying vegetation and under bark and debris.

Pauropods have a small head with branched antennae and no true eyes (Figure 19.24A). Their 12 trunk segments usually bear nine pairs of legs (none on the first or the last two segments). They have only one tergal (dorsal) plate covering every two segments.

Tracheae, spiracles, and circulatory system are lacking. Pauropods are probably most closely related to diplopods.

Class Symphyla

Symphyla (Gr. *sym,* together, + *phylon,* tribe) are small (2 to 10 mm) and have centipede-like bodies (Figure 19.24B). They live in humus, leaf mold, and debris. *Scutigerella* (L. dim. of *Scutigera)* are often pests on vegetables and flowers, particularly in greenhouses. They are soft bodied, with 14 segments, 12 of which bear legs and one a pair of spinnerets. The antennae are long and unbranched.

Symphylans are eyeless but have sensory pits at the bases of the antennae. Their tracheal system is limited to a pair of spiracles on their head and tracheal tubes to anterior segments only. Only 160 species are described.

> The mating behavior of *Scutigerella* (Figure 19.24B) is unusual. Males place a spermatophore at the end of a stalk. When a female finds it, she takes it into her mouth, storing sperm in special buccal pouches. Then she removes eggs from her gonopore with her mouth and attaches them to moss or lichen, or to walls of crevices, smearing them during handling with some of the semen, thereby fertilizing them. Nymphs hatch from eggs, adding pairs of legs as they grow toward adulthood.

19.5 PHYLOGENY AND ADAPTIVE DIVERSIFICATION

Phylogeny

Extant arthropods are divided among four subphyla. Relationships among subphyla are subject to debate, but the taxon Pancrustacea, which contains hexapods and crustaceans, is well supported. Which subphylum is the sister taxon to Pancrustacea? According to the mandibulate hypothesis, Myriapoda is grouped with Pancrustacea.

Biologists assume that the ancestral arthropod had a segmented body with one pair of appendages per segment. During evolution, adjacent segments fused to make body regions (tagmata). How many

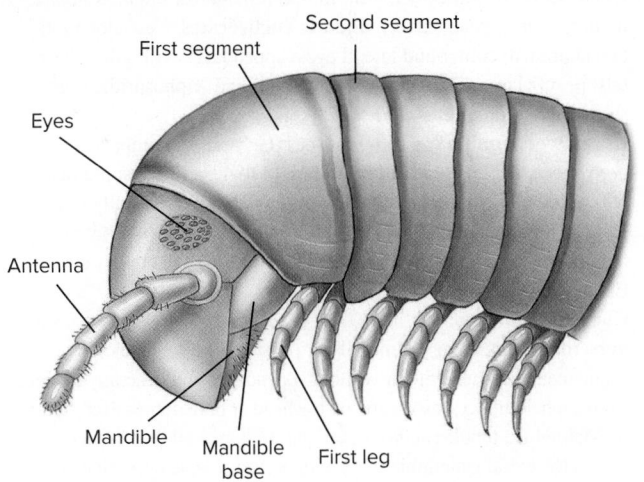

First segment
Second segment
Eyes
Antenna
Mandible
Mandible base
First leg

Figure 19.23 **A,** A pair of mating millipedes from South Africa. **B,** Head of millipede.

A   B

Figure 19.24 **A,** Pauropod. Pauropods are minute, whitish myriapods with three-branched antennae and nine pairs of legs. They live in leaf litter and under stones. They are eyeless but have sense organs that resemble eyes. **B,** *Scutigerella,* a symphylan, is a minute whitish myriapod that is sometimes an agricultural and greenhouse pest.

segments contributed to a head in each group of arthropods? *Hox* gene studies indicate that the first five segments, at least, fused to form the head tagma in all four extant subphyla. It is surprising to find the same pattern of fusion in chelicerates as in other subphyla because a head is not immediately obvious in a chelicerate. Spider bodies have two tagmata: prosoma, or cephalothorax, and opisthosoma, or abdomen. Is the head part of the prosoma? *Hox* gene comparisons indicate that the entire prosoma corresponds to the head of other arthropods.

Studies of the heads of pycnogonids were used to detect the phylogenetic position of these odd animals. Sea spiders have spindly bodies and unusual chelicerae. There was speculation that pycnogonids were not chelicerates at all, but instead were the sister taxon of all other arthropods. In the earliest fossil arthropods, appendages emerge from the first head segment, but in spiders and horseshoe crabs, chelicerae and the nerves that control them originate from the second segment during early development. Initial studies of nerve patterns in larval sea spiders indicated that their chelicerae arose, and were controlled, from the first segment. If this result had been confirmed, pycnogonids would be considered the sister taxon to all other arthropods. However, subsequent studies using *Hox* gene expression to define segment boundaries did not support this result. Sea spiders remain within subphylum Chelicerata. They and all living arthropods have head appendages that arise from the region of the head that corresponds to the second segment.

Another controversial area of arthropod biology where genetic studies have proved helpful lies in the evolution and antiquity of uniramous and biramous appendages. Hexapods and myriapods have uniramous appendages, but trilobites and some crustaceans have biramous appendages. If the ancestral appendage were biramous, then the switch to uniramous appendages might have occurred in one lineage whose descendants now carry this trait. Such reasoning led biologists to group hexapods with myriapods, but phylogenies using molecular characters repeatedly placed hexapods with crustaceans. Is it likely that the uniramous limb evolved more than once? This question would be more easily answered if the genetic basis of limb structure were understood. Studies of the genetic determination of limb branching show that modulation of expression of one gene (*Distal-less,* or *Dll*) determines the number of limb branches (see pp. 437–439). Gene expression can be modified within a lineage, so the number of limb branches present is unlikely to be homologous.

The number of appendages per segment is another variable character within Arthropoda. The ancestral arthropod is assumed to have had one pair per segment. Millipedes, in class Diplopoda, have two pairs of appendages on most trunk segments. Did the millipede pattern originate by the repeated fusion of two ancestral segments? Perhaps it did, but expression of the *Distal-less* gene might also have a role here. Larval millipedes have only one pair of appendages per segment.

Adaptive Diversification

Arthropods demonstrate multiple evolutionary trends toward pronounced tagmatization by differentiation or fusion of segments, giving rise to such combinations of tagmata as head and trunk; head, thorax, and abdomen; or cephalothorax (fused head and thorax) and abdomen. The ancestral arthropod condition is to have similar appendages on each segment. More derived forms have appendages specialized for specific functions, or some segments that lack appendages entirely.

Much of the amazing diversity in arthropods seems to have evolved because of modification and specialization of their cuticular exoskeleton and their jointed appendages, yielding a wide variety of locomotor and feeding adaptations.

Taxonomy of Phylum Anthropoda

Subphylum Trilobita (trī′lō-bi′ta) (Gr. *tri,* three, + *lobos,* lobe): **trilobites**. All extinct forms; Cambrian to Carboniferous; body divided by two longitudinal furrows into three lobes; distinct head, trunk, and abdomen; biramous (two-branched) appendages.

Subphylum Chelicerata (ke-lis′e-ra′ta) (Gr. *chēlē,* claw, + *keras,* horn, + *ata,* group suffix): **eurypterids, horseshoe crabs, spiders, ticks**. First pair of appendages modified to form chelicerae; pair of pedipalps and four pairs of legs; no antennae; no mandibles; cephalothorax and abdomen usually unsegmented.

Subphylum Myriapoda (mir-ē-a′pōda) (Gr. *myrias,* a myriad, + *pous, podus,* foot); **myriapods**. All appendages uniramous; head appendages consisting of one pair of antennae, one pair of mandibles, and one or two pairs of maxillae.

Subphylum Crustacea (crus-ta′she-a) (L. *crusta,* shell, + *acea,* group suffix): **crustaceans**. Mostly aquatic, with gills; cephalothorax usually with dorsal carapace; biramous appendages, modified for various functions; head appendages consisting of two pairs of antennae, one pair of mandibles, and two pairs of maxillae; development primitively with nauplius stage (see the Taxonomy of Crustacea box on, p. 440).

Subphylum Hexapoda (hex′-a pōda) (Gr. *hex,* six, + *pous, podus,* foot): **hexapods**. Body with distinct head, thorax, and abdomen; pair of antennae; mouthparts modified for different food habits; head of six fused segments; thorax of three segments; abdomen with variable number, usually 11 somites; thorax with two pairs of wings (sometimes one pair or none) and three pairs of jointed legs; separate sexes; usually oviparous; gradual or abrupt metamorphosis.

Taxonomy of Subphylum Chelicerata

Class Merostomata (mer′ō-sto′ma-ta) (Gr. *mēros,* thigh, + *stoma,* mouth, + *ata,* group suffix): **aquatic chelicerates**. Cephalothorax and abdomen; compound lateral eyes; appendages with gills; sharp telson; subclasses Eurypterida (all extinct) and Xiphosurida, horseshoe crabs. Example: *Limulus.*

Class Pycnogonida (pik′no-gon′i-da) (Gr. *pyknos,* compact, + *gony,* knee, angle): **sea spiders**. Small (3 to 4 mm), but some reach 500 mm; body chiefly cephalothorax; tiny abdomen; usually four pairs of long walking legs (some with five or six pairs); mouth on long proboscis; four simple eyes; no respiratory or excretory system. Example: *Pycnogonum.*

Class Arachnida (ar-ack′ni-da) (Gr. *arachnē,* spider): **scorpions, spiders, mites, ticks, harvestmen**. Four pairs of legs; segmented or unsegmented abdomen with or without appendages and generally distinct from cephalothorax; respiration by tracheae or book lungs; excretion by Malpighian tubules and/or coxal glands; dorsal bilobed brain connected to ventral ganglionic mass with nerves, simple eyes; chiefly oviparous; no true metamorphosis. Examples: *Argiope, Centruroides.*

Taxonomy of Subphylum Myriapoda

Class Diplopoda (dip-lō′ pōda) (Gr. *diploos*, double, + *pous, podos*, foot): **millipedes**. Body almost cylindrical; head with short antennae and simple eyes; body with variable number of segments; short legs, usually two pairs of legs to a segment; oviparous. Examples: *Julus, Spirobolus.*

Class Chilopoda (kī-lō′-pōda) (Gr. *cheilos*, lip, + *pous, podos*, foot): **centipedes**. Dorsoventrally flattened body; variable number of segments, each with one pair of legs; one pair of long antennae; oviparous. Examples: *Cermatia, Lithobius, Geophilus.*

Class Pauropoda (pau-ro+pōda) (Gr. *pauros*, small, + *pous, podos*, foot): **pauropods**. Minute (1 to 1.5 mm); cylindrical body consisting of double segments and bearing 9 or 10 pairs of legs; no eyes. Examples: *Pauropus, Allopauropus.*

Class Symphyla (sym′fy-la) (Gr. *syn*, together, + *phylē*, tribe): **garden centipedes**. Slender (1 to 8 mm) with long, filiform antennae; body consisting of 15 to 22 segments with 10 to 12 pairs of legs; no eyes. Example: *Scutigerella.*

While adaptations and specializations made possible by the cuticular exoskeleton of arthropods and other morphological and behavioral features may have fostered high diversity, another important factor ensuring the incredible evolutionary success of the arthropods was undoubtedly their small body size, which allowed them many more types of specialized niches than would be available for larger organisms.

SUMMARY

Arthropoda is the largest, most abundant and diverse phylum of animals. Arthropods are segmented, coelomate, ecdysozoan protostomes with well-developed organ systems. Most show marked tagmatization. They occur in virtually all habitats capable of supporting life. Perhaps more than any other single factor, prevalence of arthropods is explained by adaptations made possible by their cuticular exoskeleton and small size. Other important elements are jointed appendages, tracheal respiration, efficient sensory organs, complex behavior, and metamorphosis.

Trilobites were a dominant Paleozoic subphylum, now extinct. Members of subphylum Chelicerata have no antennae, and their main feeding appendages are chelicerae. In addition, they have a pair of pedipalps (which may be similar to the walking legs) and four pairs of walking legs. Class Merostomata includes the extinct eurypterids and the ancient, although still extant, horseshoe crabs. Class Pycnogonida contains the sea spiders, which are odd little animals with a large suctorial proboscis and vestigial abdomen. The great majority of living chelicerates are in class Arachnida: spiders (order Araneae), scorpions (order Scorpiones), harvestmen (order Opiliones), ticks and mites (order Acari), and others.

Tagmata of most spiders (cephalothorax and abdomen) show no external segmentation and are joined by a waistlike pedicel. Spiders are predaceous, and their chelicerae are provided with venom glands for paralyzing or killing prey. They breathe by book lungs, tracheae, or both. Spiders can spin silk, which they use for a variety of purposes, including in some cases webs for trapping prey.

Distinctive characters of scorpions are their large, clawlike pedipalps and their clearly segmented abdomen, which bears a terminal stinging apparatus. Harvestmen have small, ovoid bodies with very long, slender legs. Their abdomen is segmented and broadly joined to their cephalothorax.

The cephalothorax and abdomen of ticks and mites are completely fused, and mouthparts are borne on an anterior capitulum. Like spiders, some mites can spin silk. They are the most numerous of any arachnids; some are important carriers of disease, and others are serious plant pests.

Members of subphylum Myriapoda have a head followed by a series of trunk segments. The most familiar myriapods are predatory centipedes and herbivorous millipedes.

REVIEW QUESTIONS

1. What are important distinguishing features of arthropods?
2. Name the subphyla of arthropods, and give a few examples of each.
3. Briefly discuss the contribution of the cuticle to the success of arthropods, and name some other factors that have contributed to their success.
4. How are tagmata formed?
5. What is a trilobite?
6. What appendages characterize chelicerates?
7. Briefly describe the distinguishing morphological features of each of the following: eurypterids, horseshoe crabs, pycnogonids.
8. Why are horseshoe crabs in the same subphylum as spiders?
9. What are the tagmata of arachnids, and which tagmata bear appendages?
10. Describe the mechanism of each of the following with respect to spiders: feeding, excretion, sensory reception, web-spinning, reproduction.
11. Is it true that spider silk is stronger than steel?
12. What are the most important spiders in the United States that are dangerous to humans? How does their venom work?
13. How are spiders beneficial to humans?
14. Distinguish each of the following orders from the others: Araneae, Scorpiones, Opiliones, Acari.
15. Discuss the economic and medical importance of members of order Acari to human well-being.
16. What kind of arthropods are chiggers?
17. How do centipedes capture and subdue prey?
18. Which myriapods occur in moist soil or leaf litter?

For Further Thought Filter-feeding is a very common way to collect food from water, but this method is rarely used on land. Why might webspinning spiders be considered filter feeders?

SELECTED REFERENCES

Abe, F. R., and B. S. Lieberman. 2009. The nature of evolutionary radiations: a case study involving Devonian trilobites. Evol. Biol. **36:** 225–234. *An interesting study of speciation involving vicariance and changing sea level.*

Averof, M. 1998. Evolutionary biology: origin of the spider's head. Nature **395:** 436–437. *Summary of research on homology of the head across arthropod subphyla.*

Bowman, A. S., J. W. Dillwith, and J. R. Sauer. 1996. Tick salivary prostaglandins: presence, origin and significance. Parasitol. Today **12:** 388–396. *Tick prostaglandins act as immunosuppressants, anticoagulants, and analgesics. They allow the tick to feed over an extended time without the blood clotting, an inflammatory reaction occurring, or the host dislodging them.*

Eldridge, B. F., and J. D. Edman (eds). 2004, rev. ed. Medical entomology. Kluwer Academic Publishers, Dordrecht. *The best reference currently available.*

Foelix, R. F. 2011, ed. 3. Biology of spiders. New York, Oxford University Press. *Attractive, comprehensive book with extensive references; of interest to both amateurs and professionals.*

Jager, M., J. Murienne, C. Clabaut, J. Deutsch, H. Le Guyader, and M. Manuel. 2006. Homology of arthropod anterior appendages revealed by *Hox* gene expression in a sea spider. Nature **441:** 506–508. *Segment boundaries in sea spider heads show that chelifores (chelicerae) originate from the second head segment.*

Legg, D. A., M. D. Sutton, and G. D. Edgecombe. 2013. Arthropod fossil data increase congruence of morphological and molecular phylogenies. Nature Communications **4:** 2485. *This work supports the mandibulate hypothesis for arthropod phylogeny and indicates that hexapods arose within Crustacea.*

Luoma, J. R. 2001. The removable feast. Audubon **103** (3): 48–54. *During May and June large numbers of horseshoe crabs ascend the shores of U.S. Atlantic states to breed and lay eggs. Since the 1980s they have been heavily harvested to be chopped up and used for bait. This practice has led to serious declines in* Limulus *populations, with accompanying declines in populations of migrating shore birds that feed on* Limulus *eggs.*

Mallatt, J., J. R. Garey, and J. W. Shultz. 2004. Ecdysozoan phylogeny and Bayesian inference: first use of nearly complete 28s and 18s rRNA gene sequences to classify arthropods and their kin. Mol. Phylogen. Evol. **31:** 178–191. *Results indicate that the Crustacea is paraphyletic without hexapods, but that Pancrustacea is a monophyletic group, that chelicerates and myriapods are sister taxa, and that Panarthropoda is a monophyletic group. There was no support for a mandibulate clade.*

McDaniel, B. 1979. How to know the ticks and mites. Dubuque, Iowa, William C. Brown Publishers. *Useful, well-illustrated keys to genera and higher categories of ticks and mites in the United States.*

Ostfeld, R. S. 1997. The ecology of Lyme-disease risk. Am. Sci. **85:** 338–346. *Lyme disease, caused by a bacterium transmitted by ticks, has been reported in 48 of the 50 United States and seems to be increasing in frequency and geographic range.*

Polis, G. A. (ed). 1990. The biology of scorpions. Stanford, California, Stanford University Press. *The editor brings together a readable summary of what is known about scorpions.*

Rehm, P., K. Meusemann, J. Borner, B. Misof, and T. Burmester. 2014. Phylogenetic position of Myriapoda revealed by 454 transcriptome sequencing. Molecular Phylogenetics and Evolution **77:**25–33. *Data support the mandibulate hypothesis and Diplopoda as sister taxon to Pauropoda.*

Rota-Stabelli, O., L. Campbell, H. Brinkmann, G. D. Edgecombe, S. J. Longhorn, K. J. Peterson, D. Pisani, H. Philippe, and M. J. Telford. 2011. A congruent solution to arthropod phylogeny: phylogenomics, microRNAs and morphology support monophyletic Mandibulata. Proc. R. Soc. B **278:** 298–306. *Several lines of evidence suggest myriapods are the sister taxon to Pancrustacea resulting in a clade of arthropods sharing mandibles.*

Rota-Stabelli, O., E. Kayal, D. Gleeson, J. Daub, J. L. Boore, M. J. Telford, D. Pisani, M. Blaxter, and D. V. Lavrov. 2010. Ecdysozoan mitogenomics: evidence for a common origin of the legged invertebrates, the Panarthropoda. Genome Biol. Evol. **2** :425–440. *Support for the mandibulata hypothesis for arthropod relationships is provided here.*

Suter, R. B. 1999. Walking on water. Am. Sci. **87:** 154–159. *Fishing spiders* (Dolomedes) *depend on surface tension to walk on water.*

Villalobos, E. M. 2016. The mite that jumped, the bee that traveled, the disease that followed. Science **351:** 554–556. *A clear description of how mites and humans spread a deadly honey bee virus.*

Weaver, D. C. 1999. Mysterious fevers. Discover **20:** 37–40. *Ehrlichiosis is caused by a bacterial parasite of white blood cells transmitted by ticks.*

20

CHAPTER

Crustaceans

- **PHYLUM ARTHROPODA**
  - Subphylum Crustacea

A Sally Lightfoot crab, Grapsus grapsus, *from the Galápagos Islands.*

© Cleveland P. Hickman, Jr.

"Insects of the Sea"

The Crustacea (L. *crusta,* shell) are named after the hard shell that most crustaceans bear. Over 67,000 species have been described, and several times that number probably exist. Most familiar to people are the edible lobsters, crayfishes, shrimps, and crabs, but there is an astonishing array of less familiar forms, such as copepods, ostracods, water fleas, whale lice, tadpole shrimp, and krill. They fill a wide variety of ecological roles and show enormous variation in morphological characteristics, making a satisfactory description of the group as a whole very difficult.

We live in the Age of Arthropods, notwithstanding our anthropocentric attachment to our tradition of calling the current era the Age of Mammals. Together, insects and crustaceans compose more than 80% of all named animal species. Just as insects pervade terrestrial habitats (more than a million named species and countless trillions of individuals), crustaceans abound in oceans, lakes, and rivers. Some walk, some burrow, and some (such as barnacles) are sessile. Some swim upright, others swim upside down, and many are delicate microscopic forms that drift as plankton in oceans or in lakes. Indeed, it is probable that some of the most abundant animals in the world are members of the copepod genus *Calanus.* In recognition of their dominance of marine habitats, crustaceans have been called "insects of the sea."

Extant arthropods are divided among four subphyla (see Figure 19.2, p. 405). Crustacea and Hexapoda share five derived features and are united in clade Pancrustacea (Figure 20.1). We depict crustaceans and hexapods as sister taxa, but some phylogenies using molecular characters support the hypothesis that hexapods arose from *within* the crustacean branch, perhaps within the Remipedia. If the same pattern emerges from studies using other genes, it will be phylogenetically correct to refer to insects as "terrestrial crustaceans." Our description of crustaceans as

"insects of the sea" in the prologue to this chapter describes only the ecological role of these animals.

Crustaceans are divided among three subgroups (Figure 20.1). One of these, Oligostraca, contains members of the former phylum Pentastomida. Pentastomids, often called tongue worms, are parasites of vertebrates, living in lungs or nasal cavities. They are closely related to fish lice in subclass Branchiura.

Features of crustacean classes and subclasses are discussed following a general introduction to crustacean biology.

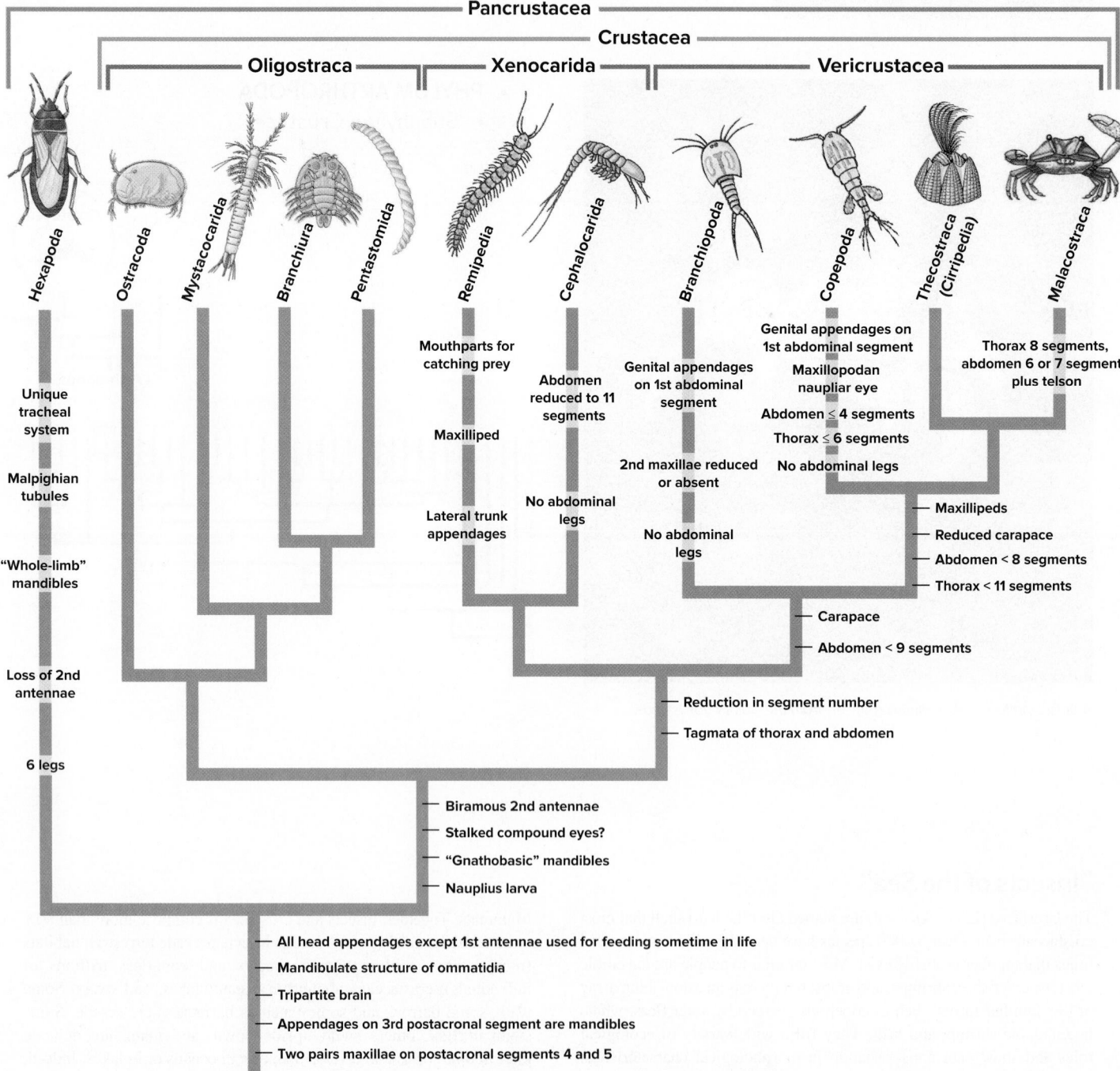

Figure 20.1   Cladogram showing hypothetical relationships of hexapods and classes of crustaceans. Hexapods and crustaceans form a clade diagnosed by numerous shared derived characteristics. Characters followed by a question mark may be ancestral rather than shared derived features. The acron is the anterior region of the head and is not counted as a segment.

20.1 SUBPHYLUM CRUSTACEA

General Nature of a Crustacean

Crustacea are mainly marine; however, there are many freshwater and a few terrestrial species. Crustaceans differ from other arthropods in a variety of ways, but the distinguishing characteristic is that crustaceans are the only arthropods with **two pairs of antennae.** In addition to two pairs of antennae and a pair of **mandibles,** crustaceans have two pairs of **maxillae** on the head, followed by a pair of appendages on each body segment. In some crustaceans not all segments bear appendages. All appendages, except perhaps the first antennae, are ancestrally **biramous** (two main branches), and at least some appendages of present-day adults show that condition. Organs specialized for respiration, if present, function as gills in aquatic species.

Most crustaceans have between 16 and 20 segments, but some forms have 60 segments or more. A larger number of segments is an ancestral feature. The more derived condition is to have fewer segments and increased tagmatization (see p. 406). Major tagmata are head, thorax, and abdomen. In most Crustacea one or more thoracic segments are fused with the head to form a **cephalothorax.** Tagmata are not homologous throughout the subphylum (or even within some classes) because in different groups different segments may have fused to form what we now call, for example, a head or a cephalothorax.

By far the largest class of crustaceans is the class Malacostraca, which includes lobsters, crabs, shrimps, beach hoppers, sow bugs, and many others. These species show a surprisingly constant arrangement of body segments and tagmata, which is considered the ancestral plan of this class (Figure 20.2). This typical body plan has a head of five (six embryonically) fused segments, a thorax of eight segments, and an abdomen of six segments (seven in a few species). At the anterior end is a nonsegmented **rostrum** and at the posterior end

is a nonsegmented **telson,** which with the last abdominal segment and its **uropods** forms a tail fan in many forms.

In many crustaceans the dorsal cuticle of the head may extend posteriorly and around the sides of the animal to cover or be fused with some or all of the thoracic and abdominal segments. This covering is called a **carapace** (Figure 20.2). In some groups the carapace forms clamshell-like valves that cover most or all of the body. In decapods (including lobsters, shrimp, crabs, and others), the carapace covers the entire cephalothorax but not the abdomen.

Form and Function

Because of their size and easy availability, large crustaceans such as crayfishes have been well studied and are commonly included in introductory laboratory courses. Hence many of the comments here apply specifically to crayfishes and their relatives.

External Features

Bodies of crustaceans are covered with a secreted cuticle composed of chitin, protein, and calcareous material. The harder, heavy plates of larger crustaceans are particularly high in calcareous deposits. The hard protective covering is soft and thin at the joints between segments, allowing flexibility of movement. The carapace, if present, covers much or all of the cephalothorax; in decapods such as crayfishes, all head and thoracic segments are enclosed dorsally by the carapace. Each segment not enclosed by the carapace is covered by a dorsal cuticular plate, or **tergum** (Figure 20.3A), and two lateral covering plates called **pleurites** that form the pleuron. A ventral transverse bar, or **sternum,** lies between the segmental appendages (Figure 20.3B). The abdomen terminates in a telson that bears the anus.

Position of the **gonopores** varies according to sex and group of crustaceans. They may be on or at the base of a pair of appendages, at the terminal end of the body, or on segments without legs. For example, in crayfishes openings of the vasa deferentia are on the median side at the base of the fifth pair of walking legs, and those of the oviducts are at the base of the third pair. In females an opening to the seminal receptacle is usually located in the midventral line between the fourth and fifth pairs of walking legs.

Appendages

Members of classes Malacostraca (for example, crayfishes) and Remipedia typically have a pair of jointed appendages on each segment (Figure 20.3B), although abdominal segments in the other classes typically do not bear appendages. Considerable specialization occurs as a derived condition in some crustaceans, such as crayfishes. The basic, biramous plan is illustrated by a crayfish appendage such as a maxilliped, a thoracic limb modified to become a feeding appendage (Figure 20.4). The basal portion, or **protopod,** bears a lateral **exopod** and a medial **endopod.** The protopod is composed of two parts (**basis** and **coxa**), whereas the exopod and endopod have from one to several parts each. Variations on the basic form exist (Figure 20.5). Some appendages, such as the walking legs of crayfishes, have become secondarily uniramous. Medial or lateral processes, called **endites** and **exites,** respectively, sometimes occur on crustacean limbs. An exite on the protopod is called an **epipod,** which is often modified as a gill. Table 20.1 shows how various

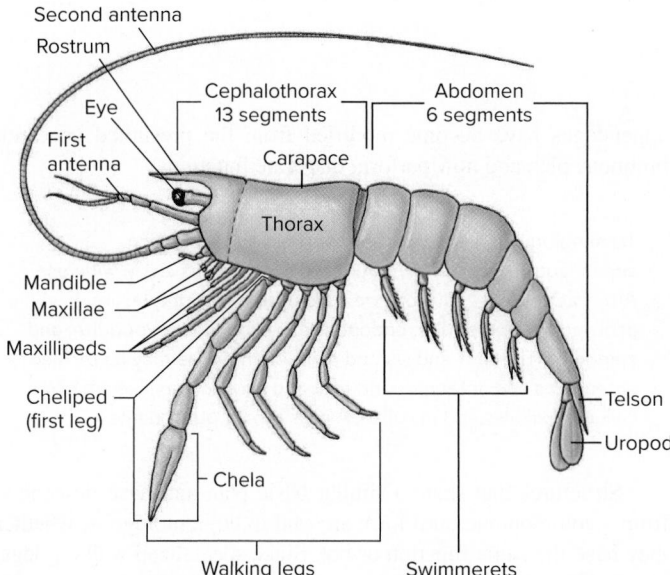

Figure 20.2 Archetypical plan of Malacostraca. The two maxillae and three maxillipeds have been separated diagrammatically to illustrate the general plan.

Labels on figure:
Second antenna
Rostrum
Eye
First antenna
Cephalothorax 13 segments
Abdomen 6 segments
Carapace
Thorax
Mandible
Maxillae
Maxillipeds
Cheliped (first leg)
Chela
Walking legs
Swimmerets
Telson
Uropod

Second
antenna

Chela

First
antennae

Major
Cheliped

Second
antenna

Chela

Rostrum

Eye

Second walking
leg (a minor
cheliped)

Third walking
leg

Carapace

Cervical groove

Fourth
walking leg

Fifth
walking leg

Tergum

Telson

Uropod

Abdomen

First
antennae

Exopod of first
maxilliped

Second maxilliped

Mouth

Third maxilliped

Second
walking leg

Third walking leg

Fourth walking leg

First (copulatory)
swimmeret of male

Second
swimmeret

Swimmerets 3 to 5

Sternum

Anus

Telson

Uropod

A   **B**

Figure 20.3   External structure of crayfishes. **A,** Dorsal view. **B,** Ventral view.

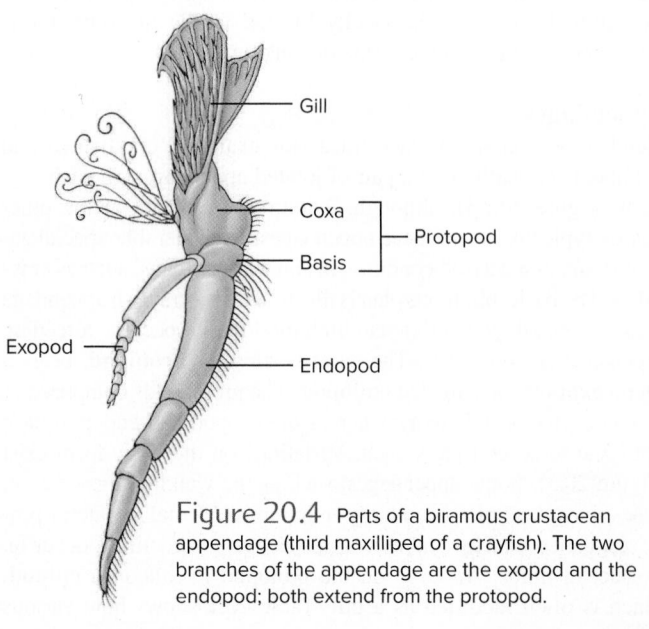

Gill

Coxa

Basis

Protopod

Exopod

Endopod

Figure 20.4   Parts of a biramous crustacean appendage (third maxilliped of a crayfish). The two branches of the appendage are the exopod and the endopod; both extend from the protopod.

appendages have become modified from the presumed ancestral biramous plan and now perform disparate functions.

> Terminology applied by various workers to crustacean appendages is not uniform. At least two systems are in wide use. Alternative terms to those we have used, for example, are protopodite, exopodite, endopodite, basipodite, coxopodite, and epipodite. The first and second pairs of antennae may be termed antennules and antennae, and first and second maxillae are often called maxillules and maxillae. A rose by any other name . . .

Structures that share a similar basic plan and have descended from a common ancestral form are said to be homologous, whether they have the same function or not. Since specialized walking legs, mouthparts, chelipeds, and swimmerets have all developed from a common biramous appendage (that has become modified to perform different functions), they are all homologous to each other—a condition called **serial homology.** Limbs ancestrally were all very similar,

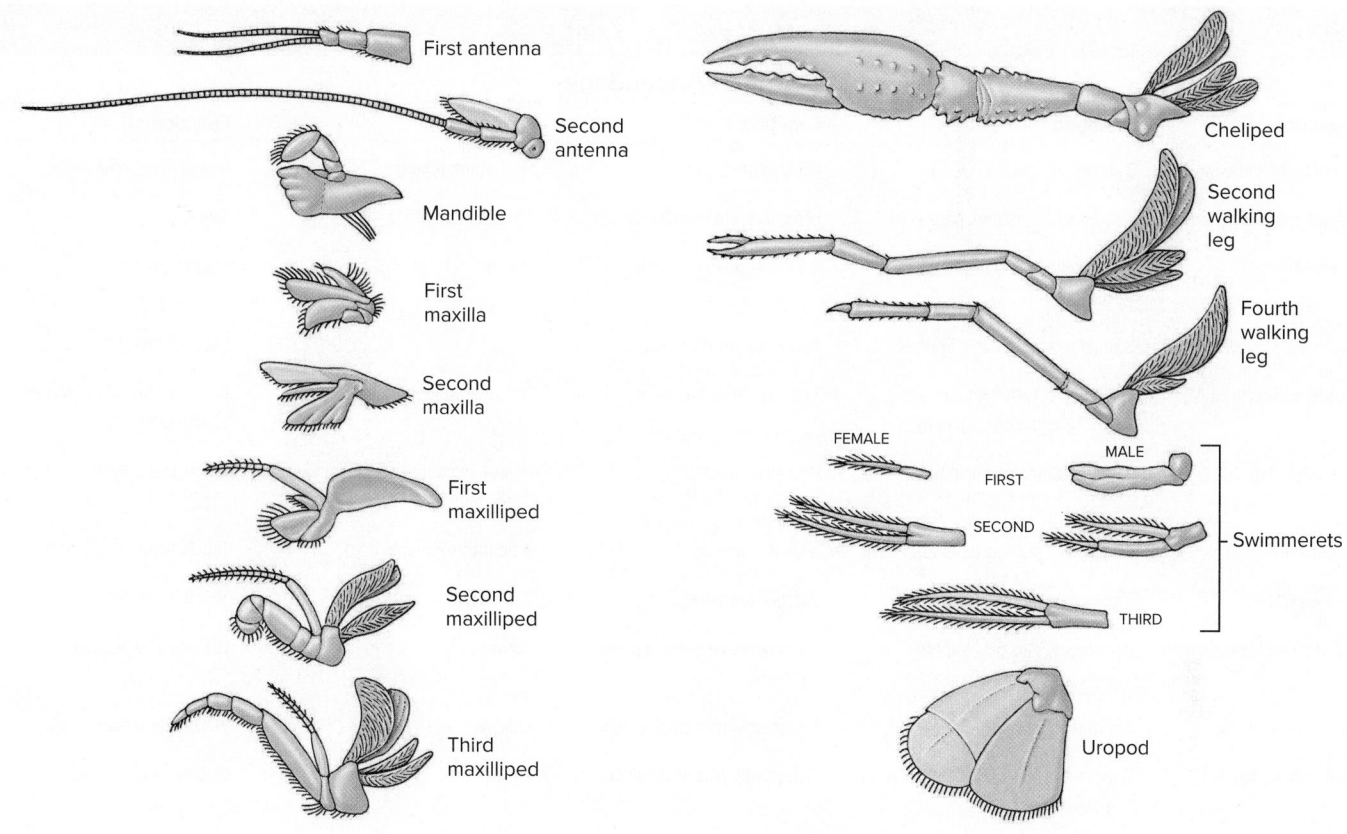

Figure 20.5 Appendages of a crayfish showing variations on the basic biramous plan, as found in a swimmeret. Protopod, brown; endopod, blue; exopod, yellow.

but during structural evolution some branches have been reduced, some lost, and some greatly altered; some have had new parts added. Crayfishes and their allies possess the most elaborate serial homology in the animal kingdom, with 17 distinct but serially homologous types of appendages (Table 20.1). For example, compare the size of the chela of the cheliped to the tiny claw (chela) of the second walking leg in Figure 20.5 .

Internal Features

The muscular and nervous systems in the thorax and abdomen clearly show segmentation, but there are marked modifications in other systems. Most changes involve concentration of parts in a particular region or else reduction or complete loss of parts.

Hemocoel The major body space in arthropods is not a coelom, but a persistent blastocoel that becomes a blood-filled **hemocoel** (see p. 397 and p. 685). In crustaceans the only coelomic compartments remaining are the end sacs of excretory organs and the space around the gonads.

Muscular System Striated muscles form a considerable part of the body of most Crustacea. Muscles are usually arranged in antagonistic groups: **flexors,** which draw a part toward the body, and **extensors,** which extend the part outward. The abdomen of a crayfish

has powerful flexors (Figure 20.6), which are used when the animal swims backward with a burst of speed to escape from predators.

Respiratory System Gas exchange in smaller crustaceans occurs across thin areas of cuticle (for example, in the appendages) or the entire body, and specialized structures for gas exchange may be absent. Larger crustaceans have gills, which are delicate, featherlike projections with very thin cuticle (see Chapter 31, p. 693). In decapods the sides of the carapace enclose the gill cavity, which is open anteriorly and ventrally (Figure 20.7). Gills may project from the pleural wall into the gill cavity, from the articulation of thoracic legs with the body, or from thoracic coxae. The latter two types are typical of crayfishes. The "bailer," a part of the second maxilla, draws water over the gill filaments, into the gill cavity at the bases of the legs, and out of the gill cavity at the anterior.

Circulatory System Crustaceans and other arthropods have an "open" or lacunar type of circulatory system (see Chapter 31, pp. 685–686). This means that there are no veins and no separation of blood from interstitial fluid, as there is in animals with closed systems. Hemolymph (blood plus interstitial fluid) leaves the heart through arteries, circulates through the hemocoel, and returns to venous **sinuses,** or spaces, instead of veins before it enters the heart.

TABLE 20.1
Crayfish Appendages

Appendage	Protopod	Endopod	Exopod	Function(s)
First antenna (antennule)	3 segments, statocyst in base	Many-jointed feeler	Many-jointed feeler	Touch, taste, equilibrium
Second antenna (antenna)	2 segments, excretory pore in base	Long, many-jointed feeler	Thin, pointed blade	Touch, taste
Mandible	2 segments, heavy jaw and base of palp	2 distal segments of palp	Absent	Crushing food
First maxilla (maxillule)	2 segments, with 2 thin endites	Small unjointed lamella	Absent	Food handling
Second maxilla (maxilla)	2 segments, with 2 endites and 1 scaphognathite (epipod)	1 small pointed segment	Part of scaphognathite (bailer)	Drawing currents of water into gills
First maxilliped	2 medial plates and epipod	2 small segments	1 basal segment, plus many-jointed filament	Touch, taste, food handling
Second maxilliped	2 segments plus gill (epipod)	5 short segments	2 slender segments	Touch, taste, food handling
Third maxilliped	2 segments plus gill (epipod)	5 larger segments	2 slender segments	Touch, taste, food handling
First walking leg (cheliped)	2 segments plus gill (epipod)	5 segments with heavy pincer (chela)	Absent	Offense and defense
Second walking leg	2 segments plus gill (epipod)	5 segments plus small pincer	Absent	Walking and prehension
Third walking leg	2 segments plus gill (epipod); genital pore in female	5 segments plus small pincer	Absent	Walking and prehension
Fourth walking leg	2 segments plus gill (epipod)	5 segments, no pincer	Absent	Walking
Fifth walking leg	2 segments; genital pore in male; no gill	5 segments, no pincer	Absent	Walking
First swimmeret	In female reduced or absent; in male fused with endopod to form tube			In male, transferring sperm to female
Second swimmeret				
Male	Structure modified for transfer of sperm to female	Structure modified for transfer of sperm to female		
Female	2 segments	Jointed filament	Jointed filament	Creating water currents; carrying eggs and young
Third, fourth, and fifth swimmerets	2 short segments	Jointed filament	Jointed filament	Creating water currents; in female carrying eggs and young
Uropod	1 short, broad segment	Flat, oval plate	Flat, oval plate; divided into 2 parts with hinge	Swimming; egg protection in female

A dorsal heart is the chief propulsive organ. It is a single-chambered sac of striated muscle. Hemolymph enters the heart from the surrounding **pericardial sinus** through paired ostia, with valves that prevent backflow into the sinus (Figure 20.7). From the heart hemolymph enters one or more arteries. Valves in the arteries prevent a backflow of hemolymph. Small arteries empty into tissue sinuses, which in turn often discharge into a large **sternal sinus** (Figure 20.7).

From there, afferent sinus channels carry hemolymph to the gills, if present, for oxygen and carbon dioxide exchange. Hemolymph then returns to the pericardial sinus by efferent channels (Figure 20.7).

Hemolymph in arthropods may be colorless, reddish, or bluish, as in many Crustacea. Hemocyanin, a copper-containing respiratory pigment, or hemoglobin, an iron-containing pigment, may be carried in solution. Hemolymph has the property of clotting, which prevents

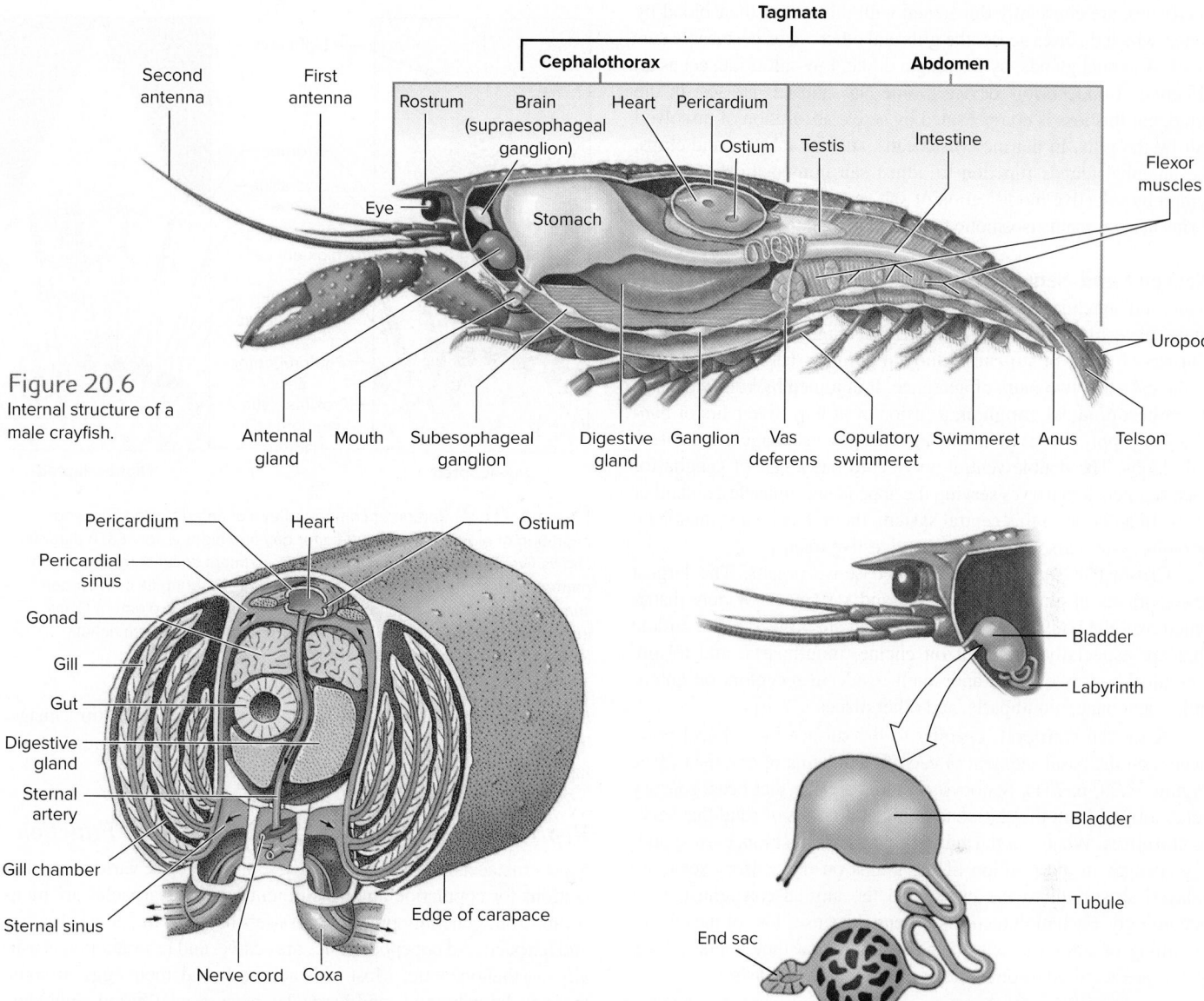

Figure 20.6
Internal structure of a male crayfish.

Figure 20.7 Diagrammatical cross section through heart region of a crayfish showing direction of blood flow in this "open" blood system. Heart pumps blood to body tissues through arteries, which empty into tissue sinuses. Returning blood enters sternal sinus, then goes through gills for gas exchange, and finally back to pericardial sinus by efferent channels. Note absence of veins.

Figure 20.8 Scheme of antennal gland (green gland) of crayfishes. (In natural position organ is much folded.) Some crustaceans lack a labyrinth, and the excretory tubule (nephridial canal) is a muchcoiled tube.

its loss in minor injuries. Some ameboid cells release a thrombinlike coagulant that precipitates clotting.

Excretory System   Excretory organs of adult crustaceans are a pair of tubular structures located in the ventral part of their head anterior to the esophagus (Figure 20.6). They are called **antennal glands** or **maxillary glands,** depending on whether they open at the base of the antennae or at the base of the second maxillae (see Chapter 30, pp. 665–666). A few adult crustaceans have both. Excretory organs of decapods are antennal glands, also called green glands in this group. Crustaceans do not have Malpighian tubules, the excretory organs of spiders and insects.

The **end sac** of the antennal gland consists of a small vesicle (**saccule**) and a spongy mass called a **labyrinth.** The labyrinth connects by an **excretory tubule** to a dorsal **bladder,** which opens to the exterior by a pore on the ventral surface of the basal antennal segment (Figure 20.8). Hydrostatic pressure within the hemocoel provides force for filtration of fluid into the end sac. Filtrate is excreted as urine after the resorption of salts, amino acids, glucose, and some water.

Excretion of nitrogenous wastes (mostly ammonia) takes place by diffusion across thin areas of cuticle, especially gills. The so-called "excretory organs" function principally to regulate ionic and osmotic composition of body fluids. Freshwater crustaceans, such as crayfishes, are constantly threatened with dilution of their blood by water, which diffuses across the gills and other water-permeable surfaces. Antennal glands, by forming a dilute, low-salt urine, act as an effective "floodcontrol" device. Some Na^+ and Cl^- are lost in the urine, but this loss is compensated by active absorption of dissolved salt by the gills. In marine crustaceans, such as lobsters and crabs, the antennal glands function to adjust salt composition of hemolymph by selective modification of salt content from urine. In these forms urine remains isosmotic to the blood.

Nervous and Sensory Systems   Nervous systems of crustaceans and annelids have much in common, although those of crustaceans have more fusion of ganglia (see Figure 20.6). The brain consists of a pair of **supraesophageal ganglia** that supplies nerves to the eyes and two pairs of antennae. It is joined by connectives to the **subesophageal ganglion,** a fusion of at least five pairs of ganglia that supply nerves to mouth, appendages, esophagus, and antennal glands. The double ventral nerve cord has a pair of ganglia for each segment and nerves serving the appendages, muscles, and other parts. In addition to this central system, there may be a sympathetic nervous system associated with the digestive tract.

Crustaceans have well-developed sense organs. The largest sense organs of crayfishes are eyes and statocysts. Widely distributed over the body are **tactile hairs,** delicate projections of cuticle that are especially abundant on chelae, mouthparts, and telson. Chemical senses of taste and smell reside in receptors on antennules, antennae, mouthparts, and other places.

A saclike **statocyst,** opening to the surface by a dorsal pore, occurs on the basal segment of each first antenna of crayfishes (see Figure 33.28, p. 741). Statocysts contain a ridge that bears sensory setae formed from the chitinous lining and grains of sand that serve as **statoliths.** Whenever the animal changes its position, corresponding changes in the position of the grains on the sensory setae are relayed as stimuli to the brain, and the animal can adjust itself accordingly. Each molt (ecdysis) of cuticle causes loss of the cuticular lining of statocysts and the sand grains that they contain. New grains are acquired through the dorsal pore after ecdysis.

Eyes in many crustaceans are compound, composed of many photoreceptor units called **ommatidia** (Figure 20.9). Covering the rounded surface of each eye is a transparent area of cuticle, the **cornea,** which is divided into many small squares or hexagons called facets. These facets form the outer faces of the ommatidia. Each ommatidium behaves like a tiny eye and contains several kinds of cells arranged in a columnar fashion (Figure 20.9). Black pigment cells are located between adjacent ommatidia, and movement of pigment in an arthropod compound eye permits it to adjust to different amounts of light. There are three sets of pigment cells in each ommatidium: distal retinal, proximal retinal, and reflecting; these are so arranged that they can form a collar or sleeve around each ommatidium. For strong light (day adaptation) the distal retinal pigment moves inward and meets the outward-moving proximal retinal pigment so that a complete pigment sleeve forms around the ommatidium (Figure 20.9). In this condition only rays that strike the cornea directly reach the photoreceptor cells (retinuli), for each ommatidium is shielded from others. Thus each ommatidium sees only a limited area of the field of vision (a mosaic, or **apposition,** image). In dim light distal and proximal pigments separate so that light rays, with the aid of reflecting pigment cells, have a chance to spread to

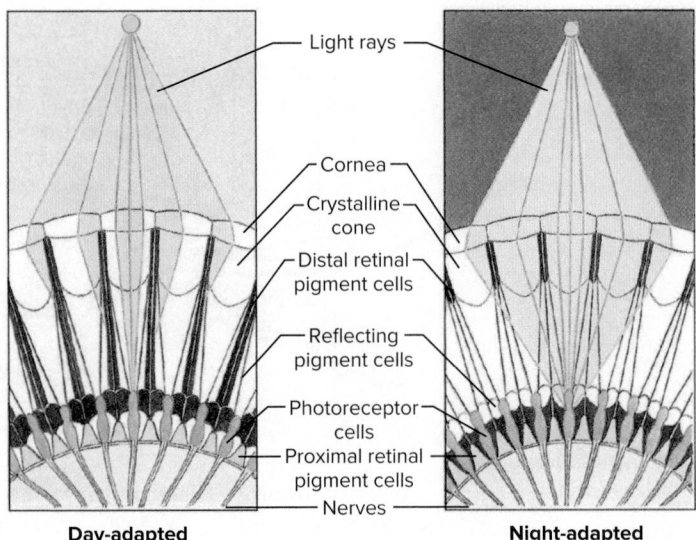

Day-adapted   **Night-adapted**

Figure 20.9   Portion of compound eye of an arthropod showing migration of pigment in ommatidia for day and night vision. Each diagram shows five ommatidia. In daytime, a dark pigment collar surrounds each ommatidium so that only the light rays directly entering its own cornea stimulate the ommatidium (mosaic vision); in nighttime, pigment forms incomplete collars and light rays can spread to adjacent ommatidia (continuous, or superposition, image).

adjacent ommatidia and to form a continuous, or **superposition,** image. This second type of vision is less precise but takes maximum advantage of the limited amount of light received.

Reproduction, Life Cycles, and Endocrine Function

Most crustaceans have separate sexes, and there are various specializations for copulation among different groups. Barnacles are monoecious but generally practice cross-fertilization. In some ostracods and harpacticoid copepods males are scarce, and reproduction is usually parthenogenetic. Most crustaceans brood their eggs in some manner: branchiopods and barnacles have special brood chambers, copepods have brood sacs attached to the sides of their abdomen (see Figure 20.18), and many malacostracans carry eggs and young attached to their abdominal appendages.

Crayfishes have direct development: there is no larval form. A tiny juvenile with the same form as the adult and a complete set of appendages and segments hatches from the egg. However, development is indirect in most crustaceans, and larvae, quite unlike adults in structure and appearance, hatch from eggs. Larvae undergo **metamorphosis** to become adults. The ancestral and most widely occurring larva in Crustacea is the **nauplius** (Figures 20.10 and 20.20). Nauplii bear only three pairs of appendages: uniramous first antennules, biramous antennae, and biramous mandibles. All function as swimming appendages at this stage. Subsequent development may involve a gradual change to adult body form, and appendages and segments are added through a series of molts. However, transformation to the adult form may involve more abrupt changes. For example, metamorphosis of a barnacle proceeds from a free-swimming nauplius to a larva with a bivalve carapace called a cyprid and finally to a sessile adult with calcareous plates.

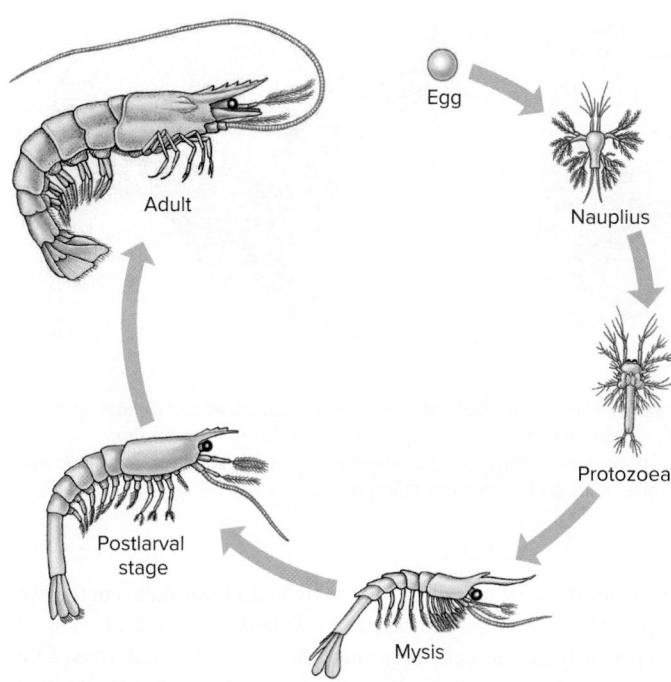

Figure 20.10 Life cycle of a Gulf shrimp, *Farfantepenaeus*. Penaeids spawn at depths of 40 to 90 m. Young larval forms are planktonic and move inshore to water of lower salinity to develop as benthic juveniles and adults. Older shrimp return to deeper water offshore.

Molting and Ecdysis   Molting, the physiological process of making a larger cuticle, and ecdysis (ek′dī-sis) (Gr. *ekdyein,* to strip off), the shedding of the cuticle, are necessary for the body to increase in size because the exoskeleton is nonliving and does not grow as the animal grows. Much of a crustacean's functioning, including its reproduction, behavior, and many metabolic processes, is directly affected by the physiology of the molting cycle.

Cuticle, which is secreted by underlying epidermis, has several layers (Figure 20.11; see also Figure 19.3, p. 405). The outermost is **epicuticle,** a very thin layer of lipid-impregnated protein. The bulk of cuticle is the several layers of **procuticle:** (1) **exocuticle,** which lies just beneath the epicuticle and contains protein, calcium salts, and chitin; (2) **endocuticle,** which itself is composed of (3) a **principal layer,** which contains more chitin and less protein and is heavily calcified, and (4) an uncalcified **membranous layer,** a relatively thin layer of chitin and protein.

Molting animals grow in the **intermolt** phases, or instars, with soft tissues increasing in size until there is no free space within the cuticle. When the body fills the cuticle, the animal enters the **premolt** phase. Growth occurs over a much longer time period than is apparent from examining the external size of the animal.

During the molting process and some time before actual ecdysis, epidermal cells enlarge considerably. They separate from the membranous layer, secrete a new epicuticle, and begin secreting a new exocuticle (Figure 20.11). Enzymes are released into the area above the new epicuticle. These enzymes begin to dissolve old endocuticle, and soluble products are resorbed and stored within the body of the crustacean.

Intermolt condition

— Epicuticle
— Exocuticle

— Endocuticle

— Epidermis

STEP 1:
In the premolt stage, the old procuticle separates from the epidermis, which secretes a new epicuticle.

— New epicuticle

STEP 2:
Still in the premolt stage, new exocuticle is secreted as molting fluid dissolves the old endocuticle. Solution products are reabsorbed.

— Dissolving endocuticle

— New exocuticle

— Discarded old epicuticle and exocuticle

STEP 3:
At ecdysis, the old epicuticle and exocuticle are discarded.

STEP 4:
In postecdysis, new cuticle is stretched and unfolded, and endocuticle is secreted.

— New endocuticle

Figure 20.11 Cuticle secretion and resorption in ecdysis.

Some calcium salts are stored as **gastroliths** (mineral accretions) in the walls of the stomach. Finally, only exocuticle and epicuticle of the old cuticle remain, underlain by new epicuticle and new exocuticle. The animal swallows water, which it absorbs through its gut, and

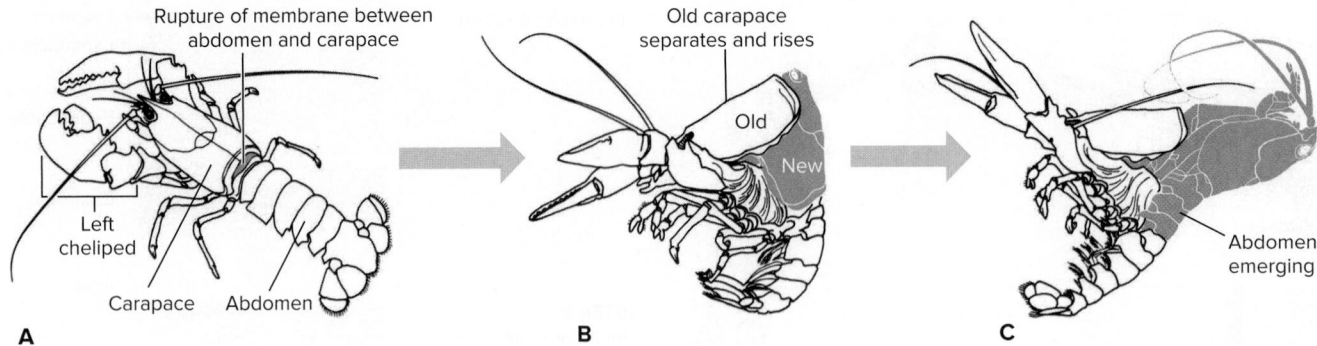

Figure 20.12   Molting sequence in a lobster, *Homarus americanus*. **A,** Membrane between carapace and abdomen ruptures, and carapace begins slow elevation. This step may take up to 2 hours. **B** and **C,** Head, thorax, and finally abdomen withdraw. This process usually takes no more than 15 minutes. Immediately after ecdysis, chelipeds are desiccated and body is very soft. Lobster continues rapid absorption of water so that within 12 hours the body increases about 20% in length and 50% in weight. Tissue water will be replaced by protein in succeeding weeks.

its blood volume increases greatly. Internal pressure causes the cuticle to split along preformed lines of weakness in the cuticle, and the animal pulls itself out of its old exoskeleton (Figure 20.12). Following this is a stretching of the still soft new cuticle, deposition of new endocuticle, redeposition of salvaged inorganic salts and other constituents, and hardening of the new cuticle. During the period of molting, the animal is defenseless and remains hidden and quiescent.

When a crustacean is young, ecdysis must occur frequently to allow growth, and the molting cycle is relatively short. As the animal approaches maturity, intermolt periods become progressively longer, and in some species molting ceases entirely. During intermolt periods, increase in tissue mass occurs as living tissue replaces water.

Hormonal Control of the Ecdysis Cycle   Although ecdysis is hormonally controlled, the cycle is often initiated by environmental stimuli perceived by the central nervous system. Such stimuli may include temperature, day length, and humidity (in the case of land crabs) or a combination of environmental signals. The signal from the central nervous system decreases production of a **molt-inhibiting hormone** by the **X-organ.** (The X-organ is a group of neurosecretory cells in the medulla terminalis of the brain.) In crayfishes and other decapods, the medulla terminalis is found in the eyestalk. The hormone is carried in axons of the X-organ to the **sinus gland** (which itself is probably not glandular in function), and also in the eyestalk, where it is released into the hemolymph.

A drop in level of molt-inhibiting hormone promotes release of a **molting hormone** (see p. 457 and pp. 749–750) from the **Y-organs.** Y-organs lie beneath the epidermis near the adductor muscles of the mandibles, and they are homologous to prothoracic glands of insects, which produce the hormone ecdysone. Action of molting hormone is to initiate processes leading to ecdysis. Once initiated, the cycle proceeds automatically without further action of hormones from either X- or Y-organs.

Other Endocrine Functions   Removal of eyestalks accelerates molting; in addition, crustaceans whose eyestalks have been removed can no longer adjust their body coloration to match the background conditions. It was discovered long ago that this defect

was caused not by loss of vision but by loss of hormones in the eyestalks. The body color of crustaceans is largely a result of pigments in special branched cells (chromatophores) in the epidermis. Concentration of pigment granules in the center of the cells causes lightening, and dispersal of pigment throughout the cells causes darkening. Pigment behavior is controlled by hormones from neurosecretory cells in the eyestalk, as is migration of retinal pigment for light and dark adaptation in the eyes (see Figure 20.9).

Neurosecretory cells are nerve cells that are modified for secretion of hormones. They are widespread in invertebrates and also occur in vertebrates. Cells in the vertebrate hypothalamus and posterior pituitary are good examples (see pp. 749–753).

Release of neurosecretory material from the pericardial organs in the wall of the pericardium causes an increase in the rate and amplitude of the heartbeat.

Androgenic glands, first discovered in an amphipod (*Orchestia*, a common beach hopper), occur in male malacostracans. Unlike most other endocrine organs in crustaceans, these are not neurosecretory organs. Their secretion stimulates expression of male sexual characteristics. Young malacostracans have rudimentary androgenic glands, but in females these glands fail to develop. If they are artificially implanted in a female, her ovaries transform to testes and begin to produce sperm, and her appendages begin to acquire male characteristics at the next molt. In isopods the androgenic glands are in the testes; in all other malacostracans they are between muscles of the coxopods of the last thoracic legs and partly attached near ends of the vasa deferentia. Although females do not possess organs similar to androgenic glands, their ovaries produce one or two hormones that influence secondary sexual characteristics.

Hormones that influence other body processes in Crustacea also may be present, and evidence suggests that a neurosecretory substance produced in the eyestalk regulates the level of blood sugar.

Feeding Habits   Feeding habits and adaptations for feeding vary greatly among crustaceans. Many forms can shift from one type

Figure 20.13 Malacostracan stomach showing gastric "mill" and directions of food movements. Mill has chitinous ridges, or teeth, for mastication, and setae for straining food before it passes into the pyloric stomach.

of feeding to another depending on environment and food availability, but all use the same fundamental set of mouthparts. Mandibles and maxillae function to ingest food; maxillipeds hold and crush food. Walking legs, particularly chelipeds, capture food in predaceous forms.

Many crustaceans, both large and small, are predatory, and some have interesting adaptations for killing prey. For example, one kind of mantis shrimp carries, on one of its walking legs, a specialized digit that can be drawn into a groove and released suddenly to pierce passing prey. Pistol shrimps (*Alpheus* spp.) have an enormously enlarged chela that can be cocked like the hammer of a gun and snapped shut at great speed, forming a cavitation bubble that implodes with a force sufficient to stun their prey.

Food sources of **suspension feeders** range from plankton and detritus to bacteria. **Predators** consume larvae, worms, crustaceans, snails, and fishes. **Scavengers** eat dead animal and plant matter. Suspension feeders, such as fairy shrimps, water fleas, and barnacles, use their legs, which bear a thick fringe of setae, to create water currents that sweep food particles through the setae. Mud shrimps (*Upogebia* spp.) use long setae on their first two pairs of thoracic appendages to strain food from water circulated through their burrow by movements of their swimmerets. Yeti crabs from hydrothermal vents harbor sulfur-oxidizing bacteria on special setae, as well as farming methane- and sulfide-binding chemosynthetic proteobacteria. This novel feeding method relies on **symbiosis**.

Crayfishes have a two-part stomach (Figure 20.13). The first part contains a **gastric mill** in which food, already shredded by their mandibles, can be further ground by three calcareous teeth into particles fine enough to pass through a filter of setae in the second part; food particles then pass into the intestine for chemical digestion.

20.2 A BRIEF SURVEY OF CRUSTACEANS

Crustaceans are an extensive group of over 67,000 species worldwide with many subdivisions. They have many structures, habitats, and modes of living. Some are much larger than crayfishes; others are smaller, even microscopic. Some are highly developed and specialized; others have simpler organization.

Readers should realize that the following summary of crustaceans is misleadingly brief. Taxonomy within Crustacea is in flux

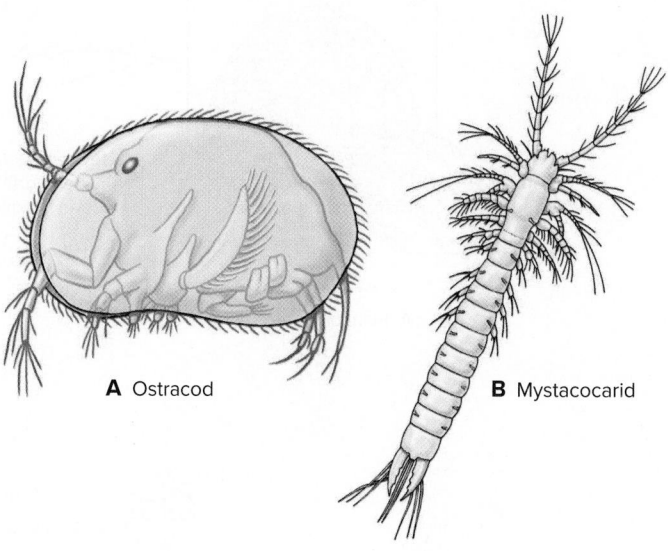

Figure 20.14 **A**, An ostracod. **B**, A mystacocarid.

with traditional classes and subclasses no longer supported by molecular phylogenies. Crustaceans are divided among three groups: Oligostraca, Xenocarida, and Vericrustacea. We present taxa in the order shown in Figure 20.1. Within Vericrustacea, the largest and most familiar group of crustaceans, Malacostraca, is described on page 435. The taxonomy on page 440 provides a concise summary of the groups.

Oligostraca

Ostracoda

Members of Ostracoda are, like diplostracans, enclosed in a bivalve carapace and resemble tiny clams, 0.25 to 10 mm long (Figure 20.14A). They are commonly called mussel shrimp or seed shrimp; they have a worldwide distribution and are important in aquatic food webs. Ostracods show considerable fusion of trunk segments, obscuring division between the thorax and the abdomen. The trunk has one to three pairs of limbs, with the number of thoracic appendages reduced to two or none. Feeding and locomotion are principally by use of the head appendages. Most ostracods are benthic or climb on plants, but some are planktonic or burrowing, and a few are parasitic. Feeding habits are diverse; there are particle, plant, and carrion feeders and predators. They are widespread in both marine and freshwater habitats. Most of the 6000 known species are dioecious, but some are parthenogenetic. Some bizarre male mussel shrimps emit light and may synchronize their flashing to attract females. Development is by gradual metamorphosis. There are thousands of extant species and over 10,000 ostracod fossil species whose presence in certain rock strata often serve as important indicators of oil deposits.

Mystacocarida

Mystacocarida is a class of tiny crustaceans (less than 0.5 mm long) that live in interstitial water between sand grains of marine

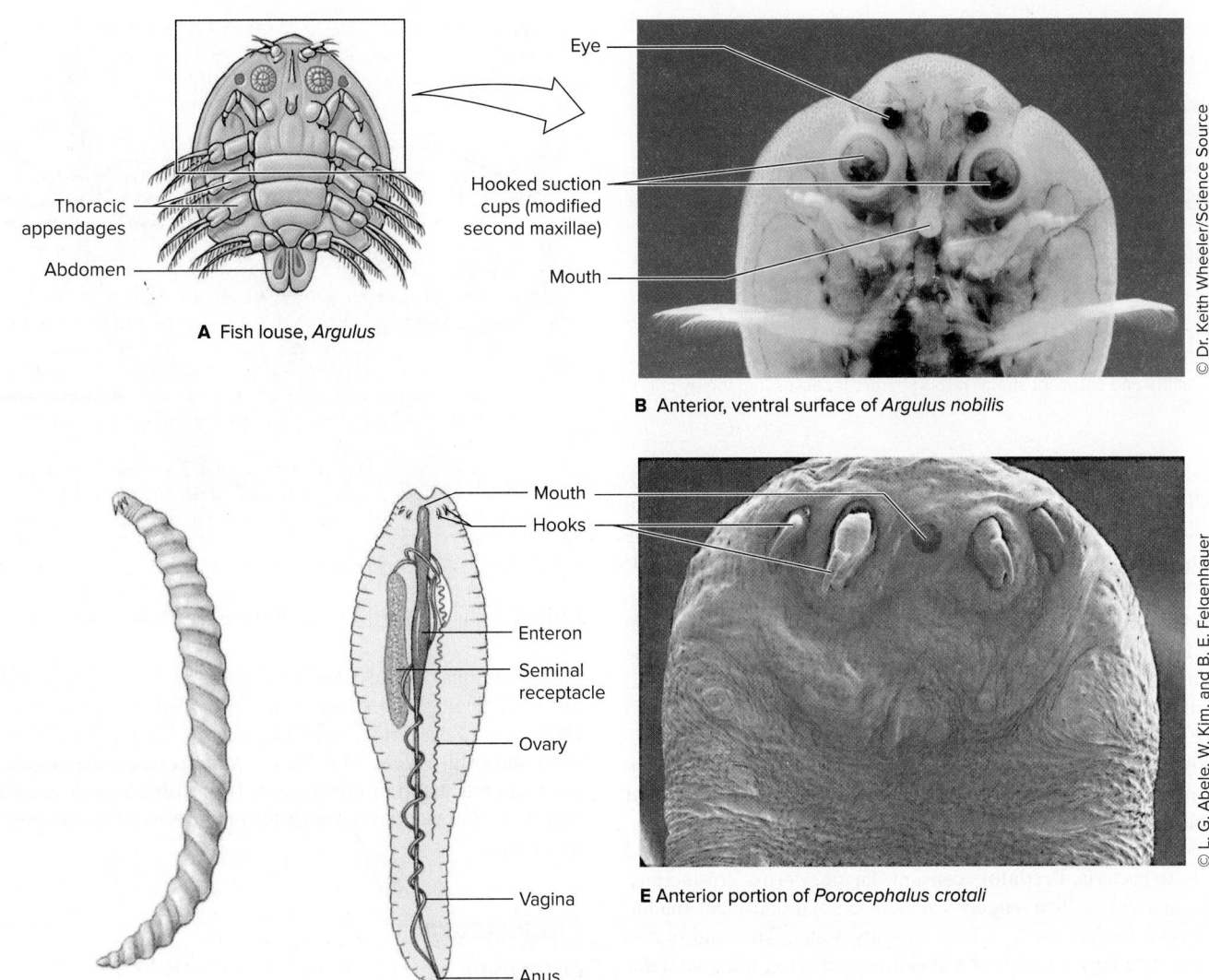

A Fish louse, *Argulus*

Eye

Hooked suction cups (modified second maxillae)

Mouth

B Anterior, ventral surface of *Argulus nobilis*

© Dr. Keith Wheeler/Science Source

Mouth

Hooks

Enteron

Seminal receptacle

Ovary

Vagina

Anus

Thoracic appendages

Abdomen

C *Armillifer*   **D** *Linguatula*

E Anterior portion of *Porocephalus crotali*

© L. G. Abele, W. Kim, and B. E. Felgenhauer

Figure 20.15   **A**, Fish lice, in Branchiura, are closely related to pentastomids (**C** and **D**). Images of the anterior regions of both animals (**B** and **E**) demonstrate morphological similarities.

beaches (Figure 20.14B). Only 10 species have been described, but mystacocarids are widely distributed through many parts of the world.

Branchiura

Branchiurans are a small group of primarily fish ectoparasites whose mouthparts are modified for sucking. Members of this group are usually between 5 and 10 mm long and may occur on marine or freshwater fish. They typically have a broad, shieldlike carapace, compound eyes, four biramous thoracic appendages for swimming, and a short, unsegmented abdomen (Figure 20.15A). Second maxillae have become modified as suction cups, enabling the parasites to move on their fish host or even from fish to fish. Heavily infested fish may get fungal infections and die. There is no nauplius, and young resemble adults except in size and degree of development of appendages.

Pentastomida

Members of former phylum Pentastomida (pen-ta-stō-mʹi-da) (Gr. *pente,* five, + *stoma,* mouth), or tongue worms, comprise about 130 species of wormlike parasites of the respiratory system of vertebrates. Adult pentastomids live mostly in lungs of reptiles, such as snakes, lizards, and crocodiles, but one species, *Reighardia sternae,* lives in air sacs of terns and gulls, and another, *Linguatula serrate* (Gr. *lingua,* tongue), lives in the nasopharynx of canines and felines (and occasionally humans). Although more common in tropical areas, they also occur in North America, Europe, and Australia.

Adults range from 1 to 13 cm in length. Their body is covered with a nonchitinous and highly porous cuticle in the form of transverse rings (Figure 20.15C and D) that is molted periodically during larval stages. The anterior end may bear five short protuberances (hence the name Pentastomida). Four of these bear chitinous claws, and the fifth bears the mouth (Figure 20.15). A simple, straight digestive system is adapted for processing blood from the host. The nervous system, similar to that of other arthropods, has paired ganglia along the ventral nerve cord. The only sense organs appear to be papillae. There are no circulatory, excretory, or respiratory organs.

Sexes are separate, and females are usually larger than males. A female may produce several million eggs, which pass up the trachea of the host, are swallowed, and exit with feces. Larvae hatch as oval, tailed creatures with four stumpy legs. Most pentastomid life cycles require an intermediate vertebrate host such as a fish, a reptile, or, rarely, a mammal, that is eaten by the definitive vertebrate host. After ingestion by an intermediate host, larvae penetrate the intestine, migrate randomly in the body, and finally metamorphose into nymphs. After growth and several molts, a nymph finally becomes encapsulated and dormant. When eaten by a final host, a juvenile finds its way to a lung, feeds on blood and tissue, and matures.

Several species have been found encysted in humans, the most common being *Armillifer armillatus* (L. *armilla,* ring, bracelet, + *fero,* to bear), but usually they cause few symptoms. *Linguatula serrata* is a cause of nasopharyngeal pentastomiasis, or "halzoun," a disease of humans in the Middle East and India.

Xenocarida

Remipedia

Remipedia (Figure 20.16A) is a very small group within Crustacea. The 10 species described so far have come from caves with connections to the sea. A blind cave-dwellling remipedian was recently identified as the first venomous crustacean. Remipedians have 25 to 38 trunk segments (thorax and abdomen), all bearing paired, biramous, swimming appendages that are essentially alike. Antennules are biramous. Both pairs of maxillae and a pair of maxillipeds, however, are prehensile and apparently adapted for feeding. The shape of the swimming appendages is similar to that of Copepoda, but unlike copepods and cephalocarids, swimming legs are directed laterally rather than ventrally.

Cephalocarida

Cephalocarida (Figure 20.16B) is also a small group, with only nine species known. Cephalocarids occur along both coasts of the United States, in the West Indies, and in Japan. They are 2 to 3 mm long and live in bottom sediments from the intertidal zone to a depth of 300 m. Some features are ancestral: thoracic limbs are very similar to each other, and second maxillae are similar to thoracic limbs.

Cephalocarids have no eyes, carapace, or abdominal appendages. True hermaphrodites, they are unique among Arthropoda in discharging both eggs and sperm through a common duct.

Vericrustacea

Branchiopoda

There are over 10,000 species of Branchiopoda, which represent a crustacean type with some ancestral characters. Three orders are recognized: **Anostraca** (fairy shrimp and brine shrimp, Figure 20.17B), with no carapace; **Notostraca** (tadpole shrimp, Figure 20.17A), whose carapace forms a large dorsal shield; and **Diplostraca** (water fleas, Figure 20.17C), which typically have a carapace that encloses the body but not the head, or a carapace that encloses the entire body (clam shrimps). Branchiopods have flattened and leaflike **phyllopodia**—legs that serve as the chief respiratory organs (hence the name branchiopods). Most branchiopods also use their legs for suspension feeding, and in groups other than cladocerans, they use their legs for locomotion as well.

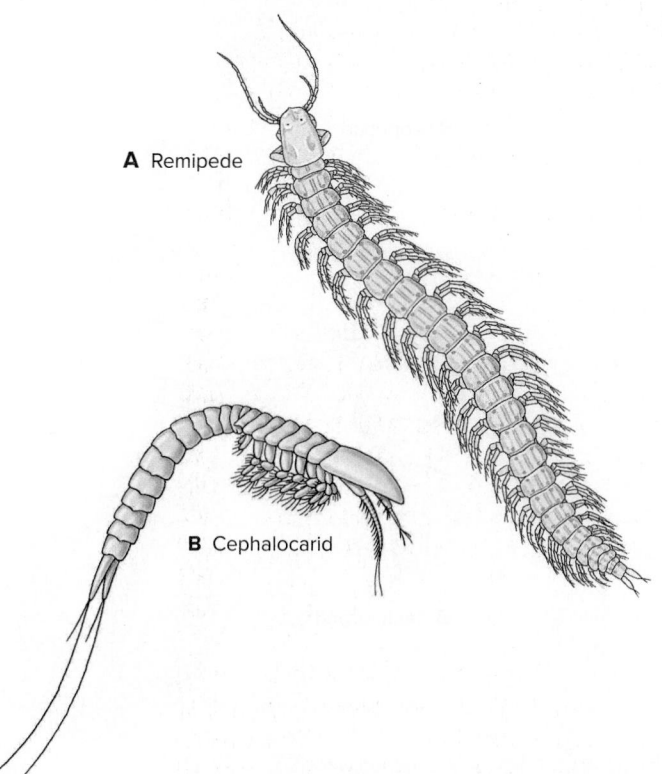

Figure 20.16   **A**, A remipede crustacean of class Remipedia. **B**, A cephalocarid crustacean of class Cephalocarida.

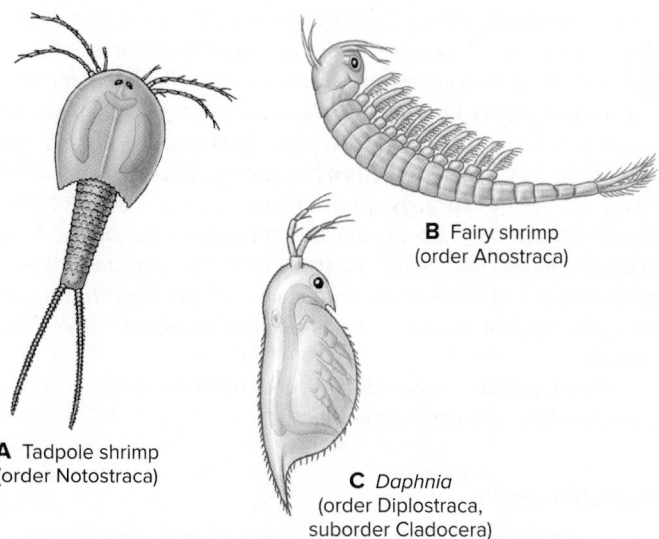

Figure 20.17   Animals in **A, B,** and **C** are members of class Branchiopoda.

Most branchiopods are freshwater forms. Most important and abundant are water fleas (cladocerans), which often form a large proportion of freshwater zooplankton. Their interesting means of reproduction is reminiscent of that of some rotifers (see Chapter 14). During summer cladocerans often produce only females, by parthenogenesis, rapidly increasing the population. With onset of unfavorable conditions, some males are produced, and eggs that must be fertilized are produced by normal meiosis (fertilized eggs that overwinter are called ephipia). Fertilized eggs are highly resistant to cold and desiccation, and they are very important to the survival of the overwintering population and for passive transfer to new habitats. Most cladocerans have direct development, whereas other branchiopods have gradual metamorphosis.

Copepoda

This group is third only to Malacostraca in number of species, and their collective biomass exceeds billions of metric tons throughout marine and freshwaters of the world. Copepods are small (usually a few millimeters or less in length) and rather elongate, tapering toward the posterior. They lack a carapace and retain a simple, median, nauplius (maxillopodan) eye in adults (Figure 20.18A). They have a single pair of uniramous maxillipeds and four pairs of rather flattened, biramous, thoracic swimming appendages. The fifth pair of legs is reduced. The posterior part of the body is usually separated from the anterior, appendage-bearing portion by a major articulation. Antennules are often longer than other appendages and used in swimming. Copepoda have become very diverse and evolutionarily enterprising, with large numbers of symbiotic as well as free-living species. Many parasites are highly modified, and adults may be so highly modified (and may depart so far from the description just given) that they can hardly be recognized as arthropods, let alone crustaceans.

Ecologically, free-living copepods are of extreme importance, often dominating the consumer level (p. 827) in aquatic communities. In many marine localities the copepod *Calanus* is the most abundant organism in zooplankton and has the greatest proportion of total biomass (p. 828). In other localities it may be surpassed in biomass only by euphausids (p. 437). *Calanus* is an important dietary component of such economically and ecologically important fish as herring, menhaden, and sardines. This genus is also important to the larvae of larger fish and (along with euphausids) is an important food item for some whales and sharks that are filter feeders. Other genera commonly occur in marine zooplankton, and some forms such as *Cyclops* and *Diaptomus* may form an important component of freshwater plankton. Many species of copepods are parasites of a wide variety of other marine invertebrates and marine and freshwater fish, and can be of economic importance. Some species of free-living copepods serve as intermediate hosts for parasites of humans, such as *Diphyllobothrium* (a tapeworm) and *Dracunculus* (a nematode), as well as of other animals.

Development in copepods is indirect, and some highly modified parasites show striking metamorphoses.

Tantulocarida

Tantulocarida (Figure 20.18B) is a group of crustaceans described first in 1983. Only about 12 species are known so far. They are tiny (0.15 to 0.2 mm) copepod-like ectoparasites of other deep-sea benthic crustaceans. They have no recognizable head appendages except one pair of antennae on sexual females. Their life cycle is not known with certainty, but present evidence suggests that there is a parthenogenetic cycle and a bisexual cycle with fertilization. Tantulus larvae penetrate the cuticle of their hosts by a mouth tube. Then their abdomen and all thoracic limbs are lost during metamorphosis to an adult. Alone among maxillopodans, juveniles bear six to seven abdominal segments, but other evidence supports inclusion in this group.

Thecostraca (Cirripedia)

Thecostraca includes barnacles (order Thoracica), which are usually enclosed in a shell of calcareous plates, as well as three smaller orders of burrowing or parasitic forms. Barnacles are sessile as adults and may be attached to the substrate by a stalk (gooseneck barnacles) (Figure 20.19B) or directly (acorn barnacles) (Figure 20.19A). Typically their carapace (mantle) surrounds their body and secretes a shell of calcareous plates. The head is reduced, they have no abdomen, and the thoracic legs are long, many-jointed cirri with hairlike setae. The cirri are extended through an opening between the calcareous plates to filter small particles on which the animal feeds (Figure 20.19). Although all barnacles are marine, they are often in the intertidal zone and are therefore exposed to drying and sometimes freshwater for some periods of time. For example, *Semibalanus balanoides* can tolerate below-freezing temperatures in the Arctic tidal zone and can survive exposed on its rocky substrate for up to nine hours in the summer. During these periods the aperture between the plates closes to a very narrow slit.

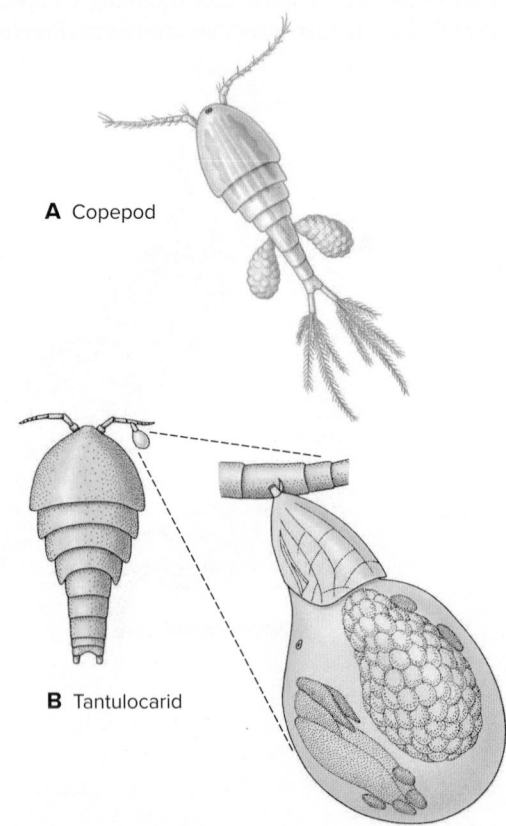

A Copepod

B Tantulocarid

Figure 20.18 A, A copepod with attached eggsacs. **B,** A tantulocarid. This odd little parasite is shown attached to the first antenna of its copepod host at left.

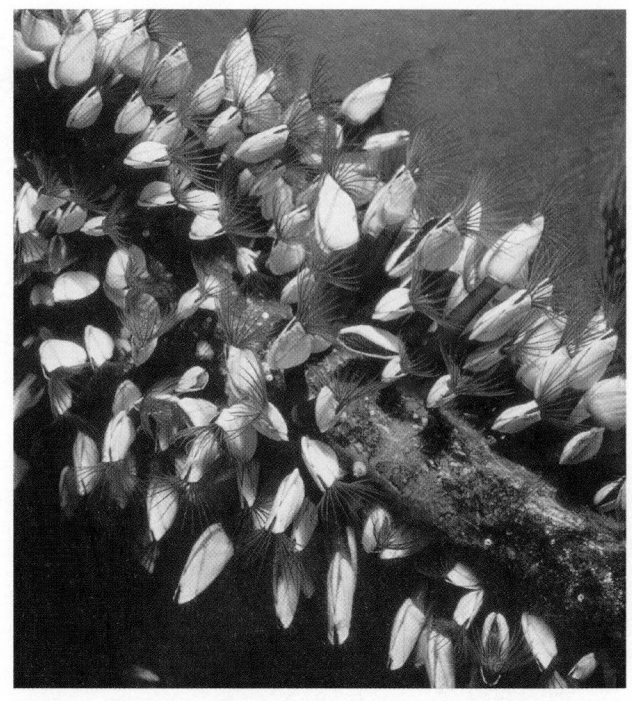

A   **B**

Figure 20.19   Barnacles; order Thoracica, subclass Cirripedia, class Maxillopoda. **A,** Acorn barnacles, *Balanus balanoides,* on an intertidal rock await the return of the tide. **B,** Gooseneck barnacles, *Lepas* sp. Note the feeding legs, or cirri, on *Lepas.* Barnacles attach themselves to a variety of firm substrates, including rocks, pilings, and boat bottoms.

Barnacles frequently foul ship bottoms by settling and growing there. So great may be their number that the speed of the ship may be reduced by 30% to 40%, necessitating drydocking the ship to remove them. They may also live atop whales (see Figure 20.24A).

Most nonparasitic barnacles are hermaphroditic and undergo a striking metamorphosis during development. Most hatch as nauplii, which soon become cyprid larvae, so called because of their resemblance to an ostracod genus *Cypris.* They have a bivalve carapace and compound eyes. Cyprids attach to the substrate by means of their first antennae, which have adhesive glands, and begin their metamorphosis. This involves several dramatic changes, including secretion of calcareous plates, loss of eyes, and transformation of swimming appendages to cirri. Because adult barnacles are sessile and do not resemble typical crustaceans, they were originally classified as molluscs and their body parts were given molluscan names. This is why the tissue inside the barnacle plates is called a "mantle" when this term normally is limited to the molluscan tissue that lines and makes the shell. The discovery of barnacle nauplius larvae led to the correct taxonomy.

Members of order Rhizocephala, such as *Sacculina,* are highly modified parasites of crabs. These barnacles are dioecious. Like other cirripedes, they start life as nauplii and then become cyprid larvae, but when they find a host, females of most species metamorphose into a **kentrogon** (Gr. *kentron,* a point, spine, + *gonos,*

progeny), which injects cells of the parasite into the hemocoel of its crab host (Figure 20.20). Eventually, rootlike absorptive processes grow throughout the crab's body, and the parasite's reproductive structures become externalized between the cephalothorax and reflexed abdomen of the crab. Males at the cyprid stage attach to the external female brood chamber.

The exact position at which reproductive structures become externalized from the crab's body is of great adaptive value for rhizocephalan parasites. Because a crab's egg mass would be borne in this position, a crab treats the parasite as if it were a mass of the crab's own eggs. It protects, ventilates, and grooms its parasite and actually assists in the parasite's reproduction by performing spawning behavior at the appropriate time. The crab's grooming is necessary for continued good health of the parasite. But what if the rhizocephalan's larva is so unlucky as to infect a male crab? During the parasite's internal growth in the male crab, it castrates its host, and the crab behaves like a female in terms of caring for the parasite. Gonads do not develop in parasitized crabs of either gender.

Malacostraca

Malacostraca, with over 20,000 species worldwide, shows great diversity. We confine our coverage to a few of the most important groups. We described the characteristic body plan of malacostracans on page 423.

Seta on crab

Kentrogon larva
penetrates at base
of seta, injects cells

Nauplius larvae
of parasite
released

Web of Sacculina
tissue spreads
throughout
internal body
cavity

Male cyprid
enters
externalized
female

Embryonic cell
mass attaches
to midgut and
begins to grow

External portion
of *Sacculina* on
host crab

Sacculina tissue growing
on the crab midgut
spreads internally to
cover organs

Figure 20.20 Life cycle of *Sacculina* (order Rhizocephala, subclass Cirripedia, class Maxillopoda), parasite of crabs (*Carcinus*).

Order Isopoda Isopods are one of the few crustacean groups to have successfully invaded terrestrial habitats in addition to freshwater and seawater habitats and the only crustaceans to have become truly terrestrial.

They typically are dorsoventrally flattened, lack a carapace, and have sessile compound eyes. Maxillipeds are the first pair of thoracic limbs; other thoracic limbs lack exopods and are similar. Abdominal appendages bear gills or lunglike organs called pseudotracheae and, except for uropods, also are similar to each other (hence the name isopods). Many species have the ability to roll into a tight ball for protection.

Common land forms are the sow bugs, or pill bugs (*Porcellio* and *Armadillidium,* Figure 20.21A), which live under stones and in damp places. Although they are terrestrial, they lack an efficient cuticular covering and other adaptations possessed by insects to conserve water; therefore they must live in moist environments (for example, under wet logs or rocks). *Caecidotea* (Figure 20.21 B) is a common freshwater form found under rocks and among aquatic plants. *Ligia* is a common marine form that scurries across the beach or rocky shore. Some isopods are parasites of fishes (Figure 20.22A) or crustaceans.

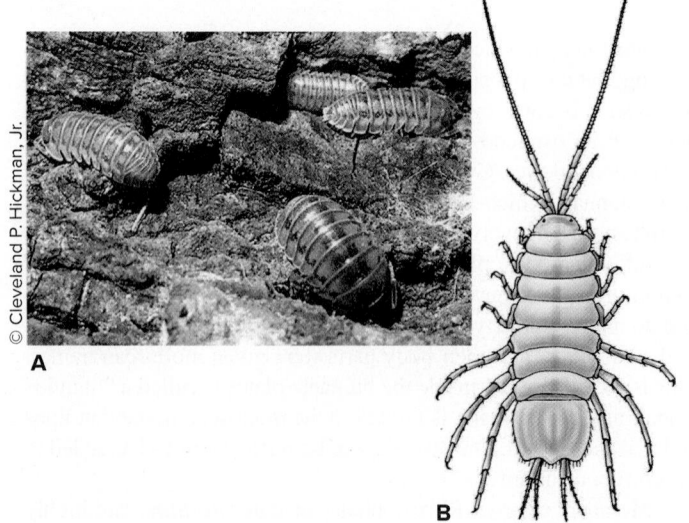

© Cleveland P. Hickman, Jr.

A

B

Figure 20.21 **A,** Four pill bugs, *Armadillidium vulgare* (order Isopoda, Malacostraca), common terrestrial forms. **B,** Freshwater sow bug, *Caecidotea* sp., an aquatic isopod.

Some giant isopods from the deep sea approach 40 cm in length (Figure 20.22B). They live at depths up to 2100 m and are common at 300 to 600 m. They feed on the bodies of dead animals that sink into their habitat or capture smaller prey.

Development is essentially direct but may be highly metamorphic in specialized parasites.

Order Amphipoda   Amphipods resemble isopods in that they lack a carapace, and have sessile compound eyes and only one pair of maxillipeds (Figure 20.23). However, they are usually compressed laterally, and their gills are in the typical thoracic position. Furthermore, their thoracic and abdominal limbs are each arranged in two or more groups that differ in form and function. For example, one group of abdominal legs may be for swimming and another group for jumping. There are many marine amphipods, including some beach-dwelling forms (for example, *Orchestia,* a beach hopper), numerous freshwater genera (*Hyalella* and *Gammarus),* and a few parasites (Figure 20.24B). Development is direct and without a true metamorphosis.

Order Euphausiacea   Euphausiacea is a group of only about 90 species, but they are important as oceanic plankton called "krill" (Figure 20.25). They are about 3 to 6 cm long and have a carapace that is fused with all thoracic segments but does not entirely enclose their gills. They have no maxillipeds, but have thoracic limbs with exopods. Most are bioluminescent, with a light-producing substance in an organ called a **photophore.** Some species may occur in enormous swarms, covering up to 45 m² and extending up to 500 m in one direction. They form a major portion of the diet of baleen whales and many fishes. Eggs hatch as nauplii, and development is indirect and metamorphic.

Order Decapoda   Decapods have three pairs of maxillipeds and five pairs of walking legs. In crabs the first pair of walking legs is modified to form pincers **(chelae),** but the second and third pairs may also be chelate, as in crayfishes, lobsters, and most shrimps. Decapods range in size from a few millimeters to the largest of all

arthropods, Japanese spider crabs, whose chelae span 4 m from tip to tip. Crayfishes, lobsters, crabs, and "true" shrimp belong in this group (Figures 20.26 and 20.27). There are about 18,000 species of decapods, and the order is extremely diverse. They are very important ecologically and economically, and numerous species are relished as food.

Crabs, especially, exist in a great variety of forms. Although resembling crayfishes, they differ from the latter in having a broader cephalothorax and reduced abdomen. Familiar examples along the seashore are hermit crabs (Figure 20.26B), which live in snail shells (because their abdomens are not protected by the same heavy exoskeleton as are the anterior parts); fiddler crabs, *Uca* (Figure 20.26C), which burrow in sand just below the high-tide level and emerge to run over the sand while the tide is out; spider crabs such as *Libinia;* interesting decorator crabs *Dromidia,* and others, which cover their carapaces with sponges and sea anemones for protective camouflage (Figure 20.27).

20.3 PHYLOGENY AND ADAPTIVE DIVERSIFICATION
Phylogeny
Crustacean phylogenies using molecular characters differ considerably from those based on morphological characters. Figure 20.1 depicts an arrangement of crustacean taxa with pentastomids, branchiurans, mystacocarids, and ostracods forming a clade that diverged from other crustaceans at the base of the tree. Morphologists had assumed that members of Remipedia occupied this position. Remipedians have a long body, with no tagmatization behind the head, a double ventral nerve cord, and serially arranged digestive ceca. Fossils of a puzzling arthropod from the Mississippian period seem to be the sister group of remipedians, and their morphology suggested one mechanism for the origin of biramous appendages. They have *two pairs* of uniramous limbs on each segment. Thus, it was suggested that each crustacean segment represents two ancestral segments that fused ("diplopodous condition," as seen in Diplopoda,

A

B

Figure 20.22   **A,** An isopod parasite (*Anilocra* sp.) on a coney (*Cephalopholis fulvus*) inhabiting a Caribbean coral reef (order Isopoda, Malacostraca). **B,** A giant isopod, *Bathynomus giganteus,* from the deep Atlantic Ocean.

© Steve Trewhella/FLPA/Science Source

© NHPA/Science Source

© Andrey Nekrasov/Alamy Stock Photo

Figure 20.23   Marine amphipods. **A,** An amphipod or sandhopper, *Orchestia gammarellus,* from Kimmeridge Bay, England. **B,** A skeleton shrimp, *Caprella linearis,* from the Arctic. **C,** *Phronima,* a pelagic marine amphipod, feeds on the inner portion of a salp (subphylum Urochordata, see Chapter 23), taking over its tunic. The amphipod protrudes its abdominal swimmerets from the barrel-shaped tunic to maneuver its new home through the water and to capture prey. Only the amphipod is shown in this photo.

© Larry S. Roberts

© Larry S. Roberts

Figure 20.24   **A,** Head and mouth of a healthy California grey whale, *Eschrichtius robustus,* bearing its characteristic heavy load of barnacles (Thecostraca) and cyamid parasites (order Amphipoda, Malacostraca) (*arrows*). Note yellowish plates of baleen in mouth (p. 704). **B,** Cyamid parasites of grey whale. Unlike most amphipods, these are dorsoventrally flattened. They have sharp, grasping claws on their legs.

p. 415), and that biramous appendages derived from fusion of both limbs on an ancestral diplopodous segment. However, it is now known that modulation in expression of the *Distal-less (Dll)* gene determines location of distal ends of arthropod limbs. In each primordial (embryonic) biramous appendage, the gene product of *Dll* can be observed in two groups of cells, each of which will become a branch of the limb. In a uniramous limb primordium, there is only one such group of cells, and in primordia of phyllopodous

Figure 20.25 *Meganyctiphanes* (order Euphausiacea, Malacostraca), "northern krill."

Figure 20.27
Sponge crab, *Dromidia antillensis.* This crab is one of several species that deliberately mask themselves with material from their immediate environment (order Decapoda, Malacostraca).

A

B

C

D

E

Figure 20.26 Decapod crustaceans. **A,** A bright orange tropical rock crab, *Grapsus grapsus,* is a conspicuous exception to the rule that most crabs bear cryptic coloration. **B,** A hermit crab, *Dardanus megistos,* which has a soft abdominal exoskeleton, lives in a snail shell into which it can withdraw for protection. **C,** A male fiddler crab, *Uca* sp., waves its enlarged cheliped to make territorial displays and in threat and combat. **D,** A red night shrimp, *Rhynchocinetes rigens,* prowls caves and coral reefs, but only at night. **E,** A spiny lobster, *Panulirus argus* (shown here), and the northern lobster, *Homarus americanus,* are consumed with gusto by many people (order Decapoda, Malacostraca).

limbs (as in class Branchiopoda), there are as many groups expressing *Dll* as there are limb branches. Thus, uniramous limbs are not necessarily ancestral to biramous limbs. Biramous crustaceans occur on both sides of the deepest phylogenetic split within the group, suggesting that the ancestral crustacean was probably biramous.

The wormlike pentastomids were placed in Ecdysozoa near arthropods because their larval form resembles tardigrade larvae, their cuticle is molted, and there are other similarities in sperm morphology and larval appendages. Phylogenies based on sequences of ribosomal RNA genes indicate that pentastomids are crustaceans.

Taxonomy of Subphylum Crustacea

Higher taxonomy of Crustacea is complex and subject to change as new data become available. This listing relies on several sources, omitting many smaller taxa.

Ostracoda (ästrə′kōdə) (Gr. *ostrakodes,* having a shell): **ostracods.** Bivalve carapace entirely encloses body; body unsegmented or indistinctly segmented; no more than two pairs of trunk appendages. Examples: *Cypris, Cypridina, Gigantocypris.*

Mystacocarida (mistäkō-′karədə) (Gr. *mystax,* mustache, + *karis,* shrimp, + *ida,* pl. suffix): **mustache shrimps.** No carapace; body of cephalon and 10-segmented trunk; telson with clawlike caudal rami; cephalic appendages nearly identical, but antennae and mandibles biramous, other head appendages uniramous; second through fifth trunk segments with short, single-segment appendages. Example: *Derocheilocaris.*

Branchiura (bran-′kēyurə) (Gr. *branchia,* gills, + *ura,* tail): **fish lice.** Body oval, head and most of trunk covered by flattened carapace, incompletely fused to first thoracic segment; thorax with four pairs of appendages, biramous; abdomen unsegmented, bilobed; eyes compound; antennae and antennules reduced; maxillules often forming suctoral discs. Examples: *Argulus, Chonopeltis.*

Pentastomida (pen-ta-stōm′i-da) (Gr. *pente,* five, + *stoma,* mouth): **pentastomids.** Wormlike unsegmented body with five short anterior protuberances, four bear claws and the fifth bearing the sucking mouth. Examples: *Armillifer, Linguatula.*

Remipedia (remə-′pədēə) (L. *remipedes,* oar-footed). No carapace; one-segmented protopods; biramous antennules and antennae; all trunk appendages similar; cephalic appendages large and raptorial; maxilliped segment fused to head; trunk unregionalized. Example: *Speleonectes.*

Cephalocarida (se′fə-′lō ka′rədə) (Gr. *kephalē,* head, + *karis,* shrimp, + *ida,* pl. suffix). No carapace; phyllopodia, one-segmented protopods; uniramous antennules and biramous antennae; compound eyes lacking; no abdominal appendages; maxilliped similar to thoracic leg. Example: *Hutchinsoniella.*

Branchiopoda (bran-kēā′pōdə) (Gr. *branchia,* gills, + *pous, podos,* foot). Phyllopodia; carapace present or absent; no maxillipeds; antennules reduced; compound eyes present; no abdominal appendages; maxillae reduced.

    Order Anostraca (ənästrəkə) (Gr. *an-,* prefix meaning without, + *ostrakon,* shell): **fairy shrimp and brine shrimp.** No carapace; no abdominal appendages; uniramous antennae. Examples: Artemia, *Branchinecta.*

    Order Notostraca (nō′tästrəkə) (Gr. *nōtos,* the back, + *ostrakon,* shell): **tadpole shrimp.** Carapace forming large dorsal shield; abdominal appendages present, reduced posteriorly; antennae vestigial. Examples: *Triops, Lepidurus.*

    Order Diplostraca (diplōstrəkə) (Gr. *diploos,* double, + *ostrakon,* shell): **water fleas** (cladocerans) and **clam shrimps** (conchostracans). Carapace folded, usually enclosing trunk but not head (cladocerans) or enclosing entire body (conchostracans); biramous antennae. Examples: *Daphnia, Leptodora, Lynceus.*

Copepoda (kōpe-′pədə) (Gr. *kōpē,* oar, + *pous, podos,* foot): **copepods.** No carapace; thorax typically of seven segments, of which first and sometimes second fuse with head to form cephalothorax; antennules uniramous; antennae bi- or uniramous; four to five pairs swimming legs; parasitic forms often highly modified. Examples: *Cyclops, Diaptomus, Calanus, Ergasilus, Lernaea, Salmincola, Caligus.*

Tantulocarida (tan′tü′lōkarədə) (L. *tantulus,* so small, + *caris,* shrimp). No recognizable cephalic appendages except antennae on sexual female; solid median cephalic stylet; six free thoracic segments, each with pair of appendages, anterior five biramous; six abdominal segments; minute copepod-like ectoparasites. Examples: *Basipodella, Deoterthron.*

Thecostraca (Cirripedia) (sirə-′pēdēə) (L. *cirrus,* curl of hair, + *pes, pedis,* foot): **barnacles.** Sessile or parasitic as adults; head reduced and abdomen rudimentary; paired compound eyes absent; body segmentation indistinct; usually hermaphroditic; in free-living forms carapace becomes mantle, which secretes calcareous plates; antennules become organs of attachment, then disappear. Examples: *Balanus, Policipes, Sacculina.*

Malacostraca (malə-′kä-strəkə) (Gr. *malakos,* soft, + *ostrakon,* shell). Usually with eight segments in thorax and six plus telson in abdomen; all segments with appendages; antennules often biramous; first one to three thoracic appendages often maxillipeds; carapace covering head and part or all of thorax, sometimes absent; gills usually thoracic epipods.

    Order Isopoda (īso-′pōdə) (Gr. *isos,* equal, + *pous, podos,* foot): **isopods.** No carapace; antennules usually uniramous, sometimes vestigial; eyes sessile (not stalked); gills on abdominal appendages; body commonly dorsoventrally flattened; second thoracic appendages usually not prehensile. Examples: *Armadillidium, Caecidotea, Ligia, Porcellio.*

    Order Amphipoda (am-′fi-′pōdə) (Gr. *amphis,* on both sides, + *pous, podos,* foot): **amphipods.** No carapace; antennules often biramous; eyes usually sessile; gills on thoracic coxae; second and third thoracic limbs usually prehensile; typically bilaterally compressed body form. Examples: *Orchestia, Hyalella, Gammarus.*

    Order Euphausiacea (yü-foz-ē-āshēə) (Gr. *eu,* well, + *phausi,* shining bright, + L. *acea,* suffix, pertaining to): **krill.** Carapace fused to all thoracic segments but not entirely enclosing gills, no maxillipeds; all thoracic limbs with exopods. Example: *Meganyctiphanes.*

    Order Decapoda (də′ka-′pōdə) (Gr. *deka,* ten, + *pous, podos,* foot): **shrimps, crabs, lobsters.** All thoracic segments fused with and covered by carapace; eyes on stalks; first three pairs of thoracic appendages modified to maxillipeds. Examples: *Farfantepenaeus* (= *Penaeus*), *Cancer, Pagurus, Grapsus, Homarus, Panulirus.*

A study of gene arrangements and base sequences of mitochondrial DNA confirmed this result. Pentastomids are now considered highly derived crustaceans, placed near fish lice (Branchiura).

Molecular phylogenies sometimes place insects (Hexapoda) within Crustacea, but there is no general agreement as to where they belong within this group. In some cases they are placed near Branchiopoda, but several recent studies place them within Remipedia.

Adaptive Diversification

The level of adaptive diversification demonstrated by the crustaceans is great, with the exploitation of virtually all aquatic resources. They are unquestionably the dominant arthropod group in marine environments, and they share dominance of freshwater habitats with insects. Invasions of terrestrial environments have been much more limited, with isopods being the only notable success. There are a few other terrestrial examples, such as land crabs. The most diverse class is Malacostraca, and the most abundant groups are Copepoda and Ostracoda.

Members of both taxa include planktonic suspension feeders and numerous scavengers. Tiny herbivorous copepods sit at the base of nearly all marine food webs and are critical to ocean ecology. Copepods have been particularly successful as parasites of both vertebrates and invertebrates, and it is clear that present parasitic copepods are products of numerous invasions of such niches.

SUMMARY

Crustacea is a large, primarily aquatic subphylum. In addition to a pair of mandibles, crustaceans have two pairs of antennae and two pairs of maxillae. Their tagmata are a head and trunk or a head, thorax, and abdomen. Many have a carapace. Crustacean appendages are ancestrally biramous.

All arthropods must periodically cast off their old cuticle (ecdysis) and grow in size before the newly secreted cuticle hardens. Premolt and postmolt periods are hormonally controlled, as are several other processes, such as change in body color and expression of sexual characteristics.

Feeding habits vary greatly in Crustacea, and there are many forms of predators, scavengers, suspension feeders, and parasites. Some even obtain nutrition from chemosynthetic bacteria. Respiration is through the body surface or by gills, and excretory organs take the form of maxillary or antennal glands. Circulation, as in other arthropods, is through an open system of sinuses (hemocoel), and a dorsal, tubular heart is the chief pumping organ. Most crustaceans have compound eyes composed of units called ommatidia. Sexes are usually separate.

Branchiopoda is characterized by phyllopodia and contains, among others, order Diplostraca, which is ecologically important as zooplankton. Copepods lack a carapace and abdominal appendages. They are abundant and are among the most important of the primary consumers in many freshwater and marine ecosystems. Many are parasitic. Most members of Thecostraca (barnacles) are sessile as adults, secrete a shell of calcareous plates, and filter-feed by means of their thoracic appendages. Branchiura contains fish lice. Closely related to fish lice are tongue worms (former members of phylum Pentastomida); they are parasitic in the lungs and nasal cavities of vertebrates.

Malacostraca is a diverse and familiar group whose important orders are Isopoda, Amphipoda, Euphausiacea, and Decapoda. All have both abdominal and thoracic appendages. Isopods lack a carapace and are usually dorsoventrally flattened. Amphipods also lack a carapace but are usually laterally flattened. Euphausiaceans are important oceanic plankton called krill. Decapods include crabs, shrimps, lobsters, crayfishes, and others; they have five pairs of walking legs (including chelipeds) on their thorax.

REVIEW QUESTIONS

1. List the appendages on the head of a crustacean. What other important characteristics of Crustacea distinguish them from other arthropods?
2. Define each of the following: tergum, sternum, telson, protopod, exopod, endopod, epipod, endite, and exite.
3. What is meant by homologous structures? What is meant by serial homology, and how do crustaceans show serial homology?
4. What is a carapace?
5. Briefly describe respiration and circulation in crayfishes.
6. Briefly describe the function of antennal and maxillary glands in Crustacea.
7. How does a crayfish detect changes in position?
8. What is the photoreceptor unit of a compound eye? How does this unit adjust to varying amounts of light?
9. What is a nauplius? What is the difference between direct and indirect development in Crustacea?
10. Describe the molting process in Crustacea, including the action of hormones and the process of ecdysis.
11. What characteristics distinguish Branchiopoda, Ostracoda, Copepoda, Thecostraca, and Malacostraca?
12. Explain why remipedians were once thought to resemble ancestral crustaceans.
13. Explain the ecological importance of copepods to marine and freshwater ecosystems.
14. Briefly explain the genetic determination of biramous and uniramous appendages.
15. What is a tongue worm, and where would it be found?

For Further Thought The flexibility inherent in a segmented body plan was clearly important to crustacean evolution, but why are crustaceans so much more diverse than annelids, another segmented taxon?

SELECTED REFERENCES

Bliss, D. E. (editor-in-chief). 1982–1985. The biology of Crustacea, vols. 1–10. New York, Academic Press, Inc. *This series is a standard reference for all aspects of crustacean biology.*

Boore, J. L., D. V. Lavrov, and W. M. Brown. 1998. Gene translocation links insects and crustaceans. Nature **392**: 667–668. *A single mitochondrial gene translocation, indicative of a recent common ancestor, is shared by insects and crustaceans, but not present in chelicerates or myriapods.*

Chang, E. S., S. A. Chang, and E. P. Mulder. 2001. Hormones in the lives of crustaceans: an overview. Am. Zool. **41**: 1090–1097. *Summary of research into hormone function in the American lobster.*

Galant, R., and S. B. Carroll. 2002. Evolution of a transcriptional repression domain in an insect *Hox* protein. Nature **415**: 910–913. *There are levels of a protein (Ultrabithorax, Ubx), encoded by a* Hox *gene, in abdomens of insects, where they repress expression of another gene,* Distal-less (Dll), *which is required for limb information. Crustacean abdomens and onychophorans have high Ubx but can form limbs on their abdomen, showing that Ubx is a conditional repressor in those groups.*

Giribet, G., G. D. Edgecombe, and W. C. Wheeler. 2001. Arthropod phylogeny based on eight molecular loci and morphology. Nature **413:** 157–161. *Support for Crustacea and Insecta as sister taxa in a mandibulate clade.*

Gould, S. J. 1996. Triumph of the root-heads. Nat. Hist. **105:** 10–17. *An informative essay on parasite-host coevolution using* Sacculina *as an example.*

Huys, R., G. A. Boxhall, and R. J. Lincoln. 1993. The tantulocaridan life cycle: the circle closed? J. Crust. Biol. **13:** 432–442. *The current hypothesis of a parthenogenetic cycle alternating with a cycle that includes fertilization in these bizarre little creatures.*

Laufer, H., and W. J. Biggers. 2001. Unifying concepts learned from methyl farnesoate for invertebrate reproduction and postembryonic development. Am. Zool. **41:** 442–457. *In crustaceans, methyl farnesoate has the same role in development that juvenile hormone has in insects.*

Lavrov, D. L., W. M. Brown, and J. L. Boore. 2004. Phylogenetic position of Pentastomida and (pan)crustacean relationships. Proc. R. Soc. Lond. Ser. B. **271:** 537–544. *Pentastomids are maxillopod crustaceans, probably closely related to fish lice.*

Panganiban, G., A. Sebring, L. Nagy, and S. Carroll. 1995. The development of crustacean limbs and the evolution of arthropods. Science **270:** 1363–1366. *Probing for particular homeotic gene products suggests that all arthropods derive from a common ancestor and that biramous and uniramous limbs derive from modulation of* Distalless (Dll) *gene expression.*

Regier, J. C., J. W. Shultz, A. Zwick, A. Hussey, B. Ball, R. Wetzer, J. W. Martin, and C. W. Cunningham. 2011. Arthropod relationships revealed by phylogenomic analysis of nuclear protein-coding sequences. Nature **463** :1079–1083. *A crustacean phylogeny based on molecular characters does not show* Remipedia *as a basal form.*

Thatje, S., L. Marsh, C. N. Roterman, M. N. Mavrogordato, and K. Linse. 2015. Adaptations to hydrothermal vent life in *Kiwa tyleri,* a new species of Yeti crab from the East Scotia Ridge, Antarctica. PLos One **10**: e0127621. doi: 10.1371/journal. *A new addition to the chemoautotrophic animals.*

Versluis, M., B. Schmitz, A. von der Heydt, and D. Lohse. 2000. How snapping shrimp snap: through captivating bubbles. Science **289:** 2114–2117. *Snapping of their chela is strong enough to cause cavitation bubbles. Implosion of the bubbles stuns prey.*

von Reumont, B. M., A. Blanke, S. Richter, F. Alvarez, C. Bleidorn, and R. A. Jenner. 2014. The first venomous crustacean revealed by transcriptomics and functional morphology: Remipede venom glands express a unique toxin cocktail dominated by enzymes and a neurotoxin. Mol. Biol. Evol. **31**: 45–58. *Venomous species are now found in all major branches of arthropods.*

Zill, S. N., and E.-A. Seyfarth. 1996. Exoskeletal sensors for walking. Sci. Am. **275:** 86–90 (July). *Cockroaches, crabs, and spiders have sensors in the exoskeleton of their legs that act as biological strain gauges.*

21
CHAPTER

Hexapods

Most animal species are insects.

© Helen I'Anson

- **PHYLUM ARTHROPODA**
  - Subphylum Hexapoda

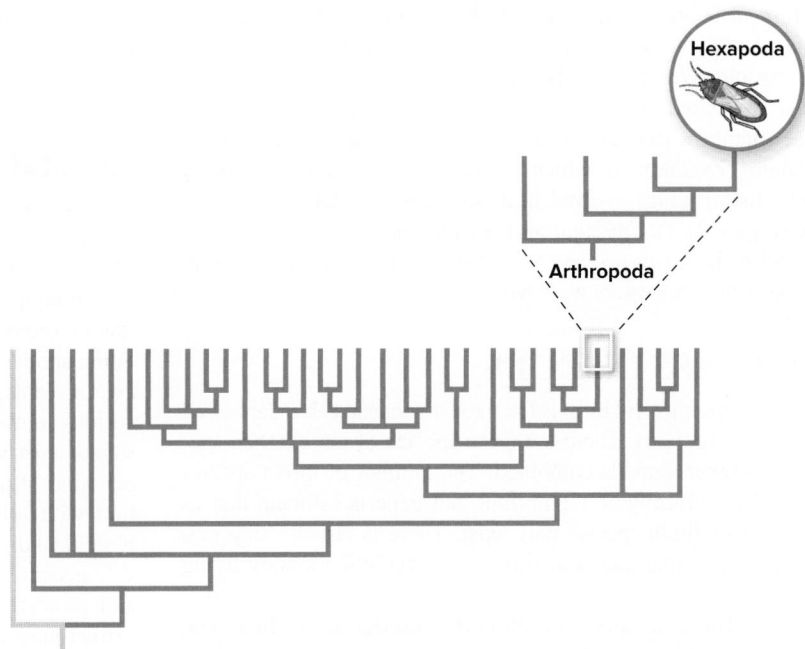

Insect Outbreaks

Humans suffer staggering economic losses to insects. Locust outbreaks in Africa seem a thing of the past to many today, but this is far from true. Locust populations fluctuate between quiet phases, where they cover only 16 million square kilometers in 30 African countries, and plague phases, where they cover 29 million square kilometers of land in 60 countries. A swarm of locusts, *Schistocerca gregaria,* contains 40 to 80 million insects per square kilometer. In peak phases, they cover 20% of the earth's land surface and affect the livelihood of one-tenth of the earth's population. The last plague phase was 1986 to 1989, but the Food and Agriculture Organization (FAO) of the United Nations monitors and maps population sizes continuously to respond quickly to outbreaks (http://www.fao.org/ag/locusts/en/info/info/faq/index.html).

In the western United States and Canada, outbreaks of mountain pine beetles in the 1980s and 1990s killed pines on huge acreages, and an ongoing outbreak has harmed forests in Wyoming and Colorado. Recently spruce trees in the same area have been damaged by a new outbreak of the spruce budworm that killed millions of conifers in the 1973 to 1985 outbreak. Since its introduction in the 1920s, a fungus that causes Dutch elm disease, mainly transmitted by European bark beetles, has virtually obliterated American elm trees in North America. Since 2004, another alien invader, the emerald ash borer, a beetle whose larvae consume the wood of maples and other trees, threatens the continent's ash trees. Recently, the equally destructive Asian Longhorn beetle established populations in Massachusetts forests.

Insects far outnumber all other species of animals in the world combined, and numbers of individuals are equally enormous. Some scientists have estimated that there are 200 million insects for every single human alive today! Fortunately, many of these insects are beneficial, including those that consume harmful insects. Ladybug beetles consume aphids and can prevent garden outbreaks. Bees and moths pollinate many flowers including important crop plants. Many birds and other wildlife depend upon insects for food. Despite the potential damage from outbreaks, our world would be unrecognizable without insects.

Subphylum Hexapoda is named for the presence of **six legs** in members of the group. All legs are **uniramous.** Hexapods have **three tagmata**—head, thorax, and abdomen—with appendages on the head and thorax. Abdominal appendages are greatly reduced or absent. There are two classes within Hexapoda: Entognatha and Insecta (Figure 21.1).

Entognatha is a small group whose members have the bases of mouthparts enclosed within the head capsule. There are three orders of entognathans. Members of Protura and Diplura are tiny, are eyeless, and inhabit soils or dark, damp places where they are rarely noticed. Members of Collembola are commonly called springtails because of their ability to leap; an animal 4 mm long may leap 20 times its body length. Collembolans live in soil, in decaying plant matter, on freshwater pond surfaces, and along the seashore. They can be very abundant, reaching millions per hectare in some soils, but like other entognathans, their small size makes them less visible to the casual observer.

Insecta is an enormous class whose members have ectognathous mouthparts, where the bases of mouthparts lie outside the head capsule. Winged insects are called pterygotes, and wingless insects are called apterygotes. Class Insecta contains one group whose members diverged from ancestors of the wingless order Thysanura (Zygentoma), which forms the sister taxon to all other insects. Insect wings evolved in a common ancestor of the latter clade (Figure 21.1). Thysanurans are called primitively wingless to distinguish them from orders whose members do not have wings now, but whose ancestors were winged.

21.1 CLASS INSECTA

Insecta (L. *insectus,* cut into) is the most diverse and abundant of all groups of arthropods. There are more species of insects than species of all other animals combined. The number of insect species classified is currently at 1.1 million, but experts estimate that as many as 30 million species may exist. There is also striking evidence of continuing and sometimes rapid evolution among living insects.

It is difficult to appreciate fully the ecological, medical, and economic significance of this extensive group. The study of insects **(entomology)** occupies the time and resources of skilled men and women all over the world. The struggle between humans and their insect competitors seems to be endless, yet paradoxically insects have so interwoven themselves into the economy of nature in so many useful roles that most terrestrial ecosystems would collapse without them. In particular, the role of bees as pollinators of world food crops cannot be overestimated (see p. 462).

Insects differ from other arthropods in having ectognathous mouthparts and usually **two pairs of wings** on the thoracic region of the body, although some have one pair of wings or none. Insects range in size from less than 1 mm to 20 cm in length, the majority being less than 2.5 cm long. Some of the largest insects live in tropical areas.

Distribution

Insects are among the most abundant and widespread of all land animals. They have spread into practically all habitats that sup-

port life except the sea. Relatively few are truly marine, but some are common in intertidal zones. Marine water striders (*Halobates*), which live on the *surface* of the ocean, are the only marine invertebrates that live on the sea-air interface. Insects are common in brackish water, in salt marshes, and on sandy beaches. They are abundant in freshwater, in soil, in forests (especially the tropical forest canopy), and they are found even in deserts and wastelands, on mountaintops, and as parasites in and on plants and animals.

Their wide distribution is made possible by their powers of flight and their highly adaptable nature. Insects evolved wings and invaded the air 250 million years before flying reptiles, birds, or mammals. In most cases they can easily surmount barriers that are virtually impassable to many other animals. Their small size allows them to be carried by currents of both wind and water to far regions. Their well-protected eggs can withstand rigorous conditions and can be carried long distances by birds and other animals. Their agility and ecological aggressiveness enable them to occupy every possible niche in a habitat. No single pattern of biological adaptation can be applied to them.

Adaptability

Insects have shown an amazing evolutionary adaptability, as evidenced by their wide distribution and enormous diversity of species. Most of their structural modifications are in their wings, legs, antennae, mouthparts, and alimentary canals. Such wide diversity enables this vigorous group to use all available food and shelter resources. Some are parasitic, some suck the sap of plants, some chew the foliage of plants, some are predaceous, and some live on the blood of various animals. Within these different groups, specialization occurs, so that a particular kind of insect will eat, for instance, leaves of only one kind of plant. This specificity of eating habits lessens competition with other species and to a great extent accounts for their biological diversity.

Insects are well adapted to dry and desert regions. Their hard and protective exoskeleton limits evaporation. Some insects also extract most of the water from food, fecal material, and by-products of cell metabolism.

External Form and Function

Insects show a remarkable variety of morphological characteristics, but, as in other arthropods, the exoskeleton is formed of a complex system of plates called **sclerites,** connected by concealed, flexible hinge joints. Muscles between sclerites enable insects to make precise movements. Rigidity of their exoskeleton is attributable to unique scleroproteins and not to its chitin component. It is waterproof and its lightness makes flying possible. By contrast, the cuticle of crustaceans is stiffened mostly by minerals.

Insects are much more homogeneous in tagmatization than are Crustacea. Insect tagmata comprise a head, thorax, and abdomen. The cuticle of each body segment typically is composed of four plates (sclerites): a dorsal notum (tergum), a ventral sternum, and a pair of lateral pleura. Pleura of abdominal segments are often partially membranous rather than sclerotized. Some insects

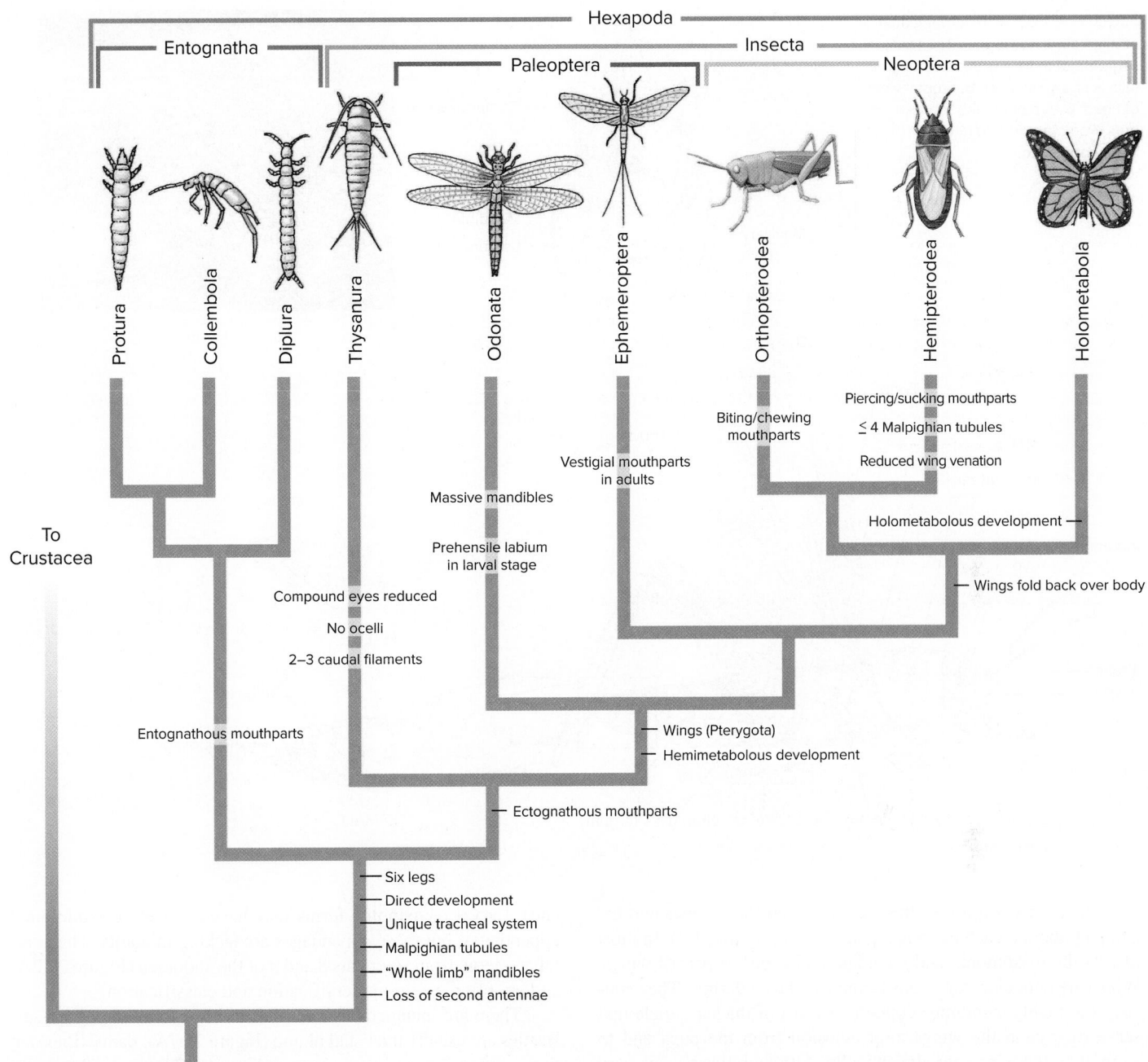

Figure 21.1 Cladogram showing hypothetical relationships among hexapods. Many synapomorphies have been omitted. Orders Protura, Collembola, and Diplura are entognathous. These orders, plus Thysanura, originated before the earliest winged ancestors. Orders Odonata and Ephemeroptera form Paleoptera, where wings are outspread. The remaining orders have wings that can fold back over the abdomen (Neoptera). Superorder Orthopterodea includes orders Orthoptera, Blattodea, Phasmatodea, Mantodea, Isoptera, Plecoptera, Embiidina, and Dermaptera. Hemipterodea includes orders Zoraptera, Psocoptera, Hemiptera, Thysanoptera, and Phthiraptera; and superorder Holometabola encompasses all holometabolous orders. Somewhat different relationships have been proposed from a phygogenomic data set (see Misof et. al., 2014; reference on p. 470).

are fairly generalized in body structure; some are highly special-ized. Grasshoppers, or locusts, are a generalized type often used in laboratories to demonstrate general features of insects (Figure 21.2).

The head usually bears a pair of relatively large compound eyes, a pair of antennae, and typically three ocelli (Figure 21.2). Antennae, which vary greatly in size and form (Figure 21.3),

act as tactile organs, as olfactory organs, and in some cases as auditory organs. Mouthparts, formed from specially hard-ened cuticle, typically consist of a labrum, a pair each of mandi-bles and maxillae, a labium, and a tonguelike hypopharynx. The type of mouthparts that an insect possesses determines how it feeds. We discuss some of these modifications later in this chapter.

Figure 21.2 **A,** External features of a female grasshopper. The terminal segment of a male with external genitalia is shown in inset. **B,** Frontal view of head. **C,** A pair of lubber grasshoppers, *Romalea guttata* (order Orthoptera), copulating.

The thorax comprises three segments: prothorax, mesothorax, and metathorax, each bearing a pair of legs (Figure 21.2). In most insects the mesothorax and metathorax also bear a pair of wings. Wings are cuticular extensions formed by the epidermis. They consist of a double membrane containing veins of thicker cuticle that serve to expand the wings after eclosion from the pupa and to strengthen the wings aerodynamically. Although these veins vary in their patterns among different taxa, they are relatively constant within a family, genus, or species and serve as one means of classification and identification.

Legs of insects often are modified for special purposes. Many terrestrial forms have walking legs with terminal pads and claws. These pads may be sticky for walking upside down, as in house flies. The hindlegs of grasshoppers and crickets are adapted for jumping (Figure 21.4). Mole crickets have the first pair of legs modified for burrowing in the ground. Water bugs and many beetles have paddle-shaped appendages for swimming. For grasping prey, the forelegs of a praying mantis are long and strong (Figure 21.5). The legs of honey bees show complex adaptations for collecting pollen (Figure 21.6).

The insect abdomen comprises 9 to 11 segments; the eleventh, when present, bears a pair of **cerci** (appendages at the posterior end). Larval or nymphal forms may have a variety of abdominal appendages, but these appendages are lacking in adults. The genitalia emerge from segments 8 and 9 of the abdomen (Figure 21.2A) and are often useful in identification and classification.

There are innumerable variations in body form among insects. Beetles are usually thick and plump (Figure 21.7A); damselflies, ant lions, and walkingsticks are long and slender (Figure 21.7B); many aquatic beetles are streamlined; butterflies have the broadest wings of all; and cockroaches are flat, adapted to living in crevices. The ovipositor of female ichneumon wasps is extremely long (Figure 21.8), whereas the anal cerci form forceps in earwigs but are long and many-jointed in stoneflies and mayflies. The hymenopteran stinger is a modified ovipositor. Antennae are long in cockroaches and katydids, short in dragonflies and most beetles, knobbed in butterflies, and plumed in some moths. Many other dramatic variations exist (Figure 21.3). Perhaps most amazing is the fact that mouthparts, antennae, legs, cerci, and ovipositors are all modified appendages.

Locomotion

Walking   When walking, most insects use a triangle of legs involving the first and last leg of one side together with the middle leg of the opposite side. In this way, terrestrial insects keep at least three of their six legs on the ground at all times—a tripod arrangement that bestows stability.

Some insects, such as water striders, *Gerris* (L. *gero,* to carry), can walk on the surface of water. A water strider has on its footpads nonwetting hairs that do not break the surface film of water but merely indent it. As it skates along on its two pairs of posterior legs, *Gerris* uses its reduced and toothed pair of prothoracic legs to capture and hold prey. Water striders exhibit unusual cleaning behavior and may do complete flips on the water surface in an attempt to dislodge debris from their thoracic terga (Figure 21.9). Bodies of marine water striders, *Halobates* (Gr. *halos,* the sea, + *bātes,* one that treads), excellent surfers on rough ocean waves, are further protected by a water-repellent coat of close-set hairs shaped like thick hooks.

Power of Flight Insects are the only invertebrates that can fly, and they share the power of flight with birds and flying mammals. However, their wings evolved in a manner different from limb buds of birds and mammals and are not homologous to them. Insect wings are formed by outgrowths from the body wall of the mesothoracic and metathoracic segments and are composed of cuticle. Fossil evidence suggests that insects may have evolved fully functional wings over 400 million years ago.

Most insects have two pairs of wings, but Diptera (true flies) have only one pair (Figure 21.10), the hindwings being represented by a pair of tiny **halteres** (balancers) that vibrate and are responsible for equilibrium during flight. Males of order Strepsiptera have only a hind pair of wings and an anterior pair of "halteres." Males of scale insects also have one pair of wings but no halteres. Some insects are ancestrally (for example, silverfish) or secondarily (for example, fleas) wingless. Reproductive female ants shed their wings after their nuptial flight (males die), and reproductive male and female termites have wings, but workers in both cases are wingless. Lice and fleas are always wingless.

Wings may be thin and membranous, as in flies and many other groups (Figure 21.10); thick and hard, such as the front wings of beetles (Figure 21.7A); parchmentlike, such as the front wings of grasshoppers; covered with fine scales, as in butterflies and moths; or covered with hairs, as in caddisflies.

Wing movements are controlled by a complex of muscles in the thorax. **Direct flight muscles** are attached to a part of the wing itself. **Indirect flight muscles** are not attached to the wing and cause wing movement by altering the shape of the thorax. The wing is hinged at the thoracic tergum and also slightly laterally on a pleural process, which acts as a fulcrum (Figure 21.11). In most insects, the upstroke of the wing is effected by contracting indirect muscles that pull the tergum down toward the sternum (Figure 21.11A). Dragonflies and cockroaches accomplish the downstroke by contracting direct muscles attached to the wings lateral to the pleural fulcrum. In Hymenoptera and Diptera (see pp. 466–467) all major flight muscles are indirect. The downstroke occurs when the sternotergal muscles (muscles inserted on sternum and tergum) relax and longitudinal muscles of the thorax contract and arch the tergum (Figure 21.11B), pulling up the tergal articulations relative to the pleura. The downstroke in beetles and grasshoppers involves both direct and indirect muscles.

Insectan flight-muscle contraction has two basic types of neural control: **synchronous** and **asynchronous.** Larger insects such as

Silkworm larva   Silkworm moth

Japanese beetle   Checkered beetle

Yellow jacket   Syrphid fly

Figure 21.3 A few types of insect antennae.

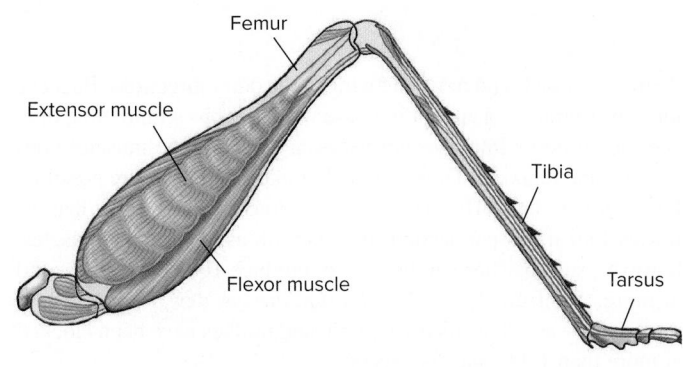

Femur

Extensor muscle

Flexor muscle

Tibia

Tarsus

Figure 21.4 Hindleg of grasshopper. Muscles that operate the leg are found within a hollow cylinder of exoskeleton. Here they are attached to the internal wall, from which they manipulate segments of limb on the principle of a lever. Note pivot joint and attachment of tendons of extensor and flexor muscles, which act reciprocally to extend and flex the limb.

dragonflies and butterflies have wings with synchronous muscles, in which a single nerve impulse stimulates a muscle contraction and thus one wing stroke. Wings with asynchronous muscles occur in Hymenoptera, Diptera, Coleoptera, and some Hemiptera (see pp. 466–467). Their mechanism of action is complex and depends on the storage of potential energy in resilient parts of the thoracic cuticle. As one set of muscles contracts (moving the wing in one direction), they also stretch the antagonistic set of muscles, causing

A

B

Figure 21.5 **A**, Praying mantis (order Mantodea) feding on an insect. **B**, Praying mantis laying eggs.

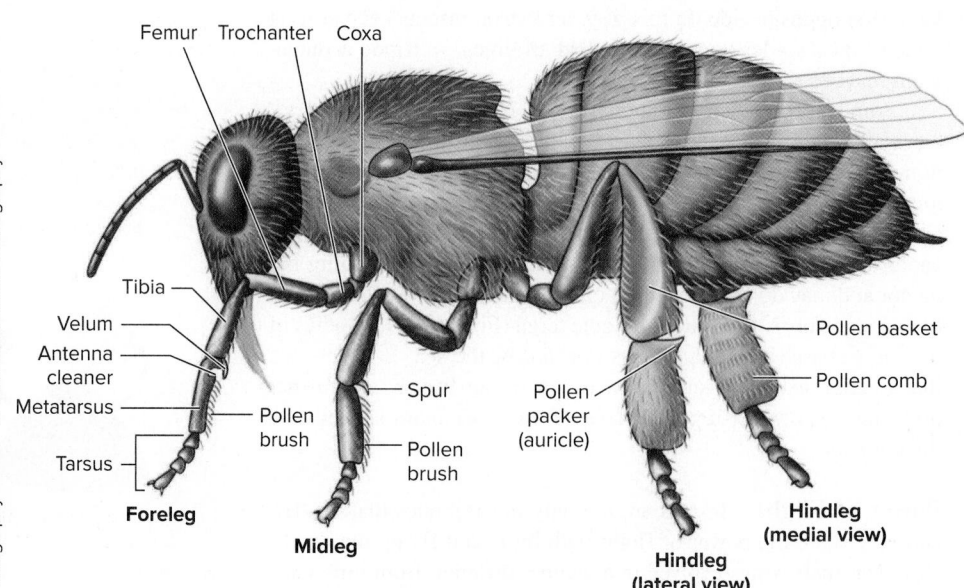

Figure 21.6 Adaptive legs of worker honey bee. In the foreleg, the toothed indentation covered with the velum combs out the antennae. The spur on the middle leg removes wax from wax glands on the abdomen. Pollen brushes on the front and middle legs comb off pollen picked up on body hairs and deposit it on the pollen brushes of the hindlegs. Long hairs of the pollen comb on the hindleg remove pollen from the comb of the opposite leg; then the auricle (pollen packer) presses it into a pollen basket when the leg joint is flexed back. A bee carries her load in both baskets to the hive and pushes pollen into a cell, to be cared for by other workers.

them to contract (and move the wing in the other direction). Because muscle contractions are not in phase with nervous stimulation, only occasional nerve impulses are necessary to keep the muscles contracting and relaxing. Thus extremely rapid wing beats are possible. For example, butterflies (with synchronous muscles) may beat as few as four times per second. Insects with asynchronous muscles, however, such as flies and bees, may produce 100 beats per second or more. The fruit fly, *Drosophila* (Gr. *drosos*, dew, + *philos*, loving), can fly at 300 beats per second, and midges have been clocked at more than 1000 beats per second.

Obviously flying entails more than a simple flapping of wings; a forward thrust is necessary. As indirect flight muscles alternate rhythmically to raise and to lower the wings, direct flight muscles alter the angle of the wings so that they act as lifting airfoils during both upstroke and downstroke, twisting the leading edge of the wings downward during the downstroke and upward during the upstroke. A figure-eight movement (Figure 21.11C) results, spilling air from the trailing edges of the wings. The quality of forward thrust depends, of course, on several factors, such as variations in wing venation, wing load (grams of body weight divided by total wing area), how much the wings are tilted, and wing length and shape.

Flight speeds vary tremendously. The fastest flyers usually have narrow, fast-moving wings with a strong tilt and a strong figure-eight component. Sphinx moths and horse flies achieve approximately 48 km (30 miles) per hour and dragonflies approximately 40 km (25 miles) per hour. Some insects are capable of long, continuous

A

B

Figure 21.7

A, A stag beetle on a branch. Though the processes from the head and thorax might appear to be for pinching or stabbing an opponent, they actually are used to lift or pry up a rival of the same species away from resources. **B**, A walkingstick in genus *Diapheromera*. Walkingsticks are common in much of North America. They are wingless, and despite their camouflage as a twig, they are eaten by numerous predators.

flights. Migrating monarch butterflies, *Danaus plexippus* (Gr. After Danaus, mythical king of Arabia) (see Figure 21.27A), travel hundreds to thousands of miles south in the fall, flying at a speed of approximately 10 km (6 miles) per hour, to reach their overwintering roosts in Mexico and California.

Internal Form and Function

Nutrition

The digestive system (Figure 21.12; see also Figure 32.9, p. 708) consists of a foregut (mouth with salivary glands, esophagus, crop for storage, and gizzard for grinding in some); a midgut (stomach and gastric ceca); and a hindgut (intestine, rectum, and anus). Some digestion may occur in the crop as food mixes with enzymes

from the saliva, but no absorption occurs there. The main site for digestion and absorption is the midgut, and the ceca may increase the digestive and absorptive area. Little absorption of nutrients occurs in the hindgut (with certain exceptions, such as wood-eating termites), but this is a major area for resorption of water and some ions (see p. 454).

Most insects feed on plant juices and plant tissues (**phytophagous** or **herbivorous**). Some insects feed on specific plants; others, such as grasshoppers, eat almost any plant. Caterpillars of many moths and butterflies eat foliage of only certain plants. Certain species of ants and termites cultivate fungus gardens as a source of food.

Many beetles and larvae of many insects live on dead animals (**saprophagous**). Some insects are **predaceous,** catching and eating other insects as well as other types of animals (see Figure 21.5). However, the so-called predaceous diving beetle, *Cybister fimbriolatus* (Gr. *kybistēr,* diver), is not as predaceous as once supposed, but is largely a scavenger.

Many insects are **parasitic** as adults or as larvae; in some cases, both juveniles and adults are parasites. For example, fleas (Figure 21.13) live on blood of mammals as adults, but their larvae are free-living scavengers. Lice (Figures 21.14 and 21.15) are parasitic throughout their life cycle. Many parasitic insects are themselves parasitized by other insects, a condition called **hyperparasitism.** Larvae of many varieties of wasps live and complete much of their metamorphosis inside the bodies of spiders or other insects (Figure 21.16), consuming their hosts and eventually killing them. Because they always kill their hosts, they are known as **parasitoids** (a lethal type of parasite). Parasitoid insects are enormously important in controlling populations of other insects.

For each type of feeding, mouthparts are adapted in a specialized way. **Sucking mouthparts** usually form a tube and can easily pierce the tissues of plants or animals. Mosquitos (order Diptera) demonstrate this arrangement well. Their mandibles, maxillae, hypopharynx, and labrum-epipharynx are elongated into needlelike stylets, together forming a **fascicle** (Figure 21.17C), which pierces the skin of

Figure 21.8 An ichneumon wasp with the end of the abdomen raised to thrust her long ovipositor into wood to find a tunnel made by the larva of a wood wasp or wood-boring beetle. She can bore 13 mm or more into the wood to lay her eggs in the larva of a wood-boring beetle, which will become host for the ichneumon larvae. Other ichneumon species attack spiders, moths, flies, crickets, caterpillars, and other insects.

© Cleveland P. Hickman, Jr.

NPS Photo by Rosalie LaRue

Figure 21.9 Water strider, *Gerris* sp. (order Hemiptera). The animal is supported on its long, slender legs by the water's surface tension.

© INSADCO Photography/Alamy Stock Photo

Figure 21.10 House fly, *Musca domestica* (order Diptera). House flies can become contaminated with over 100 human pathogens, which may be transferred to human and animal food by direct contact, regurgitated food, and feces.

Figure 21.11 **A,** Flight muscles of insects such as cockroaches, in which upstroke is by indirect muscles and downstroke is by direct muscles. **B,** In insects such as flies and bees, both upstroke and downstroke are by indirect muscles. **C,** The figure-eight path followed by the wing of a flying insect during the upstroke and downstroke.

their prey to enter a blood vessel. The hypopharynx bears a salivary duct, and the labrum-epipharynx forms a food channel. The labrum forms a sheath for the fascicle that bends back during feeding (Figure 21.17C). In honey bees the labium forms a flexible and contractile "tongue" covered with many hairs. When a bee plunges its proboscis into nectar, the tip of the tongue bends upward and moves back and forth rapidly. Liquid enters the tube by capillary action and is drawn inside continuously by a pumping pharynx. In adult butterflies and moths, mandibles are usually absent (they are always present in larvae), and the maxillae form a long sucking proboscis (Figure 21.17D) for drawing nectar from flowers. At rest the proboscis coils into a flat spiral. While feeding the proboscis extends and fluid is pumped inside by pharyngeal muscles.

House flies, blow flies, and fruit flies have **sponging** and **lapping mouthparts** (Figure 21.17E). At the apex of the labium is a pair of large, soft lobes with grooves on the lower surface that serve as food channels. These flies lap liquid food or liquefy food first with salivary secretions. Horse flies not only sponge surface liquids but pierce the skin with slender, tapering mandibles, and then absorb blood.

Chewing mouthparts such as those of grasshoppers and many other herbivorous insects are adapted for seizing and crushing food (Figure 21.17A); those of most carnivorous insects are sharp and pointed for piercing their prey. Mandibles of chewing insects are strong, toothed plates whose edges can bite or tear while the maxillae hold the food and pass it toward the mouth. Enzymes secreted by the salivary glands provide chemical action to aid the chewing process.

Circulation

A tubular heart creates a peristaltic wave (Figure 21.12) that moves hemolymph (blood) forward through the only blood vessel, a dorsal aorta. Accessory pulsatory organs help move hemolymph into the wings and legs, and flow is also facilitated by body movements. The heart is part of an open circulatory system (see Chapter 31); it possesses

ostia through which blood is returned from the hemolymph. Hemolymph consists of plasma and amebocytes and apparently has little role in oxygen transport in most insects, but hemoglobin occurs in the hemolymph of some species (especially aquatic immatures occupying environments of low oxygen tension) and functions in oxygen transport.

Gas Exchange

Terrestrial animals require efficient respiratory systems that permit rapid oxygen and carbon dioxide exchange but at the same time restrict water loss. In insects this is the function of the **tracheal system,** an extensive network of thin-walled tubes that branch into every part of the body (Figure 21.18). The tracheal system of insects evolved independently of that of other arthropodan groups such as spiders. The tracheal trunks open to the outside by **spiracles,** usually two pairs on the thorax and seven or eight pairs on the abdomen. A spiracle may be merely a hole in the integument, as in primitively wingless insects, but there is usually a valve or some other closing mechanism that reduces water loss. The evolution of a tracheal system with valves must have been very important in enabling insects to move into drier habitats. Spiracles may also possess a filtering device such as a sieve plate or a set of interlocking bristles that prevents the entrance of water, parasites, or dust into the tracheae.

Tracheae are composed of a single layer of cells and are lined with cuticle that is shed during molts along with the outer cuticle. Spiral thickenings of cuticle (called **taenidia**) support the tracheae and prevent their collapse. Tracheae branch out into smaller tubes, ending in very fine, fluid-filled tubules called **tracheoles** (lined with cuticle, but not shed at ecdysis), which branch into a fine network over the cells (see Figure 31.19). Large insects may have tracheae several millimeters in diameter that taper down to 1 to 2 μm. Tracheoles then taper to 0.5 to 0.1 μm in diameter. In one stage of silkworm larvae it is estimated that there are 1.5 million tracheoles! Some lepidopterous (moth and butterfly) larvae have an abdominal mass of tracheoles that forms the structural and physiological equivalent of a

Figure 21.12   Internal structure of female grasshopper.

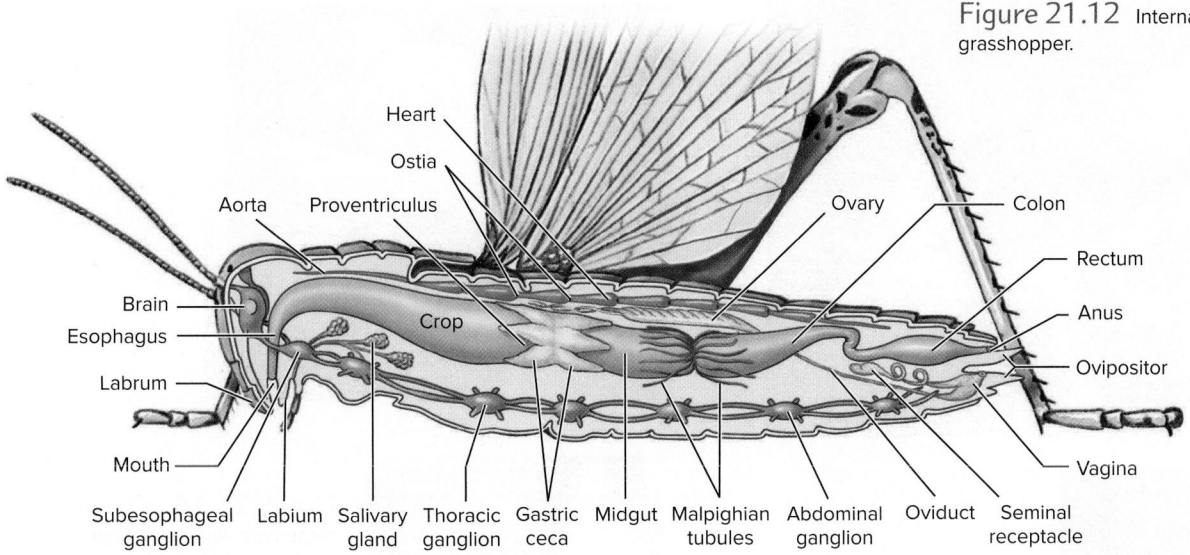

The labels in the figure are: Heart, Ostia, Aorta, Proventriculus, Ovary, Colon, Rectum, Brain, Esophagus, Crop, Anus, Labrum, Ovipositor, Mouth, Vagina, Subesophageal ganglion, Labium, Salivary gland, Thoracic ganglion, Gastric ceca, Midgut, Malpighian tubules, Abdominal ganglion, Oviduct, Seminal receptacle

Figure 21.13   Female human flea, *Pulex irritans* (order Siphonaptera).

© E. Pollard/PhotoLink/Getty Images

© Dennis Kunkel Microscopy, Inc.

Figure 21.14   *Linognathus vituli,* the long-nosed cattle louse, attaches to hairs. Five egg cases, cemented to hairs, are also visible. Larvae are emerging from two egg cases.

vertebrate lung. Scarcely any living cell is more than a few micrometers away from a tracheole. In fact, the ends of some tracheoles actually indent the membranes of cells that they supply, so that they terminate close to mitochondria. The tracheal system affords efficient transport usually without use of oxygen-carrying pigments in hemolymph, although hemoglobin is present in some.

The tracheal system may also include **air sacs,** which are apparently dilated tracheae without taenidia (Figure 21.18A). They are thin walled and flexible and are most common in the body cavity, although they sometimes occur in appendages. Air sacs may allow internal organs to change in volume during growth without changing the shape of the insect, and they reduce the weight of large insects. However, in many insects the air sacs increase the volume of air inspired and expired. Muscular movements in the abdomen draw air into the tracheae and expand the sacs, which collapse on expiration. In some insects—locusts, for example—additional pumping is provided by telescoping the abdomen, pumping with the prothorax, or thrusting the head forward and backward.

Studies of insect respiration using X-rays have shown that tracheal expansion and compression also occur in response to movements of jaw or limb muscles. Contraction of these muscles increases pressure inside the exoskeleton, and this elevated pressure causes contraction of tracheae, effectively permitting insects to exhale. When muscles involved in tracheal compression relax, tracheae expand due to the recoil by taenidial rings. If tracheae contract when spiracles are closed, increased internal pressure enhances oxygen diffusion to cells.

In some very small insects, gas transport occurs entirely by diffusion along a concentration gradient. Consumption of oxygen causes a reduced pressure in their tracheae that pulls air inward through the spiracles.

The tracheal system is an adaptation for air breathing, but many insects (nymphs, larvae, and adults) live in water. In small, soft-bodied aquatic nymphs, gaseous exchange may occur by diffusion through the body wall, usually into and out of a tracheal network just under the integument. Aquatic nymphs of stoneflies, mayflies, and damselflies have a variety of **tracheal gills,** which are thin extensions of the body wall containing a rich tracheal supply. Gills of dragonfly nymphs are ridges in the rectum (rectal gills) where gas exchange occurs as water enters and leaves.

James Gathany/CDC

A

B

James Gathany/CDC

Figure 21.15 *Pediculus humanus var. corporis* is the human body louse; it lives on the body, clothing, or bedding of some people and is spread by human contact. **A,** Female louse with ingested blood. **B,** Male louse showing grasping claws

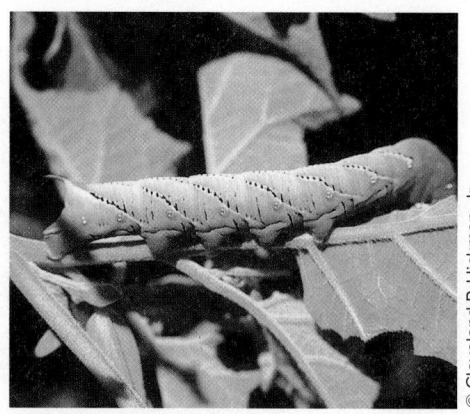

© Cleveland P. Hickman, Jr.

A

© Paparazzi by Appointment/Alamy Stock Photo

B

Figure 21.16 **A,** Larval stage of the tomato hornworm, *Manduca sexta* (order Lepidoptera). The more than 100 species of North American sphinx moths are strong fliers and mostly nocturnal feeders. Their larvae are called hornworms because of the large, fleshy posterior spine. **B,** Hornworm parasitized by a tiny wasp, *Apanteles* (a parasitoid), which laid its eggs inside the caterpillar. The wasp larvae have emerged, and their pupae are formed on the caterpillar's skin. Young wasps emerge in 5 to 10 days, but the caterpillar dies.

Although diving beetles, *Dytiscus* (Gr. *dytikos*, able to swim), can fly, they spend most of their life in water as excellent swimmers. How do they, and other aquatic insects, respire? They use an "artificial gill" in the form of a bubble of air (a plastron) held under the first pair of wings. The bubble is kept stable by a layer of hairs on top of the abdomen and is in contact with spiracles on the abdomen. Oxygen from the bubble diffuses into their tracheae and is replaced by diffusion of oxygen from the surrounding water. However, nitrogen from the bubble diffuses into the water, slowly decreasing the size of the bubble; therefore, diving beetles must surface every few hours to replace the air. Mosquito larvae are not good swimmers but live just below the surface, protruding short breathing tubes like snorkels to the surface for air (Figure 21.23B). Spreading oil on the water, a favorite method of mosquito control, clogs their tracheae with oil and suffocates the larvae. "Rat-tailed maggots" of certain syrphid flies have an extensible tail that can stretch as much as 15 cm to the water surface.

Excretion and Water Balance

Insects and spiders have each evolved independently a unique excretory system consisting of **Malpighian tubules** that operate in conjunction with specialized glands in the wall of the rectum. Malpighian tubules, variable in number, are thin, elastic, blind tubules attached to the juncture between the midgut and hindgut (Figures 21.12 and 21.19 A). Free ends of the tubules lie in the hemocoel and are bathed in hemolymph.

The mechanism of urine formation in Malpighian tubules of herbivorous insects appears to depend on a proton pump that adds hydrogen ions to the lumen of the tubule. Hydrogen ions are then exchanged for potassium ions (Figure 21.19B). This primary secretion of ions pulls water with it by osmosis to produce a potassium-rich fluid. Other solutes and waste materials also are secreted or diffuse into the tubule. The predominant waste product of nitrogen metabolism in most insects is uric acid, which is virtually insoluble in water (see p. 666). Uric acid enters the upper end of tubules, where the pH is slightly alkaline, as relatively soluble potassium and urate (abbreviated KHUr in Figure 21.19). As forming urine passes into the lower end of tubules, potassium combines with carbon dioxide and is reabsorbed as potassium bicarbonate ($KHCO_3$). As a result the pH of the fluid becomes acidic (pH 6.6), and insoluble uric acid (HUr) precipitates. As urine drains into the intestine and passes through the hindgut, specialized rectal glands reabsorb chloride, sodium (and in some cases potassium), and water.

Since water requirements vary among different types of insects, this ability to cycle water and salts is very important. Insects living in dry environments may resorb nearly all water from the rectum, producing a nearly dry mixture of urine and feces. However, freshwater larvae need to excrete water as well as to conserve salts. Insects that feed on dry grains need to conserve water and to excrete salt. In contrast, leaf-feeding insects ingest and excrete quantities of fluid. For example, aphids (p. 466) pass the excess fluid in the form of a sweetish material called **honeydew,** which is relished by other insects, especially ants (see Figure 21.33A, p. 462). Honeydew promotes growth of sooty mold (fungus) on leaves of infested plants and "rains" onto cars parked beneath infested trees.

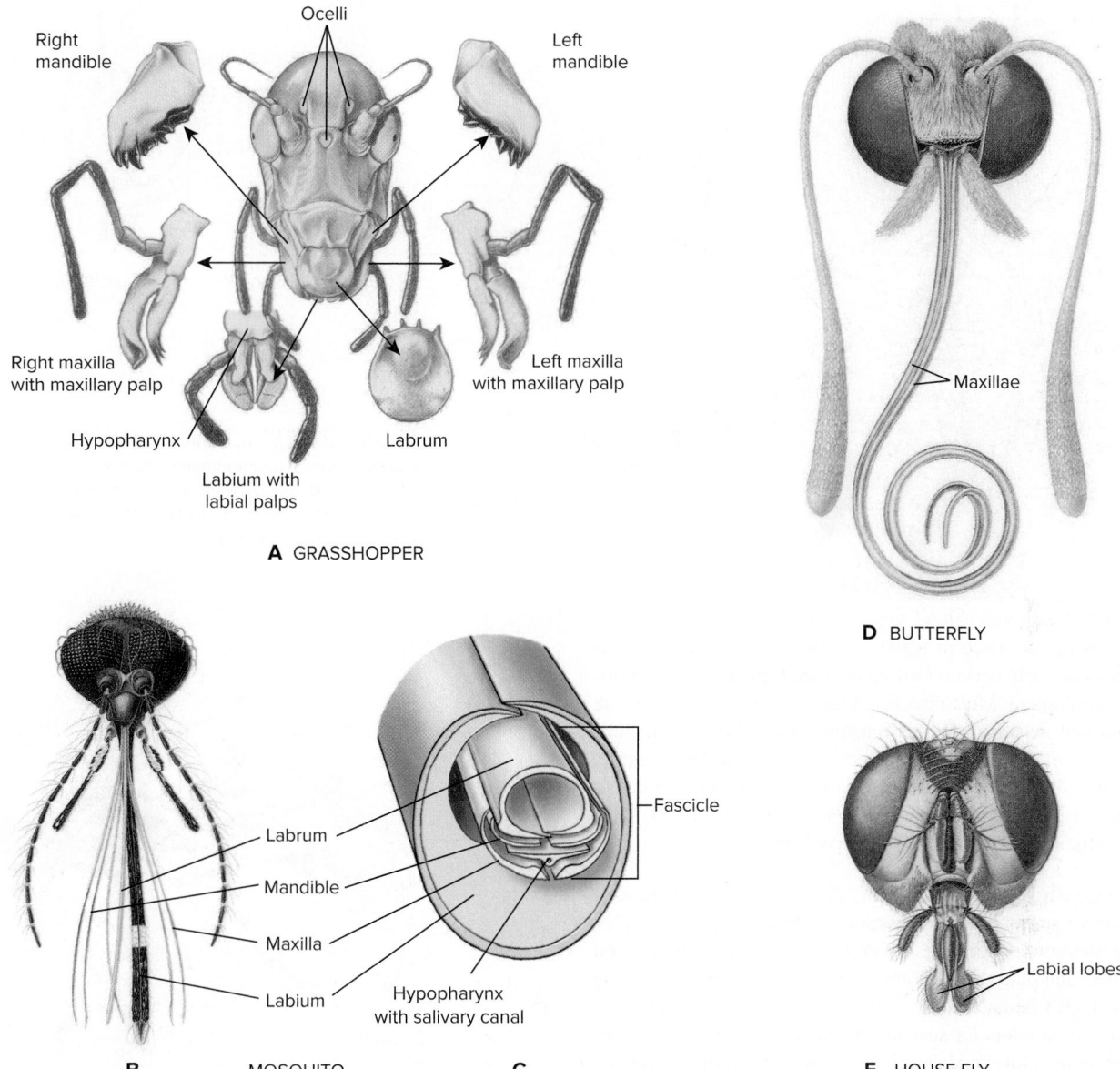

Figure 21.17   Four types of insect mouthparts. **A,** Chewing mouthparts of a grasshopper. **B** and **C,** Sucking mouthparts of a mosquito. Parts of piercing fascicle are shown in cross section **(C)**. **D,** Sucking mouthparts of a butterfly. Mandibles are absent, and maxillae form a long proboscis. **E,** Sponging mouthparts of a house fly. A pair of large lobes with grooves on their lower surface are on the end of the labium.

Nervous System

The nervous system in general resembles that of larger crustaceans, with a similar tendency toward the fusion of ganglia (see Figure 21.12). A number of insects have a giant fiber system. There is also a stomodeal nervous system that corresponds in function to the autonomic nervous system of vertebrates. Neurosecretory cells located in various parts of the brain have an endocrine function, but, except for their role in molting and metamorphosis, little is known of their activity.

Sense Organs

Along with neuromuscular coordination, insects have unusually keen sensory perception. Their sense organs are mostly microscopic and are located chiefly in the body wall. Each type usually responds to a specific stimulus, including mechanical, auditory, chemical, visual, and other stimuli.

Mechanoreception   Mechanical stimuli (those involving touch, pressure, and vibration) are detected by **sensilla.** A sensillum may be simply a **seta,** or hairlike process, connected with a nerve cell; a nerve ending just under the cuticle and lacking a seta; or a more complex organ (scolophorous organ) consisting of sensory cells with their endings attached to the body wall. Such organs are widely distributed over the antennae, legs, and body.

Auditory Reception   Very sensitive setae (hair sensilla) or tympanal organs may detect specific frequencies of airborne sounds. In tympanal organs a number of sensory cells (ranging

Figure 21.18 **A,** Generalized arrangement of insect tracheal system (diagrammatic). Air sacs and tracheoles not shown. **B,** Relationship of spiracle, tracheae, taenidia (chitinous bands that strengthen the tracheae), and tracheoles (diagrammatic).

from a few to hundreds) extend to a very thin tympanic membrane that encloses an air space in which vibrations are detected. Tympanal organs occur in certain Orthoptera (see Figure 21.2), Hemiptera, and Lepidoptera. Most insects are fairly insensitive to airborne sounds but can detect vibrations reaching them through the substrate. Organs on the legs usually detect vibrations of the substrate. Some nocturnal moths (for example, family Noctuidae) can detect ultrasonic pulses emitted by bats for echolocation (pp. 623, 738) and dive to the ground when they perceive bats.

Chemoreception   Chemoreceptors (for taste or smell) are usually bundles of sensory cell processes often located in sensory pits. These are often on mouthparts, but in many insects they are also on the antennae, and butterflies, moths, and flies also have them on the tarsi of the legs. Chemical sense is generally keen, and some insects can detect certain odors for several kilometers. Pheromones are detected at extremely low concentrations (see p. 735). Many patterns of insect behavior such as feeding, mating, habitat selection, caste production, and host-parasite relations are mediated through chemical senses. These senses also play a crucial role in responses of insects to artificial repellents and attractants. For example, an increase in carbon dioxide concentration, such as would be caused by a potential host nearby, causes a resting mosquito to begin flying; then it follows gradients of warmth and moisture and other cues to find its host. Diethyl toluamide (DEET), a repellent, apparently blocks the mosquito's ability to sense lactic acid, thus preventing host location.

Visual Reception   Insect eyes are of two types, simple and compound. **Simple eyes** are found in some nymphs and larvae and in many adults. Most insects have three ocelli on their head. Honey bees probably use ocelli to monitor light intensity and photoperiod (length of day) but not to form images.

Most adult insects have **compound eyes,** which may cover much of the head. They consist of thousands of ommatidia—6300 in the eye of a honey bee, for example. The structure of the compound eye is similar to that of crustaceans (Figure 21.20). An insect such as a honey

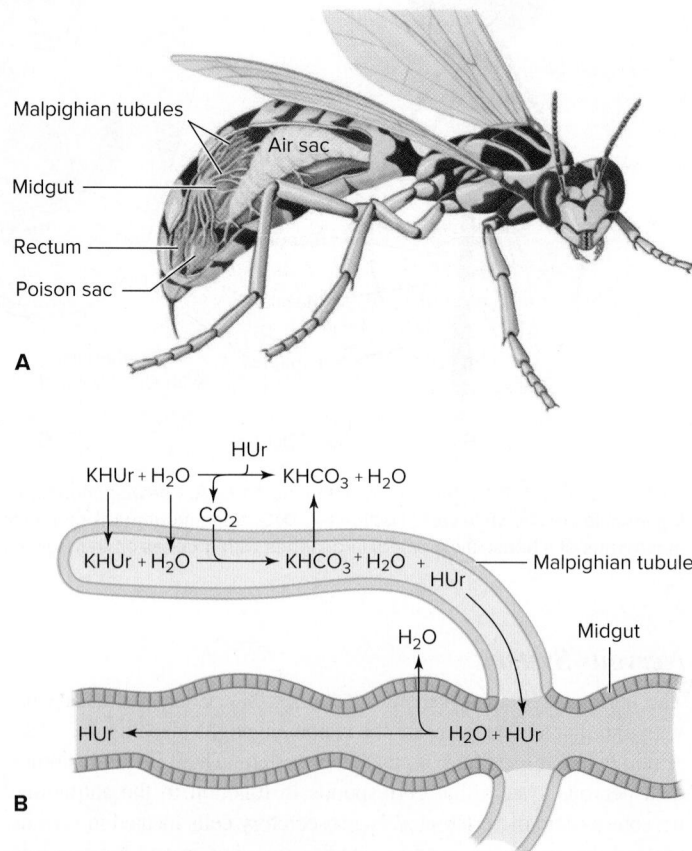

Figure 21.19   Malpighian tubules of insect. **A,** Malpighian tubules are located at the juncture of the midgut and hindgut (rectum) as shown in the cutaway view of a wasp. **B,** Function of Malpighian tubules. Hydrogen ions are actively exchanged for potassium ions in the upper tubules. Water and potassium acid urate (KHUr) follow. Potassium is resorbed with water and other solutes in the rectum.

bee can see simultaneously in al-most all directions around its body, but it is more myopic than humans, and images, even of nearby objects, are probably fuzzy. However, most flying insects rate much higher than humans in flicker-fusion tests. Flick-ers of light become fused in human eyes at a frequency of 45 to 55 per second, but bees and blow flies can distinguish as many as 200 to 300 separate flashes of light per second. This is undoubtedly advantageous in analyzing a fast-changing landscape during flight.

A bee can distinguish colors, but its sensitivity begins in the ultraviolet range, which human eyes cannot see. Although uniform in color to our perception, bee-pollinated flowers often have petals with lines and angular shapes that differ in ultraviolet (UV) light absorption and reflection. The lines and shapes of UV absorption act as a "nectar guide," leading bees to nectar in the flower. Many in-sects, such as butterflies, also have vision sensitive to red wave-lengths, but honey bees are red-blind.

Figure 21.20 Compound eye of an insect. A single ommatidium is shown enlarged to the right.

Other Senses Insects also have well-developed senses for tem-perature, especially on the antennae and legs, and for humidity, as well as for proprioception (sensation of muscle stretch and body po-sition), gravity, and other physical properties.

Neuromuscular Coordination

Insects are active creatures with excellent neuromuscular coordina-tion. Arthropod muscles are typically cross-striated, just as vertebrate skeletal muscles are. A flea can leap a distance of 100 times its own length, and an ant can carry in its jaws a load greater than its own weight. This sounds as though insect muscle were stronger than that of other animals. Actually, however, the force that a particular muscle can exert is related directly to its cross-sectional area, not its length. Based on maximum load moved per square centimeter of cross sec-tion, the strength of insect muscle is relatively the same as that of vertebrate muscle. The illusion of great strength of insects (and other small animals) is simply a consequence of small body size.

In terms of proportionate body length, a flea's jump would be the equivalent of a 6-foot human executing a standing high jump of 600 feet! Actually, a flea's muscles are not entirely responsible for its jump; they cannot contract rapidly enough to reach the re-quired acceleration. Fleas depend on pads of *resilin,* a protein with unusual elastic properties, which is also found in wing-hinge liga-ments of many other insects. Resilin releases 97% of its stored en-ergy on returning from a stretched position, compared with only 85% in most commercial rubber. When a flea prepares to jump, it rotates its hind femurs and compresses the resilin pads, then en-gages a "catch" mechanism. In effect, it has cocked itself. To take off, the flea needs to exert a relatively small muscular action to un-hook the catches, allowing the resilin to expand.

Reproduction

Parthenogenesis occurs prominently in life cycles of some Hemip-tera and Hymenoptera (see insect orders, pp. 465–467), but sexual reproduction is the norm for insects. Sexes are separate and various means are used to attract mates. A female moth releases a powerful pheromone that males can detect for a great distance. Fireflies use flashes of light; some insects find each other by sounds or color sig-nals and by arious kinds of courtship behavior.

Once a mate has been attracted, fertilization is usually internal. Sperm may be released directly or packaged into spermatophores. During the evolutionary transition of ancestral insects from marine to terrestrial life, spermatophores were widely used. Spermatophores may be transferred without copulation, as in silverfish, where a male deposits a spermatophore on the ground, then spins signal threads to guide a female to it. Alternatively, spermatophores may be deposited in the female vagina (see Figure 21.12) during copulation; in many cases, especially in butterflies, nutrients are also passed to the female via the spermatophore. Copulation (Figure 21.21) evolved much later than did indirect sperm transfer using spermatophores.

Usually sperm are stored in the seminal receptacle of a female in numbers sufficient to fertilize more than one batch of eggs. Many insects mate only once during their lifetime, but others, such as male damselflies, copulate several times per day.

Insects usually lay a great many eggs. A queen honey bee, for example, may lay more than 1 million eggs during her lifetime. On the other hand, some flies are viviparous and bring forth only a sin-gle offspring at a time. Insects that make no provision for care of their young may lay many more eggs than do insects that provide for their young or those that have a very short life cycle.

Most species lay their eggs in a particular habitat to which visual, chemical, or other cues guide them. Butterflies and moths lay their eggs on the specific kind of plant on which the caterpillar must feed. For example, a tiger moth may look for a pigweed, a sphinx moth for a tomato or tobacco plant, and a monarch butterfly for a milkweed plant (Figures 21.22 and 21.27 A). Insects whose imma-ture stages are aquatic characteristically lay their eggs in water (Figure 21.23). A tiny braconid wasp lays her eggs on the caterpillar of the sphinx moth, where the larvae will feed internally. After feeding and growing within the caterpillar, the braconid larvae emerge from the host and pupate externally within tiny cocoons (see Figure 21.16). An ichneumon wasp, with unerring accuracy, seeks a certain kind of larva in which her young will live as parasitoids. Her long ovipositor may have to penetrate 1 to 2 cm of wood to find a larva of a wood wasp or a wood-boring beetle in which she will deposit her eggs (see Figure 21.8).

Metamorphosis and Growth

Early development occurs within an egg. Once out of the egg, a larval or young ametabolous insect can grow only until it reaches the limits of its exoskeleton. At this point, it must make a larger cuticle, in a process called molting, and shed its existing cuticle in a process called ecdysis (see p. 457). Because molting produces an incremental increase in size, insects must molt many times; a stage that occurs between molts on the way to pupation or adulthood is called an **instar**.

In ametabolous insects, the young have the same morphology as adults, but larvae bear little resemblance to adults in insects that undergo **metamorphosis** (Figure 21.22). Although metamorphosis occurs in many animals, insects illustrate it more dramatically than any other group. The transformation, for instance, of a hickory horned devil caterpillar into a beautiful royal walnut moth represents an astonishing morphological change. In insects metamorphosis is associated with the evolution of wings, which are restricted to the reproductive stage. Adults, in effect, have become mating and dispersal stages for those species possessing flight.

Ametabolous (Direct) Development

A few insects, such as silverfish and springtails, undergo direct development. The young, or juveniles, are similar to adults except in size and sexual maturation. Stages are egg, juvenile, and adult. Primitively wingless insects are among those with this type of development.

Hemimetabolous (Incomplete) Metamorphoss

Some insects undergo a **hemimetabolous** (Gr. *hemi,* half, + *metabolē,* change), or gradual (incomplete), metamorphosis. These include grasshoppers, cicadas, mantids, and terrestrial bugs, which have terrestrial young, and mayflies, stoneflies, dragonflies, and aquatic bugs that lay their eggs in water and whose young are aquatic. The young are called **nymphs,** and their wings develop externally as budlike outgrowths in early instars and increase in size as the animal grows by successive molts and becomes a winged **adult** (Figures 21.24 and 21.25). Aquatic nymphs in some orders have tracheal gills or other modifications for aquatic life (Figure 21.26). The stages of hemimetabolous insects are egg, nymph (several instars), and adult (Figure 21.25).

> The *biological* meaning of the word "bug" is a great deal more restrictive than in common English usage. People often refer to all insects as "bugs," even extending the word to include such nonanimals as bacteria, viruses, and glitches in computer programs. Strictly speaking, however, a bug is a member of order Hemiptera and nothing else.

Figure 21.21   Copulation in insects (see also Figure 21.2C). **A,** *Omura congrua* (order Orthoptera) is a grasshopper found in Brazil. **B,** Bluet damselflies, *Enallagma* sp. (order Odonata), are common throughout North America. Here, a male continues to grasp a female after copulation. The female (white abdomen) lays eggs in the water.

Holometabolous (Complete) Metamorphosis

Approximately 88% of insects undergo a complete or **holometabolous** (Gr. *holo,* complete, + *metabolē,* change) metamorphosis, which separates physiological processes of growth (larva) from those of differentiation (pupa) and reproduction (adult) (Figure 21.22). In effect, each stage functions without competition with the other stages, for larvae often live in entirely different surroundings and eat different foods from adults. The wormlike **larvae,** which usually have chewing mouthparts, are called caterpillars, maggots, bagworms, fuzzy worms, or grubs. After a series of instars, a larva molts to enter a transitional stage called a **pupa.** Pupae are usually inactive and enveloped by a case, which may take several different forms. Pupae are nonfeeding and are the stage in which many insects also pass the winter. The **adult** insect emerges (ecloses) from the overwintering pupa in the spring, revealing an insect with crumpled,

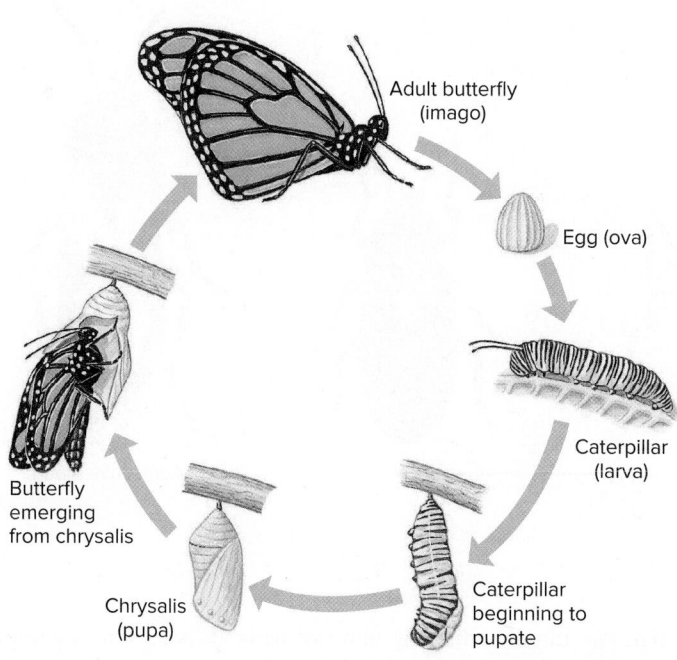

Figure 21.22 Complete (holometabolous) metamorphosis in a butterfly, *Danaus plexippus*. Eggs hatch to produce first of several larval instars. Last larval instar molts to become a pupa. Adult emerges at pupal molt.

Figure 21.23 **A,** Mosquito *Culex* (order Diptera) lays her eggs in small packets or rafts on the surface of standing or slowly moving water. **B,** Mosquito larvae are the familiar wrigglers of ponds and ditches. To breathe, they hang head down, with respiratory tubes projecting through the surface film of water. Motion of vibratile tufts of fine hairs on the head brings a constant supply of food to these filter feeders.

miniature wings. In a short time its wings expand and harden, and the insect flies. The stages of a holometabolous insect, then, are egg, larva (several instars), pupa, and adult (Figure 21.22). Adults undergo no further molting.

Physiology of Metamorphosis

Hormones regulate metamorphosis in insects (see Chapter 34, pp. 749–751). Major endocrine organs that are involved in development are the **brain, prothoracic (ecdysial) glands, corpora cardiaca,** and **corpora allata** (Figure 34.4, p. 750).

The intercerebral part of the brain and the ganglia of the nerve cord contain several groups of neurosecretory cells that produce a brain hormone called **prothoracicotropic hormone (PTTH).** These neurosecretory cells send their axons to paired organs behind the brain, called corpora cardiaca, which serve as storage and release organs for PTTH (and also produce other hormones). PTTH is carried via the hemolymph to the prothoracic gland, a glandular organ in the head or prothorax that produces **molting hormone,** or **ecdysone** (ek-dī´sōn) in response to PTTH. Ecdysone sets in motion certain processes that lead to molting and discarding the old cuticle (ecdysis).

Larval molting persists as long as **juvenile hormone** from the corpora allata is present in sufficient amounts, along with molting hormone in the hemolymph. Under these conditions, each molt produces a larger larva (see Figure 34.4, p. 750).

In later instars, the corpora allata release progressively less juvenile hormone. When juvenile hormone is at a very low level, a larva molts to become a pupa (instead of a *larger* larva), and likewise, the cessation of juvenile hormone production in a pupa leads to an adult at the next molt (metamorphosis). Control of development is the same in hemimetabolous insects, except that there is no pupa, and cessation of juvenile hormone production occurs in the final nymphal instar. The corpora allata again become active in adult insects, in which juvenile hormone is important in normal sexual reproduction and gamete formation. The prothoracic glands degenerate in adults of most insects, and adults do not molt.

Insect hormones have been the focus of fascinating experiments. For example, if the corpora allata (and thus juvenile hormone) are removed surgically from a larva, the following molt is a metamorphosis. Conversely, if the corpora allata from a young larva are transplanted into a final larval instar, the latter can be converted into a giant larva, because metamorphosis to a pupa cannot occur.

Diapause

Many animals, including many types of insects, undergo a period of dormancy in their annual life cycle. In temperate zones there may be a

Figure 21.24 **A,** Ecdysis in a cicada *Magicicada septendecim* (order Hemiptera). The old cuticle splits along a dorsal midline as a result of increased blood pressure and of air forced into the thorax by muscle contraction. The emerging insect is pale, and its new cuticle is soft. The wings will be expanded by blood pumped into veins, and the insect enlarges by taking in air. **B,** An adult cicada.

Figure 21.25
Life history of a hemimetabolous insect.Image: © Helen I'Anson.

Egg

Nymphs

Adult chinch bug

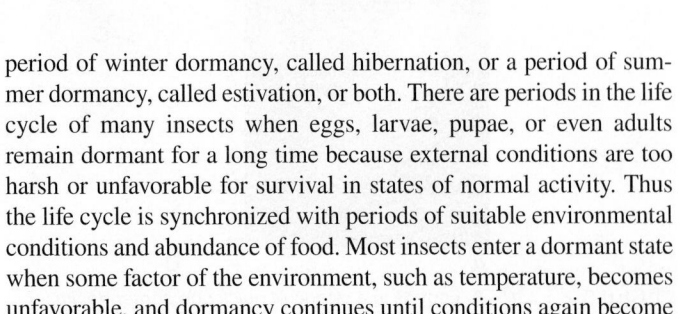

© Cleveland P. Hickman, Jr.

© Creatas/PunchStock

© Vitalii Hulai/Getty Images

A

B

C

Figure 21.26 A, Stonefly, *Perla* sp. (order Plecoptera). **B,** A dragonfly, order Odonata. **C,** Nymph of dragonfly. Both stoneflies and dragonflies have aquatic nymphs that undergo gradual metamorphosis.

(Gr. *dia,* through, dividing into two parts, + *pausis,* a stopping), and it is an important adaptation to survive adverse environmental conditions. Diapause is internally controlled in each species and sometimes varies between subspecies within a species, but it is usually initiated by a particular signal. In the environment of an insect, such signals forecast adverse conditions to come, for example, lengthening or shortening of days. Thus photoperiod, or day length, is often a signal that initiates diapause. After diapause is initiated, another environmental signal is usually required to end it. Such signal may be return of favorable temperature after a prolonged period of cold or an occasion of rain after a dry period, as in a desert.

Diapause always occurs at the end of an active growth stage of the molting cycle so that, when the diapause period ends, an insect is ready for another molt. One species of the ant *Myrmica* reaches the third instar stage in late summer. Many larvae do not develop beyond this point until the following spring, even if temperatures are mild or if the larvae are kept in a warm laboratory. Depending on the species, insects may enter diapause at any stage of their life cycle.

Defense

Insects have many means of protecting themselves. The cuticular exoskeleton affords good protection for many. Some, such as stink bugs, have repulsive odors and tastes; others protect themselves by a good offense, for many are very aggressive and fight (for example, bees and ants); and still others are swift in running for cover when danger threatens.

Many insects practice chemical warfare in a variety of ingenious ways. Some repel an assault by virtue of their bad taste, odor, or poisonous properties; others use chemical exudates that mechanically prevent a predator from attacking. Caterpillars of some monarch butterflies (Figure 21.27A) assimilate cardiac glycosides from certain species of milkweed (family Asclepiadaceae); this substance confers unpalatability on larvae and adults and induces vomiting in some (but not all) of their avian predators. Bombardier beetles, on the other hand, produce an irritating spray, which they aim accurately at attacking ants or other enemies.

period of winter dormancy, called hibernation, or a period of summer dormancy, called estivation, or both. There are periods in the life cycle of many insects when eggs, larvae, pupae, or even adults remain dormant for a long time because external conditions are too harsh or unfavorable for survival in states of normal activity. Thus the life cycle is synchronized with periods of suitable environmental conditions and abundance of food. Most insects enter a dormant state when some factor of the environment, such as temperature, becomes unfavorable, and dormancy continues until conditions again become favorable.

However, some species have a prolonged arrest of growth that occurs regardless of environment, whether or not favorable conditions prevail. This type of dormancy is called **diapause** (di´a-poz)

Luna moths are lovely insects with long extensions on their hindwings. They are active at night, as are their bat predators, who use echolocation to find prey. The wing extensions produce an acoustic diversion, directing predatory strikes toward the extensions and away from the moth body, enabling many moths to survive an attack.

A

B

Figure 21.27 Mimicry in butterflies. **A,** Monarch butterfly is distasteful to, and avoided by, some birds because as a caterpillar it fed on the acrid milkweed. **B,** The monarch is mimicked by the smaller viceroy butterfly, *Limenitis archippus,* which feeds on willows and may (or may not) be tasteful to birds, but is not eaten because it so closely resembles the monarch in color and markings. This kind of mimicry is called Batesian mimicry, although some evidence argues for Müllerian mimicry in this case.

Some color patterns and body forms in insects are highly adaptive in evasion of predation, such as those for **aposematic** coloration (warning coloration to advertise noxious qualities) and **crypsis** (camouflage in shape or coloration to escape notice, Figure 21.28). Aposematic coloration is subject to **mimicry** (Figure 21.27), in which two or more noxious species share aposematic coloration to avoid a common predator (Müllerian mimicry) or a tasteful species adopts the aposematic coloration of a noxious species to fool a common predator (Batesian mimicry).

Behavior and Communication

The keen sensory perceptions of insects make them extremely responsive to many stimuli. Stimuli may be internal (physiological) or external (environmental), and responses are governed by both the physiological state of the animal and the pattern of nerve pathways traveled by the impulses. Many responses are simple, such as orientation toward or away from a stimulus, for example, avoidance of light by a cockroach, or attraction of carrion flies to the odor of dead flesh.

Much behavior of insects, however, is not a simple matter of orientation but involves a complex series of responses. A pair of dung beetles chew a bit of dung, roll it into a ball, and roll the ball laboriously to where they intend to bury it after laying their eggs in it (Figure 21.29). Cicadas slit the bark of a twig and then lay an egg in each of the slits. The female potter wasp *Eumenes* scoops clay into pellets, carries them one by one to a building site, and fashions them into dainty little narrow-necked clay pots, into each of which the wasp lays an egg. Then the mother wasp hunts and paralyzes a number of caterpillars, pokes them into the opening of a pot, and seals the opening with clay. Each egg, in its own protective pot, hatches to find a well-stocked larder of food.

Much of such behavior is innate; however, learning is more important than was once believed. The potter wasp, for example, must learn where she has left her pots if she is to return to fill them with

A © Creatas Images/PictureQuest   **B** © Kjell Sandved/Alamy Stock Photo   **C** © Francesco Tomasinelli/Science Source

Figure 21.28 Crypsis (camouflage) in insects. **A,** The dead leaf butterfly, *Kallima paralekta*, resembles a dead leaf. It is native to Indonesia and Malaysia. **B,** Bizarre processes from the thorax of a treehopper from Mexico, *Sphongophorus* sp. (order Hemiptera), masquerade as parts of the twig on which it feeds. **C,** Broken outlines and color of a lichen katydid, *Dysonia* sp. (order Orthoptera) allow it to blend with leaves in the Peruvian cloud forest.

Figure 21.29
This African dung beetle, *Kheper aegyptiorum purpurascens,* is a generalist when it comes to selecting dung, but here it is rolling elephant dung. Typically, dung is formed into a ball and rolled away by males, by females, or by both individuals together. Eggs are laid in the dung and young beetles develop in the brood ball, sometimes with extended care by the mother beetle.

© Paul Souders/The Image Bank/Getty Images

caterpillars one at a time. Social insects, which have been studied extensively, are capable of most basic forms of learning used by mammals. An exception is insight learning. Apparently insects, when faced with a new problem, cannot reorganize their memories to construct a new response.

Some insects can memorize and perform in sequence tasks involving multiple signals in various sensory areas. Worker honey bees have been trained to walk through mazes that involved five turns in sequence, using such clues as color of a marker, distance between two spots, or angle of a turn. The same is true of ants. Workers of one species of *Formica* learned a six-point maze at a rate only two or three times slower than that of laboratory rats. Foraging trips of ants and bees often wind and loop in a circuitous route, but once a forager has found food, the return trip is relatively direct. How an insect does this is unknown.

Insects communicate with each other by means of chemical, visual, auditory, and tactile signals. **Chemical** signals take the form of **pheromones,** which are substances secreted by one individual that affect behavior or physiological processes of another individual. Many pheromones have been described. Like hormones, pheromones are effective in minute quantities. Pheromones are used in a wide range of situations. They may attract opposite sex, signal an alarm, or initiate behaviors such as overwintering in lady beetles, or aggregation in bark beetles. They may also be used to mark trails, to define territories, or to stimulate defensive responses. Social insects, such as bees, ants, wasps, and termites, can recognize a nestmate—or an alien in the nest—by means of identification pheromones. Social parasites escape detection—and certain destruction—by imitating or duplicating pheromones produced by members of their host colony. Pheromones determine caste in termites and to some extent in ants and bees. They are a primary integrating force in populations of social insects.

Many insect pheromones have been isolated and identified. Traps baited with pheromones have been used for many years to monitor insects of economic importance. They can be used to detect presence of an insect, such as a new arrival from a neighboring area (tracking the spread of European gypsy moth in the United States or presence of European corn ear worms in a field), or to monitor changes in population levels. Use of pheromone traps has become an important tool to detect potential outbreaks, allowing sufficient time to plan remedial action.

Sound production and **reception** in insects have been studied extensively, and although a sense of hearing is not present in all insects, this means of communication is meaningful to those insects that use it. Sounds serve as warning devices, advertisement of territorial claims, or courtship songs. Sounds of crickets and grasshoppers seem to be concerned with courtship and aggression. Male crickets scrape the modified edges of their forewings together to produce their characteristic chirping. The long, drawn-out sound of male cicadas, a call to attract females, is produced by vibrating membranes in a pair of organs on the ventral side of the basal abdominal segment.

There are many forms of **tactile communication,** such as tapping, stroking, grasping, and antennae touching, which evoke responses varying from recognition to recruitment and alarm. Certain kinds of flies, springtails, and beetles manufacture their own **visual signals** in the form of **bioluminescence.** Best known of luminescent beetles are fireflies, or lightning bugs (which are neither flies nor bugs, but beetles), in which a flash of light helps to locate a prospective mate. Each species has its own characteristic flashing rhythm produced on the ventral side of the last abdominal segments. Females flash an answer to the species-specific pattern to attract males. Female *Photuris* sometimes flash the patterns of other species to attract males of those species as prey (Figure 21.30).

Figure 21.30 A lightning "bug" is actually a beetle in the genus *Photuris*, family Lampyridae. A female emits flashes of light to attract males for mating. However, females of certain species prey on males of other species. They flash a pattern characteristic of the other species, but then consume the males arriving to mate.

Figure 21.31 Queen bee surrounded by her court. The queen is the only egg layer in the colony. The attendants, attracted by her pheromones, constantly lick her body. As food and the queen's pheromones are transferred from these bees to others, the queen's presence is communicated throughout the colony.

Social Behavior

Some community groups are temporary and uncoordinated, as are hibernating associations of carpenter bees or feeding clusters of aphids. Some are coordinated for only brief periods, and some cooperate more fully, such as tent caterpillars, *Malacosoma,* which join in building a home web and feeding net. However, all of these are open communities with limited social behavior.

In the **eusocial** Hymenoptera (honey bees and ants) and Isoptera (termites), a complex social life is regulated by chemical and tactile communication. Social interactions occur at all stages of the life cycle, communities are usually ongoing, all activities are collective, and there is reciprocal communication and division of labor. The society usually demonstrates polymorphism, or **caste** differentiation.

Honey bees have one of the most complex organizations in the insect world. As many as 60,000 to 70,000 honey bees may occupy a single hive. There are three castes: a single sexually mature female, or **queen;** a few hundred **drones,** which are sexually mature males; and the rest, **workers,** which are sexually inactive genetic females (Figure 21.31).

Workers take care of young, secrete wax with which they build the six-sided cells of honeycomb, gather nectar from flowers, manufacture honey, collect pollen, and ventilate and guard the hive. Several drones fertilize a queen during her nuptial (mating) flight, at which time enough sperm is stored in her seminal receptacle to last her a lifetime. Drones die after mating, and those remaining in the hive at summer's end are driven out by workers and starve.

Castes are determined partly by fertilization and partly by what is fed to larvae. Drones develop parthenogenetically from unfertilized eggs (and consequently are haploid); queens and workers develop from fertilized eggs (and thus are diploid; see haplodiploidy, p. 137 and Figure 36.20). Female larvae that will become queens are fed royal jelly, a secretion from the salivary glands of nurse workers. Both the amount of royal jelly fed and the duration of feeding affect queen formation. Diet-induced metabolites affect gene expression

and signal transduction within the developing larva, thus directing the phenotype toward workers or queens. Queens live one to two years whereas the lifespan of workers is only months. Honey and pollen are added to worker diet about the third day of larval life. Pheromones in "queen substance," which is produced by a queen's mandibular glands, prevent female workers from maturing sexually. Workers produce royal jelly only when the level of "queen substance" pheromone in the colony drops, most typically due to overcrowding. This change also occurs when the queen becomes too old, dies, or is removed. Then workers' ovaries develop, and they start enlarging a larval cell and feeding a larva royal jelly to produce a new queen. Production of a new queen may be followed by swarming, where the old queen leaves with part of the colony.

Honey bees have evolved an efficient system of communication by which, through certain body movements, their scouts inform workers of the location and quantity of food sources (Figure 36.23, p. 793).

Termite colonies contain several castes, consisting of fertile individuals, both males and females, and immature individuals (Figure 21.32). Some fertile individuals may have wings and may

Figure 21.32 **A,** Workers of the western subterranean termite, *Reticulitermes hesperus* (order Isoptera). Workers are wingless immature animals that tend the nest. **B,** Termite queen becomes a distended egg-laying machine. The queen and several workers and soldiers are shown here.

leave the colony, mate, lose their wings, and as **king** and **queen** start a new colony. Wingless fertile individuals may under certain conditions substitute for the king or queen. Immature members are wingless and become **workers** and **soldiers.** Soldiers have large heads and mandibles and serve to defend the colony. As in bees and ants, extrinsic factors cause caste differentiation. Reproductive individuals and soldiers secrete inhibiting pheromones that pass throughout the colony to nymphs through a mutual feeding process, called **trophallaxis,** so that they become sterile workers. Workers also produce pheromones, and if the level of "worker substance" or "soldier substance" falls, as might happen after an attack by marauding predators, for example, the next generation produces compensating proportions of the appropriate caste.

Ants are closely related to bees and wasps (order Hymenoptera) and also have highly organized societies. Superficially, they resemble termites, but they can be distinguished easily. In contrast to termites, ants are usually dark in color, are hard bodied, and have a constriction posterior to their first abdominal somite. Antennae of ants are elbowed, while those of termites are threadlike or resemble a string of beads (moniliform).

In ant colonies males die soon after mating and the queen either starts her own new colony or joins some established colony and does the egg laying. Sterile females are wingless workers and soldiers that do work of the colony: gather food, care for young, and protect the colony. In many larger colonies there may be two or three types of individuals within each caste.

Ants have evolved some striking patterns of "economic" behavior, such as making slaves, farming fungi, herding "ant cows" (aphids or other members of Sternorrhyncha) (Figure 21.33A), sewing their nests together with silk (Figure 21.33B), and using tools.

21.2 INSECTS AND HUMAN WELFARE

Beneficial Insects

Although most of us think of insects primarily as pests, all terrestrial life, including human life, would have great difficulty surviving if all insects were suddenly to disappear. Some produce useful materials: honey and beeswax from bees, silk from silkworms, and shellac from a wax secreted by lac insects. More important, however, insects are necessary for the cross-fertilization of many crops. Bees pollinate almost $14 billion worth of food crops per year in the United States alone, and this figure does not include pollination of forage crops for livestock or pollination by other insects. Bee populations are in decline in many parts of the world, including the United States. Bee populations have fluctuated in ancient and modern times, but the recent disappearance of entire colonies that began about 2006 has had a significant impact on agriculture. In Colony Collapse Disorder, worker bees disappear and leave the queen stranded with little or no care for her offspring. The workers likely fall victim to the "deformed wing virus" DWV, which is carried by *Varroa* mites and likely spread by human traffic in bee colonies (see p. 414 and associated references, p. 420 and p. 470).

The evolution of flowering plants led to a great insect radiation. Insects and plants coevolved an astonishing array of adaptations involving both reward and deception. Floral morphology, color, and scent attract specific pollinators and guide their movements; for example, see the nectar guides on p. 455.

A

B

Figure 21.33 **A,** A red carpenter ant (*Camptonotus ferrugineus*, order Hymenoptera) tending a group of aphids (order Hemiptera). The aphids feed copiously on plant juices and excrete the excess as a clear liquid rich in carbohydrates ("honeydew"), which is cherished as a food by ants. **B,** A weaver ant nest in Australia.

Many predaceous insects, such as tiger beetles, aphid lions, ant lions, praying mantids, and ladybug beetles, destroy harmful insects (Figure 21.34). In turn, insects serve as important sources of food for many other animals. Parasitoid insects are very important in controlling populations of many harmful insects. Dead animals are quickly consumed by maggots hatched from eggs laid in carcasses. Forensic entomology uses the succession of insects in a dead body to estimate the age of a corpse and thus provides valuable information to investigators about time of death.

Harmful Insects

Harmful insects include those that eat and destroy plants and fruits, such as grasshoppers, chinch bugs, corn borers, boll weevils, grain weevils, San Jose scale, and hundreds of others (Figure 21.35). Nearly every cultivated crop has several insect pests. People expend enormous resources in all agricultural activities, in forestry, and in the food industry to counter insects and the damage they cause. Outbreaks of bark beetles or defoliators such as spruce budworms,

Figure 21.34 Some beneficial insects. **A,** A predaceous spined soldier bug (order Hemiptera) feeds on a larva of the Mexican bean beetle. Note the sucking proboscis of the bug. Mexican bean beetles are pests of snap beans and of soybeans. **B,** A ladybug beetle (order Coleoptera). Adults (and larvae of most species) feed voraciously on plant pests such as mites, aphids, scale insects, and thrips. **C,** A parasitic wasp (*Aleiodes indiscretus*) inserting fertilized eggs into a gypsy moth (*Lymantria dispar*) caterpillar. After the wasp deposits her eggs, the caterpillar recovers and resumes an active life—until it is killed by developing wasp larvae.

Figure 21.35 Some insect pests. **A,** Japanese beetles, *Popillia japonica* (order Coleoptera), are serious pests of fruit trees and ornamental shrubs. They were introduced into the United States from Japan in 1917. **B,** Citrus mealybugs, *Planococcus citri* (order Hemiptera, suborder Sternorrhyncha). Many mealybugs are pests of commercially valuable plants. **C,** A corn ear worm, *Heliothis zea* (order Lepidoptera). An even more serious pest of corn is the infamous corn borer, an import from Europe in 1908 or 1909.

hemlock woolly adelgids, and gypsy moths have caused tremendous economic losses and have become a major element in determining the composition of forests in the United States. Gypsy moths, introduced into the United States in 1869 in an ill-advised attempt to breed a better silkworm, have spread throughout the northeast as far south as Virginia and as far west as Minnesota. They defoliate forests in years when there are outbreaks. In 2013, the U.S. Forest Service estimated that 574,150 acres were defoliated in northeastern states.

Ten percent of all arthropod species are parasitic insects, or insects that in essence are "micropredators" because they attack but do not remain on their hosts. Lice, bloodsucking flies, warble flies, bot flies, and many others attack humans or domestic animals or both. Malaria, carried by *Anopheles* mosquitos (Figure 21.36), is still one of the world's major diseases, infecting hundreds of millions of people each year and causing millions of deaths. Mosquitos also transmit yellow fever, lymphatic filariasis, and most recently Zika virus. Fleas carry plague, which at times in history has killed significant portions of human populations. House flies are vectors of typhoid, as are lice for typhus fever; tsetse flies carry African sleeping sickness; and bloodsucking bugs, *Rhodnius* and related genera, transmit Chagas's disease. The newest viral plague to hit North America, the West Nile virus, is carried by over 40 species of mosquito, especially *Culex,* and infects humans, some other mammals, and over 75 species of birds, some of which serve as reservoirs for the virus.

There is tremendous destruction of food, clothing, and property by weevils, cockroaches, ants, clothes moths, termites, and carpet beetles. Not the least of insect pests are bedbugs, *Cimex* (Figure 21.36B), bloodsucking hemipterous insects whose association with humans originated from bats that shared their caves early in human evolution. Bedbug infestation is rising throughout the developed world for unknown reasons. Proposed explanations include increased importation of bugs by travelers and reluctance to use insecticides.

A

© Digital Vision/Getty Images

B

© Science Source

Figure 21.36 **A,** A mosquito in genus *Aedes* (order Diptera) from South Africa. Members of this genus spread dengue fever and yellow fever, among other illnesses. **B,** A bedbug, *Cimex lectularius,* is a common human parasite.

West Nile virus, a disease agent spread by mosquitos, affects mammals and birds throughout the world. First identified in Uganda in 1937, it spread to North America in 1999. Birds serve as a reservoir for the virus: a bird bitten by an infected mosquito plays host to the virus for 1 to 4 days, during which the virus can be picked up by other mosquitos and spread to new hosts. Human response to infection varies, with about 80% showing no symptoms, close to 20% exhibiting flulike symptoms such as fever and achiness, and less than 1% developing potentially fatal encephalitis or other possibly permanent neurological effects. Preventing mosquito bites is the best way to avoid infection, so studies of mosquito behavior are useful. For example, researchers wonder whether mosquitos that initially feed on infected birds tend to choose another bird for the next bite, or simply feed opportunistically on any available animal. Understanding virus transmission makes possible mathematical models to predict how and when the disease will spread.

Taxonomy of Subphylum Hexapoda

Class Entognatha (en´tog-natha) (Gr. *entos,* within, inside, + *gnathos,* jaw): **entognaths.** Base of mouthparts lies within head capsule; mandibles have one articulation. Example: *Entomobrya.*

Class Insecta (in-sek´ta) (L. *insectus,* cut into): **insects.** Bases of mouthparts exposed and exiting head capsule; mandibles usually have two regions of articulation. Examples: *Drosophila, Bombus, Anopheles* (insect orders listed on pp. 465–467).

Control of Insects

Because all insects are an integral part of the ecological communities to which they belong, their total destruction would obviously do more harm than good. All terrestrial food webs would be seriously disturbed or destroyed. The beneficial role of insects in our environment is often overlooked, and in our zeal to control the pests we spray the landscape indiscriminately with extremely effective "broad-spectrum" insecticides that eradicate good, as well as harmful, insects. We have also found, to our dismay, that many chemical insecticides persist in the environment and accumulate as residues in the bodies of animals higher in the food chain, including ourselves. Furthermore, many insects have evolved a resistance to insecticides in common use. Honey bees are especially susceptible to insecticides, and resistance mainly is developed by harmful insects.

For several years, methods of control other than chemical insecticides have been under intense investigation, experimentation, and development. Economics, concern for the environment, and consumer demand are causing thousands of farmers across the United States to use alternatives to strict dependence on chemicals.

Many organisms useful in the biological control of insects are currently being used or are under scientific investigation. These organisms include bacterial, viral, and fungal pathogens. For example, a bacterium, *Bacillus thuringiensis,* produces a toxin that is quite effective in control of lepidopteran pests (cabbage looper, imported cabbage worm, tomato worm, gypsy moth). Other strains of *B. thuringiensis* have toxins that kill insects in other orders, and the species diversity of target insects is being widened by techniques of genetic engineering. Genes coding for the toxin produced by *B. thuringiensis* also have been introduced into other bacteria and even into the plants themselves, which makes the plants resistant to insect attack. Many of our grain crops, especially corn, now possess genes that express protein toxic to specific insect pests, thereby obviating the need for pesticides. However, some insects have now evolved resistance to the *B. thuringiensis* toxin.

A number of viruses and fungi that have potential as insecticides have been isolated. Difficulties and expense in rearing these agents have been overcome in certain cases, and some are now commercially produced.

Introduction of natural predators or parasitoids of insect pests has met with some success. In the United States vedalia beetles from Australia help to control cottony-cushion scale on citrus plants, and numerous instances of control by use of insect parasitoids are recorded.

Another approach to biological control is to interfere with reproduction or behavior of insect pests with sterile males or with naturally occurring organic compounds that act as hormones or pheromones.

Taxonomy of Class Entognatha and Class Insecta

Entomologists do not all agree on names of orders or on the limits of each order. Some choose to combine and others to divide the groups. However, this synopsis of orders is one that is rather widely accepted.

I. Class Entognatha

Order Protura (pro-tu´ra) (Gr. *protos,* first, + *oura,* tail): **proturans.** Minute (1 to 1.5 mm); no eyes or antennae; appendages on abdomen as well as thorax; live in soil and dark, humid places; direct development.

Order Diplura (dip-lu´ra) (Gr. *diploos,* double, + *oura,* tail): **japygids** and **campodeids.** Usually less than 10 mm; pale, eyeless; a pair of long terminal filaments or pair of caudal forceps; live in damp humus or rotting logs; development direct.

Order Collembola (col-lem´bo-la) (Gr. *kolla,* glue, + *embolon,* peg, wedge): **springtails, jumping bristletails,** and **snow fleas.** Small (5 mm or less); compound eyes lacking, eye patches of 1 to several lateral ocelli; respiration by trachea or body surface; a springing organ folded under the abdomen for leaping; abundant in soil; sometimes swarm on pond surface film or on snowbanks in spring; development direct.

II. Class Insecta

Subclass Apterygota

Order Thysanura (thy-sa-nu´ra) (Gr. *thysanos,* tassel, + *oura,* tail): **silverfish** (Figure 21.37) and **bristletails.** Small to medium size; large eyes; long antennae; three long terminal cerci; live under stones and leaves and around human habitations; development direct.

Subclass Pterygota

1. Infraclass Paleoptera

Order Ephemeroptera (e-fem-er-op´ter-a) (Gr. *ephēmeros,* lasting but a day, + *pteron,* wing): **mayflies** (Figure 21.38). Wings membranous; forewings larger than hindwings; adult mouthparts vestigial; nymphs aquatic, with lateral tracheal gills.

Order Odonata (o-do-na´ta) (Gr. *odontos,* tooth, + *ata,* characterized by): **dragonflies, damselflies** (see Figures 21.21B and 21.26B). Large; membranous wings are long, narrow, net veined, and similar in size; long and slender body; aquatic nymphs with gills and prehensile labium for capture of prey.

2. Infraclass Neoptera

Order Orthoptera (or-thop´ter-a) (Gr. *orthos,* straight, + *pteron,* wing): **grasshoppers** (see Figure 21.2), **locusts, crickets, katydids.** Wings, when present, with forewings thickened and hindwings folded like a fan under forewings; chewing mouthparts.

Order Blattodea (blə-´tō dēə) (L. *blatta,* cockroach, + Gr. *eidos,* form, + *ea,* characterized by): **cockroaches.** Common insects in tropical areas; often in houses in northern areas; oval, flattened bodies may exceed 5 cm in length; tarsi with 5 segments; wings typically present, often reduced.

Order Phasmatodea (faz-mə-´tō-dēə) (Gr. *phasma,* apparition, + *eidos,* form, + *ea,* characterized by): **walkingsticks** (see Figure 21.7B) and **leaf insects.** Bodies elongated and sticklike or flattened and laterally expanded; herbivorous; tropical forms may be very large (up to 30 cm).

Order Mantodea (man-´tō-dēə) (Gr. *mantis,* soothsayer, + *eidos,* form, + *ea,* characterized by): **mantids** (see Figure 21.5).
Bodies elongated with raptorial front legs; predatory; may reach 10 cm in length.

Order Mantophasmatodea (man-´tō-faz-mə-´tō-dēə) (an amalgamation of the order names for praying mantids [Mantodea] and walkingsticks [Phasmatodea]): **gladiators.** Secondarily wingless; chewing mouthparts; resemble a combination of a praying mantis and a walkingstick; nocturnal predators, on insects and spiders; described in 2002; rare, found in Africa; 6 to 8 species.

A

Figure 21.37 Silverfish *L episma* (order Thysanura) is often found in homes.

B

Figure 21.38 Mayfly (order Ephemeroptera). **A,** Nymph. **B,** Adult.

Taxonomy of Class Entognatha and Class Insecta

Order Dermaptera (der-map´ter-a) (Gr. *derma,* skin, + *pteron,* wing): **earwigs.** Very short forewings; large and membranous hindwings folded under forewings when at rest; chewing mouthparts; forcepslike cerci.

Order Plecoptera (ple-kop´ter-a) (Gr. *plekein,* to twist, + *pteron,* wing): **stoneflies** (Figure 21.26A). Membranous wings; larger and fanlike hindwings; aquatic nymph with tufts of tracheal gills.

Order Isoptera (i-sop´ter-a) (Gr. *isos,* equal, + *pteron,* wing): **termites** (Figure 21.32). Small; membranous, narrow wings similar in size with few veins; wings shed at maturity; erroneously called "white ants"; distinguishable from true ants by broad union of thorax and abdomen; complex social organization.

Order Embiidina (em-bē-ə´di-nə) (Gr. *embios,* lively, + *eidos,* form, + *ina,* resembling): **webspinners.** Small; male wings membranous, narrow, and similar in size; wingless females; chewing mouthparts; colonial; make silk-lined channels in tropical soil.

Order Psocoptera (so-cop´ter-a) (Gr. *psoco,* rub away, + *pteron,* wing) **(Corrodentia): psocids, book lice, bark lice.** Body usually small, may be as large as 10 mm; membranous, narrow wings with few veins, usually held rooflike over abdomen when at rest; some wingless species; found in books, bark, bird nests, on foliage.

Order Zoraptera (zo-rap´ter-a) (Gr. *zōros,* pure, + *apteryos,* wingless): **zorapterans.** As large as 2.5 mm; membranous, narrow wings usually shed at maturity; colonial and termitelike.

Order Phthiraptera (thī–rap´-ter-a) (Gr. *phteir,* louse, + *apteros,* wingless): **lice.** Wingless ectoparasites adapted for clinging to warm-blooded hosts. **Sucking lice** (Figure 21.15) in former order Anoplura now constitute suborder Anoplura, mouthparts adapted for piercing and sucking, includes head lice, crab lice, and body lice. **Chewing lice** (Figure 21.14) in former order Mallophaga now divided among three suborders.

Order Thysanoptera (thy-sa-nop´ter-a) (Gr. *thysanos,* tassel, + *pteron,* wing): **thrips.** Length 0.5 to 5 mm (a few longer); wings, if present, long, very narrow, with few veins, and fringed with long hairs; sucking mouthparts; destructive plant-eaters, but some feed on insects.

Order Hemiptera (he-mip´ter-a) (Gr. *hemi,* half, + *pteron,* wing). Members have unique mouthparts specialized for piercing and sucking. Hemiptera is divided into three suborders: Heteroptera, Auchenorrhyncha, and Sternorrhyncha. Heteroptera contains **true bugs;** size 2 to 100 mm; wings present or absent; forewings with basal portion thickened and partly sclerotized; apical portion membranous; hindwings membranous; at rest, wings held flat over abdomen; many with odorous scent glands; includes water scorpions, water striders (see Figure 21.9), bedbugs, squash bugs, assassin bugs, chinch bugs, stink bugs, plant bugs, lace bugs, and many others. Auchenorrhyncha contains **hoppers** (Figures 21.28 B and 21.39) and **cicadas** (see Figure 21.24); four wings typical if wings are present. Sternorrhyncha contains **whiteflies, psyllids, aphids, mealybugs** (Figure 21.35 B), and **scale insects;** four wings typical if wings are present; often with complex life histories; many species are plant pests.

Order Neuroptera (neu-rop´ter-a) (Gr. *neuron,* nerve, + *pteron,* wing): **dobsonflies, ant lions** (Figure 21.40), **lacewings.** Medium to large size; similar, membranous wings with many cross veins; chewing mouthparts; dobsonflies with greatly enlarged mandibles in males, and with aquatic larvae; ant lion larvae (doodlebugs) make craters in sand to trap ants.

Order Coleoptera (ko-le-op´ter-a) (Gr. *koleos,* sheath, + *pteron,* wing): **beetles** (see Figures 21.7A, 21.29, 21.35A), including **fireflies** (see Figure 21.30) and **weevils** (Figure 21.41D). The largest order of animals in the world with 250,000 described species; front wings (elytra) thick, hard, opaque; membranous hindwings folded under front wings at rest; mouthparts for biting and chewing; includes ground beetles, carrion beetles, whirligig beetles, darkling beetles, stag beetles, dung beetles, diving beetles, boll weevils, and many others.

Order Strepsiptera (strep-sip´ter-a) (Gr. *strepsis,* a turning, + *pteron,* wing): **stylops** or **twisted wing parasites.** Females wingless, without eyes or antennae; males with vestigial forewings and fan-shaped hindwings; females and larvae parasites of bees, wasps, and other insects.

Order Mecoptera (me-kop´ter-a) (Gr. *mekos,* length, + *pteron,* wing): **scorpionflies.** Small to medium size; wings long, slender, with many veins; at rest, wings held rooflike over back; males have scorpion-like clasping organ at end of abdomen; carnivorous; live in most woodlands.

Order Lepidoptera (lep-i-dop´ter-a) (Gr. *lepidos,* scale, + *pteron,* wing): **butterflies** (Figure 21.41A) and **moths.** Membranous wings covered with overlapping scales; wings coupled or overlapping; mouthparts a sucking tube, coiled when not in use; larvae (caterpillars) with chewing mandibles for plant eating, stubby prolegs on abdomen, and silk glands for spinning cocoons; antennae knobbed in butterflies, and usually filamentous (sometimes plumed) in moths (Figure 21.41B).

Order Diptera (dip´ter-a) (Gr. *di,* two, + *pteron,* wing): **true flies** (Figure 21.10). Single pair of wings, membranous and narrow; hindwings reduced to inconspicuous balancers (halteres); sucking mouthparts or adapted for sponging, lapping, or piercing; legless larvae (maggots); includes crane flies, mosquitos, moth flies, midges, fruit flies, flesh flies, house flies, horse flies, bot flies, blow flies, and many others.

Order Trichoptera (tri-kop´ter-a) (Gr. *trichos,* hair, + *pteron,* wing): **caddisflies.** Small, soft bodies; wings well veined and partially scaled, hairy, folded rooflike over hairy body; chewing mouthparts, mandibles much reduced; aquatic larvae of many species construct cases of leaves, sand, gravel, bits of shell, or plant matter, bound together with secreted silk or cement; some make silk feeding nets attached to rocks in stream.

Order Siphonaptera (si-fon-ap´ter-a) (Gr. *siphon,* a siphon, + *apteros,* wingless): **fleas** (see Figure 21.13). Small; wingless; bodies laterally compressed; legs adapted for leaping; ectoparasites or micropredators of birds and mammals; larvae legless, maggotlike scavengers.

Figure 21.39 An oak treehopper, *Platycotis vittata* (order Hemiptera).

James Carmichael, Jr./NHPA

Taxonomy of Class Entognatha and Class Insecta

Order Siphonaptera (si-fon-ap´ter-a) (Gr. *siphon,* a siphon, + *apteros,* wingless): **fleas** (See Figure 21.13). Small; wingless; bodies laterally compressed; legs adapted for leaping; ectoparasites or micropredators of birds and mammals; larvae legless, maggotlike scavengers.

Order Hymenoptera (hi-men-op´ter-a) (Gr. *hymen,* membrane, + *pteron,* wing): **ants, bees, wasps** (Figure 21.41C). Very small to large; membranous, narrow wings coupled distally; subordinate hindwings; mouthparts for chewing and lapping liquids; ovipositor sometimes modified into stinger, piercer, or saw (see Figure 21.8); both social and solitary species, most larvae legless, blind, and maggotlike.

A

B

Figure 21.40   A, Adult ant lion (order Neuroptera). **B,** Note the long, prominent, clubbed antennae on the ant lion head.

A **B** **C** **D**

Figure 21.41   A, Citrus swallowtail butterfly (*Papilio demodocus*). Members of the Papilionidae grace many areas of the world, both tropical and temperate, including North America. Compare knobbed antennae with plumed antennae in **B,** the North American polyphemus moth (*Antheraea polyphemus*). **C,** Paper wasps (order Hymnoptera) attending their pupae and larvae. **D,** This black vine weevil *Otiorhynchus sulcatus* has spread through North America via ornamental potted plants. It was introduced from Europe. Weevils belong to the largest family, Curculionidae, in the largest insect order, Coleoptera (beetles).

Such research, although very promising, is slow because of our limited understanding of insect behavior and the problems of isolating and identifying complex compounds that are produced in such minute amounts. Nevertheless, pheromones will probably play an important role in biological pest control in the future.

A systems approach called **integrated pest management** is practiced with many crops. This approach involves mixing all possible, practical techniques to contain pest infestations at a tolerable level; for example, cultivating techniques (resistant plant varieties, crop rotation, timing of sowing, planting or harvesting, and others), use of biological controls, and sparse use of insecticides.

The sterile-male approach has been used effectively in eradicating screwworm flies, a livestock pest. Large numbers of male insects, sterilized by irradiation in the pupal stage, are introduced into the natural population; females that mate with the sterile flies lay infertile eggs.

21.3 PHYLOGENY AND ADAPTIVE DIVERSIFICATION

Our understanding of evolutionary relationships among the arthropod subphyla has changed greatly over the last decade. Subphylum Uniramia, which united myriapods and hexapods, was predicated on the assumption that uniramous appendages were a shared derived character (synapomorphy) that united these groups to the exclusion of all other arthropods. However, Uniramia did not emerge as a monophyletic taxon in phylogenies based on molecular characters, which grouped hexapods and crustaceans as a clade. Crustaceans have biramous appendages, so such a large difference in appendage form was initially considered strong evidence against the hypothesis of a sister-taxon relationship of hexapods and crustaceans. However, once the genetic basis of branched appendages was better understood, it was clear that changes in the number of branches on an appendage could evolve by a relatively simple change in gene expression. Members of former subphylum Uniramia were then divided between subphylum Myriapoda and subphylum Hexapoda.

Hexapods are now united with members of subphylum Crustacea in clade Pancrustacea. Both taxa have mandibles, and several recent phylogenies indicate that pancrustaceans are more closely related to myriapods, a mandibulate taxon, than they are to chelicerates, which lack mandibles.

While clade Pancrustacea is well supported, the nature of the relationship between crustacean and hexapod subgroups is subject to debate. Some phylogenies indicate a sister-taxon relationship between crustaceans and hexapods, as we show in Figure 19.2, but others indicate that hexapods arose *within* Crustacea. If this result is supported by future studies, subphylum Crustacea will be paraphyletic unless it is redefined to include hexapods. In phylogenies where hexapods originate within Crustacea, they appear closest to branchiopod, cephaplocarid, or remipedian crustaceans. The next decade of research should clarify the evolutionary position of Hexapoda within Pancrustacea.

Within Hexapoda, class Entognatha is shown as the sister taxon to class Insecta in Figure 21.1, but some research indicates that entognathous mouthparts may have evolved several times, and that some entognathans may be closer to insects than they are to other entognathans.

Insect fossils date the origin of insects to the late Devonian period, but molecular and phylogenomic studies suggest that insects evolved in the early Ordovician period, about 479 MYA. The phylogenomic study supports remipedians as the closest living relatives of hexapods. The phylogeny from this study differs from Figure 21.1 in that Ephemeroptera and Odonata are sister taxa, and in the relative positions of Psocoptera, Phthirapera, Thysanoptera, and Hemiptera, among others. Some new higher order taxa are also developed.

It is likely that ancestral insects had a head and trunk of many similar segments, most or all of which bore limbs. Early fossil insects had small abdominal appendages (and apparently some multiramous appendages, Figure 21.42), and some modern, apterygote (primitively wingless) orders have abdominal styli that are considered vestigial legs. We now understand that absence of abdominal legs in most insects results from a pattern of expression of certain *Hox* genes that prevents expression of the *Distal-less* gene in the abdomen of insects but not crustaceans or onychophorans (p. 439).[1,2]

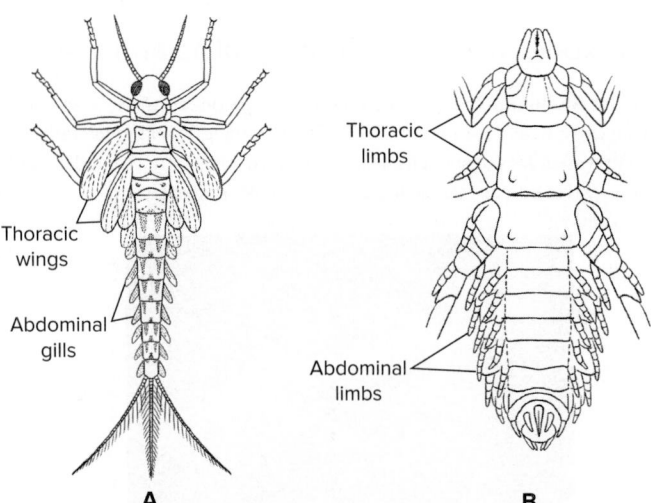

Figure 21.42 **A,** A Paleozoic "mayfly" nymph with thoracic winglets and abdominal gills. The thoracic winglets could have been precursors of wings. **B,** A Paleozic insect with multiramous thoracic legs and vestigial, multiramous abdominal limbs.

The evolutionary origin of insect wings has long been a puzzle. Evidence based on fossil insect mandibles suggests that winged insects were in existence about 400 million years ago. By the Carboniferous period, several orders of winged insects (Paleoptera), most of which are now extinct, had appeared. The adaptive value of wings for flight is clear, but such structures did not spring into existence fully developed. One hypothesis was that wings developed from lateral thoracic expansions that were useful in gliding. However, this hypothesis did not explain an origin or function of articulations and neuromusculature in the protowings that would provide raw material for selection and eventual evolution of flapping wings to support flight. An alternative hypothesis is that ancestral flying insects were derived from aquatic insects or insects with aquatic juveniles that bore external gills on their thorax from which wings could have been derived. Thoracic and abdominal gills on Paleozoic insects were apparently articulated and movable, capable of ventilation and swimming movements. They may have provided the morphological structures for "prowings." The evolution of broadly attached thoracic prowings (incapable of providing flight) in semi-aquatic insects would have increased the body temperature of such insects during basking. The subsequent expansion of these thoracic prowings for behavioral temperature regulation (basking) could easily have provided the morphological stage required for the evolution of truly functional wings (Figure 21.43) large enough to support flight.

The ancestral winged insect gave rise to three lines, which differed in ability to flex their wings. Two of these (Odonata and Ephemeroptera) have outspread wings or hold their wings vertically over the abdomen. The other line has wings that can fold back horizontally over the abdomen; this line branched into three groups, all of which

[1]Galant, R., and S. B. Carroll. 2002. Evolution of a transcriptional repression domain in an insect Hox protein. Nature **415:** 910–913.

[2]Ronshaugen, M., N. McGinnis, and W. McGinnis. 2002. Hox protein mutation and macroevolution of the insect body plan. Nature **415:** 914–917.

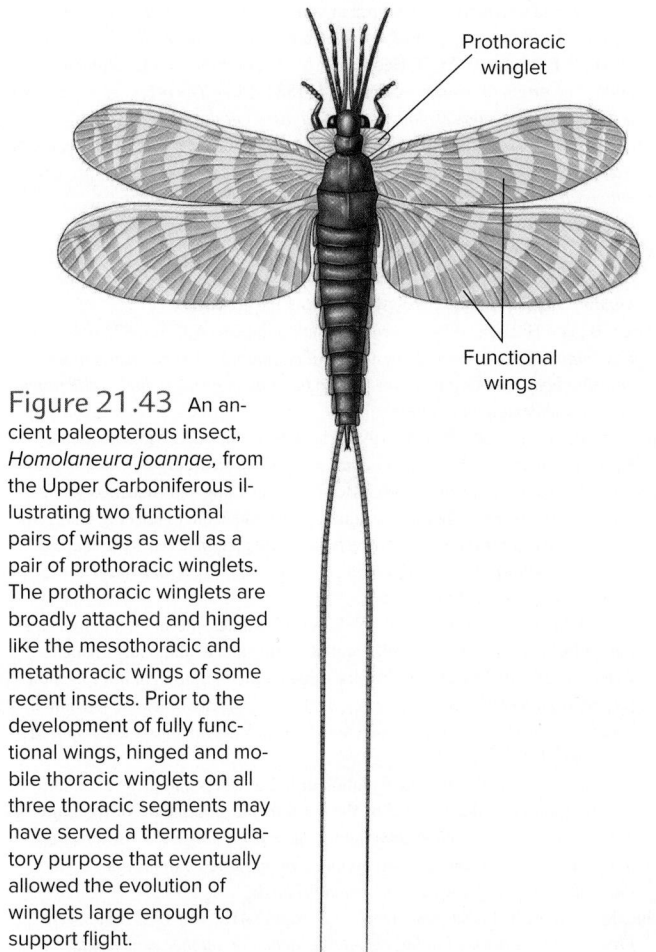

Figure 21.43 An ancient paleopterous insect, *Homolaneura joannae,* from the Upper Carboniferous illustrating two functional pairs of wings as well as a pair of prothoracic winglets. The prothoracic winglets are broadly attached and hinged like the mesothoracic and metathoracic wings of some recent insects. Prior to the development of fully functional wings, hinged and mobile thoracic winglets on all three thoracic segments may have served a thermoregulatory purpose that eventually allowed the evolution of winglets large enough to support flight.

Labels on figure: Prothoracic winglet; Functional wings

were present by the Permian period. One group with hemimetabolous metamorphosis, chewing mouthparts, and cerci includes Orthoptera, Dermaptera, Isoptera, and Embiidina; another group with hemimetabolous metamorphosis and a tendency toward sucking mouthparts includes Thysanoptera, Hemiptera, and perhaps also Psocoptera, Zoraptera, and Phthiraptera, although there is some disagreement among authorities about the last group. Insects with holometabolous metamorphosis have the most specialized life history, and these apparently form a clade that includes the remaining neopterous orders (for example, Lepidoptera, Diptera, Hymenoptera).

Adaptive properties of insects have been stressed throughout this chapter. The directions and ranges of their adaptive radiation, both structurally and physiologically, have been amazingly varied. Whether it be in the area of habitat, feeding adaptations, means of locomotion, reproduction, or general mode of living, the adaptive achievements of insects are truly remarkable.

S U M M A R Y

Members of subphylum Hexapoda have uniramous appendages and bear one pair of antennae, a pair of mandibles, and two pairs of maxillae on the head. Tagmata are head, thorax, and abdomen. Hexapods have three pairs of jointed thoracic legs. There are two classes: Entognatha and Insecta. Entognatha contains three orders of tiny animals commonly inhabiting soils.

Insecta is the largest class of the world's largest phylum. Insects have mouthparts whose bases extend out of the head capsule.

The evolutionary success of insects in terrestrial habitats is largely explained by features such as wings (present in most), waterproofing of their cuticle and other mechanisms to minimize water loss, and the ability to become dormant during adverse conditions.

Most insects bear two pairs of wings on their thorax, although some have one pair and some are primitively or secondarily wingless. Wing movements in some insects are controlled by direct flight muscles, which insert directly on the base of the wings in the thorax, whereas others have indirect flight muscles, which move the wings by changing the shape of the thorax. Each contraction of synchronous flight muscles requires a nerve impulse, while asynchronous flight muscles contract many times for each nerve impulse.

Feeding habits vary greatly among insects, and there is an enormous variety of specialization of mouthparts reflecting the particular feeding habits of a given insect. Insects breathe by means of a tracheal system, which is a system of tubes that opens by spiracles on the thorax and abdomen. Excretory organs are Malpighian tubules floating freely in the hemocoel.

Sexes are separate in insects, and fertilization is usually internal. Almost all insects undergo metamorphosis during development. In hemimetabolous (incomplete) metamorphosis, larval instars are called immatures or nymphs, and adults emerge at the last nymphal molt. In holometabolous (complete) metamorphosis, the last larval molt gives rise to a nonfeeding stage (pupa). A winged adult emerges at the final, pupal, molt. Both types of metamorphosis are hormonally controlled.

Insects are important to human welfare, particularly because they pollinate food and forage crop plants; control populations of other, harmful insects by predation and parasitism; and serve as food for other animals. Forensic entomology makes use of insect succession in dead bodies to estimate time of death. Many insects are harmful to human interests because they inflict great damage on crops, food, forests, clothing, and property, and many are carriers of important diseases affecting humans and domestic animals.

Hexapods are united with crustaceans in clade Pancrustacea. Entognathans either form or include the sister taxon of insects. They—like some insects—are wingless. Legs have been lost from the abdomen of insects, being now confined to each of the three thoracic segments. Wings of ancestral winged insects may have been derived from external gills of aquatic nymphs or adults, and there may have been a stage in which tergal expansions (or prowings), whether articulated or not, served to enhance the efficiency of behavioral thermoregulation until a size was reached to form efficient aerofoils for truly functional wings.

Adaptive diversity and the numbers of both species and individuals in Insecta are enormous.

R E V I E W   Q U E S T I O N S

1. What characteristics distinguish hexapods from *all* other arthropods?
2. How are insects distinguished from other hexapods?
3. Explain why indirect flight muscles can beat much more rapidly than direct flight muscles.
4. How do insects walk?
5. What major parts form an insect's gut, and how do these parts function?
6. Describe three different types of mouthparts found in insects, and tell how they are adapted for feeding on different foods.
7. Describe the tracheal system of a typical insect and explain why it can function efficiently without oxygen-carrying pigments in the hemolymph. Why would a tracheal system not be suitable for humans?

8. Describe the unique excretory system of insects. How is uric acid formed?

9. Describe sensory receptors that insects have to various stimuli.

10. Explain the difference between holometabolous and hemimetabolous metamorphosis in insects, including stages of each.

11. Describe hormonal control of metamorphosis in insects, including the action of each hormone and where each is produced.

12. What is diapause, and what is its adaptive value?

13. Briefly describe three features that insects have evolved to avoid predation.

14. Describe and give an example of each of four ways in which insects communicate with each other.

15. What castes occur in honey bees and in termites, and what is the function of each?

16. What mechanisms determine castes in honey bees and termites?

17. What is trophallaxis? What function(s) does it serve in termites?

18. Name several ways in which insects are beneficial to humans and several ways they are detrimental.

19. In what ways are detrimental insects controlled? What is integrated pest management?

20. What are the most probable characteristics of the most recent common ancestor of insects? What major lineages descended from it?

21. What is a plausible scenario for evolution of wings and flying insects?

For Further Thought Under what circumstances should natural selection favor possession of a holometabolous life cycle over an ametabolous or hemimetabolous life cycle? Consider food availability, specialized body parts, and dispersal ability in your answer.

SELECTED REFERENCES

Arnett, R. H., Jr., and M. C. Thomas, eds. 2000. American beetles, vol. 1. Boca Raton, Florida, CRC Press.

Arnett, R. H., Jr., M. C. Thomas, P. E. Skelley, and J. H. Franks, eds. 2002. American beetles, vol. 2. Polyphaga: Scarabaeoidea through Curculionoidea. Boca Raton, Florida, CRC Press. *These two volumes present recent detailed keys to families of American beetles; the second edition of volume 1 (2001) includes additions and corrections.*

Barber, J. R., B. C. Leavell, A. L. Keene, J. W. Breinholt, B. A. Chadwell, C. J. W. McClure, G. M. Hill, and A. Y. Kawahara. 2015. Moth tails divert bat attack: evolution of acoustic deflection. PNAS **112:** 2812–2816. *The long tails on the hindwings of luna moths confuse echo–locating bats and enhance moth survivorship.*

Beckage, N. E. 1997. The parasitic wasp's secret weapon. Sci. Am. **277:** 82–87 (Nov.). *This parasitoid wasp carries a virus that invades the host insect when the wasp lays its eggs, then paralyzes the host.*

Douglas, M. M. 1981. Thermoregulatory significance of thoracic lobes in the evolution of insect wings. Science **211:** 84–86. *Evolution of broadly attached thoracic winglets could have increased the body temperature excess of ancient wingless insects by 55% over that of forms without winglets. Subsequent expansion of winglets for behavioral thermoregulation could have provided the morphological stage required for evolution of functional wings.*

Engel, M. S., and D. A. Grimaldi. 2004. New light shed on the oldest insect. Nature **427:** 627–630. *Silurian fossil mandibles indicate that insects and entognathous hexapods had evolved by this period.*

Foret, S., R. Kucharski, M. Pellegrini, S. Feng, S. E. Jacobsen, G. E. Robinson, and R. Maleszka. 2012. DNA methylation dynamics, metabolic fluxes, gene splicing, and alternative phenotypes in honey bees. PNAS **109:** 4968–4973. *An exploration of the role of royal jelly in determining the queen bee phenotype.*

Glenner, H., P. F. Thomsen, M. B. Hebsgaard, M. V. Sorensen, and E. Willerslev. 2006. The origin of insects. Science **314:** 1883–1884. *The origin of insects as terrestrial crustaceans explains a puzzling aspect of the fossil record.*

Gullan, P. J., and P. S. Cranston. 2005. The insects: an outline of entomology, ed. 3. Malden, Massachusetts, Blackwell Publishing. *An easy-to-use text for general entomology with good figures, modern phylogenies, and data on biogeography*

Hayashi,. M. 1999. Attack of the fire ants. Sci. Am. **280:** 26, 28 (Feb.). *Fire ants have invaded Galápagos, Melanesia, and West Africa, where they may be blinding elephants and otherwise disrupting the ecosystem.*

Heinrich, B., and H. Esch. 1994. Thermoregulation in bees. Am. Sci. **82:** 164–170. *Fascinating behavioral and physiological adaptations for increasing and decreasing body temperature allow bees to function in a surprisingly wide range of environmental temperatures.*

Hölldobler, B. H., and E. O. Wilson. 1990. The ants. Cambridge, Massachusetts, Harvard University Press. *The fascinating story of social organization in ants.*

Johnson, N. F., and C. Triplehorn. 2005. Borror and DeLong's introduction to the study of insects, ed. 7. Belmont, California, Brooks/Cole Publishing Company. *A reference text for the study of insects; keys are included, but some require specialized knowledge of morphological characters. See Arnett and Thomas for keys to the American beetle families.*

Kingsolver, J. G., and M. A. R. Koehl. 1985. Aerodynamics, thermoregulation, and the evolution of insect wings: differential scaling and evolutionary change. Evolution **39 (3):** 488–504. *This work supports the thermoregulatory hypothesis proposed by Douglas by showing that as winglets approach the maximum benefit for temperature regulation they also reach the minimum size required for either gliding and/or flapping flight.*

Kukalova-Peck, J. 1978. Origin and evolution of insect wings and their relation to metamorphosis, as documented by the fossil record. Journal of Morphology **156:** 53–126. *A thorough examination of the paleontological evidence of the paleopterous insects and a re-examination of the stages that may have led to the evolution of insect wings and metamorphosis.*

Levine, M. 2002. How insects lose their limbs. Nature **415:** 848–849. *A Hox gene product in the abdomen of insects inhibits action of another gene product that is necessary for limb formation.*

Misof, B. et al. 2014. Phylogenomics resolves the timing and pattern of insect evolution. Science **346:** 763–767. *Insects are inferred to be much older than fossils indicate and a new phylogeny is presented.*

Raff, R. A. 1996. The shape of life: genes, development, and the evolution of animal form. Chicago, University of Chicago Press. *Includes a good account of how insect wings may have evolved.*

Regier, J. C., J. W. Shultz, and R. E. Kambic. 2005. Pancrustacean phylogeny: hexapods are terrestrial crustaceans and maxillopods are not monophyletic. Proc. R. Soc. Lond. Ser. B. **272:** 395–401. *Hexapods are closely related to Branchiopoda, Cephalocarida, and Remipedia.*

Topoff, H. 1990. Slave-making ants. Am. Sci. **78:** 520–528. *An amazing type of social parasitism in which certain species of ants raid the colonies of related species, abduct their pupae, then exploit them to do all the work in the host colony.*

Westneat, M. W., O. Betz, R. W. Blob, K. Fezzaa, W. J. Cooper, and W. Lee. 2003. Tracheal respiration in insects visualized with synchrotron X-ray imaging. Science **299:** 558–560. *Tracheae compress actively, exhaling air, but expand passively.*

Whiting, R. 2004. Phylogenetic relationships and evolution of insects, pp. 330–344. In J. Cracraft and M. J. Donoghue, eds., Assembling the tree of life. New York, Oxford University Press. *A detailed discussion of current hypotheses for insect evolution, including the suggestion that entognathous hexapods do not form a monophyletic group, unlike ectognathous forms.*

Wilfert, L., G. Long, C. Leggett, P. Schmid-Hempel, R. Butlin, S. J. M. Martin, and M. Boots. 2016. Deformed wing virus is a recent global epidemic in honeybees driven by Varroa mites. Science **351:** 594–597. *This paper and a more general companion article by Villalobos cited in Chapter 19 outline causes of Colony Collapse Disorder.*

22
CHAPTER

Chaetognaths, Echinoderms, and Hemichordates

- **PHYLUM CHAETOGNATHA**
- **PHYLUM XENOTURBELLIDA**
- **CLADE AMBULACRARIA**
  - Phylum Echinodermata
  - Phylum Hemichordata

Chaetognatha
Xenoturbellida
Echinodermata
Hemichordata

Sea *stars* (Pisaster ochraceus) *surrounded by green algae at low tide.*

A Design to Puzzle the Zoologist

Libbie Hyman, a distinguished American zoologist, once described echinoderms as a "noble group especially designed to puzzle the zoologist." With a combination of characteristics that should delight the most avid reader of science fiction, echinoderms would seem to confirm Lord Byron's observation that

> *Tis strange—but true;*
> *for truth is always strange;*
> *Stranger than fiction.*

The echinoderm fossil record contains bilaterally symmetrical forms. This, in conjunction with their bilaterally symmetrical larvae, indicates that the current radial symmetry is derived, despite the adaptive value of bilaterality for free-moving animals, and the merits of radial symmetry for sessile animals.

Most echinoderm larvae develop a rudimentary calcareous endoskeleton and use ciliary tracts for feeding and swimming, but once they metamorphose into sessile forms their bodies change radically. One coelomic compartment develops into a unique water-vascular system that uses hydraulic power to operate a multitude of tiny tube feet used in food gathering and locomotion. The endoskeleton may be reduced, as in a soft-bodied sea cucumber, or rigid, as in a sea star or a sea urchin. Many echinoderms have miniature jawlike pincers (pedicellariae) scattered on their body surface, often stalked and sometimes equipped with poison glands.

This constellation of characteristics is unique in the animal kingdom. Despite the vast amount of research that has been devoted to them, we are still far from understanding many aspects of echinoderm biology. Even the position of the anterior-posterior body axis, a feature so obvious in almost all other animals, is difficult to discern in echinoderms.

Triploblastic animals are divided into two large clades: Protostomia and Deuterostomia. Clades are diagnosed by a combination of molecular and morphological features, with suites of morphological features visible in early development. Classical developmental characters associated with protostomes are spiral mosaic cleavage; formation of the mouth from the embryonic blastopore (protostomy); and formation of a coelom by schizocoely, when a coelom is present (see Chapter 8, pp. 170–171). Classical deuterostome developmental features are radial regulative cleavage, formation of the mouth from a second opening (deuterostomy), and coelom formation by enterocoely (see Chapter 8, pp. 171–172). All deuterostomes other than *Xenoturbella* are coelomate.

Members of some phyla possess all the developmental characters in each suite: marine annelids and molluscs are classical protostomes, and echinoderms are classical deuterostomes. However, readers of the preceding chapters will be aware that some taxa, such as the problematic lophophorates, do not exhibit all protostome features.

Despite the progress made in charting the path of animal evolution, there are some phyla whose body plans seem to represent a "mix-and-match" strategy; they have both protostome and deuterostome features. The lophophorates have already been mentioned, and here we include another phylum, Chaetognatha, whose evolutionary position is much debated. The chaetognaths are pelagic marine predators commonly called arrow worms. They are an ancient group with representatives among the Burgess Shale fossils, so they are more than 500 million years old. Fossil forms are strikingly similar to modern forms in external appearance. This fascinating phylum has been placed in Deuterostomia, in Protostomia, and outside the two groups.

We depict Chaetognatha as outside the protostome and deuterostome clades (see the Cladogram of All Animal Phyla on the inside front cover and Figure 22.1), pending new studies. Chaetognaths are

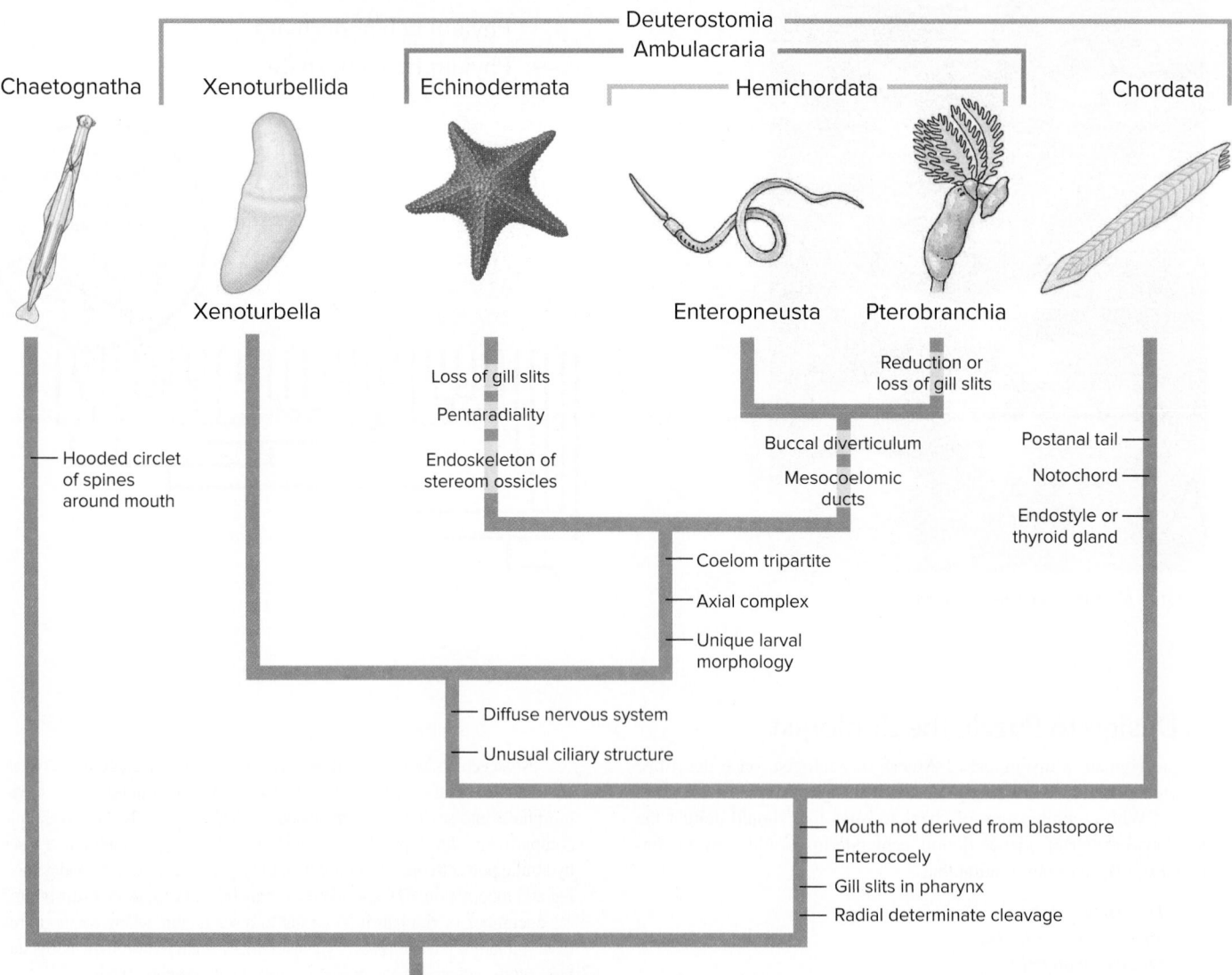

Figure 22.1 Cladogram showing hypothetical relationships among deuterostome phyla. Based on this hypothesis, the ancestral deuterostome was a marine benthic form with gill slits, perhaps similar to enteropneusts. Gill slits apparently were present in stem echinoderms, but lost in modern echinoderms. Although this relationship is now widely accepted, some prefer an alternative hypothesis that hemichordates and chordates form a clade, based on shared gill slits and a dorsal, hollow nerve cord.

discussed in this chapter, followed by three deuterostome phyla, Xenoturbellida, Echinodermata, and Hemichordata. The latter two are now considered sister taxa in clade Ambulacraria (Figure 22.1). The status of Xenoturbellida is much debated.

22.1 PHYLUM CHAETOGNATHA

A common name for chaetognaths is arrow worms. They are all marine animals and most are highly specialized for their planktonic existence. The name Chaetognatha (ke-tog′na-tha) (Gr. *chaitē,* long flowing hair, + *gnathos,* jaw) denotes the sickle-shaped bristles on each side of their mouth. This is not a large group, with about 130 known species. Their small, straight bodies resemble miniature torpedoes, or darts, ranging from less than 1 to about 12 cm in length (Figure 22.2).

With the exception of *Spadella* (Gr. *spadix,* palm frond, + *ella,* dim. suffix), a benthic genus, and a few species that live close to the deep-ocean floor, arrow worms are all adapted for a planktonic existence. They usually swim to the surface at night and descend during daytime. Much of the time they drift passively, but they can dart forward in swift spurts, using their caudal fin and longitudinal muscles—a fact that no doubt contributes to their success as planktonic predators. Horizontal fins bordering the trunk serve largely as stabilizers and are used in flotation rather than in active swimming.

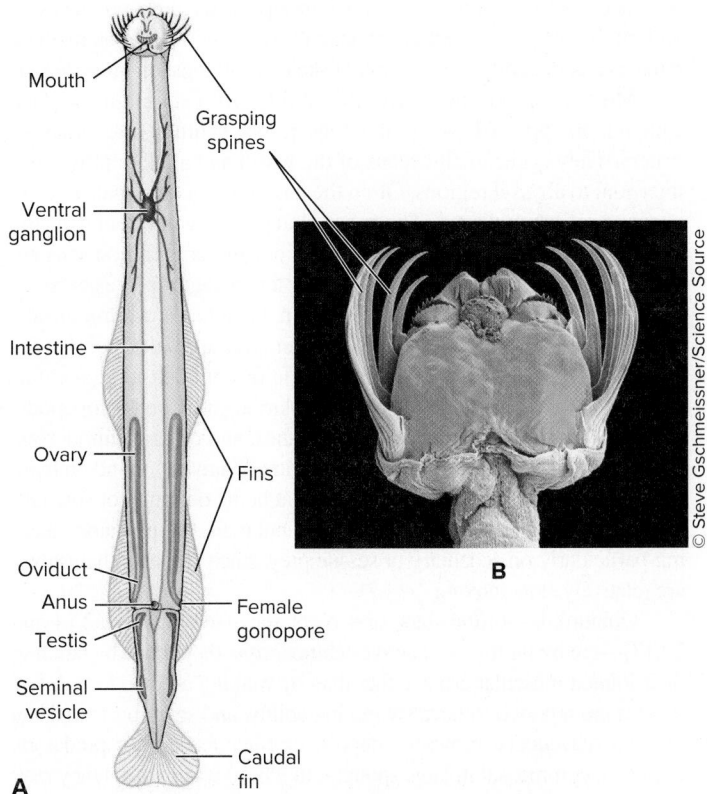

Figure 22.2 Arrow worms. **A,** Internal structure of *Sagitta.*
B, Scanning electron micrograph of the anterior end of an arrow worm.

22.2 FORM AND FUNCTION

Arrow worms are unsegmented, and their body includes a head, trunk, and postanal tail (Figure 22.2A). Beneath the head is a large depression, called a vestibule, leading to the mouth. The vestibule contains teeth and is flanked on both sides by curved chitinous spines used in seizing prey (Figure 22.2B). A pair of eyes is located dorsally. A peculiar hood formed from a fold of the neck can be drawn forward over the head and spines. When a chaetognath captures prey, it retracts its hood, and its teeth and raptorial spines spread apart and then snap shut with startling speed. Arrow worms are voracious predators, living on planktonic forms, especially copepods, but including a variety of other planktonic crustaceans, small fish, and even other chaetognaths. When they are abundant, as they often are, they may have a substantial ecological impact. Most species are highly mobile and nearly transparent, characteristics likely of adaptive value in their role as planktonic predators.

A thin cuticle covers the body, and their epidermis is single layered except along the sides, where it is stratified in a thick layer. These are the only invertebrates with a many-layered epidermis.

Arrow worms have a complete digestive system and a well-developed coelom. The coelom is used as a hydrostatic skeleton. There is a nervous system with a nerve ring connecting a cerebral ganglion above the esophagus to a number of lateral ganglia and a large ventral ganglion. Sense organs include eyes, sensory bristles, and possibly a unique U-shaped ciliary loop that extends over their neck from the back of the head. The exact function of this loop remains unknown, but it may detect vibrations or water currents, or may be chemosensory. Chaetognaths use vibrations of the sensory bristles to detect prey. Respiratory and excretory systems are entirely lacking and these processes appear to be accomplished solely by diffusion. A loosely organized hemal system has been described for chaetognaths.

Arrow worms are hermaphroditic with either reciprocal or self-fertilization. Eggs of *Sagitta* (L. arrow) bear a coat of jelly and are planktonic. Eggs of other arrow worms may be released to sink to the substratum during development, attached to stationary objects, or attached to the parent and carried while they develop. Juveniles develop directly without metamorphosis.

Chaetognath embryogenesis suggests deuterostome affinities. The mouth does not arise from the blastopore, and the coelom develops by enterocoely. However, the exact nature of coelom formation remains contentious. Some authorities state that embryonic coelom formation is clearly enterocoelous, while others argue that it differs from that of typical deuterostomes because the chaetognath coelom is formed by a backward extension from the archenteron rather than by pinched-off coelomic sacs. There is no true peritoneum lining the coelom.

Chaetognaths were described as having a tripartite coelom like that present in members of Ambulacraria, but apparently the third section of the coelom is merely a partition between male and female gonads in these hermaphrodites.

Early descriptions of chaetognath cleavage describe it as radial, complete, and equal, but more recent studies dispute this description, finding instead that cleavage planes in four-cell embryos are similar to those of crustaceans and nematodes (see Chapter 8 for a discussion of cleavage).

Chaetognaths and nematodes both lack circular muscles and have a similar arrangement of longitudinal muscles. Some phylogenies based on nucleotide sequences placed chaetognaths within Ecdysozoa, but there have been no reports that the thin cuticle is molted.

One phylogenetic study using *Hox* genes as characters suggested that chaetognaths branched from the animal lineage before the protostome/deuterostome split. Some phylogenies place them within protostomes as the sister taxon to all other members. Protostome features present in adult animals include the ventral nerve cord. Adult characters and nucleotide sequence data are at odds with developmental data, so the evolutionary position of chaetognaths remains in doubt despite continued study. A recent phylogeny indicated that the modern taxa share many convergent features, making it difficult to identify the ancestral body plan.

22.3 PHYLUM XENOTURBELLIDA

It is surprising that a fourth deuterostome phylum would be erected now, because all other deuterostomes have an extensive fossil record and are quite well studied. The yellowish, wormlike, ciliated animals in this phylum were first described in 1949, but they have been slow to find a taxonomic home. They have been placed with turbellarian flatworms (hence the name), and with molluscs.

Xenoturbellida contains a single genus, *Xenoturbella*, with two species: *Xenoturbella bocki* and *X. westbladi*. The animals reach lengths of 3 cm and widths near 0.5 cm, and live in North Sea mud. They feed on bivalves and bivalve eggs.

The body has few distinct morphological features, but two furrows are visible externally: one runs around the middle of the body and a second occurs laterally. The furrows may have a sensory function because they overlie thickenings in the diffuse nerve net. The body is not cephalized. The mouth opens in the middle of the flattened ventral body surface; it leads into a blind gut. There are four tissue layers, including a muscular layer with radial, circular, and longitudinal muscles. There are no structured gonads, but sexual reproduction occurs. *Xenoturbella* has direct development similar to that of acoelomorph flatworms; there is no feeding larval stage.

Does this animal belong within Deuterostomia? Studies of mitochondrial genes and gene order say that it does, but in some molecular phylogenies, acoelomorph flatworms (see p. 290) and xenoturbellids form a clade called Xenacoelomorpha. In some instances this clade is placed at the base of Bilateria and in others, the entire clade is placed within the deuterostomes. The placement of *Xenoturbella* as the sister taxon to Ambulacraria (see below) is tentative and based on the shared presence of unusual ciliary structures and the diffuse nervous system. If this position is supported by future studies, we must assume that many ancestral features have been lost in *Xenoturbella*.

22.4 CLADE AMBULACRARIA

Ambulacraria is a superphylum that contains two deuterostome phyla: Echinodermata and Hemichordata (Figure 22.1). Echinoderms, including sea stars, brittle stars, and sea cucumbers, are familiar to many people, but hemichordates, including acorn worms and pterobranchs, are much less familiar. Aside from classical deuterostome features, members of Ambulacraria share a three-part (tripartite) coelom, similar larval forms, and an axial complex (a highly specialized metanephridium).

22.5 PHYLUM ECHINODERMATA

Echinoderms are marine animals such as sea stars (also called starfishes), brittle stars, sea urchins, sea cucumbers, and sea lilies. They represent a bizarre group sharply distinguished from all other animals. The name Echinodermata (e-ki′nō-der′ma-ta) (L. *echinatus*, prickly, + Gr. *derma*, skin, + *ata*, characterized by) is derived from their external spines or protuberances. Echinoderms have a combination of characteristics found in no other phylum: (1) an endoskeleton of large plates or small scattered ossicles, (2) a water-vascular system, (3) pedicellariae, (4) dermal branchiae, and (5) basic pentaradial symmetry in the adults. No other group with such complex organ systems has radial symmetry.

Echinoderms are an ancient group of animals extending back to at least the Cambrian period. The recent discovery of bilaterally symmetrical adult echinoderms from the early middle Cambrian (Europe) indicates that pentaradial symmetry is derived. Biologists had long assumed a bilateral ancestor because echinoderm larvae are bilaterally symmetrical, but fossil evidence had been lacking. The newly discovered ancient bilateral forms were benthic deposit feeders. Most echinoderm fossils are attached (sessile) forms with radial symmetry. Hence, the body plan of present-day echinoderms seems to have been derived from one that was attached to the seafloor, had radial symmetry and radiating grooves (ambulacra) for food gathering, and had an upward-facing oral side. Attached forms likely were once plentiful, but only about 80 species, all in class Crinoidea, still survive (Figure 22.3). Instead, their free-moving descendants have become quite abundant and most have maintained radial symmetry. Nevertheless, in the exception that proves the rule (that bilaterality is adaptive for free-moving animals), at least three groups of echinoderms (sea cucumbers and two groups of sea urchins) have secondarily evolved a superficial bilateral organization (although there remains a pentaradial organization of skeletal and most organ systems).

Most echinoderms have no ability to osmoregulate (see Chapter 30, pp. 661–662), and thus rarely venture into brackish waters. They occur in all oceans of the world and at all depths, from intertidal to abyssal regions. Often the most common animals in deep oceans are echinoderms. The most abundant species found in the Philippine Trench (10,540 m) was a sea cucumber. Echinoderms are virtually all bottom dwellers, although a few pelagic species exist.

No parasitic echinoderms are known, but a few are commensals. On the other hand, a wide variety of other animals make their homes in or on echinoderms, including parasitic or commensal algae, unicellular eukaryotes, ctenophores, turbellarians, cirripeds, copepods, decapods, snails, clams, polychaetes, fishes, and other echinoderms.

Asteroids, or sea stars (Figure 22.4), are commonly found on hard, rocky surfaces, but numerous species are at home on sandy or soft substrates. Some species are particle feeders, but many are predators, feeding particularly on sedentary or sessile prey, since sea stars themselves are relatively slow moving.

Ophiuroids—brittle stars, or serpent stars (see Figures 22.14 and 22.17)—are by far the most active echinoderms; they move by bending their jointed muscular arms, rather than by walking on tube feet. A few species are reported to have swimming ability, and some burrow. They may be scavengers, browsers, deposit or filter feeders, or predators. Some are commensal in large sponges, in whose water canals they may live in great numbers.

Holothurians, or sea cucumbers (see Figure 22.24), are prevalent in all seas. Many inhabit sandy or mucky substrates, where they

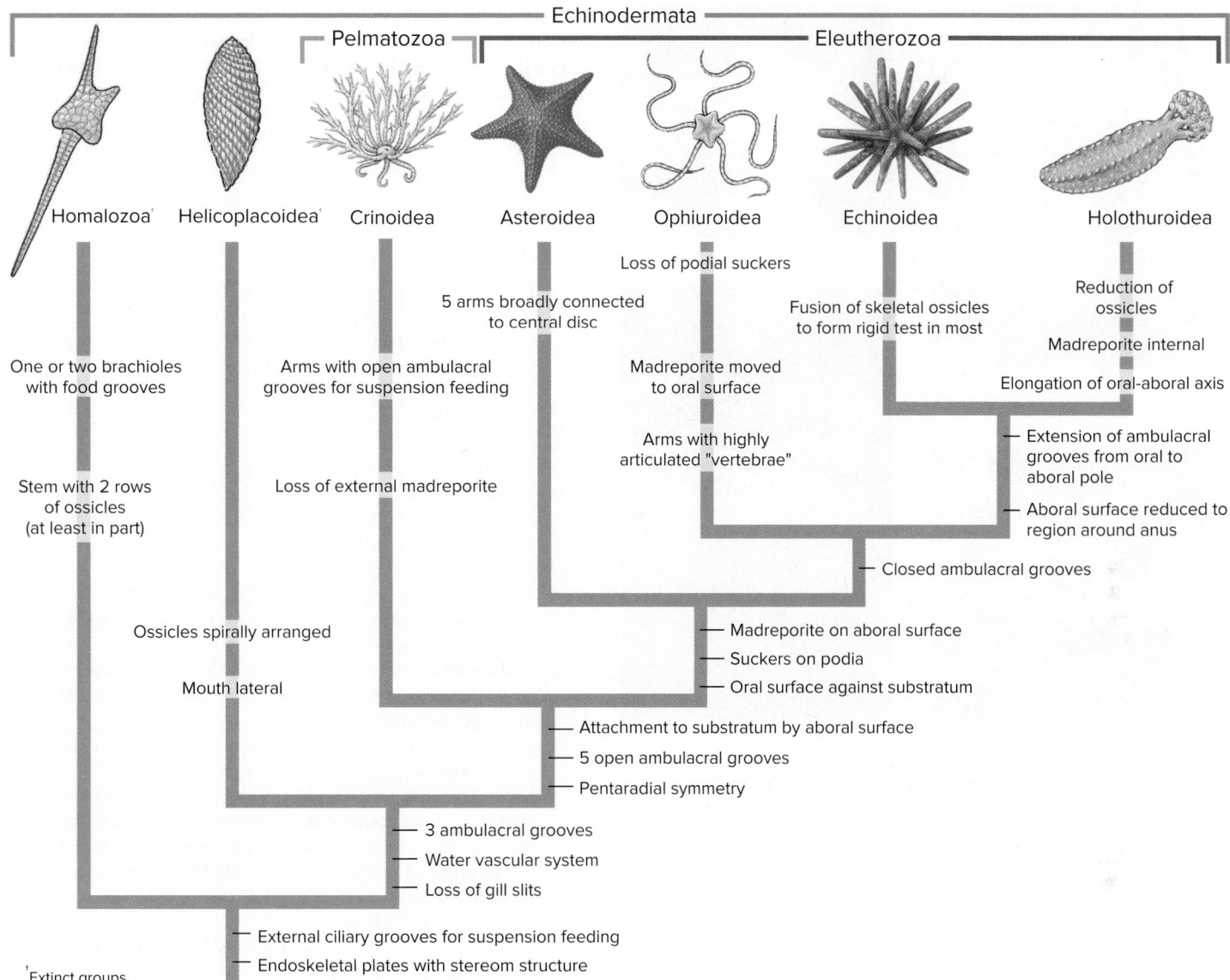

Figure 22.3 Cladogram showing hypothetical relationships among echinoderm groups. The extinct Homalozoa (carpoids), which were not radial in symmetry but had stereom endoskeletal plates, represent an early split from echinoderm ancestry. The extinct helicoplacoids had three ambulacral grooves that wound around their bodies in spiral fashion, and are hypothesized to be the sister group of modern echinoderms. Evolution of pentaradial symmetry was an adaptation to sessile existence and is a synapomorphy for modern echinoderms. The cladogram presented here depicts ophiuroids as having arisen separately from asteroids, after evolution of closed ambulacral grooves; possession of five arms would have arisen independently in these groups. An alternate scenario, also well supported, unites Asteroidea and Ophiuroidea in a clade, with five arms being a synapomorphy and independent evolution of closed ambulacral grooves in ophiuroids and the common ancestor of echinoids and holothuroids.

lie concealed. Compared with other echinoderms, holothurians are greatly extended in the oral-aboral axis. They are oriented with that axis more or less parallel to the substrate and lying on one side. Most are suspension or deposit feeders.

Echinoids, or sea urchins (see Figure 22.19), are adapted for living on the ocean floor and always keep their oral surface in contact with the substratum. Regular urchins, which are radially symmetrical, feed chiefly on algae or detritus, while irregulars, which are secondarily bilateral, feed on small particles. "Regular" sea urchins prefer hard substrates, but sand dollars and heart urchins ("irregular" urchins) are usually found on sand.

Crinoids (see Figure 22.29) stretch their arms outward and up likea flower's petals and feed on plankton and suspended particles. Most living species spend most of their time on the substrate, fastened by aboral appendages called cirri.

Because of their spiny bodies, echinoderms are not often prey of other animals—except other echinoderms (sea stars). Some fishes have strong teeth and other adaptations that enable them to feed on echinoderms. A few mammals, such as sea otters, feed on sea urchins. Some people relish sea urchin gonads, either raw or roasted on the half shell. Trepang, the cooked, protein-rich body wall of certain large sea cucumbers, is a delicacy in many east Asian countries. Unfortunately, the

Figure 22.4 Some sea stars (class Asteroidea) from the Pacific. **A,** Pincushion star, *Culcita navaeguineae,* preys on coral polyps and also eats other small organisms and detritus. **B,** *Choriaster granulatus* scavenges dead animals on shallow Pacific reefs. **C,** *Tosia queenslandensis* from the Great Barrier Reef System browses encrusting organisms. **D,** Crown-of-thorns star, *Acanthaster planci,* is a major coral predator (see note, p. 480).

biology of sea urchin development than that of almost any other embryonic system. Artificial parthenogenesis was first discovered in sea urchin eggs, when it was found that, by treating eggs with hypertonic seawater or subjecting them to a variety of other stimuli, development would proceed without sperm.

Characteristics of Phylum Echinodermata

1. Unique **water-vascular system** of coelomic origin extends from body surface as series of tentacle-like projections (**podia,** or **tube feet**) protracted by increase of fluid pressure within them; opening to exterior (**madreporite** or **hydropore**) usually present
2. Living in marine habitats
3. Free-living taxa
4. Body unsegmented (nonmetameric) with **pentaradial symmetry;** body rounded, cylindrical, or star-shaped, with five or more radiating areas, or **ambulacra,** alternating with interambulacral areas; no head
5. Triploblastic body
6. Coelom extensive, forming perivisceral cavity and cavity of water-vascular system; coelom of enterocoelous type; coelomic fluid with amebocytes
7. **Endoskeleton** of **dermal calcareous ossicles** with **spines** or of calcareous **spicules** in dermis; covered by epidermis (ciliated in most); **pedicellariae** (in some)
8. Digestive system usually complete; axial or coiled; anus absent in ophiuroids
9. Skeletal elements connected by ligaments of mutable collagenous tissue under neural control, ligaments can be "locked" into rigid posture or relaxed to allow free movement at will; locomotion by **tube feet,** which project from **ambulacral areas,** by movement of spines, or by movement of arms, which project from central disc of body
10. Nervous system with circumoral ring and radial nerves; usually two or three systems of networks located at different levels in the body, varying in degree of development according to group
11. **No brain;** few specialized sensory organs; sensory system of tactile and chemoreceptors, podia, terminal tentacles, photoreceptors, and statocysts
12. Autotomy and regeneration of lost parts conspicuous; asexual reproduction by fragmentation in some
13. Sexes separate (except a few hermaphroditic) with large gonads, single in holothuroids but multiple in most; simple ducts with no elaborate copulatory apparatus or secondary sexual structures; fertilization usually external; eggs brooded in some; development through **free-swimming, bilateral, larval stages** (some with direct development); metamorphosis to radial adult or subadult form; radial cleavage and regulative development
14. **Excretory organs absent**
15. Respiration by **papulae, tube feet, respiratory tree** (holothuroids), and **bursae** (ophiuroids)
16. Blood-vascular system (**hemal system**) much reduced, playing little if any role in circulation, and surrounded by extensions of coelom (**perihemal sinuses);** main circulation of body fluids (coelomic fluids) by peritoneal cilia

intense, often illegal, fishery for sea cucumbers has severely depleted their populations in many areas of the tropical world. However, aquaculture methods are being developed and sea cucumber ranching is becoming more common throughout the world. There are ranches in Alaska, Australia, China, Japan, the Philippines, New Caledonia, New Zealand, and Mexico, as well as other places.

Sea stars feed on a variety of molluscs, crustaceans, and other invertebrates. In some areas they may perform an important ecological role as a top carnivore in a community. Their chief economic impact is on clams and oysters. A single starfish can eat as many as a dozen oysters or clams in a day. To rid shellfish beds of these pests, quicklime is sometimes spread over areas where they abound. Quicklime damages the delicate epidermal membrane, destroying the dermal branchiae and ultimately the animal itself. Unfortunately, other softbodied invertebrates are also damaged. However, oysters remain with their shells tightly closed until the quicklime is degraded.

Echinoderms have been widely used in developmental studies because their gametes are usually abundant and easy to collect and raise in a laboratory. Investigators can follow embryonic developmental stages with great accuracy. We know more about molecular

Class Asteroidea

Sea stars, previously called starfishes, illustrate the basic features of echinoderm structure and function. Thus we will consider them first, then comment on major differences shown by other groups.

There are about 1500 living species. Sea stars are familiar along the shoreline where large numbers may aggregate on rocks. Sometimes they cling so tenaciously that they are difficult to dislodge without tearing off some tube feet. They also live on muddy or sandy substrates and among coral reefs. They are often brightly colored and range in size from a centimeter in greatest diameter to about a meter across. *Asterias* (Gr. *asteros,* a star) is a common genus of the east coast of the United States and is commonly studied in zoology laboratories. *Pisaster* (Gr. *pisos,* a pea, + *asteros,* a star, see photo p. 471) is common on the west coast of the United States, as is *Dermasterias* (Gr. *dermatos,* skin, leather, + *asteros,* a star), the leather star.

Form and Function

External Features   Sea stars are composed of a central disc that merges gradually with the tapering arms (rays). The body is somewhat flattened, flexible, and covered with a ciliated, pigmented epidermis. Their mouth is centered on the under, or oral, side, surrounded by a soft peristomial membrane. An **ambulacrum** (pl., **ambulacra,** L. *ambulacrum,* a covered way, an alley, a walk planted with trees), or **ambulacral area,** runs from the mouth on the oral side of each arm to the tip of the arm. Sea stars typically have five arms, but they may have more (Figure 22.4D), and there are as many ambulacral areas as there are arms. An **ambulacral groove** is found along the middle of each ambulacral area, and the groove is bordered by rows of **tube feet (podia)** (Figure 22.5). These in turn are usually protected by movable **spines.** A large **radial nerve** can be seen in the center of each ambulacral groove (Figure 22.6C), between the rows of tube feet. The nerve is very superficially located, covered only by thin epidermis. Under the nerve is an extension of the coelom and the radial canal of the water-vascular system, all of which are external to the underlying ossicles (Figure 22.6C). In all other classes of living echinoderms

except crinoids, these structures are covered by ossicles or other dermal tissue; thus ambulacral grooves in asteroids and crinoids are said to be *open,* and those of the other groups are *closed.*

The aboral surface is usually rough and spiny, although spines of some species are flattened, so that the surface appears smooth (Figure 22.4C). Around the bases of spines are groups of minute, pincerlike **pedicellariae,** bearing tiny jaws manipulated by muscles (Figure 22.7). These jaws keep the body surface free of debris, protect papulae, and sometimes aid in food capture. **Papulae (dermal branchiae** or **skin gills)** are soft, delicate projections of the coelomic cavity, covered only with epidermis and lined internally with peritoneum; they extend out through spaces between ossicles and are involved with respiration (Figures 22.6C and 22.7F). Also on the aboral side are an inconspicuous anus and a conspicuous circular **madreporite** (Figure 22.5A), a calcareous sieve plate leading to the water-vascular system.

Endoskeleton   Beneath the epidermis of sea stars is a mesodermal endoskeleton of small calcareous plates, or **ossicles,** bound together with connective tissue. This connective tissue is an unusual form of mutable collagen, termed **catch collagen,** that is under neurological control. Catch collagen can be changed from a "liquid" to a "solid" form very quickly when stimulated by the nervous system. This characteristic provides echinoderms with some unique mechanical properties; perhaps the most important is the ability to hold various postures without muscular effort. From the ossicles, spines and tubercles project to form the spiny surface. Ossicles are penetrated by a meshwork of spaces, usually filled with fibers and dermal cells. This internal meshwork structure is described as **stereom** and is unique to echinoderms.

Muscles in the body wall move the rays and can partially close the ambulacral grooves by drawing their margins together.

Coelom, Excretion, and Respiration   Coelomic compartments of larval echinoderms give rise to several structures in adults, one of which is a spacious body coelom filled with fluid. This fluid contains amebocytes (coelomocytes), bathes internal organs, and projects into papulae. The ciliated peritoneal lining of the coelom circulates the fluid

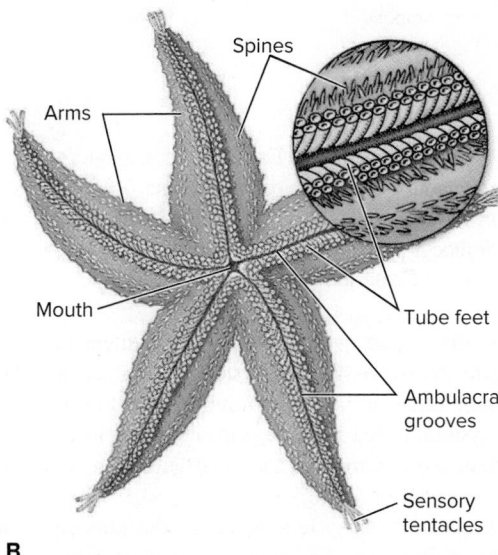

Figure 22.5
External anatomy of asteroid.
A, Aboral view.
B, Oral view.

A   B

Figure 22.6 **A,** Internal anatomy of a sea star. **B,** Water-vascular system. Podia penetrate between ossicles. (Polian vesicles are not present in *Asterias.*) **C,** Cross section of arm at level of gonads, illustrating open ambulacral grooves.

around the body cavity and into the papulae. Exchange of respiratory gases and excretion of nitrogenous waste, principally ammonia, occur by diffusion through the thin walls of papulae and tube feet. Some wastes may be engulfed by coelomocytes, which migrate through the epithelium of papulae or tube feet to the exterior.

Water-Vascular System
A second coelomic compartment in echinoderms gives rise to the water-vascular system. It is a set of canals and specialized tube feet that, together with the dermal ossicles, forms a hydraulic system. In sea stars the primary functions of the water-vascular system are locomotion and food gathering, in addition to respiration and excretion.

Structurally, the water-vascular system opens to the outside through small pores in the madreporite. The madreporite of asteroids is on the aboral surface (Figure 22.5A) and leads into a **stone canal,** which descends toward a **ring canal** around the mouth (Figure 22.6B). **Radial canals** diverge from the ring canal, one into the ambulacral groove of each ray. Also attached to the ring canal are four or five pairs of folded, pouchlike **Tiedemann's bodies** and from one to five **polian vesicles** (polian vesicles are absent in some sea stars, such as *Asterias*). Tiedemann's bodies may produce coelomocytes, and polian vesicles are apparently for fluid storage and regulation of the internal pressure within the water-vascular system.

A series of small **lateral canals,** each with a one-way valve, connects the radial canal to the cylindrical podia, or tube feet, along the sides of the ambulacral groove in each ray. Each podium is a hollow, muscular tube, the inner end of which is a muscular sac, or **ampulla,** which lies within the body coelom (Figure 22.6A and 22.6C), and

Figure 22.7 Pedicellariae of sea stars
A, Close-up view of the aboral surface of the sea star *Pycnopodia helianthoides*. Note the large pedicellariae, as well as groups of small pedicellariae around the spines. Many thin-walled papulae can be seen. **B,** Forceps-type pedicellaria with associated muscles. **C,** A rosette of small pedicellariae are shown at the base of a spine; note the relative size of the spine. **D,** Structure of a small pedicellaria.

D Small pedicellaria

C Rosette

Spine

Pedicellariae

© Jeff Rotman/Science Source

A Close-up view of the aboral surface of the sea star *Pycnopodia helianthoides*.

Valve

Adductor muscles

Abductor muscle

Stalk

B Large Forceps-type pedicellaria

the outer end of which usually bears a **sucker.** Some species lack suckers. Podia pass to the outside between ossicles in the ambulacral groove.

The water-vascular system operates hydraulically and is an effective locomotor mechanism. The basic design applies muscular pressure to coelomic fluid in tube feet to stiffen them for walking. The ampulla at the top of the tube foot serves as a fluid reservoir. Each tube foot has in its walls connective tissue that maintains the cylinder at a relatively constant diameter. Contraction of muscles in the ampulla forces fluid into the podium, extending it. Valves in the lateral canals prevent backflow of fluid into the radial canals. Conversely, contraction of the longitudinal muscles in the tube foot retracts the podium, forcing fluid back into the ampulla. Contraction of muscles in one side of the podium bends the organ toward that side. Small muscles at the end of the tube foot can raise the middle of the disclike end, creating suction when the end is applied to a firm substrate. It has been estimated that by combining mucous adhesion with suction, a single podium can exert a pull equal to 0.25 to 0.3 newtons. Coordinated action of all or many of the tube feet is sufficient to draw the animal up a vertical surface or over rocks. The ability to move while firmly adhering to the substrate is a clear advantage to an animal living in a sometimes wave-churned environment.

On a soft surface, such as muck or sand, suckers are ineffective (numerous sand-dwelling species have no suckers), so the tube feet are employed as legs. Locomotion becomes mainly a stepping process. Most sea stars can move only a few centimeters per minute, but some very active ones—*Pycnopodia* (Gr. *pyknos,* compact, dense,

+ *pous, podos,* foot), for example—can move 75 to 100 cm per minute. When inverted, a sea star bends its rays until some tubes reach the substratum and attach as an anchor; then it slowly rolls over.

Tube feet are innervated by the central nervous system (ectoneural and hyponeural systems). Nervous coordination enables tube feet to move in a single direction, although not in unison, so that the sea star may progress. If the radial nerve in an arm is cut, podia in that arm lose coordination, although they can still function. If the circumoral nerve ring is cut, podia in all arms become uncoordinated, and movement ceases.

Feeding and Digestive System   The mouth on the oral side leads through a short esophagus to a large stomach in the central disc. The lower (cardiac) part of the stomach can be everted through the mouth during feeding (Figure 22.9), and excessive eversion is prevented by gastric ligaments. The upper (pyloric) part is smaller and connects by ducts to a pair of large **pyloric ceca (digestive glands)** in each arm (Figure 22.6A). Digestion is mostly extracellular, although some intracellular digestion may occur in the ceca. A short intestine leads aborally from the pyloric stomach, and there are usually a few small, saclike **intestinal ceca** (Figure 22.6A). The anus is inconspicuous, and some sea stars lack an intestine and anus.

Many sea stars are carnivorous and feed on molluscs, crustaceans, polychaetes, echinoderms, other invertebrates, and sometimes small fishes. Sea stars consume a wide range of food items, but many show particular preferences. Some select brittle stars, sea urchins, or

A   B

© Mark Conlin/Alamy Stock Photo

© Paul Nicklen/National Geographic/Getty Images

Figure 22.8 **A,** *Orthasterias koehleri* eating a clam. **B,** A sea star captures red sea urchins as prey.

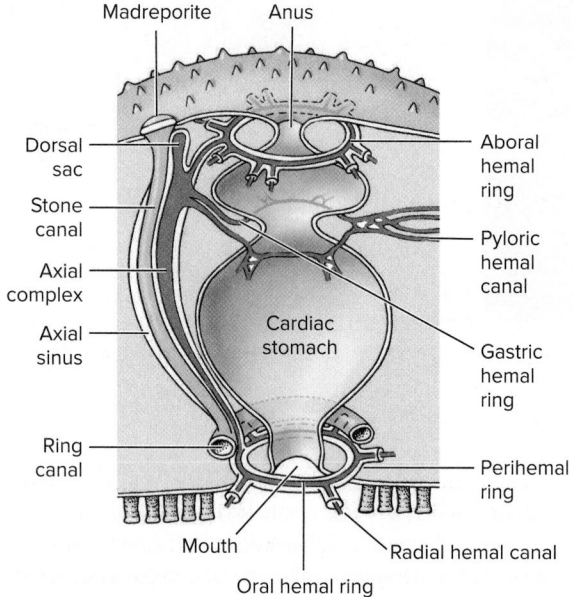

Figure 22.9 Hemal system of asteroids. The main perihemal channel is the thin-walled axial sinus, which encloses both the axial organ and the stone canal. Other features of the hemal system are shown.

Labels on Figure 22.9:
Madreporite
Anus
Dorsal sac
Stone canal
Axial complex
Axial sinus
Ring canal
Mouth
Aboral hemal ring
Pyloric hemal canal
Cardiac stomach
Gastric hemal ring
Perihemal ring
Radial hemal canal
Oral hemal ring

Since 1963 there have been numerous reports of increasing numbers of the crown-of-thorns starfish (*Acanthaster planci* [Gr. *akantha,* thorn, + *asteros,* star]) (see Figure 22.4D) that were damaging large areas of coral reef in the Pacific Ocean. Crown-of-thorns stars feed on coral polyps, and they sometimes occur in large aggregations, or "herds." There is some evidence that outbreaks have occurred in the past, but an increase in frequency during the past 40 years suggests that some human activity may be affecting the sea stars. Of reefs surveyed in 2002, 12% had outbreaks, compared with 1988 when 10% had outbreaks, causing extensive damage. Efforts to control these organisms are very expensive and of questionable effectiveness. The controversy continues, especially in Australia, where it is exacerbated by extensive media coverage.

sand dollars, swallow them whole, and later regurgitate undigestible ossicles and spines (Figure 22.8B). Some attack other sea stars, and if they are small compared with their prey, they may attack and begin eating at the end of one arm.

Some asteroids feed heavily on molluscs (Figure 22.8A); *Asterias* is a significant predator on commercially important clams and oysters. When feeding on a bivalve, a sea star will wrap itself around its prey, attaching its podia to the valves, and then exert a steady pull, using its feet in relays. A force of some 12.75 newtons can thus be exerted. This would be roughly equivalent to a human attempting to lift a 1000-lb weight with one hand. In half an hour or so the bivalve's adductor muscles fatigue and relax. With a very small gap available, the star inserts its soft, everted stomach into the space between the valves and wraps it around the soft parts of the bivalve to begin digestion. After feeding, the sea star draws its

stomach inward by contraction of its stomach muscles and relaxation of body-wall muscles.

Some sea stars feed on small particles, either entirely or in addition to carnivorous feeding. Plankton and other organic particles coming in contact with an animal's surface are carried by epidermal cilia to the ambulacral grooves and then to the mouth.

Hemal System   The so-called hemal system is not very well developed in asteroids, and its function in all echinoderms is unclear. The hemal system has little to do with circulation of body fluids. It is a system of tissue strands enclosing unlined sinuses and is itself enclosed in another coelomic compartment, or **perihemal channels** (Figure 22.9). Research with at least one sea star species shows that absorbed nutrients appear in the hemal system within a few hours of feeding, and eventually concentrate in the gonads and podia. Thus, the hemal system appears to play a role in distributing digested nutrients. It also includes the **axial complex,** which pressure-filters blood vascular fluids (Figure 22.9).

Nervous System   Echinoderms do not have a brain or distinct ganglia. The nervous system consists of three units lying within the disc and arms. Chief of these is an **oral (ectoneural)** system composed of a **nerve ring** around the mouth and a main **radial nerve** into each arm. It appears to coordinate the tube feet. A **deep (hyponeural)** system lies aboral to the oral system, and an **aboral** system consists of a ring around the anus and radial nerves along the roof of each ray. An **epidermal nerve plexus** or nerve net freely connects these systems with the body wall and related structures. The epidermal plexus coordinates responses of the dermal branchiae to tactile stimulation—the only instance known in echinoderms in which coordination occurs through a nerve net.

Sense organs are not well developed, but echinoderms do respond to changes in temperature, chemical composition of their surroundings, and light intensity. Tactile organs and other sensory cells are scattered over the surface, and an ocellus is at the tip of each arm. Sea stars are usually more active at night.

Reproductive System, Regeneration, and Autotomy   Most sea stars have separate sexes. A pair of gonads lies in each interradia space (see Figure 22.6A). Fertilization is external and occurs in early summer when eggs and sperm are shed into the water. A secretion from neurosecretory cells located on the radial nerves stimulates maturation and shedding of asteroid eggs.

Echinoderms can regenerate lost parts. Sea-star arms can regenerate readily, even if all are lost. Sea stars also have the power of autotomy and can cast off an injured arm near its base. Regeneration of a new arm may take several months.

Some species can regenerate a complete new sea star (Figure 22.10) from a detached arm. For most asteroids to regenerate, the detached arm must contain a part (about one-fifth) of the central disc. However, in some species, such as *Linckia*, no trace of the central disc is needed for regeneration. Some sea stars reproduce asexually under normal conditions by cleaving their central disc, each part regenerating the rest of the disc and missing arms.

Development Development is quite varied among different lineages of sea stars. Some species produce benthic egg masses in which juveniles develop. Other species produce eggs that are brooded, either under the oral side of the animal or in specialized aboral structures, and development is direct. Some species even have viviparous brooding of young in the gonads of the adults. However, most sea stars produce free-swimming planktonic larvae. Even here there is variation, and some species provision their young with sufficient yolk to develop without feeding in the water column, whereas others require a prolonged period of feeding to gain sufficient energy to metamorphose into the adult body form.

Early embryogenesis shows a typical ancestral deuterostome pattern (see Figures 8.8A and 8.19B). Gastrulation is by invagination, and the anterior end of the archenteron pinches off to become a coelomic cavity, which expands in a U-shape to fill the blastocoel. Each leg of the U, at the posterior, constricts to become a separate vesicle, and these eventually give rise to the main coelomic compartments of the body (metacoels, called **somatocoels** in echinoderms). The anterior portion of the U undergoes subdivision to form protocoels and mesocoels (called **axocoels** and **hydrocoels** in echinoderms) (Figure 22.11). The left hydrocoel will become the water-vascular system, and the left axocoel will give rise to the stone canal and perihemal channels. The right axocoel and hydrocoel will disappear. The free-swimming larva has cilia arranged in bands and is called a **bipinnaria** (Figure 22.12A). These ciliated tracts become extended into larval arms. Soon the larva grows three adhesive arms and a sucker at its anterior end and is then called a **brachiolaria** (Figure 22.12B). At that time it attaches to the substratum, forms a temporary attachment stalk, and undergoes metamorphosis.

Metamorphosis involves a dramatic reorganization of a bilateral larva into a radial juvenile. The anteroposterior axis of the larva is lost, and *what was the left side becomes the oral surface, and the larval right side becomes the aboral surface* (Figure 22.11). Correspondingly, larval mouth and anus disappear, and a new mouth and anus form on what were originally the left and right sides, respectively. The portion of the anterior coelomic compartment from the left side expands to form the ring canal of the water-vascular system around the mouth, and then it grows branches to form radial canals. As the short, stubby arms and the first podia appear, the animal detaches from its stalk and begins life as a young sea star. A number of regulatory genes found in bilateral animals are conserved in echinoderms and have surprisingly similar functions. For example, in vertebrates *Distal-less* and its homolog regulate outgrowth of limbs, whereas in echinoderms, its homolog is active in development of tube feet.

Sea Daisies Strange little (less than 1 cm diameter) disc-shaped animals (Figure 22.13) were discovered in water over 1000 m deep off New Zealand. Sometimes called sea daisies, they were originally described (1986) as a new class of echinoderms called Concentricycloidea. Only three species are known. Most zoologists now agree that they are highly derived spinulosid asteroids. Phylogenetic analysis of rDNA places them within Asteroidea.

Sea daisies are pentaradial in symmetry but have no arms. Their tube feet are located around the periphery of the disc, rather than along ambulacral areas, as in other echinoderms. Their water-vascular system includes two concentric ring canals; the outer ring may represent radial canals since podia arise from it. A hydropore, homologous to the madreporite, connects the inner ring canal to the aboral surface. One species has no digestive tract; its oral surface is covered by a membranous **velum,** by which it apparently absorbs nutrients. Another species has a shallow, saclike stomach but no intestine or anus.

Class Ophiuroidea

Brittle stars and basket stars form the largest group of echinoderms with over 2000 extant species. They abound in all types of benthic marine habitats, even carpeting the abyssal seafloor in many areas. Brittle stars have only 5 arms, but basket stars such as *Astrophyton* (Gr. *asteros,* star + *phyton,* creature, animal) (Figure 22.14B) and *Gorgonocephalus* (Gr. *Gorgo,* name of a female monster of terrible aspect + *kephalē,* a head) (see Figure 22.14B) have arms that branch repeatedly.

Form and Function

Although both have five arms, brittle stars are surprisingly different from asteroids. Their arms stars are slender and sharply set off from the central disc (Figure 22.14). They have no pedicellariae or papulae, and their ambulacral grooves are closed and covered with arm

Figure 22.10 Pacific sea star, *Echinaster luzonicus,* can reproduce itself by splitting across the disc, then regenerating missing arms. The one shown here has evidently regenerated six arms from the longer one at top left.

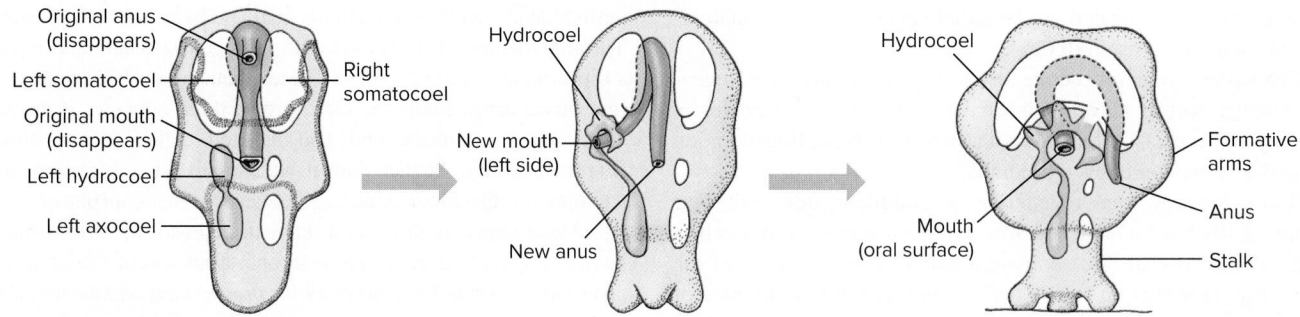

Figure 22.11   Asteroid metamorphosis. The left somatocoel becomes the oral coelom, and the right somatocoel becomes the aboral coelom. The left hydrocoel becomes the water-vascular system and the left axocoel the stone canal and perihemal channels. The right axocoel and hydrocoel are lost.

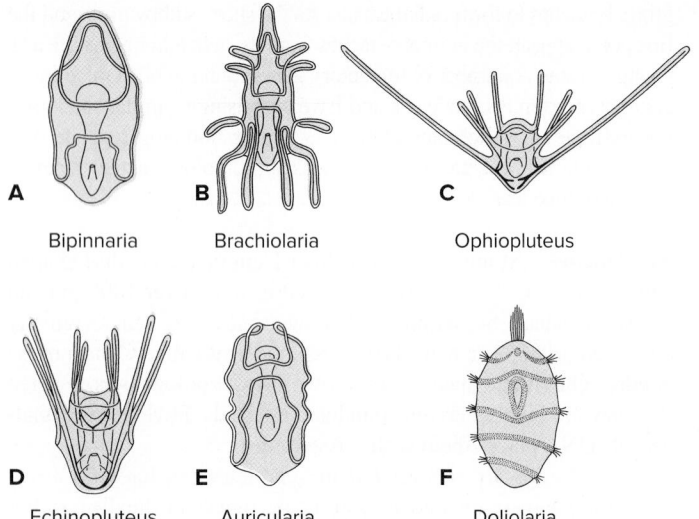

Figure 22.12   Larvae of echinoderms. **A,** Bipinnaria of asteroids. **B,** Brachiolaria of asteroids. **C,** Ophiopluteus of ophiuroids. **D,** Echinopluteus of echinoids. **E,** Auricularia of holothuroids. **F,** Doliolaria of crinoids.

ossicles. Their tube feet are without suckers; they aid in feeding but are of limited use in locomotion. In contrast to asteroids, the madreporite of ophiuroids is located on the oral surface, on one of the oral shield ossicles (Figure 22.15). Ampullae are absent from podia, and force for protrusion of a podium is generated by a proximal muscular portion of the podium.

Each jointed arm consists of a column of articulated ossicles (so-called **vertebrae**), connected by muscles and covered by plates. Locomotion is by arm movement. Arms are moved forward in pairs and are placed against the substratum, while one (any one) is extended forward or trailed behind, and the animal is pulled or pushed along in a jerky fashion.

Five movable plates that serve as jaws surround the mouth (Figure 22.15). There is no anus. Their skin is leathery, with dermal plates and spines arranged in characteristic patterns. Surface cilia are mostly lacking.

Visceral organs are confined to the central disc, since the rays are too slender to contain them (Figure 22.16). The stomach is sac-like, and there is no intestine. Indigestible material is cast out of the mouth.

Five pairs of invaginations called **bursae** open toward the oral surface by genital slits at the bases of the arms. Water circulates in and out of these sacs for exchange of gases. On the coelomic wall of each bursa are small gonads that discharge their ripe sex cells into the bursa. Gametes pass through the genital slits into the water for fertilization (Figure 22.17A).

Sexes are usually separate; a few ophiuroids are hermaphroditic. Some brood their young in the bursae; the young escape through the genital slits or by rupturing the aboral disc. Most species produce a free-swimming larva called an ophiopluteus, and its ciliated bands extend onto delicate, beautiful larval arms (Figure 22.12C). During metamorphosis to a juvenile, there is no temporarily attached phase, as there is in asteroids.

Water-vascular, nervous, and hemal systems are similar to those of sea stars. Each arm contains a small coelom, a radial nerve, and a radial canal of the water-vascular system.

Behavior and Ecology

Brittle stars tend to be secretive, living on hard substrates in locations where little or no light penetrates. They are often negatively phototropic and insinuate themselves into small crevices between rocks, becoming more active at night. They are commonly fully exposed only in the permanent darkness of deep seas. Ossicles in arms of at least some photosensitive ophiuroids show a remarkable adaptation for photoreception. Tiny, rounded structures on their aboral surface serve as microlenses, focusing light on nerve bundles just beneath them. Related species that are indifferent to light have no such structures.

Ophiuroids feed on a variety of small particles, either browsing food from the seafloor or suspension-feeding. Podia are important in transferring food to the mouth. Some brittle stars extend arms into the water and catch suspended particles in mucous strands between arm spines. Basket stars perch on corals, extending their branched arms to capture plankton (Figure 22.14B).

Some ophiuroids are carnivorous, and at least one species is a fish specialist that assumes an ambush posture with the central disc held off the substrate. When an unsuspecting fish enters the "shelter" beneath the central disc, the star twists abruptly to trap the fish in a spiral cylinder formed by the spiny arms.

Regeneration and autotomy are even more pronounced in brittle stars than in sea stars. Many seem very fragile, releasing an arm or even part of the disc at the slightest provocation. Some can reproduce asexually by cleaving the disc; each new individual then regenerates missing parts.

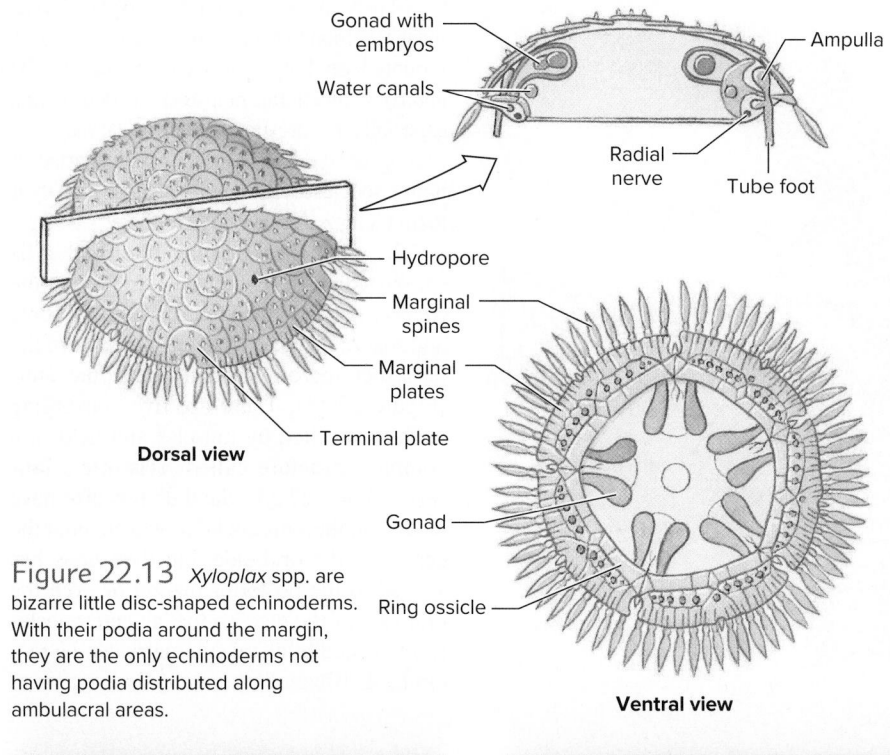

Figure 22.13 *Xyloplax* spp. are bizarre little disc-shaped echinoderms. With their podia around the margin, they are the only echinoderms not having podia distributed along ambulacral areas.

Figure 22.14 **A,** Brittle star, *Ophiothrix suensonii* (class Ophiuroidea) on a red rope sponge from Belize. Brittle stars do not use their tube feet for locomotion but can move rapidly (for an echinoderm) by means of their arms. **B,** Oral view of a giant basket star, *Astrophyton muricatum*. Basket stars extend their many-branched arms to filter-feed, usually at night.

Some common ophiuroids along the coast of the United States are *Amphipholis* (Gr. *amphi*, both sides of, + *pholis*, horny scale) (viviparous and hermaphroditic), *Ophioderma* (Gr. *ophis*, snake, + *dermatos*, skin), *Ophiothrix* (Gr. *ophis*, snake, + *thrix*, hair), and *Ophiura* (Gr. *ophis*, snake, + *oura*, tail) (Figure 22.14). Most ophiuroids are drab, but some are attractive, with bright color patterns (Figure 22.17A).

Class Echinoidea

There are about 950 living species of echinoids, which generally have a compact body enclosed in an endoskeletal shell, also called

a test, composed of tightly fitted dermal ossicles. Echinoids lack arms, but their tests bear a typical pentamerous plan of echinoderms in their five ambulacral areas. These areas can be hard to see on urchins with dense spines, but Figure 22.18A depicts four of the five rows of tube feet, and the external pores for tube feet can be seen in Figure 22.18B. In echinoids, the most notable modification of the ancestral body plan is that the oral surface, which bears the tube feet and faces the substratum in sea stars, has expanded around to the aboral side, so that ambulacral areas extend up to an area close to the anus **(periproct)**.

Most living species of sea urchins are "regular"; they have a hemispherical shape, radial symmetry, and medium to long spines (Figures 22.19 and 22.20). Sand dollars (Figure 22.21) and heart urchins (Figure 22.22) are "irregular" because members of their orders have become secondarily bilateral. Regular urchins move by means of their tube feet, with some assistance from their spines, and irregular urchins move chiefly by their short spines (Figure 22.21). Some echinoids are quite colorful, and some have greatly reduced tests, making them soft-bodied. These soft urchins often have bright warning coloration and their pedicellariae deliver painful toxins.

Echinoids have a wide distribution in all seas, from intertidal regions to deep oceans. Regular urchins often prefer rocky or hard substrates, whereas sand dollars and heart urchins burrow into a sandy substrate. Distributed along one or both coasts of North America are common genera of regular urchins (*Arbacia* [Gr. *Arbakēs*, first king of Media], *Strongylocentrotus* [Gr. *strongylos*, round, compact, + *kentron*, point, spine] [Figure 22.20], *Lytechinus* [Gr. *lytos*, dissolvable, broken, + *echinos*, sea urchin]) sand dollars (*Dendraster* [Gr. *dendron*, tree, stick, + *asteros*, star] and *Echinarachnius* [Gr. *echinos*, sea urchin, + *arachnē*, spider]). The West Indies–Florida region is rich in echinoderms, including echinoids, of which *Diadema* (Gr. *diadeō*, to bind around), with its long, needle-sharp spines, is a notable example (Figure 22.19D; see note p. 485).

Form and Function

In general, an echinoid test is a compact skeleton of 10 double rows of plates that bear movable, stiff spines (Figure 22.18). Collagen

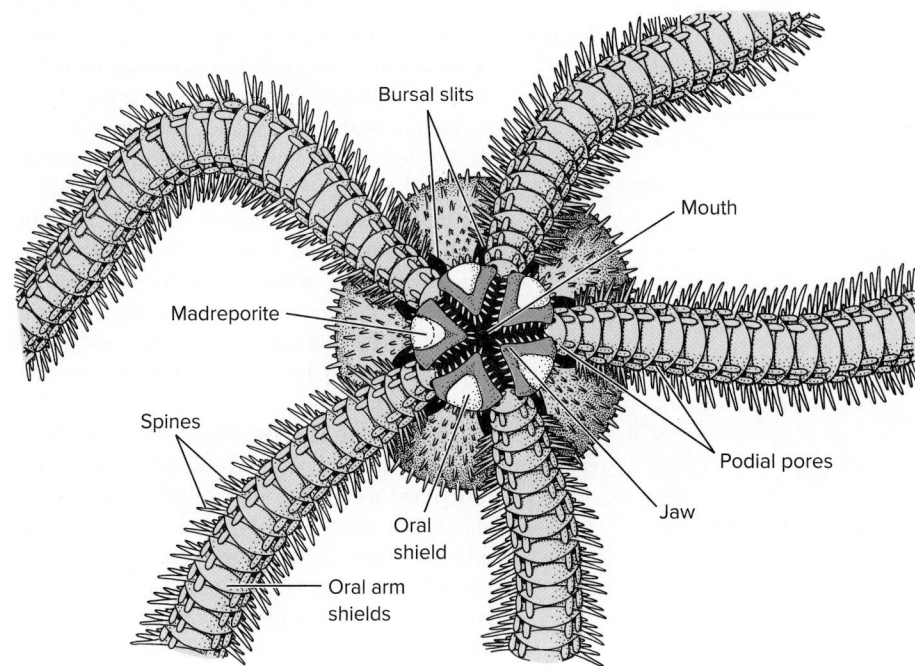

Figure 22.15   Oral view of spiny brittle star, *Ophiothrix*.

There are several kinds of pedicellariae, most common of which are three-jawed forms mounted on long stalks (see Figure 22.7D and E). Pedicellariae help keep the body clean, especially by preventing marine larvae from settling on the body surface. Pedicellariae of many species bear venom glands, and their toxin paralyzes small prey.

In some sea urchins branched gills (modified podia) encircle the peristome (region around the mouth). Genital pores and madreporite are located aborally, in the periproct region that surrounds the anus (Figure 22.18). There are five converging teeth, controlled by muscles and held in a complex structure called **Aristotle's lantern** (Figure 22.23). Sand dollars also have teeth, and their mouth is located at about the center of the oral side, but their anus has shifted to the posterior margin or even the oral side of the disc, so that an anteroposterior axis and bilateral symmetry can be recognized. Bilateral symmetry is even more

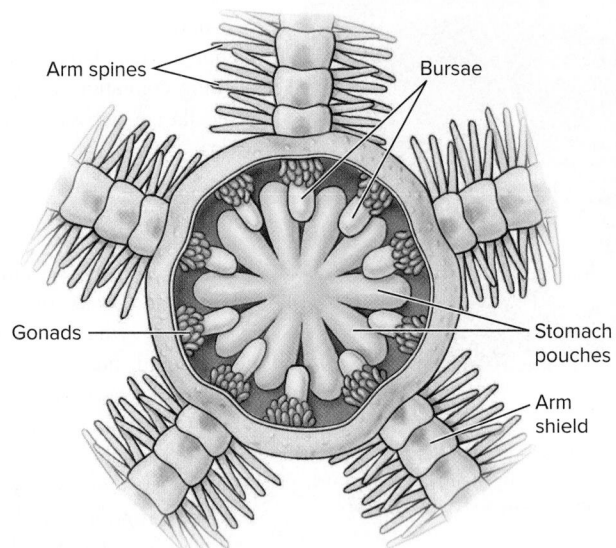

Figure 22.16   Ophiuroid with aboral disc wall cut away to show principal internal structures. Bursae are fluid-filled sacs in which water constantly circulates for respiration. They also serve as brood chambers. Bases of arms are shown.

Diadema antillarum is not nearly as prominent as it once was. In January 1983, an epidemic swept through the Caribbean and along the Florida Keys. Its cause has never been determined, but it decimated the *Diadema* population, leaving less than 5% of the original numbers. Other species of sea urchins were unaffected. However, various types of algae formerly grazed heavily by *Diadema* have increased greatly on the reefs, and *Diadema* populations have not recovered. This abundance of algae has had a disastrous effect on coral reefs around Jamaica. Herbivorous fishes around that island had been chronically overharvested, leaving nothing to control algal overgrowth following the decline of *Diadema*. Coral reefs around Jamaica have been largely destroyed as a result. A modest recovery in *Diadema* populations is occurring in parts of the Caribbean.

© Sanamyan/Alamy Stock Photo

Figure 22.17   This brittle star, *Ophiopholis aculeata*, broods eggs in its bursae. The arms have been broken and are regenerating.

fibers suture the plates together forming a firm endoskeleton, but during periods of growth the sutures loosen so that each plate can add new calcium at its edges. The five ambulacral areas are homologous to the five arms of sea stars and have pores (Figure 22.18B) through which long tube feet extend. The plates bear small tubercles on which the round ends of spines articulate as ball-and-socket joints. Spines are moved by small muscles around the bases.

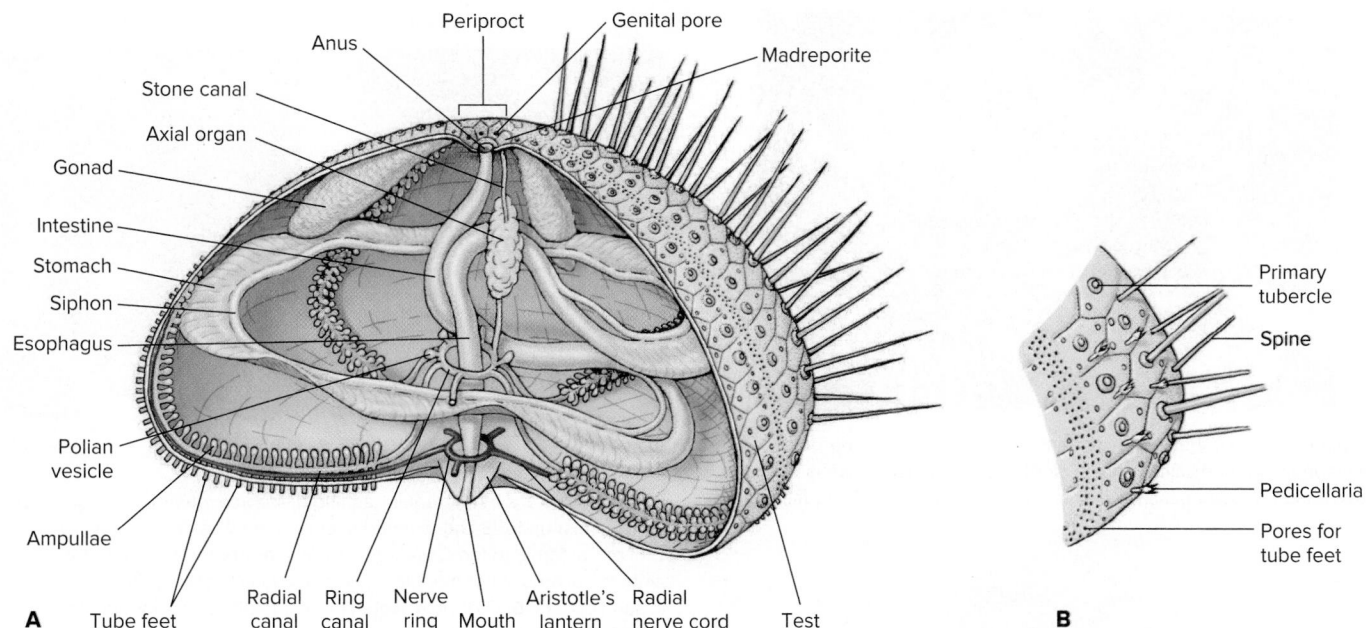

Figure 22.18 **A,** Internal structure of a sea urchin; water-vascular system in tan. **B,** Detail of portion of endoskeleton.

Figure 22.19 Diversity among regular sea urchins (class Echinoidea). **A,** Ten-lined pencil urchin, *Eucidaris metularia*, in the Red Sea. Members of this order have many ancestral characters and have survived since the Paleozoic era. They may be closest in resemblance to the common ancestor of all other extant echinoids. **B,** Slate-pencil urchin, *Heterocentrotus mammilatus.* The large, triangular spines of these urchins were formerly used for writing on slates. **C,** Aboral spines of the intertidal urchin, *Colobocentrotus atratus,* are flattened and mushroom shaped, while the marginal spines are wedge shaped, giving the animal a streamlined form to withstand pounding surf. **D,** *Diadema antillarum* is a common species in the West Indies and Florida. **E,** *Astropyga magnifica* is one of the most spectacularly colored sea urchins.

accentuated in heart urchins, with their anus near the posterior on the oral side and their mouth moved away from the oral pole toward the anterior (Figure 22.22).

Inside the test (Figure 22.18) is the coiled digestive system. A ciliated **siphon** connects the esophagus to the intestine and enables water to bypass the stomach to concentrate food for digestion in the intestine. Sea urchins are largely omnivorous, but their primary diet often consists mostly of algae and other organic material, which they graze with their teeth. Sand dollars have short, club-shaped spines that move sand and its organic contents

Figure 22.20 Purple sea urchins, *Strongylocentrotus purpuratus,* are common along the Pacific coast of North America where there is heavy wave action. These animals are from Santa Cruz Island in California.

Figure 22.21 Two sand dollar species. **A,** *Mellita sp.* burrowing in the sand underwater on Sanibel Island, Florida Gulf Coast. **B,** Removed from the sand, the short spines and petaloids on the aboral surface of this *Echinarachnius parma* are easily seen. This sand dollar is from the Gulf of Mexico.

Aboral pole   Periproct

A   **B**   Mouth

Figure 22.22 An irregular echinoid, *Meoma,* one of the largest heart urchins (test up to 18 cm). *Meoma* occurs in the West Indies and from the Gulf of California to the Galápagos Islands. **A,** Aboral view. Anterior ambulacral area is not modified as a petaloid in heart urchins, although it is in sand dollars. **B,** Oral view. Note curved mouth at anterior end and periproct at posterior end.

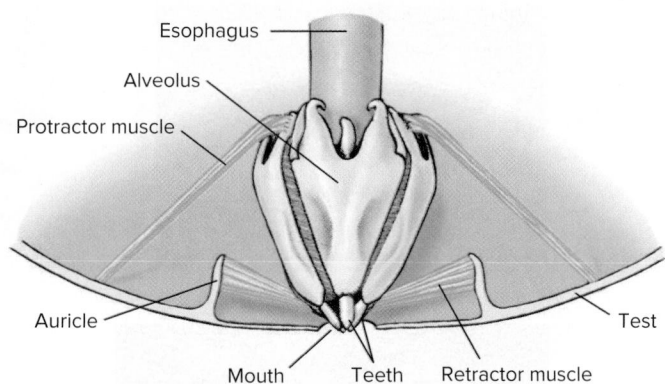

Esophagus

Alveolus

Protractor muscle

Auricle   Test

Mouth   Teeth   Retractor muscle

Figure 22.23 Aristotle's lantern, a complex mechanism used by sea urchins for masticating their food. Five pairs of retractor muscles draw the lantern and teeth up into the test; five pairs of protractors push the lantern down and expose the teeth. Other muscles produce a variety of movements. Only major skeletal parts and muscles are shown in this diagram.

over the aboral surface and down the sides. Fine food particles drop between the spines, and ciliated tracts on the oral side carry the particles to their mouth.

Hemal and nervous systems are basically similar to those of asteroids. Ambulacral grooves are closed, and radial canals of the water-vascular system run just beneath the test, one in each ambulacral radius (Figure 22.18). Ampullae for the podia are within the test, and each ampulla usually communicates with its podium by *two* canals through pores in the ambulacral plate; consequently, such pores in the plates are in pairs.

Peristomial gills, where present, are of little importance in respiratory gas exchange, this function being performed principally by other podia. Although the gills appear to provide some oxygen to the muscles associated with Aristotle's lantern, they seem to function primarily to accommodate pressure changes in the pharyngeal coelom during feeding movements of the lantern complex. In irregular urchins respiratory podia are thin walled, flattened, or lobulate and are arranged in ambulacral fields called **petaloids** on the aboral surface. Petaloids form what appears to be a flower pattern atop sand dollars (Figure 22.21). Irregular urchins also have short, suckered, single-pored podia in ambulacral and sometimes interambulacral areas; these podia function in food handling.

Sexes are separate, and both eggs and sperm are shed into the sea for external fertilization. Some, such as certain pencil urchins, brood their young in depressions between the spines. **Echinopluteus larvae** (see Figure 22.12D) of nonbrooding echinoids may live a planktonic existence for several months and then metamorphose into young urchins (see Figure 8.10, p. 161).

Class Holothuroidea

In a phylum characterized by odd animals, class Holothuroidea (sea cucumbers) contains members that both structurally and physiologically are among the strangest. These animals bear a remarkable resemblance to the vegetable after which they are named (Figure 22.24). Compared with other echinoderms, holothurians are greatly elongated in the oral‐aboral axis, and ossicles are much reduced in most; consequently, these animals are soft bodied. Some species crawl on the seafloor, others are found beneath rocks, and some are burrowers.

There are approximately 1150 living species of holothuroids. Common species along the east coast of North America are *Cucumaria frondosa* (L. *cucumis,* cucumber), *Sclerodactyla briareus* (Gr. *skleros,* hard, + *daktylos,* finger) (Figure 22.26), and the translucent, burrowing *Leptosynapta* (Gr. *leptos,* slender, + *synapsis,* joining together). Along the Pacific coast there are several species of *Cucumaria* (Figure 22.24C) and the striking reddish brown *Parastichopus* (Gr. *para,* beside, + *stichos,* line or row, + *pous, podos,* foot) (Figure 22.24A), with very large papillae.

Form and Function

A sea cucumber's body wall is usually leathery, with tiny ossicles embedded in it (Figure 22.25), although a few species have large ossicles forming a dermal armor (Figure 22.24B). Because of the elongate body form of sea cucumbers, they characteristically lie on one side. The body wall contains circular and longitudinal muscles along the ambulacra.

In some species locomotor tube feet are restricted to the five ambulacral areas (Figure 22.24C) or occur all over the body, but most sea cucumbers have well-developed tube feet only in the ambulacra normally applied to the substratum (Figure 22.24A and B). Thus a secondary bilaterality is present, albeit of quite different origin from that of irregular urchins. The side applied to the substratum has three ambulacra and is called a **sole;** tube feet in the dorsal ambulacral areas, if present, are usually without suckers and may be modified as sensory papillae. All tube feet, except oral tentacles, may be absent in burrowing forms. Oral tentacles are 10 to 30 retractile, modified tube feet around the mouth.

A sea cucumber's coelomic cavity is spacious and fluid filled and has many coelomocytes. The fluid-filled coelom now serves as a hydrostatic skeleton. Dermal ossicles are small and not connected to each other, so they no longer form an endoskeleton.

The digestive system empties posteriorly into a muscular **cloaca** (Figure 22.26). A **respiratory tree** composed of two long, many-branched tubes also empties into the cloaca, which pumps seawater into it. The respiratory tree serves for both respiration and excretion

A

B

C

Figure 22.24   Sea cucumbers (class Holothuroidea). **A,** *Stichopus horrens*, a sea cucumber from the Galapagos, is nocturnal. Before dawn, it slips under a stone or overhang, retracts to empty seawater from its respiratory tree, and rests throughout the day. **B,** In sharp contrast to most sea cucumbers, the surface ossicles of *Psolus chitonoides* are developed into a platelike armor. The ventral surface is a flat, soft, creeping sole, and the mouth (surrounded by tentacles) and anus are turned dorsally. **C,** Tube feet cover all ambulacral areas of *Cucumaria miniata* but are better developed on its ventral side, shown here.

Figure 22.25 Ossicles of sea cucumbers are usually microscopic bodies buried in leathery dermis. They can be extracted from this tissue with commercial bleach and are important taxonomic characteristics. The ossicles shown here, called tables, buttons, and plates, are from *Holothuria difficilis*. They illustrate the meshwork (stereom) structure observed in ossicles of all echinoderms at some stage in their development (×250).

and is not present in any other group of living echinoderms. Gas exchange also occurs through skin and tube feet.

The hemal system is better developed in holothurians than in other echinoderms. Their water-vascular system is peculiar in that the madreporite lies free in the coelom.

Sexes are usually separate, but some holothurians are hermaphroditic. Among echinoderms, only sea cucumbers have a single gonad. The gonad is usually in the form of one or two clusters of tubules that join at the gonoduct. Fertilization is external, and the free-swimming larva is called an **auricularia** (see Figure 22.12E). Some species brood their young either inside their body or somewhere on the body surface.

Behavior and Ecology

Sea cucumbers are sluggish, moving partly by means of their ventral tube feet and partly by waves of contraction in the muscular body wall. Most sedentary species trap suspended food particles in mucus of their outstretched oral tentacles or pick up particles from the surrounding surface. They then stuff their tentacles into the pharynx, one by one, ingesting captured food (Figure 22.27A). Others crawl along the substrate grazing the seafloor with their tentacles (Figure 22.27B).

Sea cucumbers have a peculiar power of what appears to be self-mutilation but is in reality a defense mechanism. When irritated or subjected to unfavorable conditions, many species can cast out a part of their viscera by a strong muscular contraction that may either rupture the body wall or evert its contents through their anus. Lost parts

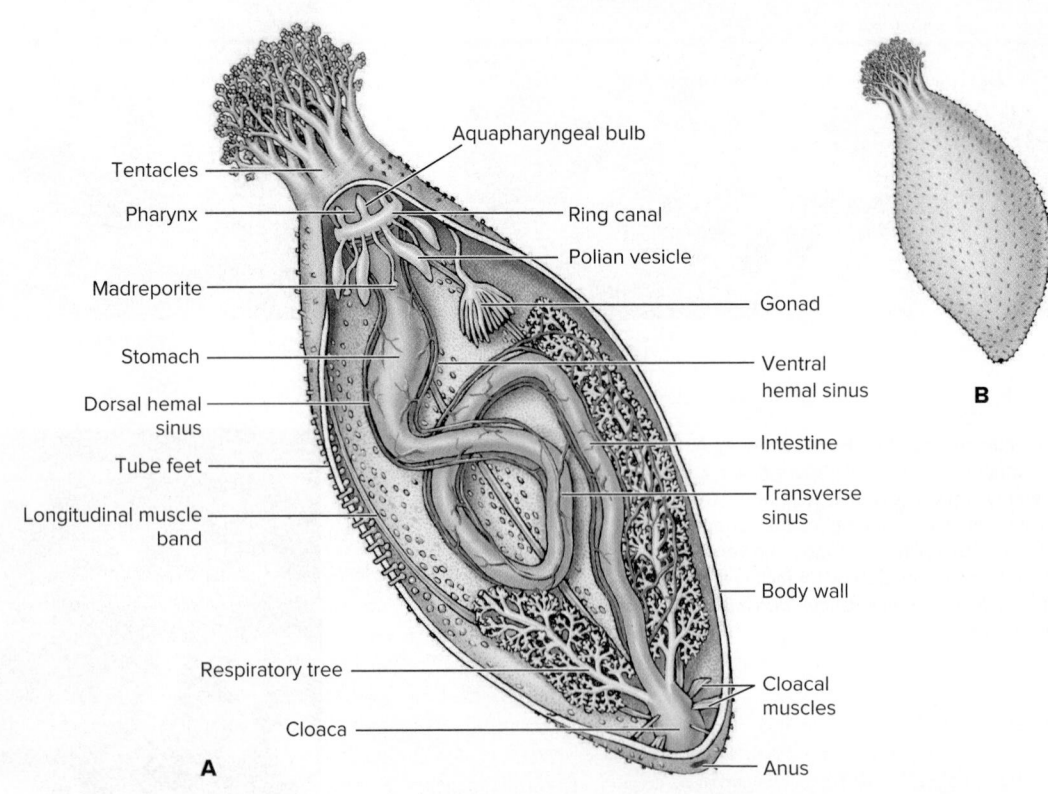

Figure 22.26 Anatomy of a sea cucumber *Sclerodactyla*. **A,** Internal. *Red,* hemal system. **B,** External.

A　　　　　**B**　　　　　**C**

Figure 22.27　**A,** *Eupentacta quinquesemita* extends its tentacles to collect particulate matter in the water, then puts them one by one into its mouth and cleans the food from them. **B,** Moplike tentacles of *Parastichopus californicus* are used for deposit-feeding on the bottom. **C,** *Bohadschia argus* expels its Cuvierian tubules, modified parts of its respiratory tree, when it is disturbed. These sticky strands, containing a toxin, discourage potential predators.

are soon regenerated. Certain species have Cuvierian tubules, which are attached to the posterior part of the respiratory tree and when expelled may entangle an enemy (Figure 22.27C). These tubules become long and sticky after expulsion, and some contain toxins.

An interesting commensal relationship exists between certain sea cucumbers and the pealfish, *Carapus,* which uses the cloaca and respiratory tree of the sea cucumber as shelter.

Class Crinoidea

Crinoids include about 625 species of sea lilies and feather stars. As fossil records reveal, crinoids were once far more numerous than they are now. They differ from other echinoderms by being attached during a substantial part of their lives. Sea lilies have a flower-shaped body at the tip of an attached stalk (Figure 22.28). Feather stars have long, many-branched arms, and adults are free-moving, though they may remain in the same spot for long periods (Figure 22.29). During metamorphosis feather stars become sessile and stalked, but after several months they detach and become free-moving. Many crinoids are deep-water forms, but feather stars may inhabit shallow waters, especially in Indo-Pacific and West-Indian–Caribbean regions, where the largest numbers of species occur.

Form and Function

Their body disc, or **calyx,** is covered with a leathery skin **(tegmen)** containing calcareous plates. Epidermis is poorly developed. Five flexible arms branch to form many more arms, each with many lateral **pinnules** arranged like barbs on a feather (Figure 22.28). Calyx and arms together form the **crown.** Sessile forms have a long, jointed **stalk** attached to the aboral side of the body. This stalk is composed of plates, appears jointed, and may bear **cirri.** Madreporite, spines, and pedicellariae are absent.

Their upper (oral) surface bears a mouth, which opens into a short esophagus, from which the long intestine with diverticula proceeds aborally for a distance and then makes a complete turn to an **anus,** which may be on a raised cone (Figure 22.28B). **Ambulacral**

grooves are open and ciliated and serve to carry food to their mouth (Figure 22.28B). Simple tube feet without suckers line the ambulacral grooves, which extend into the pinnules. With the aid of tube feet and mucous strands, crinoids collect small organisms from the surrounding waters.

Their water-vascular system follows the basic echinoderm plan. However, the system functions entirely on existing coelomic fluid. There is no madreporite to allow exchange of fluid with the surrounding environment. The nervous system has an oral ring and a radial nerve that runs to each arm. The aboral or entoneural system is more highly developed in crinoids than in most other echinoderms. This system innervates the podia, which proliferate along the lateral pinnules. Podia collect food and act as sensory organs. Additional sensory organs are scanty and simple.

Sexes are separate. Gonads are simply masses of cells in the genital cavity of arms and pinnules. Gametes escape without ducts through a rupture in pinnule walls. Brooding occurs in some forms. **Doliolaria** larvae (Figure 22.12F) are free-swimming for a time before they become attached and metamorphose.

22.6 PHYLOGENY AND ADAPTIVE DIVERSIFICATION

Phylogeny

Echinoderms left an extensive fossil record and evolved about 26 anatomically distinct body forms, which account for 20 currently recognized classes. Members of most classes were extinct by the end of the Paleozoic era, and only five classes survive today. Based on their bilateral larvae, and newly discovered bilaterally symmetrical fossil forms, it appears that ancestral echinoderms were bilateral and that their coelom had three pairs of spaces (trimeric or tripartite). Two major clades of ancient echinoderms developed, one bilateral and deposit-feeding and the other radially symmetrical and suspension-feeding.

Ancestors of the extant echinoderms were sessile, became radial as an adaptation to that existence, and then gave rise to the free-moving groups. Figure 22.3 is consistent with this hypothesis for the

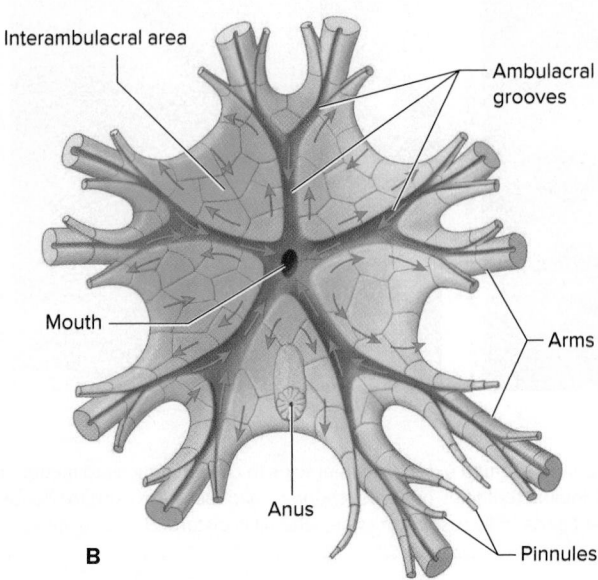

Figure 22.28 Crinoid structure. **A,** Sea lily (stalked crinoid) with portion of stalk. Modern crinoid stalks rarely exceed 60 cm, but fossil forms were as much as 20 m long. **B,** Oral view of calyx of a crinoid, *Antedon,* showing direction of ciliary food currents. Ambulacral grooves with podia extend from mouth along arms and branching pinnules. Food particles touching podia are tossed into ambulacral grooves and carried, tangled in mucus, by strong ciliary currents toward the mouth. Particles falling on interambulacral areas are carried by cilia first toward the mouth and then outward and finally dropped off the edge, thus keeping the oral disc clean.

Figure 22.29 *Comantheria briareus* are crinoids found on Pacific coral reefs. They extend their arms into the water to catch food particles both during the day and at night.

taxa most closely related to modern forms. It depicts endoskeletal plates with stereom structure and the presence of external ciliary grooves for feeding as early echinoderm (or pre-echinoderm) characteristics. Extinct carpoids (Figure 22.30A) had stereom ossicles but were not radially symmetrical, and the status of their water-vascular system, if any, is uncertain. Fossil helicoplacoids (Figure 22.30B) show evidence of three true ambulacral grooves, and their mouth was on the side of their body. We show both groups as early diverging echinoderms (Figure 22.3).

Attachment to a substratum by their aboral surface would have selected for radial symmetry, explaining the origin of subphylum Pelmatozoa, whose living members are crinoids. Both Cystoidea (extinct) and Crinoidea were primitively attached to a substratum by an aboral stalk. An ancestor that became free-moving and applied its oral surface to the substratum would have given rise to subphylum Eleutherozoa.

Phylogeny within Eleutherozoa is controversial. Most investigators agree that echinoids and holothuroids form a clade, but opinions diverge on the relationship of ophiuroids and asteroids. Figure 22.3 illustrates the view that ophiuroids arose after closure of their

Taxonomy of Phylum Echinodermata

There are about 7000 living and 20,000 extinct or fossil species of Echinodermata. The traditional taxonomy placed all free-moving forms that were oriented with oral side down in subphylum Eleutherozoa, containing most living species. The other subphylum, Pelmatozoa, contained mostly forms with stems and oral side up; most extinct classes and living Crinoidea belong to this group. Although alternative schemes have strong supporters, cladistic analysis provides evidence that the two traditional subphyla are monophyletic. This list includes only groups with living members.

Subphylum Pelmatozoa (pel-ma'to-zo'a) (Gr. *pelmatos*, a stalk, + *zēon*, animal). Body in form of cup or calyx, borne on aboral stalk during part or all of life; oral surface directed upward; open ambulacral grooves; madreporite absent; both mouth and anus on oral surface; several fossil classes plus living Crinoidea.

Class Crinoidea (krin-oi'de-a) (Gr. *krinon*, lily, + *eidos*, form, + *ea*, characterized by): **sea lilies** and **feather stars.** Five arms branching at base and bearing pinnules; ciliated ambulacral grooves on oral surface with tentacle-like tube feet for food gathering; spines, madreporite, and pedicellariae absent. Examples: *Antedon, Comantheria* (Figure 22.29).

Subphylum Eleutherozoa (e-lu'ther-o-zo'a) (Gr. *eleutheros*, free, not bound, + *zōon*, animal). Body form star-shaped, globular, discoidal, or cucumber-shaped; oral surface directed toward substratum or oral-aboral axis parallel to substratum; body with or without arms; ambulacral grooves open or closed.

Class Asteroidea (as'ter-oy'de-a) (Gr. *aster*, star, + *eidos*, form, + *ea*, characterized by): **sea stars (starfish).** Star-shaped, with arms not sharply marked off from central disc; ambulacral grooves open, with tube feet on oral side; tube feet often with suckers; anus and

madreporite aboral; pedicellariae present. Examples: *Orthasterias, Pisaster* (p. 471). This group includes former members of **class Concentricycloidea** (kon-sen'tri-sy-kloy'de-a) (L. *cum*, together, + *centrum*, center [having a common center], + Gr. *kyklos*, circle, + *eidos*, form, + *ea*, characterized by): **sea daisies.** Disc-shaped body, with marginal spines but no arms; concentrically arranged skeletal plates; ring of suckerless podia near body margin; hydropore present; gut present or absent, no anus. Example: *Xyloplax* (Figure 22.13).

Class Ophiuroidea (o'fe-u-roy'de-a) (Gr. *ophis*, snake, + *oura*, tail, + *eidos*, form, + *ea*, characterized by): **brittle stars** and **basket stars.** Star-shaped, with arms sharply marked off from central disc; ambulacral grooves closed, covered by ossicles; tube feet without suckers and not used for locomotion; pedicellariae absent; anus absent. Examples: *Ophiura* (Figure 22.14A), *Gorgonocephalus* (Figure 22.14B).

Class Echinoidea (ek'i-noy'de-a) (Gr. *echinos*, sea urchin, hedgehog, + *eidos*, form, + *ea*, characterized by): **sea urchins, sea biscuits,** and **sand dollars.** More or less globular or disc-shaped, with no arms; compact skeleton or test with closely fitting plates; movable spines; ambulacral grooves closed; tube feet with suckers; pedicellariae present. Examples: *Arbacia, Strongylocentrotus* (Figure 22.20), *Lytechinus, Mellita.*

Class Holothuroidea (hol'o-thu-roy'de-a) (Gr. *holothourion*, sea cucumber, + *eidos*, form, + *ea*, characterized by): **sea cucumbers.** Cucumber-shaped, with no arms; spines absent; microscopic ossicles embedded in thick muscular wall; anus present; ambulacral grooves closed; tube feet with suckers; circumoral tentacles (modified tube feet); pedicellariae absent; madreporite internal. Examples: *Sclerodactyla, Parastichopus, Cucumaria* (Figure 22.24C).

---

ambulacral grooves, but this scheme treats evolution of five ambulacral rays (arms) in ophiuroids and asteroids as independently evolved. Alternatively, if ophiuroids and asteroids are a single clade, as recent phylogenies suggest, then closed ambulacral grooves must have been lost in the asteroids, or evolved separately in ophiuroids and in the common ancestor of echinoids and holothuroids.

Adaptive Diversification

Diversification of echinoderms has been limited by their most important characters: radial symmetry, water-vascular system, and dermal endoskeleton. If their ancestors had a brain and specialized sense organs, these were lost in adoption of radial symmetry. Only now are gene expression studies beginning to help researchers identify structures such as the anterior-posterior axis in adult echinoderms. The best vidence currently available suggests that the oral surface is anterior and the aboral surface is posterior. Under this hypothesis, the arms represent lateral growth zones.

Echinoderms have diversified within the benthic habitat. There are large numbers of creeping forms with filter-feeding, deposit-feeding, scavenging, and herbivorous habits, and very rare pelagic forms. In this light the relative success of asteroids as predators is impressive and probably attributable to the extent to which they have exploited the hydraulic mechanism of their tube feet.

The basic body plan of echinoderms appears to have severely limited their evolutionary opportunities to become parasites. Indeed, the most mobile of echinoderms, ophiuroids, which are also the ones most able to insert their bodies into small spaces, are the only group with significant numbers of commensal species.

22.7 PHYLUM HEMICHORDATA

Hemichordata (hem'i-kor-da'ta) (Gr. *hemi*, half, + *chorda*, string, cord) are marine animals formerly considered a subphylum of chordates, based on their possession of gill slits and a rudimentary notochord. However, the so-called hemichordate notochord is really a buccal diverticulum (called a **stomochord,** meaning "mouth-cord") and not homologous to a chordate notochord, so hemichordates are ranked as a separate phylum.

Hemichordates are vermiform bottom dwellers, living usually in shallow waters. Some colonial species live in secreted tubes. Most are sedentary or sessile. Their distribution is almost cosmopolitan, but their secretive habits and fragile bodies make collecting them difficult.

There are two classes. Members of class Enteropneusta (Gr. *enteron*, intestine, + *pneustikos*, of, or for, breathing), called acorn worms, range from 20 mm to 2.5 m in length. Members of class Pterobranchia (Gr. *pteron*, wing, + *branchia*, gills) are

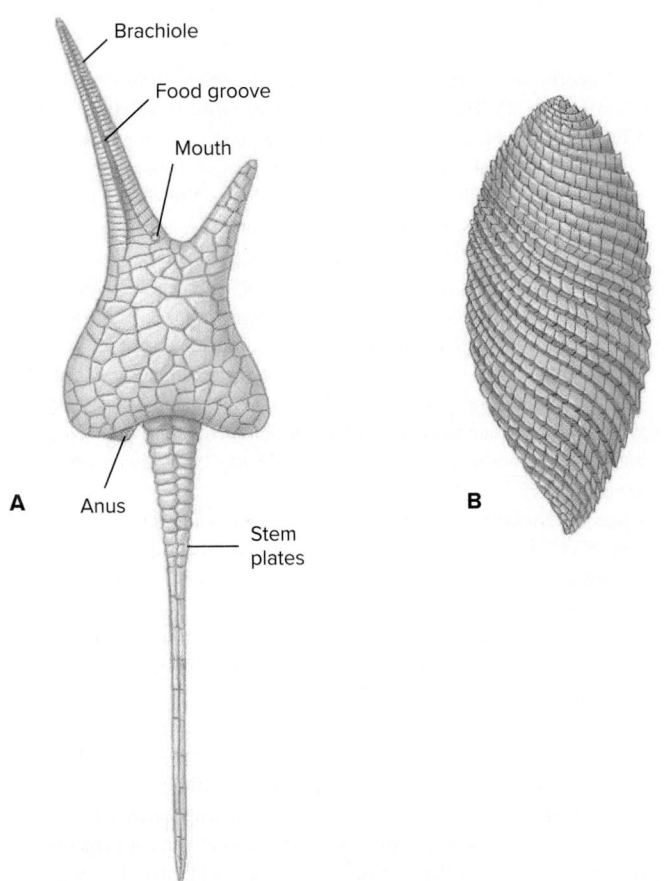

A, Brachiole
Food groove
Mouth
A   Anus
Stem plates
B

Figure 22.30   **A,** *Dendrocystites,* a carpoid (subphylum Homalozoa) with one brachiole. Brachioles are so called to distinguish them from the heavier arms of asteroids, ophiuroids, and crinoids. This group bore some characters interpreted as chordate in nature. **B,** *Helicoplacus,* a helicoplacoid, had three ambulacral areas and apparently a water-vascular system. It is the sister group to modern echinoderms.

smaller, usually 1 to 12 mm, not including their stalk. About 75 species of enteropneusts and three small genera of pterobranchs are recognized.

Hemichordates have a tripartite coelom.

Class Enteropneusta

Enteropneusts, or acorn worms, are sluggish, wormlike animals that live in burrows or under stones, usually in mud or sand flats of intertidal zones. *Balanoglossus* (Gr. *balanos,* acorn, + *glōssa,* tongue) and *Saccoglossus* (Gr. *sakkos,* sac, strainer, + *glōssa,* tongue) (Figure 22.31) are common genera.

Form and Function

Their mucus-covered body is divided into three distinct regions: a tonguelike proboscis, a short collar, and a long trunk (protosome, mesosome, and metasome).

Characteristics of Phylum Hemichordata

1. Body divided into **proboscis, collar,** and **trunk; buccal diverticulum** in posterior part of proboscis
2. Enteropneusta free-moving and of burrowing habits; pterobranchs sessile, mostly colonial, living in secreted tubes
3. Free-living
4. Bilaterally symmetrical, soft bodied; wormlike or short and compact with stalk for attachment
5. Triploblastic
6. Single coelomic pouch in proboscis, but paired pouches in collar and trunk
7. Ciliated epidermis
8. Digestive system complete
9. Longitudinal and circular muscles in body wall in some
10. A subepidermal nerve plexus thickened to form dorsal and ventral nerve cords, with a ring connective in the collar; some species with hollow **dorsal nerve cord**
11. Sensory neurons in proboscis likely function in chemoreception
12. Colonies form by asexual budding in pterobranchs; asexual reproduction by fragmentation in enteropneusts
13. Sexes separate in Enteropneusta, with gonads projecting into body cavity; tornaria larva in some Enteropneusta
14. A single **glomerulus** connected to blood vessels may have excretory function and is considered a metanephridium
15. Respiratory and filter-feeding system of **gill slits** (few or none in pterobranchs) connecting the pharynx with outside
16. Circulatory system of dorsal and ventral vessels and dorsal heart

Proboscis   Their proboscis is the active part of the animal. It probes its surroundings, collecting food in mucous strands on its surface. Cilia carry particles to the groove at the edge of the collar and direct them to the mouth on the underside, and then the particles are swallowed. Large particles can be rejected by covering the mouth with the edge of the collar (Figure 22.32). In addition to the deposit-feeding just described, enteropneusts also use a pre-oral ciliary organ and gill slits to filter-feed. They can simultaneously use filter- and deposit-feeding.

Burrow dwellers use their proboscis to excavate, thrusting it into mud or sand and allowing cilia and mucus to move sand backward. They also may ingest sand or mud as they go, extracting its organic contents. They build U-shaped, mucus-lined burrows, usually with two openings 10 to 30 cm apart and with the base of the U 50 to 75 cm below the surface. They can thrust their proboscis out the front opening for feeding. Defecation at the back opening builds characteristic spiral mounds of feces that leave a telltale clue to locations of burrows.

In the posterior end of their proboscis is a small coelomic sac (protocoel) into which extends a **buccal diverticulum,** a slender, blindly ending pouch of the gut that reaches forward into the buccal region and was formerly considered a notochord. A slender canal connects the protocoel with a **proboscis pore** to the outside (Figure 22.31C). Paired coelomic cavities in the collar also open by pores. By pulling water through these pores into the coelomic sacs, their proboscis and collar can be stiffened to aid in burrowing.

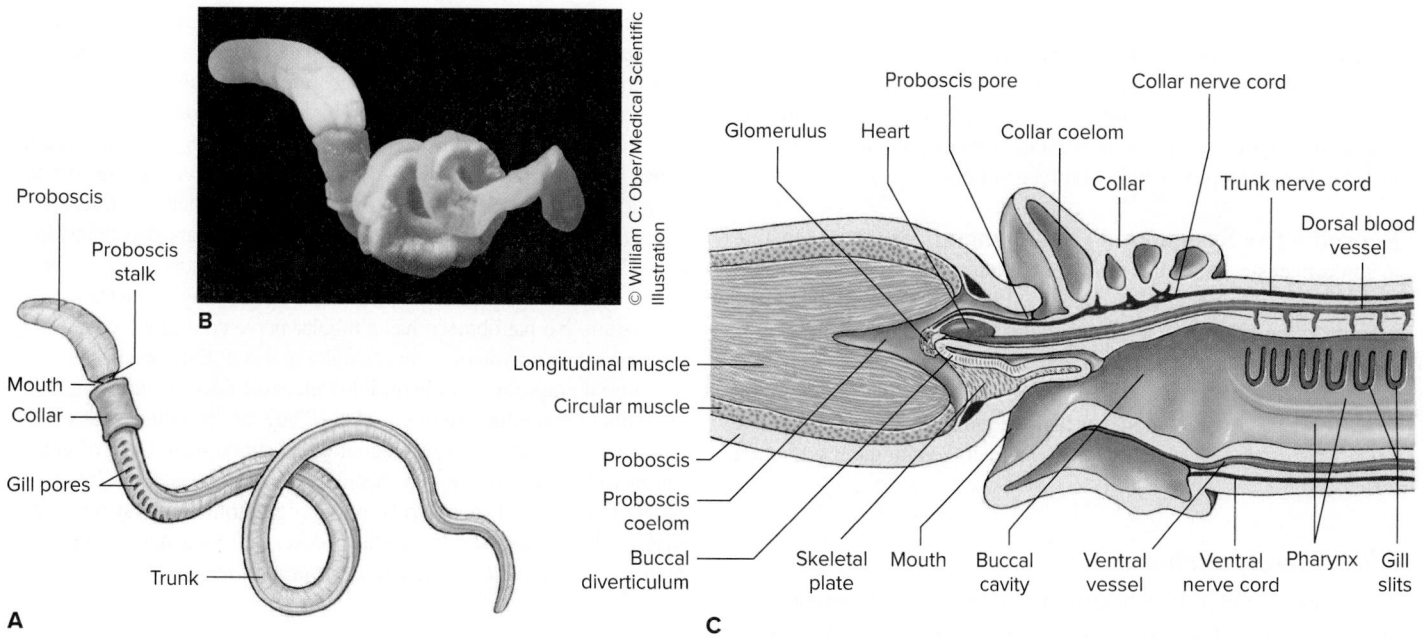

Figure 22.31 Acorn worm, *Saccoglossus* (Hemichordata, class Enteropneusta). **A,** External lateral view. **B,** Live animal. **C,** Longitudinal section through anterior end.

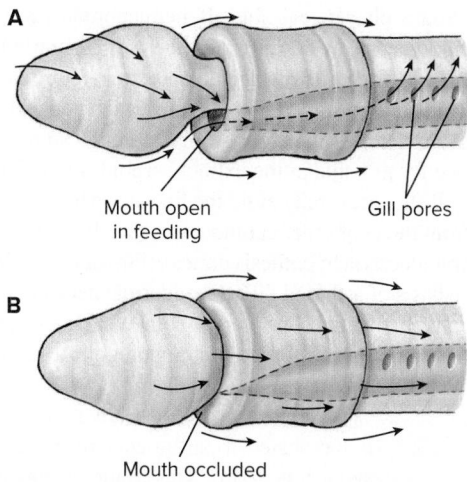

Figure 22.32 Food currents of enteropneust hemichordate. **A,** Side view of acorn worm with mouth open, showing direction of currents created by cilia on proboscis and collar. Food particles are directed toward the mouth and digestive tract. Rejected particles move outside the collar. Water leaves through gill pores. **B,** When the mouth is occluded, all particles are rejected and passed onto the collar. Nonburrowing and some burrowing hemichordates use this feeding method.

Contraction of body musculature then forces excess water out through the gill slits, reducing the hydrostatic pressure and allowing the animal to move forward.

Branchial System   A row of **gill pores** is located dorsolaterally on each side of the trunk just behind the collar (Figure 22.32A). Pores open from a series of gill chambers that in turn connect with a series of U-shaped **gill slits** in the sides of the pharynx (Figure 22.31C). There are no gills on the gill slits, but some respiratory gaseous exchange occurs in the vascular branchial epithelium, as well as in the body surface. Ciliary currents keep a fresh supply of water moving from their mouth through the pharynx and out the gill slits and branchial chambers to the outside.

Feeding and Digestive System   Hemichordates are largely ciliary-mucus feeders. Food is caught in mucus on the proboscis and collar. Cilia move the food to the ventral part of the pharynx and esophagus, then to the intestine where digestion and absorption occur.

Circulatory and Excretory Systems   A middorsal vessel carries the colorless blood forward above the gut. In the collar the vessel expands into a sinus and a heart vesicle above the buccal diverticulum. Blood then enters a network of blood sinuses called the **glomerulus,** which partially surrounds these structures (Figure 22.31C). The glomerulus has an excretory function and is homologous to the echinoderm axial complex. Blood travels posteriorly through a ventral vessel below the gut, passing through extensive sinuses to the gut and body wall.

Nervous and Sensory Systems   Their nervous system comprises a subepithelial network, or plexus, of nerve cells and fibers to which processes of epithelial cells are attached. Thickenings of this net form dorsal and ventral nerve cords that are united posterior to the collar by a ring connective. The dorsal cord continues into the collar and furnishes many fibers to the plexus of the proboscis. The dorsal nerve cord (**neurochord**) is formed by an invagination of the ectoderm and is hollow in some species. This striking similarity to the pattern in chordates is taken as evidence for homology with the dorsal nerve cord of the Chordata, but not all researchers accept this interpretation. The neurochord contains giant nerve cells with processes running to nerve trunks. This nerve plexus system is quite reminiscent of that of cnidarians and echinoderms.

Sensory receptors include neurosensory cells throughout the epidermis (especially in the proboscis, a preoral ciliary organ that may be chemoreceptive) and photoreceptor cells.

Reproductive System and Development   Sexes are separate in enteropneusts. Although most species reproduce only sexually, at least one species undergoes asexual reproduction. A dorsolateral row of gonads runs along each side of the anterior trunk. Fertilization is external, and in some species a ciliated **tornaria** larva develops and at certain stages is so similar to an echinoderm bipinnaria that it was once considered an echinoderm larva (Figure 22.33). The familiar *Saccoglossus* of American waters has direct development without a tornaria stage.

Class Pterobranchia

The basic plan of class Pterobranchia is similar to that of Enteropneusta, but certain structural differences reflect the sedentary lifestyle of pterobranchs. The first pterobranch ever reported was obtained by the famed *Challenger* expedition of 1872 to 1876. Although it was first placed among Polyzoa (Entoprocta and Ectoprocta), its affinities to hemichordates were later recognized. Only three genera (*Atubaria, Cephalodiscus,* and *Rhabdopleura*) are known.

Pterobranchs are small animals, usually within a range of 1 to 7 mm in length, although the stalk may be longer. Many individuals of *Cephalodiscus* (Gr. *kephalē,* head, + *diskos,* disc) (Figure 22.34) live together in collagenous tubes, which often form an anastomosing system. Zooids are not connected, however, and live independently in the tubes. Through apertures in these tubes, they extend their crown of tentacles. They are attached to the tube walls by extensible stalks that can retract the owners back into the tubes when necessary.

The body of *Cephalodiscus* is divided into the three regions—proboscis, collar, and trunk—characteristic of hemichordates. There is only one pair of gill slits, and the alimentary canal is U-shaped, with their anus near their mouth. The proboscis is shield-shaped. At the proboscis base are five to nine pairs of branching arms with

tentacles containing an extension of the coelomic compartment of the mesosome, as in a lophophore. Ciliated grooves on the tentacles and arms collect food. Some species are dioecious, and others are monoecious. Asexual reproduction by budding also may occur.

In *Rhabdopleura* (Gr. *rhabdos,* rod, + *pleura,* a rib, the side), which is smaller than *Cephalodiscus,* individuals remain together to form a colony of zooids connected by a stolon and enclosed in secreted tubes (Figure 22.35). The collar in these forms bears two branching arms. No gill slits or glomeruli are present. New individuals are produced by budding from a creeping basal stolon, which branches on a substratum. No pterobranch has a tubular nerve cord in the collar, but otherwise their nervous system is similar to that of Enteropneusta.

Fossil graptolites of the middle Paleozoic era often are placed as an extinct class under Hemichordata. They are important index fossils of Ordovician and Silurian geological strata. Alignment of graptolites with hemichordates has been very controversial, but discovery of an organism that seems to be a living graptolite lends strong support to the hypothesis. It has been described as a new species of pterobranch, called *Cephalodiscus graptolitoides.*

22.8 PHYLOGENY AND ADAPTIVE DIVERSIFICATION

Hemichordate phylogeny has long been puzzling. Hemichordates share characters with both echinoderms and chordates. With chordates they share pharyngeal slits. If hemichordates are the sister taxon to echinoderms, as described by the Ambulacraria hypothesis (see Figure 22.1), then gill slits are an ancestral feature for deuterostomes. Gill slits are presumed lost in pterobranch hemichordates and in a lineage ancestral to all echinoderms, although some researchers find evidence for gill slits in the extinct carpoid echinoderms. Thus, the loss of gill slits occurred before the lineages with extant members branched from those of other echinoderms (see Figure 22.3).

The Ambulacraria hypothesis unites echinoderms and hemichordates on the basis of a shared diffuse epidermal nervous system. It is therefore unlikely that the short dorsal, somewhat hollow, nerve cord in the collar zone of hemichordates is homologous to the dorsal hollow nerve cord of chordates.

Another phylogenetically important character is the shared tripartite coelom in hemichordates and echinoderms. This feature is now assumed to characterize members of Ambulacraria. Reexamination of development of the coelomic compartments in lophophorates and chaetognaths indicates that compartments in these animals are not homologous to those of Ambulacraria. The buccal diverticulum in the hemichordate mouth cavity, long thought homologous to the notochord of chordates, is now considered a synapomorphy of hemichordates themselves.

Early embryogenesis of hemichordates is remarkably like that of echinoderms, and early tornaria larvae are almost identical to bipinnaria larvae of asteroids, suggesting that echinoderms form the sister group of hemichordates (see Figure 22.1). Recent phylogenetic work supports the monophyly of echinoderms and

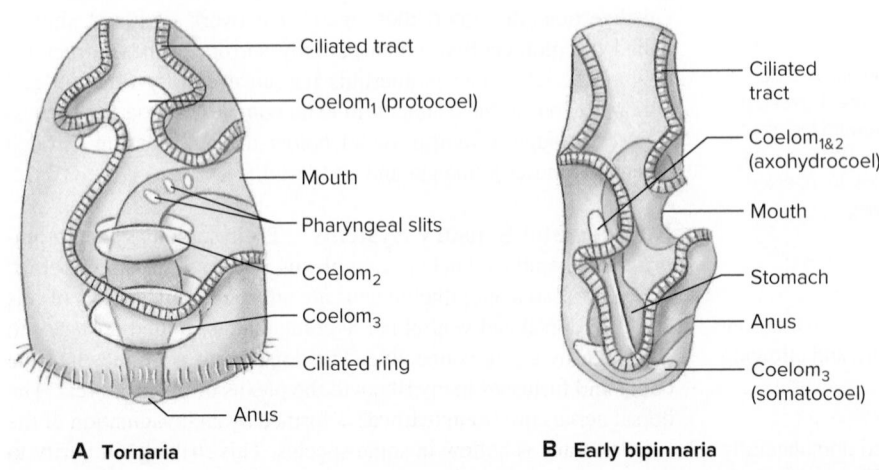

A Tornaria

- Ciliated tract
- Coelom$_1$ (protocoel)
- Mouth
- Pharyngeal slits
- Coelom$_2$
- Coelom$_3$
- Ciliated ring
- Anus

B Early bipinnaria

- Ciliated tract
- Coelom$_{1\&2}$ (axohydrocoel)
- Mouth
- Stomach
- Anus
- Coelom$_3$ (somatocoel)

Figure 22.33   Comparison of **A**, a hemichordate tornaria to **B**, an echinoderm bipinnaria.

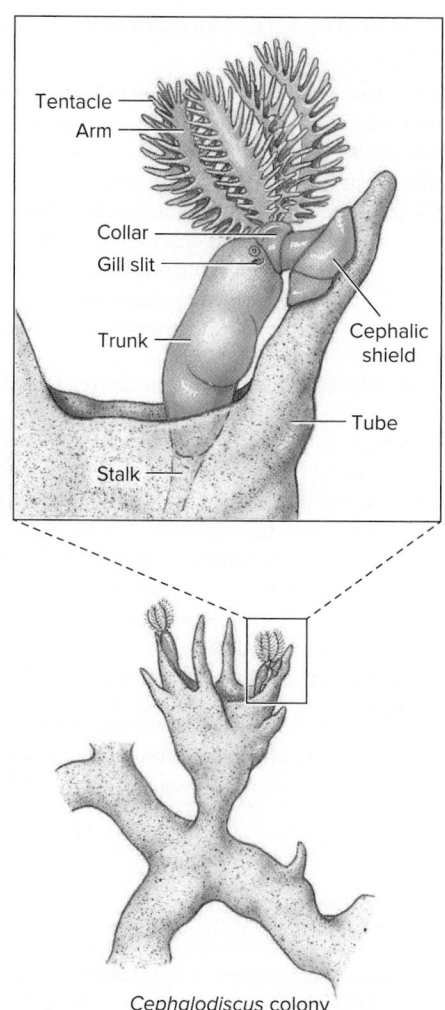

Figure 22.34 *Cephalodiscus*, a pterobranch hemichordate. These tiny (5 to 7 mm) forms live in tubes in which they can move freely. Ciliated tentacles and arms direct currents of food and water toward mouth.

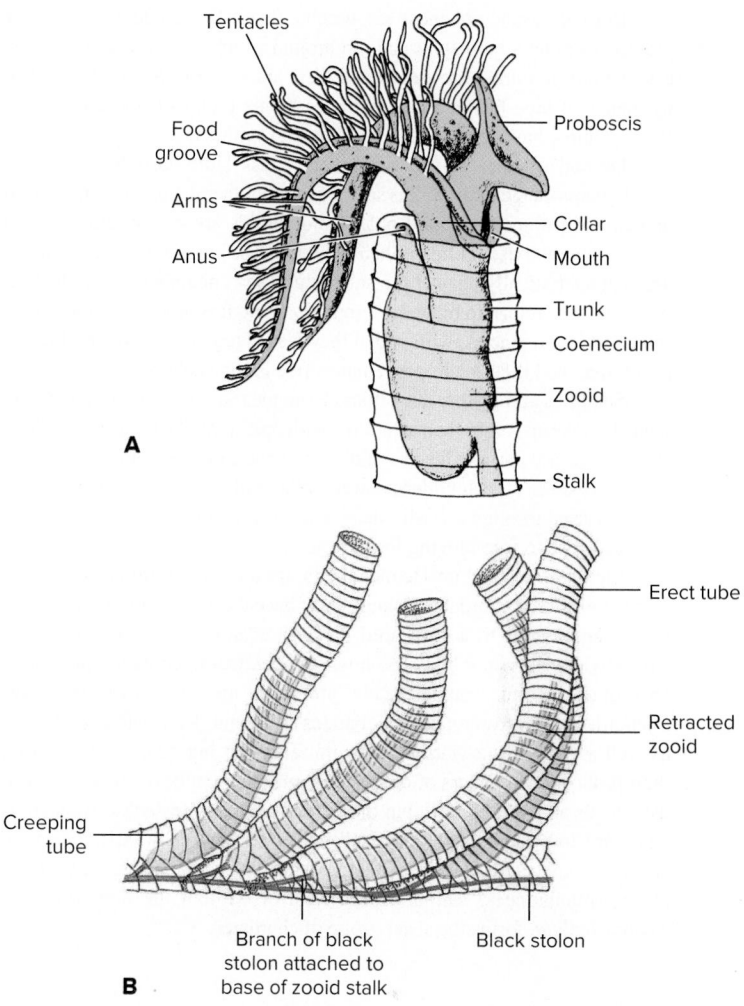

Figure 22.35 **A,** *Rhabdopleura*, a pterobranch hemichordate in its tube. Individuals live in branching tubes connected by stolons, and protrude ciliated tentacles for feeding. **B,** Portion of a colony.

hemichordates, as well as their placement as sister taxa within Ambulacraria. Gill slits in the pharynx are presumed to be a deuterostome character; one reconstruction of the ancestral feeding mode indicates that gill slits were used in filter-feeding, not respiration.

 S U M M A R Y

Arrow worms (phylum Chaetognatha) are a small group but an important component of marine plankton. They have a well-developed coelom and are effective predators, catching other planktonic organisms with the teeth and chitinous spines around their mouth. Phylum Xenoturbellida houses two species of wormlike deuterostomes with relatively simple bodies, including a blind gut. Some workers unite Xenoturbelliida with Acoelomorpha as Xenacoelomorpha, but there is no agreement on the evolutionary position of this group.

Phylum Echinodermata possesses many typical characteristics of Deuterostomia. Echinoderms are an important marine group sharply distinguished from other phyla of animals. They have pentaradial symmetry but were derived from bilateral ancestors.

Sea stars (class Asteroidea) can be used to illustrate echinoderms. Sea stars usually have five arms, which merge gradually with a central disc. Like other echinoderms, they have no head and few specialized sensory organs. Their mouth is directed toward the substratum. They have stereom dermal ossicles, respiratory papulae, and open ambulacral grooves. Many sea stars have pedicellariae. Their water-vascular system is an elaborate hydraulic system derived embryonically from one of their coelomic compartments. Along the ambulacral areas, branches of the water-vascular system (tube feet) are important in locomotion, food gathering, respiration, and excretion. Many sea stars are predators, whereas others feed on small particles. Sexes are separate, and reproductive systems are simple. The bilateral, free-swimming larva becomes attached, transforms to a radial juvenile, then detaches and becomes a motile sea star.

Sea daisies (former class Concentricycloidea) are an enigmatic group now placed within class Asteroidea. They are circular in shape, have marginal tube feet, and have two concentric ring canals in their water-vascular system.

Arms of brittle stars (class Ophiuroidea) are slender and sharply set off from the central disc. Ophiuroids have no pedicellariae or ampullae and their ambulacral grooves are closed. Their tube feet have no suckers, and their madreporite is on the oral side. They crawl by means of arm movements, and their tube feet function in food gathering.

Dermal ossicles of most sea urchins (class Echinoidea) fit together closely, forming a hard spherical test around which lie five ambulacral areas. Ambulacral areas are closed and there are no arms. Sea urchins move by means of tube feet or by their spines. Some urchins (sand dollars and heart urchins) have returned to adult bilateral symmetry.

Dermal ossicles in sea cucumbers (class Holothuroidea) are very small; therefore the body wall is soft. Their ambulacral areas also are closed and extend toward the aboral pole. Holothuroids are greatly elongated in the oral-aboral axis and lie on their side. Because certain ambulacral areas are characteristically against the substratum, sea cucumbers have also undergone some return to bilateral symmetry. Tube feet around the mouth are modified into tentacles, with which they feed. They have an internal respiratory tree, and their madreporite hangs free in the coelom.

Sea lilies and feather stars (class Crinoidea) are the only group of living echinoderms, other than asteroids, with open ambulacral grooves. They are mucociliary particle feeders and lie with their oral side up.

Ancestors of echinoderms were bilaterally symmetrical, but they likely evolved through a sessile stage that became radially symmetrical and then gave rise to free-moving forms.

Members of phylum Hemichordata are marine wormlike forms formerly considered chordates because their buccal diverticulum was thought to be homologous to a notochord. Like chordates, most have paired gill slits. Divisions of their body (proboscis, collar, trunk) contain typical deuterostome coelomic compartments (protocoel, mesocoel, metacoel). Hemichordate class Enteropneusta contains filter and deposit-feeding forms, as well as burrowing worms that capture food using mucus and cilia on their proboscis. Members of class Pterobranchia are tube dwellers that also use mucus and cilia to feed, but they capture food on tentacles. Hemichordates are important phylogenetically because they show affinities with chordates and echinoderms. Together with the echinoderms, they form clade Ambulacraria. Ambulacraria, with or without the odd wormlike *Xenoturbellida*, forms the sister group of chordates.

REVIEW QUESTIONS

1. How do chaetognaths feed?
2. Molecular sequence data indicate that *Xenoturbella* is a deuterostome. Are there corresponding morphological characters?
3. What constellation of characteristics possessed by echinoderms occurs in no other phylum?
4. How do we know that echinoderms were derived from an ancestor with bilateral symmetry?
5. Distinguish the following groups of echinoderms from each other: Crinoidea, Asteroidea, Ophiuroidea, Echinoidea, Holothuroidea.
6. What is an ambulacrum, and what is the difference between open and closed ambulacral grooves?
7. Trace or make a rough copy of Figure 22.6B without labels; then from memory label the parts of the water-vascular system of sea stars.
8. Briefly explain the mechanism of action of a sea star's tube foot.
9. What structures are involved in the following functions in sea stars? Briefly describe the action of each: respiration, feeding and digestion, excretion, reproduction.
10. Compare the structures and functions in question 9 as they occur in brittle stars, sea urchins, sea cucumbers, and crinoids.
11. Briefly describe development in sea stars, including metamorphosis.

12. Match groups in the left column with *all* correct answers in the right column.

   _____ Crinoideaa   a. Closed ambulacral grooves
   _____ Asteroidea   b. Oral surface generally upward
   _____ Ophiuroidea   c. With arms
   _____ Echinoidea   d. Without arms
   _____ Holothuroidea   e. Approximately globular or disc-shaped
                               f. Elongated in oral-aboral axis
                               g. With pedicellariae
                               h. Madreporite internal
                               i. Madreporite on oral plate

13. Define the following: pedicellariae, madreporite, respiratory tree, Aristotle's lantern, papulae, Cuvierian tubules.
14. What evidence suggests that ancestral echinoderms were sessile?
15. Give four examples of how echinoderms are important to humans.
16. What is a major difference in function of the coelom in holothurians compared with other echinoderms?
17. Describe a reason for the hypothesis that the ancestor of eleutherozoan groups was a radial, sessile organism.
18. What characteristics do Hemichordata share with Echinodermata, and how do the two phyla differ?
19. Distinguish Enteropneusta from Pterobranchia.

For Further Thought The phylogenetic placement of chaetognaths is very uncertain. What are the two other possible placements for this group and what would you need to know to decide on the best placement?

SELECTED REFERENCES

Aizenberg, J., A. Tkachenkoo, S. Weiner, L. Addadi, and G. Hendler. 2001. Calcitic microlenses as part of the photoreceptor system in brittle stars. Nature **412:** 819–822. *Tiny bumps on stereom ossicles in arms serve as microlenses to focus light on nerve photoreceptors.*

Baker, A. N., F. W. E. Rowe, and H. E. S. Clark. 1986. A new class of Echinodermata from New Zealand. Nature **321:** 862–864. *Describes the strange sea daisies.*

Ball, E. E., and D. J. Miller. 2006. Phylogeny: the continuing classificatory conundrum of chaetognaths. Curr. Biol. **16:** R593–R596. *Nice summary of chaetognath biology and two phylogenetic studies in the same issue.*

Bourlat, S. J., O. Rota-Stabelli, R. Lanfear, and M. J. Telford. 2009. The mitochondrial genome structure of *Xenoturbella bocki* (phylum Xenoturbellida) is ancestral within the deuterostomes. BMC Evol. Biol. **9:** 107–120. *These worms are clearly deuterostomes according to their mitochondrial genes and gene order.*

Cannon, J. T., K. M. Kocot, D. S. Waits, D. A. Weese, B. J. Swalla, S. R. Santos, and K. M. Halanych. 2014. Phylogenomic resolution of the echinoderm and hemichordate clade. Current Biology **24:** 2827–2832. *Data support the Asterozoa hypothesis whereby asteroids and brittle stars are sister taxa. Enteropneust hemichordates are the sister taxon of the pterobranchs.*

Gasmi, S., G. Neve, N. Pech, S. Tekaya, A. Gilles, and Y. Perez. 2014. Evolutionary history of Chaetognatha inferred from molecular and morphological data: a case study for body plan simplification. Frontiers in Zoology **11:** 84–108. *Chaetognath body plans evolved independently several times.*

Gilbert, S. F. 2013. Developmental biology, ed. 10. Sunderland, Massachusetts, Sinauer Associates. *Any modern text in developmental biology, such as this one, provides a multitude of examples in which studies on echinoderms have contributed (and continue to contribute) to our knowledge of development.*

Gonzalez, P., and C. B. Cameron. 2009. The gill slits and pre-oral ciliary organ of *Protoglossus* (Hemichordata: Enteropneusta) are filter-feeding structures. Biol. J. Linn. Soc. **98:** 898–906. *A clear description, with nice drawings and photos, of filter- and deposit-feeding in these unusual animals.*

Hendler, G., J. E. Miller, D. L. Pawson, and P. M. Kier. 1995. Sea stars, sea urchins, and allies: echinoderms of Florida and the Caribbean. Washington, Smithsonian Institution Press. *An excellent field guide for echinoderm identification.*

Hickman, C. P., Jr. 1998. A field guide to sea stars and other echinoderms of Galápagos. Lexington, VA, Sugar Spring Press. *Provides descriptions and nice photographs of members of classes Asteroidea, Ophiuroidea, Echinoidea, and Holothuroidea in the Galápagos Islands.*

Holland, N. D., L. Z. Holland, and P. W. H. Holland. 2015. Scenarios for the making of vertebrates. Nature **520**: 450–455. *The enteropneust and annelid theories for vertebrate evolution are nicely illustrated.*

Israelsson, O., and G. E. Budd. 2005. Eggs and embryos in *Xenoturbella* (phylum uncertain) are not ingested prey. Dev. Genes Evol. **215**: 358–363. *Eggs inside Xenoturbella were produced in situ and are not prey items.*

Lane, D. J. W. 1996. A crown-of-thorns outbreak in the eastern Indonesian Archipelago, February 1996. Coral Reefs **15**: 209–210. *This is the first report of an outbreak of Acanthaster planci in Indonesia. It includes a good photograph of an aggregation of these sea stars.*

Mooi, R., and B. David. 1998. Evolution within a bizarre phylum: homologies of the first echinoderms. Am. Zool. **38**: 965–974. *These authors contend that "The familiarity of a sea star or a sea urchin belies their overall weirdness." They describe the Etraxial/Axial Theory (EAT) of echinoderm skeletal homologies.*

Nielsen, C. 2010. After all: *Xenoturbella* is an acoelomorph! Evolution and Development **12**: 241–243. *The first morphological descriptions of Xenoturbella described it as an acoelomorph flatworm, and now some molecular research supports this placement.*

Philippe, H., H. Brinkmann, R. R. Copley, L. L. Moroz, H. Nakano, A. J. Poustka, A. Wallberg, K. J. Peterson, and M. J. Telford. 2011. Acoelomorph flatworms are deuterostomes related to *Xenoturbella*. Nature **470**: 255–258. *Although acoelomorph "flatworms" are close relatives to Xenoturbella, both are placed as deuterostomes, not basal bilaterians.*

Purcell, S. W., C. A. Hair, and D. J. Mills. 2012. Sea cucumber culture, farming and sea ranching in the tropics: Progress, problems and opportunities. Aquaculture **368–369**: 68–81. *Up-to-date information on the profitability and challenges of rearing sea cucumbers, especially the sandfish.*

Rahman, I. A., S. Zamora, P. L. Falkingham, and J. C. Phillips. 2015. Cambrian cinctan echinoderms shed light on feeding in the ancestral deuterostome. Proc. R. Soc. B. **282**: 20151964. *The ancestral deuterostome used pharyngeal gill slits to filter feed.*

Smith, A. B., K. J. Peterson, G. Wray, and D. T. J. Littlewood. 2003. From bilateral symmetry to pentaradiality: the phylogeny of hemichordates and echinoderms, pp. 365–383. In J. Cracraft and M. J. Donohue, eds., Assembling the tree of life. New York, Oxford University Press. *An excellent discussion of changes in deuterostome phylogeny.*

Woodley, J. D., P. M. H. Gayle, and N. Judd. 1999. Sea-urchins exert top-down control of macroalgae on Jamaican coral reefs (2). Coral Reefs **18**: 193. *In areas where Tripneustes (another sea urchin) have invaded fore reefs there is much less macroalgae, and such areas present a better chance that corals can recolonize. Diadema recovery has been slow.*

Zamora, S., I. A. Rahman, and A. B. Smith. 2012. Plated Cambrian bilaterians reveal the earliest stages of echinoderm evolution. PLoS ONE **7**: e38296. *Reconstructions of the first bilaterally symmetrical echinoderm fossil are described, along with inferences about echinoderm evolution and the origin of radial symmetry.*

23

CHAPTER

Chordates

- **PHYLUM CHORDATA**
  - Subphylum Urochordata
  - Subphylum Cephalochordata
  - Subphylum Vertebrata

Two amphioxus in feeding posture.

It's a Long Way from Amphioxus

Along the southern coasts of North America, half buried in sand on the seafloor, lives a small fishlike translucent animal quietly filtering organic particles from seawater. Inconspicuous, of no commercial value and largely unknown, this creature is nonetheless one of the famous animals of classical zoology. It is amphioxus, an animal that wonderfully exhibits the five distinctive hallmarks of the phylum Chordata: (1) a dorsal, tubular nerve cord overlying (2) a supportive notochord; (3) pharyngeal pouches or slits and (4) an endostyle for filter-feeding; and (5) a postanal tail for propulsion. Amphioxus is an animal that might have been designed by a zoologist for the classroom. During the nineteenth century, with interest in vertebrate ancestry running high, many zoologists thought amphioxus closely resembled the earliest vertebrates. Its exalted position was later acknowledged by Philip Pope in a poem sung to the tune of "Tipperary." It ends with the refrain:

It's a long way from amphioxus
*   It's a long way to us.*
It's a long way from amphioxus
*   To the meanest human cuss.*
Well, it's good-bye to fins and gill slits
*   And it's welcome lungs and hair,*
It's a long, long way from amphioxus
*   But we all came from there.*

Amphioxus's place in the sun did not endure. Amphioxus lacks one of the most important of vertebrate characteristics: a distinct head with special sense organs, an adaptation for shifting to an active predatory mode of life. Absence of a head, together with several specialized features, suggests to today's zoologists that amphioxus represents an early departure from vertebrate ancestry. We are a very long way indeed from amphioxus. Nevertheless, amphioxus probably resembles the chordate condition immediately preceding the origin of vertebrates more closely than does any other living animal.

23.1 THE CHORDATES

Animals most familiar to most people belong to phylum Chordata (kor-dät′ ə). Humans are members and share the characteristic from which the phylum derives its name—the **notochord** (Gr. *noton,* back, + L. *chorda,* cord) (Figure 23.1). All members of the phylum possess this structure, either restricted to early development or present throughout life. The notochord is a rodlike, semirigid body of fluid-filled cells enclosed by a fibrous sheath, which extends, in most cases, the length of the body just ventral to the central nervous system. Thus, the notochord is a hydrostatic organ, similar to the hydrostatic skeletons of nematodes (p. 387). Its primary purpose is to stiffen the body, providing skeletal scaffolding for the attachment of swimming muscles.

The structural plan of chordates shares features of many nonchordate invertebrates, such as bilateral symmetry, anteroposterior axis, coelom, tube-within-a-tube arrangement, metamerism, and cephalization. However, only members of the echinoderm-hemichordate assemblage (Deuterostomia branch) deserve serious consideration as the chordate sister group. Chordates share with other deuterostomes several important characteristics: radial cleavage (p. 171), an anus derived from the first embryonic opening (blastopore), a mouth derived from an opening of secondary origin, and a coelom formed by fusion of enterocoelous pouches (although in most vertebrates coelom formation is schizocoelus, but independently derived from that of protostomes, as an accommodation for their large yolks). These uniquely shared characteristics indicate a natural unity among Deuterostomia.

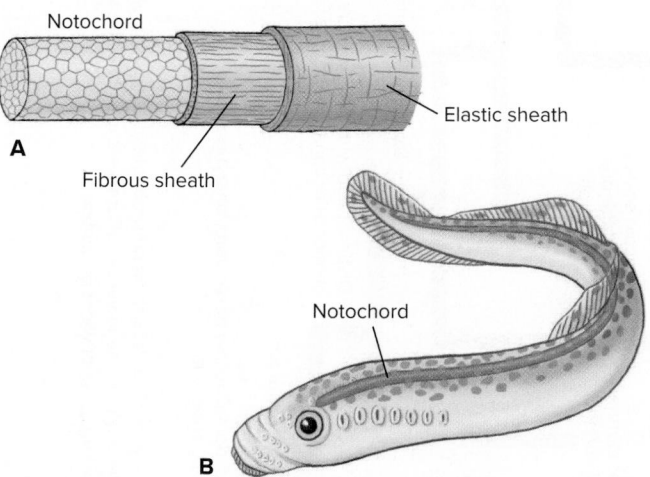

Figure 23.1 **A,** Structure of the notochord and its surrounding sheaths. Cells of the notochord proper are thick walled, pressed together closely, and filled with semifluid. Stiffness is caused mainly by turgidity of fluid-filled cells and surrounding connective tissue sheaths. This type of endoskeleton is characteristic of all chordates at some stage of life. The notochord provides longitudinal stiffening of the main body axis, a base for trunk muscles, and an axis around which the vertebral column develops. **B,** In hagfishes and lampreys it persists throughout life, but in other vertebrates it is largely replaced by vertebrae. In mammals, remnants are found in intervertebral discs.

Ecologically, chordates are among the most adaptable of animals and are able to occupy most kinds of habitat. They illustrate perhaps better than any other animal group the basic evolutionary processes of origin of new structures, adaptive strategies, and adaptive diversification.

Traditional and Cladistic Classification of the Chordates

Traditional Linnaean classification of chordates provides a convenient way to indicate the taxa included in each major group. However, in cladistic usage, some of the traditional taxa, such as Agnatha and Reptilia, are no longer recognized or have been redefined. Such taxa do not satisfy the requirement of cladistics that only **monophyletic** groups (clades), those groups that contain all known descendants of a single common ancestor, are taxonomically valid. For example, the animals traditionally called reptiles (turtles, lizards, snakes, and crocodilians) form a **paraphyletic** group, because this group does not contain all of the descendants of their most recent common ancestor. The common ancestor of reptiles as traditionally recognized is also the ancestor of birds (Figure 23.2). Reptilia is a clade if birds are included with those animals traditionally called reptiles. The reasons why nonmonophyletic groups are excluded from cladistic taxonomy are explained in Chapter 10 (p. 209).

The cladogram of chordates (Figure 23.2) shows a nested hierarchy of taxa grouped by their shared derived characters. These characters may be morphological, physiological, embryological, behavioral, chromosomal, or molecular. By contrast, the branches of a phylogenetic tree are intended to represent real lineages that occurred in the evolutionary past (Figure 23.3). Geological information regarding ages of lineages is added to information from the cladogram to generate a phylogenetic tree for the same taxa.

In our treatment of chordates, (p. 510) we have used a cladistic classification because such usage is necessary to reconstruct the evolution of characters in chordates. At present, because of numerous nested levels of clades, use of Linnaean ranks are incompatible with some taxa.

Several traditional divisions of phylum Chordata used in Linnaean classifications are shown in Table 23.1. A fundamental separation is Protochordata from Vertebrata.

The vertebrates may be variously subdivided into groups based on shared characteristics. Two such subdivisions shown in Table 23.1 are these: (1) Agnatha, vertebrates lacking jaws (hagfishes, lampreys, and jawless fossil fishes (pp. 511–512)), and Gnathostomata, vertebrates having jaws (all other vertebrates); and (2) Amniota, vertebrates whose embryos develop within a fluid-filled sac, the amnion ("reptiles", birds, and mammals), and Anamniota, vertebrates lacking this adaptation (fishes and amphibians). Gnathostomata in turn can be subdivided into Pisces, jawed vertebrates with appendages, if any, in the form of fins; and Tetrapoda (Gr. *tetras,* four, + *podos,* foot), jawed vertebrates with appendages, if any, in the form of limbs. Note that several of these groupings are paraphyletic (Protochordata, Agnatha, Anamniota, Pisces) and consequently are not accepted in cladistic classifications. Accepted monophyletic taxa are shown at the top of the cladogram in Figure 23.2 as a nested hierarchy of increasingly more inclusive groupings.

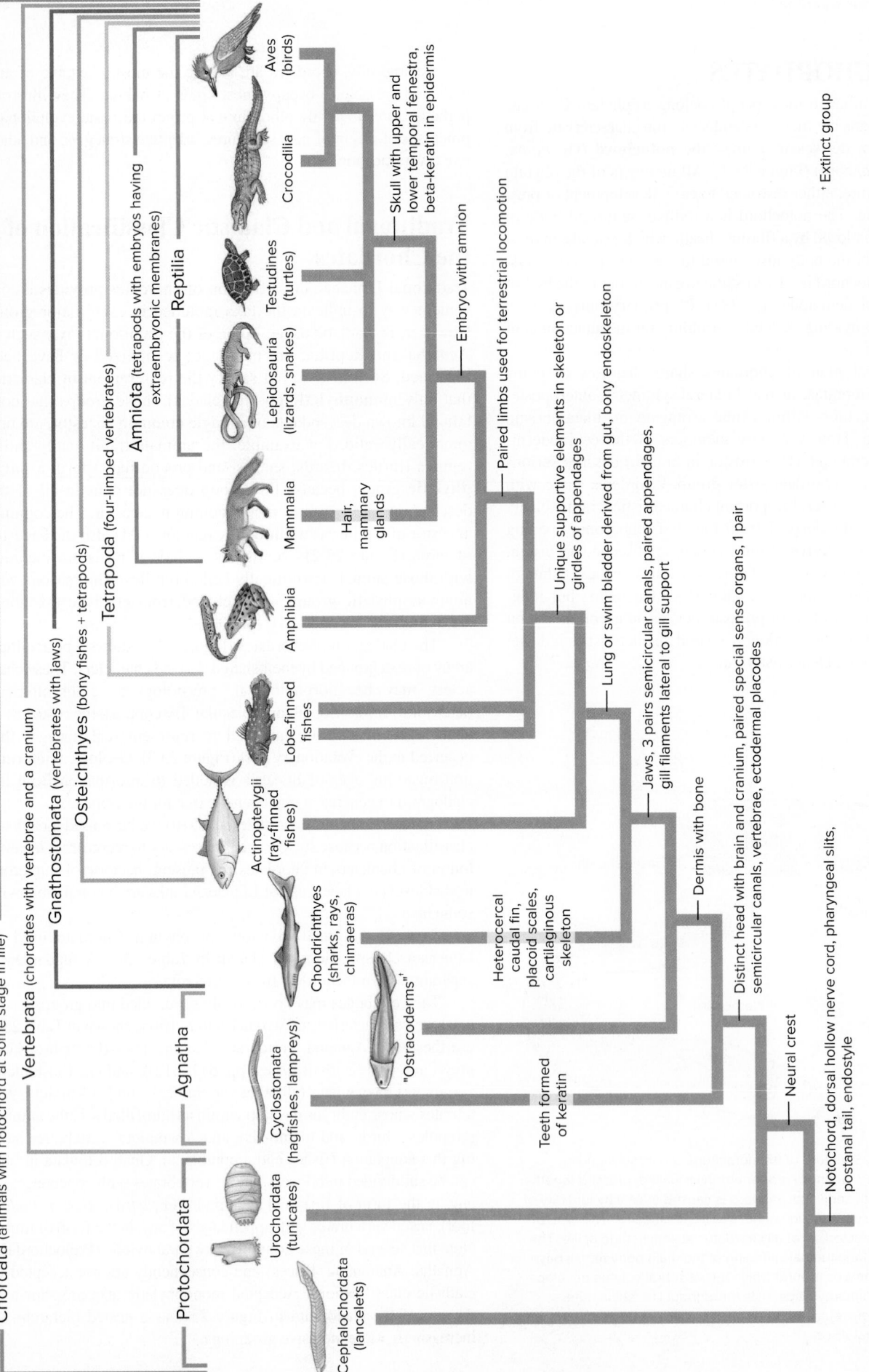

Figure 23.2 Cladogram of phylum Chordata showing probable relationships of monophyletic groups composing the phylum. Nesting brackets across the top of the cladogram identify monophyletic groupings within the phylum. The lower set of brackets identifies the traditional groupings Protochordata and Agnatha. These paraphyletic groups are not recognized in cladistic treatments, but are shown because of widespread use.

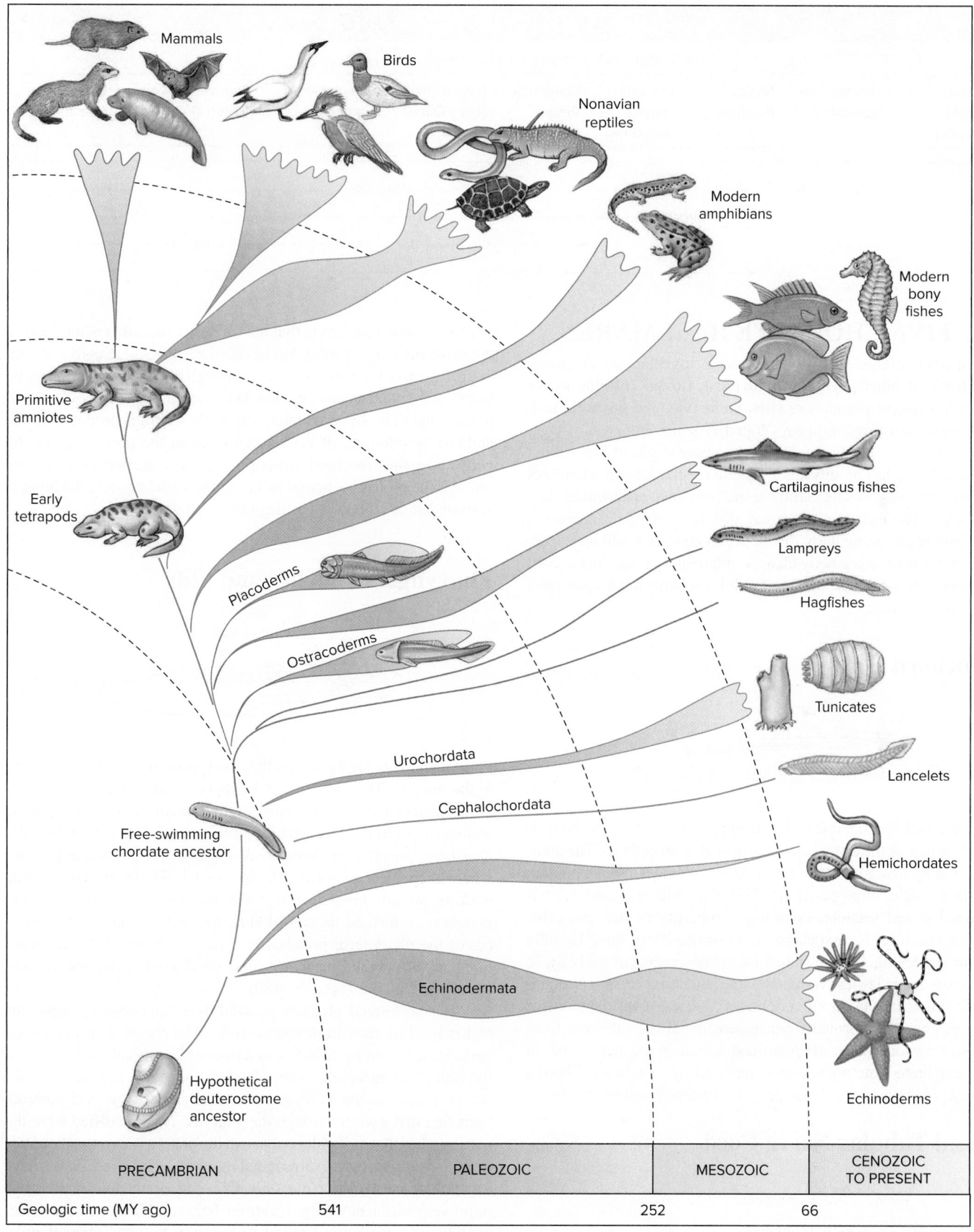

Figure 23.3 Phylogenetic tree of the chordates, suggesting probable origin and relationships. Other schemes have been suggested and are possible. The relative abundance in numbers of species of each group through geological time, as indicated by the fossil record, is suggested by the bulging and thinning of that group's line of descent.

TABLE 23.1									
Traditional Divisions of the Phylum Chordata									
Cephalo-chordata (lancelets)	Urochordata (tunicates)	Myxini (hagfishes)	Petromy-zontida (lampreys)	Chondrich-thyes (sharks)	Osteichthyes (bony fishes)	Amphibia (amphibians)	Reptilia (reptiles)	Aves (birds)	Mammalia (mammals)

23.2 FIVE CHORDATE HALLMARKS

Five distinctive characteristics that, taken together, set chordates apart from all other phyla are **notochord, dorsal tubular nerve cord, pharyngeal pouches** or **slits, endostyle,** and **postanal tail.** These characteristics are always found at some embryonic stage, although they may change or disappear in later stages of life. All but pharyngeal pouches or slits are unique to chordates; hemichordates also have pharyngeal slits and these are presumed ancestral to deuterostomes. We include pharyngeal slits as a chordate character, because they are particularly distinctive in chordates and an important part of the chordate body plan. A rudimentary dorsal nerve cord is present in some hemichordates, but it is probably not homologous to that of chordates.

Notochord

Notochord

The notochord is a flexible rod, extending the length of the body. It is the first part of the endoskeleton to appear in an embryo. The notochord is a hydrostatic organ, but unlike nematodes, which contain fluid in a single, large cavity (p. 387), the fluid in a notochord is contained in cell vacuoles or in tiny compartments between cells. Muscles attach to the notochord, and because it can bend laterally without shortening, it permits undulatory movements of the body. In amphioxus and in jawless vertebrates, the notochord persists throughout life (Figure 23.1). In all vertebrates a series of cartilaginous or bony vertebrae are formed from mesenchymal cells derived from blocks of mesodermal cells (**somites**) lateral to the notochord. In most vertebrates, the notochord is replaced by vertebrae, although remnants of the notochord may persist between or within vertebrae.

Dorsal Tubular Nerve Cord

Brain   Dorsal nerve cord

In most invertebrate phyla that have a nerve cord, it is ventral to the digestive tract and is solid, but in chordates the single cord is dorsal to the digestive tract and is a tube (although the hollow center may be nearly obliterated during growth). The anterior end becomes enlarged to form the brain in vertebrates. The hollow cord is produced in the embryo by infolding of ectodermal cells on the dorsal side of the body above the notochord. Among vertebrates, the nerve cord passes through the protective neural arches of the vertebrae, and the brain is surrounded by a bony or cartilaginous cranium.

Pharyngeal Pouches and Slits

Pharyngeal slits

Pharyngeal slits are openings that lead from the pharyngeal cavity to the outside. They are formed by inpocketing of the outside ectoderm (pharyngeal grooves) and evagination, or outpocketing, of the endodermal lining of the pharynx (pharyngeal pouches). In aquatic chordates, the two pockets break through the pharyngeal cavity where they meet to form the pharyngeal slit. In amniotes some pockets do not break through the pharyngeal cavity and only pouches are formed instead of slits. In tetrapod (terrestrial) vertebrates the pharyngeal pouches give rise to several different structures, including the Eustachian tube, middle ear cavity, tonsils, and parathyroid glands (pp. 179–180).

The perforated pharynx evolved as a filter-feeding apparatus and is used as such in protochordates. Water with suspended food particles is drawn by ciliary action through the mouth and flows out through pharyngeal slits where food is trapped in mucus. In vertebrates, ciliary action is replaced by muscular pharyngeal contractions that drive water through the pharynx. Also modified were the aortic arches that carry blood through the pharynx. In protochordates these are simple vessels surrounded by connective tissue. Fishes have a capillary network having only thin, gas-permeable walls, thus improving efficiency of gas transfer between blood and the water outside. These adaptations led to the evolution of **internal gills,** completing conversion of the pharynx from a filter-feeding apparatus in protochordates to a respiratory organ in aquatic vertebrates.

Characteristics of Phylum Chordata

1. **Postanal tail; notochord; endostyle** or **thyroid gland; bone** and **cartilage** in vertebrates
2. Living in marine, freshwater, and terrestrial habitats; many capable of flight
3. Free-living, but a very few fishes are ectoparasitic
4. Bilateral symmetry; segmented, but segmentation inconspicuous in many
5. Triploblastic
6. **Coelom well developed**
7. Epidermis present in all; dermis in vertebrates; keratinized or bony structures often present in vertebrate integument; glands often diverse and abundant in vertebrates
8. Digestive system complete; muscular gut in vertebrates; **pharyngeal pouches** present early in development, erupting to outside as gill slits in aquatic forms
9. Smooth, skeletal, and cardiac muscle tissue present; segmented myomeres in fishes and amphibians
10. **Nerve cord hollow and dorsal;** distinct, **three-lobed brain** present in vertebrates
11. Protochordates with simple, unpaired photoreceptors and statocysts; vertebrates with well-developed paired sensory organs for vision, chemoreception, hearing, balance, electroreception, and vibration sensitivity
12. Asexual reproduction by parthenogenesis in some fishes, amphibians, and lizards
13. Sexes usually separate; hermaphroditism in sea squirts and some fishes; fertilization internal or external; oviparous or viviparous; distinct larval stage in some; crocodilians, birds, mammals, and some fishes and amphibians with parental care of young
14. Paired, glomerular kidneys and ducts in vertebrates
15. Respiration primarily via gills, lungs, and skin; swim bladder present in many fishes, functioning in buoyancy
16. Closed circulation; **chambered hearts** and red blood cells in vertebrates; distinct aortic arches in all except sea squirts

Endostyle or Thyroid Gland

Endostyle

Until recently, the endostyle was not recognized as a chordate character. However, it or its derivative, the thyroid gland, occurs in all chordates but in no other animals. The endostyle, in the pharyngeal floor, secretes mucus that traps small food particles brought into the pharyngeal cavity. An endostyle occurs in protochordates and lamprey larvae. Some cells in the endostyle secrete iodinated proteins. These cells are homologous with the iodinated hormone-secreting thyroid gland of adult lampreys and all other vertebrates. In protochordates and lamprey larvae, an endostyle and a perforated pharynx work together to create an efficient filter-feeding apparatus.

Postanal Tail

Muscular, postanal tail

A postanal tail, together with somatic musculature and the stiffening notochord, provides the motility that larval tunicates and amphioxus need for their free-swimming existence. As a structure added to the body behind the end of the digestive tract, the tail clearly has evolved specifically for propulsion in water. Its efficiency is later increased in fishes with the addition of fins. The tail is evident in humans only as a vestige (the coccyx, a series of small vertebrae at the end of the spinal column), but most other mammals have a waggable tail as adults.

23.3 ANCESTRY AND EVOLUTION

Since the mid–nineteenth century when Darwin's theory of common descent became the focal point for recognizing relationships among groups of living organisms, zoologists have debated the question of chordate origins. It has been difficult to reconstruct the evolutionary history of the earliest chordates because they were probably soft-bodied creatures that had little chance of being preserved as fossils. Although Cambrian chordates have been discovered (pp. 509 and 511), the fossil record is sparse. Consequently, such reconstructions largely come from the study of living organisms, especially from an analysis of early developmental stages, which often tend to be more evolutionarily conserved than the differentiated adult forms they become.

Most early efforts to identify chordate relationships were based on similarities due to analogy rather than homology (p. 107). Analogous structures perform similar functions but have different origins (such as wings of birds and butterflies). Homologous structures, on the other hand, share a common origin but can look different (at least superficially) and perform different functions. For example, all vertebrate forelimbs are homologous because they are derived from a pentadactyl limb of the same ancestor, even though they may be modified as differently as a human's arm and a bird's wing. Homologous structures share a genetic heritage; analogous structures do not. Obviously, only homologous similarities reveal common ancestry.

Zoologists at first speculated that chordates evolved within the protostome clade (annelids and arthropods) but rejected such ideas when they realized that supposed morphological similarities were not homologous. Early in the twentieth century, further theorizing became rooted in developmental patterns of animals, and it became apparent that chordates are deuterostomes. Deuterostomia, which also includes the echinoderms and hemichordates, has several important embryological features, as well as shared gene sequences, that clearly separate it from Protostomia and establish its monophyly (Figure 8.18). Thus, deuterostomes are almost certainly a natural grouping of interrelated animals that have their common origin in ancient Precambrian seas. Several lines of anatomical, developmental, and molecular evidence suggest that somewhat later, at the base of the Cambrian period some 540 million years ago, the first distinctive chordates arose (Figure 23.3). Evidence from phylogenetic analysis of gene sequences, development, and morphology strongly suggest that a clade containing both echinoderms and hemichordates is the sister group of chordates (see Figure 22.1). Information about the biology of the earliest chordates can be gleaned from examination of the two living chordate groups that are not vertebrates, Urochordata and Cephalochordata.

23.4 SUBPHYLUM UROCHORDATA (TUNICATA)

The urochordates ("tail-chordates"), more commonly called tunicates, include about 3000 species. They live in all seas from near shoreline to great depths. Most are sessile as adults, although some are free-living. The name "tunicate" describes the usually tough, nonliving **tunic** that surrounds the animal and contains cellulose (Figure 23.4). As adults, tunicates are highly specialized chordates, for in most species only the larval form, which resembles a microscopic tadpole, bears all the chordate hallmarks. During adult metamorphosis, the notochord (which, in the larva, is restricted to the tail, hence the group name Urochordata) and tail disappear, while the dorsal nerve cord becomes reduced to a single ganglion.

Urochordata is divided into three classes: **Ascidiacea** (Gr. *askiolion*, little bag, + *acea*, suffix), **Appendicularia** (L. *appendic*, hang to), and **Thaliacea** (Gr. *thalia*, luxuriance, + *acea*, suffix). Members of Ascidiacea are by far the most common, diverse, and best known. They are often called "sea squirts" because some species forcefully discharge a jet of water from the excurrent siphon when irritated. All but a few ascidian species are sessile animals, attached to rocks or other hard substrates such as pilings or bottoms of ships. In many areas, they are among the most abundant of intertidal animals.

Ascidians may be solitary, colonial, or compound. Each of the solitary and colonial forms has its own test, but among compound forms many individuals may share the same tunic (Figure 23.5). In some compound ascidians each member has its own incurrent siphon, but the excurrent opening is common to the group.

Solitary ascidians (Figure 23.4) are usually spherical or cylindrical forms. Lining the tunic is an inner membrane, the **mantle.** On the outside are two projections: an **incurrent siphon,** or oral siphon, which corresponds to the anterior end of the body, and an **excurrent siphon,** or atrial siphon, that marks the dorsal side. Water enters the incurrent siphon and passes into a ciliated **pharynx** that is minutely perforated by slits to form an elaborate basketwork. Water passes through the slits into an **atrium** and out through the excurrent siphon.

Feeding depends on formation of a mucous net that is secreted by a glandular groove, the **endostyle,** located along the midventral side of the pharynx. Cilia on bars of the pharynx pull the

mucus into a sheet that spreads dorsally across the inner face of the pharynx. Food particles brought in the incurrent opening are trapped on the mucous net, which is then worked into a rope and carried posteriorly by cilia into the esophagus. Nutrients are absorbed in the midgut and indigestible wastes are discharged from the anus, located near the excurrent siphon.

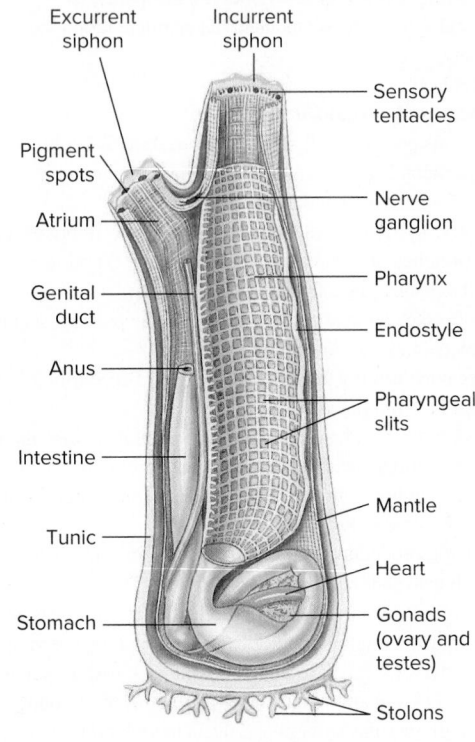

Figure 23.4   Structure of a common tunicate, *Ciona.*

B

Figure 23.5   **A,** Two yellow-edged sea squirts, *Rhopalaea,* on a Philippine reef. Note the large incurrent siphon and the smaller excurrent siphon for each animal. **B,** Seven colonies of compound tunicates, *Atriolum robustum,* on a Pacific reef. Individuals in a colony share a common tunic (yellow), but each has a separate incurrent (oral) siphon. Each colony has a single, large excurrent (atrial) siphon on top. Class Ascidiacea.

A

Figure 23.6   Metamorphosis of a solitary ascidian from a free-swimming larval stage.

The circulatory system consists of a ventral heart and two large vessels, one on either side of the heart; these vessels connect to a diffuse system of smaller vessels and spaces serving the pharyngeal basket (where respiratory exchange occurs), the digestive organs, gonads, and other structures. An odd feature, found in no other chordate, is that the heart drives the blood first in one direction for a few beats, then pauses, reverses its action, and drives the blood in the opposite direction for a few beats. Another remarkable feature is the presence of strikingly high amounts of rare elements in the blood, such as vanadium and niobium. The vanadium concentration in the sea squirt *Ciona* may reach 2 million times its concentration in seawater. The function of these rare metals in the blood is a mystery.

The nervous system is restricted to a **nerve ganglion** and plexus of nerves that lie on the dorsal side of the pharynx. Beneath the nerve ganglion is located the **subneural gland,** connected by a duct to the pharynx.

Sea squirts are hermaphroditic, with usually a single ovary and a single testis in the same animal. Gametes are carried by ducts into the atrial cavity, and then into the surrounding water where fertilization occurs.

Of the five chief characteristics of chordates, adult sea squirts have only two: pharyngeal slits and endostyle. However, the larval form reveals the secret of their true relationship. The "tadpole" larva (Figure 23.6) is an elongate, transparent form with all five chordate characteristics: notochord, hollow dorsal nerve cord, propulsive postanal tail, and large pharynx with endostyle and pharyngeal slits. The larva does not feed but swims for several hours or days before fastening itself vertically by adhesive papillae to a solid object. It then undergoes a dramatic metamorphosis (Figure 23.6) to become a sessile adult, so modified as to become almost unrecognizable as a chordate.

Tunicates of class Thaliacea, the thaliaceans or salps, are barrel- or lemon-shaped pelagic forms with transparent, gelatinous bodies that, despite the considerable size that some species reach, are nearly invisible in sunlit surface waters. They occur singly or in colonial chains that may reach several meters in length (Figure 23.7). The cylindrical thaliacean body is typically surrounded by bands of circular muscle, with incurrent and excurrent siphons at opposite ends. Water pumped through the body by muscular contraction (rather than by cilia as in ascidians) is used for locomotion by a sort of jet propulsion, for respiration, and as a source of particulate food that is filtered on mucous surfaces. Many are provided with luminous organs, which give a brilliant light at night. Most of the body is hollow, with the viscera forming a compact mass on the ventral side.

The life histories of thaliaceans are often complex and are adapted to respond to sudden increases in their food supply. The appearance of a phytoplankton bloom, for example, is met by an explosive population increase leading to extremely high density of thaliaceans. Common forms include *Doliolum* and *Salpa,* both of which reproduce by an alternation of sexual and asexual generations.

The third tunicate class, the Appendicularia (Larvacea in some taxonomies) contains tiny, pelagic creatures shaped like a bent tadpole. The name Larvacea refers to their resemblance to the larval stages of other tunicates. They feed by a method unique in the animal world. Each builds a delicate house, a transparent hollow sphere of mucus interlaced with filters and passages

Figure 23.7   Salps. The transparent individuals of this delicate, planktonic species are grouped in a chain. Visible within each individual is an opaque gonad, an opaque gut, and a long serrated gill bar. Class Thaliacea.

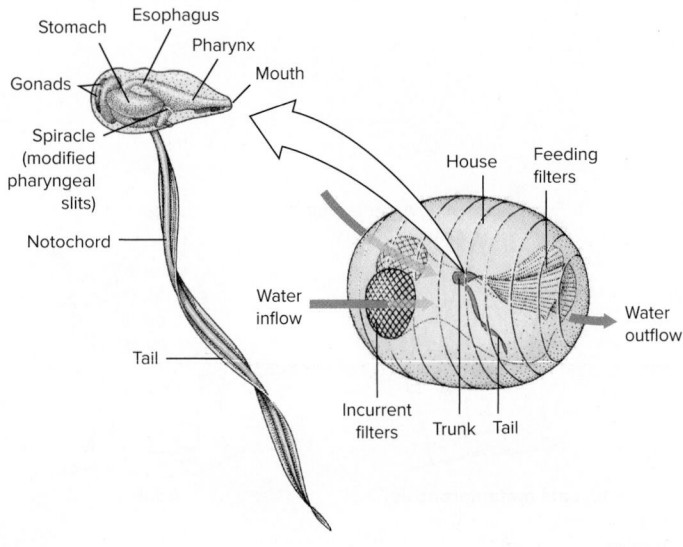

Figure 23.8 Appendicularian (larvacean) adult (*left*) and as it appears within its transparent house (*right*), which is about the size of a walnut. When the feeding filters become clogged with food, the tunicate abandons its house and builds a new one.

through which water enters (Figure 23.8). Tiny phytoplankton and bacteria trapped on a feeding filter inside the house are drawn into the animal's mouth through a strawlike tube. When the filters become clogged with waste, which happens about every 4 hours, the appendicularian abandons its house and builds a new house, a process that takes only a few minutes. Like thaliaceans, appendicularians can quickly build dense populations when food is abundant. At such times scuba diving among the houses, which are about the size of walnuts, is likened to swimming through a snowstorm!

23.5 SUBPHYLUM CEPHALOCHORDATA

Cephalochordates are lancelets: slender, laterally compressed, translucent animals about 3 to 7 cm in length (Figure 23.9) that inhabit sandy sediments of coastal waters around the world. Lancelets originally bore the generic name *Amphioxus* (Gr. *amphi*, both ends, + *oxys*, sharp), later surrendered by priority to *Branchiostoma* (Gr. *branchia*, gills, + *stoma*, mouth). Amphioxus is still used, however, as a convenient common name for all of approximately 32 species in this diminutive subphylum. Five species of amphioxus occur in North American coastal waters.

Amphioxus is especially interesting because it has the five distinctive characteristics of chordates in simple form. Water enters the mouth, driven by cilia in the buccal cavity and pharynx, then passes through numerous pharyngeal slits. The endostyle secretes a mucous sheet on the inner surface of the pharynx. Food is trapped on the mucous sheet and then moved by cilia into the gut. Here the smallest food particles are separated from the mucus and passed

into the **hepatic cecum** where they are phagocytized and digested intracellularly. Food is moved through the gut by means of cilia, which are concentrated in a darkly staining area called the ileocolic ring (Figure 23.9), rather than by muscular contractions as in vertebrates. As in tunicates, filtered water passes first into an **atrium,** then leaves the body by an **atriopore** (equivalent to the excurrent siphon of tunicates).

The closed circulatory system is complex for so simple a chordate. The flow pattern is remarkably similar to that of fishes, although there is no heart. Blood is pumped forward in the **ventral aorta** by peristaltic-like contractions of the vessel wall, then passes upward through branchial arteries (aortic arches) in the pharyngeal bars to paired **dorsal aortas,** which join posteriorly to become a single dorsal aorta. From there blood is distributed to body tissues by microcirculation and then is collected in veins, which return it to the ventral aorta. Lacking both erythrocytes and hemoglobin, their blood is thought to circulate nutrients but not respiratory gases. There are no gills specialized for respiration in the pharynx; gas exchange occurs over the surface of the body.

The nervous system is centered around a hollow nerve cord lying above the notochord. Pairs of **spinal nerve roots** emerge at each trunk myomeric (muscle) segment. Sense organs are simple, including an anterior, unpaired **ocellus** that functions as a photoreceptor. Although the anterior end of the nerve cord is not enlarged into the characteristic vertebrate brain, it is apparently homologous to parts of the vertebrate brain.

Sexes are separate. Gametes are released in the atrium, then pass through the atriopore to the outside where fertilization occurs. Cleavage is holoblastic and a gastrula is formed by invagination. Larvae hatch soon after egg deposition and gradually assume the shape of adults.

No other chordate shows the basic diagnostic chordate characteristics as clearly as does amphioxus. In addition to the five chordate anatomical hallmarks, amphioxus possesses several structural features that resemble the vertebrate plan. Among these are a hepatic cecum, a diverticulum that resembles the vertebrate pancreas in secreting digestive enzymes and the liver in storing glycogen, **segmented trunk musculature,** and the basic circulatory plan of vertebrates.

23.6 SUBPHYLUM VERTEBRATA

The third subphylum of chordates is the large and diverse Vertebrata, the subject of Chapters 24 through 28. This monophyletic group shares the basic chordate characteristics with the other two subphyla, but in addition it exhibits a number of novel characters that the others do not share. An alternative name of the subphylum, Craniata, refers to the braincase (**cranium**) of bone or cartilage, which occurs in all members.

Adaptations That Have Guided Early Vertebrate Evolution

The earliest vertebrates were substantially larger and considerably more active than the protochordates. Modifications of skeletal structures and muscles permitted increased speed and mobility. The

Figure 23.9 Amphioxus. This bottom-dwelling cephalochordate illustrates the five distinctive chordate characteristics (notochord, dorsal nerve cord, pharyngeal slits, endostyle, and postanal tail). This body plan is considered ancestral for chordates. **A,** Living amphioxus in typical position for filter-feeding. Note the oral hood with tentacles surrounding the mouth. **B,** Internal structure.

© Cleveland P. Hickman, Jr.

A

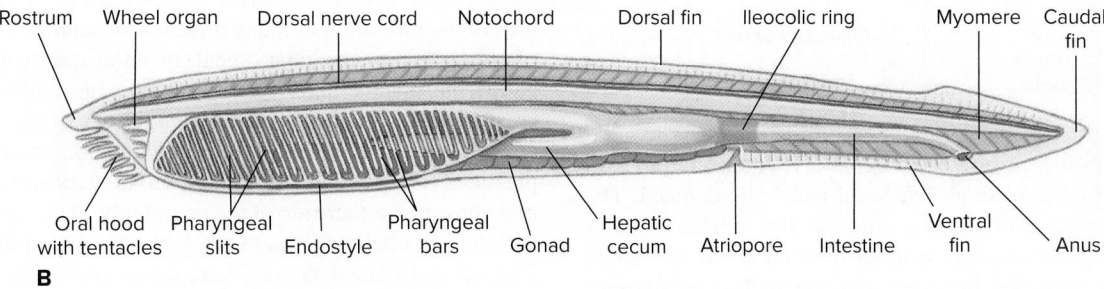

Rostrum   Wheel organ   Dorsal nerve cord   Notochord   Dorsal fin   Ileocolic ring   Myomere   Caudal fin

Oral hood with tentacles   Pharyngeal slits   Endostyle   Pharyngeal bars   Gonad   Hepatic cecum   Atriopore   Intestine   Ventral fin   Anus

B

higher activity level and size of vertebrates also required structures specialized in the location, capture, and digestion of food and adaptations designed to support a high metabolic rate.

Musculoskeletal Modifications

Centrum of vertebra
Notochord   Neural spine

Most vertebrates possess both an an exoskeleton and an endoskeleton of cartilage or bone. The endoskeleton permits almost unlimited body size with much greater economy of building materials than does the exoskeleton of arthropods. Some

vertebrates have become the most massive organisms on earth. In most vertebrates, disclike centra replace the notochord, and dorsal projections called **neural spines** are present, providing more area for attachment of segmented muscles. The segmented body muscles (myomeres) changed from the V-shaped muscles of ancestral chordates to the W-shaped muscles of vertebrates. This increased complexity of folding in the myomeres provides powerful control over an extended length of the body. Also unique to vertebrates are fin rays of dermal origin, supporting fins used in swimming.

The endoskeleton probably was composed initially of cartilage and later of bone. Cartilage, with its fast growth and flexibility, is ideal for constructing the first skeletal framework of all vertebrate embryos. The endoskeleton of living hagfishes, lampreys, sharks and their kin, and even some "bony" fishes, such as sturgeons, is mostly cartilage. Bone may have been adaptive in early vertebrates in several ways. Plates of bone in the skin of ostracoderms and other ancient fishes certainly provided protection from predators. Bone is stronger than cartilage, making it ideal for muscle attachment in areas of high mechanical stress. One of the most interesting ideas is that bone's original function was mineral storage and homeostasis. Phosphorus and calcium, the primary mineral components of bone, are used for many physiological processes and are in particularly high demand in organisms with high metabolic rates.

Most vertebrates possess an extensive exoskeleton (one that develops from the skin). Some of the earliest fishes, including ostracoderms and placoderms (see Figures 23.14 and 23.17), were partly covered in a bony, dermal armor. This armor is modified as scales in later fishes. Most terrestrial vertebrates are further protected with keratinized structures derived from the epidermis, such as reptilian scales, hair and feathers.

Physiology Upgrade

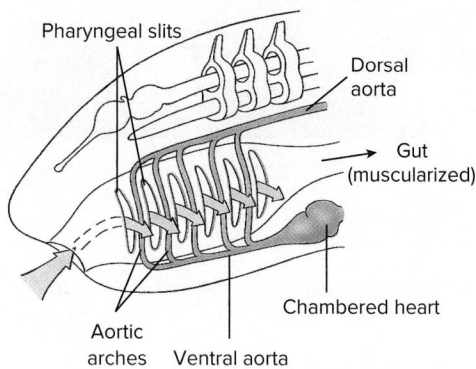

Vertebrate digestive, respiratory, circulatory, and excretory systems are modified to meet an increased metabolic demand. The perforated pharynx evolved as a filter-feeding device in early chordates, using cilia and mucus to move water and to trap small suspended food particles. In vertebrates, the addition of muscles to the pharynx created a powerful pump for moving water. With the origin of highly vascularized gills, the function of the pharynx shifted primarily to gas exchange. Early chordates moved food in the gut with cilia, but vertebrates move food in the gut with muscles. This shift, along with addition of two distinct digestive glands, the liver and pancreas, enabled vertebrates to manage the increased amount of food ingested. A chambered, ventral heart and erythrocytes with hemoglobin enhanced transportation of gases, nutrients, and other substances. Protochordates have no distinct kidneys, but vertebrates possess paired, glomerular kidneys that remove metabolic waste products and regulate body fluids and ions.

New Head, Brain, and Sensory Systems

When vertebrate ancestors shifted from filter-feeding to active predation, new sensory, motor, and integrative controls became essential for locating and capturing larger prey. The anterior end of the nerve cord became enlarged as a **tripartite brain** (forebrain, midbrain, and hindbrain) and protected by a cartilaginous or bony cranium. Paired special sense organs designed for distance reception evolved. These included eyes with lenses and inverted retinas; pressure receptors, such as paired inner ears designed for equilibrium and sound reception; chemical receptors including taste and exquisitely sensitive olfactory organs; lateral-line receptors for detecting water vibrations; and electroreceptors for detecting electrical currents that signal prey (see Chapter 33).

Neural Crest, Ectodermal Placodes, and Hox Genes

Development of the vertebrate head and special sense organs was largely the result of two embryonic innovations present only in vertebrates: **neural crest** and **ectodermal placodes.** The neural crest, a population of ectodermal cells lying along the length of the embryonic neural tube (see Figure 8.28 and p. 178), contributes to the formation of many different structures, including most of the cranium, the pharyngeal skeleton, tooth dentine, some endocrine glands, and other structures. In addition, the neural crest regulates the development of adjacent tissue, such as tooth enamel and pharyngeal muscles (branchiomeres). Ectodermal placodes (Gr. *placo,* plate) are platelike ectodermal thickenings that appear on either side of the neural tube. These give rise to the olfactory epithelium, lens of the eye, inner ear epithelium, some ganglia and cranial nerves, lateral-line mechanoreceptors, and electroreceptors. Thus the vertebrate head, with its sensory structures located adjacent to the mouth (later equipped with preycapturing jaws), stemmed from new embryonic tissues.

Studies of the distribution of homeobox (*Hox*) genes that control the body plan of chordate embryos (p. 167) suggest that *Hox* genes were duplicated at about the time of the origin of vertebrates. Amphioxus and invertebrates have only one set of *Hox* genes, whereas most living gnathostomes have four sets. Proliferation of *Hox* genes might have played a pivotal role in the evolution of the vertebrate body plan.

The Search for the Ancestral Vertebrate

Most of the early Paleozoic vertebrate fossils, the jawless ostracoderms (p. 511), share many novel features of organ system development with living vertebrates. These organ systems must have originated in an early vertebrate or invertebrate chordate lineage. Fossil invertebrate chordates are rare and known primarily from two fossil beds—the middle Cambrian Burgess Shale of Canada (p. 104) and the early Cambrian fossil beds of Chengjiang and Haikou, China. An ascidian tunicate and *Yunnanozoon,* a probable cephalochordate, are known from Chengjiang. Slightly better known is *Pikaia,* a ribbon-shaped, somewhat fishlike creature about 5 cm in length discovered in the Burgess Shale (Figure 23.10). The presence of myomeres and a notochord clearly identifies *Pikaia* as a chordate. The superficial resemblance of *Pikaia* to living amphioxus suggests that it may be an early cephalochordate.

A wealth of information about the origin of vertebrates is provided by *Haikouella lanceolata,* a small fishlike creature known from over 300 fossil specimens recently discovered in 530-million-year-old sediments near Haikou. It possessed several characters that identify it as a chordate, including notochord, pharynx, and dorsal nerve cord. Additional features of the fossils, interpreted as pharyngeal muscles, paired eyes, and an enlarged brain, are characteristic of vertebrates (Figure 23.11). However, it is not a vertebrate, because the fossils lack evidence of several diagnostic vertebrate traits, including a cranium, an ear, and a distinct telencephalon (anterior region of the forebrain). Jon Mallatt, Jun-yuan Chen, and colleagues, who have studied these fossils extensively, hypothesized *Haikouella* to be the sister taxon of vertebrates, although this is not universally accepted. Despite recent fossil discoveries of early chordates, many speculations regarding vertebrate ancestry have focused on the living protochordates, in part because they are much better known than are the fossil forms.

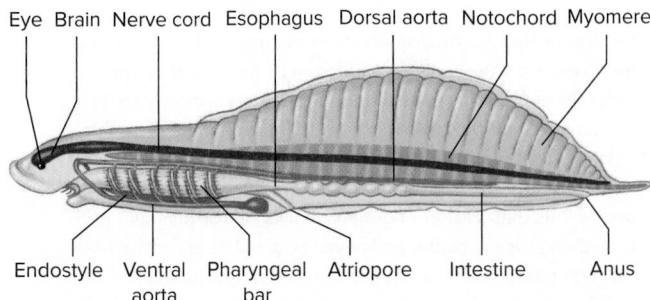

Figure 23.11   *Haikouella,* a chordate with several vertebrate features from early Cambrian shales of Haikou, China. It has been hypothesized to be the sister taxon of vertebrates.

Butler and Hodos provided an explanation for how the paired eyes of vertebrates evolved from the unpaired, median ocellus of an amphioxus-like ancestor. The homeotic gene *Pax-6* is responsible for the formation of an eye-producing region near the midbrain. Products of another gene, *sonic hedgehog,* suppress *Pax-6* expression at the midline, thus forming paired, lateral eyes. Mice with a mutated *sonic hedgehog* gene have unpaired, median eyes.

Chordate Evolution and the Position of Amphioxus

Some chordates, sea squirts, are sedentary, whereas others, amphioxus and vertebrates, are active and mobile. Which was the ancestral form? In 1928 Walter Garstang of England postulated that the ancestral chordate was a sedentary filter feeder, like adult ascidian tunicates. Garstang hypothesized that the ancestral life cycle of chordates resembled that of tunicates, and that an ancestor of vertebrates lost the ability to metamorphose

Figure 23.12   Garstang's hypothesis of larval evolution. According to this hypothesis (now rejected), the ancestral chordate was a sedentary filter feeder, like most adult tunicates. More than 540 million years ago some larvae became paedomorphic, attaining reproductive maturity in the larval body form. These became cephalized, evolving to become the first vertebrates.

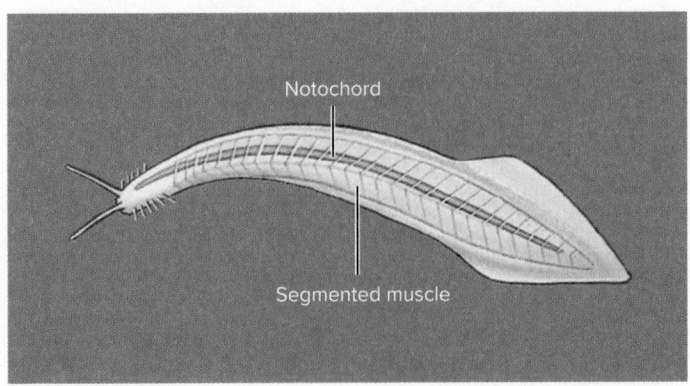

Figure 23.10   *Pikaia,* an early chordate from the Burgess Shale of British Columbia, Canada.

into a sessile adult, instead developing gonads and reproducing while retaining an otherwise larval morphology. This form, now free-swimming as an adult, would have been the ancestor of cephalochordates and vertebrates. Continued evolution, emphasizing cephalization, would have produced the first vertebrates (Figure 23.12). Garstang called this process **paedomorphosis** (Gr. *pais,* child + *morphe,* form), a term describing the evolutionary retention of juvenile or larval traits in the adult body. Paedomorphosis is a well-known phenomenon in several different animal groups (paedomorphosis in amphibians is described on pp. 550–551).

Paedomorphosis, the displacement of ancestral larval or juvenile features into a descendant adult, can be produced by three different evolutionary–developmental processes: neoteny, progenesis, and postdisplacement. In neoteny, the growth rate of body form is slowed so that the animal does not attain the ancestral adult form when it reaches maturity. Progenesis is a precocious maturation of gonads in a larval (or juvenile) body that then stops growing and never attains the ancestral body form. In postdisplacement, the onset of a developmental process is delayed relative to reproductive maturation, so that the ancestral adult form is not attained by the time of reproductive maturation. Neoteny, progenesis, and postdisplacement thus describe different ways in which paedomorphosis can happen. Biologists use the inclusive term paedomorphosis to describe results of these evolutionary–developmental processes.

Recently collected genetic and developmental evidence allows testing of Garstang's hypothesis. Numerous phylogenetic reconstructions, along with fossil evidence, place cephalochordates as the sister taxon to a clade consisting of urochordates plus vertebrates, suggesting that cephalochordates and vertebrates retain the ancestral chordate condition and that the sessile ascidians represent a derived condition (see Figure 23.2). In addition, incipient neural crest tissue recently has been identified in urochordates, supporting a sister-group relationship with vertebrates. Thus, most zoologists currently reject Garstang's hypothesis and envision the ancestral chordate as a free-swimming creature (see Figure 23.3) perhaps similar to the modern amphioxus.

Although urochordates are now considered the closest living relative of vertebrates, their sessile body form evolved in a lineage ancestral only to urochordates and not in vertebrate ancestry. Most zoologists view amphioxus as retaining much of the prevertebrate and ancestral

Taxonomy of Living Members of Phylum Chordata

Phylum Chordata

Subphylum Urochordata (yur′ō-kor-dä′tə) (Gr. *oura*, tail, + L. *chorda*, cord, + *ata*, characterized by) **(Tunicata): tunicates.** Notochord and nerve cord in free-swimming larva only; ascidian adults sessile, encased in tunic. About 3000 species.

Subphylum Cephalochordata (sef′ə-lō-kor-dä′tə) (Gr. *kephalē*, head, + L. *chorda*, cord): **lancelets (amphioxus).** Notochord and nerve cord found along entire length of body and persist throughout life; fishlike in form. 32 species.

Subphylum Vertebrata (ver′te-brä′tə) (L. *vertebratus*, backboned): **vertebrates.** Bony or cartilaginous cranium surrounding tripartite brain; well-developed head with paired sense organs, usually with vertebrae; heart with multiple chambers; muscularized digestive tract; paired kidneys.

  Cyclostomata (sī′klō-stō-mä′tä) (Gr. *cyclos*, circle, *stoma*, mouth): **hagfishes, lampreys.** Without true jaws or paired appendages.

    Myxini (mik-sē′nē) (Gr. *myxa*, slime): **hagfishes.** Four pairs of tentacles around mouth; buccal funnel absent; 1 to 16 pairs of gill openings; slime glands present; vertebrae vestigial. About 70 species.

    Petromyzontida (pet′trō-mī-zon′ti′də) (Gr. *petros*, stone, + *myzon*, sucking): **lampreys.** Buccal funnel with keratinized teeth, nasal sac not connected to pharynx; vertebrae present only as neural arches. 41 species.

  Gnathostomata (na′thō-stō′mä-tə) (Gr. *gnathos*, jaw, + *stoma*, mouth): **jawed fishes, tetrapods.** With jaws and (usually) paired appendages

    Chondrichthyes (kon-drik′thē-ēz) (Gr. *chondros*, cartilage, + *ichthys*, a fish): **sharks, skates, rays, chimaeras.** Cartilaginous skeleton, intestine with spiral valve; claspers present in males; no swim bladder. About 1000 species.

    Actinopterygii (ak′ti-nop′te-rij′ē-ī) (Gr. *aktis*, ray, + *pteryx*, fin, wing): **ray-finned fishes.** Skeleton ossified; single gill opening covered by operculum; paired fins supported primarily by dermal rays; appendage musculature within body; swim bladder mainly a hydrostatic organ, if present; atrium and ventricle not divided. About 31,000 species.

    Sarcopterygii (sär-cop-te-rij′ē-ī) (Gr. *sarkos*, flesh, + *pteryx*, fin, wing): **lobe-finned fishes.** Skeleton ossified; paired appendages with sturdy internal bones and musculature; atrium and ventricle at least partly divided. 8 species.

    Dipnoi (dip′nōi) (Gr. *di*, double + *pnoi*, breathing): **lungfishes.** Freshwater habitat; platelike, crushing teeth; caudal fin joined with anal and dorsal fins; swim bladder modified as 1 or 2 lungs. 6 species.

    Actinistia (ak′tin-is-tē′ə) (Gr. *actino*, ray + *istia*, a superlative ending): **coelacanths.** Marine habitat; three-lobed caudal fin; swim bladder filled with fat. 2 species.

    Amphibia (am-fib′ē-ə) (Gr. *amphi*, both or double, + *bios*, life): **amphibians.** Ectothermic tetrapods; respiration by lungs, gills, or skin; development through larval stage; skin moist, with mucous glands and lacking scales. About 6800 species.

    Reptilia (rep-til′ē-ə) (L. *repere*, to creep): **birds, turtles, lizards, snakes, crocodilians.** Tetrapods with amniotic membrane in embryos; lungs ventilated by aspiration; skin dry and covered with scales or feathers composed of beta-keratin.

      Lepidosauria (le-pi-dō-sor′ē-ə) (Gr. *lepidos*, scale, + *saurus*, lizard): **snakes, lizards, and tuataras.** Sprawling posture; diapsid skull often modified by loss of one or both temporal arches; transverse cloacal slit; skin shed in one piece; about 9,200 species.

      Testudines (tes-tū′di-nēz) (L. *testudo*, tortoise): **turtles.** Body enclosed in fused, bony plates forming a shell; jaws with keratinized plates instead of teeth; temporal openings of skull lost; about 325 species.

      Crocodilia (krok′ə-dil′ē-ə) (L. *crocodilus*, crocodile): **crocodilians.** Skull diapsid, elongate, and massive; secondary palate present; gizzard present; ventricle completely divided; 25 species.

      Aves (ā-vēz) (L. pl. of *avis*, bird): **birds.** Endothermic vertebrates with front limbs modified for flight; body covered with feathers; about 10,500 species.

    Mammalia (ma-mā′lē-ə) (L. *mamma*, breast): **mammals.** Endothermic vertebrates possessing mammary glands; body more or less covered with hair; brain large, with neocortex; three middle ear bones. About 5400 species.

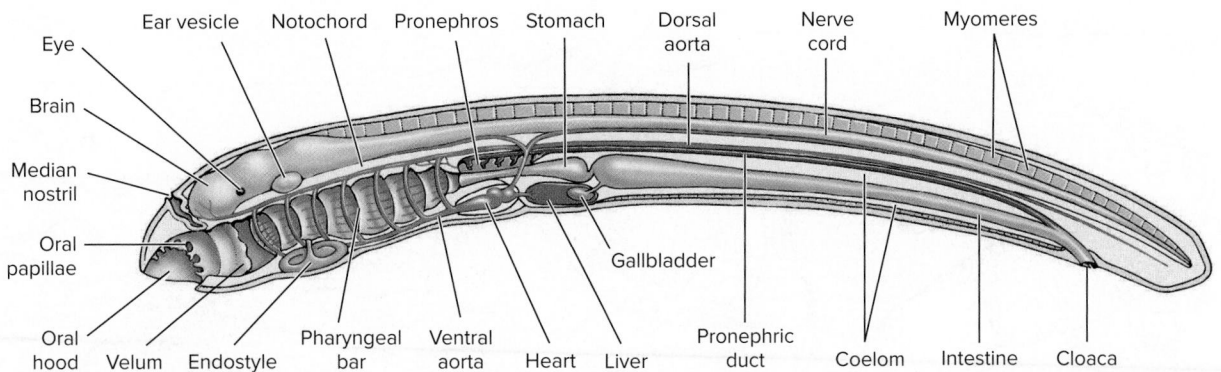

Figure 23.13 Ammocoete, freshwater larval stage of a lamprey. Although they resemble amphioxus in many ways, ammocoetes have a well-developed brain, paired eyes, distinct pronephric kidneys, heart, and other features lacking in amphioxus but characteristic of the vertebrate body plan.

chordate body structure. Cephalochordates share several characteristics with vertebrates that are absent from tunicates, including segmented myomeres, dorsal and ventral aortas, branchial or aortic arches, and podocytes, specialized excretory cells. Because of amphioxus' similarity to the presumed archetypal vertebrate form, it remains a popular reference organism in studies of early vertebrate evolution.

The Ammocoete Larva of Lampreys as a Model for the Ancestral Vertebrate Body Plan

Lampreys (jawless fishes of the clade Petromyzontida, discussed in Chapter 24) have a freshwater larval stage called an **ammocoete** Figure 23.13). In body form, appearance, life habit, and many anatomical details, an ammocoete larva resembles amphioxus. In fact, lamprey larvae were given the genus name *Ammocoetes* (Gr. *ammos*, sand, + *koite*, bed, referring to the preferred larval habitat) in the nineteenth century when it was erroneously thought to be an adult cephalochordate, closely allied with amphioxus. Ammocoete larvae are so different from adult lampreys that the mistake is understandable; the relationship was not understood until metamorphosis into an adult lamprey was observed.

An ammocoete larva has a long, slender body with an oral hood surrounding the mouth, much like an amphioxus (Figure 23.13). Ammocoetes are filter feeders, but instead of drawing water by ciliary action into the pharynx as amphioxus do, ammocoetes produce a feeding current by muscular pumping action much like modern fishes. The arrangement of body muscle into myomeres, the presence of a notochord serving as chief skeletal axis, and the plan of the circulatory system all resemble these features in amphioxus.

Ammocoetes have several characteristics lacking in amphioxus that are homologous to those of vertebrates. These include a chambered heart; tripartite brain; paired special sense organs derived, in part, from ectodermal placodes; pituitary gland; and distinct kidneys. Ammocoetes also have a true liver replacing the hepatic cecum of amphioxus, a gallbladder, and pancreatic tissue (but no distinct pancreatic gland). Instead of the numerous pharyngeal slits of amphioxus, there are only seven pairs of pharyngeal pouches and slits in ammocoetes. From pharyngeal bars

separating the pharyngeal slits project gill filaments bearing secondary lamellae much like the more extensive gills typical of bony fishes (see Figure 24.28). Thus, the function of the pharynx evolved from one of primarily filter-feeding in protochordates to one that also includes gas exchange in ammocoetes and other aquatic vertebrates.

Ammocoetes display the most ancestral condition for these characteristics of any living vertebrate. They clearly illustrate many shared derived characters of vertebrates that are obscured in the development of other vertebrates. They may approach most closely the body plan of the ancestral vertebrate.

The Earliest Vertebrates

The earliest known vertebrate fossils, until recently, were armored jawless fishes called **ostracoderms** (os-trak'o-derms) (Gr. *ostrakon*, shell, + *derma*, skin) from late Cambrian and Ordovician deposits. In the late 1990s, researchers unearthed a number of 530-million-year-old fossils in the amazing Chengjiang deposits belonging to two (possibly the same species) fishlike vertebrates: *Myllokunmingia* (Gr. *myllo*, sea fish, + Kunming, a city in China) and *Haikouichthys* (Haikou, a city in China, + Gr. *ichthy*, fish). These fossils push back the origin of vertebrates to at least the early Cambrian period. They show many vertebrate characteristics, including a heart, paired eyes, otic (ear) capsules, and what have been interpreted as rudimentary vertebrae.

The earliest ostracoderms were armored with bone in their dermis and lacked paired fins that later fishes found so important for stability (Figure 23.14). The swimming movements of one of the early groups, the **heterostracans** (Gr. *heteros*, different, + *ostrakon*, shell), must have been imprecise, although sufficient to propel them along the ocean floor where they searched for food. With fixed circular or slitlike mouth openings, they may have filtered small food particles from the water or ocean sediments. However, unlike the ciliary filter-feeding protochordates, ostracoderms pulled water into the pharynx by muscular pumping, an important innovation that suggests to some authorities that ostracoderms may have been active predators that fed on soft-bodied animals.

Figure 23.14   Three ostracoderms, jawless fishes of Silurian and Devonian times. They are shown as they might have appeared while searching for food on the floor of a Devonian sea. They employed a strong pharyngeal pump to circulate water rather than the much more limiting mode of ciliary action used by their protochordate ancestors (presumably resembling amphioxus in this way).

The term "ostracoderm" does not denote a natural evolutionary assemblage but rather is a term of convenience for describing several groups of heavily armored extinct jawless fishes.

During the Devonian period, heterostracans underwent a major diversification, producing numerous peculiar-looking forms. Without ever evolving paired fins or jaws, these earliest vertebrates flourished for 150 million years until becoming extinct near the end of the Devonian period.

Coexisting with heterostracans throughout much of the Devonian period were **osteostracans** (Gr. *osteon,* bone, + *ostrakon,* shell). Osteostracans had paired pectoral fins, an innovation that functioned to improve swimming efficiency by controlling yaw, pitch, and roll. A typical osteostracan, such as *Cephalaspis* (Gr. *kephale,* head, + *aspis,* shield) (Figure 23.14), was a small animal, seldom exceeding 30 cm in length. It was covered with a heavy, dermal armor of bone, including a single-piece head shield. Examination of internal features of the braincase reveals a sophisticated nervous system and sense organs, similar to those of modern lampreys.

Another group of ostracoderms, the **anaspids** (Figure 23.14), were more streamlined than other ostracoderms. These and other ostracoderms underwent an impressive diversification in the Silurian and Devonian periods. However, all ostracoderms became extinct by the end of the Devonian period.

For decades geologists used strange, microscopic toothlike fossils called **conodonts** (Gr. *kōnos,* cone, + *odontos,* tooth) to date Paleozoic marine sediments without knowing what kind of creature originally possessed these elements. The discovery in the early 1980s of fossils of complete conodont animals changed this situation. With their phosphatized toothlike elements, myomeres, notochord, and paired eye and otic capsules, conodonts clearly belong to the vertebrate clade (Figure 23.15), although their exact position is unclear.

The Swedish paleozoologist Erik Stensiö (1891–1984) was the first to approach fossil anatomy with the same painstaking attention to minute detail that morphologists have long applied to the anatomical study of living fishes. He developed novel and exacting methods for gradually grinding away a fossil, a few micrometers at a time, to reveal internal features. He was able to reconstruct not only bone anatomy, but nerves, blood vessels, and muscles in numerous groups of Paleozoic and early Mesozoic fishes. His innovative methods continue to be used by today's paleozoologists, but are supplemented by new imaging technologies, including computerized tomography (CT) scanning.

Early Jawed Vertebrates

All jawed vertebrates, whether extinct or living, are collectively called **gnathostomes** ("jaw mouth") in contrast to the jawless vertebrates, the **agnathans** ("without jaw"). Gnathostomes are a monophyletic group; presence of jaws is a derived character state shared by all jawed fishes and tetrapods. Agnathans, however, are defined principally by the absence of jaws, a character that is not unique to jawless fishes because jaws are lacking in vertebrate ancestors. Thus, "Agnatha" is paraphyletic.

The origin of jaws was one of the most important events in vertebrate evolution. The utility of jaws is obvious: they allow predation on large and active forms of food not available to jawless vertebrates. Ample evidence suggests that jaws arose through modifications of the first or second of the serially repeated cartilaginous gill arches. How did this mandibular arch change from a function of gill support and ventilation to one of feeding as jaws? Expansion of this arch and

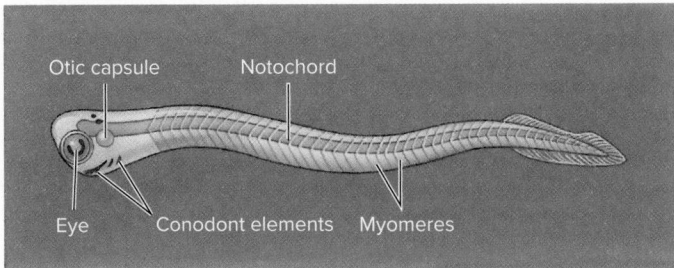

Figure 23.15 Restoration of a conodont. Conodonts superficially resembled amphioxus, but they possessed a much greater degree of encephalization (paired eyes and possibly otic [ear] capsules) and bonelike mineralized elements—all indicating that conodonts were probably vertebrates. Conodont elements are considered to be part of a food-handling apparatus.

Figure 23.16 How vertebrates got their jaws. Resemblance between jaws and gill supports of early fishes, such as this Carboniferous shark, suggests that the upper jaw (palatoquadrate) and lower jaw (Meckel's cartilage) evolved from structures that originally functioned as gill supports. The gill supports immediately behind the jaws are hinged like jaws and served to link the jaws to the braincase. Relics of this transformation are seen during the development of modern sharks.

evolution of new, associated muscles may have first assisted gill ventilation, perhaps to meet the increasing metabolic demands of early vertebrates. Once enlarged and equipped with more powerful muscles, the first pharyngeal arch could easily have been modified to serve as jaws. Evidence for this remarkable transformation includes, first, that both gill arches and jaws form from upper and lower bars that bend forward and are hinged in the middle (Figure 23.16). Second, both gill arches and jaws are derived from neural crest cells. Third, the jaw musculature is homologous to the original gill support musculature as evidenced by cranial nerve distribution. Nearly as remarkable as this drastic morphological remodeling is the subsequent evolutionary fate of jaw skeletal elements—their transformation into ear ossicles of the mammalian middle ear (p. 739).

> A growing group of zoologists use developmental characters to uncover the evolutionary history of structures or organisms, an area of research called "evo-devo." Recent studies document that expression of various homeotic genes, including *Hox* and *Dlx*, establishes dorsal-ventral and anterior-posterior boundaries during development of various head structures. The dorso-ventral patterning of mandibular arch, hyoid arch, and posterior arches are all guided by similar patterns of *Dlx* expression.

An additional feature characteristic of all gnathostomes is the presence of paired pectoral and pelvic appendages in the form of fins or limbs. These likely originated as stabilizers to check yaw, pitch, and roll generated during active swimming. According to the fin-fold hypothesis, paired fins arose from paired continuous, ventrolateral folds or fin-forming zones. The addition of skeletal supports in the fins enhanced the fins' ability to provide stability during swimming. Evidence for this hypothesis is found in the paired flaps of *Myllokunmingia* and anaspids and in the multiple paired fins of acanthodians, also described in this section. However, pectoral fins appear in the fossil record before pelvic fins do, suggesting a more complex evolutionary scenario. In one fish lineage the muscle and skeletal supports in the paired fins became strengthened, allowing them to become adapted for locomotion on land as limbs. Developmental biologists discovered that differential expression of various homeotic genes defines the boundaries of flank and paired appendages in living

gnathostomes. The origin of jaws and paired appendages may be linked to a second *Hox* duplication, near the origin of the gnathostomes. Both jaws and paired fins were major innovations in vertebrate evolution, among the most important reasons for the subsequent major diversifications of vertebrates that produced the modern fishes and all tetrapods, including you, the reader of this book.

Among the first jawed vertebrates were the heavily armored **placoderms** (plak′o-derm) (Gr. *plax*, plate, +o-*derma*, skin). These first appear in the fossil record in the early Silurian period (Figure 23.17). Placoderms evolved a great variety of forms, some very large (one was 10 m in length!). They were armored fish covered with diamond-shaped scales or with large plates of bone. All became extinct by the end of the Devonian period. Placoderms are probably paraphyletic; one placoderm group apparently gave rise to all other gnathostomes (see Figure 24.2). **Acanthodians**, a paraphyletic group of early jawed fishes known from the Silurian to Permian periods, and characterized by large, anteriorly set eyes and fins with large spines (Figure 23.17), are included in a clade that underwent a great diversification to form the cartilaginous fishes and bony fishes that dominate the waters of the world today.

SUMMARY

Phylum Chordata is named for the rodlike notochord that forms a stiffening body axis at some stage in the life cycle of every chordate. All chordates share four additional distinctive hallmarks that set them apart from all other phyla: dorsal hollow nerve cord, pharyngeal pouches or slits, endostyle, and postanal tail. Two of the three chordate subphyla are invertebrates and lack a well-developed head. They are Urochordata (tunicates), most of which are sessile as adults but all of which have a free-swimming larval stage, and Cephalochordata (lancelets), fishlike forms that include the famous amphioxus.

Figure 23.17 Early jawed fishes of the Devonian period, 400 million years ago. Jaws and gill supports from which the jaws evolved develop from neural crest cells, a nearly diagnostic character of vertebrates. Most placoderms were bottom dwellers that fed on benthic animals, although some were active predators. Acanthodians carried less armor than placoderms and had large, anteriorly placed eyes and prominent spines on paired fins. Most were marine but several species lived in freshwater.

Chordates are deuterostomes, like echinoderms and hemichordates. Most zoologists now consider the chordate ancestor to have been a small, free-swimming, filter-feeding creature. Although recent evidence suggests urochordates are the living sister group of vertebrates, the sessile lifestyle of their adults is likely derived, and amphioxus (Cephalochordata) has a body form more similar to the first vertebrates.

Subphylum Vertebrata includes the backboned members of the animal kingdom (vertebrae are transient or reduced in some, the hagfishes and lampreys). As a group vertebrates are characterized by a well-developed head, a comparatively large size, a high degree of motility, and a distinctive body plan with several distinguishing features that permitted their exceptional adaptive diversification. Most important of these are the living endoskeleton that allows continuous growth and provides a sturdy framework for efficient muscle action and attachment, a muscular pharynx with slits and gills (lost or greatly modified in terrestrial vertebrates) with vastly increased respiratory efficiency, a muscularized gut and chambered heart, which meets higher metabolic demands, and an advanced nervous system with a distinct brain and paired sense organs. Insight into the early evolution of vertebrates is provided by examination of several fossil forms, including *Haikouella,* conodonts, and ostracoderm fishes, and of ammocoetes, the larvae of living lampreys. Evolution of jaws and paired appendages likely contributed to the incredible success of one group of vertebrates, the gnathostomes.

REVIEW QUESTIONS

1. What characteristics shared by the three deuterostome phyla indicate a monophyletic group of interrelated animals?
2. Explain how a cladistic classification for the vertebrates produces important regroupings of the traditional vertebrate taxa (refer to Figure 23.2). Why are Agnatha and Reptilia, as traditionally recognized, inconsistent with cladistic principles?
3. Name five characteristics shared by all chordates, and explain the primary function of each.
4. In debating the question of chordate origins, zoologists eventually agreed that chordates must have evolved within the deuterostome assemblage rather than from a protostome group as earlier argued. What embryological evidences support this view?
5. Offer a description of an adult tunicate that would identify it as a chordate, yet distinguish it from any other chordate group.
6. Amphioxus long has been of interest to zoologists searching for a vertebrate ancestor. Explain why amphioxus captured such interest and how it is different from the hypothesized most recent common ancestor of all vertebrates.
7. Both sea squirts (urochordates) and lancelets (cephalochordates) are filter-feeding organisms. Describe the filter-feeding apparatus of a sea squirt and explain in what ways its mode of feeding is similar to, and different from, that of amphioxus.
8. Explain why it is necessary to know the life history of a tunicate to understand why tunicates are chordates.
9. List three groups of adaptations that guided vertebrate evolution, and explain how each has contributed to the success of vertebrates.
10. In 1928 Walter Garstang hypothesized that tunicates resemble the ancestral stock of the vertebrates. Explain this hypothesis and evaluate its validity with recent phylogenetic, fossil, and developmental data.
11. What is the phylogenetic position of *Haikouella*, and what evidence supports its placement there?
12. Distinguish between ostracoderms and placoderms. What important evolutionary characters first appeared in each group? What are conodonts?
13. What is the currently favored explanation for the evolution of vertebrate jaws?

For Further Thought Urochordates, instead of amphioxus, are now considered the sister group of vertebrates. Why is amphioxus, instead of urochordates, still considered a better model for the early vertebrate body plan?

SELECTED REFERENCES

Ahlberg, P. E. 2001. Major events in early vertebrate evolution. London, Taylor & Francis. *Evolution of vertebrates up to the split of major jawed fish groups, incorporating molecular, fossil, and embryological data. This book makes important contributions, but some conclusions are controversial.*

Alldredge, A. 1976. Appendicularians. Sci. Am. **235:** 94–102 (July). *Describes the biology of larvaceans, which build delicate houses for trapping food.*

Donoghue, P. C. J., P. L. Forey, and R. J. Aldridge. 2000. Conodont affinity and chordate phylogeny. Biol. Rev. **75:** 191–251. *In this summary of early chordate evolution, the authors provide evidence that conodonts are vertebrates and that lampreys and hagfishes do not form a monophyletic group.*

Forey, P., and P. Janvier. 1994. Evolution of the early vertebrates. Am. Sci. **82:** 554–565. *Summarizes the biology and evolution of many groups of ostracoderms and other primitive craniates.*

Gee, H. 1996. Before the backbone: views on the origin of the vertebrates. New York, Chapman & Hall. *Outstanding review of the many vertebrate origin hypotheses. Gee links much of the genetic, developmental, and molecular evidence in his discussion.*

Gillis, J. A., M. S. Modell, and C. V. H. Baker. 2013. Developmental evidence for serial homology of the vertebrate jaw and gill arch skeleton. Nature Commun. **4:** 1436. *Patterning of jaw and pharyngeal arches is determined by nested, dorsoventral expression of* Dlx *genes.*

Gould, S. J. 1989. Wonderful life: the Burgess Shale and the nature of history. New York, W. W. Norton & Company. *In this book describing the marvelous Cambrian fossils of the Burgess Shale, Gould "saves the best for last" by inserting an epilogue on* Pikaia, *the first known chordate.*

Jeffrey, W. R., A. G. Strickler, and Y. Yamamoto. 2004. Migratory neural crestlike cells from body pigmentation in a urochordate embryo. Nature **431:** 696–699. *Evidence that supports a sister-group relationship between urochordates and vertebrates.*

Long, J. A. 2010. The rise of fishes: 500 million years of evolution, 2nd ed. Baltimore, The Johns Hopkins University Press. *An authoritative, liberally illustrated evolutionary history of fishes.*

Maisey, J. G. 1996. Discovering fossil fishes. New York, Henry Holt & Company. *Handsomely illustrated chronology of fish evolution with cladistic analysis of evolutionary relationships.*

Mallatt, J., and J.-Y. Chen. 2003. Fossil sister group of craniates: predicted and found. J. Morph. **258:** 1–31. *Reevaluation of* Haikouella *fossils revealed several features that suggest it is the sister taxon to craniates.*

Shimeld, S. M., and P. W. Holland. 2000. Vertebrate innovations. Proc. Natl. Acad. Sci. **97:** 4449–4452. *Focuses on developmental characters, including neural crest, ectodermal placodes, and* Hox *genes.*

Shubin, N. 2008. Your inner fish. New York, Pantheon Books. *An engaging and easy read. You look the way you do because of evolutionary events that happened to fishes (and other animals) millions of years ago.*

Stokes, M. D., and N. D. Holland. 1998. The lancelet. Am. Sci. **86** (6):552–560. *Describes the historical role of amphioxus in early hypotheses of vertebrate ancestry and summarizes molecular data that has rekindled interest in amphioxus.*

Zhu, M., X. Yu, P. E. Ahlberg, B. Choo, J. Lu, J. Qiao, Q. Qu, W. Zhao, L. Jia, H. Blom, and Y. Zhu. 2013. A Silurian placoderm with osteichthyan-like marginal jaw bones. Nature **502:** 188–193. *Phylogenetic analysis of new fossils indicated new relationships of acanthodians and placoderms.*

24

CHAPTER

Fishes

Hammerhead shark, Sphyrna, *near Cocos Island, southwest of Costa Rica.*

- **PHYLUM CHORDATA**
  - Myxini
  - Petromyzontida
  - Chondrichthyes
  - Actinopterygii
  - Sarcopterygii

What Is a Fish?

In common (and especially older) usage, the term "fish" denotes a mixed assortment of water-dwelling animals. We speak of jellyfish, cuttlefish, starfish, crayfish, and shellfish, knowing full well that when we use the word "fish" in such combinations, we are not referring to a true fish. In earlier times, even biologists did not make such a distinction. Sixteenth-century natural historians classified seals, whales, amphibians, crocodiles, even hippopotamuses, as well as a host of aquatic invertebrates, as fishes. Later biologists were more discriminating, eliminating first the invertebrates and then the amphibians, reptiles, and mammals from the narrowing concept of a fish. Today we recognize a fish as an aquatic vertebrate with gills, appendages, if present, in the form of fins, and usually a skin with scales of dermal origin.

Even this modern concept of the term "fish" is used for convenience, not as a taxonomic unit. Fishes do not form a monophyletic group, because the ancestor of land vertebrates (tetrapods) is found within one group of fishes (the sarcopterygians). Thus, fishes can be defined in an evolutionary sense as all vertebrates that are not tetrapods. Because fishes live in habitats that are less accessible to humans than terrestrial habitats, people have rarely appreciated the remarkable diversity of these vertebrates. Nevertheless, the world's fishes have diversified to produce over 32,000 living species—more than all other species of vertebrates combined—with adaptations for almost every conceivable aquatic environment. No other animal group matches their domination of the world's seas, lakes, and streams.

The body of a fish is streamlined for movement through water. Suspended in a medium that is 800 times more dense than air, a trout or pike can remain motionless, varying its neutral buoyancy by adding or removing air from its swim bladder. It may dart forward or at angles, using its fins as brakes and tilting rudders. With excellent organs for salt and water exchange, fishes can finely tune their body fluid composition in their freshwater or seawater environment. Their gills are the most effective respiratory devices in the animal kingdom for extracting oxygen from a medium that contains less than 1/20 as much oxygen as air. Fishes have excellent olfactory and visual senses and a lateral-line system, which has an exquisite sensitivity to water currents and vibrations. Thus in mastering the physical problems of their element, early fishes evolved a basic body plan and set of physiological strategies that both shaped and constrained the evolution of their descendants.

24.1 ANCESTRY AND RELATIONSHIPS OF MAJOR GROUPS OF FISHES

Fishes are a vast array of distantly related gill-breathing aquatic vertebrates with fins. They are the most ancient and most diverse of the clade Vertebrata, constituting five of the nine living vertebrate classes and about half of the approximately 64,000 vertebrate species.

Fishes first appeared in the early Paleozoic era, about 550 million years ago (hypotheses of chordate and vertebrate origins are discussed in Chapter 23). The earliest vertebrates were an assemblage of jawless **agnathan** fishes, including the ostracoderms (see Figure 23.14). One group of ostracoderms gave rise to the jawed **gnathostomes.**

> The use of *fishes* as the plural form of *fish* may sound odd to most people accustomed to using *fish* in both the singular and the plural. *Fish* refers to one or more individuals of the same species; *fishes* refers to more than one species.

Along with the extinct ostracoderms, the jawless agnathans include the living **hagfishes** and **lampreys,** fishes adapted as scavengers or parasites. Hagfishes and lampreys superficially look much alike, but they diverged from each other at least 450 mya, and are placed in different groups, Myxini and Petromyzontida, respectively.

All remaining living fishes have paired appendages and jaws and are included, along with tetrapods (land vertebrates), in the monophyletic group Gnathostomata. They appear in the fossil record in the late Silurian period with fully formed jaws, and no forms intermediate between agnathans and gnathostomes are known. By the Devonian period, the "age of fishes," several distinct groups of jawed fishes were common. One of these, the placoderms (p. 513), became extinct in the subsequent Carboniferous period, although one group of placoderms apparently gave rise to all other gnathostomes. A second group, **cartilaginous fishes** of the clade Chondrichthyes (sharks, rays, and chimaeras), lost the heavy dermal armor of early jawed fishes and adopted cartilage for the endoskeleton; most are active predators with sharklike or raylike body forms that changed only slightly over the ages. As a group, chondrichthyans flourished during the Devonian and Carboniferous periods of the Paleozoic era but declined dangerously close to extinction at the end of the Paleozoic

era. They recovered in the early Mesozoic era and diversified to form modern sharks and rays (Figure 24.1).

The other two groups of gnathostomes, **acanthodians** (p. 513) and Osteichthyes (bony fishes and tetrapods), were abundant and diverse in the Devonian period. Acanthodians were distinctive in having heavy spines on all fins except the caudal fin. They became extinct in the lower Permian period. Although phylogenetic affinities of the acanthodians are much debated, the group is likely paraphyletic with respect to Chondrichthyes (Figure 24.2). There are two distinct clades of Osteichthyes. Clade Actinopterygii comprise the ray-finned fishes, which includes nearly all of our familiar bony fishes. Clade Sarcopterygii comprise the lobe-finned fishes and tetrapods. Lobe-finned fishes are represented today by **lungfishes** and **coelacanths**—meager remnants of important lineages that flourished in the Devonian period (Figure 24.1). A taxonomy of the major fish taxa is on page 541.

24.2 LIVING JAWLESS FISHES

Living jawless fishes include approximately 111 species divided between two classes: Myxini (hagfishes) with about 70 species and Petromyzontida (lampreys) with 41 species. Members of both groups lack jaws, internal ossification, scales, and paired fins, and both groups share porelike gill openings and an eel-like body form.

Based on morphological similarity of hagfishes and lampreys, early zoologists united these two groups under the name **Cyclostomata**. This grouping fell out of favor by the 1990s, when cladistics analysis of morphological characters placed lampreys with gnathostomes, which would make Cyclostomata paraphyletic. Numerous characters, including a cerebellum, eye lens, and apparently, vertebrae, are shared in lampreys and gnathostomes, but absent in hagfishes. However, subsequent phylogenetic analyses of molecular data consistently support the Cyclostomata grouping of hagfishes and lampreys, suggesting that the apparently "simple" morphology of hagfishes represents considerable degeneracy (loss of characters), rather than an ancestral vertebrate condition. In 2006, Kinya Ota and Shigeru Kuratani developed a method to rear embryos of hagfishes, and numerous papers on hagfish development and evolution soon followed. These studies also support the Cyclostomata hypothesis and are providing a wealth of new information about early vertebrate evolution. We now know that vertebrae, formerly considered absent in hagfishes, appear as rudimentary structures in some hagfish embryos, suggesting vertebrae were present in the common ancestor of all living vertebrates! We treat Cyclostomata as a monophyletic group of vertebrates (Figure 24.2).

Myxini: Hagfishes

Hagfishes are an entirely marine group that feeds on annelids, molluscs, crustaceans, and dead or dying animals. There are about 70 species of hagfishes, of which the best known in North America are the Atlantic hagfish, *Myxine glutinosa* (Gr. *myxa*, slime) (Figure 24.3), and the Pacific hagfish, *Eptatretus stoutii* (N. L. *ept,* Gr. *hepta,* seven, + *tretos,* perforated). Although almost completely blind, hagfishes are quickly attracted to food, especially dead or dying fishes, by their keenly developed senses of smell and touch. A whale that dies and falls to the bottom of the ocean attracts thousands of hagfishes, who feed on the carcass for several years.

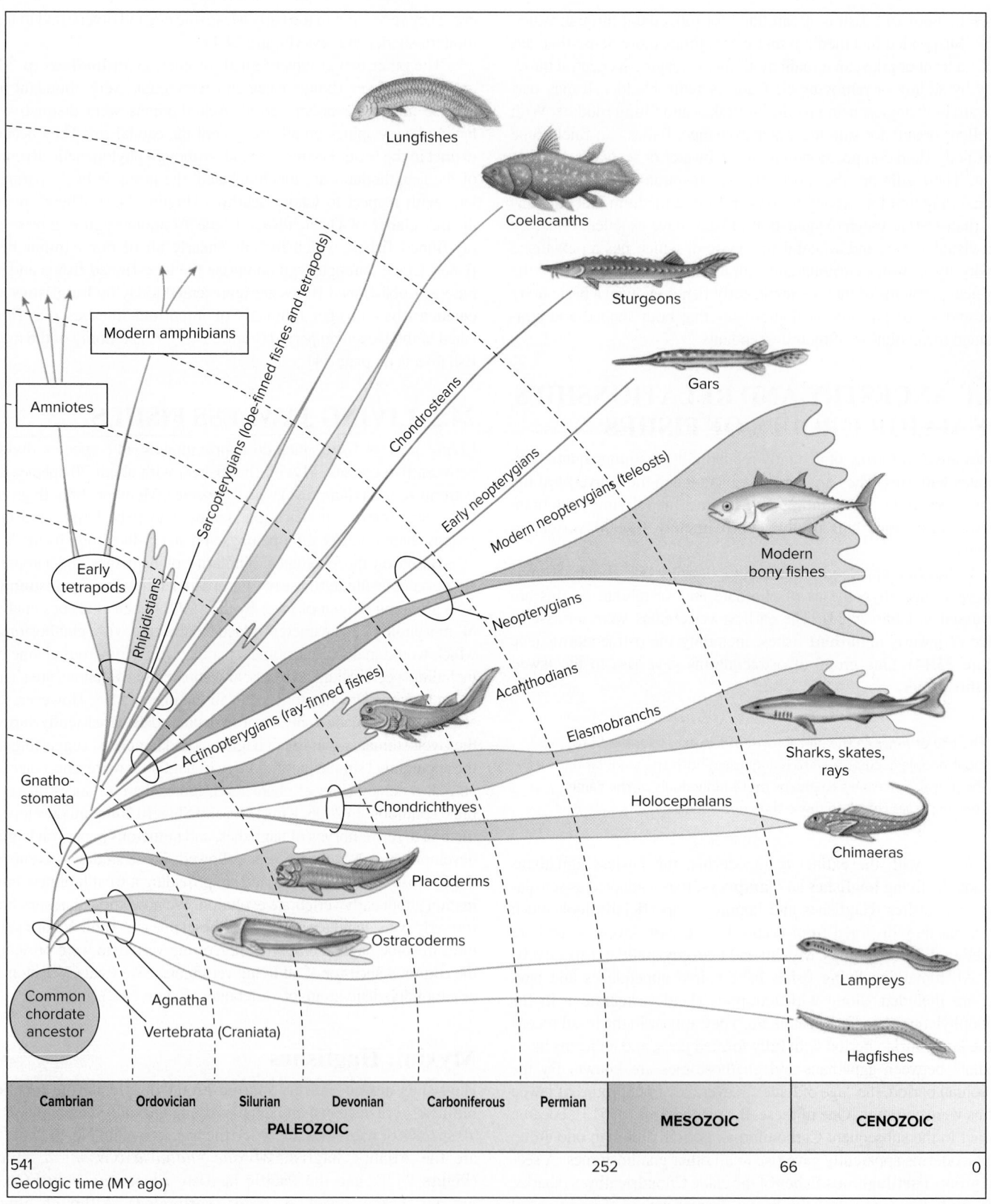

Figure 24.1   Graphic representation of the family tree of fishes, showing evolution of major groups through geological time. Numerous lineages of extinct fishes are not shown. Widened areas in the lines of descent indicate periods of adaptive diversification and relative number of species in each group. The lobe-finned fishes (sarcopterygians), for example, flourished in the Devonian period, but declined and are today represented by only four surviving genera (lungfishes and coelacanths). Sharks and rays diversified during the Carboniferous period, declined in the Permian period, then diversified again in the Mesozoic era. Johnny-come-latelies in fish evolution are the spectacularly diverse modern fishes, or teleosts, which include most living fishes.

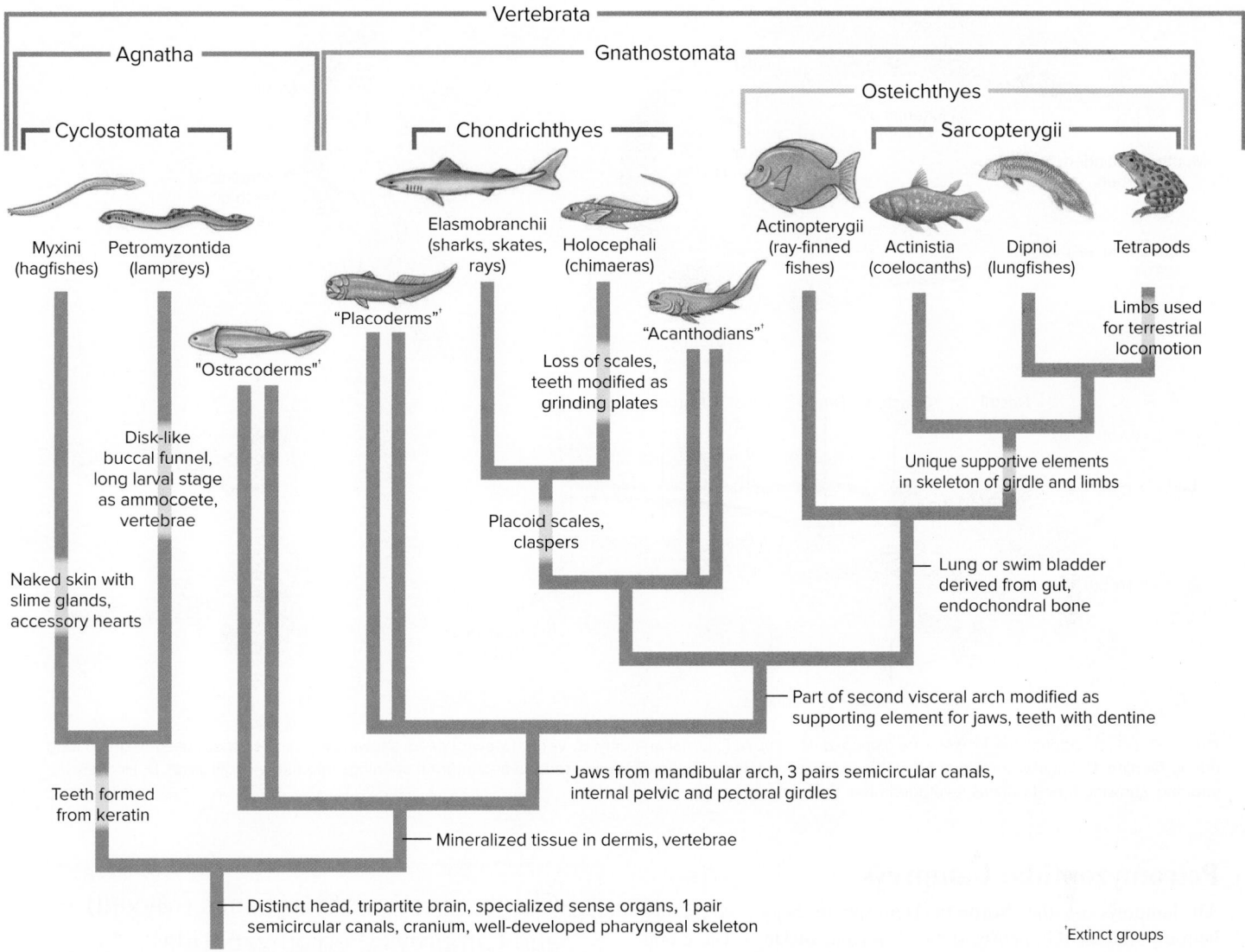

Figure 24.2 Cladogram of the fishes, showing the probable relationships of major fish taxa. Agnatha is paraphyletic with respect to gnathostomes. The placement of acanthodians is unclear; we show them as paraphyletic with respect to Chondrichthyes.

A hagfish enters a dead or dying animal through an orifice or by digging into the body. Using two toothed, keratinized plates on its tongue that fold together in a pincerlike action, the hagfish rasps bits of flesh from its prey. For extra leverage, the hagfish often ties a knot in its tail, then passes the knot forward along its body until it is pressed securely against the side of its prey (Figure 24.3D).

While the strange features of hagfishes fascinate many people, hagfishes have not endeared themselves to commercial fishermen. In earlier days, when commercial fishing was done mainly by gill nets and set lines, hagfishes often bit into the bodies of captured fish and devoured the contents, leaving behind a useless sack of skin and bones. But as large and efficient trawls came into use, hagfishes ceased to be important pests. The commercial fishing industry "turned the tables" and began targeting hagfishes as a source of leather for golf bags and boots. Fishing pressure has been so intense that some species have greatly declined.

Hagfishes are renowned for their ability to generate enormous quantities of slime. If disturbed or roughly handled, a hagfish exudes a milky fluid from special glands positioned along its body. On contact with seawater, the fluid forms a slime so slippery that the animal is almost impossible to grasp.

Unlike the body fluids of any other vertebrate, those of hagfishes are in osmotic equilibrium with seawater, as are body fluids of most marine invertebrates (see Chapter 30, p. 661). Hagfishes have several other anatomical and physiological peculiarities, including a low-pressure circulatory system served by three accessory hearts in addition to the main heart positioned behind the gills.

Much of the reproductive biology of hagfishes remains a mystery. Reproduction has been rarely documented; in these cases females produce small numbers of large, yolky eggs, which take as long as five months to hatch. There is apparently no larval stage. Almost nothing is known of spawning times, places, and behaviors or age at maturity.

Figure 24.3 Atlantic hagfish, *Myxine glutinosa* (Myxini). **A,** External anatomy. **B,** Ventral view of head, showing keratinized teeth used to grasp food during feeding. **C,** Sagittal section of head region (note retracted position of rasping tongue and internal openings into a row of gill sacs). **D,** Hagfish knotting, showing how it obtains leverage to tear flesh from prey.

Petromyzontida: Lampreys

All lampreys of the Northern Hemisphere belong to family Petromyzontidae (Gr. *petros,* stone, + *myzon,* sucking). The group name refers to a lamprey's habit of grasping a stone with its mouth to hold its position in current. The well-known, destructive, 1-m-long sea lamprey, *Petromyzon marinus,* occurs on both sides of the Atlantic Ocean (in America and Europe) (Figure 24.4). There are 24 species of lampreys in North America, of which about half are parasitic; the rest, often called "brook lampreys," are species that never feed after metamorphosis and die soon after spawning. Most parasitic species do not have a major influence on the populations of their host species.

All lampreys ascend freshwater streams to breed. Marine forms are **anadromous** (Gr. *anadromos,* running upward); they leave the sea where they spend their adult lives to swim up streams to spawn. In North America all lampreys spawn in spring or early summer. Males begin building a nest and are joined later by females. Using their oral discs to lift stones and pebbles and vigorous body vibrations to sweep away light debris, they form an oval depression (Figure 24.4). At spawning, with the female attached to a rock to maintain her position over the nest, a male attaches to the dorsal side of her head. As eggs are shed into the nest, they are fertilized by the male. The sticky eggs adhere to pebbles in the nest and become lightly covered with sand. Adults die soon after spawning.

Characteristics of Hagfishes (Myxini) and Lampreys (Petromyzontida)

1. Body slender, eel-like; no paired appendages
2. **Skin naked** (no scales)
3. **Fibrous and cartilaginous skeleton;** notochord persistent; **vertebrae reduced or absent**
4. **Jaws absent;** mouth with keratinized plates (hagfishes) or teeth (lampreys); no distinctive stomach
5. Brain small, but distinct; 10 pairs of cranial nerves
6. Eyes poorly developed (hagfishes) or moderately developed (lampreys); one pair (hagfishes) or two pairs (lampreys) of **semicircular canals**
7. Sexes separate; external fertilization
8. Large yolky eggs and no larval stage in hagfishes; small eggs and long larval stage (**ammocoete**) in lampreys
9. Excretory system of pronephric and mesonephric (hagfishes) or opisthonephric (lampreys) kidneys; kidneys drain via **archinephric duct** to cloaca; **ammonia** main nitrogenous waste
10. Hagfishes with 5–16 pairs of gills; lampreys with 7 pairs of gills
11. Heart with a sinus venosus, atrium, and ventricle; **single circulation;** accessory hearts in hagfishes

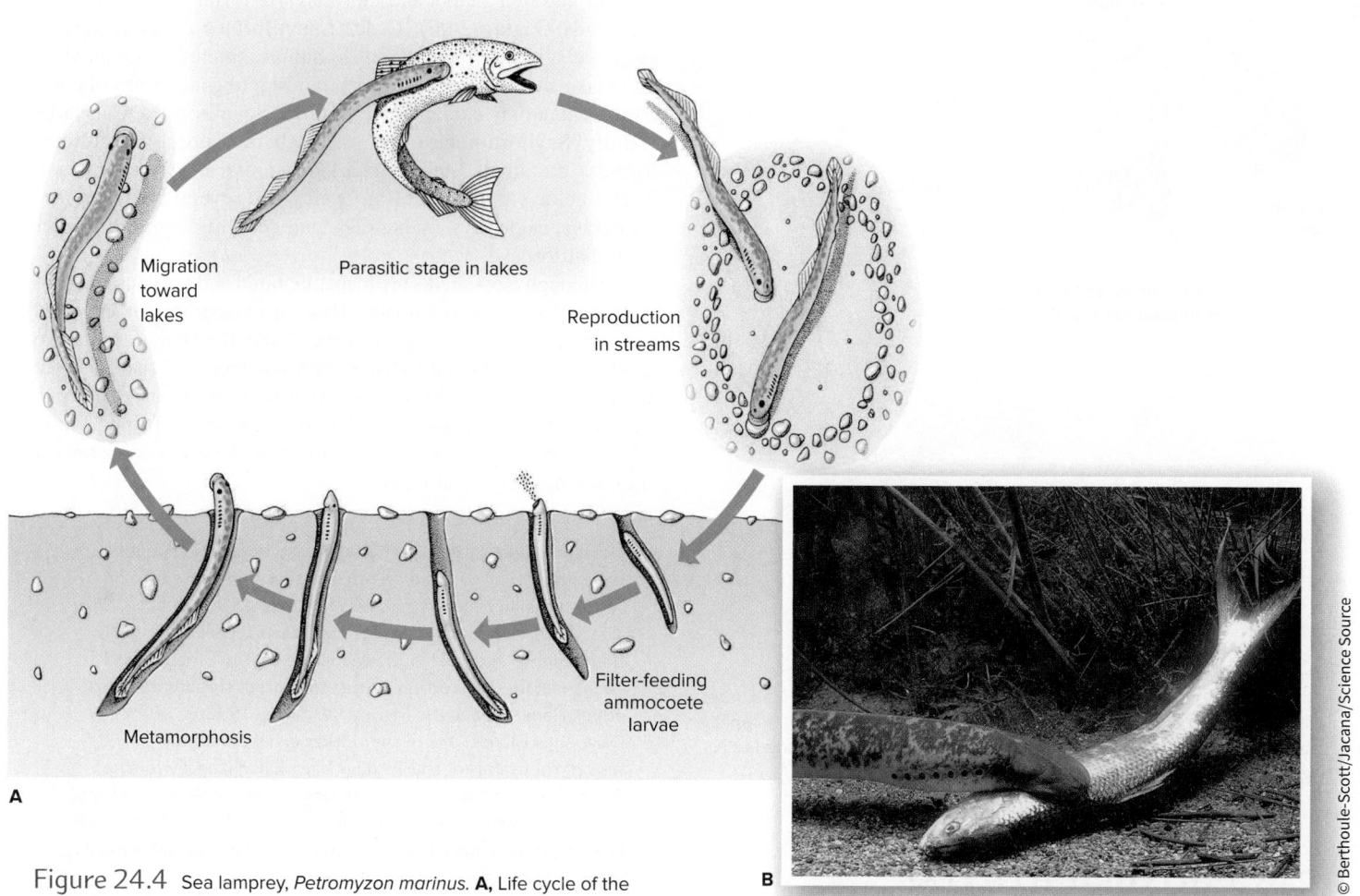

Migration toward lakes

Parasitic stage in lakes

Reproduction in streams

Filter-feeding ammocoete larvae

Metamorphosis

A

B

© Berthoule-Scott/Jacana/Science Source

Figure 24.4 Sea lamprey, *Petromyzon marinus*. **A,** Life cycle of the "landlocked" form; **B,** feeding on body fluids of a dying fish.

Eggs hatch in about 2 weeks, releasing small larvae called **ammocoetes,** which are so unlike their parents that early biologists thought they were a separate species. The larva bears a remarkable resemblance to amphioxus and possesses the basic chordate characteristics in such simplified and easily visualized form that it has been considered a chordate archetype (p. 511). When the larvae reach about 1 cm long, they leave the nest gravel and drift downstream to burrow in a suitable sandy, low-current area. The larvae filter-feed on tiny organisms and fine organic matter while growing slowly for 3 to 7 or more years, then rapidly metamorphose into adults. This change involves the eruption of eyes, replacement of the hood by a buccal funnel with keratinized teeth, enlargement of fins, maturation of gonads, and modification of the gill openings.

Parasitic lampreys either migrate to the sea, if marine, or remain in freshwater, where they attach themselves by their suckerlike mouth to fish and, with their sharp keratinized teeth, rasp through the flesh and consume body fluids or flesh (Figure 24.5). To promote the flow of blood, a lamprey injects an anticoagulant into the wound. When gorged, the lamprey releases its hold but leaves the fish with a large, gaping wound that can be fatal. Parasitic

freshwater adults live 1 to 2 years before spawning and then die; anadromous forms live 2 to 3 years.

Nonparasitic lampreys do not feed after metamorphosis, and their digestive tract degenerates to a nonfunctional strand of tissue. Within a few months they spawn and die.

Invasion of the Great Lakes by the sea lamprey, *Petromyzon marinus,* in the previous century devastated fisheries. No sea lampreys were present in the Great Lakes upstream of Niagara Falls until the Welland Ship Canal was deepened in the 1910s, allowing lampreys to bypass the Falls. Moving first through Lake Erie and then to Lakes Huron, Michigan, and Superior, sea lampreys, accompanied by overfishing, caused a total collapse of a multimillion-dollar lake trout fishery in the early 1950s. Rainbow trout, lake whitefish, lake herring, chubs, and other species were destroyed in turn. After reaching peak abundance in 1951 in Lakes Huron and Michigan and in 1961 in Lake Superior, sea lampreys began to decline, due in part to depletion of their food and in part to expensive control measures (mainly chemical larvacides placed in selective spawning streams). Lake trout, aided by a restocking program, are now recovering, but wounding rates are still high in some lakes.

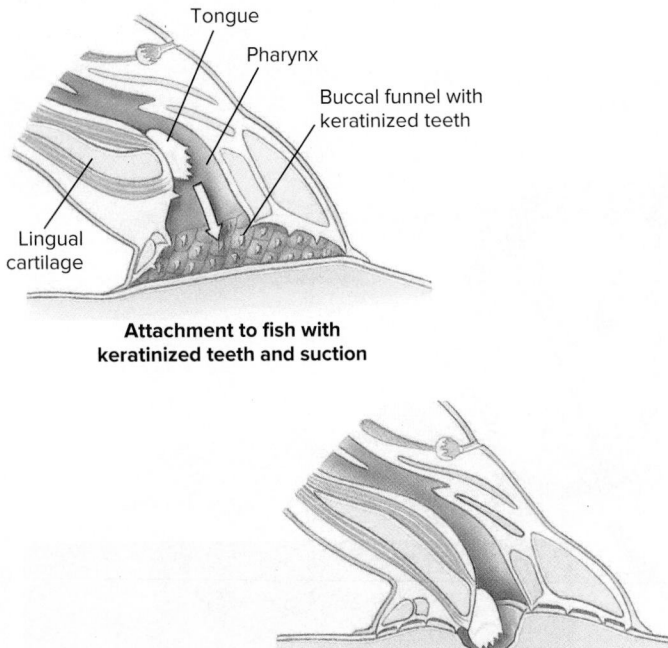

Tongue

Pharynx

Buccal funnel with
keratinized teeth

Lingual
cartilage

**Attachment to fish with
keratinized teeth and suction**

**Tongue protruded for
rasping flesh**

Figure 24.5 How a lamprey uses its keratinized tongue to feed. After firmly attaching to a fish by its buccal funnel, the protrusible tongue rapidly rasps an opening through the fish's integument. Body fluid, abraded skin, or muscle are eaten.

24.3 CHONDRICHTHYES: CARTILAGINOUS FISHES

There are about 1000 living species in Chondrichthyes, an ancient group that appeared in the Devonian period. Although a much smaller and less diverse assemblage than bony fishes, their impressive combination of well-developed sense organs, powerful jaws and swimming musculature, and predaceous habits ensures them a secure and lasting place in the aquatic community. One of their distinctive features is a cartilaginous skeleton. The cartilage in their skeletons is strengthened with calcium salts. Bone differs in having phosphate in the mineral matrix. However, phosphatized mineral tissues are retained in teeth, scales, and spines of Chondrichthyes. Almost all chondrichthyans are marine; only 28 species live primarily in freshwater. Chondrichthyes is composed of two subgroups, Elasmobranchii, the sharks, skates, and rays, and Holocephali, the chimaeras.

Elasmobranchii: Sharks, Skates, and Rays

The 12 living orders of elasmobranchs number about 960 species. Except for whales, sharks include the largest living vertebrates. The largest shark, the plankton-feeding whale shark, may reach 15 m in length. Spiny dogfish sharks commonly studied in zoological laboratories rarely exceed 1 m.

Coastal waters are dominated by ground sharks, order Carcharhiniformes, which contains typical-looking sharks such

as tiger and bull sharks and more bizarre forms, including hammerheads (Figure 24.6). Order Lamniformes contains several large, pelagic sharks dangerous to humans, including great white and mako sharks. Dogfish sharks, familiar to generations of comparative anatomy students, are in order Squaliformes. Catsharks (family Scyliorhinidae) and carpetsharks (order Orectolobiformes) are small, slender sharks that live near the seafloor. Skates belong to the order Rajiformes, and several groups of rays (stingrays, eagle rays, manta rays, and devil rays) form the order Myliobatiformes.

Although most sharks are by nature timid and cautious, some of them are dangerous to humans. There are numerous authenticated cases of shark attacks by great white sharks (reaching 6 m), mako sharks, tiger sharks, bull sharks, and hammerhead sharks. More shark casualties have been reported from tropical and temperate waters of the Australian region than from any other. During World War II there were several reports of mass shark attacks on victims of ship sinkings in tropical waters.

The worldwide shark fishery is experiencing unprecedented pressure, driven by the high price of shark fins for shark-fin soup, an Asian delicacy that sells for up to $100 per bowl. Coastal shark populations in general have declined so rapidly that "finning" has been outlawed in the United States; other countries, too, are setting quotas to protect threatened shark populations. Even in the Marine Resources Reserve of the Galápagos Islands, one of the world's exceptional wild places, tens of thousands of sharks have been killed illegally for the Asian shark-fin market. Contributing to the threatened collapse of shark fisheries worldwide is the low fecundity of sharks and the long time required by most sharks to reach sexual maturity; some species take as long as 35 years.

Form and Function

Although to many people sharks have a sinister appearance and fearsome reputation, they are at the same time among the most gracefully streamlined of all fishes. The body of a dogfish shark (Figure 24.7) is fusiform (spindle-shaped). The asymmetrical **heterocercal tail,** in which the vertebral column turns upward and extends into the dorsal lobe of the caudal fin, provides thrust and some lift as it sweeps back and forth. There are paired **pectoral** and **pelvic** fins supported by appendicular skeletons, one or two median **dorsal** fins (each with a spine in *Squalus* [L. a kind of sea fish]), and a median **caudal** fin. A median **anal** fin is present in most sharks, including the smooth dogfish, *Mustelus* (L. *mustela,* weasel). In males, the medial part of the pelvic fin is modified to form a **clasper,** which is used in copulation. Paired **nostrils** are associated with olfaction (Figure 24.8). The lateral eyes are lidless, and behind each eye a spiracle (remnant of the first gill slit) is usually present. Five (rarely, 6 or 7) gill slits are found anterior to each pectoral fin. The tough, leathery skin is covered with toothlike, dermal **placoid scales,** which reduce the turbulence of water flowing along the body surface during swimming.

Sharks are well equipped for their predatory life. They track their prey using highly sensitive senses in an orderly sequence. Sharks may initially detect prey from a kilometer or more away with

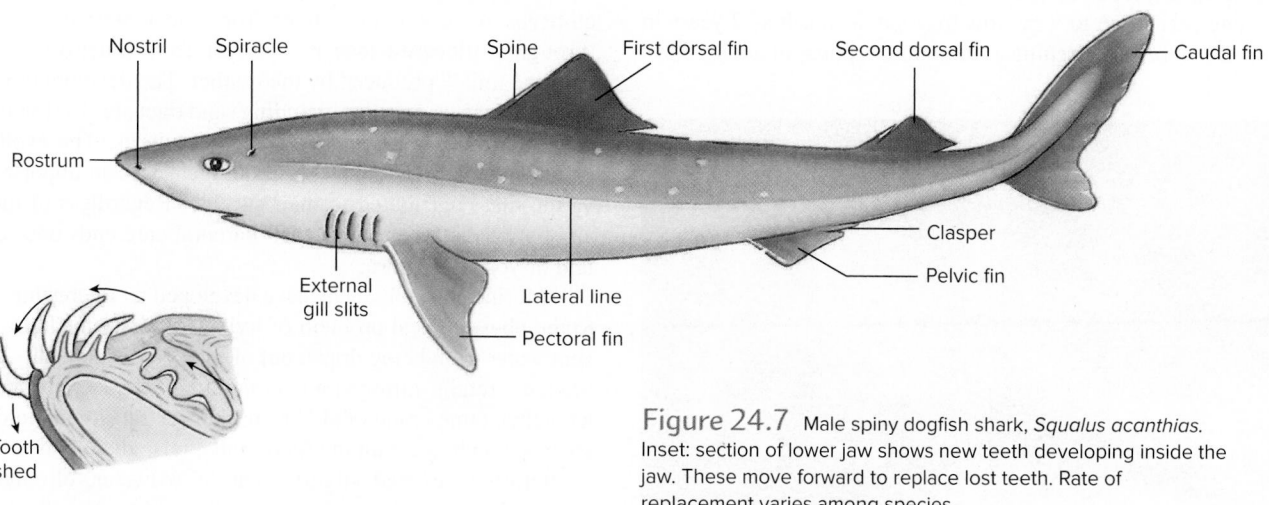

Figure 24.6 Diversity in sharks of the clade Elasmobranchii: hammerhead shark, *Sphyrna;* bull shark, *Carcharhinus leucas;* shortfin mako shark, *Isurus oxyrinchus;* great white shark, *Carcharodon carcharias;* whale shark, *Rhincodon typus;* tiger shark, *Galeocerdo cuvier;* thresher shark, *Alopias vulpinus;* spined pygmy shark, *Squaliolus laticaudus;* and collared carpetshark, *Parascyllium collare.*

Figure 24.7 Male spiny dogfish shark, *Squalus acanthias.* Inset: section of lower jaw shows new teeth developing inside the jaw. These move forward to replace lost teeth. Rate of replacement varies among species.

their large olfactory organs, capable of detecting some chemicals as low as 1 part per 10 billion. The laterally placed nostrils of hammerhead sharks (Figure 24.6) probably enhance odor localization by improving stereo-olfaction. Prey also can be located from

moderate distances by sensing low-frequency vibrations with mechanoreceptors in the **lateral-line system.** This system is composed of special receptor organs (**neuromasts**) in interconnected tubes and pores extending along the sides of the body and over the head

(Figure 24.9). At closer range a shark switches to vision as its primary method of tracking prey. Contrary to popular belief, most sharks have excellent vision, even in dimly lit waters. During the final stage of attack, sharks are guided to their prey by the bioelectric fields that surround all animals. Electroreceptors, the **ampullae of Lorenzini** (Figures 24.8 and 24.9), are located primarily on the shark's head. Sharks also can use electroreception to find prey buried in the sand.

Both upper and lower jaws of sharks are provided with many sharp teeth. The front row of functional teeth on the edge of the jaw is backed by rows of developing teeth that replace worn teeth throughout the life of the shark (Figures 24.7 and 24.8). The mouth cavity opens into a large **pharynx,** which contains openings to separate gill slits and spiracles. A short, wide esophagus runs to the J-shaped stomach. A **liver** and **pancreas** discharge their contents into a short, straight **intestine,** which contains the **spiral valve** that slows passage of food and increases the absorptive surface (Figure 24.10). Attached to a short rectum is a **rectal gland,** unique to chondrichthyans, which secretes a colorless fluid containing a high concentration of sodium chloride. The rectal gland assists the **opisthonephric kidney** (see Figure 30.9) in regulating salt concentration of the blood. Chambers of the **heart** are arranged in tandem formation, and blood circulates in the same pattern seen in other gill-breathing vertebrates (Figure 24.10). Blood leaving the heart via the ventral aorta enters capillary beds in the gills where oxygen is absorbed, then circulated to the rest of the body via the dorsal aorta, without first reentering the heart (see Figure 31.10).

All chondrichthyans have internal fertilization, but maternal support of embryos is highly variable. Some sharks and all skates lay large, yolky eggs soon after fertilization; these species are termed **oviparous.** Some deposit their eggs in a keratinized capsule called a "mermaid's purse," which often is provided with tendrils that wrap around the first firm object it contacts, much like tendrils of grape vines. Embryos are nourished from the yolk for a long period—6 to 9 months in some, as much as 2 years in one species—before hatching as miniature replicas of adults. Rays

Figure 24.8 Head of sand tiger shark, *Carcharias taurus.* Note the series of successional teeth. Also visible below the eye are nostrils and ampullae of Lorenzini.

© Wolfgang Pölzer/Alamy Stock Photo

Characteristics of Chondrichthyes

1. Body fusiform or dorsoventrally compressed; caudal fin **heterocercal** (sharks and rays) or diphycercal (chimaeras) (Figure 24.15); **paired pectoral** and **pelvic fins**
2. **Skin with placoid scales** of dermal origin or **naked**
3. **Skeleton cartilaginous;** notochord present but reduced; **vertebrae** distinct
4. **Jaws present** with **polyphyodont teeth;** stomach large (absent in chimaeras); intestine with **spiral valve** (Figure 24.10); liver often large and oil filled
5. Brain well developed; 10 pairs of cranial nerves
6. Senses of smell, vibration reception (lateral-line), vision and **electroreception** well developed; **three pairs of semicircular canals**
7. Sexes separate; **internal fertilization with claspers**
8. Oviparous or viviparous; embryos of viviparous species nourished by **placenta, yolk sac (ovoviviparity), or cannibalism;** no larval stage
9. Excretory system of **opisthonephric kidneys,** which drain via **archinephric duct** to cloaca; **high concentration of urea and trimethylamine oxide in blood; rectal gland** present
10. **Five to seven pairs of gills** leading to gill slits in rays and sharks or covered by operculum in chimaeras; **no swim bladder or lung**
11. Heart with a sinus venosus, atrium, ventricle, and conus arteriosus; **single circulation**

and many sharks are **viviparous,** retaining developing young in the uterus, and giving birth to fully formed young. Many are **ovoviviparous** (lecithotrophic viviparous) species, which retain developing young in the uterus while they are nourished by contents of their yolk sac until born. In other viviparous species, embryos receive nourishment from the maternal bloodstream through a **placenta** (see p. 176), or from nutritive secretions, "uterine milk," produced by the mother. The developing young of a few viviparous species, including sand tiger sharks (Figure 24.8), consume siblings or eggs while in the uterus. The evolution of embryo retention by many elasmobranchs was an important innovation that contributed to their success. Regardless of the initial amount of maternal support, all parental care ends once eggs are laid or young are born.

Marine elasmobranchs have developed an interesting solution to the physiological problem of living in a salty medium. To prevent water from being drawn out of the body osmotically, elasmobranchs retain nitrogenous compounds, especially urea and trimethylamine oxide (TMAO), in their extracellular fluid. These solutes, combined with the blood salts, raise the blood solute concentration to exceed slightly that of seawater, eliminating an osmotic inequality between their bodies and surrounding seawater (see p. 663).

More than half of all elasmobranchs are rays, a group that includes skates, electric rays, sawfishes, stingrays, and manta rays. Most are specialized for bottom dwelling, with a dorsoventrally flattened body and greatly enlarged pectoral fins, which

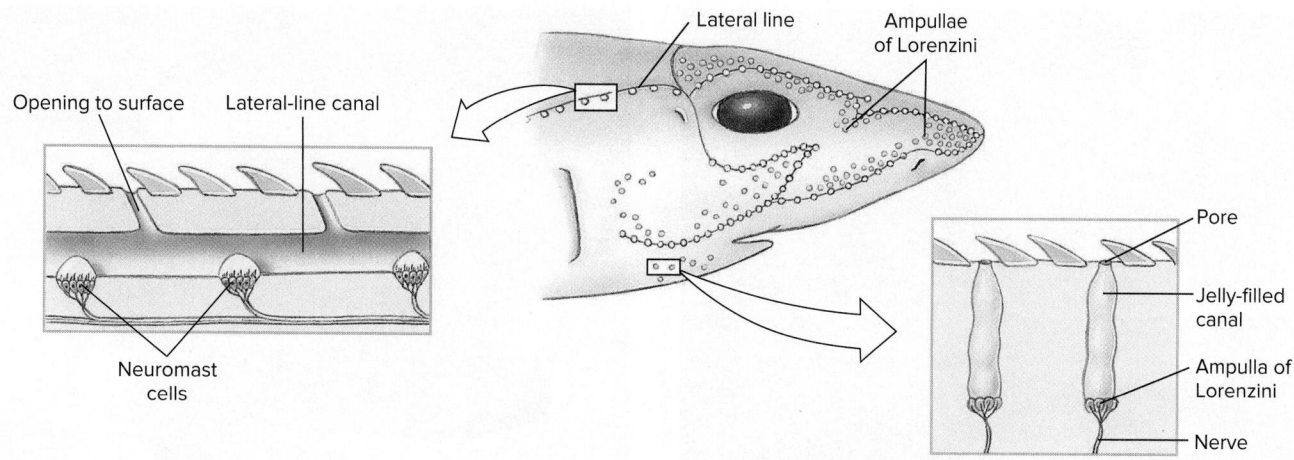

Figure 24.9 Sensory canals and receptors in a shark. The ampullae of Lorenzini respond to weak electric fields. The lateral-line sensors, called neuromast cells, are sensitive to disturbances in the water, enabling a shark to detect nearby objects by reflected waves in the water.

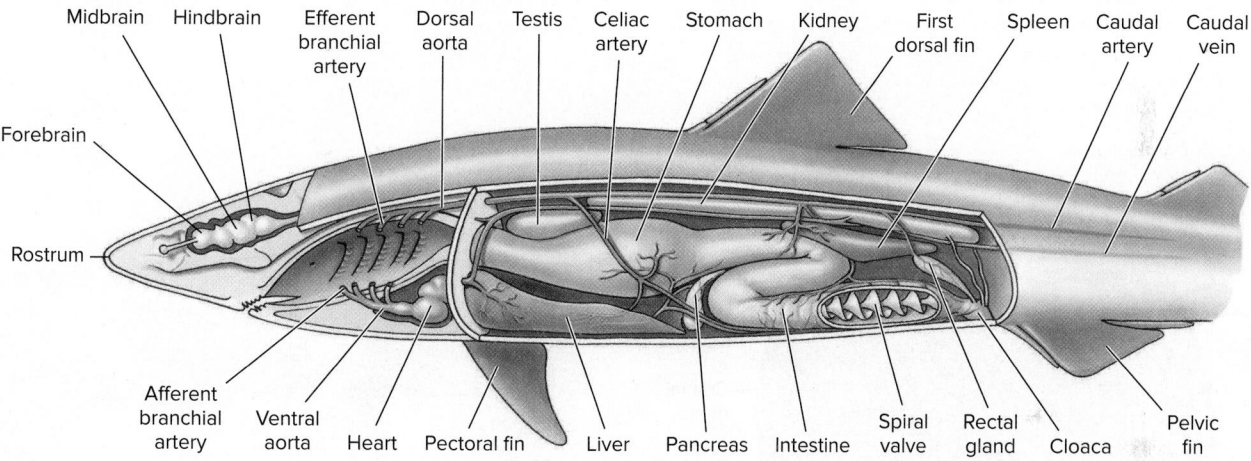

Figure 24.10 Internal anatomy of a dogfish shark, *Squalus acanthias*.

they move in a wavelike fashion to propel themselves (Figure 24.11). Gill openings are on the underside of the head, but the large spiracles are on top. Respiratory water enters through these spiracles to prevent clogging the gills, because the mouth is often buried in sand. Teeth are adapted for crushing prey: molluscs, crustaceans, and an occasional small fish.

Stingrays have a slender, whiplike tail armed with one or more serrated spines with venom glands at the base. Wounds from the spines are excruciatingly painful, and may heal slowly and with complications. Electric rays are sluggish fish with large electric organs on each side of their head (Figure 24.12). Each organ is composed of numerous vertical stacks of disclike cells connected in parallel so that when all cells discharge simultaneously, a high-amperage current is produced that flows out into the surrounding water. The voltage produced is relatively low (50 volts) but power output may be almost one kilowatt—quite sufficient to stun prey or to discourage predators. Electric rays were used by ancient Egyptians for a form of electrotherapy in the treatment of afflictions such as arthritis and gout.

Holocephali: Chimaeras

Members of the small clade Holocephali, also called ratfishes (Figure 24.13), are remnants of a line that diverged from the shark lineage at least 380 million years ago. Chimaeras (kī-mer′uz) first appeared in the Mississippian period, reached their zenith in the Cretaceous and early Tertiary periods (120 million to 50 million years ago), and then declined. Today there are only about 40 species extant.

Anatomically chimaeras have several features linking them to elasmobranchs, but they possess a suite of unique characters, too. Instead of distinct teeth, their jaws bear large flat plates. The upper jaw is completely fused to the cranium, a most unusual feature in fishes. Their gills are covered by a cartilaginous operculum, creating a single external respiratory opening. Their food includes molluscs, echinoderms, crustaceans, and fishes—a surprisingly mixed diet for such a specialized grinding dentition. Chimaeras are not commercial species and are seldom caught. Despite their bizarre shape, they are beautifully colored with a pearly iridescence.

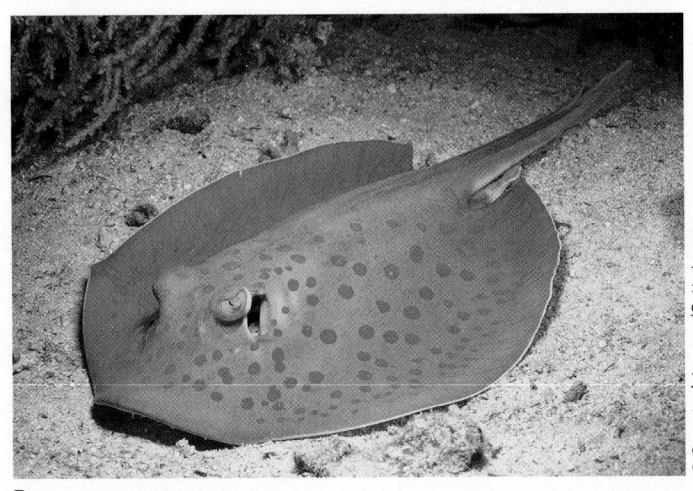

A

B

Figure 24.11 Skates and rays are specialized for life on the seafloor. Both clearnose skates, *Raja eglanteria* **(A),** and bluespotted ribbontail rays, *Taeniura lymma* **(B),** are flattened dorsoventrally and move by undulations of winglike pectoral fins.

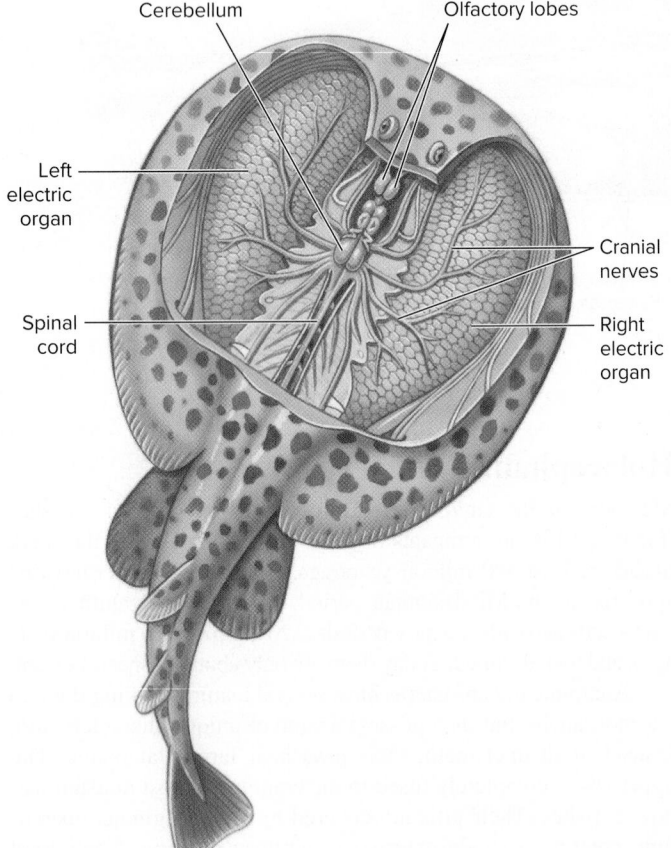

Figure 24.12 Electric ray, *Torpedo,* with electric organs exposed. Organs are built of disclike, multinucleated cells called electrocytes. When all cells are discharged simultaneously, a high-amperage current flows into the surrounding water to stun prey or to discourage predators. Electrosensory information is processed in the large cerebellum.

Figure 24.13 Spotted ratfish, *Hydrolagus collei,* of the North American west coast. This species is one of the most handsome of chimaeras, which tend to be bizarre in appearance.

24.4 OSTEICHTHYES: BONY FISHES AND TETRAPODS

Origin, Evolution, and Diversity

In the early to middle Silurian period, a lineage of fishes with bony endoskeletons gave rise to a clade of vertebrates that contains 96% of living fishes and all living tetrapods. Fishes of this clade have traditionally been called "bony fishes" **(Osteichthyes).** Bony fishes and tetrapods are united by the presence of **endochondral bone** (bone that replaces cartilage developmentally, p. 645), presence of lungs or a swim bladder derived from the gut, and several cranial and dental characters. The traditional use of "Osteichthyes" included only fishes, making it paraphyletic.

Here we treat it as a clade by also including tetrapods (see Figure 24.2).

By the middle Devonian period, two clades of bony fishes had already diversified extensively, with adaptations that fitted them for every aquatic habitat except the most inhospitable. One of these clades, ray-finned fishes (Actinopterygii), includes modern bony fishes (Figure 24.14), the most species-rich group of living vertebrates. A second clade (Sarcopterygii) is represented today by only eight fishlike species, the lungfishes and coelacanths (see Figures 24.21 and 24.22), and the land vertebrates (tetrapods).

Several key adaptations contributed to the diversification of bony fishes. They have an **operculum** over the gill composed of bony plates and attached to a series of muscles. This feature increases respiratory efficiency because outward rotation of the operculum creates a negative pressure so that water is drawn across the gills, as well as pushed across by the mouth pump (see Figure 31.20). A gas-filled diverticulum of the esophagus provides an additional means of gas exchange in hypoxic waters and an efficient means for achieving neutral buoyancy. In fishes that use these pouches primarily for gas exchange, the pouches are called lungs, while in fishes that use these pouches primarily for buoyancy, the pouches are called swim bladders (p. 533). Progressive specialization of jaw musculature and skeletal elements involved in feeding is another key feature of bonyfish evolution.

Actinopterygii: Ray-Finned Fishes

Ray-finned fishes are an enormous assemblage containing all our familiar bony fishes—over 31,000 species. The earliest actinopterygians, known as **palaeoniscids** (Gr. *palae,* ancient, + *oniskos,* sea fish), were small fishes, with large eyes, a heterocercal caudal fin (Figure 24.15), and thick, interlocking scales with an outer layer of a form of enamel called **ganoin** (Figure 24.16). These fishes had a single dorsal fin and numerous bony rays derived from scales stacked end to end, distinctively different in appearance from the lobe-finned fishes with which they shared the Devonian waters. Palaeoniscids are represented by fossil fragments as early as the late Silurian period, and flourished throughout the late Paleozoic era, during the same period that ostracoderms, placoderms, and acanthodians disappeared and sarcopterygian fishes declined in abundance (Figure 24.1).

From those earliest ray-finned fishes arose several clades. Bichirs, in the clade Cladistia, have lungs, heavy ganoid scales, and other characteristics similar to those of palaeoniscids (Figure 24.17). The 16 species of bichirs live in freshwaters of Africa. A second group are the **chondrosteans** (Gr. *chondros,* cartilage, + *osteon,* bone), represented by 27 species of freshwater and anadromous sturgeons and paddlefishes (Figure 24.17). Nearly all chondrosteans have experienced severe population declines from dam construction, overfishing, and pollution.

The third major group of ray-finned fishes to emerge from palaeoniscid stock were the **neopterygians** (Gr. *neos,* new, + *pteryx,* fin). Neopterygians appeared in the late Permian period and diversified extensively during the Mesozoic era (see Figure 24.1) when one lineage gave rise to modern bony fishes, the teleosts. There are three surviving genera of early neopterygians: the bowfin, *Amia* (Gr. tunalike fish), which inhabits shallow, weedy waters of the Great Lakes and the Mississippi River basin; and gars, *Lepisosteus* (Gr. *lepidos,* scale, + *osteon,* bone), and *Atractosteus* (Gr. *atraktos,* spindle, + *osteon,* bone), both native to eastern and southern North America (Figure 24.18). The seven species of gars are large, ambush predators with elongate bodies and jaws filled with needlelike teeth. Gars and bowfin can surface to gulp air, filling their vascularized swim bladder with air to supplement oxygen obtained in the gills.

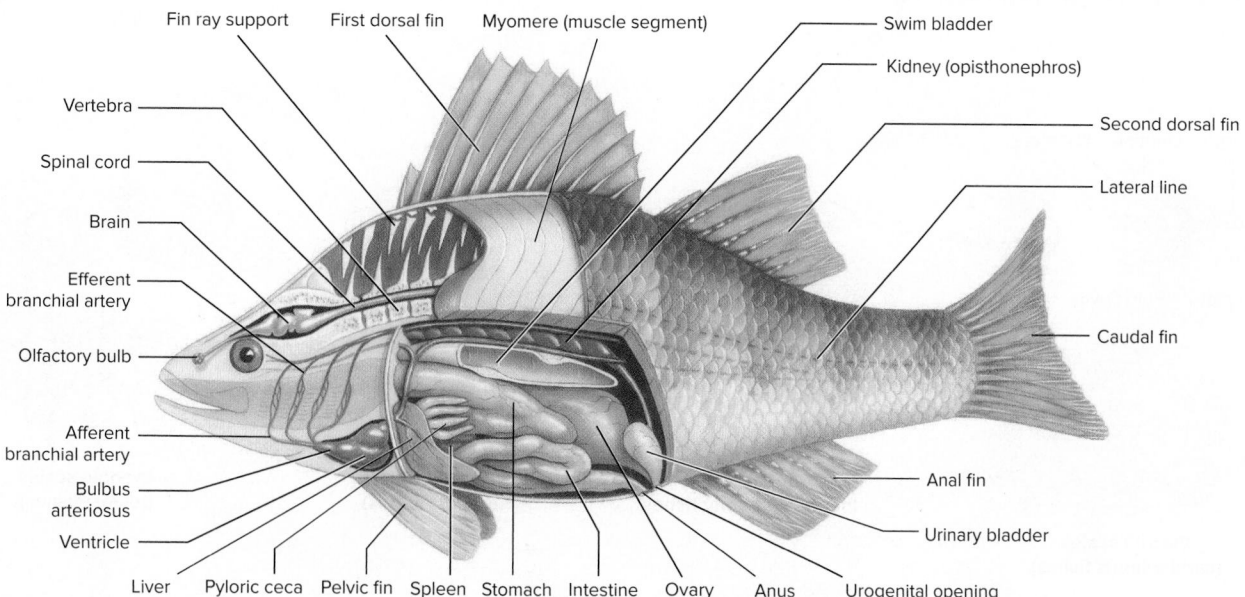

Figure 24.14 Anatomy of a yellow perch, *Perca flavescens,* a freshwater teleost fish.

Teleosts (Gr. *teleos,* perfect, + *osteon,* bone), the modern bony fishes (Figure 24.14), are the major clade of neopterygians. Teleost diversity is astounding, with about 31,000 described species and up to 10,000 undescribed species anticipated, representing about 96% of all living fishes and about half of all vertebrates (Figure 24.19). Although most of the 200 or so new species of teleosts described each year are from poorly sampled areas such as South America or deep oceanic waters, several new species are described each year from areas as well known as the freshwaters of North America! Teleosts range in size from 7 mm adult minnows to 17 m oarfish and 900 kg, 4.5 m blue marlin (Figure 24.19). These fishes occupy almost every conceivable habitat, from elevations up to 5200 m in Tibet to 8000 m below the surface of the ocean. Some species live in hot springs at 44°C, while others live under the Antarctic ice at −2°C. They may live in lakes with salt concentrations three times that of seawater, caves of total darkness, or swamps low in oxygen, or even make extended excursions onto land, as do mudskippers (Figure 24.19).

Several morphological trends in the teleost lineage allowed it to diversify into this truly incredible variety of habitats and forms. Heavy dermal armor of primitive ray-finned fishes was replaced by light, thin, flexible **cycloid** and **ctenoid** scales (Figure 24.16). Some teleosts, such as most eels and catfishes, completely lack scales. Increased mobility and speed enabled by loss of heavy armor improved predator avoidance and feeding efficiency. Changes in the fins of teleosts increased maneuverability and speed and allowed fins to serve a variety of other functions. The symmetrical shape of the **homocercal** tail (Figure 24.15) of most teleosts focused muscu-lature contractions on the tail, permitting greater speed. Other fins primarily function as fixed keels in sharks, but they are more flexi-ble and functionally versatile in teleosts (see Figure 24.14). In addi-tion to providing stability and control of complex movements,

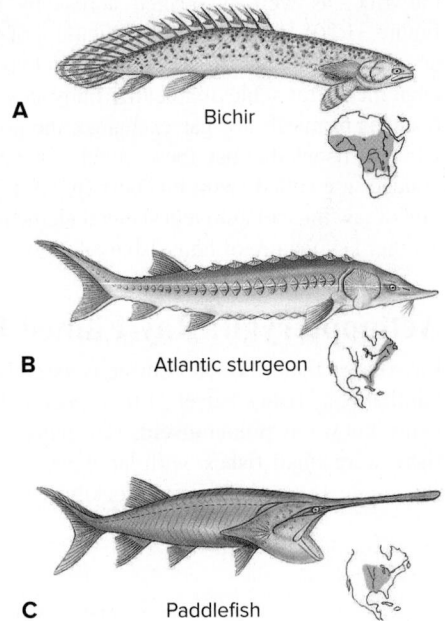

Figure 24.17 Nonteleost, ray-finned fishes of class Actinopterygii. **A,** Bichir, *Polypterus bichir,* of equatorial West Africa, is a nocturnal predator. **B,** Atlantic sturgeon, *Acipenser oxyrhynchus* (now uncommon), of Atlantic coastal rivers. **C,** Paddlefish, *Polyodon spathula,* of the Mississippi River basin, reaches 2 m and 80 kg.

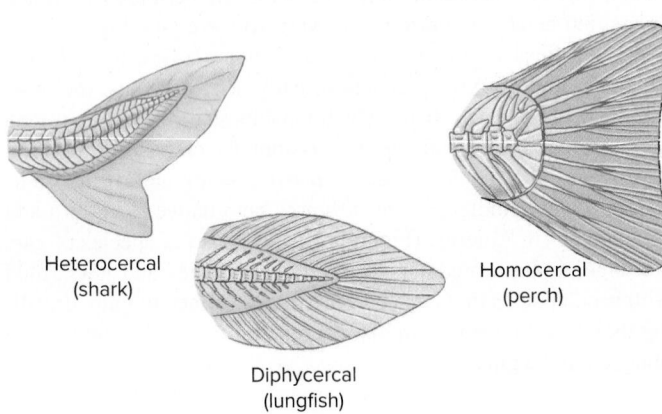

Heterocercal (shark)

Diphycercal (lungfish)

Homocercal (perch)

Figure 24.15   Types of caudal fins among fishes.

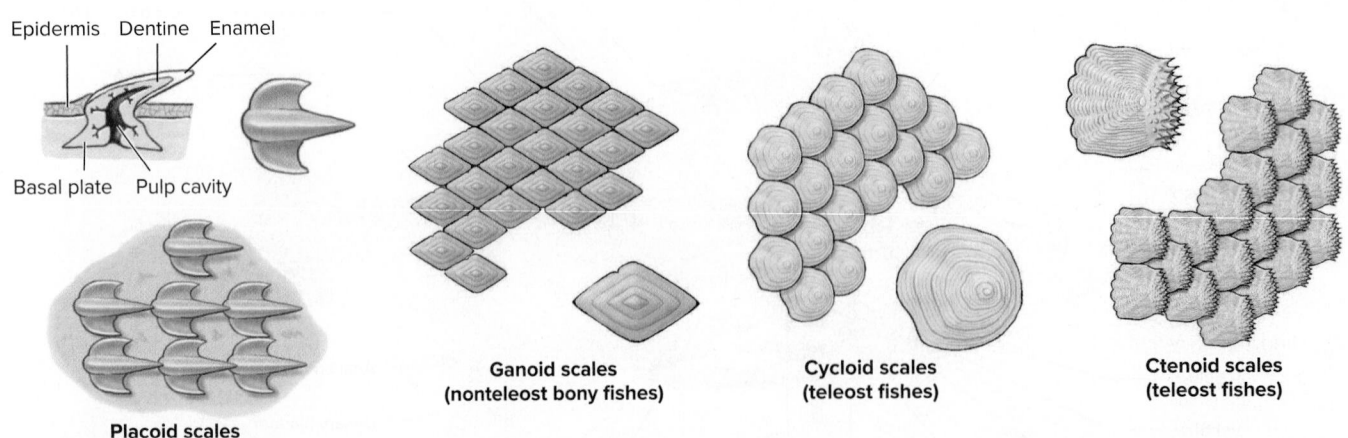

Epidermis   Dentine   Enamel

Basal plate   Pulp cavity

Placoid scales (cartilaginous fishes)

Ganoid scales (nonteleost bony fishes)

Cycloid scales (teleost fishes)

Ctenoid scales (teleost fishes)

Figure 24.16   Types of fish scales. Placoid scales are small, conical, toothlike structures characteristic of Chondrichthyes. Diamond-shaped ganoid scales, present in a few bony fishes such as gar, are composed of thick layers of silvery enamel (ganoin) on the upper surface and bone on the lower. Teleosts have either cycloid or ctenoid scales. They contain bone, but are thin and flexible and are arranged in overlapping rows.

Figure 24.18 Nonteleost neopterygian fishes. **A,** Bowfin, *Amia calva.* **B,** Longnose gar, *Lepisosteus osseus.* They frequent slow-moving streams and swamps of eastern North America where they can suspend motionless in the water, ready to snatch passing fishes.

teleost fins can function in camouflage or social communication. Bizarre modifications of the dorsal fin include the lure of anglerfishes, venom-delivering spines of scorpionfishes, and suctorial disc of sharksuckers (Figure 24.19). Teleost lineages demonstrate an increasingly fine control of gas resorption and secretion in the swim bladder. Control of buoyancy likely coevolved with fin

Characteristics of Actinopterygii

1. Caudal fin heterocercal (ancestral condition) or **homocercal** (Figure 24.15); paired pectoral and pelvic fins usually present, supported by **bony rays;** muscles controlling fin movements within trunk
2. Skin with **ganoid (ancestral condition), cycloid, or ctenoid scales** of dermal origin (Figure 24.20) or **naked**
3. **Skeleton of bone;** notochord present but reduced; **vertebrae** distinct
4. Jaws present, usually with **enameloid, polyphyodont teeth;** spiral valve present (ancestral condition) or absent
5. Brain well developed, but relatively small; 10 pairs of cranial nerves
6. Development of senses variable; three pairs of semicircular canals
7. Sexes usually separate; many hermaphroditic; some reproduce asexually by parthenogenesis; fertilization usually external, internal in some
8. Oviparous or viviparous; embryos of viviparaous species nourished by placenta or yolk sac (ovoviviparity); larval stage often greatly different from adult
9. Excretory system of opisthonephric kidneys, which drain via archinephric duct to cloaca; ammonia usually main nitrogenous waste
10. **Gills covered by a bony operculum; swim bladder present** usually functioning for buoyancy, sometimes used for respiration
11. Heart with a sinus venosus, atrium, and ventricle; **single circulation;** nucleated red blood cells

modifications to refine locomotion. Several anatomical modifications improved feeding efficiency. Changes in jaw suspension enable the orobranchial cavity to expand rapidly, creating a highly sophisticated suction device. Rapid jaw protrusion by sliding the upper jaw forward increases final attack velocity. Gill arches of many teleosts are modified into powerful **pharyngeal jaws** for chewing, grinding, and crushing. With so many changes, it is not surprising that teleosts have become the most diverse of fishes.

Sarcopterygii: Lobe-Finned Fishes and Tetrapods

The ancestor of tetrapods is found within a group of extinct sarcopterygian fishes called **rhipidistians,** which included several lineages that flourished in freshwaters and shallow coastal areas in the late Paleozoic era. Rhipidistians, such as *Eusthenopteron* (see Figure 25.3), were cylindrical, large-headed fishes with fleshy fins, and presumably lungs. The evolution of tetrapods from rhipidistians is discussed in Chapter 25.

Early sarcopterygians had lungs as well as gills, and a tail of the **heterocercal** type. However, during the Paleozoic era the orientation of the vertebral column changed so that the tail became symmetrical and **diphycercal** (see Figure 24.15). These fishes had powerful jaws; heavy, enameled scales; and strong, fleshy, paired lobed fins that might have been used like legs to support the body or to move the body over benthic substrates. The sarcopterygian clade today is represented by only eight fish species: six species of lungfishes and two species of coelacanths (Figures 24.21 and 24.22.

Of the three surviving genera of **lungfishes**, most similar to early forms is *Neoceratodus* (Gr. *neos,* new, + *cerato,* horn, + *odes,* form), the living Australian lungfish, which may attain a length of 1.5 m (Figure 24.21). This lungfish, unlike its living relatives, normally relies on gill respiration, and cannot survive long out of water. The South American lungfish, *Lepidosiren* (Gr. *lepido,* scaly, + *siren,* mythical mermaid), and the African lungfishes, *Protopterus* (Gr. *prōtos,* first, + *pteron,* wing), can live out of water for long

A © Image Source/Alamy Stock Photo
B © CP Cheah/Flickr/Getty Images
C © Brand X Pictures/PunchStock
D © Fred McConnaughey/Science Source

Figure 24.19 Diversity among teleosts. **A,** This striped marlin, *Kajikia audax*, chasing sardines near the Baja California peninsula, is one of the fastest teleosts. **B,** Mudskippers, *Periophthalmus*, make extensive excursions on land to graze on algae and capture insects; they build nests in which the young hatch and are guarded by the mother. **C,** Protective coloration of the flamboyant lionfish, *Pterois volitans*, advises caution; the dorsal spines are venomous. **D,** The suctorial disc on the sharksucker, *Echeneis naucrates,* is a modification of the dorsal fin.

coelacanths were considered to have become extinct 70 million years ago, the scientific world was astonished when a coelacanth was found on a trawl off the coast of South Africa in 1938. An intensive search to locate more specimens was successful off the coast of the Comoro Islands. There fishermen occasionally catch them at great depths with hand lines, providing specimens for research. This was the only known population of *Latimeria* until 1998, when the scientific world again was surprised by the capture of a new species of coelacanth in Sulawesi, Indonesia, 10,000 km from the Comoros!

"Modern" marine coelacanths are descendants of the Devonian freshwater stock. The tail is diphycercal (see Figure 24.15) but possesses a small lobe between the upper and lower caudal lobes, producing a three-pronged structure (Figure 24.22). Their swim bladders are fat-filled and not used for respiration.

Coelacanths are a deep metallic blue with irregular white or brassy flecks, providing camouflage against the dark lava-cave reefs they inhabit. Young are born fully formed after hatching internally from eggs 9 cm in diameter—the largest among bony fishes.

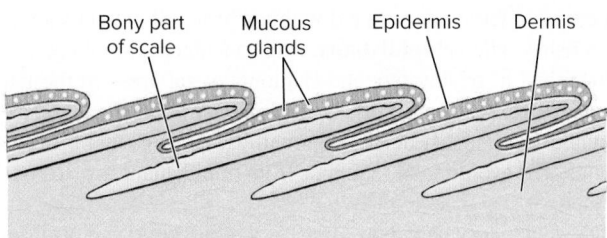

Bony part of scale   Mucous glands   Epidermis   Dermis

Figure 24.20 Section through the skin of a bony fish, showing the overlapping scales (yellow). The scales lie in the dermis and are covered by epidermis.

periods of time. *Protopterus* lives in African streams and ponds that regularly dry, with their mud beds baked hard by the hot tropical sun. The fish burrows down at the approach of the dry season and secretes a copious slime that mixes with mud to form a hard cocoon in which it remains dormant until rains return. It acquires oxygen from the atmosphere with its lungs and efficiently delivers it to tissues with partially separated pulmonary and systemic cardiovascular circuits.

Coelacanths also arose in the Devonian period, diversified somewhat, and reached their peak of diversity in the Mesozoic era. At the end of that era they nearly disappeared but left one remarkable surviving genus, *Latimeria* (Figure 24.22). Because the last

Characteristics of Sarcopterygian Fishes

1. Caudal fin heterocercal (fossil forms) or **diphycercal** in living forms (Figure 24.15); paired pectoral and pelvic fins usually present, supported by **stout bones and bony rays;** muscles controlling fin movements within fin
2. Skin with dermal scales composed of two layers of bone, a layer of **cosmine** (form of dentin) and a thin layer of **enamel** in fossil species
3. **Skeleton of bone;** vertebrae distinct
4. Jaws present, usually with **polyphyodont teeth covered with true enamel** that typically are crushing plates restricted to the palate; intestine with spiral valve
5. Brain well developed, but relatively small; 10 pairs of cranial nerves
6. Development of senses variable; three pairs of semicircular canals
7. Sexes separate; fertilization external (lungfishes) or internal (coelacanths)
8. Oviparous
9. Excretory system of opisthonephric kidneys, which drain via archinephric duct to cloaca; ammonia and urea main nitrogenous wastes
10. **Gill covered by a bony operculum; swim bladder** present, used primarily for respiration (fat-filled in coelacanths)
11. Heart with a sinus venosus, an atrium, a partly divided ventricle, and a conus arteriosus; **pulmonary and systemic circuits incompletely separated;** nucleated red blood cells

Figure 24.21 **A,** An African lungfish, *Protopterus annectens*, of Sarcoptergyi. **B,** An African lungfish burrows into the sediments of a pool as the water dries. There, it secretes a mucous cocoon and enters a state of dormancy (aestivation) for several months to 4 years, when rains return water to its stream or pond.

Figure 24.22 The coelacanth genus *Latimeria* is a surviving marine relict of a group of lobe-finned fishes that flourished some 350 million years ago.

24.5 STRUCTURAL AND FUNCTIONAL ADAPTATIONS OF FISHES

Locomotion in Water

To the human eye, some fishes appear capable of swimming at extremely high speeds, but our judgment is unconsciously tempered by our own experience that water is a highly resistant medium through which to move. Most fishes, such as a trout or a minnow, can swim maximally about 10 body lengths per second, obviously an impressive performance by human standards. When these speeds are translated into kilometers per hour it means that a 30 cm (1 foot) trout can swim only about 10.4 km (6.5 miles) per hour. As a general rule, the larger the fish, the faster it can swim.

Measuring fish cruising speeds is best done in a "fish wheel," a large ring-shaped channel filled with water that is turned at a speed equal and opposite to that of the fish. Much more difficult to measure are the sudden bursts of speed that most fish can make to capture prey or to avoid being captured. A hooked bluefin tuna was once "clocked" at 66 km per hour (41 mph); swordfish and marlin might be capable of incredible bursts of speed approaching, or even exceeding, 110 km per hour (68 mph). Such high speeds can be sustained for no more than 1 to 5 seconds.

The propulsive mechanism of a fish is its trunk and tail musculature. The axial, locomotory musculature is composed of zigzag bands, called **myomeres.** Muscle fibers in each myomere are relatively short and connect the tough connective tissue partitions that separate each myomere from the next. On the surface myomeres take the shape of a W lying on its side (Figure 24.23), but internally the bands are complexly folded and nested so that the pull of each myomere extends over several vertebrae. This arrangement produces more power and finer control of movement because many myomeres are involved in bending a given segment of the body.

Understanding how fishes swim can be approached by studying the motion of a very flexible fish such as an eel (Figure 24.24). The movement is serpentine, not unlike that of a snake, with waves of contraction moving backward along its body by alternate contraction of myomeres on either side. The anterior end of the body bends less than the posterior end, so that each undulation increases in amplitude as it travels along the body. While undulations move backward, bending of the body pushes laterally against the water, producing a **reactive force** that is directed forward, but at an angle. It can be analyzed as having two components: **thrust,** which is used to overcome drag and propels the fish forward, and **lateral force,** which tends to make the fish's head "yaw," or deviate from the course in the same direction as its tail. This side-to-side head movement is very obvious in a slender fish, like an eel or catshark, but many fishes have a large, rigid head with enough surface resistance to minimize yaw.

The movement of an eel is reasonably efficient at low speed, but its body shape generates too much frictional drag for rapid swimming. Fishes that swim rapidly, such as trout, are less flexible and limit body undulations mostly to the caudal region (Figure 24.24). Muscle force generated in the large anterior muscle mass is transferred through tendons to the relatively nonmuscular caudal peduncle and caudal fin where thrust is generated. This form of swimming reaches its highest development in tunas, whose bodies do not flex at all. Virtually all thrust is derived from powerful sweeps of the caudal fin (Figure 24.25). Many fast oceanic fishes such as marlin, swordfish, amberjacks, and wahoo have swept-back caudal fins shaped like a sickle. Such fins are the aquatic counterpart of the high-aspect wings of the swiftest birds (p. 598).

The body temperature of most fishes is the same as their environment, because any heat generated internally quickly is lost into the surrounding water. However, some fishes, such as tunas (Figure 24.25) and mako and great white sharks (see Figure 24.6), maintain a high temperature in their swimming muscles and viscera—as much as 10° C warmer than surrounding water. Marlins (see Figure 24.19A) and other billfishes elevate the temperature of their brain and retina. Research by F. G. Carey and others explains how these fishes accomplish this kind of thermoregulation, called **regional endothermy.** Heat is generated as a by-product of various activities, including digestion and swimming, or for billfishes, by a specialized heat-generating organ beneath the brain. This heat is conserved with a rete mirabile, a parallel bundle of blood vessels arranged to provide a countercurrent flow of blood (p. 675). High temperatures apparently promote powerful swimming and speed the digestive and nervous systems. Fishes with regional endothermy are the fastest in the world.

Swimming is the most economical form of animal locomotion, largely because aquatic animals are almost perfectly supported by their medium and expend little energy to overcome the force of gravity. If we compare the energy cost per kilogram of body weight of traveling 1 km by different forms of locomotion, we find swimming costs only 0.39 kcal (salmon) as compared with 1.45 kcal for flying (gull) and 5.43 kcal for walking (ground squirrel).

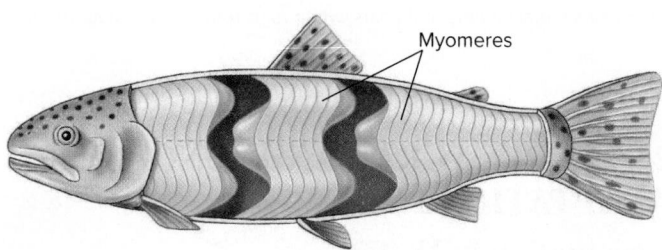

Figure 24.23 Trunk musculature of a teleost fish, partly dissected to show internal arrangement of the muscle bands (myomeres). The myomeres are folded into a complex, nested grouping, an arrangement that favors stronger and more controlled swimming.

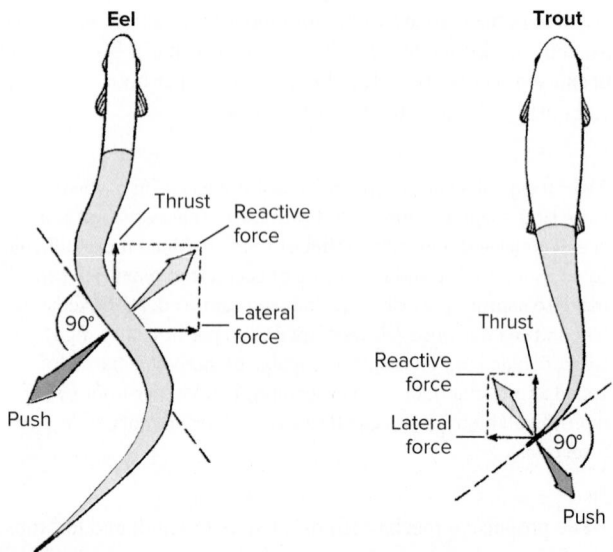

Figure 24.24 Movements of swimming fishes, showing the forces developed by an eel-shaped and by a spindle-shaped fish.

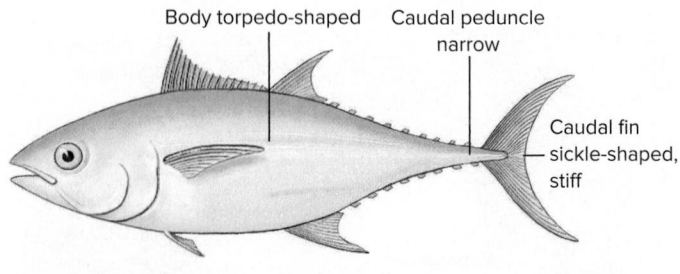

Figure 24.25 Bluefin tuna, *Thunnus thynnus*, showing adaptations for fast swimming. Powerful trunk muscles pull on the slender caudal peduncle. Since the body does not bend, all of the thrust comes from beats of the stiff, sickle-shaped caudal fin.

Neutral Buoyancy and the Swim Bladder

All fishes are slightly heavier than water because their skeletons and other tissues contain heavy elements present only in trace amounts in natural waters. To keep from sinking, sharks, which lack a swim bladder, must always keep moving forward in the water. The asymmetrical (heterocercal) tail of a shark provides lift as it sweeps through the water, and its broad head and flat pectoral fins (see Figure 24.7) act as angled planes to provide additional lift. Sharks also are aided in buoyancy by having large livers containing a lipid called **squalene** with a density of only 0.86 g per milliliter. The liver thus acts like a large sack of buoyant oil that helps to compensate for a shark's heavy body.

By far the most efficient flotation device is a gas-filled space. The **swim bladder** serves this purpose in bony fishes (Figure 24.26). It arose from the paired lungs of primitive Devonian bony fishes. Lungs were probably a ubiquitous feature of Devonian freshwater bony fishes when warm, swampy habitats would have made such an accessory respiratory structure advantageous (p. 544). Swim bladders are present in most pelagic bony fishes but are absent in tunas, most abyssal fishes, and most bottom dwellers, such as flounders and sculpins. Although its name suggests otherwise, a swim bladder is not used to swim.

Without a swim bladder, bony fishes sink because their tissues are denser than water. To achieve neutral buoyancy, they displace

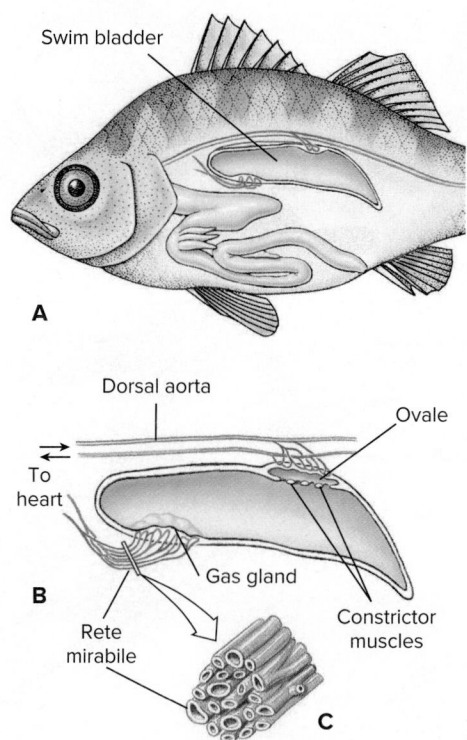

Figure 24.26 **A,** Swim bladder of a teleost fish. The swim bladder lies in the coelom just beneath the vertebral column. **B,** Gas (primarily oxygen) diffuses into the swim bladder at the gas gland. Gas from the blood is moved into the gas gland by the rete mirabile, a complex array of tightly packed capillaries that act as a countercurrent multiplier to increase oxygen concentration. The arrangement of venous and arterial capillaries in the rete is shown in **C.** To release gas during ascent, a muscular valve opens, allowing gas to enter the ovale and then diffuse into the blood.

additional water by a volume of gas in a swim bladder, thus adjusting their total density to match that of the surrounding water. This adjustment allows fishes with a swim bladder to remain suspended indefinitely at any depth with no muscular effort. Unlike bone, blood, and other tissues, gas is compressible and changes volume as a fish changes its depth. If a fish swims to a greater depth, the greater pressure exerted by the surrounding water compresses the gas in the swim bladder, so that the fish becomes less buoyant and begins to sink. The volume of gas in the swim bladder must be increased to establish a new equilibrium buoyancy. When a fish swims upward, gas in the bladder expands because of the reduced surrounding water pressure, making the fish more buoyant. Unless gas is removed, the fish continues to ascend with increasing speed as the swim bladder continues to expand.

Gas may be removed from the swim bladder in two ways. The more ancestral **physostomous** (Gr., *phys,* bladder, + *stoma,* mouth) condition (trout, for example) features a pneumatic duct that connects the swim bladder to the esophagus, through which a fish may expel air. The derived state is the **physoclistous** (Gr., *phys,* bladder, + *clist,* closed) condition in which the pneumatic duct is lost in adults. In physoclistous fishes, gas is diffused into blood at the **ovale,** a vascularized area of the swim bladder (Figure 24.26). Both types of fishes require gas to be secreted into the swim bladder from the blood, although a few shallow-water-inhabiting physostomes may gulp air to fill their swim bladder.

Physiologists who at first were baffled by the secretion mechanism now understand how it operates. In brief, the **gas gland** secretes lactic acid, which enters the blood, causing a localized high acidity in the **rete mirabile** (L., wonderful net) that forces hemoglobin to release its load of oxygen. The capillaries in the rete are arranged in parallel, creating a countercurrent multiplier (p. 675), allowing oxygen to reach high concentrations in the gas gland and to diffuse into the swim bladder. The final gas pressure attained in the swim bladder depends on the length of the rete capillaries; they are relatively short in fishes living near the surface but are extremely long in deep-sea fishes.

The amazing effectiveness of this device is exemplified by a fish living at a depth of 2400 m (8000 feet). To keep the bladder inflated at that depth, the gas inside (mostly oxygen) must have a pressure exceeding 240 atmospheres, which is much greater than the pressure in a fully charged steel gas cylinder. Yet the oxygen pressure in the fish's blood cannot exceed 0.2 atmosphere—in equilibrium with the oxygen pressure in the atmosphere at the sea surface.

Hearing and Weberian Ossicles

Fishes, like other vertebrates, detect sounds as vibrations in the inner ear. Detecting these vibrations is difficult for aquatic vertebrates because their bodies are nearly the same density as the surrounding water, letting sound waves pass through a fish's body nearly undetected.

A particularly elegant solution to this problem is found in the ostariophysans, a group of teleosts that includes minnows, characins, suckers, and catfishes. The ostariophysans include over 8000 species and are usually the dominant fishes in diversity and abundance in freshwater habitats. Their success may be partly due to their **Weberian ossicles,** a set of small bones that allow them to hear faint sounds over a much broader range of frequency than do other

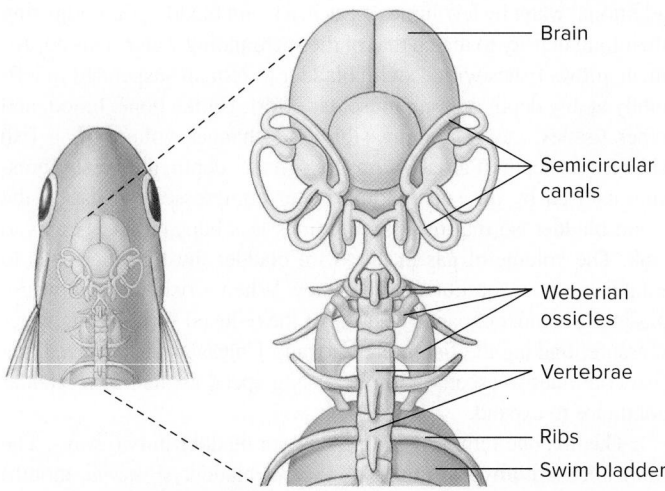

Figure 24.27 Weberian ossicles are small bones that transmit sound vibrations received in the swim bladder to the inner ear. Teleosts with this apparatus can detect faint sounds over a much broader range of frequency than do other fishes.

elasmobranchs and bony fishes the branchial mechanism is arranged to pump water continuously and smoothly over the gills, although to an observer it appears that fish breathing is pulsatile. The flow of water is opposite the direction of blood flow (countercurrent flow), the best arrangement for extracting the greatest possible amount of oxygen from water. Some bony fishes can remove as much as 85% of the dissolved oxygen from water passing over their gills. Very active fishes, such as herring and mackerel, can obtain sufficient water for their high oxygen demands only by swimming forward continuously to force water into their open mouth and across their gills. This process is called ram ventilation. Such fishes would be asphyxiated if placed in an aquarium that restricted free-swimming movements, even if the water were saturated with oxygen.

A surprising number of fishes can live out of water for varying lengths of time by breathing air. We already have discussed the lungs of lungfishes, gars, and the extinct rhipidistians. Freshwater eels often make overland excursions during rainy weather, using the skin as a respiratory surface. Electric eels, *Electrophorus* (Gr. *ēlektron,* something bright, + *phoros,* to bear), have degenerate gills and must supplement gill respiration by diffusing oxygen through the vascular

teleosts. Reception of sound begins at the swim bladder, which can be easily vibrated because it is air-filled. Sound vibrations are transmitted from the swim bladder to the inner ear by the Weberian ossicles (Figure 24.27). This system has some similarity to the tympanum and middle ear bones of mammals (p. 738), but evolved independently. Adaptations to improve hearing are not confined to ostariophysans. For example, herrings and anchovies have anterior expansions of the swim bladder that directly contact the skull. The importance of the swim bladder in these fishes is demonstrated by experiments in which the swim bladder is artificially deflated, reducing sensitivity to sounds.

Respiration

Fish gills are composed of thin filaments, each covered with a thin epidermal membrane folded repeatedly into platelike **lamellae** (Figure 24.28). These are richly supplied with blood vessels. The gills are located inside the pharyngeal cavity and are covered with a movable bony flap, the **operculum.** This arrangement protects the delicate gill filaments, streamlines the body, and makes possible a pumping system for moving water through the mouth, across the gills, and out the single pair of gill slits. Instead of opercular flaps as in bony fishes, elasmobranchs have a series of **gill slits** (see Figure 24.7) out of which the water flows. In both

Figure 24.28 Gills of fish. Muscles attached to the operculum **(A)** pump water over gills and out the gill slit. The bony, protective flap covering the gills (operculum) has been removed **(B)** to reveal branchial chamber containing the gills. There are four gill arches on each side, each bearing numerous filaments. A portion of a gill arch **(C)** shows gill rakers that project forward to strain out food and debris, and gill filaments that project to the rear. A single gill filament **(D)** is dissected to show the blood capillaries within the platelike lamellae. Direction of water flow (*blue arrows*) is opposite the direction of blood flow.

lining of the mouth cavity. One of the best air breathers of all is the Indian climbing perch, *Anabas* (Gr. *anabainō*, to go up), which spends most of its time on land near the water's edge, breathing air through special air chambers above much-reduced gills.

Osmotic Regulation

Freshwater is an extremely dilute medium with a salt concentration (0.001 to 0.005 gram moles per liter [M]) much below that of the blood of freshwater fishes (0.2 to 0.3 M). Water therefore tends to enter their bodies osmotically, and salt is lost by diffusion outward. Although the scaled and mucous-covered body surface is almost totally impermeable to water, water gain and salt loss do occur across thin membranes of the gills. Freshwater fishes are **hyperosmotic regulators** (p. 662) with several defenses against these problems (Figure 24.29). First, excess water is pumped out by the **opisthonephric** kidneys (p. 667), which are capable of forming very dilute urine. Second, special **salt-absorbing cells** located in the gill epithelium actively move salt ions, principally sodium and chloride, from water to the blood. This absorption, together with salt present in the fish's food, replaces diffusive salt loss. These mechanisms are so efficient that a freshwater fish spends only a small part of its total energy maintaining itself in osmotic balance.

Perhaps 90% of all bony fishes are restricted to either a freshwater or a seawater habitat because they are incapable of osmotic regulation in the "wrong" habitat. Most freshwater fishes quickly die if placed in seawater, as do marine fishes placed in freshwater. However, some 10% of all teleosts can pass back and forth with ease between both habitats. These **euryhaline fishes** (Gr. *eurys*, broad, + *hals*, salt) are of two types: those such as many flounders, sculpins, and killifish that live in estuaries or certain intertidal areas where salinity fluctuates throughout the day; and those such as salmon, shad, and eels, that spend part of their life cycle in freshwater and part in seawater.

Figure 24.29 Osmotic regulation in freshwater and marine bony fishes. A freshwater fish maintains osmotic and ionic balance in its dilute environment by actively absorbing sodium chloride (NaCl) across the gills (some salt is gained with food). To flush out excess water that constantly enters the body, the glomerular kidney produces a dilute urine by reabsorbing sodium chloride. A marine fish must drink seawater to replace water lost osmotically to its salty environment. Sodium chloride and water are absorbed from the stomach. Excess sodium chloride is actively transported outward by the gills. Divalent sea salts, mostly magnesium sulfate ($MgSO_4$), are eliminated with feces and secreted by the tubular kidney.

Marine bony fishes are **hypoosmotic regulators** that encounter a completely different problem. Having a much lower blood salt concentration (0.3 to 0.4 M) than seawater around them (about 1 M), they lose water and gain salt. A marine teleost fish quite literally risks drying out, much like a desert mammal deprived of water. To compensate for water loss, a marine teleost drinks seawater (Figure 24.29). Excess salt accompanying the seawater is disposed of in multiple ways. Major sea salt ions (sodium, chloride, and potassium) are carried by the blood to the gills, where they are secreted outward by special **salt-secretory cells.** The remaining sea salt ions, mostly magnesium, sulfate, and calcium, are voided with feces or excreted by the kidneys. Unlike a freshwater fish's kidneys, which form urine by the usual filtration-resorption sequence typical of most vertebrate kidneys (pp. 669–670), a marine fish's kidneys excrete divalent ions by tubular secretion. Because very little if any filtrate is formed, the glomeruli have lost their importance and disappeared in some marine teleosts. Pipefishes and frogfishes (Figure 24.31) are examples of "aglomerular" marine fishes.

Feeding Behavior

For any fish, feeding is one of its main concerns in day-to-day living. Although many a luckless angler would swear otherwise, a fish devotes more time and energy to eating, or searching for food to eat, than to anything else. Throughout the long evolution of fishes, there has been unrelenting selective pressure for those adaptations that enable a fish to win the eat-or-be-eaten contest. Certainly the most far-reaching single event was the evolution of jaws (pp. 512–513). Jaws freed fishes from a largely passive filter-feeding existence, enabling them to adopt a predatory mode of life. Improved means of capturing larger prey demanded stronger muscles, more agile movement, better balance, and improved special senses. More than any other aspect of its life habit, feeding behavior shapes a fish.

Most fishes are **carnivores** and prey on a myriad of animal foods, from zooplankton and insect larvae to large vertebrates. Some deep-sea fishes are capable of eating victims nearly twice their own size—an adaptation for life in a world where meals are infrequent. Most advanced ray-finned fishes cannot masticate their food as we can because doing so would block the current of water across the gills. Some, however, such as wolf eels (Figure 24.30), have molarlike teeth in their jaws for crushing prey, which may include hard-bodied crustaceans. Others that do grind their food use powerful pharyngeal teeth in their throat. Most carnivorous fishes swallow their prey whole, using sharp-pointed teeth in their jaws and on the roof of the mouth to seize prey. The incompressibility of water assists many large-mouthed predators in capturing prey. When the mouth is suddenly opened, water rushes in, sweeping the victim inside (Figure 24.31).

A second group of fishes are **herbivores** that eat plants and macroalgae. Plant eaters are relatively uncommon among fishes, but are crucial intermediates in the food chain in some habitats. Plant eaters are most common in coral reefs (parrotfishes, damselfishes, and surgeonfishes) and in tropical freshwater habitats (some minnows, characins, and catfishes).

Suspension feeders that crop the abundant microorganisms of the sea form a third and diverse group of fishes ranging from fish larvae to basking sharks. However, the most characteristic group of plankton feeders are herringlike fishes (menhaden, herring, anchovies, and others), mostly **pelagic** (open-sea dwellers) fishes that travel in large

Figure 24.30 Wolf eel, *Anarrhichthys ocellatus,* feeding on a sea cucumber it has captured and pulled to the opening of its den.

© Daniel W. Gotshall

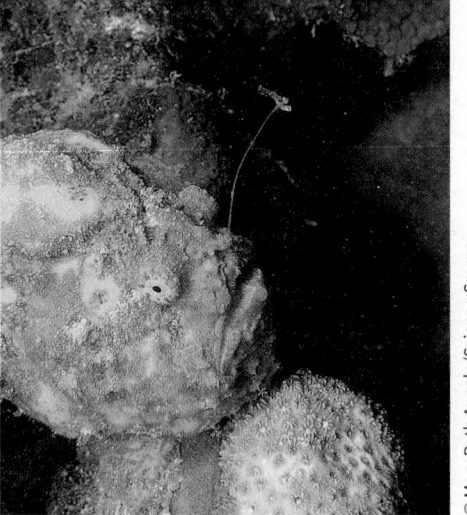

Figure 24.31 Longlure frogfish, *Antennarius multiocellatus,* awaits its meal. Above its head swings a modified dorsal fin spine ending in a fleshy tentacle that contracts and expands in a convincing wormlike manner. When a fish approaches the alluring bait, the huge mouth opens suddenly, creating a strong inward current that sweeps the prey inside. The entire process takes only 4 milliseconds!

© Mary Beth Angelo/Science Source

schools. Both phytoplankton and smaller zooplankton are strained from the water with sievelike gill rakers (see Figure 32.1). Because plankton feeders are the most abundant of all marine fishes, they are important food for numerous larger but less abundant carnivores. Many freshwater fishes also depend on plankton for food.

Other groups of fishes include **scavengers,** such as hagfishes, that consume dead and dying animals, and **detritivores,** such as some suckers and minnows, that consume fine, particulate organic matter. Some fishes use a **parasitic** mode of feeding in which they consume parts of other live fishes. Examples of these include lampreys (p. 521) and the candiru, *Vandellia* , a tiny elongate catfish that feeds on the gill epithelia of host fishes. Finally it should be noted that most fishes, though specialized for a narrow diet, may use other foods when available.

Digestion in most fishes follows the vertebrate plan (see Figure 32.9). Except in several fishes that lack distinct stomachs, food proceeds from stomach to tubular intestine, which tends to be short in carnivores (see Figure 24.14) but may be extremely long and coiled in herbivorous and detritivorous forms. In herbivorous grass carp, for example, the intestine may be nine times the body length, an adaptation for the lengthy digestion required for plant carbohydrates. In carnivores, some protein digestion may be initiated in the acid medium of the stomach, but the principal function of the stomach is to store often large and infrequent meals while awaiting their reception by the intestine.

Digestion and absorption proceed simultaneously in the intestine. A feature of ray-finned fishes, especially teleosts, is the presence of numerous **pyloric ceca** (see Figure 24.14) found in no other vertebrate group. Their primary function appears to be lipid absorption, although all classes of digestive enzymes (protein-, carbohydrate-, and lipid-splitting) are secreted there.

Migration

Freshwater Eels
For centuries naturalists had been puzzled about the life history of freshwater eels, *Anguilla* (an-gwil′lə) (L., eel), common and commercially important in coastal streams of the North Atlantic. Eels are **catadromous** (Gr. *kata,* down, +

dromos, running), meaning that they spend most of their lives in freshwater but migrate to the sea to spawn. Each fall, large numbers of eels were seen swimming down rivers toward the sea, but no adults ever returned. Each spring countless numbers of young eels, called "elvers" (Figure 24.32), each about the size of a wooden matchstick, appeared in coastal rivers and began swimming upstream. Beyond the assumption that eels must spawn somewhere at sea, the location of their breeding grounds was completely unknown.

The first clue was provided by two Italian scientists, Grassi and Calandruccio, who in 1896 reported that elvers were not larval eels but rather were advanced juveniles. True larval eels, they discovered, were tiny, leaf-shaped, completely transparent creatures that bore little resemblance to an adult eel. They had been called **leptocephali** (Gr. *lepto,* slender, + *cephal,* head) by early naturalists, who never suspected their true identity. In 1905 Johann Schmidt began a systematic study of eel biology, examining thousands of leptocephali caught in plankton nets supplied to captains of commercial vessels plying the Atlantic Ocean. By noting where larvae in different stages of development were captured, Schmidt and his colleagues eventually reconstructed the spawning migrations.

When adult eels leave the coastal rivers of Europe and North America, they swim steadily and apparently at great depth for 1 to 2 months until they reach the Sargasso Sea, a vast area of warm oceanic water southeast of Bermuda (Figure 24.32). Here, at depths of 300 m or more, the eels spawn and die. Minute larvae then begin an

Figure 24.32   Life histories of American eels, *Anguilla rostrata,* and European eels, *Anguilla anguilla.* Migration patterns of American eels are shown in blue, while those of European eels are shown in red. Circled numbers refer to stages of development. Note that American eels complete their larval metamorphosis and sea journey in 1 year, while 3 years are required for European eels to complete their much longer journey.

incredible journey back to the streams of Europe and North America. Because the Sargasso Sea is much closer to the American coastline, American eel larvae make their journey in only about 1 year, compared to 3 years for European eel larvae. Males typically remain in brackish water of coastal rivers, while females migrate as far as several hundred kilometers upstream. After 6 to 15 years of growth, females, now 1 m long, return to the ocean to join the smaller males in the journey back to the spawning grounds in the Sargasso Sea.

Enzyme electrophoretic analysis of eel larvae confirm not only the existence of separate European and American species but also Schmidt's hypothesis that European and American eels spawn in partially overlapping areas of the Sargasso Sea.

Homing Salmon

The life history of salmon is nearly as remarkable as that of freshwater eels and certainly has received far more popular attention. Most salmon are **anadromous** (Gr. *anadromos,* running upward); they spend their adult lives at sea but return to freshwater to spawn. Atlantic salmon, *Salmo salar* (L. *salmo,* salmon, + *sal,* salt), can make repeated upstream spawning runs. The seven Pacific salmon (*Oncorhynchus*) [on-kō-rēnk′us] [Gr. *onkos,* hook, + *rhynchos,* snout] species (sockeye, coho, pink, Chinook, chum, steelhead, and masu) each make a single spawning run (Figure 24.33), after which they die.

The virtually infallible homing instinct of the Pacific species is legendary. After migrating downstream as a smolt (a juvenile stage; Figure 24.34), a sockeye salmon ranges many hundreds of kilometers over the Pacific for nearly 4 years, grows to 2 to 5 kg in weight, and then returns almost unerringly to spawn in the headwaters of its parent stream. Some straying does occur and is an important means of increasing gene flow and populating new streams.

Figure 24.33 Sockeye salmon, *Oncorhynchus nerka*, spawning in Power Creek, Alaska. Females build nests called rudds, using their caudal fins to excavate shallow depressions in gravel riffles. After a female deposits her eggs, a male, distinguished by hooked jaws and large teeth, fertilizes the eggs. The female covers the eggs with gravel and the young begin their development.

Salmon runs along the Pacific coast of North America have been devastated by a lethal combination of increased siltation from logging, pollution, and especially, the presence of more than 50 hydroelectric dams, which obstruct upstream migration of adult salmon and kill downstream migrants as they pass through the dams' power-generating turbines. The result is that the annual run of wild salmon is today less than 10% of the 10 to 16 million fishes that ascended the rivers 150 years ago. Some hope comes from dam removals, which are occurring at an increasing rate in the United States. The "new" streams formed are quickly recovering. Only days after removal of a 62-m-high dam on the Elwha River in Washington in 2012, Chinook salmon were detected moving upstream of the old dam.

Experiments by A. D. Hasler and others show that homing salmon are guided upstream by the characteristic odor of their parent stream. When salmon finally reach the spawning beds of their parents (where they themselves were hatched), they spawn and die. The following spring, newly hatched fry transform into smolts before and during the downstream migration. At this time they are imprinted (p. 783) with the distinctive odor of the stream, which is apparently a mosaic of compounds released by the characteristic vegetation and soil in the watershed of the natal stream. They also seem to imprint on odors of other streams they pass while migrating downriver and use these odors in reverse sequence as a map during the upriver migration as returning adults.

How do salmon find their way to the mouth of a coastal river from the trackless miles of open ocean? Salmon move hundreds of kilometers away from the coast, much too far to be able to detect the odor of their natal stream. Experiments suggest that some migrating fish, like birds, can navigate by orienting to the position of the sun. However, migrant salmon can navigate on cloudy days and at night, indicating that sun navigation, if used at all, cannot be a salmon's only navigational cue. Fish also (again, like birds) appear able to detect and to navigate to the earth's magnetic field. Finally, fishery biologists concede that salmon may not require precise navigational abilities at all, but instead may use ocean currents, temperature gradients, and food availability to reach the general coastal area where "their" river is located. From this point, they navigate by their imprinted odor map, making correct turns at each stream junction until they reach their natal stream.

Reproduction and Growth

In a group as diverse as fishes, it is no surprise to find extraordinary variations on the basic theme of sexual reproduction. Most fishes favor a simple theme: they are **dioecious,** with external fertilization and external development of their eggs and embryos (oviparity). However, as tropical fish enthusiasts are well aware, the ever-popular ovoviviparous guppies and mollies of home aquaria, and several other teleosts, develop in the ovarian cavity of the mother, then are born (Figure 24.35). As described earlier in this chapter (p. 524), some viviparous sharks develop a kind of placental attachment through which the young are nourished during gestation.

Oviparity is the most common mode of reproduction in fishes. Many marine fishes are extraordinarily profligate egg producers. Males and females come together in great schools and release vast

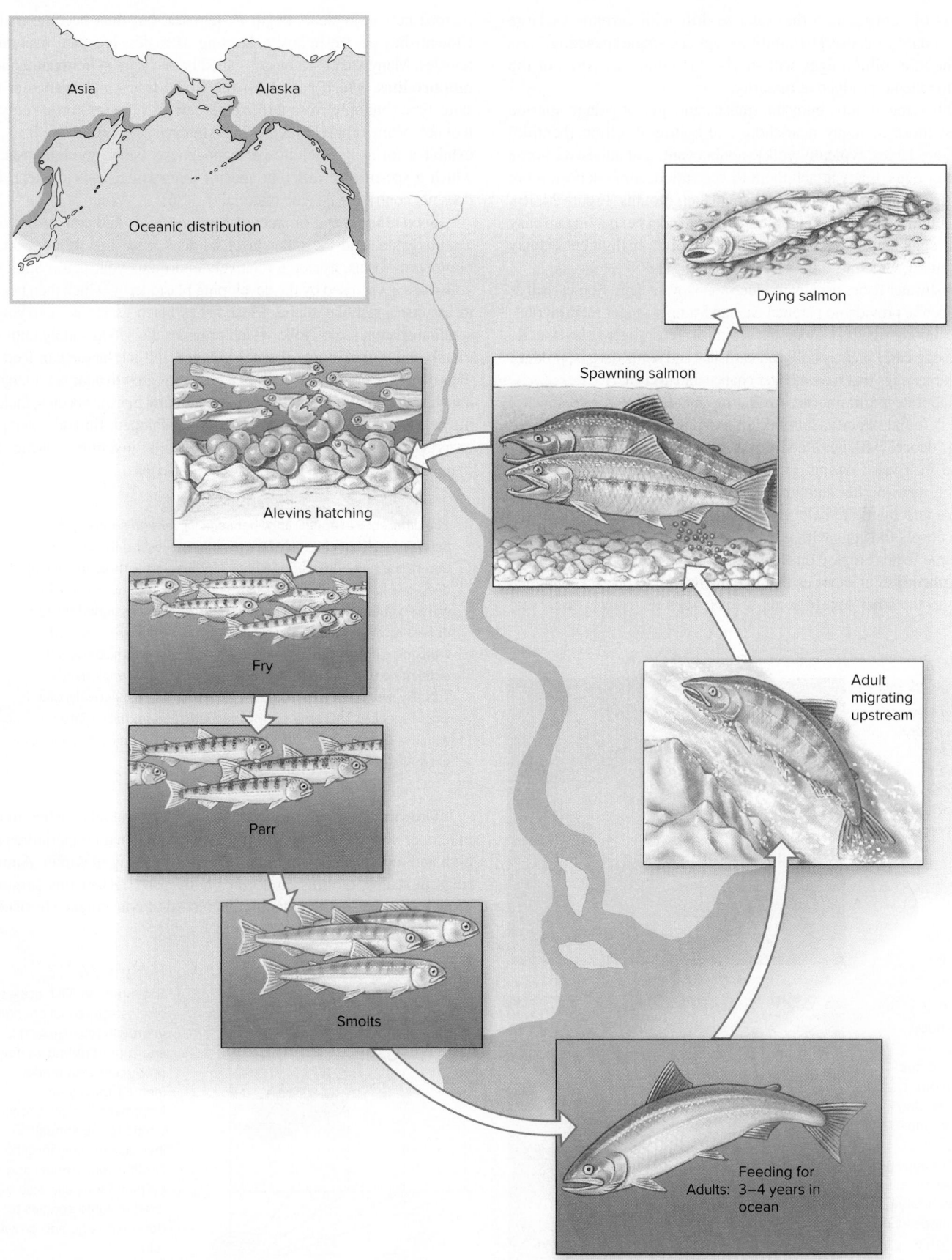

Figure 24.34   Life history of Pacific salmon, *Oncorhynchus*.

numbers of gametes into the water to drift with currents. A large female cod may release 4 to 6 million eggs at a single spawning. Less than one in a million eggs will survive the numerous perils of the ocean to reach reproductive maturity.

Unlike the minute, buoyant, transparent eggs of pelagic marine teleosts, those of many near-shore and bottom-dwelling (benthic) species are larger, typically yolky, nonbuoyant, and adhesive. Some bury their eggs, many attach them to vegetation, some deposit them in nests, and some even incubate them in their mouths (Figure 24.36). Many benthic spawners guard their eggs. Intruders expecting an easy meal of eggs may be met with a vivid and often belligerent display by the guard, which is almost always male.

Freshwater fishes usually produce nonbuoyant eggs. Some, such as yellow perch, provide no parental care and simply scatter their myriads of eggs among weeds or along the sediment. Freshwater fishes that do provide egg care, such as bullhead catfishes and some darters, produce fewer, larger eggs that have a better chance for survival.

Elaborate preliminaries to mating are the rule for freshwater fishes. A female Pacific salmon, for example, performs a ritualized mating "dance" with her breeding partner after arriving at the spawning bed in a fast-flowing, gravel-bottomed stream (Figure 24.34). She then turns on her side and scoops a nest hole with her tail. As eggs are laid by the female, they are fertilized by a male. After the female covers the eggs with gravel, the exhausted fish dies.

Some fishes employ unusual reproductive strategies. **Sequential hermaphrodites** are species that initially mature as one sex, and then change to the other sex. In some species, such as many wrasses and parrotfishes, individuals begin as females, but then become males. Clownfishes do the opposite, starting as males, but then becoming females. Many small sea basses called hamlets are **synchronous hermaphrodites,** which have both functional testes and ovaries at the same time (but only cross-fertilize). A few fish species consist only of females. Many of these, such as the Amazon molly, *Poecillia formosa,* exhibit a form of ameiotic parthenogenesis called **gynogenesis,** in which a sperm of a different species initiates egg development, but does not contribute genetic material (p. 136).

Soon after an egg of an oviparous species is laid and fertilized, it absorbs water and the outer layer hardens. Cleavage follows, and a blastoderm forms, astride a relatively enormous yolk mass. Soon the yolk mass is enclosed by the developing blastoderm, which then begins to assume a fishlike shape. Most fishes hatch as larvae, carrying a semitransparent sac of yolk, which provides their food supply until the mouth and digestive tract have developed and the larvae can feed on their own (see Figure 8.24). After a period of growth a larva undergoes a metamorphosis, especially dramatic in many marine species, including eels (Figure 24.32). Body shape is refashioned, fin and color patterns change, and the animal becomes a juvenile bearing the unmistakable definitive body form of its species.

The difference in appearance between larvae and adults and between males and females sometimes makes it difficult to determine taxonomic boundaries. Until recently, three groups of deep-sea fishes, called tapetails, bignose fishes, and whalefishes, were so different in morphology that they were assigned three different taxonomic families. Comparisons of newly collected, transforming specimens and genetic data allowed scientists to determine that the tapetails, bignose fishes, and whalefishes actually represent larvae, males, and females, respectively, of a single species! This amazing ontogenetic and sexual variation, which is correlated to different feeding adaptations, is the most extreme among vertebrates.

Growth is temperature dependent. Consequently, fishes living in temperate regions grow rapidly in summer when temperatures are high and food is abundant but nearly stop growing in winter. Annual rings in scales, otoliths, and other bony parts reflect this seasonal growth (Figure 24.37), a distinctive record of convenience to fishery

Figure 24.35
Rainbow seaperch, *Hypsurus caryi,* giving birth. All of the West Coast surfperches (family Embiotocidae) are ovoviviparous.

© Daniel W. Gotshall

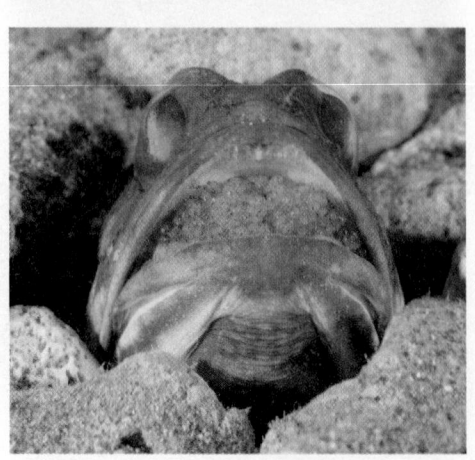

Figure 24.36
Male banded jawfish, *Opistognathus macrognathus,* orally brooding its eggs. The male retrieves the female's eggs and incubates them until they hatch. During brief periods when the jawfish is feeding, the eggs are left in the burrow.

© Andrew J. Martinez/Science Source

Figure 24.37
Scale growth. Fish scales disclose seasonal changes in growth rate. Growth is interrupted during winter, producing year marks (annuli). Each year's increment in scale growth is a ratio to the annual increase in body length. Otoliths (ear stones) and certain bones can also be used in some species to determine age and growth rate.

Taxonomy of Living Fishes

The following taxonomy primarily follows that of Nelson (2006). The probable relationships of these traditional groupings, together with the major extinct groups of fishes, are shown in the cladogram in Figure 16.2. Other schemes of classification have been proposed. Because of the difficulty of determining relationships among the numerous living and fossil species, we can appreciate why fish classification has undergone, and will continue to undergo, continuous revision. As in the previous chapter, we do not show Linnaean ranks below subphylum. Nested hierarchies of monophyletic groups are shown by indenting.

Subphylum Vertebrata

   Cyclostomata (sī´klō-stō-mä´tä) (Gr. *cyclos*, circle, + *stoma*, mouth). Without true jaws or paired appendages; teeth of keratin; vertebrae reduced or absent in adults.

      Myxini (mik-sin´ē) (Gr. *myxa*, slime): **hagfishes.** Mouth with four pairs of tentacles; buccal funnel absent; 1–16 pairs of external gill openings; vertebrae rudimentary, only in embryos; slime glands present. Examples: *Myxine*, *Epaptretus*; about 70 species, marine.

      Petromyzontida (pet´trō-mī-zon´ti-də) (Gr. *petros*, stone, + *myzon*, sucking): **lampreys.** Mouth surrounded by keratinized teeth but no barbels; buccal funnel present; seven pairs of external gill openings; vertebrae present only as neural arches. Examples: *Petromyzon*, *Ichthyomyzon*, *Lampetra*; 41 species, freshwater and anadromous.

   Gnathostomata (na´thō-stō´mä-tə) (Gr. *gnathos*, jaw, + *stoma*, mouth). Jaws present; paired appendages present (secondary lost in a few forms); three pairs of semicircular canals; notochord partly or completely replaced by centra.

      Chondrichthyes (kon-drik´thē-ēz´) (Gr. *chondros*, cartilage, + *ichthys*, fish): **cartilaginous fishes.** Cartilaginous skeleton; teeth not fused to jaws and usually replaced; no swim bladder; intestine with spiral valve; claspers present in males.

         Elasmobranchii (ē-laz´mō-brănk´ē-ī) (Gr. *elasmos*, plated, + *branchia*, gills): **sharks, skates,** and **rays.** Placoid scales or derivatives (scutes and spines) usually present; five to seven gill arches and gill slits in separate clefts along pharynx; upper jaw not fused to cranium. Examples: *Squalus*, *Raja*, *Sphyrna*; about 960 species, mostly marine.

         Holocephali (hō-lō-sef´ə-lī) (Gr. *holos*, entire, + *kephale*, head): **chimaeras** and **ratfishes.** Scales absent; four gill slits covered by operculum; jaws with tooth plates; accessory clasping organ (tentaculum) in males; upper jaw fused to cranium. Examples: *Chimaera*, *Hydrolagus*; 40 species, marine.

      Actinopterygii (ak´ti-nop´te-rij´ē-ī) (Gr. *aktis*, ray, + *pteryx*, fin, wing): **ray-finned fishes.** Skeleton ossified; gills covered by bony operculum; paired fins supported primarily by dermal rays; limb musculature within body; swim bladder mainly for buoyancy, if present; atrium and ventricle not divided; teeth with enameloid covering.

         Cladistia (clə-dis´tē-ə) (Gr. *cladi*, branch): **bichirs.** Rhombic ganoid scales; lungs; spiracle present; dorsal fin consisting of 5–18 finlets. Examples: *Polypterus*; about 16 species, freshwater.

         Chondrostei (kon-dros´tē-ī) (Gr. *chondros*, cartilage, + *osteon*, bone): **paddlefishes, sturgeons.** Skeleton primarily cartilage; caudal fin heterocercal; large scutes or tiny ganoid scales present. Examples: *Polyodon*, *Acipenser*; 29 species, freshwater and anadromous.

         Neopterygii (nē´op-te-rij´ē-ī) (Gr. *neo*, new, + *pteryx*, fin, wing): **gars, bowfin,** and **teleosts.** Skeleton primarily bone; caudal fin usually homocercal; scales cycloid, ctenoid, absent, or rarely, ganoid. Examples: *Amia*, *Lepisosteus*, *Anguilla*, *Oncorhynchus*, *Perca*; about 31,000 species, nearly all aquatic habitats.

      Sarcopterygii (sär-cop-te-rij´-ē-ī) (Gr. *sarkos*, flesh, + *pteryx*, fin, wing): **lobe-finned fishes, tetrapods.** Fish-like members with skeleton ossified; gills covered by bony operculum; paired fins with sturdy internal skeleton and musculature within appendage; diphycercal tail; usually with lungs; atrium and ventricle at least partly divided; teeth with enamel covering. Fish examples: *Latimeria* (coelacanths); *Neoceratodus*, *Lepidosiren*, *Protopterus* (lungfishes); 8 fish species, marine and freshwater.

---

biologists who wish to determine a fish's age. Unlike birds and mammals, which stop growing after reaching maturity, most reproductively mature fishes continue to grow, though more slowly, for as long as they live.

SUMMARY

Fishes are gill-breathing aquatic vertebrates with fins for appendages. They include the oldest vertebrates, having originated from an ancestor in the Cambrian period or possibly earlier. Five clades with living fishes are recognized. The jawless hagfishes (Myxini) and lampreys (Petromyzontida), which together make up the clade Cyclostomata, have an eel-like body form without paired fins, a cartilaginous skeleton, a notochord that persists throughout life, and a disclike mouth adapted for sucking or biting. All other vertebrates have jaws, a major development in vertebrate evolution. Members of Chondrichthyes (sharks, rays, and chimaeras) have a cartilaginous skeleton, paired fins, well-developed sensory organs, and an active, characteristically predaceous habit. The fishlike members of Osteichthyes have bony endoskeletons and an operculum and are divided into the clades Sarcopterygii

and Actinopterygii. Lobe-finned fishes of Sarcopterygii include lungfishes and coelacanths. Terrestrial vertebrates also are a part of this clade. Actinopterygii is composed of the ray-finned fishes, a huge and diverse modern assemblage containing nearly all familiar freshwater and marine fishes. Success of this group may be linked to adaptations of skeletal, muscular, and buoyancy systems, which increased locomotion and feeding efficiency.

One group of Actinopterygii, the teleosts, has diversified into about 31,000 species that reveal an enormous diversity of adaptations, body form, behavior, and habitat preference. Most fishes swim by undulatory contractions of the body muscles, which generate thrust (propulsive force) and lateral force. Eel-like fishes oscillate the whole body, but in more rapid swimmers undulations are limited to the caudal region or caudal fin alone.

Most pelagic bony fishes achieve neutral buoyancy in water using a gas-filled swim bladder, the most effective gas-secreting device known in the animal kingdom. Sensitivity to sounds is enhanced by Weberian ossicles that transmit sounds from the swim bladder to the inner ear. Gills of fishes, having efficient countercurrent flow between water and blood, facilitate high rates of oxygen exchange. All fishes show well-developed osmotic and ionic regulation, achieved principally by kidneys and gills. With the exception of agnathans, all fishes have jaws and teeth that are variously modified for carnivorous, herbivorous, planktivorous, and other feeding modes.

Many fishes are migratory, and some, such as catadromous freshwater eels and anadromous salmon, make remarkable journeys of great length and precision.

Fishes reveal an extraordinary range of sexual reproductive strategies. Most fishes are oviparous, but ovoviviparous and viviparous fishes are not uncommon. Reproductive investment may be in large numbers of eggs with low survival (many marine fishes) or in fewer eggs with greater parental care for better survival (many freshwater fishes).

REVIEW QUESTIONS

1. Provide a brief description of fishes, citing characteristics that would distinguish them from all other animals.
2. What evidence suggests that hagfishes and lampreys form a monophyletic group?
3. Describe feeding behavior in hagfishes and lampreys. How do they differ?
4. Describe the life cycle of sea lampreys, *Petromyzon marinus,* and the history of their invasion of the Great Lakes.
5. In what ways are sharks well equipped for a predatory life habit?
6. What function does the lateral-line system serve? Where are receptors located?
7. Explain how bony fishes differ from sharks and rays in the following systems or features: skeleton, scales, buoyancy, respiration, and reproduction.
8. Match the ray-finned fishes in the right column with the group to which each belongs in the left column:

_____ Chondrosteans	a. Perch
_____ Nonteleost	b. Sturgeon
neopterygians	c. Gar
_____ Teleosts	d. Salmon
	e. Paddlefish
	f. Bowfin

9. Make a cladogram that includes the following groups of fishes: chondrosteans, elasmobranchs, hagfishes, holocephalans, lampreys, lungfishes, teleosts. Add the following synapomorphies to the diagram: claspers, cranium, endochondral bone, fins with stout bones, jaws, keratinized teeth.
10. List four characteristics of teleosts that contributed to their incredible diversity and abundance.
11. Only eight species of lobe-finned fishes are alive today, remnants of a group that flourished in the Devonian period of the Paleozoic era. What morphological characteristics distinguish lobe-finned fishes? What is the literal meaning of Sarcopterygii, the taxon comprising lobe-finned fishes and tetrapods?
12. Explain how lungfishes are adapted to survive out of water.
13. Describe the discoveries of living coelacanths. What is the evolutionary significance of the group to which they belong?
14. Compare the swimming movements of eels with those of trout, and explain why the latter are more efficient for rapid locomotion.
15. Sharks and bony fishes approach or achieve neutral buoyancy in different ways. Describe the methods evolved in each group. Why must a teleost fish adjust the gas volume in its swim bladder when it swims upward or downward? How is gas volume adjusted?
16. What is meant by "countercurrent flow" as it applies to fish gills?
17. How do Weberian ossicles increase a fish's sensitivity to sounds?

18. Compare the osmotic problem and the mechanism of osmotic regulation in freshwater and marine bony fishes.
19. Two principal groups of fishes, with respect to feeding behavior, are carnivores and suspension feeders. How are these two groups adapted for their feeding behavior?
20. Describe the life cycle of freshwater eels. How does the life cycle of American eels differ from that of European eels?
21. How do adult Pacific salmon find their way back to their parent stream to spawn?
22. What mode of reproduction in fishes is described by each of the following terms: oviparous, ovoviviparous, viviparous?
23. Reproduction in marine pelagic fishes and in benthic freshwater fishes is distinctively different. How and why do they differ?

For Further Thought Regarding osmoregulation, what physiological and behavioral changes would occur as a fish migrates from a freshwater stream to the ocean?

SELECTED REFERENCES

Carey, F. G. 1973. Fishes with warm bodies. Sci. Am. **228:** 36–44 (Feb.). *Classic paper about how fishes with regional endothermy keep warm.*

Heimberg, A. M., R. Cowper-Sal-lari, M. Semon, P. C. J. Donoghue, and K. J. Peterson. 2010. MicroRNAs reveal the relationship of hagfish, lampreys, and gnathostomes and the nature of the ancestral vertebrate. Proc. Nat. Acad. Sci. **107:** 19379–19383. *Analysis that supports the cyclostome hypothesis and suggests that miRNA evolution was important in the formation of the vertebrate body plan.*

Helfman, G. J., B. B. Collette, D. E. Facey, and B. W. Brown. 2009. The diversity of fishes: biology, evolution, and ecology, ed. 2. Malden, Massachusetts, Wiley-Blackwell. *This information-packed textbook focuses on adaptation and diversity, and is particularly strong in evolution and ecology.*

Johnson, G. D., J. R. Paxton, T. T. Sutton, T. P. Satoh, M. Nishida, and M. Miya. 2009. Deep-sea mystery solved: astonishing larval transformations and extreme sexual dimorphism unite three fish families. Biology Letters **5:** 235–239. *Morphologically different fishes formerly allocated to three families are now known to represent different life stages in the same species.*

Kuratani, S., and K. G. Ota. 2008. Hagfish (Cyclostomata, Vertebrata): searching for the ancestral developmental plan of vertebrates. BioEssays **30:** 167–172. *The engaging story of the discovery of a method to obtain hagfish embryos and its scientific importance.*

Long, J. A. 2010. The rise of fishes: 500 million years of evolution, ed. 2. Baltimore, The Johns Hopkins University Press. *A lavishly illustrated evolutionary history of fishes.*

Martini, F. H. 1998. Secrets of the slime hag. Sci. Am. **279:** 70–75 (Oct.). *Biology of our most distant living craniate relative.*

Nelson, J. S. 2006. Fishes of the world, ed. 4. New York, John Wiley & Sons, Inc. *Authoritative classification of all major groups of fishes.*

Ota, K. G., S. Fujimoto, Y. Oisi, and S. Kuratani. 2013. Late development of hagfish vertebral elements. J. Exp. Zool. B. Mol. Dev. Ecol. **320:** 129–139. *Hagfish embryos have vertebrae.*

Paxton, J. R., and W. N. Eschmeyer. 1998. Encyclopedia of fishes, ed. 2. San Diego, Academic Press. *Excellent authoritative reference that focuses on diversity and is spectacularly illustrated.*

Robison, B. H., and K. R. Resenbichler. 2008. *Macropinna microstoma* and the paradox of its tubular eyes. Copeia **2008:** 780–784. *A cnidarian-eating fish with tubular eyes located beneath a transparent dome covering its head.*

Springer, V. G., and J. P. Gold. 1989. Sharks in question. Washington, Smithsonian Institution Press. *Morphology, biology, and diversity of sharks, richly illustrated.*

Wegner, N. C., O. E. Snodgrass, H. Dewar, and J. R. Hyde. 2015. Whole-body endothermy in a mesopelagic fish, the opah, *Lampris guttatus.* Science **348:** 786–789. *Unlike other "regionally endothermic" fishes, the opah, a large, bony fish of the open ocean, heats its whole body.*

Weinberg, S. 2000. A fish caught in time: the search for the coelacanth. London, Fourth Estate. *The exciting history of the coelacanth discoveries.*

Early Tetrapods and Modern Amphibians

- PHYLUM CHORDATA
- CLASS AMPHIBIA

A frog tadpole undergoing metamorphosis.

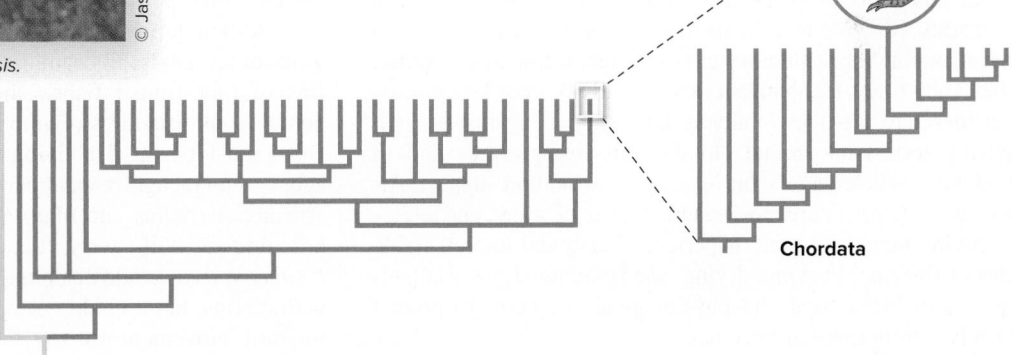

Amphibia

Chordata

From Water to Land in Ontogeny and Phylogeny

A chorus of frogs beside a pond in spring announces the start of a new life cycle. Mating frogs produce masses of eggs, which later hatch into limbless, gill-breathing, fishlike tadpole larvae that feed and grow. Then a remarkable transformation occurs. Hindlegs appear and gradually lengthen. The tail shortens and eventually disappears. Larval teeth and gills are lost. Eyelids develop. Forelimbs emerge. In a few weeks the aquatic tadpole has completed its metamorphosis to an adult frog.

The evolutionary transition from water to land occurred not in weeks but over millions of years. A lengthy series of alterations cumulatively fitted the vertebrate body plan for life on land. The origin of

land vertebrates is no less a remarkable feat for this fact—a feat that would not occur again because well-established terrestrial competitors would exclude poorly adapted transitional forms.

Amphibians include the only living vertebrates that have a transition from water to land in both their ontogeny and their phylogeny. Even after 350 million years of evolution, amphibians remain quasi-terrestrial, hovering between aquatic and land environments. This double life is expressed in their name. Even the amphibians best adapted for a terrestrial existence cannot stray far from moist conditions. Many, however, have developed ways to keep their eggs out of open water where their larvae would be exposed to enemies.

Life on land is a major theme of the remaining vertebrate groups, which form a clade called the superclass Tetrapoda. Amphibians and amniotes (including nonavian reptiles, birds, and mammals) are the two major extant branches of **tetrapod** phylogeny, which originates in the Devonian period. Many tetrapod lineages have lost one or both of the two pairs of limbs that give the group its name; amphibian examples include sirens, a kind of salamander, and caecilians (p. 548). Amphibians are ectothermic (p. 672), relying on environmental context rather than metabolic energy to regulate body temperature. Many amphibians depend on freshwater streams or pools for their reproduction. In this chapter, we review the origins of terrestrial vertebrates and discuss the amphibian branch in detail. We discuss the major amniote groups in Chapters 26 through 28.

Movement from water to land is perhaps the most dramatic event in animal evolution because it involves invasion of a physically hazardous habitat. Life originated in water. Animals are mostly water in composition, and all cellular activities occur in water. Nevertheless, organisms invaded land, carrying their watery composition with them. Vascular plants, pulmonate snails, and tracheate arthropods made this transition much earlier than did vertebrates, and thus generated a food supply that terrestrial vertebrates eventually would use. Although invasion of land required modification of almost every organ system, aquatic and terrestrial vertebrates retain many structural and functional similarities. We see a transition between aquatic and terrestrial vertebrates most clearly today in the many living amphibians that make this transition during their own life histories.

Important physical differences that animals must accommodate when moving from water to land include those of (1) oxygen content, (2) density, (3) temperature regulation, and (4) habitat diversity. Oxygen is at least 20 times more abundant in air, and it diffuses much more rapidly through air than through water. It is thus readily available to animals that possess lungs and/or a skin surface suitable for respiratory gas exchange. Compared to water, air has 1000 times less buoyant density and 50 times less viscosity. It therefore provides relatively little support against gravity, requiring terrestrial animals to develop strong limbs and to remodel their skeleton to achieve adequate structural support. Air fluctuates in temperature more readily than does water, and terrestrial environments therefore experience harsh and unpredictable cycles of freezing, thawing, drying, and flooding. Terrestrial animals require behavioral and physiological strategies to protect themselves from thermal extremes.

25.1 DEVONIAN ORIGIN OF TETRAPODS

By the Devonian period, beginning about 416 million years ago, bony fishes had diversified to include many freshwater forms. An important combination of characteristics that evolved originally in aquatic habitats provided access to terrestrial habitats (Figure 25.1). An air-filled cavity, which functioned as a swim bladder, and paired internal nares (nostrils; evolutionary origin shown on Figure 25.2), which functioned in chemoreception, connected to the pharynx. On land, this combination of structures would draw oxygen-rich air through the nares and into the air-filled cavity, whose surface would exchange oxygen and carbon dioxide with body fluids. The bony elements of paired fins, modified for support and movement on underwater surfaces (evolutionary origin shown on Figure 25.2), gained sufficient strength to support and to move the body on land.

The internal nares, air-filled cavity, and paired limbs of an aquatic tetrapod ancestor therefore made possible evolution of terrestrial breathing and support. The airfilled cavity illustrates the important evolutionary principle of exaptation (p. 121), in which a structure that has evolved by natural selection for an initial utility or role is later recruited or "coopted" for a new role. Note that the air-filled cavities called "lungs" and "swim bladders" are homologous, with the specific terms distinguishing the structure's role for air breathing in lungfishes and tetrapods versus buoyancy during swimming in ray-finned fishes. Zoologists continue to debate the question of whether the lung or the swim bladder was the original role of the air-filled cavity.

Freshwater habitats are inherently unstable, being prone to evaporation or depletion of the dissolved oxygen needed to support vertebrate life. It is therefore not surprising that multiple fish groups, given a combination of structures that could be coopted for terrestrial breathing and locomotion, evolved some degree of terrestriality. Mudskippers and lungfish are two familiar examples of evolution of terrestriality by fishes; however, only one such transition, occurring in the early Devonian period, provided the ancestral lineage of all tetrapod vertebrates. This lineage ultimately evolved the characteristic tetrapod adaptations to air breathing, including increased vascularization of the air-filled cavity with a rich capillary network to form an efficient lung, and a **double circulation** to direct the deoxygenated blood into the lungs for oxygenation, and the oxygenated blood from the lungs to other body tissues (see Chapter 31, p. 686).

Tetrapods evolved limbs in an ancestral aquatic habitat during the Devonian period prior to their evolutionary movement onto land. Although fish fins may seem very different from the jointed limbs of tetrapods, an examination of bony elements of the paired fins of lobe-finned fishes shows that they broadly resemble homologous structures of amphibian limbs. In *Eusthenopteron,* a Devonian lobe-fin fish living approximately 385 million years ago, we can recognize an upper arm bone (humerus) and two forearm bones (radius and ulna) as well as other elements that we homologize with wrist bones of tetrapods (Figure 25.3). *Eusthenopteron* could push itself through the bottom mud of pools with its fins, but it could not walk upright because backward and forward movement of the fins was limited to about 20 to 25 degrees. The fossil genus *Tiktaalik,* which lived approximately 375 million years ago, is morphologically intermediate between lobe-finned fishes and tetrapods. *Tiktaalik* probably inhabited shallow, oxygen-depleted streams or swamps, using its appendages to support the body while placing its snout above water to breathe air. This form also might have traversed land.

Acanthostega, one of the earliest known Devonian tetrapods, had well-formed tetrapod limbs with clearly formed digits on both foreand hindlimbs, but these limbs were too weakly constructed to hoist the animals' bodies off the surface for walking on land. *Acanthostega* and *Ichthyostega,* both approximately 365 million years old, reveal that the earliest tetrapods had more than five digits per limb and that the five-digit pattern more characteristic of living forms was stabilized later in tetrapod evolution.

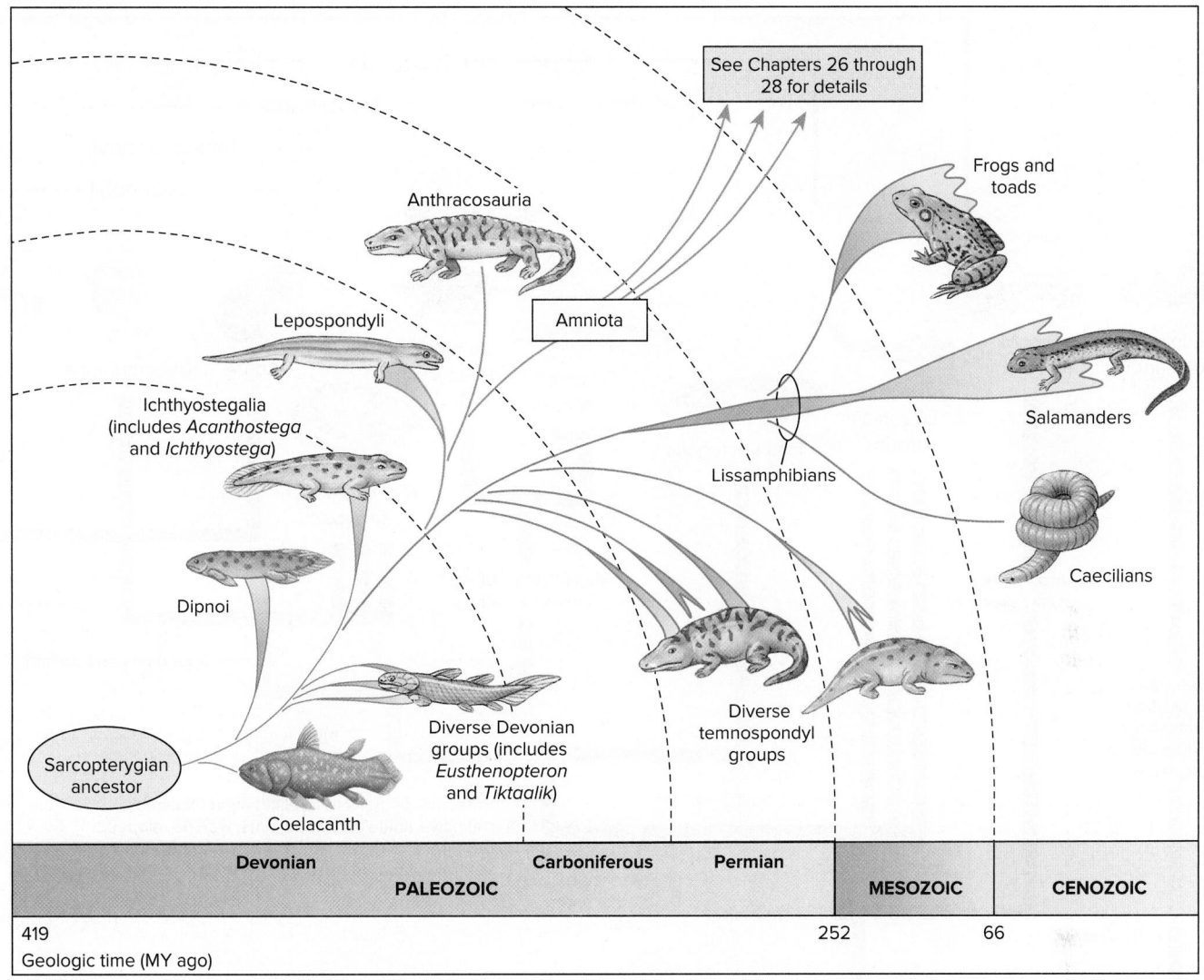

Figure 25.1 Early tetrapod evolution and the descent of amphibians. Tetrapods share most recent common ancestry with diverse Devonian groups. Amphibians share most recent common ancestry with diverse temnospondyls of the Carboniferous and Permian periods of the Paleozoic era, and the Triassic period of the Mesozoic era.

Ichthyostega (Gr. *ichthys,* fish, + *stegē,* roof, or covering, in reference to the roof of the skull, which was shaped like that of a fish) possessed several adaptations, in addition to jointed limbs, for life on land. These include stronger vertebrae and associated muscles to support the body in air, new muscles to elevate the head, strengthened shoulder and hip girdles, protective rib cage, and ears modified for detecting airborne sounds. *Ichthyostega* nonetheless resembled aquatic forms in retaining a tail complete with fin rays and in having opercular (gill-covering) bones.

Early Diversification of Tetrapods

Evolutionary relationships of early tetrapod groups remain controversial. We present a tentative cladogram (Figure 25.2), which

Bones of *Ichthyostega,* the most thoroughly studied of all early tetrapods, were first discovered on an East Greenland mountainside in 1897 by Swedish scientists looking for three explorers lost two years earlier during an ill-fated attempt to reach the North Pole by a hot-air balloon. Later expeditions by Gunnar Säve-Söderberg uncovered skulls of *Ichthyostega,* but Säve-Söderberg died at age 38 before he could study the skulls. After Swedish paleontologists returned to the Greenland site where they found the remainder of *Ichthyostega's* skeleton, Erik Jarvik, one of Säve-Söderberg's assistants, examined the skeleton in detail. This research became his life's work, producing a description of *Ichthyostega* that remains the most detailed of any Paleozoic tetrapod. Jarvik suffered a crippling stroke at age 88 in 1994, but had by then virtually completed an extensive monograph on *Ichthyostega,* which was published in 1996.

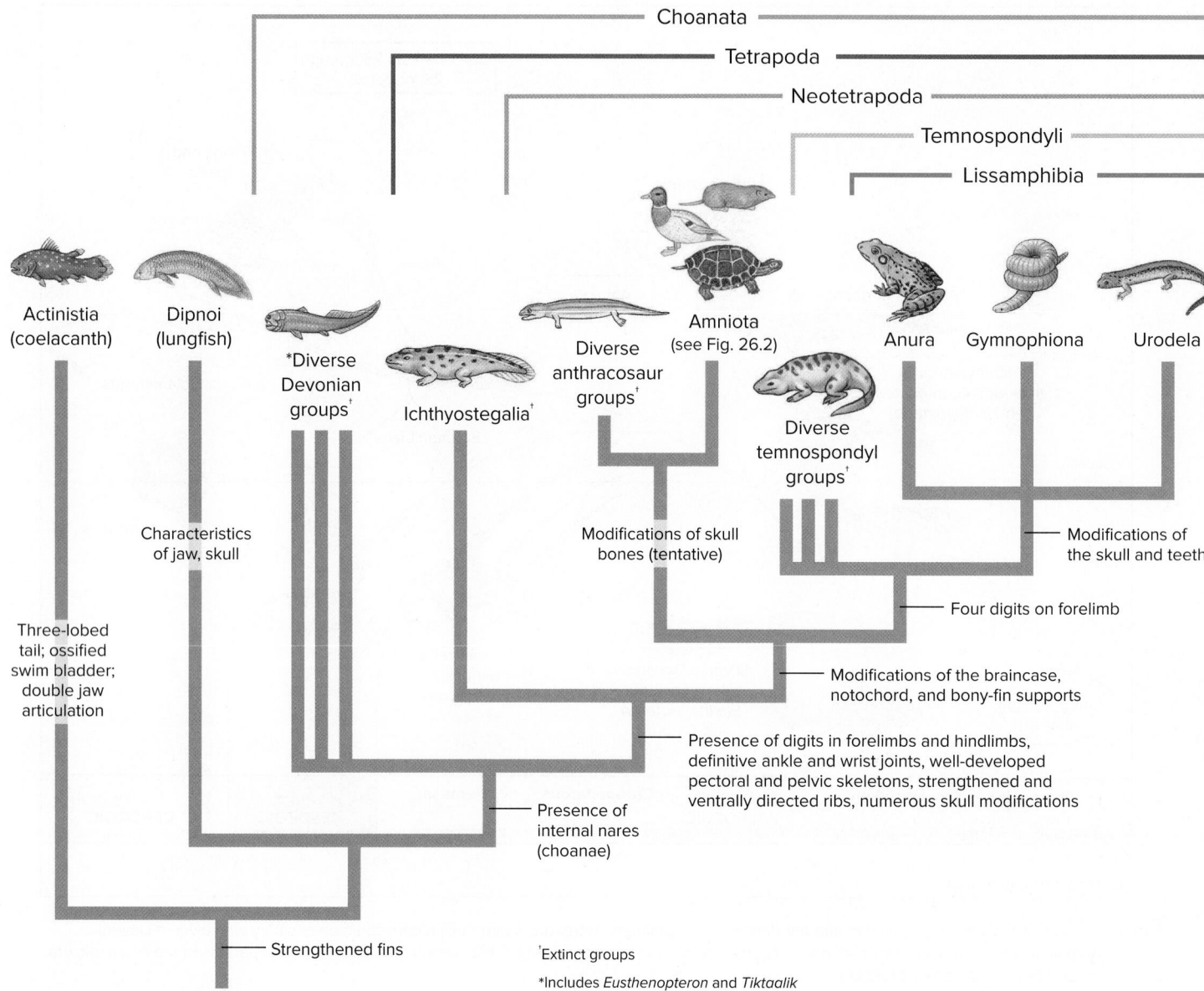

Figure 25.2 Tentative cladogram of the Tetrapoda with emphasis on descent of the amphibians. Ichthyostegalia includes *Acanthostega* and *Ichthyostega*.

almost certainly will undergo revision as new data are reported. Several extinct lineages plus **Lissamphibia,** which contains modern amphibians, form a group called **temnospondyls.** This group generally has only four digits on the forelimb rather than the five typical of most tetrapods. Lissamphibians arose during the Carboniferous period and later diversified, probably by the early Triassic period, to produce ancestors of the three major groups of amphibians alive today, **frogs** (Anura or Salientia), **salamanders** (Caudata or Urodela), and **caecilians** (Apoda or Gymnophiona).

Two additional generally recognized but nonetheless controversial groupings of Carboniferous and Permian tetrapods, **lepospondyls** and **anthracosaurs,** are judged from skull structure to be closer to amniotes than to temnospondyls (Figure 25.2). We cover the amniote branch of tetrapod phylogeny in Chapters 26 through 28.

25.2 MODERN AMPHIBIANS

The three living amphibian orders comprise approximately 7,500 species. Most share general adaptations for life on land, including skeletal strengthening.

In the ancestral amphibian life history, eggs are aquatic and hatch to produce an aquatic larval form that uses gills for respiration. A metamorphosis follows in which gills are lost. Metamorphosed amphibians use cutaneous respiration on land, and many have lungs, which exist throughout larval life and are activated for breathing air at metamorphosis. Many amphibians retain this general pattern, but important exceptions include some salamanders that lack a complete metamorphosis and retain a permanently aquatic, larval morphology throughout life. Some caecilians, some frogs, and some other

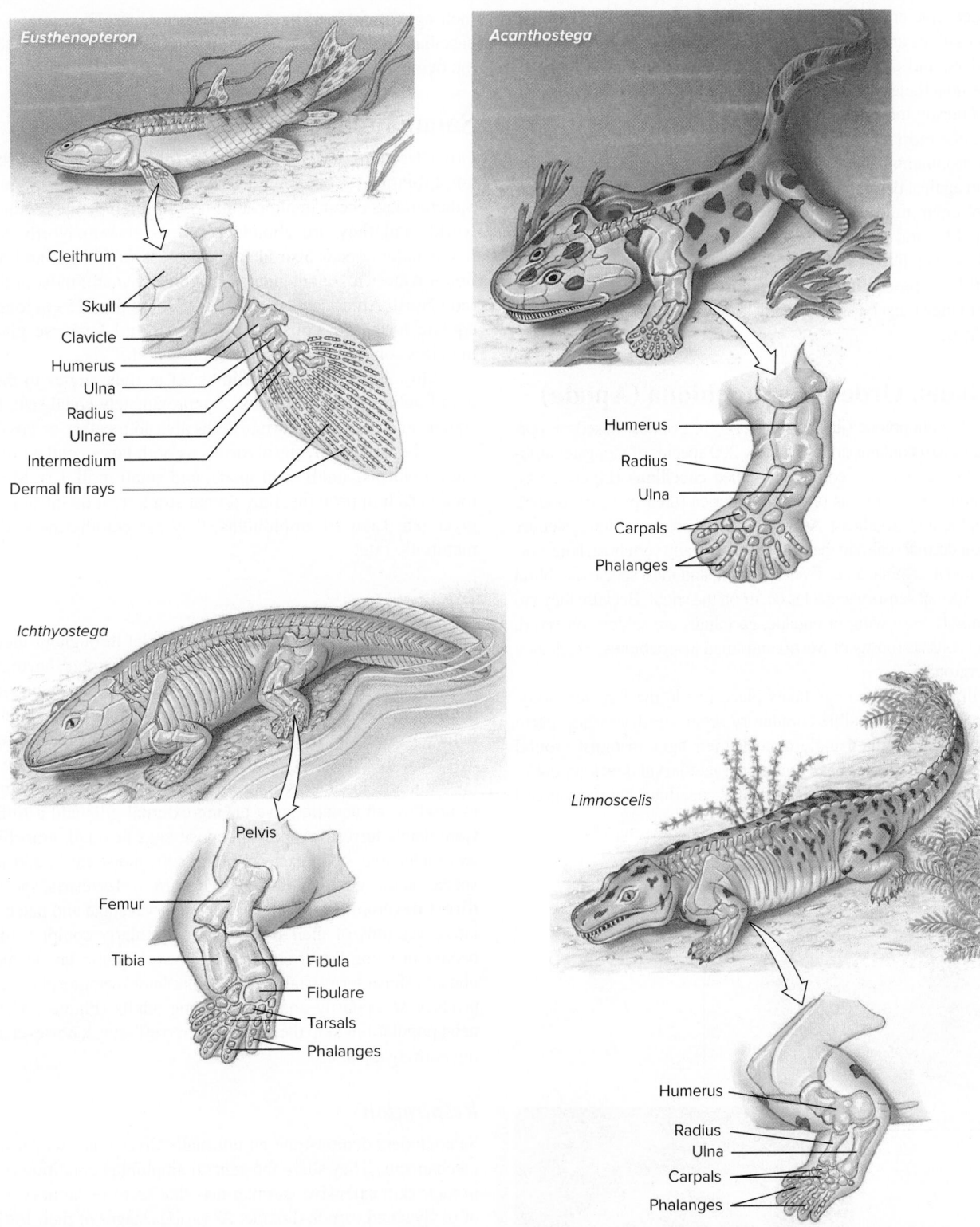

Figure 25.3 Evolution of tetrapod limbs. Limbs of tetrapods evolved from fins of Paleozoic fishes. *Eusthenopteron,* a late Devonian lobe-finned fish (rhipidistian) had paired muscular fins supported by bony elements that foreshadowed the bones of tetrapod limbs. The anterior fin contained an upper arm bone (humerus), two forearm bones (radius and ulna), and smaller elements homologous to wrist bones of tetrapods. As typical of fishes, the pectoral girdle, consisting of the cleithrum, clavicle, and other bones, was firmly attached to the skull. In *Acanthostega,* one of the earliest known Devonian tetrapods (appearing about 365 million years BP), dermal fin rays of the anterior appendage were replaced by eight fully evolved fingers. *Acanthostega* was probably exclusively aquatic because its limbs were too weak for travel on land. *Ichthyostega,* a contemporary of *Acanthostega,* had fully formed tetrapod limbs and must have been able to walk on land. The hindlimb bore seven toes (the number of front limb digits is unknown). *Limnoscelis,* an anthracosaur of the Carboniferous period (about 300 million years BP), had five digits on both front and hindlimbs, the basic pentadactyl model that became the tetrapod standard.

salamanders live entirely on land and have no aquatic larval phase. Both alternatives are evolutionarily derived conditions. Some frogs, salamanders, and caecilians that undergo a complete metamorphic life cycle nonetheless remain in water as adults rather than moving onto land during their metamorphosis.

Even the most terrestrial amphibians remain dependent on very moist environments. Their skin is thin, and it requires moisture for protection against desiccation in air. Amphibians also require moderately cool environments. As ectotherms, their body temperature is determined by and varies with the environment, greatly restricting where they can live. Cool and wet environments are especially important for reproduction. Eggs are not well protected from desiccation, and they must be shed directly into water or onto moist terrestrial surfaces.

Caecilians: Order Gymnophiona (Apoda)

The order Gymnophiona (jim´nō-fy´ō-na) (Gr. *gymnos,* naked, + *opineos,* of a snake) contains approximately 200 species of elongate, limbless, burrowing creatures commonly called **caecilians** (Figure 25.4). They inhabit tropical forests of South America (their principal home), Africa, India, and Southeast Asia. Caecilians possess a long, slender body, small dermal scales in the skin of some, many vertebrae, long ribs, no limbs, and a terminal anus. Eyes are small, and most species are blind as adults. Special sensory tentacles occur on the snout. Because they are almost entirely burrowing or aquatic, caecilians are seldom observed. Their food consists mostly of worms and small invertebrates, which they find underground.

Fertilization of the eggs takes place inside the female's body, and males have a protrusible copulatory organ for delivering sperm to the eggs. Caecilians usually deposit their eggs in moist ground near water. Some species have aquatic larvae; larval development in other species occurs within the egg. Some caecilians carefully guard

A

B

Figure 25.4 **A,** Female caecilian coiled around eggs in burrow. **B,** Pink-head caecilian (*Herpele multiplicata*), native to western Africa.

their eggs in folds of the body. Viviparity also is common in some caecilians, with embryos obtaining nourishment by eating the wall of the oviduct.

Salamanders: Order Urodela (Caudata)

Order Urodela (Gr. *oura,* tail, + *delos,* evident) comprises tailed amphibians, approximately 700 species of salamanders. Salamanders occur in almost all northern temperate regions of the world, and they are abundant and diverse in North America. Salamanders occur also in tropical areas of Central and northern South America. Salamanders are typically small; most of the common North American salamanders are less than 15 cm long. Some aquatic forms are considerably longer, and Japanese giant salamanders sometimes exceed 1.5 m in length.

Most salamanders have limbs set at right angles to the trunk, with forelimbs and hindlimbs of approximately equal size. In some aquatic and burrowing forms, limbs are rudimentary or missing.

Salamanders are carnivorous as both larvae and adults, preying on worms, small arthropods, and small molluscs. Since their food is rich in proteins, they do not store great quantities of fat or glycogen. Like all amphibians, they are ectotherms with a low metabolic rate.

Life Cycles

Some salamanders are aquatic or terrestrial throughout their entire life cycle, but the ancestral condition is metamorphic, having aquatic larvae and terrestrial adults that occupy moist places under stones and rotten logs. Eggs of most salamanders are fertilized internally; a female recovers in her cloaca a packet of sperm (**spermatophore**) deposited by a male on a leaf or stick (Figure 25.5). Aquatic species lay their eggs in clusters or stringy masses in water. Their eggs hatch to produce an aquatic larva having external gills and a finlike tail. Completely terrestrial species deposit eggs in small, grapelike clusters under logs or in excavations in soft, moist earth, and in many species adults guard their eggs (Figure 25.6). Terrestrial species have **direct development:** they bypass the larval stage and hatch as miniature versions of their parents. A particularly complex life cycle occurs in some American newts, whose aquatic larvae metamorphose to form terrestrial juveniles that later metamorphose again to produce secondarily aquatic, breeding adults (Figure 25.7). Many newt populations skip the terrestrial "red eft" stage, however, remaining entirely aquatic.

Respiration

Salamanders demonstrate an unusually diverse array of respiratory mechanisms. They share the general amphibian condition of having in their skin extensive vascular nets that serve respiratory exchange of oxygen and carbon dioxide. At various stages of their life history, salamanders also may have external gills, lungs, both, or neither gills nor lungs. Salamanders that have an aquatic larval stage have gills as larvae but lose them later if a metamorphosis occurs. Several lineages of salamanders have evolved permanently aquatic forms that fail to complete metamorphosis and retain their gills and finlike tail throughout life. Lungs, the most widespread respiratory organ of terrestrial vertebrates, are present from birth in salamanders that have them, and lungs become active as the primary means of respiration following metamorphosis.

Characteristics of Modern Amphibians

1. **Limbs usually four (quadrupedal)** in two pairs with associated shoulder/hip girdle, although some salamanders have forelimbs only and caecilians have no limbs; no true nails; feet often webbed; **forelimb usually with four digits** and hindlimb with five. **Skeleton mostly bony** with varying numbers of vertebrae; salamanders usually have distinct head, neck, trunk, and tail; adult frogs have a fused head and trunk and usually no tail; caecilians have an elongated trunk not strongly demarcated from the head and a terminal anus; **ectothermic**

2. **Skin smooth, moist, and glandular**; integument modified for **cutaneous respiration**; pigment cells (chromatophores) common and varied; **granular glands** secrete defensive compounds

3. Skull relatively lighter, less ossified, flattened in profile and with fewer bones than other vertebrates

4. Mouth usually large with small teeth in upper or both jaws and on the vomer/palate

5. **Tripartite brain** includes forebrain (telencephalon) coordinating olfaction, midbrain (mesencephalon) coordinating vision, and hindbrain (rhombencephalon) coordinating hearing and balance; ten pairs of cranial nerves

6. Ear with **tympanic membrane** (eardrum) and **stapes** (columella) for transmitting vibrations to inner ear; for vision in air, cornea rather than lens is principal refractive surface for bending light; **eyelids** and **lachrymal glands** protect and wash eyes; paired **internal nostrils** open into a nasal cavity lined with **olfactory epithelium** at anterior part of mouth cavity

7. Separate sexes; fertilization mostly external in frogs and toads but internal via a spermatophore in most salamanders and caecilians; predominantly oviparous, but some ovoviviparous or viviparous

8. **Eggs moderately yolky (mesolecithal) with jellylike membrane coverings**; aquatic larva often present with metamorphosis to a more terrestrial adult form

9. Excretory system of paired mesonephric or opisthonephric kidneys; urea main nitrogenous waste

10. Respiration by skin and in some forms by gills and/or lungs; presence of gills and lungs varies among species and by developmental state of some species; forms with aquatic larvae lose gills at metamorphosis in frogs; many salamanders retain gills and an aquatic habitat throughout life; paired nostrils enable breathing in lung-breathing forms; vocal cords present between lungs and vocal sacs, especially in frogs

11. **Heart with a sinus venosus,** two atria, one ventricle, a conus arteriosus; **double circulation** through the heart in which pulmonary arteries and veins supply lungs (when present) and return oxygenated blood to heart; skin abundantly supplied with blood vessels

Female   Male

Spermatophore

A

Base of spermatophore beneath female's vent

B

© Dr. Lynne Houck

Figure 25.5 Courtship and sperm transfer in pygmy salamanders, *Desmognathus wrighti*. After judging the female's receptivity by the presence of her chin on his tail base, the male deposits a spermatophore on the ground, then moves forward a few paces. **A,** The white mass of the sperm atop a gelatinous base is visible at the level of the female's forelimb. The male moves ahead, the female following until the spermatophore is at the level of her vent. **B,** The female has recovered the sperm mass in her vent, while the male arches his tail, tilting the female upward and presumably facilitating recovery of the sperm mass. The female later uses the sperm stored in her body to fertilize eggs internally before laying them.

Although we normally associate lungs with terrestrial organisms and gills with aquatic ones, salamander evolution has produced aquatic forms that breathe primarily with lungs and terrestrial forms that lack them completely.

Amphiumas of the salamander family Amphiumidae have evolved a completely aquatic life history with a greatly reduced metamorphosis. Amphiumas nonetheless lose their gills before adulthood and then breathe primarily by lungs. They must raise their nostrils above the surface of the water to breathe.

In contrast to amphiumas, all species of family Plethodontidae are lungless, and many of these species are strictly terrestrial. This large family contains more than 450 species, including many familiar

Figure 25.6 Female dusky salamander (*Desmognathus* sp.) attending eggs. Some salamanders exercise parental care of eggs, which includes rotating eggs and protecting them from fungal infections and predation by various arthropods and other salamanders.

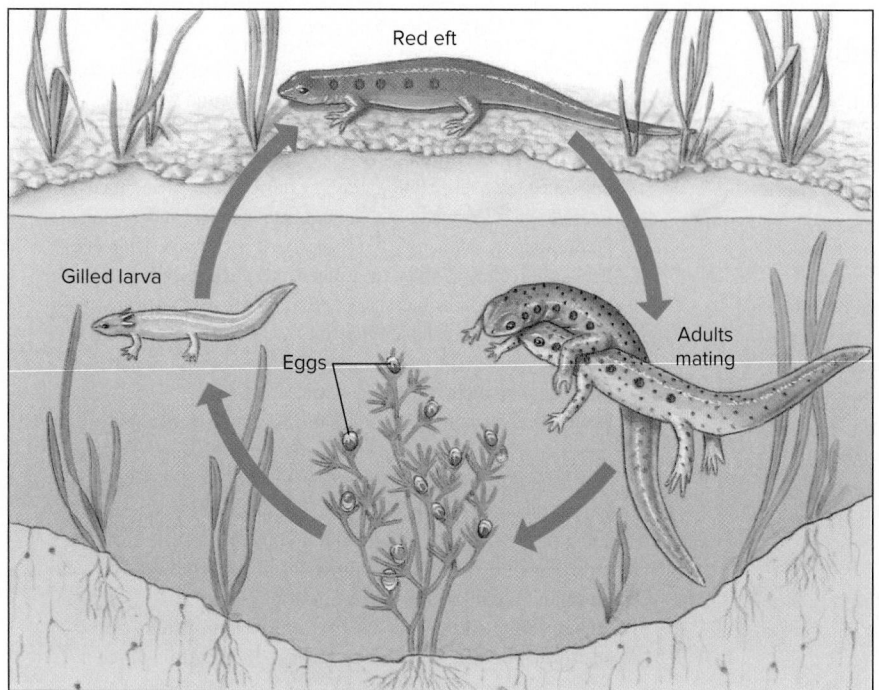

Figure 25.7 Life history of a red-spotted newt, *Notophthalmus viridescens* of the family Salamandridae. In many populations the aquatic larva metamorphoses into a brightly colored "red eft" stage, which remains on land from 1 to 3 years before transforming into a secondarily aquatic adult.

North American salamanders (Figures 25.5, 25.6, and 25.8). The efficiency of cutaneous respiration is increased by penetration of a capillary network into the epidermis or by thinning of the epidermis over superficial dermal capillaries. Cutaneous respiration is supplemented by pumping air through the mouth, where respiratory gases are exchanged across the vascularized membranes of the buccal (mouth) cavity (buccopharyngeal breathing). Evolutionary loss of lungs probably occurred in an ancestral plethodontid lineage that occupied swift streams where lungs would have conferred too much buoyancy, and where water is so cool and well oxygenated that cutaneous respiration suffices. Some plethodontids have aquatic larvae whose gills are lost at metamorphosis. Others retain a permanently larval form with gills throughout life. Many others are completely terrestrial and bear the distinction of being the only vertebrates to have neither lungs nor gills at any stage of their life history. It is odd that the most completely terrestrial salamanders evolved in a group that lacks lungs.

Paedomorphosis

A persistent phylogenetic trend observed in salamander evolution is for descendants to retain into adulthood features that occurred only in preadult stages of their ancestors. Some characteristics of ancestral adult morphology are consequently eliminated. This condition is called **paedomorphosis** (Gr. "child form"; see Chapter 6, p. 112). The most dramatic form of paedomorphosis occurs in species that become sexually mature while retaining their gills, aquatic life habit, and other larval characteristics.

These nonmetamorphic species are called **perennibranchiate** ("permanently gilled"). Mud puppies of genus *Necturus* (Figure 25.9), which inhabit submerged substrates in ponds and lakes, are an extreme example. These and many other salamanders are obligately perennibranchiate; they do not metamorphose under any conditions.

Some other species of salamanders reach sexual maturity with larval morphology but, unlike *Necturus*, metamorphose to terrestrial forms under certain environmental conditions. *Ambystoma* species from Mexico and the United States include naturally transforming types, nontransforming types, and nontransforming types that can be made to transform by experimental treatment with the thyroid hormone, thyroxine (T_4). Gilled individuals are called **axolotls** (Figure 25.9). Their typical habitat includes small ponds that can disappear through evaporation in dry weather. When its pond evaporates, an axolotl metamorphoses to a terrestrial form, losing its gills and breathing with lungs. It can then travel across land in search of new ponds, to which it must return to reproduce. Axolotls are forced to metamorphose artificially when they are treated with thyroxine (T_4). Thyroid hormones (T_3 and T_4) are essential for amphibian metamorphosis. The pituitary gland appears not to become fully active in non-metamorphosing forms, thereby failing to release thyroid-stimulating hormone (TSH, p. 753), which causes the thyroid gland to produce thyroid hormones.

Paedomorphosis takes many different forms in different salamanders. It alters the life history and shape of the whole body in some species, whereas in other species it alters only one or a few structures. The amphiumas lose their gills and activate their lungs before maturity, but they retain many general features of larval body form. Paedomorphosis is important even in terrestrial

Figure 25.8 Longtail salamander, *Eurycea longicauda*, a common plethodontid salamander.

A Common mud puppy
(*Necturus maculosus*)

B Axolotl (*Ambystoma mexicanum*)

Figure 25.9   Paedomorphosis in salamanders. **A,** The mud puppy
Necturus sp., is a permanently gilled (perennibranchiate) aquatic form.
B, An axolotl (*Ambystoma mexicanum*) may remain permanently gilled, or,
should its pond habitat evaporate, metamorphose to a terrestrial form that
loses its gills and breathes by lungs. The axolotl shown is an albino form
commonly used in laboratory experiments but uncommon in natural
populations.

plethodontids, which never have an aquatic larval stage. One sees
the effects of paedomorphosis, for example, in the shapes of the
hands and feet of tropical plethodontid salamanders of genus
Bolitoglossa (Figure 25.10). Ancestral morphology of *Bolitoglossa*
features well-formed digits that grow outward from the pad of the
hand or foot during development. Some species have enhanced their
ability to climb smooth vegetation, such as banana trees, by halting
growth of the digits and retaining throughout life a padlike foot.
This padlike foot can produce adhesion and suction to attach the
salamander to smooth vertical surfaces, and it thereby serves an
important adaptive function.

Frogs and Toads: Order Anura (Salientia)

The approximately 6600 species of frogs and toads that compose
order Anura (Gr. *an,* without, + *oura,* tail) are for most people the
most familiar amphibians. Anura is an old group, known from the
Jurassic period, 190 million years ago. Frogs and toads occupy a
great variety of habitats. Their aquatic mode of reproduction and
water-permeable skin prevent them from wandering too far from
sources of water, however, and their ectothermy bars them from
polar and subarctic habitats. The name of the order, Anura, denotes
an obvious group characteristic, absence of tails in adults. Although
all have a tailed larval stage during embryonic or larval development,
only genus *Ascaphus* contains a tail-like structure in adults. Frogs
and toads are specialized for jumping, as suggested by the alternative
order name, Salientia, which means leaping.

We see in the appearance and life habit of their larvae further
distinctions between Anura and Caudata. Eggs of most frogs hatch
into a tadpole having a long, finned tail, both internal and external
gills, no legs, specialized mouthparts for herbivorous feeding (salamander larvae and some tadpoles are carnivorous), and a highly specialized internal anatomy. They scarcely resemble adult frogs.
Metamorphosis of a frog tadpole to an adult frog is thus a striking
transformation. The perennibranchiate condition never occurs in
frogs and toads as it does in salamanders.

Taxonomists recognize 45 families of frogs and toads. The best-
known frog families in North America are Ranidae, which contains
most of our familiar frogs (Figure 25.11A), and Hylidae, tree frogs
(Figure 25.11B). True toads, belonging to family Bufonidae, have
short legs, stout bodies, and thick skins, usually with prominent
warts (Figure 25.12). However, the term "toad" is used informally
also for some terrestrial members of several other families.

The largest anuran is the West African *Conraua goliath,* which
is more than 30 cm long from tip of nose to anus (Figure 25.13). This
giant eats animals as big as rats and ducks. The smallest frogs
recorded are *Eleutherodactylus iberia* and *Psyllophryne didactyla,*

Figure 25.10   Foot structure of
representatives of three different species of
the tropical plethodontid salamander genus
Bolitoglossa. These specimens have been
treated chemically to clear the skin and
muscles and to stain the bone red/pink and
cartilage blue. The species having the most
fully ossified and distinct digits **(A, C)** live
primarily on the forest floor. The species
having the padlike foot caused by restricted
digital growth **(B)** climbs smooth leaves and
stems using the foot surface to produce
suction or adhesion for attachment. The
padlike foot evolved by paedomorphosis; it
was derived evolutionarily by truncating
development of the body, which prevents full
digital development.

© Allan Larson

Figure 25.11 Two common North American frogs. **A,** Bullfrog, *Lithobates catesbeianus,* largest American frog and mainstay of the frogleg epicurean market (family Ranidae). **B,** Green tree frog, *Hyla cinerea,* a common inhabitant of swamps of the southeastern United States (family Hylidae). Note adhesive pads on the feet.

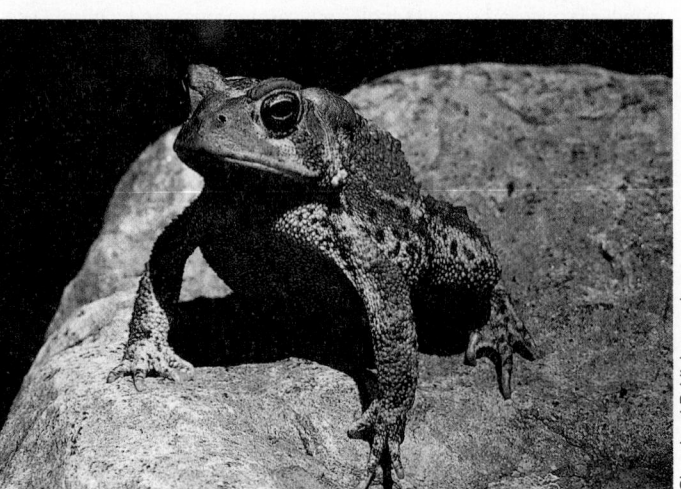

Figure 25.12 American toad, *Bufo americanus* (family Bufonidae). This principally nocturnal yet familiar amphibian feeds on large numbers of insect pests and on snails and earthworms. The rough skin contains numerous glands that produce a surprisingly poisonous milky fluid, providing excellent protection from a variety of potential predators.

Figure 25.13 *Conraua (Gigantorana) goliath* (family Petropedetidae) of West Africa, the world's largest frog. This specimen weighed 3.3 kg (approximately 7½ pounds).

measuring less than 1 cm in length; they are also the smallest known tetrapods. These tiny frogs, which can be covered by a dime, live respectively in Cuba and in the Brazilian rain forest. The largest American frog is the bullfrog, *Lithobates catesbeianus* (Figure 25.11A), which reaches a head and body length of 20 cm.

Habitats and Distribution

Probably the most familiar frogs are the species of family Ranidae (Gr. frog), found throughout temperate and tropical regions of the world except in New Zealand, the oceanic islands, and southern South America. They usually live near water, although some, such as wood frogs, *Lithobates sylvatica,* spend most of their time on damp forest floors. Wood frogs probably return to pools only for breeding in early spring. The larger bullfrogs, *Lithobates catesbeianus,* and

green frogs, *Lithobates clamitans,* are nearly always in or near permanent water or swamps. Leopard frogs, *Lithobates pipiens* and related species, have a wider variety of habitats and are the most widespread of North American frogs; they are commonly used in biology laboratories and for classical electrophysiological research. They occur in some form in nearly every state, and extend far into northern Canada and as far south as Panama.

Most larger frogs are solitary except during breeding season. During breeding periods most of them, especially males, are very noisy. Each male usually occupies a particular perch near water, where he may remain for hours or even days, trying to attract a female to that spot.

At times frogs are mainly silent, and their presence is not detected until they are disturbed. When they enter the water, they dart swiftly to the bottom of the pool and kick the substrate to conceal themselves in a cloud of muddy water. In swimming, they hold their forelimbs near their body and kick backward with their webbed hindlimbs, propelling themselves forward. When they surface to breathe, only the head and foreparts are exposed, since they conceal themselves behind any available vegetation.

Amphibian populations are declining in numbers in various parts of the world, although many species continue to thrive. Epidemic infections by chytrid fungi and losses of habitat explain many declines. Climatic changes that reduce water depth at oviposition sites increase ultraviolet exposure of embryos and make them more susceptible to fungal infection. Declines in population survival may include an increased incidence of malformed individuals, such as frogs with extra limbs. Malformed limbs are often associated with infection by trematodes (p. 299).

In contrast to these declines, at least two introduced amphibian populations have prospered to the point of being considered pests. African clawed frogs *Xenopus laevis* (Figure 25.14), following release from biomedical laboratories, are now well established in southern California. The large marine toad *Rhinella marina*, introduced to Queensland, Australia, to control agricultural pests, has now spread beyond agricultural fields and is considered a pest.

During winter months most frogs in temperate climates hibernate in the oxygen-rich water of pools and streams. Their life processes are at a very low ebb during hibernation, sustained by diffusion of oxygen across the skin and the energy derived from glycogen and fat stored in their bodies during spring and summer. More terrestrial frogs, such as tree frogs, hibernate in humus of the forest floor. They tolerate low temperatures, and many actually survive freezing all extracellular fluid, representing 35% of the body water. Such frost-tolerant frogs prepare for winter by accumulating glucose and glycerol in body fluids, thereby protecting tissues from the normally damaging formation of ice crystals.

Adult frogs have numerous enemies, including snakes, aquatic birds, turtles, raccoons, and humans; fish prey on tadpoles, and only a few tadpoles survive to maturity. Although most frogs and toads are defenseless, many in the tropics and subtropics are aggressive, jumping and biting at predators. Some defend themselves by feigning death. Most anurans can inflate their lungs so that they are difficult to swallow. When disturbed along the margin of a pond or brook,

a frog either remains quite still or jumps into concealing vegetation. When held, a frog may cease its struggles for an instant and then suddenly leap, voiding its urine. A frog's best protection is its ability to leap and, in some species, to use poison glands. Dendrobatid frogs use potent toxins for defense.

Integument and Coloration

A frog's skin is thin and moist, and it is attached loosely to the body only at certain points. Histologically the skin contains two layers: an outer stratified **epidermis** and an inner spongy **dermis** (Figure 25.15). The epidermal layer, which is shed periodically when a frog or toad "molts," contains deposits of **keratin,** a tough, fibrous protein that limits abrasion and loss of water from the skin. More terrestrial amphibians such as toads have especially heavy deposits of keratin, although amphibian keratin is soft, unlike the hard keratin that forms scales, claws, feathers, horns, and hair of amniotes.

The epidermis produces two types of integumentary glands that grow into the loose dermal tissues below. Small mucous glands secrete a protective mucous waterproofing onto the skin surface, and large granular glands produce a whitish, watery poison highly irritating to predators. All amphibians produce a skin poison, but its effectiveness varies among species and with different predators. A western Colombian tribe uses the extremely toxic poison of three species of *Phyllobates,* a genus of small South American dendrobatid frogs, to poison points of blowgun darts. Most species of the family Dendrobatidae produce toxic skin secretions, some of which are among the most lethal animal secretions known, more dangerous even than venoms of sea snakes or any of the most venomous arachnids.

A frog's skin color is produced, as in other amphibians, by special pigment cells, **chromatophores,** located mainly in the dermis. Amphibian chromatophores, like those of many other vertebrates, are branched cells containing pigment that may be concentrated in a small area or dispersed throughout the branching processes to control skin coloration (Figure 25.16; see also p. 642). Most amphibians have three types of chromatophores: uppermost in the dermis are **xanthophores,** containing yellow, orange, or red pigments; beneath these lie **iridophores,** containing a silvery, light-reflecting pigment; and lowermost are **melanophores,** containing black or brown melanin. Iridophores act like tiny mirrors, reflecting light back through the xanthophores to produce the bright colors of many tropical frogs. Surprisingly perhaps, green hues so common in North American

Figure 25.14 African clawed frog, *Xenopus laevis.* The claws, an unusual feature in frogs, are on the hind feet. This frog has been introduced into California, where it is considered a serious pest.

© Cleveland P. Hickman, Jr.

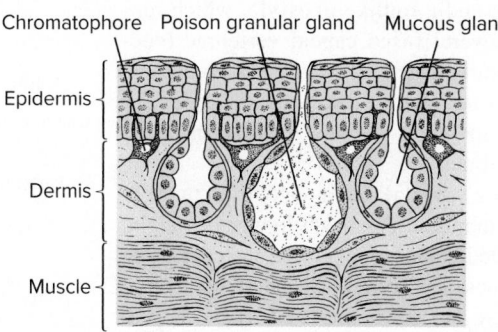

Figure 25.15 Section through frog skin.

Figure 25.16   Pigment cells (chromatophores). **A,** Pigment dispersed. **B,** Pigment concentrated. The pigment cell does not contract or expand; color effects are produced by streaming of cytoplasm, carrying pigment granules into cell branches for maximum color effect or to the center of the cell for minimum effect. Control over dispersal or concentration of pigment is mostly by light stimuli acting through a pituitary hormone.

frogs are produced not by green pigment but by an interaction of xanthophores containing a yellow pigment and underlying iridophores that, by reflecting and scattering light (Tyndall scattering), produce a blue color. Blue light is filtered by the overlying yellow pigment and thus appears green. Many frogs can adjust their color to match their background and thus to camouflage themselves (Figure 25.17).

Skeletal and Muscular Systems

Amphibians, like other vertebrates, have a well-developed **endoskeleton** of bone and cartilage to support the body and its muscular movements. Movement onto land required limbs capable of supporting the body's weight and introduced a new set of stress and leverage forces. The entire musculoskeletal system of an adult frog is specialized for jumping and swimming by simultaneous extensor thrusts of the hindlimbs.

The amphibian vertebral column assumes a new role as a support from which the abdomen is suspended and to which limbs attach. It is a rigid frame for transmitting force from the hindlimbs to the body. Anurans are further specialized by an extreme shortening of the body. Typical frogs have only nine trunk vertebrae and a rodlike **urostyle,** which represents several fused caudal vertebrae (coccyx) (Figure 25.18). The limbless caecilians, which do not share these specializations for tetrapod locomotion, may have as many as 285 vertebrae.

A frog's skull differs greatly from those of other vertebrates; it is much lighter in weight, less ossified, flattened in profile, and containing fewer bones. The front of the skull, containing the nose, eyes, and brain, is better developed, whereas the back of the skull, which contains the gill apparatus in fishes, is much reduced (Figure 25.18).

Figure 25.17   Cryptic coloration of the gray tree frog, *Hyla versicolor.* Camouflage is so good that presence of this frog usually is disclosed only at night by its resonant, flutelike call.

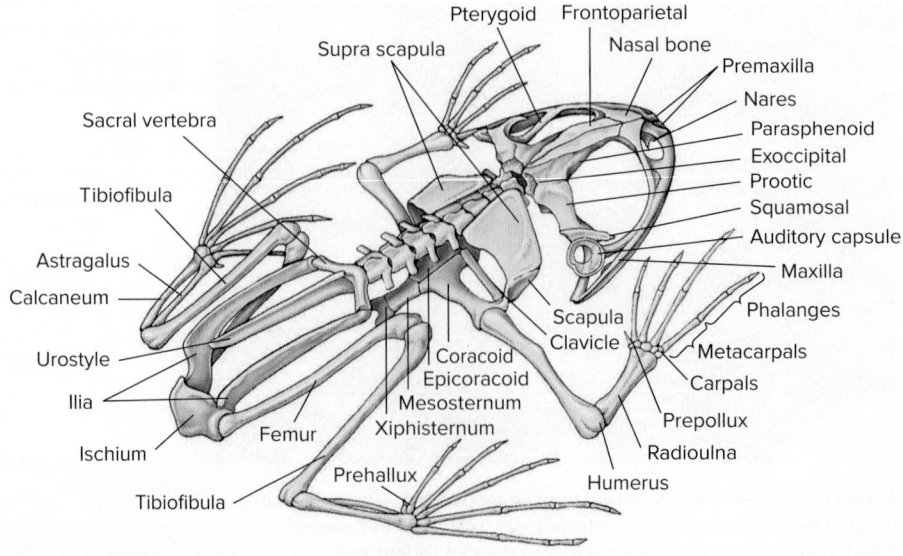

Figure 25.18   Skeleton of a bullfrog, *Lithobates catesbeiana.*

Limbs are of typical tetrapod pattern in having three main joints (hip, knee, and ankle; or shoulder, elbow, and wrist). The foot is typically five-rayed (pentadactyl) and the hand is four-rayed, with both foot and hand having several joints in each digit (Figure 25.18). It is a repetitive system resembling the bone structure of lobefins, which are distinctly suggestive of amphibian limbs (see Figure 25.3). One can easily imagine how selective pressures generated by terrestrial locomotion remodeled ancestral lobefins into limbs.

Limb muscles are presumably homologous to radial muscles that move the fins of fishes, but the muscular arrangement has become so complex in tetrapod limbs that its exact correspondence with fin musculature is unclear. Despite this complexity, we can identify two major groups of muscles on any limb: an anterior and ventral group that pulls the limb forward and toward the midline (protraction and adduction), and a second set of posterior and dorsal muscles that draws the limb back and away from the body (retraction and abduction).

Trunk musculature, which in fishes is segmentally organized into powerful muscular bands (myomeres, p. 531) for locomotion by lateral flexion, has been much modified during amphibian evolution. Dorsal (epaxial) muscles are arranged to support the head and to brace the vertebral column. Ventral (hypaxial) muscles are used almost exclusively for locomotion in fishes, caecilians, and salamanders, but in frogs they serve mainly to compress the abdomen for breathing and for flexing the trunk.

Respiration and Vocalization

Amphibians use three respiratory surfaces for gas exchange in air: skin (cutaneous breathing), mouth (buccal breathing), and lungs. Frogs and toads depend more heavily than do salamanders on lung breathing; nevertheless, skin provides an important supplementary avenue for gas exchange in anurans, especially during hibernation in winter. Even when lung breathing predominates, carbon dioxide is lost primarily across the skin while oxygen is absorbed primarily across the lungs.

Lungs are supplied by pulmonary arteries (derived from the sixth aortic arches) and blood returns directly to the left atrium by pulmonary veins. Frog lungs are ovoid, elastic sacs with their inner surfaces divided into a network of septa subdivided into small terminal air chambers called faveoli. Faveoli of frog lungs are much larger than alveoli of amniote vertebrates, and consequently frog lungs have a smaller relative surface available for gas exchange: the respiratory surface of *Lithobates pipiens* is about 20 cm^2 per cubic centimeter of air contained, compared with 300 cm^2 for humans. The major challenge in lung evolution was not development of a good internal vascular surface, but rather the mechanism of moving air. A frog is a positive-pressure breather that fills its lungs by forcing air into them; this system contrasts with the negative-pressure system of amniotes (see Chapter 31, p. 694). A frog's breathing is explained in Figure 25.19. One can easily follow this sequence in a living frog at rest: rhythmical throat movements of mouth breathing occur continuously before flank movements indicate that the lungs are being emptied and refilled.

Vocal cords, located in the **larynx,** or voice box, are much more developed in male frogs than in females. A frog produces sound by passing air back and forth over the vocal cords between the lungs and a large pair of sacs (vocal pouches) in the floor of the mouth. The vocal sacs also serve as effective resonators in males, which use their voices to attract mates. Most species have unique sound patterns.

Figure 25.19 Breathing in a frog. Frogs are positive-pressure breathers that fill their lungs by forcing air into them. **A,** Floor of mouth is lowered, drawing air in through nostrils. **B,** With nostrils closed and glottis open, the frog forces air into its lungs by elevating the floor of its mouth. **C,** With the glottis closed, the mouth cavity may rhythmically ventilate for a period. **D,** Lungs are emptied by contraction of body-wall musculature and by elastic recoil of lungs.

Circulation

Amphibian circulation is a closed system of arteries and veins serving a vast peripheral network of capillaries through which blood is forced by a single pressure pump, the heart (see Chapter 31, p. 686). The principal differences in circuitry relative to fishes involve the shift from gill to lung breathing. Elimination of gills removes a major obstacle to blood flow from the arterial circuit, but lung breathing introduces two new evolutionary challenges. The first challenge is to provide a blood circuit to the lungs. As we have seen, this challenge was met by converting the sixth aortic arch into pulmonary arteries to serve the lungs and by developing new pulmonary veins for returning oxygenated blood to the heart (p. 687). The second evolutionary challenge was to separate pulmonary circulation from the rest of the body's circulation, so that oxygenated blood from the lungs is sent to the body (the systemic circuit) and deoxygenated venous blood returning from the body is sent to the lungs (the pulmonary circuit). Tetrapods have solved the problem by evolving a partition down the center of the heart, creating a double pump, one for each circuit. However, partitioning is only partial in amphibians and most nonavian reptiles compared to birds and mammals, which have the most extensively divided hearts containing two atria and two ventricles.

Frog hearts (Figure 25.20) have two separate atria and a single undivided ventricle. Blood from the body (systemic circuit) first enters a large receiving chamber, the sinus venosus, from which

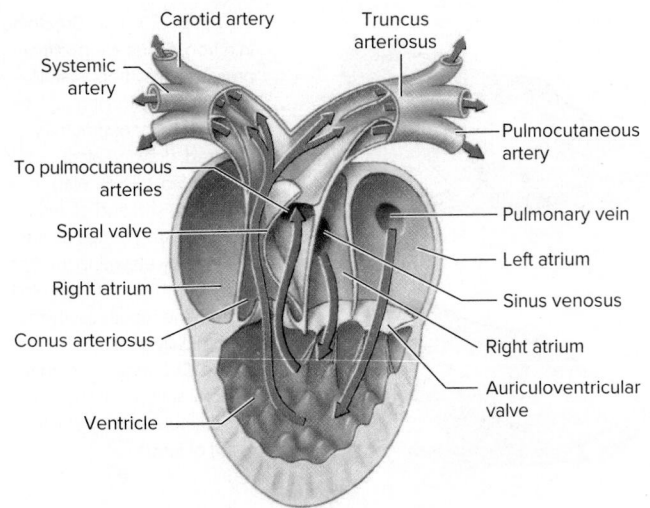

Figure 25.20 Structure of a frog heart. *Red arrows*, oxygenated blood. *Blue arrows*, deoxygenated blood.

blood is aspirated into the right atrium. The left atrium receives freshly oxygenated blood from the lungs and skin. The right and left atria contract asynchronously so that although the **ventricle** is undivided, blood remains mostly separated when it enters this chamber. When the ventricle contracts, oxygenated pulmonary blood enters the systemic circuit and deoxygenated systemic blood enters the pulmonary circuit. This separation is aided by a **spiral valve,** which divides the systemic and pulmonary flows in the **conus arteriosus** (Figure 25.20), and by different blood pressure in the pulmonary and systemic blood vessels leaving the conus arteriosus.

Feeding and Digestion

Frogs are carnivorous as adults, as are most other amphibians, and they feed on insects, spiders, worms, slugs, snails, millipedes, and nearly anything else that moves and is small enough to swallow whole. They snap at moving prey with their protrusible tongue, which is attached to the front of the mouth and is free at the posterior end. The highly glandular free end of the tongue produces a sticky secretion that adheres to prey. When teeth are present on the premaxillae, maxillae, and vomers (roof of the mouth), they are used to grasp prey, not for chewing. The digestive tract is relatively short in adult amphibians, a characteristic of most carnivores, and it produces a variety of enzymes for digesting proteins, carbohydrates, and fats.

Larval stages of anurans (tadpoles) are usually herbivorous, feeding on pond algae and other vegetable matter; they have a relatively long digestive tract because digestion of plant matter requires time-consuming fermentation before useful products can be absorbed.

Nervous System and Special Senses

Three fundamental parts of the brain—forebrain (telencephalon) coordinating olfaction, midbrain (mesencephalon) coordinating vision, and hindbrain (rhombencephalon) coordinating hearing and balance—show dramatic evolutionary trends in tetrapods (pp. 729–732).

Cephalization increases with emphasis on information processing by the brain and a corresponding loss of independence of the spinal ganglia, which are capable only of stereotyped reflexive behavior. Nonetheless, a frog with only the spinal cord and not the brain intact maintains normal body posture and can accurately raise its leg to wipe an irritant from its skin. It even uses the opposite leg if the closer leg is held.

The forebrain (Figure 25.21) contains the olfactory center, which assumes greatly increased importance on land for detecting dilute airborne odors. The sense of smell is one of the dominant special senses in frogs. The remainder of the forebrain, or cerebrum, has little importance in amphibians. Instead, complex integrative activities of frogs occur in the midbrain optic lobes. The hindbrain is divided into an anterior cerebellum and a posterior medulla. The cerebellum (Figure 25.21), which coordinates equilibrium and movement, is not well developed in amphibians. All sensory neurons except those of vision and smell pass through the medulla, located at the anterior end of the spinal cord. Here are located centers for auditory reflexes, respiration, swallowing, and vasomotor control.

Evolution of a semiterrestrial life by amphibians features a reordering of sensory receptor priorities on land. The pressure-sensitive

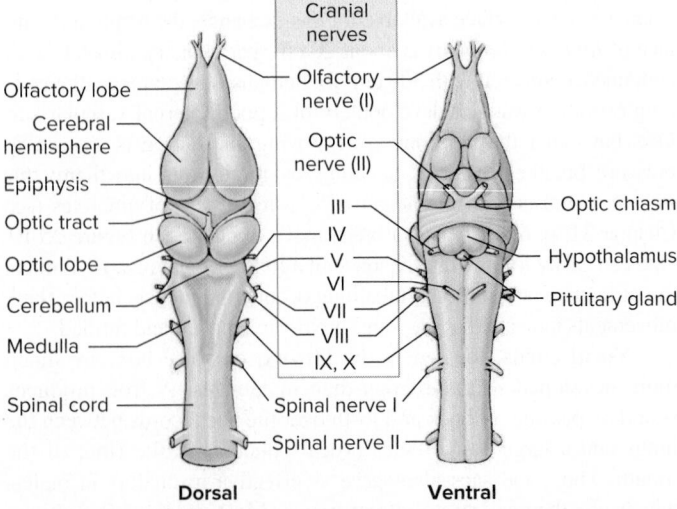

Figure 25.21 Brain of a frog, dorsal and ventral views.

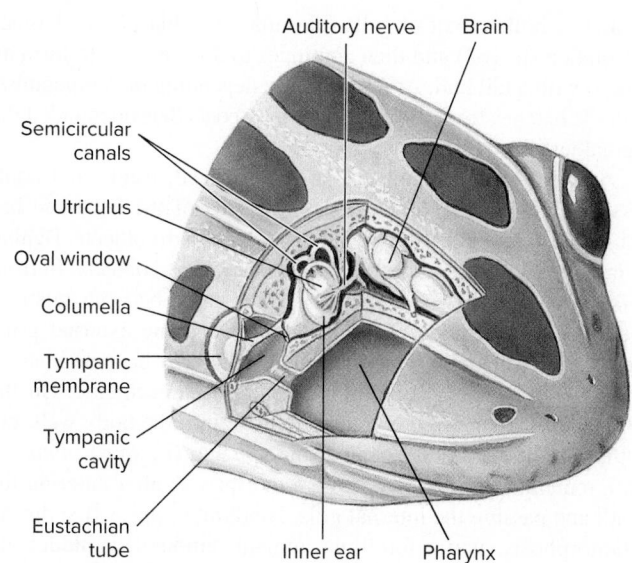

Figure 25.22 Cutaway of frog head showing ear structure. Sound vibrations are transmitted from the tympanic membrane by way of the columella to the inner ear. The eustachian tube allows pressure equilibration between the tympanic cavity and the pharynx.

lateral-line (acousticolateral) system of fishes remains only in aquatic larvae of amphibians and in a few strictly aquatic adult amphibian species. This system serves no purpose on land, because it relies on reflected pressure waves to detect and to localize objects in water.

A frog's ear is by amniote standards a simple structure: a middle ear closed externally by a large **tympanic membrane** (eardrum) and containing a **columella** (stapes) that transmits vibrations to the inner ear (Figure 25.22). The latter contains a **utricle,** from which arise three semicircular canals, and a **saccule** bearing a diverticulum, or **lagena.** The lagena is partly covered with a **tectorial membrane,** which in its fine structure resembles the much more complex mammalian cochlea (see Chapter 33, p. 739). In most frogs this structure senses low-frequency sound energy not greater than 4000 Hz (cycles per second); in bullfrogs the main frequency response is in the 100 to 200 Hz range, which matches the energy of a male frog's low-pitched call.

Vision is the dominant special sense in many amphibians (the mostly blind caecilians are obvious exceptions). Several modifications of ancestral aquatic eyes serve to adapt them for use in air. Lachrymal glands and eyelids keep eyes moist, wiped free of dust, and shielded from injury. Since the cornea is exposed to air, it is an important refractive surface, removing much of the burden from the lens of bending light rays and focusing an image on the retina. As in fishes, accommodation (adjusting focus for near and distant objects) occurs by moving the lens. Unlike eyes of most fishes, amphibian eyes at rest are adjusted for distant objects, and the lens is moved forward to focus on nearby objects.

Keeping a sharp image on the retina for approaching or receding objects requires accommodation, which occurs in several different ways among vertebrates. Eyes of bony fishes and lampreys are set for near vision in the resting stage; to focus on distant objects, the lens is moved backward. In amphibians,

sharks, and snakes, the relaxed eye is focused on distant objects and the lens is moved *forward* to focus on nearby objects. In birds, mammals, and all nonavian reptiles except snakes, the lens accommodates by changing its *curvature* rather than by moving forward or backward. The resting eye in these forms is set for distant vision, and to focus on nearby objects the lens curvature is increased by being squeezed (or, in some, allowed to relax) into a rounded shape.

Figure 25.23 Amphibian eye.

A **retina** contains both **rods and cones,** the latter providing color vision (see Chapter 33, p. 743). The iris contains well-developed circular and radial muscles and can rapidly expand or contract the aperture (pupil) to accommodate changing illumination. The upper lid of the eye is fixed, but the lower one is folded into a transparent **nictitating membrane** capable of moving across the eye surface (Figure 25.23). Frogs and toads generally possess good vision, which is crucial for animals that rely on quick escape to avoid their numerous predators and on accurate movements to capture rapidly moving prey.

Other sensory receptors include tactile and chemical receptors in skin, taste buds on the tongue and palate, and a well-developed olfactory epithelium lining the nasal cavity.

Reproduction

Because frogs and toads are ectothermic, they breed, feed, and grow only during warm seasons. One of the first drives after the dormant period is breeding. In spring males call vociferously to attract females.

When their eggs are mature, females enter water and males clasp them in a process called **amplexus** (Figure 25.24) in which eggs are fertilized externally (after leaving the female's body). As a female lays eggs, a male discharges sperm over the eggs to fertilize them. After fertilization, the jelly layers absorb water and swell. Eggs are laid in large masses, usually anchored to vegetation.

A fertilized egg (zygote) begins development almost immediately (Figure 25.25). By repeated division (cleavage) a zygote

Figure 25.24 A male green tree frog, *Hyla cinerea,* clasps a larger female during breeding season in a South Carolina swamp. Clasping (amplexus) lasts until the female deposits her eggs. Like most tree frogs, these can change color rapidly; the male here, normally green, has darkened during amplexus.

© Cleveland P. Hickman, Jr.

becomes a hollow ball of cells (blastula). The blastula undergoes gastrulation (p. 162) and then continues to differentiate to form an embryo with a tail bud. At 2 to 21 days, depending on temperature, a tadpole hatches from the protective jelly coats that surrounded the original fertilized egg.

At hatching, a tadpole has a distinct head, trunk, and compressed tail. Its mouth is on the ventral side of the head and has keratinized jaws for scraping vegetation from hard objects. Behind the mouth is a ventral adhesive disc for clinging to objects. In front of the mouth are two deep pits, which later become nostrils. Swellings on each side of the head later become external gills. There are three pairs of external gills, which later become internal gills covered by a flap of skin (operculum) on each side. On the right side the operculum completely fuses with the body wall, but on the left side a small opening, the spiracle (L. *spiraculum,* air hole), remains. Water flows through the spiracle after entering the mouth and passing the internal gills. Hindlimbs appear first during metamorphosis, while forelimbs remain temporarily hidden by

Tail shortens by reabsorption, metamorphosis nearing completion

Sexually mature frog at 3 years

Male clasps female (amplexus); eggs fertilized as they are shed

Hindlimbs, then forelimbs emerge

Spiracle

Skinfold grows over exernal gill, water exits through spiracle

External gills

Tail bud

Eggs surrounded by jelly coats

Tadpole begins feeding on algae

Embryo nourished by yolk

Cleavage

Figure 25.25 Life cycle of a leopard frog, *Lithobates pipiens.* Cleavage and embryonic stages occur within the jelly-coated eggs within the egg masses as shown at the right side of the figure. The cleavage and embryonic stages are shown free and disproportionately large to reveal structural detail.

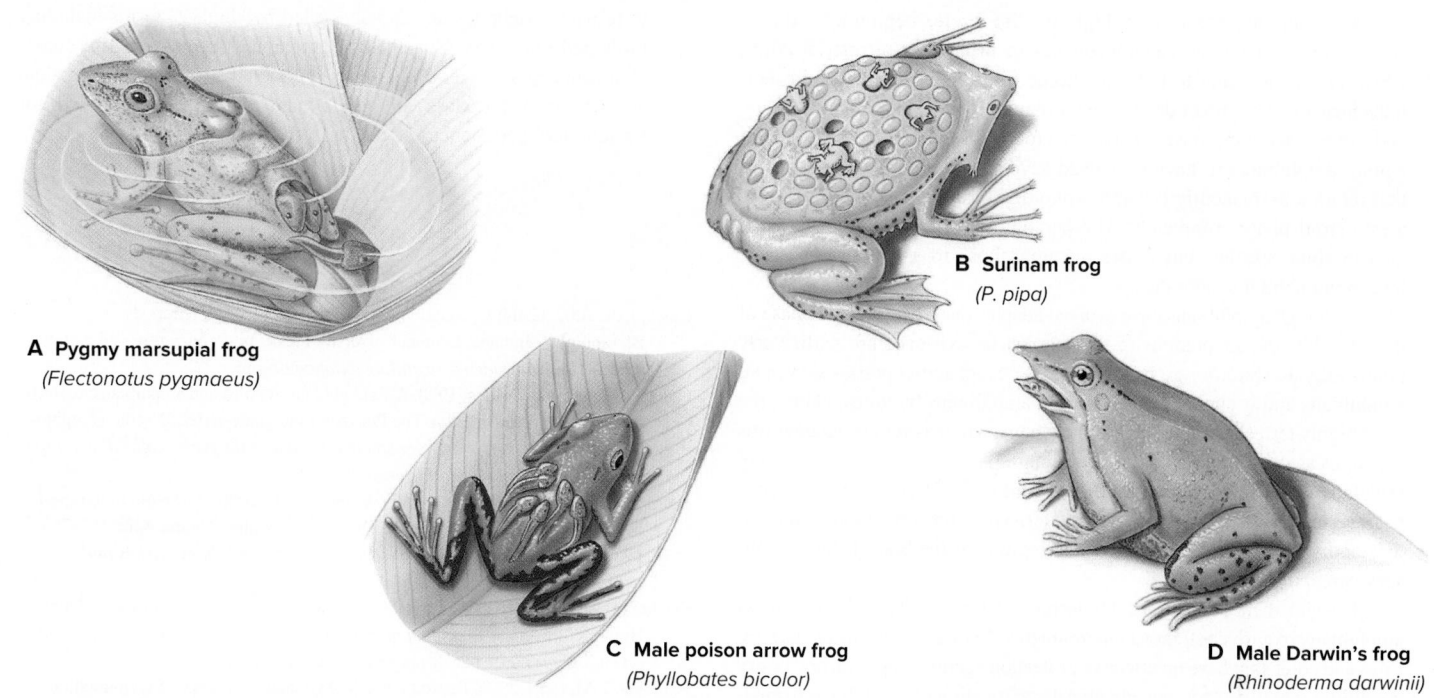

Figure 25.26 Unusual reproductive strategies of anurans. **A,** Female South American pygmy marsupial frog, *Flectonotus pygmaeus,* carries developing larvae in a dorsal pouch. **B,** Female Surinam frog, *Pipa pipa,* carries eggs embedded in specialized brooding pouches on the dorsum; froglets emerge and swim away when development is complete. **C,** Male poison arrow frog, *Phyllobates bicolor,* carries tadpoles adhering to its back. **D,** Tadpoles of a male Darwin's frog, *Rhinoderma darwinii,* develop into froglets in its vocal pouch. When ready to emerge, a froglet crawls into the parent's mouth, which the parent opens to allow the froglet's escape.

folds of the operculum. The tail is resorbed. The intestine becomes much shorter. The mouth undergoes a transformation into the adult condition. Lungs develop, and gills are resorbed (Figure 25.25). Leopard frogs usually complete metamorphosis within 3 months, whereas bullfrogs take 2 or 3 years to complete the process.

The life history just described is typical of most temperate-zone anurans but only one of many alternative patterns in tropical anurans. Some remarkable reproductive strategies of tropical anurans are illustrated in Figure 25.26. Some species lay their eggs in foam masses that float on the surface of the water; some deposit their eggs on leaves overhanging ponds and streams into which the emerging tadpoles drop; some lay their eggs in damp burrows; and others place their eggs in water trapped in tree cavities or in water-filled chambers of some bromeliads (epiphytic plants in the tropical forest canopy).

While most frogs abandon their eggs, some, such as the tropical dendrobatids (a family that includes the poison-dart frogs), tend their eggs. When the tadpoles hatch, they squirm onto the parent's back to be carried for varying lengths of time (Figure 25.26C). Tadpoles of Darwin's frog develop into froglets in their father's vocal pouch (Figure 25.26D), whereas Australian gastric-brooding frogs develop in their mother's stomach. Marsupial frogs carry their developing eggs in a pouch on the back (Figure 25.26A). In Surinam frogs (Figure 25.26B), the male and female do backward somersaults during mating, and eggs and sperm slide into the space between the mating pair. The male presses the fertilized eggs into the female's back, which develops a spongy incubating layer that eventually sloughs once young are hatched.

In the species-rich tropical genus *Eleutherodactylus,* mating occurs on land, and eggs hatch directly into froglets; the aquatic larval stage is eliminated, freeing these frogs from an obligatory association with pools or streams. A Puerto Rican species *E. jasperi* has evolved internal fertilization and live birth of offspring.

Migration of frogs and toads depends upon their breeding habits. Males usually return to a pond or stream before females, which they then attract by calling. Some salamanders also have a strong homing instinct, returning each year to reproduce in the same pool, to which they are guided by olfactory cues. The initial stimulus for migration in many cases is attributable to a seasonal cycle in the gonads plus hormonal changes that increase frogs' sensitivity to changes in temperature and humidity.

 S U M M A R Y

Amphibians are ectothermic, primitively quadrupedal vertebrates that have glandular skin and breathe by lungs, gills, and/or skin. They form one of two major branches of tetrapod phylogeny, the other one being amniotes. Modern amphibians comprise three major evolutionary groups. Caecilians (order Gymnophiona) are a small tropical group of limbless, elongate forms. Salamanders (order Urodela) are tailed amphibians, most of which have retained the generalized four-limbed body plan of their Paleozoic ancestors. Frogs and toads (order Anura) are the largest group of modern amphibians, adults of which are specialized for jumping on land or in water.

Most amphibians have a biphasic life cycle, beginning with an aquatic larva that later metamorphoses to produce a terrestrial adult, which returns to water to lay eggs. Some frogs, salamanders, and caecilians have evolved direct development that omits the aquatic larval stage, and some caecilians have evolved viviparity. Salamanders are unique among amphibians in having evolved several perennibranchiate species that retain a permanently larval morphology throughout life, eliminating a terrestrial phase completely. The perennibranchiate condition is obligate in some species, but others metamorphose to a terrestrial form if their pond habitat evaporates.

Although amphibians have evolved adaptations to the aquatic phase of their life history, adaptations to their terrestrial existence are particularly noteworthy. Respiratory exchange of gases occurs across porous skin in all amphibians and is supplemented in most amphibians by lungs. Oddly, the most highly terrestrial salamanders lack lungs, whereas some aquatic forms use lungs as their major respiratory structure. Terrestrial forms have strong skeletons with terrestrial adaptations evident in the structuring of the ribs, pectoral and pelvic girdles, and limbs. Derived features of amphibian auditory and visual systems and associated regions of the brain facilitate sensory perception on land.

Despite their adaptations for terrestrial life, adults and eggs of all amphibians require cool, moist environments if not actual pools or streams. Eggs and adult skin have no effective protection against very cold, hot, or dry conditions, greatly restricting adaptive diversification of amphibians to environments that have moderate temperatures and abundant water.

REVIEW QUESTIONS

1. How did the characteristic differences between aquatic and terrestrial environments influence the early evolution of tetrapods?
2. Describe the different modes of respiration used by amphibians. What paradox do amphiumas and terrestrial plethodontids present regarding the association of lungs with life on land?
3. Evolution of the tetrapod limb was one of the most important events in vertebrate history. Describe the supposed sequence in its evolution.
4. Compare the general life-history patterns of salamanders with those of frogs. Which group shows a greater variety of evolutionary changes of the ancestral biphasic amphibian life cycle?
5. Give the literal meaning of the name Gymnophiona. What animals are included in this amphibian order, what do they look like, and where do they live?
6. What is the literal meaning of the order names Urodela and Anura? What major features distinguish members of these two orders from each other?
7. Describe the breeding behavior of a typical woodland salamander.
8. How has paedomorphosis been important to evolutionary diversification of salamanders?
9. Describe the integument of a frog. How are the various skin colors produced in frogs?
10. Describe amphibian circulation.
11. Explain how the forebrain, midbrain, hindbrain, and their associated sensory structures have developed to meet sensory requirements for amphibian life on land.
12. Briefly describe the reproductive behavior of frogs. In what important ways do leopard frogs (*Lithobates pipiens*) and pygmy salamanders (*Desmognathus wrighti*) differ in their reproduction?

For Further Thought On the nineteenth-century notion of a "scale of nature," living amphibians were considered remnants of archaic terrestrial vertebrates largely superseded by "higher" forms including birds and mammals. Nonetheless, amphibian species are often more abundant and of longer evolutionary duration than avian or mammalian species. In what ways are amphibians unusually well suited for evolutionary persistence?

SELECTED REFERENCES

Clack, J. A. 2002. Gaining ground: the origin and evolution of tetrapods. Bloomington, Indiana, Indiana University Press. *An authoritative account of paleontological evidence regarding tetrapod origins.*

Conant, R., and J. T. Collins. 1998. A field guide to reptiles and amphibians: eastern and central North America. The Peterson field guide series. Boston, Houghton Mifflin Company. *Updated version of a popular field guide; color illustrations and distribution maps for all species.*

Daeschler, E. B., N. H. Shubin, and F. A. Jenkins, Jr. 2006. A Devonian tetrapod-like fish and the evolution of the tetrapod body plan. Nature **440:** 757–763. *Describes Tiktaalik, a fossil intermediate between lobe-fin fishes and tetrapods.*

Duellman, W. E., and L. R. Trueb. 1994. Biology of amphibians. Baltimore, Johns Hopkins University Press. *Important comprehensive sourcebook of information on amphibians, extensively referenced and illustrated.*

Exbrayat, J. M. (ed.). 2006. Reproductive biology and phylogeny of Gymnophiona (caecilians). Enfield, New Hampshire, Science Publishers, Inc. *Authoritative coverage of caecilian biology.*

Frost, D. 2015. Amphibian species of the world 6.0, an online reference. New York, NY. American Museum of Natural History. *A continuously updated online catalogue of all amphibian species and higher taxa.*

Frost, D. R., T. Grant, J. Faivovich, R. H. Bain, A. Haas, C. F. B. Haddad, R. O. De Sá, A. Channing, M. Wilkinson, S. C. Donnellan, C. J. Raxworthy, J. A. Campbell, B. L. Blotto, P. Moler, R. C. Drewes, R. A. Nussbaum, J. D. Lynch, D. M. Green, and W. C. Wheeler. 2006. The amphibian tree of life. Bull. Am. Mus. Nat. Hist. **297:** 1–370. *A phylogeny for living amphibians derived from a large compilation of morphological and molecular characters.*

Halliday, T. R., and K. Adler (eds.). 2002. Firefly encyclopedia of reptiles and amphibians. Toronto, Canada, Firefly Books. *Excellent authoritative reference work with high-quality illustrations.*

Heatwole, H., ed. 1995–2015. Amphibian biology. Baulkham Hills, Australia. Surrey Beatty and Sons. *The first 11 volumes of a continuing series covering amphibian biology in great detail.*

Jamieson, B. G. M. (ed.). 2003. Reproductive biology and phylogeny of Anura. Enfield, New Hampshire, Science Publishers, Inc. *Provides detailed coverage of reproductive biology and early evolutionary diversification of frogs and toads.*

Lannoo, M. (ed.). 2005. Amphibian declines: the conservation status of United States species. Berkeley, California, University of California Press. *A survey of conservation status of American amphibians.*

Lewis, S. 1989. Cane toads: an unnatural history. New York, Dolphin/Doubleday. *Based on an amusing and informative film of the same title, this book describes the introduction of cane toads to Queensland, Australia, and the unexpected consequences of their population explosion there. "If Monty Python teamed up with National Geographic, the result would be* Cane Toads.*"*

Petranka, J. W. 1998. Salamanders of the United States and Canada. Washington, DC, Smithsonian Institution Press. *A comprehensive coverage of life history and ecology of American and Canadian salamanders.*

Pough, F. H., R. M. Andrews, J. E. Cadle, M. L. Crump, A. H. Savitsky, and K. D. Wells. 2004. Herpetology, ed. 3. Upper Saddle River, New Jersey, Prentice-Hall. *A current general textbook of herpetology.*

Savage, J. M. 2002. The amphibians and reptiles of Costa Rica. Chicago, University of Chicago Press. *Costa Rica hosts a diversity of anuran, caecilian, and salamander species. Courses offered by the Organization for*

Tropical Studies give college students an opportunity to study this amphibian fauna.

Sever, D. M. (ed.). 2003. Reproductive biology and phylogeny of Urodela (Amphibia). Enfield, New Hampshire, Science Publishers, Inc. *A thorough review of reproductive biology and evolutionary relationships among salamanders.*

Stebbins, R. C., and N. W. Cohen. 1995. A natural history of amphibians. Princeton, New Jersey, Princeton University Press. *Worldwide treatment of amphibian*

biology, emphasizing physiological adaptations, ecology, reproduction, behavior, and a concluding chapter on amphibian declines.

Vitt, L. J., and J. P. Caldwell. 2013. Herpetology: an introduction to the biology of amphibians and reptiles, ed. 4. San Diego, Academic Press. *A current general textbook of herpetology.*

Wake, D. B. 2015. AmphibiaWeb. Berkeley, CA. University of California Digital Library. *A continuously updated online source for detailed information on amphibian species and higher taxa.*

26
CHAPTER

Amniote Origins and Nonavian Reptiles

- **PHYLUM CHORDATA**
  - Reptilia

Hatching Madagascar day gecko, Phelsuma madagascariensis.

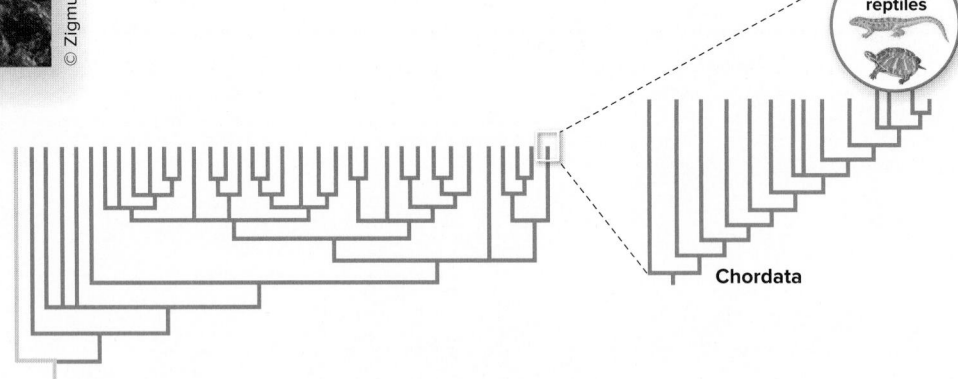

Enclosing the Pond

Amphibians, with well-developed limbs, redesigned sensory and respiratory systems, and modifications of the postcranial skeleton for supporting the body in air, have made a notable conquest of land. However, their shell-less eggs, thin, moist skin, and often gilled larvae keep their development hazardously tied to freshwater or moist terrestrial habitats. An ancestor of a clade containing nonavian reptiles, birds, and mammals evolved an egg better adapted for dry terrestrial conditions. This shelled egg enclosed the formerly aquatic early stages of development. In fact, the "pond-dwelling" stages were not eliminated but enclosed within a series of extraembryonic membranes that provided complete support for embryonic development. One membrane, the amnion, encloses a fluid-filled cavity, an aquatic environment for the developing

embryo. Another membranous sac, the allantois, serves both as a respiratory surface and as a chamber for storing nitrogenous wastes. Enclosing these membranes is a third membrane, the chorion, through which oxygen and carbon dioxide freely pass. Finally, surrounding and protecting everything is a porous shell.

With the last ties to aquatic reproduction severed, conquest of land by vertebrates was ensured. Paleozoic tetrapods that developed this reproductive pattern were ancestors of a single, monophyletic assemblage called Amniota, named after the innermost of the three extraembryonic membranes, the amnion. Before the end of the Paleozoic era, amniotes had diverged into multiple lineages that gave rise to all nonavian reptiles, birds, and mammals.

The nonavian members of the clade Reptilia (rep-til′ē-ə) (L. *repto,* to creep) include nearly 9500 species (approximately 320 native species in the United States and Canada) occupying a great variety of aquatic and terrestrial habitats, in many of which they are diverse and abundant. The four living reptile groups are turtles, lepidosaurs (snakes, lizards, and tuataras), crocodilians, and birds (Chapter 27). Some of the best known members of Reptilia are the awe-inspiring dinosaurs, reptiles of giant stature that diversified during the Mesozoic era. Dinosaurs form a clade with birds, to the exclusion of other reptiles.

Reptilia and Mammalia (Chapter 28) are the two clades of living amniotes. The morphology, physiology, and behavior of the first amniotes were probably most similar to those of lizards, among living amniotes. In the next section we discuss the origin of amniotes, their diversification into various groups, and their adaptations for life on dry land.

26.1 ORIGIN AND EARLY EVOLUTION OF AMNIOTES

As mentioned in the prologue to this chapter, amniotes are a monophyletic group that appeared and diversified in the late Paleozoic era. Most zoologists agree that amniotes are most closely related to anthracosaurs, a group of **anamniotes** (vertebrates lacking an amnion) of the early Carboniferous period. Anthracosaurs were better adapted for terrestrial life than most other anamniotes, and sometimes initially mistaken for early reptiles. Their diet probably consisted mostly of insects, which had greatly diversified by the Carboniferous period. *Diadectes* is the most likely sister taxon of amniotes, and interestingly enough, may be the only completely herbivorous anamniotic tetrapod, past or present. The first amniotes were small and lizardlike, but by the early Permian period they evolved into numerous forms diverse in morphology, feeding biology, and use of habitats (Figure 26.1).

Early diversification of amniotes produced three patterns of holes (fenestra) in the temporal region of the skull. **Anapsid** (Gr. *an,* without, + *apsis,* arch) skulls have no openings in the temporal area of the skull behind the **orbit** (opening in the skull for the eye); thus, the temporal region of the skull is completely roofed by dermal bones (Figure 26.2). This skull morphology was present in the earliest amniotes. It also occurs in one living group, the turtles, although the anapsid condition in turtles likely evolved secondarily, from ancestors having temporal fenestra. Two other amniote clades, Diapsida and Synapsida, represent separate evolutionary derivations from the ancestral anapsid condition.

The **diapsid** (Gr. *di,* double, + *apsis,* arch) skull has two temporal openings: one pair located low over the cheeks, and a second pair positioned above the lower pair, in the roof of the skull, and separated from the first by a bony arch (Figure 26.2). Diapsid skulls characterize birds and all amniotes traditionally considered "reptiles," except turtles (Figure 26.1). In many living diapsids (lizards, snakes, and birds) one or both of the bony arches and openings have been lost, perhaps facilitating skull kinesis (see Figure 26.13). The earliest diapsids gave rise to five morphologically distinct clades. The **lepidosaurs** include lizards, snakes, and tuataras. The **archosaurs** include dinosaurs, pterosaurs, crocodilians, and birds. We discuss the origin and early evolution of birds in Chapter 27 (pp. 585–588). A third, smaller clade, the **sauropterygians,** includes

several extinct aquatic groups, the most familiar of which are the large, long-necked plesiosaurs (Figure 26.1). **Ichthyosaurs,** represented by extinct, aquatic, dolphinlike forms (Figure 26.1), form a fourth clade of diapsids. Placement of the last clade, the **turtles,** with Diapsids has been controversial, in part because of the "anapsid" form of their skull. All turtles, including the earliest fossil forms, lack temporal fenestra and were considered the only living descendants of parareptiles, an early anapsid group. However, other morphological and genetic evidence published over the past 20 years places turtles within the diapsid clade, suggesting that the two pairs of temporal fenestra characteristic of diapsids were lost early in turtle evolution. The relationship of turtles to other diapsids is still being investigated, although the bulk of genetic evidence suggests affinities with archosaurs.

The third skull fenestration condition is **synapsid** (Gr. *syn,* together, + *apsis,* arch), characterized by a single pair of temporal openings located low on the cheeks and bordered by a bony arch (Figure 26.2). The synapsid condition occurs in a clade that includes mammals and their extinct relatives, the therapsids and pelycosaurs (Figure 26.1). Synapsids were the first amniote group to undergo extensive adaptive diversification and were the dominant large amniotes of the late Paleozoic era. Mammal evolution is discussed in Chapter 28 (pp. 611–613).

What was the functional significance of the temporal openings in early amniotes? In living forms these openings are occupied by large muscles that elevate (adduct) the lower jaw. Changes in jaw musculature might reflect a shift from suction feeding in aquatic vertebrates (p. 529) to terrestrial feeding, which required larger muscles to produce more static pressure, for doing activities such as nipping plant material with the anterior teeth, or for grinding food with the posterior teeth. Amniotes exhibit considerably more variation in feeding biology than do anamniotes, and herbivory is common in many amniote lineages. Although the functional significance of the evolution of temporal openings in amniotes is not fully understood, it is clear that expansion of the jaw musculature of amniotes must have been important. Even in fenestra-lacking turtle skulls, emarginations (notches) in the temporal area of the skull provide room for large jaw muscles.

Adaptations of Amniotes

Derived characters of amniotes include an amniotic egg, rib ventilation, and desiccation-resistant skin, in addition to skeletal characteristics of the head, shoulder, and ankle. These and other morphological and physiological characters allowed amniotes to be more energetic and successful at exploiting dry, terrestrial habitats than are anamniotes (especially modern amphibians).

1. **Amniotic egg.** All amniotes have eggs with four extraembryonic membranes, the **amnion, allantois, chorion,** and **yolk sac** (Figure 26.3). The amnion encloses the embryo in fluid, cushioning the embryo and providing an aqueous medium for growth. Metabolic wastes are stored in a sac formed by the allantois. The chorion surrounds the entire contents of the egg, and like the allantois, is highly vascularized. Thus, the chorion and allantois form an efficient respiratory organ for removing carbon dioxide and acquiring oxygen. Most amniotic eggs are surrounded by a mineralized but often flexible shell, although many lizards and snakes and most mammals lack shelled eggs. The shell forms an important mechanical support, and especially

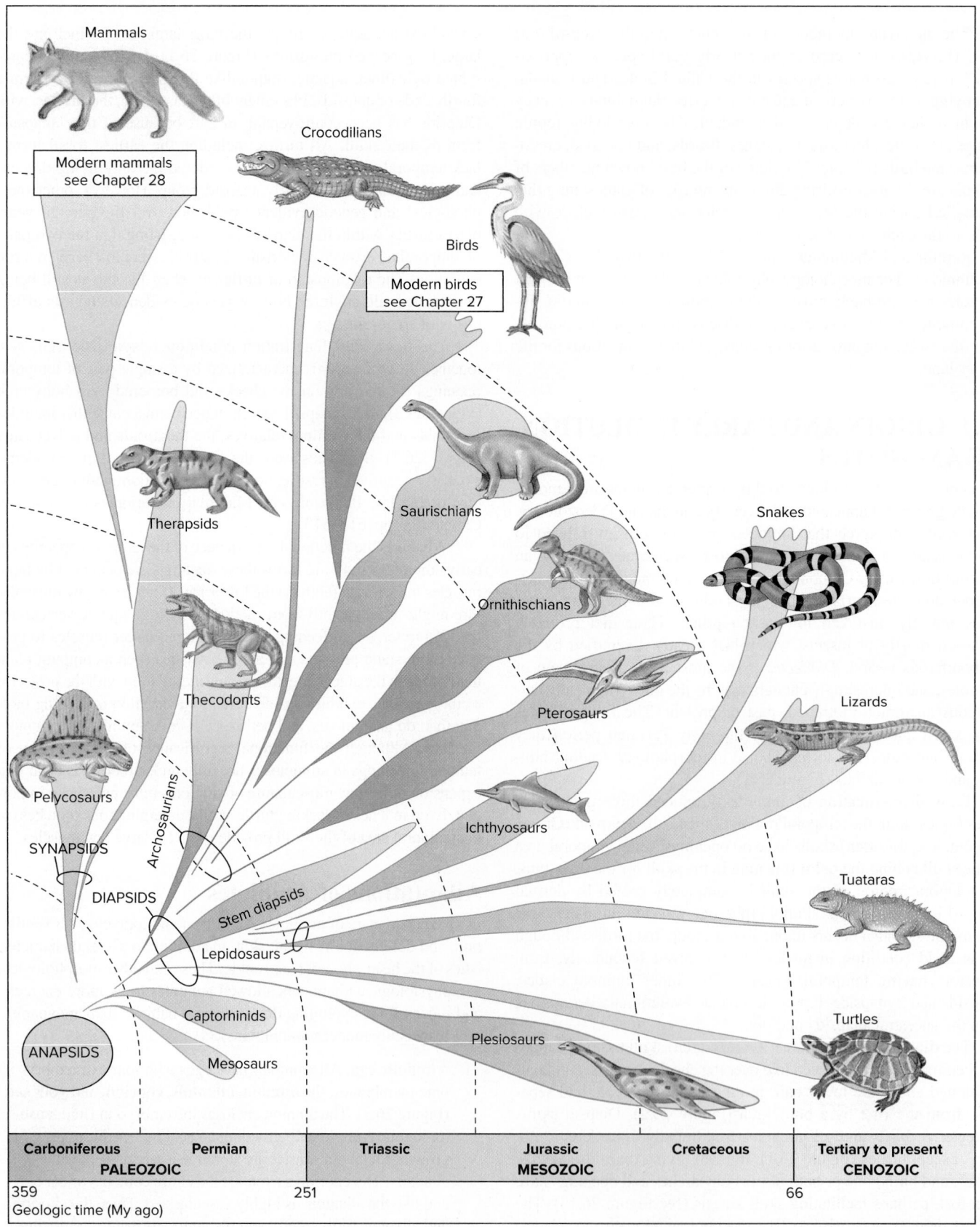

Figure 26.1 Evolution of amniotes. The earliest amniotes evolved an amniotic egg, which allowed amniotes to exploit drier habitats. The living amniotes, which include nonavian reptiles, birds, and mammals, evolved from a lineage of small, lizardlike forms that retained the anapsid skull pattern of early, anamniote tetrapods. One lineage that descended from the early amniotes had a synapsid skull pattern and gave rise to modern mammals. Birds, squamates, and crocodilians have a diapsid skull pattern. Turtles have an anapsid skull, although it is probably independently evolved.

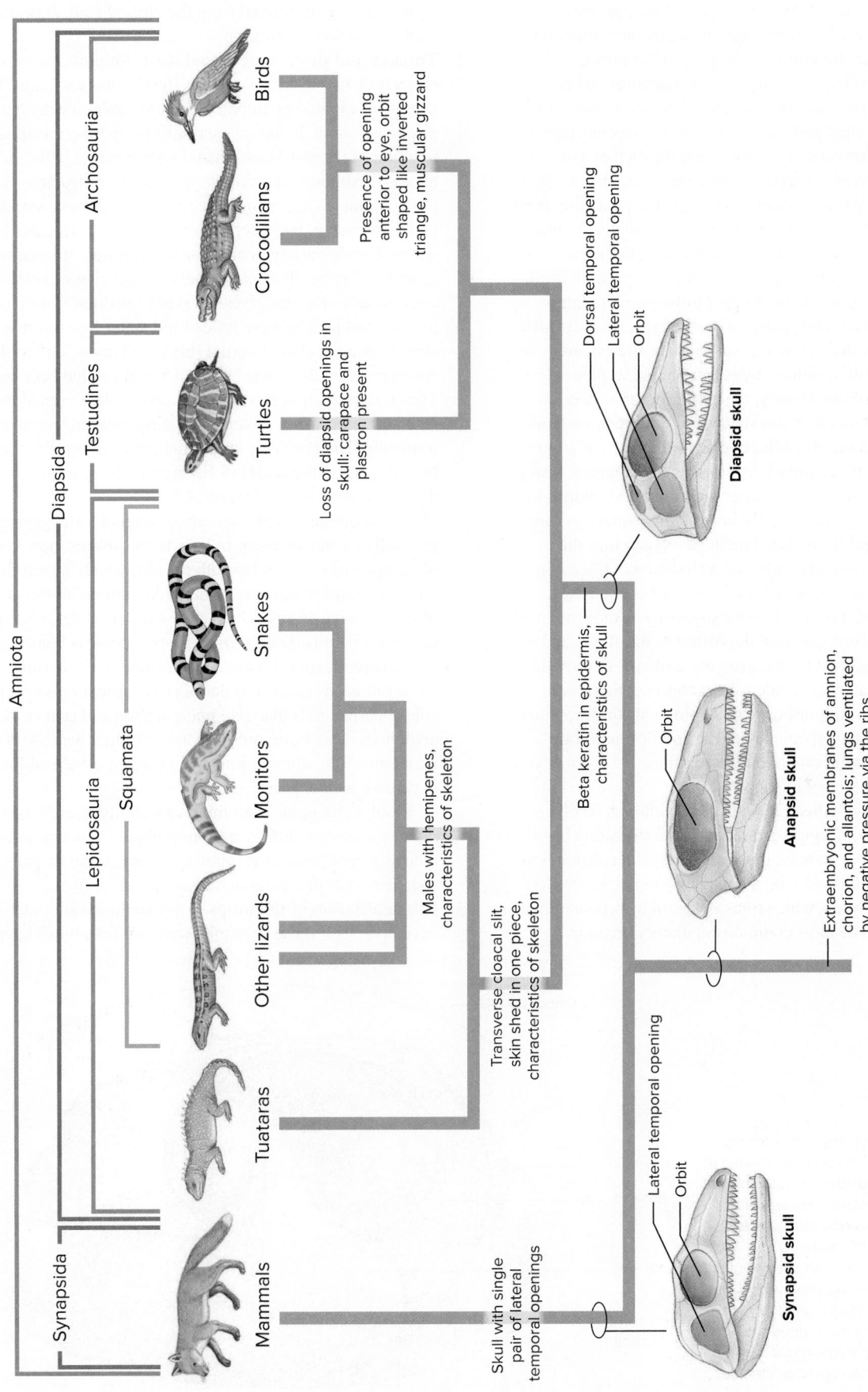

Figure 26.2   Cladogram of living Amniota showing monophyletic groups. The skulls represent the ancestral condition of the three groups. Skulls of modern diapsids and synapsids are often highly modified by loss or fusion of skull bones that obscures the ancestral condition. Representative skulls for anapsids are *Nyctiphruetus* of the upper Permian period; for diapsids, *Youngina* of the upper Permian period; for synapsids, *Aerosaurus*, a pelycosaur of the lower Permian period. The relationship of turtles to other reptiles is controversial, although most recent genetic evidence indicates a close relationship with archosaurs, as shown here.

for birds, a semipermeable barrier, which allows passage of gases but limits water loss. Like eggs of anamniotes, amniotic eggs have a yolk sac for nutrient storage (p. 174), although the yolk sac tends to be larger in amniotes. In marsupial and placental mammals, the yolk sac does not store yolk, but it may form a temporary or persistent yolk-sac placenta to transfer nutrients, gases, and wastes between the embryo and the mother. For many amniote species, embryonic development occurs in a female's reproductive tract, providing even greater protection from predators and dehydration, and potential for the mother to manage the embryo's nutritional and other physiological needs.

How did the amniotic egg evolve? It is tempting to think of the amniotic egg as *the* land egg. However, many amphibians lay eggs on land and many amniotic eggs, such as those of turtles, must be buried in wet soil or deposited in areas of high humidity. Still, amniotic eggs can be laid in places too dry for any amphibian; clearly, the evolution of amniotic eggs was a major factor in the success of tetrapods on land. Perhaps a more important selective advantage of the amniotic egg was that it permitted development of a larger, faster-growing embryo. Anamniotic eggs are supported mainly by a thick, jellylike layer. This jelly layer is inadequate to support large eggs and limits movement of oxygen into the eggs. One hypothesis suggests that a first step in the evolution of the amniotic egg was replacement of the jelly layer with a shell, which provided better support and movement of oxygen. Furthermore, calcium deposited in the shell can be dissolved and absorbed by the growing embryo, to provide one of the raw materials needed for skeleton construction. This hypothesis is supported by physiological studies, which show that embryos of species with the smallest amniotic eggs have a metabolic rate about three times that of embryos of anamniotes with eggs of the same size.

All amniotes lack gilled larvae and have internal fertilization. Shelled amniotic eggs require internal fertilization because sperm cannot penetrate the shell. Internal fertilization in amniotes is accomplished with a copulatory organ, except for tuataras and most birds, which transfer sperm by pressing their cloacas together. The most common copulatory organ in amniotes, a penis derived from the cloacal wall, appears to be another amniote innovation.

2. **Thicker and more waterproof skin.** Amphibians must maintain a thin, moist skin to permit effective gas exchange. However, this skin makes amphibians vulnerable to dehydration and physical trauma. In amniotes, a shift away from a respiratory function of the skin is associated with changes in the skin morphology. Although the skin varies widely in structure among living amniotes and anamniote tetrapods, amniote skin tends to be much thicker, more keratinized, and less permeable to water. A wide variety of structures composed of **keratin,** such as scales (Figure 26.4), hair, feathers, and claws, project from amniote skin. Keratin gives the skin protection from physical trauma, and hydrophobic lipids limit water loss through the skin. Unique to clade Reptilia (birds and nonavian reptiles), the epidermis contains a hard form of keratin called beta keratin. The characteristic scales of nonavian reptiles, formed mostly of beta keratin, provide protection against wear in terrestrial environments. The primarily epidermal scales of reptiles are not homologous to the scales of fishes, which are mostly bony, dermal structures (see Figure 29.2).

In crocodilians, scales remain throughout life, growing gradually to replace wear. In lizards and snakes, new keratinized epidermis grows beneath the old, which is then shed at intervals. Turtles add new layers of keratin under the old layers of the platelike scutes, which are modified scales. Crocodiles and many lizards (skinks, for example) possess bony plates called **osteoderms** (Figure 26.4) located in the dermis, beneath the keratinized scales. The dermis has chromatophores (p. 642), color-bearing cells that give many lizards and snakes their colorful hues. This layer, unfortunately for their bearers, is converted into alligator and snakeskin leather, esteemed for expensive handbags and shoes.

Keratin and lipids limit the skin's ability to exchange respiratory gases—so, unlike most amphibians, few amniotes use their skin as a primary respiratory organ. Amniote gas exchange occurs primarily in the lungs.

3. **Rib ventilation of the lungs.** Amniote lungs are much better developed than those of amphibians; amniote lungs have much

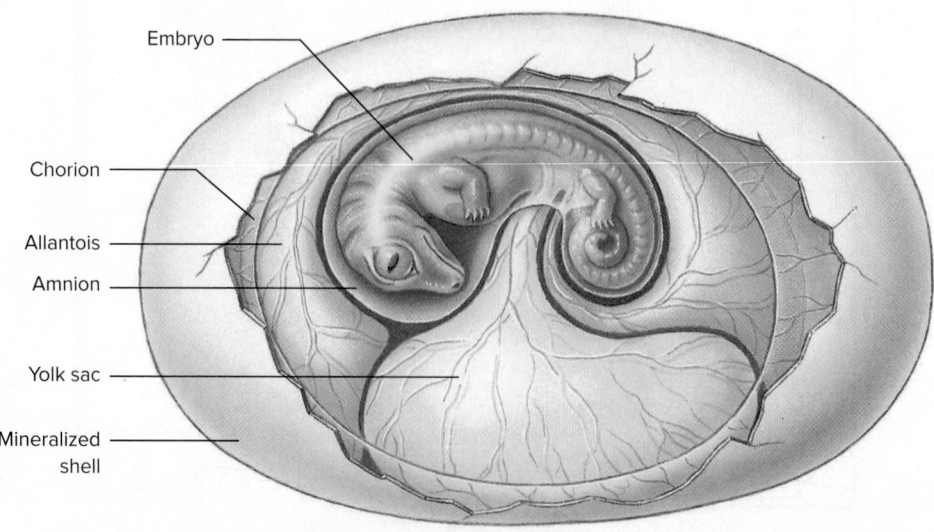

Figure 26.3 Amniotic egg. The embryo develops within the amnion and is cushioned by amniotic fluid. Food is provided by yolk from the yolk sac and metabolic wastes are deposited within the allantois. As development proceeds, the allantois fuses with the chorion, a membrane lying against the inner surface of the shell; both membranes are supplied with blood vessels that assist in the exchange of oxygen and carbon dioxide across the porous shell. Because this kind of egg is an enclosed, self-contained system, it is often called a "cleidoic" egg (Gr. *kleidoun,* to lock in).

Embryo

Chorion

Allantois

Amnion

Yolk sac

Mineralized shell

Figure 26.4 Section of the skin of a nonavian reptile showing overlapping, keratinized scales in the epidermis and bony osteoderms in the dermis.

more surface area (see Figure 31.21) and are ventilated by a different mechanism. These changes reflect both the increased metabolic demands of amniotes and the reduced ability of the amniote skin to serve as a gas-exchange organ.

Amphibians, like air-breathing fishes, fill their lungs by *pushing* air from the oral and pharyngeal cavities into the lungs (see Figure 31.22). In contrast, amniotes *draw* air into their lungs (**aspiration**) by expanding the thoracic cavity using costal (rib) muscles or pulling the diaphragm or liver (with other muscles) posterior. This shift from positive to negative ventilation probably influenced the evolution of amniote limbs. Early tetrapods used their rib muscles to make lateral undulations, causing a wriggling motion not unlike that of eels (p. 532). Enlargement and mechanical rearrangement of the limbs enabled more efficient locomotion of amniotes, particularly for those with large bodies, while allowing rib muscles to serve a primary function in lung ventilation. However, we note that limbs are not required for terrestrial locomotion as some amniotes, such as snakes and amphisbaenians, are very successful with no limbs at all! Although lungs are the primary gas exchange organ for nearly all amniotes, other areas of the body can be used. Many aquatic turtles supplement pulmonary respiration with gas exchange at their pharynx or cloaca, and most gas exchange in sea snakes occurs across their skin.

4. **Stronger jaws.** Most fish jaws are designed for suction and quick closure, but once prey is seized, little static force can be applied. Suction feeding is not possible with terrestrial vertebrates, and the skeleton and muscles of the jaws of early tetrapods became adapted to seize prey. In amniotes, expansion of the jaw musculature, often into temporal openings (Figure 26.2) or notches, provided better mechanical advantage. In contrast to fishes, the tongue of tetrapods is muscular and mobile, functioning to move food in the mouth for mastication and swallowing.

5. **High-pressure cardiovascular systems.** In all amniotes the right atrium, which receives deoxygenated blood from the body, is completely partitioned from the left atrium, which receives oxygenated blood from the lungs. Mammals, birds, and crocodilians have two completely separated ventricles as well (Figure 26.2); other reptiles have a single ventricle

incompletely partitioned into multiple chambers. Even in species with incomplete ventricular septa, flow patterns within the heart limit mixing of oxygen-rich blood and oxygen-poor blood; all amniotes therefore have functionally separate circulations. This better separation of circuits permits higher systemic blood pressure in amniotes; fishes and amphibians typically have systemic pressures of 15 to 40 mm Hg, compared with about 80 mm Hg in varanids (a group of large, active lizards; see page 573). Higher pressure is adaptive for active, terrestrial organisms because of their higher metabolic needs and because the heart must overcome gravity to pump blood "uphill."

Is the incomplete separation of hearts of many nonavian reptiles merely a transitional stage on the evolutionary path to the "advanced" heart of mammals and birds? No, incomplete separation is adaptive for these vertebrates, as it allows blood to bypass the lungs when pulmonary respiration is not occurring (for example, diving or estivation). Thus, during lung bypass, energy is not expended pumping blood through the capillary beds of the lungs.

6. **Water-conserving nitrogen excretion.** Most amphibians excrete their metabolic waste primarily as ammonia or urea. Ammonia is toxic at relatively low concentrations and must be removed in a dilute solution. Because of the water required for excretion of ammonia, it is not adaptive for vertebrates occupying dry, terrestrial habitats. Mammals excrete their nitrogenous waste as urea, which is concentrated in their kidneys (see pp. 671–672), reducing water loss through excretion. Birds and nonavian reptiles excrete their nitrogenous wastes primarily as uric acid. Because it is relatively nontoxic, uric acid can be concentrated, requiring very little water for excretion. Birds and nonavian reptiles have limited abilities to concentrate their urine in the kidneys, so their urinary bladder receives dilute urine. Water (and many salts) are resorbed in the bladder, and "urine" is voided as a semisolid mass of uric acid. Salts are removed in many seabirds and nonavian reptiles by salt glands—located near the nose, eyes, or tongue—which secrete a salty fluid that is strongly hyperosmotic to body fluids.

7. **Expanded brain and sensory organs.** The brain has a relatively large cerebrum and cerebellum in all amniotes, but especially so in birds and mammals. Enlargement of the cerebrum is correlated with integration of sensory information and control of muscles during locomotion. Nonavian reptiles and birds have particularly good vision, processed in the optic lobe (Figure 26.5), and many species display brilliant coloration. Smell is not as developed in many reptiles, but snakes and some lizards use a highly sensitive sense of smell to find prey and mates. In lizards and snakes, olfaction is assisted by well-developed Jacobson's organs, specialized olfactory chambers in the roof of the mouth (p. 575). Some lizards and snakes can detect ultraviolet or infrared light.

Changes in Traditional Classification of Reptiles

With increasing use of cladistic methodology in zoology, and its insistence on hierarchical arrangement of monophyletic groups (see p. 209), important changes have been made in the classification of

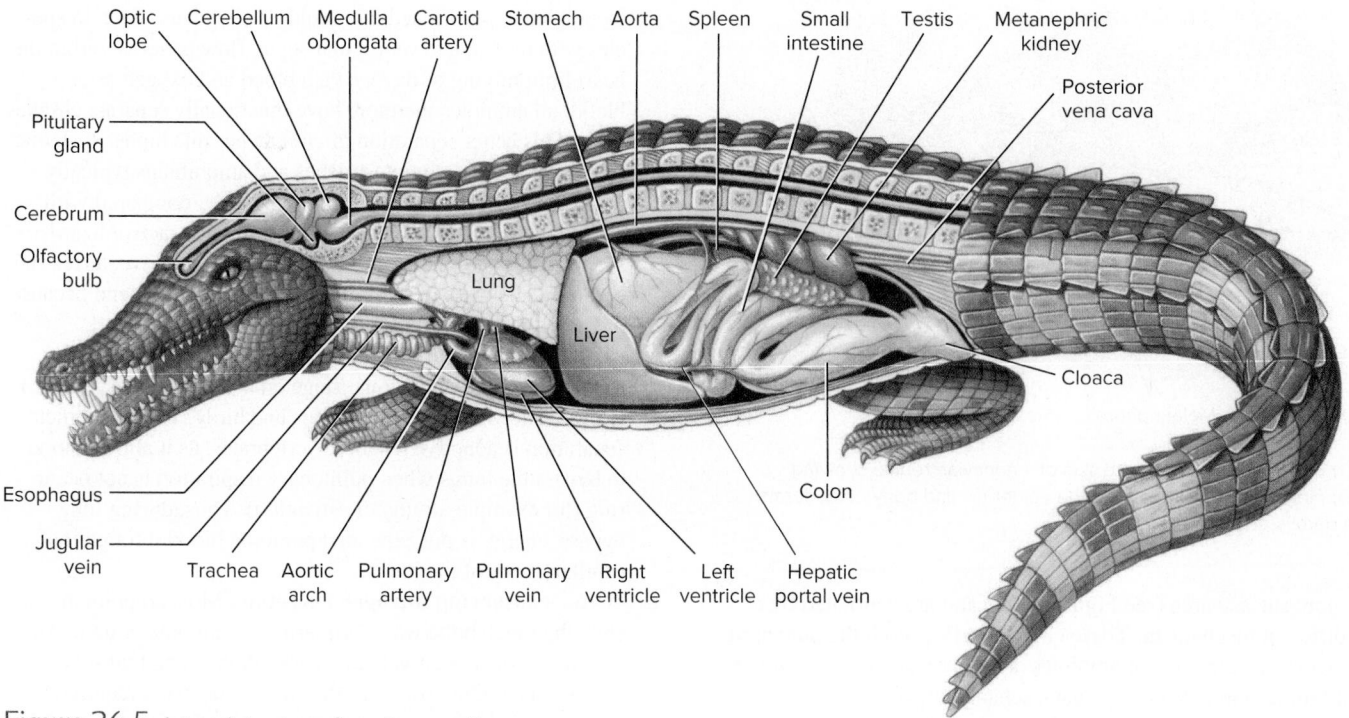

Figure 26.5 Internal structure of a male crocodile.

amniotes. As traditionally defined, "reptiles" are snakes, lizards, tuataras, crocodilians, and turtles, and several extinct groups, such as dinosaurs, plesiosaurs, pterosaurs, and many other early amniotes. However, birds and reptiles (excluding certain early amniotes called "mammal-like reptiles") share several derived characters, including certain skull and ankle characteristics along with presence of a special type of keratin in the skin called beta keratin, which unites them as a monophyletic group (see Figure 26.2). Thus, "reptiles" as traditionally defined form a **paraphyletic** group because they do not include all descendants of their most recent ancestor.

Crocodilians and birds are sister groups; they are more recently descended from a common ancestor than either is from any other living reptilian lineage. In other words, birds and crocodilians belong to a monophyletic group apart from other reptiles and, according to the rules of cladistics, should be assigned to a clade that separates them from the remaining reptiles. This clade is in fact recognized; it is Archosauria (see Figures 26.1 and 26.2), a grouping that also includes the extinct dinosaurs and pterosaurs. Archosaurs, along with their sister group, the lepidosaurs (tuataras, lizards, and snakes), and turtles, form a monophyletic group that cladists call Reptilia. Here we use Reptilia and reptiles in a cladistic fashion to include living amniote groups traditionally termed "reptiles" together with birds and all extinct groups more closely related to these than to mammals. The term "nonavian reptiles" is used informally to refer to a paraphyletic group that includes the living turtles, lizards, snakes, tuataras, and crocodilians, and a number of extinct groups, including plesiosaurs, ichthyosaurs, pterosaurs, and dinosaurs. Nonavian reptiles are the subject of most of the remainder of this chapter; birds, which form the remainder of clade Reptilia, are discussed in Chapter 27. Four clades of living nonavian reptiles are recognized: (1) Testudines, turtles; (2) Squamata, lizards and snakes; (3) Sphenondonta, tuataras; (4) Crocodilia, crocodilians.

26.2 CHARACTERISTICS AND NATURAL HISTORY OF REPTILIAN GROUPS

Testudines (Chelonia): Turtles

Turtles appear in the fossil record in the Triassic period, some 240 million years ago. The earliest turtles had teeth and a reduced shell, but were otherwise similar to those of today. Lacking teeth, a modern turtle's jaw has tough, keratinized plates for gripping food (Figure 26.6). Turtles are enclosed in shells consisting of a dorsal **carapace** (Fr. from Sp., *carapacho*, covering) and a ventral **plastron** (Fr. breastplate). The shell is bone, overlain by enlarged scales of keratin. The bony part forms from expansion and fusion of ribs, vertebrae, and many dermally ossifying elements (Figure 26.7). Unique among vertebrates, turtle limbs and limb girdles are located *inside* the ribs! Recently discovered fossils clarify early turtle evolution. A 260-million-year old amniote, *Eunotosaurus*, had broadened ribs, but no shell; a 240-million-year-old amniote, *Pappochelys*, had broadened ribs and a partial plastron, but no carapace. These fossils indicate the first step in the evolution of the turtle's shell was broadening of the ribs. Then a plastron evolved, followed by a carapace. Also, *Pappochelys* had two temporal fenestra, providing further evidence of the turtle's place in clade Diapsida.

How did the turtle shell evolve? The carapace of turtles is partly formed from ribs, which enclose the scapula. This is in contrast to all other amniotes, which have the scapula outside the ribs. Recent developmental studies revealed that turtles follow the ancestral amniote pattern early in rib development, but then rib growth is redirected laterally (instead of ventrally) toward the carapacial ridge (which outlines a disc that is the precursor to the carapace) (Figure 26.8). Furthermore, the lateral body wall folds inward just below the

Characteristics of Nonavian Reptiles

1. Two paired limbs, usually with five toes each; limbs vestigial or absent in many; **ectothermic**
2. Body covering of **keratinized epidermal scales** and sometimes bony dermal plates; integument with few glands
3. Skull with **one occipital condyle** (bony bump that connects to the first vertebra); lower jaws of several bones; distinct atlas and axis; usually two sacral vertebrae
4. Teeth **polyphyodont** (replaced many times) or absent (turtles); when present, teeth usually **homodont** (all similar in function and shape) with a single point; **gizzard** in crocodilians
5. Brain moderately well developed with expanded cerebrum; 12 pairs of cranial nerves
6. Eyes with color vision in some; snakes and some lizards with highly developed chemoreception using olfactory epithelia and **Jacobson's organ;** some snakes with **heat-sensitive pit organs**; middle ear with a single bone
7. Usually separate sexes, but some lizards reproduce asexually by parthenogenesis; internal fertilization; copulatory organ a **penis, hemipene,** or (rarely) absent; sex determined by chromosomes or by environment
8. Embryonic membranes of **amnion, chorion, and allantois**; oviparous or viviparous; eggs with leathery or calcareous shells; embryos of viviparous species nourished by **placenta** or **yolk sac** (ovoviviparity); parental care absent, except in crocodilians; no larval stage
9. Excretory system of **metanephric kidneys** and ureters that open into a cloaca; uric acid usually main nitrogenous waste
10. Lungs filled by **aspiration** (negative ventilation); **no gills**; some supplement gas exchange with cloaca, skin, or pharynx
11. Heart with a sinus venosus, two atria, and ventricle incompletely divided into three chambers; crocodilian heart with sinus venosus, two atria, and two ventricles; **pulmonary and systemic circuits incompletely separated**; nucleated red blood cells

Figure 26.6 Common snapping turtle, *Chelydra serpentina,* showing the absence of teeth. Instead, the jaw edges are covered with a keratinized plate.

© Cleveland P. Hickman, Jr.

Figure 26.7 Skeleton and shell of a turtle, showing fusion of vertebrae and ribs with the carapace. The long and flexible neck allows the turtle to withdraw its head into its shell for protection.

carapacial ridge, carrying the scapula to a position underneath the ribs. At this stage the carapace is still incomplete, similar to the 220-million-year-old fossil turtle, *Odontochelys.* Subsequent ossification completes the carapace in most modern turtles.

The terms "turtle," "tortoise," and "terrapin" are applied variously to different members of the turtle order. In North American usage, they are all correctly called turtles. The term "tortoise" is frequently applied to land turtles, especially large forms. British usage of the terms is different: "tortoise" is the inclusive term, whereas "turtle" is applied only to the aquatic members.

Because its ribs are fused to the shell, a turtle cannot expand its chest to breathe as do other amniotes. Instead, turtles employ certain abdominal and pectoral muscles as a "diaphragm." Air is drawn in by contracting limb flank muscles to make the body cavity larger. Exhalation is also active and is accomplished by drawing the limbs back into the shell, thus compressing the viscera and forcing air out of the lungs. Breathing is visible as the bellowslike movements of the turtle's "limb pockets": folds of skin between the limbs and the rigid shell. Movements of the limbs during walking also help to ventilate the lungs. Many aquatic turtles can acquire oxygen by pumping water in and out of a vascularized mouth cavity or cloaca; this activity enables them to remain submerged for long periods when inactive.

Although a turtle's brain is small, never exceeding 1% of body weight, the cerebrum is larger than that of an amphibian, and a turtle is able to learn a maze about as quickly as a rat. Turtles have a middle and an inner ear, but sound perception is poor. Not

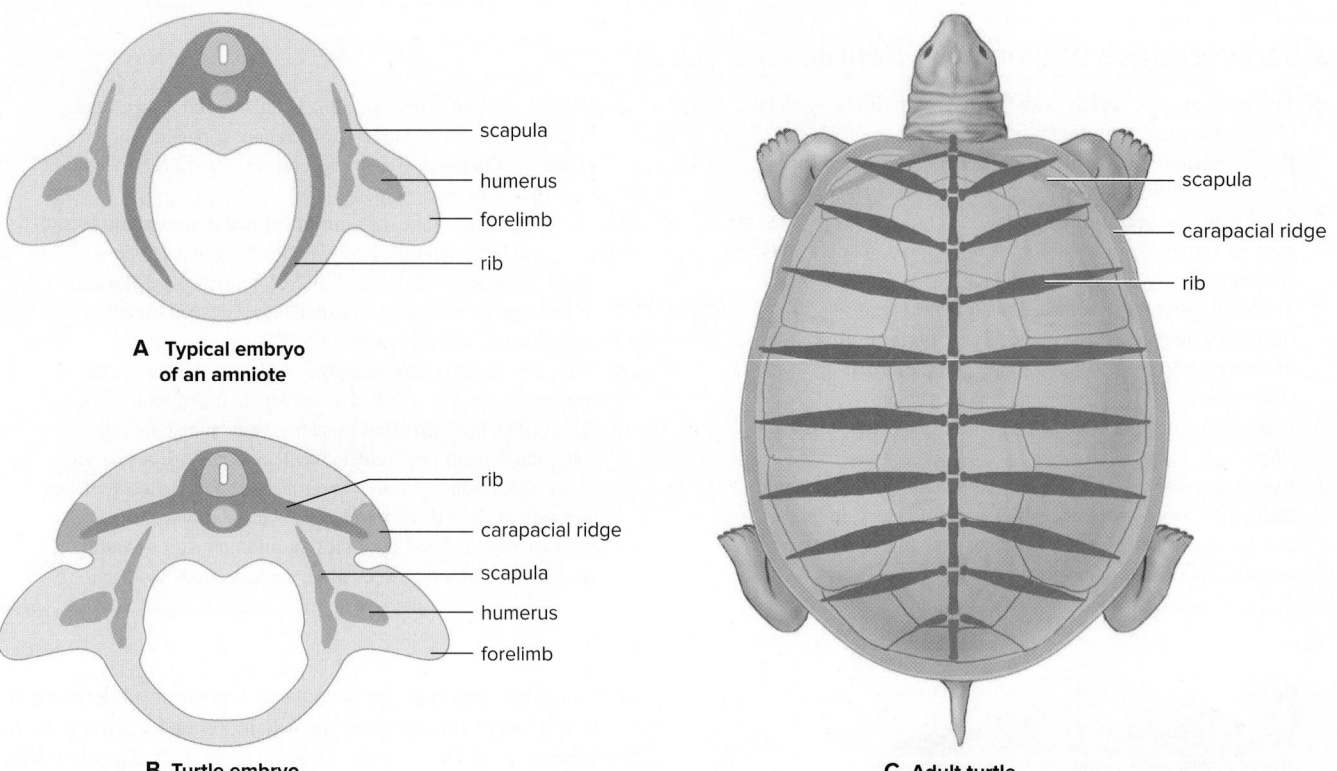

scapula
humerus
forelimb
rib

A Typical embryo of an amniote

rib
carapacial ridge
scapula
humerus
forelimb

B Turtle embryo

scapula
carapacial ridge
rib

C Adult turtle

Figure 26.8   Development of the turtle shell. **A,** In most amniotes, rib growth is directed ventrally, leaving the scapula outside (lateral to) the ribs. **B,** In turtles, rib growth is directed laterally, toward a carapacial ridge (pink), ultimately enclosing the scapula within the ribs. **C,** In most modern turtles, the ribs expand and fuse, to form a solid, bony carapace.

Figure 26.9   Mating Galápagos tortoises, *Geochelone elephantopus*. The male has a concave plastron that fits over the highly convex carapace of the female, helping to provide stability during mating. Males utter a roaring sound during mating, the only time they are known to emit vocalizations.

unexpectedly, therefore, most turtles are mute, although many tortoises utter grunting or roaring sounds during mating (Figure 26.9). Compensating for poor hearing are a good sense of smell and color vision.

Turtles are oviparous, and fertilization is internal, using a penis as the copulatory organ. All turtles, even marine forms, bury their shelled, amniotic eggs in the ground. Usually they exercise considerable care in constructing their nest, but once eggs are deposited and covered, the female deserts them. An interesting feature of turtle reproduction is that in some turtle families, as in all crocodilians and some lizards, nest temperature determines sex of the hatchlings. In turtles, low temperatures during incubation produce males and high temperatures produce females (Figure 26.10).

Marine turtles, buoyed by their aquatic environment, may reach great size. Leatherbacks are largest, attaining a length of 2 m and weight of 725 kg. Green turtles (Figure 26.11), so named because of their greenish body fat, may exceed 360 kg, although most individuals of this economically valuable and heavily exploited species seldom live long enough to reach anything approaching this size. Some land tortoises may weigh several hundred kilograms, such as the giant tortoises (Figure 26.9) of the Galápagos Islands that so intrigued Darwin during his visit there in 1835. Most tortoises are rather slow moving; an hour of determined trudging carries a large Galápagos tortoise approximately 300 m (although they may move much more rapidly for short distances). Their low metabolism probably explains their longevity, for some live more than 150 years.

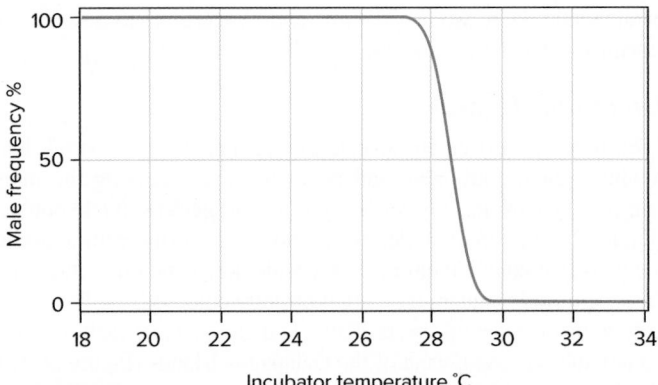

Figure 26.10 Temperature-dependent sex determination in the European pond turtle, *Emys orbicularis.* Eggs incubated at high temperatures produce females, while eggs incubated at low temperatures produce males.

Figure 26.11 Green sea turtle, *Chelonia mydas.* Green turtles are herbivores that subsist on marine grasses and algae. Sea turtles range widely in the oceans, returning to land only to deposit their eggs. Sea turtles are found in all tropical oceans.

The shell, like a medieval coat of armor, offers obvious advantages. In many species, the head and appendages can be drawn in for protection. The familiar common box turtle, *Terrapene carolina,* has a plastron that is hinged, forming two movable parts that can be pulled up against the carapace so tightly that one can hardly force a knife blade between the shells. Some aquatic turtles, such as the large leatherback turtle and common snapping turtle (Figure 26.6), have reduced shells, making complete withdrawal of limbs and head impossible. Snappers, however, have another formidable defense, as their name implies. They are carnivorous, living on carrion, fishes, frogs, crayfishes, or almost anything that comes within reach of their powerful jaws. An alligator snapper lures unwary fish into its mouth with a pink, wormlike extension of its tongue that serves as "bait" (Figure 26.12). Alligator snappers are wholly aquatic and come ashore only to lay their eggs.

Figure 26.12 An alligator snapping turtle, *Macroclemys temminckii,* of the southeastern United States lies underwater, mouth agape, luring fishes and other unwary prey by undulating a pink, wormlike protrusion from its tongue. Any prey attempting to eat the bait is instantly captured in powerful jaws.

Squamata: Lizards and Snakes

Squamates are the most diverse living, nonavian reptiles, comprising approximately 95% of known species of nonavian reptiles. Lizards appeared in the fossil record in the Jurassic period, but they did not diversify until the Cretaceous period of the Mesozoic era when dinosaurs were at the climax of their diversity. Snakes appeared during the late Jurassic period, probably evolving from a group of lizards whose descendants include the Gila monster, iguanas, and monitor lizards. Two specializations in particular characterize snakes: (1) extreme elongation of their body and accompanying displacement and rearrangement of internal organs, and (2) skull specializations for swallowing large prey.

Viviparity in living nonavian reptiles is limited to squamates and has evolved in that clade about 100 separate times. Viviparity typically occurs by increasing the length of time that eggs are kept within the oviduct. Developing young obtain nutrients from yolk sacs (**lecithotrophy** or ovoviviparity) or via the mother (**placentotrophy**), or some combination of both. Viviparous snakes and lizards are more commonly associated with cool climates. Viviparity may be an adaptation that permits squamates to regulate the temperature of their developing embryos, using **ectothermy** (see p. 674), for optimal development. Gravid mothers can move to favorable environments to keep their bodies, and their young, warm and stable in temperature, which ensures rapid development. Also, young retained in the mother's body are better protected from predation than are deposited eggs.

Skulls of squamates are modified from the ancestral diapsid condition by loss of dermal bone ventral and posterior to the lower temporal opening. This modification allowed evolution, in most lizards and snakes, of a **kinetic skull** having movable joints. The quadrate, which in other reptiles is fused to the skull, has a joint at its dorsal end, as well as its usual articulation with the lower jaw. In addition, joints in the palate and across the roof of the skull allow the snout to be tilted (Figure 26.13). Specialized mobility of the skull enables squamates to seize and to manipulate their prey; it also increases the effective closing force of the jaw musculature. The skull of snakes is even more kinetic than that of lizards. Such exceptional skull mobility is considered a major factor in diversification of lizards and snakes.

Traditionally, Squamata was divided into three subgroups: Lacertilia (lizards), Serpentes (snakes), and Amphisbaenia (amphisbaenians). Amphisbaenians now are considered modified lizards, and we discuss them with lizards. Snakes form a monophyletic group but evolved within a lizard subgroup, thus making Lacertilia paraphyletic. Here we use "lizard" and "Lacertilia" to describe all squamates that are not snakes.

Lacertilia: Lizards

Lizards are an extremely diverse group, including terrestrial, burrowing, aquatic, arboreal, and aerial members. Among the more familiar groups in this varied group are **geckos** (Gekkonidae) (Figure 26.14), small, agile, mostly nocturnal forms with adhesive toe pads that enable them to walk upside down and on vertical surfaces; **iguanids** (Iguanidae), often brightly colored New World lizards with ornamental crests, frills, and throat fans, including the remarkable marine iguana of the Galápagos Islands (Figure 26.15); **skinks** (Scincidae), with elongate bodies, an armor of tight-fitting osteoderms (see Figure 26.4), and reduced limbs in many species;

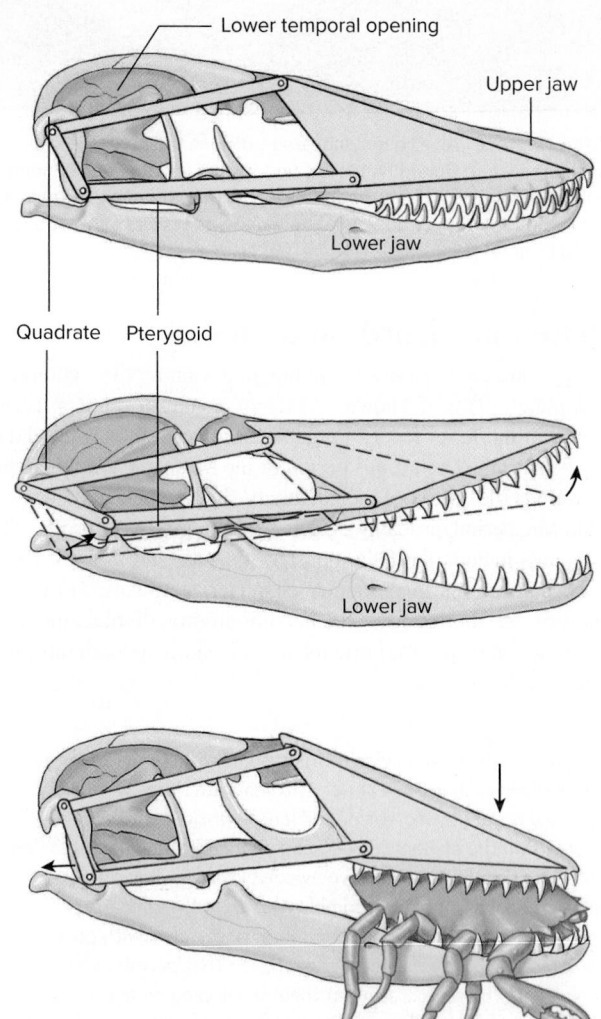

Figure 26.13 Kinetic diapsid skull of a monitor lizard, *Varanus*, showing the joints that allow the snout and upper jaw to move on the rest of the skull. The quadrate can move at its dorsal and ventral ends and at both the lower jaw and the pterygoid bone. Also, the snout can be raised to open the mouth wide, or depressed to ensure parallel bite force between the jaws. Note that the lower temporal opening is very large with no lower border; this modification of the diapsid condition, common in modern lizards, provides space for expansion of large jaw muscles. The upper temporal opening lies dorsal and medial to the lower opening and is not visible in this drawing.

Figure 26.14 Tokay, *Gekko gecko*, of Southeast Asia has a true voice and is named after the strident repeated *to-kay, to-kay* call.

Figure 26.15 A large male marine iguana, *Amblyrhynchus cristatus*, of the Galápagos Islands, feeding underwater on algae. This is the only marine lizard in the world. It has special salt-removing glands in the eye orbits and long claws that enable it to cling to a hard surface while feeding on small red and green algae, its principal diet. It may dive to depths exceeding 10 m (33 feet) and remain submerged more than 30 minutes.

monitors or varanids (Varanidae), large, active predators that include the largest lizard, the Komodo dragon, *Varanus komodensis* (Figure 26.16); and **chameleons** (Chamaeleonidae), a group of arboreal lizards, mostly of Africa and Madagascar. Chameleons are entertaining creatures that catch insects with a sticky-tipped tongue that can be projected accurately and rapidly to a distance greater than the length of their body (Figure 26.17). Most lizards have four limbs and relatively short bodies, but in many the limbs are reduced, and a few, such as glass lizards (Figure 26.18), are completely limbless.

Most lizards have movable eyelids, whereas a snake's eyes are permanently covered with a transparent cap. Lizards have keen vision for daylight (retinas rich in both cones and rods; see pp. 743–744 for discussion of color vision), although one group, the nocturnal geckos, has retinas composed entirely of rods. Most lizards have an external ear that snakes lack. The inner ear of lizards is variable in structure, but as with other nonavian reptiles, hearing does not play an important role in the lives of most lizards. Geckos are exceptions because males are strongly vocal (to announce territory and to discourage approach of other males) and they must, of course, hear their own vocalizations.

Many lizards live in the world's hot, arid regions, aided by adaptations that make desert life possible. Lipids in their thick skin minimize water loss. Little water is lost in their urine because they primarily excrete uric acid, as do other groups that are successful in arid habitats (birds, insects, and pulmonate snails). Some, such as Gila monsters of southwestern United States deserts, store fat in their tails, which they use during drought to provide energy and metabolic water (Figure 26.19).

Lizards, like nearly all nonavian reptiles, are **ectotherms**, adjusting their body temperature by moving among different microclimates (see Chapter 30, p. 674). Because cold climates provide limited opportunities for ectotherms to raise their body temperature, relatively few nonavian reptile species live in cold climates. Ectotherms use considerably less energy than do endotherms; therefore, nonavian reptiles are successful in ecosystems with low productivity and warm climates, such as tropical deserts, dry, open forests, and grasslands. Thus, ectothermy is not an "inferior" characteristic of nonavian reptiles but a successful strategy for coping with specific environmental challenges.

The amphisbaenians, or "worm lizards," are lizards highly specialized for a **fossorial** (burrowing) life. Amphisbaenia means "double walk," in reference to their peculiar ability to move backward nearly as effectively as forward. Amphisbaenians have elongate, cylindrical bodies of nearly uniform diameter, and most lack any trace of external limbs (Figure 26.20). Eyes usually are hidden underneath the skin, and there are no external ear openings. Their skull is solidly built and either conical or spade-shaped, which aids in tunneling through the soil. The skin is formed into numerous, independently moving rings, which can grip the soil, creating a movement not unlike that of earthworms. Amphisbaenians have an

Figure 26.16   The Komodo dragon, *Varanus komodoensis*, is the largest lizard, reaching up to 3 m in length. It feeds on pigs, deer, and carrion, and like other monitors, is mildly venomous.

Figure 26.17   A chameleon snares a dragonfly. After cautiously edging close to its target, the chameleon suddenly lunges forward, anchoring its tail and feet to the branch. A split second later, it launches its sticky-tipped, foot-long tongue to trap the prey. The eyes of this panther chameleon, *Fucifer pardalis*, are swiveled forward to provide binocular vision and excellent depth perception.

Figure 26.18   A European glass lizard or sheltopusik, *Pseudopus apodus*. This legless lizard is stiff and has an extremely long, fragile tail that readily fractures when the animal is struck or seized. Most specimens have only a partly regenerated tip to replace a much longer tail previously lost. Glass lizards can be readily distinguished from snakes by a deep, flexible groove running along each side of the body. They feed on worms, insects, spiders, birds' eggs, and small nonavian reptiles.

Figure 26.19 A Gila monster, *Heloderma suspectum,* of southwestern United States desert regions. It and the related Mexican beaded lizard are the only highly venomous lizards known. These brightly colored lizards feed principally on birds' eggs, nesting birds, mammals, and insects. Unlike venomous snakes, the Gila monster secretes venom from glands in its lower jaw. The chewing bite is painful to humans but seldom fatal.

Figure 26.20 An amphisbaenian, or "worm lizard." Amphisbaenians burrow, using their solid skull as a digging tool. The species pictured, the Iberian worm lizard, *Blanus cinereus,* occurs in Portugal and Spain.

extensive distribution in South America and tropical Africa. In the United States, one species, *Rhineura floridana,* occurs in Florida.

Serpentes: Snakes

Snakes are a monophyletic group of squamates. Although the earliest, fossil snakes had limbs, all living snakes are limbless and usually lack both pectoral and pelvic girdles (the latter persists as a vestige in pythons, boas, and a few other snakes). The numerous vertebrae of snakes, shorter and wider than those of tetrapods, permit quick lateral undulations through grass and over rough terrain. Ribs increase rigidity of the vertebral column, providing more resistance to lateral stresses. Elongation of the **neural spine** gives the numerous muscles more leverage.

Many lineages of lizards exhibit reduction or loss of limbs, but none of these lineages experienced the remarkable diversification of snakes. The skull of snakes is more kinetic than that of lizards, enabling snakes to swallow much larger prey. Unlike lizard jaws, the two halves of the lower jaw (mandibles) are joined only by muscles and skin, allowing them to spread widely apart. Many skull bones are so loosely articulated that the entire skull can flex asymmetrically to accommodate oversized prey (Figure 26.21). Prey is slowly engulfed as the jaws and palatal bones (palatines and pterygoids), all of which bear posteriorly curved teeth, are moved forward over the prey. While one side of the jaws and palate are anchored in the prey, the other side is advanced, ratcheting the prey deeper into the mouth. Because a snake must keep breathing during the slow process of swallowing, its tracheal opening (glottis) is thrust forward between the two mandibles.

Figure 26.21 **A,** Lateral view of a python skull. Each side of the extremely kinetic skull has several movable bones (labeled) which permit extraordinary movements of the jaw in feeding. The halves of the lower jaw are united by flexible soft tissues, permitting wide separation and independent movement of each side. **B,** The great mobility of a snake's jaw and skull elements is evident in this snake, *Dayspeltis inornata,* swallowing an egg.

The cornea of a snake's eye, lacking a movable eyelid, is permanently protected with a transparent membrane called a spectacle, which, together with reduced eyeball mobility, gives snakes the unblinking stare that many people find unnerving. Most snakes have relatively poor vision, but arboreal snakes possess excellent binocular vision that helps them to track prey through branches where scent trails are difficult to follow (Figure 26.22).

Snakes have no external ears or tympanic membrane. This condition, together with absence of any obvious response to aerial sounds, led to a widespread opinion that snakes are totally deaf. Nonetheless, snakes do have internal ears, and recent work has shown quite clearly that within a limited range of low frequencies (100 to 700 Hz), hearing in snakes compares favorably with that of most lizards. Snakes are also quite sensitive to vibrations conducted through the ground.

Nevertheless, most snakes employ chemical senses rather than vision and hearing to hunt their prey. In addition to the usual olfactory areas in the nose, which are not well developed, snakes have a pair of pitlike **Jacobson's organs** (vomeronasal organs; p. 736) in the roof of the mouth. These are lined with richly innervated chemosensory epithelium. The forked tongue, flicked through the air, collects scent molecules (Figure 26.23). The tongue is then drawn past Jacobson's organs and information is transmitted to the brain, where scents are identified.

Boids (pythons and boas) and pit vipers have special heat-sensitive **pit organs** on their heads, located between their nostrils and their eyes (Figure 26.24). These respond to radiant energy (long-wave infrared, 5000 to 15,000 nm) and are especially sensitive to heat emitted by warm-bodied birds and mammals that are their food (infrared wavelengths of about 10,000 nm). Some measurements suggest that pit organs can distinguish temperature differences of only 0.003° C from a radiating surface. Pit organs serve to track warm prey and to aim strikes, which are as effective in total darkness as in daylight. The anatomy of the heat receptors is quite different between boids and pit vipers, suggesting that their heat receptors evolved independently.

Snakes have evolved several solutions to the obvious problem of movement without limbs. The most typical pattern of movement is **lateral undulation** (Figure 26.25A). Movement follows an S-shaped path, with a snake propelling itself by exerting lateral force against surface irregularities. A snake seems to "flow," because the moving loops appear stationary with respect to the ground. Lateral undulatory movement is fast and efficient under most but not all circumstances. **Concertina movement** (Figure 26.25B) enables a snake to move in a narrow passage, as when climbing a tree by using irregular channels in the bark. A snake extends forward while bracing S-shaped loops against the sides of the channel. To advance in a straight line,

Figure 26.23 A copperhead, *Agkistrodon contortrix*, flicks its tongue to smell its surroundings. Scent particles trapped on the tongue's surface are transferred to Jacobson's organs, olfactory organs in the roof of the mouth. Note the heat-sensitive pit organ between the nostril and the eye.

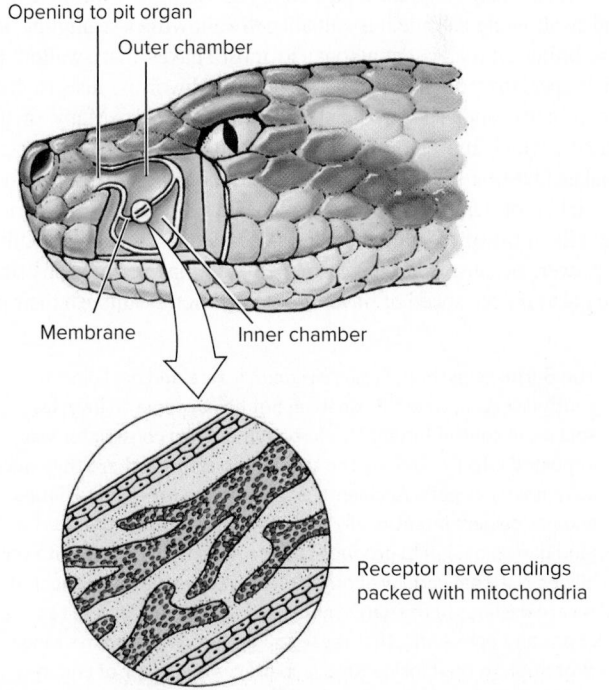

Opening to pit organ
Outer chamber
Membrane
Inner chamber
Receptor nerve endings packed with mitochondria

Figure 26.24 Pit organ of rattlesnake, a pit viper. Cutaway shows the location of a deep membrane that divides the pit into inner and outer chambers. Heat-sensitive nerve endings are concentrated in the membrane.

Figure 26.22 A parrot snake, *Leptophis ahaetulla*. The slender body of this Central American tree snake is an adaptation for traveling among branches.

A Lateral undulation   **B Concertina motion**   Move Contract Fix Stretch Move **C Rectilinear motion**   **D Sidewinder motion**

0 sec   1 sec   2 sec

Figure 26.25 Snake locomotion. **A,** Lateral undulation. **B,** Concertina motion. **C,** Rectilinear motion. **D,** Sidewinder motion. Refer to text for explanation.

as when stalking prey, many heavy-bodied snakes employ **rectilinear movement.** Two or three sections of the body rest on the ground to support the snake's weight. Intervening sections are lifted free of the ground and pulled forward by muscles (shown in red in Figure 26.25C) that originate on ribs and insert on the ventral skin. Rectilinear movement is a slow but effective way of moving inconspicuously toward prey, even when there are no surface irregularities. **Side-winding** is a fourth form of movement that enables desert vipers to move with surprising speed across loose, sandy surfaces with minimum surface contact (Figure 26.25D). The sidewinder rattlesnake moves by throwing its body forward in loops with its body lying at an angle of about 60 degrees to its direction of travel.

Most snakes capture their prey by grasping it with their mouth and swallowing it while it is still alive. Swallowing a struggling, kicking, biting animal is dangerous, so most snakes that swallow prey alive specialize on small prey, such as earthworms, insects, fishes, amphibians, and, less frequently, small mammals. Many of these snakes, which may be quite fast, locate prey by actively foraging. Snakes that first kill their prey by constriction (Figure 26.26) usually specialize on large, often mammalian prey. The largest constrictors can kill and swallow prey as large as deer, leopards, and crocodilians. However, because muscle rearrangements that permit constricting prey also reduce speed of travel, most constrictors ambush their prey.

The Burmese python, *Python bivittatus*, is rare and declining in southeast Asia, where it is native, but has become an invasive species in south Florida. This large (up to 5 m) constrictor was imported into the U.S. by the tens of thousands, where they were sold mostly as pets. Accidental escapes from breeding facilities and pet owners intentionally releasing their difficult-to-care-for (and dangerous) pets produced a reproducing population in Everglades National Park by about 2006. A 2011 study documented severe declines in the native mid-sized mammals, such as raccoons and opossums, that these snakes eat. The python's range expansion to the Florida Keys jeopardizes a number of endangered native species, such as the Key Largo wood rat. Through intensive efforts, thousands of Burmese pythons have been removed from south Florida, but the population continues to grow.

© Cleveland P. Hickman, Jr.

Figure 26.26 Nonvenomous African house snake, *Boaedon fuluginosus,* constricting a mouse before swallowing it.

Other snakes kill their prey by injecting it with venom. Less than 20% of all snakes are venomous, although venomous species outnumber nonvenomous species 4 to 1 in Australia. Venomous snakes are usually divided into five families, based in part on type of fangs.

Vipers (family Viperidae) have highly developed, movable, tubular fangs at the front of the mouth (Figure 26.27). The fangs lie in a membranous sheath when the mouth is closed. When a viper strikes, a special muscle and bone lever system erects the fangs as the mouth opens. Fangs are driven into the prey by the thrust of the strike, and venom is injected into the wound through a canal in the fangs. A viper immediately releases its prey after the bite and follows it until it is paralyzed or dead. Vipers of the Old World lack facial heat-sensing pits. Among the latter are the puff adders (Africa), the saw-scaled vipers (Africa and South Asia), and the

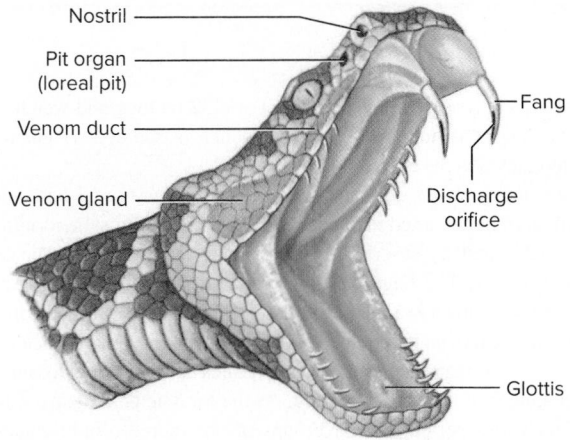

Figure 26.27 Head of rattlesnake showing the venom apparatus. The venom gland, a modified salivary gland, is connected by a duct to the hollow fang.

Nostril
Pit organ (loreal pit)
Venom duct
Venom gland
Fang
Discharge orifice
Glottis

Figure 26.28 Monocled cobra, *Naja kaouthia*. Cobras erect the front of the body and flatten the neck as a threat display. Although a cobra's strike range is limited, all cobras are dangerous because of the extreme toxicity of their venom.

© Mark Carwardine/Getty Images

common European adder. The Gaboon viper, of Africa, has the longest fangs of any snake, measuring about 5 cm (2 inches) long. Snakes of subfamily Crotalinae within family Viperidae are called **pit vipers** because they possess special heat-sensitive pit organs on their heads (Figure 26.24). All best-known North American venomous snakes are pit vipers, including cottonmouths, copperheads (Figure 26.23), and rattlesnakes. Each year in the United States approximately 3500 bites from pit vipers are reported, causing about 5 deaths.

A second, large family of venomous snakes (family Elapidae) has short, permanently erect fangs in the front of the mouth. In this group are cobras (Figure 26.28), mambas, coral snakes, and kraits. All of the terrestrial venomous snakes in Australia are elapids. Most elapids have extremely toxic venom. The highly venomous sea snakes are usually placed in a third family (Hydrophiidae). Mole vipers and stiletto snakes (Atractaspididae) are small, fossorial venomous species that vary in fang type. The large family Colubridae, which contains most familiar nonvenomous snakes, also includes numerous mildly venomous species, all of which have small, fixed fangs at the rear of the mouth. A few, including the boomslang and the savanna twigsnake of Africa, are very dangerous and have caused human fatalities.

Even the saliva of harmless snakes possesses limited toxic qualities, which is likely the ancestral state from which highly toxic venom evolved. Snake venoms have traditionally been divided into two types. The **neurotoxic** type acts mainly on the nervous system, affecting the optic nerves (causing blindness) or the phrenic nerve of the diaphragm (causing paralysis of respiration). The **hemorrhagin** type destroys red blood cells and blood vessels, causing extensive hemorrhaging of blood into tissue spaces. Actually, most snake venoms are complex mixtures of substances that are not easily assigned to one of these types. In addition, all venoms possess enzymes that accelerate digestion.

The toxicity of a venom is measured by the median lethal dose on laboratory animals (LD_{50}). By this standard the venoms of the inland taipan of Australia and some of the sea snakes appear to be the most deadly. However, several larger snakes inject more venom during a bite and are more dangerous. Black mambas are large and

fast, and inject a large amount of fast-acting toxin during a bite. This is one of the most feared snakes in Africa, as untreated bites are almost always fatal, sometimes in as little as 1 hour. It is estimated worldwide that at least 20,000 (possibly up to 90,000) people die from snakebite each year. Most deaths occur in India, Pakistan, Bangladesh, and nearby countries, where poorly shod people frequently encounter venomous snakes and frequently do not get immediate medical attention following a snakebite. The snakes primarily responsible for deaths in these areas are Russell's viper, sawscaled viper, and several species of cobras.

The LD_{50} (median lethal dose) has been the standardized procedure for assaying toxicity of drugs; it was originally developed in the 1920s by pharmacologists. In practice, small samples of laboratory animals, usually mice, are exposed to a graded series of doses of the drug or toxin. The dose that kills 50% of the animals in the test period is recorded as the LD_{50}. Expensive and time consuming, this classical procedure is being replaced by alternative methods that greatly reduce the number of animals needed. Among these alternatives are cytotoxicity tests that evaluate the ability of test substances to kill cells, and toxikinetic procedures that measure interaction of a drug or toxin with a living system.

Most snakes are **oviparous** species that lay their shelled, elliptical eggs beneath rotten logs, under rocks, or in holes in the ground. The remainder are **viviparous**, giving birth to well-formed young. Many of these, including all American pit vipers except for the tropical bushmaster, are **ovoviviparous**, in which the young

The Mesozoic World of Dinosaurs

In 1842, when the English anatomist Richard Owen coined the term *dinosaur* ("fearfully-great lizard") to describe fossil Mesozoic reptiles of gigantic size, only three poorly known dinosaur genera were distinguished. New and marvelous fossil discoveries quickly followed, and by 1887 zoologists were able to distinguish two groups of dinosaurs based on differences in structure of the pelvic girdles. The Saurischia ("lizard-hipped") had a simple, three-pronged pelvis with hip bones arranged much as they are in other nonavian reptiles. The large, bladelike ilium is attached to one or two sacral vertebrae. The pubis and ischium extend anteriorly and ventrally, respectively. The Ornithischia ("bird-hipped") had a somewhat more complex pelvis. The ilium and ischium were arranged similarly in ornithischians and saurischians, but the ornithischian pubis was a narrow, rod-shaped bone with anteriorly and posteriorly directed processes lying alongside the ischium. Oddly, while the ornithischian pelvis, as the name suggests, was similar to that of modern birds, birds are of the saurischian clade.

Dinosaurs and their living descendants, the birds, are archosaurs ("ruling lizards"), a group that includes crocodiles and pterosaurs (refer to the taxonomy of amniotes on p. 580). As traditionally recognized, dinosaurs are a paraphyletic group because they do not include birds.

From among the various archosaurian radiations of the Triassic period, there emerged a lineage with limbs drawn under the body to provide an upright posture. This lineage gave rise to the earliest dinosaurs of the late Triassic period. *Herrerasaurus,* a bipedal dinosaur from Argentina, has one of the most distinctive characteristics of dinosaurs: walking upright on pillarlike legs, rather than on legs splayed outward like those of modern amphibians and nonavian reptiles. This arrangement allowed the legs to support the great weight of the body while providing an efficient and rapid stride.

Two groups of saurischian dinosaurs have been proposed based on differences in feeding habits and locomotion: the carnivorous and bipedal theropods, which include birds, and the herbivorous and quadrupedal sauropods. *Coelophysis* was an early theropod with a body form typical of all theropods: powerful hindlegs with three-toed feet; long, heavy counterbalancing tail; slender, grasping forelimbs; flexible neck; and a large head with jaws armed with daggerlike teeth. Large predators such as *Allosaurus,* common during the Jurassic period, were replaced by even more massively built carnivores of the Cretaceous period, such as *Tyrannosaurus,* which reached a length of 12.2 m (40 ft), stood nearly 6 m high, and weighed more than 7200 kg (8 tons). Not all predatory saurischians were massive; several were swift and nimble, such as *Velociraptor* ("speedy predator") of the late Cretaceous period.

Herbivorous saurischians, the quadrupedal sauropods, appeared in the late Triassic period. Although early sauropods were small- and medium-sized dinosaurs, those of the Jurassic and Cretaceous periods attained gigantic proportions, the largest terrestrial vertebrates ever to have lived. *Brachiosaurus* reached 25 m (82 ft) in length and may have weighed in excess of 30,000 kg (33 tons). Even larger sauropods have

been discovered; *Argentinosaurus* was 40 m (132 ft) long and weighed about 80,000 kg. With long necks and long front legs, sauropods were the first vertebrates adapted to feed on trees.

The second group of dinosaurs, the Ornithischia, were all herbivorous. Although more varied in appearance than saurischians, the ornithischians are united by several derived skeletal features that indicate common ancestry. The huge back-plated *Stegosaurus* of the Jurassic period is a well-known example of armored ornithischians, which comprised two of the five major groups of ornithischians. Even more shielded with bony plates than stegosaurs were the heavily built ankylosaurs, "armored tanks" of the dinosaur world. As the Jurassic period gave way to the Cretaceous period, several groups of unarmored ornithischians appeared, although many bore impressive horns. The steady increase in ornithiscian diversity in the Cretaceous period paralleled a concurrent gradual decline in giant sauropods, which had flourished in the Jurassic period. *Triceratops* is representative of horned dinosaurs that were common in the late Cretaceous period. Even more prominent in the late Cretaceous period were the hadrosaurs, such as *Parasaurolophus,* which probably lived in large herds. Many hadrosaurs had skulls elaborated with crests that probably functioned as vocal resonators to produce species-specific calls. The fifth group, the bipedal pachycephalosaurs of the late Cretaceous period, had thick skulls possibly used in fighting.

Dinosaurs likely provided considerably more complex parental care than did most other nonavian reptiles. Because both crocodilians (p. 580) and birds, members of Archosauria, share well-developed parental care, it is likely that dinosaurs exhibited similar behavior. Fossil nests of dinosaurs have been discovered for several groups. In one case, a fossil adult of the small theropod *Oviraptor* was found with a nest of eggs. Originally, it was hypothesized that the adult was a predator on the eggs (*Oviraptor* means "egg seizer"). Later, an embryo in similar eggs was found and identified as *Oviraptor,* indicating that the adult was probably with its own eggs! Examination of baby *Maiasaura* (a hadrosaur) found in a nest revealed considerable wear on their teeth. This suggests that the babies had remained in a nest and possibly were fed by adults during part of their early life.

Sixty-five million years ago, the last Mesozoic dinosaurs became extinct, leaving birds and crocodilians as the only surviving lineages of archosaurs. The demise of dinosaurs coincided with a large asteroid impact on the Yucatan peninsula that would have produced worldwide environmental upheaval. Although the impact event is generally accepted as the primary cause of the extinctions, other events, including a massive volcanic eruption in the Deccan Plateau of India, lowering of sea level, and climate change, have been suggested to contribute to the demise of dinosaurs and other animals. We continue to be fascinated by the awe-inspiring, often staggeringly large creatures that dominated the Mesozoic era for 165 million years—an incomprehensibly long period of time. Today, inspired by clues from fossils and footprints from a lost world, scientists continue to piece together the puzzle of how various dinosaur groups arose, behaved, and diversified.

---

receive nourishment only from their yolk sacs. Other viviparous snakes, including garter snakes (*Thamnophis*), nourish their young with **placentae,** permitting exchange of materials between the embryonic and maternal bloodstreams. Snakes are able to store sperm and can lay several clutches of fertile eggs at long intervals after a single mating.

Sphenodonta: Tuataras

Sphenodonta is represented by two living species of the genus *Sphenodon* (Gr. *sphenos,* wedge, + *odontos,* tooth) of New Zealand (Figure 26.29). Tuataras are sole survivors of the sphenodontid lineage that diversified modestly during the early Mesozoic but declined

SAURISCHIANS

ORNITHISCHIANS

66 MY ago

CRETACEOUS

Titanosaurus
12 m (40 ft)

Parasaurolophus
10 m (33 ft)

Velociraptor
1.8 m (6 ft)

Triceratops
9 m (30 ft)

145 MY ago

JURASSIC

Brachiosaurus
25 m (82 ft)

Stegosaurus
9 m (30 ft)

Ilium
Ischium
Pubis

Allosaurus
11 m (35 ft)

201 MY ago

TRIASSIC

Coelophysis
3 m (10 ft)

Ilium
Pubis
Ischium

Herrerasaurus 4 m (13 ft)
One of the oldest known
dinosaurs. Has characteristics
of both saurischians and
ornithischians.

252 MY ago

Taxonomy of Early Amniotes and Living Nonavian Reptiles

The following cladistic taxonomy lists the living members of Amniota and several extinct groups of the clade Reptilia. There is no consensus of a system assigning Linnean ranks consistent with a cladistic taxnomy for these groups and thus ranks are not assigned here. However, it should be noted that the four major clades of living reptiles (Testudines, Squamata, Sphenodonta, and Crocodilia) are ranked as orders in many references. Relationships of turtles to other diapsids are controversial. Extinct groups are indicated by a dagger.

Anapsida[†] (an-ap′si-də) (Gr. *a n*, without, + *a psis,* arch): **anapsids.** Amniotes having a skull with no temporal opening.

Diapsida (dī-ap′si-də) (Gr. *di,* double, + *apsis,* arch): **diapsids.** Amniotes having a skull with two temporal openings.

Testudines (tes-tū′di-nēz) (L. *testudo,* tortoise) (**Chelonia**): **turtles.** Body in a bony case of dorsal carapace and ventral plastron; jaws with keratinized plates instead of teeth; vertebrae and ribs fused to overlying carapace; temporal openings lost; about 325 species.

Lepidosauria (le-pi-dō-sor′ē-ə) (Gr. *lepidos,* scale, + *sauros,* lizard). Characterized by sprawling posture; no bipedal specializations; diapsid skull usually modified by loss of one or both temporal arches; transverse cloacal slit, skin shed in one piece.

Squamata (skwä-mä′tə) (L. *squamatus,* scaly, + *ata,* characterized by): **snakes and lizards.** Skin of keratinized epidermal scales or plates, which are shed; quadrate movable; skull kinetic (except amphisbaenians); vertebrae usually concave in front; paired copulatory organs; about 5810 lizard species and 3370 snake species.

Sphenodonta (sfē′nō-don′tə) (Gr. *sphen,* wedge, + *odontos,* tooth): **tuataras.** Skull retains both pairs of temporal openings; vertebrae biconcave; quadrate immovable; parietal eye prominent; copulatory organ absent; two extant species in *Sphenodon.*

Ichthyosauria[†] (ik′thē-ō-sor′e-ə) (Gr. *ichthys,* fish, + *sauros,* lizard). Mesozoic marine dolphin-shaped diapsids with large eyes and vertical tails.

Sauropterygia[†] (sor-op-te-rij′ē-ə) (Gr. *sauros,* lizard, + *pteryginos,* winged). Carnivorous, Mesozoic marine diapsids with large, paddle-like limbs. This includes long-necked "plesiosaurs" and short-necked "pliosaurs."

Archosauria (ärk′ō-sor′ē-ə) (Gr. *archo,* ruling, + *sauros,* lizard). Orbit shaped like an upside-down triangle; anteorbital fenestra (opening in the skull anterior to the orbit) and gizzard present; heart ventricle fully divided; parental care of young; many forms bipedal.

Crocodilia (krok′ə-dil′ē-ə) (L. *crocodilus,* crocodile): **alligators, caimans, crocodiles, and gharials.** Skull elongate and massive; nares terminal; secondary palate present; vertebrae usually concave in front; forelimbs usually with five digits; hindlimbs with four digits; quadrate immovable; 25 species.

Pterosauria[†] (ter-ə-sor′ē-ə) (Gr. *pteron,* winged, + *sauros,* lizard). Flying, Mesozoic archosaurs with membranous wings.

Saurischia (sor-is′kē-ə) (Gr. *sauros,* lizard, + *ischion,* hip). Mesozoic dinosaurs; bipedal carnivores and mostly quadrupedal herbivores; ancestral (lizardlike) hip structure.

Sauropodomorpha[†] (sor′ō-pod-ō-morf′ə) (Gr. *sauros,* lizard, + *podos,* foot, + *morphē,* form). Herbivorous saurischians including Mesozoic giants such as *Brachiosaurus, Apatosaurus,* and *Diplodocus.*

Theropoda (ther-ə-pō-d′ə) (Gr. *thēr,* wild beast, + *podos,* foot). Carnivorous saurischians including huge predators such as *Tyrannosaurus* and small, agile predators such as *Deinonychus* and *Velociraptor.* Birds also are a member of this clade.

Ornithischia[†] (or-ni-this′kē-ə) (Gr. *ornis,* bird, + *ischion,* hip). Mesozoic dinosaurs; bipedal and quadrupedal herbivores such as *Stegosaurus, Triceratops,* and *Parasaurolophus;* derived (birdlike) hip structure.

Synapsida (sin-ap′si-də) (Gr. *syn,* together, + *apsis,* arch). Amniotes having skull with one pair of lateral temporal openings. This clade contains numerous extinct forms from the Carboniferous to Triassic periods called "pelycosaurs" and "therapsids," and the living mammals.

---

toward the end of that era. Tuataras were once widespread throughout the two main islands of New Zealand but are now restricted to small islets of Cook Strait and off the northeast coast of North Island. On some of these islands, under protection by the New Zealand government, they are prospering. Loss of the tuatara populations on the main islands of New Zealand was caused by humans' intentional or accidental introducing of nonnative animals, including rats, cats, dogs, and goats, which preyed upon tuataras and their eggs or destroyed their habitat. Tuataras are particularly vulnerable because they have slow growth and low reproductive rates.

Tuataras are lizardlike forms, up to 80 cm in length, that live in burrows often shared with sea birds called petrels. They have one of the slowest reproductive rates of reptiles: they take 10–20 years to become sexually mature and typically produce eggs only once every four years, which take seven months to hatch. They are slow-growing animals with long lives; one captive male was 118 years old in 2015.

Tuataras have captured the interest of zoologists because of numerous features that are almost identical to those of early Mesozoic diapsids that lived 200 million years ago. These features include a diapsid skull with two temporal openings. Tuataras also bear a well-developed median parietal eye complete with elements of cornea, lens, and retina (although because it is buried beneath opaque skin, this "third eye" can register only changes in light intensity). Parietal eyes, which also occur in many anamniotes, play a role in regulating daily and seasonal rhythms of behavior. *Sphenodon* represents one of the slowest rates of morphological evolution known among vertebrates.

Crocodilia: Crocodiles, Alligators, Caimans, and Gharials

Modern crocodilians and birds are the only survivors of the archosaurian lineage that gave rise to the great Mesozoic diversification of dinosaurs and their kin. Although modern crocodiles belong to a clade that began its diversification in the late Cretaceous period, they differ little in anatomy from crocodilians of the early Mesozoic era. Having remained mostly unchanged for nearly 200 million years,

A

Figure 26.29 Tuatara, *Sphenodon punctatus,* a living representative of Sphenodonta. This "living fossil" reptile has on top of its head a well-developed parietal eye with retina, lens, and nervous connections to the brain. Although covered with scales, this third eye is sensitive to light. Tuataras occur today only on certain islands off the coast of New Zealand.

B

Figure 26.30 Crocodilians. **A,** Nile crocodile, *Crocodylus niloticus,* basking. The fourth tooth of the lower jaw fits outside the slender upper jaw; alligators lack this feature. **B,** American alligator, *Alligator mississippiensis,* an increasingly noticeable resident of rivers, bayous, and swamps of the southeastern United States.

crocodilians face an uncertain future in a world dominated by humans. Modern crocodilians are divided into three families: alligators and caimans, mostly a New World group; crocodiles, which are widely distributed and include the saltwater crocodile, one of the largest living nonavian reptiles; and gharials, represented by a single species that now persists only in parts of India and Nepal.

All crocodilians have an elongate, robust, well-reinforced skull and massive jaw musculature arranged to provide a wide gape and rapid, powerful closure. Teeth are set in sockets, a type of dentition called **thecodont** that was typical of Mesozoic archosaurs, including the earliest birds. Another adaptation, found in no other vertebrate except mammals, is a complete secondary palate. This innovation pushed the internal nares to the posterior, allowing a crocodilian to breathe when its mouth is filled with water or food (or both). Crocodilians, along with birds and mammals, have a four-chambered heart with completely divided atria and ventricles.

Estuarine crocodiles, *Crocodylus porosus,* of southern Asia, and Nile crocodiles, *C. niloticus* (Figure 26.30A), grow to great size (adults weighing 1000 kg have been reported) and are swift and aggressive. Crocodiles are known to attack animals as large as cattle, deer, and people. Alligators are usually less aggressive than these crocodiles and far less dangerous to people. In the United States, *Alligator mississippiensis* (Figure 26.30B) is the only species of alligator; *Crocodylus acutus,* restricted to extreme southern Florida, is the only species of crocodile. Large crocodilians are powerful animals, and adults have almost no nonhuman enemies. However, during early life, they are vulnerable to many predators. Nests left unguarded by the mother are almost certain to be discovered and raided by any of several mammals that relish eggs, and young hatchlings may be devoured by large fishes or wading birds.

Male alligators make loud bellows during the mating season. Crocodilians are oviparous. A female usually lays 20 to 50 eggs in a mass of dead vegetation or buried in the sand, and then guards them. Unlike other living nonavian reptiles, crocodilians provide extensive parental care. The mother hears vocalizations from hatching young and responds by opening the nest to allow the hatchlings to escape. Young are guarded by their mother for two years or more after hatching. Although the young are capable of catching their own food after hatching, they also feed on small pieces that fall from food the mother is eating. As with many turtles and some lizards, incubation temperature of the eggs determines sex ratio of the offspring. However, unlike turtles (p. 570), low nest temperatures produce only females, whereas high nest temperatures produce only males.

Crocodiles and alligators can be distinguished on the basis of head morphology. Crocodiles have a relatively narrow snout, and when their mouths are closed, the fourth lower jaw tooth is visible where it fits into a notch on the upper jaw (Figure 26.30A). Alligators (and caimans) generally have a broader snout (Figure 26.30B), and their fourth lower jaw tooth is hidden when the mouth is closed. Gharials have very narrow snouts filled with slender teeth and eat mostly fishes.

SUMMARY

Amniotes diverged from a group of early tetrapods during the late Paleozoic era, about 300 million years ago, and diversified into diverse forms occupying a range of aquatic and terrestrial habitats. Their success as terrestrial vertebrates can be attributed to several adaptations, including the amniotic egg. The amniotic egg, with its shell and four extraembryonic membranes—the amnion, allantois, chorion, and yolk sac—permits rapid development of embryos in terrestrial environments. Additional amniote adaptations that support occupation of dry environments and a relatively active lifestyle include a thick, water-resistant skin, excretion of urea or uric acid, high-surface-area lungs ventilated by trunk muscles, expanded jaw musculature, and an efficient cardiovascular system.

Before the end of the Paleozoic era, amniotes diversified to form three groups distinguished by skull structure: anapsids, which lack temporal fenestrae; synapsids, which have one pair of temporal fenestrae; and diapsids, which have two pairs of temporal fenestrae. Mammals evolved from early synapsids. Early diapsids gave rise to all living nonavian reptiles (tentatively including turtles) and to birds. No living amniotes retain the Paleozoic anapsid condition. One clade of diapsids, the archosaurs, underwent a great, worldwide diversification during the Mesozoic era into large and morphologically diverse forms, including the ichthyosaurs, plesiosaurs, pterosaurs, and dinosaurs. Although these are now extinct, descendants of some dinosaurs survived the great extinction at the end of the Mesozoic era, and underwent their own diversification as birds.

Turtles (Testudines), with their distinctive shells, have changed little in anatomy since the Triassic period. Turtles are a small group of long-lived terrestrial, semiaquatic, and aquatic species. They lack teeth, and instead bear keratinized plates on their jaws. All are oviparous, and all, including marine forms, bury their eggs.

Lizards and snakes (Squamata) represent 95% of all living nonavian reptiles. Both are diversified and successful groups, particularly in hot climates. Lizards are distinguished from snakes by having united lower jaw halves, movable eyelids, and external ear openings. Lizards and snakes are ectothermic, regulating their temperature by moving among different microenvironments. Most lizards and snakes are oviparous, although viviparity is not uncommon, especially in cooler climates. Amphisbaenians are a small group of tropical lizards specialized for burrowing. They have ringed, usually limbless, bodies and a solid skull.

Snakes, which evolved from one group of lizards, are characterized by elongate, limbless bodies and a highly kinetic skull that allows them to swallow prey several times larger than the snake's diameter. Most snakes rely on chemical senses, including Jacobson's organ, to hunt prey, rather than on visual and auditory senses. Two groups of snakes (pit vipers and boids) have unique infrared-sensing organs for tracking prey. Some snakes swallow their prey alive; others kill their prey by constriction or with venom. Different groups of venomous snakes are distinguished by the anatomy and placement of their fangs.

The tuataras of New Zealand (Sphenodonta) are relics and sole survivors of a group that otherwise disappeared 100 million years ago. They bear several features, including two pairs of temporal openings and a prominent parietal eye, that are almost identical to those of early Mesozoic diapsids. These rare reptiles are particularly vulnerable because they have slow growth and low reproductive rates.

Crocodiles, alligators, caimans, and gharials (Crocodilia) are the only living nonavian representatives of the archosaurian clade, which includes the extinct dinosaurs and the living birds. Crocodilians have several adaptations for a carnivorous, semiaquatic life, including a massive skull with powerful jaws, and a secondary palate. They have the most complex parental care of any living nonavian reptile.

REVIEW QUESTIONS

1. What are the four membranes associated with amniotic eggs? What is the function of each of these membranes?
2. How do the skin and respiratory systems of amniotes differ from those of their early tetrapod ancestors?
3. Amniotes are divided into three groups based on their skull morphologies. What are these three groups, and how do their skulls differ? Which living amniotes, if any, originated from each of these three groups?
4. Why are "reptiles," as traditionally defined, a paraphyletic group? How has cladistic taxonomy revised Reptilia to make it monophyletic?
5. Describe ways in which amniotes are more functionally or structurally suited for terrestriality than are anamniotes.
6. Describe the principal structural features of turtles that distinguish them from other nonavian reptiles.
7. How might nest temperature affect egg development in turtles? In crocodilians?
8. What is meant by a "kinetic" skull and what benefit does it confer? How are snakes able to eat large prey?
9. Most lizards and snakes are oviparous, but some are ovoviviparous or have placental viviparity. Compare these methods of reproduction in squamates. In what climate is viviparity most common?
10. In what ways are the special senses of snakes similar to those of lizards, and in what ways have they evolved for specialized feeding strategies?
11. What are amphisbaenians? What morphological adaptations do they have to aid in burrowing?
12. What is the function of Jacobson's organ of snakes?
13. What is the function of the "pit" of pit vipers?
14. Describe how a snake moves by lateral undulation. Why might this form of locomotion be inefficient on an unstable surface (such as sand) or a surface lacking irregularities? What forms of locomotion would work for a snake under these conditions?
15. What is the difference in structure or location of the fangs of a rattlesnake, a cobra, and an African boomslang?
16. How does snake venom kill prey? How is venom toxicity measured?
17. Why are tuataras (Sphenodon) of special interest to biologists? Why are they rare?
18. Distinguish ornithischian and saurischian dinosaurs based on their hip anatomy.
19. Was the parental care of a dinosaur more like that of a lizard or that of a crocodile? Explain.
20. From which diapsid subgroup have crocodilians descended? What other major fossil and living vertebrate groups belong to this same subgroup? In what structural and behavioral ways do crocodilians differ from other living nonavian reptiles?

For Further Thought How could changes in the environment affect populations of species with temperature-dependent sex determination?

SELECTED REFERENCES

Alvarez, W., and F. Asaro. 1990. An extraterrestrial impact. Sci. Am. **263:** 78–84 (Oct.). *This article and an accompanying article by V. E. Courtillot, "A volcanic eruption," present opposing interpretations of the cause of the Cretaceous mass extinction that led to the demise of the dinosaurs.*

Cogger, H. G., and R. G. Zweifel (eds.). 1998. Encyclopedia of reptiles and amphibians. San Diego, Academic Press. *This comprehensive and lavishly illustrated volume was written by some of the best-known herpetologists in the field.*

Crews, D. 1994. Animal sexuality. Sci. Am. **270:** 108–114 (Jan.). *The reproductive strategies of reptiles, including nongenetic sex determination, provide insights into the origins and functions of sexuality.*

Erickson, G. M. 1999. Breathing life into *Tyrannosaurus rex*. Sci. Am. **281:** 42–49 (Sept.). *Present evidence suggests that* T. rex *was gregarious and acquired food by scavenging and predation.*

Gans, C., and 15 other authors, eds. 1969–2010. Biology of the Reptilia. *This monumental work includes 22 edited volumes that provide a detailed resource about nonavian reptiles.*

Greene, H. W. 1997. Snakes: The evolution of mystery in nature. Berkeley, University of California Press. *Beautiful photographs accompany a well-written volume for the scientist or novice.*

Halliday, T. R., and K. Adler(eds.). 2002. The encyclopedia of reptiles and amphibians. Richmond Hill, Ontario, Firefly Books Ltd. *Comprehensive and beautifully illustrated treatment of reptilian groups with helpful introductory sections on origins and characteristics.*

King, G. 1996. Reptiles and herbivory. London, Chapman & Hall. *Explains the adaptations reptiles use in obtaining nutrients from a herbivorous diet.*

Lohmann, K. J. 1992. How sea turtles navigate. Sci. Am. **266:** 100–106 (Jan.). *Evidence suggests that sea turtles use the earth's magnetic field and the direction of ocean waves to navigate back to their natal beaches to nest.*

Martill, D. M., H. Tischlinger, and N. R. Longrich. 2015. A four-legged snake from the early Cretaceous of Gondwana. Science **349:** 416–419. *The first known snake with four limbs. It fed on small vertebrates, probably killing them by constriction.*

Nagashima, H., F. Sugahara, M. Takechi, R. Ericsson, Y. Kawashima-Ohya, Y. Narita, and S. Kuratani. 2009. Evolution of the turtle body plan by the folding and creation of new muscle connections. Science **325:** 193–196. *Evo-devo research offers insight into how the turtle shell evolved.*

Paul, G. S. 2000. The Scientific American book of dinosaurs. New York, St. Martin's Press. *Essays emphasizing functional morphology, behavior, evolution, and extinction of dinosaurs.*

Pianka, E. R., and L. J. Vitt. 2003. Lizards: windows to the evolution of diversity. Berkeley, University of California Press. *Hundreds of color photographs highlight a treatment of the behavior and evolution of lizards.*

Schoch, R. R., and H. -D. Sues. 2015. A middle Triassic stem-turtle and the evolution of the turtle body plan. Nature **523:** 584–587. *The fossil turtle* Pappochelys *has two pairs of temporal openings, confirming that turtles are diapsids.*

Sumida, S. S., and K. L. M. Martin (eds.). 1997. Amniote origins: completing the transition to land. San Diego, Academic Press. *Discusses the diversity, evolution, ecology, and adapative morphology of early amniotes, with special emphasis on the origin of the amniotic egg.*

Vitt, L. J., and J. P. Caldwell. 2013. Herpetology: an introductory biology of amphibians and reptiles, ed. 4. San Diego, Academic Press. *A general textbook of herpetology.*

Vonk, F. J., and 16 other authors. 2013. Evolutionary origin and development of snake fangs. Nature **454:** 630–633. *Despite their incredible morphological variation, all snake fangs are homologous.*

27

CHAPTER

Birds

Snow geese, Chen caerulescens, *during migration.*

- **PHYLUM CHORDATA**
  - Aves

Aves

Chordata

Long Trip to a Summer Home

Some birds, having mastered flight, use this power to make long seasonal migrations. Moving between southern wintering regions and northern summer breeding regions with long days and an abundance of insects provides parents with ample food to rear their young. Predators of birds are not so abundant in the far North, and a brief once a-year appearance of vulnerable young birds does not encourage buildup of predator populations. Migration also vastly increases the amount of space available for breeding and reduces aggressive territorial behavior. Finally, migration favors homeostasis—the balancing of physiological

processes that maintains internal stability—by allowing birds to avoid climatic extremes.

The migratory pageant inspires wonder, and the physiological mechanisms of migration are equally challenging to researchers. What determines the timing of migration, and how does each bird store sufficient fuel for the journey? How did the sometimes difficult migratory routes originate, and what cues do birds use in navigation? How does instinct drive the migratory waves in spring and fall, carrying most birds successfully to their northern nests, while countless others fail and die, winnowed by this ever-challenging task?

Of the vertebrates, birds (Aves, ā′vēz; L. pl. of *avis,* bird) are the most noticeable, the most melodious, and, many think, the most beautiful. With more than 10,500 species collectively distributed nearly worldwide, birds outnumber any other vertebrate group except fishes. Birds inhabit forests and deserts, mountains and prairies, and all oceans. Four species are known to have visited the North Pole, and one, a skua, was seen at the South Pole. Some birds live in total darkness in caves, finding their way by echolocation, and others dive to depths greater than 45 m to prey on aquatic life. The "bee" hummingbird of Cuba, weighing only 1.8 g, is one of the smallest vertebrate endotherms.

The single unique feature that distinguishes birds from other living animals is their feathers. If an animal has feathers, it is a bird; if it lacks feathers, it is not a bird. However, we note that feathers were not so diagnostic in the past; some theropod dinosaurs that were not birds had feathers.

There is great uniformity of structure among birds. Despite approximately 150 million years of evolution, during which they proliferated and adapted to specialized ways of life, we have no difficulty recognizing a living bird. In addition to feathers, all birds have forelimbs modified into wings (although not always used for flight); all have hindlimbs adapted for walking, swimming, or perching; all have keratinized beaks lacking teeth; and all lay eggs. The reason for this great structural and functional uniformity is that birds evolved into flying animals, which forces them to retain these diagnostic characteristics.

A bird's entire anatomy is designed around flight. An airborne life for a large vertebrate is a highly demanding evolutionary challenge. A bird must, of course, have wings for lift and propulsion. Bones must be light yet serve as a rigid airframe. The respiratory system must be highly efficient to meet the intense metabolic demands of flight. A bird must have a rapid and efficient digestive system to process an energy rich diet; it must have a high metabolic rate; and it must have a high pressure circulatory system. Above all, birds must have a finely tuned nervous system and acute senses, especially superb vision, to handle the complex problems of headfirst, high-velocity flight.

27.1 ORIGIN AND RELATIONSHIPS

Approximately 147 million years ago, a flying animal died and settled to the bottom of a shallow marine lagoon in what is now Bavaria, Germany. It was rapidly covered with fine silt and eventually fossilized. There it remained until discovered in 1861 by a workman splitting slate in a limestone quarry. The fossil was approximately the size of a crow, with a skull not unlike that of modern birds except that the beaklike jaws bore small bony teeth set in sockets like those of dinosaurs (Figure 27.1). The skeleton was decidedly reptilian, with a long, bony tail, clawed fingers, and abdominal ribs. It might have been considered a typical theropod dinosaur except that it carried an unmistakable imprint of **feathers.** Named *Archaeopteryx lithographica* (är-kē-op′ter-iks lith-ō-graf′i-kə, Gr., meaning "ancient wing inscribed in stone"), the fossil was an especially fortunate discovery because the fossil record of birds was disappointingly meager. The finding was also dramatic because it demonstrated beyond reasonable doubt the phylogenetic relatedness of birds and extinct theropod dinosaurs.

Zoologists had long recognized the similarity of birds and nonavian reptiles. The skulls of birds and nonavian reptiles abut the first neck vertebra by a single occipital condyle (a small bony knob; mammals have two such knobs). Birds and nonavian reptiles have a single middle ear bone, the stapes (mammals have three middle ear bones). Birds and nonavian reptiles have a lower jaw consisting of

B

Figure 27.1 *Archaeopteryx,* a 147-million-year-old early bird. **A,** Cast of the second and most nearly perfect fossil of *Archaeopteryx,* which was discovered in a Bavarian stone quarry. Twelve specimens of *Archaeopteryx* have been discovered, the most recent one in 2011. **B,** Reconstruction of *Archaeopteryx.*

© Jason Edwards/Getty Images

A

five or six bones, whereas the lower jaw of mammals has one bone, the dentary. Birds and nonavian reptiles excrete their nitrogenous wastes as uric acid, whereas mammals excrete theirs as urea. Birds and most nonavian reptiles lay large, yolked eggs with the early embryo developing on the surface by shallow cleavage divisions.

The distinguished English zoologist Thomas Henry Huxley was so impressed with these and many other anatomical and physiological affinities that he called birds "glorified reptiles" and classified them with a group of dinosaurs called theropods (p. 578) that displayed several birdlike characteristics (Figures 27.2 and 27.3).

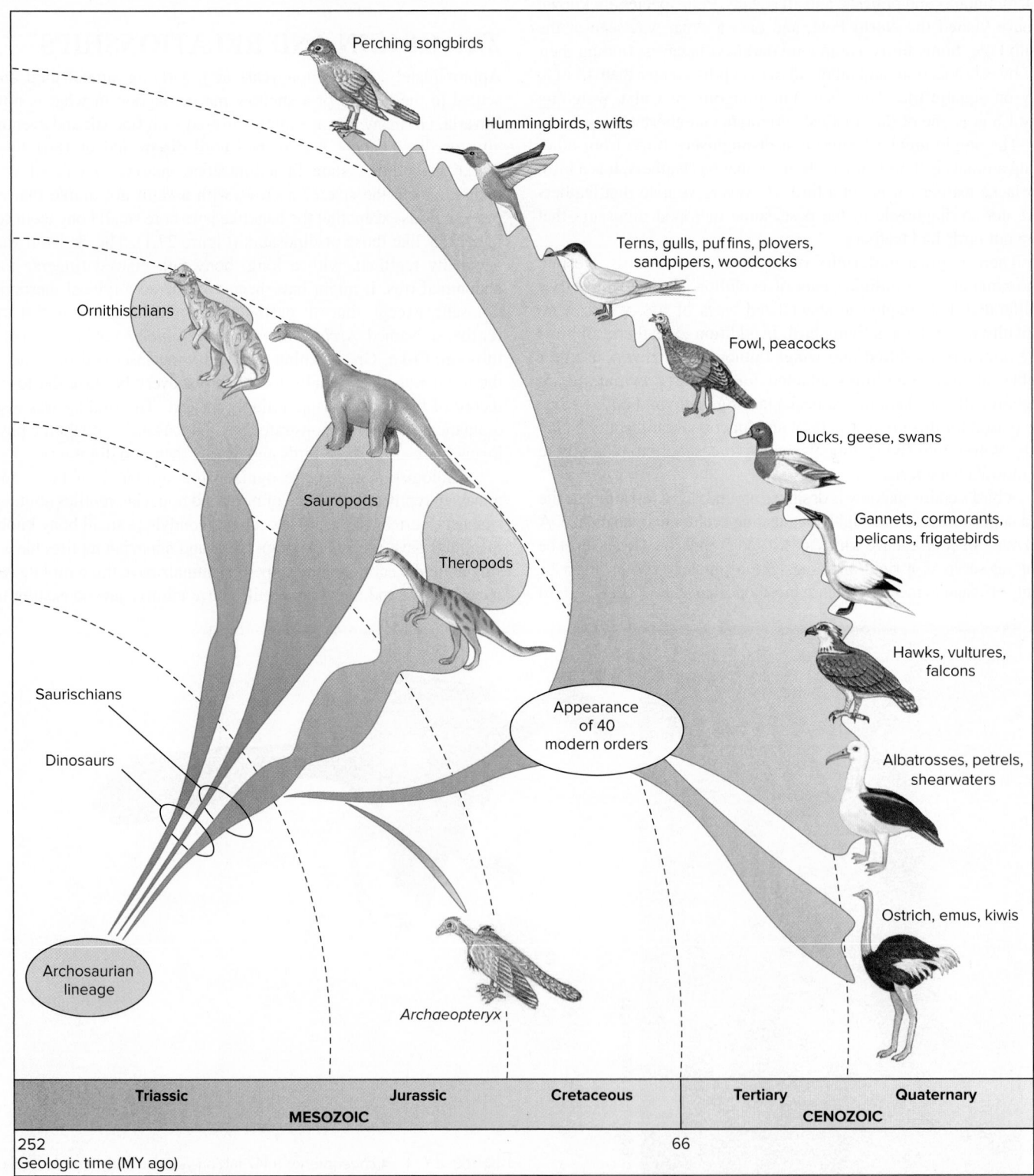

Figure 27.2 Evolution of modern birds. Of 40 living bird orders, 9 of the more important are shown. The earliest known bird, *Archaeopteryx,* lived in the Upper Jurassic period about 147 million years ago. *Archaeopteryx* uniquely shares many specialized aspects of its skeleton with the smaller theropod dinosaurs and is considered to have evolved within the theropod clade. Evolution of modern bird orders occurred rapidly during the Cretaceous and early Tertiary periods.

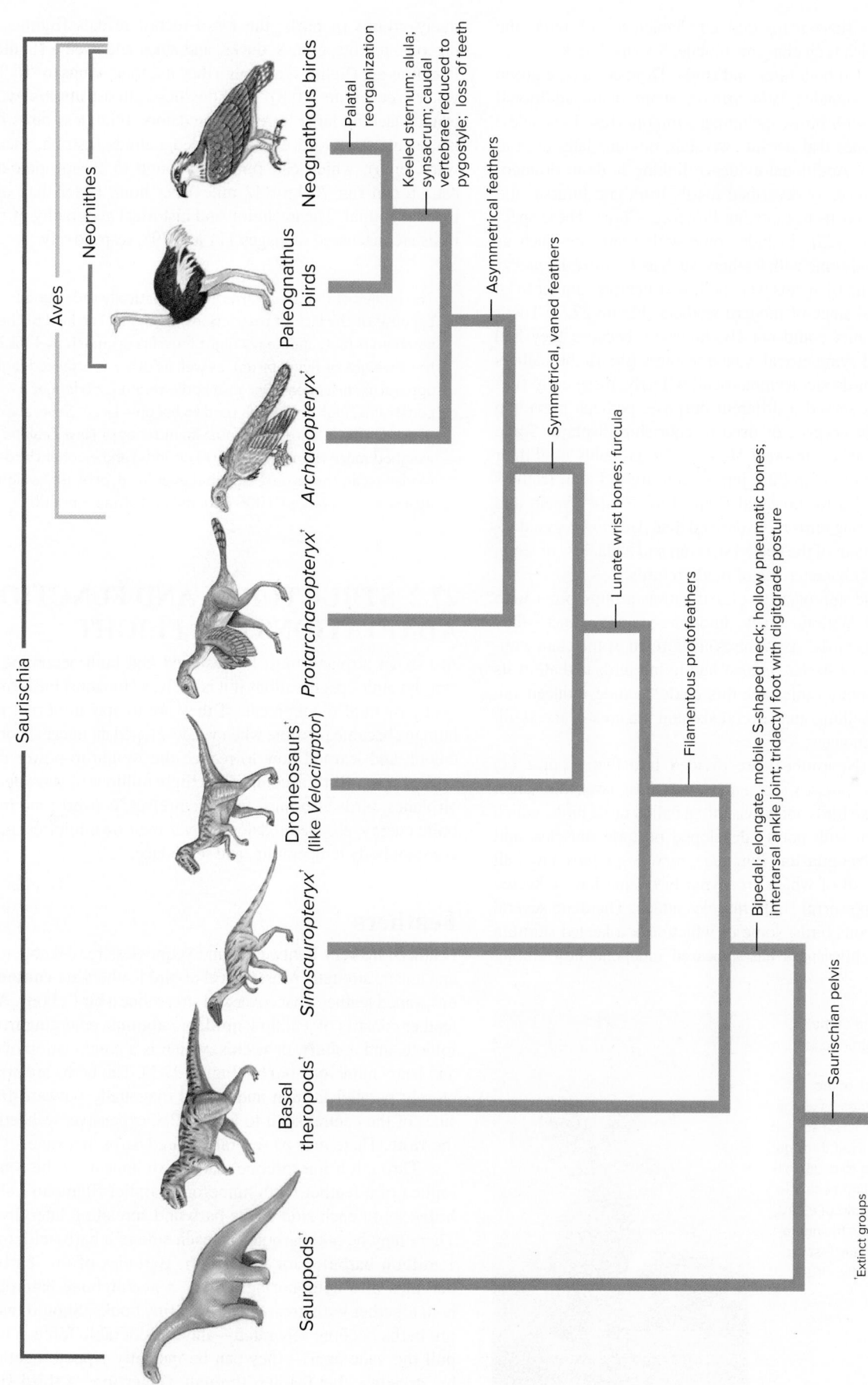

Figure 27.3 Cladogram of Saurischia, showing the relationship of several taxa to modern birds. Shown are a few shared derived characters, mostly related to flight. The ornithischians are the sister group to the saurischians and all are members of the clade Archosauria (see Figures 26.1 and 26.2).

Theropod dinosaurs share many derived characters with birds, the most obvious of which is an elongate, mobile, S-shaped neck.

Feathers preceded both birds and flight. Dromeosaurs, a group of theropods that includes *Velociraptor,* share many additional derived characters with birds, including a furcula (fused clavicles) and lunate wrist bones that permit swiveling motions later used in flight (Figure 27.3). Additional evidence linking birds to dromeosaurs comes from recently described fossils from late Jurassic and early Cretaceous deposits in Liaoning Province, China. These spectacular dromeosaur fossils include some with filaments, such as *Sinosauropteryx,* and some with feathers, such as *Protarchaeopteryx* and *Caudipteryx.* The filaments were hollow structures, similar to an early developmental stage of modern feathers (Figure 27.5). These feathered dromeosaurs could not fly, however, because they had short forelimbs and symmetrical, vaned feathers (the flight feathers of modern flying birds are asymmetrical). Clearly, these early filaments and feathers served a different purpose, perhaps providing thermoregulation or crypsis, or used in courtship displays. Some dromeosaurs, such as the arboreal *Microraptor,* probably used their feathers for gliding or controlling jumps from trees. Later, feathers were exapted (p. 121) for powered flight. Fossils from Spain and Argentina, representing birds more derived than *Archaeopteryx,* document the development of the keeled sternum and alula, loss of teeth, and fusion of bones characteristic of modern birds.

A phylogenetic approach to classification groups birds with theropod dinosaurs. With this view, dinosaurs are not extinct—they are with us today as birds! Aves (birds) is defined as the clade containing the ancestor of *Archaeopteryx* and living birds and all of its descendants. Synapomorphies for this clade include reduced tail length, larger cerebellum, and several skeletal features of the skull, vertebrae, and appendages.

Living birds (Neornithes) are divided into two groups: (1) **Paleognathae** (Gr. *palaios,* ancient, + *gnathos,* jaw), the large, flightless, ostrichlike birds and kiwis, often called ratite birds, which have a flat sternum with poorly developed pectoral muscles, and tinamous, and (2) **Neognathae** (Gr. *neos,* new, + *gnathos,* jaw), all other birds, nearly all of which are flying birds that have a keeled sternum to which powerful flight muscles attach. There are several flightless neognathous birds, some of which lack a keeled sternum (Figure 27.4). Flightlessness has appeared independently among

many groups of birds; the fossil record reveals flightless wrens, pigeons, parrots, cranes, ducks, and auks, and even a flightless owl. Penguins are flightless although they use their wings to "fly" through water (see Figure 10.8). Flightlessness almost always evolved on islands lacking large terrestrial predators. Flightless birds living on continents today are the large paleognathids (ostrich, rhea, cassowary, emu), which can run fast enough to escape predators. An ostrich can run 70 km (42 miles) per hour, faster than any other bipedal animal. The evolution and historical geography of flightless birds are discussed on pages 111 and 798, respectively.

> The bodies of flightless birds are dramatically redesigned because of the lack of restrictions for flight. The keel of the sternum is lost, and heavy flight muscles (as much as 17% of the body weight of flying birds), as well as other specialized flight apparatus, disappear. Because body weight is no longer a constraint, flightless birds tend to become large. Several extinct flightless birds were enormous: giant moas of New Zealand weighed more than 225 kg (500 pounds) and elephantbirds of Madagascar, the largest bird that ever lived, probably weighed nearly 450 kg (about 1000 pounds) and stood 3 m tall.

27.2 STRUCTURAL AND FUNCTIONAL ADAPTATIONS FOR FLIGHT

Just as an airplane must be designed and built according to rigid aerodynamic specifications if it is to fly, so too must birds meet stringent structural requirements if they are to stay airborne. Flight by humans became possible when we developed an internal combustion engine and learned how to reduce the weight-to-power ratio to a critical point. Birds accomplished flight millions of years ago. Unlike airplanes, birds also must feed themselves, convert food into metabolic energy, escape predators, repair their own injuries, maintain a constant body temperature, and reproduce.

Feathers

Feathers are very lightweight and yet possess remarkable toughness and tensile strength. Most typical of bird feathers are **contour feathers,** vaned feathers that cover and streamline a bird's body. A contour feather consists of a hollow **quill,** or **calamus,** emerging from a skin follicle, and a **shaft,** or **rachis,** which is a continuation of the quill and bears numerous **barbs** (Figure 27.5). The barbs are arranged in closely parallel fashion and spread diagonally outward from both sides of the central shaft to form a flat, expansive, webbed surface, the **vane.** There may be several hundred barbs in a vane.

Through a microscope, each barb appears to be a miniature replica of a feather, with numerous parallel filaments called **barbules** set in each side of the barb and spreading laterally from it. There may be 600 barbules on each side of a barb, with more than 1 million barbules for the feather. Barbules of one barb overlap barbules of a neighboring barb in a herringbone pattern and are held together with great tenacity by tiny hooks. Should two adjoining barbs become separated—and considerable force is needed to pull the vane apart—they can be instantly zipped together again by drawing the feather through fingertips. A bird does this

Figure 27.4 One of the strangest birds in a strange land, a flightless cormorant, *Phalacrocorax harrisi,* of the Galápagos Islands, dries its wings after a fishing forage. It is a superb swimmer, propelling itself through the water with its feet to catch fish and octopuses. The flightless cormorant is an example of a bird with carinate ancestors (having a keeled sternum) that has lost the keel and the ability to fly. Order Suliformes.

© Cleveland P. Hickman, Jr.

Figure 27.5

Types of bird feathers and their development. **A** to **E,** Successive stages in development of a vaned, or contour, feather. Growth occurs within a protective sheath, **D,** that splits open when growth is complete, allowing the mature feather to spread flat. **F** to **H,** Other feather types, including a pheasant feather with aftershaft, **F;** filoplumes, **G;** and down feathers, **H.**

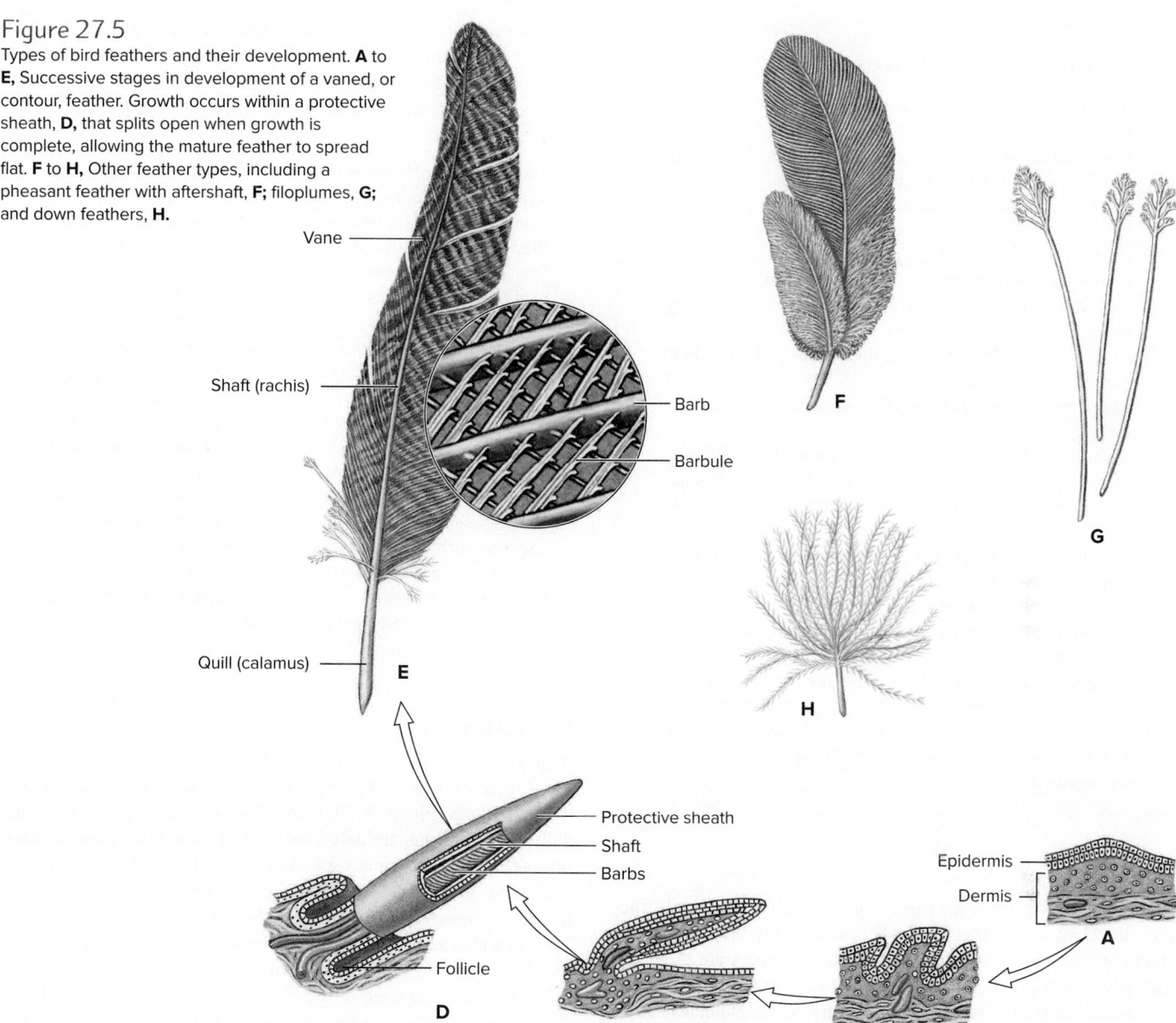

preening with its beak, and it spends much time preening to keep its feathers in perfect condition.

Types of Feathers

Different types of bird feathers serve different functions. **Contour feathers** (Figure 27.5E) give the bird its outward form and are the type we have already described. Contour feathers that extend beyond the body and are used in flight are called **flight feathers. Filoplume feathers** (Figure 27.5G) are hairlike, degenerate feathers; each is a weak shaft with a tuft of short barbs at the tip. They are the "hairs" of a plucked fowl. They have no known function. Bristles around the mouths of flycatchers and whippoorwills are probably modified filo-plumes. **Down feathers** (Figure 27.5H) are soft tufts without a prominent rachis, hidden beneath contour feathers. They are soft because their barbules lack hooks. They are especially abundant on the breast and abdomen of water birds and on young quail and grouse and function principally to conserve heat. A fourth type of highly

modified feather, the **powder-down feather,** characterizes herons, bitterns, hawks, and parrots. Tips of these disintegrate as they grow, releasing a talclike powder that helps to waterproof the feathers and to give them metallic luster.

Origin and Development

Like a reptile's scale, to which it is homologous, a feather develops from an epidermal thickening overlying a nourishing dermal core (Figure 27.5A). However, rather than flattening like a scale, a feather bud forms a hollow cylinder and partly sinks into the follicle from which it is growing. The hollow cylinder has two layers of epidermis, an outer layer that forms a protective sheath, and an inner layer that forms a ridge destined to become the rachis and barbs. As a feather enlarges and nears the end of its growth, the soft rachis and barbs are transformed into hard structures by deposition of **keratin**. The protective sheath splits apart, allowing the end of a feather to protrude and barbs to unfold.

Characteristics of Aves

1. **Neck elongate and S-shaped;** forelimbs modified as **wings; endothermic**
2. Epidermal converting of **feathers** and **leg scales;** thin epidermis and dermis sweat glands; oil gland at base of tail
3. Skull with **one occipital condyle; many bones with air cavities;** ribs with strengthening, uncinate processes; tail short, caudal vertebrate reduced to a **pygostyle;** pelvic girdle a **synsacrum; sternum usually large and keeled**
4. **No teeth;** each jaw covered with a keratinized sheath, forming a **beak; gizzard** present
5. Brain well developed, with **large optic lobes and cerebellum;** 12 pairs of cranial nerves
6. Eyes large, with **pecten** (Figure 27.15); middle ear with a single bone
7. Separate sexes; internal fertilization; copulatory organ (penis) only in paleognathids, ducks, geese, and a few others; **females with functional left ovary and oviduct only;** sex determined by chromosomes (female is heterogametic)
8. Fetal membranes of **amnion, chorion, and allantois; oviparous; eggs with much yolk and hard, calcareous shells;** extensive parental care of young
9. Excretory system of **metanephric kidneys** and ureters that open into cloaca; uric acid main nitrogenous waste
10. Lungs of **parabronchi** with **continous air flow; syrinx** (voice box) present; **air sacs** among visceral organs and skeleton
11. Heart with two atria and two ventricles; **separate pulmonary and systemic circuits; persistent right aortic arch;** nucleated red blood cells

Molting

When fully grown, a feather, like a mammalian hair, is a dead structure. Shedding, or molting, of feathers is a highly orderly process. Except in penguins, which molt all at once, feathers are discarded gradually, which avoids the appearance of bare spots. Flight and tail feathers are lost in exact pairs, one from each side, maintaining balance (Figure 27.6). Replacements emerge before the next pair is lost, and most birds can continue to fly unimpaired during the molting period; however, many water birds (ducks, geese, loons, and others) lose all their primary feathers at once and are grounded during the molt. Many prepare for molting by moving to isolated bodies of water where they can find food and more easily escape enemies. Nearly all birds molt at least once a year, usually in late summer after nesting season.

> The vivid color of feathers is of two kinds: pigmentary and structural. Red, orange, and yellow feathers are colored by pigments, called lipochromes, deposited in feather barbules as they are formed. Black, brown, red-brown, and gray colors are from a different pigment, melanin. Blue feathers of blue jays, indigo buntings, and bluebirds depend not on pigment but on scattering of shorter wavelengths of light by particles within the feather; these are structural colors. Blue feathers are usually underlain by melanin, which absorbs certain wavelengths, thus intensifying the blue. Such feathers look the same from any angle of view. Green colors are almost always a combination of yellow pigment and blue feather structure. Another kind of structural color is the beautiful iridescent color of many birds, which ranges from red, orange, copper, and gold to green, blue, and violet. Iridescent color is based on interference that causes light waves to reinforce, to weaken, or to cancel each other. Iridescent colors may change with the angle of view; quetzals, for example, look blue from one angle and green from another. Among vertebrates, only tropical reef fishes can vie with birds for intensity and vividness of color.

Skeleton

A major structural requirement for flight is a light, yet sturdy, skeleton (Figure 27.7A). Compared with those of the earliest known bird, *Archaeopteryx* (Figure 27.7B), bones of modern birds are phenomenally light, delicate, and laced with air cavities. Such **pneumatized** bones (Figure 27.8) are nevertheless strong. The skeleton of a frigate bird with a 2.1 m (7-foot) wingspan weighs only 114 grams (4 ounces), less than the weight of all its feathers.

As archosaurs, birds evolved from ancestors with diapsid skulls (p. 563). However, skulls of modern birds are so specialized that it is difficult to see any trace of the original diapsid condition. A bird's skull is built lightly and most bones are fused into one piece. The braincase and orbits are large to accommodate a bulging brain and large eyes needed for quick motor coordination and superior vision. Yet, a pigeon skull weighs only 0.21% of its body weight; by comparison, a rat's skull weighs 1.25% of its body weight. As a whole, however, the skeleton of a bird is not lighter than that of a mammal of similar size. The difference is in distribution of mass: whereas the skull and pneumatized wing bones are especially light, the leg bones are heavier than those of mammals. This lowers the bird's center of gravity, which improves aerodynamic stability.

In *Archaeopteryx*, both jaws contained teeth set in sockets, an archosaurian characteristic. Modern birds are completely toothless, having instead a keratinous beak molded around the bony jaws. The mandible is a complex of several bones hinged to provide a double-jointed action that permits the mouth to gape widely. Most birds have kinetic skulls (kinetic skulls of lizards are described on pp. 571–572) with a flexible attachment between upper jaw and skull.

The most distinctive feature of the vertebral column is its rigidity. Most vertebrae except the **cervicals** (neck vertebrae) are fused together. Most caudal vertebrae are fused into a **pygostyle** (Figure 27.7A),

Figure 27.6 Osprey, *Pandion haliaetus*, lands while holding a freshly captured fish. Feathers are molted in sequence in exact pairs so that balance is maintained during flight. Order Accipitriformes.

© Corbis

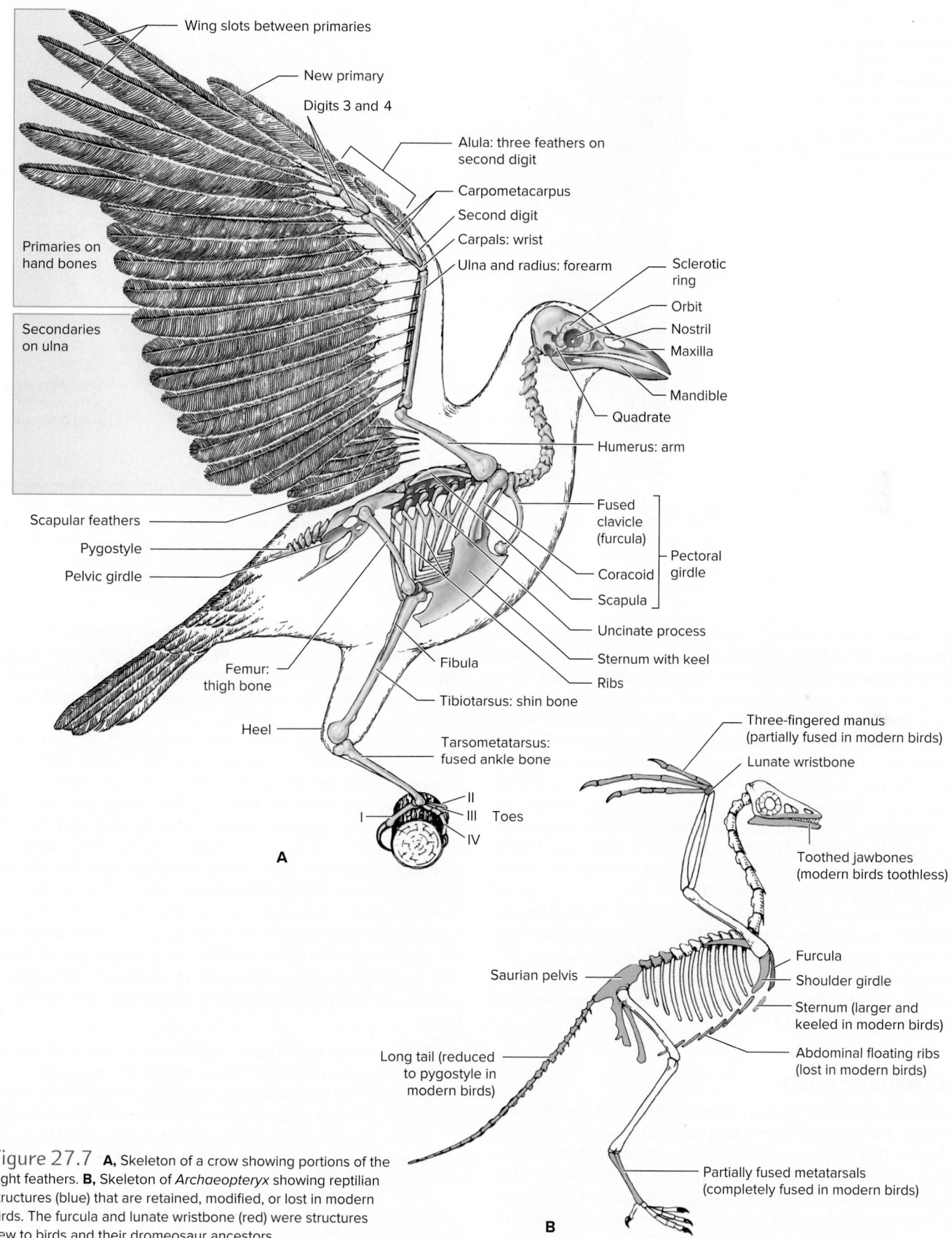

Figure 27.7 **A,** Skeleton of a crow showing portions of the flight feathers. **B,** Skeleton of *Archaeopteryx* showing reptilian structures (blue) that are retained, modified, or lost in modern birds. The furcula and lunate wristbone (red) were structures new to birds and their dromeosaur ancestors.

Wing slots between primaries

New primary

Digits 3 and 4

Alula: three feathers on second digit

Carpometacarpus

Second digit

Carpals: wrist

Ulna and radius: forearm

Sclerotic ring

Orbit

Nostril

Maxilla

Mandible

Quadrate

Humerus: arm

Primaries on hand bones

Secondaries on ulna

Scapular feathers

Pygostyle

Pelvic girdle

Fused clavicle (furcula)

Coracoid

Scapula

Pectoral girdle

Uncinate process

Sternum with keel

Ribs

Femur: thigh bone

Fibula

Tibiotarsus: shin bone

Heel

Tarsometatarsus: fused ankle bone

I

II

III

IV

Toes

A

Three-fingered manus (partially fused in modern birds)

Lunate wristbone

Toothed jawbones (modern birds toothless)

Furcula

Shoulder girdle

Sternum (larger and keeled in modern birds)

Abdominal floating ribs (lost in modern birds)

Saurian pelvis

Long tail (reduced to pygostyle in modern birds)

Partially fused metatarsals (completely fused in modern birds)

B

Figure 27.8
Hollow wing bone of a songbird showing stiffening struts and air spaces that replace bone marrow. Such pneumatized bones are remarkably light and strong.

Struts

Compact bone

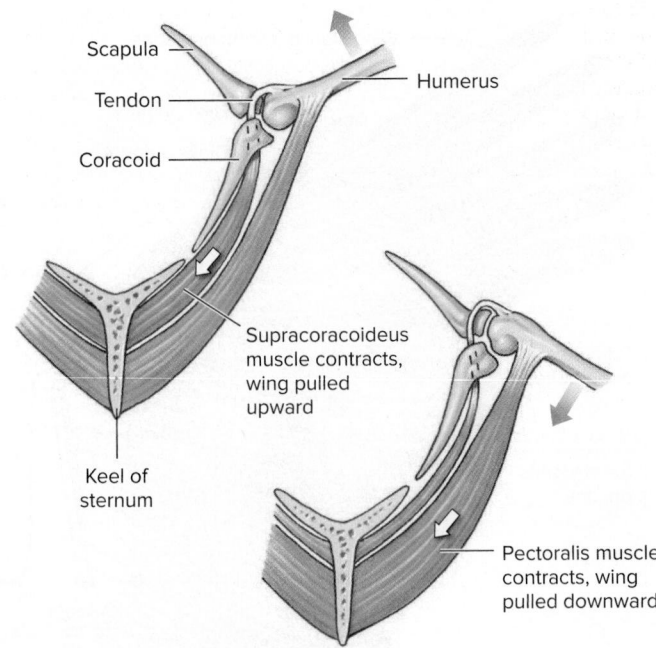

Figure 27.9
Flight muscles of a bird are arranged to keep the center of gravity low in the body. Both major flight muscles are anchored on the sternum keel. Contraction of the pectoralis muscle pulls the wing downward. Then, as the pectoralis relaxes, the supracoracoideus muscle contracts and, acting as a pulley system, pulls the wing upward.

while many of the remaining vertebrae in the trunk are fused as the **synsacrum.** These fused vertebrae and the pelvic girdle form a stiff but light framework to support the legs and to provide rigidity for flight. To assist in this rigidity, ribs are braced against each other with uncinate processes (Figure 27.7A). Except in flightless birds, the sternum bears a large, thin keel called a carina that provides an attachment for powerful flight muscles. Fused clavicles form an elastic **furcula** that apparently stores energy as it flexes during wing beats. The asymmetrical feathers and large furcula of *Archaeopteryx,* along with anatomy of its brain and inner ear, suggest that it had some flight abilities. However, it would have been a weak flier, because its small sternum offered little area for attachment of flight muscles (Figure 27.7B).

Bones of the forelimbs are highly modified for flight. They are reduced in number, and several are fused together. Despite these alterations, a bird's wing is clearly a rearrangement of the basic vertebrate tetrapod limb from which it arose (see Figure 25.3), and all the elements—arm, forearm, wrist, and fingers—are represented in modified form (Figure 27.7).

Muscular System

Locomotor muscles of wings are relatively massive to meet demands of flight. The largest of these is the **pectoralis,** which depresses the wings in flight. Its antagonist is the **supracoracoideus** muscle, which raises the wing (Figure 27.9). Surprisingly, perhaps, this latter muscle is not located on the backbone (anyone who has been served the back of a chicken knows that it offers little meat) but is positioned under the pectoralis on the breast. It is attached by a tendon to the upper side of the humerus of the wing so that it pulls from below by an ingenious "rope-and-pulley" arrangement. Both pectoralis and supracoracoideus are anchored to the keel of the sternum. Positioning the main muscle mass low in the body improves aerodynamic stability.

From the main leg muscle mass in the thigh, thin but strong tendons extend downward through sleevelike sheaths to the toes. Consequently the feet are nearly devoid of muscles, explaining the thin, delicate appearance of a bird's leg. This arrangement places the main muscle mass near a bird's center of gravity and at the same time allows great agility to the slender, lightweight feet. Because the feet are composed mostly of bone, tendon, and tough, scaly skin, they are highly resistant to damage from freezing. When a bird perches on a branch, an ingenious toe-locking mechanism (Figure 27.10) is activated, which prevents the bird from falling off its perch when asleep. The same mechanism causes the talons of a hawk or owl automatically to sink deeply into its prey as the legs bend under the impact of a strike. The powerful grip of a bird of prey was described by L. Brown:[1]

> When an eagle grips in earnest, one's hand becomes numb, and it is quite impossible to tear it free, or to loosen the grip of the eagle's toes with the other hand. One just has to wait until the bird relents, and while waiting one has ample time to realize that an animal such as a rabbit would be quickly paralyzed, unable to draw breath, and perhaps pierced through and through by the talons in such a clutch.

---

[1]From Brown, L. 1970. *Eagles.* New York, Arco Publishing.

Figure 27.10 Perching mechanism of a bird. When a bird settles on a branch, tendons automatically tighten, closing the toes around the perch.

Birds have lost the long, ancestral tail, still fully evident in *Archaeopteryx,* and have substituted a pincushion-like muscle mound into which the tail feathers are rooted. It contains an array of tiny muscles, as many as 1000 in some species, which control the crucial tail feathers. The most complex muscular system is in the neck of birds; the thin and stringy muscles, elaborately interwoven and subdivided, provide the bird's neck with great vertebral flexibility.

Food, Feeding, and Digestion

The first birds were carnivorous, feeding principally on insects, already well established on the earth's surface in both variety and numbers long before birds made their appearance. With the advantage of flight, birds could hunt insects on the wing and carry their assault to insect refuges mostly inaccessible to their earthbound tetrapod peers. Today, there is a bird to hunt nearly every insect; they probe the soil, search the bark, scrutinize every leaf and twig, and drill into insect galleries hidden in tree trunks.

Other animal foods (worms, molluscs, crustaceans, fishes, frogs, reptiles, and mammals, as well as other birds) all found their way into the diet of birds. A very large group, nearly one-fifth of all birds, feeds on nectar. Some birds are omnivores (often termed **euryphagous,** or "wide-eating" species) that will eat whatever is seasonally abundant. Others are specialists (called **stenophagous,** or "narrow-eating" species) that have the pantry to themselves—but at a price. Should the food source be reduced or destroyed for some reason (disease or adverse weather, for example), their survival may be jeopardized.

The **beaks,** or bills, of birds are strongly adapted to their food habits—from generalized types such as strong, pointed beaks of ravens, to highly specialized ones in flamingoes, pelicans, and avocets (Figure 27.11). The beak of a woodpecker is a straight, hard, chisel-like structure. Anchored to a tree trunk with its tail serving as a brace, a woodpecker delivers powerful, rapid blows to excavate nest cavities or to expose burrows of wood-boring insects. It then uses its long, flexible, barbed tongue to seek insects in their galleries. A woodpecker's skull is especially thick to absorb shock.

How much do birds eat? By a peculiar twist of reality, the commonplace "to eat like a bird" is supposed to signify a diminutive appetite. Yet birds, because of their intense metabolism, are voracious feeders. Small birds with their high metabolic rate eat more food per body mass than large birds. This happens because oxygen consumption increases only about three-fourths as rapidly as body weight. For example, the resting metabolic rate (oxygen consumed per gram of body weight) of a hummingbird is 12 times that of a pigeon and 25 times that of a chicken. A 3 g hummingbird may eat 100% of its body weight in food each day; an 11 g blue tit, about 30%; and a 1880 g domestic chicken, 3.4%. Obviously the weight of food consumed also depends on water content of the food, since water has no nutritive value. A 57 g Bohemian waxwing was estimated to eat 170 g of watery *Cotoneaster* berries in one day— three times its body weight! Seed-eaters of equivalent size might eat only 8 g of dry seeds per day.

Birds process their food rapidly and thoroughly with efficient digestive equipment. A shrike can digest a mouse in 3 hours, and berries will pass completely through the digestive tract of a thrush in just 30 minutes. Because birds lack teeth, foods that require grinding are reduced in the gizzard. Many birds have an enlargement (**crop**) at the lower end of the esophagus, which serves as a storage chamber. In pigeons, doves, and some parrots, the crop not only stores food but also produces a lipid- and protein-rich fluid composed of sloughed epithelial cells of the crop lining. For a few days after hatching, the helpless young are fed regurgitated crop fluid by both parents.

The stomach proper consists of two compartments, a **proventriculus,** which secretes gastric juice, and a muscular **gizzard,** which has a heavily keratinized lining and serves to grind food. To assist in the grinding process, birds swallow coarse, gritty objects or pebbles, which lodge in the gizzard. A turkey's gizzard is especially powerful and can break up acorns, hickory nuts, and walnuts. Certain birds of prey, such as owls, form pellets of indigestible materials, mainly bones and fur, in the proventriculus, ejecting them through the mouth. At the junction of the small intestine with the colon are paired **ceca;** these are well developed in herbivorous birds, in which they serve as fermentation chambers. The terminal part of the digestive system is the **cloaca,** which also receives genital ducts and ureters. In young birds the dorsal wall of the cloaca bears the **bursa of Fabricius,** which processes the B lymphocytes that are important in the immune response (p. 767).

Circulatory System

The general plan of bird circulation is not greatly different from that of mammals, although it evolved independently. Their four-chambered heart is large, with strong ventricular walls. Birds share with mammals a complete separation of respiratory and systemic circulations. However, the right aortic arch, instead of the left as in the mammals, leads to the dorsal aorta. The two jugular veins in the neck

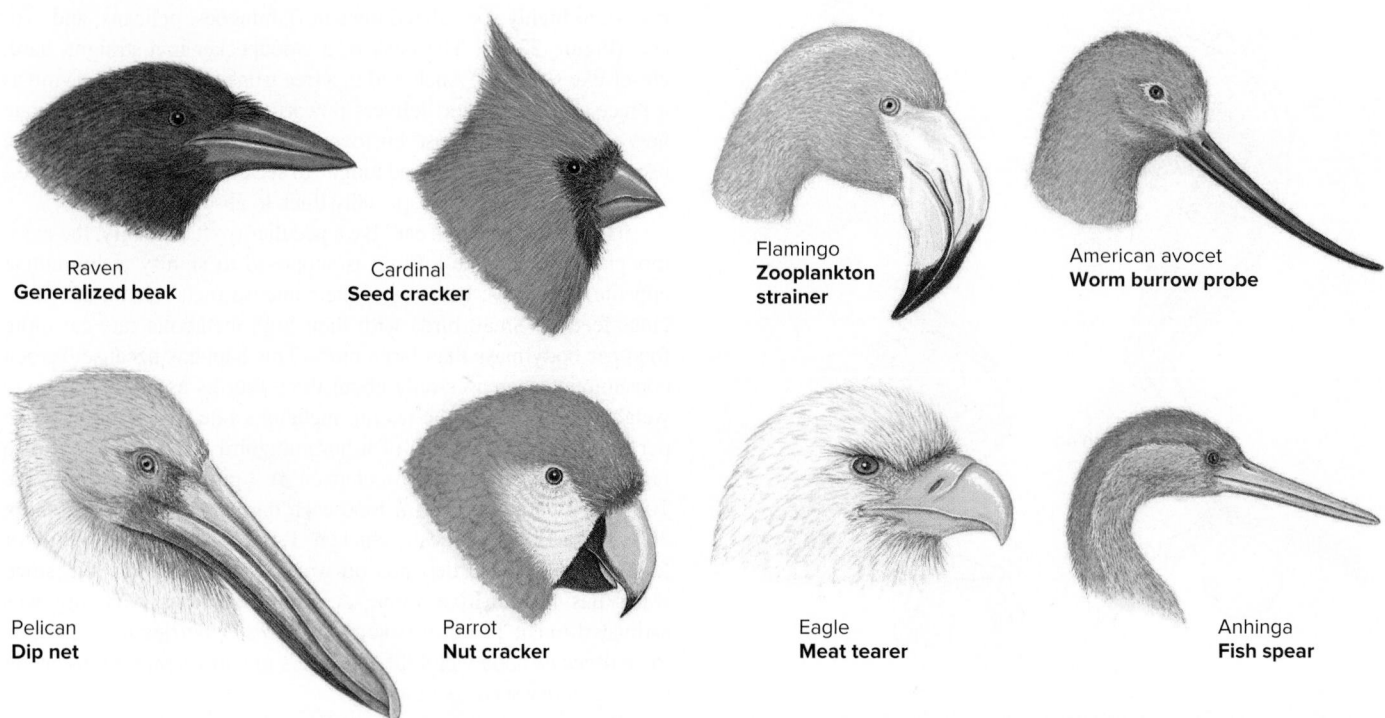

Figure 27.11   Some beaks of birds showing variety of adaptations.

are connected by a cross vein, an adaptation for shunting blood from one jugular to the other as the head rotates. The brachial and pectoral arteries to the wings and breast are unusually large.

A bird's heartbeat is extremely fast, and, as in mammals, there is an inverse relationship between heart rate and body weight. For example, a turkey has a heart rate at rest of about 93 beats per minute, a chicken has a resting rate of 250 beats per minute, and a black-capped chickadee has 500 beats per minute when asleep, which may increase to a phenomenal 1000 beats per minute during exercise. Blood pressure in birds is roughly equivalent to that in mammals of similar size.

A bird's blood contains **nucleated, biconvex erythrocytes.** (Mammals, the only other endothermic vertebrates, have enucleated, biconcave erythrocytes that are somewhat smaller than those of birds.) **Phagocytes,** or mobile ameboid cells of the blood, are particularly efficient in birds in repairing wounds and destroying microbes.

Respiratory System

The respiratory system of birds differs radically from the lungs of nonavian reptiles and mammals and is marvelously adapted for meeting the high metabolic demands of flight. In birds the finest branches of the bronchi, rather than ending in saclike alveoli as in mammals, are developed as tubelike **parabronchi** through which air flows continuously. The parabronchi form the lungs of birds. Also unique is an extensive system of nine interconnecting **air sacs** located in pairs in the thorax and abdomen and even extended by tiny tubes into the centers of the long bones (Figure 27.12). Air sacs connect to the lungs in such a way that most of the inspired air bypasses the lungs and flows directly into the posterior air sacs, which serve as reservoirs for fresh air. On expiration, this oxygenated air is passed through the

lungs and collected in the anterior air sacs. From there it flows directly to the outside. Thus, it takes two respiratory cycles for a single breath of air to pass through the respiratory system (Figure 27.12). The advantage of such a system is that an almost continuous stream of oxygenated air is passed through the richly vascularized parabronchi. It is clearly the most efficient respiratory system of any terrestrial vertebrate.

> The remarkable efficiency of a bird's respiratory system is emphasized by bar-headed geese, *Anser indicus*, that routinely migrate over the Himalayan mountains at altitudes up to 7300 m (24,000 ft), under conditions that are hypoxic to humans. They reach these altitudes in less than a day, without the acclimatization that is absolutely essential for a person even to approach the upper reaches of the Himalayas.

In addition to performing its principal respiratory function, the air sac system helps to cool the bird during vigorous exercise. A pigeon, for example, produces about 27 times more heat when flying than when at rest. Air sacs have numerous diverticula that extend inside the larger pneumatic bones (Figure 27.8) of the pectoral and pelvic girdles, wings, and legs. Because they contain warmed air, they provide considerable buoyancy to the bird.

Excretory System

Urine is formed in the relatively large, paired metanephric kidneys (p. 669) by glomerular filtration followed by selective modification of the filtrate in the tubule. Urine passes by way of **ureters** to the **cloaca.** There is no urinary bladder.

Figure 27.12 Respiratory system of a bird. **A,** Lung and air sacs. One side of the bilateral air sac system is shown. **B,** Movement of a single volume of air through a bird's respiratory system. Two full respiratory cycles are required to move air through the system. The first breath of air is shown in dark blue, while the second breath of air is shown in light blue. Note that air actually moves continually through the parabronchi.

Like other reptiles, birds excrete their nitrogenous wastes as uric acid. In shelled eggs, all excretory products remain within the egg-shell with the growing embryo. Uric acid crystallizes from solution and can be stored harmlessly within the allantoic sac (p. 563). Because of uric acid's low solubility, a bird can excrete 1 g of uric acid in only 1.5 to 3 ml of water, whereas a mammal may require 60 ml of water to excrete 1 g of urea. Concentration of uric acid occurs almost entirely in the cloaca, where it is combined with fecal material, and the water is reabsorbed.

Bird kidneys are much less efficient than those of mammals in removing salts, especially sodium, potassium, and chloride. Most mammals can concentrate solutes to 4 to 8 times that of blood concentration, and some desert rodents can concentrate urine to nearly 25 times that of blood. By comparison, most birds concentrate solutes only slightly greater than that of blood (the best that any bird can concentrate is about 6 times that of blood).

To compensate for weak solute-concentrating ability of the kidney, some birds, especially marine birds, use extrarenal mechanisms to excrete salts gained from the food they eat and seawater they drink. **Salt glands,** one located above each eye of sea birds (Figure 27.13), excrete highly concentrated solutions of sodium chloride, up to twice the concentration of seawater. The salt solution runs out the internal or external nostrils, giving gulls, petrels, and other sea birds a perpetual runny nose. The size of the salt gland in some birds depends on how much salt the bird consumes. For example, a population of mallard ducks living a semimarine life in Greenland has salt glands 10 times larger than those of ordinary freshwater mallards.

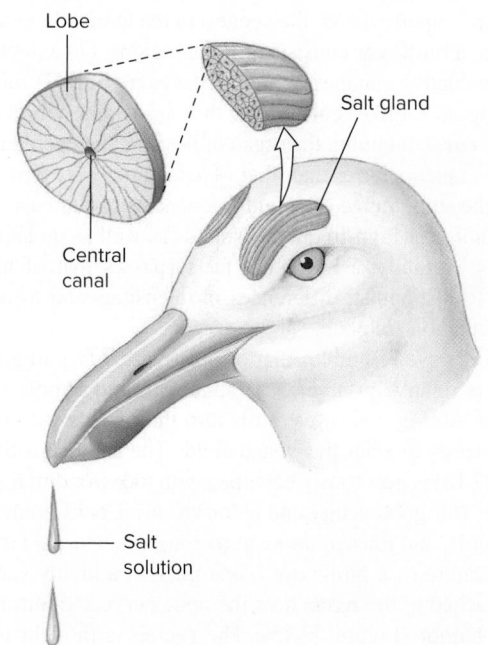

Figure 27.13 Salt glands of a marine bird (gull). One salt gland is located above each eye. Each gland consists of several lobes arranged in parallel. One lobe is shown in cross section, much enlarged. Salt is secreted into many radially arranged tubules, then flows into a central canal that leads into the nose.

Nervous and Sensory Systems

The design of a bird's nervous and sensory system reflects the complex problems of flight and a highly visible existence, in which it must gather food, mate, defend territory, incubate and rear young, and correctly distinguish friend from foe. The brain of a bird has well-developed **cerebral hemispheres, cerebellum,** and **optic lobes** (Figure 27.14). The **cerebral cortex**—chief coordinating center of a mammalian brain—is thin, unfissured, and poorly developed in birds. Instead, the core of the cerebrum, the **dorsal ventricular ridge,** has enlarged into the principal integrative center, controlling such activities as eating, singing, flying, and complex reproductive behavior. Relatively intelligent birds, such as crows and parrots, have larger cerebral hemispheres than do less intelligent birds such as chickens and pigeons. The **cerebellum,** much larger in birds than in nonavian reptiles, coordinates muscle position,

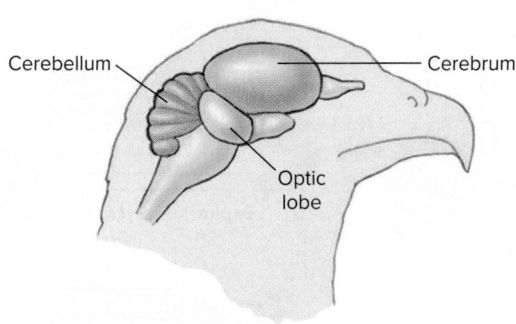

Figure 27.14 Bird brain showing principal divisions.

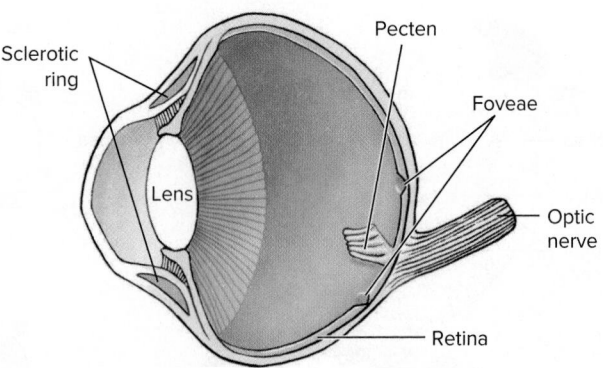

Figure 27.15 A hawk's eye has all the structural components of a mammalian eye, plus a pleated structure, the pecten, hypothesized to provide nourishment to the retina. The extraordinarily keen vision of hawks is attributed to the extreme density of cone cells in the foveae: 1.5 million per fovea compared to 0.2 million for humans.

equilibrium, and visual information used in movement and balance. The **optic lobes,** laterally bulging structures of the midbrain comparable to the visual cortex of mammals, organize visual information.

The senses of smell and taste are poor in some birds, but are relatively well developed in many other birds, such as carnivorous birds, flightless birds, oceanic birds, and waterfowl. Birds have good hearing and superb vision, the keenest in the animal kingdom. As in mammals, a bird's ear consists of three regions: (1) **external ear,** a sound-conducting canal extending to the **eardrum,** (2) **middle ear,** containing a rodlike **columella** that transmits vibrations, and (3) **inner ear,** containing the organ of hearing, the **cochlea.** A bird's cochlea is much shorter than that of a mammal, yet birds can hear roughly the same range of sound frequencies as humans. However, birds do not hear high-frequency sounds as well as do similar-sized mammals. Actually, a bird's ear far surpasses that of humans in capacity to distinguish differences in intensities and to respond to rapid fluctuations in pitch.

A bird's eye resembles that of other vertebrates in gross structure but is relatively larger, less spherical, and almost immobile; instead of turning their eyes, birds turn their heads with their long, flexible necks to scan the visual field. The light-sensitive **retina** (Figure 27.15) is generously equipped with rods (for dim light vision) and cones (for good acuity and color vision). Cones predominate in diurnal birds, and rods are more numerous in nocturnal birds. A distinctive feature of a bird's eye is the **pecten,** a highly vascularized organ attached to the retina near the optic nerve and jutting into the vitreous humor (Figure 27.15). The pecten is thought to provide nutrients and oxygen to the eye. On the anterior side of the eye is a **sclerotic ring** of platelike bones that serve to strengthen and to focus the large eye (see Figure 27.7).

The position of a bird's eyes in its head is correlated with its life habits. Herbivores that must avoid predators have eyes placed laterally to give a wide view of the world; predaceous birds such as hawks and owls have eyes directed to the front, allowing more binocular vision for better depth perception. In birds of prey and some others, the **fovea,** or region of keenest vision on the retina, is placed in a deep pit, which requires the bird to focus exactly on the source of the image. Many birds, moreover, have two foveae on the retina (Figure 27.15): the central one for sharp monocular views and the posterior one for binocular vision. Woodcocks can see binocularly both forward and backward. The visual acuity of a hawk is about eight times that of a human (enabling it to see clearly a crouching

rabbit more than a mile away), and an owl's ability to see in dim light is more than 10 times that of a human. Birds have good color vision, especially toward the red end of the spectrum.

Many birds can see at ultraviolet wavelengths, enabling them to view environmental features inaccessible to us but accessible to insects (such as flowers with ultraviolet-reflecting "nectar guides" that attract pollinating insects). Several species of ducks, hummingbirds, kingfishers, and passerines (songbirds) can see in the near ultraviolet (UV) wavelengths down to 370 nm (the human eye filters out ultraviolet light below 400 nm). For what purpose do birds use their UV-sensitivity? Some, such as hummingbirds, may be attracted to nectar-guiding flowers, like insects. But, for others, the benefit derived from UV-sensitivity is unknown.

27.3 FLIGHT

What prompted evolution of flight, a bird's ability to rise free of earthbound concerns, as almost every human has dreamed of doing? Two competing hypotheses of the origin of bird flight have been offered: birds began to fly either by climbing to a high place and gliding down, or by flapping their wings to launch into the air from the ground. The first hypothesis, termed arboreal, or "trees down," has been long favored. Proponents of this view envision an arboreal ancestor of *Archaeopteryx* gliding from tree to tree, or perhaps "pouncing" on prey below using wings to control its attack. Modifications permitting lift and powered flight would be highly advantageous for this kind of life. Indeed, there are many arboreal squirrels and lizards that use gliding to move among trees. The kind of locomotion envisioned by proponents of the arboreal hypothesis is perhaps best exhibited in the kakapo, a living species of flightless New Zealand parrot, which climbs trees using its hindlimbs, and glides down, sometimes refining its glides by flapping its wings. A weakness of this hypothesis is that few feathered dromeosaurs were arboreal, although some of the smaller ones, such as *Microraptor*, were probably climbers and arboreal.

Proponents of the cursorial, or "ground-up," hypothesis suggest that the feathered wings of bipedal, ground-dwelling ancestors of the first flying birds may have been used as snares to capture insects or to refine aerodynamic control during leaps to capture flying insects. Thus, as the wings became larger, they would have been capable of powered flight. However, a ground launch requires working against gravity rather than enlisting its help! No living gliders launch from the ground. A slightly more convincing scenario is suggested by studies of chukar partridge chicks, which use wingbeats to assist running up steep inclines. Although the evidence is weighted toward the arboreal hypothesis, the debate about origin of flight has not been settled. Interestingly, feathers were certainly a requirement for bird flight, but were not required for powered flight in two other vertebrate lineages, bats and extinct pterosaurs, which lack feathers.

Bird Wing as a Lift Device

Bird flight requires that a bird becomes airborne and moves forward. To become airbone it must generate lift forces greater than its own mass and to move forward it must generate thrust to move against resistive forces of drag. It uses its wings to do both. A wing is streamlined in cross section, with a slightly concave lower surface (**cambered**) with small, tight-fitting feathers where the leading edge meets the air. Air slips smoothly over the wing, creating lift with minimal drag. In general, the outer part of the wing, the modified hand bones with the attached **primary feathers** (see Figure 27.7A), provides the thrust needed to move the bird forward against the resistive forces of friction. The inner, highly cambered, part of the wing, with the **secondary feathers** and associated forearm, has less vertical motion than the outer wing, and mainly acts as an airfoil, generating lift.

Wings create lift from (1) a reaction from air deflected downward and (2) higher air pressure below the wing than above the wing. A flying bird holds its wings at an angle, so that the leading edge of the wing is higher than the trailing edge (Figure 27.16A). Air moving over the wing, from front to back, deflects downward, and slightly forward. In accordance with Newton's Third Law (for every action there is an equal but opposite reaction), the wing (and the bird) deflects upward. You can demonstrate this by moving your hand, held at an angle, through a pool of water. During flight, a wing's angle or its cambered shape creates a high pressure area in front of and below the wing and a low pressure area above and behind the wing. Because fluids move from high to low pressure, air moves

toward the trailing edge of the wing's top. The wing's angle deflects air under the wing slightly forward, causing air to move more slowly under than above the wing. The inertia of the fast-moving air above the wing keeps the air moving toward the trailing edge and pushing less against the top surface of the wing. Thus, additional lift occurs because air exerts more pressure on the underside of the wing than on the top of the wing.

Flapping Flight

During the downstroke, a bird's outer wing moves down and slightly forward, and it rotates so that the leading edge is below the trailing edge (Figure 27.16B). As the wing moves downward, the primaries bite through the air like a propeller, displacing the air backward and propelling the bird forward. Air movement over the wing creates the same forces as in gliding flight, but the wing's angled position rotates the resultant net forces forward. The wing folds slightly during the upstroke, and it returns to its original position with a minimum of drag (Figure 27.17). Little lift is generated during the upstroke for most birds, but some larger birds can rotate their wings to create additional lift during the upstroke. Birds that hover have an especially powerful upstroke and orient their wing motions in a way that produces lift, but not thrust (Figure 27.18).

Wing Dynamics at Low and High Speeds

The lift-to-drag ratio of an airfoil is determined by the angle of attack (angle of tilt) and the airspeed (Figure 27.19A). At high speeds sufficient lift is generated so that the wing is held at a low angle of attack, creating less drag. As speed decreases, lift can be increased by increasing the angle of attack, but drag forces also increase. When the angle of attack becomes too steep, usually around 15 degrees, turbulence appears on the upper surface, lift is destroyed, and stalling occurs (Figure 27.19B). Stalling can be delayed or prevented by **wing slots,** which direct a layer of rapidly moving air across the upper wing surface (Figure 27.19C). Wing slots are used in aircraft traveling at a low speed. In birds, two kinds of wing slots occur: (1) the **alula,** or group of small feathers on the thumb (Figure 27.7), which provides a midwing slot, and (2) **gaps between the primary feathers,** which create wing-tip slots. Eddies of air that form at the tips of wings, called **wing-tip**

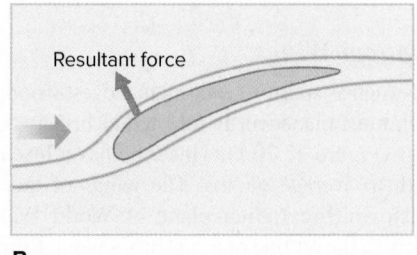

Figure 27.16   Cross sections of a bird's wing. **A,** Lift is generated when air is deflected downward and higher air pressure occurs below the wing than above the wing. **B,** Forward thrust is produced during a wing's downstroke. The wing's leading edge is below the trailing edge, rotating the resultant force forward.

Figure 27.17 In normal flapping flight of strong fliers like ducks, the wings sweep downward and forward fully extended. Primary feathers at the wing tips provide thrust. To begin the upbeat, the wing is bent, bringing it upward and backward. The wing then extends, ready for the next downbeat.

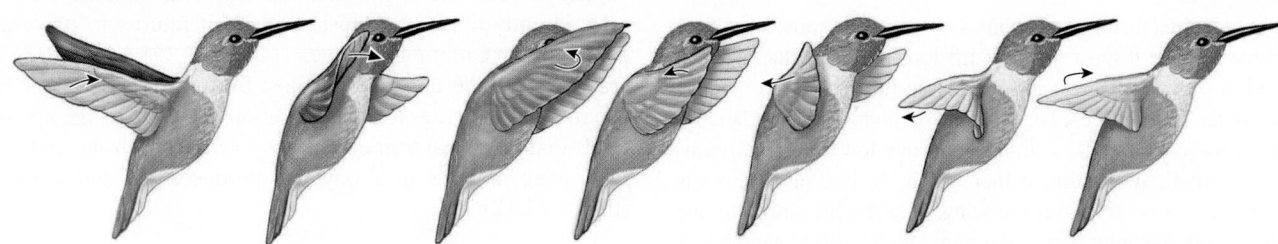

Figure 27.18 The secret of a hummingbird's ability to change direction instantly, or hang motionless in the air while sipping nectar from a flower, lies in its wing structure. The wing is nearly rigid, but hinged at the shoulder by a swivel joint and powered by a supracoracoideus muscle (p. 592) that is unusually large for the bird's size. When hovering the wing moves in a sculling motion. The leading edge of the wing moves forward on the forward stroke, then swivels nearly 180 degrees at the shoulder to move backward on the backstroke. The effect is to provide lift without propulsion on *both* forward and backward strokes.

vortexes, are particularly problematic at high speeds because they create drag (Figure 27.19D). This drag is reduced in wings with pointed tips and effectively reduced in long wings with widely separated tips **(high-aspect ratio wings),** which increase the area of the wings free from wing-tip vortexes.

Basic Forms of Bird Wings

Bird wings vary in size and form because successful exploitation of different habitats has imposed special aerodynamic requirements. Four types of bird wings are recognized.

Elliptical Wings

Birds such as sparrows, warblers, doves, woodpeckers, and magpies, which must maneuver in forested or brushy habitats, have elliptical wings (Figure 27.20A). This type has a **low aspect ratio** (ratio of length to average width). The wings of the highly maneuverable British Spitfire fighter plane of World War II fame conformed closely to the outline of a sparrow's wing. Elliptical wings have both an alula and gaps between the primary feathers; this arrangement helps to prevent stalling during sharp turns, low-speed flight, and frequent landing and takeoff. Each separated primary feather behaves as a narrow wing with a high angle of attack, providing

high lift at low speed. High maneuverability of elliptical wings is exemplified by the tiny chickadee, which can change course within 0.03 second.

High-Speed Wings

Birds that feed during flight, such as swallows, falcons, and swifts, or that make long migrations, such as plovers, sandpipers, terns, and gulls (Figure 27.20B), have wings that sweep back and taper to a slender tip. They are rather flat in section, have a high aspect ratio, and lack wing-tip slotting characteristic of elliptical wings. Sweepback and wide separation of wing tips reduce "tip vortex" (Figure 27.19D). This type of wing is aerodynamically efficient for high-speed flight but cannot easily keep a bird airborne at low speeds, except for hummingbirds, which rapidly move their wings in a specialized way to hover (Figure 27.18). The fastest birds, such as sandpipers, clocked at 175 km (109 miles) per hour, belong to this group.

Active Soaring Wings

Oceanic soaring birds, including albatrosses, shearwaters, and gannets (Figure 27.20C), also have wings with high aspect ratios, resembling those of a sailplane. Such long, narrow wings lack slots and are adapted for **active (dynamic) soaring.** Active soaring can

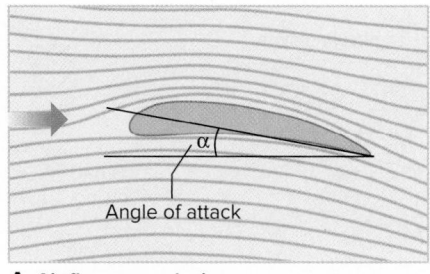

A Air flow around wing

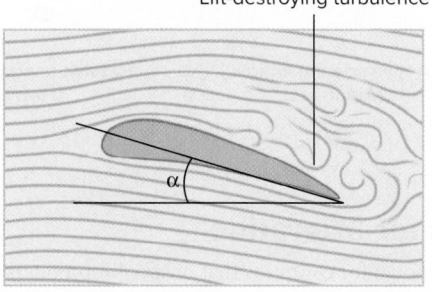

Lift-destroying turbulence

B Stalling at low speed

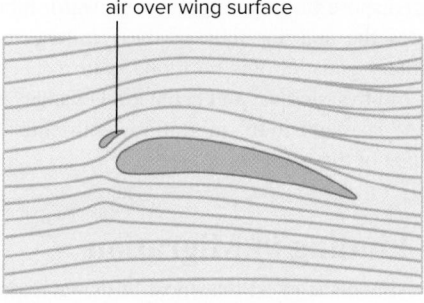

Wing slot directs fast-moving
air over wing surface

C Preventing stall with wing slots

D Formation of wing tip vortex

Figure 27.19 **A,** Air patterns formed by an airfoil, or wing, moving from right to left. **B,** At low speed the angle of attack (a) must increase to maintain lift but this increases the threat of stalling. **C,** Low-speed stalling can be prevented with wing slots. **D,** Wing tip vortex (*bottom*), a turbulence that tends to develop at high speeds, reduces flight efficiency. The effect is reduced in wings that sweep back and taper to a tip.

be done only over seas with strong, reliable winds, and exploits different wind speeds near the ocean surface (slow) and well above the surface (fast). Passive soaring, by contrast, occurs when a bird uses rising columns of warm air for lift, as described in the next

section. A bird that uses active soaring begins a downwind glide from an elevated position, gaining speed as it descends. Near the surface of the ocean, it turns into the wind and rises into stronger winds. Although its velocity relative to the ocean winds slows, the strong winds over its wings provide the lift to keep it aloft.

Passive Soaring Wings

Vultures, hawks, eagles, owls, and ospreys (Figure 27.20D)—predators that carry heavy loads— have large wings with high camber, promoting high lift. Their wings have alulas and primary-feather slots, which prevent stalling at low speeds. Wings of these birds have an aspect ratio intermediate between those of elliptical wings and high-speed wings. Many of these birds are land soarers, with broad, slotted wings that provide the sensitive response and maneuverability required for soaring in capricious air currents over land.

27.4 MIGRATION AND NAVIGATION

We described advantages of migration in the prologue to this chapter. Not all birds migrate, of course, but most North American and European species do, and the biannual journeys of some are truly extraordinary undertakings.

Migration Routes

Most migratory birds have well-established routes trending north and south. Because most birds breed in the Northern Hemisphere, where most of the earth's landmass occurs, most of the 4000 species of migrant birds migrate south for the northern winter and north to nest during the northern summer. Exceptions include many shearwaters, seabirds which breed in the southern hemisphere, and migrate north for the southern winter. Some use different routes in the fall and spring (Figure 27.21). Some complete their migratory routes in a very short time. Bar-tailed godwits, *Limosa lapponica,* fly non-stop 11,000 km from Alaska to New Zealand, relying on large stores of body fat to fuel their nine-day journey. Others, however, make a leisurely trip, often stopping along the way to feed. Some warblers are known to take 50 to 60 days to migrate from their winter quarters in Central America to their breeding areas in Canada. Many smaller species migrate at night and feed by day; others migrate chiefly in daytime; and many waterfowl and wading birds migrate by either day or night.

Many birds follow landmarks, such as rivers and coastlines, but others do not hesitate to fly directly over large bodies of water in their routes. Some birds have very wide migration lanes, whereas others, such as certain sandpipers, are restricted to very narrow ones, keeping to the coastlines because of their food requirements.

Some species have extremely long-distance migrations. Arctic terns, greatest globe spanners of all, breed north of the Arctic Circle during the northern summer, then migrate to Antarctic regions for the northern winter. This species takes a circuitous route in migrations from North America, passing over to the coastlines of Europe and Africa and then to their winter quarters, a trip that may exceed 18,000 km (11,200 miles).

Many small songbirds, such as warblers, vireos, thrushes, flycatchers, and sparrows, also make great migratory treks (Figure 27.21). Migratory birds that nest in Europe or Central Asia spend the northern winter in Africa.

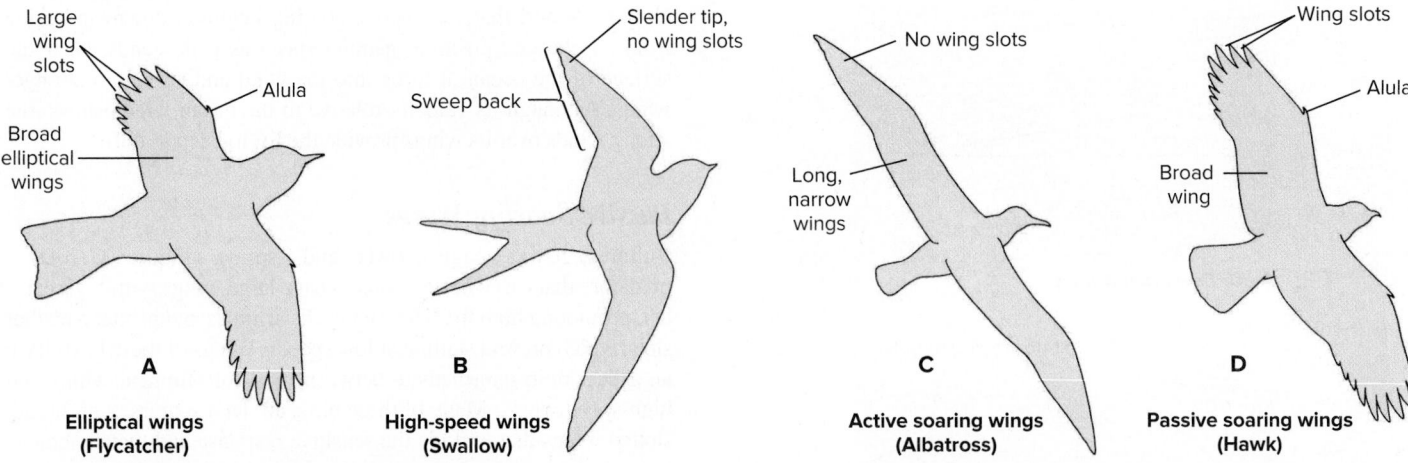

A
Large wing slots
Alula
Broad elliptical wings
Elliptical wings (Flycatcher)

B
Slender tip, no wing slots
Sweep back
High-speed wings (Swallow)

C
No wing slots
Long, narrow wings
Active soaring wings (Albatross)

D
Wing slots
Alula
Broad wing
Passive soaring wings (Hawk)

Figure 27.20   Four basic forms of bird wings.

Summer nesting range

Range during northern winter

Bobolink

Summer nesting range

Range during northern winter

American golden plover

Figure 27.21   Migrations of bobolinks, *Dolichonyx oryzivorus,* and American golden plovers, *Pluvialis dominica.* Bobolinks commute 22,500 km (14,000 miles) each year between nesting sites in North America and their range in Argentina, where they spend the northern winters, a phenomenal feat for such a small bird. Although the breeding range has extended to colonies in western areas, these birds take no shortcuts but adhere to the ancestral seaboard route. American golden plovers fly a loop migration, striking out across the Atlantic in their southward autumnal migration but returning in the northern spring by way of Central America and the Mississippi Valley because ecological conditions are more favorable at that time.

Stimulus for Migration

People have known for centuries that onset of reproductive cycles of birds is closely related to season. It has been demonstrated that lengthening days of late winter and early spring stimulate development of the gonads and accumulation of fat—both important internal changes that predispose birds to migrate northward. Increasing day length stimulates the anterior lobe of the pituitary into activity. Release of pituitary gonadotropic hormone in turn sets in motion a complex series of physiological and behavioral changes, stimulating gonadal growth, fat deposition, migration, courtship and mating behavior, and care of the young.

Direction Finding in Migration

Numerous experiments suggest that most birds navigate chiefly by sight. Birds recognize topographical landmarks and follow familiar migratory routes—a behavior assisted by flock migration, during which navigational resources and experience of older birds can be pooled. In addition to visual navigation, birds use a variety of orientation cues. Birds have a highly accurate sense of time. Numerous studies support a hypothesis that birds can detect and navigate by the earth's magnetic field. The navigational abilities of birds are primarily instinctive, although they may require calibration with existing navigational landmarks. In addition, learning can play a role because a bird's navigational abilities may improve with experience.

In the early 1970s W. T. Keeton showed that flight bearings of homing pigeons were significantly disturbed by magnets attached to the birds' heads, or by minor fluctuations in the geomagnetic field. Deposits of a magnetic substance called magnetite (Fe_3O_4) have been discovered in the beaks of pigeons. Recent experiments showed that a pigeon could discriminate between the presence and absence of magnetic anomaly, but not when its upper beak was anesthetized, nor when its trigeminal nerve, which innervates the upper beak, was severed.

Experiments by German ornithologists G. Kramer and E. Sauer and American ornithologist S. Emlen demonstrated convincingly that birds can navigate by celestial cues: the sun by day and the stars by night. Using special circular cages, Kramer concluded that birds maintain compass direction by referring to the sun (Figure 27.22). This is called **sun-azimuth orientation** (*azimuth,* compass bearing of the sun). To use the sun as a compass, birds must know the time of day, because the sun's position changes throughout the day. By exposing birds to altered light cycles to shift their perception of daybreak, researchers showed that birds do use an internal clock in this fashion. Sauer's and Emlen's ingenious planetarium experiments strongly suggest that some birds, probably many, can detect and navigate by the North Star axis around which the constellations appear to rotate.

A

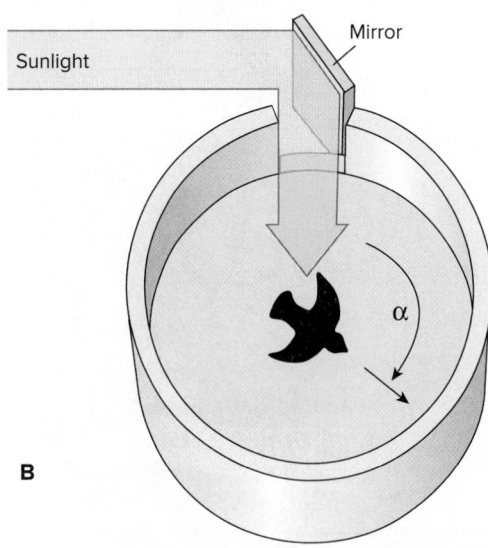

B

Figure 27.22   Gustav Kramer's experiments with sun-compass navigation in starlings. **A,** In a windowed, circular cage, the bird fluttered to align itself in the direction it would normally follow if it were free. **B,** When the true angle of the sun is deflected with a mirror, the bird maintains the same position relative to the sun. This shows that these birds use the sun as a compass. The bird navigates correctly throughout the day, changing its orientation to the sun as the sun moves across the sky.

In an elegant set of experiments designed to determine whether nocturnal migrants have an innate sense of direction or learn direction as nestlings, Stephen Emlen raised indigo buntings under three sets of conditions in a planetarium in which star patterns could be modified. One group of nestlings was allowed to see stars in a normal night sky rotating around the North Star. A second group of nestlings saw an equivalent pattern that rotated around Betelgeuse, a bright star in the constellation Orion, as if Betelgeuse were the North Star. A third group of nestlings was raised seeing at night only points of light that did not rotate.

When the birds were old enough to migrate, they were placed in cages under a normal night sky that allowed recording of the direction in which they tried to migrate. Birds that had seen only points of light during their development, with no rotation of the sky, showed no ability to detect direction and moved randomly. Birds that had developed seeing the normal sky rotated around the North Star oriented correctly for migration; and the group that developed seeing the sky rotated about Orion showed consistent orientation as if Betelgeuse were the North Star, even though now exposed to a normal night sky with stars rotating around the North Star. Thus, Emlen elegantly showed that these birds do not hatch with an innate sense of direction but must learn direction by seeing the sky rotate around a "pole" star!

27.5 SOCIAL BEHAVIOR AND REPRODUCTION

The adage says "birds of a feather flock together," and many birds are indeed highly social creatures. Especially during the breeding season, sea birds gather, often in enormous colonies, to nest and to rear young (Figure 27.23). Land birds, with some conspicuous exceptions, such as starlings and rooks, tend to be less gregarious than sea birds during breeding and to seek isolation for rearing their brood. Species that separate from their kind during breeding may aggregate for migration or feeding. Togetherness offers advantages: mutual protection from enemies, greater ease in finding mates, less opportunity for individual straying during migration, and mass huddling for protection against low night temperatures during migration. Certain species, such as pelicans (Figure 27.24), use cooperative behavior to feed. At no time are the highly organized social interactions of birds more evident than during the breeding season, as they establish territorial claims, select mates, build nests, incubate and hatch their eggs, and rear their young.

Figure 27.23   Part of a colony of northern gannets, *Morus bassanus,* showing extremely close spacing between pairs in this highly social bird. Order Suliformes.

© DeAgostini/Getty Images

A

B

Figure 27.24 Cooperative feeding behavior by white pelicans, *Pelecanus onocrotalus*. **A,** Pelicans form a horseshoe to drive fish together. **B,** Then they plunge simultaneously to scoop fish in their huge beaks. Order Pelecaniformes.

Figure 27.25 Copulation in waved albatrosses, *Diomeda irrorata*. In most bird species males lack a penis. A male passes sperm by standing on the back of a female, pressing his cloaca against that of the female. Order Procellariiformes.

Reproductive System

During most of the year the **testes** of males are tiny, bean-shaped bodies. At the breeding season they enlarge greatly, as much as 300 times their nonbreeding size. Because males of most species lack a penis, copulation is a matter of bringing cloacal surfaces into contact, usually while the male stands on the back of the female (Figure 27.25). Some swifts and hawks copulate in flight.

In females of most birds, only the **left ovary and oviduct** develop; those on the right dwindle to vestigial structures (Figure 27.26). Eggs discharged from the ovary enter the expanded end of the oviduct, where fertilization occurs. Several hours later, while eggs are passing down the oviduct, **albumin,** or egg white, from special glands is added to them; farther down the oviduct, shell membrane, shell, and shell pigments are also secreted around the egg. Sperm remain alive in a female's oviduct for many days after a single mating. Hen eggs show good fertility for 5 or 6 days after mating, but then fertility drops rapidly. However, an egg occasionally can be fertile as long as 30 days after separation of hens from a rooster.

Mating Systems

Two types of mating systems in animals are **monogamy,** in which an individual has only one mate, and **polygamy,** in which an individual has more than one mate during a breeding period. Monogamy is rare in most animal groups but common in birds; more than 90% are monogamous. A few bird species, such as swans and geese, choose partners for life and often remain together throughout the year. Seasonal monogamy is more common as most migrant birds pair only during the breeding season, living independent lives the rest of the year and perhaps choosing a different mate the next breeding season.

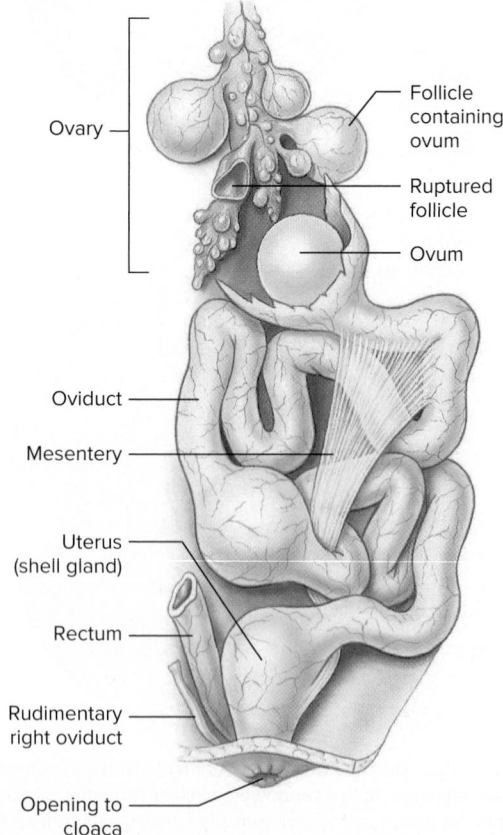

Figure 27.26 Reproductive system of a female bird. For most birds, only the left ovary and reproductive tract are functional. Structures on the right dwindle to vestiges.

Upper-level carnivores, species at the top of the food chain (p. 826), are vulnerable to decline through biomagnification of toxins. After World War II, DDT was widely used to control insect populations, especially mosquitos, which transmit malaria, in the United States. Brown pelicans, hawks, eagles, ospreys, and other raptorial birds consumed prey containing DDT, which then became concentrated in the birds' bodies. A primary effect on these birds was thinning of eggshells, possibly because DDT (or its metabolite, DDE) interferes with movement of calcium from blood into the shell glands. The fragile eggs often broke before hatching and these birds were in great decline by the middle of the twentieth century. Public outcry led to banning of DDT use in 1972, in part championed by Rachel Carson's *Silent Spring*, warning of the dangers of DDT. Subsequently, most raptor populations rebounded, including bald eagles, leading to their removal from the Endangered Species List. However, DDT is persistent in the environment and continues to affect some birds. In 2013, shell thinning was still a problem in certain California condor populations that feed on sea lions that forage in the highly contaminated Palo Verdes Shelf.

Figure 27.27 Dominant male greater sage grouse, *Centrocercus urophasianus,* with a hen that has been attracted by his "booming" display. Order Galliformes.

One reason that monogamy is much more common among birds than among mammals is that male and female birds are equally adept at most aspects of parental care. Because female mammals gestate the young and feed them by lactation, they provide important kinds of parental care that males do not. Female and male birds can alternate care of the nest and young, which permits one parent to be at the nest at all times. For some species, a female remains on the nest for months at a time, and is fed by the male. This constant attendance to the nest can be particularly important in species that would experience high loss of eggs or young to predators or rival birds if a nest were left unguarded. For many bird species, the high demands on a male to care for the young or his mate preclude the establishment of nests with additional females.

Although most birds have a monogamous mating system (social monogamy), either member of a pair may also mate with an individual that is not its partner. Recent **DNA** analyses have shown that most songbird species frequently are "unfaithful," engaging in extra-pair copulations. As a result, nests of many of these monogamous species contain a sizeable portion (30% or more) of young with fathers other than the attendant male. One possible benefit of extra-pair copulations is increased genetic diversity of the young. Second, extra-pair copulations permit a male to increase the number of sired young, increasing his reproductive success. Third, by mating with an individual of better genetic quality than the present mate, fitness of offspring could be increased. This can be particularly beneficial for females, who cannot easily increase the number of their offspring because the large size of a bird's eggs limits the number of eggs (and offspring) she can produce. Thus, extra-pair copulations allow males to increase the number of their offspring and could allow females to improve the genetic quality of their offspring. Genetic diversity of offspring produced is increased by extra-pair copulation for both males and females.

The most common form of polygamy in birds, when it occurs, is **polygyny** ("many females"), in which a male has more than one female mate. In many species of grouse, males gather in a collective display ground, the **lek,** which is divided into individual territories, each vigorously defended by a displaying male (Figure 27.27). There is nothing of value in a lek to the female except the male, and all he can offer are his genes, for only females care for the young. Usually a dominant male and several subordinate males occur in a lek. Competition among males for females is intense, but females appear to choose the dominant male for mating because, presumably, social rank correlates with genetic quality.

Polyandry ("many males"), in which a female mates with several males and the male incubates the eggs, is relatively rare in birds. It is practiced by several shorebird species, including spotted sandpipers, *Actitis macularia.* Female spotted sandpipers defend territories and mate with multiple males. Males incubate eggs within the female's territory and provide most parental care. This unusual reproductive strategy and clustering of individuals may be in response to high predation on spotted sandpiper nests.

Nesting and Care of Young

Most birds build some form of nest in which to rear their young. Some birds simply lay their eggs on bare ground or rocks. Others build elaborate nests such as pendant nests constructed by orioles, delicate lichen-covered mud nests of hummingbirds (Figure 27.28) and flycatchers, chimney-shaped mud nests of cliff swallows, the floating nests of red-necked grebes, and the huge sand and vegetation mounds of Australian brush turkeys. Most birds take considerable pains to conceal their nests from enemies. Woodpeckers, chickadees, bluebirds, and many others place their nests in tree hollows or other cavities; kingfishers excavate tunnels in the banks of streams for their nests; and birds of prey build high in lofty trees or on inaccessible cliffs. Nest parasites such as brown-headed cowbirds and European cuckoos build no nests at all but simply lay their eggs in the nests of birds smaller than themselves. When the eggs hatch, the foster parents care for the cowbird or cuckoo young, which outcompete the host's own hatchlings.

The developmental state of newly hatched birds varies among species. **Precocial** young, such as quail, fowl, ducks, and most water birds, are covered with down when hatched and can run or swim as soon as their plumage is dry (Figure 27.29). The most precocial birds are brush turkeys or megapodes of Australia, which can fly at hatching. However, most precocial young, even those able to leave the nest soon after hatching, are still fed or protected from predators by their parents for some time. **Altricial** hatchlings, which are naked and unable to see or to walk at birth, remain in the nest for a week or more. Parents of altricial species must supply food to their young almost constantly, for young birds may eat more than their weight each day (Figure 27.28). Many young are not easily categorized as precocial or altricial because they are intermediate in development at birth. For example, gulls and terns are born covered with down and with eyes open, but are unable to leave the nest for some time.

Although it may seem that precocial chicks have all the advantages, with their greater ability to find food and to escape predation, altricial birds have some advantages of their own. Because altricial birds lay relatively small eggs with minimal yolk supplies, a mother has a relatively small investment in her eggs and can more easily replace eggs lost to predation or extreme weather conditions. Altricial chicks also grow faster, perhaps due to the higher growth potential of immature tissue.

27.6 BIRD POPULATIONS AND THEIR CONSERVATION

Bird populations, like those of other animals, vary in size from year to year. Snowy owls, for example, are subject to population cycles that closely follow cycles in their food supply, mainly rodents. Voles, mice, and lemmings in the north have a fairly regular 4-year cycle of abundance; at population peaks, predator populations of foxes, weasels, and buzzards, as well as snowy owls, increase because there is abundant food for rearing their young. After a crash in the rodent population, snowy owls move south, seeking alternative food supplies. They occasionally appear in large numbers in southern Canada and the northern United States, attracting groups of birders eager to see an unusual and fascinating bird.

Activities of people may cause spectacular changes in bird distribution. Both common starlings (Figure 27.30) and house sparrows have been accidentally or deliberately introduced into numerous countries, to become the two most abundant bird species on earth, with the exception of domestic fowl.

Humans also are responsible for the extinction of many bird species. More than 140 species of birds have, since 1681, followed the last dodo to extinction. Most were victims of changes in their habitat or competition with introduced species. Overhunting contributed to the extinction of some species, among them passenger pigeons, which 150 years ago darkened the skies over North America in incredible numbers estimated in the billions (Figure 27.31).

Today, game birds are a well-managed renewable resource in the United States and Canada, and while hunters kill millions of

Figure 27.28 Anna's hummingbird, *Calypte anna,* feeding its young in its nest of fibers and down from plants, bound together with spiderwebs and camouflaged with lichens. A female builds the nest, incubates two pea-sized eggs, and rears the young with no assistance from a male. Anna's hummingbird, a common resident of California, is the only hummingbird to winter in the United States. Order Apodiformes.

© Richard R. Hansen/Science Source

Altricial
One-day-old meadowlark

Precocial
One-day-old ruffed grouse

Figure 27.29 Comparison of one-day-old altricial and precocial young. The altricial meadowlark (*left*) hatches nearly naked, blind, and helpless. The precocial ruffed grouse (*right*) is covered with down, alert, strong legged, and able to feed itself.

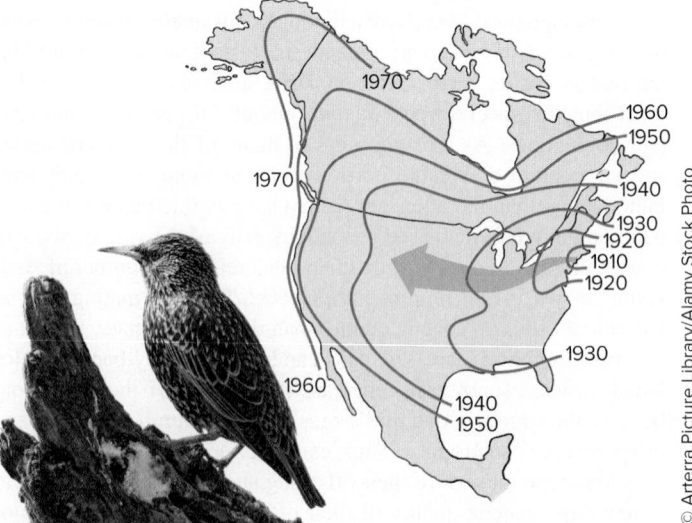

Figure 27.30 Colonization of North America by common starlings, *Sturnus vulgaris,* after the introduction of 60 birds into Central Park in New York City in 1890. There are now an estimated 150 million starlings in the United States alone, testimony to the great reproductive potential of birds. Starlings are omnivorous, eating mostly insects in spring and summer and shifting to wild fruits in the fall. Order Passeriformes.

© Arterra Picture Library/Alamy Stock Photo

Figure 27.31   Sport-shooting of passenger pigeons in Louisiana during the nineteenth century, before establishment of state and federal hunting regulations. In addition to clearing of the hardwood forests that served as nesting habitats, relentless sport and market hunting eventually dropped the population too low to sustain colonial breeding. The last passenger pigeon died in captivity in 1914.

game birds each year, none of the bird species legally hunted are endangered. Hunting interests, by acquiring large areas of wetlands for migratory bird refuges and sanctuaries, have contributed to the recovery of both game and nongame birds.

Of particular concern is the sharp decline of songbirds in the United States and southern Canada. Amateur birdwatchers and ornithologists have recorded that many songbird species that were abundant as recently as 40 years ago are now scarce. There are several reasons for the decline. Intensification of agriculture, permitted by the use of herbicides, pesticides, and fertilizers, has deprived ground-nesting birds of fields that were formerly left fallow. Excessive fragmentation of forests throughout much of the United States has increased exposure of nests of forest-dwelling species to nest predators such as blue jays, raccoons, and opossums, and to nest parasites such as brownheaded cowbirds. House cats also kill millions of small birds every year. From a study of radio-collared farm cats in Wisconsin, researchers estimated that in that state alone, cats may kill 19 million songbirds in a single year.

The rapid loss of tropical forests—approximately 120,000 square kilometers each year, an area about the size of the state of Louisiana—is depriving some 390 species of neotropical migrant birds of their wintering homes. Studies indicate that stressors on the wintering grounds are seriously decreasing the physiological condition of birds, particularly songbirds, prior to northward migration. Of all long-term threats facing songbird populations, tropical deforestation is the most serious and most intractable to change.

Some birds, such as robins, house sparrows, and starlings, can accommodate these changes and may even thrive on them, but for most birds the changes are adverse. Unless we take leadership in managing our natural resources wisely we soon could be facing the "silent spring" that Rachel Carson envisioned in 1962.

Lead poisoning of waterfowl is a side effect of hunting and fishing. Before federal regulations went into effect in 1991, requiring the use of nonlead shot for all inland and coastal waterfowl hunting, shotguns scattered more than 3000 tons of lead each year in the United States alone. When waterfowl eat the pellets (which they mistake for seeds or grit), the pellets are ground and eroded in their gizzards, facilitating absorption of lead into their blood. Lead poisoning paralyzes or weakens birds, leading to death by starvation. Although poisoning of birds from lead shot has declined, lead sinkers used in fishing still poison large numbers of waterfowl. Recently, several states have banned lead sinkers, requiring anglers to use nontoxic alternatives.

SUMMARY

The more than 10,500 species of living birds are feathered, egg-laying, endothermic vertebrates with forelimbs modified as wings. Birds are in a clade with dromeosaurs, a group of small, bipedal theropod dinosaurs. Feathers, which were present in some dromeosaurs, probably were originally used for thermoregulation or courtship, and later were adapted for flight. The oldest known fossil bird, *Archaeopteryx* from the Jurassic period of the Mesozoic era, had numerous reptilian characteristics and was almost identical to certain theropod dinosaurs except that it had asymmetrical feathers. It is probably the sister taxon of modern birds.

Adaptations of birds for flight are of two basic kinds: those reducing body weight and those promoting more power for flight. Feathers combine lightness with strength, water repellency, and high insulative value. Body weight is further reduced by elimination of some bones, fusion of others (also providing rigidity for flight), and presence of hollow, air-filled spaces in many bones. The light, keratinized beak, replacing the heavy jaws and

Taxonomy of Living Members of Aves

The clade Aves contains about 10,500 species distributed among 40 orders of living birds. Understanding the relationships of living birds, and consequently placing them in a classification, has been difficult because of the apparent rapid diversification of birds in the Cretaceous and early Tertiary periods. Relationships among bird groups are still being evaluated, leading to refinements in classification schemes. We present a taxonomy of the 40 bird orders, which follows the 2015 International Ornithologists Union World Bird List. The Linnean rank of order is used here for birds, because of its routine usage.

Aves (L. *avis,* bird)

Paleognathae (Gr. *palaios,* ancient, + *gnathos,* jaw). Modern birds with ancestral archosaurian palate. Ratites, which include ostriches, rheas, cassowaries, emus, and kiwis, (with unkeeled sternum) and tinamous (with keeled sternum).

Order Struthioniformes (strū′thē-on-i-for′mēz) (L. *struthio,* ostrich, + *forma,* form): **ostriches.** The ostrich, *Struthio camelus* (Figure 27.32), which reaches 2.4 m tall and 135 kg, is the largest living bird. The feet have only two toes of unequal size covered with pads, which enable the birds to travel rapidly over sandy ground. 2 species in Africa.

Order Rheiformes (rē′i-for-mēz) (Gr. *rhea,* rhea, + form): **rheas.** Two species of large, flightless birds found in grasslands of South America.

Order Casuariiformes (kas′u-er-i-for′mēz) (Mal. *casuar,* cassowary, + form): **cassowaries** and **emu.** The four species of cassowaries occupy forests of northern Australia and New Guinea. The emu is the second largest living bird species, and is confined to Australia. All are flightless.

Order Apterygiformes (ap-te-rij′i-for′mēz) (Gr. *a,* without + *pteryg,* wing, + form): **kiwis.** Kiwis, about the size of a domestic fowl, are unusual in having only the mere vestige of a wing. 5 species, all in New Zealand.

Order Tinamiformes (tin-am′i-for′mēz) (N.L. *Tinamus,* type genus, + form): **tinamous.** Ground-dwelling, grouselike birds of Central and South America. 47 species.

Neognathae (Gr. *neos,* new, + *gnathos,* jaw). Modern birds with flexible palate.

Order Anseriformes (an′ser-i-for′mēz) (L. *anser,* goose, + form): **swans, geese,** and **ducks.** The members of this order have broad beaks with filtering ridges at their margins, a foot web restricted to the front toes, and a long breastbone with a low keel. 177 species, worldwide distribution.

Order Galliformes (gal′li-for′mēz) (L. *gallus,* cock, + form): **quail, grouse, pheasants, ptarmigan, turkeys,** and **domestic fowl.** Chickenlike ground-nesting herbivores with strong beaks and heavy feet. The bobwhite quail, *Colinus virginianus,* occurs across the eastern half of the United States. The ruffed grouse, *Bonasa umbellus,* occupies the same region, but in woods instead of the open pastures and grain fields the bobwhite frequents. 298 species, worldwide distribution.

Order Sphenisciformes (sfē-nis′i-for′mēz) (Gr. *Sphēniskos,* dim. of *sphen,* wedge, from the shortness of the wings, + form): **penguins.** Web-footed marine swimmers of southern seas from Antarctica north to the Galápagos Islands. Although penguins

Figure 27.32 Ostrich, *Struthio camelus,* of Africa, the largest of all living birds. Order Struthioniformes.

© Cleveland P. Hickman, Jr.

have a keeled sternum, they use their wings as paddles for swimming rather than for flight. 18 species.

Order Gaviiformes (gā′vēi-for′mēz) (L. *gavia,* bird, probably sea mew, + form): **loons.** The five species of loons are remarkable swimmers and divers with short legs and heavy bodies. They live exclusively on fish, supplemented with amphibians and crayfishes. The familiar great northern loon, *Gavia immer,* occurs mainly in northern waters of North America and Eurasia.

Order Podicipediformes (pōd′i-si-ped′i-for′mez) (L. *podex,* rump, + *pedi,* foot): **grebes.** Short-legged divers with lobate webbed toes. The pied-billed grebe, *Podilymbus podiceps,* is widely distributed in North America. Grebes are most common in old ponds where they build their raftlike floating nests. 23 species, worldwide distribution.

Order Phoenicopteriformes (fē′ni-cop-ter′i-for′mēz) (Gr. *phoenico,* reddish-purple, + *pter,* wing, + form): **flamingos** (Figure 27.33). Large, colorful, wading birds that use lamellae in their beaks to strain zooplankton from the water. 6 species.

Order Procellariiformes (prō-sel-lar′ē-i-for′mēz) (L. *procella,* tempest, + form): **albatrosses, petrels, fulmars,** and **shearwaters.** All are marine birds with hooked beaks and tubular nostrils. In wingspan (more than 3.6 m in some), albatrosses are the largest flying birds. 143 species, worldwide distribution.

Order Pelecaniformes (pel-e-can′i-for′mēz) (Gr. *pelekan,* pelican, + form): **pelicans, ibises,** and **herons.** Mostly colonial fish-eaters inhabiting coasts, lakes, wetlands, and streams. 118 species, worldwide distribution, especially in the tropics.

Taxonomy of Living Members of Aves

Figure 27.33 Greater flamingos, *Phoenicopterus ruber,* on an alkaline lake in East Africa. Order Phoenicopteriformes.

Figure 27.34 Laughing gulls, *Larus atricilla,* in flight. Order Charadriiformes.

Order Phaethontiformes (fã-thon´ti-for´mēz) (Gr. *phaethont,* shining + form): **tropicbirds.** The three species in this order are large, tropical seabirds with white plumage.

Order Suliformes (sū´li-for´mēz) (Ice. *sul,* gannet + form): **frigatebirds, gannets, boobies,** and **cormorants.** Medium to large diving seabirds that feed mostly on fishes. 60 species, worldwide distribution.

Order Ciconiiformes (si-ko´nē-i-for´mēz) (L. *ciconia,* stork, + form): **storks.** Long-necked, long-legged, waders with long, stout beaks. 19 species, worldwide, mostly tropical distribution.

Order Accipitriformes (as-sip´i-tri-for´mēz) (L. *accipiter,* hawk + form): **vultures, condors, eagles, buzzards,** and **hawks.** Most are diurnal birds of prey with keen vision and sharp, curved talons. 265 species, worldwide distribution.

Order Falconiformes (fal´ko-ni-for´mēz) (L. *falco,* falcon, + form): **falcons.** Very fast birds of prey that primarily eat other birds. The peregrine falcon, *Falco peregrinus,* dives at speeds up to 320 km/h (200 mph). 67 species, worldwide distribution.

Order Otidiformes (o-tid´i-for´mēz) (Gr. *otid,* bustard + form): **bustards.** Large, long-legged, terrestrial birds of mostly arid habitats. 26 species in Africa, Asia, and Europe.

Order Mesitornithiformes (mēs-it´or-nith´i-for´mēz) (Gr. *mesit,* mediator + *ornith,* bird + form): **mesites.** Small, terrestrial birds inhabiting forests and brushlands. 3 species, restricted to Madagascar.

Order Cariamiformes (car-ē-am´i-for´mēz) (Tupi *cariama,* seriema + form): **seriemas.** Small, long-legged, terrestrial birds inhabiting grasslands and open forests. 2 species, restricted to South America.

Order Eurypygiformes (yur-ē-pij´i-for´mēz) (Gr. *eury,* broad + *pyg,* rump + form): **kagu** and **sunbittern.** The two birds in this order inhabit tropical forests, the kagu in New Caledonia and the sunbittern in Central and South America. 2 species.

Order Gruiformes (gru´i-for´mēz) (L. *grus,* crane, + form): **cranes, rails, coots,** and **gallinules.** Mostly prairie and marsh breeders. 190 species, worldwide distribution.

Order Charadriiformes (kā-ra-drē´i-for´mēz) Gr. *charadri,* curlew (a sandpiper) + form): **gulls** (Figure 27.34), **plovers, sandpipers, terns, woodcocks, snipes, avocets, phalaropes, skuas, skimmers, auks, puffins.** All are shorebirds. They are strong fliers and are usually colonial. 384 species, worldwide distribution.

Order Pterocliformes (ter-o´cli´or´mēz) (Gr. *ptero,* wing + *clid,* key + form): **sandgrouse.** Seed-eating, gregarious pigeonlike birds. 16 species, occurring in Africa, Asia, and southern Europe.

Order Columbiformes (kō-lum´bē-i-for´mēz) (L. *columba,* dove, + form): **pigeons** and **doves.** All have short necks, short legs, and a short, slender beak. The flightless dodo, *Raphus cucullatus,* of the Mauritius Islands became extinct in 1681. 334 species, worldwide distribution.

Order Psittaciformes (sit´ta-sē-for´mēz) (L. *psittacus,* parrot, + form): **parrots** and **parakeets.** Birds with hinged and movable upper beak, fleshy tongue. 395 species, pantropical distribution.

Order Opisthocomiformes (o-pis´th´-co-mi-for´mēz) (Gr. *opistho,* back, + L. *comos,* with long hair, + form): **hoatzin.** The young of this herbivorous bird use the large claws on their wings to climb trees. 1 species, in South America.

Order Musophagiformes (myū´sō-fa´ji-for´mēz) (L. *musa,* banana, + Gr. *phagō,* to eat, + form): **turacos.** Medium to large birds of dense forest or forest edge having a conspicuous patch of crimson on the spread wing, brightly colored bill, and short and rounded wings. 23 species restricted to Africa.

Order Cuculiformes (ku-ku´li-for´mēz) (L. *cuculus,* cukoo, + form): **cuckoos** and **roadrunners.** Common cuckoos, *Cuculus canorus,* lay their eggs in nests of smaller birds, which rear the young cuckoos. American cuckoos usually rear their own young. 149 species, worldwide distribution.

Order Strigiformes (strij´i-for´mēz) (L. *strix,* screech owl, + form): **owls.** Nocturnal predators with large eyes, powerful beaks and feet, and silent flight. 240 species, worldwide distribution.

Order Caprimulgiformes (kap-ri-mul´ji-for´mēz) (L. *caprimulgus,* goatsucker, + form): **nightjars** and **nighthawks.** Night and twilight feeders with small, weak legs and wide mouths fringed with bristles. Whippoorwills, *Antrostomus vociferus,* are common in the woods of the eastern states, and nighthawks, *Chordeiles minor,* are often seen and heard during evening flights. 122 species, worldwide distribution.

Taxonomy of Living Members of Aves

Order Apodiformes (ə-pōd′i-for′mēz) (Gr. *apous,* footless, + form): **swifts** and **hummingbirds.** Small birds with short legs and rapid wingbeat. The familiar chimney swift, *Chaetura pelagia,* fastens its nest in chimneys by means of saliva. A swift in China builds a nest of saliva used for soup making. Most species of hummingbirds occur in the tropics, but there are 24 species in the United States, of which only one, the ruby-throated hummingbird, occurs in the eastern part of the country. 469 species, worldwide distribution.

Order Coliiformes (kä-lā′i-for′mēz) (Gr. *kolios,* green woodpecker, + form): **mousebirds.** Small crested birds. 6 species restricted to southern Africa.

Order Trogoniformes (trō-gon′i′-for′mēz) (Gr. *tr ogon,* gnawing, + form): **trogons.** Richly colored, long-tailed birds. 43 species, pantropical distribution.

Order Leptosomiformes (lep-t o-som′i-for′mēz) (Gr. *lepto,* slender + *som,* body + form): **cuckoo roller.** The single species in this order inhabits Madagascar forests.

Order Coraciiformes (kä-rä′sē-′i′for′mēz) (N.L. *coracii* from Gr. *korakias,* a kind of raven, + form): **kingfishers** and **bee-eaters.** Birds that have strong, prominent beaks and nest in cavities. In the eastern half of the United States, belted kingfishers, *Megaceryle alcyon,* are common along waterways. 156 species, worldwide distribution.

Order Bucerotiformes (byū-ser′ot-i-for′mēz) (Gr. *bu,* ox + *cerat,* horn, referring to the bill shape of hornbills + form): **hoopoes** and **hornbills.** Most nest in tree cavities and have a large, curved beak. 74 species, mostly tropical Asia and Africa.

Order Piciformes (pis′i-for′mēz) (L. *picus,* woodpecker, + form): **woodpeckers, toucans, puffbirds,** and **honeyguides.** Birds with highly specialized beaks and two toes extending forward and two backward. All nest in cavities. The largest species in North America is the pileated woodpecker, usually found in mature forests. 443 species, worldwide distribution.

Order Passeriformes (pas-ser′i-for′mēz) (L. *passer,* sparrow, + form): **perching songbirds.** This is the largest order of birds, containing 106 families and 60% of all birds. Most have a highly developed syrinx (voice box). Their feet are adapted for perching on thin stems and twigs. The young are altricial. To this order belong many birds with beautiful songs such as thrushes, warblers, finches (Figure 27.35), mockingbirds, meadowlarks, sparrows, vireos, chickadees, and hosts of others. Others of this order, such as swallows, magpies, crows, ravens, jays, nuthatches, and creepers, have no songs worthy of the name. 6390 species, worldwide distribution.

Figure 27.35 Ground finch, *Geospiza fuliginosa,* one of the famous Darwin's finches of the Galápagos Islands. Order Passeriformes.

© Cleveland P. Hickman, Jr.

---

teeth of nonavian reptiles, serves as both hand and mouth for birds and is variously adapted for different feeding habits.

Adaptations that provide power for flight include high metabolic rate and body temperature coupled with an energy-rich diet; a highly efficient respiratory system consisting of a system of air sacs arranged to provide a constant, one-way flow of air through the lungs; powerful flight and leg muscles arranged to place muscle weight near the bird's center of gravity; and an efficient, high-pressure circulation.

Birds have keen eyesight, good hearing, and superb coordination for flight. The kidneys produce uric acid as the principal nitrogenous waste.

Birds fly by applying the same aerodynamic principles as an airplane and using similar equipment: wings for lift, support, and propulsion; a tail for steering and landing control; and wing slots for control at low flight speed. Flightlessness in birds is not uncommon and has evolved independently in several bird orders, usually on islands where terrestrial predators are absent; all are derived from flying ancestors. Arboreal and cursorial hypotheses for the origin of flight have been proposed. The arboreal hypothesis, currently favored by most zoologists, proposes that wings were first used to glide down from trees and later modified for powered flight.

Bird migration refers to regular movements between summer nesting places and wintering regions. Because land mass is concentrated in the northern hemisphere, most birds migrate north in spring and south in fall. Many cues are used for navigation during migration, including using geographic landmarks, the position of the sun, the stars, or the earth's magnetic field.

The highly developed social behavior of birds is manifested in vivid courtship displays, mate selection, territorial behavior, and incubation of eggs and care of the young. Most birds have a social monogamous mating system, although extra-pair copulations are common. Young hatch at various levels of development; altricial young are naked and helpless, while precocial young are feathered and able to walk and to feed themselves.

REVIEW QUESTIONS

1. Explain the significance of the discovery of *Archaeopteryx*. Why did this fossil demonstrate beyond reasonable doubt that birds are grouped phylogenetically with dinosaurs?
2. The special adaptations of birds contribute to two essentials for flight: more power and less weight. Explain how each of the following contributes to one or both of these two essentials: feathers, skeleton, muscle distribution, digestive system, circulatory system, respiratory system, excretory system, reproductive system.
3. How do marine birds rid themselves of excess salt?
4. In what ways are a bird's ears and eyes specialized for the demands of flight?
5. Explain how a bird's wing produces lift. What features help to prevent stalling at low flight speeds? What features help to reduce drag?
6. Describe four basic forms of bird wings. How does wing shape correlate with flight speed and maneuverability?
7. Contrast the arboreal and cursorial hypotheses for the origin of flight in birds.
8. What advantages does seasonal migration provide for birds?
9. Describe different navigational resources birds may use in long-distance migration.
10. What are some advantages of social aggregation among birds?
11. More than 90% of all bird species are monogamous. Explain why monogamy is much more common among birds than among mammals.
12. Briefly describe examples of polygyny and polyandry among birds.
13. Why might a "monogamous" bird seek extra-pair copulations?
14. Define the terms precocial and altricial as they relate to birds.
15. Offer some examples of how human activities have affected bird populations.

For Further Thought Reproductive behavior and strategies are better known in birds than in any other vertebrate group. Why?

SELECTED REFERENCES

Ackerman, J. 1998. Dinosaurs take wing. Nat. Geog. **194(1):** 74–99. *Beautifully illustrated synopsis of dinosaur-to-bird evolution.*

Bennet, P. M., and I. E. F. Owens. 2002. Evolutionary ecology of birds: life histories, mating systems, and extinction. Oxford, UK, Oxford University Press. *A phylogenetic approach to understanding how natural and sexual selection have led to the incredible diversity of bird mating systems.*

Elphick, J. (ed.). 1995. The atlas of bird migration: tracing the great journeys of the world's birds. New York, Random House. *Lavishly illustrated collection of maps of birds' breeding and wintering areas, migration routes, and many facts about each bird's migration journey.*

Emlen, S. T. 1975. The stellar-orientation system of a migratory bird. Sci. Am. **233:** 102–111 (Aug.). *Describes fascinating research with indigo buntings, revealing their ability to navigate by the center of celestial rotation at night.*

Feduccia, A. 1999. The origin and evolution of birds, ed. 2. New Haven, Yale University Press. *An updated successor to the author's* The Age of Birds *(1980) but more comprehensive; rich source of information on evolutionary relationships of birds.*

Gill, F. B. 2006. Ornithology, ed. 3. New York, W. B. Freeman and Company: *Popular, comprehensive, and accurate ornithology text.*

Mora, C. V., M. Davison, J. M. Wild, and M. W. Walker. 2004. Magnetoreception and its trigeminal mediation in the homing pigeon. Nature **432:** 508–511. *Pigeons can discriminate between the presence and the absence of a magnetic anomaly.*

Padian, K., and L. M. Chiappe. 1998. The origin of birds and their flight. Sci. Am. **279:** 38–47 (Feb.). *The authors argue that birds evolved from small, predatory dinosaurs that lived on the ground.*

Perrins, C. 2004. The Princeton encyclopedia of birds. Princeton University Press. *A comprehensive guide to all bird orders of the world, richly illustrated with color art and photographs.*

Prum, R. O., and A. H. Brush. 2003. Which came first, the feather or the bird? Sci. Am. **288(3):** 84–93. *Evolution of feathers is remarkably similar to their development.*

Sibley, C. G., and J. E. Ahlquist. 1990. Phylogeny and classification of birds: a study in molecular evolution. New Haven, Yale University Press. *A comprehensive application of DNA annealing experiments to the problem of resolving avian phylogeny.*

Videler, J. J. 2006. Avian flight. Oxford, U.K., Oxford Univ. Press. *An up-to-date introduction to all aspects of bird flight and scientific controversies regarding its origin.*

Waldvogel, J. A. 1990. The bird's eye view. Am. Sci. **78:** 342–353 (July–Aug.). *Birds possess visual abilities unmatched by humans. So how can we know what they really see?*

Wellnhofer, P. 1990. *Archaeopteryx*. Sci. Am. **262:** 70–77 (May). *Description of perhaps the most important fossil ever discovered.*

28

CHAPTER

Mammals

Juvenile grizzly bear, *Ursus arctos horribilis*.

- **PHYLUM CHORDATA**
  - Mammalia

Mammalia

Chordata

The Tell-Tale Hair

Hair evolved in a common ancestor of all mammals and has been retained to varying degrees in all species descended from that ancestor. Hair is therefore diagnostic for mammals; except in some pathological conditions, all mammals have hair at some point in their lives, and hair occurs in no other living organisms. Even those living mammals apparently without hair, such as whales, usually have a few hairs on their bodies. Mammalian hair has undergone numerous adaptive modifications for diverse uses. Mammals use hair for concealment, behavioral

signaling, waterproofing, and buoyancy; their hair may serve as sensitive vibrissae on their snouts or as prickly quills. Perhaps the most important use of their hair is thermal insulation, helping to maintain a high, constant body temperature in all climates, and thus support a high level of activity.

Mammals are among the most active animals, exhibiting speed and endurance in aquatic, aerial, and terrestrial habitats. They maintain this activity in nearly all environmental conditions, including the cool of night, baking deserts, frigid polar seas, and icy winters. Although hair is perhaps the most obvious feature of mammals, many other evolutionary innovations support their diversification. These unique innovations include a set of middle ear bones for transmitting sounds to the inner ear, mammary glands for nourishing newborns, a large brain with a unique covering

of the cerebrum (the neocortex), a diaphragm for efficient ventilation of lungs, and a highly developed sense of smell. Most mammals have an intrauterine, vascular placenta for feeding the embryo, specialized teeth and jaw musculature for processing food, and an erect gait for rapid and efficient locomotion.

Mammals occupy almost every environment on earth that supports life. Although not a large group (about 5700 species as compared with more than 10,000 species of birds, 32,000 species of fishes, and 1,100,000 species of insects), Mammalia (ma-mā´lē-ə; L. *mamma,* breast) is among the most biologically differentiated groups in the animal kingdom. Mammals are exceedingly diverse in size, shape, form, and function. They range in size from the Kitti's hognosed bat in Thailand, weighing only 2 g, to blue whales, exceeding 170 metric tons.

More than any other animals, mammals are targets of human activity. We have domesticated numerous mammals for food and clothing, as beasts of burden, and as pets. We use millions of mammals each year in biomedical research. We have introduced alien mammals into new habitats and extinguished mammalian populations from their native habitats. In 2016, 643 species of mammals were listed as "critically endangered" or "endangered" by the International Union for the Conservation of Nature and Natural Resources (IUCN), including many bats, cetaceans, cats, and primates. Because our welfare continues to be closely tied to that of other mammals, we must strive to preserve their natural populations and environments.

28.1 ORIGIN AND EVOLUTION OF MAMMALS

Mammalia is one of the two clades of living amniotes; the second clade, Reptilia, includes turtles, lepidosaurs (snakes, lizards, and tuataras), crocodilians (Chapter 26), and birds (Chapter 27). The evolutionary descent of mammals from their earliest amniote ancestors is perhaps the most fully documented transition in vertebrate history. From the fossil record, we can trace the evolution over 150 million years of endothermic, furry mammals from their small, ectothermic, hairless ancestors. Skull structures and especially teeth are the most abundant fossils, and it is largely from these structures that we identify the evolutionary descent of mammals.

Mammals and their closet extinct relatives have one pair of openings in the temporal region of the skull. This condition identifies them as **synapsids**, one of three major groups of amniotes that diversified during the late Paleozoic era (Figure 26.2). Synapsids were the first amniote group to diversify widely into terrestrial habitats.

The earliest synapsids diversified extensively into diverse herbivorous and carnivorous forms often collectively called **pelycosaurs** (Figures 28.1 and 28.2). These early synapsids were the most common and largest amniotes in the early Permian period. Pelycosaurs share a general outward resemblance to lizards, but this resemblance is misleading. Pelycosaurs are not closely related to lizards, which are diapsids (p. 563), nor are they a monophyletic group. From one group of early carnivorous pelycosaurs arose the **therapsids** (Figure 28.2), the only synapsid group to survive beyond the Paleozoic era. Therapsids evolved an efficient erect gait with upright limbs positioned beneath the body, rather than sprawled to the sides of the body, as in lizards and early pelycosaurs. Stability was reduced by raising the animal from the ground, and the muscular coordination

center of the brain, the cerebellum, assumed an expanded role. Changes in the morphology of the therapsid skull and mandibular adductor muscles increased feeding efficiency. Therapsids diversified into numerous herbivorous and carnivorous forms; however, most early forms disappeared during the great extinction event at the end of the Permian period. Previously, pelycosaurs and therapsids have been called "mammal-like reptiles," but use of this term is inappropriate, because they are not part of the clade Reptilia (p. 568).

One therapsid group to survive into the Mesozoic era was the **cynodonts.** Cynodonts evolved several features that supported a high metabolic rate: powerful and specialized jaw musculature, permitting a stronger bite; **heterodont** teeth, permitting better food processing and use of more diverse foods (Figure 28.3); **turbinate bones** in the nasal cavity, aiding retention of body heat (Figure 28.4); and a **secondary palate** (Figure 28.4), enabling an animal to breathe while holding prey in its mouth or chewing food. The secondary palate would be important to subsequent mammalian evolution by permitting young to breathe while suckling. Loss of lumbar ribs in cynodonts is correlated with the evolution of a **diaphragm** and also may have provided greater dorsoventral flexibility of the spinal column. Within the diverse cynodont clade (Figure 28.2), a small carnivorous group called trithelodontids most closely resembles the mammals, sharing with them several derived features of the skull and teeth.

The earliest mammals of the late Triassic period were small mouse- or shrew-sized animals with enlarged crania, redesigned jaws, and a new type of dentition, called **diphyodont,** in which teeth are replaced only once (deciduous and permanent teeth). This contrasts with the ancestral amniote pattern of continual tooth replacement throughout life (polyphyodont teeth). One of the more amazing evolutionary transformations involved the three middle ear bones, the malleus, incus, and stapes, which function to transmit sound vibrations in mammals (pp. 738–739). The stapes, homologous to the columella or hyomandibula of other vertebrates, already functioned in hearing in early synapsids. The malleus and the incus originated from the articular and the quadrate, respectively, two bones that previously served as the jaw joint but became reduced in size (better at transmitting sound vibrations) and relocated to the middle ear (Figure 28.3). A new jaw joint was formed between the dentary and squamosal (temporal) bones. This dentary-squamosal joint is the defining characteristic for fossil mammals.

The earliest mammals were almost certainly endothermic, although their body temperatures would have been rather lower than those of modern placental mammals. Hair was essential for insulation, and the presence of hair implies that sebaceous and sweat glands must also have evolved at this time to condition hair and to facilitate thermoregulation. The fossil record is silent on the appearance of mammary glands, but they must have evolved before the end of the Triassic period. The young of early mammals probably hatched from eggs in an immature condition, totally dependent on maternal milk, warmth, and protection. This mode of reproduction occurs today only in **monotremes** (a mammalian subgroup containing echidnas and platypus).

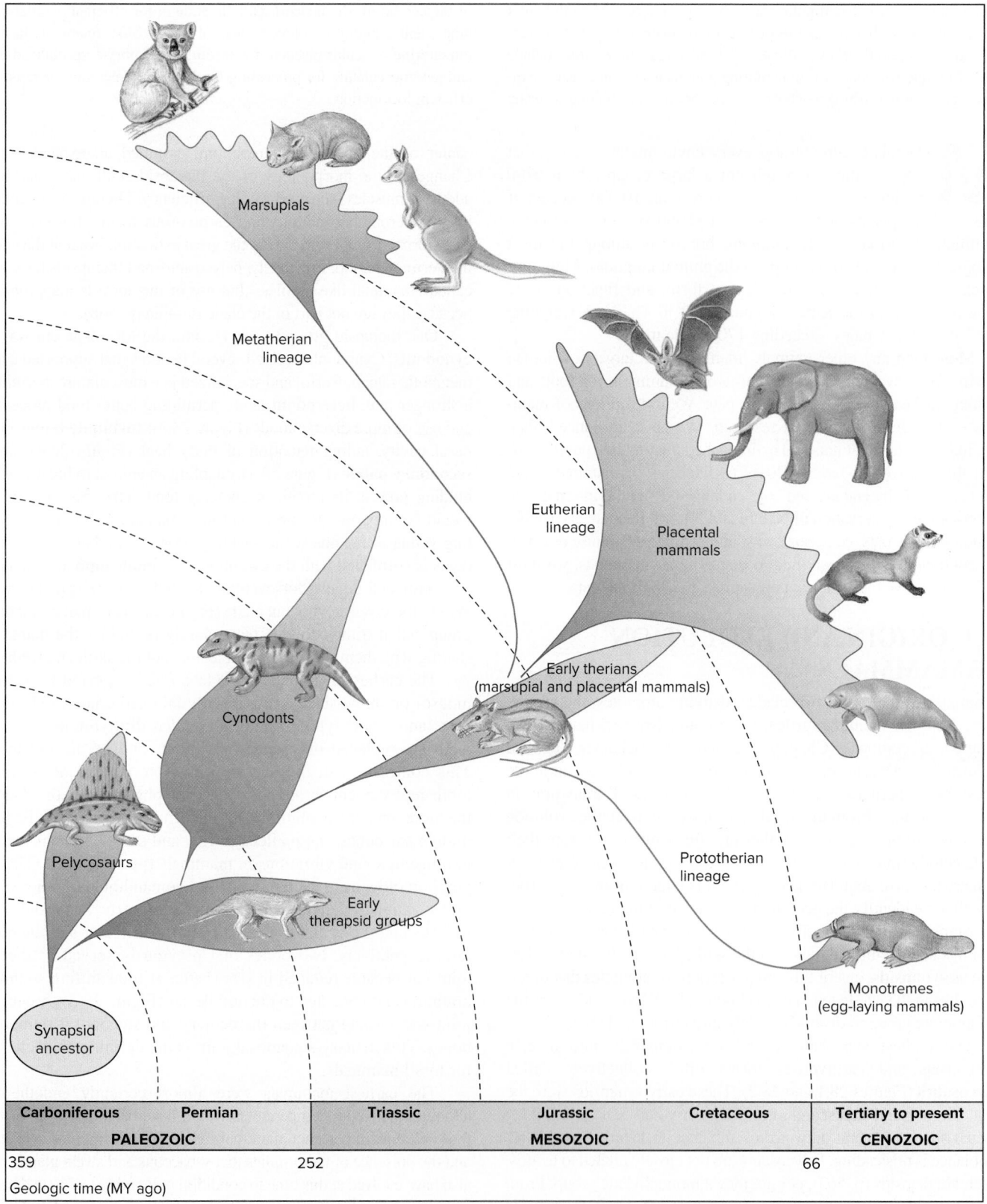

Figure 28.1 Evolution of major groups of synapsids. The synapsid lineage, characterized by lateral temporal openings in the skull, began with pelycosaurs, early amniotes of the Permian period. Pelycosaurs diversified extensively and evolved changes in jaws, teeth, and body form that presaged several mammalian characteristics. These trends continued in their successors, the therapsids, especially in cynodonts. One lineage of cynodonts gave rise in the Triassic period to early mammals. Fossil evidence indicates that all three groups of living mammals—monotremes, marsupials, and placentals—are derived from the same cynodont lineage. The great diversification of modern placental orders occurred during the Cretaceous and Tertiary periods.

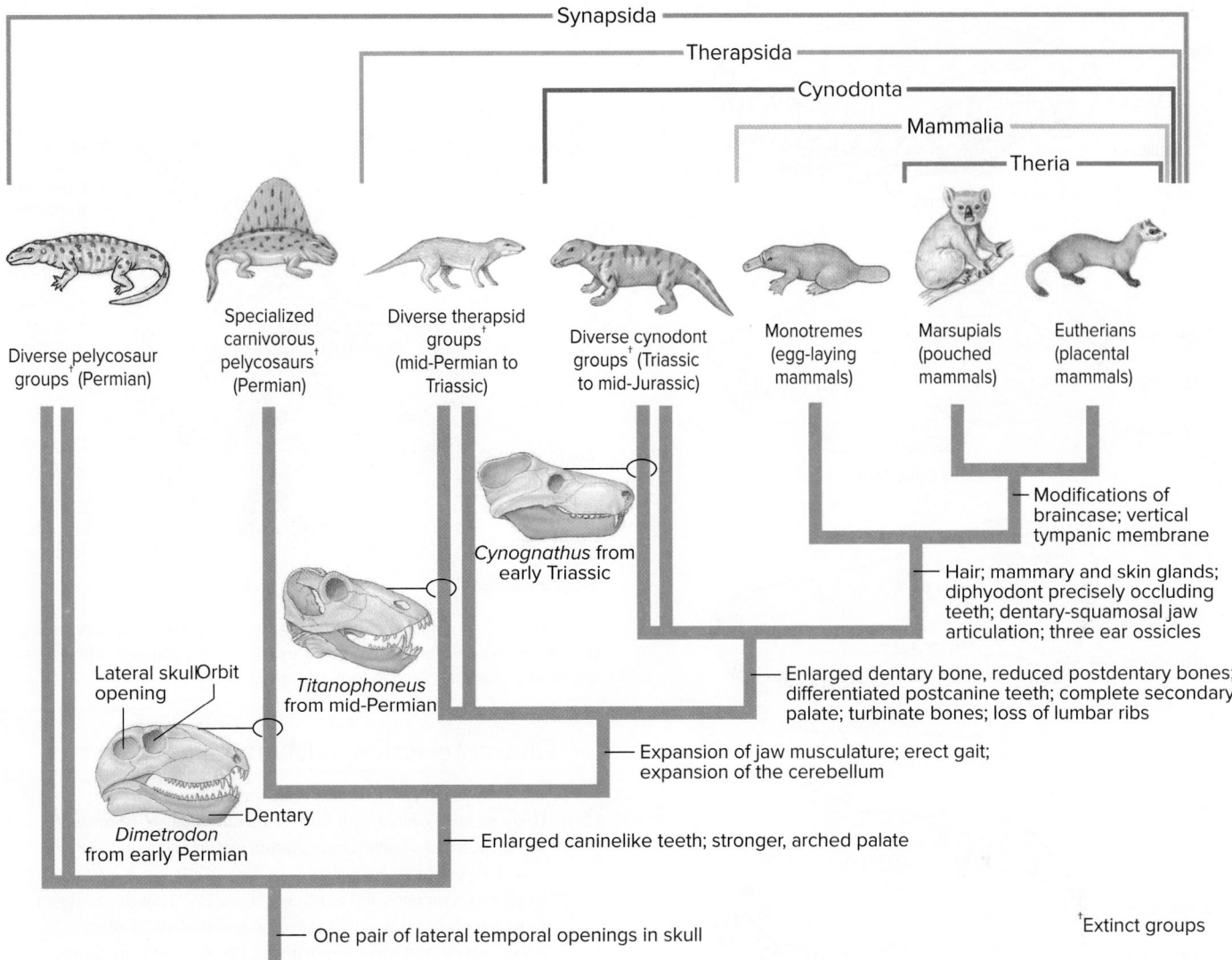

Figure 28.2 Abbreviated cladogram of synapsids emphasizing origins of important characteristics of mammals. The skulls show progressive increase in size of the dentary relative to other bones in the lower jaw and increasing heterodonty.

Oddly, early mammals of the late Triassic period, having developed nearly all novel attributes of modern mammals, had to wait for another 150 million years before they could achieve their great diversity. While dinosaurs became diverse and abundant, all nonmammalian synapsid groups became extinct. Mammals survived, first as shrewlike, probably nocturnal, creatures. Then, in the Cretaceous period, but especially during the Eocene epoch that began about 58 million years ago, modern mammals began to diversify rapidly. The great Cenozoic diversification of mammals is partly attributed to numerous habitats vacated by extinction of many diapsid groups at the end of the Cretaceous period. Mammalian diversification was almost certainly promoted by the facts that mammals were agile, endothermic, intelligent, and adaptable, and gave birth to living young, which they protected and nourished from their own milk supply, thus dispensing with vulnerable eggs laid in nests.

Living mammals are divided into two clades, the monotremes and the therians (Figure 28.2). Theria consists of the marsupials (metatherians) and placentals (eutherians). There are 29 orders of living mammals:

one order of monotremes, seven orders of marsupials, and 21 orders of placentals. A complete taxonomy is on pages 633–636.

Recent fossil discoveries and cladistic analyses have shed light on the origin of whales (Cetacea), and illustrate the importance of using fossil and molecular evidence when answering phylogenetic questions. Although the traditional view linked whales with an extinct group of wolflike creatures called mesonychids, molecular analysis of living species placed whales as the sister group to the hippopotamuses, within the order of even-toed hoofed mammals (Artiodactyla). Recent fossil discoveries in Pakistan and elsewhere provide an almost unbroken record of early whale evolution. Particularly important are remains of the ankle bones, which are diagnostic for artiodactyls. The earliest whales have a pulleylike astragulus (an ankle bone), which clearly links whales to artiodactyls. Recent cladistic analyses of combined fossil and DNA data support placement of whales as the sister group of hippos.

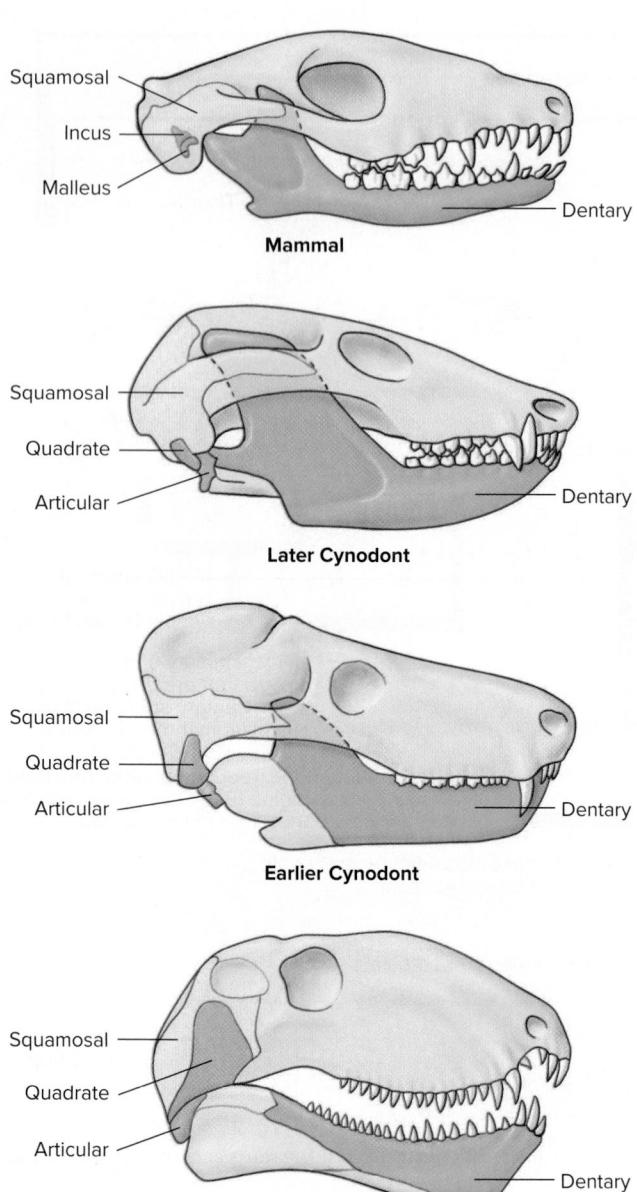

Mammal

Squamosal
Incus
Malleus
Dentary

Later Cynodont

Squamosal
Quadrate
Articular
Dentary

Earlier Cynodont

Squamosal
Quadrate
Articular
Dentary

Pelycosaur

Squamosal
Quadrate
Articular
Dentary

Figure 28.3 Evolution of the jaw joint and middle ear bones in the ancestors of mammals. The jaw joint in the earliest synapsids, the pelycosaurs, was between the articular and quadrate bones. A new joint between the dentary and squamosal bones evolved in the cynodont lineage related to mammals. In mammals, the articular and quadrate no longer function as a jaw joint, but instead transmit sound vibrations in the middle ear as the malleus and incus.

28.2 STRUCTURAL AND FUNCTIONAL ADAPTATIONS OF MAMMALS

Integument and Its Derivatives

Mammalian skin and especially its modifications distinguish mammals as a group. In general, skin is thicker in mammals than in other classes of vertebrates, although as in all vertebrates it is composed of

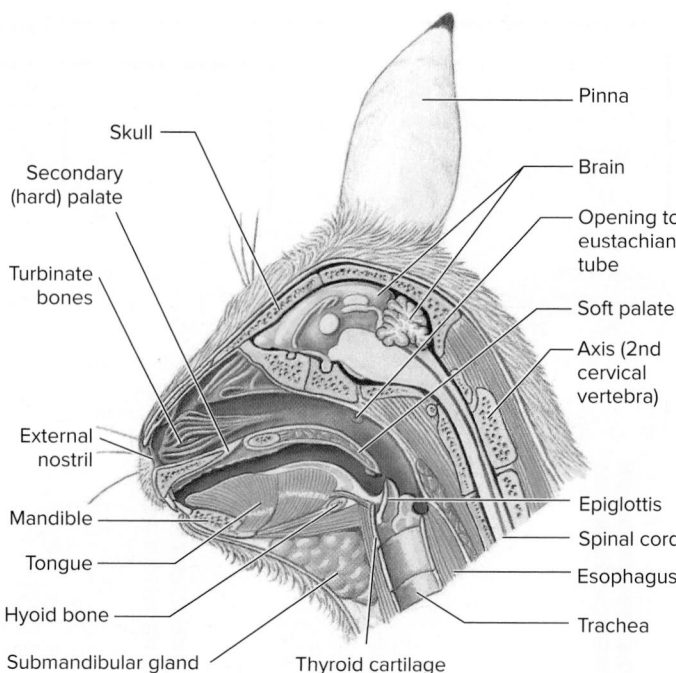

Figure 28.4   Sagittal section of the head of a rabbit. The secondary palate, which consists of bony (hard) and nonbony (soft) regions, separates the routes of air (dorsal) and food (ventral).

Characteristics of Mammalia

1. Fleshy external ears **(pinna); endothermic**
2. Body mostly covered with **hair**, but reduced coverage in some; **sweat, scent, and sebaceous glands** present; skin underlain by a thick layer of fat
3. Skull with **two occipital condyles**; lower jaw a **single enlarged bone (dentary)**; jaw joint between squamosal and dentary bones; **seven cervical vertebrae** (except in some xenarthrans [edentates] and manatees); **pelvic bones fused**
4. Mouth with **diphyodont teeth**; teeth **heterodont** in most
5. Brain highly developed, especially **cerebral cortex** (a superficial layer of the cerebrum); 12 pairs of cranial nerves
6. Olfactory sense highly developed; middle ear with **three bones** (malleus, incus, and stapes)
7. Separate sexes; internal fertilization; copulatory organ a **penis**; testes usually in a **scrotum**; sex determined by chromosomes (male is heterogametic)
8. Fetal membranes of **amnion, chorion, and allantois**; most **viviparous** with embryos developing in a **uterus** with **placental attachment**, except monotremes, which are **oviparous**; young nourished by **milk from mammary glands**
9. Excretory system of **metanephric kidneys** and ureters that usually open into a bladder; urea main nitrogenous waste
10. Lungs with high surface area from **alveoli** and ventilated by **aspiration; larynx present; secondary palate** separates air and food passages (Figure 28.4); muscular **diaphragm** ventilates lungs; convoluted **turbinate bones** in the nasal cavity warm and moisten inspired air
11. Heart with two atria and two ventricles; **separate pulmonary and systemic circuits; persistent left aortic arc;** nonnucleated, biconcave red blood cells

epidermis and **dermis** (Figure 28.5). The epidermis is thinner where it is well protected by hair, but in places that are subject to much contact and use, such as palms or soles, its outer layers become thickened with **keratin,** a fibrous protein that also constitutes nails, claws, hooves, and hair.

Hair

Hair is especially characteristic of mammals, although humans are not very hairy creatures and, in whales, hair is reduced to only a few sensory bristles on the snout. A hair grows from a hair follicle that, although epidermal in origin, is sunk into the dermis of the skin (Figure 28.5). A hair grows continuously by rapid proliferation of cells in the follicle. As a hair shaft is pushed upward, new cells are carried away from their source of nourishment and die, filled with keratin. Thus, true hair, found only in mammals, is composed of dead, keratin-packed epidermal cells.

Mammals characteristically have two kinds of hair forming their **pelage** (fur coat): (1) dense and soft **underhair** for insulation and (2) coarse and longer **guard hair** for protection against wear and to provide coloration. Underhair traps a layer of insulating air. In aquatic mammals, such as fur seals, otters, and beavers, it is so dense that it is almost impossible to wet. In water, guard hairs become wet and adhere to each other, forming a protective blanket over the underhair (Figure 28.6).

When a hair reaches a certain length, it stops growing. Normally it remains in its follicle until a new growth starts, whereupon it falls out. Most mammals experience periodic molts of the entire coat. In humans, hair is shed and replaced throughout life (although balding males confirm that replacement is not assured!).

Figure 28.6 American beaver, *Castor canadensis,* feeding on tree bark. This second largest rodent (the South American capybara is larger) has a heavy waterproof pelage consisting of long, tough guard hairs overlying the thick, silky underhair so valued in the fur trade. Order Rodentia, family Castoridae.

A hair is more than a strand of keratin. It consists of three layers: the medulla or pith in the center of the hair, the cortex with pigment granules next to the medulla, and the outer cuticle composed of imbricated scales. Hair of different mammals shows a considerable range of structure. It may be deficient in cortex, such as the brittle hair of deer, or it may be deficient in medulla, such as the hollow, air-filled hairs of wolverines. Hairs of rabbits and some others are scaled to interlock when pressed together. Curly hair, such as that of sheep, grows from curved follicles.

Figure 28.5 Structure of human skin (epidermis and dermis) and hypodermis, showing hair and glands.

Labels: Hair; Epidermis; Dermis; Hypodermis (subcutaneous layer); Sweat gland; Hair follicle; Sebaceous gland; Arrector pili muscle; Nerve endings

Some mammals, such as foxes and seals, shed their coat every summer. Most mammals have two annual molts, one in the spring and one in the fall. Summer coats are always much thinner than winter coats and in some mammals may be a different color. Several northern mustelid carnivores such as weasels have white winter coats and brown summer coats. It was once hypothesized that the white inner pelage of arctic animals conserved body heat by reducing radiation loss; in fact, dark and white pelages radiate heat equally well. The winter white pelage of arctic animals is simply camouflage in a land of snow. The snowshoe hare of North America has three annual molts: the white winter coat is replaced by a brownish gray summer coat, and this is replaced in autumn by a grayer coat, which is soon shed to reveal the winter white coat beneath (Figure 28.7). White fur of arctic mammals in winter (leukemism) is not to be confused with albinism, caused by a recessive gene that blocks pigment (melanin) formation. Albinos have red eyes and pinkish skin, whereas arctic animals in their winter coats have dark eyes and often dark-colored ear tips, noses, and tail tips.

Most mammals have somber colors that conceal their presence. Often a species is marked with "salt-and-pepper" coloration or a disruptive pattern that helps make it inconspicuous in its natural surroundings. Examples are spots of leopards and fawns

Figure 28.7 Snowshoe hare, *Lepus americanus,* in **A,** brown summer coat and **B,** white winter coat. In winter, extra hair growth on the hind feet broadens the animal's support in snow. Snowshoe hares are common residents of the taiga and are an important prey for lynxes, foxes, and other carnivores. Order Lagomorpha, family Leporidae.

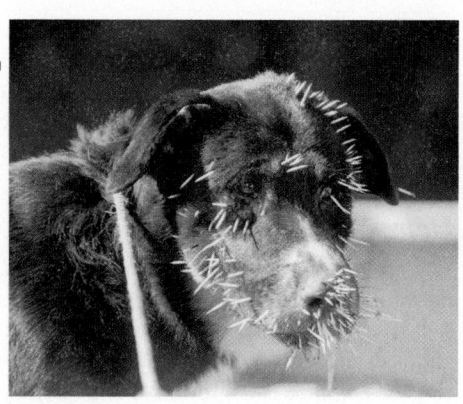

Figure 28.8 Dogs are frequent victims of a porcupine's impressive quills. Unless removed (usually by a veterinarian) quills will continue to work their way deeper in the flesh, causing great distress, and sometimes death.

and stripes of tigers. Skunks advertise their presence with conspicuous warning coloration.

The hair of mammals has become modified to serve many purposes. Bristles of hogs, spines of porcupines and their kin, and vibrissae on the snouts of most mammals are examples. **Vibrissae,** commonly called "whiskers," are really sensory hairs that provide a tactile sense to many mammals. The slightest movement of a vibrissa generates impulses in sensory nerve endings that travel to special sensory areas in the brain. Vibrissae are especially long in nocturnal and burrowing animals.

Porcupines, hedgehogs, echidnas, and a few other mammals have developed an effective and dangerous spiny armor. When cornered, the common North American porcupine turns its back toward its attacker and lashes out with its barbed tail. The lightly attached quills break off at their bases when they enter the skin and, aided by backward-pointed hooks on their tips, work deeply into tissues. Dogs are frequent victims (Figure 28.8) but fishers, wolverines, and bobcats are able to flip a porcupine onto its back to expose vulnerable underparts.

Horns and Antlers

Several kinds of horns or hornlike structures occur in mammals. **True horns** of the family Bovidae (antelopes, sheep, and cattle), are hollow sheaths of keratinized epidermis that embrace a core of bone arising from the skull (see Figure 29.3). True horns are not shed, are

not branched (although they may be greatly curved), grow continuously, and occur in both sexes.

Antlers of the deer family Cervidae are branched and composed of solid bone when mature. During their annual spring growth, antlers develop beneath a covering of highly vascular soft skin called **velvet** (Figure 28.9). Except for caribou (see Figure 28.16A), only males of the species produce antlers. When growth of antlers is complete just before the fall breeding season, blood vessels constrict and a male removes the velvet by rubbing its antlers against trees. Antlers are shed after the breeding season. New buds appear a few months later to form the next set of antlers. For several years each new pair of antlers is larger and more elaborate than the previous set. Annual growth of antlers strains mineral metabolism, because during the growing season an older moose or elk must accumulate 50 or more pounds of calcium salts from its vegetable diet.

Horns of the pronghorn (family Antilocapridae) are similar to true horns of bovids except that the keratinized portion is forked and shed annually. Giraffe horns are similar to antlers but retain their integumentary covering and are not shed. Rhinoceros horn consists of keratinized filaments that arise from dermal papillae cemented together, but not attached to the skull.

Trade in rhino products, especially the horn, has pushed Asian and African rhinos to the brink of extinction. Rhino horn was used in traditional Chinese medicine to treat fever, gout, and other ailments. Such supposed medicinal values are totally without pharmacological basis. Until recently, the principal use of rhino horns was to fashion handles for ceremonial daggers called jambiyas in the Middle East. Between 1970 and 1997, horns from 22,350 rhinos were imported into northern Yemen alone. Thanks to education efforts, use of rhino horn in Yemen has nearly ceased; the current destination of most illegal rhino horn is Vietnam. An international ban prohibiting the trade of rhino horn has reduced, but not eliminated, the now illegal trade, and populations continue to suffer.

Glands

Of all vertebrates, mammals have the greatest variety of integumentary glands. Most fall into one of four classes: sweat, scent, sebaceous, and mammary. All are derivatives of epidermis (Figure 28.5).

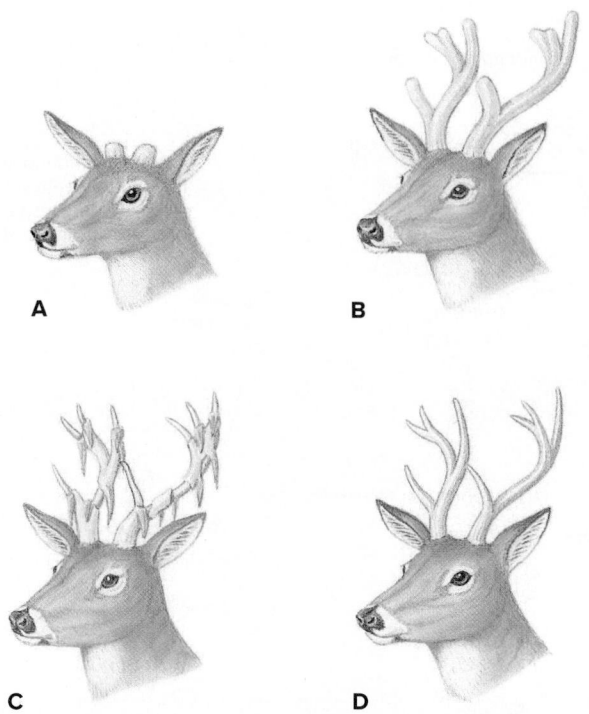

Figure 28.9 Annual growth of male deer antlers. **A,** Antlers begin growth in late spring, stimulated by pituitary gonadotropins. **B,** Bone grows very rapidly until halted by a rapid rise in testosterone (p. 138) production by the testes. **C,** The skin (velvet) dies and rubs off. **D,** Testosterone levels peak during the fall breeding season. Antlers are shed in January as testosterone levels subside.

Sweat glands are tubular, highly coiled glands that occur over much of the body surface in most mammals (Figure 28.5). They are absent from other vertebrates. There are two kinds of sweat glands: eccrine and apocrine. **Eccrine glands** secrete a watery fluid that, if evaporated on the skin's surface, draws heat away from the skin and cools it. Eccrine glands occur in hairless regions, especially foot pads, in most mammals, although in horses and most primates they are scattered over the body. They are either reduced or absent in rodents, rabbits, and whales. **Apocrine glands** are larger than eccrine glands and have longer and more convoluted ducts. Their secretory coil is in the dermis and extends deep into the hypodermis. They always open into a hair follicle or where a hair once was. Apocrine glands develop near puberty and are restricted (in humans) to the axillae (armpits), mons pubis, breasts, prepuce, scrotum, and external auditory canals. In contrast to the watery secretions of eccrine glands, apocrine secretions are milky fluids, whitish or yellow in color, that dry on the skin to form a film. Apocrine glands are not involved in heat regulation. Their activity is correlated with the reproductive cycle.

Scent glands occur in nearly all mammals. Their location and functions vary greatly. They are used for communication with members of the same species, for marking territorial boundaries, for warning, or for defense. Scent-producing glands are located in orbital, metatarsal, and interdigital regions (deer); behind the eyes and on the cheek (pica and woodchuck); on the penis (muskrats, beavers, and many canines); at the base of the tail (wolves and foxes); at the back of the head (dromedary); and in the anal region (skunks, minks, and weasels). The latter, the most odoriferous of all glands,

open by ducts into the anus; their secretions can be discharged forcefully for 2 to 3 meters. During mating season many mammals emit strong scents for attracting the opposite sex. Humans also are endowed with scent glands. However, we tend to dislike our own scent, a concern that has stimulated a lucrative deodorant industry to produce an endless output of soaps and odormasking concoctions.

Sebaceous glands (Figure 28.5) are usually associated with hair follicles, although some are free and open directly onto the surface. Gland cells are secreted in their entirety and are continuously renewed by cell division. These cells become distended with a fatty accumulation, then die, and are expelled as a greasy mixture called **sebum** into the hair follicle. Called a "polite fat" because it does not turn rancid, it serves as a dressing to keep skin and hair pliable and glossy. Most mammals have sebaceous glands over their entire body; in humans they are most numerous in the scalp and on the face.

Mammary glands, for which mammals are named, occur on all female mammals and in a rudimentary form on all male mammals. In the embryo, they develop by thickening the epidermis to form a milk line along each side of the abdomen. On certain parts of these lines the mammae appear while the intervening parts of the ridge disappear. Mammary glands increase in size at maturity, becoming considerably larger during pregnancy and subsequent nursing of young. In human females, adipose tissue begins to accumulate around the mammary glands at puberty to form the breast. In most mammals milk is secreted from mammary glands via nipples, but monotremes lack nipples and simply secrete milk onto the fur of the mother's belly, where it is lapped by the young.

Food and Feeding

Mammals exploit an enormous variety of food sources; some mammals require highly specialized diets, whereas others are opportunistic feeders that thrive on diversified diets. A mammal's anatomical and physiological adaptations for finding, capturing, chewing, swallowing, and digesting food are inextricably linked with its diet.

Teeth, perhaps more than any other single physical characteristic, reveal the life habit of a mammal (Figure 28.10). Mammals have teeth (with a few exceptions, such as monotremes, anteaters, and baleen whales), and their modifications are correlated with what a mammal eats.

As mammals evolved during the Mesozoic era, major changes occurred in teeth and jaws. Unlike the uniform **homodont** dentition of the first synapsids, mammalian teeth became differentiated, performing specialized functions such as cutting, seizing, gnawing, tearing, grinding, and chewing. Teeth differentiated in this manner are called **heterodont.** Mammalian dentition is differentiated into four types: **incisors (I),** with simple crowns and sharp edges, used mainly for snipping; **canines (C),** with long, conical crowns, specialized for piercing; **premolars (PM)** and **molars (M),** with compressed crowns and one or more cusps, suited for shearing, slicing, crushing, or grinding. The ancestral tooth formula, which expresses the number of each tooth type in one-half of the upper and lower jaw, was I 3/3, C 1/1, PM 4/4, M 3/3 = 44. Shrews, some omnivores, and carnivores come closest to this ancestral pattern (Figure 28.10).

Unlike most other vertebrates, mammals do not continuously replace their teeth throughout their lives. Most mammals grow just two sets of teeth: a temporary set, called **deciduous,** or milk, teeth, which is replaced by a permanent set when the skull has grown large

Figure 28.10 Feeding specializations of major trophic groups of placental (eutherian) mammals. Early placentals were insectivores; all other types are descended from them.

enough to accommodate a full set. Only incisors, canines, and premolars are deciduous; molars are never replaced and the single permanent set must last a lifetime.

Feeding Specializations

The feeding, or trophic, apparatus of a mammal—teeth and jaws, tongue, and digestive tract—is adapted to its particular feeding habits. Mammals are customarily divided among four basic trophic categories—insectivores, carnivores, omnivores, and herbivores—but many other feeding specializations have evolved in mammals, as in other living organisms, and feeding habits of many mammals defy exact classification. The principal feeding specializations of mammals are shown in Figure 28.10.

Insectivorous mammals, such as shrews, moles, anteaters, and most bats, are usually small. They feed on insects, as well as a variety of other small invertebrates. Insectivores eat little fibrous vegetable matter that requires prolonged fermentation, so their intestinal tract tends to be short (Figure 28.11). They have teeth with pointed cusps, permitting them to puncture the exoskeleton or skin of prey. A few insectivorous mammals, such as anteaters, completely lack teeth (Figure 28.10). Because many omnivores and carnivores also consume insects, the insectivorous diet is distinctive mainly by its lack of plant material and vertebrates.

Herbivorous mammals, which feed on grasses and other vegetation, form two main groups: (1) **browsers** and **grazers,** including ungulates (hooved mammals including horses, deer, antelope, cattle, sheep, and goats); and (2) **gnawers,** including many rodents, and

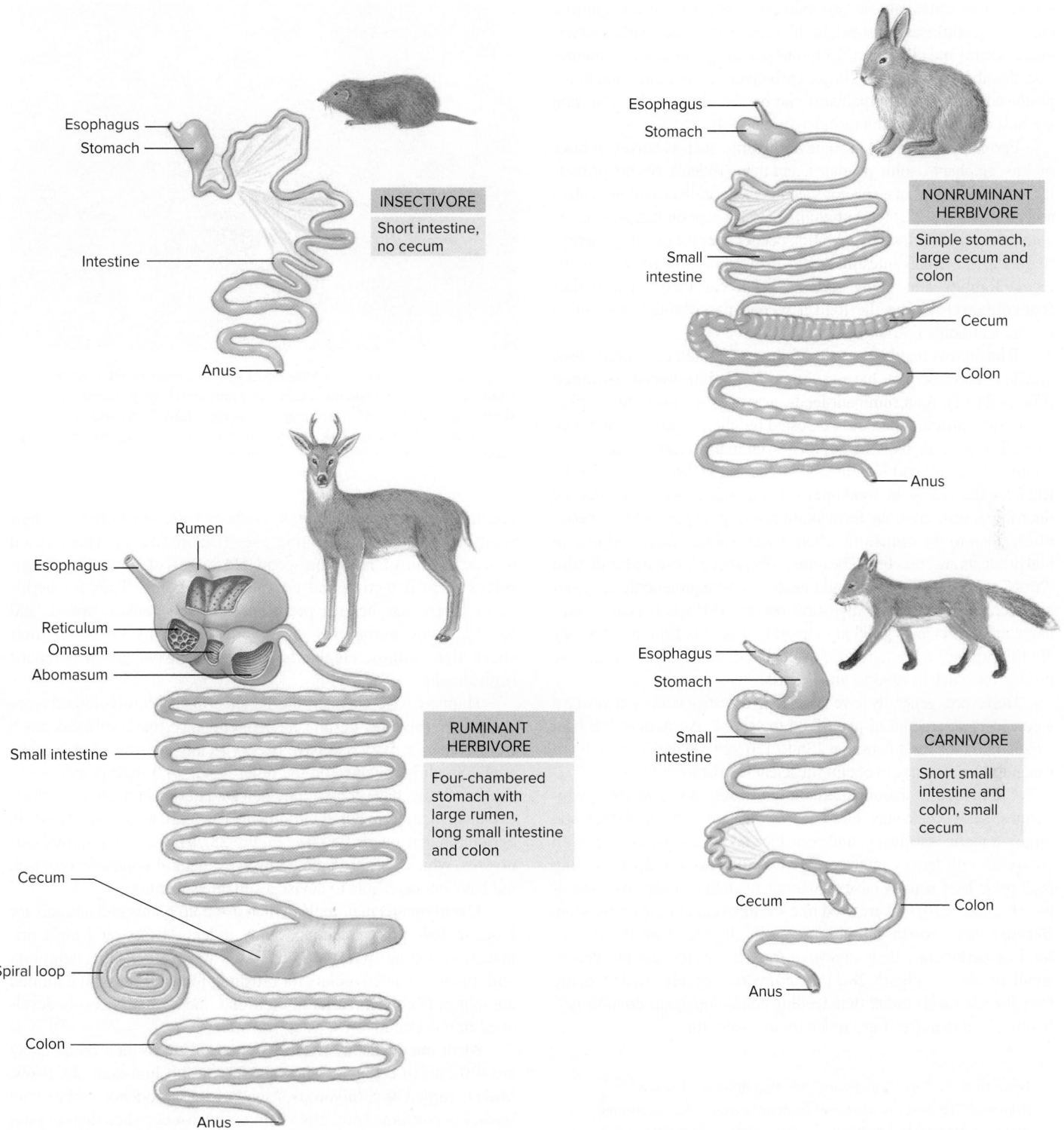

Figure 28.11 Digestive systems of mammals, showing different morphology with different diets.

rabbits and hares. In herbivores, canines are absent or reduced in size, whereas molars and premolars, which are adapted for grinding, are broad and usually high-crowned. Rodents (for example, beavers) have chisel-sharp incisors that grow throughout life and must be worn away to keep pace with their continual growth (Figure 28.10).

Herbivorous mammals have a number of adaptations for processing their fibrous diet of plant food. **Cellulose,** the structural carbohydrate of plants, is composed of long chains of glucose molecules, and therefore is a potentially nutritious food resource. However, the glucose molecules in cellulose are linked by chemical bonds that few enzymes can break. No vertebrates synthesize cellulose-splitting enzymes (**cellulases**). Instead, herbivorous vertebrates harbor anaerobic bacteria and ciliates (p. 230) that produce cellulase in fermentation chambers in their gut. Simple carbohydrates, proteins, and lipids produced by the microorganisms can be absorbed by the host, and the host can digest the microorganisms as well.

Fermentation (p. 67) in some herbivores, such as horses, zebras, rabbits, elephants, some primates, and many rodents, occurs primarily in the colon and in a spacious side pocket, or diverticulum, called a **cecum** (Figure 28.11). Although some absorption happens in the colon and cecum, most fermentation occurs after the primary absorptive area (the small intestine), and many nutrients are lost in the feces. Rabbits and many rodents often eat their fecal pellets (**coprophagy**), giving the food a second pass through the gut to extract additional nutrients.

Ruminants (cattle, bison, buffalo, goats, antelopes, sheep, deer, giraffes, and okapis) have a huge **four-chambered stomach** (Figure 28.11). As a ruminant feeds, grass passes down the esophagus to the **rumen,** where it is digested by microorganisms and then formed into small balls of cud. The ruminant returns a cud to its mouth where the cud is deliberately chewed at length to crush the fiber. Swallowed again, food returns to the rumen, where cellulolytic microorganisms continue fermentation. The pulp passes to the **reticulum,** then to the **omasum,** where water, soluble food, and microbial products are absorbed. The remainder proceeds to the **abomasum** ("true" acid stomach) and small intestine, where proteolytic enzymes are secreted and normal digestion occurs. Perhaps because ruminants are particularly good at extracting nutrients from forage, they are the primary large herbivores in ecosystems with low vegetative production, such as tundras and deserts.

Herbivores generally have large, long digestive tracts and must eat a considerable amount of plant food to survive. An African elephant weighing 6 tons must consume 135 to 150 kg (300 to 400 pounds) of rough fodder each day to obtain sufficient nourishment for life.

Carnivorous mammals, including foxes, dogs, weasels, wolverines, seals, cetaceans, and cats, feed mainly on vertebrates and large, aquatic molluscs and crustaceans. Carnivores are well equipped with large canines and powerful clawed limbs for killing their prey. They usually have carnassial, bladelike molars and premolars (Figure 28.10) that are used like shears to cut muscle and tendon. Because their protein diet is more easily digested than the fibrous food of herbivores, their digestive tract is shorter and the cecum small or absent (Figure 28.11). Carnivores organize their feeding into discrete meals rather than feeding continuously (as do most herbivores) and therefore have much more leisure time.

Note that the term "carnivore" has two different uses in mammals: to describe diet and to denote a specific taxonomic order of mammals. For example, not all carnivores belong to order Carnivora (many marsupials and cetaceans are carnivorous) and not all members of order Carnivora are carnivorous. Order Carnivora contains many omnivorous feeders and some, such as pandas, are strict vegetarians.

In general, carnivores lead more active—and by human standards more interesting—lives than do herbivores. Because a carnivore must find and catch its prey, there is a premium on

Figure 28.12 Lions, *Panthera leo,* eating a wildebeest. Lacking stamina for a long chase, lions stalk prey and then charge suddenly, surprising their prey. After gorging themselves, lions sleep and rest for periods as long as one week before eating again. Order Carnivora, family Felidae.

intelligence; many carnivores, such as cats, are noted for their stealth and cunning in hunting prey (Figure 28.12). This has led to a selection of herbivores capable either of defending themselves or of detecting and escaping carnivores. Thus for herbivores, there has been a premium on keen senses, speed, and agility. Some herbivores, however, survive by virtue of their sheer size (rhinos, elephants) or by defensive group behavior (muskoxen).

Humans have changed the balance in the carnivore-herbivore contest. Carnivores, despite their intelligence, have suffered much from human activity and have been virtually exterminated in some areas. Small herbivores, on the other hand, with their potent reproductive ability, have consistently defeated our most ingenious efforts to banish them from our environment. The problem of rodent pests in agriculture has intensified (see Figure 28.29); we have removed carnivores, which served as the herbivores' natural population control, but have not been able to devise a suitable substitute.

Omnivorous mammals, which use both plants and animals for food, include pigs, raccoons, many rodents, bears, and most primates, including humans. Omnivores have a versatile dentition, with broad, rounded molars for crushing food. Because of a limited amount of fibrous material in their diet, the cecum is poorly developed in most omnivores.

Most mammals are notably opportunistic in their diets; many are difficult to categorize in a feeding guild. For example, foxes, often regarded as carnivorous, usually feed on rodents, rabbits, and birds, but consume fruit, grains, nuts, and insects when their normal food is scarce.

Many mammals cache food stores during periods of plenty. This habit is most pronounced in rodents, such as squirrels, chipmunks, gophers, and certain mice. Tree squirrels collect nuts, conifer seeds, and fungi and store these in caches for winter use. Often each item is hidden in a different place (scatter hoarding) and marked by a scent to assist relocation in the future. Some caches of chipmunks and red squirrels can be quite large (Figure 28.13).

Figure 28.13 Eastern chipmunk, *Tamias striatus*, with cheek pouches stuffed with seeds to be carried to a hidden cache. It will try to store several liters of food for the winter. It hibernates but awakens periodically to eat some of its cached food. Order Rodentia, family Sciuridae.

Figure 28.15 A northern shorttail shrew, *Blarina brevicauda*. This tiny but fierce mammal, with a prodigious appetite for insects, mice, snails, and worms, spends most of its time underground and is seldom seen. Shrews resemble the insectivorous ancestors of placental mammals. Order Soricomorpha, family Soricidae.

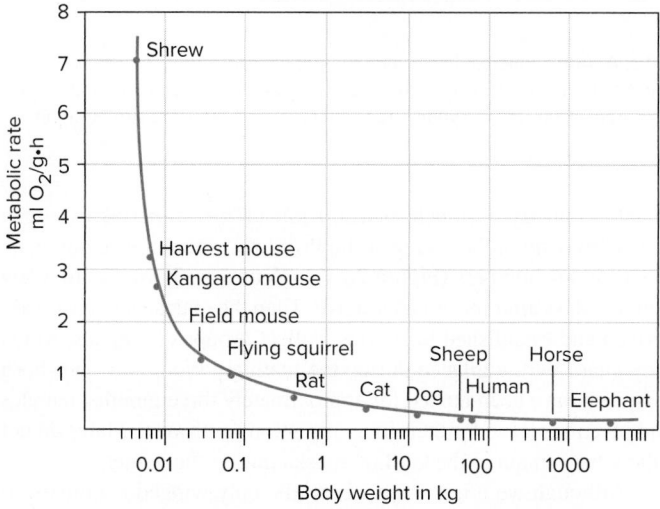

Figure 28.14 Relationship between body weight and metabolic rate for mammals. This relationship, often called the "mouse-to-elephant" curve, shows that metabolic rate is intense for small mammals like shrews and mice, and declines with increasing body weight of the species.

Source: Data from Eckert Animal Physiology: Mechanisms and Adaptations, 4/e by D. Randall, W. Burggren, K. French. © 1978, 1983, 1988, 1997, and 2000 by W. H. Freeman and Company.

Body Weight and Food Consumption

As in birds (p. 593), the smaller the mammal, the greater is its metabolic rate and the more it must consume relative to its body size (Figure 28.14). This happens because the metabolic rate of a mammal—and therefore the amount of food it must eat to sustain this metabolic rate—varies in rough proportion to the relative surface area rather than to body weight. Surface area is proportional to approximately 0.7 power of body weight, and the amount of food a mammal (or bird) eats also is roughly proportional to 0.7 power of its body weight. For example, a 3 g mouse will consume *per gram of body weight* five times more food than does a 10 kg dog and about 30 times more food than does a 5000 kg elephant. Thus small mammals (shrews, bats, and mice) must spend much more time hunting and eating food than do large mammals. The smallest shrews, weighing

only 2 g, may eat more than their body weight each day and starve to death in a few hours if deprived of food (Figure 28.15). In contrast, large carnivores can remain healthy with only one meal every few days. A cougar kills an average of one deer a week.

Migration

Migration is a more difficult undertaking for mammals than for birds or fishes, because terrestrial locomotion is more energetically expensive than swimming or flying. Not surprisingly, few terrestrial mammals make regular seasonal migrations, preferring instead to center their activities in a defined and limited home range. Nevertheless, there are some striking terrestrial mammalian migrations, especially in northern North America.

The barren-ground caribou of Canada and Alaska undertake direct and purposeful mass migrations spanning 160 to 1100 km (100 to 700 miles) twice annually (Figure 28.16). From winter ranges in boreal forests (taiga), they migrate rapidly in late winter and spring to calving ranges on the barren grounds (tundra). Calves are born in mid-June. As summer progresses, caribou are increasingly harassed by warble and nostril flies that bore into their flesh, by mosquitos that drink their blood (estimated at a liter per caribou each week during the height of the mosquito season), and by wolves that prey on their calves. They move southward in July and August, feeding little along

Caribou have suffered a drastic decline in numbers since early times when their population reached several million. By 1958 less than 200,000 remained in Canada. The decline is attributed to several factors, including habitat alteration from exploration and development in the North, but especially to excessive hunting. For example, the western Arctic herd in Alaska numbered about 242,000 caribou in 1970. Following five years of heavy unregulated hunting, a 1976 census revealed only about 75,000 animals remaining. After restricting hunting, the herd increased to about 490,000 in 2003, then declined to about 348,000 in 2011. However, this recovery is threatened by proposed expansion of oil extraction in several Alaskan wildlife refuges. In addition, caribou populations are threatened by widespread extraction of oil from tar sands, which produces large pits and tailing ponds; Sierra Club Canada reports a recent decline of woodland caribou in Alberta from such activity.

Figure 28.16 Barren-ground caribou, *Rangifer tarandus,* of Canada and Alaska. **A,** Adult male caribou in autumn pelage and antlers in velvet. **B,** Summer and winter ranges of some major caribou herds in Canada and Alaska (other herds not shown occur on Baffin Island and in western and central Alaska). Principal spring migration routes are indicated by arrows; routes vary considerably from year to year. The same species is known as reindeer in Europe. Order Artiodactyla, family Cervidae.

Figure 28.17 Annual migrations of northern fur seals, *Callorhinus ursinus,* showing the separate wintering grounds of males and females. Both males and females of the large Pribilof population migrate in early summer to the Pribilof Islands, where females give birth to their pups and then mate with males. Order Carnivora, family Otariidae.

the way. In September they reach the taiga and feed there almost continuously on low ground vegetation. Mating (rut) occurs in October.

Plains bison, before their deliberate near extinction by humans, made huge circular migrations to separate summer and winter ranges.

Oceanic seals and whales make the longest mammalian migrations. Gray whales, for example, migrate between Alaska in summer and Baja California, Mexico, in winter, an annual migration of over 18,000 km (11,250 miles). One of the most remarkable migrations is that of northern fur seals, which breed on the Pribilof Islands approximately 300 km (185 miles) off the coast of Alaska and north of the Aleutian Islands. From wintering grounds off southern California

females journey as much as 2800 km (1740 miles) across open ocean, arriving in the spring at the Pribilofs, where they congregate in enormous numbers (Figure 28.17). Young are born within a few hours or days after the females arrive. Then the males, having already arrived and established territories, collect groups of females, which they guard with vigilance during the mating period. After newborn offspring have been nursed for approximately three months, females and juveniles leave for their long migration southward. Males do not follow but remain in the Gulf of Alaska during the winter.

Although we might expect bats, the only winged mammals, to use their gift of flight to migrate, few of them do. Most spend winters in hibernation. Four species of American bats that migrate spend their summers in northern or western states and their winters in the southern United States or Mexico.

Flight and Echolocation

Many mammals scamper through trees with amazing agility; some can glide from tree to tree (Figure 28.18); and one group, the bats, has full flight. Gliding or flying evolved independently in several groups of mammals, including marsupials, rodents, flying lemurs, and bats. Flying squirrels (Figure 28.18) actually glide rather than fly, using the gliding skin (patagium) that extends from the sides of the body.

Bats are mostly nocturnal or crepuscular (active at twilight) and thus hold a niche unoccupied by most birds. Their achievement is attributed to two features: flight and capacity to navigate by echolocation. Together these adaptations enable bats to fly and to avoid obstacles in absolute darkness, to locate and to catch insects with precision, and to find their way deep into caves (a habitat largely unexploited by other mammals and birds) where they sleep during the daytime hours.

Research has concentrated on members of the family Vespertilionidae, which includes most common North American bats. When in flight, bats emit short pulses 5 to 10 msec in duration in a

Figure 28.18 Southern flying squirrel, *Glaucomys volans*, gliding in for a landing. Its undersurface is nearly tripled in area when gliding skin is spread. Glides of 40 to 50 m are possible. Good maneuverability during flight is achieved by adjusting position of the gliding skin with special muscles. Flying squirrels are nocturnal and have superb night vision. Order Rodentia, family Sciuridae.

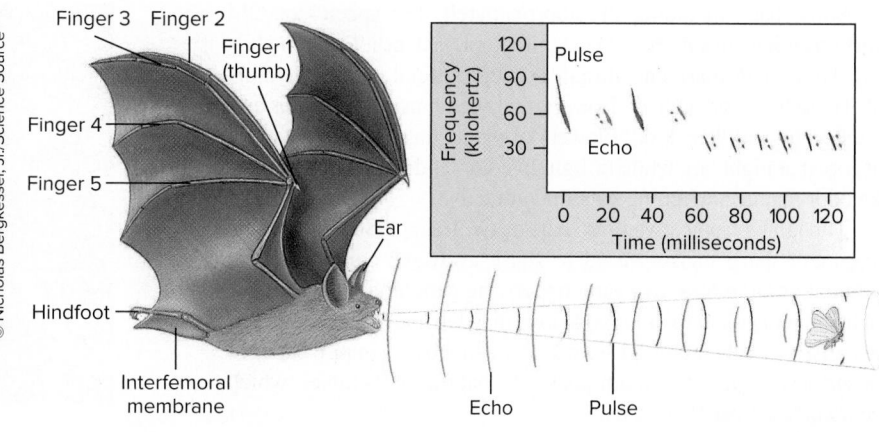

Figure 28.19 Echolocation of an insect by a little brown bat, *Myotis lucifugus*. Frequency modulated pulses are directed in a narrow beam from the bat's mouth. As the bat nears its prey, it emits shorter, lower signals at a faster rate. Order Chiroptera, family Vespertilionidae.

narrow directed beam from the mouth or nose (Figure 28.19). Each pulse is frequency modulated, being highest at the beginning, up to 100,000 Hz (hertz, cycles per second), and sweeps down to perhaps 30,000 Hz at the end. Sounds of this frequency are ultrasonic to human ears, which have an upper limit of about 20,000 Hz. When bats are searching for prey, they produce about 10 pulses per second. If prey is detected, the rate increases rapidly up to 200 pulses per second in the final phase of approach and capture. Pulses are spaced so that the echo of each is received before the next pulse is emitted, an adaptation that

prevents jamming. Because transmission-to-reception time decreases as a bat approaches an object, it can increase pulse frequency to obtain more information about an object. A bat shortens pulse length as it nears an object. Some nocturnal moths have evolved ultrasonic detectors used to detect and to avoid approaching bats (p. 738).

External ears of bats are large, like hearing trumpets, and shaped variously in different species. A bat builds a mental image of its surroundings from echo scanning that approaches the visual resolution of a diurnal animal's eyes.

Many insectivores (for example, shrews and tenrecs) use echolocation, but it is crudely developed compared with that of bats. Toothed whales, however, have a highly developed capacity to locate objects by echolocation. Totally blind sperm whales in perfect health have been captured with food in their stomachs. Although the mechanism of sound production and reception remains imperfectly understood, it is thought that clicks are produced in the nasal passages as air is moved through nasal valves and associated air sacs while the blowhole is closed. The clicks are focused into a narrow beam by the melon, a lens-shaped, fatty body in the forehead. Because of certain physical

properties of water, it is necessary for toothed whales to emit very high frequency pulses—as high as 220,000 Hz. Returning echos are primarily received through the lower jaw, and channeled through oil-filled sinuses in the dentary (lower jawbone) to the inner ear. The inner ear is surrounded by a capsule of bone enclosing a fatty mixture that blocks sounds except for those transmitted through the lower jaw. Toothed whales can apparently determine size, shape, speed, direction, and density of objects in the water and know the position of every whale in their pod.

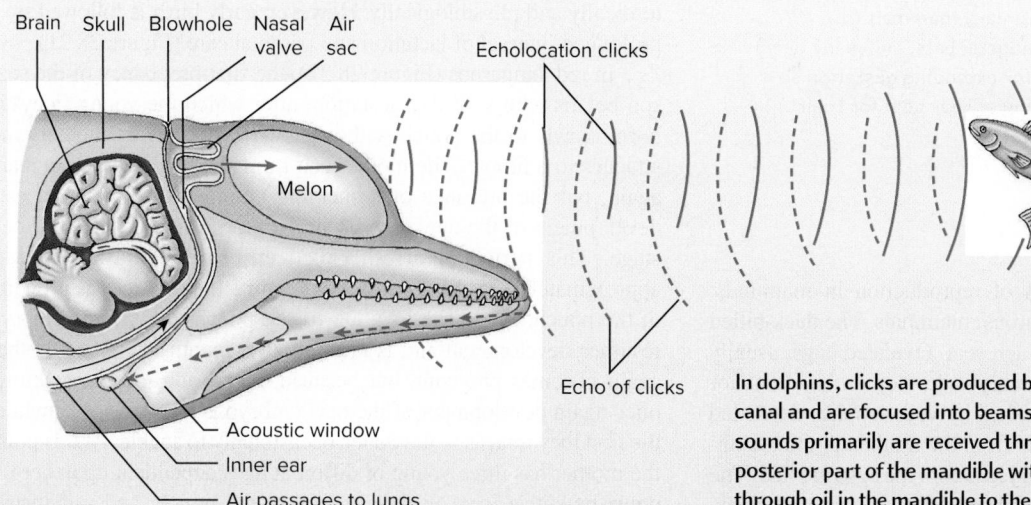

In dolphins, clicks are produced by moving air through the nasal canal and are focused into beams through the melon. Returning sounds primarily are received through the acoustic window, a posterior part of the mandible with very thin bone, and channeled through oil in the mandible to the middle and inner ears.

Some bats, including the approximately 185 species of Old World fruit bats in suborder Megachiroptera, lack echolocation abilities. They are primarily nocturnal, although several species are diurnal. These bats feed on fruit, flowers, and nectar, using large eyes and olfaction to find their food. Flowers of plants that are pollinated by bats open at night, are white or light in color, and emit a musky, batlike odor that nectarfeeding bats find attractive.

The famed tropical vampire bats of Central and South America have razor-sharp incisors used to shave away the epidermis of a mammal or large bird, exposing underlying capillaries. After infusing an anticoagulant to aid blood flow, they lap and store their blood meal in a specially modified stomach. Feeding by a vampire bat does not seriously injure the victim unless the bat transmits rabies, which occurs in less than 1% of bats.

Reproduction

Reproductive Cycle

Most mammals have definite mating seasons, usually in winter or spring and timed so that birth and the rearing of young occur at the most favorable time of the year. Many male mammals can copulate at any time, but females are receptive only during a relatively brief period of the **estrous cycle** called **estrus** or heat (Figure 28.20).

How often females are in estrus varies greatly among different mammals. Animals that have only a single estrous event during their breeding season are called **monestrous;** those having a recurrence of estrous events during their breeding season are called **polyestrous.** Dogs, foxes, and bats belong to the first group; field mice and squirrels are all polyestrous, as are many mammals living in tropical regions (Figure 28.20). Monkeys and apes have a somewhat different cycle in which the postovulation period is terminated by **menstruation,** during which the endometrium (lining of the uterus) is shed. In most primates it is reabsorbed, but in humans and chimpanzees, most of it is discharged with some blood through the vagina. This **menstrual cycle** is described in Chapter 7 (pp. 146–148).

> Delayed implantation lengthens the gestation period of many mammals. The blastocyst remains dormant while its implantation in the uterine wall is postponed for periods of a few weeks to several months. For many mammals (for example, bears, seals, weasels, badgers, bats, and many deer) delayed implantation is a device for extending gestation so that the young are born at the time of year best for their survival.

Reproductive Patterns

There are three different patterns of reproduction in mammals. **Monotremes** are egg-laying (oviparous) mammals. The duck-billed platypus has one breeding season each year. Ovulated eggs, usually two, are fertilized in the oviduct. Embryos develop in the uterus for 10 to 12 days, where they are nourished by yolk supplies deposited prior to ovulation, and by secretions from the mother. A thin, leathery shell is secreted around the embryos before the eggs are laid. The platypus lays its eggs in a burrow, where they hatch at a relatively

Figure 28.20 African lions, *Panthera leo,* mating. Lions breed at any season, although predominantly in spring and summer. During the short period that a female is receptive, she may mate repeatedly. Three or four cubs are born after gestation of 100 days. Once the mother introduces the cubs into the pride, they are treated with affection by both adult males and females. Once mature, male cubs are driven from the pride, but maturing females frequently are tolerated and remain with the pride. Order Carnivora, family Felidae.

undeveloped state after about 12 days. Echidnas incubate their eggs in an abdominal pouch. After hatching, young feed on milk produced by the mother's mammary glands. Because monotremes have no nipples, young lap milk secreted onto the belly fur of the mother.

Marsupials are pouched, viviparous mammals that exhibit a second pattern of reproduction. Although only eutherians are called "placental mammals," marsupials do have a transient type of placenta, called a **choriovitelline** (or yolk sac) **placenta.** An embryo (blastocyst) of a marsupial is at first encapsulated by shell membranes and floats free for several days in the uterine fluid. After "hatching" from the shell membranes, embryos of most marsupials do not implant (p. 176), or "take root" in the uterus, as they would in eutherians, but erode shallow depressions in the uterine wall in which they lie and absorb nutrient secretions from the mucosa through the vascularized yolk sac. Gestation (the intrauterine period of development) is brief in marsupials, and therefore all marsupials give birth to tiny young that are effectively still embryos, both anatomically and physiologically. However, early birth is followed by a prolonged interval of lactation and parental care (Figure 28.21).

In red kangaroos (Figure 28.22), the first pregnancy of the season begins with a 33-day gestation, after which the young (joey) is born, crawls to the pouch without assistance from its mother, and attaches to a nipple. The mother can immediately become pregnant again, but the presence of a suckling young in the pouch arrests development of the new embryo in the uterus at about the 100-cell stage. This period of arrest, called **embryonic diapause,** lasts approximately 235 days, during which time the first joey is growing in the pouch. When the joey leaves the pouch, the uterine embryo resumes development and is born about a month later. The mother again becomes pregnant, but because the second joey is suckling, once again development of the new embryo is arrested. Meanwhile, the first joey returns to the pouch occasionally to suckle. At this point the mother has three young of different ages dependent upon her for nourishment: a joey on foot, a joey in the pouch, and a diapause

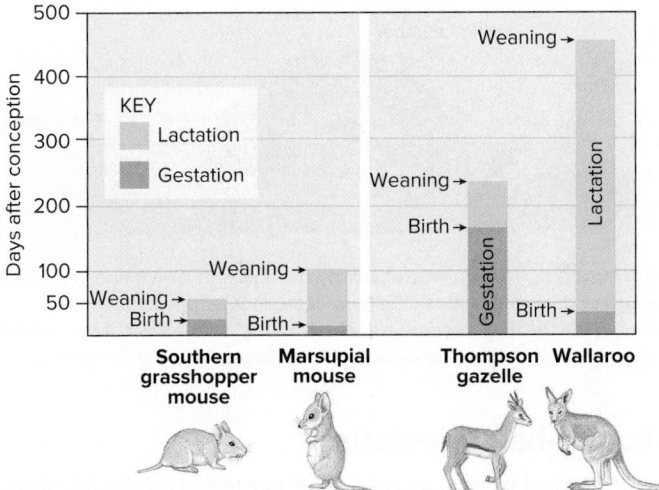

Figure 28.21   Comparison of gestation and lactation periods between matched pairs of ecologically similar species of marsupial and placental mammals. The graph shows that marsupials have shorter intervals of gestation and longer intervals of lactation than do similar species of placentals.

Joey on nipple, drinking high-protein, low-fat milk, arresting development of embryo in uterus

Embryo in diapause in uterus

Young kangaroo returning to drink low-protein, high-fat milk

True gestation

Birth

Joey leaves pouch

Dependency ends

Joey suckling in pouch 235 days   Joey suckling on foot   FIRST YOUNG

Diapause (arrested development)   True gestation 33 days   Joey suckling in pouch 235 days   SECOND YOUNG

Mating

Birth   Diapause (arrested development)   THIRD YOUNG

Mating

Figure 28.22   Kangaroos have a complicated reproductive pattern in which the mother may have three young in different stages of development dependent on her at once. Order Diprotodontia, family Macropodidae.

in prolonged gestation, unlike marsupials, in which reproductive investment is in prolonged lactation (Figure 28.21). Embryos remain in the uterus, nourished by food supplied initially by a choriovitelline placenta and later by a **chorioallantoic placenta** (described on pp. 176–177). Length of gestation is longer in placentals than in marsupials, and in large mammals it is much longer (Figure 28.21). For example, mice have a gestation period of 21 days; rabbits and hares, 30 to 36 days; cats and dogs, 60 days; cattle, 280 days; and elephants, 22 months (the longest). Important exceptions include baleen whales, the largest mammals, whose gestation period is only 12 months, and small bats no larger than mice, whose gestation period extends 4 to 5 months. The condition of the young at birth also varies. Most antelopes bear **precocial** young that are well furred, with eyes open, and able to run. Newborn mice, however, are **altricial**: blind, naked, and helpless. We all know how long it takes a human baby to learn to walk. Human growth is in fact slower than that of any other mammal, and this is one of the distinctive attributes that sets us apart from other mammals.

Is the placental mode of reproduction superior to that of marsupials? The traditional view holds this to be true, citing the low species diversity and small geographic range of marsupials, and the success of introduced placentals in Australia at the expense of some marsupials. Clearly, placentals have the advantage of higher reproductive rate, and retaining young in a pouch is not possible for fully aquatic forms. However, marsupials may have some advantages of their own. Because marsupials invest less energy in newborns, more energy would be available to a marsupial for replacement of lost young. This mode of reproduction may be advantageous in highly unpredictable climates, such as those of Australia. However, the earliest placentals and marsupials were very different from those of today, and arguments citing adaptive features for living species are not likely to apply to the early evolution of these groups. Placental and marsupial reproductive patterns are two successful methods that evolved independently. Marsupials have succeeded alongside placentals in South and Central America, where they have experienced a modest diversification into about 80 living species, and who could doubt the tenacity of the North American opossum?

The number of young that mammals produce in a season depends on mortality rate, which, for some mammals such as mice, may be high at all age levels. Usually, the larger the animal, the smaller the number of young in a litter. Small rodents, which serve as prey for many carnivores, usually produce more than one litter of several young each season. Meadow voles, *Microtus*

embryo in the uterus. There are variations on this remarkable sequence—not all marsupials have developmental delays like kangaroos, and some do not even have pouches—but in all, the young are born at an extremely early stage of development and undergo prolonged development while dependent on a mammary gland (Figure 28.23).

The third pattern of reproduction is that of viviparous **placental mammals,** the eutherians. In placentals, reproductive investment is

Figure 28.23
Virginia opossums, *Didelphis virginiana*, 15 days old, fastened to nipples in mother's pouch. When born after a gestation period of only 12 days, they are the size of honey bees. They remain attached to the nipples for 50 to 60 days. Order Didelphimorphia, family Didelphidae.

© Gary Meszaros/Science Source

Figure 28.24   Each beaver colony constructs its own lodge in a pond created by damming a stream. Each year the mother bears four or five young; when the third litter arrives, 2-year-olds are driven out of the colony to establish new colonies elsewhere. Order Rodentia, family Castoridae.

pennsylvanicus, produce as many as 17 litters of four to nine young in a year. Most carnivores have one litter of three to five young per year. Large mammals, such as elephants and horses, give birth to a single young with each pregnancy. An elephant produces, on average, four calves during her reproductive life of perhaps 50 years.

Territory and Home Range

Many mammals have territories—areas from which individuals of the same species are excluded. Many wild mammals, like some humans, are basically unfriendly to their own kind, especially so to others of their own sex during the breeding season. If a mammal dwells in a burrow or den, this area forms the center of its territory. Territories vary greatly in size depending on the size of the animal and its feeding habits. Grizzly bears have territories of several square kilometers, which they guard zealously against all other grizzlies.

Mammals mark the boundaries of their territories with secretions from scent glands or by urinating or defecating. When an intruder knowingly enters another's marked territory, it is immediately placed at a psychological disadvantage. Should a challenge follow, the intruder almost invariably breaks off the encounter in a submissive display characteristic for the species. Territoriality and aggressive and submissive displays are described in more detail in Chapter 36, pages 786–788.

A beaver colony is a family unit, and beavers are among several mammalian species in which males and females form strong monogamous bonds that last a lifetime. Because beavers invest considerable time and energy in constructing a lodge and dam and storing food for winter (Figure 28.24), the family, especially the adult male, vigorously defends its real estate against intruding beavers. Most of the work of building dams and lodges is performed by male beavers, but females assist when not occupied with their young.

An interesting exception to the strong territorial nature of most mammals are prairie dogs, which live in large, friendly communities called "towns" (Figure 28.25). When a new litter has been reared, adults relinquish the old home to the young and move to the edge of the community to establish a new home. Such a practice is opposite that of most mammals, which drive off the young when they are self-sufficient.

A **home range** of a mammal is a much larger foraging area surrounding a defended territory. Home ranges are not defended in the same way as are territories; home ranges may, in fact, overlap, producing a neutral zone that owners of several territories use for seeking food.

Mammalian Populations

A population of animals includes all members of a species that share a particular space and potentially interbreed (see Chapter 38). All mammals (like other organisms) live in ecological communities, each composed of numerous populations of different animal and plant species. Each species is affected by the activities of other species and by the physical environment, especially climate. Populations of mammals are lowest before the breeding season and greatest just after addition of new members. Beyond these expected changes in population size, mammalian populations may fluctuate from other causes.

Irregular fluctuations are commonly produced by variations in climate, such as unusually cold, hot, or dry weather, or by natural catastrophes, such as fires, hailstorms, and hurricanes. These are **density-independent** factors because they affect a population whether it is crowded or dispersed. However, the most spectacular fluctuations are **density dependent;** they correlate with population crowding, and include factors such as predation, parasitism, and competition for food and other resources. These extrinsic limits to growth are discussed in Chapter 38, page 821.

In his book *The Arctic* (1974, Montreal, Infacor, Ltd.), Canadian naturalist Fred Bruemmer describes the growth of lemming populations in arctic Canada:

"After a population crash one sees few signs of lemmings; there may be only one to every 10 acres. The next year, they are evidently numerous; their runways snake beneath the tundra vegetation, and frequent piles of rice-sized droppings indicate the lemmings fare well. The third year one sees them everywhere. The fourth year, usually the peak year of their cycle, the populations explode. Now more than 150 lemmings may inhabit each acre of land and they honeycomb it with as many as 4000 burrows. Males meet frequently and fight instantly. Males pursue females and mate after a brief but ardent courtship. Everywhere one hears the squeak and chitter of the excited, irritable, crowded animals. At such times they may spill over the land in manic migrations."

Cycles of abundance are common among many rodent species. Among the best-known examples are mass migrations of Scandinavian and arctic North American lemmings following population peaks. Lemmings (Figure 28.26) breed all year, although more in summer than in winter. The gestation period is

Figure 28.25   Adult and juvenile black-tailed prairie dogs, *Cynomys ludovicianus.* These highly social prairie dwellers are herbivores that serve as an important prey for many animals. They live in elaborate tunnel systems so closely interwoven that they form "towns" of as many as 1000 individuals. Towns are subdivided into family units, each with one or two males, several females, and their litters. Although prairie dogs display ownership of burrows with territorial calls, they are friendly with inhabitants of adjacent burrows. The name "prairie dog" derives from the sharp, doglike bark they make when danger threatens. Order Rodentia, family Sciuridae.

Figure 28.26   Northern collared lemming, *Dicrostonyx groenlandicus,* a small rodent of the far north. Populations of lemmings fluctuate widely. Order Rodentia, family Cricetidae.

Figure 28.27   Changes in population size of snowshoe hare and lynx in Canada as indicated by pelts received by the Hudson's Bay Company over a 90-year period. Abundance of lynx (predator) follows that of the hare (prey).

only 21 days; young born at the beginning of summer are weaned in 14 days and can reproduce by end of summer. At the peak of their population density, having devastated the vegetation by tunneling and grazing, lemmings begin long, mass migrations to find new, undamaged habitats for food and space. They swim across streams and small lakes as they go but cannot distinguish these from large lakes, rivers, and the sea, in which they drown. Because lemmings are the main diet of many carnivorous mammals and birds, any change in lemming population density affects their predators as well.

Snowshoe hares (see Figure 28.7) of North America show 10-year cycles in abundance. The well-known fecundity of hares enables them to produce litters of three or four young as many as five times per year. Their density may increase to 4000 hares competing for food in each square mile of northern forest. Predators (owls, minks, foxes, and especially lynxes) also increase (Figure 28.27). Then the population crashes precipitously for reasons that have long puzzled scientists. The best available evidence suggests that depletion of winter forage might explain these declines. Hare populations follow patterns of sunspot activity, which may affect vegetative production (p. 824). Whatever the causes, population crashes that follow superabundance, although harsh, permit vegetation to recover, providing survivors with a much better chance for successful breeding.

28.3 HUMANS AND MAMMALS

At least 15,000 years ago, dogs were domesticated, the first of many species to be domesticated by people. Like cats, dogs probably entered voluntarily into a mutually beneficial relationship with people. Dogs are an extremely adaptable and genetically plastic species derived from wolves. Domestic cats were domesticated from an African population of the wildcat, *Felis sylvestris,* which still occurs in Africa and Asia. Domestication of sheep, pigs, goats, cattle, donkeys, horses, camels, and llamas occurred 10,000–2,500 years ago, when agriculture was being developed throughout the world. Certain domestic species no longer exist as wild animals, for example, one-humped dromedary camels of North Africa and llamas and alpacas of South America. All truly domestic animals breed in captivity; many have been molded by selective breeding to yield characteristics desirable for human purposes.

Asian elephants are considered semidomesticated because they seldom breed in captivity. Adults are captured and submit to a life of toil with astonishing docility. Reindeer of northern Scandinavia are domesticated only in the sense that they are "owned" by nomadic peoples who follow them in their seasonal migrations (Figure 28.28). Recent attempts at domestication of other large mammals, including the giant eland of Africa, bison, and elk, have been modestly successful.

Mammals can be harmful to people by devouring our food and by carrying diseases. Rodents, rabbits, and other mammals can inflict staggering damage to growing crops and stored food (Figure 28.29). Various rodents, including brown rats and prairie dogs, carry typhus and bubonic plague. Wood ticks transmit tularemia (rabbit fever) to people from rabbits, woodchucks, muskrats, and other rodents. Ticks also transmit Rocky Mountain spotted fever to people from ground squirrels and dogs, and Lyme disease to people from white-tailed deer. *Trichinella* worms and tapeworms can infect people who eat infected beef, pork, and other mammalian meat.

28.4 HUMAN EVOLUTION

Darwin devoted an entire book, *The Descent of Man and Selection in Relation to Sex* (1871), largely to human evolution. The idea that humans share common descent with apes and other animals was repugnant to the Victorian world, which responded with predictable outrage (see Figure 6.15). When Darwin's views were first debated, few human fossils had been unearthed, but the current accumulation of fossil and DNA evidence has strongly vindicated Darwin's hypothesis that humans descend from other apes.

The search for fossils, especially for a "missing link" that would provide a connection between apes and humans, began when two Neandertal[1] skeletons were collected in the 1880s. Then in 1891, Eugene Dubois discovered the famous Java man (*Homo erectus*). Some of the most spectacular discoveries, however, occurred in Africa, especially between 1967 and 1977, and then again from 1995 to 2013. During the last 30 years, comparative biochemical studies have demonstrated that humans and chimpanzees are as similar genetically as are many sibling species. Comparative cytology provided evidence that chromosomes of humans and apes are homologous (p. 110). Darwin's hypothesis that humans descend from other apes has been confirmed.

Evolutionary Diversification of Primates

Humans are primates, a fact that even pre-evolutionist Linnaeus recognized. All primates share certain significant characteristics: grasping fingers on all four limbs, flat fingernails instead of claws, and forward-pointing eyes with binocular vision and excellent depth perception.

The earliest primate was probably a small, nocturnal animal similar in appearance to a tree shrew. This ancestral primate lineage split into two lineages, one of which gave rise to lemurs and lorises (Strepsirrhini); and the other to tarsiers (Figure 28.30), monkeys

Figure 28.28 Herd of reindeer, *Rangifer tarandus,* during annual roundup by Laplanders in northern Sweden. The same species is called caribou in North America. Order Artiodactyla, family Cervidae.

© Cleveland P. Hickman, Jr.

© MikeLane45/iStock/Getty Images

Figure 28.29 Brown rat, *Rattus norvegicus,* living all too successfully beside human habitations. Brown rats not only cause great damage to food stores but also spread disease, including bubonic plague (a disease, carried by infected fleas, that greatly influenced human history in medieval Europe), typhus, infectious jaundice, *Salmonella* food poisoning, and rabies. Order Rodentia, family Muridae.

(Figure 28.31), and apes (Figure 28.32). Traditionally lemurs, lorises, and tarsiers have been called **prosimians,** a paraphyletic group, and apes and monkeys have been called **simians** or **anthropoids.** Prosimians and many simians are arboreal (tree-dwellers), which is probably the ancestral lifestyle for both groups. Flexible limbs are essential for active animals moving through trees. Grasping hands and feet, in contrast to clawed feet of squirrels and other rodents, enable primates to grip limbs, to hang from branches, to seize and to manipulate food, and, most significantly, to use tools. Primates have highly developed special senses, especially acute, binocular vision, and proper coordination of limb and finger muscles to assist their active arboreal life. Of course, sense organs are no better than the brain processing sensory information. A large cerebral cortex supports precise timing of movements, judgment of distance, and perception of the environment.

The earliest simian fossils appeared in Africa in late Eocene deposits, some 55 million years ago. Many of these primates became diurnal rather than nocturnal, making vision the dominant special sense, now enhanced by color perception. We recognize three major simian clades. These are (1) New World monkeys of Central and South America, including howler monkeys, spider monkeys (Figure 28.31A), and tamarins; (2) Old World monkeys, including baboons (Figure 28.31B), mandrills, and colobus monkeys; and (3) apes (Figure 28.32). Old World monkeys and apes (including humans) are sister taxa, and together form the sister group of New World monkeys. In addition to their geographic separation, Old World monkeys differ from New World monkeys in lacking a grasping tail, while having close-set, downward-facing nostrils, better opposable, grasping thumbs, and only two premolars in each jaw half.

---

[1]The traditional spelling "Neanderthal" has been replaced by "Neandertal" in many recent publications. In old German, *thal* (valley) is pronounced with a hard *t*, rather than with a soft *th*. Modern spellings of many German words were changed to reflect their pronunciation, resulting in *thal* becoming *tal*.

Figure 28.30   A prosimian, the Mindanao tarsier, *Tarsius syrichta carbonarius,* of Mindanao Island in the Philippines. Order Primates, family Tarsiidae.

Figure 28.32   Western lowland gorillas, *Gorilla gorilla gorilla,* order Primates, family Hominidae, are apes.

A

B

Figure 28.31   Monkeys. **A,** Geoffroy's spider monkey, *Ateles geoffroyi,* order Primates, family Atelidae, is a New World monkey. **B,** Olive baboons, *Papio homadryas,* order Primates, family Cercopithecidae, are Old World monkeys.

Apes, which differ from Old World monkeys in having a larger cerebrum and loss of a tail, include gibbons, orangutans, gorillas, chimpanzees, bonobos, and humans. Except for gibbons, all apes are in the family Hominidae and are here called **hominids.** Chimpanzees and bonobos form the living sister group to humans (see Figure 10.9). All fossil hominid species that are phylogenetically placed closer to living humans than to chimpanzees are called here **humans** or **hominins.** The earliest known fossils of apes are from 23-million-year-old rocks in eastern Africa. These early, forest-dwelling hominids subsequently diversified into numerous forms that spread across Africa and Eurasia.

The First Humans and the Origin of Bipedalism

Trends in the evolution of skeletal differences between humans and other hominids often are associated with changes in diet and posture. Human jaws are less robust and bear smaller teeth than

those of other hominids, reflecting a more omnivorous diet. Position of the **foramen magnum** (a hole in the skull through which the spinal cord passes) shifted to be directly underneath the braincase in humans, indicative of bipedalism and upright posture. Additional skeletal changes in the human lineage associated with bipedalism include shorter pelvic bones, an S-shaped vertebral column, longer hindlimb bones, and parallel alignment of all five digits of the feet (see Figure 29.9). Bipedalism provided a better view of the landscape and freed the hands for using tools, defending against predators, carrying young, and gathering food.

Most genetic evidence suggests that humans diverged from chimpanzees about 6 million years ago, but some studies date the divergence to as recent as 4 million years ago or as long as 10 million years ago. Fossil evidence of humans from this period is sparse and controversial. In 2001 the desert sands of Chad yielded a remarkably complete skull of a hominid, *Sahelanthropus tchadensis,* dated at about 6.5 million years ago (Figure 28.33). Although its brain was no larger than that of a chimpanzee (320 to 380 cm^3), its relatively small canine teeth and ventral position of the foramen magnum suggest that the skull could be human.

The earliest well-known human was *Ardipithecus ramidus* from Ethiopia, dated at about 4.4 million years ago. Until recently this species was known only from teeth, but in 2009 numerous additional fossils were described, including a 45% complete skeleton, named "Ardi." *Ardipithecus* was about 120 cm tall and bipedal, although it retained many ancestral adaptations for arboreal life, including long arms, fingers, and toes. Both *Sahelanthropus* and *Ardipithecus* were probably woodland dwellers, based on associated fossil vertebrates and invertebrates from the same deposits. Earlier hypotheses suggested that bipedality arose as an adaptation to drying African environments, as woodlands were replaced with open savannas. However, because formation of the savannas of Africa did not occur until about 3 million years ago, it is now clear that bipedality first arose in woodland-dwelling hominins.

Another celebrated early human fossil is a 40% complete skeleton of a female *Australopithecus afarensis* (Figure 28.34). Unearthed in 1974 and named "Lucy" by its discoverer Donald Johanson, *A. afarensis* was a short, bipedal human with face and brain size (380 to 450 cm^3) slightly larger than that of a chimpanzee. This species was sexually dimorphic in size; females stood about 1 m (Figure 28.35) and males stood about 1.5 m. Their teeth suggest that they were primarily fruit- and leaf-eaters, but they likely incorporated some meat into their diet. The numerous fossils of *A. afarensis* date from 3.7 to 3 million years ago.

An explosion of australopithecine fossil finds during the last four decades documents at least eight species. Most of the early forms are considered gracile australopithecines because of their relatively light build, especially in skull and teeth (although all were more robust than modern humans). The oldest of these is *Australopithecus anamensis* of Kenya and Ethiopia, which lived from 4.2 to 3.9 million years ago. This species is morphologically intermediate between *Ardipithecus* and *A. afarensis*; some researchers consider it the ancestor or sister group of modern humans and all other australopithecines. One of the most recent of the gracile australopithecines was *Australopithecus africanus,* dated at 3 to 2.3 million years ago. In 2010 two hominid skeletons were found in South Africa from 2-million-year-old rocks. These were described as *Australopithecus sediba* and suggested to be phylogenetically close to *Homo.*

Coexisting with the earliest species of *Homo* and the latest species of gracile *Australopithecus* were at least three robust australopithecines. One of these was *Paranthropus robustus* (Figure 28.33), about 1–1.2 m tall and 40–55 kg. The "robust" australopithecines, which lived from 2.5 to 1.2 million years ago, were specialized with skull crests, heavy jaws, and large molars. Their diet included coarse seeds, roots, and nuts, in addition to the fruit, soft plants, and invertebrates typical of other early hominins. They are an extinct branch in hominid evolution and not part of our own ancestry.

Early *Homo*: Tool-Making and Migration Out of Africa

The earliest *Homo* fossils are dated about 2.8 million years ago. Because the human fossil record during this time is sparse, the evolution of *Homo* is unclear, although many researchers consider *Homo habilis* ("handy man") to be representative of the earliest *Homo.* This species was similar in form to *Australopithecus,* about 111 to 135 cm tall (3.7 to 4.5 feet), with long arms and short legs. Like its ancestors, *H. habilis* was adapted for both arboreal life and bipedality. However, it had a larger brain (500 to 600 cm^3) than *Australopithecus,* which might have been used in a novel cultural feature of *H. habilis*—the making of stone tools. This capacity to make stone tools may be the defining feature of genus *Homo.* Chimpanzees and bonobos do not make stone tools, and they cannot be taught to make simple stone tools, despite efforts by patient researchers. *Homo habilis* shared the early Pleistocene drying African landscape with many other hominids, including species of *Paranthropus, Austalopithecus, Homo rudolfensis,* and later, *Homo erectus. Homo rudolfensis,* a slightly more robust species than *H. habilis,* is of uncertain relationship with other hominids. Although the expansion of savannas is no longer linked to the origin of bipedality, perhaps it contributed to the spread of savanna-dwelling *Homo.*

About 1.9 million years ago *Homo erectus* appeared, a tall hominid standing 150 to 190 cm (5 to 6 feet) tall, with a low but distinct forehead and strong brow ridges. This species is known from Africa and Eurasia; those from Africa are considered to be a different species, *H. ergaster,* by some researchers. Early fossils had a cranial capacity of 850 cm^3, slightly larger than that of *H. habilis,* but later *H. erectus* had a cranial capacity of about 1100 cm^3, only slightly smaller than that of modern humans (Figure 28.33). The technology of *H. erectus* is characterized by more advanced tools and possibly by use of fire, as indicated by charcoal deposits and burned bones. It spread throughout southern Europe and in Asia as far east as China and Java, where it survived until about 150,000 years ago. One of the most unusual hominid finds was announced in 2004: *Homo floresiensis,* a species only 1 m tall, from the island of Flores, Indonesia. This species probably diverged from *H. erectus* and became extinct only about 13,000 years ago.

Modern Humans

Modern humans diverged from African *H. erectus* at least 800,000 years ago. These early humans, formerly designated as "archaic *Homo sapiens,*" are now usually assigned to *H. antecessor* and *H. heidelbergensis.* The better known of the pair, *H. heidelbergensis,* spread throughout Africa, Europe, and the Middle East. The brain of *H. heidelbergensis* (about 1250 cc^3) was larger than those of its ancestors, and it exhibited reduced brow ridges and teeth (but not so reduced as in *Homo sapiens*). During the time of *H. heidelbergensis* the Earth's climate cooled, with long glacial conditions predominating. About 300,000 to 200,000 years ago, *Homo heidelbergensis* was replaced by two hominins, *H. neanderthalensis* in Europe and *H. sapiens* in Africa. It has been hypothesized that northern (European) populations of *H. heidelbergensis* became more robust, adapting to cold conditions, forming *H. neanderthalensis,* while southern (African) populations retained their more gracile appearance, forming *H. sapiens.* **Neandertals** (*H. neanderthalensis*) occupied most of Europe and the Middle East. They had a brain size similar to that of modern humans and made more sophisticated stone tools than did *H. erectus.* Their robust, heavily muscled bodies allowed them to survive the cold climates of the Ice Age and to hunt large Pleistocene mammals, including woolly rhinoceros, bison, and mammoth. Judging by their numerous upper-body injuries, Neandertals attacked large game in close quarters, probably stabbing, rather than throwing, spears at prey. Burial of the dead is first known in Neandertals, and they possibly had complex rituals or religion. However, they lacked the complex art, technology, and culture of their successors.

Fossil evidence indicates that characteristics of *H. sapiens,* as presently defined, arose in Africa about 200,000 years ago (Figure 28.33). About 39,000 years ago Neandertals disappeared, approximately 3000 years after the first appearance of *H. sapiens* in Europe and Asia. The first modern humans were tall people with a culture very different from that of Neandertals. Implement crafting developed rapidly, and human culture became enriched with aesthetics, artistry, and sophisticated language.

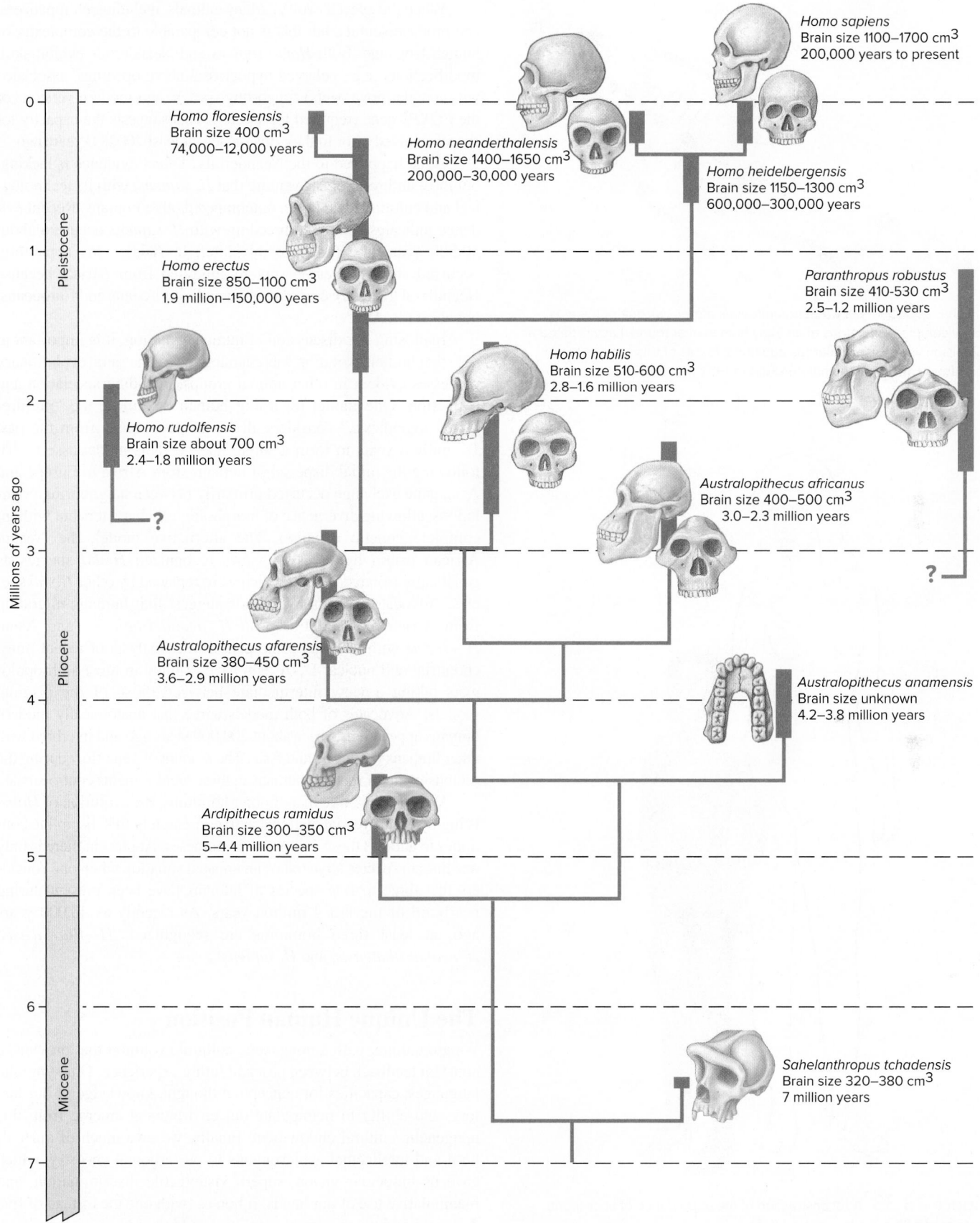

Figure 28.33 Evolution of hominins. Shown is one possible phylogenetic reconstruction. Many relationships are controversial. Note that in this reconstruction *Australopithecus* is paraphyletic. Thick vertical lines denote known durations of fossil forms in the stratigraphic record.

© John Reader/Science Source

Figure 28.34 Lucy (*Australopithecus afarensis*), one of the most nearly complete skeletons of an early human ever found. Lucy is dated at 3.2 million years old. Numerous additional fossils of this species have been found, including complete skulls in 1992 and 2000.

Figure 28.35 A reconstruction of the appearance of Lucy (*right*) compared with a modern human (*left*).

When did speech evolve? Many animals, including chimpanzees, can produce sounds, but this is not comparable to the complexity of human language. Both *Homo sapiens* and Neandertals exhibit skull modifications (e.g., enlarged hypoglossal nerve openings, associated with tongue movements) supporting speech, and modern versions of the *FOXP2* gene (required for speech). This suggests the capacity for speech evolved prior to their divergence, at least 200,000 years ago.

What happened to the Neandertals? Direct evidence is lacking, but most anthropologists venture that *H. sapiens,* with its technological and cultural advantages, outcompeted other humans. Recent evidence indicates some interbreeding with *H. sapiens* occurred about 50,000 years ago (and possibly later, in Asia). This apparently occurred shortly after *H. sapiens* migrated from Africa, because Neandertal genes are common in many living Asians and Europeans, but absent in Africans.

In closing our discussion of human evolution, it is important to note that human evolution was charaterized by the same evolutionary processes evident in other animal groups, including speciation and extinction. One model regarding human evolution, the "multiregional hypothesis," considers all *Homo* populations from the past 1.7 million years to form a single lineage. This model asserts that following the initial dispersal of humans from Africa to Europe and Asia, gene exchange occurred primarily between neighboring populations, allowing divergence of morphological characters but without complete genetic isolation. The alternative model, the "recent African origin hypothesis," views recognized *Homo* species as genetically isolated species which were replaced by other *Homo* species. Advocates of this hypothesis suggest that humans migrating from Africa replaced European *H. neanderthalensis* and Asian *H. erectus* with minimal or no gene flow. Analysis of recent mitochondrial and nuclear DNA data have resulted in most anthropologists taking a view intermediate between those of the original models. Advocates of both models agree that anatomically modern humans appeared in Africa about 200,000 years ago and interbred with other humans in Europe and Asia. The amount of gene flow during the establishment of modern humans in these areas remains controversial.

We note that, like that of other hominins, the evolution of *Homo* is highly controversial, and that active research in this discipline continues to address these competing hypotheses. At present there is only one human species alive today, an unusual situation when one considers that three to five species of humans have been present during nearly all of the last 4 million years. As recently as 40,000 years ago, at least three hominins are recognized: *H. floresiensis, H. neanderthalensis,* and *H. sapiens*!

The Unique Human Position

We are unique, with a nongenetic cultural evolution that provides a constant feedback between past and future experience. Our symbolic languages, capacities for conceptual thought, knowledge of our history, and ability to manipulate our environment emerge from this nongenetic cultural endowment. Finally, we owe much of our cultural and intellectual achievements to our arboreal ancestry, which gave us binocular vision, superb visuotactile discrimination, and manipulative use of our hands. If horses (with one toe instead of five fingers) had human mental capacity, could they have accomplished what humans have?

Taxonomy of Living Mammalian Orders

The taxonomy follows Wilson and Reeder (2005). The 29 recognized living mammalian orders include one monotreme order, seven marsupial orders, and 21 placental orders. Recent phylogenetic analysis of DNA sequences has led to many changes in taxonomy. For example, the former order Insectivora has been shown to be polyphyletic and its members are now placed in three orders, Afrosoricida, Soricomorpha, and Erinaceomorpha.

Mammalia

Prototheria (prō´tō-thir´ē-ə) (Gr. *prōtos*, first, + *thēr*, wild animal). Monotreme mammals.

Order Monotremata (mon´ō-trē-mä´tä) (Gr. *monos*, single, + *trēma*, hole): **egg-laying (oviparous) mammals: duck-billed platypus, echidnas.** Five species in this order are from Australia, Tasmania, and New Guinea. The most noted member of the order is the duck-billed platypus, *Ornithorhynchus anatinus.* Spiny anteaters, or echidnas, *Tachyglossus*, have a long, narrow snout adapted for feeding on ants, their chief food.

Theria (thir´ē-a) (Gr. *thēr*, wild animal).

Metatheria (met´ə-thir´ē-ə) (Gr. *meta*, after, + *thēr*, wild animal). Marsupial mammals.

Order Didelphimorphia (di´del-fi-mor´fē-ə) (Gr. *di*, two, + *delphi*, uterus, + *morph*, form): **American opossums.** These mammals, like other marsupials, are characterized by an abdominal pouch, or marsupium, in which they rear their young. Most species are in Central and South America, but one species, the Virginia opossum, *Didelphis virginiana,* is widespread in North America; 87 species.

Order Paucituberculata (pa´se-tū-ber-cū-la´te) (L. *pauci,* few, + *tuberculum,* knob): **shrew opossums.** Tiny, shrew-sized marsupials that occur in western South America; six species.

Order Microbiotheria (mī´crō-bī-ō-ther´ē-ə) (Gr. *micro,* small, + *bio,* life, + *thər,* wild animal): **Monito del Monte.** A South American mouse-sized marsupial that may be more closely related to Australian marsupials; one species.

Order Dasyuromorphia (das-ē-yur´ō-mor´fē-ə) (Gr. *dasy,* hairy, + *uro,* tail, + *morph,* form): **Australian carnivorous mammals.** In addition to a number of larger carnivores, this order includes a number of marsupial "mice," all of which are carnivorous. Confined to Australia, Tasmania, and New Guinea; 71 species.

Order Peramelemorphia (per´ə-mel-e-mor´fē-ə) (Gr. *per,* pouch, + *mel,* badger, + *morph,* form): **bandicoots.** Like placentals, members of this group have a chorioallantoic placenta and a relatively high rate of reproduction. Confined to Australia, Tasmania, and New Guinea; 22 species.

Order Notoryctemorphia (nō´tō-rik-te-mor´fē-a) (Gr. *not,* back, + *oryct,* digger, + *morph,* form): **marsupial moles.** Bizarre, semifossorial marsupials in Australia; two species.

Order Diprotodontia (di´prō-tō-don´tē-ə) (Gr. *di,* two, + *pro,* front, + *odont,* tooth): **koalas, wombats, possums, wallabies, kangaroos.** Diverse marsupial group containing some of the largest and most familiar marsupials. Present in Australia, Tasmania, New Guinea, and many islands of the East Indies; 143 species.

Eutheria (yū-ther-ē-ə) (Gr. *eu,* true, + *ther,* wild animal). Placental mammals.

Order Afrosoricida (af´rō-sor-i´si-də) (*afro,* of Africa + L. *soric,* shrew): **tenrics, golden moles.** Small, insectivorous mammals of Africa and Madagascar; 51 species.

Order Macroscelidea (mak´rō-se-lid´ē-ə) (Gr. *makros,* large, + *skelos,* leg): **elephant shrews.** Secretive mammals with long legs, a snoutlike nose adapted for foraging for insects, large eyes. Widespread in Africa; 15 species.

Order Tubulidentata (tū´byū-li-den´tä-tä) (L. *tubulus,* tube, + *dens,* tooth): **aardvark.** "Aardvark" is Dutch for earth pig, a peculiar animal with a piglike body found in Africa; one species.

Order Proboscidea (prō´bo-sid´ē-ə) (Gr. *proboskis,* elephant's trunk, from *pro,* before, + *boskein,* to feed): **elephants.** Largest of living land animals, with two upper incisors elongated as tusks, and well-developed molar teeth. Asian elephants, *Elephas maximus,* have long been partly domesticated and trained to do heavy tasks. Taming of African elephants, *Loxodonta africana,* is more difficult but was done extensively by ancient Carthaginians and Romans, who employed them in their armies; three species.

Order Hyracoidea (hī-ra-koi´dē-ə) (Gr. *hyrax,* shrew): **hyraxes** (coneys). Hyraxes are herbivores restricted to Africa and Syria. They have some resemblance to short-eared rabbits but have teeth like rhinoceroses, with hooves on their toes and pads on their feet. They have four toes on the front feet and three toes on the back; four species.

Order Sirenia (sī-rē´nē-ə) (Gr. *seiren,* sea nymph): **sea cows, manatees.** Large, aquatic mammals with large head, no hindlimbs, and forelimbs modified into flippers. The sea cow (dugong) of tropical coastlines of East Africa, Asia, and Australia and three species of manatees of the Caribbean area and Florida, Amazon River, and West Africa are the only living species. A fifth species, the large Steller's sea cow, was hunted to extinction by humans in the mid-eighteenth century; four species.

Order Cingulata (sin´gyū-lät´ə) (L. *cingul,* belt); **armadillos** (Figure 28.36). Insectivorous mammals with small, peglike teeth and beltlike bands of armor. Inhabit South and Central America; the nine-banded armadillo is expanding its range northward in the United States; 21 species.

Order Pilosa (pī-lō´sə) (L. *pilos,* hairy): **sloths, anteaters.** Anteaters are toothless mammals that use long tongues to feed on ants and termites; sloths are arboreal leaf-eaters. Both are restricted to Central and South America; 10 species.

Order Dermoptera (der-mop´ter-ə) (Gr. *derma,* skin, + *pteron,* wing): **flying lemurs.** These are related to true bats and consist of a single genus *Galeopithecus.* They are not lemurs (which are primates) and cannot fly but glide like flying squirrels. They occur in the Malay peninsula in the East Indies; two species.

Order Scandentia (scan-den´tē-ə) (L. *scandentis,* climbing): **tree shrews.** Small, squirrel-like mammals of the tropical rain forests of southern and southeastern Asia. Despite their name, many are not especially well-adapted for life in trees, and some are almost completely terrestrial; 20 species.

Taxonomy of Living Mammalian Orders

Order Primates (prī-ā´tēz or prī´māts) (L. *prima*, first): **prosimians, monkeys, apes.** First among animals in brain development, with especially large cerebral cortex. Most species are arboreal, with large eyes, binocular vision, grasping hands, and five digits (usually with flat nails) on both forelimbs and hindlimbs. It is hypothesized that their tree-dwelling habits of agility in capturing food or avoiding enemies were largely responsible for their advances in brain structure. There are two suborders; 376 species.

Suborder Strepsirrhini (strep´sə-rī-nē) (Gr. *strepsō*, to turn, twist, + *rhinos*, nose): **lemurs, lorises, pottos, bush babies.** Seven families of arboreal primates concentrated in Madagascar, but with species in Africa, southeast Asia, and Malay peninsula. All have a wet, naked region (rhinarium) surrounding comma-shaped nostrils, a long nonprehensile tail, and a second toe provided with a claw; 88 species.

Suborder Haplorhini (hap´lō-rī-nē) (Gr. *haploos*, single, simple, + *rhinos*, nose): **tarsiers, New and Old World monkeys, apes.** Six families, all except tarsiers are in the clade Anthropoidea. Haplorhine primates have dry, hairy noses, ringed nostrils and differences in skull morphology that distinguish them from strepsirrhine primates. Family **Tarsiidae** contains crepuscular and nocturnal tarsiers (Figure 28.30), with large, forward-facing eyes and reduced snout (five species). New World Monkeys, sometimes called platyrrhine monkeys because the nostrils are widely separated, are contained in five families: **Callitrichidae** (marmosets and tamarins; 42 species), **Cebidae** (capuchins and squirrel monkeys; 14 species), **Atelidae** (howler and spider monkeys; 29 species), **Pithecidae** (titis, sakis, and ukaris; 54 species), and **Aotidae** (night monkeys; 11 species). Howler and spider monkeys have prehensile tails (Figure 28.31A) used like an additional hand.

Old World monkeys, termed catarrhine monkeys because their nostrils are set close together and open to the front, are placed in family **Cercopithecidae,** with 122 species. They include mandrills, *Mandrillus*, baboons, *Papio*, macaques, *Macaca*, and langurs, *Presbytis*. The thumb and large toe are opposable. Some have internal cheek pouches; none have prehensile tails. Family **Hylobatidae** contains gibbons (17 species) with arms much longer than legs, prehensile hands with fully opposable thumb, and locomotion by true brachiation. Family **Hominidae** contains four living genera and six species: *Gorilla* (two species), *Pan* (two species, chimpanzee and bonobo), *Pongo* (one species, orangutan), and *Homo* (one species, human).

Order Lagomorpha (lag´ō-mor´fə) (Gr. *lagos*, hare, + *morphē*, form): **rabbits, hares, pikas** (Figure 28.37). With long, constantly growing incisors, like rodents, but unlike rodents, they have four upper incisors instead of two. All lagomorphs are herbivores with cosmopolitan distribution; 92 species.

Order Rodentia (rō-den´tē-ə) (L. *rodere*, to gnaw): **gnawing mammals: squirrels, rats, woodchucks.** Most numerous of all mammals in both numbers and species. Characterized by two pairs of chisel-like incisors that grow throughout life and are adapted for gnawing. With their impressive reproductive rate, adaptability, and capacity to invade nearly all terrestrial habitats, they are of great ecological importance. Important families of this order are **Sciuridae** (squirrels and woodchucks), **Muridae** (rats and house mice), **Castoridae** (beavers), **Erethizontidae** (porcupines), **Geomyidae** (pocket gophers), and **Cricetidae** (hamsters, deer mice, gerbils, voles, lemmings); 2277 species.

Order Soricomorpha (sor´i-cō-mor´fə) (L. *soric*, shrew, + *morph*, form): **shrews, moles** Small, sharp-snouted animals, which feed principally on small invertebrates and live most of their lives underground or undercover. Shrews are among the smallest mammals. Worldwide except for New Zealand and Australia; 428 species.

Order Erinaceomorpha (er´i-nās-ē-ō-morf´ə) (Gr. *erin*, hedgehog, + *morph*, form): **hedgehogs, gymnures.** Hedgehogs, of Eurasia and Africa, are nocturnal, omnivorous burrowers with numerous hairs modified as spines. Gymnures are similar to shrews in appearance, but are larger; they live in southeast Asia; 24 species.

Order Chiroptera (kī-rop´ter-ə) (Gr. *cheir*, hand, + *pteron*, wing): **bats.** Wings of bats, the only true flying mammals, are modified forelimbs in which the second to fifth digits are elongated to support a thin integumental membrane for flying. The first digit (thumb) is short with a claw. Most bats, including all North American species, are nocturnal insect-eaters. In Old World tropics, fruit bats, or "flying foxes," *Pteropus*, are the largest of all bats, with a wingspread up to 1.7 m; they live chiefly on fruits; 1116 species.

Order Pholidota (fōli-dō-´tə) (Gr. *pholis*, horny scale): **pangolins.** Pangolins are ant- and termite-eaters whose bodies are covered with overlapping keratinized scales formed from fused

Figure 28.36 Nine-banded armadillo, *Dasypus novemcinctus*. During the day this nocturnal species occupies long tunnels, which it digs with powerful, clawed forelimbs. Order Cingulata, family Dasypodidae.

© Joel Zatz/Alamy Stock Photo

Taxonomy of Living Mammalian Orders

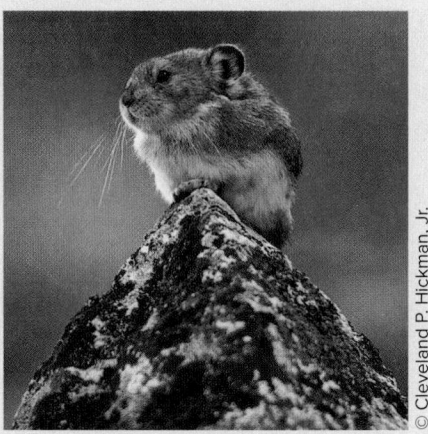

Figure 28.37 A pika, *Ochotona princeps,* atop a rockslide in Alaska. This rat-sized mammal does not hibernate but prepares for winter by storing dried grasses beneath boulders. Order Lagomorpha, family Ochotonidae.

Figure 28.39 A Galápagos sea lion bull, *Zalophus californianus,* barks to indicate his territorial ownership. Order Carnivora, family Otariidae.

Figure 28.38 Wolverine, *Gulo gulo,* in Yellowstone National Park. Wolverines, the largest member of the weasel family, are residents in boreal forests and alpine and tundra habitats in North America, Europe, and Asia. Order Carnivora, family Mustelidae.

bundles of hair. Their home is in tropical Asia and Africa; eight species.

Order Carnivora (car-niv´or-ə) (L. *caro,* flesh, + *vorare,* to devour): **flesh-eating mammals: dogs, wolves, cats, bears, weasels, seals, sea lions** (Figure 28.39), **walruses.** All except the giant panda have predatory habits, and their teeth are especially adapted for consuming flesh. They are distributed worldwide, but only seals occur in Australian and Antarctic regions. Among more familiar families are **Canidae** (dogs), containing wolves, foxes, and coyotes; **Felidae** (cats), whose members include tigers, lions, cougars, and lynxes; **Ursidae** (bears); **Procyonidae** (raccoons); **Mustelidae** (weasels), containing martens, skunks, weasels, otters, badgers, minks, and wolverines

(Figure 28.38); and **Otariidae** (ear ed seals), containing fur seals and sea lions; 280 species.

Order Perissodactyla (pe-ris´sō-dak´til-ə) (Gr. *perissos,* odd, 1 *dactylos,* toe): **odd-toed hoofed mammals: horses, asses, zebras, tapirs, rhinoceroses.** Odd-toed hoofed mammals have an odd number of toes (one or three), each with a keratinized hoof (Figure 28.40). Both Perissodactyla and Artiodactyla are often called **ungulates** (L. *ungula,* hoof), or hoofed mammals, with teeth adapted for grinding plants. The horse family (Equidae), which also includes asses and zebras, has only one functional toe. Tapirs have a short proboscis formed from the upper lip and nose. *Rhinoceros* includes several species found in Africa and Southeast Asia. All are herbivorous; 17 species.

Order Artiodactyla (är´tē-ō-dak´til-ə) (Gr. *artios,* even, 1 *daktylos,* toe): **even-toed hoofed mammals: swine, camels, deer and their allies, giraffes, hippopotamuses, antelopes, cattle, sheep, goats.** Most have two toes, although the hippopotamus and some others have four (Figure 28.40). Each toe is sheathed in a keratinized hoof. Many, such as cattle, deer, and sheep, have horns or antlers. Many are ruminants. Most are strictly herbivores, but some species, such as pigs, are omnivorous. The group includes some of the most valuable domestic animals; 240 species.

Order Cetacea (se-tā´shē-ə) (L. *cetus,* whale): **whales** (Figure 28.41), **dolphins, porpoises.** Anterior limbs of cetaceans are modified into broad flippers; posterior limbs are absent. Some have a fleshy dorsal fin and the tail is divided into transverse fleshy flukes. Nostrils are represented by a single or double blowhole on top of the head. Hairs are limited to a few on the muzzle, and skin glands are absent except for those of the mammary and the eye. The order is divided into **toothed whales**

Taxonomy of Living Mammalian Orders

Figure 28.40 Odd-toed and even-toed ungulates. Rhinoceroses and horses (order Perissodactyla) are odd-toed; hippopotamuses and deer (order Artiodactyla) are even-toed. Lighter, faster mammals run on only one or two toes.

© John Pitcher/Design Pics/Corbis

Figure 28.41 Humpback whale, *Megaptera novaeangliae,* breaching. Among the most acrobatic of whales, humpbacks appear to breach to stun fish schools or to communicate information to other pod members. Order Cetacea, family Balaenopteridae.

(suborder Odontoceti), represented by dolphins, porpoises, and sperm whales; and **baleen whales** (suborder Mysticeti), represented by rorquals, right whales, and gray whales. The blue whale, a rorqual, is the heaviest animal that has ever lived.

Rather than teeth, baleen whales have a straining device made of keratin, called baleen, used to filter plankton. Cetaceans are the sister group to hippos; thus, some taxonomists place cetaceans within the order Artiodactyla; 84 species.

SUMMARY

Mammals are endothermic vertebrates whose bodies have hair and who nurse their young with milk. The approximately 5700 species of mammals descend from the synapsid lineage of amniotes that arose during the Carboniferous period of the Paleozoic era. Their evolution is traced from pelycosaurs of the Permian period to therapsids of the late Permian and Triassic periods of the Mesozoic era. One group of therapsids, the cynodonts, gave rise during the Triassic period to mammals. Mammalian evolution was accompanied by the appearance of many important derived characteristics, among these an enlarged brain with greater sensory integration, high metabolic rate, endothermy, heterodont teeth, and many changes in the skeleton that supported a more active life. Mammals diversified rapidly during the Tertiary period of the Cenozoic era.

Mammals are named for the glandular milk-secreting organs of females (rudimentary in males), a unique adaptation that, combined with prolonged parental care, buffers infants from demands of foraging for themselves and eases the transition to adulthood. Hair, the integumentary outgrowth that covers most mammals, serves variously for mechanical protection, thermal insulation, protective coloration, and waterproofing.

Mammalian skin is rich in glands: sweat glands that function in evaporative cooling, scent glands used in social interactions, and sebaceous glands that secrete lubricating skin oil.

Placental mammals have deciduous teeth that are replaced by permanent teeth (diphyodont dentition). Four kinds of teeth—incisors, canines, premolars, and molars—are highly modified in different mammals for specialized feeding tasks, or they may be absent. Food habits of mammals strongly influence their body form and physiology. Insectivores have pointed teeth for piercing the exoskeletons of insects and the integument of other small invertebrates. Herbivorous mammals have specialized molars and premolars for grinding cellulose and silica-rich plants and have specialized regions of the gut for harboring microorganisms that digest cellulose. Carnivorous mammals have adaptations, including specialized jaw muscles and teeth, for killing and removing flesh from their prey, mainly herbivorous mammals. Omnivores feed on both plant and animal foods and have a variety of tooth types.

Some marine, terrestrial, and aerial mammals migrate; some migrations, such as those of fur seals and caribou, are extensive. Migrations are usually made toward climatic conditions favorable for finding food, mating, or rearing young.

Mammals with true flight, the bats, are mainly nocturnal and thus avoid direct competition with birds. Most employ ultrasonic echolocation to navigate and to feed in darkness.

Monotremes, of the Australian region, are the only mammals that retain the ancestral trait of egg-laying. After hatching, the young are nourished with their mother's milk. All other mammals are viviparous. Embryos of marsupials have brief gestation periods, are born underdeveloped, and complete their early growth in the mother's pouch, nourished by milk. The remaining mammals are eutherians, mammals that develop a sophisticated placental attachment between mother and embryos through which embryos are nourished for a prolonged period.

Mammal populations fluctuate due to both density-dependent and density-independent causes and some mammals, particularly rodents, may experience extreme cycles of abundance in population density. The unqualified success of mammals as a group reflects their impressive overall adaptability.

Humans are primates, a mammalian group that descended from a shrewlike ancestor. The common ancestor of all modern primates was arboreal and had grasping fingers and forward-facing eyes capable of binocular vision. Primates diversified to form (1) lemurs and lorises and (2) tarsiers, monkeys, and apes (including humans). Chimpanzees and bonobos together form the sister group of humans. The earliest humans appeared in Africa about 6.5 million years ago and gave rise to several genera of australopithecines, which persisted for about 3 million years. Australopithecines were shorter and smaller-brained than modern humans, but were bipedal. They gave rise to, and coexisted with, *Homo habilis,* the first maker of stone tools. *Homo erectus* appeared about 1.9 million years ago and spread throughout Africa, Europe, and Asia. They were eventually replaced by modern humans, *Homo sapiens.*

REVIEW QUESTIONS

1. Describe the evolution of mammals, tracing their synapsid lineage from early amniote ancestors to true mammals. How would you distinguish pelycosaurs, early therapsids, cynodonts, and mammals?
2. Describe structural and functional adaptations that appeared in early amniotes that foreshadowed the mammalian body plan. Which mammalian attributes do you think were especially important to successful diversification of mammals?
3. Hair is hypothesized to have evolved in therapsids as an adaptation for insulation, but modern mammals have adapted hair for several other purposes. Describe these.
4. What is distinctive about each of the following: horns of bovids, antlers of deer, and horns of rhinos? Describe the growth cycle of antlers.
5. Describe location and principal function(s) of each of the following skin glands: sweat glands, scent glands, sebaceous glands, and mammary glands.
6. Define "diphyodont" and "heterodont" and explain why both terms apply to mammalian dentition.
7. Describe food habits of insectivorous, herbivorous, carnivorous, and omnivorous mammals. List common names of some mammals belonging to each group.
8. Most herbivorous mammals depend on cellulose as their main energy source, yet no mammal synthesizes cellulose-splitting enzymes. How are the digestive tracts of mammals specialized for symbiotic digestion of cellulose?
9. How does fermentation differ between horses and cattle?
10. What is the relationship of body mass to metabolic rate of mammals?
11. Describe the annual migrations of barren-ground caribou and fur seals.
12. Explain what is distinctive about the life habit and mode of navigation in bats.
13. Describe and distinguish patterns of reproduction in monotremes, marsupials, and placental mammals. What aspects of mammalian reproduction characterize *all* mammals but no other vertebrates?
14. Distinguish between territory and home range for mammals.
15. What is the difference between density-dependent and densityindependent causes of fluctuations of the size of mammalian populations?
16. Describe the hare-lynx population cycle, considered a classic example of a prey-predator relationship (Figure 28.27). From your examination of the cycle, formulate a hypothesis to explain the oscillations.
17. What do the terms Theria, Metatheria, Eutheria, Monotremata, and Marsupialia mean? List mammals that are grouped under each taxon.
18. What anatomical characteristics set primates apart from other mammals?
19. What role do the fossils named "Ardi" and "Lucy" play in the reconstruction of human evolutionary history?
20. In what ways do the genera *Australopithecus* and *Homo* differ?
21. When did the different species of *Homo* appear and how did their cultures differ?

For Further Thought Many zoologists consider mammalian teeth, rather than endothermy, hair, or lactation, to be the characteristic most responsible for success of mammals. Explain why this could be true.

SELECTED REFERENCES

Cameron, D. W., and C. P. Groves. 2004. Bones, stones, and molecules. Burlington, Massachusetts, Elsevier Academic Press. *Comprehensive synthesis of human evolution that highlights speciation and extinction in the human lineage.*

DeSalle, R., and I. Tattersall. 2008. Human origins. College Station, Texas A & M University Press. *The human genome contribution to our understanding of human evolution.*

Feldhamer, G. A., L. C. Drickamer, S. H. Vessey, J. F. Merritt, and C. Krajewski. 2015. Mammalogy: adaptation, diversity, and ecology, ed. 4. Baltimore, Johns Hopkins University Press. *Modern, well-illustrated textbook.*

Gibbons, A. 2009. A new kind of ancestor: *Ardipithecus* unveiled. Science **326:** 36–40. *This, and eleven other articles in this issue, describe the anatomy and life of "Ardi."*

Green, R. E., and 55 others. 2010. A draft sequence of the Neandertal genome. Science **328:** 701–722. *Neandertal genes are in living Europeans and Asians, evidence of limited interbreeding between neandertals and modern humans.*

Johanson, D. C., and M. A. Edey. 1981. Lucy, the beginnings of humankind. New York, Simon & Schuster. *Entertaining account of Johanson's discovery of the famous, nearly complete* Australopithecus afarensis *skeleton.*

Macdonald, D. (ed.). 2009. The Princeton encyclopedia of mammals. Princeton, NJ, Princeton University Press. *This large volume covers nearly all mammals of the world, enhanced with fine photographs and color artwork.*

Nowak, R. M. 1999. Walker's mammals of the world, ed. 6. Baltimore, The Johns Hopkins University Press. *An illustrated reference work on mammals, with descriptions of all extant and recently extinct species.*

Rismiller, P. D., and R. S. Seymour. 1991. The echidna. Sci. Am. **294:** 96–103 (Feb.). *Studies of this fascinating monotreme have revealed many secrets of its natural history and reproduction.*

Suga, N. 1990. Biosonar and neural computation in bats. Sci. Am. **262:** 60–68 (June). *How the bat nervous system processes echolocation signals.*

Villmoare, B., W. H. Kimbel, C. Seyoum, C. J. Campisano, E. N. DiMaggio, J. Rowan, D. R. Braun, J. R. Arrowsmith, and K. E. Reed. 2015. Early *Homo* at 2.8 Ma from Ledi-Geraru, Afar, Ethiopia. Science **347:** 1352–1359. *New fossils extend origins of* Homo *by 400,000 years; drying climates created environments for diversification of hominid taxa.*

Wilson, D. E., and D. Reeder (eds.). 2005. Mammal species of the world: a taxonomic and geographic reference, ed. 3. Baltimore, Johns Hopkins University Press. *An authoritative compilation of the world's mammal species.*

Wong, K. 2002. The mammals that conquered the seas. Sci. Am. **286:** 70–79 (May). *Fossil and DNA evidence help to unravel the evolutionary history of whales.*

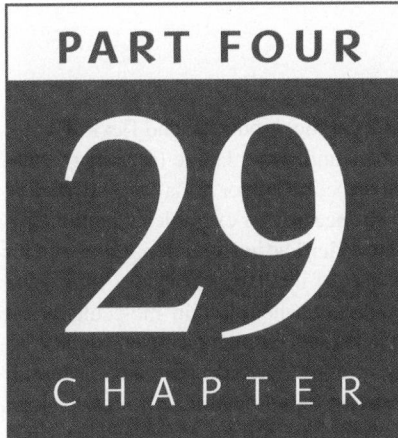

PART FOUR

29

CHAPTER

Support, Protection, and Movement

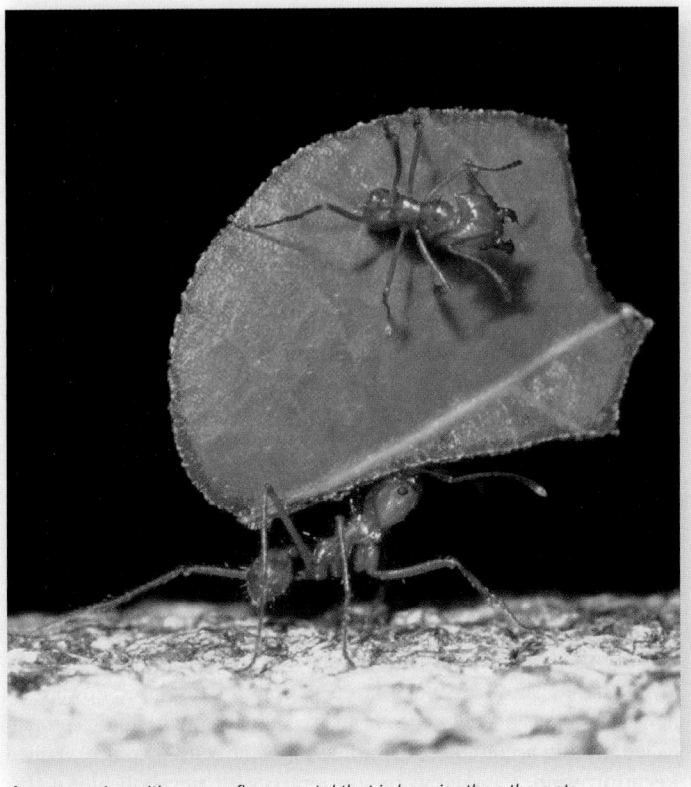

An ant carries with ease a flower petal that is heavier than the ants body weight.

© Eric Soder/Science Source

Of Grasshoppers and Superman

"A dog," remarked Galileo in the seventeenth century, "could probably carry two or three such dogs upon his back; but I believe that a horse could not carry even one of its own size." Galileo was expressing the principle of scaling, a principle that allows us to understand the physical consequences of changing body size. A grasshopper can jump to a height of 50 times the length of its body, yet a man in a standing jump cannot clear an obstacle of his own height. Without an understanding of scaling, this comparison could easily lead us to expect something extraordinary about the muscles of insects. To the authors of a nineteenth-century entomology text it seemed that "this wonderful strength of insects is doubtless the result of something peculiar in the structure and arrangement of their muscles, and principally their extraordinary power of contraction." But grasshopper muscles are no more powerful than human muscles because *muscles of small and large animals exert the same force per cross-sectional area.* Grasshoppers leap high in proportion to their size because they are small, not because they possess extraordinary muscles.

The authors of this nineteenth-century text further suggested that if vertebrates had the powers of insects, they would surely have "caused the early desolation of the world." Earthly mortals would need more than superhuman muscles were they to leap in the proportions of a grasshopper. They would require superhuman tendons, ligaments, and bones to withstand the stresses of mighty contractions, not to mention the crushing strains of landing at terminal velocity. The feats of Superman would be quite impossible were he built of the structural materials available to earthbound animals, rather than of the wondrous materials available to inhabitants of the mythical planet Krypton.

29.1 INTEGUMENT

The integument is the outer covering of the body, a protective wrapping that includes the skin and all structures derived from or associated with skin, such as hair, setae, scales, feathers, and horns. In most animals it is tough and pliable, providing mechanical protection against abrasion and puncture and forming an effective barrier against bacterial invasion. It may also provide moisture proofing against fluid loss or gain. The integument helps to protect underlying cells against the damaging action of the sun's ultraviolet rays. In addition to being a protective cover, the integument serves a variety of important regulatory functions. For example, in endothermic animals, most of the body's heat is lost through the integument; the integument contains mechanisms that cool the body when it is too hot and slow heat loss when the body is too cold. The integument contains sensory receptors that provide essential information about the immediate environment. It has excretory functions and in some animals respiratory functions as well. In some animals integument can also absorb nutrients, for example, many endoparasites. Through integument pigmentation some organisms can make themselves more or less conspicuous. Integument secretions can make an animal sexually attractive or repugnant, or provide olfactory and/or pheromonal cues that influence behavioral interactions between individuals.

Invertebrate Integument

Many unicellular eukaryotes have only the delicate plasma membrane for external covering; others, such as *Paramecium,* have developed a protective pellicle (see Chapter 11, p. 230). Most invertebrates, however, have more complex tissue coverings. The principal covering is a single-layered **epidermis.** Some invertebrates have added a secreted noncellular **cuticle** over the epidermis for additional protection.

Members of the large parasitic clade within Phylum Platyhelminthes (flatworms) have a **syncytial tegument** that is resistant to immune responses of the host and to digestion by host enzymes. It has been hypothesized that the lack of individual cell boundaries at the surface provide this protection (see Chapter 14, p. 293).

The molluscan epidermis is delicate and soft and contains mucous glands, some of which secrete the calcium carbonate of the shell (see Chapter 16, p. 337). Some cephalopod molluscs (squids and octopuses) have developed a more complex integument, consisting from the surface inward of cuticle, simple epidermis, layer of connective tissue, layer of reflecting cells (iridocytes), and thicker layer of connective tissue.

Arthropods have the most complex of invertebrate integuments, providing not only protection but also skeletal support. Development of a firm exoskeleton and jointed appendages suitable for attachment of muscles has been a key feature in the extraordinary diversity of this phylum, largest of animal groups. The arthropod integument consists of a single layered epidermis (more precisely called **hypodermis**), which secretes a complex cuticle of two zones (Figure 29.1A). The thicker inner zone, the **procuticle,** is composed of protein and chitin (a polysaccharide) deposited in layers (lamellae). The outer zone of cuticle, lying on the external surface above the procuticle, is the thin **epicuticle.** The epicuticle is a nonchitinous complex of proteins and lipids that provides a protective moisture-proofing barrier to the integument.

An arthropod cuticle may be a tough but soft and flexible layer, as it is in many microcrustaceans and insect larvae (see Chapter 20, p. 429 and Chapter 21, p. 456, respectively), or it may be hardened in one of two ways. In decapod crustaceans, for example, crabs and lobsters, the cuticle is stiffened by **calcification,** the deposition of calcium carbonate in the outer layers of the procuticle (see Chapter 20, p. 429). In insects hardening occurs when protein molecules bond together with stabilizing cross-linkages within and between adjacent lamellae of the procuticle. The result of this process, called **sclerotization,** is formation of a highly resistant and insoluble protein, **sclerotin.** The arthropod cuticle is one of the toughest materials synthesized by animals; it is strongly resistant to pressure and tearing and can withstand boiling in concentrated alkali, yet it is light, having a specific mass of only 1.3 (1.3 times the weight of water).

Because of the tough cuticle, arthropods must molt to grow. When they molt, epidermal cells first divide by mitosis. Enzymes secreted by the epidermis digest most of the procuticle. Digested materials are then absorbed and consequently not lost by the body. Then in the space beneath the old cuticle a new epicuticle and procuticle are formed. After the old cuticle is shed, the new cuticle is thickened and calcified or sclerotized (see Figure 20.11, p. 429).

Vertebrate Integument and Derivatives

The basic plan of the vertebrate integument, as exemplified by frog and human skin (Figure 29.1B and C), includes a thin, outer stratified epithelial layer, the epidermis, derived from ectoderm and an inner, thicker layer, the **dermis,** or true skin, which is of mesodermal origin. (Ectoderm and mesoderm are embryonic germ layers, described in Figure 8.27, p. 178.)

Although the epidermis is thin and appears simple in structure, it gives rise to most derivatives of the integument, such as hair, feathers, claws, and hooves. The epidermis is a stratified squamous epithelium (p. 190 and Figure 9.9, p. 191) consisting of several layers of cells. Cells of the basal epidermal layer undergo frequent mitosis to renew layers that lie above. As outer layers of cells are displaced upward by new generations of cells beneath, an exceedingly tough, fibrous protein called **keratin** accumulates in the interior of the cells—a process called **keratinization.** Gradually, keratin replaces all metabolically active cytoplasm and the cells die as they become **cornified.** Cornified cells, highly resistant to abrasion and water diffusion, form the outermost **stratum corneum.** This epidermal layer becomes especially thick in areas exposed to persistent pressure or wear, such as calluses, foot pads of mammals, and the scales of reptiles and birds. Eventually the cells of the stratum corneum die and are shed, lifeless and scalelike. Such is the origin of dandruff as well as a significant fraction of household dust.

The dermis is a dense connective-tissue layer (p. 191) containing blood vessels, collagenous fibers, nerves, pigment cells, fat cells, and connective-tissue cells called fibroblasts. These elements support, cushion, and nourish the epidermis, which is devoid of blood vessels. In addition, other cells present in this connective-tissue layer (macrophages, mast cells, and lymphocytes; see Chapter 35), provide a first line of defense if the outer epidermal layer is broken.

The dermis may also contain true bony structures of dermal origin. Heavy bony plates were common in ostracoderms and placoderms of the Paleozoic era (see Figure 23.17, p. 514) and persist in some living fishes, such as sturgeons (see Figure 24.17B, p. 528). Scales of contemporary fishes are bony dermal structures that have

Figure 29.1 Integumentary systems of animals, showing the major layers. **A,** Structure of arthropod (crustacean) body wall showing cuticle and epidermis. **B,** Structure of amphibian (frog) integument. **C,** Structure of human integument.

Lizards, snakes, turtles, and crocodilians were among the first to exploit the adaptive possibilities of the remarkably tough protein keratin. A reptilian epidermal scale that develops from keratin is a much lighter and more flexible structure than the bony, dermal scale of a fish, yet it provides excellent protection from abrasion and desiccation (Figure 29.2). Scales may be overlapping structures, as in snakes and some lizards, or develop into plates, as in turtles and crocodilians. In birds, keratin found new uses. Feathers, beaks, and claws, as well as scales, are all epidermal structures composed of dense keratin. Mammals continued to capitalize on keratin's virtues by turning it into hair, hooves, claws, and nails. As a result of its keratin content, hair is by far the strongest material in the body. It has a tensile strength comparable to that of rolled aluminum and is nearly twice as strong, weight for weight, as the strongest bone.

evolved from the bony armor of the Paleozoic fishes but are much smaller and more flexible. Fish scales are thin bony slivers covered with a mucus-secreting epidermis (Figure 29.2). Most amphibians lack dermal bones in their skin, except for vestiges of dermal scales found in a few tropical caecilian species. In reptiles dermal bones provide the armor of crocodilians and the beaded skin of many lizards, and contribute to the shell of turtles. Dermal bone also gives rise to antlers, as well as the bony core of horns.

Structures such as claws, beaks, nails, and horns contain combinations of epidermal (keratinized) and dermal components. Their basic structure is the same, with a central bony core covered by a vascularized nutritive layer of the dermis, and an outer epithelial layer. This epithelial layer has a germinative component responsible for the continual growth of horns, hooves, claws, and beaks. The outer epithelial layer is keratinized. Overgrowth of these structures is prevented by constant wear and abrasion (Figure 29.3).

Animal Coloration

The colors of animals tend to be vivid and dramatic when serving as important recognition marks (to attract a mate, for example), or as warning coloration, or they are subdued or cryptic when used for camouflage. Integumentary color is usually produced by pigments, but in many insects and in some vertebrates, especially birds, certain colors are produced by the physical structure of the surface tissue, which reflects certain light wavelengths and eliminates others. Colors produced this way are called **structural color,** and they are responsible for the most beautifully iridescent and metallic hues in the animal kingdom. Many butterflies, beetles, and a few fishes share with birds the distinction of being the earth's most resplendent animals. Certain structural colors of feathers are caused by minute, air-filled spaces or pores that reflect white light (white feathers) or some portion of the spectrum (for example, Tyndall blue coloration produced by scattering of light [see note, p. 590]). Iridescent colors that change hue as an animal's angle shifts with respect to an observer are produced when light is reflected from several layers of thin, transparent film. By phase interference, light waves reinforce, weaken, or eliminate each other to produce some of the purest and most brilliant colors we know.

More common than structural colors in animals are **pigments** (biochromes), an extremely varied group of large molecules that reflect light rays. In crustaceans and ectothermic vertebrates these pigments are contained in large cells with branching processes, called **chromatophores** (Figures 29.1B and 29.4A). The pigment may concentrate in the center of the cell in an aggregate too small to be visible, or it may disperse throughout the cell and its processes, providing maximum display. Chromatophores of cephalopod molluscs are entirely different (Figure 29.4B). Each is a small saclike cell filled with pigment granules and surrounded by muscle cells that, when contracted, stretch the whole cell into a pigmented sheet. When muscles relax, the elastic chromatophore quickly shrinks to a small sphere. With such pigment cells squids and octopuses can alter their color more rapidly than any other animal.

Most widespread of animal pigments are the **melanins,** a group of black or brown polymers responsible for various earth-colored shades that most animals wear; melanins are contained within pigment cells called **melanophores** or **melanocytes.** Mammals are a somber-colored group (pp. 616, 620, and 621), and most largely lack color vision, a deficiency associated with a lack of bright colors in the group. Exceptions are the brilliantly colored skin patches of some baboons and mandrills. Significantly, primates have color vision and thus can appreciate such eye-catching ornaments, which are

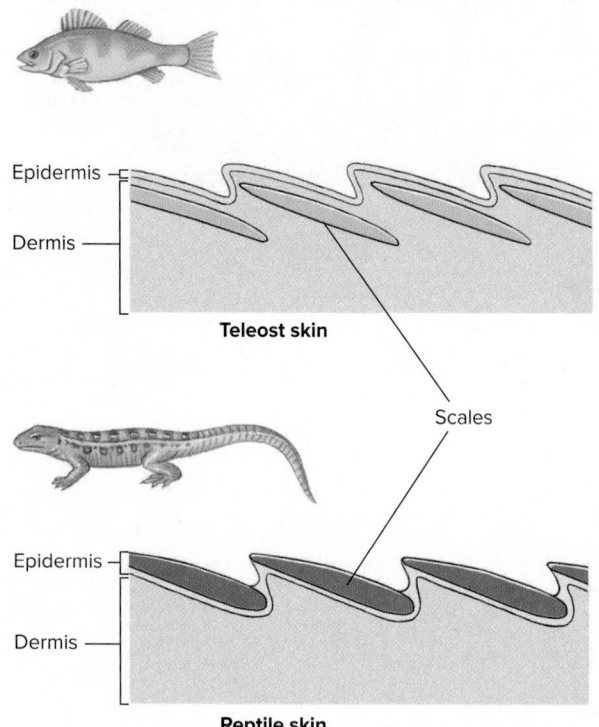

Teleost skin

Reptile skin

Epidermis

Dermis

Scales

Figure 29.2 Integument of bony fishes and lizards. Bony (teleost) fishes have bony scales from dermis, and lizards have keratinized scales from epidermis. Thus they are not homologous structures. Dermal scales of fishes are retained throughout life. Since a new growth ring is added to each scale each year, fishery biologists use scales to determine the age of fishes. Epidermal scales of reptiles are shed periodically.

Claw

Beak

Horn

KEY:
- Bone
- Vascular layer of dermis
- Germinative epidermis
- Keratinized epidermis

Figure 29.3 Similarity of structure of integumentary derivatives. Claws, beaks, and horns are all built of similar combinations of epidermal (keratinized) and dermal components. A central bony core is covered by a vascularized nutritive layer of dermis. An outer epithelial layer has a basal germinative component, which proliferates to allow these structures to grow continually. The thickened surface epithelium is keratinized or cornified. Note that the relative thickness of each component is not drawn to scale.

A   Crustacean chromatophores

B   Cephalopod chromatophores

Figure 29.4 Chromatophores. **A,** The crustacean chromatophore showing the pigment dispersed (*left*) and concentrated (*right*). Vertebrate chromatophores are similar. **B,** The cephalopod chromatophore is an elastic capsule surrounded by muscle fibers that, when contracted (*left*), stretch the capsule to expose the pigment.

————— Direction of movement —————▶

Figure 29.5 Earthworm movement. When circular muscles contract, longitudinal muscles relax and that portion of the body stretches by internal fluid pressure and the worm elongates. By alternate contraction of longitudinal and circular muscles, a wave of contraction passes from anterior to posterior. Bristlelike setae are extended to anchor the animal and to prevent slippage as it moves forward.

associated with reproductive fitness in these species. The muted colors of mammals are caused by melanin, which is deposited in growing hair by dermal melanophores.

Yellow and red colors are often caused by **carotenoid** pigments, which are frequently contained within special pigment cells called **xanthophores.** Most vertebrates are incapable of synthesizing their own carotenoid pigments but must obtain them directly or indirectly from plants. Two entirely different classes of pigments called ommochromes and pteridines are usually responsible for the yellow pigments of molluscs and arthropods. Green colors are rare; when they occur, they are usually produced by yellow pigment overlying blue structural color. **Iridophores,** a third type of chromatophore, contain crystals of guanine or some other purine, rather than pigment. They produce a silvery or metallic effect by reflecting light.

Injurious Effects of Sunlight

The familiar vulnerability of human skin to sunburn reminds us of the potentially damaging effects of ultraviolet radiation on protoplasm. Many animals, such as flatworms, if exposed to the sun in shallow water, are damaged or killed by ultraviolet radiation. Most land animals are protected from such damage by the screening action of special body coverings, for example, the cuticle of arthropods, scales of reptiles, and feathers and fur of birds and mammals, respectively. Humans, however, are "naked apes" that lack the furry protection of most other mammals. We must depend on thickening of the epidermis (particularly the outermost stratum corneum) and on epidermal pigmentation for protection. Most ultraviolet radiation is absorbed in the epidermis, but about 10% penetrates the dermis. Damaged cells in both epidermis and dermis release histamine and other vasodilator substances that cause blood-vessel enlargement in the dermis and the characteristic red coloration of sunburn. Light skins suntan through the formation of the pigment **melanin** in the deeper epidermis and by "pigment darkening," that is, the

photooxidative blackening of bleached pigment already present in the epidermis. Unfortunately, tanning does not bestow protection. Sunlight still ages the skin prematurely, and tanning itself causes the skin to become dry and leathery. There is now strong evidence that genetic mutations caused by high doses of sunlight received during the pre-adult years are responsible for skin cancers that appear later in life. Each year from 2002 to 2011 in the United States, there has been an increased incidence rate of malignant melanoma between 1.5% and 1.1% in males and females, respectively.[1]

29.2 SKELETAL SYSTEMS

Skeletons are supportive systems that provide rigidity to the body, surfaces for muscle attachment, and protection for vulnerable body organs. The familiar bone of the vertebrate skeleton is only one of several kinds of supportive and connective tissues serving various binding and weight-bearing functions, which are described in this section.

Hydrostatic Skeletons

Not all skeletons are rigid; invertebrates of many groups use their body fluids as an internal hydrostatic skeleton. Muscles in the body wall of an earthworm, for example, have no firm base for attachment but develop muscular force by contracting against incompressible coelomic fluids, enclosed within a limited space.

Alternate contractions of the circular and longitudinal muscles of the body wall enable a worm to become thin and then thicken, producing backward-moving waves of motion that propel the animal forward (Figure 29.5). Movement in earthworms and other annelids is helped by septa that separate their body into more or less independent compartments (see Figure 17.3, p. 366), and by tiny bristles called setae that anchor segments as the worm moves (see Figure 17.19, p. 375).

[1] Kohler, B. A., R. L. Sherman, N. Howlader, A. Jemal, A. B. Ryerson, K. A. Henry, F. P. Boscoe, K. A. Cronin, A. Lake, A. -M. Noone, S. J. Henley, C. R. Eheman, R. N. Anderson, L. Penberthy. *Annual report to the nation on the status of cancer, 1975–2011, featuring incidence of breast cancer subtypes by race/ethnicity, poverty, and state. Journal of the National Cancer Institute* 2015; 107 (6): djv048.

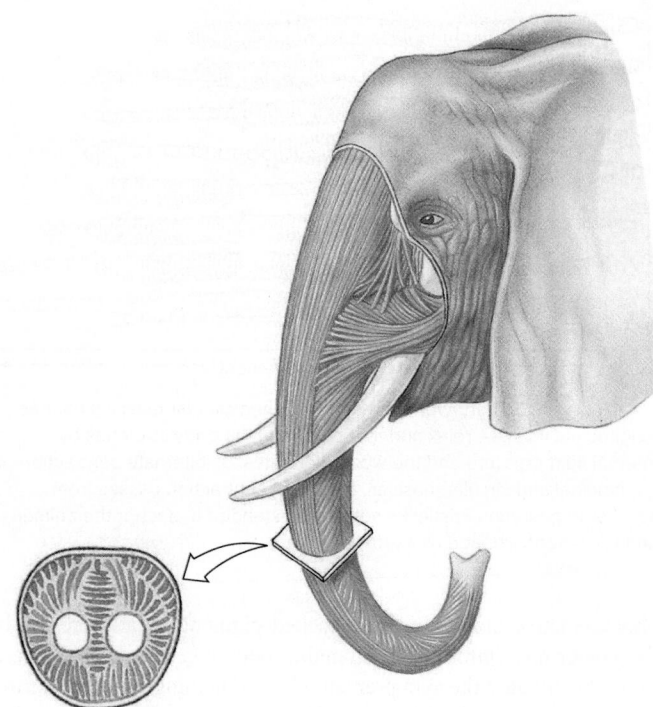

Figure 29.6 Muscular trunk of an elephant, an example of a muscular hydrostat.

An obvious advantage of a compartmentalized body plan is that if a worm is punctured or even cut into pieces, each part can still develop pressure and move. Worms that lack internal compartments, for example, the lugworm *Arenicola* (see Figure 17.11, p. 370), are rendered helpless if body fluid is lost through a wound.

Many muscles not only produce movement but also provide a unique form of skeletal support. The elephant's trunk is an excellent example of a structure that lacks any obvious form of skeletal support, yet is capable of bending, twisting, elongating, and lifting heavy weights (Figure 29.6). An elephant's trunk, tongues of mammals and reptiles, and tentacles of cephalopod molluscs are examples of **muscular hydrostats.** Like hydrostatic skeletons of worms, muscular hydrostats work because they are composed of incompressible tissues that remain at constant volume. The remarkably diverse movements of muscular hydrostats depend on muscles arranged in complex patterns.

Rigid Skeletons

Rigid skeletons differ from hydrostatic skeletons in one fundamental way: rigid skeletons consist of rigid elements, usually jointed, to which muscles can attach. Since muscles can only contract and not actively lengthen, rigid skeletons provide the anchor points required by opposing sets of muscles, such as flexors and extensors, that allow movement in more than one direction.

There are two principal types of rigid skeletons: **exoskeleton,** typical of molluscs, arthropods, and many other invertebrates; and **endoskeleton,** characteristic of echinoderms, vertebrates, and some cnidarians. An invertebrate's exoskeleton may be protective, but it

may also perform a vital role in locomotion. An exoskeleton may take the form of a shell, a spicule, or a calcareous, proteinaceous, or chitinous plate. It may be rigid, as in molluscs, or jointed and movable, as in arthropods. Unlike an endoskeleton, which grows with the animal, an exoskeleton usually does not grow and must be periodically molted to make way for an enlarged replacement (molting and ecdysis in crustaceans are described on p. 429). Some invertebrate exoskeletons, such as the shells of snails and bivalves, grow with the animal.

The vertebrate endoskeleton is formed inside the body and is composed of bone and cartilage, which are forms of specialized connective tissue (see p. 192 and Figure 9.11C, D). Bone not only supports and protects but is also the major body reservoir for calcium and phosphorus. In amniote vertebrates red blood cells, platelets, and white blood cells are formed in the marrow found within bone.

> The arthropod-type exoskeleton is perhaps a better arrangement for small animals than a vertebrate-type endoskeleton because a hollow cylindrical tube can support much more weight without collapsing than can a solid cylindrical rod of the same material and weight. Arthropods can thus enjoy both protection and structural support from their exoskeleton. For larger animals the hollow cylinder would be completely impractical. If made thick enough to support the body weight, it would be too heavy to lift; but if kept thin and light, it would be extremely sensitive to buckling or shattering on impact. Finally, can you imagine the sad plight of an animal the size of an elephant when it shed its exoskeleton to molt?

Notochord and Cartilage

The **notochord** (see Figure 23.1, p. 499) is a semirigid supportive axial rod of protochordates and all vertebrate larvae and embryos. It is composed of large, vacuolated cells and is surrounded by layers of elastic and fibrous collagen sheaths. It is a stiffening device, preserving body shape during locomotion. Except in jawless vertebrates (lampreys and hagfishes, see p. 520 and p. 517, respectively), the notochord is surrounded or replaced by vertebrae during embryonic development.

Cartilage is a major skeletal element of some vertebrates. Jawless fishes (for example, lampreys) and elasmobranchs (sharks, skates, and rays, see p. 522) have purely cartilaginous skeletons. Other vertebrates as adults have principally bony skeletons with some cartilage interspersed. Cartilage is a soft, pliable tissue that resists compression. The basic form, **hyaline cartilage,** has a clear, glassy appearance (see Figure 9.11, p. 192). It is composed of cartilage cells **(chondrocytes)** surrounded by a firm, complex protein-carbohydrate gel interlaced with a meshwork of collagen fibers. Blood vessels are virtually absent—the reason that sports injuries involving cartilage heal slowly. In addition to forming the cartilaginous skeleton of some vertebrates and that of all vertebrate embryos, hyaline cartilage forms articulating surfaces of many bone joints of most adult vertebrates and supporting tracheal, laryngeal, and bronchial rings of the respiratory system (see p. 695). Two other types of cartilage, elastic and fibrous, are similar to hyaline cartilage, except that in the case of elastic cartilage the fiber type is predominantly elastic, whereas in the case of fibrous cartilage numerous collagen fiber bundles are present, often arranged in herringbone patterns.

Cartilage similar to hyaline cartilage occurs in some invertebrates, for example, the radula of gastropod molluscs (p. 335) and lophophore of brachiopods (p. 325). The cartilage of cephalopod molluscs is of a special type with long, branching processes that resemble cells of vertebrate bone.

Bone

Bone is a living tissue that differs from other connective and supportive tissues by having significant deposits of inorganic calcium salts laid down in an extracellular matrix composed of collagen fibers in a protein-carbohydrate gel. Unlike cartilage, it is highly vascular and therefore is capable of rapid healing and growth. Its structural organization is such that bone has nearly the tensile strength of cast iron, yet is only one-third as heavy.

Bone is never formed in vacant space but is always laid down by replacement in areas occupied by some form of connective tissue. Most bone develops from cartilage and is called **endochondral** ("within-cartilage") or replacement bone. Embryonic cartilage is gradually eroded, leaving it extensively honeycombed; bone-forming cells then invade these areas and begin depositing extracellular bone matrix, which becomes calcified, around strandlike remnants of the cartilage. A second type of bone is **intramembranous bone,** which develops directly from sheets of embryonic cells. Dermal bone, mentioned on page 640, is a type of intramembranous bone. In tetrapod vertebrates intramembranous bone is restricted mainly to bones of the face, cranium, and clavicle; the remainder of the skeleton is endochondral bone. Whatever the embryonic origin, once fully formed, endochondral and intramembranous bone look similar.

Fully formed bone, however, may vary in density. **Spongy** (or cancellous) **bone** consists of an open, interlacing framework of bony tissue, oriented to give maximum strength under normal stresses and strains that bone receives. Bone usually develops first as spongy bone, but some bones, through further deposition of bone matrix, become **compact.** Compact bone is dense, appearing solid to the unaided eye. Both spongy and compact bone occur in typical long bones of tetrapods (Figure 29.7).

Microscopic Structure of Bone   Compact bone is composed of a calcified bone matrix arranged in concentric rings. Between these rings are cavities **(lacunae)** containing bone cells **(osteocytes),** whose processes extend through many minute passages **(canaliculi).** These processes interconnect bone cells and allow

Figure 29.7 Structure of compact bone. **A,** Adult long bone with a cut into the medullary cavity. **B,** Enlarged section showing osteons, the basic histological unit of bone. **C,** Enlarged view of an osteon showing the concentric lamellae and the osteocytes (bone cells) arranged within lacunae. **D,** An osteocyte within a lacuna. Bone cells receive nutrients and growth factors from the circulatory system via tiny canaliculi that interlace the calcified matrix. Bone cells are called osteoblasts when they are building bone, but, in mature bone shown here, they become resting osteocytes. Bone is covered with dense connective tissue called periosteum.

communication between them, via gap junctions (p. 44), and serve to distribute nutrients and growth factors throughout the bone. This entire organization of lacunae and canaliculi is arranged into an elongated cylinder called an **osteon** (also called **haversian system**) (Figure 29.7). Bone consists of bundles of osteons cemented together and interconnected with blood vessels and nerves, although noncellular matrix predominates. As a result of the presence of blood vessels and nerves, bone breaks can heal rapidly, and bone diseases can be as painful as any other tissue disease.

Bone is a dynamic tissue, and bone remodeling and growth are complex restructuring processes, involving both destruction by bone-resorbing cells **(osteoclasts)** and deposition by bone-building cells **(osteoblasts).** Both processes occur simultaneously so that new osteons form as old ones are resorbed. The marrow cavity inside grows larger by bone resorption of the inner surface of the surrounding bone while new bone is laid down on the outer bone surface by bone deposition. Bone growth responds to several hormones, in particular **parathyroid hormone** from the parathyroid gland, which stimulates bone resorption, and **calcitonin** from the thyroid gland (or the ultimobrachial gland in nonmammalian vertebrates), which inhibits bone resorption. These two hormones, together with vitamin D$_3$, **1,25- dihydroxyvitamin D$_3$,** are responsible for maintaining a constant level of calcium in the blood. The effect of hormones on bone growth and resorption is described on page 756.

Like muscle, bone is subject to "use and disuse." When we exercise our muscles, our bones respond by producing new bone tissue to give added strength. In fact, the bumps and processes to which muscles attach are produced by bone in response to the action of muscle forces. Conversely, when bones are not subject to stress, as in space flight, the body resorbs the mineral, and the bones become weak. Astronauts who spend many months in space must exercise to a greater degree than on earth to prevent such resorption and bone weakness.

Plan of the Vertebrate Skeleton

The vertebrate skeleton is composed of two main divisions: **axial skeleton,** which includes skull, vertebral column, sternum, and ribs, and **appendicular skeleton,** which includes the limbs (or fins or wings) and pectoral and pelvic girdles (Figures 29.8 and 29.9). Not surprisingly, the skeleton has undergone extensive remodeling in the course of vertebrate evolution. The move from water to land forced dramatic changes in body form. With increased cephalization, further concentration of brain, sense organs, and food-gathering apparatus in the head occurred and the skull became the most intricate portion of the skeleton. Some early fishes had as many as 180 skull bones (a source of frustration to paleontologists) but through loss of some bones and fusion of others, skull bones became greatly reduced in number during evolution of the tetrapods. Amphibians and lizards have 50 to 95, and mammals, 35 or fewer. Humans have 29.

The vertebral column is the main stiffening axis of the postcranial skeleton. In fishes it serves much the same function as the notochord, providing stiffness as well as points for muscle attachment, preserving body shape during muscle contraction. With evolution of amphibious and terrestrial tetrapods, the vertebrate body was no longer buoyed by the aquatic environment. The vertebral column became structurally adapted to withstand new regional stresses transmitted to the column by the two pairs of appendages. In amniote tetrapods (reptiles, birds, and mammals), the vertebrae are differentiated into **cervical** (neck), **thoracic** (chest), **lumbar** (back), **sacral** (pelvic), and **caudal** (tail) vertebrae. In frogs, birds, and also humans, caudal vertebrae are reduced in number and size, and sacral vertebrae are fused. The number of vertebrae varies among vertebrates. Pythons lead the list with more than 400. In humans (Figure 29.9) there are 33 in a young child, but in adults 5 are fused to form the **sacrum** and 4 to form the **coccyx.** Besides the sacrum and coccyx, humans have 7 cervical, 12 thoracic, and 5 lumbar vertebrae. The number of cervical vertebrae (7) is constant in nearly all mammals, whether the neck is short as in dolphins, or long as in giraffes.

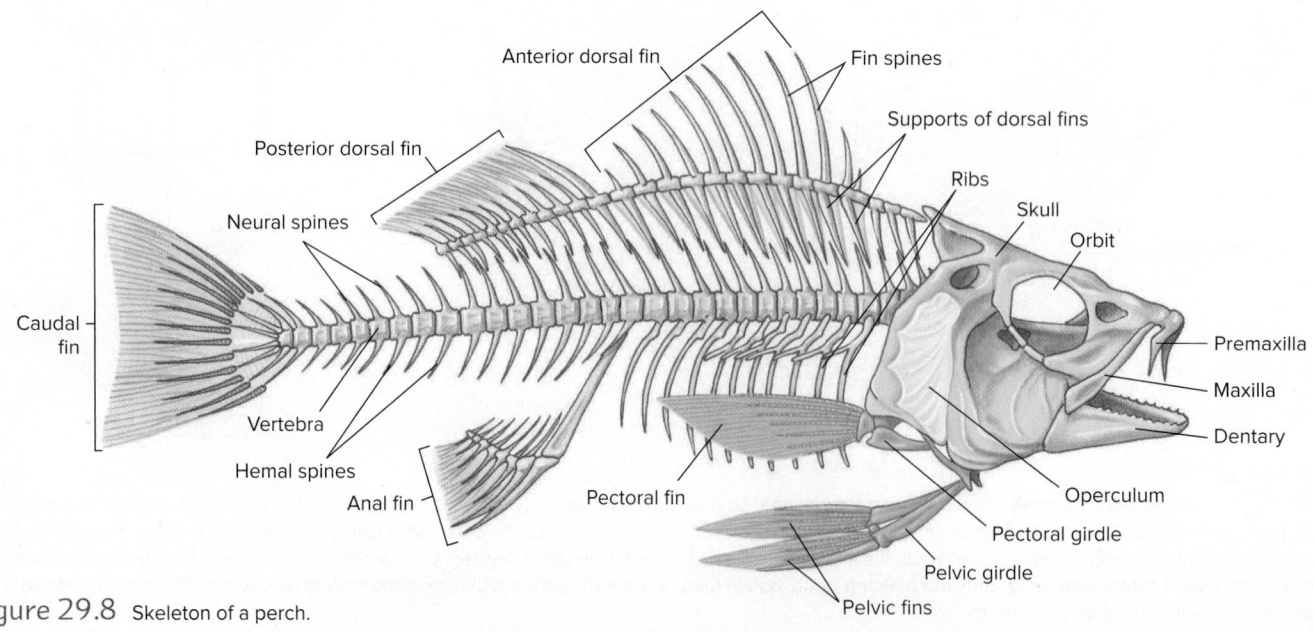

Figure 29.8 Skeleton of a perch.

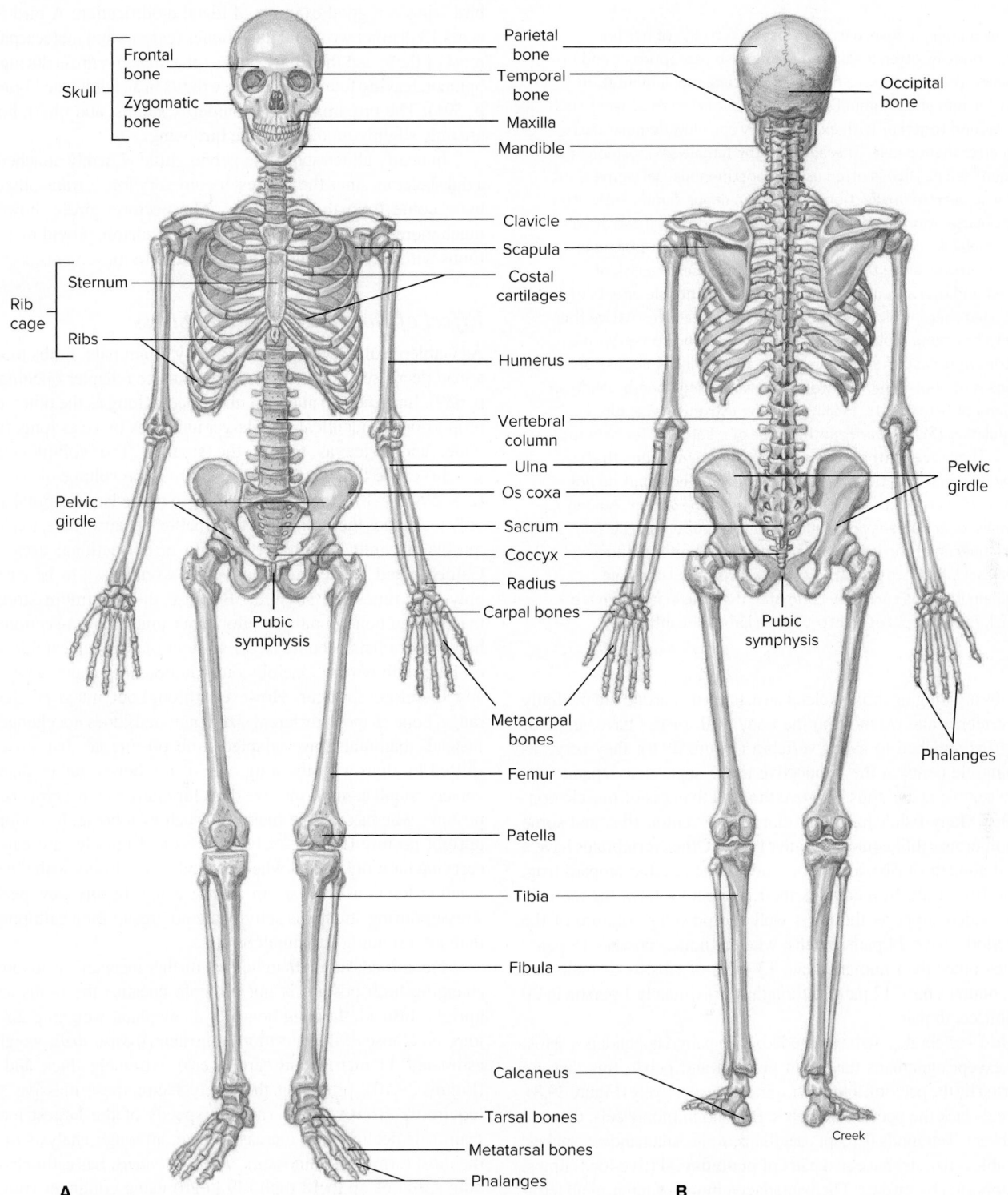

Figure 29.9   Human skeleton. **A,** Ventral view. **B,** Dorsal view. In comparison with those of other mammals, the human skeleton is a patchwork of primitive and specialized parts. Erect posture, produced by specialized changes in legs and pelvis, enabled the primitive arrangement of arms and hands (arboreal adaptation of human ancestors) to be used for manipulation of tools. Development of the skull and brain followed as a consequence of the premium natural selection put on dexterity and ability to appraise the environment.

The first two cervical vertebrae, **atlas** and **axis,** are present in all amniotes, but are modified in mammals to support the skull and to permit pivotal movements. The atlas bears the globe of the head much as the mythological Atlas bore the earth on his shoulders. The axis, the second vertebra, permits the head to turn from side to side.

Following menopause, a woman loses 5% to 6% of her bone mass annually, often leading to the disease osteoporosis and increasing the risk of bone fractures. Dietary supplementation with calcium and vitamin D_3 has been advocated to prevent such losses, and together with exercise they can slow demineralization after menopause. Therapy with the female sex hormone estrogen (see p. 146) is often used in postmenopausal women, because ovarian production of estrogen drops significantly after menopause. More frequently, however, low doses of estrogen are usually accompanied by low doses of the female hormone progesterone, since this combination decreases the risk of breast and uterine cancer, which are significant side effects of estrogen therapy alone. **Bisphosphonates** are an alternative therapy to hormone replacement therapy (HRT) for women with a history of breast or uterine cancer in the family. This class of drugs is nonhormonal and works by decreasing bone-breakdown activity of osteoclasts. Finally, **selective estrogen-receptor modulators (SERMs)** are another type of treatment for osteoporosis. These are synthetic hormone replacement drugs that work by mimicking the effects of estrogen on bone but do not appear to increase the risk of breast or uterine cancer. Among animals, only humans are troubled with osteoporosis, perhaps a consequence of the long postreproductive life of the human species. Osteoporosis is traditionally considered a female problem, but it is currently estimated that one in five men are at risk for osteoporosis (http://www.iofbonehealth.org/).

Ribs are long or short skeletal structures that articulate medially with vertebrae and extend into the body wall. Fishes have single or paired ribs attached to every vertebra (Figure 29.8); they serve as stiffening elements in the connective-tissue septa that separate the muscle segments and thus improve the effectiveness of muscle contractions. Many fishes have both dorsal and ventral ribs, and some have numerous riblike intermuscular bones. Other vertebrates have a reduced number of ribs, and some, such as the familiar leopard frog, have no ribs at all. In mammals the ribs together form the thoracic basket, which supports the chest wall and prevents collapse of the lungs. Sloths have 24 pairs of ribs, whereas horses possess 18 pairs. Primates other than humans have 13 pairs of ribs; both male and female humans have 12 pairs, although approximately 1 person in 20 has a thirteenth pair.

Most vertebrates, fishes included, have paired appendages. Most fishes except agnathans have thin pectoral and pelvic fins that are supported by the pectoral and pelvic girdles, respectively (Figure 29.8). Some eels lack the pectoral or pelvic fins and in moray eels, they are both absent. Tetrapods (except caecilians, some salamanders, snakes, and limbless lizards) have two pairs of **pentadactyl** (five-toed) limbs, also supported by girdles. The pentadactyl limb is similar in all tetrapods, alive and extinct; even when highly modified for various modes of life, the elements are rather easily homologized (the evolution of the pentadactyl limb is illustrated in Figure 25.3).

Modifications of the basic pentadactyl limb for life in different environments often involve bone loss or fusion rather than addition of new bone. The ends of the appendages are more likely to be modified, such as the bony structures of the feet and hands. Horses and their relatives evolved a foot structure for fleetness by elongation of the third toe. In effect, a horse stands on its third fingernail (hoof). A

bird wing is a good example of distal modification. A bird embryo bears 13 distinct wrist and hand bones (carpals and metacarpals), but most of these and the finger bones (phalanges) regress during development, leaving four bones in three digits in an adult (see Figure 27.7, p. 591). The proximal bones (humerus, radius, and ulna), however, are only slightly modified in the bird wing.

In nearly all tetrapods the pelvic girdle is firmly attached to the axial skeleton, since the greatest locomotory forces transmitted to the body come from the hindlimbs. The pectoral girdle, however, is much more loosely attached to the axial skeleton, providing the forelimbs with greater freedom for manipulative movements.

Effect of Body Size on Bone Stress

As Galileo realized in 1638, the ability of animals' limbs to support a load decreases as animals increase in size (chapter opening essay, p. 639). Imagine two animals, one twice as long as the other, that are proportionally identical. The larger animal is twice as long, twice as wide, and twice as tall as the smaller. The volume (and the weight) of the larger animal is eight times the volume of the smaller ($2 \times 2 \times 2 = 8$). However, the strength of the larger animal's legs is only four times the strength of the smaller, because bone, tendon, and muscle strength are proportional to cross-sectional area. So, as Galileo noted, eight times the weight would have to be carried by only four times the strength. Because the maximum strength of mammalian bone is rather uniform per unit of cross-sectional area, how can animals become larger without placing unbearable stresses on long limb bones? One obvious solution is to make bones stouter and therefore stronger. However, throughout much of their size range, bone shape in different sized mammals does not change much. Instead, mammals have adapted limb posture so that stresses are shifted to align with the long axis of the bones, rather than transversely. Small animals the size of a chipmunk run in a crouched-limb posture, whereas a large mammal, such as a horse, has adopted an upright posture (Figure 29.10). Bones and muscles are capable of carrying far more weight when aligned more closely with the ground reaction force, as they are in a horse's leg. In this way, peak bone stresses during strenuous activity are no greater for a galloping horse than for a running chipmunk or dog.

For animals larger than horses, further mechanical advantage by changing limb posture is not possible because the limbs are fully upright. Instead, the long bones of an elephant weighing 2.5 metric tons, and those of the enormous dinosaur *Apatosaurus,* weighing an estimated 34 metric tons, are (were) extremely thick and robust (Figure 29.10), providing the safety factor these massive animals require(d). However, top running speeds of the largest terrestrial mammals decline with increasing size, although analysis of one of the most formidable dinosaurs, *Tyrannosaurus,* has estimated a running speed of up to 18 mph (29 km/h) using computer simulations (Sellers and Manning, 2007).

29.3 ANIMAL MOVEMENT

Movement is an important characteristic of animals. Animal movement occurs in many forms in animal tissues, ranging from barely discernible streaming of cytoplasm to extensive movements of powerful skeletal muscles. Most animal movement depends on a single fundamental mechanism: **contractile proteins,** which allow

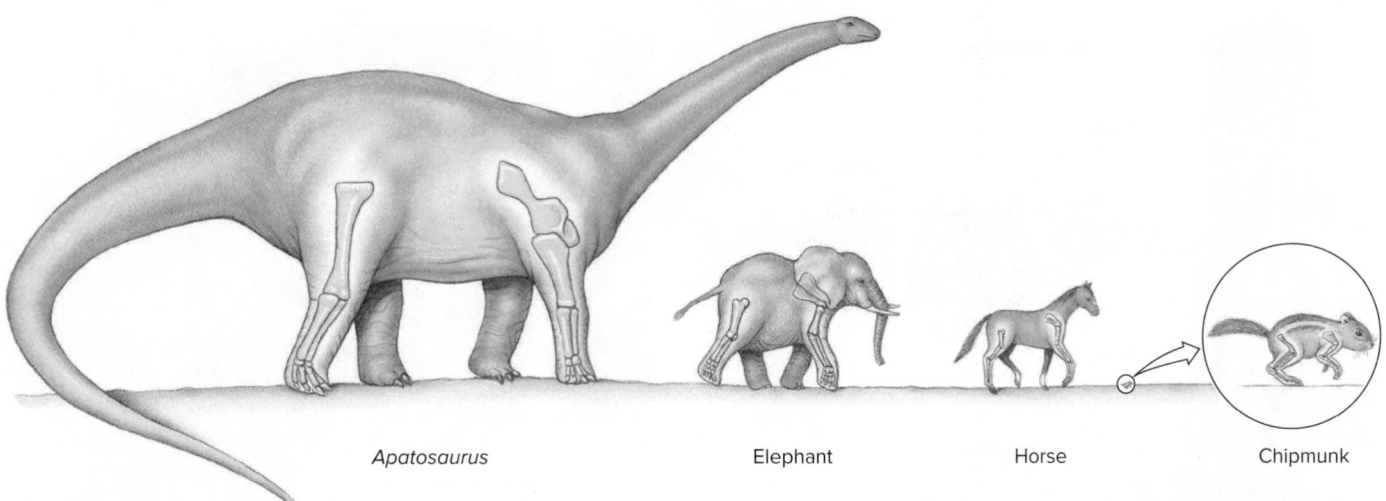

Figure 29.10   Comparison of postures in small and large mammals, showing the effect of scale. Because of its more upright posture, bone stresses in the horse are similar to those in the chipmunk. In mammals larger than horses (above about 300 kg), greatly increased stresses require that bones become exceedingly robust and that the animal lose agility.

relaxation and contraction. Contractile machinery is composed of ultrafine protein fibrils arranged to contract when powered by **ATP.** By far the most important protein contractile system is the **actomyosin system,** composed of two proteins, **actin** and **myosin.** This is an almost universal biomechanical system found from unicellular eukaryotes to vertebrates; it performs a long list of diverse functional roles. Cilia and flagella, however, are composed of proteins other than actin and myosin, and thus are exceptions to the rule. In this discussion we examine the three principal kinds of animal movement: ameboid, ciliary and flagellar, and muscular.

Ameboid Movement

Ameboid movement is a form of movement especially characteristic of amebas and other unicellular forms; it occurs also in many wandering cells of animals, such as macrophages, white blood cells, embryonic mesenchyme, and numerous other mobile cells that move through tissue spaces.

Research with a variety of ameboid cells, including pathogen-fighting phagocytes in blood (see p. 000), has produced a consensus model to explain extension and withdrawal of **pseudopodia** (false feet) and ameboid crawling. Optical studies of an ameba in movement suggest that the outer layer of nongranular, gel-like **ectoplasm** surrounds a more fluid core of **endoplasm** (see Figure 11.9, p. 222). Movement depends on actin, actin-binding proteins (ABPs), and other regulatory proteins. According to this hypothesis (Stossel, 1994), as the pseudopod extends, hydrostatic pressure forces actin subunits in the flowing endoplasm into the pseudopod, where they dissociate from ABPs that inhibit actin polymerization and are then able to assemble into a polymer network of actin filaments to form the gel-like ectoplasm. At the trailing edge of the gel, where the network disassembles, actin filaments interact, in the presence of calcium ions, with myosin to create a contractile force that pulls the cell forward behind the extending pseudopod. Locomotion is assisted by membrane-adhesion proteins that attach temporarily to the substrate to provide traction, enabling the cell to crawl steadily forward (see Figure 11.11, p. 223).

Ciliary and Flagellar Movement

Cilia are minute, hairlike, motile processes that extend from surfaces of cells of many animals. They are a particularly distinctive feature of ciliate unicellular eukaryotes, but except for nematodes, in which motile cilia are absent, and arthropods, in which they are rare, cilia occur in all major groups of animals. Cilia perform many roles either in moving small organisms such as unicellular ciliates and ctenophores (Figure 29.12B) through their aquatic environment or in propelling fluids and materials over epithelial surfaces of larger animals.

Research suggests that within the human body the presence of cilia on cells is the rule, rather than the exception. Cilia are implicated in cellular communication, particularly during development, and appear to play a role in organ placement within the developing embryo.

Cilia are of remarkably uniform diameter (0.2 to 0.5 μm) wherever they occur. Electron microscopy reveals that each cilium has at its base a **basal body** (kinetosome) that is structurally similar to a centriole (see Figure 3.14, p. 44). Each basal body gives rise to a peripheral circle of nine double microtubules arranged around two single microtubules in the center (Figure 29.11) forming the structural support and machinery for movement within each cilium. Each microtubule is composed of a spiral array of protein subunits called **tubulin** (see Figure 3.13B, p. 43). The microtubule doublets around the periphery are connected to each other and to the central pair of microtubules by a complex system of **microtubule-associated proteins** (MAPs). Also extending from each doublet is a pair of arms composed of the MAP, **dynein.** The dynein arms, which act as cross bridges between doublets, operate to produce a sliding force between microtubules. During ciliary movement, microtubules behave as "sliding filaments" that move past one another much like the sliding filaments of vertebrate skeletal muscle described in the discussion on the sliding-filament hypothesis, page 654. During ciliary flexion, dynein arms link to adjacent microtubules, then

A

B

Figure 29.11 **A,** Longitudinal and cross section of a cilium showing the microtubules and microtubule-associated proteins (MAPs) of the 9 + 2 arrangement typical of both cilia and flagella. The central pair of microtubules end near the level of the cell surface. The peripheral microtubules continue inward for a short distance to compose two of each of the triplets in the basal body (kinetosome). **B,** Electron micrograph of section through several cilia. (×133,000).

A Cilium

B

Figure 29.12 **A,** Flagellum beats in wavelike undulations, propelling water parallel to its main axis. Cilium propels water parallel to the cell surface. **B,** Movement of cilia in comb plates of a ctenophore. Note how the waves of beating comb plates pass down a comb row, opposite the direction of the power stroke of individual cilia. The movement of one comb plate lifts the plate below it and so triggers the next lower plate and others below it in sequence.

swivel and release in repeated cycles, causing microtubules on the concave side to slide outward past microtubules on the convex side. This process increases curvature of the cilium. During the recovery stroke microtubules on the opposite side slide outward to bring the cilium back to its starting position.

A **flagellum** is a whiplike structure longer than a cilium and usually present singly or in small numbers at one end of a cell. Flagella occur in many single-celled eukaryotes, in animal spermatozoa, and in sponges. Flagella have the same basic internal structure as cilia, although several exceptions to the 9 + 2 arrangement have

been noted; for example, sperm tails of flatworms have but one central microtubule, and sperm tails of mayflies have no central microtubule. The main difference between a cilium and a flagellum is in their beating pattern rather than in their structure. A flagellum beats symmetrically with snakelike undulations so that water is propelled parallel to the long axis of the flagellum. A cilium, in contrast, beats asymmetrically with a fast power stroke in one direction followed by a slow recovery during which the cilium bends as it returns to its original position (Figure 29.12A). Water is propelled parallel to the ciliated surface (Figure 29.12A and B).

Muscular Movement

Contractile tissue is most highly developed in muscle cells called **fibers.** Although muscle fibers themselves can do work only by contraction and cannot actively lengthen, they can be arranged in so many different configurations and combinations that almost any movement is possible.

Types of Vertebrate Muscle

Vertebrate muscle is broadly classified on the appearance of muscle cells (fibers) when viewed with a light microscope. Both **skeletal muscle** and **cardiac muscle** appear transversely striped **(striated)**, with alternating dark and light bands (Figure 29.13), although unlike skeletal muscle, cardiac muscle is uninucleate and with branching cells. A third type of vertebrate muscle is **smooth (or visceral) muscle,** which lacks the characteristic alternating bands of the striated type.

Skeletal muscle is typically organized into sturdy, compact bundles or bands (Figure 29.13A). It is called skeletal muscle because it is attached to and interacts with skeletal elements and is responsible for movements of the trunk, appendages, respiratory organs, eyes, mouthparts, and other structures. Skeletal muscle fibers are extremely long, cylindrical, multinucleate cells that usually reach from one end of the whole muscle to the other. They are packed into bundles called **fascicles** (L. *fasciculus,* small bundle), which are enclosed by tough connective tissue (Figure 29.14). The fascicles are in turn grouped into a discrete **muscle** surrounded by a thick connective tissue layer. Most skeletal muscles taper at their ends, where they connect to bones by tendons. Other muscles, such as the ventral abdominal muscles, are flattened sheets.

In most fishes, amphibians, and to some extent lizards and snakes, muscles are organized into segments that alternate with the vertebrae. Skeletal muscles of other vertebrates, by splitting, fusion, and shifting, have developed into specialized muscles best suited for manipulating jointed appendages that have evolved for locomotion on land. Skeletal muscle contracts powerfully and quickly but fatigues more rapidly than does smooth muscle. Skeletal muscle is sometimes called **voluntary muscle** because it is stimulated by motor neurons (see p. 727) under conscious control.

> Muscles can only contract or shorten themselves and require external forces to restore their original length following each contraction; they provide movement only in one direction, and are therefore often grouped as an antagonistic set of muscles. Antagonistic muscles are functional opposites that oppose each other's action. For example, the biceps brachii on one side of the upper arm is opposed in its action by the triceps brachii on the opposite side of the arm. By contracting against each other, they balance and smooth rapid movements in two different directions.

Cardiac muscle, the seemingly tireless muscle of the vertebrate heart, possesses some characteristics of skeletal muscle (Figure 29.13B). It is fast acting and striated like skeletal muscle and the mechanism of contraction is similar to that of skeletal muscle; however, contraction is under involuntary autonomic and hormonal control (like smooth muscle). External control mechanisms serve only to modulate the intrinsic rate and strength of contraction; the heartbeat originates within specialized cardiac muscle, and the

Nuclei

Intercalated discs

© Ed Reschke

B

Nuclei

© Ed Reschke

A

Nuclei

© Ed Reschke

C

Figure 29.13 Photomicrographs of types of vertebrate muscle. **A,** Skeletal muscle (human) showing several striated fibers (cells) lying side by side. Note that nuclei are peripheral (× 250). **B,** Cardiac muscle (monkey) is striated, similar to skeletal muscle, although the striations are less obvious due to fewer contractile fibrils per fiber. Note the vertical bars, called intercalated discs, joining separate fibers end to end, and the centrally placed nuclei within the fibers (×250). **C,** Smooth muscle (human) showing absence of striations. Note elongate nuclei in the long fibers (× 250).

Figure 29.14 Organization of skeletal muscle from gross to molecular level. A skeletal muscle (*left*) is composed of thousands of multinucleated muscle fibers or cells (*center*), each containing thousands of myofibrils (*right*). Each myofibril contains numerous myosin and actin filaments that interact to slide past one another during contraction to shorten the muscle. The sarcoplasmic reticulum is a network of modified endoplasmic reticulum tubules surrounding the myofibrils and serves as a reservoir for stored calcium that is released during each membrane depolarization, and initiates sliding of action and myosin filaments during muscle contraction.

heart continues to beat even if removed from the body (heart excitation is described on p. 688). Cardiac muscle is composed of closely opposed, but separate, uninucleate cell fibers joined to each other end to end by junctional complexes (see Chapter 3, p. 43) called **intercalated discs.**

Smooth muscle lacks the striations typical of skeletal and cardiac muscle (Figure 29.13C). The cells are much smaller, tapering strands, each containing a single, central nucleus. The cells interdigitate with each other such that the tapered end of one cell lies close to the central nuclear region of the next. Smooth muscle cells are organized into sheets of muscle encircling cavities and tubular structures of the body, such as the walls of the alimentary canal, blood vessels, respiratory passages, and urinary and reproductive ducts and cavities. Smooth muscle is typically slow acting and can maintain prolonged contractions with very little energy expenditure. Like cardiac muscle, it is controlled by the autonomic nervous system (see p. 732), and hormones, in addition to having local regulatory mechanisms; thus, unlike skeletal muscle, its contractions are involuntary and unconscious. Smooth muscle functions by sustained contraction or relaxation. For example, smooth muscles push material through a tube, such as the intestine, by active contractions, or they change the diameter of a tube to regulate fluid or air flow, as in a blood vessel or airway.

Types of Invertebrate Muscle

Smooth and striated muscles also characterize invertebrate animals, as does another type called oblique striated muscle. There are many variations of all three types and even instances in which structural and functional features of vertebrate smooth and striated muscle are combined. Striated and smooth muscle appear in invertebrate groups as diverse as cnidarians and arthropods. The evolution of striated and smooth muscle in cnidarians and ctenophores relative to bilaterians

is discussed in Chapter 13 (p. 286). The thickest muscle fibers known, approximately 3 mm in diameter and 6 cm long, are those of giant barnacles and of Alaska king crabs living along the Pacific coast of North America. Such large muscle cells lend themselves well to physiological studies and are understandably popular with muscle physiologists.

We illustrate the range of invertebrate muscle types with two functional extremes: the specialized adductor muscles of molluscs and the fast flight muscles of insects.

Bivalve molluscan muscles contain fibers of two types. One kind is striated muscle that can contract rapidly, enabling the bivalve to snap shut its valves when disturbed. Scallops use these "fast" muscle fibers to swim (see Figure 16.25B, p. 347). The second muscle type is smooth muscle, capable of slow, long-lasting contractions. Using these fibers, a bivalve can keep its valves tightly shut for hours or even days (see Figure 16.26). Such adductor muscles use little metabolic energy and receive remarkably few neural signals to maintain the activated state. The contracted state has been likened to a "catch mechanism" involving a low rate of cross-bridge cycling (see p. 655) between contractile proteins within the muscle fiber with a low energy expenditure. Similar mechanisms have been discovered in some types of vertebrate smooth muscle.

Insect flight muscles are virtually the functional antithesis of the slow, long-lasting contractions observed in adductor muscles of bivalves. The wings of some small flies operate at frequencies greater than 1000 beats per second. The **fibrillar muscle,** which contracts at these frequencies—far greater than even the most active of vertebrate muscles—shows unique characteristics. It has very limited extensibility; the wing leverage system is arranged so that the muscles shorten only slightly during each downbeat of the wings. Furthermore, muscles and wings operate as a rapidly oscillating system in an elastic thorax (see Figure 21.11, p. 450). Since the muscles recoil elastically and are activated by stretch during flight, they receive excitatory neural signals only periodically rather than one signal per contraction; one reinforcement signal for every 20 or 30 contractions is enough to keep the system active. Insect flight muscles are described in Chapter 21 (pp. 447–449).

Structure of Striated Muscle

As mentioned on page 651, striated muscle is so named because of periodic bands, plainly visible under the light microscope, that pass across the widths of muscle cells. Each cell, or **fiber,** is multinucleated and tube shaped, containing numerous **myofibrils,** packed together and invested by a plasma membrane, the **sarcolemma** (Figure 29.14). The myofibril contains two types of **filaments,** composed of the proteins **myosin** and **actin.** These are the contractile proteins of the muscle. Actin extends in parallel filaments from a dense protein complex called the Z line. The functional unit of the myofibril, the **sarcomere,** extends between successive Z lines. Figure 29.14 shows these anatomical relationships.

Each myosin filament is composed of many myosin molecules packed together in an elongate bundle (Figure 29.15). Each myosin molecule contains two polypeptide chains, each having a club-shaped head (Figure 29.15A), and aligned in two bundles to form a myosin filament. The two myosin bundles are held end to end at the center of each sarcomere so that the double heads of each myosin molecule face outward from the center of the filament and point toward the Z lines to which the actin filaments attach (Figure 29.15B). The myosin heads act as binding sites for high-energy ATP and during muscle contraction they form molecular cross bridges that interact with the actin filaments.

Actin filaments are composed of a backbone of a double strand of the protein actin, twisted into a double helix. In addition, two actin-binding proteins, tropomyosin and troponin, form part of the actin filament complex. They are important in regulating the

Human muscle tissue develops before birth, and a newborn child's complement of skeletal muscle fibers is similar to that of an adult. Although an adult male weight lifter and a young boy have a similar number of muscle fibers, the weight lifter may be several times the boy's strength because repeated high-intensity, short-duration exercise has induced synthesis of additional actin and myosin filaments. Each fiber has hypertrophied, becoming larger and stronger. This type of exercise favors hypertrophy of fast glycolytic fibers (see p. 657) that fatigue quickly. Endurance exercise such as long-distance running produces a very different response. Fast oxidative and intermediate fiber types are stimulated (see p. 657), and develop more mitochondria and myoglobin and become adapted for a high rate of oxidative phosphorylation. These changes, together with the development of more capillaries serving the fibers, increase capacity for longduration activity rather than strength of contraction.

interaction of actin and myosin during muscle contraction. Two thin strands of **tropomyosin** lie near the grooves between the actin strands. Each tropomyosin strand is itself a double helix as shown in Figure 29.15C. **Troponin,** a complex of three globular proteins, is located at intervals along the actin filament. Troponin acts as a calcium-dependent switch that controls the contraction process.

The actin filament complexes extend outward from both sides of the Z line and overlap myosin bundles toward the center of each sarcomere (Figures 29.15B and 29.16). Nebulin and titin are structural proteins in vertebrate striated muscle that interact with actin (nebulin) and myosin (titin) (Figure 29.15B & C). Nebulin regulates actin filament length and titin, an elastic protein, provides support for myosin and anchors it to the middle of the sarcomere at the M line.

Figure 29.15 Molecular structure of actin and myosin filaments of skeletal muscle. **A,** The myosin molecule is composed of two polypeptides coiled together and expanded at their ends into two globular heads. **B,** The myosin filament is composed of a bundle of myosin molecules with the globular heads extended outward toward the actin filaments on both sides. Titin protein runs through each myosin filament and attaches at the M line in the middle of the sacromere and at the Z line. **C,** The actin filament consists of a double strand of actin surrounded by two tropomyosin strands. Nebulin protein forms a rodlike core around which the actin filament aligns. A globular protein complex, troponin, occurs in pairs at every seventh actin unit. Troponin is a calcium-dependent switch that controls the interaction between actin and myosin.

Figure 29.16 Sliding-filament hypothesis, showing how actin and myosin filaments interact during contraction. **A,** Muscle relaxed. **B,** Muscle contracted. Nebulin, titin, tropomyosin, and troponin have been omitted from this figure so that the contraction mechanism can be more easily viewed.

Sliding-Filament Hypothesis of Muscle Contraction

In the 1950s the English physiologists A. F. Huxley and H. E. Huxley independently proposed the **sliding-filament hypothesis** to explain striated muscle contraction. According to this hypothesis, the actin and myosin filaments become linked together by molecular cross bridges, which act as levers to pull the filaments past each other. During contraction, the club-shaped heads on the myosin filaments form cross bridges that move rapidly back and forth, alternately attaching to and releasing from receptor sites on the actin filaments, and drawing actin filaments past the myosin filaments in a ratchetlike action. As contraction continues, the Z lines are pulled closer together (Figure 29.16). Thus the sarcomere shortens. Because all sarcomere units shorten together, the whole muscle contracts. Relaxation is a passive process. When cross bridges between the actin and myosin filaments release, the sarcomeres are free to lengthen. This requires some force, which is supplied by recoil of elastic fibers within the connective tissue layers of the muscle (see p. 728) and by antagonistic muscles or the force of gravity.

Control of Contraction

Muscle contracts in response to nerve stimulation. If the nerve supply to a muscle is severed, the muscle **atrophies,** or wastes away. Skeletal muscle fibers are innervated by motor neurons whose cell bodies are located in the central nervous system (brain and spinal cord) (see p. 000). Each cell body gives rise to a motor axon that leaves the central nervous system to travel by way of a peripheral nerve trunk to a muscle, where it branches repeatedly into many terminal branches. Each terminal branch innervates a single muscle fiber. Depending on the type of muscle, a single motor axon may innervate as few as three or four muscle fibers (where very precise control is needed, such as the muscles that control eye movement) or as many as 2000 muscle fibers (where precise control is not required, such as large leg muscles). A motor neuron and all muscle fibers it innervates is called a **motor unit.** The motor unit is the functional unit of skeletal muscle control. When a motor neuron fires, the action potential passes to all fibers of the motor unit and each is stimulated to contract simultaneously. Total force exerted by a muscle depends on the number of motor units activated. Precise control of movement is achieved by varying the number of motor units activated at any one time. Increasing the number of motor units brought into play produces a smooth and steady increase in muscle tension; this is called motor unit **recruitment.**

The Neuromuscular Junction

The place where a motor axon terminates on a muscle fiber is called the **neuromuscular (or myoneural) junction** (Figure 29.17). At the junction is a tiny gap, or **synaptic cleft,** that thinly separates a nerve terminal and muscle fiber. Within each nerve terminal and close to the synaptic cleft, the neuron stores a chemical, **acetylcholine,** in minute vesicles called **synaptic vesicles.** Vesicles of acetylcholine are released into the synaptic cleft when a nerve signal or action potential (see Chapter 33, p. 725) reaches a synapse. Acetylcholine is a chemical mediator or **neurotransmitter** that diffuses across the synaptic cleft and acts on the muscle fiber membrane, or sarcolemma, by binding to receptor sites and generating an electrical depolarization (Figure 29.17 inset and p. 725). The depolarization spreads rapidly over the musclefiber sarcolemma and induces muscle contraction. Thus the synapse is a special chemical bridge that couples together the electrical activities of nerve and muscle fibers. The mechanism of transmission of an electrical signal from nerve fiber to muscle is very similar to signal transmission between two nerve fibers described on page 725 and in Figures 33.7 and 33.8 (p. 726 and p. 727).

Built into vertebrate skeletal muscle is an elaborate conduction system that serves to carry the depolarization from the neuromuscular junction to the densely packed filaments within the fiber. Along the surface of the sarcolemma are numerous invaginations of the membrane that project into the muscle fiber as a system of tubules,

Figure 29.17 Section of vertebrate skeletal muscle showing a nerve-muscle synapse (neuromuscular or myoneural junction), sarcoplasmic reticulum, and connecting transverse tubules (T-tubules). Arrival of a nerve impulse or action potential at the synapse triggers a release of acetylcholine into the synaptic cleft (*inset at left*). Transmitter molecules binding to receptors generate membrane depolarization. This depolarization spreads across the sarcolemma, into the T-tubules, and to the sarcoplasmic reticulum, where the sudden release of calcium sets in motion the contractile machinery of the myofibril.

called **T-tubules** (Figure 29.17). The membrane depolarization from the neuromuscular junction passes down these T-tubules and into the muscle fiber. The T-tubules are closely associated with the **sarcoplasmic reticulum,** a system of modified endoplasmic reticulum (p. 40) that runs parallel to the actin and myosin filaments. The sarcoplasmic reticulum stores calcium, and its release around the actin and myosin filaments enables the muscle fiber to contract.

Excitation-Contraction Coupling

How does electrical depolarization of the sarcolemma and T-tubules activate the contractile machinery? In resting, unstimulated muscle, shortening does not occur because thin tropomyosin strands surrounding the actin filaments lie in a position that prevents the myosin heads from attaching to actin. When muscle is stimulated and the action potential is transmitted down the T-tubules, the electrical depolarization stimulates the sarcoplasmic reticulum surrounding the fibrils to release calcium ions (Figure 29.18). The calcium binds to the actin-binding protein, troponin. Troponin immediately undergoes changes in shape that cause tropomyosin to leave its blocking position, exposing active sites on the actin filaments. Myosin heads then bind to these sites, forming cross bridges between adjacent myosin and actin filaments. This sets in motion an attach-pull-release cycle, or **cross-bridge cycling,** that occurs in a series of steps as shown in Figure 29.18. Release of bond energy from ATP hydrolysis (see Chapter 4, p. 60) activates the myosin head, which swings 45 degrees, at the same time releasing a molecule of ADP. This is the power

stroke that pulls the actin filament a distance of about 10 nm, and it comes to an end when inorganic phosphate (Pi) is released. Another ATP molecule now binds to the myosin head, freeing myosin from the active site on actin. Thus each cycle requires expenditure of energy in the form of ATP (Figure 29.18).

Shortening continues as long as nerve action potentials arrive at the neuromuscular junction and free calcium remains available around the actin and myosin filaments. The cross-bridge cycling can repeat again and again, 50 to 100 times per second, pulling actin and myosin filaments past each other. While the distance that each sarcomere can shorten is very small, this distance is multiplied by the thousands of sarcomeres lying end to end in a muscle fiber. Consequently, a strongly contracting muscle may shorten by as much as one-third its resting length.

When stimulation stops, calcium is quickly pumped back into the sarcoplasmic reticulum. Troponin resumes its original configuration; tropomyosin moves back into its blocking position on actin, and the muscle relaxes.

Energy for Contraction

Muscle contraction requires large amounts of energy. ATP is the immediate source of energy, and its levels within muscle are held almost constant due to immediate replenishment from three main sources: glucose, glycogen, and creatine phosphate. Glucose is transported to muscle in the blood where it is catabolized during **aerobic metabolism** (see Chapter 4, p. 61) to produce ATP.

T-tubule

Sarcoplasmic
reticulum

Sarcolemma

Myofibril

Action potential

Calcium attaches
to troponin

Troponin and tropomyosin
move off active site

Ca⁺⁺

Troponin

Tropomyosin

Actin

ADP
+ Pᵢ

Myosin

Step 1: An action potential spreads along the sarcolemma and is conducted inward to the sarcoplasmic reticulum by way of T-tubules. Calcium ions are released from the sarcoplasmic reticulum and diffuse rapidly into the myofibrils, binding to troponin molecules on the actin filament. Troponin and tropomyosin molecules move away from the active sites.

Figure 29.18 Excitation-contraction coupling in vertebrate skeletal muscle.

Glycogen store within muscle can also supply glucose molecules for ATP production. Glycogen is a polysaccharide chain of glucose molecules (p. 22) stored in both liver and muscle. Muscle has by far the larger store—three-fourths of all glycogen in the body is stored in muscle. As a supply of energy for contraction, glycogen has three important advantages: it is relatively abundant, it can be mobilized quickly, and it can provide energy during anaerobic, as well as aerobic, conditions. Enzymes convert glycogen to glucose-6-phosphate molecules, the first stage of glycolysis that leads to generation of ATP (p. 62).

Finally, muscles have an energy reserve in the form of **creatine phosphate.** Creatine phosphate is a high-energy phosphate compound that stores bond energy during periods of rest. As ADP is produced from ATP during contraction, creatine phosphate releases its stored bond energy to convert ADP to ATP. This reaction can be summarized as follows:

$$\text{Creatine Phosphate} + \text{ADP} \rightarrow \text{ATP} + \text{Creatine}$$

Some muscle types (**slow** and **fast oxidative fibers,** next section) rely heavily on glucose and oxygen supplies transported to muscle via the circulatory system. If muscular contraction is not too vigorous or too prolonged, glucose from the blood or released from glycogen can be completely oxidized to carbon dioxide and water by aerobic metabolism. During prolonged or heavy exercise, however, blood flow to the muscles, although greatly increased above resting levels, cannot supply oxygen to the mitochondria rapidly enough to complete oxidation of glucose. The contractile machinery then receives its energy largely by **anaerobic glycolysis,** a process that does not require oxygen (p. 67). This anaerobic pathway, although not nearly as efficient as the aerobic one, is of great importance; without it, all forms of heavy muscular exertion would be impossible. Indeed, **fast glycolytic fibers** (next section) rely almost exclusively on anaerobic glycolysis to produce energy for contraction.

During anaerobic glycolysis, glucose is degraded to lactic acid with release of energy. Lactic acid accumulates in the muscle and

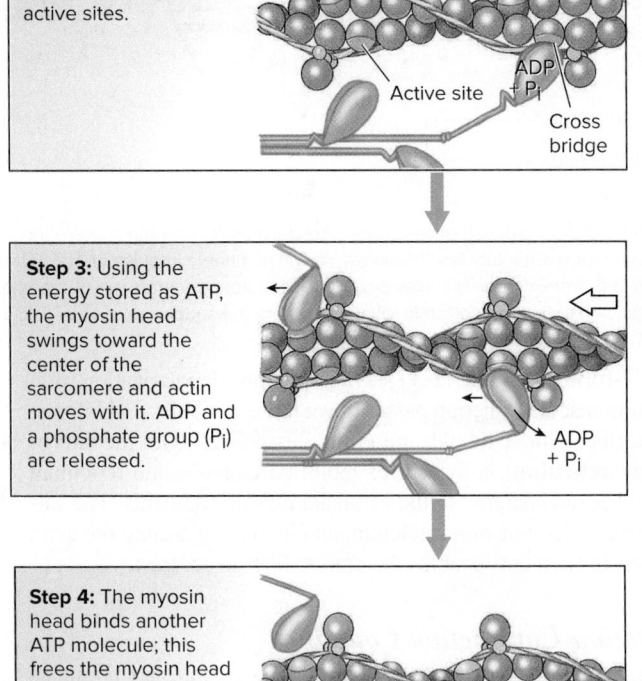

Step 2: Myosin forms cross bridges as it binds to the exposed active sites.

Active site

ADP
+ Pᵢ

Cross
bridge

Step 3: Using the energy stored as ATP, the myosin head swings toward the center of the sarcomere and actin moves with it. ADP and a phosphate group (Pᵢ) are released.

ADP
+ Pᵢ

Step 4: The myosin head binds another ATP molecule; this frees the myosin head from the active site on actin.

ATP

Step 5: The myosin head splits ATP, retaining the energy released as well as the ADP and the phosphate group. The cycle can now repeat as long as calcium is present to open active sites on the actin molecules.

ADP
+ Pᵢ

diffuses rapidly into the general circulation. If muscular exertion continues, muscle fatigue occurs. This was originally attributed to the buildup of lactic acid, and a resultant decrease in pH, causing enzyme inhibition. However, data suggest that accumulation of inorganic phosphate also may cause muscle fatigue, at least in muscle types that rely heavily on creatine phosphate stores. Thus the anaerobic pathway is a self-limiting one, since continued heavy exertion leads to exhaustion. The muscles incur an **oxygen debt** because accumulated lactic acid must be converted to pyruvic acid, which can be fed into the Krebs cycle via conversion to Acetyl-CoA (see Chapter 4, p. 63). Thus lactic acid is oxidized by extra oxygen as the oxygen debt is repaid. After a period of exertion, oxygen consumption remains elevated until all lactic acid has been oxidized or consumed in resynthesis of glycogen.

Muscle Performance

Fast and Slow Fibers

Skeletal muscles of vertebrates consist of more than one type of fiber. **Slow oxidative fibers,** which are specialized for slow, sustained contractions without fatigue, are important in maintaining posture in terrestrial vertebrates. Such muscles are often called **red muscles** because they contain an extensive blood supply, a high density of mitochondria for supplying ATP via aerobic metabolism, and abundant stored myoglobin which supplies additional stored oxygen, all of which give the muscle a red color.

Two kinds of **fast fibers,** capable of fast, powerful contractions, are known. One kind of fast fiber (**fast glycolytic fiber**) lacks an efficient blood supply and has a low density of mitochondria and myoglobin. Muscles composed of these fibers (often called **white muscles**) are usually pale in color, function anaerobically, and fatigue

rapidly. The "white meat" of chicken breast is a familiar example. Members of the cat family have running muscles formed almost entirely of fast glycolytic fibers that operate anaerobically. During a chase, such muscles accumulate a substantial oxygen debt, which is replenished after the chase. For example, a cheetah after a high-speed chase lasting less than a minute will pant heavily for 30 to 40 minutes before its oxygen debt is relieved. Weight lifters favor activation and development of these muscle fibers and thus cannot sustain lifting heavy weights for long periods. The other kind of fast fiber (**fast oxidative fiber**) has an extensive blood supply and a high density of mitochondria and myoglobin, and functions largely aerobically. Some animals use these muscles for rapid, sustained activities. Most migratory birds, such as geese and swans, as well as dogs and ungulates (hoofed mammals), for example, have limb (or wing) muscles with a high percentage of fast oxidative fibers, and are capable of active locomotion for long periods of time. Most muscles possess a mixture of these different fiber types, which permits a range of activity.

Importance of Tendons in Energy Storage

When mammals walk or run, much kinetic energy is stored from step to step as elastic strain energy in the tendons. For example, during running the Achilles tendon is stretched by a combination of downward force of the body on the foot and contraction of the calf muscles. The tendon then recoils, extending the foot while the muscle is still contracted, propelling the leg forward (Figure 29.19). An extreme example of this bouncing-ball principle is the bounding of a kangaroo (Figure 29.19). This type of movement uses far less energy than would be required if every step relied solely on alternate muscle contraction and relaxation.

There are many examples of elastic storage in the animal kingdom. It is used in the ballistic jumps of grasshoppers and fleas,

Figure 29.19 Energy storage in the Achilles tendon of human and kangaroo legs. During running, stretching of the Achilles tendon when the foot strikes the ground stores kinetic energy that is released to propel the leg forward.

in the wing hinges of flying insects, in the hinge ligaments of bivalve molluscs, and in the highly elastic large dorsal ligament (ligamentum nuchae) that helps to support the head of hoofed mammals.

Skeletal muscle has been shown to secrete a hormone called **irisin.** Studies showed that during exercise in humans and mice, blood irisin levels rise and cause many of the beneficial effects of exercise on the body. Major benefits included an increase in energy expenditure with no change in food intake, and improved glucose control. Interestingly, white adipose tissue, which stores energy, was converted to brown adipose tissue, which consumes energy. Thus, the benefits of irisin on obesity in humans may be considerable.

SUMMARY

An animal is wrapped in a protective covering, the integument, which may be as simple as the single-layered epidermis of many invertebrates or as complex as the skin of a mammal. The arthropod exoskeleton is the most complex of invertebrate integuments, consisting of a two-layered cuticle secreted by a single-layered epidermis. It may be hardened by calcification or sclerotization and must be molted at intervals to permit body growth. Vertebrate integument consists of two layers: the epidermis, which gives rise to various derivatives such as hair, feathers, and claws; and the dermis, which supports and nourishes the epidermis. It also is the origin of bony derivatives such as fish scales and deer antlers.

Integument color is of two kinds: structural color, produced by refraction or scattering of light by particles in the integument; and pigmentary color, produced by pigments that are usually confined to special pigment cells (chromatophores).

Skeletons are supportive systems that may be hydrostatic or rigid. The hydrostatic skeletons of several soft-walled invertebrate groups depend on body-wall muscles that contract against a noncompressible internal fluid of constant volume. In a similar manner, muscular hydrostats, such as the tongue of mammals and reptiles, and the trunk of elephants, rely on muscle bundles arranged in complex patterns to produce movement without either skeletal support or a liquid-filled cavity. Rigid skeletons have evolved with attached muscles that act with the supportive skeleton to produce movement. Arthropods have an external skeleton, which must be shed periodically to make possible an enlarged replacement. Vertebrates evolved an internal skeleton, a framework formed of cartilage and/or bone, that can grow with the animal, while, in the case of bone, additionally serving as a reservoir of calcium and phosphate.

Animal movement, whether in the form of ameboid movement of wandering cells such as white blood cells, or in the form of contraction of an organized muscle mass, depends on specialized contractile proteins. The most important of these is the actomyosin system, which is usually organized into elongate actin and myosin filaments that slide past each other during contraction. When a muscle is stimulated, an electrical depolarization is conducted into the muscle fibers through T-tubules to the sarcoplasmic reticulum, causing the release of calcium. Calcium binds to a troponin complex associated with the actin filament. This binding causes tropomyosin to shift from its blocking position and allows the myosin heads to form cross bridges with the actin filament. Powered by ATP, the myosin heads swivel back and forth to pull the actin and myosin filaments past one another. Phosphate bond energy for contraction is ultimately supplied by carbohydrate fuels.

Vertebrate skeletal muscle consists of variable percentages of both slow fibers, used principally for sustained postural contractions, and fast fibers, used in locomotion. Tendons are important in locomotion because the kinetic energy stored in stretched tendons at one stage of a locomotory cycle is released at a subsequent stage.

REVIEW QUESTIONS

1. The arthropod exoskeleton is the most complex of invertebrate integuments. Describe its structure, and explain the difference in the way cuticle is hardened in crustaceans and in insects.
2. Distinguish between epidermis and dermis in vertebrate integument, and describe structural derivatives of these two layers.
3. What is the difference between structural colors and colors based on pigments? How do the chromatophores of vertebrates and cephalopod molluscs differ in structure and function?
4. As "naked apes" humans lack the protective investment of fur that shields other mammals from the damaging effects of sunlight. How does human skin respond to ultraviolet radiation in the short term and with continued exposure?
5. Hydrostatic skeletons have been defined as a mass of fluid enclosed within a muscular wall. How would you modify this definition to apply it to a muscular hydrostat? Offer examples of both a hydrostatic skeleton and a muscular hydrostat.
6. One of the special qualities of vertebrate bone is that it is a living tissue that permits continuous remodeling. Explain how the structure of bone allows this remodeling to happen.
7. What is the difference between endochondral and membranous bone? Between spongy and compact bone?
8. Discuss the role of osteoclasts, osteoblasts, parathyroid hormone, and calcitonin in bone growth.
9. The laws of scaling tell us that doubling the length of an animal will increase its weight eightfold, while the force its bones can bear increases only fourfold. What solutions to this problem have evolved that allow animals to become large, while maintaining bone stresses within margins of safety?
10. Name the major skeletal components included in the axial and in the appendicular skeleton.
11. An unexpected discovery from studies of ameboid movement is that the same proteins found in the contractile system of animal muscle—actin and myosin—are present in ameboid cells. Explain how these and other proteins interact during ameboid movement.
12. A "9 + 2" arrangement of microtubules is typical of both cilia and flagella. Explain how this system produces a bending motion. What is the difference between a cilium and a flagellum?
13. What functional features of molluscan smooth muscle and insect fibrillar muscle set them apart from typical vertebrate muscle?
14. The sliding-filament model of skeletal-muscle contraction assumes a sliding or slipping of interdigitating filaments of actin and myosin. Electron micrographs show that during contraction the actin and myosin filaments remain of constant length while the distance between Z lines shortens. Explain how this happens in terms of the molecular structure of the muscle filaments. What is the role of regulatory proteins (troponin and tropomyosin) in contraction?
15. While the sarcoplasmic reticulum of muscle was first described by nineteenth-century microscopists, its true significance was not appreciated until its intricate structure was revealed much later by the electron microscope. What could you tell a nineteenth-century microscopist to enlighten him or her about the structure of the

sarcoplasmic reticulum and its role in the coupling of excitation and contraction?

16. The filaments of skeletal muscle are moved by free energy derived from the hydrolysis of ATP. During sustained muscle contraction ATP levels remain fairly constant, while levels of creatine phosphate fall. Explain why this is so. Under what circumstances is an oxygen debt incurred during muscle contraction?

17. During evolution, skeletal muscle became adapted to functional demands ranging from sudden, withdrawal movements of a startled worm, to the sustained contractions required to maintain mammalian posture, to supporting a long, fast chase across an African savanna. What are some of the fiber types in vertebrate muscle that evolved to support these kinds of activities?

For Further Thought   A weight lifter decides to go jogging with a friend. Explain why the weight lifter can't keep jogging for long in terms of her/his leg muscle structure and function.

SELECTED REFERENCES

Ainsworth, C. 2007. Tails of the unexpected. Native **448**: 638–641. *Exciting new roles for cilia are coming to light that appear to have implications in human disease and regulation of embryonic development.*

Anderson, J. I., P. Schjerling, and B. Saltin. 2000. Muscle, genes and athletic performance. Sci. Am. **283**: 48–55 (Sept.). *Good discussion of muscle structure and function related to human athletic performance.*

Bostrom, P., J. Wu, M. P. Jedrychowski, A. Korde, L. Ye, J. C. Lo, K. A. Rasbach, E. A. Bostrom, J. H. Choi, J. Z. Long, S. Kajimura, M. C. Zingaretti, B. F. Vind, H. Tu, S. Cinti, K. Hojlund, S. P. Gygi, and B. S. Spiegelman. 2012. A PGC1-alpha-dependent myokine that drives brown-fat-like development of white fat and thermogenesis. Nature **481**: 463–469. *This research article outlines original research on the newly discovered skeletal muscle hormone, irisin.*

Hadley, N. F. 1986. The arthropod cuticle. Sci. Am. **255**: 104–112 (July). *Describes properties of this complex covering that account for much of the adaptive success of arthropods.*

Le Clainche, C., and M. E. Carlier. 2008. Regulation of actin assembly associated with protrusion and adhesion in cell migration. Physiol. Rev. **88**: 489–513. *An in-depth discussion of actin and actin-binding proteins in the regulation of cell crawling.*

Marx, J. 2004. Coming to grips with bone loss. Science **305**: 1420–1422. *An interesting review of new therapies for prevention and treatment of osteoporosis.*

Novelle, M. G., C. Contreras, A. Romero-Pick, M. Lopez, and C. Diequez. 2013. Irisin, two years later. Int. J. Endocrinol. **1013**: 1–8. *A review article that describes the most important actions of irisin on energy and metabolic homeostasis.*

Randall, D., W. Burggren, and K. French. 2002. Eckert animal physiology: mechanisms and adaptations, ed. 5. New York, W. H. Freeman & Company. *A comprehensive and comparative treatment of animal physiology.*

Sellers, W. I., and P. L. Manning. 2007. Estimating dinosaur maximum running speeds using evolutionary robotics. Proc. R. Soc. B (The Royal Society) **274**: 2711–6. *Computer simulations used to estimate the running speed of dinosaurs including* Tyrannosaurus.

Sparrow, J. C., and F. Schock. 2009. The initial steps of myofibril assembly: integrins pave the way. Nature Reviews. Molecular Cell Biology **10**: 203–208. *A thorough discussion of sarcomere structure and assembly in striated muscle.*

Stossel, T. P. 1994. The machinery of cell crawling. Sci. Am. **271**: 54–55, 58. *Describes the process of ameboid movement using assembly and disassembly of protein polymers in the cytoplasm.*

Westerblad, H., D. G. Allen, and J. Lännergren. 2002. Muscle fatigue: lactic acid or inorganic phosphate the major cause? News Physiol. Sci. **17**: 17–21. *Well-written review of recent data that provides an alternate explanation for muscle fatigue.*

Willmer, P., G. Stone, and I. Johnston. 2005. Environmental physiology of animals, ed. 2. Oxford, U.K., Blackwell Science Ltd. *Well-written information regarding environmental adaptations in both vertebrates and invertebrates.*

30
CHAPTER

Homeostasis: Osmotic Regulation, Excretion, and Temperature Regulation

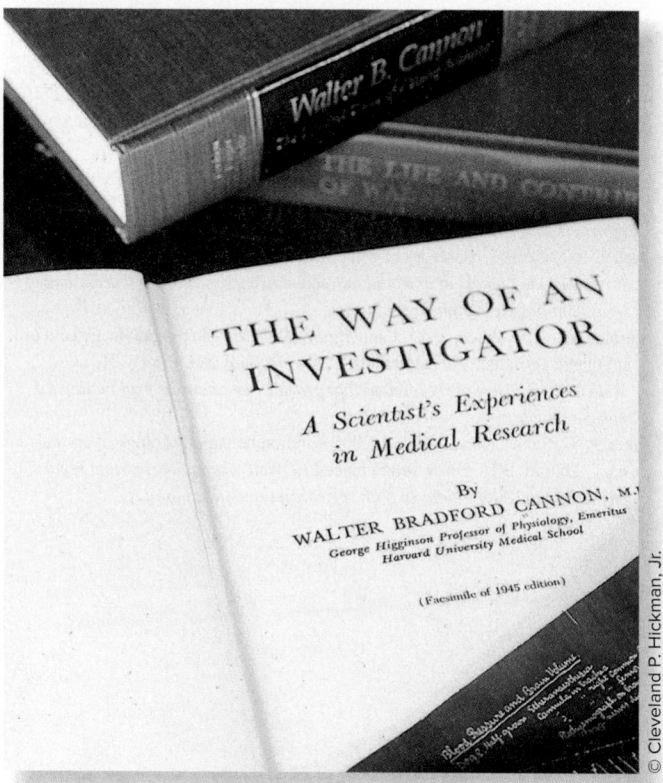

Title page of Walter B. Cannon's autobiography.

Homeostasis: Birth of a Concept

The tendency toward internal stabilization of the animal body was first recognized by Claude Bernard, great French physiologist of the nineteenth century who, through his studies of blood glucose and liver glycogen, first discovered internal secretions. After a lifetime of study and experimentation, Bernard gradually developed the principle for which he is best remembered, that of constancy of the internal environment, a principle that in time would pervade physiology and medicine.

Years later, at Harvard University, American physiologist Walter B. Cannon (Figure 30.1) reshaped and restated Bernard's idea. From his studies of the nervous system and reactions to stress, he described the ceaseless balancing and rebalancing of physiological processes that maintain stability and restore the normal internal state of the body when it has been disturbed. He also gave it a name: **homeostasis.** The term became pervasive in the medical literature of the 1930s. Physicians spoke of getting their patients back into homeostasis. Even politicians and sociologists saw what they considered deep nonphysiological implications. Cannon enjoyed this broadened application of the concept and later suggested that democracy was a form of government that took a homeostatic middle course. Despite the enduring importance of the homeostasis concept, Cannon never received the Nobel Prize—one of several acknowledged oversights of the Nobel Committee.

Figure 30.1 Walter Bradford Cannon (1871 to 1945), Harvard professor of physiology who coined the term "homeostasis" and developed the concept originated by French physiologist Claude Bernard (see Figure 31.2, p. 682).

© Science Source

The concept of homeostasis, described in the chapter opening essay, permeates all physiological thinking and is the theme of this chapter and Chapter 31. Although this concept was first developed from studies with mammals, it applies to all organisms. Potential changes in the internal environment arise from two sources. Materials from the environment surrounding cells, such as oxygen, nutrients, and salts, that are required for metabolic activities, must be continuously replenished. Cellular activity also produces waste products that must be expelled. Second, the internal environment changes as it responds to the organism's external environment. Changes from either source must be stabilized by the physiological mechanisms of homeostasis.

In animals, cellular homeostasis is maintained by the coordinated activities of all systems of the body, except for the reproductive system. The various homeostatic activities are coordinated by the circulatory, nervous, and endocrine systems, and also by organs that serve as sites of exchange with the external environment. These last include kidneys, lungs or gills, digestive tract, and integument. Through these organs oxygen, nutrients, minerals, and other constituents of body fluids enter, water is exchanged, heat is lost, and metabolic wastes are eliminated.

Thus, systems within an organism function in an integrated way to maintain a constant internal environment for all cells around a setpoint. Small deviations from this setpoint in pH, temperature, osmotic pressure, metabolic fuels (for example, glucose or fatty acids), carbon dioxide, and oxygen levels activate physiological mechanisms that return the variable to its setpoint by a process called **negative feedback** regulation (see Chapter 34, p. 749).

We look first at the problems of controlling the internal fluid environment of aquatic animals. Next we briefly examine how terrestrial animals regulate their internal state. Finally we look at strategies for regulating body temperature.

30.1 WATER AND OSMOTIC REGULATION

Water and osmotic regulation provide a means for maintaining internal solute concentrations within a range that allows cellular function to proceed. As discussed in Chapter 3 (p. 44), the selective permeability of plasma membranes means that changes in ion concentrations inside or outside the cell will dramatically alter ion and water flow through the membrane. Cell volume will rise and fall if cells are exposed to a hypoosmotic (hypotonic) or hyperosmotic (hypertonic) environment, respectively, and both changes will produce negative

effects on cellular metabolism. The concept of water and osmotic regulation applies to single-celled eukaryotes and animals alike; however, multicellular animals can control ion and water balance inside cells by regulating the ion and water content of the fluids that bathe them.

How Marine Invertebrates Maintain Salt and Water Balance

Most marine invertebrates are in osmotic equilibrium with their seawater environment. With body surfaces permeable to salts and water, their body-fluid concentration rises or falls in conformity with changes in concentrations of seawater. Because such animals are incapable of regulating osmotic pressure of their body fluid, they are called **osmoconformers.** Invertebrates living in the open sea are seldom exposed to osmotic fluctuations because the ocean is a highly stable environment. Oceanic invertebrates have, in fact, very limited abilities to withstand osmotic change. Should they be exposed to dilute seawater, they absorb water by osmosis and die quickly because their body's cells cannot tolerate dilution and are helpless to prevent it. These animals are restricted to living in a narrow salinity range and are called **stenohaline** (Gr. *stenos,* narrow, + *hals,* salt). An example is the marine spider crab, *Maia* (Figure 30.2).

Conditions along coasts and in estuaries and river mouths are much less constant than those of the open ocean. Here animals must cope with large and often abrupt changes in salinity as the

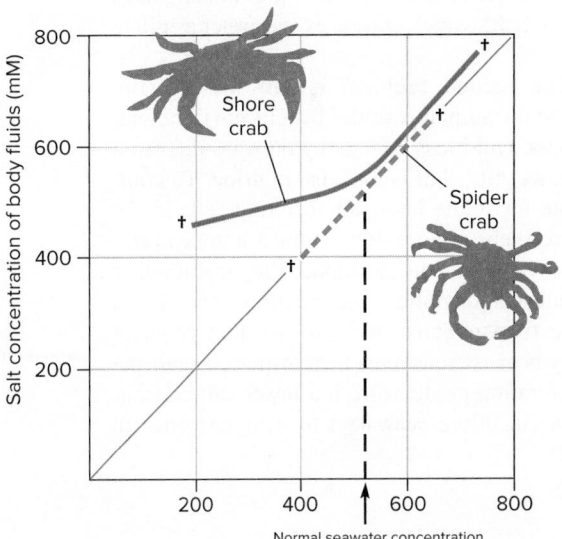

Figure 30.2 Salt concentration of body fluids of two crabs as affected by variations in seawater concentration. The 45-degree line represents equal concentration between body fluids and seawater. Since the spider crab, *Maia* sp., cannot regulate the salt concentration of its fluids, it conforms to whatever changes happen in the seawater (*dashed red line*). The shore crab, *Eriocheir* sp., however, can regulate osmotic concentration of its fluids to some degree in dilute seawater (*solid blue line*). Daggers at ends of lines indicate limits of tolerance for each species beyond which they will die.

tides ebb and flow and mix with freshwater draining from rivers. These animals are termed **euryhaline** (Gr. *eurys,* broad, + *hals,* salt), meaning that they can survive a wide range of salinity changes, mainly by demonstrating varying powers of **osmoregulation.** For example, a brackish-water shore crab, *Eriocheir,* can resist dilution of body fluids by dilute (brackish) seawater (Figure 30.2). Although the concentration of salts in the body fluids falls, it does so less rapidly than the fall in seawater concentration. This crab is a **hyperosmotic regulator,** meaning that it maintains its body fluids more concentrated (hence *hyper-*) than the surrounding water.

By preventing excessive dilution, and thus protecting the cells from extreme changes, brackish-water shore crabs can live successfully in the physically unstable but biologically rich coastal environment. Nevertheless, with only a limited capacity for osmotic regulation, they will die if exposed to greatly diluted seawater. Because a crab's body fluids are osmotically more concentrated than the dilute seawater outside, water flows into its body, especially across the thin, permeable membranes of the gills (see Chapter 20, Figure 20.7, p. 427, showing crustacean gills). As with the red blood cells placed in pure water (p. 47), water diffuses inward because of higher solute concentration inside. For the crab, were this inflow of water allowed to continue unchecked, its body fluids would soon become diluted. The problem is solved by the kidneys (antennal glands located in the crab's head; see Figure 30.7 showing similar glands of a crayfish), which excrete excess water as dilute urine.

The second problem is salt loss. Again, because the animal is saltier than its environment, it cannot avoid loss of ions by outward diffusion across its gills. Salt is lost also in urine. To compensate for solute loss, salt-secreting cells in the gills actively remove ions from dilute seawater and move them into the blood, thus maintaining the internal osmotic concentration. This is an **active transport** (p. 47) process that requires energy because ions must be transported against a concentration gradient from a lower salt concentration (in dilute seawater) to a higher one (in blood).

Salt and Water Balance in Freshwater

During the Silurian and Lower Devonian periods, the major groups of jawed fishes began to penetrate into brackish-water estuaries and then gradually into freshwater rivers. Before them lay a new, unexploited habitat already stocked with food in the form of insects and other invertebrates, which had preceded them into freshwater. However, advantages that this new habitat offered were

balanced by a tough physiological challenge: the necessity for osmoregulation.

Freshwater animals face challenges similar to but more extreme than those of the shore crab just described. They must keep the salt concentration of their body fluids higher than that of the water in which they live. Water enters their bodies osmotically, and salt is lost by outward diffusion. Unlike the habitat of the brackish-water shore crab, freshwater is much more dilute than are coastal

A FRESHWATER FISH

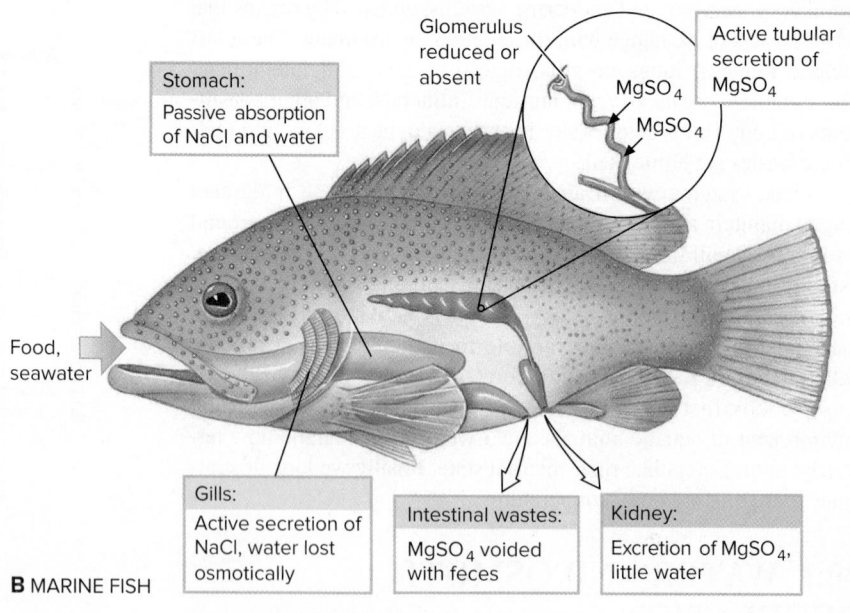

B MARINE FISH

Figure 30.3 Osmotic regulation in freshwater and marine bony fishes. **A,** A freshwater fish maintains osmotic and ionic balance in its dilute environment by actively absorbing sodium chloride across the gills (some salt enters with food). To flush out excess water that constantly enters the body, the glomerular kidney produces dilute urine by reabsorbing sodium chloride. **B,** A marine fish must drink seawater to replace water lost osmotically to its salty environment. Sodium chloride and water are absorbed from the stomach. Excess sodium chloride is secreted outward by the gills. Divalent sea salts, mostly magnesium sulfate, are eliminated with feces and secreted by the tubular kidney.

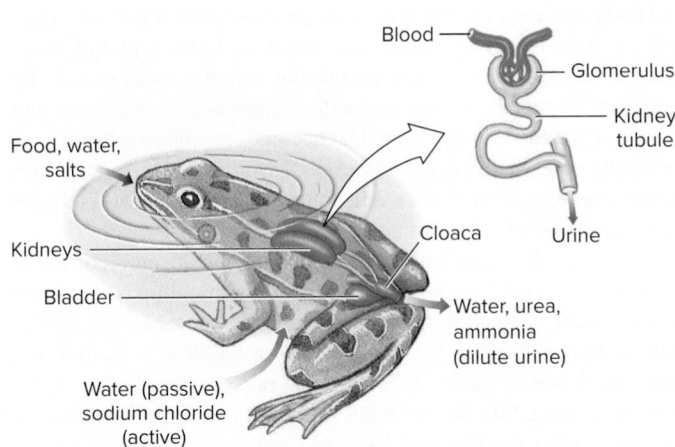

Figure 30.4 Exchange of water and solute in a frog. Water enters the highly permeable skin and is excreted by kidneys. The skin also actively transports ions (sodium chloride) from the environment. Kidneys form dilute urine by reabsorbing sodium chloride. Urine flows into the urinary bladder, where, during temporary storage, most of the remaining sodium chloride is removed and returned to the blood.

estuaries, and there is no retreat, no salty sanctuary into which a freshwater animal can retire for osmotic relief. Freshwater organisms maintain their internal osmotic concentrations lower than marine organisms. This represents a balance between solute requirements and the amount of work it takes to keep water out. Thus, freshwater organisms have become permanent and highly efficient hyperosmotic regulators.

The scaled and mucus-covered body surface of a fish is nearly waterproof, yet remains flexible. In addition, freshwater fishes have several defenses against the problems of water gain and salt loss. First, water that inevitably enters the body by osmosis is pumped out by the kidney, which is capable of forming very dilute urine by reabsorbing sodium chloride (Figure 30.3A). Second, salt-absorbing cells located in the gills transport salt ions, principally sodium and chloride (present in small quantities even in freshwater), from the water to the blood. This process, together with salt present in the fish's food, replaces diffusive salt loss. These mechanisms are so efficient that a freshwater fish devotes only a small part of its total energy expenditure to osmoregulation.

Crayfishes, aquatic insect larvae, clams, and other freshwater animals are also hyperosmotic regulators and face the same hazards as freshwater fishes; they tend to gain too much water and lose too much salt. Like freshwater fishes, they excrete excess water as urine and replace lost salt by active ion transport across the gills.

Amphibians living in water also must compensate for salt loss by actively absorbing salt from the water (Figure 30.4). They use their skin for this purpose. Pieces of frog skin continue to transport sodium and chloride actively for hours when removed and placed in a specially balanced salt solution, and frog skin became a favorite membrane system for studies of ion-transport phenomena.

Salt and Water Balance in Marine Fishes

Bony fishes living in the oceans today are descendants of earlier freshwater bony fishes that moved back into the sea during the Triassic period approximately 230 million years ago. Over millions of years freshwater fishes established an ionic concentration in the body fluid equivalent to approximately one-third that of seawater. Body fluid of terrestrial vertebrates is remarkably similar to that of dilute seawater too, a fact that is undoubtedly related to the lower osmotic concentrations of the body fluids of their freshwater ancestors.

By expressing concentration of salt in seawater or body fluids in molarity, we are saying that the osmotic strength is equivalent to the molar concentration of an ideal solute having the same osmotic strength. In fact, seawater and animal body fluids are not ideal solutions because they contain electrolytes that dissociate in solution. A 1 M solution of sodium chloride (which dissociates in solution) has a much greater osmotic strength than a 1 M solution of glucose, an ideal solute (nonelectrolyte) that does not dissociate in solution. Consequently, biologists usually express osmotic strength of a biological solution in osmolarity rather than molarity. A 1 osmolar solution exerts the same osmotic pressure as a 1 M solution of a nonelectrolyte.

When some freshwater bony fishes of the Triassic period ventured back to the sea, they would have had a much lower internal osmotic concentration than the seawater around them, causing them to lose water and to gain salt. Indeed, a marine bony fish literally risks desiccation, much like a desert mammal deprived of water.

Marine bony fishes living today maintain the salt concentration of their body fluids similar to that of freshwater fishes, at approximately one-third that of seawater (body fluids = 0.3 to 0.4 gram mole per liter [M]; seawater = 1 M). They are **hypoosmotic regulators** because they maintain their body fluids at a lower concentration (hence *hypo*) than their seawater environment. To compensate for water loss a marine bony fish drinks seawater (Figure 30.3B). This seawater is absorbed from the intestine, and the major sea salt, sodium chloride, is carried by the blood to the gills, where salt-secreting cells transport it back into the surrounding sea. Ions remaining in the intestinal residue, especially magnesium, sulfate, and calcium, are voided with feces or excreted by kidneys. In this indirect way, marine fishes rid themselves of excess sea salts and replace water lost by osmosis. Samuel Taylor Coleridge's ancient mariner, surrounded by "water, water, everywhere, nor any drop to drink" undoubtedly would have been tormented even more had he known of the marine fishes' ingenious solution to thirst. A marine fish regulates the amount of seawater it drinks, consuming only enough to replace water loss and no more.

Sharks and rays (elasmobranchs) are almost totally marine, and achieve osmotic balance differently. The salt composition of shark's blood is similar to that of bony fishes, but the blood also carries large amounts of the organic compounds urea and trimethylamine oxide (TMAO). Urea is a metabolic waste product that most animals quickly excrete. The shark kidney, however, conserves urea, allowing it to accumulate in the blood and body tissues and raising the blood osmolarity to equal or slightly exceed that of seawater. With the osmotic difference between blood and seawater eliminated, water balance is not a problem for sharks and other elasmobranchs; they are in osmotic equilibrium with their environment.

The high concentration of urea in the blood and body tissues of sharks and rays—more than 100 times as high as in mammals—could not be tolerated by most other vertebrates. In the latter, such high concentrations of urea would disrupt peptide bonds of proteins, altering protein configuration. This would also occur in sharks and rays except that TMAO has the opposite effect and serves to stabilize proteins in the presence of high urea levels. So accommodated are elasmobranchs to urea that their tissues cannot function without it, and their heart will stop beating in its absence.

Salt and Water Balance in Terrestrial Animals

The problems of living in an aquatic environment seem small indeed compared with problems of life on land. Although animal bodies are mostly water, all metabolic activities proceed in water, and life itself was conceived in water, many animals, like the plants preceding them, moved onto land. Once on land, terrestrial animals evolved physiological mechanisms to avoid desiccation, eventually becoming abundant even in some of the most arid parts of the earth.

TABLE 30.1

Labels in figure: Animal remains cool in burrow during daytime; Respiratory moisture condensed in nasal passages; Metabolic water from dry seeds; Produces dry feces; Urine concentrated by countercurrent exchange in extra long loop of Henle

Water Balance in a Human and a Kangaroo Rat, a Desert Rodent

	Human (%)	Kangaroo Rat (%)
Gains		
Drinking	48	0
Free water in food	40	10
Metabolic water	12	90
Losses		
Urine	60	25
Evaporation (lungs and skin)	34	70
Feces	6	5

Source: Some data from K. Schmidt-Nielsen, How animals work Cambridge University Press, 1972.

Terrestrial animals lose water by evaporation from respiratory and body surfaces, excretion in urine, and elimination in feces. They replace such losses by consuming water in food, drinking water when available, and retaining **metabolic water** formed in cells by oxidation of metabolic fuel molecules, such as carbohydrates and fats (see Chapter 4, p. 62). Stored fat becomes an important source of metabolic water in diving mammals. Certain arthropods—for example, desert roaches, certain ticks and mites, and mealworms—absorb water vapor directly from atmospheric air.

In some desert rodents, metabolic water gain may constitute most of the animals' water intake. Particularly revealing is a comparison of water balance in humans with that of kangaroo rats, desert rodents that may drink no water at all. In addition to conserving water as shown in Table 30.1, kangaroo rats acquire all their water from their food: 90% as metabolic water derived from oxidation of fuel molecules (see Figure 4.10, p. 61) and 10% as free moisture in food. Although we eat foods with a much higher water content than the dry seeds of a kangaroo rat's diet, we still must drink half our total water requirement.

Given ample water to drink, humans can tolerate extremely high environmental temperatures while preventing a rise in body temperature. Our ability to keep cool by evaporation was impressively demonstrated more than 200 years ago by a British scientist who remained for 45 minutes in a room heated to 260° F (126° C). A steak he carried in with him was thoroughly cooked, but he remained uninjured and his body temperature did not rise. Sweating rates may exceed 3 liters of water per hour under such conditions and cannot be tolerated unless the lost water is replaced by drinking. Without water, a human continues to sweat unabatedly until the water deficit exceeds 10% of the body weight, when collapse occurs. With a water deficit of 12% a person cannot swallow even if offered water, and death occurs when the water deficit reaches about 15% to 20%. Few people can survive more than a day or two in a desert without water. Thus humans are not physiologically well adapted for desert climates.

Excretion of waste products presents a special problem in water conservation. The primary end product of protein breakdown is ammonia, a highly toxic material. Fishes easily excrete ammonia by diffusion across their gills, since there is an abundance of water to wash it away. Terrestrial insects, nonavian reptiles, and birds have no convenient way to rid themselves of toxic ammonia; instead, they convert it into uric acid, a nontoxic, almost insoluble compound. This conversion enables them to excrete a semisolid urine with little water loss. Uric acid as an end product has another important benefit. Nonavian reptiles and birds lay amniotic eggs enclosing their embryos (see Figure 26.3, p. 566), together with their stores of food and water, and waste products that accumulate during development. By converting ammonia to uric acid, a developing embryo's waste can be precipitated into solid crystals, which are stored harmlessly within the amniotic sac, inside the egg until hatching.

Marine birds and turtles have evolved an effective solution for excreting large loads of salt eaten with their food. Located above each eye is a **salt gland** capable of excreting a highly concentrated solution of sodium chloride—up to twice the concentration of seawater. In birds the salt solution runs out the nares (see p. 595

and Figure 27.13). Marine lizards and turtles, like *Alice in Wonderland*'s Mock Turtle, shed their salt-gland secretion as salty tears. Salt glands are important accessory organs of salt excretion in these animals because their kidneys cannot produce a concentrated urine, as can mammalian kidneys.

30.2 INVERTEBRATE EXCRETORY STRUCTURES

Many unicellular eukaryotes and some freshwater sponges have special excretory organelles called contractile vacuoles. Other invertebrates have excretory organs that are simple tubular structures forming urine by first producing an ultrafiltrate or fluid secretion from the blood. This ultrafiltrate enters the proximal end of the tubule and is modified continuously as it flows down the tubule. The final product is urine.

Contractile Vacuole

The tiny, spherical, intracellular vacuole of unicellular eukaryotes and freshwater sponges is not a true excretory organ, since ammonia and other nitrogenous wastes of metabolism readily diffuse into the surrounding water. The contractile vacuole of freshwater unicellular eukaryotes is an organ of water balance that expels excess water gained by osmosis. As water enters the cell, a vacuole grows and finally contracts and empties its contents through a pore on the surface. The cycle is repeated rhythmically. Although the mechanism for filling the vacuole is not fully understood, recent research suggests that contractile vacuoles have numerous proton pumps located within their membrane (proton pumps are described in connection with the electron transport chain in Chapter 4, p. 64 and following). Proton pumps create H^+ and HCO^- gradients that draw water into the vacuole, forming an isoosmotic solution. These ions are excreted when the vacuole empties (see Figure 11.15, p. 226).

Contractile vacuoles are common in freshwater unicellular eukaryotes, sponges, and radiate animals (such as hydra), but rare or absent in marine forms of these groups, which are isoosmotic with seawater and consequently neither lose nor gain too much water.

Nephridium

The most common type of invertebrate excretory organ is the nephridium, a tubular structure designed to maintain appropriate osmotic balance. One of the simplest arrangements is the flame cell system (or **protonephridium**) of acoelomates (for example, flatworms) and some pseudocoelomates (for example, rotifers).

In planaria and other flatworms the protonephridial system comprises two highly branched duct systems distributed throughout the body, each branch terminating in a "flame cell" (Figure 30.5). Fluid enters the system through the flame cells, where rhythmical beating of the flagellar tuft, suggestive of a tiny flickering flame, creates a negative pressure that draws fluid into the tubular portion of the system. As the fluid moves down the tubule, water and metabolites valuable to the body are recovered by reabsorption, leaving wastes behind to be expelled through excretory pores that open at intervals on the body surface. Nitrogenous wastes (mainly ammonia) diffuse across the entire body surface.

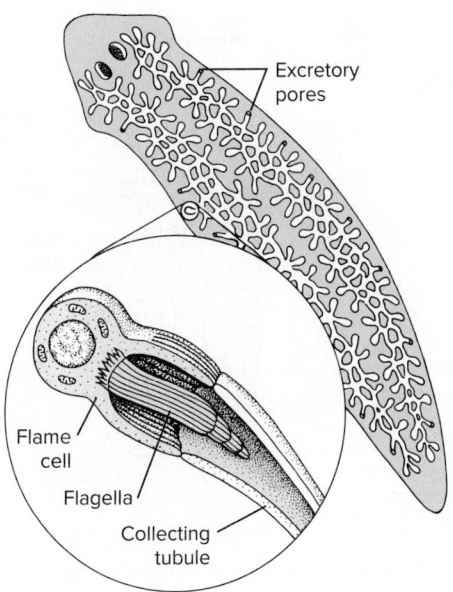

Figure 30.5   Flame-cell system of a flatworm. Body fluids collected by flame cells (protonephridia) are passed down a system of ducts to excretory pores on the body surface.

The flame-cell system is extensively branched throughout a flatworm's body. Thus, these animals have no necessity for a circulatory system to deliver wastes to a centralized excretory system (such as the kidneys of vertebrates and many invertebrates).

The protonephridium just described is a **closed** system. The tubules are closed on the inner end and urine is formed from a fluid that must first enter the tubules by being transported across flame cells. A more advanced type of nephridium is the **open,** or "true," nephridium (**metanephridium**) found in several eucoelomate phyla such as annelids (Figure 30.6), molluscs, and several smaller phyla. A metanephridium differs from a protonephridium in two important ways. First, the tubule is open at *both* ends, allowing fluid to be swept into the tubule through a ciliated funnel-like opening, the **nephrostome.** Second, a metanephridium is surrounded by a network of blood vessels that assists in reabsorption from the tubular fluid of water and valuable materials such as salts, sugars, and amino acids.

Despite these differences, the basic process of urine formation is the same in protonephridia and metanephridia: fluid enters and flows continuously through a tubule where the fluid is selectively modified by (1) reabsorbing valuable solutes from it and returning these to the body, and (2) adding waste solutes to it (secretion). The sequence ensures removal of wastes from the body without loss of useful materials. Kidneys of vertebrates operate in basically the same way (pp. 688–672).

Arthropod Kidneys

The paired **antennal glands** of crustaceans, located in the ventral part of the head (Figure 30.7), are an elaboration of the basic nephridial organ. However, they lack open nephrostomes. Instead, blood hydrostatic pressure forms a protein-free filtrate of the blood

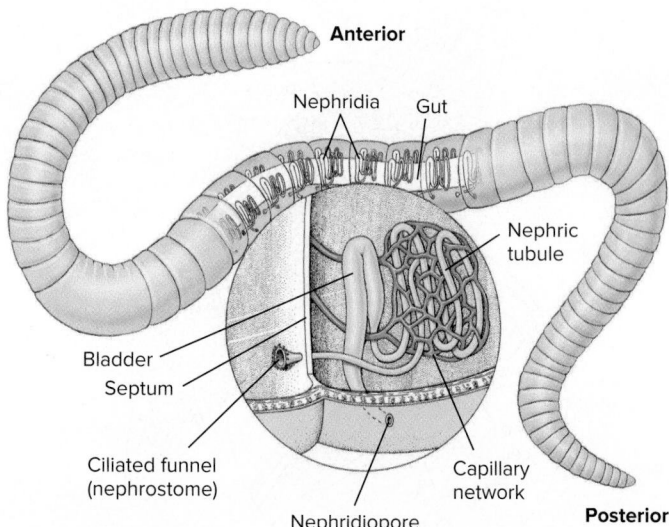

Figure 30.6 Excretory system of an earthworm. Each segment has a pair of large nephridia suspended in a fluid-filled coelom. Each nephridium occupies two segments because the ciliated funnel (nephrostome) drains the segment anterior to the segment containing the rest of the nephridium. The nephric tubule reabsorbs valuable materials from the tubular fluid into the capillary network.

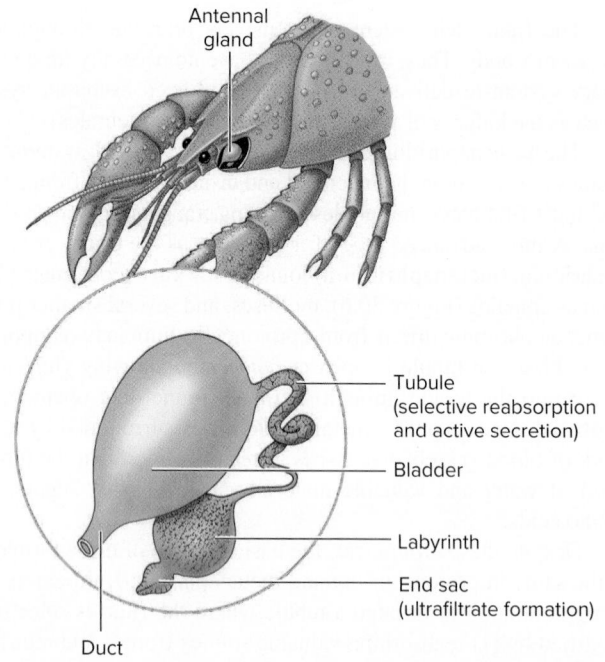

Figure 30.7 Crustacean antennal glands shown in a crayfish. These are filtration kidneys in which a filtrate of the blood is formed in the end sac. The ultrafiltrate is converted into urine as it passes down the tubule toward the bladder.

Insects and spiders have a unique excretory system consisting of **Malpighian tubules** that operate in conjunction with specialized glands in the wall of the rectum (Figure 30.8). These thin, elastic, blind Malpighian tubules are closed and lack an arterial supply. Urine formation does not occur by filtration of body fluids as is the case in the nephridium, but is produced by tubular secretion mechanisms by the cells lining the Malpighian tubules that are bathed in hemolymph (blood). This process is initiated by active transport (p. 47) of hydrogen ions into the Malpighian tubular lumen. These hydrogen ions are then moved back into the cells lining the tubule using protein transporters (p. 46) that move hydrogen ions in exchange for sodium or potassium ions, and chloride ions follow passively. Herbivorous and omnivorous insects secrete primarily potassium into the tubular lumen. Carnivorous insects, such as blood-sucking mosquitos, initially secrete a fluid high in sodium that reflects the high salt content of a blood plasma meal (see Chapter 31, p. 682). As the red blood cells are digested, the fluid content of sodium falls and becomes rich in potassium ions. The secretion of ions creates an osmotic pressure that draws water, solutes, and nitrogenous wastes, especially uric acid, into the tubule. Uric acid enters the blind-ending distal segment of the tubule as soluble potassium urate, which precipitates as insoluble uric acid in the proximal end of the tubule. Once the formative urine drains into the rectum, water and salts may be reabsorbed by specialized rectal glands, leaving behind uric acid, excess water, salts, and other wastes, which are expelled in feces. Rectal glands of aquatic insect larval stages absorb

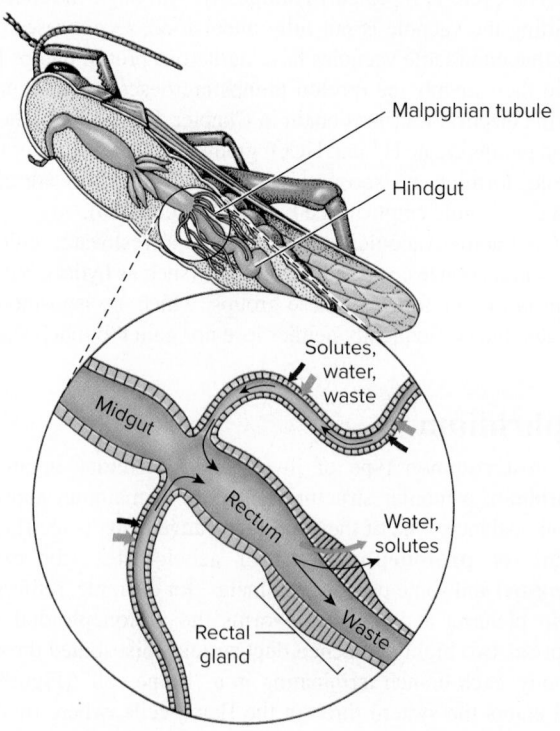

Figure 30.8 Malpighian tubules of insects. Malpighian tubules are located at the juncture of the midgut and the hindgut (rectum). Solutes, especially potassium, are actively secreted into the tubules from the surrounding hemolymph (blood). Water, uric acid, and other wastes follow. This fluid drains into the rectum, where solutes (including potassium) and water are actively reabsorbed, leaving wastes to be excreted.

(ultrafiltrate) in the end sac. In the tubular portion of the gland, selective reabsorption of certain salts and active secretion of others modifies the filtrate as it moves toward the bladder. Thus crustaceans have excretory organs that are basically vertebrate-like in the functional sequence of urine formation.

solute, but little water, whereas blood-sucking insects can alter the amount of salt and water reabsorption during and between meals. The feces of bloodsucking insects will be high in water and excess salts during and immediately after a blood meal, but low in salts and water between meals. The Malpighian tubule excretory system is ideally suited for life in dry environments and has contributed to the adaptive diversification of insects on land.

30.3 VERTEBRATE KIDNEY

Ancestry and Embryology

From comparative studies of development, biologists infer that the kidney of the earliest vertebrates extended the length of the coelomic cavity and was composed of segmentally arranged tubules, each resembling an invertebrate nephridium. Each tubule opened at one end into the coelom by a nephrostome and at the other end into a common **archinephric duct.** This ancestral kidney is called an **archinephros** ("ancient kidney"), and we find a segmented kidney very similar to an archinephros in embryos of hagfishes and caecilians (Figure 30.9). Almost from the beginning, the reproductive system, which develops beside the excretory system from the same segmental blocks of trunk mesoderm, used the nephric ducts as a convenient conducting system for reproductive products. Thus even though the two systems have nothing functionally in common, they are closely associated in their use of common ducts (Figure 30.9).

Kidneys of living vertebrates developed from this ancient plan. During embryonic development of amniote vertebrates, kidneys develop in three successive stages: **pronephros, mesonephros,** and **metanephros** (Figure 30.9). Some, but not all, of these stages are observed in other vertebrate groups. In all vertebrate embryos, the pronephros is the first kidney to appear. It is located anteriorly in the body and becomes part of the persistent kidney only in adult hagfishes and a few bony fish species. In all other vertebrates the pronephros degenerates during development and is replaced by a more centrally located mesonephros. The mesonephros is the functional kidney of embryonic amniotes (nonavian reptiles, birds, and mammals). The mesonephros and metanephros, together called the **opisthonephros** (Figure 30.9), form the adult kidney of most fishes and amphibians.

The metanephros, characteristic of adult amniotes, is distinguished in several ways from the pronephros and mesonephros. It is more caudally located and it is a much larger, more compact

Figure 30.9 Comparative development of male vertebrate kidney. *Red,* functional structures. *Light red,* degenerative or undeveloped parts.

structure containing a very large number of nephric tubules. It is drained by a new duct, the **ureter,** which developed when the old archinephric duct was relinquished to the reproductive system of the male for sperm transport. Thus three successive kidney types—pronephros, mesonephros, metanephros—succeed each other embryologically, and to some extent phylogenetically, in amniotes.

Vertebrate Kidney Function

The vertebrate kidney is part of many interconnected mechanisms that maintain cellular homeostasis, and it is the principal organ that regulates volume and composition of the internal fluid environment. While we commonly describe the vertebrate kidney as an organ of excretion, the removal of metabolic wastes is incidental to its osmoregulatory function.

The organization of kidneys differs somewhat in different groups of vertebrates, but in all the basic functional unit is the **nephron,** and urine is formed by three well-defined physiological processes: **filtration, reabsorption,** and **secretion.** This discussion focuses mainly on the mammalian kidney, which is the most completely understood osmoregulatory organ.

The two human kidneys are small organs comprising less than 1% of the body weight, yet they receive 20% to 25% of the total cardiac output—1440 to 1800 liters of blood each day. This vast blood flow is channeled to approximately 2 million nephrons, which form the bulk of the two kidneys. Each nephron begins with an expanded chamber, the **Bowman's capsule,** containing a ball of intertwined capillaries, the **glomerulus** (glo-mer'yoo-lus), which together form

the **renal corpuscle.** Blood pressure in the glomerular capillaries forces an almost protein-free **filtrate** into a Bowman's capsule and along a **renal tubule,** consisting of several segments that perform the functions of reabsorption and secretion in the process of urine formation. The filtrate passes first into a **proximal convoluted tubule,** then into a long, thin-walled **loop of Henle,** which may extend deep into the inner portion of the kidney (the **medulla**) before returning to the outer portion (the **cortex**) where it becomes a **distal convoluted tubule.** From the distal tubule the fluid empties into a **collecting duct,** which drains into the **renal pelvis.** Here the urine collects before being carried by the **ureter** to the **urinary bladder.** These anatomical relationships are shown in Figure 30.10.

Urine that leaves the collecting duct is very different from the filtrate produced in the renal corpuscle. During its travels through the renal tubule and collecting duct, both the composition and the concentration of the original filtrate change. Some solutes such as all glucose and amino acids, and most sodium, have been reabsorbed while others, such as hydrogen ions and urea, have been concentrated in urine.

The nephron, with its pressure filter and tubule, is intimately associated with blood circulation (Figures 30.10 and 30.11). Blood from the aorta enters each kidney through a large **renal artery,** which divides into a branching system of smaller arteries. The arterial blood reaches the glomerulus through an **afferent arteriole** and leaves by an **efferent arteriole.** From the efferent arteriole the blood travels to an extensive capillary network that surrounds and supplies the proximal and distal convoluted tubules and the loop of Henle. This capillary network provides a means for the pickup and delivery of materials that are reabsorbed or secreted by the kidney tubules.

Figure 30.10 Urinary system of humans, with enlargements showing detail of the kidney and a single nephron.

Figure 30.11 Scanning electron micrograph of a cast of the microcirculation of a mammalian kidney, showing several glomeruli (pink) and associated, larger blood vessels (gold). The Bowman's capsule, which normally surrounds each glomerulus, has been removed in preparing the cast.

From these capillaries blood collects in veins that unite to form the **renal vein.** This vein returns blood to the posterior vena cava.

Glomerular Filtration

The process of urine formation begins in the glomerulus. The glomerulus acts as a specialized mechanical filter, which produces an almost protein-free filtrate of the plasma in the fluid-filled space of the Bowman's capsule as a result of high blood pressure across glomerular capillary walls. The diameter of the afferent arteriole entering the glomerulus is greater than that of the exiting efferent arteriole, providing the high hydrostatic pressure that allows formation of the glomerular filtrate. Solute molecules small enough to pass through the filtration slits of the capillary wall are carried through by the filtrate in which they are dissolved. Red blood cells and almost all plasma proteins, however, are withheld because they are too large to pass through these pores (Figure 30.12).

The filtrate continues through the renal tubular system, where it undergoes extensive modification before becoming urine. Human kidneys form approximately 180 liters (nearly 50 gallons) of filtrate each day, a volume many times exceeding total blood volume. If this volume of water and the valuable nutrients and salts it contains were lost, the body would soon be depleted of these compounds. Depletion does not happen because nearly all of the filtrate is reabsorbed. The final urine volume in humans averages 1.2 liters per day.

Conversion of filtrate into urine involves two processes: (1) modification of the composition of the filtrate through tubular reabsorption and secretion, and (2) changes in the total osmotic concentration of urine through regulation of water excretion.

Tubular Reabsorption

Approximately 60% of the filtrate volume and virtually all of the glucose, amino acids, vitamins, and other valuable nutrients are reabsorbed in the proximal convoluted tubule. Much of this reabsorption is by **active transport,** in which cellular energy is used to

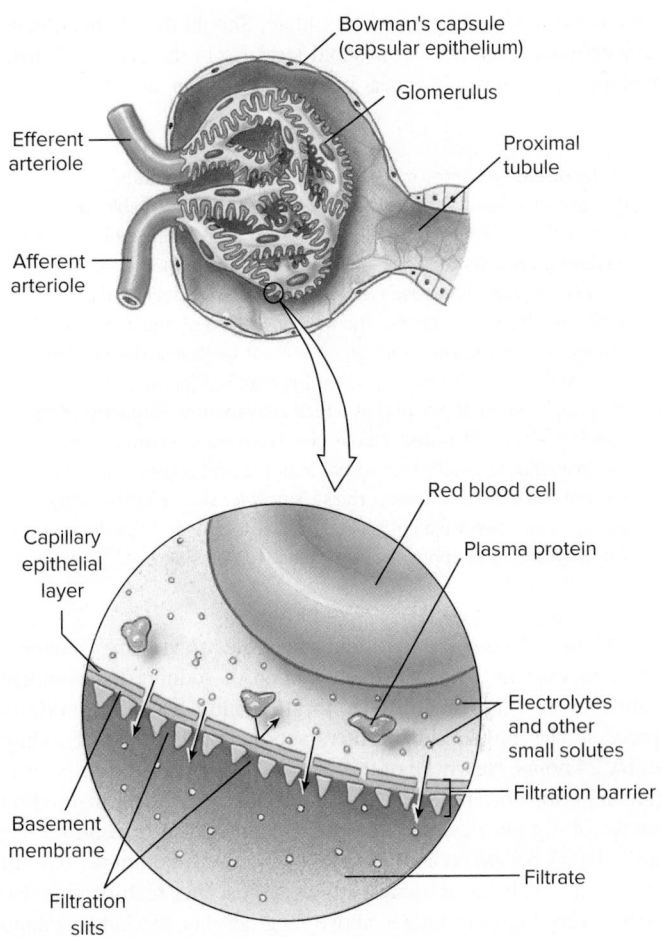

Figure 30.12 Renal corpuscle, showing (*enlargement*) the filtration of fluid through the glomerular capillary membrane. Water, electrolytes, and other small molecules pass the porous filtration barrier, but almost all plasma proteins are too large to pass the barrier. The filtrate is thus almost protein-free.

transport materials from tubular fluid to the surrounding capillary network and thus into the blood circulation. Electrolytes such as sodium, potassium, calcium, bicarbonate, and phosphate are reabsorbed by ion pumps, which are carrier proteins driven by the hydrolysis of ATP (ion pumps are described on p. 48). Because an essential function of the kidney is to regulate plasma concentrations of electrolytes, all are individually reabsorbed by ion pumps specific for each electrolyte. The degree of reabsorption depends on the body's ability to conserve each mineral. Some materials are passively reabsorbed. Negatively charged chloride ions, for example, passively flow by diffusion through protein channels (p. 45) specific for chloride ions, drawn by the active reabsorption of positively charged sodium ions in the proximal convoluted tubule. Water, too, passively diffuses from the tubule through open protein channels, as it osmotically follows the active reabsorption of solutes.

For most substances there is an upper limit to the amount of substance that can be reabsorbed. This upper limit is termed the **transport maximum** (renal threshold) for that substance. For example, glucose normally is reabsorbed completely by the kidney because the transport maximum for glucose is well above the amount of glucose usually present in the glomerular filtrate. Should the plasma glucose concentration exceed this threshold level, as in the disease diabetes mellitus, glucose would appear in the urine (Figure 30.13).

In the disease diabetes mellitus ("sweet running through"), glucose rises to abnormally high concentrations in the blood plasma (hyperglycemia) because the hormone insulin, which enables body cells to acquire glucose, is deficient (see Chapter 34, p. 760). As blood glucose rises above a normal level of about 100 mg/100 ml of plasma, the concentration of glucose in the filtrate also rises, and more glucose must be reabsorbed by the proximal tubule. Eventually a point is reached (about 300 mg/100 ml of plasma) at which reabsorptive capacity of the tubular cells is saturated. Above this transport maximum for glucose, glucose spills into urine. In untreated diabetes mellitus the patient's urine is sweet, thirst is unrelenting, and the body wastes away despite a large food intake. In England the disease for centuries was appropriately called the "pissing evil."

Unlike glucose or amino acids, most electrolytes are excreted in urine in variable amounts. Reabsorption of sodium, the dominant cation in the plasma, illustrates the flexibility of the reabsorption process. The human kidney filters approximately 600 g of sodium every 24 hours. Nearly all of this sodium is reabsorbed, but the exact amount is matched precisely to sodium intake. With a normal sodium intake of 4 g per day, the kidney excretes 4 g and reabsorbs 596 g each day. A person on a low-salt diet of 0.3 g of sodium per day still maintains salt balance because only 0.3 g escapes reabsorption. But with a very high salt intake, above 20 g per day, the kidney cannot excrete sodium as fast as it enters. The unexcreted sodium chloride causes retention of additional water in the body fluids, and the person gains weight. (The salt intake of an average North American is about 6 to 18 g per day, approximately 20 times more than the body needs, and three times more than is considered acceptable for those predisposed to high blood pressure.)

Distal convoluted tubules make additional adjustments of filtrate composition. Sodium reabsorbed by the proximal convoluted tubule—some 85% of the total filtered—is obligatory reabsorption; this amount is reabsorbed independent of sodium intake. In the distal convoluted tubule, however, sodium reabsorption is controlled by **aldosterone,** a steroid hormone secreted by the adrenal gland (p. 759). Aldosterone

A human kidney can adapt to excrete large quantities of salt (sodium chloride) under conditions of high salt intake. In societies accustomed to widespread use of foods heavily salted for preservation (for example, salted pork and salt herring) daily intakes may approach or even exceed 100 g. Body weight remains normal under such conditions. However, the acute ingestion of 20 to 40 g/day by volunteers unadapted to such large intakes of salt caused swelling of tissues, increase in body weight, and an increase in blood pressure.

increases both active reabsorption of sodium and secretion of potassium by the distal tubules; the hormone thus decreases loss of sodium and increases loss of potassium in the urine. The secretion of aldosterone is regulated (1) by the enzyme **renin,** produced by the **juxtaglomerular apparatus,** a complex of cells located at the junction of the afferent arteriole with the glomerulus (see Figure 30.10), and (2) by elevated blood potassium levels. Renin is released in response to a low blood sodium level, to low blood pressure (which can occur if the blood volume drops too low), or to low sodium in the glomerular filtrate. Renin then initiates a series of enzymatic events culminating in the production of **angiotensin,** a blood-borne hormone that has several related effects. First, it stimulates the release of aldosterone, which acts in turn to increase sodium reabsorption and potassium secretion by the distal tubule. Second, it increases the secretion of **antidiuretic hormone** (vasopressin, discussed on p. 671), which promotes water conservation by the kidney. Third, it increases blood pressure. Finally, it stimulates thirst, which also is stimulated by decreased blood volume or increased blood osmolarity. These actions of angiotensin tend to reverse the circumstances (low blood sodium and low blood pressure and/or blood volume) that triggered the secretion of renin. Sodium and water are conserved, and blood volume and blood pressure are restored to normal.

The flexibility of distal reabsorption of sodium varies considerably in different animals: it is restricted in humans but very broad in many rodents. These differences have appeared because selective pressures during evolution have adapted rodents for dry environments. They must conserve water and at the same time excrete considerable sodium. Humans, however, have not evolved to accommodate the large salt appetites many have. Our closest relatives, the great apes, are vegetarians with an average salt intake of less than 0.5 g per day.

Glomerulus

Glucose

Urine   Blood   Urine   Blood   Urine   Blood

A Low glucose level   **B** Maximum transport capacity   **C** High glucose level

Figure 30.13 The mechanism for the tubular reabsorption of glucose can be likened to a conveyor belt running at constant speed. **A,** When glucose concentrations in the glomerular filtrate are low, all is reabsorbed. **B,** When glucose concentrations in the filtrate have reached the transport maximum, all carrier sites for glucose are occupied. If glucose levels rise further, **C,** as in the disease diabetes mellitus, some glucose escapes the carriers and appears in urine.

Tubular Secretion

In addition to reabsorbing materials from glomerular filtrate, the nephron can secrete materials across the tubular epithelium and *into* the filtrate. In this process, the reverse of tubular reabsorption, carrier proteins in the tubular epithelial cells selectively transport substances from blood in capillaries outside the tubule to the filtrate inside the tubule. Tubular secretion enables a kidney to increase the urine concentrations of materials to be excreted, such as hydrogen and potassium ions, drugs, and various foreign organic materials. The tubular epithelium is able to recognize foreign organic materials, such as ingested pharmaceutical drugs, because they are metabolized by the liver to form cationic or anionic molecules. These molecules are transported by the tubular epithelium, which has both cationic and anionic transporters in its membrane. The distal convoluted tubule is the site of most tubular secretion.

In the kidneys of bony marine fishes, nonavian reptiles, and birds, tubular secretion is a different process than it is in mammalian kidneys. Marine bony fishes actively secrete large amounts of magnesium and sulfate, seawater salts that are by-products of their mode of osmoregulation (see p. 662). Nonavian reptiles and birds excrete uric acid instead of urea as their major nitrogenous waste (see Figure 4.19, p. 68). The material is actively secreted by the tubular epithelium. Since uric acid is nearly insoluble, it forms crystals in the urine and requires little water for excretion. Thus excretion of uric acid is an important adaptation for water conservation.

Water Excretion

Kidneys closely regulate the osmotic pressure of blood. When fluid intake is high, the kidney excretes dilute urine, saving salts and excreting water. When fluid intake is low, kidneys conserve water by forming concentrated urine. A dehydrated person can concentrate urine to approximately four times blood osmotic concentration. This important ability to concentrate urine enables us to excrete wastes with minimal water loss.

The capacity of mammalian and some avian kidneys to produce a concentrated urine involves an interaction between the loop of Henle and the collecting ducts. This interplay produces an osmotic gradient in the kidney, as shown in Figure 30.14. In the cortex, interstitial fluid is isoosmotic with blood, but deep in the medulla the osmotic concentration is four times greater than that of blood (in rodents and desert mammals that can produce highly concentrated urine the osmotic gradient is much greater than in humans). The high osmotic concentrations in the medulla are produced by an exchange of ions in the loop of Henle by **countercurrent multiplication.** "Countercurrent" refers to the opposite directions of fluid movement in the two limbs of the loop of Henle: down in the descending limb and up in the ascending limb. "Multiplication" describes the increasing osmotic concentration in the medulla surrounding the loops of Henle and the collecting ducts resulting from ion exchange between the two limbs of the loop.

The functional characteristics of this system are as follows. The descending limb of the loop of Henle is permeable to water but impermeable to solutes. The ascending limb is nearly impermeable to water. Sodium chloride passively moves from the lower ascending limb and is actively transported from the thick portion of the ascending limb into the surrounding tissue fluid (Figure 30.14). As the

interstitium surrounding the loop becomes more concentrated with solute, water is withdrawn from the descending limb by osmosis. The tubular fluid in the base of the loop, now more concentrated, moves up the ascending limb, where still more sodium chloride diffuses or is pumped out. In this way the effect of ion movement in the ascending limb is multiplied as more water is withdrawn from the descending limb and more concentrated fluid is presented to the ascending limb (Figures 30.14 and Figure 30.15). The blood capillaries surrounding the loops of Henle, the **vasa recta,** are also arranged in a countercurrent fashion. Thus salt that diffuses into the blood of the vasa recta close to the ascending limb of the loop of Henle will not leave the medulla, but will diffuse from here into the blood entering the medulla in the vasa recta, so that very little salt is lost from this region. This arrangement of blood vessels is important in maintaining the osmotic concentration gradient of the medulla and cortex.

Final adjustment of urine concentration occurs not in the loops of Henle but in the collecting ducts. Formative urine that enters the distal tubule from the loop of Henle is dilute (because of salt withdrawal) and may be diluted still more by active reabsorption of more sodium chloride in the distal tubule. The formative urine, low in solutes but carrying urea, now flows down into the collecting duct. Because of the high concentration of solutes in the interstitial fluid surrounding the collecting duct, water is withdrawn from the urine. As the urine becomes more concentrated, urea also diffuses out of it. The lower regions of the collecting duct are permeable to urea, and thus far, four different urea transporters have been discovered. Some of this urea flows back into the lower portion of the ascending limb of the loop of Henle, but since the loop of Henle is less permeable to urea, its concentrations build within the medulla tissue fluid. This buildup of urea contributes significantly to the high osmotic concentration of the medulla and to the countercurrent multiplication mechanism (Figure 30.15).

The amount of water reabsorbed and the final concentration of the urine depend on permeability of the walls of the distal convoluted tubule and the collecting duct. This is controlled by **antidiuretic hormone** (ADH, or vasopressin), which is released by the posterior pituitary gland (neurohypophysis, see p. 751). Special receptors in the brain that constantly sense the osmotic pressure of body fluids govern the release of this hormone. When the blood osmotic pressure increases or blood volume decreases, as during dehydration, the pituitary gland releases more ADH. ADH increases permeability of the collecting duct by increasing the number of water channels in the collecting duct epithelial cells. Then, as the fluid in the collecting duct passes through the hyperosmotic region of the kidney medulla, water diffuses through the channels into the surrounding interstitial fluid and is carried away by the blood circulation. Thus, urine loses water and becomes more concentrated. Given this sequence of events for dehydration, it is not difficult to anticipate how the system responds to overhydration: the pituitary stops releasing ADH, the water channels in the collecting duct epithelial cells decrease in number, and a large volume of dilute urine is excreted.

The varying ability of different mammals to form a concentrated urine correlates closely with length of the loops of Henle. Beavers, which have no need to conserve water in their aquatic environment, have short loops and can concentrate their urine only to about twice the osmolarity of blood. Humans, with relatively longer loops, can concentrate urine 4.2 times that of blood. As we would anticipate, desert mammals have much greater urine-concentrating

Figure 30.14 Mechanism of urine concentration in mammals. Sodium and chloride diffuse or are pumped from the ascending limb of the loop of Henle, and water is withdrawn passively from the descending limb, which is impermeable to sodium chloride. Sodium chloride reabsorbed from the ascending limb of the loop of Henle and urea reabsorbed from the collecting duct raise the osmotic concentration in the kidney medulla, creating an osmotic gradient for controlled reabsorption of water from the collecting duct.

KEY:
- Water diffusion
- Variable water movement
- NaCl diffusion
- NaCl pump
- Na and K exchange
- Urea reabsorption

Within figure:

mosm

Cortex

Medulla

Tissue fluid osmolarity

Bowman's capsule

Proximal tubule

Glomerulus

Distal tubule: Permeability to sodium and potassium controlled by aldosterone

Collecting duct: Permeability to water controlled by ADH

Water permeable

Descending limb of loop of Henle

Impermeable to sodium chloride

Water impermeable

Low permeability to urea

Permeable to urea

Loop of Henle

NaCl, H_2O, Na/K, Urea

300, 300, 300, 500, 750, 1000, 1200, 100, 400, 600, 1200

powers. A camel can produce a urine 8 times the plasma concentration, a gerbil 14 times, and an Australian hopping mouse 22 times. In this creature, the greatest urine concentrator of all, the long loops of Henle extend to the tip of a long renal papilla that pushes out into the mouth of the ureter.

30.4 TEMPERATURE REGULATION

We have seen that a fundamental problem facing an animal is keeping its internal environment in a state that permits normal cell function. Biochemical activities are sensitive to the chemical environment and our discussion thus far has examined how the chemical environment is stabilized. Biochemical reactions are also extremely sensitive to temperature. All enzymes have an optimum temperature; at temperatures above or below this optimum, enzyme function is impaired. Temperature therefore is a severe constraint for animals, all of which must maintain biochemical stability. When body

temperature drops too low, metabolic processes slow, reducing the amount of energy the animal can muster for activity and reproduction. If body temperature rises too high, metabolic reactions become unbalanced and enzymatic activity is hampered or even destroyed. Thus animals can succeed only in a restricted range of temperature, usually between 0° and 40°C. Animals either must find a habitat where they do not have to contend with temperature extremes, or must develop means of stabilizing their metabolism independent of temperature extremes.

Ectothermy and Endothermy

The terms "cold-blooded" and "warm-blooded" have long been used to divide animals into two groups: invertebrates and vertebrates that feel cold to the touch, and those, such as humans, other mammals, and birds, that do not. It is true that body temperature of mammals and birds is usually (though not always) warmer than the air temperature,

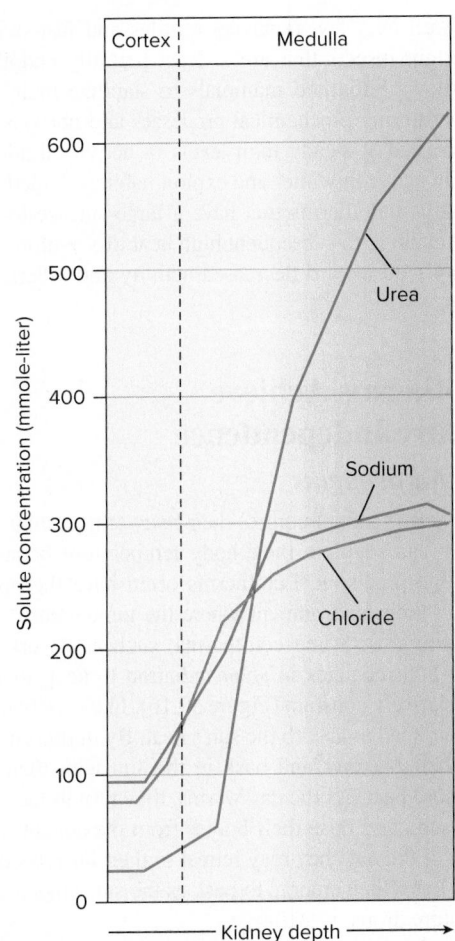

Figure 30.15 Osmotic concentration of tissue fluid in the mammalian kidney. Tissue fluid is isosmotic in the kidney cortex (*to left in diagram*) but osmotic concentration increases continuously through the medulla, reaching a maximum at the papilla where the urine drains into the ureter.

A temperature difference of 10° C has become a standard used to measure the temperature sensitivity of a biological function. This value, called the Q_{10}, is determined (for temperature intervals of exactly 10° C) simply by dividing the value of a rate function (such as metabolic rate or rate of an enzymatic reaction) at the higher temperature by the value of the rate function at the lower temperature. In general, metabolic reactions have Q_{10} values of about 2.0 to 3.0. Purely physical processes, such as diffusion, have much lower Q_{10} values, usually close to 1.0.

but a "cold-blooded" animal is not necessarily cold. Tropical fishes, insects, and nonavian reptiles basking in the sun may have body temperatures equaling or surpassing those of mammals. Conversely, many "warm-blooded" mammals hibernate, allowing their body temperature to approach the freezing point of water. Thus the terms "warm-blooded" and "cold-blooded" are hopelessly subjective and nonspecific but are so firmly entrenched in our vocabulary that most biologists find it easier to accept the usage than to try to change it.

Zoologists often use **poikilothermic** (body temperature that fluctuates with environmental temperature) and **homeothermic** (constant body temperature, regulated independent of environmental temperature) as alternatives to "cold-blooded" and "warm-blooded,"

respectively. These terms, which refer to variability of body temperature, are more precise and more informative, but still offer difficulties. For example, deep-sea fishes live in an environment having no perceptible temperature change. Even though their body temperature is absolutely stable, to call such fishes homeotherms would imply that they possess the ability to regulate their body temperature if environmental temperature were to change—but this would be unlikely. Furthermore, many homeothermic birds and mammals allow their body temperature to change between day and night or during seasonally induced hibernation.

Physiologists prefer yet another way to contrast thermoregulatory mechanisms, one that reflects the fact that an animal's body temperature is a balance between heat gain and heat loss. All animals produce heat from cellular metabolism, but in most the heat is conducted away as fast as it is produced. In these animals, the **ectotherms** —and the overwhelming majority of animals belong to this group—body temperature is determined solely by environmental context. Many ectotherms exploit their environment behaviorally to select areas of more favorable temperature (such as basking in the sun), but the source of energy used to increase body temperature comes from the environment, not from within the body. Alternatively, some animals can generate and retain enough metabolic heat to elevate their own body temperature to a high but stable level. Because the source of their body heat is internal, they are called **endotherms.** These favored few in the animal kingdom are the birds and mammals, as well as a few nonavian reptiles and fast-swimming fishes, and certain insects that are at least partially endothermic. Endothermy allows birds and mammals to stabilize their internal temperature, permitting biochemical processes and nervous system functions to proceed at steady high levels of activity. Endotherms may thus remain active in winter and exploit habitats denied to ectotherms. In reality, endotherms that have a large surface-to-volume ratio (small animals) with subsequent high heat loss and/or a limited food availability tend toward decreased activity and hibernation in colder climates.

How Ectotherms Achieve Temperature Independence

Behavioral Adjustments

Although ectotherms cannot control their body temperature physiologically, many can regulate their body temperature behaviorally with considerable precision. Ectotherms often have the option of seeking areas in their environment where the temperature is favorable to their activities. Some ectotherms, such as desert lizards, exploit hour-to-hour changes in solar radiation to keep their body temperatures relatively constant (Figure 30.16). In the early morning they expose their head to absorb the sun's heat. By midmorning they emerge from their burrows and bask in the sun with their bodies flattened to absorb heat. As the day warms, they turn to face the sun to reduce exposure, and raise their bodies from the hot substrate. In the hottest part of the day they may retreat to their burrows or move into the shade. Later they emerge to bask as the sun sinks lower and the air temperature drops.

These behavioral patterns help to maintain a relatively steady body temperature of 36° to 39° C while the air temperature varies between 29° and 44° C. Some lizards can tolerate intense midday heat without shelter. The desert iguana of the southwestern United States prefers a body temperature of 42° C when active and can tolerate a rise to 47° C, a temperature that is lethal to all birds and mammals and most other lizards. The term "cold-blooded" clearly does not apply to these animals!

Metabolic Adjustments

Even without the behavioral adjustments just described, most ectotherms can adjust their metabolic rates to the prevailing temperature such that the intensity of metabolism remains mostly unchanged. This is called **temperature compensation** and involves complex biochemical and cellular adjustments. These adjustments enable a fish or a salamander, for example, to benefit from almost the same level of activity in both warm and cold environments. Thus, whereas endotherms achieve metabolic homeostasis by maintaining their body temperature independent of environmental temperature, ectotherms accomplish much the same by directly maintaining their metabolism independent of body temperature. Such metabolic regulation also is a form of homeostasis.

Temperature Regulation in Endotherms

Most mammals have body temperatures between 36° and 38° C, somewhat lower than those of birds, which range between 40° and 42° C. Constant temperature is maintained by a delicate balance between heat production and heat loss—not a simple matter when these animals are alternating between periods of rest and bursts of heat-producing activity.

Heat is produced by the animal's metabolism. This includes oxidation of foods, basal cellular metabolism, and muscular contraction.

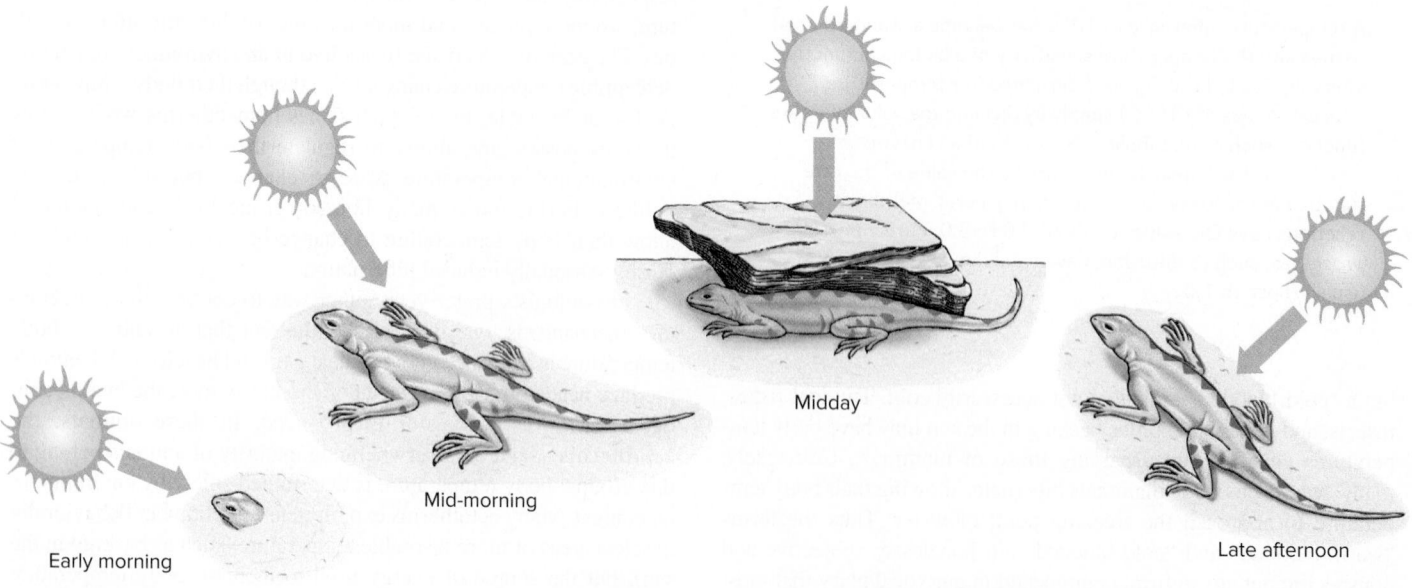

Midday

Mid-morning

Early morning

Late afternoon

Figure 30.16 How a lizard regulates its body temperature behaviorally. In the morning the lizard absorbs the sun's heat through its head while keeping the rest of its body protected from cool morning air. Later it will emerge to bask. At noon, with its body temperature high, it seeks shade from the hot sun. When the air temperature drops in the late afternoon, it emerges to absorb the final rays of the sun.

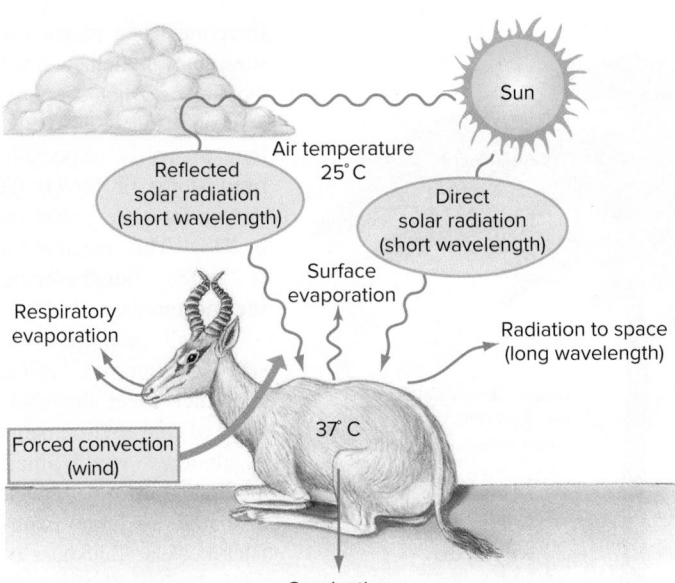

Figure 30.17 Exchange of heat between a large mammal and its environment on a warm day. Red arrows indicate sources of net heat gain by the animal (all radiation); blue arrows are avenues of net heat loss (evaporative cooling, conduction to the ground, long-wave radiation into space, and forced convection by the wind). If air and ground temperatures were warmer than the animal, the arrows for forced convection, conduction, and radiation would be reversed. Then the animal could lose heat only by evaporative cooling.

Because much of an endotherm's daily caloric intake is required to generate heat, especially in cold weather, the endotherm must eat more food than an ectotherm of the same size. Heat is lost by radiation, conduction, and convection (air movement) to a cooler environment and by evaporation of water (Figure 30.17). A bird or mammal can control both processes of heat production and heat loss within rather wide limits. If the animal becomes too cool, it can generate heat by increasing muscular activity (exercise or shivering) and can decrease heat loss by increasing its insulation. If it becomes too warm, it decreases heat production and increases heat loss. We examine these processes with examples.

Adaptations for Hot Environments

Despite the harsh conditions of deserts—intense heat during the day, cold at night, and scarcity of water, vegetation, and cover—many kinds of animals live there successfully. The smaller desert mammals are mostly **fossorial** (living mainly in the ground) and/or **nocturnal** (active at night). The lower temperature and higher humidity of burrows help to reduce water loss by evaporation. As explained on page 664, desert animals such as kangaroo rats and American desert ground squirrels can, if necessary, derive the water they need from metabolism of their dry food, drinking no water at all. Such animals produce a highly concentrated urine and form almost completely dry feces.

Large desert ungulates (hooved mammals that regurgitate and chew their partially digested food) obviously cannot escape desert heat by living in burrows. Animals such as camels and desert antelopes (gazelle, oryx, and eland) possess a number of adaptations for coping with heat and dehydration. Figure 30.18 shows those of the

eland. Mechanisms for controlling water loss and preventing overheating are closely linked. The glossy, pallid color of fur reflects direct sunlight, and fur itself is an excellent insulation that resists heat. Heat is lost by convection and conduction from the underside of elands where the fur is very thin. Fat tissue, an essential food reserve, is concentrated in a single hump on the back, instead of being uniformly distributed under the skin where it would impair loss of heat by radiation. Elands avoid evaporative water loss—the only means an animal has for cooling itself when the environmental temperature is higher than that of the body—by permitting their body temperature to drop during the cool night and then to rise slowly during the day as the body stores heat. Only when the body temperature reaches 41° C must elands prevent further rise through **evaporative cooling** by sweating and panting. Respiratory moisture is condensed and reabsorbed in nasal passages as air is breathed out. They conserve water by producing a concentrated urine and dry feces. Camels have all of these adaptations developed to a similar or even greater degree; they are perhaps the most perfectly adapted of all large desert mammals.

Adaptations for Cold Environments

In cold environments mammals and birds use two major mechanisms to maintain homeothermy: (1) **decreased conductance,** reduction of heat loss by increasing the effectiveness of the insulation, and (2) **increased heat production.**

In all mammals living in the cold regions of the earth, fur thickness increases in winter, sometimes by as much as 50%. Thick underhair is the principal insulating layer, whereas the longer and more visible guard hair serves as protection against wear and for protective coloration. In birds, down feathers similarly function to conserve heat (see Chapter 27, p. 589). Unlike the well-insulated trunk of the body, the body extremities (legs, tail, ears, nose) of arctic birds and mammals are thinly insulated and exposed to rapid cooling. To prevent these parts from becoming major avenues of heat loss, they are allowed to cool to low temperatures, often approaching freezing point. The heat in the warm arterial blood is not lost from these regions, however. Instead, a **countercurrent heat exchange** between the outgoing warm blood and the returning cold blood prevents heat loss. Arterial blood in the leg of an arctic mammal or bird passes in close contact with a network of small veins. Because arterial blood flow is opposite to that of returning venous blood, heat is exchanged very efficiently from artery to veins. When arterial blood reaches the foot it has transferred nearly all of its heat to the veins returning blood to the body core (Figure 30.19). Thus little heat is lost to the surrounding cold air from poorly insulated distal regions of the leg. Countercurrent heat exchangers in appendages also are common in aquatic mammals such as seals and whales, which have thinly insulated flippers that would be avenues of excessive heat loss in the absence of this heat-salvaging arrangement.

A consequence of peripheral heat exchange is that legs and feet of mammals and birds living in cold environments must function at low temperatures. Temperatures of the feet of arctic foxes and barren-ground caribou are just above freezing point; in fact, the temperature may be below 0°C in footpads and hooves. To keep feet supple and flexible at such low temperatures, fats in the extremities have very low melting points, perhaps 30°C lower than ordinary body fats.

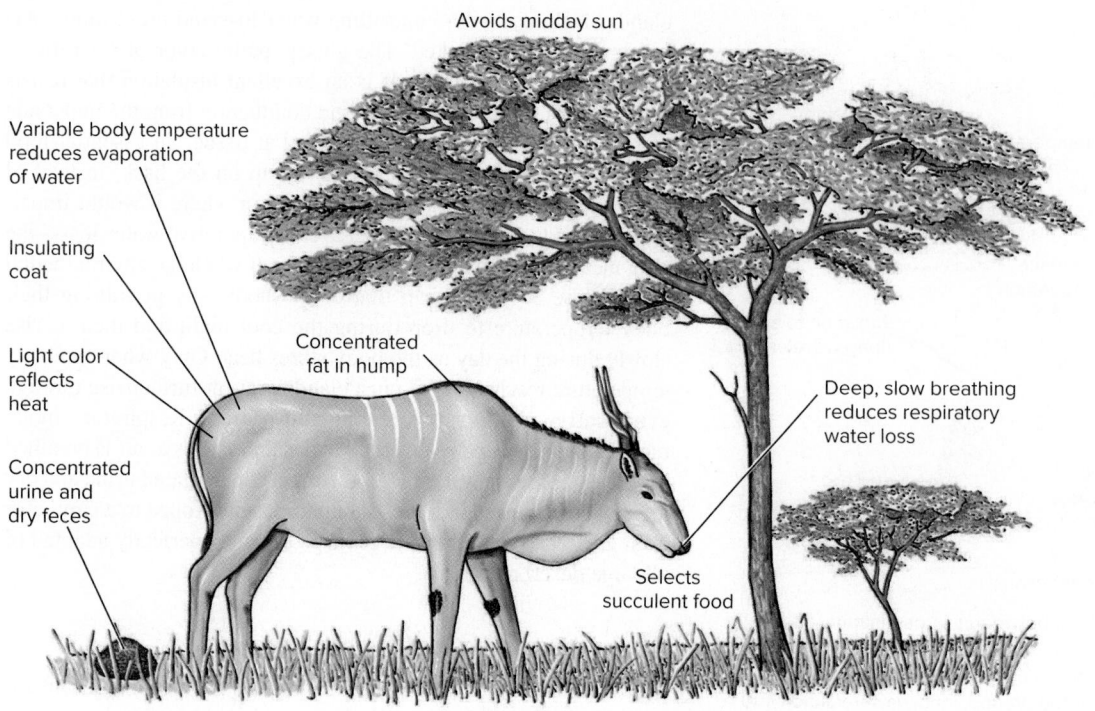

Avoids midday sun

Variable body temperature reduces evaporation of water

Insulating coat

Light color reflects heat

Concentrated fat in hump

Concentrated urine and dry feces

Deep, slow breathing reduces respiratory water loss

Selects succulent food

Figure 30.18 Physiological and behavioral adaptations of the common eland for regulating temperature in the hot, arid savanna of Central Africa.

23°

Core body temperature 37°

7.5°

5°

37°

Air temperature –30°

35°

14°

8°

0°

36°   37°

31° ← 32°

25° ← 26°

19° ← 20°

14° ← 15°

10° ← 11°

7°   8°

Figure 30.19 Countercurrent heat exchange in the leg of an arctic wolf. The diagram to the left shows how the extremities cool when the animal is exposed to low air temperatures. The diagram to the right depicts a portion of the front leg artery and vein, showing how heat is exchanged between arterial and venous blood. Thus, heat is shunted back into the body and conserved.

In severely cold conditions all mammals can produce more heat by **augmented muscular activity** through exercise or shivering. A person can increase heat production as much as 18-fold by violent shivering when maximally stressed by cold, though not indefinitely. Another source of heat is increased oxidation of foods, especially from stores of brown fat (brown fat is described on p. 714). This mechanism is called **nonshivering thermogenesis.**

Small mammals the size of lemmings, voles, and mice meet the challenge of cold environments in another way. Small mammals are not as well insulated as large mammals because thickness of fur is limited by the need to maintain mobility. So in addition to the mechanisms just mentioned, these forms exploit the excellent insulating qualities of snow by living under it in runways on the forest floor, where, incidentally, their food also is located. In this **subnivean environment** the temperature seldom drops below –5° C even though the air temperature above may fall to –50° C. Snow insulation decreases thermal conductance from small mammals just as thick pelage does for large mammals. Living beneath the snow is really a type of avoidance response to cold.

Adaptive Hypothermia in Birds and Mammals

Endothermy is energetically expensive. Whereas an ectotherm can survive for weeks in a cold environment without eating, an endotherm must always have energy resources to supply its high metabolic rate. The problem is especially acute for small birds and mammals which, because of their intense metabolism, may require a daily intake of food approaching their own body weight to maintain homeothermy (food consumption by birds is discussed on p. 593, and by mammals on p. 617). It is not surprising, then, that a few small birds and mammals have evolved ways to abandon homeothermy for periods ranging from a few hours each day to several months, allowing their body temperature to fall until it approaches or equals the temperature of surrounding air.

Some very small mammals, such as bats, maintain high body temperatures when active but allow their body temperature to drop profoundly when inactive and asleep. This is called **daily torpor,** an adaptive hypothermia that provides enormous saving of energy to small endotherms that are never more than a few hours away from starvation at normal body temperatures. Hummingbirds also may drop their body temperature at night when food supplies are low (Figure 30.20).

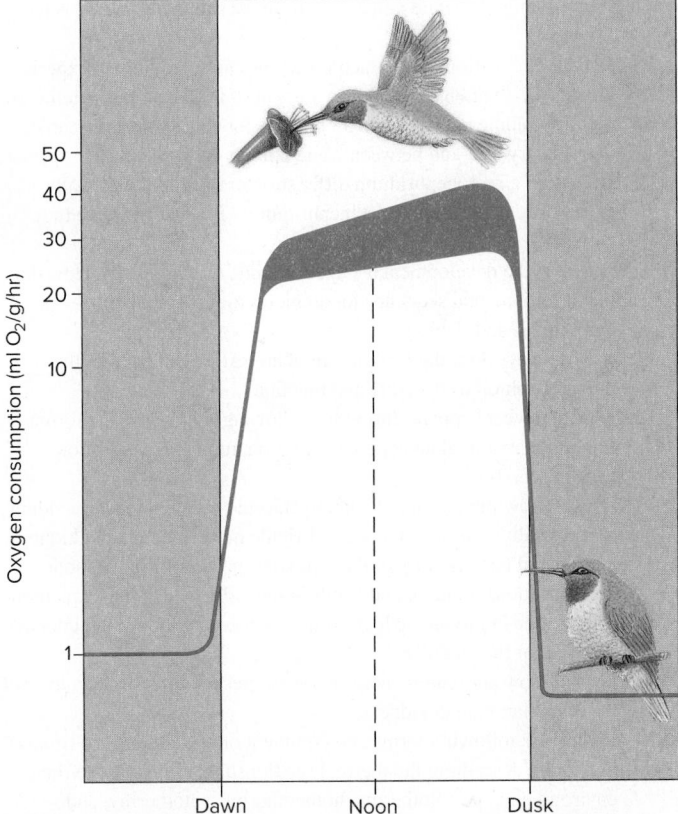

Figure 30.20 Torpor in hummingbirds. Body temperature and oxygen consumption (red line) are high when hummingbirds are active during the day but may drop to one-twentieth these levels during periods of food shortage. Torpor vastly lowers demands on the bird's limited energy reserves.

Figure 30.21 Hibernating hazel dormice, *Muscardinus avellanarius* (order Rodentia), in a woven nest. They begin hibernating in October and sleep until April or May.

this state breathing rates and metabolism decrease when temperatures are high, food is scarce, or dehydration threatens. Some examples of animals that estivate are land snails, blue land crabs, African lungfishes, desert tortoises, pigmy mice, and Columbian ground squirrels.

SUMMARY

Throughout life, matter and energy pass through the body, potentially disturbing the internal physiological state. Homeostasis, the ability of an organism to maintain internal stability despite such challenges, is a characteristic of all living systems. Homeostasis involves the coordinated activity of several physiological and biochemical mechanisms, and it is possible to relate some major events in animal evolution to increasing internal independence from the consequences of environmental change. In this chapter we have examined two aspects of homeostasis: (1) the varying ability of animals to stabilize the osmotic and chemical composition of the body, and (2) the capacity of animals to regulate their temperatures in thermally challenging environments.

Most marine invertebrates must either depend on the osmotic stability of the ocean to which they conform, or be able to tolerate wide fluctuations in environmental salinity. Some of the latter show limited powers of osmotic regulation, the capacity to resist internal osmotic change through the evolution of specialized regulatory organs. All animals living in freshwater are hyperosmotic to their environment and have developed mechanisms for recovering salt from the environment and eliminating excess water that enters the body osmotically.

All vertebrate animals show excellent osmotic homeostasis. Marine bony fishes maintain their body fluids distinctly hypoosmotic to their environment by drinking seawater and physiologically distilling it. Elasmobranchs (sharks and rays) have adopted a strategy of osmotic homeostasis by retaining urea and trimethylamine oxide (TMAO) in the blood.

The kidney is the most important organ for regulating the chemical and osmotic composition of the blood or interstitial fluid. All animal kidneys are variations on a basic theme: a tubular structure that forms urine by introducing a fluid secretion or filtrate of the blood or interstitial fluid into a tubule in which it is selectively modified to form urine. Terrestrial vertebrates have especially sophisticated kidneys, since they must be able to regulate closely the water content of the blood by balancing gains and expenditures.

Many small- and medium-sized mammals in northern temperate regions solve the problem of winter scarcity of food and low temperature by entering a prolonged and controlled state of dormancy: **hibernation.** True hibernators, such as ground squirrels, jumping mice, marmots, woodchucks, and dormice (Figure 30.21), prepare for hibernation by storing body fat. Entry into hibernation is gradual. After a series of "test drops" during which body temperature decreases a few degrees and then returns to normal, the animal cools to within a degree or less of the ambient temperature. Metabolism decreases to a fraction of normal. In ground squirrels, for example, the respiratory rate decreases from a normal rate of 200 breaths per minute to 4 or 5 breaths per minute, and the heart rate from 150 to 5 beats per minute. When an animal awakens from hibernation, it both shivers violently and employs nonshivering thermogenesis to produce heat.

Some mammals, such as bears, badgers, raccoons, and opossums, enter a state of prolonged sleep in winter with little or no decrease in body temperature. Prolonged sleep is not true hibernation. Bears of the northern forest sleep for several months. A bear's heart rate may decrease from 40 to 10 beats per minute, but body temperature remains normal and the bear is awakened if sufficiently disturbed.

Some invertebrates and vertebrates will enter a state of dormancy during the summer called **estivation** or "summer sleep." In

The basic excretory unit is the nephron, composed of a glomerulus, in which an ultrafiltrate of the blood is formed, and a long nephric tubule, in which the formative urine is selectively modified by the tubular epithelium. Water, salts, and other valuable materials pass by reabsorption to the circulation, and certain wastes pass by secretion from the circulation to the tubular urine. All mammals and some birds can produce urine more concentrated than blood by means of a countercurrent multiplier system localized in the loops of Henle, a specialization not found in other vertebrates.

Temperature has a profound effect on the rate of biochemical reactions and, consequently, on the metabolism and activity of all animals. Animals may be classified according to whether body temperature varies with environmental temperature (poikilothermic) or is stable (homeothermic), or by the source of body heat, whether external (ectothermic) or internal (endothermic).

Ectotherms partially free themselves from thermal constraints by seeking habitats with favorable temperatures, by behavioral thermoregulation, or by adjusting their metabolism to the prevailing temperature through biochemical alterations.

Endothermic birds and mammals differ from ectotherms in having a much higher production of metabolic heat and a much lower conductance of heat from the body. They maintain constant body temperature by balancing heat production with loss.

Small mammals in hot environments for the most part escape intense heat and reduce evaporative water loss by burrowing. Large mammals employ several strategies for dealing with direct exposure to heat, including reflective insulation, heat storage by the body, and evaporative cooling.

Endotherms in cold environments maintain body temperature by decreasing heat loss with thickened pelage or plumage, by peripheral cooling, and by increasing heat production through shivering or nonshivering thermogenesis. Small endotherms also may avoid exposure to low temperatures by living under the snow.

Adaptive hypothermia is a strategy used by small mammals and birds to blunt energy demands during periods of inactivity (daily torpor) or periods of prolonged cold and minimal food availability (hibernation). Some vertebrates and invertebrates enter a similar state during the summer when temperatures are high, food is scarce, or dehydration threatens (estivation).

REVIEW QUESTIONS

1. Define homeostasis. What evolutionary advantages for a species might result from successful maintenance of internal homeostasis?
2. Describe the physiological challenges confronting marine invertebrates entering freshwater and, using crustaceans as an example, suggest solutions to these challenges.
3. Distinguish the terms in the following pairs: osmotic conformity and osmotic regulation; stenohaline and euryhaline; hyperosmotic and hypoosmotic.
4. Young downstream salmon migrants moving from their freshwater natal streams into the sea leave an environment nearly free of salt to enter one containing three times as much salt as their body fluids. Describe osmotic challenges of each environment and suggest physiological adjustments salmon must make in moving from freshwater to the sea.
5. Most marine invertebrates are osmotic conformers. How does their body fluid differ from that of sharks and rays, which are also in near osmotic equilibrium with their environment?
6. What strategy does a kangaroo rat use that allows it to exist in the desert without drinking any water?

7. In what animals would you expect to find a salt gland? What is its function?
8. Relate the function of contractile vacuoles to the following experimental observations: to expel an amount of fluid equal in volume to its own volume required 4 to 53 minutes for some freshwater unicellular eukaryotes, and between 2 and 5 hours for some marine species.
9. How does a protonephridium differ structurally and functionally from a true nephridium (metanephridium)? In what ways are they similar?
10. Describe the developmental stages of kidneys in amniotes. How does the developmental sequence for amniotes differ from that of amphibians and fishes?
11. In what ways does the nephridium of an earthworm parallel the human nephron in structure and function?
12. Describe what happens during the following stages in urine formation in the mammalian nephron: filtration, tubular reabsorption, tubular secretion.
13. Explain how the cycling of sodium chloride between the descending and ascending limbs of the loop of Henle in the mammalian kidney, and special permeability of these tubules, produces high osmotic concentrations in interstitial fluids in the kidney medulla. Explain the role of urea in producing high osmotic concentrations in the interstitial fluid of the medulla.
14. Explain how antidiuretic hormone (vasopressin) controls excretion of water in mammalian kidneys.
15. Define the following terms and comment on the limitations (if any) of each in describing the thermal relationships of animals to their environments: poikilothermy, homeothermy, ectothermy, endothermy.
16. Large mammals live successfully in deserts and in the arctic. Describe different adaptations mammals use to maintain homeothermy in each environment.
17. Explain why it is advantageous for certain small birds and mammals to abandon homeothermy during brief or extended periods of their lives.

For Further Thought If daytime body temperature of some ectotherms were measured, the values would be fairly constant. Provide an explanation for such observations, giving some possible mechanisms that might be involved.

SELECTED REFERENCES

Beyenbach, K. W. 2003. Transport mechanisms of diuresis in Malpighian tubules of insects. J. Exp. Biol. **206:** 3845–3856. *An excellent review of Malpighian tubular secretion in blood-sucking insects compared with aquatic and terrestrial insects.*

Heinrich, B. 1996. The thermal warriors: strategies of insect survival. Cambridge, Massachusetts, Harvard University Press. *Describes the many fascinating ways that insects respond to their temperature environment.*

Ianowski, J. P., and M. J. O'Donnell. 2004. Basolateral ion transport mechanisms during fluid secretion by *Drosophila* Malpighian tubules: Na$^+$ recycling, Na$^+$: K$^+$: 2Cl$^-$ cotransport and Cl$^-$ conductance. J. Exp. Biol. **207:** 2599–2609. *Research paper that presents data in support of a revised model for Malpighian tubular secretion in a terrestrial insect.*

Louw, G. N., and D. Mitchell. 1996. Physiological animal ecology. New York, Longman Scientific & Technical. *Clearly presented survey with emphasis on thermoregulation and water relations in animals.*

Randall, D., W. Burggren, and K. French. 2001. Eckert animal physiology: mechanisms and adaptations. New York, W. H. Freeman & Company. *A comprehensive and comparative treatment of animal physiology.*

Sands, J. M. 1999. Urea transport: it's not just "freely diffusible" anymore. News Physiol. Sci. **14:** 46–47. *Summarizes studies that have reported urea transport proteins in the collecting duct of the nephron.*

Schmidt-Nielsen, K. 1981. Countercurrent systems in animals. Sci. Am. **244:** 118–128 (May). *Explains how countercurrent systems transfer heat, gases, or ions between fluids moving in opposite directions.*

Schultz, S. G. 1996. Homeostasis, Humpty Dumpty and integrative biology. News Physiol. Sci. **11:** 238–246. *Describes the central role that homeostasis plays in the study of physiological systems.*

Smith, H. W. 1953. From fish to philosopher. Boston, Little, Brown & Company. *Classic account of vertebrate kidney evolution.*

Storey, K. B., and J. M. Storey. 1990. Frozen and alive. Sci. Am. **263:** 92–97 (Dec.). *Explains how many animals have evolved strategies for surviving complete or almost complete freezing during the winter months.*

Willmer, P., G. Stone, and I. Johnston. 2004. Environmental physiology of animals, ed. 2. Oxford, Blackwell Science Ltd. *Well-written information regarding excretory systems in both vertebrates and invertebrate.*

Homeostasis: Internal Fluids and Respiration

© Andrew Syred/Science Source

Colored scanning electron micrograph of blood cells.

William Harvey's Discovery

Ceaselessly, during a human life, the heart pumps blood through arteries, capillaries, and veins: about 5 liters per minute, until by the end of a normal life the heart has contracted some 2.5 billion times and pumped 300,000 tons of blood. When the heart stops its contractions, life also ends.

The crucial role of heart contractions for human life has been known since antiquity; however, the circuit flow of blood, the notion that the heart pumps blood into arteries through the circulation and receives it back in veins, became known only a few hundred years ago. The first correct description of blood flow by the English physician William Harvey was vigorously opposed when first published in 1628.

Centuries earlier, the Greek anatomist Galen had taught that air enters the heart from the windpipe and that blood was able to pass from one ventricle to the other through "pores" in the interventricular septum of the heart. Galen also thought that blood first flowed from the heart into all vessels, then returned—a kind of ebb and flow of blood. Even though there was almost nothing correct about this concept, it was still doggedly trusted at the time of Harvey's publication.

Harvey's conclusions were based on sound experimental evidence. He used a variety of animals for his experiments and chided human anatomists, saying that if only they had acquainted themselves with anatomy of other vertebrates, they would have understood the blood's circuit. By tying ligatures on arteries, he noticed that the region between the heart and the ligature swelled. When veins were tied off, the swelling occurred beyond the ligature. When blood vessels were cut, blood flowed in arteries from the cut end nearest the heart; the reverse happened in veins. By means of such experiments, Harvey discovered the correct scheme of blood circulation, even though he could not see the capillaries that connected the arterial and venous flows.

ingle-celled organisms live with their cell surface in direct contact with their environment, and they obtain nutrients and oxygen and release wastes directly across it. These organisms are so small that no special internal system of transport, beyond normal cytoplasmic streaming, is required. Except for some simple multicellular forms, such as sponges, cnidarians, and flatworms, most multicellular organisms, because of their increased size, activity, and complexity, need a specialized circulatory system to transport nutrients, waste products, and respiratory gases to and from all tissues of their body. Circulatory systems also transport hormones (see Chapter 34) from glands or cells that produce them to target tissues where they act in concert with the nervous system (see Chapter 33) to integrate organismal function. Water, electrolytes, and the many other constituents of body fluids are distributed and exchanged between different organs and tissues. An effective response to disease and injury is vastly accelerated by an efficient circulatory system. Homeothermic birds and mammals depend heavily on blood circulation to conserve or to dissipate heat as required for maintenance of constant body temperature.

Gas exchange solely by diffusion across surface membranes is possible only for very small organisms less than 1 mm in diameter. For example, in single-celled organisms, oxygen is acquired and carbon dioxide is liberated in this way, because diffusion paths are short and the surface area of the organism is large relative to volume. As animals became larger and evolved a waterproof covering, specialized devices, such as lungs and gills, evolved to increase the effective surface for gas exchange. Because gases diffuse so slowly through living tissue, a circulatory system was necessary to distribute gases to and from deep tissues of the body. Even these adaptations were inadequate for complex animals with high rates of cellular respiration. Because the solubility of oxygen in the blood plasma is so low, plasma alone cannot carry enough oxygen to support metabolic demands. The oxygen-carrying capacity of blood increased greatly with the evolution of special oxygen-transporting blood proteins, such as hemoglobin, which seems to have evolved in conjunction with the circulatory system.

31.1 INTERNAL FLUID ENVIRONMENT

The body fluid of a single-celled organism is cytoplasm, a liquid-gel substance surrounding the various membrane systems and organelles. In animals, body fluids are divided into **intracellular** and **extracellular** compartments. The intracellular compartment (also called intracellular fluid) is the collective fluid inside all the body's cells. The extracellular compartment (or extracellular fluid) is the fluid outside and surrounding the cells (Figure 31.1A). The extracellular fluid buffers cells from physical and chemical changes occurring outside the body. The importance of extracellular fluid was first emphasized by the great French physiologist Claude Bernard (Figure 31.2). In animals having closed circulatory systems (vertebrates, annelids, and a few other invertebrate groups; see p. 685), extracellular fluid is further subdivided into **blood plasma** and **interstitial (intercellular) fluid** (Figure 31.1A). Blood vessels of a closed circulatory system contain plasma, whereas interstitial fluid, also called tissue fluid, surrounds body cells. Nutrients and gases passing between vascular plasma and cells must traverse this narrow fluid separation. Interstitial fluid is constantly formed from plasma by movement of fluid to and from microscopic vessels that are in close proximity to every cell (capillaries; see p. 690).

Figure 31.1 Fluid compartments of an animal's body. **A,** All body cells can be represented as belonging to a single large fluid compartment that is completely surrounded and protected by extracellular fluid (*milieu intérieur*). In animals with a closed circulatory system, this fluid is further subdivided into plasma and interstitial fluid. All exchanges with the environment occur across the plasma compartment. **B,** Typical electrolyte composition of extracellular and intracellular fluids. Total equivalent concentration of each major constituent is shown. Equal amounts of anions (negatively charged ions) and cations (positively charged ions are in each fluid compartment. Note that sodium, calcium, and chloride, abundant in the extracellular fluid, are present only in low concentrations in the intracellular fluid). Note the much higher concentration of protein inside cells.

Figure 31.2 French physiologist Claude Bernard (1813 to 1878), one of the most influential of nineteenth-century physiologists. Bernard believed in the constancy of the *milieu intérieur* ("internal environment"), which is the extracellular fluid bathing the cells. He emphasized that it is through the *milieu intérieur* that foods, wastes, and gases are exchanged and through which chemical messengers are distributed. He wrote, "The living organism does not really exist in the external environment (the outside air or water) but in the liquid *milieu intérieur* . . . that bathes the tissue elements."

Composition of the Body Fluids

All these fluid spaces—plasma, interstitial, and intracellular—differ from one another in solute composition, but all have one feature in common: they are mostly water. Despite their firm appearance, animals are 70% to 90% water. Humans, for example, are approximately 70% water by weight. Of this, 50% is cell water, 15% is interstitial-fluid water, and the remaining 5% is in blood plasma. Plasma serves as the pathway of exchange between the cells of the body and the outside world. This exchange of respiratory gases, nutrients, and wastes is accomplished by specialized organs (for example, kidney, lung, gill, alimentary canal), as well as by skin (Figure 31.1A).

Body fluids contain many inorganic and organic substances in solution. Principal among these are inorganic electrolytes and proteins. Sodium, chloride, and bicarbonate ions are the chief extracellular electrolytes, whereas potassium, magnesium, and phosphate ions and **proteins** are the major intracellular electrolytes (Figure 31.1B). These differences are dramatic; they are maintained by active transport ion pumps (see Chapter 3, p. 48) despite continuous flow of materials into and out of cells of the body. The two subdivisions of extracellular fluid—plasma and interstitial fluid—have similar compositions except that plasma has more proteins, which are mostly too large to move from capillaries into interstitial fluid.

31.2 COMPOSITION OF BLOOD

Invertebrates that lack a circulatory system (such as flatworms and cnidarians) also lack true "blood," but they possess a clear, watery tissue fluid containing some phagocytic cells, a little protein, and a mixture of salts similar to seawater. The "blood" of invertebrates with open circulatory systems (p. 685) is more complex and is often called hemolymph (Gr. *haimo,* blood, 1 L. *lympha,* water).

Invertebrates with closed circulatory systems (p. 685), maintain a clear separation between blood contained within blood vessels and tissue (interstitial) fluid surrounding blood vessels and cells.

In vertebrates, blood is a complex liquid tissue composed of plasma and cellular components suspended in plasma.

The composition of mammalian blood follows:

Plasma—55% of blood

1. Water 90%
2. Dissolved solids, consisting of plasma proteins (albumin, globulins, fibrinogen), glucose, amino acids, electrolytes, various enzymes, antibodies, hormones, metabolic wastes, and traces of many other organic and inorganic materials
3. Dissolved gases, especially oxygen, carbon dioxide, and nitrogen

Cellular components (Figure 31.3)—45% of blood

1. Red blood cells (erythrocytes), containing hemoglobin for transport of oxygen and carbon dioxide
2. White blood cells (leukocytes), serving as scavengers and as defensive cells
3. Cell fragments (platelets in mammals) or cells (thrombocytes in other vertebrates) that function in blood coagulation

Plasma proteins are a diverse group of large and small proteins that perform numerous functions. The major protein groups are (1) **albumins,** the most abundant group, constituting 60% of the total, which help to keep plasma in osmotic equilibrium with the cells of the body; (2) **globulins,** a diverse group of high-molecular-weight protein (35% of total) including immunoglobulins, which function in acquired immunity (see p. 769), and various metal-binding proteins; and (3) **fibrinogen,** a very large protein that functions in blood coagulation. Blood **serum** is plasma minus the proteins involved in clot formation.

Red blood cells, or **erythrocytes,** occur in enormous numbers in blood, approximately 5.4 million per milliliter of blood in adult men and 4.8 million in adult women. In mammals and birds, red cells form continuously from large nucleated **erythroblasts** in red bone marrow (in other vertebrates, kidneys and spleen are the principal sites of red blood cell production). During erythrocyte formation hemoglobin is synthesized and the precursor cells divide several times. In mammals the nucleus shrinks during development to become nonfunctional and eventually is lost from the cell by exocytosis (p. 49). The majority of cellular organelles, such as ribosomes, mitochondria, and most enzyme systems, are also lost. What is left is a biconcave disc consisting of a baglike membrane packed with about 280 million molecules of the blood-transporting pigment hemoglobin. Approximately 33% of an erythrocyte by weight is hemoglobin. The biconcave shape (Figures 31.3 and 31.4A) is a mammalian innovation that provides a larger surface for gas diffusion than would a flat or spherical shape. Other vertebrates typically have nucleated erythrocytes that are usually ellipsoidal in shape (Figure 31.4B).

Figure 31.3 Cellular components of human blood. Hemoglobin-containing erythrocytes, or red blood cells of humans and other mammals lack nuclei, but those of all other vertebrates have nuclei (Figure 31.4B). Various leukocytes, or white blood cells provide a system of protection for the body. Platelets are cell fragments that participate in the blood's clotting mechanism.

Figure 31.4 Mammalian and amphibian erythrocytes, or red blood cells. **A,** Erythrocytes of a gerbil are biconcave discs containing hemoglobin and no nucleus. **B,** Frog erythrocytes are convex discs, each containing a nucleus, which is plainly visible in the scanning electron micrograph as a bulge in the center of each cell. (Magnifications: mammalian erythrocytes, × 6300; frog erythrocytes, × 2400.)

An erythrocyte enters the circulation for an average lifespan of approximately 4 months. During this time it may journey 11,000 km, squeezing repeatedly through the smallest blood vessels, capillaries, which are sometimes so narrow that the erythrocyte must bend to pass through. At last it fragments and is quickly engulfed by large scavenger cells called **macrophages** located in the liver, bone marrow, and spleen. Iron from the heme component of hemoglobin is salvaged to be used again; the rest of the heme is converted to **bilirubin,** a bile pigment. It is estimated that a human body produces 10 million erythrocytes and destroys another 10 million every second.

White blood cells, or **leukocytes,** form part of the immune system of the body (see Chapter 35). In adults they number only approximately 50,000 to 100,000 per milliliter of blood, a ratio of 1 white cell to 500 to 1000 red cells. There are several kinds of white blood cells: **granulocytes** (subdivided into **neutrophils, basophils,** and **eosinophils**) and **agranulocytes,** the **lymphocytes** and **monocytes** (Figure 31.3).

Hemostasis: Prevention of Blood Loss

It is essential that animals have ways of preventing rapid loss of body fluids after an injury. In animals with closed circulatory systems (p. 685), blood flows under considerable hydrostatic pressure. Thus, they are especially vulnerable to hemorrhagic loss.

When a blood vessel is damaged, smooth muscle (p. 193) in the wall of the vessel contracts, which causes the vessel lumen to narrow, sometimes so strongly that blood flow is completely stopped. This simple but highly effective means of preventing hemorrhage is used by invertebrates and vertebrates alike. Beyond this first defense against blood loss, all vertebrates, as well as some larger, active invertebrates with high blood pressures, have in the blood special cellular components and proteins that can form plugs, or clots, at an injury site.

In crustaceans (see Chapter 20, p. 425), the hemolymph contains a clotting factor, similar to fibrinogen, and ameboid cells that release clotting factors explosively and initiate coagulation and subsequent clotting in response to minor injuries.

In vertebrates, **blood coagulation** occurs as a complex series of chemical reactions that produce a tangled network of fibers from one of the plasma proteins, **fibrinogen.** The enzyme thrombin catalyzes the transformation of fibrinogen into a **fibrin** meshwork (Figure 31.5) that entangles blood cells to form a gel-like clot. Thrombin is normally present in blood in an inactive form called **prothrombin,** which must be activated for coagulation to occur.

In this process, blood platelets (Figure 31.3) and the damaged cells of blood vessels play a vital role. Platelets form in red bone marrow from large, multinucleated cells that regularly pinch off bits of their cytoplasm; thus platelets are fragments of cells. There are 150,000 to 300,000 platelets per cubic millimeter of blood. When the normally smooth inner surface of a blood vessel is disrupted, either by a break or by deposits of a cholesterol-lipid material, platelets rapidly adhere to the surface and release **thromboplastin** and other clotting factors. These platelet-derived factors, in addition to calcium ions, and tissue thromboplastin and other factors released from damaged cells of blood vessels and underlying tissue, initiate conversion of prothrombin to active thrombin (Figure 31.6).

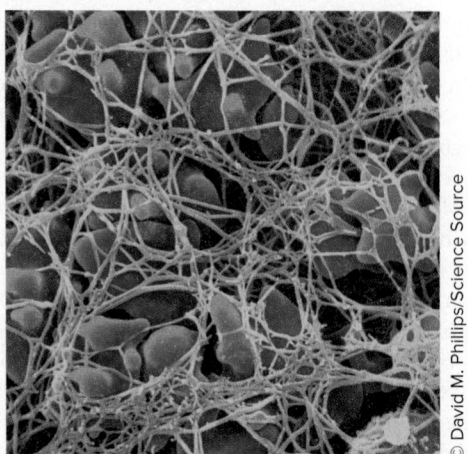

© David M. Phillips/Science Source

Figure 31.5 Scanning electron micrograph of human red blood cells trapped in a fibrin clot. Clotting is initiated after tissue damage by disintegration of platelets in blood, resulting in a complex series of intravascular reactions that end with conversion of a plasma protein, fibrinogen, into long, tough, insoluble polymers of fibrin. Fibrin and entangled red blood cells form the blood clot, which arrests bleeding.

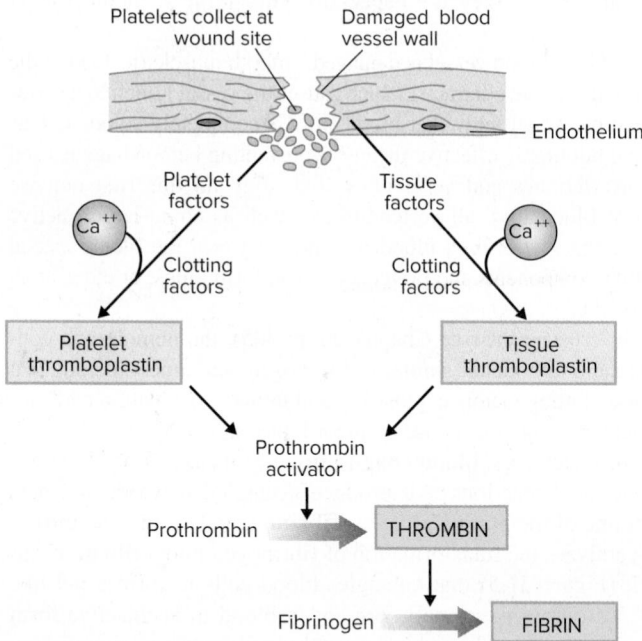

Figure 31.6 Stages in formation of fibrin.

The catalytic sequence in this scheme involves a series of plasma protein factors, each normally inactive until activated by a previous factor in the sequence. The sequence forms an amplified "cascade" with each reactant in the sequence leading to a large increase in the amount of the next reactant. At least 13 different plasma clotting factors have been identified. A deficiency of only a single factor can delay or prevent the clotting process. Why has such a complex clotting mechanism evolved? One advantage is that amplification of the signal occurs at each step of the cascade,

so even small injuries cause a rapid and full-blown response. Probably these proteins evolved by multiple gene mutations since the plasma proteins in the cascade are all closely related serine proteases.

Several kinds of clotting abnormalities occur in humans. One of these, hemophilia, is a condition characterized by failure of blood to clot, so that even insignificant wounds can cause continuous severe bleeding and death. A rare mutation (the condition occurs in about 1 in 10,000 males) on the X sex chromosome (see Chapter 5, p. 82) causes a lack of one of the platelet factors in males and in homozygous females. Called the "disease of kings," it once ran through several interrelated royal families of Europe, apparently having originated from a mutation in one of Queen Victoria's parents.

Hemophilia is one of the best-known cases of sex-linked inheritance in humans (p. 82). Actually two different loci on the X chromosome are involved. Classic hemophilia (hemophilia A) accounts for about 80% of persons with the condition, and the remainder are caused by Christmas disease (hemophilia B). The recessive allele at each locus causes a deficiency of a different platelet factor.

31.3 CIRCULATION

Most animals have evolved mechanisms for transporting materials throughout their body. For sponges (see Figure 12.5, p. 250) and diploblasts (see Chapter 13), the water in which they live provides the medium for transport. Water, propelled by ciliary, flagellar, or body movements, passes through channels or compartments to facilitate the movement of food, respiratory gases, and wastes. An animal's shape obviously is important. Flattened and leaflike acoelomate flatworms (see Chapter 14) survive because the distance of any body part from the surface is short; respiratory gases and metabolic wastes transfer by simple diffusion even though many are relatively large animals. True circulatory systems—containing vessels through which blood moves—are essential to animals so large or so active that diffusional processes alone cannot supply their oxygen needs.

A circulatory system having a full set of components—pump, arterial distribution system, capillaries that interface with cells, venous reservoir, and return system—is fully recognizable in annelid worms (see Chapter 17). In earthworms (Figure 31.7) there are two main vessels, a dorsal vessel carrying blood anteriorly, and a ventral vessel delivering blood posteriorly throughout the body by way of segmental vessels and a dense tissue capillary network. The dorsal vessel drives the blood forward by peristalsis (see p. 708) and thus serves as a heart. Five aortic arches that connect the dorsal and ventral vessels laterally are contractile and serve as accessory hearts to maintain a steady flow of blood into the ventral vessel and to the head region, which has its own afferent and efferent blood vessels (Figure 31.7). Many smaller segmental vessels that deliver blood to tissue capillaries are actively contractile as well. There is no localized pump pushing the blood through a system of passive tubes; instead, the power of contraction is widely distributed throughout the vascular system.

Open and Closed Circulations

The system just described is a **closed circulation** because the circulating fluid, **blood,** is confined to vessels throughout its journey through the vascular system. Many invertebrates have an **open circulation** in which there are no small blood vessels or capillaries interfacing with cells or connecting arteries with veins. In insects and other arthropods (see Chapter 20, p. 425, and Chapter 21, p. 450), in most molluscs, and in many smaller invertebrate groups, blood sinuses, collectively called a **hemocoel,** replace capillary beds. During development of the body cavity in these groups, the blastocoel is not completely obliterated by the mesoderm (Figure 31.8) and fuses with embryonic coelomic cavities. Thus, the hemocoel comprises the primary body cavity (persistent blastocoel) and secondary coelomic cavities through which blood (also called **hemolymph**) freely circulates (bottom diagrams in Figure 31.8). Since there is no separation of the extracellular fluid into blood plasma and lymph (as there is in a closed circulation), blood volume is large and may constitute 20% to 40% of body volume. By contrast,

blood volume in animals with closed circulations (vertebrates, for example) is only about 5% to 10% of body volume.

In arthropods, the heart and all viscera lie in the hemocoel, bathed by blood (Figure 31.8). Blood enters the heart through valved openings, the ostia, and the heart's contractions, which resemble a forward-moving peristaltic wave, propel blood into a limited arterial system. Blood travels to the head and other organs, then flows into the hemocoel. It is routed through the body and appendages by a system of baffles and longitudinal membranes (septa) before returning to the heart (Figure 31.9). Because blood pressure is very low in open systems, seldom exceeding 4 to 10 mm Hg, many arthropods have auxiliary hearts or contractile vessels to boost blood flow.

Insects and many other terrestrial arthropods do not use their circulatory system for respiratory gas transport. Instead, a separate respiratory system has evolved for this purpose: a tracheal system in insects (see Figure 31.19) and some other terrestrial arthropods (centipedes, millipedes, and some spiders, p. 403), pseudotracheae in terrestrial pill bugs, and book lungs in some spiders.

In animals with closed circulatory systems (most annelids, cephalopod molluscs, and all vertebrates), the coelom increases in size during embryonic development to obliterate the blastocoel and forms a secondary body cavity (top diagrams in Figure 31.8). A system of continuously connected blood vessels develops within the mesoderm. Closed systems have certain features in common. A **heart** pumps blood into **arteries** that branch and narrow into **arterioles** and then into a vast system of **capillaries** that interfaces with cells in body tissues. Blood leaving capillaries flows into **venules** and then **veins** that return the blood to the heart. Capillary walls are thin (only one cell thick), permitting rapid transfer of materials between blood and tissues. Closed systems are more suitable for large and active animals because blood moves rapidly to the most active tissues. In addition, flow to various organs can be readjusted to meet changing needs by varying the diameters of blood vessels.

Because blood pressures are much higher in closed than in open systems, fluid is constantly moving through capillary walls into the

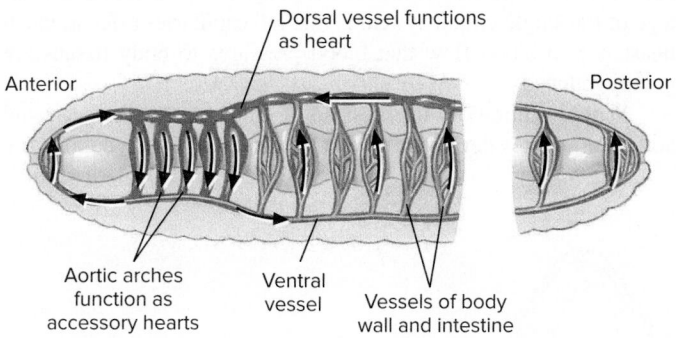

Figure 31.7 Blood flow through the closed vascular system of an earthworm.

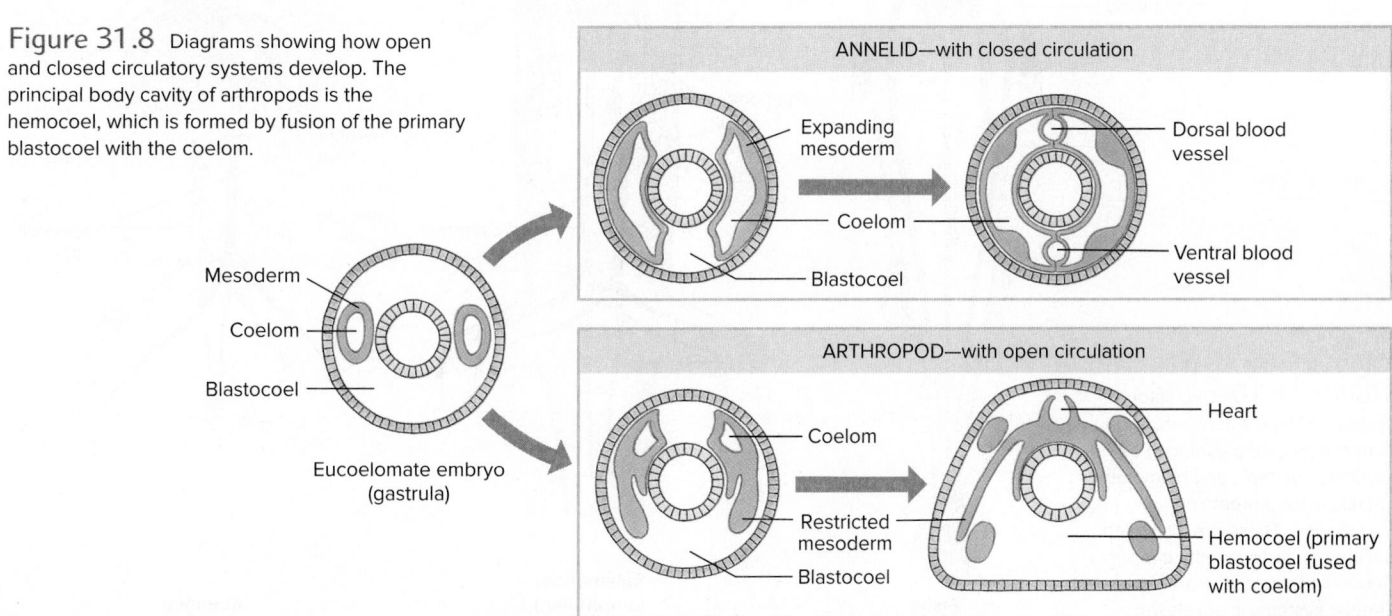

Figure 31.8 Diagrams showing how open and closed circulatory systems develop. The principal body cavity of arthropods is the hemocoel, which is formed by fusion of the primary blastocoel with the coelom.

surrounding tissue spaces. Most of this fluid is drawn back into capillaries by osmosis (see p. 46). The remainder is recovered by a **lymphatic system** (p. 691), which has evolved separately but in conjunction with the high-pressure system of vertebrates.

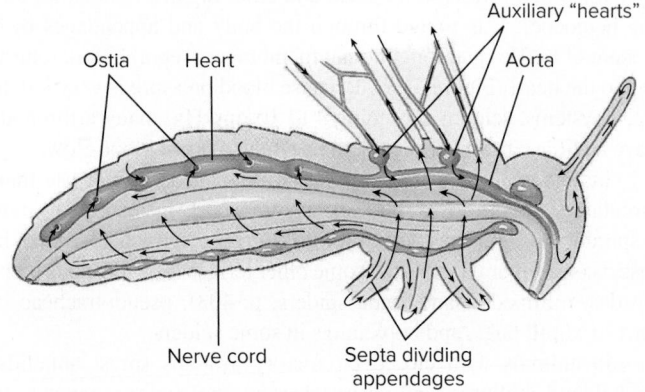

Figure 31.9 Circulatory system of an insect. Although the circulatory system is open, blood is directed through the appendages in channels formed by longitudinal septa. Arrows indicate the course of circulation.

Plan of Vertebrate Circulatory Systems

In vertebrates the principal differences in the blood-vascular system involve the gradual separation of the heart into two separate pumps as vertebrates evolved from aquatic life with gill breathing to fully terrestrial life with lung breathing. These changes are shown in Figure 31.10, which compares the circulation of fish, amphibians, and mammals.

A fish heart contains two main chambers in series, an **atrium** and a **ventricle.** The atrium is preceded by an enlarged chamber, the **sinus venosus,** which collects blood from the venous system and assures a smooth delivery of blood to the heart. Elasmobranchs have a fourth heart chamber, the **conus arteriosus,** which dampens blood-pressure oscillations before blood flows into delicate blood capillaries. Teleost fish have a **bulbus arteriosus** that serves the same function (see Figure 24.14, p. 527). Blood makes a single circuit through a fish's vascular system; it is pumped from the heart to the gills, where it is oxygenated, then flows into the dorsal aorta to be distributed to body organs, and finally returns by veins to the heart. In this circuit the heart must provide sufficient pressure to push the blood through two sequential capillary systems, first that of the gills, and then that of the remainder of the body. The principal disadvantage of the single-circuit system is that gill capillaries offer so much resistance to blood flow that blood pressures to body tissues are greatly reduced.

With evolution of lungs instead of gills between the heart and aorta, vertebrates developed a high-pressure **double circulation:** a

Figure 31.10 Circulatory systems of fish, amphibian, and mammal, showing evolution of separate systemic and pulmonary circuits in lung-breathing vertebrates. These are all closed circulatory systems—the capillary network that joins the arterial and venous systems is not shown.

Fish

Salamander (amphibian)

Mammal

systemic circuit that provides oxygenated blood to the capillary beds of the body organs; and a **pulmonary circuit** that serves the lungs. The beginning of this major evolutionary change probably resembled the condition seen in lungfishes and amphibians. In modern amphibians (frogs, toads, salamanders), the atrium is completely separated by a partition into two atria (Figure 31.11). The right atrium receives venous blood from the body while the left atrium receives oxygenated blood from the lungs and skin. The ventricle is undivided, but venous and arterial blood remain mostly separate due to the spiral fold of the conus arteriosus, the arrangement of septa or folds in vessels leaving the heart (Figure 31.11), and differential blood pressures in the exiting vessels. A septum partially divides the ventricle in most nonavian reptiles, and it is complete in crocodilians, birds, and mammals (Figures 31.10 and 31.12). Systemic and pulmonary circuits are now separate circulations, each served by one half of a dual heart (Figures 31.10 and 31.12).

Mammalian Heart

The four-chambered mammalian heart (Figure 31.12) is a muscular organ located in the thorax and covered by a tough, fibrous sac, the **pericardium.** Blood returning from the lungs flows through the **pulmonary veins** and collects in the **left atrium,** passes into the **left ventricle,** and is pumped into the body (systemic) circulation via the **aorta.** Blood returning from the body flows through the **inferior** (posterior) and **superior** (anterior) **vena cava** into the **right atrium,** and passes into the **right ventricle,** which pumps it into the lungs via the **pulmonary arteries.** Backflow of blood in the heart is prevented by two sets of valves, formed as extensions of the inner wall of the heart, that open and close passively in response to pressure differences between the heart chambers. The **left atrioventricular** (bicuspid) and the **right atrioventricular** (tricuspid) valves separate the cavities of the atrium and ventricle in each half of the heart. Where the great arteries, the **pulmonary** from the right ventricle and the **aorta** from the left ventricle, leave the heart, **semilunar valves** prevent backflow from these arteries into the ventricles.

Contraction is called **systole** (sis′to-lee), and relaxation, **diastole** (dy-as′to-lee) (Figure 31.13). When the atria contract (atrial systole), the ventricles relax (ventricular diastole) and fill with blood. Ventricular systole is accompanied by atrial filling during atrial diastole, although a large volume of blood returns to the heart due to negative thoracic pressure during inspiration and ventricular systole that draws blood into the heart. The rate of contraction or heartbeat depends on age, sex, and especially exercise. Exercise can increase **cardiac**

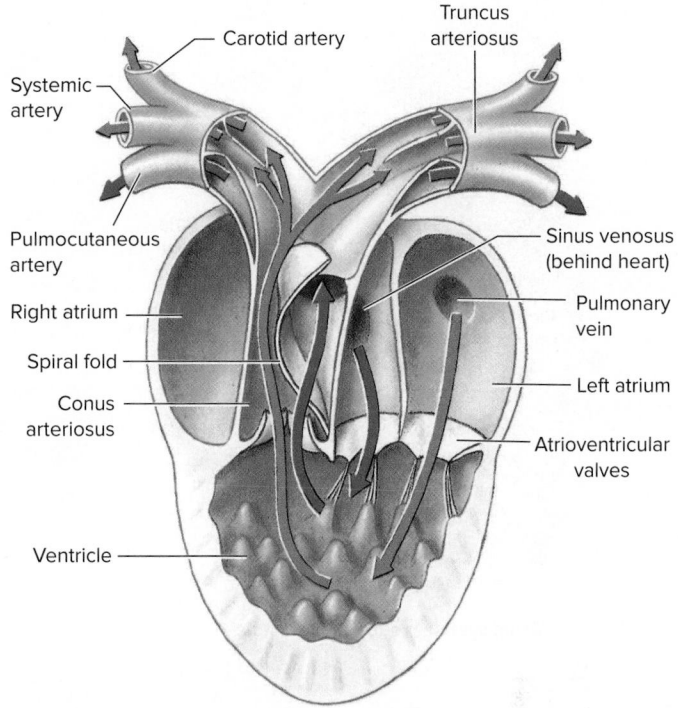

Figure 31.11   Route of blood through a frog heart. Atria are completely separated, and the spiral fold in the conus arteriosus helps to route blood to lungs and systemic circulation.

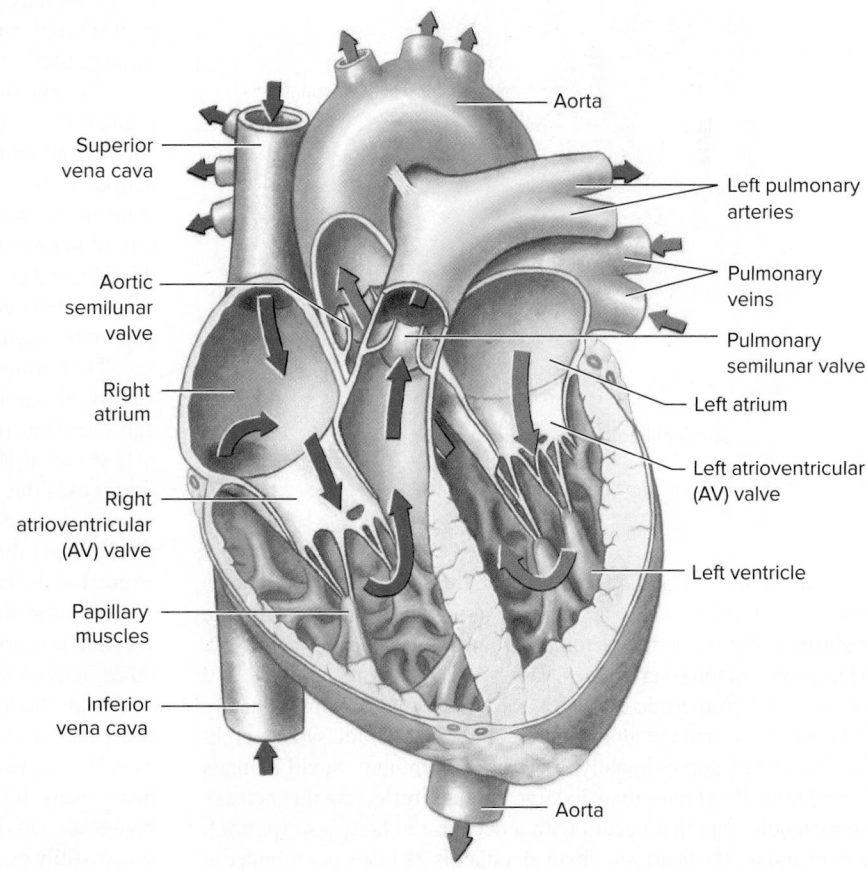

Figure 31.12   Human heart. Deoxygenated blood enters the right side of the heart and is pumped to the lungs. Oxygenated blood returning from the lungs enters the left side of the heart and is pumped to the body. The right ventricular wall is thinner than that of the left ventricle, a result of less muscular force generation to pump blood to such close structures as the lungs.

Atrial systole/ventricular diastole

Atrial diastole/ventricular systole

Figure 31.13 Human heart in systole and diastole.

output (volume of blood forced from either ventricle each minute) more than fivefold due to an increase in both heart **rate** and **stroke volume** (volume of blood forced from either ventricle per beat). Heart rates among vertebrates vary with level of metabolism and body size. Ectothermic codfish have a heart rate of approximately 30 beats per minute; endothermic rabbits of about the same weight have a rate of approximately 200 beats per minute. Small animals have higher heart rates than do large animals, reflecting the increase in metabolic rate that occurs with a decrease in body size (p. 621). For example, the heart rate in an elephant is 25 beats per minute, in

a human 70 per minute, in a cat 125 per minute, and in a mouse 400 per minute; in the tiny 4 g shrew, the smallest mammal, the heart rate approaches 800 beats per minute.

Excitation and Control of the Heart

The vertebrate heart is a muscular pump composed of **cardiac muscle.** Cardiac muscle resembles skeletal muscle—both are types of striated muscle—but cardiac cells are branched and joined end-to-end by junctional complexes (intercalated discs) to form a branching network (see Figure 29.13, p. 651). Unlike skeletal muscle, vertebrate cardiac muscle does not depend on nerve activity to initiate a contraction. Instead, regular contractions are established by specialized cardiac muscle cells, called **pacemaker cells.** In a nonavian reptile, bird, or mammal heart, the pacemaker is in the **sinoatrial (SA) node,** a remnant of the ancestral sinus venosus found in fish and amphibians (Figure 31.11). Electrical activity initiated in the pacemaker spreads over the muscle of the two atria and then, after a slight delay, to the secondary pacemaker, the atrioventricular (AV) node, at the top of the ventricles. At this point electrical activity is conducted rapidly through the **bundle of His** and left and right branching bundles to the apex of the ventricle and then continues through specialized fibers (**Purkinje fibers**) to the apex or "tip" of the ventricles (Figure 31.14). Excitation and contraction of cardiac muscle cells thus begins within cardiac cells in the apex of the ventricles and spreads upward to propel blood in the most efficient way; it also ensures that both ventricles contract simultaneously and with enough of a delay to allow the atria to fill before electrical activity is again initiated at the SA node. Structural specializations in Purkinje fibers, such as well-developed intercalated discs (see Figure 29.13, p. 651) and numerous gap junctions, facilitate rapid conduction through these fibers.

The **cardiac center** in the brain's medulla provides external regulation to the heart (p. 730). It connects to the heart by two sets of nerves, the parasympathetic **vagus** nerves, which apply a braking action to the heart rate when activated by the brain, and the sympathetic nerves, which increase heart rate when activated. Both sets of nerves terminate in the SA node, thus regulating the rate of the pacemaker. The sympathetic nerves also innervate ventricular muscle fibers and cause an increase in stroke volume in addition to heart rate when activated.

The cardiac center in turn receives sensory information about a variety of stimuli. Pressure receptors (sensitive to blood pressure) and chemical receptors (sensitive primarily to carbon dioxide and pH) occur at strategic points in the vascular system. The cardiac center uses this information to adjust heart rate and stroke volume in response to activity or changes in body position. Feedback mechanisms thus control the heart and keep its activity constantly attuned to the body's needs.

Because the heartbeat is initiated in specialized muscle cells, vertebrate hearts, together with the hearts of molluscs and several other invertebrates, are called **myogenic** ("muscle origin") hearts. Although the nervous system does alter pacemaker activity to vary heart rate, a myogenic heart beats spontaneously and involuntarily even if completely removed from the body. An isolated turtle or frog heart beats for hours if placed in a balanced salt solution. The myogenic activity of a human heart allows heart transplants to occur successfully even if a heart has been removed from a body for several

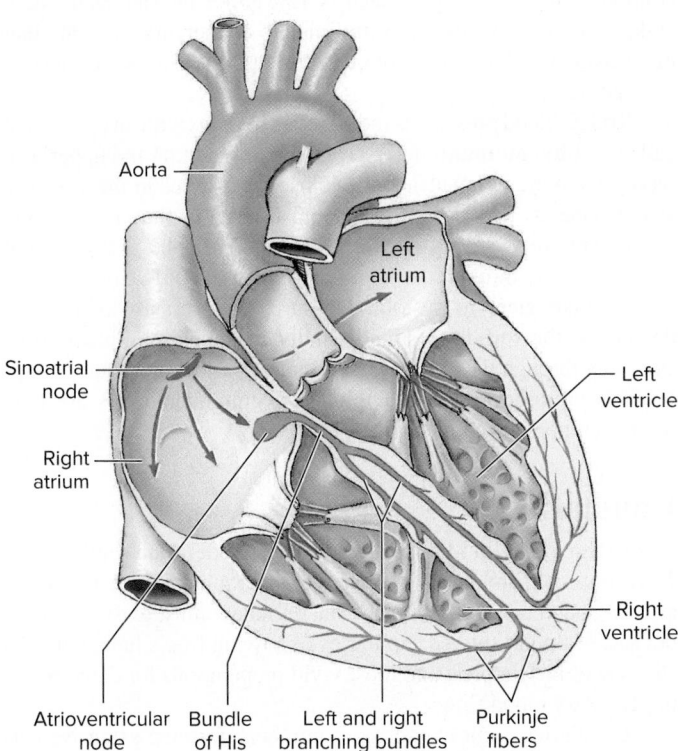

Figure 31.14 Mechanisms controlling heartbeat. Arrows indicate spread of excitation from the sinoatrial node, across the atria, to the atrioventricular node. The wave of excitation is then conducted very rapidly to ventricular muscle over the specialized conducting bundles and Purkinje fiber system.

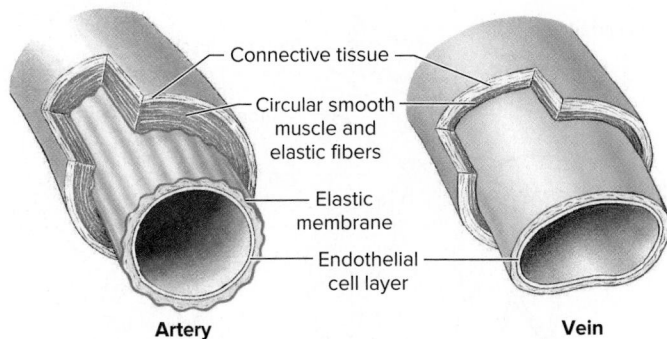

Figure 31.15 Artery and vein, showing tissue layers. Note greater thickness of the muscle layer in the artery. This layer has more elastic fibers in elastic arteries and more smooth muscle in muscular arteries.

strenuous exercise. Any reduction in coronary circulation due to partial or complete blockage (coronary artery disease) may lead to a heart attack (myocardial infarction), in which heart cells die from lack of oxygen.

Coronary artery disease (CAD) is currently the number one killer in the United States. Risk factors can be divided into those that cannot be modified and those that can be modified. Risk factors that cannot be modified include a family history of heart disease, being a male, being a postmenopausal female, and being over 45 years old. Risk factors that can be modified include smoking, high blood cholesterol levels, high blood pressure, uncontrolled diabetes, being overweight or obese, stress, high saturated fat and cholesterol diets, and a sedentary lifestyle. Reducing modifiable risk factors can significantly decrease the risk of CAD.

Arteries

All vessels leaving the heart are called arteries, whether they carry oxygenated blood (aorta) or deoxygenated blood (pulmonary artery). To withstand high, pulsing pressures generated during ventricular systole, the largest arteries closest to the heart (**elastic arteries**) are invested with thick layers of elastic fibers, very little smooth muscle (p. 193), and tough inelastic connective tissue (Figure 31.15). The elasticity of these arteries allows them to stretch as the surge of blood leaves the heart during ventricular systole and then to recoil, compressing the fluid column, during ventricular diastole. This elasticity maintains the high blood pressure created by each heartbeat, and blood moves in a forward direction because semilunar valves at the entrances to these arteries impede backward blood flow (Figure 31.12). Thus the normal arterial pressure in humans varies only between 120 mm Hg (systole) and 80 mm Hg (diastole) (usually expressed as 120/80), rather than dropping to zero during diastole, as we might expect in a fluid system with an intermittent pump. Arteries farther away from the heart possess more smooth muscle and fewer elastic fibers. These arteries, called **muscular arteries,** can increase or decrease their diameter, which dampens the heartbeat-associated high pressure and flow oscillations before the blood reaches body organs.

hours, especially if the heart's activity is slowed by cooling. Some invertebrates, for example, decapod crustaceans, have **neurogenic** ("nerve origin") hearts. In these hearts a cardiac ganglion (group of nerve cell bodies) located on the heart serves as pacemaker. If this ganglion is separated from the heart, the heart stops beating, even though the ganglion itself remains rhythmically active.

Coronary Circulation

In most vertebrates, an organ as active as the heart needs a generous blood supply of its own. The heart of small fishes and frogs is small, however, and the heart's own pumping action provides sufficient oxygenated blood to the muscle through extensive spaces between the muscle fibers. In larger fishes and frogs, as well as in nonavian reptiles, birds, and mammals, however, the larger heart size and the thickness of the heart muscle require that the heart have its own blood vessels, the **coronary circulation.** Coronary arteries are the first branches from the aorta. They divide to form the right and left coronary arteries that feed an extensive capillary network surrounding the muscle fibers and provide them with oxygen and nutrients. Heart muscle has an extremely high oxygen demand. Even at rest the heart removes 70% of oxygen from the blood, in contrast to most other body tissues, which remove only about 25%. Therefore, an increase in the work of the heart must be met by a massive increase in coronary blood flow—up to nine times the resting level during

Figure 31.16
Capillary bed. Precapillary sphincters (smooth muscles encircling an entrance to a capillary) control blood flow through capillaries.

Labels on figure: Small arteriole; Smooth muscle cells; Neuron; Sphincter contracted; Sphincter open; Capillaries; Small venule

Arteriosclerosis is a thickening and loss of elasticity in arteries. When arteriosclerosis is characterized by fatty deposits of cholesterol in artery walls, the condition is *atherosclerosis.* Inflammation precedes accumulation of fat in the arteries. Irregularities in the walls of blood vessels often cause blood to clot around them, forming a *thrombus.* When a bit of the thrombus breaks off and is carried by the blood to lodge elsewhere, it is an *embolus.* If the embolus blocks one of the coronary arteries, the person has a heart attack (a "coronary"). The portion of the heart muscle served by the blocked branch of the coronary artery is starved for oxygen. Scar tissue replaces dead muscle cells if the person survives. A thrombus also can form within a coronary artery and this is a more common cause of heart attack.

As arteries branch and narrow into **arterioles,** the walls are composed primarily of only one or two layers of smooth muscle surrounding an epithelial layer (Figure 31.16). Contraction of this muscle narrows the arterioles and reduces blood flow to some body organs, while simultaneous relaxation of arteriolar smooth muscle increases blood flow to other organs, diverting blood to where it is most needed. Blood must be pumped with a hydrostatic pressure sufficient to overcome resistance of the narrow passages through which it must flow. Consequently, large animals tend to have higher blood pressure than do small animals.

Blood pressure was first measured in 1733 by Stephan Hales, an English clergyman with unusual inventiveness and curiosity. He tied his mare, which was "to have been killed as unfit for service," on her back and exposed the femoral artery. This he cannulated with a brass tube, connecting it to a tall glass tube with the windpipe of a goose. The use of the windpipe was both imaginative and practical; it gave the apparatus flexibility "to avoid inconveniences that might arise if the mare struggled." The blood rose 8 feet in the glass tube and bobbed up and down with systolic and diastolic beats of the heart. The weight of the 8-foot column of blood was equal to the blood pressure. We now express blood pressure as the height of a column of mercury (Hg), which is 13.6 times heavier than water. Hales's figures, expressed in millimeters of mercury, indicate that he measured a blood pressure of 180 to 200 mm Hg, about normal for a horse.

Today, blood pressure is measured in humans with an instrument called a **sphygmomanometer.** A cuff is inflated on the upper arm with air to a pressure sufficient to close the arteries in the arm. If a stethoscope is held over the brachial artery (in the crook of the elbow) and air is slowly released from the cuff, the first turbulent spurts of blood through the artery as it opens slightly can be heard. This is equivalent to systolic pressure. As pressure in the cuff decreases, the turbulent sound finally disappears as blood runs smoothly through the artery. The pressure at which the sound disappears is diastolic pressure. Digital sphygmomanometers are now more commonly used.

Capillaries

The Italian Marcello Malpighi was the first to describe capillaries in 1661, thus confirming the existence of the minute links between the arterial and venous systems that Harvey knew must exist but could not see. Malpighi studied capillaries of a living frog's lung, which is still one of the simplest and most vivid preparations for demonstrating capillary blood flow.

Capillaries occur in enormous numbers, forming extensive networks in nearly all tissues (Figure 31.16). In muscle there are more than 2000 per square millimeter (1,250,000 per square inch), but not all are open at the same time. Indeed, perhaps fewer than 1% are open in resting skeletal muscle. But when muscle is active, all capillaries may open to bring oxygen and nutrients to the working muscle fibers and to carry away metabolic wastes.

Capillaries are extremely narrow, averaging in mammals about 8 µm in diameter, which is only slightly wider than the red blood cells that must pass through them. Their walls are formed by a single layer of thin **endothelial** cells, held together by a delicate basement membrane and a few connective-tissue fibers.

Capillary Exchange

Capillaries are permeable to small ions, nutrients, and water. Blood pressure within a capillary tends to force fluids out through or between the capillary endothelial cells and into the surrounding interstitial space (p. 681). Fluid may pass between the endothelial cells via water-filled clefts (approximately 4 nm wide) or through endothelial cells in pinocytotic vesicles (p. 49) that shuttle material from one side of an epithelial cell to the other. Lipid-soluble substances can diffuse easily through the plasma membranes of endothelial cells and into the interstitial fluid. Larger molecules within blood, such as plasma proteins, contribute to the colloid osmotic pressure of blood. Because these proteins cannot pass through the capillary endothelial cell clefts, an almost protein-free filtrate is forced outward. This fluid movement is important in irrigating the interstitial space, and in providing tissue cells with oxygen, glucose, amino acids, and other nutrients. For capillary exchange to be effective, fluids that leave the capillaries must at some point reenter the circulation and bring with them cellular metabolic wastes. If they did not, fluid would quickly accumulate in tissue spaces, causing swelling or edema. The delicate balance of

Figure 31.17 Fluid movement across the capillary endothelial cell clefts. At the arterial end of the capillary, hydrostatic (blood) pressure exceeds colloid osmotic pressure contributed by plasma proteins, and a plasma filtrate is forced out of the capillary. At the venous end, colloid osmotic pressure exceeds the hydrostatic pressure, and fluid is drawn into the capillary. In this way plasma nutrients are carried into the interstitial space where they can enter cells, and metabolic wastes from the cells are drawn into the plasma and carried away.

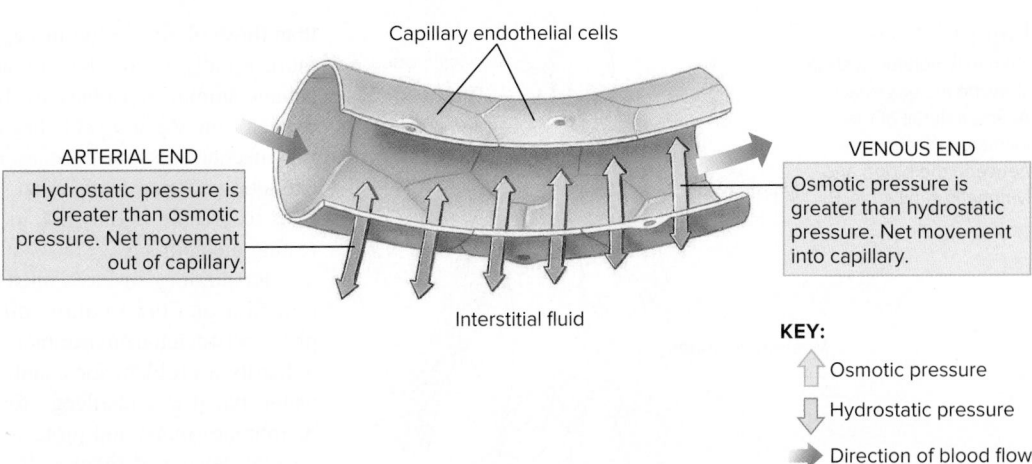

fluid exchange across the capillary wall is achieved by the two opposing forces of hydrostatic (blood) pressure and colloid osmotic pressure (Figure 31.17).

In a capillary, the blood pressure that pushes water molecules and solutes through the capillary endothelial cell clefts is greatest at the arteriolar end of the capillary and declines along its length as blood pressure falls (Figure 31.17). Opposing blood hydrostatic pressure is an osmotic pressure (p. 46) created by the proteins that cannot pass through the capillary endothelial cell clefts. This colloid osmotic pressure, which is about 25 mm Hg in mammalian plasma, tends to draw water back into the capillary from the tissue fluid. The result of these two opposing forces is that water and solutes tend to be filtered out of the arteriolar end of the capillary, where hydrostatic pressure exceeds colloid osmotic pressure, and to be drawn in again at the venous end, where colloid osmotic pressure exceeds hydrostatic pressure.

The amount of fluid filtered across the capillary endothelial cells fluctuates greatly among different capillaries. Usually outflow exceeds inflow, and some excess fluid remains in the interstitial spaces between cells, where it is called interstitial (intercellular) fluid. This excess interstitial fluid is collected and removed by **lymph capillaries** of the **lymphatic system** and eventually this fluid, now called **lymph,** returns to the circulatory system via larger lymph vessels.

Veins

Venules and veins into which the capillary blood drains for its return journey to the heart are thinner walled, less elastic, and of considerably larger diameter than their corresponding arteries and arterioles (Figure 31.15). Blood pressure in the venous system is low, varying from approximately 10 mm Hg, where capillaries drain into venules, to almost zero in the right atrium. Because pressure is so low, venous return must be assisted by valves in the veins, body skeletal muscles surrounding the veins, suction created during diastole of the heart, and the rhythmical action of the lungs during breathing. Without these mechanisms, blood would pool in the lower extremities of a standing animal. Veins that lift blood from the extremities to the heart contain valves that divide the long column of blood into segments. Valves are formed as infoldings of the endothelial cell layer and underlying connective tissue. When skeletal muscles contract, as

in even slight activity, veins are squeezed, and blood within them moves toward the heart because valves within the veins keep blood from flowing backward. The well-known risk of fainting while standing at stiff attention in hot weather usually can be prevented by deliberately pumping leg muscles. Negative pressure in the thorax created by inspiratory movements of the lungs and by ventricular systole also speeds venous return by drawing blood up the large vena cava into the heart.

Lymphatic System

The lymphatic system of vertebrates is an extensive network of thin-walled vessels that arise as blind-ended lymph capillaries in most tissues of the body. These unite to form a treelike structure of increasingly larger lymph vessels, which finally drain into veins in the lower neck (Figure 31.18). One function of the lymphatic system is to return to the blood the excess interstitial fluid filtered across capillary endothelial cells into interstitial spaces. This fluid-filtrate, called lymph once it has moved into the lymphatic vessels, is similar to plasma but has a much lower concentration of protein. Large molecules, especially fats absorbed from the gut, also reach the circulatory system by way of the lymphatic system when they pass into lacteals, lymphatic capillaries that occur within the villi of the small intestine (see Chapter 32, p. 713, and Figure 32.13). The rate of lymph flow is very low, a minute fraction of blood flow, and larger lymph vessels possess valves similar in structure and function to those found in veins.

The lymphatic system also plays a central role in the body's defenses. Located at intervals along the lymph vessels are **lymph nodes** (Figure 31.18) that have several defense-related functions (see Chapter 35). Cells in the lymph glands such as macrophages remove foreign particles, especially bacteria, which might otherwise enter the general circulation. They are also centers (together with bone marrow and thymus gland) for production, maintenance, and distribution of lymphocytes—essential components of a body's defense mechanisms.

31.4 RESPIRATION

Energy from food is released by oxidative processes, usually with molecular oxygen as the terminal electron acceptor. Physiologists distinguish two separate but interrelated respiratory processes:

Figure 31.18

Human lymphatic system, showing major vessels, **A,** and a detail of the intimate association between the blood and lymphatic capillaries, **B.**

Cervical nodes

Thoracic duct

Right lymphatic duct

Axillary nodes

Mesenteric nodes

Inguinal nodes

Lymphatic vessel

Arteriole

Valve

Venule

Lymphatic capillary

B

A

cellular respiration, the oxidative processes that occur within cells (p. 61), and **external respiration,** the exchange of oxygen and carbon dioxide between the organism and its environment across a respiratory surface. In this section we describe external respiration and transport of gases from respiratory surfaces to body tissues.

Problems of Aquatic and Aerial Breathing

An animal's mechanism of external respiration is determined largely by the nature of its environment. The two great arenas of animal evolution—water and land—are vastly different in their physical characteristics. The most obvious difference is that air contains far more oxygen—at least 20 times more—than does water. For example, water at 5°C (41°F) fully saturated with air contains approximately 9 ml of oxygen per liter (0.9%); by comparison, air contains 209 ml of oxygen per liter (21%). The density and viscosity of water are approximately 800 and 50 times greater, respectively,

than those of air. Furthermore, gas molecules diffuse 10,000 times more rapidly in air than in water. These differences mean that aquatic animals must have evolved very efficient ways of removing oxygen from water—yet fishes with highly efficient gills and pumping mechanisms may use as much as 20% of their energy just extracting oxygen from water. By comparison, the cost for mammals to breathe is only 1% to 2% of the energy produced during resting metabolism.

Respiratory surfaces must be thin and always kept wet with a thin film of fluid to allow diffusion of gases across an aqueous phase between the environment and the underlying circulation. This is hardly a problem for aquatic animals, immersed as they are in water, but it is a challenge for air breathers. To keep respiratory membranes moist and protected from injury, air breathers have in general developed them as invaginations of the body surface and then added pumping mechanisms to move air in and out of the respiratory region. The lung is the best example of a structure adapted for breathing on land. In general, **evaginations** of the body surface, such as gills, are most suitable for aquatic respiration; **invaginations,** such as lungs and tracheae, are best for air breathing.

Respiratory Organs

Gas Exchange by Direct Diffusion

Single-celled eukaryotes, sponges, cnidarians, and many worms respire by direct diffusion of gases between organism and environment. As noted at the beginning of this chapter, this kind of **cutaneous respiration** is not adequate when the cellular mass exceeds approximately 1 mm in diameter. However, by greatly increasing the surface of the body relative to its mass, many multicellular animals can supply part or all of their oxygen requirements by direct diffusion. Flatworms illustrate this strategy.

Cutaneous respiration frequently supplements gill or lung breathing in larger animals such as amphibians and fishes. For example, an eel can exchange 60% of its oxygen and carbon dioxide through its highly vascular skin. During winter hibernation, frogs and even turtles exchange all respiratory gases through their skin while submerged in ponds or springs. Most species of salamanders are lungless. Some lungless salamanders have larvae with gills, and gills persist in the adults of some (see Figure 25.9, p. 551), but adults of most species have neither lungs nor gills.

Gas Exchange Through Tubes: Tracheal Systems

Insects and certain other terrestrial arthropods (centipedes, millipedes, and some spiders) have a highly specialized type of respiratory system, in many respects the simplest, most direct, and most efficient respiratory system of active animals. It is a branching system of tubes **(tracheae)** that extends to all parts of the body (Figure 31.19). The smallest end channels are fluid-filled **tracheoles,** less than 1 μm in diameter, which terminate in close association with the plasma membranes of cells. Air enters and leaves the tracheal system through valvelike openings **(spiracles)** that may be closed to reduce water loss. A filter may also decrease entrance of water, debris, or parasites (see Figure 21.18, p. 454). Some insects can ventilate the tracheal system with body movements; the telescoping movement of the abdomen of bees on hot summer days is an example. Respiratory pigments occur in insect blood, but because the cells have direct access through the tracheal system to incoming oxygen and can release carbon dioxide from cells through this system an insect's respiration is independent of its circulatory system. Consequently, insect blood plays a minor role in oxygen transport.

Efficient Exchange in Water: Gills

Gills or **branchia** of various types are effective respiratory devices for life in water. Gills may be simple **external** extensions of the body surface, such as **dermal papulae** of sea stars (p. 477) or **branchial**

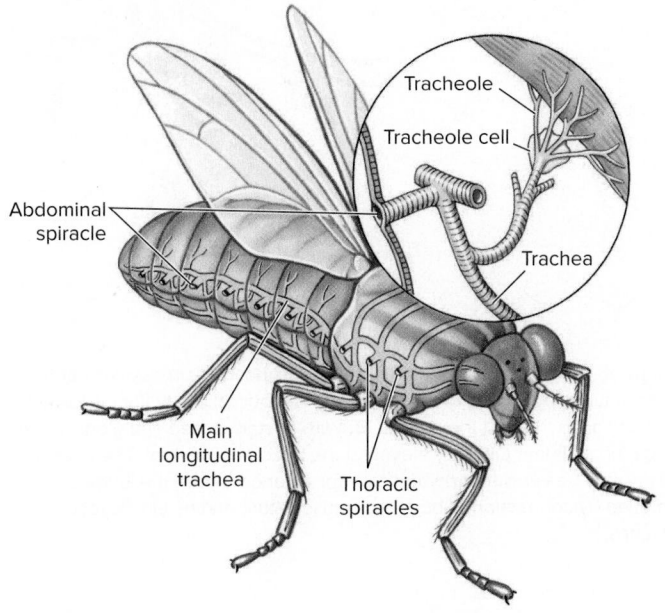

Figure 31.19　Tracheal system of insects. Air enters through spiracles, then travels through tracheae to reach tissues at tracheoles.

tufts (gills) of marine worms (p. 370) and aquatic amphibians (p. 548). The dorsal lobe of paddlelike appendages called **parapodia** also serves as an external respiratory surface for some marine polychaete worms whose blood vessels branch through the parapodia close to the surface for enhanced gas exchange (p. 366). Most efficient are **internal gills** of fishes (p. 534), molluscs (see Figures 16.5 and 16.18), and arthropods. Fish and mollusc gills are thin filamentous structures, richly supplied with blood vessels arranged so that blood flow is opposite to the flow of water across the gills. This arrangement, called **countercurrent flow** (pp. 534 and 671), provides the greatest possible extraction of oxygen from water. In molluscs, water is moved over the gill filaments by cilia. In fish, water flows over the gills in a steady stream, pulled and pushed by an efficient, two-valved, branchial pump composed of the mouth and opercular cavities (Figure 31.20). Gill ventilation is often assisted by the fish's forward movement through the water with its mouth open (ram ventilation).

Lungs

Gills are unsuitable for life in air because, when removed from the buoying water medium, gill filaments collapse, dry, and stick together; a fish out of water rapidly asphyxiates despite the abundance of oxygen around it. Consequently most air-breathing vertebrates possess highly vascularized internal cavities called lungs. Structures called lungs occur in invertebrates, such as pulmonate snails (see p. 344), scorpions, some spiders (see Figure 19.9), and some small crustaceans, but these structures are not homologous to vertebrate lungs and are usually not efficiently ventilated.

Lungs that can be ventilated by muscle movements to produce a rhythmic exchange of air characterize most terrestrial vertebrates. Most rudimentary of vertebrate lungs are those of lungfishes (pp. 529–530), which use them to supplement, or even to replace, gill respiration during periods of drought. Although of simple construction, a lungfish lung has a capillary network in its largely unfurrowed walls, a tubelike connection to the pharynx, and a primitive ventilating system for moving air into and out of the lung.

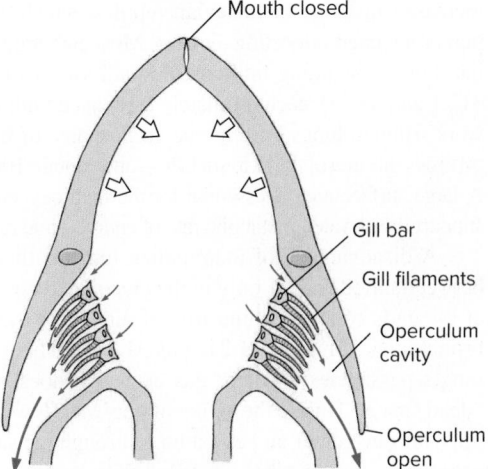

Figure 31.20　How a fish ventilates its gills. Through the action of two skeletal muscle pumps, one in the mouth cavity, the other in the opercular cavity, water is drawn into the mouth, passes over the gills, and exits through the gill covers (opercular clefts).

Mouth cavity expands, operculum is closed, opercular cavity expands, water is pulled into the mouth and across the gills.

Mouth closed, mouth cavity is compressed, operculum opens, water is pushed across gills.

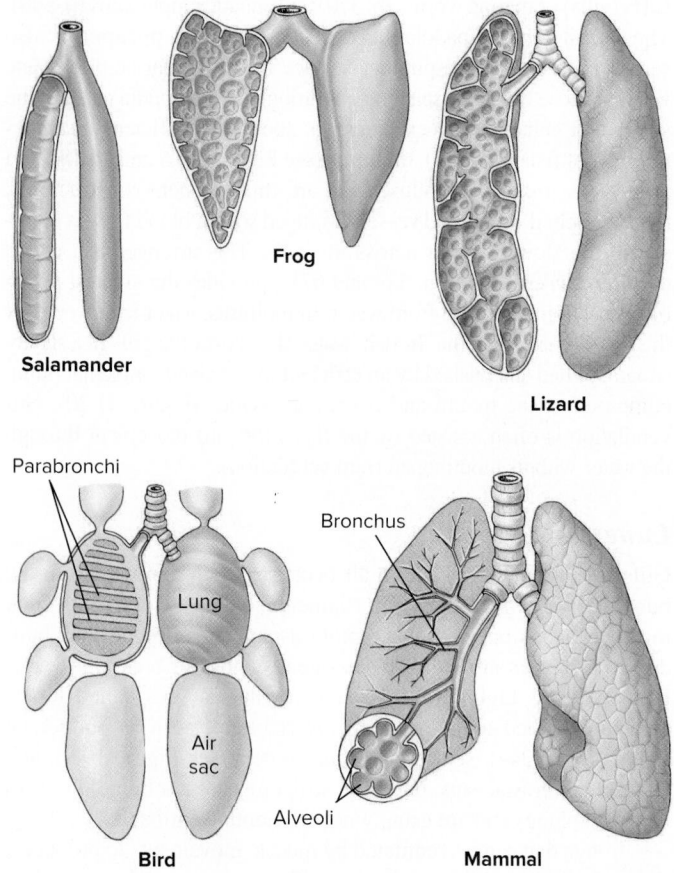

Figure 31.21 Variations among internal structures of lungs among vertebrate groups ranging from simple sacs with little exchange surface between blood and air spaces in amphibians to lobulated structures, each with complex divisions and extensive exchange surfaces, in birds and mammals.

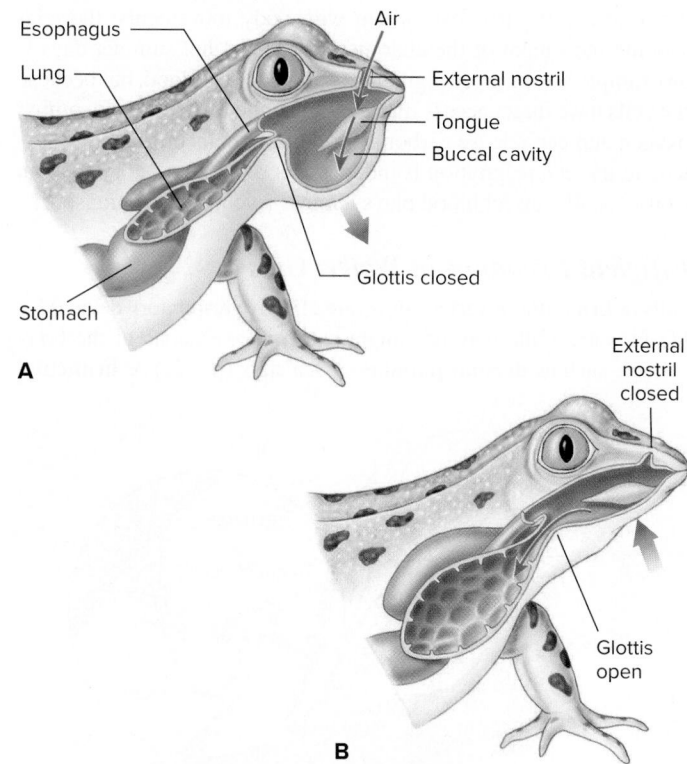

Figure 31.22 Breathing in frogs. Frogs, positive-pressure breathers, fill their lungs by forcing air into them. **A,** The buccal cavity floor is lowered, drawing air in through the nostrils. **B,** With nostrils closed and glottis open, frogs force air into lungs by elevating the buccal cavity floor. The buccal cavity may be ventilated rhythmically for a period before the lungs are emptied by contraction of body-wall musculature and by elastic recoil of lungs.

Amphibian lungs vary from simple, smooth-walled, baglike lungs of some salamanders to the subdivided lungs of frogs and toads (Figure 31.21). Total surface available for gas exchange is much increased in lungs of nonavian reptiles, which are subdivided into numerous interconnecting air sacs. Most elaborate of all are mammalian lungs containing millions of small sacs, called **alveoli** (Figures 31.21 and 31.23), each intimately associated with a rich vascular network. Human lungs have a total surface area of from 50 to 90 m^2—50 times the area of the skin surface—and contain 1000 km of capillaries. A large surface area is essential for the high oxygen uptake required to support the elevated metabolic rate of endothermic mammals.

A disadvantage of mammalian lungs is that gas is exchanged between blood and air only in the alveoli and alveolar ducts, located at the ends of a branching tree of air tubes (trachea, bronchi, and bronchioles [Figures 31.21 and 31.23]). The volume of air in a lung's passageways where gas exchange does not occur is called "dead space." Unlike the efficient one-way flow of water across fish gills, air must enter and exit a lung through the same channel. After exhalation, the air tubes are filled with "used" air from the alveoli which, during the following inhalation, is pulled back into the lungs with fresh air. This air shuttles back and forth with each breath, decreasing the efficiency of gas exchange. Lung ventilation in humans is so inefficient that in normal breathing only approximately one-sixth of the air in the lungs is replenished with each inspiration. Even after forced expiration, 20% to 35% of the air remains in the lungs.

In birds, lung efficiency is improved vastly over that of other terrestrial vertebrates by adding an extensive system of air sacs (Figure 31.21 and p. 594) that serve as air reservoirs during ventilation. On inspiration, about 25% of incoming air passes over the lung **parabronchi** (one-cell-thick air capillaries), where gas exchange occurs. The remaining 75% of incoming air bypasses the lungs to enter the air sacs (gas exchange does not occur here). At expiration some of this fresh air passes directly through the lung passages and into the lung parabronchi. Thus the parabronchi receive nearly fresh air during both inspiration and expiration. The highly evolved bird lung meets the extraordinary metabolic demands of flight.

Amphibians and lungfishes employ a **positive pressure** action to force air into their lungs, unlike most nonavian reptiles, birds, and mammals, which ventilate their lungs by **negative pressure,** in which expansion of the thoracic cavity pulls air into lungs. Frogs ventilate the lungs by first drawing air into the mouth through the **nostrils** (external nares). Then, closing the nostrils and raising the floor of the mouth, or buccal cavity, they drive air into the lungs (Figure 31.22). Much of the time, however, frogs rhythmically ventilate only the buccal cavity, a well-vascularized respiratory surface that supplements cutaneous and pulmonary respiration.

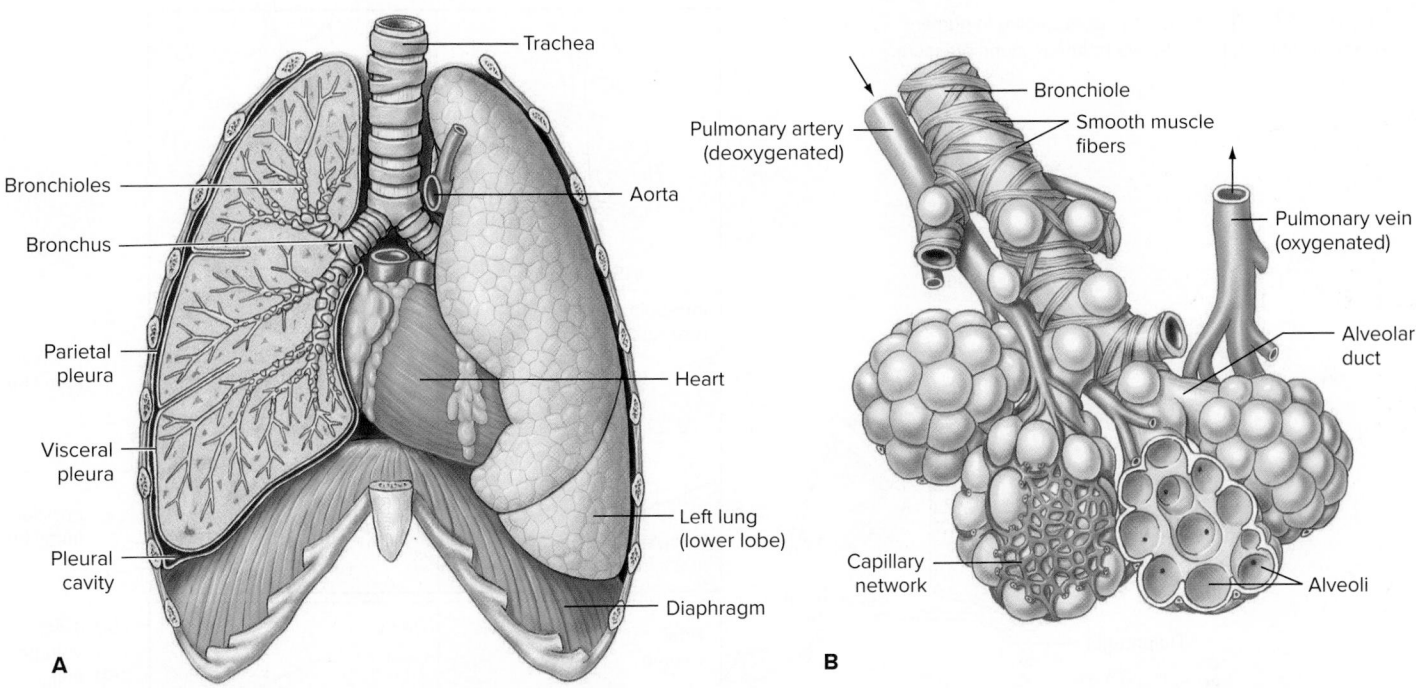

Figure 31.23   **A,** Human lungs with right lung shown in section. **B,** Terminal portion of bronchiole showing air sacs with their blood supply. Arrows show direction of blood flow.

Structure and Function of the Mammalian Respiratory System

Air enters a mammalian respiratory system through nostrils (external nares) and passes through a **nasal chamber,** lined with mucus-secreting epithelium, and then through **internal nares,** nasal openings connected to the **pharynx.** Here, where pathways of digestion and respiration cross, inhaled air leaves the pharynx by passing into a narrow opening, the **glottis,** while food enters the esophagus to pass to the stomach (see Figure 32.10, p. 709). The glottis opens into the **larynx,** or voice box, and then into the **trachea,** or windpipe. The trachea branches into two **bronchi,** one to each lung (Figure 31.23A). Within the lungs each bronchus divides and subdivides into small tubes (**bronchioles**) that lead via **alveolar ducts** to the air sacs (**alveoli**) (Figure 31.23B). The single-layered endothelial walls of the alveoli and alveolar ducts are thin and moist to facilitate gas exchange between air and adjacent blood capillaries. Air passageways are lined with both mucus-secreting and ciliated epithelial cells, which play an important role in conditioning the air before it reaches the alveoli. Partial cartilage rings in the walls of the tracheae, bronchi, and even some of the larger bronchioles provide flexible support and prevent those structures from collapsing.

In its passage to the air sacs, air is filtered free from most dust and other foreign substances, warmed to body temperature, and saturated with moisture.

Lungs contain much elastic connective tissue. They are covered by a thin layer of tough epithelium called the **visceral pleura.** Continuous with this layer is a similar layer, the **parietal pleura,** that lines the inner surface of the walls of the chest (Figure 31.23A). The two layers of the pleura are in close contact, are lubricated by tissue fluid, and slide over one another as the lungs expand and contract. The "space" between the pleura, called the **pleural cavity,** maintains a partial vacuum, or negative **intrapleural pressure,** which helps to keep the lungs expanded. The chest cavity is bounded by the spine, ribs, and breastbone, and floored by the **diaphragm,** a dome-shaped, muscular partition between the chest cavity and the abdomen. A muscular diaphragm occurs only in mammals.

Ventilating the Lungs

The chest cavity is an air-tight chamber. During inspiration the ribs are pulled upward by contraction of the external intercostal muscles, and the diaphragm contracts and flattens. The resultant increase in volume of the chest cavity (Figure 31.24) causes intrapleural pressure to fall to a more negative value and air pressure in the lungs, **intrapulmonary pressure,** to fall below atmospheric pressure: air rushes in through passageways to equalize the pressure. **Tidal volume** is the amount of air (ml) moved during this process. Normal **expiration** is a less active process than inspiration, primarily occurring by elastic recoil of connective tissues and relaxation of muscles. When this occurs, the ribs and diaphragm return to their original position, the chest cavity decreases in size, intrapulmonary and intrapleural pressures increase, the elastic lungs deflate, and air exits (Figure 31.24). During forced expiration, which occurs during exercise, the ribs are pulled down and inward to a greater extent than during normal expiration by contraction of an additional set of muscles located between the ribs (internal intercostal muscles). Simultaneous contraction of abdominal muscles forces the diaphragm upward to a greater degree as it relaxes due to upward pressure from abdominal organs, especially the liver, below it. Together these mechanisms expel more air and also enhance inspiratory volume during the next breath.

Figure 31.24 Mechanism of breathing in humans. Pressures (mm Hg) are relative to atmospheric pressure.

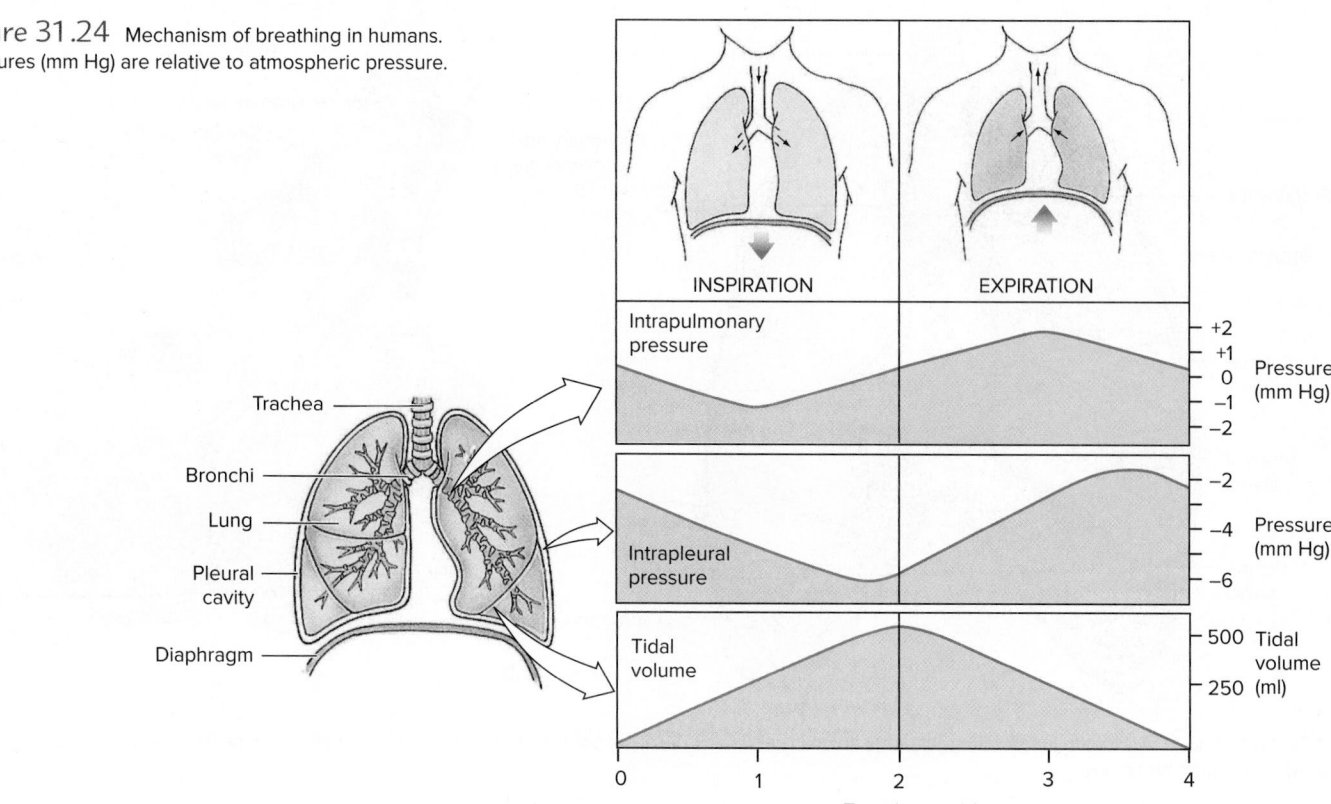

Coordination of Breathing

Breathing is normally involuntary and automatic but can be voluntarily controlled. Neurons in the medulla of the brain (p. 730) regulate normal, quiet breathing. They spontaneously produce rhythmical bursts that stimulate contraction of the diaphragm and external intercostal muscles during inspiration. However, respiration must adjust itself to changing requirements of the body for oxygen. Carbon dioxide rather than oxygen has the greatest effect on respiratory rate because chemical receptors (**chemoreceptors**) located centrally in the medulla of the brain and peripherally in the vascular system are more sensitive to elevated carbon dioxide levels than to low oxygen levels. Even a small rise in carbon dioxide level in the blood has a powerful effect on respiratory activity. Actually, the stimulatory effects of carbon dioxide are due in part to an increase in hydrogen-ion concentration in cerebrospinal fluid bathing the brain or in blood plasma bathing peripheral receptors.

$$CO_2 + H_2O \leftrightarrow H_2CO_3 \leftrightarrow H^+ + HCO_3^-$$

The above reaction shows that carbon dioxide combines with water to form carbonic acid. Carbonic acid then dissociates to release hydrogen ions, making the cerebrospinal fluid more acidic, and stimulating respiratory chemoreceptors in the medulla of the brain. Both rate and depth of respiration increase. Peripheral vascular chemoreceptors, located close to the heart and in the neck region, monitor peripheral changes in blood levels of carbon dioxide and hydrogen ions and send stimulatory signals to the medulla respiratory centers if these levels rise.

Swimmers can remain submerged much longer without breathing if they vigorously hyperventilate first to blow off carbon dioxide from the lungs, lowering circulating carbon dioxide levels in the blood and thereby delaying the overpowering urge to breathe. The practice is dangerous because blood oxygen is depleted while the individual is swimming just as rapidly as without prior hyperventilation, but this will not stimulate respiratory centers, and the swimmer may lose consciousness when the oxygen supply to the brain drops below a critical point. Several documented drownings among swimmers attempting long underwater swimming records have been caused by this practice.

Gaseous Exchange in Lungs and Body Tissues: Diffusion and Partial Pressure

Air (the atmosphere) is a mixture of gases: about 71% nitrogen and 20.9% oxygen, in addition to fractional percentages of other gases, such as carbon dioxide (0.03%). Gravity attracts the atmospheric gases to the earth. At sea level the atmosphere exerts a pressure due to gravity equal to 760 mm Hg (atmospheric pressure; 1 atm). Because air is a mixture of gases, *part* of the 760 mm Hg pressure is due to each component gas and is termed **partial pressure.** For example, the partial pressure of oxygen is $0.209 \times 760 = 159$ mm, and that for carbon dioxide is $0.0003 \times 760 = 0.23$ mm, in dry air. (Atmospheric air is never completely dry, and the varying amount of water vapor present exerts a pressure in proportion to its concentration, like other gases.)

TABLE 31.1

Partial Pressures and Gas Concentrations in Air and Body Fluids

	Nitrogen (N_2)	Oxygen (O_2)	Carbon Dioxide (CO_2)	Water Vapor (H_2O)
Inspired air (dry)	600 (79%)	159 (20.9%)	0.2 (0.03%)	—
Alveolar air (saturated)	573 (75.4%)	100 (13.2%)	40 (5.2%)	47 (6.2%)
Expired air (saturated)	569 (74.8%)	116 (15.3%)	28 (3.7%)	47 (6.2%)
Arterial blood	573	100	40	
Peripheral tissues	573	0–30	45–68	
Venous blood	573	40	46	

Note: Values expressed in millimeters of mercury (mm Hg). Percentages indicate proportion of total atmospheric pressure at sea level (760 mm Hg). Inspired air is shown as dry, although atmospheric air always contains variable amounts of water. If, for example, atmospheric air at 20° C were half saturated (relative humidity 50%), the partial pressures and percentages would be N_2 593.5 (78.1%); O_2 157 (20.6%); CO_2 0.2 (0.03%); and H_2O 8.75 (1.1%).

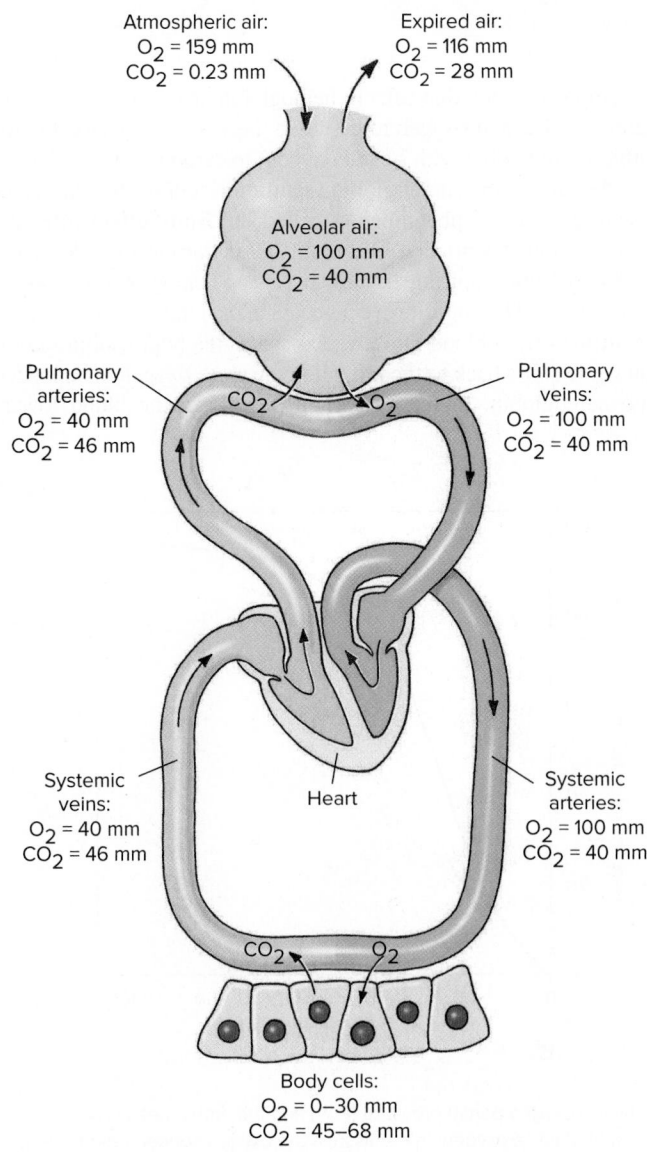

Figure 31.25 Exchange of respiratory gases in lungs and tissue cells. Numbers are partial pressures in millimeters of mercury (mm Hg).

As soon as air enters the respiratory tract, its composition changes (Table 31.1, Figure 31.25). Inspired air becomes saturated with water vapor as it travels through the air-filled passageways toward the alveoli. When inspired air reaches the alveoli, it mixes with residual air remaining from the previous respiratory cycle. Partial pressure of oxygen drops and that of carbon dioxide rises. Upon expiration, air from the alveoli mixes with air in the dead space to produce still a different mixture (Table 31.1). Although no significant gas exchange occurs in the dead space, the air it contains is the first air to leave the body when expiration begins.

Because the partial pressure of oxygen in lung alveoli is greater (100 mm Hg) than it is in blood entering lung capillaries (40 mm Hg), oxygen diffuses into the capillaries. In a similar manner carbon dioxide in blood of lung capillaries has a higher concentration (46 mm Hg) than is present in lung alveoli (40 mm Hg), so carbon dioxide diffuses from the blood capillaries into the alveoli.

In tissues respiratory gases also move along their concentration gradients (Figure 31.25). Cells continuously use and produce oxygen and carbon dioxide, respectively, so the partial pressure of oxygen in the arterial blood (100 mm Hg) entering the tissue bed is greater than that in the tissues (0 to 30 mm Hg), and partial pressure of carbon dioxide in tissues (45 to 68 mm Hg) is greater than that in blood (40 mm Hg). In each case gases diffuse from a location of higher concentration to one of lower concentration.

Because of the weight of water, pressure increases the equivalent of 1 atmosphere for every 10 m of depth in seawater, and the pressure of the air supplied to a diver must be increased correspondingly so that it can be drawn into the lungs. Under the increased pressure, additional air dissolves in the blood, the amount depending on depth and time at depth of a dive. If a diver ascends slowly, the gas comes out of solution imperceptibly and is breathed out from the lungs. However, if the ascent is too rapid, the air comes out of solution and forms bubbles in the blood and other tissues, a condition called *decompression sickness* or *the bends*. The result is painful and, if severe, can cause paralysis or death.

How Respiratory Gases Are Transported

In some invertebrates respiratory gases are simply carried dissolved in body fluids. However, solubility of oxygen is so low in water that this mode of transport is adequate only for animals with low metabolic rates. For example, only approximately 1% of a human's oxygen requirement can be transported in this way. Consequently in many invertebrates and in virtually all vertebrates, nearly all oxygen and a significant amount of carbon dioxide are transported in the blood by colored proteins called **respiratory pigments**. In all vertebrates these respiratory pigments are packaged into red blood cells.

The most widespread respiratory pigment in the animal kingdom is **hemoglobin,** a red, iron-containing protein present in all vertebrates and many invertebrates. Each molecule of hemoglobin is 5% **heme,** an iron-containing compound giving the red color to blood, and 95% **globin,** a colorless protein. The heme portion of hemoglobin has a great affinity for oxygen; each gram of hemoglobin can carry a maximum of approximately 1.3 ml of oxygen. Because there are approximately 15 g of hemoglobin in each 100 ml of blood, fully oxygenated blood contains approximately 20 ml of oxygen per 100 ml. Of course, for hemoglobin to be of value it must hold oxygen in a reversible chemical combination so that it can release oxygen to tissues. The actual amount of oxygen that combines with hemoglobin depends on the shape or conformation of the hemoglobin molecule, which is affected by several factors, including the concentration of oxygen itself. When the oxygen concentration is high, as it is in the capillaries of the lung alveoli, hemoglobin binds oxygen; in tissues where the prevailing oxygen partial pressure is low, hemoglobin releases its oxygen reserves (Figure 31.26).

We express the oxygen-carrying capacity of hemoglobin relative to the surrounding oxygen concentration as **hemoglobin saturation curves** (also called oxygen dissociation curves [Figure 31.26]). As these curves show, the lower the surrounding oxygen partial pressure, the greater the quantity of oxygen released from hemoglobin. This important characteristic of hemoglobin allows more oxygen to be released to those tissues that have a high level of aerobic cellular respiration (p. 61) and therefore need it most (those having the lowest partial pressure of oxygen).

> Sickle cell anemia is an up-to-now incurable, inherited condition (p. 95) in which a single amino acid (glutamic acid) in normal hemoglobin (HbA) is replaced by a valine in sickle cell hemoglobin (HbS). The ability of HbS to carry oxygen is severely impaired, and erythrocytes tend to crumple during periods of oxygen stress (for example, during exercise). Capillaries become clogged with misshapen red cells; the affected area is very painful, and the tissue may die. About 1 in 10 African-Americans carry the trait (heterozygous). Heterozygotes do not have sickle cell anemia and live normal lives, but if both parents are heterozygous, each of their offspring has a 25% chance of inheriting the disease.

Another factor that affects hemoglobin saturation curves and therefore release of oxygen to tissues is the sensitivity of **oxyhemoglobin** (hemoglobin with bound oxygen) to carbon dioxide. Carbon dioxide shifts the hemoglobin saturation curve to the right (Figure 31.26B), a phenomenon called the **Bohr effect** after the Danish scientist who first described it. As carbon dioxide enters the blood from respiring tissues, it causes hemoglobin to unload more oxygen. The opposite event occurs in the lungs; as carbon dioxide diffuses from blood into alveolar space, the hemoglobin saturation curve shifts back to the left, allowing more oxygen to be loaded onto hemoglobin. Increased carbon dioxide in the blood lowers

Figure 31.26 Hemoglobin saturation curves. Curves show the relationship between oxygen partial pressure (PO_2) and the amount of oxygen that can bind to hemoglobin. **A,** At the higher partial pressure in the lungs, hemoglobin can bind more oxygen. In the tissues the oxygen concentration is less, so hemoglobin can bind less and unloads more oxygen. **B,** Hemoglobin is also sensitive to carbon dioxide partial pressure (Bohr effect) and pH. As carbon dioxide enters blood from the tissues, it shifts the curve to the right, decreasing affinity of hemoglobin for oxygen. Thus, hemoglobin unloads more oxygen in the tissues where carbon dioxide concentration is higher or when pH is lower.

blood pH, as does addition of acid to blood (for example, lactic acid from exercising muscles, p. 656). A lower pH also shifts the hemoglobin saturation curve to the right and causes release of oxygen to active tissues (Figure 31.26B).

Although hemoglobin is the only vertebrate respiratory pigment, several other respiratory pigments occur among invertebrates. *Hemocyanin,* a blue, copper-containing protein, occurs in crustaceans and most molluscs. Among other pigments is *chlorocruorin* (klor-a-cru9-o-rin), a green, iron-containing pigment found in four families of polychaete tubeworms. Its structure and oxygen-carrying capacity are very similar to those of hemoglobin, but it is carried free in the plasma rather than being enclosed in blood cells. *Hemerythrin* is a red pigment found in some polychaete worms. Although it contains iron, this metal is not present in a heme group (despite the name of the pigment!), and its oxygen-carrying capacity is poor compared with hemoglobin.

The same blood that transports oxygen to the tissues from the lungs must carry carbon dioxide back to the lungs on its return trip. However, unlike oxygen, which is transported almost exclusively in combination with hemoglobin, carbon dioxide is transported in three different forms. A small fraction of the blood-borne carbon dioxide, only about 5%, occurs as gas physically dissolved in the plasma. The remainder diffuses into red blood cells. In red blood cells, most carbon dioxide, approximately 70%, becomes carbonic acid through action of the enzyme carbonic anhydrase. Carbonic acid immediately dissociates into hydrogen ion and bicarbonate ion. We can summarize the entire reaction as follows:

$$\overset{\text{carbonic anhydrase}}{CO_2 + H_2O \rightleftarrows H_2CO_3 \rightleftarrows H^+ + HCO_3^-}$$

The hydrogen ions combine with hemoglobin to form **deoxyhemoglobin,** thus preventing a severe decrease in blood pH, and releasing oxygen simultaneously. Biocarbonate ions are transported out of the red blood cells in exchange for chloride ions (the **chloride shift**). Bicarbonate ions remain in solution in plasma because unlike carbon dioxide, bicarbonate is extremely soluble (Figure 31.27).

Another fraction of the carbon dioxide, approximately 25%, combines reversibly with hemoglobin. Carbon dioxide combines not with the heme group but with amino groups of several amino acids to form a compound called **carbaminohemoglobin.**

All of these reactions are reversible. When blood reaches the lungs, bicarbonate is transported back into red blood cells (the chloride shift transporter reverses direction), it reacts with the hydrogen ions that are released from hemoglobin as oxygen is bound preferentially, and it is converted back into carbon dioxide. Carbon dioxide diffuses from red blood cells into plasma, along with carbon dioxide released from hemoglobin as it combines with oxygen. This carbon dioxide ultimately diffuses from plasma into alveolar air.

Unfortunately for humans and many other animals, hemoglobin has an affinity for carbon monoxide, an atmospheric contaminant due to industrialization, that is about 200 times greater than its affinity for oxygen. Consequently, even when carbon monoxide occurs in the atmosphere at lower concentrations than oxygen, it tends to displace oxygen from hemoglobin to form a stable compound called carboxyhemoglobin. Air containing only 0.2% carbon monoxide may be fatal. Because of their higher respiratory rate, children and small animals are poisoned more rapidly than are adults.

Figure 31.27 Transport of carbon dioxide in the blood. **A,** Carbon dioxide produced by cellular respiration diffuses from the tissues into plasma and red blood cells. Carbonic anhydrase in red blood cells catalyzes conversion of carbon dioxide into carbonic acid, then bicarbonate and hydrogen ions. The bicarbonate diffuses out of the cells, and inward diffusion of chloride ions maintains electrical balance. Hydrogen ions associate with hemoglobin. **B,** The lower partial pressure of carbon dioxide in the alveoli of the lungs favors reversal of these reactions.

SUMMARY

Fluid in the body, whether intracellular, plasma, or interstitial, is mostly water, but contains many dissolved substances, including electrolytes and proteins. Vertebrate blood consists of fluid plasma and cellular components, including red and white blood cells and platelets. Plasma has many dissolved solids, as well as dissolved gases. Mammalian red blood cells lose their nucleus during development and contain the oxygen-carrying pigment, hemoglobin. White blood cells are important in immune function. Platelets are vital in the process of clotting, necessary to prevent excess blood loss when a blood vessel is damaged. Platelets release a series of factors that convert prothrombin to thrombin, an enzyme that causes fibrinogen to change to an insoluble form, fibrin, forming the fibers of a blood clot.

In open circulatory systems, such as those of arthropods and most molluscs, blood moves from arteries into a hemocoel, which is a body cavity derived from the blastocoel and embryonic coelomic cavities. In closed circulatory systems, such as those of annelids, vertebrates, and cephalopod molluscs, the heart pumps blood into arteries, then into arterioles of smaller diameter, through a bed of fine capillaries, through venules, and finally through veins, which lead back to the heart. In fishes, which have a single atrium and a single ventricle, blood is pumped to gills and then directly to systemic capillaries throughout the body without first returning to the heart. With evolution of lungs, vertebrates developed a double circulation consisting of a systemic circuit serving the body, and a pulmonary circuit serving the lungs. To be fully efficient, this change required partitioning of both atrium and ventricle to form a double pump; partial partitioning occurs in lungfishes and amphibians, which have two atria but an undivided ventricle, and is complete in crocodilians, birds, and mammals, which have two ventricles.

One-way flow of blood during the heart's contraction (ventricular systole) and relaxation (ventricular diastole) is assured by valves between the atria and the ventricles and between the ventricles and the pulmonary arteries and aorta. Although the heart can beat spontaneously, due to the presence of pacemaker cells, its rate is controlled by hormones and by nerves from the autonomic system. Heart muscle uses much oxygen and has a well-developed coronary blood circulation. The walls of arteries are thicker than those of veins, and the elastic connective tissue in the walls of larger arteries allows them to expand during ventricular systole and to recoil during ventricular diastole. Normal arterial blood pressure (hydrostatic) of humans in systole is 120 mm Hg; in diastole, it is 80 mm Hg. Because capillary endothelial cells possess narrow water-filled clefts between them, a protein-free filtrate crosses capillary walls, its movement determined by a balance between opposing forces of hydrostatic and protein osmotic pressure. Substances also leave and enter the blood via pinocytotic vesicles and diffusion (lipid-soluble molecules) through the endothelial cells. Tissue fluid that does not reenter the capillary system is collected by the lymphatic system (now called lymph) and returned to blood by lymph ducts.

Very small animals can depend on diffusion between the external environment and their tissues or cytoplasm for transport of respiratory gases, but larger animals require specialized organs, such as gills, tracheae, or lungs, for this function. Gills and lungs provide an increased surface area for exchange of respiratory gases between blood and the environment. Many animals have special respiratory pigments and other mechanisms to help transport oxygen and carbon dioxide in blood. The most widespread respiratory pigment in the animal kingdom, hemoglobin, has a high affinity for oxygen at high oxygen concentrations but releases it at lower concentrations. Vertebrate hemoglobin, which is packaged in red blood cells, combines readily with oxygen in gills or lungs, then releases it in respiring body tissues where the oxygen partial pressure is low. Blood carries carbon dioxide from tissues to lungs as bicarbonate ion and dissolved gas in the plasma, and, in combination with hemoglobin, in red blood cells.

REVIEW QUESTIONS

1. Name the chief intracellular electrolytes and the chief extracellular electrolytes.
2. What is the fate of spent erythrocytes in the body?
3. Outline or briefly describe the sequence of events that leads to blood coagulation.
4. Two distinctly different styles of circulatory systems have evolved among animals: open and closed. What is "open" about an open circulatory system? Closed systems sometimes are cited as adaptive for actively moving animals with (at least at times) high metabolic demand. Can you suggest possible reasons for this assertion?
5. Place the following in correct order to describe the circuit of blood through the vascular system of a fish: ventricle, gill capillaries, sinus venosus, body tissue capillaries, atrium, dorsal aorta.
6. Trace the flow of blood through a mammalian heart, naming the four chambers and their valves, and explaining where the blood entering each atrium comes from and where blood leaving each ventricle goes. When the ventricles contract, what prevents blood from reentering the atria? What factors cause blood to move forward at high pressure in the aorta?
7. Explain the origin and conduction of the excitation that leads to a heart contraction. Why is the vertebrate heart said to be myogenic? If the heart is myogenic, how do you account for alterations in rate of the heartbeat?
8. Define the terms systole and diastole. Distinguish atrial and ventricular systole and diastole.
9. Explain the movement of fluid across capillary endothelial cells. How does balance of hydrostatic pressure and colloid osmotic pressure determine direction of net fluid flow?
10. Blood pressure at the arterial end of capillaries is about 40 mm Hg in humans. If blood pressure at the venous end is about 15 mm Hg, and colloid osmotic pressure is 25 mm Hg throughout, what is the net effect on fluid movement between capillaries and tissue spaces?
11. Provide a brief description of the lymphatic system. What are its principal functions? Why is movement of lymph through the lymphatic system very slow?
12. Describe an advantage of a fish's gills for breathing in water and a disadvantage for breathing on land.
13. Describe the tracheal system of insects. What is the advantage of such a system for a small animal?
14. Trace the route of inspired air in humans from the nostrils to the smallest chamber of the lungs. What is the "dead air space" of a mammalian lung and how does it affect the partial pressure of oxygen reaching the alveoli? How is this problem partially solved by bird respiratory systems?
15. The amount of time that scuba divers can spend underwater is limited by several factors, including time required to deplete the air supply in their tanks. To make their air last longer, novice divers may be instructed to breathe slowly and exhale as much as possible on each breath. Can you suggest a reason why this behavior would lengthen a diver's air supply?
16. How does a frog ventilate its lungs? Contrast an amphibian's positive-pressure breathing with a mammal's negative-pressure breathing.
17. What is the role of carbon dioxide in the control of rate and depth of mammalian breathing?
18. The air pressure supplied to a scuba diver must equal that exerted by the surrounding seawater, and for each 10 m increase in depth,

pressure of the surrounding seawater increases one full atmosphere. Assuming the partial pressure of oxygen in air at sea level (one atmosphere) is 0.209 × 760 mm Hg (= 159 mm Hg), what partial pressure of oxygen would a diver be breathing at a depth of 30 m?

19. Explain how oxygen is carried in blood, including specifically the role of hemoglobin. Answer the same question with regard to carbon dioxide transport.

20. The ability of hemoglobin to bind oxygen decreases with decreasing oxygen concentration and also decreases with increasing carbon dioxide concentration. What effect do these phenomena have on the delivery of oxygen to tissues?

For Further Thought At high altitudes, the partial pressure of oxygen is lower. How might you expect this lower oxygen level to affect the ability of the blood to transport respiratory gases, and what would be the effect on respiring tissues?

SELECTED REFERENCES

Bartecchi, C. E. 1998. If you don't have a defibrillator. Sci. Am. **278:** 91. *Describes "cough" and "thump" techniques that can be used instead of cardiopulmonary resuscitation (CPR).*

Brusca, R., and G. Brusca. 2003. Invertebrates, ed. 2. Sunderland, MA, Sinauer Associates. *Good discussion of origins of body cavities in invertebrates.*

Burggren, W. W. 1997. Identifying and evaluating patterns in cardiorespiratory physiology. Am. Zool. **37:** 109–115. *One of several papers in this issue that covers cardiorespiratory physiology.*

Eisenberg, M. S. 1998. Defibrillation: The spark of life. Sci. Am. **278:** 86–90. *When the pacemaker loses its rhythm, heart muscle commences uncoordinated contractions. Application of a brief electrical shock from a defibrillator often can "reset" the pacemaker. Defibrillators have saved many lives.*

Glass, M. L., and S. C. Wood (eds.). 2009. Cardio-respiratory control in vertebrates. Berlin, Germany, Springer. *Reviews many aspects of vertebrate respiration from an evolutionary perspective.*

Hardison, R. 1999. The evolution of hemoglobin. Am. Sci. **87:** 126–137. *Comparison of amino-acid sequences in hemoglobins from animals, plants, unicellular eukaryotes, and eubacteria suggests that all share a common ancestor early in organismal evolution.*

Hill, R. W., G. A. Wyse, and A. Anderson. 2012. Animal Physiology, ed. 3. Sunderland, MA, Sinauer Associates, Inc. Publishers. *Well-written chapters on respiratory and circulatory systems.*

Jain, R. K., and P. F. Carmeliet. 2001. Vessels of death or life. Sci. Am. **285:** 38–43. *Overview of blood vessel development (angiogenesis) and its role in many serious diseases.*

Kiberstis, P., and J. Marx. 1996. Cardiovascular medicine. Science **272:** 663. *Introduction to a series of articles on heart development, genetics of blood pressure, genetics of cardiovascular disease, mouse models of atherosclerosis, molecular therapies for vascular diseases, and drugs for stroke.*

Libby, P. 2002. Atherosclerosis: the new view. Sci. Am. **286:** 47–55 (May). *Describes the data on development of atherosclerosis.*

Lillywhite, H. B. 1988. Snakes, blood circulation and gravity. Sci. Am. **259:** 92–98 (Dec.). *How a snake's vascular system is designed to counter the effects of gravity.*

Nucci, M. L., and A. Abuchowski. 1998. The search for blood substitutes. Sci. Am. **278:** 73–77. *Shortage of blood supplies and risk of contamination have made the search for substitutes more urgent.*

Randall, D., W. Burggren, and K. French. 2002. Eckert animal physiology: mechanisms and adaptations, ed. 5. New York, W. H. Freeman & Company. *A readable, comparative approach to animal physiology.*

Williams, T. M., P. Bengtson, D. L. Steller, D. A. Croll, and R. W. Davis. 2015. The healthy heart: lessons from nature's elite athletes. Physiology **30:** 349–357. *This article provides insight into the reasons cardiovascular disease is three times more likely in humans than in wild mammals, even though heart structure and function are similar.*

32
CHAPTER

Digestion and Nutrition

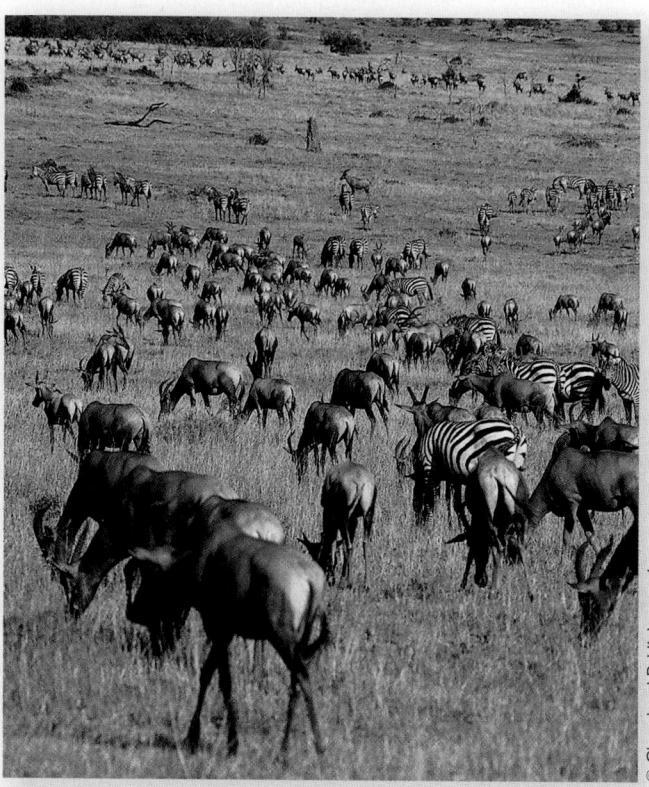

Topi and zebras on the African savannah.

© Cleveland P. Hickman, Jr.

A Consuming Cornucopia

Sir Walter Raleigh observed that the difference between a rich man and a poor man is that the former eats when he pleases while the latter eats when he can get food. Unlike the affluent for whom food acquisition requires only the selection of prepackaged foods at a well-stocked supermarket, the world's poor can appreciate that for them, as for the rest of the animal kingdom, procuring food is the main challenge for survival.

Potential food is everywhere and little remains unexploited. Animals bite, chew, nibble, crush, graze, browse, shred, rasp, filter, engulf, enmesh, suck, and absorb foods of incredible variety. What an animal eats and how it eats profoundly affect an animal's feeding specialization, its behavior, its physiology, and its internal and external anatomy—in

short, both its body form and its role in the web of life. The endless evolutionary jostling between predator and prey has produced adaptations for eating and adaptations for avoiding being eaten. By whatever means food may be secured, there is far less variation among animals in the subsequent digestion or breakdown of foods into useful, simpler products. Vertebrates and invertebrates alike use similar digestive enzymes. Even more uniform are the final biochemical pathways for nutrient use and energy transformation (see Chapter 4). The nourishment of animals is like a cornucopia in which the food flows in rather than out. A great diversity of foods procured by countless feeding adaptations enters the mouth of the horn, is simplified, and finally is used for organismal survival and reproduction.

All organisms require energy to maintain their highly ordered and complex structures. This energy is chemical bond energy released by transforming complex compounds acquired from an organism's environment into simpler ones.

The ultimate source of energy for life on earth is the sun. Sunlight is captured by chlorophyll molecules in green plants, which transform a portion of this energy into chemical bond energy (food energy). Green plants are **autotrophic** organisms, which require only inorganic compounds absorbed from their surroundings to provide the raw material for synthesis and growth. Most autotrophic organisms are the chlorophyll-bearing **phototrophs,** although some, the chemosynthetic bacteria, are **chemotrophs** that gain energy from inorganic chemical reactions.

Almost all animals are **heterotrophic** organisms that depend on already synthesized organic compounds of plants and other animals to obtain the materials they will use for growth, maintenance, and reproduction. An animal's food is normally the complex tissues of other organisms that must be digested into soluble molecules small enough to be absorbed as nutrients by its body's cells.

Animals are divided into categories based on dietary habits. **Herbivorous** animals feed mainly on plant life. **Carnivorous** animals feed mainly on other animals. **Omnivorous** animals eat both plants and animals. **Saprophagous** animals feed on decaying organic matter.

Ingestion of foods and their simplification by **digestion** are only initial steps in nutrition. Foods reduced by digestion to more simple, soluble, molecular forms are **absorbed** and **transported** to the body's tissues. There they are **assimilated** into the structures of cells, **oxidized** to yield energy and heat, or if not immediately used, **stored** for future use. Wastes produced by oxidation must be **excreted.** Food products that remain undigested and are not absorbed are **egested** in feces.

32.1 FEEDING MECHANISMS

Few animals can absorb nutrients directly from their external environments. Exceptions are some blood parasites (p. 236), certain intestinal amoebozoan parasites (p. 241), and tapeworms (p. 304) and acanthocephalans (p. 313) that nourish themselves on primary organic molecules absorbed directly across their body surfaces. These nutrients have already been digested by their host organism. Most animals must work for their meals. They are active feeders that have evolved numerous specializations for obtaining food. With food procurement as one of the most potent driving forces in animal evolution, natural selection has placed a high priority on adaptations for exploiting new sources of food and the means of food capture and intake.

Feeding on Particulate Matter

Drifting microscopic particles abound in the upper hundred meters of ocean water. Most of this multitude is **plankton,** organisms that are nonmotile or too small or weak to do anything but drift with the ocean's currents. The rest is organic debris, disintegrating remains of dead plants and animals. Although this oceanic swarm of plankton forms a rich life domain, it is unevenly distributed. The heaviest plankton growth occurs in estuaries and areas of upwelling, where there is an abundant nutrient supply. It is consumed by numerous larger animals, invertebrates and vertebrates, using a variety of feeding mechanisms.

One of the most important and widely employed methods for feeding is **suspension feeding** (Figure 32.1). Most suspension feeders use ciliated surfaces to produce currents that draw drifting food particles into their mouths. Most suspension-feeding invertebrates, such as sedentary polychaete worms, bivalve molluscs, hemichordates (see Figures 22.34 and 22.35), and most protochordates (see Figures 23.4 and 23.8), entrap particulate food on mucous sheets that convey the food into the digestive tract. Others, such as fairy shrimps, (see Figure 20.17), water fleas, and barnacles, use sweeping movements of their setae-fringed legs to create water currents and to entrap food, which is transferred to their mouth. Freshwater developmental stages of certain insect orders use fanlike arrangements of setae or spin silk nets to entrap food.

One form of suspension feeding, often called **filter-feeding,** has evolved frequently as a secondary modification among representatives of groups that are primarily selective feeders. These animals possess filtering devices that strain food from water as it passes through them. Examples include many microcrustaceans, fishes such as herring, menhaden, and basking sharks, certain birds such as flamingos, and the largest of all animals, baleen (whalebone) whales (Figure 32.1). The vital importance of one component of plankton, diatoms, in supporting a great pyramid of filter-feeding animals is stressed by N. J. Berrill:[1]

> A humpback whale . . . needs a ton of herring in its stomach to feel comfortably full—as many as five thousand individual fish. Each herring, in turn, may well have 6000 or 7000 small crustaceans in its own stomach, each of which contains as many as 130,000 diatoms. In other words, some 400 billion golden-brown diatoms sustain a single medium-sized whale for a few hours at most.

Another type of particulate feeding exploits deposits of disintegrated organic material (detritus) that accumulates on and in the substratum; this type is called **deposit feeding.** Some deposit feeders, such as many annelids and some hemichordates, simply pass the substrate through their bodies, removing nutrients from it (see p. 374). Others, such as scaphopod molluscs, certain bivalve molluscs, and some sedentary, tube-dwelling polychaete worms, use appendages to gather organic deposits some distance from the body and move them toward the mouth (Figure 32.2).

Feeding on Food Masses

Among the most interesting animal adaptations are those that have evolved for procuring and manipulating solid food. Such adaptations and animals bearing them are largely shaped by what the animal eats.

Predators must locate, capture, hold, and swallow prey. Many carnivorous animals simply seize food and swallow it intact, while some employ toxins that paralyze or kill prey upon capture. Although no true teeth appear among invertebrates, many have beaks or toothlike structures for biting and holding. A familiar example is the errant, carnivorous polychaete *Nereis,* which possesses a muscular pharynx armed with chitinous jaws that can be

---

[1]Berrill, N. J. 1958. You and the universe. New York, Dodd, Mead & Co.

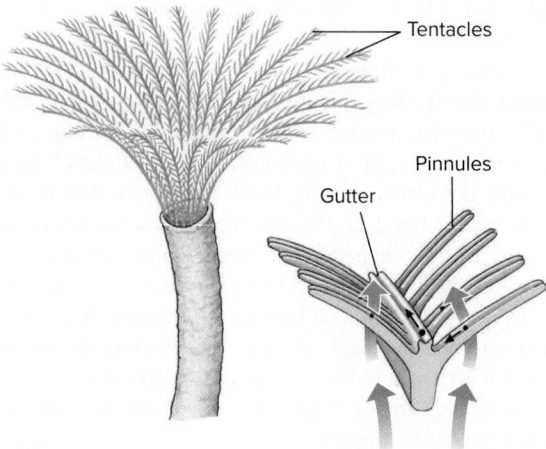

A Marine fan worms (class Sedentaria, phylum Annelida) have a crown of tentacles. Numerous cilia on the edges of the tentacles draw water (*blue arrows*) between pinnules where food particles are entrapped in mucus; particles are then carried down a "gutter" in the center of the tentacle to the mouth (*black arrows*).

B Bivalve molluscs (class Bivalvia, phylum Mollusca) use their gills as feeding devices, as well as for respiratory gas exchange. Water currents created by cilia on the gills carry food particles into the incurrent siphon and between slits in the gills where they are entangled in a mucous sheet covering the gill surface. Ciliated food grooves then transport the particles to the mouth (not shown). Arrows indicate direction of water movement.

C Barnacles (class Maxillopoda, subphylum Crustacea, phylum Arthropoda) sweep their thoracic appendages (cirri) through the water to trap plankton and other organic particles on fine bristles that fringe the cirri. Food is transferred to the barnacle's mouth by the first, short cirri.

D Herring and other filter-feeding fishes (class Actinopterygii, phylum Chordata) use gill rakers that project forward from the gill arches into the pharyngeal cavity to strain plankton. Herring swim almost constantly, forcing water and suspended food into their mouth; food is strained by their gill rakers, and water passes through the gill openings.

E Baleen whales (class Mammalia, phylum Chordata) filter plankton, principally large crustaceans called krill, with baleen. Water enters the swimming whale's open mouth by force of the animal's forward motion and is strained through more than 300 keratinized baleen plates that hang like a curtain from the roof of the mouth. Krill and other plankton caught in the baleen are periodically collected by the huge tongue and swallowed.

Figure 32.1   Some suspension and filter feeders and their feeding mechanisms.

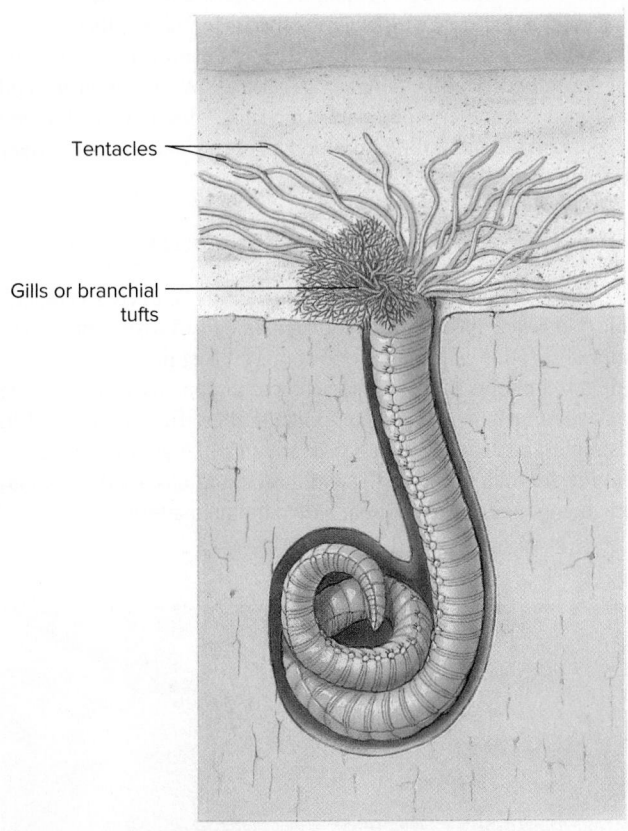

Figure 32.2   The annelid *Amphitrite* is a deposit feeder that lives in a mucus-lined burrow and extends long feeding tentacles in all directions across the surface. Food trapped on mucus is conveyed along the tentacles to the mouth.

Figure 32.3   This African egg-eating snake, *Dasypeltis,* subsists entirely on hard-shelled birds' eggs, such as this quail egg, which it swallows whole. Its special adaptations are reduced size and number of teeth, enormously expansible jaw provided with elastic ligaments, and teethlike vertebral spurs that puncture the shell. Subsequently, the snake will puncture and collapse the egg, swallow its contents, and regurgitate the crushed shell.

everted with great speed to seize prey (see Figure 17.2, p. 366). Once a capture is made, the pharynx is retracted and prey swallowed. Fishes, amphibians, and nonavian reptiles use their teeth principally to grip prey and to prevent its escape until they can swallow it whole. Snakes and some fishes can swallow enormous meals. An absence of limbs means that alternate methods of prey capture and swallowing have evolved. Recurved teeth seize and hold prey and distensible jaws and stomachs accommodate large and infrequent meals (Figure 32.3). Birds lack teeth, but their beaks often possess serrated edges, or the upper beak is hooked for seizing and tearing prey (see Figure 27.11, p. 594).

Many invertebrates reduce food size by shredding devices (such as shredding mouthparts of many crustaceans) or by tearing devices (such as beaklike jaws of cephalopod molluscs; see Figure 16.38, p. 355, and Figure 16.40, p. 356). Insects have three pairs of appendages on their heads that serve variously as jaws, chitinous teeth, chisels, tongues, or sucking tubes (see Figure 21.17, p. 453). Usually the first pair serves as crushing teeth; the second, as grasping jaws; and the third, as a probing and tasting tongue.

True mastication, chewing or crushing of food as opposed to tearing, is nearly restricted to mammals. Mammals usually have four different types of teeth, each adapted for specific functions. **Incisors** are designed for biting, cutting, and stripping; **canines** are for

seizing, piercing, and tearing; **premolars** and **molars,** at the back of the jaw, are for grinding and crushing (Figure 32.4). This basic pattern is often greatly modified in animals having specialized food habits (Figure 32.5; see also Figure 28.10, p. 618). Herbivores usually lack canines but have well-developed molars with enamel ridges for grinding. Well-developed, self-sharpening incisors of rodents grow throughout life and must be worn away by gnawing to keep pace with growth. Some teeth have become so highly modified that they are no longer useful for biting or chewing food. An elephant's tusk (Figure 32.6) is a modified upper incisor used for defense, attack, and rooting, and a male wild boar has modified canines used as weapons. Many feeding specializations of mammals are described on pages 618–620.

Herbivorous, or plant-eating, animals have evolved special devices for crushing and cutting plant material. Some invertebrates have scraping mouthparts, such as the radula of snails (see Figure 16.4, p. 336). Insects such as locusts have grinding and cutting mandibles; herbivorous mammals such as horses and cattle use wide, corrugated molars for grinding. All such mechanisms disrupt the tough cellulose plant cell wall to accelerate its digestion by intestinal microorganisms, as well as to release cell contents for direct enzymatic breakdown. Thus herbivores digest food that carnivores cannot, and in doing so, convert plant material into organic compounds for their own use and for ultimate consumption by carnivores and omnivores.

Feeding on Fluids

Fluid-feeding is especially characteristic of parasites, but it is practiced among many free-living forms as well. Some internal parasites (endoparasites) simply absorb the nutrient surrounding them, unwittingly provided by the host. Others bite and rasp host tissue, suck blood, and feed on contents of the host's intestine. External parasites (ectoparasites) such as leeches, lampreys (see Figure 24.5, p. 521), parasitic crustaceans, and insects use a variety of efficient piercing and sucking mouthparts to feed on blood or other body fluid. There are numerous arthropods that feed on fluids, for example, fleas, mosquitos, sucking lice, bedbugs, ticks, and mites. Many are vectors of serious diseases of humankind and thus qualify as far more than pesky annoyances.

Unfortunately for humans and other warm-blooded animals, the ubiquitous mosquito excels in its blood-sucking habit. Alighting gently, the mosquito punctures its prey with an array of six needlelike mouthparts (see Figure 21.17B and C, p. 453). One of these injects an anticoagulant saliva, which causes the irritating itch that follows the "bite" and serves as a vector for microorganisms causing malaria, yellow fever, encephalitis, dengue, Zika virus, West Nile virus, and other diseases. Another mouthpart is a channel through which blood is sucked. Only females dine on blood to obtain nutrients necessary for forming their eggs.

32.2 DIGESTION

In the process of digestion, which means literally "carrying asunder," organic foods are mechanically and chemically broken into small units for absorption. Although food solids consist principally of carbohydrates, proteins, and fats, the very components that form the body of the consumer, these components must first be reduced to their simplest molecular units before they can be assimilated. Each animal reassembles some of these digested and absorbed units into organic compounds of the animal's own unique pattern.

Figure 32.4 Structure of human molar tooth. A tooth is built of three layers of calcified tissue covering: enamel, which is 98% mineral and the hardest material in the body; dentine, which composes the mass of the tooth and is approximately 75% mineral; and cementum, which forms a thin covering over the dentine in the root of the tooth and is very similar to dense bone in composition. The pulp cavity contains loose connective tissue, blood vessels, nerves, and toothbuilding cells.

Figure 32.6 An African elephant loosening soil from a salt lick with its tusk. Elephants use their powerful modified incisors in many ways to search for food and water: plowing the ground for roots, prying apart branches to reach the edible cambium, and drilling into dry riverbeds for water.

© Cleveland P. Hickman, Jr.

Figure 32.5 Mammalian dentition. **A,** Teeth of gray fox, a carnivore, showing the four types of teeth. **B,** Woodchuck, a rodent, with chisel-like incisors that continue to grow throughout life as they wear. **C,** White-tailed deer, a browsing ungulate, with flat premolars and molars bearing complex ridges suited for grinding.

In unicellular eukaryotes and sponges digestion is entirely **intracellular** (Figure 32.7). A food particle is enclosed within a food vacuole by phagocytosis (see p. 48). Digestive enzymes are added and the products of digestion, the simple sugars, amino acids, and other molecules, are absorbed into the cell cytoplasm, where they may be used directly or, in the case of multicellular animals, may be transferred to other cells. Food wastes are simply extruded from the cell by exocytosis (p. 49).

There are important limitations to intracellular digestion. Only particles small enough to be phagocytized can be engulfed, and every cell must be capable of secreting all necessary enzymes, and of absorbing the products into its cytoplasm. These limitations were resolved with evolution of an **alimentary system** in which **extracellular** digestion of large food masses occurs. In extracellular digestion certain cells lining the **lumen** (cavity) of an alimentary canal specialize in forming various digestive secretions, such as enzymes, whereas other cells function largely, or entirely, in absorption. Many simpler animals, such as radiates, turbellarian flatworms, and ribbon worms (nemerteans), practice both intracellular and extracellular digestion. With evolution of greater complexity and appearance of complete mouth-to-anus alimentary systems, extracellular digestion became emphasized, together with increasing regional specialization of the digestive tract. For arthropods and vertebrates, digestion is almost entirely extracellular. Ingested food is exposed to various mechanical, chemical, and bacterial treatments, to different acidic and alkaline regions, and to digestive juices that are added at appropriate stages as food passes through the alimentary canal.

Action of Digestive Enzymes

Mechanical processes of cutting and grinding by teeth and muscular mixing by an intestinal tract are important in digestion. However, reduction of foods to small, absorbable units relies principally on chemical breakdown by **enzymes,** discussed in Chapter 4 (p. 57). Digestive enzymes are **hydrolytic** enzymes, or **hydrolases,** so called because food molecules are split by a process of **hydrolysis,** breaking of a chemical bond by adding the components of water across it:

$$R-R + H_2O \xrightarrow{\text{digestive enzyme}} R-OH + H-R$$

In this general enzymatic reaction, R—R represents a larger food molecule split into two products, R—OH and H—R. Usually these reaction products must in turn be split repeatedly before the original molecule is reduced to its numerous subunits. Proteins, for example, are composed of hundreds, or even thousands, of interlinked amino acids, which must be separated before individual amino acids or dipeptides (two linked amino acids) can be absorbed. Similarly, complex carbohydrates must be reduced to simple sugars. Fats (lipids) are reduced to molecules of glycerol, fatty acids, and monoglycerides (see Digestion in the Vertebrate Small Intestine for specific enzyme examples, p. 711). There are specific enzymes for digestion of each class of organic compounds. These enzymes are located in specific regions of an alimentary canal in an "enzyme chain," in which one enzyme may complete digestion that was begun by preceding enzymes in the chain. Some fats, unlike proteins and

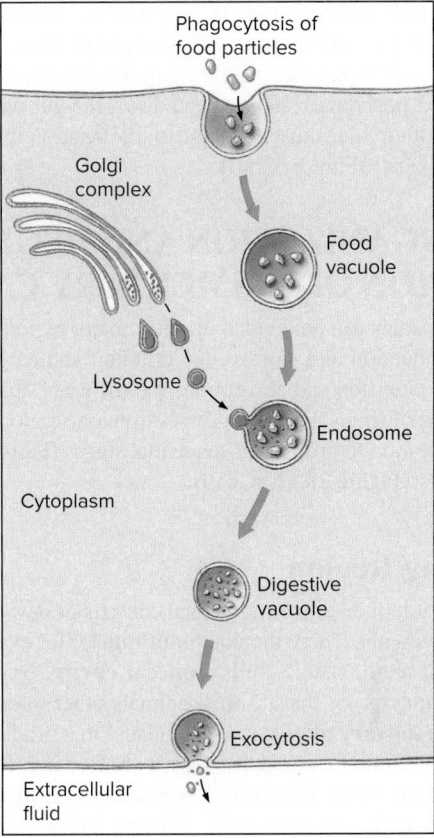

Figure 32.7 Intracellular digestion. Lysosomes containing digestive enzymes (lysozymes) are produced within a cell by the Golgi complex. Lysosomes fuse with food vacuoles and release enzymes that digest the enclosed food. Usable products of digestion are absorbed into the cytoplasm, and indigestible wastes are expelled.

carbohydrates, may be absorbed without first being completely hydrolyzed, since they can diffuse through the plasma membrane of the cells lining the alimentary canal.

Motility in Alimentary Canal

Food is moved through digestive tracts by **cilia** or by specialized **musculature,** and often by both. Movement is usually by cilia in acoelomate and pseudocoelomate animals that lack the mesodermally derived gut musculature of true coelomates (see Figure 9.4, p. 187 for a reminder of these body plans). Cilia move intestinal fluids and materials also in some eucoelomates, such as most molluscs, in which the coelom is weakly developed. In animals with well-developed coeloms, the gut is usually lined with two opposing layers of smooth muscle: a longitudinal layer, in which smooth muscle fibers run parallel with the length of the gut, and a circular layer, in which muscle fibers embrace the circumference of the gut (see Figure 32.13). A characteristic gut movement is **segmentation,** the alternate constriction of rings of smooth muscle of the intestine that constantly divide and squeeze contents back and forth (Figure 32.8A). Walter B. Cannon of homeostasis fame (p. 661), while still a medical student at Harvard in 1900, was the first to use

X-rays to watch segmentation in experimental animals that had been fed suspensions of barium sulfate. Segmentation serves to mix food but does not move it through the gut. Another kind of muscular action, called **peristalsis,** moves food down the gut with waves of contraction of circular muscle behind and relaxation in front of the food mass (bolus) (Figure 32.8B).

32.3 ORGANIZATION AND REGIONAL FUNCTION OF ALIMENTARY CANALS

Alimentary canals can be divided into five major regions: (1) reception, (2) conduction and storage, (3) grinding and early digestion, (4) terminal digestion and absorption, and (5) water absorption and concentration of solids. Food progresses from one region to the next, allowing digestion to proceed in sequential stages (Figures 32.9 and 32.10; see also Figure 28.11, p. 619).

Receiving Region

The first region of an alimentary canal consists of devices for feeding and swallowing. These include **mouthparts** (for example, mandibles, jaws, teeth, radula, bills), **buccal cavity,** or mouth, and muscular **pharynx,** or throat. Most animals other than suspension feeders have **salivary glands** (buccal glands) that produce lubricating secretions containing mucus to assist swallowing (Figure 32.9). Salivary glands often have other specialized functions, such as secretion of toxic chemicals for quieting struggling prey and secretion of salivary enzymes to begin digestion. Salivary secretions of leeches, for example, are complex mixtures containing an anesthetic substance (making their bite nearly painless) and several enzymes that prevent blood coagulation and increase blood flow by dilating veins and dissolving cell adhesion molecules that bind cells together (see Figure 17.28, p. 380).

Salivary **amylase** is a carbohydrate-splitting enzyme that begins hydrolysis of plant and animal starches as soon as they have been ingested. It occurs only in certain herbivorous molluscs, some insects, and primate mammals, including humans. Starches are long polymers of glucose, and salivary amylase does not completely hydrolyze it but breaks it mostly into two-glucose fragments called **maltose.** Some free glucose and longer fragments of starch are also produced. When the food bolus is swallowed, salivary amylase continues to act for some time, digesting perhaps half of the starch before the enzyme is inactivated by the acidic environment of the stomach. Further starch digestion resumes beyond the stomach in the intestine as the pH of the intestinal lumen increases.

A tongue is a vertebrate innovation, usually attached to the floor of the mouth, that assists in food manipulation and swallowing. Tongues are also used as chemosensors and possess taste buds that are used to determine palatability of foods (see Chapter 33, p. 735). They may be used for other purposes, however, such as food capture (for example, frogs, chameleons, woodpeckers, anteaters; see Figure 26.17, p. 573) or as olfactory sensors (many lizards and snakes; see Figure 26.23, p. 575).

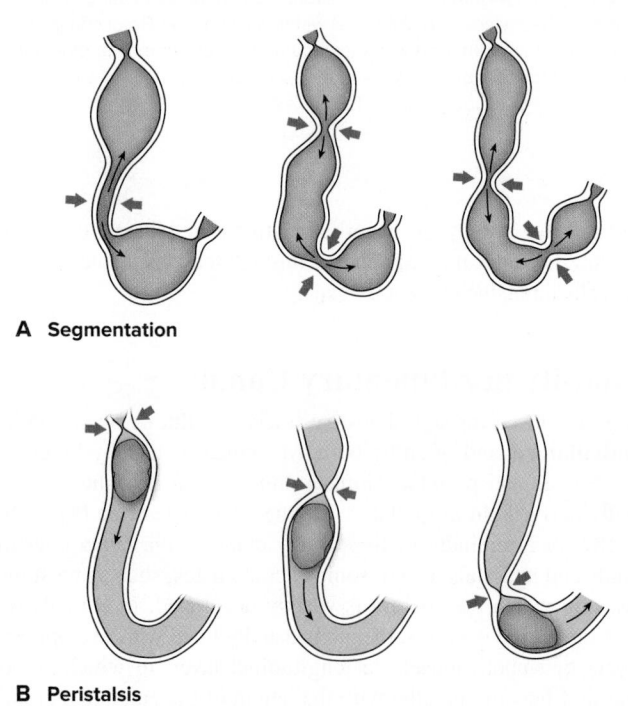

Figure 32.8 Movement of intestinal contents by segmentation and peristalsis. **A,** Segmentational movements of food showing how constrictions squeeze food back and forth, mixing it with enzymes. The sequential mixing movements occur at about 1-second intervals. **B,** Peristaltic movement, showing how food is propelled through the alimentary canal by a traveling wave of contraction behind the food mass (bolus).

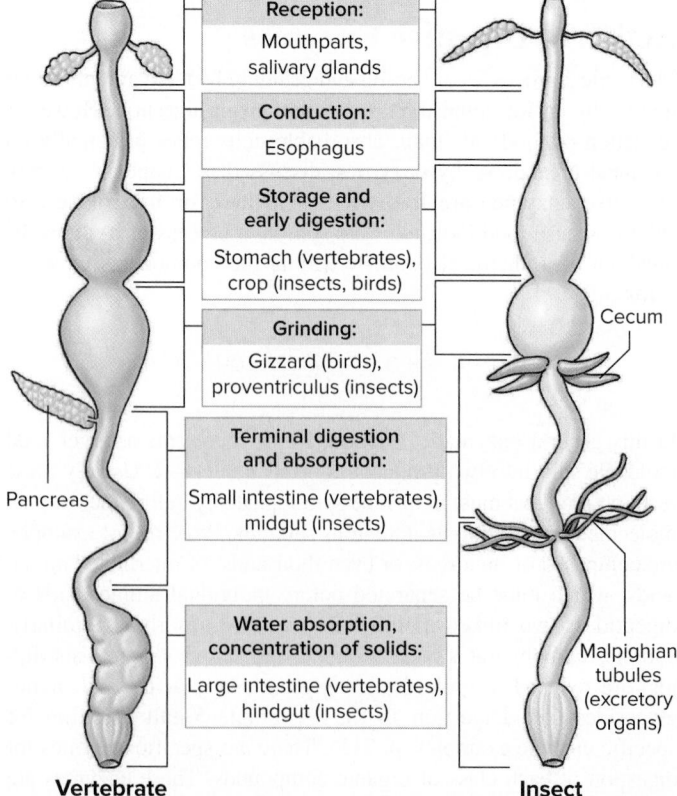

Figure 32.9 Generalized digestive tracts of a vertebrate and an insect, showing the major functional regions of animal digestive systems.

In humans, swallowing begins with the tongue pushing food, moistened with saliva, toward the pharynx. The nasal cavity closes reflexively by raising the soft palate. As food slides into the pharynx, the epiglottis tips down over the trachea, nearly closing it (Figure 32.10). Some particles of food may enter the opening of the trachea but contraction of laryngeal muscles prevents them from going farther. Once food is in the esophagus, peristaltic contraction of esophageal muscles forces it smoothly toward the stomach. The top one-third of the esophagus is surrounded by skeletal muscle as well as smooth muscle, so the act of swallowing is voluntary until the food has traveled past this upper region.

Conduction and Storage Region

The **esophagus** of vertebrates and many invertebrates serves to transfer food to the digestive region. In many invertebrates (annelids, insects, octopods) the esophagus is expanded into a **crop** (Figure 32.9), used for food storage before digestion. Among vertebrates, only birds have a crop, which serves to store and to soften food (grain, for example) before it passes to the stomach, or to allow mild fermentation of food before it is regurgitated to feed nestlings.

Region of Grinding and Early Digestion

In most vertebrates, and in some invertebrates, the **stomach** provides initial digestion as well as storage and mixing of food with digestive juices. Mechanical breakdown of food, especially plant food with its tough cellulose cell walls, often continues in herbivorous animals by grinding and crushing devices in the stomach. The muscular **gizzard**

of terrestrial oligochaete worms and birds is assisted by stones and grit swallowed along with food or, in arthropods, by hardened linings (for example, chitinous teeth of an insect's proventriculus [Figure 32.9], and calcareous teeth of gastric mills of crustaceans [see Figure 20.13, p. 431]).

Digestive diverticula—blind tubules or pouches arising from the main passage—often supplement stomachs of many invertebrates. They are usually lined with a multipurpose epithelium having cells specialized for secreting mucus or digestive enzymes, or absorption or storage. Examples include ceca of polychaete worms and insects, digestive glands of bivalve molluscs (see Figure 16.31, p. 351), hepatopancreas of crustaceans, and pyloric ceca of sea stars.

Herbivorous vertebrates have evolved several strategies for exploiting cellulose-splitting microorganisms to derive maximal nutrition from plant food. Despite its abundance on earth, cellulose that encloses plant cells can be digested only by an enzyme, **cellulase,** that has limited distribution in the living world. No animals can produce intestinal cellulase for direct digestion of cellulose. However, many herbivorous animals harbor microorganisms (bacteria and unicellular eukaryotes) in their gut that do produce cellulase. These microorganisms ferment cellulose under anaerobic conditions in the gut, producing fatty acids and sugars that herbivores can use. While the ultimate fermentation machine is the multichambered stomach of the ruminant ungulates described on page 620, many other animals harbor microorganisms in other parts of the gut, such as their intestine proper or their cecum (see Figure 28.11, p. 619).

The stomach of carnivorous and omnivorous vertebrates is typically a J-shaped muscular tube whose inner epithelial layer

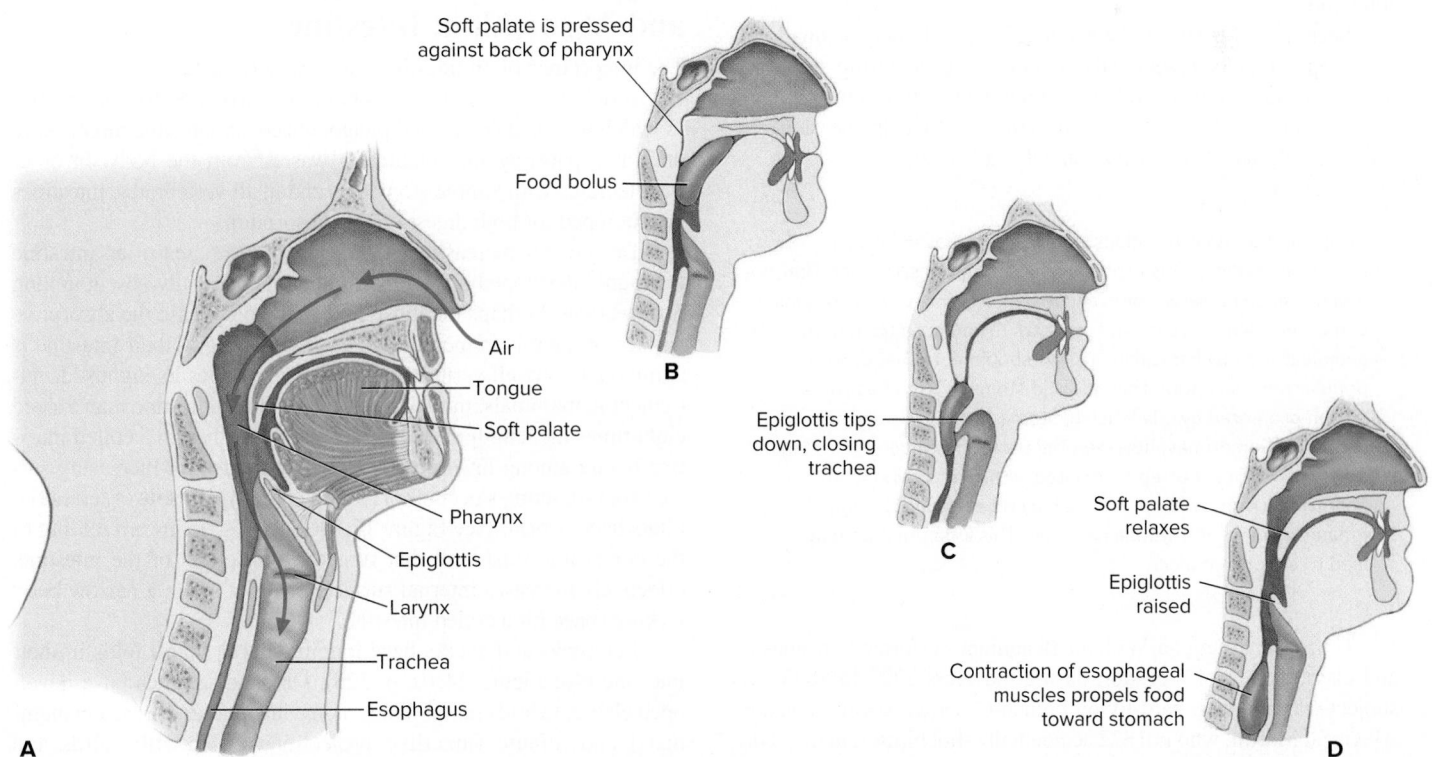

Figure 32.10 Oral cavity and throat in humans in sagittal section (A), and sequence of swallowing (B–D).

contains glands that produce proteolytic enzymes and strong acids. When food arrives at a stomach, a **cardiac sphincter** opens reflexively to allow food to enter, then closes to prevent regurgitation back into the esophagus. In humans, peristaltic waves pass over the filled stomach at a rate of approximately three each minute. Churning is most vigorous at the intestinal end, where food is steadily released into the **duodenum,** the first region of the small intestine. A **pyloric sphincter** regulates flow of food into the intestine and prevents regurgitation into the stomach. Deep tubular glands in the stomach wall secrete **gastric juice,** in humans approximately 2 liters each day. Three types of cells line these glands: **goblet cells,** which secrete mucus; **chief cells,** which secrete **pepsinogen;** and **parietal** or **oxyntic cells,** which secrete **hydrochloric acid.** Pepsinogen is the precursor of **pepsin,** a **protease** (protein-splitting enzyme) produced from pepsinogen only in an acid medium (pH 1.6 to 2.4). The inability of pepsin to function until acidification of gastric juice helps to protect the stomach lining from digestion by this protease. This highly specific enzyme splits large proteins by preferentially hydrolyzing certain peptide bonds scattered along the peptide chain of a protein molecule. Although pepsin, because of its specificity, cannot completely degrade proteins, it effectively hydrolyzes them into smaller polypeptides. Other proteases complete digestion of protein in the intestine. Pepsin occurs in stomachs of nearly all vertebrates.

Rennin (not to be confused with renin, an enzyme produced by vertebrate kidneys, p. 670) is a milk-curdling enzyme found in the stomachs of ruminant mammals. It probably occurs in many other mammals. By clotting and precipitating milk proteins, it slows movement of milk through the stomach. Rennin extracted from stomachs of calves is used in making cheese. Human infants, lacking rennin, digest milk proteins with acidic pepsin, just as adults do.

Secretion of gastric juice is intermittent. Although a small volume of gastric juice is secreted continuously, even during prolonged periods of starvation, secretion normally increases when stimulated by sight and smell of food, by presence of food in the stomach, and by emotional states such as anxiety and anger.

Stomach mucosa is not digested by its own powerful acid secretions because it is protected by another gastric secretion, mucus, made of water, salts, and mucin, a highly viscous organic compound. Mucus coats and protects the mucosa from both chemical and mechanical injury. We should note that despite a popular misconception that an "acid stomach" is unhealthy, a notion promoted by advertising, stomach acidity is normal and essential. Sometimes, however, the protective mucous coating fails. This failure is often associated with an infection from a bacterium (*Helicobacter pylori*) that secretes toxins causing inflammation of the stomach's lining. This inflammation may lead to a stomach ulcer.

U.S. Army surgeon William Beaumont performed an unusual and classic study of digestion during the years 1825 to 1833. His subject was a young, hard-living French-Canadian voyageur named Alexis St. Martin, who in 1822 accidentally shot himself in the abdomen with a musket. Miraculously, the wound healed, but a permanent opening, or fistula, formed that permitted Beaumont to see

Figure 32.11   Dr. William Beaumont at Fort Mackinac, Michigan Territory, collecting gastric juice from Alexis St. Martin.

directly into St. Martin's stomach (Figure 32.11). St. Martin became a permanent patient in Beaumont's care. Over a period of eight years, Beaumont observed and recorded how the lining of the stomach changed under different psychological and physiological conditions, how foods changed during digestion, effects of emotional states on stomach motility, and many other facts about the digestive process of his famous patient.

Region of Terminal Digestion and Absorption: Intestine

The importance of an intestine varies widely among animal groups. In invertebrates whose stomachs have extensive digestive diverticula in which food is digested and phagocytized, an intestine might serve only as a pathway for conducting wastes from the body. In other invertebrates with simple stomachs, and in all vertebrates, intestines are equipped for both digestion and absorption.

Devices for increasing the internal surface area of an intestine are highly developed in vertebrates, but are generally absent among invertebrates. Perhaps the most direct way to increase the absorptive surface of a gut is to increase its length. Coiling of their intestine is common among all vertebrate groups and reaches its highest development in mammals, in which the length of the intestine may exceed eight times the length of the animal's body. Although a coiled intestine is rare among invertebrates, other strategies for increasing surface area sometimes occur. For example, the **typhlosole** of terrestrial oligochaete worms (see Figure 17.19C, p. 375), an inward folding of the dorsal intestinal wall that runs the full length of the intestine, effectively increases internal surface area of a gut in a narrow body lacking space for a coiled intestine.

Lampreys and sharks have longitudinal or spiral folds in their intestine (see Figure 24.10, p. 525). Other vertebrates have developed elaborate folds (amphibians, nonavian reptiles, birds, and mammals) and minute fingerlike projections called **villi** (birds and mammals; Figure 32.12). Electron microscopy reveals that each cell lining the intestinal cavity additionally is bordered by hundreds of

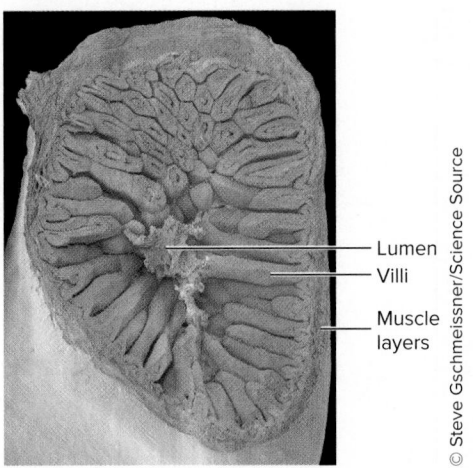

Figure 32.12 Scanning electron micrograph of a rat intestine showing numerous fingerlike villi that project into the lumen and vastly increase effective absorptive and secretory surface of the intestine. (×35)

Lumen
Villi
Muscle layers

© Steve Gschmeissner/Science Source

short, delicate processes called **microvilli** (Figure 32.13C and D; see also Figure 3.15B, p. 45). These processes, together with larger villi and intestinal folds, may increase the internal surface area of an intestine more than a thousand times as compared to a smooth cylinder of the same diameter. This elaborate surface greatly facilitates absorption of food molecules.

Digestion in the Vertebrate Small Intestine

Food is released into the small intestine through a **pyloric sphincter,** which relaxes at intervals to allow entry of acidic stomach contents into the initial segment of the small intestine, or **duodenum.** Two secretions pour into this region: **pancreatic juice** and **bile** (Figure 32.14). Both secretions have a high bicarbonate content, especially pancreatic juice, which effectively neutralizes gastric acid, raising the pH of the liquefied food mass, now called **chyme,** from 1.5 to 7 as it enters the duodenum. This change in pH is essential because all intestinal enzymes are effective only in a neutral or slightly alkaline medium.

Mesentery

Serosa

Mucosa
Submucosa
Circular muscle
Longitudinal muscle

A

Digestive gland

Duct

Villus

Lacteal

Capillary network

Lymphatic vessel

Blood vessels

Intestinal crypt

B

© Don W. Fawcett/Science Source

D

Endoplasmic reticulum

Mitochondrion

Nucleus

C

Figure 32.13 Organization of vertebrate digestive tract, showing **A,** the successive layers of mucosa, submucosa, muscle, and enveloping serosa; an enzyme-secreting digestive gland (e.g., pancreas); and the thin mesentery that positions the intestine within the body cavity. **B,** Portion of mucosal lining of intestine, showing fingerlike villi. **C,** Section of a single cell from the mucosal lining. **D,** Microvilli on surface of mucosal cell, rat intestine. (×16,400)

Region	Secretion	pH	Composition
Salivary glands	Saliva	6.5	Amylase Bicarbonate
Stomach	Gastric juice	1.5	Pepsinogen HCl Rennin in ruminant mammals
Liver and gallbladder	Bile	7–8	Bile salts and pigments Cholesterol
Pancreas	Pancreatic juice	7–8	Trypsin, Chymotrypsin Carboxypeptidase Lipase, Amylase Nucleases Bicarbonate
Small intestine	Membrane enzymes	7–8	Aminopeptidase Maltase Lactase Sucrase Alkaline Phosphatase Nucleotidase Nucleosidase

Figure 32.14 Secretions of a mammalian alimentary canal with the principal components and pH of each secretion.

Cells of intestinal mucosa (Figure 32.13A), like those of the stomach mucosa, are subjected to considerable wear and are constantly undergoing replacement. Cells deep in the crypt between adjacent villi divide rapidly and migrate up the villus. In mammals the cells reach the tip of the villus in about two days. There they are shed, along with their membrane enzymes, into the lumen at a rate of some 17 billion a day along the length of the human intestine. Before they are shed, however, these cells differentiate into absorptive cells that function to transport nutrients from the intestinal lumen into the network of blood and lymph vessels, once digestion is complete.

Bile   The liver secretes bile into the **bile duct,** which drains into the upper intestine (duodenum). Between meals bile collects in the **gallbladder,** an expansible storage sac that releases bile when stimulated by fatty food in the duodenum. Bile contains water, bile salts, and pigments, but no enzymes (Figure 32.14). **Bile salts** (mainly sodium taurocholate and sodium glycocholate) are essential for digestion of fats. Fats, because of their tendency to remain in large, water-insoluble globules, are especially resistant to enzymatic digestion. Bile salts reduce surface tension of fat globules, allowing the churning action of the intestine to break fats into tiny droplets (emulsification). With total surface exposure of fat particles greatly increased, fat-splitting lipases can reach and hydrolyze the triglyceride molecules. The yellow-green color of bile is produced by **bile pigments,** breakdown products of hemoglobin from aged red blood cells. Bile pigments also give feces its characteristic color.

Bile production is only one of the liver's many functions. This highly versatile organ is a storehouse for glycogen, production center for plasma proteins, site of protein synthesis and detoxification of protein wastes, site for destruction of aged red blood cells, and center for metabolism of fat, amino acids, and carbohydrates.

Pancreatic Juice   Pancreatic secretion of vertebrates contains several enzymes of major importance in digestion (Figure 32.14). Two powerful proteases, **trypsin** and **chymotrypsin,** continue enzymatic digestion of proteins begun by pepsin, which is now inactivated by the alkalinity of the intestine. Trypsin and chymotrypsin, like pepsin, are highly specific proteases that split peptide bonds of a protein molecule. Hydrolysis of a peptide linkage may be shown as follows:

Pancreatic juice also contains **carboxypeptidase,** which removes amino acids from carboxyl ends of polypeptides; **pancreatic lipase,** which hydrolyzes fats into fatty acids and glycerol; **pancreatic amylase,** a starch-splitting enzyme identical to salivary amylase in its action; and **nucleases,** which degrade RNA and DNA to nucleotides.

Membrane Enzymes   Cells lining the intestine have digestive enzymes embedded as transmembrane proteins in their plasma membrane that continue digestion of carbohydrates, proteins, and phosphate compounds (Figure 32.14). These enzymes of microvillus membranes (Figure 32.13D) include **aminopeptidase,** which splits terminal amino acids from the amino end of short peptides, and several **disaccharidases,** enzymes that split 12-carbon sugar molecules into 6-carbon units. The disaccharidases include **maltase,** which splits maltose into two molecules of glucose; **sucrase,** which splits sucrose into fructose and glucose; and **lactase,** which breaks lactose (milk sugar) into glucose and galactose. Also present are **alkaline phosphatase,** an enzyme that attacks a variety of phosphate compounds, and **nucleotidases** and **nucleosidases,** which continue the breakdown of nucleotides into nucleosides and finally ribose and deoxyribose sugars, and purines and pyrimidines.

Although milk is the universal food of newborn mammals and one of the most complete human foods, many adult humans cannot digest milk because they are deficient in lactase, the enzyme that hydrolyzes lactose (milk sugar). Lactose intolerance is genetically determined. It is characterized by abdominal bloating, cramps, flatulence, and watery diarrhea, all appearing within 30 to 90 minutes after ingesting milk or its unfermented by-products. (Fermented dairy products, such as yogurt and cheese, create no intolerance problems.)

Northern Europeans and their descendants, which include a majority of European-Americans, are most tolerant of milk. Many other ethnic groups are generally intolerant to lactose, including Japanese, Chinese, Inuit (Eskimos), South American Indians, and most Africans. Only about 30% of African-Americans are tolerant; those who are tolerant trace their ancestry to east and central Africa where dairying is traditional and tolerance to lactose is high.

Absorption

Little food is absorbed in the stomach because digestion is still incomplete and because of limited absorptive surface area. However, some materials, such as glucose and lipid-soluble drugs and alcohol, are absorbed mostly there, which contributes to their rapid appearance in the bloodstream. Most digested food is absorbed from the small intestine, where numerous finger-shaped villi and microvilli provide an enormous surface area through which materials can pass from the intestinal lumen into the circulation.

Carbohydrates are absorbed almost exclusively as simple sugars (monosaccharides, for example, glucose, fructose, and galactose) because the intestine is virtually impermeable to polysaccharides. Proteins are absorbed principally as their amino acid subunits, although a limited amount of small proteins or peptide fragments may be absorbed. Both active and passive transport processes transfer simple sugars and amino acids across the intestinal epithelium (see p. 45).

Immediately after a meal these materials are in such high concentration in the gut that they readily diffuse by facilitated transport into the blood (p. 47), where their concentration is initially lower. However, if absorption were passive only, we would expect transfer to cease as soon as concentrations of a substance became equal on both sides of the intestinal epithelium. Passive transfer alone would permit valuable nutrients to be lost in feces. In fact, very little is lost because passive transfer is supplemented by an **active transport** (p. 47) mechanism located in epithelial cells that transfers digested food molecules into the blood. Materials thus are moved *against*

their concentration gradient, a process requiring energy. Although not all food products are actively transported, those that are, such as glucose, galactose, and most amino acids, are carried by specific protein transporters for each kind of molecule.

As mentioned previously, fat droplets are emulsified by bile salts and then digested by pancreatic lipase. Triglycerides are broken into fatty acids and monoglycerides, which complex with bile salts to form minute droplets called **micelles.** When micelles contact the microvilli of the intestinal epithelium, fatty acids and monoglycerides are absorbed by simple diffusion. They then enter the endoplasmic reticulum of absorptive cells, where they are resynthesized into triglycerides before passing into **lacteals** (Figure 32.13B). From the lacteals, fat droplets enter the lymphatic system (see Figure 31.18, p. 692) and eventually pass into the blood circulation through the thoracic duct. After a fatty meal, even a peanut butter sandwich, the presence of numerous fat droplets in the blood imparts a milky appearance to blood plasma. Triglyceride digestion, absorption, and metabolism are currently a major research focus and have led to the development of a number of antiobesity drugs used to control our body weight.

Region of Water Absorption and Concentration of Solids

The large intestine consolidates the indigestible remnants of digestion by reabsorption of water to form solid or semisolid feces for removal from the body by defecation. Reabsorption of water is of special significance in insects, especially those living in dry environments, which must (and do) conserve nearly all water entering the rectum. Specialized **rectal glands** absorb water and ions as needed, leaving behind fecal pellets that are almost completely dry. In nonavian reptiles and birds, which also produce nearly dry feces, most water is reabsorbed in the cloaca. A white, pastelike feces is formed containing both indigestible food wastes and uric acid.

The colon of most vertebrates contains enormous numbers of bacteria, which first enter the sterile colon of a newborn infant with its food. In adult humans approximately one-third of the dry weight of feces is bacteria; these include harmless bacteria as well as bacteria that can cause serious illness should they escape into the abdomen or bloodstream. Normally the body's defenses prevent invasion of such bacteria. Bacteria degrade organic wastes in feces and provide some nutritional benefit by synthesizing certain vitamins (vitamin K and small quantities of some B vitamins), which are absorbed by the body.

32.4 REGULATION OF FOOD INTAKE

Most animals unconsciously adjust intake of food to balance energy expenditure. If energy expenditure is increased by greater physical activity, more food is consumed. Most vertebrates, from fish to mammals, eat for calories rather than bulk because if their diet is diluted with fiber, they respond by eating more. Similarly, intake is adjusted downward following a period of several days when caloric intake is too high.

A brain's **hunger center,** located in the hypothalamus and brain stem (pp. 730 and 731), regulates the intake of food. A drop in blood glucose level stimulates a craving for food. While most animals seem able to stabilize their weight at normal levels with ease, many humans cannot. Obesity is rising throughout the industrial world and is a major health problem in many countries today. According to recent surveys, two-thirds of adults and one-third of children (2011–2012) in the United States are either overweight or obese.[2] In Canada, 24% of children and adolescents were categorized as overweight or obese in 2011 (Statistics Canada). Assessment of excess weight relies on body mass index (BMI, weight in kilograms divided by the square of height in meters), waist circumference, and risk factor for diseases associated with obesity, such as type 2 diabetes, cardiovascular disease, and some cancers. A BMI of 25 or greater is considered overweight, whereas a BMI of 30 or greater is determined obese.

Many obese people do not eat significantly more food than thinner people, but rather have begun to defend a higher setpoint for body weight. The increase in fastfood meals, larger portion sizes, and more sedentary lifestyle is associated with the prevalence of obesity in developed countries. There is also a high correlation between the increase in fructose consumption (found in processed food), along with high-fat diets, and the rise in obesity. Ingestion of liquid calories from fructose appears to lead to decreased regulation of food intake and therefore more weight gain than ingestion of solid calories from fructose. Evidence suggests that increased dietary fructose and fat consumption leads to insulin insensitivity or resistance (see p. 760), subsequent weight gain, and type 2 diabetes. Some obese people may also have a reduced capacity to burn excess calories by "diet-induced thermogenesis." Placental mammals are unique in having a dark adipose tissue called **brown fat,** specialized for generation of heat. Newborn mammals, including human infants, have much more brown fat per body weight than do adults. In human infants brown fat is located in the chest and upper back, and near the kidneys. Abundant mitochondria in brown fat contain a protein called **uncoupling protein,** which acts to uncouple the production of ATP during oxidative phosphorylation (p. 6). Thermogenesis in brown fat is stimulated by excess food and by cold temperatures (**nonshivering thermogenesis,** see p. 676) and is activated by the sympathetic nervous system (p. 733), which responds to signals from the hypothalamus and brain stem. In people of average weight, increased caloric intake induces brown fat to dissipate excess energy as heat through the action of uncoupling protein. The Pima Indians of Arizona have low sympathetic nervous system activity, and this may contribute to the prevalence of obesity in this population. Research on this link between diet-induced thermogenesis and obesity may provide new therapies to obese people.

Many mammals have two kinds of adipose tissue that perform completely different functions. **White adipose tissue,** the bulk of body fat, is adapted for storage of fat derived mainly from surplus fats and carbohydrates in the diet. It is distributed throughout the body, particularly in the deep layers of skin. **Brown adipose tissue** is highly specialized for mediating nonshivering and diet-induced thermogenesis rather than for the storage of fat. Brown fat, unique to placental mammals, is especially well developed in hibernating species of bats and

rodents, but it occurs also in many nonhibernating species, such as rabbits, artiodactyls, carnivores, and primates (including humans). It is brown because it is packed with mitochondria containing large quantities of iron-bearing cytochrome molecules. In ordinary body cells, ATP is generated by a flow of electrons down the electron transport chain (p. 64). This ATP then powers various cellular processes. In brown fat cells heat is generated instead of ATP. A recently discovered hormone, called irisin, is secreted from skeletal muscle during exercise. This hormone appears to stimulate conversion of white adipose tissue to brown adipose tissue and thus enhances energy consumption.

There are other reasons for obesity in addition to the fact that many people initially eat too much and get too little exercise. Fat stores are supervised by the hypothalamus and brain stem, which may be set at a point higher or lower than the norm. A high setting can be lowered somewhat by exercise, but as dieters are painfully aware, the body defends its fat stores with remarkable tenacity. In 1995, a hormone produced by fat cells was found to cure obesity in mutant mice lacking the gene that produces the hormone. The hormone, called **leptin,** appears to operate through a feedback system that tells the hypothalamus and brain stem how much fat the body carries (see Chapter 34, p. 761). If levels are high, release of leptin by fat cells diminishes appetite and increases thermogenesis. The discovery of leptin initiated a flurry of research on obesity and a resurgence of commercial interest in producing a weight-loss drug based on leptin. Unfortunately, most obese people do not respond to infusions of leptin and actually produce higher than normal amounts themselves. Their brain appears to have become resistant to these high leptin levels and does not respond by decreasing appetite. Fat cells have now been shown to secrete a variety of hormones that induce an inflammatory response in various tissues associated with metabolism. This response appears to be at least partly responsible for the change in regulation of food intake when an individual becomes overweight. Much research focuses on understanding how these multiple signals interact with short-term satiety signals produced during the digestive process (discussed in the next section) to regulate food intake and, subsequently, body weight.

Regulation of Digestion

The digestive process is coordinated by a family of hormones (see Chapter 34) produced by the body's most diffuse endocrine tissue, called the enteroendocrine system, found within the gastrointestinal (GI) tract. The hormones of this system are examples of the many substances produced by a vertebrate's body that have hormonal function, yet they are not necessarily produced by discrete endocrine glands.

Among the principal GI hormones are gastrin, cholecystokinin (CCK), and secretin (Figure 32.15). **Gastrin** is a small polypeptide hormone produced by endocrine cells in the pyloric portion of the stomach. Gastrin is secreted in response to stimulation by parasympathetic nerve endings (vagus nerve), and when protein food enters the stomach. Its main actions are to stimulate hydrochloric acid secretion from parietal or oxyntic cells and to increase gastric motility. Gastrin is an unusual hormone because it exerts its action on the same organ from which it is secreted. **CCK** is also a

[2]Ogden, C. L., M. D. Carroll, B. K. Kit, and K. M. Flegal. Prevalence of childhood and adult obesity in the United States, 2011–2012. 2014. JAMA. **311:** 806–814.

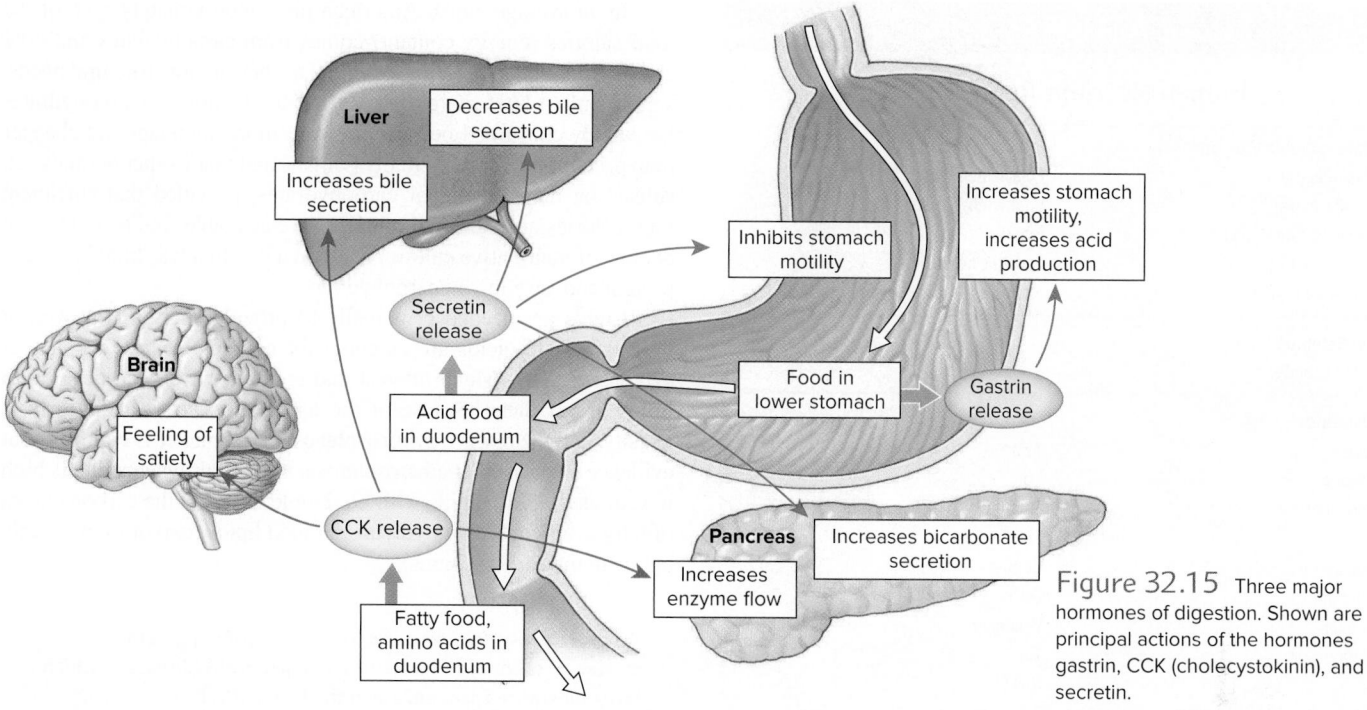

Figure 32.15 Three major hormones of digestion. Shown are principal actions of the hormones gastrin, CCK (cholecystokinin), and secretin.

polypeptide hormone, and it has a striking structural resemblance to gastrin, suggesting that the two arose by duplication of ancestral genes. CCK is secreted by endocrine cells in the walls of the upper small intestine in response to the presence of fatty acids and amino acids in the duodenum. It has at least three distinct functions. It stimulates gallbladder contraction and thus increases the flow of bile salts into the intestine; it stimulates an enzyme-rich secretion from the pancreas; and it acts on the brain stem to contribute a feeling of satiety after a meal, particularly one rich in fats. The first hormone to be discovered, **secretin** (see the opening essay for Chapter 34 on p. 746), is produced by endocrine cells in the duodenal wall. It is secreted in response to food and strong acid in the stomach and small intestine, and its principal action is to stimulate release of an alkaline pancreatic fluid that neutralizes stomach acid as it enters the intestine. It also aids fat digestion by inhibiting gastric motility and increasing production of an alkaline bile secretion from the liver.

GI hormones continue to be isolated and their structures determined. So far, all are peptides, and many occur in both the GI tract and the central nervous system. One of these is CCK, which occurs in high concentrations in the cerebral cortex and hypothalamus of mammals. By providing a feeling of satiety after eating (just mentioned), it may play a role in regulating appetite. Several other GI peptides, for example, vasoactive intestinal peptide (VIP), glucagon-like peptide 1 (GLP-1), pancreatic polypeptide (PP), gastric inhibitory peptide (GIP), ghrelin, and peptide YY (PYY), appear to be neurotransmitters in the brain. For example, ghrelin, PP, and PYY also appear to be short-term regulators of food intake. Ghrelin levels rise before a meal and appear to stimulate appetite, while PP and PYY levels rise during a meal and induce satiety. Much research focuses on newly discovered peptides in the hope of finding a "magic bullet" to solve the current obesity crisis.

32.5 NUTRITIONAL REQUIREMENTS

Food of animals must include **carbohydrates, proteins, fats, water, mineral salts,** and **vitamins.** Carbohydrates and fats are necessary fuels for energy and for the synthesis of various substances and structures. Proteins (actually the amino acids of which they are composed) are needed for synthesis of species-specific proteins and other nitrogen-containing compounds. Water is required as the solvent for body chemistry and as a major component of all fluids of the body. Inorganic salts are required as anions and cations of body fluids and tissues and form important structural and physiological components throughout the body. Vitamins are accessory factors from food that are often built into the structure of many enzymes.

A **vitamin** is a relatively simple organic compound that is required in very small amounts in the diet for some specific cellular function. Vitamins are not sources of energy but function as coenzymes (see p. 58) that are often associated with activity of important enzymes that serve vital metabolic roles. Plants and many microorganisms synthesize all the organic compounds they need; animals, however, have lost certain synthetic abilities during their evolution and depend ultimately on plants to supply these compounds. Vitamins therefore represent synthetic gaps in the metabolic machinery of animals.

Vitamins are usually classified as fat soluble (soluble in fat solvents such as ether) or water soluble. The **water-soluble vitamins** include the B complex and vitamin C (Table 32.1). Vitamins of the B complex, so grouped because the original B vitamin was subsequently found to comprise several distinct molecules, tend to occur together in nature. Almost all animals, vertebrate and invertebrate, require B vitamins; they are "universal" vitamins. A dietary need for vitamin C and the **fat-soluble vitamins** A, D_3, E, and K is mostly restricted to vertebrates, although some are required by certain invertebrates. Even within groups of close relationship, requirements for vitamins are relative, not

TABLE 32.1

Human Nutrient Requirements

Water-Soluble Vitamins

Thiamine (B$_1$)	Folacin (folic acid)
Riboflavin (B$_2$)	Vitamin B$_{12}$ (cobalamin)
Niacin (nicotinic acid)	Biotin
Pyridoxine (B$_6$)	Ascorbic acid (C)
Pantothenic acid	

Fat-Soluble Vitamins

A, D$_3$, E, and K

Minerals

Major	*Trace*
Calcium	Iron
Phosphorus	Fluorine
Sulfur	Zinc
Potassium	Copper
Chlorine	Silicon
Sodium	Vanadium
Magnesium	Tin
	Nickel
	Selenium
	Manganese
	Iodine
	Molybdenum
	Chromium
	Cobalt

Amino Acids

Phenylalanine	Methionine
Lysine	Tryptophan
Isoleucine	Threonine
Leucine	Arginine*
Valine	Histidine*

Polyunsaturated Fatty Acids

Arachidonic
Linoleic
Linolenic

*Required for normal growth of children.

In an average North American diet, approximately 50% of the total calories (energy content) comes from carbohydrates and 40% comes from lipids. Proteins, essential as they are for structural needs, supply only a little more than 10% of total calories. Carbohydrates are widely consumed because they are more abundant and cheaper than proteins or lipids. Actually humans and many other animals can subsist on diets devoid of carbohydrates, provided that sufficient total calories and essential nutrients are available. Inuits, before the decline of their native culture, lived on a diet that was high in fat and protein and very low in carbohydrate.

Lipids are needed principally to provide energy. However, at least three fatty acids are essential for humans because we cannot synthesize them. Much interest and research have been devoted to lipids in our diets because of the association between fatty diets, obesity, and the disease **atherosclerosis.** The matter is complex, but evidence suggests that atherosclerosis may occur when a diet is high in saturated lipids (lipids with no double bonds in the carbon chains of fatty acids) but low in polyunsaturated lipids (two or more double bonds in the carbon chains).

> Atherosclerosis (Gr. *atheroma,* tumor containing gruel-like matter, + *sclerosis,* to harden) is a degenerative disease in which fatty substances accumulate in the lining of arteries, causing narrowing of the passage and eventual hardening and loss of elasticity. Current evidence suggests that inflammation of the artery wall precedes deposition of fat. Elevated cholesterol levels can fuel such inflammation. Cholesterol-lowering drugs are used by many people in an effort to decrease the risk of cardiovascular diseases such as atherosclerosis and related conditions.

Proteins are expensive foods and limited in the diet. Proteins, of course, are not themselves the essential nutrients but rather contain essential amino acids. Of the 20 amino acids commonly found in proteins, eight and possibly 10 are essential to humans (Table 32.1). We can synthesize the rest from other amino acids. Generally, animal proteins have more essential amino acids than do proteins of plant origin. All eight essential amino acids must be present simultaneously in the diet for protein synthesis. If one or more is missing, use of the other amino acids is reduced proportionately; they cannot be stored and are metabolized for energy. Thus heavy reliance on a single plant source as a diet inevitably leads to protein deficiency. This problem can be corrected if two kinds of plant proteins having complementary strengths in essential amino acids are ingested together. For example, a balanced protein diet can be prepared by mixing wheat flour, which is deficient only in lysine, with a legume (peas or beans), which is a good source of lysine but deficient in methionine and cysteine. Each plant complements the other by having adequate amounts of those amino acids that are deficient in the other.

Because animal proteins are rich in essential amino acids, they are in great demand in all countries. North Americans eat far more animal proteins than do Asians and Africans. In 2011 in North America, 13% of calories came from meat. By comparison, in Somalia only 7% of calories came from meat.[3]

absolute. A rabbit does not require vitamin C, but guinea pigs and humans do. Some songbirds require vitamin A, but others do not.

The recognition years ago that many human diseases and those of domesticated animals were caused by or associated with dietary deficiencies led biologists to search for specific nutrients that would prevent such diseases. These studies eventually yielded a list of **essential nutrients** for people and other animal species studied. Essential nutrients are those needed for normal growth and maintenance and that *must* be supplied in the diet. It is "essential" that these nutrients be in the diet because an animal cannot synthesize them from other dietary constituents. Nearly 30 organic compounds (amino acids and vitamins) and 21 elements are essential for humans (Table 32.1). Considering that the body contains thousands of different organic compounds, the list in Table 32.1 is remarkably short. Animal cells can synthesize compounds of enormous variety and complexity from a small, select group of raw materials.

---

[3]http://www.nationalgeographic.com/what-the-world-eats/

Two different types of severe food deficiency are recognized: marasmus, general undernourishment from a diet low in both calories and protein, and kwashiorkor, protein malnourishment from a diet adequate in calories but deficient in protein. Marasmus (Gr. *marasmos,* to waste away) is common in infants weaned too early and placed on low-calorie–low-protein diets; these children are listless, and their bodies waste away. Kwashiorkor is a West African word describing a disease a child gets when displaced from the breast by a newborn sibling. This disease is characterized by retarded growth, anemia, weak muscles, a bloated body with typical pot belly, acute diarrhea, susceptibility to infection, and high mortality.

Undernourishment and malnourishment rank as two of the world's oldest problems and remain major health problems today, afflicting an eighth of the human population. Growing children and pregnant and lactating women are especially vulnerable to the devastating effects of malnutrition. Cell proliferation and growth in the human brain are most rapid in the terminal months of gestation and the first year after birth. Adequate protein for neuron development is a requirement during this critical time to prevent neurological dysfunction. Brains of children who die of protein malnutrition during the first year of life have 15% to 20% fewer brain cells relative to normal children (Figure 32.16). Malnourished children (Figure 32.17) who survive this period may suffer permanent brain damage and cannot be helped by later corrective treatment. Studies suggest that poverty, with attendant lack of educational and medical resources, and lowered expectations, exacerbates effects of malnutrition by delaying intellectual development.[4]

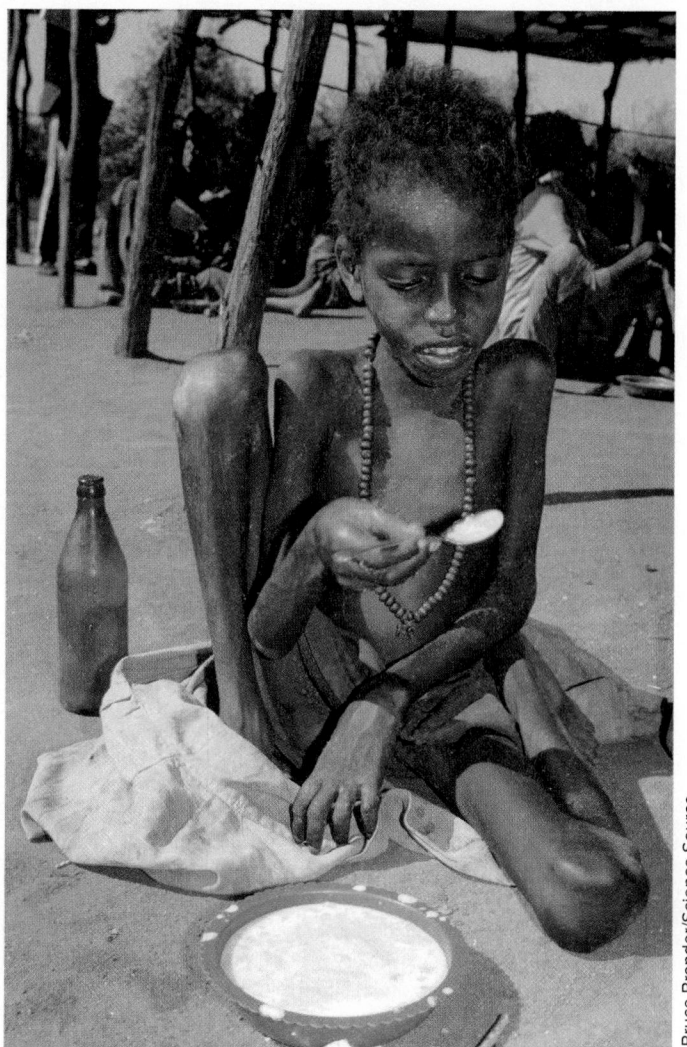

Figure 32.17 Refugee child suffering severe malnutrition.

Figure 32.16 Effect of early malnutrition on cell number (measured as total DNA content) in a human brain. This graph shows that malnourished infants (*purple oval*) have far fewer brain cells than do normal infants (*green growth curve*).

[4]Brown, J. L. and E. Pollitt. 1996. Malnutrition, poverty and intellectual development. Sci. Am. **274**: 38–43 (Feb.).

SUMMARY

Autotrophic organisms (mostly green plants and cyanobacteria), using inorganic compounds as raw materials, capture the energy of sunlight through photosynthesis and produce complex organic molecules. Heterotrophic organisms (bacteria, fungi, and animals) use organic compounds synthesized by plants, and chemical bond energy stored therein, for their own nutritional and energy needs.

A large group of animals with very different levels of complexity feed by filtering minute organisms and other particulate matter suspended in water. These are either suspension feeders, that ingest all food particles, or filter-feeders, that are more selective in straining food particles of a particular size from water. Others feed on organic detritus deposited in the substrate. Selective feeders have evolved mechanisms for manipulating larger food masses, including various devices for seizing, scraping, boring, tearing, biting, and chewing. Fluid-feeding characterizes endoparasites, which may absorb food across the general body surface, and ectoparasites, herbivores, and predators that have developed specialized mouthparts for piercing and sucking.

Digestion is a process of breaking food mechanically and chemically into molecular subunits for absorption. Digestion is intracellular in unicellular eukaryotic groups and sponges. In more complex animals it is supplemented, and finally replaced entirely, by extracellular digestion, which occurs in sequential stages in a tubular cavity, or alimentary canal. The mouth receives food, mixes it with lubricating saliva, then passes it down the esophagus to regions where food may be stored (crop), ground (gizzard), or acidified and subjected to early digestion (vertebrate stomach). Among vertebrates, most digestion occurs in the small intestine. Enzymes from the pancreas and intestinal mucosa hydrolyze proteins, carbohydrates, fats, nucleic acids, and various phosphate compounds. The liver secretes bile, containing salts that emulsify fats. Once foods are digested, their products are absorbed as molecular subunits (monosaccharides, amino acids, fatty acids, and glycerol) into blood or lymph vessels of villi of the small intestine. The large intestine (colon) serves mainly to absorb water and minerals from the food wastes as they pass through it. It also contains symbiotic bacteria that produce certain vitamins.

Most animals balance food intake with energy expenditure. Food intake is regulated primarily by hunger centers located in the hypothalamus and brain stem. In mammals, should caloric intake exceed requirements for energy, excess calories normally are dissipated as heat in specialized brown fat tissue.

Several gastrointestinal hormones coordinate digestive functions. They include gastrin, which stimulates acid secretion by the stomach; CCK, which stimulates gallbladder and pancreatic secretion and leads to satiety; and secretin, which stimulates bicarbonate secretion from the pancreas and inhibits gastric motility. New GI hormones are being added to this list as they are discovered. For example, ghrelin stimulates appetite, while PP and PYY induce satiety.

All animals require a balanced diet containing both fuels (mainly carbohydrates and lipids) and structural and functional components (proteins, minerals, and vitamins). For every animal, certain amino acids, lipids, vitamins, and minerals are "essential" dietary factors that cannot be produced by an animal's own synthetic machinery. Animal proteins are better-balanced sources of amino acids than are plant proteins, which tend to lack one or more essential amino acids. Undernourishment and protein malnourishment are among the world's major health problems, afflicting millions of people. Ironically, obesity and diseases associated with this condition are major health problems in developed countries.

REVIEW QUESTIONS

1. Distinguish the terms in the following pairs: autotrophic and heterotrophic; phototrophic and chemotrophic; herbivores and carnivores; omnivores and insectivores.
2. Suspension feeding is one of the most important methods of feeding among animals. Explain the characteristics, advantages, and limitations of suspension feeding, and name three different groups of animals that are suspension feeders.
3. An animal's feeding adaptations are an integral part of its behavior and usually shape the appearance of the animal itself. Discuss contrasting feeding adaptations of carnivores and herbivores.
4. Explain how food is propelled through the digestive tract.
5. Compare intracellular with extracellular digestion and suggest why there has been a phylogenetic trend in some animals from intracellular to extracellular digestion.
6. Which structural modifications vastly increase the internal surface area of the intestine (both invertebrate and vertebrate), and why is this large surface area important?

7. Trace digestion and final absorption of a carbohydrate (starch) in the vertebrate gut, naming the carbohydrate-splitting enzymes, where they are found, the breakdown products of starch digestion, and in what form they are finally absorbed.
8. As in question 7, trace digestion and final absorption of a protein.
9. Explain how fats are emulsified and digested in the vertebrate gut. Explain how bile aids the digestive process even though it contains no enzymes. Provide an explanation for the following observation: fats are broken into fatty acids and monoglycerides in the intestinal lumen, but appear later in the blood as fat droplets.
10. Explain the phrase "diet-induced thermogenesis" and relate it to the problem of obesity in some people. What other factors may contribute to human obesity?
11. Name three hormones of the gastrointestinal tract and explain how they assist in the coordination of gastrointestinal function.
12. Name the basic classes of foods that serve mainly as (a) fuels and (b) structural and functional components.
13. If vitamins are neither biochemically similar compounds nor sources of energy, what characteristics distinguish vitamins as a distinct group of nutrients? What are the water-soluble and the fat-soluble vitamins?
14. Why are some nutrients considered "essential" and others "nonessential" even though both types of nutrients are used in growth and tissue repair?
15. Explain the difference between saturated and unsaturated lipids, and comment on the current interest in these compounds as they relate to human health.
16. What is meant by "protein complementarity" among plant foods?

For Further Thought   Explain why a balanced diet of vegetables, a starchy food (rice, pasta, or potatoes), and a smaller meat portion would be appropriate for human health.

SELECTED REFERENCES

Bachman, E. S., H. Dhillon, C-Y. Zhang, S. Cinti, A. C. Bianco, B. K. Kobilka, and B. B. Lowell. 2002. βAR signalling required for diet-induced thermogenesis and obesity resistance. Science **297:** 843–845. *This paper describes an important study performed in mice showing that diet-induced thermogenesis is regulated by the sympathetic nervous system.*

Blaser, M. J. 1996. The bacteria behind ulcers. Sci. Am. **274:** 104–107 (Jan). *We now know that most cases of stomach ulcers are caused by acid-loving microbes. At least one-third of the human population are infected although most do not become ill.*

Bostrom, P., J. Wu, M. P. Jedrychowski, A. Korde, L. Ye, J. C. Lo, K. A. Rasbach, E. A. Bostrom, J. H. Choi, J. Z. Long, S. Kajimura, M. C. Zingaretti, B. F. Vind, H. Tu, S. Cinti, K. Hojlund, S. P. Gygi, and B. S. Spiegelman. 2012. A PGC1-alpha-dependent myokine that drives brown-fat-like development of white fat and thermogenesis. Nature **481:** 463–469. *This research article outlines original research on the newly discovered skeletal muscle hormone, irisin.*

Bray, G. A., and G. M. Popkin. 2013. Calorie sweetened beverages and fructose: what have we learned 10 years later? Pediatric Obesity. **8:** 242–248. *Summarizes the adverse effects of fructose in human diets.*

Hill, J. O., H. R. Wyatt, G. W. Reed, and J. C. Peters. 2003. Obesity and the environment: where do we go from here? Science **299:** 853–855. *Read this and related articles on the obesity crisis in a special issue of* Science.

Martins, I. J., and T. G. Redgrave. 2004. Obesity and post-prandial lipid metabolism. Feast or famine? J. Nut. Biochem. **15:** 130–141. *An excellent review of the disease of obesity and current research in lipid metabolism.*

Milton, K. 1993. Diet and primate evolution. Sci. Am. **269:** 86–93 (Aug). *Studies with primates suggest that modern human diets often diverge greatly from those to which the human body may be adapted.*

Morrison, S. F. 2004. Central pathways controlling brown adipose tissue thermogenesis. News Physiol. Sci. **19:** 67–74. *A readable review relating thermogenesis to energy expenditure.*

Randall, D., W. Burggren, and K. French. 2002. Eckert animal physiology: mechanisms and adaptations, ed. 5. New York, W. H. Freeman & Company. *A comprehensive and comparative treatment of animal physiology including an excellent section on digestion in animals.*

Ronti, T., G. Lupattelli, and E. Mannarino. 2006. The endocrine function of adipose tissue: an update. Clin. Endocrinol. **64:** 355–365. *Summary of the signaling molecules secreted by fat stores and their role in energy balance.*

Sanderson, S. L., and R. Wassersug. 1990. Suspension-feeding vertebrates. Sci. Am. **262:** 96–101 (Mar). *A variety of vertebrates, some enormous in size, eat by filtering out small organisms from massive amounts of water passed through a feeding apparatus.*

Weindrach, R. 1996. Caloric restriction and aging. Sci. Am. **274:** 46–52 (Jan). *Organisms from single-celled protists to mammals live longer on well-balanced but low-calorie diets. The potential benefits for humans are examined.*

Willmer, P., G. Stone, and I. Johnston. 2005. Environmental physiology of animals, ed. 2. Oxford, U.K., Blackwell Science Ltd. *Well-written information regarding environmental adaptations of both vertebrates and invertebrates.*

33

CHAPTER

Nervous Coordination: Nervous System and Sense Organs

James Gathany/CDC

Female Rocky Mountain wood tick on a grass stem awaits its host.

The Private World of the Senses

Information from the senses of vision, hearing, taste, olfaction, and touch continually assaults us. These classic five senses are supplemented by sensory inputs of cold, warmth, vibration, and pain, as well as by information from numerous internal sensory receptors that operate silently and automatically to keep our body working smoothly.

The world revealed by our senses is uniquely human even if our individual senses are not. We share this exclusive world with no other animal, nor can we venture into the sensory world of any other animal except as an abstraction through our imagination.

The idea that each animal enjoys an unshared sensory world was first conceived by Jakob von Uexküll, a seldom-cited German biologist of the early part of the last century. Von Uexküll asks us to try to enter the world of a tick through our imagination, supplemented by what we know of tick biology. It is a world of temperature,

of light and dark, and of the odor of butyric acid, a chemical common to all mammals. Insensitive to all other stimuli, the tick climbs up a blade of grass to wait, for years if necessary, for cues that will betray the presence of a potential host onto which she will jump. Later, swollen with blood, she drops to the earth, lays her eggs, and dies. The tick's impoverished sensory world, fine-tuned by natural selection for the world she will encounter, has ensured her single goal, reproduction.

A bird and a bat may share for a moment precisely the same environment. The worlds of their perceptions, however, are vastly different, structured by the limitations of the sensory receptors each employs and by the brain that receives and processes what it needs for survival. For one it is a world dominated by vision; for the other, by echolocation. The world of each is alien to the other, just as their worlds are to us.

The nervous system originated in a fundamental property of life: **irritability,** the ability to respond to environmental stimuli (see Chapter 1, p. 8). The response may be simple, such as a unicellular eukaryote moving to avoid a noxious substance, or quite complex, such as a vertebrate animal responding to elaborate signals of courtship. A unicellular eukaryote receives and responds to a stimulus, all within the confines of a single cell. Evolution of multicellularity and more complex levels of animal organization required increasingly complex mechanisms for communication between cells and organs. Relatively rapid communication is by **neural mechanisms** and involves propagated electrochemical signaling along and between cell membranes. The basic plan of a nervous system is to receive information from external and internal environments, to encode this information, and to transmit and to process it for appropriate action. We examine these functions in this chapter. Relatively less rapid or long-term adjustments in animals are governed by **hormonal mechanisms,** the subject of Chapter 34.

33.1 NEURONS: FUNCTIONAL UNITS OF NERVOUS SYSTEMS

A **neuron,** or nerve cell, may assume many shapes, depending on its function and location; a typical kind is shown diagrammatically in Figure 33.1. From the nucleated cell body extend cytoplasmic processes of two types: one or more **dendrites** in all but the simplest neuron, and a single **axon.** As the name dendrite suggests (Gr. *dendron,* tree), these processes are often highly branched. They, and the entire cell body surface, are the nerve cell's receptive apparatus, designed to receive information from multiple sources at once. Some of these inputs are excitatory, causing a signal to be generated and propagated; others are inhibitory, making signal generation and propagation less likely.

The single axon (Gr. *axon,* axle), often a long fiber that may be meters in length in the largest mammals, is relatively uniform in diameter, and typically carries signals away from the cell body. In vertebrates and some complex invertebrates, the axon is often covered with an insulating sheath of **myelin,** which accelerates signal propagation.

Neurons are commonly classified as **afferent,** or **sensory; efferent,** or **motor;** and **interneurons,** which are neither sensory nor motor but interconnect neurons. Afferent and efferent neurons lie mostly outside the **central nervous system** (brain and nerve cord) in the **peripheral nervous system,** while interneurons, which in humans are 99% of all neurons in the body, lie entirely within the central nervous system. Afferent neurons are connected to **receptors.** Receptors function to convert external and internal environmental stimuli into nerve signals, which are carried by afferent, sensory neurons into the central nervous system. Some of these signals may be perceived as conscious sensation, but many are not. Nerve signals also move to efferent, motor neurons, which carry them via the peripheral nervous system to **effectors,** such as muscles or glands, which respond to the signals.

In vertebrates, nerve processes (usually axons) in the peripheral nervous system are often bundled together in a wrapping of connective tissue to form a **nerve** (Figure 33.2). Cell bodies of these nerve processes are located either in the central nervous system or in

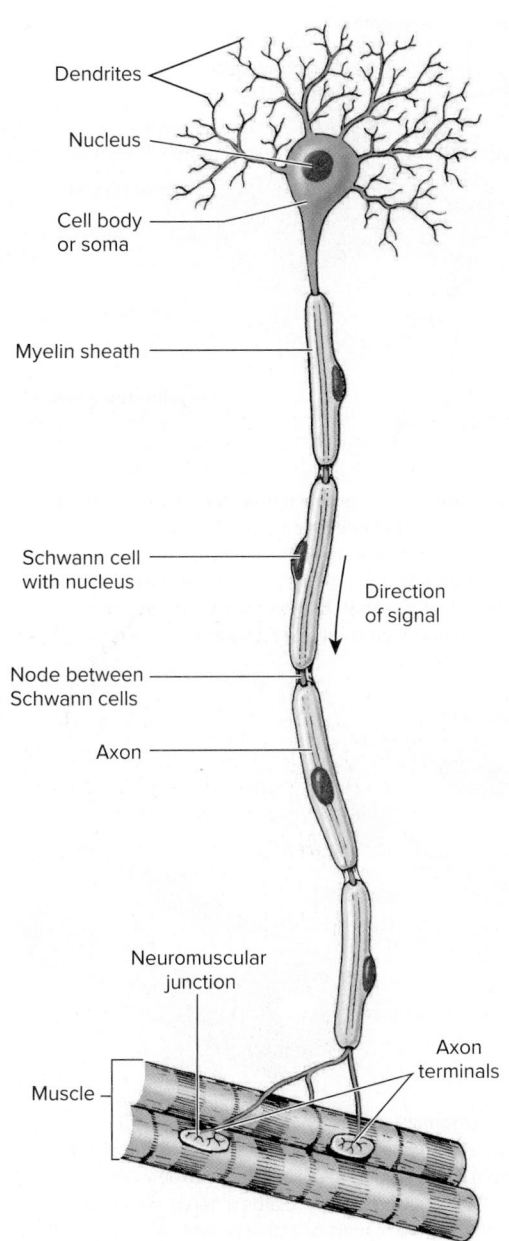

Figure 33.1
Structure of a motor (efferent) neuron.

Dendrites

Nucleus

Cell body or soma

Myelin sheath

Schwann cell with nucleus

Direction of signal

Node between Schwann cells

Axon

Neuromuscular junction

Muscle

Axon terminals

ganglia, which are discrete bundles of nerve-cell bodies located outside the central nervous system.

Surrounding neurons are non-nervous **neuroglial cells** (often simply called **"glial"** cells) that have a special relationship to neurons. Glial cells are extremely numerous in the vertebrate brain, where they outnumber neurons 10 to 1 and may form almost half the volume of the brain. Some glial cells form intimate insulating sheaths of lipid-containing **myelin** around nerve fibers. Myelin is essentially composed of multiple layers of glial cell plasma membrane that wrap around the axon in concentric rings as it forms the myelin sheath (Figure 33.3). Vertebrate nerves are often enclosed by myelin, produced by special glial cells called **Schwann cells** (Figure 33.3) in the peripheral nervous system, and **oligodendrocytes** in the central nervous system. Certain glial cells, called **astrocytes** because of their radiating, starlike shape, serve as nutrient and ion reservoirs for neurons,

Myelinated neuron
Schwann cell
Axon

Schwann cell
Axons
Unmyelinated neurons
(surrounded by a single Schwann cell)

Connective tissue sheath
Fascicle (connective tissue bundle)
Artery and vein

Figure 33.2 Structure of a peripheral nerve showing nerve fibers surrounded by various layers of connective tissue. A nerve may contain thousands of both efferent and afferent fibers. Multiple unmyelinated neurons are held in place within the nerve by a single Schwann cell providing support. Individual myelinated neurons are surrounded by multiple layers of myelin produced by single Schwann cells wrapping many times around each neuron.

Nucleus
Schwann cell
Axon
Myelin sheath

Figure 33.3 Development of the myelin sheath in a myelinated neuron of the peripheral nervous system. The whole Schwann cell grows around an axon, then rotates around it, enclosing the axon in a tight, multilayered sheath. The myelin sheath insulates a nerve axon and facilitates transmission of nerve signals or action potentials.

as well as providing a scaffold during brain development, enabling migrating neurons to find their destinations from points of origin. Astrocytes, and smaller **microglial** cells, are essential for the regenerative process that follows brain injury. Microglial cells are members of the **mononuclear phagocyte system** (see Chapter 35, p. 766) and are important in brain immune function. Following brain injury, they activate to phagocytose cellular debris. Unfortunately, astrocytes also participate in several diseases of the nervous system, including Parkinsonism, multiple sclerosis, and brain tumor development.

Nature of a Nerve Action Potential

A nerve signal or **action potential** is an electrochemical message of neurons, the common functional denominator of all nervous-system activity. Despite the incredible complexity of nervous systems of many animals, action potentials are basically alike in all neurons and

in all animals. An action potential is an "all-or-none" phenomenon; either the fiber is conducting an action potential, or it is not. Because all action potentials are alike, the only way that a nerve fiber can vary its signal is by changing the frequency of signal conduction. Frequency change is the language of a nerve fiber. A fiber may conduct no action potentials at all or very few per second up to a maximum approaching 1000 per second. The higher the frequency (or rate) of conduction, the greater is the level of excitation.

Resting Membrane Potential

Membranes of neurons, like all plasma membranes, have a selective permeability that creates ionic imbalances across the membrane. This ion imbalance is the basis of the resting membrane potential and a change in this imbalance generates a signal. Therefore, it is important to consider the ionic basis of the resting membrane potential, prior to considering signal generation. The interstitial fluid surrounding neurons contains relatively high concentrations of sodium (Na^+) and chloride (Cl^-) ions, but a low concentration of potassium ions (K^+) and large impermeable anions with negative charge, such as proteins. Inside the neuron, the ratio is reversed: K^+ and impermeable anion concentrations are high, but Na^+ and Cl^- concentrations are low (Figure 33.4; see also Figure 31.1B, p. 681). These differences are pronounced; there is approximately 10 times more Na^+ outside than in and 25 to 30 times more K^+ inside than out.

When at rest, the membrane of a neuron is selectively permeable to K^+, which can traverse the membrane through potassium channels that are open in the resting membrane (see Chapter 3, p. 45). The permeability to Na^+ is nearly zero because Na^+ channels are closed in a resting membrane. Potassium ions tend to diffuse outward through the membrane, following the gradient of potassium concentration. Very quickly the positive charge outside reaches a level that prevents any more K^+ from diffusing out of the axon (because like charges repel each other), and because large anions cannot pass through the membrane, positively charged potassium ions are drawn back into the cell. Now the membrane is at equilibrium for K^+ (the equilibrium potential) with an electrical gradient that exactly balances the concentration gradient (approximately −90 mV [millivolts]). The **resting membrane potential** is usually −70 mV with the inside of the membrane negative with respect to the outside. This is somewhat higher than the K^+ equilibrium potential. The difference in the resting membrane potential (−70 mV) and the potassium equilibrium potential (−90 mV) is due to a very low permeability to Na^+, which leaks through the open potassium channels into the cell, even in the resting condition. This Na^+ inward leak is due to the high Na^+ concentration gradient and electrical attraction.

Sodium Pump

With Na^+ leaking into the cell, there is also more of a tendency of K^+ ions to flow out of the cell at rest, because it is not in equilibrium. Thus, the ion gradients that occur across a resting membrane tend to decay over time. This decay is prevented by **sodium pumps,** each a complex of protein subunits embedded in the plasma membrane of the axon (see Figure 3.19, p. 48). Each sodium pump uses energy from the hydrolysis of ATP to transport sodium from the inside to the outside of the membrane. The sodium pump in nerve axons, as in many other cell membranes, also moves K^+ into the axon while it is moving Na^+ out. Thus, it is a **sodium-potassium exchange pump**

Figure 33.4 Ionic composition inside and outside a resting nerve cell. An active sodium-potassium exchange pump located in the plasma membrane drives sodium to the outside, keeping its concentration low inside and potassium concentration high inside. The membrane is permeable to potassium, but this ion is held inside by the repelling positive charge outside the membrane, by its attraction to large negatively charged anions, such as proteins, which cannot leave the cell, and by a constant influx of potassium due to the sodium-potassium exchange pump.

that helps to restore the ion gradients of both Na^+ and K^+. In addition, in the central nervous system, astrocytes help to maintain the correct balance of ions surrounding neurons by storing excess potassium produced during neuronal activity.

Action Potential

A nerve **action potential** is a very rapid and brief moving change in electrical membrane potential (Figure 33.5) that results from a **depolarization** of the membrane of the nerve fiber. This means that the membrane potential changes from rest (approximately -70 mV) in a positive direction and overshoots 0 mV to reach about $+35$ mV. The membrane potential thus reverses for an instant so that the outside becomes negative compared with the inside. Then, as the action potential moves ahead, the membrane returns to its normal resting membrane potential, ready to conduct another signal. The entire event occupies approximately a millisecond. Perhaps the most significant property of the nerve action potential is that it is **self-propagating;** once started the action potential moves along the nerve fiber automatically and unchanged in intensity, much like the burning of a fuse.

What causes the reversal of polarity in the plasma membrane during passage of an action potential? We have seen that the resting membrane potential depends on the high membrane permeability to K^+, some 50 to 70 times greater than the permeability to Na^+. When the action potential arrives at a given point in a neuron membrane, the change in membrane potential causes **voltage-gated Na^+ channels** (p. 45) to open suddenly, permitting Na^+ to diffuse into the axon from the outside, moving down the concentration gradient for Na^+. The voltage-gated Na^+ channels remain open for less than a millisecond. Only a very minute amount of Na^+ moves across the membrane— less than one-millionth of the Na^+ outside—but this sudden rush of positive ions cancels the local resting membrane potential and the membrane is **depolarized.** Then, as the Na^+ channels are inactivated

and close, the membrane quickly regains its resting properties as K^+ ions quickly diffuse out through voltage-gated K^+ channels that open briefly in response to the membrane depolarization. The membrane becomes once again practically impermeable to Na^+, and permeable primarily to K^+ as the resting membrane potential is reestablished. The continuing activity of the sodium-potassium pumps also helps to reestablish Na^+ and K^+ ion gradients.

Thus, the rising phase of an action potential is associated with rapid influx (inward movement) of Na^+ (Figure 33.5). When the action potential reaches its peak, Na^+ permeability is restored to normal, and K^+ permeability briefly increases above the resting level and K^+ ions exit. Increased potassium permeability causes the action potential to drop rapidly toward the resting membrane potential, during the **repolarization** phase. Voltage-gated K^+ channels are slow to close during repolarization and so most action potentials exhibit an **afterhyperpolarization** phase where the membrane potential falls below the resting membrane potential as more K^+ ions flow out of the cell. Once these K^+ channels finally close, the resting membrane potential is rapidly achieved. The membrane is now ready to transmit another action potential. These events occur at each point along the neuron membrane, as the action potential is conducted from the axon hillock, where the axon potential originated, to the axon terminals (Figures 33.5 and 33.6A).

High-Speed Conduction

Although ionic and electrical events associated with action potentials are much the same throughout the animal kingdom, conduction velocities vary enormously from neuron to neuron and from animal to animal—from as slow as 0.1 m/sec in sea anemones to as fast as 120 m/sec in some mammalian motor axons. The speed of conduction is highly correlated with diameter of the axon. Small axons conduct slowly because internal resistance to current flow is high.

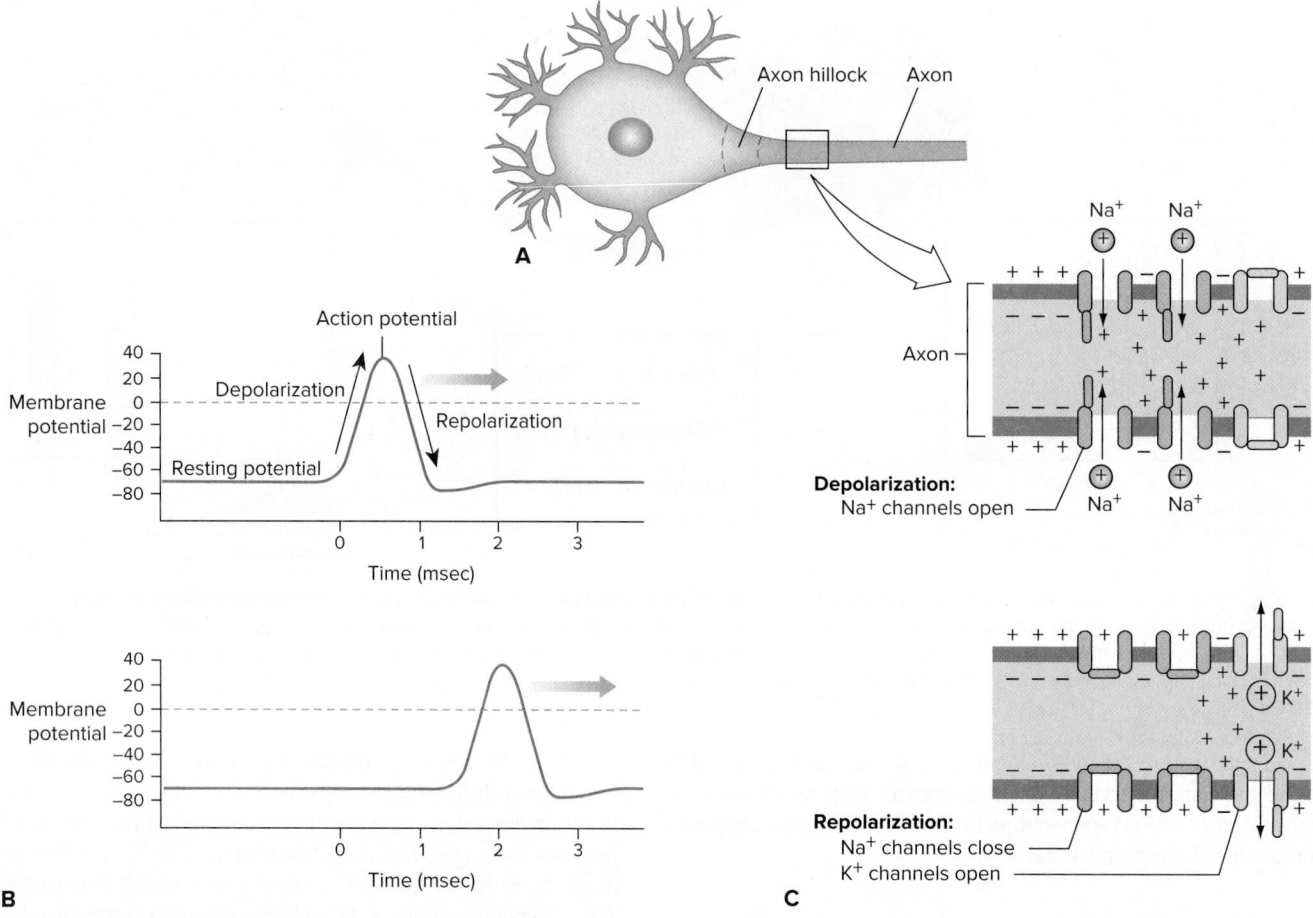

Figure 33.5 Conduction of an action potential or nerve signal. The action potential originates in the axon hillock of the neuron **(A)** and moves toward the right. **B** and **C** show the electrical event and associated changes in localized membrane permeability to sodium and potassium measured at two different points as the action potential moves along the membrane. The second action potential in **B** is recorded about 4 millimeters farther along the axon from the first one. **C,** During depolarization, when the action potential arrives at a point in the membrane, voltage-gated sodium channels open, allowing sodium ions to enter. Sodium inflow reverses the membrane polarity, making the inner surface of the axon positive and the outside negative. During repolarization, sodium channels close and voltage-gated potassium channels open. Potassium ions move out and restore the normal resting potential after a brief after-hyperpolarization phase. These events occur repeatedly along a membrane until the axon terminals are reached.

In most invertebrates, where fast conduction velocities are important for quick response, such as in locomotion to capture prey or to avoid capture, axon diameters are larger. Giant axons of squids are nearly 1 mm in diameter and carry impulses 10 times faster than ordinary axons in the same animal. A squid's giant axon innervates the animal's mantle musculature, which is used for powerful mantle contractions when it swims by jet propulsion. Similar giant axons enable earthworms, which are normally slow-moving animals, to withdraw almost instantaneously into their burrows when startled.

Although vertebrates do not possess giant axons, they achieve high conduction velocities by a cooperative relationship between axons and the investing layers of myelin laid down by the Schwann cells or oligodendrocytes described on page 721. Insulating myelin sheaths are interrupted at intervals of about 1 mm or less by nodes (called **nodes of Ranvier**) where the plasma membrane of the axon is exposed to interstitial fluid surrounding the nerve (see Figure 33.2). In these **myelinated fibers** action potentials depolarize the axon membrane only at nodes because the myelin sheath prevents depolarization elsewhere (Figure 33.6B). Ion pumps and

channels that move ions across a membrane are concentrated in each node. Once an action potential starts down an axon, depolarization of the first node initiates an electrical current that flows to the neighboring node, causing it to depolarize and to trigger an action potential. Thus the action potential leaps from node to node, a kind of conduction called **saltatory** (L. *salto,* to dance, leap). The gain in efficiency as compared with non-myelinated axons is impressive. For example, a frog myelinated axon only 12 μm in diameter conducts nerve impulses at the same speed as a squid unmyelinated axon 350 μm in diameter.

Some invertebrates, including prawns and insects, also have fast axons invested with multiple layers of a myelinlike substance that is interrupted by nodes at intervals much like myelinated axons of vertebrates. Conduction rates, though not as fast as in vertebrate saltatory conduction, are much faster than unmyelinated axons of the same diameter in other invertebrates.

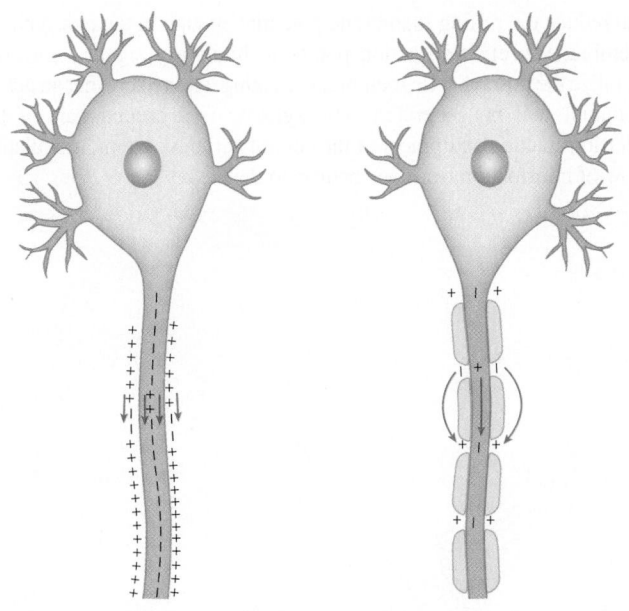

A Unmyelinated neuron   **B Myelinated neuron**

Figure 33.6 Action potential conduction in unmyelinated and myelinated fibers. In unmyelinated fibers **A,** the action potential spread is continuous, and must depolarize the entire length of the axon membrane. In myelinated fibers **B,** the action potential leaps from node to node, bypassing the insulated portions of the fiber. This is saltatory conduction, which is much faster than continuous conduction.

Temperature also alters conduction velocity in animals. Endotherms usually have a high conduction velocity because they maintain a constant high body temperature (37° C in humans), whereas the conduction velocity in ectotherms fluctuates with environmental temperatures.

33.2 SYNAPSES: JUNCTIONS BETWEEN NERVES

When an action potential passes down an axon to its terminal, it must cross a small gap, or **synapse** (Gr. *synapsis,* contact, union), separating it from another neuron or an effector organ. Two distinct kinds of synapses are known: electrical and chemical.

Electrical synapses, although much less common than chemical synapses, have been demonstrated in both invertebrate and vertebrate groups. Electrical synapses are points at which ionic currents flow through narrow gap junctions (see Figure 3.15, p. 45) from one neuron to another. Electrical synapses show no time lag and consequently are important for escape reactions. Signals are bidirectional at many electrical synapses, but unidirectional ones have been demonstrated in Crustacea. Electrical synapses also have been observed in other excitable cell types, and form an important method of communication in the heart between cardiac muscle cells (p. 679) and between cells in smooth muscle tissue (for example, the uterus, p. 144).

Much more complex than electrical synapses are **chemical synapses,** which contain packets or vesicles of specialized chemicals called **neurotransmitters.** Neurons bringing action potentials toward chemical synapses are called **presynaptic neurons;** those carrying action potentials away are **postsynaptic neurons.** At a synapse, membranes are separated by a narrow gap, or **synaptic cleft,** having a width of approximately 20 nm.

The axon of most neurons divides at its end into many branches, each of which bears a synaptic knob or terminal that interacts with a dendrite or the cell body of the next neuron (Figure 33.7A). Because a single action potential coming down a nerve axon is transmitted along these many branches and synaptic endings, many impulses may converge on the cell body at one instant or may diverge onto more than one postsynaptic neuron. In addition, axon terminals of many neurons may almost cover a nerve cell body and its dendrites with thousands of synapses.

The 20-nm interstitial fluid-filled gap of chemical synapses between presynaptic and postsynaptic membranes prevents action potentials from spreading directly to a postsynaptic neuron. Instead, synaptic knobs secrete one or more specific neurotransmitters that communicate chemically with the postsynaptic cell. One of the most common neurotransmitters of the peripheral nervous system is **acetylcholine,** which illustrates typical synaptic transmission. Inside synaptic terminals of presynaptic neurons are numerous tiny **synaptic vesicles,** each containing several thousand molecules of acetylcholine. When an action potential arrives at a synaptic terminal a sequence of events occurs as portrayed in Figures 33.7 and 33.8. An action potential causes an inward movement of calcium (Ca^+) ions through voltage-gated channels in the synaptic terminal membrane and this induces exocytosis of some neurotransmitter-filled synaptic vesicles. Acetylcholine molecules diffuse into the gap in a fraction of a millisecond and bind briefly to receptor molecules on ion channels in the postsynaptic membrane. These **chemically-gated** channels (p. 45) open and ions flow through the channels while they remain open. This flow of ions creates a voltage change in the postsynaptic membrane. Whether this postsynaptic excitatory potential is large enough to trigger an action potential depends on how many acetylcholine molecules are released and how many channels are opened. Acetylcholine is rapidly destroyed by the enzyme **acetylcholinesterase,** which converts acetylcholine into acetate and choline. As long as the neurotransmitter is present in the synaptic cleft it continues to stimulate the postsynaptic membrane by binding to ion channel receptor sites, indefinitely stimulating the ion channels to open. Organophosphate insecticides (such as malathion) and certain military nerve gases are poisonous for precisely this reason; they block acetylcholinesterase. The final step in the sequence is reabsorption of choline into the presynaptic terminal, resynthesis of acetylcholine, and its storage in synaptic vesicles, ready to respond to another action potential.

Vertebrate and invertebrate nervous systems both have many different chemical neurotransmitters. Those that depolarize postsynaptic membranes are released at **excitatory synapses,** while those that move the resting membrane potential in a more negative direction **(hyperpolarization),** thereby stabilizing them against depolarization, are released at **inhibitory synapses.** Whether a neurotransmitter causes a postsynaptic excitatory or inhibitory potential depends on what specific ions flow through the chemically-gated channels to which they bind. Thus, neurotransmitters can be both excitatory and inhibitory. Such examples include acetylcholine, norepinephrine, dopamine, and serotonin. However, some neurotransmitters always appear to be inhibitory (for example, glycine and gamma aminobutyric acid [GABA]), while others seem always to be excitatory (for example, glutamate). Neurons in the central nervous system have both excitatory and inhibitory synapses among the hundreds or thousands of synaptic terminals on the dendrites and cell body of each neuron.

The net balance of all excitatory and inhibitory inputs received by a postsynaptic cell determines whether it generates an action potential (Figure 33.8). If many excitatory signals are received at one time, they can reduce the resting membrane potential enough in the postsynaptic membrane to elicit an action potential. Inhibitory signals, however, stabilize the postsynaptic membrane, making it less likely that an action potential will be generated. The synapse is a crucial part of the decision-making equipment of the central nervous system, modulating flow of information from one neuron to the next.

Figure 33.7 Transmission of action potentials across nerve synapses. **A,** A cell body of a motor nerve is shown with the axon terminals of interneurons. Each terminal ends in a synaptic knob or terminal; thousands of synaptic terminals may interact with a single nerve cell body and its dendrites. **B,** A synaptic terminal enlarged 60 times more than in **A.** An action potential traveling down the axon causes movement of synaptic vesicles to the presynaptic membrane where exocytosis occurs, releasing neurotransmitter molecules into the cleft. **C,** Diagram of a synaptic cleft at the ultrastructural level. Upon vesicular exocytosis, neurotransmitter molecules move rapidly across the gap to bind briefly with chemically-gated ion channels in the postsynaptic membrane. Binding of neurotransmitter to its receptor produces a change in the potential of the postsynaptic membrane, in this case, caused by opening of the ion channels.

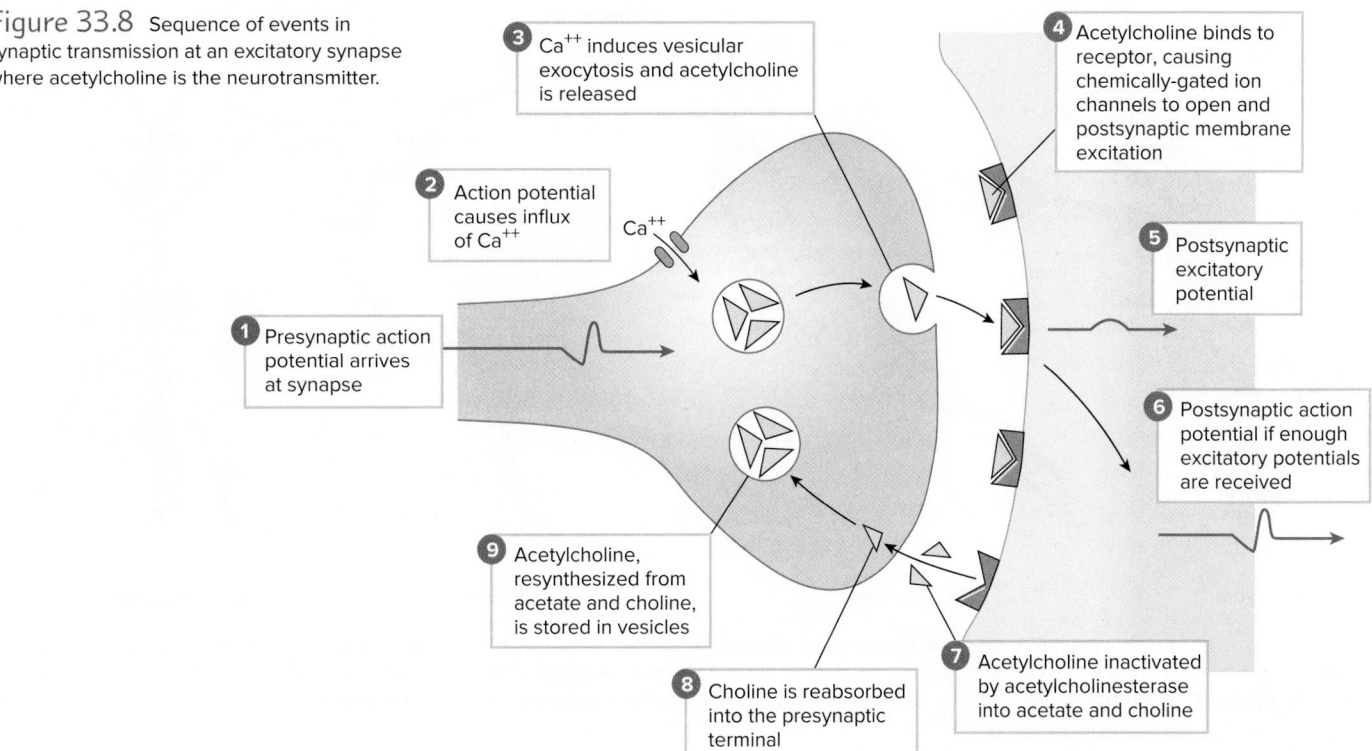

Figure 33.8 Sequence of events in synaptic transmission at an excitatory synapse where acetylcholine is the neurotransmitter.

3 Ca⁺⁺ induces vesicular exocytosis and acetylcholine is released

4 Acetylcholine binds to receptor, causing chemically-gated ion channels to open and postsynaptic membrane excitation

2 Action potential causes influx of Ca⁺⁺

5 Postsynaptic excitatory potential

1 Presynaptic action potential arrives at synapse

6 Postsynaptic action potential if enough excitatory potentials are received

9 Acetylcholine, resynthesized from acetate and choline, is stored in vesicles

8 Choline is reabsorbed into the presynaptic terminal

7 Acetylcholine inactivated by acetylcholinesterase into acetate and choline

33.3 EVOLUTION OF NERVOUS SYSTEMS

Invertebrates: Development of Centralized Nervous Systems

Various animal phyla reveal a progressive increase in complexity of nervous systems that probably reflects in a general way stages in evolution of nervous systems. The simplest pattern of invertebrate nervous systems is the nerve net of radially symmetrical animals, such as sea anemones, jellyfishes, hydras, and comb jellies (Figure 33.9A). A nerve net forms an extensive network in and under the epidermis over all the body. A signal starting in one part of this net spreads in all directions, since synapses in most radially symmetrical animals allow bidirectional signal transmission. There is evidence of organization into **reflex arcs** (p. 728) with branches of a nerve net connecting to sensory receptors in the epidermis and to epithelial cells that have contractile properties. Although most responses tend to be generalized, many are astonishingly complex for so simple a nervous system. Part of the nerve net is concentrated into two nerve rings in medusan forms of Cnidaria (see Figure 13.11, p. 269) and receives sensory input from **statocysts,** organs of balance (p. 268), and **ocelli,** light-sensitive organs, as well as sensory cells detecting chemical and tactile stimuli. In vertebrates, nerve nets occur in nerve plexuses located, for example, in the intestinal wall; such nerve plexuses govern generalized intestinal movements such as peristalsis and segmentation (p. 707).

Bilateral nervous systems, the simplest of which occur in flatworms, represent a distinct increase in complexity over the nerve net of radiate animals. Evidence from the study of genetic mechanisms controlling brain development in insect and mouse embryos shows homology of regulatory gene families. These data suggest that a common ancestral brain utilizing these genes may have evolved prior to the protostome-deuterostome divergence. Flatworms have two anterior ganglia, composed of groups of nerve cell bodies from which two main nerve trunks run posteriorly, with lateral branches extending throughout the body (Figure 33.9B). This is the simplest nervous system showing differentiation into a **peripheral nervous system** (a communication network extending to all parts of the body) and a **central nervous system** (a concentration of nerve cell bodies), which coordinates everything. More complex invertebrates exhibit a more centralized nervous system (brain), with two longitudinal fused nerve cords and many ganglia. Elaborate nervous systems of annelids consist of a bilobed brain, a double nerve cord with segmental ganglia, and distinctive **afferent** (sensory) and **efferent** (motor) neurons (Figure 33.9C). Segmental ganglia are relay stations for coordinating regional activity.

The basic plan of molluscan nervous systems is a series of three pairs of well-defined ganglia, but cephalopods (such as octopuses and squids) have ganglia burgeoned into textured nervous centers of great complexity; those of octopuses contain more than 160 million cells. Sense organs, too, are highly developed. Consequently, cephalopod behavior far outstrips that of any other invertebrate, and they are capable of learning (p. 356).

The basic plan of arthropod nervous systems (Figure 33.9D) resembles that of annelids, but ganglia are larger and sense organs are much better developed. Social behavior is often elaborate, particularly in hymenopteran insects (bees, wasps, and ants), and most arthropods are capable of considerable manipulation of their environment. Despite the small size of an insect's brain, examples of learning have been documented in bees, wasps, ants, flies, locusts, and grasshoppers. The brain region associated with learning appears

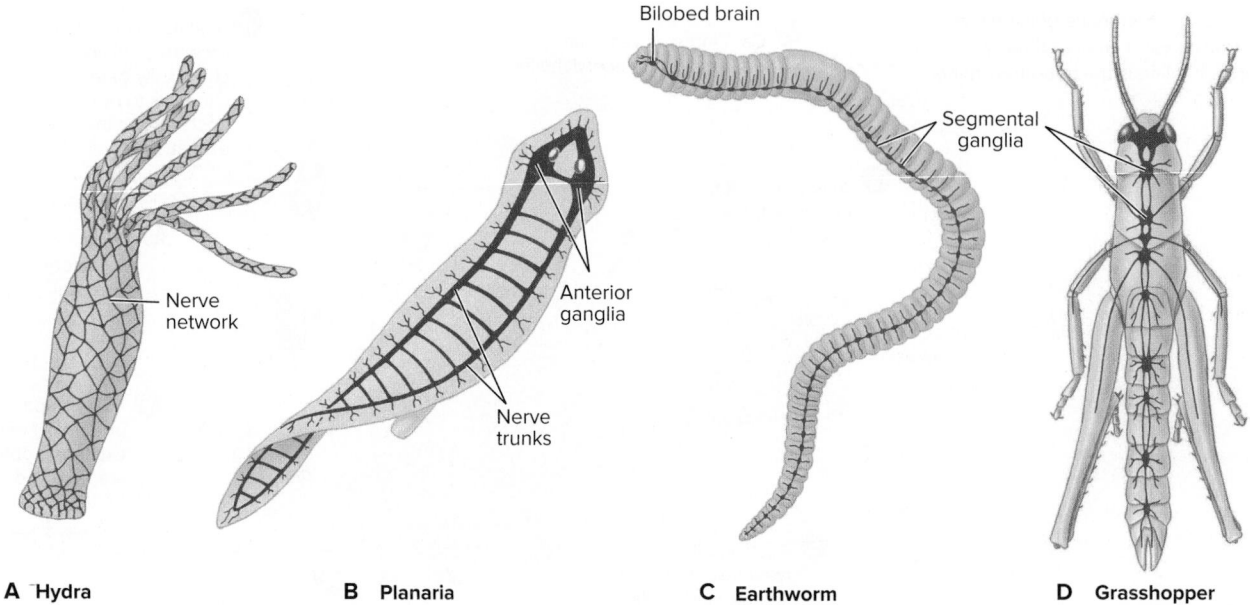

Figure 33.9 Invertebrate nervous systems. **A,** Nerve net of radiates, the simplest neural organization. **B,** Flatworm system, the simplest linear-type nervous system of two nerves connected to a complex neuronal network. **C,** Annelid nervous system, organized into a bilobed brain and ventral cord with segmental ganglia. **D,** Arthropod nervous system, also segmental, with large ganglia and more elaborate sense organs.

to be areas called **mushroom bodies,** which are largest in social insects, and experiments have shown changes in these structures with age and experience.

Vertebrates: Fruition of Encephalization

The basic plan of the vertebrate nervous system is a hollow, *dorsal* nerve cord terminating anteriorly in a large neural mass, or brain. This pattern contrasts with the nerve cord of bilateral invertebrates outside of the phylum Chordata, which is solid and ventral to their alimentary canal. By far the most important trend in evolution of vertebrate nervous systems is the increase in size, configuration, and functional capacity of the brain, a process called **encephalization.** Vertebrate encephalization has brought to full fruition several functional capabilities, including fast responses, great capacity for storage of information, and enhanced complexity and flexibility of behavior. Another consequence of encephalization is the ability to form associations between past, present, and (at least in humans) future events.

Spinal Cord

The **brain** and **spinal cord** compose the central nervous system. During early embryonic development, the spinal cord and brain begin as an ectodermal neural groove, which by folding and enlarging becomes a long, hollow neural tube (see Figure 8.28, p. 179). The cephalic end enlarges to form brain vesicles, and the rest becomes the spinal cord. Unlike any invertebrate nerve cord, segmental nerves of vertebrate spinal cords (31 pairs in humans) are separated into dorsal sensory roots and ventral motor roots. Sensory nerve cell bodies are gathered together into dorsal root (spinal) ganglia. Both dorsal (sensory) and ventral (motor) roots meet beyond the spinal cord to form a mixed spinal nerve (Figure 33.10).

The spinal cord encloses a central spinal canal and is additionally wrapped in three layers of membranes called **meninges** (men-injeez; Gr. *meningos,* membrane). In cross section the cord shows two zones (Figure 33.10). An inner zone of gray matter, resembling in shape the wings of a butterfly, contains the cell bodies of motor neurons and interconnecting interneurons. An outer zone of white matter contains bundles of axons and dendrites linking different levels of the cord with each other and with the brain.

Reflex Arc

Many neurons work in groups called **reflex arcs,** a fundamental unit of neural operation that has remained conserved during evolution of the nervous system. A reflex arc contains at least two neurons, but usually there are more. Parts of a typical reflex arc are (1) a **receptor,** a sense organ in skin, muscle, or another organ; (2) an **afferent,** or sensory, neuron, which carries impulses toward the central nervous system; (3) the **central nervous system,** where synaptic connections are made between sensory neurons and interneurons; (4) an **efferent,** or motor, neuron, which makes a synaptic connection with the interneuron and carries impulses from the central nervous system; and (5) an **effector,** by which an animal responds to environmental changes. Examples of effectors are muscles, glands, ciliated cells, cnidocytes of cnidarians, electric organs of fishes, and pigmented cells called chromatophores (p. 642).

A reflex arc in vertebrates in its simplest form contains only two neurons—a sensory (afferent) neuron and a motor (efferent) neuron (for example, the "knee-jerk" or stretch reflex, Figure 33.11A). Usually, however, interneurons are interposed between sensory and motor neurons (Figure 33.11B). An interneuron may connect afferent and efferent neurons on the same side of the spinal cord or on opposite sides, or it may connect them on different anterior-posterior levels of the spinal cord, on either the same or opposite sides.

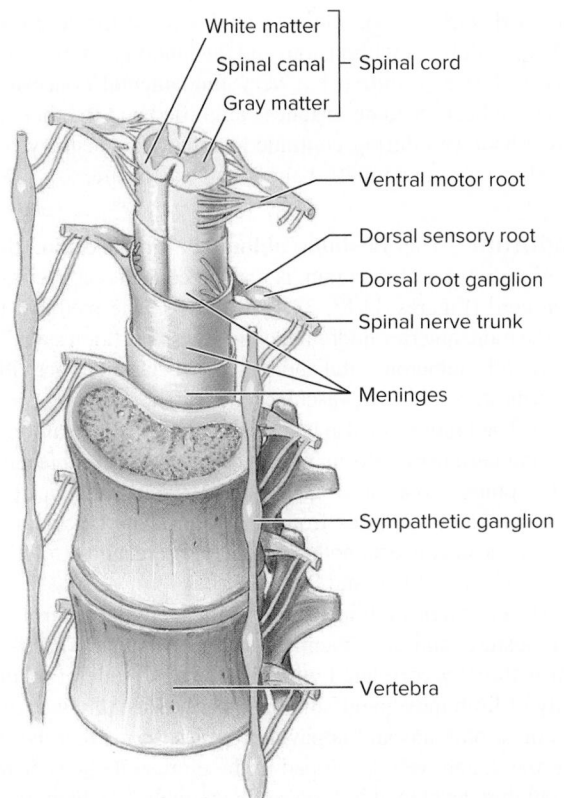

Figure 33.10 Human spinal cord and its protection. Two vertebrae show the position of the spinal cord, emerging spinal nerves, and the sympathetic ganglion chain. The cord is wrapped by three layers of membrane (meninges), between two of which lies a protective bath of cerebrospinal fluid.

A **reflex act** is a response to a stimulus acting over a reflex arc. It is involuntary, meaning that it is often not under the control of the will. For example, many vital processes of the body, such as control of breathing, heartbeat, diameter of blood vessels, and sweat secretion, are reflex acts. Some reflex acts are innate; others are acquired through learning.

In almost any reflex act, a number of reflex arcs are involved. For instance, a single afferent sensory neuron may make synaptic connections with many efferent motor neurons. In a similar way an efferent motor neuron may receive signals from many afferent sensory neurons. Afferent neurons also make connections with ascending sensory neurons, which travel in the white matter of the spinal cord, bringing information about peripheral reflexes to the brain. Reflex activity may then be modified by signals from the brain traveling along descending motor neurons, which impinge on the final efferent motor neurons before they leave the spinal cord for the periphery.

Brain

Unlike the spinal cord, which has changed little in structure during vertebrate evolution, the brain has changed dramatically. The ancestral vertebrate brain of fishes and early tetrapods expanded to form a deeply fissured and enormously intricate brain in the lineage leading to mammals (Figure 33.12). It reaches its greatest complexity in the human brain, which contains some 35 billion neurons, each of which may receive information from tens of thousands of synapses at one time. The ratio between weight of the brain and that of the spinal cord affords a fair criterion of an animal's intelligence. In fishes and amphibians this ratio is approximately 1:1; in humans the ratio is 55:1—in other words, the brain is 55 times heavier than the spinal cord. Although the human brain is not the largest (the sperm whale's brain is seven times heavier) nor the most convoluted (that of the porpoise is even more folded), it is the best in overall performance.

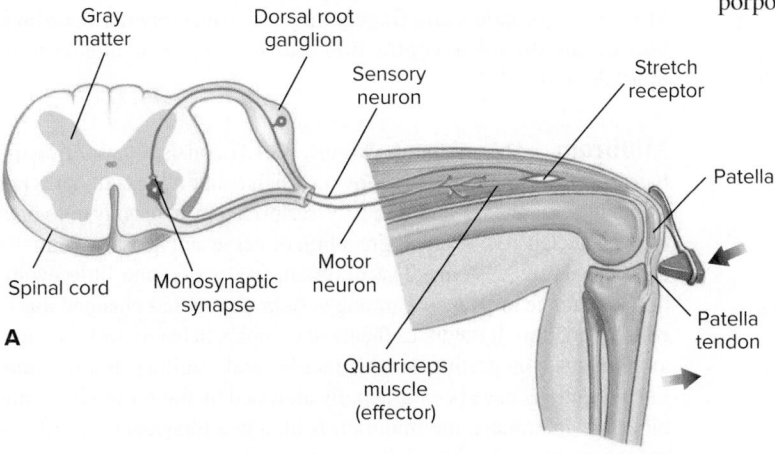

Figure 33.11 The reflex arc. **A,** The "knee-jerk" or stretch reflex, a simple reflex arc. Sudden pressure on the patellar ligament stretches muscles in the upper leg. Action potentials generated in stretch receptors are conducted along afferent (sensory) neurons to the spinal cord and relayed directly to an efferent (motor) nerve cell body. Action potentials pass along efferent neurons to leg muscles (effectors), stimulating them to contract. **B,** Multisynaptic reflex arc. A more common reflex arc includes interneurons between the sensory and the motor neuron. For example, a tack puncture is sensed by pain receptors in the skin and the signal is conducted along afferent fibers to the spinal cord, where synaptic connections are made with interneurons. Here, an interneuron is shown making connections with motor neurons on both sides of the spinal cord, such that stimulation of muscle fibers in more than one part of the body (both legs, for example) allows coordination of muscle responses to the tack puncture.

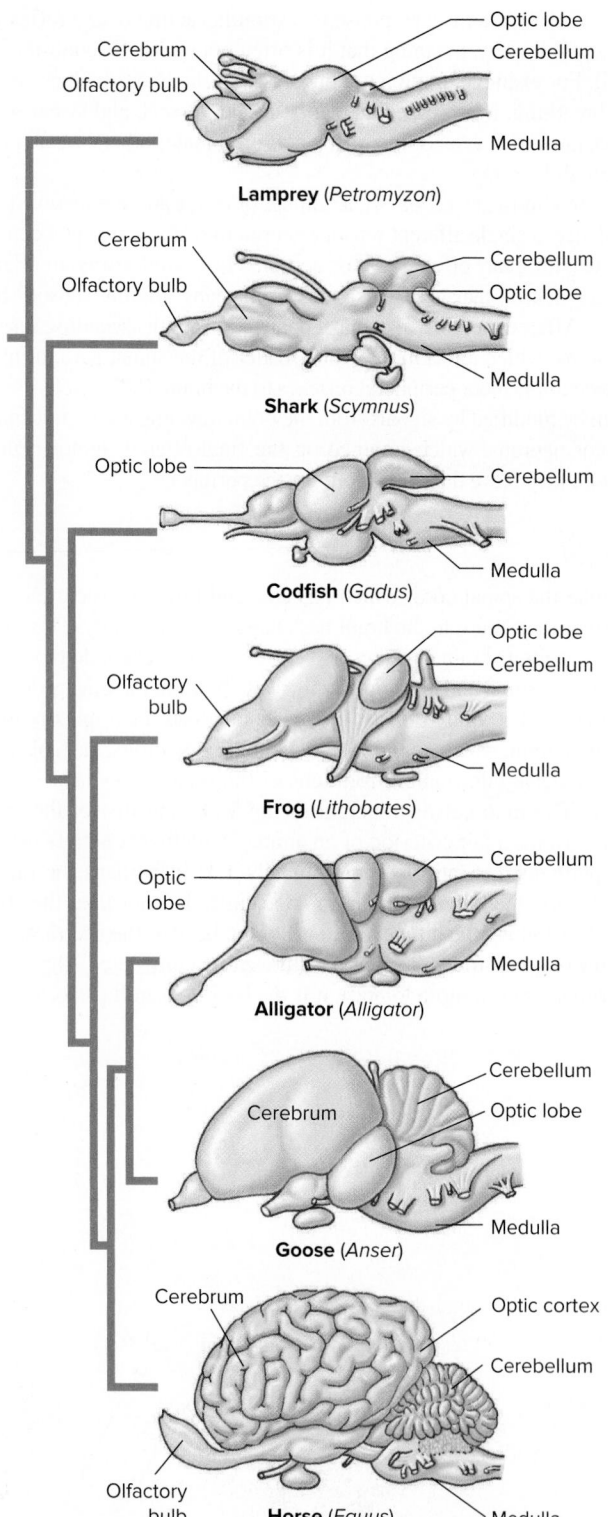

Figure 33.12 Evolution of the vertebrate brain. Note the progressive increase in size of the cerebrum. The cerebellum, concerned with equilibrium and motor coordination, is largest in animals whose balance and precise motor movements are well developed (fishes, birds, and mammals).

Brains of early vertebrates had three principal divisions: a forebrain, or **prosencephalon;** a midbrain, or **mesencephalon;** and a hindbrain, or **rhombencephalon** (Figure 33.13). Each part was

concerned with one or more special senses: the forebrain with smell, the midbrain with vision, and the hindbrain with hearing and balance. These primitive but very fundamental concerns of the brain have been in some instances amplified and in others reduced or overshadowed during continued evolution as sensory priorities were shaped by an animal's habitat and way of life.

Hindbrain   The **medulla oblongata or medulla,** the most posterior division of the brain, is really a conical continuation of the spinal cord (Figures 33.13, 33.14A and B). The medulla, together with the more anterior midbrain, constitutes the "brain stem," an area that controls numerous vital and largely subconscious activities such as heartbeat, respiration, vascular tone, gastric secretions, and swallowing. The brain stem also contains centers that appear to integrate incoming peripheral information regarding satiety and feeding stimuli. The **pons,** also a part of the hindbrain, contains a thick bundle of fibers that carry impulses from one side of the cerebellum to the other, and also connects both medulla and cerebellum to other brain regions (Figure 33.14A and B).

The **cerebellum,** lying dorsal to the medulla, controls equilibrium, posture, and movement (Figure 33.14A and B). Its development is directly correlated with an animal's mode of locomotion, agility of limb movement, and balance. It is usually weakly developed in amphibians and nonavian reptiles, forms that live close to the ground, and well developed in the more agile bony fishes. It is most highly developed in birds and mammals, in which it is greatly expanded and folded. The cerebellum does not initiate movement but operates as a precision error-control center, or servomechanism, that programs a movement initiated somewhere else, such as the motor cortex of the cerebrum (Figure 33.14A). Primates and especially humans, who possess a manual dexterity far surpassing that of other animals, have the most complex cerebellum. Movements of hands and fingers may involve cerebellar coordination of simultaneous contraction and relaxation of hundreds of individual muscles.

Midbrain   The midbrain (Figure 33.13) consists mainly of the **tectum** (including the **optic lobes**), which contains nuclei that serve as centers for visual and auditory reflexes. (In neurophysiological usage a nucleus is a small aggregation of nerve cell bodies within the central nervous system.) The midbrain has undergone little evolutionary change in structure among vertebrates but has changed markedly in function. It mediates the most complex behavior of fishes and amphibians, integrating visual, tactile, and auditory information. Such functions have been gradually assumed by the forebrain in amniotes. In mammals, the midbrain is mainly a relay center for information on its way to higher brain centers.

Forebrain   Just anterior to the midbrain lie the **thalamus** and **hypothalamus,** the most posterior elements of the forebrain (Figure 33.14B). The egg-shaped thalamus is a major relay station that analyzes and passes sensory information to higher brain centers. In the hypothalamus are several "housekeeping" centers that regulate body temperature, water balance, appetite, and thirst—all functions concerned with maintenance of internal constancy (homeostasis). Neurosecretory cells located in the hypothalamus produce several neurohormones (described in Chapter 34). The hypothalamus also

Development of the Vertebrate Brain

Embryonic vesicle		Main components in adults	Function
Early embryo	Late embryo		
Forebrain (Prosencephalon)	Telencephalon	Cerebrum	Motor area controls voluntary muscle movements; sensory cortex is center of conscious perception of touch, pressure, vibration, pain, temperature, and taste; association areas integrate and process sensory data
	Diencephalon	Thalamus	Part of limbic system; integrates sensory information arriving at thalamus, projects to cerebral frontal lobes
		Hypothalamus	Controls autonomic functions; sets appetitive drives (thirst, hunger, sexual desire) and reproductive behavior; participates in emotional responses; secretes ADH, oxytocin; secretes releasing hormones for anterior pituitary regulation
Midbrain (Mesencephalon)	Mesencephalon	Optic lobes (tectum)	Integrates visual information with other sensory inputs; relays auditory information
		Midbrain nuclei	Involuntary control of muscle tone; processing of incoming sensations and outgoing motor commands
Hindbrain (Rhombencephalon)	Metencephalon	Cerebellum	Involuntary coordination and control of outgoing movements for equilibrium, muscle tone, posture
		Pons	Links cerebellum with other brain centers and with medulla and spinal cord; modifies output of respiratory centers in medulla
	Myelencephalon	Medulla oblongata	Regulates heart rate and force of contraction; vasomotor control; sets rate of respiration; relays information to the cerebellum; integrates feeding and satiety stimuli

Figure 33.13   Divisions of the vertebrate brain.

contains centers for regulating reproductive function and sexual behavior, and it participates in emotional behaviors.

The anterior portion of the forebrain, or **cerebrum** (Figure 33.14 A and B), can be divided into two anatomically distinct areas, the **paleocortex** and the **neocortex.** Originally concerned with smell, it became well developed in advanced fishes and early terrestrial vertebrates, which depend on this special sense. In mammals and especially in primates the paleocortex is a deep-lying area called a rhinencephalon ("nose brain"), because many of its functions depend on olfaction. Better known as the **limbic system,** it mediates several species-specific behaviors that relate to fulfilling needs, such as feeding and sex. One region of the limbic system, the **hippocampus,** has been extensively studied as a site of spatial learning and memory. The hippocampus is now well-known because its neurons have mitotic capabilities in adults, a previously unknown occurrence in mammalian neurons.

Although a late arrival in vertebrate evolution, the neocortex completely overshadows the paleocortex and has become so expanded that it envelops much of the forebrain and all of the midbrain (Figure 33.14). Almost all integrative activities primitively assigned to the midbrain are transferred to the neocortex, or **cerebral cortex** that forms the surface gray matter of the cerebrum.

Functions in the cerebrum have been localized by direct stimulation of exposed brains of people and experimental animals, postmortem examination of persons suffering from various lesions, and surgical removal of specific brain areas in experimental animals. The cortex contains discrete motor and sensory areas (Figures 33.14 and 33.15). The motor areas control voluntary muscle movements, while the sensory cortex is the center of conscious perception of touch, pressure, pain, temperature, and taste. Vision, smell, hearing, and speech are purely sensory or motor regions located in specific areas on the cerebral lobes. In addition, there are large "silent" regions, called **association areas,** concerned with memory, judgment, reasoning, and other integrative functions. These regions are not directly connected to sense organs or muscles.

Thus in mammals, and especially in humans, separate parts of the brain mediate conscious and unconscious functions. The unconscious mind, all of the brain except the cerebral cortex, governs numerous vital functions that are removed from conscious control: respiration, blood pressure, heart rate, hunger, thirst, temperature balance, salt balance, sexual drive, and basic (sometimes irrational) emotions. The brain is also a complex endocrine gland that regulates and receives feedback from the body's endocrine system (see Chapter 34). The conscious mind, or cerebral cortex, is the site of higher mental activities (for example, planning and reasoning), memory, and integration of sensory information. Memory appears to transcend all parts of the brain rather than being a property of any particular part of the brain as once believed.

The right and left hemispheres of the cerebral cortex are bridged through the corpus callosum (Figure 33.14B), a neural connection through which the two hemispheres are able to transfer information and coordinate mental activities. In humans, the two hemispheres of the brain are specialized for different functions: the left hemisphere for language development, mathematical and learning capabilities, and sequential thought processes; and the right hemisphere for spatial, musical, artistic, intuitive, and perceptual activities. In addition,

Figure 33.14 A, External view of the human brain, showing lobes of the cerebrum and localization of major function of the cerebrum and cerebellum. **B,** Section through midline of the human brain showing one cerebral hemisphere of the cerebrum, the thalamus and hypothalamus of the forebrain, and the pons, medulla, and cerebellum of the hindbrain.

each hemisphere controls the opposite side of the body. It has been known for a long time that even extensive damage to the right hemisphere may cause varying degrees of left-sided paralysis but has little effect on intellect and speech. Conversely, damage to the left hemisphere usually causes loss of speech and may have disastrous effects on intellect. Because these differences in brain symmetry and function exist at birth, they appear to be inborn rather than the result of developmental or environmental effects as once believed.

Hemispheric specialization has long been considered a unique human trait, but it was recently discovered in the brains of songbirds, in which one side of the brain is specialized for song production.

Although the large size of their brain undoubtedly makes humans the most intelligent of animals, it is apparent that they can do without much of it and still remain intelligent. Brain scans of persons with hydrocephalus (enlargement of the head as a result of pressure disturbances that cause the brain ventricles [fluid-filled cavities within the brain] to enlarge many times their normal size) show that although many such persons are functionally disabled, others are nearly normal. The cranium of one person with hydrocephalus was nearly filled with cerebrospinal fluid and the only remaining cerebral cortex was a thin layer of tissue, 1 mm thick, pressed against his cranium. Yet this young man, with only 5% of his brain, had achieved first-class honors in mathematics at a British university and was socially normal. This and other similarly dramatic observations suggest that there is enormous redundancy and spare capacity in corticocerebral function. It also suggests that the deep structures of the brain, which are relatively spared in hydrocephalus, may perform functions once attributed to the cortex.

Peripheral Nervous System

The peripheral nervous system includes all nervous tissue outside the central nervous system. It has two functional divisions: **sensory** or **afferent division,** which brings sensory information to the central nervous system, and **motor** or **efferent division,** which conveys motor commands to muscles and glands. The efferent division has two components: (1) **somatic nervous system,** which innervates skeletal muscle (see Figure 33.11), and (2) **autonomic nervous system,** which innervates smooth muscle, cardiac muscle, and glands.

Autonomic Nervous System The autonomic system governs involuntary, internal functions of the body that do not ordinarily affect consciousness, such as movements of alimentary canal and heart, contraction of smooth muscle of blood vessels, urinary bladder, iris of the eye, and others, plus secretions of various glands.

Autonomic nerves originate in the brain or spinal cord as do nerves of the somatic nervous system, but unlike the latter,

Figure 33.15 Arrangement of sensory and motor cortices shown in transverse section (see Figure 33.14 for surface view). These regions form part of the cerebral cortex of the cerebrum. Localizations of sensory terminations from different parts of the body are shown at right; origins of descending motor pathways are shown at left. The motor cortex lies in front of the sensory cortex, so the two are not superimposed. These maps were produced from the 1930s work of Canadian neurosurgeon Wilder Penfield. Research shows that the motor cortex is not as orderly as the map suggests; rather, correspondence between cortical areas and areas of the body they control is more diffuse.

Figure 33.16 General organization of the autonomic nervous system.

Parasympathetic Ganglion

(close to effector organ) Acetylcholine is neurotransmitter released by both pre- and postganglionic neuron endings

Sympathetic Ganglion

(close to spinal cord) Acetylcholine released from preganglionic neuron ending; norepinephrine released from postganglionic neuron ending

autonomic fibers consist of not one but two motor neurons (Figure 33.16). They synapse once after leaving the cord and before arriving at the effector organ. These synapses are located outside the spinal cord in ganglia. Axons passing from the cord to the ganglia are called *preganglionic* neurons; those passing from the ganglia to the effector organs are called *postganglionic* neurons.

Subdivisions of the autonomic system are the **parasympathetic** and **sympathetic** systems. Most organs in the body are innervated by both sympathetic and parasympathetic neurons, whose actions are antagonistic (Figure 33.17). If one neuron stimulates an activity, the other inhibits it. However, neither kind of nerve is exclusively excitatory or inhibitory. For example, parasympathetic neurons inhibit heartbeat but excite peristaltic movements of the intestine; sympathetic neurons increase heartbeat but inhibit intestinal peristaltic movement.

Parasympathetic neurons emerge from the central nervous system either in brain stem cranial nerves or in spinal nerves emerging from the sacral (pelvic) region of the spinal cord (Figures 33.16 and 33.17). In the sympathetic division nerve cell bodies of all preganglionic neurons are located in thoracic and upper lumbar areas of the spinal cord. Their neurons exit through the ventral roots of spinal nerves, separate from these, and go to sympathetic ganglia (Figures 33.10 and 33.17), which are paired and form a chain on each side of the spinal column.

Ganglia are usually remote from the effector organ in the sympathetic system (for example, the sympathetic ganglion shown in Figure 33.10) and are often embedded in tissue layers close to effector organs in the parasympathetic system (Figure 33.16).

All preganglionic neurons, whether sympathetic or parasympathetic, release acetylcholine at their synapse with postganglionic cells. However, parasympathetic postganglionic neurons release acetylcholine at their endings, whereas sympathetic postganglionic neurons with a few exceptions release norepinephrine (also called noradrenaline). This difference is another important characteristic distinguishing the two parts of the autonomic nervous system.

As a general rule the parasympathetic division is associated with nonstressful activities, such as resting, eating, digestion, and urination. The sympathetic division is active under conditions of physical or emotional stress. Under such conditions heart rate increases, blood vessels to the skeletal muscles dilate, blood vessels in the viscera constrict, activity of the intestinal tract decreases, and metabolic rate increases. The importance of these responses in emergency reactions (sometimes called the fright, fight or flight response) are described in Chapter 34 (p. 759). It should be noted, however, that the sympathetic division is active to some degree also during resting conditions in maintaining normal blood pressure and body temperature.

33.4 SENSE ORGANS

Animals require a constant inflow of information from the environment to regulate their lives. Sense organs are specialized sensory receptors that have evolved to detect environmental status and change. An animal's sense organs are its first level of environmental perception—they bring information to the central nervous system.

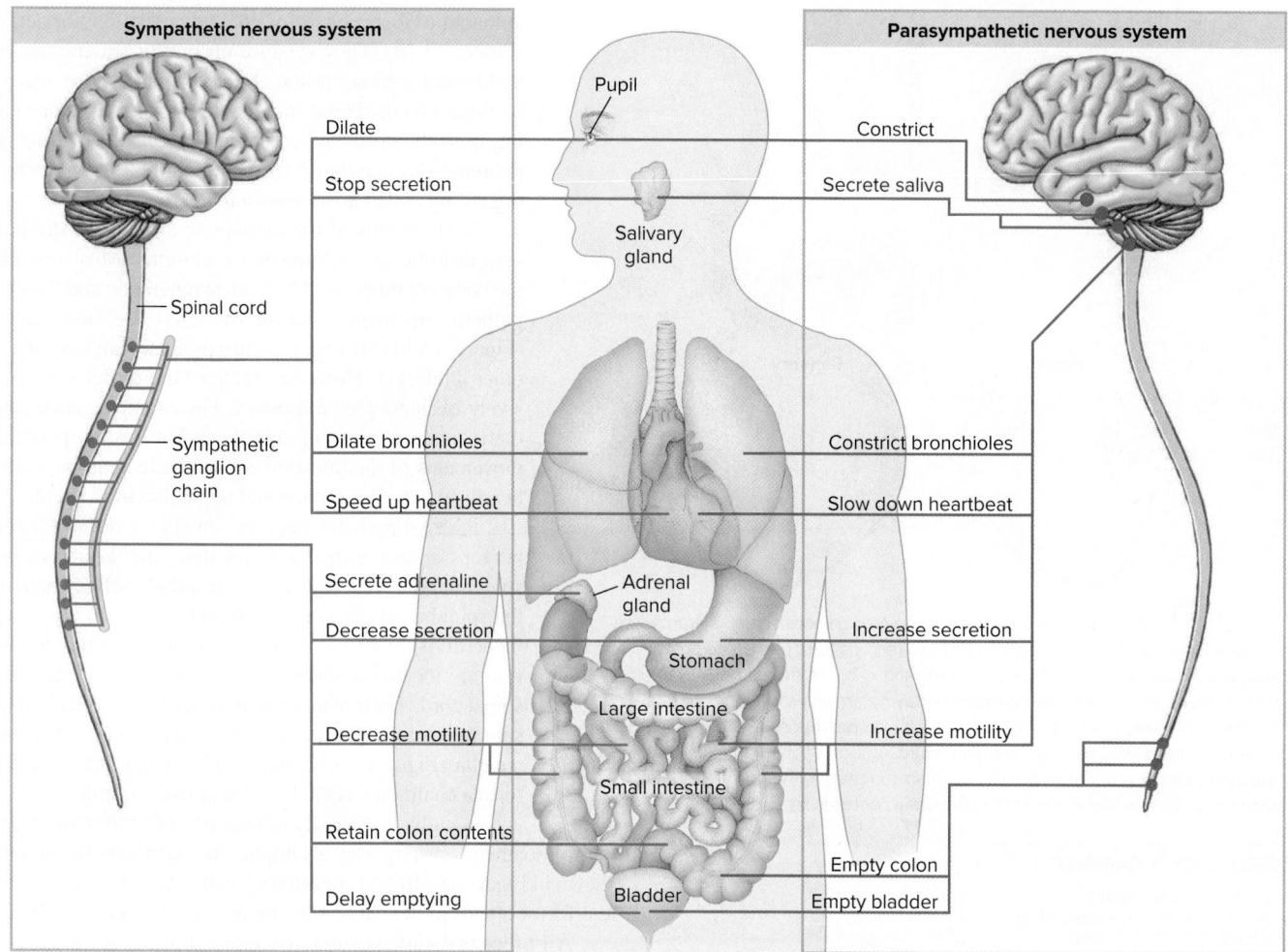

Figure 33.17 Autonomic nervous system in humans. Outflow of autonomic nerves from the central nervous system is shown. Sympathetic (*red, on left*) outflow is from the thoracic and lumbar areas of the spinal cord by way of a chain of sympathetic ganglia. Parasympathetic (*blue, on right*) outflow is from the cranial and sacral regions of the central nervous system; parasympathetic ganglia (not shown) are located in or adjacent to the organs innervated. Most organs are innervated by fibers from both sympathetic and parasympathetic divisions.

A **stimulus** is some form of energy—electrical, mechanical, chemical, or radiant. Sensory receptors within a sense organ transform energy from a stimulus into nerve action potentials, the common language of the nervous system. In a very real sense, then, sense organs are biological transducers. A microphone, for example, is a transducer that converts mechanical (sound) energy into electrical energy. Like a microphone, which is sensitive only to sound, sensory receptors are, as a rule, specific for one kind of stimulus. Thus eyes respond only to light, ears to sound, pressure receptors to pressure, and chemoreceptors to chemicals, converting all forms of energy into nerve action potentials that can be transmitted to the central nervous system. A response occurs either via the previously described reflex arc that is fundamental to all nervous systems, or by a much more complex response involving multiple brain regions and effectors.

Since all nerve action potentials are qualitatively alike, how do animals perceive and distinguish different sensations of varying stimuli? The answer is that real perception of sensation is performed by localized regions of the brain, where sensory receptors from each sense organ have their own hookup. This concept of "labeled lines" of communication to specific brain regions was first described in the 1830s by Johannes Müller, who called this the **law of specific nerve energies.** Action potentials arriving at a particular sensory area of the brain can be interpreted in only one way. For example, pressure on the eye causes us to see "stars" or other visual patterns; mechanical distortion of the eye initiates action potentials in the optic nerve fibers that are perceived as light sensations.

Classification of Receptors

Receptors are traditionally classified by their location. Those near the external surface, called **exteroceptors,** keep an animal informed about its external environment. Internal parts of the body are provided with **interoceptors,** which receive stimuli from internal organs. Muscles, tendons, and joints have **proprioceptors,** which are sensitive to changes in tension of muscles and provide an organism with a sense of body position. Sometimes receptors are classified by the form of energy to which the receptors respond, such as **chemical, mechanical, light,** or **thermal.**

Chemoreception

Chemoreception is the oldest and most universal sense in the animal kingdom and in unicellular eukaryotes. It probably guides organismal behavior more than does any other sense. Unicellular forms use **contact chemical receptors** to locate food and adequately oxygenated water and to avoid harmful substances. These receptors elicit an orientation behavior, called **chemotaxis,** toward or away from a chemical source. Most animals have specialized **distance chemical receptors,** which are often developed to a remarkable degree of sensitivity. Distance chemoreception, usually called smell or olfaction, guides feeding behavior, location and selection of sexual mates, territorial and trail marking, and alarm reactions of numerous animals.

In all vertebrates and in insects, the senses of **taste** and **smell** are clearly distinguishable. Although there are similarities between taste and smell receptors, in general taste is more restricted in response and is less sensitive than smell. Central nervous system centers for taste and smell are located in different parts of the brain.

Insect chemoreceptors are located in sensory hairs called sensilla. Taste sensilla occur on the mouthparts, legs, wing margins, and ovipositor in females. They have a single pore at the tip and recognize four classes of compounds: sugar (attractive), bitter (repelling), salts, and water. Olfactory sensilla occur on the head on two pairs of olfactory organs: the antennae and the maxillary palps (see Figure 21.17, p. 453). Pores on the cuticular walls of these sensilla allow odorant and pheromone molecules from the environment to contact the olfactory receptor neurons.

Figure 33.18 Pheromone-producing glands of an ant (*shown in orange*).

Social insects and many other animals, including mammals, produce species-specific compounds, called **pheromones,** that constitute a highly developed chemical language. Pheromones are a diverse group of organic compounds that an animal releases to affect the physiology or behavior of another individual of the same species. Information regarding territory, societal hierarchy, sex, and reproductive state are transmitted via this system. Ants, for example, are walking batteries of glands (Figure 33.18) that produce numerous chemical signals. These include releaser pheromones, such as alarm and trail pheromones, and primer pheromones, which alter endocrine and reproductive systems of different castes in a colony.

In vertebrates, taste receptors occur in the mouth cavity and especially on the tongue (Figure 33.19), where they provide a means for judging foods before they are swallowed. A **taste bud** consists of a cluster of receptor cells surrounded by supporting cells; it has a small external pore through which slender tips of sensory cells project. Chemicals being tasted interact with specific receptor sites on microvilli of receptor cells. Taste sensations are categorized as sweet, salty, sour, bitter, and umami (Japanese, roughly translated to mean "meaty" or "savory"). Although the mechanisms are different for each basic taste sensation, receptor cells are depolarized by the specific chemical, and action potentials are generated. Contrary to what was originally thought, taste receptors can respond to different types of taste categories, although they may respond more strongly to one particular type. Action potentials are transmitted across chemical synapses (p. 725) and travel along sensory neurons to specific brain regions. Taste discrimination depends on assessment by the brain of the relative activity of the five different taste receptor subtypes. This assessment is similar to color vision in vertebrates, where a whole rainbow of colors can be distinguished by relative excitation of only three types of color photoreceptors (see p. 743). Because receptor cells are subject to abrasion by foods, taste buds have a short life (5 to 10 days in mammals) and are continually being replaced.

Although olfaction is a primal sense for many animals, used for identification of food, sexual mates, and predators, it is most highly developed in mammals. Even humans, although a species not celebrated for detecting smells, can discriminate perhaps 20,000 different odors. A human nose can detect 1/25 of one-millionth of 1 mg of mercaptan, the odoriferous substance of skunks. Nonetheless, our olfactory abilities compare poorly with those of other mammals that rely on olfaction for survival. A dog explores new surroundings with

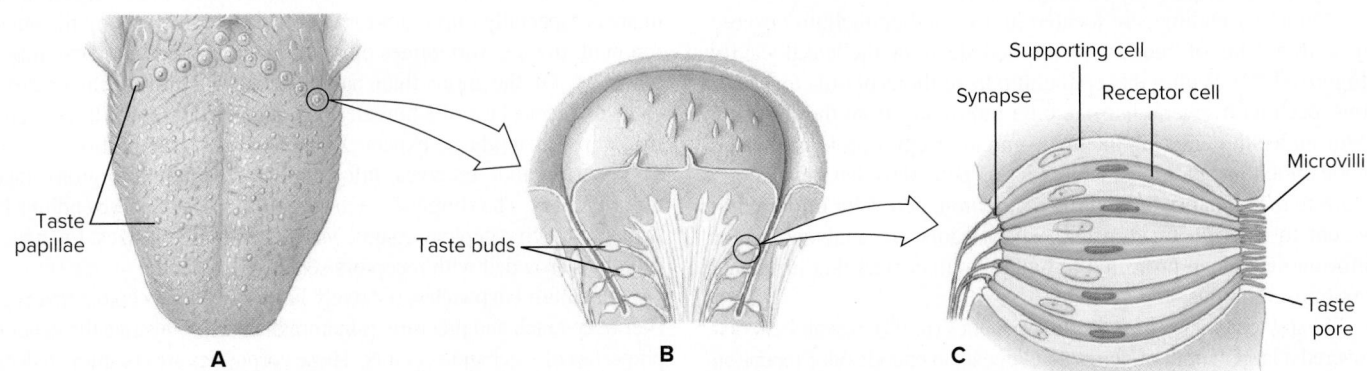

Figure 33.19 Taste receptors. **A,** Surface of human tongue showing location of taste papillae. **B,** Position of taste buds on a taste papilla. **C,** Structure of a taste bud.

Figure 33.20
Human olfactory epithelium. **A,** The epithelium is a patch of tissue positioned in the roof of the nasal cavity. **B,** It is composed of supporting cells, basal cells, and olfactory receptor cells with cilia protruding from their free ends.

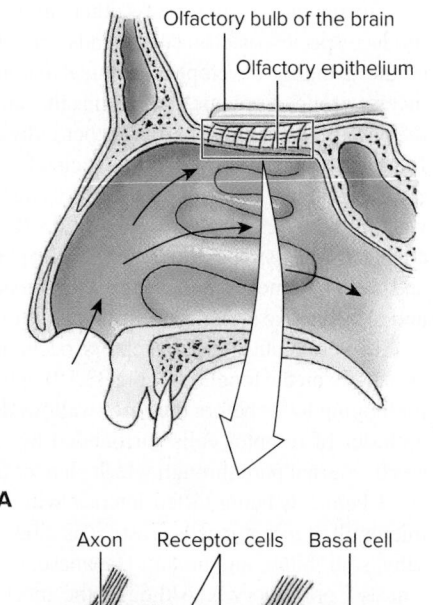

Olfactory bulb of the brain

Olfactory epithelium

A

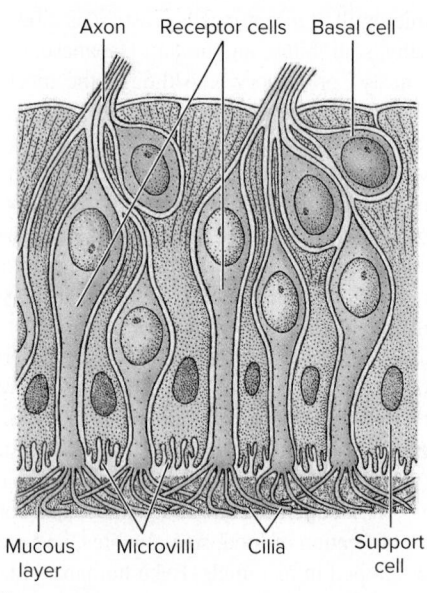

Axon Receptor cells Basal cell

Mucous layer Microvilli Cilia Support cell

B

ancient and highly conserved through evolution. Each of the 1000 or so genes discovered in mammals encodes a separate type of odor receptor. Since mammals can detect at least 20,000 different odors, each receptor must respond to several odor molecules, and each odor molecule must bind several types of receptors, each of which responds to a part of the molecule's structure. Brain-mapping techniques have shown that each olfactory neuron projects to a characteristic location on the olfactory bulb, providing a two-dimensional map that identifies which receptors have been activated in the nose. In addition, olfactory neurons expressing the same odor-receptor gene converge to a fixed olfactory bulb region, which might provide an explanation for the extremely high sensitivity of smell. Projected to the brain, odor information is recognized as a unique scent.

Because flavor of food depends on odors reaching the olfactory epithelium through the throat passage, taste and smell are easily confused. All "tastes" other than the five basic ones (sweet, sour, bitter, salty, and umami) result from flavor molecules reaching the olfactory epithelium in this manner. Food loses its appeal during a common cold because a stuffy nose blocks odors rising from the mouth.

Many terrestrial vertebrates possess an additional olfactory organ, the **vomeronasal organ (VNO** or **Jacobson's organ).** The VNO is also lined with olfactory epithelium and is located in paired, blindending passages that open into the nasal or oral cavity. The VNO olfactory receptor cells respond to chemical signals carried by a variety of chemicals, including food odors (snakes) and pheromones (mammals). The best established example of pheromonal communication in humans is that of menstrual cycle synchronization that occurs in women living together in close association, such as in dormitories.

Mechanoreception

Mechanoreceptors are sensitive to quantitative forces such as touch, pressure, stretching, sound, vibration, and gravity—in short, they respond to motion. Animals interact with their environments, feed themselves, maintain normal postures, and walk, swim, or fly, using a steady flow of information from mechanoreceptors.

Touch

Invertebrates, especially insects, have many kinds of receptors sensitive to touch. Such receptors are well endowed with tactile hairs sensitive to both touch and vibrations. Superficial touch receptors of vertebrates are distributed over the body but tend to be concentrated in areas especially important for exploring and interpreting the environment. In most vertebrates these areas are the face and extremities of limbs. Of the more than half-million separate touch-sensitive spots on the surface of a human body, most occur on the tongue and fingertips, as might be expected based on the large portion of sensory cortex that receives information from these regions (see Figure 33.15). The simplest touch receptors are bare nerve endings in skin, but touch receptors assume varying shapes and sizes. Each hair follicle is crowded with receptors sensitive to touch.

Pacinian corpuscles, relatively large mechanoreceptors that register deep touch and pressure in mammalian skin, illustrate the general properties of mechanoreceptors. These corpuscles are common in deep layers of skin, connective tissue surrounding muscles and tendons, and abdominal mesenteries. Each corpuscle consists of a neuron terminus surrounded by a capsule of numerous concentric, onionlike layers of

its nose much as we do with our eyes. A dog's nose is justifiably renowned; with some odorous sources a dog's nose is at least a million times more sensitive than ours. Dogs are assisted in their proficiency by having a nose located close to the ground, where odors from passing creatures tend to linger.

Olfactory endings are located in a special epithelium covered by a thin film of mucus, positioned deep in the nasal cavity (Figure 33.20). Within this epithelium lie millions of olfactory neurons, each with several hairlike cilia protruding from the free end. Odor molecules entering the nose bind to receptor proteins located in the cilia; this binding generates action potentials that travel along axons to the olfactory bulb of the brain. From here odor information is sent to the olfactory cortex, where odors are analyzed. Odor information is then projected to higher brain centers that influence emotions, thoughts, and behavior.

Using techniques of molecular genetics (p. 93), researchers discovered a large family of genes that appears to encode odor reception in mammals (including humans). About 70 genes from the same family have been identified in the fruit fly, *Drosophila,* and some also in the nematode, *C. elegans.* The gene family for olfaction is thus

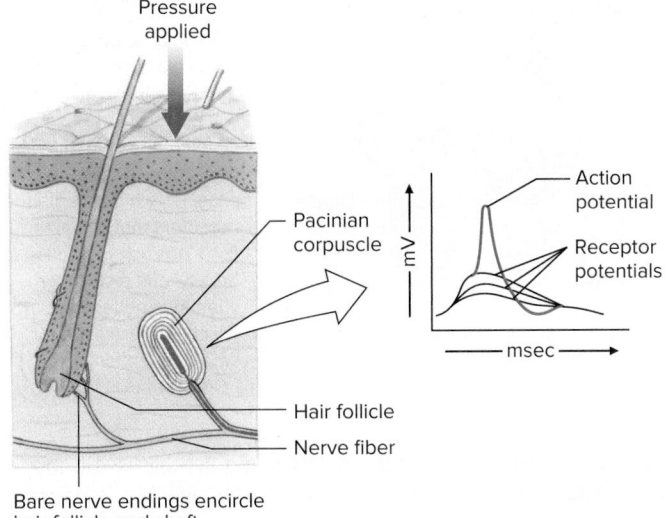

Figure 33.21 Response of pacinian corpuscle to applied pressure. Progressively stronger pressure produces stronger receptor potentials. When the threshold stimulus is reached, an all-or-none action potential is generated in the afferent nerve fiber.

connective tissue (Figure 33.21). Pressure at any point on a capsule distorts its nerve ending, producing a graded **receptor potential,** a local flow of electric current similar to a postsynaptic excitatory potential (p. 725). Progressively stronger stimuli lead to correspondingly stronger receptor potentials until a **threshold current** is produced; this current initiates an action potential in a sensory nerve fiber. A second action potential is initiated at removal of the pressure, but not during the pressure. This response is called adaptation (not to be confused with the evolutionary meaning of this term [see Chapter 6]) and characterizes many touch receptors, which are admirably suited to detecting a sudden mechanical change but readily adapt to new conditions. We are aware of new pressures when we put on our shoes and clothing in the morning, but we are not reminded of these pressures all day.

Pain

Pain receptors are relatively unspecialized nerve fiber endings that respond to a variety of stimuli signaling possible or real damage to tissues. These free nerve endings also respond to other stimuli, such as mechanical movement of a tissue and temperature changes. Pain fibers respond to small peptides, such as substance P and bradykinins, which are released by injured cells. This type of response is termed *slow pain. Fast pain* responses (for example, a pin prick, cold or hot stimuli) are a more direct response of the nerve endings to mechanical or thermal stimuli.

> Pain is a distress call from the body signaling some noxious stimulus or internal disorder. Although there is no cortical pain center, discrete areas have been located in the brain stem where pain messages from the periphery terminate. These areas contain two kinds of small peptides, endorphins and enkephalins, that are endogenous opiates with morphinelike or opium-like activity. When released, they bind with specific opiate receptors in the midbrain. They are the body's own analgesics.

Just as pain is a sign of danger, sensory pleasure is a sign of a stimulus useful to the subject. Pleasure depends on the internal state of an animal and is judged with reference to homeostasis and some physiological set point. Pleasure states may be produced by release of endogenous opioids within the central nervous system.

Lateral-Line System of Fishes and Amphibian

A lateral line is a distant touch reception system for detecting wave vibrations and currents in water. Receptor cells, called **neuromasts,** are located on the body surface in aquatic amphibians and some fishes, but in many fishes they are located within canals running beneath the epidermis; these canals open at intervals to the surface (Figure 33.22). Each neuromast is a collection of **hair cells** with sensory endings, or cilia, embedded in a gelatinous, wedge-shaped mass, the **cupula.** The cupula projects into the center of the lateral-line canal so that it bends in response to any disturbance of water on the body surface. The lateral-line system is one of the principal sensory systems that guide fishes in their movements and in location of predators, prey, and social partners (pp. 522 and 525).

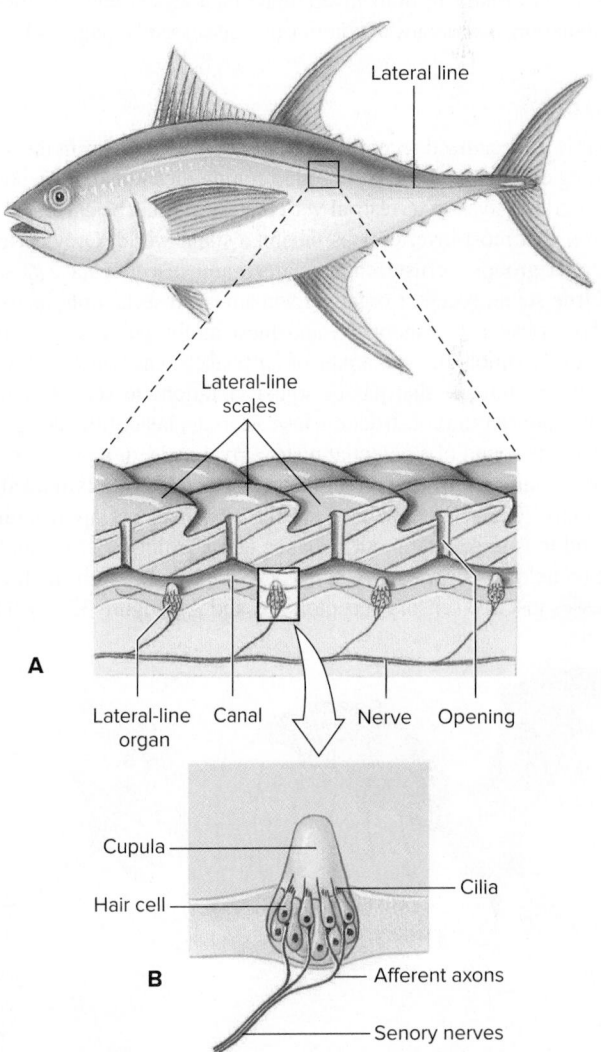

Figure 33.22 Lateral-line system. **A,** Lateral line of a bony fish with exposed neuromasts. **B,** Structure of a neuromast (lateral-line organ).

The lateral line serves another function in some fish, the reception of small bioelectric signals (produced during heart and muscle activity) from other members of their species or from intruder or prey species. The **electroreceptor cells** are found in pores closely associated with the lateral-line system and in some species, such as the shark, are concentrated primarily on the head (see Figure 24.9, p. 525). In addition to receiving electric signals, some fish are able to generate weak or strong electric fields produced by **electric organs,** which are modified muscles located near the tail (for example, in some freshwater fish, such as the catfish and electric eel). Intruders or prey can be located as they produce a disturbance in the electric field. Potential mates are recognized in some species since the discharge frequency is different between the sexes. Fish with strong electric fields can use electrolocation both to locate and to stun their prey (electric eels). Others have electric organs on either side of the head (for example, in marine electric rays; see p. 526) and use an electric current to stun prey.

Hair cells form an important sensory component of several mechanoreceptors found in both invertebrate (statocysts) and vertebrate (vestibular organs) organs of equilibrium, discussed on page 740.

Hearing

An ear is a specialized receptor for detecting sound waves in the surrounding environment. Because sound communication and reception are integral to lives of terrestrial vertebrates, we may be surprised to discover that most invertebrates inhabit a silent world. Only certain arthropod groups—crustaceans, spiders, and insects—have developed true sound-receptor organs. Even among insects, only locusts, cicadas, crickets, grasshoppers, and most moths possess ears, and these are of simple design: a pair of air pockets, each enclosed by a tympanic membrane that passes sound vibrations to sensory cells. Despite their spartan construction, insect ears are beautifully designed to detect the sound of a potential mate, a rival male, or a predator.

Especially interesting are ultrasonic detectors of certain nocturnal moths. These have evolved specifically to detect approaching bats and thus lessen the moth's chance of becoming a bat's evening meal (echolocation in bats is described on p. 622). Each moth ear possesses just two sensory receptors, A_1 and A_2 (Figure 33.23). The

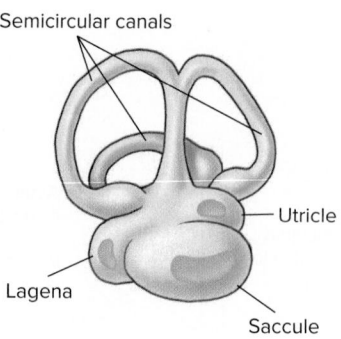

Figure 33.24 Vestibular apparatus of a teleost fish, containing three semicircular canals, which respond to angular acceleration; two balance organs (utricle and saccule), which are static receptors that signal the fish's position in relation to gravity; and a small chamber, the lagena, which is specialized for sound reception.

A_1 receptor responds to ultrasonic cries of a bat that is still too far away to detect the moth. As the bat approaches and its cries increase in intensity, the receptor fires more rapidly, informing the moth that the bat is coming nearer. Since the moth has two ears, its nervous system can determine the bat's position by comparing firing rates from the two ears. The moth's strategy is to fly away before the bat detects it. But if the bat continues its approach, the second (A_2) receptor in each ear, which responds only to high-intensity sounds, will fire. The moth responds immediately with an evasive maneuver, usually making a power dive to a bush or the ground, where it is safe because the bat cannot distinguish a moth's echo from those of its surroundings. (See page 458 to read how a luna moth survives attack by redirecting the bat to nonessential wing tails.)

In their evolution, vertebrate ears originated as a balance organ, or **labyrinth,** whose function of maintaining equilibrium is described in the next section. In all jawed vertebrates, from fishes to mammals, the labyrinth has a similar structure, consisting of two small chambers called the **saccule** and the **utricle,** and three **semicircular canals** (Figure 33.24). In fishes the base of the saccule is extended into a tiny pocket (**lagena**) that, during evolution of vertebrates, developed into the hearing receptor of tetrapods. With continued elaboration and elongation in birds and mammals, the fingerlike lagena was modified to form a **cochlea.**

A human ear (Figure 33.25) is representative of mammalian ears. The outer, or external, ear collects sound waves and funnels them through an **auditory canal** to an eardrum or **tympanic membrane** lying next to the middle ear. The middle ear is an air-filled chamber containing a chain of three tiny bones, or ossicles, the **malleus** (hammer), **incus** (anvil), and **stapes** (stirrup), named because of their fancied resemblance to these objects. These bones conduct sound waves across the middle ear (Figure 33.25B). They are arranged so that the force of sound waves pushing against the tympanic membrane is amplified as much as 90 times where the stapes contacts the **oval window** of the inner ear. Muscles attached to the middle ear bones contract when the ear receives very loud noises, providing the inner ear some protection from damage. The middle ear connects with the pharynx by means of a **eustachian tube,** which permits pressure equalization on both sides of the tympanic membrane.

Figure 33.23 Ear of a moth used to detect approaching bats. See text for explanation.

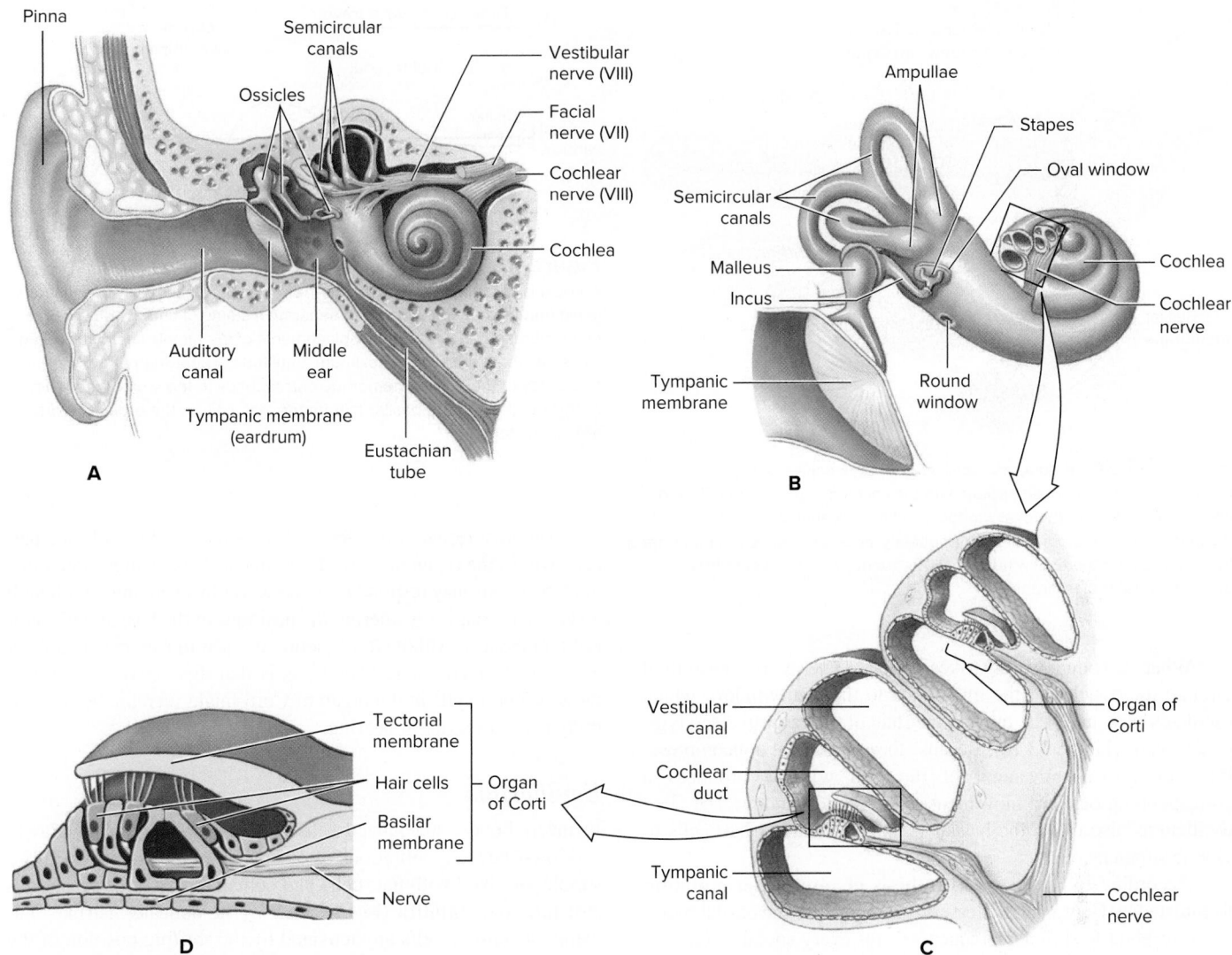

Figure 33.25 Human ear. **A,** Longitudinal section showing external, middle, and inner ear. **B,** Enlargement of middle ear and inner ear. The cochlea of the inner ear has been opened to show the arrangement of canals within. **C,** Enlarged cross section of cochlea showing the organ of Corti. **D,** Detail of ultrastructure of the organ of Corti.

The origin of the three tiny bones of the mammalian middle ear—the malleus, incus, and stapes—is one of the most extraordinary and well-documented transitions in vertebrate evolution. Amphibians, nonavian reptiles, and birds have a single rodlike ear ossicle, the stapes (also called the columella), which originated as a jaw support (hyomandibular) as seen in fishes (see Figure 23.16, p. 513). With evolution of the earliest tetrapods, the braincase became firmly sutured to the skull, and the hyomandibular, used to brace the jaw, was converted into the stapes. In a similar way, the two additional ear ossicles of the mammalian middle ear—the malleus and the incus— originated from parts of the jaw of early vertebrates. The quadrate bone of nonavian reptilian upper jaws became the incus, and the articular bone of the lower jaw became the malleus. Homology of nonavian reptilian jaw bones to mammalian ear bones is clearly documented in the fossil record and in embryological development of mammals.

Within the inner ear is the organ of hearing, or **cochlea** (Gr. *cochlea*, snail's shell), which is coiled in mammals, making two

and one half turns in humans (Figure 33.25B). The cochlea is divided longitudinally into three fluid-filled, tubular canals running parallel with one another. This relationship is indicated in Figure 33.26. These canals become progressively smaller from the base of the cochlea to the apex. One of these canals is called the **vestibular canal;** its base is closed by the oval window. The **tympanic canal,** which is in communication with the vestibular canal at the tip of the cochlea, has its base closed by the **round window.** Between these two canals is a **cochlear duct,** which contains the **organ of Corti,** the actual sensory apparatus (Figure 33.25C and D). Within the organ of Corti are fine rows of hair cells that run lengthwise from the base to the tip of the cochlea. At least 24,000 hair cells are present in a human ear. The 80 to 100 "hairs" on each cell are actually microvilli and a single large cilium (see Chapter 3, p. 43, and Chapter 29, p. 649), which project into the endolymph of the cochlear canal. Each cell is connected with neurons of the auditory-nerve. Hair cells rest on the **basilar membrane,** which separates the tympanic canal and the cochlear duct, and they are covered by the **tectorial membrane,** lying directly above them (Figure 33.25D).

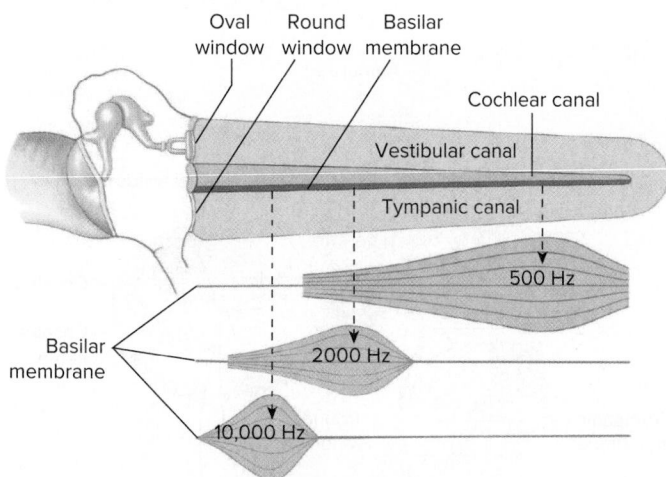

Oval window   Round window   Basilar membrane

Cochlear canal

Vestibular canal

Tympanic canal

500 Hz

Basilar membrane

2000 Hz

10,000 Hz

Figure 33.26 Frequency localization in the cochlea of the mammalian ear as it would appear with the cochlea stretched out. Sound waves transmitted to the oval window produce vibration waves that travel down the basilar membrane. High-frequency vibrations cause the membrane to resonate near the oval window. Low-frequency tones travel farther down the basilar membrane.

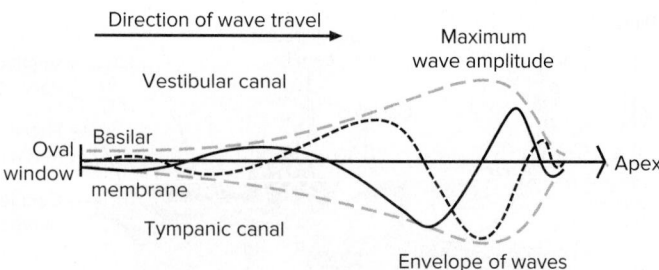

Direction of wave travel

Maximum wave amplitude

Vestibular canal

Oval window   Basilar membrane

Tympanic canal

Envelope of waves

Apex

Figure 33.27 Traveling waves along the basilar membrane. The oval window is at left, and the cochlear apex at right. The two wave formations (*solid* and *dashed lines*) occur at separate instants of time. The curves in color represent the extreme displacements of the membrane by traveling waves as they reach their maximum amplitude, where the natural frequency of the basilar membrane corresponds to the sound frequency. At this point along the basilar membrane, hair cells in the organ of Corti are stimulated.

When a sound wave strikes the ear, energy is transmitted through the ossicles of the middle ear to the oval window, which oscillates back and forth, moving the fluid of the vestibular and tympanic canals (Figure 33.26). Because these fluids are noncompressible, an inward movement of the oval window produces a corresponding outward movement of the round window. The fluid oscillations also cause the basilar membrane with its hair cells to vibrate simultaneously.

According to the **place hypothesis of pitch discrimination** formulated by Georg von Békésy, different areas of the basilar membrane respond to different frequencies; for every sound frequency, there is a specific "place" on the basilar membrane where hair cells respond to that frequency (Figure 33.26). Initial displacement of the basilar membrane starts a wave traveling down the membrane, much as flipping a rope at one end starts a wave moving down the rope (Figure 33.27). The displacement wave increases in amplitude as it moves from the oval window toward the apex of the cochlea, reaching a maximum at the region of the basilar membrane where the natural frequency of the membrane corresponds to the sound frequency. Here, the membrane vibrates with such ease that the energy of the traveling wave is completely dissipated. Hair cells within the organ of Corti in that region are stimulated, and graded receptor potentials (similar to a postsynaptic excitatory potential, p. 725) are generated with enough amplitude to cause action potentials to fire in the axons of the auditory nerve. Isolated hair cells have been shown to respond to a particular band of frequencies depending on their location within the cochlea. Thus, action potentials that are carried by certain axons of the auditory nerve are interpreted by the hearing center as particular tones. **Loudness** of a tone depends on the number of hair cells stimulated, whereas **timbre,** or quality, of a tone is produced by the pattern of hair cells stimulated by vibration. This latter characteristic of tone enables us to distinguish different human voices and different musical instruments, although the notes in each case may be of the same pitch and loudness.

Auditory research has focused on a more active role for hair cells within the organ of Corti. Experiments have demonstrated that outer hair cells may respond to sound waves by changing their length and thus mechanically altering the positions of the basilar and tectorial membranes. Although a function of such movements is not yet established in vivo, one hypothesis is that this active response of these receptor cells in the organ of Corti might increase both sensitivity and selectivity of hearing.

Equilibrium

In invertebrates, specialized sense organs for monitoring gravity and low-frequency vibrations often appear as **statocysts.** Each is a simple sac lined with hair cells and containing a heavy calcareous structure, the **statolith** (Figure 33.28). The delicate, hairlike filaments of sensory cells are activated by the shifting position of the statolith when the animal changes position. Statocysts occur in many invertebrate phyla from cnidarians to arthropods. All are built on similar principles.

The vertebrate organ of equilibrium is the **labyrinth,** or **vestibular organ.** It consists of two small chambers (**saccule** and **utricle**) and three **semicircular canals** (Figure 33.25B). The utricle and the saccule are static balance organs that, like invertebrate statocysts, give information about position of the head or body with respect to the force of gravity. As the head tilts in one direction or another, statoliths press on different groups of hair cells; these cells send nerve action potentials to the brain, which interprets this information with reference to head position.

The semicircular canals of vertebrates respond to **rotational acceleration** but are relatively insensitive to linear acceleration. The three semicircular canals are at right angles to each other, one for each axis of rotation. They are filled with fluid (endolymph), and within each canal is a bulblike enlargement, the **ampulla,** which contains hair cells. The hair cells are embedded in a gelatinous membrane, the **cupula,** which projects into the fluid. The cupula is similar in structure to the cupula of the lateral-line system of fishes (p. 737). When the head rotates, fluid in the canal at first tends not to move because of inertia. Since the cupula is attached, its free end is pulled in the direction opposite the direction of rotation (Figure 33.29). Bending of the cupula distorts and excites the hair cells embedded in it, increases the discharge rate of afferent sensory nerve fibers leading from the ampulla to the brain, and produces a sensation of rotation. Since the three canals of each ear are in different planes, acceleration in any direction stimulates at least one ampulla.

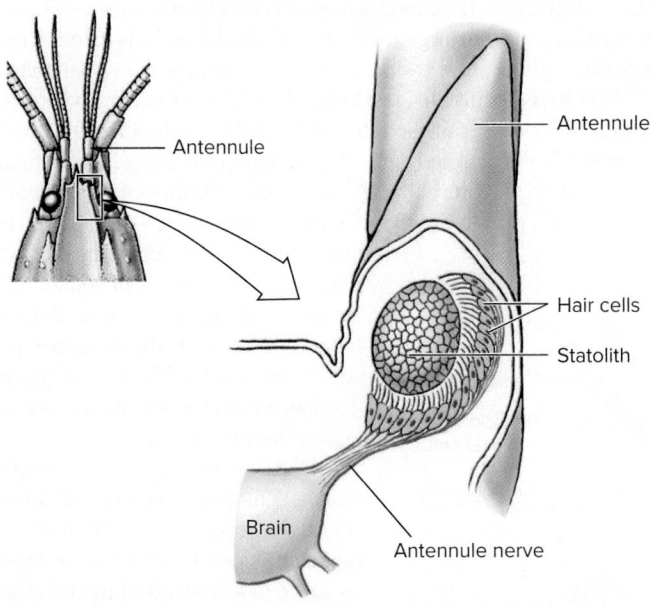

Figure 33.28   Statolith of the crayfish, a static balance organ of invertebrates.

Photoreception: Vision

Light-sensitive receptors are called **photoreceptors.** These receptors range from simple light-sensitive cells scattered randomly on the body surface of many invertebrates (dermal light sense) to the exquisitely developed camera-type eye of vertebrates and cephalopods. Eyespots of astonishingly advanced organization appear even in some unicellular forms. That of a dinoflagellate, *Nematodinium,* bears a lens, a light-gathering chamber, and a photoreceptive pigment cup—all developed within a single-celled organism (Figure 33.30). The dermal light receptors of many invertebrates are of much simpler design. They are far less sensitive than optic receptors, but they are important in locomotory orientation, pigment distribution in chromatophores, photoperiodic adjustment of reproductive cycles, and other behavioral changes.

More highly organized eyes, many capable of excellent image formation, are based on one of two different principles: either a single-lens, camera-type eye such as those of cephalopod molluscs and vertebrates; or a multifaceted (compound) eye as in arthropods. Arthropod **compound eyes** are composed of many independent visual units called **ommatidia** (Figure 33.31). Light enters each corneal lens and is absorbed by visual pigments in the rhabdomere of the retinular cells. These receptor cells depolarize and generate action potentials in the axon leaving each ommatidium. Eyes of bees contain about 15,000 of these units, each of which views a separate narrow sector of the visual field. Such eyes form a mosaic of images of varying brightness from the separate units. Many insects have color vision; honeybees can use ultraviolet light to see nectar guides in flowers. Many flying insects also detect polarized light and use it to navigate through their environments. Resolution (the ability to see objects sharply) is poor compared with that of a vertebrate eye. A fruit fly, for example, must be closer than 3 cm to see another fruit fly as anything but a single spot. However, a compound eye is especially well suited to detecting motion, as anyone who has tried to swat a fly knows.

Eyes of certain annelids and molluscs and all vertebrates are built like a camera—or rather we should say that a camera is modeled somewhat after these eyes. A camera-type eye contains in the front a light-tight chamber and lens system, which focuses an image of the visual field on a light-sensitive surface (the retina) in the back (Figure 33.32; also see Figure 16.41 p. 357).

The spherical eyeball is built of three layers: (1) a tough outer white **sclera** for support and protection, (2) a middle **choroid coat,** containing blood vessels for nourishment, and (3) a light-sensitive **retina** (Figure 33.32). The **cornea** is a transparent anterior modification of the sclera. A pigmented region, or **iris,** regulates the size of the light opening, or **pupil.** The pupil is usually a round or vertical slit-shaped opening in vertebrates, but

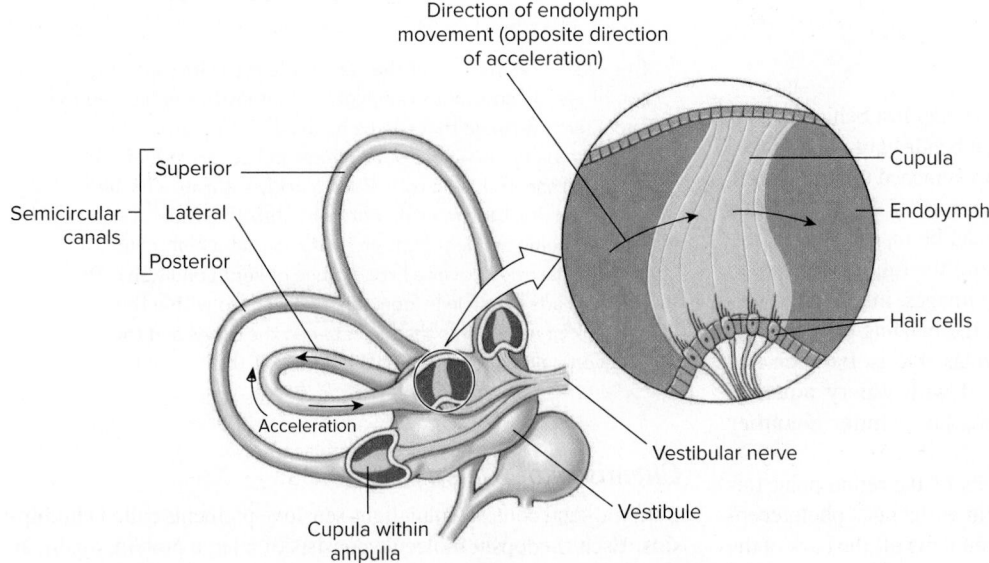

Figure 33.29   How the semicircular canals respond to angular acceleration. Because of inertia, endolymph in the semicircular canal corresponding to the plane of motion moves past the cupula in a direction opposite to that of angular acceleration. Movement of the cupula stimulates hair cells.

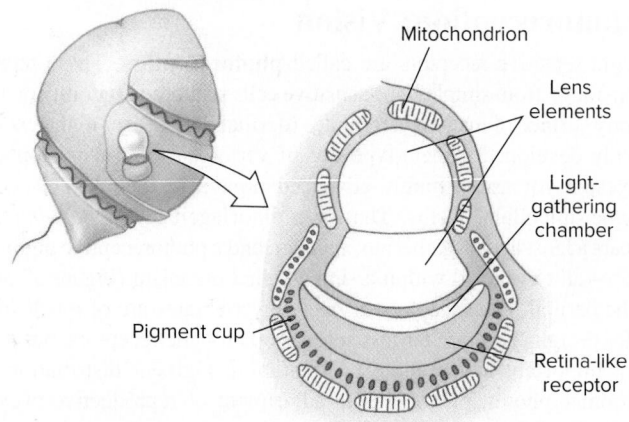

Figure 33.30 Eyespot of the dinoflagellate, *Nematodinium.*

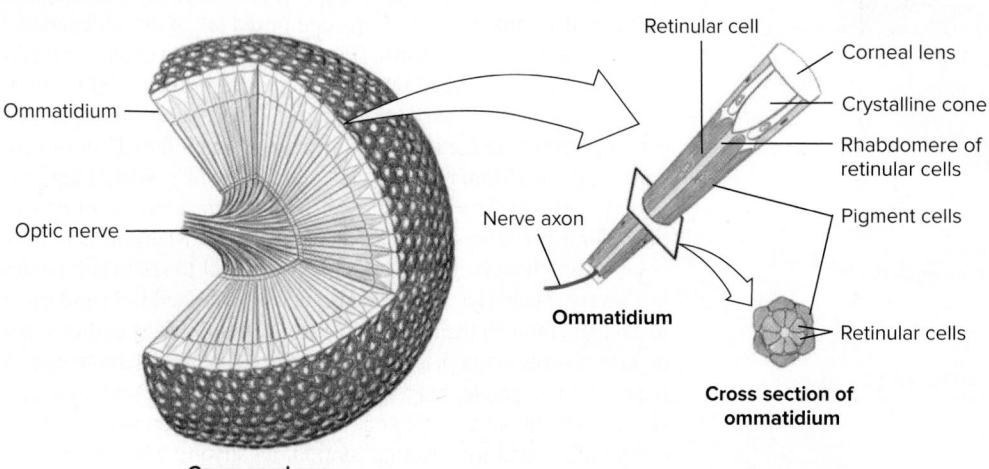

Figure 33.31 Compound eye of an insect. A single ommatidium is shown enlarged at right.

and 1 million cones are present in each human eye. Cones are primarily concerned with color vision in ample light and rods with colorless vision in dim light. Next is a network of **intermediate neurons** (bipolar, horizontal, and amacrine cells) that process and relay visual information from the photoreceptors to ganglion cells whose axons form the optic nerve. The network permits much convergence, especially for rods. Information from several hundred rods may converge on a single ganglion cell, an adaptation that greatly increases the effectiveness of rods in dim light. Cones show very little convergence. By coordinating activities between different ganglion cells, and adjusting sensitivities of bipolar cells, horizontal and amacrine cells improve overall contrast and quality of a visual image.

The **fovea centralis,** or **fovea,** the region of keenest vision, is located in the center of the retina (Figure 33.32), in direct line with the center of the lens and cornea. It contains only cones, a vertebrate specialization for diurnal (daytime) vision. Acuity of an animal's eyes depends on the density of cones in the fovea. A human fovea and that of a lion contain approximately 150,000 cones per square millimeter, but many water and field birds have up to 1 million cones per square millimeter. Their eyes are as good as our eyes would be if aided by eight-power binoculars.

Rods alone occupy the peripheral parts of the retina. Rods are high-sensitivity receptors for dim light. At night, the cone-filled fovea is unresponsive to low levels of light and we become functionally color blind ("at night all cats are gray"). Under nocturnal conditions, the position of greatest visual acuity is not at the center of the fovea but at its edge. Thus it is easier to see a dim star at night by looking slightly to one side of it.

a horizontal slit-shaped opening in cephalopods. Just behind the iris is a **lens,** a transparent, elastic oval disc that bends light rays to focus an image on the retina. **Ciliary muscles** are attached to and encircle the lens. In vertebrates they can alter curvature of the lens so that images at varying distances from the eye can be focused on the retina. In cephalopods the ciliary muscles move the rigid lens closer to or farther away from the retina to focus on images. In terrestrial vertebrates the cornea actually does most of the bending of light rays, whereas the lens adjusts focus for near and far objects. Between cornea and lens is an **outer chamber** filled with watery **aqueous humor;** between lens and retina is a much larger **inner chamber** filled with viscous **vitreous humor.**

In cephalopods the photoreceptor cells of the retina point forward and directly absorb incoming light; in vertebrates photoreceptor cells point backward and absorb light bouncing off the back of the eye. In vertebrates the retina is composed of several cell layers (Figure 33.33). The outermost layer, closest to the sclera, consists of pigment cells or **chromatophores.** Adjacent to this layer are the photoreceptors, **rods** and **cones.** Approximately 125 million rods

One of several marvels of the vertebrate eye is its capacity to compress the enormous range of light intensities presented to it into a narrow range that can be handled by optic nerve fibers. Light intensity between a sunny noon and a starlit night differs by more than 10 billion to 1. Rods quickly saturate with high light intensity, but cones do not; they shift their operating range with changing ambient light intensity so that a high-contrast image is perceived over a broad range of light conditions. This shift is made possible by complex interactions within the network of nerve cells that lie between the cones and the ganglion-cells that generate retinal output to the brain.

Chemistry of Vision

Both rods and cones contain light-sensitive pigments called **rhodopsins.** Each rhodopsin molecule consists of a large protein, **opsin,** an enzyme, and a small carotenoid molecule, **retinal,** a derivative of vitamin A. When a quantum of light strikes the photopigment and is absorbed by the rhodopsin molecule, retinal is isomerized, changing the shape of the molecule. This molecular change triggers the

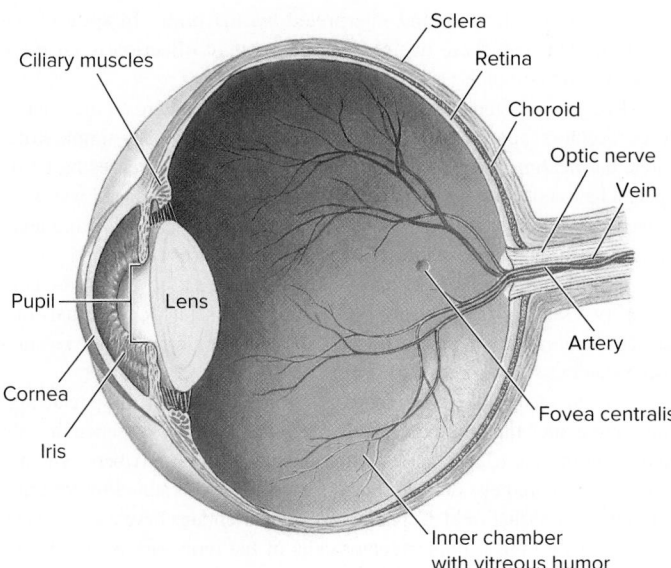

Figure 33.32 Structure of the human eye.

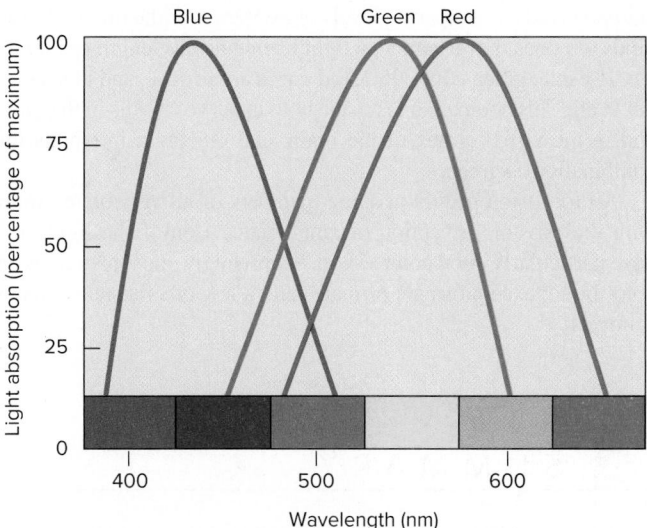

Figure 33.34 The absorption spectrum of human vision. Three types of visual pigments in cones absorb maximally at 430 nm (blue cones), 540 nm (green cones), and 575 nm (red cones).

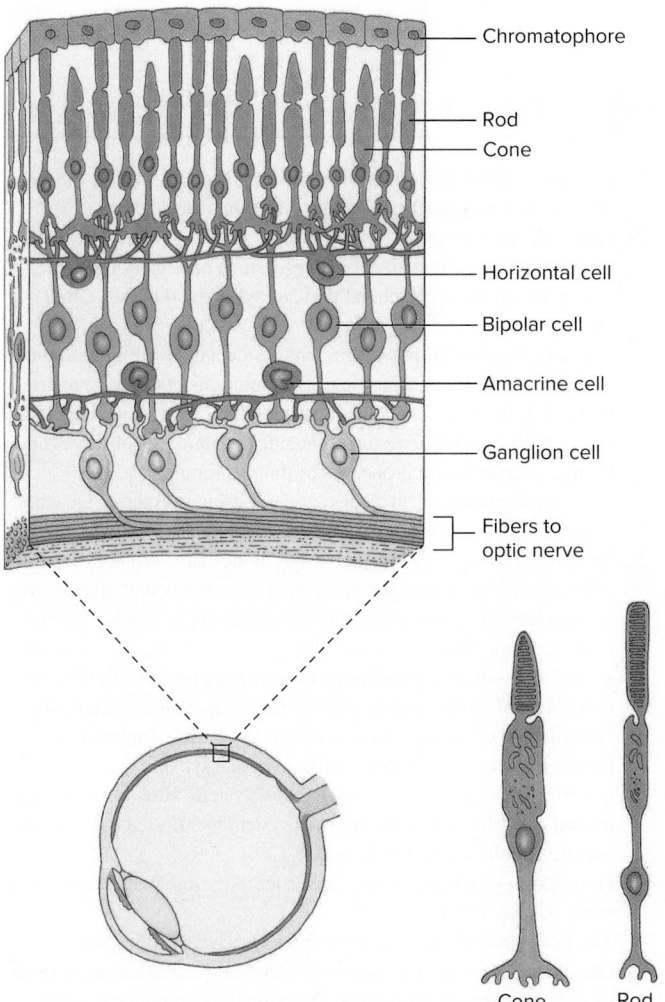

Figure 33.33 Structure of a primate retina, showing organization of intermediate neurons that connect photoreceptor cells to ganglion cells of the optic nerve.

enzymatic activity of opsin, which sets in motion a biochemical sequence of several steps. This complex sequence behaves as an excitatory cascade that vastly amplifies the energy of a single photon to cause hyperpolarization (p. 725) of a rod or cone. This hyperpolarization signal is transmitted through the intermediate neurons and leads to depolarization and generation of action potentials in a ganglion cell. It is interesting to note that light reception in invertebrate eyes leads to *depolarization* of receptor cells, while similar light signals induce *hyperpolarization* in vertebrate receptor cells.

The amount of intact rhodopsin in a retina depends on the intensity of light reaching the eye. A dark-adapted eye contains much rhodopsin and is very sensitive to weak light. Conversely, in a light-adapted eye, most of the rhodopsin is split into retinal and opsin. It takes approximately half an hour for a light-adapted eye to accommodate to darkness, while the rhodopsin level gradually increases.

Color Vision

Cones function to perceive color and require 50 to 100 times more light for stimulation than do rods. Consequently, night vision is almost totally rod vision. Unlike humans, who have both day and night vision, some vertebrates specialize for one or the other. Strictly nocturnal animals, such as bats and owls, have pure rod retinas. Purely diurnal forms, such as common gray squirrels and some birds, have only cones; they are virtually blind at night.

In 1802 an English physician and physicist Thomas Young speculated that we see color by relative excitation of three kinds of photoreceptors: one each for red, green, and blue. In the 1960s Young's prescient hypothesis was eventually supported through the combined work of several groups of researchers. Humans have three types of cones, each containing a visual pigment that responds to a particular wavelength of light (Figure 33.34). Blue cones absorb the most light at 430 nm, green cones at 540 nm, and red cones at 575 nm. Variation in the structure of opsin produces the different visual pigments found in rods and the three types of cones. Colors are perceived by comparing levels of excitation of the three different kinds of cones. For example, a light having a wavelength of 530 nm would excite green cones 95%, red cones about 70%, and blue cones not at all. This comparison is made both in nerve circuits in the retina and in the visual cortex of the brain, and our brain interprets this combination as green.

Color vision occurs in some members of all vertebrate groups with the possible exception of amphibians. Bony fishes and birds have particularly good color vision. Surprisingly, most mammals are color blind; exceptions are primates and a few other species such as squirrels.

SUMMARY

The nervous system is a rapid-communication system that interacts continuously with the endocrine system to control coordination of body function. The basic unit of nervous integration in all animals is the neuron, a highly specialized cell designed to conduct self-propagating electrical events, called action potentials, to other cells. Action potentials are transmitted from one neuron to another across synapses, which may be either electrical or chemical. The narrow gap between neurons at chemical synapses is bridged by a chemical neurotransmitter molecule, which is released from the synaptic knob and can be either stimulatory or inhibitory.

The simplest organization of neurons into a system is the nerve net of cnidarians, basically a plexus of nerve cells that, with additions, is the basis of nervous systems of several invertebrate phyla. With the appearance of ganglia (nerve centers) in bilateral flatworms, nervous systems differentiated into central and peripheral divisions. Molluscs and arthropods possess a higher level of brain complexity than do other invertebrates and are capable of learning. In vertebrates, the central nervous system consists of a brain and spinal cord. Fishes and amphibians have a three-part brain, whereas in mammals, the cerebral cortex has become a vastly enlarged multicomponent structure that has assumed the most important integrative activities of the nervous system. It completely overshadows the ancient brain, which is consigned to the role of relay center and to monitoring numerous unconscious but nonetheless vital functions such as breathing, blood pressure, and heart rate.

In humans the left cerebral hemisphere is usually specialized for language and mathematical skills while the right hemisphere is specialized for visual-spatial and musical skills.

The peripheral nervous system connects the central nervous system to receptors and effector organs. It is divided broadly into an afferent system, which conducts sensory signals to the central nervous system, and an efferent system, which conveys motor impulses to effector organs. This efferent system is subdivided into a somatic nervous system, which innervates skeletal muscle, and an autonomic nervous system, which innervates smooth and cardiac muscles and glands. The autonomic nervous system is subdivided into anatomically distinct sympathetic and parasympathetic systems, each of which sends fibers to most body organs. Generally the sympathetic system governs excitatory activities and the parasympathetic system governs maintenance and restoration of body resources.

Sensory organs are receptors designed especially to respond to internal or environmental change. The most primitive and ubiquitous sense is chemoreception. Chemoreceptors may be contact receptors such as the insect and vertebrate sense of taste, or distance receptors such as olfaction, which detects airborne molecules. In both cases, a specific chemical interacts with a particular receptor and produces

impulses transmitted to, and interpreted by, the brain. In spite of the similarity between these two senses, the sense of olfaction is far more sensitive and complex.

Receptors for touch, pain, equilibrium, and hearing are all mechanical force receptors. Touch and pain receptors are characteristically simple structures, but hearing and equilibrium are highly specialized senses based on special hair cells that respond to mechanical deformation. Sound waves received by the ear are mechanically amplified and transmitted to the inner ear, where different areas of the cochlea respond to different sound frequencies. Equilibrium receptors, also located in the inner ear in vertebrates, consist of two saclike static balance organs and three semicircular canals that detect rotational acceleration. Invertebrates monitor gravity and position using statocysts.

Vision receptors (photoreceptors) are associated with special pigment molecules that photochemically decompose in the presence of light and, in doing so, trigger nerve action potentials in optic fibers. The advanced compound eye of arthropods is especially well suited to detecting motion in the visual field. Cephalopods and vertebrates have a camera eye with focusing optics. Photoreceptor cells of the retina are of two kinds: rods, designed for high sensitivity with dim light, and cones, designed for color vision in daylight. Cones predominate in the fovea centralis of human eyes, the area of keenest vision. Rods are more abundant in peripheral areas of the retina.

REVIEW QUESTIONS

1. Define the following terms: neuron, axon, dendrite, myelin sheath, afferent neuron, efferent neuron, association neuron.
2. Glial cells far outnumber neurons and contribute roughly half the weight of the mammalian nervous system. What functions do glial cells perform in the peripheral nervous system and in the central nervous system?
3. The concentration of potassium ions inside a nerve cell membrane is higher than the concentration of sodium ions outside the membrane, yet the inside of the membrane (where the cation concentration is higher) is negative to the outside. Explain this observation in terms of permeability properties of the membrane.
4. What ionic and electrical changes occur during passage of an action potential along an axon?
5. Explain different ways in which invertebrates and vertebrates have achieved high velocities for conduction of action potentials. Can you suggest why the invertebrate solution would not be suitable for the homeothermic birds and mammals?
6. Why is the sodium-potassium pump *indirectly* important to the action potential and to maintaining the resting membrane potential?
7. Describe the microstructure of a chemical synapse. Summarize what happens when an action potential arrives at a synapse.
8. Describe the cnidarian (radiate) nervous system. How is a tendency toward centralization of the nervous system manifested in flatworms, annelids, molluscs, and arthropods?
9. How does a vertebrate spinal cord differ morphologically from nerve cords of invertebrates?
10. The knee-jerk reflex is often called a stretch reflex because a sharp tap on the patellar ligament stretches the quadriceps femoris, the extensor muscle of the leg. Describe the components and sequence of events that lead to a "knee jerk." Why is this reflex simpler than most reflex arcs? What is the difference between a reflex arc and a reflex act?

11. Name major functions associated with the following brain structures: medulla oblongata, cerebellum, tectum, thalamus, hypothalamus, cerebrum, limbic system.

12. What functional activities are associated with the left and the right hemispheres of the cerebral cortex?

13. What is the autonomic nervous system and what activities distinguish it from the somatic nervous system? Why can the autonomic nervous system be described as a "two-neuron" system?

14. Give the meaning of the statement, "The idea that all sense organs behave as biological transducers is a unifying concept in sensory physiology."

15. Chemoreception in vertebrates and insects is mediated through clearly distinguishable senses of taste and smell. Contrast these two senses in humans in terms of anatomical location and nature of the receptors and sensitivity to chemical molecules. Describe chemoreception in insects.

16. What is the vomeronasal organ and what activity does it perform? Why is its functioning often considered distinct from the sense of smell, but a component of the olfactory system of vertebrates?

17. Explain how ultrasonic detectors of certain nocturnal moths are adapted to help them escape an approaching bat.

18. Outline the place theory of pitch discrimination as an explanation of the human ear's ability to distinguish sounds of different frequencies.

19. Explain how the semicircular canals of the ear are designed to detect rotation of a human head in any directional plane.

20. Contrast the structure and functioning of the compound eye of arthropods with the camera-type eye of cephalopod molluscs and vertebrates.

21. Explain what happens when light strikes a dark-adapted rod that leads to the generation of a nerve impulse. What is the difference between rods and cones in their sensitivity to light?

22. In 1802 Thomas Young hypothesized that we see color because the retina contains three kinds of receptors. What evidence substantiates Young's hypothesis? How can we perceive any color in the visible spectrum when our retinas contain only three classes of color cones?

For Further Thought Animals are continuously bombarded by a vast array of environmental stimuli. Consider how animals with the simplest and most complex bilateral nervous systems (flatworms and mammals, respectively) might filter and preferentially respond to these environmental stimuli.

SELECTED REFERENCES

Axel, R. 1995. The molecular logic of smell. Sci. Am. **273:** 154–159 (Oct.). *Describes a surprisingly large family of genes that encodes odor-detecting molecules. This and other findings help to illuminate how the nose and brain may perceive scents.*

Behmer, S. T. 2005. Learning in insects. Encyclopedia of entomology. Dordrecht, The Netherlands, Kluwer Academic Publishers, pp. 1278–1283. *Provides an overview of insect learning, their abilities, and their limitations.*

Changeux, J-P. 1993. Chemical signaling in the brain. Sci. Am. **269:** 58–62 (Nov.). *Studies of the electric organs of fishes provide insights into how neurons in the human brain transmit information from one to the next.*

Fettiplace, R., and C. M. Hackney. 2006. Nature Reviews Neuroscience **7:** 19–29. *Describes the motor role of the sensory hair cells within the sensory system of mammals.*

Haddad, D., F. Schaupp, R. Brandt, G. Manz, R. Menzel, and A. Haase. 2004. NMR imaging of the honeybee brain. J. Insect Sci. **4:** 7–13. *NMR microscopy reveals the 3-D structure of the honeybee brain in a noninvasive way. Excellent images illustrate the structures of this tiny brain that is capable of rapid neural processing, memory retention, and communication of those memories.*

Hill, R. W., G. A. Wyse, and A. Anderson. 2012. Animal Physiology, ed. 3. Sunderland, MA, Sinauer Associates, Inc. Publishers. *Well-written chapters on neural mechanisms, nervous system organization, and sensory systems.*

Kammermeier, L., and H. Reichert. 2001. Evolution of the nervous system: common developmental genetic mechanisms for patterning invertebrate and vertebrate brains. Br. Res. Bull. **55:** 675–682. *Discussion of the discovery of homology of genes regulating embryonic development of an insect* (Drosophila melanogaster) *and a mammalian (mouse) brain.*

McClintock, M. K. 2000. Human pheromones: primers, releasers, signalers, or modulates? In K. Wallen and J. E. Schneider (eds.), Reproduction in context. Social and environmental influences on reproduction. Cambridge, Massachusetts, The MIT Press, pp. 355–420. *A discussion of pheromones with particular reference to equivocal information in humans.*

Nathan, P. 1997. The nervous system, ed. 4. London, Whurr Publications Ltd. *One of the best of several semipopular accounts of the nervous system.*

Nef, P. 1998. How we smell: the molecular and cellular bases of olfaction. News Physiol. Sci. **13:** 1–5 (Feb.). *Describes three models for odor perception, each based on experimental data.*

Randall, D., W. Burggren, and K. French. 2001. Eckert animal physiology: mechanisms and adaptations. New York, W. H. Freeman & Company. *A comprehensive and comparative treatment of animal physiology, with particularly good information on the nervous systems and sense organs of animals.*

Smith, D. V., and R. F. Margolskee. 2001. Making sense of taste. Sci. Am. **284:** 32–39 (March). *Describes the mechanism of taste reception in a very readable way.*

34
CHAPTER

Chemical Coordination: Endocrine System

An endocrine islet of Langerhans, site of insulin and glucagon synthesis, within the human pancreas.

© Ed Reschke

The Birth of Endocrinology

The birth date of endocrinology as a science is usually given as 1902, when two English physiologists, W. H. Bayliss and E. H. Starling (Figure 34.1), demonstrated the action of a hormone in a classic experiment that is still considered a model of the scientific method. Bayliss and Starling were interested in determining how the pancreas secreted its digestive juice into the small intestine at the proper time of the digestive process. They tested the hypothesis that acidic food entering the intestine triggered a nervous reflex that released pancreatic juice. To test this hypothesis, Bayliss and Starling cut away all nerves serving a tied-off loop of the small intestine of an anesthetized dog, leaving the isolated loop connected to the body only by its circulation. Injecting acid into the nerveless loop, they saw a pronounced flow of pancreatic juice. Thus, rather than a nervous reflex, some chemical messenger had circulated through the blood from the intestine to the pancreas, causing the pancreas to secrete. Yet acid itself could not be the factor because it had no effect when injected directly into the circulation.

Bayliss and Starling then designed the crucial experiment that started the new science of endocrinology. Suspecting that the chemical messenger originated in the mucosal lining of the intestine, they prepared an extract of scrapings from the mucosa, injected it into the dog's circulation, and were rewarded with an abundant flow of pancreatic juice. They named the messenger present in the intestinal mucosa *secrtin.* Later Starling coined the term **hormone** to describe all such chemical messengers, since he correctly surmised that secretin was only the first of many hormones awaiting discovery.

A   **B**

Figure 34.1   Founders of endocrinology. **A,** Sir William H. Bayliss (1860 to 1924). **B,** Ernest H. Starling (1866 to 1927).

The endocrine system, the second great integrative system controlling an animal's activities, communicates by chemical messengers called **hormones** (Gr. *hormōn,* to excite). The classical definition of hormones states that they are chemical compounds released into the blood in small amounts and transported by the circulatory system throughout the body to distant **target cells,** where they initiate physiological responses.

Many hormones are secreted by **endocrine glands,** small, well-vascularized ductless glands composed of groups of cells arranged in cords or plates. Since endocrine glands have no ducts, their only connection with the rest of the body is by blood or other body fluid; in vertebrates they receive their raw materials from their extensive blood supply and secrete their finished hormonal products into it. **Exocrine glands,** in contrast, have ducts for discharging their secretions onto a free surface. Examples of exocrine glands are sweat glands and sebaceous glands of skin, salivary glands, and the various enzyme-secreting glands in the walls of the stomach and intestine (see Chapter 32).

The classical definitions of hormones and endocrine glands just given, like so many other generalizations in biology, are changing as new information appears. Some hormones may never enter the general circulation. Furthermore, many hormones, such as those of digestion (see Chapter 32, p. 714), are synthesized in minute amounts by highly diffuse cells of the **enteroendocrine system,** some are transported into the brain where they act on neurons and glial cells (for example, insulin and leptin), and some, such as cytokines, are secreted by cells of the immune system (p. 771). Hormones may function as **neurotransmitters** in the brain or as local tissue factors **(parahormones),** which stimulate cell growth or some biochemical process. Most hormones, however, are blood borne and therefore diffuse into every tissue space in the body.

Compared with nervous systems, endocrine systems are slow acting because of the time required for a hormone to reach the appropriate tissue, cross the capillary endothelium, and diffuse through tissue fluid to, and sometimes into, cells. The minimum response time is seconds, but it can be much longer. Hormonal responses in general are long lasting (minutes to days) whereas those under nervous control are short term (milliseconds to minutes). We expect endocrine control where a sustained effect is required, as in many metabolic, growth, and reproductive processes. Despite such differences, nervous and endocrine systems function without sharp separation as a single, interdependent system. Endocrine glands often receive directions from the brain. Conversely, many hormones act on the nervous system and significantly affect a wide array of animal behaviors.

The first formal experiment in endocrinology was performed in 1849 by Professor Arnold Adolph Berthold at the University of Gottingen. He conclusively demonstrated that a blood-borne signal was produced by the testes, and that this chemical was responsible for producing both physical and behavioral characteristics that distinguished an adult male rooster from immature chickens and adult, castrated male chickens (capons). Berthold castrated male chicks and divided them into three groups. He left one group of controls to grow normally without their testes, and he reimplanted the testes into the second group. The third group was implanted with testes from different chicks. As the chicks grew, he observed that the castrated group developed into capons, with no interest in hens, lacking rooster plumage and male aggressive behavior. The second and third groups of birds were indistinguishable from each other, with full male plumage, normal aggressive behavior, and interest in hens. Berthold then killed the birds and dissected them. He found that the transplanted testes had developed their own blood supply and were functioning normally. From this classic experiment, Berthold concluded that because there was no nerve supply to the testes, the testes must secrete a blood-borne signal that produced all characteristics of maleness.

All hormones are low-level signals. Even when an endocrine gland is secreting maximally, its hormone is so greatly diluted by the large volume of blood it enters that its plasma concentration seldom exceeds 10^{-9} M (or one billionth of a 1 M concentration). Some target cells respond to plasma concentrations of hormone as low as 10^{-12} M. Since hormones have far-reaching and often powerful influences on cells, it is evident that their effects are vastly amplified at the cellular level.

34.1 MECHANISMS OF HORMONE ACTION

Widespread distribution of hormones in an animal permits certain hormones, such as growth hormone from the vertebrate pituitary gland, to affect most, if not all, cells during specific stages of cellular differentiation. Whether hormones produce widespread responses or highly specific responses only in certain cells and at certain times depends on the presence of **receptor molecules** on or in target cells. A hormone engages only those cells that display the receptor that, by virtue of its specific molecular shape, binds with the hormone molecule. Other cells are insensitive to the hormone's presence because they lack specific receptors. Hormones act through **membrane-bound receptors: nuclear receptors** and **cytoplasmic receptors.**

Membrane-Bound Receptors and the Second-Messenger Concept

Many hormones, such as most amino acid derivatives, and peptide hormones that are too large, or too polar, to pass through plasma membranes, bind to transmembrane proteins (see Figure 3.6, p. 40) that act as receptor sites on the surfaces of target-cell membranes. The hormone and receptor form a complex that triggers a cascade of molecular events within a cell. The hormone thus behaves as a **first messenger** that causes activation of a **second-messenger** system in the cytoplasm. At least six different molecules have been identified as second messengers. Each works via a specific **kinase,** which causes activation or inactivation of rate-limiting enzymes (see Chapter 4, p. 59) that modify the direction and rate of cytoplasmic processes (Figure 34.2). Since many molecules are activated at each level in the second-messenger system cascade after a single hormone molecule has been bound, the message is amplified, perhaps many thousands of times.

Second-messenger systems known to participate in hormone actions are **cyclic AMP (cAMP), cyclic GMP (cGMP), Ca^{++}/calmodulin, inositoltrisphosphate (IP$_3$),** and **diacylglycerol (DAG).** Cyclic AMP was the first to be investigated, and it has been shown to mediate actions of many peptide hormones, including parathyroid hormone, glucagon, adrenocorticotropic hormone (ACTH), thyrotropic hormone (TSH), melanophore-stimulating hormone (MSH), and vasopressin. It also mediates action of epinephrine (also called adrenaline), an amino acid derivative.

Interestingly, the same hormone may activate different second-messenger systems in each type of target cell, such that a single hormone produces multiple actions within an animal.

Other membrane-bound receptors possess kinase activity themselves and are activated when the hormone binds to the receptor, for example, the insulin and insulin-like growth factor membrane receptors.

The 2012 Nobel Price in Chemistry was awarded to Brian Kobilka (Stanford University) and Robert Lefkowitz (Duke University) for their work on yet another class of membrane-bound receptors called **G Protein Coupled Receptors** (GPCRs). This is a large class of important receptors not only for many protein and peptide hormones, but also for neurotransmitters (see Chapter 33). GPCRs may be coupled with any one of the second-messenger systems mentioned above leading to a diverse array of responses from one hormone interacting with multiple cellular targets. GPCRs may produce both excitatory and inhibitory responses in target cells depending on the GPCR subtype present in the membrane. GPCRs are becoming extremely important as drug targets in medicine.

Nuclear Receptors

Unlike peptide hormones and epinephrine, which are much too large to pass through plasma membranes, **steroid hormones** (for example, estrogen, testosterone, and aldosterone) are lipid-soluble molecules that readily diffuse through plasma membranes. Once inside the cytoplasm of target cells, steroid hormones bind selectively to receptor molecules. While these receptor molecules may be located in either cytoplasm or nucleus, their ultimate site of activity is nuclear. The hormone-receptor complex, now called a **gene regulatory protein,** then activates or inhibits specific genes. As a result, gene transcription is altered (see Chapter 5, p. 88), since messenger RNA molecules are synthesized on specific sequences of DNA. Stimulation or inhibition of mRNA formation modifies production of key enzymes, thus setting in motion a hormone's observed effect (Figure 34.2). Thyroid hormones and the insect-molting hormone, ecdysone (a steroid, p. 24), also act through nuclear receptors. Thyroid hormones first bind to a transmembrane protein-transport molecule (p. 46) that uses ATP to move the hormones into the cell.

Compared with peptide hormones that act *indirectly* through second-messenger systems, steroids and thyroid hormones have a *direct* effect on protein synthesis because they bind a nuclear receptor that modifies specific gene activity.

Cytoplasmic Receptors

Lipid-soluble hormones, such as estrogen, are now known to interact not only with nuclear receptors but also with cytoplasmic receptors that are either membrane bound or free within the cytoplasm. Once activated, these hormone-receptor complexes interact with second-messenger systems within the cytoplasm (much like peptide hormones) or through a cascade of events activate factors that enter the nucleus either to stimulate or to inhibit transcription processes (p. 88). Thus, lipid-soluble hormones provide multiple and complex control of target cells.

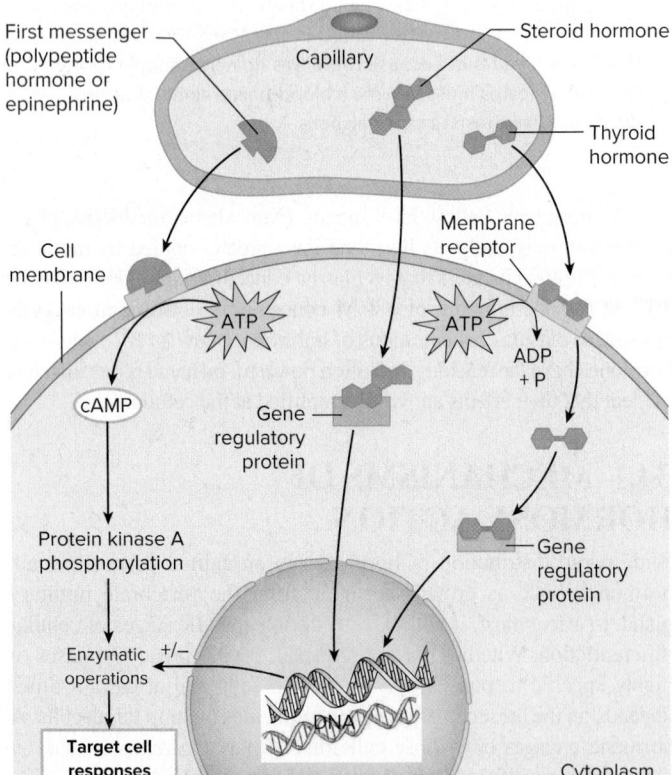

Figure 34.2 Mechanisms of hormone action shown in a stylized cell. The cell is enlarged relative to the capillary so that more than one mechanism can be illustrated. Peptide hormones and epinephrine act through second-messenger systems, as, for example, cyclic AMP, shown here. The combination of hormone with a membrane receptor stimulates the enzyme adenylate cyclase to catalyze formation of cyclic AMP (second messenger). Thyroid hormones bind to a membrane receptor and are transported into the cell by active transport. There they bind to cytoplasmic receptors that are transported into the nucleus to alter gene transcription. Steroid hormones diffuse through the cell membrane to combine with cytoplasmic or nuclear receptors that alter gene transcription.

Control of Secretion Rates of Hormones

Hormones influence cellular functions by altering rates of many different biochemical processes. Many affect enzymatic activity and thus alter cellular metabolism, some change membrane permeability, some regulate synthesis of cellular proteins, and some stimulate release of hormones from other endocrine glands. Because these are all dynamic processes that must adapt to changing metabolic demands, they must be regulated, not merely activated, by appropriate hormones. This regulation is achieved by precisely controlled release of a hormone into the blood or surrounding tissue. Concentration of a hormone in body fluid depends on two factors: its rate of secretion and the rate at which it is inactivated and removed from the circulation. Consequently, if secretion is to be correctly controlled, an endocrine gland requires information about the level of its own hormone(s) in the plasma.

Most hormones are controlled by **negative feedback systems** that operate between the glands secreting hormones and the outputs or products of target cells (Figure 34.3). A feedback pattern is one in which output is constantly compared with a setpoint, like a thermostat. For example, CRH (corticotrophin-releasing hormone), secreted by the hypothalamus, stimulates the pituitary (containing target cells) to release ACTH. ACTH stimulates the adrenal gland (containing target cells) to secrete cortisol. As the level of ACTH rises in the plasma, it acts on, or "feeds back" on, the hypothalamus to inhibit release of CRH. Similarly, as cortisol levels rise in the plasma, it "feeds back" on the hypothalamus and pituitary to inhibit release of both CRH and ACTH, respectively. Thus any deviation from the setpoint (a specific plasma level of each hormone) leads to corrective action in the opposite direction (Figure 34.3). Such a negative feedback system is highly effective in preventing extreme oscillations in hormonal output. However, hormonal feedback systems are more complex than a rigid "closed-loop" system such as a thermostat that controls a house's central heating system, because hormonal feedback may be altered by input from the nervous system or by metabolites or other hormones.

Extreme oscillations in hormone output do sometimes occur under natural conditions. However, because they have the potential to disrupt finely tuned homeostatic mechanisms, such extreme oscillations, as a result of **positive feedback,** are highly regulated and possess an obvious shutoff mechanism. During positive feedback, the signal (or output of the system) feeds back to the control system and causes an increase in the initial signal. In this way the initial signal becomes progressively amplified to produce an explosive event. For example, hormones controlling parturition (birth) become elevated from their normal setpoint and are shut off by birth of offspring from the uterus (see Chapter 7).

34.2 INVERTEBRATE HORMONES

All invertebrate taxa produce hormones, and cnidarians, nematodes, and annelids possess endocrine cells that act autonomously and are not organized into glands. Endocrine glands appear in molluscs and arthropods, with the latter being more complex. In most invertebrate phyla these chemical signals must move from source to target cells via body fluids other than circulating blood. Invertebrate hormones are peptides (often neuropeptides), steroids, or terpenoids (lipid-soluble organic molecules), but peptides and neuropeptides are the most common among different invertebrate groups. Some hormones are vertebratelike in structure and function (for example, steroids), but there is much greater diversity in invertebrate endocrine function than in that of vertebrates. Invertebrate hormones regulate color changes, growth, reproduction, and internal homeostatic mechanisms, such as metabolism, metabolic fuel levels, and osmoregulation.

In many animal phyla, the principal source of hormones is **neurosecretory cells,** specialized nerve cells capable of synthesizing and secreting hormones. Their products, called neurosecretory hormones or neurohormones, are discharged directly into body fluids, and serve as a crucial link between the nervous and endocrine systems.

Peptides and neuropeptides have been shown to regulate many physiological processes in invertebrates. In Crustacea, **cardioactive peptide** increases heart rate. Hormones that regulate metabolism of carbohydrates, fats, and amino acids belong to the **crustacean hyperglycemic hormone** family (CHH) in Crustacea and to the **adipokinetic hormone** family in insects. **Diuretic hormones** stimulate the secretion of fluid in insect Malpighian tubules (p. 666). A family of small neuropeptides called **FMRFamide-related peptides** (FaRPs) appears to have evolved as early as bilateral symmetry, and their functions seem to have been conserved across phyla. Peptides of this family are known to regulate muscular tissues of the body, and digestive and reproductive processes in many invertebrates, as well as osmoregulatory processes in nematodes, annelids, molluscs, and insects, and arterial hemolymph flow in crustaceans. They are now being isolated and characterized in vertebrates. An extensively studied neurosecretory process in invertebrates is control of development and metamorphosis of insects. In insects, as in other arthropods, growth is a series of steps in which the rigid, nonexpansible exoskeleton is periodically discarded and replaced with a new, larger one. Most insects undergo a process of metamorphosis (p. 456), in which a series of juvenile stages, each requiring formation of a new exoskeleton, end with a molt.

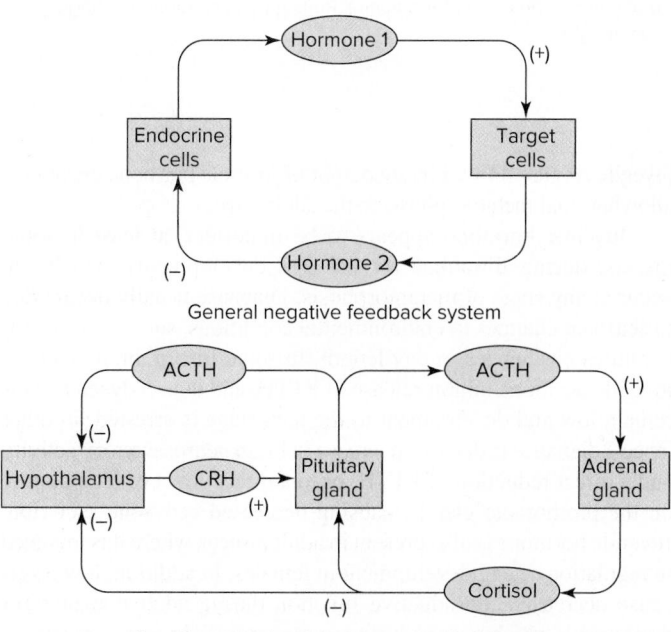

General negative feedback system

Specific example of a negative feedback system

Figure 34.3 Negative feedback systems.

Endocrine Control of Molting in a Moth

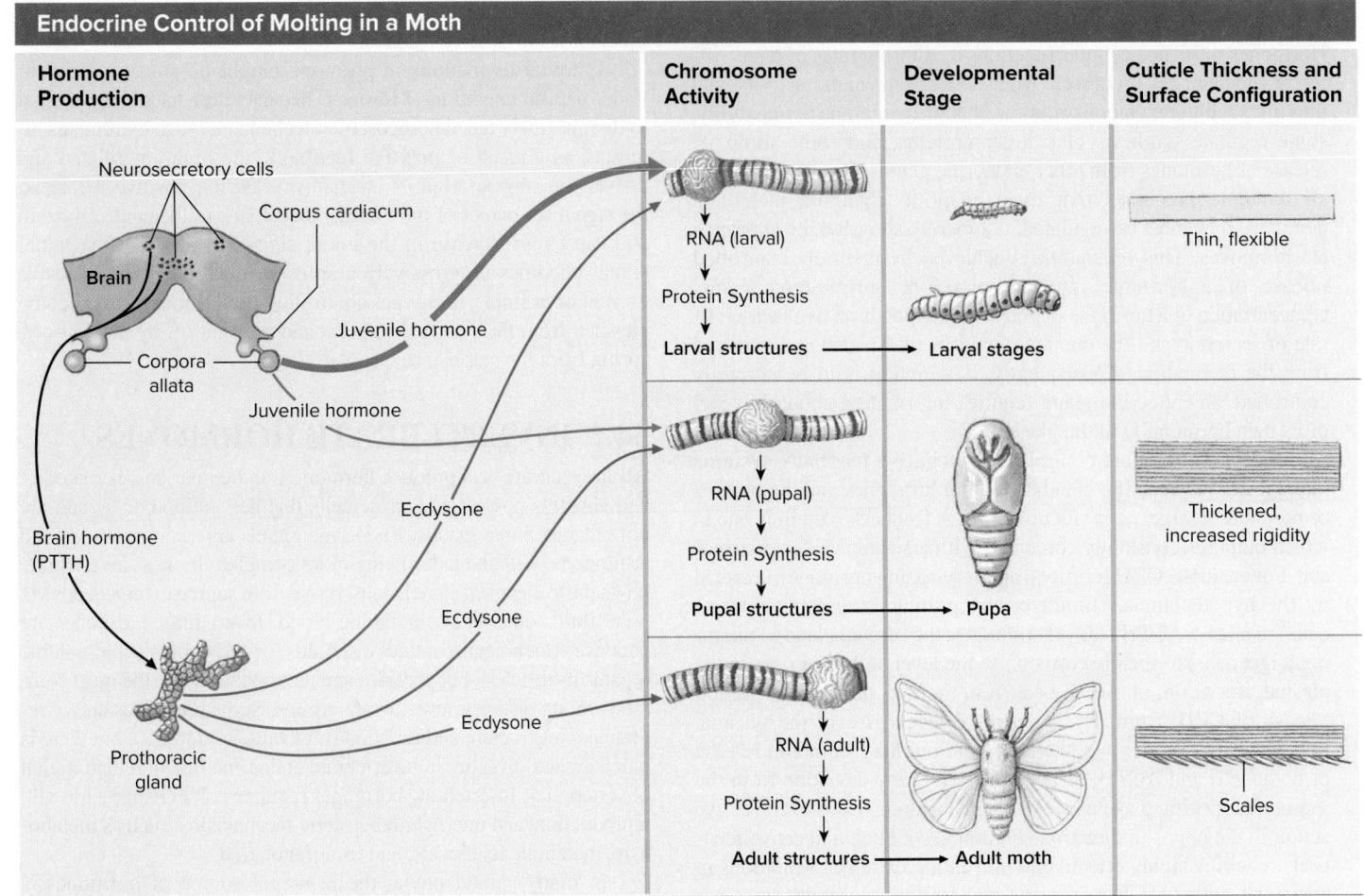

Hormone Production	Chromosome Activity	Developmental Stage	Cuticle Thickness and Surface Configuration

Figure 34.4 Endocrine control of molting in a moth, typical of insects having complete metamorphosis. Many moths mate in spring or summer, and eggs soon hatch into the first of several larval stages, called instars. After the final larval molt, the last and largest larva (caterpillar) spins a cocoon in which it pupates. The pupa overwinters, and an adult emerges in the spring to start a new generation. Juvenile hormone and ecdysone interact to control molting and pupation. Many genes are activated during metamorphosis, as seen by puffing of chromosomes (center column). Puffs form in sequence during successive molts. Changes in cuticle thickness and surface characteristics are shown at right.

Insect physiologists discovered that molting and metamorphosis (see Chapter 21, p. 456) are primarily controlled by interaction of two hormones, one favoring growth and differentiation of adult structures and another favoring retention of juvenile structures. These two hormones are **molting hormone** or **ecdysone,** a steroid produced by the prothoracic gland, and **juvenile hormone,** a terpenoid produced by the corpora allata (Figure 34.4).

Ecdysone is controlled by **prothoracicotropic hormone** or **PTTH.** This hormone is a polypeptide (molecular weight about 5000) produced by neurosecretory cells of the brain, and transported by axons to the corpora cardiaca where it is stored. At intervals during juvenile growth, release of PTTH into the blood stimulates the prothoracic gland to secrete ecdysone. Ecdysone bound to its nuclear receptor acts directly on the chromosomes as a gene regulatory protein (Figure 34.2) to cause a molt, and subsequent development of adult structures. It is held in check, however, by juvenile hormone, which favors retention of juvenile characteristics. During juvenile life, juvenile hormone predominates and each molt yields another larger

juvenile (Figure 34.4). Finally output of juvenile hormone decreases, allowing final metamorphosis to the adult stage.

Juvenile hormone appears to be important, at least in some insects, during **diapause** (or arrested development), which can occur at any stage of metamorphosis. Diapause usually occurs due to seasonal changes in environmental conditions, such as cold temperatures or changes in day length. In some insects high levels of juvenile hormone inhibit release of PTTH, and thus ecdysone levels remain low and development to the next stage is arrested. In other insects diapause is due to a decrease in brain neurosecretory activity and a direct reduction of PTTH, or to a direct effect of temperature on the prothoracic glands, causing decreased ecdysone secretion. Juvenile hormone is also present in adult insects, where it is involved in regulation of egg development in females. In addition, low levels cause decreased reproductive function during adult diapause (or dormancy), which occurs during winter months in some insects.

Based on the extensive study of all aspects of endocrine regulation in insects, scientists have synthesized many different endocrine disruptors designed to regulate insect populations, such as potent analogs of juvenile hormone, which induce abnormal final molts or prolong or block development. Unlike chemical insecticides, they are highly specific, but given the conservation of hormone functions across invertebrate groups, they are less ecologically benign than originally thought.

34.3 VERTEBRATE ENDOCRINE GLANDS AND HORMONES

We describe some of the best understood and most important vertebrate hormones, including a brief overview of mammalian hormonal mechanisms (because laboratory mammals and humans have always been the objects of the most intensive research) and discussion of some important differences in functional roles of hormones among different vertebrate groups.

Hormones of the Hypothalamus and Pituitary Gland

The pituitary gland, or **hypophysis,** is a small gland (0.5 g in humans) lying in a depression in the floor of the cranium (Figure 34.5). It is a two-part gland having a double embryological origin. The **anterior pituitary** (adenohypophysis) is derived embryologically from the

Figure 34.5 Human hypothalamus and pituitary gland. The posterior lobe is connected directly to the hypothalamus by axons of neurosecretory cells. The anterior lobe is indirectly connected to the hypothalamus by a portal circulation (shown in red) beginning in the base of the hypothalamus and ending in the anterior pituitary.

Figure 34.6 Relationship of hypothalamic, pituitary, and target-gland hormones. The hormone sequence controlling the release of cortisol from the adrenal cortex is used as an example.

roof of the mouth. The **posterior pituitary** (neurohypophysis) arises from a ventral portion of the brain, the **hypothalamus,** and is connected to it by a stalk, the **infundibulum.** Although the anterior pituitary lacks any anatomical connection to the brain, it is functionally connected to it by a special portal circulatory system. A portal circulation is one that delivers blood from one capillary bed to another (Figures 34.5 and 34.6). In this case, the portal circulation provides a link between neurosecretory cells of the hypothalamus and endocrine cells of the anterior pituitary gland.

Hypothalamus and Neurosecretion

Because of the strategic importance of the pituitary in influencing most hormonal activities in the body, the pituitary was once called the "master gland." This description is not appropriate, however, because anterior pituitary hormones are regulated by a higher council, the neurosecretory centers of the hypothalamus. The hypothalamus is itself under ultimate control by inputs from other regions of the brain. The hypothalamus contains groups of neurosecretory cells, which are specialized nerve cells (Figures 34.5 and 34.6) that manufacture neurohormones. Those that regulate the anterior pituitary are called **releasing hormones** or **release-inhibiting hormones** (or "factors"). These neurohormones travel down nerve axons to their endings in the median eminence. Here they enter a capillary network to complete their journey to the anterior pituitary by way of the hypothalamic-pituitary portal system. The hypothalamic hormones then stimulate or inhibit release of various anterior pituitary hormones. Several hypothalamicreleasing and release-inhibiting hormones have been discovered, characterized chemically, and isolated in pure state (Table 34.1),

although the identification and action of some of the hypothalamic hormones listed in Table 34.1 is still tentative. Other hypothalamic neurosecretory cells possess nerve axons that travel down the infundibulum and have their nerve endings in the posterior pituitary. These secrete neurohormones directly into the general blood circulation and not into the hypothalamic-pituitary portal system (Figure 34.5).

Anterior Pituitary

The anterior pituitary consists of an **anterior lobe** (pars distalis) as shown in Figure 34.5, and an **intermediate lobe** (pars intermedia), which is absent in some animals (including humans). The anterior pituitary produces seven hormones, and in those animals with an intermediate lobe, all but one are released by the anterior lobe.

TABLE 34.1
Hormones of the Vertebrate Pituitary

	Hormone	Chemical Nature	Principal Action (s)	Hypothalamic Control(s)
Adenohypophysis (Anterior lobe)	Thyroid-stimulating hormone (TSH)	Glycoprotein	Stimulates thyroid hormone synthesis and secretion	TSH-releasing hormone (TRH)
	Follicle-stimulating hormone (FSH)	Glycoprotein	Female: follicle maturation and estrogen synthesis Male: stimulates sperm production	Gonadotropin-releasing hormone (GnRH)[1]
	Luteinizing hormone (LH)	Glycoprotein	Female: stimulates ovulation, corpus luteum formation, estrogen and progesterone synthesis Male: testosterone secretion	Gonadotropin-releasing hormone(GnRH)[1] Gonadotropin release-inhibiting hormone (GnIH)[2]
	Prolactin (PRL)	Protein	Mammary gland growth, milk synthesis in females, immune response and angiogenesis in mammals, parental behavior, electrolyte and water balance in lower vertebrates	Dopamine (prolactin release-inhibiting hormone or PIH) Prolactin-releasing factor (PRF)?
	Growth hormone (GH) (somatotropin)	Protein	Stimulates soft tissue and bone growth, protein synthesis, mobilization of glycogen and fat stores	Growth hormone-releasing hormone (GHRH) Growth hormone release-inhibiting hormone (GHIH) or somatostatin
	Adrenocorticotropic hormone (ACTH)	Polypeptide	Stimulates glucocorticoid synthesis by adrenal cortex	Corticotropin-releasing hormone (CRH)
Intermediate lobe[3]	Melanocyte-stimulating hormone (MSH)	Polypeptide	Increased melanin synthesis by melanocytes in epidermis of ectotherms; function not clear in endotherms	Melanocyte-stimulating hormone-inhibiting hormone (MSHIH)
Neurohypophysis (Posterior lobe)	Oxytocin	Octapeptide	Milk ejection and uterine contractions in female, sexual behavior and pair bonding in monogamous species	
	Vasopressin[4] (antidiuretic hormone or ADH)	Octapeptide	Water reabsorption in mammalian kidneys	
	Vasotocin[5]	Octapeptide	Increases water reabsorption	

[1]One GnRH hormone regulates both FSH and LH, but some studies suggest a separate FSH-releasing hormone (FSH-RH).
[2]GnIH has been discovered in birds and mammals.
[3]Birds and some mammals lack an intermediate lobe. In these forms, MSH is produced by the anterior lobe.
[4]In mammals.
[5]In all vertebrate classes except mammals, although additional, related hormones have been identified.

The anterior pituitary produces four hormones that regulate other endocrine glands, collectively called **tropic hormones** (from the Greek *tropē,* to turn toward) (Table 34.1). **Thyroid-stimulating hormone (TSH)** or **thyrotropin** stimulates production of thyroid hormones by the thyroid gland. Two tropic hormones, commonly called **gonadotropins,** act on the gonads (ovaries of females, testes of males). These are **follicle-stimulating hormone (FSH)>** and **luteinizing hormone (LH).** FSH promotes egg production and secretion of estrogen in females, and supports sperm production in males. LH induces ovulation, corpus luteum production, and secretion of female sex steroids, progesterone and estrogen. In males, LH promotes production of male sex steroids (primarily testosterone). It once was called interstitial cell stimulating hormone (ICSH) in males, before it was discovered to be identical to LH in females. The hormonal control of reproduction is extensively discussed in Chapter 7. The fourth tropic hormone, **adrenocorticotropic hormone (ACTH),** increases production and secretion of steroid hormones from the adrenal cortex.

Prolactin and the structurally related **growth hormone (GH)** are proteins. Prolactin is essential for preparing mammary glands for lactation; after birth it is required for production of milk. Prolactin also is implicated in parental behavior in a wide variety of vertebrates. Beyond its more traditional role in reproductive processes, prolactin regulates water and electrolyte balance in many species. Prolactin has also been shown to be a chemical mediator of the immune system and is important in formation of new blood vessels (angiogenesis). Unlike tropic hormones, prolactin acts directly on its target tissues rather than through other hormones.

GH (also called somatotropin) performs a vital role in governing body growth through its stimulatory effect on cellular mitosis, on synthesis of messenger RNA and protein, and on metabolism, especially in new tissue of young vertebrates. Growth hormone acts directly on growth and metabolism, as well as indirectly through a polypeptide hormone, **insulin-like growth factor (IGF)** or somatomedin, produced by the liver.

The only anterior pituitary hormone produced by the intermediate lobe is **melanocyte-stimulating hormone (MSH).** In cartilaginous and bony fishes, amphibians, and nonavian reptiles, MSH is a directacting hormone that promotes dispersion of the pigment melanin within melanocytes, causing darkening of the skin. In birds and mammals, MSH is produced by cells in the anterior pituitary rather than the intermediate lobe, but its physiological function remains unclear. MSH appears unrelated to pigmentation in endotherms, although it causes darkening of the skin in humans if injected into the circulation. MSH has been isolated from specific regions of the hypothalamus, where it has been linked to regulation of ingestive behaviors and metabolism in adult mammals. MSH and ACTH are derived from a precursor molecule (pro-opiomelanocortin or POMC) that is transcribed and translated from a single gene.

Posterior Pituitary

The hypothalamus is the source of two hormones of the posterior lobe of the pituitary (Table 34.1). They are formed in neurosecretory cells in the hypothalamus, whose axons extend down the infundibular stalk and into the posterior lobe. The hormones are secreted from axon terminals ending close to blood capillaries, which the

Figure 34.7 Posterior pituitary hormones of mammals. Both oxytocin and vasopressin consist of eight amino acids (the two sulfur-linked cysteine molecules are considered a single amino acid, cystine). Oxytocin and vasopressin are identical except for amino-acid substitutions in the blue positions. Abbreviations represent amino acids.

hormones enter when released (Figure 34.5). In a sense the posterior lobe is not a true endocrine gland, but a storage and release center for hormones manufactured entirely in the hypothalamus. The two posterior-lobe hormones of mammals, oxytocin and vasopressin, are chemically very much alike. Both are polypeptides consisting of eight amino acids (octapeptides, Figure 34.7). These hormones are among the fastest-acting hormones, since they can produce a response within seconds of their release from the posterior lobe.

Oxytocin has two important specialized reproductive functions in adult female mammals. It stimulates contraction of uterine smooth muscles during parturition (birth of young). In clinical practice, oxytocin is used to induce delivery during prolonged labor and to prevent uterine hemorrhage after birth. A second action of oxytocin is that of milk ejection by the mammary glands in response to suckling. A role for oxytocin in pair-bonding behavior in both sexes in some monogamous vole species has now been established.

Vasopressin, the second posterior-lobe hormone, acts on collecting ducts of the kidney to increase water reabsorption and thus restrict urine flow, as already described on page 671. It is therefore often called **antidiuretic hormone** or **ADH.** Vasopressin also increases blood pressure through its generalized constrictor effect on smooth muscles of arterioles, and acts centrally to increase thirst, and therefore, drinking behavior. Finally, vasopressin has been linked to pair-bonding behavior in some monogamous vole species.

All jawed vertebrates secrete two posterior-lobe hormones that are quite similar to those of mammals. All are octapeptides, but their structures vary because of amino-acid substitutions in three of eight aminoacid positions in the molecule.

Of all posterior-lobe hormones, **vasotocin** (Table 34.1) has the widest phylogenetic distribution and is considered the parent hormone from which other octapeptides evolved. It occurs in all vertebrate classes except mammals. It is a water-balance hormone in amphibians, especially toads, in which it acts to conserve water by (1) increasing permeability of skin (to promote water absorption from the environment), (2) stimulating water reabsorption from the urinary bladder, and (3) decreasing urine flow. Action of vasotocin is best understood in amphibians, but it appears to play some water-conserving role in birds and nonavian reptiles as well.

Pineal Gland

In all vertebrates the dorsal part of the brain, the diencephalon (Figure 33.13, p. 731), gives rise to a saclike evagination called the pineal complex, which lies just beneath the skull in a midline position. In ectothermic vertebrates the pineal complex contains glandular tissue and a photoreceptive sensory organ involved in pigmentation responses and in light-dark biological rhythms. In lampreys, many amphibians, lizards, and tuataras (*Sphenodon*, p. 578), the median photoreceptive organ is so well developed, containing structures analogous to the lens and cornea of lateral eyes, that it is often called a third eye. In birds and mammals, the pineal complex is an entirely glandular structure called the **pineal gland.** The pineal gland produces the hormone **melatonin.** Melatonin secretion is strongly affected by exposure to light. Its production is lowest during daylight hours and highest at night. In nonmammalian vertebrates, the pineal gland is responsible for maintaining **circadian rhythms**—self-generated (endogenous) rhythms that are about 24 hours in length. A circadian rhythm serves as a biological clock for many physiological processes that follow a regular pattern.

In mammals, an area of the hypothalamus called the **suprachiasmatic nucleus** has become the primary circadian pacemaker, although the pineal gland still produces melatonin nightly and serves to reinforce the circadian rhythm of the suprachiasmatic nucleus. In birds and mammals in which seasonal rhythms in reproduction are regulated by **photoperiod,** melatonin plays a critical role in timing gonadal activity. In long-day breeders, such as horses, ferrets, hamsters, and deer mice, reduced light stimulation with shortening day length in the autumn increases melatonin secretion, and in these species reproductive activity is suppressed during winter months. Lengthening days in the spring have the opposite effect, and reproductive activities are resumed. Short-day breeders, such as white-tailed deer, silver fox, spotted skunk, and sheep, are stimulated by reduced day length in the autumn; increasing melatonin levels at this time are associated with increased reproductive activity. The role of melatonin is an indirect one in both cases because melatonin itself does not stimulate or inhibit the reproductive axis.

The pineal gland produces subtle effects on circadian and annual rhythms in nonphotoperiodic mammals (such as humans). For example, melatonin secretion has been linked to a sleeping and eating disorder in humans called seasonal affective disorder (SAD). Some people living in northern latitudes, where day lengths are short in winter and when melatonin production is elevated, become depressed in winter, sleep long periods, and may have eating binges. Often this wintertime disorder can be treated by exposure to sunlamps with full-spectrum light; such exposure depresses melatonin secretion by the pineal gland. Disturbed physiological rhythms associated with jet lag, shift work, and aging also are linked to inappropriate melatonin rhythms. Circadian-rhythm problems may also be linked to psychiatric disorders, metabolic syndrome (simultaneous occurrence of high blood pressure, elevated insulin levels, excess body fat around the waist, or elevated cholesterol levels), and increased cancer risks.

Brain Neuropeptides

The blurred distinction between endocrine and nervous systems is nowhere more evident than in the growing list of hormonelike neuropeptides that have been discovered in the central and peripheral nervous systems of vertebrates and invertebrates. Neuropeptides and associated GPCRs (p. 748) have been identified in Cnidaria as well as in protostomes and deuterostomes. In mammals, approximately 40 neuropeptides (short chains of amino acids) have been located using immunological labeling with antibodies and visualized in histological sections under a microscope, and the list is still growing. Many lead double lives—behaving both as hormones, carrying signals from gland cells to their targets, and as neurotransmitters, relaying signals between nerve cells. For example, both oxytocin and vasopressin have been located at widespread sites in the brain by immunological methods. Related to this discovery is the fascinating observation that people and experimental animals injected with minute quantities of vasopressin experience enhanced learning and improved memory. This effect of vasopressin in brain tissue is unrelated to its well-known antidiuretic function in the kidney (p. 671). Several hormones, such as gastrin and cholecystokinin (p. 714) (formerly considered only as part of the enteroendocrine system of the gastrointestinal tract), have been discovered in the cerebral cortex, hippocampus, and hypothalamus. In addition to its gastrointestinal actions, cholecystokinin modulates feeding and satiety and may serve other functions as a brain neuroregulator. Endorphins and enkephalins, neuropeptides that bind with opiate receptors and influence perception of pleasure and pain (see note on p. 737), also function in brain circuits that modulate several other functions unrelated to pleasure and pain, such as control of blood pressure, body temperature, body movement, feeding, and reproduction. Even more intriguing, endorphins are derived from the same prohormone (POMC) that gives rise to the anterior pituitary hormones ACTH and MSH.

> The radioimmunoassay technique developed by Solomon Berson and Rosalyn Yalow about 1960 revolutionized endocrinology and neurochemistry. First, antibodies to the hormone of interest (insulin, for example) are prepared by injecting a mammal, such as guinea pigs or rabbits, with the hormone, and then collecting the antibodies that are produced in the blood. Then, a fixed amount of radioactively labeled insulin and unlabeled insulin antibodies is mixed with the sample of blood plasma to be measured. The native insulin in the blood plasma and the radioactive insulin compete for insulin antibodies. The more insulin present in the sample, the less radioactive insulin will bind to the antibodies. Bound and unbound insulin are then separated, and their radioactivities are measured together with those of appropriate standard solutions containing known amounts of insulin to determine the amount of insulin present in the blood sample.

Prostaglandins and Cytokines

Prostaglandins

Prostaglandins are derivatives of long-chain unsaturated fatty acids that were discovered in seminal fluid in the 1930s. At first they were thought to be produced only by the prostate gland (hence the name) but now they have been found in virtually all mammalian tissues. Prostaglandins often act as local hormones and have diverse actions on many different tissues, making generalizations about their effects difficult. Many of their effects, however, involve smooth muscle. In some tissues prostaglandins regulate vasodilation or vasoconstriction by their action on smooth muscle in walls of blood vessels. They are known to stimulate contraction of uterine smooth muscle during childbirth. There also is evidence that overproduction of uterine prostaglandins is responsible for painful symptoms of menstruation (dysmenorrhea) experienced by many women, and inhibitors of prostaglandins may provide relief from these symptoms. Among other actions of prostaglandins is their intensification of pain in damaged tissues, mediation of the inflammatory response (see Chapter 35, pp. 767–769), and involvement in fever.

Cytokines

For some years we have known that cells of the immune system communicate with one another and that this communication was crucial to the immune response. Now we understand that a large group of polypeptide hormones called **cytokines** (p. 771) mediate communication between a wider variety of cells and can affect the cells that secrete them, can affect nearby cells, and like other hormones, can affect cells in distant locations. Their target cells bear specific receptors for the cytokine bound to the surface membrane. Cytokines coordinate a complex network, with some target cells being activated, stimulated to divide and often to secrete their own cytokines. The same cytokine that activates some cells may suppress division of other target cells. Cytokines also are involved in formation of blood, and in regulation of energy balance by the central nervous system. Many cytokines that are secreted by adipose tissue, termed **adipokines,** have become a focus of research on the regulation of energy balance and obesity. It appears that secretion of these hormones elicits an inflammatory response (see Chapter 35, p. 767) within adipose tissue and other organs that mediates many of the symptoms associated with obesity.

Hormones of Metabolism

An important group of hormones adjusts the delicate balance of metabolic activities. Rates of chemical reactions within cells are often regulated by long sequences of enzymes (see Chapter 4, p. 57). Hormones may alter activity of crucial enzymes in a metabolic process, thus accelerating or inhibiting the entire process. The most important hormones of metabolism are those of the thyroid, parathyroid, adrenal glands, pancreas, and white adipose tissue, as well as the previously mentioned growth hormone of the anterior pituitary. Most recently, skeletal muscle has been added to this list.

Thyroid Hormones

The thyroid gland is a large endocrine gland located in the neck of all vertebrates. It is composed of thousands of tiny spherelike units, called follicles, where two thyroid hormones, **triiodothyronine** and **thyroxine** (T_3 and T_4, respectively), are synthesized, stored, and released into the bloodstream as needed. The size of the follicles, and amount of stored T_3 and T_4 they contain, depends on the activity of the gland (Figure 34.8). A third hormone, **calcitonin,** is secreted by C cells located between the follicles in the connective tissue of the thyroid gland in mammals; this hormone is discussed in the section on calcium metabolism (p. 756).

A unique characteristic of the thyroid is its ability to concentrate high levels of **iodine;** in most animals this single gland contains well over half the body store of iodine. Epithelial cells of thyroid follicles actively trap iodine from the blood and combine it with the amino

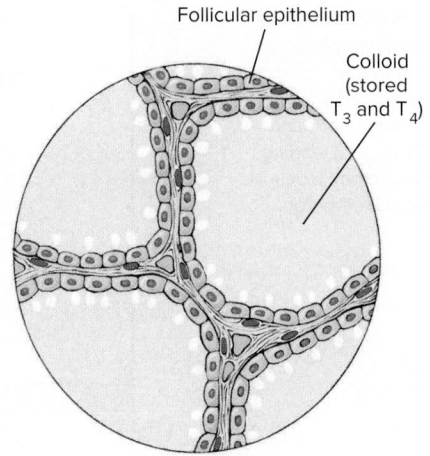

Follicular epithelium

Colloid (stored T_3 and T_4)

Inactive follicles

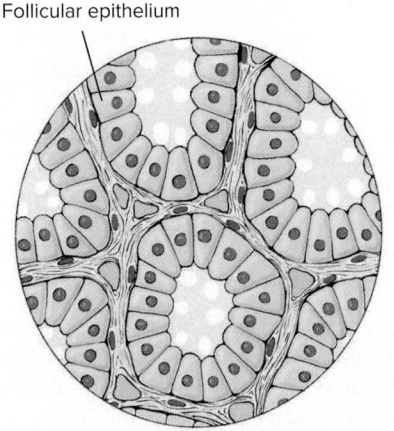

Follicular epithelium

Active follicles

Figure 34.8
Appearance of thyroid gland follicles viewed through the microscope (approximately × 350). When inactive, follicles are distended with colloid, the storage form of thyroid hormones, and epithelial cells are flattened. When active, the colloid disappears as thyroid hormones are secreted into the circulation, and epithelial cells become greatly enlarged.

acid tyrosine, creating the two thyroid hormones. T_3 contains three iodine atoms, whereas T_4 contains four iodine atoms. T_4 is formed in much greater amounts than T_3, but in many animals T_3 is the more physiologically active hormone, and T_4 is now considered a precursor to T_3. The most important actions of T_3 and T_4 are to (1) promote normal growth and development of the nervous system of growing animals, and (2) stimulate metabolic rate.

Undersecretion of thyroid hormones in fishes, birds, and mammals dramatically impairs growth, especially of the nervous system. Thyroid malfunction from a very early age produces a human **cretin,** a mentally disabled dwarf. Conversely, oversecretion of thyroid hormones causes precocious development in all vertebrates, although its effect is particularly prominent in fishes and amphibians. In frogs and toads, transformation from an aquatic herbivorous tadpole without lungs or legs to a semiterrestrial or terrestrial carnivorous adult with lungs and four legs occurs when the thyroid gland becomes active at the end of larval development. Stimulated by rising thyroid hormone levels of the blood, metamorphosis and climax occur (Figure 34.9). Growth of frogs after metamorphosis is directed by growth hormone. Permanently nontransforming salamanders, such as axolotl, result from defects in thyroid hormone production or in receptors (see Chapter 6, p. 113 and Chapter 25, pp. 550–551).

Figure 34.9 Effect of thyroid hormones (T_3 and T_4) on growth and metamorphosis of a frog. The release of TRH from the hypothalamus at the end of premetamorphosis sets in motion hormonal changes (increased TSH, T_3, and T_4) leading to metamorphosis. Thyroid-hormone levels are maximal at the time forelimbs emerge.

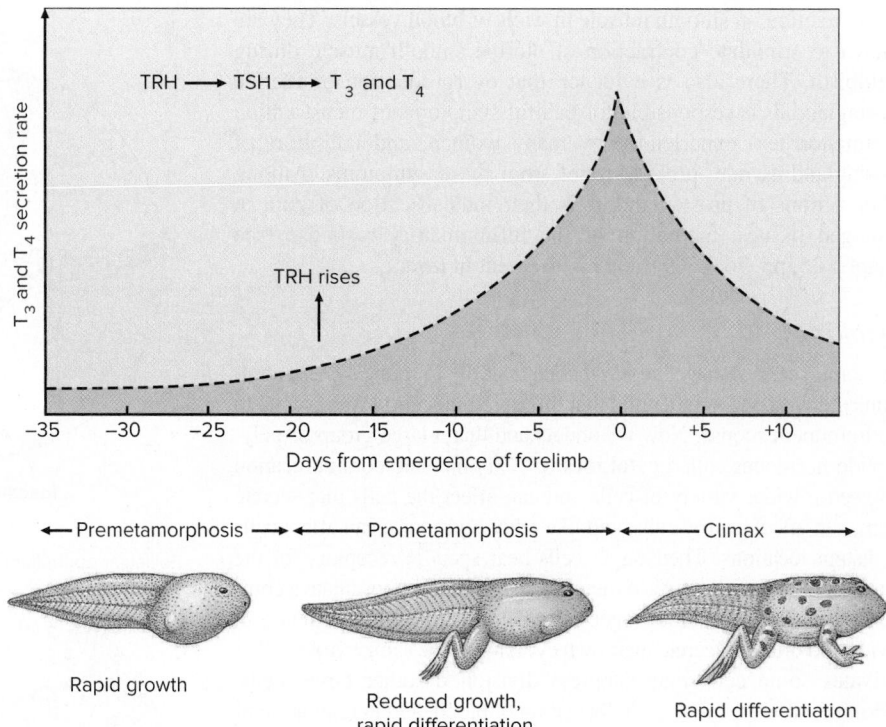

Rapid growth

Reduced growth, rapid differentiation

Rapid differentiation

In birds and mammals, the best-known action of thyroid hormones is control of oxygen consumption and heat production. The thyroid maintains metabolic activity of homeotherms (birds and mammals) at a normal level. Oversecretion of thyroid hormones accelerates body processes by up to 50%, causing irritability, nervousness, fast heart rate, intolerance of warm environments, and loss of body weight despite increased appetite. Undersecretion of thyroid hormones slows metabolic activities, causing loss of mental alertness, slowing of heart rate, muscular weakness, increased sensitivity to cold, and weight gain. An important function of the thyroid gland is to promote adaptation to cold environments by increasing heat production. Thyroid hormones stimulate cells to produce more heat and store less chemical energy (ATP); in other words, thyroid hormones *reduce* efficiency of cellular oxidative phosphorylation (p. 64). Consequently, many cold-adapted mammals have heartier appetites and eat more food in winter than in summer, although their activity is about the same in both seasons. In winter, a larger portion of the food is being converted directly into body-warming heat.

Synthesis and release of thyroid hormones are governed by thyrotropic hormone (TSH) from the anterior pituitary gland (Table 34.1). TSH is regulated in turn by thyrotropin-releasing hormone (TRH) of the hypothalamus. As noted on page 751, TRH is part of a higher regulatory council that controls the tropic hormones of the anterior pituitary. TRH and TSH control thyroid activity in an excellent example of negative feedback (p. 749). It can be overridden, however, by neural stimuli, such as exposure to cold, which directly stimulates increased release of TRH and thus TSH.

In the 1930s, a condition called goiter was common among people living in the Great Lakes region of the United States and Canada, as well as some other parts of the world, such as the Swiss Alps. This type of goiter is an enlargement of the thyroid gland caused by deficiency of iodine in the diet. Thus, TSH levels rise because of a decrease in thyroid-hormone negative feedback. Overstimulation of the thyroid gland by TSH to produce thyroid hormone with insufficient iodine causes the gland to hypertrophy, sometimes so much that the entire neck region becomes swollen (Figure 34.10). Goiter caused by iodine deficiency seldom occurs in North America because of the widespread use of iodized salt. However, it is estimated that even today 200 million people worldwide experience varying degrees of goiter, mostly in high mountains of South America, Europe, and Asia.

Hormonal Regulation of Calcium Metabolism

Closely associated with the thyroid gland and, in some animals, buried within it are **parathyroid glands.** These tiny glands occur as two pairs in humans but vary in number and position in other vertebrates. In birds and mammals, including humans, removal of the parathyroid glands rapidly decreases the level of calcium in the blood, leading to a serious increase in nervous-system excitability, severe muscular spasms, tetany, and finally death. Parathyroid glands secrete a hormone, **parathyroid hormone (PTH),** which is essential to maintenance of calcium homeostasis. Calcium ions are extremely important for formation of healthy bones. In addition, they are needed for a multitude of functions, such as neurotransmitter and hormone release, muscle contraction, intracellular signaling, and blood clotting.

Before considering how hormones maintain calcium homeostasis, it is helpful to summarize mineral metabolism in bone, a densely packed storehouse of both calcium and phosphorus (see Chapter 29, p. 645, for bone structure and function). Bone contains approximately

Figure 34.10
A large goiter caused by iodine deficiency. Overstimulation by excess TSH causes the gland to enlarge enormously, as the thyroid gland is stimulated to extract enough iodine from the blood to synthesize the body's requirement for thyroid hormones.

© Alison Wright/Science Source

Figure 34.11 How rate of secretion of parathyroid hormone (PTH) and calcitonin respond to changes in blood calcium level in a mammal.

98% of the calcium and 80% of the phosphorus in humans. Although bone is second only to teeth as the most durable material in the body (as evidenced by survival of fossil bones for millions of years), it is in a state of constant turnover in living vertebrates. Bone-building cells **(osteoblasts)** synthesize organic fibers and glycoproteins of bone matrix, which becomes mineralized with a form of calcium phosphate called hydroxyapatite. Bone-resorbing cells **(osteoclasts)** are giant, multinucleated cells that dissolve the bony matrix, releasing calcium and phosphate into the blood. These opposing activities allow bone constantly to remodel itself, especially in a growing animal, producing structural improvements to counter new mechanical stresses on the body. They additionally provide a vast and accessible reservoir of minerals that can be withdrawn as needed for general cellular requirements. One of the more prominent ideas for the evolution of bone in fishes was that it might have served as a reservoir for calcium to support heart muscle contractions in fish that colonized freshwater.

The level of calcium in the blood is maintained by three hormones that coordinate absorption, storage, and excretion of calcium ions. If blood calcium should decrease slightly, the parathyroid gland increases its secretion of PTH. This increase stimulates osteoclasts to dissolve bone adjacent to them, thus releasing calcium and phosphate into the bloodstream and returning blood calcium level to normal. PTH also decreases the rate of calcium excretion by the kidney and increases production of the hormone 1,25-dihydroxyvitamin D_3. PTH levels vary inversely with blood calcium level, as shown in Figure 34.11.

A second hormone involved in calcium metabolism in all tetrapods is derived from vitamin D_3. Vitamin D_3, like all vitamins, is a dietary requirement. But unlike other vitamins, vitamin D_3 may also be synthesized in the skin from a precursor by irradiation with ultraviolet light from the sun. Vitamin D_3 is then converted in a two-step oxidation to a hormonal form, **1,25-dihydroxyvitamin D_3.** This steroid hormone is essential for active calcium absorption by the gut (Figure 34.12). Production of 1,25-dihydroxyvitamin D_3 is stimulated by low plasma phosphate as well as by an increase in PTH secretion.

In humans, a deficiency of vitamin D_3 causes rickets, a disease characterized by low blood calcium and weak, poorly calcified bones that tend to bend under postural and gravitational stresses. Rickets has been called a disease of northern winters, when sunlight

Figure 34.12 Regulation of blood calcium in birds and mammals.

is minimal. It once was common in the smoke-darkened cities of England and continental Europe.

A third calcium-regulating hormone, **calcitonin,** is secreted by specialized cells (C cells) in the thyroid gland of mammals and in the ultimobranchial glands of other vertebrates. Calcitonin is released in response to elevated levels of calcium in the blood. It rapidly suppresses calcium withdrawal from bone, decreases intestinal absorption of calcium, and increases excretion of calcium by the kidneys. Calcitonin therefore protects the body against an increase in level of calcium in the blood, just as parathyroid hormone protects it from a decrease in blood calcium (Figure 34.12). Calcitonin has been identified in all vertebrate groups, but its importance is uncertain in humans because replacement of calcitonin is not required for

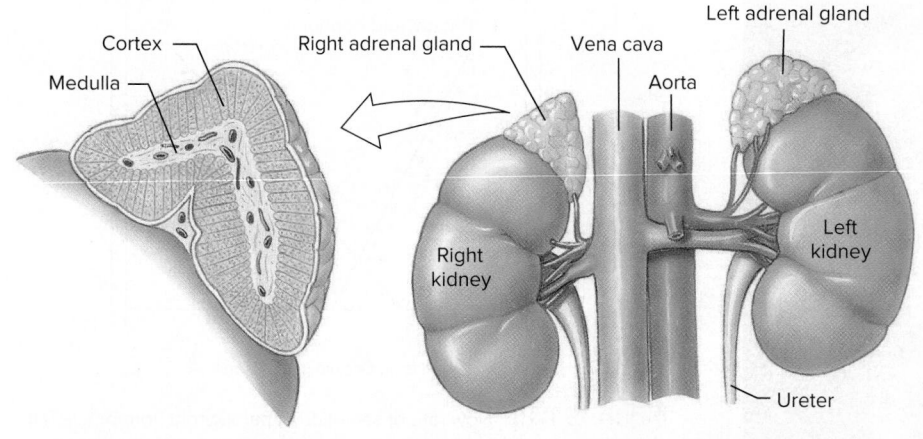

Figure 34.13 Paired adrenal glands of humans, showing gross structure and position on the upper poles of the kidneys. Steroid hormones are produced by the cortex. The sympathetic hormones, epinephrine and norepinephrine, are produced by the medulla.

Cholesterol   **Aldosterone**   **Cortisol**

Figure 34.14 Hormones of the adrenal cortex. Cortisol (a glucocorticoid) and aldosterone (a mineralocorticoid) are two of several steroid hormones synthesized from cholesterol in the adrenal cortex.

maintenance of calcium homeostasis, if the thyroid gland is surgically removed (also removing the C cells). It seems to be fairly important in fishes, because fishes do not have discrete parathyroid glands and thus use calcitonin more for calcium regulation.

Hormones of the Adrenal Cortex

The mammalian adrenal gland is a double gland composed of two unrelated types of glandular tissue: an outer region of adrenocortical cells, or **cortex,** and an inner region of specialized cells, the **medulla** (Figure 34.13). In nonmammalian vertebrates homologs of adrenocortical and medullary cells are organized quite differently; they may be intermixed or distinct, but never arranged in a cortexmedulla relationship as in mammals.

At least 30 different compounds have been isolated from adrenocortical tissue, all of them closely related steroids. Only a few of these compounds are true steroid hormones; most are various intermediates in synthesis of steroid hormones from **cholesterol** (Figure 34.14). Corticosteroid hormones are commonly classified into two groups, according to their function: glucocorticoids and mineralocorticoids.

Glucocorticoids, such as **cortisol** (Figure 34.14) and **corticosterone,** influence food metabolism, inflammation, and stress. They promote synthesis of glucose from compounds other than carbohydrates, particularly amino acids and fats. The overall effect of this process, called **gluconeogenesis,** is to increase the level of glucose in the blood, thus providing a quick energy source for muscle and nervous tissue. Glucocorticoids are also important in diminishing the immune response to various inflammatory conditions. Because several diseases of humans are inflammatory diseases (for example,

allergies, hypersensitivity, and rheumatoid arthritis), these corticosteroids have important medical applications.

Synthesis and secretion of glucocorticoids are controlled principally by ACTH of the anterior pituitary (see Figure 34.6), while ACTH is controlled in turn by corticotropin-releasing hormone (CRH) of the hypothalamus (Table 34.1). As with pituitary control of the thyroid, negative feedback exists between CRH, ACTH, and the adrenal cortex (see Figure 34.3). An increase in release of glucocorticoids suppresses output of CRH and ACTH; the resulting decline in blood level of CRH and ACTH then inhibits further release of glucocorticoids from the adrenal cortex. An opposite sequence of events happens should the blood level of glucocorticoids drop: CRH and ACTH output increases, which in turn stimulates secretion of glucocorticoids. CRH is known to mediate stressful stimuli through the adrenal axis.

Adrenal steroid hormones, especially the glucocorticoids, are remarkably effective in relieving symptoms of rheumatoid arthritis, allergies, and various disorders of connective tissue, skin, and blood. Following the report in 1948 by P. S. Hench and his colleagues at the Mayo Clinic that cortisone dramatically relieved pain and crippling effects of advanced arthritis, steroid hormones were hailed by the media as "wonder drugs." Optimism soon diminished, however, when it became apparent that severe side effects always attended long-term administration of antiinflammatory steroids. Steroid therapy lulls the adrenal cortex into inactivity and may permanently impair the body's capacity to produce its own steroids. Today steroid therapy is applied with caution, because the inflammatory response (see p. 767) is a necessary part of the body's defenses.

Mineralocorticoids, the second group of corticosteroids, are those that regulate salt balance. **Aldosterone** (Figure 34.14) is by far the most important steroid of this group. Aldosterone promotes tubular reabsorption of sodium and tubular secretion of potassium by the kidneys. Since sodium usually is in short supply in diets of many animals and potassium is in excess, mineralocorticoids play vital roles in preserving the correct balance of blood electrolytes. The salt-regulating action of aldosterone is controlled by the renin-angiotensin system, and by blood levels of potassium ions described on page 670.

Adrenocortical tissue also produces **androgens** (Gr. *andros,* man, + *genesis,* origin), which, as the name implies, are similar in effect to the male sex hormone, testosterone. Adrenal androgens promote some developmental changes, such as the growth spurt, that occur just before puberty in human males and females. Development of **anabolic steroids,** synthetic hormones related to testosterone, has led to widespread abuse of steroids among athletes.

Despite almost universal condemnation by Olympic, medical, and college sports authorities, an unscientific and clandestine program of experimentation with anabolic steroids has become popular with many amateur and professional athletes in many countries. These synthetics (and testosterone and its precursors) cause hypertrophy of skeletal muscle and improve performance that depends on strength. Unfortunately, they also have serious side effects, including testicular atrophy (and infertility), periods of irritability, abnormal liver function, and cardiovascular disease. Recent data suggest that steroid abuse among adolescents has declined since 2008. Most users are male and in 2010, self-reporting estimates of use were 0.5% in eighth-graders, 0.7% in tenth-graders, and 1.7% in twelfth-graders (http://www.drugabuse.gov/drugs-abuse/steroids-anabolic). Use among professional athletes is well documented in the media, even though steroid use is banned from many sports.

Hormones of the Adrenal Medulla

Adrenal medullary cells secrete two structurally similar hormones: **epinephrine (adrenaline)** and **norepinephrine (noradrenaline).** The adrenal medulla is derived embryologically from the same tissue that gives rise to postganglionic sympathetic neurons of the autonomic nervous system (p. 732). Norepinephrine serves as a neurotransmitter at the endings of sympathetic nerve axons. Thus functionally, as well as embryologically, the adrenal medulla can be considered a very large sympathetic ganglion.

It is not surprising, then, that adrenal medullary hormones and the sympathetic nervous system have the same general effects on the body. These effects center on responses to emergencies, such as fear and strong emotional states, flight from danger, fighting, lack of oxygen, blood loss, and exposure to pain. Walter B. Cannon, of homeostasis fame (p. 661), termed these "fight or flight" responses that are appropriate for survival. We are familiar with the increased heart rate, tightening of the stomach, dry mouth, trembling muscles, general feeling of anxiety, and increased awareness that attends sudden fright or other strong emotional states. These effects are attributable to increased activity of the sympathetic nervous system and to rapid release into the blood of epinephrine and norepinephrine from the adrenal medulla.

Activation of the adrenal medulla by the sympathetic nervous system prolongs the effects of sympathetic-system activation.

Epinephrine and norepinephrine have many other effects of which we are not as aware, including constriction of arterioles (which, together with increased heart rate, increases blood pressure), mobilization of liver glycogen and fat stores to release glucose and fatty acids for increased energy availability, increased oxygen consumption and heat production, hastening of blood coagulation, and inhibition of the gastrointestinal tract. These changes prepare the body for emergencies and are activated in stressful conditions.

Hormones from Islet Cells of the Pancreas

The pancreas is both an exocrine and an endocrine organ (Figure 34.15). The exocrine portion produces pancreatic juice, a mixture of digestive enzymes and bicarbonate ions conveyed by a duct (or ducts) to the digestive tract (see Chapter 32). Scattered within the extensive exocrine portion of the pancreas are numerous small islets of tissue, called **islets of Langerhans** (Figure 34.15 and photograph on p. 746). This endocrine portion of the pancreas is only 1% to 2% of the total weight of the organ. The islets are without ducts and secrete their hormones directly into blood vessels that extend throughout the pancreas.

Several polypeptide hormones are secreted by different cell types within the islets: **insulin** and **amylin,** produced by **beta cells; glucagon,** produced by **alpha cells; somatostatin,** produced by **delta cells;** and **pancreatic polypeptide (PP),** produced by

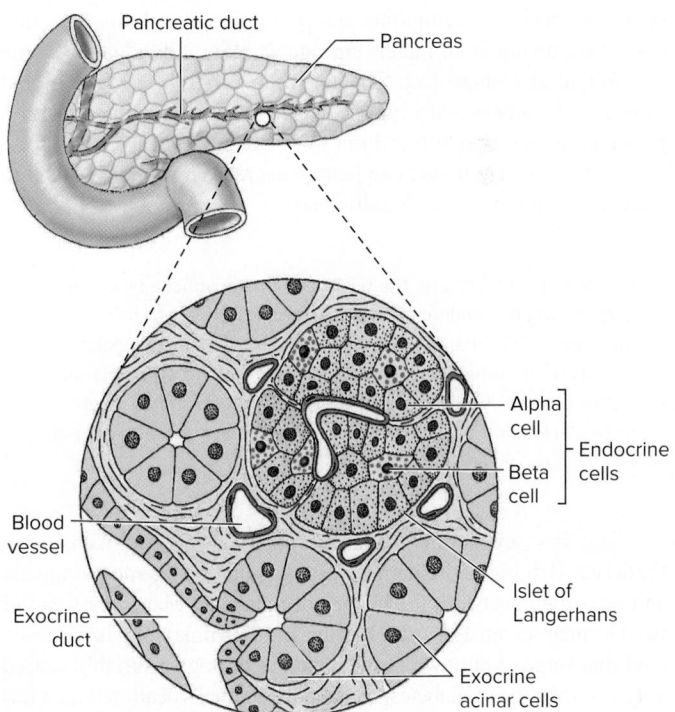

Figure 34.15 The pancreas is composed of two kinds of glandular tissue: exocrine acinar cells that secrete digestive juices that enter the intestine through the pancreatic duct, and endocrine islets of Langerhans. The islets of Langerhans secrete the hormones insulin, amylin, glucagon, somatostatin, and pancreatic polypeptide directly into the blood circulation.

gamma cells. Insulin and glucagon have antagonistic actions of great importance in metabolism of carbohydrates and fats. Meals rich in carbohydrates stimulate insulin and amylin release as blood glucose levels rise following digestion and absorption of the meal (p. 713). Insulin is essential for uptake of blood-borne glucose by cells, especially skeletal-muscle cells. Insulin promotes entry of glucose into body cells through its action on a glucose-transporter molecule found in cell membranes. Amylin's actions appear to enhance insulin's actions. Although insulin dependent glucose-transporter molecules have been demonstrated on neurons of the central nervous system, neurons do not require insulin for glucose uptake. This independence from insulin is very important because unlike other cells of the body, neurons almost exclusively use glucose as an energy source. The exact role of insulin-dependent glucose transporters in the brain is not clear, but insulin is important in central regulation of food intake and body weight. Cells of the rest of the body, however, require insulin to use glucose; without insulin the level of glucose in the blood rises to abnormally high levels, a condition called hyperglycemia. When this level exceeds the transport maximum of the kidney (see note on p. 670), sugar (glucose) appears in urine. Insulin deficiency also inhibits uptake of amino acids by skeletal muscle, and fats and proteins in muscle are metabolized to provide energy. Body cells starve while the urine abounds in the very substance the body craves. The disease of insulin deficiency, called **diabetes mellitus type 1,** afflicts nearly 5% of humans in varying degrees of severity. If left untreated, it can lead to severe damage to kidneys, eyes, and blood vessels, and it can greatly shorten life expectancy. Humans can also develop insulin insensitivity or **diabetes mellitus type 2,** with similar symptoms to type 1 diabetes mellitus. This disease is occurring at an increasing rate as more individuals become overweight and obese (see Chapter 32, p. 714). In 2000, a global estimate of humans with type 2 diabetes was 171 million, with a projected increase to 366 million by 2030. An increase in exercise levels and a change in diet can help to decrease insulin levels and to alleviate symptoms in such individuals.

> In 1982, insulin became the first hormone produced by genetic engineering (recombinant DNA technology, p. 93) to be marketed for human use. Recombinant insulin has the exact structure of human insulin and therefore will not stimulate an immune response (see Chapter 35, p. 769), which has often been a problem for diabetics receiving insulin purified from pig or cow pancreas.

The first extraction of insulin in 1921 by two Canadians, Frederick Banting and Charles Best, was one of the most dramatic and important events in the history of medicine. Many years earlier two German scientists, J. Von Mering and O. Minkowski, had discovered that surgical removal of the pancreas of dogs invariably caused severe symptoms of diabetes, causing the animal's death within a few weeks. Many attempts were made to isolate the diabetes preventive factor, but all failed because powerful protein-splitting digestive enzymes in the exocrine portion of the pancreas destroyed the hormone during extraction procedures. Following a hunch, Banting, in collaboration with Best and his physiology professor J. J. R. Macleod, tied shut the pancreatic ducts of several dogs. This caused the exocrine portion of the gland with its hormone-destroying enzyme to

Figure 34.16   Charles H. Best and Sir Frederick Banting in 1921 with the first dog to be kept alive by insulin.

degenerate, but it left the islets' tissues healthy long enough for Banting and Best to extract insulin successfully from these glands. Injected into another dog, the insulin immediately lowered the level of sugar in the blood (Figure 34.16). Their experiment paved the way for commercial extraction of insulin from slaughterhouse animals. It meant that millions of people with diabetes, previously fated to disability or death, could look forward to more normal lives.

Glucagon, another hormone of the pancreas, has several effects on carbohydrate and fat metabolism that are opposite to the effects of insulin. Low blood glucose levels and absorption of amino acids into the blood following digestion (p. 714) stimulate glucagon secretion. For example, glucagon raises the blood glucose level (by converting liver glycogen to glucose), whereas insulin lowers blood glucose. Glucagon and insulin do not have the same effects in all vertebrates, and in some, glucagon is lacking altogether.

Somatostatin, from pancreatic delta cells, inhibits secretion of other pancreatic hormones, reduces the rate of gastric emptying, and inhibits pancreatic exocrine secretion. Somatostatin is also secreted from the hypothalamus (here termed growth hormone release–inhibiting hormone) and inhibits growth hormone release from the anterior pituitary (see p. 752).

Pancreatic hormone, PP, is released after a meal and reduces appetite. It appears to have a physiological role in regulating gastric secretions, and it decreases secretions from the pancreatic exocrine gland and gallbladder, and inhibits intestinal motility. But because its administration to mice and humans reduces food intake, it has become a focus of research in the fight against obesity.

Growth Hormone and Metabolism

Growth hormone (GH) is a particularly important metabolic hormone in young animals during growth and development. It acts directly on long bones to promote cartilaginous growth and bone formation by cell division and protein synthesis, thus producing an increase in length and density of bone. GH also acts indirectly on growth via stimulation of **insulin-like growth factor (IGF)** or somatomedin release from the liver. This polypeptide hormone promotes mobilization of glycogen from liver stores and release of fat from adipose tissue stores necessary for growth processes. Thus, GH is considered a **diabetogenic hormone,** since oversecretion leads to an increase in blood glucose and can cause insulin insensitivity or diabetes mellitus type 2. If produced in excess, GH causes giantism. A deficiency of this hormone in a human child leads to dwarfism.

White Adipose Tissue as an Endocrine Organ

The discovery in 1994 of the *ob* gene, which codes for the hormone **leptin** (Gr. *leptos,* thin), produced by adipose tissue, began a period of intense research on adipose tissue as a possible regulator of human obesity (p. 714). So far many adipose-derived hormones with autocrine, paracrine, and endocrine functions have been described.

Leptin is an important hormone that regulates eating behavior and long-term energy balance as part of a feedback system that informs the brain, particularly the hypothalamus and brain stem, of the energy status of the periphery. Evidence suggests that leptin is more important during times of lower food and energy availability, since decreased fat stores secrete less leptin. At such times, the brain responds by directing available energy away from nonessential processes, such as reproduction, and stimulates increased foraging and feeding behaviors. Blood plasma levels of leptin mirror those of insulin, which also provides an important feedback signal to the brain regarding adipose tissue stores.

Of the long list of recently discovered adipose-derived hormones, there are several, in addition to leptin, that appear to be involved in the regulation of energy balance. **Adiponectin** tends to lower blood glucose levels by increasing the effects of insulin on liver and skeletal muscle. Several studies also suggest that low blood adiponectin levels are linked with obesity and type 2 diabetes incidence. In addition, high blood adiponectin levels are associated with a decreased risk of coronary artery disease, which is due in part to adiponectin decreasing the likelihood of fatty deposits of cholesterol in the artery walls (p. 689). **Tumor necrosis factor-α (TNF-α),** as well as other cytokines, are secreted by adipose tissue (termed adipokines), and high levels are associated with obesity-related insulin insensitivity (p. 714).

Skeletal Muscle as an Endocrine Tissue

Skeletal muscle has most recently been added to the ever-growing number of body tissues that secrete hormones. The newly discovered hormone, **irisin,** is secreted when skeletal muscle is active during exercise. Irisin appears to cause an increase in energy expenditure without increasing food intake, as well as improved glucose-regulatory mechanisms. In addition, irisin converts white adipose tissue, which stores energy, into brown adipose tissue, which releases energy as heat (p. 714). These effects have thus far been shown in mice and humans. It appears that the beneficial effects of exercise on the body may be mediated in part by irisin.

S U M M A R Y

Hormones are chemical messengers synthesized by special endocrine and other cells and transported by blood or other body fluid to target cells where they affect cell function by altering specific biochemical processes. Specificity of response is ensured by the presence on or in target cells of protein receptors that bind only selected hormones. Hormone effects are vastly amplified in target cells by acting through one of two basic mechanisms. Many hormones, including epinephrine, glucagon, vasopressin, and some hormones of the anterior pituitary, cause production of a "second messenger," such as cyclic AMP, that relays the hormone's message from a surface receptor to a cell's biochemical machinery. Steroid hormones and thyroid hormones operate through mainly cytoplasmic or nuclear receptors. A hormone-receptor complex is formed that alters protein synthesis by stimulating or inhibiting gene transcription.

Most invertebrate hormones are products of neurosecretory cells, although endocrine cells occur in cnidarians, nematodes, and annelids, and endocrine glands occur in molluscs and arthropods. Peptide, neuropeptide, steroid, and terpenoid hormones regulate many physiological processes. The best understood invertebrate endocrine system is that controlling molting and metamorphosis in insects. A juvenile insect grows by passing through a series of molts under control of two hormones, one (ecdysone) favoring molting to an adult and the other (juvenile hormone) favoring retention of juvenile characteristics. Ecdysone is controlled by a neurosecretory hormone (PTTH) from the brain. Juvenile hormone, ecdysone, and PTTH all play an important role in the regulation of diapause (arrested development) that can occur at any stage of metamorphosis, as well as in the adult (dormancy).

The vertebrate endocrine system is orchestrated by the hypothalamus. Release of all anterior pituitary hormones is primarily regulated by hypothalamic neurosecretory products called releasing hormones (or release-inhibiting hormones). The hypothalamus also produces two neurosecretory hormones, which are stored in and released from the posterior lobe of the pituitary. In mammals these two hormones are oxytocin, which stimulates milk production and uterine contractions during parturition; and vasopressin (antidiuretic hormone), which acts on the kidney to restrict urine production, causes vasoconstriction of blood vessels, and increases thirst. In amphibians, nonavian reptiles, and birds, vasotocin replaces vasopressin as the water-balance hormone.

The anterior lobe of the pituitary produces seven well-characterized hormones. Four of these are tropic hormones that regulate subservient endocrine glands: thyrotropic hormone (TSH), which controls secretion of thyroid hormones; adrenocorticotropic hormone (ACTH), which stimulates release of steroid hormones by the adrenal cortex, primarily the glucocorticoids, cortisol or corticosterone; and follicle-stimulating hormone (FSH) and luteinizing hormone (LH), which act on ovaries and testes. Three direct-acting hormones are (1) prolactin, which plays several diverse roles, including stimulation of milk production during lactation; (2) growth hormone, which governs body growth and metabolism; and (3) melanocyte-stimulating hormone (MSH), which controls melanocyte dispersion in ectothermic vertebrates.

The pineal gland, a derivative of the pineal complex of the diencephalon of the brain, produces the hormone melatonin. In many vertebrates, melatonin, which is released in response to darkness, maintains circadian rhythms. In birds and mammals that are seasonal breeders, melatonin level provides information regarding day length, and thereby indirectly regulates seasonal reproductive activity.

Application of ultrasensitive radioimmunochemical techniques has revealed many neuropeptides in the brain, several of which behave as neurotransmitters in the brain but as hormones elsewhere in the body. The classical definition of a hormone has been modified to include other chemical messengers, such as prostaglandins and cytokines, which originate in sources other than clearly defined endocrine glands.

Several hormones play important roles in regulating cellular metabolic activities. Two thyroid hormones, triiodothyronine (T_3) and thyroxine (T_4), control growth, development of the nervous system, and cellular metabolism. Calcium metabolism is regulated principally by three hormones: parathyroid hormone from the parathyroid glands; a hormonal derivative of vitamin D_3, 1,25-dihydroxyvitamin D_3; and calcitonin from the C cells of the thyroid gland or ultimobranchial glands. Parathyroid hormone and 1,25- dihydroxyvitamin D_3 increase plasma calcium levels; calcitonin decreases plasma calcium levels.

The principal steroid hormones of the adrenal cortex are glucocorticoids, which stimulate formation of glucose from nonglucose sources (gluconeogenesis), and mineralocorticoids, which regulate blood electrolyte balance. The adrenal medulla is the source of epinephrine and norepinephrine, which have many effects, including assisting the sympathetic nervous system in emergency responses. They also increase energy substrates in the blood for use in emergency situations.

Glucose metabolism is regulated by the antagonistic action of three pancreatic hormones. Insulin is needed for cellular use of blood glucose; it also increases storage of fats in adipose tissue and uptake of amino acids in muscle. Amylin augments insulin actions. Glucagon opposes the action of insulin and amylin. Pancreatic somatostatin inhibits pancreatic exocrine and endocrine secretion and inhibits gastric emptying. Pancreatic polypeptide is released after meals and reduces appetite.

White adipose tissue is now considered an endocrine organ and secretes many peptides. Leptin feeds back to the hypothalamus to modulate food intake and long-term energy balance. Adiponectin tends to lower blood glucose levels and decreases the risk of cardiovascular disease, while tumor necrosis factor-α appears to increase the risk of obesity-related insulin insensitivity.

Skeletal muscle has recently been shown to secrete the hormone irisin, which appears to regulate energy metabolism in a beneficial way.

REVIEW QUESTIONS

1. Outline Berthold's first endocrine experiment. What was the hypothesis?
2. Provide definitions for the following: hormone, endocrine gland, exocrine gland, hormone receptor molecule.
3. Hormone receptor molecules are the key to understanding specificity of hormone action on target cells. Describe and distinguish between receptors located on the cell surface and those located in the nucleus of target cells. Name two hormones whose action is mediated through each type of receptor.
4. What is the importance of feedback systems in the control of hormonal output? Offer an example of a hormonal feedback pattern.
5. Give two examples of invertebrate hormones that regulate metabolism.
6. Explain how the three hormones involved in insect growth—ecdysone, juvenile hormone, and PTTH—interact in molting and metamorphosis.
7. Name seven hormones produced by the anterior pituitary gland. Why are four of these seven hormones called "tropic hormones"? Explain how neurosecretory cells in the hypothalamus control secretion of hormones by the anterior pituitary gland.
8. Describe the chemical nature and function of two posterior-lobe hormones, oxytocin and vasopressin. What is distinctive about the way these neurosecretory hormones are secreted compared with the neurosecretory release and release-inhibiting hormones that control the anterior pituitary hormones?
9. What is the evolutionary origin of the pineal gland of birds and mammals? Explain the role of the pineal hormone, melatonin, in

regulating seasonal reproductive rhythms in birds and mammals. Does melatonin have any function in humans?
10. What are endorphins and enkephalins? What are prostaglandins?
11. What are some functions of hormones called cytokines?
12. What are the two most important functions of the thyroid hormones?
13. Explain how you would interpret the graph in Figure 34.11 to show that PTH and calcitonin act in a complementary way to control blood calcium level.
14. Describe the principal functions of the two major groups of adrenocortical steroids, the glucocorticoids, and the mineralocorticoids. To what extent do these names provide clues to their function?
15. Where are the hormones epinephrine and norepinephrine produced and what is their relationship to the sympathetic nervous system and its response to emergencies?
16. Explain the actions of two hormones of the islets of Langerhans on the level of glucose in the blood. What is the consequence of insulin insufficiency or insensitivity as in the disease diabetes mellitus?
17. What is the function of the hormone leptin? Why has its discovery proved important in the area of regulation of feeding?
18. Why is adipose tissue considered to be an endocrine organ?

For Further Thought Endocrine regulation of the body is generally significantly slower than neural regulatory mechanisms. Consider the benefits produced by such slow regulation that might have promoted development of this system.

SELECTED REFERENCES

Bostrom, P., J. Wu, M. P. Jedrychowski, A. Korde, L. Ye, J. C. Lo, K. A. Rasbach, E. A. Bostrom, J. H. Choi, J. Z. Long, S. Kajimura, M. C. Zingaretti, B. F. Vind, H. Tu, S. Cinti, K. Hojlund, S. P. Gygi, and B. S. Spiegelman. 2012. A PGC1-alpha-dependent myokine that drives brown-fat-like development of white fat and thermogenesis. Nature **481:** 463–469. *This research article outlines original research on the newly discovered skeletal muscle hormone, irisin.*

Cornelis, J. P., Grimmelikhuijzen, and F. Hauser. 2012. Mini-review: the evolution of neuropeptide signaling. Regulatory Peptides **177:** S6–S9. *This short article looks at the evolution of G protein-coupled receptors in neural and endocrine signaling pathways from Cnidaria to vertebrates.*

Dockray, G. J. 2004. The expanding family of –RFamide related peptides and their effects on feeding behavior. Exp. Physiol. **89:** 229–235. *This review describes an ever-expanding family of neuropeptides (FaRPs) that appear to regulate a diverse number of endocrine functions in invertebrates and vertebrates.*

Hadley, M. E., and J. E. Levine. 2007. Endocrinology, ed. 6. Englewood Cliffs, New Jersey, Prentice-Hall, Inc. *Undergraduate-level textbook in vertebrate endocrinology.*

Kahn, S. E., R. L. Hull, and K. M. Utzschneider. 2006. Mechanisms linking obesity to insulin resistance and type 2 diabetes. Nature **444:** 840–846. *This review discusses the endocrine mechanisms that can lead to type 2 diabetes in obese humans.*

Kovacs, W. J., and S. R. Ojeda. 2011. Textbook of endocrine physiology, ed. 6. Oxford, Oxford University Press. *Excellent upper-undergraduate and graduate text of human endocrinology.*

Krajniak, K. G. 2005. Annelid endocrine disruptors and a survey of invertebrate FMRFamide-related peptides. Integr. Comp. Biol. **45:** 88–96. *A brief overview of functions of FMRFamide-related peptides (FaRPs) in invertebrates can be found in the second part of this review.*

LaFont, R. 2000. The endocrinology of invertebrates. Ecotoxicology **9:** 41–57. *A well-written review of invertebrate endocrinology compared with vertebrate endocrine function.*

Oehlmann, J., and U. Schulte-Oehlmann. 2003. Endocrine disruption in invertebrates. Pure Appl. Chem. **75:** 2207–2218. *Provides an overview of* *invertebrate hormones and the effects of endocrine disruptors in invertebrate groups.*

Phillips, M. L. 2009. Of owls, larks and alarm clocks. Nature **458:** 142–144. *This news feature highlights the progress of research to determine the role of circadian rhythms in contributing to human disease.*

Ronti, T., G. Lupattelli, and E. Mannarino. 2006. The endocrine function of adipose tissue: an update. Clin. Endocrinol. **64:** 355–365. *This review discusses evidence for white adipose tissue as an endocrine organ.*

Rosen, E. D., and B. M. Spiegelman. 2006. Nature **444:** 847–853. *Provides a discussion of the biology of white adipose tissue and its role in endocrine regulation of energy balance.*

Tarrant, A. M. 2005. Endocrine-like signaling in cnidarians: current understanding and implications for ecophysiology. Integr. Comp. Biol. **45:** 201–214. *Reviews endocrine function in cnidarians in comparison to other invertebrate and vertebrate endocrinology.*

Van Gaal, L. F., I. L. Mertens, and C. E. De Block. 2006. Mechanisms linking obesity with cardiovascular disease. Nature **444:** 875–880. *Provides an excellent summary of the mechanisms by which obesity can lead to an elevated risk of cardiovascular disease.*

Woods, S. C., P. A. Rushing, and R. J. Seeley. 2001. Neuropeptides and the control of energy homeostasis. Nutrition and Brain **5:** 93–115. *A readable discussion of food intake regulation and body weight control, incorporating our current hypotheses regarding leptin.*

35
CHAPTER

Immunity

A leukocyte (orange) is attaching to and engulfing Bacillus *cells (blue, rod-shaped). It will use enzymes to digest the bacteria. The whole process is called phagocytosis.*

© Dr. Kari Lounatmaa/Science Source

The Language of Cells in Immunity

Cells of the immune system communicate with each other by mechanisms similar to those used by a peptide hormone (see Chapter 34). Target cells have receptors in their plasma membrane that specifically bind the signal molecules and only those molecules. Binding of a signal causes changes in the receptor molecule (or in an associated membrane protein), and this initiates a cascade of events involving protein kinases and phosphorylases (enzymes that transfer phosphate groups). As a result, transcription factors are mobilized, and these initiate transcription of formerly inactive genes in the nucleus, leading to synthesis of proteins needed for an immune response (see Chapter 5).

Signaling molecules of the immune system include *cytokines*. Cytokines and their receptors orchestrate cellular activation and regu-

lation, causing some cells to proliferate, suppressing proliferation of others, and stimulating secretion of additional cytokines or other defense molecules. Precise signaling among cells and exact performance of their duties are essential to maintenance of health and defense against invading viruses, bacteria, and parasites and for prevention of unrestrained cell division, as in cancer. Successful establishment of invaders in our bodies depends on evasion or subversion of our immune system, and inappropriate response of immune cells may itself produce disease. We have learned to manipulate the immune response so that we can transplant organs between individuals, but progressive failure in immune cell communication causes profound disease, such as AIDS.

The immune system is distributed throughout an animals body, and it is as crucial to survival as any other body system. Every animals environment is filled with parasites and potential parasites: flatworms, nematodes, arthropods, single-celled eukaryotes, bacteria, and viruses. Whether a parasite can infect an animal, the **host,** and the resulting severity of disease depends largely on the hosts defense system.

35.1 SUSCEPTIBILITY AND RESISTANCE

A host is **susceptible** to a parasite if the host cannot eliminate a parasite before the parasite becomes established. The host is **resistant** if its physiological status prevents establishment and survival of the parasite. Corresponding terms from the viewpoint of a parasite would be **infective** and **noninfective.**

These terms denote only the success or failure of infection, not the underlying mechanisms. Mechanisms that increase resistance (and correspondingly reduce susceptibility and infectivity) may involve either attributes of a host not related to active defense mechanisms or specific defense mechanisms mounted by a host in response to a foreign invader. It is important to remember that these terms are relative, not absolute; for example, an organism may be more or less resistant than another, and its resistance can vary, depending on age, health, and environmental exposure.

The term **immunity** is often used as a synonym for resistance. A more precise statement is that *an animal demonstrates immunity if it possesses cells or tissues capable of recognizing and protecting the animal against nonself invaders.* Most animals show some degree of **innate immunity,** defense that does not depend on prior exposure to the invader. In addition to having innate immunity, animals may develop **acquired immunity,** which is specific to a particular nonself material, requires time for its development, and occurs more quickly and vigorously on subsequent exposure.

Frequently, resistance conferred by immune mechanisms is not complete. In some instances a host may recover clinically and be resistant to a specific challenge, but some parasites may remain and reproduce slowly, as in toxoplasmosis (p. 235), Chagas disease (p. 229), and malaria (p. 237). This condition is called **premunition.**

35.2 INNATE DEFENSE MECHANISMS

Innate defense mechanisms fall into several categories: (1) physical barriers and chemical defenses, (2) cellular defenses, such as phagocytosis, and (3) the inflammatory response. We discuss these categories in turn.

Physical Barriers and Chemical Defenses

The unbroken surface of most animals provides a barrier to invading organisms. It may be tough and cornified, as in many terrestrial vertebrates, or sclerotized, as in arthropods (see Figure 29.1, p. 641). Soft outer surfaces are usually protected by a layer of mucus, which lubricates the surface and helps to dislodge particles or potential parasites from it.

A variety of antimicrobial substances occur in secretions of animals. In insects, pathogens such as parasitic wasp eggs induce release of chemicals that form melanin to encapsulate the eggs. Chemical defenses present in many vertebrates include a low pH in the stomach and vagina and hydrolytic enzymes in secretions of the alimentary tract. Mucus is produced by mucous membranes lining the digestive and respiratory tracts of vertebrates and contains parasiticidal substances such as **IgA** and **lysozyme.** IgA is a class of antibody (p. 770) that can cross cellular barriers easily and is an important protective agent in mucus of the intestinal epithelium. It is secreted onto the surface of cells lining the alimentary canal in response to invasion by specific bacteria. Thus, it is part of an animals acquired immune response (p. 769). IgA is present also in saliva and sweat. Lysozyme is an enzyme that attacks the cell wall of many bacteria.

Various cells, including those involved in the acquired immune response, liberate protective compounds. A family of low-molecular-weight glycoproteins, the **interferons,** are released by a variety of eukaryotic cells in response to invasion by intracellular parasites (including viruses) and other stimuli. **Tumor necrosis factor (TNF)** is a member of the protein-signaling molecule family called **cytokines** (see Table 35.3) and is produced by many cell types, including **macrophages** (p. 766), some **T lymphocytes** (p. 767), and white adipose tissue (p. 714). TNF is a major mediator of inflammation (p. 767) and in sufficient concentration causes **fever.** Fever in mammals is one of the most common symptoms of infection. The protective role of fever, if any, remains unclear, but high body temperature may destabilize certain viruses and bacteria.

The intestine of most animals harbors a population of bacteria that seem not to be harmed by host defenses, and inhibit establishment of pathogenic microbes. Data suggest that these bacteria are maintained within the intestine by cytokine secretions from innate lymphoid cells (p. 767). This gut microbiome is being intensely studied in humans and animal models as it appears to be altered by obesity. These changes may even contribute to the progression of obesity and associated disorders.

Substances in normal human milk can kill intestinal parasites such as *Giardia lamblia* and *Entamoeba histolytica* (see Chapter 11), and these substances may be important in protecting infants against these and other infections. Antimicrobial elements in human breast milk include lysozyme, defensins (see Cellular Defenses: Phagocytosis section), IgA, IgG (another class of antibody), interferons, and leukocytes (white blood cells; see also Chapter 31).

Some species of mammals are susceptible to infection with parasites such as *Schistosoma mansoni* (see Chapter 14), and others are partially or completely resistant. Macrophages (p. 766) of more resistant species (rats, guinea pigs, rabbits) kill schistosome juveniles, but macrophages of susceptible species do not.

Complement is a series of proteins that are activated in sequence as a host response to invading organisms. Activation of complement by the **classical pathway** (so called only because it was discovered first) depends on antibody (p. 770) bound to the surface of the invading organism and so is an effector mechanism in the acquired immune response (p. 769). Complement activated by the **alternative pathway** is an important innate defense against invasion by bacteria and some fungi. Activation of this pathway is by interaction of complement proteins produced early in the cascade sequence with polysaccharides in the outer coating of the microorganism. Both pathways rely on activation of the third component in the complement cascade (C3), and from this point on both pathways are the same. Active C3 initiates a cascade that ultimately lyses the invading cell. The hosts own cells are not lysed because regulatory proteins rapidly inactivate the first active component of complement when it binds host cells but not foreign cells. The active C3 component also binds invading target cells, effectively tagging them for **phagocytosis** (p. 48). The process of tagging pathogens for subsequent phagocytosis is called **opsonization;** the tagging molecules are called **opsonins.** Finally, active C3 attracts lymphocytes (p. 767) to the site of infection and enhances inflammation (p. 767). Complement-like proteins called Teps (thioester-containing proteins) have been discovered in insects and appear to function in a way similar to the alternative pathway of the complement system.

Insects tend to be resistant to infection by many microbial pathogens. In the 1980s experiments showed that inoculation of moth larvae with bacteria caused a release of a barrage of antimicrobial agents that killed the bacteria, even without prior exposure to the invader. Since that time, hundreds of **antimicrobial peptides** have been described from a broad spectrum of animals, invertebrates and vertebrates. They are especially important at surfaces where organism meets environment, such as skin or mucous membranes. For example, epithelial glands in the frog skin secrete high concentrations of antimicrobial peptides at sites of irritation or injury. Antimicrobial peptides do not have such high specificity as does the acquired immune response of vertebrates, but rather each peptide is effective against a different category of microbe, for example, gram-positive bacteria (bacteria that stain with Gram stain), gram-negative bacteria, and fungi. Release of the peptides is immediate upon contact with the foreign organism and is not subject to prior immunizing experience with the microbe. Conventional antibiotics usually work by blocking a critical protein in an invading microbe, but these peptides interfere with internal signaling of a microbe or perforate its surface with holes.

Antimicrobial peptides in mammals have been called **defensins.** They do not harm cells of the organism from which they originate. Defensins are secreted by macrophages, neutrophils, eosinophils (Table 35.2), and epithelial cells lining intestinal, respiratory, and urogenital tracts. Secretions occur in response to stimulation by molecules on the surface of microbes or, in some cases, to their metabolic products. Such molecules are conserved over a range of microbes but do not occur on host cells. Defensins may be chemoattractants to neutrophils and monocytes, or they may enhance the inflammatory response (p. 767) or the acquired immune response (p. 769). Several neuropeptides (p. 754) and cytokines (Table 35.3) demonstrate antimicrobial activity.

Release of peptides begins when receptors on a cells surface recognize a microbial molecule. Many of these receptors are **Toll** proteins or **Toll-like receptors (TLRs),** so called because they occur in a cell membrane where they receive signals from the outside. At least nine TLRs have been described in humans, each of which recognizes a specific pattern of molecules from a class of microbes. Activation of a particular TLR signals the nucleus to synthesize a peptide against that particular microbe. New discoveries include an ever-increasing list of non-TLRs that also function as innate immunity receptor systems.

Cellular Defenses: Phagocytosis

For defense against an invader an animals cells must recognize when a substance does not belong; they must recognize *nonself* as well as

self (see section titled Basis of Self and Nonself Recognition on p. 48). Phagocytosis illustrates the process of nonself recognition, and it also serves as a process for removing senescent cells and cellular debris from the host. Phagocytosis occurs in almost all animals and is a feeding mechanism in many single-celled eukaryotes (p. 224). A cell that has this ability is a **phagocyte.** Phagocytes engulf a particle within an invagination of the phagocytes plasma membrane (see Figure 3.20, p. 49). The invagination becomes pinched off, and the particle becomes enclosed within an intracellular vesicle. Other cytoplasmic vesicles called **lysosomes** (see Chapter 3, p. 40) join with the particle-containing vesicle and provide digestive enzymes to destroy the particle. Lysosomes of many phagocytes also contain enzymes that catalyze production of cytotoxic **reactive oxygen intermediates (ROIs)** and **reactive nitrogen intermediates (RNIs).** ROIs include superoxide radical (O_2^-), hydrogen peroxide (H_2O_2), singlet oxygen ($^1O_2$), and hydroxyl radical ($OH \cdot$). RNIs include nitric oxide (NO) and its oxidized forms, nitrite (NO_2^-) and nitrate (NO_3^-). All such intermediates are potentially toxic to invasive microorganisms or parasites.

Phagocytes and Other Defense Cells

Many invertebrates have specialized cells that function as itinerant troubleshooters within the body, acting to engulf or to encapsulate foreign material (Table 35.1) and to repair wounds. Such cells are variously called amebocytes, hemocytes, or coelomocytes in different animals. If a foreign particle is small, it is engulfed by phagocytosis; but if it is larger than about 10 μm, it is usually encapsulated. Arthropods can encapsulate a foreign object by deposition of melanin around it, either from cells of the capsule or by precipitation from the hemolymph (blood).

In vertebrates several categories of cells perform phagocytosis (Table 35.2). **Monocytes** arise from stem cells in the bone marrow (Figure 35.1) and give rise to **macrophages.** These cells are members of the **mononuclear phagocyte system** (formerly called the **reticuloendothelial system),** which are phagocytic cells in the macrophage family stationed around the body. The mononuclear phagocyte system includes macrophages in connective tissue, lymph nodes, spleen, and lung; **Kupffer cells** in sinusoids of the liver; **osteoclasts** of bone; **Langerhans cells (dendritic cells)** of skin; and **microglial cells** in the central nervous system. Macrophages also have important roles in the acquired immune response of vertebrates (p. 769) because in addition to phagocytosis, they are **antigen-presenting cells (APCs).** APCs are cells that activate acquired immunity responses, and include macrophages, Kupffer cells, Langerhans cells, microglial cells, and B lymphocytes (p. 767).

Some **polymorphonuclear leukocytes (PMNs),** named for the highly variable shape of their nucleus (see Figures 31.3, p. 683, and 35.3), are circulating phagocytes in blood (Table 35.2). A more common name for these leukocytes is **granulocytes,** which alludes to the many small granules visible in their cytoplasm. Granulocytes are further subdivided into **neutrophils, eosinophils,** and **basophils;** the latter two are named according to the staining properties of their granules. Neutrophils are the most abundant (60% to 70% of the total leukocytes), and they provide the first line of phagocytic defense in an infection. Their lysosomes catalyze production of ROIs that are toxic to invading organisms. Eosinophils in normal blood account for about 2% to 5% of the total leukocytes, and basophils are the least numerous

TABLE 35.1

Some Invertebrate Leukocytes and Their Functions

Group	Cell Types and Functions	Phagocytosis	Encapsulation	Allograft Rejection	Xenograft Rejection
Sponges	Archaeocytes (wandering cells that differentiate into other cell types and can act as phagocytes)	+	+	+*	+*
Cnidarians	Amebocytes: "lymphocytes"	+		+	+
Nemertines	Agranular leukocytes; granular macrophage-like cells	+		−	±
Annelids	Basophilic amebocytes (accumulate as "brown bodies"), acidophilic granulocytes	+	+	+	+
Sipunculids	Several types	+	+	±	+
Insects	Several types, depending on family; e.g., plasmatocytes, granulocytes, spherule cells, coagulocytes (blood clotting)	+	+	−	±
Crustaceans	Granular phagocytes; refractile cells that lyse and release contents	+	+	−	+
Molluscs	Amebocytes	+	+	−	+
Echinoderms	Amebocytes, spherule cells, pigment cells, vibratile cells (blood clotting)	+	+	+	+
Tunicates	Many types, including phagocytes; "lymphocytes"	+	+	+	+

Data from Lackie, A. M. 1980. Parasitology 80:393–412. (See Lackies article for references.)
*Transplantation reactions occur, but the extent to which the leukocytes are involved is unknown.

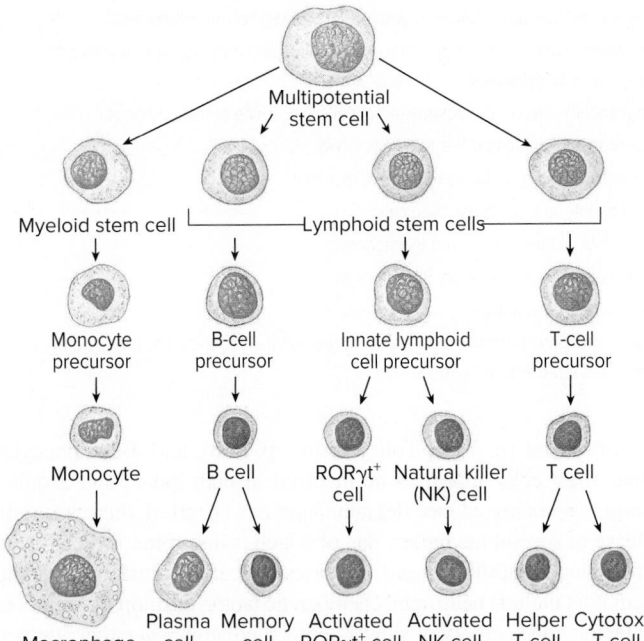

Figure 35.1 Lineage of some cells active in immune response. These cells, as well as red blood cells and other white blood cells, are derived from multipotential stem cells in the bone marrow. B cells mature in bone marrow and are released into blood or lymph. Precursors of T cells complete their development in the thymus gland. Precursors of macrophages circulate in blood as monocytes.

at about 0.5%. Eosinophils actively destroy invading parasites by releasing their granules at the site of such an infection. They also limit the inflammatory response (see Inflammation section). Basophils release their granules during an inflammatory response.

Several other kinds of cells, including basophils and eosinophils, are not important as phagocytes but are important cellular components of the defense system. **Mast cells** are basophil-like cells found in the dermis and other connective tissue (see Chapter 9, p. 191), and like basophils, they release their granules (degranulation) during inflammation (see Inflammation section). **Lymphocytes** (Figure 35.1), including **T lymphocytes (T cells)** and **B lymphocytes (B cells),** are crucial in the acquired immune response of vertebrates (Table 35.2). **Natural killer (NK) cells** are lymphocyte-like cells that belong to a class of immune cells called innate lymphoid cells (ILCs). NK cells can kill virus-infected and tumor cells. They release interferons (Table 35.3), cytokines that activate other defense cells, as well as substances that lyse the target cell. Another ILC, called an **RORγt⁺ ILC,** occurs within the intestinal lumen, on the epithelial surface, or within underlying lymphatic tissue of the intestine. These cells secrete the cytokine called **interleukin 22 (IL-22)** (Table 35.3), which restricts commensal bacteria within the intestine and stimulates antimicrobial peptide production in response to intestinal pathogen invasion.

Inflammation

Inflammation is a vital process in the innate immune response involved in mobilization of body defenses against an invading organism or other tissue damage and in repair of damage thereafter. The course of events in the inflammatory process is greatly influenced by prior exposure of the body to the invader and by the duration of the invaders presence or its persistence in the body. The innate processes by which an invader is actually destroyed, however, are themselves nonspecific but usually lead to activation of the acquired immune response (p. 769).

During an inflammatory response there is an early phase in which damaged tissue releases chemicals, such as histamine, which cause dilation of blood capillaries and increased permeability, allowing

TABLE 35.2

Cells of the Vertebrate Immune System

Innate Immune Cells	Functions
Mononuclear phagocyte system:	
Macrophages	Monocyte is precursor cell; phagocytize foreign substancesbacteria, antigen-presenting cells (APCs) in connective tissue; secrete cytokines
Kupffer cells	Phagocytize foreign substances/bacteria, APCs in liver; secrete cytokines
Microglial cells	Phagocytize foreign substances/bacteria, APCs in the central nervous system; secrete cytokines
Langerhans cells (dendritic cells)	Phagocytize foreign substances/bacteria, APCs in skin; secrete cytokines
Polymorphonuclear leukocytes (PMNs) or Granulocytes:	
Neutrophils	Phagocytize foreign substances/bacteria
Eosinophils	Defend against parasitic worms and limit the inflammatory response
Basophils	Participate in the inflammatory response and allergic reactions
Mast cells	Participate in the inflammatory response and allergic reactions
Natural Killer (NK) cells	Innate lymphoid cells (ILCs) that become lymphocyte-activated killer (LAK) cells and kill virus-infected cells and tumors by lysing them; secrete cytokines
RORγt$^+$ ILCs	ILCs residing in the intestine that limit commensal bacteria within the intestine and stimulate antimicrobial peptides when pathogens invade the intestine
Acquired Immune Cells	
T lymphocytes (T cells):	Work in cell- and humoral-mediated immunity
T-Helper lymphocytes:	
T$_H$1 cells	Work in cell-mediated immunity and activate cytotoxic T lymphocytes (CTLs); secrete cytokines
T$_H$2 cells	Work in humoral-mediated immunity and activate B lymphocytes to divide and produce plasma cells that secrete specific antibodies; secrete cytokines
Cytotoxic T lymphocytes (CTLs)	Killer cells of cell-mediated immunity; attach and destroy target cells expressing certain antigens
T-suppressor lymphocytes	Eventually inhibit immune response by inhibiting T- and B-cell activity
T-memory lymphocytes	Provide antigen memory for activation during future immune responses
B lymphocytes (B cells):	Work in humoral-mediated immunity, antigen-presenting cells
Sensitized B lymphocyte	Present antigen to T$_H$2 cells, which activate sensitized B lymphocyte
Activated B lymphocyte	Grow and differentiate into plasma cells and memory B lymphocytes
Plasma cells	Secrete antibodies specific to a particular antigen
Memory B lymphocytes	Long-lived B lymphocytes with antibody specific to a particular antigen on their surface that can multiply rapidly to produce plasma cells if the antigen invades later

leukocytes to move from capillaries to reach the damaged region. Blood plasma also leaks into the region and the area swells (and becomes red), helping to isolate the region from potential foreign invaders. Bradykinins are also released from the damaged cells and this causes a pain sensation. Leukocytes are attracted to the region by chemotactic chemicals secreted by damaged cells, degranulated basophils, and mast cells. Mast cells and basophils bear receptors for the Fc portions of IgE antibodies (p. 770). A more classical explanation for their activation involves B lymphocytes, which upon first exposure to an antigen, usually a specific foreign substance (p. 770), release specific IgEs from activated B lymphocytes (**plasma cells,** see p. 771). These IgEs bind to mast cell and basophil Fc receptors and the cells become sensitized. If subsequent exposure to the specific antigen occurs, then the antigen binds the Fc-IgE receptor complex on the mast cell or basophil membrane and elicits degranulation of the cells. More recent evidence suggests that mast cells (and basophils to a lesser extent, due to their limited numbers) are activated by a myriad of immune chemicals. They express receptors for a variety of cytokines (p. 771), complement proteins (p. 765), superoxides,

neuropeptides (p. 754), Toll proteins (p. 766), and T lymphocytes. Thus, mast cells appear to be involved in both innate and acquired immune responses. Once degranulation has occurred, there is a rapid release of several mediators, one of which is histamine, which further dilates local capillaries and increases vascular permeability. Other mediators include neutrophil chemotactic factor, eosinophil chemotactic factor, and a variety of slower-acting mediators that either augment the effects of histamine or begin to limit the inflammatory response. The first line of phagocytic defense is neutrophils, which may remain active for several days. Then macrophages, either those already present in the tissue or newly differentiated monocytes, begin phagocytosis and antigen presentation to lymphocytes, which have been attracted to the region. The local inflammatory response increases adhesion molecules for leukocytes on the surface of capillary endothelial cells (p. 690) such that circulating leukocytes adhere to these cells and leave the blood by passing between endothelial cells into the damagedinfected tissue region. Once macrophages begin to phagocytize foreign substances, they express a portion of the digested antigen (termed an epitope) on their surface, which they present to T lymphocytes—hence the term **antigen presenting cells (APCs).** Thus begins an acquired immune response (see Acquired Immune Response in Vertebrates section), which is a secondary cellular response that occurs about 24 to 72 hours after the immediate, nonspecific reaction.

Some degree of cell death (**necrosis**) always occurs in inflammation, but necrosis may not be prominent if the inflammation is minor. When necrotic debris is confined within a localized area, pus (spent leukocytes and tissue fluid) may cause an increase in hydrostatic pressure, forming an **abscess.** An area of inflammation that opens out to skin or a mucous surface is an **ulcer.**

Immediate hypersensitivity can occur when mast cells and basophils become sensitized by foreign substances, such as pollen or bee venom, and their response is exaggerated. In humans this is the basis for **allergies** and **asthma,** which are quite undesirable conditions, leading one to wonder why they evolved. Some scientists propose that the allergic response originally evolved to help the body ward off parasites, because only allergens and parasite antigens stimulate production of large quantities of IgE. Avoidance of or reduction in effects of parasites would have conferred a selective advantage in human evolution. The hypothesis is that in absence of heavy parasitic challenge, the immune system is free to react against other substances, such as ragweed pollen. People now living where parasites remain abundant are less troubled with allergies than are those living in relatively parasite-free areas. If immediate hypersensitivity becomes more widespread in the body, it is called **anaphylaxis,** which may be fatal if not treated rapidly.

35.3 IMMUNITY IN INVERTEBRATES

One principal test of the ability of tissues to recognize nonself, and by inference potential pathogens, is by grafting a piece of tissue from another individual of the same species (**allograft**) or a different species (**xenograft**) onto the host. If the graft grows in place with no host response, the host tissue is treating it as self, but if cell response and rejection of the graft occur, the host exhibits immune recognition. Most invertebrates, even simple sponges (p. 247), tested in this way reject xenografts; and almost all can reject allografts to some degree (Table 35.1). Interestingly, nemertines and molluscs apparently do not reject allografts. Even some animals with quite loose body organization, such as Porifera and Cnidaria, can reject allografts; this response may be an adaptation to avoid loss of integrity of an individual sponge or colony under conditions of crowding, with attendant danger of overgrowth or fusion with other individuals. Interestingly, sponges, cnidarians, annelids, and insects (for example, American cockroaches, *Periplaneta americana)* reject allografts from the same source more quickly upon second exposure; thus they show at least short-term immunological memory.

Hemocytes of molluscs release degradative enzymes during phagocytosis and encapsulation, and antimicrobial substances occur in body fluids of a variety of invertebrates. Substances functioning as opsonins (p. 766) occur in annelids, insects, crustaceans, echinoderms, and molluscs.

Bacterial, viral, and fungal infections in some insects stimulate production of antimicrobial peptides (p. 766), but these peptides show broad-spectrum activity and are not specific for a single infective agent. Specific, induced responses that demonstrate memory upon challenge, previously considered a hallmark of acquired immunity of vertebrates, occur in copepods and American cockroaches. Injection of prawns with a specific virus coat protein produces protection against the virus from which the coat protein was isolated. In addition, studies in water fleas (a microcrustacean) and bumble bees show that immunity can be passed from one generation to the next. Social insects, such as ants, promote resistance to fungal infections in the entire colony by heavy grooming of infected ants and subsequent transfer of the fungal pathogen to healthy colony members. Experiments suggest that low-grade infection with the fungus induces up-regulation of antimicrobial peptide secretions of a colony member as opposed to sharing of antimicrobial peptides between nestmates.

Contact with infectious organisms can bring the defense systems of snails into enhanced levels of readiness that last for up to two months or more. Susceptibility of snail hosts of the trematode *Schistosoma mansoni* (p. 301) depends heavily on genotype of the snail. Excretory/secretory products of the trematode stimulate motility of hemocytes from resistant snails but inhibit motility of hemocytes from susceptible hosts. Hemocytes from resistant snails encapsulate the trematode larva and apparently kill it with superoxide and H_2O_2 and then destroy it by phagocytosis. It appears that the cytokine interleukin-1 occurs in resistant snails and is responsible for activating hemocytes.

Evidence from studies of innate immunity in invertebrates has begun to blur the lines between acquired and innate immune systems. Although many of the mechanisms are quite different, analogous phenomena of memory and specificity of response have now been found in invertebrates. These criteria have always been used to distinguish acquired and innate immunity. One key difference that is still recognized, however, is amplification of the immune response by secondary exposures through immune-cell proliferation specific to the pathogen challenge. This phenomenon is central to acquired immune responses of vertebrates.

35.4 ACQUIRED IMMUNE RESPONSE IN VERTEBRATES

The specialized system of nonself recognition possessed by vertebrates produces increased resistance to *specific* foreign substances or invaders on repeated exposures.

The immune response is stimulated by a specific foreign substance called an **antigen,** and an antigen is any substance that can stimulate an immune response. Antigens may be any of a variety of substances with a molecular weight over 3000. They are most commonly proteins and are usually (but not always) foreign to the host. The acquired immune response has two defense mechanisms, called **humoral** and **cell-mediated.** Humoral immunity is based on **antibodies,** which are both on cell surfaces and released into blood and lymph, whereas cellular immunity is entirely associated with cell surfaces. There is extensive communication and interaction among cells of these two defense mechanisms.

The following section first considers how self and nonself are recognized, the types of recognition molecules involved, and the chemical mediators, cytokines, that provide communication between cells of humoral and cellular immunity. Next we discuss generation of a humoral response and a cell-mediated response, and the consequences of destruction of acquired immune cells in AIDS (Acquired Immune Deficiency Syndrome).

Basis of Self and Nonself Recognition

Major Histocompatibility Complex

We have known for many years that nonself recognition is very specific. If tissue from one individual is transplanted into another individual of the same species, the graft will grow for a time and then die as immunity against it rises. In the absence of drugs that modify the immune response, tissue grafts grow successfully only if they are between identical twins or between individuals of highly inbred strains of animals. The molecular basis for this nonself recognition involves a specific group of proteins embedded in the cell surface. In vertebrates these proteins are coded by certain genes, now called the **major histocompatibility complex (MHC)**. MHC proteins are among the most variable known, and unrelated individuals almost always have different genes. There are two types of MHC proteins: class I and class II. Class I proteins occur on the surfaces of virtually all cells, whereas class II MHC proteins occur only on certain cells participating in immune responses, such as lymphocytes and macrophages.

An acquired immune response develops over a period of time in the early development of an organism. All substances present at the time the capacity develops are recognized as self in later life. Unfortunately, the system of self and nonself recognition sometimes fails, and an animal may begin to produce antibodies against some part of its own body. This condition leads to one of several known *autoimmune diseases,* such as rheumatoid arthritis, multiple sclerosis, lupus, and diabetes mellitus type 1.

Recognition Molecules

MHC proteins are not themselves the molecules that recognize foreign substances. This task falls to two basic types of molecules: **antibodies** and **T-cell receptors,** the genes for which probably evolved from a common ancestor. Each vertebrate individual has an enormous variety of antibodies and T-cell receptors, *each of which binds specifically* to one particular antigen (or part of an antigen), even though that antigen may have never been present in the body previously.

Antibodies

Antibodies are proteins called **immunoglobulins.** They are borne in the plasma membrane of B lymphocytes or secreted by cells **(plasma cells)** derived from B cells. The basic antibody molecule consists of four polypeptide strands: two identical light chains and two identical, longer heavy chains held together in a Y-shape by disulfide bonds and hydrogen bonds (Figure 35.2). The amino-acid sequence toward the two top ends of the Y varies in both the heavy and the light chains, according to the specific antibody molecule (the **variable region**), and this variation determines with which antigen the antibody can bind. Each of the ends of the Y forms a cleft that acts as the antigen-binding site (Figure 35.2), and specificity of the molecule depends on the shape of the cleft and properties of the chemical groups that line its walls. The remainder of an antibody is called the **constant region.** The variable end of the antibody molecule is often called **Fab,** for **a**ntigen-**b**inding **f**ragment, and the constant end is called **Fc,** for **c**rystallizable **f**ragment (Figure 35.2). The so-called constant region is not really constant:

the light chains can be either of two types, and the heavy chains may be any of five types. The type of heavy chain determines the **class** of the antibodies: **IgM, IgG** (commonly called *gamma globulin*), **IgA, IgD,** and **IgE.** The class of the antibody determines the specific role of the antibody in the immune response (for example, whether the antibody is secreted or held on a cell surface), but not its antigen.

A major problem of immunology is understanding how the mammalian genome could contain information needed to produce at least a million different antibodies. The answer seems to be that antibody genes occur in pieces, rather than as continuous stretches of DNA, and that the antigen-recognizing sites (variable regions) of heavy and light chains of antibody molecules are pieced together from information supplied by separate DNA sequences, which can be shuffled by recombination activating genes (RAGs) to increase diversity of the gene products. The immense repertoire of antibodies is achieved in part by complex gene rearrangements and in part by frequent somatic mutations that produce additional variation in protein structure of the variable regions of the heavy and light antibody chains. Analogous processes occur in the production of genes for T-cell receptors.

Functions of Antibody in Host Defense   Antibodies can mediate destruction of an invader (antigen) in a number of ways. A foreign antigen, for example, becomes coated with antibody molecules as their Fab regions become bound to it, effectively immobilizing the antigen. Macrophages recognize the projecting Fc regions and are stimulated to phagocytize the antigen-antibody complex (opsonization). Antibodies also may be able to neutralize toxins that are secreted by an invader.

Another important process, particularly in destruction of bacterial cells, is interaction of antibodies with complement activated by the classical pathway. As noted on page 765, the first component in the classical pathway is activated by antibody bound to the surface of the invading organism. The end result in both classical and alternative complement pathways is the same, lysis of a foreign cell. Both pathways also lead to opsonization or enhancement of inflammation (p. 767). Binding of complement to antigen-antibody complexes can facilitate clearance of these potentially harmful masses by phagocytic cells.

Antibody bound to the surface of an invader may trigger contact killing of the invader by host natural killer cells in what is called **antibody-dependent, cell-mediated cytotoxicity (ADCC).** Receptors for Fc of bound antibody on a microorganism or tumor cell cause natural killer cells to adhere to them and to secrete the cytotoxic contents of their vesicles.

T-Cell Receptors

T-cell receptors are transmembrane proteins on the surfaces of T cells. Like antibodies, T-cell receptors have a constant region and a variable region. The constant region extends slightly into the cytoplasm and the variable region, which binds with specific antigens, extends outward. Most T cells also bear other transmembrane proteins closely linked to the T-cell receptors, which serve as **accessory** or **coreceptor** molecules. There are about 200 known **CD** (for

Figure 35.2 **A,** Antibody molecule is composed of two shorter polypeptide chains (light chains) and two longer chains (heavy chains) held together by hydrogen and covalent disulfide bonds. These are further subdivided into variable and constant regions that have independent folding units, or domains, of about 110 amino acids. The folding pattern is more complex than that shown here. Interchain disulfide bonds at the hinge region give the molecule flexibility at that point. The variable domains of both the light and heavy chains have hypervariable ends, which serve as the antigenbinding sites. **B,** Molecular model of antibody molecule.

clusters of **d**ifferentiation proteins) **molecules** or **markers** (see Figure 35.5), one of which, CD3, associates with the constant region of T-cell receptors. The other CD molecules bind to specific ligands on target cells.

Subsets of T Cells

Lymphocytes are **activated** when they are stimulated to move from their recognition phase, in which they simply bind with particular antigens, to a phase in which they proliferate and differentiate into cells that function to eliminate the antigens. We also speak of activation of effector cells, such as macrophages, when they are stimulated to perform their protective function.

Communication between cells in the immune response, regulation of the response, and certain effector functions are performed by different kinds of T cells (Figure 35.3 and Table 35.2). Although morphologically similar, subsets of T cells can be distinguished by characteristic proteins in their surface membranes. For example, cells with the coreceptor protein CD4 and CD28 are **T-helper cells** (or T_H). These cells secrete cytokines that modulate the activity of other types of lymphocytes and macrophages during an immune response. Some T_H cells (designated $T_H 1$) activate cell-mediated immunity against bacterial and viral attack, and others (called $T_H 2$) activate humoral immunity and antibody release (Figure 35.3 and Table 35.2).

Cytotoxic T lymphocytes (CTLs) are cells with the coreceptor protein $CD8^+$; they kill target cells expressing certain antigens. A CTL binds tightly to the target cell and secretes a protein that causes pores to form in the cell membrane, causing lysis. **T-suppressor cells** eventually suppress an immune response by inhibiting other T- and B-cell activity and **T-memory cells** provide antigen memory for activation during future immune responses.

Cytokines

The 1980s saw rapid advances in our knowledge of how cells of immunity communicate with one another. They do this by means of protein hormones called **cytokines** (Table 35.3; see also Chapter 34, p. 755). Cytokines include the interleukins (IL), interferons (IFN), and Tumor Necrosis Factor (TNF). They can act on the same cells that produce them, on cells nearby, or on cells distant in the body from their production site. Several cytokines including interferon have been shown to possess antimicrobial activity.

> Interleukins (ILs) were originally so called because they are synthesized by leukocytes and have their effect on leukocytes. We now know that other kinds of cells can produce interleukins, and interleukins produced by leukocytes can affect other kinds of cells.

Generation of a Humoral Response: $T_H 2$ Arm

When an antigen is introduced into a body, it binds to a specific antibody on the surface of an appropriate B cell, but this binding is usually not sufficient to activate the B cell to multiply. The B cell internalizes the antigen-antibody complex and then incorporates portions of the

Figure 35.3 Major pathways involved in cell-mediated (T_H1) and humoral (T_H2) immune responses as mediated by cytokines. Solid arrows indicate positive signals and broken arrows indicate inhibitory signals. Broken lines without arrows indicate path of cellular activation.

antigen into its own cell surface, bound in the cleft of MHC II protein (Figures 35.4 and 35.5). That portion of the antigen presented on the surface of the B cell (**antigen-presenting cell** or **APC**) is called the **epitope** (or **determinant**). T_H2 cells with a specific T-cell receptor for that particular epitope recognize the epitope bound to the MHC II protein. Binding of the T-cell receptor to the epitope–MHC II complex is enhanced by the coreceptor CD4, which itself binds to the constant portion of the MHC II protein (Figure 35.5). Bound CD4 also transmits a stimulation signal to the interior of the T cell. Activation of the T cell further requires interaction of additional costimulatory signals (for example, the CD40 molecule and receptor) from other proteins on the surface of the B and T cells. CD8 coreceptors function in a similar way on cytotoxic T lymphocytes (CTLs); they enhance binding of the T-cell receptor and transmit a stimulatory signal into the T cell.

Activated T_H2 cells secrete IL-4, IL-5, IL-6, and IL-10 (Table 35.3), which activate the B cell that has the same epitope and class II MHC protein on its surface (Figures 35.3 and 35.4). The B cell multiplies rapidly and many differentiate into plasma cells, which secrete large quantities of antibody. The antibody binds to antigen, and macrophages recognize this complex and are stimulated to engulf it (opsonization) (Figure 35.4). The antibody is secreted for a period of time, then the plasma cells die. Thus if we measure the concentration of antibody (**titer**) soon after the antigen is injected, we can detect little or none. Titer then rises rapidly as plasma cells secrete the antibody, and it may decrease somewhat as they die and antibody is degraded

(Figure 35.6). However, if we give another dose of antigen (**challenge**), there is little or no lag, and antibody titer rises more quickly to a higher level than after the first dose. This is the **secondary** or **anamnestic response**, and it occurs because some of the activated B cells gave rise to long-lived **memory cells.** There are many more memory cells present in the body than the original B lymphocyte with the appropriate antibody on its surface, and they rapidly multiply to produce additional plasma cells. Existence of an anamnestic response has great practical value because it is the basis for protective vaccines.

Many aspects of immunology have been greatly assisted by the discovery of a method for producing stable clones of cells that produce only one kind of antibody. Such monoclonal antibodies bind only to *one kind* of antigenic epitope (most proteins bear many different antigenic epitopes and thus stimulate the body to produce complex mixtures of antibodies). *Monoclonal* antibodies are made by fusing normal antibody-producing plasma cells with a continuously growing plasma cell line, producing a hybrid of the normal cell with one that can divide indefinitely in culture. This cell line is called a *hybridoma*. Clones are selected from among the hybrids and are grown to become "factories" that produce almost unlimited quantities of one specific antibody. Hybridoma techniques discovered in 1975 have become one of the most important research tools for producing antibodies used by immunologists.

TABLE 35.3

Some Important Cytokines

Cytokine	Principal Source	Functions
Interleukin-1 (IL-1)	Activated macrophages	Mediates inflammation; activates T cells, B cells, and macrophages
Interleukin-2 (IL-2)	T_H1 cells	Major growth factor for T and B cells; enhances cytolytic activity of natural killer cells, causing them to proliferate and become **lymphocyte-activated killer (LAK) cells**
Interleukin-3 (IL-3)	Activated T and B cells	Multilineage colony-stimulating factor; promotes growth and differentiation of all cell types in bone marrow
Interleukin-4 (IL-4)	Mostly by T_H2 cells	Growth factor for B cells, some T cells, and mast cells; promotes antibody secretion
Interleukin-5 (IL-5)	$T_H2$cells	Activates eosinophils; acts with IL-2 and IL-4 to stimulate growth and differentiation of B cells
Interleukin-6 (IL-6)	Macrophages, endothelial cells, fibroblasts, and T_H2 cells	Important growth factor for B cells late in their differentiation; activates CTLs
Interleukin-8 (IL-8)	Antigen-activated T cells, macrophages, endothelial cells, fibroblasts, and platelets	Activating and chemotactic factor for neutrophils, and to a lesser extent, other PMNs
Interleukin-10 (IL-10)	T_H2 cells	Inhibits T_H1, NK, and macrophage cytokine synthesis; promotes B-cell proliferation
Interleukin-12 (IL-12)	Macrophages and B cells	Activates NK cells and T cells; potently induces production of IFN-; shifts immune response to T_H1
Interleukin-22 (IL-22)	RORγt$^+$ ILCs	Contains commensal bacteria within the intestine and increases the production of microbial peptides in response to intestinal pathogen invasions
Transforming growth factor-β (TGF-β)	Macrophages, lymphocytes, and other cells	Inhibits lymphocyte proliferation, CTL and LAK cell generation, and macrophage cytokine production
Interferon-α (IFN-α)	Body cells under viral attack	Activates NK cells and macrophages
Interferon-β (IFN-β)	Body cells under viral attack	Activates NK cells and macrophages
Interferon-γ (IFN-γ)	T_H1 cells and LAK cells	Strong macrophage-activating factor; causes a variety of cells to express class II MHC molecules; promotes T- and B-cell differentiation; activates neutrophils; activates endothelial cells to allow lymphocytes to pass through walls of vessels; antimicrobial activity
Tumor necrosis factor (TNF)	Activated macrophages and T_H1 cells	Major mediator of inflammation; low concentrations activate endothelial cells, activate PMNs, stimulate macrophages and cytokine production (including IL-1, IL-6, IL-12, and TNF itself); higher concentrations cause increased synthesis of prostaglandins, resulting in fever

Figure 35.4 Humoral immune response. (1) Surface antibodies of the B cell bind antigen. (2) The B cell internalizes the antigen-antibody complex, partially digests it, and displays portions on its surface (epitope), along with class II MHC protein. (3) The T_H2 cell recognizes antigen and class II protein on the B cell, is activated, and secretes interleukins (Table 35.3). T_H2 then stimulates proliferation of the B cell, which carries antigen and class II protein on its surface. (4) Interleukins promote activation and differentiation of B cells into (5) many plasma cells that secrete antibodies. (6) Some B-cell progeny become memory cells. (7) Antibody produced by plasma cells binds to antigen and stimulates macrophages to consume antigen (opsonization).

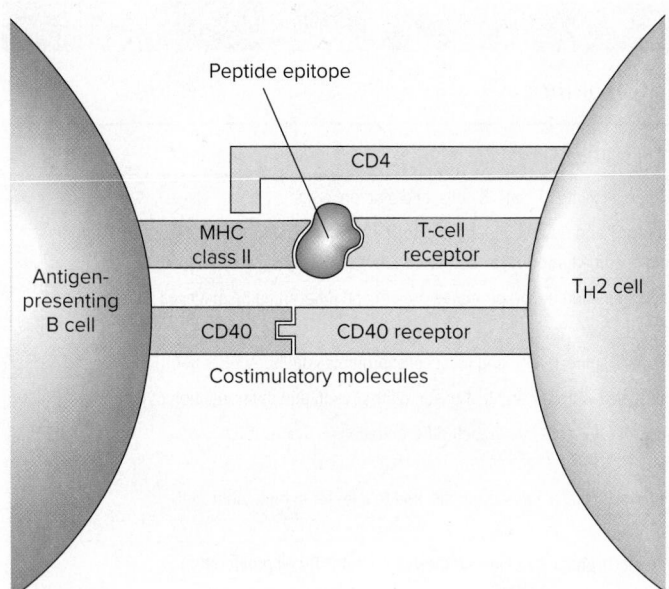

Figure 35.5 Interacting molecules during activation of a T$_H$2-helper cell by an antigen-presenting B cell.

Figure 35.6 Typical antibody response after primary and challenge immunizations. The secondary response is a result of the large numbers of memory cells produced after the primary B-cell activation.

Cell-Mediated Response: T$_H$1 Arm

Many immune responses involve little, if any, antibody and depend on the action of cells only. In cell-mediated immunity the epitope of an antigen is also presented by APCs, but the T$_H$1 arm of the immune response is activated (Figure 35.3). In this case APCs may be virally infected cells, tumor cells, or infected macrophages that have phagocytosed bacteria. T$_H$1 cells recognize the epitope-MHC II complex and become activated to release IL-2, TNF, and interferon-γ (INF-γ) (Table 35.3). IL-2 promotes the activity of activated B and T cells and enhances the cytotoxic activity of natural killer cells, causing them to proliferate and to become **lymphocyte-activated killer (LAK) cells** (Table 35.2). LAK cells also release INF-γ. INF-γ is a potent macrophage-activating factor. It promotes B- and T-cell differentiation and B-cell proliferation, and

activates capillary endothelial cells so that lymphocytes may pass through blood vessels into infected tissue areas. INF-γ is also a major inducer of inflammation. TNF activates PMNs and stimulates macrophages and cytokine production. Cytotoxic T cells also interact with APC surface receptors and this, together with stimulation by IL-2 and INF-γ (secreted from activated T$_H$1 cells), causes them to proliferate and to secrete proteins that cause pores to form in the infected cells, causing lysis.

Transplantation of organs from one person to another requires immunosuppressing the recipient enough that the new organ is not rejected without leaving the patient defenseless against infection. Since discovery of a fungus-derived drug called cyclosporine, many organs, for example, kidneys, hearts, lungs, and livers, can be transplanted. Cyclosporine inhibits IL-2 and affects CTLs more than T$_H$2 lymphocytes. It has no effect on other white blood cells or on healing mechanisms, so that a patient can still mount an immune response but not reject the transplant. However, the patient must continue to take cyclosporine because if the drug is stopped, the body recognizes the transplanted organ as foreign and rejects it.

Like humoral immunity, cell-mediated immunity shows a secondary response due to large numbers of memory T cells produced from the original activation. For example, a second tissue graft (challenge) between the same donor and host is rejected much more quickly than the first.

Acquired Immune Deficiency Syndrome (AIDS)

AIDS is an extremely serious disease in which the ability to mount an immune response is disabled severely. It is caused by **human immunodeficiency virus (HIV).** HIV preferentially invades and destroys T$_H$ lymphocytes due to the virus expressing CD4 protein as a major surface receptor. Normally, T$_H$ cells form 60% to 80% of the T-cell population; in AIDS their levels plummet to undetectable, so the humoral immune response is destroyed and the cell-mediated response is compromised.

The first case of AIDS was recognized in 1981. At the end of 2011, about 34 million people were living with HIV.[1] Only 10% of the worlds population live in sub-Saharan Africa, but about two-thirds of all people infected with HIV/AIDS live in this region. Worldwide, 2.5 million were newly infected in 2011, a 20% lower infection rate than in 2001. The sharpest declines in infection rates have occurred in the Caribbean (42%) and sub-Saharan Africa (25%). HIV infection virtually always progresses to AIDS after a latent period of some years. Since T$_H$ cell populations are virtually destroyed by HIV, AIDS patients are continuously plagued by infections with microbes and parasites that cause insignificant problems in persons with normal immune responses. If left untreated, AIDS is a terminal disease. There are now highly effective, though costly, drugs that can slow progression of the disease. Some drugs (for example, AZT) attack the enzyme needed by the virus to produce its DNA, whereas others (for example, protease inhibitors) inhibit the enzymes required to assemble a new

[1]UNAIDS 2012 Report on Global AIDS epidemic. All data in this section are taken from this report.

virus. Because the virus mutates (p. 95) to produce many different strains during progress of an infection, efforts to produce a vaccine have been fruitless thus far.

35.5 BLOOD GROUP ANTIGENS

ABO Blood Types

Blood cells differ chemically from person to person, and when two incompatible blood types are mixed, **agglutination** (clumping together) of erythrocytes results. The basis of these chemical differences is naturally occurring antigens on the membranes of red blood cells. The best known of these inherited immune systems is the ABO blood group (p. 122). Antigens A and B are inherited as codominant alleles of a single gene. Homozygotes for a recessive allele at the same gene have type O blood, which lacks A and B antigens. Thus, as shown in Table 35.4, an individual with, for example, genes I^A/I^A or I^A/i develops A antigen (blood type A). The presence of an I^B gene produces B antigens (blood type B), and for the genotype I^A/I^B, both A and B antigens develop on the erythrocytes (blood type AB). Epitopes of A and B also occur on the surfaces of many epithelial and most endothelial cells.

There is an odd feature about the ABO system. Normally we would expect that a type A individual would develop antibodies against type B cells only if cells bearing B epitopes were first introduced into the body. In fact, type A persons acquire anti-B antibodies soon after birth, even without exposure to type B cells. Similarly, type B individuals come to carry anti-A antibodies at a very early age. Type AB blood has neither anti-A nor anti-B antibodies (because if it did, it would destroy its own blood cells), and type O blood has both anti-A and anti-B antibodies. There is evidence that the antibodies develop as a response to A and B epitopes on intestinal microorganisms when the intestine becomes colonized with bacteria after birth. Presumably, small and unnoticed infections with the bacteria occur. The antibodies thus produced cross-react with the A and B epitopes on erythrocytes.

We see then that the blood-group names identify their *antigen* content. Persons with type O blood are called universal donors because, lacking antigens, their blood can be infused into a person with any blood type. Even though it contains anti-A and anti-B antibodies, these are so diluted during transfusion that they do not react with A or B antigens in a recipients blood. Persons with AB blood are universal recipients because they lack antibodies to A and B antigens. In practice, however, clinicians insist on matching blood types to prevent any possibility of incompatibility.

Rh Factor

Karl Landsteiner, an Austrian—later American—physician, discovered ABO blood groups in 1900. In 1940, 10 years after receiving a Nobel Prize, he made still another famous discovery. This was a blood group called the Rh factor, named after rhesus monkeys, in which it was first found. Approximately 85% of white individuals in the United States have the factor (positive) and the other 15% do not (negative). The Rh factor is encoded by a dominant allele at a single gene. Rh-positive and Rh-negative bloods are incompatible; shock and even death may follow their mixing when Rh-positive blood is introduced into an Rh-negative person who has been sensitized by an earlier transfusion of Rh-positive blood. Rh incompatibility accounts for a peculiar and often fatal **hemolytic disease of the newborn (erythroblastosis fetalis).** If an Rh-negative mother has an Rh-positive baby (father is Rh-positive), she can become immunized by fetal blood during the birth process. Anti-Rh antibodies are predominately IgG and can cross the placenta during a subsequent pregnancy and agglutinate fetal blood. Erythroblastosis fetalis normally is not a problem in cases of ABO incompatibility between mother and fetus because antibodies to ABO antigens are primarily IgM and cannot cross the placenta. Erythroblastosis fetalis can now be prevented by giving an Rh-negative mother anti-Rh antibodies just after the birth of her first child. These antibodies remain long enough to neutralize any Rh-positive fetal blood cells that may have entered her circulation, thus preventing her own antibody machinery from being stimulated to produce Rh-positive antibodies. Active, permanent immunity is blocked. The mother must be treated after every subsequent pregnancy (assuming the father is Rh-positive). If the mother has already developed an immunity, however, the baby may be saved by an immediate, massive transfusion of blood free of antibodies.

SUMMARY

A plethora of viral, prokaryotic, and eukaryotic parasites exists in every animals environment, and a defense (immune) system is crucial to survival. Immunity can be defined concisely as possession of tissues capable of recognizing and protecting the animal against nonself invaders. Most animals have some amount of innate immunity, and vertebrates develop acquired immunity. The surface of most animals provides a physical barrier to invasion, and contains antimicrobial substances in its body secretions.

Exposure of many animals, vertebrate and invertebrate, to many microorganisms stimulates their innate immune response. This response is

TABLE 35.4

Major Blood Groups

Blood Type	Genotype	Antigens on Red Blood Cells	Antibodies in Serum	Can Give Blood To	Can Receive Blood From	Frequency in United States (%)		
						Whites	Blacks	Asians
O	*i/i*	None	Anti-A and anti-B	All	O	45	48	31
A	*I^A/I^A*, *I^A/i*	A	Anti-B	A, AB	O, A	41	27	25
B	*I^B/I^B*, *I^B/i*	Anti-A	B, AB	O, B	10	21	34	AB
AB	*I^A/I^B*	None	AB	All	4	4	10	

based on release of antimicrobial peptides, is immediate, not requiring prior immunizing exposure, and is nonspecific but related to the category of invading microbe.

Phagocytes engulf particles and usually digest or kill them with enzymes and cytotoxic secretions. Many invertebrates have specialized cells that can perform defensive phagocytosis. Several kinds of vertebrate cells, especially macrophages and neutrophils, are important phagocytes, and cells of the mononuclear phagocyte system reside in various sites in the body. Eosinophils are important in allergies and many parasitic infections. Basophils, mast cells, T and B lymphocytes, and ILCs (natural killer cells and RORγt⁺ ILCs) are not phagocytic but play vital roles in defense.

Inflammation is an important part of the body's defense; it is greatly influenced by prior immunizing experience with an antigen.

Many invertebrates show nonself recognition by rejection of xenografts, allografts, or both. In some cases they may show enhanced response on repeated exposure. Bacterial, viral, and fungal infections stimulate the release of degradative enzymes and anti-microbial peptides, and they activate phagocytes. Acquired responses that demonstrate memory and specificity have been demonstrated in some invertebrate phyla.

An acquired immune response is elicited by an antigen. Vertebrates demonstrate increased resistance to *specific* foreign substances (antigens) on repeated exposure, and the resistance is based on a vast number of specific recognition molecules: antibodies and T-cell receptors. Nonself recognition depends on markers in cell surfaces called major histocompatibility (MHC) proteins. Antibodies are borne on the surfaces of B lymphocytes (B cells) and in solution in the blood after secretion by the progeny of B cells, plasma cells. T-cell receptors occur only on the surfaces of T lymphocytes (T cells).

The cells of immunity communicate with each other and with other cells in the body by means of protein hormones called cytokines such as interleukins, tumor necrosis factor, and interferon-γ. The two defense mechanisms of the vertebrate immune response are the humoral response (T_H2), involving antibodies, and the cell-mediated response (T_H1), involving cell surfaces only. When one defense mechanism is activated or stimulated, its cells produce cytokines that tend to suppress activity in the other defense mechanism. Activation of either defense mechanism requires that the antigen be phagocytosed by an APC (antigen-presenting cell, for example, a macrophage or a B cell), which partially digests the antigen and presents its determinant (epitope) on the surface of the APC along with an MHC class II protein. Virally infected cells and tumor cells also present antigen to T_H1 cells. Extensive communication by cytokines and activation (and suppression) of various cells in the response lead to production of specific antibody or proliferation of T cells with specific receptors that recognize an antigenic epitope. After the initial response, memory cells remain in the body and are responsible for enhanced response on the next exposure to the antigen.

Damage to the immune response caused by HIV (human immunodeficiency virus) in production of AIDS (acquired immune deficiency syndrome) is due primarily to destruction of a crucial set of helper T cells, bearing the CD4 coreceptor on their surface.

People have genetically determined antigens on the surfaces of their red blood cells (ABO blood groups and others); blood types must be compatible in transfusions, or transfused blood is agglutinated by antibodies in the recipient.

REVIEW QUESTIONS

1. Distinguish susceptibility from resistance, and innate from acquired immunity. Why are these traditionally recognized types of immunity now more difficult to distinguish?

2. What are some examples of innate defense mechanisms that are chemical in nature? What is complement?

3. After a phagocyte has engulfed a particle, what usually happens to the particle?

4. What are some important phagocytes in vertebrates?

5. What is the molecular basis of self and nonself recognition in vertebrates?

6. What is the difference between T cells and B cells?

7. What is a cytokine? What are some functions of cytokines?

8. Outline the sequence of events in a humoral immune response, from the introduction of antigen to the production of antibody.

9. Define the following: plasma cell, secondary response, memory cell, opsonization, titer, challenge, cytokine, natural killer cell, interleukin-2.

10. What are the functions of CD4 and CD8 proteins on the surface of T cells?

11. In general, what are consequences of activation of the T_H1 arm of the immune response? Activation of the T_H2 arm?

12. Distinguish between class I and class II MHC proteins.

13. Describe a typical inflammatory response.

14. What is a major mechanism by which HIV damages the immune system in AIDS?

15. Give the genotypes of each of the following blood types: A, B, O, AB. What happens when a person with type A gives blood to a person with type B? With type AB? With type O?

16. What causes hemolytic disease of the newborn (erythroblastosis fetalis)? Why does the condition not arise in cases of ABO incompatibility?

17. Give some evidence that cells of many invertebrates bear molecules on their surface that are specific to the species and even to a particular individual animal.

18. Give an example of immunological memory in invertebrates.

For Further Thought   Traditionally, acquired immunity is considered a characteristic of vertebrate immune systems. Provide an argument that it is probably also found among invertebrates.

SELECTED REFERENCES

Aderem, A., and R. J. Ulevitch. 2000. Toll-like receptors in the induction of the innate immune response. Nature **406:** 782–787. *A good review of the role of Toll-like receptors in the innate immune response.*

Alberts, B., A. Johnson, J. Lewis, M. Raff, K. Roberts, and P. Walter. 2007. Molecular biology of the cell, ed. 5. New York, Garland Publishing. *Concise discussion of complement system with clear figures.*

Beaven, M. A. 2009. Our perception of the mast cell from Paul Ehrlich to now. European J. of Immunology **39:** 11–25. *An update on information regarding mast cell function.*

Cherry, S., and N. Silverman. 2006. Host-pathogen interactions in *Drosophila*: new tricks from an old friend. Nature Immunol. **7:** 911–917. *Using an array of genetic tools and* Drosophila *as a model system, this review describes exciting new advances in the study of insect innate immune function.*

Devereux, G. 2006. The increase in the prevalence of asthma and allergy: food for thought. Nature Rev. **6:** 869–874. *An interesting article that reviews evidence for dietary changes causing an increase in asthma and allergies in developed countries.*

Flajnik, M. F., and L. Du Pasquier. 2004. Evolution of innate and adaptive immunity: can we draw a line? Trends Immunol. **25:** 640–644. *Discusses how sequencing animal genomes provides evidence of adaptive immunity genes and molecules in invertebrates, and how the line between innate and adaptive immune systems is becoming blurred.*

Gartner, L. P., and J. L. Hiatt. 2010. Color textbook of histology, ed. 5 Secaucus, NJ. Lippincott, Williams and Wilkins. *Excellent chapters describing cells of the mammalian immune system and their functions.*

Gura, T. 2001. Innate immunity: ancient system gets new respect. Science **291:** 2068–2071. *Innate immunity evolved early in animal evolution.*

Klotman, M. E., and T. L. Chang. 2006. Defensins in innate antiviral immunity. Nature Rev. **6:** 447–456. *This review describes data that suggest an antiviral role for mammalian defensins against both the virus and the host cell.*

Konrad, D., and S. Wueest. 2014. The gut-adipose-liver axis in the metabolic syndrome. Physiology **29:** 304–313. *An excellent review of the role of the gut microbiome in development of obesity related insulin resistance and Type 2 diabetes.*

Konrad, M., M. L. Vyleta, F. J. Thels, M. Stock, S. Tragust, M. Klatt, V. Drescher, C. Marr, L. V. Ugelvig, and S. Cremer. 2012. PLoS Biol **10(4):** e100130. dol:10.1371journal.pbio.1001300. *This research article provides evidence that social transfer of fungal pathogens by grooming infected ants within the colony promotes innate immune responses in colony nestmates.*

Kurtz, J. 2005. Specific memory within innate immune systems. Trends Immunol. **26:** 186–192. *An excellent review of evidence for memory within the innate immune system of invertebrates.*

Kurtz, J., and K. Franz. 2003. Evidence for memory in invertebrate immunity. Nature **425:** 37–38. *Copepods show memory when infected with their natural parasite, a tapeworm.*

Letvin, N. L. 2006. Progress and obstacles in the development of an AIDS vaccine. Nature **6:** 930–939. *Provides an excellent review of HIV biology and immune regulatory mechanisms, as well as a good discussion of progress toward an effective vaccine against AIDS.*

Litman, G. W., J. P. Cannon, and L. J. Dishaw. 2005. Reconstructing immune phylogeny: new perspectives. Nature Rev. Immunol. **5:** 866–879. *Describes the interrelatedness of adaptive and innate immunity from an evolutionary perspective and provides a clear visualization of the similar selection pressures that have influenced pathogenicity and host protection mechanisms.*

Little, T. J., D. Hultmark, and A. F. Read. 2005. Invertebrate immunity and the limits of mechanistic immunology. Nature Immunol. **6:** 651–654. *A review of some exciting data showing anticipatory behavior in the invertebrate innate immune system.*

Medzhitov, R., and C. A. Janeway, Jr. 2002. Decoding the patterns of self and nonself by the innate immune system. Science **296:** 298–300. *The several strategies employed by innate immune systems for discrimination of self and nonself.*

Roberts, J. P. 2004. Are HIV vaccines fighting fire with gasoline? The Scientist **18:** 26–27. *Clear and concise summary of why a vaccine for HIV/AIDS is proving difficult to produce. Associated material includes information on vaccines that are being tested.*

Spits, H., and T. Cupedo. 2012. Innate lymphoid cells: emerging insights in development, lineage relationships, and function. Annu. Rev. Immunol. **30:** 647–675. *An article that reviews current evidence on the subtypes and functions of innate lymphoid cells.*

Vizioli, J., and M. Salzet. 2002. Antimicrobial peptides from animals: focus on invertebrates. Trends Pharmacol. Sci. **23:** 494–496. *A short review of invertebrate antimicrobial peptides and their possible use as antibiotics.*

Zasloff, M. 2002. Antimicrobial peptides of multicellular organisms. Nature **415:** 389–395. *This short review discusses the molecules and mechanisms behind antimicrobial peptide activity in innate immune function and their possible use as antiinfective drugs.*

36

CHAPTER

Animal Behavior

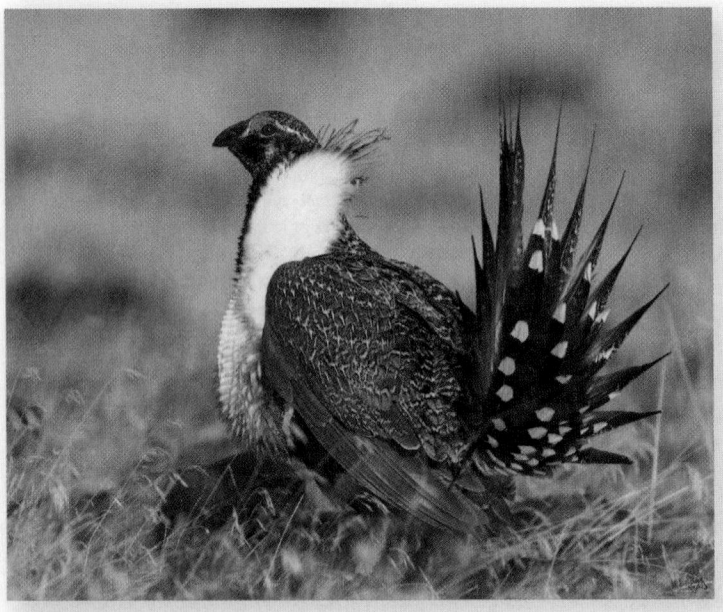

Male sage grouse, Centrocercus urophasianus, *displaying at its lek.*

Jeanne Stafford/U.S. Fish and Wildlife Service

The Lengthening Shadow of One Person

Ralph Waldo Emerson said that an institution is the lengthening shadow of one person. Charles Darwin's shadow is a long one covering the entire fields of evolution, ecology, and finally, after a long gestation, animal behavior. Each of these fields alters how we think about ourselves, the earth we inhabit, and the animals that share it with us.

Charles Darwin showed with great insight how natural selection would favor various behavioral patterns for survival and reproduction. Darwin's pioneering book, *The Expression of the Emotions in Man and Animals,* published in 1872, mapped a strategy for behavioral research still used today. However, science in 1872 was unprepared for Darwin's

central insight that behavioral patterns, no less than bodily structures, have histories of evolution by natural selection. Another 60 years would pass before such concepts would flourish within behavioral science.

In 1973, the Nobel Prize in Physiology or Medicine was awarded to three pioneering zoologists, Karl von Frisch, Konrad Lorenz, and Niko Tinbergen (Figure 36.1). The citation stated that these three were the principal architects of the new science of ethology, the scientific study of animal behavior, particularly under natural conditions. It was the first time that any contributor to the behavioral sciences was so honored. The discipline of animal behavior had arrived.

Behavioral biologists ask *how* animals behave and *why* they behave as they do. "How" questions concern mechanisms of immediate or **proximate causation,** and are studied experimentally (p. 12). For example, a biologist might explain the singing of a male white-throated sparrow in the spring by hormonal or neural mechanisms. Such physiological or mechanistic causes of behavior are proximate factors. We examine the proximate factors underlying animal behavior in Chapter 33 and 34, emphasizing the functioning of nervous systems and hormonal coordination. Alternatively, a biologist might ask what function singing serves the sparrow, and then identify events in the ancestry of birds that led to springtime singing. These are "why" questions that focus on **ultimate causation,** the evolutionary origin and purpose of a behavior. Questions of ultimate causation are answered using comparative methodology (p. 12), applying phylogenetic analysis (p. 205) to identify evolutionary changes in behavior and their associated morphological and environmental contexts. This chapter presents evolutionary explanations of animal behavior and the challenges of behavioral evolution for Darwin's theory of natural selection (p. 119).

The study of animal behavior has several different historical roots, and there is no universally accepted term for the whole subject. **Comparative psychology** emerged from efforts to find general laws of behavior that apply to many species, including humans. Psychologists compared results of laboratory experiments on white rats, pigeons, dogs, and occasionally primates to identify neurological and behavioral similarities and differences. Following criticisms that the discipline lacked an evolutionary perspective and focused too narrowly on white rats as a model for other organisms, many comparative psychologists developed phylogenetically based investigations, some of these conducted under natural conditions.

The aim of a second approach, **ethology,** is to describe the behavior of an animal in its *natural habitat.* Most ethologists have been zoologists who gather data by field observations and experiments. Nature provides the variables, which are manipulated using such approaches as presenting animal models, playing recordings of animal vocalizations, and altering habitats. Ethologists also test their hypotheses under controlled laboratory conditions. Laboratory results often direct further testing of hypotheses using observations of free-ranging animals in undisturbed natural environments.

Ethology emphasizes the importance of ultimate factors influencing behavior. A great contribution of von Frisch, Lorenz, and Tinbergen was to demonstrate that behavioral traits are measurable entities like anatomical or physiological traits. The central theme of ethology is that behavioral traits can be identified and measured, homologies determined, and their evolutionary histories investigated to provide causal explanations.

Work by comparative psychologists and ethologists gave rise to the discipline of **behavioral ecology.** Behavioral ecologists study the evolutionary and environmental contexts of animal behaviors and how particular behaviors serve to maximize reproductive and evolutionary success. Behavioral ecologists often concentrate on a particular aspect of behavior, such as mate choice, foraging, or parental investment. **Sociobiology,** the ethological study of social behavior, was formalized with the 1975 publication of E. O. Wilson's *Sociobiology: The New Synthesis.* Wilson describes social behavior as reciprocal communication of a cooperative nature (transcending mere sexual activity) that permits a group of organisms of the same species to become organized in a cooperative manner. In a complex

A

B

C

Figure 36.1   Pioneers of the science of ethology. **A,** Konrad Lorenz (1903 to 1989). **B,** Karl von Frisch (1886 to 1982). **C,** Niko Tinbergen (1907 to 1988).

system of social interactions, individuals are highly dependent on others for daily living. While social behavior appears in many groups of animals, Wilson identified four "pinnacles" of complex social behavior. These are (1) colonial invertebrates, such as the Portuguese man-of-war (p. 271), which is a composite of interdependent individual organisms; (2) social insects, such as ants, bees, and termites, which have developed sophisticated systems of communication; (3) nonhuman mammals, such as dolphins, elephants, and primates, that have highly developed social systems; and (4) humans.

Wilson's inclusion of human behavior in sociobiology and his assumptions of genetic foundations for variation among many human social behaviors have been strongly criticized. Complex systems of human social interactions, including religion, economic systems, and such objectionable characteristics as racism, sexism, and war, are explained more easily as emergent properties (p. 4) of human culture and its history. Is it meaningful to search for a specific genetic basis or justification for such phenomena? Many answer "no," and look instead to the field of sociology, rather than sociobiology, to explain complex, emergent properties of human societies. For this reason, most researchers studying animal social behavior have rejected "sociobiology" as a title of their discipline in favor of "behavioral ecology."

36.1 DESCRIBING BEHAVIOR: PRINCIPLES OF CLASSICAL ETHOLOGY

Early ethologists sought to identify and to explain relatively invariant components of behavior shared by diverse animal species. From such studies emerged several concepts first popularized in Tinbergen's influential book, *The Study of Instinct* (1951).

The egg-retrieval response of greylag geese, *Anser anser* (Figure 36.2), described by Lorenz and Tinbergen in a famous paper in 1938, illustrates an instinctual behavior. If Lorenz and Tinbergen presented a female greylag goose with an egg a short distance from her nest, she would rise, extend her neck until the beak was just over the egg and then bend her neck to pull the egg into the nest.

Although this behavior appeared rational, Tinbergen and Lorenz noticed that if they removed the egg after the goose had begun her retrieval, or if the egg being retrieved slipped away, the goose would continue the retrieval movement without the egg until she was again settled on her nest. Then, seeing the egg outside the nest, she would repeat the egg retrieval pattern.

Thus, the bird performed egg-retrieval behavior as a program that, once initiated, had to be completed in a standard way. Lorenz and Tinbergen viewed egg retrieval as a "fixed" pattern of behavior: a motor pattern mostly invariable in its performance. A behavior of this type, performed in an orderly, predictable sequence, is called **stereotypical behavior.** Of course, stereotyped behavior may not occur identically on all occasions, but it should be recognizable, even when performed inappropriately. Further experiments by Tinbergen disclosed that the greylag goose was not particularly discriminating about what she retrieved. Almost any smooth and rounded object placed outside the nest would trigger the retrieval behavior; even a small toy dog and a large yellow balloon were dutifully retrieved. When the goose settled down on such objects, they obviously did not feel right, and she discarded them.

Lorenz and Tinbergen realized that an egg outside the nest must act as a stimulus, or trigger, to release egg-retrieval behavior. Lorenz termed the triggering stimulus a **releaser,** a simple stimulus in the environment that would trigger a certain innate behavior. Because the animal usually responded to some specific aspect of the releaser (sound, shape, or color, for example), the effective stimulus was called a **sign stimulus.** Ethologists have described hundreds of sign stimuli. In every case the response is highly predictable. For example, the alarm call of adult herring gulls always releases a crouching freeze response in their chicks. Certain nocturnal moths take evasive maneuvers or drop to the ground when they hear ultrasonic cries of bats that feed on them (p. 738); most other sounds do not release this response.

These examples illustrate the predictable and programmed nature of much animal behavior. This is even more evident when stereotyped behavior is released inappropriately. In spring a male three-spined stickleback, *Gasterosteus aculeatus,* a small fish, selects a territory that it defends vigorously against other males. The underside of the male becomes bright red, and an approach of another red-bellied male releases a threat posture or even an aggressive attack. Tinbergen's suspicion that a male's red belly served as a releaser for aggression was reinforced when a passing red postal truck evoked attack behavior from males in his aquarium. Tinbergen then performed experiments using a series of models presented to the males. He found that they vigorously attacked any model bearing a red stripe, even a plump lump of wax with a red underside, yet a model that closely resembled a male stickleback but lacked the red belly was attacked less frequently (Figure 36.3). Tinbergen discovered other examples of stereotyped behavior released by simple sign stimuli. Male European robins, *Erithacus rubecula,* furiously

Figure 36.2 Egg-retrieval behavior of the greylag goose, *Anser anser,* as studied by Lorenz and Tinbergen. In this stereotypical behavior, the egg outside the nest (*1*) is a sign stimulus for the goose to approach it (*2*) and to pull the egg toward the nest (*3–4*). The position shown in frame 4 is used to roll the egg toward the nest. The goose completes its return to the nest in this fashion even if the egg being retrieved rolls away.

Figure 36.3 Stickleback models used to study territorial behavior. The carefully made model of a stickleback, *Gasterosteus aculeatus (left)*, without a red belly, is attacked much less frequently by a territorial male stickleback than are the four simple red-bellied models.

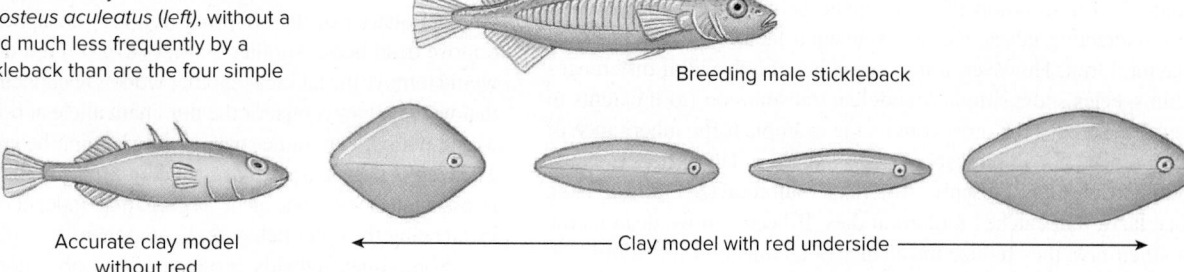

Breeding male stickleback

Accurate clay model without red

Clay model with red underside

Figure 36.4 Two models of a European robin, *Erithacus rubecula*. The bundle of red feathers is attacked by male robins, whereas the stuffed juvenile bird *(right)* without a red breast is ignored.

Source: Modified from D. Lack, *The Life of the Robin,* H. F. & G. Witherby Ltd., London, England, 1943.

attacked a bundle of red feathers placed in their territory but ignored a stuffed juvenile robin withoutred feathers (Figure 36.4).

Under conditions that are relatively consistent and predictable, automatic preprogrammed responses may be most efficient. Releasers have the advantage of focusing an animal's attention on the relevant signal, and release of a preprogrammed stereotyped behavior enables an animal to respond rapidly when speed is essential for survival or reproductive success.

Behavioral ecologists studying behavioral variation within species observe that individuals often differ from each other in a suite of correlated behaviors, and that these differences are expressed consistently in different situations. For example, in *D. melanogaster* flies (p. 83), some larvae are "rovers" in their foraging behavior, traveling long path distances and exploring different patches of food, whereas others are "sitters," moving only short distances and concentrating their foraging within a local patch of food. Rovers and sitters do not differ in overall level of foraging activity, only in the path distances that they travel during

foraging. Rovers and sitters illustrate contrasting behavioral types within a **behavioral syndrome,** defined by Andrew Sih and colleagues as "a suite of correlated behaviors reflecting between-individual consistency in behavior across multiple (two or more) situations." Behavioral syndromes can have a genetic basis, and thus may be subject to evolution by natural selection. The rover larval type in *D. melanogaster,* for example, is genetically dominant over the sitter type as alternative alleles at an autosomal locus.

36.2 CONTROL OF BEHAVIOR

Stereotyped behavior suggested to ethologists that they were observing inherited, or **innate,** behavior. Many kinds of preprogrammed behavior appear suddenly in an animal's ontogeny and are indistinguishable from similar behavior performed by older, experienced individuals. Orb-weaving spiders build their webs without practice, and male crickets court females without lessons from more experienced crickets or by learning from trial and error. We call such behaviors innate or instinctive.

Instinctive behaviors are, like organismal morphology, dependent on interactions between an organism and its environments during ontogeny. Although an instinct might appear rigid and fixed, instincts are products of evolutionary change and remain subject to further evolutionary change by selection. Dog breeds, for example, often have characteristic behavioral tendencies that differ among breeds according to the selection imposed by past breeders. Breeders of sheepdogs, for example, have constructed genetic combinations that reinforce behaviors useful in herding sheep and suppress behaviors destructive to this endeavor. Training individual sheepdogs enhances behavioral performance. Dogs of another breed might only rarely express behaviors conducive to sheep herding, yet all dogs share very recent common ancestry on evolutionary standards. Evolution of instincts is perhaps best conceived as a narrowing of behavioral repertoires so that the nervous system reinforces a particular subset of behaviors and limits use of alternatives. As environments and selective forces change, different behavioral modes are genetically stabilized.

Many complex sequences of behavior in invertebrate animals are largely invariate in their execution and appear to follow precise rules without learning. Programmed behavior is important for survival, especially for animals that never know their parents. They must respond to stimuli immediately and correctly as they emerge. Animals with long lives and with parental care or other opportunities for social interactions nonetheless may improve or change their behavior by learning.

Genetics of Behavior

Hereditary transmission of most innate behavior is complex, with many interacting genes and environmental factors influencing each behavioral trait. However, a few examples of behavioral differences within species show simple Mendelian transmission from parents to offspring. Perhaps the most convincing example is the inheritance of hygienic behavior in honeybees, *Apis mellifera*. Honeybees are susceptible to a bacterial disease, American foulbrood (*Bacillus larvae*). A bee larva that catches foulbrood dies. If bees remove dead larvae from their hive they reduce the chance of an infection spreading.

Some strains of bees, called "hygienic," uncap hive cells containing rotting larvae and remove them from the hive. W. C. Rothenbuhler described two components to this behavior: first, removal of cell caps, and second, removal of larvae. Hygienic bees have homozygous recessive genotypes for two different genes. Uncapping behavior is performed by individuals homozygous for a recessive allele, *u,* at one gene, and removal behavior is performed by individuals homozygous for a recessive allele, *r,* at a second gene. (Because the behaviors being studied are expressed only by diploid females, we characterize strains using diploid genotypes; males are haploid, as indicated in Figure 36.5.) When Rothenbuhler crossed hygienic bees (*u/u r/r*) with a nonhygienic strain (*U/U R/R*), he found that all hybrids (*U/u R/r*) were nonhygienic. Thus only workers having both genes in the homozygous recessive condition show the complete behavior. Next, Rothenbuhler performed a "backcross" between the hybrids and the hygienic parental strain. As we expect if hygienic behavior is transmitted by allelic variation of two genes, four different kinds of bees resulted (Figure 36.5). Approximately one-quarter of the bees were

homozygous recessive for both *u* and *r* and showed the complete behavior: they both uncapped the cells and removed infected larvae. Another quarter of the offspring (*u/u R/r*) uncapped cells but did not remove dead bees. Another quarter (*U/u r/r*) did not uncap cells, but would remove the larvae if another worker uncapped the cells. Workers that were heterozygous for the dominant allele at both genes (*U/u R/r*) would not perform either part of the cleaning behavior (Figure 36.5). The results show clearly that each component of the cleaning behavior is associated with one gene segregating independently of the gene influencing the other behavioral component.

Sometimes hybrids produce behaviors intermediate between those of their parents. A classic study by W. C. Dilger on nest-building behavior in different species of lovebirds revealed such an outcome. Lovebirds are small parrots of the genus *Agapornis* (Figure 36.6). Each species has its own method of gathering material used to build a nest. Fischer's lovebirds, *A. personata fischeri,* cut long strips of nesting material from vegetation, then carry them to the nest one strip at a time. Peach-faced lovebirds, *A. roseicollis,* carry several such strips of nesting material simultaneously by tucking them into feathers of the lower back and rump. Dilger, who crossed the two species successfully, found that hybrids attempted both feather-tucking and beak-carrying behaviors but performed neither behavior correctly (Figure 36.6). Hybrids inherited a behavior intermediate between those of the parents. With experience, hybrids improved their carrying ability by tending to carry the material in their beaks, like Fischer's lovebird. In this case, disruption of genetically programmed behavior by hybridization caused birds to rely on learning by trial and error how to gather nesting material.

Homozygous females   **Hemizygous males**
hygienic strain   **nonhygienic strain**

u/u r/r ✕ *U/- R/-*

↓

U/u R/r   Nonhygienic hybrids

⇣

U/u R/r ✕ *u/- r/-*   Testcross of hybrid bees

u/u r/r	*U/u r/r*	*u/u R/r*	*U/u R/r*
Hygienic	Nonhygienic, does not uncap but can remove dead larvae	Nonhygienic, uncaps, leaves dead larvae inside cells	Nonhygienic, neither uncaps nor removes dead larvae

KEY: Worker females

u/u = uncaps cells
U/U or *U/u* = does not uncap cells
r/r = removes dead larvae
R/R or *R/r* = does not remove dead larvae

Figure 36.5   The genetics of hygienic behavior in female worker honeybees, as demonstrated by W. C. Rothenbuhler. Results reveal two independently assorting genes, one associated with uncapping cells containing diseased larvae, and the other associated with removing diseased larvae from cells. Hemizygous male progeny are not shown. See text for further explanation.

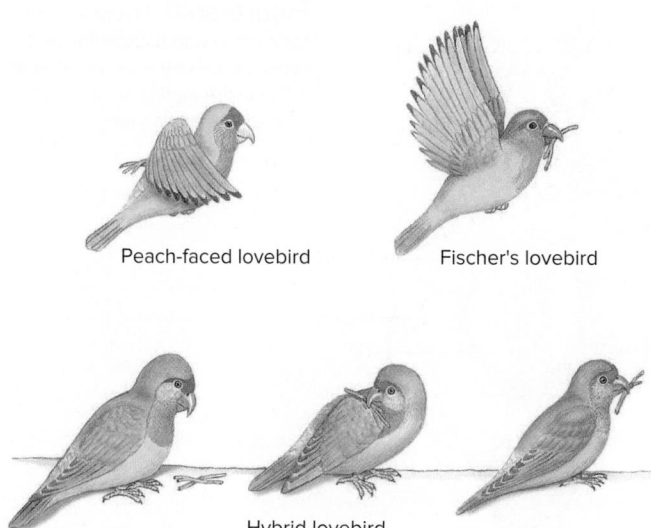

Figure 36.6 Intermediate behavior in hybrid lovebirds, *Agapornis* sp. The peach-faced lovebird carries nest-building material tucked into its feathers; Fischer's lovebird carries nest-building material in its beak. The hybrids attempted both carrying methods; neither method was initially accomplished successfully, but birds eventually learned to carry material in their beaks, like Fischer's lovebird.

Peach-faced lovebird

Fischer's lovebird

Hybrid lovebird

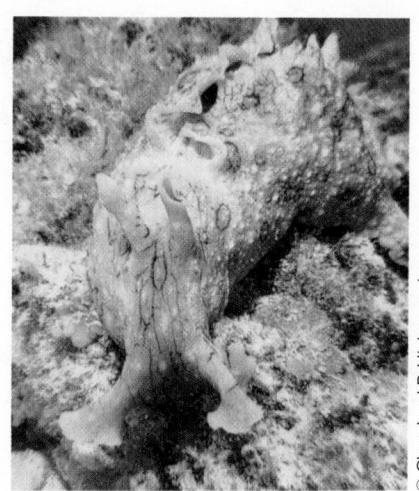

Figure 36.7 The sea hare, *Aplysia* sp., an opisthobranch gastropod used in many neurophysiological and behavioral studies.

Learning and Diversity of Behavior

Another aspect of behavior is learning, which we define as modification of behavior through experience. An excellent model system for studying learning processes is the marine opisthobranch snail, *Aplysia* (Figure 36.7), a subject of intense experimentation by E. R. Kandel and his associates. Gills of *Aplysia* are partly covered by the mantle cavity and open to the outside by a siphon (Figure 36.8). If one prods the siphon, *Aplysia* withdraws its siphon and gills and folds them in the mantle cavity. This simple protective response, called gill-withdrawal reflex, is repeated when *Aplysia* extends its siphon again. If the siphon is touched repeatedly, however, *Aplysia* decreases the gill-withdrawal response and ignores the stimulus. This behavioral modification illustrates a widespread form of learning called **habituation.** If *Aplysia* is given a noxious stimulus (for example, an electric shock) to the head at the same time the siphon is touched, it becomes **sensitized** to the stimulus and withdraws its gills as completely as it did before habituation occurred. Sensitization, then, can reverse previous habituation.

The nervous pathways of habituation and sensitization in *Aplysia* are known. Receptors in the siphon connect through sensory neurons (black pathways in Figure 36.8) to motor neurons (blue pathway in Figure 36.8) that control the gill-withdrawal muscles and muscles of the mantle cavity. Kandel found that repeated stimulation of the siphon diminished the release of synaptic neurotransmitter from sensory neurons. Sensory neurons continue to fire when the siphon is probed but, with less neurotransmitter being released into the synapse, the system becomes less responsive.

Sensitization requires action of a different kind of neuron called a facilitating interneuron. These interneurons make connections between sensory neurons in the snail's head and motor neurons that control muscles of the gill and mantle (Figure 36.8). When

sensory neurons in the head are stimulated by an electric shock, they fire facilitating interneurons, which end on the synaptic terminals of sensory neurons (red pathways in Figure 36.8). These endings in turn *increase* the amount of neurotransmitter released by the siphon sensory neurons. This release increases excitation of the excitatory interneurons and motor neurons leading to the gill and mantle muscles. Motor neurons now fire more readily than before. The system is now sensitized because any stimulus to the siphon produces a strong gill-withdrawal response.

Kandel's studies indicate that strengthening or weakening of the gill-withdrawal reflex involves changes in levels of neurotransmitter in existing synapses. However, more complex kinds of learning may involve formation of new neural pathways and connections, as well as changes in existing circuits.

Imprinting

Another kind of learned behavior is **imprinting,** the imposition of a stable behavior in a young animal by exposure to particular stimuli during a critical period in the animal's development. As soon as a newly hatched gosling or duckling is strong enough to walk, it follows its mother when away from the nest. After it has followed the mother for some time it follows no other animal. However, if the eggs are hatched in an incubator or if the mother is separated from the eggs as they hatch, the goslings follow the first large object that they see. As they grow, the young geese prefer the artificial "mother" to anything else, including their true mother. The goslings are said to be imprinted on the artificial mother.

Imprinting was observed at least as early as the first century A.D., when the Roman naturalist Pliny the Elder wrote of "a goose which followed Lacydes as faithfully as a dog." Konrad Lorenz was the first to study imprinting systematically. When Lorenz hand-reared greylag goslings, they formed an immediate and permanent attachment to him and waddled or swam after him wherever he went (Figure 36.9). They could no longer be induced to follow their own mother or another human being. Lorenz found that imprinting occurs in a brief sensitive period early in an individual's life and that once established the imprinted bond usually persists throughout life.

Figure 36.8 Neural circuitry concerned with habituation and sensitization of the gill-withdrawal reflex in the marine snail, *Aplysia*. See text for explanation.

What imprinting shows is that a goose's brain (or the brain of numerous other birds and mammals that show imprinting-like behavior) accommodates the imprinting experience. Natural selection in some species favors evolution of a brain that imprints in this way, in which following the mother and obeying her commands are important for survival. The fact that a gosling can be made to imprint to a mechanical toy duck or a person under artificial conditions is a cost that is tolerated because goslings seldom encounter these stimuli in their natural environments. The disadvantages of the system's simplicity are outweighed by the advantages of its reliability.

One final example completes our consideration of learning. Songbirds demonstrate robust sex differences in many behaviors. Males of many species of birds have characteristic territorial songs that identify the singers to other birds and announce territorial rights to other males of that species. Like many other songbirds, a male whitecrowned sparrow, *Zonotrichia leucophrys,* must learn the song of its species by hearing the song of its father. If a sparrow is hand-reared in acoustic isolation in a laboratory, it develops an abnormal song (Figure 36.10); if an isolated bird is allowed to hear recordings of normal white-crowned sparrow songs during a critical period of 10 to 50 days after hatching, it learns to sing normally. It even imitates the local dialect that it hears.

Characteristics of the song are not determined by learning alone. If, during the critical learning period, an isolated male white-crowned sparrow hears a recording of another species of sparrow, even a closely related one, it does not learn the song. It learns only the song appropriate to its own species. Thus, although the song must be learned, the brain is constrained to recognize and to learn vocalizations produced by males of its species alone. Learning a wrong song would produce behavioral chaos, and natural selection favors a system that eliminates such errors. Another example of complex interactions between learned and innate factors is illustrated by navigation of seasonally migratory birds (p. 599).

Figure 36.9 Greylag goslings imprinted on Konrad Lorenz follow him as faithfully as they would a natural mother.

36.3 SOCIAL BEHAVIOR

When we think of "social" animals we tend to think of highly structured honeybee colonies, herds of antelope grazing on the African plains (Figure 36.11), schools of herring, or flocks of starlings. Social behavior of animals *of the same species* living together is by no means limited to such obvious examples. Any interaction resulting from a response of one animal to another of the same species is a social behavior. Even a pair of rival males fighting for possession of a female displays a social interaction.

Social aggregations are only one kind of social behavior, and indeed not all aggregations of animals are social. Clouds of moths attracted to a light at night, barnacles attracted to a common float, or trout gathering in the coolest pool of a stream are nonsocial groupings

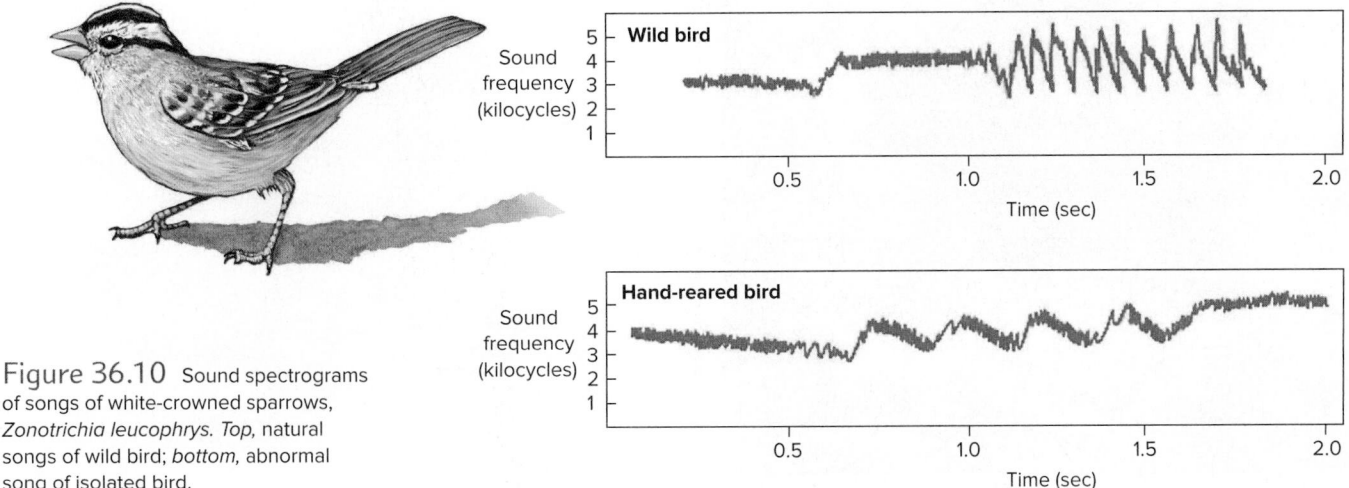

Figure 36.10 Sound spectrograms of songs of white-crowned sparrows, *Zonotrichia leucophrys. Top,* natural songs of wild bird; *bottom,* abnormal song of isolated bird.

of animals responding to environmental signals. Social aggregations depend on signals from the animals themselves, causing the animals to remain together and to influence one another.

Animal species vary in degree of sociality. Males and females of sexually reproducing species must bond at least temporarily in pairs to achieve fertilization, and some animals limit their adult sociality to breeding. Alternatively, swans, geese, albatrosses, and beavers, to name just a few, form strong monogamous bonds that last a lifetime. The most persistent social bonds usually form between mothers and their young and, for birds and mammals, these bonds usually terminate at fledging or weaning.

Selective Consequences of Sociality

Ethologist Tim Clutton-Brock distinguishes **socially coordinated behavior,** in which an individual adjusts its actions to the presence of others to increase its own reproductive success directly, from **cooperative behavior,** in which an individual performs activities that benefit others because such behavior ultimately benefits that individual's genetic contributions to future generations. The former category includes agonistic and competitive behaviors, territoriality, and formation of various mating systems. The latter category includes cooperative foraging and breeding behaviors, and especially behaviors that may benefit an individual's close relatives or cause others to reciprocate a beneficial behavior. If an individual's presence in a group benefits its survival and reproductive success, selection should favor evolution of cooperative strategies.

Social aggregations are beneficial for defense, both passive and active, from predators. Musk-oxen form a passive defensive circle when threatened by a wolf pack and are much less vulnerable than an individual facing wolves alone.

As an example of active defense, a breeding colony of black-headed gulls, *Chroicocephalus ridibundus,* alerted by alarm calls of a few, attack predators *en masse;* this collective attack discourages a predator more effectively than would individual attacks. Members of a town of prairie dogs, *Cynomys ludovicianus* (p. 627), although divided into social units called coteries, cooperate by warning one another with a special bark when danger threatens. Thus every individual in a social organization benefits from the eyes, ears, and noses of all other members of the group. Experimental tests using a wide

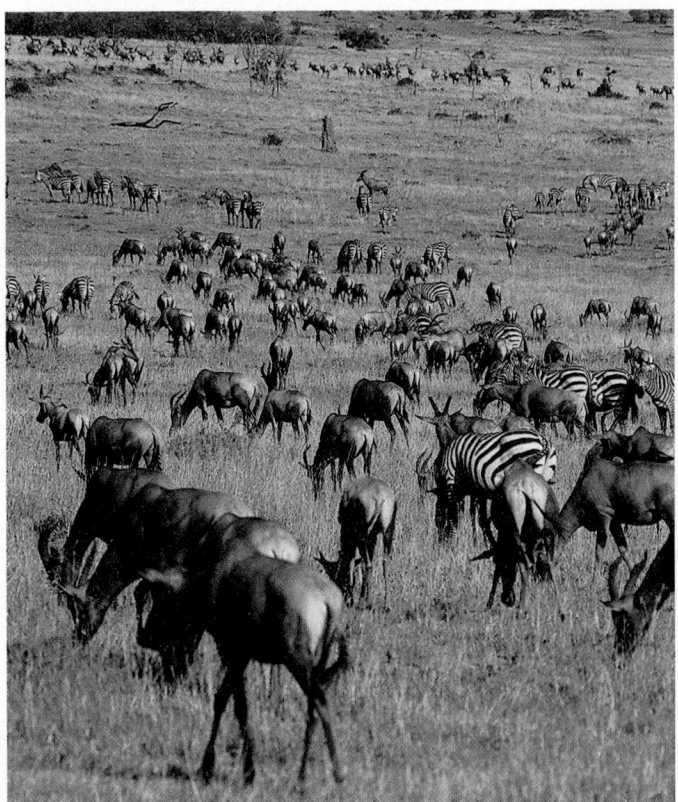

Figure 36.11 Mixed herd of topi and common zebra grazing on a savanna of tropical Africa.

variety of predators and prey support the notion that the larger the group is, the less likely is an individual within the group to be eaten.

Sociality offers several potential benefits to reproduction. It facilitates encounters between males and females, which can consume much time and energy for solitary animals. Sociality also synchronizes reproductive behavior through the mutual stimulation that individuals give one another. Furthermore, parental care that social animals provide their offspring increases survival of the brood (Figure 36.12). Social living provides opportunities for individuals to give aid and to share food with young other than their own. Such

Figure 36.12   An infant yellow baboon, *Papio cyanocephalus,* "jockey rides" its mother. Later, as the infant is weaned, the mother-infant bond weakens and the infant will be refused rides.

Figure 36.13   Japanese macaque washing sweet potatoes. The tradition began when a young female named Imo began washing sand from the potatoes before eating them. Younger members of the troop quickly learned the behavior.

interactions within a social network have produced some intricate cooperative behavior among parents, their young, and their kin.

Of the many other advantages of social organization noted by ethologists, we mention only a few in this brief treatment: cooperation in hunting for food; huddling for mutual protection from severe weather; opportunities for division of labor, which is especially well developed in social insects; and the potential for learning and transmitting useful information through a society.

Observers of a seminatural colony of macaque monkeys in Japan recount an interesting example of acquiring and passing tradition in a society. The macaques were given sweet potatoes and wheat at a feeding station on the beach of an island colony. One day a young female named Imo was observed washing the sand off a sweet potato in seawater. The behavior was quickly imitated by Imo's playmates and later by Imo's mother. Still later, when the young members of the troop became mothers, they waded into the sea to wash their potatoes; their offspring imitated them without hesitation. The tradition became firmly established in the troop (Figure 36.13).

Some years later, Imo, as an adult, discovered that she could separate wheat from sand by tossing a handful of sandy wheat in the water; allowing the sand to sink, she could retrieve the floating wheat to eat. Again, within a few years, wheat-sifting became a tradition in the troop.

Imo's peers and social inferiors copied her innovations most readily. The adult males, her superiors in the social hierarchy, would not adopt the practice but continued laboriously to pick wet sand grains off their sweet potatoes and to scout the beach for single grains of wheat.

Acquisition of food-cleaning skills by Imo and her peers shows that a social setting provides opportunities for acquisition and sharing of complex learned behaviors, ones that transcend simple imprinting and habituation. Imo's food-cleaning behaviors demonstrate a conditioned response, learning that certain methods lead repeatedly to a desired result, plus reasoning and insight for evaluating which methods are useful for cleaning various food items.

Social living also can have disadvantages compared with a solitary existence. Species that survive by camouflage from potential predators profit by being dispersed. Large predators benefit from a solitary existence for a different reason: their requirement for a large supply of prey. Thus there is no overriding adaptive advantage to sociality that inevitably disfavors a solitary way of life. Advantages and disadvantages of sociality depend upon the ecological situation.

Agonistic or Competitive Behavior

Animals may fight for access to food, water, sexual mates, or shelter when such resources are limited. Much of what animals do to resolve conflicts is called **aggression,** which we define as an offensive physical action, or threat, to force others to abandon something they own or might attain. Many ethologists consider aggression part of a somewhat more inclusive category called **agonistic** (Gr. contest) behavior, any activity related to fighting, whether it be aggression, defense, submission, or retreat.

Most aggressive encounters lack the violence that we usually associate with fighting. Many species possess specialized weapons such as sharpened teeth, beaks, claws, or horns used for protection from, or predation on, other species. Although potentially dangerous, such weapons are seldom used in any severely damaging way against members of *their own species.*

Animal aggression within a species seldom produces injury or death because animals have evolved many symbolic **ritualized threat displays** that carry mutually understood meanings to establish a dominance hierarchy within the population. A ritualized display is a behavior that has been modified through evolution to make it increasingly effective in serving a communicative function. Through **ritualization,** simple movements or traits become more intensive, conspicuous, or precise, and acquire the function of a signal to reduce misunderstanding. Fights over mates, food, or territory become ritualized jousts rather than bloody, no-holds-barred battles. When fiddler crabs (see Figure 20.26C) spar for reproductive territory on intertidal sands, their large claws usually are only slightly opened. Even in intense fighting when claws are used, the crabs grasp each other in a way that prevents reciprocal injury. Male venomous black mambas, *Dendroaspis polylepis,* competing for a female engage in stylized bouts by winding themselves together; each attempts to butt the other's head with its own until one snake becomes pinned to the ground and retreats. The rivals do not bite each other. Males of many fish species contest territorial boundaries with lateral threat displays, puffing themselves to look as threatening as possible. The encounter is usually settled when either animal perceives itself obviously inferior in the social hierarchy, withdraws, and swims away. Rival giraffes engage in largely symbolic "necking" matches in which two males standing side by side wrap and unwrap their necks (Figure 36.14). Neither uses its potentially lethal hooves on the other, and neither is injured.

Thus animals fight as though programmed by rules that prevent serious injury. Fights between rival bighorn rams are spectacular to watch, and the sound of clashing horns may be heard for hundreds of meters (Figure 36.15), but the skull is so well protected by massive horns that injury occurs only by accident. Nevertheless, despite these constraints, aggressive encounters on occasion can be true fights to the death of a rival. If male African elephants, *Loxodonta africana,* are unable to resolve dominance conflicts painlessly with ritual postures, they resort to incredibly violent battles, with each trying to plunge its tusks into the most vulnerable parts of the opponent's body.

More commonly, however, the loser of a ritualized encounter may simply run away, or signal defeat by a specialized subordination ritual. A likely loser profits by communicating submission as quickly as possible, thereby avoiding a thrashing. Such submissive displays that signal the end of a fight can be almost the opposite of threat displays (Figure 36.16). In his book, *The Expression of the Emotions in Man and Animals* (1872), Charles Darwin described the seemingly opposite nature of threat and appeasement displays as the "principle of antithesis." The principle remains accepted by ethologists today.

The winner of an aggressive competition is dominant to the loser, the subordinate. For the victor, dominance means enhanced access to all contested resources that contribute to reproductive success: food, mates, and territory. In a social species, dominance interactions often form a **dominance hierarchy.** One animal at the top wins encounters with all other members in the social group; the second in rank wins all but those with the top-ranking individual.

A simple dominance hierarchy is often called a "peck order," following Schjelderup-Ebbe's original description of social dominance mediated by pecking in chickens, *Gallus gallus.* Chickens typically live in flocks, and young chicks aggressively peck one another when competing for access to food or a water dish. Chickens learn their relative strengths within the flock, and pecking from a stronger individual causes a weaker one to surrender a contested resource. Hens high in the peck order drive weaker hens from a favored nesting site. The top rooster dominates calling and breeding. Top hens and roosters get unquestioned access to food and water. The system works because it reduces social tensions that would constantly surface if animals had to fight constantly over resources.

Not all dominance hierarchies have clear-cut dominant and subordinate individuals. In some hierarchies, dominant animals are frequently challenged by subordinates. Subordinates in any social order may never reproduce and are often the first to die during times of food scarcity.

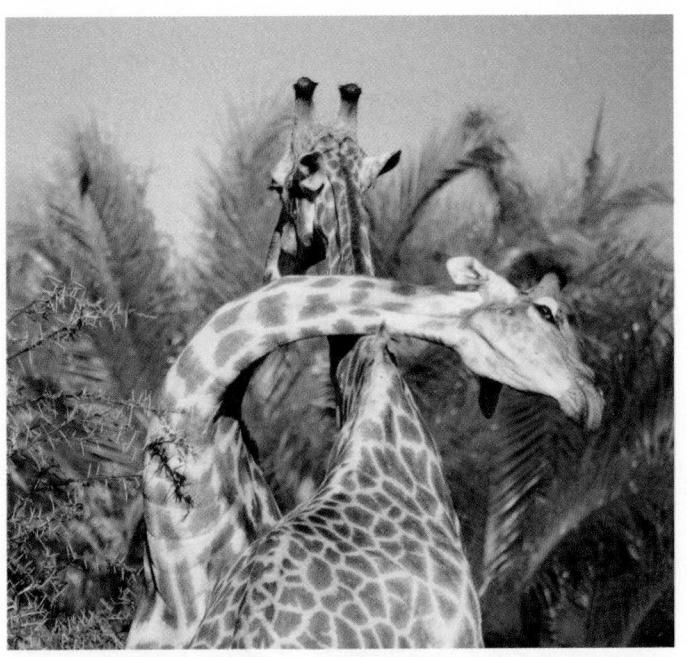

Figure 36.14 Male Masai giraffes, *Giraffa camelopardalis,* fight for social dominance. Such fights are largely symbolic, seldom causing injury.

Figure 36.15 Male bighorn sheep, *Ovis canadensis,* fight for social dominance during the breeding season.

Territoriality

Territorial ownership is another facet of sociality in animal populations. A territory is a fixed area whose occupant excludes intruders *of the same species* and sometimes other species. This exclusion involves defending the area from intruders and being conspicuous on the site. Territorial defense occurs in numerous animals: insects, crustaceans, fishes, amphibians, lizards, birds, and mammals, including humans. Sometimes the space defended moves with the individual. Individual distance, as one example, can be observed in the spacing between swallows or pigeons on a wire, gulls on a beach, or people waiting for a bus.

Territoriality is generally an alternative to dominance behavior, although both systems can operate in the same species. A territorial

Figure 36.16 Darwin's illustration of his principle of antithesis as shown by the postures of dogs. **A,** A dog approaches another dog with hostile, aggressive intentions. **B,** The same dog is in a humble and conciliatory state of mind. The signals of aggressive display have been reversed.

Source: C. Darwin, *Expression of the Emotions in Man and Animals,* Appleton and Co., Newyork, 1872.

Figure 36.17 Northern gannet, *Morus bassanus* nesting colony. Note precise spacing of nests with each occupant just beyond pecking distance of its neighbors.

system may work well when the population is low, but it may fail with increasing population density and be replaced with dominance hierarchies when all animals occupy a common space.

Like every other competitive endeavor, territoriality carries both costs and advantages. It is beneficial when it ensures access to limited resources, *unless* territorial boundaries cannot be maintained with little effort. Presumed benefits of a territory are, in fact, numerous: uncontested access to a foraging area; enhanced attractiveness to females, thus reducing problems of pair bonding, mating, and rearing young; reduced disease transmission; reduced vulnerability to predators. Advantages of holding a territory wane if an individual must spend most of its time in boundary disputes with neighbors and other intruders.

Most of the time and energy required for territoriality are expended when the territory is first established. Once boundaries are located they tend to be respected, and aggressive behavior diminishes as territorial neighbors come to recognize each other. Indeed, neighbors may look so peaceful that an observer not present when the territories were established may conclude incorrectly that the animals are not territorial. In Gálapagos sea lions, *Zalophus wollebaeki,* a dominant male with many females seldom quarrels with neighbors, who have their own territories to defend. However, he must be constantly vigilant against bachelor bulls who challenge his mating privileges, and one of which ultimately will replace him, usually within a few months.

Birds are conspicuously territorial. A male song sparrow, *Melospiza melodia,* for example, has a territory of approximately three-fourths of an acre. In any given area, the number of song

sparrows remains approximately the same each year. The population remains stable because the young occupy territories of adults that die. Any surplus in the song sparrow population is excluded from territories and thus not able to mate.

Sea birds such as gulls, gannets, boobies, and albatrosses occupy colonies that are divided into very small territories just large enough for nesting (Figure 36.17). Territories of these birds cannot include their fishing grounds, since they all forage at sea, where their food is always shifting in location and shared by all.

Territorial behavior is not as prominent with mammals as it is with birds. Mammals are less mobile than birds, making it more difficult for them to patrol a territory for trespassers. Instead, many mammals have **home ranges** (p. 626). A home range is the total area that an individual traverses in its activities. It is not an exclusive, defended preserve but overlaps home ranges of other individuals of the same species.

For example, home ranges of baboon troops overlap extensively, although a small part of each range becomes the recognized territory of each troop for its exclusive use. Home ranges may shift considerably among the seasons. A baboon troop may have to shift to a new range during the dry season to obtain water and better grass. Elephants, before their movements were restricted by humans, made long seasonal migrations across the African savanna to new feeding ranges. However, home ranges established for each season are remarkably consistent in size.

Mating Systems

Animals display diverse mating systems. Behavioral ecologists generally classify mating systems by the ways that males and females associate for mating. **Monogamy** is an association between one male and one female at a time. **Polygamy** is a general term that

Figure 36.18 Two elephant seals, *Mirounga angustirostris,* fight to establish dominance. Males are much larger than females in this highly polygynous society.

Figure 36.19 Sage grouse *(Centrocercus urophasianus)* male displaying to females, North America.

incorporates all multiple mating systems where females or males have more than one mate. **Polygyny** refers to a male that mates with more than one female. **Polyandry** is a system in which a female mates with more than one male. There are specific types of polygyny. **Resource-defense polygyny** occurs when males gain access to females indirectly by holding critical resources. For example, female bullfrogs, *Lithobates catesbeianus* (p. 552), prefer to mate with males who are larger and older. These males defend territories of higher quality than those of smaller males because their territories have favored temperatures or are free of predatory leeches. **Female-defense polygyny** occurs when females aggregate and, consequently, are defendable. Thus, when female elephant seals occupy a small island, dominant males can defend and mate with them relatively easily (Figure 36.18). This situation was previously called a "harem." **Male-dominance polygyny** occurs when females select mates from aggregations of males competing for an opportunity to mate. For example, some animals form **leks.** A lek is a communal display ground where males congregate to attract and to court females. Females choose and mate with the male having the most attractive qualities (Figure 36.19). Leks characterize some birds, including prairie chickens and sage grouse. In these systems, sexual selection (p. 125) is often intense, producing evolution of bizarre courtship rituals and exaggerated morphological traits.

Cooperative Behavior, Altruism, and Kin Selection

If, as Darwin suggested, animals should behave selfishly and strive to produce as many offspring as possible, why do some animals help others at some risk to themselves? As noted earlier (p. 785), general advantages of a group context may explain selection for cooperative behavior; however, some forms of cooperative behavior are so extreme that they would seem to require an additional explanation. Why do some individuals forgo breeding to serve the reproductive success of others? Why do some individuals appear to sacrifice themselves so that other members of their group can survive? Until the mid-1960s, scientists had trouble explaining in Darwinian terms how such **altruistic behavior** might persist in a population.

Most altruistic behaviors formerly were explained using a **group-selection** argument. Group selectionists suggested that animals that helped others or that failed to mate did so for the benefit of the group as a whole. Therefore, such behaviors produced increased survivorship of groups whose members behaved altruistically. According to proponents of this argument, selection occurs at the level of the group, not at the level of the individual as Darwin suggested. However, the group-selection argument as originally proposed by V. C. Wynne-Edwards in 1962 has been rejected by most behavioral ecologists for several reasons.

For example, if alleles associated with a risky altruistic behavior, such as giving calls to warn others of predators, were distributed randomly in a social group, those lacking such genes would flourish. They would be warned with no risk to themselves; their chances of reproduction would be greater and, in time, the "selfish" alleles would eliminate the altruistic ones from the group's gene pool.

In 1964, W. D. Hamilton, based largely on his studies of insects, proposed a new way to explain altruistic behavior by modifying Ronald Fisher's concept of the fitness of an allele (p. 125). In Fisher's original formulation of the genetic theory of natural selection, a particular allele has high fitness if its possessors on average contribute more offspring to the next generation than do individuals having contrasting alleles at the locus. Fisher's model assumed additive inheritance, in which an individual homozygous for a favorable allele would have more offspring on average than would an individual homozygous for a different allele, and fitness of heterozygous individuals would be the average of the contrasting homozygous genotypes. Selection would increase the frequency of the favorable allele in the gene pool based on the allele's direct effects on fitness of its possessors. Hamilton argued that an allele also might have indirect effects that would increase its frequency in the gene pool. The indirect effects occur if the possessor of an allele acts to increase fitness of other individuals that carry the same allele. An allele that promotes social behavior directed preferentially toward the welfare of close relatives could increase its frequency in the gene pool through indirect effects, because close relatives likely have the same allele. Parental behaviors that benefit survival of offspring, as studied in birds (p. 603), are an obvious example. Thus, if everything else were equal, brothers who on average share half their alleles would be more likely to aid one another than they would a cousin who shares on average only 25% of their alleles. Hamilton's hypothesis based on this genetic explanation for altruism and cooperation is called **kin selection.** Essentially, kin selection operates by individuals assisting the survival and reproduction of other individuals who possess the same genes by common descent. In an extreme case, an individual who dies in an act that saves a large number of siblings might favor its alleles' presence in the gene pool despite self-sacrifice. Indirect effects on fitness of an allele promoting such behavior outweigh its direct effects on fitness in this case.

Hamilton's hypothesis revolutionized evolutionary and behavioral biology. The main criterion of Darwinian fitness in genetic models of natural selection is the relative number of an individual's alleles that are passed to future generations. Hamilton, however, developed the concept of **inclusive fitness,** which is the relative number of an individual's alleles that are passed to future generations as a result of either an individual's own reproductive success *or that of related individuals.* Thus, kin selection and inclusive fitness may explain many altruistic behaviors that have perplexed biologists for many years.

A good example of altruism and kin selection in nature is the remarkable cooperation and coordination among social insects such as ants, bees, and wasps. Through haplodiploidy (p. 137), where males are haploid and females are diploid, full sisters are genetically related on average by 75% rather than 50% (Figure 36.20). Sisters are more closely related to each other than to their own daughters! Therefore, they cooperate with other members of their social group, forego breeding themselves, and aid their queen to produce more sisters who are more closely related (75% related) than potential offspring (50% related). This explanation is challenged by molecular genetic discoveries that in many haplodiploid insects, reproductive females have multiple mates. The nonreproductive females who tend their mother's younger offspring indiscriminately are therefore not likely to be tending only their full siblings. An average relatedness between nonreproductive workers and their tended siblings would need to exceed 50% for kin selection to provide a convincing explanation for reproductive division of labor (often called **eusociality**) in Hymenoptera.

Female Belding's ground squirrels, *Spermophilus beldingi,* found in the High Sierra of California, give alarm calls when a predator approaches (Figure 36.21). Alarm calling warns other members of the social group and is risky to the alarm caller. However, benefits to alarm calling outweigh risks because alarm callers are warning *related* individuals. Thus, alarm-calling behavior, even if it is risky, may be favored by selection if it increases inclusive fitness of a caller.

Through experimental studies we now know that animals of many kinds discriminate kin and nonkin: isopods, insects, fishes, frog and toad tadpoles, birds, squirrels, and monkeys. Individuals of some species can even discriminate between full siblings and half-siblings and between cousins and unrelated individuals. Thus, some species have a finely tuned ability to identify relatives of various degrees of relatedness. Cues used in kin recognition vary among species. Birds often use vocalizations, whereas chemical cues are common in other groups.

Because altruistic behavior occurs among unrelated individuals in many natural populations, kin-selection theory does not explain all altruistic behavior. The theory of **reciprocal altruism,** formulated initially by Robert Trivers, provides an additional Darwinian foundation for explaining altruistic behaviors among individuals, including ones who are not close relatives. According to this theory, an individual is selected to perform altruistic acts if

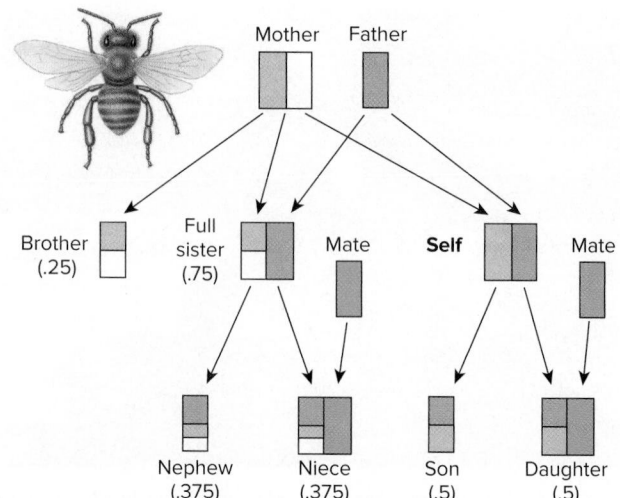

Figure 36.20 Haplodiploidy in honeybees, showing degrees of relatedness of a female worker bee (labeled Self) to individuals she might raise. In honeybees, as in other haplodiploid animals, diploid females develop from fertilized eggs, and males develop from unfertilized eggs. Each daughter of a male gets all his genes (purple bar) and *full* sisters receive an identical one-half of their genome from the same father. Open bars represent other, unrelated alleles. Because full sisters also share half the genes they receive from their common mother (orange bar), the relatedness of Self to a full sister is 0.75, the average of 0.5 and 1.0. (In a diploid–diploid system as in humans, the relatedness of siblings is 0.5 because both paternally and maternally inherited genes have a 50% chance of being present in a sibling.) Note that relatedness of female workers to a brother is only 0.25, because brothers are fatherless. If the mother bee has mated with more than one male, average relatedness of a worker bee to her mother's immature daughters will vary between 0.5 and 0.75.

Figure 36.21 Belding's ground squirrel, *Spermophilus beldingi,* gives an alarm call to warn of an approaching predator. This risky behavior endangers the callers to benefit noncallers. Females whose young are nearby give more such alarm calls than do other individuals.

doing so increases that individual's chances of receiving equal or more valuable favors from others. Reciprocal altruism is most likely to evolve in species that have stable social groupings whose members are mutually interdependent for defense, nutrition, or reproduction, with frequent opportunities for altruistic interaction.

Darwinian selective explanations for social behavior have been investigated extensively using the mathematical theory of games, models constructed initially to evaluate the economic consequences of cooperation among individuals in business. Behavioral ecologists have adapted these models to determine which behaviors may qualify as an **evolutionary stable strategy** or **ESS**. An ESS is expected to persist over long periods of evolutionary time because it prevails in competition with alternative strategies that might arise. An altruistic behavior would not be an ESS if it were subject to cheaters who could preferentially obtain altruistic behaviors from unrelated others without giving in return. Because altruistic behavior is hard to study in nature, results of ESS theory help to focus investigations on populations and behaviors that are most likely to show evolutionary stability. Ritualized displays of animal aggression (p. 787) are considered strong examples of evolutionary stable strategies because they prevent individuals from escalating conflicts to the point of serious injury and thereby enhance survivorship. Selection should favor ritualized displays over alternative behaviors that would subject an individual to violence. An important research topic is to determine how truthful are the displays used by animals to avoid violent conflict or to procure mates. Would it be an evolutionary stable strategy for organisms to evolve displays that deceptively overstate their actual strength or desirability for mating?

Gerald Wilkinson's studies of food sharing by vampire bats show that individuals do reciprocate altruistic behavior. Vampire bats, *Desmodus rotundus,* aggregate in resting areas and leave those areas at night to obtain blood meals from large mammals, which are often difficult to find. A vampire bat that obtains a blood meal may perform the altruistic behavior of regurgitating blood to other hungry members of its group. Wilkinson used laboratory experiments to show that hungry bats were not fed at random, but that individual bats were more likely to be fed by those to whom they had previously provided the same service. These results confirm that vampire bats recognize each other as individuals, remember which individuals have performed altruistic behaviors, and then return favors. Despite such encouraging results, reciprocal altruism is difficult to study in nature because it generally requires long-term observations of marked individuals and may occur simultaneously with kin selection.

Animal Communication

Only through communication can one animal influence behavior of another. Compared to the enormous communicative potential of human speech, however, nonhuman communication is severely restricted. Animals might communicate by sounds, scents, touch (including electric and thermal signals), pheromones (as sensed by antennae in insects and vomeronasal glands in mammals), and movement. Indeed any sensory channel or combination of channels might be used, giving animal communication richness and variety.

Phylogenetic studies are important for testing hypotheses of the evolution of mating-behavioral and morphological characters by sexual selection. Phylogenetic and behavioral studies of swordtail fish species by Alexandra Basolo show that female preference for the male sword evolved prior to evolution of the sword in genus *Xiphophorus,* consistent with the hypothesis that initial formation and elongation of the sword in males occurred through sexual selection by female preference. Another important phylogenetic study of mating behavior reveals evolution of male displays in leks of neotropical manakins, family Pipridae. Richard Prum identified 44 behavioral characters that have been used to evolve species-specific behavioral displays in these birds, and he reconstructed the historical sequence by which these displays evolved. His results show a general evolutionary trend toward increased complexity of displays and a tendency for behavioral changes to precede changes in plumage. A new behavioral display that highlights a particular area of plumage subjects that plumage to sexual selection for morphological elaboration. These studies reveal that behavioral evolution may be a major factor determining the action of selection on morphological characters. Some evolutionists have proposed that behavioral evolution generally accelerates morphological evolution, and that a change in behavior is often a critical factor permitting evolution of new adaptive zones.

Unlike our language, which is composed of words with definite meanings that may be rearranged to generate an almost infinite array of new meanings and images, communication of other animals consists of a limited repertoire of signals. Typically, each signal conveys one and only one message. These messages are not divided or rearranged to construct *new kinds* of information. A single message from a sender may, however, contain several bits of relevant information for a receiver.

A cricket's call announces to an unfertilized female the species of the sender (males of different species have different songs), his sex (only males sing), his location (source of the song), and his social status (only a male able to defend the area around his burrow calls from one location). This information is crucial to a female and accomplishes a biological function. However, there is no way for a male to alter his song to provide additional information concerning food, predators, or habitat, which might improve his mate's chances for survival and thus enhance his own fitness.

Chemical Sex Attraction in Moths

Mate attraction in silkworm moths illustrates an extreme case of stereotyped, single-message communication that has evolved to serve a single biological function: mating. Virgin female silkworm moths have special glands that produce a chemical sex attractant to which males are sensitive. Adult males use their large bushy antennae, covered with thousands of sensory bristles that function as receptors (Figure 36.22), to detect the chemical attractant **bombykol** (a complex alcohol named after the silkworm, *Bombyx mori*).

To attract males, females sit quietly and emit a minute amount of bombykol, which is carried downwind. When a few molecules reach a male's antennae, he is stimulated to fly upwind in search of the female. His search is at first random, but, when by chance he approaches within a few hundred yards of the female, he encounters a concentration gradient of the attractant. Guided by the gradient, he flies toward the female, finds her, and copulates with her.

In this example of chemical communication the attractant bombykol, a pheromone (p. 735), serves as a signal to bring the sexes together. Its effectiveness is ensured because natural selection favors evolution of males with antennal receptors sensitive enough to detect the attractant at great distances (several kilometers). Males whose genotype produces a less sensitive sensory system fail to locate a female and thus are reproductively eliminated from the population.

Language of Honeybees

Among the most sophisticated and complex of all nonhuman communication systems is the symbolic language of honeybees. Bee colonies construct a honeycomb, a vertical surface of hexagonal wax cells containing larvae and stores of honey and pollen brought to the hive by worker females. If nectar and pollen are abundant around the hive, workers do not need to communicate to others the locations of food that they have found. When food is scarce, a forager bringing nectar or pollen to the hive directs others to the food resource by walking ritualized patterns on the surface of the honeycomb among the other workers.

If food is nearby (less than 50 m from the hive), the forager performs what researchers call a **round dance**: the forager walks a complete clockwise circle, then turns and walks a counterclockwise circle, and repeats this sequence of movements many times. The round dance signals other workers to fly from the hive, searching nearby areas for nectar and pollen. If the food resource is farther away, the forager uses a more complex set of walking movements called a **waggle dance** (Figure 36.23). The waggle dance provides both distance and directional information to guide other workers

toward the food. The waggle dance takes a figure-eight pattern with three parts per cycle: (1) a circle whose diameter is about three times the length of the bee, (2) a straight run while waggling the abdomen from side to side and emitting a pulsed, low-frequency sound, and (3) another circle, turning in the opposite direction from the first. This sequence is repeated many times. The straight, waggle run (part 2 of the cycle) communicates direction of the food source using the sun as a reference point. Waggle dances thus require clear weather. If the food is directly toward the sun, the waggle run moves straight upward over the vertical surface of the honeycomb. If the food lies 60 degrees to the right of the sun, the waggle run is 60 degrees right of vertical. The waggle run thus points at the same angle relative to the vertical as the food is located relative to the sun. The waggle dance communicates the food's distance by varying its tempo. If food is about 100 m away, each figure-eight cycle lasts about 1.25 seconds; if 1000 m away, it lasts about 3 seconds; and if about 8 km (5 miles) away, it lasts 8 seconds.

Studies of foraging communication in honeybees provide an outstanding example of how active controversy drives scientific discovery. The precise information communicated by bee dances on location of food sources was discovered in 1943 by Karl von Frisch. This discovery was strongly challenged beginning in 1967 by Adrian Wenner and Patrick Wells, whose experiments indicated that a bee's dance was not needed to communicate position of food sources, and that odors could best explain a bee's foraging communications. The resulting conflict stimulated many researchers to design new experiments to test these hypotheses, ultimately revealing that each side of the original controversy had revealed a different strategy among a large repertoire of communication mechanisms that bees use in different environmental contexts. Among the experiments demonstrating that bees communicate precise information through dances are ones in which a robot bee with vibrating metal wings was moved through the waggle dance and its associated sounds. When operated in a hive, the computer-directed robot successfully recruited attending bees to visit preselected food dishes outside the hive that had not been visited previously. This work showed that the sounds associated with the dance were as important as its visual appearance. The dance is used, however, mainly when food near the hive is limited and distant food sources are ephemeral; when a scented food source is readily and continuously available nearby, bees communicate primarily using the odor cues reported by Wenner and Wells. A major message of James and Carol Gould's work on honeybees is that bees use several sensory systems to navigate, and that a primary system has backups in case it fails. Discovery of this full range of sensory communication mechanisms, and the circumstances conducive to a bee's use of each one, owes a great debt to the controversy sparked by the apparently conflicting, but ultimately consolidated, findings of von Frisch, Wenner, and Wells.

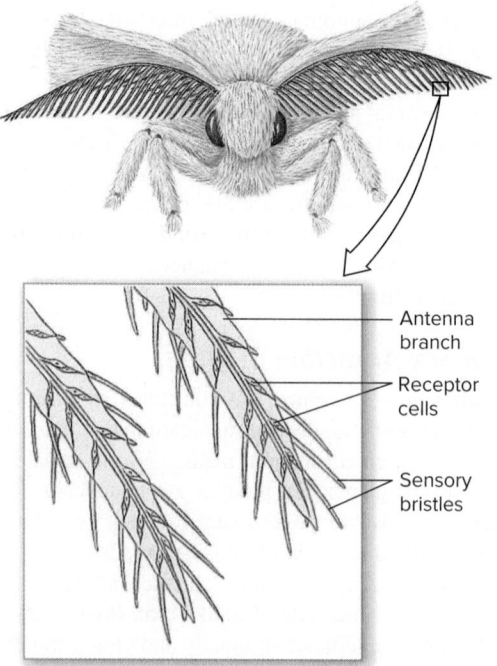

Figure 36.22 Large antennae of a male silkworm moth, *Bombyx mori*; these are especially sensitive to the sex attractant (pheromone) released by the female moth.

Antenna branch

Receptor cells

Sensory bristles

Communication by Displays

A display is a kind of behavior or series of behaviors that serves communication. The release of sex attractant by a female moth and the dances of bees just described are examples of displays; so are

Figure 36.23 Waggle dance of honeybees used to communicate both the direction and the distance of a food source. The straight run of the waggle dance indicates direction according to the position of the sun (angles X and Y).

alarm calls of herring gulls, songs of the white-crowned sparrow, courtship dances of the sage grouse, and "eyespots" on the hind wings used by certain moths to startle potential predators.

The elaborate pair-bonding displays of blue-footed boobies, *Sula nebouxii* (Figure 36.24), are performed with maximum intensity when the birds come together after a period of separation. The male at right in the illustration is sky pointing: the head and tail are pointed skyward and the wings are swiveled forward to display their glossy upper surfaces to the female. This display is accompanied by a high-piping whistle. The female at left is parading. She steps with exaggerated slow deliberation, lifting each brilliant blue foot in turn, as though holding it aloft momentarily for the male to admire. Such highly personalized displays, performed with droll solemnity, appear comical, even inane, to an observer. Indeed the boobies, whose name is derived from the Spanish word "bobo" meaning clown, presumably were so designated for their amusing antics.

The exaggerated nature of the displays ensures that the message is not missed or misunderstood. Such displays are essential to establish and to maintain a strong pair bond between male and female. This requirement also explains the repetitious nature of displays that follow one another throughout courtship and until laying of eggs. Repetitious displays maintain a state of mutual stimulation between male and female, ensuring the cooperation necessary for copulation and subsequent incubation and care of young.

Animal Cognition

One of the most fascinating subjects in animal behavior is animal intelligence and awareness. Animal cognition is a general term for mental function, including perception, thinking, and memory. Many biologists consider some mental processes of animals similar to those of humans. Studies on animal cognition with nonhuman primates and African grey parrots have yielded fascinating results.

In the late 1960s Beatrix and Allen Gardner of the University of Nevada in Reno began using American Sign Language (ASL) to train a chimpanzee named Washoe to communicate with her hands as deaf people do. By age five Washoe could sign 132 words, which she could put into strings forming sentences and phrases. She could answer questions, make suggestions, and convey moods. In one session, when asked what a swan was, Washoe answered "water bird." Washoe also taught signs to other chimpanzees. At first, signs were used as play but soon the chimpanzees used them to make spontaneous requests to trainers such as "drink," "tickle," and "hug." Similar work has been done with other primates, including gorillas, orangutans, and bonobos.

Irene Pepperberg of the University of Arizona worked for years with an African grey parrot named Alex. Because parrots can vocalize like humans, Pepperberg could communicate with Alex using human vocal language. Over the years Alex learned a number of attributes, including colors, shapes, and materials for more than 100 objects. Alex could identify objects by colors, size, and shape. Thus, if Alex was given two objects of the same color but one larger than the other, he could state that the difference between them was "size." Alex also could count and relate to the trainer how many objects of each particular category were present.

Conscious awareness is also part of cognition. Donald Griffin wrote two books suggesting that many animals are self-aware and can think and reason. The ability of apes, parrots, and other animals to use language-related skills is significant because it shows us their cognitive abilities so that we can begin to communicate with them. In a recent book, animal behaviorist Marc Bekoff and bioethicist

Figure 36.24 A pair of Galápagos blue-footed boobies, *Sula nebouxii,* display to each other. A male *(right)* is sky pointing; a female *(left)* is parading. Such vivid, stereotyped, communicative displays serve to maintain reciprocal stimulation and cooperative behavior during courtship, mating, nesting, and care of the young.

Jessica Pierce discuss evidence that animals have recognizable emotions and moral intelligence. Interpretations of animal cognition remain highly controversial.

The animal behaviorist Irven DeVore reported how choosing the proper channel for dialogue can have more than academic interest:[1]

> One day on the savanna I was away from my truck watching a baboon troop when a young juvenile came and picked up my binoculars. I knew if the glasses disappeared into the troop they'd be lost, so I grabbed them back. The juvenile screamed. Immediately every adult male in the troop rushed at me—I realized what a cornered leopard must feel like. The truck was 30 or 40 feet away. I had to face the males. I started smacking my lips very loudly, a gesture that says as strongly as a baboon can, "I mean you no harm." The males came charging up, growling, snarling, showing their teeth. Right in front of me they halted, cocked their heads to one side—and started lip-smacking back to me. They lip-smacked. I lip-smacked, "I mean you no harm." "I mean *you* no harm."
> It was, in retrospect, a marvelous conversation. But while my lips talked baboon, my feet edged toward the truck until I could leap inside and close the door.

 ## SUMMARY

Animal behavior has emerged as a scientific discipline from several different approaches. Comparative psychology emphasizes the identification of mechanisms controlling behavior, using relatively few species, with the intent that these mechanisms might have wide applicability among animals. Ethology is the study of behavior, both innate and learned, of animals in their natural habitats. Behavioral ecologists have shown that behavioral traits have evolutionary histories and may evolve by natural selection. Behavioral ecologists distinguish between studies that focus on the mechanisms of behavior

(proximate causation) and those that focus on function or evolution of behavior (ultimate causation).

Students of animal behavior have observed and cataloged many behavioral patterns of animals that are highly predictable and almost invariable in performance. Often these patterns are triggered, or "released," by specific, and usually simple, environmental stimuli, called sign stimuli. Although such formalized behaviors might be released inappropriately at times, they are efficient and enable the animal to respond rapidly. Development of behavioral patterns depends on an interaction between an organism and its environment. Calling formalized behaviors "instinctive" or "innate" implies stability of critical environmental components of the behaviors.

Behavior can be modified by learning through experience. Two simple kinds of learning behavior are habituation, which is reduction or elimination of a behavioral response in the absence of any reward or punishment; and sensitization, in which a repeated stimulus increases the strength of a behavioral response. We describe the gill-withdrawal reflex of the marine mollusc, *Aplysia,* as a protective response that can be modified experimentally to show either habituation or sensitization. Modification of the alarm response of herring-gull chicks is another example of habituation. Another form of learning is imprinting, the lasting recognition bond that forms early in life between the young of many social animals and their mothers.

Social behavior is behavior arising from interactions of members of a species with one another. In social organizations, animals tend to remain together, communicate with each other, and usually resist intrusions by "outsiders." Advantages of sociality include cooperative defense from predators, cooperative searching for food, improved reproductive performance and parental care of young, and transmission of useful information through the society. Because social animals compete with one another for resources (such as food, sexual mates, and shelter), conflicts are often resolved by a form of overt hostility called aggression. Most aggressive encounters between conspecifics are stylized bouts involving more bluff than intent to injure or to kill. Dominance hierarchies, in which a priority of access to common resources is established by aggression, are common in social organizations. Territoriality is an alternative to dominance. A territory is a defended area from which intruders of the same species are excluded.

Mating systems include monogamy, the mating of an individual with only one partner of the opposite sex each breeding season, and polygamy, the mating of an individual with two or more partners in a breeding season. Two

---

[1]DeVore, Irven. The marvels of animal behavior. 1972. Washington, D.C., National Geographic Society.

forms of polygamy are polygyny, the mating of a male with more than one female, and polyandry, the mating of a female with more than one male. Polygyny may occur directly, by males defending their access to females, or indirectly, by males defending a resource.

A behavior in which one animal may risk its own fitness to increase fitness of others is called altruistic behavior. Examples of risky behaviors include one member of a social group warning others of a predator, and cooperative behavior among social insects in which an individual may forgo reproduction to raise its younger siblings. A common explanation of altruism is kin selection, in which the recipient of an altruistic act is sufficiently closely related to the altruist that the recipient's survival would benefit the genes shared with the altruist. An alternative explanation is reciprocal altruism, in which individuals perform altruistic behaviors toward others from whom they have received altruistic acts.

Communication, often considered the essence of social organization, is the means by which animals influence behavior of other animals, using sounds, chemical signals, visual displays, touch, or other sensory signals. Compared to the richness of human language, animals communicate with a very limited repertoire of signals. One of the most famous examples of animal communication is that of the symbolic dances of honeybees. Birds communicate by calls and songs and, especially, by visual displays. By ritualization, simple movements have evolved to become conspicuous signals having definite meanings.

REVIEW QUESTIONS

1. How do experimental approaches of comparative psychology and behavioral ecology differ? Comment on the aims and methods employed by each.
2. Egg-retrieval behavior of greylag geese is an excellent example of a highly predictable behavior. Interpret this behavior within the framework of classical ethology, using these terms: releaser, sign stimulus, and stereotyped behavior. Interpret the territorial defense behavior of male three-spined sticklebacks in the same context.
3. The idea that behavior must be *either* innate or learned has been called a "nature versus nurture" controversy. What reasons are there for believing that such a strict dichotomy does not exist?
4. Two kinds of simple learning are habituation and imprinting. Distinguish these two types of learning, and offer an example of each.
5. Some strains of bees show hygienic behavior by uncapping cells containing larvae infected with a bacterial disease called foulbrood and removing the dead larvae from the hive. What evidence shows that this behavior is transmitted by two independently segregating genes?
6. Discuss some advantages of sociality for animals. If social living has so many advantages, why do many animals successfully live alone?
7. Suggest why aggression, which might seem counterproductive, exists among social animals.
8. What is the selective advantage to winners, as well as to losers, that aggressive encounters within species for social dominance are usually ritualized displays or symbolic fights rather than unrestrained fights to death?
9. Of what use is a territory to an animal, and how is a territory established and kept? What is the difference between territory and home range?
10. Polygyny is a form of polygamy in which a male mates with more than one female. Explain how three forms of polygyny differ from each other: resource-defense polygyny, female-defense polygyny, and male-dominance polygyny.

11. Give an example of an altruistic behavior and explain how such behavior conflicts with Darwin's expectation that animals should act selfishly to produce as many offspring as possible.
12. Earlier explanations of altruistic behavior as a form of group selection have been supplanted by Hamilton's hypothesis of kin selection. What distinguishes kin selection and how does it accord with the notion of inclusive fitness, the relative number of an individual's alleles that pass to the next generation?
13. How does reciprocal altruism contrast with kin selection as an explanation of apparently altruistic behavior? What evidence from vampire bats supports reciprocal altruism?
14. Comment on limitations of animal communication compared with those of human communication.
15. Dance language used by returning forager honeybees to specify location of food is an example of remarkably complex communication among "simple" animals. How is direction and distance information coded into the waggle dance of the bees? What is the purpose of the round dance?
16. What is meant by "ritualization" in display communication? What is the adaptive significance of ritualization?
17. Early efforts by humans to communicate vocally with chimpanzees were almost total failures; however, researchers have learned how to communicate successfully with apes. How was this task accomplished?

For Further Thought In earlier chapters, we applied the critical concept of homology to molecular, chromosomal, and morphological characters. How is homology applied to studies of animal behavior?

SELECTED REFERENCES

Alcock, J. 2013. Animal behavior: an evolutionary approach, ed. 10. Sunderland, Massachusetts, Sinauer Associates, Inc. *Clearly written and well-illustrated discussion of genetics, physiology, ecology, and history of behavior in an evolutionary perspective.*

Attenborough, D. 1990. The trials of life: a natural history of animal behavior. Boston, Little, Brown and Company. *Superb photographs and flowing text describe the life cycles of organisms, often focusing on unusual and exciting patterns of behavior.*

Basolo, A. L. 1996. The phylogenetic distribution of a female preference. Syst. Biol. **45:** 290–307. *A phylogentic analysis of behavioral evolution.*

Bekoff, M., and J. Pierce. 2009. Wild justice: the moral lives of animals. Chicago, The University of Chicago Press. *A search for emotions and moral intelligence in animals.*

Bradbury, J. W., and S. L. Vehrencamp. 2011. Principles of animal communication, ed. 2. Sunderland, Massachusetts, Sinauer Associates. *A comprehensive text on animal communication.*

Clutton-Brock, T. 2002. Breeding together: kin selection and mutualism in cooperative vertebrates. Science **296:** 69–72. *A good summary of selective explanations of cooperative behavior.*

Drickamer, L. C., S. H. Vessey, and E. Jakob. 2002. Animal behavior: mechanisms, ecology, and evolution, ed. 5. Boston, McGraw-Hill Publishers. *Comprehensive, with helpful discussions on the methods and experimental approaches used to answer behavioral questions.*

Gould, J. L., and C. G. Gould. 1994. The animal mind. New York, Scientific American Library. *Attractively illustrated, engagingly written exploration of animal behavior and efforts of researchers to determine what happens inside the minds of animals.*

Houck, L. D., and L. C. Drickamer (eds.). 1996. Foundations of animal behavior. Chicago, The University of Chicago Press. *Classic papers in animal behavior with commentaries.*

Kendler, K. S., and R. J. Greenspan. 2006. The nature of genetic influences on behavior: lessons from "simpler" organisms. American Journal of Psychiatry **163:** 1683–1694. *Reviews genetic studies of behavior in fruit flies and other genetic model systems.*

Kirchner, W. H., and W. F. Towne. 1994. The sensory basis of the honeybee's dance language. Sci. Am. **270:** 74–80 (June). *Experiments with a robotic honeybee that can dance and emit sounds similar to living bees show conclusively that the dance language successfully recruits foragers to food outside the hive.*

Lorenz, K. Z. 1952. King Solomon's ring. New York, Thomas Y. Crowell Company, Inc. *One of the most delightful books ever written about behavior of animals.*

Manning, A., and M. S. Dawkins. 2012. An introduction to animal behaviour, ed. 6. Cambridge, England, Cambridge University Press. *Survey of animal behavior, drawing from ethology, physiology, and comparative psychology.*

Preston-Mafham, R., and K. Preston-Mafham. 1993. The encyclopedia of land invertebrate behavior. Cambridge, Massachusetts, The MIT Press. *Numerous examples of fascinating invertebrate behavior in a series of informative and beautifully illustrated essays. Highly recommended.*

Prum, R. O. 1990. Phylogenetic analysis of the evolution of display behavior in the neotropical manakins (Aves: Pipridae). Ethology **84:** 202–231. *A phylogenetic analysis of behavioral evolution.*

Queller, D. C., and J. E. Strassmann. 1998. Kin selection and social insects. BioScience **48** (3):165–175. *How kin selection operates in social insects, and why most cases of altruism are found in social insects.*

Ridley, M. 1995. Animal behavior: an introduction to behavioral mechanisms, development, and ecology, ed. 2. Oxford, Blackwell Scientific Publications. *The principles of animal behavior presented with well-chosen examples and clear illustrations.*

Savage-Rumbaugh, S., S. G. Shanker, and T. J. Taylor. 1998. Apes, language, and the human mind. New York, Oxford University Press. *Details the general area of ape language.*

Sih, A., A. Bell, and J. C. Johnson. 2004. Behavioral syndromes: an ecological and evolutionary overview. Trends in Ecology and Evolution **19:** 372–378. *Provides theory and examples of behavioral syndromes.*

PART FIVE

37

Animal Distributions

Spaceship earth.

© StockTrek/Getty Images

Spaceship Earth

All life is confined to the **biosphere,** a thin veneer on the earth's surface. From the first remarkable photographs of earth taken from the *Apollo* spacecraft, revealing a beautiful blue and white globe lying against a limitless backdrop of space, the phrase "spaceship earth" became part of our vocabulary. All resources for sustaining life except solar energy are restricted to a thin layer of land and sea and a narrow veil of atmosphere above it. If we could shrink the earth and all of its dimensions to a 1 m sphere, we no longer would perceive vertical dimensions on the earth's surface. The highest mountains would not penetrate a thin coat of paint applied to our shrunken earth; a fingernail's scratch would exceed the depth of an ocean's deepest trenches.

Earth's biosphere and the organisms in it evolve together. Approximately 5 billion years ago, the earth was barren, stormy, and volcanic with a reducing atmosphere of ammonia, methane, and water, but it supported the prebiotic syntheses that led to life's beginnings. Appearance of free oxygen in the atmosphere, produced largely if not almost entirely by life, illustrates the reciprocity between organism and environment. Although oxygen was poisonous to early forms of life, its gradual accumulation from photosynthesis led some forms to evolve oxygen metabolism, on which most organisms now depend. As living organisms adapt and evolve, they change their environments. In so doing, they must themselves change.

37.1 PRINCIPLES OF HISTORICAL BIOGEOGRAPHY

Zoogeographers describe patterns of animal distribution and species diversity and seek to explain why species and species diversity are distributed as they are. Zoogeography is a subset of biogeography, the evolutionary and ecological theories that explain the spatial distributions of all living forms. Most species occupy limited geographic areas. Why animals are distributed as they are is not always obvious, because similar habitats on separate continents may contain quite different kinds of animals. A particular species may be absent from a region that supports similar animals because barriers to dispersal prevent it from getting there or because established populations of other animals prevent it from colonizing. Thus, we aim to discover why animals occur where they are and not where one thinks they should be.

Explanations for geographical distributions of animals lie in the history of nature. The fossil record plainly shows that animals once flourished in regions that they no longer occupy. Extinction has a major role, but many groups left descendants that migrated to other regions and survived. For example, camels originated in North America, where their oldest fossils occur. Camels spread during the Pleistocene epoch by way of Alaska to Eurasia and Africa, where true camels live today, and to South America, where their living descendants include llamas, alpacas, guanacos, and vicuñas. (The Pleistocene began about 2.6 million years BP and ended about 11 thousand years BP; see the geological timetable, titled Timeline of Major Biological Events, on the back inside cover.) Then camels went extinct in North America about 10,000 years BP at the end of the Ice Age. Thus the history of an animal species and its ancestors must be documented before one can understand why it lives where it does. Earth's surface undergoes enormous change. Many land areas were once covered by seas. Fertile plains have been claimed by advancing desert; impassable mountain barriers have formed where none existed before; and inhospitable ice fields have retreated in a warming climate to be replaced by forests. Geological change is responsible for much of the alteration in animal (and plant) distribution and is powerful in shaping organic evolution.

Phylogenetic systematics permits us to reconstruct histories of animal distributions (see Chapter 10). A cladogram presents the structure of common evolutionary descent among species. Geographical distributions of closely related species are mapped onto a cladogram, in the same manner that one would map characteristics of organismal morphology such as presence of vertebrae, to generate hypotheses of those species' geographic histories. The large aquatic hellbenders of eastern North America form a clade with two species of East Asia. Molecular calibration of their phylogeny (p. 206) suggests that the hellbenders separated from their East Asian relatives about 28 million years ago, when forests and streams of eastern Asia and eastern North America were connected across Alaska and northern Canada, areas that climatic changes later rendered inhospitable for salamanders. Because the closest living relatives of hellbenders and Asian giant salamanders are the smaller hynobiid salamanders of Asia, the best hypothesis is that the giant salamanders originated in Asia and that a hellbender lineage dispersed to North America 28 million years ago. Many other forest-dwelling plant and animal populations moved between eastern Asia and eastern North America at about the same time, and subsequently became isolated on different continents when colder climates destroyed the northern forests.

Disjunct Distributions

Zoogeographers are challenged to explain numerous discontinuous or **disjunct distributions:** closely related species living in widely separated areas of a continent, or even the world (Figure 37.1). How could a group of animals become so dispersed geographically? Either a population moves from its place of origin to a new location (**dispersal),** traversing intervening territory unsuited for long-term colonization, or the environment changes, breaking a once continuously distributed species into geographically separated populations (**vicariance).** Climatic changes might contract and fragment areas of habitat favorable for a species, or physical movement of landmasses or waterways might carry different populations of a species away from each other.

Distribution by Dispersal

By dispersal, animals enter new localities from their places of origin. Dispersal involves *emigration* from one region and *immigration* into another. Dispersal is a *one-way,* outward movement distinguished from *periodic* movement back and forth between two localities, such as seasonal migration of many birds (p. 599). Dispersing animals may move actively by their own power or be passively dispersed by wind, by floating or rafting on rivers, lakes, or the sea, or by hitching rides on other animals. Animal species should expand their geographic distributions in this manner across all areas of favorable habitat accessible to them. As the last Pleistocene glaciers retreated northward, habitats favorable for many temperate species became available on formerly glaciated territory in North America, Europe, and Asia. Species that originated immediately south of the glaciated territory prior to glacial retreat then expanded northward as new habitats appeared. Because the reproductive rate of animal populations is great, a continuous pressure drives populations to expand across all favorable habitats.

Dispersal easily explains movement of animal populations into favorable habitats that are geographically adjacent to their places of origin. This movement produces an expanded but geographically continuous distribution. Can dispersal also explain the origins of geographically disjunct distributions? For example, species of flightless ratite birds (p. 111) collectively inhabit disjunct landmasses primarily of the Southern Hemisphere, including Africa, Australia, Madagascar, New Guinea, New Zealand, and South America. Each species occupies only one of these landmasses; no sharing of ratite species occurs among these areas. These landmasses are separated from each other by ocean, a very strong barrier to ratite dispersal. Could one explain the origin and geographic distribution of ratite birds by dispersal from a **center of origin?** Because ratite birds form a clade, one of the landmasses would be the site of origin of the group, and new populations on the other landmasses would be established by rare events of overseas dispersal, initially from the group's center of origin. Populations isolated on different landmasses would then evolve separately, eventually producing different species. Because ratites do not fly, a dispersalist hypothesis requires intermittent, passive rafting of individuals across the ocean. Is this hypothesis reasonable? We know from studies of the Galápagos Islands and Hawaii (see Chapter 6) that occasional, long-distance

Figure 37.1   Disjunct distributions in North America. **A,** Moles of the family Talpidae probably entered North America across the Bering land bridge that once joined North America and Asia during the Tertiary period. Eastern and western populations are now separated by the Rocky Mountains. **B,** Gopher tortoises of the genus *Gopherus* occupy three fully isolated areas.

dispersal of terrestrial animals and plants across oceans does occur. This is the only way that terrestrial animals could colonize islands produced by oceanic volcanoes. For flightless birds and many other discontinuously distributed animals, however, an alternative explanation of disjunct distributions is the hypothesis of vicariance (L. *vicarius,* a substitute).

Distribution by Vicariance

Disjunct distributions of animals can be created by physical changes in the environment that cause formerly continuous habitats to become disjunct. Areas once joined may become separated by barriers impenetrable for many animals. The study of fragmentation of biotas in this manner is called **vicariance biogeography.** Recall that "vicariance" is an example of "allopatry," which is simply a distribution of populations in geographically separated areas (p. 114). Lava flows from a volcano may cause a formerly continuous forest to separate into geographically discontinuous patches, thereby breaking many species of plants and animals into geographically isolated populations.

Perhaps the most dramatic vicariant phenomenon is continental drift, through which a once continuous landmass is sequentially broken into continents and islands separated by ocean. All terrestrial and freshwater animal species that had spread across the initially continuous landmass would be sequentially fragmented into many populations on different continents and islands separated by ocean.

Vicariance by continental drift gives us another hypothesis for explaining the disjunct distribution of flightless birds; they may descend from an ancestral species that was widespread in the Southern Hemisphere when Africa, Australia, Madagascar, New Guinea, New Zealand, and South America were physically connected (Figure 37.2). When these landmasses moved apart across the ocean, ancestral species would have fragmented into disjunct populations carried away from one another, each one diverging from the others to form a distinct species unique to its landmass.

Suppose that different species of flightless birds evolved allopatrically as continental drifting sequentially broke their terrestrial environment into isolated pieces. If we construct a cladogram (p. 205) or phylogenetic tree of these birds as shown in Figure 37.3, the earliest divergence should correspond to the first geological splitting of the ancestral landmass into two parts, each one carrying part of the ancestral ratite species. All subsequent branching events on the tree should correspond sequentially to subsequent splittings of landmasses that fragmented major lineages further. Our tree hypothetically reconstructs the history of splitting of landmasses that carry ratite populations. If we erase the names of species from terminal branches of the tree and replace them with geographic areas in which each species is found, we have a hypothesis for the sequential separation of different geographic areas.

We can test this vicariant hypothesis further by identifying other groups of terrestrial organisms that have different species in each of the same geographic areas as flightless birds. If our

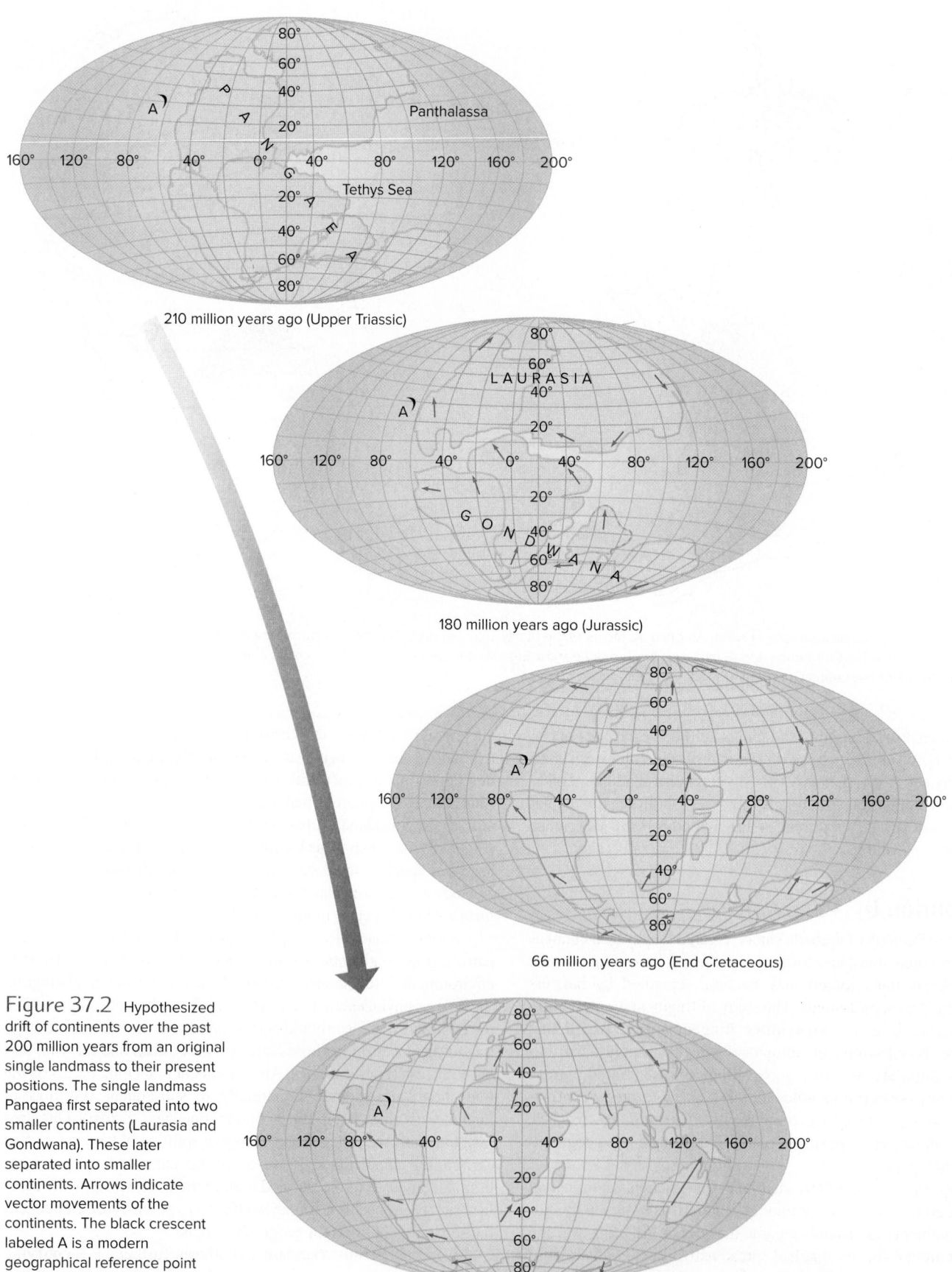

210 million years ago (Upper Triassic)

180 million years ago (Jurassic)

66 million years ago (End Cretaceous)

Today

Figure 37.2 Hypothesized drift of continents over the past 200 million years from an original single landmass to their present positions. The single landmass Pangaea first separated into two smaller continents (Laurasia and Gondwana). These later separated into smaller continents. Arrows indicate vector movements of the continents. The black crescent labeled A is a modern geographical reference point representing the Antilles arc in the West Indies.

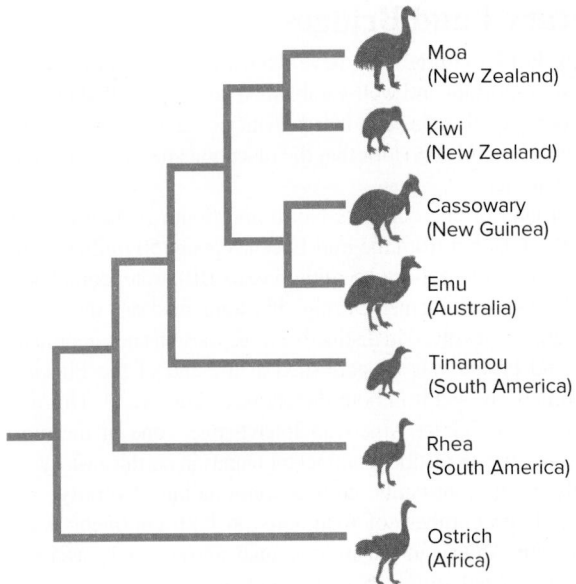

Figure 37.3 Phylogenetic relationships inferred for flightless birds using molecular genetic data (and morphology for the extinct moa) (see Chapter 6, p. 111). Vicariance biogeography proposes that these flightless species descended from an ancestral species that was widespread in the Southern Hemisphere when Africa, Australia, New Guinea, New Zealand, and South America were connected. By moving apart, these landmasses fragmented both themselves and the flightless populations they contained. If the vicariance hypothesis is correct, the phylogenetic branching sequence inferred for allopatric flightless species reflects the sequence by which their landmasses broke apart from each other. Replace species names with names of their geographic areas to produce an area cladogram. Area cladograms for other groups of animals and plants whose ancestral populations would have been fragmented by the same geological events should match the area cladogram for flightless birds.

hypothesis is correct, these groups were fragmented geographically by the same geological splitting events that fragmented flightless birds. We therefore predict that the cladograms or phylogenetic trees constructed for species in the other groups will show the same branching pattern as the flightless-bird tree when we replace the species names with those of the areas they inhabit. If this hypothesis is confirmed, we have a **general area cladogram** that depicts the history of fragmentation of the different geographic areas studied. This general hypothesis of vicariance can be investigated further using geological and climatic studies to identify the mechanisms of geographic fragmentation.

In most groups of organisms, recurring vicariance and dispersal have contributed to evolution of disjunct distributional patterns. Methods of vicariance biogeography are useful for finding such cases. Indeed, the cladogram of flightless birds is not just a simple grouping of birds that inhabit nearby areas. We can ask whether any branch on the cladogram for a particular group of species is inconsistent with the general area cladogram for geographic areas that these species inhabit. Suppose that the cladogram for a particular taxon is consistent with the area cladogram except for placement of a single branch. We explain most geographic disjunctions within the taxon by vicariance but look for dispersal to explain the single branch not compatible with the general area cladogram. In this way, we can

focus our study of dispersal on specific cases in which it is most likely to have occurred.

Continental-Drift Theory

It is no accident that enthusiasm for vicariance biogeography accompanied acceptance of continental-drift theory by geologists. Continental-drift theory is not new (it was proposed in 1912 by German meteorologist Alfred Wegener), but it remained largely disfavored until the theory of **plate tectonics** provided a mechanism to explain drifting continents. According to the theory of plate tectonics (tectonics means "deforming movement"), earth's crust is composed of 6 to 10 rocky plates, about 100 km thick, which shift their positions across an underlying surface. Wegener proposed that earth's continents had been drifting like rafts following breakup of a single great landmass called Pangaea ("all land"). The original breakup of Pangaea occurred approximately 200 million years ago. Two great supercontinents formed: a northern Laurasia and a southern Gondwana, separated from each other by the Tethys Sea (Figure 37.2). At the end of the Jurassic period, some 146 million years ago, the supercontinents began to fragment and to drift apart. Laurasia split into North America, most of Eurasia, and Greenland. Gondwana split into South America, Africa, Madagascar, Arabia, India, Australia, New Guinea, Antarctica, and numerous smaller fragments that now form Southeast Asia. The Arabian, Indian, and Southeast Asian fragments gradually crossed the Tethys Sea and eventually collided with Laurasia, to which they are now attached. This theory is supported by the appearance of fit between the continents, by airborne paleomagnetic surveys, by seismographic studies, by mid-ocean ridges located where the tectonic plates arise, and by a wealth of biological data.

Continental drift explains several otherwise puzzling distributions of animals, such as the similarity of invertebrate fossils in Africa and South America, as well as certain similarities in present-day faunas at the same latitudes on the two continents. However, the continents have been separated for all of the Cenozoic era and probably for much of the Mesozoic era as well, much too long to explain the distributions of some modern organisms such as placental mammals. Continental-drift theory is, nevertheless, enormously useful in explaining interconnections between floras and faunas of the past.

The present distribution of marsupial mammals is an excellent example of the influence of continental breakup. Marsupials appeared in the Middle Cretaceous period, about 100 million years bp, probably in South America. Because South America was then connected to Australia through Antarctica (then much warmer than it is today), marsupials spread through all three continents. They also moved into North America, but there they encountered placental mammals, which had dispersed to that continent from Asia. Marsupials evidently could not coexist with placentals, and so became extinct in North America. (North American marsupials today, opossums, are relatively recent arrivals from South America.) Placentals followed marsupials into South America, but by that time marsupials had expanded and were too firmly established to be displaced. In the meantime, about 50 million years BP, Australia drifted apart from Antarctica, barring entrance to placentals. Australia remained in isolation, allowing marsupials to diversify into the present rich and varied faunas.

Many people consider Alfred Russel Wallace (see Figure 6.1) the founder of modern historical biogeography. Wallace conducted extensive field studies in the Malay Archipelago, where he discovered an abrupt faunal change between Asian faunal elements and Australian/New Guinean ones. Pheasants, parrots, monkeys, numerous lizard groups, and even marine invertebrates are among the faunal elements whose geographic distributions show abrupt boundaries at this line. This biogeographic boundary is called "Wallace's Line," and it bisects the current country of Indonesia (Figure 37.4). Wallace's Line has been a biogeographic mystery since he described it because there are no obvious environmental changes or barriers that would explain the abrupt displacement of faunas across the line.

Plate tectonics offers the best explanation of Wallace's Line. Although currently close, the Southeast Asian plates separated from the Australian/New Guinean plate during the breakup of Gondwana, and these different plates spent many millions of years crossing the Tethys Sea in isolation from each other before finding their present locations. Wallace's Line marks the approximate boundary between the Southeast Asian and Australian/New Guinean plates, whose faunas diverged greatly during their long evolutionary separation. Molecular phylogenetic studies comparing lizard groups distributed on opposite sides of Wallace's Line support the interpretation that these groups were formerly isolated by a large expanse of ocean and only recently became geographic neighbors in the Malay Archipelago.

Figure 37.4 The enigmatic borderline described by Alfred Russel Wallace as marking the geographic separation of Asian (upper left) and Australian (center to lower right) faunas in the Malay Archipelago. A collision of tectonic plates brought formerly distant landmasses and their faunas in close geographic proximity, thereby forming "Wallace's Line."

Temporary Land Bridges

Temporary land bridges also have been important pathways of dispersal. An important and well-established land bridge that no longer exists connected Asia and North America across the Bering Strait. It was across this corridor that the placentals moved from Asia into North America.

Today a land bridge connects North and South America at the Isthmus of Panama, but from the mid-Eocene epoch (50 million years BP) to the late Pliocene epoch (3.5 million years BP), water completely separated the two continents. During this long interval, the major groups of mammals evolved in distinctive directions on each continent. When the land bridge was reestablished at the end of the Pliocene epoch, mammals crossed it in both directions (Figure 37.5). This dispersal is called the "Great American Interchange," one of the most important minglings of distinct continental faunas in earth's history.

Initially both continents gained mammalian diversity, but extinction of large numbers of mammals on both continents soon followed. North American carnivores such as raccoons, weasels, foxes, dogs, cats (including sabercats), and bears began preying on South American mammals. Other North American invaders included hoofed mammals (horses, tapirs, peccaries, camelids, deer, pronghorns, and mastodonts), rabbits, and several families of rodents. These mammals displaced many South American residents occupying similar habitats. Today nearly half of South America's mammals descend from North American invaders that arrived within the past 3.5 million years. Only a few South American invaders survived in North America: porcupines, armadillos, and opossums. Several other South American groups, including giant ground sloths, glyptodonts, anteaters, giant aquatic capybaras, toxodonts (rhino-sized plant eaters), and giant armadillos, entered North America but subsequently went extinct there.

Climatic Cycles and Vicariance

Within the past 3 million years, global cycles of average world temperature have occurred with durations typically of 20,000 to 100,000 years. Such cycles undoubtedly occurred also in the more distant evolutionary past, but are more difficult to measure. These cycles are called glacial cycles, because as global temperatures drop to a low point, glaciers expand geographically from the north and south poles and from the highest mountain peaks outward. Sea levels typically drop worldwide when large amounts of water are frozen in glaciers. At the warmest phases of each cycle, glaciers retreat to uncover habitable land, and sea levels rise, sometimes inundating formerly exposed land. Earth is currently at a relatively warm phase, between glacial maxima.

Climatic cycles have an enormous influence on geographic distributions of animals. Cycling temperatures interact with the habitat requirements of animals to generate alternating episodes in which favorable habitat is geographically continuous and widespread versus fragmented into isolated patches. For example, salamanders of the *Plethodon jordani* group of species (including *Plethodon shermani*, see Figure 6.20) are cold adapted and occupy only the top peaks of the southern Appalachian Mountains and not the intervening lower, warmer areas (Figure 37.6). At the coldest parts of the glacial cycles, the environmental conditions favorable for these salamanders spread into lower elevations, allowing a continuous distribution of this group of species among the mountain peaks and

Figure 37.5 The Great American Interchange. The Isthmus of Panama emerged approximately 3.5 million years ago, permitting the extensive interchange of many families of mammals. At top are representatives of 38 South American genera that walked north across the isthmus. At bottom are representatives of 47 North American genera that migrated to South America. The North American immigrants diversified rapidly after entering South America. South American immigrants to North America diversified little and most became extinct. Adapted from Marshall 1988, American Scientist, 76, 380–88.

Figure 37.6 Geographic distributions of 7 species of salamanders of the *Plethodon jordani* group in the southern Appalachian Mountains. These cold-adapted species are restricted to isolated mountaintop populations and do not occupy the lower, warmer areas. Gray shading shows the larger, geographically continuous distribution of habitat favorable to these salamanders at the last glacial maximum, when global climate was cooler than it is now. Climatic cycles in the history of these species cause episodic expansions of populations across the gray region followed by fragmentation into mountaintop isolates.

Source: *Weisrock & Larson. 2006.* Biological Journal of the Linnean Society *89:25–51.*

intervening areas. As a glacial episode occurs, species expand and make geographic contact with one another. When climate warms in the next cycle, these salamanders once again retreat to geographically isolated populations on mountain peaks. If mountaintop populations evolve species-level differences during geographic isolation, they do not merge when they make geographic contact at the next cold episode. Neighboring species typically block each other's expansion when they meet geographically; each species holds its territory and prevents its closest relatives from entering. Species of the *Plethodon jordani* group, which occupy different mountaintops (Figure 37.6), illustrate this process.

Should other groups of species, such as cold-adapted cave spiders, whose distributions match those of the *P. jordani* group, show matching area cladograms for these mountaintop isolates? Vicariance-biogeographic theory predicts that the order of phylogenetic branching for both groups should match the order in which the various mountaintops became disconnected "islands" of habitat favorable for cold-adapted species.

The cyclic aspect of climatic changes "throws a wrench" into the expectation of a general area cladogram. Clades of salamander and spider species that share the same fragmented mountaintop distribution today likely entered the southern Appalachian Mountains at different times in the past and from different geographic directions. A cold point in any of the past cycles could have permitted a cold-adapted species to spread across these mountains, only to be fragmented into mountaintop isolates at the next global warming. This issue was intitially called the "deep-history problem" for vicariance biogeography. We now recognize that taxa sharing the same areas of endemism, such as mountaintops, nonetheless typically differ greatly in evolutionary age and in phylogenetic branching pattern. Expectation of a general area cladogram is reduced even further if one mountaintop population goes extinct in a particularly warm episode, and its mountaintop is recolonized from another mountain at a later cold episode.

Although the hypothesis of a general area cladogram has failed in most biogeographic studies, speciation by vicariance (p. 114) remains a central principle in historical biogeography. When biogeographers claim that the vicariance biogeographic paradigm has failed, they refer specifically to rejecting hypotheses of general area cladograms, which does not diminish the evolutionary importance of vicariance.

37.2 DISTRIBUTION OF LIFE ON EARTH

We proceed from the general principles of historical biogeography to a description of the major physical boundaries that delimit animal evolution, including dimensions of the biosphere and major discontinuities between terrestrial environments, inland waters, and oceanic environments.

Biosphere and Its Subdivisions

The biosphere as usually defined is the thin outer layer of the earth capable of supporting life. It is a global system that includes all life on earth and the physical environments in which living organisms exist and interact. The physical subdivisions of the biosphere include the lithosphere, hydrosphere, and atmosphere.

The **lithosphere** is the rocky material of the earth's outer shell, the ultimate source of all mineral elements required by living organisms. The **hydrosphere** is the water on or near the earth's surface, and it extends into the lithosphere and atmosphere. A global hydrological cycle of evaporation, precipitation, and runoff distributes water across the earth. Five-sixths of the evaporation is from oceans, and more water evaporates from oceans than returns to them by precipitation. Oceanic evaporation therefore provides much of the rainfall that supports life on land. The gaseous

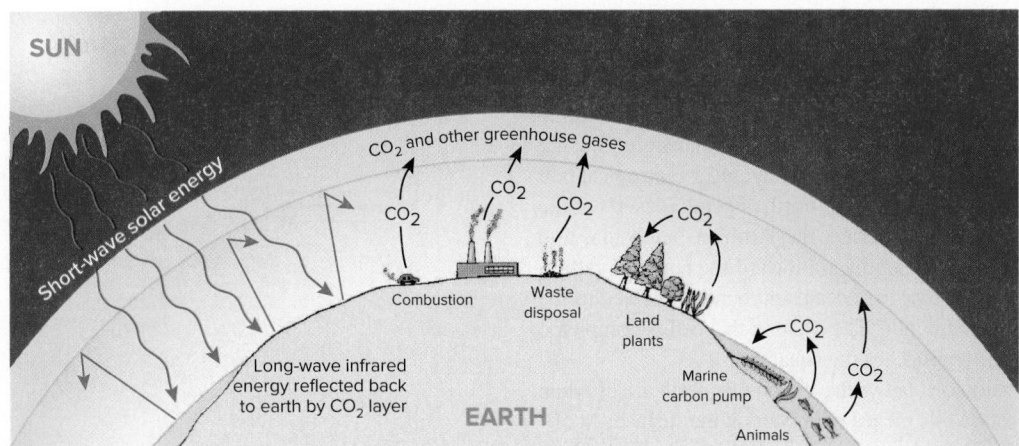

Figure 37.7 "Greenhouse effect." Carbon dioxide and water vapor in the atmosphere are transparent to sunlight but absorb heat energy reradiated from the earth, leading to warming of atmospheric air. In the marine carbon pump, carbon dioxide fixed by marine plants (especially phytoplankton) and carried in ionic form (HCO_3^-) by cold water sinks to great depths and is thereby removed from the atmosphere until deep-sea currents move it to the ocean surface in tropical regions.

component of the biosphere, the **atmosphere,** extends to some 3500 km above the earth's surface, but all life is confined to the lowest 8 to 15 km (troposphere). The screening layer in the atmosphere of oxygen-ozone is concentrated mostly between 20 and 25 km. The main gases in the troposphere are (by volume) nitrogen, 78%; oxygen, 21%; argon, 0.93%; carbon dioxide, 0.03%; and water vapor (variable amounts).

Atmospheric oxygen originates almost entirely from photosynthesis. Since the mid-Paleozoic era, oxygen consumed by living organisms for respiration has approximately equaled oxygen production. Earth's surplus of free oxygen is unlikely to be depleted because oxygen reserves in the atmosphere and in the oceans are so enormous that the supply could last thousands of years even if photosynthetic replenishment ceased.

The rapid input of carbon dioxide into the atmosphere from burning fossil fuels may significantly affect earth's heat budget. Much of the sun's shortwave light energy absorbed by earth's surface reradiates as long-wave infrared heat energy (Figure 37.7). Materials in the atmosphere, especially carbon dioxide and water vapor, impede this heat loss and raise atmospheric temperature. This heating of the atmosphere is called the "greenhouse effect," since the atmosphere traps reradiated heat from the earth much as the glass of a greenhouse traps heat reradiated by plants and soil inside. Although the greenhouse effect provides conditions essential for all life on earth, the gradual accumulation of carbon dioxide could increase the temperature of the biosphere as a whole and raise sea level by melting polar ice (Figure 37.8).

Global Climatic Influences on Animal Environments

Earth's environments vary greatly as determined by climate, oceanic currents, characteristic rainfall and temperature of each region, and

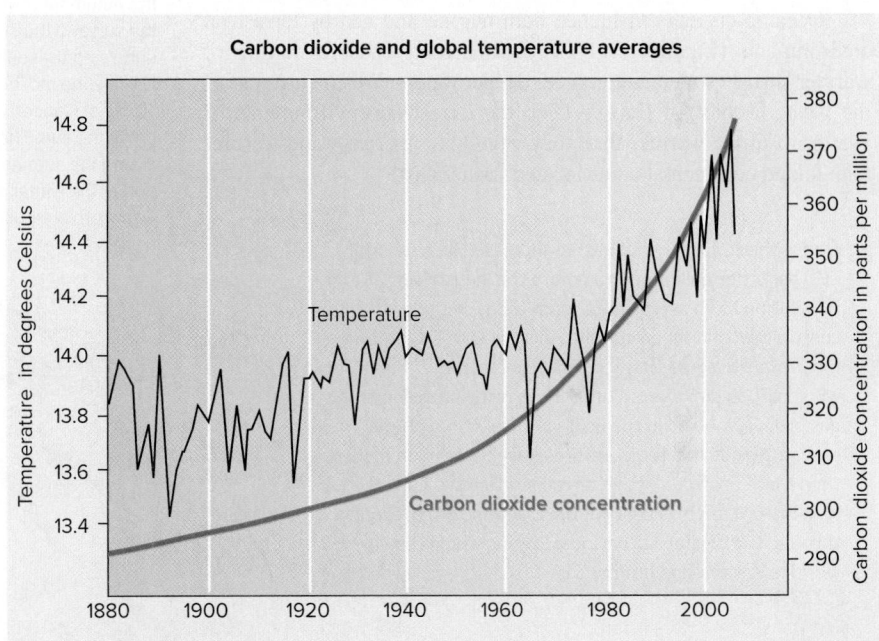

Figure 37.8 Rise in global atmospheric carbon dioxide and global temperature averages for the past 126 years. Data points before 1958 come from analysis of air trapped in bubbles in glacial ice from sites around the world. Atmospheric carbon dioxide has climbed on average for more than a century while earth's temperature has followed a more erratic upward trend.

amount of solar radiation. These factors impose strong boundaries on the geographic areas suitable for occupation by each animal species. Global variation in climate arises from uneven heating of the atmosphere by sunlight. Because the sun's rays strike higher latitudes at a lower angle, atmospheric heating is less there than at tropical latitudes (Figure 37.9). Air warmed in the tropics (latitudes between 23 degrees north and 23 degrees south) rises and moves toward the poles. It is replaced by cold air moving away from the poles at lower levels. Earth's rotation complicates this pattern, producing a Coriolis effect that deflects moving air to the right in the Northern Hemisphere and to the left in the Southern Hemisphere. Air circulation in each hemisphere forms three latitudinal zones, called cells (Figure 37.10). In the

Northern Hemisphere, hot, moist air in the tropics cools and condenses as it rises, providing rainfall for the lush vegetation of tropical rainforests. Warm air then flows northward at high levels, cools, and sinks at 20° to 30° latitude. This air is very dry, having lost its moisture in the tropics. As air heats it absorbs even more moisture, causing intense evaporation at earth's surface and producing a subtropical belt of deserts centered between 15° and 30° north or south (deserts of the American southwest, Saharan Africa, the Arabian peninsula, and Australia). The air then flows southward toward the equator, gathering moisture as it moves across the ocean, and being deflected to the right as the northeast trade winds. The cycle in this cell is completed when the now moist air approaches the equator.

A second circulation cell between 30° and 60° north arises when cool air sinking at around 30° moves northward at the surface. At 50° to 60° north it encounters cold air moving south from the North Pole, producing an unstable stormy area with abundant precipitation. Warmer air from the south deflects upward and turns south at high altitude to complete the second cell. A third polar cell forms when cold, southward-moving Arctic air returns to the pole at high altitude.

Oceanic currents influence both marine and nearby terrestrial environments (Figure 37.11). Global currents cause a mixing of warmer surface waters with colder deeper waters. Warm currents off the western coast of Europe keep the terrestrial environments of Western Europe warmer than they would be otherwise and warmer than inland continental areas located farther east.

Atmospheric carbon dioxide has increased from about 280 parts per million (ppm) before the Industrial Revolution to an average 400 ppm today. In the past century global temperature has increased 0.4° C and most experts agree that it will have increased 2° to 6° C when carbon dioxide and other heat-trapping greenhouse gases have doubled in the next century. Atmospheric carbon dioxide and temperatures have been much higher than these projected levels at various times in earth's long history, such as parts of the Paleozoic era (Devonian through Carboniferous periods), during which the entire world was warm and humid.

Terrestrial Environments: Biomes

A biome is a major biotic unit bearing a characteristic and easily recognized array of plant life. Botanists long ago recognized that the earth's terrestrial environment could be divided into large units having a distinctive vegetation, such as forests, prairies, and deserts. Animal distribution has always been more difficult to map, because plant and animal distributions do not exactly coincide. Zoogeographers use plant distributions as the basic biotic units and recognize biomes as distinctive combinations of plants and animals. A biome is therefore identified by its dominant plant formation (Figure 37.12), but because animals depend on plants, each biome supports a characteristic fauna.

Each biome is distinctive, but plant communities blend into one another over broad areas. As one moves westward across North America, moist deciduous

Figure 37.9 Earth's climate is determined by differential solar radiation between the higher latitudes and tropical latitudes (between 23 degrees north and 23 degrees south). Solar energy at high latitudes is spread across a much larger, slanting surface area than is an equivalent amount of energy at the tropics. This diagram shows the sun's rays as they would hit the earth during the autumnal or vernal equinox, when the equator is most perpendicular to the sun and thus receives the most concentrated radiation. Because the earth's north–south axis is not perpendicular to the sun, the latitudes that receive the most concentrated rays vary between 23 degrees north (the Tropic of Cancer) and 23 degrees south (the Tropic of Capricorn) as the earth orbits the sun. The sun's rays are most concentrated on the Tropic of Cancer during the summer solstice, and most concentrated on the Tropic of Capricorn during the winter solstice. Latitudes between the tropics of Cancer and Capricorn are collectively called the tropics.

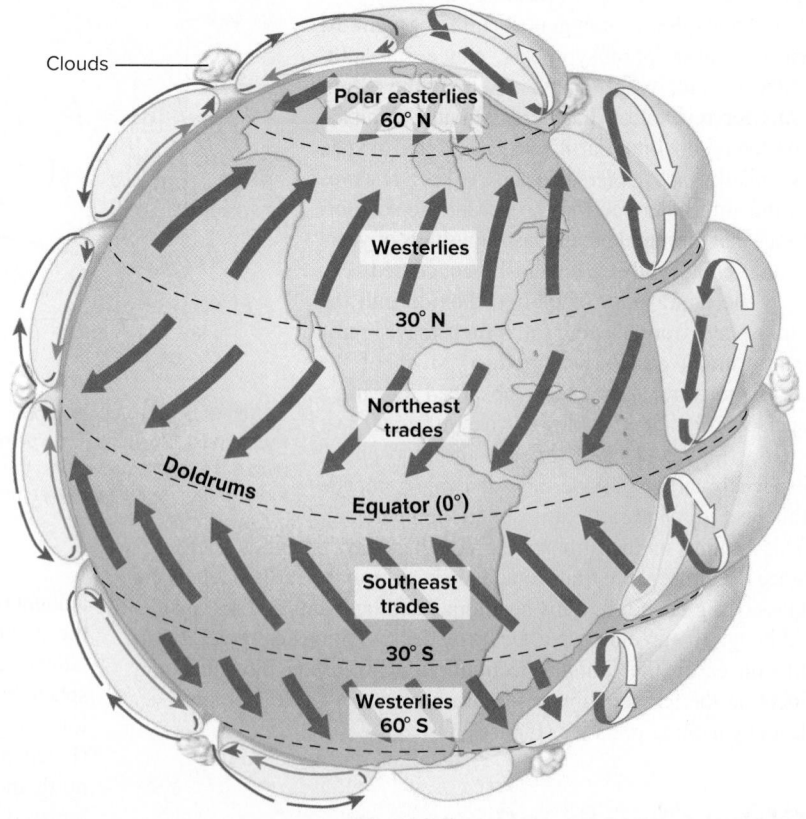

Figure 37.10 Earth as a heat engine. As the result of unequal heating across earth's surface, together with other factors such as earth's rotation, circulation of the oceans, and presence of landmasses, the earth acts like a giant heat engine that imposes a complex patchwork of climates on earth. See text for explanation.

Figure 37.11 Oceanic currents and locations of nutrient upwellings that support abundant animal life in the pelagic realm.

Upwelling by currents and/or wind
Surface flow
Deep or bottom flow

forests of the Appalachians change gradually to drier oak forests of the upper Mississippi River valley, and then to oak woodlands with grassy understory, which yield to tall and mixed prairies (now corn and wheatlands), then to desert grasslands, and finally to desert shrublands. The indistinct boundaries where the dominant plants of adjacent biomes mix together form an almost continuous gradient called an **ecocline.** Thus biomes are abstractions, a convenient way to organize our concepts about different communities. Nevertheless, anyone can distinguish a grassland, deciduous forest, coniferous forest, or shrub desert by the dominant plants in each, and we can predict from the vegetation which animals live in each biome.

The principal terrestrial biomes are temperate deciduous forest, temperate coniferous forest, tropical forest, grassland, tundra, and desert. In this brief survey, we refer especially to biomes of North America and consider predominant features of each.

Temperate Deciduous Forest

Temperate deciduous forest, best developed in eastern North America, encompasses several forest types that change gradually from northeast to south. Deciduous, broad-leaved trees such as oak, maple, and beech predominate. Seasons are most distinct from each other in this biome. The deciduous habit is an adaptation of dormancy for low energy levels from the sun in winter and freezing winter temperatures. In summer, the relatively dense forests form a closed canopy and deep shade below. Consequently, understory plants grow rapidly in spring and flower early before the canopy develops. Mean annual precipitation is relatively high (75 to 125 cm, or 30 to 50 inches), and rain falls episodically throughout the year. Mean annual temperatures range between 5° and 18° C (41° to 65° F).

Animal communities in deciduous forests respond to the onset of winter in various ways. Some, such as insect-eating warblers, migrate. Others, such as woodchucks, hibernate during winter months. Others survive by using available food (as do deer) or stored food supplies (as do squirrels). Hunting and habitat loss have largely eliminated large carnivores (mountain lions, bobcats, and wolves) from eastern forests.

Figure 37.12 Major biomes of North America. Boundaries between biomes are not distinct but grade into one another over broad areas.

Deer, however, thrive in second-growth forests. Insect and invertebrate communities are abundant in deciduous forests where decaying logs and forest-floor debris provide excellent shelter.

Heavy exploitation of deciduous forests of North America began in the seventeenth century and peaked in the nineteenth century. Logging removed the once-magnificent stands of temperate hardwoods. With the opening of the prairie for agriculture, many eastern farms were abandoned and returned gradually to deciduous forests.

Coniferous Forest

In North America coniferous forests form a broad, continuous, continent-wide belt stretching across Canada and Alaska, and south through the Rocky Mountains into Mexico. This biome continues across northern Eurasia, making it one of the largest plant formations on earth. Evergreen trees, including pines, firs, spruces, and cedars, dominate this biome, being adapted to withstand freezing winter temperatures and short summer growing seasons. Conical trees with flexible branches shed snow easily. The northern area is the **boreal** (northern) **forest,** often called **taiga** (a Russian word, pronounced "tie-ga"). Taiga is dominated by white and black spruce, balsam, subalpine fir, larch, and birch trees. Mean annual precipitation is less than 100 cm (40 inches), and average temperatures range from −5° to +3° C (23° to 37° F).

In central North America, the taiga merges into temperate deciduous forest through a **mixed temperate forest,** which combines coniferous and deciduous trees. It is dominated by sugar maple, white pine, red pine, and eastern hemlock. Much of this forest was destroyed by logging and replaced by second growth, which has a higher proportion of deciduous trees, including hard maples, beech, and yellow birch. Oaks and hickories dominate the mixed temperate forests farther south and occur less frequently in the pine-dominated **southern evergreen forests,** which cover much of the southeastern United States. The last old-growth coniferous forests of the Pacific northwest are rapidly falling to commercial logging.

Mammals of boreal forests include deer, moose (Figure 37.13), elk, snowshoe hares, various rodents, carnivores such as wolves, foxes, wolverines, lynxes, weasels, and martens, and omnivorous bears. They are adapted physiologically or behaviorally for long, cold, snowy winters, as illustrated by hibernation of bears during the coldest months. Common birds include chickadees, nuthatches, warblers, and jays. One bird, the red crossbill, has a beak specialized for picking seeds from cones. Mosquitos and flies pester animals and people in this biome. Southern coniferous forests lack many mammals found in the north, but they have more snakes, lizards, and amphibians.

Tropical Forest

A worldwide equatorial belt of tropical forests has high rainfall (more than 200 cm [80 inches] per year), high humidity, relatively high and constant temperatures averaging over 17° C (63° F), and little seasonal variation in day length. These conditions nurture luxurious, uninterrupted growth that reaches its greatest intensity in rainforests. In sharp contrast to temperate deciduous forests, dominated by relatively few tree species, tropical forests contain thousands of species, none of which dominates. A single hectare typically contains 50 to 70 tree species versus 10 to 20 tree species in an equivalent area of hardwood forest in the eastern United States. Climbing plants and epiphytes occur among the trunks and limbs.

A distinctive feature of tropical forests is stratification of life into six to eight feeding strata (Figure 37.14). Insectivorous birds and bats

Figure 37.13   A bull moose browses on dwarf birch in the boreal coniferous forest or taiga biome. Note shedding of antler skin ("velvet"), signifying that antler growth is complete and that breeding season is approaching.

Meters	
50 —	
45 —	*Emergent species level:* Exposed to full sunlight, wind, temperature changes
40 —	
35 —	Animal life, mostly birds and insects
30 —	
25 —	*Canopy level:* Main habitat of forest animals
20 —	Animals seldom descend to forest floor
15 —	Animals descend to forest floor for short periods
10 —	
5 —	*Ground level:* Protected, little light, poor growth, soil fauna
0 —	

Figure 37.14   Profile of tropical forest, showing vertical differentiation of animal and plant life into six feeding strata. The animal biomass is small compared with the plant biomass.

fly above the canopy; below the canopy, birds, fruit bats, and other mammals feed on leaves and fruit. In the middle zones are arboreal mammals (such as monkeys and sloths), numerous birds, insectivorous bats, insects, and amphibians. Climbing animals, such as squirrels and civets, move along tree trunks, feeding from all strata. On the ground are large mammals such as the large rodents of South America (for example, capybara, paca, and agouti) and members of the pig family. Finally, small insectivorous, carnivorous, and herbivorous animals search the litter and lower tree trunks for food. No other biome matches a tropical forest's incredible variety of animal species. Food webs (p. 826) are intricate and notoriously difficult to unravel.

Tropical forests, especially the enormous expanse centered in the Amazon Basin, are the most seriously threatened forest ecosystems. Large areas are cleared for agriculture by "slash-and-burn" methods, but farms are soon abandoned because soil fertility is low. It may seem paradoxical that a biome as luxuriant as a tropical forest should have poor soil; however, nutrients released by decomposition are rapidly recycled by plants, microbes, and fungi, leaving no reservoir of humus. In many areas, the soil rapidly becomes a hard, bricklike crust called **laterite** after plants are removed. Tropical plants do not recolonize such areas. Other pressures on tropical forests include logging by multinational timber companies and clearing of land for cattle ranching.

Grassland

The North American prairie biome is among the world's most extensive grasslands, ranging from the Rocky Mountain edge on the west to eastern deciduous forest on the east, and from northern Mexico in the south to the Canadian provinces of Alberta, Saskatchewan, and Manitoba in the north. Original grassland associations of plants and animals have been widely transformed into the world's most productive agricultural region, dominated by monocultures of cereal grains. In grazing lands, virtually all major native grasses have been replaced by alien species. Of the once dominant herbivore, bison (Figure 37.15), very few survive, but jackrabbits, prairie dogs, ground squirrels, and pronghorns remain. Mammalian predators include coyotes, ferrets, and badgers, although only coyotes are common. Vast tracts of open tall-grass prairie remain in the Flint Hills of Kansas and northern Oklahoma, and large areas of short-grass prairie occur in western Kansas and Nebraska. These areas maintain native vegetation and predatory animals, including raptors, cougars, and bobcats. Rainfall on the North American prairie ranges from about 80 cm (31 inches) in the east to 40 cm (16 inches) in the west. Average annual temperatures range between 10° and 20° C (50° and 68° F).

Tundra

Tundra characterizes severely cold regions, especially treeless Arctic regions and high mountaintops. Plant life must adapt to a short growing season of about 60 days and to a soil that is usually frozen. Average annual precipitation is less than 25 cm (10 inches) and annual temperature averages about −10° C (14° F).

Most tundra regions are covered with bogs, marshes, ponds, and a spongy mat of decayed vegetation, although high tundras may be covered only with lichens and grasses. Despite thin soil and a short growing season, dwarf woody plants, grasses, sedges, and lichens may be profuse. Characteristic animals of the Arctic tundra are lemmings, caribou (Figure 37.16), musk-oxen, arctic foxes, arctic hares, ptarmigans, and (during the summer) many migratory birds.

Desert

Deserts are arid regions where rainfall is low (less than 25 cm [10 inches] a year), and water evaporation is high. The North American desert has two parts, the hot deserts of the southwest (Mojave, Sonoran, and Chihuahuan) and the cool, high desert in the rain shadow of the High Sierras and Cascade Mountains. Desert plants, such as thorny shrubs and cacti, have reduced foliage, drought-resistant seeds, and other adaptations for conserving water. Many large desert animals have remarkable anatomical and physiological adaptations for keeping cool and conserving water (p. 675). Most smaller animals avoid severe conditions by living in burrows or using nocturnal habits. Mammals include mule deer, peccaries, cottontails, jackrabbits, kangaroo rats, and ground squirrels. Physiological mechanisms that permit a mammal to conserve body water in arid conditions are particularly well studied in kangaroo rats (p. 664). Typical birds are roadrunners, cactus wrens, turkey vultures, and burrowing owls. Lizards, snakes, and tortoises are numerous, and a few species of toads are common. Arthropods include a great variety of insects and arachnids.

Inland Waters

Only 2.5% of the world's water is freshwater. Most freshwater occurs in polar ice caps or underground in aquifers and soil, leaving only 0.01% of the world's inland waters as habitat for aquatic life. A quarter of the world's vertebrates and nearly half of its fishes live in these fragile "islands" of water, which must also supply human needs for irrigation, drinking water, hydroelectric power, and waste disposal.

Inland waters exist as running-water, or **lotic** (L. *lotus,* action of washing) habitats, and standing-water, or **lentic** (L. *lentus,* slow) habitats. Lotic habitats follow a gradient from mountain brooks to streams and rivers. Brooks and streams with high-velocity water

Figure 37.15 Bison grazing on a short-grass prairie.

© F. Gohier/Science Source

Figure 37.16 A large male caribou on the Alaskan tundra. The gregarious caribou travel in large herds, feeding in summer on grasses, dwarf willow, and birch, but in winter almost exclusively on lichen.

© Cleveland P. Hickman, Jr.

flow contain much dissolved oxygen because of their turbulence. Nutrients are chiefly from organic detritus washed from adjacent terrestrial areas. More slowly moving rivers have less dissolved oxygen and more floating algae and plants. Their fauna tolerates lower oxygen concentration.

Lentic habitats, such as ponds and lakes, have still lower concentrations of oxygen, particularly in the deeper areas. Animals living on underwater substrates or submerged vegetation (**benthos**) include snails and mussels, crustaceans, and a wide variety of insects. Many swimming forms, called **nekton,** occur in lakes and larger ponds. Depending on the nutrients available, a large contingent of small, floating plants and animals (**plankton**) may occur. Ponds and lakes have short lifespans, a few hundred to many thousands of years depending on size and rate of sedimentation, and undergo great physical change as they age. For example, the Great Lakes of North America, which occupy depressions gouged by glacial advances of the Pleistocene epoch, became ice free about 5000 years ago.

> A striking exception to the short lifetimes of most lakes is Lake Baikal in southern Siberia. This enormous lake, 1741 m deep (more than 1 mile), is by far the oldest lake in the world, dating from at least the Paleocene—more than 60 million years BP. The speciation of sculpins in Lake Baikal is illustrated in Figure 6.21, page 116.

Many freshwater habitats are severely damaged by human pollution such as dumping of toxic industrial wastes and enormous quantities of sewage. Of the Great Lakes, Lake Erie is the most seriously affected by inflow of nitrates and phosphates. These nutrients fertilize the lake, creating huge blooms of algae that sink and decompose, producing anoxic conditions that harm aquatic life.

Oceanic Environments

Oceans occupy by far the largest portion of earth's biosphere, covering 71% of earth's surface to an average depth of 3.75 km (2.3 miles), with their greatest depths reaching more than 11.5 km (7.2 miles) below sea level. The monotony of the ocean's surface belies a variety of life below. Oceans are the cradle of life, as reflected by the variety of organisms living there—more than 200,000 species of unicellular forms, plants, and animals. About 98% of these forms live on the seafloor (**benthic**); only 2% live freely in the open ocean (**pelagic**). Of the benthic forms, the largest biomass occurs in the intertidal zone or shallow depths of oceans, but species diversity increases from shallow waters to a maximum at 2000 to 3000 m depth and then declines at greater depths.

The most productive areas are concentrated along continental margins and a few areas where waters are enriched by organic nutrients and debris lifted by upwelling currents into the sunlit, or **photic,** zone, where photosynthetic activity occurs. With certain notable exceptions (see the box "Life Without the Sun" in Chapter 38, p. 830), all life below the photic zone is supported by a light "rain" of organic particles from above.

An **estuary** is a semienclosed transition zone where freshwater enters the sea. Despite unstable salinity caused by the variable entry of freshwater, an estuary is a nutrient-rich habitat that supports a diverse fauna.

Benthic communities of the ocean floor occupy geological provinces categorized by topography, substrate, and distance from shore (Figure 37.17). Closest to shore are an ocean's **continental margins,** which contain (1) a **continental shelf** extending from shallow nearshore waters outward to a depth of 120 to 400 m, (2) a **continental slope** marking a sharp descent from the outer shelf to a depth of 3000 to 5000 m, and (3) a **continental rise,** thick sediments piled at the base of the continental slope. Beyond this continental margin is the deep-sea basin or **abyssal plain,** an offshore plain with submarine channels and hills averaging 4000 m in depth but descending to 11,000 m. The abyssal plain has little seasonal variation in temperature and illumination and therefore is relatively stable despite considerable spatial heterogeneity.

Rocky Intertidal Zone

The **intertidal zone** is the portion of the continental shelf exposed to air during low tides; animals of intertidal communities experience daily fluctuations between marine and terrestrial environments. Attached to rocky intertidal substrates are periwinkles, barnacles, mussels, and other forms whose exoskeleton protects them from desiccation and physical abrasion by waves (Figure 37.18). Marine gastropods and sea stars consume these attached forms. Interactions between physical stresses, predation, and interspecific competition (p. 822) often produce visibly distinct bands (Figure 37.18B); periwinkles dominate the most exposed rocks, barnacles intermediate areas, and mussels the most submerged surfaces. Depressions in the rocky surface

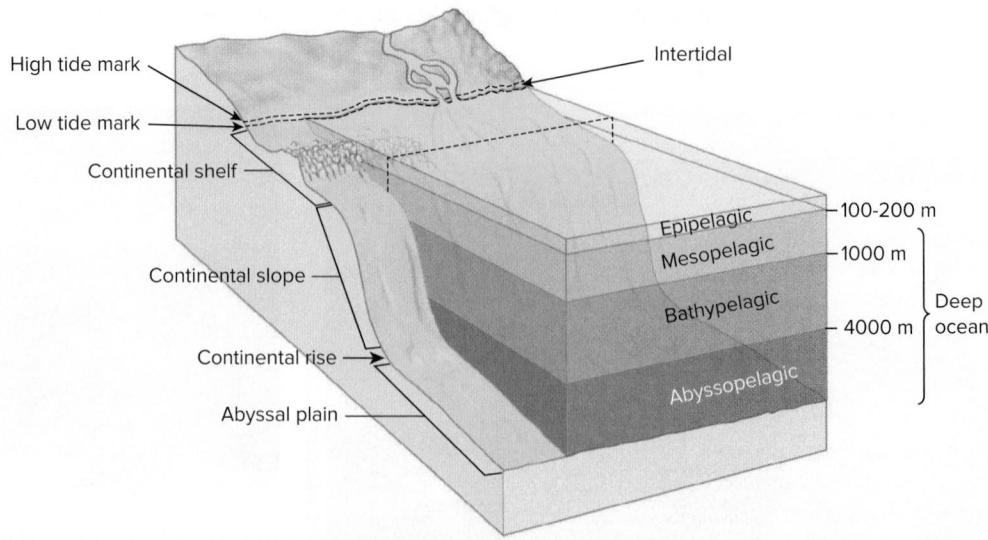

Figure 37.17 Major marine zones. The continental shelf, slope, and rise collectively form the continental margin.

often produce tidal pools isolated on an otherwise exposed shore. These tidal pools support anemones, corals, tunicates, and other forms ill-suited to the fully exposed surfaces. Attached seaweeds are often interspersed among rocky intertidal faunas. Rocky intertidal faunas are abundant on the northern Atlantic and Pacific coasts of North America.

Rocky Subtidal Zone

Kelp forests (Figure 37.19) dominated by brown seaweeds occupy shallow subtidal waters worldwide, reaching even the Arctic and Antarctic Circles. Kelps attach to a firm substrate by holdfasts and grow upward, some reaching the sea surface and forming a canopy analogous to the canopy of a tropical forest. Grazing by sea urchins and damage by storms greatly alter the structure of a kelp forest. Numerous species of abalones, sea urchins, and limpets graze kelp forests of the Pacific Coast of North America. These kelp forests support diverse animal life including suspension-feeding mussels and their crustacean predators. Populations of sea otters, which consume molluscs, sea urchins, and fishes of kelp forests, increase a kelp forest's density by removing sea urchins, a powerful grazer.

Coral reefs occur off the coasts of continents and volcanic islands and include atolls, a series of reefs encircling a submerged volcanic island. Reefs protect from wave damage a particularly diverse subtidal community (Figure 37.20), whose substrate is a topographically complex structure formed by mutualistic growths of corals and single-celled algae (pp. 280–282). A single reef may contain 50 or more species of coral with different species dominant at different depths. The complex topography of a reef divides its surface into numerous subcommunities associated with varying levels of illumination and physical orientation, thereby supporting hundreds of species of fishes and snails, plus other cnidarians, crustaceans, sponges, polychaetes, molluscs, echinoderms, tunicates, and other invertebrates. Because complex symbiotic relationships characterize a coral-reef community, no single species dominates its structure. Reefs are therefore less disturbed by changes in single species than are Pacific kelp forests, whose community structure changes greatly with local density of otter populations. Reef communities nonetheless are very susceptible to damage by chemical pollution and increases in water temperature.

Many species competing for limited space on a coral reef show agonistic interactions. Clonal ectoprocts (p. 322) compete by overgrowing others, causing some groups to evolve rapid clonal budding and erect structures resistant to overgrowth. Some slowly growing corals destroy their neighbors with stinging tentacles and digestive secretions. Mutualistic relationships also exist, including protection of clown fishes by otherwise predatory sea anemones attached to reef corals (p. 277). Thus, the apparent stability of coral reefs is achieved by dynamic interactions among many species.

Nearshore Soft Sediments

Intertidal and nearshore subtidal environments with soft sediments support a number of marine biomes including beaches, mudflats, **salt marshes, sea-grass beds,** and **mangrove communities.** A sandy substrate that is exposed to air during low tides can be colonized by grasses, followed by marsh mussels, burrowing crabs and shrimp, and deposit-feeding polychaetes, forming the characteristics of a salt

A

B

© jatrax/iStock/Getty Images

© Jean-Louis Klein & Marie-Luce Hubert / Science Source

Figure 37.18 A, A rocky intertidal community typically contains attached bivalve molluscs and barnacles, whose exoskeleton provides protection from desiccation and wave action, and predatory sea stars. **B,** At low tide, intertidal animal communities sometimes display themselves as visibly distinct bands, as seen on this exposed shore.

marsh. Small creeks in salt marshes are particularly favorable habitats for many polychaetes, mussels, periwinkles, crustaceans, and fishes. Small fishes, such as killifishes, attract predatory terns and kingfishers. These marshes are important sources of organic matter and provide nurseries for many marine fish species.

Shallow coastal subtidal environments include beds of sea grasses, which often colonize newly deposited sediments and become dense along the Atlantic coasts of Europe and North America. Hydroids, sponges, and ectoprocts occur among the grasses, which also support larvae of scallops.

In calm waters of tropical and subtropical marine coasts, mangrove trees rooted in submerged soft sediments form thick forests along the shoreline. Submerged mangrove roots support a rich marine community of detritus feeders, including oysters, crabs, and shrimp. Fishes also abound. Mangroves are perhaps unique in supporting a marine community at their roots and simultaneously a terrestrial community in their exposed branches (Figure 37.21).

© Gregory Ochocki/Science Source

A

© David Courtenay/Getty Images

B

Figure 37.19 Kelp forests of the rocky subtidal zone (**A**) are grazed by sea urchins, which are consumed by sea otters (**B**). A large population of sea otters keeps a kelp forest dense by removing sea urchins.

Deep-Sea Sediments

The deep sea includes the continental slope, continental rise, and abyssal plain (Figure 37.17). Soft sediments prevail: clean sands where strong currents flow and fine muds where currents are weak. Suspension-feeding invertebrates (p. 703) dominate sandy substrates but are rare on muddy ones. Experiments on suspension-feeding clams show that high turbidity in muddy areas damages suspension-feeding systems. Deposit feeders abound in muddy substrates, producing a patchiness of animal communities on the abyssal plain corresponding to substrate type. Deposit-feeding sea cucumbers, polychaetes, and echiurid worms produce fecal mounds that provide local substrates for smaller suspension-feeding bivalves,

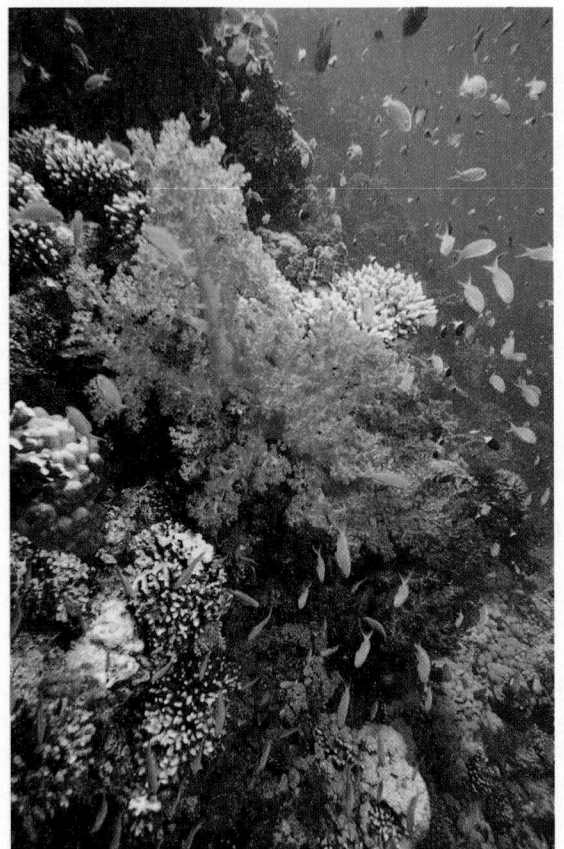

© Frank & Joyce Burek/Getty Images

Figure 37.20 A coral reef's topographic complexity permits it to support diverse subtidal communities with complex symbiotic relationships among species.

polychaetes, and crustaceans in an area otherwise unsuited for suspension feeding. Dead fish and plants fall to the seafloor, supporting bacteria and deposit feeders.

Hydrothermal Vents

Sporadic occurrence of hydrothermal vents (p. 830) contributes further patchiness to the environments and animal communities of the deep sea. Hydrothermal vents occur on the abyssal plain where submarine volcanic activity forces hot, sulfide-rich water upward through a rock crevice on the seafloor. Archaebacteria that derive their energy by oxidizing sulfide form mats on the rocky surfaces near the vents, and these mats are grazed by bivalves, limpets, and crabs. Other bivalves have sulfur-oxidizing archaebacteria as symbionts in their gills. Giant pogonophoran or siboglinid worms (p. 371) also harbor symbiotic archaebacteria to obtain nutrition. Hydrothermal vents are ephemeral; repeated colonization of newly formed vents propagates these communities.

Pelagic Realm

The vast open ocean is called the **pelagic** realm (Figure 37.22). Despite its size (90% of the total oceanic area), the pelagic realm is relatively impoverished biologically because, as organisms die, they sink from the photic zone, carrying nutrients into the bathypelagic zone.

Figure 37.21 Nearshore marine environments with soft sediments include mangrove forests, which support different animal life below (**A**) and above water (**B**).

Figure 37.22 Life in pelagic zones. Each zone supports a distinct community of organisms. Animals in zones below the mesopelagic depend on the meager rain of food that sinks from the epipelagic and mesopelagic zones.

Areas of **upwelling** and convergence of ocean currents (see Figure 37.11) are vital sources of nutrient renewal for the surface photic zone. Phytoplankton abound in upwellings at the Antarctic Sea, and photosynthesis by these phytoplankton introduces an extensive supply of organic nutrients into these waters. Major predators of the phytoplankton include a small crustacean called krill (see Figure 20.25), which is the major food source for baleen whales. It is estimated that at the height of their population numbers, baleen whale populations could have consumed over 70 million tons of krill per year.

The world's most productive fisheries are centered on upwelling regions. Before its collapse in 1972, the Peruvian anchoveta fishery, which depended on the Peru Current, provided 22% of all fish caught in the world! Earlier, the California sardine fishery and the Japanese herring fishery, both fisheries of upwelling regions, were intensively harvested to collapse and never recovered. The world's fisheries today are seriously imperiled by overfishing, degradation of fish habitats by trawling, wasteful fishing methods, and marine pollution. Some of the world's greatest fishing grounds, such as the Grand Banks and Georges Banks of eastern North America, have been destroyed.

Below the surface, or **epipelagic** zone, lie the great ocean depths, characterized by enormous pressure, perpetual darkness, and a constant temperature near 0° C. The deep sea remained a world unknown to humans until baited cameras, bathyscaphs, and deep-water trawls were lowered to view and to sample the ocean floor. There are several distinct habitats in the ocean depths (Figure 37.22). The **mesopelagic** zone is the "twilight zone," which receives dim light and supports a varied community of animals. Below the mesopelagic zone is a world of perpetual darkness, divided into three depth zones as shown in Figure 37.22: bathypelagic, abyssopelagic, and hadopelagic. Deep-sea forms depend on a meager rain of organic debris from above that escapes consumption by organisms in the water column.

The biosphere comprises lithosphere, the earth's rocky shell; hydrosphere, the global distribution of water; and atmosphere, the blanket of gas surrounding the earth.

Earth's terrestrial environment comprises various biomes, each one bearing a distinctive array of plant and animal species. Eastern deciduous forest has distinct seasons and autumn leaf fall. North of the deciduous forest is coniferous forest, whose northern range is called taiga, an area dominated by needleleafed trees adapted for heavy snowfall. Animals of the taiga are adapted for long, snowy winters.

The tropical forest is the richest biome, characterized in part by a great diversity of plant species and the vertical stratification of animal habitats. Most tropical forest soils rapidly deteriorate when forest is removed.

The most modified biome is grassland, or prairie, which has been converted largely to agriculture and grazing. The tundra biome of the far north and the desert biome are both severe environments for animal life, but they are populated nevertheless with organisms that have evolved appropriate adaptations.

Freshwater habitats include rivers and streams (lotic habitats) and ponds and lakes (lentic habitats). All are geologically ephemeral habitats strongly influenced by nutrient input.

Oceans occupy 71% of earth's surface. The photic, or sunlit, zone supports photosynthetic activity by phytoplankton. Oceanic animal communities are classified according to topographic features, substrate, and distance from the shore. Benthic communities with rocky substrates include intertidal regions with tide pools, subtidal coral reefs and kelp forests, and hydrothermal vent communities of the deep sea. Benthic communities on soft sediments include nearshore salt marshes, sea-grass beds, and mangrove communities. Benthic deep-sea communities form a patchwork of suspension feeders on sandy substrates and deposit feeders on muddy ones. Coral reefs are the most ecologically diverse of these benthic communities. Pelagic communities include a shallow-water zone overlying the continental shelf. This zone is the locus of the world's great fisheries, which are especially productive in areas of upwelling where nutrients are constantly renewed. The deeper waters of open ocean occupy most of the ocean's area but have low biological productivity.

SUMMARY

Zoogeography is the study of animal distribution on earth and its history using evidence from current distributions of animal species, fossils, and phylogenetic systematic analyses. Animals have become distributed by dispersal, the spread of populations from their centers of origin, and by vicariance, the separation of populations by barriers. Continental drift, now strongly supported by plate tectonic theory, explains how some animal groups became geographically separated so that evolutionary diversification could occur. It also explains how certain groups, such as marsupial mammals, became isolated from others. Temporary land bridges have served as important pathways for animal dispersal. Global climatic cycles interact with local topographic features to generate complex histories of population expansion and vicariance for many animal taxa.

The biosphere is a thin, life-containing blanket surrounding the earth. Life on earth is possible because of a steady supply of energy from the sun, the presence of water, a suitable range of temperatures, the correct proportion of major and minor elements, and the screening of lethal ultraviolet radiation by atmospheric ozone. Earth's environment and living organisms have evolved together, each deeply marking the other.

REVIEW QUESTIONS

1. What are some reasons a species may be absent from a habitat or region to which it should adapt well?
2. Define and distinguish the alternative explanations for disjunct distributions among animals: dispersal and vicariance.
3. Who first proposed the continental-drift theory? What three sources of evidence convinced geologists that the theory was correct?
4. How does continental-drift theory help to explain the disjunct distribution of marsupial mammals in Australia and South America?
5. What was the Great American Interchange? When did it occur, and what were the results?
6. What are the consequences of climatic cycles for vicariance? What difficulties do they introduce for the traditional methodology of vicariance biogeography?
7. How have the earth and life on it evolved together with each deeply influencing the other?
8. What is the biosphere? How would you distinguish between the following subdivisions of the biosphere: lithosphere, hydrosphere, atmosphere?

9. What is the origin of oxygen on earth? What would happen to earth's supply of oxygen if photosynthesis suddenly were to cease?

10. What is the evidence that increasing carbon dioxide levels in the atmosphere are responsible for an increase in the "greenhouse effect"?

11. What is a biome? Briefly describe six examples of biomes.

12. Describe three kinds of benthic marine communities that use hard substrates and three that use soft substrates. What are the main physical factors that separate community types within each substrate category?

13. What are some very productive marine environments, and why are they so productive?

14. What is the source of nutrients for animals living in the deep-sea habitat?

For Further Thought What consequences would climatic cycles have for vicariance on an oceanic island that has both mountains and valleys, with some valleys being inundated when sea level is high?

SELECTED REFERENCES

Berner, E. K., and R. A. Berner. 2012. Global environment: water, air, and geo-chemical cycles, ed. 2. Princeton, New Jersey, Princeton University Press. *A geochemistry textbook with good coverage of global water and air circulation, greenhouse effect, acid rain, and geochemistry of rivers, lakes, and oceans.*

Castro, P., and M. E. Huber. 2015. Marine biology, ed. 10. Boston, McGraw-Hill Higher Education. *A global coverage of marine biology designed for nonmajors.*

Cox, C. B., and P. D. Moore. 2016. Biogeography: an ecological and evolutionary approach, ed. 9. Hoboken, New Jersey, John Wiley and Sons. *Highly readable account with a strong ecological emphasis.*

Heaney, L. R. 2007. Is a new paradigm emerging for oceanic island biogeography? Journal of Biogeography **34:** 753–757. *A review of the roles of dispersal and vicariance in generating animal species on oceanic islands.*

Levinton, J. S. 2013. Marine biology, ed. 4. Oxford, U.K., Oxford University Press. *A thorough coverage of oceanic ecosystems with photographs of diverse communities.*

Lieberman, B. S. 2000. Paleobiogeography. New York, Kluwer Academic/Plenum Publishers. *A coverage of historical biogeography with emphasis on use of fossils to study global change, plate tectonics, and evolution.*

Lomolino, M. V., B. R. Riddle, R. J. Whittaker, and J. H. Brown. 2010. Biogeography, ed. 4. Sunderland, Massachusetts, Sinauer Associates. *A current general textbook of biogeography.*

Lomolino, M. V., D. F. Sax, and J. H. Brown. 2004. Foundations of biogeography. Chicago, University of Chicago Press. *Collection of classic papers of biogeography from 1700s to 1975 with new commentary.*

MacDonald, G. 2003. Biogeography: introduction to space, time, and life. New York, John Wiley and Sons, Inc. *An introductory textbook of biogeography.*

Marshall, R. 2005. Alaska wilderness: exploring the central Brooks Range, ed. 3. Berkeley, California, University of California Press. *A chronicle of the original exploration of Alaskan tundra, with updates on its conservation status.*

Paillard, D. 2006. What drives the ice age cycle? Science **313:** 455–456. *A concise review of climatic cycles of the past three million years and their causes.*

Rothschild, L. J., and A. M. Lister. 2003. Evolution on planet earth: the impact of the physical environment. San Diego, Academic Press. *A geological and climatic perspective on life's evolutionary history.*

Van Oosterzee, P. 1997. Where worlds collide: the Wallace Line. Ithaca, New York, Cornell University Press. *An entertaining account of the founding of biogeography by Alfred Russel Wallace with emphasis on the paradox of an abrupt faunal change in the Malay Archipelago.*

Whitfield, P., P. D. Moore, and B. Cox. 2002. Biomes and habitats. New York, Macmillan Reference USA. *A survey of the earth's biomes.*

38
CHAPTER

Animal Ecology

The ecological niche of a praying mantis includes other insects as a food resource.

Ecological Niche and Animal Species Diversity

Life's diversity forms a hierarchy of interacting units: an individual organism, a population, a community, and most inclusively an ecosystem. Central to ecological study of animal diversity is habitat, the spatial location where an animal lives. What an animal does in its habitat is its niche: how it gets its food, how it achieves reproductive perpetuity—in short, how it survives and stays adapted in a Darwinian sense.

The "competitive exclusion principle" states that no two species can occupy the same niche if they are to live together in a stable ecological community. Different species are therefore able to form an ecological community in which each has a different role in their shared environment.

In the mid–nineteenth century, German zoologist Ernst Haeckel introduced the term **ecology,** defined as the relation of the animal to its organic as well as inorganic environment. Environment here includes everything external to the animal but, most important, its niche. Although we no longer restrict ecology to animals alone, Haeckel's definition remains basically sound. Animal ecology is now a highly synthetic science that incorporates the behavior, physiology, genetics, and evolution of animals to study interactions between populations of animals and their environments.

The major goal of ecological studies is to explain how these diverse interactions influence the geographical distributions and abundances of animal populations. Measuring the niche requirements of each species is crucial for understanding animal diversity and for ensuring the continued survival of many populations.

38.1 THE HIERARCHY OF ECOLOGY

We study ecology as a hierarchy of biological systems in interaction with their environments. At the base of the ecological hierarchy is an **organism,** such as an individual animal. To understand why animals live where they do, ecologists must examine the varied physiological and behavioral mechanisms that animals use to survive, to grow, and to reproduce. Endothermic animals, including birds and mammals, must maintain a near-perfect physiological balance between production and loss of heat under extreme temperatures, as occur in the Arctic or a desert. Other species succeed in these situations by escaping the most extreme conditions by migration or hibernation. Insects, fishes, and other ectotherms (animals whose body temperature depends on environmental heat) respond to fluctuating temperatures by altering behavior and biochemical and cellular processes. Thus an animal's physiological capacities permit it to live in changing and often adverse environments. Behavioral responses are important also for obtaining food, finding shelter, escaping enemies and unfavorable environments, finding a mate, and tending offspring. Physiological mechanisms and behaviors that improve adaptability to environments assist survival of organisms. Ecologists who focus their studies at the organismal level are called physiological ecologists or behavioral ecologists.

Animals in nature coexist with others of the same species; animals of the same species form reproductive communities called **populations** (p. 818). Measurable properties of populations include genetic variability among individuals (polymorphism), growth in numbers over time, and factors that limit the density of individuals in each area. Ecological studies of these properties help us to predict the future success of an endangered species and to discover controls for pest species.

Just as individuals do not live alone in nature, populations of different species co-occur in more complex associations called **ecological communities.** One measure of a community's complexity is **species diversity,** the number of different species that coexist to form the community. Populations of species in a community interact with one another in many ways, the most prevalent of which are **predation, parasitism,** and **competition. Predators** obtain energy and nutrients by killing and eating other animals, called prey. **Parasites** derive similar benefits from their host organisms, but they live on or in the host and usually do not kill the host. A **parasitoid** lives on or in the host but eventually kills its host organism. Competition occurs when food or space are limited, and members of the same or different species interfere with each other's use of shared resources. Communities are complex because all of these interactions occur simultaneously, and their individual effects on the community often cannot be isolated.

Ecological communities are biological components of even larger, more complex entities called **ecosystems.** An ecosystem consists of all populations in an ecological community together with their physical environments. The study of ecosystems reveals two key processes in nature, the flow of energy and the cycling of materials through biological channels. The largest ecosystem is the **biosphere,** the thin veneer of land, water, and atmosphere that envelops the earth and supports all life (see Chapter 37).

Environment and Niche

An animal's environment comprises all conditions that directly affect its survival and reproduction. These factors include space; forms of energy such as sunlight, heat, and wind and water currents; and also materials such as soil, air, water, and numerous chemicals. Environment also includes other organisms, which can be an animal's food, or its predators, competitors, hosts, or parasites. Environment thus includes both abiotic (nonliving) and biotic (living) factors. **Resources,** such as space and food, are environmental factors that an animal uses directly.

An investigator categorizes a particular resource as being either expendable or nonexpendable, depending on how an animal uses it. Food is expendable, because once eaten it is no longer available. Food therefore must be continuously replenished. Space, whether total living area or a subset, such as the number of suitable nesting sites, is not exhausted by being used, and thus is nonexpendable.

The physical space where an animal lives is its **habitat.** Size of a habitat varies. A rotten log is a normal habitat for carpenter ants. Such logs occur in larger habitats called forests. Deer also inhabit forests, but because they forage in open meadows, their habitat is larger than the forest. On a larger scale, some migratory birds occupy forests of the north temperate region during summer and move to the tropics during winter. Thus, we define habitat by an animal's normal activity rather than by arbitrary physical boundaries.

Animals of any species experience environmental limits of temperature, moisture, and food within which they can grow, reproduce, and survive. A suitable environment therefore must meet all requirements for life. A freshwater clam inhabiting a tropical lake could tolerate the temperature of a tropical ocean, but the ocean's salinity would kill the clam. A brittle star living in the Arctic Ocean could tolerate the salinity of the tropical ocean but not its temperature. Thus temperature and salinity are two separate dimensions of an animal's environmental limits. If we add another variable, such as pH (p. 22), we increase our description to three dimensions (Figure 38.1).

If we consider all environmental conditions that permit members of a species to survive and to multiply, we define the role of that species in nature as distinct from all others. This unique, multidimensional relationship of a species with its environment is called its **niche** (see the chapter-opening essay, p. 816). Niche dimensions vary among members of a species, making niche subject to evolution by natural selection. The niche of a species undergoes evolutionary changes over successive generations.

One can categorize a particular animal species as being either a generalist or a specialist with respect to a particular environmental resource or condition. For example, most fish species are specialized for either freshwater or seawater, but not both. However, those that occupy marshes, such as the killifish *Fundulus heteroclitus,* are generalists regarding salinity, and easily tolerate changes in salinity as freshwater from land mixes with seawater during tidal cycles. Similarly, although most snakes eat a wide variety of animal prey, others are narrow dietary specialists; for example, the African snake *Dasypeltis scaber* is specialized to eat bird eggs (see Figure 32.3, p. 705).

However broad may be the tolerance limits of an animal, it experiences only a single set of conditions at a time. An animal probably

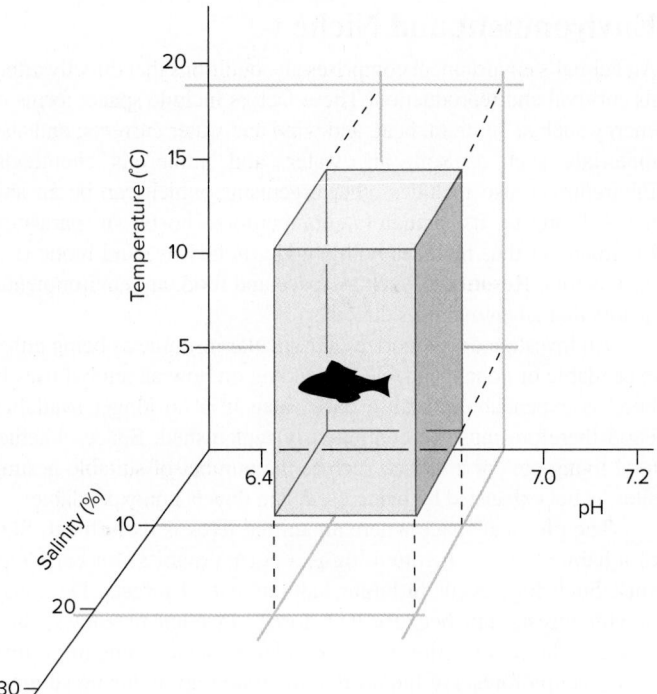

Figure 38.1 Three-dimensional niche volume of a hypothetical animal showing three tolerance ranges. This graphic representation is one way to show a portion of the multidimensional nature of environmental relations.

does not experience in its lifetime all environmental conditions that it might tolerate. Thus, we must distinguish an animal's **fundamental niche,** which describes its potential role, and its **realized niche,** the subset of potentially suitable environments that an animal actually experiences. Likewise, we must distinguish fundamental niche from realized niche at the population and species levels. For example, competition within a community may limit the realized niche of a species to a much smaller range of conditions than those predicted by its fundamental niche. For example, the lungless slimy salamander *Plethodon teyahalee* occupies forest-floor habitats in the southern Appalachian Mountains. It occupies the coolest, mountaintop forests only where species of the *Plethodon jordani* group are absent (see Figure 6.20). Climatic requirements restrict the *P. jordani*-group species to cool mountaintop isolates, from which they exclude *P. teyahalee* through aggressive encounters. Competition with the *P. jordani*-group species thereby constrains the realized niche of some *P. teyahalee* populations to a smaller and higher average range of temperatures than its fundamental niche would permit.

Populations

An animal exists in nature as a member of a population, a reproductively interactive group of animals of a single species (p. 113). A species may be a single, cohesive population or may contain many geographically disjunct populations, often called **demes.** Because members of a deme interbreed, they share a common gene pool (p. 122). A deme of cichlid fishes in Cuatro Cienegas, Mexico, was mistakenly diagnosed as two separate species because some individuals have strong "molariform" jaws capable of crushing snails,

whereas others have weaker "papilliform" jaws capable of processing only soft food items. Molecular genetic studies show that despite their morphological differences, these fishes interbreed and share a common gene pool, thereby forming a single deme of the species *Cichlasoma minckleyi.*

To study the spatial ecology of a species that has multiple geographic populations, ecologists use a model called the **metapopulation.** A metapopulation comprises two or more geographically distinct demes, each of which maintains continuity in time and space. Most interactions among individual organisms, including reproduction, occur within a deme, but some individuals move from one deme to a different one. This movement between demes is called migration if the individual's change of demes is a permanent one. Movement of individuals among demes within a species can impart some evolutionary cohesion to the species as a whole. Local environments can change unpredictably, sometimes severely depleting or eliminating a local deme. Migration is therefore a crucial source of replacement among demes within a region. A species can avoid extinction if risk of extinction is spread among many demes, because simultaneous destruction of all demes in this manner is unlikely unless a catastrophe is geographically widespread. Interaction among demes in this manner is called **metapopulation dynamics.** In some species, migration and recolonization among demes are nearly symmetrical. If some demes are stable and others more susceptible to extinction, the more stable ones, termed **source demes,** differentially supply emigrants to the less stable ones, called **sink demes.**

Suppose that *Cichlasoma minckley* is subdivided geographically into multiple demes that differ from one another in the relative proportions of molariform and papilliform individuals. A deme that contains mostly molariform individuals should be more stable in times of food scarcity than demes having mostly "papilliform" individuals, because the molariform jaws permit a fish to exploit snails as food when the preferred softer food items are depleted. Such a deme could function as a source deme following a decline in the fish's favored prey species.

Demography

Each population or deme has a characteristic **age structure, sex ratio,** and **growth rate.** The study of these properties and the factors that influence them is called demography. Demographic characteristics vary according to lifestyles of the species under study. For example, some animals (and many plants) are **modular.** Modular animals, such as sponges, corals, and ectoprocts, consist of colonies of genetically identical organisms. Reproduction is by asexual **cloning,** as described for hydrozoans in Chapter 13 (pp. 267, 270). Colonies propagate also by fragmentation, as seen on coral reefs during severe storms. Pieces of coral may scatter by wave action on a reef, forming propagules for new reefs. For these modular animals, age structure and sex ratio are difficult to determine. Changes in the area that a colony occupies are used to measure growth rate, but counting individuals is difficult and less meaningful than in **unitary** animals, which live independently.

Populations of unitary animals that reproduce by **parthenogenesis** (p. 136) share some characteristics with asexual clones. Strictly parthenogenetic species contain only females, which lay unfertilized eggs that hatch into daughters whose genotype comes entirely from their mothers. The praying mantid, *Bruneria borealis,* common in the southeastern United States, is a parthenogenetic unitary animal, as are some fishes, salamanders, and lizards.

Most animals are biparental (p. 135), and reproduction follows a period of organismal growth and maturation. Each new generation begins with a **cohort** of individuals born at the same time. Of course, individuals of any cohort do not all survive to reproduce. For a population to retain constant size from generation to generation, each adult female must replace herself on average with one daughter that survives to reproduce. If females produce on average more than one viable daughter, the population grows; if fewer than one, the population declines.

Animal species have different characteristic patterns of **survivorship** from birth until death of the last member of a cohort. The three principal types of survivorship are illustrated in Figure 38.2. In curve I, most individuals die at old age through senescence. Human populations that have the resources to support expanding population size and to prevent infant mortality approximate curve I. Curve II, in which rate of mortality as a proportion of survivors is constant over all ages, characterizes some animals that care for their young, as do many birds, and thus avoid high infant mortality. In contrast to curve I, however, populations at curve II have important sources of mortality other than old age. Predation or lack of food, for example, would kill individuals at any age. Curve III characterizes populations whose infant or juvenile mortality is very high relative to that of young adults.

Survivorship of most invertebrates, and of vertebrates like fish that produce great numbers of offspring, resembles curve III. For example, a mature female marine prosobranch snail, *Ilyanassa obsoleta,* produces thousands of eggs each reproductive period. Zygotes become free-floating planktonic veliger larvae (p. 337), which scatter far from the mother's location in oceanic currents. They experience high mortality from numerous animals that consume plankton. Furthermore, the larvae require a specific, sandy substrate on which to settle and to metamorphose into adult snails. The probability of a larva surviving long enough to find a suitable habitat is very low, and most of the cohort dies during the veliger stage. We therefore see a rapid drop in survivorship in the first part of the curve. The few larvae that do survive to become snails have improved odds of surviving further, as shown by the more gentle slope of the curve for older snails. Thus, high reproductive output balances high juvenile mortality.

Most animals do not survive to reach reproductive age, and those that do survive might reproduce only once before they die, as do many insect species of the temperate zone. Here, adults reproduce before the onset of winter and die, leaving only their eggs to overwinter and to repopulate their habitat the following spring. Similarly, Pacific salmon after several years return from the ocean to freshwater to spawn only once, after which all adults of a cohort die. However, other animals survive long enough to produce multiple cohorts of offspring that may mature and reproduce while their parents are still alive and reproductively active.

Populations of animals containing multiple cohorts, such as robins, box turtles, and humans, exhibit **age structure.** Analysis of age structure reveals whether a population is actively growing, stable, or declining. Figure 38.3 shows age profiles for the people of Mexico as measured in 1975 and 2005. The former profile shows an actively growing population, whereas the latter one shows a population maintaining a stable size.

Population Growth and Intrinsic Regulation

Population growth is the difference between rates of birth and death. As Darwin recognized from an essay by Thomas Malthus (p. 103), all populations have an inherent ability to grow exponentially. This ability is called the **intrinsic rate of increase,** denoted by the

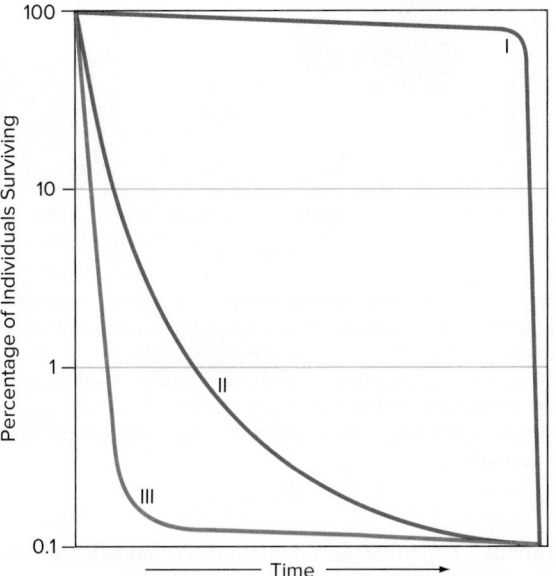

Figure 38.2 Three principal types of theoretical survivorship curves. See text for explanation.

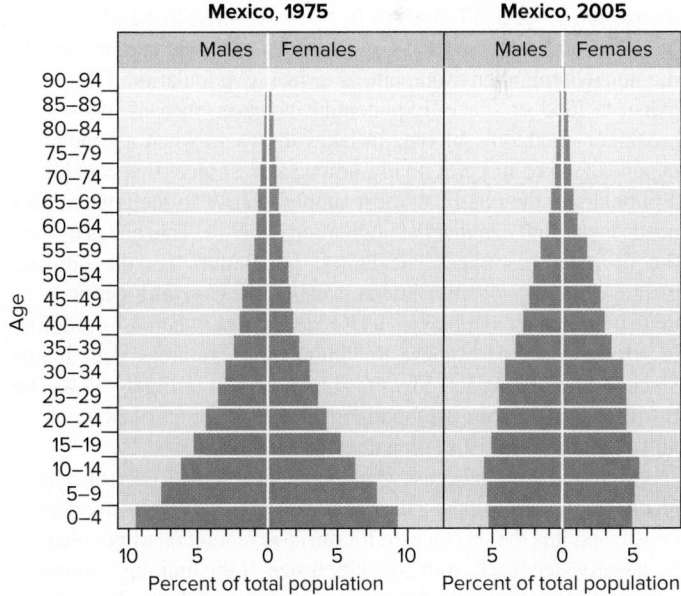

Figure 38.3 Age structure profiles of the human populations of Mexico in 1975 and 2005 contrast the rapidly growing, youthful population of 1975 with the stable population of 2005, where the fertility rate approximates replacement This **demographic transition** from high birth and death rates to low ones often occurs as a country becomes increasingly industrial in its economic development. Popular references to the "demographic transition" in human societies are specifically to this kind of change.

Figure 38.4 Population growth, showing exponential growth of a species in an unlimited environment, and logistic growth in a limited environment. K = population size at carrying capacity; r = intrinsic rate of population growth.

symbol **r.** The steeply rising curve in Figure 38.4 shows this kind of growth. If species continually grew in this fashion, earth's resources soon would be exhausted. A bacterium dividing three times per hour could produce a colony a foot deep over the entire earth after 38 hours, and this mass would be over our heads only one hour later. Potential growth rates of bacterial populations greatly exceed those of animals, but animal populations could achieve the same kind of result over a longer period of time, given unlimited resources. Many insects lay thousands of eggs each year. A single Atlantic codfish, *Gadus morhua,* may spawn 6 million eggs in a season, and a field mouse can produce 17 litters of five to seven offspring each year.

Obviously, unrestricted growth is uncommon in nature. Even in the most benign environment, a growing population eventually exhausts food or space. Exponential increases such as locust out-breaks or planktonic blooms in lakes must end when a population expends food or space. Actually, among all resources that could limit a population, the one in shortest supply relative to the population's needs is the first one that the population depletes. This one is termed the **limiting resource.** The largest population that the limiting resource can support in a habitat is called the **carrying capacity** of that environment, symbolized as **K.** Ideally, a population would slow its growth rate in response to diminishing resources until it just reached K, as represented by the sigmoid curve in Figure 38.4. The box on page 822 compares mathematical expressions of exponential and sigmoid (or logistic) growth curves.

Sigmoid growth occurs when population density exerts negative feedback on growth rate. This phenomenon is called density depen-dence, and it is the mechanism for intrinsic regulation of populations by negative feedback from population size. If the limiting resource is expendable, as is food, carrying capacity is reached when rate of resource replenishment equals rate of depletion by the population; the population is then at K for that limiting resource. According to the sigmoid model, when population density reaches K, rates of birth and death are equal and growth of the population ceases. If food is replaced at a rate that supports the current population, but no more, a population of grasshoppers in a green meadow, for example, may be at carrying capacity even though we see plenty of unconsumed food.

A

B

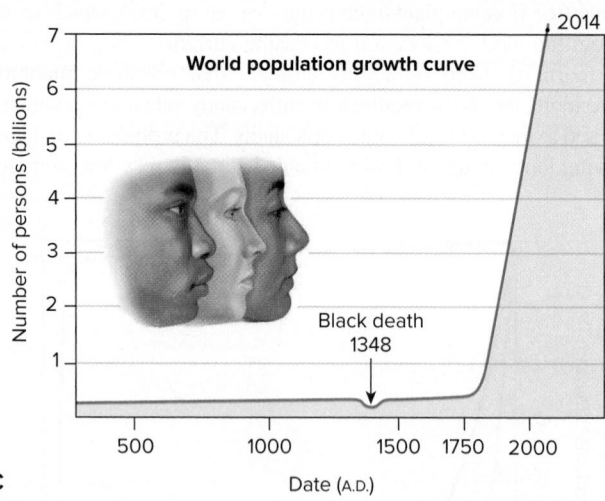

C

Figure 38.5 Growth curves for sheep, *Ovis aries* (**A**), ring-necked pheasant, *Phasianus colchicus* (**B**), and world human populations (**C**) throughout history. Note that the sheep population on an island is stable because people control the population, but the ring-necked pheasant population oscillates greatly, probably caused by large changes in carrying capacity. Where would you place the carrying capacity for the human population?

Although experimental populations of unicellular eukaryotes often fit the sigmoid growth curve closely, most populations in nature fluctuate above and below carrying capacity. For example, after sheep were introduced to Tasmania around 1800, their numbers changed logistically with small oscillations around an average population size of about 1.7 million; we thereby infer the carrying capacity of the environment to be 1.7 million sheep (Figure 38.5A). Ring-necked pheasants introduced on an island in Ontario, Canada, exhibited wider oscillations (Figure 38.5B).

Why do intrinsically regulated populations oscillate this way? First, the carrying capacity of an environment can change over time, causing a change of population density as dictated by a limiting resource. Second, animals always experience a lag between the time that a resource becomes limiting and the time that the population responds by reducing its rate of growth. Third, extrinsic factors occasionally can limit a population's growth below carrying capacity. We consider extrinsic factors in the next section.

On a global scale, humans have the longest record of exponential population growth (Figure 38.5C). Although famine and war have restrained growth of populations locally, global human growth declined only when bubonic plague ("black death") decimated much of Europe during the fourteenth century. What, then, is the carrying capacity for the human population? The answer is unclear, because technological advances have increased our ability to extract useful resources from our environments.

Agriculture increased the carrying capacity of the environment, and the human population grew steadily from 5 million when agriculture was introduced around 8000 B.C. to 16 million around 4000 B.C. Despite the toll of terrible famines, disease, and war, the population reached 500 million by 1650. With the Industrial Revolution in Europe and England in the eighteenth century, followed by a medical revolution, discovery of new lands for colonization, and better agricultural practices, the human carrying capacity increased dramatically. The population doubled to 1 billion around 1850. It doubled again to 2 billion by 1927, reached 4 billion in 1974, passed 6 billion in October 1999, and is expected to reach 8 billion by the year 2040. Thus, the growth has been exponential and remains high (Figure 38.5C). What would constitute the maximum sustainable agricultural output is uncertain, although food production cannot increase to sustain exponential population growth indefinitely. The United Nations has taken the stance that K for humans is approximately 7–11 billion people.

> Between 1970 and 2004 the annual growth rate of the human population decreased from 1.9% to 1.23%. At 1.23%, it will take nearly 57 years for the world population to double, rather than 36.5 years at the higher annual growth rate. The decrease is credited to better education for women and family planning. Nevertheless, half the global population is under 25 years old. Thus, despite a drop in growth rate, the greatest surge in population lies ahead, with a projected 3 billion people added within the next five decades, the most rapid increase ever in human numbers.

Extrinsic Limits to Growth

We have seen that the intrinsic carrying capacity of a population for an environment prevents unlimited exponential growth of the population. Population growth also can be limited by extrinsic biotic factors, including predation, parasitism (including disease-causing pathogens), and interspecific competition, or by abiotic influences such as floods, fires, and storms. Although abiotic factors certainly reduce populations in nature, they cannot truly regulate population growth because their effect is wholly independent of population size; abiotic limiting factors are thus called **density-independent.** A single hailstorm may kill most young wading birds, and a forest fire may eliminate entire populations of many animals, regardless of how many individuals exist.

In contrast, biotic factors act in a **density-dependent** manner. Predators and parasites respond to changes in density of their prey and host populations, respectively, to maintain those populations at fairly constant sizes. These sizes are below carrying capacity, because populations regulated by predation or parasitism are not limited by their resources. Competition between species for a common limiting resource lowers the effective carrying capacity for each species below that of either one alone. Salamanders of the species *Plethodon teyahalee* occur together with single species of the *Plethodon jordani* group at some scattered intermediate elevations in the southern Appalachian Mountains; where the species overlap, both species have lower densities than are typical for them outside the small area of co-occurrence. Experimental removal of either species from a site of co-occurrence increases population density of the remaining species.

Community Ecology
Interactions Among Populations in Communities

Populations of animals that form a **community** interact in ways that can be detrimental (−), beneficial (+), or neutral (0) to each species, depending on the interaction. For instance, we consider a predator's effect on prey as (−), because survival of the prey animal is reduced. However, the same interaction benefits the predator (+) because food obtained from prey increases a predator's ability to survive and to reproduce. Thus, the predator-prey interaction is + −. Ecologists use this shorthand notation to characterize interspecific interactions because it shows the direction in which the interaction affects each species.

We see other kinds of + − interactions. One is **parasitism,** in which the parasite benefits by using the host as a home and source of nutrition, and the host is harmed. **Herbivory,** in which an animal eats a plant, is another + − relationship. **Commensalism** is an interaction that benefits one species and neither harms nor benefits the other (+0). Most bacteria that normally inhabit our intestinal tracts do not affect us (0), but the bacteria benefit (+) by having food and a place to live. A classic example of commensalism is the association of pilot fishes and remoras with sharks (Figure 38.6). These fishes get the "crumbs" remaining when the host shark makes its kill.

Species engaged in mutualism both benefit from their ecological interaction (+ +; Figure 38.7). A distinction between commensalism and mutualism is sometimes difficult to make in practice. For example, if the harmless bacteria residing in the human gut prevent entry of harmful bacteria, the apparent commensalism becomes a mutualism. Likewise, if remoras remove parasites from their shark hosts (Figure 38.6), then this commensalism likewise becomes a mutualism.

Some mutualistic relationships are not only beneficial but necessary for survival of one or both species. An example is the relationship

© Noble Proctor/Science Source

Figure 38.6 Four remoras, *Remora* sp., attached to a shark. Remoras feed on fragments of food left by their shark host, as well as on pelagic invertebrates and small fishes. Although they actually are good swimmers, remoras prefer to be pulled through the water by marine creatures or boats. The shark host may benefit by having embedded copepod skin parasites removed by remoras.

between a termite and parabasalids (p. 228) inhabiting its gut. Bacterial symbionts of the parabasalids digest wood eaten by the termite because they produce an enzyme, lacking in the termite, that digests cellulose; the termite lives on waste products of parabasalid-bacterial metabolism. In return, the parabasalids and their bacteria gain a habitat and food supply.

Competition between species reduces abundances of both (− −). Many biologists, including Darwin, considered competition the most common and important interaction in nature. Ecologists have constructed most of their theories of community structure from the premise that competition is the chief organizing factor in species assemblages. Sometimes the effect on one species in a competitive relationship is negligible. This condition is called **amensalism,** or **asymmetric competition** (0−). For example, two species of barnacles in rocky intertidal habitats, *Chthamalus stellatus* and *Balanus balanoides,* compete for space. A famous experiment by

---

Exponential and Logistic Growth

We describe the sigmoid growth curve (see Figure 38.4) by a simple model called the logistic equation. The slope at any point on the growth curve is the growth rate, how rapidly population size changes with time. If **N** represents the number of organisms and **t** the time, we can express growth in the language of calculus as an instantaneous rate:

$$d\mathbf{N}/d\mathbf{t} = \text{rate of change in number of organisms per time}$$
at a particular instant in time

When populations experience unlimited resources (unlimited food and space, and no competition from other organisms), growth is limited only by the inherent capacity of the population to reproduce itself. Under these ideal conditions growth is expressed by the symbol **r,** defined as the intrinsic rate of population growth per capita. The index **r** is actually the difference between the birth rate and death rate per individual in the population per unit time. The growth rate of the population as a whole is then:

$$d\mathbf{N}/d\mathbf{t} = r\mathbf{N}$$

This expression describes the rapid **exponential growth** illustrated by the early upward-curving portion of the sigmoid growth curve (see Figure 38.4).

Growth rate for populations in the real world slows as the upper limit is approached, and eventually stops. At this point **N** has reached its maximum density because the space being studied is "saturated" with animals. This limit is called the carrying capacity of the environment and is expressed by the symbol **K.** The sigmoid population growth curve is described by the logistic equation, written as:

$$d\mathbf{N}/d\mathbf{t} = r\mathbf{N}[\mathbf{K} - \mathbf{N}/\mathbf{K}]$$

This equation states that the rate of increase per unit of time ($d\mathbf{N}/d\mathbf{t}$) = rate of growth per capita (**r**) × population size (**N**) × unutilized freedom for growth ([**K** − **N**]/**K**). We see from the equation that when the population approaches carrying capacity, **K** − **N** approaches 0, $d\mathbf{N}/d\mathbf{t}$ also approaches 0, and the curve flattens.

Populations occasionally overshoot the carrying capacity of the environment so that **N** exceeds **K.** The population then exhausts a resource (usually food or shelter). The rate of growth, $d\mathbf{N}/d\mathbf{t}$, then becomes negative and the population declines.

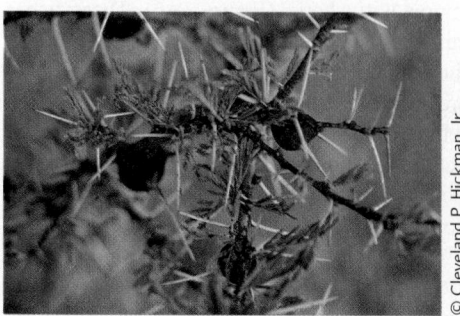

© Cleveland P. Hickman, Jr.

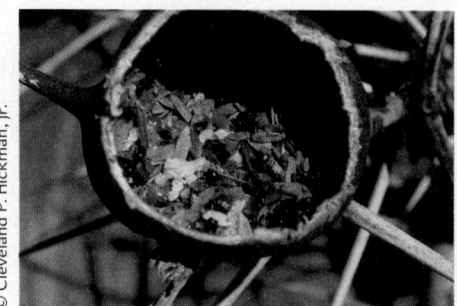

© Cleveland P. Hickman, Jr.

Figure 38.7 Among the many mutualisms that abound in nature is that of the whistling thorn acacia, *Acacia drepanolobium,* of the African savanna and the ants, *Crematogaster mimosae* and other species, that make their homes in the acacia's swollen galls. The acacia provides both protection for the ants' larvae (*photograph of opened gall, right*) and honeylike secretions used by the ants as food. In turn, the ants protect the tree from herbivores by swarming out as soon as the tree is touched. Giraffes, *Giraffa camelopardalis,* which love the tender acacia leaves, seem immune to the ants' fiery stings.

Joseph Connell[1] demonstrated that *B. balanoides* excluded *C. stellatus* from a portion of the habitat, whereas *C. stellatus* had no effect on *B. balanoides.*

We have treated interactions as occurring between pairs of species. However, in natural communities containing populations of many species, a predator may have more than one prey, and several animals may compete for the same resource (p. 817). Thus, ecological communities are complex and dynamic, a challenge to ecologists who study this level of natural organization.

Competition and Resource Partitioning

Competition occurs when two or more species share a **limiting resource.** Simply sharing food or space with another species does not produce competition unless the resource is in short supply relative to the needs of the species that share it. Thus, we cannot demonstrate competition in nature simply by showing that two species share a resource. However, we find evidence of competition by investigating the different ways that species exploit a resource.

Competing species may reduce conflict by reducing overlap of their niches. **Niche overlap** is the portion of resources shared by the niches of two or more species. For example, if two species of birds eat seeds of exactly the same size, competition eventually excludes the species less able to exploit the resource. This example illustrates the principle of **competitive exclusion:** strongly competing species cannot coexist indefinitely. This principle was discovered in 1932 by the Soviet microbiologist G. F. Gause, who performed experiments on the mechanism of competition between two ciliate species (*Paramecium aurelia* and *P. caudatum*) and two yeast species (*Saccharomyces cerevisiae* and *S. kefir*). To coexist in the same habitat, species must specialize by partitioning a shared resource and using different portions of it. Specialization of this kind is called **resource partitioning.**

Resource partitioning usually appears as differences in organismal morphology or behavior related to exploitation of a resource. For example, in his classic study of Galápagos finches (p. 117), English ornithologist David Lack noticed that beak sizes of these birds depended on whether they occurred together on the same island (Figure 38.8). On the islands Daphne and Los Hermanos, where *Geospiza fuliginosa* and *G. fortis* occur separately and therefore do not compete with each other, beak sizes are nearly identical; on the island Santa Cruz, where both *G. fuliginosa* and *G. fortis* coexist, their beak sizes do not overlap. These results suggest resource partitioning, because beak size determines the size of seeds eaten. Work by American ornithologist Peter Grant confirms what Lack suspected: *G. fuliginosa* with its smaller beak selects smaller seeds than does *G. fortis* with its larger beak. Where these two forms coexist, competition between them has led to evolutionary displacement of beak sizes to diminish competition. Absence of competition today has been called "the ghost of competition past."

Ecological character displacement is the mechanism of resource partitioning illustrated by these Galápagos finches; co-occurrence in the same community causes species to evolve niche dimensions more restricted than those of their ancestors. An alternative

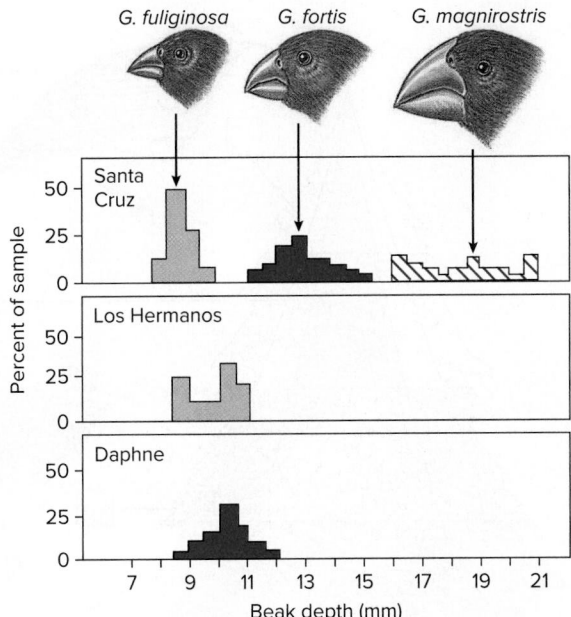

Figure 38.8 Displacement of beak sizes in Darwin's finches from the Galápagos Islands. Beak depths are given for the ground finches *Geospiza fuliginosa* (pink bars) and *G. fortis* (blue bars) where they occur together (sympatric) on Santa Cruz Island and where they occur alone on the islands Daphne and Los Hermanos. *Geospiza magnirostris* is another large ground finch that lives on Santa Cruz.

and perhaps more common means of evolving resource partitioning is through **microhabitat selection,** formation of a community by species whose preferred microhabitats differ sufficiently to prevent them from competing for what otherwise would be a shared limiting resource.

When several species share the same general resources, they form a **guild.** Just as a guild in medieval times constituted a brotherhood of men sharing a common trade, species in an ecological guild share a common livelihood. The term guild was introduced to ecology by Richard Root in his 1967 paper on niche patterns of the blue-gray gnatcatcher.[2]

The theory of competitive exclusion predicts that when different species of a guild occur in the same geographic area, detailed study will reveal resource partitioning among the guild's species. A classic test of resource partitioning among species in a bird guild is Robert MacArthur's study of a feeding guild comprising five species of warblers in spruce woods of the northeastern United States.[3]

At first glance, we might ask how five species of birds, very similar in size and appearance, could coexist by feeding on insects in the same tree. However, on close inspection MacArthur found subtle differences among these birds in sites of foraging (Figure 38.9). One species searched only on outer branches of spruce crowns; a second species used the top 60% of the tree's outer and inner branches away from the trunk; a third species concentrated on inner branches near the trunk; a fourth species used the midsection from the periphery to

[1]Connell, J. H. 1961. The influence of interspecific competition and other factors on the distribution of the barnacle *Chthamalus stellatus.* Ecology **42:** 710–723.

[2]Root, R. B. 1967. The niche exploitation pattern of the blue-gray gnatcatcher. Ecological Monographs 37: 317–350.

[3]MacArthur, R. H. 1958. Population ecology of some warblers of northeastern coniferous forests. Ecology 39: 599–619.

Cape May
warbler

Blackburnian
warbler

Black-throated
green warbler

Bay-breasted
warbler

Yellow rumped
(Myrtle) warbler

Figure 38.9 Distribution of foraging effort among five species of wood warblers in a northeastern North American spruce forest. The warblers form a feeding guild. Resource partitioning by selection of different microhabitats among species permits them to coexist in the same trees despite sharing a limited number of insect species as prey.

the trunk; and a fifth species foraged in the bottom 20% of the tree. These observations suggest that structural differences in habitat separate each warbler's niche within this guild.

Predators and Parasites

Ecological warfare waged by predators against their prey causes coevolution: predators get better at catching prey, and prey get better at escaping predators. This is an evolutionary race that predators cannot afford to win. If a predator became so efficient that it exterminated its prey, the predator species would need to find alternative food or become extinct. Because most predators feed on more than a single species, specialization on a single prey species to the point of extermination is uncommon.

When a predator does rely primarily on a single prey species, both populations tend to fluctuate cyclically. First prey density increases, then that of the predator, until prey become scarce. At that point, predators must adjust their population size downward by leaving the area, lowering reproduction, or dying. When density of the predator population falls enough to allow reproduction by prey to outpace mortality from predation, the cycle begins again. Thus, populations of both predators and prey show cycles of abundance, but increases and decreases in predator abundance are slightly delayed relative to those of prey because of the time lag in a predator's response to changing prey density. Laboratory experiments with ciliates reveal this process (Figure 38.10). Perhaps the longest documented natural

example of a predator-prey cycle is between Canadian populations of snowshoe hares and lynxes (see Figure 28.27, p. 627). Abundance of lynx (predator) follows that of showshoe hares (prey) on a 10-year

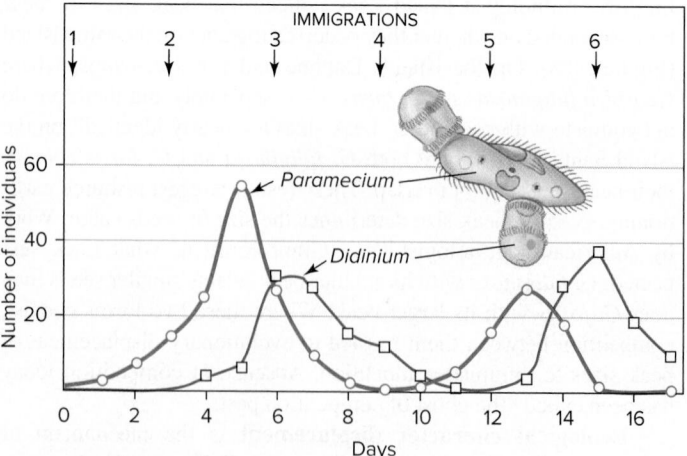

Figure 38.10 Classic predator-prey experiment by Russian biologist G. F. Gause in 1934 shows the cyclic interaction between predator (*Didinium*) and prey (*Paramecium*) in laboratory culture. When the *Didinium* find and eat all the *Paramecium*, the *Didinium* themselves starve. Gause could keep the two species coexisting only by occasionally introducing one *Didinium* and one *Paramecium* to the culture (*arrows*); these introductions simulated migration from an outside source.

Figure 38.11 Artful guises abound in the tropics. Batesian mimicry: **A,** A harmless yellowjacket clearwing moth *Pennisetia marginata* (top) mimics a yellowjacket wasp (*Vespula maculifrons,* bottom), which has a powerful stinger. **B,** Müllerian mimicry: Neotropical butterflies *Heliconis ismenius* (top) and *Lycorea cleobaea* share distastefulness and matching coloration to warn common predators. **C,** Mimicry between the viceroy butterfly (*Limenitis archippus,* top) and monarch butterfly (*Danaus plexippus)* is often presented as a case of Batesian mimicry in which the presumably tasteful viceroy mimics warning coloration of the distasteful monarch; evidence that the viceroy also is distasteful would make this a case of Müllerian mimicry.

cycle. Interestingly, the timing of these predator-prey cycles appears to track 10-year cycles in abundance of sunspots, which increase the solar energy that reaches earth and possibly also growth of vegetable matter on which hares feed. A global climatic variable thus influences timing of the density-dependent predator-prey cycles.

The war between predators and prey reaches high art in the evolution of defenses by potential prey. Potential prey can escape detection by matching their background, or by resembling some inedible feature of the environment (such as a stick). Such defenses are called cryptic. In contrast to cryptic defenses, toxic or distasteful animals advertise their strategy with bright colors and conspicuous behavior. Such defenses are termed **aposematic.** These species are protected because predators learn to recognize and to avoid them after distasteful encounters.

When distasteful prey adopt warning coloration, advantages of deceit arise for palatable prey. Palatable prey can deceive potential predators by mimicking distasteful prey, a phenomenon called **Batesian mimicry.** Coral snakes and yellowjacket wasps are both brightly colored, harmful prey. Coral snakes have a venomous bite, and yellowjacket wasps have a powerful sting. Both species serve as **models** for other species, called **mimics,** that are harmless but resemble the harmful model species (Figure 38.11A). By matching another organism that attracts its prey, a predator can use mimicry aggressively. The crab spider shown in Figure 19.13C (p. 412) matches flower petals; it hides among the petals and eats insects attracted to the flowers in search of nectar or pollen.

In another form of mimicry termed **Müllerian mimicry,** two or more species that are toxic or otherwise harmful resemble each other (Figure 38.11B). What may an animal that has its own poison gain by evolving resemblance to another poisonous animal? The answer is that a predator needs only to experience the toxicity

of one species to avoid all similar prey. A predator can learn one warning signal more easily than many! The benefits that two distasteful species derive from mutual mimicry are not always equal, for example, when a mildly distasteful species mimics a highly poisonous one. Such cases illustrate a continuum between Batesian and Müllerian mimicry. For example, mimicry between viceroy and monarch butterflies (Figure 38.11C) is often presented as Batesian mimicry of the distasteful monarch by the presumably tasteful viceroy; however, some data suggest that the viceroy also is distasteful, which would make the shared warning coloration of monarchs and viceroys a case of Müllerian mimicry.

Sometimes the influence of one population on others is so pervasive that its absence drastically changes an entire community. We call such a population a **keystone species.**[4] On the rocky intertidal shores of western North America, the sea star, *Pisaster ochraceus,* is a keystone species. Sea stars are a major predator of the mussel, *Mytilus californianus.* When sea stars were removed experimentally from a patch of Washington State coastline, mussels expanded in numbers, occupying all space previously used by 25 other invertebrate and algal species (Figure 38.12). Keystone predators act by reducing prey populations below the level at which resources, such as space, are limiting. The original notion that all keystone species were predators has been broadened to include any species whose removal causes extinction of others.

By reducing competition, keystone species may allow more species to coexist on the same resource. Consequently they contribute to maintaining diversity in a community. Keystone

[4]Paine, R. T. 1969. A note on trophic complexity and community stability. American Naturalist **103:** 91–93.

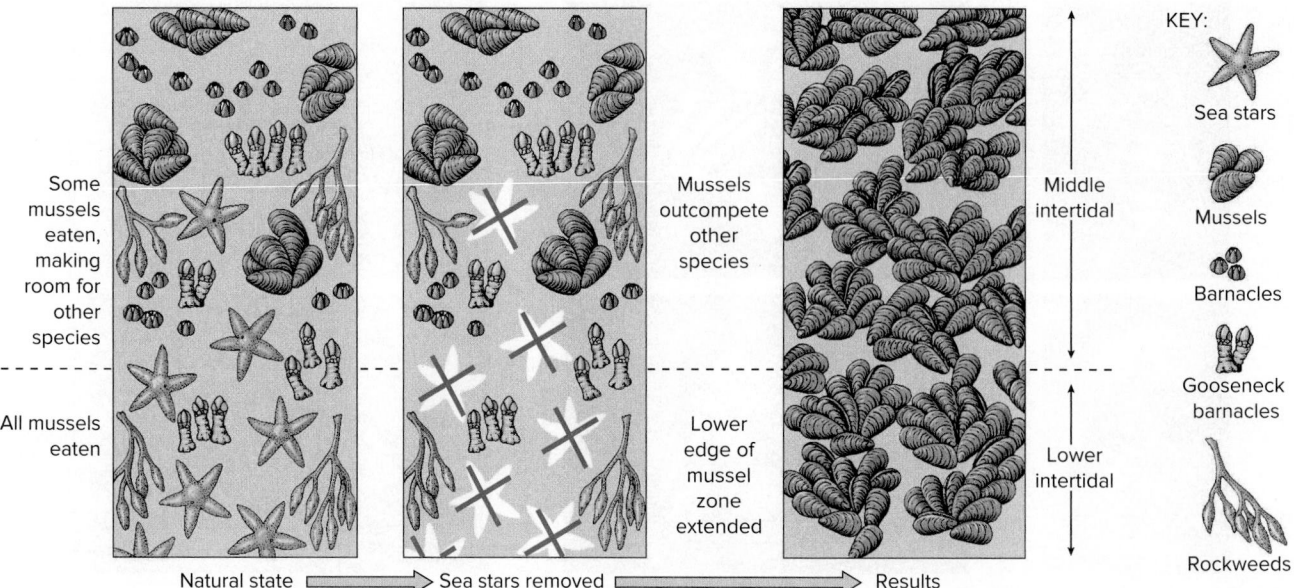

KEY:

Sea stars

Mussels

Barnacles

Gooseneck barnacles

Rockweeds

Some mussels eaten, making room for other species

All mussels eaten

Mussels outcompete other species

Lower edge of mussel zone extended

Middle intertidal

Lower intertidal

Natural state ⟹ Sea stars removed ⟹ Results

Figure 38.12 The experimental removal of a keystone species, the predatory sea star, *Pisaster ochraceus*, from an intertidal community completely changes the structure of the community. With their principal predator missing, mussels form dense beds by outcompeting and replacing other intertidal species.

species illustrate a more general phenomenon, disturbance. Periodic natural disturbances such as fires and hurricanes also can prevent monopolization of resources and competitive exclusion by a few broadly adapted competitors. Disturbances can permit more species to coexist in such highly diverse communities as coral reefs and rainforests.

Parasites are often considered freeloaders because they appear to benefit from their hosts at no expense. **Ectoparasites,** such as ticks and lice, infest many different kinds of animals. The host provides nutrition from its body and aids dispersal of the parasite. However, we must consider that the evolutionary pathway to parasitism from free-living forms often has costs as well as benefits. **Endoparasites,** such as tapeworms (see Chapter 14), have lost their ability to choose habitats. Also, because they must move among hosts to complete their life cycle, the chance that a single individual will live to reproduce is very low. The higher the number of intermediate hosts involved in a parasite's life cycle, the lower the likelihood of success, and the greater reproductive output must be to balance mortality.

Biologists often are puzzled by the complexity of parasite-host relationships. For example, a trematode parasite (p. 299) of the marine gastropod *Ilyanassa obsoleta* actually changes its host's behavior to complete its life cycle. These snails live in sandy intertidal substrates in eastern North America. If the snails are exposed to air when the tide recedes, they normally burrow into sand to avoid desiccation. If, however, a snail is infected with the trematode *Gynaecotyla adunca,* it moves shoreward on high tides preceding low night tides to be left on the beach on the receding tide. Then, as in the legend of the Trojan Horse, the snail sheds the larval parasites, called cercariae, into the sand, where they can infect the next intermediate host, a beach-living crustacean. The definitive host for this trematode, a gull or other shorebird, acquires it by eating an infected crustacean. The life cycle is completed when the bird defecates into water, releasing the trematode's eggs, from which hatch larvae that will infect more snails.

Virulence of a parasite is correlated, at least in part, with availability of new hosts. Coevolution between parasite and host is expected to generate an increasingly benign, less virulent relationship if host organisms are uncommon and/or difficult for a parasite to infect. Selection favors a benign relationship, because a parasite's chances of survival are diminished if its host dies. A highly virulent parasite should evolve only if the parasite can easily move to a new host after killing a previous one.

Ecosystems

Transfer of energy and materials among organisms within ecosystems is the ultimate level of organization in nature. Energy and materials are required to construct and to maintain life, and their incorporation into biological systems is called **productivity.** Ecologists subdivide productivity into component **trophic levels** based on how organisms obtain energy and materials. Trophic levels are linked together into **food chains,** which denote movement of energy from plant compounds to organisms that eat plants, then to other organisms that eat the plant feeders, and possibly further through a linear series of organisms that feed and are then eaten by others. A **food web** shows the branching pathways for transfer of energy and materials among species in an ecosystem. Figure 38.13 summarizes a food web for a salt marsh, with each group of species ordered vertically according to its level in the corresponding food chain, shown at the left of the figure.

Primary producers are organisms that begin productivity by fixing carbon and nitrogen, and storing energy acquired usually from sunlight as chemical energy in covalent carbon-carbon bonds. "Fixing" consists in acquiring carbon dioxide and nitrogen gasses from air and using them to assemble organic molecules, such as

Figure 38.13 Midwinter food web in *Salicornia* salt marsh of San Francisco Bay area. The connected trophic levels shown at the left constitute a food chain for this community. The connections between the species that occupy these trophic levels, summarized in the rest of the figure, constitute a food web for this community.

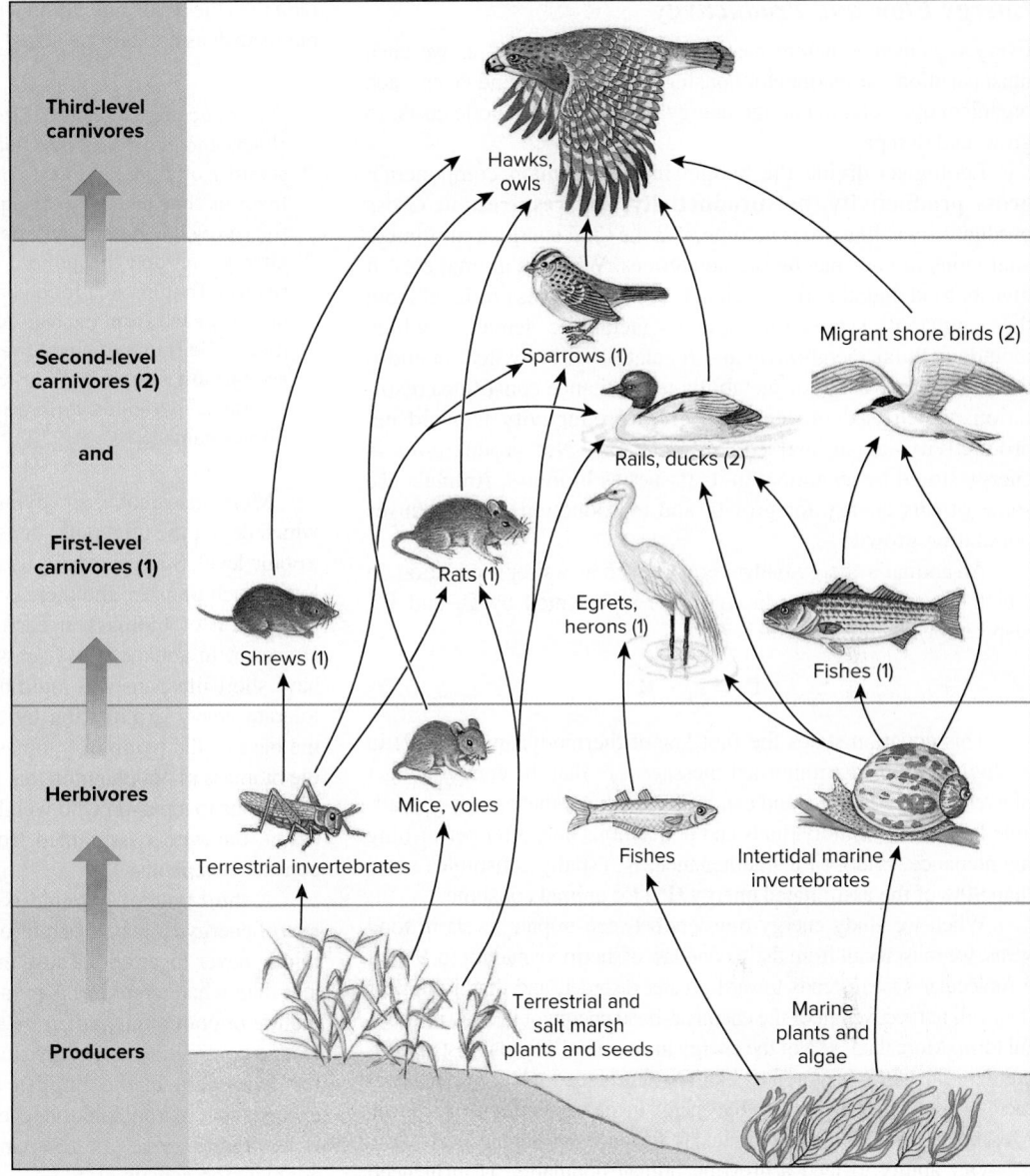

sugars (p. 22) and amino acids (p. 24). Primary producers usually are green plants that capture solar energy through **photosynthesis** (but see an exception in the box titled "Life Without the Sun," p. 830). Powered by solar energy, plants assimilate and organize minerals, water, and carbon dioxide into living tissue. All other organisms survive by consuming this tissue, or by consuming organisms that consumed this tissue. **Consumers** include **herbivores,** which eat plants directly, and **carnivores,** which eat other animals. The most important consumers are **decomposers,** mainly bacteria and fungi, that break dead organic matter into its mineral components, returning it to a soluble form again available to plants at the base of the nutrient cycle (p. 828).

Earth's ecosystems have a finite supply of matter, which must be recycled for life to continue. Important chemicals such as carbon and nitrogen are endlessly reused as organisms die, and their decomposition returns these chemicals to the environment, from which plants construct new living matter. Unlike matter, energy ultimately dissipates from the ecosystem as heat and is not recycled. Thus, no ecosystems are truly closed; all require new energy input from the sun or hydrothermal vents (p. 812).

Energy Flow and Productivity

Every organism in nature has an **energy budget.** Just as we each must partition our income for housing, food, utilities, and taxes, each organism must obtain enough energy to meet its metabolic costs, to grow, and to reproduce.

Ecologists divide the budget into three main components: **gross productivity, net productivity,** and **respiration.** Gross productivity is like gross income; it is the total energy assimilated, analogous to your pay before deductions. When an animal eats, it digests food and absorbs nutrients. Most energy assimilated from these nutrients serves the animal's metabolic demands, which include cellular metabolism and regulation of body heat in endotherms. Energy used for metabolic maintenance constitutes respiration, which is deducted from gross productivity to yield net productivity, an animal's take-home pay. Net productivity is energy stored by an animal in its tissues as **biomass.** Animals use some of this energy for growth and reproduction, which causes population growth.

An animal's energy budget is expressed by a simple equation, in which gross and net productivity are represented by P_g and P_n, respectively, and respiration is **R:**

$$P_n = P_g - R$$

This equation states the first law of thermodynamics (p. 9) in ecological terms. Its important messages are that the energy budget of every animal is finite and can be limiting, and that energy is available for growth of individuals and populations only after performing maintenance. Energy for maintenance, **R,** usually constitutes more than 90% of the assimilated energy (P_g) for animal consumers.

When we study energy transfers between trophic levels in food webs, we must recall from the second law of thermodynamics (p. 9) that a molecular system tends toward greater disorder, and that a predator thus will not recover all of the chemical-bond energy of its prey in a useful form. More than 90% of the energy in an animal's food is lost as heat, and less than 10% is stored as biomass. Each succeeding trophic level therefore contains only 10% of the energy in the next lower trophic level. Most ecosystems are thereby limited to five or fewer trophic levels.

Our ability to feed a growing human population is influenced profoundly by the second law of thermodynamics (p. 9). People, who are at the end of the food chain, may eat the grain, fruits, and vegetables of plants that fix the sun's energy in chemical bonds; this very short chain represents an efficient use of potential energy. People also may eat beef from animals that eat grass that fixes the sun's energy; the addition of a trophic level decreases available energy by a factor of 10. Nonetheless, cattle and other agricultural animals can convert to meat and/or milk or eggs vegetable matter, such as grass, that is unsuited for direct human consumption. Because grass is abundant and otherwise inaccessible for human nutrition, farm animals that convert it to meat or other food give us access to an important source of energy despite adding a trophic level to the food chain.

When we examine the food chain in terms of biomass at each level, we can construct **ecological pyramids** of either numbers, biomass, or energy. A pyramid of numbers (Figure 38.14A), also called an **Eltonian pyramid,** depicts numbers of organisms transferred between each trophic level. This pyramid provides a vivid impression of the great difference in numbers of organisms involved in each step of the chain, and it supports the observation that large predatory animals are rarer than the small animals on which they feed. However, a pyramid of numbers does not indicate actual mass of organisms at each level.

Oxford ecologist Charles Elton invented the concepts of food chains and ecological pyramids in 1923. Working for a summer on a treeless arctic island, Elton watched arctic foxes as they roamed, noting what they ate and, in turn, what their prey had eaten, until he traced the complex cycling of nitrogen in food throughout the animal community. Elton realized that life in a food chain comes in discrete sizes, because each form evolved to be much bigger than the thing it eats. He thus explained the common observation that large animals are rare while their smaller prey are common; the ecological pyramids illustrating this phenomenon now bear Elton's name.

More instructive are pyramids of biomass (Figure 38.14B), which depict the total bulk, or "standing crop," of organisms at each trophic level. Such pyramids usually slope upward because energy is lost at each transfer, and there is thus less energy available to sustain production of biomass at each successively higher trophic level. However, in aquatic ecosystems whose producers are algae, which have short lifespans and rapid turnover, the pyramid inverts. Algae tolerate heavy exploitation by zooplankton consumers. Therefore, the base of the pyramid (biomass of phytoplankton) is smaller than the biomass of zooplankton that it supports. This inverted pyramid is analogous to a person who weighs far more than the food in a refrigerator, but who is sustained from the refrigerator because food is constantly replenished.

A third type of pyramid is a pyramid of energy, which shows rate of energy flow between levels (Figure 38.14C). An energy pyramid is never inverted because energy transferred from each level is less than what entered it. A pyramid of energy gives the best overall picture of community structure because it depicts the loss of energy from a community as energy flows from primary producers through the food web. In the English Channel, energy of phytoplankton exceeds that of zooplankton, even though biomass of phytoplankton is less than biomass of zooplankton (because of heavy grazing by zooplankton consumers).

Nutrient Cycles

All elements essential for life come from environmental air, soil, rocks, and water. When plants and animals die and their bodies decay, or when organic substances are burned or oxidized, elements and inorganic compounds essential for life (nutrients) return to the environment. Decomposers fulfill an essential role in this process by feeding on remains of plants and animals and on fecal material. The result is that nutrients flow in a perpetual cycle between biotic and abiotic components of an ecosystem. Nutrient cycles are often called **biogeochemical cycles** because they involve exchanges between living organisms (bio-) and rocks, air, and water of the earth's crust (geo-). Continuous input of energy from the sun keeps nutrients flowing and the ecosystem functioning (Figure 38.15).

Many industrially produced organic chemicals that enter the ecosystem as waste challenge nature's nutrient cycling because decomposers have not evolved ways to degrade these compounds. Probably most harmful to ecosystemic processes are chemical

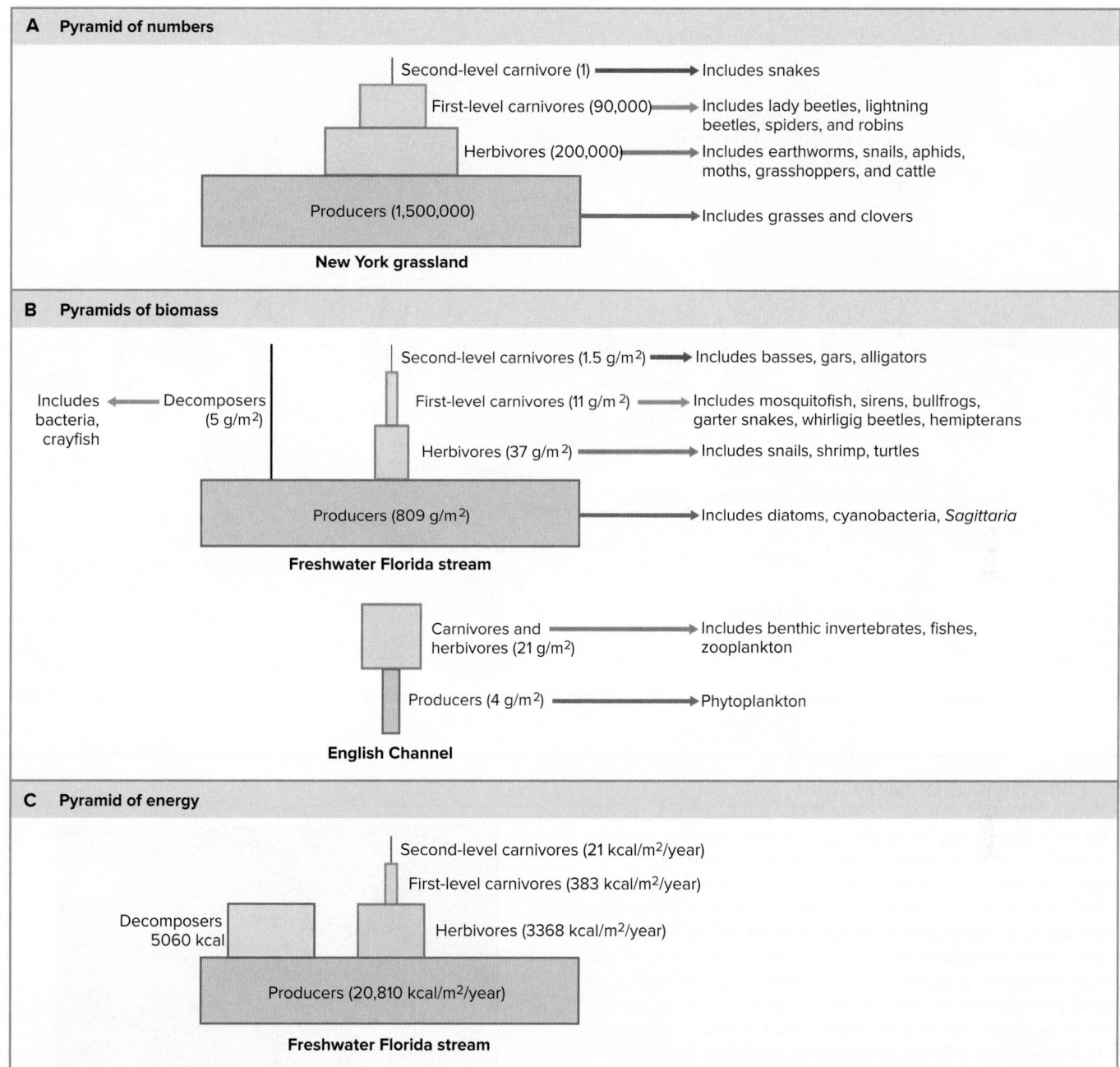

Figure 38.14 Ecological pyramids with representative organisms from each trophic level. **A,** Pyramid of numbers of organisms counted for 100 square feet of grassland in New York. **B,** Pyramids of biomass for a freshwater Florida stream (top) and plankton of the English Channel (inverted pyramid at bottom). **C,** Pyramid of energy for the same Florida stream shown in **B.** (*From the work of E. P. Odum and H. T. Odum.*)

pesticides. Pesticides in natural food webs can be insidious for three reasons. First, many pesticides become concentrated as they travel up succeeding trophic levels. The highest concentrations occur in the biomass of top carnivores such as hawks and owls, diminishing their ability to reproduce. Second, many species killed by pesticides are not pests but merely innocent bystanders, called nontarget species. Nontarget effects happen when pesticides move outside the agricultural field to which they were applied, through rainwater runoff, leaching through soil, or being dispersed by wind. The third problem is persistence; some chemical pesticides have great longevity in the environment, so that nontarget effects persist long after pesticide application. Genetic engineering of crop plants aims to improve their resistance toward pests to lessen the need for chemical pesticides.

38.2 EXTINCTION AND BIODIVERSITY

Biodiversity exists because rates of speciation on average slightly exceed rates of extinction in earth's evolutionary history. Approximately 99% of all species that have lived are extinct. Speciation rates

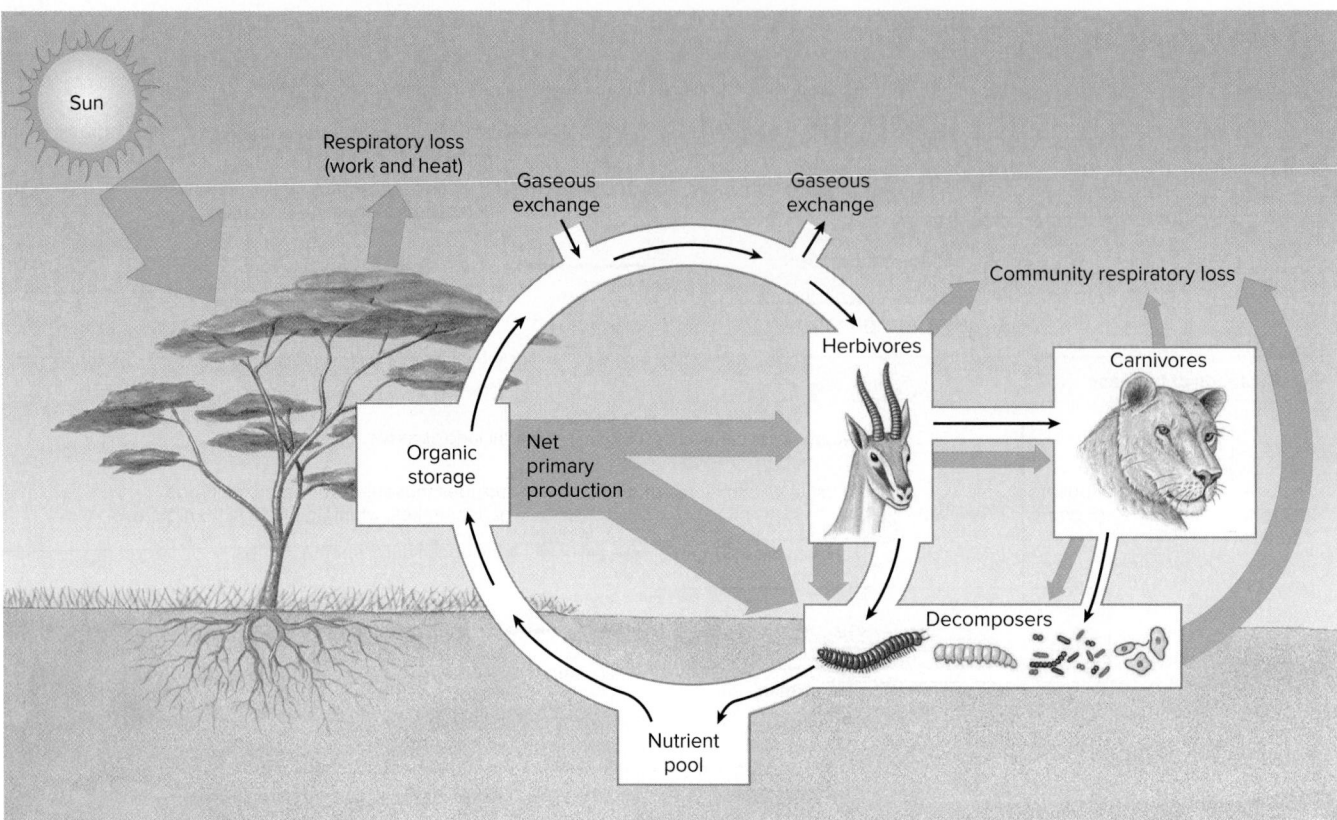

Figure 38.15 Nutrient cycles and energy flow in a terrestrial ecosystem. Note that nutrients are recycled, whereas energy flow *(orange)* is one way.

Life Without the Sun

For many years, ecologists thought that all animals depended directly or indirectly on primary production from solar energy. However, in 1977 and 1979, dense communities of animals were discovered living on the seafloor adjacent to vents of hot water issuing from rifts (Galápagos Rift and East Pacific Rise) where tectonic plates on the seafloor slowly spread apart. These communities (see photo) include several species of molluscs, some crabs, polychaete worms, enteropneusts (acorn worms), and giant pogonophoran or siboglinid worms. The temperature of seawater above and immediately around vents is 7° to 23° C where it is heated by basaltic intrusions, whereas the surrounding normal seawater is 2° C.

The producers in these vent communities are chemoautotrophic bacteria that derive energy from oxidation of hydrogen sulfide abundant in the vent water and fix carbon dioxide into organic carbon. Some animals in the vent communities—for example, bivalve molluscs—are filter feeders that ingest bacteria. Others, such as giant pogonophoran tubeworms (see p. 371), which lack mouths and digestive tracts, harbor colonies of symbiotic bacteria in their tissues and use the organic carbon that these bacteria synthesize.

NOAA PMEL Vents Program

A population of giant pogonophoran or siboglinid tubeworms grows in dense profusion near Sully Vent, a hydrothermal vent in the Main Endeavour Vent Field of the Juan de Fuca Ridge, NE Pacific Ocean, photographed at 2250 m (about 7400 feet) below sea level.

represent an ongoing process of geographic expansion of populations by dispersal followed by geographic fragmentation (p. 799), producing a multiplication of species. Speciation rates vary enormously among animal taxa and geographic areas; typical rates include 0.2 to 0.4 speciation events per species per million years, as measured for Cretaceous marine gastropods of the Atlantic coast; average duration of these gastropod species was 2 to 6 million years.

Extinction rates show episodic peaks and valleys throughout earth's evolutionary history. Paleontologist David Raup analyzed episodicity of extinction peaks by dividing the 600-million-year

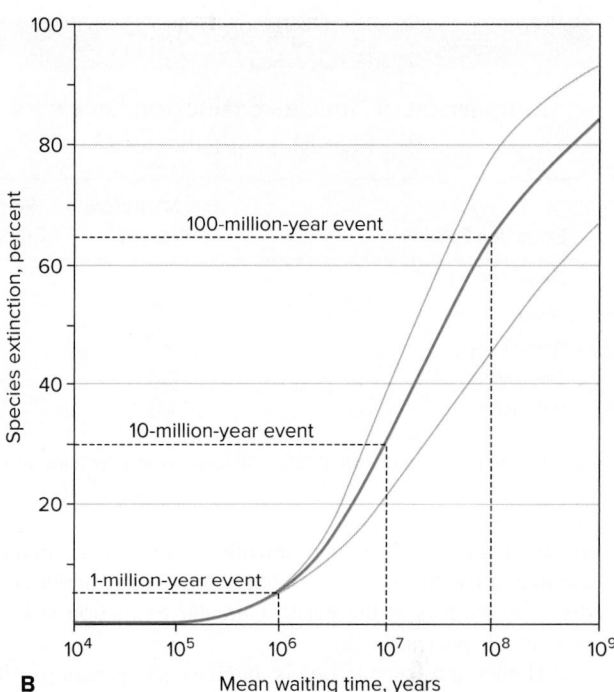

A Percent species extinction per 1-million-year intervals

B Mean waiting time, years

Figure 38.16 **A,** Variation in extinction rates for species in the marine fossil record. In this analysis, David Raup divided the fossil record into 600 consecutive intervals, each one having a duration of 1 million years (Myr), beginning 600 Myr ago. Percent species extinction was measured for each interval. Almost 20% of the intervals had extinction rates of 10 to 15%, the highest bar on this graph. The mean species extinction rate is 25% per million years, and the mean species duration is 4 Myr. **B,** Species "kill curve" for the data summarized in **A.** Waiting time is the average interval between events equal to or greater than a given intensity of species extinction. The kill curve demonstrates an episodic distribution of extinction peaks over the past 600 Myr. If extinction rates were distributed randomly through time, the kill curve would be indistinguishable from the x axis. The lighter curves on either side of the dark one show the statistical measurement error; the actual curve could lie anywhere between these lighter curves, but the dark line provides the best estimate. Source: Data from *David Raup (1995)*.

history[5] of the marine fossil record into successive intervals of 1 million years' duration; he then measured the percentage of existing species that suffered extinction during each interval. Rates of species extinction in the 600 intervals range from near zero to 96%, averaging about 25% (Figure 38.16A). We measure episodicity of extinction by asking, How long must we wait on average for an extinction peak killing at least 30% of existing species, or perhaps 65%? Answers to these questions are shown by Raup's "kill curve" (Figure 38.16B). At least 5% of the existing species go extinct during any million-year interval of geological time. Extinction events killing at least 30% of existing species occur every 10 million years on average, whereas extinction events killing at least 65% of existing species occur on average every 100 million years; the latter episodes clearly qualify as mass extinctions (p. 129).

Figure 38.16A reveals a continuous distribution of extinction rates from very high to very low. Extinction rates intermediate between mass extinction and background extinction occur primarily in the Paleozoic era, with the contrast between extinction peaks and valleys being more pronounced in the post-Paleozoic fossil record.

The contrast between "mass extinction" and "background extinction" thus remains useful despite the apparent continuity of high and low extinction rates in Raup's analysis.

Fossil studies show that species whose geographic ranges are large have lower average rates of extinction than do those with small geographic ranges, although mass extinctions can erase this contrast. Species of Atlantic Cretaceous gastropods differed greatly in geographic range depending on modes of larval feeding. Some species had pelagic, plankton-feeding ("planktotrophic") larvae, which were carried long distances by oceanic currents; these species maintained large geographic ranges averaging 2000 km along the Atlantic Coast. Other species had heavy larvae that settled to the ocean floor as benthic feeders immediately upon hatching; these nonplanktotrophic species average less than one-fourth the geographic range of their plankton-feeding counterparts. A nonplanktotrophic species is about three times more likely to suffer extinction than a plankton-feeding one, but their geographic fragmentation also makes nonplanktotrophic species twice as likely to undergo speciation.

A paradox of biodiversity is that geographic **habitat fragmentation** of a species simultaneously increases rates of both local extinction and speciation. The African antelopes shown in Figure 6.11 (p. 106) illustrate a similar contrast; in the past 6 million years, one group (hartebeests, blesboks, and wildebeests) experienced multiple events of speciation and extinction with seven species alive today, whereas the other lineage (impalas) persisted as a single species through the same time period. Impalas

---

[5]Following Raup's analysis, the estimated age of the Phanerozoic fossil record (see Timeline of Major Biological Events, on inside back cover) was decreased from 600 million years to 542 million years. This revision should not alter the major conclusions of Raup's analysis, but the numbers are no longer consistent with geological dating. When describing Raup's results, we continue to refer to the 600-million-year figure used in his analyses while using the newer estimates elsewhere in the book.

TABLE 38.1

Comparison of Species Extinction Levels for the Big Five Mass Extinctions*

Extinction Episode	Age, MY Before Present	Percent Extinction
Cretaceous	66	76
Triassic	200	76
Permian	251	96
Devonian	359	82
Ordovician	443	85

*Source: Data from David Raup (1995), with revised dates for these extinctions.

are about equal to the sum of the other seven species in number of animals alive today. This contrast shows that increased species diversity evolves at the peril of greater extinction risk for each individual species.

Higher taxa, such as orders, families, and genera (p. 198), also gain some protection from extinction by having large geographic ranges. Raup notes that higher taxa containing many species collectively distributed across a large geographic area are unlikely to go extinct. When such extinctions do happen, as seen for many dinosaur and marine ammonoid taxa at the end of the Cretaceous era (Table 38.1), unusually catastrophic conditions appear responsible. An asteroid bombardment at the end-Cretaceous extinction (p. 129) appears to have caused within a short time widespread fires and extreme darkness and cold followed by extreme heat, all conditions well outside the evolutionary tolerances of many animal taxa whose members were formerly abundant. Only by chance would a taxon likely contain species able to withstand a challenge unprecedented in the group's evolutionary history.

Darwin explained extinctions of higher taxa by interspecific competition, but paleontological studies now refute this claim. Paleontologist Michael Benton estimates that less than 15% of the extinct tetrapod families could have been killed by competition with other families. Ecological and fossil studies of ectoprocts demonstrate that species of one order (Cheilostomatida) have outcompeted those of another order (Tubuliporata) ecologically by overgrowing their colonies for many millions of years without driving Tubuliporata to extinction.

Decline or extinction of one taxon often frees resources to which another taxon becomes adapted at a much later time, leading to evolutionary proliferation of ecologically diverse species in the latter. Resources freed by extinction of dinosaur taxa at the end of the Cretaceous period are considered important for the later proliferation of mammalian adaptive and species diversity in the Cenozoic era (p. 129).

Fossil studies of extinction help us to put into evolutionary perspective the consequences of human-mediated ecological changes for biodiversity. Fragmentation of populations as seen especially on islands (p. 117) produces locally high rates of species formation and endemism (p. 200), but these young species are unusually prone to extinction because their geographic ranges are small. About half of all areas in the world that contain at least two endemic bird species are islands, even though islands form less than

10% of the earth's terrestrial habitats. Island species are often particularly prone to destruction by introduction of invasive exotic species. For example, land snails of the genus *Partula* on the Tahitian island of Moorea were a major study system for island speciation until introduction of exotic snails displaced these native species. Mainland habitats, such as forests, are fragmented into virtual islands when development clears vast areas of habitat and when introduced species invade these areas. Because tropical regions have high species endemism, human fragmentation of these environments is particularly likely to produce species extinction.

A major challenge for animal species conservation is to obtain an inventory of earth's species diversity. Estimates of earth's total number of species are typically as large as 10 million, and this number may be too low by an order of magnitude. Taxonomists make these estimates by calculating ratios of described versus undescribed species in a sample from a geographic region, from rates of description of new species through time, and by judging how thoroughly species diversity has been measured in various taxa. Current estimates of species diversity are least satisfactory for microorganisms. Vagueness of these estimates reflects both practical and conceptual problems. A thorough survey of geographic genetic variation among natural populations requires expensive molecular-genetic analyses (see Chapter 6) and is critical for applying any of the current species concepts (see Chapter 10). This analysis is feasible only for taxa with relatively small numbers of large individuals, such as those at the top of a food chain or ecological pyramid. Beetles and nematodes constitute two taxa whose large numbers of small organisms challenge hopes for a comprehensive taxonomic survey. Even with appropriate data in hand, conflicting perspectives on what should constitute a species (discussed in Chapter 10) preclude an unambiguous tally of species numbers. Such conflicts would be particularly intense in groups of animals that do not show primarily a simple bisexual mode of reproduction. Clearly, conservation efforts cannot await a thorough taxonomic inventory of all animal populations. Maintenance of diverse ecosystems as identified in Chapter 37 is an initial priority for preventing widespread species extinction.

One hopes that human disturbance of natural environments will not make the present time a competitor with the big five mass extinctions of Table 38.1. Human activity clearly has induced numerous species extinctions, and we must avoid making the present time one that climbs too high on Raup's kill curve (Figure 38.16B). Evolutionary studies suggest, however, that widely distributed higher taxa are unlikely to go extinct even during episodes of high species extinction. Lacking an inventory of species diversity in animals, we must avoid creating conditions that would selectively destroy any particular higher taxon.

 S U M M A R Y

Ecology is the study of relationships between organisms and their environments to explain the distribution and abundance of species on earth. The physical space that contains an animal's environment is its habitat. Within the habitat are physical and biological conditions suitable to survival and reproduction, which constitute the niche of an animal, population, or species.

Animal populations consist of demes of interbreeding members sharing a common gene pool. Cohorts of animals have characteristic patterns of survivorship that represent adaptive trade-offs between parental care and numbers of offspring. Animal populations consisting of overlapping cohorts have age structure that indicates whether they are growing, declining, or at equilibrium.

Every species in nature has an intrinsic rate of increase that provides a potential for exponential growth. Population growth may be regulated intrinsically, by the carrying capacity of the environment; or extrinsically, by competition between species for a limiting resource, or by predators or parasites. Density-independent abiotic factors can limit, but not truly regulate, population growth.

Communities consist of populations that interact with one another through competition, predation, parasitism, commensalism, and mutualism. These relationships produce coevolution among populations within communities. Guilds of species avoid competitive exclusion by partitioning limited resources, often featuring morphological specialization. Keystone predators are those that control community structure and reduce competition among prey, which increases species diversity. Parasites and their hosts evolve a benign relationship that ensures their coexistence.

Ecosystems consist of communities and their abiotic environments. Animals occupy the trophic levels of herbivorous and carnivorous consumers within ecosystems. All organisms have an energy budget consisting of gross and net productivity, and respiration. For animals, respiration usually is at least 90% of this budget. Thus, transfer of energy from one trophic level to another is limited to about 10%, which in turn limits the number of trophic levels in an ecosystem. Ecological pyramids of energy depict how productivity decreases in successively higher trophic levels of food webs.

Ecosystem productivity is described by measuring energy flow and material cycles within ecosystems. All energy is lost as heat, but nutrients and other materials are recycled. No ecosystem is closed because all depend upon acquiring energy from outside sources.

Biodiversity exists because rates of speciation on average slightly exceed rates of extinction in earth's evolutionary history. Approximately 99% of all species that have lived are now extinct. Extinction rates in the geologic past are highly episodic, with species extinction ranging from near zero to 96% in different million-year intervals. Species having large geographic ranges experience lower average rates of extinction than do species with smaller ranges, and the same relationship holds for higher taxa. Paleontological studies of extinction provide an important perspective for evaluating potential evolutionary consequences of human-mediated species extinction.

REVIEW QUESTIONS

1. The term "ecology" is derived from the Greek meaning "house" or "place to live." However, as used by scientists, the term "ecology" is not the same as "environment." How do these terms differ?
2. Distinguish the concepts of ecosystem, community, and population.
3. What is the distinction between habitat and environment?
4. Define the *niche* concept. How does the "realized niche" of a population differ from its "fundamental niche"? How does the concept of niche differ from the concept of guild?
5. Populations of independently living (unitary) animals have a characteristic age structure, sex ratio, and growth rate. However, these properties are difficult to measure for modular animals. Why?
6. Explain which of the three survivorship curves in Figure 38.2 best fits the following: (a) a population in which mortality as a proportion of survivors is constant; (b) a population in which there is little early death and most individuals live to old age; (c) a population

that experiences heavy mortality of the very young but with the survivors living to old age. Offer an example from the real world of each survivorship pattern.
7. Contrast exponential and logistic (= sigmoid) growth of a population. Under what conditions might you expect a population to exhibit exponential growth? Why can exponential growth not be perpetuated indefinitely?
8. Growth of a population can be hindered by either density-dependent or density-independent mechanisms. Define and contrast these two mechanisms. Offer examples of how growth of the human population might be curbed by either agent.
9. Herbivory is beneficial for the animal (+) but harmful to the plant it eats (−). What are some + − interactions among animal populations? What is the difference between commensalism and mutualism?
10. Explain how resource partitioning involving ecological character displacement can ease competition between coexisting species.
11. Define *predation*. How does the predator-prey relationship differ from the parasite-host relationship? Why is the evolutionary race between predator and prey one that the predator cannot afford to win?
12. Mimicry of yellowjacket wasps by clearwing moths is an example of a harmless species resembling a dangerous one. What is the advantage to the moth of this form of mimicry? What is the advantage to a dangerous species of mimicking another dangerous species?
13. A keystone species is one whose removal from a community causes extinction of other species. How does this extinction happen?
14. What is a trophic level, and how does it relate to a food web?
15. Define *productivity* as the word is used in ecology. What is a primary producer? What is the distinction between gross productivity, net productivity, and respiration? How is net productivity related to biomass (or standing crop)?
16. What is a food chain? How does a food chain differ from a food web?
17. What conditions produce an inverted pyramid of biomass in which the consumers have a greater biomass than the producers? Can you think of an example of an inverted pyramid of *numbers* in which there are, for example, more herbivores than plants on which they feed?
18. The pyramid of energy has been offered as an example of the second law of thermodynamics (p. 9). Why?
19. Animal communities surrounding deep-sea thermal vents apparently exist in total independence of solar energy. How is this existence possible?
20. What do paleontological studies show about the relationship between the geographic range of a species and its probability of undergoing speciation or extinction? How does this relationship present a paradox for biodiversity?

For Further Thought Natural processes of species formation and extinction emerge on a timescale measured in millions of years, whereas human-mediated extinctions are measured on a scale of years to decades. How does this discrepancy complicate our attempts to conserve biodiversity?

SELECTED REFERENCES

Benton, M. J. 1996. On the nonprevalence of competitive replacement in the evolution of tetrapods. In Jablonski, D., H. Erwin, and J. H. Lipps (eds.), Evolutionary paleobiology, pp. 185–210. Chicago, University of Chicago Press. *Shows that ecological competition does not explain extinction of tetrapod taxonomic families.*

Chase, J. M., and M. A. Leibold. 2003. Ecological niches: linking classical and contemporary approaches. Chicago, University of Chicago Press. *An insightful coverage of niche concepts in community ecology.*

Krebs, C. J. 2009. Ecology: the experimental analysis of distribution and abundance, ed. 6. San Francisco, Benjamin-Cummings. *Important treatment of population ecology.*

Levin, S. A. (ed.). 2009. The Princeton guide to ecology. Princeton, New Jersey, Princeton University Press. *A comprehensive review of ecological topics with contributions from many authorities.*

Molles, M. C., Jr. 2012. Ecology: concepts and applications, ed. 6. New York, McGraw-Hill. *A concise and well-illustrated survey of ecology.*

Monbiot, G. 2014. Feral: rewilding the land, the sea, and human life. Chicago, IL, University of Chicago Press. *A journalist and environmentalist explores efforts to restore vanishing wildlife, such as wolves, and their ecological consequences. The associated website includes accompanying videos.*

Pianka, E. R. 2000. Evolutionary ecology, ed. 6. San Francisco, Benjamin-Cummings. *An introduction to ecology written from an evolutionary perspective.*

Raup, D. M. 1995. The role of extinction in evolution. In W. M. Fitch and F. J. Ayala (eds.), Tempo and mode in evolution, pp. 109–124. Washington, D.C., National Academy Press. *A paleontologist's insights on extinction and biodiversity.*

Ricklefs, R. E., and G. L. Miller. 2000. Ecology, ed. 4. New York, W. H. Freeman. *Clearly written, well-illustrated general ecology text.*

Sax, D. F., J. J. Stachowicz, and S. D. Gaines. 2005. Species invasions: insights into ecology, evolution, and biogeography. Sunderland, Massachusetts, Sinauer Associates. *The establishment of alien species in an area can reveal ecosystem properties and the structuring of ecological communities.*

Sibly, R. M., J. Hone, and T. H. Clutton-Brock. 2003. Wildlife population growth rates. Cambridge, U.K., Cambridge University Press. *Contributions by many ecologists illustrate diverse methods for measuring growths of natural populations.*

Sinclair, A. R. R., J. M. Gosline, G. Holdsworth, C. J. Krebs, S. Boutin, J. N. M. Smith, R. Boonstra, and M. Dale. 1993. Can the solar cycle and climate synchronize the snowshoe hare cycle in Canada? Evidence from tree rings and ice cores. American Naturalist **141:** 173–198. *An example of how a global climatic variable can influence the timing of a density-dependent, predator-prey cycle*

Smil, V. 1997. Global population and the nitrogen cycle. Sci. Am. **277:** 78–81 (July). *The sudden growth of the human population worldwide in the 20th century parallels global consumption of synthetically produced nitrogen-rich fertilizers upon which humans now heavily depend for food production, but there are adverse consequences for the environment.*

Smith, T. M., and R. L. Smith. 2015. Elements of ecology, ed. 9. San Francisco, Benjamin-Cummings. *Clearly written, well-illustrated general ecology text.*

This glossary lists definitions, pronunciations, and derivations of the most important recurrent technical terms, units, and names (excluding taxa) used in the text.

A

abdomen (ab′-dō-mən). Portion of a vertebrate body between the thorax (chest) and pelvis; portion of an arthropod body posterior to the thorax or cephalothorax.

abiotic (ā′bī-ät′ik) (Gr. *a*, without, + *biōtos*, life, livable). Characterized by the absence of life.

abomasum (ab′ō-mā′səm) (L. *ab*, from, + *omasum*, paunch). Fourth and last chamber of the stomach of ruminant mammals.

aboral (ab-o′rəl) (L. *ab*, from, + *os*, mouth). A region of an animal opposite the mouth.

aboral sense organ In ctenophores, a balance organ (statocyst) at the opposite side of the body from the mouth.

abscess (ab′ses) (L. *abscessus*, a going away). Dead cells and tissue fluid confined in a localized area, causing swelling.

absorb In a cell or tissue, acquire a nutrient moved from blood through pores.

abyssal plain (ə-bis′-əl plān′). An offshore seafloor with submarine channels and hills averaging 4000 m in depth but descending to 11,000 m below the sea surface.

acanthodians (a′kan-thō′dē-əns) (Gr. *akantha*, prickly, thorny). Group of jawed fishes, characterized by large spines in their fins, from Lower Silurian to Lower Permian.

acanthor (ə-kan′thor) (Gr. *akantha*, spine or thorn, + *or*). First larval form of acanthocephalans in the intermediate host.

accessory (branchial) heart In cephalopod molluscs, an arterial muscle that increases blood pressure into the capillaries of a gill.

accessory (coreceptor) molecule On a T cell, a transmembrane protein that enables a **T-cell receptor** to bind a specific antigen.

accessory organ or accessory sex organ Any structure, other than a gonad (= primary organ), that serves primarily to produce or to deliver a germ cell (e.g., yolk gland, oviduct, vas deferens), or to provide nourishment of the developing offspring by the mother (e.g., placenta, uterus).

acclimatization (ə-klī′mə-tə-za′-shən) (L + Gr. *klima*, climate). Gradual physiological adjustment of an organism in response to relatively long-lasting environmental changes.

acetylcholine (as-e′-til-kō′-lēn). A chemical neurotransmitter in animal nervous systems.

acetylcholinesterase (as-e′-til-kō′-lən-es′tə-rās). Enzyme that degrades the neurotransmitter acetylcholine and thereby terminates signal transmission across neural synapses and neuromuscular junctions.

acetyl coenzyme A (= acetyl-CoA) (as-e′-til kō′-en′- zīm ā). In cellular respiration, a molecule, formed in the mitochondrion, through which breakdown products of glycolysis enter the citric acid cycle for further oxidation; also a chemical precursor to synthesis of the neurotransmitter acetylcholine.

acid A molecule that dissociates in solution to produce a hydrogen ion (H^+).

acinus (as′ə-nəs), pl. **acini** (as′ə-nī) (L. grape). A small lobe of a compound gland or a saclike cavity at the termination of a passage.

acoelomate (ā-sēl′ə-māt′) (Gr. *a*, not, + *koilōma*, cavity). Without a coelom, as in flatworms and proboscis worms.

acontium (ə-kän′tē-əm), pl. **acontia** (Gr. *akontion*, dart). Threadlike structure bearing nematocysts located on mesentery of sea anemone.

acrocentric (ak′rō-sen′trək) (Gr. *akros*, tip, + *kentron*, center). Chromosome with centromere near the end.

acron (a′crän) (Gr. *akron*, mountaintop, fr. *akros*, tip). Preoral region of an insect.

acrosome (ak′-rō-sōm). Organelle at the head of a sperm cell that contains digestive enzymes permitting the sperm to penetrate the cell membrane of an ovum to cause fertilization.

actin (Gr. *aktis*, ray, beam). A protein in all cells that forms part of the cytoskeleton and is involved in the processes of endocytosis and exocytosis. In contractile tissues it forms the thin filaments of striated muscle.

actinotroch (ək-tin′ə-trōk) (Gr. *aktis*, ray, beam, + *trochos*, wheel). Larval form in Phoronida.

action potential A transitory difference in voltage occurring across a cell membrane; in neurons, it triggers opening of a membrane channel during neurotransmission.

activation energy Minimum energy required to start a chemical reaction.

active transport Mediated transport in which a transmembrane protein transports a molecule across a plasma membrane against a concentration gradient; requires expenditure of energy; contrasts with **facilitated diffusion.**

actomyosin system (ak′-tō-mī′ō-sin). A mechanism of animal movement based on alternating contraction and relaxation of structures caused by movements of the proteins actin and myosin.

adaptation (L. *adaptatus*, fitted). An anatomical structure, physiological process, or behavioral trait that evolved by natural selection and improves an organism's ability to survive and to leave descendants; contrasts with **exaptation.**

adaptive radiation Evolutionary diversification that produces numerous ecologically disparate lineages from a single ancestral one, especially when this diversification occurs within a short interval of geological time.

adaptive zone A characteristic reaction and mutual relationship between environment and organism ("way of life") demonstrated by members of a taxonomic group.

adductor (ə-duk′tər) (L. *ad*, to, + *ducere*, to lead). A muscle that draws a part toward a median axis, or a muscle that draws the two valves of a mollusk shell together.

adenine (ad′nēn, ad′ə-nēn) (Gr. *adēn*, gland, + *ine*, suffix). A purine base; component of nucleotides and nucleic acids.

adenosine (ə-den′ə-sēn) **(di-, tri) phosphate** (ADP and ATP). A nucleotide composed of adenine, ribose sugar, and two (ADP) or three (ATP) phosphate units; ATP is an energy-rich compound that, with ADP, serves as a phosphate bond-energy transfer system in cells.

adhesion junction Transmembrane proteins serving as an intercellular, anchoring connection.

adhesive sac Structure in a larval ectoproct whose secretions attach the larva to a

bat / āpe / ärmadillo / herring / fēmale / finch / līce / crocodile / crōw / còin / cōre / duck / düke / ūnicorn / ə indicates unaccented vowel sound "uh" as in mammal, fishes, cardinal, heron, vulture / stress as in bi-ol′o-gy, bi′o-log′i-cal

G-1

substrate for metamorphosis to an adult form.

adipokine (ad′ə-pō′-kīn). A peptide-hormone **cytokine** secreted by white adipose tissue.

adiponectin Hormone secreted by adipose tissue that tends to lower blood glucose levels by increasing the effects of insulin on glucose uptake by the liver and the skeletal muscle.

adipose (ad′ə-pōs) (L. *adeps*, fat). Fatty tissue; fatty.

adrenaline (ə-dren′ə-lən) (L. *ad*, to, + *renalis*, pertaining to kidneys). A hormone produced by the adrenal, or suprarenal, gland; epinephrine.

adsorption (ad-sorp′shən) (L. *ad*, to, + *sorbeo*, to absorb). The adhesion of molecules to solid bodies.

adult Stage in an animal's life cycle from initial attainment of reproductive maturity to death.

aerobe (a′-rōb). An organism that uses oxygen for respiration.

aerobic (a-rō′bik) (Gr. *aēr*, air, + *bios*, life). Oxygen-dependent form of respiration.

afferent (af′ə-rənt) (L. *ad*, to, + *ferre*, to bear). Adjective meaning leading or bearing toward some organ, for example, nerves conducting impulses toward the brain or blood vessels carrying blood toward an organ; contrasts with **efferent.**

afterbirth Placenta and fetal membranes discarded from the uterus following mammalian birth.

age structure An accounting of the ages of individuals in a population at a particular time and place.

agglutination (ag-lu′-ti-nā′shən). Clumping of erythrocytes, as occurs when incompatible blood types mix.

aggression (ə-gre′-shən) (L. *aggressus*, attack). An offensive action or procedure.

agnathan (ag-nā′-thən) (Gr. *a*, without, + *gnathos*, jaw). A jawless fish of the paraphyletic superclass Agnatha of the phylum Chordata.

agonistic behavior (Gr. *agōnistēs*, combatant). An offensive action or threat directed toward another organism.

agranulocyte (a-gran′ū-lō-cīt). White blood cell that lacks granules in its cytoplasm; includes lymphocytes and monocytes.

air sac Air-filled space that connects a bird's lungs and bone cavities and facilitates breathing and thermoregulation.

alate (ā′lāt) (L. *alatus*, wing). Winged.

albumin (al-byū′mən) (L. *albumen*, white of egg). Any of a large class of simple proteins that are important constituents of vertebrate blood plasma and tissue fluids and also present in milk, whites of eggs, and other animal substances.

alcoholic fermentation Respiration without oxygen in which sugars are degraded to carbon dioxide and ethanol; occurs in yeasts.

aldosterone (al-dos′-te-rōn). Hormone that induces kidneys to reabsorb sodium and water and to release potassium, thus increasing blood volume and pressure.

alimentary (al′ə-men′tə-rē) (L. *alimentum*, food, nourishment). Pertaining to nutrition or nourishment.

alimentary canal Digestive tract.

alkaline phosphatase An enzyme in the plasma membrane of cells lining the intestine that attacks a variety of phosphate compounds (for example, proteins and nucleotides), removing phosphate groups from them during the digestive process.

allantois (a-lan′tō-is) (Gr. *allas*, sausage, + *eidos*, form). One of the extraembryonic membranes of the amniotes that functions in respiration and excretion in birds and nonavian reptiles and plays an important role in the development of the placenta in most mammals.

allele (ə-lēl′) (Gr. *allēlōn*, of one another). Alternative forms of genes coding for the same trait and situated at the same locus in homologous chromosomes.

allelic frequency An estimation of the proportion of gametes produced in a population (gene pool) that contains a particular allelic form of a particular gene.

allergy Acquired immune-system disorder in which a normally harmless environmental substance (allergen) causes an extreme inflammatory response.

allograft (a′lō-graft) (Gr. *allos*, other, + graft). A piece of tissue or an organ transferred from one individual to another individual of the same species, not identical twins; also called homograft.

allometry (ə-lom′ə-trē) (Gr. *allos*, other, + *metry,* measure). Relative growth of a part in relation to the whole organism.

allopatric (a′lō-pat′-rik) (Gr. *allos*, other, + *patra*, native land). In separate and mutually exclusive geographical regions.

allopatric speciation The hypothesis that new species are formed by dividing an ancestral species into geographically isolated subpopulations that evolve **reproductive barriers** between them through independent evolutionary divergence from their common ancestor.

alpha-helix (Gr. *alpha*, first, + L. *helix*, spiral). Spiral arrangement of the genetic DNA molecule; regular coiled arrangement of polypeptide chain in proteins; a secondary structure of proteins.

alternative pathway Innate immune response activated by interaction of complement proteins with polysaccharides on the outer coating of an invading microorganism; contrasts with classical pathway, which is activated by the binding of antibodies.

altricial (al-tri′shəl) (L. *altrices*, nourishers). Referring to young animals (especially birds) having the young hatched or born in an immature, dependent condition.

altruistic behavior A term used initially by Darwin to denote a behavior performed by an individual to aid others while apparently increasing its own risk. Unless such behaviors have at least indirect benefits favoring the individual performing them, evolution of such behaviors could not be explained by natural selection. Neo–Darwinian explanations of such behaviors include kin selection and reciprocal altruism.

alula (al′yə-lə) (L. dim. of *ala*, wing). The first digit or thumb of a bird's wing, much reduced in size.

alveolus (al-vē′ə-ləs) (L. dim. of *alveus*, cavity, hollow). A small cavity or pit, such as a microscopic air sac of the lungs, terminal part of an alveolar gland, or bony socket of a tooth. Membranous sacs under the plasma membrane of certain protozoa.

ambulacra (am′byə-lak′rə) (L. *ambulare*, to walk). In echinoderms, radiating grooves where podia of the water-vascular system characteristically project outside the organism.

ameboid movement (ə-mē′bóid) (Gr. *amoibē,* change, + *oid*, like). Cellular locomotion by protrusion of cytoplasm to form pseudopodia.

ameiotic parthenogenesis (ā′mī-at′ik pär′thə-nōgen′ə-sis). Unisexual reproduction by females in which the egg is produced by mitosis from diploid precursors with no haploid stage.

amensalism (ā-men′səl-iz′əm). An asymmetric competitive interaction between two species in an ecological community in which only one of the species is affected.

amictic (ā-mik′tic) (Gr. *a*, without, + *miktos*, mixed or blended). Pertaining to female rotifers, which produce only diploid eggs that cannot be fertilized, or to the eggs produced by such females; contrasts with **mictic.**

amino acid (ə-mē′nō) (amine, an organic compound). An organic acid with an amino group (–NH₂). A fundamental subunit of proteins and peptides.

aminopeptidase An enzyme found in the plasma membrane of cells lining the intestine that splits terminal amino acids from the amino end of short peptides.

amitosis (ā′mī-tō′səs) (Gr. *a,* not, + *mitos,* thread). A form of cell division in which the nucleus and cytoplasm divide without formation of a spindle or condensation of chromosomes.

ammocoetes (am-mō-cēt′ēz) (Gr. *ammo,* sand, + *coet,* bed). The filter-feeding, larval stage of lampreys.

amniocentesis (am′nē-ō-sin-tē′səs) (Gr. *amnion,* membrane around the fetus, + *centes,* puncture). Procedure for withdrawing a sample of fluid around the developing embryo for examination of chromosomes in the embryonic cells and other tests.

amnion (am′nē-än) (Gr. *amnion,* membrane around the fetus). The innermost of the extraembryonic membranes forming a fluid-filled sac around the embryo in amniotes.

amniote (am′nē-ōt). Having an amnion; as a noun, an animal that develops an amnion in embryonic life; refers collectively to nonavian reptiles, birds, and mammals.

amniotic egg (am′nē-ot′ək). A vertebrate egg containing three membranes that surround the embryo (amnion, allantois, chorion).

amphiblastula (am′fə-blas′chə-lə) (Gr. *amphi,* on both sides, + *blastos,* germ, + L. *ula,* small). Free-swimming larval stage of certain marine sponges; blastula-like but with cells of the animal pole flagellated and those of the vegetal pole unflagellated.

amphid (am′fəd) (Gr. *amphidea,* anything that is bound around). One of a pair of anterior sense organs in certain nematodes.

amphipathic (am-fi-pa′thək) (Gr. *amphi,* on both sides, + *pathos,* suffering, passion). Adjective to describe a molecule with one part soluble in water (polar) and another part insoluble in water (nonpolar). Such compounds have a natural tendency to assemble themselves into semipermeable membranes.

amphiphilic (am-fi-fi′-lik) (Gr. *amphi,* on both sides, + *philia,* love, friendship). Synonymous with **amphipathic.**

amplexus (am-plek′səs) (L. embrace). The copulatory embrace of frogs or toads in which the male fertilizes eggs as they leave the female's body.

ampulla (am-pūl′ə) (L. flask). Membranous vesicle; dilation at one end of each semicircular canal containing sensory epithelium; muscular vesicle above tube foot in water-vascular system of echinoderms.

ampullae of Lorenzini (*Lorenzini,* 17th century physician and ichthyologist) Tiny, gel-filled canals that function as electroreceptors on the skin surface of members of Chondrichthyes.

amylase (am′ə-lās′) (L. *amylum,* starch, + *ase,* suffix meaning enzyme). An enzyme that breaks starch into smaller units.

anabolism (ə-na′bə-li′zəm) (Gr. *ana,* up, + *bol,* to throw, + *ism,* suffix meaning state of condition). Constructive metabolism.

anadromous (an-ad′rə-məs) (Gr. *anadromos,* running upward). Refers to fishes that migrate up streams from the sea to spawn.

anaerobic (an′ə-rō′bik) (Gr. *an,* not, + *aēr,* air, + *bios,* life). Not dependent on oxygen for respiration.

analogy (L. *analogus,* ratio). Similarity of function but not of origin.

anamniote (an′am′-nē-ōt). A vertebrate that lacks an amniotic membrane around the embryo. Includes fishes and amphibians.

anaphylaxis (an′ə-fə-lax′əs) (Gr. *ana-,* up, + *phylax,* guard). A systemic (whole body) immediate hypersensitivity reaction.

anapsid (ə-nap′səd) (Gr. *an,* without, + *apsis,* arch). Extinct amniotes in which the skull lacks temporal openings.

anastomosis (ə-nas′tə-mō′səs) (Gr. *ana,* again, + *stoma,* mouth). A union of two or more blood vessels, fibers, or other structures to form a branching network.

ancestral Denotes a character state inferred to have been present in the most recent common ancestral population of a group of organisms.

ancestrula (an-ses′trə-lə) (NL, *ancestor*). First individual (zooid) of a bryozoan colony that arises from metamorphosis of a free-swimming larva (in marine forms) or a statoblast (in freshwater forms).

androgen (an′drə-jən) (Gr. *anēr, andros,* man, + *genēs,* born). Any of a group of steroid male sex hormones.

androgenic gland (an′drō-jen′ək) (Gr. *anēr,* male, + *gennaein,* to produce). Gland in Crustacea that causes development of male characteristics.

aneuploidy (an′ū-ploid′ē) (Gr. *an,* without, not, + *eu,* good, well, + *ploid,* multiple of). Loss or gain of a chromosome; cells of the organism have one fewer than normal chromosome number, or one extra chromosome; for example, trisomy 21 (Down syndrome).

angiotensin (an′jē-o-ten′sən) (Gr. *angeion,* vessel, + L. *tensio,* to stretch). Blood protein formed from the interaction of renin and a circulating liver protein, angiotensinogen, causing increased blood pressure and stimulating release of aldosterone and ADH.

Angstrom (after Ångström, Swedish physicist). A unit of one ten-millionth of a millimeter (one ten-thousandth of a micrometer); it is represented by the symbol Å.

anhydrase (an-hī′drās) (Gr. *an,* not, + *hydōr,* water, + *ase,* enzyme suffix). An enzyme involved in the removal of water from a compound. Carbonic anhydrase promotes the conversion of carbonic acid into water and carbon dioxide.

anisogametes (an′īs-ō-gam′ēts) (Gr. *anisos,* unequal, + *gametēs,* spouse). Gametes of a species that differ in form or size.

anlage (än′lä-gə) (Ger. laying out, foundation). Rudimentary form; primordium.

annulus (an′yəl-əs) (L. ring). Any ringlike structure, such as superficial rings on leeches.

antenna (L. sail yard). A sensory appendage on the head of arthropods, or the second of the two such pairs of structures in crustaceans.

antennal gland Excretory gland of Crustacea located in the antennal metamere.

anterior (L. comparative of *ante,* before). The head end of an organism, or (as an adjective) toward that end.

anthracosaurs (an-thrak′ə-sors) (Gr. *anthrax,* coal, carbon, + *sauros,* lizard). A group of Paleozoic labyrinthodont tetrapods.

anthropoid (an′thrə-poyd) (Gr. *anthrōpos,* man, + *eidos,* form). Resembling humans: especially the great apes.

antibodies (an′te-bä-dēs). Protein (immunoglobulins) in cell surfaces and dissolved in blood or tissue fluid, capable of combining with the antigens that stimulated their production. Formation of antibody-antigen complexes immobilizes invading organisms, which can then be removed by phagocytosis of the complexes **(opsonization).**

anticodon (an′tī-kō′don). A sequence of three nucleotides in transfer RNA that is complementary to a codon in messenger RNA.

antigen (an′ti-jən). Any substance capable of stimulating an immune response, most often a protein.

---

bat / āpe / ärmadillo / herring / fēmale / finch / līce / crocodile / crōw / cȯin / cōre / duck / dūke / ūnicorn / ə indicates unaccented vowel sound "uh" as in mammal, fishes, cardinal, heron, vulture / stress as in bi-ol′o-gy, bi′o-log′i-cal

antigenic determinant See **epitope.**

antigen-presenting cell Immune-system cell that phagocytizes antigen and activates acquired immune responses.

antimicrobial peptides Peptides secreted during an innate immune response in animals and plants. See **defensins.**

antiparallel In the double helix of DNA, paired strands are oriented so that the 3′ end of one strand is opposite the 5′ end of the other strand.

antler (ant′lur) (Fr. *andouiller*, antler). Bony skull projection of members of the deer family; usually found only in males and often shed annually.

aperture (ap′ər-chər) (L. *apertura* from *aperire*, to uncover). An opening; the opening into the first whorl of a gastropod shell.

apex (ā′peks) (L. tip). Highest or uppermost point; the lower pointed end of the heart. Adj., **apical** (ā′pə-kl).

apical complex A certain combination of organelles found in the protozoan phylum Apicomplexa.

apocrine (ap′ə-krən) (Gr. *apo*, away, + *krinein*, to separate). Applies to a type of mammalian sweat gland that produces a viscous secretion by detaching part of the cytoplasm of secreting cells.

apodeme A protrusion on the inside of the cuticle of certain arthropods (insects and crustaceans) to which the muscles attach.

apoptosis (a′pə-tō′səs) (Gr. *apo-*, away from, + *ptōsis*, a falling). Genetically determined cell death, "programmed" cell death.

apopyle (ap′ə-pīl) (Gr. *apo*, away from, + *pylē*, gate). In sponges, opening of the radial canal into the spongocoel.

aposematic (ap′-ə-si-mat′ik). A conspicuous condition that serves as a warning; for example, the bright colors of a monarch butterfly's wings warn potential predators that the butterfly is distasteful.

appendicular (L. *ad*, to, + *pendere*, to hang). Pertaining to appendages; pertaining to vermiform appendix.

aquaporins (aw′kwə-pōr′ins). Water channels or pores composed of transmembrane proteins that allow water to move through plasma membranes. They may be open at all times or gated to open or to close by a specific signal.

arboreal (är-bōr′ē-əl) (L. *arbor*, tree). Living in trees.

archaeocytes (är′-kē-ō-sīt′z) (Gr. *archaios*, beginning, + *kytos*, hollow vessel). Ameboid cells of varied function in sponges.

archenteron (är-ken′tə-rän) (Gr. *archē*, beginning, + *enteron*, gut). The main cavity of an embryo in the gastrula stage; it is lined with endoderm and represents the future digestive cavity.

archinephros (är′kə-nəf′rōs) (Gr. *archaois*, ancient, + *nephros*, kidney). Ancestral vertebrate kidney, existing today only in the embryo of hagfishes.

archosaur (är′kə-sor) (Gr. *archōn*, ruling, + *sauros*, lizard). A clade of diapsid vertebrates that includes the living birds and crocodiles and the extinct pterosaurs and dinosaurs.

arginine phosphate Phosphate storage compound (phosphagen) found in many invertebrates and used to regenerate stores of ATP.

Aristotle's lantern Masticating apparatus of some sea urchins.

arteriole (är-tir′ē-ōl) (L. *arteria*, artery). A small arterial branch that delivers blood to a capillary network.

artery (är′tə-rē) (L. *arteria*, artery). A blood vessel that carries blood away from the heart and toward a peripheral cavity.

asconoid (as′kə-noid) (Gr. *askos*, bladder). Simplest form of sponges, with canals leading directly from the outside to the interior.

asexual Without distinct sexual organs; not involving formation of gametes.

aspiration (as′pir-ā-shun) Drawing a fluid into a cavity by suction, as when air is drawn into lungs of amniotes.

assimilation (L. *assimilatio*, bringing into conformity). Absorption of digested nutrients to form complex organic protoplasmic materials.

asymmetric competition See **amensalism.**

athecate (ā′thē-kāt′) (Gr. *a-*, an absence of something, + *thēkē*, a case, box). Denotes an organism without a theca.

atherosclerosis (a′thə-rō-sklə-rō′səs) (Gr. *athērōma*, tumor full of gruel-like material, + *sklērōs*, hard). Disease characterized by fatty plaques forming in the inner linings of arteries.

atmosphere (at′-mə-sfi(ə)r). The gaseous component of the biosphere, extending from the earth's surface to an altitude of 3500 km.

atoke (ā′tōk) (Gr. *a*, without, + *tokos*, offspring). Anterior, nonreproductive part of a marine polychaete, as distinct from the posterior, reproductive part (epitoke) during the breeding season.

atoll (ə-tōl′) (Maldivian, *atolu*). A coral reef or island surrounding a lagoon.

atom The smallest unit of an element, composed of a dense nucleus of protons and (usually) neutrons surrounded by a system of electrons.

ATP Adenosine triphosphate. In biochemistry, an ester of adenosine and triphosphoric acid.

atrioventricular valve Structure in the heart between each **atrium** and **ventricle** that allows one-way flow of blood.

atrium (ā′trē-əm) (L. *atrium*, vestibule). One of the less muscular chambers of the heart; also, the tympanic cavity of the ear; also, the large cavity containing the pharynx in tunicates and cephalochordates.

auricle (aw′ri-kl) (L. *auricula*, a small ear). One of the less muscular chambers of the heart; atrium; the external ear, or pinna; any earlike lobe or process.

auricularia (ə-rik′u-lar′e-ə) (L. *auricula*, a small ear). A type of larva found in Holothuroidea.

autogamy (aw-täg′ə-mē) (Gr. *a utos*, self, + *g amos*, marriage). Condition in which the gametic nuclei produced by meiosis fuse within the same organism that produced them to restore the diploid number.

autosome (aw′tə-sōm) (Gr. *autos*, self, + *sōma*, body). Any chromosome that is not a sex chromosome.

autotomy (aw-tät′ə-mē) (Gr. *autos*, self, + *tomos*, a cutting). Detachment of a part of the body by an organism.

autotroph (aw′tə-trōf) (Gr. *autos*, self, + *trophos*, feeder). An organism that makes its organic nutrients from inorganic raw materials.

autotrophic nutrition (Gr. *autos*, self, + *trophia*, denoting nutrition). Nutrition characterized by the ability to use simple inorganic substances for the synthesis of more complex organic compounds, as in green plants and some bacteria.

average effect A quantitative-genetic parameter estimating the incremental contribution of each copy of a particular allele to the mean value of a particular organismal phenotype (such as height or weight) in the population being studied. Average effect is calculated from measurements of the population frequencies of all genotypes containing the allele, and the mean deviations of each genotypic class from the mean value of the phenotype in the population as a whole.

avicularium (a-vik′yə-ler′-ē-əm) (L. *avicula*, small bird, + *aria*, like or connected with). Modified zooid that is attached to the surface of the feeding zooid in Ectoprocta and resembles a bird's beak.

axial (L. *axis*, axle). Relating to the axis, or stem; on or along the axis.

axial complex A pressure-filtering system for circulatory fluids in echinoderms.

axocoel (ak′sə-sēl) (Gr. *axon*, an axle, + *koilos*, hollow). The most anterior of three coelomic spaces that appear during larval echinoderm development.

axolotl (ak′sə-lot′l) (Nahuatl, *atl*, water, + *xolotl*, doll, servant, spirit). Salamanders of the species *Ambystoma mexicanum,* which do not metamorphose and which retain aquatic larval characteristics throughout adulthood.

axon (ak′sän) (Gr. *axōn*). Elongate extension of a neuron that conducts signals away from the cell body and toward the synaptic terminals.

axoneme (aks′ə-nēm) (L. *axis*, axle, + Gr. *nēma*, thread). The microtubules in a cilium or flagellum, usually arranged as a circlet of nine pairs enclosing one central pair; also, the microtubules of an axopodium.

axopodium (ak′sə-pō′dē-um) (Gr. *axon*, an axis, + *podion*, small foot). Long, slender, more or less permanent pseudopodium found in certain amebas. (Also **axopod.**)

axostyle (aks′ō-stīl). Tubelike organelle in some flagellate protozoa, extending from the area of the kinetosomes to the posterior end, where it often protrudes.

B

barrier reef A coral reef that runs approximately parallel to the shore and is separated from the shore by a lagoon.

basal body Also called kinetosome and blepharoplast, a cylinder of nine triplets of microtubules found at the base of a flagellum or cilium; same structure as a centriole.

basal disc Aboral attachment site on a cnidarian polyp.

base A molecule that dissociates in solution to produce a hydroxide ion.

basis, basipodite (bā′səs, bā-si′pə-dīt) (Gr. *basis*, base, + *pous, podos,* foot). The distal or second joint of the protopod of a crustacean appendage.

Batesian mimicry A condition in which a prey species evolves close resemblance to the aposematic phenotype (such as warning coloration) of an inedible species to deceive a potential predator into avoiding it.

bathypelagic (bath′ə-pe-laj′ik) (Gr. *bathys*, deep, + *pelagos*, open sea). Relating to or inhabiting the deep sea.

B cell A type of lymphocyte that is most important in the humoral immune response.

behavioral ecology The study of animal behaviors as used to promote survival and reproduction in a population's natural habitat.

behavioral syndrome A contrasting set of correlated behaviors reflecting interindividual consistency in behavior across multiple situations; for example, the contrast between fruit-fly larvae that take long paths in their foraging versus others that concentrate their feeding in a small area.

benthos (ben′thäs) (Gr. depth of the sea). Organisms that live on the floor of a sea or lake; adj., **benthic.** Also, the submerged substrate itself.

bilateral cleavage In an early embryo, the first cell-division plane divides the zygote into left and right sides, which are maintained throughout subsequent cleavages.

Bilateria (bī′lə-tir′-ē-ə) (L. *bi* -, two, + *latus,* side). Bilaterally symmetrical animals.

bilirubin (bil′ə-rü-bən) (L. *bilis*, bile, + *rubeo,* to be red). A breakdown product of the heme group of hemoglobin, excreted in bile.

binary fission A mode of asexual reproduction in which an animal splits into two approximately equal offspring.

binomial nomenclature The Linnaean system of naming species, in which the first word is the name of the genus (first letter capitalized) and the second word is the specific epithet (uncapitalized), usually an adjective modifying the name of the genus. Both of these words are written in italics.

biogenetic law See **recapitulation.**

biogeochemical cycle A description of the flow of elementary matter, such as carbon or phosphorus, through the component parts of an ecosystem and its **abiotic** environment, including the amount of an element present at the various stages of a **food web.**

biological species concept Defines species as a reproductive community of populations (reproductively isolated from others) that occupies a specific niche in nature.

bioluminescence Method of light production by living organisms in which certain proteins (luciferins), in the presence of oxygen and an enzyme (luciferase), usually are converted to oxyluciferins with the liberation of light.

biomass (Gr. *bios*, life, + *maza*, lump or mass). The weight of total living organisms or of a species population per unit of area.

biome (bī′-ōm) (Gr. *bios*, life, + *ōma*, abstract group suffix). Complex of plant and animal communities characterized by climatic and soil conditions; the largest ecological unit.

biosphere (Gr. *bios*, life, + *sphaira*, globe). That part of earth containing living organisms.

biotic (bī-ät′ik) (Gr. *biōtos*, life, livable). Of or relating to life.

bipinnaria (L. *bi*, double, + *pinna*, wing, + *aria*, like or connected with). Free-swimming, ciliated, bilateral larva of the asteroid echinoderms; develops into the brachiolaria larva.

biradial symmetry A type of radial symmetry in which only two planes passing through the oral-aboral axis yield mirror images because some structure is paired.

biramous (bī-rām′əs) (L. *bi*, double, + *ramus*, a branch). Adjective describing appendages with two distinct branches; contrasts with uniramous, unbranched.

bivalent (bī-vāl′ənt) (L. *bi*, double, + *valen,* strength, worth). The pairs of homologous chromosomes at synapsis in the first meiotic division, a tetrad.

blastocoel (blas′tə-sēl) (Gr. *blastos*, germ, + *koilos,* hollow). Cavity of the blastula.

blastocyst (blast′ō-sist) (Gr. *blastos*, germ, + *kystis*, bladder). Mammalian embryo in the blastula stage.

blastomere (Gr. *blastos*, germ, + *meros*, part). An early cleavage cell.

blastopore (Gr. *blastos*, germ, + *poros*, passage, pore). External opening of the archenteron in the gastrula.

blastula (Gr. *blastos*, germ, + L. *ula*, dim.). Early embryological stage of many animals; consists of a hollow mass of cells.

blending See **polygenic inheritance.**

blepharoplast (blə-fä′rə-plast) (Gr. *blepharon,* eyelid, + *plastos*, formed). See **basal body.**

blood plasma The liquid, noncellular fraction of blood, including dissolved substances.

blood type Characteristic of human blood given by the particular antigens on the membranes of the erythrocytes, genetically determined, causing agglutination when incompatible groups are mixed; the blood types are designated A, B, O, AB, Rh negative, Rh positive, and others.

Bohr effect A characteristic of hemoglobin that causes it to dissociate from oxygen in greater degree at higher concentrations of carbon dioxide.

bombykol (bäm′-bə-käl). A volatile chemical produced by virgin female silkworm moths to attract males.

book gill Respiratory structure of aquatic chelicerates (Arthropoda) where many thin

---

bat / āpe / ärmadillo / herring / fēmale / finch / līce / crocodile / crōw / côin / cōre / duck / düke / ūnicorn / ə indicates unaccented vowel sound "uh" as in mammal, fishes, cardinal, heron, vulture / stress as in bi-ol′o-gy, bi′o-log′i-cal blood-filled gills are layered like the pages of a book. Gas exchange occurs as seawater passes between each pair of gills.

book lung Respiratory structure of terrestrial chelicerates (Arthropoda) where many thin-walled air pockets extend into a blood-filled chamber in the abdomen.

boreal (bōr′ē-əl) (L. *boreas,* north wind). Relating to a northern biotic area characterized by a predominance of coniferous forests and tundra.

BP Before the present.

brachial (brāk′ē-əl) (L. *brachium,* forearm). Referring to the arm.

brachiation (brāk′ē-ā′shən) (L. *brachium,* arm). Locomotion by swinging by the arms from one hold to another.

brachiolaria (brāk′ē-ō-lār′ē-ə) (L. *brachiola,* little arm, + *aria,* pertaining to). This asteroid larva develops from the bipinnaria larva and has three preoral holdfast processes.

brackish Water having a salinity intermediate between freshwater and seawater, ranging from 0.5 to 30 parts per thousand.

bradyzoite (brā′dē-zō′īt). An individual coccidian (a single-celled parasite) such as *Toxoplasma gondii* that is encased in a tissue cyst and divides slowly.

branchial (brank′ē-əl) (Gr. *branchia,* gills). Referring to gills.

bronchiole (brän′kē-ōl) (Gr. *bronchion,* dim. of *bronchos,* windpipe). Small, thin-walled branch of the bronchus.

bronchus (brän′kəs), pl. **bronchi** (Gr. *bronchos,* windpipe). Either of two primary divisions of the trachea that lead to the right and left lungs.

brown bodies Remnants of the lophophore and digestive tract of a degenerating adult ectoproct left behind in the chamber as a new lophophore and digestive tract are formed.

brown fat Mitochondria-rich, heat-generating adipose tissue of endothermic vertebrates.

bryophytes (brī′ə-fīts) (Gr. *bry ō,* to sprout up, + *phyta,* plants). Nonvascular plants comprising mosses, liverworts, and hornworts.

buccal (buk′əl) (L. *bucca,* cheek). Referring to the mouth cavity.

budding Reproduction in which the offspring arises as an outgrowth from the parent and is initially smaller than the parent. Failure of the offspring to separate from the parent leads to colony formation.

buffer Any substance or chemical compound that tends to keep pH levels constant when acids or bases are added.

bursa, pl. **bursae** (M.L. *bursa,* pouch, purse made of skin). A saclike cavity. In ophiuroid echinoderms, pouches opening at bases of arms and functioning in respiration and reproduction (genitorespiratory bursae).

C

caecilian (si-sil′-yən). Denotes any member of the amphibian order Gymnophiona (also called Apoda).

caecum See **cecum.**

calciferous glands (kal-si′fə-rəs). Glands in an earthworm that secrete calcium ions into the gut.

calcitonin Hormone secreted by C cells in the thyroid gland of mammals and in the ultimobranchial glands of other vertebrates. It is released in response to elevated levels of calcium in the blood and causes decreased intestinal calcium absorption, increased calcium excretion by the kidneys, and decreased release of calcium from bone.

calorie (kal′ə-rē) (L. *calere,* to be warm). Unit of heat defined as the amount of heat required to heat 1 g of water from 14.5° to 15.5° C; 1 cal = 4.184 joules in the International System of Units.

calyx (kā-′liks) (L. bud cup of a flower). Any of various cup-shaped zoological structures.

capillary Extremely narrow (average in mammals is about 8 μm in diameter) blood vessels linking arterial and venous systems within tissues, and composed of a single layer of endothelial cells. Capillaries provide an interface between the circulatory system and cells allowing filtration of oxygen, nutrients, and signaling molecules (for example, hormones) to cells and uptake of metabolic wastes from cells.

capitulum (ka-pi′chə-ləm) (L. small head). Term applied to small, headlike structures of various organisms, including the projection from the body of ticks and mites carrying mouthparts.

captacula (kap-tak′ū-lə) (L. *captare,* to lie in wait for). Tentacles extending from the head of scaphopod molluscs, used in feeding.

carapace (kar′ə-pās) (F. from Sp. *carapacho,* shell). Shieldlike plate covering the cephalothorax of certain crustaceans; dorsal part of the shell of a turtle.

carbaminohemoglobin Form of **hemoglobin** produced when carbon dioxide combines with amino acid groups of hemoglobin in vertebrate red blood cells or invertebrate blood plasma.

carbohydrates (L. *carbo,* charcoal, + Gr. *hydōr,* water). Compounds of carbon, hydrogen, and oxygen having the generalized formula $(CH_2O)_n$; aldehyde or ketone derivatives of polyhydric alcohols, with hydrogen and oxygen atoms attached in a 2:1 ratio.

carboxyl (kär-bäk′səl) (carbon + oxygen + yl, chemical radical suffix). The acid group of organic molecules (–COOH).

carboxypeptidase An enzyme in pancreatic juices that removes amino acids from the carboxyl end of polypeptides.

cardiac (kär′dē-ak) (Gr. *kardia,* heart). Belonging or relating to the heart.

carinate (kar′ə-nāt) (L. *carina,* keel). Having a keel, in particular the flying birds with a keeled sternum for the insertion of flight muscles; contrasts with ratite.

carnivore (kar′nə-vōr′) (L. *carnivorous,* flesh eating). A mammal of the order Carnivora. Also, any organism that eats animals. Adj., **carnivorous.**

carotene (kär′ə-tēn) (L. *carota,* carrot, + *ene,* unsaturated straight-chain hydrocarbons). A red, orange, or yellow pigment belonging to the group of carotenoids; precursor of vitamin A.

carrier An individual heterozygous for a recessive allele, such as an allele for a genetic disease, who is phenotypically normal but may transmit the recessive allele to offspring.

carrying capacity The maximum number of individuals that can persist under specified environmental conditions.

cartilage (L. *cartilago;* akin to L. *cratis,* wickerwork). A translucent, specialized connective tissue that forms most of the skeleton of embryos, very young vertebrates, and adult cartilaginous fishes, such as sharks and rays; in other adult vertebrates much of it is converted into bone.

caste (kast) (L. *castus,* pure, separated). One of the polymorphic forms within an insect society, each caste having its specific duties, such as queen, worker, and soldier.

catabolism (Gr. *kata,* downward, + *bol,* to throw, + *ism,* suffix meaning state of condition). Destructive metabolism; process in which complex molecules are reduced to simpler ones.

catadromous (kə-tad′rə-məs) (Gr. *kata,* down, + *dromos,* a running). Refers to fishes that migrate from freshwater to the ocean to spawn.

catalyst (kat′ə-ləst) (Gr. *kata,* down, + *lysis,* a loosening). A substance that accelerates a chemical reaction but does not become a part of the end product.

catastrophic species selection Differential survival among species during a time of mass extinction based on character variation that permits some species but not others to withstand severe environmental disturbances, such as those caused by an asteroid impact.

catch collagen In echinoderms a mutable collagenous tissue under neuronal control that can be changed from a "liquid" to a "solid" form very quickly.

caudal (käd′l) (L. *cauda*, tail). Constituting, belonging to, or relating to a tail.

caveolae (ka-vē′ə-lē) (L. *cavea*, a cave, + dim. suffix). The invaginated vesicles and pits in pinocytosis.

cDNA See **complementary DNA.**

cecum, caecum (sē′kəm) (L. *caecus*, blind). A blind pouch at the beginning of the large intestine; any similar pouch.

cell-mediated immune response Acquired immune response involving cell surfaces only, not antibody production, specifically the T_H1 arm of the immune response. Contrasts with **humoral immune response.**

cellulase (sel′ū-lās) (L. *cella*, small room). An enzyme that cleaves cellulose; synthesized only by bacteria and some protists.

cellulose (sel′ū-lōs) (L. *cella*, small room). Chief polysaccharide constituent of the cell wall of green plants and some fungi; an insoluble carbohydrate $(C_6H_{10}O_5)_n$ that is converted into glucose by hydrolysis.

center of origin The geographic area occupied by a species or higher taxon during its initial evolution; contrasts to areas colonized by dispersal following a species' evolutionary origin.

centriole (sen′trē-ōl) (Gr. *kentron*, center of a circle, + L. *ola*, small). A minute cytoplasmic organelle usually found in the centrosome and considered the active division center of the animal cell; organizes spindle fibers during mitosis and meiosis. Same structure as basal body or kinetosome.

centrolecithal (sen′tro-les′ə-thəl) (Gr. *kentron*, center, + *lekithos*, yolk, + Eng. *al*, adjective). Pertaining to an insect egg with the yolk concentrated in the center.

centromere (sen′trə-mēr) (Gr. *kentron*, center, + *meros*, part). A localized constriction in a characteristic position on a given chromosome, bearing the kinetochore.

centrosome (sen′trə-sōm) (Gr. *kentron*, center, + *sōma*, body). Microtubule organizing center in nuclear division in most eukaryotic cells, and also the organizing center for the microtubule portion of the cytoskeleton; in animals and many unicellular organisms it surrounds the centrioles.

cephalization (sef′ə-li-zā-shən) (Gr. *kephale*, head). The evolutionary process by which sensory organs and specialized appendages become localized in the head end of animals.

cephalothorax (sef′ə-lä-thō′raks) (Gr. *kephale*, head, + thorax). A body division found in many Arachnida and higher Crustacea, in which the head is fused with some or all of the thoracic segments.

cerata (sə-ra′tə) (Gr. *keras*, a horn, bow). Dorsal processes on some nudibranchs for gaseous exchange.

cercaria (ser-kar′ē-ə) (Gr. *kerkos*, tail, + L. *aria*, like or connected with). Tadpolelike larva of trematodes (flukes).

cerebellum (ser′ə-bel′əm) (L. *cerebrum*, brain). Region of the hindbrain that controls muscle coordination.

cervix (L. *cervix*, neck). Opening or neck of the uterus or womb of a mammal. Adj., cervical, refers also to vertebrae located in the neck.

channels Pores created by transmembrane proteins that allow ions and water to move through plasma membranes by diffusion. They may be open at all times or gated by specific signals to open or to close them (for example, chemically-gated or voltage-gated channels).

character (kar′ik-tər). A component of phenotype (including specific molecular, morphological, behavioral, or other features) used by systematists to diagnose species or higher taxa, or to evaluate phylogenetic relationships among different species or higher taxa, or relationships among populations within a species.

character displacement Differences in morphology or behavior within a species caused by competition with another species; characteristics typical of one species differ according to whether the other species is present in or absent from a local community.

charging In protein synthesis, a reaction catalyzed by tRNA synthetase, in which an amino acid is attached to its particular tRNA molecule.

chela (kēl′ə) (Gr. *chēlē*, claw). Pincerlike claw.

chelicera (kə-lis′ə-rə), pl. **chelicerae** (Gr. *chēlē*, claw, + *keras*, horn). One of a pair of the most anterior head appendages on the members of the subphylum Chelicerata.

chelifore (kə′-li-fōr) (Gr. *chēlē*, claw, + OE *fore*, before). First pair of appendages on a

pycnogonid; sometimes absent; if present, chelate or achelate.

chelipeds (kēl′ə-peds) (Gr. *chēlē*, claw, + L. *pes*, foot). Pincerlike first pair of legs in most decapod crustaceans; specialized for seizing and crushing.

chemoautotroph (ke′mō-aw′tō-trōf) (Gr. *chemeia*, transmutation, + *autos*, self, + *trophos*, feeder). An organism utilizing inorganic compounds as a source of energy.

chemotaxis (kē′mō-tak′səs) (Gr. *chēmeia*, an infusion, + *taxō* from *tassō*, to put in order). Orientation movement of cells or organisms in response to a chemical stimulus.

chemotroph (kem′ə-trōf) (Gr. *chēmeia*, an infusion, + *trope*, to turn). An organism that derives nourishment from inorganic substances without using chlorophyll.

chiasma (kī-az′mə), pl. **chiasmata** (Gr. cross). An intersection or crossing, as of nerves; a connection point between homologous chromatids where crossing over has occurred at synapsis.

chitin (kī′tən) (Fr. *chitine*, from Gr. *chitōn*, tunic). A hard substance that forms part of the cuticle of arthropods and occurs sparingly in certain other invertebrates; a nitrogenous polysaccharide insoluble in water, alcohol, dilute acids, and digestive juices of most animals.

chloragogen cells (klōr′ə-gog-ən) (Gr. *chlōros*, light green, + *agōgos*, a leading, a guide). Modified peritoneal cells, greenish or brownish, clustered around the digestive tract of certain annelids; apparently they aid in elimination of nitrogenous wastes and in food transport.

chloride shift Chloride ions are exchanged for bicarbonate ions in vertebrate red blood cells during carbon dioxide transport.

chlorocruorin (klō′rō-krü′ə-rən) (Gr. *chlōros*, light green, + L. *cruor*, blood). A greenish iron-containing respiratory pigment dissolved in the blood plasma of certain marine polychaetes.

chlorophyll (klō′rə-fil) (Gr. *chlōros*, light green, + *phyllōn*, leaf). Green photosynthetic pigment of cyanobacteria, plants, and some unicellular eukaryotes.

chloroplast (klō′rə-plast) (Gr. *chlōros*, light green, + *plastos*, molded). An organelle in plant cells that contains chlorophyll and is the physical location of photosynthesis; occurs also in photosynthetic unicellular eukaryotes.

choanoblast (kō-an′ə-blast) (Gr. *choanē*, funnel, + *blastos*, germ). One of several cellular elements within the syncytial tissue of a hexactinellid sponge; choanoblasts bear flagellated extensions called collar bodies.

choanocyte (kō-an′ə-sīt) (Gr. *choanē*, funnel, + *kytos*, hollow vessel). One of the flagellated collar cells that line cavities and canals of sponges.

choanoflagellate (kō-an′ə-fla-jel′āt). Any member of a protozoan clade having a single flagellum surrounded by a column of microvilli; some form colonies, and all are included within the larger clade of opisthokonts.

cholinergic (kōl′i-nər′jik) (Gr. *chōle*, bile, + *ergon*, work). A nerve fiber that releases acetylcholine from its axon terminal.

chorioallantoic membrane (kō′rē-ō-al′an-tō′-ic) (Gr. *chorion*, skin, + *allas*, sausage, + *eidos*, form). A vascular envelope around some amniote embryos formed by fusion of mesoderm from the chorion and allantois.

chorioallantoic placenta (kō′rē-ō-al′an-tō′-ic) (Gr. *chorion*, skin, + *allas*, sausage). A type of placenta that occurs in placental mammals and some marsupials in which materials are exchanged between the embryo and mother through a structure modified from the embryonic chorionic and allantoic membranes.

chorion (kō′rē-on) (Gr. *chorion*, skin). The outer of the double membrane that surrounds the embryo of nonavian reptiles, birds, and mammals; in mammals it contributes to the placenta.

chorionic villi (kō′rē-än′ic vil′ē) (Gr. *chorion*, skin, + L. *villi*, pl. of *villus*, shaggy hair, tuft of hair). Fingerlike extensions containing blood vessels that lie on the outer surface of a vertebrate chorionic membrane.

choriovitelline placenta (kor′ē-ō-vi′təl-ən) (Gr. *chorion*, skin, + *vittel*, yolk of an egg). An often transitory placenta that forms during the early developmental stages of marsupials and placental mammals. Also called the "yolk sac placenta," it is formed from the yolk sac and chorionic membrane of the embryo.

choroid (kōr′oid) (Gr. *chorion*, skin, + *eidos*, form). Delicate, highly vascular membrane in vertebrate eye; the layer between the retina and the sclera.

chromatid (krō′mə-tid) (Gr. *chromato*, from *chrōma*, color, + L. *id*, feminine stem for particle of specified kind). Either of the paired strands of a replicated chromosome joined by the centromere; at anaphase of mitosis or anaphase of the second meiotic division, the centromere divides and each of the formerly attached chromatids becomes a separate, unreplicated chromsome.

chromatin (krō′mə-tin) (Gr. *chrōma*, color). The nucleoprotein material of a chromosome; the hereditary material containing DNA.

chromatophore (krō-mat′ə-fōr) (Gr. *chrōma*, color, + *pherein*, to bear). Pigment cell, usually in the dermis, in which the pigment can be dispersed or concentrated.

chromosomal theory of inheritance The general theory synthesizing results of Mendelian genetics and cytology to propose particulate inheritance of hereditary factors carried on chromosomes in eukaryotes.

chromosome (krō′mə-sōm) (Gr. *chrōma*, color, + *sōma*, body). A complex body, spherical or rod-shaped, that becomes visible following condensation of the nuclear material during mitosis, and carries a part of the organism's genetic information as genes composed of DNA associated with proteins; comprises a single centromere and all chromatin physically connected to that centromere.

chrysalis (kris′ə-lis) (L. from Gr. *chrysos*, gold). The pupal stage of a butterfly.

chyme (kīm) (Gr. *chymos*, juice). Semifluid mass of partly digested food in stomach and small intestine as digestion proceeds.

chymotrypsin An enzyme in pancreatic juices that splits peptide bonds of a protein molecule.

cilium (sil′ē-əm), pl. **cilia** (L. eyelid). A hairlike, vibratile organelle process found on many animal cells. Cilia may be used in moving particles along the cell surface or, in ciliated unicellular forms, for locomotion. Some cilia are sensory, as are those in the inner ear and lateral-line organs.

cinclides (sing′klid-əs), sing. **cinclis** (sing′kləs) (Gr. *kinklis*, latticed gate or partition). Small pores in the external body wall of sea anemones for extrusion of acontia.

circadian (sər-kād′ē-ən) (L. *circa*, around, + *dies*, day). Occurring at a period of approximately 24 hours.

cirrus (sir′əs), pl. **cirri** (L. curl). A hairlike tuft on an insect appendage; locomotor organelle of fused cilia; male copulatory organ of some invertebrates. Cirri can refer to tufts of fused cilia in members of the Ciliophora.

cis-regulation Control of the rate of transcription of a gene that encodes mRNA and rRNA by untranscribed sequences physically adjacent to it on the same DNA molecule; cis-regulatory sequences influence the expression only of sequences physically linked to them and not of those present on a homologous chromosome in the same diploid nucleus.

cisternae (sis-ter′nē) (L. *c ista*, box). Spaces between membranes of the endoplasmic reticulum within cells.

cistron (sis′tron) (L. *cista*, box). A series of codons in DNA that code for an entire polypeptide chain.

clade (klād) (Gr. *klados*, branch, sprout). A taxon or other group consisting of an ancestral species and all of its descendants, forming a distinct branch on a cladogram or phylogenetic tree.

cladistics (klad-is′-təks) (Gr. *klados*, branch, sprout). A system of arranging taxa by analysis of evolutionarily derived characteristics so that the arrangement reflects phylogenetic relationships.

cladogram (klād′ə-gram) (Gr. *klados*, branch, + *gramma*, letter). A branching diagram showing the pattern of sharing of evolutionarily derived characters among species or higher taxa.

clasper (klas′-pər). Digitiform projection on the medial side of pelvic fins of male chondrichthians and some placoderms; used as an intromittent organ to transfer sperm to female reproductive tract.

clathrin (kla′thrən) (L. *chathri*, latticework). A protein forming a lattice structure lining the invaginated pits during receptor-mediated endocytosis.

cleavage (O.E. *cleofan*, to cut). Process of nuclear and cell division in animal zygote.

cleavage furrow In a developing animal embryo, a groove marking a future plane of cell division.

climax (klī′maks) (Gr. *klimax*, ladder). Stage of relative stability attained by a community of organisms, often the culminating development of a natural succession. Also, orgasm.

climax community (Gr. *klimax*, ladder, staircase, climax). A self-perpetuating, more-or-less stable community of organisms that continues as long as environmental conditions under which it developed prevail.

clitellum (klī-tel′əm) (L. *clitellae*, packsaddle). Thickened saddlelike portion of certain midbody segments of many oligochaetes and leeches. Produces the cocoon in which eggs are deposited during sexual reproduction.

cloaca (klō-ā′kə) (L. sewer). Posterior chamber of digestive tract in many vertebrates, receiving feces and urogenital products. In certain invertebrates, a terminal portion of digestive tract that serves also as respiratory, excretory, or reproductive tract.

clone (klōn) (Gr. *klōn*, twig). All descendants derived by asexual reproduction from a single individual.

cloning (klō′ning). Production of genetically identical organisms by asexual reproduction.

cnida (nī′də) (Gr. *knidē*, nettle). Stinging or adhesive organelles formed within cnidocytes in phylum Cnidaria; nematocysts are a common type.

cnidoblast (nī′də-blast) (Gr. *knidē*, nettle, + *blastos*, germ). See **cnidocyte.**

cnidocil (nī′də-sil) (Gr. *knidē*, nettle, + L. *cilium*, hair). Modified cilium on nematocyst-bearing cnidocytes in cnidarians; triggers nematocyst.

cnidocyte (nī′də-sīt) (Gr. *knidē*, nettle, + *kytos*, hollow vessel). Modified interstitial cell that holds the nematocyst; during development of the nematocyst, the cnidocyte is a cnidoblast.

coacervate (kō′a-sər′vāt) (L. *coacervatus*, to heap up). An aggregate of colloidal droplets held together by electrostatic forces.

coagulation (kō-ag′ū-lā-shən). Process in which enzymes are activated to produce clotting of blood.

coccidian (kok-sid′ē-ən) (Gr. *kokkis*, kernel, grain). Intracellular protozoan parasite belonging to a class within phylum Apicomplexa; the organism causing malaria is an example.

cochlea (kōk′lē-ə) (L. snail, from Gr. *kochlos*, a shellfish). A tubular cavity of the inner ear containing the essential organs of hearing; occurs in crocodiles, birds, and mammals; spirally coiled in mammals.

cocoon (kə-kün′) (Fr. *cocon*, shell). Protective covering of a resting or developmental stage, sometimes used to refer to both the covering and its contents; for example, the cocoon of a moth or the protective covering for the developing embryos in some annelids.

codominance A condition in which each allele maintains its distinctive homozygous expression in the heterozygous condition, not a blending of the separate homozygous phenotypes (contrasts with intermediate inheritance). Genes for A and B blood types show codominance.

codon (kō′dän) (L. code, + on). In messenger RNA, a sequence of three adjacent nucleotides that code for one amino acid.

coelacanth (sē′la-canth) (Gr. *coel*, hollow + a*canth*, spine, in reference to hollow caudal fin rays). A bony fish with lobelike fins, of the clade Sarcopterygii.

coelenteron (sē-len′tər-on) (Gr. k *oilos*, hollow, + *enteron*, intestine). Internal cavity of a cnidarian; gastrovascular cavity; archenteron.

coelom (sē′lōm) (Gr. *koilōma*, cavity). The body cavity in triploblastic animals, lined with mesodermal peritoneum.

coelomate (sē′lōm-āt). Animals that possess a **coelom;** also called eucoelomate.

coelomocyte (sē′lō′mə-sīt) (Gr. *koilōma*, cavity, + *kytos*, hollow vessel). Another name for amebocyte; primitive or undifferentiated cell of the coelom and the water-vascular system of echinoderms.

coelomoduct (sē-lō′mə-dukt) (Gr. *koilos*, hollow, + L. *ductus*, a leading). A duct that carries gametes or excretory products (or both) from the coelom to the exterior.

coenecium, coenoecium (sə-nēs[h] ′ē-əm) (Gr. *koinos*, common, + *oikion*, house). The common secreted investment of an ectoproct colony; may be chitinous, gelatinous, or calcareous.

coenenchyme (sēn′ən-kīm) (Gr. *koinos*, shared in common, + *enchyma*, something poured in). Extensive mesogleal tissue between the polyps of an octocorallian (phylum Cnidaria) colony.

coenosarc (sē′nə-särk) (Gr. *koinos*, shared in common, + *sarkos*, flesh). The inner, living part of hydrocauli in hydroids.

coenzyme (kō-en′zīm) (L. prefix, *co*, with, + Gr. *enzymos*, leavened, from *en* in, + *zymē*, leaven). A required substance in the activation of an enzyme; a prosthetic or nonprotein constituent of an enzyme.

cohesion species concept Defines species as the most inclusive population of individuals having the potential for phenotypic cohesion through intrinsic cohesion mechanisms; a refinement of the **evolutionary species concept** emphasizing population-genetic processes.

cohort (kō′hōrt). All organisms of a population born within a specified interval of time.

collagen (käl′ə-jən) (Gr. *kolla*, glue, + *genos*, descent). A structural protein of connective tissue, the most abundant protein in the animal kingdom, characterized by high content of the amino acids glycine, alanine, proline, and hydroxyproline.

collar bodies Extensions of choanoblasts bearing flagellated collars in hexactinellid sponges.

collar cells Cells having a single flagellum surrounded by a ring of microvilli. Sponge choanocytes are collar cells, as are choano-flagellates, but collar cells also occur in other taxa.

collenchyme (käl′ən-kīm) (Gr. *kolla*, glue, + *enchyma*, infusion). A gelatinous mesenchyme containing undifferentiated cells; occurs in cnidarians and ctenophores.

collencyte (käl′lən-sīt) (Gr. *kolla*, glue, + *en*, in, + *kytos*, hollow vessel). A type of cell in sponges that is star-shaped and apparently contractile.

colloblast (käl′ə-blast) (Gr. *kolla*, glue, + *blastos*, germ). A glue-secreting cell on the tentacles of ctenophores.

colloid (kä′löid) (Gr. *kolla*, glue, + *eidos*, form). A two-phase system in which particles of one phase are suspended in the second phase.

columella (kä′lū-mel′ə) (L. small column). Central pillar in gastropod shells; also, a sound-transmitting bone of amphibians.

comb plate One of the plates of fused cilia arranged in rows for ctenophore locomotion.

commensalism (kə-men′səl-iz′əm) (L. *cum*, together with, + *mensa*, table). A relationship in which one individual lives close to or on another and benefits, and the host is unaffected; often symbiotic.

common descent Darwin's theory that all forms of life are derived from a shared ancestral population through a branching of evolutionary lineages.

community (L. *communitas*, community, fellowship). See **ecological community** and **reproductive community.**

comparative biochemistry Studies of the structures of biological macromolecules, especially proteins and nucleic acids, and their variation within and among species to reveal homologies of macromolecular structure.

comparative cytology Studies of the structures of chromosomes within and among species to reveal homologies of chromosomal structure.

comparative method Use of patterns of similarity and dissimilarity among species or populations to test hypotheses of character **homology** and to infer phylogenetic relationships; use of phylogeny to examine evolutionary processes and history.

comparative morphology Studies of organismal form and its variation within and among species to reveal homologies of organismal characters.

comparative psychology A field of study used to identify general rules of behavior that apply to humans and other animals.

bat / āpe / ärmadillo / herring / fēmale / finch / līce / crocodile / crōw / cóin / cōre / duck / dūke / ūnicorn / ə indicates unaccented vowel sound "uh" as in mammal, fishes, cardinal, heron, vulture / stress as in bi-ol′o-gy, bi′o-log′i-cal

competition Some degree of overlap in ecological niches of two populations in the same community, such that both depend on the same food source, shelter, or other resources, and negatively affect each other's survival.

competitive exclusion An ecological principle stating that two species whose niches are very similar cannot coexist indefinitely in the same community; one species is driven to extinction by competition between them.

complement Collective name for a series of proteins in the blood. Proteins of the complement series are activated by antibody bound to invading organisms and lead to rupture of foreign cells. Some complement proteins also bind to antibody-antigen complexes enhancing phagocytosis by phagocytic cells of the immune system.

complementary Descriptive of a relationship between two single strands of DNA or RNA whose base sequences permit formation of a double-stranded hybrid molecule by hydrogen bonding between the bases of the two strands.

complementary DNA (cDNA) DNA prepared by transcribing the base sequence from mRNA into DNA by reverse transcriptase; also called **copy DNA.**

compound A substance whose molecules are composed of atoms of two or more elements.

condensation reaction A chemical reaction in which reactant molecules are combined by the removal of a water molecule (a hydrogen from one and a hydroxyl from the other reactant).

conditional specification During embryonic cleavage, diffusion of molecules from neighboring cells, providing positional information to specify cell fate.

condyle (kän′dīl) (Gr. *kondylos*, bump). A rounded process on a bone used for articulation.

cone (kōn′). One of a group of cells on the retina of a vertebrate eye used to perceive color and to form images in well-illuminated settings.

conjugation (kon′ju-ga′shən) (L. *conjugare*, to yoke together). Temporary union of two ciliate protozoa while they are exchanging chromatin material and undergoing nuclear phenomena resulting in binary fission. Also, formation of cytoplasmic bridges between bacteria for transfer of plasmids.

conodonts (kon′-ə-donts) (Gr. *con*, cone + *odont*, tooth). Microscopic, toothlike fossils belonging to an extinct vertebrate-like animal known from the Cambrian to Triassic periods.

conspecific (L. *com*, together, + *species*). A member of the same species.

consumer (kən-sū′-mer). An organism whose energy and matter are acquired by eating other organisms, which may be **primary producers, herbivores,** or **carnivores**.

continental margin A portion of the ocean floor located adjacent to the shore; comprises a continental shelf, continental slope, and continental rise.

continental rise Thick sediments piled at the base of the submarine continental slope.

continental shelf Portion of the continental margin of the ocean floor extending from shallow nearshore waters outward to a depth of 120 to 400 m below the sea surface.

continental slope A sharp descent of the ocean floor from the outer continental shelf to a depth of 3000 to 4000 m below the sea surface.

contractile vacuole A clear fluid-filled cell vacuole in unicellular eukaryotes and a few animals; collects water and releases it to the outside in a cyclical manner, for osmoregulation and some excretion.

control That part of a scientific experiment to which the experimental variable is not applied but which is similar to the experimental group in all other respects.

conus arteriosus (kōn′-əs är-tir′-ē-ō′-səs). An extension of the ventricle through which blood exits the heart in amphibians and some fishes; in mammals, an extension of the right ventricle serving the pulmonary circuit.

convexity The property of a taxonomic group that a path can be drawn between any two members on a cladogram or phylogenetic tree without leaving the group. **Monophyletic** and **paraphyletic** groups are convex, whereas **polyphyletic** groups are not.

cooperative behavior Participation by an organism in a group activity that enhances each participant's genetic contribution to future generations. Includes collective foraging and breeding behaviors.

coprophagy (ko-prō′fā′-jē) (Gr. *kopros*, dung, + *phagein*, to eat). Feeding on dung or excrement as a normal behavior among animals; reingestion of feces.

copulation (Fr. from L. *copulare*, to couple). Sexual union to facilitate the reception of sperm by the female.

copy DNA See **complementary DNA.**

coral bleaching Normally bright-colored corals lose or expel their algal symbionts (zooxanthellae), becoming white and brittle. Due to rising ocean temperatures, the photosynthetic systems of the symbionts release damaging oxidants, which disrupt the algal-cnidarian symbiosis.

coralline algae Algae that precipitate calcium carbonate in their tissues; important contributors to coral reef mass.

coral reef Marine ecosystem of great species diversity associated with calcium carbonate deposits secreted by anthozoan cnidarians.

corium (kō′rē-um) (L. *corium*, leather). The deep layer of the skin; dermis.

cornea (kor′nē-ə) (L. *corneus*, horny). The outer transparent coat of the eye.

corneum (kor′nē-əm) (L. *corneus*, horny). Epithelial layer of dead, keratinized cells. Stratum corneum.

cornified (kor′nə-fīd) (L. *corneus*, horny). Adjective for conversion of epithelial cells into nonliving, keratinized cells.

corona (kə-rō′nə) (L. crown). Head or upper portion of a structure; ciliated disc on anterior end of rotifers.

corpora allata (kor′pə-rə əl-la′tə) (L. *corpus*, body, + *allatum*, aided). Endocrine glands in insects that produce juvenile hormone.

corpora cardiaca (kor′pə-rə kar-dī′ə-kə) (L. *corpus*, body, + Gr. *kardiakos*, belonging to the heart). Paired organs behind the brain of insects serving as storage and release organs for prothoracicotropic hormone (PTTH).

cortex (kor′teks) (L. bark). The outer layer of a structure.

cosmopolitan Used to describe a species or higher taxon that has a very large geographical range, such as the worldwide distribution of humans.

covalent bond A chemical bond in which electrons are shared between atoms.

coxa, coxopodite (kox′ə, kox-ä′pō-dīt) (L. *coxa*, hip, + Gr. *pous, podos,* foot). The proximal joint of an insect or arachnid leg; in crustaceans, the proximal joint of the protopod.

cranium (krā′nē-um) (Gr. *cranium*, skull). A protective skeletal covering of the brain, usually of cartilage or bone.

creatine phosphate High-energy phosphate compound in the muscle of vertebrates and some invertebrates, used to regenerate stores of ATP.

Cretaceous extinction A mass extinction that occurred 65 million years ago in which 76% of existing species, including all dinosaurs, became extinct, marking the end of the Mesozoic era.

cretin (krēt′n) (Fr. *crétin*, [dialect], fr. L. *christianus*, Christian, to indicate idiots so afflicted were also human). A human with severe mental, somatic, and sexual disability resulting from hypothyroidism during early stages of development.

crista (kris′ta), pl. **cristae** (L. *crista*, crest). A crest or ridge on a body organ or organelle; a platelike projection formed by the inner membrane of a mitochondrion.

crop (kräp). A region of the esophagus specialized for storing food.

crossing over Exchange of parts of nonsister chromatids at synapsis in the first meiotic division.

crystalline style A unique rodlike structure composed of enzymes inside the stomach of a bivalve.

ctenidia (te-ni′dē-ə) (Gr. *kteis*, comb). Comblike structures, especially gills of molluscs; also applied to comb plates of Ctenophora.

ctenoid scales (ten′ȯ id) (Gr. *kteis, ktenos*, comb). Thin, overlapping dermal scales of some fishes; exposed posterior margins have fine, toothlike spines.

cupula (kū′pū-lə) (L. little tub). Small inverted cuplike structure housing another structure; gelatinous matrix covering hair cells in lateral-line and equilibrium organs.

cutaneous respiration Use of the integument by amphibians for gas exchange between the blood and air. It is the primary respiratory structure of terrestrial lungless forms, including plethodontid salamanders.

cuticle (kū′ti-kəl) (L. *cutis*, skin). A protective, noncellular, organic layer secreted by the external epithelium (hypodermis) of many invertebrates. In vertebrates the term refers to the epidermis or outer skin.

Cuvierian tubules (*Cuvier*, 19th-century French comparative vertebrate anatomist). Sticky, often toxic elongate internal organs of holothurians expelled to entangle potential predators; these can be regenerated.

cyanobacteria (sī-an-ō-bak-ter′ē-ə) (Gr. *kyanos*, a dark-blue substance, + *bakterion*, dim. of *baktron*, a staff). Photosynthetic prokaryotes, also called blue-green algae or cyanophytes.

cyclin A protein important in the control of the cell-division cycle and mitosis.

cycloid scales (sī′-kloid) (Gr. *kyklos*, circle). Thin, overlapping dermal scales of some fishes; posterior margins are smooth.

cydippid larva (sī-dip′pid) (Gr. *kydippe*, mythological Athenian maiden). Free-swimming larva of most ctenophores; superficially similar to the adult.

cynodonts (sīn′ə-dänts) (Gr. *kynodōn*, canine tooth). A group of mammal-like carnivorous synapsids of the Upper Permian and Triassic periods.

cyrtocyte (ser′tō-sīt) (Gr. *kyrtē*, a fish basket, cage, + *kytos*, hollow vessel). A protonephridial cell with a single flagellum enclosed in a cylinder of cytoplasmic rods.

cyst (sist) (Gr. *kystis*, a bladder, pouch). A resistant, quiescent stage of an organism, usually with a secreted wall.

cystacanth (sis′tə-kanth) (Gr. *kystis*, bladder, pouch, + *akantha*, thorn). Juvenile stage of an acanthocephalan that is infective to the definitive host.

cysticercoid (sis′tə-ser′kȯid) (Gr. *kystis*, bladder, + *kerkos*, tail, + *eidos*, form). A type of juvenile tapeworm composed of a solid-bodied **cyst** containing an invaginated scolex; contrasts with **cysticercus.**

cysticercus (sis′tə-ser′kəs) (Gr. *kystis*, bladder, + *kerkos*, tail). A type of juvenile tapeworm in which an invaginated and introverted scolex is contained in a fluid-filled bladder; contrasts with **cysticercoid.**

cystid (sis′tid) (Gr. *kystis*, bladder). In an ectoproct, the dead secreted outer parts plus the adherent underlying living layers.

cytochrome (sī′tə-krōm) (Gr. *kytos*, hollow vessel, + *chrōma*, color). Several iron-containing pigments that serve as electron carriers in aerobic respiration.

cytokine (sī′tə-kīn) (Gr. *kytos*, hollow vessel, + *kinein*, to move). A polypeptide hormone that mediates communication between cells participating in an immune response and is secreted as a neurotransmitter by the brain. Many different cytokines have been isolated, and these can affect cells that secrete them, nearby cells, or cells at distant locations. Includes **adipokines.**

cytokinesis (sī′tə-kin-ē′sis) (Gr. *kytos*, hollow vessel, + *kinesis*, movement). Division of the cytoplasm of a cell.

cytopharynx (Gr. *kytos*, hollow vessel, + *pharynx*, throat). Short tubular gullet in ciliated unicellular eukaryotes.

cytoplasm (sī′tə-plasm) (Gr. *kytos*, hollow vessel, + *plasma*, mold). The living matter of the cell, excluding the nucleus.

cytoplasmic specification During embryonic cleavage, process by which molecules within the cytoplasm of each cell specify cell fate; same as autonomous specification.

cytoproct (sī′tə-prokt) (Gr. *kytos*, hollow vessel, + *prōktos*, anus). Site on a unicellular eukaryote where undigestible matter is expelled.

cytopyge (sī′tə-pīj) (Gr. *kytos*, hollow vessel, + *pyge*, rump or buttocks). In some unicellular eukaryotes, localized site for expulsion of wastes.

cytosol (sī′tə-sol) (Gr. *kytos*, hollow vessel, + L. *sol*, from *solutus*, to loosen). Unstructured portion of the cytoplasm in which the organelles are suspended.

cytosome (sī′tə-sōm) (Gr. *kytos*, hollow vessel, + *sōma*, body). The cell body inside the plasma membrane.

cytostome (sī′tə-stōm) (Gr. *kytos*, hollow vessel, + *stoma*, mouth). The cell mouth in many unicellular eukaryotes.

cytotoxic T cells (Gr. *kytos*, hollow vessel, + toxin). A special T cell activated during cell-mediated immune responses that recognizes and destroys virus-infected cells.

D

dactylozooid (dak-til′ə-zō-id) (Gr. *dakos*, bite, sting, + *tylos*, knob, + *zōon*, animal). A polyp of a colonial hydroid specialized for defense or killing food.

Darwinism Theory of evolution emphasizing common descent of all living organisms, gradual change, multiplication of species, and natural selection.

data, sing. **datum** (Gr. *dateomai*, to divide, cut in pieces). The results in a scientific experiment, or descriptive observations, upon which a conclusion is based.

deciduous (də-sij′ə-wəs) (L. *decidere*, to fall off). Shed or falling off at end of a growing period.

decomposer (dē′-kəm-pō′zər). A **consumer** that breaks organic matter into soluble components available to plants at the base of the food web; most are bacteria or fungi.

deduction (L. *deductus*, led apart, split, separated). Reasoning from the general to the particular from given premises to their necessary conclusion.

defensins Antimicrobial peptides produced abundantly by cells in and around the linings of mammalian intestine, urogenital tract, and respiratory tract and by neutrophils.

definitive host The host in which sexual reproduction of a symbiont occurs; if no sexual reproduction, then the host in which the symbiont becomes mature and reproduces; contrasts with **intermediate host.**

delayed type hypersensitivity A pathological manifestation of the inflammatory reaction based primarily on cell-mediated immunity.

---

bat / āpe / ärmadillo / herring / fēmale / finch / līce / crocodile / crōw / cȯin / cōre / duck / düke / ūnicorn / ə indicates unaccented vowel sound "uh" as in mammal, fishes, cardinal, heron, vulture / stress as in bi-ol′o-gy, bi′o-log′i-cal

deletion Chromosomal material is excised and lost from a chromosome.

deme (dēm) (Gr. populace). A local population of closely related animals.

demographic transition A change in the characteristic age structure and rate of growth of a population; specifically, a change in human populations coincident with industrialization from having a higher proportion of young individuals and a net increase in population size to a more even distribution of old and young individuals and a more stable population size.

demography (də-mäg′rə-fē) (Gr. *demos*, people, + *graphy*, field of study). The properties of the rate of growth and the **age structure** of populations.

dendrite (den′drīt) (Gr. *dendron*, tree). Any nerve-cell process that conducts impulses toward the cell body.

dendritic cell A member of the mononuclear phagocyte system that performs phagocytosis and is an antigen-presenting cell of the skin; secretes cytokines. Also called Langerhans cell.

density-dependent Refers to biotic environmental factors, such as predators and parasites, whose effects on a population vary according to the number of organisms in the population.

density-independent Refers to abiotic environmental factors, such as fires, floods, and temperature changes, whose effects on a population are unaffected by the number of organisms in the population.

deoxyhemoglobin Form of **hemoglobin** produced when hydrogen ions combine with the heme group of hemoglobin in vertebrate red blood cells or invertebrate blood plasma.

deoxyribonucleic acid (DNA) The genetic material of all organisms, characteristically organized into linear sequences of genes.

deoxyribose (dē-äk-sē-rī′bōs) (L. *deoxy*, loss of oxygen, + *ribose*, a pentose sugar). A five-carbon sugar similar to ribose but lacking an oxygen atom at the 2′ carbon. It forms part of the fundamental structure of DNA nucleotides.

depolarization A change in the voltage in a positive direction recorded across a plasma membrane (see **membrane potential**). This allows a signal to be transmitted in excitable cells, such as nerve, muscle, and sensory cells.

deposit feeders Aquatic organisms that consume detritus and small organisms in submerged sediments.

derived character state Condition of a taxonomic character inferred by cladistic analysis to have arisen within a taxon being examined cladistically rather than having

been inherited from the most recent common ancestor of all members of the taxon.

dermal (Gr. *derma*, skin). Pertaining to the skin; cutaneous.

dermal ostia (Gr. *derma*, skin; L. *ostium*, door). Incurrent pores in a sponge.

dermis The inner, sensitive mesodermal layer of skin; corium.

desmosome (dez′mə-sōm) (Gr. *desmos*, bond, + *sōma*, body). Buttonlike plaque serving as an intercellular connection.

determinate cleavage The type of cleavage, usually spiral, in which the fate of the blastomeres is determined very early in development; mosaic cleavage.

detritus (də-trī′tus) (L. that which is rubbed or worn away). Any fine particulate debris of organic or inorganic origin.

Deuterostomia (dü′tə-rō-stō′mē-ə) (Gr. *deuteros*, second, secondary, + *stoma*, mouth). A group of phyla in which cleavage is indeterminate (regulative) and ancestrally radial. The endomesoderm is enterocoelous, and the mouth is derived away from the blastopore. Includes Echinodermata, Chordata, and Hemichordata; contrasts with **Protostomia.**

dextral (dex′trəl) (L. *dexter*, right-handed). Pertaining to the right; in gastropods, shell is dextral if opening is to right of columella when held with spire up and facing observer.

diapause (dī′ə-pawz) (Gr. *diapausis*, pause). A period of arrested development in the life cycle of insects and certain other animals in which physiological activity is very low and the animal is highly resistant to unfavorable external conditions.

diaphragm (dī′ə-fram) (Gr. *dia*, separate + *phragm*, partition). A sheetlike muscle that separates the thoracic and abdominal cavities of mammals. Contraction of this muscle draws air into the lungs.

diapsids (dī-ap′səds) (Gr. *di*, two, + *apsis*, arch). Amniotes in which the skull bears two pairs of temporal openings; includes nonavian reptiles (possibly excluding turtles) and birds as living representatives.

diastole (dī-as′tə-lē) (Gr. *diastolē*, dilation). Relaxation and expansion of the heart during which the chambers are filled with blood.

dictyosome (dik′tē-ə-sōm) (Gr. *diction*, to throw, + *sōma*, body). A part of the secretory system of endoplasmic reticulum in unicellular eukaryotes, also called Golgi bodies.

diffusion (L. *diffusus*, dispersion). The movement of particles or molecules from

area of high concentration of the particles or molecules to area of lower concentration.

digestion Reduction of food by mechanical and chemical means to simple, soluble molecules that can be absorbed and transported to body cells.

digitigrade (dij′ə-tə-grād) (L. *digitus*, finger, toe, + *gradus*, step, degree). Walking on the digits with the heel of the foot raised; contrasts with **plantigrade** in describing mammalian locomotion.

dihybrid (dī-hī′brəd) (Gr. *dis*, twice, + L. *hibrida*, mixed offspring). A hybrid whose parents differ in two distinct characters; an offspring having two different alleles at two different loci, for example, *A/a B/b.*

1,25-dihydroxyvitamin D$_3$ The hormonally active form of vitamin D, which increases uptake of calcium into the bloodstream from the digestive tract and from bone, and decreases loss of calcium from blood to urine.

dimorphism (dī-mōr′fizm) (Gr. *di*, two, + *morphē*, form). Existence within a species of two distinct forms according to color, sex, size, organ structure, or behavior. Occurrence of two kinds of zooids in a colonial organism.

dioecious (dī-ē′shəs) (Gr. *di*, two, + *oikos*, house). Having male and female gonads in separate individuals.

diphycercal (dif′i-ser′kəl) (Gr. *diphyēs*, twofold, + *kerkos*, tail). A tail that tapers to a point, as in lungfishes; vertebral column extends to tip without upturning.

diphyodont (di′fi-ə-dänt) (Gr. *diphyēs*, twofold, + *odous*, tooth). Having deciduous and permanent sets of teeth successively.

diploblastic (di′plə-blas′tək) (Gr. *diploos*, double, + *blastos*, bud). Organism with two germ layers, endoderm and ectoderm.

diploid (dip′lòid) (Gr. *diploos*, double, + *eidos*, form). Having the somatic (double, or 2*n*) number of chromosomes or twice the number characteristic of a gamete of a given species.

direct development A life cycle sequence from zygote to adult without intervening larval stages.

directional selection A selective process favoring an extreme value of a quantitative phenotype in a population, potentially causing a change in the mean value of the phenotype of the population.

disaccharidase An enzyme in the plasma membrane of cells lining the intestine that splits 12-carbon sugar molecules into 6-carbon sugar molecules.

disaccharides (dī-sak′ə-rīds) (Gr. *dis*, twice, + L. *saccharum*, sugar). A class of sugars (such as lactose, maltose, and sucrose) that yield two monosaccharides on hydrolysis.

discoidal cleavage In an early embryo, cell division occurs within a small disc of cytoplasm atop a large mass of yolk.

disjunct distribution Denotes the geographic range of a species or group of closely related species being separated into two or more geographically isolated areas.

dispersal (dis-pər'-səl). Movement of organisms from their place of birth to a new geographic area for permanent residence. Founder events are special, rare cases of dispersal in which dispersing individuals cross a geographic barrier unsuited to permanent residence and start a new population beyond the barrier.

disruptive selection A selective process in which a population's mean value of a quantitative phenotype is disfavored over extreme values, potentially causing a bimodal distribution of phenotypes to evolve.

distal (dis'təl). Farther from the center of the body than a reference point.

disulfide bond (dī-səl'-fīd bänd). A covalent bond between the sulfur atoms of two cysteine amino acids. Formation of such bonds between nonadjacent cysteines in a polypeptide stabilizes the tertiary structure of a protein; disulfide bonds between cysteines of different polypeptides contribute to the quaternary structure of a protein.

DNA See **deoxyribonucleic acid.**

DNA barcoding A technique for identifying organisms to species using sequence information of a standard gene present in all animals. The mitochondrial gene encoding cytochrome *c* oxidase I (*COI*) is often used.

DNA ligase (lī'gāse). An enzyme that joins the ends of two separate pieces of DNA.

domain (dō-mān'). An informal taxonomic rank above the Linnaean kingdom; Archaea, Bacteria, and Eucarya are ranked as domains.

dominance hierarchy A social ranking, formed through agonistic behavior, in which individuals are associated with one another so that some have greater access to resources than do others.

dominant An allele that is expressed regardless of the nature of the corresponding allele on the homologous chromosome.

dorsal (dōr'səl) (L. *dorsum,* back). Toward the back, or upper surface, of an animal.

double circulation A system in tetrapod vertebrates for distributing blood through two separate circuits. One circuit carries deoxygenated blood from the heart through the lungs for oxygenation and back to the heart as oxygenated blood; a separate circuit carries oxygenated blood to tissues where oxygen is released and then back to the heart in deoxygenated form. Double circulation features various mechanisms for separating the alternative circuits.

double helix The fundamental structure of a DNA molecule consisting of two paired strands held together by complementary pairing of bases and forming the three-dimensional structure of an alpha helix. The paired strands are antiparallel because the 3′ end of one strand is opposite the 5′ end of the other one.

Down syndrome A congenital syndrome including mental disability, caused by the cells in a person's body having an extra chromosome 21; also called trisomy 21.

dual-gland adhesive organs Organs in the epidermis of most turbellarians, with three cell types: viscid and releasing gland cells and anchor cells.

duodenum (dü-ə-dēn'əm) (L. *duodeni,* twelve each, for its length, about 12 fingers' width). The first and shortest portion of the small intestine lying between the pyloric end of the stomach and the jejunum.

duplication An extra copy of chromosomal material is produced and inserted within a chromosome.

dyad (dī'ad) (Gr. *dyas,* two). A product of the division of a tetrad during the first meiotic division.

E

eccrine (ek'rən) (Gr. *ek,* out of, + *krinein,* to separate). Applies to a type of mammalian sweat gland that produces a watery secretion.

ecdysiotropin (ek-dī'-sē-ə-trō'pən) (Gr. *ekdysis,* to strip off, escape, + *tropos,* a turn, change). Hormone secreted in brain of insects that stimulates prothoracic gland to secrete molting hormone. Prothoracicotropic hormone; brain hormone.

ecdysis (ek-dī'-sis) (Gr. *ekdysis,* to strip off, escape). Shedding of outer cuticular layer; molting, as in insects or crustaceans.

ecdysone (ek-dī'sōn) (Gr. *ekdysis,* to strip off). Molting hormone of arthropods, stimulates

growth and ecdysis, produced by prothoracic glands in insects and Y organs in crustaceans.

ecdysozoan protostome (ek'dī-sō-zō'ən prō'təstōm) (Gr. *ekdysis,* to strip off, escape, + *zōon,* animal. Gr. *protos,* first, + *stoma,* mouth). Any member of a clade within Protostomia whose members shed the cuticle as they grow; includes arthropods, nematodes, and several smaller phyla.

ecocline (ēk'ō-klīn) (Gr. *oikos,* home, + *klino,* to slope, recline). The gradient between adjacent biomes; a gradient of environmental conditions.

ecological character displacement Differences in morphology or behavior within a species caused by competition with another species; characteristics typical of one species differ according to whether the other species is present in or absent from a local community.

ecological community An assemblage of species that are associated in a common area and interact with one another in a self-sustaining and self-regulating relationship.

ecological pyramid A quantitative measurement of a **food web** in terms of amount of **biomass,** numbers of **organisms,** or energy at each of the different **trophic** levels present (**producers, herbivores,** first-level **carnivores,** higher-level carnivores).

ecology (Gr. *oikos,* house, + *logos,* discourse). Part of biology that concerns the relationship between organisms and their environment.

ecosystem (ēk'ō-sis-təm) (eco[logy] from Gr. *oikos,* house, + system). An ecological unit consisting of both the biotic communities and the nonliving (abiotic) environment, which interact to produce a stable system.

ecotone (ēk'ō-tōn) (eco[logy] from Gr. *oikos,* home, + *tonos,* stress). The transition zone between two adjacent ecological communities.

ectoderm (ek'tō-derm) (Gr. *ektos,* outside, + *derma,* skin). Outer layer of cells of an early embryo (gastrula stage); one of the germ layers, also sometimes used to include tissues derived from ectoderm.

ectognathous (ek'tə-nā'thəs) (Gr. *ektos,* outside, without, + *gnathos,* jaw). A derived character shared by most insects: mandibles and maxillae not in pouches.

ectolecithal (ek'tō-les'ə-thəl) (Gr. *ektos,* outside, + *lekithos,* yolk). Yolk for nutrition of the embryo contributed by cells that are

---

bat / āpe / ärmadillo / herring / fēmale / finch / līce / crocodile / crōw / cȯin / cōre / duck / düke / ūnicorn / ə indicates unaccented vowel sound "uh" as in mammal, fishes, cardinal, heron, vulture / stress as in bi-ol′o-gy, bi′o-log′i-cal separate from the egg cell and are combined with the zygote by envelopment within the eggshell.

ectomesoderm In a developing animal, a middle layer of cells derived from ectoderm, in contrast to the more typical derivation from endoderm.

ectoneural (ek'tə-nu-rəl) (Gr. *ektos*, outside, without, + *neuron*, nerve). Oral (chief) nervous system in echinoderms.

ectoparasite (ek'tō-par'ə-sīt). A **parasite** that resides on the outside surface of its host organism; contrasts with **endoparasite.**

ectoplasm (ek'tō-plazm) (Gr. *ektos*, outside, + *plasma*, form). The cortex of a cell or that part of cytoplasm just under the cell surface; contrasts with **endoplasm.**

ectothermic (ek'tō-therm'ic) (Gr. *ektos*, outside, + *thermē*, heat). Having a body temperature derived from heat acquired from the environment; contrasts with **endothermic.**

edema (ē-dē'mə) (Gr. *oidēma*, swelling). Buildup of fluid in interstitial spaces, causing swelling.

effect macroevolution Differential rates of speciation and/or extinction among evolving lineages attributed to interactions between their differing organismal-level emergent properties and the environments shared by the lineages; contrasts with **species selection.**

effector (L. *efficere*, bring to pass). An organ, tissue, or cell that becomes active in response to stimulation.

efferent (ef'ə-rənt) (L. *ex*, out, + *ferre*, to bear). Leading or conveying away from some organ, for example, action potentials conducted away from the brain, or blood conveyed away from an organ; contrasts with **afferent.**

egestion (ē-jes'chən) (L. *egestus*, to discharge). Act of casting out indigestible or waste matter from the body by any normal route.

electron A subatomic particle with a negative charge and a mass of 9.1066×10^{-28} gram.

elephantiasis (el-ə-fən-tī'ə-səs). Disfiguring condition caused by chronic infection with filarial worms *Wuchereria bancrofti* and *Brugia malayi.*

Eltonian pyramid An **ecological pyramid** showing numbers of **organisms** at each of the **trophic** levels.

embryogenesis (em'brē-ō-jen'ə-səs) (Gr. *embryon*, embryo, + *genesis*, origin). The origin and development of the embryo; embryogeny.

emergence (L. *e*, out, + *mergere*, to plunge). The appearance of properties in a biological system (at the molecular, cellular, organismal, or species levels) that cannot be deduced from knowledge of the component parts taken separately or in partial combinations; such properties are termed **emergent properties.**

emulsion (ē-məl'shən) (L. *emulsus*, milked out). A colloidal system in which both phases are liquids.

enamel (ē-nam'əl) (Anglo-French *enamailler*). The hard, outer layer of teeth and fish scales.

encystment (en-sist'-mənt). Process of cyst formation.

endemic (en-dem'ik) (Gr. *en*, in, + *demos*, populace). Peculiar to a certain region or country; native to a restricted area; not introduced.

endergonic (en-dər-gän'ik) (Gr. *endon*, within, + *ergon*, work). Used in reference to a chemical reaction that requires energy; energy absorbing.

endite (en'dīt) (Gr. *endon*, within). Medial process on an arthropod limb.

endochondral (en'dō-kän'drəl) (Gr. *endon*, within, + *chondros*, cartilage). Occurring with the substance of cartilage, especially bone formation.

endocrine (en'də-krən) (Gr. *endon*, within, + *krinein*, to separate). Refers to a gland that is without a duct and that releases its product directly into the blood or lymph.

endocytosis (en'dō-sī-tō'-səs) (Gr. *endon*, within, + *kytos*, hollow vessel). The engulfment of matter by phagocytosis, pinocytosis, receptor-mediated endocytosis, and bulk-phase (nonspecific) endocytosis.

endoderm (en'də-dərm) (Gr. *endon*, within, + *derma*, skin). Innermost germ layer of an embryo, forming the primitive gut; also may refer to tissues derived from endoderm.

endognathous (en'də-nā-thəs) (Gr. *endon*, within, + *gnathous*, jaw). An ancestral character in insects, found in orders Diplura, Collembola, and Protura, in which the mandibles and maxillae are located in pouches.

endolecithal (en'də-les'ə-thəl) (Gr. *endon*, within, + *lekithos*, yolk). Yolk for nutrition of the embryo incorporated into the egg cell itself.

endolymph (en'də-limf) (Gr. *endon*, within, + *lympha*, water). Fluid that fills most of the membranous labyrinth of the vertebrate ear.

endometrium (en'də-mē'trē-əm) (Gr. *endon*, within, + *mētra*, womb). The mucous membrane lining the uterus.

endoparasite (en'dō-par'ə-sīt). A **parasite** that resides inside the body of its host organism; contrasts with **ectoparasite.**

endoplasm (en'də-pla-zm) (Gr. *endon*, within, + *plasma*, mold or form). The portion of cytoplasm that immediately surrounds the nucleus; contrasts with **ectoplasm.**

endoplasmic reticulum (ER) A complex of membranes within a cell; may bear ribosomes (rough ER) or not (smooth ER).

endopod, endopodite (en'də-päd, en-dop'ə-dīt) (Gr. *endon*, within, + *pous, podos*, foot). Medial branch of a biramous crustacean appendage.

endopterygote (en'dəp-ter'i-gōt) (Gr. *endon*, within, + *pteron*, feather, wing). Insect in which the wing buds develop internally; has holometabolous metamorphosis.

endorphin (en-dōr'fin) (contraction of endogenous morphine). Group of opiate-like brain neuropeptides that modulate pain perception and are implicated in many other functions.

endoskeleton (Gr. *endon*, within, + *skeletos*, hard). A skeleton or supporting framework within the living tissues of an organism; contrasts with **exoskeleton.**

endostyle (en'də-stīl) (Gr. *endon*, within, + *stylos*, a pillar). Mucus-secreting, ciliated groove(s) in the floor of the pharynx of tunicates, cephalochordates, and larval jawless fishes useful for accumulating and moving food particles to the stomach.

endosymbiosis (en'dō-sim-bī-ō-səs) (Gr. *endon*, within, + *syn*, with, + *bios*, life). Association between organisms of different species whereby one lives inside another.

endothelium (en-də-thē'lē-əm) (Gr. *endon*, within, + *thēlē*, nipple). Squamous epithelium lining internal body cavities such as heart and blood vessels. Adj., **endothelial.**

endothermic (en'də-therm'ic) (Gr. *endon*, within, + *thermē*, heat). Having a body temperature determined by heat derived from an animal's own oxidative metabolism; contrasts with **ectothermic.**

energy budget An economic analysis of the energy used by an organism, partitioned into **gross productivity, net productivity,** and **respiration.**

enkephalin (en-kef'ə-lin) (Gr. *endon*, within, + *kephale*, head). Group of small brain neuropeptides with opiate-like qualities.

enterocoel (en'tər-ō-sēl') (Gr. *enteron*, gut, + *koilos*, hollow). A type of coelom formed by the outpouching of a mesodermal sac from the endoderm of the primitive gut.

enteroendocrine system Endocrine cells dispersed within the lining of the digestive tract. They secrete hormones that may act locally to regulate the digestive processes or be transported to the brain, where they act on neurons and glial cells.

enteron (en′tə-rän) (Gr. intestine). The digestive cavity.

entocodon (en′tə-kō′dän) (L. *entos,* within, + L. *codex,* tablet). One of three layers, with ectoderm and endoderm, of a developing medusa bud on a hydrozoan colony; the entocodon derives from ectoderm and produces smooth and striated muscles in the medusa.

entognathy (en′tə-nā-thē). Bases of mouthparts enclosed within the head capsule in insect orders Collembola, Diplura, and Protura.

entomology (en′tə-mol′ə-jē) (Gr. *entoma,* an insect, + *logos,* discourse). Study of insects.

entozoic (en-tə-zō′ic) (Gr. *entos,* within, + *zōon,* animal). Living within another animal; internally parasitic (chiefly parasitic worms).

entropy (en′trə-pē) (Gr. *en,* in, on, + *tropos,* turn, change in manner). The proportion of a system's energy that is not available for doing work.

enzyme (en′zīm) (Gr. *enzymos,* leavened, from *en,* in, + *zyme,* leaven). A substance, produced by living cells, that is capable of accelerating specific chemical transformations, such as hydrolysis, oxidation, or reduction, but is unaltered itself in the process; a biological catalyst.

ephyra (ef′ə-rə) (Gr. *Ephyra,* Greek city). Refers to castlelike appearance. Medusa bud from a scyphozoan polyp.

epidermis (ep′ə-dər′məs) (Gr. *epi,* on, upon, + *derma,* skin). The outer, nonvascular layer of skin of ectodermal origin; in invertebrates, a single layer of ectodermal epithelium.

epididymis (ep′ə-did′ə-məs) (Gr. *epi,* on, upon, + *didymos,* testicle). Part of the sperm duct that is coiled and lying near the testis. Plural: epididymides (ep′ə-did′ə-mə-dēz′).

epigenesis (ep′ə-jen′ə-sis) (Gr. *epi,* on, upon, + *genesis,* birth). The embryological (and generally accepted) view that an embryo is a new creation that develops and differentiates step by step from a zygote undergoing cell division; the progressive production of new parts that were nonexistent as such in the original zygote.

epigenetics (ep′ə-je-net′iks) (Gr. *epi,* on, upon, + *genesis,* birth). Study of the relationship between genotype and phenotype as mediated by developmental processes.

epipelagic (ep′-ə-pə-laj′-ik). Refers to the upper zone of the marine pelagic realm, which receives greater illumination than lower zones.

epipod, epipodite (ep′ē-päd, e-pi-pō′-dīt) (Gr. *epi,* on, upon, + *pous, podos,* foot). A lateral process on the protopod of a crustacean appendage, often modified as a gill.

epistasis (e-pis′tā-səs) (Gr. *epi,* on, upon, + *stasis,* standing). Prevention of expression of an allele at one locus by an allele at another locus.

epistome (ep′i-stōm) (Gr. *epi,* on, upon, + *stoma,* mouth). Flap over the mouth in some lophophorates bearing the protocoel.

epithelium (ep′ə-thē′lē-əm) (Gr. *epi,* on, upon, + *thēlē,* nipple). A cellular tissue covering a free surface or lining a tube or cavity.

epitoke (ep′ə-tōk) (Gr. *epitokos,* fruitful). Posterior part of a marine polychaete when swollen with developing gonads during the breeding season; contrasts with **atoke.**

epitope (ep′ə-tōp) That portion of an antigen to which an antibody or T-cell receptor binds. Also called **antigenic determinant.**

erythroblastosis fetalis (ə-rith′rə-blas-tō′səs fə-tal′əs) (Gr. *erythros,* red, + *blastos,* germ, + *osis,* a disease; L. *fetalis,* relating to a fetus). A disease of newborn infants caused when Rh-negative mothers develop antibodies against the Rh-positive blood of the fetus. See **blood type.**

erythrocyte (ə-rith′rə-sīt) (Gr. *erythros,* red, + *kytos,* hollow vessel). Red blood cell; has hemoglobin to carry oxygen from lungs or gills to tissues. During formation in mammals, erythrocytes lose their nuclei; those of other vertebrates retain the nuclei.

esthete (es-thēt′) (Gr. *esthēs,* a garment). Light sensory receptor on a shell of a chiton (phylum Mollusca).

estivation (es′tə-vā′shen) (L. *aestivates,* to spend the summer). A state of dormancy during the summer when temperatures are high, food is scarce, and/or dehydration threatens. Metabolism and breathing rate decrease.

estrous cycle Periodic episodes of estrus, or "heat," when females of most mammalian species become sexually receptive.

estrus (es′trəs) (L. *oestrus,* gadfly, frenzy). The period of heat of a female mammal associated with ovulation. Time of maximum sexual receptivity. Adj., **estrous.**

estuary (es′chə-we′rē) (L. *aestuarium,* estuary). An arm of the sea where the tide meets the current of a freshwater drainage.

ethology (e-thäl′-ə-jē) (Gr. *ethos,* character, + *logos,* discourse). The study of animal behavior in natural environments.

euchromatin (ū′krō′-mə-tən) (Gr. *eu,* good, well, + *chrōma,* color). Part of the chromatin that absorbs less stain than heterochromatin; contains active genes.

eukaryote, eukaryotic, eucaryotic (ū′ka′-rē-ōt′, ū′ka′-rē-ot′ik) (Gr. *eu,* good, true, + *karyon,* nut, kernel). Organisms whose cells characteristically contain a membrane-bound nucleus or nuclei; contrasts with **prokaryotic.**

eumetazoan (ū′-met′-ə-zō′ən) (Gr. *eu,* good, true, + *meta,* after, + *zōon,* animal). Any multicellular animal with distinct germ layers that form true tissues; animals at or beyond the cellular grade of organization.

euploidy (ū′ploid′ē) (Gr. *eu,* good, well, + *ploid,* multiple of). Presence of one or more complete sets of chromosomes and no partial sets in a cell nucleus; includes **haploidy, diploidy,** and **polyploidy.**

euryhaline (ū′-rə-hā′līn) (Gr. *eurys,* broad, + *hals,* salt). Able to tolerate wide ranges of saltwater concentrations.

euryphagous (ū-rif′ə-gəs) (Gr. *eurys,* broad, + *phagein,* to eat). Eating a large variety of foods.

eurytopic (ū-rē-tōp′ik) (Gr. *eurys,* broad, + *topos,* place). Refers to an organism with a wide environmental range.

eusociality (ū′sō-shē-al′-ət-ē). A reproductive division of labor among members of a population or species. Generations overlap, and nonreproductive individuals help to raise younger relatives that are not the helpers' offspring. Ants, many bees, and some wasps are eusocial.

eutely (ū′tē′-lē) (Gr. *euteia,* thrift). Condition of a body composed of a constant number of cells or nuclei in all adult members of a species, as in rotifers, acanthocephalans, and nematodes.

evagination (ē-vaj′ə-nā′shən) (L. *e,* out, + *vagina,* sheath). An outpocketing from a hollow structure.

evolution (L. *evolvere,* to unfold). Organic evolution encompasses all changes in the characteristics and diversity of life on earth throughout its history.

evolutionary duration The length of time that a species or higher taxon exists in geological time.

evolutionary sciences Empirical investigation of ultimate causes in biology using the comparative method.

evolutionary species concept Defines species as a single lineage of ancestral-descendant populations that maintains its identity from other such lineages and has its own evolutionary tendencies and historical fate; differs from the biological species concept by explicitly including a time dimension and including asexual lineages.

evolutionary stable strategy (ESS) Application of the mathematical theory of games to evaluate whether a system of social behaviors is robust to the evolution of "cheating" behaviors that would threaten its stability; denotes a social system that should persist over long periods of evolutionary time because it prevails in competition with others that might arise.

evolutionary taxonomy A system of classification, formalized by George Gaylord Simpson, that groups species into Linnaean higher taxa representing a hierarchy of distinct **adaptive zones;** such taxa may be **monophyletic** or **paraphyletic** but not **polyphyletic.**

evolvability Refers to the opportunities for morphological evolution conferred on an evolving lineage that possesses a "toolkit" of semiautonomous developmental modules that can be expressed at various stages of ontogeny and at multiple physical locations in the body. For example, evolution of tetrapod limbs featured ectopic expression of developmental genetic modules normally expressed in the vertebral column.

exaptation (ek-sap′-tā′shən). Evolutionary cooption of an organismal or molecular character for a biological role unrelated to the character's evolutionary origin. Bird feathers are considered an exaptation for flight because they originated prior to avian flight but provided utility for flight following its origin; contrasts with **adaptation;** bird feathers are considered an adaptation for the biological role of thermoregulation.

excision repair Means by which cells repair certain kinds of damage (dimerized pyrimidines) in their DNA.

exergonic (ek′sər-gon′ik) (Gr. *exō,* outside of, + *ergon,* work). An energy-yielding reaction.

exite (ex′īt) (Gr. *exō,* outside). Process from lateral side of an arthropod limb.

exocrine (ek′sə-krən) (Gr. *exō,* outside, + *krinein,* to separate). A type of gland that releases its secretion through a duct; contrasts with **endocrine.**

exocytosis (eks′ə-sī-tō′səs) (Gr. *exo,* outside, + *kytos,* hollow vessel). Transport of a substance from inside a cell to the outside.

exon (ex′on) (Gr. *exō,* outside). Part of the mRNA as transcribed from the DNA that contains a portion of the information necessary for final gene product.

exopod, exopodite (ex′ə-päd, ex-äp′ə-dīt) (Gr. *exō,* outside, + *pous, podos,* foot). Lateral branch of a biramous crustacean appendage.

exopterygote (ek′sop-ter′i-gōt) (Gr. *exō,* without, + *pteron,* feather, wing). Insect in which the wing buds develop externally during nymphal instars; has hemimetabolous metamorphosis.

exoskeleton (ek′sō-skel′ə-tən) (Gr. *exō,* outside, + *skeletos,* hard). A supporting structure secreted by ectoderm or epidermis; external, not enveloped by living tissue; contrasts with **endoskeleton.**

experiment (L. *experiri,* to try). A trial made to support or to disprove a hypothesis.

experimental method A general procedure for testing hypotheses by predicting how a biological system will respond to a treatment, making the treatment under controlled conditions, and then comparing the observed results with the predicted ones.

experimental sciences Empirical investigation of proximate causes in biology using the **experimental method.**

exponential growth Refers to an increase in the numbers of individuals in a population by at least a factor of 2 each generation.

exteroceptor (ek′stər-ō-sep′tər) (L. *exter,* outward, + *capere,* to take). A sense organ excited by stimuli from the external world.

extrinsic factor An environmental variable that influences the biological properties of a population, such as observed number of individuals or rate of growth.

extrusome (eks′trə-sōm) (L. *extrusus,* driven out, + *soma,* body). Any membrane-bound organelle used to extrude something from a cell.

eyelid (ī′-lid). A thin surface of skin and muscle that can be closed to protect an eye from light, abrasion, and/or desiccation. Occurs in many but not all terrestrial vertebrates.

F

facilitated diffusion Mediated transport in which a transmembrane protein makes possible diffusion of a molecule across a plasma membrane in the direction of a concentration gradient; contrasts with **active transport.**

facilitated transport See **facilitated diffusion.**

FAD Abbreviation for flavin adenine dinucleotide, an electron acceptor in the respiratory chain.

fascicle (fas′ə-kəl) (L. *fasciculus,* small bundle). A small bundle, usually referring to a collection of muscle fibers or nerve axons.

fatty acid Any of a series of saturated organic acids having the general formula $C_nH_{2n}O_2$,

or unsaturated molecules in which the number of hydrogen atoms is less than 2n; occurs in natural fats of animals and plants.

female-defense polygyny A male obtains more than one mate because multiple females aggregate and can be guarded as a unit from other males.

fermentation (L. *fermentum,* ferment). Enzymatic catalysis, without oxygen, of organic substrates, especially carbohydrates, yielding simple alcohols, acids, and carbon dioxide and converting ADP to ATP.

fertilization cone A volcano-like structure forming on the surface of some animal eggs into which the fertilizing sperm will be drawn. The egg surface responds to chemicals released by the incoming sperm.

fiber (L. *fibra,* thread). A thin cell or strand of protoplasmic material produced or secreted by a cell and lying outside the cell.

fibril (L. *fibra,* thread). A strand of protoplasm produced by a cell and lying within the cell.

fibrillar (fī′brə-lər) (L. *fibrilla,* small fiber). Composed of or pertaining to fibrils or fibers.

fibrin Protein that forms a meshwork, trapping erythrocytes, to become a blood clot. Precursor is fibrinogen.

fibrosis (fī-brō′səs). Deposition of fibrous connective tissue in a localized site, during process of tissue repair or to isolate a source of antigen.

filipodium (fī′li-pō′de-əm) (L. *filum,* thread, + Gr. *pous, podos,* a foot). A type of pseudopodium that is very slender and may branch but does not rejoin to form a mesh.

filter-feeding Any feeding process by which particulate food is filtered from water in which it is suspended.

first law of thermodynamics Energy is neither created nor destroyed but may be converted from one form to another.

fission (L. *fissio,* a splitting). Asexual reproduction by a division of the body into two or more parts.

fitness Degree of suitability for a particular environment. Genetic fitness is relative contribution by organisms of a particular genotype to the next generation; organisms with high genetic fitness are naturally selected and their genetic characteristics become prevalent in a population.

5′ end (five-prime end). The terminus of a nucleic acid molecule that consists of phosphate attached to the 5′ carbon of the terminal sugar (the opposite terminus is the 3′ end).

flagellum (flə-jel′em), pl. **flagella** (L. a whip). Whiplike organelle of locomotion.

flame cell Specialized hollow excretory or osmoregulatory structure of one or several small cells containing a tuft of flagella (the "flame") and situated at the end of a minute tubule; connected tubules ultimately open to the outside. See **solenocyte, protonephridium.**

fluke (O.E. *flōc,* flatfish). A member of class Trematoda or class Monogenea. Also, certain of the flatfishes (order Pleuronectiformes).

food chain Movement of energy from plant compounds to organisms that eat plants, then to other organisms that eat the plant feeders, and possibly further through a linear series of organisms that feed and are then eaten by others. Food chains connect and branch to form food webs.

food vacuole A digestive organelle in the cell.

food web An analysis relating species in an ecological community according to how they acquire nutrition, such as fixing atmospheric carbon **(producers),** consuming producers **(herbivores),** consuming herbivores (first-level **carnivores),** or consuming carnivores (higher level carnivores).

foraminiferan (for'əm-i-nif'-ər-ən) (L. *foramin,* hole, perforation, + *fero,* to bear). Amebas with netlike pseudopodia and bearing a test with many openings.

fossil (fos'əl). Any remains or impression of an organism from a past geological age that has been preserved by natural processes, usually by mineralization in the earth's crust.

fossorial (fä-sōr'ē-əl) (L. *fossor,* digger). Characterized by digging or burrowing.

fouling Contamination of feeding or respiratory areas of an organism by excrement, sediment, or other matter. Also, accumulation of sessile marine organisms on the hull of a boat or ship so as to impede its progress through the water.

founder event Establishment of a new population by a small number of individuals (sometimes a single female carrying fertile eggs) that disperse from their parental population to a new location geographically isolated from the parental population.

fovea (fō'vē-ə) (L. small pit). A small pit or depression; especially the fovea centralis, a small pit containing only cones in the retina of some vertebrates, a point of acute vision.

free energy The energy available for doing work in a chemical system.

free-living Any organism not intimately associated with a host organism.

fringing reef A type of coral reef close to a landmass having a small lagoon or no lagoon between it and the shore.

frog (fräg'). Denotes any member of the amphibian order Anura (also called Salientia).

frontal plane A plane parallel to the main axis of the body and at right angles to the sagittal plane.

fundamental niche A variety of roles potentially performed by an organism or population in an ecological community; limits on such roles are set by the intrinsic biological attributes of an organism or population. See also **niche** and **realized niche.**

funnel The tube from which a jet of water exits the mantle cavity of a cephalopod mollusc.

furcula (fur'cū-la) (L. *furc,* fork). Fused clavicles of birds and some dinosaurs.

fusiform (fū'zə-form) (L. *fusus,* spindle, + *forma,* shape). Spindle-shaped; tapering toward each end.

G

gamete (ga'mēt, gə-mēt') (Gr. *gamos,* marriage). A mature haploid sex cell; usually, male and female gametes can be distinguished. An egg or a sperm.

gametic meiosis Meiosis that occurs during formation of the gametes, as in humans and other metazoa.

gametocyte (gə-mēt'ə-sīt) (Gr. *gametēs,* spouse, + *kytos,* hollow vessel). The mother cell of a gamete; an immature gamete.

ganglion (gang'lē-ən), pl. **ganglia** (Gr. little tumor). An aggregation of nerve tissue containing nerve-cell bodies located outside the central nervous system.

ganoid scales (ga'nȯid) (Gr. *ganos,* brightness). Thick, bony, rhombic scales of some bony fishes; not overlapping.

gap genes Genes expressed in broad regions along the anterior-posterior axis of a developing embryo (e.g., produce head, thorax, and abdomen in *Drosophila);* mutations cause gaps in formation of segments.

gap junction Pores formed by a ring of transmembrane proteins connecting the cytoplasm between two cells.

gastrocoel (gas'trə-sēl) (Gr. *gastēr,* stomach, + *koilos,* hollow). An embryonic cavity forming in gastrulation that becomes the adult gut; also called an **archenteron.**

gastrodermis (gas'trə-dər'mis) (Gr. *gastēr,* stomach, + *derma,* skin). Lining of the digestive cavity of cnidarians.

gastrolith (gas'trə-lith) (Gr. *gastēr,* stomach, + *lithos,* stone). Calcareous body in the wall of the cardiac stomach of crayfish and other Malacostraca, preceding the molt.

gastrovascular cavity (Gr. *gastēr,* stomach, + L. *vasculum,* small vessel). Body cavity in certain lower invertebrates that functions in both digestion and circulation and has a single opening serving as both mouth and anus.

gastrozooid (gas'trə-zō-id) (Gr. *gastēr,* stomach, + *zōon,* animal). The feeding polyp of a hydroid; a hydranth.

gastrula (gas'trə-lə) (Gr. *gastēr,* stomach, + L. *ula,* dim.). Embryonic stage, usually cap- or sac-shaped, with walls of two layers of cells surrounding a cavity (archenteron) with one opening (blastopore).

gastrulation (gas'trə-lā'shən) (Gr. *gastēr,* stomach). Process by which an early animal embryo becomes a gastrula, acquiring first two and then three layers of cells.

gated channels Pores in the plasma membrane created by transmembrane proteins that open or close in response to a specific signal, such as binding of a chemical **(chemically-gated),** a change in membrane potential **(voltage-gated),** or a membrane distortion **(mechanically gated).** When open, these channels allow ions and water to move through the plasma membrane by diffusion.

gel (jel) (from gelatin, from L. *gelare,* to freeze). That state of a colloidal system in which the solid particles form the continuous phase and the fluid medium forms the discontinuous phase.

gemmule (je'mūl) (L. *gemma,* bud, + *ula,* dim.). Asexual, cystlike reproductive unit in freshwater sponges; formed in summer or autumn and capable of overwintering.

gene (Gr. *genos,* descent). A nucleic acid sequence (usually DNA) that encodes a functional polypeptide or RNA sequence.

gene pool A collection of all of the alleles of all of the genes in a population.

general area cladogram A cladogram depicting the sequence of historical fragmentation among various geographic areas of endemism; represents the shared branching patterns of the individual cladograms of many taxa that share common areas of endemism.

---

bat / āpe / ärmadillo / herring / fēmale / finch / līce / crocodile / crōw / cȯin / cōre / duck / dūke / ūnicorn / ə indicates unaccented vowel sound "uh" as in mammal, fishes, cardinal, heron, vulture / stress as in bi-ol'o-gy, bi'o-log'i-cal

general lineage concept of species Claim by Kevin de Queiroz that all contemporary concepts of species share the primary definition that a species is a segment of a population-level lineage. The alternative concepts (**biological species concept, cohesion species concept, evolutionary species concept, phylogenetic species concept**, and others not covered in this text) differ in secondary properties used to guide taxonomic recognition of species (reproductive community, genetic and demographic exchangeability, unique evolutionary role and tendencies, diagnosable lineage with no branching).

genetic code The correspondence between the sequence of bases in a DNA or messenger RNA molecule and the sequence of amino acids in an encoded protein.

genetic drift Random change in allelic frequencies in a population. In small populations, genetic variation at a locus may be lost by chance fixation of a single allelic variant.

genome (jē′nōm) (Gr. *genos,* offspring, + *ōma,* abstract group). All the DNA in a haploid set of chromosomes (nuclear genome), organelle (mitochondrial genome, chloroplast genome) or virus (viral genome, which in some viruses consists of RNA rather than DNA).

genomics (jē-nō′miks). Mapping and sequencing of genomes (= structural genomics). Functional genomics is the development and application of experimental approaches to assess gene function. Functional genomics uses information derived from structural genomics.

genotype (jēn′ō-tīp) (Gr. *genos,* offspring, + *typos,* form). The genetic constitution, expressed and latent, of an organism; the total set of genes present in the cells of an organism; contrasts with **phenotype.**

genus (jē-nus), pl. **genera** (L. race). A group of related species with taxonomic rank between family and species.

geographic range The specific geographic area occupied by members of a population, species, or higher taxon.

germ cell A haploid cell (ovum or sperm) whose fertilization by one of the opposite kind produces a diploid zygote; also called a **gamete.**

germinal vesicle Mature nucleus of a primary oocyte, enlarged and filled with RNA.

germinative zone The site immediately following the scolex on the body of a mature tapeworm where new proglottids are produced.

germ layer In an animal embryo, one of three basic layers (ectoderm, endoderm, mesoderm) from which the various organs and tissues arise in the multicellular animal.

germovitellarium (jer′mō-vit-ə-lar′ē-əm) (L. *germen,* a bud, offshoot, + *vitellus,* yolk). Closely associated ovary (germarium) and yolk-producing structure (vitellarium) in rotifers.

germ plasm Cell lineages giving rise to the germ cells of a multicellular organism, as distinct from the somatoplasm.

gestation (jes-tā′shən) (L. *gestare,* to bear). The period in which offspring are carried in the uterus.

gizzard (giz′ərd) (L. *gigeria,* bird entrails). The heavy muscled stomach of birds and other animals, used for grinding food.

globulins (glo′bū-lənz) (L. *globus,* a globe, ball, + *-ulus,* ending denoting tendency). A large group of compact proteins with high molecular weight; includes immunoglobulins (antibodies).

glochidium (glō-kid′ē-əm) (Gr. *glochis,* point, + *idion,* dim.). Bivalved larval stage of freshwater mussels.

glomerulus (glä-mer′ū-ləs) (L. *glomus,* ball). A tuft of capillaries projecting into a renal corpuscle in a kidney. Also, a small spongy mass of tissue in the proboscis of hemichordates, presumed to have an excretory function. Also, a concentration of nerve fibers situated in the olfactory bulb.

glucocorticoid A corticosteroid hormone produced and secreted from the cortex of the adrenal gland. Two such hormones are cortisol and corticosterone, which have roles in metabolism, inflammation, and stress.

gluconeogenesis (glü-cō-nē-ō-gən′ə-səs) (Gr. *glykys,* sweet, + *neos,* new, + *genesis,* origin). Synthesis of glucose from protein or lipid precursors.

glucose A 6-carbon sugar particularly important for cellular metabolism in living organisms (= dextrose).

glycogen (glī′kə-jən) (Gr. *glykys,* sweet, + *genēs,* produced). A polysaccharide constituting the principal form in which carbohydrate is stored in animals; animal starch.

glycolysis (glī-kol′ə-səs) (Gr. *glykys,* sweet, + *lysis,* a loosening). Enzymatic breakdown of glucose (especially) or glycogen into phosphate derivatives with release of energy.

gnathobase (nath′ō-bās′) (Gr. *gnathos,* jaw, + base). A median basic process on certain appendages in some arthropods, usually for biting or crushing food.

gnathostomes (nath′ō-stōms) (Gr. *gnathos,* jaw, + *stoma,* mouth). Vertebrates with jaws.

Golgi complex (gōl′jē) (after Golgi, Italian histologist). A part of the cellular endomembrane system that serves as a collecting and packaging center for secreted polypeptides and proteins. Also called **Golgi apparatus.**

gonad (gō′nad) (N.L. *gonas,* primary sex organ). An organ that produces gametes (ovary in the female and testis in the male).

gonadotropin A peptide hormone secreted by the anterior pituitary gland that acts on the gonads (ovaries of females and testes of males). There are two gonadotropins: follicle-stimulating hormone (FSH) and luteinizing hormone (LH). FSH promotes egg production in the female and sperm production in the male. LH induces ovulation, corpus luteum production, and sex steroid secretion in females, and sex steroid secretion in males.

gonangium (gō-nan′jē-əm) (N.L. *gonas,* primary sex organ, + *angeion,* dim. of vessel). Reproductive zooid of hydroid colony (Cnidaria).

gonoduct (gän′ə-dukt) (Gr. *gonos,* seed, progeny, + duct). Duct leading from a gonad to the exterior.

gonophore (gän′ə-for) (Gr. *gonos,* seed, progeny, + *phoros,* bearer). Sexual reproductive structure developing from reduced medusae in some hydrozoans; it may be retained on the colony or released.

gonopore (gän′ə-por) (Gr. *gonos,* seed, progeny, + *poros,* an opening). A genital pore in many invertebrates.

grade (L. *gradus,* step). A level of organismal complexity or adaptive zone characteristic of members of a taxonomic group.

gradualism (graj′ə-wal-iz′əm). A component of Darwin's evolutionary theory postulating that evolution occurs by the temporal accumulation of small, incremental changes in populations, usually across very long periods of geological time; it opposes claims that evolution can occur by large, discontinuous or macromutational changes.

granular glands Integumentary structures of modern amphibians associated with secretion of defensive compounds.

granulocytes (gran′ū-lə-sīts) (L. *granulus,* small grain, + Gr. *kytos,* hollow vessel). White blood cells (neutrophils, eosinophils, and basophils) bearing in their cytoplasm "granules" (vesicles) that stain histochemically.

green gland Excretory gland of certain Crustacea; the antennal gland.

gregarine (gre-ga-rin′) (L. *gregarious,* belonging to a herd or flock). Unicellular eukaryotic parasites belonging to class Gregarinea within phylum Apicomplexa; these organisms infect guts or body cavities of invertebrates.

gregarious (L. *grex,* herd). Living in groups or flocks.

gross productivity A measurement of the total energy assimilated by an organism.

ground substance The matrix in which connective tissue fibers are embedded.

group selection A hypothesis that selection sometimes acts on an association of individuals in a population rather than directly on the individuals themselves; proposed to explain evolution of individual behaviors that do not benefit an individual directly but may benefit an association that includes the individual performing the behavior. Critical analyses have largely discredited group-selectional hypotheses in favor of alternatives, such as kin selection and reciprocal altruism.

growth rate The proportion by which a population changes in numbers of individuals at a given time by reproduction and possibly immigration.

guanine (gwä′nēn) (Sp. from Quechua, *huanu,* dung). A white crystalline purine base, $C_5H_5N_5O$, occurring in various animal tissues and in guano and other animal excrements.

guild (gild) (M.E. *gilde,* payment, tribute). Species of a local community that partition resources through character displacement to avoid niche overlap and competition, such as warbler communities whose component species forage at different characteristic locations within a conifer tree.

gynandromorph (ji-nan′drə-mōrf) (Gr. *gyn,* female, + *andr,* male, + *morphē,* form). An abnormal individual exhibiting characteristics of both sexes in different parts of the body; for example, the left side of a bilateral organism may show female characteristics, while the right side shows male characteristics.

gynecophoric canal (gī′nə-kə-fōr′ik) (Gr. *gynē,* woman, + *pherein,* to carry). Groove in male schistosomes (certain trematodes) that carries the female.

H

habitat (L. *habitare,* to dwell). The place where an organism normally lives or where individuals of a population live.

habitat fragmentation Emergence of geographic barriers that disrupt a formerly continuous geographic distribution of populations of a species. Evolutionary rates of species formation and species extinction are increased by this occurrence.

habituation A kind of learning in which continued exposure to the same stimulus produces diminishing responses; contrasts with **sensitization.**

hair cell An important sensory component of several kinds of mechano- and auditory receptors found in both invertebrate (statocyst) and vertebrate (vestibular organ, organ of Corti) organs of equilibrium and hearing. The "hairs" are stereocilia, or sensory endings that project from the cell surface, and when bent by mechanical stimuli they generate nerve impulses or action potentials, communicating a signal to the central nervous system.

haltere (hal′tēr), pl. **halteres** (hal′tērz) (Gr. leap). In Diptera, small club-shaped structure on each side of the metathorax representing the hindwings; thought to be sense organs for balancing; also called balancer.

haplodiploidy (Gr. *haploos,* single, + *diploos,* double, + *eidos,* form). Reproduction in which haploid males are produced parthenogenetically, and diploid females are from fertilized eggs.

haploid (Gr. *haploos,* single). The reduced, or *n,* number of chromosomes, typical of gametes, as opposed to the diploid, or *2n,* number found in somatic cells. In certain groups, mature organisms may have a haploid number of chromosomes.

Hardy-Weinberg equilibrium Mathematical demonstration that the Mendelian hereditary process does not change the populational frequencies of alleles or genotypes across generations, and that change in allelic or genotypic frequencies requires factors such as natural selection, genetic drift in finite populations, recurring mutation, migration of individuals among populations, and nonrandom mating.

hectocotylus (hek-tə-kät′ə-ləs) (Gr. *hekaton,* hundred, + *kotylē,* cup). Specialized, and sometimes autonomous, arm that serves as a male copulatory organ in cephalopods.

heliozoan (hēl′ē-ō-zō′ən) (Gr. *hēlios,* sun, + *zōon,* animal). Descriptive term for a freshwater ameba, naked or testate.

hemal system (hē′məl) (Gr. *haima,* blood). System of small vessels in echinoderms; function is probably distribution of nutrients to specific body regions.

hemerythrin (hē′mə-rith′rin) (Gr. *haima,* blood, + *erythros,* red). A red, iron-containing respiratory pigment in the blood of some polychaetes, sipunculids, priapulids, and brachiopods.

hemidesmosome (he-mē-dez′mə-sōm) (Gk. *hēmi,* half, + *desmos,* bond, + *sōma,* body). Buttonlike plaque composed of transmembrane proteins that anchor a cell to underlying connective-tissue layers.

hemimetabolous (he′mē-mə-ta′bə-ləs) (Gr. *hēmi,* half, + *metabolē,* change). Refers to gradual metamorphosis during development of insects, without a pupal stage.

hemizygous (he′mē-zī′gəs) (Gr. *hēmi,* half, + *zygōtos,* yoked). For animals having chromosomal sex determination in which one sex (termed the heterogametic sex) has only one copy of a particular sex chromosome, genotypes of heterogametic individuals consist of a single copy of all genes located on that sex chromosome.

hemocoel (hēm′ə-sēl) (Gr. *haima,* blood, + *koiloma,* cavity). Major body space in arthropods formed by fusion of embryonic coelom with blastocoel; contains blood (hemolymph).

hemoglobin (Gr. *haima,* blood, + L. *globulus,* globule). An iron-containing respiratory pigment occurring in vertebrate red blood cells and in blood plasma of many invertebrates; a compound of an iron porphyrin heme and globin proteins.

hemolymph (hē′mə-limf) (Gr. *haima,* blood, + L. *lympha,* water). Fluid in the coelom or hemocoel of some invertebrates that functions similarly to the blood and lymph of vertebrates.

hemozoin (hē-mə-zō′ən) (Gr. *haima,* blood, + *zōon,* an animal). Insoluble digestion product of malarial parasites produced from hemoglobin.

hepatic (hə-pat′ik) (Gr. *hēpatikos,* of the liver). Pertaining to the liver.

herbivore ([h]ərb′ə-vōr′) (L. *herba,* green crop, + *vorare,* to devour). Any organism subsisting on plants. Adj., **herbivorous.**

herbivory ([h]ər-biv′-ə-rē). Denotes the condition of animals feeding on plants or the destruction of plant biomass by such feeding.

heredity (L. *heres,* heir). The faithful transmission of biological traits from parents to their offspring.

---

hermaphrodite (hə[r]-maf′rə-dīt) (Gr. *hermaphroditos,* containing both sexes; from Greek mythology, Hermaphroditos, son of Hermes and Aphrodite). An organism with both male and female functional reproductive organs. **Hermaphroditism** usually refers to an aberrant individual in a dioecious species; **monoecy** implies that this is the normal condition for the species.

hermatypic (hər-mə-ti′pik) (Gr. *herma,* reef, + *typos,* pattern). Relating to reef-forming corals.

heterocercal (het′ə-rō-sər′kəl) (Gr. *heteros,* different, + *kerkos,* tail). In some fishes, a tail with the upper lobe larger than the lower, and the end of the vertebral column somewhat upturned in the upper lobe, as in sharks.

heterochromatin (het′ə-rō-krōm′ə-tən) (Gr. *heteros,* different, + *chrōma,* color). Chromatin that stains intensely and appears to represent genetically inactive areas.

heterochrony (het′ə-rō-krō′-nē) (Gr. *heteros,* different, + *chronos,* time). Evolutionary change in the relative time of appearance or rate of development of characteristics from ancestor to descendant.

heterodont (het′ə-rō-dänt) (Gr. *heteros,* different, + *odous,* tooth). Having teeth differentiated into incisors, canines, and molars for different purposes.

heterokont (het′ə-rō-känt) (Gr. *heteros,* other, different, + *kont,* pole). Refers to flagellated cells with two different anterior flagella: one long, hairy, and forward directed, and the other short, smooth, and trailing posteriorly.

heterolobosea (het′ə-rō-lō-bō′sē-ə) (Gr. *heteros,* other, different, + *lobos,* lobe). A unicellular eukaryotic clade in which most members can assume both ameboid and flagellate forms.

heterostracans (het′ə-rō-strā′-kənz) (Gr. *heteros,* different, + *ostrakon,* shell). A group of extinct fishes with dermal armor and no jaws or paired fins; known from the Ordovician to Devonian periods.

heterotopy (het′ə-rō-tō′-pē). An evolutionary change in the physical location of a structure or developmental process in the organismal body plan.

heterotroph (het′ə-rō-trōf) (Gr. *heteros,* different, + *trophos,* feeder). An organism that must obtain both organic and inorganic raw materials from the environment in order to live; includes most animals, fungi, and those plants and microorganisms that do not have photosynthesis.

heterozygote (het′ə-rō-zī′gōt) (Gr. *heteros,* different, + *zygōtos,* yoked). An organism in which homologous chromosomes contain different allelic forms (often dominant and recessive) of a locus; derived from a zygote formed by union of gametes of dissimilar allelic constitution.

hexamerous (hek-sam′ər-əs) (Gr. *hex,* six, + *meros,* part). Six parts, specifically, symmetry based on six or multiples thereof.

hibernation (L. *hibernus,* wintry). Condition, especially of mammals, of passing the winter in a torpid state in which the body temperature drops nearly to freezing and the metabolism drops close to zero.

hierarchical system A scheme arranging organisms into a series of taxa of increasing inclusiveness, as illustrated by Linnaean classification.

high heat of vaporization Denotes the large amount of energy needed to convert liquid water at 100° C to gaseous phase (more than 500 kilocalories per gram).

high specific heat capacity Denotes the large amount of energy needed to raise the temperature of liquid water (1° C per gram).

high surface tension Denotes the large pressure needed to break the surface of liquid water relative to all other liquids except mercury.

histogenesis (his-tō-jen′ə-sis) (Gr. *histos,* tissue, + *genesis,* descent). Formation and development of tissue.

histology (hi-stäl′-ə-jē) (Gr. *histos,* web, tissue, + *logos,* discourse). The study of the microscopic anatomy of tissues.

histone (hi′stōn) (Gr. *histos,* tissue). Any of several simple proteins found complexed with DNA in cell nuclei. Histones are involved in regulation of gene expression and DNA packaging. Histones yield a high proportion of basic amino acids on hydrolysis; characteristic of eukaryotes.

holoblastic cleavage (Gr. *holo,* whole, + *blastos,* germ). Complete and approximately equal division of cells in an early embryo; occurs in mammals, amphioxus, and many aquatic invertebrates that have eggs with a small amount of yolk.

holometabolous (hō′lō-mə-ta′bə-ləs) (Gr. *holo,* complete, + *metabolē,* change). Complete metamorphosis during development.

holophytic nutrition (hōl′ō-fit′ik) (Gr. *holo,* whole, + *phyt,* plant). Occurs in green plants and certain protozoa and involves synthesis of carbohydrates from carbon dioxide and water in the presence of light, chlorophyll, and certain enzymes.

holozoic nutrition (hōl′ō-zō′ik) (Gr. *holo,* whole, + *zoikos,* of animals). Type of nutrition involving ingestion of liquid or solid organic food particles.

homeobox (hō′mē-ō-box) (Gr. *homoios,* like, resembling, + L. *buxus,* boxtree [used in the sense of enclosed, contained]). A highly conserved 180-base-pair sequence found in homeotic genes, which encode polypeptides that function as transcription factors to activate expression of other genes at critical times in development. The homeobox encodes the DNA-binding domain of a transcription factor.

homeostasis (hō′mē-ō-stā′sis) (Gr. *homeo,* alike, + *stasis,* state or standing). Maintenance of an internal steady state by means of self-regulation.

homeothermic (hō′mē-ō-thər′mik) (Gr. *homeo,* alike, + *thermē,* heat). Having a nearly uniform body temperature, regulated independent of the environmental temperature.

homeotic genes (hō-mē-ät′ik) (Gr. *homoios,* like, resembling). Genes, identified through mutations, that give developmental identity to specific body segments. These genes encode transcription factors required to activate gene expression at critical times during development.

home range The area over which an animal moves in its activities. Unlike territories, home ranges are not defended.

hominid (häm′ə-nid) (L. *homo, hominis,* man). A member of the family Hominidae, which includes chimpanzees, gorillas, humans, orangutans, and extinct forms descended from their most recent common ancestor.

hominoid (häm′ə-nȯid). Relating to the Hominoidea, a superfamily of primates to which apes and humans are assigned.

homocercal (hō′mə-ser′kəl) (Gr. *homos,* same, common, + *kerkos,* tail). A tail with the upper and lower lobes symmetrical and the vertebral column ending near the middle of the base, as in most teleost fishes.

homodont (hō′mō-dänt) (Gr. *homos,* same, + *odous,* tooth). Having all teeth similar in form.

homograft See **allograft.**

homolog, homologue (hōm′ə-log) One member of a set of homologous structures or one of a pair of homologous chromosomes.

homology (hō-mäl′ə-jē) (Gr. *homologos,* agreeing). Equivalence of parts or organs of different organisms caused by evolutionary derivation from a corresponding part or organ in a remote ancestor, and usually having a similar embryonic origin. May also refer to molecular (DNA, RNA, protein) sequences or chromosomes whose equivalence represents descent from a common ancestral molecule or chromosome, respectively. Serial homology is the correspondence in the same individual of repeated structures having the same origin and development, such as the appendages of arthropods. Adj., **homologous.**

homoplasy (hō'mō'plā'sē). Phenotypic similarity among characteristics of different species or populations (including molecular, morphological, behavioral, or other features) that does not accurately represent patterns of common evolutionary descent (= nonhomologous similarity); it is produced by evolutionary parallelism, convergence and/or reversal, and is revealed by incongruence among different characters on a cladogram or phylogenetic tree.

homozygote (hō-mō-zī'gōt) (Gr. *homos,* same, + *zygotos,* yoked). An organism having identical alleles at one or more genetic loci. Adj., **homozygous.**

humoral (hū'mər-əl) (L. *humor,* a fluid). Pertaining to an endocrine secretion.

humoral immune response Acquired immune response involving production of antibodies, specifically the T_H2 arm of the immune response. Contrasts with **cell-mediated immune response.**

hyaline (hī'ə-lən) (Gr. *hyalos,* glass). Adj., glassy, translucent. Noun, a clear, glassy, structureless material occurring, for example, in cartilage, vitreous body, mucin, and glycogen.

hybrid Refers to offspring of crosses between genetically distinct populations, such as those recognized as different varieties or species.

hybridization Natural or artificial genetic crossing between genetically distinct populations, such as ones considered different varieties or species.

hybridoma (hī-brid-ō'mə) (contraction of hybrid + myeloma). Fused product of a normal and a myeloma (cancer) cell, which has some of the characteristics of the normal cell.

hydatid cyst (hī-da'təd) (Gr. *hydatis,* watery vesicle). A type of cyst formed by juveniles of certain tapeworms (*Echinococcus*) in their vertebrate hosts.

hydranth (hī'dranth) (Gr. *hydōr,* water, + *anthos,* flower). Nutritive zooid of hydroid colony.

hydrocaulus (hī'drə-kä'ləs) (Gr. *hydōr,* water, + *kaulos,* stem of a plant). Stalks or "stems"

of a hydroid colony, the parts between the hydrorhiza and the hydranths.

hydrocoel (hī'-drə-sēl) (Gr. *hydōr,* water, + *koilos,* hollow). Second or middle coelomic compartment in echinoderms; left hydrocoel gives rise to water-vascular system.

hydrocorals Members of phylum Cnidaria, class Hydrozoa, with massive calcareous skeletons.

hydrogen bond A relatively weak chemical bond resulting from unequal charge distribution within molecules, in which a hydrogen atom covalently bonded to another atom is attracted to the electronegative portion of another molecule.

hydrogenosome (hī-drō-gen'ə-sōm) (Gr. *hydōr,* water, + *genos,* kind, + *sōma,* body). An anaerobic cellular organelle presumed to be derived from a mitochondrion.

hydroid The polyp form of a cnidarian as distinguished from the medusa form. Any cnidarian of the class Hydrozoa, order Hydroidea.

hydrolysis (Gr. *hydōr,* water, + *lysis,* a loosening). The decomposition of a chemical compound by the addition of water; the splitting of a molecule into its groupings so that the split products acquire hydrogen and hydroxyl groups.

hydrorhiza (hī'drə-rī'zə) (Gr. *hydōr,* water, + *rhiza,* a root). Rootlike stolon that attaches a hydroid to its substrate.

hydrosphere (Gr. *hydōr,* water, + *sphaira,* ball, sphere). Aqueous envelope of the earth.

hydrostatic pressure The pressure exerted by a fluid (gas or liquid), defined as force per unit area. For example, the hydrostatic pressure of one atmosphere (1 atm) is 14.7 lb/in^2.

hydrostatic skeleton A mass of fluid or plastic parenchyma enclosed within a muscular wall to provide the support necessary for antagonistic muscle action; for example, parenchyma in acoelomates and perivisceral fluids in pseudocoelomates serve as hydrostatic skeletons.

hydrothermal vent A submarine hot spring; seawater seeping through the seafloor is heated by magma and expelled back into the sea through the hydrothermal vent.

hydroxyl (hydrogen + oxygen, + yl). Containing an OH^- group, a negatively charged ion formed by alkalies in water.

hyomandibular (hī-ō-mən-dib'yə-lər) (Gr. *hyoeides* [shaped like the Gr. letter upsilon γ, + *eidos,* form], + L. *mandere,* to chew). Skeleton derived from the hyoid gill arch,

forming part of articulation of the lower jaw of fishes, and forming the stapes of the ear of amniotic vertebrates.

hyperosmotic (Gr. *hyper,* over, + *ōsmos,* impulse). Refers to a solution whose osmotic pressure is greater than that of another solution to which it is compared; contains a greater concentration of dissolved particles and gains water through a selectively permeable membrane from a solution containing fewer particles; contrasts with **hypoosmotic.**

hyperparasitism (hī'pər-par'ə-sit-iz'-əm) (Gr. *hyper,* over, + *para,* beside, + *sitos,* food). Parasitism of a parasite by another parasite.

hyperpolarization A change in the voltage in a negative direction recorded across a plasma membrane (see **membrane potential**). This allows a signal to be transmitted in excitable cells, such as nerve, muscle, and sensory cells.

hypertonic Refers to a solution that contains a greater concentration of dissolved particles than that of another solution with which it is compared; contrasts with **hypotonic.**

hypertrophy (hī-pər'trə-fē) (Gr. *hyper,* over, + *trophē,* nourishment). Abnormal increase in size of a part or organ.

hypodermis (hī'pə-dər'mis) (Gr. *hypo,* under, + L. *dermis,* skin). The cellular layer lying beneath and secreting the cuticle of annelids, arthropods, and certain other invertebrates.

hypoosmotic (Gr. *hypo,* under, + *ōsmos,* impulse). Refers to a solution whose osmotic pressure is less than that of another solution with which it is compared or which is taken as a standard; contains a lesser concentration of dissolved particles and loses water during osmosis; contrasts with **hyperosmotic.**

hypophysis (hī-pof'ə-sis) (Gr. *hypo,* under, + *physis,* growth). Pituitary body.

hypostome (hī'pə-stōm) (Gr. *hypo,* under, + *stoma,* mouth). Name applied to a structure in various invertebrates (such as mites and ticks), located at the posterior or ventral area of the mouth.

hypothalamus (hī-pō-thal'ə-mis) (Gr. *hypo,* under, + *thalamos,* inner chamber). A ventral part of the forebrain beneath the thalamus; one of the centers of the autonomic nervous system and neuroendocrine regulation.

hypothesis (Gr. *hypothesis,* foundation, supposition). A statement or proposition

---

bat / āpe / ärmadillo / herring / fēmale / finch / līce / crocodile / crōw / cóin / cōre / duck / dūke / ūnicorn / ə indicates unaccented vowel sound "uh" as in mammal, fishes, cardinal, heron, vulture / stress as in bi-ol'o-gy, bi'o-log'i-cal that can be tested by observation or experiment.

hypothetico-deductive method (Gr. *hypotithenai,* to suppose, + L. *deducere,* to lead). Scientific process of making a conjecture and then seeking empirical tests that potentially lead to its rejection.

hypotonic Refers to a solution that contains a lesser concentration of dissolved particles than that of another solution with which it is compared; contrasts with **hypertonic.**

I

ichthyosaur (ik′thē-ō-sōr) (Gr. *ichthyo,* fish, + *saur,* lizard). Aquatic, Mesozoic reptiles characterized by a porpoiselike body, but with a vertical tail and large eyes.

imago (i-mä′gō). An adult and sexually mature insect.

immediate cause See **proximate cause.**

immediate hypersensitivity Over-responsive inflammatory response in a sensitized human based primarily on humoral immunity.

immunity Ability of tissues in an organism to recognize and to defend against nonself invaders. **Innate immunity** is a mechanism of defense that does not depend on prior exposure to the invader; **acquired immunity** is specific to a nonself material, requires time for development, and occurs more quickly and vigorously on secondary response.

immunoglobulin (im′yū-nō-glä′byū-lən) (L. *immunis,* free, + *globus,* globe). Any of a group of plasma proteins, produced by B cells and plasma cells, that participates in the immune response by combining with the antigen that stimulated its production. Antibody.

imprinting (im′prin-ting) (L. *imprimere,* to impress, imprint). Rapid and usually stable learning pattern appearing early in the life of a member of a social species and involving recognition of its own species; may involve attraction to the first moving object seen.

inbreeding The tendency among members of a population to mate preferentially with close relatives.

inclusive fitness A modification of the genetic concept of fitness to consider not only the average numbers of offspring produced by organisms of a particular genotype but also the consequences of those organisms for the reproductive outputs of their close relatives. Because close relatives have genes that are identical by descent, an organism that forgoes reproduction to enhance the reproductive success of numerous close relatives nonetheless contributes copies of its genes to future generations; such an organism could have high inclusive fitness despite a fitness of zero.

incomplete dominance See **intermediate inheritance.**

incus (in′kəs) (L. *incus,* anvil). The middle of a chain of three bones of the mammalian middle ear; homologous with the quadrate bone of early vertebrates.

indeterminate cleavage A type of embryonic development in which the fate of the blastomeres is not determined very early as to tissues or organs, for example, in echinoderms and vertebrates; regulative cleavage.

indigenous (in-dij′ə-nəs) (L. *indigena,* native). Pertains to organisms that are native to a particular region; not introduced.

indirect development A life-cycle sequence from zygote to adult with intervening larval stages.

induction (L. *inducere, inductum,* to lead). Reasoning from the particular to the general; deriving a general statement (hypothesis) based on individual observations. In embryology, a developmental response resulting from interaction with neighboring cells.

inductor (in-duk′ter) (L. *inducere,* to introduce, lead in). In embryology, a tissue or organ that causes the differentiation of another tissue or organ.

inflammation (in′fləm-mā′shən) (L. *inflammare,* from *flamma,* flame). A part of the innate immune response at a site of antigen invasion or tissue injury. A cascade of events causes phagocytic immune cell activation and damage repair in the affected region. The area becomes swollen and red, and can be painful during this process.

infraciliature (in-frə-sil′ē-ə-tūr) (L. *infra,* below, + *cilia,* eyelashes). The organelles just below the cilia in ciliated unicellular eukaryotes.

infundibulum (in′fun-dib′ū-ləm) (L. funnel). Stalk of the neurohypophysis linking the pituitary to the diencephalon.

ingression (in-gre′shən) (L. *ingressus,* to go into, enter). Migration of individual cells from the embryo surface to the interior of the embryo during development.

inheritance of acquired characteristics The discredited Lamarckian notion that organisms, by striving to meet the demands of their environments, obtain new adaptations and pass them by heredity to their offspring.

innate (i-nāt′) (L. *innatus,* inborn). A characteristic based partly or wholly on genetic or epigenetic constitution.

instar (inz′tär) (L. form). Stage in the life of an insect or other arthropod between molts.

instinct (L. *instinctus,* impelled). Stereotyped, predictable, genetically programmed behavior. Learning may or may not be involved.

insulin-like growth factor (IGF) A polypeptide hormone secreted by the liver, and also called somatomedin. This hormone is stimulated by growth hormone and causes similar effects, such as body growth and enhanced cellular metabolism.

integument (in-teg′ū-mənt) (L. *integumentum,* covering). An external covering or enveloping layer.

interbreeding population The most inclusive grouping of organisms for which sexual reproduction and associated genetic recombination occur freely across generations; implies an absence of biological barriers to fertilization of gametes between females and males included in the grouping. A criterion of the biological species concept.

intercellular (in-tər-sel′yə-lər) (L. *inter,* among, + *cellula,* chamber). Occurring between body cells.

interferons (in-tər-fēr′-änz). Several cytokines encoded by different genes, important in mediation of innate immunity and inflammation.

interleukin (in-tər-lü′-kin). One of a series of cytokines produced primarily by cells of the immune system, such as macrophages, mast cells, T and B cells, and also by endothelial cells and fibroblasts. Target cells are various leukocytes and other cells primarily involved with enhancing an immune response. The name "interleukins" derives from a mistaken early notion that they were produced only by leukocytes and that their target cells were limited to leukocytes.

interleukin-1 A cytokine produced by macrophages that stimulates T cells, B cells, and macrophages.

interleukin-2 A cytokine produced by T helper lymphocytes that leads to proliferation of B and T cells and enhanced activity of natural killer cells.

intermediary meiosis Meiosis that occurs neither during gamete formation nor immediately after zygote formation, producing both haploid and diploid generations, such as in foraminiferans.

intermediate filaments Components of the cytoskeleton in all cells that are important in resisting cell stretching. They also help to hold adjacent cells together in association with **desmosomes.**

intermediate host A host in which some development of a symbiont occurs, but in which maturation and sexual reproduction do not occur.

intermediate inheritance The alternative alleles of a gene do not have a simple dominance relationship, so that the heterozygote shows a condition intermediate between or otherwise different from homozygotes for each allele.

internal nostrils Palatal structures connecting the nasal cavity and throat in lungfishes and tetrapod vertebrates; used for olfaction and/or breathing when the mouth is closed.

interstitial (in-tər-sti′shəl) (L. *inter,* among, + *sistere,* to stand). Situated in the interstices or spaces between structures such as cells, organs, or grains of sand.

intertidal zone A portion of an ocean's continental shelf that gets exposed to air during low tides and submerged at high tides.

intracellular (in-trə-sel′yə-lər) (L. *intra,* inside, + *cellula,* chamber). Occurring within a body cell or within body cells.

intrinsic growth rate Exponential growth rate of a population; the difference between the density independent components of the birth and death rates of a natural population with stable age distribution.

intrinsic rate of increase See **intrinsic growth rate.**

intron (in′trän) (L. *intra,* within). Portion of mRNA as transcribed from DNA that will not form part of mature mRNA, and therefore does not encode an amino-acid sequence in the protein product.

introvert (L. *intro,* inward, + *vertere,* to turn). The anterior narrow portion that can be withdrawn (introverted) into the trunk of a sipunculid worm.

invagination (in-vaj′ə-nā′shən) (L. *in,* in, + *vagina,* sheath). An infolding of a layer of tissue to form a saclike structure.

inversion (L. *invertere,* to turn upside down). A turning inward or inside out, as in embryogenesis of sponges; also, reversal in order of genes or reversal of a chromosome segment.

ion An atom or group of atoms with a net positive or negative electrical charge because of the loss or gain of electrons.

ionic bond A chemical bond formed by transfer of one or more electrons from one atom to another; characteristic of salts.

iridophore (i-rid′ə-fōr) (Gr. *iris,* rainbow, or iris of eye). Iridescent or silvery chromatophores containing crystals or plates of guanine or other purine.

irritability (L. *irritare,* to provoke). A general property of all organisms involving the ability to respond to stimuli or changes in the environment.

isogametes (ī′sō-gam′ēts) (Gr. *isos,* equal, + *gametēs,* spouse). Gametes of a species in which gametes of both sexes are alike in size and appearance.

isolecithal (ī′sō-les′ə-thəl) (Gr. *isos,* equal, + *lekithos,* yolk, + *al*). Pertaining to a zygote (or ovum) with yolk evenly distributed. Homolecithal.

isoosmotic Refers to a liquid having the same osmotic pressure as another, reference liquid.

isotonic (Gr. *isos,* equal, + *tonikos,* tension). Pertaining to solutions having the same or equal osmotic pressure; isoosmotic.

isotope (Gr. *isos,* equal, + *topos,* place). One of several different forms (species) of a chemical element, differing from each other in atomic mass but not in atomic number.

iteroparity (i′tər-ō-pa′ri-tē′). A life history in which individual organisms of a population normally reproduce more than one time before dying; contrasts with **semelparity.**

J

Jacobson's organ (*Jacobson,* nineteenth-century Danish surgeon and anatomist). Also called the vomeronasal organ. A chemosensory organ in the roof of the mouth of many terrestrial vertebrates; the tongue transfers pheromones and food derived chemicals to this organ.

juvenile hormone Hormone produced by the corpora allata of insects; its effects include maintenance of larval or nymphal characteristics during development.

juxtaglomerular apparatus (jək′stə-glä-mer′yə-lər) (L. *juxta,* close to, + *glomus,* ball). Complex of three kinds of sensory cells formed from specialization of the afferent arteriole (granular cells), the glomerulus (mesangial cells), and the distal tubule (macula densa cells) and located close to the glomerulus, where these regions of the nephron are in close apposition. The granular cells of this apparatus produce the enzyme renin.

K

kelp forest Marine ecosystem characterized by high density of seaweeds of order Laminariales of the class Phaeophyceae (brown algae).

kentrogon (ken′trə-gən) (Gr. *kentron,* a point, spine, + *gonos,* progeny, generation). A larva of the cirripede order Rhizocephala (subphylum Crustacea) that functions to inject the parasite cells into the host hemocoel.

keratin (ker′ə-tən) (Gr. *kera,* horn, + *in,* suffix of proteins). A scleroprotein found in epidermal tissues and modified into hard structures such as horns, hair, nails, and reptilian scales.

keystone species A species (typically a predator) whose removal leads to reduced species diversity within the ecological community.

kinesis (kə-nē′səs) (Gr. *kinēsis,* movement). Movements by an organism in response to stimulus.

kinetochore (ki-ne′tə-kor) (Gr. *kinein,* to move, + *choris,* asunder, apart). A disc of proteins located on the centromere, specialized to interact with the spindle fibers during mitosis.

kinetodesma (ki-nē′tə-dez′mə), pl. **kinetodesmata** (Gr. *kinein,* to move, + *desma,* bond). Fibril arising from the kinetosome of a cilium in a ciliate and passing along the kinetosomes of cilia in that same row.

kinetoplast (ki-nē′tə-plast) (Gr. *kinətos,* moving, + *plastos,* molded, formed). Cellular organelle that functions in association with a kinetosome at the base of a flagellum; presumed to be derived from a mitochondrion.

kinetosome (ki-ne′tə-sōm) (Gr. *kinētos,* moving, + *sōma,* body). The self-duplicating structure at the base of the flagellum or cilium; similar to centriole, also called basal body or blepharoplast.

kinety (kī-nē′tē) (Gr. *kinein,* to move). All the kinetosomes and kinetodesmata of a row of cilia.

kinin (kī′nin) (Gr. *kinein,* to move, + *in,* suffix of hormones). A type of local hormone that acts near its site of origin; also called **parahormone** or **tissue hormone.**

kin selection An extension of the genetical theory of natural selection to explain altruistic behaviors that preferentially benefit close relatives; an individual's genetic contribution to future generations is enhanced by promoting survival of relatives because their shared genes are identical by descent.

Kupffer cells Phagocytic cells and antigen-presenting cells in the liver; part of the mononuclear phagocyte system.

kwashiorkor (kwäsh-ē-or′kər) (from Ghana). Malnutrition caused by diet high in carbohydrate and extremely low in protein.

L

labium (lā′bē-əm) (L. a lip). The lower lip of the insect formed by fusion of the second pair of maxillae. Also, part of human female external genitalia.

labrum (lā′brəm) (L. a lip). The upper lip of insects and crustaceans situated above or in front of the mandibles; also refers to the outer lip of a gastropod shell.

labyrinth (L. *labyrinthus,* labyrinth). Vertebrate internal ear, composed of a series of fluid-filled sacs and tubules (membranous labyrinth) suspended within bone cavities (osseous labyrinth).

labyrinthodont (lab′ə-rin′thə-dänt) (Gr. *labyrinthos,* labyrinth, + *odous, odontos,* tooth). A group of Paleozoic tetrapods containing the temnospondyls and the anthracosaurs.

lachrymal glands (lak′rə-məl) (L. *lacrimia,* tear). Structures in terrestrial vertebrates that secrete tears to lubricate the eyes.

lactase An enzyme in the plasma membrane of cells lining the intestine that splits lactose (milk sugar) into glucose and galactose.

lacteal (lak′tē-əl) (L. *lacteus,* of milk). Noun, one of the lymph vessels in the villus of the intestine. Adj., relating to milk.

lacuna (lə-kū′nə), pl. **lacunae** (L. pit, cavity). A sinus; a space between cells; a cavity in cartilage or bone.

lacunar system A netlike set of circulatory canals filled with fluid in an acanthocephalan.

lagena (lə-jē′nə) (L. large flask). Portion of the primitive ear in which sound is translated into nerve impulses; evolutionary beginning of cochlea.

Lamarckism Hypothesis, as expounded by Jean Baptiste de Lamarck, of evolution by the acquisition during an organism's lifetime of characteristics that are transmitted to offspring.

lamella (lə-mel′ə) (L. dim. of *lamina,* plate). One of the two plates forming a gill in a bivalve mollusc. One of the thin layers of bone laid concentrically around an osteon (Haversian canal). Any thin, platelike structure.

lappets Lobes around the margin of scyphozoan medusae (phylum Cnidaria).

larva (lar′və), pl. **larvae** (L. a ghost). An immature stage that is quite different from the adult.

larynx (lar′inks) (Gr. the larynx, gullet). Modified upper portion of respiratory tract of air-breathing vertebrates, bounded by the glottis above and the trachea below; voice box. Adj., **laryngeal** (lə-rin′j(ē)əl), relating to the larynx.

lateral (L. *latus,* the side, flank). Of or pertaining to the side of an animal; a *bilateral* animal has two sides.

lateral-line system Sensory organ, consisting of neuromast organs in canals and grooves on the head and sides of the body of fishes and some amphibians, that detects water vibrations.

laterite (lat′ə-rīt) (L. *later,* brick). Group of hard, red soils from tropical areas that show intense weathering and leaching of bases and silica, leaving aluminum hydroxides and iron oxides. Adj., **lateritic.**

law of independent assortment Also called **Mendel's second law.** Segregation of allelic forms of a gene into gametes occurs at random with respect to the segregation of alleles of a second gene located on a different pair of homologous chromosomes. Genes located very far apart on a single chromosome also sometimes show independent assortment because crossovers occur between them at a high rate.

law of segregation Mendel's first law of inheritance, in which the paired particulate factors influencing variation of a trait separate from each other in gamete formation so that each gamete carries only one of the factors.

lecithotrophy (le′si-thə-trō′fē) (Gr. *lekithos,* yolk of egg, + *trophos,* one who feeds). Nutrition of an embryo directly from the yolk of an ovum.

leishmaniasis (līsh′mə-nī′ə-səs) (Fr. Sir W. B. Leishman, 1926, British medical officer). Disease caused by infection with protozoans of the genus *Leishmania.*

lek (lek) (Sw. play, game). An area where animals assemble for communal courtship display and mating.

lemniscus (lem-nis′kəs) (L. ribbon). One of a pair of internal projections of the epidermis from the neck region of Acanthocephala; functions in fluid control in the protrusion and invagination of the proboscis.

lentic (len′tik) (L. *lentus,* slow). Of or relating to standing water, such as a swamp, pond, or lake.

lepidosaurs (lep′ə-dō-sorz) (L. *lepidos,* scale, + *sauros,* lizard). A lineage of diapsid reptiles that appeared in the Permian period and that includes the modern snakes, lizards, amphisbaenids, and tuataras, and the extinct ichthyosaurs.

lepospondyls (lep′ə-spänd′lz) (Gr. *lepos,* scale, + *spondylos,* vertebra). A group of Paleozoic tetrapods distinguished by the possession of spool-shaped vertebral centra.

leptocephalus (lep′tə-sef′ə-ləs), pl. **leptocephali** (Gr. *leptos,* thin, + *kephalē,* head). Transparent, ribbonlike migratory larva of eels and related teleosts.

leuconoid (lü′kə-nóid) (Gr. *leukon,* white, + *eidos,* like). A type of canal system in sponges where choanocytes reside in chambers.

leukemism (lü′kə-mi′-zəm) (Gr. *leukos,* white, + *ismos,* condition of). Presence of white pelage or plumage in animals with normally pigmented eyes and skin.

leukocyte (lü′kə-sīt) (Gr. *leukos,* white, + *kytos,* hollow vessel). Any of several kinds of white blood cells (for example, granulocytes, lymphocytes, monocytes), so called because, unlike red blood cells, they bear no hemoglobin.

library In molecular biology, a set of clones containing recombinant DNA. Obtained from and representing the genome of an organism.

ligament (lig′ə-mənt) (L. *ligamentum,* bandage). A tough, dense band of connective tissue connecting one bone to another.

ligand (lī′gənd) (L. *ligo,* to bind). A molecule that specifically binds to a receptor; for example, a hormone (ligand) binds specifically to its receptor on the cell surface.

ligation (lī-gā′shən) (L. *ligo,* to bind). End-to-end joining of two pieces of DNA.

limax form (lī′məx) (L. *limax,* slug). Form of pseudopodial movement occurring after ameboid cells of a cellular slime mold unite to make a sluglike body; no discrete pseudopodia are extended.

limiting resource A particular source of nutrition, energy, or living space whose scarcity is causally associated with a population having fewer individuals than otherwise expected in a particular environment.

lineage (lin′-ē-əj, lin′-əj). An unbranched series of ancestral-descendant populations extending through evolutionary time. Lineages related to others through branching of ancestral lineages form a phylogenetic tree. In molecular evolution, a series of ancestral-descendant DNA molecules traced through an organismal genealogy or phylogeny.

linkage, linked Genes carried on the same chromosome whose allelic variants tend to sort together in gamete formation (violates Mendel's second law of independent assortment).

lipase (lī′pās) (Gr. *lipos,* fat, + *ase,* enzyme suffix). An enzyme that accelerates the hydrolysis or synthesis of fats.

lipid, lipoid (li′pid) (Gr. *lipos,* fat). Certain fatlike substances such as triglycerides composed of fatty acids and glycerol, often containing other groups such as phosphoric acid. Lipids combine with proteins and carbohydrates to form principal structural components of cells such as the plasma membrane.

Lissamphibia (lis′-am-fi′-bē-ə). A tetrapod clade comprising modern amphibians (caecilians, frogs, and salamanders) and all descendants of their most recent common ancestor.

lithosphere (lith′ə-sfēr) (Gr. *lithos,* rock, + *sphaira,* ball). The rocky component of the earth's surface layers.

littoral (lit′ə-rəl) (L. *litoralis,* seashore). Adj., pertaining to the shore. Noun, that portion of the seafloor between the extent of high and low tides, intertidal; in lakes, the shallow part from the shore to the lakeward limit of aquatic plants.

lobopodium (lō′bə-pō′dē-əm) (Gr. *lobos,* lobe, + *pous, podos,* foot). Blunt, lobelike pseudopodium.

lobosea (lō-bō′sē-ə) (Gr. *lobos,* lobe). A clade comprising amebas with lobopodia.

locus (lō′kəs), pl. **loci** (lō′sī) (L. place). Position of a gene in a chromosome.

logistic equation A mathematical expression describing an idealized sigmoid curve of population growth.

lophocyte (lō′fə-sīt) (Gr. *lophos,* crest, + *kytos,* hollow vessel). Type of sponge amebocyte that secretes bundles of fibrils.

lophophore (lōf′ə-fōr) (Gr. *lophos,* crest, + *phoros,* bearing). Tentacle-bearing ridge or arm within which is an extension of the coelomic cavity in lophophorate animals (ectoprocts, brachiopods, and phoronids).

lophotrochozoan protostome (lō′fə-trō′kə-zō′ən) (Gr. *lophos,* crest, + *trochos,* wheel, + *zōon,* animal). Any member of a clade within Protostomia whose members generally possess either a trochophore larva or a lophophore; examples are annelids, molluscs, and ectoprocts.

lorica (lo′rə-kə) (L. corselet). Protective external case found in some unicellular eukaryotes, rotifers, and others.

lotic (lō′tik) (L. *lotus,* action of washing or bathing). Of or pertaining to running water, such as a brook or river.

low viscosity Refers to the ability of liquid water to flow quickly and easily through containers of varying spatial dimensions, such as vessels of animal circulatory systems.

lumbar (lum′bär) (L. *lumbus,* loin). Relating to or near the loins or lower back.

lumen (lü′mən) (L. light). The cavity of a tube or organ.

lymph (limf) (L. *lympha,* water). The fluid in the lymphatic system formed from excess fluid that would otherwise accumulate in interstitial spaces between cells during capillary exchange.

lymphocyte (lim′fō-sīt) (L. *lympha,* water, goddess of water, + Gr. *kytos,* hollow vessel). Cell in blood and lymph that has a central role in immune responses. See **T cell** and **B cell.**

lysosome (lī′sə-sōm) (Gr. *lysis,* loosing, + *sōma,* body). Intracellular organelle consisting of a membrane enclosing several digestive enzymes that are released when the lysosome fuses with vescicles or endosomes produced by endocytosis.

M

macroevolution (L. *makros,* long, large, + *evolvere,* to unfold). Evolutionary change on a grand scale, encompassing the origin of novel designs, evolutionary trends, adaptive radiation, and mass extinction.

macrogamete (mak′rə-gam′ēt) (Gr. *makros,* long, large, + *gamos,* marriage). The larger of the two gamete types in a heterogametic organism, considered the female gamete.

macromere (mak′rə-mēr′) (Gr. *makros,* long, large, + *meros,* part). The largest size class of blastomeres in a cleaving embryo when the blastomeres differ in size from one another.

macronucleus (ma′krō-nü′klē-əs) (Gr. *makros,* long, large, + *nucleus,* kernel). The larger of the two kinds of nuclei in ciliates; controls all cell functions except reproduction.

macrophage (mak′rə-fāj) (Gr. *makros,* long, large, + *phagō,* to eat). A cell of the mononuclear phagocyte system in vertebrates that performs crucial functions in the immune response and inflammation

because it is an antigen-presenting cell that also produces cytokines.

madreporite (ma′drə-pōr′īt) (Fr. *madrépore,* reef-building coral, + *ite,* suffix for some body parts). Sievelike structure providing the intake for the water-vascular system of echinoderms.

major histocompatibility complex (MHC) Complex of genes coding for proteins inserted in the cell membrane; the proteins are the basis of self-nonself recognition by the immune system.

malaria (mə-lar′-ē-ə) (It. *malaria,* bad air). A disease marked by periodic chills, fever, anemia, and other symptoms, caused by *Plasmodium* spp.

male-dominance polygyny A male obtains more than one mate because females choose him over other males in an aggregation.

malleus (mal′ē-əs) (L. hammer). The ossicle attached to the tympanum in middle ears of mammals.

Malpighian tubules (mal-pig′ē-ən) (Marcello Malpighi, Italian anatomist, 1628–1694). Blind tubules opening into the hindgut of nearly all insects and some myriapods and arachnids, and functioning primarily as excretory organs.

maltase An enzyme in the plasma membrane of cells lining the intestine that splits maltose into two molecules of glucose.

mandible (L. *mandibula,* jaw). One of the lower jawbones in vertebrates; one of the head appendages in arthropods.

mangrove community A rich submarine community of deposit feeders (crabs, oysters, and shrimp) associated with the submerged roots of mangrove trees along some tropical coasts.

mantle Soft extension of the body wall in certain invertebrates, for example, brachiopods and molluscs, which usually secretes a shell; thin body wall of tunicates.

manubrium (man-ü′brē-əm) (L. handle). The portion projecting from the oral side of a jelly medusa, bearing the mouth; oral cone; presternum or anterior part of sternum; handlelike part of malleus of ear.

marasmus (mə-raz′məs) (Gr. *marasmos,* to waste away). Malnutrition, especially of infants, caused by a diet deficient in both calories and protein.

marsupial (mär-sü′pē-əl) (Gr. *marsypion,* little pouch). One of the pouched mammals of the subclass Metatheria.

mass extinction A relatively short interval of geological time in which a large portion

bat / āpe / ärmadillo / herring / fēmale / finch / līce / crocodile / crōw / còin / cōre / duck / dūke / ūnicorn / ə indicates unaccented vowel sound "uh" as in mammal, fishes, cardinal, heron, vulture / stress as in bi-ol′o-gy, bi′o-log′i-cal

(75% to 95%) of existing species or higher taxa are eliminated nearly simultaneously.

mast cells Immune-function cells found primarily in connective tissue. Upon activation by an antigen they release chemicals important in the inflammatory response of innate immunity. They also activate cells of the acquired immune response.

mastax (mas′təx) (Gr. jaws). Pharyngeal mill of rotifers.

matrix (mā′triks) (L. *mater,* mother). The intercellular substance of a tissue, or that part of a tissue into which an organ or process is set.

maxilla (mak-sil′ə) (L. dim. of *mala,* jaw). One of the upper jawbones in vertebrates; one of the head appendages in arthropods.

maxilliped (mak-sil′ə-ped) (L. *maxilla,* jaw, + *pes,* foot). One of the pairs of head appendages located just posterior to the maxilla in crustaceans, a thoracic appendage that has become incorporated into the feeding mouthparts.

medial (mē′dē-əl). Situated, or occurring, in the middle.

mediated transport Transport of a substance across a cell membrane mediated by a protein carrier molecule in the membrane.

medulla (mə-dül′ə) (L. marrow). The inner portion of an organ in contrast to the cortex or outer portion. Also, part of hindbrain.

medusa (mə-dū-sə) (Gr. mythology, female monster with snake-entwined hair). A jellyfish, or the free swimming stage in the life cycle of cnidarians.

Mehlis′ gland (me′ləs). Glands of uncertain function surrounding the ootype of trematodes and cestodes.

meiosis (mī-ō′səs) (Gr. from *mieoun,* to make small). The nuclear and cellular divisions that reduce the cellular chromosome content from the diploid to the haploid number; in animals, usually occurs in the last two divisions in the formation of the mature egg or sperm.

melanin (mel′ə-nin) (Gr. *melas,* black). Black or dark-brown pigment found in plant or animal structures.

melanophore (mel′ə-nə-fōr, mə-lan′ə-fōr) (Gr. *melania,* blackness, + *pherein,* to bear). Black or brown chromatophore containing melanin.

membranelle (mem-bra-nel′). A tiny membranelike structure; may be formed by fused cilia.

membrane potential The voltage recorded across a plasma membrane due to an unequal distribution of ions and charged molecules on either side of the membrane.

Such an unequal distribution of charge is caused by selective permeability of plasma membranes to specific ions and charged molecules

memory cells Population of long-lived T and B lymphocytes remaining after an initial immune response and providing the secondary response.

Mendel's second law See **law of independent assortment.**

meninges (mə-nin′jez), sing. **meninx** (Gr. *mēninx,* membrane). Any of three membranes (arachnoid, dura mater, pia mater) that envelop the vertebrate brain and spinal cord. Also, solid connective tissue sheath enclosing the central nervous system of some vertebrates.

menopause (men′ō-päz) (Gr. *men,* month, + *pauein,* to cease). In the human female, that time of life when ovulation ceases; cessation of the menstrual cycle.

menstruation (men′strü-ā′shən) (L. *menstrua,* the menses, from *mensis,* month). The discharge of blood and uterine endometrial tissue from the vagina at the beginning of a menstrual cycle. It occurs during the first few days of the ovarian cycle.

meroblastic (mer-ə-blas′tik) (Gr. *meros,* part, + *blastos,* germ). Partial cleavage occurring in zygotes having a large amount of yolk at the vegetal pole; cleavage restricted to a small area on the surface of the egg.

merozoite (me′rə-zō′īt) (Gr. *meros,* part, + *zōon,* animal). A very small trophozoite at the stage just after cytokinesis has been completed in multiple fission of a unicellular eukaryote.

mesenchyme (me′zən-kīm) (Gr. *mesos,* middle, + *enchyma,* infusion). Embryonic connective tissue; irregular or amebocytic cells often embedded in gelatinous matrix.

mesentery (me′zən-ter′ē) (L. *mesenterium,* mesentery). Peritoneal fold serving to hold the viscera in position.

mesocoel (me′zō-sēl) (Gr. *mesos,* middle, + *koilos,* hollow). Middle body coelomic compartment in some deuterostomes, anterior in lophophorates, corresponds to hydrocoel in echinoderms.

mesoderm (me′zə-dərm) (Gr. *mesos,* middle, + *derm,* skin). The third germ layer, formed in the gastrula between the ectoderm and endoderm; gives rise to connective tissues, muscle, urogenital and vascular systems, and the peritoneum.

mesoglea (me′zō-glē′ə) (Gr. *mesos,* middle, + *glia,* glue). The layer of jellylike or cement material between the epidermis and gastrodermis in cnidarians and ctenophores.

mesohyl (me′zō-hīl) (Gr. *mesos,* middle, + *hylē,* a wood). Gelatinous matrix surrounding sponge cells; mesoglea, mesenchyme.

mesolecithal (me′zō-les′ə-thəl) (Gr. *mesos,* middle, + *lekithos,* yolk). Pertaining to a zygote (or ovum) having a moderate amount of yolk concentrated in the vegetal pole.

mesonephros (me′zō-nef′rōs) (Gr. *mesos,* middle, + *nephros,* kidney). The middle of three pairs of embryonic renal organs in vertebrates. Functional kidney of embryonic amniotes; its collecting duct is a Wolffian duct. Adj., **mesonephric.**

mesopelagic (me′zō-pə-laj′-ik). Refers to the "twilight zone" of open oceanic waters marking the transition from the well-lit epipelagic zone above it to a zone of total darkness below it.

mesosome (me′zə-sōm) (Gr. *mesos,* middle, + *sōma,* body). The portion of the body in lophophorates and some deuterostomes that contains the mesocoel.

messenger RNA (mRNA) A form of ribonucleic acid that carries genetic information from the gene to the ribosome, where it determines the order of amino acids as a polypeptide is formed.

metabolism (Gr. *metabolē,* change). A group of processes that includes digestion, recovery of energy (respiration), and synthesis of molecules and structures by organisms; the sum of the constructive (anabolic) and destructive (catabolic) processes.

metacentric (me′tə-sen′trək) (Gr. *meta,* between, among, after, + *kentron,* center). Chromosome with centromere at or near the middle.

metacercaria (me′tə-sər-ka′rē-ə) (Gr. *meta,* between, among, after, + *kerkos,* tail, + L. *aria,* connected with). Fluke juvenile (cercaria) that has lost its tail and has become encysted.

metacoel (me′tə-sēl) (Gr. *meta,* between, among, after, + *koilos,* hollow). Posterior coelomic compartment in some deuterostomes and lophophorates; corresponds to somatocoel in echinoderms.

metamere (me′tə-mēr) (Gr. *meta,* after, + *meros,* part). A repeated body unit along the longitudinal axis of an animal, a somite, or segment.

metamerism (mə-ta′-mə-ri′zəm) (Gr. *meta,* between, after, + *meros,* part). Condition of being composed of serially repeated parts (metameres); serial segmentation.

metamorphosis (Gr. *meta,* between, among, after, + *morphē,* form, + *osis,* state of). Sharp change in form during postembryonic development, for example, tadpole to frog or larval insect to adult.

metanephridium (me′tə-nə-fri′dē-əm) (Gr. *meta,* between, among, after, + *nephros,* kidney). A type of tubular nephridium with the inner open end draining the coelom and the outer open end discharging to the exterior.

metanephros (me′tə-ne′frōs) (Gr. *meta,* between, among, after, + *nephros,* kidney). Vertebrate kidney formed from the most posterior of three embryonic regions capable of forming renal organs; functional kidney of adult amniotes; drained by ureter. Adj., **metanephric.**

metapopulation A large population that comprises numerous semiautonomous subpopulations, termed demes, with some limited movement of individuals among demes. Demes of a metapopulation are often geographically distinct. Patterns of movement of individuals among versus within demes constitute **metapopulation dynamics.**

metasome (me′tə-sōm) (Gr. *meta,* after, behind, + *sōma,* body). The portion of the body in lophophorates and some deuterostomes that contains the metacoel.

metazoa (me′tə-zō′ə) (Gr. *meta,* after, + *zōon,* animal). Animals.

MHC See **major histocompatibility complex.**

microevolution (mī-krō-ev-ə-lü′shən) (L. *mikros,* small, + *evolvere,* to unfold). A change in the gene pool of a population across generations.

microfilament (mī′krō-fil′ə-mənt) (Gr. *mikros,* small, + L. *filum,* a thread). A thin, linear structural protein forming a part of the cytoskeleton in all cells. Part of the contractile protein machinery of muscle cells.

microfilariae (mīk′rə-fil-ar′ē-ē) (Gr. *mikros,* small, + L. *filum,* a thread). Partially developed juveniles born alive in filarial worms (phylum Nematoda).

microgamete (mīk′rə-gam′ēt) (Gr. *mikros,* small, + *gamos,* marriage). The smaller of the two gamete types in a heterogametic organism, considered the male gamete.

microglial cells Phagocytic and antigen-presenting cells in the central nervous system that secrete cytokines; part of the mononuclear phagocyte system.

microhabitat selection When species having similar ecological niches occur in the same community, they partition shared resources by specializing on different aspects of the shared resource, such as insectivorous birds favoring different characteristic foraging sites.

micromere (mīk′rə-mēr′) (Gr. *mikros,* small, + *meros,* part). The smallest size class of blastomeres in a cleaving embryo when the blastomeres differ in size from one another.

micron (μ) (mī′krän) (Gr. neuter of *mikros,* small). One one-thousandth of a millimeter; about 1/25,000 of an inch. Now largely replaced by micrometer (μm).

microneme (mī′krə-nēm) (Gr. *mikros,* small, + *nēma,* thread). One type of structure forming the apical complex in the phylum Apicomplexa, slender and elongate, leading to the anterior and thought to function in host-cell penetration.

micronucleus The smaller of two nuclei found in ciliates; controls the reproductive functions of these organisms.

micropyle (mīk′rə-pīl) (Gr. *mikros,* small, + *pileos,* a cap). The small opening through which the cells emerge from a gemmule (phylum Porifera).

microsporidian (mī′krə-spō-rid′ē-ən) (Gr. *micros,* small, + *spora,* seed, + *idion,* dim. suffix). Any member of a unicellular eukaryotic clade comprising intracellular parasites with a distinctive spore morphology.

microthrix See **microvillus.**

microtubule (Gr. *mikros,* small, + L. *tubule,* pipe). A long, tubular cytoskeletal element composed of tubulin dimers, with an outside diameter of 20 to 27 nm. Microtubules influence cell shape and play important roles during cell division.

microvillus (Gr. *mikros,* small, + L. *villus,* shaggy hair). Narrow, cylindrical cytoplasmic projection from epithelial cells; microvilli form the brush border of several types of epithelial cells. Also, microvilli with unusual structure cover the surface of cestode tegument (also called **microthrix** [pl. **microtriches**]).

mictic (mik′tik) (Gr. *miktos,* mixed or blended). Pertaining to haploid eggs of rotifers or the females that lay such eggs.

mimic (mim′ik) (Gr. *mimicus,* imitator). A species whose morphological or behavioral characteristics copy those of another species because those characteristics deter shared predators.

mimicry Evolution by natural selection of similar forms by different species, such as the sharing of warning signals that discourage common predators. In Batesian mimicry, a species tasteful to a predator deceptively copies warning signals of a distasteful species. In Müllerian mimicry,

two or more distasteful species evolve common warning signals to avoid a common predator.

mineralocorticoids (min′ (ə)rəl-ō-kort′ə-kȯids) (M. E. *minerale,* ore, + L. *cortex,* bark, + *oid,* suffix denoting likeness of form). Steroid hormones of the adrenal cortex, especially aldosterone, that regulate salt balance.

miracidium (mīr′ə-sid′ē-əm) (Gr. *meirakidion,* youthful person). A minute ciliated larval stage in the life of flukes.

mitochondrion (mī′tə-kän′drē-ən) (Gr. *mitos,* a thread, + *chondrion,* dim. of *chondros,* corn, grain). An eukaryotic cellular organelle in which aerobic metabolism occurs.

mitosis (mī-tō′səs) (Gr. *mitos,* thread, + *osis,* state of). Nuclear division in which there is an equal qualitative and quantitative division of the chromosomal material between the two resulting nuclei; ordinary cell division.

mixed temperate forest Central North American woodlands containing evergreen trees such as white pine, red pine, and hemlock mixed with various deciduous trees, including maples, oaks, and hickories.

model (mod′l) (Fr. *modèle,* pattern). A species whose morphological or behavioral characteristics are copied by another species because those characteristics deter shared predators.

modular (mäj′ə-lər). Describes the structure of a colony of genetically identical organisms that are physically associated and produced asexually by cloning.

molecule A configuration of atomic nuclei and electrons bound together by chemical bonds.

molting (mōl′-ting). Shedding of the outer cuticular layer; see **ecdysis.**

monocyte (mon′ə-sīt) (Gr. *monos,* single, + *kytos,* hollow vessel). A type of leukocyte that becomes a phagocytic and antigen-producing cell (macrophage) after moving into tissues.

monoecious (mə-nē′shəs) (Gr. *monos,* single, + *oikos,* house). Having both male and female gonads in the same organism, usually denoting that this is the typical condition of a species; hermaphroditic.

monogamy (mə-näg′ə-mē) (Gr. *monos,* single, + *gamos,* marriage). The condition of having a single mate at any one time. Adj., **monogamous.**

monohybrid cross (Gr. *monos,* single, +
L. *hybrida,* mongrel). Production of hybrid
offspring from parents different in one
specified character.

monomer (mä′nə-mər) (Gr. *monos,* single, +
meros, part). A molecule of simple
structure, but capable of linking with others
to form polymers.

mononuclear phagocyte system The fixed
phagocytic cells in the tissues, especially
the liver, lymph nodes, spleen, and others;
also called the reticuloendothelial system.

monophyly (mo′nə-fī′-lē′) (Gr. *monos,* single,
+ *phyle,* tribe). The condition that a taxon
or other group of organisms contains the
most recent common ancestor of the group
and all of its descendants; contrasts with
polyphyly and **paraphyly.**

monosaccharide (mo′nə-sa′kə-rīd) (Gr. *monos,*
one, + *sakcharon,* sugar, from Sanskrit
sarkarā, gravel, sugar). A simple sugar that
cannot be decomposed into smaller sugar
molecules; the most common are pentoses
(such as ribose) and hexoses (such as
glucose).

monosomy, monosomic (mo′nə-sō′mē) (Gr.
monos, one, + *sōmē,* body). The
chromosomal constitution of an otherwise
diploid organism in which a single
chromosome is missing (chromosome
number = 2*n* – 1).

monotreme (mo′nō-trēm) (Gr. *monas,* single,
+ *trēmatos,* hole). Any member of an order
of egg-laying (oviparous) mammals; duck-
billed platypus and echidnas.

monozoic (mo′nə-zō′ik) (Gr. *monos,* single,
+ *zōon,* animal). Tapeworms that have a
single proglottid, do not undergo
strobilation to form chain of proglottids.

morphogen (mor′fə-jən) (Gr. *morphē,* form,
+ *genesis,* origin). Soluble molecule acting
on target cells, or forming a gradient from
producing cells to target cells, to specify
cell fates; an agent of embryonic induction
and epigenesis.

morphogenesis (mor′fə-je′nə-səs) (Gr. *morphē,*
form, + *genesis,* origin). Development of
the architectural features of organisms;
formation and differentiation of tissues and
organs.

morphogenetic determinant Certain protein or
messenger RNA in egg cytoplasm
distributed among descendant cells during
cleavage to direct later gene expression and
to specify cell fate; the basis of mosaic
development.

morphological species concept See
typological species concept.

morphology (Gr. *morphē,* form, + L. *logia,*
study, from Gr. *logos,* work). The science of

structure. Includes cytology, the study of
cell structure; histology, the study of tissue
structure; and anatomy, the study of gross
structure.

mosaic development Embryonic development
characterized by independent differentiation
of each part of the embryo; determinate
cleavage.

mucus (mū′kəs) (L. *mucus,* nasal mucus).
Viscid, slippery secretion rich in mucins
produced by secretory cells such as those in
mucous membranes. Adj., **mucous.**

Müllerian mimicry A condition in which two
inedible species evolve similar aposematic
phenotypes (such as warning coloration) to
discourage potential predators from
attempting to eat either one.

Müller's larva Free-swimming ciliated larva
that resembles a modified ctenophore;
characteristic of certain marine polyclad
turbellarians.

multiple fission A mode of asexual
reproduction in some single-celled
eukaryotes in which the nuclei divide more
than once before cytokinesis occurs.

multiplication of species The Darwinian
theory that the evolutionary process
generates new species through a branching
of evolutionary lineages derived from an
ancestral species.

mushroom bodies Region of the
protocerebrum of an insect brain
associated with learning.

mutation (mū-tā′shən) (L. *mutare,* to change).
A stable and abrupt change of a gene; the
heritable modification of a characteristic.

mutualism (mū′chə-wə-li′zəm) (L. *mutuus,*
lent, borrowed, reciprocal). A type of
interaction in which two different species
derive benefit from their association, which
may be necessary to both; often symbiotic.

mycetozoa (mī-sē′-tə-zō′ə) (Gr. *mykētos,* a
fungus, + *zōon,* animal). A eukaryotic clade
containing cellular, acellular, and protostelid
slime molds.

myelin (mī′ə-lən) (Gr. *myelos,* marrow). A fatty
material forming the medullary sheath of
nerve fibers.

myocyte (mī′ə-sīt) (Gr. *mys,* muscle, + *kytos,*
hollow vessel). Contractile cell (pinacocyte)
in sponges.

myofibril (mī′ō-fī′-bril) (Gr. *mys,* muscle, + L.
dim. of *fibra,* fiber). A contractile filament
within muscle or muscle fiber.

myogenic (mī′o-jen′ik) (Gr. *mys,* muscle, +
N.L., *genic,* giving rise to). Originating in
muscle, such as heartbeat arising in
vertebrate cardiac muscle because of
inherent rhythmical properties of muscle
rather than because of neural stimuli.

myomere (mī′ə-mēr) (Gr. *mys,* muscle, +
meros, part). A muscle segment of
successive segmental trunk musculature.

myosin (mī′ə-sin) (Gr. *mys,* muscle, + *in,* suffix,
belonging to). A large, actin-binding protein
of all cells. In contractile tissue it is
organized to form the thick filaments of
striated muscle, for example. During
contraction it combines with actin to form
actomyosin.

myotome (mī′ə-tōm) (Gr. *mys,* muscle, + *tomos,*
cutting). That part of a somite destined to
form muscles; the muscle group innervated
by a single spinal nerve.

N

nacre (nā′kər) (F. mother-of-pearl). Innermost
lustrous layer of mollusc shell, secreted by
mantle epithelium. Adj., **nacreous.**

NAD Abbreviation of nicotinamide adenine
dinucleotide, an electron acceptor in many
metabolic reactions; the reduced form
NADH is an electron donor.

nares (ner′ēz), sing. **naris** (L. nostrils).
Openings into the nasal cavity, both
internally and externally, in the head of a
vertebrate.

natural killer cells Lymphocyte-like cells that
can kill virus-infected cells and tumor cells
in the absence of antibody.

natural selection A nonrandom reproduction of
varying organisms in a population that
causes the survival of those best adapted to
their environment and elimination of those
less well adapted; leads to evolutionary
change if the variation is heritable,
accumulating the most favorable traits in the
population and discarding less
favorable ones.

nauplius (nä′plē-əs) (L. a kind of shellfish). A
free swimming microscopic larval stage of
certain crustaceans, with three pairs of
appendages (antennules, antennae, and
mandibles) and median eye. Characteristic
of ostracods, copepods, barnacles, and some
others.

nekton (nek′tən) (Gr. neuter of *nēktos,*
swimming). Term for actively swimming
organisms, essentially independent of wave
and current action; contrasts with **plankton.**

nematocyst (ne-ma′-tə-sist′) (Gr. *nēma,* thread,
+ *kystis,* bladder). Stinging organelle of
cnidarians.

neo-Darwinism (nē′ō′ där′wə-niz′əm). A
modified version of Darwin's evolutionary
theory that eliminates elements of the
Lamarckian inheritance of acquired
characteristics and pangenesis that were
present in Darwin's formulation; this theory

originated with August Weismann in the late nineteenth century and, after incorporating Mendelian genetic principles, has become the currently favored version of Darwinian evolutionary theory.

neodermis See **syncytial tegument.**

neoplastic growth (nē′-ō-plas′-tik grōth′). Proliferation of cells at an uncharacteristically high rate within the body of a multicellular organism, leading to cancerous tumors and metastases.

neopterygian (nē-äp′tə-rij′ē-ən) (Gr. *neos,* new, + *pteryx,* fin). Any of a large group of bony fishes that includes most modern species.

neotenine See **juvenile hormone.**

neoteny (nē-ot′ə-nē) (Gr. *neos,* new, + *teinein,* to extend). An evolutionary process by which somatic development is retarded relative to sexual maturation; produces a descendant that reaches sexual maturity while retaining a morphology characteristic of the preadult or larval stage of an ancestor.

nephridiopore (nə-frid′ē-ə-pōr) (Gr. *nephros,* kidneys, + *porus,* pore). An external excretory opening in invertebrates.

nephridium (nə-frid′ē-əm) (Gr. *nephridios,* of the kidney). One of the segmentally arranged, paired excretory tubules of many invertebrates, notably the annelids. In a broad sense, any tubule specialized for excretion and/or osmoregulation; with an external opening and with or without an internal opening.

nephron (ne′frän) (Gr. *nephros,* kidney). Functional unit of a vertebrate kidney, comprising a Bowman's capsule, an enclosed glomerulus, and the attached uriniferous tubule.

nephrostome (nef′rə-stōm) (Gr. *nephros,* kidney, + *stoma,* mouth). Ciliated, funnel-shaped opening of a nephridium.

neritic (nə-rit′ik) (Gr. *nērites,* a mussel). Portion of the sea overlying the continental shelf, specifically from the subtidal zone to a depth of 200 m.

nested hierarchy A pattern in which species are ordered into a series of increasingly more inclusive clades according to the taxonomic distribution of synapomorphies.

net productivity The energy stored by an organism, equal to the energy assimilated (**gross productivity**) minus the energy used for metabolic maintenance (**respiration**).

neural crest Populations of ectodermally derived embryonic cells that differentiate into many skeletal, neural, and sensory structures unique to vertebrates.

neural spine A dorsal projection of a vertebra that functions as a muscle-attachment site.

neurochord (nur′ə-kord) (L. *nervus,* sinew, tendon, + Gr. *chorda,* cord). Longitudinal nerve chord of hemichordates that lies above the notochord.

neurogenic (nur′ə-jen′ik) (Gr. *neuron,* nerve, + N.L. *genic,* give rise to). Originating in nervous tissue, as does the rhythmical beat of some arthropod hearts.

neuroglia (nur′ə-glē′-ə) (Gr. *neuron,* nerve, + *glia,* glue). Tissue supporting and filling the spaces between the nerve cells of the central nervous system.

neurolemma (nur′ə-lem′ə) (Gr. *neuron,* nerve, + *lemma,* skin). Delicate nucleated outer sheath of a nerve cell; sheath of Schwann.

neuromast (Gr. *neuron,* sinew, nerve, + *mastos,* knoll). Cluster of sense cells on or near the surface of a fish or amphibian that is sensitive to water movements.

neuron (Gr. nerve). A nerve cell.

neuropodium (nur′ə-pō′dē-əm) (Gr. *neuron,* nerve, + *pous, podos,* foot). Lobe of parapodium nearer the ventral side in polychaete annelids.

neurosecretory cell (nur′ə-se′-kre-to′-rē). Any cell (neuron) of the nervous system that produces a hormone.

neurotransmitter Specialized chemical that is stored and released from vesicles found at chemical synapses of the nervous system. These chemicals transmit signals between neurons, between receptors and neurons, and between neurons and effectors such as muscles and glands.

neutron A particle in an atomic nucleus lacking an electrical charge and having a mass 1839 times that of an electron.

niche (nich′). The role of an organism, population, or species in an ecological community comprising its usage of resources; its unique way of life and its relationship to other biotic and abiotic factors.

niche overlap A comparison of two species quantifying the proportion of each species' resources that are utilized also by the other species.

nictitating membrane (nik′tə-tā-ting) (L. *nicto,* to wink). Third eyelid, a transparent membrane of birds and many nonavian reptiles and mammals, that can be pulled across the eye.

nitrogen fixation (Gr. *nitron,* soda, + *gen,* producing). Reduction of molecular nitrogen to ammonia by some bacteria and

cyanobacteria, often followed by **nitrification,** the oxidation of ammonia to nitrites and nitrates by other bacteria.

nitrogenous base The molecular subunit of a nucleotide attached at the 1′ carbon of deoxyribose or ribose and which participates in hydrogen bonding between nucleotide strands. Includes adenine, cytosine, guanine, thymine, and uracil.

nondisjunction Failure of a pair of homologous chromosomes to separate during meiosis, leading to one gamete with $n + 1$ chromosomes (see **trisomy**) and another gamete with $n - 1$ chromosomes.

notochord (nō′tə-kōrd′) (Gr. *nōtos,* back, + *chorda,* cord). An elongated cartilaginous cellular cord, enclosed in a sheath, which forms the axial skeleton of chordate embryos, jawless fishes, and adult cephalochordates.

notopodium (nō′tə-pō′dē-əm) (Gr. *nōtos,* back, + *pous, podos,* foot). Lobe of parapodium nearer the dorsal side in polychaete annelids.

nucleic acid (nü′klā′ik) (L. *nucleus,* kernel). One of a class of molecules composed of joined nucleotides; chief types are deoxyribonucleic acid (DNA), found in cell nuclei (chromosomes) and mitochondria, and ribonucleic acid (RNA), found both in cell nuclei (chromosomes and nucleoli) and in cytoplasmic ribosomes.

nucleoid (nü′klē-ōid) (L. *nucleus,* kernel, + *oid,* like). The region in a prokaryotic cell that contains the genome.

nucleolus (nü-klē′ə-ləs) (dim. of L. *nucleus,* kernel). A deeply staining body within the nucleus of a cell and containing ribosomal RNA; nucleoli are specialized portions of certain chromosomes that carry multiple copies of the genes encoding ribosomal RNA and where ribosomal RNA is actively synthesized.

nucleoplasm (nü′klē-ō-pla′-zm) (L. *nucleus,* kernel, + Gr. *plasma,* mold). Protoplasm of a nucleus, as distinguished from cytoplasm.

nucleoprotein Any protein structurally associated with DNA or RNA.

nucleosidase An enzyme in the plasma membrane of cells lining the intestine that splits nucleosides into ribose and deoxyribose sugars and purines and pyrimidines.

nucleosome (nü′klē-ə-sōm′) (L. *nucleus,* kernel, + *sōma,* body). A repeating subunit of chromatin in which one and three-quarter turns of the double-helical DNA are wound around eight molecules of histones.

nucleotidase An enzyme in the plasma membrane of cells lining the intestine that splits nucleotides into nucleosides.

nucleotide (nü′klē-ə-tīd′). A molecule consisting of phosphate, 5-carbon sugar (ribose or deoxyribose), and a purine or a pyrimidine; the purines are adenine and guanine, and the pyrimidines are cytosine, thymine, and uracil.

nucleus (nü′klē-əs) (L. *nucleus*, a little nut, the kernel). The organelle in eukaryotes that contains the chromatin and which is bounded by a double membrane (nuclear envelope).

nuptial flight (nup′shəl). The mating flight of insects, especially that of a queen with a male or males.

nurse cells Single cells or layers of cells surrounding or adjacent to other cells or structures for which the nurse cells provide nutrient or other molecules (for example, for insect oocytes or *Trichinella* spp. juveniles).

nymph (L. *nympha*, nymph, bride). An immature stage (following hatching) of a hemimetabolous insect that lacks a pupal stage.

O

ocellus (ō-sel′əs) (L. dim. of *oculus*, eye). A simple eye or eyespot in many types of invertebrates.

octomerous (ok-tom′ər-əs) (Gr. *oct*, eight, + *meros*, part). Eight parts, specifically, symmetry based on eight.

odontophore (ō-don′tə-fōr′) (Gr. *odous*, tooth, + *pherein*, to carry). Tooth-bearing organ in molluscs, including the radula, radular sac, muscles, and cartilages.

olfactory (ōl-fakt′ (ə)-rē) (L. *olor*, smell, + *factus*, to bring about). Pertaining to the sense of smell.

olfactory epithelium A specialized chemosensory surface tissue inside the nasal cavities of aquatic and terrestrial vertebrates.

omasum (ō-ma′səm) (L. paunch). The third compartment of the stomach of a ruminant mammal.

ommatidium (ä′mə-tid′ē-əm) (Gr. *omma*, eye, + *idium*, small). One of the optical units of the compound eye of arthropods.

omnivore (äm′nə-vōr) (L. *omnis*, all, + *vorare*, to devour). An animal that uses a variety of animal and plant material in its diet.

oncogene (än′kə-jēn) (Gr. *onkos*, protuberance, tumor, + *genos*, descent). Any of a number of genes associated with neoplastic growth

(cancer). The gene in its benign state, either inactivated or performing its normal role, is a **proto-oncogene.**

oncomiracidium (än′kō-mir′ə-sid′ē-əm) (Gr. *onkos*, barb, hook, + *meirakidion*, youthful person). A ciliated larva of a monogenetic trematode.

oncosphere (än′kəs-fēr) (Gr. *onkinos*, a hook, + *sphaira*, ball). Rounded larva common to all cestodes, bears hooks.

ontogeny (än-tä′jə-nē) (Gr. *ontos*, being, + *geneia*, act of being born, from *genēs*, born). The course of development of an individual from egg to senescence.

oocyst (ō′ə-sist) (Gr. *ōion*, egg, + *kystis*, bladder). Cyst formed around a zygote of malarial parasites and related organisms.

oocyte (ō′ə-sīt) (Gr. *ōion*, egg, + *kytos*, hollow). Stage in formation of ovum, just preceding first meiotic division (primary oocyte) or just following first meiotic division (secondary oocyte).

ooecium (ō-ēs′ē-əm) (Gr. *ōion*, egg, + *oikos*, house, + L. *ium*, from). Brood pouch; compartment for developing embryos in ectoprocts.

oogenesis (ō-ə-jen′ə-səs) (Gr. *ōion*, egg, + *genesis*, descent). Formation, development, and maturation of a female gamete or ovum.

oogonium (ō′ə-gōn′ē-əm) (Gr. *ōion*, egg, + *gonos*, offspring). A cell that, by continued division, gives rise to oocytes; an ovum in a primary follicle immediately before the beginning of maturation.

ookinete (ō-ə-kī′nēt) (Gr. *ōion*, egg, + *kinein*, to move). The motile zygote of malarial parasites.

ootid (ō-ə-tid′) (Gr. *ōion*, egg, + *idion*, dim.). Stage of formation of ovum after second meiotic division following expulsion of second polar body.

ootype (ō′ə-tūp) (Gr. *ōion*, egg, + *typos*, mold). Part of oviduct in flatworms that receives ducts from vitelline glands and Mehlis′ gland.

operculum (ō-per′kū-ləm) (L. cover). The gill cover in bony fishes; protective plate in some snails.

operon (äp′ə-rän) A genetic unit consisting of a cluster of genes under the control of other genes; occurs in prokaryotes.

ophthalmic (äf-thal′mik) (Gr. *ophthalamos*, an eye). Pertaining to the eye.

opisthaptor (ä′pis-thap′tər) (Gr. *opisthen*, behind, + *haptein*, to fasten). Posterior attachment organ of a monogenetic trematode.

opisthokont (ō-pis′thə-känt) (G. *opisthen*, behind, + *kontos*, a pole). Any member of a eukaryotic clade comprising fungi,

microsporidians, choanoflagellates, and animals; if present, flagellated cells possess a single posterior flagellum.

opisthonephros (ō-pis′-thō-nef′-rōs) (Gr. *opisth*, back + *nephros*, kidney). A kidney that develops from the middle and posterior portions of the nephrogenic region of vertebrates and is drained by the Wolffian duct or accessory ducts. Functional kidney of most adult anamniotes (fishes and amphibians). Adj., **opisthonephric.**

opisthosoma (ō-pis′thə-sō′mə) (Gr. *opisthe*, behind, + *sōma*, body). Posterior body region in arachnids and pogonophorans.

opsonization (op′sən-i-zā′shən) (Gr. *opsonein*, to buy victuals, to cater). The facilitation of phagocytosis of antigens by phagocytes in the blood or tissues. Mediated by antibody bound to the particles to form antibody-antigen complexes, or by complement proteins (vertebrates) or complement-like proteins (invertebrates) bound to antigen.

oral disc The end of a cnidarian polyp bearing the mouth.

oral lobe A flaplike extension of the mouth of a scyphozoan medusa that aids in feeding.

orbit (L. circle). The cavity of the skull housing the eyeball.

organelle (Gr. *organon*, tool, organ, + L. *ella*, dim.). Specialized part of a cell; a subcellular structure that performs functions analogous to organs of animals.

organism (ōr′-gə-niz′-əm). A biological individual composed of one or more cells, tissues, and/or organs whose parts are interdependent in producing a collective physiological system. Organisms of the same species may form **populations.**

organizer (ōr′gan-ī′-zer) (Gr. *organos*, fashioning). Area of an embryo that directs subsequent development of other parts.

orthogenesis (ōr′thō-jen′ə-səs). A unidirectional trend in the evolutionary history of a lineage as revealed by the fossil record; also, a now discredited, anti-Darwinian evolutionary theory, popular around 1900, postulating that genetic momentum forced lineages to evolve in a predestined linear direction that was independent of external factors and often led to decline and extinction.

osculum (os′kū-ləm) (L. *osculum*, a little mouth). Excurrent opening in a sponge.

osmole (oz′-mōl). Molecular weight of a solute, in grams, divided by the number of ions or particles into which it dissociates in solution. Adj., **osmolar.**

osmoregulation Maintenance of proper internal salt and water concentrations in a cell or in the body of a living organism; active regulation of internal osmotic pressure.

osmosis (oz-mō′sis) (Gr. *ōsmos*, act of pushing, impulse). A flow of solvent (usually water) through a semipermeable membrane.

osmotic pressure The pressure that resists the flow of water into cytoplasm.

osmotroph (oz′mə-trōf) (Gr. *ōsmos*, a thrusting, impulse, + *trophē*, to eat). A heterotrophic organism that absorbs dissolved nutrients.

osphradium (äs-frā′dē-əm) (Gr. *osphradion*, small bouquet, dim. of *osphra*, smell). A chemoreceptive sense organ in aquatic snails and bivalves that tests incoming water.

ossicles (L. *ossiculum*, small bone). Small separate pieces of echinoderm endoskeleton. Also, tiny bones of the middle ear of vertebrates.

osteoblast (os′tē-ō-blast′) (Gr. *osteon*, bone, + *blastos*, bud). A bone-forming cell.

osteoclast (os′tē-ō-clast′) (Gr. *osteon*, bone, + *klan*, to break). A large, multinucleate cell that functions in bone dissolution.

osteocyte (os′tē-ə-sīt′) (Gr. *osteon*, bone, + *kytos*, hollow). A bone cell characteristic of adult bone. It has developed from an osteoblast, and is located in a lacuna of the bone substance.

osteoderm (os′tē-ə-dərm′) (Gr. *osteon*, bone, + *derma*, skin). A bony, dermal plate located under and supporting an epidermal scale.

osteon (os′tē-on) (Gr. bone). Unit of bone structure; haversian system.

osteostracans (os′tē-os′trə-kəns) (Gr. *osteon*, bone, + *ostrakon*, shell). A group of jawless, extinct fishes with dermal armor and pectoral fins from the Silurian and Devonian periods.

ostium (L. door). Opening.

ostracoderm (os-trak′ō-derm) (Gr. *ostrakon*, shell, + *derma*, skin). A paraphyletic group of extinct, jawless fishes with dermal armor known from the late Cambrian to Devonian periods.

otolith (ō′-tō-lith) (Gr. *ous, otos*, ear, + *lithos*, stone). Calcareous concretions in the membranous labyrinth of the inner ear of some vertebrates, or in the auditory organ of certain invertebrates.

outgroup In phylogenetic systematic studies, a species or group of species closely related to but not included within a taxon whose phylogeny is being studied, and used to polarize variation of characters and to root the phylogenetic tree.

outgroup comparison A method for determining the polarity of a character in cladistic analysis of a taxonomic group. Character states found within the group being studied are judged ancestral if they occur also in related taxa outside the study group (= outgroups); character states that occur only within the taxon being studied but not in outgroups are judged to have been derived evolutionarily within the group being studied.

oviger (ō′vi-jər) (L. *ovum*, egg, + *gerere*, to bear). Leg that carries eggs in pycnogonids.

oviparity (ō′və-pa′rə-tē) (L. *ovum*, egg, + *parere*, to bring forth). Reproduction in which eggs are released by the female; development of offspring occurs outside the maternal body. Adj., **oviparous** (ō-vip′ə-rəs).

ovipositor (ō′və-päz′ə-tər) (L. *ovum*, egg, + *positor*, builder, placer, + *or*, suffix denoting agent or doer). In many female insects a structure at the posterior end of the abdomen for laying eggs.

ovoviviparity (ō′vō-vī-və-par′ə-tē) (L. *ovum*, egg, + *vivere*, to live, + *parere*, to bring forth). A reproduction mode in which young develop (often in an egg) in the uterus, but rely only on their yolk sacs for nutrition, with no placental connection. Also called lecithotropic viviparity. Adj., **ovoviviparous** (ō′vō-vī-vip′ə-rəs).

ovum (L. *ovum*, egg). Mature female germ cell (egg).

oxidation (äk′sə-dā-shən) (Fr. *oxider*, to oxidize, from Gr. *oxys*, sharp, + *ation*). The loss of an electron by an atom or molecule; sometimes addition of oxygen chemically to a substance. Opposite of reduction, in which an electron is accepted by an atom or molecule.

oxidative (aerobic) metabolism Cellular respiration using molecular oxygen as the final electron acceptor.

oxidative phosphorylation (äk′sə-dā′-tiv fäs′fər-i-lā′shən). The conversion of inorganic phosphate to energy-rich phosphate of ATP, involving electron transport through a respiratory chain to molecular oxygen.

P

p53 protein A tumor-suppressor protein with critical functions in normal cells. A mutation in the gene that encodes it, *p53*, can cause loss of control over cell division and thus may lead to cancer.

paedomorphosis (pē-dō-mor′fə-səs) (Gr. *pais*, child, + *morphē*, form). Retention of ancestral juvenile features in later stages of the ontogeny of descendants.

pair bond An affiliation between an adult male and an adult female for reproduction. Characteristic of monogamous species.

pancreatic amylase A carbohydrate-splitting enzyme in pancreatic juices that hydrolyzes starch into shorter starch polymers, and into maltose and glucose.

pancreatic lipase An enzyme in pancreatic juices that hydrolyzes fats into fatty acids and glycerol.

papilla (pə-pil′ə) (L. nipple). A small nipplelike projection. A vascular process that nourishes the root of a hair, feather, or developing tooth.

papula (pa′pū-lə) (L. pimple). Respiratory processes on skin of sea stars; also, pustules on skin.

parabasal bodies Cellular organelles similar to Golgi bodies, presumed to function as part of the secretory system in endoplasmic reticulum.

parabasalid (pe′rə-bā′sa-lid) (Gr. *para*, beside, + *basis*, body). Any member of a unicellular eukaryotic clade having a flagellum and parabasal bodies.

parabiosis (pe′rə-bī-ō′sis) (Gr. *para*, beside, + *biosis*, mode of life). The fusion of two individuals, causing mutual physiological intimacy.

parabronchi (pe′rə-brong′kī) (Gr. *para*, beside, + *bronchos*, windpipe). Fine air-conduction pathways of a bird lung.

paradigm (pe′rə-dīm′). A powerful scientific theory that explains diverse observations and guides active research, such as Darwin's theory of common descent of life.

parahormone Local tissue factor that stimulates cell growth and some biochemical processes.

paramylon granules (pe′rə-mī-lən) (Gr. *para*, beside, + *mylos*, mill, grinder). Organelles containing the starchlike substance paramylon; in some algae and flagellates.

parapatric speciation A branching of population lineages to form separate species in which the diverging lineages are mostly nonoverlapping in geographic distribution but make contact along a narrow borderline. This controversial mode of speciation contrasts with allopatric speciation and sympatric speciation.

---

bat / āpe / ärmadillo / herring / fēmale / finch / līce / crocodile / crōw / cóin / córe / duck / dūke / ūnicorn / ə indicates unaccented vowel sound "uh" as in mammal, fishes, cardinal, heron, vulture / stress as in bi-ol′o-gy, bi′o-log′i-cal

paraphyly (pe′rə-fī-lē) (Gr. *para*, beside, + *phyle*, tribe). The condition that a taxon or other group of organisms contains the most recent common ancestor of all members of the group but excludes some descendants of that ancestor; contrasts with **monophyly** and **polyphyly.**

parapodium (pe′rə-pō′dē-əm) (Gr. *para*, beside, + *pous, podos*, foot). One of the paired lateral processes on each side of most segments in polychaete annelids; variously modified for locomotion, respiration, or feeding.

parasite (per′ə-sīt). An organism that lives physically on or in and at the expense of another organism.

parasitism (per′ə-si′ti-zm) (Gr. *parasitos*, from *para*, beside, + *sitos*, food). The condition of an organism living in or on another organism (host) at whose expense the parasite is maintained; destructive symbiosis.

parasitoid (per′ə-si′-tȯid). An organism that is a typical parasite early in its development but that finally kills the host during or at the completion of development; used to reference many insect parasites of other insects.

parasympathetic (pe′rə-sim-pə-thet′ik) (Gr. *para*, beside, + *sympathes*, sympathetic, from *syn*, with, + *pathos*, feeling). One of the subdivisions of the autonomic nervous system, whose neurons originate in the brain and enter the periphery through the brain stem and posterior part of the spinal cord.

parenchyma (pä-ren′kə-mə) (Gr. anything poured in beside). In lower animals, a spongy mass of vacuolated mesenchyme cells filling spaces between viscera, muscles, or epithelia; in some, cell bodies of muscle cells. Also, the specialized tissue of an organ as distinguished from the supporting connective tissue.

parenchymula (pä′ren-ki′mū-lə) (Gr. *para*, beside, + *enchyma*, infusion). Flagellated, solid-bodied larva of some sponges.

parietal (pä-rī-ə-təl) (L. *paries*, wall). Something next to, or forming part of, a wall of a structure.

parietal eye An eye on the top of the head of some vertebrates, functioning primarily as a photoreceptor; also called **pineal eye.**

parsimony (pär′sə-mō′nē′) (L. *parsus*, to spare). A general methodological principle that the simplest hypothesis capable of explaining observations is the best working hypothesis and should be tested first before investigating more complex hypotheses. In phylogenetic systematics, this principle involves using the phylogenetic tree that requires the smallest amount of evolutionary change as the best working hypothesis of phylogenetic relationships.

parthenogenesis (pär′thə-nō-gen′ə-sis) (Gr. *parthenos*, virgin, + L. from Gr. *genesis*, origin). Unisexual reproduction involving the production of young by females not fertilized by males; common in rotifers, cladocerans, aphids, bees, ants, and wasps. A parthenogenetic egg may be diploid or haploid.

particulate inheritance Theories of heredity in which hereditary factors are discrete entities that do not blend when transmitted through the same organism, such as the paired factors identified in Mendel's experiments.

pathogenic (path′ə-jen′ik) (Gr. *pathos*, disease, + N.L. *genic*, giving rise to). Producing or capable of producing disease.

PCR See **polymerase chain reaction.**

peck order A hierarchy of social privilege in a flock of birds.

pecten (L. comb). Any of several types of comblike structures on various organisms, for example, a pigmented, vascular, and comblike process that projects into the vitreous humor from the retina at a point of entrance of the optic nerve in the eyes of all birds and many reptiles.

pectines (pek′tīnz) (L. comb, pl. of **pecten**). Sensory appendage on abdomens of scorpions.

pectoral (pek′tə-rəl) (L. *pectoralis*, from *pectus*, the breast). Of or pertaining to the breast or chest; to the pectoral girdle; or to a pair of keratinized shields of the plastron of certain turtles.

pedal laceration Asexual reproduction in sea anemones, a form of fission.

pedalium (pə-dal′ē-əm) (L. *pedalis*, of or belonging to the foot). Flattened blade at the base of the tentacles in cubozoan medusae (Cnidaria).

pedicel (ped′ə-sel) (L. *pediculus*, little foot). A small or short stalk or stem. In insects, the second segment of an antenna or the waist of an ant.

pedicellaria (ped′ə-sə-ler′ē-ə) (L. *pediculus*, little foot, + *aria*, like or connected with). One of many minute pincerlike organs on the surface of certain echinoderms.

pedipalps (ped′ə-palps′) (L. *pes, pedis*, foot, + *palpus*, stroking, caress). Second pair of appendages of arachnids.

peduncle (pē-dun′-kəl) (L. *pedunculus*, dim. of *pes*, foot). A stalk. Also, a band of white matter joining different parts of the brain.

pelage (pel′ij) (Fr. fur). Hairy covering of mammals.

pelagic (pə-laj′ik) (Gr. *pelagos*, the open sea). Occupying or moving through water rather than the underlying substrate; contrasts with **benthic** (see **benthos**).

pellicle (pel′ə-kəl) (L. *pellicula*, dim. of *pellis*, skin). Thin, translucent, secreted envelope covering many protozoa.

pelvic (pel′vik) (L. *pelvis*, a basin). Situated at or near the pelvis, as applied to girdle, cavity, fins, and limbs.

pelycosaur (pel′ə-kō′-sȯr) (Gr. *pelyx*, basin, + *sauros*, lizard). Any of a group of Permian synapsids characterized by homodont dentition and sprawling limbs.

pen A flattened flexible internal support in a squid; a remnant of the ancestral shell.

pentadactyl (pen′-tə-dak′təl) (Gr. *pente*, five, + *daktylos*, finger). With five digits, or five fingerlike parts, to the hand or foot.

pentamerous symmetry (pen-tam′ər-əs) (Gr. *pente*, five, + *meros*, part). A radial symmetry based on five or multiples thereof.

peptidase (pep′tə-dās) (Gr. *peptein*, to digest, + *ase*, enzyme suffix). An enzyme that splits peptides, releasing smaller peptides or amino acids.

peptide bond A bond that binds amino acids together into a polypeptide chain, formed by removing an OH from the carboxyl group of one amino acid and an H from the amino group of another to form an amide bond –CO–NH–.

perennibranchiate (pə-ren′ə-brank′ē-āt) (L. *perennis*, throughout the year, + Gr. *branchia*, gills). Having permanent gills, relating especially to certain paedomorphic salamanders.

pericardium (per′-i-kär′dē-əm) (Gr. *peri*, around, + *kardia*, heart). Area around heart; membrane around heart.

perihemal channels Coelomic compartments that may aid in distributing nutrients throughout the bodies of echinoderms.

periostracum (per′-ē-äs′trə-kəm) (Gr. *peri*, around, + *ostrakon*, shell). Outer layer of a mollusc shell.

peripheral (pə-ri′fər-əl) (Gr. *peripherein*, to move around). Structure or location distant from center, near outer boundaries.

periproct (per′ə-präkt) (Gr. *peri*, around, + *prōktos*, anus). Region of aboral plates around the anus of echinoids.

perisarc (per′ə-särk) (Gr. *peri*, around, + *sarx*, flesh). Sheath covering the stalk and branches of a hydroid.

perissodactyl (pə-ris′ə-dak′təl) (Gr. *perissos*, odd, + *daktylos*, finger, toe). Pertaining to an order of ungulate mammals with an odd number of digits.

peristalsis (per′ə-stal′səs) (Gr. *peristaltikos*, compressing around). The series of alternate relaxations and contractions that force food through an alimentary canal.

peristomium (per′ə-stō′mē-əm) (Gr. *peri*, around, + *stoma*, mouth). Foremost true segment of an annelid; it bears the mouth.

peritoneum (per′ə-tə-nē′əm) (Gr. *peritonaios*, stretched around). The membrane that lines the coelom and covers the coelomic viscera.

Permian extinction A **mass extinction** that occurred 245 million years ago in which 96% of existing species became extinct, marking the end of the Paleozoic era.

perpetual change The most basic theory of evolution, that the living world is neither constant nor cycling, but is always undergoing irreversible modification through time.

petaloids (pe′tə-lóids) (Gr. *petlon*, leaf, + *eidos*, form). Describes flowerlike arrangement of respiratory podia in irregular sea urchins.

pH (*potential of hydrogen*). A symbol denoting the relative concentration of hydrogen ions in a solution; pH values are from 0 to 14, and the lower the value, the more acid or hydrogen ions in the solution. Equal to the negative logarithm of the hydrogen ion concentration.

phagocyte (fag′ō-sīt) (Gr. *phagein*, to eat, + *kytos*, hollow vessel). Any cell that engulfs and devours microorganisms or other particles.

phagocytosis (fag′ō-sī-tō-səs) (Gr. *phagein*, to eat, + *kytos*, hollow vessel). Engulfment of a particle by a phagocyte or a unicellular eukaryote.

phagosome (fa′gō-sōm) (Gr. *phagein*, to eat, + *sōma*, body). Membrane-bound vesicle in cytoplasm containing food material engulfed by phagocytosis.

phagotroph (fag′ō-trōf) (Gr. *phagein*, to eat, + *trophē*, food). A heterotrophic organism that ingests solid particles for food.

pharyngeal pouches Paired evaginations, or pockets, that develop on the inside of the pharynx in chordates. In fishes and protochordates, they open to the outside of the body as pharyngeal slits.

pharynx (far′inks), pl. **pharynges** (Gr. *pharynx*, gullet). The part of the digestive tract between the mouth cavity and the esophagus that, in chordates, is common to both digestive and respiratory tracts.

phasmid (faz′mid) (Gr. *phasma*, apparition, phantom, + *id*). One of a pair of glands or sensory structures in the posterior end of certain nematodes.

phenetic taxonomy (fə-ne′tik) (Gr. *phaneros*, visible, evident). Refers to the use of a criterion of overall similarity to classify organisms into taxa; contrasts with classifications based explicitly on a reconstruction of phylogeny.

phenotype (fē′nō-tīp) (Gr. *phainein*, to show). The visible or expressed characteristics of an organism, influenced by the genotype, although not all genes in the genotype are expressed.

phenotypic gradualism The hypothesis that new traits, even those that are strikingly different from ancestral ones, evolve by a long series of small, incremental steps.

pheromone (fer′ō-mōn) (Gr. *pherein*, to carry, + *hormōn*, exciting, stirring up). Chemical substance released by one organism that influences the behavior or physiological processes of another organism.

phosphagen (fäs′fə-jən) (phosphate + gen). A term for creatine phosphate and arginine phosphate, which store and may be sources of high-energy phosphate bonds.

phosphate group $\mathrm{HO{-}\!\!\!\!\overset{\displaystyle OH}{\underset{\displaystyle \underset{O}{\parallel}}{P}}\!\!\!\!{-}O{-}}$,

chemical compound present in up to three copies in series on the 5′ carbon of the sugar moiety of a nucleotide; in a nucleic acid, a single copy connects the 5′ carbon of one nucleotide to the 3′ carbon of an adjacent nucleotide.

phosphorylation (fäs′fō-rə-lā′shən). The addition of a phosphate group, that is, —PO₃ , to a compound.

photic (fō′-tik). Sunlit portions of oceanic waters inhabited by photosynthetic organisms.

photoautotroph (fōt-ō-aw′-tō-trōf) (Gr. *photōs*, light, + *autos*, self, + *trophos*, feeder). An organism requiring light as a source of energy for making organic nutrients from inorganic raw materials.

photosynthesis (fōt-ō-sin′thə-sis) (Gr. *phōs*, light, + *synthesis*, action or putting together). The synthesis of carbohydrates from carbon dioxide and water in chlorophyll-containing cells exposed to light.

phototaxis (fōt′ō-tak′sis) (Gr. *phōs*, light, + *taxis*, arranging, order). A taxis in which light is the orienting stimulus. An involuntary tendency for an organism to turn toward (positive) or away from (negative) light.

phototrophs (fōt′-ō-trōfs) (Gr. *phōs*, *photōs*, light, + *trophē*, nourishment). Organisms

capable of using CO₂ in the presence of light as a source of metabolic energy.

phyletic gradualism A model of evolution in which morphological evolutionary change is continuous and incremental and occurs mainly within unbranched species or lineages over long periods of geological time; contrasts with **punctuated equilibrium.**

phyllopodium (fī′lə-pō′dē-əm) (Gr. *phyllon*, leaf, + *pous, podos*, foot). Leaflike swimming appendage of branchiopod crustaceans.

phylogenetic species concept Defines species as an irreducible (basal) cluster of organisms, diagnosably distinct from other such clusters, and within which there is a parental pattern of ancestry and descent.

phylogenetic systematics See **cladistics.**

phylogenetic tree A tree diagram whose branches represent current or past evolutionary lineages and which shows the hypothesized patterns of common descent among those lineages.

phylogeny (fī-loj′ə-nē) (Gr. *phylon*, tribe, race, + *geneia*, origin). The origin and diversification of any taxon, or the evolutionary history of its origin and diversification, usually presented in the form of a dendrogram.

phylum (fī′ləm), pl. **phyla** (N.L. from Gr. *phylon*, race, tribe). A chief category, between kingdom and class, of taxonomic classifications into which are grouped organisms of common descent that share a fundamental pattern of organization.

physiology (L. *physiologia*, natural science). A branch of biology covering the organic processes and phenomena of an organism or any of its parts or of a particular bodily process.

phytoflagellates (fī-tə-fla′jə-lāts). Members of the former class Phytomastigophorea, plantlike flagellates.

phytophagous (fī-tof′ə-gəs) (Gr. *phyton*, plant, + *phagein*, to eat). Organisms that feed on plants.

pinacocyte (pin-a′-kō-sīt′) (Gr. *pinax*, tablet, + *kytops*, hollow vessel). Flattened cells composing dermal epithelium in sponges.

pinacoderm (pin-a′-kō-dərm′) (Gr. *pinax*, plank, tablet, + *derma*, skin). The layer of pinacocytes in sponges.

pinna (pin′ə) (L. feather, sharp point). The external ear. Also a feather, wing, or fin or similar part.

bat / āpe / ärmadillo / herring / fēmale / finch / līce / crocodile / crōw / cóin / cõre / duck / dūke / ūnicorn / ə indicates unaccented vowel sound "uh" as in mammal, fishes, cardinal, heron, vulture / stress as in bi-ol′o-gy, bi′o-log′i-cal

pinocytosis (pin′ō-sī-tō′sis, pīn′o-sī-tō′sis) (Gr. *pinein*, to drink, + *kytos*, hollow vessel, + *osis*, condition). Fluid acquisition by a cell in which specific receptors bind ions/molecules present on plasma membranes, which are invaginated and pinch off to form small vesicles. See **caveolae.**

pit organ A heat-sensing organ on the snout of some vipers, pythons, and boas.

placenta (plə-sen′tə) (L. flat cake). The vascular structure, embryonic and maternal, through which the mammalian embryo and fetus are nourished while in the uterus.

placentotrophy (plə-sent′ō-trō′fē) (L. *placenta*, flat cake, + *trophos*, one who feeds). Nutrition of an embryo from a placenta.

placode (pla′kōd) (Gr. *plakos*, flat round plate). Localized, platelike thickening of vertebrate head ectoderm from which a specialized structure develops; such structures include eye lens, special sense organs, and certain neurons.

placoderms (plak′ə-dərmz) (Gr. *plax*, plate, + *derma*, skin). A group of heavily armored jawed fishes of the Silurian and Devonian periods.

placoid scale (pla′kȯid) (Gr. *plax, plakos*, tablet, plate). Type of scale found in cartilaginous fishes, with basal plate of dentin embedded in the skin and a backward-pointing spine tipped with enamel.

plankton (plank′tən) (Gr. neuter of *planktos*, wandering). The passively floating animal and plant life of a body of water; contrasts with **nekton.**

plantigrade (plan′tə-grād′) (L. *planta*, sole, + *gradus*, step, degree). Pertaining to mammals that walk on the whole surface of the foot (for example, humans and bears); contrasts with **digitigrade.**

planula (plan′yə-lə) (N.L. dim. from L. *planus*, flat). Free-swimming, ciliated larval type of cnidarians; usually flattened and ovoid, with an outer layer of ectodermal cells and an inner mass of endodermal cells.

plasma cell (plaz′mə) (Gr. *plasma*, a form, mold). Produced from a B cell that has been activated during a humoral immune response; functions to secrete antibodies.

plasma membrane (plaz′mə) (Gr. *plasma*, a form, mold). A living, external, limiting, protoplasmic structure that functions to regulate exchange of nutrients across the cell surface.

plasmalemma (plaz′mə-lem-ə) (Gr. *plasma*, a form, mold, + *lemma*, rind, sheath). The cell membrane or plasma membrane.

plasmid (plaz′məd) (Gr. *plasma*, a form, mold). A small circle of DNA that may be carried by a bacterium in addition to its genomic DNA.

plasmodium (plaz-mō′dē-əm) (Gr. *plasma*, a form, mold, + *eidos*, form). Multinucleate ameboid mass, syncytial.

plastid (plas′təd) (Gr. *plast*, formed, molded, + L. *id*, feminine stem for particle of specified kind). A membranous organelle in plant cells functioning in photosynthesis and/or nutrient storage, for example, chloroplast.

plastron (plast′tron) (Fr. *plastron*, breast plate). Ventral body shield of turtles; structure in corresponding position in certain arthropods; thin film of gas retained by epicuticle hairs of aquatic insects.

platelet (plāt′lət) (Gr. dim. of *plattus*, flat). A tiny cell fragment in the blood that releases substances initiating blood clotting.

plate tectonics Geological shifting of position by rocky plates of the earth's crust relative to underlying layers. This theory explains the changing positions of continents through geological time, formation of mountain ranges, and patterns in the formation of volcanic-island archipelagoes.

pleiotropy, pleiotropic (plī-ə-trō′pē, plī-ə-trō′pic) (Gr. *pleiōn*, more, + *tropos*, to turn). Pertaining to a gene producing more than one effect; affecting multiple phenotypic characteristics.

pleopod (plē′ə-päd) (Gr. *plein*, to sail, + *pous, podos*, foot). One of the swimming appendages on the abdomen of a crustacean.

plesiomorphic (plē′sē-ə-mȯr′fik). An ancestral condition of a variable character.

pleura (plu′rə) (Gr. side, rib). The membrane that lines each half of the thorax and covers the lungs.

plexus (plek′səs) (L. network, braid). A network, especially of nerves or blood vessels.

pluteus (plü′tē-əs), pl. **plutei** (L. *pluteus*, movable shed, reading desk). Echinoid or ophiuroid larva with elongated processes like the supports of a desk; originally called "painter's easel larva."

pneumatophore (nü-ma′-tə-fȯr) (Gr. *pneumatos*, wind, breathing, + *phōros*, bearing). A gas-filled float of the Portuguese man-of-war and some other siphonophores, which are specialized hydrozoan colonies (phylum Cnidaria).

pneumostome (nü′mə-stōm) (Gr. *pneuma*, breathing, + *stoma*, mouth). The opening of the mantle cavity (lung) of pulmonate gastropods to the outside.

podium (pō′dē-əm) (Gr. *pous, podos*, foot). A footlike structure, for example, the tube foot of echinoderms.

poikilothermic (pȯi-ki′lə-thər′mik) (Gr. *poikilos*, variable, + thermal). Pertaining to animals whose body temperature is variable and fluctuates with that of the environment; cold blooded; contrasts with **ectothermic.**

polarity (Gr. *polos*, axis). In systematics, the ordering of alternative states of a taxonomic character from evolutionarily ancestral to derived conditions. In developmental biology, the tendency for the axis of an ovum to orient corresponding to the axis of the mother. Also, condition of having opposite poles; differential distribution of gradation along an axis.

polarization (L. *polaris*, polar, + Gr. *iz*, make). The arrangement of positive electrical charges on one side of a surface membrane and negative electrical charges on the other side (in nerves and muscles).

polian vesicles (pō′lē-ən) (from G. S. Poli, Italian naturalist). Vesicles opening into ring canal in most asteroids and holothuroids.

polyandry (pol′ē-an′drē) (Gr. *polys*, many, + *anēr*, man). Condition of having more than one male mate at one time.

polyembryony Asexual proliferation of a single fertilized egg to produce many embryos.

polygamy (pə-lig′ə-mē) (Gr. *polys*, many, + *gamos*, marriage). Condition of having more than one mate at a time.

polygenic inheritance Inheritance of traits influenced by multiple alleles; traits show continuous variation between extremes; offspring are usually intermediate between the two parents; also called **blending** and **quantitative inheritance.**

polygyny (pə-lij′ə-nē) (Gr. *polys*, many, + *gynē*, woman). Condition of having more than one female mate at one time.

polymer (pä′lə-mər) (Gr. *polys*, many, + *meros*, part). A chemical compound composed of repeated structural units called monomers.

polymerase chain reaction (PCR) A technique for preparing large quantities of DNA from tiny samples, amplifying a specific gene using a pair of synthetic oligonucleotides complementary to its flanking sequences.

polymerization (pə-lim′ər-ə-zā′shən). The process of forming a polymer or polymeric compound.

polymorphism (pä′lē-mor′fī-zəm) (Gr. *polys*, many, + *morphē*, form). The presence in a species of more than one structural type of individual; genetic variation in a population.

polynucleotide (poly + nucleotide). A nucleic acid having many nucleotides combined in a linear chain.

polyp (pä′lip) (Gr. *polypous*, many-footed). Individual of the phylum Cnidaria, generally adapted for attachment to the substratum at the aboral end; often forms colonies.

polypeptide (pä′lē-pep′tīd) (Gr. *polys*, many, + *peptein*, to digest). A molecule consisting of a single linear chain of amino acids.

polyphyletic (pä′lē-fi-let′-ik) (Gr. *polys*, many, + *phylon*, tribe). Derived from more than one ancestral source; contrasts with monophyletic and paraphyletic.

polyphyly (pä′lē-fī′lē) (Gr. *polys*, many, + *phylon*, tribe). The condition that a taxon or other group of organisms does not contain the most recent common ancestor of all members of the group, implying that it has multiple evolutionary origins; such groups are not valid as formal taxa and are recognized as such only through error. Contrasts with **monophyly** and **paraphyly.**

polyphyodont (pä′li-fi′ə-dänt) (Gr. *polyphyes*, manifold, + *odous*, tooth). Having several sets of teeth in succession.

polypide (pä′li-pīd) (L. *polypus*, polyp). In an ectoproct colony, an individual or zooid that has a lophophore, a digestive tract, muscles, and nerve centers.

polyploid (pä′lē-plóid′) (Gr. *polys*, many, + *ploidy*, number of chromosomes). An organism possessing more than two full homologous sets of chromosomes.

polysaccharide (pä′lē-sak′ə-rīd) (Gr. *polys*, many, + *sakcharon*, sugar, from Sanskrit *sarkarā*, gravel, sugar). A carbohydrate composed of many monosaccharide units, for example, glycogen, starch, and cellulose.

polysome (polyribosome) (Gr. *polys*, many, + *sōma*, body). Two or more ribosomes connected by a molecule of messenger RNA.

polyspermy (pä′lē-spər′mē) (Gr. *polys*, many, + *sperma*, seed). Entrance of more than one sperm during fertilization of an egg.

polytene chromosomes (pä′li-tēn) (Gr. *polys*, many, + *tainia*, band). Chromosomes in some somatic cells of some insects in which the chromatin replicates repeatedly without undergoing mitosis.

polytypic (pol-ē-tip′-ik). Refers to a species that contains two or more taxonomically designated subspecies; taxonomic recognition of subspecies is controversial and rejected by many taxonomists.

population (L. *populus*, people). A group of organisms of the same species inhabiting a specific geographical locality.

populational gradualism The observation that new genetic variants become established in a population by increasing their frequencies across generations incrementally, initially from one or a few individuals and eventually characterizing a majority of the population.

porocyte (pō′rə-sīt) (Gr. *porus*, passage, pore, + *kytos*, hollow vessel). A type of cell in asconoid sponges through which water enters the spongocoel.

portal system (L. *porta*, gate). Circulatory structure in which a capillary bed drains into another capillary bed through veins; for example, hepatic portal and renal portal systems in vertebrates.

positive assortative mating A tendency for individuals to mate with others that resemble themselves for one or more varying traits in a population.

posterior (L. latter). Situated at or toward the rear of the body; in bilateral forms, the end of the main body axis opposite the head.

preadaptation Possession of a trait that coincidentally predisposes an organism for survival in an environment different from those encountered in its evolutionary history.

prebiotic synthesis The chemical synthesis that occurred before the emergence of life.

precocial (prē-kō′shəl) (L. *praecoquere*, to ripen beforehand). An organism capable of independent living at or shortly after birth, requiring little parental care; contrasts with **altricial.**

predaceous, predacious (prə-dā′shəs) (L. *praedator*, a plunderer, *praeda*, prey). Living by killing and consuming other animals; predatory.

predation (prə-dā′shən). An interaction between species in an ecological community in which members of one species (prey) serve as food for another species (**predator).**

predator (pred′ə-tər) (L. *praedator*, a plunderer, *praeda*, prey). An organism that preys on other organisms for its food.

preformation Discredited concept that gametes contained already-formed young that unfolded or expanded during development.

prehensile (prē-hen′səl) (L. *prehendere*, to seize). Adapted for grasping.

premunition A resistance to reinfection by an animal (host) when some infective organisms remain in the host's body.

primary bilateral symmetry Usually applied to a radially symmetrical organism descended from a bilateral adult ancestor and developing from a bilaterally symmetrical larva.

primary heterotroph Denotes the hypothesis that the earliest microorganisms to evolve obtained nutrients from an environment that did not contain autotrophic organisms.

primary organizer Region of an embryo near the dorsal lip of the blastopore that is self-differentiating and induces development of the neural plate and main body axis.

primary producer A species whose members begin **productivity** by acquiring energy and matter from **abiotic** sources, such as plants that synthesize sugars from water and carbon dioxide using solar energy (see **photosynthesis**).

primary radial symmetry Usually applied to a radially symmetrical organism that did not have a bilateral adult ancestor or larva, in contrast to a secondarily radial organism.

primary structure For a protein, the linear sequence of amino acids in a polypeptide chain. For a nucleic acid, the linear sequence of bases in the molecule.

primitive (L. *primus,* first). Primordial; ancient; little evolved; characteristics closely approximating those possessed by an early ancestor.

prion (prī′-än). An infectious protein that causes proteins of the host organism to assume a nonstandard and often pathological spatial conformation, as in "mad cow" disease.

proboscis (prō-bäs′əs) (Gr. *pro*, before, + *boskein*, feed). A snout or trunk. Also, tubular sucking or feeding organ with the mouth at the end as in planarians, leeches, and insects. Also, the sensory and defensive organ at the anterior end of certain invertebrates.

producers (L. *producere*, to bring forth). Organisms, such as plants, able to produce their own food from inorganic substances.

production In ecology, the energy accumulated by an organism that becomes incorporated into new biomass.

productivity (prō′dək-tiv′-ət-ē). A property of a biological system measured by the amount of energy and/or materials that it incorporates.

product rule The probability of independent events occurring together is the product of the probabilities of the events occurring separately.

progesterone (prō-jes′tə-rōn′) (L. *pro*, before, + *gestare*, to carry). Mammalian steroid hormone secreted by the corpus luteum and the placenta; prepares the uterus for the fertilized egg and maintains the capacity of the uterus to hold the embryo and fetus.

proglottid (prō-glät′əd) (Gr. *proglōttis*, tongue tip, from *pro*, before, + *glōtta*, tongue, + *id*, suffix). Boxlike portion of a tapeworm containing a set of reproductive organs; usually corresponds to a segment.

prohormone (prō′hor-mōn) (Gr. *pro*, before, + *hormaein*, to excite). A precursor of a hormone, especially a peptide hormone.

prokaryote, prokaryotic, procaryotic (prō-kar′ē-ōt′, prō-kar′ē-ät′ik) (Gr. *pro*, before, + *karyon*, kernel, nut). Not having a membrane-bound nucleus or nuclei. Prokaryotic cells characterize the bacteria and archaea; contrasts with **eukaryotic.**

promoter A region of DNA to which the RNA polymerase must have access for transcription of a structural gene to begin.

pronephros (prō-nef′rōs) (Gr. *pro*, before, + *nephros*, kidney). Most anterior of three pairs of embryonic renal organs of vertebrates, functional only in adult hagfishes and larval fishes and amphibians, and vestigial in amniote embryos. Adj., **pronephric.**

proprioceptor (prō′prē-ə-sep′tər) (L. *proprius*, own, particular, + receptor). Sensory receptor located deep within the tissues, especially muscles, tendons, and joints, that is responsive to changes in muscle stretch, body position, and movement.

prosimian (prō-sim′ē-ən) (Gr. *pro*, before, + L. *simia*, ape). Any member of a group of arboreal primates including lemurs, tarsiers, and lorises, but excluding monkeys, apes, and humans.

prosoma (prō-sōm′ə) (Gr. *pro*, before, + *sōma*, body). Anterior part of an invertebrate in which primitive segmentation is not visible; fused head and thorax of arthropod; cephalothorax.

prosopyle (prōs′ō-pīl) (Gr. *prosō*, forward, + *pyle*, gate). Connections between the incurrent and radial canals in some sponges.

prostaglandins (präs′tə-glan′dənz). A family of fatty-acid hormones, originally discovered in semen, known to have powerful effects on smooth muscle, nerves, circulation, and reproductive organs.

prostomium (prō-stōm′ē-əm) (Gr. *protos*, first, + *stoma*, mouth, + *-idion*, dim. ending). Region of a segmented animal anterior to the mouth.

protandrous (prō-tan′drəs) (Gr. *prōtos*, first, + *anēr*, male). Condition of hermaphroditic animals and plants in which male organs and their products appear before the corresponding female organs and products, thus preventing self-fertilization.

protease (prō′tē-ās) (Gr. *protein*, + *ase*, enzyme). An enzyme that digests proteins; includes proteinases and peptidases.

protein (prō′tēn, prō′tē-ən) (Gr. *protein*, from *proteios*, primary). A macromolecule of carbon, hydrogen, oxygen, nitrogen, and usually sulfur; composed of chains of amino acids joined by peptide bonds; present in all cells.

protein polymorphism Occurrences of allelic variants in the amino acid sequences of proteins within a population or species. Separation of allelic variants by protein electrophoresis provided an early method for quantifying the amount of genetic variation present in natural populations.

proteome (prō′tē-ōm) (Gr. *protein*, + L. suffiix *-oma*, group). The set of protein molecules produced by an organism during its lifetime. Scientific study of this phenomenon is called **proteomics.**

prothoracic glands Glands that secrete the hormone ecdysone, or molting hormone, in the prothorax of insects.

prothoracicotropic hormone (PTTH) Hormone secreted by the brain of insects that stimulates the prothoracic gland to secrete ecdysone, or molting hormone.

prothrombin (prō-thräm′bən) (Gr. *pro*, before, + *thrombos*, clot). A constituent of blood plasma that is changed to thrombin by a complex sequence of chemical reactions involving locally derived factors and circulating plasma proteins; involved in blood clotting.

protist (prō′tist) (Gr. *protos*, first). A member of the paraphyletic kingdom Protista, generally considered to include the unicellular eukaryotes.

protocoel (prō′tə-sēl) (Gr. *protos*, first, + *koilos*, hollow). The anterior coelomic compartment in some deuterostomes; corresponds to the axocoel in echinoderms.

protocooperation A mutually beneficial interaction between organisms in which the interaction is not physiologically necessary to the survival of either.

proton A subatomic particle with a positive electrical charge and having a mass of 1836 times that of an electron; located in the nucleus of atoms.

protonephridium (prō′tə-nə-frid′ē-əm) (Gr. *protos*, first, + *nephros*, kidney). Primitive osmoregulatory or excretory organ consisting of a tubule terminating internally with flame-bulb or solenocyte; the unit of a flame-bulb system.

proto-oncogene See **oncogene.**

protoplasm (prō′tə-plazm) (Gr. *protos*, first, + *plasma*, form). Organized living substance; cytoplasm and nucleoplasm of the cell.

protopod, protopodite (prō′tə-päd, prō-top′ə-dīt) (Gr. *protos*, first, + *pous, podos*, foot).

Basal portion of crustacean appendage, containing coxa and basis.

Protostomia (prō′tə-stō′mē-ə) (Gr. *protos*, first, + *stoma*, mouth). A group of phyla in which cleavage is mosaic, the coelom (in coelomate forms) is formed by proliferation of mesodermal bands (schizocoelic formation), the mesoderm is formed from a particular blastomere (called 4d), and the mouth is derived from or near the blastopore. Includes the Annelida, Arthropoda, Mollusca, and a number of minor phyla; contrasts with **Deuterostomia.**

proventriculus (prō′ven-trik′ū-ləs) (L. *pro*, before, + *ventriculum*, ventricle). In birds the glandular stomach between the crop and the gizzard. In insects, a muscular dilation of the foregut armed internally with chitinous teeth.

proximal (L. *proximus*, nearest). Situated toward or near the point of attachment; opposite of distal, distant.

proximate cause (L. *proximus*, nearest, + *causa*). Explanations of the functioning of a biological system at a particular time and place, such as how an animal performs its metabolic, physiological, and behavioral activities. (= immediate cause)

pseudocoelom (sü′do-sēl-əm) (Gr. *pseudēs*, false, + *koilōma*, cavity). A body cavity not lined with peritoneum and not a part of the blood or digestive systems; embryonically derived from the blastocoel.

pseudocoelomate (sü′də-sē′lə-māt) (Gr. *pseudēs*, false, + *koilōma*, cavity, + *ate*, suffix). Having a body cavity formed from a persistent blastocoel and lined with mesoderm on only one side.

pseudopodium (sü′də-pō′dē-əm) (Gr. *pseudēs*, false, + *podion*, small foot, + *eidos*, form). A temporary cytoplasmic protrusion extended out from an ameboid cell, and serving for locomotion or for engulfing food.

pulmonary (pul′mə-ner′-ē) (L. *pulmo*, lung, + *aria*, suffix denoting connected to). Relating to or associated with lungs.

punctuated equilibrium A model of evolution in which morphological evolutionary change is discontinuous, being associated primarily with discrete, geologically instantaneous events of speciation leading to phylogenetic branching; morphological evolutionary stasis characterizes species between episodes of speciation; contrasts with **phyletic gradualism.**

pupa (pū′pə) (L. girl, doll, puppet). Inactive quiescent stage of the holometabolous insects. It follows the larval stages and precedes the adult stage.

purine (pū′rēn) (L. *purus,* pure, + *urina,* urine). Organic base with carbon and nitrogen atoms in two interlocking rings. The parent substance of adenine, guanine, and other naturally occurring bases.

pygidium (pī-jid′ē-əm) (Gr. *pygē,* rump, buttocks, + *-idion,* dim. ending). Posterior closure of a segmented animal, bearing the anus.

pygostyle (pī-′gō-stīl) (Gr. *pygo,* rump, + *styl,* pillar). A bone at the end of a bird's vertebral column, formed from fused caudal vertebrae.

pyrimidine (pī-rim′ə-dēn) (alter. of pyridine, from Gr. *pyr,* fire, + *id,* adj. suffix, + *ine*). An organic base composed of a single ring of carbon and nitrogen atoms; parent substance of several bases found in nucleic acids.

Q

quaternary structure For a protein that contains more than one polypeptide chain, the three-dimensional configuration formed by bonds between side groups of amino acids located on different polypeptide chains.

quantitative inheritance See **polygenic inheritance.**

queen In entomology, the single fully developed female in a colony of social insects such as bees, ants, and termites, distinguished from workers, nonreproductive females, and soldiers.

R

radial canals Canals along the ambulacra radiating from the ring canal of echinoderms; also choanocyte-lined canals in syconoid sponges.

radial cleavage Embryonic development in which early cleavage planes are symmetrical to the polar axis, each blastomere of one tier lying directly above the corresponding blastomere of the next layer; typically occurs with regulative cleavage.

radial symmetry A morphological condition in which the parts of an animal are arranged concentrically around an oral-aboral axis, and more than one imaginary plane through this axis yields halves that are mirror images of each other.

Radiata (rä′dē-ä′tə) (L. *radius,* ray). Phyla showing radial symmetry, specifically Cnidaria and Ctenophora.

radiolarian (rā′dē-ə-la′rē-ən) (L. *radiolus,* small sunbeam). Amebas with actinopodia and beautiful tests.

radioles (rā′dē-ōlz) (L. *radius,* ray, spoke of a wheel). Featherlike processes from the head of many tubicolous polychaete worms (phylum Annelida), used primarily for feeding.

radula (ra′jə-lə) (L. scraper). Rasping tongue found in most molluscs.

Ras protein A protein that initiates a cascade of reactions leading to cell division when a growth factor is bound to the cell surface. The gene encoding Ras becomes an oncogene when a mutation produces a form of Ras protein that initiates the cascade even in the absence of the growth factor.

ratite (ra′tīt) (L. *ratis,* raft). Referring to birds having an unkeeled sternum; contrasts with **carinate.**

realized niche The role actually performed by an organism or population in its ecological community at a particular time and place as constrained by both its intrinsic biological attributes and its particular environmental conditions. See also **niche** and **fundamental niche.**

recapitulation Summarizing or repeating; hypothesis that an individual repeats its phylogenetic history in its development.

receptor-mediated endocytosis Endocytosis of large molecules, which are bound to surface receptors in clathrin-coated pits.

recessive An allele that must be homozygous to influence a phenotype.

recombinant DNA DNA from two different species, such as a virus and a mammal, combined into a single molecule.

reciprocal altruism Evolution of a behavioral repertoire in which an organism performs behaviors benefiting other members of the population, possibly at its own increased peril, because such behaviors are returned by individuals that receive its benefits.

redia (rē′dē-ə), pl. **rediae** (rē′dē-ē) (from Redi, Italian biologist). A larval stage in the life cycle of flukes; it is produced by a sporocyst larva, and in turn gives rise to many cercariae.

reduction In chemistry, the gain of an electron by an atom or molecule of a substance; also the addition of hydrogen to, or the removal of oxygen from, a substance.

regional endothermy Regulation of only part of the body using its own metabolism, as when tunas maintain high temperature in the swimming muscles and the brain, but not in other organs.

regulative cleavage See **radial cleavage.**

regulative development Embryonic development determined by interactions among neighboring cells; cell fates not determined by cytoplasmic content; same as indeterminate cleavage.

regulatory gene A gene that influences the rate of transcription of another gene. A cis-acting regulatory gene binds transcription factors necessary for expression of another gene located adjacent to it on the same DNA molecule. A trans-acting regulatory gene encodes a transcription factor that influences expression of genes located elsewhere in the genome.

relative fitness A comparison of two or more different genotypes for the average numbers of offspring produced per individual in a population. Based on measurements of relative fitnesses of alternative diploid genotypes, relative fitnesses can be assigned analytically to individual alleles.

releaser (L. *relaxare,* to unloose). Simple stimulus that elicits an innate behavior pattern.

renin (rē′nin) (L. *ren,* kidney). An enzyme produced by the kidney juxtaglomerular apparatus that initiates changes causing increased blood pressure and increased sodium reabsorption.

rennin (re′nən) (M.E. *renne,* to run). A milk-clotting endopeptidase secreted by the stomach of some young mammals, including bovine calves and human infants.

replication (L. *replicatio,* a folding back). In genetics, the duplication of one or more DNA molecules from the preexisting molecule.

reproductive barrier (L. *re,* + *producere,* to lead forward; M.F. *barriere,* bar). The factors that prevent one sexually propagating population from interbreeding and exchanging genes with another population.

reproductive community A general criterion for the species category shared to some degree by all formal species concepts is that species constitute a reproductively bounded population or lineage of populations that does not freely merge with others in nature.

repugnatorial glands (L. *repugnare,* to resist). Glands secreting a noxious substance for defense or offense, for example, as in the millipedes.

---

bat / āpe / ärmadillo / herring / fēmale / finch / līce / crocodile / crōw / côin / cõre / duck / düke / ūnicorn / ə indicates unaccented vowel sound "uh" as in mammal, fishes, cardinal, heron, vulture / stress as in bi-ol′o-gy, bi′o-log′i-cal

resource (rē′so(ə)rs). An available source of nutrition, energy, or space in which to live.

resource-defense polygyny A male gains reproductive access to multiple females indirectly by holding a critical resource.

resource partitioning To coexist in the same habitat, two or more species specialize on different portions of a shared resource; for example, different warbler species forage in different characteristic parts of a conifer tree when they are part of the same ecological community.

respiration (L. *respiratio*, breathing). Gaseous interchange between an organism and its surrounding medium. In the cell, the release of energy by the oxidation of food molecules.

restriction endonuclease An enzyme that cleaves a DNA molecule at a particular base sequence.

rete mirabile (rē′tē mə-rab′ə-lē) (L. wonderful net). A network of small blood vessels so arranged that the incoming blood runs countercurrent to the outgoing blood and thus makes possible efficient exchange between the two bloodstreams. Such a mechanism serves to maintain the high concentration of gases in the fish swim bladder.

reticular (rə-tik′ū-lər) (L. *reticulum*, small net). Resembling a net in appearance or structure.

reticuloendothelial system (rə-tic′ū-lō-en-dō-thēl′ē-əl) (L. *reticulum*, dim. of net, + Gr. *endon*, within, + *thele*, nipple). The fixed phagocytic cells in the tissues, especially the liver, lymph nodes, spleen, and others; more commonly called the mononuclear phagocyte system.

reticulopodia (rə-tik′ū-lə-pō′dē-ə) (L. *reticulum*, dim. of *rete*, net, + *podos, pous*, foot). Pseudopodia that branch and rejoin.

reticulum (rə-tik′ū-ləm) (L. *rete*, dim. *reticulum*, a net). Second stomach of ruminants; a netlike structure.

retina (ret′nə, ret′ən-ə) (L. *rete*, net). The posterior sensory cell layer of the eye. Sensory cells in this layer (rods and/or cones) receive light signals and transmit them to the brain, where images are formed.

retortamonad (rē-tort′ə-mō′nəd) (L. *retro*, bend backward, + *monas*, single). Any member of a protozoan clade comprising certain heterotrophic flagellates.

rhabdite (rab′dīt) (Gr. *r habdos*, rod). Rodlike structures in the cells of the epidermis or underlying parenchyma in certain turbellarians. They are discharged in mucous secretions.

rheoreceptor (rē′ō-rē-cep′tər) (Gr. *rheos*, a flowing, + receptor). A sensory organ of aquatic animals that responds to water current.

rhinarium (rī-na′rē-əm) (Gr. *rhis*, nose). Hairless area surrounding the nose of a mammal.

rhinophore (rī′nə-fōr) (Gr. *rhis*, nose, + *pherein*, to carry). Chemoreceptive tentacles in some molluscs (opisthobranch gastropods).

rhipidistian (rip-ə-dis′tē-ən) (Gr. *rhipis*, fan, + *histion*, sail, web). Member of a group of Paleozoic lobe-finned fishes.

rhizopodia (rī′zə-pō′dē-ə) (Gr. *rhiza*, root, + *podos*, foot). Branched filamentous pseudopodia made by some amebas.

rhopalium (rō-pā′lē-əm) (N.L. from Gr. *rhopalon*, a club). One of the marginal, club-shaped sense organs of certain jellies; tentaculocyst.

rhoptries (rop′trēz) (Gr. *rhopalon*, club, + *tryō*, to rub, wear out). Club-shaped bodies in Apicomplexa composing one of the structures of the apical complex; open at anterior and apparently functioning in penetration of host cell.

rhynchocoel (ring′kō-sēl) (Gr. *rhynchos*, snout, + *koilos*, hollow). In nemerteans, the dorsal tubular cavity that contains the inverted proboscis. It has no opening to the outside.

ribonucleic acid (RNA) A linear polymer of nucleotides containing the sugar ribose, often bending to form complex spatial structures stabilized by hydrogen bonds between bases of nonadjacent nucleotides. It functions in gene expression and protein synthesis in all living forms and is hypothesized to have been the basis of early, precellular life.

ribose (rī′bōs). A five-carbon sugar forming part of the fundamental structure of RNA nucleotides, including ATP molecules used to store chemical energy in cellular metabolism.

ribosomal RNA (rRNA) Ribonucleic acids that form the physical structures of ribosomes in association with ribosomal proteins.

ribosome (rī′bə-sōm). Subcellular structure composed of protein and ribonucleic acid. Occurs free in the cytoplasm or attached to the membranes of the endoplasmic reticulum; functions in protein synthesis.

ritualization In ethology, the evolutionary modification, usually intensification, of a behavior pattern to serve communication. Includes symbolic **ritualized threat displays** that carry mutually understood meanings to establish a dominance hierarchy within a population.

RNA Ribonucleic acid, of which there are several different kinds, such as messenger RNA, ribosomal RNA, and transfer RNA (mRNA, rRNA, tRNA), as well as many structural and regulatory RNAs.

RNA polymerase One of three kinds of enzymes that synthesize RNA using ribonucleotide triphosphates (ATP, CTP, GTP, UTP) and a DNA template. In eukaryotes, RNA polymerase I synthesizes ribosomal RNA; RNA polymerase II synthesizes messenger RNA; and RNA polymerase III synthesizes transfer RNA.

RNA world Hypothetical stage in the evolution of life on earth in which both catalysis and replication were performed by RNA, not protein enzymes and DNA.

rod One of a group of cells in the retina of a vertebrate eye serving vision in low light.

rostellum (ros′tel′ləm) (L. small beak). Projecting structure on scolex of tapeworm, often with hooks.

rostrum (ros′trəm) (L. ship's beak). A snoutlike projection on the head.

rotational cleavage Unique pattern of early embryo formation in most mammals whereby cells at the second cell division appear to have turned relative to each other.

round dance A behavior performed by a hive bee to notify other members of the hive population that a food source is nearby.

rumen (rü′mən) (L. cud). The large first compartment of the stomach of ruminant mammals. Serves as a fermentation chamber in which bacteria degrade cellulose.

ruminant (rüm′ə-nənt) (L. *ruminare*, to chew the cud). Artiodactyl mammals with a complex four-chambered stomach including a foregut that contains bacteria and serves as a fermentation chamber.

S

saccule (sa′kūl) (L. *sacculus*, small bag). Small chamber of the membranous labyrinth of the inner ear.

sacrum Adj., **sacral** (sā′krəm, sā′krəl) (L. *sacer*, sacred). Bone formed by fused vertebrae to which pelvic girdle is attached; pertaining to the sacrum.

sagittal (saj′ə-təl) (L. *sagitta*, arrow). Pertaining to the median anteroposterior plane that divides a bilaterally symmetrical organism into right and left halves.

salamander (sal′-ə-man′-der). Denotes any member of the amphibian order Urodela (also called Caudata, sometimes with a different content of fossil forms).

salt (L. *sal*, salt). The reaction product of an acid and a base; dissociates in water solution to negative and positive ions, but not H$^+$ or OH$^-$.

salt marsh A marine community located on intertidal sand flats typically containing grasses, marsh mussels, burrowing crabs and shrimp, and deposit-feeding polychaetes; provides nursery areas for many marine fishes.

saprophagous (sə-prof′ə-gəs) (Gr. *sapros,* rotten, + *phagos,* from *phagein,* to eat). Feeding on decaying matter; saprobic; saprozoic.

saprozoic nutrition (sap-rə-zō′ik) (Gr. *sapros,* rotten, + *zōon,* animal). Animal nutrition by absorption of dissolved salts and simple organic nutrients from surrounding medium; also refers to feeding on decaying matter.

sarcolemma (sär′kə-lem′ə) (Gr. *sarx,* flesh, + *lemma,* rind). The thin, noncellular sheath that encloses a striated muscle fiber.

sarcomere (sär′kə-mir) (Gr. *sarx,* flesh, + *meros,* part). Transverse segment of striated muscle forming the fundamental contractile unit.

sarcoplasm (sär′kə-plaz′əm) (Gr. *sarx,* flesh, + *plasma,* mold). The clear, semifluid cytoplasm between the fibrils of muscle fibers.

saturated fatty acid A subunit of fats containing a carboxyl group and a carbon chain in which all carbons are joined by single bonds and each carbon in the chain is bonded to two hydrogen atoms.

sauropterygians (so-rop′tə-rij′ē-əns) (Gr. *sauros,* lizard, + *pteryginos,* winged). Marine Mesozoic reptiles often with paddle-like limbs and elongate necks; includes plesiosaurs, pliosaurs, and placodonts.

scalids (skä′-lədz) (Gr. *skalis,* hoe, mattock). Recurved spines on the head of kinorhynchs.

schistosomiasis (shis′tō-sō-mī′-ə-sis) (Gr. *schistos,* divided, + *soma,* body, + *lasis,* a diseased condition). Infection with blood flukes of the genus *Schistosoma.*

schizocoel (skiz′ō-sēl) (Gr. *schizo,* from *schizein,* to split, + *koilōma,* cavity). A coelom formed by the splitting of embryonic mesoderm. Noun, **schizocoelomate,** an animal with a schizocoel, such as an arthropod or mollusc. Adj., **schizocoelous,** the process of forming a coelom by mesodermal splitting.

schizocoelous mesoderm formation (skiz′ō-sēl-ləs). Embryonic formation of the mesoderm as cords of cells between ectoderm and endoderm; splitting of these cords produces the coelomic space.

schizogony (ski-zo′gə-nē) (Gr. *schizein,* to split, + *gonos,* seed). Multiple asexual fission.

scientific revolution Philosopher Thomas Kuhn's term for a phase of scientific discovery in which research reveals flaws in an existing paradigm causing it to be discarded in favor of an alternative one.

sclerite (skler′īt) (Gr. *sklēros,* hard). A hard chitinous or calcareous plate or spicule; one of the plates forming the exoskeleton of arthropods, especially insects.

scleroblast (skler′ə-blast) (Gr. *sklēros,* hard, + *blastos,* germ). An amebocyte specialized to secrete a spicule; occurs in sponges.

sclerocyte (skler′ə-sīt) (Gr. *sklēros,* hard, + *kytos,* hollow vessel). An amebocyte in sponges that secretes spicules.

sclerotic (skler-ot′ik) (Gr. *sklēros,* hard). Pertaining to the tough outer coat of the eyeball.

sclerotin (skler′-ə-tən) (Gr. *sklērotēs,* hardness). Insoluble, tanned protein permeating the cuticle of arthropods.

sclerotization (sklər′ə-tə-zā′shən). Process of hardening the cuticle of arthropods by formation of stabilizing cross linkages between peptide chains of adjacent protein molecules.

scolex (skō′leks) (Gr. *skōlēx,* worm, grub). An attachment organ or holdfast of a tapeworm; bears suckers and, in some, hooks, and posterior to it new proglottids are differentiated.

scrotum (skrō′təm) (L. bag). The pouch that contains the testes in most mammals.

scyphistoma (sī-fis′tə-mə) (Gr. *skyphos,* cup, + *stoma,* mouth). The polyp form of a scyphozoan.

sea-grass bed A coastal subtidal marine community comprising a thicket of sea grasses and associated invertebrates, including hydroids, sponges, ectoprocts, and larvae of sea scallops.

sebaceous (sə-bāsh′əs) (L. *s ebaceus,* made of tallow). A type of mammalian epidermal gland that produces a fatty substance.

sebum (sē′bəm) (L. grease, tallow). Oily secretion of the sebaceous glands of the skin.

secondary induction Specification of cell fates due to interactions with cells not belonging to the primary organizer region of the embryo.

secondary structure For a protein, the three-dimensional configuration formed by bond angles between adjacent amino acids in the linear polypeptide chain. A common secondary structure is an alpha-helix, which makes helical turns in a clockwise direction like a screw.

second law of thermodynamics Physical systems tend to proceed to a state of increased disorder termed entropy.

sedentary (sed′ən-ter-ē). Stationary, sitting, inactive; staying in one place.

segment-polarity genes Genes active during development to specify anterior-posterior structures within a segment.

segmentation Division of the body into discrete segments or metameres; also called **metamerism.**

selectively permeable Permeable to small particles, such as water and certain inorganic ions, but not to larger molecules.

semelparity (se′məl-pe′ri-tē′). A life history in which individual organisms of a population normally reproduce only one time before dying, although numerous offspring may be produced at the time of reproduction; contrasts with **iteroparity.**

seminiferous tubules (sem-ə-nif′rəs) (L. *semen,* semen, + *ferre,* to bear). Tubes within the testes where sperm develop and that carry the sperm from the testes to become a component of semen.

semipermeable (L. *semi,* half, + *permeabilis,* capable of being passed through). Permeable to small particles, such as water and certain inorganic ions, but not to larger molecules.

sensillum, pl. sensilla (sen-si′ləm) (L. *sensus,* sense). A small sense organ, especially in arthropods.

sensitization (sen′-sət-ə-zā′-shən). A kind of learning in which an animal acquires a characteristic response to a particular stimulus; contrasts with **habituation.**

septal filament Unattached edge of an internal partition (septum) in a sea anemone gastrovascular cavity that extends into the cavity and is armed with nematocysts and gland cells.

septum, pl. septa (L. fence). A wall between two cavities.

sequential hermaphrodite A species in which individuals initially mature as one sex, but later can change to the opposite sex.

serial homology See **homology.**

serosa (sə-rō′sə) (N.L. from L. *serum,* serum). The outer embryonic membrane of birds and reptiles; chorion. Also, the peritoneal lining of the body cavity.

serotonin (ser′ə-tōn′ən) (L. *serum*, serum). A phenolic amine serving as a neurotransmitter in the central nervous system; also occurs in the serum of clotted blood and in many other tissues; 5-hydroxytryptamine.

serous (sir′əs) (L. *serum*, serum). Watery, resembling serum; applied to glands, tissue, cells, fluid.

serum (sir′əm) (L. whey, serum). The liquid that separates from the blood after coagulation; blood plasma from which fibrinogen has been removed. Also, the clear portion of a biological fluid separated from its particulate elements.

sessile (ses′əl) (L. *sessilis*, low, dwarf). Attached at the base; fixed to one spot.

set [of chromosomes] All chromosomal material present in a haploid egg or sperm; exact composition of a set of chromosomes varies among species.

seta (sē′tə), pl. **setae** (sē′tē) (L. bristle). A needlelike chitinous structure of the integument of annelids, arthropods, and others.

sex chromosomes Chromosomes that determine gender of an animal. They may bear a few or many other genes.

sex ratio An accounting of the proportion of males versus females in a population at a particular time and place.

sexual selection Differential propagation among varying organisms caused by the greater success of some forms in the reproductive process (mating success and fertility). A characteristic favored by sexual selection may be detrimental to survival and disfavored by natural selection.

sibling species Reproductively isolated species that are so similar morphologically that they are difficult or impossible to distinguish using morphological characters.

sickle cell anemia A condition in individuals homozygous for hemoglobin S (HbS) that causes the red blood cells to collapse (sickle) under oxygen stress.

sign stimulus An ethological term denoting an entity (such as a particular sound, shape, or color) whose perception by an animal elicits a stereotypical behavioral pattern.

siliceous (sə-li′shəs) (L. *silex*, flint). Containing silica.

simian (sim′ē-ən) (L. *simia*, ape). Pertaining to monkeys or apes.

sinistral (si′nə-strəl, sə-ni′strəl) (L. *sinister*, left). Pertaining to the left; in gastropods, shell is sinistral if opening is to left of

columella when held with spire up and facing observer.

sink deme A subpopulation (deme) whose members are drawn disproportionately from other subpopulations of the same species (see **metapopulation dynamics**); for example, a deme occupying an environmentally unstable area whose members are periodically destroyed by climatic changes and then replenished by colonists from other demes when favorable conditions are restored.

sinoatrial (SA) node (si′nō-āt′rē-əl) (L. *sinus*, curved, + *atrium*, hall). Specialized cardiac muscle cells in the wall of the right atrium that serve as the pacemaker in a tetrapod heart.

sinus (sī′nəs) (L. curve). A cavity or space in tissues or in bone.

siphon (sī′-fən). A tube for directing water flow.

siphonoglyph (sī′fän′ə-glif′) (Gr. *siphōn*, reed, tube, siphon, + *glyphē*, carving). Ciliated furrow in the gullet of sea anemones.

siphuncle (sī′fun-kəl) (L. *siphunculus*, small tube). Cord of tissue running through the shell of a nautiloid, connecting all chambers with the animal's body.

sister group (= "sister taxon"). The relationship between a pair of species or higher taxa that are each other's closest phylogenetic relatives.

smallest distinct groupings A general criterion for the species category shared to some degree by all formal species concepts is that species be the least inclusive population or lineage of populations having a unique history of common descent. Violation of this criterion would blur the distinction between species and higher taxa.

socially coordinated behavior Any activity in which an organism adjusts its actions to the presence of others to increase its own reproductive success. Includes cooperative behaviors as well as agonistic ones.

sociobiology Ethological study of social behavior in humans or other animals.

solenia (sō-len′ē-ə) (Gr. *sōlēn*, pipe). Channels through the coenenchyme connecting the polyps in an octocorallian colony (phylum Cnidaria).

solenocyte (sō-len′ə-sīt) (Gr. *sōlēn*, pipe, + *kytos*, hollow vessel). Special type of flame bulb in which the bulb bears a flagellum instead of a tuft of flagella. See **flame cell, protonephridium.**

solvent (sol′vent). A liquid into which a substance is dissolved.

soma (sō′mə) (Gr. body). The whole of an organism except the germ cells (germ plasm).

somatic (sō-mat′ik) (Gr. *sōma*, body). Refers to the body, for example, somatic cells in contrast to germ cells.

somatic embryogenesis An asexual reproductive process in sponges whereby isolated cells or bits of tissue reorganize into an incipient sponge.

somatocoel (sə-mat′ə-sēl) (Gr. *sōma*, the body, + *koilos*, hollow). Posterior coelomic compartment of echinoderms; left somatocoel gives rise to oral coelom, and right somatocoel becomes aboral coelom.

somatoplasm (sō′ma-tō-pla′zm) (Gr. *sōma*, body, + *plasma*, anything formed). The living matter that forms the mass of the body as distinguished from germ plasm, which comprises the reproductive cells. The protoplasm of body cells.

somite (sō′mīt) (Gr. *soma*, body). One of the blocklike masses of mesoderm arranged segmentally (metamerically) in a longitudinal series beside the neural tube of the embryo; metamere.

sorting Differential survival and reproduction among varying individuals; often confused with natural selection, which is one possible cause of sorting.

southern evergreen forests Pine-dominated woodlands of the southeastern United States.

source deme A stable subpopulation (deme) that serves differentially as a source of colonists for establishing, joining, or replacing other such subpopulations of the same species (see **metapopulation dynamics**); for example, a deme inhabiting an environmentally stable area whose members routinely establish transitory populations in environmentally unstable nearby areas.

speciation (spē′sē-ā′shən) (L. *species*, kind). The evolutionary process or event by which new species arise.

species (spē′shez, spē′sēz) sing. and pl. (L. particular kind). A group of interbreeding individuals of common ancestry that are reproductively isolated from all other such groups; a taxonomic unit ranking below a genus and designated by a binomen consisting of its genus and the species name.

species diversity The number of different **species** that coexist at a given time and place to form an **ecological community.**

species epithet The second (uncapitalized) word in the Linnaean binomial nomenclature of species, used to separate an individual species from other members of the same genus.

species selection Differential rates of speciation and/ or extinction among varying evolutionary lineages caused by interactions among species-level emergent characteristics and the environment; contrasts with **effect macroevolution.**

Spemann organizer A region of an embryo acting as the primary organizer (see **primary organizer**).

spermatheca (spər′mə-thē′kə) (Gr. *sperma*, seed, + *thēkē*, case). A sac in the female reproductive organs for the reception and storage of sperm.

spermatid (spər′mə-tid) (Gr. *sperma*, seed, + *eidos*, form). A growth stage of a male reproductive cell arising by division of a secondary spermatocyte; gives rise to a spermatozoon.

spermatocyte (spər-ma′-tō-sīt) (Gr. *sperma*, seed, + *kytos*, hollow vessel). A growth stage of a male reproductive cell; gives rise to a spermatid.

spermatogenesis (spər-ma′-tō-jen′-ə-səs) (Gr. *sperma*, seed, + *genesis*, origin). Formation and maturation of spermatozoa.

spermatogonium (spər′ma′-tō-gō′nē-əm) (Gr. *sperma*, seed, + *gonē*, offspring). Precursor of mature male reproductive cell; gives rise directly to a spermatocyte.

spermatophore (spər-ma′-tō-fōr′) (Gr. *sperma*, *spermatos*, seed, + *pherein*, to bear). Capsule or packet enclosing sperm, produced by males of several invertebrate groups and a few vertebrates.

sphincter (sfingk′tər) (Gr. *sphinkter*, band, sphincter, from *sphingein*, to bind tight). A ring-shaped muscle capable of closing a tubular opening by constriction.

spicule (spi′kyul) (L. dim. *spica*, point). One of the minute calcareous or siliceous skeletal bodies found in sponges, radiolarians, soft corals, and sea cucumbers.

spiracle (spi′rə-kəl) (L. *spiraculum*, from *spirare*, to breathe). External opening of a trachea in arthropods. One of a pair of openings on the head of elasmobranchs for passage of water. Exhalent aperture of tadpole gill chamber.

spiral cleavage A type of embryonic cleavage in which cleavage planes are diagonal to the polar axis, and unequal cells are produced by the alternate clockwise and counterclockwise cleavage around the axis of polarity; typically occurs with mosaic cleavage.

spiral valve A thin surface within the conus arteriosus of an amphibian heart that directs oxygenated blood toward the lungs and deoxygenated blood to other body organs.

spongin (spun′jin) (L. *spongia*, sponge). Fibrous, collagenous material forming the skeletal network of demosponges.

spongioblast (spun′jē-o-blast) (Gr. *spongos*, sponge, + *blastos*, bud). Cell in a sponge that secretes spongin, a protein.

spongocoel (spun′jō-sēl) (Gr. *spongos*, sponge, + *koilos*, hollow). Central cavity in sponges.

spongocyte (spun′jō-sīt) (Gr. *spongos*, sponge, + *kytos*, hollow vessel). A cell in sponges that secretes spongin.

sporocyst (spō′rə-sist) (Gr. *sporos*, seed, + *kystis*, pouch). A larval stage in the life cycle of flukes; it originates from a miracidium.

sporogony (spor-og′ə-nē) (Gr. *sporos*, seed, + *gonos*, birth). Multiple fission to produce sporozoites after zygote formation.

sporozoite (spō′rə-zō′īt) (Gr. *sporos*, seed, + *zōon*, animal, + *ite*, suffix for body part). A stage in the life history of many sporozoans; released from oocysts.

squalene (skwā′lēn) (L. *squalus*, a kind of fish). A liquid acyclic triterpene hydrocarbon found especially in the liver oil of sharks; because of its low density, this lipid helps to maintain buoyancy in sharks.

squamous epithelium (skwä′məs) (L. *squama*, scale, + *osus*, full of). Simple epithelium of flat, nucleated cells.

stabilizing selection A selective process in which the mean value of a quantitative phenotype is favored over extreme values in a population, potentially stabilizing the mean value.

stapes (stā′pēz) (L. stirrup). Stirrup-shaped innermost bone of the middle ear; homologous to the hymandibula of fishes.

statoblast (stat′ə-blast) (Gr. *statos*, standing, fixed, + *blastos*, germ). Asexually produced overwintering stage of many freshwater ectoprocts.

statocyst (Gr. *statos*, standing, + *kystis*, bladder). Sense organ of equilibrium; a fluid-filled cellular cyst containing one or more granules (statoliths) used to sense direction of gravity.

statolith (Gr. *statos*, standing, + *lithos*, stone). Small calcareous body resting on tufts of cilia in the statocyst.

stenohaline (sten-ə-hā′līn, -lən) (Gr. *stenos*, narrow, + *hals*, salt). Pertaining to aquatic organisms that have restricted tolerance to changes in environmental saltwater concentration.

stenophagous (stə-nof′ə-gəs) (Gr. *stenos*, narrow, + *phagein*, to eat). Eating few kinds of foods.

stenotopic (sten-ə-tō′pik) (Gr. *stenos*, narrow, + *topos*, place). Refers to an organism with a narrow range of adaptability to environmental change; having a restricted environmental distribution.

stereogastrula (ste′rē-ə-gas′trə-lə) (Gr. *stereos*, solid, + *gastēr*, stomach, + L. *ula*, dim). A solid type of gastrula, such as the planula of cnidarians.

stereom (ster′ē-ōm) (Gr. *stereos*, solid, hard, firm). Meshwork structure of endoskeletal ossicles of echinoderms.

stereotyped behavior A pattern of behavior repeated with little variation in performance.

sternum (ster′nəm) (L. breastbone). Ventral plate of an arthropod body segment; breastbone of vertebrates.

sterol (ste′rol), **steroid** (ste′roid) (Gr. *stereos*, solid, + L. *ol*, from *oleum*, oil). One of a class of organic compounds containing a molecular skeleton of four fused carbon rings; it includes cholesterol, sex hormones, adrenocortical hormones, and vitamin D.

stigma (Gr. *stigma*, mark, tattoo mark). Eyespot in certain unicellular eukaryotes. Spiracle of certain terrestrial arthropods.

stolon (stō′lən) (L. *stolō, stolonis*, a shoot, or sucker of a plant). A rootlike extension of the body wall giving rise to buds that may develop into new zooids, thus forming a compound animal in which the zooids remain united by the stolon. Occurs in some colonial anthozoans, hydrozoans, ectoprocts, and ascidians.

stoma (stō′mə) (Gr. mouth). A mouthlike opening.

stomochord (stō′mə-kord) (Gr. *stoma*, mouth, + *chordē*, cord). Anterior evagination of the dorsal wall of the buccal cavity into the proboscis of hemichordates; the buccal diverticulum.

stramenopile (strə-men′ə-pīl) (L. *stramen*, straw, + *pilus*, a hair). Any member of a clade of unicellular eukaryotes with tubular mitochondrial cristae, and typically having 3-part tubular hairs on a long anterior flagellum.

strobila (strō′bə-lə) (Gr. *strobilē*, lint plug like a pine cone [*strobilos*]). A stage in the development of the scyphozoan jellyfish. Also, the chain of proglottids of a tapeworm.

strobilation (strō′bə-lā′shən) (Gr. *strobilos,* a pine cone). Repeated, linear budding of individuals, as in scyphozoans (phylum Cnidaria), or sets of reproductive organs, as in tapeworms (phylum Platyhelminthes).

stroma (strō′mə) (Gr. *strōma,* bedding). Supporting connective tissue framework of an animal organ; filmy framework of red blood corpuscles and certain cells.

structural gene A gene carrying the information to construct a protein.

subnivean (sub-ni′vē-ən) (L. *sub,* under, below, + *nivis,* snow). Applied to environments beneath snow, in which snow insulates against a colder atmospheric temperature.

substrate The substance upon which an enzyme acts; also, a base or foundation (substratum); and the substance or base on which an organism grows.

sucrase An enzyme in the plasma membrane of cells lining the intestine that splits sucrose into glucose and fructose.

sugar The fundamental subunit of carbohydrates having carbon, hydrogen, and oxygen in a usual ratio of 1:2:1, respectively. Two sugars (deoxyribose, ribose) form part of the nucleotide structure of nucleic acids.

superficial cleavage Cleavage in which nuclear division occurs without cytoplasmic division in the early stages. Later, the cell membrane moves inward to enclose nuclei in cells.

survivorship (sər-vī′vər-ship). The proportion of individuals of a cohort or population that persist from one point in their life history (such as birth) to another one (such as reproductive maturity or a specified age).

suspension feeder Aquatic organisms that collect suspended food particles from the surrounding water; particles are filtered or taken by other methods.

swim bladder Gas-filled sac of many bony fishes used in buoyancy and in some cases, respiratory gas exchange.

sycon (sī′kon) (Gr. *sykon,* fig). A type of canal system in certain sponges. Sometimes called syconoid.

symbiosis (sim′bī-ōs′əs, sim′bē-ōs′əs) (Gr. *syn,* with, + *bios,* life). The living together of two different species in an intimate relationship. At least one species benefits; the other species may benefit, may be unaffected, or may be harmed (mutualism, commensalism, and parasitism, respectively).

sympatric (sim′pa′-trik) (Gr. *syn,* with, + *patra,* native land). Having the same or overlapping regions of geographical distribution. Noun, **sympatry.**

sympatric speciation A branching of population lineages to form separate species during which the diverging lineages co-occupy a geographic area. A controversial mode of speciation that contrasts with allopatric speciation and parapatric speciation.

symplesiomorphy (sim-plē′zē-ə-mōr′fē). Sharing among species of ancestral characteristics, not indicative that the species form a monophyletic group.

synapomorphy (sin-ap′o-mōr′fē) (Gr. *syn,* together with, + *apo,* of, + *morphe,* form). Shared, evolutionarily derived character states that are used to recover patterns of common descent among two or more species.

synapse (si′naps, si-naps′) (Gr. *synapsis,* contact, union). The place at which an action potential passes between neuronal processes, typically from an axon of one nerve cell to a dendrite of another nerve cell.

synapsids (si-nap′sədz) (Gr. *synapsis,* contact, union). An amniote group comprising the mammals and the ancestral "mammal-like reptiles," having a skull with a single pair of temporal openings.

synapsis (si-nap′səs) (Gr. *synapsis,* contact, union). The time when paired homologous chromosomes lie alongside one another in close association in the first meiotic division.

synaptonemal complex (sin-ap′tə-nē′məl) (Gr. *synapsis,* a joining together, + *nēma,* thread). The structure that holds homologous chromosomes together during synapsis in prophase of meiosis I.

synchronous hermaphrodite A species which has individuals with functional ovaries and testes at the same time. Also called simultaneous hermaphrodite.

syncytial epidermis A multinucleate epidermal layer occurring beneath the cuticle in rotifers.

syncytial specification During embryonic development, diffusion of molecules within syncytial cytoplasm provides positional information to specify cell fate after cytokinesis.

syncytial tegument A unique multinucleate nonciliated epidermis occurring in many parasitic flatworms; also called a neodermis.

syncytium (sən-sish′-əm) (Gr. *syn,* with, + *kytos,* hollow). A multinucleated cell. Adj., **syncytial.**

syndrome (sin′drōm) (Gr. *syn,* with, + *dramein,* to run). A group of symptoms characteristic of a particular disease or abnormality.

syngamy (sin′gə-mē) (Gr. *syn,* with, + *gamos,* marriage). Fertilization of one gamete with another individual gamete to form a zygote; occurs in most animals with sexual reproduction.

synkaryon (sin-ker′ē-on) (Gr. *syn,* with, + *karyon,* nucleus). Zygote nucleus resulting from fusion of pronuclei.

synsacrum (sin-sa′krəm) (Gr. *syn,* together + *sacrum*). A bone of birds, consisting of fused lumbar, sacral, and caudal vertebrae.

syrinx (sir′inks) (Gr. shepherd's pipe). The vocal organ of birds located at the base of the trachea.

systematics (sis-tə-mat′iks). Science of taxonomy and reconstruction of phylogeny.

systematization (sis-tem′ə-tī-zā′shən). Taxonomic grouping of species to represent patterns of descent from common ancestral lineages. Taxa are defined as comprising the most recent common ancestor of a designated pair of species and all descendants of that ancestor. Contrasts with classification, in which taxa are defined as all organisms that possess an essential organismal characteristic.

systole (sis′tə-lē) (Gr. *systolē,* drawing together). Contraction of heart.

T

tactile (tak′til) (L. *tactilis,* able to be touched, from *tangere,* to touch). Pertaining to touch.

taenidia (tə′nid′-ē-ə) (Gr. *tainia,* ribbon). Spiral thickenings of the cuticle that support tracheae (phylum Arthropoda).

tagma, pl. **tagmata** (Gr. *tagma,* arrangement, order, row). A compound body section of an arthropod resulting from embryonic fusion of two or more segments; for example, head, thorax, abdomen.

tagmatization, tagmosis Organization of the arthropod body into tagmata.

taiga (tī′gä) (Russ.). Habitat zone characterized by large tracts of coniferous forests, long, cold winters, and short summers; most typical in Canada and Siberia.

tantulus (tan′tə-ləs) (Gr. *tantulus,* so small). Larva of a tantulocaridan (subphylum Crustacea).

taxis (tak′sis), pl. **taxes** (Gr. *taxis,* arrangement). An orientation movement by a (usually) simple organism in response to an environmental stimulus.

taxon (tak′son), pl. **taxa** (Gr. *taxis,* arrangement). Any taxonomic group or entity.

taxonomic rank The Linnaean category (kingdom, phylum, class, order, family, genus, species, and variations thereof) into which a recognized taxon is placed.

taxonomy (tak-son′ə-mē) (Gr. *taxis,* arrangement, + *nomos,* law). Study of the principles of scientific classification; systematic ordering and naming of organisms.

T cell A type of lymphocyte important in cellular immune response and in regulation of most immune responses.

T-cell receptors Receptors borne on surfaces of T cells. The variable region of a T-cell receptor binds with a specific antigen.

tectorial membrane (tek-to′-rē-əl mem′-brān). A structure in the anuran inner ear that detects low-frequency sounds.

tectum (tek′təm) (L. roof). A rooflike structure, for example, dorsal part of capitulum in ticks and mites.

tegmen (teg′mən) (L. *tegmen,* a cover). External epithelium of crinoids (phylum Echinodermata).

tegument (teg′ū-ment) (L. *tegumentum,* from *tegere,* to cover). An integument: specifically, external covering in cestodes and trematodes.

telencephalon (tel′en-sef′ə-lon) (Gr. *telos,* end, + *encephalon,* brain). The most anterior vesicle of the brain; the anterior-most subdivision of the prosencephalon that becomes the cerebrum and associated structures.

teleology (tē′lē-äl′ə-jē) (Gr. *telos,* end, + L. *logia,* study of, from Gr. *logos,* word). The philosophical view that natural events are goal directed and are preordained, as opposed to the scientific view of mechanical determinism.

teleost (tē′lē-ost). A clade of ray-finned fishes characterized by a homocercal caudal fin.

telocentric (tē′lō-sen′trək) (Gr. *telos,* end, + *kentron,* center). Chromosome with centromere at the end.

telolecithal (te-lō-les′ə-thəl) (Gr. *telos,* end, + *lekithos,* yolk, + *al*). Having the yolk concentrated at one end of an egg.

telson (tel′sən) (Gr. *telson,* extremity). Posterior projection of the last body segment in many crustaceans.

temnospondyls (tem-nō-spän′dəls) (Gr. *temnō,* to cut, + *spondylos,* vertebra). A large group of early tetrapods that lived from the Carboniferous to the Triassic periods.

template (tem′plāt). A pattern or mold guiding the formation of a duplicate; often used with reference to gene replication and transcription.

temporary land bridge A connection of two terrestrial areas that is inundated at times of high sea level but which emerges as dry land at times of low sea level.

tendon (ten′dən) (L. *tendo,* tendon). Fibrous band connecting muscle to bone or other movable structure.

tentaculocyst (ten-tak′u-lō-sist) (L. *tentaculum,* feeler, + *kystis,* pouch). One of the sense organs along the margin of medusae; a rhopalium.

tergum (ter′gəm) (L. back). Dorsal part of an arthropod body segment.

territory (L. *territorium,* from *terra,* earth). A restricted area preempted by an animal or pair of animals, usually for breeding purposes, and guarded from other individuals of the same species.

tertiary structure For a protein, the three-dimensional configuration formed by bonds between side groups of amino acids located in different regions of a polypeptide chain. The disulfide bond between two cysteine amino acids is a common example.

test (L. *testa,* shell). A shell or hardened outer covering.

testate The condition of having a test.

testcross A genetic cross used to determine the genotype (homozygous versus heterozygous) of an individual showing a genetically dominant phenotype. The individual tested is crossed to a homozygous recessive individual. Homozygous individuals tested produce only offspring with the dominant phenotype, whereas heterozygous individuals produce offspring with approximately even numbers of dominant and recessive phenotypes.

tetrad (te′trad) (Gr. *tetras,* four). A pair of replicated homologous chromosomes (two chromatids each, four total) physically aligned and held together in synapsis at prophase I of meiosis; a bivalent.

tetrapods (te′trə-pods) (Gr. *tetras,* four, + *pous, podos,* foot). Four-footed vertebrates; the group includes amphibians, nonavian reptiles, birds, and mammals, including species that have lost paired limbs, such as caecilians and snakes.

theca (thē′kə) (Gr. *thēkē,* a case for something, a box). A protective covering for an organism or an organ.

thecate (thē′kāt) (Gr. *thēkē,* a case, box). An organism bearing a theca.

theory (thē′-o-rē). A scientific hypothesis or set of related hypotheses that offer very powerful explanations for a wide variety of related phenomena and serve to organize scientific investigation of those phenomena.

therapsids (thə-rap′sidz) (Gr. *theraps,* an attendant). Extinct amniotes, from the Permian to Triassic periods from which true mammals evolved; traditionally a paraphyletic group that can be rendered monophyletic by including mammals.

thermocline (thər′mō-klīn) (Gr. *thermē,* heat, + *klinein,* to swerve). Layer of water separating upper warmer and lighter water from lower colder and heavier water in a lake or sea; a stratum of abrupt change in water temperature.

thoracic (thō-ra′sək) (L. *thōrax,* chest). Pertaining to the thorax or chest.

3′ end (three-prime end). The terminus of a nucleic acid molecule that consists of a hydroxyl group attached to the 3′ carbon of the terminal sugar. Nucleic-acid synthesis consists of adding nucleotides to this end of the molecule. (The opposite terminus is the 5′ end.)

thrombin Enzyme catalyzing fibrinogen transformation into fibrin during the process of blood clotting. Percursor is **prothrombin.**

thyrotropin A peptide hormone that is also called thyroid-stimulating hormone. It stimulates production and secretion of thyroid hormones by the thyroid gland.

thyroxine One of two thyroid hormones that is also called T4 because it contains four iodine atoms. It is considered the precursor for triiodothyronine, or T3, because it is less active than T3.

Tiedemann's bodies (tēd′ə-mənz) (from F. Tiedemann, German anatomist). Four or five pairs of pouchlike bodies attached to the ring canal of sea stars, apparently functioning in production of coelomocytes.

tight junction Region of extremely close apposition of cell membranes between two adjacent cells. The junction is formed by bonding between rows of transmembrane proteins in each closely linked membrane.

tissue (ti′shu) (M.E. *tissu,* tissue). An aggregation of cells, usually of the same kind, organized to perform a common function.

titer (tī′tər) (Fr. *titrer,* to titrate). Concentration of a substance in a solution as determined by titration.

Toll-like receptors (TLRs) Named for the Toll protein family discovered in *Drosophila,* Toll-like receptors occur on cell membranes of vertebrates. When activated by binding a microbe, they signal the cell to synthesize

bat / āpe / ärmadillo / herring / fēmale / finch / līce / crocodile / crōw / cŏin / cōre / duck / dūke / ūnicorn / ə indicates unaccented vowel sound "uh" as in mammal, fishes, cardinal, heron, vulture / stress as in bi-ol′o-gy, bi′o-log′i-cal an appropriate antimicrobial peptide. Recognizing patterns rather than specific molecular configurations, they are a vital part of innate immune defenses.

tornaria (tor-na′rē-ə) (L. *tornare*, to turn). A free swimming larva of enteropneusts that rotates as it swims; resembles somewhat the bipinnaria larva of echinoderms.

torsion (L. *torquere*, to twist). A twisting phenomenon in gastropod development that alters the position of the visceral and pallial organs by 180 degrees.

toxicyst (tox′i-sist) (Gr. *toxikon*, poison, + *kystis*, bladder). Structures possessed by predatory ciliate protozoa, which on stimulation expel a poison to subdue prey.

trabecular reticulum (trə-bek′ū-lər rə-tik′ū-ləm) (L. *trabecula*, a small beam; *reticulum*, a net). A bilayered, syncytial tissue forming the main body structure of hexactinellid sponges (phylum Porifera).

trachea (trā′kē-ə) (M.L. windpipe). The cartilage-lined tube used to conduct air between the pharynx and lungs of tetrapods. Also, any of the air tubes of insects.

tracheal system (trāk′ē-əl) (L. *trachia*, windpipe). A network of thin-walled tubes that branch throughout the entire body of terrestrial insects; used for respiration.

tracheole (trāk′ē-ōl) (L. *trachia*, windpipe). Fine branches of the tracheal system, filled with fluid, but not shed at ecdysis.

trachyline (trak′i-līn) (Gr. trachys, rough, + *linum*, thread). Descriptive term for an unusual hydrozoan life cycle (phylum Cnidaria) where a larva metamorphoses directly to a medusa stage without an intervening polyp stage.

transcription Formation of messenger RNA from the coded DNA.

transcription factor A steroid or protein molecule that binds to a chromosome at the locus of a gene, either to activate or to repress synthesis of RNA complementary to the gene's sense strand.

transfer RNA (tRNA) A form of RNA of about 70 or 80 nucleotides that is an adapter molecule in the synthesis of proteins. A specific amino-acid molecule is carried by transfer RNA to a ribosome-messenger RNA complex for incorporation into a nascent polypeptide.

transformational theory of evolution Any evolutionary hypothesis in which change occurs via the restructuring of individual organisms during their ontogenies and transmission of such phenotypic alterations to offspring, as in Lamarck's theory. Contrasts with variational theories of

evolution, such as Darwinian natural selection.

translation (L. a transferring). The process in which the genetic information present in messenger RNA is used to direct the order of specific amino acids during protein synthesis.

translocation Chromosomal material moved from one chromosome to a nonhomologous chromosome, often in a reciprocal exchange.

transporter A transmembrane protein transport molecule in the plasma membrane that permits otherwise impermeable ions and/or molecules to be transported across the membrane, called carrier-mediated transport.

trans-regulatory Pertains to control of the rate of transcription of a gene that encodes mRNA or rRNA by a protein transcription factor that binds DNA at or near the gene whose transcription it influences; contrasts with cis-regulation in typically influencing both copies of the target gene present on homologous chromosomes in a diploid nucleus.

transverse plane (L. *transversus*, across). A plane or section that lies or passes across a body or structure.

trend A directional change in the characteristic features or patterns of diversity in a group of organisms when viewed over long periods of evolutionary time in the fossil record.

trichinosis (trik-in-ō′səs). Disease caused by infection with the nematode *Trichinella spiralis.*

trichocyst (trik′ə-sist) (Gr. *thrix*, hair, + *kystis*, bladder). Saclike protrusible organelle in the ectoplasm of ciliates, which discharges as a threadlike weapon of defense.

triglyceride (trī-glis′ə-rīd) (Gr. *tria*, three, + *glykys*, sweet, + *ide*, suffix denoting compound). A triester of glycerol with one, two, or three kinds of fatty acids.

triiodothyronine One of two thyroid hormones that is also called T3 because it contains three iodine atoms. T3 promotes normal growth and development of the nervous system in growing animals, and also stimulates metabolism.

trimerous (trī′mə-rəs) (Gr. *treis*, three, + *meros*, a part). Body in three main divisions, as in lophophorates and some deuterostomes.

tripartite (trī-par′tīt). See **trimerous.**

triploblastic (trip′lō-blas′tik) (Gr. *triploos*, triple, + *blastos*, germ). Pertaining to animals in which the embryo has three

primary germ layers—ectoderm, mesoderm, and endoderm.

trisomy The chromosomal constitution of an otherwise diploid organism in which a single extra chromosome is present (chromosome number + 2 *n* + 1).

trisomy 21 See **Down syndrome.**

trochophore (trōk′ə-fōr) (Gr. *trochos*, wheel, + *pherein*, to bear). A free-swimming ciliated marine larva characteristic of most molluscs and certain ectoprocts, brachiopods, and marine worms; an ovoid or pyriform body with preoral circlet of cilia and sometimes a secondary circlet behind the mouth.

trophallaxis (trōf′ə-lak′səs) (Gr. *trophē*, food, + *allaxis*, barter, exchange). Exchange of food between young and adults, especially certain social insects.

trophi (trō′fī) (Gr. *trophos*, one who feeds). Jawlike structures in the mastax of rotifers.

trophic (trō′fək) (Gr. *trophē*, food). Pertaining to feeding and nutrition.

trophic level Position of a species in a **food web,** such as producer, herbivore, first-level carnivore, or higher-level carnivore.

trophoblast (trōf′ə-blast) (Gr. *trephein*, to nourish, + *blastos*, germ). Outer ectodermal nutritive layer of blastodermic vesicle; in mammals it is part of the chorion and attaches to the uterine wall.

trophosome (trōf′ə-sōm) (Gr. *trophē*, food, + *sōma*, body). Organ in pogonophorans or siboglinids bearing mutualistic bacteria, derived from midgut.

trophozoite (trōf′ə-zō′īt) (Gr. *trophē*, food, + *zōon*, animal). Adult stage in the life cycle of a unicellular eukaryote in which it is actively absorbing nourishment.

tropic (tro′pic) (Gr. *tropē*, to turn toward). Related to the tropics (tropical); in endocrinology, a hormone that influences the action of another hormone or endocrine gland (usually pronounced trō′pic).

tropomyosin (trōp′ə-mī′ə-sən) (Gr. *tropos*, turn, + *mys*, muscle). Low-molecular-weight, actin-binding protein surrounding the actin filaments of striated muscle. It works with troponin to regulate muscle contraction.

troponin (trə-pōn′in). Complex of globular, actin-binding proteins positioned at intervals along the actin filament of skeletal muscle; serves as a calcium-dependent switch in muscle contraction.

true horn A pointed projection on the head of some mammals, consisting of a hollow sheath of keratin that encloses a bony core, as in sheep, cattle, and goats (Bovidae). Contrast with **antler.**

tube feet (podia) Numerous small, muscular, fluid-filled tubes projecting from an echinoderm; part of water-vascular system; used in locomotion, clinging, food handling, and respiration.

tubercle (tü′bər-kəl) (L. *tuberculum,* small hump). Small protuberance, knob, or swelling.

tubulin (tü′bū-lən) (L. *tubulus,* small tube, + *in,* belonging to). Globular protein, which usually occurs as a dimer, forming the hollow cylinder of microtubules.

tumor necrosis factor A cytokine, the most important source of which is macrophages and T-helper lymphocytes, that is a major mediator of inflammation.

tumor-suppressor gene A gene whose products act as restraints on cell division by triggering apoptosis, controlling transcription of other genes, restraining progression in phases of the cell cycle, or by other means.

tundra (tun′drə) (Russ. from Lapp, *tundar,* hill). Terrestrial habitat zone, located between taiga and polar regions; characterized by absence of trees, short growing season, and mostly frozen soil during much of the year.

tunic (L. *tunica,* tunic, coat). In tunicates, a cuticular, cellulose-containing covering of the body secreted by the underlying body wall.

turbinates (tər′bin-āts) (L. *turbin,* whirling). Highly convoluted bones covered in mucous membrane in the nasal cavity of endotherms; serve to reduce amount of heat and water lost during respiration.

tympanic membrane (tim-pan′ik) (Gr. *tympanon,* drum). The surface that separates the outer and middle ear (eardrum).

type specimen A specimen deposited in a museum that formally defines the name of the species that it represents.

typhlosole (tif′lə-sōl′) (Gr. *typhlos,* blind, + *sōlēn,* channel, pipe). A longitudinal fold projecting into the intestine in certain invertebrates such as the earthworm.

typological species concept The discredited, pre-Darwinian notion that species are classes defined by the presence of fixed, unchanging characters (= "essence") shared by all members.

typology (tī-pol′ə-jē) (L. *typus,* image). A classification of organisms in which members of a taxon are perceived to share intrinsic, essential properties, and variation among organisms is considered uninteresting and unimportant.

U

ulcer (ul′-sər) (L. *ulcus,* ulcer). An abscess that opens through the skin or a mucous surface.

ultimate cause (L. *ultimatus,* last, + *causa*). The evolutionary factors responsible for the origin, state of being, or role of a biological system.

umbilical (L. *umbilicus,* navel). Refers to the navel, or umbilical cord.

umbo (um′bō), pl. **umbones** (əm-bō′nēz) (L. boss of a shield). One of the prominences on either side of the hinge region in a bivalve mollusc shell. Also, the "beak" of a brachiopod shell.

undulating membrane A membranous structure on a unicellular eukaryote associated with a flagellum; may be formed from fused cilia.

ungulate (un′gū-lət) (L. *ungula,* hoof). Hooved (noun); any hooved mammal.

uniformitarianism (ū′nə-fōr′mə-ter′ē-ə-niz′əm). Methodological assumptions that the laws of chemistry and physics have remained constant throughout the history of the earth, and that past geological events occurred by processes that can be observed today.

unique density behavior Refers to greater density of water in liquid than in solid phase.

uniramous (ū′nə-rām′əs) (L. *unus,* one, + *ramus,* a branch). Adjective describing unbranched appendages (phylum Arthropoda).

unitary (ū′nə-ter′ē). Describes the structure of a population in which reproduction is strictly sexual and each organism is genetically distinct from others.

unsaturated fatty acid A subunit of fat molecules containing a carboxyl group and a carbon chain in which two or more carbons are joined by double bonds and therefore capable of forming additional bonds with other atoms, such as hydrogen.

upwelling Oceanic region at which cool and nutrient-rich water rises to the ocean surface, replacing the warmer, usually nutrient-depleted surface water; such regions have high ecological productivity, supporting many fishes.

ureter (ūr′ə-tər) (Gr. o *uētēr,* ureter). Duct carrying urine from metanephric kidney to bladder or cloaca.

urethra (ū-rē′thrə) (Gr. *ourethra,* urethra). The tube from the urinary bladder to the exterior in both sexes.

uric acid (yur′ik) (Gr. *uri,* urine). Nitrogenous waste product requiring little water to excrete; the main waste product in insects, birds, and nonavian reptiles.

uropod (ū′rə-pod) (Gr. *oura,* tail, + *pous, podos,* foot). Posteriormost appendage of many crustaceans.

urostyle (ū′-rō-stīl). A rodlike structure comprising fused vertebrae at the posterior end of the anuran vertebral column; homologous to tail vertebrae of other tetrapods.

utricle (ū′trə-kəl) (L. *utriculus,* little bag). That part of the inner ear containing the receptors for dynamic body balance; the semicircular canals lead from and to the utricle.

V

vacuole (vak′yə-wōl) (L. *vacuus,* empty, + Fr. *ole,* dim). A membrane-bound, fluid-filled space in a cell.

valve (L. *valva,* leaf of a double door). One of the two shells of a typical bivalve mollusc or brachiopod. In cardiovascular and lymphatic systems, valves allow one-way flow of blood or lymph.

variation (L. *varius,* various). Differences among individuals of a group or species that cannot be ascribed to age, sex, or position in the life cycle.

variational theory of evolution An evolutionary hypothesis, such as Darwinian natural selection, in which change occurs in the frequencies of alternative genetic characteristics across generations in a population rather than by modification of heredity by characteristics acquired by an organism during its ontogeny. Contrasts with transformational theories of evolution, such as Lamarckism.

vector (L. a bearer, carrier, from *vehere, vectum,* to carry). Any agent that carries and transmits pathogenic microorganisms from one host to another host. Also, in molecular biology, an agent such as bacteriophage or plasmid that carries recombinant DNA.

vegetal plate Region formed by flattening the embryonic vegetal pole at the beginning of gastrulation.

vegetal pole Region of an egg with a high concentration of yolk; this region is

---

bạt / āpe / ärmadillo / herring / fēmale / fĭnch / līce / crocodile / crōw / cŏin / cōre / duck / dūke / ūnicorn / ə indicates unaccented vowel sound "uh" as in mammal, fishes, cardinal, heron, vulture / stress as in bi-ol′o-gy, bi′o-log′i-cal opposite the animal pole where cytoplasm is concentrated.

veins (vānz) (L. *vena*, a vein). Blood vessels that carry blood toward the heart; in insects, fine extensions of the tracheal system that support the wings.

velarium (və-la′rē-əm) (L. *velum*, veil, covering). Shelflike extension of the subumbrellar edge in cubozoans (phylum Cnidaria).

veliger (vel′ə-jər, vel′ə-gər) (L. *velum*, veil, covering). Larval form of certain molluscs; develops from the trochophore and has the beginning of a foot, mantle, and shell.

velum (vē′ləm) (L. veil, covering). A membrane on the subumbrellar surface of jellyfish of class Hydrozoa. Also, a ciliated swimming organ of the veliger larva.

ventral (ven′trəl) (L. *venter*, belly). Situated on the lower or abdominal surface.

ventricle (ven′-tri-kəl). A chamber of the vertebrate heart that receives blood from an atrium (a separate heart chamber) and pumps blood from the heart.

venule (ven′ūl) (L. *venula*, dim. of *vena*, vein). Small vessel conducting blood from capillaries to vein; small vein of insect wing.

vermiform (ver′mə-form) (L. *vermis*, worm, + *forma*, shape). Adjective to describe any wormlike organism; an adult (nematogen) rhombozoan (phylum Mesozoa).

vesicular (ve-sik′yə-ler) (L. *vesicula*, small bladder, blister). Descriptive term for a granular appearance of the nucleus in many unicellular eukaryotes due to clumping of chromatin; also composed of vesicle-like cavities; also bladderlike.

vestige (ves′tij) (L. *vestigium*, footprint). A rudimentary organ that may have been well developed in some ancestor or in the embryo.

vibrissa (vī-bris′ə), pl. **vibrissae** (L. nostril hair). Stiff hairs that grow from the nostrils or other parts of the face of many mammals and that serve as tactile organs; "whiskers."

vicariance (vī-kar′ē-ənts) (L. *vicarius*, a substitute). Geographical separation of populations, especially as imposed by discontinuities in the physical environment that fragmented formerly geographically continuous populations.

vicariance biogeography An approach to historical biogeography that emphasizes locating physical barriers that simultaneously fragment codistributed species into local areas of geographic endemism; explains shared patterns of cladogenesis among geographically codistributed taxa.

vicariant speciation Formation of species in allopatry initiated by intrusion of a physical barrier that fragments a species into geographically isolated populations. Contrasts with speciation by a founder event, which requires establishment of a new population by a rare movement of individuals across an otherwise strong geographic barrier into territory not occupied by the parent population.

villus (vil′əs), pl. **villi** (L. tuft of hair). A small fingerlike process on the wall of the small intestine that increases the surface area for absorption of digested nutrients. Also one of the branching, vascular processes on the embryonic portion of the placenta.

virus (vī′rəs) (L. slimy liquid, poison). A submicroscopic noncellular particle composed of a nucleoprotein core and a protein shell; parasitic; grows and reproduces only in a host cell.

viscera (vis′ər-ə) (L. pl. of *viscus*, internal organ). Internal organs in the body cavity.

visceral (vis′ər-əl). Pertaining to viscera.

vitalism (L. *vita*, life). The discredited viewpoint that natural processes are controlled by supernatural forces and cannot be explained through the laws of physics and chemistry and mechanistic processes alone.

vitamin (L. *vita*, life, + *amine*, from former supposed chemical origin). An organic substance required in small amounts for normal metabolic function; must be supplied in the diet or by intestinal flora because the organism cannot synthesize it. An exception is vitamin D_3, which is manufactured in the skin in the presence of sunlight.

vitellaria (vi′təl-lar′ē-ə) (L. *vitellus*, yolk of an egg). Structures in many flatworms that produce vitelline cells, which provide eggshell material and nutrient for the embryo.

vitelline gland See **vitellaria.**

vitelline membrane (vi-tel′ən, vī′təl-ən) (L. *vitellus*, yolk of an egg). The noncellular membrane that encloses the egg cell.

viviparity (vi′və-par′ə-tē) (L. *vivus*, alive, + *parere*, to bring forth). Reproduction in which eggs develop within the female body, which supplies nutritional aid as in therian mammals, many nonavian reptiles, and some fishes; offspring are born as juveniles. Adj., **viviparous** (vī-vip′ə-rəs).

vocal cords Paired mucous membrane folds whose vibration in the larynx (voice box) of many terrestrial vertebrates produces sound.

W

waggle dance A complex behavior performed by a hive bee to direct other members of the hive population toward a distant food source.

water-vascular system System of fluid-filled closed tubes and ducts peculiar to echinoderms; used to move tentacles and tube feet that serve variously for clinging, food handling, locomotion, and respiration.

weir (wēr) (Old English *wer*, a fence placed in a stream to catch fish). Interlocking extensions of a flame cell and a collecting tubule cell in some protonephridia.

X

xanthophore (zan′thə-fōr) (Gr. *xanthos*, yellow, + *pherein*, to bear). A chromatophore containing yellow pigment.

xenograft (zē′nə-graft). Graft of tissue from a species different from the recipient.

X-organ Neurosecretory organ in eyestalk of crustaceans that secretes molt-inhibiting hormone.

Y

Y-organ Gland in the antennal or maxillary segment of some crustaceans that secretes molting hormone.

Z

zoecium, zooecium (zō-ē′shē-əm) (Gr. *zōon*, animal, + *oikos*, house). Cuticular sheath or shell of Ectoprocta.

zooflagellates (zō′ə-fla′jə-lāts). Members of the former Zoomastigophorea, animal-like flagellates (former phylum Sarcomastigophora).

zooid (zō′-id) (Gr. *zōon*, life). An individual member of a colony of animals, such as colonial cnidarians and ectoprocts.

zooxanthella (zō′-zan-thəl′ə) (Gr. *zōon*, animal, + *xanthos*, yellow). A minute dinoflagellate alga living in the tissues of many types of marine invertebrates.

zygote (Gr. *zygōtos*, yoked). A fertilized egg.

zygotic meiosis Meiosis that occurs within the first few divisions after zygote formation; thus, all stages in the life cycle other than the zygote are haploid.

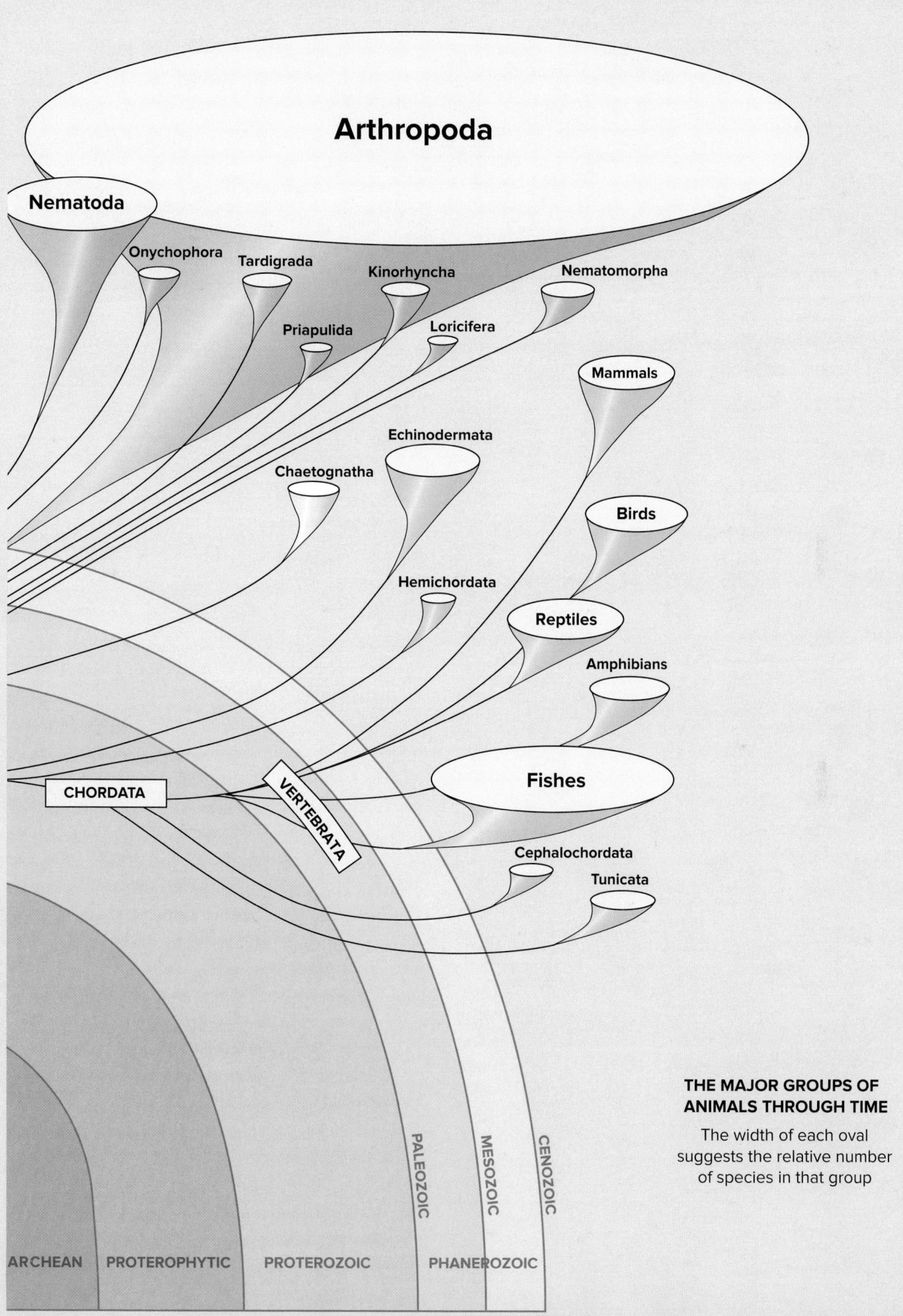

Arthropoda

Nematoda

Onychophora

Tardigrada

Priapulida

Kinorhyncha

Loricifera

Nematomorpha

Mammals

Echinodermata

Chaetognatha

Birds

Hemichordata

Reptiles

Amphibians

Fishes

CHORDATA

VERTEBRATA

Cephalochordata

Tunicata

THE MAJOR GROUPS OF ANIMALS THROUGH TIME

The width of each oval suggests the relative number of species in that group

PALEOZOIC

MESOZOIC

CENOZOIC

ARCHEAN   PROTEROPHYTIC   PROTEROZOIC   PHANEROZOIC

EONS

PHANEROZOIC

PROTEROZOIC

541 MY BP

ARCHEAN

Multi-cellular organisms

2500 MY BP

Appearance of oxygen in atmosphere

Photosynthesis

Earliest known fossils

4000 MY BP

Oldest known rocks

Origin of life

Formation of Earth

Formation of the Sun

Origin of Solar System

ERAS	PERIOD	Millions of Years Before Present	BIOLOGICAL EVENT
CENOZOIC	QUATERNARY	2.6	
	TERTIARY		Diversification of mammals; continental drift isolating many
	Great Extinction	66	
MESOZOIC	CRETACEOUS	145	Climax of dinosaurs and marine reptiles followed by extinction; radiation of marsupial and plac mammals; first flowering plants. decline of gymnosperms
	JURASSIC	201	First birds; dinosaurs abundant forest growth
	TRIASSIC	252	First dinosaurs; first true mamm lush forest growth, conifers dor
	Great Extinction		
PALEOZOIC	PERMIAN	299	Radiation of reptiles; early syna common; displacement of ampl period ends with mass extinctio
	CARBONIFEROUS — PENNSYLVANIAN	323	First amniotes; giant insects; gr conifer forests
	CARBONIFEROUS — MISSISSIPPIAN	359	Radiation of amphibians; shark abundant; scale trees and seed
	DEVONIAN	419	First tetrapods on land; first bo finned and lobe-finned fishes; w and swampy environments
	SILURIAN	443	First jawed fishes; first land-bas invertebrates
	ORDOVICIAN	485	Fishes begin radiation; abundant marine invertebrates land plants
	CAMBRIAN	541	Origin of many invertebrate ph classes; Burgess Shale fauna; e chordates; earliest vertebrates fishes); shelled molluscs; trilobi abundant